The Michigan Historical Reprint Series

Reprints from the collection of the University of Michigan University Library

The Scholarly Publishing Office
the University of Michigan
University Library

SUBJECT-MATTER INDEX

OF

PATENTS FOR INVENTIONS

ISSUED BY THE

UNITED STATES PATENT OFFICE

FROM

1790 TO 1873, INCLUSIVE.

VOLUME II.

COMPILED AND PUBLISHED UNDER THE DIRECTION OF

M. D. LEGGETT,

COMMISSIONER OF PATENTS.

WASHINGTON:
GOVERNMENT PRINTING OFFICE.
1874.

INDEX OF PATENTS

ISSUED FROM

THE UNITED STATES PATENT-OFFICE

FROM 1790 TO 1873, INCLUSIVE.

VOLUME II.

Invention.	Inventor.	Residence.	Date.	No.
H.				
Hackling-machine	L. F. Lannay and J. Webb	Baltimore County, Md	Mar. 12, 1872	124, 497
Hackling-machine	W. Montgomery	West Roxbury, Mass	July 2, 1872	128, 647
Hæmagatactiphorus	D. Houghtaling and A. Meneely	Watervliet, N. Y	June 11, 1829	....
Hair and cutting fur from peltry, Machine for extracting.	E. Cutter	Brattleborough, Vt	Mar. 1, 1809	....
Hair and grass cloth, Machine for preparing woof for the manufacture of.	J. Downie	Paterson, N. J	July 18, 1865	48, 796
Hair and oakum, Picking	R. B. Lewis	Hallowell, Me	June 25, 1836	....
Hair, Apparatus for arranging	G. Lieberknecht	Wiemar, North German Confederation.	Aug. 1, 1871	117, 648
Hair, Application for restoring the	B. Harris	New Orleans, La	Mar. 1, 1859	23, 086
Hair, Barber's apparatus for brushing	C. P. Kroll	New York, N. Y	Sept. 12, 1865	49, 893
Hair-clasp	C. Rowland	Washington, D. C	Feb. 16, 1869	87, 072
Hair-clipping device	F. C. Richardson	New York, N. Y	May 2, 1871	114, 477
Hair-clipping device	J. C. Wilson	New York, N. Y	Nov. 22, 1870	109, 479
Hair-clipping machine	G. H. Pratt	Boston Highlands, Mass	Feb. 6, 1872	123, 508
Hair-cloth	I. Lindsley	Pawtucket, R. I	Apr. 7, 1868	76, 476
Hair-cloth	J. Noblit	Philadelphia, Pa	Jan. 28, 1868	73, 920
Hair-cloth	W. Rossnagel	Newark, N. J	Oct. 24, 1871	120, 330
Hair-cloth, Imitation	W. Rossnagel	Newark, N. J	Oct. 24, 1871	120, 331
Hair-cloth, Imitation	C. Sherriff	Newark, N. J	Dec. 5, 1871	121, 674
Hair-cloth, Imitation	I. N. Tichenor	Newark, N. J	Jan. 16, 1872	122, 868
Hair-cloth, Machine for trimming and shearing	O. Arnold and I. Lindsay	North Providence, R. I	Aug. 9, 1870	106, 106
Hair-cloth, Making	W. Shotwell and A. Kinder	New York	July 23, 1813	....
Hair-cloth, Manufacture of imitation	J. J. Comstock and J. Aborn	Providence, R. I	May 17, 1870	103, 018
Hair-cloth, Process of separating the hair from	D. Whiteley	Providence, R. I	Dec. 6, 1870	109, 857
Hair-comb	G. Hooker	Bristol, Conn	Apr. 3, 1835	....
Hair-comb	M. Moss	Farmington, Conn	Jan. 10, 1809	....
Hair-comb	T. Stanley	Southington, Conn	July 6, 1809	....
Hair-comb, Metal	L. North	Otsego, N. Y	May 28, 1818	....
Hair-comb, Metallic	N. Bushnell	Middletown, Conn	Oct. 31, 1835	....
Hair-comb, Socket	E. Parsons	Bristol, Conn	Aug. 16, 1810	....
Hair-combs, Machinery for pointing the teeth of	C. B. Rogers	Deep River, Conn	Jan. 4, 1859	22, 513
Hair-combs, Manufacturing	I. Ives	Bristol, Conn	Apr. 5, 1823	....
Hair, Composition for coloring	J. C. Ayer and E. Haeffely	Lowell, Mass	Jan. 28, 1868	73, 865
Hair, Composition for preserving and curing the	H. Rose	Milford, Mass	Apr. 12, 1864	42, 311
Hair, Composition for the	J. Mackay	New York	Apr. 6, 1842	2, 541
Hair-compound	B. F. Atwood	New York, N. Y	Oct. 27, 1868	83, 440
Hair, Compound for cleansing	A. Miles	Toledo, Ohio	Dec. 5, 1871	121, 536
Hair, Compound for coloring	G. Smith	Groton Junction, Mass	Mar. 23, 1869	88, 224
Hair, Compound for coloring and preserving the	G. Smith	Groton Junction, Mass	Feb. 1, 1870	99, 487
Hair, Compound for curling	E. L. Parsons	Grand Ledge, Mich	June 20, 1871	116, 215
Hair, Compound for treating human	W. R. Lount	Austin, Tex	Oct. 26, 1869	96, 245
Hair-crimper	H. H. Adams	New York, N. Y	Mar. 24, 1863	38, 010
Hair-crimper	H. Christian	New York, N. Y	Mar. 27, 1866	53, 413
Hair-crimper	S. F. Conant	Skowhegan, Me	Apr. 7, 1868	76, 401
Hair-crimper	W. F. George	New York, N. Y	Jan. 29, 1861	31, 234
Hair-crimper	E. Ivins	Waterbury, Conn	May 3, 1859	23, 842
Hair-crimper	B. Mannon	Newport, Ky	Aug. 21, 1866	57, 441
Hair-crimper	J. B. Siccardi	New York, N. Y	Feb. 19, 1867	62, 294
Hair-crimper	G. W. Wood	Philadelphia, Pa	Apr. 14, 1868	76, 683
Hair-crimping pin	E. A. Tyler	Buffalo, N. Y	Mar. 6, 1866	53, 060
Hair-curler	C. H. Barney	Providence, R. I	Oct. 17, 1871	119, 960
Hair-curler	S. A. Early	Philadelphia, Pa	July 9, 1867	66, 476
Hair-curler	J. W. Kenny and J. H. Adams	Albany, N. Y	Oct. 4, 1870	107, 920
Hair-curler	C. H. Lavis and J. McMillan	Philadelphia, Pa	July 9, 1867	66, 599
Hair-curler	I. S. Marcy	Nashua, N. H	July 4, 1871	116, 731
Hair-curler	C. Markley	New York, N. Y	July 25, 1871	117, 439
Hair curler and crimper, Combined	E. Matteson	Jersey City, N. J	July 26, 1870	105, 822
Hair, Curling	E. T. Crain	Kansas City, Mo	May 20, 1873	138, 995
Hair-curling apparatus	M. Adkins	Oswego, N. Y	July 27, 1869	92, 926
Hair-curling clamp	F. Arnold	Middle Haddam, Conn	Feb. 2, 1858	19, 228
Hair-curling device	S. L. Tibbals	Dutch Flat, Cal	July 19, 1870	105, 519

Index of patents issued from the United States Patent Office from 1790 *to* 1873, *inclusive*—Continued.

Invention.	Inventor.	Residence.	Date.	No.
Hair-curling fluid	R. Clark	Chicago, Ill	Apr. 24, 1866	54, 115
Hair-curling instrument	M. M. Lewis	Albany, N. Y	Apr. 21, 1857	17, 103
Hair-curling iron	H. S. Maxim	Boston, Mass	Aug. 21, 1866	57, 354
Hair-curling pin	A. Vogels and F. Krebs	New York, N. Y., and Georgetown, D. C.	Apr. 26, 1870	102, 338
Hair curls and apparatus for manufacturing, Detachable.	J. Mayer	Philadelphia, Pa	Dec. 26, 1871	122, 183
Hair-cutter	G. A. Harley	New York, N. Y	Sept. 14, 1869	94, 820
Hair-cutter	G. A. Harley	New York, N. Y	Dec. 28, 1869	98, 256
Hair-cutting comb and shears, Combined	J. H. Atkinson	San Francisco, Cal	May 18, 1869	90, 063
Hair-cutting machine	R. Maynard and J. J. Purkiss	Cambridge and London, Great Britain.	Dec. 17, 1867	72, 214
Hair-cutting shears	L. D. Craig	Nevada City, Cal	Dec. 15, 1868	84, 860
Hair, Device for curling and dressing	P. Ceredo	Düsseldorf, Prussia	Sept. 20, 1870	107, 596
Hair-dressing	F. R. Taylor	Waverly, N. Y	Apr. 25, 1871	114, 223
Hair dressing and coloring compound	J. N. Fallis	Newport, Ky	Jan. 31, 1871	111, 443
Hair-dressing composition	J. C. Tilton	Sanbornton Bridge, N. H	Sept. 29, 1868	82, 660
Hair-dressing compound	F. Stearns	Detroit, Mich	Mar. 24, 1868	75, 806
Hair-dressing machine	W. Chubb	New York	July 9, 1829	....
Hair-dressing roll	E. Schnautz	New York, N. Y	Jan. 31, 1871	111, 388
Hair-drying machine	T. Malley	Allegheny City, Pa	Jan. 19, 1869	85, 944
Hair-dye	D. Duprat	New York, N. Y	Dec. 15, 1863	40, 918
Hair-dye	J. Lory	Memphis, Tenn	Mar. 16, 1869	88, 793
Hair-dye	G. S. G. Saur	Washington, D. C	July 8, 1862	38, 840
Hair-dye and cosmetic	J. L. Wild	Bay City, Mich	Aug. 1, 1871	117, 583
Hair dye and dressing compound	G. F. Peckham	Grafton, Ohio	May 12, 1868	77, 911
Hair-dyeing	L. S. Stimson	Lowell, Mass	Dec. 24, 1867	72, 558
Hair-dyeing apparatus	C. Merritt	South Weymouth, Mass	Nov. 19, 1867	71, 036
Hair-dyeing apparatus	W. Patton	Springfield, Mass	Oct. 13, 1868	82, 982
Hair-dyeing composition	A. Grandjean	New York, N. Y	Feb. 28, 1844	3, 453
Hair-dyeing compound	J. S. Letord	Sedalia, Mo	June 27, 1871	116, 462
Hair, Fibers to imitate	W. Stauf	Bonn, Prussia	Mar. 12, 1872	124, 652
Hair for ladies' head-dress, Imitation	L. F. Shaw	New York, N. Y	Oct. 26, 1869	96, 275
Hair for mattress, Picking, twisting, curling, &c	H. Green and J. C. Osgood	Hamilton, N. Y	Dec. 18, 1832	....
Hair for stuffing cushions, &c	W. Adamson	Philadelphia, Pa	Apr. 2, 1872	125, 246
Hair for weaving, Treating	J. Gledhill	New York, N. Y	Feb. 21, 1854	10, 555
Hair for wigs, &c., Machine for weaving	L. A. Seago	Jerseyville, Ill	Feb. 9, 1869	86, 872
Hair from fur, Machine for separating	J. MacDonald	New York, N. Y	Sept. 11, 1827	....
Hair from fur-skins, Extracting	J. Hollinghead and D. Baker	Trenton, N. J	Aug. 19, 1813	....
Hair from peltry, Machine for extracting	E. Cutter	Walpole, N. H	Feb. 12, 1810	....
Hair from skins, Instrument for extracting	E. Flint	New York, N. Y	June 30, 1837	244
Hair, grass, &c., Machine for cutting	C. F. Harlow and E. H. Perry	Boston and Roxbury, Mass	May 28, 1867	65, 077
Hair, Imitation of braided human	S. E. Cook	Philadelphia, Pa	Apr. 24, 1866	54, 116
Hair, Instrument for parting lady's	J. L. Meek	New York, N. Y	Mar. 13, 1866	53, 163
Hair, Machine for beating and cleaning	G. P. Holloway and W. J. Huey	Portland, Ind	Dec. 6, 1870	109, 903
Hair, Machine for carding or picking curled	F. Harding	Cleveland, Ohio	Aug. 25, 1840	1, 736
Hair, Machine for heading	E. Hough	Brooklyn, N. Y	Dec. 2, 1873	145, 060
Hair, Machine for heading	R. Turneaure	Rockford, Ill	Dec. 2, 1873	145, 259
Hair, Machine for picking curled	W. Adamson	Philadelphia, Pa	May 29, 1860	28, 447
Hair, Machine for picking curled	N. L. Cole	Norwich, Conn	Mar. 8, 1870	100, 604
Hair, Machine for picking curled	L. F. Lannay and J. Webb	Baltimore County, Md	May 28, 1872	127, 354
Hair, Machine for picking curled	J. Thompson, 3d	Essex County, Mass	July 8, 1834	....
Hair, Machine for removing burs from	F. Walpert	Baltimore, Md	Mar. 31, 1868	76, 122
Hair, Machinery for cleaning	J. Radebaugh and J. A. Matlack.	Lancaster, Ohio	Apr. 17, 1849	6, 331
Hair, Manufacture of crimped	R. F. Donisthorpe and T. A. W. Clarke.	Leicester, England	Dec. 2, 1873	145, 056
Hair, Medical compound for the	P. Schlicher	Louisville, Ky	Sept. 19, 1871	119, 052
Hair, Medical compound for treating the	W. F. Staten	Jasper, Fla	Dec. 9, 1873	145, 313
Hair-medicine	A. J. Fletcher	Red Bluff, Cal	Apr. 12, 1870	101, 722
Hair or wool from animals, Machine for clipping	C. W. Emery	Dorchester, Mass	Jan. 10, 1865	45, 821
Hair or wool from animals, Machine for clipping	C. W. Emery	Dorchester, Mass	Feb. 7, 1865	46, 226
Hair-picker	F. Calvert	Lowell, Mass	Mar. 31, 1868	76, 158
Hair-picking machine	C. W. Couillard	Bath, Me	Oct. 1, 1872	131, 853
Hair-picking machine	F. Frey	Liberty, Ill	Nov. 19, 1867	70, 987
Hair-picking machine	E. Hoffstaetter	New York, N. Y	Apr. 7, 1868	76, 451
Hair-picking machine	D. M. Varney	Burlington, Vt	Aug. 13, 1872	130, 550
Hair-pin	C. M. Gormly	Pittsburgh, Pa	Feb. 7, 1871	111, 635
Hair-pin	E. Hewitt and J. McAuliffe	New York, N. Y	June 13, 1871	115, 855
Hair-pin	J. C. Howells	Washington, D. C	Aug. 23, 1864	43, 914
Hair-pin	A. T. Thayer	New York, N. Y	July 5, 1864	43, 439
Hair-pin	W. Wickersham	Boston, Mass	Aug. 2, 1870	106, 009
Hair-pin box	J. C. Howells	New York, N. Y	Feb. 2, 1869	86, 544
Hair, Preparation for the	V. Clirehugh	New York, N. Y	Apr. 11, 1842	2, 551
Hair, Preparation for the	N. T. Ormsby	Chicago, Ill	June 18, 1872	127, 983
Hair, Preparation of artificial skins with natural	J. R. and F. C. Tussaud	London, England	Apr. 30, 1872	126, 166
Hair, Process for treating	W. Adamson	Philadelphia, Pa	July 12, 1864	43, 466
Hair puff or roll	J. D. Openheimer	Philadelphia, Pa	Jan. 3, 1871	110, 782
Hair-renewing compound	A. L. Baker	Newark, N. J	Mar. 9, 1869	87, 618
Hair-restorative	E. J. Balcear	Martinez, Cal	Nov. 10, 1868	83, 820
Hair-restorative	E. W. Barnes	Syracuse, N. Y	Mar. 12, 1872	124, 533
Hair-restorative	P. Bearsch	Baltimore, Md	July 16, 1872	129, 310
Hair-restorative	A. K. Benson	Allegheny City, Pa	Dec. 21, 1869	98, 142
Hair-restorative	R. W. Carr	Baltimore, Md	Nov. 14, 1865	50, 901
Hair-restorative	W. H. Harris	Corry, Pa	Feb. 26, 1867	62, 328
Hair-restorative	M. Howard	Virginia City, Nev	Dec. 5, 1865	51, 319
Hair-restorative	J. Johnson	Brooklyn, N. Y	Aug. 20, 1872	130, 720
Hair-restorative	J. Loucke	Indianapolis, Ind	Jan. 5, 1864	41, 076
Hair-restorative	A. C. Maxfield	Biddeford, Me	Dec. 13, 1870	110, 058
Hair-restorative	S. M. McNett	Topeka, Kans	Mar. 26, 1872	125, 066
Hair-restorative	J. Moye	Medina, N. Y	July 21, 1863	39, 304
Hair-restorative	J. N. Smith	Colemansville, Ky	July 16, 1872	129, 608
Hair-restorative	J. Sprink	Council Bluffs, Iowa	Oct. 9, 1866	58, 690
Hair-restorative	W. P. Thomas and J. F. Boardman.	Elko, Nev	Mar. 14, 1871	112, 749
Hair-restorative	P. Trautwein	Brooklyn, N. Y	May 7, 1872	126, 591
Hair-restorer	C. Smith	Buffalo, N. Y	Mar. 1, 1864	41, 791
Hair restoring and coloring compound	G. Smith	Ayer's, Mass	Aug. 6, 1872	130, 160

Index of patents issued from the United States Patent Office from 1790 *to* 1873, *inclusive*—Continued.

Invention.	Inventor.	Residence.	Date.	No.
Hair-roll	T. Schionbeck	Saint Louis, Mo	Nov. 26, 1872	133, 490
Hair-roll	J. H. Vogt and G. Dietzel	New York, N. Y	Sept. 6, 1870	107, 134
Hair, Roll for ladies'	J. Edwards	New York, N. Y	May 19, 1863	38, 574
Hair-rope, Machine for opening	P. Wisdom	New York, N. Y	Mar. 24, 1868	76, 025
Hair-rope picker	H. R. Hildreth	Lynn, Mass	Apr. 27, 1869	89, 400
Hair-seating, Weaving	C. R. Harvey	Poughkeepsie, N. Y	Nov. 25, 1837	490
Hair spinning and curling machine	P. Wisdom	Brooklyn, N. Y	Dec. 21, 1869	98, 135
Hair spinning, twisting, and kinking machine	H. Burnham	New York	June 22, 1842	2, 680
Hair-switch	B. F. Burgess, jr	Boston, Mass	Dec. 26, 1871	122, 153
Hair-switch	G. H. Cutter	Arlington, Mass	Apr. 9, 1872	125, 382
Hair-tonic	S. A. Reybert	Plainfield, N. J	Mar. 14, 1871	112, 632
Hair tonic and restorer	C. Stock and W. B. Ross	Pawling, N. Y	Feb. 6, 1872	123, 370
Hair-turning hatchel	L. Wilbur	Watertown, N. Y	May 14, 1872	126, 661
Hair twisting and curling machine	P. Wisdom and J. H. Wilcox	Brooklyn and New York, N. Y.	Apr. 12, 1870	101, 800
Hair, Vegetable fiber as a substitute for	W. Staufen	Paris, France	Sept. 1, 1868	81, 702
Hair, Vegetable fiber to imitate	A. Couturier	Trinidad, Cuba	May 11, 1869	89, 855
Hair-wash	R. Travis	Elkton, Ky	Sept. 2, 1873	142, 532
Hair weaving and mounting apparatus	N. Demongeot	Washington, D. C	Dec. 19, 1871	121, 997
Hair-whipping machinery	I. Davis	Erie, Pa	Dec. 25, 1855	13, 981
Hall, church, &c., Seats for	J. L. Kapple	Mechanicsburgh, Ohio	June 11, 1872	127, 891
Halter	J. Carpenter	Wilmington, Ohio	Dec. 12, 1871	121, 704
Halter	S. K. Dexter	Lowell, Mass	June 25, 1872	128, 215
Halter	W. Garrison and C. H. Stevens	Syracuse, N. Y	Jan. 28, 1868	73, 711
Halter	W. M. Harris	Dixon, Ill	Feb. 22, 1870	100, 031
Halter	R. Porter	Ottumwa, Iowa	Dec. 6, 1870	109, 937
Halter	C. H. Sawyer	Buxton, Me	Oct. 30, 1866	59, 277
Halter	C. A. Steinbrocke	Louisville, Ill	Aug. 25, 1868	81, 552
Halter	J. Thornton	Wellsville, N. Y	Apr. 4, 1871	113, 468
Halter	J. Thornton and E. G. Latta	Genesee, N. Y	Aug. 20, 1872	130, 004
Halter	L. Whitehead	Nunda, N. Y	May 1, 1860	28, 121
Halter and bridle for horses	S. C. Hawkins	Patchogue, N. Y	Oct. 5, 1858	21, 674
Halter apparatus, Tethering	E. Howe	Marlborough, Mass	May 15, 1866	54, 729
Halter-clasp	J. H. Plumstead	Lynn, Mass	June 20, 1865	48, 307
Halter-clasp	G. H. Stewart	Philadelphia, Pa	Oct. 26, 1869	96, 164
Halter for hitching horses, Pastern	J. C. Ford	Montreal, Canada	Nov. 4, 1873	144, 198
Halter-hitch	A. T. Boon and L. Mills	Galesburgh, Ill	June 1, 1869	90, 812
Halter-hitch	G. Race	Norwich, N. Y	Feb. 2, 1869	86, 587
Halter-holder	G. F. Jerome	Hempstead, N. Y	July 26, 1864	43, 643
Halter, Horse	E. H. Le Baron	Mattepoisett, Mass	Aug. 9, 1864	43, 781
Halter, Horse	A. Le Plongeon	San Francisco, Cal	Nov. 22, 1870	109, 431
Halter, Horse	J. E. Marshall	West Chester, Pa	July 16, 1861	32, 837
Halter, Horse	I. P. Osborn and W. A. Bayhan	Wilmington, Ohio	Sept. 12, 1871	118, 877
Halter, Horse	M. Wesson	Springfield, Mass	Feb. 13, 1872	123, 653
Halter-loop	S. F. Cross	Canton, Ohio	Feb. 23, 1869	87, 246
Halter, Neck-rope	K. Gibbs	Berwick, Me	Dec. 19, 1865	51, 582
Halter-ring	L. C. Chase	Boston, Mass	Apr. 30, 1861	32, 180
Halter-rings, &c., Fastening for	E. M. Kinne	Cuba, N. Y	Feb. 18, 1873	136, 071
Halters, Mode of manufacturing	H. B. Ware	Burlington, Iowa	Sept. 12, 1865	49, 939
Hame	J. W. Briggs	Cleveland, Ohio	May 22, 1849	6, 465
Hame	T. G. Brooks	Oneida, Ill	Aug. 25, 1868	81, 469
Hame	R. D. Brown	Covington, Ind	Feb. 16, 1864	41, 600
Hame	W. H. Bustin	Watertown, Mass	Apr. 27, 1869	89, 287
Hame	H. Cogswell	Greenwich, N. Y	Feb. 15, 1859	22, 937
Hame	B. Crawford	Allegeny City, Pa	Mar. 10, 1868	75, 250
Hame	C. D. and R. Hazard	Cleveland Ohio	Apr. 16, 1872	125, 675
Hame	R. Hazard	New London, Ohio	Sept. 10, 1861	33, 280
Hame	F. X. Kaffer	Champaign City, Ill	Apr. 10, 1866	53, 830
Hame	E. F. Lacy and D. K. Woodbury	Danville, Ill	May 5, 1868	77, 625
Hame	G. J. Letchworth	Auburn, N. Y	Jan. 11, 1870	98, 834
Hame	G. J. Letchworth	Auburn, N. Y	July 4, 1871	116, 725
Hame	D. M. Nixon	Danville, Ill	May 5, 1868	77, 645
Hame	I. B. Smith and H. C. Burr	Springfield, Vt	Sept. 15, 1868	82, 168
Hame	J. Thornton and E. G. Latta	Wellsville and Friendship, N. Y.	Nov. 4, 1873	144, 371
Hame	L. Whitehead	Nunda, N. Y	June 28, 1870	104, 806
Hame, Adjustable	D. M. Cummings	Enfield, N. H	Oct. 28, 1862	36, 767
Hame and collar	J. G. Smith	Oregon, Wis	Dec. 28, 1869	98, 311
Hame and strap fastener	J. B. Waterman	Summit, Mich	Sept. 22, 1868	82, 367
Hame-attachment	C. H. Easte	South Boston, Mass	Apr. 19, 1870	102, 100
Hame-attachment	J. Thornton and E. G. Latta	Wellsville, N. Y	Apr. 4, 1871	113, 469
Hame-bending apparatus	A. Gardner	Buffalo, N. Y	Dec. 11, 1849	6, 934
Hame cap, Harness	L. Bonine	Vandalia, Mich	Aug. 9, 1870	106, 111
Hame-clasp	J. H. Snyder	Rockford, Ill	Aug. 13, 1867	67, 681
Hame, Coach and chaise	H. Aiken	Brighton, Mass	Mar. 12, 1823	....
Hame-coupling	G. W. Heckart	Columbiana, Ohio	June 23, 1868	79, 224
Hame, Extension	F. M. Schaeffer	Blooming Grove, Kans	Aug. 3, 1869	93, 233
Hame-fastener	H. W. Austin and E. C. Perry	Portage Township, Mich	Oct. 5, 1869	95, 410
Hame-fastener	E. Bradley	Farm Village, N. Y	Mar. 5, 1872	124, 323
Hame-fastener	T. L. Booker	Shady Grove, Va	Aug. 12, 1873	141, 690
Hame-fastener	J. Clendening	Rockford, Ill	Oct. 5, 1869	95, 430
Hame-fastener	E. A. Cooper	Lancaster, N. Y	Mar. 24, 1868	75, 864
Hame-fastener	E. A. Cooper	Buffalo, N. Y	Jan. 26, 1869	86, 137
Hame-fastener	R. A. Daniels	Wayne, Ohio	Feb. 11, 1862	34, 351
Hame-fastener	A. M. Dorman	Philadelphia, Pa	Aug. 24, 1869	94, 090
Hame-fastener	M. L. Drake	Rockford, Ill	Sept. 21, 1869	95, 007
Hame-fastener	J. G. Fleming	Cochranton, Pa	Dec. 2, 1873	145, 163
Hame-fastener	A. J. Foster	Lake Mills, Wis	Sept. 17, 1867	68, 866
Hame-fastener	A. Gale and H. R. Johnson	Shelby County, Ky	Dec. 20, 1870	110, 226
Hame-fastener	J. Harding	Schoolcraft, Mich	June 23, 1868	79, 223
Hame-fastener	J. V. Hutschler	Keyport, N. J	Feb. 9, 1869	86, 750
Hame-fastener	J. Koch and D. Seachrist	Columbiana, Ohio	Sept. 1, 1868	81, 650
Hame-fastener	J. T. McDivitt	Fayetteville, Ohio	Aug. 31, 1869	93, 230
Hame-fastener	R. R. McDonald	Syracuse, N. Y	Sept. 15, 1868	82, 142
Hame-fastener	J. H. McKinley	New York, N. Y	Dec. 29, 1868	85, 464
Hame-fastener	C. H. Miller	Buffalo, N. Y	Apr. 5, 1870	101, 642
Hame-fastener	J. D. Miller	Enon, Ohio	Aug. 4, 1868	80, 757
Hame-fastener	A. Palmer	Hudson, Mich	Nov. 19, 1867	71, 209

Index of patents issued from the United States Patent Office from 1790 *to* 1873, *inclusive*—Continued.

Invention.	Inventor.	Residence.	Date.	No.
Hame-fastener	W. W. Palmer	Hudson, Mich	Apr. 16, 1867	63, 933
Hame-fastener	W. H. Payne	Janesville, Wis	Oct. 13, 1868	83, 087
Hame-fastener	A. J. Preston	Dryden, N. Y	Nov. 7, 1865	50, 842
Hame-fastener	J. R. Richardson	Newcastle, Pa	Sept. 3, 1867	68, 575
Hame-fastener	W. A. Robinson	Grand Rapids, Mich	Jan. 21, 1868	73, 651
Hame-fastener	M. R. Shalters	Alliance, Ohio	July 24, 1866	56, 621
Hame-fastener	T. Skelton	Rockford, Ill	Dec. 22, 1868	85, 250
Hame-fastener	A. Sweetland	Syracuse, N. Y	Mar. 3, 1868	75, 074
Hame-fastener	A. Sweetland	Syracuse, N. Y	June 13, 1871	115, 908
Hame-fastener	W. S. Thompson and R. V. Love	Montgomery, Ala	Sept. 8, 1868	82, 045
Hame-fastener	W. H. Tiffon	Lee Roy, N. Y	Dec. 28, 1869	98, 315
Hame-fastener	J. Tingley	Potter County, Pa	Nov. 16, 1858	22, 096
Hame-fastener	A. J. Tompkins and J. M. Wegand.	Clarksville, Iowa	Oct. 12, 1869	95, 855
Hame-fastener	A. B. and S. A. Woodard	Alfred Centre and Hornellsville, N. Y.	July 14, 1868	79, 930
Hame-fastener	I. B. Woolsey	Bloomfield, Iowa	May 16, 1865	47, 762
Hame-fastening	W. J. Alexander	Manchester, Iowa	Dec. 11, 1866	60, 319
Hame-fastening	E. Covert	Farmer Village, N. Y	July 1, 1873	140, 475
Hame-fastening	W. Fawcett	New York, N. Y	Mar. 2, 1869	87, 402
Hame-fastening	H. A. Fowler	East Guilford, N. Y	Jan. 27, 1857	16, 475
Hame-fastening	H. B. Grumling	Grant, Pa	Jan. 10, 1871	110, 848
Hame-fastening	W. W. Kittleman	Bloomfield, Iowa	May 16, 1865	47, 735
Hame-fastening	J. McLain	Saint Mary's, Ohio	Mar. 15, 1864	41, 962
Hame-fastening	C. Morgan	Waumandee, Wis	Nov. 10, 1868	83, 987
Hame-fastening	L. D. G. Niles	Helena, Ark	Sept. 20, 1870	107, 528
Hame-fastening	M. R. Shalters and T. Catern	Alliance, Ohio	Dec. 18, 1866	60, 574
Hame-fastening	T. Taylor	Loudoun County, Va	Mar. 11, 1835	....
Hame-fastening	J. B. Tinkelpaugh	Hastings, Minn	May 10, 1864	42, 651
Hame-fastening	E. Turner	Baltimore, Md	Aug. 8, 1854	11, 501
Hame fastening, Harness	H. Compton	Wells' Corner, Pa	Feb. 3, 1857	16, 530
Hame fastening, Harness	T. Taylor	Loudoun County, Va	Jan. 15, 1850	7, 030
Hame guard, Harness	C. H. Allen	Saint Louis, Mo	Mar. 4, 1873	136, 403
Hame, Harness	M. E. Abbey	Sardis, Miss	Oct. 8, 1872	131, 986
Hame, Harness	J. E. Brown	Lansingburgh, N. Y	Nov. 28, 1865	51, 256
Hame, Harness	W. H. Bustin	Watertown, Mass	Apr. 27, 1869	89, 288
Hame, Harness	D. Clemons	Scranton, Pa	Oct. 13, 1868	83, 041
Hame, Harness	A. Dietz	New York, N. Y	Apr. 2, 1850	7, 235
Hame, Harness	C. H. Drury	Oceola, Ill	Oct. 10, 1871	119, 697
Hame, Harness	W. Duncan	Spring Hill, Ind	Aug. 16, 1870	106, 341
Hame, Harness	D. Foreman	Milton, Ill	Apr. 2, 1872	125, 278
Hame, Harness	K. Frazer	Fayetteville, N. Y	Sept. 26, 1846	4, 773
Hame, Harness	J. Holt and S. G. Cheever	Chelsea and Boston, Mass	Apr. 30, 1867	64, 222
Hame, Harness	R. W. Jones	Syracuse, N. Y	Feb. 6, 1872	123, 401
Hame, Harness	J. Letchworth	Buffalo, N. Y	Apr. 19, 1870	102, 138
Hame, Harness	J. Low	New Britain, Conn	Apr. 9, 1850	7, 265
Hame, Harness	C. Pope	Syracuse, N. Y	Nov. 6, 1849	6, 850
Hame, Harness	C. Robinson	Eau Claire, Wis	Mar. 25, 1873	137, 241
Hame, Harness	J. Thornton and E. G. Latta	Wellsville, N. Y	Apr. 18, 1871	113, 945
Hame, Harness	S. G. Tufts	Mainville, Ohio	July 30, 1867	67, 379
Hame, Harness	P. B. Watson	Belvidere, N. J	Aug. 22, 1871	118, 310
Hame hook and clevis	A. Strever	Albany, N. Y	Jan. 26, 1869	86, 188
Hame, Horse	N. Post	Madrid, N. Y	June 15, 1844	3, 634
Hame, Horse	J. K. Slater and S. G. Pratt	Boston, Mass	Sept. 20, 1844	3, 754
Hame, Horse-collar	J. E. Brown	Lansingburgh, N. Y	Apr. 10, 1866	53, 916
Hame, Horse-collar	M. Drew	Saint Paul, Minn	Sept. 18, 1860	30, 052
Hame-lock	J. A. Stansbury	Marion, Iowa	Apr. 7, 1868	76, 356
Hame-making machine	H. Burt and J. T. Hedden	Newark City, N. J	Feb. 17, 1857	16, 638
Hame-ring	J. Letchworth	Buffalo, N. Y	Mar. 9, 1869	87, 686
Hame-staple	J. Letchworth	Buffalo, N. Y	Feb. 6, 1872	123, 350
Hame-staple socket	R. J. Alger	Kalamazoo, Mich	Mar. 18, 1873	136, 803
Hame-strap	M. T. Briggs	Schoolcraft, Mich	June 4, 1867	65, 334
Hame-strap	G. Paddington and W. F. Crew	Waubeek, Iowa	June 15, 1869	91, 258
Hame-tug	J. C. Covert	Townsendville, N. Y	Dec. 10, 1867	71, 993
Hame-tug	L. Hall	Henrietta, Mich	July 7, 1863	39, 143
Hame-tug	J. Hovey	Bedford, Mich	May 8, 1860	28, 178
Hame-tug	J. McInnes	Waverly, Ill	Sept. 11, 1866	57, 944
Hame-tug	R. B. Whipple	Cleveland, Ohio	Apr. 20, 1852	8, 896
Hame-tug and breast-collar	A. McMullen	Sterling, Ill	Nov. 27, 1866	60, 031
Hame tug and buckle	L. W. Heelan	Petersburgh, Ill	Sept. 11, 1866	57, 899
Hame tug and buckle	J. S. Topham	Washington, D. C	June 12, 1860	28, 700
Hame tug fastener	E. D. Lockwood	Churchville, N. Y	May 18, 1858	20, 278
Hame-tug fastening	J. E. Ball	Newark, N. J	May 18, 1858	20, 246
Hame-tug fastening	J. W. Church	Cold Water, Mich	Dec. 18, 1860	30, 911
Hame-tug fastening	W. J. Lockwood	Sturgis, Mich	Sept. 29, 1857	18, 290
Hame tug, Harness	N. Botsford	Somerset, N. Y	June 14, 1870	104, 106
Hame tug, Harness	P. R. Dawson	Brenham, Tex	Aug. 15, 1871	117, 994
Hame tug, Harness	S. A. Summers	Trappe, Md	Aug. 27, 1872	130, 955
Hame-tug, Metallic	J. M. Curran and J. C. Baxter	Washington, D. C	Oct. 1, 1867	69, 413
Hames and collars, Stretching and blocking	M. Eddy	Adams, N. Y	Apr. 22, 1835	....
Hames and means of hitching horses to vehicles, Appliance to.	J. L. Kreider	Chestnut Level, Pa	June 1, 1869	90, 669
Hames, Breast-loop for	W. B. Hayden	Columbus, Ohio	Sept. 7, 1869	94, 597
Hames, Mode of attaching hold-back rings to	W. B. Hayden	Columbus, Ohio	July 26, 1870	105, 679
Hames to supersede use of collars, Wooden	S. Norton	East Bloomfield, N. Y	June 12, 1835	....
Hammer	S. Anderson	Garrettsville, N. Y	Aug. 20, 1845	4, 155
Hammer	C. Ball	Augusta, Mich	Sept. 9, 1862	36, 389
Hammer	R. Boeklen	Brooklyn, N. Y	Aug. 28, 1860	29, 760
Hammer	R. Boeklen and G. W. Schramm	Brooklyn and New York, N. Y.	Nov. 4, 1862	36, 829
Hammer	H. Cheney	Little Falls, N. Y	July 2, 1867	66, 298
Hammer	J. P. Clark	Portland, Me	Oct. 19, 1858	21, 820
Hammer	T. S. Coffin	Harrington, Me	Oct. 27, 1868	83, 463
Hammer	D. T. Crockett	Newark, N. J	Dec. 18, 1866	60, 482
Hammer	R. Devereux	Buffalo, N. Y	Mar. 26, 1867	63, 228
Hammer	J. H. Godwin	Scotland Neck, N. C	June 23, 1868	79, 066
Hammer	C. Hammond	Philadelphia, Pa	Jan. 19, 1847	4, 934
Hammer	P. C. Havely and W. W. Coggshall.	Rensselaerville, N. Y	June 30, 1868	79, 346

Index of patents issued from the United States Patent Office from 1790 *to* 1873, *inclusive*—Continued.

Invention.	Inventor.	Residence.	Date.	No.
Hammer	G. W. Hubbard	Lowville, N. Y	Sept. 29, 1868	82, 621
Hammer	J. H. Littlefield	Cambridge, Mass	Sept. 12, 1865	49, 898
Hammer	G. H. Mills and J. M. Hanscom	Boston, Mass	July 15, 1862	35, 885
Hammer	C. Monson	New Haven, Conn	Oct. 3, 1865	50, 262
Hammer	J. P. Radley	Albany, N. Y	Feb. 8, 1870	99, 705
Hammer	O. Rock	Hudson, Mass	May 31, 1870	103, 660
Hammer	O. Shepard	Alton, Ill	Feb. 18, 1868	74, 615
Hammer	S. B. Smith	New Haven, Conn	Dec. 7, 1869	97, 715
Hammer	H. H. W. Wright	South Boston, Mass	Mar. 17, 1868	75, 615
Hammer	G. M. Young	El Paso, Ill	May 16, 1871	115, 008
Hammer	W. Zimmerman	Quincy, Ill	Nov. 10, 1868	83, 897
Hammer and anvil-stock, Trip	J. C. Higgins	Skowhegan, Me	Oct. 4, 1870	108, 023
Hammer and anvil, Trip	D. A. Morris	Pittsburgh, Pa	Oct. 5, 1858	21, 691
Hammer and carpet-stretcher, Tack	A. Hicks	Factoryville, N. Y	June 28, 1864	43, 309
Hammer and drop, Adjustable	T. P. Keeler	Worcester, Mass	May 26, 1868	78, 291
Hammer and mallet	W. S. McNeil	Springfield, Mass	Sept. 8, 1868	82, 017
Hammer and nail-holder combined	R. W. Green	Bradford, Pa	Nov. 24, 1868	84, 351
Hammer and turning nipple locks, Concealed	A. Wurfflein	Philadelphia, Pa	Dec. 18, 1849	6, 964
Hammer, Atmospheric	W. D. Grimshaw	Birmingham, England	Jan. 10, 1865	45, 896
Hammer, Atmospheric	M. Peck	New Haven, Conn	Apr. 17, 1860	27, 925
Hammer, Atmospheric	J. Robertson	New York, N. Y	May 1, 1866	54, 413
Hammer, Atmospheric	H. Shattuck	Hamden, Conn	Aug. 22, 1865	49, 593
Hammer, Atmospheric trip	B. Hotchkiss	New Haven, Conn	Sept. 15, 1863	39, 924
Hammer, Automatic steam	C. Sellers	Philadelphia, Pa	Oct. 22, 1872	132, 374
Hammer, Automatic steam	W. Sellers	Philadelphia, Pa	Oct. 22, 1872	132, 375
Hammer, Blacksmith's	G. B. Cubberley	Milwaukee, Wis	Feb. 25, 1873	136, 139
Hammer, Blacksmith's spring	J. Rainey	Orange County, N. C	Mar. 7, 1827	....
Hammer, Bush	J. W. Maloy	Boston, Mass	Dec. 7, 1869	97, 663
Hammer, Bush	W. C. Peckham	Troy, Ohio	Oct. 29, 1872	132, 539
Hammer, Bush	A. Wheeler	Gloucester, Mass	Nov. 6, 1866	59, 491
Hammer, Bush	E. J. Worcester	Worcester, Mass	May 13, 1873	138, 975
Hammer, Carpenter's	J. O. Montignani	Albany, N. Y	Feb. 28, 1865	46, 574
Hammer, Cast-iron	T. Jones	Glastenbury, Conn	July 31, 1828	....
Hammer, Claw	L. P. Edmeston	New York	Aug. 1, 1848	5, 697
Hammer claw-die	H. Hammond	Hartford, Conn	Nov. 19, 1867	70, 998
Hammer, Compressed-air forge	C. Vogel	New York, N. Y	Aug. 30, 1870	106, 892
Hammer, Steam-cylinder	R. Mitchell	Wolverhampton, Great Britain.	Nov. 5, 1867	70, 595
Hammer, Dental	J. C. Dean	Chicago, Ill	July 11, 1865	48, 708
Hammer-die	H. W. Bailey	Springfield, Mass	Mar. 17, 1868	75, 662
Hammer-die	J. Yerkes	Philadelphia, Pa	Aug. 19, 1873	142, 070
Hammer, Driect-action steam	J. H. Towne	Philadelphia, Pa	Sept. 3, 1850	7, 623
Hammer, Drop	J. Aughe	Dayton, Ohio	July 3, 1866	55, 976
Hammer, Drop	J. Blackadder	New Orleans, La	Aug. 6, 1872	130, 183
Hammer, Drop	J. E. Crisp	Charlestown, Mass	June 11, 1872	127, 853
Hammer, Drop	J. Evans	New Haven, Conn	Apr. 25, 1865	47, 405
Hammer, Drop	B. Hotchkiss	New Haven, Conn	Mar. 1, 1864	41, 775
Hammer, Drop	E. Kaylor	Pittsburgh, Pa	Oct. 12, 1869	95, 693
Hammer, Drop	J. McGeorge	Bellaire, Ohio	July 8, 1873	140, 717
Hammer, Drop	L. H. Olmsted	Stamford, Conn	July 16, 1867	66, 879
Hammer, Drop	E. K. Root	Hartford, Conn	Aug. 16, 1853	9, 941
Hammer, Drop	N. C. Stiles	Middletown, Conn	Aug. 27, 1872	130, 877
Hammer, Drop	J. Tobin	Newark, N. J	Dec. 16, 1873	145, 601
Hammer, Drop	W. H. Waters	Springfield, Mass	Oct. 1, 1867	69, 518
Hammer, Drop	J. Wool	Boston, Mass	Apr. 21, 1868	77, 155
Hammer, Fly or trip	J. Hines	Providence, R. I	June 16, 1810	....
Hammer, Foot-trip	E. Pierce and J. Hathaway	Pultney, N. Y	Oct. 19, 1827	....
Hammer, Foot-trip	C. V. Statler and G. W. Wilson.	Walnut Grove, Ill	Jan. 11, 1859	22, 589
Hammer for bending couplings	J. T. Bruen	New York, N. Y	June 26, 1866	55, 813
Hammer for cutting and dressing stone	J. Richards	Braintree, Mass	Feb. 20, 1828	....
Hammer for forging bloom	G. D. Manley	Cogan Station, Pa	Nov. 6, 1866	59, 425
Hammer for pecking or working mill and other stones, Peck.	B. Gardner	Ashfield, Mass	Aug. 3, 1838	873
Hammer for planishing sheet-metal	W. D. Wood	McKeesport, Pa	Sept. 9, 1873	142, 754
Hammer for plating out scythes, Gage to be attached to the anvil of trip.	B. Richardson	Western, Mass	May 17, 1814	....
Hammer, Forge	J. Comstock	New London, Conn	Apr. 24, 1855	12, 770
Hammer, Forge	E. Paye	New York, N. Y	Oct. 9, 1860	30, 343
Hammer, Forge	B. Shiverick	Pittsburgh, Pa	Nov. 16, 1858	22, 092
Hammer, Forge	E. Spaulding	Worcester, Mass	Nov. 22, 1864	45, 187
Hammer, Forge	B. Walker	Green Point, N. Y	Dec. 3, 1867	71, 823
Hammer, Forge and trip	J. Goulding	Leicester, Mass	Mar. 1, 1811	....
Hammer, Forge arm	J. Sharp	Colerain, Pa	June 14, 1834	....
Hammer, Forging	S. B. Dodge	Roslyn, N. Y	Jan. 1, 1869	90, 642
Hammer, Forging	W. D. Grimshaw	Newark, N. J	May 20, 1862	35, 309
Hammer, Forging	L. Wetherell	Boston, Mass	Mar. 12, 1867	62, 793
Hammer forging-die	T. Daffin	Philadelphia, Pa	Oct. 7, 1873	143, 406
Hammer, Furrowing	L. Sauers	Mount Joy, Pa	Apr. 24, 1866	54, 258
Hammer, Hand	A. Gregory	Washington, D. C	Nov. 16, 1858	22, 073
Hammer-handle	T. Phillips	Ann Arbor, Mich	Apr. 9, 1861	31, 997
Hammer-handle	D. Weiser	Philadelphia, Pa	May 19, 1868	78, 164
Hammer, Hatchet	J. Howe	Princeton, Mass	July 8, 1834	....
Hammer-head	R. Black	Holyoke, Mass	July 16, 1867	66, 781
Hammer-head	R. Dawes	Washington, D. C	Oct. 19, 1858	21, 823
Hammer, Lasting	J. W. Warner	Dover, N. H	Oct. 19, 1869	96, 061
Hammer lifter, Drop	F. M. Hodge	Shelburne Falls, Mass	Jan. 3, 1871	110, 655
Hammer lifter, Drop	C. Peck	New Haven, Conn	Sept. 9, 1873	142, 723
Hammer, Machine	D. Noyes	Abington, Mass	Oct. 25, 1853	10, 170
Hammer, Meat	A. H. Brown	Springfield, Vt	Apr. 30, 1867	64, 279
Hammer, Nail	C. Carlisle	Woodstock, Vt	Oct. 16, 1860	30, 386
Hammer, Nail	H. Cheney	Little Falls, N. Y	July 4, 1871	116, 553
Hammer, Nail	C. G. Dodge, jr	Providence, R. I	July 10, 1866	56, 191
Hammer, Nail	E. S. Morton	Plymouth, Mass	June 7, 1870	103, 914
Hammer, Nail	G. Selsor	Philadelphia, Pa	Mar. 19, 1867	63, 106
Hammer, Nail	H. M. Stocum	Painted Post, N. Y	Dec. 5, 1871	121, 682
Hammer, Nail	W. G. Ward	Savona, N. Y	July 31, 1866	56, 838
Hammer, Portable trip	S. Kilburn	Sterling, Md	Aug. 18, 1829	....
Hammer, Power	I. Althouse	Columbus, Ohio	Oct. 17, 1871	120, 017

Index of patents issued from the United States Patent Office from 1790 *to* 1873, *inclusive*—Continued.

Invention.	Inventor.	Residence.	Date.	No.
Hammer, Power	J. C. Butterfield	Chicago, Ill	Oct. 28, 1873	144, 058
Hammer, Power	J. C. Butterfield	Chicago, Ill	Oct. 28, 1873	144, 059
Hammer, Power	A. Cunningham and A. Sharp	Salem, Ohio	July 2, 1867	66, 222
Hammer, Power	T. B. Harrison	Maquoketa, Iowa	Sept. 10, 1867	68, 740
Hammer, Power	M. Hunkley	Rochester, N. Y	Aug. 27, 1867	68, 301
Hammer, Power	J. M. Long	Hamilton, Ohio	Feb. 25, 1873	136, 253
Hammer, Power	J. Palmer	Concord, N. H	Jan. 9, 1872	122, 647
Hammer, Power	S. Pennock	Kennett's Square, Pa	Mar. 25, 1873	137, 092
Hammer, Power	T. F. Preston	Pawtucket, R. I	May 7, 1867	64, 567
Hammer, Power	T. T. Prosser	Chicago, Ill	Apr. 13, 1869	88, 980
Hammer, Power	T. Scott and J. Clarridge	Madison Mills and Pancoastburgh, Ohio.	Sept. 10, 1867	68, 797
Hammer, Power	T. Shaw	Philadelphia, Pa	Feb. 27, 1866	52, 894
Hammer, Power	T. Shaw	Philadelphia, Pa	Nov. 17, 1868	84, 221
Hammer, Power	A. Strube	Frederick, Md	Mar. 24, 1868	75, 808
Hammer, Revolving	G. Stacy	Nanuet, N. Y	Nov. 18, 1873	144, 638
Hammer, Riveting	H. A. Seymour and W. H. Nettleton.	Bristol, Conn	May 10, 1870	102, 871
Hammer saw-set, Trip	W. Duesler	Saint Joseph, Mo	Sept. 28, 1869	95, 331
Hammer, screw-driver, and wrench, Combined	W. D. Gold	Philadelphia, Pa	May 14, 1867	64, 662
Hammer shafts or stamps, Device for lubricating, cooling, and washing vertical.	Z. E. Coffin	Newton, Mass	Sept. 6, 1870	107, 005
Hammer, Shoemaker's	A. Clarke	Boston, Mass	Mar. 21, 1871	112, 902
Hammer, Shoemaker's	A. and A. Clarke	Philadelphia, Pa	July 2, 1872	128, 463
Hammer, Shoemaker's	J. Vigeant	Marlborough, Mass	Jan. 7, 1868	73, 141
Hammer, Spring-balance	J. Collins	Parma, Ohio	Apr. 27, 1869	89, 292
Hammer standards, Construction of steam	O. C. Ferris and F. B. Miles	Philadelphia, Pa	Apr. 4, 1871	113, 508
Hammer, Steam	T. Beach	Freeport, Pa	Sept. 4, 1860	29, 854
Hammer, Steam	P. Danvers	New York, N. Y	Sept. 14, 1858	21, 489
Hammer, Steam	O. C. Ferris and F. B. Miles	Philadelphia, Pa	July 5, 1870	104, 948
Hammer, Steam	D. Joy	Middlesboro', Great Britain.	Aug. 4, 1868	80, 550
Hammer, Steam	E. L. Kinsley	Cambridge, Mass	May 8, 1866	54, 560
Hammer, Steam	L. Kirk	Reading, Pa	Sept. 19, 1848	5, 774
Hammer, Steam	J. L. L. Morris	Reading, Pa	Jan. 31, 1854	10, 479
Hammer, Steam	R. Morrison	Newcastle-upon-Tyne, England.	Oct. 29, 1861	33, 595
Hammer, Steam	D. R. Quick and G. A. Gardner	New York, N. Y	Aug. 12, 1873	141, 667
Hammer, Steam	W. Sellers	Philadelphia, Pa	Oct. 22, 1872	132, 373
Hammer, Steam	T. Sumner	Paterson, N. J	June 27, 1854	11, 182
Hammer, Steam	R. R. Taylor	Reading, Pa	Jan. 1, 1861	31, 041
Hammer, Steam	P. L. Weimer	Reading, Pa	Jan. 31, 1854	10, 486
Hammer, Steam	C. W. and J. P. Willard	Chicago, Ill	Oct. 4, 1864	44, 573
Hammer, Steam drop	F. B. Miles	Philadelphia, Pa	Apr. 1, 1873	137, 313
Hammer, Steam trip	L. Kirk	Reading, Pa	Apr. 3, 1847	5, 044
Hammer, Stone	G. H. Pierce	Middlebury, Mass	Dec. 17, 1834	....
Hammer-strap	W. J. Lewis	Pittsburgh, Pa	Oct. 4, 1870	107, 929
Hammer-straps for wagons, Blank for	W. J. Lewis	Pittsburgh, Pa	Oct. 4, 1870	107, 931
Hammock-support	F. Park	Deerfield, Mass	Nov. 4, 1873	144, 219
Hammock-support	O. Tufts	Boston, Mass	Oct. 28, 1873	144, 168
Hammer, Swage spring	W. Field	North Providence, R. I	Sept. 14, 1833	....
Hammer-swaging die	H. Hammond	Hartford, Conn	Nov. 19, 1867	70, 997
Hammer, Tack	G. J. Capewell	Cheshire, Conn	Nov. 25, 1873	144, 953
Hammer, Tack	J. Crandell and N. P. Braman	Chicopee Falls, Mass., and Bridgeport, Conn.	Jan. 5, 1869	85, 568
Hammer, Tack	T. Evans	Newark, N. J	Nov. 26, 1867	71, 290
Hammer, Tack	A. Pond	Southington, Conn	June 19, 1866	55, 704
Hammer, Tack	T. B. Stout	Keyport, N. J	June 11, 1867	65, 706
Hammer, Tilt	H. Baines	Toronto, Canada	Aug. 22, 1871	118, 181
Hammer, Tilt	P. Breen	Auburn, N. Y	Oct. 31, 1871	120, 412
Hammer, Tilt	J. F. Gould	....	Mar. 8, 1806	....
Hammer, Trip	O. Ames	Plymouth, Mass	Apr. 25, 1810	....
Hammer, Trip	J. Beal	Ontario County, N. Y	Oct. 5, 1811	....
Hammer, Trip	J. Briggs	Louisville, Ky	Oct. 27, 1863	40, 390
Hammer, Trip	H. Bushnell	New Haven, Conn	Mar. 3, 1857	16, 714
Hammer, Trip	J. C. Butterfield and J. Hay	Chicago, Ill	Oct. 18, 1870	108, 326
Hammer, Trip	E. Downer	De Ruyter, N. Y	Apr 11, 1810	....
Hammer, Trip	H. E. Fessel and F. Krautwadel	Chicago, Ill	Feb. 7, 1865	46, 229
Hammer, Trip	J. C. Forrest and G. Baker	Schenectady, N. Y	Dec. 14, 1852	9, 462
Hammer, Trip	B. Hershey	Erie, Pa	Feb. 21, 1871	112, 036
Hammer, Trip	B. Hotchkiss	New Haven, Conn	June 14, 1859	24, 428
Hammer, Trip	B. Hotchkiss	New Haven, Conn	May 3, 1864	42, 581
Hammer, Trip	D. Howell	Louisville, Ky	Apr. 10, 1860	27, 804
Hammer, Trip	B. Hughes	Rochester, N. Y	May 16, 1854	10, 923
Hammer, Trip	L. Kinsley	Cambridgeport, Mass	Feb. 10, 1863	37, 628
Hammer, Trip	J. Kreshbiel	Williamsville, N. Y	July 25, 1871	117, 432
Hammer, Trip	S. Morrill	Readfield, Mass	Sept. 28, 1812	....
Hammer, Trip	J. Neuberger	Buffalo, N. Y	Oct. 15, 1872	132, 173
Hammer, Trip	E. Osborn	Berkshire County, Mass	Jan. 8, 1810	....
Hammer, Trip	U. Palmer	Spencer, N. Y	July 31, 1815	....
Hammer, Trip	M. Peck	New Haven, Conn	July 3, 1855	13, 179
Hammer, Trip	J. W. Peer	Schenectady, N. Y	Nov. 29, 1853	10, 274
Hammer, Trip	P. P. Read	Mercer, Me	Dec. 7, 1826	....
Hammer, Trip	H. Redfield	Grafton, Mass	Mar. 5, 1835	....
Hammer, Trip	T. J. Root	Galena, Ill	Feb. 7, 1865	46, 267
Hammer, Trip	T. J. and R. L. Root	Andover, Ohio	Aug. 20, 1867	67, 913
Hammer, Trip	L. Rosenkrans	Big Flats, N. Y	Mar. 17, 1828	....
Hammer, Trip	J. Smith	Otsego County, N. Y	Oct. 14, 1809	....
Hammer, Trip	J. Tandler	Grand Rapids, Mich	Aug. 27, 1867	68, 295
Hammer, Trip	W. Van Anden	Poughkeepsie, N. Y	Aug. 16, 1853	9, 942
Hammer, Trip	C. Vogel	New York, N. Y	Feb. 25, 1868	74, 868
Hammer, Trip	P. L. Weimer	Reading, Pa	Sept. 25, 1855	13, 608
Hammer, Vertical trip	P. Stebbins and J. Holmes	Schenectady, N. Y	June 1, 1852	8, 989
Hammer, &c., with craw and peen, Nail	C. A. Strong	South Glastonbury, Conn	Mar. 3, 1829	....
Hammers and hatchets, Mode of manufacturing socket.	P. Eastman	Canaan, N. H	July 17, 1839	1, 247
Hammers and stamps, Operating	C. K. James	Jersey City, N. J	June 19, 1866	55, 772
Hammers, Arrangement of valves, ports, and passages for operating steam.	R. R. Taylor	Reading, Pa	Nov. 29, 1853	10, 276

Index of patents issued from the United States Patent Office from 1790 *to* 1873, *inclusive*—Continued.

Invention.	Inventor.	Residence.	Date.	No.
Hammers, Cylinder for steam	T. R. Morgan	Pittsburgh, Pa	Apr. 23, 1872	125, 976
Hammers, Machine for forging	W. Evans	Long Island City, N. Y	Nov. 25, 1873	144, 969
Hammers, Machine for making	R. B. Perkins	Meriden, Conn	Oct. 13, 1857	18, 408
Hammers, Making	J. Yerkes	Fox Chase, Pa	Feb. 12, 1867	62, 105
Hammers, Manner of working forge	G. E. Sellers	Cincinnati, Ohio	Jan. 10, 1845	3, 882
Hammers, Manufacture of tack	T. A. Conklin	New Britain, Conn	Dec. 10, 1867	71, 986
Hammers, Method of adjusting the stroke of trip	L. Briggs, jr	Braintree, Mass	Mar. 11, 1851	7, 965
Hammers, Mode of hanging forge	J. McNeel	Coleman Township, Pa	Jan. 21, 1829	....
Hammers, Moving trip	E. Crowell	Gardiner, Me	Aug. 4, 1826	....
Hammers, Operating blacksmith's	J. W. Kerr	Rochester, N. Y	Apr. 20, 1858	19, 997
Hammers, Operating steam trip	J. S. Bonney and C. W. Willard	Hanson and Bridgewater, Mass.	Aug. 17, 1858	21, 183
Hammers, spikes, gimlets, &c., Machine for shaping	E. L'Hommedieu	Saybrook, Conn	Apr. 28, 1830	....
Hammers to helves, Attachment of	D. Hicks	Duncansville, Pa	Apr. 2, 1850	7, 242
Hammers, Valve arrangement for steam	J. Watt	South Boston, Mass	Dec. 6, 1853	10, 297
Hammers, Valve-gear for steam	F. B. Miles	Philadelphia, Pa	July 21, 1868	80, 082
Hammers with perpendicular handles, Machine for working trip.	J. Fox, jr	Middlesex, Mass	Feb. 25, 1807	....
Hammock	W. H. Forbush	Buffalo, N. Y	Nov. 5, 1861	33, 678
Hammock	A. Woods	Liverpool, England	Sept. 17, 1867	68, 927
Hammock, Knapsack	G. Abbott	United States Army	Feb. 2, 1864	41, 418
Hammock, Knapsack	A. W. Süs	New York, N. Y	Mar. 8, 1864	41, 870
Hammock-supporter	J. M. K. Southwick	Newport, R. I	May 23, 1871	115, 128
Hammocks, Device for suspending	O. Tufts	Boston, Mass	Sept. 30, 1873	143, 392
Hams, Apparatus for extracting marrow from	W. W. Macqueen	Saint Louis, Mo	Mar. 26, 1872	125, 062
Hams, Composition for covering	H. A. Amelung	New York, N. Y	Dec. 8, 1863	40, 803
Hams, Composition for covering	H. Billings	Beardstown, Ill	Apr. 9, 1850	7, 256
Hams, Holder for slicing	J. Baumgartner and L. Angster	Newark, N. J	Nov. 12, 1867	70, 780
Hams, Putting up	S. E. Kelly	Philadelphia, Pa	Mar. 7, 1871	112, 467
Hand and arm, Artificial	J. S. Drake	Boston, Mass	July 22, 1856	15, 372
Hand, Apparatus for development of the muscles of the.	D. P. D. Kops	New York, N. Y	Nov. 28, 1871	121, 289
Hand, Artificial	O. Lindsay and I. Vance	Washington, D. C	Sept. 19, 1865	50, 014
Hand, Artificial	J. F. Maguire	East Boston, Mass	July 17, 1866	56, 427
Hand, Artificial	W. Selpho	New York, N. Y	Aug. 18, 1857	18, 021
Hand, Artificial	W. Selpho and J. Walber	New York, N. Y	Dec. 6, 1859	26, 378
Hand clamp	S. L. Thompson	Lowell, Mass	May 7, 1872	126, 425
Hand-drill	E. Beeman	Owego, N. Y	Aug. 12, 1862	36, 135
Hand-drill	W. Bushnell	New York, N. Y	Dec. 2, 1851	8, 554
Hand-drill	W. C. Burch	Gloucester, N. J	May 26, 1868	78, 183
Hand-drill	L. Hillebrand	Philadelphia, Pa	Mar. 19, 1872	124, 681
Hand-drill	J. Johnson	Somerville, Mass	Oct. 31, 1848	5, 894
Hand-drill	F. McNair	Fultonham, Ohio	Nov. 16, 1858	22, 085
Hand-drill	S. H. Ostrom	Schenectady, N. Y	May 18, 1869	90, 292
Hand-drill	H. H. Packer	Boston, Mass	June 29, 1858	20, 728
Hand-drill	J. H. Parker	Boston, Mass	May 15, 1860	28, 327
Hand-drill	J. H. Parker	Boston, Mass	May 15, 1860	28, 328
Hand-drilling machine	J. E. Hunter	North Adams, Mass	Apr. 5, 1870	101, 466
Hand-engine	C. T. Ulmann and M. Beckman	New York and Brooklyn, N. Y.	Mar. 17, 1868	75, 657
Hand-fork	F. W. Palmer	West Richmondville, N. Y	May 12, 1868	77, 906
Hand griping-tool	E. A. Alpress	Bristol, Conn	Dec. 31, 1867	72, 712
Hand-hole plate, Crab for	J. S. Hooton	New Carlisle, Ind	July 12, 1870	105, 208
Hand-indicator for showing course of vessels	H. P. Tuttle	Brooklyn, N. Y	Mar. 15, 1870	100, 821
Hand-mill	E. Alsop	New York, N. Y	Sept. 1, 1868	81, 574
Hand-motor, Mechanical	H. Fowler	Washington, D. C	Sept. 20, 1870	107, 474
Hand-motor, Mechanical	H. Fowler	Washington, D. C	Sept. 20, 1870	107, 475
Hand-preserving compound	J. W. Osborn	Brooklyn, N. Y	Jan. 31, 1871	111, 376
Hand press	D. F. Day	Philadelphia, Pa	Aug. 15, 1854	11, 590
Hand-press, Jeweler's	J. McWilliams	Providence, R. I	Sept. 23, 1873	143, 023
Hand-press, Self-inking	D. Zuern and L. L. Bevan	Shamokin, Pa	Nov. 2, 1858	21, 997
Hand-presses, Rack-motion for	C. Wells	Cincinnati, Ohio	Dec 4, 1866	60, 301
Hand-protector	J Turnbull	Simsbury, Conn	Dec. 19, 1871	121, 974
Hand-rake	C. Bissell	Aurora, Ohio	May 7, 1867	64, 475
Hand-rake	H. M. Clark	Brewer, Me	Sept. 15, 1868	82, 207
Hand-rake	M. Ellis	Greenfield, Mass	Sept. 6, 1870	107, 018
Hand-rake	D. Johnston	Sulphur Springs, Ohio	Aug. 24, 1869	94, 118
Hand-rake	G. W. Stearns	Lebanon, N. H	Feb. 6, 1872	123, 369
Hand-rake	A. Winters	Washington, Pa	Dec. 22, 1868	85, 154
Hand-rake for hay, &c	A. Foster	Auburn, N. Y	Dec. 19, 1827	....
Hand-rake, Revolving	S. C. Rundlett	Portland, Me	Oct. 25, 1864	44, 824
Hand-rest	I. Belford	Caldwell, Ohio	Mar. 4, 1873	136, 357
Hand-shears	J. N. Wallis	Fleming, N. Y	Dec. 27, 1864	45, 659
Hand-shears	C. Witte	Brooklyn, N. Y	Dec. 29, 1868	85, 500
Hand shears and nippers	T. Wallis and T. Witbeck	Scipio, N. Y	Mar. 21, 1865	46, 960
Hand-spring for machinery	P. Howe	Saffordville, Conn	Apr. 30, 1867	64, 318
Hand-support, Penman's	D. A. Sanborn	Brooklyn, N. Y	July 5, 1870	105, 001
Hand-wrench	G. B. Phillips	Albany, N. Y	July 21, 1857	17, 844
Hands, Substitute for artificial	J. Reichenbach	Pittsburgh, Pa	June 27, 1865	48, 440
Handcuff	W. H. Kimball	Augusta, Me	Aug. 7, 1860	29, 495
Handcuff, Adjustable	O. C. Phelps	New York, N. Y	July 17, 1866	56, 463
Handcuff-lock	H. H. Cheney	East Saginaw, Mich	Apr. 4, 1871	113, 499
Handcuff-lock	J. A. Wisner and M. Hoyt	East Saginaw, Mich	July 6, 1869	92, 414
Handkerchief and fan holder	G. D. Stevens	New York, N. Y	July 23, 1872	129, 871

Handle:

See Augur-handle.
Awl-handle.
Ax-handle.
Basket-handle.
Boat-hook handle.
Bone handle.
Bottle-handle.
Brad-awl handle.
Broom-handle.
Brush-handle.
Brush and broom handle.
Burial-casket handle.
Can-handle.
Cane-handle.
Carriage-handle.
Casket-handle.
Chisel-handle.
Coffee and spice mill handle.
Coffin-handle.
Comb-handle.
Cross-cut-saw handle.
Crutch-handle.
Cutlery-handle.
Dipper-handle.

Index of patents issued from the United States Patent Office from 1790 *to* 1873, *inclusive*—Continued.

Invention.	Inventor.	Residence.	Date.	No.
Handle—Continued. *See* Drawer-handle. File-handle. File-griping handle. Fork and spoon handle. Gimlet-handle. Hammer-handle. Implement-handle. Kettle-handle. Knife-handle. Knife and fork handle. Lamp-chimney handle. Latch-handle. Lifting-handle. Milk-can handle. Mop-handle. Mucilage-brush handle. Organ-handle. Pick-handle. Plow-handle. Plow-beam handle. Pocket-knife handle. Pump-handle. Rake-handle. Sadiron-handle. Satchel-handle. Sauce-pan handle. Saw-handle. Screw-driver handle. Sheet-metal-ware handle. Shovel-handle. Smoothing-iron handle. Spade-handle. Tea and coffee pot handle. Tool-handle. Traveling-bag handle. Trunk-handle. Umbrella-handle. Whip-handle.				
Handles, &c., Finishing and bleaching	D. H. Taylor	Westfield, N. Y	Nov. 1, 1870	108, 846
Handles, molds, &c., Composition for making	C. Branwhite	New York, N. Y	Jan. 23, 1846	4, 362
Hanger: *See* Adjustable hanger. Bracket-hanger. Broom-hanger. Car-hanger. Clothes-hanger. Door-hanger. Eave-trough hanger. Garment-hanger. Grindstone-hanger. Harness-check-hook hanger. Journal or shaft hanger. Lamp-hanger. Mirror-hanger. Picture-hanger. Pulley-hanger. Shaft-hanger. Shafting-hanger. Sword-hanger. Tobacco-hanger. Whip-hanger.				
Hanger	P. P. Lane and E. Myers	Cincinnati, Ohio	Oct. 12, 1869	95, 813
Hanger	J. M. Stone	North Andover, Mass	Sept. 17, 1867	69, 041
Hanger-bar socket	T. A. Summers	Rochester, N. Y	Dec. 22, 1863	41, 027
Hanging bracket	H. Polley	San Francisco, Cal	Sept. 10, 1872	131, 298
Hanging bracket or shelf	A. C. Brown	Chicago, Ill	Mar. 18, 1873	136, 900
Hanging gate	E. Secor	Niles, Mich	Sept. 2, 1873	142, 522
Harbor and river channels, Means of defending	R. H. Jewett	Mount Sterling, Ill	Sept. 8, 1863	39, 816
Harbor defense, Submarine	J. S. Gilbert	New York, N. Y	Apr. 7, 1863	38, 106
Harbor obstructions, Method of removing	J. D. Hall	Philadelphia, Pa	Dec. 20, 1864	45, 562
Harbors, &c., Mud-machine for deepening	J. Watchman and J. Bratt	Baltimore, Md	Feb. 18, 1825	....
Hard-rubber molds, Apparatus for filling	J. C. Robie	Binghamton, N. Y	Jan. 2, 1866	51, 865
Harmonicons, Glass or metal	J. Koppe	New York, N. Y	Oct. 2, 1860	30, 230
Harmonicons, Musical glass called the grand	F. H. Smith	Northampton, Va	Apr. 7, 1825	....
Harmonium	J. Gilmour	Glasgow, North Britain	July 18, 1865	48, 881
Harmonium	H. W. Smith	Boston, Mass	July 3, 1860	29, 041
Harmonium	C. F. Uhlig	Chemnitz, Saxony	Aug. 14, 1866	57, 270
Harness	M. E. Abbey	Sardiz, Miss	Oct. 8, 1872	131, 987
Harness	E. S. Basnett	Morgantown, W. Va	Apr. 5, 1870	101, 569
Harness	S. L. Bond	Greenwood, S. C	Oct. 23, 1860	30, 457
Harness	J. Calkins	Hudson, Mich	July 18, 1865	48, 790
Harness	S. G. Cheever	Boston, Mass	June 8, 1869	91, 087
Harness	E. Covert	Farmer Village, N. Y	Aug. 12, 1873	141, 698
Harness	C. Daniels	Fremont, Ohio	Feb. 19, 1867	62, 257
Harness	G. B. Durkee	Alden, N. Y	Oct. 8, 1867	69, 550
Harness	G. W. Dutton	Tomales, Cal	Oct. 17, 1871	120, 049
Harness	I. Ellis	Tyler, Tex	June 25, 1872	128, 377
Harness	J. Fassnacht	New Milltown, Pa	Sept. 13, 1859	25, 401
Harness	E. E. Hardy	New York, N. Y	May 30, 1865	47, 949
Harness	G. M. Harnisch	Chicago, Ill	July 6, 1869	92, 188
Harness	J. K. Harris	Springfield, Ohio	May 26, 1868	78, 282
Harness	J. K. Harris	Springfield, Ohio	June 8, 1869	91, 127
Harness	A. Harroun, jr	South Onondaga, N. Y	Mar. 17, 1868	75, 632
Harness	N. T. Healy	Medina, N. Y	Nov. 5, 1867	70, 434
Harness	R. Huyck	Sheboygan Falls, Wis	Aug. 14, 1866	57, 143
Harness	W. A. Jordan	New Orleans, La	Sept. 1, 1868	81, 788
Harness	A. D. Kendig	Safe Harbor, Pa	Oct. 30, 1866	59, 316
Harness	F. D. Ladenberger	Glenbeulah, Wis	Nov. 21, 1865	51, 062
Harness	A. W. Lawton	Rochester, N. Y	Oct. 31, 1871	120, 445
Harness	J. B. Lindeman	Manor Township, Pa	Mar. 19, 1867	63, 064
Harness	A. McCracken	South Byron, N. Y	May 2, 1871	114, 318
Harness	A. McMullen and J. H. Rock	Sterling, Ill	May 5, 1868	77, 508
Harness	B. Mirick	Rutland County, N. Y	Apr. 21, 1809	....
Harness	F. Monroe	Romeo, Mich	Aug. 24, 1858	21, 267
Harness	R. O'Brien	Fowlersville, N. Y	Apr. 7, 1868	76, 507
Harness	E. C. Patterson	Rochester, N. Y	Nov. 5, 1872	132, 855
Harness	W. Peabody	Dixmont Centre, Me	July 17, 1866	56, 462
Harness	G. Renton	Boston, Mass	May 6, 1873	138, 530
Harness	H. B. Richmond	New Britain, Conn	Aug. 20, 1872	130, 747
Harness	H. T. Robbins	Hyde Park, Mass	Oct. 22, 1872	132, 413
Harness	J. J. Smokey	Natchez, Miss	Feb. 18, 1868	74, 623
Harness	J. C. Spooner	Houlton, Me	Oct. 17, 1871	120, 118
Harness	J. Stanbrough	Newark, N. J	Mar. 22, 1853	9, 628
Harness	D. Waldhauer	New York, N. Y	Nov. 23, 1869	97, 251
Harness	L. Whitehead	Nunda, N. Y	May 17, 1870	103, 264
Harness	W. S. Wood	Haltborough, Pa	Apr. 7, 1868	76, 369
Harness	G. M. Zell	Waynesville, Ohio	Apr. 14, 1868	76, 874
Harness adapted to horse-rakes	W. Parker	Putney, Vt	June 26, 1849	6, 551
Harness and carriage to prevent accident from runaway horses, Making safety.	E. Eastlack	Greenwich, N. J	Jan. 10, 1840	1, 467
Harness and other articles of leather, Composition for dressing.	M. T. Boyd	New York, N. Y	June 1, 1869	90, 813

Index of patents issued from the United States Patent Office from 1790 *to* 1873, *inclusive*—Continued.

Invention.	Inventor.	Residence.	Date.	No.
Harness and saddle pad	G. Elsey	Sterling, Ill	Feb. 11, 1873	135, 790
Harness and thills of vehicles, Safety-apparatus to be applied to.	J. H. Wilson, jr	Nashville, Tenn	Mar. 18, 1856	14, 484
Harness, Attaching rosettes to	W. L. Denio	Rochester, N. Y	Oct. 13, 1868	82, 927
Harness-attachment	W. W. Beebee	Dubuque, Iowa	Jan. 14, 1868	73, 286
Harness-attachment	W. A. Blundell	Saint Louis, Mo	Mar. 21, 1871	112, 889
Harness-attachment	E. Croughwell, A. Tellman, and E. L. Vanderburg.	San Francisco, Cal	Aug. 26, 1873	142, 085
Harness-attachment for hitching and unhitching horses.	T. J. Dobbs	Weehawken, N. J	Oct. 14, 1873	143, 678
Harness-attachment for supporting driving-lines	T. D. Brown	Montville, Ohio	Mar. 1, 1859	23, 073
Harness back-band	A. J. Rutledge	Jefferson Parish, La	July 16, 1872	129, 498
Harness back-band hook	S. Reynolds	Pittsburgh, Pa	July 11, 1871	116, 993
Harness back-pad	J. Umbach	Kankakee, Ill	July 28, 1868	80, 433
Harness-bell	C. T. Liebold	New York, N. Y	Jan. 16, 1866	52, 058
Harness-bell	R. Schmidt	New York, N. Y	July 10, 1866	56, 277
Harness-bit	B. S. Roberts	New Haven, Conn	Nov. 8, 1870	109, 145
Harness, Bitting	B. F. Brewster	Norwich, Conn	July 27, 1869	93, 047
Harness, Blacking for	S. Sherwood	New York, N. Y	May 8, 1866	54, 616
Harness-blinds, Die for shaping	E. Ward	Newark, N. J	July 16, 1872	129, 630
Harness breast-collar	A. B. Morrison	Le Roy, N. Y	Apr. 22, 1873	138, 039
Harness breast-plate	O. Ramsdell	Westfield, Vt	Feb. 19, 1850	7, 106
Harness, Breast-plate for the breast-collar of double	C. Graham	New York, N. Y	Mar. 1, 1870	100, 283
Harness breast-strap attachment	J. Letchworth	Buffalo, N. Y	Oct. 19, 1869	96, 016
Harness, Breast-strap fastening for	J. H. Martin	Columbus, Ohio	June 21, 1870	104, 475
Harness, Breast-strap protector for	L. R. Ward	Ward's Corners, Iowa	Sept. 6, 1870	107, 134
Harness breast-straps, Shield for	M. W. Pond and H. E. Mussey	Elyria, Ohio	Sept. 6, 1864	44, 113
Harness breast-straps, Slide for	D. Pettengill	Delhi, N. Y	Nov. 25, 1862	37, 006
Harness breast-straps, Slide for	M. W. Pond and L. W. Darling	Elyria, Ohio	Jan. 19, 1864	41, 320
Harness breeching and breast plate, Brace for	E. Brickett	Minot, Me	Feb. 14, 1860	27, 098
Harness-breeching holder	F. L. Churchill and R. C. Beach	Minneapolis, Minn	May 16, 1871	114, 925
Harness, Breeching-stay for	W. R. Knowles	Columbiana, Ohio	Apr. 23, 1872	125, 963
Harness-breeching to wagon-thills, Mode of attaching.	A. Parker	Coventry, N. Y	May 3, 1859	23, 854
Harness, bridles, &c., Machine for the manufacture of.	S. Wilson	Dansville, N. Y	May 9, 1846	4, 509
Harness-buckle	J. McLean	Meadville, Pa	Apr. 8, 1873	137, 558
Harness-buckle	I. Roraback	South Bend, Ind	July 7, 1868	79, 779
Harness, Carriage	E. Wilder	South Hingham, Mass	Jan. 7, 1868	73, 149
Harness, Cart	P. K. Curll	Elkridge Landing, Md	Jan. 14, 1868	73, 236
Harness, Cart	E. E. Hardy	New York, N. Y	Sept. 13, 1864	44, 185
Harness-check hook	C. H. Bassett	Derby, Conn	Nov. 17, 1868	84, 080
Harness-check hook	A. B. Buell	Westmoreland, N. Y	Mar. 13, 1844	3, 484
Harness-check hook	H. A. Carlton	Chelsea, Mass	Apr. 22, 1873	138, 003
Harness-check hook and terret	W. V. Kay	Chicago, Ill	May 23, 1871	115, 061
Harness, Check-hook guard for	W. C. Shipherd	Cleveland, Ohio	Oct. 1, 1872	131, 793
Harness-check loop	W. Parsons	Palmyra, N. Y	Mar. 4, 1873	136, 454
Harness-clamp	T. B. Hodge	Francistown, N. H	Feb. 12, 1867	61, 939
Harness-clamp	A. W. Porter and S. A. Drake	Farmer Village, N. Y	Aug. 27, 1872	130, 942
Harness-clasp	M. A. Penn	Washington, D. C	Apr. 15, 1873	137, 954
Harness-clip	J. S. Hays	Williamsport, Pa	July 12, 1870	105, 202
Harness-clip	C. M. Tripp	Pittsburgh, Pa	Feb. 8, 1870	99, 731
Harness-clips, Mode of attaching	R. S. B. Milliken	South Bend, Ind	June 4, 1872	127, 631
Harness coach-pad	A. Gilliam	Pittsburgh, Pa	Aug. 19, 1873	141, 873
Harness, Cock-eye for	S. D. Bingham	Maumee City, Ohio	Jan. 12, 1869	85, 784
Harness, Cock-eye for	J. W. Briggs	Cleveland, Ohio	June 5, 1849	6, 490
Harness, Cock-eye for	T. J. Magruder	Marion, Ohio	Dec. 7, 1869	97, 662
Harness, Cock-eye for	L. Worster	Syracuse, N. Y	Aug. 17, 1869	93, 788
Harness, Crupper for	J. J. Flack	Joliet, Ill	July 20, 1852	9, 133
Harness, Draft	G. Chamberlin	Olean, N. Y	Sept. 13, 1870	107, 335
Harness, Elastic draft-attachment for single and double.	J. Barrow	Cincinnati, Ohio	Sept. 15, 1868	82, 072
Harness-fastening	A. and G. Bernd and A. White	Macon, Ga	Sept. 14, 1869	94, 861
Harness-fastening	E. Covert	Farmer Village, N. Y	July 1, 1873	140, 474
Harness-fastening	W. Hayden	New Milford, Pa	Jan. 21, 1837	111
Harness-fastening	T. Henderson	Harford County, Md	July 27, 1852	9, 149
Harness-fastening	S. S. Little and H. P. Westcott	Macon, Ga	Mar. 22, 1870	101, 141
Harness-fastening	J. Shepard	New Britain, Conn	Apr. 25, 1865	47, 500
Harness for breaking horses	T. Hammond	Petersburgh, Ind	Apr. 14, 1868	76, 626
Harness for carriages and saddlery, Making	B. Foster, D. Washburn, N. Rockwell, and J. O. Walker.	Castleton, Vt	Aug. 16, 1810	....
Harness for horses	J. Palen	Lockport, N. Y	Dec. 14, 1869	97, 799
Harness for horses	J. Rouse	Port Gibson, N. Y	Sept. 27, 1859	25, 587
Harness for horses	H. C. Smith	Dublin, Ind	July 10, 1866	56, 283
Harness for horses	J. Smith	Delaware, Ohio	Feb. 17, 1857	16, 660
Harness for horses, Breast-plate	A. Post	Henrietta, N. Y	Apr. 8, 1840	1, 545
Harness for horses in shafts	R. Beale	Washington, D. C	June 15, 1837	237
Harness for preventing the fore legs of horses from interfering.	D. G. Kettell	Worcester, Mass	Jan. 22, 1861	31, 174
Harness for shoeing horses	J. P. Reynolds	Mirabile, Mo	Jan. 22, 1861	31, 213
Harness for shoeing horses	W. P. Thomas	Hillsborough, Ind	Mar. 18, 1856	14, 477
Harness for vicious horses	S. L. Gray	Chillicothe, Ohio	May 19, 1868	78, 082
Harness-frame, Guide or clamp for	A. S. Villee	Lancaster, Pa	Nov. 12, 1867	70, 922
Harness from horses, Apparatus for detaching	S. Hunt	Baltimore, Md	July 11, 1854	11, 262
Harness from horses, Detaching	G. Yellott	Bel Air, Md	June 15, 1852	9, 044
Harness-hook	S. P. Babcock	Jordan, N. Y	May 10, 1864	42, 634
Harness-hook	G. and A. Bernd	Macon, Ga	Feb. 22, 1870	100, 233
Harness-hook	J. U. Fiester	Winchester, Ohio	Feb. 6, 1872	123, 340
Harness-hook	M. Reilly	Covington, Ky	June 21, 1870	104, 646
Harness-hook	I. C. Richmond	West Meriden, Conn	July 2, 1867	66, 391
Harness-hook	J. Seaman	New York, N. Y	Feb. 15, 1870	99, 957
Harness-hook	W. H. Taylor	Baldwinsville, N. Y	Oct. 25, 1870	108, 739
Harness-hook	A. White	Macon, Ga	Feb. 22, 1870	100, 226
Harness-hook	J. J. Wilkins	Virden, Ill	June 26, 1866	55, 942
Harness hook and bit	A. A. Hotchkiss and E. Garnsey.	Sharon, Conn	Jan. 23, 1834	....
Harness hook and loop shield	W. H. Morgan	Alliance, Ohio	Sept. 6, 1870	107, 087
Harness-hook, Safety	J. L. Dickinson	Dubuque, Iowa	Feb. 4, 1868	73, 958

Index of patents issued from the United States Patent Office from 1790 *to* 1873, *inclusive*—Continued.

Invention.	Inventor.	Residence.	Date.	No.
Harness, Horse yoke	B. F. Baker	Milton, N. Y	Apr. 2, 1867	63, 357
Harness-leather-cutting machine	J. Wehr	Roanoke, Ind	Aug. 15, 1865	49, 459
Harness, Leather-fastening for	T. G. Moore	Albia, Iowa	Jan. 4, 1870	98, 514
Harness-loop	G. L. Du Laney and J. S. Huston.	Mechanicsburgh, Pa	May 24, 1870	103, 311
Harness-loop	F. Hickman	Reading, Pa	Dec. 9, 1873	145, 299
Harness-loop	A. M. Osborn	Girard, Pa	Aug. 15, 1871	118, 046
Harness-loop	G. W. Roland	Salem, Ohio	May 12, 1868	77, 918
Harness-loop	A. Worster	Hannibal, N. Y	June 22, 1869	91, 809
Harness-loops, Clamp for embossing	O. H. Morris	New Haven, Conn	Dec. 7, 1869	97, 675
Harness-maker's bench-iron	J. Thornton and E. G. Latta	Genesee and Friendship, N. Y.	Sept. 16, 1873	142, 967
Harness-maker's clamp	D. Eighme	Chicago, Ill	Sept. 2, 1873	142, 384
Harness-maker's clamp	J. F. Johnson	Monrovia, Ind	Sept. 1, 1868	81, 787
Harness-maker's gage	D. G. Garlock	White Cloud, Kans	Feb. 4, 1873	135, 472
Harness, Metallic layer loops for	R. J. Algeo	Kalamazoo, Mich	Mar. 26, 1872	124, 996
Harness, Mode of dressing side-strap for	J. F. Valentine	Union County, Ohio	June 18, 1867	65, 845
Harness-mounting	T. Fawcett	New York, N. Y	Oct. 21, 1873	143, 817
Harness-mounting	J. F. Knorr	New York, N. Y	Feb. 18, 1873	136, 073
Harness-mounting	S. S. Sargent	Newark, N. J	July 16, 1872	129, 254
Harness-mounting	C. M. Theberath	Newark, N. J	Dec. 3, 1867	71, 819
Harness-mounting	C. H. Thornton	Newark, N. J	Nov. 14, 1871	121, 022
Harness-mountings, Mode of covering	W. Fawcett	New York, N. Y	Dec. 26, 1871	122, 163
Harness-nails, Die for forming metal heads on	F. Reynold and F. L. Hilbright.	Newark, N. J	June 12, 1866	55, 536
Harness-nails, Mode of manufacturing	F. Reynold	Newark, N. J	July 24, 1866	56, 611
Harness-operating mechanism	R. Blakie	Hyde Park, Mass	May 31, 1870	103, 708
Harness or vehicles, Breaking the surge on	F. P. Hart	Strasburgh, Pa	Dec. 29, 1868	85, 307
Harness-ornament	W. Ulrich and C. Hackmeister.	Newark, N. J	May 25, 1869	90, 610
Harness, Ornamenting	W. W. Kitch	Crestline, Ohio	Feb. 13, 1872	123, 709
Harness, Ornamenting	F. Meinberg	New York, N. Y	Mar. 12, 1872	124, 446
Harness-pad	C. Drew	Newark, N. J	Feb. 2, 1869	86, 376
Harness-pad	C. Drew	Newark, N. J	Nov. 9, 1869	96, 563
Harness-pad	J. H. Garrett	Greencastle, Ind	Dec. 12, 1871	121, 863
Harness pad	A. Gilliam	Cincinnati, Ohio	Apr. 30, 1867	64, 215
Harness-pad	G. W. Graves	Chicago, Ill	Sept. 13, 1870	107, 361
Harness-pad	J. D. Groff and K. H. Wilson	Oxford, Pa	Feb. 25, 1873	136, 155
Harness-pad	J. Hosford	Monroeville, Ohio	June 25, 1867	66, 086
Harness-pad	J. Hughes	Newark, N. J	Feb. 18, 1868	74, 691
Harness-pad	J. Hughes	Newark, N. J	June 22, 1869	91, 747
Harness-pad	J. Ives	Mount Carmel, Conn	Apr. 24, 1860	27, 984
Harness-pad	J. F. Knorr	New York, N. Y	Oct. 22, 1872	132, 472
Harness-pad	J. P. Lovett and F. P. Lefevre.	Oxford, Pa	Dec. 10, 1872	133, 786
Harness pad	J. Maclure	Newark, N. J	Dec. 24, 1867	72, 514
Harness-pad	S. A. Marker	Indianapolis, Ind	Nov. 25, 1873	144, 853
Harness-pad	M. W. Pond	Elyria, Ohio	Feb. 21, 1871	112, 073
Harness-pad	M. W. Pond	Elyria, Ohio	July 1, 1873	140, 384
Harness-pad	M. W. Pond	Elyria, Ohio	July 1, 1873	140, 385
Harness-pad	W. A. Reddick	Niles, Mich	Sept. 30, 1873	143, 253
Harness-pad	H. R. Ridgley	Mansfield, Ohio	Jan. 14, 1868	73, 389
Harness-pad	W. H. Rutan	Newark, N. J	Nov. 26, 1872	133, 338
Harness-pad	R. M. Selleck	New York, N. Y	June 15, 1858	20, 588
Harness-pad	H. C. Swift	Fond du Lac, Wis	Feb. 16, 1869	87, 078
Harness-pad	J. H. Van Riper	Chelsea, Mich	Mar. 26, 1872	125, 101
Harness-pad	P. H. Wiedersum	New York, N. Y	Feb. 27, 1872	124, 185
Harness-pad block	W. H. Rannels	Oakland Mills, Pa	June 4, 1867	65, 433
Harness-pad block	P. Shaw	Syracuse, N. Y	Dec. 3, 1867	71, 653
Harness-pad block	J. Zimmer, jr	Half Day, Ill	Nov. 16, 1869	96, 860
Harness-pad brace or stiffener	S. Samson	Sterling, Mass	Jan. 5, 1833	....
Harness-pad clamp	J. H. Garrett	Greencastle, Ind	Feb. 13, 1872	123, 689
Harness-pad crimp and last	R. W. Calvert and W. Michael.	Mansfield, Ind	July 30, 1872	129, 931
Harness-pad, Elastic water-proof	A. Deitz	Albany, N. Y	Oct. 14, 1836	....
Harness-pad former or mold	T. H. Girard	Batavia, N. Y	Nov. 3, 1863	40, 473
Harness-pad plate	C. Gahr	Newark, N. J	May 24, 1870	103, 321
Harness-pad plate	P. Shaw and E. S. Dawson	Syracuse, N. Y	Mar. 19, 1867	63, 108
Harness-pad press	G. W. Lawbaugh	Geneseo, Ill	Feb. 25, 1868	74, 921
Harness-pad press	G. W. Lawbaugh	Geneseo, Ill	Dec. 15, 1868	85, 013
Harness-pad tree	J. W. Hinman	Berlin, Wis	Feb. 25, 1868	74, 823
Harness-pad tree	W. V. Kay	Chicago, Ill	May 23, 1871	115, 062
Harness-pad tree	P. Shaw and E. S. Dawson	Syracuse, N. Y	Nov. 5, 1867	70, 633
Harness-pad tree	W. H. Taylor	Baldwinsville, N. Y	Aug. 12, 1873	141, 835
Harness-pads, Manufacturing	J. Maclure	Newark, N. J	Feb. 11, 1868	74, 390
Harness-pads, Method of making	C. Gahr	Newark, N. J	Nov. 28, 1871	131, 352
Harness-padding	J. Carmac and A. Stobbs	Hector, N. Y	Aug. 16, 1870	106, 321
Harness, Padding or stuffing for	H. Hauer	Philadelphia, Pa	Nov. 17, 1868	84, 116
Harness-receptacle	H. E. Pond	Franklin, Mass	Aug. 13, 1867	67, 672
Harness-rein guide	L. Richmond	Derby, Vt	Dec. 14, 1869	97, 814
Harness, &c., Renovating	D. McCleary	Hollidaysburgh, Pa	July 23, 1867	66, 982
Harness-ring	S. Aldud	Avoca, Wis	July 19, 1870	105, 406
Harness-ring	G. H. Buckius	Canton, Ohio	July 7, 1868	79, 725
Harness-ring	A. Dietz	New York, N. Y	Apr. 9, 1850	7, 259
Harness-ring	H. N. Eames	Newport, N. Y	Mar. 1, 1870	100, 384
Harness-ring	W. Yates	Canton, Ohio	July 7, 1868	79, 797
Harness, Riveting	W. Dukehart	Baltimore, Md	Mar. 3, 1836	....
Harness round-knife	J. H. Quackenbush	Springfield, Mass	Aug. 4, 1868	80, 765
Harness-rosette	G. S. Caldwell	Syracuse, N. Y	Nov. 19, 1867	71, 132
Harness-rosette	W. D. Davis, C. W. Blakeslee, and J. C. Peck.	Watertown, Conn	Mar. 21, 1871	112, 906
Harness-rosette	J. W. Dayton	Waterbury, Conn	Apr. 23, 1872	125, 885
Harness-rosette	F. F. Reynold	Newark, N. J	Nov. 12, 1872	132, 983
Harness-rosette	C. F. Richers	New York, N. Y	Aug. 18, 1868	81, 110
Harness-rosette	L. C. Wilson	Albany, N. Y	Mar. 16, 1869	87, 896
Harness-saddle	M. E. Abbey	Sardis, Miss	Oct. 8, 1872	131, 988
Harness-saddle	V. Borst	New York, N. Y	June 25, 1867	65, 995
Harness-saddle	V. Borst	New York, N. Y	Apr. 20, 1869	89, 196
Harness-saddle	V. Borst	New York, N. Y	May 17, 1870	103, 132
Harness-saddle	V. Borst	New York, N. Y	Nov. 8, 1870	109, 108
Harness-saddle	V. Borst	New York, N. Y	Mar. 28, 1871	113, 136

Index of patents issued from the United States Patent Office from 1790 *to* 1873, *inclusive*—Continued.

Invention.	Inventor.	Residence.	Date.	No.
Harness-saddle	V. Borst	New York, N. Y	Sept. 19, 1871	119, 112
Harness-saddle	V. Borst	New York, N. Y	Mar. 26, 1872	125, 011
Harness-saddle	P. Böttyer	Newark, N. J	Apr. 18, 1865	47, 276
Harness-saddle	J. W. Briggs	Cleveland, Ohio	June 12, 1849	6, 518
Harness-saddle	A. D. Brown	New York, N. Y	Dec. 28, 1847	5, 408
Harness-saddle	A. D. Brown	New York, N. Y	Oct. 3, 1848	5, 822
Harness-saddle	G. H. Buckins	Canton, Ohio	Aug. 25, 1868	81, 336
Harness-saddle	J. T. Denniston	Lynde, N. Y	Nov. 20, 1846	4, 860
Harness-saddle	E. Dixon	Emporia, Kans	June 18, 1872	128, 025
Harness-saddle	G. W. Dutton	Tomales, Cal	Aug. 27, 1872	130, 906
Harness-saddle	J. Fye	Hamilton, Ohio	Dec. 4, 1866	60, 168
Harness-saddle	A. H. Gaylay	Saratoga Springs, N. Y	Mar. 14, 1848	5, 476
Harness-saddle	A. Gilliam	Cincinnati, Ohio	Apr. 30, 1867	64, 214
Harness-saddle	A. Gilliam	Pittsburgh, Pa	Oct. 19, 1869	95, 895
Harness-saddle	A. Gilliam	Pittsburgh, Pa	Jan. 17, 1871	110, 969
Harness-saddle	A. Gilliam	Pittsburgh, Pa	July 11, 1871	116, 946
Harness-saddle	A. Gilliam	Pittsburgh, Pa	Nov. 18, 1873	144, 613
Harness-saddle	E. E. Hardy	Joliet, Ill	June 7, 1870	104, 023
Harness-saddle	W. G. Hull	Sing Sing, N. Y	Apr. 13, 1869	88, 961
Harness-saddle	W. Leonard	Boston, Mass	Dec. 31, 1867	72, 743
Harness-saddle	W. and W. E. Leonard	Chelsea, Mass	Aug. 15, 1871	118, 137
Harness-saddle	C. K. Marshall	New Orleans, La	May 10, 1870	102, 953
Harness-saddle	G. Marshall	Indiana, Pa	Nov. 5, 1867	70, 589
Harness-saddle	J. H. Martin	Columbus, Ohio	Apr. 9, 1872	125, 589
Harness-saddle	J. McLain	Circleville, Ohio	Dec. 23, 1851	8, 610
Harness-saddle	O. B. North	New Haven, Conn	Apr. 11, 1865	47, 244
Harness-saddle	C. Pope and K. Frazer	Syracuse, N. Y	Feb. 5, 1847	4, 959
Harness-saddle	H. Sanders	Utica, N. Y	Jan. 5, 1858	19, 048
Harness-saddle	A. V. Sargeant	Newark, N. J	Mar. 18, 1873	136, 870
Harness-saddle	R. M. Selleck	New York, N. Y	Nov. 7, 1854	11, 911
Harness-saddle	P. Shaw	Syracuse, N. Y	Apr. 21, 1857	17, 117
Harness-saddle	R. Spencer	Brooklyn, N. Y	Aug. 6, 1850	7, 551
Harness-saddle	R. Spencer	New York, N. Y	Aug. 22, 1854	11, 576
Harness-saddle	R. Swift	New Haven, Conn	Dec. 29, 1857	18, 996
Harness-saddle	J. C. Tobias	Middleport, Ill	June 17, 1862	35, 656
Harness-saddle	S. E. Tompkins	Newark, N. J	Nov. 17, 1863	40, 653
Harness-saddle	S. E. Tompkins	Sing Sing, N. Y	Apr. 16, 1872	125, 770
Harness-saddle	S. E. Tompkins	Sing Sing, N. Y	July 15, 1873	140, 968
Harness-saddle	S. E. Tompkins	Sing Sing, N. Y	July 15, 1873	140, 969
Harness-saddle	S. E. Tompkins	Sing Sing, N. Y	July 15, 1873	140, 970
Harness-saddle	J. L. Van Wert	Tolland, Mass	Jan. 18, 1870	98, 899
Harness-saddle	J. Waite	Palmer, Mass	Mar. 9, 1869	87, 606
Harness saddle, Cart	H. A. Rains	Nashville, Tenn	Nov. 13, 1866	59, 651
Harness-saddle check-hook	A. V. Sargeant	Newark, N. J	Sept. 2, 1873	142, 415
Harness-saddle check-hook	P. H. Wiedersum	New York, N. Y	Mar. 5, 1872	124, 408
Harness-saddle hook	R. Hart	Dunkirk, N. Y	Jan. 25, 1870	99, 185
Harness-saddle hook and terret	S. E. Tompkins	Newark, N. J	Aug. 19, 1862	36, 248
Harness-saddle loop	J. M. Roe	Worcester, Mass	Mar. 16, 1869	87, 973
Harness-saddle mounting	J. J. Carrel	Petersburgh, Va	Oct. 17, 1848	5, 860
Harness-saddle pad	H. Lerew	Panora, Iowa	May 27, 1873	139, 403
Harness-saddle pad	R. C. Sturges	Boston, Mass	Jan. 19, 1869	86, 112
Harness-saddle pad or housing	C. B. Hogg	Boston, Mass	Nov. 30, 1869	97, 295
Harness-saddle plate	J. H. Garrett	Mount Pleasant, Iowa	Nov. 1, 1870	108, 900
Harness-saddle seat	P. Burns	Syracuse, N. Y	May 10, 1870	103, 004
Harness-saddle seat	O. B. North	New Haven, Conn	June 5, 1866	55, 348
Harness-saddle tree	J. C. Dickey	Saratoga Springs, N. Y	Mar. 6, 1855	12, 481
Harness-saddle tree	J. Fonda	Albany, N. Y	Apr. 7, 1863	38, 104
Harness-saddle tree	G. D. Gillett	Meridian, N. Y	July 20, 1869	92, 717
Harness-saddle tree	J. F. Knorr	Orange, N. J	Oct. 13, 1868	82, 961
Harness-saddle tree	A. Koehler	Holyoke, Mass	May 9, 1865	47, 647
Harness-saddle tree	J. A. Lawrence	New Haven, Conn	July 22, 1851	8, 238
Harness-saddle tree	T. Mardock and W. C. Kellar	Cincinnati, Ohio	Oct. 12, 1852	9, 328
Harness-saddle tree	J. H. Martin	Columbus, Ohio	Feb. 21, 1871	112, 061
Harness-saddle tree	J. H. Martin	Columbus, Ohio	Apr. 4, 1871	113, 541
Harness-saddle tree	J. H. Martin	Columbus, Ohio	July 25, 1871	117, 309
Harness-saddle tree	T. G. Moore	Albia, Iowa	Aug. 29, 1871	118, 547
Harness-saddle tree	O. B. North	New Haven, Conn	Feb. 21, 1865	46, 489
Harness-saddle tree	C. J. Paine	Young America, Ill	Apr. 23, 1872	126, 080
Harness-saddle tree	G. Pennoyer	New York, N. Y	May 24, 1870	103, 495
Harness-saddle tree	R. Spencer	Southport, Conn	Aug. 22, 1854	11, 573
Harness-saddle tree	S. E. Tompkins	Newark, N. J	Jan. 10, 1865	45, 880
Harness-saddle tree	S. E. Tompkins and J. Maclure	Newark, N. J	Feb. 1, 1859	22, 841
Harness-saddle tree	P. Washburn	Taunton, Mass	Mar. 8, 1832	......
Harness-saddle tree	P. H. Wiedersum	New York, N. Y	Sept. 27, 1870	107, 843
Harness-saddle tree	P. H. Wiedersum	New York, N. Y	Jan. 31, 1871	111, 502
Harness-saddle tree	P. H. Wiedersum	New York, N. Y	June 27, 1871	116, 518
Harness-saddle tree	P. H. Wiedersum	New York, N. Y	June 27, 1871	116, 519
Harness-saddle tree	P. H. Wiedersum	New York, N. Y	Apr. 30, 1872	126, 364
Harness-saddle tree, Metallic	S. E. Tompkins	New York, N. Y	Nov. 27, 1855	13, 858
Harness-saddle tree, Safety	J. S. Hall	Pittsburgh, Pa	Sept. 1, 1868	81, 629
Harness-saddle tree, Wooden	F. P. Ambler, jr	Trumbull, Conn	June 1, 1858	20, 463
Harness-saddles, Back-band for	S. F. Marshall	Wilmington, Del	Feb. 11, 1873	135, 831
Harness-saddles, Buckle-attachment to	J. B. Welpton	Tabor, Iowa	Apr. 1, 1873	137, 519
Harness-saddles, Fastening of terrets in	P. B. Cool	Columbus, Ohio	Mar. 12, 1850	7, 157
Harness-saddles, Revolving back-band hook for	G. Elbet	Pittsburgh, Pa	May 31, 1870	103, 588
Harness, Safety	J. L. Arnold	Lowndesville, S. C	June 4, 1872	127, 403
Harness, Safety	I. Ellis	Tyler, Tex	Oct. 1, 1872	131, 747
Harness, Safety	C. Rogers	Baltimore, Md	July 22, 1833	......
Harness safety-hook	H. Beagle, jr	Philadelphia, Pa	Apr. 9, 1861	31, 946
Harness, Safety-line for	A. H. Rockwell	Harpersville, N. Y	Dec. 4, 1866	60, 253
Harness safety-loop	C. H. Trumbull	Marion, N. Y	Mar. 5, 1872	124, 298
Harness, Safety-trap for	C. W. Nichols	Chicago, Ill	Nov. 25, 1873	145, 006
Harness, Securing buckles and rings to	R. B. Anderson	Oneida, Ill	June 23, 1868	79, 180
Harness shaft-loop	B. J. Aurand	Mount Gilead, Ohio	July 16, 1867	66, 771
Harness shaft-tug	J. L. Hoyt	Port Jervis, N. Y	June 19, 1847	5, 160
Harness, Shaft-tug lug for	T. J. Magruder	Marion, Ohio	Dec. 14, 1869	97, 786
Harness-shield	M. W. Pond	Elyria, Ohio	Feb. 23, 1869	87, 110
Harness, Single	O. V. Flora	Madison, Ind	July 7, 1868	79, 745
Harness, Single	J. S. Reid	Orange, Ind	Jan. 28, 1868	73, 835

Index of patents issued from the United States Patent Office from 1790 *to* 1873, *inclusive*—Continued.

Invention.	Inventor.	Residence.	Date.	No.
Harness, Single	C. R. Stuart	Winslow, Me	Oct. 24, 1871	120, 344
Harness-slide	J. S. Topham	Washington, D. C	Apr. 22, 1862	35, 049
Harness-slide snap	J. and D. C. Kerr	Ames, Iowa	June 9, 1868	78, 807
Harness-snap	P. Beckman	Naperville, Tenn	Oct. 6, 1863	40, 152
Harness-snap	E. A. Cooper	Lancaster, N. Y	Apr. 2, 1867	63, 478
Harness-snap	E. A. Cooper	Buffalo, N. Y	June 23, 1868	79, 107
Harness-snap	W. F. Davidson, O. A. Bates, S. M. Wilson, and A. P. Russell.	Janesville, Wis	Dec. 24, 1867	72, 462
Harness-snap	G. W. Devin	Ottumwa, Iowa	May 15, 1866	54, 700
Harness-snap	G. B. Durkee	Alden, N. Y	Oct. 8, 1867	69, 548
Harness-snap	G. B. Durkee	Alden, N. Y	Oct. 8, 1867	69, 549
Harness-snap	H. Harris	Newark, N. J	Feb. 21, 1865	46, 468
Harness-snap	H. Harris	Newark, N. J	Mar. 14, 1865	46, 796
Harness-snap	B. B. Hotchkiss	Sharon, Conn	Dec. 14, 1858	22, 290
Harness-snap	P. W. Jones	Syracuse, N. Y	Nov. 3, 1863	40, 486
Harness-snap	C. H. Palmer	Newark, N. J	Nov. 29, 1864	45, 299
Harness-snap	S. Reynolds	Allegheny, Pa	Mar. 12, 1872	124, 628
Harness-snap	J. B. Tibbits	Palmyra, N. Y	Aug. 11, 1863	39, 506
Harness-snap	A. B. Woodward and W. Bruen	Alfred Centre, N. Y., and Newark, N. J.	Feb. 7, 1871	111, 710
Harness steel spring-worm	G. Pritchard	Clarksburgh, Va	May 3, 1832	....
Harness-tree	T. Dempsey	Newark, N. J	Jan. 12, 1858	19, 078
Harness-tree	F. B. Kuehnhold and D. B. Sturges.	Newark, N. J	Feb. 16, 1858	19, 371
Harness-tree	C. M. Sturgess	Washington, Iowa	Feb. 5, 1867	61, 891
Harness-tree	J. H. Whissemore	Mansfield, Ohio	Aug. 11, 1868	81, 041
Harness-tree pad	W. A. Sharp and J. A. Shannon	Tama City, Iowa	Nov. 17, 1868	84, 071
Harness-trimming	J. Bauer	Newark, N. J	July 4, 1871	116, 536
Harness-trimming	J. Bauer	Newark, N. J	Oct. 17, 1871	119, 914
Harness-trimming	M. A. Fisk	Springfield, Mass	Mar. 18, 1868	75, 257
Harness-trimming	T. J. Magruder	Marion, Ohio	Feb. 11, 1868	74, 391
Harness-trimming	S. Wiener	Newark, N. J	July 11, 1871	116, 904
Harness-trimmings and carriage-ornaments, Manufacture of.	W. S. Robinson	Taunton, Mass	Sept. 22, 1829	....
Harness-trimmings, Manufacture of rubber-coated.	A. Albright	Newark, N. J	Feb. 13, 1872	123, 603
Harness-trimmings, Manufacture of rubber-coated.	A. Albright	Newark, N. J	Apr. 15, 1873	137, 873
Harness-trimmings, Mode of covering	C. M. Theberath	Newark, N. J	Jan. 18, 1870	99, 032
Harness-tug	I. H. Alexander	Newfield, N. Y	Oct. 10, 1871	119, 684
Harness-tug	N. Botsford	Lockport, N. Y	June 11, 1872	127, 840
Harness-tug	J. S. H. Dickinson	Jackson, Pa	June 15, 1869	91, 425
Harness-tug clasp	L. D. Cowles	Armada, Mich	Feb. 17, 1863	37, 678
Harness-tug clasp	L. D. Cowles	Armada, Mich	Dec. 15, 1863	40, 911
Harness-tug clasp	L. D. Cowles	Armada, Mich	Dec. 15, 1863	40, 912
Harness-tug clasp	J. S. Gilman	Tecumseh, Mich	July 29, 1862	35, 995
Harness-tug or trace-supporter	C. W. Terpening	Geneseo, Ill	June 29, 1869	91, 986
Harness, Water-hook bolt for	J. W. Bishop	New Haven, Conn	Mar. 16, 1869	87, 755
Harness, Wire	A. R. Reese	Phillipsburgh, N. J	Mar. 8, 1864	41, 860
Harnessing horses to vehicles, Mode of	R. McHardy	Edinburgh, Great Britain	Oct. 3, 1871	119, 526
Harp, Keyed	A. Kuhn	Baltimore, Md	Jan. 27, 1857	16, 489
Harpoon	M. Adams	Chilmark, Mass	Apr. 21, 1863	28, 207
Harpoon	W. Carsley	New Bedford, Mass	July 29, 1841	2, 195
Harpoon	D. H. Chamberlain	Boston, Mass	Aug. 17, 1835	....
Harpoon	G. Doyle	Provincetown, Mass	Nov. 2, 1858	21, 949
Harpoon	J. Holmes and A. West	Tisbury, Mass	Nov. 24, 1846	4, 865
Harpoon	J. Q. Kelley	Sag Harbor, N. Y	Oct. 20, 1857	18, 458
Harpoon	Z. Kelley	New Bedford, Mass	Dec. 3, 1867	71, 763
Harpoon	A. Moor	Hampden, Me	Mar. 16, 1844	3, 490
Harpoon	C. Randall	Palmyra, Ga	Dec. 3, 1846	4, 872
Harpoon	J. Sizer, 2d	New London, Conn	Oct. 16, 1829	....
Harpoon	J. D. B. Stillman	New York, N. Y	Apr. 6, 1852	8, 862
Harpoon and lance	H. W. Harkness	Bristol, Conn	Feb. 16, 1858	19, 363
Harpoon and lance, Gun	R. Brown	New London, Conn	Aug. 20, 1850	7, 572
Harpoon, Bomb	C. Freeman	Brewster, Mass	May 7, 1872	126, 388
Harpoon, Exploding	C. Burt	Belfast, Me	May 6, 1851	8, 073
Harpoon, Explosive	T. Briggs	Philadelphia, Pa	Dec. 11, 1860	30, 869
Harpoon, Gun	O. Allen	Norwich, Conn	Dec. 5, 1848	5, 949
Harpoon, Gun	R. Brown	New London, Conn	June 4, 1850	7, 410
Harpoon, Gun	J. P. Rechten	New York, N. Y	Dec. 7, 1869	97, 693
Harpoon, Gun	R. E. Smith	Provincetown, Mass	Apr. 23, 1867	64, 045
Harpoon, Gun	T. W. Roys	Southampton, N. Y	Jan. 22, 1861	31, 190
Harpoon, Hinged gun	W. Albertson	New London, Conn	Nov. 19, 1850	7, 777
Harpoon, Rocket	T. W. Roys	Southampton, N. Y	June 3, 1862	35, 474
Harpoon, Rocket	T. W. Roys and G. A. Lilliendahl.	New York, N. Y	Apr. 24, 1866	54, 211
Harpoons, Method of attaching lines to	C. F. Brown	Warren, R. I	Sept. 3, 1850	7, 610
Harrow	J. Adams	Monroe, Mich	Sept. 24, 1861	33, 327
Harrow	J. Aiken	Warner, N. H	Dec. 17, 1867	72, 260
Harrow	J. Allen	Union, N. Y	Dec. 18, 1860	30, 905
Harrow	W. Anderson	Jacksonville, N. Y	Aug. 1, 1854	11, 452
Harrow	M. Atwood	New Sharon, Iowa	Nov. 24, 1868	84, 339
Harrow	J. Ayers	Canterbury, N. H	July 10, 1866	56, 159
Harrow	D. C. Ayres	Lumberland, N. Y	Sept. 7, 1858	21, 403
Harrow	A. W. Ball	Delaware Grove, Pa	June 8, 1869	90, 980
Harrow	A. H. Ballagh	Westport, Md	Mar. 16, 1869	87, 819
Harrow	W. H. Barry	Rabbit River, Mich	Oct. 20, 1868	83, 238
Harrow	L. Beckelshymen	Leavenworth, Kans	Aug. 20, 1867	67, 940
Harrow	R. N. Bennett	Union Mills, Ind	Oct. 26, 1869	96, 191
Harrow	J. Benson	Belle Plains, Iowa	Aug. 1, 1871	117, 595
Harrow	A. Berdan	Macon, Mich	Aug. 10, 1858	21, 113
Harrow	W. Berlin	Berryville, Va.	Apr. 12, 1853	9, 657
Harrow	A. W. Bohaker	Napa, Cal	Dec. 3, 1872	133, 623
Harrow	M. Böshenz	Chili, Ill	Feb. 11, 1868	74, 194
Harrow	E. Bradley	Clyde, N. Y	Sept. 11, 1866	57, 851
Harrow	T. E. C. Brinly	Louisville, Ky	May 14, 1872	126, 779
Harrow	T. S. and T. A. Brown	Brooklyn, N. Y	Aug. 24, 1869	93, 959
Harrow	W. Brown	Butte des Morts, Wis	Feb. 18, 1873	136, 027
Harrow	R. W. Buckles	Grayville, Ill	May 24, 1859	24, 095

Index of patents issued from the United States Patent Office from 1790 *to* 1873, *inclusive*—Continued.

Invention.	Inventor.	Residence.	Date.	No.
Harrow	M. Bucklin	Grafton, N. H	June 4, 1861	32, 462
Harrow	J. G. Burcham	Noblesville, Ind	Mar. 29, 1870	101, 350
Harrow	W. J. Burdick	Alfred, N. Y	July 26, 1870	105, 775
Harrow	S. P. Campbell	Rochester, Minn	Feb. 9, 1864	41, 479
Harrow	A. C. Carnes	Smithville, Tenn	July 1, 1873	140, 403
Harrow	H. Cartwright	Owatonna, Minn	Feb. 4, 1873	135, 464
Harrow	J. C. Center	Mason City, Ill	July 25, 1871	117, 258
Harrow	V. M. Chafee	Grayville, Ill	May 25, 1858	20, 325
Harrow	J. Chase	Farmington, Pa	Nov. 24, 1868	84, 342
Harrow	J. Chellew	Glasford, Ill	July 18, 1871	117, 149
Harrow	C. Clareni	New York, N. Y	Apr. 3, 1855	12, 659
Harrow	J. Click	Springfield, Ohio	Aug. 27, 1867	68, 169
Harrow	O. Coe	Port Washington, Wis	Mar. 2, 1858	19, 489
Harrow	D. C. Colby	Newport, N. H	June 26, 1860	28, 938
Harrow	L. Coleman	New Orleans, La	Aug. 13, 1867	67, 634
Harrow	T. B. Collins	Noank, Conn	June 29, 1869	91, 914
Harrow	J. C. Concklin	Peekskill, N. Y	Apr. 21, 1836	....
Harrow	J. C. Concklin	Yorktown, N. Y	June 3, 1862	35, 487
Harrow	J. C. Conkey	Washington, Ohio	May 11, 1858	20, 195
Harrow	D. B. Conover	New Bedford, N. J	Dec. 20, 1870	110, 344
Harrow	G. Cooke	Bristol Station, Ill	June 10, 1862	35, 563
Harrow	W. J. Cordill	Blue Earth City, Minn	Apr. 30, 1872	126, 186
Harrow	A. J. Craig	Ashmore Station, Ill	Oct. 13, 1868	82, 923
Harrow	I. Crum	West Chester, Ohio	Mar. 24, 1868	75, 867
Harrow	I. Crum	West Chester, Ohio	June 22, 1869	91, 525
Harrow	J. S. Davis	Washington, Ohio	June 1, 1858	20, 410
Harrow	L. P. Davis and O. P. Beagle	Altona, Ill	Nov. 26, 1872	133, 308
Harrow	N. A. Davis	Sutton, N. H	Apr. 10, 1866	53, 792
Harrow	J. Dawson	Greenwood, Ill	Oct. 25, 1870	108, 690
Harrow	J. Dawson	Greenwood, Ill	July 2, 1872	128, 601
Harrow	J. Delong	Covert, N. Y	Oct. 14, 1862	36, 644
Harrow	A. B. Depeny	Hamburgh, Iowa	Nov. 19, 1872	133, 212
Harrow	W. De Witt and O. D. Barrett	Cleveland, Ohio	Mar. 2, 1858	19, 494
Harrow	D. L. Dickson	Durham, Ill	Feb. 18, 1868	74, 673
Harrow	J. Dingman	Decatur, Ill	Sept. 13, 1870	107, 344
Harrow	S. Donaldson	Fenton, Mich	Jan. 16, 1872	122, 711
Harrow	M. Easterbrook, jr	Geneva, N. Y	Sept. 12, 1865	49, 867
Harrow	O. W. Edmunds	Bluffdale, Ill	Dec. 8, 1868	84, 807
Harrow	G. M. Ellis	Huntertown, Ind	Aug. 23, 1870	106, 677
Harrow	J. H. Elward	Ottawa, Ill	Nov. 17, 1863	40, 614
Harrow	H. L. Eshelman	Elizabethtown, Pa	July 30, 1867	67, 178
Harrow	T. H. Enlass	Mason City, Ill	Jan. 26, 1869	86, 218
Harrow	D. S. Fisher	Cedar Spring, Ind	May 28, 1867	65, 196
Harrow	J. Foltz	Valley Mills, Ind	June 8, 1869	90, 939
Harrow	J. N. Fordyce	Cambridge, Ohio	Aug. 13, 1867	67, 835
Harrow	E. Forney	Fishersville, Pa	Jan. 28, 1868	73, 792
Harrow	J. T. Foster	Jersey City, N. J	June 18, 1861	32, 599
Harrow	W. Fox	Beaver Dam, Wis	Apr. 4, 1871	113, 286
Harrow	J. T. Frankeberger	Hensly, Ill	Dec. 24, 1867	72, 623
Harrow	J. H. French	Syracuse, N. Y	May 10, 1859	23, 914
Harrow	A. Friedemann	Waverly, Ohio	Aug. 2, 1870	106, 046
Harrow	A. Friedemann	Waverly, Iowa	Jan. 31, 1871	111, 445
Harrow	D. L. Garver	Hart Township, Mich	Oct. 5, 1869	95, 458
Harrow	J. F. Gazley	Canyonville, Oreg	May 27, 1873	139, 309
Harrow	L. C. Gifford	Monticello, Ill	Mar. 11, 1873	136, 717
Harrow	L. C. Gillaspie	Denmark, Tenn	July 10, 1860	29, 073
Harrow	E. A. Goodes	Philadelphia, Pa	July 20, 1869	92, 719
Harrow	W. Gourley	Clarke County, Va	July 17, 1855	13, 257
Harrow	W. Grange	Augusta, Ky	Sept. 2, 1862	36, 345
Harrow	F. Granger	Lockport, Ill	July 17, 1866	56, 405
Harrow	P. S. Graves and P. B. Parcell	Ashmore, Ill	Oct. 5, 1869	95, 467
Harrow	O. C. Green	Dublin, Ind	Nov. 22, 1859	26, 179
Harrow	M. W. Gunn	La Salle, Ill	Feb. 11, 1868	74, 348
Harrow	E. L. Hagan	Frankfort, N. Y	Dec. 12, 1854	12, 059
Harrow	C. Hairgrove	Jacksonville, Ill	Jan. 16, 1872	122, 765
Harrow	D. Haldeman	Morgantown, Va	Feb. 20, 1855	12, 415
Harrow	S. W. Hamsher	Decatur, Ill	Oct. 18, 1859	25, 824
Harrow	C. Hanson	Owatonna, Minn	Dec. 1, 1868	84, 547
Harrow	J. M. Harper	El Paso, Ill	Dec. 13, 1870	110, 133
Harrow	J. Harris	San Francisco, Cal	Aug. 27, 1872	130, 016
Harrow	G. Heffner	Homer, Iowa	Mar. 16, 1869	87, 845
Harrow	J. H. Hendee	Jackson, Mich	June 21, 1864	43, 207
Harrow	J. Herald and C. B. Tomkins	Trumansburgh, N. Y	June 7, 1859	24, 303
Harrow	E. W. Herendeen	Geneva, N. Y	Oct. 24, 1871	120, 195
Harrow	E. S. Herrington	Emmett, Ohio	Nov. 14, 1871	120, 873
Harrow	H. M. Hickman and B. G. Devoe.	Vandalia, Ill	Nov. 10, 1868	83, 962
Harrow	A. Hockstein	Williamsville, N. Y	July 14, 1868	79, 829
Harrow	S. S. Hogle	York, Ohio	Mar. 17, 1857	16, 866
Harrow	W. R. Hollingsworth	Mount Pleasant, Iowa	Sept. 28, 1869	95, 348
Harrow	T. C. Hooker	Kendall, N. Y	June 18, 1861	32, 564
Harrow	T. C. Hooker	Brockport, N. Y	Nov. 12, 1872	132, 964
Harrow	M. W. House	Cleveland, Ohio	Sept. 4, 1860	29, 885
Harrow	A. Jacobson	Logan, Nebr	Apr. 9, 1872	125, 395
Harrow	J. Jacobson	Austin, Minn	Nov. 1, 1870	108, 910
Harrow	D. L. Jaques	Hudson, Mich	Dec. 13, 1870	110, 142
Harrow	D. L. Jaques	Hudson, Mich	Apr. 15, 1873	137, 926
Harrow	J. Jay and J. Coppock	Jonesborough, Ind	Aug. 3, 1869	93, 306
Harrow	C. Jillson	Worcester, Mass	Oct. 17, 1865	50, 478
Harrow	B. Johnston	Sterling, Ill	July 4, 1871	116, 716
Harrow	A. Jones	Randolph, Wis	Mar. 12, 1872	124, 586
Harrow	S. G. Jones	Niantic, Ill	June 8, 1869	91, 024
Harrow	W. P. Jones	Arcade, N. Y	Sept. 6, 1870	107, 057
Harrow	A. and N. Kane	Newport, N. Y	July 13, 1869	92, 614
Harrow	J. Kelsey	Yardleyville, Pa	Dec. 23, 1862	37, 234
Harrow	T. M. King	Murfreesborough, Tenn	May 20, 1873	139, 065
Harrow	J. Kinhart	Athens, Ill	Jan. 23, 1872	123, 025
Harrow	J. Knickerbocker	Dunning, Pa	July 6, 1869	92, 320

Index of patents issued from the United States Patent Office from 1790 *to* 1873, *inclusive*—Continued.

Invention.	Inventor.	Residence.	Date.	No.
Harrow	C. Lane and J. M. Healy	Jamestown, N. Y	Mar. 23, 1869	88, 046
Harrow	O. J. Leabo	Forest Grove, Oreg	Mar. 19, 1872	124, 752
Harrow	A. Lewis	Hastings, Minn	Sept. 13, 1870	107, 269
Harrow	J. S. Lewis	Elkader, Iowa	Oct. 26, 1869	96, 244
Harrow	H. C. Lezott	Osage, Iowa	July 20, 1869	92, 849
Harrow	S. H. Lintan	Burrows, Ind	Nov. 30, 1869	97, 416
Harrow	P. H. Lang	Palmyra, Me	Dec. 31, 1872	134, 433
Harrow	I. Low	East Fairfield, Ohio	Aug. 31, 1869	94, 326
Harrow	L. Lupton	Winchester, Va	May 24, 1853	9, 747
Harrow	C. R. Macy	Bedminster, N. J	Dec. 28, 1869	98, 281
Harrow	M. R. Marcell	Dansville, N. Y	June 29, 1869	92, 071
Harrow	N. B. Marsh	Marengo, Ill	Mar. 9, 1869	87, 577
Harrow	B. T. Martin	Charlotte, Mich	Dec. 15, 1868	84, 956
Harrow	J. Mathison	Fremont, Nebr	Oct. 24, 1871	120, 204
Harrow	D. B. Maze	West Buffalo, Ohio	Oct. 7, 1873	143, 520
Harrow	J. G. McCauley	Stone Bridge, Va	Aug. 29, 1854	11, 614
Harrow	N. McCuen	South Potsdam, N. Y	Feb. 25, 1862	34, 516
Harrow	W. McFishback	Union, Ohio	Aug. 14, 1866	57, 166
Harrow	J. Mellinger	Greensburgh, Pa	Oct. 11, 1870	108, 279
Harrow	S. Mendenhall	Muncy Station, Pa	Mar. 23, 1869	88, 195
Harrow	J. Mercier	Detroit, Mich	Jan. 7, 1868	73, 024
Harrow	M. D. Meriweather	Denmark, Tenn	Aug. 28, 1860	29, 804
Harrow	A. Mero	New Haven, Conn	Apr. 10, 1866	53, 850
Harrow	J. H. Miller and F. A. Pickering	Niantic, Ill	Nov. 23, 1869	97, 104
Harrow	N. B. Miller	Orion, Mich	Jan. 21, 1868	73, 626
Harrow	S. Miller	Winchester, Ohio	Mar. 19, 1861	31, 731
Harrow	J. E. Morgan	Deerfield, N. Y	Mar. 31, 1857	16, 933
Harrow	J. E. Morgan	Deerfield, N. Y	July 16, 1867	66, 869
Harrow	S. A. and C. C. Morgan	Auburn, N. Y	June 26, 1860	28, 890
Harrow	D. C. Myers	South Bend, Ind	Mar. 12, 1867	62, 769
Harrow	J. Myers	Powhatan Point, Ohio	Aug. 15, 1854	11, 528
Harrow	D. B. Neal	Mount Gilead, Ohio	Aug. 24, 1858	21, 269
Harrow	S. Neill	Chillicothe, Ill	Sept. 4, 1860	29, 903
Harrow	S. J. Newell	South China, Me	Nov. 5, 1872	132, 853
Harrow	A. A. Nuquist	Oneida, Ill	July 27, 1869	93, 115
Harrow	W. S. O'Brien	Brimfield, Ill	June 5, 1866	55, 351
Harrow	G. Ogg	Lacon, Ill	Mar. 12, 1867	62, 875
Harrow	R. W. O'Neal	Punta Arenas, Cal	Jan. 30, 1872	123, 193
Harrow	S. J. Orange	Grayville, Ill	Feb. 2, 1858	19, 259
Harrow	S. J. Orange and G. Beidelman	Grayville, Ill	Sept. 7, 1858	21, 439
Harrow	W. F. Osborn	Mount Pleasant, Pa	Nov. 10, 1868	83, 878
Harrow	G. Paddington	Springville, Iowa	Dec. 21, 1869	98, 100
Harrow	O. D. Padrick	Shelbyville, Ind	July 3, 1866	56, 088
Harrow	W. F. Pagett	Stone Bridge, Va	May 2, 1854	10, 863
Harrow	D. A. Parkman	Union City, Tenn	Sept. 2[illegible], 1871	119, 397
Harrow	J. A. B. Patterson	Springfield, Ill	Apr. 20, 1869	89, 238
Harrow	N. A. Patterson	Kingston, Tenn	Feb. 14, 1860	27, 150
Harrow	J. M. Payne	Benton, Ill	Sept. 26, 1871	119, 398
Harrow	G. W. Pense and C. E. Lykke	Franklin Grove, Ill	Mar. 16, 1869	87, 872
Harrow	W. H. Platt	Dayton, Ohio	Apr. 8, 1873	137, 569
Harrow	E. J. Preston	Eureka, Mo	Nov. 16, 1869	96, 936
Harrow	J. Rankin	Taunton, Mass	Apr. 21, 1868	77, 091
Harrow	G. Reed	Greencastle, Ind	June 18, 1872	128, 067
Harrow	W. Rennyson	Norristown, Pa	Sept. 12, 1871	118, 973
Harrow	W. Rennyson	Norristown, Pa	Nov. 14, 1871	121, 003
Harrow	J. W. Richardson	Sligo, Ohio	Jan. 15, 1867	61, 258
Harrow	W. E. Robbins and G. Enderton	Sterling, Ill	May 30, 1871	115, 525
Harrow	J. Robins	Leominster, Mass	Feb. 2, 1858	19, 281
Harrow	H. W. Robinson	Woodstock, Ill	Jan. 23, 1872	122, 911
Harrow	L. C. Rumberger	Cass, Pa	Nov. 25, 1873	145, 016
Harrow	J. Russell	Grampian Hills, Pa	Jan. 17, 1860	26, 864
Harrow	J. Ruth	Defiance, Ohio	Mar. 3, 1868	75, 202
Harrow	J. Ruth and A. Vaughn	Grayville, Ill	Aug. 10, 1858	21, 153
Harrow	A. Safley	Mount Vernon, Iowa	Apr. 19, 1864	42, 408
Harrow	S. C. Schofield	Freeport, Ill	Mar. 13, 1860	27, 477
Harrow	D. W. Shares	Hamden, Conn	Jan. 27, 1857	16, 498
Harrow	M. Sheldon, jr	Calais, Vt	July 1, 1862	35, 784
Harrow	B. F. Shellabarger	Mifflintown, Pa	Aug. 1, 1848	5, 687
Harrow	H. and C. Shirk	Lebanon, Pa	Feb. 8, 1870	99, 719
Harrow	V. Sippel	Niagara City, N. Y	June 29, 1869	91, 977
Harrow	C. Smith	Irwin's Station, Pa	Feb. 4, 1873	135, 448
Harrow	J. Smith	Woodland, Mich	Mar. 25, 1873	137, 254
Harrow	J. Snowdin and J. Kent	Westford and Beaver Dam, Wis.	Mar. 23, 1869	88, 225
Harrow	N. Starr, jr	Homer, N. Y	Dec. 17, 1867	72, 334
Harrow	C. E. Steller	Genesee, Wis	Dec. 2, 1862	37, 061
Harrow	C. E. Steller	Chicago, Ill	Feb. 18, 1868	74, 626
Harrow	A. J. Stewart	Chillicothe, Mo	Nov. 4, 1873	144, 236
Harrow	W. D. Summers	Hopkinsville, Ky	Aug. 19, 1873	142, 055
Harrow	S. Svanson	Sweede Point, Iowa	Mar. 12, 1867	62, 904
Harrow	C. Svenzen	Blair, Nebr	Oct. 21, 1873	143, 941
Harrow	H. N. Swift	Matteawan, N. Y	Aug. 13, 1872	130, 451
Harrow	F. Y. Tavenner, J. W. Galbraith, and A. Smith.	Sedalia, Mo	Aug. 24, 1869	94, 044
Harrow	R. L. Taylor	Golden Prairie, Iowa	Sept. 9, 1873	142, 593
Harrow	J. Temple	Bellefonte, Pa	Nov. 27, 1860	30, 789
Harrow	C. and G. P. Tharp	Bryan, Ohio	Nov. 14, 1871	121, 021
Harrow	C. Thayer and L. L. Thomas	Otego, N. Y	Aug. 23, 1870	106, 742
Harrow	J. J. Thomas	Union Springs, N. Y	Sept. 22, 1868	82, 451
Harrow	G. W. Tolhurst	Cleveland, Ohio	Apr. 21, 1857	17, 121
Harrow	W. Tuttle	Fayette, Miss	Apr. 5, 1870	101, 547
Harrow	M. Vanbibber	Tipton, Ind	Jan. 4, 1870	98, 531
Harrow	J. E. Van Riper	Dearborn, Mich	Aug. 6, 1867	67, 466
Harrow	J. A. Walker	Nashville, Tenn	July 22, 1873	141, 021
Harrow	J. N. and T. Wallis	Fleming, N. Y	Jan. 7, 1868	73, 061
Harrow	J. Walsh	Stark County, Ill	July 30, 1867	67, 387
Harrow	J. N. Wamsley	Galena, Ohio	Jan. 21, 1873	135, 022
Harrow	E. S. Warner	Union Township, Ohio	Aug. 24, 1869	94, 153

Index of patents issued from the United States Patent Office from 1790 *to* 1873, *inclusive*—Continued.

Invention.	Inventor.	Residence.	Date.	No.
Harrow	C. Watson	Cascade, Va	Oct. 2, 1860	30, 280
Harrow	G. Watt	Richmond, Va	July 1, 1873	140, 563
Harrow	H. Wendover	Norway, N. Y	Apr. 1, 1873	137, 520
Harrow	M. D. Wells	Morgantown, W. Va	Jan. 23, 1866	52, 233
Harrow	S. White	Penfield, Ohio	Sept. 14, 1858	21, 542
Harrow	J. Wigle	West Point, Ill	Oct. 17, 1871	120, 140
Harrow	J. P. Wile	Nashville, Mich	June 1, 1869	90, 713
Harrow	B. B. Williams	Laclede, Mo	Apr. 13, 1869	88, 931
Harrow	J. H. Williams	Tontzville, Kan	Sept. 14, 1869	94, 855
Harrow	J. L. Willoughby	Bowling Green, Ky	Aug. 29, 1871	118, 570
Harrow	F. R. Willson	Columbus, Ohio	June 16, 1868	79, 042
Harrow	L. L. Willson	Denmark, Mich	Nov. 5, 1872	132, 885
Harrow	S. and A. Woodard	Saratoga, Ind	Aug. 30, 1870	106, 908
Harrow	P. N. Woodworth	Stony Point, Cal	June 17, 1873	139, 987
Harrow	G. Workman	Rochester, N. Y	Dec. 21, 1869	98, 137
Harrow	M. K. Young	Glen Haven, Wis	Oct. 28, 1873	144, 044
Harrow, Adjustable	J. C. Center	Bath, Ill	Apr. 28, 1868	77, 254
Harrow, Adjustable	A. Havens	Trenton, N. J	Oct. 18, 1870	108, 478
Harrow, Adjustable	J. Kinhart	Athens, Ill	Sept. 21, 1869	94, 964
Harrow, Adjustable	H. Pulse	Saint Paul, Ind	Dec. 26, 1865	51, 750
Harrow and clod-crusher, Combined	A. C. Tower	Troy Grove, Ill	Aug. 20, 1872	130, 606
Harrow and cultivator	W. A. Estes	China, Me	Aug. 3, 1869	93, 187
Harrow and cultivator	J. F. Jaquess	Commerce, Miss	Jan. 24, 1871	111, 125
Harrow and cultivator	J. Lerch	Uhlersville, Pa	Aug. 16, 1870	106, 376
Harrow and cultivator	E. S. Rice	Paw-Paw, Mich	Apr. 6, 1869	88, 740
Harrow and cultivator	A. S. White	Malone, N. Y	Sept. 29, 1868	82, 573
Harrow and cultivator, Combined	G. W. Bressler	La Fayette, Iowa	Mar. 30, 1869	88, 363
Harrow and cultivator, Combined	H. Culver	Dansville, N. Y	Mar. 9, 1869	87, 642
Harrow and cultivator, Combined	W. R. Cummings	McCutchanville, Ind	June 29, 1869	91, 917
Harrow and cultivator, Combined	W. J. Funk	Portland, Oreg	July 20, 1869	92, 714
Harrow and cultivator, Combined	D. M. Harkrader	Chili, Ill	Oct. 13, 1868	82, 942
Harrow and cultivator, Combined	J. Lerch	Uhlersville, Pa	Nov. 23, 1869	97, 205
Harrow and cultivator, Combined	J. K. Minich	Mount Joy, Pa	May 8, 1866	54, 648
Harrow and cultivator, Combined	J. R. Ross and W. D. Mitchell	Centralia, Ill	Apr. 6, 1869	88, 669
Harrow and cultivator, Combined	A. H. Sheffer	West Donegal, Pa	May 1, 1866	54, 421
Harrow and cultivator, Combined	N. W. Wheeler	Ripon, Wis	Oct. 6, 1868	82, 775
Harrow and cultivator combined, Revolving	S. D. Carpenter	Madison, Wis	Dec. 29, 1868	85, 282
Harrow and cultivator, Revolving	A. Newell	New York, N. Y	July 2, 1867	66, 245
Harrow and cultivator, Sulky	W. and J. Whait	Independence, Iowa	July 23, 1867	67, 009
Harrow and earth-cutter	J. Schroeffel and W. Dell	Allegheny City, Pa	Apr. 5, 1870	101, 522
Harrow and field-roller	A. S. Keagy	Harristown, Ill	Feb. 28, 1871	112, 149
Harrow and marker, Combined	W. Addleton	Mottville, N. Y	July 6, 1869	92, 139
Harrow and marker, Combined	B. F. Barney	Pontiac, Ill	July 27, 1869	93, 038
Harrow and marker, Combined	S. Hutchinson	Griggsville, Ill	July 18, 1871	117, 079
Harrow and marker, Combined	A. Pierce	Olmsted, Ohio	Sept. 11, 1866	57, 961
Harrow and plow-hoe	I. Cheatham	Providence Inn, Va	July 31, 1827	....
Harrow and roller, Combined	T. Bailey	Wyoming, Ill	Aug. 30, 1870	106, 765
Harrow and roller, Combined	J. M. Blankenbeker	Power's Station, Ind	Aug. 23, 1870	106, 639
Harrow and roller, Combined	W. H. Converse	New Castle, Me	Apr. 4, 1865	47, 090
Harrow and roller, Combined	F. A. Dann and J. McKibben	Wellsville, Mo	Sept. 20, 1870	107, 461
Harrow and roller, Combined	J. Steiner and J. P. Miller	Harrisburgh, Ohio	May 15, 1866	54, 790
Harrow and roller, Revolving	E. K. Harvey	Quincy, Ohio	May 26, 1868	78, 283
Harrow and seed-drill	J. A. Underwood	Grand River, Iowa	Oct. 21, 1862	36, 739
Harrow and seed-sower, Combined	E. A. Barton	Boonville, Ind	Dec. 28, 1869	98, 335
Harrow and seeder	D. L. and J. M. Barlow	Cohoctah, Mich	Jan. 17, 1865	45, 903
Harrow and seeder, Combined	L. G. Jackson	Marion, Wis	May 13, 1873	138, 806
Harrow and seeder, Combined	B. Randall	Adams, N. Y	Nov. 12, 1867	70, 898
Harrow and seeder, Combined	E. Stoner	Greencastle, Pa	Nov. 15, 1870	109, 269
Harrow and seeder combined, Revolving sulky	A. L. Taveau	Chaptico, Md	May 4, 1869	89, 812
Harrow and seeder combined, Revolving sulky	A. L. Taveau	Chaptico, Md	Apr. 5, 1870	101, 540
Harrow and seeder, Combined wheel	F. Bramer	Little Falls, N. Y	June 24, 1873	140, 182
Harrow and sower, Combined	J. Brewer	Albany, Ill	Oct. 1, 1861	33, 377
Harrow and stone-remover, Combined	E. Babcock	Scott, N. Y	Feb. 4, 1868	74, 035
Harrow, Carriage	T. J. Marinus, and S. and W. Whait.	Independence, Iowa	Oct. 16, 1866	58, 854
Harrow, Circular	G. W. Hills	Dodge Centre, Minn	Nov. 25, 1873	144, 978
Harrow, Corn, &c	P. Clark	Aurora, N. Y	Aug. 20, 1835	....
Harrow, Corn	G. Hill	Galva, Ill	May 23, 1871	115, 055
Harrow, Corn and grain cultivator	N. Stanton, jr	Florida, N. Y	July 9, 1832	....
Harrow, Cultivating	J. Slocum	Syracuse, N. Y	Nov. 27, 1860	30, 767
Harrow, cultivator, and planter combined	J. G. S. Garwood	Vermillion, Ill	Jan. 21, 1868	73, 593
Harrow, cultivator, wheelbarrow, and sled combined	J. R. Wagner	Manada Hill, Pa	Aug. 2, 1870	106, 004
Harrow, Double-rotary	G. W. Hall	New Haven, Mich	Oct. 1, 1867	69, 340
Harrow, drill, grass-seeder, and roller, Combined	J. P. Long	Osage, Iowa	May 12, 1863	38, 495
Harrow, drill, planter, and roller combined	D. B. Platt	Madison, Ind	June 16, 1868	78, 998
Harrow, Expanding	I. J. Halsted	Springfield, Ill	Jan. 5, 1869	85, 582
Harrow, Flexible	C. D. Blim	Port Huron, Mich	May 5, 1868	77, 574
Harrow, Flexible	W. B. and G. M. Ramsay	South Strabane, Pa	May 2, 1854	10, 865
Harrow for clearing ground, Cleave	L. Tam	Sussex County, Del	July 11, 1826	....
Harrow-frame	J. Brainerd, E. F. Olds, and A. W. Olds.	Cleveland, Ohio, and Lyon and Green Oak, Mich.	Apr. 2, 1861	31, 865
Harrow, Hoe	A. Newbern	Bristol Township, Pa	June 1, 1824	....
Harrow, Plow	S. Vinton	Wellington, Conn	Nov. 18, 1807	....
Harrow, Portable field	J. D. Johnson	Tylersville, Pa	Mar. 30, 1869	88, 301
Harrow, Reciprocating	J. Densmore	Holley, N. Y	July 16, 1867	66, 684
Harrow, Revolving	G. S. Bartholomew and A. R. Chapman.	Reading, Mich	May 18, 1869	90, 222
Harrow, Revolving	C. Bates	Kingston, Mass	June 1, 1869	90, 628
Harrow, Revolving	M. G. Cass	Utica, N. Y	Sept. 10, 1839	1, 317
Harrow, Revolving	J. Hill and J. G. Stewart	Jonesville, Ind	Dec. 26, 1871	122, 251
Harrow, Revolving	W. A. Horrall and R. G. Sirwell	Grayville, Ill	Feb. 16, 1858	19, 365
Harrow, Revolving	M. W. House	Cleveland, Ohio	Jan. 18, 1859	22, 651
Harrow, Revolving	B. Lyon	Mount Pleasant, Ill	Sept. 29, 1868	82, 623
Harrow, Revolving	H. Mittendorf	Stony Hill, Mo	Feb. 25, 1873	136, 169
Harrow, Revolving	H. H. Monroe	Thomaston, Me	Mar. 8, 1870	100, 547
Harrow, Revolving	G. J. Olendorf	Middlefield, N. Y	June 17, 1856	15, 151
Harrow, Revolving	S. Rugg	Lancaster, Mass	Jan. 11, 1831	....
Harrow, Revolving	H. C. Stoll	Mokena, Ill	Feb. 26, 1867	62, 379

Index of patents issued from the United States Patent Office from 1790 *to* 1873, *inclusive*—Continued.

Invention.	Inventor.	Residence.	Date.	No.
Harrow, Revolving	W. R. Toby and M. J. Barcalo	Nunda and Mount Morris, N. Y.	Dec. 1, 1868	84, 656
Harrow, Revolving cylindrical	J. D. Woodside	Washington, D. C	July 28, 1832	....
Harrow, Revolving disk	J. F. Pond	Cleveland, Ohio	May 9, 1871	114, 707
Harrow, roller, and drill, Combined	S. Bradbury	Dresden, Mo	Aug. 24, 1869	93, 956
Harrow, roller, and seed-planter, Combined	R. J. Colvin	Lancaster, Pa	May 23, 1871	115, 026
Harrow, roller, and seed-sower combined	C. E. Pierce	New York, N. Y	Feb. 15, 1870	99, 946
Harrow, roller, and seeder, Combined	A. Powell	New Baltimore, N. Y	Jan. 4, 1870	98, 627
Harrow, Rotary	A. H. Acken	Griggstown, N. J	Apr. 23, 1867	64, 055
Harrow, Rotary	O. D. Barrett	Cleveland, Ohio	Oct. 4, 1859	25, 620
Harrow, Rotary	J. W. Barton and T. Holmes	Emporia, Kans	Mar. 14, 1871	112, 676
Harrow, Rotary	C. Bates	Kingston, Mass	Mar. 26, 1867	63, 199
Harrow, Rotary	M. L. Bawder	Cleveland, Ohio	Dec. 4, 1860	30, 795
Harrow, Rotary	E. Bennett	Naukin, Mich	Apr. 9, 1872	125, 523
Harrow, Rotary	L. Brainard and L. Newton	Attica, N. Y	Apr. 10, 1855	12, 667
Harrow, Rotary	J. Brainerd	Cleveland, Ohio	Oct. 2, 1860	30, 198
Harrow, Rotary	M. B. Champion	Sturgis, Mich	July 13, 1869	92, 422
Harrow, Rotary	J. F. Chase	Westbrook, Me	Mar. 28, 1871	113, 141
Harrow, Rotary	G. Collins	Fremont, Nebr	Sept. 13, 1870	107, 337
Harrow, Rotary	G. Cook	Paris, Ill	Sept. 13, 1859	25, 390
Harrow, Rotary	C. Daniel	Sigel, Mo	Dec. 1, 1863	40, 741
Harrow, Rotary	C. P. Fisher	Leesville, Ohio	June 1, 1869	90, 652
Harrow, Rotary	H. A. Gaston	Stockton, Cal	Oct. 12, 1869	95, 788
Harrow, Rotary	P. Gevin and E. Foreman	Summer Hill, Pa	Nov. 27, 1860	30, 728
Harrow, Rotary	C. and J. K. Gingrich	Annville, Pa	June 28, 1859	24, 554
Harrow, Rotary	W. P. Goolman	Dublin, Ind	Aug. 30, 1859	25, 301
Harrow, Rotary	W. T. Hildrup	Harrisburgh, Pa	Feb. 1, 1859	22, 805
Harrow, Rotary	W. T. Hildrup	Harrisburgh, Pa	Apr. 12, 1859	23, 578
Harrow, Rotary	S. S. Hogle	Cleveland, Ohio	Oct. 11, 1859	25, 736
Harrow, Rotary	S. S. Hogle	Cleveland, Ohio	Oct. 30, 1860	30, 554
Harrow, Rotary	C. Hood	Seneca Falls, N. Y	Dec. 5, 1871	121, 621
Harrow, Rotary	C. Howell	Cleveland, Ohio	Jan. 4, 1859	22, 502
Harrow, Rotary	A. C. Jenson and J. Mathison	Fremont, Nebr	Apr. 9, 1872	125, 572
Harrow, Rotary	M. C. Kilgore	Washington, Iowa	Oct. 4, 1859	25, 651
Harrow, Rotary	S. Lubolt and J. Trout	Lykens, Pa	Feb. 9, 1869	86, 768
Harrow, Rotary	W. H. Main	Liverpool, Ohio	Aug. 24, 1858	21, 265
Harrow, Rotary	W. H. Main	Liverpool, Ohio	Nov. 9, 1858	22, 026
Harrow, Rotary	W. H. Main	Liverpool, Ohio	Mar. 26, 1861	31, 813
Harrow, Rotary	J. W. McLean	Lebanon, Ind	June 7, 1859	24, 322
Harrow, Rotary	H. H. Monroe	Thomaston, Me	Aug. 2, 1864	43, 700
Harrow, Rotary	D. Morris	Bartlett, Ohio	Jan. 1, 1867	60, 776
Harrow, Rotary	A. W. Olds	Green Oak, Mich	Sept. 3, 1861	33, 212
Harrow, Rotary	A. W. Olds	Green Oak, Mich	Dec. 24, 1861	34, 007
Harrow, Rotary	A. W. Olds	Green Oak Station, Mich	Oct. 1, 1872	131, 893
Harrow, Rotary	J. F. Ostrander	New York, N. Y	Aug. 26, 1851	8, 313
Harrow, Rotary	J. D. Parrot	Morristown, N. J	Oct. 31, 1865	50, 729
Harrow, Rotary	R. Rakestraw	Wyoming, Ill	July 11, 1871	116, 991
Harrow, Rotary	R. Rakestraw	Wyoming, Ill	June 10, 1873	139, 814
Harrow, Rotary	F. Raymond	Sandusky, Ohio	Nov. 27, 1860	30, 759
Harrow, Rotary	J. Robins	Boston, Mass	Sept. 21, 1858	21, 577
Harrow, Rotary	H. W. Robinson	Woodstock, Ill	Aug. 20, 1872	130, 748
Harrow, Rotary	P. B. B. Stiles	Galesburgh, Ill	July 30, 1867	67, 372
Harrow, Rotary	H. Thompson	Rockland, Me	May 22, 1866	55, 024
Harrow, Rotary	S. S. Thompson	Heller's Corners, Ind	Sept. 21, 1858	21, 580
Harrow, Rotary	G. W. Toleman	Augusta, Ky	May 10, 1859	23, 962
Harrow, Rotary	L. S. Tyler	Lineville, Pa	Feb. 5, 1861	31, 341
Harrow, Rotary	S. M. Wade	Andover, Ohio	Apr. 12, 1859	23, 627
Harrow, Rotary	B. A., and M. H. Welds and H. W. Strong.	Reading, Mich	Feb. 2, 1869	86, 478
Harrow, Rotating	S. M. Garver	Monticello, Ill	Sept. 8, 1863	39, 806
Harrow, Rotating	J. B. Glascock	Fancy Creek, Ill	July 21, 1857	17, 831
Harrow, Rotating	S. Grenell	Mokena, Ill	Sept. 18, 1866	58, 191
Harrow, Seed-sowing	J. Davis	Allegheny City, Pa	Nov. 11, 1862	36, 898
Harrow, seeder, and roller combined	J. Vessot, sr., and S. Vessot, jr	Joliette, Canada	Mar. 21, 1871	112, 869
Harrow, Seeding	E. Badlam	Ogdensburgh, N. Y	June 18, 1861	32, 554
Harrow, Seeding	W. Finlay	Schoolcraft, Mich	Apr. 10, 1860	27, 789
Harrow, Seeding	H. Hewitt	San Francisco, Cal	Mar. 27, 1860	27, 668
Harrow, Seeding	A. E. Jerome	Monrooville, Ohio	Nov. 15, 1859	26, 110
Harrow, Seeding	M. S. Root	Medina, Ohio	May 22, 1860	28, 411
Harrow, Self-sharpening	H. Hopkins and J. Pratt	Pompey, N. Y	Nov. 9, 1824	....
Harrow, Sulky	W. C. Bye	Collette, Ind	Feb. 25, 1873	136, 131
Harrow, &c., Sulky	J. A. Casey	Maysville, Ky	June 21, 1870	104, 423
Harrow, Sulky	J. E. Cheasebro	Marilla, N. Y	Feb. 9, 1869	86, 731
Harrow, Sulky	W. H. H. Frye	North Fryeburgh, Me	Sept. 7, 1869	94, 589
Harrow, Sulky	E. W. Hewitt	Pecatonica, Ill	Dec. 15, 1868	84, 879
Harrow, Sulky	B. Randall	Adams, N. Y	Apr. 14, 1868	76, 657
Harrow, Sulky	F. P. L. Reimer and H. Asbahr	Davenport, Iowa	July 25, 1871	117, 330
Harrow, Sulky	P. Speelmon	Knox, Ind	Nov. 4, 1873	144, 231
Harrow, Sulky	G. W. Van Goeder	Warren, Ohio	Mar. 4, 1873	136, 564
Harrow, Sward-cutting	D. Rice	Rowe, Mass	May 17, 1844	3, 589
Harrow-teeth	S. W. Corbin	Bainbridge, N. Y	Nov. 29, 1870	109, 719
Harrow-teeth	D. M. Cummings	Enfield, N. H	Nov. 22, 1859	26, 167
Harrow-teeth	D. France	Lakeville, Minn	Dec. 24, 1872	134, 137
Harrow-teeth	E. N. Higley	Lake Village, N. H	July 12, 1864	43, 500
Harrow-teeth	P. V. Hixson	Tioga, Pa	Oct. 6, 1868	82, 835
Harrow-teeth	A. Jones	Westford, Wis	May 12, 1868	77, 737
Harrow-teeth	H. R. Kinney	Portsmouth, Ohio	Sept. 20, 1870	107, 506
Harrow-teeth	W. McConaughey	New Garden, Pa	Dec. 8, 1823	....
Harrow-teeth	A. Patterson	Birmingham, Pa	Nov. 3, 1868	83, 656
Harrow-teeth	W. H. Platt	Dayton, Ohio	Oct. 14, 1873	143, 713
Harrow-teeth	P. Prettyman	Georgetown, Del	Apr. 3, 1834	....
Harrow-teeth	F. R. Willson	Columbus, Ohio	Sept. 24, 1867	69, 290
Harrow-teeth, Apparatus for making	W. McConaughey	New Garden, Pa	Feb. 16, 1827	....
Harrow-teeth, Die for forging	J. Pedder and G. Abel	Pittsburgh, Pa	Mar. 12, 1872	124, 619
Harrow-teeth, Fluke	L. C. Tann	Milton, Del	Mar. 30, 1830	....
Harrow-teeth, Machine for making	J. Morgan	Pittsburgh, Pa	June 14, 1870	104, 186
Harrow-teeth, Machine for rolling	A. H. Holmes	Allegheny, Pa	Feb. 27, 1872	124, 062
Harrow-teeth, Machine for rolling	S. C. Murdoch	Pittsburgh, Pa	Dec. 5, 1871	121, 653

Index of patents issued from the United States Patent Office from 1790 *to* 1873, *inclusive*—Continued.

Invention.	Inventor.	Residence.	Date.	No.
Harrow-teeth, Mode of fastening	G. Fry	Newaygo, Mich	Feb. 28, 1871	112, 236
Harrow-teeth, Reversible	G. W. Hurst	Avon, Ohio	June 24, 1873	140, 198
Harrow, Wheel	T. M. Brintnall	Medina, Ohio	Aug. 27, 1872	130, 839
Harrow, Wheel	E. R. Powell	Jeffersonville, Vt	Nov. 29, 1870	109, 754
Harrow, Wheeled	A. C. Baker and N. O. Hoyt	La Fayette, N. Y	Oct. 29, 1867	70, 314
Harrow, Wheeled	A. L. P. Vairin	Ripley, Miss	Jan. 4, 1870	98, 647
Harrow with cultivator-attachment	A. M. Bakewell	Normal, Ill	Apr. 20, 1869	89, 192
Harrows, Combined	C. Wickoff, jr	Fairview, Ill	Apr. 25, 1871	114, 244
Harrows, Riding-attachment for	J. M. Freeman	Belleville, N. Y	Aug. 27, 1867	68, 179
Harrows, Riding-attachment to	A. D. Shaver	Belleville, N. Y	July 7, 1868	79, 604
Harrows to land-rollers, Attachment of	D. Hill	Bartonia, Ind	Oct. 11, 1853	10, 109
Harrows to plows, Attachment of	J. Stroop	Philadelphia, Pa	June 26, 1849	6, 562
Harrows to seed-planters, Attachment of	M. and S. S. Sage	Windsor, N. Y	Apr. 16, 1850	7, 293
Harvester	A. S. Acker	Albion, N. Y	Feb. 27, 1866	52, 809
Harvester	H. A. Adams	Sandwich, Ill	Nov. 4, 1873	144, 179
Harvester	J. H. Adamson	Auburn, South Australia	Aug. 29, 1871	118, 574
Harvester	H. Adkins	Plymouth, Ill	Sept. 2, 1856	15, 638
Harvester	R. L. Allen	New York, N. Y	Sept. 7, 1858	21, 401
Harvester	S. S. Allen	Bristol, Pa	Apr. 7, 1857	16, 957
Harvester	N. Allstatter	Hamilton, Ohio	Sept. 28, 1869	95, 301
Harvester	J. Anthony	Ledyard, N. Y	Oct. 8, 1867	69, 606
Harvester	S. Augspurger	Trenton, Ohio	Feb. 16, 1864	41, 593
Harvester	E. M. Awrey	Caistor, Canada	July 15, 1873	140, 804
Harvester	A. P. Ayers	Chicago, Ill	June 21, 1870	104, 407
Harvester	D. Babcock	Warsaw, Ill	June 30, 1868	79, 300
Harvester	G. Bailey	Wiscotta, Iowa	Sept. 6, 1864	44, 064
Harvester	N. A. Baker	Covington, Ky	May 21, 1872	126, 919
Harvester	M. and W. P. Bales	London, Ohio	Sept. 13, 1870	107, 214
Harvester	E. Ball	Canton, Ohio	Oct. 18, 1859	25, 797
Harvester	E. Ball and M. L. Ballard	Canton, Ohio	Mar. 20, 1860	27, 511
Harvester	V. S. Barber	Alliance, Ohio	Aug. 23, 1870	106, 535
Harvester	S. and J. H. Barley	Longwood, Mo	June 19, 1860	28, 727
Harvester	A. B. Barnard, S. R. Nye, and R. L. Hewett.	West Fitchburgh, Mass	Aug. 18, 1868	81, 129
Harvester	J. Barnes	Rockford, Ill	Dec. 13, 1864	45, 383
Harvester	J. Barnes	Rockford, Ill	Jan. 12, 1869	85, 722
Harvester	J. Barnes	Rockford, Ill	Sept. 28, 1869	95, 304
Harvester	J. Barnes	Rockford, Ill	May 16, 1871	114, 747
Harvester	L. Barnes	Islip Township, N. Y	Jan. 8, 1856	14, 046
Harvester	T. J. Barnes	Corry, Pa	Jan. 17, 1871	110, 949
Harvester	J. A. Barrington	Fredericktown, Ohio	Dec. 21, 1858	22, 341
Harvester	S. S. Bartlett	Providence, R. I	Feb. 25, 1862	34, 545
Harvester	S. S. Bartlett	Providence, R. I	Nov. 10, 1863	40, 585
Harvester	S. S. Bartlett	Providence, R. I	May 16, 1865	47, 691
Harvester	S. O. Bartow	Bethel, Conn	Feb. 4, 1868	74, 036
Harvester	S. D. Bates	Lewisburgh, Pa	Nov. 9, 1869	96, 660
Harvester	L. M. Batty	Canton, Ohio	Dec. 28, 1869	98, 220
Harvester	L. M. Batty	Canton, Ohio	Jan. 31, 1871	111, 424
Harvester	L. M. Batty	Canton, Ohio	Mar. 14, 1871	112, 520
Harvester	J. Beach	De Ruyter, N. Y	Mar. 19, 1872	124, 781
Harvester	J. H. Beam	Springfield, Ill	Mar. 28, 1871	113, 134
Harvester	W. G. Beels	Independence, Iowa	Jan. 18, 1870	98, 910
Harvester	S. Bell	Marseilles, Ill	Feb. 28, 1854	10, 581
Harvester	O. Billings	Elyria, Ohio	Sept. 12, 1871	118, 837
Harvester	J. Birch, A. Crosby, and T. Birch.	Westfield, N. Y., and Meadville, Pa.	June 28, 1870	104, 695
Harvester	V. W. Blanchard	Bridgeport, Vt	Dec. 8, 1863	40, 812
Harvester	V. W. Blanchard	Bridgeport, Vt	Oct. 17, 1865	50, 441
Harvester	V. W. Blanchard	Bridgeport, Vt	Jan. 1, 1867	60, 676
Harvester	J. Blue	Covert, N. Y	Dec. 18, 1860	30, 953
Harvester	J. Blue	Covert, N. Y	Mar. 26, 1861	31, 774
Harvester	J. Blue	Trumansburgh, N. Y	Nov. 24, 1868	84, 349
Harvester	O. Bonney, jr	San Francisco, Cal	July 21, 1868	80, 124
Harvester	O. Bonney, jr	San Francisco, Cal	Sept. 28, 1869	95, 188
Harvester	J. W. Bosse	Saint Louis, Mo	Jan. 10, 1865	45, 810
Harvester	J. W. Bosse	Saint Louis, Mo	Jan. 17, 1865	45, 905
Harvester	J. W. Bosse	Saint Louis, Mo	Jan. 17, 1865	45, 906
Harvester	J. Bardwell	Brockport, N. Y	Jan. 2, 1872	122, 432
Harvester	J. B. Bowen and J. E. Baker	Madison, Wis	Apr. 22, 1862	35, 009
Harvester	W. F. Brabrook	South Hardwick, Vt	Aug. 6, 1867	67, 490
Harvester	C. J. Brackebush and C. E. Merrifield.	Indianapolis, Ind	Sept. 14, 1869	94, 865
Harvester	H. Brackett	Valley Falls, N. Y	May 4, 1869	89, 730
Harvester	C. Bradfield	Philadelphia, Pa	Aug. 21, 1855	13, 480
Harvester	I. W. Bragg	Kidder, Mo	July 8, 1873	140, 675
Harvester	F. Bramer	Fabius, N. Y	Dec. 19, 1865	51, 546
Harvester	T. Brett	Geneva, Ohio	Nov. 17, 1863	40, 601
Harvester	H. H. Bridenthall, jr	New Derby, Pa	Oct. 17, 1871	120, 027
Harvester	A. D. Briggs	Springfield, Mass	Feb. 16, 1858	19, 344
Harvester	A. B. Briggs	South Deerfield, Mass	Feb. 9, 1864	41, 566
Harvester	C. R. Brinkerhoff	Batavia, N. Y	July 10, 1860	29, 053
Harvester	C. R. Brinkerhoff	Rochester, N. Y	Jan. 11, 1870	98, 737
Harvester	A. Brown	Worcester, Mass	Aug. 24, 1869	94, 167
Harvester	A. Brown, L. G. Kniffen, and T. H. Dodge.	Worcester, Mass., and Nashua, N. H.	Nov. 17, 1863	40, 658
Harvester	C. B. Brown	Alton, Ill	May 11, 1858	20, 191
Harvester	L. D. Brown	Saint Louis, Mo	Nov. 20, 1860	30, 664
Harvester	R. Brown	Frederick, Md	June 18, 1861	32, 556
Harvester	R. Brown	Frederick, Md	Oct. 27, 1863	40, 395
Harvester	R. D. Brown	Covington, Ky	Feb. 9, 1864	41, 477
Harvester	T. S. Brown	Poughkeepsie, N. Y	Aug. 25, 1868	81, 472
Harvester	T. S. Brown	Poughkeepsie, N. Y	July 2, 1872	128, 584
Harvester	A. C. Brownlich	Buffalo, N. Y	Jan. 4, 1859	22, 483
Harvester	C. Brownlich	Buffalo, N. Y	Mar. 1, 1859	23, 075
Harvester	S. T. Bruce	Marshall, Mo	May 1, 1860	28, 061
Harvester	R. Bryson	Schenectady, N. Y	June 5, 1860	28, 557
Harvester	R. Bryson	Schenectady, N. Y	Mar. 31, 1863	38, 030
Harvester	R. Bryson	Schenectady, N. Y	July 31, 1866	56, 698

Index of patents issued from the United States Patent Office from 1790 *to* 1873, *inclusive*—Continued.

Invention.	Inventor.	Residence.	Date.	No.
Harvester	A. Bullock	Busti, N. Y	Nov. 11, 1856	16, 079
Harvester	C. Bullock	Jamestown, N. Y	Dec. 21, 1858	22, 345
Harvester	O. H. Burdick	Auburn, N. Y	Feb. 27, 1866	52, 927
Harvester	J. Burke	Sycamore, Ill	Nov. 12, 1867	70, 798
Harvester	W. H. Burkhart	Bucyrus, Ohio	May 30, 1865	47, 926
Harvester	W. H. Burkhart	Bucyrus, Ohio	Feb. 13, 1866	52, 526
Harvester	H. K. Burnett	Poughkeepsie, N. Y	July 28, 1868	80, 272
Harvester	J. P. Burnham	Rockford, Ill	Nov. 29, 1859	26, 251
Harvester	G. E. Burt	Harvard, Mass	Dec. 29, 1868	85, 274
Harvester	S. and J. H. Buser	Warren, Ill	Feb. 14, 1860	27, 105
Harvester	J. Butter	Buffalo, N. Y	Nov. 15, 1859	26, 142
Harvester	J. Butter	Buffalo, N. Y	June 16, 1863	38, 882
Harvester	J. S. Butterfield	Philadelphia, Pa	Mar. 2, 1858	19, 483
Harvester	C. Cadwell	Waukegan, Ill	Jan. 16, 1866	52, 026
Harvester	C. Cadwell	Waukegan, Ill	Jan. 16, 1866	52, 027
Harvester	C. Cadwell	Waukegan, Ill	Aug. 14, 1866	57, 082
Harvester	C. Cadwell	Waukegan, Ill	May 12, 1868	77, 714
Harvester	J. M. Canfield, A. T. Still, and E. P. Wheeler.	Lawrence, Kans	Mar. 27, 1866	53, 406
Harvester	C. W. Cardot	Fredonia, N. Y	May 14, 1867	64, 749
Harvester	S. D. Carpenter	Madison, Wis	Nov. 4, 1873	144, 189
Harvester	W. D. Carpenter	South Berwick, Me	Jan. 21, 1868	73, 501
Harvester	H. Carter	Greene, N. Y	July 19, 1859	24, 793
Harvester	H. M. Carter	La Fayette, Ind	Feb. 4, 1873	135, 403
Harvester	J. A. Cauldwell	Watkins, N. Y	Aug. 8, 1871	117, 862
Harvester	W. Chapman	Salisbury, Conn	Feb. 23, 1869	87, 244
Harvester	C. H. Charlesworth and J. H. Short.	Avoca, N. Y	May 17, 1870	103, 136
Harvester	G. E. Chenoweth	Baltimore, Md	Mar. 30, 1858	19, 749
Harvester	G. E. Chenoweth	Baltimore, Md	Feb. 8, 1859	22, 855
Harvester	G. E. Chenoweth	Baltimore, Md	Mar. 1, 1859	23, 077
Harvester	G. E. Chenoweth	Baltimore, Md	Nov. 15, 1859	26, 091
Harvester	G. E. Chenoweth	Baltimore, Md	Mar. 13, 1860	27, 617
Harvester	W. L. Childs	Piermont, N. Y	Mar. 2, 1858	19, 486
Harvester	C. Clapp	Trumansburgh, N. Y	May 10, 1870	102, 776
Harvester	C. Clapp	Trumansburgh, N. Y	Aug. 1, 1871	117, 514
Harvester	H. Clark	Rochester, N. Y	Mar. 10, 1857	16, 788
Harvester	S. A. Clemens	Rockford, Ill	Mar. 3, 1857	16, 721
Harvester	E. H. Clinton	Iowa City, Iowa	Dec. 2, 1873	145, 050
Harvester	D. Clow	Janesville, Wis	Feb. 15, 1859	22, 936
Harvester	D. Clow	Janesville, Wis	July 22, 1862	35, 923
Harvester	N. Clute	Dunnsville, N. Y	June 16, 1857	17, 555
Harvester	N. Clute	Dunnsville, N. Y	Nov. 30, 1358	22, 163
Harvester	W. F. Cochrane	Springfield, Ohio	Jan. 31, 1865	46, 178
Harvester	W. F. Cochrane	Springfield, Ohio	Jan. 31, 1865	46, 179
Harvester	W. F. Cochrane	Springfield, Ohio	Jan. 31, 1865	46, 180
Harvester	W. F. Cochrane	Springfield, Ohio	Jan. 31, 1865	46, 181
Harvester	W. F. Cochrane	Springfield, Ohio	Jan. 31, 1865	46, 182
Harvester	W. F. Cochrane	Springfield, Ohio	Jan. 31, 1865	46, 183
Harvester	W. F. Cochrane	Springfield, Ohio	Jan. 1, 1867	60, 692
Harvester	W. F. Cochrane	Springfield, Ohio	Jan. 1, 1867	60, 693
Harvester	J. F. Coddington	Boundbrook, N. J	July 7, 1868	79, 551
Harvester	J. T. Coddington	Newark, N. J	Mar. 9, 1869	87, 637
Harvester	W. Cogswell	Ottawa, Ill	Aug. 29, 1865	49, 607
Harvester	W. Cogswell	Ottawa, Ill	Aug. 29, 1865	49, 608
Harvestor	W. Cogswell	Ottawa, Ill	Nov. 2, 1869	96, 311
Harvester	W. and I. Cogswell, jr	Ottawa, Ill	Dec. 6, 1859	26, 338
Harvester	L. H. Colborn	Baltimore, Md	Apr. 12, 1859	23, 552
Harvester	I. H. Caller	Poughkeepsie, N. Y	Sept. 1, 1863	39, 719
Harvester	I. H. Caller	Poughkeepsie, N. Y	Nov. 24, 1863	40, 675
Harvester	J. Collins	Guelph, Canada	Feb. 19, 1867	62, 252
Harvester	R. Conarroe	Camden, Ohio	Dec. 14, 1869	97, 884
Harvester	I. H. Conklin	Rockford, Ill	May 25, 1858	20, 334
Harvester	J. M. Connel	Newark, Ohio	Sept. 14, 1869	94, 871
Harvester	R. F. Cooke	Newark, N. J	Dec. 31, 1867	72, 722
Harvester	G. T. Coolman and C. M. Young	Corry, Pa	Mar. 30, 1869	88, 453
Harvester	G. T. Coolman and C. M. Young	Corry, Pa	May 11, 1869	89, 973
Harvester	G. T. Coolman and C. M. Young	Corry, Pa	Feb. 15, 1870	99, 850
Harvester	G. E. Cooper	Baltimore, Md	Oct. 12, 1858	21, 741
Harvester	F. C. Coppage	Terre Haute, Ind	June 19, 1866	55, 763
Harvester	F. C. Coppage	Terre Haute, Ind	July 10, 1866	56, 320
Harvester	F. C. Coppage	Terre Haute, Ind	Dec. 10, 1867	71, 989
Harvester	W. R. Cory	Springfield, Ill	Jan. 3, 1871	110, 742
Harvester	J. F. Crawford	New York Mills, N. Y	Jan. 11, 1870	98, 671
Harvester	S. Crawford	London, Canada	May 21, 1872	126, 932
Harvester	T. Crumbling	Hellam, Pa	Sept. 27, 1859	25, 559
Harvester	G. S. Curtis	Chicago, Ill	Feb. 23, 1858	19, 411
Harvester	G. S. and H. Curtis	Chicago, Ill	May 31, 1864	42, 930
Harvester	J. Curtis	Hackettstown, N. J	Nov. 29, 1864	45, 228
Harvester	T. Curtis	New Hudson, Mich	May 3, 1864	42, 564
Harvester	J. D. Custer	Norristown, Pa	Sept. 27, 1859	25, 561
Harvester	R. Daniels	Woodstock, Vt	Oct. 13, 1857	18, 390
Harvester	A. L. Darby	White Creek, N. Y	Dec. 23, 1862	37, 253
Harvester	J. S. Davis	Tiffin, Ohio	Oct. 31, 1865	50, 693
Harvester	J. S. Davis	Tiffin, Ohio	Mar. 10, 1868	75, 383
Harvester	J. S. Davis	Toledo, Ohio	May 20, 1873	139, 122
Harvester	R. M. Davis	Eaton, N. Y	July 22, 1862	35, 970
Harvester	T. D. Davis and J. M. Waldron.	Syracuse and South Otselic, N. Y.	Jan. 7, 1862	34, 097
Harvester	C. Denton	Pekin, Ill	Oct. 9, 1866	58, 720
Harvester	C. Denton	Decatur, Ill	June 30, 1868	79, 452
Harvester	W. De Witt	Cleveland, Ohio	Jan. 7, 1862	34, 090
Harvester	J. Dick, jr	Canton, Ohio	Oct. 9, 1866	58, 617
Harvester	C. G. Dickinson	Poughkeepsie, N. Y	Feb. 1, 1859	22, 786
Harvester	C. G. Dickinson	Poughkeepsie, N. Y	Jan. 29, 1861	31, 229
Harvester	S. P. Doane	San Francisco, Cal	Feb. 2, 1869	86, 375
Harvester	J. A. Dodge	Auburn, N. Y	May 23, 1865	47, 807
Harvester	J. A. Dodge	Auburn, N. Y	June 26, 1866	55, 837

Index of patents issued from the United States Patent Office from 1790 *to* 1873, *inclusive*—Continued.

Invention.	Inventor.	Residence.	Date.	No.
Harvester	J. A. Dodge	Auburn, N. Y	Aug. 20, 1867	67, 852
Harvester	J. A. Dodge	Auburn, N. Y	Dec. 22, 1868	85, 168
Harvester	J. A. Dodge	Auburn, N. Y	Aug. 6, 1872	130, 285
Harvester	T. H. Dodge	Worcester, Mass	Jan. 2, 1872	122, 445
Harvester	A. G. Donnelly	Breesport, N. Y	Nov. 17, 1868	84, 051
Harvester	J. W. H. Doubler	West Alden, N. Y	Sept. 16, 1873	142, 903
Harvester	W. Dripps	Coatesville, Pa	Oct. 7, 1856	15, 843
Harvester	J. J. Duchesne	Lacon, Ill	July 7, 1868	79, 740
Harvester	I. A. Dufield	McHenry, Ill	Nov. 1, 1859	25, 960
Harvester	R. Dutton	Dayton, Ohio	Apr. 27, 1858	20, 050
Harvester	R. Dutton	New York, N. Y	Feb. 11, 1868	74, 208
Harvester	R. Dutton	New York, N. Y	Feb. 11, 1868	74, 209
Harvester	R. Dutton	New York, N. Y	Feb. 11, 1868	74, 210
Harvester	P. Dyer, jr	Unadilla, Mich	May 18, 1869	90, 247
Harvester	J. F. Earl	San Francisco, Cal	Nov. 10, 1868	83, 839
Harvester	M. Easterbrook, jr	Geneva, N. Y	May 22, 1866	54, 876
Harvester	M. Easterbrook, jr	Geneva, N. Y	June 11, 1867	65, 732
Harvester	J. Ebner and F. Leuthy	Lancaster, Pa	Oct. 18, 1859	25, 815
Harvester	R. Eickemeyer	Yonkers, N. Y	Apr 12, 1870	101, 719
Harvester	G. S. Ellard	Westerly, R. I	July 6, 1869	92, 294
Harvester	M. E. Ellsworth	Hudson, Ohio	Aug. 10, 1858	21, 125
Harvester	J. H. Elward	Ottawa, Ill	Apr. 26, 1864	42, 532
Harvester	J. H. Elward	Ottawa, Ill	Nov. 1, 1864	44, 857
Harvester	J. H. Elward	Polo, Ill	June 14, 1870	104, 290
Harvester	J. H. Elward	Polo, Ill	Apr. 25, 1871	113, 990
Harvester	J. H. Elward	Polo, Ill	May 28, 1872	127, 324
Harvester	J. H. Elward	Saint Paul, Minn	Aug. 19, 1873	141, 863
Harvester	J. H. Elward	Polo, Ill	Aug. 19, 1873	141, 864
Harvester	J. H. Elward	Saint Paul, Minn	Aug. 19, 1873	141, 865
Harvester	J. H. Elward	Saint Paul, Minn	Aug. 19, 1873	141, 866
Harvester	J. H. Elward	Saint Paul, Minn	Aug. 19, 1873	141, 867
Harvester	J. H. Elward	Saint Paul, Minn	Aug. 19, 1873	141, 868
Harvester	P. Embree	West Chester, Pa	May 19, 1863	38, 626
Harvester	D. L. Emerson	Rockford, Ill	Dec. 10, 1861	33, 918
Harvester	D. L. Emerson	Rockford, Ill	Aug. 11, 1863	39, 526
Harvester	D. L. Emerson	Rockford, Ill	Aug. 18, 1863	39, 559
Harvester	D. L. Emerson	Rockford, Ill	Sept. 6, 1864	44, 084
Harvester	D. L. Emerson	Rockford, Ill	Sept. 10, 1867	68, 615
Harvester	R. Emerson, jr., and F. Graham	Rockford, Ill	Jan. 14, 1862	34, 181
Harvester	H. L. Emery	Albany, N. Y	Oct. 25, 1859	25, 888
Harvester	J. Emery	Cedar Falls, Iowa	June 25, 1867	66, 135
Harvester	E. Emmert	Franklin Grove, Ill	Jan. 19, 1858	19, 137
Harvester	E. Emmert	Franklin Grove, Ill	Jan. 28, 1868	73, 788
Harvester	E. Emmert	Franklin Grove, Ill	July 13, 1869	92, 435
Harvester	G. Engle	Bunker Hill, Wis	Mar. 3, 1863	37, 813
Harvester	D. W. Entrikin and L. H. Davis	West Chester, Pa	Apr. 13, 1858	19, 919
Harvester	G. Esterly	White Water, Wis	Jan. 22, 1861	31, 158
Harvester	G. Esterly	White Water, Wis	Jan. 22, 1861	31, 159
Harvester	G. Esterly	White Water, Wis	Jan. 13, 1863	37, 391
Harvester	G. Esterly	White Water, Wis	Feb. 18, 1868	74, 676
Harvester	F. Ewer	Mendon Centre, N. Y	Sept. 8, 1863	39, 804
Harvester	E. W. Fairman	Orfordville, Wis	Mar. 10, 1868	75, 401
Harvester	J. A. Falk, A. Johnson, and G. A. Erickson.	Altona, Ill	Aug. 30, 1859	25, 251
Harvester	G. C. Fanckboner	Schoolcraft, Mich	Nov. 28, 1865	51, 163
Harvester	G. Farmer	Oceola, Ill	Aug. 21, 1860	29, 685
Harvester	J. Farrington	Corry, Pa	June 20, 1871	116, 037
Harvester	J. Farrington	Corry, Pa	Jan. 7, 1873	134, 531
Harvester	D. E. Fisher	Cedar Spring, Ind	Jan. 14, 1868	73, 314
Harvester	H. Fisher	Alliance, Ohio	July 26, 1859	24, 860
Harvester	H. Fisher	Alliance, Ohio	Jan. 26, 1864	41, 411
Harvester	R. H. Fisher	Claremont, N. H	May 4, 1858	20, 152
Harvester	R. H. Fisher	Claremont, N. H	Oct. 19, 1858	21, 827
Harvester	B. G. Fitzhugh	Baltimore, Md	May 1, 1866	54, 466
Harvester	P. Flickinger	Hanover, Pa	Feb. 14, 1860	27, 120
Harvester	A. B. J. Flowers	Greenfield, Ind	Sept. 22, 1857	18, 240
Harvester	A. B. J. Flowers	Greenfield, Ind	June 1, 1858	20, 416
Harvester	A. Foot	Earlville, Ill	Apr. 13, 1869	88, 951
Harvester	E. T. Ford	Stillwater, N. Y	Jan. 6, 1863	37, 371
Harvester	E. T. Ford	Stillwater, N. Y	July 17, 1866	56, 397
Harvester	T. N. Foster	Watertown, N. Y	Oct. 2, 1860	30, 215
Harvester	J. L. and H. K. Fountain	Rockford, Ill	May 15, 1849	6, 450
Harvester	H. E. Fowler	North Branford, Conn	Aug. 18, 1868	81, 158
Harvester	J. Fox	Baltimore, Md	Oct. 15, 1867	69, 908
Harvester	J. Fox and S. C. Ridgway	Baltimore, Md	Apr. 17, 1866	53, 969
Harvester	H. H. Faye	Ottawa, Ill	Sept. 25, 1860	30, 135
Harvester	J. Q. A. Frazier	Piqua, Ohio	July 16, 1861	32, 828
Harvester	W. Gage	Buffalo, N. Y	Sept. 16, 1856	15, 735
Harvester	J. L. Garber	Hamilton, Ohio	Dec. 19, 1865	51, 579
Harvester	C. B. and G. B. Garlinghouse	Allensville, Ind	July 19, 1859	24, 803
Harvester	G. B. and C. B. Garlinghouse	Allensville, Ind	Oct. 28, 1862	36, 775
Harvester	J. Geiss, J. Brosius, and W. P. Penn.	Belleville, Ill	July 8, 1862	35, 815
Harvester	D. D. Gitt	Arendtsville, Pa	Apr. 12, 1864	42, 282
Harvester	D. D. Gitt	Arendtsville, Pa	Jan. 10, 1865	45, 823
Harvester	D. D. Gitt	Arendtsville, Pa	Feb. 27, 1866	52, 843
Harvester	C. W. Glover	Roxbury, Conn	Oct. 14, 1856	15, 882
Harvester	R. Glover	Grayville, Ill	May 12, 1863	38, 533
Harvester	W. H. Goble and A. Stuart	Urbana, Ohio	Nov. 21, 1848	5, 933
Harvester	W. F. Goodwin	Washington, D. C	Mar. 13, 1866	53, 137
Harvester	W. F. Goodwin	East New York, N. Y	Jan. 14, 1868	73, 424
Harvester	W. F. Goodwin	East New York, N. Y	June 9, 1868	78, 800
Harvester	W. F. Goodwin	East New York, N. Y	Mar. 30, 1869	88, 381
Harvester	W. F. Goodwin	Metuchen, N. J	Mar. 22, 1870	101, 120
Harvester	W. F. Goodwin	Metuchen, N. J	Feb. 7, 1871	111, 633
Harvester	W. F. Goodwin	Metuchen, N. J	Aug. 29, 1871	118, 526
Harvester	W. F. Goodwin and A. M. Smith	East New York, N. Y., and Washington, D. C.	Mar. 17, 1868	75, 677

Index of patents issued from the United States Patent Office from 1790 *to* 1873, *inclusive*—**Continued.**

Invention.	Inventor.	Residence.	Date.	No.
Harvester	H. S. Gondon	Steel's Mills, Ill	Oct. 8, 1872	132,003
Harvester	J. F. Gordon	Rochester, N. Y	Aug. 27, 1872	130,852
Harvester	J. Gore	Brattleborough, Vt	Dec. 27, 1859	26,582
Harvester	A. B. Graham	Waukegan, Ill	July 23, 1867	67,041
Harvester	A. B. Graham	Waukegan, Ill	Feb. 11, 1868	74,342
Harvester	A. B. Graham	Waukegan, Ill	Apr. 7, 1868	76,436
Harvester	A. B. Graham	Waukegan, Ill	Mar. 9, 1869	87,561
Harvester	G. S. Grier	Milford, Del	Oct. 31, 1871	120,431
Harvester	J. M. Groh	Shaefferstown, Pa	Oct. 28, 1862	36,780
Harvester	C. P. Gronberg	Aurora, Ill	Mar. 16, 1869	87,770
Harvester	S. Gumaer	Chicago, Ill	Aug. 4, 1857	17,927
Harvester	S. B. Haines	Lewistown, Pa	Mar. 8, 1864	41,842
Harvester	S. B. Haines	Lewistown, Pa	Jan. 7, 1868	72,999
Harvester	J. C. Hall	Monroe, Wis	Mar. 2, 1869	87,488
Harvester	L. Hall	Metamora, Ill	Sept. 6, 1870	107,039
Harvester	M. Hallenbeck	Albany, N. Y	May 18, 1858	20,271
Harvester	M. Hallenbeck	Albany, N. Y	May 18, 1858	20,272
Harvester	M. Hallenbeck	Albany, N. Y	Nov. 3, 1868	83,627
Harvester	M. Hallenbeck	Phillipsburgh, N. Y	July 23, 1872	129,817
Harvester	M. Hallenbeck	Albany, N. Y	July 30, 1872	130,041
Harvester	G. Hamel	Jenkintown, Pa	Sept. 3, 1872	130,993
Harvester	H. D. Hammond	Batavia, N. Y	June 30, 1857	17,676
Harvester	R. Hance	Piscatonica, Ill	Sept. 16, 1862	36,460
Harvester	W. H. Harman	Westminster, Md	July 4, 1871	116,588
Harvester	D. S. Harner	Xenia, Ohio	Nov. 19, 1867	71,164
Harvester	J. Harris	Janesville, Wis	July 28, 1868	80,479
Harvester	J. Harris	Janesville, Wis	July 9, 1872	128,726
Harvester	J. K. Harris	Allensville, Ind	Oct. 12, 1858	21,804
Harvester	L. W. Harris	Waterville, N. Y	Mar. 3, 1857	16,730
Harvester	W. Harrison	Linneus, Mo	June 10, 1873	139,787
Harvester	W. O. Harrison	Chittenden, Vt	July 21, 1868	80,170
Harvester	A. J. Haswell and J. W. Irwin	Circleville, Ohio	Sept. 20, 1870	107,487
Harvester	B. G. H. Hathaway	Rock Stream, N. Y	May 28, 1867	65,218
Harvester	B. G. H. Hathaway	Rock Stream, N. Y	Feb. 4, 1868	74,085
Harvester	B. G. H. Hathaway	Rock Stream, N. Y	Feb. 18, 1868	74,685
Harvester	B. G. H. and G. M. Hathaway	Rock Stream, N. Y	May 28, 1867	65,219
Harvester	J. S. and R. Hawkins	Greenfield, Ind	Aug. 23, 1859	25,194
Harvester	G. F. Hawley	Grand Rapids, Mich	Dec. 27, 1870	110,568
Harvester	A. A. Heath	West Greenville, Pa	May 9, 1865	47,639
Harvester	B. Hess	Manor Township, Pa	Feb. 27, 1866	52,851
Harvester	J. C. Henermann and J. Reeves	Camden, N. J	July 1, 1856	15,236
Harvester	A. Hilts, jr	Springdale, Ohio	Oct. 1, 1867	69,342
Harvester	I. I. Hite	White Post, Va	July 10, 1855	13,226
Harvester	P. F. Hodges	Massillon, Ohio	Aug. 26, 1873	142,231
Harvester	R. Hoffheins	Dover, Pa	May 20, 1862	35,315
Harvester	R. Hoffheins	Dover, Pa	Nov. 3, 1863	40,481
Harvester	L. B. Hoit and M. Laflin	Cedar Falls, Iowa, and Chicago, Ill.	July 20, 1869	92,723
Harvester	C. Holloway	Petersburgh, Va	Mar. 17, 1857	16,836
Harvester	S. T. Holly	Rockford, Ill	Mar. 25, 1862	34,804
Harvester	S. T. Holly	Rockford, Ill	Apr. 29, 1862	35,096
Harvester	S. T. Holly	Rockford, Ill	June 23, 1863	39,014
Harvester	C. C. Holman	Clayville, N. Y	Jan. 23, 1866	52,168
Harvester	G. W. Holmes	Council Bluffs, Iowa	Feb. 13, 1872	123,631
Harvester	H. L. Hopkins	Lebanon, N. Y	Dec. 17, 1861	33,943
Harvester	H. L. Hopkins	Eaton, N. Y	Dec. 12, 1871	121,784
Harvester	J. Howard and E. T. Bonsfield	Bedford, England	Nov. 19, 1867	71,175
Harvester	H. Howe	Oneonta, N. Y	Oct. 20, 1868	83,164
Harvester	C. Howell	Cleveland, Ohio	Feb. 23, 1858	19,422
Harvester	M. G. Hubbard	Penn Yan, N. Y	Jan. 20, 1857	16,441
Harvester	M. G. Hubbard	Penn Yan, N. Y	Jan. 27, 1857	16,484
Harvester	M. G. Hubbard	Penn Yan, N. Y	Mar. 17, 1857	16,840
Harvester	M. G. Hubbard	Penn Yan, N. Y	May 12, 1857	17,277
Harvester	M. G. Hubbard	Penn Yan, N. Y	May 12, 1857	17,280
Harvester	M. G. Hubbard	Penn Yan, N. Y	Feb. 9, 1858	19,298
Harvester	M. G. Hubbard	Penn Yan, N. Y	Feb. 15, 1859	22,953
Harvester	M. G. Hubbard	Penn Yan, N. Y	Feb. 22, 1859	23,029
Harvester	M. G. Hubbard	New York, N. Y	Nov. 15, 1859	26,105
Harvester	M. G. Hubbard	Syracuse, N. Y	Sept. 13, 1864	44,192
Harvester	M. G. Hubbard	Syracuse, N. Y	Sept. 22, 1868	82,412
Harvester	M. G. Hubbard	Syracuse, N. Y	Sept. 22, 1868	82,413
Harvester	M. G. Hubbard	Syracuse, N. Y	Sept. 22, 1868	82,414
Harvester	M. G. Hubbard	Syracuse, N. Y	Sept. 22, 1868	82,415
Harvester	S. Hull	Poughkeepsie, N. Y	Nov. 16, 1858	22,077
Harvester	S. Hull	Poughkeepsie, N. Y	Dec. 2, 1862	37,073
Harvester	S. Hull	Poughkeepsie, N. Y	Dec. 22, 1863	41,040
Harvester	S. Hull	Poughkeepsie, N. Y	June 14, 1864	43,116
Harvester	J. P. Humphreys	Nashville, Tenn	Apr. 16, 1872	125,816
Harvester	S. R. Hunter	Cortland, N. Y	July 22, 1856	15,377
Harvester	T. S. Hunter	Cross Plains, Iowa	Jan. 5, 1864	41,071
Harvester	O. Hussey	Baltimore, Md	Aug. 23, 1859	25,201
Harvester	J. H. Irwin	Beardstown, Ill	Aug. 21, 1860	29,698
Harvester	G. M. Jackson	North Hector, N. Y	Dec. 10, 1867	72,046
Harvester	G. M. Jackson	North Hector, N. Y	Mar. 30, 1869	88,483
Harvester	J. Jann	New Windsor, Md	June 5, 1866	55,304
Harvester	B. A. Jenkins	White Water, Wis	July 10, 1860	29,084
Harvester	G. T. and M. Jerome	Mineola, N. Y	Oct. 5, 1858	21,681
Harvester	I. Jewell	Wheaton, Ill	May 21, 1861	32,400
Harvester	W. B. Johns and W. J. Read	Georgetown, D. C., and Cumberland, Md.	June 28, 1870	104,852
Harvester	I. A. Johnson	Rockford, Ill	Feb. 28, 1871	112,147
Harvester	L. H. Johnson	Rochester, Minn	Feb. 16, 1869	87,048
Harvester	S. and L. Johnson, jr	Avon, N. Y	Dec. 8, 1857	18,813
Harvester	W. N. Johnson	Oxford, Wis	Aug. 1, 1871	117,639
Harvester	S. Johnston	Buffalo, N. Y	Oct. 28, 1862	36,843
Harvester	S. Johnston	Buffalo, N. Y	Jan. 31, 1865	46,190
Harvester	S. Johnston	Syracuse, N. Y	July 7, 1868	79,575
Harvester	S. Johnston and C. H. Jenner	Brockport, N. Y	June 20, 1871	116,063

Index of patents issued from the United States Patent Office from 1790 *to* 1873, *inclusive*—Continued.

Invention.	Inventor.	Residence.	Date.	No.
Harvester	E. Jones	Chester Cross Roads, Ohio	Mar. 27, 1860	27, 636
Harvester	J. H. Jones	Rockford, Ill	Feb. 23, 1869	87, 263
Harvester	J. H. Jones	Rockford, Ill	Feb. 23, 1869	87, 264
Harvester	J. H. Jones	Rockford, Ill	June 1, 1869	90, 666
Harvester	J. H. Jones	Rockford, Ill	Mar. 12, 1872	124, 587
Harvester	L. A. and G. J. Jones	Barrington, N. Y	May 21, 1867	64, 984
Harvester	W. Jones	Saint Louis, Mo	Dec. 8, 1863	40, 883
Harvester	A. and H. Kane	Newport, N. Y	Apr. 24, 1860	27, 989
Harvester	H. R. Keese	Bridport, Vt	Mar. 29, 1859	23, 376
Harvester	J. M. and M. L. Kellar	Buckeye, Iowa	Oct. 8, 1867	69, 679
Harvester	M. A. Keller	Littletown, Pa	Oct. 22, 1867	70, 001
Harvester	D. A. Kellogg	Valparaiso, Ind	July 4, 1871	116, 601
Harvester	P. H. Kells	Hudson, N. Y	Mar. 21, 1854	10, 671
Harvester	W. A. Kerby	Auburn, N. Y	Apr. 19, 1870	102, 131
Harvester	J. Kershaw	Kent, Ohio	June 15, 1869	91, 347
Harvester	W. F. Ketchum	Buffalo, N. Y	June 29, 1858	20, 719
Harvester	W. F. Ketchum	Buffalo, N. Y	Jan. 18, 1859	22, 655
Harvester	M. C. Kilgore	Washington, Iowa	Nov. 8, 1864	44, 956
Harvester	O. H. King	Salem, Iowa	Aug. 2, 1859	24, 941
Harvester	J. F. Kingwill	Chicago, Ill	Feb. 25, 1873	136, 325
Harvester	D. P. Kinyon	Raritan, N. J	Mar. 1, 1859	23, 090
Harvester	W. A. Kirby	Buffalo, N. Y	Nov. 15, 1859	26, 114
Harvester	W. A. Kirby	Buffalo, N. Y	May 15, 1860	28, 284
Harvester	W. A. Kirby	Auburn, N. Y	Apr. 5, 1864	42, 255
Harvester	W. A. Kirby	Buffalo, N. Y	Apr. 19, 1864	42, 440
Harvester	W. A. Kirby	Auburn, N. Y	May 18, 1869	90, 273
Harvester	W. A. Kirby	Auburn, N. Y	June 25, 1872	128, 402
Harvester	J. Kline	Mechanicsburgh, Pa	May 10, 1870	102, 835
Harvester	G. S. Knapp	Dubuque, Iowa	Dec. 24, 1861	33, 997
Harvester	I. Knauer	Valley Forge, Pa	Apr. 28, 1857	17, 157
Harvester	J. M. Knepley	Jersey Shore, Pa	Nov. 10, 1868	83, 975
Harvester	L. G. Kniffen	North Salem, N. Y	Dec. 13, 1859	26, 437
Harvester	L. G. Kniffen	Worcester, Mass	Apr. 12, 1864	42, 296
Harvester	L. G. Kniffen	Worcester, Mass	Apr. 12, 1864	42, 331
Harvester	L. G. Kniffen	Worcester, Mass	Aug. 20, 1867	67, 882
Harvester	L. G. Kniffen	Worcester, Mass	Aug. 20, 1867	67, 883
Harvester	L. G. Kniffen	Worcester, Mass	Aug. 20, 1867	67, 884
Harvester	L. G. Kniffen	Worcester, Mass	Aug. 20, 1867	67, 885
Harvester	A. E. Kroger	Norwalk, Conn	Sept. 4, 1855	13, 524
Harvester	E. M. Krum	Nassau, N. Y	May 1, 1866	54, 372
Harvester	M. Laflin	Chicago, Ill	July 2, 1872	128, 638
Harvester	S. I. Lamb and W. N. Whiteley	Springfield, Ohio	July 12, 1870	105, 343
Harvester	I. Lancaster	Baltimore, Md	Feb. 18, 1868	74, 552
Harvester	I. Lancaster	Baltimore, Md	Mar. 24, 1868	75, 773
Harvester	I. Lancaster	Baltimore, Md	Mar. 24, 1868	75, 774
Harvester	I. Lancaster	Baltimore, Md	Aug. 10, 1869	93, 627
Harvester	F. Landon	Brockport, N. Y	Nov. 13, 1860	30, 655
Harvester	F. Landon	Brockport, N. Y	May 28, 1861	32, 446
Harvester	C. Lee	Sandy, Ohio	June 25, 1867	66, 154
Harvester	G. W. Leisher	New Wilmington, Pa	June 17, 1873	139, 900
Harvester	E. E. Lewis	Geneva, N. Y	Feb. 25, 1868	74, 835
Harvester	E. I. Leyburn	Lexington, Va	May 12, 1868	77, 897
Harvester	C. Lidren	La Fayette, Ind	Sept. 29, 1868	82, 535
Harvester	C. Lidren	La Fayette, Ind	May 28, 1872	127, 252
Harvester	C. Lidren	La Fayette, Ind	Feb. 4, 1873	135, 480
Harvester	S. K. Lighter	Hamilton, Ohio	Nov. 3, 1863	40, 488
Harvester	S. K. Lighter and J. Curtis	Hamilton, Ohio	Dec. 15, 1868	84, 887
Harvester	S. A. Lindsay	Unionville, Md	Aug. 2, 1859	24, 944
Harvester	S. D. Locke	Janesville, Wis	Aug. 6, 1872	130, 227
Harvester	J. Long	Massillon, Ohio	Dec. 29, 1857	18, 981
Harvester	J. M. Long	Hamilton, Ohio	Sept. 4, 1860	29, 938
Harvester	J. M. Long	Hamilton, Ohio	Apr. 21, 1863	38, 232
Harvester	J. M. Long, P. Black, and R. Allstatter.	Hamilton, Ohio	Mar. 23, 1858	19, 703
Harvester	G. Lord	Watertown, N. J	Apr. 12, 1859	23, 592
Harvester	I. S. Love	Beloit, Wis	Oct. 7, 1856	15, 855
Harvester	W. R. Low	Sandwich, Ill	Nov. 12, 1872	132, 970
Harvester	W. R. Low and A. Adams	Sandwich, Ill	Feb. 1, 1870	99, 451
Harvester	A. Lowmiller	Jewett, Ohio	Feb. 20, 1866	52, 724
Harvester	S. Luce	Syracuse, N. Y	Aug. 19, 1873	141, 938
Harvester	D. B. Luckey	Bloomingburgh, N. Y	Jan. 6, 1863	37, 351
Harvester	D. B. Luckey	Bloomingburgh, N. Y	Sept. 13, 1864	44, 259
Harvester	C. M. Lufkin	Acworth, N. H	Sept. 8, 1857	18, 173
Harvester	G. G. Lyman	Independence, Iowa	Apr. 21, 1868	77, 058
Harvester	G. G. Lyman	Independence, Iowa	June 22, 1869	91, 757
Harvester	A. J. Manny	Freeport, Ill	Nov. 28, 1865	51, 203
Harvester	F. H. Manny	Rockford, Ill	Aug. 28, 1860	29, 800
Harvester	F. H. Manny	Rockford, Ill	Oct. 2, 1860	30, 235
Harvester	F. H. Manny	Rockford, Ill	June 23, 1863	38, 969
Harvester	F. H. Manny	Rockford, Ill	June 23, 1863	38, 970
Harvester	J. H. Manny	Waddam's Grove, Ill	Nov. 23, 1852	9, 423
Harvester	J. H. Manny	Rockford, Ill	Jan. 1, 1856	14, 026
Harvester	J. P. Manny	Rockford, Ill	July 14, 1857	17, 779
Harvester	J. P. Manny	Rockford, Ill	July 6, 1858	20, 806
Harvester	J. P. Manny	Rockford, Ill	Mar. 25, 1862	34, 760
Harvester	J. P. Manny	Rockford, Ill	Mar. 25, 1862	34, 761
Harvester	J. P. Manny	Rockford, Ill	Mar. 25, 1862	34, 762
Harvester	J. P. Manny	Rockford, Ill	Mar. 31, 1863	38, 053
Harvester	J. P. Manny	Rockford, Ill	Mar. 31, 1863	38, 054
Harvester	J. P. Manny	Rockford, Ill	June 15, 1869	91, 246
Harvester	J. P. Manny	Rockford, Ill	Mar. 7, 1871	112, 361
Harvester	J. P. Manny	Rockford, Ill	Mar. 7, 1871	112, 362
Harvester	J. P. Manny	Rockford, Ill	Mar. 21, 1871	112, 942
Harvester	J. P. Manny	Rockford, Ill	Apr. 16, 1872	125, 746
Harvester	J. P. Manny	Rockford, Ill	Jan. 21, 1873	135, 041
Harvester	P. Manny	Stephenson County, Ill	June 26, 1849	6, 560
Harvester	P. Manny	Waddam's Grove, Ill	Oct. 21, 1856	15, 927
Harvester	P. Manny	Waddam's Grove, Ill	Jan. 20, 1857	16, 445

Index of patents issued from the United States Patent Office from 1790 *to* 1873, *inclusive*—Continued.

Invention.	Inventor.	Residence.	Date.	No.
Harvester	P. Manny	Waddam's Grove, Ill	Apr. 7, 1857	16,984
Harvester	P. Manny	Waddam's Grove, Ill	Apr. 7, 1857	16,985
Harvester	P. Manny	Waddam's Grove, Ill	Mar. 27, 1860	27,642
Harvester	H. Marcellus	Amsterdam, N. Y	Apr. 20, 1858	19,999
Harvester	C. W. and W. W. Marsh	Clinton, Ill	Feb. 14, 1865	46,373
Harvester	C. W. and W. W. Marsh	Shabbona, Ill	Nov. 12, 1867	70,730
Harvester	C. W. and W W. Marsh	Sycamore, Ill	June 18, 1872	127,981
Harvester	J. S. Marsh	Lewisburgh, Pa	Nov. 16, 1858	22,084
Harvester	J. S. Marsh	Lewisburgh, Pa	May 21, 1861	32,375
Harvester	J. S. Marsh	Lewisburgh, Pa	Feb. 10, 1863	37,630
Harvester	J. S. Marsh	Lewisburgh, Pa	Feb. 12, 1867	61,944
Harvester	J. S. Marsh	Lewisburgh, Pa	Feb. 28, 1871	112,263
Harvester	J. S. Marsh, E. and C. C. Shorkley, and P. Beaver.	Lewisburgh, Pa	Feb. 10, 1863	37,631
Harvester	C. E. Mason	Elgin, Ill	Sept. 21, 1869	94,966
Harvester	N. Maxson	Wilmington, Ohio	Nov. 13, 1860	30,636
Harvester	J. McAleer	Chambersburgh, Pa	Oct. 11, 1859	25,748
Harvester	J. McCafferty	Waterloo, Iowa	Feb. 22, 1870	100,051
Harvester	C. N. McCandlish and J. C. Naginey.	West Rushville, Ohio	Apr. 8, 1873	137,614
Harvester	R. W. McClelland	Springfield, Ill	Mar. 22, 1864	42,009
Harvester	S. S. McColloch	Wheeling, W. Va	Apr. 22, 1873	138,172
Harvester	J. A. McCormick	Versailles, Ky	Apr. 27, 1858	20,080
Harvester	J. B. McCormick	Saint Louis, Mo	May 19, 1863	38,596
Harvester	J. B. McCormick	Dayton, Ohio	Feb. 15, 1870	99,929
Harvester	J. B. McCormick	Dayton, Ohio	Mar. 22, 1870	101,031
Harvester	L. J. McCormick and W. R. Baker.	Chicago, Ill	June 21, 1870	104,618
Harvester	L. J. McCormick and W. R. Baker.	Chicago, Ill	Nov. 1, 1870	108,922
Harvester	L. J. McCormick and W. R. Baker.	Chicago, Ill	Oct. 10, 1871	119,712
Harvester	L. J. McCormick and W. R. Baker.	Chicago, Ill	Oct. 17, 1871	119,937
Harvester	L. J. McCormick, W. R. Baker, and L. Erpelding.	Chicago, Ill	Oct. 27, 1868	83,520
Harvester	L. J. McCormick, W. B. Baker, and L. Erpelding.	Chicago, Ill	Apr. 27, 1869	89,324
Harvester	L. J. McCormick, W. R. Baker, and L. Erpelding.	Chicago, Ill	June 14, 1870	104,331
Harvester	L. J. McCormick and L. Erpelding.	Chicago, Ill	Jan. 15, 1867	61,228
Harvester	L. J. McCormick, L. Erpelding, and W. R. Baker.	Chicago, Ill	Dec. 28, 1869	98,394
Harvester	J. C. McDougal	Black Rock, N. Y	Jan. 26, 1864	41,384
Harvester	W. McIntosh	Wilmington, Ill	Oct. 22, 1861	33,533
Harvester	W. McLagan	Cuylerville, N. Y	June 15, 1852	9,031
Harvester	D. S. McNamara	North Hoosick, N. Y	June 30, 1857	17,691
Harvester	D. S. McNamara	North Hoosick, N. Y	Sept. 28, 1858	21,612
Harvester	W. Meigs	Waynesville, Ohio	July 22, 1862	35,945
Harvester	E. W. Merrill	West Buxton, Me	Mar. 17, 1868	75,640
Harvester	Q. F. Messinger	Easton, Pa	Jan. 9, 1872	122,630
Harvester	H. Merves	Binghamton, N. Y	July 5, 1870	104,978
Harvester	W. Michael	Murrysville, Pa	Feb. 9, 1869	86,774
Harvester	J. C. Miles	Bloomington, Wis	June 21, 1870	104,482
Harvester	C. G. Miller	Springfield, Ohio	Sept. 10, 1867	68,775
Harvester	C. G. Miller	Springfield, Ohio	Mar. 10, 1868	75,444
Harvester	C. G. Miller and B. Kersting	Springfield, Ohio	Jan. 11, 1870	98,700
Harvester	L. Miller	Canton, Ohio	May 4, 1858	20,180
Harvester	L. Miller	Canton, Ohio	May 4, 1858	20,181
Harvester	L. Miller	Canton, Ohio	May 4, 1858	20,182
Harvester	L. Miller	Akron, Ohio	Nov. 24, 1868	84,432
Harvester	L. and J. Miller	Canton, Ohio	July 5, 1859	24,700
Harvester	L. and J. Miller	Canton, Ohio	July 8, 1862	35,830
Harvester	W. K. Miller	Canton, Ohio	Apr. 6, 1858	19,864
Harvester	W. K. Miller	Canton, Ohio	Feb. 8, 1859	22,885
Harvester	W. K. Miller	Canton, Ohio	July 2, 1861	32,707
Harvester	W. K. Miller	Canton, Ohio	Sept. 10, 1867	68,642
Harvester	W. K. Miller	Canton, Ohio	June 3, 1873	139,596
Harvester	W. K. Miller	Canton, Ohio	June 3, 1873	139,597
Harvester	W. T. Miller	New Geneva, Pa	Oct. 25, 1870	108,716
Harvester	J. Mitchell	Gosport, N. Y	July 6, 1858	20,813
Harvester	J. A. Moore and A. H. Patch	Louisville, Ky	Mar. 29, 1869	23,383
Harvester	S. M. Moore	Beloit, Wis	Nov. 15, 1864	45,067
Harvester	J. Moran and C. D. Wallace	Auburn, N. Y., and Corry, Pa.	Feb. 7, 1871	111,557
Harvester	D. S. Morgan	Brockport, N. Y	Jan. 22, 1861	31,183
Harvester	R. Morris	Salem, Ind	Nov. 9, 1869	96,606
Harvester	W. Morrison	Carlisle, Pa	Nov. 22, 1859	26,198
Harvester	C. Moul	Hanover, Pa	Jan. 28, 1868	73,914
Harvester	D. Mulock	Mount Hope, N. Y	Feb. 28, 1871	112,166
Harvester	W. Needham	Rockford, Ill	Aug. 28, 1866	57,552
Harvester	W. Needham and J. Nelson	Rockford, Ill	Feb. 1, 1865	46,486
Harvester	W. Neff	Centre Hall, Pa	Feb. 18, 1868	74,577
Harvester	H. H. Nestestu	Deerfield, Wis	Sept. 6, 1870	107,193
Harvester	C. Newhall	Hinsdale, N. H	July 29, 1862	36,017
Harvester	P. Nicola	Massillon, Ohio	Nov. 21, 1871	121,061
Harvester	F. Nishwitz	Brooklyn, N. Y	Feb. 16, 1858	19,377
Harvester	F. Nishwitz	Brooklyn, N. Y	Feb. 21, 1865	46,488
Harvester	F. Nishwitz	Brooklyn, N. Y	Jan. 28, 1868	73,826
Harvester	F. Nishwitz	Brooklyn, N. Y	Jan. 28, 1868	73,827
Harvester	F. Nishwitz	Brooklyn, N. Y	Sept. 28, 1869	95,372
Harvester	F. Nishwitz	Brooklyn, N. Y	Sept. 28, 1869	95,373
Harvester	D. G. Norris	Saint Johnsbury, Vt	June 19, 1866	55,780
Harvester	H. Opp	Belleville, Ill	Dec. 7, 1858	22,237
Harvester	D. M. Osborne	Auburn, N. Y	June 19, 1866	55,781
Harvester	C. N. Owen	Salem, Ohio	May 24, 1870	103,490
Harvester	E. F. Page	Brooklyn, N. Y	July 25, 1865	48,988

Index of patents issued from the United States Patent Office from 1790 *to* 1873, *inclusive*—Continued.

Invention.	Inventor.	Residence.	Date.	No.
Harvester	D. R. Paiste	Williamstown, Pa	June 5, 1866	55, 427
Harvester	A. Palmer	Brockport, N. Y	May 1, 1866	54, 396
Harvester	I. H. Palmer	Lodi, Wis	Dec. 22, 1863	41, 016
Harvester	I. H. Palmer	Lodi, Wis	Oct. 20, 1868	83, 304
Harvester	I. H. Palmer	Lodi, Wis	Aug. 2, 1870	105, 973
Harvester	J. and E. Panton	London, Canada	Oct. 22, 1872	132, 408
Harvester	L. F. Parker	Davenport, Iowa	Dec. 1, 1868	84, 507
Harvester	H. A. Parkhurst	Fairfield, N. Y	Feb. 23, 1858	19, 442
Harvester	E. G. Passmore, jr	Philadelphia, Pa	Apr. 10, 1866	53, 862
Harvester	E. G. Passmore, jr	Philadelphia, Pa	Sept. 15, 1868	82, 156
Harvester	J. W. Patterson and L. H. Colborn.	Baltimore, Md	Mar. 8, 1859	23, 190
Harvester	N. A. Patterson	Kingston, Tenn	Oct. 13, 1857	18, 405
Harvester	J. L. Paxson	Norristown, Pa	Apr. 10, 1860	27, 826
Harvester	H. Pease	Brockport, N. Y	July 30, 1867	67, 211
Harvester	E. Peck	San José, Cal	Feb. 14, 1860	27, 151
Harvester	E. A. Peck	Sycamore, Ill	July 16, 1872	128, 974
Harvester	H. W. Pell	Rome, N. Y	Dec. 10, 1867	71, 906
Harvester	H. W. Pell	Rome, N. Y	Jan. 21, 1868	73, 641
Harvester	H. W. Pell	Rome, N. Y	Jan. 21, 1868	73, 642
Harvester	W. P. Penn	Belleville, Ill	Sept. 10, 1861	33, 263
Harvester	W. P. Penn	Belleville, Ill	Sept. 27, 1864	44, 452
Harvester	W. P Penn, J. Geiss, and J. Brosius.	Belleville, Ill	July 8, 1862	35, 836
Harvester	W. P. Penn, J. Geiss, and J. Brosius.	Belleville, Ill	Oct. 27, 1863	40, 427
Harvester	S. Pennock	Kennett's Square, Pa	Sept. 22, 1857	18, 267
Harvester	J. G. Perry	South Kingston, R. I	May 22, 1866	54, 946
Harvester	J. G. Perry	Kingston, R. I	Feb. 2, 1869	86, 584
Harvester	J. G. Perry	Kingston, R. I	May 11, 1869	90, 020
Harvester	J. G. Perry	Kingston, R. I	June 21, 1870	104, 489
Harvester	J. G. Perry	Kingston, R. I	June 21, 1870	104, 490
Harvester	J. G. Perry	Kingston, R. I	Aug. 13, 1872	130, 386
Harvester	J. G. Perry	Kingston, R. I	Sept. 16, 1873	142, 811
Harvester	S. Persons and A. M. Cone	Panama, N. Y	Feb. 1, 1859	22, 824
Harvester	G. M. Peters, jr	Granville, Ohio	Sept. 17, 1867	69, 018
Harvester	C. J. C. Petersen	Port Chester, N. Y	Sept. 19, 1871	119, 172
Harvester	J. J. Piggott and C. L. Crowell	Belleville and Peoria, Ill	Apr. 7, 1863	38, 123
Harvester	J. Pine	Troy, N. Y	May 7, 1861	32, 267
Harvester	J. Pine	Troy, N. Y	Oct. 22, 1861	33, 557
Harvester	J. Pine	Troy, N. Y	Oct. 27, 1863	40, 448
Harvester	J. Pine	Troy, N. Y	Apr. 26, 1864	42, 506
Harvester	J. Pine	Troy, N. Y	Mar. 5, 1867	62, 681
Harvester	N. Platt	Ottawa, Ill	June 12, 1849	6, 517
Harvester	T. E. Platt	Newtown, Conn	Dec. 31, 1867	72, 756
Harvester	J. C. and L. C. Pluche	Cape Vincent, N. Y	June 17, 1856	15, 146
Harvester	T. Plumleigh	Dundee, Ill	Jan. 30, 1872	123, 291
Harvester	J. Powers and E. M. Smith	New York, N. Y	Jan. 14, 1862	34, 186
Harvester	J. W. Prentiss	Pultney, N. Y	Nov. 28, 1865	51, 011
Harvester	J. W. Prentiss and E. M. Birdsall.	Penn Yan, N. Y	June 21, 1864	43, 225
Harvester	J. B. Pressey, M. A. Wheaton, and D. Sheets.	Dubuque County, Iowa, Suisun City, Cal., and Pike County, Mo.	July 12, 1864	43, 525
Harvester	K. H. C. Preston	Manlius, N. Y	Dec. 17, 1861	33, 975
Harvester	K. H. C. Preston	Manlius, N. Y	Oct. 15, 1867	69, 838
Harvester	K. H. C. Preston	Manlius, N. Y	Sept. 1, 1868	81, 686
Harvester	A. Pritz	Dayton, Ohio	Apr. 16, 1861	32, 080
Harvester	S. N. Purse	Ashley, Mo	Dec. 27, 1859	26, 616
Harvester	G. Pye	Boston, Mass	June 15, 1869	91, 482
Harvester	O. Qwick, W. S. Opie, and A. J. Farrand.	Raritan, N. J	Dec. 24, 1867	72, 678
Harvester	W. K. Rairigh	Plumville, Pa	June 3, 1873	139, 611
Harvester	W. K. Rairigh	Plumville, Pa	June 3, 1873	139, 612
Harvester	A. Ralston	West Middletown, Pa	Mar. 8, 1859	23, 194
Harvester	H. B. Ramsey	Indianapolis, Ind	Oct. 11, 1859	25, 761
Harvester	J. Ramsey	Milford, Tex	May 5, 1868	77, 524
Harvester	A. Rank	Salem, Ohio	Apr. 4, 1865	47, 128
Harvester	A. Rank	Salem, Ohio	Feb. 12, 1867	61, 952
Harvester	A. Rank	Salem, Ohio	Feb. 12, 1867	61, 953
Harvester	A. Rank	Salem, Ohio	Nov. 5, 1867	70, 614
Harvester	A. Rank	Salem, Ohio	May 5, 1868	77, 525
Harvester	A. Rank	Salem, Ohio	May 5, 1868	77, 526
Harvester	A. Rank	Salem, Ohio	June 30, 1868	79, 500
Harvester	A. Rank	Salem, Ohio	Mar. 23, 1869	88, 209
Harvester	A. Rank	Salem, Ohio	Nov. 2, 1869	96, 350
Harvester	A. Rank	Salem, Ohio	Nov. 2, 1869	96, 351
Harvester	A. Rank	Salem, Ohio	Nov. 2, 1869	96, 352
Harvester	A. Rank	Salem, Ohio	July 19, 1870	105, 597
Harvester	A. Rank	Salem, Ohio	Nov. 7, 1871	120, 669
Harvester	A. Rank and J. H. Cox	Salem, Ohio	Oct. 13, 1868	82, 987
Harvester	A. Rank and J. H. Cox	Salem, Ohio	Feb. 2, 1869	86, 449
Harvester	D. Rauck	Intercourse, Pa	Mar. 1, 1859	23, 115
Harvester	S. and I. Rawson	Almond, N. Y	Nov. 29, 1870	109, 755
Harvester	S. and I. Rawson	Almond, N. Y	July 25, 1871	117, 460
Harvester	B. F. Ray	Baltimore, Md	Feb. 5, 1856	14, 205
Harvester	B. F. Ray	Baltimore, Md	Feb. 15, 1859	22, 977
Harvester	B. F. Ray	Baltimore, Md	Jan. 21, 1862	34, 215
Harvester	S. Ray and E. Grant	Alliance, Ohio	Nov. 14, 1865	50, 956
Harvester	H. W. Read	West Windsor, Vt	Nov. 9, 1858	22, 032
Harvester	W. T. B. Read	Alton, Ill	June 2, 1857	17, 451
Harvester	G. Rector	Sodus, Mich	June 2, 1868	78, 482
Harvester	J. H. Redfield and W. J. Cox	Salem, Ind	Feb. 25, 1868	74, 847
Harvester	O. Redmond	Rochester, N. Y	Apr. 10, 1866	53, 872
Harvester	J. H. and A. E. Redstone	Indianapolis, Ind	Aug. 19, 1862	36, 258
Harvester	C. A. Reed and J. M. Campbell	Madison and Beaver Dam, Wis.	Aug. 10, 1869	93, 641
Harvester	A. R. Reese	Phillipsburgh, N. J	May 14, 1867	64, 797

Index of patents issued from the United States Patent Office from 1790 *to* 1873, *inclusive*—Continued.

Invention.	Inventor.	Residence.	Date.	No.
Harvester	L. C. Reese	Phillipsburgh, N. J	Apr. 10, 1860	27, 832
Harvester	J. Reily	Hart Prairie, Wis	Jan. 8, 1856	14, 079
Harvester	B. W. S. and C. H. Ressler	Tipton, Ind	June 10, 1873	139, 817
Harvester	S. Rex	Orefield, Pa	Apr. 7, 1868	76, 522
Harvester	G. S. Reynolds	Tunbridge, Vt	July 23, 1861	32, 915
Harvester	I. Reynolds	Republic, Ohio	June 13, 1854	11, 086
Harvester	E. B. Rice	Madison, Wis	Apr. 5, 1870	101, 659
Harvester	E. B. Rice	Madison, Wis	Nov. 14, 1871	121, 005
Harvester	G. W. Richardson	Grayville, Ill	June 25, 1861	32, 669
Harvester	G. W. Richardson	Grayville, Ill	Mar. 4, 1862	34, 614
Harvester	G. W. Richardson	Grayville, Ill	Dec. 6, 1864	45, 345
Harvester	G. W. Richardson	Grayville, Ill	Nov. 7, 1865	50, 844
Harvester	G. W. Richardson and R. Glover.	Grayville, Ill	Jan. 25, 1859	22, 772
Harvester	A. Rickart	Schoharie, N. Y	Feb. 11, 1873	135, 729
Harvester	G Riemer	Fayette, N. Y	Nov. 12, 1867	70, 744
Harvester	M. H. Ripley	Minneapolis, Minn	Jan. 4, 1870	98, 635
Harvester	M. H. Ripley and W. N. Temple.	Minneapolis, Minn	Mar. 30, 1869	88, 511
Harvester	C. Roberts	Livonia, N. Y	Feb. 23, 1858	19, 447
Harvester	R. A. Roberts	Salisbury, Mo	Mar. 18, 1873	136, 940
Harvester	S. Rockafellow	Muscatine, Iowa	May 26, 1868	78, 236
Harvester	J. L. Rohrer and C. E. Roper	Upper Leacock, Pa., and Canton, Ohio.	Dec. 19, 1871	122, 062
Harvester	W. Rose	Lake, Ill	Feb. 18, 1868	74, 600
Harvester	A. M. Ross	Ilion, N. Y	Jan. 2, 1872	122, 492
Harvester	N. Rowe	Emmittsburgh, Md	Dec. 11, 1866	60, 427
Harvester	J. S. Royce	Cuylerville, N. Y	Apr. 1, 1862	34, 849
Harvester	G. H. Rugg	South Ottawa, Ill	June 8, 1852	9, 005
Harvester	C. Russell and W. K. Miller	Massillon, Ohio	Mar. 24, 1868	75, 797
Harvester	E. P. Russell	Manlius, N. Y	Nov. 12, 1861	33, 712
Harvester	E. P. Russell	Manlius, N. Y	Dec. 24, 1861	34, 014
Harvester	E. P. Russell	Manlius, N. Y	Mar. 4, 1862	34, 594
Harvester	E. P. Russell	Manlius, N. Y	Nov. 14, 1865	50, 959
Harvester	F. Russell	Boston, Mass	Aug. 14, 1855	13, 438
Harvester	H. R. Russell	New Market, Md	Sept. 1, 1863	39, 780
Harvester	I. S. and H. R. Russell	New Market, Md	Mar. 29, 1859	23, 399
Harvester	I. S. and H. R. Russell	New Market, Md	Aug. 20, 1867	67, 914
Harvester	I. S. and H. R. Russell	New Market, Md., and Woodbury, N. J.	Aug. 18, 1868	81, 215
Harvester	E. H. Rurve	Walcott, Iowa	Feb. 13, 1872	123, 645
Harvester	J. Rynearson	Farmington, Ill	Oct. 28, 1862	36, 800
Harvester	A. Saltsman, C. H. Charlesworth, and R. F. Osgood.	Avoca and Rochester, N. Y	Apr. 20, 1869	89, 082
Harvester	D. Sandford	Taylor, Ill	Dec. 6, 1859	26, 376
Harvester	W. and T. Schnebly	New York, N. Y	June 15, 1852	9, 038
Harvester	W. and T. Schnebly	Hackensack, N. J	Nov. 30, 1858	22, 203
Harvester	J. W. Schuckers	Philadelphia, Pa	Jan. 31, 1871	111, 482
Harvester	J. W. Schuckers	Philadelphia, Pa	Feb. 28, 1871	112, 288
Harvester	H. H. Scoville	Syracuse, N. Y	Apr. 12, 1859	23, 613
Harvester	J. Scoville	Buffalo, N. Y	Jan. 31, 1860	27, 010
Harvester	J. Scuder and E. Miller	Litiz, Pa	May 30, 1871	115, 376
Harvester	T. C. Sebring	Milford, Mich	Sept. 15, 1868	82, 253
Harvester	J. Seibel	Manlius, Ill	Dec. 27, 1864	45, 642
Harvester	J. Seibel	Manlius, Ill	Feb. 21, 1865	46, 502
Harvester	J. Seibel	Manlius, Ill	May 15, 1866	54, 784
Harvester	J. Seibel	Manlius, Ill	Nov. 10, 1868	84, 006
Harvester	J. Seibel	Manlius, Ill	June 7, 1870	104, 067
Harvester	J. F. Seiberling	Doylestown, Ohio	Oct. 15, 1861	33, 496
Harvester	J. F. Seiberling	Doylestown, Ohio	Nov. 1, 1864	44, 893
Harvester	J. F. Seiberling	Doylestown, Ohio	Nov. 1, 1864	44, 894
Harvester	J. F. Seiberling	Doylestown, Ohio	Dec. 5, 1865	51, 359
Harvester	J. F. Seiberling	Doylestown, Ohio	Jan. 30, 1866	52, 327
Harvester	J. F. Seiberling	Doylestown, Ohio	May 8, 1866	54, 611
Harvester	J. F. Seiberling	Akron, Ohio	Aug. 28, 1866	57, 583
Harvester	J. F. Seiberling	Akron, Ohio	June 8, 1869	91, 172
Harvester	J. F. Seiberling	Akron, Ohio	Aug. 1, 1871	117, 692
Harvester	J. F. Seiberling	Akron, Ohio	Jan. 21, 1873	135, 013
Harvester	J. F. Seiberling	Akron, Ohio	Apr. 1, 1873	137, 490
Harvester	J. F. Seiberling	Akron, Ohio	May 27, 1873	139, 268
Harvester	W. H. Seymour	Brockport, N. Y	Mar. 28, 1854	10, 707
Harvester	W. H. Seymour and D. S. Morgan.	Brockport, N. Y	Feb. 11, 1873	135, 731
Harvester	W. H. Seymour and H. Pease	Brockport, N. Y	Sept. 9, 1856	15, 721
Harvester	W. H. Seymour and H. Pease	Brockport, N. Y	Sept. 9, 1856	15, 722
Harvester	W. H. Seymour and H. Pease	Brockport, N. Y	May 25, 1858	20, 394
Harvester	M. R. Shalters	Alliance, Ohio	Jan. 9, 1866	51, 975
Harvester	W. A. Sharpe	Syracuse, N. Y	July 27, 1869	93, 013
Harvester	J. W. Shaw	Concord, N. H	Apr. 27, 1869	89, 509
Harvester	M. and S. Shawver	Bellefontaine, Ohio	May 27, 1862	35, 397
Harvester	A. Shebanck	Cleveland, Ohio	Mar. 31, 1868	76, 255
Harvester	A. Sheline and E. Burke	Edon, Ohio	Dec. 7, 1869	97, 556
Harvester	J. A. Shephard	Portsmouth, Ohio	Sept. 10, 1872	131, 307
Harvester	N. C. Sherman and S. Lightcap	Hazle Green, Wis	Aug. 4, 1857	17, 942
Harvester	A. Sherwood	Auburn, N. Y	Oct. 5, 1869	95, 522
Harvester	A. and C. Sherwood	Auburn, N. Y	Nov. 8, 1870	109, 062
Harvester	J. H. Shireman	New York, N. Y	Aug. 5, 1873	141, 597
Harvester	H. J. Silvernale	Golden Lake, Wis	May 20, 1873	139, 199
Harvester	W. H. H. Sisum	Cherry Valley, N. Y	Aug. 24, 1869	94, 141
Harvester	E. W. Skinner	Madison, Wis	Aug. 18, 1868	81, 221
Harvester	W. D. Slack	Lewis urgh, Pa	Mar. 30, 1869	88, 334
Harvester	W. D. Slack	Lewisburgh, Pa	Oct. 5, 1869	95, 526
Harvester	J. Smalley	Boundbrook, N. J	Apr. 12, 1859	23, 618
Harvester	A Smith	Vienna Cross Roads, Ohio	Aug. 25, 1868	81, 423
Harvester	A. Smith	Springfield, Ohio	Oct. 12, 1869	95, 738
Harvester	A. B. Smith	Rochester, Pa	Sept. 1, 1868	81, 832
Harvester	A. L. Smith	Bristol Centre, N. Y	Dec. 21, 1869	98, 113
Harvester	B. H. Smith and G. W. Archer	Ipswich, Mass	June 24, 1862	35, 712

Index of patents issued from the United States Patent Office from 1790 *to* 1873, *inclusive*—Continued.

Invention.	Inventor.	Residence.	Date.	No.
Harvester	E. Smith	Cold Spring Harbor, N. Y	Jan. 21, 1862	34,219
Harvester	E. Smith	Cold Spring Harbor, N. Y.	Apr. 22, 1862	35,044
Harvester	E. H. Smith and H. G. Smith	Saint Albans, Vt., and Goshen, Conn.	May 19, 1868	78,144
Harvester	E. M. Smith	New York, N. Y	Feb. 23, 1864	47,746
Harvester	H. C. Smith	Cleveland, Ohio	Mar. 9, 1858	19,590
Harvester	H. C. Smith	Cleveland, Ohio	May 11, 1858	20,225
Harvester	J. B. Smith	Winfield, N. Y	Feb. 19, 1861	31,486
Harvester	J. D. Smith	Lancaster, Ohio	Oct. 19, 1858	21,854
Harvester	S. H. Smith	Magnolia, Ill	June 15, 1858	20,593
Harvester	D. Snell	Little Falls, N. Y	July 29, 1862	36,034
Harvester	L. D. Snook	Barrington, N. Y	June 2, 1868	78,490
Harvester	J. H. Snyder	Killbuck, Ill	Mar. 29, 1864	42,148
Harvester	W. W. Sowles	Manlius, N. Y	Aug. 7, 1866	57,005
Harvester	G. H. Spaulding	Rockford, Ill	Dec. 17, 1867	72,238
Harvester	G. H. Spaulding	Rockford, Ill	Feb. 21, 1871	111,987
Harvester	G. H. Spaulding	Rockford, Ill	July 8, 1873	140,738
Harvester	W. Sprague	Farnham, N. Y	July 14, 1868	79,871
Harvester	P. H. Standish	Pacheco, Cal	Feb. 19, 1861	31,490
Harvester	C. T. Stetson	Amherst, Mass	June 30, 1857	17,705
Harvester	C. T. Stetson	Amherst, Mass	Nov. 2, 1858	21,993
Harvester	W. S. Stetson	Baltimore, Md	Apr. 5, 1859	23,508
Harvester	W. S. Stetson	Baltimore, Md	Aug. 14, 1860	29,632
Harvester	W. H. Stevenson	Auburn, N. Y	Mar. 3, 1868	75,070
Harvester	G. W. Stewart	Adairsville, Ga	Dec. 22, 1868	85,254
Harvester	L. B. Stilson	Woodland, Minn	Oct. 6, 1868	82,890
Harvester	O. Stoddard	Busti, N. Y	May 11, 1858	20,221
Harvester	O. Stoddard	Busti, N. Y	Dec. 14, 1858	22,312
Harvester	A. Stoler and S. A. Sisson	Bristol, Pa	May 15, 1860	28,313
Harvester	C. S. Stone	Indianapolis, Ind	Nov. 4, 1873	144,367
Harvester	G. Stone and J. P. Bullock	Beloit, Wis	Mar. 14, 1865	46,830
Harvester	J. W. Stout	Raritan, N. J	June 3, 1862	35,479
Harvester	J. W. Street	Salem, Ohio	July 15, 1862	35,901
Harvester	H. Stuckey	Bucyrus, Ohio	Jan. 24, 1871	111,272
Harvester	J. M. Swain	Howard, Ind	Feb. 26, 1867	62,452
Harvester	D. M. Swartz and J. Kreamer	Millheim, Pa	May 12, 1863	38,514
Harvester	A. J. Sweeney	Wheeling, W. Va	Sept. 3, 1872	131,134
Harvester	W. A. Sweet	Syracuse, N. Y	Dec. 8, 1863	40,864
Harvester	L. Swenson	North Cape, Wis	Aug. 17, 1869	93,765
Harvester	Z. Swope	Lancaster, Pa	Feb. 25, 1873	136,279
Harvester	C. R. and J. O. Taber	Salem, Ohio	June 30, 1868	79,518
Harvester	E. Taylor	Memphis, Tenn	Apr. 1, 1873	137,506
Harvester	T. H. Taylor	Jeffersonville, Ill	Oct. 5, 1869	95,537
Harvester	D. H. Thayer	Ludlowville, N. Y	Mar. 24, 1868	76,001
Harvester	D. H. Thayer	Ludlowville, N. Y	Feb. 13, 1872	123,650
Harvester	J. B. Thomison	Lynchburgh, Tenn	Dec. 26, 1871	122,201
Harvester	J. W. Thompson	Greenfield, Mass	Feb. 4, 1868	74,016
Harvester	T. J. Tindall	New York, N. Y	Oct. 11, 1864	44,679
Harvester	C. Tinker and J. A. Sprague	Mantua, Ohio	Aug. 4, 1857	17,945
Harvester	J. B. Tinker	Plymouth, N. Y	Feb. 5, 1861	31,339
Harvester	W. Tinker	Kelloggsville, Ohio	Aug. 19, 1856	15,582
Harvester	W. Tinker	Kelloggsville, Ohio	Dec. 9, 1856	16,194
Harvester	B. Titcomb	Baltimore, Md	Apr. 19, 1864	42,416
Harvester	J. Toay	Mineral Point, Wis	Oct. 27, 1868	83,568
Harvester	P. A. Tobey	Caton, N. Y	Mar. 15, 1870	100,817
Harvester	C. Tompkins	Troy, N. Y	Mar. 6, 1866	53,057
Harvester	J. S. Troxel	Mount Pleasant, Pa	May 11, 1858	20,227
Harvester	I. V. Trump	Somerville, N. Y	Aug. 3, 1858	21,093
Harvester	J. S. Truxell	Greensburgh, Pa	Aug. 30, 1870	106,891
Harvester	J. S. Truxell	Greensburgh, Pa	Nov. 14, 1871	120,911
Harvester	J. S. Truxell	Greensburgh, Pa	Mar. 11, 1873	136,628
Harvester	W. O. Tubbs	Spring, Pa	Apr. 19, 1864	42,420
Harvester	A. W. Tucker	Waxahachie, Tex	Nov. 12, 1867	70,918
Harvester	B. H. Turner	Fremont, Nebr	May 2, 1871	114,493
Harvester	S. W. Tyler	Greenwich, N. Y	Jan. 26, 1858	19,218
Harvester	S. W. Tyler	Greenwich, N. Y	Mar. 29, 1859	23,413
Harvester	S. W. Tyler	Greenwich, N. Y	Nov. 13, 1860	30,651
Harvester	J. Urmy	Wilmington, Del	Mar. 10, 1863	37,881
Harvester	W. Van Anden	Poughkeepsie, N. Y	Mar. 30, 1858	19,803
Harvester	W. Van Anden	Poughkeepsie, N. Y	Dec. 24, 1861	34,019
Harvester	W. Van Anden	Poughkeepsie, N. Y	July 1, 1862	35,792
Harvester	W. Van Anden	Poughkeepsie, N. Y	Sept. 9, 1862	36,430
Harvester	W. Van Anden	Poughkeepsie, N. Y	Aug. 11, 1863	39,511
Harvester	J. Van De Water	Whitewater, Wis	Mar. 28, 1871	113,225
Harvester	J. Van Doren	Somerville, N. J	Mar. 2, 1858	19,522
Harvester	J. Van Doren	Somerville, N. J	Apr. 6, 1858	19,884
Harvester	D. Van Kleeck	Cohocton, N. Y	Mar. 20, 1860	27,583
Harvester	J. P. Van Sickle	Geneva, N. Y	Oct. 26, 1869	96,170
Harvester	N. T. Veatch	Camden, Ill	July 16, 1872	129,501
Harvester	H. C. Velie	Poughkeepsie, N. Y	July 19, 1870	105,612
Harvester	H. C. Velie	Poughkeepsie, N. Y	Jan. 2, 1872	122,501
Harvester	W. A. Vertrees	Winchester, Mo	Feb. 28, 1860	27,323
Harvester	A. H. Wagner	Chicago, Ill	Oct. 4, 1870	108,069
Harvester	A. H. Wagner	Chicago, Ill	Mar. 14, 1871	112,750
Harvester	C. B. Wagner	Philadelphia, Pa	June 24, 1856	15,204
Harvester	D. B. Waite	Spring Water, N. Y	June 1, 1858	20,456
Harvester	L. Wallace	Leavenworth, Kans	Mar. 17, 1868	75,606
Harvester	W. Wallace	Syracuse, N. Y	Aug. 19, 1873	141,970
Harvester	W. J. Wallis and W. E. Huttmann	Chicago, Ill	June 15, 1869	91,501
Harvester	T. and I. W. Ward	Lane Depot, Ill	July 7, 1863	39,186
Harvester	R. Warner	Brattleborough, Vt	Mar. 1, 1859	23,132
Harvester	D. Warren	Gettysburgh, Pa	Dec. 16, 1862	37,187
Harvester	D. Watson	Petersburgh, Va	July 17, 1855	13,285
Harvester	D. Watson	New York, N. Y	Mar. 3, 1857	16,763
Harvester	W., jr., and J. Webber	Rockton, Ill	Dec. 22, 1857	18,938
Harvester	G. Weiland	Buffalo, N. Y	May 26, 1868	78,245

Index of patents issued from the United States Patent Office from 1790 *to* 1873, *inclusive*—Continued.

Invention.	Inventor.	Residence.	Date.	No.
Harvester	T. Welch	Churchville, N. Y	Aug. 1, 1865	49, 183
Harvester	T. Welch	Churchville, N. Y	June 16, 1868	78, 904
Harvester	H. Wells	Erieville, N. Y	Apr. 16, 1872	125, 711
Harvester	J. V. A. Wemple	Chicago, Ill	Apr. 19, 1859	23, 730
Harvester	J. Werner, jr	Prairie du Lac, Wis	June 17, 1873	139, 986
Harvester	P. Werni	Chicago, Ill	Nov. 19, 1867	71, 097
Harvester	C. Wheeler, jr	Poplar Ridge, N. Y	May 26, 1863	38, 708
Harvester	C. Wheeler, jr	Auburn, N. Y	Oct. 8, 1867	69, 732
Harvester	C. Wheeler, jr	Auburn, N. Y	Apr. 7, 1868	76, 571
Harvester	C. Wheeler, jr	Auburn, N. Y	Apr. 21, 1868	77, 145
Harvester	C. Wheeler, jr	Auburn, N. Y	Nov. 17, 1868	84, 151
Harvester	C. Wheeler, jr., and C. Young	Auburn, N. Y	Apr. 4, 1871	113, 475
Harvester	J. Whitehead	Manchester, Va	Jan. 5, 1858	19, 055
Harvester	A. Whiteley	Springfield, Ohio	Apr. 24, 1855	12, 769
Harvester	A. Whiteley	Springfield, Ohio	Aug. 30, 1859	25, 297
Harvester	A. Whiteley	Springfield, Ohio	Mar. 3, 1868	75, 229
Harvester	W. N. Whiteley	Springfield, Ohio	Jan. 30, 1866	52, 351
Harvester	W. N. Whiteley	Springfield, Ohio	Aug. 13, 1867	67, 828
Harvester	W. N. Whiteley	Springfield, Ohio	Nov. 19, 1867	71, 255
Harvester	W. N. Whiteley	Springfield, Ohio	May 5, 1868	77, 691
Harvester	W. N. Whiteley	Springfield, Ohio	July 21, 1868	80, 258
Harvester	W. N. Whiteley	Springfield, Ohio	Aug. 11, 1868	81, 046
Harvester	W. N. Whiteley	Springfield, Ohio	Nov. 24, 1868	84, 399
Harvester	W. N. Whiteley	Springfield, Ohio	Dec. 15, 1868	85, 045
Harvester	W. N. Whiteley	Springfield, Ohio	June 14, 1870	104, 386
Harvester	W. N. Whiteley	Springfield, Ohio	July 12, 1870	105, 394
Harvester	W. N. Whiteley	Springfield, Ohio	July 12, 1870	105, 395
Harvester	W. N. Whiteley	Springfield, Ohio	Sept. 12, 1871	118, 996
Harvester	W. N. Whiteley, jr	Springfield, Ohio	Feb. 23, 1864	47, 734
Harvester	W. N. Whiteley, jr	Springfield, Ohio	Dec. 5, 1865	51, 374
Harvester	W. N. Whiteley, jr	Springfield, Ohio	Jan. 30, 1866	52, 349
Harvester	W. N. Whiteley, jr	Springfield, Ohio	Jan. 30, 1866	52, 350
Harvester	W. N. Whiteley, jr	Springfield, Ohio	Aug. 21, 1866	57, 419
Harvester	W. N. Whiteley, jr	Springfield, Ohio	Oct. 30, 1866	59, 302
Harvester	W. N. Whiteley, jr	Springfield, Ohio	Oct. 30, 1866	59, 303
Harvester	W. N. Whiteley, jr	Springfield, Ohio	Mar. 26, 1867	63, 192
Harvester	W. N. Whiteley, jr	Springfield, Ohio	May 14, 1867	64, 818
Harvester	W. N. Whiteley and S. T. Lamb.	Springfield, Ohio	Aug. 22, 1871	118, 315
Harvester	W. N. and A. Whiteley	Springfield, Ohio	Jan. 18, 1859	22, 684
Harvester	W. N., jr., and A. Whiteley	Springfield, Ohio	Dec. 17, 1867	72, 349
Harvester	J. H. Whitenack	Somerville, N. J	Mar. 6, 1866	53, 069
Harvester	D. D. Whitker	New York, N. Y	Aug. 30, 1870	106, 903
Harvester	J. A. Whitney	Maryland, N. Y	July 8, 1862	35, 852
Harvester	J. H. Whitney	Rochester, N. Y	Jan. 16, 1872	122, 750
Harvester	B. Wieland	Orangeville, Ill	Apr. 18, 7865	47, 351
Harvester	J. D. Wilber	Poughkeepsie, N. Y	Oct. 22, 1867	70, 059
Harvester	J. D. Wilber	Poughkeepsie, N. Y	Oct. 31, 1871	120, 407
Harvester	J. D. Wilber	Poughkeepsie, N. Y	May 13, 1873	138, 830
Harvester	J. D. Wilber	Poughkeepsie, N. Y	Sept. 16, 1873	142, 828
Harvester	D. H. Wilgus and S. M. Denny	Harveysburgh, Ohio	July 13, 1869	92, 500
Harvester	H. Willard and R. Ross	Vergennes, Vt	Nov. 3, 1857	18, 562
Harvester	J. S. Williams	Chicago, Ill	Dec. 20, 1864	45, 550
Harvester	R. M. Williams	Rockville, Md	Oct. 12, 1869	95, 864
Harvester	S. Williams	Stockton, Cal	June 15, 1858	20, 600
Harvester	J. C. and T. G. Wilson	Cedar Hill, Tex	Apr. 21, 1857	17, 123
Harvester	T. Windell	New Albany, Ind	June 8, 1858	20, 525
Harvester	M. J. Wine	Long Glade, Va	Oct. 27, 1868	83, 578
Harvester	C. P. Wing	Fayetteville, N. Y	May 28, 1861	32, 443
Harvester	C. P. Wing	Fayetteville, N. Y	Apr. 5, 1864	42, 248
Harvester	C. P. Wing	Fayetteville, N. Y	June 28, 1864	43, 361
Harvester	J. Winters and C. C. Gapen	Lacon, Ill	Dec. 10, 1867	71, 937
Harvester	C. B. Withington	Rock, Wis	Feb. 21, 1860	27, 252
Harvester	C. W. and B. F. Witt	Indianapolis, Ind	Aug. 4, 1868	80, 694
Harvester	C. W. and B. F. Witt	Indianapolis, Ind	July 12, 1870	105, 399
Harvester	D. Wolf	Lebanon, Pa	May 7, 1867	64, 615
Harvester	C. B. Wolgemuth	Mount Joy, Pa	Apr. 9, 1872	125, 512
Harvester	A. Wood	Henrietta, N. Y	Jan. 31, 1865	46, 170
Harvester	W. A. Wood	Hoosick Falls, N. Y	Dec. 29, 1857	19, 001
Harvester	W. A. Wood	Hoosick Falls, N. Y	Dec. 29, 1857	19, 002
Harvester	W. A. Wood	Hoosick Falls, N. Y	Feb. 22, 1859	23, 056
Harvester	W. A. Wood	Hoosick Falls, N. Y	Dec. 3, 1861	33, 862
Harvester	W. A. Wood	Hoosick Falls, N. Y	Apr. 15, 1862	34, 995
Harvester	W. A. Wood	Hoosick Falls, N. Y	June 10, 1862	35, 558
Harvester	W. A. Wood	Hoosick Falls, N. Y	Dec. 3, 1867	71, 832
Harvester	W. A. Wood	Hoosick Falls, N. Y	Feb. 25, 1873	136, 295
Harvester	W. A. Wood and J. M. Rosebrooks.	Hoosick Falls, N. Y	July 19, 1859	24, 836
Harvester	W. A. and W. A. Wood and J. M. Rosebrooks.	Hoosick Falls, N. Y	Jan. 3, 1871	110, 713
Harvester	W. A. and W. A. Wood and J. M. Rosebrooks.	Hoosick Falls, N. Y	Jan. 3, 1871	110, 714
Harvester	W. A. and W. A. Wood and J. M. Rosebrooks.	Hoosick Falls, N. Y	Jan. 3, 1871	110, 715
Harvester	J. Woodruff	Rahway, N. J	June 5, 1860	28, 625
Harvester	J. Woody	Mount Vernon, Ind	Oct. 12, 1858	21, 792
Harvester	J. Woody	Mount Vernon, Ind	Aug. 20, 1861	33, 120
Harvester	T. Y. Woolford	Romney, W. Va	Feb. 11, 1873	135, 741
Harvester	H. M. Yale	Cold Water, Mich	Mar. 25, 1873	137, 162
Harvester	T. Yates	Milwaukee, Wis	Sept. 23, 1873	143, 110
Harvester	B. Yeakel	Allentown, Pa	Feb. 23, 1858	19, 463
Harvester	G. W. N. Yost	Corry, Pa	June 9, 1868	78, 708
Harvester	G. W. N. Yost	Corry, Pa	June 9, 1868	78, 709
Harvester	G. W. N. Yost	Corry, Pa	June 9, 1868	78, 710
Harvester	G. W. N. Yost	Corry, Pa	Aug. 18, 1868	81, 241
Harvester	G. W. N. Yost	Corry, Pa	Aug. 25, 1868	81, 569
Harvester	G. W. N. Yost	Corry, Pa	Oct. 6, 1868	82, 782
Harvester	G. W. N. Yost	Corry, Pa	Dec. 15, 1868	84, 927

Index of patents issued from the United States Patent Office from 1790 *to* 1873, *inclusive*—Continued.

Invention.	Inventor.	Residence.	Date.	No.
Harvester	G. W. N. Yost	Corry, Pa	Dec. 22, 1868	85, 159
Harvester	G. W. N. Yost	Corry, Pa	Dec. 29, 1868	85, 349
Harvester	G. W. N. Yost	Corry, Pa	Jan. 5, 1869	85, 628
Harvester	G. W. N. Yost	Corry, Pa	Jan. 12, 1869	85, 887
Harvester	G. W. N. Yost	Corry, Pa	Mar. 23, 1869	88, 254
Harvester	G. W. N. Yost	Corry, Pa	Mar. 30, 1869	88, 533
Harvester	G. W. N. Yost	Corry, Pa	May 25, 1869	90, 621
Harvester	G. W. N. Yost	Corry, Pa	June 1, 1869	90, 909
Harvester	G. W. N. Yost	Corry, Pa	Apr. 15, 1873	137, 812
Harvester	G. W. N. Yost	Corry, Pa	Apr. 15, 1873	137, 813
Harvester	G. W. N. Yost	Corry, Pa	Apr. 15, 1873	137, 814
Harvester	G. W. N. Yost	Corry, Pa	Apr. 15, 1873	137, 815
Harvester	C. M. Young	Meadville, Pa	Jan. 3, 1871	110, 718
Harvester	C. M. Young	Meadville, Pa	Jan. 28, 1873	135, 394
Harvester	E. Young	Fayetteville, Mo	Oct. 1, 1867	69, 529
Harvester	M. Young, jr	Frederick, Md	Sept. 21, 1858	21, 587
Harvester	M. Young, jr	Frederick, Md	July 9, 1861	32, 807
Harvester	W. Zimmerman	Oskaloosa, Iowa	June 19, 1866	55, 758
Harvester	D. Zug	Shaefferstown, Pa	Oct. 4, 1859	25, 697
Harvester	D. Zug	Shaefferstown, Pa	June 3, 1862	35, 485
Harvester and binder	J. E. Nesen	Buffalo, N. Y	Dec. 13, 1853	10, 314
Harvester and binder, Combined	J. M. McKesson	Lincoln, Nebr	Feb. 28, 1871	112, 267
Harvester and binder, Combined grain	J. H. Whitney	Rochester, Minn	Dec. 6, 1870	109, 985
Harvester and binder, Grain	L. B. Stilson	Minneapolis, Minn	Feb. 15, 1870	99, 968
Harvester and binder, Grain	P. H. Watson and S. Renwick	Washington, D. C	Aug. 16, 1853	9, 930
Harvester and binder, Grain	W. Watson, E. S. Renwick, and P. H. Watson.	Chicago, Ill., and Washington, D. C.	May 13, 1851	8, 083
Harvester and husker combined, Corn	L. Devore	Victor, Iowa	Aug. 22, 1871	118, 217
Harvester and husker, Combined corn	J. McLeish	Chicago, Ill	Apr. 2, 1872	125, 318
Harvester and husker, Combined corn	S. Patton	Chatsworth, Ill	May 24, 1870	103, 362
Harvester and husker, Combined corn	A. Smith	Pontiac, Ill	Oct. 11, 1870	108, 197
Harvester and husker, Combined corn	T. F. Vincent	Rock Island, Ill	June 18, 1872	127, 999
Harvester and shocker, Corn	E. Culp	Hilliard Station, Ohio	Oct. 24, 1871	120, 249
Harvester and shocker, Corn	W. H. Karicofe	Harrisonburgh, Va	Aug. 3, 1869	93, 311
Harvester and thrasher, Combined	J. H. Adamson	Auburn, Australia	July 1, 1873	140, 396
Harvester and thrasher, Combined	W. G. and L. T. Davis	McMinnville, Oreg	Nov. 16, 1869	96, 896
Harvester and thrasher, Combined	J. D. Field	Keokuk, Iowa	Apr. 12, 1870	101, 849
Harvester and thrasher combined	D. Howell	Saint Helena, Cal	Jan. 25, 1870	99, 089
Harvester and thrasher combined	W. B. Hubard	Arrington Depot, Va	Dec. 31, 1872	134, 476
Harvester and thrasher, Combined	J. H. Robbins	Bethel, Oreg	July 16, 1872	129, 171
Harvester-attachment for dumping and shocking sheaves.	W. A. Sharp and C. L. McClung.	Tama City, Iowa	May 6, 1873	138, 536
Harvester-attachment for raking and binding grain.	H. A. Reid	Beaver Dam, Wis	June 28, 1870	104, 883
Harvester, Bean	N. Chappell	East Avon, N. Y	Feb. 7, 1865	46, 216
Harvester, Bean	J. Hinds	Hindsburgh, N. Y	Apr. 1, 1873	137, 370
Harvester, Bean	A. Hutchinson and S. I. Lynd	Albion, N. Y	Oct. 15, 1872	132, 210
Harvester, Bean	J. Ketcham and J. Waterman	Orangeport, N. Y	Nov. 24, 1857	18, 697
Harvester, Bean	D. B. Munger	Mumford, N. Y	Apr. 18, 1865	47, 320
Harvester, Bean	D. C. Rosier	Clarkson, N. Y	Sept. 18, 1866	58, 140
Harvester, Bean	J. A. and T. Wood	Chemung, Ill	Mar. 28, 1871	113, 125
Harvester-binder	J. Barta	La Crosse, Wis	May 16, 1871	114, 749
Harvester binder-attachment	J. H. Garnhart	Madison, Wis	Sept. 2, 1873	142, 331
Harvester binder-attachment	J. H. Garnhart	Madison, Wis	Oct. 7, 1873	143, 505
Harvester binder-attachment	S. Mills	Madison, Wis	Nov. 26, 1872	133, 473
Harvester binder-attachment	H. Porter	Polo, Ill	Nov. 18, 1873	144, 789
Harvester binder-attachment	C. G. Price	Amana, Iowa	Aug. 27, 1872	130, 865
Harvester binder-attachment	W. L. Sanford	Ashton, Ill	Sept. 16, 1873	142, 870
Harvester binder-attachment	T. Urdahl	Cross Plains, Wis	Nov. 18, 1873	144, 715
Harvester binder attachment	T. Webb	Freeport, Ill	May 13, 1873	138, 771
Harvester binder, Grain	C. A. McPhetridge	Saint Louis, Mo	Nov. 18, 1856	16, 097
Harvester-binder stand and table	W. R. Low	Sandwich, Ill	Feb. 4, 1873	135, 481
Harvester binding-apparatus	C. H. McAleer	Chambersburgh, Pa	Aug. 16, 1859	25, 131
Harvester binding-apparatus	D. W. Travis and C. M. Clinton	Enfield and Ithaca, N. Y	Sept. 6, 1870	107, 126
Harvester binding-attachment	C. Alvord	Westford, Wis	Mar. 5, 1861	31, 584
Harvester binding-attachment	J. W. Bates	Saint Paul, Minn	Apr. 20, 1869	89, 194
Harvester binding-attachment	J. Bebel and W. Hedges	Earlville, Ill	Sept. 6, 1864	44, 139
Harvester binding-attachment	R. D. Brown	Covington, Ind	June 27, 1865	48, 363
Harvester binding-attachment	H. M. and W. W. Burson	Atkinson, Ill	Mar. 10, 1863	37, 852
Harvester binding-attachment	A. Freed and J. Snook	La Porte, Ind	Jan. 10, 1871	110, 841
Harvester binding-attachment	W. Grey	Nicholsville, Ohio	Nov. 16, 1858	22, 074
Harvester binding-attachment	L. P. Harris	Mansfield, Ohio	Feb. 26, 1861	31, 540
Harvester binding-attachment	F. B. Isett	Hollidaysburgh, Pa	Aug. 2, 1870	105, 946
Harvester binding-attachment	C. Powers and P. Lancaster	Bronson, Mich	Oct. 29, 1861	33, 601
Harvester binding-attachment	B. W. Squires and C. M. Clinton	Ithaca, N. Y	Jan. 25, 1870	99, 256
Harvester binding-attachment	A. Underwood	Kenosha, Wis	Aug. 11, 1863	39, 510
Harvester binding-device	G. Notman	Deerfield, Ohio	May 11, 1858	20, 215
Harvester, Broom-corn	J. O. Norton	Wilton, Ill	Oct. 14, 1862	36, 668
Harvester, Cane	W. B. Robertson	West Baton Rouge, La	Feb. 12, 1861	31, 406
Harvester, Cane	A. St. Dizier	Plaquemine, La	Oct. 9, 1860	30, 361
Harvester caster-wheel	J. S. and H. R. Russell	New Market, Md., and Woodbury, N. J.	Sept. 28, 1869	95, 383
Harvester, Clover	J. R. Bruton	Concord, N. C	May 13, 1873	138, 786
Harvester, Clover	F. Decker	Ostrander, Ohio	May 28, 1861	32, 416
Harvester, Clover	F. Decker	Ostrander, Ohio	Feb. 28, 1865	46, 547
Harvester, Clover	P. Dismukes	Gallatin, Tenn	Nov. 24, 1868	84, 416
Harvester, Clover	P. Dismukes	Gallatin, Tenn	Mar. 23, 1869	88, 144
Harvester, Clover	J. S. Gage	Dowagiac, Mich	Dec. 19, 1854	12, 095
Harvester, Clover	M. Garretson	Bermudian, Pa	Jan. 6, 1852	8, 628
Harvester, Clover	D. Hinkle	New Pittsburgh, Ohio	July 2, 1861	32, 692
Harvester, Clover	J. Hinton	Monroe County, Va	May 22, 1849	6, 475
Harvester, Clover	J. W. Hull and A. G. Stiffler	Alquina, Ind	Feb. 7, 1871	111, 544
Harvester, Clover	E. Kramer	Alvira, Pa	May 1, 1866	54, 370
Harvester, Clover	J. Krauser	Reading, Pa	June 22, 1852	9, 055
Harvester, Clover	S. Krauser	Reading, Pa	Dec. 18, 1849	6, 954
Harvester, Clover	J. Lamburn	Boundary City, Ind	Sept. 26, 1871	119, 369
Harvester, Clover	C. McCombie	Carrolltown, Pa	May 14, 1872	126, 726
Harvester, Clover	T. S. Steadman	Murray, N. Y	May 23, 1854	10, 967

Index of patents issued from the United States Patent Office from 1790 *to* 1873, *inclusive*—Continued.

Invention.	Inventor.	Residence.	Date.	No.
Harvester, Clover	S. L. Stockstill and W. H. H. Scarff.	Medway, Ohio	Feb. 25, 1868	74, 861
Harvester, Clover	J. A. Wagener	Pultney, N. Y	May 24, 1853	9, 750
Harvester, Clover	J. Westhafer	Quincy, Ohio	Jan. 12, 1869	85, 880
Harvester, Clover-seed	C. B. Wheeler and A. Bascom	Steuben, Ohio	July 29, 1856	15, 449
Harvester, Corn	I. V. Adair	Varick, N. Y	Apr. 6, 1858	19, 822
Harvester, Corn	M. and W. P. Bales	London, Ohio	May 12, 1868	77, 708
Harvester, Corn	M. and W. P. Bales	London, Ohio	Jan. 26, 1869	86, 203
Harvester, Corn	B. F. Barney	Pontiac, Ill	May 4, 1869	89, 546
Harvester, Corn	W. Beach	Baltimore, Md	Oct. 4, 1859	25, 699
Harvester, Corn	J. H. Beam	Woodside, Ill	May 11, 1869	89, 968
Harvester, Corn	G. W. S. Bell	Tallula, Ill	Sept. 7, 1869	94, 549
Harvester, Corn	J. H. Besant and M. B. Atkinson	Point of Rocks, Md., and Georgetown, D. C.	Mar. 19, 1872	124, 661
Harvester, Corn	W. M. Bonwill	Camden, Del	Mar. 4, 1856	14, 344
Harvester, Corn	J. W. Bope	Saint Louis, Mo	Jan. 10, 1865	45, 811
Harvester, Corn	E. Boswell	Highland, Ohio	May 3, 1868	75, 119
Harvester, Corn	J. Bowers	Iola, Kans	Nov. 19, 1872	133, 141
Harvester, Corn	T. Boyd	Des Moines, Iowa	Feb. 27, 1866	52, 819
Harvester, Corn	I. V. Brower	Millstone, N. J	Aug. 13, 1861	33, 026
Harvester, Corn	A. R. Burdick and C. D. Read	Racine, Wis., and Elgin, Ill	July 22, 1862	35, 921
Harvester, Corn	J. Burke	Sycamore, Ill	Jan. 9, 1872	122, 561
Harvester, Corn	T. Butterworth	Shelbyville, Mo	Feb. 6, 1866	52, 384
Harvester, Corn	J. F. Byland	Walton, Ky	Dec. 7, 1869	97, 599
Harvester, Corn	J. and W. Carrothers	Newton, Iowa	Dec. 24, 1872	134, 251
Harvester, Corn	J. and W. Carrothers	Newton, Iowa	May 27, 1873	139, 367
Harvester, Corn	J. L. Chapman	Kinmundy, Ill	Mar. 1, 1859	23, 076
Harvester, Corn	O. N. Chase	Boston, Mass	Oct. 13, 1863	40, 243
Harvester, Corn	W. Cogswell and C. A. Mathewson.	Ottawa, Ill	Aug. 9, 1859	24, 992
Harvester, Corn	R. B. Corbin and J. Morris	Saint Augustine, Ill	June 22, 1858	20, 628
Harvester, Corn	C. U. Crandall	Galesburgh, Ill	Dec. 31, 1867	72, 811
Harvester, Corn	J. Crossland	Spencer Grove, Iowa	July 25, 1871	117, 264
Harvester, Corn	J. H. Crip	Quincy, Ohio	Nov. 12, 1867	70, 816
Harvester, Corn	B. T. Currier	Bath, Me	Aug. 9, 1859	24, 994
Harvester, Corn	T. E. Curtis	Titusville, Pa	Aug 22, 1871	118, 207
Harvester, Corn	J. Dodenhoff	Bloomington, Ill	July 7, 1857	17, 729
Harvester, Corn	A. Dyson and W. N. Macqueen.	Saint Louis, Mo	June 26, 1866	55, 840
Harvester, Corn	E. J. Eno	Jacksonville, Ill	Dec. 15, 1863	40, 979
Harvester, Corn	E. J. Eno	Springfield, Ill	Sept. 28, 1869	95, 214
Harvester, Corn	J. H. Fisher and C. Holcomb	Mendota, Ill	May 11, 1869	89, 979
Harvester, Corn	A. W. Fleming	Springfield, Ill	Apr. 27, 1869	89, 396
Harvester, Corn	H. Flescher	Springfield, Ill	Apr. 12, 1870	101, 721
Harvester, Corn	D. C. Flint	Bardolph, Ill	May 23, 1871	115, 046
Harvester, Corn	A. Ford	Toledo, Ohio	June 8, 1869	91, 005
Harvester, Corn	B. M. Fowler	Brooklyn, N. Y	July 12, 1864	43, 487
Harvester, Corn	F. F. Fowler	Upper Sandusky, Ohio	Aug. 14, 1866	57, 113
Harvester, Corn	W. N. Gates	Manchester Centre, N. Y	July 30, 1867	67, 287
Harvester, Corn	G. Geer	Douglas, Ill	June 2, 1863	38, 738
Harvester, Corn	S. Gesley	Beloit, Wis	Aug. 6, 1872	130, 211
Harvester, Corn	M. A. Getzendaner	Polo, Ill	May 3, 1870	102, 532
Harvester, Corn	H. Gortner and J. McCann	Nashport, Ohio	Mar. 6, 1860	27, 360
Harvester, Corn	L. Hamilton	Panora, Iowa	July 22, 1873	141, 138
Harvester, Corn	L. B. Hamilton	Boston, Mass	Dec. 31, 1867	72, 848
Harvester, Corn	S. H. Hamilton	Macomb, Ill	Aug. 27, 1861	33, 142
Harvester, Corn	G. B. Hamlin	Willimantic, Conn	June 28, 1870	104, 846
Harvester, Corn	J. D. Hampshire	Paper Mills Post-Office, Md	Oct. 20, 1868	83, 155
Harvester, Corn	H. Harris	Circleville, Ohio	Jan. 5, 1869	85, 585
Harvester, Corn	E. K. Harvey	Quincy, Ohio	Nov. 11, 1868	83, 851
Harvester, Corn	G. D. Haworth	Mechanicsburgh, Ill	July 21, 1857	17, 832
Harvester, Corn	A. Henlings	Philadelphia, Pa	Mar. 31, 1857	16, 921
Harvester, Corn	W. Hill and J. A. Harpham	Havana, Ill	Oct. 29, 1867	70, 335
Harvester, Corn	N. Hospers and M. Sellers	Pella and Keokuk, Iowa	Jan. 12, 1869	85, 828
Harvester, Corn	W. B. Hubard	Arrington Depot, Va	June 15, 1869	91, 444
Harvester, Corn	A. Humberger	Somerset, Ohio	Dec. 1, 1857	18, 747
Harvester, Corn	A. Humberger	Somerset, Ohio	Apr. 27, 1858	20, 067
Harvester, Corn	A. Humberger	Somerset, Ohio	Apr. 3, 1860	27, 720
Harvester, Corn	J. Johnson	Mott Haven, N. Y	Dec. 26, 1871	122, 176
Harvester, Corn	S. Johnston	Buffalo, N. Y	Oct. 28, 1862	36, 848
Harvester, Corn	W. H. Kenyon	North Providence, R. I	Jan. 7, 1873	134, 678
Harvester, Corn	J. Kilborn	Chicago, Ill	Sept. 16, 1873	142, 920
Harvester, Corn	J. H. Kite	Conrad's Store, Va	Mar. 8, 1859	23, 174
Harvester, Corn	D. Landon	Wyandot, Ohio	June 22, 1858	20, 645
Harvester, Corn	C. M. Lightner	Harrisburgh, Pa	July 10, 1866	56, 236
Harvester, Corn	R. B. Linthicum	Lexington, Ill	Aug. 21, 1866	57, 345
Harvester, Corn	C. B. Maclay	Delavan, Ill	Nov. 9, 1869	96, 602
Harvester, Corn	J. Mains	Olean, Ill	Feb. 4, 1868	74, 104
Harvester, Corn	G. H. Manlove and J. P. Green.	Chicago, Ill	Sept. 25, 1866	58, 348
Harvester, Corn	G. H. Manlove and J. P. Green.	Chicago, Ill	Sept. 25, 1866	58, 349
Harvester, Corn	W. M. Mason	Polo, Ill	Mar. 29, 1864	42, 097
Harvester, Corn	R. C. Mauck	Conrad's Store, Va	Jan. 4, 1859	22, 508
Harvester, Corn	R. C. Mauck and W. T. McGahey.	Conrad's Store and McGaheysville, Va.	Apr. 22, 1856	14, 730
Harvester, Corn	J. McLeish	Chicago, Ill	Sept. 21, 1869	95, 031
Harvester, Corn	G. Meader	Prairie Centre, Ill	Apr. 28, 1868	77, 394
Harvester, Corn	T. Merrell	Dixon, Ill	Mar. 19, 1872	124, 757
Harvester, Corn	T. Merrell	Dixon, Ill	Nov. 26, 1872	133, 327
Harvester, Corn	Q. F. Messinger	Easton, Pa	Feb. 25, 1868	74, 928
Harvester, Corn	J. I. Mettler	Mendota, Ill	July 4, 1871	116, 735
Harvester, Corn	H. Molby	Davisburgh, Mich	Jan. 21, 1868	73, 629
Harvester, Corn	A. Moravek	Rosnyo, Hungary	Feb. 5, 1867	61, 852
Harvester, Corn	C. G. Moremen	Brandenburgh, Ky	Aug. 21, 1866	57, 360
Harvester, Corn	B. Murray and J. Van Doren	Ottawa and Farm Ridge, Ill	Dec. 7, 1858	22, 259
Harvester, Corn	W. Murray	Alexandria, Va	Feb. 1, 1870	99, 463
Harvester, Corn	R. L. Nelson	Orange Court-House, Va	Apr. 25, 1871	114, 182
Harvester, Corn	R. L. Nelson	Orange Court-House, Va	June 13, 1871	115, 978
Harvester, Corn	R. L. Nelson	Orange Court-House, Va	July 30, 1872	130, 065
Harvester, Corn	N. Newman	Springfield, Ill	Sept. 15, 1868	82, 148

Index of patents issued from the United States Patent Office from 1790 *to* 1873, *inclusive*—**Continued.**

Invention.	Inventor.	Residence.	Date.	No.
Harvester, Corn	N. Newman	Springfield, Ill	Nov. 30, 1869	97, 429
Harvester, Corn	N. Newman	Springfield, Ill	July 26, 1870	105, 716
Harvester, Corn	J. Oadhoudt	Saint Anthony's Falls, Minn	May 14, 1867	64, 789
Harvester, Corn	S. Patton	Chatsworth, Ill	Aug. 8, 1868	81, 202
Harvester, Corn	S. Patton	Chatsworth, Ill	Apr. 22, 1873	138, 181
Harvester, Corn	J. and I. M. Poffenberger	Urbana, Ohio	June 29, 1869	91, 964
Harvester, Corn	W. E. Prall	Knoxville, Tenn	May 22, 1866	54, 951
Harvester, Corn	E. W. Quincy	Lacon, Ill	July 7, 1868	79, 775
Harvester, Corn	E. W. Quincy	Peoria, Ill	June 24, 1873	140, 305
Harvester, Corn	I. Reamer	Conrad's Store, Va	Jan. 11, 1859	22, 583
Harvester, Corn	I. Reamer and H. Miller	Conrad's Store, Va	Sept. 14, 1858	21, 516
Harvester, Corn	I. Reamer and H. Miller	Conrad's Store, Va	Apr. 26, 1859	23, 783
Harvester, Corn	J. H. Rible	Somerset, Ohio	Dec. 1, 1857	18, 769
Harvester, Corn	J. E. Rice	Moline, Ill	May 31, 1870	103, 779
Harvester, Corn	G. W. Richardson and J. W. White.	Grayville, Ill	June 14, 1859	24, 434
Harvester, Corn	T. A. Risher	Circleville, Ohio	Mar. 23, 1858	19, 716
Harvester, Corn	T. A. Risher	Oskaloosa, Iowa	Nov. 17, 1863	40, 641
Harvester, Corn	A. F. Roberts	Lexington, Ky	Sept. 19, 1871	119, 181
Harvester, Corn	J. A. C. Rose	Carrollton, Ill	Feb. 16, 1869	87, 071
Harvester, Corn	D. Sarvar and R. Coons	Greensburgh, Pa	May 5, 1868	77, 660
Harvester, Corn	A. Scott	Rochester, N. Y	Aug. 5, 1873	141, 466
Harvester, Corn	S. Secrist	West Liberty, Ohio	Sept. 18, 1866	58, 143
Harvester, Corn	J. W. Sharrard, S. Bryan, and H. Hunt.	Janesville, Wis	Aug. 31, 1869	94, 446
Harvester, Corn	J. Shobe	Principio, Md	Mar. 19, 1867	63, 111
Harvester, Corn	H. L. Smith	Watkins, N. Y	Dec. 17, 1867	72, 236
Harvester, Corn	J. W. Smith	Iowa Point, Kans	Aug. 2, 1864	43, 716
Harvester, Corn	J. W. Smith	Iowa Point, Kans	May 30, 1865	47, 995
Harvester, Corn	M. A. Smith	Middlefield, N. Y	Oct. 4, 1870	107, 972
Harvester, Corn	J. H. Spears and K. and R. Wells.	Piper City, Ill	June 3, 1873	139, 623
Harvester, Corn	A. Sprague	Cold Water, Mich	Aug. 12, 1856	15, 533
Harvester, Corn	L. Stadler	Bowensburgh, Ill	Sept. 7, 1869	94, 662
Harvester, Corn	A. Stoddard	Tecumseh, Mich	July 27, 1858	21, 031
Harvester, Corn	T. H. Storms and J. C. Poffenberger.	Jacksonville, Ill	Nov. 29, 1864	45, 282
Harvester, Corn	C. S. Stull	Poolesville, Md	Feb. 28, 1871	112, 191
Harvester, Corn	J. B. Sweetland	Pontiac, Mich	Dec. 3, 1867	71, 815
Harvester, Corn	J. B. Sweetland	Pontiac, Mich	Feb. 11, 1868	74, 447
Harvester, Corn	M. Thorp	Warterloo, Iowa	Sept. 12, 1871	118, 984
Harvester, Corn	W. S. Tilton	Boston, Mass	June 17, 1856	15, 152
Harvester, Corn	D. J. Tittle	Albany, N. Y	Sept. 25, 1866	58, 320
Harvester, Corn	G. M. Tope	Leesville, Mo	July 15, 1873	140, 971
Harvester, Corn	J. H. L. Tuck	Ottawa, Ill	Feb. 1, 1870	99, 501
Harvester, Corn	E. Turk	Manchester, Ill	Jan. 18, 1870	99, 034
Harvester, Corn	J. M. Wallis and E. Miller	Milton, Iowa	Apr. 3, 1866	53, 710
Harvester, Corn	S. Ward	Lane, Ill	July 11, 1865	48, 750
Harvester, Corn	J. H. Whitney	Rochester, Minn	Nov. 7, 1871	120, 686
Harvester, Corn	L. C. Wilder	Lexington, N. C	Jan. 26, 1858	19, 221
Harvester, Corn	C. W. Williams	Wyandotte, Mich	Oct. 8, 1867	69, 600
Harvester, Corn	H. Williams	Chicago, Ill	July 29, 1862	36, 046
Harvester, Corn	J. S. Williams	Chicago, Ill	Dec. 12, 1865	51, 505
Harvester, Corn	J. F. Winchell	Springfield, Ohio	Oct. 9, 1866	58, 706
Harvester, Corn	J. F. Winchell	Springfield, Ohio	May 5, 1868	77, 696
Harvester, Corn	A. N. Woodard	Fentonville, Mich	Apr. 25, 1871	114, 076
Harvester, Corn	J. Wright and J. J. Johnson	Cold Water, Mich	May 28, 1867	65, 324
Harvester, Corn	T. Yates	Dubuque, Iowa	Dec. 4, 1866	60, 318
Harvester, Corn	R. Yeilding	Ypsilanti, Mich	July 15, 1862	35, 907
Harvester, Corn	G. W. N. Yost	Port Gibson, Miss	Jan. 8, 1856	14, 076
Harvester, Corn	P. C. Yost	Hamilton, Ill	May 5, 1868	77, 562
Harvester, Corn and cane	E. H. Clinton, W. Prather, and H. O. Hutchinson.	Iowa City, Iowa	Aug. 20, 1867	67, 955
Harvester, Corn and cane	D. P. Flynn and R. S. Hayes	Le Roy, N. Y	Mar. 6, 1860	27, 356
Harvester, Corn and cane	W. M. Mason	Polo, Ill	Sept. 3, 1861	33, 204
Harvester, Corn and cane	H. D. McGeorge and D. S. Greer.	Morgantown, Va	June 21, 1859	24, 477
Harvester, Corn and sugar-cane	G. W. N. Yost	Cincinnati, Ohio	July 10, 1870	29, 131
Harvester, Cotton	W. Apperly and C. P. Johnson	Memphis, Tenn	Dec. 11, 1860	30, 862
Harvester, Cotton	L. Bishop	Talladega, Ala	July 5, 1859	24, 609
Harvester, Cotton	J. Davis, jr., and D. Scott, jr.	Greensborough, N. C	Apr. 5, 1870	101, 439
Harvester, Cotton	J. Griffin	Louisville, Ky	Mar. 8, 1859	23, 168
Harvester, Cotton	J. Griffin	Louisville, Ky	Nov. 22, 1859	26, 180
Harvester, Cotton	M. Hosford and J. C. Avery	Macon, Miss	Apr. 27, 1858	20, 066
Harvester, Cotton	D. C. Hubbard	Point Coupee Parish, La	July 12, 1870	105, 211
Harvester, Cotton	W. H. Irving	Philadelphia, Pa	Feb. 13, 1872	123, 704
Harvester, Cotton	J. W. Leigh	Norfolk, Va	Mar. 18, 1873	136, 924
Harvester, Cotton	O. P. Myers	Canton, Ohio	Dec. 24, 1872	134, 157
Harvester, Cotton	W. H. Pedrick	Richmond, Ind	Nov. 18, 1873	144, 629
Harvester, Cotton-stalk	S. Bowerman	Detroit, Mich	Oct. 1, 1850	7, 677
Harvester, Cradling	M. Barlow	Lexington, Ky	Apr. 29, 1856	14, 761
Harvester-crank	H. L. Wanzer	Lanesville, Conn	Nov. 9, 1869	96, 643
Harvester crank and pin	J. Kline	Mechanicsburgh, Pa	Mar. 22, 1870	101, 134
Harvester-crank motion	A. Palmer	Brockport, N. Y	May 1, 1866	54, 397
Harvester-cutter	E. M. Allen	Darlington, Mich	Mar. 31, 1868	76, 135
Harvester-cutter	J. Atkins	Mokena, Ill	Feb. 11, 1768	74, 272
Harvester-cutter	H. C. Aydelot	Carthage, Ind	Aug. 16, 1870	106, 305
Harvester-cutter	T. D. Aylesworth	Ilion, N. Y	Sept. 6, 1859	25, 312
Harvester-cutter	E. M. Birdsall	Penn Yan, N. Y	Oct. 12, 1869	95, 638
Harvester-cutter	J. A. Bonham and A. J. Harrington.	Lovely Dale, Ind	Jan. 16, 1870	100, 801
Harvester-cutter	H. Bonholtzer and J. S. Shopp	Cumberland County, Pa	Sept. 14, 1869	94, 705
Harvester-cutter	C. Cadwell	Waukegan, Ill	Jan. 22, 1867	61, 394
Harvester-cutter	G. W. Chapman, jr	Iowa Falls, Iowa	Aug. 4, 1868	80, 598
Harvester-cutter	T. J. Christy	Olney, Ill	June 2, 1868	78, 515
Harvester-cutter	T. J. and G. M. Clark	Higganum, Conn	Oct. 25, 1870	108, 686
Harvester-cutter	J. M. Connel	Newark, Ohio	July 19, 1870	105, 550
Harvester-cutter	H. Cutler	Central Village, Conn	Mar. 14, 1871	112, 692

Index of patents issued from the United States Patent Office from 1790 *to* 1873, *inclusive*—Continued.

Invention.	Inventor.	Residence.	Date.	No.
Harvester-cutter	J. W. H. Doubler	West Alden, N. Y	July 29, 1873	141, 210
Harvester-cutter	G. L. Du Laney	Mechanicsburgh, Pa	May 18, 1869	90, 086
Harvester-cutter	G. L. Du Laney	Mechanicsburgh, Pa	June 8, 1869	91, 102
Harvester-cutter	G. L. Du Laney	Mechanicsburgh, Pa	Nov. 23, 1869	97, 062
Harvester-cutter	G. Fetter	Philadelphia, Pa	Apr. 3, 1860	27, 757
Harvester-cutter	S. B. Fisk	Rising Sun, Ind	Oct. 1, 1872	131, 869
Harvester-cutter	J. L. Fountain	Rockford, Ill	Nov. 17, 1857	18, 659
Harvester-cutter	T. Garrick	Providence, R. I	Mar. 14, 1871	112, 582
Harvester-cutter	S. Gillam	New York, N. Y	May 29, 1866	55, 084
Harvester-cutter	J. Gore	Fredonia, N. Y	Feb. 16, 1858	19, 360
Harvester-cutter	P. Gregg	Brownsville, Mich	Apr. 25, 1871	114, 003
Harvester-cutter	M. Hallenbeck	Albany, N. Y	Nov. 3, 1868	83, 628
Harvester-cutter	M. Harrison	Laclede, Mo	Sept. 19, 1871	119, 029
Harvester-cutter	M. P. Hathaway	Mankato, Minn	Aug. 14, 1866	57, 130
Harvester-cutter	I. A. Hebbard	Rochester, N. Y	Apr. 5, 1870	101, 460
Harvester-cutter	H. L. Hervey	Quincy, Ill	Mar. 18, 1856	14, 453
Harvester-cutter	M. G. Hubbard	Penn Yan, N. Y	Jan. 20, 1857	16, 442
Harvester-cutter	M. G. Hubbard	Penn Yan, N. Y	June 16, 1857	17, 575
Harvester-cutter	S. Hull	Poughkeepsie, N. Y	Aug. 4, 1863	39, 401
Harvester-cutter	S. Hull	Poughkeepsie, N. Y	Sept. 14, 1869	94, 826
Harvester-cutter	J. Irwin	Frankfort, Ohio	July 7, 1857	17, 739
Harvester-cutter	W. G. Kenyon	Wakefield, R. I	Mar. 28, 1871	113, 175
Harvester-cutter	T. R. and S. Knowles	Jersey City, N. J	Feb. 18, 1868	74, 550
Harvester-cutter	A. G. Laughlin and K. G. Rice	Providence, Ky	Oct. 22, 1872	132, 513
Harvester-cutter	M. Lewis	Odell, Ill	June 2, 1868	78, 460
Harvester-cutter	H. A. Link	Columbus, Ohio	Aug. 3, 1869	93, 319
Harvester-cutter	I. S. Love	Beloit, Wis	Mar. 11, 1856	14, 402
Harvester-cutter	J. H. Manny	Waddam's Grove, Ill	June 21, 1853	9, 800
Harvester-cutter	J. H. Manny	Rockford, Ill	June 26, 1855	13, 149
Harvester-cutter	J. H. Manny	Rockford, Ill	Mar. 25, 1856	14, 544
Harvester-cutter	J. P. Manny	Rockford, Ill	July 7, 1857	17, 745
Harvester-cutter	H. W. Mason	Hagerstown, Md	Aug. 30, 1870	106, 948
Harvester-cutter	W. McKeever	Staunton, Va	Apr. 8, 1873	137, 705
Harvester-cutter	H. Mewes	Binghamton, N. Y	Dec. 17, 1872	134, 085
Harvester-cutter	R. J. Morrison	Richmond, Ind	Jan. 13, 1857	16, 303
Harvester-cutter	B. Murray	Farm Ridge, Ill	June 13, 1854	11, 076
Harvester-cutter	C. K. Myers	Pekin, Ill	Feb. 8, 1870	99, 695
Harvester-cutter	T. Neys	Menomonee, Wis	Dec. 21, 1869	98, 182
Harvester-cutter	J. T. Norris	Tiffin, Ohio	June 16, 1868	78, 889
Harvester-cutter	J. H. Owen	Houston Township, Ill	Apr. 20, 1869	89, 236
Harvester-cutter	W. J. Oxer	Williamsport, Ind	Sept. 19, 1871	119, 170
Harvester-cutter	J. T. Polson	Laclede, Mo	Oct. 3, 1871	119, 640
Harvester-cutter	C. Pomeroy	Mattoon, Ill	Apr. 22, 1873	138, 192
Harvester-cutter	G. F. Quick	Philadelphia, Pa	July 19, 1864	43, 602
Harvester-cutter	S. H. Reed and T. Thompson	Fredericktown, Ohio	Feb. 25, 1873	136, 266
Harvester-cutter	F. E. Rogers	Paw Paw, Ill	June 28, 1870	104, 773
Harvester-cutter	B. T. Roney	Philadelphia, Pa	Apr. 29, 1856	14, 777
Harvester-cutter	B. T. Roney	Philadelphia, Pa	Sept. 11, 1860	29, 999
Harvester-cutter	D. Russell	Drewersburgh, Ind	Jan. 30, 1855	12, 327
Harvester-cutter	L. Russell	Otsego, Mich	Feb. 13, 1872	123, 731
Harvester-cutter	J. Schneider	Canton, Ohio	Dec. 28, 1869	98, 303
Harvester-cutter	H. F. Shaw	West Roxbury, Mass	Sept. 14, 1869	94, 918
Harvester-cutter	W. E. Shoales	Sherburne Four Corners, N. Y.	Sept. 2, 1873	142, 417
Harvester-cutter	J. S. Smith and J. Coder	Swanton, Ohio	Aug. 31, 1869	94, 251
Harvester-cutter	W. G. Smith	Elizabethport, N. J	Nov. 6, 1860	30, 595
Harvester-cutter	L. E. Stillwell	Franklinville, N. Y	July 16, 1872	129, 068
Harvester-cutter	D. Stukey	Lancaster, Ohio	Sept. 6, 1870	107, 118
Harvester-cutter	F. R. and W. O. Sutton	Wellington, Ill	July 29, 1873	141, 298
Harvester-cutter	J. M. Taft	Arcadia, Wis	Nov. 2, 1869	96, 503
Harvester-cutter	P. Thayer	Lansingburgh, N. Y	Mar. 11, 1856	14, 422
Harvester-cutter	H. Unger	Germantown, Ohio	Sept. 24, 1872	131, 580
Harvester-cutter	C. B. Wagner	Philadelphia, Pa	June 24, 1856	15, 205
Harvester-cutter	W. Wasson, P. F. Powers, and G. W. Dungan.	Genoa, Nev	Feb. 4, 1873	135, 500
Harvester-cutter	J. M. Wehrly	Somerville, N. J	Aug. 7, 1866	57, 023
Harvester-cutter	S. H. Wellings and S. Soules	Bridgeville, Mich	June 10, 1873	139, 695
Harvester-cutter	W. N. Whiteley	Springfield, Ohio	Aug. 11, 1868	81, 044
Harvester-cutter	W. N. Whiteley, jr., J. Fassler, and O. S. Kelly.	Springfield, Ohio	Aug. 13, 1867	67, 829
Harvester-cutter	F. Wittram	San Francisco, Cal	Dec. 28, 1869	98, 456
Harvester-cutter	W. A. Wood	Hoosick Falls, N. Y	Dec. 2, 1862	37, 066
Harvester-cutter	E. L. Yancy	Batavia, N. Y	Aug. 6, 1872	130, 266
Harvester cutter and raker, Grain and grass	H. Knowles and H. C. Bevington.	Washington, D. C., and Holmes County, Ohio.	July 2, 1850	7, 475
Harvester-cutter bar	R. Allstatter	Hamilton, Ohio	May 31, 1870	103, 699
Harvester-cutter bar	B. Johnson and W. Johnson	Carrollton and Hanover, Ohio.	Aug. 10, 1869	93, 448
Harvester-cutter bar	J. H. Manny	Rockford, Ill	Jan. 22, 1856	14, 149
Harvester-cutter bar	W. J. Oxer	Williamsport, Ind	May 24, 1870	103, 361
Harvester-cutter bar	E. G. Short and C. Oberly	Carthage, N. Y	Apr. 12, 1870	101, 930
Harvester-cutter bar	T. Welch	Churchville, N. Y	Mar. 19, 1867	62, 985
Harvester-cutter-bar connection	G. D. Culp and W. J. Keeney	Allensville and Florence, Ind.	Aug. 25, 1863	39, 633
Harvester-cutter-bar heads, Manufacture of	A. Padgham	Syracuse, N. Y	June 20, 1871	116, 088
Harvester cutter-bars, Machine for grinding	J. H. and J. S. Griffin	Harvard, Ill	Feb. 20, 1872	123, 890
Harvester-cutter blades to sickle-bars, Attaching	W. H. Hovey	Springfield, Mass	Apr. 29, 1856	14, 768
Harvester cutter, Corn	J. J. De Frietas	Springfield, Ill	July 26, 1870	105, 784
Harvester-cutter finger	J. H. Manny	Waddam's Grove, Ill	Apr. 19, 1853	9, 675
Harvester cutter, Grain and grass	M. G. Hubbard	Penn Yan, N. Y	Nov. 11, 1856	16, 057
Harvester cutter, Grain and grass	W. Pierpoint	Salem, N. J	Nov. 22, 1853	10, 258
Harvester-cutter grinder	J. P. Barker	Vienna, Ohio	Sept. 19, 1871	119, 108
Harvester-cutter grinder	W. B. Denel	Ithaca, N. Y	Nov. 2, 1869	96, 402
Harvester-cutter grinder	E. F. Keeling	Milton, Ohio	Aug. 13, 1861	33, 043
Harvester-cutter grinder	E. R. McCall	Simcoe, Canada	Oct. 24, 1871	120, 205
Harvester-cutter grinder	S. O. and P. W. Vaughn	De Kalb, Ill	Oct. 4, 1870	107, 985
Harvester-cutter holder	S. H. Wilson	Auburn, N. Y	Jan. 5, 1869	85, 627
Harvester-cutter holder for grinders	T. E. King and G. C. Dolph	Cleveland and West Andover, Ohio.	Sept. 27, 1870	107, 784

Index of patents issued from the United States Patent Office from 1790 *to* 1873, *inclusive*—Continued.

Invention.	Inventor.	Residence.	Date.	No.
Harvester-cutter sharpener	E. L. Bushnell	Poughkeepsie, N. Y	July 14, 1868	79, 948
Harvester-cutter sharpener	I. H. Coller	Poughkeepsie, N. Y	Mar. 8, 1864	41, 831
Harvester-cutter sharpener	J. McKnight	Pomeroy, Ohio	July 24, 1866	56, 589
Harvester-cutter sharpener	J. K. Staman	Mifflin, Ohio	Sept. 8, 1863	39, 866
Harvester-cutters, Apparatus for grinding	J. H. Curran	Rochester, N. Y	Jan. 30, 1872	123, 088
Harvester-cutters, Apparatus for grinding	J. F. and R. G. Kirkwood	Ellicott City, Md	July 2, 1872	128, 634
Harvester-cutters, Hand-grinder for	H. C. Fisk	Wellsville, N. Y	Oct. 24, 1871	120, 258
Harvester-cutters, Machine for grinding	J. Weichhart	San Francisco, Cal	Oct. 12, 1869	95, 751
Harvester-cutters, Machine for grinding	H. Whitall	Woodbury, N. J	May 14, 1867	64, 817
Harvester-cutters, Machine for sharpening	W. S. Ingraham	Evanston, Ill	Aug. 9, 1870	106, 167
Harvester-cutters, Rest for grinding	T. Brett	Geneva, Ohio	Apr. 21, 1868	76, 883
Harvester-cutters, Sharpening-bench for	J. R. Cliffton	West Unity, Ohio	Nov. 26, 1872	133, 409
Harvester-cutters while being sharpened, Holder for.	W. H. Daniels	Bryan, Ohio	Feb. 27, 1872	124, 121
Harvester cutting-apparatus	J. L. Abell	Cummington, Mass	Nov. 29, 1870	109, 569
Harvester cutting-apparatus	S. W. Beach	Ypsilanti, Mich	July 2, 1872	128, 456
Harvester cutting-apparatus	S. Comfort, jr	Morrisville, Pa	Apr. 7, 1857	16, 968
Harvester cutting-apparatus	G. W. Cook and F. M. Duncan	Geneseo and Aledo, Ill	Mar. 14, 1871	112, 689
Harvester cutting-apparatus	J. M. Connel	Newark, Ohio	Aug. 20, 1872	130, 568
Harvester cutting-apparatus	R. Dutton	Brooklyn, N. Y	Feb. 11, 1868	74, 211
Harvester cutting-apparatus	R. Dutton	Brooklyn, N. Y	Feb. 11, 1868	74, 212
Harvester cutting-apparatus	R. Dutton	New York, N. Y	Feb. 11, 1868	74, 213
Harvester cutting-apparatus	T. C. Hargrave	Schenectady, N. Y	Dec. 17, 1861	33, 941
Harvester cutting-apparatus	C. Howell	Cleveland, Ohio	Sept. 14, 1858	21, 499
Harvester cutting-apparatus	W. G. Kenyon	Wakefield, R. I	July 18, 1871	117, 178
Harvester cutting-apparatus	L. G. Kniffen	Worcester, Mass	Dec. 24, 1861	33, 999
Harvester cutting-apparatus	P. Manny	Waddam's Grove, Ill	Aug. 25, 1857	18, 052
Harvester cutting-apparatus	H. Mewes	Binghamton, N. Y	July 5, 1870	105, 111
Harvester cutting-apparatus	H. Mewes	Binghamton, N. Y	July 16, 1872	129, 357
Harvester cutting-apparatus	R. J. Morrison	Richmond, Va	Aug. 14, 1855	13, 433
Harvester cutting-apparatus	C. D. Read	Ayer's, Mass	June 25, 1872	128, 326
Harvester cutting-apparatus	A. R. Reese	Phillipsburgh, N. J	Apr. 22, 1862	35, 040
Harvester cutting-apparatus	A. Stoler and S. A. Sisson	Bristol, Pa	Dec. 17, 1861	33, 966
Harvester cutting-apparatus	D. H. Thayer	Lansing, N. Y	Mar. 26, 1861	31, 854
Harvester cutting-apparatus	W. A. Wood	Hoosick Falls, N. Y	Dec. 28, 1858	22, 468
Harvester cutting-apparatus, Attachment for	A. Anderson and L. Johnston	London, Canada	May 30, 1871	115, 411
Harvester cutting-apparatus, Corn and cane	J. W. Batson	Triadelphia, Md	July 29, 1856	15, 409
Harvester cutting-apparatus, Grain and grass	J. W. Baltzly and W. Hobson	Pana, Ill	Sept. 22, 1857	18, 229
Harvester cutting-device	D. W. Entrikin and L. H. Davis	West Chester, Pa	Sept. 13, 1858	19, 920
Harvester cutting-device	C. W. Glover	Roxbury, Conn	July 15, 1856	15, 334
Harvester cutting-device	C. P. Gronberg	Montgomery, Ill	Sept. 7, 1858	21, 414
Harvester cutting-device	C. Wheeler, jr	Poplar Ridge, N. Y	Sept. 2, 1856	15, 677
Harvester discharging-apparatus	A. J. Cook	Enon, Ohio	Mar. 28, 1854	10, 691
Harvester discharging-apparatus	J. S. Fowler	Davenport, Iowa	July 14, 1868	79, 967
Harvester-divider	J. J. Lurvey	North Prairie, Wis	Oct. 20, 1868	83, 185
Harvester-divider	J. H. Shireman	East Berlin, Pa	Oct. 4, 1859	25, 679
Harvester divider-attachment	J. H. Keller and D. F. Luse	Boalsburgh and Centre Hall, Pa.	Apr. 2, 1872	125, 198
Harvester drive-wheel	W. A. Wood	Hoosick Falls, N. Y	Apr. 13, 1869	88, 938
Harvester driving-wheel	D. L. Emerson	Rockford, Ill	Oct. 25, 1864	44, 834
Harvester driving-wheel	E. P. Russell	Manlius, N. Y	Aug. 15, 1865	49, 439
Harvester driving-wheel	W. N. Whiteley and T. Harding	Springfield, Ohio	June 23, 1868	79, 171
Harvester dropper	W. Allstatter and F. Schurger	Hamilton, Ohio	Mar. 5, 1872	124, 311
Harvester-dropper	J. J. and D. N. Barnhill	Vincennes, Ind	Aug. 8, 1871	117, 726
Harvester-dropper	W. G. Beels	Independence, Iowa	Jan. 18, 1870	98, 909
Harvester-dropper	C. Bickel and J. F. Leibold	Delaware, Ohio	Aug. 22, 1871	118, 331
Harvester-dropper	H. Brackett	Valley Falls, N. Y	May 4, 1869	89, 732
Harvester-dropper	H. Brackett	Valley Falls, N. Y	May 4, 1869	89, 733
Harvester-dropper	J. Case	La Fayette, Ind	Feb. 9, 1869	86, 730
Harvester-dropper	G. R. Clements	Prescott, Wis	Oct. 13, 1868	83, 040
Harvester-dropper	O. Dorsey	Newark, Ohio	July 23, 1872	129, 798
Harvester-dropper	W. H. Forker	Meadville, Pa	Dec. 17, 1872	133, 933
Harvester-dropper	J. S. Fowler	Davenport, Iowa	Dec. 16, 1873	145, 640
Harvester-dropper	J. B. Gathright	Louisville, Ky	Mar. 4, 1873	136, 370
Harvester-dropper	A. Goodyear, 2d	Hamden, Conn	Nov. 12, 1872	133, 025
Harvester-dropper	O. C. Green	Dublin, Ind	Jan. 11, 1870	98, 689
Harvester-dropper	D. S. Harner	Xenia, Ohio	July 5, 1870	104, 953
Harvester-dropper	O. M. Harrison	Glasgow, Mo	Aug. 22, 1871	118, 236
Harvester-dropper	I. Hedges	Radnor, Ohio	May 23, 1871	115, 199
Harvester-dropper	A. J. Hodges	Peoria, Ill	Dec. 9, 1873	145, 419
Harvester-dropper	J. W. Irwin	Circleville, Ohio	Jan. 23, 1872	122, 892
Harvester-dropper	C. D. Jeffries	Wooster, Ohio	Feb. 1, 1870	99, 444
Harvester-dropper	N. Johnson	Newburgh, Minn	May 27, 1873	139, 395
Harvester-dropper	N. S. Ketchum	Marshalltown, Iowa	Mar. 29, 1870	101, 273
Harvester-dropper	T. W. S. Kidd	Springfield, Ill	Dec. 20, 1870	110, 246
Harvester-dropper	T. F. Lippincott	Conemaugh, Pa	Dec. 21, 1869	98, 075
Harvester-dropper	J. Miller	Canton, Ohio	Mar. 27, 1866	53, 551
Harvester-dropper	J. Miller	Canton, Ohio	Dec. 24, 1867	72, 523
Harvester-dropper	L. Miller	Akron, Ohio	Nov. 12, 1867	70, 735
Harvester-dropper	T. C. Moore	Dublin, Ind	Sept. 20, 1870	107, 522
Harvester-dropper	E. Myers	Creagerstown, Md	Dec. 21, 1869	98, 090
Harvester-dropper	E. Myers	Creagerstown, Md	Feb. 22, 1870	100, 181
Harvester-dropper	H. Nisen	Brighton, Wis	June 4, 1872	127, 636
Harvester-dropper	A. L. and G. M. Peters	Lancaster, Ohio	Nov. 17, 1868	84, 133
Harvester-dropper	G. M. Peters	Lancaster, Ohio	Sept. 8, 1868	81, 941
Harvester-dropper	W. J. Plecker	Bushnell, Ill	May 6, 1873	138, 526
Harvester-dropper	A. Rank	Salem, Ohio	May 4, 1869	89, 792
Harvester-dropper	R. A. Roberts	Salisbury, Mo	July 2, 1872	128, 564
Harvester-dropper	B. Seneff	Chillicothe, Ohio	Feb. 27, 1872	124, 166
Harvester-dropper	B. W. Steschult	Glandorf, Ohio	Dec. 29, 1868	85, 487
Harvester-dropper	G. Stevenson	Zionsville, Ind	July 6, 1869	92, 392
Harvester-dropper	J. O. Taber	Salem, Ohio	Feb. 2, 1869	86, 602
Harvester-dropper	J. O. Taber	Salem, Ohio	May 31, 1870	103, 796
Harvester-dropper	J. S. Truxell	Greensburgh, Pa	Sept. 30, 1873	143, 391
Harvester-dropper	J. N. and T. Wallis	Fleming, N. Y	June 20, 1871	116, 120
Harvester-dropper	A. Ward	Dublin, Ind	May 10, 1870	102, 890
Harvester-dropper	W. N. Whiteley	Springfield, Ohio	Sept. 28, 1869	95, 397
Harvester-dropper	W. N. Whiteley	Springfield, Ohio	July 12, 1870	105, 396

Index of patents issued from the United States Patent Office from 1790 *to* 1873, *inclusive*—Continued.

Invention.	Inventor.	Residence.	Date.	No.
Harvester dropping-platform	E. and M. Ball	Canton, Ohio	July 21, 1868	80, 112
Harvester dropping-platform	E. P. Cady	Trenton, Wis	May 10, 1870	102, 914
Harvester dropping-platform	J. Miller	Canton, Ohio	Nov. 24, 1868	84, 368
Harvester dropping-platform	G. M. Peters	Lancaster, Ohio	Jan. 14, 1868	73, 380
Harvester dropping-platform	B. T. Reynolds	Centreville, Ind	May 24, 1870	103, 370
Harvester dropping-platform	J. B. Sawyer	Templeton, Mass	Apr. 26, 1870	102, 439
Harvester dropping-platform	A. H. Shreffler	Joliet, Ill	Nov. 3, 1868	83, 735
Harvester dropping-platform	C. R. and J. O. Taber	Salem, Ohio	Mar. 31, 1868	76, 269
Harvester dropping-platform	P. Warner	Gilead, Mich	Nov. 4, 1873	144, 239
Harvester dropping-platform	G. Wellhouse	Akron, Ohio	Nov. 10, 1868	84, 033
Harvester dumping-platform	J. W. Bope	Saint Louis, Mo	Apr. 14, 1868	76, 703
Harvester dumping-platform	D. M. Swartz	Lewisburgh, Pa	Aug. 20, 1867	67, 929
Harvester dumping-reel	E. P. Russell	Manlius, N. Y	Sept. 3, 1867	68, 461
Harvester-finger	J. P. Manny	Rockford, Ill	July 6, 1858	20, 808
Harvester-finger	J. Reily	Heart Prairie, Wis	Apr. 29, 1856	14, 790
Harvester-finger	E. P. Russell	Manlius, N. Y	Mar. 19, 1861	31, 745
Harvester-finger	E. P. Russell	Manlius, N. Y	Oct. 13, 1863	40, 283
Harvester-finger	H. C. Smith	Cleveland, Ohio	Mar. 2, 1858	19, 518
Harvester finger-bar	J. J. Barnes	Monticello, Ind	Feb. 11, 1868	74, 191
Harvester finger-bar	R. Dutton	New York, N. Y	May 29, 1866	55, 074
Harvester finger-bar	R. Dutton	New York, N. Y	May 26, 1868	78, 197
Harvester finger-bar	E. F. Ford	Stillwater, N. Y	Nov. 3, 1868	83, 619
Harvester finger-bar	J. M. Long, P. Black, and R. Allstatter.	Hamilton, Ohio	Dec. 1, 1857	18, 754
Harvester finger-bar	W. H. Seymour	Brockport, N. Y	Dec. 16, 1856	16, 253
Harvester finger-bar	C. A. Smith	Napa, Cal	Feb. 4, 1873	135, 495
Harvester finger-bars, Manufacture of	R. Dutton	New York, N. Y	May 26, 1868	78, 196
Harvester finger-beam	A. Rank	Salem, Ohio	Mar. 14, 1871	112, 629
Harvester finger-beam	A. R. Reese	Phillipsburgh, N. J	Apr. 22, 1862	35, 041
Harvester finger-guard	C. T. Bush	Rensselaerville, N. Y	Sept. 5, 1865	49, 713
Harvester finger-guard	R. Dutton	New York, N. Y	May 15, 1866	54, 702
Harvester finger-guards, Machine for slotting	J. Fassler	Springfield, Ohio	May 28, 1867	65, 193
Harvester finger-guards, Making	M. L. Ballard	Canton, Ohio	Jan. 29, 1861	31, 218
Harvester finger-guards, Making	M. L. Ballard	Canton, Ohio	Jan. 29, 1861	31, 219
Harvester finger-guards, Making	M. L. Ballard	Canton, Ohio	Jan. 29, 1861	31, 220
Harvester finger-guards, Making	M. L. Ballard	Canton, Ohio	Jan. 29, 1861	31, 284
Harvester finger-guards, Making	L. Miller	Canton, Ohio	Jan. 29, 1861	31, 285
Harvester finger or guard	L. Miller	Canton, Ohio	May 11, 1858	20, 243
Harvester, Flax	S. W. Tyler	Greenwich, N. Y	Sept. 1, 1863	39, 764
Harvester, Flax	S. W. Tyler	Troy, N. Y	June 13, 1871	115, 910
Harvester-frame	M. G. Hubbard	Penn Yan, N. Y	Apr. 28, 1857	17, 151
Harvester-frame	H. F. Mann	Westville, Ind	June 3, 1856	15, 013
Harvester frame, Grass	A. Palmer	Brockport, N. Y	Jan. 30, 1855	12, 323
Harvester friction-gear	H. F. and G. F. Shaw	West Roxbury, Mass	Dec. 3, 1872	133, 675
Harvester gaveling-attachment	J. W. Harvey	Marshalltown, Iowa	Dec. 27, 1864	45, 604
Harvester gaveling-attachment	C. C. Shults	Waverly, Iowa	May 7, 1872	126, 582
Harvester gaveling-fork	O. Webster	Murray, N. Y	Nov. 7, 1871	120, 602
Harvester-gearing	J. Farrington	Corry, Pa	June 20, 1871	116, 038
Harvester-gearing	M. G. Hubbard	Syracuse, N. Y	Sept. 22, 1868	82, 410
Harvester-gearing	M. G. Hubbard	Syracuse, N. Y	Sept. 22, 1868	82, 411
Harvester-gearing	J. P. Manny	Rockford, Ill	Dec. 12, 1871	121, 881
Harvester-gearing	A. H. Wagner	Chicago, Ill	Feb. 22, 1870	100, 221
Harvester-gearing	A. Warner	Ontario, N. Y	Nov. 14, 1865	50, 969
Harvester-gearing for changing speed	M. G. Hubbard	Syracuse, N. Y	Feb. 1, 1870	99, 439
Harvester, Grain	N. T. Allen	Ludlowville, N. Y	June 10, 1851	8, 157
Harvester, Grain	T. D. Burrall	Geneva, N. Y	Apr. 5, 1853	9, 644
Harvester, Grain	E. Danford, jr	Geneva, Ill	Sept. 17, 1850	7, 649
Harvester, Grain	B. Densmore	Sweden, N. Y	Feb. 10, 1852	8, 720
Harvester, Grain	A. Elliott	San Francisco, Cal	Apr. 1, 1856	14, 556
Harvester, Grain	D. Fitzgerald	New York County, N. Y	Sept. 7, 1852	9, 247
Harvester, Grain	B. G. Fitzhugh	Frederick, Md	Mar. 28, 1854	10, 693
Harvester, Grain	J. Haines	Pekin, Ill	May 22, 1855	12, 907
Harvester, Grain	S. S. Hurlbut	Racine, Wis	Feb. 4, 1851	7, 928
Harvester, Grain	T. N. Lupton	Winchester, Va	May 8, 1855	12, 824
Harvester, Grain	C. Marston	Viroqua, Wis	Aug. 14, 1860	29, 610
Harvester, Grain	D. S. Middlekauff	Hagerstown, Md	Mar. 14, 1854	10, 652
Harvester, Grain	J. E. Newcomb	Whitehall, N. Y	Jan. 9, 1855	12, 215
Harvester, Grain	J. F. Nicholson	Davidsonville, Md	May 15, 1855	12, 888
Harvester, Grain	F. Nishwitz	Williamsburgh, N. Y	Aug. 30, 1853	9, 975
Harvester, Grain	A. Palmer and S. G. Williams	Brockport, N. Y	July 1, 1851	8, 192
Harvester, Grain	A. Palmer and S. G. Williams	Brockport, N. Y., and Janesville, Wis.	Jan. 24, 1854	10, 459
Harvester, Grain	J. Phillips	Harrisburgh, Pa	Feb. 12, 1856	14, 250
Harvester, Grain	W. H. Start	Smyrna, Del	June 24, 1851	8, 182
Harvester, Grain	M. Vanderpool	Polk County, Oreg	Oct. 6, 1868	82, 896
Harvester, Grain	T. Van Fossen	Lancaster, Ohio	Jan. 20, 1852	8, 667
Harvester, Grain and grass	O. Billings	La Grange, Ohio	Nov. 12, 1861	33, 691
Harvester, Grain and grass	C. B. Brown	Griggsville, Ill	Dec. 7, 1852	9, 446
Harvester, Grain and grass	J. E. Brown and S. S. Bartlett	Woonsocket, R. I	Dec. 20, 1853	10, 320
Harvester, Grain and grass	J. E. Brown and S. S. Bartlett	Woonsocket, R. I	Jan. 2, 1855	12, 121
Harvester, Grain and grass	R. Bryson	Schenectady, N. Y	Aug. 3, 1858	21, 063
Harvester, Grain and grass	M. Burnett and C. Vander Woerd.	Boston, Mass	Jan. 2, 1855	12, 123
Harvester, Grain and grass	T. D. Burrall	Geneva, N. Y	Mar. 18, 1856	14, 441
Harvester, Grain and grass	J. Case	Springfield, Ohio	Apr. 17, 1855	12, 714
Harvester, Grain and grass	S. Colburn	Ansonia, Conn	July 3, 1855	13, 160
Harvester, Grain and grass	A. Dietz and J. G. Dunham	Raritan, N. J	Mar. 27, 1855	12, 584
Harvester, Grain and grass	D. Fitzgerald and J. H. Smith	New York, N. Y	Aug. 10, 1852	9, 182
Harvester, Grain and grass	E. B. Forbush	Buffalo, N. Y	July 20, 1852	9, 134
Harvester, Grain and grass	E. B. Forbush	Buffalo, N. Y	Apr. 17, 1855	12, 721
Harvester, Grain and grass	E. B. Forbush	Buffalo, N. Y	Mar. 18, 1856	14, 448
Harvester, Grain and grass	U. H. Goble	Springfield, Ohio	Dec. 20, 1853	10, 323
Harvester, Grain and grass	H. Green	Ottawa, Ill	Mar. 21, 1854	10, 657
Harvester, Grain and grass	J. H. Heyser and E. M. Mobley	Hagerstown, Md	May 19, 1857	17, 328
Harvester, Grain and grass	W. H. Hovey	Springfield, Mass	July 3, 1855	13, 173
Harvester, Grain and grass	W. H. Hovey	Springfield, Mass	Apr. 15, 1856	14, 661
Harvester, Grain and grass	M. G. Hubbard	New York, N. Y	June 5, 1855	13, 004
Harvester, Grain and grass	M. G. Hubbard	Penn Yan, N. Y	Dec. 7, 1858	22, 251

Index of patents issued from the United States Patent Office from 1790 *to* 1873, *inclusive*—**Continued.**

Invention.	Inventor.	Residence.	Date.	No.
Harvester, Grain and grass	W. G. Huyett	Williamsburgh, Pa	June 14, 1853	9,779
Harvester, Grain and grass	O. B. Judd	Little Falls, N. Y	Jan. 16, 1855	12,252
Harvester, Grain and grass	W. A. Kerby	Buffalo, N. Y	Apr. 15, 1856	14,694
Harvester, Grain and grass	W. F. Ketchum	Buffalo, N. Y	Dec. 19, 1854	12,113
Harvester, Grain and grass	W. F. Ketchum	Buffalo, N. Y	Jan. 15, 1856	14,102
Harvester, Grain and grass	J. H. Manny	Freeport, Ill	Oct. 17, 1854	11,810
Harvester, Grain and grass	J. H. Manny	Rockford, Ill	May 8, 1855	12,825
Harvester, Grain and grass	J. H. Manny	Rockford, Ill	Jan. 22, 1856	14,148
Harvester, Grain and grass	J. H. Manny and H. Marcellus.	Rockford, Ill., and Amsterdam, N. Y.	Mar. 6, 1855	12,499
Harvester, Grain and grass	H. Marcellus	Amsterdam, N. Y	Apr. 13, 1858	19,938
Harvester, Grain and grass	W. P. Maxson	Albion, Wis	Sept. 9, 1856	15,701
Harvester, Grain and grass	L. Miller	Canton, Ohio	Dec. 10, 1861	33,895
Harvester, Grain and grass	R. T. Osgood	Orland, Me	Feb. 17, 1852	8,743
Harvester, Grain and grass	B. T. Roney	Philadelphia, Pa	Mar. 11, 1856	14,409
Harvester, Grain and grass	G. Sanford and T. and S. Hull.	Poughkeepsie, N. Y	Jan. 15, 1856	14,127
Harvester, Grain and grass	W. and T. Schnebly	New York, N. Y	Dec. 20, 1853	10,326
Harvester, Grain and grass	W. H. Seymour	Brockport, N. Y	Dec. 14, 1852	9,476
Harvester, Grain and grass	O Stoddard	Busti, N. Y	Sept. 2, 1856	15,672
Harvester, Grain and grass	J. Swartz	Buffalo, N. Y	Nov. 14, 1854	11,951
Harvester, Grain and grass	P. Sylla	Elgin, Ill	Apr. 17, 1855	12,745
Harvester, Grain and grass	P. Sylla and A. Adams	Elgin, Ill	Sept. 20, 1873	10,038
Harvester, Grain and grass	J. Thompson	Clifton, N. Y	Aug. 28, 1855	13,508
Harvester, Grain and grass	J. Urmy	Wilmington, Del	July 24, 1855	13,330
Harvester, Grain and grass	A. Van Duzer	Goshen, N. Y	Feb. 9, 1858	19,319
Harvester, Grain and grass	J. J. Weeks	Oyster Bay, N. Y	Sept. 26, 1854	11,735
Harvester, Grain and grass	C. Wheeler, jr	Poplar Ridge, N. Y	Dec. 5, 1854	12,044
Harvester, Grain and grass	C. Wheeler, jr	Venice, N. Y	Feb. 6, 1855	12,367
Harvester, Grain and grass	A. Whiteley	Springfield, Ohio	Sept. 19, 1854	11,710
Harvester, Grain and grass	A. Whiteley	Clark County, Ohio	July 10, 1855	13,246
Harvester, Grain and grass	A. Whiteley	Springfield, Ohio	Feb. 5, 1856	14,212
Harvester, Grain and grass	A. Whiteley	Springfield, Ohio	Mar. 11, 1856	14,428
Harvester, Grain and grass	A. Whiteley	Springfield, Ohio	Mar. 25, 1856	14,541
Harvester, Grain and grass	A. P. Wilson	Waterbury, Conn	June 3, 1856	15,029
Harvester, Grain and grass	W. A. Wood	Hoosick Falls, N. Y	Mar. 20, 1855	12,570
Harvester, Grain and grass	G. W. N. Yost	Pittsburgh, Pa	Feb. 12, 1856	14,266
Harvester, Grain and grass	G. W. N. Yost	Pittsburgh, Pa	Apr. 1, 1856	14,582
Harvester, Grain and grass	J. T. Youart	Troy, Ohio	May 27, 1856	14,980
Harvester, Grain and maize	E. Quincy	Lacon, Ill	Oct. 8, 1850	7,705
Harvester, Grain-binding	H. H. Bridenthall, jr	Youngstown, Pa	Dec. 10, 1872	133,744
Harvester, Grain binding	C. F. Goddard	Saint Ansgar, Iowa	Feb. 18, 1873	135,985
Harvester grain-platform	A. E. Gleason	Newton, Mich	Mar. 17, 1868	75,674
Harvester, Grass	G. Esterly	Heart Prairie, Wis	June 27, 1854	11,155
Harvester, Grass	J. Haines	Pekin, Ill	Sept. 4, 1855	13,523
Harvester, Grass	W. K. Hall	Philippi, Va	Nov. 22, 1853	10,267
Harvester, Grass	M. Hallenbeck	Albany, N. Y	Apr. 18, 1854	10,802
Harvester, Grass	W. F. Ketchum	Buffalo, N. Y	Feb. 10, 1852	8,724
Harvester, Grass	J. S. and D. Lake	Smith's Landing, N. J	July 20, 1852	9,137
Harvester, Grass	W. Manning	South Trenton, N. J	July 20, 1852	9,138
Harvester, Grass	J. H. Maydole and A. W. Morse	Eaton, N. Y	Feb. 6, 1855	12,363
Harvester, Grass	R. J. Morrison	Richmond, Va	Feb. 13, 1855	12,393
Harvester, Grass	E. Neely	Savannah, Mo	Jan. 7, 1851	7,888
Harvester, Grass	F. Peabody	Salem, Mass	May 29, 1855	12,963
Harvester-guard	J. Birch	Corry, Pa	Sept. 28, 1869	95,187
Harvester-guard	J. H. Jones and M. S. Prentice.	Rockford, Ill	Sept. 7, 1869	94,608
Harvester guard-finger	W. Allen and L. Russ	Worcester, Mass	Oct. 29, 1867	70,311
Harvester guard-finger	R. Beans	Johnsville, Pa	Mar. 20, 1860	27,515
Harvester guard-finger	J. W. Brokaw	Springfield, Ohio	Sept. 14, 1858	21,533
Harvester guard-finger	A. Brown, H. D. Worcester, A. M. Griswold.	Ganier, Ill	Oct. 22, 1867	70,071
Harvester guard-finger	S. Copeland	Worcester, Mass	May 16, 1865	47,702
Harvester guard-finger	A. Crosby	Westfield, N. Y	July 12, 1870	105,311
Harvester guard-finger	G. Fyfe and G. Hard	Ottawa, Ill	Mar. 7, 1871	112,440
Harvester guard-finger	A. A. Hotchkiss	Sharon, Conn	Mar. 15, 1864	41,922
Harvester guard-finger	A. A. Hotchkiss and J. P. Adriance.	Sharon, Conn., and New York, N. Y.	June 21, 1859	24,461
Harvester guard-finger	W. F. Ketchum	Buffalo, N. Y	Apr. 25, 1854	10,841
Harvester guard-finger	J. H. Manny	Rockford, Ill	June 26, 1855	13,150
Harvester guard-finger	K. H. C. Preston	Manlius, N. Y	Sept. 10, 1861	33,265
Harvester guard-finger	A. R. Reese	Phillipsburgh, N. J	July 28, 1857	17,894
Harvester guard-finger	M. B. Riggs	New York, N. Y	Jan. 28, 1862	34,267
Harvester guard-finger	M. B. Riggs	New York, N. Y	Oct. 28, 1862	36,796
Harvester guard-finger	W. A. Wood	Hoosick Falls, N. Y	July 1, 1856	15,264
Harvester guard-finger	W. A. Wood	Hoosick Falls, N. Y	Nov. 19, 1867	71,257
Harvester guard-fingers, Blank for	J. Fassler	Springfield, Ohio	Sept. 28, 1869	95,334
Harvester guard-fingers, Machine for drilling	J. Fassler	Springfield, Ohio	May 28, 1867	65,191
Harvester guard-fingers, Machine for making	J. Fassler	Springfield, Ohio	Mar. 15, 1870	100,878
Harvester guard-fingers, Machine for milling	J. Fassler	Springfield, Ohio	May 28, 1867	65,192
Harvester guard-fingers, Machine for sawing the slot in.	J. Fassler	Springfield, Ohio	July 12, 1870	105,323
Harvester guard-fingers, Making	L. Miller	Akron, Ohio	Nov. 24, 1868	84,433
Harvester guard-fingers, Making	G. W. Slough	Canton, Ohio	May 8, 1860	28,206
Harvester guard-fingers, Manufacture of	W. N. Whiteley, jr., J. Fassler, and O. S. Kelly.	Springfield, Ohio	June 18, 1867	65,977
Harvester guard or finger	A. Wissler	Clay Township, Pa	Dec. 5, 1865	51,380
Harvester-guards, Method of renovating the cutting-edge of.	J. Rummel, jr	New Middletown, Ohio	Nov. 23, 1869	97,232
Harvester guiding-wheel	W. A. and W. A. Wood	Hoosick Falls, N. Y	Jan. 3, 1871	110,712
Harvester, Hemp	T. Berry	Louisburgh, Ky	June 22, 1858	20,618
Harvester, Hemp	W. B. Coates	Big Lick, Va	Oct. 15, 1850	7,719
Harvester, Hemp	O. Farrar	Jessamine County, Ky	May 14, 1872	126,793
Harvester, Hemp	P. S. Fitch	Hanly, Ky	Jan. 7, 1873	134,656
Harvester, Hemp	R. C. Wrenn	Waverly, Mo	Mar. 3, 1868	75,103
Harvester-knife	G. J. Wardwell	Rutland, Vt	Feb. 15, 1870	99,797
Harvester-knife grinder	J. Y. Hoagland	Auburn, N. Y	Mar. 17, 1868	75,633
Harvester-knife grinder	T. Loring	Blackwoodtown, N. J	May 31, 1870	103,633
Harvester-knife grinder	J. A. Thompson	Auburn, N. Y	Mar. 17, 1868	75,709
Harvester-knife grinder	P. W. Vaughan	De Kalb, Ill	Dec. 19, 1871	122,083

Index of patents issued from the United States Patent Office from 1790 *to* 1873, *inclusive*—Continued.

Invention.	Inventor.	Residence.	Date.	No.
Harvester-knife grinder	E. L. Yancey	Batavia, N. Y	Dec. 14, 1869	97, 847
Harvester-knife head	H. A. Soliday	Doylestown, Ohio	Mar. 23, 1869	88, 091
Harvester-knife sharpener	H. Fisher	Canton, Ohio	July 16, 1872	128, 957
Harvester-knives, Apparatus for sharpening	I. C. Saunders	Trenton, Mich	Dec. 19, 1871	122, 064
Harvester-knives, Machine for grinding	J. M. Connel	Newark, Ohio	July 15, 1873	140, 810
Harvester-knives, Machine for grinding	J. M. Connel	Newark, Ohio	July 15, 1873	140, 811
Harvester-knives, Machine for grinding	J. Murphy	New York, N. Y	Feb. 25, 1873	136, 171
Harvester-knives, Machine for sharpening	J. S. Elliott	Chelsea, Mass	May 13, 1873	138, 743
Harvester-knives, Sharpening	A. P. Taylor	Darien, N. Y	Mar. 22, 1870	101, 061
Harvester-knives, Tempering	J. F. Shippey	Valley Falls, N. Y	Apr. 29, 1873	138, 349
Harvester, Maize	G. A. Bruce	Mechanicsburgh, Ill	Aug. 29, 1854	11, 595
Harvester, Maize	J. S. Burnham	West Jefferson, Ohio	Sept. 19, 1854	11, 688
Harvester, Maize	S. Lapham	Salem, Ohio	Apr. 18, 1854	10, 811
Harvester, Maize	C. B. Matthews	Oquawka, Ill	Oct. 19, 1858	21, 840
Harvester, Maize	J. L. Ream	Mount Pulaski, Ill	Dec. 21, 1852	9, 487
Harvester, Maize	W. Watson	Chicago, Ill	Oct. 15, 1850	7, 725
Harvester, &c., pitman	J. W. Doty	Lockport, N. Y	Oct. 30, 1866	59, 192
Harvester-pitman	O. P. Drury	Niles, Mich	Aug. 4, 1868	80, 614
Harvester-pitman	B. E. J. Eils	Washington, D. C	Sept. 24, 1872	131, 671
Harvester-pitman	J. R. Finley	Delphi, Ind	Apr. 13, 1869	88, 863
Harvester-pitman	G. B. and C. B. Garlinghouse	North Madison, Ind	June 23, 1868	79, 112
Harvester-pitman	G. W. Harrison	Lansing, Mich	Sept. 26, 1871	119, 356
Harvester-pitman	B. J. Hunter	Ledyard, N. Y	Aug. 14, 1866	57, 140
Harvester-pitman	W. J. Keeney	Florence, Ind	Aug. 27, 1867	68, 207
Harvester-pitman	J. L. Kintner	Harrison County, Ind	May 21, 1867	64, 985
Harvester-pitman	J. M. Mourer	Millheim, Pa	Jan. 30, 1872	123, 283
Harvester-pitman	C. E. Roper	Canton, Ohio	Apr. 27, 1869	89, 348
Harvester-pitman	T. C. Sebring	Rochester, N. Y	June 25, 1867	66, 047
Harvester-pitman	G. L. Squier	Buffalo, N. Y	Sept. 3, 1867	68, 581
Harvester-pitman connection	J. Dixon and M. B. Sampson	Eddyville, Iowa	Aug. 16, 1870	106, 339
Harvester-pitman connection	W. Ferris	Pleasant Plain, Ohio	Nov. 26, 1872	133, 363
Harvester-pitman connection	S. J. Green	Syracuse, N. Y	July 25, 1871	117, 409
Harvester-pitman connection	H. Howe	Houston, Minn	May 28, 1872	127, 240
Harvester-pitman connection	W. Loucks	Lowville, N. Y	Apr. 30, 1872	126, 220
Harvester-pitman connection	D. M. Osborne	Auburn, N. Y	Nov. 7, 1865	50, 879
Harvester-pitman connection	D. M. Osborne	Auburn, N. Y	Nov. 7, 1865	50, 880
Harvester-pitman connection	H. L. Wanzer	Clyde, Ohio	July 2, 1867	66, 430
Harvester-pitman connection	B. F. Waters	Garrettsville, Ohio	Jan. 14, 1873	134, 954
Harvester-pitman connection	C. Wheeler, jr	Auburn, N. Y	Feb. 11, 1868	74, 463
Harvester-pitman connection	R. C. Wood	Le Roy, Kans	Oct. 5, 1869	95, 548
Harvester-pitman coupling	H. Bishop	Batavia, Iowa	Mar. 20, 1866	53, 261
Harvester-pitman head	S. H. Wilson	Auburn, N. Y	June 5, 1866	55, 435
Harvester-pitman joint	S. T. Lamb	New Albany, Ind	Sept. 28, 1869	95, 359
Harvester-pitman joint	S. T. Lamb	New Albany, Ind	Feb. 22, 1870	100, 158
Harvester-pitman joint	S. L. Lamb	New Albany, Ind	Feb. 22, 1870	100, 159
Harvester-platform	J. P. Adriance and T. S. Brown	Poughkeepsie, N. Y	Feb. 8, 1870	99, 617
Harvester-platform	G. B. Deardorff	Canal Dover, Ohio	July 18, 1871	117, 055
Harvester-platform	C. Lidren	La Fayette, Ind	May 10, 1870	102, 839
Harvester platform, Grain	J. Atkins	Chicago, Ill	Apr. 24, 1855	12, 756
Harvester-platforms, Adjustable grain-wheel for	G. W. Patten	Auburn, N. Y	Jan. 7, 1873	134, 610
Harvester, Potato	J. E. Hardenbergh	Fultonville, N. Y	Nov. 15, 1859	26, 102
Harvester, Potato	J. J. Hill	Xenia, Ohio	May 15, 1866	54, 724
Harvester, Potato	J. D. Otstot	Springfield, Ohio	Nov. 15, 1859	26, 118
Harvester-rake	A. Adams	Boston Station, Ky	June 15, 1869	91, 401
Harvester-rake	N. Allstatter	Hamilton, Ohio	Apr. 26, 1864	42, 454
Harvester-rake	E. Ames	Austin, Minn	Mar. 16, 1869	87, 817
Harvester-rake	P. Ammerman	Cynthiana, Ky	Oct. 27, 1868	83, 355
Harvester-rake	C. J. Arlington	Auburn, N. Y	July 7, 1868	79, 539
Harvester-rake	J. K. Augspurger	Trenton, Ohio	Aug. 26, 1873	142, 139
Harvester-rake	C. Aultman	Canton, Ohio	Dec. 31, 1867	72, 775
Harvester-rake	J. Bacon	Medina, Wis	Oct. 17, 1865	50, 529
Harvester-rake	J. Baldwin	Saint Paris, Ohio	June 28, 1864	43, 282
Harvester-rake	T. P. Barbour	Louisville, Ky	Jan. 14, 1873	134, 722
Harvester-rake	J. Barnes	Rockford, Ill	Apr. 17, 1866	53, 942
Harvester-rake	J. Barnes	Rockford, Ill	Dec. 10, 1867	71, 952
Harvester-rake	J. Barnes	Rockford, Ill	Jan. 12, 1869	85, 723
Harvester-rake	J. Barnes	Rockford, Ill	Apr. 25, 1871	114, 094
Harvester-rake	C. Barns	Oskaloosa, Iowa	Mar. 30, 1869	88, 358
Harvester-rake	C. L. Barritt	Richland, Mich	Apr. 28, 1868	77, 244
Harvester-rake	T. Baylis and D. Williams	Tecumseh, Mich	Jan. 11, 1853	9, 528
Harvester-rake	E. R. Bergstresser	Hublersburgh, Pa	Feb. 5, 1867	61, 704
Harvester-rake	G. Blake	Whitby, Canada	Sept. 8, 1868	81, 978
Harvester-rake	J. B. Bowen, C. A. Reed, and C. A. Whelan.	Madison, Wis	Aug. 18, 1868	81, 130
Harvester-rake	J. B. Bowen, C. A. Whelan, and C. A. Reed.	Madison, Wis	July 20, 1869	92, 783
Harvester-rake	H. Brackett	Valley Falls, N. Y	May 4, 1869	89, 731
Harvester-rake	C. B. Brown	Upper Alton, Ill	July 11, 1854	11, 249
Harvester-rake	C. B. Brown	Springfield, Mo	July 5, 1870	104, 927
Harvester-rake	J. O. Brown, A. Ingham, and F. T. Lomont.	Massillon, Ohio	July 31, 1866	56, 696
Harvester-rake	R. D. Brown	Covington, Ind	July 24, 1866	56, 521
Harvester-rake	R. D. Brown	Covington, Ind	Dec. 24, 1867	72, 599
Harvester-rake	T. S. Brown	Poughkeepsie, N. Y	Aug. 25, 1868	81, 473
Harvester-rake	T. S. Brown	Poughkeepsie, N. Y	Sept. 28, 1869	95, 191
Harvester-rake	F. Brua	Gordonsville, Pa	Jan. 8, 1867	60, 993
Harvester-rake	R. Bryson	Schenectady, N. Y	Apr. 8, 1862	34, 875
Harvester-rake	R. Bryson	Schenectady, N. Y	Mar. 3, 1868	74, 983
Harvester-rake	F. M. Buckles	Altona, Ill	July 27, 1869	93, 049
Harvester-rake	C. Buckwalter	Davenport, Iowa	Feb. 1, 1870	99, 397
Harvester-rake	C. Bullock	Jamestown, N. Y	Apr. 23, 1861	32, 160
Harvester-rake	O. H. Burdick	Auburn, N. Y	June 7, 1864	43, 007
Harvester-rake	O. H. Burdick	Auburn, N. Y	Dec. 8, 1868	84, 730
Harvester-rake	C. T. Burgess	Brentwood, England	Nov. 3, 1868	83, 599
Harvester-rake	I. P. Cadman	Mendota, Ill	Dec. 17, 1867	72, 163
Harvester-rake	R. Carkhuff	Lewisburgh, Pa	Dec. 15, 1868	84, 857
Harvester-rake	C. L. Carter	Union City, Ind	May 1, 1866	54, 294
Harvester-rake	H. J. Case	Auburn, N. Y	Feb. 16, 1869	86, 904

Index of patents issued from the United States Patent Office from 1790 *to* 1873, *inclusive*—Continued.

Invention.	Inventor.	Residence.	Date.	No.
Harvester-rake	W. J. and R. Case	Pittstown, N. J	Sept. 19, 1865	49, 976
Harvester-rake	W. J. and R. Case	Pittstown, N J	May 29, 1866	55, 059
Harvester-rake	P. C. Chipron	Highland, Ill	May 29, 1866	55, 060
Harvester-rake	S. Clevenger	Vibbard, Mo	Aug. 12, 1873	141, 695
Harvester-rake	D. Clow	Janesville, Wis	Sept. 6, 1870	107, 004
Harvester-rake	H. Clymo	Galena, Ill	May 1, 1866	54, 298
Harvester-rake	W. F. Cochrane	Springfield, Ohio	Sept. 19, 1865	50, 066
Harvester-rake	O. B. Colcord	Greenville, Ill	Apr. 4, 1871	113, 632
Harvester-rake	J. Collins	Guelph, Canada	Feb. 11, 1873	135, 691
Harvester-rake	S. Comfort, jr	Morrisville, Pa	Oct. 20, 1857	18, 437
Harvester-rake	S. Comfort, jr	Morrisville, Pa	Jan. 5, 1858	19, 019
Harvester-rake	F. E. Cook	Seville, Ohio	May 28, 1867	65, 059
Harvester-rake	A. G. Crane and W. T. Johnston.	Ottumwa, Iowa	Nov. 15, 1870	109, 301
Harvester-rake	M. Crossman and P. A. Spicer.	Marengo and Marshall, Mich.	Oct. 1, 1867	69, 322
Harvester-rake	J. S. Crump	Williamsburgh, Mo	Dec. 10, 1867	71, 994
Harvester-rake	J. D. Custer	Norristown, Pa	Feb. 2, 1869	86, 511
Harvester-rake	C. F. Davis	Auburn, N. Y	May 22, 1866	54, 871
Harvester-rake	N. H. Dederer	Greene, N. Y	Feb. 6, 1866	52, 396
Harvester-rake	C. C. Dennis	Auburn, N. Y	Apr. 21, 1863	38, 219
Harvester-rake	H. F. W. Deterding	Alton, Ill	June 23, 1868	79, 057
Harvester-rake	J. Dick, jr	Canton, Ohio	Jan. 28, 1868	73, 878
Harvester-rake	J. Dick, jr	Oshawa, Ontario	Nov. 3, 1868	83, 610
Harvester-rake	W. P. Dillman	Joliet, Ill	Apr. 17, 1866	53, 958
Harvester-rake	W. P. Dillman	Joliet, Ill	Nov. 19, 1867	70, 976
Harvester-rake	J. A. Dodge	Auburn, N. Y	Aug. 20, 1867	67, 850
Harvester-rake	J. A. Dodge	Auburn, N. Y	Aug. 20, 1867	67, 851
Harvester-rake	J. A. Dodge and G. Perry	Auburn, N. Y	Dec. 3, 1867	71, 718
Harvester-rake	J. A. Dodge and W. H. and H. S. Stevenson.	Auburn, N. Y	Aug. 20, 1867	67, 853
Harvester-rake	O. Dorsey	Howard County, Md	Mar. 4, 1856	14, 350
Harvester-rake	J. C. Durborow	Ellicott City, Md	June 9, 1868	78, 654
Harvester-rake	J. C. Durborow	Ellicott City, Md	June 15, 1869	91, 315
Harvester-rake	W. T. Eastes	Summitville, Ind	Aug. 16, 1870	106, 342
Harvester-rake	W. T. Eastes	Summitville, Ind	July 22, 1873	141, 127
Harvester-rake	B. Ellingworth	Le Roy, Minn	May 2, 1871	114, 441
Harvester-rake	R. Emerson	Sycamore, Ill	Sept. 19, 1871	119, 130
Harvester-rake	J. Farrington	Corry, Pa	Mar. 29, 1870	101, 246
Harvester-rake	V. H. Felt	Rochester, N. Y	Jan. 7, 1868	72, 990
Harvester-rake	V. H. Felt	Kendall, N. Y	May 3, 1870	102, 525
Harvester-rake	J. R. Finley	Delphi, Ind	July 20, 1869	92, 713
Harvester-rake	E. T. Ford	Batavia, N. Y	Nov. 24, 1857	18, 686
Harvester-rake	E. T. Ford	Stillwater, N. Y	Feb. 20, 1866	52, 700
Harvester-rake	J. S. Fowler	Davenport, Iowa	Mar. 23, 1869	88, 156
Harvester-rake	J. French	Independence, Iowa	Jan. 18, 1870	98, 861
Harvester-rake	M. Gibbs	Homer, Mich	Dec. 8, 1868	84, 818
Harvester-rake	J. H. and A. J. Glass	McGregor, Iowa	Dec. 10, 1867	72, 016
Harvester-rake	W. F. Goodwin	Washington, D. C	Mar. 13, 1866	53, 136
Harvester-rake	W. F. Goodwin	Washington, D. C	June 12, 1866	55, 487
Harvester-rake	W. F. Goodwin	Washington, D. C	Mar. 12, 1867	62, 838
Harvester-rake	W. F. Goodwin	Washington, D. C	May 7, 1867	64, 520
Harvester-rake	W. F. Goodwin	Washington, D. C	May 7, 1867	64, 521
Harvester-rake	W. F. Goodwin	East New York, N. Y	Dec. 31, 1867	72, 839
Harvester-rake	W. F. Goodwin	East New York, N. Y	Dec. 31, 1867	72, 840
Harvester-rake	W. F. Goodwin	East New York, N. Y	Dec. 31, 1867	72, 841
Harvester-rake	W. F. Goodwin and A. W. Browne.	Washington, D. C., and Brooklyn, N. Y.	May 7, 1867	64, 522
Harvester-rake	W. F. Goodwin and A. W. Browne.	Washington, D. C., and Brooklyn, N. Y.	May 7, 1867	64, 523
Harvester-rake	W. F. Goulding	Providence, R. I	Mar. 8, 1870	100, 615
Harvester-rake	O. C. Green	Belleville, Ill	Aug. 28, 1855	13, 490
Harvester-rake	G. S. Grier	Milford, Del	Feb. 27, 1872	124, 130
Harvester-rake	G. S. Grier	Milford, Del	July 9, 1872	128, 875
Harvester-rake	C. P. Gronberg	Geneva, Ill	Dec. 1, 1857	18, 736
Harvester-rake	C. P. Gronberg	Elgin, Ill	Aug. 6, 1872	130, 295
Harvester-rake	G. Hall	Baltimore, Md	June 2, 1863	38, 796
Harvester-rake	J. C. Hall	Monroe, Wis	July 9, 1867	66, 584
Harvester-rake	J. C. Hall	Monroe, Wis	Sept. 1, 1868	81, 628
Harvester-rake	T. Harding	Springfield, Ohio	Mar. 10, 1868	75, 263
Harvester-rake	J. H. Harnly	Penn Township, Pa	Dec. 9, 1862	37, 133
Harvester-rake	H. A. M. Harris	Philadelphia, Pa	Dec. 17, 1867	72, 198
Harvester-rake	H. A. M. Harris	Philadelphia, Pa	Dec. 17, 1867	72, 199
Harvester-rake	J. Harris	Janesville, Wis	May 23, 1871	115, 052
Harvester-rake	G. Heuerman, W. Sternberg, and J. Stuhr.	Davenport, Iowa	Nov. 28, 1871	121, 247
Harvester-rake	G. W. Hines	Brookfield, Wis	Jan. 31, 1871	111, 344
Harvester-rake	J. F. Hirschy and W. M. McDonald.	Wooster, Ohio	Apr. 16, 1867	63, 890
Harvester-rake	S. T. Holly	Rockford, Ill	Mar. 7, 1871	112, 341
Harvester-rake	S. Hull	Poughkeepsie, N. Y	Aug. 18, 1868	81, 090
Harvester-rake	S. Hull	Poughkeepsie, N. Y	May 11, 1869	89, 867
Harvester-rake	W. R. G. Humphrey	Chillicothe, Mo	Feb. 13, 1872	123, 701
Harvester-rake	W. H. H. Hunter	Versailles, Ind	Apr. 26, 1864	42, 488
Harvester-rake	W. H. Hurlbut	Mirabile, Mo	May 14, 1872	126, 705
Harvester-rake	J. Irvine	Parkersburgh, Iowa	Sept. 9, 1873	142, 632
Harvester-rake	W. B. Johns	Cumberland, Md	Dec. 24, 1867	72, 644
Harvester-rake	I. A. Johnson	Middletown, Conn	Nov. 22, 1870	109, 415
Harvester-rake	S. Johnston	Buffalo, N. Y	Dec. 22, 1863	41, 009
Harvester-rake	S. Johnston	Syracuse, N. Y	Feb. 2, 1869	86, 554
Harvester-rake	J. H. Jones	Rockton, Ill	Aug. 22, 1865	49, 530
Harvester-rake	J. H. Jones	Rockford, Ill	Mar. 27, 1866	53 450
Harvester-rake	R. V. Jones and H. Fessler	Canton, Ohio	Feb. 12, 1867	61, 035
Harvester-rake	M. A. Keller	York County, Pa	Feb. 28, 1865	46, 565
Harvester-rake	D. A. Kellogg	Valparaiso, Ind	May 16, 1871	114, 827
Harvester-rake	W. King	Richmond, Va	July 16, 1872	129, 035
Harvester-rake	W. W. Kingsbury	Kalamazoo, Mich	June 29, 1869	92, 058
Harvester-rake	W. E. Kinnear	Bucyrus, Ohio	Oct. 29, 1872	132, 669

Index of patents issued from the United States Patent Office from 1790 *to* 1873, *inclusive*—Continued.

Invention.	Inventor.	Residence.	Date.	No.
Harvester-rake	D. R. Kinyon	Raritan, N. J	Mar. 14, 1871	112, 605
Harvester-rake	W. A. Kirby	Auburn, N. Y	Dec. 8, 1868	84, 833
Harvester-rake	W. A. Kirby	Auburn, N. Y	Apr. 2, 1872	125, 308
Harvester-rake	I. Lancaster	Baltimore, Md	Aug. 20, 1867	67, 888
Harvester-rake	I. Lancaster	Baltimore, Md	July 21, 1868	80, 188
Harvester-rake	I. Lancaster	Baltimore, Md	Apr. 27, 1869	89, 414
Harvester-rake	C. Ledren	La Fayette, Ind	Feb. 25, 1868	74, 836
Harvester-rake	E. J. Leyburn	Lexington, Va	June 26, 1866	55, 877
Harvester-rake	E. J. Leyburn	Lexington, Va	Oct. 22, 1867	70, 007
Harvester-rake	E. J. Leyburn	Lexington, Va	Dec. 10, 1867	71, 890
Harvester-rake	C. Lidren	La Fayette, Ind	Mar. 30, 1869	88, 394
Harvester-rake	S. K. Lighter	Hamilton, Ohio	Dec. 11, 1866	60, 392
Harvester-rake	S. A. Lindsay	Unionville, Md	Dec. 11, 1860	30, 882
Harvester-rake	E. Lippoldt	Brighton, Ill	Nov. 18, 1873	144, 683
Harvester-rake	R. Little and L. Gibbs	Canton, Ohio	Dec. 17, 1867	72, 208
Harvester-rake	S. D. Locke	Hoosick Falls, N. Y	Sept. 16, 1873	142, 708
Harvester-rake	J. M. Long	Hamilton, Ohio	May 21, 1867	64, 888
Harvester-rake	J. D. Loveland	Wheatfield, N. Y	Apr. 3, 1866	53, 740
Harvester-rake	F. H. Manny	Rockford, Ill	Dec. 28, 1869	98, 284
Harvester-rake	J. P. Manny	Rockford, Ill	Mar. 7, 1871	112, 363
Harvester-rake	J. P. Manny	Rockford, Ill	Mar. 21, 1871	112, 940
Harvester-rake	J. P. Manny	Rockford, Ill	Mar. 21, 1871	112, 941
Harvester-rake	J. P. Manny	Rockford, Ill	Apr. 25, 1871	114, 165
Harvester-rake	J. P. Manny	Rockford, Ill	Apr. 25, 1871	114, 166
Harvester-rake	J. P. Manny	Rockford, Ill	Apr. 16, 1872	125, 747
Harvester-rake	J. S. Marsh	Lewisburgh, Pa	Jan. 21, 1868	73, 619
Harvester-rake	A. J. Martin	Rockford, Ill	Apr. 30, 1867	64, 345
Harvester-rake	A. McArthur	Boonville, Mo	Apr. 4, 1871	113, 321
Harvester-rake	E. R. McCall	Simcoe, Canada	May 17, 1870	103, 217
Harvester-rake	J. B. McCormick	Saint Louis, Mo	Jan. 21, 1868	73, 623
Harvester-rake	L. J. McCormick, W. R. Baker, and L. Erpelding.	Chicago, Ill	Sept. 15, 1868	82, 141
Harvester-rake	J. B. McMillan	North Vernon, Ind	Sept. 2, 1873	142, 490
Harvester-rake	J. H. Mears	Oshkosh, Wis	Mar. 4, 1862	34, 584
Harvester-rake	J. H. Mears	Oshkosh, Wis	June 17, 1862	35, 653
Harvester-rake	D. Mendenhall	Fairfield, Iowa	Apr. 7, 1868	76, 339
Harvester-rake	W. G. Merrell	Auburn, N. Y	Jan. 21, 1868	73, 452
Harvester-rake	J. Miller	Canton, Ohio	July 9, 1867	66, 609
Harvester-rake	J. Miller	Canton, Ohio	July 23, 1867	66, 983
Harvester-rake	L. Miller	Akron, Ohio	May 28, 1867	65, 106
Harvester-rake	L. Miller	Akron, Ohio	Feb. 25, 1868	74, 929
Harvester-rake	L. Miller	Akron, Ohio	May 28, 1872	127, 181
Harvester-rake	S. Miller	Urbana, Ohio	July 15, 1851	8, 225
Harvester-rake	W. W. Miller	Zionsville, Ind	June 21, 1870	104, 483
Harvester-rake	J. B. Morse and L. L. Carter	La Fayette, Ind	Mar. 9, 1869	87, 699
Harvester-rake	C. Moul	Hanover, Pa	Apr. 3, 1866	53, 653
Harvester-rake	J. Mumma	Middletown, Ohio	Mar. 13, 1866	53, 171
Harvester-rake	T. Murphy	Pine Bluffs, Wis	Aug. 26, 1873	142, 171
Harvester-rake	M. L. Nix	Chetopah, Kans	July 22, 1873	141, 164
Harvester-rake	G. Oerllein	Utica, Minn	Nov. 22, 1870	109, 441
Harvester-rake	E. Ogden	Lynchburgh, Va	Mar. 17, 1868	75, 570
Harvester-rake	J. L. Owens	Cambria, Wis	Sept. 9, 1873	142, 644
Harvester-rake	A. and C. W. Palmer	Brockport, N. Y	Mar. 16, 1869	87, 963
Harvester-rake	A. and C. W. Palmer	Brockport, N. Y	Feb. 1, 1870	99, 340
Harvester-rake	A. and C. W. Palmer	Brockport, N. Y	Mar. 15, 1870	100, 922
Harvester-rake	W. Partridge	Philadelphia, Pa	Mar. 22, 1870	101, 156
Harvester-rake	H. Pease	Brockport, N. Y	Dec. 20, 1870	110, 279
Harvester-rake	W. P. Penn	Belleville, Ill	Apr. 24, 1860	28, 010
Harvester-rake	G. M. Peters	Granville, Ohio	July 9, 1867	66, 625
Harvester-rake	G. M. Peters	Granville, Ohio	Oct. 1, 1867	69, 476
Harvester-rake	H. F. Phillips	Auburn, N. Y	Jan. 26, 1869	86, 178
Harvester-rake	W. Pimlott	Springfield, Ohio	Jan. 25, 1870	99, 108
Harvester-rake	W. Pimlott	Brockport, N. Y	Nov. 22, 1870	109, 449
Harvester-rake	J. Pine	Troy, N. Y	May 7, 1861	32, 266
Harvester-rake	J. Poulson, jr	Pittstown, N. J	Sept. 26, 1865	50, 159
Harvester-rake	D. J. Powers	Madison, Wis	Oct. 6, 1868	82, 872
Harvester-rake	K. H. C. Preston	Manlius, N. Y	Sept. 1, 1868	81, 685
Harvester-rake	A. Pritz	Dayton, Ohio	Dec. 4, 1860	30, 835
Harvester-rake	A. Quick, W. S. Opie, and A. J. Farrand.	Raritan, N. J	Feb. 25, 1868	74, 845
Harvester-rake	J. M. Randle	Brighton, Ill	Oct. 10, 1865	50, 388
Harvester-rake	A. Rank	Salem, Ohio	Aug. 18, 1868	81, 208
Harvester-rake	A. Rank	Salem, Ohio	Nov. 2, 1869	96, 353
Harvester-rake	A. Rank	Salem, Ohio	Sept. 27, 1870	107, 721
Harvester-rake	A. Rank	Salem, Ohio	Aug. 12, 1873	141, 668
Harvester-rake	C. D. Read	Hamilton, Ohio	Aug. 11, 1863	39, 497
Harvester-rake	O. Redmond	Rochester, N. Y	Mar. 10, 1868	75, 297
Harvester-rake	A. R. Reese	Phillipsburgh, N. J	Feb. 16, 1864	41, 641
Harvester-rake	A. R. Reese	Phillipsburgh, N. J	May 1, 1866	54, 406
Harvester-rake	A. R. Reese, W. Gould, and N. Lake.	Phillipsburgh, N. J	Nov. 5, 1861	33, 663
Harvester-rake	T. S. Reeve and C. D. Smith	Chicago, Ill	Oct. 13, 1868	82, 991
Harvester-rake	G. H. Reister	Washington, Iowa	Aug. 31, 1869	94, 440
Harvester-rake	L. M. Rice	Oregon, Wis	Apr. 2, 1867	63, 427
Harvester-rake	S. C. Ridgaway	Baltimore, Md	Nov. 26, 1867	71, 327
Harvester-rake	S. C. Ridgaway	Baltimore, Md	Jan. 14, 1868	73, 388
Harvester-rake	C. Roberts	Belleville, Ill	May 16, 1854	10, 904
Harvester-rake	A. B. Rodman	Lyons, Iowa	Oct. 18, 1864	44, 749
Harvester-rake	J. L. Rohrer	Upper Leacock Township, Pa.	Dec. 3, 1867	71, 649
Harvester-rake	J. M. Rosebrook	Hoosick Falls, N. Y	Aug. 1, 1871	117, 687
Harvester-rake	J. M. Rosebrooks	Hoosick Falls, N. Y	Nov. 7, 1871	120, 777
Harvester-rake	J. W. Schuckers	Philadelphia, Pa	Aug. 16, 1870	106, 513
Harvester-rake	F. Schürger and N. Allstatter	Hamilton, Ohio	Aug. 11, 1868	80, 879
Harvester-rake	F. B. Scott	Lancaster, N. Y	Feb. 28, 1871	112, 289
Harvester-rake	J. Seibel	Manlius, Ill	Apr. 16, 1872	125, 696
Harvester-rake	W. H. Seymour	Brockport, N. Y	July 8, 1851	8, 212
Harvester-rake	W. H. Seymour and A. Palmer	Brockport, N. Y	Apr. 15, 1862	34, 986

Index of patents issued from the United States Patent Office from 1790 *to* 1873, *inclusive*—Continued.

Invention.	Inventor.	Residence.	Date.	No.
Harvester-rake	W. H. Seymour and A. Palmer	Brockport, N. Y	Mar. 24, 1868	75, 801
Harvester-rake	S. S. and J. G. Sherman	McHenry, Ill	Mar. 6, 1866	53, 051
Harvester-rake	S. S. and J. G. Sherman	McHenry, Ill	Dec. 22, 1868	85, 139
Harvester-rake	A. Sherwood	Auburn, N. Y	Aug. 20, 1872	130, 597
Harvester-rake	D. B. Shirk	Brunersville, Pa	Feb. 22, 1870	100, 078
Harvester-rake	D. B. Shirk	Brunersville, Pa	Nov. 1, 1870	108, 839
Harvester-rake	E. W. Skinner	Madison, Wis	Feb. 22, 1870	100, 200
Harvester-rake	E. M. Smith	New York, N. Y	June 30, 1868	79, 507
Harvester-rake	E. H. Smith	Westminster, Md	Feb. 23, 1869	87, 211
Harvester-rake	E. M. Smith	New York, N. Y	May 14, 1872	126, 839
Harvester-rake	W. T. Smith	Bellefontaine, Ohio	Nov. 5, 1872	132, 870
Harvester-rake	A. D. Sprague and A. Dockum	Caledonia, Minn	May 15, 1866	54, 789
Harvester-rake	G. L. Squire	Buffalo, N. Y	Apr. 8, 1862	34, 909
Harvester-rake	E. Stewart	Fort Madison, Iowa	Dec. 10, 1867	72, 109
Harvester-rake	O. O. Storle	North Cape, Wis	June 12, 1866	55, 551
Harvester-rake	O. O. Storle	North Cape, Wis	May 5, 1868	77, 675
Harvester-rake	O. O. Storle	Norway, Wis	Jan. 12, 1869	85, 870
Harvester-rake	O. O. Storle	North Cape, Wis	July 5, 1870	105, 139
Harvester-rake	H. A. Stringer and A. F. Ward	Chatham, Canada	Nov. 22, 1870	109, 468
Harvester-rake	W. Sutliff	Burns, Wis	Nov. 22, 1870	109, 469
Harvester-rake	J. B. Sweetland	Pontiac, Mich	Dec. 3, 1867	71, 816
Harvester-rake	J. B. Sweetland	Pontiac, Mich	Mar. 24, 1868	76, 000
Harvester-rake	J. Taylor	Magnolia, Ill	May 13, 1862	35, 270
Harvester-rake	T. Taylor	Baltimore, Md	Apr. 10, 1866	53, 929
Harvester-rake	J. Tray	Mineral Point, Wis	July 19, 1870	105, 608
Harvester-rake	J. S. Truxell	Greenburgh, Pa	Apr. 1, 1873	137, 510
Harvester-rake	J. Underwood	Muscatine, Iowa	June 9, 1868	78, 700
Harvester-rake	I. Van Doren	Somerville, N. J	Sept. 22, 1857	18, 256
Harvester-rake	I. Van Doren	Somerville, N. J	Mar. 2, 1858	19, 523
Harvester-rake	J. Walmsley	London, Canada	Feb. 13, 1872	123, 752
Harvester-rake	W. H. Ward	Auburn, N. Y	June 22, 1869	91, 582
Harvester-rake	G. H. Weller	New Village, N. J	May 23, 1871	115, 141
Harvester-rake	G. Wellhouse	Akron, Ohio	Apr. 24, 1866	54, 235
Harvester-rake	H. Wells	Erieville, N. Y	Apr. 16, 1872	125, 710
Harvester-rake	P. Werni	Chicago, Ill	Nov. 19, 1867	71, 098
Harvester-rake	C. Wheeler, jr	Auburn, N. Y	Apr. 21, 1868	77, 146
Harvester-rake	C. Wheeler, jr	Auburn, N. Y	July 14, 1868	79, 881
Harvester-rake	C. Wheeler, jr	Auburn, N. Y	Dec. 15, 1868	85, 044
Harvester-rake	C. Wheeler, jr	Auburn, N. Y	Apr. 4, 1871	113, 474
Harvester-rake	W. N. Whiteley	Springfield, Ohio	Feb. 11, 1868	74, 464
Harvester-rake	W. N. Whiteley	Springfield, Ohio	June 23, 1868	79, 170
Harvester-rake	W. N. Whiteley	Springfield, Ohio	Aug. 11, 1868	81, 045
Harvester-rake	W. N. Whiteley	Springfield, Ohio	Jan. 26, 1869	86, 265
Harvester-rake	W. N. Whiteley	Springfield, Ohio	Feb. 1, 1870	99, 506
Harvester-rake	W. N. Whiteley, jr	Springfield, Ohio	May 14, 1867	64, 819
Harvester-rake	W. N. Whiteley, jr	Springfield, Ohio	May 21, 1867	65, 030
Harvester-rake	W. N. Whiteley and J. Fassler	Springfield, Ohio	Feb. 11, 1868	74, 465
Harvester-rake	T. S. Whitenack	Easton, Pa	Feb. 5, 1861	31, 345
Harvester-rake	J. R. Whittemore	Chicopee Falls, Mass	May 20, 1862	35, 343
Harvester-rake	J. Wilcke	Chicago, Ill	Aug. 3, 1869	93, 260
Harvester-rake	W. S. Wilmot	New York, N. Y	June 4, 1861	32, 493
Harvester-rake	W. A. Wood	Hoosick Falls, N. Y	Nov. 19, 1861	33, 761
Harvester-rake	W. A. Wood	Hoosick Falls, N. Y	Jan. 3, 1871	110, 716
Harvester rake and reel	A. B. Hitchcock	Juneau, Wis	Sept. 14, 1869	94, 887
Harvester rake and reel combined	P. F. Hodges	Moline, Ill	May 17, 1870	103, 185
Harvester rake and reel combined	P. F. Hodges	Moline, Ill	May 17, 1870	103, 186
Harvester rake and reel combined	S. Johnston	Buffalo, N. Y	Feb. 7, 1865	46, 300
Harvester rake and reel, Combined	W. A. Kirby	Auburn, N. Y	June 25, 1872	128, 403
Harvester rake and reel, Combined	J. D. Whiting	Ripon, Wis	May 16, 1871	115, 004
Harvester rake and reel, Combined	W. A. Wood	Hoosick Falls, N. Y	Dec. 8, 1863	40, 876
Harvester rake and reel, Combined	M. Young, jr	Frederick, Md	Sept. 18, 1860	30, 103
Harvester rake-attachment	C. Barns	West Liberty, Iowa	Aug. 9, 1870	106, 108
Harvester rake-attachment	L. M. Batty	Canton, Ohio	Sept. 19, 1865	49, 964
Harvester rake-attachment	R. D. Brown	Covington, Ind	Dec. 19, 1865	51, 550
Harvester rake-attachment	H. Fisher	Canton, Ohio	Nov. 21, 1865	51, 111
Harvester rake-attachment	L. Miller	Akron, Ohio	Nov. 21, 1865	51, 070
Harvester rake-attachment	L. Miller	Akron, Ohio	Nov. 21, 1865	51, 071
Harvester rake-attachment	W. B. Parsons	Granger, N. Y	July 18, 1865	48, 834
Harvester rake-attachment	E. Smith	Clinton, Pa	June 5, 1866	55, 386
Harvester-rake, Automatic	J. W. Brokaw	Springfield, Ohio	Aug. 25, 1857	18, 070
Harvester-rake, Automatic	S. Comfort, jr	Morrisville, Pa	Dec. 23, 1856	16, 307
Harvester-rake, Automatic	T. F. Kums	Rockford, Ill	Oct. 1, 1861	33, 418
Harvester-rake, Automatic	T. F. Kums	Rockford, Ill	Oct. 1, 1861	33, 419
Harvester-rake, Automatic	T. F. Kums	Rockford, Ill	Oct. 1, 1861	33, 420
Harvester-rake, Automatic	S. T. Lamb	New Washington, Ind	Apr. 29, 1856	14, 769
Harvester-rake, Automatic	J. S. Manning	Philadelphia, Pa	June 30, 1857	17, 687
Harvester-rake, Automatic	J. Ollis	Bloomington, Ill	June 26, 1860	28, 894
Harvester-rake, Automatic	I. C. Twining	Wrightstown, Pa	Aug. 14, 1860	29, 640
Harvester-rake, Automatic	W. N. Whiteley, jr	Springfield, Ohio	Aug. 29, 1865	49, 669
Harvester rake, Grain	J. Atkins	Chelsea, Ill	Dec. 21, 1852	9, 479
Harvester-rake, Self-acting	S. T. Lamb	New Washington, Ind	June 30, 1857	17, 685
Harvester-rake, Self-acting	J. Whitehead	Manchester, Va	Sept. 16, 1856	15, 751
Harvester-rake, Self-acting	J. Whitehead	Manchester, Va	Dec. 2, 1856	16, 156
Harvester-rakes, &c., Producing intermittent acceleration of motion in.	J. Richardson	Buckeystown, Md	June 19, 1855	13, 102
Harvester-raker	J. W. Haggard and G. Bull	Bloomington, Ill	Dec. 11, 1855	13, 933
Harvester-raker	W. A. Wood	Hoosick Falls, N. Y	Sept. 11, 1860	30, 018
Harvester raking and binding apparatus	A. Sherwood	Auburn, N. Y	Sept. 14, 1858	21, 540
Harvester raking and binding apparatus	A. B. Smith	Clinton, Pa	June 19, 1860	28, 783
Harvester raking and binding attachment	J. Barta	La Crosse, Wis	May 15, 1866	54, 672
Harvester raking and binding attachment	J. P. Manny	Rockford, Ill	July 6, 1858	20, 805
Harvester raking and binding attachment	T. Palmer	Catskill, N. Y	Jan. 13, 1863	37, 408
Harvester raking and binding device	A. Sherwood	Auburn, N. Y	Jan. 26, 1858	19, 212
Harvester raking and delivering attachment	W. A. Wood	Hoosick Falls, N. Y	Apr. 27, 1859	20, 119
Harvester raking-apparatus	G. A. Clarke	Philadelphia, Pa	Jan. 1, 1856	14, 043
Harvester raking-apparatus	I. Dodenhoff	Bloomington, Ill	Aug. 18, 1857	18, 009
Harvester raking-apparatus	S. R. Hunter	Cortland, N. Y	Dec. 2, 1856	16, 145
Harvester raking-apparatus	J. P. Manny	Rockford, Ill	July 14, 1857	17, 798

Index of patents issued from the United States Patent Office from 1790 *to* 1873, *inclusive*—Continued.

Invention.	Inventor.	Residence.	Date.	No.
Harvester raking-apparatus	D. C. Smith	Tecumseh, Mich	June 30, 1857	17, 703
Harvester raking-apparatus	J. Urmy	Wilmington, Del	Mar. 31, 1857	10, 941
Harverter raking-apparatus, Automatic	W. H. Wilson	Denton, Md	May 8, 1860	28, 228
Harvester raking-apparatus, Corn and cane	J. W. Batson	Triadelphia, Md	July 29, 1856	15, 408
Harvester raking-attachment	J. A. Barrington	Fredericktown, Ohio	June 8, 1858	20, 475
Harvester raking attachment	W. W. Burson	Yates City, Ill	Nov. 2, 1858	21, 940
Harvester raking-attachment	A. H. Caryl	Sandusky, Ohio	Feb. 5, 1856	14, 183
Harvester raking-attachment	G. E. Chenoweth	Baltimore, Md	Apr. 9, 1861	31, 954
Harvester raking-attachment	I. H. Conklin	Rockford, Ill	Apr. 14, 1857	17, 025
Harvester raking-attachment	I. C. Crane	Edgerton, Ohio	May 17, 1864	42, 752
Harvester raking-attachment	P. S. Crawford	Marengo, Ill	Sept. 21, 1858	21, 552
Harvester raking-attachment	H. Curtis	New York, N. Y	Oct. 18, 1864	44, 709
Harvester raking-attachment	D. Davis, J. Heibeler, and S. A. Porter.	Prescott, Wis	Apr. 5, 1864	42, 171
Harvester rak'ng-attachment	D. O. De Wolf	New York, N. Y	June 1, 1858	20, 411
Harvester raking-attachment	J. L. Fountain	Rockford, Ill	Jan. 12, 1858	19, 085
Harvester raking-attachment	J. Fox and J. W. Van Hook	Uniontown, D. C	Dec. 27, 1864	45, 597
Harvester raking-attachment	G. V. Griffith	Sandusky, Ohio	Apr. 27, 1858	20, 061
Harvester raking-attachment	C. P. Gronberg	Montgomery, Ill	June 28, 1859	24, 555
Harvester raking-attachment	C. P. Gronberg	Aurora, Ill	Sept. 11, 1860	29, 966
Harvester raking-attachment	D. Guptail	Elgin, Ill	June 26, 1860	28, 854
Harvester raking-attachment	W. H. Hovey	Springfield, Mass	Apr. 15, 1856	14, 693
Harvester raking-attachment	R. V. Jones	Canton, Ohio	Dec. 27, 1864	45, 614
Harvester raking-attachment	D. W. and H. A. Lafetra	Eatontown, N. J	Apr. 14, 1857	17, 045
Harvester raking attachment	A. E. McGaughey	Red Wing, Minn	Feb. 9, 1864	41, 522
Harvester raking-attachment	J. McIntosh	Geneva, Ill	June 30, 1857	17, 690
Harvester raking-attachment	W. G. Merrell	Auburn, N. Y	Apr. 26, 1864	42, 498
Harvester raking-attachment	J. Nelson	Rockford, Ill	Sept. 7, 1858	21, 437
Harvester raking-attachment	S. N. Page	Salona, Pa	Mar. 21, 1865	46, 930
Harvester raking-attachment	J. W. Patterson	Philadelphia, Pa	Feb. 16, 1858	19, 370
Harvester raking-attachment	A. R. Reese	Phillipsburgh, N. J	Oct. 19, 1858	21, 847
Harvester raking-attachment	I. S. and H. R. Russell and J. I. D. Bristol.	New Market, Md., and Detroit, Mich.	July 21, 1863	39, 329
Harvester raking-attachment	W. T. Shaw and J. Manz	Wilmington, Del	Nov. 22, 1864	45, 185
Harvester raking-attachment	A. B. Smith	Clinton, Pa	May 14, 1861	32, 317
Harvester raking-attachment	B. Smith	Batavia, Ill	July 14, 1863	39, 249
Harvester raking-attachment	I. B. Snyder	Lancaster County, Pa	Jan. 20, 1863	37, 463
Harvester raking-attachment	J. A. St. John	Janesville, Wis	May 25, 1858	20, 378
Harvester raking-attachment	O. Stoddard	Busti, N. Y	Apr. 13, 1858	19, 958
Harvester raking-attachment	A. Stoler and S. A. Sisson	Bristol, Pa	Mar. 1, 1864	41, 793
Harvester raking-attachment	H. K. Taylor	Racine, Wis	July 12, 1864	43, 531
Harvester raking-attachment	J. V. A. and A. Wemple	Chicago, Ill	Feb. 16, 1858	19, 393
Harvester raking-attachment	C. Wheeler, jr	Poplar Ridge, N. Y	July 8, 1856	15, 311
Harvester raking-attachment	C. Wheeler, jr	Poplar Ridge, N. Y	Apr. 26, 1864	42, 521
Harvester raking-attachment	W. Whiteley, jr	Springfield, Ohio	Nov. 25, 1856	16, 131
Harvester raking-attachment	J. Young	Marshallton, Pa	Dec. 14, 1858	22, 326
Harvester raking-device	S. R. Hunter	Cortland, N. Y	Sept. 1, 1857	18, 096
Harvester raking-device	C. W. and W. W. Marsh	Shabbona, Ill	Jan. 5, 1864	41, 080
Harvester raking-device	J. Nelson	Rockford, Ill	Jan. 5, 1864	41, 123
Harvester ratchet-attachment	G. E. Burt and S. B. Hildreth	Barvard, Mass	Dec. 1, 1868	84, 475
Harvester ratchet-attachment	W. Cogswell	Ottawa, Ill	May 15, 1866	54, 689
Harvester-reel	J. M. Bope	South Bend, Ind	June 7, 1870	103, 969
Harvester-reel	J. M. Bope	South Bend, Ind	June 7, 1870	103, 970
Harvester-reel	G. A. Bruce	Mechanicsburgh, Ill	July 3, 1855	13, 159
Harvester-reel	G. S. Curtis	Chicago, Ill	Sept. 27, 1859	25, 560
Harvester-reel	C. F. Goddard	Saint Ansgar, Iowa	Sept. 2, 1873	142, 389
Harvester-reel	G. S. Grier	Milford, Del	Dec. 19, 1871	122, 007
Harvester-reel	G. S. Grier	Milford, Del	July 2, 1872	128, 616
Harvester-reel	H. B. Hurford and E. Miller	Canton, Ohio	Feb. 13, 1872	123, 702
Harvester-reel	J. R. Jones	Clarksville, Iowa	Mar. 23, 1869	88, 043
Harvester-reel	J. H. Keller	Boalsburgh, Pa	Nov. 1, 1870	108, 912
Harvester-reel	J. Miller	Swan, Ind	Aug. 8, 1871	117, 911
Harvester-reel	D. C. Nutting and J. D. Allen	Bowling Green, Ky	Aug. 15, 1871	118, 045
Harvester-reel	C. N. Owen	Salem, Ohio	Sept. 27, 1870	107, 804
Harvester-reel	J. L. Rohrer	Upper Leacock Twnsp., Pa	Nov. 17, 1868	84, 138
Harvester-reel	W. F. Rundell	Genoa, N. Y	Dec. 31, 1867	72, 910
Harvester-reel	W. F. Rundell	Genoa, N. Y	June 16, 1868	79, 009
Harvester-reel	E. P. Russell	Manlius, N. Y	Nov. 12, 1861	33, 713
Harvester-reel	E. P. Russell	Manlius, N. Y	Apr. 18, 1865	47, 338
Harvester-reel	J. A. Shephard	Portsmouth, Ohio	July 30, 1872	129, 990
Harvester-reel	J. Werner, jr	Prairie du Sac, Wis	Sept. 16, 1873	142, 881
Harvester-reel	W. W. and C. C. Wright	Canton, Pa	Dec. 7, 1852	9, 458
Harvester-reel	G. W. N. Yost	Corry, Pa	June 9, 1868	78, 711
Harvester reel and rake	W. H. Seymour	Brockport, N. Y	Jan. 31, 1871	111, 483
Harvester reel and rake	W. N. Whiteley	Springfield, Ohio	Sept. 8, 1868	82, 050
Harvester reel and rake combined	S. W. Bingman	Laurelton, Pa	Feb. 15, 1870	99, 820
Harvester reel-rake	M. K. Church	Stamford, Canada	Dec. 17, 1872	133, 966
Harvester reel-rake	C. A. Werden	Waukegan, Ill	Mar. 18, 1873	136, 885
Harvester reel-rake	C. Young	Auburn, N. Y	Nov. 19, 1872	133, 285
Harvester-reels, Mode of supporting	T. I. Stealey	Middlebourne, Va	Dec. 15, 1857	18, 871
Harvester, Rice	J. J. Herndon	Marlborough District, S. C.	Oct. 1, 1850	7, 683
Harvester scroll-wheel	C. D. Rogers	Utica, N. Y	July 7, 1857	17, 749
Harvester-seat	T. S. Brown	Poughkeepsie, N. Y	Apr. 24, 1866	54, 245
Harvester-seat	W. J. Ludlow	Cleveland, Ohio	June 2, 1868	78, 531
Harvester-seat	E. A. Smith	Saint Albans, Vt	Mar. 10, 1868	75, 306
Harvester-seat	G. W. N. Yost	Corry, Pa	Apr. 15, 1873	137, 816
Harvester, Seed	F. Decker	Delaware, Ohio	Aug. 2, 1870	105, 920
Harvester sheaf-dropper	N. S. Ketchum	Marshalltown, Iowa	June 4, 1872	127, 611
Harvester-sickle	I. C. Crane	Edgerton, Ohio	Apr. 5, 1864	42, 169
Harvester-sickle	P. Manny	Waddam's Grove, Ill	Oct. 21, 1856	15, 926
Harvester sickle-guard	A. Shogren	Chicago, Ill	Aug. 2, 1859	24, 957
Harvester sickle-head	M. H. Hilburn	Wilmington, Ill	July 31, 1866	56, 749
Harvester-sickles, Tool for shaping	J. Jann	New Windsor, Md	Nov. 20, 1866	59, 770
Harvester, Steam	R. C. Parvin	Philadelphia, Pa	Apr. 16, 1872	125 843
Harvester, Sugar-cane	W. T. Johnson and J. Doyle	Philadelphia, Pa., and Wetumpka, Ala.	Oct. 2, 1860	30, 226
Harvester, Sugar-cane	R. R. Taylor	Reading, Pa	May 24, 1859	24, 164
Harvester, Thrashing	S. S. Rembert	Memphis, Tenn	Mar. 26, 1850	7, 222

Index of patents issued from the United States Patent Office from 1790 *to* 1873, *inclusive*—Continued.

Invention.	Inventor.	Residence.	Date.	No.
Harvester-tongue	M. Rohrer	Polo, Ill	Apr. 25, 1871	114, 203
Harvester-tongue	J. Wadleigh	Chebanse, Ill	Dec. 22, 1868	85, 150
Harvester track-clearer	O. Du Bois	San José, Cal	Apr. 1, 1873	137, 351
Harvester track-clearer	W. F. Ketchum	Buffalo, N. Y	May 17, 1853	9, 737
Harvester track-clearer	J. P. Manny	Rockford, Ill	July 6, 1858	20, 807
Harvester track-clearer	C. N. Owen	Salem, Ohio	Nov. 21, 1871	121, 123
Harvester track-clearer	N. Stonecipher	Cambridge City, Ind	July 7, 1868	79, 609
Harvester track-clearer	J. O. Taber	Salem, Ohio	Mar. 24, 1863	38, 007
Harvester track-clearer	G. W. N. Yost	Corry, Pa	Dec. 8, 1868	84, 786
Harvester track-clearer, Grass	A. Whiteley	Springfield, Ohio	Aug. 22, 1854	11, 579
Harvester-wheel	P. Lincoln	Cold Water, Mich	Mar. 24, 1863	37, 969
Harvester-wheel	A. J. Sweeney	Wheeling, W. Va	Sept. 3, 1872	131, 133
Harvester wheel and axle	B. G. Turner	Fremont, Nebr	Aug. 19, 1873	141, 965
Harvesters, Adjustable reel and cut-off for	J. H. Keller and D. F. Luse	Boalsburgh and Centre Hill, Pa.	Nov. 21, 1871	121, 107
Harvesters, Adjustable reel-post for	J. W. Davis	Dublin, Ohio	July 26, 1870	105, 654
Harvesters, Apparatus for removing grain from	S. Comfort, jr	Morrisville, Pa	Apr. 1, 1856	14, 553
Harvesters, Arrangement of cutters in grain and grass.	J. Peirson	Wilmington, Del	July 2, 1850	7, 481
Harvesters, Attaching connecting-bars to cutters of	J. and W. Little	Princeton, Ind	July 3, 1855	13, 176
Harvesters, Attaching raker's seat to	S. Hull	Poughkeepsie, N. Y	June 26, 1855	13, 147
Harvesters, Attaching teeth to sickle-bars of	I. C. and L. C. Pluche	Cape Vincent, N. Y	June 10, 1856	15, 084
Harvesters, &c., Attaching wheels to	A. Whiteley	Springfield, Ohio	Apr. 24, 1855	12, 768
Harvesters, Binder's platform for	C. Miller	Lincoln, Pa	June 27, 1871	116, 471
Harvesters, Binder's table for	P. M. Donohoo	Saint Rose, Wis	Nov. 21, 1871	121, 090
Harvesters, Binder's table for	J. H. Smith	Orford, Iowa	Jan. 21, 1873	135, 168
Harvesters, Clutch-mechanism for	W. Michael	Murrysville, Pa	June 7, 1870	103, 910
Harvesters, Combined rake and reel attachment to	R. Hoffheins	Dover, Pa	July 4, 1865	48, 557
Harvesters, Cut-off for	W. R. G. Humphrey	Chillicothe, Mo	Jan. 23, 1872	123, 023
Harvesters, Cutting-gear of grain and grass	S. S. Allen	Salem, N. J	Nov. 8, 1853	10, 201
Harvesters, Cutting-off apparatus for	J. S. Schoonover and P. S. Bacon.	Corry, Pa	Apr. 30, 1873	126, 236
Harvesters, Delivering-apparatus of grain	E. A. Morrison	Lawrenceville, Pa	Jan. 30, 1855	12, 339
Harvesters, Draft-device for	J. M. Smith	Jerseyville, Ill	Nov. 15, 1870	109, 353
Harvesters, Finger-bar arrangement for	J. A. Moore and A. H. Patch	Louisville, Ky	Nov. 25, 1856	16, 134
Harvesters, Finger-fan for	W. A. Wood	Hoosick Falls, N. Y	Dec. 10, 1867	72, 149
Harvesters from reapers to mowers, &c., Mechanism for changing.	R. Beans	Johnsville, Pa	Aug. 28, 1855	13, 504
Harvesters, Gleaner-attachment for	J. V. Bryson	Greenup, Ky	Oct. 1, 1872	131, 844
Harvesters, Grain-binder for	G. W. N. Yost	Port Gibson, Miss	Jan. 1, 1856	14, 036
Harvesters, Grain-binding attachment for	T. S. Minnis	Meadville, Pa	Sept. 27, 1870	107, 703
Harvesters, Grain-discharger for	J. D. Congue	Lyndon, Ill	Nov. 24, 1863	40, 676
Harvesters, Grain-lifter for	W. M. Jackson	Woodland, Cal	June 21, 1870	104, 460
Harvesters, Lifting-finger for	A. Hughes	Brampton Ash, England	Nov. 11, 1873	144, 457
Harvesters, Means of adjusting the grain-wheel of	J. D. Wright	Columbia, Tenn	Nov. 22, 1870	109, 481
Harvesters, Method of gathering grain upon, and discharging it from, the platform of.	O. Hussey	Baltimore, Md	Dec. 21, 1858	22, 368
Harvesters, Method of raising and lowering the cutters of.	J. F. Barrett	North Granville, N. Y	July 10, 1855	13, 205
Harvesters, Running-gear for	M. G. Hubbard	Syracuse, N. Y	June 4, 1867	65, 386
Harvesters, &c., Seat for	J. G. Perry	Kingston, R. I	July 22, 1873	141, 075
Harvesters, Self-rake for	R. D. Brown	Covington, Ind	Apr. 7, 1863	38, 094
Harvesters, Self-raker for	S. G. Randall	Rockford, Ill	July 22, 1856	15, 387
Harvesters, Self-raker for	J. T. Whitaker	Saint Charles, Ill	Apr. 29, 1856	14, 784
Harvesters, Self-raking attachment to	H. Foresman	Enon, Ohio	May 13, 1856	14, 861
Harvesters, Spring-seat for	T. Brett	Geneva, Ohio	Feb. 27, 1872	124, 113
Harvesters, Swathing-apparatus for	S. C. Longshore	Lahaska, Pa	Oct. 20, 1857	18, 462
Harvesters, &c., Tempering knife sections for	A. R. Reynolds	Auburn, N. Y	Aug. 28, 1866	57, 655
Harvesting and seeding machine, Frame for	J. H. Snyder	Killbuck, Ill	Dec. 20, 1864	45, 531
Harvesting grain	A. Churchill	Geneva, Ill	Mar. 16, 1841	2, 007
Harvesting grain and grass, Machinery for	W. and T. Schnebly	Hackensack, N. Y	Feb. 17, 1857	16, 658
Harvesting, husking, and shelling machine, Corn	D. A. Dickinson	Baltimore, Md	Oct. 16, 1866	58, 785
Harvesting-machine	W. Allen	Worcester, Mass	Oct. 4, 1864	44, 500
Harvesting-machine	O. Barr	Aurora, Ill	Jan. 16, 1849	6, 034
Harvesting-machine	J. F. Barrett	North Granville, N. Y	Apr. 28, 1857	17, 134
Harvesting-machine	L. M. Batty	Canton, Ohio	Sept. 19, 1865	49, 963
Harvesting-machine	E. Bayliss, J. O. Brown, and F. T. Lomont.	Massillon, Ohio	Oct. 14, 1873	143, 609
Harvesting-machine	A. Belchambers	Ripley, Ohio	Mar. 7, 1865	46, 628
Harvesting-machine	W. B. Birdsall and E. H. Coggswell.	Hudson, Mich	Nov. 1, 1864	44, 847
Harvesting-machine	J. M. Bowman	Brockport, N. Y	July 10, 1860	29, 119
Harvesting-machine	J. Bradley	Cedar Falls, Iowa	Sept. 20, 1864	44, 277
Harvesting-machine	C. R. Brinckerhoff	Batavia, N. Y	May 24, 1859	24, 093
Harvesting machine	C. R. Brinckerhoff	Rochester, N. Y	Aug. 22, 1865	49, 497
Harvesting-machine	W. H. Burkhart	Bucyrus, Ohio	Jan. 30, 1866	52, 264
Harvesting-machine	T. B. Butler	Newark, Conn	July 5, 1859	24, 613
Harvesting-machine	J. Carpenter	Yorktown, N. Y	Dec. 23, 1856	16, 313
Harvesting-machine	G. and W. Chamberlin	Olean, N. Y	Apr. 26, 1859	23, 756
Harvesting-machine	H. Chandler	New Gloucester, Me	Aug. 22, 1865	49, 501
Harvesting-machine	D. A. Church, L. H. Obert, and W. W. and O. F. Willoughby.	Friendship, N. Y., and Chicago, Ill.	Aug. 7, 1847	5, 228
Harvesting-machine	F. Clark	Charlotte, N. Y	June 18, 1861	32, 602
Harvesting-machine	W. F. Cochrane	Springfield, Ohio	July 25, 1865	49, 024
Harvesting-machine	I. H. Coller	Poughkeepsie, N. Y	July 11, 1865	48, 658
Harvesting-machine	J. C. Cox and R. Newton	Greenville, N. C	Nov. 17, 1857	18, 628
Harvesting-machine	D. Cushing	Aurora, Ill	Nov. 21, 1848	5, 934
Harvesting-machine	J. S. Davis	Tiffin, Ohio	Aug. 22, 1865	49, 506
Harvesting-machine	N. A. Dederer	Greene, N. Y	Feb. 6, 1866	52, 395
Harvesting-machine	J. Dunlap	Geneva, Wis	June 20, 1847	5, 174
Harvesting-machine	R. Dutton	Dayton, Ohio	Feb. 12, 1861	31, 378
Harvesting-machine	R. Dutton	New York, N. Y	July 17, 1866	56, 389
Harvesting-machine	G. Esterly	Heart Prairie, Wis	Oct. 22, 1844	3, 803
Harvesting-machine	G. Esterly	Heart Prairie, Wis	Mar. 24, 1857	16, 873
Harvesting-machine	G. Esterly	Whitewater, Wis	Apr. 12, 1859	23, 666
Harvesting-machine	B. G. Fitzhaugh	Frederick, Md	June 28, 1859	24, 549
Harvesting-machine	G. W. Folhurst	Cleveland, Ohio	Sept. 16, 1856	15, 748
Harvesting-machine	E. T. Ford	Stillwater, N. Y	Feb. 20, 1866	52, 701

Index of patents issued from the United States Patent Office from 1790 *to* 1873, *inclusive*—Continued.

Invention.	Inventor.	Residence.	Date.	No.
Harvesting-machine	C. Foster	La Perte, Ind	Jan. 1, 1847	4, 916
Harvesting-machine	L. J. Fountain	New Milford, Ill	June 20, 1865	48, 335
Harvesting-machine	J. Haines	Union Grove, Ill	Mar. 27, 1849	6, 245
Harvesting-machine	M. Hallenbeck	Albany, N. Y	Apr. 17, 1860	27, 901
Harvesting-machine	T. C. Hargraves	Schenectady, N. Y	Apr. 30, 1861	30, 233
Harvesting-machine	J. K. Harris	Allensville, Ind	June 30, 1857	17, 678
Harvesting-machine	I. Hawley and D. S. Stafford	Pekin and Decatur, Ill	Nov. 8, 1864	44, 996
Harvesting-machine	J. E. Heath	Warren, Ohio	Jan. 15, 1850	7, 020
Harvesting-machine	J. E. Heath	Geneva, Ohio	Sept. 11, 1855	13, 545
Harvesting-machine	M. G. Hubbard	Penn Yan, N. Y	May 10, 1859	23, 921
Harvesting-machine	A. H. Inskeep	Middleburgh, Ohio	Sept. 13, 1859	25, 415
Harvesting-machine	J. Jann	New Windsor, Md	Aug. 4, 1863	39, 403
Harvesting-machine	W. Jones	Bradford, Vt	July 8, 1851	8, 207
Harvesting-machine	W. H. Jordan	Roseville, Ind	Apr. 5, 1864	42, 197
Harvesting-machine	H. G. Kaufmann	Saint Louis, Mo	Aug. 31, 1858	21, 343
Harvesting-machine	W. A. Kirby	Buffalo, N. Y	Sept. 2, 1856	15, 659
Harvesting-machine	W. A. Kirby	Buffalo, N. Y	July 2, 1861	32, 736
Harvesting-machine	S. T. Lamb	New Washington, Ind	Oct. 9, 1860	30, 330
Harvesting-machine	J. Little	Chambersburgh, Pa	May 24, 1859	24, 128
Harvesting-machine	H. H. Luther	Warren, R. I	June 7, 1859	24, 318
Harvesting-machine	F. H. Manny	Rockport, Ill	Apr. 30, 1861	32, 197
Harvesting-machine	H. Marcellus	Amsterdam, N. Y	May 3, 1859	23, 851
Harvesting-machine	D. J. Marvin	Stockton, Cal	Nov. 15, 1864	45, 061
Harvesting-machine	J. C. and C. N. Mayberry	White Rock, Ill	Nov. 19, 1861	33, 748
Harvesting-machine	J. H. Maydole	Eaton, N. Y	Aug. 7, 1860	29, 504
Harvesting-machine	R. W. McClelland	Springfield, Ill	Aug. 2, 1864	43, 697
Harvesting-machine	J. McPherson	Pennington, N. J	May 31, 1859	24, 226
Harvesting-machine	B. Mertz	Burlington, Iowa	Sept. 11, 1860	29, 985
Harvesting-machine	L. Miller	Canton, Ohio	Dec. 3, 1861	33, 845
Harvesting-machine	L. and J. Miller	Canton, Ohio	Dec. 2, 1862	37, 049
Harvesting-machine	H. Moore and J. Hascall	Kalamazoo, Mich	June 28, 1836	......
Harvesting-machine	L. L. Moore	Petersburgh, Va	Aug. 19, 1856	15, 569
Harvesting-machine	R. J. Morrison	Richmond, Va	Dec. 16, 1856	16, 244
Harvesting-machine	G. Murphy	Lewisburgh, Pa	Sept. 26, 1865	50, 206
Harvesting-machine	W. Neff	Centre Hall, Pa	Oct. 22, 1861	33, 539
Harvesting-machine	J. M. Orfut	Malta, Ill	Aug. 27, 1861	33, 158
Harvesting-machine	S. K Paden	Volant, Pa	June 28, 1864	43, 331
Harvesting-machine	S. N. Page	Salona, Pa	Mar. 21, 1865	46, 929
Harvesting-machine	F. S. Pease	Buffalo, N. Y	Nov. 14, 1848	5, 925
Harvesting-machine	A. J. Purviance	Updegraff's, Ohio	May 22, 1849	6, 476
Harvesting-machine	S. Ray and M. R. Shalters	Alliance, Ohio	Apr. 19, 1859	23, 707
Harvesting-machine	J. Reily	Heart Prairie, Wis	Nov. 20, 1855	13, 828
Harvesting-machine	J. H. Rible	Somerset, Ohio	July 10, 1860	29, 129
Harvesting-machine	G. W. Richardson	Grayville, Ill	July 19, 1859	24, 848
Harvesting machine	G. W. Richardson and R. Glover	Grayville, Ill	Sept. 6, 1859	25, 370
Harvesting-machine	J. S. Royce	Cuylerville, N. Y	Dec. 18, 1860	30, 933
Harvesting-machine	E. P. Russell	Manlius, N. Y	Feb. 14, 1865	46, 394
Harvesting-machine	J. Y. Schelly and J. Stauffer	Hereford and Hosensack, Pa.	Sept. 2, 1856	15, 669
Harvesting-machine	W. and T. Schnebly	Hackensack, N. J	June 21, 1859	24, 495
Harvesting-machine	W. and T. Schnebly	Hackensack, N. J	June 21, 1859	24, 496
Harvesting-machine	J. Seibel	Manlius, Ill	Apr. 19, 1864	42, 410
Harvesting-machine	W. H. Seymour and D. S. Morgan.	Brockport, N. Y	June 8, 1858	20, 516
Harvesting-machine	J. B. Smith	Winfield, N. Y	Nov. 22, 1864	45, 186
Harvesting-machine	J. D. Smith	Lancaster, Ohio	May 3, 1859	23, 869
Harvesting-machine	W. S. Stetson	Baltimore, Md	May 17, 1859	24, 062
Harvesting-machine	W. S. Stetson and R. F. Maynard.	Baltimore, Md	May 17, 1859	24, 063
Harvesting-machine	A. G. Stipher	Richmond, Ind	June 28, 1859	24, 591
Harvesting-machine	S. Thomas	Burnett, Wis	May 10, 1859	23, 961
Harvesting-machine	B. Titcomb	Baltimore County, Md	June 5, 1860	28, 618
Harvesting-machine	J. A. Wagener	Pultney, N. Y	May 22, 1860	28, 427
Harvesting-machine	W., jr., and J. Webber	Rockton, Ill	Mar. 29, 1859	23, 417
Harvesting-machine	T. Welch	Churchville, N. Y	Apr. 3, 1866	53, 713
Harvesting-machine	J. Werner, jr	Prairie du Sac, Wis	Sept. 26, 1865	50, 191
Harvesting-machine	E. C. West	Bradford, Vt	June 25, 1845	4, 092
Harvesting-machine	M. A. Wheaton	Suisun City, Cal	Aug. 1, 1865	49, 186
Harvesting-machine	C. Wheeler, jr	Poplar Ridge, N. Y	Feb. 9, 1864	41, 556
Harvesting-machine	C. Wheeler, jr	Poplar Ridge, N. Y	Feb. 9, 1864	41, 557
Harvesting-machine	C. Wheeler, jr	Poplar Ridge, N. Y	Feb. 9, 1864	41, 558
Harvesting-machine	C. Wheeler, jr	Poplar Ridge, N. Y	Feb. 9, 1864	41, 559
Harvesting-machine	E. H. Wheeler	Keokuk, Iowa	Jan. 29, 1861	31, 275
Harvesting-machine	J. Whitehead	Manchester, Va	May 24, 1859	24, 170
Harvesting-machine	T. S. Whitenack	Easton, Pa	Apr. 30, 1861	32, 216
Harvesting-machine	H. Willard and R. Ross	Vergennes, Vt	May 10, 1859	23, 971
Harvesting-machine	W. Wilmington	Toledo, Ohio	July 3, 1860	29, 026
Harvesting-machine	W. H. Wilson	Denton, Md	May 10, 1859	23, 972
Harvesting-machine	B. F. Witt	Dublin, Ind	Apr. 10, 1860	27, 852
Harvesting-machine	W. A. Wood	Hoosick Falls, N. Y	Feb. 10, 1857	16, 618
Harvesting-machine	W. A. Wood	Hoosick Falls, N. Y	Feb. 10, 1857	16, 619
Harvesting-machine	W. A. Wood	Hoosick Falls, N. Y	Feb. 10, 1857	16, 620
Harvesting-machine	W. A. Wood	Hoosick Falls, N. Y	May 3, 1859	23, 878
Harvesting-machine	W. A. Wood	Hoosick Falls, N. Y	July 29, 1862	36, 047
Harvesting-machine	M. Young, jr	Frederick, Md	June 28, 1859	24 598
Harvesting-machine	M. Young, jr	Frederick, Md	Dec. 18, 1860	30, 999
Harvesting-machine, Bean	S. V. Newman	Covington, N. Y	June 21, 1859	24, 482
Harvesting-machine cutter	S. T. Lamb	New Washington Ind	Oct. 2, 1860	30, 231
Harvesting-machine for cutting, thrashing, and winnowing grain.	D. A. Church	Friendship, N. Y	May 4, 1841	2, 074
Harvesting machine, Grain	G. R. Crane	Caldwell, N. J	Mar. 10, 1857	16, 789
Harvesting machine, Grain	G. F. Foote	Buffalo, N. Y	Nov. 11, 1856	16, 052
Harvesting-machine guards, Manufacture of	J. E. Sweet	Syracuse, N. Y	Nov. 28, 1871	121, 261
Harvesting-machine header-attachment	G. Easterly	Whitewater, Wis	Jan. 13, 1863	37, 392
Harvesting machine, Hemp	J. B. McCormick	Versailles, Ky	June 2, 1857	17, 438
Harvesting-machine, Rake	L. P. Brady	Mount Joy, Pa	Dec. 18, 1860	30, 908
Harvesting-machine raking-attachment	G. Tatlock	Salem, Ind	May 10, 1859	23, 959
Harvesting-machines, Automatic rake for	J. P. Greene and I. Dodenhoff	Bloomington, Ill	Apr. 21, 1857	17, 088

Index of patents issued from the United States Patent Office from 1790 *to* 1873, *inclusive*—Continued.

Invention.	Inventor.	Residence.	Date.	No.
Harvesting-machines, Crank-wrist for	J. O. Taber	Salem, Ohio	June 7, 1870	104, 080
Harvesting-machines, Divider for	W. A. Woods	Hoosick Falls, N. Y	Mar. 24, 1863	37, 995
Harvesting-machines, Form of teeth in	E. B. Forbush	Buffalo, N. Y	Nov. 27, 1849	6, 903
Harvesting-machines, Grain-carrier for	J. J. and H. F. Mann	Clinton, Ind	June 19, 1849	6, 540
Harvesting-machines, Grain-discharging attachment to.	J. F. Black	Lancaster, Ill	Oct. 26, 1858	21, 869
Harvesting-machines, Machine for grinding the cutters of.	W. H. Stevenson	Auburn, N. Y	Feb. 18, 1868	74, 730
Harvesting-machines, Machine for grinding the cutters of.	W. H. Stevenson	Auburn, N. Y	Feb. 18, 1868	74, 731
Harvesting-machines, Method of manufacturing cutter-bars for.	G. G. Taylor	Worcester, Mass	Jan. 10, 1865	45, 879
Harvesting-machines, Method of operating the cutters of.	J. Haviland	Milton, N. Y	Sept. 1, 1857	18, 093
Harvesting-machines, Pitman-head and crank-wrist box for.	W. N. Whiteley, jr	Springfield, Ohio	July 10, 1866	56, 304
Harvesting-machines, Tongue and caster-plate for	R. Emerson, jr	Rockford, Ill	May 26, 1857	17, 367
Harvesting-rake	P. Sylla	Elgin, Ill	May 12, 1863	38, 515
Harvesting standing corn, Machine for	J. H. Frampton	Hopewell, Ohio	Mar. 3, 1857	16, 726
Hasp and staple	O. J. Smith	Wauwatosa, Wis	Dec. 24, 1872	134, 322
Hasp and staple fastening	E. P. Jones	Shell Mound, Miss	Mar 15, 1870	100, 899
Hasp-lock	J. H. Beauregard	Kingsbury, N. Y	Oct. 6, 1868	82, 681
Hasp-lock	W. N Chamberlain	Denton, Mich	Apr. 4, 1871	113, 496
Hasp-lock	E. R. Colver	New London, Conn	Dec. 21, 1869	98, 036
Hasp-lock	G. Crompton	Jersey City, N. J	Oct. 24, 1871	120, 247
Hasp-lock	J. S. Elliott	Philadelphia, Pa	May 12, 1868	77, 723
Hasp-lock	J. Fisler and G. Crompton	Jersey City, N. J	July 4, 1871	116, 581
Hasp-lock	E. L. Gaylord	Terrysville, Conn	July 27, 1869	93, 078
Hasp lock	W. D. Heister	Newton Hamilton, Pa	Dec. 19, 1871	122, 013
Hasp-lock	M. G. Imbach	Hartford, Conn	Oct. 27, 1868	83, 500
Hasp-lock	J. Kinzer	Pittsburgh, Pa	July 15, 1873	140, 926
Hasp-lock	W. C. McGill	Cincinnati, Ohio	June 27, 1871	116, 469
Hasp-lock	C. W. A. Romer	Newark, N. J	Nov. 16, 1869	96, 968
Hasp-lock	T. Slaight	Newark, N. J	May 7, 1872	126, 583
Hasp-lock	A. B. Smith	Memphis, Mich	May 27, 1873	139, 430
Hasp-lock	E. P. Stimens and T. C. Martin	Perryville, Ohio	Mar. 18, 1873	136, 877
Hasp-lock	C. Walsh, J. F. Connelly, and A. Bratt.	Newark, N. J	Dec. 21, 1869	98, 212
Hasp or latch, Safety	E. C. Robinson	Boston, Mass	Mar. 14, 1871	112, 635
Hassock-machine	C. F. Anthony	Chicago, Ill	May 26, 1868	78, 350
Hassocks or stools, Device for forming	J. G. Flagg	Philadelphia, Pa	Jan. 15, 1867	61, 183
Hat	D. K. Albright and L. H. De Lange.	Philadelphia, Pa., and Burlington, N. J.	June 13, 1865	48, 222
Hat	D. K. Albright and L. H. De Lange.	Philadelphia, Pa., and Bordentown, N. J.	Nov. 5, 1867	70, 493
Hat	J. P. Beatty	Norwalk, Conn	June 27, 1865	48, 357
Hat	J. P. Beatty	Norwalk, Conn	Mar. 23, 1869	88, 116
Hat	A. Bogardus	Matteawan, N. Y	Sept. 19, 1871	119, 073
Hat	J. W. Corey	Newark, N. J	Jan. 24, 1871	111, 178
Hat	P. Curtis	Amesbury, Mass	June 7, 1864	43, 072
Hat	F. Degen	New York, N. Y	Dec. 7, 1852	9, 442
Hat	W. E. Doubleday	Brooklyn, N. Y	Aug. 26, 1862	36, 284
Hat	R. Dunlap	New York, N. Y	June 27, 1865	48, 384
Hat	J. H. Earle	Fall River, Mass	Nov. 7, 1865	50, 809
Hat	J. R. Ender	Trenton, La	Apr. 17, 1860	27, 897
Hat	F. P. Flanagan	Newark, N. J	June 16, 1863	38, 895
Hat	D. W. Gitchell	New York, N. Y	Nov. 28, 1865	51, 172
Hat	H. Hayward	New York, N. Y	Aug. 10, 1869	93, 616
Hat	J. M. Heard	Aberdeen, Miss	May 24, 1870	103, 457
Hat	C. A. Hopkins, C. H. Reid, G. N. Raymond, and J. S. Meeker.	Danbury, Conn	July 4, 1871	116, 710
Hat	F. Howard	Boston, Mass	Jan. 12, 1864	41, 260
Hat	H. C. Hulbert and A. Tollett	Brooklyn, N. Y	Aug. 18, 1868	81, 089
Hat	F. Huot and C. Baudouin	New York, N. Y	June 9, 1868	78, 668
Hat	R. B. Hurd and W. Halladay	Paterson, N. J., and Brooklyn, N. Y.	Jan. 12, 1864	41, 261
Hat	G. Johnson	Philadelphia, Pa	Dec. 27, 1870	110, 570
Hat	J. Keith	Charlton, Mass	Jan. 10, 1871	110, 921
Hat	H. Kellogg	New Haven, Conn	July 28, 1868	80, 490
Hat	W. B. Lodge	Danbury, Conn	Oct. 15, 1867	69, 921
Hat	G. Mallory	Bridgeport, Conn	Feb. 11, 1868	74, 392
Hat	W. H. Mallory	Watertown, Conn	Sept. 8, 1863	39, 822
Hat	J. W. McGill	Washington, D. C	Nov. 29, 1870	109, 641
Hat	E. Morris	Philadelphia, Pa	Jan. 6, 1857	16, 338
Hat	E. Morris	Philadelphia, Pa	July 18, 1865	48, 828
Hat	W. S. Nelson	Saint Louis, Mo	Aug. 28, 1866	57, 554
Hat	R. S. Nickerson and J. Wallace	Philadelphia, Pa	May 30, 1865	47, 974
Hat	J. O'Donnell	Saint Louis, Mo	Apr. 29, 1873	138, 430
Hat	W. H. Plumb	New York, N. Y	Sept. 17, 1867	68, 899
Hat	C. L. Rahmer	Brooklyn, N. Y	June 6, 1865	48, 092
Hat	C. L. Rahmer	Brooklyn, N. Y	July 17, 1866	56, 447
Hat	C. E. Richards	North Attleborough, Mass	Apr. 23, 1867	64, 145
Hat	B. Sherwood	New York County, N. Y	Aug. 10, 1852	9, 190
Hat	B. B. Taggart and C. W. Brown	Watertown, N. Y	June 23, 1868	79, 278
Hat	W. H. Towers	New York, N. Y	Aug. 22, 1865	49, 570
Hat	P. W. Vail	Newark, N. J	Aug. 3, 1869	93, 251
Hat	W. F. Warburton	Philadelphia, Pa	Jan. 2, 1855	12, 173
Hat	W. F. Warburton	Philadelphia, Pa	May 31, 1864	42, 977
Hat	W. F. Warburton	Philadelphia, Pa	Aug. 7, 1866	57, 019
Hat	M. S. Watkins	Mansfield, Tex	Sept. 1, 1868	81, 715
Hat	W. H. White	Kent Island, Md	Sept. 25, 1866	58, 328
Hat	W. H. White	Kent Island, Md	Sept. 25, 1866	58, 329
Hat	W. H. White	New York, N. Y	May 19, 1868	78, 036
Hat	W. H. White	Baltimore, Md	Mar. 14, 1871	112, 661
Hat	D. Wilcox	Boston, Mass	Mar. 23, 1869	88, 107
Hat	D. Wilcox	Boston, Mass	June 8, 1869	91, 189
Hat	S. Wilks and J. E. Dow	Poland, Russia, and Boston, Mass.	July 16, 1872	129, 194

Index of patents issued from the United States Patent Office from 1790 *to* 1873, *inclusive*—Continued.

Invention.	Inventor.	Residence.	Date.	No.
Hat, Adjustable	I. Y. Cassiano	San Antonio, Tex	Dec. 2, 1873	145, 092
Hat and block for forming, Ventilated	W. F. Warburton	Philadelphia, Pa	May 14, 1872	126, 854
Hat and bonnet	J. Dennis, jr	Portsmouth, R. I	May 2, 1843	3, 065
Hat and bonnet box	O. A. Dailey	Washington, D. C	Mar. 14, 1865	46, 782
Hat and bonnet pressing apparatus	W. E. Doubleday	Brooklyn, N. Y	Sept. 20, 1864	44, 293
Hat and bonnet pressing machine	H. E. West	Attleborough, Mass	July 25, 1865	49, 048
Hat and cap	W. H. White	West River, Md	Apr. 17, 1866	54, 049
Hat and cap	W. H. White	West River, Md	May 22, 1866	54, 986
Hat and cap bands, Manufacture of India-rubber	W. H. Halsey	Hoboken, N. J	Apr. 10, 1866	53, 920
Hat and cap bracket	T. Lawrence	Peoria, Ill	Sept. 13, 1870	107, 389
Hat and cap ear covering	B. R. Morse	Galesburgh, Ill	Apr. 22, 1873	138, 040
Hat and cap holder	C. Beeny	Albany, N. Y	Mar. 21, 1871	112, 767
Hat and cap holder and bracket	J. R. Hartman and M. Kais	Peoria, Ill	Aug. 2, 1870	105, 938
Hat and cap linings, Sweat-knife for cutting	E. R. Pye	New York, N. Y	Nov. 8, 1859	26, 052
Hat and cap ventilator	W. Dale	New York, N. Y	Nov. 7, 1871	120, 629
Hat and clothes hook	T. Bowerman	Newark, N. J	Apr. 30, 1872	126, 256
Hat and coat guard	S. M. Robbins	New York, N. Y	Feb. 18, 1873	136, 005
Hat and coat holder	E. J. Post	Vienna, N. J	Jan. 29, 1867	61, 633
Hat and coat hook	C. G. Cole	Bennington, Vt	Feb. 18, 1873	135, 968
Hat and coat hook	J. Danner	Canton, Ohio	Mar. 18, 1873	136, 905
Hat and coat hook	A. J. Goodrich	Wolcottville, Conn	Oct. 24, 1871	120, 156
Hat and coat rack	G. T. Benson	Jersey City, N. J	Nov. 8, 1870	108, 961
Hat and coat rack	T. W. Brown	New York, N. Y	Oct. 1, 1867	69, 311
Hat and coat rack	G. F. J. Colburn	Newark, N. J	Feb. 7, 1865	46, 296
Hat and coat rack	J. E. Osborn	Chicago, Ill	June 22, 1869	91, 559
Hat and coat rack, Safety	F. W. Roth	Washington, D. C	Apr. 20, 1869	89, 171
Hat and dish rack	W. W. Horton	Katonah, N. Y	Apr. 28, 1868	77, 377
Hat and parcel holder	A. Clarke	Philadelphia, Pa	Mar. 11, 1873	136, 587
Hat and velvet polisher	J. A. Thompson	Auburn, N. Y	Mar. 8, 1864	41, 895
Hat bats, Machine for forming felt	W. G. Hagaman	Philadelphia, Pa	Feb. 3, 1857	16, 543
Hat-block	J. B. Brown	Middletown, N. Y	Sept. 24, 1867	69, 172
Hat-block	W. W. Cumberland	Newark, N. J	Feb. 19, 1861	31, 442
Hat-block	W. W. Jemeson	Wheeling, Ohio	Mar. 30, 1836	....
Hat-block	J. H. Masker	Newark, N. J	July 21, 1863	39, 299
Hat-block	H. A. Robison	Cleveland, Ohio	May 8, 1866	54, 601
Hat-block, Expansible	R. Eickemeyer	Yonkers, N. Y	July 29, 1873	141, 338
Hat-block to color hats	A. and S. Chichester	Wilton, Conn	May 29, 1835	....
Hat-blocks, Apparatus for holding	J. White	Cleveland, Ohio	Aug. 10, 1869	93, 571
Hat-blocks, Machine for turning	J. H. Masker	Newark, N. J	Jan. 3, 1860	26, 691
Hat blocking and shaping machine	R. Eickemeyer	Yonkers, N. Y	Dec. 31, 1867	72, 726
Hat blocking and shaping machine	G. Osterheld and R. Eickemeyer	Yonkers, N. Y	Apr. 3, 1866	53, 661
Hat blocking and stretching machine	J. De La Mar	Brooklyn, N. Y	Oct. 11, 1870	108, 116
Hat blocking and stretching machine	G. H. Hawkins	New York, N. Y	Aug. 4, 1868	80, 733
Hat blocking and stretching machine	G. A. Mandeville and W. E. Pine	Newark, N. J	Jan. 21, 1868	73, 541
Hat-blocking machine	J. A. Ball and P. B. Lawson	Cold Spring, N. Y	Nov. 3, 1868	83, 590
Hat-blocking machine	C. H. Berry	Brooklyn, N. Y	Nov. 7, 1871	120, 699
Hat-blocking machine	S. Boyden	Newark, N. J	Nov. 13, 1866	59, 707
Hat-blocking machine	J. De La Mar	Brooklyn, N. Y	Feb. 26, 1867	62, 321
Hat-blocking machine	J. Eberhardt	Newark, N. J	June 23, 1868	79, 059
Hat-blocking machine	E. C. Fales	Foxborough, Mass	May 20, 1873	139, 133
Hat-blocking machine	W. C. Griswold	New York, N. Y	Jan. 7, 1868	72, 998
Hat-blocking machine	W. C. Griswold, A. Pelisse, and A. H. Hook	Brooklyn, N. Y., Newark, N. J., and New York, N. Y.	Aug. 25, 1868	81, 364
Hat-blocking machine	S. S. Middlebrook	Sandy Hook, Conn	Apr. 2, 1867	63, 549
Hat-blocking machine	J. Sheldon	New York, N. Y	Mar. 19, 1867	63, 109
Hat-blocking machine	J. Sheldon	New York, N. Y	Apr. 2, 1867	63, 434
Hat-blocking machine	J. Sheldon	New York, N. Y	May 4, 1869	89, 601
Hat-blocking machine	J. Sheldon	New York, N. Y	May 31, 1870	103, 787
Hat-blocking machine	J. Surerus and W. H. Behrens	Newark, N. J	May 3, 1870	102, 613
Hat-blocking machines, Adjustable banding-ring for	G. Hayden	Newark, N. J	Nov. 7, 1871	120, 740
Hat-blocking machines, Adjustable crown-blocks for	W. C. Griswold	Newark, N. J	Nov. 7, 1871	120, 739
Hat-bodies, Alarm-attachment for machine for forming	W. C. Waring	Yonkers, N. Y	Jan. 3, 1871	110, 700
Hat-bodies, Apparatus for forming	G. and C. Beatty	Norwalk, Conn	May 21, 1861	32, 346
Hat-bodies, Apparatus for stretching	T. G. Oakley and W. R. Finch	Brooklyn, N. Y	Oct. 20, 1863	40, 383
Hat-bodies, Apparatus for stretching	W. H. Stevens	Danbury, Conn	Sept. 27, 1864	44, 468
Hat-bodies, Batting or web for	T. Blanchard	New York, N. Y	June 14, 1837	230
Hat-bodies by machinery at one operation, Manufacturing	G. L. Thatcher	Brooklyn, N. Y	May 31, 1828	....
Hat-bodies by machinery, Making	L. Van Hoesen	Norwalk, Conn	June 19, 1829	....
Hat bodies by machinery, Roll for felting or sizing	P. W. Somers	Danbury, Conn	Dec. 4, 1866	60, 275
Hat-bodies by sewing, Manner of cutting fabrics to be made into	I. L. Chapman	New York, N. Y	May 1, 1845	4, 021
Hat-bodies, Clearing	W. F. Young	Buffalo, N. Y	Aug. 20, 1872	130, 781
Hat-bodies, Composition for stiffening	J. M. Bottum	New York, N. Y	June 21, 1864	43, 176
Hat-bodies, Composition for stiffening	T. Trowbridge	Danbury, Conn	June 20, 1865	48, 328
Hat-bodies, Conductor in machine for forming	L. E. Hopkins	New York, N. Y	Aug. 30, 1853	9, 970
Hat-bodies, Device for printing	A. Barnes	Newark, N. J	Apr. 14, 1868	76, 696
Hat-bodies, &c., Dyeing	A. C. Brush and G. C. White	Norwalk and Danbury, Conn.	Oct. 9, 1866	58, 591
Hat-bodies, Felted	R. Johnson	Danbury, Conn	Jan. 28, 1868	73, 812
Hat-bodies, Felting	E. R. Barnes and J. B. Blakslee	Brookfield and Newton, Conn.	Aug. 12, 1856	15, 508
Hat-bodies, Felting	W. W. Cumberland	Newark, N. J	Oct. 23, 1855	13, 698
Hat-bodies, Felting	I. Searles	Newark, N. J	Dec. 18, 1855	13, 963
Hat-bodies, Forming	S. Graham	Fayette, Me	May 15, 1832	....
Hat-bodies, Forming	M. Osborne	New York, N. Y	Nov. 18, 1845	4, 275
Hat-bodies, Forming	A. H. Stevens	Richland, N. Y	Jan. 27, 1830	....
Hat-bodies, Forming	H. F. West and A. H. Stevens	Richland, N. Y	Oct. 29, 1828	....
Hat-bodies, Forming	J. Wharton	Newark, N. J	Dec. 2, 1873	145, 262
Hat-bodies, Forming and hardening	A. Spencer and A. Loeschner	New York and Brooklyn, N. Y.	Apr. 25, 1854	10, 832
Hat-bodies, Implement for alluring	J. Northrop and J. F. Emmons	Bridgeport, Conn	May 2, 1871	114, 327
Hat-bodies, Machine for blocking	W. A. Fenn	New Milford, Conn	Apr. 21, 1857	17, 083
Hat-bodies, Machine for blocking and stretching	W. Humphreys, jr	Cold Spring, N. Y	Apr. 23, 1867	64, 107

Index of patents issued from the United States Patent Office from 1790 *to* 1873, *inclusive*—Continued.

Invention.	Inventor.	Residence.	Date.	No.
Hat-bodies, Machine for felting	J. W. Blackham	Brooklyn, N. Y	Nov. 20, 1866	59, 758
Hat-bodies, Machine for felting	J. Bowden and R. Shaw	Marple, Great Britain	Feb. 4, 1873	135, 399
Hat-bodies, Machine for felting	C. Faure	New York, N. Y	Feb. 6, 1866	52, 489
Hat-bodies, Machine for felting	W. Fuzzard	Cambridgeport, Mass	July 8, 1856	15, 290
Hat-bodies, Machine for felting	W. Fuzzard	Cambridgeport, Mass	Apr. 7, 1857	16, 973
Hat-bodies, Machine for felting	S. H. Gray	Bridgeport, Conn	June 3, 1856	15, 008
Hat-bodies, Machine for felting	L. E. Hopkins	Brooklyn, N. Y	July 22, 1856	15, 375
Hat-bodies, Machine for felting	J. Kirk, S. Shelmerdine, and C. Froggart.	Stockport, Great Britain	Mar. 29, 1870	101, 276
Hat-bodies, Machine for felting	R. Northrop	New Milford, Conn	Mar. 19, 1861	31, 732
Hat-bodies, Machine for felting	H. L. Randall	Roxbury, Conn	Apr. 21, 1857	17, 115
Hat-bodies, Machine for felting	R. Smith	Danbury, Conn	June 4, 1861	32, 500
Hat-bodies, Machine for felting	P. W. Somers	Danbury, Conn	Dec. 4, 1866	60, 274
Hat-bodies, Machine for felting	J. Thomas	Brooklyn, N. Y	Aug. 26, 1856	15, 627
Hat-bodies, Machine for felting	J. Vero	Dewsbury near Leeds, England.	Feb. 28, 1871	112, 197
Hat-bodies, Machine for felting	J. Walber	Washington, D. C	Apr. 23, 1867	64, 049
Hat-bodies, Machine for felting and hardening	J. T. Earle	Danbury, Conn	Oct. 4, 1870	107, 887
Hat-bodies, Machine for forming	F. Degen	Newark, N. J	Dec. 10, 1867	71, 996
Hat-bodies, Machine for forming	R. Fitzgerald	Newark, N. J	Dec. 6, 1859	26, 395
Hat-bodies, Machine for forming	I. Gill	Walpole, Mass	Jan. 13, 1857	16, 426
Hat-bodies, Machine for forming	I. Nutt	Newark, N. J	Oct. 11, 1870	108, 175
Hat-bodies, Machine for forming	O. Root	Montgomery County, N. Y	Dec. 19, 1829	....
Hat-bodies, Machine for forming and constructing	B. Lapham and E. Cady	Canaan, N. Y	Apr. 15, 1829	....
Hat-bodies, Machine for forming and hardening	A. C. Arnold	Norwalk, Conn	Oct. 6, 1857	18, 316
Hat-bodies, Machine for fulling	A. Cattaneo	Newark, N. J	Mar. 3, 1868	75, 122
Hat-bodies, Machine for fulling and felting	W. M. Storm and G. H. Ennis	New York, N. Y., and Hudson County, N. J.	Nov. 3, 1868	83, 743
Hat-bodies, Machine for hardening	J. Booth	Newark, N. J	Sept. 15, 1857	18, 181
Hat-bodies, Machine for hardening	S. Boyden	Newark, N. J	Aug. 30, 1859	25, 300
Hat-bodies, Machine for making	D. Barnum	Philadelphia County, Pa	July 1, 1851	8, 195
Hat-bodies, Machine for making	A. Clark and H. Chase	Falmouth, Mass	June 11, 1829	....
Hat-bodies, Machine for making	E. F. Condit and A. Taylor	Springfield, N. J	Mar. 19, 1850	7, 182
Hat-bodies, Machine for making	J. F. Greene	Brooklyn, N. Y	Jan. 1, 1861	31, 053
Hat-bodies, Machine for manufacturing	J. Booth	Newark, N. J	Aug. 25, 1857	18, 034
Hat-bodies, Machine for manufacturing	L. E. Hopkins	New York, N. Y	Dec. 7, 1852	9, 450
Hat-bodies, Machine for molding, felting, and fulling	M. A. Johnson and W. Murkland.	Lowell, Mass	Oct. 23, 1866	59, 031
Hat-bodies, Machine for planking	P. Emmons	New York, N. Y	Apr. 19, 1853	9, 672
Hat-bodies, Machine for pouncing and napping	R. Eickemeyer	Yonkers, N. Y	Feb. 28, 1865	46, 552
Hat-bodies, Machine for shrinking	P. V. W. Bishop	Morristown, N. J	Nov. 7, 1871	120, 617
Hat-bodies, Machine for shrinking	J. S. Taylor	Danbury, Conn	May 3, 1853	9, 700
Hat-bodies, Machine for sizing	P. Emmons	New York, N. Y	July 24, 1855	13, 304
Hat-bodies, Machine for sizing	S. H. Gray	Bridgeport, Conn	Dec. 23, 1856	16, 305
Hat-bodies, Machine for sizing	W. E. Hatfield	Newark, N. J	Dec. 3, 1861	33, 867
Hat-bodies, Machine for sizing	W. H. Hoyt	Bethel, Conn	July 14, 1868	79, 833
Hat-bodies, Machine for sizing	S. C. Ketchum	Brooklyn, N. Y	May 27, 1856	14, 960
Hat-bodies, Machine for sizing	W. B. Lodge and H. Platner	Danbury, Conn., and Hudson, N. Y.	June 25, 1867	66, 097
Hat-bodies, Machine for sizing	A. Spencer	New York, N. Y	Mar. 25, 1856	14, 521
Hat-bodies, Machine for sizing	J. Thomas	Brooklyn, N. Y	June 17, 1856	15, 154
Hat-bodies, Machine for sizing	H. Warner	Newark, N. J	Jan. 14, 1868	73, 411
Hat-bodies, Machine for sizing	S. W. Wood	Washington, D. C	June 15, 1858	20, 602
Hat-bodies, Machine for sizing and felting	G. A. Mandeville	Newark, N. J	Oct. 25, 1870	108, 715
Hat-bodies, Machine for sizing or planking	W. B. Lodge and H. Platner	Danbury, Conn	June 25, 1867	66, 096
Hat-bodies, Machine for stretching	R. Eickemeyer	Yonkers, N. Y	Feb. 28, 1865	46, 553
Hat-bodies, Machine for stretching	R. Eickemeyer	Yonkers, N. Y	Jan. 21, 1873	135, 033
Hat-bodies, Machine for stretching	W. C. Griswold	Brooklyn, N. Y	Dec. 8, 1868	84, 690
Hat-bodies, Machine for stretching	P. Keffer	Reading, Pa	Mar. 17, 1868	75, 684
Hat-bodies, Machinery for felting	J. B. Blakslee and E. R. Barnes	Newtown and Brookfield, Conn.	Mar. 17, 1857	16, 823
Hat-bodies, Machinery for felting	J. B. Blakslee and S. S. Middlebrook.	Newtown, Conn	Jan. 10, 1860	26, 744
Hat-bodies, Machinery for felting	W. Fuzzard	Newark, N. J	Feb. 13, 1855	12, 386
Hat-bodies, Machinery for felting	M. R. Lemman	Columbus, Miss	May 8, 1860	28, 185
Hat-bodies, Machinery for felting	S. S. Middlebrook and J. B. and C. F. Blakslee.	Newtown, Conn	Jan. 30, 1855	12, 321
Hat-bodies, Machinery for felting	J. S. Taylor	Danbury, Conn	May 6, 1856	14, 845
Hat-bodies, Machinery for felting	J. Thomas	Brooklyn, N. Y	July 1, 1856	15, 261
Hat-bodies, Machinery for forming	P. Arneson	Newark, N. J	Aug. 23, 1859	25, 168
Hat-bodies, Machinery for forming	D. Barnum	New York, N. Y	Oct. 17, 1854	11, 805
Hat-bodies, Machinery for forming	S. Boyden	Newark, N. J	Jan. 25, 1859	22, 698
Hat-bodies, Machinery for forming	S. Boyden	Newark, N. J	Jan 10, 1860	26, 811
Hat-bodies, Machinery for forming	W. Fuzzard	Charlestown, Mass	Aug. 7, 1860	29, 478
Hat-bodies, Machinery for forming	M. Hardy	New York, N. Y	Dec. 14, 1858	22, 288
Hat-bodies, Machinery for forming	L. Platt and R. Wildman	Danbury, Conn	July 31, 1860	29, 441
Hat-bodies, Machinery for forming	A. B. Taylor	Newark, N. J	July 29, 1856	15, 443
Hat-bodies, Machinery for forming	J. S. Taylor	Danbury, Conn	Oct. 14, 1856	15, 903
Hat-bodies, Machinery for forming	J. S. Taylor	Danbury, Conn	Sept. 13, 1859	25, 454
Hat-bodies, Machinery for forming	T. Walber	New York, N. Y	Aug. 17, 1852	9, 205
Hat-bodies, Machinery for forming	D. G. Wells	New York, N. Y	Sept. 9, 1856	15, 715
Hat-bodies, Machinery for forming	R. Wildman	Hartford, Conn	July 20, 1846	4, 643
Hat-bodies, Machinery for hardening	G. E. Cowperthwait	Danbury, Conn	May 17, 1859	24, 007
Hat-bodies, Machinery for making	L. W. Boynton	South Coventry, Conn	Oct. 10, 1854	11, 773
Hat-bodies, Machinery for making	W. Fosket	Ware, Mass	Jan. 23, 1846	4, 363
Hat-bodies, Machinery for making	I. Searles	Newark, N. J	Nov. 7, 1854	11, 910
Hat-bodies, Machinery for making	A. B. Taylor	Newark, N. J	Mar. 18, 1856	14, 476
Hat-bodies, Machinery for making	A. B. Taylor and H. A. Burr	New York, N. Y	Feb. 9, 1847	4, 962
Hat-bodies, Machinery for making	H. A. Wells	New York, N. Y	Apr. 25, 1846	4, 472
Hat-bodies, Machinery for manufacturing	P. Arneson, J. Pedorson, and H. Rees.	New York, N. Y	Oct. 9, 1855	13, 614
Hat-bodies, Machinery for manufacturing	L. E. Hopkins	New York, N. Y	Dec. 21, 1852	9, 484
Hat-bodies, Machinery for perforating	W. F. Warburton and W. B. Aitken.	Philadelphia, Pa	Sept. 20, 1859	25, 551
Hat-bodies, Making	A. Rankin	Newark, N. J	Oct. 3, 1854	11, 756
Hat-bodies, Making a batting or web for	H. A. Wells, J. James, and R. W. Peck.	Brooklyn, N. Y	Sept. 22, 1837	399
Hat-bodies, Manufacture of	S. Lyon	Roxbury, Mass	Feb. 5, 1847	4, 951

Index of patents issued from the United States Patent Office from 1790 *to* 1873, *inclusive*—Continued.

Invention.	Inventor.	Residence.	Date.	No.
Hat-bodies, Manufacture of	H. Moore	New York, N. Y	Apr. 20, 1837	173
Hat-bodies, Manufacture of	A. B. Taylor	Newark, N. J	Aug. 12, 1856	15,534
Hat-bodies, Pouncing	W. H. Tupper	New York, N. Y	Nov. 15, 1859	26,132
Hat-bodies, Preparing material for	P. Arnesen, J. Pederson, and H. Rees.	New York, N. Y	Oct. 2, 1855	13,613
Hat-bodies, Rolling and felting	J. Crisp and V. V. Dodd	Bloomfield and Orange, N. J.	Oct. 17, 1871	119,970
Hat-bodies, Sizing	J. McCracken	Brooklyn, N. Y	Oct. 21, 1856	15,929
Hat-bodies, Steaming-on	S. Polley	Brooklyn, N. Y	Apr. 2, 1867	63,556
Hat-bodies, Stiffening	H. Blynn	Newark, N. J	May 9, 1835	....
Hat-bodies, Stiffening	J. P. Kettell and J. Wright	Worcester County, Mass	June 11, 1836	....
Hat-bodies, Stiffening	J. McCracken	Brooklyn, N. Y	Sept. 2, 1856	15,664
Hat-bodies, Stiffening	E. P. Spear	Lexington, Mass	July 2, 1836	....
Hat-bodies, Winding-machine for setting up	J. Grant	Providence, R. I	Apr. 28, 1825	....
Hat-bodies, Winding-machine for setting up	J. Grant	Providence, R. I	Apr. 10, 1827	....
Hat-bodies with fur, Process of facing wool	R. D. Hine	Matteawan, N. Y	May 19, 1868	78,090
Hat-body	G. Johnson	Camden, N. J	Dec. 24, 1872	134,206
Hat-body	W. Nunns	New York	Aug. 14, 1834	....
Hat-body	J. A. Pease	New York, N. Y	Jan. 22, 1867	61,451
Hat-body bat	S. Hurlbut	Glastenbury, Conn	Feb. 14, 1831	....
Hat-body formers, Cover for cone of	F. Degen and E. R. Parsil	Newark, N. J	Aug. 22, 1871	118,213
Hat-body machinery	A. B. Taylor	Newark, N. J	Aug. 31, 1858	21,382
Hat-body machinery	D. Tenny	Plattsburgh, N. Y	Oct. 1, 1830	....
Hat-box	C. W. Packer	Philadelphia, Pa	July 16, 1867	66,880
Hat-box and valise	Z. S. Durfee	Philadelphia, Pa	Oct. 16, 1866	58,792
Hat-box-cutting die	F. Schoettle	Philadelphia, Pa	Jan. 30, 1872	123,127
Hat-brim, Extension	I. Y. Cassiano	San Antonio, Tex	July 8, 1873	140,614
Hat-brim, Extension	I. Y. Cassiano	San Antonio, Tex	July 8, 1873	140,615
Hat brim, Soft	J. W. Corey	Newark, N. J	Nov. 1, 1870	108,884
Hat-brim spring	G. Mallory	Bridgeport, Conn	Dec. 4, 1866	60,213
Hat-brim spring	S. Peck	New Haven, Conn	Dec. 4, 1866	60,236
Hat-brims, Apparatus for forming	F. Degan	New York, N. Y	Apr. 17, 1847	5,065
Hat-brims, Block for cutting and shaping	F. R. Going	Boston, Mass	Mar. 24, 1868	75,897
Hat-brims, Block for setting	S. Billings	Spring Garden, Pa	Aug. 7, 1849	6,627
Hat-brims, Curling	F. Degen	New York, N. Y	Nov. 13, 1849	6,867
Hat-brims, Curling	F. S. Sibley	Brooklyn, N. Y	Oct. 9, 1860	30,379
Hat-brims, Curved spring for	E. S. Nichols	New Haven, Conn	Feb. 5, 1867	61,856
Hat-brims, Machine for forming	C. T. Rowe	Norwalk, Conn	Jan. 30, 1872	123,200
Hat-brims, Method of setting or ironing	F. Degen	New York	Aug. 31, 1842	2,765
Hat-brims, Pattern for trimming	C. M. Hawes	New York, N. Y	Aug. 4, 1868	80,732
Hat-brushing machine	C. Faure	New York, N. Y	June 13, 1865	48,235
Hat-brushing machine	J. W. Hopkins	Brooklyn, E. D., N. Y	Feb. 23, 1869	87,103
Hat-brushing machine	J. D. Parsons	Yonkers, N. Y	July 20, 1869	92,877
Hat brushing and napping machine	J. W. Hopkins	Brooklyn, N. Y	Mar. 2, 1869	87,495
Hats, Case for ladies' calash	H. Cook	Onondaga, N. Y	Feb. 20, 1833	....
Hat, Cassimere	O. Brooks and J. A. Sloan	Philadelphia, Pa	Nov. 9, 1842	2,847
Hat-check	J. E. Smith	Waterbury, Conn	May 10, 1870	102,982
Hat, Child's	S. B. Pratt	Boston, Mass	Sept. 17, 1872	131,371
Hat-conformator	J. Dickson, jr	Brooklyn, N. Y	June 5, 1860	28,564
Hat-conformator	B. Hogan	Albany, N. Y	Mar. 24, 1868	75,760
Hat-conformator	J. F. Klein	Trenton, N. J	Feb. 24, 1863	37,759
Hat-conformator	J. J. McKnight	Tarrytown, N. Y	Nov. 25, 1873	144,855
Hat-conformator	E. Z. Webster	Norwich, Conn	Dec. 15, 1868	85,041
Hat, Cork	A. C. Crondal	New York, N. Y	Nov. 8, 1864	44,941
Hat-cushion	A. D. Purinton	Dover, N. H	June 26, 1860	28,962
Hat, Elastic ventilating	D. Greenleaf	Vicksburgh, Miss	Nov. 26, 1836	90
Hat, Elastic water-proof	A. Buffum and J. Kelly	Smithfield, R. I	Feb. 17, 1820	....
Hat, Enameled	E. S. Cheney and G. P. Perry	Providence, R. I	July 20, 1869	92,704
Hat, Felt	J. T. Waring	Yonkers, N. Y	Aug. 6, 1872	130,342
Hat, Felt	J. W. Whittal and W. W. Pendleton.	Greenwich, Conn	Aug. 1, 1854	11,462
Hat, Felted	E. Sealy	Newark, N. J	May 18, 1869	90,198
Hat-felting apparatus	W. H. Hoyt	Bethel, Conn	Apr. 15, 1862	34,963
Hat-felting machine	J. F. Badoye	New York, N. Y	Oct. 22, 1867	70,066
Hat-felting machine	J. W. Blackham	Brooklyn, N. Y	Dec. 10, 1867	71,959
Hat-felting machine	A. Cataneo	Newark, N. J	June 21, 1870	104,424
Hat-felting machine	A. C. Fuller	Danbury, Conn	May 13, 1856	14,862
Hat-felting machinery	J. S. Taylor	Danbury, Conn	Jan. 15, 1856	14,121
Hat-finishing apparatus	J. A. Roche and J. J. Stewart	Williamsburgh, N. Y	Sept. 26, 1865	50,165
Hat-finishing machine	J. Cooper and T. Barnett	Philadelphia, Pa	May 31, 1826	....
Hat-finishing machine	J. H. La Ban	New York, N. Y	Jan. 22, 1861	31,177
Hat finishing machine	J. C. Richardson	Newark, N. J	Jan. 19, 1869	85,964
Hat-finishing machine	G. W. Stout	Newark, N. J	Nov. 8, 1870	109,072
Hat-finishing press	W. W., jr., and M. J. Walsh	Brooklyn, N. Y	Oct. 2, 1866	58,513
Hat, Fireman's	J. Brown	Newark, N. J	Nov. 1, 1845	4,246
Hat-joiner	L. Simonet	Paris, France	Apr. 2, 1867	63,435
Hat-forming die	R. T. Wilde and S. H. Lyon	Brooklyn, N. Y	Nov. 11, 1862	36,927
Hat-forming die	R. T. Wilde and S. H. Lyon	Brooklyn, N. Y	Mar. 24, 1863	37,993
Hat-forming die and block	L. Bommer	New York, N. Y	Nov. 3, 1868	83,597
Hat-forming machine	D. Dennis	Barre, Mass	Sept. 12, 1871	118,914
Hat-forming machine	H. N. Swift	Matteawan, N. Y	Mar. 13, 1866	53,196
Hat-forming mold	J. E. Ward	Bredbury, Great Britain	Dec. 22, 1868	85,191
Hat-frame	A. Komp	New York, N. Y	June 27, 1865	48,412
Hat-frame, Metallic skeleton	A. Komp	New York, N. Y	Apr. 25, 1865	47,431
Hat-frames, Machine for forming	B. Ertola and A. Caselli	New York, N. Y	Oct. 21, 1873	143,814
Hat-frames, Machine for pressing	L. P Faught	Foxborough, Mass	July 30, 1872	130,031
Hat-holder	Z. Waters	Bloomington, Ill	July 28, 1868	80,374
Hat-holder for pews and seats	J. M. Cain	La Fayette, Wis	Mar. 16, 1869	87,756
Hat-hook	J. Harvey	Scranton, Pa	Mar. 3, 1868	75,159
Hat-hook	C. L. Spencer	Providence, R. I	Oct. 27, 1868	83,562
Hat-hook for pews	R. W. Whitney and J. W. Shaw.	South Berwick, Me., and Concordia, N. H.	Aug. 6, 1867	67,620
Hat-ironing machine	W. Best	Newark, N. J	Mar. 27, 1866	53,401
Hat-ironing machine	W. Best and M. S. Drake	Newark, N. J	Mar. 27, 1866	53,518
Hat-ironing machine	M. S. Drake	Newark, N. J	Aug. 21, 1866	57,437
Hat-ironing machine	R. Eickemeyer	Yonkers, N. Y	Dec. 30, 1873	145,934
Hat-ironing machine	G. W. Stout and J. C. Richardson.	Newark, N. J	Dec. 3, 1867	71,662

Index of patents issued from the United States Patent Office from 1790 *to* 1873, *inclusive*—Continued.

Invention.	Inventor.	Residence.	Date.	No.
Hat-ironing machine	G. W. Stout and J. C. Richardson.	Newark, N. J	Dec. 15, 1868	84, 915
Hat-ironing machine	J. T. Waring	Yonkers, N. Y	Oct. 15, 1867	69, 950
Hat-ironing machinery	S. A. Kinsman and S. Field	Barre, Mass	Mar. 11, 1856	14, 401
Hat, Leather	A. Bogardus	Newburgh, N. Y	June 4, 1872	127, 550
Hat lining and tips, Ornamenting	T. W. Bracher	New York, N. Y	June 6, 1871	115, 689
Hat-linings, Embossing	T. W. Bracher	New York, N. Y	Jan. 10, 1871	110, 891
Hat-linings, Press for printing	W. Moultrie	New York, N. Y	Dec. 16, 1856	16, 245
Hat-machine	C. M. Osgood	Amherst, Mass	Jan. 17, 1871	111, 079
Hat machine, Paper	H. Kellogg	New Haven, Conn	June 4, 1867	65, 393
Hat, Man's	T. W. Adams	Baltimore, Md	Dec. 24, 1861	34, 043
Hat, Military	C. L. Pascal	Philadelphia, Pa	Dec. 10, 1861	33, 900
Hat, Military	W. F. Warburton	Philadelphia, Pa	Feb. 25, 1862	34, 538
Hat or bonnet	D. Scrymgeour	Foxborough, Mass	Dec. 5, 1871	121, 668
Hat or cap ventilator	W. Dale	New York, N. Y	Dec. 19, 1871	121, 995
Hat or hat-body felting machinery	J. B. Laville	Paris, France	Sept. 26, 1854	11, 731
Hat, Palm-leaf	F. Groening	Brooklyn, N. Y	June 25, 1836	....
Hat, Paper-pulp	J. A. Pease	New York, N. Y	Oct. 18, 1864	44, 744
Hat-planking machine	R. Bacon	Boston, Mass	Jan. 31, 1827	....
Hat-pouncing apparatus	E. Nougaret	Newark, N. J	Feb. 20, 1866	52, 739
Hat-pouncing machine	J. B. Brown	Danbury, Conn	Oct. 25, 1870	108, 562
Hat-pouncing machine	M. S. Drake	Newark, N. J	Feb. 5, 1867	61, 814
Hat-pouncing machine	R. Eickemeyer	Yonkers, N. Y	Nov. 23, 1869	97, 178
Hat-pouncing machine	J. L. Labiaux	Newark, N. J	Mar. 26, 1867	63, 261
Hat-pouncing machine	E. Mougaret	Newark, N. J	Sept. 18, 1866	58, 126
Hat-pouncing machine	A. Pelisse and G. W. Stoute	Newark, N. J	Apr. 4, 1871	113, 446
Hat-pouncing machine	J. C. Richardson	Newark, N. J	Jan. 7, 1868	73, 044
Hat-pouncing machine	J. C. Richardson	Newark, N. J	Aug. 17, 1869	93, 836
Hat-pouncing machine	J. Rosencranz	Boston, Mass	Sept. 28, 1869	95, 270
Hat-pouncing machine	P. W. Vail	Newark, N. J	Sept. 24, 1867	69, 277
Hat-pouncing machine	C. H. Van Houten and J. M. Crane.	Newark, N. J	Apr. 9, 1867	63, 674
Hat-pouncing machine	S. S. Wheeler	Danbury, Conn	Aug. 24, 1869	94, 158
Hat-pouncing machine	S. S. Wheeler and D. B. Manley	Danbury, Conn	Aug. 14, 1866	57, 232
Hat-press	G. Bartel	New York, N. Y	Aug. 29, 1871	118, 578
Hat-press	S. Beatty	Norwalk, Conn	Mar. 14, 1871	112, 677
Hat-press	R. Eickemeyer	Yonkers, N. Y	May 13, 1873	138, 871
Hat-press	R. Eickemeyer	Yonkers, N. Y	June 24, 1873	140, 335
Hat-press	G. C. Howard	Philadelphia, Pa	May 4, 1869	89, 834
Hat pressing and ironing apparatus	H. A. Robison	Cleveland, Ohio	July 22, 1862	35, 954
Hat-pressing apparatus	J. S. Giles, W. Halladay, and J. A. Rue.	New York and Brooklyn, N. Y.	Dec. 23, 1862	37, 255
Hat-pressing apparatus	M. Morse	Franklin, Mass	Mar. 6, 1866	53, 083
Hat-pressing apparatus	H. L. Sweet	Foxborough, Mass	Mar. 31, 1863	38, 067
Hat-pressing device	C. M. Osgood	Amherst, Mass	Feb. 4, 1873	135, 557
Hat-pressing die	J. Desper	Barre, Mass	June 10, 1873	139, 708
Hat-pressing die	J. Desper	Barre, Mass	Nov. 4, 1873	144, 262
Hat-pressing die	W. E. Doubleday and S. H. Lyon	Brooklyn, N. Y	Mar. 5, 1861	31, 592
Hat pressing machine	S. G. Congdon and D. C. Mowry	Mansfield and Milford, Mass.	Dec. 27, 1861	15, 528
Hat-pressing machine	J. De La Mar	Brooklyn, N. Y	Jan. 12, 1867	62, 018
Hat-pressing machine	W. E. George	Wrentham, Mass	June 30, 1868	79, 338
Hat-pressing machine	I. T. Green	Milford, Conn	Oct. 2, 1866	58, 544
Hat-pressing machine	E. B. Hastings and J. Desper	Palmer and Barre, Mass	Feb. 4, 1873	135, 428
Hat-pressing machine	N. B. Hooper	Newark, N. J	Apr. 10, 1866	53, 826
Hat-pressing machine	S. Howard	Luton, Great Britain	July 21, 1868	80, 181
Hat-pressing machine	M. Morse	Franklin, Mass	Aug. 28, 1866	57, 638
Hat-pressing machine	M. and C. H. Morse	Franklin, Mass	Oct. 1, 1867	69, 358
Hat-pressing machine	J. Stearns	Templeton, Mass	July 8, 1851	8, 213
Hat-pressing machine	G. L. Thompson	New York, N. Y	Nov. 27, 1866	60, 088
Hat-pressing machine	F. Wolfram	New York, N. Y	Mar. 9, 1869	87, 744
Hat-pressing machinery	D. Brown	Wrentham, Mass	July 22, 1873	140, 993
Hat-pressing machinery	B. Potter, jr	Templeton, Mass	Oct. 29, 1850	7, 745
Hat-protector	G. W. Bidwell	New Haven, Conn	Aug. 14, 1866	57, 074
Hat-protector	H. Hodgson	New York, N. Y	Oct. 23, 1866	59, 019
Hat-rack	N. Clarke	Boston, Mass	Feb. 18, 1873	136, 032
Hat-rack	C. H. Keener	Baltimore, Md	July 24, 1866	56, 569
Hat-rack	T. S. Lambert	Peekskill, N. Y	Oct. 6, 1863	40, 173
Hat-rack and seat	E. S. Blake	Pittsburgh, Pa	Aug. 14, 1866	57, 076
Hat-rack and umbrella-stand combined	J. W. Currier	Springfield, Mass	Mar. 12, 1872	124, 552
Hat-rack, Extension	N. Hayden	Chicago, Ill	Oct. 25, 1870	108, 590
Hat-rack, Folding	N. Hayden	Chicago, Ill	Dec. 26, 1871	122, 168
Hat-rack, Folding	M. L. Roddy	Cincinnati, Ohio	Oct. 29, 1872	132, 692
Hat-rack for cigars, &c	C. Branwhite	Williamsburgh, N. Y	Apr. 24, 1860	28, 034
Hat-rack for seats	P. Sylla	Elgin, Ill	Feb. 25, 1868	74, 953
Hat-rounder and hat-gage	S. Jones	Bridgeport, Pa	June 2, 1819	....
Hat, Safety	J. J. Giltenan	Cincinnati, Ohio	Feb. 23, 1869	87, 164
Hat-shade	M. L. Battle	Bainbridge, Ga	July 9, 1872	128, 780
Hat-shaper	A. and F. Brown	New York, N. Y	July 11, 1854	11, 302
Hat-shaping apparatus	S. H. Apple, B. Lindheim, and H. Baer.	New York, N. Y	Dec. 4, 1860	30, 791
Hat-shaping machine	G. W. Gallagiere and E. W. Ruby.	New Milford, N. Y	Apr. 13, 1869	88, 865
Hat-shaping machine	J. P. Morlat	Paris, France	Feb. 1, 1870	99, 458
Hat-shaping machine	S. Wing	Munson, Mass	Aug. 17, 1869	93, 783
Hat-shaping machine	R. Cooke	New Hartford, N. Y	Dec. 2, 1873	145, 052
Hat-shell iron	J. P. Kettell	Worcester, Mass	Apr. 1, 1862	34, 838
Hat sizing and felting machine	E. R. Gardiner	Brooklyn, N. Y	May 21, 1867	64, 966
Hat-sizing machine	J. H. Hopkins	Newark, N. J	Feb. 21, 1871	111, 935
Hat-sizing machine	A. Pelisse and F. Degen	Newark, N. J	Feb. 8, 1870	99, 700
Hat-sizing machine	J. S. Taylor	Danbury, Conn	May 28, 1867	65, 300
Hat, Spring-brim	S. Collins	New Haven, Conn	Nov. 8, 1864	44, 939
Hat-stand	J. B. Wickersham	New York, N. Y	June 2, 1857	17, 461
Hat steaming and hardening machine	N. Wildman	Danbury, Conn	Apr. 2, 1824	....
Hat-steaming apparatus	C. W. Banta	Orange, N. J	July 1, 1873	140, 341
Hat stretching and blocking machine	S. Polley	Brooklyn, N. Y	Oct. 6, 1868	82, 748
Hat-stretching apparatus	J. W. Blackham	Brooklyn, N. Y	Dec. 22, 1863	41, 036
Hat-stretching machine	R. Eickemeyer	Yonkers, N. Y	June 22, 1869	91, 730

Index of patents issued from the United States Patent Office from 1790 *to* 1873, *inclusive*—Continued.

Invention.	Inventor.	Residence.	Date.	No.
Hat-stretching machine	R. Eickemeyer	Yonkers, N. Y	July 15, 1873	140, 903
Hat-stretching machine	R. Eickemeyer	Yonkers, N. Y	July 29, 1873	141, 337
Hat-stretching machine	S. Goodman	Natick, Mass	Jan. 21, 1873	135, 112
Hat-stretching machine	A. Pelisse	Newark, N. J	Jan. 4, 1870	98, 624
Hat supporter and ventilator, Combined	J. A. Borthwick	Philadelphia, Pa	Jan. 31, 1871	111, 309
Hat-sweat	B. W. Fay	Boston, Mass	Sept. 2, 1862	36, 344
Hat-sweat	G. W. Thompson	Brooklyn, N. Y	Sept. 22, 1868	82, 364
Hat sweat-band	W. Burgyes	Chelsea, Mass	Apr. 22, 1862	35, 011
Hat-sweat from water-proof paper	W. M. Waterbury	New York, N. Y	Dec. 12, 1871	121, 829
Hat-tip	S. H. Greenbaum	New York, N. Y	Oct 26, 1869	96, 222
Hat tips and linings, Ornamentation of	T. W. Bracher	New York, N. Y	Apr. 9, 1872	125, 534
Hat-tips, Cutting	T. W. Bracher	New York, N. Y	Jan. 8, 1869	91, 077
Hat-tips, Lining	W. H. M. Pye	Brooklyn, N. Y	Aug. 12, 1873	141, 727
Hat-tips, Machine for pulling out	J. Stevens	Chicopee Falls, Mass	Feb. 5, 1867	61, 773
Hat-tips, Machinery for pulling out	W. Selpho	New York, N. Y	Apr. 24, 1847	5, 082
Hat-tips, Printing	M. J. Duffy	New York, N. Y	Oct. 1, 1872	131, 858
Hat-trimming, Machine for covering reeds for	G. A. Shephard	Bethel, Conn	May 3, 1870	102, 722
Hat, Ventilated	J. N. Genin	New York, N. Y	July 22, 1845	4, 118
Hat, Ventilated	W. F. Warburton	Philadelphia, Pa	Dec. 11, 1860	30, 900
Hat-ventilation	J. McMannus	New York, N. Y	Jan. 3, 1860	26, 692
Hat-ventilator	D. K. Albright and L. H. De Laage.	Philadelphia, Pa., and Bordentown, N. J.	Nov. 21, 1865	50, 999
Hat-ventilator	S. Beatty	Norwalk, Conn	May 31, 1870	103, 703
Hat-ventilator	S. Beatty	Norwalk, Conn	Aug. 1, 1871	117, 506
Hat-ventilator	F. H. Bell	Washington, D. C	July 10, 1860	29, 050
Hat-ventilator	T. W. Bracher	New York, N. Y	July 20, 1869	92, 785
Hat-ventilator	T. W. Bracher	New York, N. Y	Sept. 21, 1869	95, 074
Hat-ventilator	B. J. Burnett	Mount Vernon, N. Y	Mar. 27, 1866	53, 406
Hat-ventilator	C. H. Coffin	San Francisco, Cal	Oct. 15, 1867	69, 900
Hat-ventilator	W. Dale	New York, N. Y	Apr. 16, 1872	125, 725
Hat-ventilator	G. Deas	New York, N. Y	Apr. 21, 1868	77, 007
Hat-ventilator	P. N. Horsley	Jersey City, N. J	June 3, 1862	35, 446
Hat-ventilator	H. A. House	Bridgeport, Conn	Mar. 28, 1871	113, 054
Hat-ventilator	W. M. Irwin and A. H. Moses	Montgomery, Ala	Sept. 1, 1868	81, 784
Hat-ventilator	J. Jenkinson	Brooklyn, N. Y	Apr. 24, 1860	27, 985
Hat-ventilator	A. O. Knapp	Danbury, Conn	Apr. 23, 1872	125, 962
Hat-ventilator	A. Komp	New York, N. Y	Sept. 5, 1865	49, 767
Hat-ventilator	G. Munro	Philadelphia, Pa	July 6, 1869	92, 340
Hat-ventilator	E. G. Nichols	Beaufort, S. C	Oct. 24, 1871	120, 163
Hat-ventilator	J. Pollock	Morrisania, N. Y	Apr. 3, 1860	27, 736
Hat-ventilator	G. B. Smith	Detroit, Mich	July 23, 1872	129, 866
Hat-ventilator	W. Smith	Philadelphia, Pa	Oct. 3, 1865	50, 285
Hat-ventilator	A. Solmans	New York, N. Y	Apr. 18, 1871	113, 941
Hat-ventilator	C. C. Stremme	Austin, Tex	Mar. 30, 1869	88, 527
Hat-ventilator	W. F. Warburton	Philadelphia, Pa	Nov. 24, 1868	84, 323
Hat-ventilator	W. F. Warburton	Philadelphia, Pa	Nov. 22, 1870	109, 473
Hat-ventilator	A. H. Young	Milford, Mass.	Aug. 6, 1872	130, 173
Hat, Water-proof	R. Mills	Baltimore, Md	Feb. 27, 1832	....
Hat, Water-proof silk	G. B. Dexter	Boston, Mass	Jan. 6, 1836	....
Hat wiring and binding machine	H. A. Whiting	New York, N. Y	Apr. 29, 1873	138, 459
Hats and bonnets, Apparatus for forming	S. H. Lyon	New York, N. Y	Jan. 19, 1864	41, 342
Hats and bonnets, Fabric for	E. Copleston	Wrentham, Mass	Nov. 20, 1866	59, 822
Hats and bonnets, Fabric for	P. and A. Erhard	New York, N. Y	June 22, 1869	91, 614
Hats and bonnets, Fabric for	H. Loenberg	Boston, Mass	Aug. 5, 1862	36, 097
Hats and bonnets, Fabric for	H. Lowenberg	New York, N. Y	Feb. 28, 1865	46, 568
Hats and bonnets, Finishing leghorn	J. Snyder	Philadelphia, Pa	Aug. 21, 1834	....
Hats and bonnets, Manufacture of palm-leaf	D. C. Perrin	Roxbury, Mass	Jan. 13, 1863	37, 409
Hats and bonnets of horse-hair, Manufacture of	C. L. Noe	New York, N. Y	July 20, 1843	3, 188
Hats and bonnets, Pressing	S. E. Pettee	Foxborough, Mass	Feb. 6, 1855	12, 353
Hats and caps, Fabric for the manufacture of	T. Garceau and E. De La Granja	Boston, Mass	Mar. 10, 1868	75, 406
Hats and caps, Shaping and embossing	A. L. Bayley	Amesbury, Mass	Apr. 9, 1861	31, 944
Hats and caps, Ventilating	J. O. Blythe	Germantown, Pa	Aug. 20, 1861	33, 121
Hats and caps, Ventilating sweat-leather for	O. A. Taylor	San Francisco, Cal	Oct. 11, 1864	44, 697
Hats and caps, Water-proof sweat-band for	P. F. Lenhart	Brooklyn, N. Y	June 28, 1870	104, 861
Hats, and frame-block for, Dyeing and siring	N. L. Canfield	Wilton, Conn	Mar. 14, 1834	....
Hats and making cloth without yarn, Machine for planking.	J. Pitkin and T. Kimball	Hartford, Conn	Oct. 19, 1807	....
Hats, Apparatus for attaching mourning-badge to	T. H. Lowerre	New York, N. Y	June 27, 1865	48, 417
Hats, Apparatus for making brush	J. C. Raake	Brooklyn, N. Y	Apr. 26, 1864	42, 508
Hats, Attaching wire to brims of	C. F. Bosworth	Milford, Conn	Aug. 18, 1868	81, 246
Hats, Blocking	D. Beard	Guilford, N. C	May 28, 1816	....
Hats, Blocking	W. Wright	Philadelphia, Pa	Oct. 10, 1848	5, 849
Hats, bonnets, &c., Compound for coating textile fabrics for manufacture of.	J. L. Kendall and R. H. Trested	Foxborough, Mass., and Jamaica, N. Y.	Feb. 9, 1869	86, 841
Hats, bonnets, &c., Machine for ironing and pressing	R. Murdock	Baltimore, Md	June 17, 1840	1, 635
Hats, bonnets, &c., Machine to cut chips or strips of wood to make chip.	J. Roberts, A. D. Allen, and E. Kelsey.	....	Sept. 5, 1804	....
Hats, bonnets, &c., Machine to cut strips or chips of wood to make chip.	A. D. Allen	....	May 10, 1804	....
Hats, bundles, &c., Rack for	C. A. Young	Providence, R. I	Mar. 12, 1872	124, 469
Hats by application of hot water, Napping	H. Wortham and G. S. Petty	Lancaster, Ky	July 9, 1832	....
Hats by application of steam, Making	W. Harkins	Wheeling, Va	Dec. 29, 1826	....
Hats, &c., by machinery, Mode of planking wool	L. L. Macomber	Bath, Me	Dec. 31, 1825	....
Hats by steam, Manufacture of	D. Sutton	Lancaster, Ky	June 13, 1831	....
Hats, caps, &c., Application of bark for	S. G. Whipple	Hallowell, Mass	Apr. 17, 1807	....
Hats, caps, bonnets, &c., Making	E. Pratt	Cambridge, Mass	May 16, 1835	....
Hats, Carding fur and wool for	A. Buffum	....	Feb. 21, 1807	....
Hats, cloths, &c., from a fleece or sheet taken from a carding-engine without bowing, Manufacturing.	R. Pitkin	East Hartford, Conn	Apr. 23, 1808	....
Hats, &c., Coloring	J. Taylor	Danbury, Conn	Nov. 20, 1823	....
Hats, Coloring woolen	G. D. Foote	Danbury, Conn	Sept. 27, 1859	25, 564
Hats, Combined sweat-band, fastener, and size-mark for.	T. A. Lawrence and J. H. Murfey.	New York, N. Y	Mar. 17, 1868	75, 685
Hats, Composition for stiffening	C. Bent and F. Bush	Chelmsford, Mass	Nov. 3, 1831	....
Hats, Composition for stiffening wool and fur	L. L. Macomber	Gardiner, Me	Jan. 19, 1831	....

Index of patents issued from the United States Patent Office from 1790 *to* 1873, *inclusive*—Continued.

Invention.	Inventor.	Residence.	Date.	No.
Hats, Composition of materials for the manufacture of.	F. Messer	Guildhall, Vt	Jan. 23, 1812	
Hats, Cooling and coloring	R. Pike	Wilton, Conn	Jan. 23, 1834	
Hats, Dyeing	A. Gould	Washington, Conn	Jan. 11, 1836	
Hats, Embossing	R. Eickemeyer	Yonkers, N. Y	Feb. 8, 1870	99, 544
Hats, Felting	M. D. Whipple	Cambridge, Mass	Dec. 1, 1868	84, 599
Hats, Felting and napping	T. J. Cornell	Randolph, Vt	Feb. 11, 1831	
Hats, Finishing palm-leaf	D. Dennis	Barre, Mass	July 4, 1854	11, 214
Hats, Forming bell-crown	W. E. Doubleday	Brooklyn, N. Y	Feb. 17, 1863	37, 682
Hats from a mixture of clothier's flocks with sheep's wool, Manufacturing felt.	J. Carleton	Salem, N. H	Jan. 21, 1814	
Hats, fur, &c., Process of coloring	H. Hibbard	Attica, N. Y	May 25, 1838	746
Hats, Gearing of cones for bowing	T. F. Mayhew	Boston	Aug. 22, 1827	
Hats, Machine for basoning, hardening, and steaming.	J. Sizer	New London, Conn	Nov. 6, 1811	
Hats, Machine for carding, winding, and making.	I. Sanford	Blockley, Pa	Oct. 11, 1828	
Hats, Machine for coating	J. F. Mathias and D. M. Legat.	Paris, France	June 18, 1867	65, 926
Hats, Machine for coloring	G. M. Johnson	Port Deposit, Md	Dec. 31, 1839	1, 454
Hats, Machine for forming the brims of felt	W. A. Fenn	New Milford, Conn	Apr. 14, 1857	17, 033
Hats, Machine for lustering	R. Eickemeyer	Yonkers, N. Y	July 29, 1873	141, 339
Hats Machine for making bats or frames for wool.	R. Gookins	Hampshire, Mass	Jan. 24, 1806	
Hats, Machine for making bodies of fur	H. A. Wells, J. James, and R. W. Peck.	Brooklyn, N. Y	Sept. 20, 1837	383
Hats, Machine for making felt in the manufacture of fur.	H. S. Miller	Philadelphia, Pa	Mar. 5, 1839	1, 094
Hats, Machine for making leather and India-rubber.	S. Gleason	New York, N. Y	Nov. 24, 1843	3, 357
Hats, Machine for napping	A. Rankin	Newark, N. J	Sept. 20, 1839	1, 328
Hats, Machine for polishing and equalizing the nap or pile on the surface of.	J. Loudon and T. Shaw	New York, N. Y	May 7, 1845	4, 033
Hats, Machine for pressing bell-crowned	L. P. Faught	Foxborough, Mass	June 25, 1872	128, 293
Hats, Machine for pressing palm-leaf	C. Gorham	Barre, Mass	Mar. 3, 1840	1, 503
Hats, Machine for pressing straw or chip	S. Prince	New York, N. Y	Aug. 31, 1808	
Hats, Machine for printing	T. Byrne and T. Henry	New York, N. Y	May 2, 1865	47, 516
Hats, Machine for raising and drying the nap of	M. Esselen	Roxbury, Mass	Aug. 13, 1867	67, 740
Hats, Machine for stretching the brims of	J. Sheldon	Newark, N. J	May 23, 1871	115, 114
Hats, Machine for washing and cleaning	W. Carlock	Baltimore, Md	Mar. 12, 1830	
Hats, Machinery for forming brims of felt	W. A. Fenn	Brookfield, Conn	Jan. 19, 1858	19, 138
Hats, Machinery for hardening	R. Wildman	Charlestown, Mass	Feb. 26, 1856	14, 330
Hats, Machinery for setting up	J. Grant	Providence, R. I	Aug. 11, 1821	
Hats, Making	W. E. Doubleday	Brooklyn, N. Y	Oct. 2, 1866	58, 391
Hats, Making	W. Goodrich	Philadelphia, Pa	Aug. 20, 1816	
Hats, Making paper water-proof	B. Grut	New York, N. Y	Oct. 1, 1830	
Hats, Making printed felt	A. Bailey	Amesbury, Mass	Apr. 18, 1865	47, 269
Hats, Manufacturing	A. Bancker and C. F. Alvord.	New York, N. Y	Jan. 9, 1849	6, 010
Hats, Manufacturing	J. W. Beebe	Brooklyn, N. Y	Mar. 4, 1856	14, 343
Hats, Manufacturing	J. and J. Bryan	Saratoga, N. Y	Aug. 26, 1815	
Hats, Manufacturing	E. Corning, jr		June 13, 1803	
Hats, Manufacturing	L. E. Hopkins	Brooklyn, N. Y	June 5, 1855	13, 005
Hats, Manufacturing	S. Hurlbut	Glastenbury, Conn	June 13, 1831	
Hats, Manufacturing	J. Johnson	New Orleans, La	Jan. 8, 1856	14, 062
Hats, Manufacturing	H. Kellogg	New Haven, Conn	Feb. 12, 1867	61, 037
Hats, Manufacturing	J. Maguire	Washington, D. C	Dec. 7, 1844	3, 851
Hats, Manufacturing	A. Rankin	Newark, N. J	June 5, 1855	13, 013
Hats, Manufacturing	T. Sealy	Newark, N. J	Dec. 24, 1867	72, 546
Hats, Manufacturing	J. Taylor and C. Brown	Danbury, Conn	Jan. 13, 1832	
Hats, Manufacturing	S. Tibballs, jr	Tyringham, Mass	Dec. 26, 1810	
Hats, Manufacturing	W. F. Warburton	Philadelphia, Pa	Aug. 2, 1859	24, 968
Hats, Manufacturing	W. F. Warburton and C. C. Lovett.	Philadelphia, Pa	Apr. 10, 1860	27, 848
Hats, Manufacturing cemented	N. Weston	Reading, Mass	May 24, 1816	
Hats, Manufacturing felt	J. Monach	Rahway, N. J	May 10, 1859	23, 934
Hats, Manufacturing felt	G. H. Scribner	Natick, Mass	Jan. 21, 1870	104, 652
Hats, Manufacturing leather	J. S. and W. Wibirh	Eden, N. Y	Sept. 4, 1841	2, 241
Hats, Manufacturing light	H. Lainhart, jr	Baltimore, Md	July 6, 1809	
Hats, Manufacturing military	S. Hoskin	Catskill, N. Y	Apr. 30, 1810	
Hats, Manufacturing napped	L. Lyon, 2d	Needham, Mass	May 16, 1834	
Hats, Manufacturing soft	B. McNamee	Boston, Mass	July 16, 1872	129, 356
Hats, Method of extracting alcohol used in stiffening	W. McCay	Northumberland, Pa	June 22, 1832	
Hats, Method of surfacing felt	A. Hurd	Danbury, Conn	Feb. 10, 1857	16, 588
Hats, &c., Method of treating straw-braid for	G. Cornwall, 2d	Milford, Conn	May 19, 1857	17, 318
Hats, Mold for forming wool and rorams, &c., into bats for.	J. Werely	Albany, N. Y	Oct. 14, 1813	
Hats, Napping	J. Dart, jr., W. Wells, and D. Olmsted.	Buffalo, N. Y	May 12, 1832	
Hats, Napping	J. Long		Aug. 5, 1799	
Hats, Napping	L. Lyon, 2d	Needham, Mass	Apr. 13, 1836	
Hats of leather, skins, &c., Machinery for forming, &c.	R. Fish	New York, N. Y	Oct. 12, 1844	3, 790
Hats of paper, Making	J. Ford and J. White	Boston, Mass	Nov. 25, 1811	
Hats, Planking	J. Gregg	Greene County, Pa	Mar. 14, 1810	
Hats, Planking	G. and E. Page	Manchester, Conn	Nov. 9, 1832	
Hats, Planking and felting	J. Swan	Haverhill, N. H	Mar. 7, 1811	
Hats, Preparing seal's fur for	G. Cleveland		Sept. 9, 1803	
Hats, Pressing	J. F. Mathias	Paris, France	Jan. 19, 1864	41, 311
Hats, Pressing leghorn, straw, and other	R. Tyler and B. P. Coston	Philadelphia, Pa	May 23, 1826	
Hats, Process of treating muslin for sweat-linings, &c., for.	W. P. Wright	Philadelphia, Pa	Apr. 14, 1868	76, 685
Hats, Rounding-jack for trimming brims of	S. Polley	Brooklyn, N. Y	Mar. 26, 1867	63, 173
Hats, Scalding and napping	D. Baldwin	Ithaca, N. Y	Oct. 15, 1829	
Hats, Scalding and napping	A. P. Gregory	Ithaca, N. Y	May 13, 1830	
Hats, Size-mark for	L. C. Woehming	New York, N. Y	Apr. 20, 1869	89, 108
Hats, Sizing and napping	G. Henning	Ithaca, N. Y	Apr. 28, 1831	
Hats, &c., Sizing for	H. E. Pond	Franklin, Mass	Dec. 8, 1863	40, 886
Hats, Stiffening for	A. and G. V. Raymond	Baltimore, Md	Feb. 7, 1818	
Hats, Stiffening for	J. D. Wilson	New York	Aug. 6, 1829	
Hats, Stiffening for water-proof	G. W. Downs	Circleville, Ohio	Oct. 25, 1832	
Hats, Sweat-leather for	D. Shive	Philadelphia, Pa	May 14, 1867	64, 717
Hats, Sweat-leather ventilator for	H. Curtis and A. Tufts	Charlestown, Mass	June 24, 1862	35, 672

Index of patents issued from the United States Patent Office from 1790 *to* 1873, *inclusive*—Continued.

Invention.	Inventor.	Residence.	Date.	No.
Hats, Ventilating	J. W. Beebe	New York, N. Y	Mar. 16, 1858	19, 616
Hats, Ventilating	G. W. Cherry	Alexandria, D. C	July 26, 1845	4, 122
Hats, Ventilating	R. Halloran	New York, N. Y	Mar. 21, 1846	4, 427
Hats, Ventilating	T. R. Johnson	Montreal, Canada	Aug. 11, 1868	80, 963
Hats, Ventilating	A. Maginnis	Philadelphia, Pa	May 17, 1859	24, 035
Hats, Ventilating	A. Maginnis	Philadelphia, Pa	Sept. 11, 1860	29, 984
Hats, Ventilating	W. Sellers	New York, N. Y	July 31, 1855	13, 361
Hats, Ventilating sweat-leather for	J. Pollock	Morrisania, N. Y	Nov. 13, 1860	30, 643
Hats, Washing	S. Dinkhouse	Eastonborough, Pa	Jan. 30, 1832	....
Hats, Water-proof stiffening for	S. Hempstead, jr	Saint Charles County, Mo	May 25, 1827	....
Hats, Water-proof stiffening for	S. Hempstead, jr	Saint Charles County, Mo	Oct. 26, 1827	....
Hats, Water-proof stiffening for	J. H. Tilgo	Washington, D. C	Sept. 23, 1819	....
Hats, Water-proof stiffening for fur and wool	W. Buckles	Baltimore, Md	Nov. 28, 1817	....
Hats, &c., Weaving grass for making	L. Burnap	Merrimack, N. H	Feb. 16, 1823	....
Hats with rabbit's fur, Napping	D. Evans	Alexandria, Va	Oct. 2, 1820	....
Hatch	D. Talcot	New York, N. Y	May 9, 1854	10, 899
Hatch for vessels	R. C. Holmes	Cape May, N. J	Mar. 7, 1846	4, 397
Hatch for warehouses, Safety	A. D. Medlicott	Windsor Locks, Conn	Nov. 19, 1867	71, 199
Hatch of vessels, Securing	E. S. Keyser	New York, N. Y	Dec. 8, 1857	18, 816
Hatch, Safety	G. N. Creamer	Trenton, N. J	May 26, 1868	78, 265
Hatch, Safety	R. H. Martin	Staten Island, N. Y	July 12, 1868	77, 744
Hatch, Self-closing	C. H. Reynolds	Williamsburgh, N. Y	Feb. 11, 1873	135, 728
Hatch, Ship's	J. R. Rich	Tremont, Me	Mar. 30, 1869	88, 330
Hatchet	S. Daugherty	Belle Vernon, Pa	Sept. 10, 1872	131, 211
Hatchet	N. F. English	Hartland, Vt	Apr. 27, 1858	20, 052
Hatchet	J. Jenkins	Philadelphia, Pa	July 13, 1869	92, 532
Hatchet	M. E. Rudasill	Shelby, N. C	Nov. 20, 1860	30, 692
Hatchet	R. H. Thompson	Troy, N. Y	Nov. 26, 1872	133, 343
Hatchet	D. E. Weaver	Cheviot, Ohio	Nov. 12, 1872	132, 996
Hatchet-blanks, Die for shaping	J. Yerkers	Philadelphia, Pa	Mar. 4, 1873	136, 479
Hatchet, Carpenter's	J. T. Shank	Martinsburgh, W. Va	Nov. 26, 1867	71, 416
Hatchet, hammer, and scraper, Combined	A. Barbarin	New Orleans, La	Oct. 13, 1868	83, 021
Hatchet-heads, Machine for swaging	L. Dodge	Cohoes, N. Y	Sept. 1, 1857	18, 087
Hatching chickens by artificial heat	N. E. Guerin	New York, N. Y	Mar. 30, 1843	3, 019
Hatchway	G. Follett and A. Brummel	Brooklyn, N. Y	May 27, 1873	139, 307
Hatchway	C. I. Russell	Chicago, Ill	Dec. 24, 1872	134, 169
Hatchway	J. D. Sinclair	Brooklyn, N. Y	Nov. 24, 1868	84, 387
Hatchway-door, Metallic	T. Hyatt	New York, N. Y	Dec. 2, 1873	145, 184
Hatchway for elevators	J. Wayland	Jersey City, N. J	Dec. 24, 1872	134, 334
Hatchway-guard	E. H. Ball	New York, N. Y	July 2, 1872	128, 523
Hatchway-guard	H. H. Covert	New York, N. Y	Nov. 2, 1869	96, 398
Hatchway-guard	M. J. Hinden	Detroit, Mich	Feb. 25, 1873	136, 161
Hatchway-guard, Automatic	G. E. Berry and F. C. Pingree	Detroit, Mich	Nov. 11, 1873	144, 432
Hatchway opening and closing apparatus	E. C. Ford	New York, N. Y	Sept. 25, 1860	30, 134
Hatchway-protector, Automatic	J. B. Waring and J. W. Wilson	New York, N. Y	Nov. 26, 1872	133, 345
Hatchway protector, Elevator	J. W. Meaker	Detroit, Mich	May 20, 1873	139, 014
Hatchway safety-gate	J. W. Meaker	Detroit, Mich	Sept. 3, 1872	131, 109
Hatchway safety guard	Z. S. B. Weeks and C. L. Kohler	New York, N. Y	Mar. 5, 1872	124, 301
Hatchway, Self-closing	W. P. Cherrington	Boston, Mass	Dec. 16, 1873	145, 626
Hatchway, Self-closing	A. B. Lee	Yonkers, N. Y	Sept. 9, 1873	142, 738
Hatchway, Self-closing	A. Reid	New York, N. Y	July 22, 1873	141, 171
Hatchways, Apparatus for opening and closing	H. Sizer and E. Stone	New York, N. Y., and Lowell, Mass.	Oct. 2, 1855	13, 624
Hatchways, Closing	J. S. Baldwin	Newark, N. J	May 20, 1873	138, 984
Hatchways, Closing	J. M. Bradner	Williamsport, Pa	July 23, 1872	129, 646
Hatchways, Closing	E. M. Hackett	New York, N. Y	July 1, 1873	140, 499
Hatchways, Operating shutters for	J. H. McKernan	Indianapolis, Ind	May 10, 1870	102, 954
Hatchways, Safety-attachment for	J. Bridge	Augusta, Me	June 16, 1857	17, 551
Hatchways, Safety-device for	A. Fries	Cincinnati, Ohio	Feb. 7, 1871	111, 628
Hatters' cards or jacks	J. C. Seeley	Dutchess County, N. Y	Mar. 15, 1827	....
Hatters' finishing-irons, &c., Furnaces for heating	G. V. Raymond	Richmond, Va	Sept. 29, 1825	....
Hatters' iron-making die	B. Hull	Westport, Conn	May 1, 1866	54, 356
Hatters' irons, Furnace for heating	D. L. Tuthill	New York, N. Y	Nov. 8, 1831	....
Hatters, Machine for bowing wool for	H. S. Holladay and E. G. Griffen	Lyme, Conn	June 6, 1812	....
Hatters, Machine for bowing wool or fur for	E. C. Griffing	Lyme, Conn	Aug. 10, 1815	....
Hatters, Making bow-strings for	D. Nichols	New Milford, Conn	Mar. 1, 1809	....
Hatters, Mode of preparing fur for	A. Buffum	Troy, Mass	Sept. 28, 1825	....
Hatters' plank-kettle	S. Fisher	Herkimer County, N. Y	July 15, 1816	....
Haversack	T. Keech	New York, N. Y	Sept. 22, 1863	40, 047
Hawse-hole, Ship's	R. R. Osgood	Troy, N. Y	May 5, 1857	17, 228
Hawse-pipe	P. Burnham	Gloucester, Mass	Oct. 8, 1867	69, 625
Hawse-pipe	P. Moody	Gloucester, Mass	May 21, 1867	64, 997
Hawse-pipe	C. Perley	New York, N. Y	Nov. 27, 1860	30, 755
Hawse-pipe	T. J. Southard	Richmond, Me	Mar. 13, 1860	27, 482
Hawse-pipe for ships	A. S. Phillips	Boston, Mass	July 5, 1859	24, 702
Hawse-pipe stopper	O. B. Bearse	Hyannis, Mass	June 14, 1870	104, 101
Hawse-pipe stopper	C. Perley	New York, N. Y	Aug. 9, 1864	43, 793
Hawse-pipe stopper	J. Stewart	Bangor, Me	Mar. 12, 1867	62, 902
Hawser-clamp	H. H. Ellis and C. F. Gladding	Norwich, Conn	June 9, 1868	78, 794
Hay and cotton press	A. Adams	Jerseyville, Ill	Nov. 28, 1854	11, 986
Hay and cotton press	H. A. Ashley	Springfield, Ohio	Jan. 6, 1863	37, 370
Hay and cotton press	S. T. Baker	West Gorham, Me	Mar. 20, 1835	....
Hay and cotton press	J. C. Baldwin	Stanton, Va	Feb. 22, 1839	1, 086
Hay and cotton press	R. Ball	Petersburgh, Va	June 28, 1870	104, 689
Hay and cotton press	W. C. Banks	Como Depot, Miss	Oct. 12, 1869	95, 631
Hay and cotton press	H. Barnes	Blairsville, Pa	Nov. 30, 1858	22, 216
Hay and cotton press	C. J. Beasley	Petersburgh, Va	July 5, 1870	104, 922
Hay and cotton press	J. Berkeley	Washington, Tex	Oct. 5, 1869	95, 416
Hay and cotton press	B. R. Brown and J. Toone, jr	Jackson, Tenn	May 31, 1870	103, 557
Hay and cotton press	S. W. Bullock	Catskill, N. Y	Mar. 23, 1842	2, 507
Hay and cotton press	S. Q. Carey	Waxahatchie, Tex	Nov. 10, 1868	83, 926
Hay and cotton press	N. Chapman	Hopedale, Mass	Dec. 13, 1870	110, 011
Hay and cotton press	N. Chapman	Hopedale, Mass	Jan. 17, 1871	111, 043
Hay and cotton press	R. H. and G. F. Cole	Greenport, N. Y	Mar. 15, 1870	100, 861
Hay and cotton press	L. L. Cummings	Munnsville, N. Y	June 15, 1858	20, 551
Hay and cotton press	M. G. Cunningham	Corsicana, Tex	Dec. 27, 1870	110, 442
Hay and cotton press	J. K. Davis	Monticello, S. C	Dec. 27, 1870	110, 443
Hay and cotton press	E. M. Day and J. T. Noel	Lower Lake, Cal	Feb. 20, 1866	52, 786

Index of patents issued from the United States Patent Office from 1790 *to* 1873, *inclusive*—Continued.

Invention.	Inventor.	Residence.	Date.	No.
Hay and cotton press	L. Dederick	Albany, N. Y	May 16, 1854	10, 920
Hay and cotton press	L. Dederick	New York, N. Y	May 25, 1869	90, 431
Hay and cotton press	G. N. Doolittle	Louisville, Ky	June 23, 1863	38, 950
Hay and cotton press	B. F. Dunning	Galesburgh, Ill	Mar. 14, 1865	46, 785
Hay and cotton press	E. Evans	Montgomery, Ala	Jan. 12, 1869	85, 808
Hay and cotton press	C. D. Findlay and D. D. Craig	Macon, Ga	Feb. 21, 1871	111, 923
Hay and cotton press	D. Flagg and C. Peck	Gardiner, Me	Dec. 30, 1826	....
Hay and cotton press	J. Garfield	Groton, Mass	Apr. 23, 1867	64, 093
Hay and cotton press	J. Grosvenor	New York, N. Y	Aug. 31, 1837	369
Hay and cotton press	J. K. Harris	Allensville, Ind	Aug. 20, 1861	33, 083
Hay and clothes press	G. W. Hart	Aurora, Ind	Dec. 15, 1863	40, 930
Hay and cotton press	C. W. Hawkes	Brunswick, Me	May 29, 1841	2, 111
Hay and cotton press, &c	G. A. Hotchkiss	Pleasant Township, Ind	July 15, 1862	35, 913
Hay and cotton press	P. C. Ingersoll	Green Point, N. Y	July 28, 1863	39, 365
Hay and cotton press	S. Ingersoll	Green Point, N. Y	Apr. 15, 1856	14, 663
Hay and cotton press	I. James	Mattoon, Ill	Feb. 16, 1864	41, 625
Hay and cotton press	J. H. Johnson	Griffin, Ga	Oct. 25, 1870	108, 599
Hay and cotton press	D. Knowles	Philadelphia, Pa	July 11, 1871	116, 965
Hay and cotton press	N. J. Lampman	Coxsackie, N. Y	Oct. 17, 1846	4, 818
Hay and cotton press	C. Lent	Washington, D. C	Feb. 18, 1868	74, 703
Hay and cotton press	E. G. Maxey and W. R. Mason	Troy, Ind., and Lewisport, Ky.	Jan. 21, 1868	73, 542
Hay and cotton press	W. H. McBurney	Sacramento City, Cal	May 13, 1873	138, 909
Hay and cotton press	M. McElroy	Springfield, Mass	Sept. 3, 1867	68, 568
Hay and cotton press	J. A. McGillivrae and C. O. Wheeler.	Mattison, Ill	Oct. 12, 1869	95, 707
Hay and cotton press	J. W. McIntyre	Philadelphia, Pa	Apr. 9, 1872	125, 471
Hay and cotton press	D. L. Miller	Madison, N. J	Mar. 20, 1860	27, 556
Hay and cotton press	S. Miller	Mount Union, Pa	Mar. 8, 1870	100, 546
Hay and cotton press	M. D. Moore	Brooklyn, N. Y	Feb. 6, 1866	52, 495
Hay and cotton press	J. Müller	Nacogdoches, Tex	Oct. 14, 1873	143, 586
Hay and cotton press	P. L. Negley	Castleton, Ind	Oct. 3, 1871	119, 528
Hay and cotton press	C. H. Parshall	Detroit, Mich	Dec. 19, 1865	51, 615
Hay and cotton press	I. N. Patten and D. G. Marden	Memphis, Tenn	Jan. 25, 1870	99, 228
Hay and cotton press	J. Peevy	Passadumkeag, Me	Jan. 1, 1856	14, 031
Hay and cotton press	J. B. Pugh	Champaign, Ill	Dec. 17, 1872	134, 097
Hay and cotton press	W. Read	Greenwood, Cal	Sept. 22, 1863	40, 064
Hay and cotton press	T. D. and F. R. Reed and C. Dorrel.	Rising Sun, Ind	May 22, 1866	54, 956
Hay and cotton press	W. Ridemour and M. K. Biser	Springfield, Ohio	Mar. 8, 1864	41, 893
Hay and cotton press	J. Robertson	Gosport, Ind	Sept. 8, 1868	82, 036
Hay and cotton press	C. A. Robinson	Florence, Ind	Mar. 21, 1871	112, 966
Hay and cotton press	C. H. Robinson	Bath, Me	Aug. 18, 1863	39, 588
Hay and cotton press	E. Rock	Greenvale, N. Y	May 23, 1871	115, 106
Hay and cotton press	J. S. Schofield	Macon, Ga	Sept. 3, 1867	68, 462
Hay and cotton press	P. A. Shearer	Morristown, Tenn	Jan. 28, 1873	135, 291
Hay and cotton press	W. E. Sheffield	Saint Joseph, Mo	Aug. 17, 1869	93, 914
Hay and cotton press	J. Simpson	Chester, S. C	Aug. 31, 1869	94, 347
Hay and cotton press	J. J. Sivley	Clarksville, Tex	Nov. 21, 1871	121, 133
Hay and cotton press	J. N. Smith	Jersey City, N. J	Jan. 16, 1866	52, 084
Hay and cotton press	G. W. Stewart	Atlanta, Ga	Apr. 22, 1873	138, 210
Hay and cotton press	G. W. Swift	Memphis, Tenn	Mar. 8, 1870	100, 687
Hay and cotton press	G. W. Swinebroad	Bolivar, Tenn	Aug. 16, 1870	106, 427
Hay and cotton press	B. F. Taft	Groton Junction, Mass	Aug. 4, 1868	80, 681
Hay and cotton press	E. Thomas	Craigsville, Va	June 8, 1869	90, 970
Hay and cotton press	G. Utley	Charlotte, N. C	May 12, 1868	77, 852
Hay and cotton press	G. Utley	Charlotte, N. C	Aug. 23, 1870	106, 638
Hay and cotton press	M. Wallace	Little Rock, Ark	Aug. 30, 1870	106, 803
Hay and cotton press	H. R. Walton	Philadelphia, Pa	Apr. 26, 1870	102, 463
Hay and cotton press	C. O. Wheeler	Mattison, Ill	Feb. 13, 1872	123, 754
Hay and cotton press	W. H. Whetstone	Lowndesborough, Ala	June 4, 1872	127, 666
Hay and cotton press	J. L. White	Hernando, Miss	Feb. 7, 1871	111, 706
Hay and cotton presses, Mode of fastening door of	P. C. Ingersoll	Green Point, N. Y	June 16, 1863	38, 928
Hay and dung forks, &c., Making	C. Lawrence	Chatham, N. Y	July 23, 1823	....
Hay and grain, Apparatus for stacking	W. F. Browne	Washington, D. C	June 4, 1867	65, 471
Hay and grain cock cover	E. R. Whitney	Plattsburgh, N. Y	Nov. 24, 1868	84, 453
Hay and grain covering	F. Wingate	Hallowell, Me	Mar. 19, 1833	....
Hay and grain elevator	J. Dennis	Oswego, N. Y	Dec. 21, 1869	98, 042
Hay and grain loader	W. J. Webber	Hollister, Cal	Aug. 12, 1873	141, 840
Hay and grain protector	A. J. Frisbie	Saint Mary's, Ohio	Oct. 2, 1866	58, 401
Hay and grain rack	R. Cobb	Hadley, Mich	Oct. 7, 1862	36, 604
Hay and grain rake, Revolving hand	S. Coats	Shoreham, Vt	Apr. 17, 1837	163
Hay and grain roof, Portable	J. J. Naylon	Brighton, Mich	Feb. 4, 1868	74, 120
Hay and grain, Thatching for stacks of	R. McLarn	Shirland, Pa	Mar. 2, 1869	87, 425
Hay and hemp press	G. W. Penniston	North Vernon, Ind	Sept. 23, 1862	36, 525
Hay and lime elevator	A. and G. D. Thomas	Saint Thomas, Pa	May 3, 1870	102, 729
Hay and manure fork	J. K. Babcock	Shortsville, N. Y	Apr. 30, 1867	64, 272
Hay and manure fork	G. L. Barton and A. E. Roberts	Albany, N. Y	Nov. 24, 1857	18, 670
Hay and manure fork	W. Jones	Speedsville, N. Y	Aug. 25, 1857	18, 047
Hay and manure fork	A. J. Martin	Catskill, N. Y	July 7, 1868	79, 586
Hay and manure fork	E. Moore	Brooklyn, N. Y	July 27, 1869	92, 989
Hay and manure fork	R. A. Peet	Caledonia Township, Mich	July 5, 1870	104, 993
Hay and manure fork	L. D. Pitcher	Pitcherville, Ill	May 25, 1869	90, 463
Hay and manure fork	F. Villard	Mount Eaton, Ohio	Feb. 15, 1870	99, 982
Hay and manure fork, Combined	F. Villard	Mount Eaton, Ohio	Aug. 22, 1871	118, 410
Hay and manure forks, Fastening for	A. Clark	Southfield, N. Y	Mar. 5, 1850	7, 134
Hay and manure forks, Shank of	R. M. Hine	Mentz, N. Y	Jan. 3, 1854	10, 384
Hay and other forks, Manufacture of	G. B. Ely	Saint Johnsbury, Vt	Sept. 24, 1867	69, 196
Hay and pruning knife	J. Fasig	West Salem, Ohio	Jan. 30, 1866	52, 279
Hay and pruning knife combined	J. Fasig	West Salem, Ohio	May 3, 1870	102, 524
Hay and straw cutter	A. C. Both	Hesse-Cassel, Germany	May 31, 1870	103, 552
Hay and straw cutter	C. D. and W. S. Brewer	Lewisburgh, Pa	Dec. 13, 1864	45, 385
Hay and straw cutter	J. H. Dickinson	Chicopee Falls, Mass	June 4, 1867	65, 355
Hay and straw cutter	J. D. Felthousen	Michigan City, Ind	Sept. 18, 1860	30, 056
Hay and straw cutter	H. R. Hawkins	Akron, Ohio	Nov. 20, 1860	30, 677
Hay and straw cutter	T. Hazard	Wilmington, Ohio	Aug. 30, 1864	43, 988
Hay and straw cutter	W. Henshaw	Spencer, Mass	Apr. 3, 1866	53, 614
Hay and straw cutter	J. N. Neff	Strasburgh, Pa	Nov. 27, 1860	30, 750

Index of patents issued from the United States Patent Office from 1790 *to* 1873, *inclusive*—Continued.

Invention.	Inventor.	Residence.	Date.	No.
Hay and straw cutter	F. Pohlman	Coxsackie, N. Y	June 3, 1862	35, 468
Hay and straw cutter	D. H. Whittemore	Lynchburgh, Va	Sept. 4, 1860	29, 930
Hay and straw cutting machine	C. Brown	Buffalo, N. Y	May 26, 1868	78, 181
Hay and straw elevator	J. H. Gill	Mount Pleasant, Ohio	Jan. 12, 1858	19, 087
Hay and straw for feed for cattle and horses, Preparing.	C. Brown	Buffalo, N. Y	Sept. 3, 1867	68, 345
Hay and straw stacker	C. Rundell	Chicago, Ill	Apr. 24, 1866	54, 213
Hay and straw stacking apparatus	D. W. Baird	Lebanon, Tenn	Sept. 24, 1872	131, 590
Hay-binder	A. R. Clark	Onondaga, N. Y	Nov. 15, 1870	109, 176
Hay-binder	M. A. Dilley	Le Roy, Kans	Sept. 24, 1872	131, 509
Hay-carrier	C. S. Boothby	Saco, Me	May 3, 1870	102, 651
Hay-carrier	J. B. Drake	Indianapolis, Ind	Apr. 2, 1872	125, 183
Hay-carrier	T. J. Powell	Naples, N. Y	July 26, 1870	105, 723
Hay-carrier	H. C. Stouffer	Beaver Township, Ohio	Dec. 1, 1868	84, 591
Hay-carrier	E. B. Tanner	Attica, Ohio	May 13, 1873	138, 955
Hay, Carrier for unloading	G. Smith	Rochester, N. Y	June 20, 1871	116, 230
Hay, cider, and oil press	N. Whitney	Augusta, Me	Feb. 15, 1826	....
Hay-cock protector	O. R. Dinsmore	Auburn, N. H	Mar. 23, 1858	19, 689
Hay-crane	M. Simms and J. V. Chambers.	Wheeling, W. Va	Jan. 14, 1868	73, 263
Hay-crane or carrier	B. P. Barackman	Lineville, Pa	Feb. 8, 1870	99, 520
Hay-cutter	F. Gerfen	West Hempfield Township, Pa.	May 20, 1868	78, 369
Hay-cutter	H. Kinsey, F. W. Kissell, and J. E. and J. M. Smith.	Ligonier, Pa	Nov. 17, 1868	84, 063
Hay-cutter	J. G. Perry	South Kingston, R. I	July 14, 1863	39, 242
Hay, &c., Cutting and pressing	R. Wakeman and J. L. Ballance	Port Deposit, Md	May 23, 1865	47, 883
Hay, &c., cutting machine	T. H. and D. T. Willson	Harrisburgh, Pa	Nov. 6, 1860	30, 597
Hay-derrick	I. Van Voorhis	Hillsborough, Pa	Oct. 1, 1867	69, 375
Hay, Derrick for stacking	T. G. Palmer	Greenville, N. Y	Aug. 23, 1864	43, 925
Hay, Derrick for stacking	S. Turner	Onargo, Ill	Aug. 23, 1864	43, 936
Hay, Device for mowing away	L. E. Palmer	Le Roy, Pa	Apr. 6, 1869	88, 664
Hay, Double box for pressing	E. Waterhouse	Gardiner, Me	Nov. 14, 1826	....
Hay-drag	D. and A. Eddelman	Madison, Ind	Feb. 15, 1870	99, 766
Hay elevating and unloading apparatus	H. Buck	Polo, Ill	Feb. 18, 1873	136, 029
Hay-elevating apparatus	J. Bolles	Jackson, Ohio	Oct. 31, 1871	120, 365
Hay-elevating apparatus	J. F. Nugent	Cannonsburgh, Mich	Apr. 7, 1868	76, 506
Hay-elevating device	F. Calvert	Wabash City, Ind	Sept. 12, 1871	118, 791
Hay-elevating fork	L. A. Beardsley	South Edmeston, N. Y	Sept. 11, 1860	29, 948
Hay-elevating fork	L. A. Beardsley	South Edmeston, N. Y	Sept. 17, 1861	33, 288
Hay-elevating fork	T. T. Calkins and D. E. Wing	Coxsackie, N. Y	Aug. 16, 1864	43, 832
Hay-elevating fork	C. E. and J. N. Gladding	Troy, Pa	May 11, 1858	20, 241
Hay-elevating fork	J. B. Hawley	Albany, N. Y	Aug. 23, 1864	43, 946
Hay-elevating fork	G. C. Howard and I. N. Wilfong	Philadelphia, Pa	Mar. 24, 1863	39, 965
Hay-elevating fork	D. F. Neikirk	Republic, Ohio	Feb. 23, 1864	41, 715
Hay-elevating fork	W. S. Newton	Norwich, Conn	Jan. 24, 1865	46, 018
Hay-elevating fork	E. Reynolds	Corunna, Mich	Dec. 27, 1864	45, 634
Hay-elevating fork	L. Rundell	Coxsackie, N. Y	Apr. 7, 1863	38, 129
Hay-elevating fork	L. Rundell	New Baltimore, N. Y	May 5, 1863	38, 420
Hay-elevating fork	E. W. Seymour and G. W. Gregory.	Centre Lisle and Binghamton, N. Y.	June 9, 1863	38, 872
Hay-elevating fork	R. J. Stanley	Mount Morris, N. Y	Dec. 8, 1863	40, 860
Hay-elevating fork	S. Wheeler	Albany, N. Y	Feb. 23, 1864	47, 733
Hay-elevating fork	J. A. Whitney	Maryland, N. Y	Aug. 23, 1864	43, 939
Hay-elevating machine	J. C. McGrew	Smithfield, Ohio	Dec. 21, 1858	22, 372
Hay-elevator	T. H. Arnold	Troy, Pa	Aug. 10, 1869	93, 579
Hay-elevator	T. V. Bagly	Jones' Station, Ind	Apr. 18, 1871	113, 837
Hay-elevator	A. S. Brown	Delphi, Ind	Mar. 19, 1872	124, 787
Hay-elevator	E. H. Carpenter	Dexter, Mich	Dec. 10, 1867	71, 850
Hay-elevator	A. Chapman	Delta, N. Y	July 2, 1872	128, 537
Hay-elevator	F. A. Crane	Zanesville, Ohio	Oct. 27, 1868	83, 479
Hay-elevator	W. Derr	Tiffin, Ohio	Aug. 10, 1869	93, 423
Hay-elevator	J. M. Dick	Buffalo, N. Y	July 7, 1863	39, 128
Hay-elevator	F. R. Dufour	Vevay, Ind	Nov. 30, 1869	97, 282
Hay-elevator	W. E. Durkee	Fort Edward, N. Y	June 12, 1860	28, 713
Hay-elevator	P. C. Ellsworth	Venice, N. Y	Oct. 2, 1866	58, 541
Hay-elevator	F. F. Fowler	Crane Township, Ohio	Apr. 17, 1860	27, 899
Hay-elevator	R. Furnas	Friendswood, Ind	Dec. 20, 1870	110, 224
Hay-elevator	J. Gifford, jr	Watertown, N. Y	Nov. 13, 1866	59, 585
Hay-elevator	C. E. Gladding	Towanda, Pa	June 27, 1871	116, 300
Hay-elevator	E. C. Green	Plainfield, Ind	Mar. 19, 1867	63, 037
Hay-elevator	T. Harter	Ilion, N. Y	Oct. 24, 1871	120, 273
Hay-elevator	G. F. Hipp and J. B. Fast	Nova, Ohio	Sept. 3, 1867	68, 508
Hay-elevator	J. L. Hubbell and E. Sherman	Fairfield, Conn	Sept. 25, 1866	58, 259
Hay-elevator	T. T. Jarrett	Horsham, Pa	May 30, 1854	10, 989
Hay-elevator	J. H. Jenkins	Upper Sandusky, Ohio	Nov. 28, 1865	51, 190
Hay-elevator	J. L. Kintner	Harrison, Ind	Apr. 23, 1867	64, 111
Hay-elevator	C. H. Kirkpatrick	La Fayette, Ind	Sept. 9, 1873	142, 635
Hay-elevator	J. Linderman	Bullville, N. Y	Sept. 27, 1870	107, 695
Hay-elevator	W. Louden	Fairfield, Iowa	Mar. 17, 1868	75, 686
Hay-elevator	O. E. Mabie	Camden, N. Y	May 4, 1869	89, 673
Hay-elevator	H. and L. M. McCown	Enon Valley, Pa	Jan. 7, 1868	73, 190
Hay-elevator	H. and L. M. McCown	Enon Valley, Pa	Sept. 22, 1868	82, 427
Hay-elevator	M. Mitchell	Crown Point, Ind	Jan. 1, 1867	60, 773
Hay elevator	A. Mott	Scott, N. Y	Jan. 21, 1868	73, 544
Hay-elevator	M. D. Myers	Ilion, N. Y	July 15, 1862	35, 887
Hay-elevator	J. W. Odaniel	Cloverdale, Ind	Dec. 21, 1869	98, 097
Hay-elevator	W. L. Overhiser	Stockton, Cal	July 23, 1867	66, 990
Hay-elevator	N. Palmer	Greenville, N. Y	Sept. 30, 1862	36, 572
Hay-elevator	H. G. Porter	Grand Rapids, Mich	Mar. 3, 1868	75, 193
Hay-elevator	G. W. Prout	Ashland, N. Y	Feb. 16, 1864	41, 640
Hay-elevator	H. H. Quint	Brookfield, Mo	Aug. 22, 1871	118, 270
Hay-elevator	E. M. Rees	Norristown, Pa	Aug. 10, 1858	21, 150
Hay-elevator	S. Rogers	Pittsburgh, Pa	Jan. 24, 1865	46, 027
Hay-elevator	S. B. Secrist and I. Seyster	Ogle County, Ill	Aug. 3, 1869	93, 236
Hay-elevator	A. Smith	Shellsburgh, Pa	Oct. 5, 1869	95, 611
Hay-elevator	H. C. Stouffer	East Lewiston, Ohio	Nov. 16, 1869	96, 996
Hay-elevator	J. B. Summerill	Penn's Grove, N. J	Mar. 24, 1868	75, 810
Hay-elevator	J. M. Taber	Greenwich, N. Y	Apr. 22, 1862	35, 045

Index of patents issued from the United States Patent Office from 1790 *to* 1873, *inclusive*—Continued.

Invention.	Inventor.	Residence.	Date.	No.
Hay-elevator	N. C. Thomas and J. H. Coe	Brighton, Mich	Oct. 13, 1868	83, 007
Hay-elevator	E. J. Toof	Fort Madison, Iowa	May 9, 1865	47, 671
Hay-elevator	J. M. Van Demark and N. Barlow.	Phelps, N. Y	May 2, 1871	114, 496
Hay-elevator	E. L. Walker	Bendford's Store, Pa	Sept. 6, 1864	44, 129
Hay-elevator	G. Wilbur	Macedon, N. Y	July 17, 1837	271
Hay-elevator	E. L. Yancy	Batavia, N. Y	Nov. 9, 1869	96, 647
Hay elevator and carrier	W. S. Neil	Greensborough, Pa	July 4, 1871	116, 622
Hay elevator and carrier	J. H. White	Columbus City, Iowa	May 21, 1872	127, 003
Hay elevator and carrier, Track-rope	E. H. Carpenter	Dexter, Mich	Feb. 20, 1866	52, 676
Hay elevator and conveyer	T. E. Haymond	Morris, Ill	Dec. 13, 1870	110, 135
Hay elevator and conveyer	N. D. Hinman	Stepney Depot, Conn	Oct. 23, 1866	59, 018
Hay elevator and conveyer	C. A. Miller	Marengo Township, Mich	July 19, 1870	105, 477
Hay elevator and conveyer	T. I. Powell	Naples, N. Y	Jan. 26, 1869	86, 179
Hay elevator and loader	J. H. Bean	Macon, Ill	Sept. 16, 1873	142, 836
Hay-elevator and manure-drag combined	P. H. Stauffer	Lehighton, Pa	Jan. 19, 1869	86, 109
Hay elevator and stacker	A. W. Trcoker	Harvard, Ill	July 11, 1865	48, 742
Hay-elevator clutch	N. D. Hinman	Stepney Depot, Conn	Nov. 12, 1867	70, 717
Hay-elevator fork	B. G. Fox	Pricetown, Pa	Aug. 15, 1871	118, 121
Hay-elevator or stack-builder	F. Wicks	Kansas, Ill	Oct. 18, 1864	44, 761
Hay-elevator sling	G. Smith	Rochester, N. Y	June 20, 1871	116, 231
Hay-elevators, Tackle-block for	F. Wicks and F. F. Fowler	Upper Sandusky, Ohio	Aug. 28, 1866	57, 610
Hay fed to stock, Device for saving the seed from	R. A. Campbell	Salem, Ind	Nov. 16, 1858	22, 062
Hay for baling, Device for preparing	G. H. Nye	Monmouth, Ill	Jan. 10, 1865	45, 853
Hay for market, Apparatus for preparing	S. Colahan	Cleveland, Ohio	May 22, 1866	54, 864
Hay for pressing, Machine for cutting	O. and C. Waste	Cameron, Ill	Jan. 26, 1864	41, 416
Hay-fork	C. S. Ambruster	Woodstown, N. J	Sept. 1, 1868	81, 726
Hay-fork	H. F. Bernendefer and D. H. Finch.	Attica, Ohio	Apr. 21, 1868	76, 980
Hay-fork	A. Clark	Southfield, N. Y	Oct. 8, 1850	7, 697
Hay-fork	J. Dampman	Lebanon, Pa	Dec. 28, 1869	98, 236
Hay-fork	E. G. Dorchester and U. Scott	Geneva, N. Y	Dec. 8, 1868	84, 712
Hay-fork	C. L. Driesslein	Chicago, Ill	July 11, 1865	48, 665
Hay-fork	W. H. Elliot	New York, N. Y	Nov. 13, 1866	59, 570
Hay-fork	T. Foster	Coxsackie, N. Y	Sept. 29, 1863	40, 098
Hay-fork	J. G. Hitchcock	New York, N. Y	Jan. 7, 1868	73, 099
Hay-fork	W. Jones	Speedsville, N. Y	Jan. 13, 1857	16, 388
Hay-fork	J. E. Lobdell and L. H. Smith	Centre Lisle, N. Y	Feb. 1, 1870	99, 449
Hay-fork	J. A. Montgomery	Williamsport, Pa	Jan. 23, 1866	52, 186
Hay-fork	F. L. Morrison	New Albany, Ind	June 29, 1869	91, 959
Hay-fork	G. W. Robinson	New Wilmington, Pa	Aug. 25, 1868	81, 541
Hay-fork	L. Rogers	Pittsburgh, Pa	Jan. 11, 1870	98, 711
Hay-fork	W. F. Rundell	Genoa, N. Y	Mar. 28, 1865	47, 042
Hay-fork	W. F. Rundell	Genoa, N. Y	Nov. 27, 1866	60, 065
Hay-fork	J. B. Stewart	South Paris, Me	May 1, 1866	54, 435
Hay-fork	E. W. Walton and A. J. Brown	Stockton, Cal	Nov. 2, 1869	96, 517
Hay fork and carrier	J. B. Drake	Indianapolis, Ind	Aug. 6, 1872	130, 286
Hay-fork hook	T. W. Peirce	Minneapolis, Minn	Sept. 20, 1870	107, 531
Hay-fork, Elevator	J. F. H. Brown	Greenup, Ill	Apr. 25, 1871	113, 975
Hay-fork grapple	J. K. O'Neil and J. F. Thomas	Kingston and Herkimer County, N. Y.	Jan. 12, 1869	85, 844
Hay-fork, Horse	B. F. Alexander	Glen Hope, Pa	June 29, 1869	91, 998
Hay-fork, Horse	D. W. Amos	Broad Top City, Pa	Jan. 14, 1868	73, 222
Hay-fork, Horse	T. H. Arnold	Troy, Pa	May 29, 1866	55, 035
Hay-fork, Horse	T. H. Arnold	Troy, Pa	Nov. 13, 1866	59, 539
Hay-fork, Horse	T. H. Arnold	Troy, Pa	Sept. 10, 1867	68, 683
Hay-fork, Horse	G. Atherton	Sparta Township, Mich	Feb. 18, 1873	136, 016
Hay-fork, Horse	L. Atwater	Ithaca, N. Y	Jan. 14, 1868	73, 223
Hay-fork, Horse	C. N. Baldwin	Wilmington, Conn	Oct. 25, 1870	108, 551
Hay-fork, Horse	W. D. Ballard	Davisburgh, Mich	Aug. 31, 1869	94, 174
Hay-fork, Horse	C. Bean	Pawtucket, R. I	May 2, 1871	114, 391
Hay-fork, Horse	S. Bebout	Waterford, Ohio	Oct. 11, 1870	108, 094
Hay-fork, Horse	H. J. Beemer	Honesdale, Pa	Nov. 10, 1868	83, 821
Hay-fork, Horse	J. R. Benedict	Marion, N. Y	May 21, 1872	126, 866
Hay-fork, Horse	W. H. Berdan	York, Mich	Mar. 16, 1869	87, 902
Hay-fork, Horse	D. S. Bigler	Siddonstown, Pa	May 18, 1869	90, 149
Hay-fork, Horse	M. D. Birge	Grand Rapids, Mich	Sept. 10, 1867	68, 690
Hay-fork, Horse	C. C. Blodgett	Watertown, N. Y	Dec. 18, 1866	60, 467
Hay-fork, Horse	C. C. Blodgett	Watertown, N. Y	Aug. 13, 1867	67, 711
Hay-fork, Horse	D. S. Blue	Tremont, Ohio	May 7, 1867	64, 403
Hay-fork, Horse	L. L. Bond	Chicago, Ill	Dec. 3, 1867	71, 575
Hay-fork, Horse	S. B. Bowen	Stockton, Cal	Aug. 19, 1873	141, 851
Hay-fork, Horse	G. W. Bowlsby	Monroe, Mich	July 28, 1868	80, 442
Hay-fork, Horse	J. Bradley	Racine, Wis	June 8, 1869	91, 078
Hay-fork, Horse	A. C. Briggs	North Easton, N. Y	July 10, 1866	56, 170
Hay-fork, Horse	W. D. Brooks	Bethany, Pa	Sept. 29, 1868	82, 485
Hay-fork, Horse	A. S. Brown	Lebanon, Pa	May 18, 1869	90, 233
Hay-fork, Horse	B. F. Brown	Catlin, Ind	Mar. 21, 1871	112, 777
Hay-fork, Horse	J. S. Brown	Washington, D. C	July 17, 1866	56, 359
Hay-fork, Horse	J. S. Brown and W. F. Browne	Washington, D. C	Apr. 9, 1867	63, 608
Hay-fork, Horse	E. and A. Buckman	East Greenbush, N. Y	Jan. 23, 1866	52, 134
Hay-fork, Horse	J. H. Bunton	Thornbury, Pa	Nov. 17, 1868	84, 084
Hay-fork, Horse	B. S. Burgan	Congress, Ohio	Nov. 3, 1868	83, 598
Hay-fork, Horse	J. R. Cadwell	Dexter, Mich	Dec. 20, 1864	45, 472
Hay-fork, Horse	J. R. Cadwell	Dexter, Mich	Aug. 1, 1865	49, 083
Hay-fork, Horse	W. Carlton	Adrian, Mich	Nov. 26, 1867	71, 453
Hay-fork, Horse	J. B. Carothers	Pine Grove Mills, Pa	Sept. 19, 1871	119, 120
Hay-fork, Horse	W. Carroll	Hillsdale, Mich	Nov. 12, 1867	70, 695
Hay-fork, Horse	A. Y. Case	Dexter, Mich	Dec. 19, 1865	51, 555
Hay-fork, Horse	A. M. Cheney and H. B. Kimball.	Charlotte, Mich	Dec. 1, 1868	84, 477
Hay-fork, Horse	S. Clark	Howard, N. Y	Feb. 1, 1870	99, 292
Hay-fork, Horse	D. B. Clement	Brooklyn, N. Y	Feb. 9, 1864	41, 484
Hay-fork, Horse	D. B. Clement	Brooklyn, N. Y	Jan. 24, 1865	46, 047
Hay-fork, Horse	A. Coates	Watertown, N. Y	Nov. 19, 1867	71, 137
Hay-fork, Horse	M. Coffin	Milton, Ky	Nov. 5, 1867	70, 527
Hay-fork, Horse	W. S. Coffman	Coldwater, Mich	June 23, 1868	79, 106
Hay-fork, Horse	A. J. Cook	Guilford, Conn	Nov. 19, 1867	70, 962

Index of patents issued from the United States Patent Office from 1790 to 1873, inclusive—Continued.

Invention.	Inventor.	Residence.	Date.	No.
Hay-fork, Horse	S. Cook	Harrison, Ill	Sept. 17, 1872	131, 394
Hay-fork, Horse	A. J. Cooley	Chardon, Ohio	July 17, 1866	56, 375
Hay-fork, Horse	J. A. Cowles	Chicago, Ill	Feb. 21, 1865	46, 450
Hay-fork, Horse	F. Cramer	Chess Springs, Pa	Aug. 10, 1869	95, 520
Hay-fork, Horse	E. G. Crandell	Belfast, N. Y	Jan. 18, 1870	98, 930
Hay-fork, Horse	J. Crandell	Ilion, N. Y	Oct. 25, 1864	44, 788
Hay-fork, Horse	T. C. Craven	Albany, N. Y	June 26, 1866	55, 824
Hay-fork, Horse	C. N. Culver	Bowling Green, Ohio	June 5, 1866	55, 246
Hay-fork, Horse	F. Culver	Elkland, Pa	Dec. 17, 1867	72, 270
Hay-fork, Horse	J. S. Culver	Springport, N. Y	Nov. 12, 1867	70, 700
Hay-fork, Horse	J. Cummins	Perry, Mich	Mar. 9, 1869	87, 643
Hay-fork, Horse	J. and E. C. Cummins	Perry, Mich	Nov. 16, 1869	96, 894
Hay-fork, Horse	J. J. De Grummond	Knoxville, Ill	Nov. 23, 1869	97, 171
Hay-fork, Horse	M. Dennis	Barton, N. Y	Dec. 22, 1868	85, 167
Hay-fork, Horse	W. E. Derrick	Jordan, N. Y	Nov. 24, 1868	84, 269
Hay-fork, Horse	W. E. Derrick	Jordan, N. Y	Jan. 12, 1869	85, 733
Hay-fork, Horse	W. E. Derrick	Jordan, N. Y	July 6, 1869	92, 175
Hay-fork, Horse	W. E. Derrick	Jordan, N. Y	June 14, 1870	104, 124
Hay-fork, Horse	H. E. Dewey	Aurora, Ill	Jan. 21, 1868	73, 515
Hay-fork, Horse	H. L. Doane	Green Oak, Mich	June 23, 1868	79, 058
Hay-fork, Horse	L. M. Doudna	Elmira, N. Y	Apr. 23, 1867	64, 083
Hay-fork, Horse	G. H. Dow	Freeport, Ill	Nov. 24, 1868	84, 271
Hay-fork, Horse	J. B. Drake	Picture Rocks, Pa	July 31, 1866	56, 851
Hay-fork, Horse	J. Drinkwater	Adams, Ohio	June 30, 1868	79, 328
Hay-fork, Horse	A. T. Dunbar and J. H. Fellows.	Alba, Pa	Nov. 15, 1870	109, 305
Hay-fork, Horse	J. A. Eberly and H. Wechter	Reamstown Station, Pa	July 13, 1869	92, 434
Hay-fork, Horse	F. Ebert	Saxonburgh, Pa	May 11, 1869	89, 920
Hay-fork, Horse	J. T. Elliott	Grand Rapids, Mich	Dec. 29, 1868	85, 437
Hay-fork, Horse	H. A. Estes	Jersey City, N. J	Feb. 4, 1868	74, 064
Hay-fork, Horse	O. Evans	Alliance, Ohio	Sept. 24, 1867	69, 084
Hay-fork, Horse	E. J. Fenn	Medina, Ohio	July 7, 1868	79, 644
Hay-fork, Horse	E. J. Fenn	Medina, Ohio	Sept. 21, 1869	95, 012
Hay-fork, Horse	B. Field	Three Rivers, Mich	Dec. 24, 1872	134, 265
Hay-fork, Horse	D. Figge	Jenner's Cross Roads, Pa	Apr. 14, 1866	57, 110
Hay-fork, Horse	H. Fisher	Canton, Ohio	July 31, 1866	56, 739
Hay-fork, Horse	H. Fisher and M. Ball	Canton, Ohio	Mar. 20, 1866	53, 286
Hay-fork, Horse	E. Forney and J. Swab	Elizabethville, Pa	Jan. 18, 1870	98, 946
Hay-fork, Horse	F. Foster	Coxsackie, N. Y	Mar. 20, 1866	53, 288
Hay-fork, Horse	F. F. Fowler	Upper Sandusky, Ohio	Aug. 14, 1866	57, 114
Hay-fork, Horse	R. S. Frame	Washington, D. C	Dec. 1, 1868	84, 540
Hay-fork, Horse	R. S. Frame	Washington, Ohio	Mar. 8, 1870	100, 518
Hay-fork, Horse	W. B. Gabel	East Cocalico Township, Pa.	Dec. 17, 1867	72, 282
Hay-fork, Horse	J. S. Gage	Dowagiac, Mich	Feb. 9, 1864	41, 499
Hay-fork, Horse	D. M. Garrett	Shelby, Ohio	Aug. 29, 1865	49, 621
Hay-fork, Horse	S. L. Gates	Verona, N. Y	Aug. 25, 1863	39, 641
Hay-fork, Horse	S. L. Gates	Verona, N. Y	Sept. 6, 1864	44, 087
Hay-fork, Horse	C. H. Gifford	Philadelphia, Pa	Oct. 8, 1867	69, 651
Hay fork, Horse	W. M. Gillan	Mount Parnel, Pa	Nov. 3, 1868	83, 705
Hay-fork, Horse	E. Gilliam	Allegheny, Pa	June 5, 1866	55, 278
Hay-fork, Horse	J. Gilmore	Phœnixville, Pa	Dec. 31, 1867	72, 837
Hay-fork, Horse	J. Gilmore	Phœnixville, Pa	Oct. 27, 1868	83, 482
Hay-fork, Horse	B. F. Gladding	Providence, R. I	Feb. 9, 1869	86, 661
Hay-fork, Horse	C. E. Gladding	Troy, Pa	Jan. 7, 1868	73, 179
Hay-fork, Horse	J. A. Glenn	West Middlesex, Pa	Dec. 1, 1868	84, 544
Hay-fork, Horse	J. A. Glenn	West Middlesex, Pa	May 18, 1869	90, 256
Hay-fork, Horse	G. D. S. Gochnauer	Mulberry, Pa	Feb. 9, 1869	86, 831
Hay-fork, Horse	J. S. Gochnauer	York, Pa	Dec. 17, 1867	72, 286
Hay-fork, Horse	A. Gordon	Rochester, N. Y	Oct. 9, 1866	58, 635
Hay-fork, Horse	H. N. Green	Fort Wayne, Ind	Nov. 22, 1870	109, 509
Hay-fork, Horse	C. S. Grimes	Lancaster, Ky	Mar. 22, 1870	101, 122
Hay-fork, Horse	J. T. Hall	Trenton, N. Y	Apr. 9, 1867	63, 719
Hay-fork, Horse	A. M. Halsted	Rye, N. Y	Mar. 7, 1865	46, 664
Hay-fork, Horse	J. D. Halsted	Rye, N. Y	Jan. 5, 1864	41, 155
Hay-fork, Horse	W. Hannah	Middlefield Centre, N. Y	July 7, 1868	79, 655
Hay-fork, Horse	O. J. Hardgrove	Canton, Ohio	Mar. 19, 1867	63, 042
Hay-fork, Horse	S. and D. A. Harris	Shippensburgh, Pa	Aug. 7, 1866	56, 936
Hay-fork, Horse	S. and D. A. Harris	Shippensburgh, Pa	Apr. 23, 1867	64, 100
Hay-fork, Horse	A. J. Harrison	Cadiz, Ohio	Jan. 5, 1869	85, 658
Hay-fork, Horse	E. Harter	Dowagiac, Mich	Apr. 27, 1869	89, 478
Hay-fork, Horse	E. Harter	Dowagiac, Mich	Dec. 28, 1869	98, 258
Hay-fork, Horse	H. H. Hatheway	Clockville, N. Y	May 21, 1867	64, 863
Hay-fork, Horse	I. J. Hattabaugh	Santa Clara County, Cal	Nov. 19, 1867	71, 167
Hay-fork, Horse	L. Haverstick	Manor Township, Pa	Sept. 17, 1867	68, 873
Hay-fork, Horse	L. Haverstick	Manor Township, Pa	Jan. 12, 1869	85, 739
Hay-fork, Horse	L. Haverstick	Manor Township, Pa	Apr. 29, 1873	138, 249
Hay-fork, Horse	S. Z. Hawbecker and A. Thomas.	Upton and Saint Thomas, Pa.	May 18, 1869	90, 098
Hay-fork, Horse	G. W. Heath	Burlington, Pa	Mar. 26, 1867	63, 244
Hay-fork, Horse	G. W. Heath	Burlington, Pa	Apr. 9, 1867	63, 634
Hay-fork, Horse	G. W. Heath	Burlington, Pa	Aug. 11, 1868	80, 930
Hay-fork, Horse	G. W. Heath	Burlington, Pa	Mar. 30, 1869	88, 477
Hay-fork, Horse	J. S. Henry and A. H. Reist	Manheim, Pa	Aug. 27, 1867	68, 195
Hay-fork, Horse	P. Hill	Millport, N. Y	Dec. 6, 1864	45, 324
Hay-fork, Horse	N. Hinman	Sparta, Mich	Aug. 10, 1869	93, 439
Hay-fork, Horse	N. D. Hinman	Pleasant Dale, Conn	Aug. 15, 1865	49, 405
Hay-fork, Horse	B. F. Hisert	Norton Hill, N. Y	Sept. 1, 1863	39, 732
Hay-fork, Horse	B. F. Hisert	Norton Hill, N. Y	Feb. 20, 1866	52, 805
Hay-fork, Horse	S. M. Hoagland	Catawissa, Pa	Apr. 14, 1868	76, 758
Hay-fork, Horse	A. Houghton	Seville, Ohio	June 23, 1868	79, 228
Hay-fork, Horse	C. A. Howard	Pontiac, Mich	Oct. 13, 1868	83, 063
Hay-fork, Horse	C. A. Howard	Pontiac, Mich	Oct. 31, 1871	120, 440
Hay-fork, Horse	D. J. Howenstine	Marshallville, Ohio	Feb. 18, 1868	74, 538
Hay-fork, Horse	J. W. Hull	Connersville, Ind	May 17, 1870	103, 192
Hay-fork, Horse	A. B. Hunt	Matteson, Mich	Nov. 16, 1869	96, 810
Hay-fork, Horse	E. Huoncker	Bethel, Pa	Mar. 16, 1869	87, 979
Hay-fork, Horse	J. Huy	Bakerstown, Pa	Aug. 1, 1871	117, 541

Index of patents issued from the United States Patent Office from 1790 *to* 1873, *inclusive*—Continued

Invention.	Inventor.	Residence.	Date.	No.
Hay-fork, Horse	T. D. Ingersoll	Monroe, Mich	Mar. 15, 1870	100, 767
Hay-fork, Horse	T. H. and H. James	Stockport, N. Y	Jan. 5, 1864	41, 072
Hay-fork, Horse	J. Jennings	Jacksonville, Ill	Jan. 5, 1869	85, 669
Hay-fork, Horse	J. Jones	Burtonville, N. Y	July 16, 1872	129, 237
Hay-fork, Horse	R. V. Jones	Canton, Ohio	Dec. 31, 1867	72, 860
Hay-fork, Horse	H. Kauffman	York, Pa	Oct. 23, 1866	59, 033
Hay-fork, Horse	C. H. B. Kellogg	Tontogany, Ohio	Oct. 13, 1868	83, 068
Hay-fork, Horse	T. C. Kelly	West Liberty, Pa	July 27, 1869	92, 975
Hay-fork, Horse	J. H. Kendrick	Dexter, Mich	Oct. 22, 1867	70, 002
Hay-fork, Horse	J. G. Kimberlin	West Dryden, N. Y	Feb. 16, 1869	87, 051
Hay-fork, Horse	N. W. Kingsley	Swansea, Mass	Aug. 31, 1869	94, 319
Hay-fork, Horse	G. Kinney	Bristol, Ind	Mar. 10, 1868	75, 430
Hay-fork, Horse	A. Knapp	North Fairfield, Ohio	Dec. 10, 1867	72, 050
Hay-fork, Horse	L. G. Kniffen	Worcester, Mass	Oct. 25, 1864	44, 805
Hay-fork, Horse	H. Krafft	Mulberry, Pa	May 5, 1868	77, 496
Hay-fork, Horse	E. M. Krum	Nassau, N. Y	May 28, 1867	65, 237
Hay-fork, Horse	J. B. Kurtz	Davisburgh, Pa	Oct. 20, 1868	83, 175
Hay-fork, Horse	A. J. Laird	Middletown, Pa	Aug. 21, 1866	57, 337
Hay-fork, Horse	A. J. Laird	Middletown, Pa	Dec. 17, 1867	72, 304
Hay-fork, Horse	H. Laird	Mechanicsburgh, Pa	May 26, 1868	78, 294
Hay-fork, Horse	J. C. Lampman	Baltimore, Md	Oct. 3, 1871	119, 619
Hay-fork, Horse	S. F. Leavitt	North Fairfield, Ohio	Sept. 12, 1865	49, 897
Hay-fork, Horse	C. E. Lins	Ashland, Pa	Aug. 25, 1868	81, 517
Hay-fork, Horse	D. Lippy	Mansfield, Ohio	Jan. 9, 1866	51, 950
Hay-fork, Horse	D. Lippy and J. H. Palm	Mansfield, Ohio	Sept. 6, 1864	44, 102
Hay-fork, Horse	R. W. Liscomb	Smithfield, Pa	Jan. 16, 1866	52, 060
Hay-fork, Horse	T. Lloyd	Muncy, Pa	Sept. 18, 1866	58, 110
Hay-fork, Horse	J. E. Lobdell	Centre Lisle, N. Y	Mar. 10, 1868	75, 282
Hay-fork, Horse	J. W. Lowe	Ashland, Ohio	Sept. 23, 1873	143, 170
Hay-fork, Horse	A. W. Lozier	New York, N. Y	Dec. 1, 1868	84, 559
Hay-fork, Horse	A. W. Lozier	New York, N. Y	Mar. 30, 1869	88, 397
Hay-fork, Horse	J. R. Lyons	Montrose, Pa	Feb. 4, 1868	74, 103
Hay-fork, Horse	J. M. Mansfield	Watertown, N. Y	Oct. 1, 1867	69, 353
Hay-fork, Horse	H. C. Mapes	Rushville, N. Y	Apr. 28, 1868	77, 299
Hay-fork, Horse	A. M. Martin and J. C. Blocher	Bloomville, Ohio	Apr. 14, 1868	76, 787
Hay-fork, Horse	L. S. Mason	Middlefield Centre, N. Y	Feb. 4, 1868	74, 106
Hay-fork, Horse	N. F. Mathewson	Barrington, R. I	Feb. 16, 1869	86, 934
Hay-fork, Horse	D. C. Mattison and T. P. Williamson.	Stockton, Cal	Feb. 2, 1869	86, 565
Hay-fork, Horse	S. S. Mattis	Curtin, Pa	Jan. 12, 1869	85, 840
Hay-fork, Horse	D. C. McDonald	Rushford, N. Y	May 6, 1873	138, 572
Hay-fork, Horse	J. M. McDonald	McCoysville, Pa	Mar. 8, 1870	100, 650
Hay-fork, Horse	W. W. McFaddin	Ennisville, Pa	Nov. 8, 1870	109, 032
Hay-fork, Horse	E. McKenzie	Watertown, N. Y	Feb. 18, 1868	74, 565
Hay-fork, Horse	S. P. Mecay	Kilbourne, Ohio	May 24, 1870	103, 485
Hay-fork, Horse	J. Milholland	New Concord, Ohio	May 19, 1868	78, 114
Hay-fork, Horse	J. A. Miller	Shippensburgh, Pa	Mar. 9, 1869	87, 696
Hay-fork, Horse	J. A. Miller	Shippensburgh, Pa	Dec. 28, 1869	98, 397
Hay-fork, Horse	S. Miller	Mohawk, N. Y	Feb. 9, 1869	86, 775
Hay-fork, Horse	D. Morton	Mount Vernon, Ohio	Jan. 5, 1869	85, 686
Hay-fork, Horse	P. A. Mowers	Cleversburgh, Pa	Mar. 24, 1868	75, 956
Hay-fork, Horse	J. H. Mullin	Schellsburgh, Pa	Feb. 11, 1868	74, 405
Hay-fork, Horse	C. E. Murray	Sugar Valley, Pa	July 14, 1868	80, 002
Hay-fork, Horse	M. D. Myers	Dion, N. Y	Mar. 14, 1865	46, 814
Hay-fork, Horse	D. B. Neal	Mount Gilead, Ohio	May 11, 1869	89, 938
Hay-fork, Horse	D. F. Neikirk	Republic, Ohio	July 16, 1867	66, 730
Hay-fork, Horse	A. J. Nellis	Pittsburgh, Pa	Oct. 21, 1873	143, 775
Hay-fork, Horse	H. Newmeyer	Millerstown, Pa	Nov. 20, 1866	59, 859
Hay-fork, Horse	S. T. Nigh	Leitersburgh, Md	May 4, 1869	89, 683
Hay-fork, Horse	F. Nishwitz	Williamsburgh, N. Y	July 17, 1866	56, 439
Hay-fork, Horse	F. Nishwitz and P. F. Hyers	Williamsburgh, N. Y., and Pekin, Ill.	July 17, 1866	56, 440
Hay-fork, Horse	J. K. O'Neil	Kingston, N. Y	Apr. 3, 1866	53, 660
Hay-fork, Horse	J. K. O'Neil	Kingston, N. Y	Apr. 17, 1866	54, 007
Hay-fork, Horse	J. K. O'Neil	Kingston, N. Y	June 12, 1866	55, 528
Hay-fork, Horse	J. K. O'Neil	Kingston, N. Y	Nov. 13, 1866	59, 633
Hay-fork, Horse	J. K. O'Neil	Kingston, N. Y	Oct. 13, 1868	82, 979
Hay-fork, Horse	J. O'Rourke	Columbiana, Ohio	Apr. 7, 1868	76, 511
Hay-fork, Horse	O. Paddock	Watertown, N. Y	Mar. 24, 1868	75, 786
Hay-fork, Horse	O. Paddock	Watertown, N. Y	Oct. 6, 1868	82, 867
Hay-fork, Horse	O. Paddock	Watertown, N. Y	Mar. 23, 1869	88, 067
Hay-fork, Horse	S. K. Padden	Pulaski, Pa	Mar. 5, 1872	124, 381
Hay-fork, Horse	S. Page	McAllisterville, Pa	Dec. 17, 1867	72, 321
Hay-fork, Horse	S. Page	McAllisterville, Pa	Mar. 10, 1868	75, 452
Hay-fork, Horse	S. Page	McAllisterville, Pa	June 30, 1868	79, 384
Hay-fork, Horse	H. D. Palmer	Leonidas, Mich	June 9, 1868	78, 816
Hay-fork, Horse	N. Palmer	Albany, N. Y	Feb. 25, 1868	74, 934
Hay-fork, Horse	N. Palmer	Albany, N. Y	May 12, 1868	77, 907
Hay-fork, Horse	N. Palmer	Athens, N. Y	Dec. 31, 1872	134, 482
Hay-fork, Horse	J. A. Park	Lansing, Mich	Apr. 13, 1869	88, 804
Hay-fork, Horse	J. A. Park	Lansing, Mich	Sept. 14, 1869	94, 909
Hay-fork, Horse	J. A. Park	Lansing, Mich	Nov. 2, 1869	96, 474
Hay-fork, Horse	E. M. Parker	Zion, Md	June 9, 1868	78, 820
Hay-fork, Horse	J. H. Parker, J. T. Hall, and I. Pierce.	Trenton, N. Y	Mar. 19, 1867	62, 970
Hay-fork, Horse	S. W. Patterson and S. Dewey	Mainesburgh, Pa	July 9, 1867	66, 621
Hay-fork, Horse	G. C. Perry	Ortonville, Mich	June 8, 1869	91, 160
Hay-fork, Horse	J. F. Pierce	Holland Patent, N. Y	Mar. 6, 1866	53, 085
Hay-fork, Horse	M. H. Pope	Susquehanna Depot, Pa	May 26, 1868	78, 324
Hay-fork, Horse	H. G. Porter	Grand Rapids, Mich	July 26, 1870	105, 722
Hay-fork, Horse	I. T. Price	Leesville, Ohio	June 26, 1866	55, 902
Hay-fork, Horse	E. Raber	Lake, Ohio	May 12, 1868	77, 839
Hay-fork, Horse	E. Raber	Roanoke, Ind	Sept. 7, 1869	94, 644
Hay-fork, Horse	A. Ream and C. L. Bush	Reading, Pa	Apr. 2, 1872	125, 330
Hay-fork, Horse	C. W. Reed	Chagrin Falls, Ohio	Dec. 24, 1867	72, 540
Hay-fork, Horse	C. W. Reed	Chagrin Falls, Ohio	Dec. 8, 1868	84, 708
Hay-fork, Horse	A. Reynolds	Rock Spring, Md	Apr. 14, 1868	76, 661
Hay-fork, Horse	E. Reynolds	Corunna, Mich	Oct. 18, 1864	44, 748

Index of patents issued from the United States Patent Office from 1790 *to* 1873, *inclusive*—Continued.

Invention.	Inventor.	Residence.	Date.	No.
Hay-fork, Horse	R. Reynolds and C. Young	Stockport, N. Y	May 30, 1865	47, 982
Hay-fork, Horse	R. Reynolds and C. Young	Stockport, N. Y	May 30, 1865	48, 029
Hay-fork, Horse	E. Rhoades, sr	Clyde, Ohio	Jan. 21, 1868	73, 650
Hay-fork, Horse	S. H. Rhoades and W. Carroll	Clyde, Ohio, and Hillsdale, Mich.	May 12, 1868	77, 760
Hay-fork, Horse	T. Rhoades	Ottawa, Ill	Apr. 29, 1873	138, 286
Hay-fork, Horse	E. Rhoades, jr	Clyde, Ohio	Apr. 16, 1867	63, 812
Hay-fork, Horse	E. Rhoades, jr	Clyde, Ohio	Nov. 3, 1868	83, 794
Hay-fork, Horse	J. L. Ripley	Fremont, Ohio	Sept. 6, 1864	44, 116
Hay-fork, Horse	J. L. Ripley	Fremont, Ohio	Feb. 14, 1865	46, 393
Hay-fork, Horse	G. M. Robinson	New Wilmington, Pa	May 12, 1868	77, 843
Hay-fork, Horse	L. M. Roby	Leesville, Ohio	Apr. 10, 1866	53, 878
Hay-fork, Horse	B. B. Rockwell	Waterford Mills, Ind	Oct. 28, 1873	144, 143
Hay-fork, Horse	L. Rogers	Pittsburgh, Pa	Jan. 18, 1870	98, 887
Hay-fork, Horse	S. Rogers	Pittsburgh, Pa	Mar. 20, 1866	53, 345
Hay-fork, Horse	S. Rogers	Pittsburgh, Pa	Oct. 23, 1866	59, 139
Hay-fork, Horse	J. W. Row	Lewisburgh, Pa	Aug. 17, 1869	93, 750
Hay fork, Horse	E. D. Rundell	Hudson, N. Y	Dec. 13, 1864	45, 439
Hay-fork, Horse	A. V. Ryder	Germano, Ohio	Aug. 14, 1866	57, 193
Hay-fork, Horse	W. Scholl	Marion Township, Pa	Mar. 24, 1868	75, 800
Hay-fork, Horse	P. Schwitzer	Robeson Township, Pa	Apr. 7, 1868	76, 352
Hay-fork, Horse	E. U. Scoville	Manlius, N. Y	Apr. 5, 1870	101, 524
Hay-fork, Horse	O. P. Secor	Chicago, Ill	Sept. 5, 1865	49, 834
Hay-fork, Horse	G. W. Shade	Shippensburgh, Pa	Dec. 17, 1867	72, 232
Hay-fork, Horse	G. W. Shade	Shippensburgh, Pa	June 28, 1870	104, 891
Hay-fork, Horse	J. Shearer	Reading, Pa	Nov. 26, 1867	71, 334
Hay-fork, Horse	R. S. Sheldon	Chicago, Ill	Jan. 5, 1869	85, 702
Hay-fork, Horse	A. Shellenberger	Versailles, Ohio	June 14, 1870	104, 365
Hay-fork, Horse	H. L. Sheperd	Osborn, Ohio	July 28, 1868	80, 513
Hay-fork, Horse	E. Shorkley	Lewisburgh, Pa	Aug. 14, 1866	57, 199
Hay-fork, Horse	S. G. Simpson	Mill Creek, Pa	Nov. 30, 1869	97, 449
Hay-fork, Horse	S. Slack	Dowagiac, Mich	Aug. 24, 1869	94, 142
Hay-fork, Horse	A. Smith	Schellsburgh, Pa	Aug. 11, 1868	80, 839
Hay-fork, Horse	R. A. Smith	Washington Mills, N. Y	May 4, 1869	89, 603
Hay-fork, Horse	F. Snyder	Hinkletown, Pa	July 9, 1867	66, 530
Hay-fork, Horse	I. E. Snyder	Woodward, Pa	Sept. 23, 1873	143, 192
Hay-fork, Horse	I. C. Spear	New Wilmington, Pa	Jan. 14, 1868	73, 403
Hay-fork, Horse	W. S. Spratt	West Manchester, Pa	Sept. 4, 1866	57, 789
Hay-fork, Horse	A. B. Sprout	Hughesville, Pa	May 1, 1866	54, 431
Hay-fork, Horse	A. B. Sprout	Hughesville, Pa	Apr. 23, 1867	64, 163
Hay-fork, Horse	A. B. Sprout	Hughesville, Pa	May 26, 1868	78, 335
Hay-fork, Horse	N. Starr, jr	Homer, N. Y	Apr. 21, 1868	77, 120
Hay-fork, Horse	G. M. Stearns	Syracuse, N. Y	Jan. 25, 1870	99, 120
Hay-fork, Horse	H. B. Steele	West Winsted, Conn	Mar. 10, 1868	75, 481
Hay-fork, Horse	W. W. Stevens and J. Patcher, jr	Tontogany, Ohio	June 30, 1868	79, 406
Hay-fork, Horse	B. F. Stewart	Freeport, Ohio	Oct. 15, 1867	69, 857
Hay-fork, Horse	D. P. Stewart	Spruce Creek, Pa	Oct. 12, 1869	95, 743
Hay-fork, Horse	O. O. Storle	North Cape, Wis	Apr. 23, 1867	64, 164
Hay-fork, Horse	O. O. Storle	Norway, Wis	Nov. 10, 1868	84, 017
Hay-fork, Horse	H. C. Stouffer	Canfield, Ohio	Apr. 22, 1873	138, 211
Hay-fork, Horse	H. C. Stouffer, J. Heaton, and A. A. Bushong.	Columbiana, Ohio	Dec. 3, 1867	71, 661
Hay-fork, Horse	H. C. and A. Stouffer	Beaver Township, Ohio	Mar. 10, 1868	75, 486
Hay-fork, Horse	G. F. Strong	Onondaga, N. Y	Dec. 12, 1865	51, 490
Hay-fork, Horse	G. H. Strough	Watertown, N. Y	June 30, 1868	79, 512
Hay-fork, Horse	L. W. Stuart	Narrowsburgh, N. Y	Aug. 9, 1870	106, 230
Hay-fork, Horse	J. W. Summers	Sandy Hill, N. Y	Mar. 24, 1868	75, 997
Hay-fork, Horse	J. B. Sweetland	Pontiac, Mich	Apr. 21, 1868	77, 125
Hay-fork, Horse	J. B. Sweetland	Pontiac, Mich	Apr. 13, 1869	88, 991
Hay-fork, Horse	J. B. Sweetland	Pontiac, Mich	May 4, 1869	89, 810
Hay-fork, Horse	E. B. Tanner	Attica, Ohio	Sept. 27, 1864	44, 470
Hay-fork, Horse	L. H. Tears	Troy, Pa	Oct. 22, 1867	70, 044
Hay-fork, Horse	T. H. Tears	Le Roy, Pa	Apr. 7, 1868	76, 550
Hay-fork, Horse	A. G. Thomas	Sandy Spring, Md	Apr. 6, 1869	88, 754
Hay-fork, Horse	J. T. Thomas	Ilion, N. Y	July 20, 1869	92, 762
Hay-fork, Horse	F. W. Thorla	Hoskinville, Ohio	Jan. 18, 1870	99, 033
Hay-fork, Horse	L. N. Tinkham	Sylvania, Pa	Dec. 24, 1867	62, 567
Hay-fork, Horse	W. H. T. Tomlin	Mullica Hill, N. J	Apr. 23, 1867	64, 167
Hay-fork, Horse	H. Totten	Leesville, Ohio	Oct. 17, 1865	50, 516
Hay-fork, Horse	J. F. Troxel	Bloomville, Ohio	Apr. 6, 1869	88, 593
Hay-fork, Horse	M. I. Turck	Schodack, N. Y	May 7, 1867	64, 462
Hay-fork, Horse	P. Vanderbelt, jr	Hughesville, Pa	Nov. 26, 1867	71, 343
Hay-fork, Horse	F. Van Doren	Adrian, Mich	Jan. 23, 1872	123, 063
Hay-fork, Horse	O. Vanorman	Ripon, Wis	Dec. 10, 1867	72, 131
Hay-fork, Horse	G. H. Waldo	Prattsburgh, N. Y	Dec. 10, 1867	71, 928
Hay-fork, Horse	E. L. Walker	Benford's Store, Pa	Apr. 17, 1866	54, 044
Hay-fork, Horse	E. L. Walker	Benford's Store, Pa	Apr. 21, 1868	77, 134
Hay-fork, Horse	J. H. Walker	Grand Rapids, Mich	Jan. 7, 1868	73, 211
Hay-fork, Horse	J. H. Walker	Grand Rapids, Mich	Apr. 21, 1868	77, 136
Hay-fork, Horse	J. M. Walker	Rossville, Pa	July 13, 1869	92, 495
Hay-fork, Horse	C. E. Warner	Syracuse, N. Y	Mar. 16, 1869	87, 807
Hay-fork, Horse	C. E. Warner	Syracuse, N. Y	Oct. 26, 1869	96, 173
Hay-fork, Horse	J. L. Wells	Stockbridge, N. Y	Oct. 11, 1864	44, 6[illegible]2
Hay-fork, Horse	G. F. Weymouth	Dresden, Me	Mar. 1, 1870	100, 4[illegible]6
Hay-fork, Horse	S. H. Wheeler	Dowagiac, Mich	Sept. 26, 1865	50, 210
Hay-fork, Horse	E. I. White	Locke, N. Y	July 7, 1868	79, 711
Hay-fork, Horse	A. Whitmarsh	East Bridgewater, Mass	Mar. 6, 1866	53, 070
Hay-fork, Horse	A. F. Wicke and O. Evans	Alliance, Ohio	Mar. 26, 1867	63, 345
Hay-fork, Horse	F. Wicks	Kansas, Ill	Oct. 18, 1864	44, 760
Hay-fork, Horse	F. Wicks	Upper Sandusky, Ohio	Aug. 28, 1866	57, 609
Hay-fork, Horse	J. H. Wilder	Farmington, Ohio	Apr. 7, 1868	76, 573
Hay-fork, Horse	H. Willard	Grand Rapids, Mich	Dec. 3, 1867	71, 829
Hay-fork, Horse	H. Willard	Grand Rapids, Mich	June 23, 1868	79, 173
Hay-fork, Horse	C. W. Willis and A. Stephenson	Cambridge City, Ind	Apr. 17, 1866	54, 050
Hay-fork, Horse	F. M. Willson	Whitney's Point, N. Y	Apr. 14, 1868	76, 868
Hay-fork, Horse	W. D. Wilson	Watertown, N. Y	May 26, 1868	78, 347
Hay-fork, Horse	M. Winsler, W. Campbell, and L. Hardman.	Tuscarawas County, Ohio	May 21, 1867	65, 031

Index of patents issued from the United States Patent Office from 1790 *to* 1873, *inclusive*—Continued.

Invention.	Inventor.	Residence.	Date.	No.
Hay-fork, Horse	P. A. Wise	Stockbridge, N. Y	June 5, 1866	55,408
Hay-fork, Horse	J. Wolfrom	York, Pa	June 2, 1868	78,504
Hay-fork, Horse	E. B. Woodruff	Morristown, N. J	Jan. 29, 1867	61,697
Hay-fork, Horse	L. Woodworth	Troy, Pa	May 19, 1868	78,169
Hay-fork, Horse	E. Yancey	Utica, N. Y	Sept. 3, 1867	68,405
Hay-fork, Horse	E. Yeiser and J. S. Sheetz	Sheridan, Pa	Jan. 15, 1867	61,302
Hay-fork, Horse	C. Yengst	Annville, Pa	June 13, 1871	116,000
Hay-fork, Horse	J. S. Yinger	Manchester Township, Pa	Dec. 29, 1868	85,416
Hay-fork, Horse	J. S. Yinger	Manchester, Pa	June 21, 1870	104,529
Hay-fork, Horse	W. Zimmerman	Colfax, Iowa	Mar. 24, 1868	76,029
Hay-fork, Spiral	A. Neumeyer	Millerstown, Pa	Apr. 2, 1867	63,554
Hay-forks, Apparatus for operating horse	G. M. Robinson	New Wilmington, Pa	Feb. 4, 1868	74,142
Hay-forks, Clutch for	G. D. Melotte	Watertown, N. Y	Oct. 29, 1867	70,239
Hay-forks, Clutch for operating horse	D. B. Baker	Rollersville, Ohio	Dec. 31, 1867	72,778
Hay-forks, Device for suspending	W. S. Salisbury	Adams Centre, N. Y	May 28, 1867	65,283
Hay-forks, &c., Device for suspending horse	G. B. and C. Lewis	Adams Centre, N. Y	Jan. 1, 1867	60,758
Hay-forks, Elevating	J. H. Chapman	Utica, N. Y	Mar. 26, 1867	63,212
Hay-forks, Grapple for suspending	W. E. Derrick	Jordan, N. Y	Jan. 12, 1869	85,734
Hay-forks, Machine for bending	N. Brand	Leonardsville, N. Y	June 17, 1856	15,118
Hay-forks, Operating horse	H. Maycock	Verona, N. Y	Feb. 20, 1866	52,808
Hay-forks, Operating horse	A. J. Purviance	Keosauqua, Iowa	Oct. 1, 1867	69,479
Hay-forks, Self-tightening band for	J. H. Melick	Albany, N. Y	Nov. 24, 1863	40,701
Hay-forks, Shank and socket for hand	F. F. Terry	Port Gibson, N. Y	Apr. 30, 1867	64,381
Hay-forks, Shank of	D. Anthony, jr	Spring Port, N. Y	Oct. 8, 1850	7,694
Hay-forks, Suspending claw for horse	E. S. Kershaw	Sherburne, N. Y	July 30, 1867	67,199
Hay-gatherer	W. E. Phelps	Elmwood, Ill	Feb. 8, 1870	99,589
Hay-gatherer	A. Stream	Harrodsburgh, Ind	June 6, 1871	115,783
Hay-gatherer	A. Stream	Harrodsburgh, Ind	May 21, 1872	127,116
Hay-gatherer	C. Waste	Galesburgh, Ill	June 14, 1870	104,233
Hay gatherer and carrier	G. Meaders	Jeffersonville, Ind	June 21, 1870	104,480
Hay gatherer and shocker	C. M. Terrell	Oskaloosa, Iowa	Mar. 22, 1870	101,183
Hay-gathering machine	F. F. Fowler	Crane Township, Ohio	Jan. 22, 1861	31,162
Hay, grain, &c., Curing and drying	T. Heermans	Mitchellsville, Tenn	July 5, 1864	43,406
Hay, grain, &c., Mode of ventilating	A. Post	Henrietta, N. Y	June 18, 1861	32,584
Hay-grinding machine	W. P. Miller	New York, N. Y	June 17, 1873	139,911
Hay-hook, Horse	A. Kratz	Plumsteadville, Pa	Nov. 17, 1863	40,628
Hay-knife	J. S. Ball	Mishawaka, Ind	Nov. 18, 1873	144, 59
Hay-knife	C. Dittman	Upper Leacock Township, Pa.	July 5, 1870	105,050
Hay-knife	J. Fasig	Jackson, Ohio	Apr. 13, 1858	19,921
Hay-knife	J. Fasig	West Salem, Ohio	Feb. 11, 1862	34,356
Hay-knife	G. Fenton	Streetsborough, Ohio	Mar. 10, 1868	75,255
Hay-knife	F. J. Fischer	Hamilton, Ohio	June 19, 1866	55,765
Hay-knife	C. A. Fisher	Geneseo, Ill	Jan. 7, 1868	73,177
Hay-knife	C. A. Fisher	Geneseo, Ill	Sept. 29, 1868	82,505
Hay-knife	M. Merrill	Oneida County, N. Y	Jan. 18, 1870	98,876
Hay-knife	J. Offineer	Ashland, Ohio	Sept. 15, 1868	82,152
Hay-knife	G. W. Parsons and W. S. Finney.	Harrisburgh, Ohio	July 16, 1872	129,247
Hay-knife	G. Reiber	Dover, Pa	Aug. 22, 1871	118,390
Hay-knife	J. L. Ripley	Fremont, Ohio	Sept. 29, 1868	82,641
Hay-knife	H. M. Smith	Kalamazoo, Mich	Sept. 3, 1867	68,536
Hay-knife	P. O. Soper	San Francisco, Cal	Jan. 29, 1867	61,576
Hay-knife	A. N. Staley	Waynesborough, Pa	July 18, 1869	90,202
Hay-knife	G. F. Weymouth	Dresden, Me	Mar. 7, 1871	112,400
Hay-knife	S. Whalen	West Milton, N. Y	Mar. 2, 1854	10,874
Hay-knife	D. H. Wile	New Pittsburgh, Ohio	Oct. 4, 1870	108,076
Hay-knife	J. H. Wolfe	Lytle City, Iowa	Aug. 29, 1871	118,571
Hay-knife and pruner combined	D. Fasig	Rowsburgh, Ohio	Feb. 16, 1869	87,035
Hay-knife and pruning-hook, Combined	D. Fasig	Rowsburgh, Ohio	Mar. 8, 1870	100,611
Hay-loader	I. Anderson	Poland, Ohio	Oct. 5, 1869	95,408
Hay-loader	J. Bachelder	Norwich, Conn	Mar. 15, 1870	100,710
Hay-loader	J. R. Bailey	Woonsocket, R. I	Aug. 10, 1869	93,581
Hay-loader	H. Baker	Cortland, N. Y	July 18, 1871	117,033
Hay-loader	A. Barker	Camanche, Iowa	June 2, 1868	78,566
Hay-loader	J. M. Boorman	Scarborough, N. Y	Apr. 12, 1870	101,705
Hay-loader	A. Buck	Hebron, Ind	July 19, 1870	105,418
Hay loader	J. Bullis	Macedon, N. Y	May 7, 1867	64,482
Hay-loader	J. Capen	Charlton, Mass	Dec. 21, 1869	98,027
Hay-loader	D. Clagett	Hagerstown, Md	Aug. 30, 1870	106,916
Hay-loader	E. Cooper	Theresa, N. Y	Apr. 13, 1869	88,945
Hay-loader	E. Cooper	Theresa, N. Y	July 12, 1870	105,308
Hay-loader	E. N. Curtice	Spring Water, N. Y	Feb. 4, 1868	74,055
Hay-loader	T. P. Davidson and A. Parvis	Liberty, Ind	Apr. 8, 1873	137,751
Hay-loader	L. De Lacee	Springfield, Ill	May 21, 1867	64,848
Hay-loader	W. Denton	Iowa City, Iowa	Apr. 27, 1869	89,388
Hay-loader	M. A. Dilley	Mendon, Mich	Sept. 17, 1867	68,855
Hay-loader	N. B. Douglas	Cornwall, Vt	June 30, 1868	79,327
Hay-loader	W. H. Elliot	Plattsburgh, N. Y	Apr. 26, 1864	42,466
Hay-loader	W. H. Elliot	New York, N. Y	July 3, 1866	56,023
Hay-loader	W. H. Elliot	New York, N. Y	Apr. 23, 1867	64,003
Hay-loader	W. H Elliot	New York, N. Y	June 4, 1867	65,478
Hay-loader	J. E. Fell and J. H. Mattocks	Maquokita, Iowa	Feb. 14, 1871	111,735
Hay-loader	A. Fuller	Plymouth, Ind	Aug. 7, 1866	56,924
Hay-loader	C. W. Gage	Homer, N. Y	July 16, 1867	66,699
Hay-loader	C. B. Garlinghouse	Carpentersville, Ind	Jan. 25, 1870	99,181
Hay-loader	C. B. Garlinghouse and J. Dickason.	Allensville, Ind	July 14, 1863	39,264
Hay-loader	A. Garver	Lime Spring Station, Iowa	Nov. 12, 1872	132,960
Hay-loader	C. Gibbs	Pittsfield, Vt	May 29, 1866	55,083
Hay-loader	T. Giffin	erryville, Ohio	Jan. 30, 1872	123,166
Hay-loader	C. H. Gifford	Philadelphia, Pa	Oct. 8, 1867	69,652
Hay-loader	H. L. Gockley	Jackson, Ill	Sept. 6, 1870	107,038
Hay-loader	W. H. Gray	Ashfield, Mass	Aug. 16, 1870	106,355
Hay-loader	C. P. Hale	Calhoun, Ky	Oct. 12, 1869	95,680
Hay-loader	W. M. Hall and J. Johnson	Barrington, N. Y	Apr. 2, 1867	63,382
Hay-loader	H. W. Hamilton	Brandon, Vt	Dec. 12, 1871	121,777
Hay-loader	N. L. Hatch	Cape Elizabeth, Me	Dec. 22, 1868	85,092

Index of patents issued from the United States Patent Office from 1790 *to* 1873, *inclusive*—Continued.

Invention.	Inventor.	Residence.	Date.	No.
Hay-loader	S. R. Higgins	Parma, Mich	May 15, 1866	54, 723
Hay-loader	H. C. Howland and T. L. Moore	Oxford, N. Y	Aug. 26, 1873	142, 234
Hay-loader	W. L. Hubbell	Brooklyn, N. Y	Aug. 25, 1863	39, 655
Hay-loader	H. Hull	West Exeter, N. Y	Feb. 11, 1868	74, 226
Hay-loader	D. C. Jewett and A. C. Bowen	Sand Spring and Bowen's Prairie, Iowa.	Jan. 9, 1866	51, 947
Hay-loader	M. K. Lewis and J. C. Durbin	Iowa City, Iowa	Jan. 3, 1865	45, 726
Hay-loader	A. W. Lozier	New York, N. Y	June 23, 1868	79, 135
Hay-loader	A. W. Lozier	New York, N. Y	Dec. 1, 1868	84, 560
Hay-loader	A. W. Lozier	New York, N. Y	June 15, 1869	91, 242
Hay-loader	O. T. Nanny	Mamakating, N. Y	Mar. 27, 1866	53, 470
Hay-loader	A. J. and S. D. McKee	Beaver Dam, Ohio	June 8, 1869	91, 147
Hay-loader	B. J. Moore	Dresbach, Minn	Nov. 2, 1869	96, 462
Hay-loader	J. Morgan	West Springfield, Mass	Apr. 16, 1867	63, 806
Hay-loader	G. G. Park	Xenia, Nebr	Aug. 23, 1870	106, 611
Hay-loader	I. J. Parker	Buffalo Grove, Iowa	July 26, 1870	105, 837
Hay-loader	W. Platt and A. G. Burnham	Greenfield, Mass	Jan. 31, 1865	46, 139
Hay-loader	A. C. Preston	Battle Creek, Mich	Feb. 15, 1870	99, 947
Hay-loader	G. Race	Norwich, N. Y	Apr. 17, 1866	54, 084
Hay-loader	F. M. Reynolds	Mile Strip, N. Y	July 21, 1868	80, 222
Hay-loader	J. P. Rideout	Bowdoinham, Me	Jan. 24, 1871	111, 252
Hay-loader	J. G. Seborn	Iowa City, Iowa	Sept. 14, 1869	94, 846
Hay-loader	R. Shaw	Warsaw, Ind	Apr. 23, 1872	125, 992
Hay-loader	W. H. Straub	Danville, Pa	Dec. 7, 1869	97, 724
Hay-loader	E. Sweet	Whitney's Point, N. Y	Mar. 26, 1867	63, 332
Hay-loader	L. D. Taylor	Granville Centre, Pa	May 7, 1867	64, 459
Hay-loader	L. D. Taylor	Granville Centre, Pa	Apr. 4, 1871	113, 363
Hay-loader	F. Van Etten	Jackson, Mich	May 8, 1866	54, 627
Hay-loader	A. Vose	Pittsfield, Vt	Nov. 19, 1867	71, 250
Hay-loader	C. E. Warner	Burlington, Kans	Dec. 9, 1873	145, 321
Hay, Loading	R. Cobb	Hadley, Mich	Dec. 5, 1865	51, 294
Hay, Loading	W. Dixon	Chicago, Ill	June 26, 1860	28, 839
Hay, Loading	T. J. Jolly	Olean, Ind	June 26, 1860	28, 868
Hay loading and pitching machine	J. S. De Haven	Akron, Ohio	Aug. 19, 1862	36, 208
Hay loading and unloading apparatus	G. W. Long	Delaware Centre, Iowa	Feb. 11, 1873	135, 720
Hay-loading device	A. B. Clark	Grass Lake, Mich	June 7, 1870	103, 844
Hay-loading machine	J. B. Atwater	Chicago, Ill	Aug. 30, 1864	43, 964
Hay-loading machine	W. H. Bentley	Westford, N. Y	Nov. 17, 1863	40, 598
Hay-loading machine	G. Constable	Cannonsville, N. Y	July 14, 1863	39, 215
Hay-loading machine	L. De Lacie	Newark, N. Y	Dec. 13, 1864	45, 394
Hay-loading machine	S. V. Essick	Moultrie, Ohio	May 10, 1859	23, 911
Hay-loading machine	J. W. Foust	Hammondsburgh, Pa	Nov. 18, 1862	36, 947
Hay-loading machine	J. F. Hall and I. Pierce	Holland Patent, N. Y	June 28, 1864	43, 307
Hay-loading machine	J. F. Hall and I. Pierce	Holland Patent, N. Y	Dec. 13, 1864	45, 405
Hay-loading machine	S. R. Higgins	Parma, Mich	Jan. 10, 1865	45, 826
Hay-loading machine	H. Maycock	Verona, N. Y	Oct. 4, 1864	44, 540
Hay-loading machine	J. B. McIntosh	Girard, Pa	Feb. 19, 1861	31, 470
Hay-loading machine	L. Mishler	Mogadore, Ohio	Sept. 24, 1861	33, 359
Hay-loading machines, Tongue for	E. Sweet	Whitney's Point, N. Y	Jan. 21, 1868	73, 474
Hay, Machine for spreading and turning	J. Center	Hudson, N. Y	June 27, 1834	....
Hay-making machine	G. A. Brown	Middletown, R. I	Jan. 23, 1855	12, 269
Hay-making machine	T. J. Goff	Warren, R. I	June 14, 1859	24, 386
Hay-making machine	F. Peabody	Salem, Mass	Apr. 17, 1855	12, 728
Hay-making machine	J. C. Stoddard	Worcester, Mass	June 28, 1859	24, 588
Hay-making machine	J. C. Stoddard	Worcester, Mass	Dec. 6, 1859	26, 380
Hay-mow, Ventilating	G. Race	Norwich, N. Y	Dec. 10, 1867	72, 081
Hay or cotton press	F. Horton	Silver Creek, N. Y	Nov. 12, 1867	70, 720
Hay or cotton press	D. A. Nelson	Tyler, Tex	Sept. 9, 1873	142, 718
Hay or feed cutter	J. R. Whittemore	Chicopee Falls, Mass	Aug. 26, 1862	36, 316
Hay or straw cutter	S. S. Clark	Manchester, N. H	Jan. 17, 1860	26, 835
Hay, &c., Preserving	A. D. Ditman	Chester, Pa	Feb. 15, 1838	599
Hay-press	D. R. Allen	Cumberland, Me	Apr. 25, 1846	4, 484
Hay-press	J. S. Arnold	South Milan, Ind	July 3, 1860	28, 952
Hay-press	G. H. Aylworth	Brighton, Ill	May 21, 1867	64, 828
Hay-press	G. H. Aylworth	Brighton, Ill	Apr. 5, 1870	101, 566
Hay-press	H. F. Blank	Liberty, Ill	Nov. 18, 1873	144, 733
Hay-press	M. B. Bliss	Pittston, Me	Jan. 26, 1828	....
Hay-press	I. Bisbee and A. Bisbee	East Pharsalia, N. Y., and Polk Township, Mo.	Apr. 8, 1862	34, 872
Hay-press	O. Bossée	Millbrae, Cal	Aug. 29, 1871	118, 582
Hay-press	J. Briggs	Homewood, Ill	Mar. 19, 1872	124, 717
Hay-press	A. R. Chamberlin and A. Cleflin	Richmond, Me	Mar. 30, 1836	....
Hay-press	S. Colahan	Cleveland, Ohio	Mar. 11, 1862	34, 622
Hay-press	G. W. D. Culp	Allensville, Ind	July 12, 1864	43, 480
Hay-press	G. W. D. Culp	East Enterprise, Ind	Feb. 4, 1868	74, 054
Hay-press	G. W. D. Culp	East Enterprise, Ind	Oct. 22, 1872	132, 448
Hay-press	J. H. Davison and T. J. Sasher	Mound Station, Ill	Nov. 1, 1870	108, 829
Hay-press	L. Dederick	Kinderhook, N. Y	June 24, 1843	3, 148
Hay-press	L. Dederick	Albany, N. Y	June 6, 1854	11, 043
Hay-press	L. Dederick	Albany, N. Y	July 14, 1863	39, 258
Hay-press	W. Deering	Louisville, Ky	May 26, 1863	38, 659
Hay-press	E. A. Field	Sidney, Me	Dec. 26, 1865	51, 709
Hay-press	M. Fletcher	Louisville, Ky	Sept. 27, 1864	44, 413
Hay-press	F. Frey	Liberty, Ill	July 5, 1870	105, 062
Hay-press	G. Gibbs	Sugar Branch, Ind	Oct. 25, 1864	44, 795
Hay-press	J. H. Gove	San Francisco, Cal	Mar. 6, 1860	27, 361
Hay-press	F. F. Hamilton	Green Bay, Wis	Jan. 10, 1871	110, 849
Hay-press	F. F. Hamilton	Green Bay, Wis	Nov. 28, 1871	121, 244
Hay-press	C. W. Hawkes	Brunswick, Me	Apr. 18, 1839	1, 124
Hay-press	A. Hayford and A. Strout	Belfast, Me	May 30, 1865	47, 950
Hay-press	S. Hewitt	Allensville, Ind	Dec. 30, 1843	3, 394
Hay-press	H. C. Hunt	Indianapolis, Ind	Dec. 6, 1870	109, 905
Hay-press	I. H. Johnson	Long Reach, W. Va	Aug. 21, 1866	57, 331
Hay-press	G. W. Lockhart	Charlestown, Ind	Apr. 7, 1868	76, 478
Hay-press	S. Mahurin	Liberty, Ill	July 16, 1867	66, 860
Hay-press	P. Manny	Waddam's Grove, Ill	Apr. 17, 1855	12, 740
Hay-press	C. Martine	Greensburgh, N. Y	May 30, 1842	2, 657
Hay-press	W. McCord	Sing Sing, N. Y	Jan. 24, 1860	26, 917
Hay-press	M. R. Molleur	Chicago, Ill	Feb. 18, 1868	74, 571

Index of patents issued from the United States Patent Office from 1790 *to* 1873, *inclusive*—Continued.

Invention.	Inventor.	Residence.	Date.	No.
Hay-press	M. V. Northrup	Hornitos, Cal	Dec. 27, 1870	110, 585
Hay-press	G. Noyes	Pownal, Me	Dec. 17, 1867	72, 221
Hay-press	C. F. Paine	Winslow, Me	Apr. 25, 1844	3, 561
Hay-press	P. C. Pearson	Harrison County, Ind	July 19, 1864	43, 600
Hay-press	W. R. and H. M. Peavey	Swanville, Me	July 15, 1862	35, 889
Hay-press	S. G. Randall	New Britain, Conn	July 17, 1860	29, 194
Hay-press	W. H. Reynolds	New Orleans, La	July 11, 1871	116, 869
Hay-press	A. C. Richard	Point Lookout, Tenn	Dec. 24, 1867	72, 681
Hay-press	W. Ridonour	Springfield, Ohio	Mar. 10, 1863	37, 873
Hay-press	B. Roberts	Highland, N. Y	Apr. 24, 1866	54, 209
Hay-press	C. Rundlett and J. W. Drummond.	Alden and Winslow, Me	July 19, 1859	24, 824
Hay-press	J. Scott	Brooklyn, N. Y	Feb. 6, 1872	123, 517
Hay-press	A. Spanier	Saint John, Ind	Aug. 15, 1871	118, 164
Hay-press	C. S. Stevens and A. T. Boon	Galesburgh, Ill	Oct. 11, 1864	44, 673
Hay-press	S. Stevens	Sacramento, Cal	July 24, 1860	29, 320
Hay-press	R. V. W. Thorn	New York, N. Y	Apr. 29, 1813	....
Hay-press	E. A. Ward	Gallipolis, Ohio	Aug. 14, 1866	57, 227
Hay-press pawl	L. D. Rundell	South Westerlo, N. Y	Feb. 10, 1863	37, 643
Hay-press, Portable	J. Mauck	Cheshire, Ohio	May 28, 1867	65, 249
Hay-press, Wagon	H. Brocksmith	Saint Louis, Mo	Dec. 30, 1873	145, 986
Hay-presses, Capstan for working beater	P. K. Dederick	Albany, N. Y	Aug. 29, 1865	49, 678
Hay-presses, Method of securing the door of	C. Martratt	New Baltimore, N. Y	Apr. 21, 1857	17, 108
Hay, Pressing and baling	G. H. Aylworth	Brighton, Ill	Mar. 21, 1871	112, 766
Hay-rack	G. Baldwin	Bluffton, Ind	Apr. 25, 1865	47, 379
Hay-rack	A. C. Bayley	Battle Creek, Mich	May 4, 1869	89, 619
Hay-rack	C. G. Boyer	Greenfield, Ind	July 2, 1867	66, 292
Hay-rack	S. Brownell	Irving, N. Y	Sept. 22, 1868	82, 288
Hay-rack	G. Denis and G. Grassel	Oceola, Iowa	Nov. 23, 1869	97, 172
Hay-rack	W. E. Grimes	Allen, Mo	Apr. 10, 1866	53, 814
Hay-rack	W. P. Jones	Arcade, N. Y	Jan. 2, 1872	122, 387
Hay-rack	J. Mandigo	Wayland, Mich	Oct. 3, 1871	119, 627
Hay-rack	D. Matteson	Pittsfield, Mich	Aug. 17, 1869	93, 897
Hay-rack	A. Naramor	Berlin Heights, Ohio	Jan. 24, 1865	46, 017
Hay-rack	W. Nash	Corning, N. Y	Apr. 30, 1867	64, 349
Hay-rack	W. Smith	Tod Township, Ohio	Oct. 28, 1868	83, 561
Hay-rack	A. T. Soule	Savannah, N. Y	July 21, 1868	80, 095
Hay-rack	D. N. Webster	Geneva, Ohio	Sept. 18, 1866	58, 158
Hay-rack and wagon-box, Combined	F. G. McClellan	Attica, Ohio	Apr. 19, 1870	102, 023
Hay-rack, Feeding	T. Moore	New York, N. Y	Apr. 17, 1866	54, 005
Hay-rack, Hinged	M Carpenter	Wayland, Mich	June 20, 1871	116, 153
Hay-rake	D. Armel	Somerset, Pa	Jan. 20, 1863	37, 429
Hay-rake	E. P. Barton and R. W. Towle	Batavia and Bath, N. Y	July 9, 1861	32, 745
Hay-rake	H. Brown	Cape Elizabeth, Me	July 3, 1866	56, 000
Hay-rake	L. A. C. Brown	Sparta, Ill	Dec. 1, 1857	18, 731
Hay-rake	R. I. Burbank	Boston, Mass	Mar. 23, 1869	88, 127
Hay-rake	C. P. Carpenter	Saint Johnsbury, Vt	Aug. 26, 1856	15, 601
Hay-rake	C. P. Carpenter	Saint Johnsbury, Vt	Sept. 15, 1863	39, 989
Hay-rake	W. Deckman	Canton, Ohio	Apr. 16, 1861	32, 054
Hay-rake	C. J. Fay	Hammonton, N. J	Sept. 25, 1860	30, 132
Hay-rake	P. Fitzgerald	Constantia, Ohio	Oct. 5, 1858	21, 664
Hay-rake	A. J. Greene	Sterling, Mass	Nov. 5, 1867	70, 550
Hay-rake	T. C. Hance	Palmyra, N. Y	June 25, 1821	....
Hay-rake	H. Heaberlin	Scipio, Ind	Sept. 2, 1856	15, 653
Hay-rake	S. J. Homan	Walden, N. Y	Oct. 2, 1860	30, 222
Hay-rake	W. King	Springfield, Ill	Sept. 1, 1868	81, 791
Hay-rake	R. G. Lamson	Brownsville, Vt	Oct. 16, 1866	58, 841
Hay-rake	W. H. Misner and G. E. Marker.	Heyworth, Ill	Aug. 2, 1870	105, 967
Hay-rake	J. I. Monroe	Burlington, Mass	Mar. 19, 1867	63, 079
Hay-rake	F. L. Nagler	Irving Township, Mich	Jan. 24, 1871	111, 138
Hay-rake	G. Notman	Deerfield, Ohio	Aug. 8, 1871	117, 917
Hay-rake	F. B. Parker	Queensville, Ind	Aug. 23, 1853	9, 956
Hay-rake	M. Raezer	Reading, Pa	Oct. 5, 1858	21, 698
Hay-rake	T. R. Roach	West Needham, Mass	Nov. 4, 1856	16, 025
Hay-rake	I. I. Robbins	Penn's Manor, Pa	Sept. 23, 1856	15, 777
Hay-rake	D. Robinson	Reading Centre, N. Y	Apr. 1, 1862	34, 848
Hay-rake	Z. Sanders	West Windsor, Vt	June 8, 1852	9, 007
Hay-rake	D. Smith	Vincennes, Ind	June 7, 1838	773
Hay-rake	C. R. Soule	Fairfield, Vt	May 18, 1852	8, 965
Hay-rake	C. R. Soule	Fairfield, Vt	July 9, 1861	32, 791
Hay-rake	J. J. Squire	Saint Louis, Mo	Dec. 23, 1856	16, 318
Hay-rake	J. C. Stoddard	Lockport, N. Y	June 4, 1872	127, 659
Hay-rake	J. S. Sturges	Litchfield, Ohio	Mar. 9, 1852	8, 791
Hay-rake	J. Wallace and D. Carpenter	Goshen, N. Y	Dec. 1, 1863	40, 782
Hay-rake	G. Whitcomb	Port Chester, N. Y	Oct. 5, 1858	21, 712
Hay-rake	F. G. Willson	Ontario, Canada	July 30, 1861	32, 935
Hay-rake	S. W. Wood	Washington, D. C	July 7, 1857	17, 772
Hay rake and feeder, Combined	S. J. Taylor	Rome, N. Y	May 24, 1870	103, 524
Hay rake and fork	T. W. Peirce	Minneapolis, Minn	July 18, 1871	117, 201
Hay-rake and grain-harrow	G. Davis	Belmont, Ohio	July 22, 1839	1, 258
Hay rake and loader	J. Adams	Transfer, Pa	June 16, 1868	78, 912
Hay rake and loader	H. P. Burdick	Buffalo, N. Y	June 1, 1869	90, 818
Hay rake and loader	R. Chesnut	Richmond, Ind	Mar. 23, 1869	88, 009
Hay rake and loader	A. Clark	Cadiz, Ohio	May 25, 1869	90, 425
Hay rake and loader	W. A. Dean	New Lexington, Ohio	July 16, 1872	129, 009
Hay rake and loader	N. Gabel	Minneapolis, Kans	Jan. 4, 1870	98, 581
Hay rake and loader	F. W. Harlow	Hannibal, Mo	June 22, 1869	91, 623
Hay rake and loader	J. W. Henry	Pecatonica, Ill	Oct. 1, 1867	69, 432
Hay rake and loader	H. Kimmell	Waynesburgh, Ohio	Aug. 13, 1867	67, 771
Hay rake and loader	O. Luce	Virgil, N. Y	Mar. 31, 1868	76, 215
Hay rake and loader	F. Marion and N. E. Wilson	Groveland, Ill	Jan. 21, 1873	134, 997
Hay rake and loader	A. J. Preston	East Guilford, N. Y	July 18, 1871	117, 204
Hay rake and loader	A. Sheline	Edon, Ohio	Sept. 14, 1869	94, 782
Hay rake and loader	F. Terwilleger and J. R. Isdell	Wyanet, Ill	Sept. 14, 1869	94, 791
Hay rake and loader	J. J. Thompson	Richwood, Ohio	Oct. 13, 1868	83, 117
Hay rake and loader	M. Webb	Chenango Forks, N. Y	Sept. 3, 1867	68, 587
Hay rake and loader	G. A. Wing	Albany, N. Y	May 3, 1870	102, 744

Index of patents issued from the United States Patent Office from 1790 *to* 1873, *inclusive*—Continued.

Invention.	Inventor.	Residence.	Date.	No.
Hay rake and seeder combined, Horse	J. H. Bear	York, Pa	Jan. 1, 1867	60, 826
Hay rake and tedder	J. C. Mills	Palmyra, N. Y	Dec. 7, 1869	97, 543
Hay rake and tedder combined	C. La Dow	South Galway, N. Y	Mar. 28, 1871	113, 066
Hay rake and tedder combined	F. A. Thayer	Sheldonville, Mass	Apr. 25, 1871	114, 065
Hay rake and tedder, Combined horse	H. C. Varnum	Hartford, Vt	July 19, 1870	105, 611
Hay-rake fastening	O. W. Hogle	Somerset, N. Y	Sept. 3, 1850	7, 618
Hay-rake, Flexible horse	W. Buckminster	Framingham, Mass	July 12, 1838	838
Hay-rake, Horse	N. E. Allen	Trenton, Wis	Apr. 20, 1858	19, 975
Hay-rake, Horse	A. Amos	Potsdam Junction, N. Y	Nov. 4, 1873	144, 305
Hay-rake, Horse	A. Amos	Potsdam Junction, N. Y	Nov. 4, 1873	144, 306
Hay-rake, Horse	F. Andrews	Galesburgh, Ill	June 27, 1871	116, 254
Hay-rake, Horse	O. B. Austin	Potsdam Junction, N. Y	Dec. 24, 1872	134, 239
Hay-rake, Horse	L. B. Ball	Dayton, Ohio	Feb. 27, 1872	124, 110
Hay-rake, Horse	A. B. Barnard	Worcester, Mass	Feb. 22, 1870	100, 104
Hay-rake, Horse	N. M. Barnes	Tiffin, Ohio	July 19, 1870	105, 542
Hay-rake, Horse	N. M. Barnes	Tiffin, Ohio	May 28, 1872	127, 295
Hay-rake, Horse	J. H. Bean	Macon, Ill	July 8, 1873	140, 671
Hay-rake, Horse	C. D. Blinn	Port Huron, Mich	Aug. 27, 1867	68, 160
Hay-rake, Horse	O. Bonney, jr	San Francisco, Cal	Jan. 31, 1871	111, 308
Hay-rake, Horse	L. S. Bortree	Grand Rapids, Mich	June 21, 1870	104, 545
Hay-rake, Horse	C. L. and F. Brough	Delphi, Ind	Dec. 31, 1872	134, 353
Hay-rake, Horse	A. Brown	Worcester, Mass	July 12, 1870	105, 165
Hay-rake, Horse	F. Brown	Russell, Ohio	Jan. 30, 1872	123, 081
Hay-rake, Horse	J. H. Bullard	Chicopee, Mass	Nov. 12, 1872	133, 012
Hay-rake, Horse	G. E. Burt	Harvard, Mass	Dec. 17, 1872	133, 920
Hay-rake, Horse	M. Butler	Vernon, Ind	Mar. 19, 1867	63, 011
Hay-rake, Horse	G. E. Carleton	Old Town, Me	Sept. 27, 1870	107, 660
Hay-rake, Horse	A. L. Chubb	Grand Rapids, Mich	Sept. 26, 1871	119, 320
Hay-rake, Horse	J. M. Colson	Morrill, Me	Aug. 23, 1870	106, 554
Hay-rake, Horse	J. Comly	York, Pa	Dec. 26, 1871	122, 109
Hay-rake, Horse	I. N. Condra	Genoa, Iowa	Jan. 31, 1871	111, 320
Hay-rake, Horse	J. H. Cook	Hagerstown, Md	Feb. 8, 1870	99, 643
Hay-rake, Horse	A. Cowley	Harpersfield, N. Y	Mar. 30, 1858	19, 753
Hay-rake, Horse	J. Crellin	Marshalltown, Iowa	May 16, 1865	47, 703
Hay-rake, Horse	L. A. Crockett	Wythe County, Va	Nov. 21, 1871	121, 159
Hay-rake, Horse	N. W. Curtis	Johnsburgh, N. Y	Jan. 25, 1870	99, 066
Hay-rake, Horse	S. L. Denny	Christiana, Pa	Aug. 16, 1870	106, 336
Hay-rake, Horse	B. J. Downing	Mitchell, Iowa	Sept. 30, 1873	143, 278
Hay-rake, Horse	A. T. Dunbar	Alba, Pa	Dec. 15, 1868	84, 999
Hay-rake, Horse	H. Eastman	Indianapolis, Ind	Mar. 6, 1860	27, 407
Hay-rake, Horse	C. Edgar	Dayton, Ohio	Sept. 16, 1873	142, 844
Hay-rake, Horse	D. S. Emrick	Fayette, N. Y	Apr. 26, 1870	102, 238
Hay-rake, Horse	W. P. Ewing	Fancy Hill, Va	Sept. 21, 1869	95, 010
Hay-rake, Horse	J. F. Faust	Lebanon, Ohio	July 6, 1858	20, 844
Hay-rake, Horse	W. H. Fay	Camden, N. J	June 14, 1870	104, 131
Hay-rake, Horse	H. Gale	Albion, Mich	Nov. 9, 1869	96, 576
Hay-rake, Horse	O. L. Genung and W. H. Blackman.	Caroline, N. Y	Nov. 19, 1872	133, 091
Hay-rake, Horse	C. N. Goss	Claremont, N. H	Jan. 28, 1873	135, 330
Hay-rake, Horse	P. M. Gundlach	Belleville, Ill	Nov. 2, 1869	96, 425
Hay-rake, Horse	E. R. Hall	Utica, N. Y	June 9, 1868	78, 735
Hay-rake, Horse	G. N. Hall	Mamakating, N. Y	Dec. 6, 1859	26, 396
Hay-rake, Horse	J. Harris	Mount Vernon, Ind	June 11, 1872	127, 876
Hay-rake, Horse	W. H. Hartley	Philadelphia, Pa	Oct. 14, 1873	143, 572
Hay-rake, Horse	G. Hauck	Mechanicsburgh, Pa	Sept. 13, 1870	107, 256
Hay-rake, Horse	J. Hayward	Greene, N. Y	Oct. 2, 1866	58, 412
Hay-rake, Horse	W. A. Heath	Apalachin, N. Y	May 19, 1868	78, 086
Hay-rake, Horse	J. Heidy	Martinsburgh, Ohio	Nov. 12, 1872	133, 028
Hay-rake, Horse	T. N. Henderson	Jackson, Mich	Apr. 23, 1867	64, 104
Hay-rake, Horse	J. F. Henkley	Saint Louis, Mo	Mar. 28, 1871	113, 165
Hay-rake, Horse	M. S. Holman and C. S. Farrer.	Armada, Mich	Aug. 15, 1871	118, 130
Hay-rake, Horse	W. Holmes	Clarksville, N. Y	Sept. 8, 1868	82, 002
Hay-rake, Horse	B. F. Horton	Ithaca, N. Y	Feb. 15, 1870	99, 897
Hay-rake, Horse	C. Howard	West Hurley, N. Y	Feb. 1, 1870	99, 319
Hay-rake, Horse	J. Howard and E. S. Bousfield.	Bedford, England	Mar. 7, 1871	112, 461
Hay-rake, Horse	J. G. Huntington	Atkinson, Me	Oct. 19, 1869	96, 005
Hay-rake, Horse	A. K. Hurst	Chambersburgh, Pa	July 10, 1860	29, 082
Hay-rake, Horse	G. L. Ives	Rome, N. Y	Aug. 13, 1872	130, 431
Hay-rake, Horse	O. S. Jarvis	Xenia, Ill	Apr. 25, 1871	114, 150
Hay-rake, Horse	A. B. Johnson	Washington, Ind	Mar. 6, 1860	27, 366
Hay-rake, Horse	J. F. Keller	Hagerstown, Md	Mar. 25, 1873	137, 212
Hay-rake, Horse	J. G. Kimberlin	Dryden, N. Y	Apr. 30, 1867	64, 331
Hay-rake, Horse	W. S. Kincaid	Leavenworth County, Kans	Jan. 19, 1869	86, 080
Hay-rake, Horse	G. W. King	Greenville, N. Y	May 2, 1865	47, 553
Hay-rake, Horse	C. Kugler	Barnesville, Ohio	Sept. 5, 1871	118, 726
Hay-rake, Horse	G. O. Lackey	Akron, Ohio	Sept. 6, 1870	107, 183
Hay-rake, Horse	D. Z. Lantz	Salisbury Township, Pa	Dec. 17, 1872	133, 944
Hay-rake, Horse	S. Lessig	Reading, Pa	Dec. 27, 1859	26, 599
Hay-rake, Horse	L. Litchfield and J. S. Corbin	Gouverneur, N. Y	July 1, 1873	140, 513
Hay-rake, Horse	W. H. Locke	Canton, Pa	Apr. 18, 1871	113, 895
Hay-rake, Horse	J. G. Lockwood	West Davenport, N. Y	May 23, 1871	115, 071
Hay-rake, Horse	J. G. Lockwood	West Davenport, N. Y	Dec. 19, 1871	122, 039
Hay-rake, Horse	W. H. Long	Eddyville, Iowa	Feb. 1, 1870	99, 331
Hay-rake, Horse	C. M. Lufkin and C. G. Allen	Barre, Mass	Aug. 19, 1873	142, 031
Hay-rake, Horse	S. I. Lynd and E. M. Tousley	Albion, N. Y	July 23, 1872	129, 742
Hay-rake, Horse	S. I. Lynd and E. M. Tousley	Albion, N. Y	July 23, 1872	129, 743
Hay-rake, Horse	R. H. Martindale	Hobbieville, Ind	Apr. 30, 1872	126, 314
Hay-rake, Horse	W. C. Martindale	Philadelphia, Pa	July 8, 1873	140, 634
Hay-rake, Horse	W. Matthews	Vinton, Ohio	Aug. 31, 1869	94, 425
Hay-rake, Horse	R. W. McClelland	Springfield, Ill	Apr. 12, 1870	101, 752
Hay-rake, Horse	J. A. McGee	Sharpsville, Ind	May 17, 1870	103, 065
Hay-rake, Horse	G. M. L. McMillen	Drayton, Ohio	Aug. 16, 1870	106, 385
Hay-rake, Horse	J. H. Mears	Oshkosh, Wis	July 29, 1873	141, 370
Hay-rake, Horse	C. R. Merriam	Middlebury, Vt	Jan. 3, 1871	110, 771
Hay-rake, Horse	J. Mills	Milan, Ind	Aug. 23, 1870	106, 604
Hay-rake, Horse	P. J. Moose and J. Kuhn	Dansville, N. Y	Oct. 24, 1871	120, 160
Hay-rake, Horse	J. Morgan and W. Cline, jr	Clayton, Ind	Feb. 1, 1870	99, 457
Hay-rake, Horse	J. H. Morris	Maquoketa, Iowa	Apr. 5, 1870	101, 646

Index of patents issued from the United States Patent Office from 1790 *to* 1873, *inclusive*—Continued.

Invention.	Inventor.	Residence.	Date.	No.
Hay-rake, Horse	B. Morse	Ithaca, N. Y	Aug. 19, 1873	141, 884
Hay-rake, Horse	J. I. Munroe	Woburn, Mass	Apr. 12, 1870	101, 904
Hay-rake, Horse	L. Myers	Ludlowville, N. Y	July 5, 1870	105, 117
Hay-rake, Horse	W. A. Myers	York, Pa	Apr. 4, 1871	113, 552
Hay-rake, Horse	O. T. Nanny	Amity, N. Y	Nov. 22, 1870	109, 439
Hay-rake, Horse	D. B. Neal	Mount Gilead, Ohio	Aug. 17, 1869	93, 829
Hay-rake, Horse	G. Notman	Deerfield, Ohio	May 3, 1870	102, 578
Hay-rake, Horse	G. Notman	Deerfield, Ohio	May 9, 1871	114, 591
Hay-rake, Horse	S. R. Nye	Barre, Mass	July 26, 1870	105, 833
Hay-rake, Horse	J. K. O'Neil	Kingston, N. Y	Feb. 15, 1870	99, 939
Hay-rake, Horse	L. A. Paddock	Pecatonica, Ill	Feb. 6, 1872	123, 357
Hay-rake, Horse	L. A. Paddock	Pecatonica, Ill	Apr. 2, 1872	125, 216
Hay-rake, Horse	M. Pennock	East Marlborough, Pa	June 26, 1822	....
Hay-rake, Horse	P. Pfeifer	Durhamville, N. Y	Oct. 3, 1871	119, 532
Hay-rake, Horse	R. Pratt	Marple Township, Pa	Jan. 8, 1856	14, 067
Hay-rake, Horse	P. Prescott	Boonville, N. Y	Oct. 6, 1868	82, 749
Hay-rake, Horse	M. Raezer	Reading, Pa	Nov. 29, 1859	26, 294
Hay-rake, Horse	M. C. Remington	Weedsport, N. Y	Sept. 28, 1869	95, 268
Hay-rake, Horse	M. C. Remington	Weedsport, N. Y	Oct. 10, 1871	119, 791
Hay-rake, Horse	G. S. Reynolds	East Bethel, Vt	May 10, 1859	23, 943
Hay-rake, Horse	M. J. Robinson	Ashley, Ill	Sept. 13, 1870	107, 411
Hay-rake, Horse	S. Rockafellow	Moline, Ill	Jan. 24, 1871	111, 150
Hay-rake, Horse	S. Rockafellow	Moline, Ill	Dec. 26, 1871	122, 282
Hay-rake, Horse	S. Rockafellow	Moline, Ill	Mar. 19, 1872	124, 855
Hay-rake, Horse	R. S. Ryker	Canaan, Ind	Sept. 10, 1872	131, 183
Hay-rake, Horse	C. W. Sanborn	Morrill, Me	Jan. 4, 1870	98, 637
Hay-rake, Horse	W. Sharkey	Chico, Cal	Dec. 7, 1869	97, 708
Hay-rake, Horse	D. P. Sharp	Ithaca, N. Y	Mar. 5, 1872	124, 225
Hay rake, Horse	R. B. Sheldon	Canastota, N. Y	Apr. 1, 1873	137, 389
Hay-rake, Horse	J. H. Shoonmaker	Bethlehem, N. Y	Mar. 21, 1871	112, 854
Hay-rake, Horse	J. S. Shrawder	Fairview Village, Pa	May 17, 1870	103, 092
Hay-rake, Horse	A. J. Skunk	Des Moines, Iowa	July 26, 1870	105, 733
Hay-rake, Horse	S. P. Smith	Waterford, N. Y	Apr. 19, 1870	102, 172
Hay-rake, Horse	S. P. Smith	Waterford, N. Y	May 31, 1870	103, 789
Hay-rake, Horse	S. P. Smith	Waterford, N. Y	Mar. 7, 1871	112, 507
Hay-rake, Horse	S. P. Smith	Waterford, N. Y	Jan. 23, 1872	123, 053
Hay-rake, Horse	S. Stevenson	Dansville, N. Y	Mar. 11, 1873	136, 785
Hay-rake, Horse	J. C. Stoddard	Worcester, Mass	Aug. 6, 1861	33, 023
Hay-rake, Horse	J. C. Stoddard	Worcester, Mass	Sept. 27, 1870	107, 832
Hay-rake, Horse	J. C. Stoddard	Worcester, Mass	Jan. 10, 1871	110, 876
Hay-rake, Horse	O. O. Storle	North Cape, Wis	Apr. 18, 1871	113, 813
Hay-rake, Horse	G. Sweet	Dansville, N. Y	Oct. 3, 1871	119, 669
Hay-rake, Horse	G. Sweet	Dansville, N. Y	Nov. 25, 1873	145, 029
Hay-rake, Horse	B. C. Taylor	Dayton, Ohio	Sept. 1, 1868	81, 706
Hay-rake, Horse	R. M. Treat	Morris, Conn	May 2, 1871	114, 368
Hay-rake, Horse	M. W. Trescott	North Canaan, Conn	June 13, 1871	115, 909
Hay-rake, Horse	C. Tyler	Dryden, N. Y	Mar. 22, 1870	101, 189
Hay-rake, Horse	J. H. Walker	Grand Rapids, Mich	Aug. 12, 1873	141, 684
Hay-rake, Horse	C. W. Warner	New Haven, Vt	Aug. 18, 1868	81, 271
Hay-rake, Horse	D. H. Weaver	South Bend, Ind	May 31, 1870	103, 691
Hay-rake, Horse	G. Weiland	Dayton, Ohio	Mar. 22, 1870	101, 196
Hay-rake, Horse	J. R. Whittemore	Chicopee Falls, Mass	Nov. 19, 1872	133, 280
Hay-rake, Horse	J. E. Wisner	Friendship, N. Y	Oct. 25, 1870	108, 744
Hay-rake, Horse	J. E. Wisner	Friendship, N. Y	Nov. 29, 1870	109, 700
Hay-rake, Horse	J. E. Wisner	Friendship, N. Y	Nov. 14, 1871	121, 035
Hay-rake, Horse	E. Zimerman	Pamelia Four Corners, N. Y.	May 31, 1870	103, 815
Hay-rake, Revolving	E. Calderwood	Portland, Me	Jan. 21, 1865	46, 447
Hay-rake, Revolving	H. and E. Hoopes	Wilmington, Del	July 26, 1824	....
Hay-rake, Revolving	E. B. and M. D. Wells	Morgantown, Va	Sept. 20, 1839	1, 329
Hay-rake spring-teeth	Z. Breed	Weare, N. H	June 18, 1859	7, 434
Hay rake, thrasher, loader, and stacker	J. R. Hammond	Sedalia, Mo	Nov. 30, 1869	97, 395
Hay-rakes, Tooth-fastening for horse	B. Pickering	Dayton, Ohio	July 9, 1872	128, 749
Hay raker and loader	J. Armstrong and J. Jeffcoat	Onawa, Iowa	Feb. 2, 1869	86, 345
Hay raker and loader	H. Baker	Cortland, N. Y	July 3, 1866	55, 979
Hay raker and loader	J. M. Boorman	Scarborough, N. Y	Apr. 28, 1868	77, 247
Hay raker and loader	C. C. Brandt and J. C. Shillook	New Ulm, Minn	June 18, 1872	128, 012
Hay raker and loader	W. F. Brown and J. N. Smith	Washington, D. C., and Jersey City, N. J.	Dec. 3, 1867	71, 577
Hay raker and loader	A. Campbell	Peoria, Ill	Oct. 15, 1867	69, 764
Hay raker and loader	J. F. Chandler	Concord, N. H	Aug. 17, 1869	93, 806
Hay raker and loader	L. Clarke	Candor, N. Y	Feb. 11, 1868	74, 307
Hay raker and loader	J. S. Coffman and M. Graybill	Greenville, Ind	Mar. 17, 1868	75, 528
Hay raker and loader	L. D. Copeland	Chenango Forks, N. Y	Sept. 24, 1867	69, 187
Hay raker and loader	E. N. Curtice	Springwater, N. Y	Nov. 26, 1867	71, 366
Hay raker and loader	M. A. Dilly	Mendon, Mich	May 12, 1868	77, 721
Hay raker and loader	R. Elarton and E. A. Thomas	Hillsborough, Iowa	Jan. 5, 1869	85, 651
Hay raker and loader	N. Farlow and J. A. Ham	Sullivan, Ill	Oct. 5, 1869	95, 451
Hay raker and loader	J. W. Foust	Evansburgh, Pa	Oct. 16, 1866	58, 804
Hay raker and loader	M. B. Fuller	Sanborn, N. Y	Feb. 25, 1868	74, 814
Hay raker and loader	G. B. Garlinghouse and J. C. Moore.	Madison, Ind	Apr. 14, 1868	76, 743
Hay raker and loader	A. Godfrey	Dupage, Ill	Oct. 15, 1867	69, 794
Hay raker and loader	J. Harper	Hillsborough, Iowa	Mar. 10, 1868	75, 416
Hay raker and loader	S. R. Higgins	Parma, Mich	May 12, 1868	77, 817
Hay raker and loader	W. H. Hiteshew	Perrysburgh, Ind	Sept. 22, 1868	82, 316
Hay raker and loader	G. F. Holmes	Cortland, N. Y	Feb. 16, 1869	87, 045
Hay raker and loader	H. Kewley	Perry, Ohio	Dec. 3, 1867	71, 764
Hay raker and loader	N. S. Kinyon and B. F. Smith	Chenango Forks, N. Y	Sept. 17, 1867	68, 887
Hay raker and loader	J. C. Leonard, S. B. Holcomb, and W. B. Wright.	Clinton, Mo	Oct. 5, 1869	95, 493
Hay raker and loader	S. M. Livingston	Claverack, N. Y	Jan. 28, 1868	73, 731
Hay raker and loader	A. W. Lozier	New York, N. Y	June 15, 1869	91, 243
Hay raker and loader	G. C. Mead	Guilford, N. Y	Nov. 19, 1867	71, 198
Hay raker and loader	J. C. Moore and C. B. and G. B. Garlinghouse.	Madison and North Madison, Ind.	Feb. 23, 1869	87, 106
Hay raker and loader	S. K. Morse	Commerce, Mich	May 5, 1868	77, 641
Hay raker and loader	W. T. Nichols	Rutland, Vt	Jan. 28, 1868	73, 747
Hay raker and loader	W. T. Nichols	Rutland, Vt	Mar. 10, 1868	75, 450

Index of patents issued from the United States Patent Office from 1790 to 1873, inclusive—Continued.

Invention.	Inventor.	Residence.	Date.	No.
Hay raker and loader	C. Plaisted	Kenockee, Mich	Jan. 21, 1868	73, 460
Hay raker and loader	A. J. Preston	Guilford, N. Y	Mar. 24, 1868	75, 790
Hay raker and loader	G. H. Reister	Washington, Iowa	Mar. 15, 1870	100, 804
Hay raker and loader	F. M. Robinson and T. G. Springer.	Conneautville, Pa	July 16, 1867	66, 741
Hay raker and loader	A. Royse and M. K. Morris	Le Roy, Pa	Dec. 3, 1867	71, 792
Hay raker and loader	J. Ruhl and E. S. Herrington	Defiance, Ohio	May 19, 1868	78, 136
Hay raker and loader	B. Shirley	Moravia, N. Y	Oct. 22, 1867	70, 032
Hay raker and loader	G. W. Swartz	Newburgh, Pa	July 2, 1867	66, 412
Hay raker and loader	L. Underwood	Ottawa, Ill	Oct. 6, 1868	82, 770
Hay raker and loader	A. Vose	Pittsfield, Vt	Dec. 10, 1867	72, 133
Hay raker and loader	H. Willard	Vergennes, Vt	Oct. 15, 1867	69, 881
Hay raking and bunching machine	L. Wallace	Leavenworth City, Kans	Dec. 12, 1865	51, 500
Hay raking and cocking machine	F. Granger	Homer, Ill	Jan. 1, 1867	60, 880
Hay raking and cocking machine	Z. C. Steele	Pana, Ill	May 17, 1870	103, 099
Hay raking and cocking machine	S. Stewart	Lowell, Mass	May 17, 1870	103, 100
Hay raking and cocking machine	L. R. Stone	Owassa, Mich	Feb. 26, 1861	31, 568
Hay raking and cocking machine	J. Wadleigh	Chebanse, Ill	May 4, 1869	89, 611
Hay raking and loading machine	J. A. Althouse	Phillipstown, Ill	Nov. 22, 1859	26, 225
Hay raking and loading machine	W. W. Barnes	Lyme, N. H	Feb. 27, 1866	52, 814
Hay raking and loading machine	J. B. Benton, J. F. Behn, and G. Bastian.	Buffalo, N. Y	July 6, 1858	20, 772
Hay raking and loading machine	M. and D. Cole	Colebrook and Kent, Conn	Oct. 9, 1866	58, 601
Hay raking and loading machine	M. G. Couch	Odessa, N. Y	July 23, 1861	32, 860
Hay raking and loading machine	W. A. Duncan	Syracuse, N. Y	Mar. 7, 1865	46, 647
Hay raking and loading machine	W. A. Duncan	Syracuse, N. Y	Apr. 24, 1866	54, 133
Hay raking and loading machine	W. A. Duncan	Syracuse, N. Y	July 30, 1867	67, 176
Hay raking and loading machine	T. Frame	Barnesville, Ohio	Oct. 30, 1866	59, 205
Hay raking and loading machine	J. K. Harris	Allensville, Ind	Dec. 11, 1855	13, 908
Hay raking and loading machine	A. W. Heaney	Doylestown, Pa	May 29, 1866	55, 098
Hay raking and loading machine	G. Howard	Westfield Township, Ohio	July 19, 1859	24, 806
Hay raking and loading machine	M. K. Lewis, J. C. Durbin, and L. P. Lewis.	Iowa City, Iowa	Aug. 15, 1865	49, 423
Hay raking and loading machine	W. D. Mayfield	Ashley, Ill	Mar. 27, 1866	53, 464
Hay raking and loading machine	F. Nevergold and G. Stackhouse.	Pittsburgh, Pa	May 1, 1866	54, 392
Hay raking and loading machine	H. S. Palmer	Norvell, Mich	Sept. 25, 1866	58, 282
Hay raking and loading machine	J. S. Preston	Mechanicsville, Pa	Feb. 13, 1866	52, 601
Hay raking and loading machine	M. S. and C. B. Rawson	Winhall and Londonderry, Vt.	Feb. 19, 1867	62, 224
Hay raking and loading machine	J. Smith	Condit, Ohio	June 3, 1856	15, 046
Hay raking and loading machine	J. N. Smith	Jersey City, N. J	Mar. 14, 1865	46, 863
Hay raking and loading machine	J. P. Smith	Hudson, N. Y	July 10, 1866	56, 331
Hay raking and loading machine	D. H. Thompson	Fitchburgh, Mass	Mar. 25, 1856	14, 538
Hay raking and loading machine	B. M. Townsend	Quincy, Ill	Apr. 9, 1850	7, 277
Hay raking and loading machine	T. J. Wallace	Cameron, Ill	Sept. 13, 1859	25, 459
Hay raking and loading machine	M. Webb	Chenango Forks, N. Y	Oct. 23, 1866	59, 105
Hay raking and loading machine	H. Willard	Vergennes, Vt	Aug. 21, 1866	57, 421
Hay raking and loading machine	R. and J. Wright	Franklin Township, Pa	Oct. 23, 1866	59, 116
Hay raking and pitching machine	A. I. Preston	East Guilford, N. Y	Oct. 16, 1860	30, 423
Hay-raking machine	D. Ramler	Union Deposit, Pa	Apr. 26, 1859	23, 782
Hay-rigging	O. Reed	Paris, Mich	May 20, 1862	35, 352
Hay-shocker	C. M. Terrell	Oskaloosa, Iowa	Sept. 17, 1867	69, 045
Hay-shocking machine	N. W. Plymate	Freeland, Iowa	Mar. 8, 1864	41, 859
Hay, &c., Sling for loading and unloading	G. W. Long	Delaware Centre, Iowa	Sept. 9, 1873	142, 638
Hay-spreader	R. Adams and J. D. Sheetz	Heidelburgh Township, Pa	June 22, 1869	91, 699
Hay-spreader	A. B. Barnard	Worcester, Mass	Dec. 8, 1868	84, 725
Hay-spreader	H. Beers	Brookfield, Conn	May 30, 1865	47, 919
Hay-spreader	M. Bowker	Newark, N. J	Oct. 19, 1869	95, 875
Hay-spreader	N. Brand	Ilion, N. Y	Mar. 2, 1869	87, 392
Hay-spreader	H. M. Burdick	Ilion, N. Y	Mar. 30, 1869	88, 270
Hay-spreader	J. M. Burdick	Ilion, N. Y	Nov. 17, 1868	84, 087
Hay-spreader	G. E. Burt	Harvard, Mass	June 25, 1867	66, 074
Hay-spreader	G. E. Burt	Harvard, Mass	Oct. 15, 1867	69, 761
Hay-spreader	G. E. Burt and E. A. Hildreth	Harvard, Mass	Nov. 24, 1868	84, 257
Hay-spreader	W. H. Butterworth	Trenton, N. J	Nov. 24, 1868	84, 409
Hay-spreader	W. H. Butterworth	Trenton, N. J	May 18, 1869	90, 077
Hay-spreader	A. H. Caryl	Groton, Mass	Aug. 11, 1868	80, 806
Hay-spreader	N. Chapman	Milford, Mass	Sept. 22, 1868	82, 322
Hay-spreader	D. B. Clement	Brighton, Mass	July 6, 1869	92, 166
Hay-spreader	T. C. Craven	Albany, N. Y	Dec. 8, 1868	84, 800
Hay-spreader	N. Eaton	Woburn, Mass	Sept. 7, 1869	94, 480
Hay-spreader	W. H. Elliot	New York, N. Y	Feb. 2, 1869	86, 516
Hay-spreader	W. H. Elliot	New York, N. Y	June 8, 1869	91, 001
Hay-spreader	C. R. Frink	Norwich, N. Y	May 8, 1866	54, 524
Hay-spreader	C. R. Frink	Norwich, N. Y	Oct. 23, 1866	59, 001
Hay-spreader	C. R. Frink	Norwich, N. Y	Dec. 24, 1867	72, 624
Hay-spreader	C. R. Frink	Norwich, N. Y	Jan. 4, 1870	98, 486
Hay-spreader	W. H. H. Frye	North Fryeburgh, Me	May 24, 1870	103, 441
Hay-spreader	J. Garfield	Groton, Mass	May 7, 1867	64, 518
Hay-spreader	W. C. Gifford	Jamestown, N. Y	Jan. 24, 1865	45, 992
Hay-spreader	D. Kennedy	New York, N. Y	July 28, 1868	80, 412
Hay-spreader	W. H. Kenyon	Wakefield, R. I	Aug. 3, 1869	93, 313
Hay-spreader	L. L. Knight	Barre, Mass	Apr. 3, 1866	53, 739
Hay-spreader	G. G. Knowles	Wakefield, R. I	May 12, 1868	77, 739
Hay-spreader	D. Lyman	Middlefield, Mass	Apr. 25, 1865	47, 437
Hay-spreader	N. F. Matheuson	Barrington, R. I	Mar. 30, 1869	88, 314
Hay-spreader	M. D. Myers	Frankfort, N. Y	Feb. 9, 1869	86, 777
Hay-spreader	E. W. Nichols	Worcester, Mass	Dec. 28, 1869	98, 292
Hay-spreader	H. L. Perkins	Kingsman, Ohio	May 31, 1870	103, 771
Hay-spreader	E. D. and O. B. Reynolds	North Bridgewater, Mass	Feb. 16, 1869	87, 066
Hay-spreader	G. T. Savary	Byfield, Mass	Jan. 12, 1869	85, 858
Hay-spreader	J. C. Stoddard	Worcester, Mass	June 29, 1869	92, 122
Hay-spreader	B. F. Taft	Groton Junction, Mass	Apr. 13, 1869	88, 822
Hay-spreader	J. A. Talpey	Somerville, Mass	June 22, 1869	91, 790
Hay-spreader	J. F. Thomas	Ilion, N. Y	May 11, 1869	89, 899
Hay-spreader	J. F. Thomas and D. H. McLane	Ilion, N. Y	July 5, 1870	105, 141

Index of patents issued from the United States Patent Office from 1790 *to* 1873, *inclusive*—Continued.

Invention.	Inventor.	Residence.	Date.	No.
Hay-spreader	M. N. Ward	Bangor, Me	Mar. 2, 1869	87, 383
Hay-spreader	G. I. Wooster	Plymouth, Conn	Aug. 16, 1870	106, 444
Hay spreader and cocker combined	R. T. Gill	Poughkeepsie, N. Y	Mar. 31, 1868	76, 073
Hay spreader and elevator combined	T. C. Craven	Albany, N. Y	Dec. 19, 1865	51, 650
Hay-spreader fork	G. E. Burt and E. A. Hildreth	Harvard, Mass	June 8, 1869	90, 991
Hay-spreader teeth	G. A. Squier	Syracuse, N. Y	Apr. 21, 1868	76, 952
Hay spreading and turning machine	G. Moody	Falmouth, Me	Oct. 28, 1862	36, 791
Hay-spreading machine	C. Willard	Newtown, Pa	July 25, 1865	49, 021
Hay-stacker	J. T. Breneman	Springfield, Ohio	Jan. 8, 1867	61, 044
Hay-stacker	T. N. Bunnell	Reynolds, Ind	Oct. 19, 1869	95, 879
Hay-stacker	J. Forshee and J. C. McCland	Unionville Centre, Ohio	Oct. 29, 1867	70, 189
Hay-stacker	W. S. Nichols	Rutland, Vt	Jan. 28, 1868	73, 748
Hay-stacker	C. Rundell	Chicago, Ill	July 10, 1866	56, 273
Hay-stacker	S. J. Wallace	Keokuk, Iowa	Apr. 23, 1867	64, 050
Hay-stacking apparatus	C. H. Kirkpatrick	Sugar Grove, Ind	July 30, 1872	129, 967
Hay-stacking apparatus	C. H. Tryon	Greenwood, Ill	Jan. 29, 1867	61, 642
Hay-stacking device	W. Louden	Fairfield, Iowa	Aug. 28, 1866	57, 525
Hay-stacking machine	W. M. Mason	Polo, Ill	Feb. 25, 1862	34, 513
Hay-stacks, &c., Portable roof for	T. Munson	Canandaigua, N. Y	July 28, 1868	80, 301
Hay, straw, and vegetables, Machine for cutting	R. I. Burbank	Boston, Mass	Dec. 22, 1868	85, 067
Hay-tedder	O. A. Benton	Amenia, N. Y	Dec. 6, 1870	109, 863
Hay-tedder	A. Brown	Worcester, Mass	Jan. 10, 1871	110, 893
Hay-tedder	R. Bryson	Schenectady, N. Y	Feb. 15, 1870	99, 833
Hay-tedder	E. W. Bullard	Barre, Mass	Jan. 31, 1871	111, 314
Hay-tedder	H. M. Burdick	Ilion, N. Y	Oct. 12, 1869	95, 647
Hay-tedder	H. M. Burdick	Ilion, N. Y	Apr. 18, 1871	113, 848
Hay-tedder	H. M. Burdick	Ilion, N. Y	Apr. 18, 1871	113, 849
Hay-tedder	J. M. Burdick	Ilion, N. Y	Mar. 22, 1870	100, 972
Hay-tedder	W. H. Butterworth	Trenton, N. J	Jan. 17, 1871	111, 041
Hay-tedder	A. H. Caryl	Groton, Mass	Aug. 17, 1869	93, 805
Hay-tedder	T. J. and G. M. Clark	Higganum, Conn	Mar. 15, 1870	100, 856
Hay-tedder	T. J. and G. M. Clark	Higganum, Conn	Aug. 19, 1873	141, 999
Hay-tedder	J. K. Collins	Hartford, Vt	Oct. 12, 1869	95, 656
Hay-tedder	J. K. Collins	Hartford, Vt	Nov. 7, 1871	120, 626
Hay-tedder	T. C. Craven	Albany, N. Y	Nov. 16, 1869	96, 892
Hay-tedder	C. R. Frink	Norwich, N. Y	Sept. 24, 1872	131, 513
Hay-tedder	L. Hale	Hollis, N. H	Feb. 21, 1871	112, 030
Hay-tedder	M. B. Harvey	Stafford, Conn	Sept. 13, 1870	107, 365
Hay-tedder	J. B. Kelley	Brandon, Vt	Sept. 28, 1869	95, 234
Hay-tedder	W. G. Kenyon	Wakefield, R. I	Mar. 14, 1871	112, 602
Hay-tedder	J. Lee	Pottstown, Pa	Feb. 13, 1872	123, 700
Hay-tedder	W. H. Mickle	Utica, N. Y	Sept. 12, 1871	118, 810
Hay-tedder	M. D. Myers	Frankfort, N. Y	Dec. 7, 1869	97, 677
Hay-tedder	J. G. Perry	Kingston, R. I	Nov. 30, 1869	97, 437
Hay-tedder	J. G. Perry	Kingston, R. I	Nov. 30, 1869	97, 438
Hay-tedder	J. G. Perry	Kingston, R. I	Aug. 1, 1871	117, 561
Hay-tedder	J. G. Perry	Kingston, R. I	Aug. 1, 1871	117, 562
Hay-tedder	J. G. Perry	Kingston, R. I	Dec. 19, 1871	121, 958
Hay-tedder	H. F. and G. F. Shaw	West Roxbury, Mass	July 16, 1872	129, 431
Hay-tedder	F. R. Smith	Ilion, N. Y	Aug. 2, 1870	106, 087
Hay-tedder	W. Smith	Weston, Mass	Apr. 19, 1870	102, 173
Hay-tedder	J. A. Talpey	Somerville, Mass	Nov. 9, 1869	96, 741
Hay-tedder	H. Warren	Leicester, Mass	June 21, 1870	104, 669
Hay tedder and rake, Combined	S. Perry	Newport, N. Y	Jan. 2, 1872	122, 487
Hay-tedders, Device for securing the tines of	D. Kennedy	New York, N. Y	Jan. 10, 1871	110, 856
Hay, Teeth for scattering	J. C. Stoddard	Worcester, Mass	July 24, 1860	29, 327
Hay turning and spreading machine	E. W. Bullard	Barre, Mass	July 21, 1861	32, 350
Hay-unloader	H. H. Ensminger	Buffalo, N. Y	July 5, 1870	104, 947
Hay-unloader	J. Wood	Smyrna, N. Y	May 14, 1867	64, 729
Hay, Unloading	H. H. Angell	Clermont, Iowa	June 26, 1860	28, 822
Hay unloading and mowing apparatus	A. Smith	Hoosick Four Corners, N.Y	May 23, 1871	115, 120
Hay unloading and stacking device	W. R. Waldron	Webster, Mich	May 28, 1867	65, 316
Hay-unloading apparatus	E. Harrison	Mountain View, Cal	Aug. 29, 1871	118, 531
Hay-unloading apparatus	G. Smith	Providence, R. I	Oct. 26, 1869	96, 161
Hay-unloading device	J. Backus	Greenvale, Ill	Oct. 13, 1868	83, 018
Hay-unloading device	G. Smith	Providence, R. I	Nov. 3, 1868	83, 667
Hay-unloading device	G. Van Sickle	Auburn, N. Y	Sept. 29, 1868	82, 570
Head and shoulder rest	A. V. W. Van Vechten	New York, N. Y	June 9, 1868	78, 702
Head-band	D. McKinnon	Wappinger's Falls, N. Y	Aug. 26, 1873	142, 169
Head-block	A. A. Adams	Felchville, Vt	Sept. 22, 1868	82, 269
Head-block	N. F. Beckwith	Omro, Wis	Dec. 26, 1871	122, 215
Head-block	C. R. Bushnell	Saint Anthony's Falls, Minn	Nov. 10, 1868	83, 830
Head-block	T. L. Clark	Mount Vernon, Ohio	Mar. 10, 1868	75, 371
Head-block	J. M. Comins	Boscawen, N. H	Dec. 26, 1871	122, 228
Head-block	J. F. Cook	Baltimore, Md	Dec. 15, 1868	84, 937
Head-block	P. M. Cummings	Cincinnati, Ohio	Dec. 13, 1870	110, 119
Head-block	P. M. Cummings	Lyons, Iowa	Nov. 11, 1873	144, 388
Head-block	T. Douglass	Warren, Ohio	Dec. 19, 1871	121, 934
Head-block	T. Douglass	Warren, Ohio	Jan. 30, 1872	123, 091
Head-block	C. R. Ely	Northfield, Vt	Feb. 2, 1869	86, 517
Head-block	P. Estes	Leavenworth, Kans	Dec. 31, 1867	72, 828
Head-block	H. H. Gridley	Auburn, N. Y	Aug. 30, 1870	106, 932
Head-block	N. Hunt	Salem, Ohio	July 20, 1869	92, 828
Head-block	W. A. L. Kirk	Hamilton, Ohio	Sept. 22, 1868	82, 328
Head-block	M. C. Lewis	Glasgow, Mo	Dec. 24, 1867	72, 510
Head-block	A. C. Martin and W. Ritchie	Hamilton, Ohio	Nov. 10, 1868	83, 982
Head-block	H. R. Martin	Hillsborough, N. H	Aug. 30, 1871	115, 492
Head-block	D. Parkhurst	Saint Louis, Mo	Feb. 6, 1872	123, 507
Head-block	H. M. Popple	Warren, Pa	May 23, 1871	115, 237
Head-block	E. H. Stearns	Erie, Pa	Sept. 1, 1868	81, 837
Head-block	W. Tauney	Jeromesville, Ohio	July 16, 1872	129, 620
Head-block, Shoemaker's	J. H. Morton	Deering, Me	Mar. 26, 1872	124, 908
Head-comb, Lady's	D. Warring	Boston, Mass	July 16, 1872	129, 441
Head-covering	T. Bracher	New York, N. Y	July 4, 1865	48, 510
Head-covering	M. D. Cohen	Philadelphia, Pa	Dec. 24, 1861	33, 984
Head-covering	E. Copleston	Wrentham, Mass	Jan. 22, 1867	61, 403
Head-covering	T. S. Lambert	Peekskill, N. Y	Oct. 20, 1863	40, 382
Head-dress for ladies	J. Edwards	New York, N. Y	Jan. 5, 1864	41, 062

Index of patents issued from the United States Patent Office from 1790 to 1873, inclusive—Continued.

Invention.	Inventor.	Residence.	Date.	No.
Head-dress for ladies	H. S. C. Iwerson	New York, N. Y	Aug. 21, 1866	57, 328
Head-dress for ladies, Waterfall	P. Walter	New York, N. Y	Mar. 21, 1865	46, 961
Head-dress for smoke, Safety	W. N. Ball	La Porte, Ind	Apr. 9, 1872	125, 515
Head dress, Infant's	L. E. Love	New York, N. Y	Nov. 28, 1871	121, 391
Head-dresses, Band for	N. Grant and G. Downs	Providence, R. I	July 11, 1865	48, 677
Head-light	H. S. Maxim and J. Radley	New York, N. Y	Sept. 28, 1869	95, 248
Head-light	M. M. Rounds	New Haven, Conn	Apr. 17, 1866	54, 020
Head-light burner, Locomotive	A. C. Vaughan	Philadelphia, Pa	Aug. 30, 1870	106, 971
Head-light, Engine	P. Budenbach	New York, N. Y	May 22, 1866	54, 850
Head-light, Engine	S. M. Davies	Chicago, Ill	Mar. 28, 1865	46, 998
Head-light, Engine	J. Radley	New York, N. Y	June 26, 1866	55, 904
Head-light, Gasoline locomotive	J. B. Terry	Hartford, Conn	Dec. 24, 1867	72, 697
Head-light lens or glass	H. C. Felthousen	Buffalo, N. Y	Dec. 27, 1870	110, 449
Head-light, Locomotive	J., W. H., and A. E. Briggs	Newport, Ky	Aug. 26, 1873	142, 204
Head-light, Locomotive	S. E. Cleveland	Buffalo, N. Y	June 3, 1862	35, 486
Head-light, Locomotive	A. Collins and J. Hardy	Galesburgh, Ill	June 27, 1871	116, 272
Head-light, Locomotive	S. M. Davis	Chicago, Ill	Dec. 6, 1870	109, 877
Head-light, Locomotive	E. L. Hall	Utica, N. Y	July 30, 1867	67, 192
Head-light, Locomotive	R. W. Love	Windsor, Vt	June 7, 1870	103, 905
Head-light, Locomotive	H. S. Maxim and J. Radley	New York, N. Y	Oct. 5, 1869	95, 498
Head-light, Locomotive	L. Michaels	Cincinnati, Ohio	Nov. 23, 1869	97, 211
Head-light, Locomotive	L. Michaels	Covington, Ky	Apr. 29, 1873	138, 426
Head-light, Locomotive	T. J. Newland	Utica, N. Y	Feb. 14, 1865	46, 380
Head-light, Locomotive	J. Radley	New York, N. Y	Nov. 13, 1866	59, 650
Head-light, Locomotive	T. S. Ray and S. E. Cleveland	Buffalo, N. Y	Dec. 5, 1865	51, 351
Head-light, Locomotive	T. S. Ray and S. E. Cleveland	Buffalo, N. Y	Apr. 10, 1866	53, 871
Head-light, Locomotive	P. F. Stout	Philadelphia, Pa	July 22, 1873	141, 183
Head-light, Locomotive	A. C. Vaughan	Philadelphia, Pa	Nov. 26, 1867	71, 431
Head light, Locomotive	W. Westlake	Chicago, Ill	Oct. 7, 1873	143, 550
Head lights, Mechanism for adjustment of	C. D. Gibson	New York, N. Y	May 15, 1866	54, 714
Head-lights, Reflector for locomotive	C. S. Lee and W. M. Baldwin	Troy, N. Y	July 18, 1871	117, 089
Head-rest	J. S. Bartlett	Warsaw, N. Y	Nov. 8, 1870	108, 956
Head-rest	E. P. Cook	Cartersville, Ga	Feb. 22, 1870	100, 014
Head-rest	A. Dunlap	Clyde, Ohio	Dec. 24, 1867	72, 467
Head-rest	F. Fabrici	New York, N. Y	Oct. 1, 1872	131, 816
Head-rest	D. R. V. Goetchins	Little Falls, N. Y	Dec. 28, 1869	98, 250
Head-rest	R. Hale	Chicago, Ill	Nov. 5, 1867	70, 555
Head-rest	H. C. Hunt	Amboy, Ill	Apr. 21, 1868	77, 043
Head-rest	C. V. Littlepage	Austin, Tex	May 31, 1870	103, 631
Head-rest	J. W. Lockwood	New York, N. Y	Feb. 7, 1860	27, 058
Head-rest	C. B. Loveless	Syracuse, N. Y	Dec. 21, 1869	98, 077
Head-rest	B. Lyon and C. M. Curtis	Springfield, Mass	June 21, 1870	104, 473
Head-rest	D. M. Mitchell	Marathon, N. Y	Mar. 4, 1873	136, 531
Head-rest	T. B. Stout	Keyport, N. J	Mar. 16, 1869	87, 987
Head-rest	T. D. Thompson	Providence, R. I	Oct. 17, 1871	120, 128
Head-rest	M. Warne	Philadelphia, Pa	Oct. 20, 1868	83, 227
Head-rest	M. Warne	Philadelphia, Pa	Oct. 11, 1870	108, 308
Head-rest	E. Waters	Troy, N. Y	Dec. 29, 1857	18, 998
Head-rest for chairs	H. Snowden	Baltimore, Md	Sept. 24, 1867	69, 135
Head-rest for chairs, Portable	C. P. Bailey	Zanesville, Ohio	Mar. 28, 1854	10, 698
Head-rest for seats	J. R. Chiles	Richmond, Va	Jan. 28, 1868	73, 782
Head-rest for travelers in railway-cars, &c	G. Utley	Chapel Hill, N. C	Mar. 6, 1860	27, 395
Head-rest, Portable	H. E. Churchill	Portland, Me	Mar. 9, 1869	87, 543
Head-rest, Portable	J. M. Whitney	Boston, Mass	Jan. 9, 1866	51, 987
Head-rest, Portable	A. B. Wilson	Waterbury, Conn	Dec. 23, 1856	16, 300
Head-stall overcheck	J. B. Loomis	Ypsilanti, Mich	Sept. 10, 1872	131, 284
Head-supporting apparatus for travelers	H. C. Howells	New York, N. Y	Apr. 10, 1860	27, 806
Hearse	E. Allen	Norwich, Conn	Sept. 21, 1869	95, 066
Hearse	M. Jincks and F. Altmeyer	Danville, N. Y	Oct. 16, 1866	58, 835
Hearse	J. F. Kinback	Carbondale, Pa	Feb. 23, 1869	87, 175
Hearth, grate, and fender, Combined	G. Buchanan	Washington, Pa	June 28, 1870	104, 703
Hearth, Metallic	O. Keels	Steubenville, Ohio	Feb. 15, 1870	99, 777
Heat, Apparatus for accumulating and retaining	T. J. Chubb	Brooklyn, N. Y	May 21, 1867	64, 947
Heat, Apparatus for applying and conducting	J. C. Douglass	New York	Dec. 23, 1834	.....
Heat, Apparatus for utilizing	B. F. Sturtevant	Jamaica Plain, Mass	Oct. 12, 1869	95, 849
Heat by friction, Apparatus for generating	P. Vera	Bogota, United States of Colombia.	Jan. 19, 1869	86, 046
Heat by the circulation of hot water, Apparatus for transmitting.	A. M. Perkins	London, England	Sept. 4, 1840	1, 768
Heat by the combustion of fuel of various kinds, Mode of raising.	J. F. Bennett	Pittsburgh, Pa	Mar. 13, 1866	53, 104
Heat-controller attachment	M. S. Hitchcock	Springfield, Mass	July 25, 1865	49, 029
Heat, Evolution and management of	E. Nott	Schenectady, N. Y	Feb. 23, 1826	....
Heat, Evolution and management of	E. Nott	Schenectady, N. Y	Dec. 29, 1826	....
Heat, Evolution and management of	E. Nott	Schenectady, N. Y	May 30, 1827	....
Heat, Evolution and management of	E. Nott	Schenectady, N. Y	Mar. 26, 1828	....
Heat, Evolution and management of	E. Nott	Schenectady, N. Y	Sept. 17, 1832	....
Heat, Evolution and management of	E. Nott	Schenectady, N. Y	Sept. 17, 1832	....
Heat, Evolution, &c., of	E. Nott	Schenectady, N. Y	June 21, 1826	....
Heat from lime, &c., Applying	P. Wenn	Philadelphia, Pa	Mar. 8, 1836	....
Heat from smoke by basket-grate, Abstracting	D. Steinhauer	Philadelphia, Pa	Dec. 13, 1832	....
Heat-generating apparatus for cooking-purposes	O. F. Morrill	Chelsea, Mass	June 5, 1866	55, 340
Heat generator and radiator	G. Chilson	Boston, Mass	Sept. 26, 1854	11, 718
Heat-governor or draft-regulator	F. E. Chatard, jr	Baltimore, Md	May 13, 1873	138, 736
Heat in chemical and other processes, Method of conveying and using.	C. A. Seely	New York, N. Y	Feb. 23, 1869	87, 208
Heat indicator, Distant	J. A. Read	Enfield, New South Wales	Sept. 9, 1873	142, 727
Heat, Machinery for regulating	L. W. Wright	London, England	Aug. 25, 1831	....
Heat, Management of	L. Lewis	Lewistown, N. Y	Mar. 2, 1836	....
Heat, Management of combustion in producing	S. A. Bille	New York	Nov. 8, 1828	....
Heat, Material for transmitting	W. C. Baker	New York, N. Y	Dec. 10, 1867	71, 948
Heat, Mode of generating	W. Hartell and J. Lancaster	Kensington and Spring Garden, Pa.	Nov. 23, 1852	9, 419
Heat, Mode of obtaining	J. Johnson	Saco, Me	Apr. 7, 1868	76, 463
Heat, Non-conductor of	J. Chalmers	London, England	Aug. 4, 1868	80, 709
Heat-regulator	H. Boyle	London, England	Sept. 2, 1873	142, 371
Heat-regulator	S. M. Kellogg	Battle Creek, Mich	June 28, 1864	43, 313
Heat-regulator for hot-air furnaces	A. H. Tingley	Providence, R. I	Feb. 15, 1870	99, 972

Index of patents issued from the United States Patent Office from 1790 *to* 1873, *inclusive*—Continued.

Invention.	Inventor.	Residence.	Date.	No.
Heat-regulator for hot-water apparatus	H. L. McAvoy	Baltimore, Md	July 27, 1869	93, 104
Heat retainer and imparter	J. K. Mitchell	Philadelphia, Pa	Nov. 19, 1818	....
Heat to fluids, Application of	S. Applegate, T. Howard, and P. H. Wharton.	Philadelphia, Pa	Nov. 22, 1809	....
Heater:				
See Air-heater. Barrel-heater. Boiler-heater. Building-heater. Car-heater. Cask-heater. Cheese-vat heater. Chimney-heater. Coal-oil heater. Cotton-seed heater. Cream-heater. Curling-tongs heater. Cylinder-heater. Dairy-heater. Dentist's lamp-heater. Dish-heater. Feed-water heater. Fire-engine heater. Fire-place heater. Flat-iron heater. Friction-heater. Furnace-heater. Gas-heater. Gravel-heater. Gravel and sand heater. Hot-water heater. Hot-water and hot-air heater. Hot-water or steam heater. Hydrocarbon-heater. Hydraulic heater. Kerosene-heater. Lamp-heater. Lamp-chimney heater. Laundry-heater. Laundry-iron heater. Liquid-heater. Nursery-lamp heater. Parlor-heater. Peanut-heater. Portable heater. Ruling-machine heater. Sad-iron heater. Service-heater. Sleigh-heater. Smoke-consuming heater. Spirituous-liquor heater. Stamp-heater. Steam-heater. Steam-generating heater. Steam-generator heater. Stove and water heater. Sugar-sirup heater. Superheater. Tea-pot heater. Tire-heater. Topical heater. Tubular heater. Water-heater.				
Heater	O. Abbott	New York, N. Y	Feb. 9, 1864	41, 467
Heater	L. F. Betts	Albion, Mich	Jan. 5, 1864	41, 051
Heater	C. T. Boardman	Bergen Point, N. J	Apr. 5 1864	42, 161
Heater	N. A. Boynton	New York, N. Y	Nov. 25, 1862	36, 989
Heater	S. A. Briggs	Providence, R. I	May 20, 1862	35, 294
Heater	H. L. Budd	New York, N. Y	Jan. 12, 1864	41, 256
Heater	J. Carton	Utica, N. Y	May 13, 1862	35, 216
Heater	J. Chilcott	Brooklyn, N. Y	Aug. 22, 1865	49, 504
Heater	T. S. Clogston	Boston, Mass	Dec. 13, 1864	45, 389
Heater	T. Dowling	Lynn, Mass	Aug. 4, 1863	39, 385
Heater	C. H. Frost	Peekskill, N. Y	June 3, 1862	35, 440
Heater	C. C. Hare	Louisville, Ky	Oct. 10, 1865	50, 051
Heater	J. C. Henderson	Albany, N. Y	Dec. 23, 1862	37, 228
Heater	G. Joslin	Boston, Mass	Feb. 25, 1862	34, 503
Heater	J. Magee	Chelsea, Mass	Nov. 25, 1862	37, 027
Heater	J. A. Marvin	Red Wing, Minn	Nov. 26, 1867	71, 399
Heater	O. F. Morrill	Chelsea, Mass	Oct. 4, 1864	44, 548
Heater	O. F. Morrill	Chelsea, Mass	Oct. 4, 1864	44, 549
Heater	G. W. Oakeley	Reading, Pa	Dec. 3, 1861	33, 849
Heater	M. Pond	Boston, Mass	June 10, 1862	35, 541
Heater	J. S. Reid	Muncie, Ind	Sept. 1, 1863	39, 750
Heater	J. S. Reid	Muncie, Ind	Apr. 5, 1864	42, 225
Heater	E. C. Robinson	Troy, N. Y	Aug. 29, 1865	49, 652
Heater	W. H. Seymour	West Hartford, Conn	Mar. 25, 1862	34, 785
Heater	S. B. Sexton	Baltimore, Md	June 17, 1862	35, 632
Heater	T. Shipton	Newark, N. J	Oct. 29, 1867	70, 276
Heater	J. H. Simonds	New York, N. Y	Apr. 22, 1862	35, 043
Heater	E. Sprague	Allegheny City, Pa	Oct. 20, 1863	40, 366
Heater	J. C. Underwood	Richmond, Ind	Dec. 27, 1864	45, 656
Heater	A. Weller	Albany, N. Y	Oct. 20, 1863	40, 375
Heater and blower	W. M. Jackson	Wooland, Cal	Sept. 2, 1873	142, 397
Heater and boiler	S. A. Willett	Philadelphia, Pa	July 2, 1861	32, 731
Heater and condenser	P. W. Mackenzie	Blauveltville, N. Y	Feb. 13, 1872	123, 570
Heater and condenser combined	B. Holly	Lockport, N. Y	Apr. 25, 1871	114, 142
Heater and evaporator	F. Farquahar and R. E. Doan	Wilmington, Ohio	Aug. 13, 1867	67, 742
Heater and filter	L. F. Meunier	Fires-Lille, France	July 8, 1873	140, 719
Heater and filter, Combined	W. T. Bate	Norristown, Pa	Apr. 15, 1873	137, 878
Heater and ventilator	G. W. Blake	New York, N. Y	Feb. 22, 1870	100, 110
Heater and ventilator	B. F. Sturtevant	Jamaica Plain, Mass	Feb. 22, 1870	100, 211
Heater-boiler	C. R. Ellis	Brooklyn, N. Y	Jan. 7, 1868	73, 084
Heater, condenser, and filter combined	W. A. Lighthall	New York, N. Y	July 29, 1862	36, 010
Heater, cooler, and ventilator	D. E. Somes	Washington, D. C	Nov. 28, 1865	51, 236
Heater or cooler	J. C. Hoadley	Lawrence, Mass	Jan. 26, 1858	19, 197
Heater-screen	W. C. Baker	New York, N. Y	Apr. 21, 1863	28, 210
Heating and cooking apparatus	E. Conway	Dayton, Ohio	Aug. 28, 1860	29, 769
Heating and fuel-saving device	D. C. Colby	Claremont, N. H	Apr. 25, 1865	47, 489
Heating and illuminating apparatus	L. Ruel	Saint Johnsbury, Vt	Nov. 11, 1873	144, 414
Heating and puddling furnace	M. S. Ridgeway and C. Lewis	Danville and Harrisburgh, Pa.	May 7, 1867	64, 617
Heating and puddling furnace, Iron, &c.	J. Morrison	East Saint Louis, Ill	Jan. 30, 1872	123, 282
Heating and puddling furnaces, Boshes for	W. Batty	Lawrenceburgh, Pa	Sept. 24, 1867	69, 063
Heating and puddling furnaces, Slide-door for	S. M. Laughlin	Philadelphia, Pa	Oct. 18, 1870	108, 501
Heating and ventilating	G. F. Schulze	Janesville, Wis	Mar. 22, 1870	101, 168
Heating and warming furnace, Apartment	J. A. Page	Boston, Mass	Sept. 25, 1841	2, 269
Heating and welding furnace	G. Nimmo	Jersey City, N. J	May 28, 1867	65, 110
Heating and welding furnace	G. Nimmo and R. S. Stenton	Jersey City, N. J., and New York, N. Y.	June 14, 1864	43, 127
Heating apartments	T. H. Parker	York, Pa	Aug. 1, 1848	5, 692
Heating apartments, Apparatus for	S. Whitmarsh	Northampton, Mass	Feb. 20, 1849	6, 118
Heating-apparatus	J. Armstrong, jr.	San Francisco, Cal	Oct. 12, 1869	95, 756
Heating-apparatus	W. C. Baker	New York, N. Y	Nov. 18, 1873	144, 650

Index of patents issued from the United States Patent Office from 1790 *to* 1873, *inclusive*—Continued.

Invention.	Inventor.	Residence.	Date.	No.
Heating-apparatus	L. Bridges	Chicago, Ill	June 19, 1860	28, 805
Heating-apparatus	M. G. Fagan	Troy, N. Y	Feb. 28, 1871	112, 233
Heating-apparatus	E. C. Gillette	San Francisco, Cal	June 30, 1863	39, 041
Heating-apparatus	R. D. Granger	Philadelphia, Pa	Jan. 18, 1859	22, 642
Heating-apparatus	M. Greenebaum	Chicago, Ill	Oct. 15, 1861	33, 483
Heating-apparatus	F. L. Hedenberg	New York, N. Y	Mar. 30, 1858	19, 775
Heating-apparatus	J. R. Jenness	Norwich, Conn	Sept. 22, 1868	82, 322
Heating-apparatus	J. Johnson	Saco, Me	Apr. 7, 1868	76, 329
Heating-apparatus	A. Kohler	Boston, Mass	May 28, 1861	32, 427
Heating-apparatus	U. D. Mibills	Hartford, Wis	Jan. 25, 1859	22, 740
Heating-apparatus	P. Mitchell	Greenfield, Mass	June 25, 1861	32, 639
Heating-apparatus	G. R. Osbrey	Providence, R. I	Nov. 8, 1859	26, 048
Heating-apparatus	C. Pepper	Albany, N. Y	Dec. 20, 1859	26, 521
Heating-apparatus	S. Smith	Greenfield, Mass	July 31, 1866	56, 816
Heating-apparatus	G. W. Williamson	Scranton, Pa	Jan. 25, 1859	22, 760
Heating-apparatus regulator	E. L. Brown	Brooklyn, N. Y	Nov. 27, 1860	30, 716
Heating-boiler	Z. Ellis	Philadelphia, Pa	Nov. 15, 1864	45, 031
Heating-boiler	J. A. Maynard	Newtonville, Mass	Mar. 21, 1871	112, 824
Heating-boiler	J. G. Wilson	New York, N. Y	Jan. 10, 1871	110, 884
Heating-boiler for buildings	A. Gay	Rochester, N. Y	Sept. 12, 1871	118, 799
Heating-boiler for buildings	A. E. Hitchings	New York, N. Y	Nov. 24, 1857	18, 693
Heating-boiler for buildings	C. F. Hitchings	New York, N. Y	May 15, 1860	28, 272
Heating-boiler for buildings	C. F. Hitchings	New York, N. Y	July 23, 1867	67, 053
Heating-boiler for buildings	T. J. Myers	Philadelphia, Pa	July 25, 1871	117, 448
Heating buildings and evaporating fluids	A. M. Perkins	United States	Aug. 20, 1838	888
Heating buildings by the combination of and burning gas, air, and steam, Apparatus for.	C. H. Johnson	Boston, Mass	Mar. 4, 1856	14, 360
Heating buildings, &c., Warm air for	J. L. Dutton	Philadelphia, Pa	Mar. 13, 1834	....
Heating, cooking, and ventilating apparatus	C. B. Sawyer	Fitchburgh, Mass	June 28, 1859	24, 581
Heating-furnace	W. D. Bartlett	Amesbury, Mass	Mar. 22, 1870	101, 082
Heating-furnace	N. E. Cornnall	New York, N. Y	Feb. 4, 1868	74, 051
Heating-furnace	E. W. Crane	Roselle, N. J	Apr. 8, 1873	137, 598
Heating-furnace	C. Harkinson	Philadelphia, Pa	May 17, 1870	103, 176
Heating-furnace	T. Hydes and J. Bennett	Sheffield, England	Mar. 22, 1870	101, 017
Heating-furnace	J. Maunton	New York, N. Y	Nov. 12, 1867	70, 874
Heating-furnace	J. G. Porter and E. Davis	New York, N. Y., and Jersey City, N. J.	June 10, 1873	139, 812
Heating-furnace	C. R. Rand	Dubuque, Iowa	Aug. 14, 1866	57, 187
Heating furnace	C. R. Rand	Saint Louis, Mo	Dec. 17, 1867	72, 229
Heating-furnace	W. A. Sweet	Syracuse, N. Y	July 26, 1870	105, 738
Heating-furnace	J. Watts	Birmingham, Great Britain	Mar. 24, 1868	76, 010
Heating-furnace and cooking-stove combined	B. McConnell	Philadelphia, Pa	May 21, 1872	127, 082
Heating-furnace for buildings	B. W. Dunklee	Boston, Mass	May 10, 1859	23, 907
Heating-furnace for buildings	F. A. Frickhardt	Easton, Pa	Sept. 8, 1836	23
Heating-furnace for buildings	J. Leeds	Philadelphia, Pa	Nov. 7, 1854	11, 896
Heating-furnace for buildings	J. Leeds	Philadelphia, Pa	Aug. 14, 1855	13, 432
Heating-furnace for buildings	J. Leeds	Philadelphia, Pa	Apr. 2, 1861	31, 892
Heating-furnace for buildings	J. Plant	Washington, D. C	Oct. 5, 1858	21, 724
Heating-furnace for buildings	S. B. Sexton	Baltimore, Md	Aug. 14, 1855	13, 439
Heating-furnace for buildings	G. Walker	New Haven, Conn	June 10, 1844	3, 623
Heating furnaces, Consuming escape steam as an adjunct in.	T. Maskell	Franklin, La	Sept. 4, 1855	13, 526
Heating-furnaces, Device for drawing iron from	W. Stevens	New Albany, Ind	Aug. 15, 1871	118, 165
Heating-furnaces, Hot-air flue for	G. R. Barker	Philadelphia, Pa	July 22, 1873	141, 101
Heating-pipe, Jacket for	M. Ber	New York, N. Y	Aug. 23, 1870	106, 652
Heating plates for hot-pressing	M. A. Quigley	Washington County, Md	June 15, 1819	....
Heating-purposes, Metallic base and molding for	C. E. Fenkle	New York, N. Y	Apr. 21, 1868	77, 022
Heating rooms	A. McAllester and J. Iggett	Salem, N. Y	Dec. 15, 1827	....
Heating rooms, &c	A. Pollock	Boston, Mass	Apr. 24, 1807	....
Heating rooms	B. Russell	Hartford, Conn	Mar. 4, 1811	....
Heating rooms	R. B. Varden	Baltimore, Md	Feb. 6, 1832	....
Heating rooms and ovens	J. A. Pitts	Winthrop, Me	July 2, 1836	....
Heating rooms, &c., by rarefied air	D. Pettibone	Philadelphia, Pa	May 12, 1812	....
Heating rooms, Device for	L. Newson	Gallipolis, Ohio	Aug. 21, 1860	29, 710
Heddle-actuating mechanism	J. Crawshaw	Cobourg, Canada	July 23, 1872	129, 718
Heddle-actuating mechanism	W. Gminder	Reutlingen, Germany	Oct. 15, 1872	132, 279
Heddle-actuating mechanism	R. B. Goodyear	Wilmington, Del	July 16, 1872	129, 126
Heddle-eye opener	J. H. Crowell	Providence, R. I	June 17, 1873	139, 875
Heddle-frame	J. Dyson	Philadelphia, Pa	July 23, 1872	129, 722
Heddle-machine	P. Philip	Stockport, N. Y	Aug. 9, 1870	106, 277
Heddle, Metallic	J. Senneff	Philadelphia, Pa	Apr. 13, 1852	8, 662
Hedge and tree trimmer	C. E. Healy	New London, Ohio	Feb. 22, 1870	100, 145
Hedge and tree trimmer	S. McElhaney	Milledgeville, Ill	Sept. 3, 1872	131, 108
Hedge-cutting machine	D. Oliver	Carthage, Ill	Mar. 6, 1866	53, 033
Hedge-fastener	C. D. Brown	Tampico, Ill	July 2, 1867	66, 209
Hedge-hook	E. S. Turner	Palmyra, Wis	May 20, 1873	139, 215
Hedge-plant grab	H. Ferris	Galesburgh, Ill	Apr. 17, 1866	53, 963
Hedge-plants and weeds, Implement for extracting	J. L. Knick	Lexington, Ill	Apr. 28, 1868	77, 385
Hedge-planting device	C. D. Brown	Sterling, Ill	Jan. 15, 1867	61, 150
Hedge-plashing machine	D. Gore	Carlinville, Ill	June 22, 1869	91, 620
Hedge-setter	J. H. Holbert	Ottawa, Ill	July 27, 1869	92, 961
Hedge-shears	J. O. Minor	Wapello, Iowa	July 30, 1867	67, 207
Hedge-trimmer	C. W. Aikin	Decatur, Ill	Apr. 30, 1872	126, 243
Hedge-trimmer	F. Binder	Baltimore, Md	Aug. 27, 1867	68, 280
Hedge-trimmer	J. Block	Eureka, Ill	Jan. 30, 1872	123, 224
Hedge-trimmer	G. and S. P. Clark	Dover, Ill	Feb. 28, 1871	112, 221
Hedge-trimmer	J. S. Crum	Scottville, Ill	Apr. 23, 1872	126, 032
Hedge-trimmer	W. E. Horne	Decatur, Ill	July 5, 1870	104, 956
Hedge-trimmer	A. H. Hussey	Mount Pleasant, Ohio	Dec. 14, 1869	97, 925
Hedge-trimmer	S. T. Hyde	Piasa, Ill	May 4, 1869	89, 769
Hedge-trimmer	T. C. Mathews	Yates City, Ill	June 30, 1868	79, 372
Hedge-trimmer	J. McAnulty	Bentley, Ill	Sept. 20, 1870	107, 517
Hedge-trimmer	T. V. Nichols	Olena, Ill	Sept. 9, 1862	36, 420
Hedge-trimmer	D. Oliver	Galesburgh, Ill	Feb. 15, 1870	99, 938
Hedge-trimmer	R. T. Payne	Eureka, Ill	July 16, 1872	129, 493
Hedge-trimmer	A. Selover	Brooklyn, Ohio	Aug. 8, 1865	49, 311
Hedge-trimmer	W. and H. Perry	Tivoli, Ill	Aug. 28, 1866	57, 590
Hedge-trimmer	S. B. Turner	Adams County, Ill	Apr. 22, 1873	138, 106

Index of patents issued from the United States Patent Office from 1790 *to* 1873, *inclusive*—Continued.

Invention.	Inventor.	Residence.	Date.	No.
Hedge-trimmer	J. M. and O. W. Van Nosdall	Newark, Ill	Jan. 25, 1870	99, 121
Hedge-trimmer	J. M. and O. W. Van Nosdall	Newark, Ill	Feb. 28, 1871	112, 300
Hedge-trimmer	L. Woods	Quincy, Ill	July 25, 1854	11, 400
Hedge-trimmer and mower	G. Waddington	Le Roy, Ill	Mar. 15, 1870	100, 955
Hedge-trimmer and mower, Combined	S. H. Hepperly	Elmira, Ill	Feb. 26, 1867	62, 487
Hedge-trimmer and stalk-cutter	J. G. Sprague	Lexington, Ill	Feb. 1, 1870	99, 490
Hedge-trimmer and corn-cutting machine	J. W. Hull	Connersville, Ill	Dec. 17, 1867	72, 299
Hedge-trimming machine	E. M. Benfield	Maquin, Ill	July 29, 1862	36, 049
Hedge-trimming machine	S. Bradbury	Griggsville, Ill	Jan. 27, 1857	16, 462
Hedge-trimming machine	W. C. Hooker	Abingdon, Ill	July 25, 1865	48, 946
Hedge-trimming machine	A. Tweedy	Collinsville, Ohio	July 30, 1867	67, 380
Hedge-trimming machine	W. Wimmer	Billingsville, Ind	July 7, 1857	17, 764
Hedges, Device for bending down plants to form	C. D. Brown	Tampico, Ill	July 2, 1867	66, 210
Heel	S. A. Brackett	Boston, Mass	Nov. 5, 1872	132, 747
Heel	W. T. Downs	Saint Louis, Mo	Mar. 30, 1869	88, 461
Heel and sole edge trimming machine, Boot and shoe.	R. C. Lambert	Quincy, Mass	May 23, 1871	115, 219
Heel and spoke shave	J. A. Perley	Lynn, Mass	Feb. 8, 1859	22, 889
Heel and spur, Boot	C. F. Woodruff	Newbern, Tenn	Jan. 19, 1869	86, 049
Heel and toe irons for boots and shoes, Coating	H. O. Lothrop	Milford, Mass	Jan. 21, 1868	73, 451
Heel and toe plates for boots and shoes	I. E. Loughborough	Pittsford, N. Y	Apr. 4, 1865	47, 115
Heel-attachment for boots and shoes	G. C. Aiken	Worcester, Mass	Feb. 5, 1861	31, 286
Heel, Blank boot	H. H. Bigelow	Worcester, Mass	Sept. 7, 1869	94, 552
Heel blank manufacturing machine, Boot and shoe	C. W. Glidden	Lynn, Mass	May 27, 1873	139, 382
Heel blanks, Machine for compressing and punching.	R. C. Lambart	Quincy, Mass	Oct. 21, 1873	143, 914
Heel, Boot	F. Closs	New Haven, Conn	Sept. 18, 1866	58, 067
Heel, Boot	G. W. Davis	Milford Centre, Ohio	Feb. 12, 1867	62, 017
Heel, Boot	E. Dunbar	Buffalo, N. Y	June 20, 1865	48, 266
Heel, Boot	J. Fearn	Tompkinsville, N. Y	Feb. 18, 1868	74, 525
Heel, Boot	J. B. Gebbard and C. G. Schwarz.	Philadelphia, Pa	Oct. 14, 1873	143, 688
Heel, Boot	J. H. Greenleaf	New Haven, Conn	Nov. 20, 1866	59, 908
Heel, Boot	F. D. Hayward and P. Stone	Malden and Charlestown, Mass.	July 11, 1865	48, 682
Heel, Boot	G. Henning and H. P. Willie	Buffalo, N. Y	Jan. 14, 1868	73, 328
Heel, Boot	S. Hodgins	Saint Louis, Mo	July 25, 1865	48, 943
Heel, Boot	L. and A. A. Hoffman	Buffalo, N. Y	June 4, 1867	65, 487
Heel, Boot	J. Hubbell	Buffalo, N. Y	Sept. 3, 1867	68, 364
Heel, Boot	G. Lane	Hamilton, Ohio	Apr. 21, 1868	77, 055
Heel, Boot	G. A. Mitchell	Turner, Me	June 25, 1861	32, 638
Heel, Boot	E. Newhall	Lynn, Mass	Sept. 19, 1865	50, 026
Heel, Boot	A. T. Perrine	Louisville, Ky	Mar. 16, 1869	88, 000
Heel, Boot	F. Richardson	New Bedford, Mass	May 5, 1868	77, 529
Heel, Boot	F. Richardson and F. Hacker	Providence, R. I	May 7, 1872	126, 489
Heel, Boot	F. Richardson and F. Hacker	Providence, R. I	May 7, 1872	126, 490
Heel, Boot	C. Robinson and J. C. Marshall	Springfield, Mass	Aug. 7, 1866	56, 992
Heel, Boot	H. S. Snow	West Meriden, Conn	Mar. 12, 1867	62, 783
Heel, Boot	S. Thorp and W. Thorp	Portland and Turner, Me	Jan. 25, 1859	22, 755
Heel, Boot	R. Vandevort	Pittsburgh, Pa	Sept. 17, 1867	69, 051
Heel, Boot and shoe	I. Banister	Newark, N. J	May 21, 1872	127, 013
Heel, Boot and shoe	G. Beaty	Middlebury, Ohio	Jan. 7, 1868	72, 963
Heel, Boot and shoe	H. H. Bigelow	Worcester, Mass	Nov. 1, 1870	108, 870
Heel, Boot and shoe	H. H. Bigelow	Worcester, Mass	June 27, 1871	116, 400
Heel, Boot and shoe	L. R. Blake and A. S. Libby	Brooklyn, N. Y., and Lawrence, Mass.	Dec. 9, 1873	145, 389
Heel, Boot and shoe	F. Bleisewick	New York, N. Y	Mar. 22, 1870	100, 968
Heel, Boot and shoe	M. Bray	Newton, Mass	Sept. 12, 1871	118, 830
Heel, Boot and shoe	M. Bray	Newton, Mass	Aug. 26, 1873	142, 203
Heel, Boot and shoe	H. Cordtz	Chicago, Ill	Sept. 21, 1869	95, 087
Heel, Boot and shoe	A. O. Crane	Boston, Mass	Dec. 15, 1868	84, 861
Heel, Boot and shoe	A. O. Crane	Boston, Mass	June 6, 1871	115, 581
Heel, Boot and shoe	A. O. Crane	Boston, Mass	Aug. 22, 1871	118, 203
Heel, Boot and shoe	S. Dodge, jr., and B. Potter, jr.	Marblehead, Mass	Sept. 13, 1859	25, 395
Heel, Boot and shoe	C. Dyer, jr., and E. Drake	Stoughton, Mass	Aug. 6, 1867	67, 422
Heel, Boot and shoe	C. Dyer, jr., and E. Drake	Stoughton, Mass	Oct. 22, 1867	70, 078
Heel, Boot and shoe	C. H. Eggleston	Marshall, Mich	Dec. 26, 1871	122, 112
Heel, Boot and shoe	T. S. Engledow	Cedar Falls, Iowa	Mar. 3, 1868	75, 137
Heel, Boot and shoe	G. F. Fling	Portland, Me	Sept. 19, 1871	119, 021
Heel, Boot and shoe	S. Flint and R. S. Rogers	Lynn, Mass	Dec. 14, 1858	22, 328
Heel, Boot and shoe	H. F. Harris and G. P. Pinney	Chicago, Ill	Aug. 30, 1870	106, 814
Heel, Boot and shoe	C. H. Helms	Poughkeepsie, N. Y	June 7, 1864	43, 024
Heel, Boot, &c	R. Herr	Brooklyn, N. Y	June 9, 1868	78, 667
Heel, Boot and shoe	W. Hunt	New York, N. Y	June 21, 1859	24, 517
Heel, Boot and shoe	J. M. Hunter	New York, N. Y	Mar. 19, 1872	124, 682
Heel, Boot and shoe	J. M. Hunter	Morristown, N. J	Aug. 27, 1872	130, 805
Heel, Boot and shoe	G. W. Keene	Lynn, Mass	Mar. 6, 1860	27, 369
Heel, Boot and shoe	G. W. Keene	Lynn, Mass	Dec. 24, 1861	34, 037
Heel, Boot and shoe	F. Kettler	Milwaukee, Wis	May 27, 1862	35, 377
Heel, Boot and shoe	R. Lapham	Boston, Mass	Mar. 30, 1869	88, 310
Heel, Boot and shoe	G. W. Martin	Boston, Mass	Jan. 12, 1869	85, 748
Heel, Boot and shoe	J. Miller	Antwerp, N. Y	Mar. 15, 1864	41, 933
Heel, Boot and shoe	N. Molo	Bay City, Mich	Dec. 17, 1872	134, 087
Heel, Boot and shoe	G. E. Newcomb	Bucksport Me	May 10, 1870	102, 961
Heel, Boot and shoe	E. Newhall	Lynn, Mass	Dec. 10, 1867	72, 073
Heel, Boot and shoe	C. B. Norton	Crittenden, N. Y	Apr. 2, 1872	125, 214
Heel, Boot and shoe	S. Oliver, jr	Lynn, Mass	Apr. 14, 1857	17, 051
Heel, Boot and shoe	W. H. Peckham	Hoboken, N. J	Oct. 30, 1860	30, 542
Heel, Boot and shoe	A. T. Perrine	Boston, Mass	Oct. 3, 1871	119, 473
Heel, Boot and shoe	M. H. Prescott	Ottawa, Ill	May 18, 1869	90, 192
Heel, Boot and shoe	C. A. Read	Bridgeport, Conn	Sept. 26, 1871	119, 407
Heel, Boot and shoe	J. Read	Philadelphia, Pa	Mar. 22, 1859	23, 312
Heel, Boot and shoe	J. Read	Philadelphia, Pa	June 14, 1870	104, 200
Heel, Boot and shoe	J. R. Ryerson	St. Albans, Me	July 16, 1872	129, 369
Heel, Boot and shoe	P. Shaw	Boston, Mass	Jan. 3, 1860	26, 712
Heel, Boot and shoe	H. S. Smythe	Darlington, Wis	Feb. 21, 1871	111, 986
Heel, Boot and shoe	D. E. Somes	Biddeford, Me	July 24, 1860	29, 324
Heel, Boot and shoe	D. E. Somes	Biddeford, Me	Sept. 18, 1860	30, 096

Index of patents issued from the United States Patent Office from 1790 *to* 1873, *inclusive*—Continued.

Invention.	Inventor.	Residence.	Date.	No.
Heel, Boot and shoe	J. W. Tull and G. Stevenson	Zionsville, Ind	Sept. 29, 1868	82, 662
Heel, Boot and shoe	O. Underwood	Milford, Mass	Nov. 11, 1873	144, 580
Heel, Boot and shoe	A. B. Wilton	Dorchester, Mass	May 31, 1859	24, 269
Heel, Boot and shoe	J. Woodley	Quebec, Canada	Mar. 29, 1870	101, 399
Heel-breasting machine	C. H. Helms	Poughkeepsie, N. Y	Mar. 6, 1866	52, 997
Heel-breasting machine	A. B. Jaquith and J. Wolohan	Rochester, N. Y	Dec. 2, 1873	145, 061
Heel-breasting machine	V. K. Spear	Lynn, Mass	May 27, 1873	139, 273
Heel breasting machine, Boot and shoe	W. C. Butler	Baltimore, Md	Dec. 16, 1873	145, 625
Heel-breasting machinery	V. K. Spear	Lynn, Mass	Aug. 12, 1873	141, 735
Heel-burnisher	M. Burnham 2d	Wenham, Mass	Aug. 23, 1870	106, 657
Heel-burnishing machine	E. H. Downing and W. Joint	Lynn, Mass	Feb. 11, 1873	135, 786
Heel-burnishing machine	C. W. Glidden	Lynn, Mass	Oct. 7, 1873	143, 507
Heel-burnishing machine	C. W. Glidden	Lynn, Mass	Oct. 14, 1873	143, 690
Heel-burnishing machine	G. C. Hawkins	Boston, Mass	Dec. 30, 1873	145, 943
Heel-burnishing machine	R. C. Lambert	Quincy, Mass	Aug. 8, 1871	117, 955
Heel-burnishing machine	H. S. Vrooman	Boston, Mass	June 3, 1873	139, 636
Heel burnishing machine, Boot and shoe	C. W. Glidden	Lynn, Mass	Oct. 21, 1873	143, 899
Heel burnishing machine, Boot and shoe	G. C. Hawkins	Boston, Mass	June 4, 1872	127, 414
Heel burnishing machine, Boot and shoe	G. C. Hawkins, A. G. Mead, and V. K. Spear.	Boston and Lynn, Mass	Apr. 4, 1871	113, 658
Heel burnishing machine, Boot and shoe	R. C. Lambart	Quincy, Mass	Jan. 30, 1872	123, 112
Heel burnishing machine, Boot and shoe	R. C. Lambart	Quincy, Mass	Dec. 9, 1873	145, 430
Heel burnishing machine, Boot and shoe	A. C. McKnight	Philadelphia, Pa	Feb. 20, 1872	123, 785
Heel burnishing machine, Boot and shoe	J. H. Sawyer and C. Keniston	Boston and Somerville, Mass.	Mar. 22, 1870	101, 167
Heel burnishing machine, Boot and shoe	V. K. Spear	Lynn, Mass	June 6, 1871	115, 651
Heel burnishing machinery, Boot and shoe	J. Beasley	Lynn, Mass	Jan. 28, 1873	135, 259
Heel burnishing machinery, Boot and shoe	I. Van Nouhuys	Albany, N. Y	Aug. 12, 1873	141, 678
Heel-calk	G. F. Clemons	Springfield, Mass	May 19, 1868	77, 960
Heel-calk	J. F. Richardson and G. F. Morse.	Portland, Me	Jan. 16, 1866	52, 077
Heel-calk	T. Symonds	Portland, Me	Nov. 28, 1865	51, 241
Heel-calk	M. F. Walter and C. Staudinger.	Hartford, Conn	Sept. 6, 1870	107, 132
Heel-casing	J. R. Moffitt	Chelsea, Mass	Apr. 14, 1868	76, 794
Heel, Composition boot and shoe	F. Marquard	Newburyport, Mass	Nov. 23, 1869	97, 206
Heel, Composition for the manufacture of boot and shoe.	C. Brocking	Boston, Mass	Nov. 29, 1870	109, 578
Heel-cutter	J. H. Bean	Marietta, Ohio	Mar. 16, 1869	87, 753
Heel-cutter	J. H. Bean	Marietta, Ohio	June 22, 1869	91, 508
Heel-cutter	S. Keen	East Bridgewater, Mass	Oct. 26, 1869	96, 120
Heel-cutter	A. D. Kelley	Rochester, N. H	June 27, 1854	11, 196
Heel-cutter	O. H. and J. A. Marston	Centre Sandwich, N. H	Feb. 6, 1866	52, 430
Heel cutter, Boot	B. F. Goddard	Charlton Depot, Mass	Nov. 19, 1867	70, 992
Heel-cutter for cutting out heels of boots and shoes.	J. Shaw	Natick, Mass	June 2, 1857	17, 455
Heel cutting and finishing machine, Shoe	W. F. Edson	Philadelphia, Pa	Sept. 5, 1859	25, 326
Heel-cutting machine	O. G. Critchett	Stoneham, Mass	June 21, 1864	43, 183
Heel-cutting machine	R. C. Lambart	Raynham, Mass	Mar. 31, 1868	76, 207
Heel-cutting machine	R. C. Lambart	Raynham, Mass	Nov. 16, 1869	96, 817
Heel cutting machine, Boot and shoe	M. Bray and E. Edmands	Newton, Mass	July 16, 1872	129, 093
Heel cutting machine, Boot and shoe	E. T. Green	Stoneham, Mass	Oct. 9, 1860	30, 376
Heel cutting machine, Boot and shoe	R. C. Lambart	South Abington, Mass	Oct. 25, 1870	108, 605
Heel-cutting machine, Knife-holder and guide for	S. Keen	East Bridgewater, Mass	Oct. 25, 1870	108, 601
Heel, Detachable boot and shoe	C. W. Bailey	Boston, Mass	Nov. 9, 1869	96, 538
Heel, Detachable boot and shoe	C. W. Bailey	Boston, Mass	Sept. 6, 1870	106, 984
Heel, Detachable boot and shoe	A. C. Wyman	Boston, Mass	July 16, 1872	129, 447
Heel-die	B. F. Fisk and M. B. Stone	Haverhill, Mass	Apr. 25, 1871	114, 121
Heel, Elastic boot and shoe	T. P. Monzani	Columbusville, N. Y	Dec. 17, 1872	134, 088
Heel enameling, Boot and shoe	C. H. Orcutt	Leominster, Mass	May 2, 1871	114, 470
Heel-faces for boots and shoes, Machine for dressing.	T. K. Reed	East Bridgewater, Mass	Feb. 27, 1866	52, 883
Heel finishing machine, Boot and shoe	H. H. Bigelow	Worcester, Mass	Oct. 25, 1870	108, 678
Heel finishing machine, Boot and shoe	J. L. Joyce	New Haven, Conn	Feb. 9, 1864	41, 512
Heel-finishing machine, Shave and burnisher for	J. L. Joyce	New Haven, Conn	Aug. 13, 1872	130, 375
Heel finishing machinery, Boot and shoe	L. P. Hawkins	Lynn, Mass	Apr. 1, 1873	137, 304
Heel finishing machinery, Boot and shoe	L. P. Hawkins	Lynn, Mass	Aug. 12, 1873	141, 712
Heel forming, smoothing, and polishing machine	G. P. Merriam	Lynn, Mass	July 29, 1862	36, 015
Heel gage, Boot and shoe	S. Ward	Westfield, N. Y	Aug. 29, 1871	118, 499
Heel guard, Boot and shoe	W. A. Harris	Providence, R. I	June 5, 1860	28, 575
Heel, India-rubber boot and shoe	P. Boisset	Paris, France	Apr. 22, 1862	35, 056
Heel, Interchangeable boot and shoe	J. Norburn	Pittsburgh, Pa	Dec. 7, 1869	97, 681
Heel, Interchangeable boot and shoe	J. C. Woodhead	Pittsburgh, Pa	Dec. 14, 1869	98, 004
Heel-iron	W. M. Rice	Boston, Mass	Oct. 9, 1866	58, 675
Heel-iron and ice-calk	W. Weaver	Nashua, N. H	Aug. 11, 1863	39, 514
Heel, Metallic	E. P. Bray	Elizabeth, N. J	Aug. 22, 1871	118, 335
Heel, Metallic	A. S. Mann	Saint Louis, Mo	Jan. 3, 1871	110, 768
Heel, Metallic	A. S. Mann	Saint Louis, Mo	July 11, 1871	116, 973
Heel, Metallic boot	P. S. Devlan	Reading, Pa	July 24, 1849	6, 610
Heel, Metallic boot and shoe	E. P. Bray	Elizabeth, N. J	Oct. 10, 1871	119, 691
Heel, Metallic boot and shoe	J. V. Dinsmore	Auburn, Me	Apr. 3, 1860	27, 702
Heel, Metallic boot and shoe	M. C. Easterly	Antwerp, N. Y	Oct. 6, 1863	40, 162
Heel, Metallic boot and shoe	H. Green	Antwerp, N. Y	July 1, 1862	35, 751
Heel, Metallic boot and shoe	A. L. Holbrook	Fremont, Nebr	May 23, 1871	115, 206
Heel, Metallic boot and shoe	E. T. Miller	Albany, N. Y	Aug. 9, 1870	106, 186
Heel, Metallic boot and shoe	A. T. Perrine	Louisville, Ky	June 7, 1870	103, 921
Heel nailing and shaving machine	L. P. Hawkins	Lynn, Mass	Feb. 18, 1873	135, 911
Heel-nailing machine	C. W. Glidden	Lynn, Mass	July 23, 1872	129, 811
Heel of rubber boots and shoes	F. M. Shepard	New York, N. Y	Mar. 8, 1870	100, 677
Heel-pattern, Metallic	J. Brobst	Fort Wayne, Ind	Aug. 11, 1868	80, 904
Heel-plate	H. L. Drake	Lynn, Mass	June 3, 1873	139, 500
Heel plate, Boot and shoe	W. E. Hamlin, jr	Providence, R. I	Sept. 1, 1868	81, 777
Heel plate, Boot and shoe	W. H. and W. Lewis	New York, N. Y	Oct. 31, 1839	1, 391
Heel plate, Boot and shoe	G. B. Massey	New York, N. Y	Sept. 17, 1872	131, 452
Heel plate, Boot and shoe	G. Rohr	Williamsburgh, N. Y	Oct. 14, 1873	143, 719
Heel-plate, Skate	G. Havell	Newark, N. J	Apr. 8, 1873	137, 678
Heel polisher, Boot	V. K. Spear	Lynn, Mass	Dec. 10, 1867	72, 105
Heel polisher, Boot and shoe	C. H. Helms	Poughkeepsie, N. Y	Oct. 10, 1871	119, 707
Heel-polishing machine	B. Q. Budding	Milford, Mass	Aug. 18, 1863	39, 546

Index of patents issued from the United States Patent Office from 1790 *to* 1873, *inclusive*—Continued.

Invention.	Inventor.	Residence.	Date.	No.
Heel-polishing machine	J. M. Thompson and S. D. Tripp	Stoneham and Lynn, Mass	Mar. 7, 1865	46, 761
Heel-polishing machine	J. M. Thompson and S. D. Tripp	Stoneham and Lynn, Mass	Feb. 6, 1866	52, 498
Heel-polishing machine	J. M. Thompson and S. D. Tripp	Stoneham and Lynn, Mass	Feb. 6, 1866	52, 499
Heel-polishing machine	S. D. Tripp	Lynn, Mass	Feb. 6, 1866	52, 470
Heel-polishing machine, Boot	J. M. Thompson, G. P. French, and S. D. Tripp.	Stoneham and Lynn, Mass	Sept. 27, 1864	44, 489
Heel polishing machine, Boot and shoe	B. Q. Budding	Milford, Mass	May 3, 1864	42, 555
Heel polishing machine, Boot and shoe	B. Q. Budding	Worcester, Mass	Oct. 6, 1868	82, 798
Heel polishing machine, Boot and shoe	C. H. Helms	Poughkeepsie, N. Y	June 13, 1871	115, 853
Heel polishing machine, Boot and shoe	C. H. Helms	Poughkeepsie, N. Y	July 25, 1871	117, 283
Heel polishing machine, Boot and shoe	C. H. Helms	Poughkeepsie, N. Y	May 28, 1872	127, 234
Heel polishing machine, Boot and shoe	V. K. Spear	Lynn, Mass	Apr. 7, 1868	76, 543
Heel polishing machine, Shoe and boot	S. W. Chamberlin	Stoneham, Mass	July 23, 1861	32, 859
Heel-press	H. H. Bigelow	Worcester, Mass	July 5, 1870	105, 030
Heel press, Boot	C. H. Helms	Poughkeepsie, N. Y	Aug. 13, 1867	67, 650
Heel-protector	H. H. Southwick	Mahaska County, Iowa	Mar. 31, 1868	76, 262
Heel-randing machine, Boot and shoe	H. F. Packard	North Bridgewater, Mass	Aug. 22, 1871	118, 263
Heel-rands, Machine for forming	S. Holmes and J. F. Sargent	North Brookfield and Melrose, Mass.	Apr. 4, 1871	113, 432
Heel-rands, Machine for turning	J. B. Johnson and G. W. Moulton	Lynn, Mass	Apr. 29, 1873	138, 257
Heel, Reversible boot	F. Richardson and F. Hacker	Providence, R. I	Mar. 21, 1871	112, 962
Heel, Reversible boot	F. Richardson and F. Hacker	Providence, R. I	July 4, 1871	116, 758
Heel, Revolving boot	T. Walker	Birmingham, England	June 29, 1852	9, 088
Heel, Revolving boot and shoe	J. H. Roome	New York, N. Y	June 8, 1858	20, 510
Heel, Rotary boot and shoe	A. O. Crane	Boston, Mass	Mar. 19, 1872	124, 723
Heel-rounding machine	J. C. White	Auburn, N. Y	Aug. 30, 1870	106, 901
Heel, Rubber boot	J. F. Barrett	Concord, Mass	Sept. 24, 1867	69, 163
Heel screw, Boot	W. Ackerman	Flint, Mich	Aug. 19, 1873	141, 908
Heel seat, Boot and shoe	G. I. Mason	Lewistown, Ill	Mar. 24, 1868	75, 779
Heel-seat-cutting machine	A. B. Keith	North Bridgewater, Mass	May 26, 1868	78, 211
Heel-shave	J. G. Ross	Philadelphia, Pa	June 27, 1865	48, 448
Heel-shave	J. H. Sanford	North Bridgewater, Mass	July 9, 1872	128, 757
Heel-shave	D. E. Somes	Biddeford, Me	Nov. 13, 1860	30, 648
Heel shave, Boot	E. S. Snell	North Bridgewater, Mass	July 29, 1862	36, 035
Heel shave, Boot and shoe	F. S. Dawes	Hudson, Mass	Dec. 31, 1867	72, 724
Heel shave, Boot and shoe	V. Snell	North Bridgewater, Mass	July 20, 1858	20, 960
Heel-shaving guard	J. E. Wiggin and D. G. Crosby	Stoneham, Mass	Mar. 26, 1867	63, 346
Heel shells with wood, Machine for filling metallic boot.	F. Richardson, F. Hacker, and J. A. Blake.	Providence, R. I	May 7, 1872	126, 491
Heel, sole, &c., Composition boot and shoe	F. Marquard	Newburyport, Mass	May 23, 1871	115, 075
Heel-spur	M. Young, jr	Frederick, Md	Jan. 31, 1860	27, 019
Heel-stiffener	L. C. Crowell	Boston, Mass	Jan. 14, 1873	134, 736
Heel-stiffener	L. C. Crowell	Boston, Mass	May 27, 1873	139, 238
Heel-stiffener	A. B. Ely	Newton, Mass	Feb. 27, 1872	123, 985
Heel-stiffener	J. W. Hatch	Rochester, N. Y	July 16, 1872	129, 338
Heel-stiffener	J. R. Moffitt	Chelsea, Mass	Oct. 19, 1869	96, 023
Heel-stiffener	I. R. Rogers	Lynn, Mass	Jan. 14, 1873	134, 933
Heel-stiffener	E. M. Stevens	Chelsea, Mass	Apr. 2, 1867	63, 573
Heel-stiffener	J. M. Watson	Sharon, Mass	July 30, 1872	130, 089
Heel stiffener, Boot and shoe	S. Moore	Sudbury, Mass	Nov. 11, 1873	144, 556
Eeel stiffener, Boot and shoe	G. V. Sheffield	Northbridge, Mass	May 3, 1870	102, 720
Heel stiffener or counter, Boot and shoe	D. E. Hayward	Melrose, Mass	Oct. 27, 1863	40, 409
Heel-stiffener, Rubber	E. M. Stevens	Chelsea, Mass	Apr. 16, 1867	63, 958
Heel-stiffeners for boots and shoes, Manufacture of.	J. R. Moffitt	Chelsea, Mass	May 21, 1872	127, 090
Heel-stiffeners, Machine for making	S. Moore and H. Rogers	Sudbury, Mass	Nov. 5, 1872	132, 849
Heel-stiffeners, Machine for making	J. Samuels	Vineland, N. J	Nov. 25, 1873	145, 017
Heel stiffeners, Machine for shaping boot and shoe	J. Kimball	Boston, Mass	Oct. 8, 1872	131, 957
Heel-stiffening	A. B. Ely	Newton, Mass	Mar. 16, 1869	87, 916
Heel-stiffening for boots and shoes	F. F. Darrow and O. F. Wait	Rockport, Mass	Apr. 22, 1873	138, 010
Heel-stiffening for boots and shoes	E. M. Stevens	Boston, Mass	Aug. 18, 1863	39, 594
Heel-tap for boots and shoes	A. Warner	Brooklyn, E. D., N. Y	Jan. 3, 1871	110, 701
Heel-tread-leveling machine, Boot	A. B. Keith	Braintree, Mass	May 16, 1871	114, 949
Heel trimmer	C. H. Helms	Poughkeepsie, N. Y	Aug. 11, 1868	80, 953
Heel trimmer, Boot and shoe	L. Coté	Quebec, Canada	Nov. 15, 1870	109, 300
Heel trimming and burnishing machine, Attachment for.	S. H. Hodges	Lynn, Mass	Mar. 18, 1873	136, 916
Heel trimming and burnishing machine, Boot and shoe.	T. K. Reed	East Bridgewater, Mass	July 11, 1871	116, 867
Heel trimming and burnishing machines, Holding device for.	J. R. Folsom	Boston, Mass	Nov. 12, 1872	132, 957
Heel trimming and finishing machine, Boot and shoe.	G. C. Hawkins	Boston, Mass	Jan. 2, 1872	122, 462
Heel-trimming machine	C. J. Addy	Boston, Mass	Aug. 20, 1872	130, 686
Heel-trimming machine	C. J. Addy	Boston, Mass	Nov. 12, 1872	132, 944
Heel-trimming machine	C. J. Addy	Boston, Mass	Feb. 11, 1873	135, 745
Heel-trimming machine	C. C. Ballou	Albany, N. Y	Sept. 23, 1873	142, 980
Heel-trimming machine	E. U. Jones	Woodhaven, N. Y	July 1, 1873	140, 420
Heel-trimming machine	G. W. Moore	Newark, N. J	Oct. 15, 1872	132, 309
Heel-trimming machine	J. F. Sargent	Boston, Mass	Apr. 18, 1865	47, 341
Heel trimming machine, Boot and shoe	C. H. Helms	Poughkeepsie, N. Y	Mar. 12, 1861	31, 666
Heel trimming machine, Boot and shoe	C. H. Helms	Poughkeepsie, N. Y	Dec. 7, 1869	97, 638
Heel trimming machine, Boot and shoe	E. P. Richardson	Lawrence, Mass	Apr. 19, 1870	102, 159
Heel trimming machine, Boot and shoe	C. S. Stearns	Marlborough, Mass	June 25, 1867	66, 184
Heel-turning machine	J. Q. Molton	Lynn, Mass	Apr. 4, 1871	113, 684
Heels and soles for boots and shoes, Machine for cutting.	A. Warren	Jefferson, Ohio	July 19, 1859	24, 835
Heels and soles for boots and shoes, Machine for rasping and dressing.	J. P. Molliere	Lyons, France	Nov. 27, 1855	13, 854
Heels, Attaching boot and shoe	B. Giroux	Chicago, Ill	Dec. 13, 1870	110, 030
Heels, Breasting	W. H. Pitkin	Providence, R. I	Sept. 27, 1864	44, 453
Heels, Cutting boot	P. Shaw	Abington, Mass	Feb. 6, 1849	6, 095
Heels for boots and shoes, Machine for polishing and grinding the edges of.	J. M. Thompson and G. P. French.	Stoneham, Mass	May 3, 1864	42, 629
Heels for boots and shoes, Machine for preparing	J. Jenkins	Lynn, Mass	Sept. 30, 1862	36, 586
Heels for boots and shoes, Machine for shaping	J. H. Belser	Marlborough, Mass	Nov. 28, 1865	51, 254
Heels for boots and shoes, Machine for shaping	L. Hall	Boston, Mass	Sept. 27, 1859	25, 605
Heels for boots and shoes, Machine for shaping	J. Samuels	Lynn, Mass	Apr. 12, 1864	42, 334
Heels for boots and shoes, Process of forming pierced	H. H. Bigelow	Worcester, Mass	Apr. 9, 1872	125, 528

Index of patents issued from the United States Patent Office from 1790 *to* 1873, *inclusive*—Continued.

Invention.	Inventor.	Residence.	Date.	No.
Heels for shoes, Machine for finishing wooden	R. Hubler	New York, N. Y	Jan. 21, 1873	135, 038
Heels, Hand-tool for manufacture of	S. L. Ricker	New York, N. Y	Oct. 14, 1873	143, 718
Heels, Machine for attaching and finishing boot	H. Saloshinsky	Boston, Mass	Jan. 31, 1860	27, 008
Heels, Machine for cutting down boot and shoe	J. L. Joyce	New Haven, Conn	Jan. 8, 1869	91, 136
Heels, Machine for cutting leather for boot and shoe.	M. H. Hall	Philadelphia, Pa	Dec. 17, 1873	145, 647
Heels, Machine for dressing boot and shoe	J. S. Stuart and A. L. Corson	Marblehead, Mass	Oct. 23, 1860	30, 508
Heels, Machine for forming boot	C. W. Glidden	Lynn, Mass	May 6, 1873	138, 634
Heels, Machine for making boot and shoe	H. H. Bigelow	Worcester, Mass	Apr. 9, 1872	125, 529
Heels, Machine for making boot and shoe	C. W. Glidden	Lynn, Mass	Mar. 14, 1871	112, 702
Heels, Machine for making boot and shoe	A. Knowlton	Boston, Mass	Jan. 14, 1873	134, 899
Heels, Machine for making boot and shoe	A. C. McKnight	Philadelphia, Pa	June 17, 1873	140, 060
Heels, Machine for pricking and nail-loading boot and shoe.	C. W. Glidden	Lynn, Mass	Mar. 4, 1873	136, 503
Heels, Machine for punching the lifts of boot	G. W. Ellis	Lynn, Mass	Dec. 22, 1863	41, 037
Heels, Machine for shaping boot	G. W. Warfield	Hudson, Mass	Oct. 22, 1867	70, 139
Heels, Machine for shaping the edges of boot and shoe.	C. W. Glidden	Lynn, Mass	Feb. 4, 1873	135, 538
Heels, Machine for turning wooden	R. Hubler	New York, N. Y	Jan. 21, 1873	135, 037
Heels, Manufacture of boot and shoe	J. Blakey	Leeds, England	June 17, 1873	139, 859
Heels, Manufacture of boot and shoe	C. W. Glidden	Lynn, Mass	Jan. 7, 1873	134, 538
Heels, Metallic stiffener for boot and shoe	J. Barsaloux, J. S. James, and N. Lyon.	Sandy Hill and Hope Falls, N. Y.	July 9, 1872	128, 843
Heels of boots and shoes, Apparatus for trimming	L. Coté	St. Hyacinthe, Canada	Apr. 1, 1873	137, 293
Heels of boots and shoes, Constructing	S. Warner, J. Hodgkins, and W. E. Traver.	Lowell and Westfield, Vt., and Watervliet, N. Y.	July 1, 1840	1, 665
Heels of boots and shoes, Machine for dressing	W. H. Bush	New Haven, Conn	Mar. 12, 1872	124, 482
Heels of boots and shoes, Machine for dressing	H. Guild and L. Hall	Boston, Mass	Mar. 15, 1859	23, 245
Heels of boots and shoes, Machine for pricking and cutting.	E. S. Snell	North Bridgewater, Mass	Mar. 9, 1858	19, 611
Heels of boots and shoes, Machine for shaving	A. B. Keith and T. K. Reed	North Bridgewater and East Bridgewater, Mass.	June 21, 1870	104, 399
Heels of boots and shoes, Metallic brace for	G. W. Griswold	Carbondale, Pa	Sept. 23, 1856	15, 762
Heels of boots and shoes, Shaping	E. J. Beane	Providence, R. I	June 14, 1864	43, 082
Heels to boots and shoes and polishing the same, Machine for attaching.	G. W. Ellis and C. W. Glidden	Stoneham, Mass	Oct. 7, 1862	36, 607
Heels to boots and shoes, Machine for applying	J. Jenkins	Lynn, Mass	Mar. 12, 1861	31, 690
Heels to boots and shoes, Machine for nailing	G. W. Ellis	Lynn, Mass	Dec. 22, 1863	41, 038
Heels to boots and shoes, Mode of attaching composition.	C. H. Helms	Poughkeepsie, N. Y	Dec. 13, 1870	110, 037
Heels to boots, Attaching	C. D. Ulmer	Boston, Mass	Feb. 16, 1869	86, 955
Heels to India-rubber soles, Attaching metallic	S. T. Parmelee	New Brunswick, N. J	July 17, 1855	13, 272
Heels, Tool for trimming boot and shoe	S. C. Bedell	Red Bank, N. J	Apr. 3, 1866	53, 558
Heels, Vulcanized-rubber rings for boot and shoe	W. H. Towers	Boston, Mass	Jan. 16, 1872	122, 869
Heeling-machine	C. W. Glidden and H. P. Fairfield.	Lynn and Boston, Mass	May 20, 1873	139, 058
Heeling machine, Boot	C. W. Glidden	Lynn, Mass	May 31, 1870	103, 735
Heeling machine, Boot	T. K. Reed and A. B. Keith	East Bridgewater and North Bridgewater, Mass.	Nov. 29, 1870	109, 756
Heeling machine, Boot and gaiter	L. Graf	Newark, N. J	Dec. 16, 1873	145, 645
Heeling machine, Boot and shoe	H. H. Bigelow	Worcester, Mass	Oct. 25, 1870	108, 677
Heeling machine, Boot and shoe	J. Gilson	Stoneham, Mass	May 31, 1870	103, 734
Heeling machine, Boot and shoe	C. W. Glidden	Lynn, Mass	Feb. 25, 1873	136, 233
Heeling machine, Boot and shoe	W. F. Spinney	Lynn, Mass	May 31, 1870	103, 792
Heeling machine, Shoe	C. H. Trask and H. Eldridge	Lynn, Mass	Jan. 5, 1869	85, 625
Heeling machines, Bed-plate for boot and shoe	C. L. Frye	Marlborough, Mass	June 11, 1872	127, 861
Heeling-plate	N. A. Swett	Westbrook, Me	Sept. 3, 1867	68, 466
Heliogeastrum or planetarium	T. Newell	Poultney, Vt	Apr. 12, 1820	....
Heliographic instrument for taking the sun's altitude.	J. Oakes	New York, N. Y	June 8, 1858	20, 506
Helical spring	W. R. Nichols	Philadelphia, Pa	Apr. 26, 1870	102, 421
Helical spring	J. W. Peck, jr	Brooklyn, N. Y	Jan. 24, 1860	26, 954
Helical spring	R. Vose	New York, N. Y	Mar. 15, 1864	41, 950
Heliographic and photographic spectrum for producing line-engravings.	F. Von Egloffstein	New York, N. Y	Nov. 21, 1865	51, 103
Heliometer	C. F. L. Kisch	Huntingburgh, Ind	Apr. 28, 1868	77, 324
Heliometer	C. F. L. Kisch	Huntingburgh, Ind	May 19, 1868	78, 133
Heliotrope, Automatic	L.F. Morawetz and C. Volkmar	Baltimore, Md	June 12, 1866	55, 523
Helm-wheel, Spiral	R. P. Cunningham	Pomfret, Conn	Aug. 3, 1820	....
Hem-folder	L. Clark	Monticello, N. Y	Oct. 18, 1859	25, 807
Hem, Gathered	H. A. Ellis	Chicopee Falls, Mass	Apr. 30, 1872	126, 138
Hemmer	O. L. Brown	Hopkinton, Mass	Oct. 7, 1873	143, 433
Hemmer	E. P. Davis	North Attleborough, Mass	Apr. 14, 1868	76, 720
Hemmer	H. A. Ellis	Chicopee Falls, Mass	Apr. 30, 1872	126, 139
Hemmer	H. M. Hall	Philadelphia, Pa	July 9, 1872	128, 876
Hemmer	E. S. Yentzer	Ottawa, Ill	Apr. 22, 1873	138, 064
Hemmer and folder	J. and F. Stevens	New York, N. Y	Mar. 26, 1861	31, 833
Hemmer-guide	S. Perry	New York, N. Y	Oct. 8, 1872	132, 101
Hemming and cording guide	H. B. Odiorne	Philadelphia, Pa	May 8, 1855	12, 826
Hemming and tucking guide	G. W. Downes	New York, N. Y	May 26, 1863	38, 662
Hemming-guide	W. Gaskill	Cincinnati, Ohio	May 9, 1865	47, 629
Hemming-guide	W. Gaskill and G. H. Knight	Cincinnati, Ohio	May 9, 1865	47, 630
Hemming-guide	J. Howell	Sacramento, Cal	Mar. 5, 1861	31, 602
Hemming-guide	G. L. Jencks	Providence, R. I	June 11, 1861	32, 519
Hemorrhage from internal organs or cavities, Instruments for arresting.	A. B. Haile	Norwich, Conn	Oct. 16, 1849	6, 796
Hemorrhoidian	R. A. Cameron	Valparaiso, Ind	Aug. 27, 1867	68, 042
Hemostatic and anti-septic agents, Preparation to serve as.	P. A. F. Boboeuf	Paris, France	June 23, 1863	38, 940
Hemp and cotton stalks, Machine for pulling	A. W. Goddard	Clinton, Mass	Dec. 11, 1866	60, 359
Hemp and flax and reducing the length of the fibers, Machine for breaking.	J. S. Treat and S. Randall	Voluntown, Conn	Sept. 16, 1851	8, 360
Hemp and flax brake	A. F. Bruce	Marshall Post-Office, Mo	June 24, 1844	3, 641
Hemp and flax brake	J. Bryant	Akron, Ind	Aug. 9, 1864	43, 812
Hemp and flax, Breaking	J. C. Johnston	Fayette County, Ky	May 16, 1811	....
Hemp and flax, Breaking	R. Millor	Lexington, Ky	July 13, 1811	....
Hemp and flax breaking and cleaning machine	R. Medley	Bloomfield, Ky	June 11, 1829	....
Hemp and flax breaking and cleaning machine	W. N. Stewart	Mayslick, Ky	Feb. 4, 1843	2, 943

Index of patents issued from the United States Patent Office from 1790 *to* 1873, *inclusive*—Continued.

Invention.	Inventor.	Residence.	Date.	No.
Hemp and flax breaking and dressing	H. Lull	Ithaca, N. Y	June 14, 1837	235
Hemp and flax, Breaking and dressing	A. Salisbury and J. C. Langdon	Troy, N. Y	July 29, 1829	....
Hemp and flax breaking and dressing machine	C. Warner, A. T. Mixell, and E. I. Horn.	Belvidere, N. J	July 31, 1837	319
Hemp and flax, Breaking and swingling	A. Tanner	Trumbull County, Ohio	Oct. 24, 1811	....
Hemp and flax breaking and swingling machine	T. Cohoon	Cazenovia, N. Y	Apr. 13, 1808	....
Hemp and flax breaking machine	A. Kyes	Crittenden, Ky	Apr. 4, 1838	671
Hemp and flax breaking machine	O. A. Leavitt	Maysville, Ky	Aug. 30, 1853	9, 973
Hemp and flax breaking machine	A. Salisbury and J. C. Langdon	Troy, N. Y	Apr. 18, 1829	....
Hemp and flax breaking machine	F. Stith	Franklin, Tenn	Apr. 22, 1835	....
Hemp and flax breaking machine	J. Warren	Westbrook, Me	June 14, 1837	234
Hemp and flax breaking machine	J. Y. Watson, J. Blossom, and A. Burnett.	Salem, N. Y	Apr. 21, 1829	....
Hemp and flax, Breaking, swingling, and beating	W. Harper	Richmond, Va	Oct. 15, 1811	....
Hemp and flax cleaning and dressing machine	A. Salisbury	Troy, N. Y	Apr. 15, 1829	....
Hemp and flax, Drawing-frame for	O. S. Leavitt	Marcellus, N. Y	Sept. 20, 1853	10, 034
Hemp and flax dressing machine	H. S. Barnum and M. Stevenson	Cambridge, N. Y	July 8, 1829	....
Hemp and flax dressing machine	W. and R. Brittain	Amwell, N. J	Oct. 12, 1837	423
Hemp and flax dressing machine	H. Burden	Albany, N. Y	June 15, 1822	....
Hemp and flax dressing machine	J. Macdonald	New York, N. Y	Jan. 18, 1823	....
Hemp and flax dressing machine	G. H. Ricketts and J. Kinney, jr	Morristown, N. J	Oct. 24, 1814	....
Hemp and flax dressing machinery	W. W. Grant	Providence, R. I	Mar. 5, 1850	7, 139
Hemp and flax fiber for the manufacture of dusters, &c., Preparing.	E. G. Wayman	Louisville, Ky	Feb. 18, 1868	74, 736
Hemp and flax for carding, Machinery for preparing.	C. Beach	Penn Yan, N. Y	Sept. 23, 1862	36, 551
Hemp and flax for spinning and weaving, Preparation of.	L. Tibbitts	New Glasgow, Va	Aug. 1, 1829	....
Hemp and flax hackling machine	J. Rinek	Easton, Pa	Sept. 24, 1872	131, 708
Hemp and flax hatcheling and cleaning machine	G. M. Billings and J. Harrison	Glasgow, Mo	May 1, 1845	4, 022
Hemp and flax machine	R. Button	Cazenovia, N. Y	Feb. 11, 1807	....
Hemp and flax machine	T. Cohoon	Troy, N. Y	Apr. 25, 1829	....
Hemp and flax, Machine and mode of preparing	F. Hall	Tennessee	Oct. 7, 1813	....
Hemp and flax, Machinery for breaking	G. Sanford and J. E. Mallory	England and New York, N. Y.	Apr. 28, 1863	38, 340
Hemp and flax, Machinery for preparing	E. Davy	Crediton, England	Mar. 11, 1856	14, 394
Hemp and flax, Preparing and hatcheling	J. Goulding	Boston, Mass	Aug. 17, 1835	....
Hemp and flax, Process of bolting	G. W. Billings	New York, N. Y	Oct. 6, 1863	40, 154
Hemp and flax scutching and hackling machine	O. W. Grimes	Paducah, Ky	Sept. 23, 1851	8, 374
Hemp and flax softening machinery	R. Boyack	Poughkeepsie, N. Y	Oct. 24, 1854	11, 825
Hemp and making oakum of same, Machinery for tarring slivers or bands of.	W. Montgomery	Boston, Mass	Aug. 28, 1840	1, 747
Hemp and other fibrous material, Apparatus and process of rotting.	L. W. Wright	Plainfield, N. H	Dec. 25, 1849	6, 980
Hemp and parting its fibers, Machinery for drawing	O. S. Leavitt	Maysville, Ky	Sept. 24, 1850	7, 668
Hemp-brake	J. F. Adams	Lexington, Ky	Nov. 4, 1873	144, 301
Hemp-brake	W. W. Austin and F. Creasy	Carrollton, Mo	Mar. 5, 1861	31, 585
Hemp-brake	J. Barkley	Saint Louis, Mo	Apr. 7, 1857	17, 015
Hemp-brake	C. A. Biermann	Waterloo, Ill	Feb. 5, 1867	61, 707
Hemp-brake	R. W. Bewen	Marshall, Mo	June 24, 1856	15, 166
Hemp-brake	G. M. Campbell	Lewistown, Ill	Feb. 27, 1866	52, 821
Hemp-brake	L. S. Chichester	Williamsburgh, N. Y	Feb. 3, 1852	8, 700
Hemp-brake	E. Christianson	Saint Joseph, Mo	July 30, 1867	67, 268
Hemp-brake	E. M. Crandal	Alton, Ill	Oct. 12, 1869	95, 659
Hemp-brake	J. Crane and F. H. Hamilton	Schenectady, N. Y	Jan. 1, 1850	6, 984
Hemp-brake	R. Dodsworth	Saint Louis, Mo	Feb. 12, 1861	31, 375
Hemp-brake	Z. Feagan	Palmyra, Mo	Aug. 9, 1859	24, 999
Hemp-brake	P. S. Fitch	Hanly, Ky	June 18, 1872	128, 030
Hemp-brake	T. L. Fortune	Liberty, Mo	Sept. 9, 1846	4, 741
Hemp-brake	T. L. Fortune	Weston, Mo	July 26, 1864	43, 654
Hemp-brake	P. G. Gardiner	New York, N. Y	Feb. 28, 1845	3, 936
Hemp-brake	R. J. Gatling	Murfreesborough, N. C	Apr. 17, 1847	5, 073
Hemp-brake	R. J. Gatling	Indianapolis, Ind	Sept. 4, 1860	29, 875
Hemp-brake	J. T. Gillman	Walnut Fork, Iowa	May 24, 1864	42, 847
Hemp-brake	H. Guild	Cincinnati, Ohio	Sept. 26, 1848	5, 811
Hemp-brake	E. Guile	Waverly, Mo	Oct. 16, 1860	30, 402
Hemp-brake	W. W. Hampton	Winchester, Va	Apr. 21, 1857	17, 092
Hemp-brake	J. S. Hardeman	Arrow Rock, Mo	May 12, 1857	17, 274
Hemp-brake	R. Heneage	Buffalo, N. Y	Dec. 21, 1858	22, 399
Hemp-brake	R. Heneage	Buffalo, N. Y	Jan. 22, 1861	31, 167
Hemp-brake	J. Hindman	Haynesville, Mo	Jan. 25, 1859	22, 725
Hemp-brake	F. P. Holcomb	New Castle, Del	Mar. 13, 1847	5, 010
Hemp-brake	D. W. Hughes	New London, Mo	Feb. 27, 1855	12, 456
Hemp-brake	W. C. Hutchinson	Saint Joseph, Mo	Oct. 5, 1858	21, 680
Hemp-brake	W. Jones	Saint Louis, Mo	Jan. 29, 1861	31, 245
Hemp-brake	J. Kaye	Louisville, Ky	Sept. 15, 1868	82, 123
Hemp-brake	E. W. Lacy	Oak Park, Va	Dec. 23, 1856	16, 279
Hemp-brake	E. W. Lacy	Oak Park, Va	July 31, 1860	29, 388
Hemp-brake	W. S. Lawrence	Winchester, Ky	Apr. 9, 1872	125, 466
Hemp-brake	W. S. Lawrence	Winchester, Ky	Aug. 5, 1873	141, 569
Hemp-brake	O. S. Leavitt	Marcellus, N. Y	Sept. 20, 1853	10, 033
Hemp-brake	S. H. Little	Saint Louis, Mo	Jan. 6, 1857	16, 365
Hemp-brake	S. H. Little	Saint Louis, Mo	Aug. 24, 1858	21, 264
Hemp-brake	H. F. Mann	La Porte, Ind	May 31, 1859	24, 264
Hemp-brake	R. Mansley	Philadelphia, Pa	Jan. 18, 1859	22, 661
Hemp-brake	W. Mason	Washington County, Ohio	Dec. 28, 1809	....
Hemp-brake	J. C. Matherly	Irvine, Ky	June 24, 1873	140, 289
Hemp-brake	J. B. McCormick and W. R. Baker.	Saint Louis, Mo	Jan. 22, 1861	31, 181
Hemp-brake	J. R. McDonald	Fayette, Mo	Apr. 2, 1861	31, 899
Hemp-brake	H. D. McGeorge	Morgantown, Va	Sept. 14, 1858	21, 513
Hemp-brake	J. Mills, jr	Quincy, Ill	May 29, 1860	28, 497
Hemp-brake	S. P. Moore	Arrow Rock, Mo	Feb. 2, 1858	19, 255
Hemp-brake	T. H. Murphy	New Orleans, La	Oct. 14, 1862	36, 667
Hemp-brake	G. M. Newell	Lexington, Mo	July 13, 1858	20, 890
Hemp-brake	J. H. Phillips	Waverly, Mo	June 11, 1861	32, 530
Hemp-brake	J. W. Rinehart	Lexington, Mo	May 10, 1859	23, 946
Hemp-brake	G. Rynel	Paris, Ky	May 3, 1870	102, 596

Index of patents issued from the United States Patent Office from 1790 *to* 1873, *inclusive*—Continued.

Invention.	Inventor.	Residence.	Date.	No.
Hemp-brake	S. H. Sample	Fayette, Mo	Aug. 18, 1846	4, 698
Hemp-brake	I. Scoville	Chicago, Ill	Sept. 25, 1860	30, 161
Hemp-brake	W. Shelby	Waverly, Mo	Nov. 2, 1858	21, 983
Hemp-brake	S. Sherman	Weston, Mo	Mar. 26, 1867	63, 183
Hemp-brake	C. Simon	Louisville, Ky	Sept. 29, 1857	18, 303
Hemp-brake	A. Smith	Mobile, Ala	Nov. 20, 1849	6, 888
Hemp-brake	S. Stafford	Carroll County, Mo	July 28, 1857	17, 900
Hemp-brake	M. Thompson	Saint Joseph, Mo	Aug. 5, 1856	15, 498
Hemp-brake	W. A. Vertrees	Winchester, Mo	Aug. 30, 1859	25, 292
Hemp-brake	P. M. Walker	Marshall, Mo	May 27, 1851	8, 122
Hemp-brake	D. F. Wallace	Ripley, Ohio	May 5, 1868	77, 553
Hemp-brake	A. Wilson and G. C. Fletcher	Saint Thomas, Mo	Aug. 2, 1859	24, 966
Hemp-brake	C. Witt and A. Sina	Davenport, Iowa	Sept. 3, 1867	68, 477
Hemp-brake	G. F. S. Zimmerman and A. Beattie.	Saint Joseph, Mo	Nov. 17, 1857	18, 657
Hemp brake and cleaner	C. B. Butler	Petersburgh, Tenn	Jan. 6, 1844	3, 402
Hemp-breaker	S. A. Clemens	Rockford, Ill	Feb. 11, 1862	34, 349
Hemp, Breaking and cleaning	G. W. Billings and J. Harrison	Glasgow, Mo	June 7, 1845	4, 071
Hemp breaking and cleaning	H. Zellner	Columbia, Tenn	July 16, 1867	66, 767
Hemp breaking and cleaning machine	J. R. Booton	Luray, Va	Oct. 11, 1859	25, 717
Hemp, &c., breaking and cleaning machine	W. Y. Singleton	Springfield, Ill	May 7, 1845	4, 034
Hemp breaking and cleaning machine	G. T. Tate and W. English	Frankford, Mo	July 11, 1842	2, 717
Hemp breaking and dressing machinery	L. W. Colver	Glasgow, Mo	Sept. 26, 1848	5, 801
Hemp, Breaking and swingling	S. Mulliken		Mar. 11, 1791	
Hemp-breaking machine	L. W. Colver	Louisville, Ky	Mar. 29, 1853	9, 632
Hemp-breaking machine	W. Greathouse	Mason County, Ky	July 30, 1831	
Hemp-breaking machine	J. S Hoskins	Spring Hill, Mo	May 12, 1868	77, 815
Hemp-breaking machine	J. Pursell	Perryville, Ky	Mar. 18, 1836	
Hemp-breaking machine	S. Stafford	Carrollton, Mo	Dec. 13, 1859	26, 449
Hemp-breaking machine	W. Stone, jr	Williamson County, Tenn	Sept. 8, 1831	
Hemp-breaking machine	J. S. Van de Graaf	Scott County, Ky	May 12, 1828	
Hemp cleaner and broom-corn stripper, Sisal	G. D. Allen	Key West, Fla	Mar. 6, 1855	12, 472
Hemp-cradle	G. Reynolds, jr	Washington, Ky	May 30, 1844	3, 607
Hemp-cutter	J. L. Hardeman	Arrow Rock, Mo	Aug. 21, 1855	13, 460
Hemp-cutter	J. L. Hardeman	Arrow Rock, Mo	Nov. 17, 1857	18, 638
Hemp-cutting cradle	W. L. Larimore	Paris, Ky	Apr. 25, 1837	178
Hemp-cutting machine	J. L. Hardeman	Arrow Rock, Mo	Aug. 20, 1850	7, 578
Hemp-drawing machine	S. Lownds	Brooklyn, N. Y	July 14, 1857	17, 795
Hemp-drawing machine	G. W. Pittman	Dartmouth, Canada	Nov. 28, 1871	121, 414
Hemp-dresser	B. Langdon and A. Salisbury	Troy, N. Y	Apr. 25, 1846	4, 469
Hemp, flax, and other fibers, Bleaching and scouring.	L. Jarosson	Lille, France	Dec. 24, 1867	72, 500
Hemp, flax, and wool, Machine for hatcheling	D. Treadwell	Boston, Mass	Aug. 18, 1834	
Hemp, flax, &c., Belted roving-machine for	D. Treadwell	Boston, Mass	Feb. 3, 1834	
Hemp, flax, &c., breaking	S. Mulliken		Jan. 15, 1795	
Hemp, flax, &c., breaking and cleaning machine	G. Safford and J. E. Mallory	England and New York, N. Y.	Sept. 22, 1863	40, 063
Hemp, flax, &c., breaking and cleaning machine	T. Tebow	Lexington, Ky	May 30, 1871	115, 387
Hemp, flax, &c., Curing	W. D. Monk	Williamsburgh, N. Y	Nov. 6, 1866	59, 436
Hemp, flax, &c., for spinning, Preparing	J. B. Fuller and J. P. Upham	Claremont, N. H	May 2, 1865	47, 538
Hemp, flax, jute-grass, &c., Method of treating	S. M. Allen	Woburn, Mass	July 18, 1865	48, 782
Hemp, flax, &c., New manufacture from	S. M. Allen	Woburn, Mass	Apr. 12, 1864	42, 265
Hemp, flax, &c., Preparing short-staple fibers from	J. B. Fuller and J. P. Upham	Claremont, N. H	June 7, 1864	43, 015
Hemp, flax, &c., Process for separating fiber of	J. B. Fuller and J. P. Upham	Claremont, N. H	May 2, 1865	47, 539
Hemp, flax, &c., Separating fibers of	J. B. Fuller	Claremont, N. H	June 7, 1864	43, 073
Hemp, flax, &c., Separating fibers of	J. B. Fuller and J. P. Upham	Claremont, N. H	Sept. 27, 1864	44, 415
Hemp, flax, &c., to a fibrous condition, Reducing	R. Sherwood	Fort Edward, N. Y	Nov. 10, 1863	40, 577
Hemp, flax, tow, &c., Preparing and hatcheling	J. Goulding	Boston, Mass	Oct. 10, 1835	
Hemp, flax, &c., Treating	T. Gray	Wadsworth, England	Dec. 4, 1866	60, 177
Hemp from okra, Making	J. Blanc	New Orleans, La	June 24, 1851	8, 184
Hemp, &c., hackling machine	P. G. and G. Rice	Danville, Ky	July 11, 1837	261
Hemp-hatcheling machine	G. W. Pittman	Brooklyn, N. Y	June 20, 1871	116, 092
Hemp into rope, drawing and hackling, Spinning	J. Lang	Greenock, Scotland	Jan. 16, 1834	
Hemp-leaves, Machine for separating the fiber from the pulp in.	S. S. Mills	Charleston, S. C	Oct. 12, 1858	21, 771
Hemp-machine	J. Anderson	Louisville, Ky	Nov. 13, 1849	6, 860
Hemp-machine	G. Ehrsam	New York, N. Y	July 14, 1863	39, 219
Hemp-machine	C. C. Estes	Maury County, Tenn	July 14, 1845	4, 114
Hemp-machine	J. E. Neill	Key West, Fla	Sept. 23, 1873	143, 179
Hemp, Machine for dressing Manila	C. E. Potter	Portsmouth, N. H	Dec. 23, 1834	
Hemp, Machine for dressing Sisal	S. A. Clemens	Springfield, Mass	July 15, 1851	8, 218
Hemp, &c., Machine for lowering and separating the boon from the fibers of.	R. Deering, sr	Louisville, Ky	June 10, 1845	4, 075
Hemp, &c., Machine for preparing and spinning	M. Day	Roxbury, Mass	Apr. 30, 1840	1, 580
Hemp, Machinery for breaking and dressing	A. Eldred	Oppenheim, N. Y	Apr. 24, 1849	6, 371
Hemp, &c., Machinery for hackling or preparing and spinning.	W. Montgomery	Boston, Mass	Feb. 28, 1844	3, 452
Hemp, &c., Machinery for spinning	W. C. Hibbard	Boston, Mass	Apr. 24, 1849	6, 388
Hemp or flax breaking and cleaning machinery	G. Sanford and J. E. Mallory	New York, N. Y	Mar. 18, 1862	34, 698
Hemp or flax dressing machinery	G. F. Schaffer	New York, N. Y	Oct. 29, 1861	33, 625
Hemp or flax waste within a sliver of longer staple, Machine for enveloping.	J. Leinweber	Louisville, Ky	May 1, 1866	54, 375
Hemp, Preparation of	R. Deering, sr	Louisville, Ky	June 25, 1845	4, 093
Hemp, Process of preparing	T. H. Barlow	Lexington, Ky	June 25, 1845	4, 094
Hemp, Process of rotting	W. Watt	Glasgow, Great Britain	Nov. 21, 1854	11, 981
Hemp, Process of treating	L. C. Suggett	Lexington, Ky	May 23, 1854	10, 968
Hemp, ramie, &c., cleaning machine	B. Roezl	Santo Comapam, Mexico	Sept. 17, 1867	68, 905
Hemp, Rotting	G. W. Billings and J. Harrison	Glasgow, Mo	May 10, 1845	4, 041
Hemp-rotting apparatus	C. H. Van Dorn	Saint Louis, Mo	Sept. 19, 1848	5, 776
Hemp-scutchers	F. H. Hamilton and T. Bullock	Schenectady, N. Y	Jan. 1, 1850	6, 987
Hemp, Sisal, and other grass cleaning machine	C. A. Dean	Boston, Mass	June 13, 1871	115, 828
Hemp, Treatment of fiber of Tampico	W. Staufen	London, England	July 6, 1858	20, 827
Hens from scratching, Device for preventing	E. Rice	West Northfield, Mass	Nov. 24, 1868	84, 303
Herbs for cookery, &c., Preparing	J. Culin	Philadelphia, Pa	July 16, 1872	129, 323
Hernia by means of injections, Cure of	C. T. Sage	New York, N. Y	Jan. 10, 1843	2, 906
Hernia-pad	W. Pomeroy	Brooklyn, N. Y	Aug 25, 1868	81, 533
Hernia-pad	W. Pomeroy	New York, N. Y	Aug. 29, 1871	118, 483
Hernia, Treating reducible	Z. Jayne	Carrollton, Ill	Apr. 2, 1841	2, 032

Index of patents issued from the United States Patent Office from 1790 *to* 1873, *inclusive*—Continued.

Invention.	Inventor.	Residence.	Date.	No.
Hide and fulling mill	S. Hussey	Gowanda, N. Y	Apr. 2, 1872	125, 135
Hide and leather dressing machine	J. R. Bumgarner and L. White	Davenport, Iowa	July 13, 1858	20, 861
Hide and leather softening machine	L. H. Stanley and G. B. Draper	Attleborough, Mass	May 23, 1871	115, 131
Hide and leather washing machine	H. M. Meeker	Smith's Mills, N. Y	May 4, 1869	89, 590
Hide and leather working machine	E. Fitzhenry	Somerville, Mass	Apr. 2, 1872	125, 276
Hide and skin cleaner and scourer	J. Turney	Nottingham, England	Sept. 17, 1872	131, 480
Hide-drying apparatus	M. H. Merriam and E. L. Norton	Charlestown, Mass	June 5, 1866	55, 333
Hide-handling cylinders, Beater in	J. R. Innis	Easton, Pa	Mar. 19, 1850	7, 192
Hide-mill	J. P. Friend and B. A. Annable	Peabody and Salem, Mass	Mar. 8, 1870	100, 519
Hide-mill or machine for softening dry hides	W. Edwards	Northampton, Mass	Dec. 30, 1812	....
Hide-mills, Operating	J. G. Curtis	Emporium, Pa	Nov. 22, 1870	109, 393
Hide scouring or dressing machinery	C. and F. E. Holmes	Boston, Mass	May 24, 1870	103, 463
Hide-treating device	H. Royer	San Francisco, Cal	June 22, 1869	91, 772
Hide-washing machine	A. Ross	Maine, N. Y	May 7, 1867	64, 575
Hide-working, leather-scouring, &c., Machinery for	T. R. Williams	Salem, Mass	Aug. 16, 1870	106, 439
Hides, Adjustable frame for stretching	W. Dunn	Newark, N. J	Oct. 1, 1867	69, 327
Hides and furs, Dressing	C. I. Weston	Cummington, Mass	Mar. 24, 1868	76, 015
Hides and leather, Frame for stretching	T. P. Howell and C. P. Oliver	Newark, N. J	May 5, 1868	77, 615
Hides and manufacturing leather, Treating	W. H. Fuller	Brockport, N. Y	Dec. 27, 1870	110, 562
Hides and preserving wood, Depilating	J. E. Siebel	Chicago, Ill	July 4, 1871	116, 638
Hides and shaving leather, Machine for splitting	S. Boyden	Foxborough, Mass	Jan. 7, 1809	....
Hides and skins, Apparatus for fleshing and stoning	J. S. Wheat	South Wheeling, W. Va	Nov. 13, 1866	59, 692
Hides and skins, Breaking, unhairing, fleshing, and scraping.	T. E. Barker and S. S. Fordham	Hudson, N. Y	Oct. 15, 1812	....
Hides and skins, Compound for bating	C. F. Panknin	Charleston, S. C	Nov. 29, 1870	109, 656
Hides and skins, Compound for bating	L. F. Robertson	Morrisania, N. Y	Feb. 23, 1869	87, 202
Hides and skins, Compound for bating	L. F. Robertson	New York, N. Y	Dec. 13, 1870	110, 161
Hides and skins, Compound for treating	L. F. Robertson	West Farms, N. Y	Apr. 21, 1868	77, 099
Hides and skins, Curing	H. Napier	Elizabeth, N. J	Oct. 30, 1866	59, 251
Hides and skins, Depilating and tanning	J. Henry	New York, N. Y	Jan. 28, 1870	104, 734
Hides and skins, Machine for dressing	J. Schiffer	New York, N. Y	July 9, 1867	66, 640
Hides and skins, Machine for scouring	E. Fitzhenry	Boston, Mass	Oct. 3, 1871	119, 513
Hides and skins, Machine for splitting	W. Brown	Philadelphia, Pa	Nov. 20, 1809	....
Hides and skins, Machine for splitting	T. O. Harrison	....	May 19, 1800	....
Hides and skins, Process for treating	W. Maynard	Salem, Mass	Feb. 18, 1873	136, 082
Hides and skins, Process of bating	W. Zollikoffer	Middleburgh, Md	Feb. 3, 1838	592
Hides and skins, Reel for handling	W. Brown	Frankford and Philadelphia, Pa.	Apr. 17, 1828	....
Hides and skins, Removing hair from	W. Anderson	Inverkeithing, Scotland	Oct. 8, 1872	131, 927
Hides and skins, Stuffing	W. Panton	Quincy, Mass	Jan. 29, 1867	61, 560
Hides and skins, Treatment of	J. Armstrong	Woburn, Mass	Feb. 15, 1859	22, 925
Hides and working leather, Machine for breaking	E. D. Taylor and W. Rude	Hornellsville, N. Y	Sept. 20, 1870	107, 562
Hides, Apparatus for handling	C. Weston	Salem, Mass	Feb. 22, 1859	23, 053
Hides, Apparatus for liming	S. J. Miller, A. B. Barnett, and W. H. Study.	Economy, Ind	July 4, 1865	48, 738
Hides, Apparatus for removing hair and lime from	S. S. Weed	Stoneham, Mass	Sept. 5, 1865	49, 811
Hides, Apparatus for removing hair from	M. Bray	Boston, Mass	Aug. 22, 1865	49, 496
Hides, Apparatus for removing hair from	S. S. Weed	Stoneham, Mass	Sept. 5, 1865	49, 839
Hides, Apparatus for tanning	H. Liebermann	Louisville, Ky	June 21, 1864	43, 258
Hides, Apparatus for treating green	L. Mackey	Dunmore, Pa	May 8, 1866	54, 568
Hides, Apparatus for working	A. Adler	Paris, France	July 31, 1866	56, 687
Hides, Apparatus for working	H. Lampert	Rochester, N. Y	Mar. 15, 1870	100, 907
Hides, Bating and tanning	W. B. Milligan	Edinburgh, Va	Nov. 4, 1851	8, 500
Hides, Breaking and scraping	A. Smith	Cumberland Valley, Pa	Aug. 26, 1846	4, 712
Hides, Breaking, fleshing, and hairing	J. Dunaway	Woodville, Va	May 13, 1834	....
Hides, Breaking, working, and scouring	J. Bonney	Cornwall, Conn	May 24, 1834	....
Hides, Composition for unhairing	A. H. Ward, jr	Boston, Mass	Jan. 2, 1855	12, 151
Hides, Compound for bating and raising	G. W. Adler	Philadelphia, Pa	Feb. 1, 1870	99, 387
Hides, Compound for stuffing and tanning	E. England	Mossy Creek, Tenn	June 28, 1870	104, 719
Hides, Depilating composition for	P. H. Schlosser	Middletown, Md	Jan. 2, 1868	78, 543
Hides, Depilating compound for	A. K. Eaton	New York, N. Y	June 16, 1857	17, 562
Hides for belts, Preparing	J. J. Travis	Franklin, Conn	Oct. 22, 1828	....
Hides for tanning, Process of scraping flesh and hair from.	T. Williams	Rochester, N. Y	Nov. 4, 1830	....
Hides for tanning, Scraping the flesh and hair from	T. Williams	Rochester, N. Y	Oct. 31, 1831	....
Hides for the manufacture of various articles, Process of preparing raw.	W. H. Towers	New York, N. Y	Aug. 21, 1866	57, 409
Hides from animals, Apparatus for removing	P. Lull	Norwich, N. Y	Apr. 14, 1868	76, 642
Hides from cattle, Machine for stripping the	C. Brühl	Green Point, N. Y	July 9, 1867	66, 558
Hides, Gage-frame for slitting raw	J. Hoffman	Belvidere, N. J	May 19, 1868	78, 092
Hides, Handling	A. H. Beschormann	New York, N. Y	Nov. 14, 1846	4, 851
Hides in tanning, Apparatus for handling	D. A. Haviland and A. S. Phillips	Fort Dodge, Iowa	Nov. 5, 1861	33, 645
Hides in the lime, Softening, breaking, and stoning	B. Aylsworth	Masonville, N. Y	Feb. 13, 1833	....
Hides in the vats, Handling	S. Stem and D. Wierman	Mechanicstown, Md	Jan. 9, 1834	....
Hides into hard transparent material for making combs, &c., Manufacturing raw or green.	S. Pike	Providence, R. I	June 11, 1829	....
Hides, Knife for removing hair from	B. Harrington	Boston, Mass	July 3, 1866	56, 144
Hides, leather, &c., Apparatus for drying	A. W. Roberts	Hartford, Conn	June 11, 1867	65, 767
Hides, Machine for beaming	P. Lennox	Lynn, Mass	Aug. 11, 1868	80, 829
Hides, Machine for beaming	P. Lennox	Lynn, Mass	Nov. 8, 1870	109, 135
Hides, Machine for beaming	P. Lennox, H. H. Robbins, and E. Hayes.	Lynn, Mass	May 26, 1868	78, 389
Hides, Machine for breaking	C. Bauchman	North Whitehall, Pa	Apr. 16, 1850	7, 281
Hides, Machine for breaking	H. S. Clark	Randolph County, N. C	Dec. 22, 1826	....
Hides, Machine for breaking	I. S. Hershey	Hagerstown, Md	Sept. 11, 1849	6, 710
Hides, Machine for breaking and scouring	O. W. Bean	Farmington, Tex	July 20, 1869	92, 776
Hides, Machine for breaking, hairing, and fleshing	N. Kirk and S. C. Clark	Saint Clairsville, Ohio	July 18, 1812	....
Hides, Machine for cleaning hair from	A. Hasbrouck	Ithaca, N. Y	May 11, 1869	89, 864
Hides, Machine for cutting	J. C. Flint	Boston, Mass	Nov. 11, 1851	8, 510
Hides, Machines for dressing	H. L. Arnold	Elk Horn, Wis	Sept. 10, 1861	33, 229
Hides, Machines for dressing	E. Fitzhenry	Boston, Mass	Aug. 15, 1871	118, 002
Hides, Machine for handling	J. Hammond	Lattisburgh, Ohio	Aug. 11, 1868	80, 947
Hides, Machine for handling	B. B. Mereness	Georgetown, N. Y	Sept. 8, 1863	39, 824
Hides, Machine for handling	J. Snell, jr	Pottsville, Pa	Nov. 6, 1866	59, 469
Hides, Machine for handling	J. S. Wheat	South Wheeling, W. Va	Jan. 19, 1864	41, 336
Hides, Machine for preparing	T. W. Jones	Philomath, Ga	Feb. 4, 1851	7, 924
Hides, Machine for scraping	R. Shaler	Haddam, Conn	June 19, 1837	239
Hides, Machine for shaving	T. F. Weston	Salem, Mass	Mar. 26, 1867	63, 191

Index of patents issued from the United States Patent Office from 1790 *to* 1873, *inclusive*—Continued.

Invention.	Inventor.	Residence.	Date.	No.
Hides, &c., Machine for shaving and splitting	N. P. Whiting	Providence, R. I	Apr. 25, 1810	....
Hides, Machine for stretching	J. F. Coburn	Newark, N. J	Oct. 8, 1867	69, 630
Hides, Machine for stretching	T. P. Howell and C. P. Oliver	Essex County, N. J	Aug. 6, 1867	67, 431
Hides, Machine for treating	H. and L. Royer	San Francisco, Cal	May 12, 1868	77, 920
Hides, Machine for unhairing	E. Brock	Ellenville, N. Y	June 25, 1867	66, 124
Hides, Machine for unhairing	E. Brock and J. Schultz	Ellenville, N. Y	Aug. 18, 1868	81, 247
Hides, Machine for unhairing	J. Schultz	Ellenville, N. Y	June 25, 1867	66, 176
Hides, Machine for working	M. B. Gould	Buffalo, N. Y	Jan. 10, 1871	110, 847
Hides or skins, Apparatus for sweating	W. M. Mason	Buffalo, N. Y	Dec. 9, 1873	145, 436
Hides or skins, Machine for beaming	T. Roberts and P. Lennox	Lynn, Mass	Oct. 29, 1867	70, 268
Hides or skins, Machine for scouring, setting out, and finishing.	J. Taggart	Melrose, Mass	Dec. 21, 1869	98, 121
Hides, Order of applying tan-liquor to	S. W. Pingree	Methuen, Mass	Oct. 14, 1856	15, 896
Hides preparatory to tanning, Machine for smoothing out.	J. F. Flanders	Boston, Mass	Nov. 24, 1863	40, 680
Hides, Preparing	D. Aldrich	Saint Louis, Mo	Mar. 6, 1860	27, 338
Hides, Preparing	A. D. Lufkin	Cleveland, Ohio	July 31, 1860	29, 392
Hides, Preparing raw	H. Halvorson	Leicester, Mass	Sept. 25, 1847	5, 304
Hides, &c., Preventing moths from destroying	S. Storm	New York	Feb. 17, 1827	....
Hides, Process and apparatus for curing	A. Rock	New Orleans, La	Feb. 28, 1871	112, 285
Hides, Process for preparing and tanning	B. F. Taber	Buffalo, N. Y	Feb. 6, 1866	52, 464
Hides, Process for softening dry	J. Barrow	Cincinnati, Ohio	June 18, 1872	127, 947
Hides, Process of bating	W. Zollickoffer	Middlebury, Md	Aug. 18, 1842	2, 756
Hides, Process of treating and preserving	A. Rock	New Orleans, La	Sept. 5, 1871	118, 746
Hides, Process of treating raw	B. F. Wright	Winchester, Mass	Dec. 26, 1871	122, 142
Hides, Punching holes in raw	W. Angell	Providence, R. I	July 19, 1832	....
Hides, skins, and leather, Apparatus for treating	L. F. Robertson	Morrisania, N. Y	Feb. 23, 1869	87, 203
Hides, skins, furs, &c., for use, Preparing	J. A. Roth	Philadelphia, Pa	June 19, 1866	55, 715
Hides, skins, &c., Machine for breaking and softening.	E. Kendall	Newton, Mass	June 10, 1837	229
Hides, Softening	W. Berry	New Sharon, Me	Mar. 6, 1833	....
Hides, Softening	J. Robinson	Baltimore, Md	Mar. 18, 1834	....
Hides, Softening and working dry and slaughtered.	A. McMillen	Bedford, N. H	Feb. 4, 1833	....
Hides, Softening dry	J. M. Muller	North Becket, Mass	Nov. 13, 1866	59, 627
Hides, Treating raw	H. W. Southworth	Springfield, Mass	July 9, 1872	128, 938
Hides, Unhairing	J. Banks	Dixmont, Me	June 30, 1836	....
Hides, Working	G. Welty	West Newton, Pa	June 13, 1846	4, 570
High and low pressure engine, Combined	T. L. Jones	Natchez, Miss	Sept. 14, 1869	94, 891
High and low water alarm	G. N. Jones	Chicago, Ill	July 19, 1870	105, 461
High and low water alarm for steam-generators	T. P. Akers	New York, N. Y	Feb. 18, 1868	74, 477
High and low water alarm for steam-generators	J. H. Springer	Philadelphia, Pa	Feb. 11, 1868	74, 442
High and low water detector for steam-boilers	J. P. Hillard	Fall River, Mass	Mar. 4, 1862	34, 575
High and low water indicator	W. Butterfield	Madison, Wis	June 14, 1870	104, 265
High and low water indicator	G. B. Massey	New York, N. Y	Sept. 28, 1869	95, 246
High and low water indicator, Combined	L. F. Smith	Philadelphia, Pa	Dec. 6, 1870	109, 959
High and low water indicator for steam-boilers	S. W. Warren	Brooklyn, N. Y	Apr. 30, 1861	30, 237
High and low water indicator for steam-engines	F. Millward	Cincinnati, Ohio	May 30, 1871	115, 501
High chair and work-stand combined, Child's	D. O. Parker	Liverpool, Nova Scotia	Aug. 29, 1871	118, 479
High chair, Child's	J. Nichols	Gardner, Mass	Apr. 14, 1868	76, 801
Highways, Machine for filling wagon-ruts on	F. Monroe	Romeo, Mich	Jan. 14, 1862	34, 158
Hinge	F. C. Adams and J. Peckover	Cincinnati, Ohio	Nov. 6, 1860	30, 555
Hinge	J. Adt	Wolcottville, Conn	Mar. 31, 1868	76, 034
Hinge	J. W. Allen	Newark, N. J	Mar. 27, 1866	53, 389
Hinge	J. S. Arthur	Cordaville, Mass	Aug. 2, 1864	43, 660
Hinge	W. Baker	Utica, N. Y	Apr. 13, 1852	8, 869
Hinge	S. W. Barber	Heath, Mass	Aug. 29, 1871	118, 577
Hinge	J. Barrows	Hyde Park, Mass	Feb. 27, 1872	124, 111
Hinge	M. Bettinger and A. Boos	Cincinnati, Ohio	Oct. 12, 1858	21, 735
Hinge	S. Bingham	Saratoga Springs, N. Y	Jan. 21, 1873	134, 969
Hinge	E. S. Botner and J. B. Hopkins	Lock Haven, Pa	Feb. 5, 1867	61, 798
Hinge	E. Boileau	Saint Louis, Mo	Aug. 19, 1873	141, 989
Hinge	E. Boileau and C. Mesnier	Saint Louis, Mo	July 6, 1869	92, 150
Hinge	E. Boileau and C. Mesnier	Saint Louis, Mo	Aug. 13, 1872	130, 466
Hinge	H. D. Bradley	Forestville, Conn	May 27, 1873	139, 290
Hinge	J. D. Browne	Cincinnati, Ohio	Feb. 24, 1857	16, 678
Hinge	J. D. Browne	Madisonville, Ohio	Aug. 16, 1870	106, 315
Hinge	J. D. Browne	Madisonville, Ohio	Jan. 9, 1872	122, 559
Hinge	J. Buckman, jr	Williamsburgh, N. Y	Oct. 17, 1871	120, 032
Hinge	L. R. Chapman	Grand Rapids, Mich	Mar. 2, 1869	87, 396
Hinge	S. A. Chapman	Waterbury, Conn	Jan. 14, 1868	73, 297
Hinge	P. P. Child	Saint Louis, Mo	Oct. 8, 1872	132, 053
Hinge	J. Close and J. Buckman, jr	Brooklyn, N. Y	Sept. 19, 1865	49, 982
Hinge	J. Close and J. Buckman, jr	Brooklyn, N. Y	Dec. 12, 1865	51, 426
Hinge	A. A. Cluff	Worcester, Mass	July 2, 1872	128, 593
Hinge	C. Cole	Ithaca, N. Y	July 27, 1869	93, 058
Hinge	J. V. C. Crate	Waterbury, Conn	Oct. 11, 1864	44, 608
Hinge	J. J. Crooke	New York, N. Y	Jan. 12, 1864	41, 201
Hinge	J. J. L. and H. S. Crooke	Southfield and New York, N. Y.	June 22, 1869	91, 609
Hinge	G. H. Dodge	Camden, N. J	Dec. 3, 1861	33, 830
Hinge	W. C. Dodge	Washington, D. C	Oct. 15, 1872	132, 147
Hinge	R. Drahota	Philadelphia, Pa	Dec. 13, 1870	110, 020
Hinge	S. R. Dummer	New York, N. Y	Aug 22, 1865	49, 511
Hinge	S. R. Dummer	New York, N. Y	Oct. 24, 1865	50, 566
Hinge	S. Ehrman	Mount Joy, Pa	Jan. 29, 1861	31, 230
Hinge	W. H. Elliot	Plattsburgh, Pa	Aug. 10, 1858	21, 124
Hinge	G. F. Fischer and A. Whelan	Washington, D. C	June 14, 1870	104, 294
Hinge	R. B. Fouzer	Butler, Pa	Nov. 7, 1871	120, 580
Hinge	S. C. Frink	Indianapolis, Ind	Dec. 26, 1865	51, 711
Hinge	L. Frühinsfeld	Newark, N. J	Nov. 30, 1869	97, 381
Hinge	C. Gaylord	Washington, D. C	Aug. 16, 1870	106, 351
Hinge	E. W. Gilmore	North Easton, Mass	Dec. 12, 1865	51, 447
Hinge	K. Goddard	Philadelphia, Pa	Mar. 17, 1857	16, 831
Hinge	D. R. Gould	Chestertown, N. Y	Dec. 14, 1869	97, 907
Hinge	S. E. Harrington	Greenfield, Mass	May 17, 1864	42, 764
Hinge	R. Hart	Marietta, Ohio	Mar. 31, 1857	16, 920
Hinge	R. Hart	Washington County, Ohio	Oct. 26, 1858	21, 925
Hinge	H. Hayward	Paterson, N. J	Aug. 13, 1872	130, 369

Index of patents issued from the United States Patent Office from 1790 to 1873, inclusive—Continued.

Invention.	Inventor.	Residence.	Date.	No.
Hinge	L. T. Howell	Burlington, N. J	Apr. 26, 1859	23, 808
Hinge	A. Huffer	Hagerstown, Md	Oct. 11, 1870	108, 143
Hinge	G. S. Hurford	Canton, Ohio	June 14, 1870	104, 316
Hinge	W. Johnson	Milwaukee, Wis	Aug. 9, 1870	106, 169
Hinge	W. Johnston	Cincinnati, Ohio	Sept. 7, 1869	94, 606
Hinge	F. B. Jones	Louisville, Ky	Aug. 9, 1870	106, 170
Hinge	J. Jones	Rochester, N. Y	Feb. 7, 1860	27, 056
Hinge	D. C. Jordan, sr	Brooklyn, N. Y	Apr. 24, 1866	54, 171
Hinge	A. C. Kasson	Milwaukee, Wis	Dec. 31, 1872	134, 383
Hinge	C. M. Lane	Cincinnati, Ohio	June 4, 1861	32, 482
Hinge	G. W. Lane	Plantsville, Conn	Aug. 20, 1867	67, 889
Hinge	M. R. Lemman	Hamilton, Ohio	Mar. 21, 1871	112, 934
Hinge	E. Lindsley	Neenah, Wis	Aug. 4, 1868	80, 639
Hinge	J. R. Lomas	West Haven, Conn	Apr. 3, 1866	53, 637
Hinge	D. W. Long	Parkersburgh, W. Va	May 31, 1870	103, 758
Hinge	D. W. Long	Parkersburgh, W. Va	Aug. 20, 1872	130, 729
Hinge	W. I. Ludlow	Washington, D. C	Apr. 1, 1873	137, 458
Hinge	P. P. Lynch	Newark, N. J	Apr. 29, 1873	138, 336
Hinge	H. Manneck	New York, N. Y	Oct. 14, 1873	143, 704
Hinge	J. C. Mason	New Hartford Centre, Conn	Feb. 16, 1858	19, 374
Hinge	T. D. McCall and S. Bushnell	Walton, N. Y	Sept. 1, 1868	81, 805
Hinge	W. C. McGill	Cincinnati, Ohio	Dec. 19, 1865	51, 604
Hinge	J. Miller	Watkins, N. Y	Nov. 7, 1871	120, 762
Hinge	A. L. Mora	New York, N. Y	Nov. 5, 1867	70, 597
Hinge	J. A. Morrell	Chicago, Ill	Sept. 10, 1867	68, 644
Hinge	F. Müsser	Cincinnati, Ohio	Jan. 7, 1873	134, 697
Hinge	G. R. Nebringer	Lewisberry, Pa	July 16, 1867	66, 729
Hinge	E. D. Norton	Cuba, N. Y	May 30, 1871	115, 509
Hinge	A. A. Oat	Philadelphia, Pa	Oct. 17, 1871	119, 944
Hinge	J. K. Otis	East Cambridge, Mass	May 7, 1872	126, 571
Hinge	J. K. Otis	East Cambridge, Mass	May 14, 1872	126, 647
Hinge	G. F. Outten	Norfolk, Va	Apr. 12, 1870	101, 908
Hinge	J. Parker	Meriden, Conn	Feb. 19, 1867	62, 219
Hinge	J. Plant	Washington, D. C	Aug. 9, 1870	106, 278
Hinge	J. Plegar	Birmingham, Pa	July 24, 1855	13, 321
Hinge	E. N. Porter	Morrisville, Vt	Dec. 12, 1865	51, 476
Hinge	J. W. Powers	Evanston, Ill	June 11, 1867	65, 601
Hinge	G. M. Ramsay	New York, N. Y	Jan. 29, 1856	14, 167
Hinge	A. Rankin	Philadelphia, Pa	Apr. 8, 1873	137, 623
Hinge	V. Rathknecht	Chicago, Ill	Mar. 12, 1872	124, 627
Hinge	D. Reed	San Francisco, Cal	July 16, 1872	129, 591
Hinge	R. Reiber	Lebanon, Pa	Jan. 7, 1868	73, 120
Hinge	S. M. Richardson	New York, N. Y	Dec. 6, 1859	26, 375
Hinge	J. M. Riley	Newark, N. J	Oct. 10, 1865	50, 423
Hinge	J. Schafer	New York, N. Y	Feb. 5, 1867	61, 766
Hinge	W. R. Searle	East Hampton, Mass	May 14, 1872	126, 650
Hinge	N. Sehner	Hagerstown, Md	July 4, 1865	48, 630
Hinge	S. Selden	Fairview Township, Pa	Feb. 13, 1866	52, 649
Hinge	S. Selden and W. J. F. Liddell	Erie, Pa	July 16, 1867	66, 893
Hinge	A. P. Seymour, jr	Hecla Works, N. Y	June 1, 1869	90, 880
Hinge	R. Seymour	Waterford, N. Y	Aug. 6, 1872	130, 248
Hinge	W. Shannon	Allegheny City, Pa	Dec. 8, 1868	84, 771
Hinge	W. Shannon	Allegheny City, Pa	Dec. 8, 1868	84, 772
Hinge	J. D. Shepard	Buffalo, N. Y	Sept. 3, 1872	131, 032
Hinge	J. D. Shepard and R. W. English	Buffalo, N. Y	July 13, 1869	92, 662
Hinge	T. Smith	Boston, Mass	Feb. 15, 1870	99, 999
Hinge	J. Sowle	Boston, Mass	Jan. 21, 1868	73, 661
Hinge	S. S. Squire and T. Scharfenberg.	Brooklyn, N. Y	Feb. 1, 1859	22, 830
Hinge	S. Stiger	New York, N. Y	May 14, 1872	126, 850
Hinge	E. Swan	New Bedford, Mass	May 14, 1872	126, 654
Hinge	A. W. Sweeny	Washington, D. C	June 26, 1860	28, 944
Hinge	L. D. Tice	New York, N. Y	Sept. 17, 1872	131, 478
Hinge	L. Upham	Pawtucket, R. I	Nov. 7, 1865	50, 861
Hinge	S. D. Van Pelt	Anderson, Ind	July 26, 1870	105, 743
Hinge	A. Velguth	Milwaukee, Wis	Sept. 28, 1869	95, 290
Hinge	W. Wakenshaw	Newark, N. J	Apr. 2, 1872	125, 354
Hinge	F. H. Walker	Boston, Mass	Mar. 30, 1869	88, 351
Hinge	B. D. Washburn	Roxbury, Mass	Oct. 29, 1867	70, 299
Hinge	W. Webb	Waterbury, Conn	Aug. 2, 1864	43, 746
Hinge	W. T. Wells	Decatur, Ill	Oct. 23, 1866	59, 107
Hinge	W. T. Wells	Decatur, Ill	Dec. 19, 1871	122, 088
Hinge	D. Werner	Saint Louis, Mo	Jan. 11, 1870	98, 827
Hinge	S. M. Wheeler	Dowagiac, Mich	Apr. 4, 1871	113, 378
Hinge	A. Whelan	Washington, D. C	Jan. 19, 1869	85, 977
Hinge	A. Wirth	New York, N. Y	Aug. 26, 1873	142, 315
Hinge	J. W. Wood	Conshohocken, Pa	July 1, 1873	140, 449
Hinge	H. Young and M. Stachelin	Port Chester, N. Y	Oct. 10, 1865	50, 411
Hinge	H. M. Zimmerman	Washington, D. C	June 26, 1860	28, 937
Hinge	J. H. Zinn	Idaville, Pa	Mar. 5, 1867	62, 726
Hinge, Adjustable	J. G. Ralph	Aurora, Ill	Aug. 20, 1867	67, 904
Hinge and blind support	W. W. S. Orbeton	Haverhill, Mass	Aug. 20, 1867	67, 999
Hinge and fastener	N. B. Spooner	Plymouth, Mass	Oct. 29, 1867	70, 282
Hinge and fastener combined	L. Felker	Tewkesbury, Mass	Oct. 27, 1868	83, 369
Hinge and fastening, Combined shutter	W. S. Gerard	Newburgh, N. Y	Oct. 3, 1865	50, 239
Hinge and fastening, Window-blind	W. Baker	Utica, N. Y	Sept. 17, 1842	2, 776
Hinge and fastening, Window-blind	C. Reed	Cambridge, Mass	Aug. 1, 1848	5, 683
Hinge and hook	F. H. Cuypers	Newark, N. J	Mar. 4, 1862	34, 565
Hinge and k ob, Shutter	J. W. Tripp	Gallipolis, Ohio	Jan. 12, 1869	85, 772
Hinge and ring for lockets	G. Hartje	New York, N. Y	July 12, 1870	105, 201
Hinge and spring combined	J. J. Cowell	Newark, N. J	May 25, 1869	90, 501
Hinge and spring, Combined door	W. Gilfillan	Syracuse, N. Y	Sept. 30, 1873	143, 234
Hinge and spring, Combined double	A. B. Taft	New York, N. Y	Jan. 23, 1849	6, 052
Hinge, Angular strap	C. F. Hawley	Kansas City, Mo	June 13, 1871	115, 851
Hinge, Bag	G. Havell	Newark, N. J	June 25, 1867	66, 150
Hinge, Blind	M. Adler	Buffalo, N. Y	Nov. 3, 1868	83, 585
Hinge, Blind	W. L. Barnes	Kingston, N. Y	June 16, 1863	38, 876
Hinge, Blind	D. Bull	Amboy, Ill	Nov. 5, 1872	132, 748

Index of patents issued from the United States Patent Office from 1790 to 1873, inclusive—**Continued.**

Invention.	Inventor.	Residence.	Date.	No.
Hinge, Blind	J. L. Cathcart	Georgetown, D. C	Feb. 23, 1869	87, 243
Hinge, Blind	C. B. Clark	Buffalo, N. Y	Mar. 24, 1868	75, 862
Hinge, Blind	C. B. Clark	Buffalo, N. Y	Oct. 13, 1868	82, 920
Hinge, Blind	C. B. Clark	Buffalo, N. Y	Nov. 3, 1868	83, 603
Hinge, Blind	C. B. Clark	Buffalo, N. Y	May 25, 1869	90, 500
Hinge, Blind	C. B. Clark	Buffalo, N. Y	Feb. 15, 1870	99, 844
Hinge, Blind	A. Eccard	Washington, D. C	July 19, 1870	105, 441
Hinge, Blind	W. T. Freligh	Jersey City, N. J	Sept. 21, 1869	95, 101
Hinge, Blind	O. S. Garretson	Buffalo, N. Y	Apr. 9, 1872	125, 450
Hinge, Blind	O. S. Garretson	Buffalo, N. Y	Dec. 23, 1873	145, 797
Hinge, Blind	L. Gathmann	Chicago, Ill	Nov. 25, 1873	144, 898
Hinge, Blind	A. Huffer	Rochester, N. Y	Feb. 18, 1873	136, 061
Hinge, Blind	T. C. Lord	Chicago, Ill	Sept. 28, 1869	95, 363
Hinge, Blind	W. McKee and C. H. Jordan	Washington, D. C	Nov. 10, 1868	83, 869
Hinge, &c., Blind	J. Plant	Washington, D. C	Apr. 24, 1847	5, 080
Hinge, Blind	R. B. Prindle	Norwich, N. Y	Sept. 15, 1868	82, 245
Hinge, Blind	D. C. Sage	Middletown, Conn	Mar. 1, 1870	100, 453
Hinge, Blind	D. C. Sage	Middletown, Conn	June 14, 1870	104, 359
Hinge, Blind	A. L. Seabury and D. Morris	Norfolk, Va	Oct. 26, 1869	96, 273
Hinge, Blind	C. G. Shepard	Buffalo, N. Y	Sept. 24, 1872	131, 571
Hinge, Blind	E. R. Shepard	Scranton, Pa	Oct. 2, 1860	30, 281
Hinge, Blind	A. Velguth	Milwaukee, Wis	Oct 26, 1869	96, 171
Hinge, Blind-fastening	J. H. Nevins	Brooklyn, E. D., N. Y	Aug. 9, 1870	106, 272
Hinge, Box	R. Frisbie	Cromwell, Conn	Sept. 21, 1869	95, 103
Hinge, Butt	M. C. Ames	Hartford, Conn	June 11, 1867	65, 717
Hinge, Butt	M. A. Avery	Groton, N. Y	Feb. 25, 1868	74, 788
Hinge, Butt	B. F. Barker	San Francisco, Cal	Aug. 27, 1867	68, 150
Hinge, Butt	E. Boileau and C. Mesnier	Saint Louis, Mo	Mar. 23, 1869	88, 122
Hinge, Butt	M. Bush	Bloomington, Ill	Apr. 27, 1869	89, 464
Hinge, Butt	J. H. Carkeet	Montgomery, Ala	Nov. 3, 1868	83, 760
Hinge, Butt	J. W. Carleton	New Britain, Conn	Mar. 7, 1871	112, 417
Hinge, Butt	G. F. J. Colburn	Newark, N. J	Aug. 2, 1864	43, 670
Hinge, Butt	C. Cole	Ithaca, N. Y	Jan. 12, 1869	85, 728
Hinge, Butt	G. H. Fayman	Washington, D. C	Apr. 10, 1860	27, 788
Hinge, Butt	C. H. Foster	San Francisco, Cal	Nov. 2, 1869	96, 414
Hinge, Butt	C. Goodrich	Plainsville, Conn	Aug. 13, 1867	67, 747
Hinge, Butt	R. Hoadley	Derby, Conn	Feb. 16, 1864	41, 664
Hinge, Butt	H. Hockemeyer	Toledo, Ohio	Oct. 6, 1868	82, 717
Hinge, Butt	J. C. Hyde	West Haven, Conn	June 26, 1866	55, 954
Hinge, Butt	G. A. Jenks	Chicago, Ill	Dec. 1, 1868	84, 630
Hinge, Butt	J. Johnson	Geneseo, N. Y	Dec. 18, 1860	30, 970
Hinge, Butt	J. W. Jordan	Lexington, Va	Dec. 29, 1868	85, 310
Hinge, Butt	B. C. Pole	Richmond, Va	Nov. 3, 1868	83, 659
Hinge, Butt	A. Rankin	Philadelphia, Pa	Jan. 22, 1867	61, 458
Hinge, Butt	J. Rouse	Troy, N. Y	May 17, 1836	....
Hinge, Butt	D. C. Sage	Middletown, Conn	Nov. 12, 1867	70, 748
Hinge, Butt	J. E. Shields	Washington, D. C	May 29, 1860	28, 510
Hinge, Butt	C. Sholl	Mount Joy, Pa	July 30, 1861	32, 927
Hinge, Butt	I. L. Thompson	Sardis, Ohio	Jan. 7, 1873	134, 617
Hinge, Butt	J. F. Townsend	Westfield, N. Y	Oct. 28, 1862	36, 883
Hinge, Butt	S. Van Gilder	Knoxville, Tenn	June 17, 1873	139, 982
Hinge, &c., Butt	W. Whittaker	Troy, N. Y	Mar. 12, 1836	....
Hinge butts, Finishing loose	E. Luther, P. Lyon, and W. Edwards.	West Troy, N. Y	Dec. 22, 1868	85, 112
Hinge, Carriage	G. W. Beers	Bridgeport, Conn	July 3, 1866	55, 987
Hinge, Carriage	C. E. Schwind	New York, N. Y	Aug. 27, 1867	68, 121
Hinge, Carriage-door	G. W. Beers	Bridgeport, Conn	Mar. 7, 1871	112, 313
Hinge, Carriage-door	E. Wells	New Haven, Conn	Apr. 2, 1872	125, 236
Hinge, Carriage-top	H. Killam	New Haven, Conn	Oct. 12, 1869	95, 808
Hinge, Cast	N. Gates	Cincinnati, Ohio	Aug. 29, 1854	11, 606
Hinge-clamp	E. L. Seger and S. L. Smith	Yonkers, N. Y	Sept. 3, 1867	68, 533
Hinge, Compound	H. Loth	Philadelphia, Pa	July 15, 1873	140, 930
Hinge, Compound safe-door	P. J. Stuhltrager	Philadelphia, Pa	Sept. 29, 1868	82, 658
Hinge, Concealed	G. R. Cady and W. H. Cooper	New Haven, Conn	Nov. 26, 1867	71, 452
Hinge, Concealed	F. Wood	Bridgeport, Conn	Mar. 15, 1864	41, 956
Hinge, Door	S. M. Bullard	Holliston, Mass	June 2, 1857	17, 419
Hinge, Door	C. Dupré	Louisville, Ky	Jan. 7, 1868	73, 174
Hinge, Door	J. Elgar	Baltimore, Md	Apr. 11, 1854	10, 774
Hinge, Door	O. Judd	Cherry Valley, N. Y	Apr. 17, 1847	5, 066
Hinge, Door	G. Lane	New York, N. Y	Feb. 11, 1868	74, 384
Hinge, Door	C. H. Robison	Syracuse, N. Y	May 30, 1848	5, 608
Hinge, Door	A. W. Walter	Canton, Ohio	Oct. 27, 1868	83, 572
Hinge, Double	J. S. Jenness	Bangor, Me	Jan. 3, 1871	110, 659
Hinge, Double-acting	J. H. Carkeet	Montgomery, Ala	July 14, 1868	79, 950
Hinge, Double-acting	G. Dumbolton	Baltimore, Md	Apr. 7, 1868	76, 422
Hinge, Double-acting spring	H. E. Canfield	New York, N. Y	May 15, 1855	12, 853
Hinge, Double-acting spring	T. F. Engelbrecht	New York, N. Y	Feb. 25, 1851	7, 953
Hinge, Double-acting spring	T. F. Engelbrecht	New York, N. Y	Jan. 3, 1854	10, 389
Hinge, Elliptical	M. and R. Reeve	Burlington, N. J	Oct. 31, 1812	....
Hinge-fastener and shutter-opener, Combined	A. S. Pelton	Clinton, Conn	Jan. 23, 1849	6, 059
Hinge-finishing machine, Butt	J. J. Crooke	New York, N. Y	Feb. 26, 1867	62, 479
Hinge for blinds, Lock	T. Clucas, jr	Saint Louis, Mo	Sept. 9, 1873	142, 679
Hinge for carriage-doors, Concealed	E. Wells	New Haven, Conn	Jan. 23, 1872	122, 977
Hinge for daguerreotype-cases	E. G. Kingsley and S. A. W. Parker, jr.	Stoughton, Mass	June 1, 1858	20, 436
Hinge for doors, Helical-spring-joint	D. A. Hoyt and P. W. Bulkeley	Danbury, Conn	Apr. 28, 1838	721
Hinge for doors or window-frames of stoves	T. J. Coulston	Springfield, Pa	Nov. 9, 1869	96, 551
Hinge for doors, shutters, &c., Crane	E. Ripley	Troy, N. Y	May 13, 1851	8, 090
Hinge for doors, Spring	R. Adams	Borough of Southwark, Great Britain.	June 20, 1871	116, 002
Hinge for doors, &c., Spring	A. Rankin	Philadelphia, Pa	Sept. 27, 1870	107, 813
Hinge for doors, &c., Spring	A. F. Schiffling	Evansville, Ind	Sept. 27, 1870	107, 819
Hinge for fastening blinds, shutters, and doors	R. B. Varden	Baltimore, Md	Feb. 12, 1845	3, 903
Hinge for folding bedsteads	J. Binder	Chelsea, Mass	Aug. 16, 1853	9, 933
Hinge for gates and doors	D. S. Esten	Monson, Mass	Apr. 14, 1868	76, 727
Hinge for inkstand-covers	J. Nock	Philadelphia, Pa	Dec. 13, 1853	10, 310
Hinge for ladder-brackets	J. H. Balsley	Dayton, Ohio	Feb. 8, 1870	99, 622
Hinge for lamp-burners	W. Webb	Waterbury, Conn	June 18, 1867	65, 972

Index of patents issued from the United States Patent Office from 1790 *to* 1873, *inclusive*—Continued.

Invention.	Inventor.	Residence.	Date.	No.
Hinge for landau carriages, Concealed	E. Wells	New Haven, Conn	June 7, 1870	103, 948
Hinge for molders' flasks	G. Grant	Troy, N. Y	Dec. 7, 1852	9, 448
Hinge for molders' flasks	E. C. Little	Saint Louis, Mo	Oct. 16, 1866	58, 948
Hinge for piano-fortes, &c	R. Perkins	Boston, Mass	June 3, 1843	3, 122
Hinge for picture-cases	A. P. Critchlow	Florence, Mass	Oct. 14, 1856	15, 915
Hinge for reflectors of stereoscopes, &c	A. Beckers	New York, N. Y	Mar. 29, 1859	23, 342
Hinge for safe and vault doors	H. Gross	Cincinnati, Ohio	Nov. 18, 1873	144, 669
Hinge for safe-doors, Sliding	W. H. Butler	Brooklyn, N. Y	May 28, 1872	127, 147
Hinge for school-desks	A. E. Roberts	Des Moines, Iowa	Sept. 27, 1870	107, 724
Hinge for school-desks and settee-seats	N. O. Tiffany	Buffalo, N. Y	Dec. 20, 1870	110, 402
Hinge for sewing-machine tables	S. C. Brinser	Middletown, Pa	Aug. 13, 1872	130, 472
Hinge for sewing-machine tables	J. C. Gove	Cleveland, Ohio	Oct. 8, 1872	132, 069
Hinge for sheet-metal boxes	C. J. Hauck	Williamsburgh, N. Y	Mar. 24, 1868	75, 903
Hinge for show-cases	J. McAdams	Brooklyn, N. Y	Feb. 1, 1870	99, 453
Hinge for sleeping-car berths	J. I. Fogg	Chicago, Ill	Apr. 14, 1868	76, 620
Hinge for step-ladders	C. G. Udell	Chicago, Ill	Oct. 1, 1872	131, 798
Hinge for stove-lids	T. J. Coulston	Springville, Pa	Aug. 10, 1869	93, 600
Hinge for stoves, Quadrant	E. Barrows	Boston, Mass	Aug. 31, 1837	366
Hinge for sustaining, operating, closing, and fastening window-blinds.	L. T. Talbot	Taunton, Mass	Apr. 18, 1848	5, 521
Hinge for table-leaves	D. A. Garver	Bryan, Ohio	Apr. 2, 1872	125, 281
Hinge for table-leaves, &c	P. Hires	Columbus, Ky	Apr. 25, 1871	114, 009
Hinge for tables, &c., Lock	E. E. Hendrick	Carbondale, Pa	Mar. 12, 1872	124, 438
Hinge for tea-kettle covers, &c	J. Easterly	Albany, N. Y	Apr. 16, 1867	63, 874
Hinge for window-blinds	J. Loudon and H. Iversen	New York, N. Y	Aug. 31, 1858	21, 347
Hinge for window-blinds	T. E. Williams	Washington, D. C	Nov. 30, 1858	22, 214
Hinge for window-shutters	P. P. Child	Saint Louis, Mo	Dec. 24, 1867	72, 603
Hinge for window-shutters	W. H. Foulds	Henderson, Ky	Nov. 13, 1866	59, 581
Hinge-forming die	J. O'Kane	New York, N. Y	Apr. 9, 1867	63, 742
Hinge, Friction-roller door	J. Bailey, jr	Plymouth County, Mass	May 16, 1808	....
Hinge, Gate	C. E. Burnham	Binghamton, N. Y	Nov. 2, 1858	21, 939
Hinge, Gate	E. A. Bushnell	Burnett, Wis	July 16, 1872	129, 209
Hinge, &c., Gate	P. P. Child	Saint Louis, Mo	Oct. 31, 1871	120, 492
Hinge, Gate	C. B. Clark	Buffalo, N. Y	Feb. 7, 1871	111, 515
Hinge, Gate	P. Dennis	Schuylerville, N. Y	Feb. 25, 1868	74, 805
Hinge, Gate	J. W. Everham	Pitt's Grove, N. J	May 11, 1869	89, 978
Hinge, Gate	J. B. Farmer	Indianapolis, Ind	Feb. 9, 1869	86, 658
Hinge, Gate	W. Field and R. Carruthers	Bergen, N. J	Oct. 30, 1866	59, 202
Hinge, Gate	W. G. Franklin	Shelbina, Mo	Apr. 22, 1873	138, 081
Hinge, Gate	G. Garrett	Elkhart, Ill	Apr. 25, 1871	113, 998
Hinge, Gate	D. H. Gould	Troy, N. Y	Sept. 3, 1867	68, 434
Hinge, Gate	B. Greenside	Fort Dodge, Iowa	Oct. 30, 1866	59, 211
Hinge, Gate	A. T. Hendrick	Clyde, N. Y	Sept. 14, 1858	21, 496
Hinge, Gate	C. Hermance	Schuylerville, N. Y	Feb. 19, 1867	62, 196
Hinge, Gate	W. Hull	Fort Wayne, Ind	Oct. 21, 1873	143, 764
Hinge, Gate	H. Last	West Lebanon, Ind	June 12, 1866	55, 505
Hinge, Gate	A. D. McMaster	Rochester, N. Y	July 11, 1871	116, 851
Hinge, Gate	J. Miller	Penn Yan, N. Y	Dec. 10, 1872	133, 715
Hinge, Gate	E. D. Norton	Cuba, N. Y	Apr. 18, 1871	113, 788
Hinge, Gate	S. J. Olmsted	Binghamton, N. Y	Mar. 26, 1861	31, 852
Hinge, Gate	W. W. Robinson	Ripon, Wis	June 9, 1863	38, 843
Hinge, Gate	I. J. Ryerson	Princeton, Ind	June 8, 1869	91, 167
Hinge, Gate	J. H. Stone	Chapel Hill, Tex	Apr. 15, 1873	137, 972
Hinge, Gate	J. Sweet	Decatur, Mich	Sept. 28, 1869	95, 393
Hinge, Gate	D. Wadsworth, jr	Nashua, N. H	Nov. 6, 1860	30, 600
Hinge, Gate	J. T. Watson	Richmond, Ind	Oct. 13, 1863	40, 314
Hinge, Gate	J. T. Watson	Richmond, Ind	Jan. 1, 1867	60, 808
Hinge, Gate	L. E. Woodard	Cohocton, N. Y	Oct. 16, 1866	58, 930
Hinge hooks, Machine for making blind	G. Orr	Needham, Mass	Oct. 17, 1871	119, 945
Hinge-hooks, Manufacture of	C. Kanebeck	Boston, Mass	Jan. 21, 1868	73, 533
Hinge-hooks, Manufacture of	W. J. Lewis	Pittsburgh, Pa	Aug. 13, 1872	130, 508
Hinge hooks, Manufacture of blind	W. H. McIntosh	Needham Upper Falls, Mass.	Apr. 30, 1872	126, 150
Hinge-joints or coupling-shafts of wagons, Shifting	E. W. Seymour	Centre Lisle, N. Y	Feb. 25, 1862	34, 553
Hinge, Latch	J. Plant	Washington, D. C	May 3, 1859	23, 859
Hinge, Lock	C. B. Clark	Buffalo, N. Y	July 29, 1873	141, 323
Hinge, Lock	M. C. Lee	New York, N. Y	Apr. 22, 1873	138, 092
Hinge, Loose-joint butt	M. Perry	New York, N. Y	Oct. 1, 1867	69, 365
Hinge, Lounge	A. Spiegel	Indianapolis, Ind	Sept. 16, 1873	142, 957
Hinge-machine	J. H. Baird	Oakville, Conn	Mar. 29, 1870	101, 210
Hinge-machine	E. Brown	New York, N. Y	Nov. 24, 1868	84, 256
Hinge-making machine	E. Brown	Waterbury, Conn	May 16, 1854	10, 943
Hinge-making machine	J. B. Evrard and J. P. Boyer	Paris, France	June 19, 1866	55, 794
Hinge-making machine	D. Morse	....	June 10, 1803	....
Hinge-making machine	L. P. Summers	New Britain, Conn	Dec. 10, 1872	133, 903
Hinge-making machine, Butt	H. D. Blake	New Britain, Conn	Mar. 24, 1868	75, 724
Hinge-making machine, Butt	E. Brown and W. H. Van Gieson.	Waterbury, Conn., and Paterson, N. J.	Apr. 9, 1861	31, 949
Hinge-making machine, Butt	C. Miller	New York, N. Y	Feb. 20, 1855	12, 443
Hinge-making machine, Butt	A. Rais	Waterbury, Conn	June 25, 1867	66, 172
Hinge-making machine, Butt	A. Rais	Waterbury, Conn	July 9, 1867	66, 626
Hinge-making machine, Butt	A. Rais	Waterbury, Conn	Sept. 3, 1867	68, 529
Hinge-making machine, Butt	W. H. Van Gieson and J. J. Crooke.	Passaic, N. J., and New York, N. Y.	Mar. 10, 1868	75, 320
Hinge-making machine, Strap	W. J. Lewis	Pittsburgh, Pa	Apr. 25, 1871	114, 019
Hinge of rolling iron shutters	A. L. Johnson	Baltimore, Md	June 25, 1850	7, 457
Hinge, Pin	A. Lyons and I. B. Abrahams	New York, N. Y	Apr. 29, 1873	138, 418
Hinge-pintle	W. Wells	Ashtabula, Ohio	Mar. 9, 1869	87, 739
Hinge pivots, Machine for forging solid folded	J. J. and L. Crooke	Southfield and New York, N. Y.	Sept. 16, 1873	142, 774
Hinge, Plated butt	J. J. and L. Crooke	Southfield and New York, N. Y.	Apr. 4, 1871	113, 501
Hinge, Quadrant	N. Jones	Lockport, Ind	Aug. 24, 1869	94, 000
Hinge, Reversible	P. P. Child	Saint Louis, Mo	Sept. 26, 1871	119, 319
Hinge, Reversible	F. W. Marston	Philadelphia, Pa	May 9, 1871	114, 579
Hinge, Reversible	A. P. Seymour	Hecla Works, N. Y	Mar. 1, 1870	100, 332
Hinge, Reversible	H. J. Wolters	Salem, Mass	May 18, 1869	90, 330
Hinge, Reversible butt	D. F. Hale	Chicopee, Mass	May 3, 1870	102, 538

Index of patents issued from the United States Patent Office from 1790 *to* 1873, *inclusive*—Continued.

Invention.	Inventor.	Residence.	Date.	No.
Hinge, Reversible butt	W. Hancock	Saco, Me	Jan. 22, 1867	61, 421
Hinge, Reversible butt	W. S. Weed	Auburn, N. Y	Sept. 24, 1867	69, 287
Hinge-riveting machine	H. D. Blake	New Britain, Conn	Mar. 17, 1868	75, 663
Hinge-riveting machine	E. H. Lacy and P. Lyon	West Troy, N. Y	Jan. 16, 1872	122, 837
Hinge-riveting machine	H. M. Ritter	Covington, Ky	Dec. 15, 1868	84, 907
Hinge-riveting machine	I. W. Valance and H. Little-john.	Lansingburgh and New York, N. Y.	Aug. 27, 1861	33, 174
Hinge, Safe-door	J. L. Hall	Cincinnati, Ohio	Feb. 25, 1873	136, 239
Hinge, Safe-door	W. F. Stevens	Melrose, Mass	Dec. 30, 1873	145, 973
Hinge, Seat	M. W. Chase	Buffalo, N. Y	July 22, 1873	141, 115
Hinge, Seat	W. A. Slaymaker	Atlanta, Ga	Nov. 28, 1871	121, 259
Hinge, Self-closing	C. Vetter	Cincinnati, Ohio	Nov. 12, 1867	70, 921
Hinge, Self-fastening shutter	A. Nicholson	Poland, N. Y	Mar. 21, 1854	10, 673
Hinge, Self-lock window	J. R. Murphy	Pittsburgh, Pa	Sept. 29, 1863	40, 145
Hinge, Self-locking	G. E. Boisselier	Saint Louis, Mo	Sept. 20, 1870	107, 437
Hinge, Self-locking blind	O. S. Garretson	Buffalo, N. Y	Jan. 9, 1872	122, 594
Hinge, Self-locking shutter	E. H. Benjamin	Oak Hill, N. Y	June 23, 1868	79, 051
Hinge, Sheet-metal	W. H. Hart	New Britain, Conn	June 3, 1873	139, 464
Hinge, Shutter	D. G. Coppin	Cincinnati, Ohio	Sept. 26, 1865	50, 099
Hinge, Shutter	R. Covington	Washington, D. C	Nov. 6, 1866	59, 365
Hinge, Shutter	I. Davis	Groton, N. Y	Mar. 4, 1856	14, 349
Hinge, Shutter	H. F. Drott	Cumberland, Md	Oct. 11, 1859	25, 728
Hinge, Shutter	S. Drum	Allegheny City, Pa	June 20, 1865	48, 264
Hinge, Shutter	T. F. Duley	Olney, Md	Oct. 22, 1872	132, 512
Hinge, Shutter	C. H. Forbes and W. F. Rutter	Philadelphia, Pa	Apr. 26, 1870	102, 242
Hinge, Shutter	W. Fosket	Meriden, Conn	Nov. 25, 1873	144, 971
Hinge, Shutter	O. S. Garretson	Buffalo, N. Y	June 19, 1866	55, 644
Hinge, Shutter	J. Hof and P. Brenners	Baltimore, Md	Apr. 19, 1870	102, 125
Hinge, Shutter	C. F. Knauer	Pittsburgh, Pa	June 6, 1865	48, 075
Hinge, Shutter	R. Lee	Cincinnati, Ohio	Feb. 7, 1865	46, 251
Hinge, Shutter	H. Lull	South Coventry, Conn	Jan. 31, 1854	10, 477
Hinge, Shutter	H. L. Norton	Middletown, Conn	June 4, 1872	127, 425
Hinge, Shutter	J. Slusser	Cincinnati, Ohio	May 8, 1866	54, 618
Hinge, Shutter	J. Smith, jr	Allegheny City, Pa	Oct. 28, 1862	36, 865
Hinge, Shutter	E. L. Tevis	Philadelphia, Pa	Apr. 16, 1867	63, 962
Hinge, Sliding	J. M. Eveleth and G. P. Moore	Oroville, Cal	Feb. 17, 1863	37, 684
Hinge, Spiral or worm-joint	P. G. Bates	Waterbury, Conn	Jan. 17, 1854	10, 425
Hinge, Spring	A. Acker	Ramapo, N. Y	July 17, 1860	29, 212
Hinge, Spring	J. Bidwell	San Francisco, Cal	June 14, 1870	104, 252
Hinge, Spring	N. Birdsall	Port Jervis, N. Y	June 28, 1859	24, 535
Hinge, Spring	L. Bommer	New York, N. Y	Dec. 8, 1863	40, 879
Hinge, Spring	A. Crawford	Boston, Mass	Dec. 10, 1872	133, 699
Hinge, Spring	J. T. Garlick	New York, N. Y	Dec. 23, 1856	16, 272
Hinge, Spring	D. B. Hawkes	Providence, R. I	May 16, 1871	114, 813
Hinge, Spring	H. N. Hemingway	Rochester, N. Y	Sept. 3, 1872	190, 998
Hinge, Spring	F. Herrmann	Newport, Ky	Mar. 12, 1867	62, 747
Hinge, Spring	V. S. Herzog	Baltimore, Md	Apr. 6, 1869	88, 567
Hinge, Spring	W. Hoar	Floyd, Iowa	Sept. 2, 1873	142, 392
Hinge, Spring	S. Joyce	New York, N. Y	Sept. 16, 1873	142, 855
Hinge, Spring	C. W. Lawrence	Milton, Ind	Feb. 19, 1867	62, 140
Hinge, Spring	J. Maxson	De Ruyter, N. Y	Nov. 10, 1857	18, 593
Hinge, Spring	H. B. Middaugh	Mansfield, Pa	June 1, 1869	90, 769
Hinge, Spring	A. M. Olds	New York, N. Y	June 25, 1867	66, 103
Hinge, Spring	J. Palmer	Cincinnati, Ohio	Apr. 15, 1878	137, 951
Hinge, Spring	D. Renshaw	Brooklyn, N. Y	Jan. 11, 1870	98, 709
Hinge, Spring	C. I. Rooney and D. Renshaw	New York, N. Y	Sept. 27, 1859	25, 585
Hinge, Spring	H. W. Sabin and G. Drew	Canandaigua, N. Y	Feb. 25, 1851	7, 957
Hinge, Spring	N. Sehner	Hagerstown, Md	May 9, 1871	114, 719
Hinge, Spring	N. Sehner	Hagerstown, Md	Aug. 1, 1871	117, 691
Hinge, Spring	J. S. Smith	New York, N. Y	May 19, 1857	17, 354
Hinge, Spring	B. Turner	London, England	May 7, 1872	126, 593
Hinge, Spring	W. Wells	Cleveland, Ohio	Dec. 6, 1870	109, 855
Hinge, Spring	A. Wiswall	New York, N. Y	June 18, 1867	65, 980
Hinge, Spring butt	A. Acker	Ramapo, N. Y	Nov. 18, 1862	36, 976
Hinge, Spring butt	H. A. Clark	Boston, Mass	Mar. 25, 1873	137, 178
Hinge-stock, Composition	G. R. Sillibridge	New York, N. Y	Oct. 10, 1829	....
Hinge, Stop	G. F. J. Colburn	Newark, N. J	Sept. 6, 1864	44, 076
Hinge, Stop	H. Pennie	Buffalo, N. Y	Nov. 6, 1860	30, 587
Hinge, Stop	G. C. Thomas	Brooklyn, N. Y	Dec. 9, 1873	145, 315
Hinge-stop	M. Umstadter	Norfolk, Va	Jan. 12, 1869	85, 876
Hinge, Stove-door, &c	C. J. Woolson	Cleveland, Ohio	Mar. 16, 1852	8, 814
Hinge, Table	E. C. Brinser	Middletown, Pa	Aug. 26, 1873	142, 077
Hinge, Table	M. C. Brinser	Lancaster, Pa	Apr. 16, 1872	125, 657
Hinge, Table	W. F. Daly	Peru, Ind	Aug. 26, 1873	142, 210
Hinge, Table	J. W. Palmer	Pana, Ill	July 15, 1873	140, 947
Hinge, Telescopic quadrant	A. S. Blake	Waterbury, Conn	Aug. 10, 1869	93, 402
Hinge, Trunk	A. M. Darrell	Washington, D. C	Feb. 18, 1873	136, 040
Hinge, Trunk	O. Luce	Cortland, N. Y	Nov. 3, 1868	83, 647
Hinge, Trunk	W. Wakenshaw and T. R. Dunham.	Newark, N. J	Apr. 10, 1866	53, 908
Hinge, Trunk	L. Wheaton	Auburn, N. Y	Nov. 19, 1867	71, 254
Hinge, Trunk and box	B. Hogan	Albany, N. Y	Nov. 26, 1872	133, 313
Hinge, Wrought-iron blind	W. R. Goodrich	Utica, N. Y	Mar. 11, 1873	136, 650
Hinges and tubes, Machine for making	W. Shaw	Buffalo, N. Y	Oct. 22, 1835	....
Hinges, Arrangement of friction-roller in inclined-plane.	E. Woolman	Damascoville, Ohio	May 9, 1854	10, 900
Hinges, Casting	N. A. Fenner	Providence, R. I	Dec. 22, 1857	18, 896
Hinges, Casting	C. M. Lane	Cincinnati, Ohio	July 20, 1858	20, 948
Hinges, Casting	G. A. Rogers	Newburyport, Mass	July 3, 1810	....
Hinges, Casting	D. Stuart and C. C. Lloyd	Danville, Pa	Jan. 20, 1843	2, 917
Hinges, Casting auger gate	B. Roux	Cincinnati, Ohio	Feb. 8, 1870	99, 712
Hinges, Casting butt	R. W. Belson	Danville, Pa	July 26, 1843	3, 195
Hinges, Casting butt, &c	T. Shepherd	Philadelphia, Pa	Oct. 9, 1841	2, 288
Hinges, Casting iron	C. Alger	Boston, Mass	May 18, 1814	....
Hinges, Cutting out strap	S. M. Richardson	New York, N. Y	June 7, 1859	24, 330
Hinges, Device for inserting the pins in butt	L. P. Summers	New Britain, Conn	July 19, 1870	105, 513
Hinges, Fastening metal plates upon door	W. Whiting	Brooklyn, N. Y	Oct. 29, 1867	70, 304
Hinges, Flask for molding	T. Loring	Gloucester, N. J	Feb. 7, 1844	3, 427

Index of patents issued from the United States Patent Office from 1790 *to* 1873, *inclusive*—Continued.

Invention.	Inventor.	Residence.	Date.	No.
Hinges, Implement for expelling the joint-pin from	W. H. H. Barton	Uxbridge, Mass	Dec. 10, 1872	133, 821
Hinges, Machine for countersinking holes in butt	L. P. Summers	New Britain, Conn	Dec. 6, 1870	109, 851
Hinges, Machine for forming the eyes of	D. W. Lyon	West Troy, N. Y	Sept. 11, 1849	6, 704
Hinges, Machine for forming the eyes of butt	W. Whitaker	Albany, N. Y	July 8, 1843	3, 156
Hinges, Machine for grinding butt	C. Kenney and W. Gurley	Troy, N. Y	July 1, 1856	15, 241
Hinges, Machine for manufacturing	W. F. Lewis	Waterbury, Conn	Jan. 17, 1871	110, 98[illegible]
Hinges, Machine for manufacturing butt and other kinds of.	C. R. Macy	Hyde Park, N. Y	Oct. 5, 1838	966
Hinges, Machine for milling the knuckles of butt	E. Parker	New Britain, Conn	July 19, 1870	105, 487
Hinges, Machine for wiring	E. Luther	West Troy, N. Y	Oct. 12, 1869	95, 700
Hinges, Machinery for manufacturing	G. H. Horton and L. Armstrong	Hartford, Conn	Sept. 11, 1847	5, 293
Hinges made of wrought-iron, &c., Machine for bending the knuckles of butt.	C. Kenney	Troy, N. Y	Aug. 23, 1844	3, 717
Hinges, Manufacture of	L. Crooke	New York, N. Y	Jan. 9, 1872	122, 516
Hinges, Manufacturing butt	H. P. Anderson and J. Sawyer	Waterford, N. Y	June 18, 1833	......
Hinges, Manufacturing butt	D. Ellis	Attleborough, Mass	Aug. 8, 1834	....
Hinges, Manufacturing table butt	H. Treadwell	Hyde Park, N. Y	Dec. 31, 1833	....
Hinges, Material for manufacture of butt	O. S. Judd	New Britain, Conn	June 14, 1864	43, 117
Hinges, Mold for butt	B. F. Harley and J. D. Morris	Philadelphia, Pa	Feb. 12, 1844	3, 435
Hinges, Mold for casting butt	T. Shepherd and T. Loring	Philadelphia, Pa	Mar. 16, 1841	2, 005
Hinges, Mold for casting metallic butt	J. Kendall, jr	Watertown, Mass	July 21, 1831	....
Hinges of wrought-iron, Machinery for making butt	C. Kenney	West Troy, N. Y	Aug. 7, 1844	3, 690
Hinges on the inside, Machine for planing or dressing the knuckles of butt.	G. Stickney	Blackwoodtown, N. J	Dec. 19, 1844	3, 863
Hinges on their axes, Casting	S. Wilkes	Darleston, Great Britain	Apr. 10, 1841	2, 043
Hinges, Pintle of butt	L. Maschauer and W. Frankfurth.	Milwaukee, Wis	Oct. 27, 1868	83, 393
Hinges, &c., Plating iron for the manufacture of	J. J. and L. Crooke	Southfield and New York, N. Y.	Dec. 28, 1869	98, 354
Hinges, Recess-cutter for setting	D. Snow	Cleveland, Ohio	Feb. 16, 1869	87, 008
Hinges, Spike for	H. Garrett	Richmond, Mo	Aug. 7, 1860	29, 480
Hinges, &c., Spring for	A. French	Waterbury, Conn	June 19, 1855	13, 085
Hinges with tin, Coating	H. M. Myers	Allegheny City, Pa	June 15, 1869	91, 477
Hinged-claw wrench	A. Hay	Newark, N. J	Jan. 30, 1849	6, 077
Hinged, sliding, and other bodies, Fastening	L. R. Streeter	Chelsea, Mass	Feb. 2, 1864	41, 451
Hitch-hook	G. W. Chandler	Mason, N. H	Aug. 23, 1870	106, 549
Hitch, Horse	W. A. Middleton and J. A. Haller.	Harrisburgh, Pa	Aug. 11, 1868	80, 870
Hitch, Safety horse	T. Weaver	Harrisburgh, Pa	Nov. 24, 1868	84, 450
Hitching and sign post	C. F. Barnard	Victoria, British Columbia	Mar. 12, 1872	124, 532
Hitching-clamp or hold-fast	J. P. Sinclair	Mill Port, N. Y	Oct. 23, 1866	59, 085
Hitching-device	C. H. Sawyer	Hollis, Me	Oct. 30, 1866	59, 276
Hitching device, Horse	W. H. Bell	Spring Lake, Mich	May 24, 1870	103, 283
Hitching device, Horse	F. Deming	Bridgeport, Conn	Apr. 28, 1868	77, 364
Hitching device, Horse	S. Galbraith	New Orleans, La	Dec. 10, 1867	72, 012
Hitching device, Horse	C. Rogers	Bergen, N. J	Nov. 13, 1866	59, 660
Hitching device, Horse	J. B. Thornton	Madison, Wis	Sept. 17, 1867	69, 048
Hitching device, Horse	W. G. Ward	Steuben County, N. Y	Sept. 17, 1867	68, 919
Hitching device, Horse	M. Warne and W. H. Pearce	Philadelphia, Pa	Aug. 13, 1867	67, 689
Hitching hook, Cam	C. Starbuck	Philadelphia, Pa	Apr. 9, 1867	63, 762
Hitching horses	J. L. Kreider	Drumore Township, Pa	June 29, 1869	91, 854
Hitching horses, clothes-lines, &c., Apparatus for	E. S. Boynton	East Hartford, Conn	July 8, 1856	15, 279
Hitching horses to seeding-machines	D. C. Baughman	Tiffin, Ohio	Aug. 14, 1866	57, 068
Hitching horses to vehicles and plows	J. Graham	Ceresco, Mich	May 3, 1864	42, 574
Hitching-post	J. Gibson, jr	Albany, N. Y	Dec. 28, 1869	98, 247
Hitching-post	J. Melchers	Detroit, Mich	May 21, 1872	127, 085
Hitching-post	J. Oothoudt	Minneapolis, Minn	June 28, 1870	104, 875
Hitching-post	W. S. Owings	Pan Handle Post-Office, W. Va.	Dec. 13, 1870	110, 066
Hitching-post	W. W. Powers	Belleville, N. Y	Apr. 5, 1870	101, 657
Hitching-post	J. S. Rohrer	Lancaster, Pa	Apr. 5, 1870	101, 662
Hitching-post	E. F. Shoenberger	Germantown, Pa	May 16, 1871	114, 982
Hitching-post	B. and P. Walther	New York, N. Y	May 5, 1868	77, 686
Hitching-post	V. Ward	San Francisco, Cal	June 5, 1866	55, 445
Hitching post, Animal	D. Newton	Southington, Conn	Aug. 27, 1867	68, 308
Hitching-post cap	R. N. Roberts	New Britain, Conn	Feb. 18, 1868	74, 596
Hitching post, Horse	C. Bush	Newburgh, N. Y	June 18, 1861	32, 601
Hitching post, Horse	T. Garrish	Providence, R. I	Dec. 24, 1872	134, 266
Hitching-ring	J. P. Howell	New York, N. Y	Sept. 24, 1867	69, 212
Hitching-strap	A. M. Brubaker	Millersville, Pa	Mar. 18, 1873	136, 011
Hitching-strap	T. P. Chambers	Newtown, Pa	Dec. 3, 1867	71, 699
Hitching-strap	P. Conover	Kingsessing, Pa	Jan. 26, 1864	41, 366
Hitching-strap	J. Dubree	Drumore Township, Pa	July 24, 1866	56, 537
Hitching-strap	B. Hunter	Philadelphia, Pa	Apr. 30, 1872	126, 145
Hitching-strap	J. D. Miller	Russellville, Pa	Sept. 3, 1867	68, 569
Hitching-strap	A. J. Ross	Rochester, N. Y	Nov. 5, 1867	70, 473
Hitching-straps, Clasp for	M. R. Margerum and T. P. Marshall.	Trenton, N. J	Dec. 20, 1859	26, 508
Hobby-horse	L. Anderson	Chicago, Ill	Sept. 14, 1869	94, 858
Hobby-horse	J. A. Crandall	Brooklyn, N. Y	Apr. 15, 1873	137, 896
Hobby-horse	W. E. and J. A. Crandall	New York and Brooklyn, N. Y.	Nov. 5, 1872	132, 806
Hobby-horse	C. Hitzelberger	South Orange, N. J	Sept. 6, 1870	107, 049
Hobby-horse	J. Liming	Philadelphia, Pa	June 14, 1870	104, 170
Hobby-horse	P. J. Marqua	Cincinnati, Ohio	Feb. 7, 1865	46, 258
Hobby-horse	J. Reinhart	New York, N. Y	Oct. 21, 1873	143, 930
Hobby-horse and carriage-motor combined	J. Liming	Philadelphia, Pa	July 26, 1870	105, 701
Hobby-horse, Child's	W. L. Williams	New York, N. Y	Jan. 3, 1871	110, 709
Hobby-horse-seat guard	W. L. Williams	New York, N. Y	Dec. 27, 1870	110, 527
Hod	J. Short	Roxbury, Mass	Oct. 31, 1865	50, 741
Hod, Brick and mortar	C. Roehl	Chicago, Ill	Mar. 12, 1872	124, 448
Hod-elevator	L. Atwood	New York, N. Y	July 18, 1871	117, 140
Hod-elevator	G. W. Brown	New York, N. Y	Sept. 26, 1871	119, 305
Hod-elevator	P. Dehlinger	Buffalo, N. Y	Nov. 24, 1868	84, 268
Hod-elevator	E. H. Garrigues	Saint Louis, Mo	Oct. 24, 1871	120, 189
Hod, &c., elevator	E. Harlow	New York, N. Y	June 24, 1873	140, 195
Hod-elevator	T. M. Pelham	New York, N. Y	Sept. 28, 1869	95, 262
Hod-elevator	J. Powers	Chicago, Ill	July 12, 1870	105, 246

Index of patents issued from the United States Patent Office from 1790 *to* 1873, *inclusive*—Continued.

Invention.	Inventor.	Residence.	Date.	No.
Hod-elevator	W. H. Totten	Academia, Pa	Sept. 5, 1865	49, 805
Hod-iron, Mason's	L. W. Lake	Providence, R. I	Apr. 23, 1872	126, 061
Hod, Mason's	B. C. Monroe	Newburgh, N. Y	Dec. 16, 1873	145, 675
Hoe	S. W. Adams	Wethersfield, Conn	Feb. 13, 1866	52, 510
Hoe	S. W. Adams	Wethersfield, Conn	Oct. 23, 1866	58, 966
Hoe	L. J. Aderhold	Wedowee, Ala	July 19, 1870	105, 405
Hoe	A. Allen	Ramapo, N. Y	Jan. 23, 1836	....
Hoe	J. M. Baird	Wheeling, W. Va	July 29, 1873	141, 308
Hoe	H. R. Barnes	Rock Stream, N. Y	Feb. 4, 1873	135, 462
Hoe	H. R. Barnes	Rock Stream, N. Y	Apr. 15, 1873	137, 755
Hoe	W. G. Beckwith	Lowndesborough, Ala	Sept. 28, 1869	95, 306
Hoe	N. D. Beecroft	Bangor, Me	Feb. 18, 1873	136, 020
Hoe	L. Billings	Gallipolis, Ohio	Dec. 14, 1869	97, 863
Hoe	L. L. Bond	Chicago, Ill	Feb. 15, 1870	99, 821
Hoe	R. P. Buttles	Mansfield, Pa	Oct. 29, 1872	132, 520
Hoe	H. Camp	Covington, Ga	Jan. 21, 1873	135, 076
Hoe	J. S. Carroll	Covington, Ga	Apr. 9, 1872	125, 438
Hoe	I. Cook and J. T. Bever	Haynesville, Mo	May 2, 1871	114, 266
Hoe	M. Cookerly	Baxter Springs, Kans	Oct. 22, 1872	132, 444
Hoe	P. and H. E. Crapo	Bridgewater, Mass	Aug. 6, 1861	32, 982
Hoe	J. Dodge	Grass Valley, Cal	June 9, 1868	78, 724
Hoe	A. Ellis	Canton, Ind	Oct. 17, 1871	120, 050
Hoe	A. Ellis and O. Albertson	Salem, Ind	Oct. 18, 1870	108, 465
Hoe	J. Ells	Pittsburgh, Pa	Dec. 8, 1863	40, 826
Hoe	J. and A. Fairley	Birmingham, England	Dec. 13, 1870	110, 126
Hoe	W. C. Finney	Fayette County, Tenn	Nov. 30, 1852	9, 430
Hoe	S. H. Folsom	Winchester, Mass	Dec. 17, 1872	133, 932
Hoe	J. L. Fountain	New Milford, Ill	June 30, 1868	79, 462
Hoe	G. Freeborn	New York	July 7, 1829	....
Hoe	M. Gates	Gallipolis, Ohio	Dec. 12, 1854	12, 057
Hoe	J. Glasson	Brooklyn, N. Y	Aug. 14, 1866	57, 119
Hoe	J. H. Gould	Rutland, Ohio	Aug. 26, 1873	142, 159
Hoe	G. Harper	Montgomery County, Pa	Nov. 19, 1833	....
Hoe	B. Hinkley	Fayette, Me	June 23, 1832	....
Hoe	J. Hinman	Hamden, Conn	Apr. 22, 1835	....
Hoe	C. P. Howell	Covington, Ky	Dec. 29, 1868	85, 384
Hoe	G. S. Jobson	Macon, Ga	Jan. 4, 1870	98, 598
Hoe	M. Johnson	Three Rivers, Mich	Mar. 11, 1873	136, 733
Hoe	A. C. Kasson	Milwaukee, Wis	Dec. 4, 1866	60, 195
Hoe	T. B. Lockwood	Smyrna, Del	July 29, 1873	141, 365
Hoe	H. A. Lothrop	Sharon, Mass	Dec. 29, 1857	18, 979
Hoe	J. F. Lowe	Louisville, Ky	Mar. 23, 1869	88, 051
Hoe	C. Mahan	Grand Island, Cal	Sept. 3, 1867	68, 764
Hoe	E. W. McLendon	Griffin, Ga	June 24, 1873	140, 291
Hoe	P. H. Merrill	Hinsdale, N. H	Apr. 6, 1820	....
Hoe	J. C. Miller	Danville, Ky	Feb. 6, 1872	123, 502
Hoe	E. C. and J. W. Newland	Lawrence County, Ind	Jan. 13, 1863	37, 405
Hoe	G. H. Owens	Maysville, Ky	Apr. 21, 1868	77, 083
Hoe	H. Parkman	Philadelphia, Pa	Oct. 28, 1873	144, 127
Hoe	J. C. Plumer	Boston, Mass	Mar. 20, 1866	53, 332
Hoe	H. Porter	Cummington, Mass	Mar. 6, 1860	27, 382
Hoe	A. Prentiss	Prentiss Vale, Pa	May 14, 1867	64, 792
Hoe	E. Riston and W. W. Brigg	Maltaville, N. Y	June 5, 1866	55, 365
Hoe	H. C. Rogers	Scranton, Pa	June 17, 1862	35, 627
Hoe	S. Rogers	Pittsburgh, Pa	Jan. 16, 1866	52, 079
Hoe	C. W. Saladee and J. S. Hall	Newark, Ohio, and Pittsburgh, Pa.	June 2, 1868	78, 611
Hoe	S. Springstead	Westchester, N. Y	Nov. 9, 1869	96, 739
Hoe	J. Stilwell	Griffin, Ga	May 23, 1871	115, 132
Hoe	C. H. Strong and E. A. Sterry	Norwich, Conn	Jan. 5, 1833	....
Hoe	T. R. Timby	Saratoga Springs, N. Y	Mar. 5, 1867	62, 574
Hoe	E. Warren	Ceresco, Mich	May 10, 1870	102, 891
Hoe	P. Whitin	Northbridge, Mass	Feb. 22, 1825	....
Hoe	J. F. Wilson	Athens, Ga	Aug. 24, 1869	94, 054
Hoe	I. N. Wood	Fall River, Mass	June 2, 1868	78, 561
Hoe, Adjustable	T. Drake	Hartford, Conn	Jan. 28, 1868	73, 702
Hoe, Adjustable weeding	A. C. Arnold	Norwalk, Conn	Oct. 17, 1865	50, 436
Hoe and corn-planter, Combined	J. A. Burchard	Beloit, Wis	June 19, 1866	55, 612
Hoe and cotton-cultivator, Combined revolving	M. E. Davis	Rome, Ga	Nov. 30, 1869	97, 277
Hoe and cultivator, Horse	G. D. Rowell	Menamonee Falls, Wis	Apr. 18, 1871	113, 934
Hoe and fork, Combined	B. L. Tibbetts	South China, Me	Apr. 6, 1869	88, 591
Hoe and garden-rake combined	R. H. Gordon, sr	Brooklyn, Ohio	Jan. 21, 1868	73, 525
Hoe and garden-trimmer, Scuffle	I. Pardee	Vineland, N. J	Mar. 31, 1868	76, 288
Hoe and potato-digger, Combined	E. F. Morris and R. J. Greed	Cicero, N. Y	Oct. 15, 1867	69, 928
Hoe and rake	A. Ellis and O. Albertson	Salem, Ind	June 15, 1869	91, 317
Hoe and rake, Combined	I. Cook	Haynesville, Mo	Sept. 17, 1867	68, 958
Hoe and rake, Combined	W. M. McLendon	Greenville, Ga	May 14, 1872	126, 821
Hoe and rake, Combined	F. L. Senour	Eaton, Ohio	June 1, 1869	90, 691
Hoe and rake, Combined garden	G. W. Lockwood	Fairport, N. Y	May 21, 1872	127, 077
Hoe and roller combined, Garden	E. Blanchard	Poolesville, Md	Mar. 25, 1873	137, 124
Hoe and seed-dropper	W. Snyder	Ypsilanti, Mich	Aug. 20, 1872	130, 757
Hoe and seed-planter	C. H. Wolcott	Randolph, N. Y	Aug. 30, 1864	44, 032
Hoe-blank	H. Waters	Boston, Mass	Feb. 26, 1867	62, 456
Hoe-blank	H. Waters	Boston, Mass	Feb. 25, 1868	74, 960
Hoe-blank-rolling machine	S. A. Millard	Clayville, N. Y	July 21, 1868	80, 083
Hoe-blank-rolling machine	H. Waters	Boston, Mass	May 2, 1871	114, 498
Hoe-blanks, Machine for manufacture of	N. Brand	Leonardsville, N. Y	Feb. 26, 1861	31, 525
Hoe-blanks, Machine for rolling	W. T. Clement and E. V. Foster	Northampton, Mass	Aug. 17, 1869	93, 679
Hoe, Cane-stubble	G. H. Wright and A. K. Johnson	New Orleans, La	Oct. 17, 1871	120, 143
Hoe, Cast-iron	B. F. Boyden	Boston, Mass	Mar. 31, 1836	....
Hoe-cleansing attachment	R. R. and H. T. Spedden	Astoria, Oreg	Nov. 1, 1870	108, 843
Hoe, Compound adjustable garden	M. H. Fletcher	Richmond, Ind	Nov. 3, 1868	83, 771
Hoe, Cotton	J. S. Carroll	Covington, Ga	June 6, 1871	115, 568
Hoe, Cotton	Z. B. Sims	Bonham, Tex	Aug. 31, 1869	94, 353
Hoe, Cotton	Z. B. Sims	Bonham, Tex	Sept. 26, 1871	119, 418
Hoe, Cultivating	J. J. Ray and J. R. Young	New Orleans, La	Sept. 21, 1869	95, 844
Hoe, Double	A. Burchard	New Brenton, Ill	Mar. 3, 1868	74, 984
Hoe, Extension weeding	M. Renz	Naugatuck, Conn	Dec. 18, 1866	60, 641

Index of patents issued from the United States Patent Office from 1790 *to* 1873, *inclusive*—Continued.

Invention.	Inventor.	Residence.	Date.	No.
Hoe for weeding, Cast-iron horse	W. Carmichael	Sand Lake, N. Y	July 28, 1827	....
Hoe, Garden	J. R. Albertson	East Deer Township, Pa	Apr. 3, 1860	27,675
Hoe, Garden	I. W. Averill	Plymouth, Mich	June 11, 1836	....
Hoe, Garden	J. P. Avery	Norwich, Conn	Mar. 14, 1871	112,527
Hoe, Garden	C. Belden	Akron, Ohio	Dec. 17, 1872	134,028
Hoe, Garden	J. H. Brower	Atlas, Mich	May 28, 1867	65,053
Hoe, Garden	D. E. Eaton	Boston, Mass	Jan. 2, 1866	51,893
Hoe, Garden	C. R. Gilbert	High Point, N. C	Aug. 22, 1871	118,360
Hoe, Garden	C. S. Homer	Boston, Mass	May 8, 1840	1,593
Hoe, Garden	J. Howard	West Bridgewater, Mass	Apr. 9, 1861	31,977
Hoe, Garden	A. C. Judson	Grand Rapids, Ohio	Sept. 6, 1870	107,060
Hoe, Garden	S. V. Kimball	Indianapolis, Ind	Oct. 17, 1871	119,986
Hoe, Garden	A. M. Ross	Ilion, N. Y	Mar. 2, 1869	87,368
Hoe, Garden	S. Shetter	Allegheny, Pa	July 21, 1857	17,848
Hoe, Garden	L. Streeter	Chicopee, Mass	Sept. 1, 1868	81,839
Hoe, Garden	F. Trigalet	Astoria, N. Y	Dec. 5, 1871	121,691
Hoe-handle	S. Reynolds	Duquesneborough, Pa	Apr. 23, 1861	32,146
Hoe-handles, &c., Fastening	R. Dare	Greenwich, N. J	Feb. 14, 1832	....
Hoe-handles, Fastening	G. Hight	Gorham, Me	May 25, 1838	751
Hoe, Hand	C. A. Rose	Columbus, Ga	Jan. 1, 1867	60,941
Hoe, Horse	M. Chandler	East Corinth, Me	Jan. 14, 1862	34,128
Hoe, Horse	M. Chandler and J. B. Nickels	Corinth and Kenduskeag, Me.	Feb. 19, 1867	62,181
Hoe, Horse	G. W. Cooper	Ogeechee, Ga	June 11, 1867	65,728
Hoe, Horse	I. Copeland	North Bridgewater, Mass	Sept. 20, 1870	107,456
Hoe, Horse	T. P. Fish	Litchfield, N. Y	May 1, 1866	54,315
Hoe, Horse	J. Gifford, jr	Watertown, N. Y	Nov. 13, 1866	59,584
Hoe, Horse	D. Harris	Canaan, Me	July 24, 1866	56,555
Hoe, Horse	H. W. Hasslock	Nashville, Tenn	Mar. 22, 1870	101,010
Hoe, Horse	N. H. Lindley	Bridgeport, Conn	Feb. 9, 1869	86,846
Hoe, Horse	N. H. Lindley	Bridgeport, Conn	Jan. 18, 1870	98,983
Hoe, Horse	C. Lobdell	Fort Hill, Ill	Mar. 2, 1869	87,347
Hoe, Horse	D. C. Matteson and T. P. Williamson.	Stockton, Cal	Nov. 3, 1868	83,719
Hoe, Horse	W. Muir	Wauconda, Ill	Feb. 16, 1869	87,060
Hoe, Horse	T. Rice and L. R. Hitchcock	Caneadea, N. Y	Nov. 5, 1867	70,472
Hoe, Horse	G. W. Sawin	Nashua, N. H	Aug. 20, 1867	67,917
Hoe, Horse	E. W. Walton	Stockton, Cal	June 9, 1868	78,774
Hoe, Horse	C. Whitlock	Bridport, Vt	Oct. 25, 1870	108,741
Hoe-machine for land	M. Spofford	Georgetown, Mass	Sept. 11, 1847	5,290
Hoe-necks, Swage-die, &c., for manufacturing	N. Brand	Leonardsville, N. Y	Apr. 1, 1845	3,980
Hoe, Planting	C. N. Jones	Galway, N. Y	Sept. 29, 1863	40,107
Hoe, Planting	W. G. Sterling	Bridgeport, Conn	Feb. 14, 1854	10,525
Hoe, Planting	A. Williams	Sebec, Me	June 25, 1867	66,196
Hoe-plates, Die for forging	L. T. Richardson	Auburn, N. Y	Mar. 11, 1873	136,672
Hoe, Plowing	T. J. Mason	Harmony, Me	Dec. 15, 1868	84,957
Hoe, Prong	J. Wilson	Marlborough, N. H	Sept. 20, 1827	....
Hoe, rake, &c	J. S. Hall	West Manchester, Pa	Feb. 9, 1864	41,503
Hoe-rolling machine	W. F. Stillman	Ilion, N. Y	July 16, 1867	66,905
Hoe, Scuffle	T. R. Peck	Waterloo, N. Y	Sept. 24, 1872	131,627
Hoe seed-dropper	T. Nevison, jr	Morgan, Ohio	Mar. 31, 1868	76,235
Hoe, Seed-planting	S. Woodruff	Sparta, N. J	Apr. 20, 1858	20,014
Hoe, Seeding	L. B. Brown	Simsbury, Conn	Oct. 30, 1860	30,526
Hoe, Seeding	J. A. Pease	New York, N. Y	May 17, 1853	9,731
Hoe seeding-attachment	E. L. Berdstresser	Hublersburgh, Pa	Mar. 24, 1868	75,843
Hoe, Shuffle	J. D. Jenkins	Jackson, Ill	May 22, 1866	54,916
Hoe, Weed-puller	N. Ames	Saugus Centre, Mass	Aug. 16, 1864	43,869
Hoe, Weeding	A. B. Adams	Westport, Conn	Aug. 30, 1864	44,035
Hoe, Weeding	J. M. Adams	Canton, Mass	Oct. 18, 1859	25,862
Hoe, Weeding	G. P. Allen	Woodbury, Conn	Apr. 24, 1866	54,087
Hoe, Weeding	H. H. Baker	New Market, N. J	Oct. 4, 1859	25,619
Hoe, Weeding	A. Coleman	Red Bank, N. J	Apr. 28, 1868	77,258
Hoe, Weeding	E. M. Conkling	Parma, N. Y	Aug. 27, 1867	68,047
Hoe, Weeding	C. Crofut	Weston, Conn	Sept. 19, 1865	49,985
Hoe, Weeding	H. S. Crossland	Dresden, Tex	Aug. 26, 1873	142,149
Hoe, Weeding	J. G. Harriman	Gardiner, Me	Nov. 19, 1872	133,151
Hoe, Weeding	L. King	Oriskany Falls, N. Y	Oct. 20, 1868	83,290
Hoe, Weeding	A. E. Lyman	Northampton, Mass	Nov. 10, 1868	83,866
Hoe, Weeding	J. S. Munger	Olean, N. Y	Apr. 6, 1869	88,659
Hoe, Weeding	J. Nawgle	Mooresville, Ind	June 20, 1865	48,304
Hoe, Weeding	E. Wilcox	East Cleveland, Ohio	Oct. 31, 1871	120,408
Hoe, Wheel	E. D. and O. B. Reynolds.	North Bridgewater, Mass	Mar. 10, 1868	75,460
Hoes and picks, Die for making	D. Carr	Allegheny City, Pa	Feb. 13, 1872	123,552
Hoes and picks, Manufacture of grubbing	J. Lee	Cleveland, Ohio	Jan. 23, 1872	122,897
Hoes, Attaching eyes to the blades of	H. Havell	Newark, N. J	Feb. 10, 1857	16,584
Hoes by rolling cast-steel, Making	C. Bulkley	Colchester, Conn	Jan. 10, 1872	....
Hoes, chisels, &c., Socket for	G. Banister	Hartford, Vt	July 4, 1865	48,505
Hoes, Corn-dropping attachment for	E. L. Staples	Chillicothe, Ohio	Feb. 14, 1871	111,883
Hoes, Corn-dropping attachment for	C. W. Cotton and E. L. Staples	Cincinnati, Ohio	Sept. 10, 1867	68,607
Hoes, Garden and other	J. T. Sargent	Sutton, N. H	Oct. 25, 1853	10,162
Hoes, Machine for making	H. Sauerbier	Newark, N. J	Apr. 5, 1859	23,502
Hoes, Machinery for manufacturing	W. T. Clement and E. V. Foster	Northampton, Mass	Sept. 3, 1867	68,414
Hoes, Making	J. Abbott	Boston, Mass	Jan. 25, 1848	5,417
Hoes, Making	M. Depuy	Pittsburgh, Pa	Nov. 6, 1860	30,569
Hoes, Making	J. Watson	New Canaan, Conn	July 26, 1823	....
Hoes, Manufacture of	W. Acheson and W. H. Ridley	Pittsburgh, Pa	May 28, 1872	127,288
Hoes, Manufacture of	W. Baker	West Winsted, Conn	May 21, 1861	32,344
Hoes, Manufacture of	J. A. Black	Columbia, S. C	Oct. 13, 1828	....
Hoes, Manufacture of	S. Boyd	New York, N. Y	Jan 13, 1857	16,371
Hoes, Manufacture of	S. Boyd	Brooklyn, N. Y	Dec. 27, 1859	26,560
Hoes, Manufacture of	W. T. Clement	Northampton, Mass	Jan. 21, 1868	73,437
Hoes, Manufacture of	J. S. Craig	Guilford, Kans	Feb. 13, 1872	123,680
Hoes, Manufacture of	R. Harper	Philadelphia, Pa	Nov. 20, 1866	59,744
Hoes, Manufacture of	J. Knight	Newark, N. J	Mar. 30, 1858	19,812
Hoes, Manufacture of	A. Patterson	Birmingham, Pa	July 5, 1859	24,658
Hoes, Manufacture of	A. Patterson	Birmingham, Pa	Sept. 27, 1859	25,580
Hoes, Manufacture of	A. Patterson	Birmingham, Pa	Aug. 29, 1865	49,640
Hoes, Manufacture of	L. T. Richardson	Clayville, N. Y	May 26, 1868	78,320

Index of patents issued from the United States Patent Office from 1790 *to* 1873, *inclusive*—Continued.

Invention.	Inventor.	Residence.	Date.	No.
Hoes, Manufacture of	E. C. Tuttle	Naugatuck, Conn	Oct. 18, 1859	25, 856
Hoes, Manufacturing	S. Boyd	Brooklyn, N. Y	Aug. 2, 1859	24, 926
Hoes, Process of rolling	S. A. Millard	Clayville, N. Y	Feb. 18, 1868	74, 570
Hoes, Roll for the manufacture of planters'	J. T. Tyler	Pittsburgh, Pa	Mar. 21, 1871	112, 867
Hoes, Securing handles to	W. T. Clement	Northampton, Mass	May 14, 1861	32, 275
Hoes, Seed-dropping attachment for	A. T. Large	Chicago, Ill	Mar. 5, 1867	62, 642
Hoes to handles, Attaching	O. Clark and W. D. Hillis	Hudson, Ohio	Sept. 9, 1843	3, 255
Hoes to handles, Attaching	W. H. Hartzman	Big Lick, Va	Mar. 12 1867	62, 900
Hoes to handles, Attaching	J. A. Marine	Mooresville, Ind	Apr. 13, 1869	88, 884
Hoeing-machine	H. C. Briggs	West Auburn, Me	Nov. 17, 1868	84, 165
Hoeing-machine	H. C. Briggs	West Auburn, Me	Mar. 9, 1869	87, 627
Hoeing-machine	H. W. Clapp	Northampton, Mass	Nov. 9, 1869	96, 549
Hog-cleaning machine	N. Silverthorn	Prescott, Wis	Aug. 30, 1864	44, 021
Hog-elevator	A. J. Chambers and T. Jackson	New Washington, Ohio	Aug. 24, 1869	94, 076
Hog-elevator	D. Kaufman	Boiling Spring, Pa	Mar. 6, 1860	27, 368
Hog-feeder	I. S. Pope	Napoleon, Ohio	May 21, 1867	65, 004
Hog-holder	W. and C. Leffingwell	Clarksburgh, Ohio	May 21, 1867	64, 885
Hog-holding clamp	R. T. Leaverton	Holden, Mo	Aug. 20, 1872	130, 727
Hog-lifting machine	J. Dagley	Gosport, Ind	Nov. 14, 1871	120, 946
Hog-ring	J. J. Hutson	South Solon, Ohio	Aug. 13, 1867	67, 765
Hog-ring	W. S. and E. S. Shoemaker	Towsontown, Md., and Lancaster, Ohio.	May 11, 1869	90, 028
Hog-ringing device	G. W. Clark	Frankfort, Ohio	Oct. 1, 1867	69, 406
Hog-ringing device	W. S. Houston	Mansfield, Ohio	Nov. 14, 1871	120, 876
Hog-ringing implement	H. W. Hill	Decatur, Ill	Aug. 27, 1872	130, 853
Hog-scalder, Portable	A. Clarke	Philadelphia, Pa	Dec. 26, 1865	51, 783
Hog-scalding machine	M. Stricker	Vincennes, Ind	Oct. 23, 1866	59, 093
Hog-scalding tank	G. King	Eminence, Ky	Apr. 28, 1868	77, 198
Hog-scraper, Rotary	R. Fyfe	New York, N. Y	July 1, 1873	140, 409
Hog-scraping apparatus	O. McNeil and P. W. Dalton	Jersey City, N. J	July 15, 1873	140, 838
Hog-singeing apparatus	P. Kenney	Chicago, Ill	July 16, 1872	129, 481
Hog-snout cutter	S. Hair	Kankakee City, Ill	Aug. 23, 1864	43, 909
Hog-snout slitter	J. J. Gish	Milton, Ohio	Mar. 3, 1868	75, 148
Hog-snouter	J. C. Campbell and W. S. Bruce	Good Hope, Ill	Jan. 3, 1871	110, 629
Hog-stock	W. C. Hays and J. D. Scott	Sharonville, Ohio	June 20, 1871	116, 183
Hog-tamer	T. G. Telton	Lyons, Iowa	Apr. 18, 1865	47, 329
Hog-trap	M. Caywood	Farmington, Ill	June 14, 1872	127, 461
Hog-trap	J. J. Cox	Woodhull, Ill	Apr. 2, 1872	125, 120
Hog-trap	T. Darby	Duncan, Ill	Aug. 12, 1873	141, 633
Hog-trap	A. B. De Vore	Talkington Township, Ill	July 15, 1873	140, 816
Hog-trap	H. W. Hill	Forsythe, Ill	June 11, 1872	127, 769
Hog-trap	R. Keiler	Kent, Ill	July 23, 1872	129, 832
Hog-trap	C. B. Weeks	Galesburgh, Ill	June 29, 1869	91, 887
Hog-trough	E. Johnson	Manchester, Mass	July 21, 1857	17, 839
Hog-trough	W. H. Robbins	Richmond, Ind	Sept. 4, 1866	57, 772
Hog-trough	C. F. Rolfe	Laconia, N. H	July 7, 1868	79, 688
Hogs, Apparatus for cutting up	W. C. Marshall	New York, N. Y	Aug. 13, 1872	130, 515
Hogs by steam, Scalding	T. J. Godman	Baltimore, Md	Feb. 13, 1835	
Hogs, Device for catching and holding	F. Voss, J. P. Hax, and H. Krug.	Saint Joseph, Mo	Sept. 13, 1870	107, 312
Hogs, Forceps for snouting	G. Stephenson	Zionsville, Ind	Oct. 14, 1873	143, 730
Hogs from rooting, Attachment for preventing	J. Butts	Evansville, Wis	Aug. 20, 1867	67, 952
Hogs from rooting, Device for preventing	H. Judson	Galesburgh, Ill	July 3, 1866	56, 061
Hogs from rooting, Device for preventing	D. K. Nixon	Sandyville, Ohio	Dec. 18, 1860	30, 928
Hogs from rooting, Device for preventing	G. and H. L Nixon	Sandyville, Ohio	Feb. 18, 1868	74, 582
Hogs from rooting, Device to prevent	D. P. Heflebower	Champaign County, Ohio	Aug. 8, 1871	117, 883
Hogs from rooting, Device to prevent	R. Little	Middle Branch, Ohio	Mar. 5, 1861	31, 610
Hogs from rooting, Device to prevent	L. L. Rinehart and P. Phillips	Evansburgh, Ohio	Feb. 4, 1868	74, 002
Hogs from rooting, Preventing	J. D. Kirkpatrick	Urbana, Ohio	May 12, 1868	77, 891
Hogs in veterinary operations, Device for holding	A. M. Scott and T. Wilson	Clearmont, Mo	Jan. 7, 1873	134, 707
Hogs, Instrument for ringing	J. and G. Heesen and H. Nyland.	Tecumseh, Mich	June 27, 1871	116, 440
Hogs, Instrument for taming and marking	S. Long	Ogle County, Ill	Aug. 3, 1869	93, 210
Hogs, Nose-ring for	O. P. Goodman	Chillicothe, Ohio	Feb. 8, 1870	99, 559
Hogsheads, &c., Apparatus for moving and rolling	F. O. J. Burr	Brewer, Me	Mar. 4, 1873	136, 486
Hoist, Air	L. Taws and J. M. Hartman	Philadelphia, Pa	Oct. 19, 1869	95, 952
Hoist, Cellar	P. Coble	Mechanicsburgh, Pa	July 23, 1872	129, 790
Hoist or elevator, Cellar	H. L. Bowers	Harrisburgh, Pa	Feb. 22, 1870	100, 006
Hoist, Pneumatic	J. C. Kent and H. C. Rich	Phillipsburgh, N. J	Apr. 26, 1870	102, 275
Hoist, Portable vertical	J. E. Walsh	New York, N. Y	Oct. 8, 1872	131, 983
Hoist, Self-acting hatchway	G. N. Creamer	Trenton, N. J	Mar. 1, 1870	100, 265
Hoist, Vertical	W. E. Worthen	New York, N. Y	Nov. 14, 1871	120, 844
Hoists, Safety-device for	P. J. Borger	Cincinnati, Ohio	May 2, 1871	114, 399
Hoists, Safety-device for	P. J. Borger	Cincinnati, Ohio	Feb. 13, 1872	123, 672
Hoisting and conveying apparatus	O. T. McIntosh	New York, N. Y	May 13, 1873	138, 910
Hoisting and conveying apparatus	G. Stancliff	New York, N. Y	May 13, 1873	138, 950
Hoisting and conveying apparatus	C. B. Stough	Monticello, Ill	July 29, 1873	141, 297
Hoisting and dumping apparatus	W. B. Culver	Scranton, Pa	Dec. 15, 1868	84, 939
Hoisting and dumping apparatus	G. Martz	Pottsville, Pa	Sept. 7, 1858	21, 431
Hoisting and dumping apparatus	G. Martz	Pottsville, Pa	Sept. 28, 1869	95, 366
Hoisting and lowering apparatus	R. L. Fitch	Sing Sing, N. Y	May 30, 1871	115, 292
Hoisting and lowering apparatus	I. J. Lancaster	Vancouver, Wash	June 27, 1865	48, 414
Hoisting and lowering apparatus	H. E. Towle	New York, N. Y	Apr. 5, 1870	101, 681
Hoisting and lowering apparatus, Self-braking	W. Hart	Philadelphia, Pa	Aug. 19, 1873	142, 017
Hoisting and lowering goods, &c., Machinery for	G. Thompson	Cincinnati, Ohio	June 1, 1858	20, 455
Hoisting and lowering weights	W. C. McGill	Cincinnati, Ohio	May 30, 1865	47, 969
Hoisting and storing apparatus, Ice	H. Van Steenburgh and J. Egnor.	Catskill, N. Y	Nov. 30, 1858	22, 210
Hoisting and weighing machine	J. L. Dutton, sr	Philadelphia, Pa	Mar. 13, 1860	27, 623
Hoisting-apparatus	G. Ambrose	New York, N. Y	May 23, 1865	47, 782
Hoisting-apparatus	W. D. Andrews	New York, N. Y	Oct. 1, 1867	69, 384
Hoisting-apparatus	W. D. Andrews	Brookhaven, N. Y	Dec. 20, 1870	110, 331
Hoisting-apparatus	C. W. Baldwin	Boston, Mass	Jan. 25, 1870	99, 049
Hoisting-apparatus	J. Beattie	Chicago, Ill	July 6, 1869	92, 250
Hoisting-apparatus	J. R. Benedict	Williamson, N. Y	June 13, 1871	115, 810
Hoisting-apparatus	E. H. Bermier	Paris, France	June 1, 1869	90, 810
Hoisting-apparatus	H. Black	Carrollton, Ill	Sept. 11, 1866	57, 850
Hoisting-apparatus	J. F. Bodley, J. Reynolds, and S. Van Emon.	Cincinnati, Ohio	July 5, 1864	43, 451

Index of patents issued from the United States Patent Office from 1790 *to* 1873, *inclusive*—Continued.

Invention.	Inventor.	Residence.	Date.	No.
Hoisting-apparatus	J. N. B. Bond	New York, N. Y	Feb. 4, 1873	135, 513
Hoisting-apparatus	G. W. Brown	New York, N. Y	May 24, 1870	103, 292
Hoisting-apparatus	W. H. Brown	New York, N. Y	May 10, 1870	102, 763
Hoisting-apparatus	L. Burnell	Milwaukee, Wis	Oct. 4, 1870	107, 868
Hoisting-apparatus	F. P. Canfield	Brighton, Mass	Oct. 6, 1868	82, 799
Hoisting-apparatus	M. K. Carpenter	Cincinnati, Ohio	Aug. 30, 1870	106, 914
Hoisting-apparatus	D. H. Chamberlain	West Roxbury, Mass	Feb. 23, 1869	87, 143
Hoisting-apparatus	H. Chatfield	Wolcottville, Conn	Nov. 5, 1867	70, 412
Hoisting-apparatus	J. Christiansen	New York, N. Y	June 16, 1868	78, 927
Hoisting-apparatus	C. Clark	Unadilla Forks, N. Y	Feb. 25, 1873	136, 212
Hoisting-apparatus	C. S. Crane	Chicago, Ill	July 2, 1872	128, 465
Hoisting-apparatus	W. W. Crapster	Mechanicsburgh, Pa	Aug. 18, 1868	81, 256
Hoisting-apparatus	L. Cutting	San Francisco, Cal	June 1, 1869	90, 823
Hoisting-apparatus	J. A. Dayton	New London, Conn	May 28, 1867	65, 180
Hoisting-apparatus	W. Dyatt	New York, N. Y	Dec. 7, 1869	97, 487
Hoisting-apparatus	F. Farrell	Ansonia, Conn	Aug. 5, 1873	141, 551
Hoisting-apparatus	R. Finnegan	New York, N. Y	Oct. 22, 1867	69, 978
Hoisting-apparatus	H. Flad and J. B. Eads	Saint Louis, Mo	Jan. 24, 1871	111, 188
Hoisting-apparatus	O. H. Flook	Middletown, Md	June 21, 1870	104, 567
Hoisting-apparatus	D. Frisbie	New Haven, Conn	Sept. 25, 1866	58, 340
Hoisting-apparatus	E. R. Gard	Chicago, Ill	Aug. 29, 1871	118, 447
Hoisting-apparatus	S. C. Goodsell and D. Frisbie	New Haven, Conn	May 1, 1866	54, 467
Hoisting-apparatus	E. G. Green	East Gloucester, Mass	Sept. 14, 1869	94, 817
Hoisting-apparatus	F. H. Hambleton	Baltimore, Md	May 21, 1872	127, 049
Hoisting-apparatus	D. Head	Medusa, N. Y	June 23, 1868	79, 071
Hoisting-apparatus	P. Higdon	Lewisport, Ky	Nov. 5, 1867	70, 565
Hoisting-apparatus	J. Hoffman	Tewkesbury, N. J	Aug. 8, 1871	117, 778
Hoisting-apparatus	J. Hoffman	Lebanon, N. J	Dec. 24, 1872	134, 141
Hoisting-apparatus	G. L. Howland	Topsham, Me	Sept. 18, 1866	58, 103
Hoisting-apparatus	G. L. and W. M. Howland	Topsham, Me	Mar. 26, 1867	63, 251
Hoisting-apparatus	W. M. and G. L. Howland	Topsham, Me	Oct. 12, 1869	95, 691
Hoisting-apparatus	H. H. Hunt	Saratoga Springs, N. Y	Dec. 20, 1870	110, 238
Hoisting-apparatus	J. L. Isbell	Naugatuck, Conn	May 6, 1873	138, 655
Hoisting-apparatus	B. H. Jenks	Bridesburgh, Pa	Aug. 7, 1866	56, 947
Hoisting-apparatus	G. H. Kamnacher	Columbus, Ohio	Sept. 21, 1869	95, 022
Hoisting-apparatus	G. B. Keeler	Portchester, N. Y	Sept. 17, 1867	68, 992
Hoisting-apparatus	D. Knowles	Philadelphia, Pa	July 1, 1873	140, 421
Hoisting-apparatus	T. Krausch	New York, N. Y	Nov. 28, 1871	121, 386
Hoisting-apparatus	W. H. Kuntz	Mount Rock, Pa	Nov. 26, 1867	71, 393
Hoisting-apparatus	E. Learned	Boston, Mass	Feb. 6, 1849	6, 085
Hoisting-apparatus	A. R. Lemen	Kalamazoo, Mich	Mar. 5, 1867	62, 647
Hoisting-apparatus	J. Lemman	Cincinnati, Ohio	July 17, 1860	29, 180
Hoisting-apparatus	A. B. Lipsey	New York, N. Y	Jan. 31, 1871	111, 356
Hoisting-apparatus	S. M. Longley	Hudson, N. Y	Oct. 24, 1865	50, 607
Hoisting-apparatus	J. E. Marshall and J. W. Schroeder.	Baltimore, Md	June 23, 1868	79, 138
Hoisting-apparatus	W. K. Marvin	New York, N. Y	Feb. 7, 1865	46, 257
Hoisting-apparatus	L. S. Mason	Middlefield Centre, N. Y	Apr. 30, 1872	126, 222
Hoisting-apparatus	G. B. Massey	New York, N. Y	Oct. 7, 1873	143, 518
Hoisting-apparatus	G. B. Massey and A. B. Darling	New York, N. Y	Oct. 7, 1873	143, 519
Hoisting-apparatus	N. S. McFarland	Brooklyn, N. Y	Mar. 26, 1872	125, 064
Hoisting-apparatus	N. S. McFarland	Williamsburgh, N. Y	Oct. 14, 1873	143, 706
Hoisting-apparatus	J. V. and W. H. Merrick	Philadelphia, Pa	June 9, 1868	78, 675
Hoisting-apparatus	J. V. and W. H. Merrick	Philadelphia, Pa	Aug. 31, 1869	94, 429
Hoisting-apparatus	J. H. Mills	Boston, Mass	Jan. 19, 1869	85, 952
Hoisting-apparatus	J. Nicholson	Monticello, Ind	Dec. 5, 1871	121, 463
Hoisting-apparatus	D. A. Noble	Boston, Mass	June 11, 1872	127, 789
Hoisting-apparatus	C. R. Otis	Yonkers, N. Y	Feb. 28, 1865	46, 580
Hoisting-apparatus	C. R. Otis	Yonkers, N. Y	Mar. 31, 1868	76, 240
Hoisting-apparatus	C. R. Otis	Yonkers, N. Y	Dec. 30, 1873	146, 090
Hoisting-apparatus	C. R. and N. P. Otis	Yonkers, N. Y	Nov. 3, 1868	83, 725
Hoisting-apparatus	C. R. and N. P. Otis	Yonkers, N. Y	Jan. 17, 1871	110, 993
Hoisting-apparatus	C. R. and N. P. Otis	Yonkers, N. Y	Apr. 4, 1871	113, 555
Hoisting-apparatus	C. R. and N. P. Otis	Yonkers, N. Y	Jan. 16, 1872	122, 846
Hoisting-apparatus	E. G. Otis	Yonkers, N. Y	Jan. 15, 1861	31, 128
Hoisting-apparatus	S. K. Paden	Pulaski, Pa	Mar. 5, 1872	124, 380
Hoisting apparatus	H. Pennepacker	Kimberton, Pa	Sept. 1, 1863	39, 745
Hoisting-apparatus	T. Pollard	Philadelphia, Pa	Aug. 9, 1870	106, 284
Hoisting-apparatus	J. L. Pott	Pottsville, Pa	Aug. 16, 1859	25, 215
Hoisting-apparatus	J. L. Pott	Pottsville, Pa	Jan. 24, 1860	26, 921
Hoisting-apparatus	W. R. Reaney	Chester, Pa	Oct. 12, 1869	95, 837
Hoisting-apparatus	J. F. Rochow	New York, N. Y	Dec. 9, 1862	37, 110
Hoisting-apparatus	J. F. Rochow	New York, N. Y	Oct. 22, 1867	70, 122
Hoisting-apparatus	W. Rung	New York, N. Y	Sept. 10, 1867	68, 794
Hoisting-apparatus	J. Sanderson	Fredericksburgh, Ohio	May 19, 1868	78, 138
Hoisting-apparatus	H. A. Schneekloth	New York, N. Y	May 3, 1870	102, 597
Hoisting apparatus	E. U. and W. L. Scoville	Manlius and West Bloomfield, N. Y.	Sept. 24, 1867	69, 257
Hoisting-apparatus	E. U. and W. L. Scoville	Manlius, N. Y	Aug. 4, 1868	80, 569
Hoisting-apparatus	H. F. Shaw	West Roxbury, Mass	May 1, 1866	54, 419
Hoisting-apparatus	M. Shepard	Pontiac, Mich	Feb. 13, 1872	123, 737
Hoisting-apparatus	T. Silver	New York, N. Y	Apr. 4, 1871	113, 585
Hoisting-apparatus	W. Smith	Philadelphia, Pa	Nov. 1, 1864	44, 895
Hoisting-apparatus	G. Sprague	South Addison, N. Y	July 4, 1871	116, 769
Hoisting-apparatus	H. D. Stover	New York, N. Y	June 15, 1869	91, 283
Hoisting-apparatus	L. W. Stuart	Narrowsburgh, N. Y	Apr. 4, 1871	113, 362
Hoisting-apparatus	A. L. Sweet	Norwich, Conn	Sept. 6, 1864	44, 124
Hoisting-apparatus	J. A. Talpey	Somerville, Mass	May 23, 1865	47, 878
Hoisting-apparatus	T. Terrell	Yonkers, N. Y	Dec. 12, 1871	121, 910
Hoisting-apparatus	T. R. Vestal	Santa Fé, Tenn	July 18, 1871	117, 224
Hoisting-apparatus	J. H. Violett	Goshen, Ind	Nov. 15, 1870	109, 278
Hoisting-apparatus	J. D. Warner	Brooklyn, N. Y	Apr. 9, 1872	125, 507
Hoisting-apparatus	W. S. Watson	New York, N. Y	Nov. 20, 1866	59, 880
Hoisting-apparatus	A. Weide	Chicago, Ill	Aug. 8, 1871	117, 840
Hoisting-apparatus	T. A. Weston	Ridgewood, N. J	Dec. 14, 1869	98, 000
Hoisting-apparatus	J. Wheelock	Worcester, Mass	Aug. 8, 1871	117, 948
Hoisting-apparatus	W. C. Williamson	Philadelphia, Pa	June 11, 1872	127, 726
Hoisting-apparatus and derrick	E. Osgood	Boston, Mass	Nov. 23, 1869	97, 223

Index of patents issued from the United States Patent Office from 1790 *to* 1873, *inclusive*—Continued.

Invention.	Inventor.	Residence.	Date.	No.
Hoisting-apparatus and safety-hatch	J. W Osgood	New York, N. Y	Mar. 12, 1872	124, 614
Hoisting-apparatus, Animal	J. Newton	Marengo Township, Mich.	Feb. 7, 1871	111, 671
Hoisting-apparatus, Automatic stop for steam	W. H. Lavinia	Chicago, Ill	July 2, 1872	128, 495
Hoisting-apparatus, Brake-clutch for	T. A. Weston	Ridgewood, N. J	Jan. 14, 1873	134, 957
Hoisting-apparatus, Brake of	C. R. Otis	Yonkers, N. Y	Oct. 18, 1864	44, 740
Hoisting apparatus, Coal	L. S. Chichester	Brooklyn, N. Y	Mar. 1, 1870	100, 259
Hoisting apparatus, Coal	G. Martz	Pottsville, Pa	Aug. 15, 1856	14, 671
Hoisting apparatus, Coal	J. C. White and R. Hay	Tuckerville, Pa	Nov. 25, 1856	16, 133
Hoisting-apparatus for bricks, &c	J. Cranshaw	Rochester, N. Y	Nov. 10, 1857	18, 575
Hoisting-apparatus for builders	L. Atwood	New York, N. Y	Dec. 29, 1868	85, 419
Hoisting-apparatus for miners, Swivel	T. M. Martin	San Francisco, Cal	Mar. 11, 1873	136, 661
Hoisting-apparatus, Hydraulic	P. Hinkle	San Francisco, Cal	July 1, 1873	140, 502
Hoisting-apparatus, Hydraulic	A. Lucius	New York, N. Y	Aug. 15, 1871	118, 031
Hoisting-apparatus, Hydraulic	J. R. Ritter	Reading, Pa	Nov. 19, 1872	133, 256
Hoisting-apparatus, Hydraulic	J. R. Ritter	Reading, Pa	Nov. 19, 1872	133, 257
Hoisting-apparatus, Portable and adjustable	G. A. Myers	Williamsburgh, N. Y	July 1, 1873	140, 427
Hoisting apparatus, Rafter-hook for	D. Hicks and S. Doty	Pontiac, Mich	Apr. 28, 1868	77, 375
Hoisting-apparatus, Safety	C. W. Copeland	New York, N. Y	July 12, 1870	105, 177
Hoisting-apparatus, Safety-mechanism for	W. H. Merrick	Philadelphia, Pa	Mar. 8, 1870	100, 651
Hoisting-apparatus, Safety-pawl for	R. T. Crane and W. H. Lavinia	Chicago, Ill	July 2, 1872	128, 467
Hoisting-apparatus, Ship's	D. J. Wilcoxson	Milan, Ohio	Aug. 2, 1859	24, 967
Hoisting-apparatus, Steam	J. Beggs	Chicago, Ill	July 29, 1873	141, 197
Hoisting-apparatus, Steam	C. S. Crane and W. H. Lavinia	Chicago, Ill	July 2, 1872	128, 466
Hoisting-apparatus, Steam	J. S. Neal	Madison, Ind	Feb. 2, 1864	41, 446
Hoisting-apparatus, Steam	C. R. Otis	Yonkers, N. Y	May 16, 1865	47, 773
Hoisting-apparatus, Steam	C. R. Otis	Yonkers, N. Y	Oct. 1, 1872	131, 894
Hoisting-apparatus, Steam	C. R. Otis	Yonkers, N. Y	Oct. 1, 1872	131, 895
Hoisting-apparatus, Steam	C. R. and N. P. Otis	Yonkers, N. Y	Nov. 21, 1865	51, 076
Hoisting-apparatus, Steam	C. R. and N. P. Otis	Yonkers, N. Y	Jan. 7, 1873	134, 699
Hoisting-apparatus, Steam	O. Tufts	Boston, Mass	Apr. 10, 1866	53, 904
Hoisting apparatus, Water	H. Waterman	Haverhill, Mass	Mar. 6, 1860	27, 400
Hoisting-attachment for the shaft of well-augers	H. H. Russell	Maysville, Mo	Aug. 13, 1872	130, 442
Hoisting-block	J. A. Burr	Brooklyn, N. Y	Oct. 18, 1870	108, 446
Hoisting-block	W. Duchermin	Charlottetown, Prince Edward Island.	Dec. 17, 1861	33, 929
Hoisting-block	W. H. Merrill	Taunton, Mass	Dec. 25, 1855	13, 991
Hoisting-blocks, Anti-friction bearing of	W. Duchermin	Charlottetown, Prince Edward Island.	Feb. 4, 1862	34, 295
Hoisting-bucket	G. Focht	Reading, Pa	Nov. 3, 1857	18, 540
Hoisting-bucket for coal, &c	G. Focht	Reading, Pa	Apr. 14, 1857	17, 034
Hoisting-cages, Safety-attachment for	W. C. Williamson	Philadelphia, Pa	July 4, 1871	116, 656
Hoisting-crane	C. Bassett	Massillon, Ohio	Oct. 7, 1862	36, 598
Hoisting-device	J. J. Doyle	New York, N. Y	Jan. 8, 1861	31, 077
Hoisting-device	S. C. Goodsell	New Haven, Conn	June 25, 1867	66, 079
Hoisting-drum	G. W. La Baw	Jersey City, N. J	Apr. 15, 1856	14, 666
Hoisting-drum	H. Strickler	Carlisle, Pa	Feb. 11, 1868	74, 445
Hoisting drum and grain-fork, Hay	H. Strickler	Carlisle, Pa	Apr. 21, 1868	76, 954
Hoisting-engine	F. Murgatroyd	Cleveland, Ohio	Nov. 4, 1873	144, 217
Hoisting-fork	E. Magruder	Cap au Gris, Mo	Jan. 31, 1871	111, 361
Hoisting-gear	G. Duerre	Williamsburgh, N. Y	Nov. 14, 1871	120, 951
Hoisting-gear	W. O. Jones	Portland, Me	Aug. 11, 1868	80, 968
Hoisting-gear	H. F. Shaw	West Roxbury, Mass	May 21, 1867	64, 915
Hoisting-gear	H. F. Shaw	West Roxbury, Mass	May 3, 1870	102, 719
Hoisting-gear, Attachment for	G. Duerre	Williamsburgh, N. Y	Jan. 16, 1872	122, 712
Hoisting-grapple	J. A. Burgess	Plymouth, Mass	Sept. 20, 1870	107, 444
Hoisting-grapple	T. McGrath	Albany, N. Y	Mar. 2, 1869	87, 349
Hoisting ice, Apparatus for	A. Hunt	Philadelphia, Pa	Feb. 2, 1858	19, 250
Hoisting in mines, Cage for	H. Berry, H. Hochholzer, and F. Denver.	Virginia City, Nev	May 8, 1866	54, 490
Hoisting-jack	B. Beuchel	Kalamazoo, Mich	Nov. 2, 1869	96, 299
Hoisting-jack	E. A. Castellaw	Savannah, Ga	May 31, 1870	103, 563
Hoisting-jack	J. H. Churchill	Cross River, N. Y	Aug. 19, 1873	141, 919
Hoisting-jack	F. Hollenberry	Frizellburgh, Md	Aug. 16, 1870	106, 485
Hoisting-jack	W. Kearney	Newark, N. J	Oct. 19, 1858	21, 837
Hoisting-jack	S. B. Rittenhouse	Plymouth, Ind	Dec. 17, 1867	72, 329
Hoisting-jack	J. W. Steed	Minneapolis, Minn	Nov. 27, 1866	60, 080
Hoisting-machine	C. Abel	New York, N. Y	Jan. 24, 1865	45, 962
Hoisting-machine	W. C. Allison	Philadelphia, Pa	Apr. 2, 1850	7, 230
Hoisting-machine	J. Ashcroft	Brooklyn, N. Y	Nov. 26, 1872	133, 289
Hoisting-machine	D. W. Barr	Lancaster, Pa	Nov. 9, 1858	22, 008
Hoisting-machine	J. Bates	Baltimore, Md	Apr. 18, 1871	113, 836
Hoisting-machine	A. Betteley	Boston, Mass	Dec. 20, 1859	26, 469
Hoisting-machine	J. Bird	New York, N. Y	June 6, 1865	48, 044
Hoisting-machine	D. A. Bolt	Harrisburgh, Pa	Aug. 15, 1871	117, 971
Hoisting-machine	M. K. Carpenter	Cincinnati, Ohio	May 18, 1869	90, 236
Hoisting-machine	G. R. Clarke	New York, N. Y	June 9, 1868	78, 645
Hoisting-machine	S. Croghan	Flatbush, N. Y	Nov. 25, 1862	37, 026
Hoisting-machine	J. Darling	Cincinnati, Ohio	Dec. 23, 1873	145, 727
Hoisting-machine	J. Drummond	New York, N. Y	Mar. 10, 1834	....
Hoisting-machine	J. Edson	Boston, Mass	Oct. 29, 1867	70, 180
Hoisting-machine	W. Eppelsheimer	San Francisco, Cal	Oct. 5, 1869	95, 448
Hoisting-machine	D. Evans	Philadelphia, Pa	Mar. 19, 1834	....
Hoisting-machine	H. T. Goodling	York, Pa	July 21, 1868	80, 067
Hoisting-machine	C. H. Hersey	Boston, Mass	July 22, 1873	141, 051
Hoisting-machine	P. Higdon	Cropper's Depot, Ky	Dec. 26, 1865	51, 719
Hoisting-machine	J. Jewsbury	Brook Fields, England	Mar. 15, 1870	100, 769
Hoisting-machine	G. Johnson	Cincinnati, Ohio	Dec. 28, 1869	98, 388
Hoisting-machine	G. Johnson	Cincinnati, Ohio	Mar. 29, 1870	101, 367
Hoisting-machine	J. Jones	Providence, R. I	Dec. 16, 1873	145, 657
Hoisting-machine	R. A. and W. Kendall	Mineral Point, Wis	Apr. 26, 1870	102, 274
Hoisting-machine	J. Kennedy	Chicago, Ill	Dec. 1, 1868	84, 496
Hoisting-machine	G. R. Long and S. King	Lanark, Ill	Oct. 17, 1871	119, 988
Hoisting-machine	S. M. Longley	Hudson City, N. Y	Mar. 1, 1864	41, 779
Hoisting-machine	S. L. Lord	New York, N. Y	Nov. 21, 1871	121, 113
Hoisting-machine	W. M. Loyd	New York, N. Y	July 4, 1871	116, 607
Hoisting-machine	M. Lynch	New York, N. Y	Feb. 8, 1870	99, 578
Hoisting-machine	H. Malley	Chicago, Ill	Mar. 5, 1867	62, 549
Hoisting-machine	J. McCalvey	Philadelphia, Pa	Aug. 28, 1866	57, 535

Index of patents issued from the United States Patent Office from 1790 *to* 1873, *inclusive*—Continued.

Invention.	Inventor.	Residence.	Date.	No.
Hoisting-machine	D. McIntyre and G. C. Reeves	Central City, Colo	Nov. 15, 1864	45, 062
Hoisting-machine	P. W. Mellen	Saint Louis, Mo	July 15, 1873	140, 786
Hoisting-machine	W. Miller	Cincinnati, Ohio	May 12, 1863	38, 497
Hoisting-machine	W. Miller	Cincinnati, Ohio	July 4, 1865	48, 579
Hoisting-machine	W. Miller	Cincinnati, Ohio	May 14, 1867	64, 785
Hoisting-machine	W. Murray	Chicago, Ill	Nov. 19, 1867	71, 043
Hoisting-machine	W. Neal	Louisville, Ky	Apr. 23, 1872	125, 978
Hoisting-machine	R. Packard	Rockland, Me	May 4, 1858	20, 170
Hoisting-machine	J. S. Park	Montandon, Pa	June 29, 1869	92, 088
Hoisting-machine	N. Parrish	Kalamazoo, Mich	Sept. 27, 1870	107, 807
Hoisting-machine	F. B. Perkins	Roxbury, Mass	May 3, 1864	42, 595
Hoisting-machine	S. B. Phelps and C. A. Slack	Norwich, Vt	Mar. 28, 1865	47, 036
Hoisting-machine	G. H. Pittman	McConnellsburgh, Pa	Oct. 1, 1872	131, 902
Hoisting-machine	W. Purden	Baltimore, Md	Mar. 19, 1806	....
Hoisting-machine	H. J. Reedy	Cincinnati, Ohio	June 9, 1868	78, 829
Hoisting-machine	H. J. Reedy	Cincinnati, Ohio	June 20, 1871	116, 096
Hoisting-machine	H. J. and J. Reedy	Cincinnati, Ohio	Oct. 23, 1866	59, 137
Hoisting-machine	G. H. Reynolds	New York, N. Y	Aug. 18, 1868	81, 289
Hoisting-machine	S. Reynolds	Pierrepont Manor, N. Y	June 3, 1873	139, 614
Hoisting-machine	J. Scott	Pontiac, Mich	Mar. 21, 1871	112, 855
Hoisting-machine	W. Sellers	Philadelphia, Pa	Mar. 21, 1871	112, 856
Hoisting-machine	H. M. Smith	Richmond, Va	Apr. 16, 1867	63, 955
Hoisting-machine	I. Smith	Tonhannock, N. Y	Mar. 25, 1873	137, 152
Hoisting-machine	W. M. Smith	Augusta, Ga	Aug. 10, 1869	93, 490
Hoisting-machine	J. W. Tucker	New York, N. Y	May 28, 1867	65, 307
Hoisting-machine	S. Van Emon	Cincinnati, Ohio	Mar. 12, 1867	62, 788
Hoisting-machine	R. A. Wilder	Cressona, Pa	Mar. 18, 1862	34, 710
Hoisting-machine	N. J. Wilkinson	Kalamazoo, Mich	Apr. 26, 1870	102, 345
Hoisting-machine	M. Willard	Cincinnati, Ohio	May 16, 1865	47, 761
Hoisting-machine	W. C. Williamson	Philadelphia, Pa	Dec. 29, 1868	85, 415
Hoisting-machine	W. C. Williamson	Philadelphia, Pa	Aug. 23, 1870	106, 642
Hoisting-machine	I. W. Willson	Syracuse, N. Y	Oct. 3, 1848	5, 828
Hoisting-machine	N. Wonlarlarsky	St. Petersburg, Russia	Dec. 6, 1870	109, 988
Hoisting-machine	J. M. Yates	Princeville, Ill	Sept. 23, 1873	143, 214
Hoisting-machine capstan	C. S. Houck	Greenport, N. Y	Apr. 28, 1868	77, 191
Hoisting-machine for cellars	J. Ingram	New York, N. Y	Feb. 5, 1867	61, 835
Hoisting-machine for stacking hay	W. Louden	Fairfield, Iowa	Dec. 3, 1867	71, 771
Hoisting-machine for vessels	H. O. Perry	Buffalo, N. Y	July 2, 1867	66, 251
Hoisting-machine grapple	J. H. Hecker	Lebanon, Pa	June 25, 1872	128, 309
Hoisting-machine hatch	R. H. Martin and C. R. Otis	Brooklyn and Yonkers, N. Y.	Oct. 1, 1872	131, 888
Hoisting machine, Hay, &c	J. S. Lloyd	Salem, N. J	Apr. 24, 1860	27, 994
Hoisting-machine, Hydrostatic	S. Van Emon	Cincinnati, Ohio	Dec. 17, 1872	134, 019
Hoisting machine, Ice	W. G. Brower	Staatsburgh, N. Y	May 25, 1858	28, 322
Hoisting machine, Ice	J. Wagner	Philadelphia, Pa	Oct. 4, 1859	25, 691
Hoisting machine, Marl, &c	T. A. Granger	Wilson County, N. C	Nov. 8, 1859	26, 029
Hoisting-machine, Pneumatic	J. R. Anderson and P. C. Harlan	Freeport, Pa	July 11, 1871	116, 916
Hoisting-machines, Brake-clutch for	T. A. Weston	Ridgewood, N. J	Jan. 14, 1873	134, 958
Hoisting-machine, Safety-apparatus for	W. T. Sands	New York, N. Y	June 20, 1871	116, 103
Hoisting-machines, Safety-attachment for	W. D. Andrews	Brookhaven, N. Y	Mar. 25, 1873	137, 165
Hoisting-machinery	J. B. Holmes	Boston, Mass	June 7, 1841	2, 115
Hoisting-machinery	R. A. Wilder	Cressona, Pa	May 1, 1860	28, 122
Hoisting-machinery, Horse-power	A. Jackson	New York, N. Y	Aug. 1, 1854	11, 457
Hoisting persons, &c., from one story in a building to another, Machine for.	J. H. Hobbs	Philadelphia, Pa	May 15, 1860	28, 273
Hoisting-protector	J. W. Wilson	New York, N. Y	Oct. 1, 1872	131, 923
Hoisting-power, Tread-wheel	J. B. Cassel	Worcester Township, Pa	Nov. 20, 1866	59, 818
Hoisting-tackle	J. W. Norcross	Middletown, Conn	Dec. 26, 1865	51, 742
Hoisting-tackle	J. C. Price	New Philadelphia, Ohio	Jan. 1, 1867	60, 935
Hoisting wheels, Operating	J. McMurtry	Fayette County, Ky	Apr. 3, 1860	27, 760
Hoisting-whim	J. C. Marshall	San Francisco, Cal	May 13, 1873	138, 811
Holdback	A. C. Beckwith and J. H. Graham.	Oriskany, N. Y	Jan. 19, 1869	85, 898
Holdback	R. W. Carrier	Sherburne, N. Y	Dec. 20, 1859	26, 476
Holdback	G. D. Cleaveland	Flint, Ind	Oct. 28, 1873	143, 964
Holdback	J. C. Covert	Townsendville, N. Y	Oct. 1, 1867	69, 412
Holdback	G. F. De Vine	Hamilton, Canada	Mar. 26, 1872	124, 888
Holdback	W. Garrison and C. H. Stevens	Syracuse, N. Y	May 4, 1869	89, 573
Holdback	D. A. Gorham	Norway, Me	Mar. 23, 1869	88, 159
Holdback	L. Kruse	Sabula, Iowa	Sept. 1, 1868	81, 651
Holdback	G. B. Lumpkin	Lexington, Ga	Feb. 27, 1872	124, 144
Holdback	R. Moody	Monmouth, Me	Oct. 27, 1868	83, 400
Holdback	M. D. Myers	Schuyler, N. Y	Feb. 11, 1873	135, 833
Holdback	J. G. Rogers	Cass County, Mich	July 16, 1872	129, 253
Holdback	B. E. Sperry	Aurora, Ill	Sept. 3, 1872	131, 036
Holdback-hook	N. W. Robinson	Moriah, N. Y	Feb. 4, 1868	74, 143
Holdback-hook	U. B. Winchell	Oak Hill, N. Y	Dec. 10, 1861	33, 915
Holdback-iron for carriages	J. P. Simmons	Fulton, N. Y	May 24, 1864	42, 884
Holdback-irons	E. F. Howe	Dowagiac, Mich	Aug. 20, 1872	130, 640
Holdback-straps, Hook for	W. A. Bagley	Ansonia, Conn	Nov. 19, 1867	70, 938

Holder:

See Amunition-holder.
Bag-holder.
Barber's check-holder.
Bed-clothes holder.
Bill-holder.
Billiard-table ball-holder.
Billiard-table chalk-holder.
Bit-holder.
Blacking and brush holder.
Blacking-box holder.
Blacking-brush holder.
Blotter-holder.
Blower-holder.
Bolt-holder.
Book-holder.
Book and manuscript holder.
Bottle-holder.
Bouquet-holder.
Brace-bit holder.
Breeching-holder.
Broom-holder.
Brush-holder.
Brush and mop holder.
Bucket-holder.
Button-holder.
Camera plate-holder.
Candle-holder.

Index of patents issued from the United States Patent Office from 1790 *to* 1873, *inclusive*—Continued.

Invention.	Inventor.	Residence.	Date.	No.

Holder—Continued.

See Canopy-holder.
Cap-holder.
Car-stake holder.
Card-holder.
Carriage-tongue holder.
Cartridge-holder.
Carving knife and fork holder.
Chain-holder.
Chalk-holder.
Chalk-line holder.
Cigar-holder.
Cigar-ash holder.
Cigar-cap holder.
Clothes-holder.
Clothes-line holder.
Clothes-pin holder.
Coil-holder.
Coin-holder.
Colter-holder.
Comb-holder.
Copy-holder.
Cork or stopper holder.
Corn-holder.
Cow's tail holder.
Crayon-holder.
Cuff holder.
Currency and postage-stamp holder.
Curtain-holder.
Curtain-roller holder.
Daguerreotype-plate holder.
Diamond-holder.
Document-holder.
Door-holder.
Dress-goods holder.
Drill-holder.
Drinking-glass holder
Egg-holder.
Elastic-bosom holder.
Embroidery-holder.
Eraser-holder.
Eye-glass holder.
Fan and handkerchief holder.
Faucet-holder.
Fence-board-gage holder.
File-holder.
Fire-cracker holder.
Fire-set holder.
Fish-hook holder.
Flat-iron holder.
Fruit-jar holder.
Gas-holder.
Gimlet-holder.
Glass-holder.
Glass-blower's tool-holder.
Globe-holder.
Grain-bag holder.
Hammer-holder.
Handkerchief-holder.
Hat-holder.
Hat and cap holder.
Hat and coat holder.
Hat and parcel holder.
Hog-holder.
Horse-holder.
Horse-tail holder.
Hose-holder.
Ice-cream holder.
Insulator-holder.
Iron-holder.
Jar or vase holder.
Knife-holder.
Knife and fork holder.
Knife-blade holder.
Knife, fork, and spoon holder.
Knitting-holder.
Kite-string holder.
Label-holder.
Lamp-holder.
Lamp-chimney holder.
Lamp-shade holder.
Last-holder.
Last-block holder.
Lead-pencil holder.
Line and strap holder.
Lint and basket holder.
Mail-pouch holder.
Map-holder.
Match-holder.
Measure-holder.
Meat-holder.
Mirror-holder.
Mop holder.
Mucilage-holder.
Music-leaf holder.
Napkin-holder.
Neck-tie holder.
Needle and tool holder.
Oil-can holder.
Oil-stone holder.
Painter's holder.
Painter's brush-holder.
Painter's drill-holder.
Paper-holder.
Paper-ribbon holder.
Pen-holder.
Pencil-holder.
Percussion-cap holder.
Photographic-paper holder.
Photographic-plate holder.
Pipe-holder.
Pipe-coupling holder.
Pitcher-holder.
Planer's tool-holder.
Plate-holder.
Portfolio-holder.
Postage-holder.
Pounce-holder.
Press-holder.
Railway-rail-bolt holder.
Razor-strop holder.
Rein-holder.
Rosette-holder.
Rubber-block holder.
Ruche-holder.
Sad-iron holder.
Salt-holder.
Sand-paper holder.
Sash-holder.
Sash-spring holder.
Scythe-holder.
Scythe-stone holder.
Seat-holder.
Sewing-machine-bobbin holder.
Sewing-machine cloth-holder.
Sewing-machine work-holder.
Sewing-thread holder.
Sewing-work holder.
Shade-holder.
Shaft-holder.
Shaft and pole holder.
Sheep-holder.
Shoe-holder.
Shook-holder.
Shutter-holder.
Siding-holder.
Signal-holder.
Slat-holder.
Slate-pencil holder.
Sliding-holder.
Soap-holder.
Soda-water-tumbler holder.
Spice-holder.
Spinning-machine bobbin holder.
Spinning-ring holder.
Spittoon-holder.
Sponge-holder.
Spool-holder.
Spoon-holder.
Stalk-holder.
Stamp-holder.
Stereoscopic-picture holder.
Stereotype-block holder.
Stereotype-plate holder.
Stitching-work holder.
Stove-holder.
Stove-utensil holder.

Index of patents issued from the United States Patent Office from 1790 *to* 1873, *inclusive*—Continued.

Invention.	Inventor.	Residence.	Date.	No.
Holder—Continued. *See* Strap-holder. Tack-holder. Tag-holder. Tailor's pressing-board holder. Taper-holder. Thill-holder. Ticket-holder. Tongue-holder. Tool-holder. Towel-holder. Tray-holder. Twine-holder. Umbrella-holder. Vine-holder. Violin-holder. Wash-tub holder. Watch-wheel holder. Wax-candle holder. Wax candle or taper holder. Whip-holder. Whip and line holder. Wire jar-holder. Work-holder. Yoke-holder.				
Holding and soldering cans, Machine for	W. J., jr., and T. G. Redhefter, and L. A. Smith.	Kansas City, Mo., and Pawnee, Kans.	Mar. 18, 1873	137, 027
Hollow bolt	D. F. Fetter, M. D	New York, N. Y	Feb. 16, 1869	86, 976
Hollow drill, Adjustable	C. E. Butler	Hudson, N. Y	Apr. 19, 1870	102, 090
Hollow ring	K. Frazer	Syracuse, N. Y	Jan. 1, 1867	60, 712
Hollow-ware, &c., Apparatus for pressing	N. Thompson	Brooklyn, N. Y	Oct. 10, 1871	119, 725
Hollow-ware from tinned plates, sheet-iron, &c., Manufacturing.	E. M. Converse	Southington, Conn	July 20, 1831	....
Hollow-ware from zinc, &c., Manufacturing	S. Davis	New York	Apr. 18, 1831	....
Hollow-ware, Making gutta-percha	S. T. Armstrong	New York, N. Y	June 24, 1851	8, 180
Hollow-ware, Manufacture of sheet-iron	J. and J. D. Gray	Pittsburgh, Pa	May 5, 1863	38, 446
Hollow-ware, Mode of hanging covers to baled metallic.	E. Ripley	Troy, N. Y	Jan. 1, 1861	31, 035
Hollow-ware molding	J. I. Johnston	Cincinnati, Ohio	June 29, 1852	9, 074
Holster	W. Tileston	Georgetown, D. C	June 21, 1864	43, 275
Hominy and pearling mill	E. A. Duer	Decatur, Ill	Aug. 4, 1868	80, 713
Hominy and smut mill	W. Wright	Saint Louis, Mo	Dec. 8, 1868	84, 723
Hominy-machine	R. Campbell	Martinsburgh, Va	Apr. 9, 1827	....
Hominy-machine	W. Davis	Middleburgh, Md	May 24, 1859	24, 104
Hominy-machine	E. Fahrney	Mount Morris, Ill	May 15, 1855	12, 860
Hominy-machine	J. Gehr	Clear Spring, Md	July 17, 1860	29, 159
Hominy-machine	D. Haines	Union Bridge, Md	June 11, 1861	32, 516
Hominy-machine	O. F. Mayhew	Indianapolis, Ind	June 2, 1857	17, 469
Hominy-machine	J. Nesbitt, jr., and T. J. Crosley	Clear Spring, Md	Sept. 11, 1855	13, 549
Hominy-machine	S. Null	Carroll County, Md	May 25, 1852	8, 972
Hominy-machine	P. Siemers	Saint Louis, Mo	Oct. 13, 1857	18, 413
Hominy-machine	R. H. Taylor	Goose Creek, Va	Aug. 21, 1866	57, 406
Hominy, Making	W. Y. Singleton	Springfield, Mass	Apr. 25, 1846	4, 486
Hominy-mill	B. Bridendolph	Clear Spring, Md	Aug. 22, 1854	11, 547
Hominy-mill	J. M. Clark	Lancaster, Pa	May 25, 1858	20, 327
Hominy-mill	E. M. Coombes	Memphis, Ind	July 19, 1870	105, 430
Hominy-mill	W. C. Coombs and J. M. Gray	Memphis, Ind	Jan. 4, 1870	98, 565
Hominy-mill	F. E. Drake and J. W. Teal	Indianapolis, Ind	Mar. 23, 1858	19, 691
Hominy-mill	E. Fahrney	Deep River, Iowa	Jan. 5, 1858	19, 060
Hominy mill	J. B. Cowdy and J. A. Welsh	Xenia, Ohio	Aug. 11, 1857	17, 900
Hominy-mill	J. Haydon	Philadelphia, Pa	Apr. 3, 1866	53, 611
Hominy-mill	J. D. Heatwole and R. C. Mauck	Harrisonburgh, Va	May 1, 1860	28, 079
Hominy-mill	P. Honarighaus	Royalton, Ohio	Feb. 9, 1858	19, 297
Hominy-mill	T. Rudnut	Terre Haute, Ind	Dec. 26, 1871	122, 172
Hominy-mill	J. Hughes	Cambridge City, Ind	Oct. 12, 1852	9, 323
Hominy-mill	B. Kemp	Knoxville, Md	Dec. 11, 1866	60, 385
Hominy-mill	J. R. Leedy	Maurertown, Va	Oct. 4, 1870	108, 036
Hominy-mill	N. Read	Belfast, Me	May 8, 1820	....
Hominy-mill	I. Speight	Woodville, Miss	Dec. 21, 1858	22, 384
Hominy-mill	G. Strause	Boonsborough, Md	Sept. 20, 1859	25, 536
Hominy-mill	J. L. Toner	Edinburgh, Ind	Nov. 11, 1873	144, 486
Hominy-mill	W. Wright	Springfield, Ohio	Sept. 6, 1864	44, 136
Hominy-mill	W. Wright	Springfield, Ohio	June 20, 1865	48, 332
Hominy-mill burr	A. P. Jackson	Memphis, Ind	June 15, 1869	91, 343
Hominy-mills, Separator for	J. Donaldson	Rockford, Ill	Jan. 24, 1860	26, 893
Homomotive	W. S. Hall	Quincy, Mass	June 8, 1869	91, 123
Hone	I. Babbitt	Roxbury, Mass	Mar. 23, 1854	10, 954
Hone, Boxed	J. Potter and O. Abell	Whitehall, N. Y	Jan. 14, 1868	73, 383
Hone, Glass	A. Gordon and J. P. Bakewell	Pittsburgh, Pa	Dec. 8, 1832	....
Hone, Metallic	J. Houston	Williamsburgh, Va	May 4, 1805	....
Hone, Metallic	W. H. Webb, jr	Chelsea, Mass	Apr. 17, 1855	12, 748
Hones, Manufacture of artificial	T. Deming	East Hartford, Conn	Oct. 13, 1857	18, 392
Honey, Artificial	Z. Corbin and G. Marlett	Syracuse, N. Y	May 12, 1857	17, 264
Honey, Artificial	G. H. Smitson	Ripley, Ohio	July 13, 1869	92, 666
Honey-comb, Artificial	S. Wagner	York, Pa	May 7, 1861	32, 258
Honey-comb knife	H. O. Peabody	Boston, Mass	May 28, 1872	127, 187
Honey-combs, Device for making artificial	J. Williams	Bean's Station, Tenn	Dec. 31, 1872	134, 411
Honey-cutting knife	A. W. Todd	Chicago, Ill	May 17, 1864	42, 807
Honey from the comb, Apparatus for extracting	L. L. Langstroth and S. Wagner	Butler County, Ohio, and Washington, D. C.	Jan. 15, 1867	61, 216
Honey from the comb, Centrifugal apparatus for extracting.	W. Probasco	Lawrenceburgh, Ind	Jan. 7, 1873	134, 561
Honey from the comb, Centrifugal machine for extracting.	H. O. Peabody	Boston, Mass	Oct. 26, 1869	96, 142
Honey from the comb, Centrifugal machine for extracting.	H. O. Peabody	Boston, Mass	Dec. 7, 1869	97, 684
Hood, Lady's	E. Hill	Philadelphia, Pa	July 18, 1865	48, 867
Hood, Lady's	M. Landenberger	Philadelphia, Pa	July 19, 1864	43, 591
Hoof-expander	C. B. and S. Galentine and A. J. Russell.	Nunda, N. Y	Dec. 23, 1856	16, 273
Hoof-expander	J. Tipton	Malaga, Ohio	Jan. 21, 1868	73, 675
Hoof-hooks, Joining and fitting	F. Stanley	Austin, Tex	Sept. 15, 1868	82, 172
Hoof-pad, Elastic	F. B. Carleton	Cambridge, Vt	Jan. 1, 1867	60, 687
Hoof-parer	I. Baker	Long Branch, Mo	Oct. 10, 1871	119, 734
Hoof parer, Horse	A. Baker	Shenandoah County, Va	Aug. 7, 1860	29, 543
Hoof parer, Horse	A. Crafts and E. Weeks	Auburn, Ohio	Jan. 1, 1850	6, 983
Hoof parer, Horse	S. D. Freet	McCutchenville, Ohio	Mar. 5, 1867	62, 624
Hoof parer, Horse	F. Lehmann	Riverside, Ill	Nov. 16, 1869	90, 928
Hoof parer, Horse	V. N. Mitchell	Concord, N. C	June 2, 1857	17, 441

Index of patents issued from the United States Patent Office from 1790 *to* 1873, *inclusive*—Continued.

Invention.	Inventor.	Residence.	Date.	No.
Hoof parer, Horse	W. Tansley	Salisbury Centre, N. Y	Apr. 22, 1862	35, 046
Hoof parer, Horse	J. Temple	Van Buren, Ohio	June 15, 1869	91, 286
Hoof-pairing tool, Horse	J. C. Johnson	Sulphur Springs, Ind	Oct. 8, 1872	132, 010
Hoof-shears	M. C. Malone	Palmyra, Ill	Apr. 30, 1872	126, 221
Hoof-spreader	T. Armstrong	Hamilton, Canada	Mar. 11, 1873	136, 689
Hoofs, Machine for paring horses'	G. W. Schafer	Saint Charles, Mo	Apr. 22, 1873	138, 101
Hook:				

See Bale-hook.
Back-band hook.
Beadstead-hook.
Belt-hook.
Bench-hook.
Bird-cage hook.
Blind hinge-hook.
Boat-detaching hook.
Bracket-hook.
Breeching-hook.
Button-hook.
Cage-hook.
Cant-hook.
Carpenter's bench-hook.
Carriage-hook.
Cat-hook.
Ceiling-hook.
Chain-hook.
Chair-hook.
Chandelier-hook.
Check-hook.
Clevis-hook.
Clothes-hook.
Clothes-wringing hook.
Coat and hat hook.
Cooper's hook.
Coupling-hook.
Cultivating-hook.
Cultivator draft-hook.
Curtain-hook.
Davit-fall-block hook.
Detaching-hook.
Disengaging-hook.
Draft-hook.
Draft-chain hook.
Drag-hook.
Drapery-hook.
Draw-hook.
Dumping-tub hook.
Fench-hook.
File hook.
Fish-hook.
Garment-hook.
Gig-saddle hook.
Glass-blower's hook.
Grappling-hook.
Grub-hook.
Hames-hook.
Harness-hook.
Harness-check hook.
Hat-hook.
Hat and clothes hook.
Hat and coat hook.
Hay-hook.
Hedge-hook.
Hinge-hook.
Hitch-hook.
Hitching hold-back hook.
Hoof-hook.
Hop-hook.
Horse cant-hook.
Joint-hook.
Joist-hook.
Leader-hook.
Manure-hook.
Match-hook.
Meat-hook.
Milk-can hook.
Mousing-hook.
Needle-threading hook.
Pad-hook.
Painter's hook.
Picture-hook.
Pile-hook.
Pocket-hook.
Pruning-hook.
Rafter-hook.
Rein-hook.
Retaining and releasing hook.
Saddle-hook.
Safety-hook.
Scribe-hook.
Seat- hook.
Security-hook.
Self-adjusting hook.
Self-guarding hook.
Self-mousing hook.
Self-releasing hook.
Sewing-machine-cover hook.
Shade-hook.
Shipping-hook.
Shoe-hook.
Shoe-lacing hook.
Show-window-bracket hook.
Siding-hook.
Snap-hook.
Stable-hook.
Steam and gas pipe hook.
Suspension-hook.
Swivel-hook.
Tackle-hook.
Tackle-block hook.
Tag-hook.
Tanner's hook.
Tassel-hook.
Tipping-hook.
Tobacco-hook.
Tobacco-hanging hook.
Towing-hook.
Trace-hook.
Trace-supporting hook.
Traveler's hook.
Traveling-hook.
Traveling-bag hook.
Trolling-hook.
Vest-chain hook.
Wardrobe-hook.
Watch-chain hook.
Whiffletree-hook.
Whiffletree snap-hook.

Invention.	Inventor.	Residence.	Date.	No.
Hook	W. M. Knight and J. H. Orne	Marblehead, Mass	Aug. 22, 1865	49, 535
Hook	R. V. Laney	Cumberland, Md	July 19, 1870	105, 582
Hook	J. B. Sweetland	Pontiac, Mich	June 23, 1868	79, 159
Hook	J. H. Tracy	Mayville, Mich	Sept. 14, 1869	94, 926
Hook	R. Weiss	Brooklyn, N. Y	Feb. 18, 1873	136, 114
Hook and button, Combined	W. H. Shurtleff	Providence, R. I	June 5, 1866	55, 431
Hook and eye	A. Bennett	Brooklyn, N. Y	Feb. 12, 1867	61, 989
Hook and eye	J. P. Culver	New York, N. Y	Jan. 24, 1865	45, 979
Hook and eye	M. R. L. Davis	Kansas City, Mo	Feb. 13, 1872	123, 619
Hook and eye	J. H. Fairchild	Jericho, Vt	May 30, 1854	10, 984
Hook and eye	J. C. Howells	New York, N. Y	Dec. 13, 1864	45, 411
Hook and eye	W. Law	Birmingham, England	Jan. 28, 1868	73, 817
Hook and eye	J. B. Martindale	Newcastle, Ind	Nov. 27, 1866	60, 026
Hook and eye	J. B. Martindale	Newcastle, Ind	Mar. 5, 1867	62, 554
Hook and eye	A. C. Mason	Springfield, Vt	Apr. 9, 1861	32, 031
Hook and eye	H. McEroy	Birmingham, England	Apr. 17, 1849	6, 329
Hook and eye	H. Nickolds	Providence, R. I	Oct. 25, 1864	44, 815
Hook and eye	E. P. Roche	Bath, Me	Oct. 19, 1869	95, 936
Hook and eye	E. C. Savage	Hartford, Conn	Mar. 26, 1844	3, 512
Hook and eye	D. M. Smith	Springfield, Vt	May 22, 1860	28, 443
Hook and eye	I. Weinberg	Philadelphia, Pa	Dec. 18, 1866	60, 600
Hook and eye card	M. Fowler	Northport, Conn	Oct. 29, 1867	70, 190
Hook and eye for connecting cords	A. Codding, jr	North Attleborough, Mass	July 7, 1863	39, 121
Hook and eye for fastening garments	C. Atwood	Derby, Conn	Feb. 24, 1843	2, 978
Hook and eye making machine	J. S. Ford	San Francisco, Cal	Mar. 7, 1871	112, 331
Hook and eye making machine	J. Stewart	Boston, Mass	July 20, 1831	....
Hook and eye, Wire	C. Atwood	Derby, Conn	July 1, 1851	8, 198

Index of patents issued from the United States Patent Office from 1790 *to* 1873, *inclusive*—Continued.

Invention.	Inventor.	Residence.	Date.	No.
Hook and hasp, Combined	B. H. Brooks	New Haven, Conn	Aug. 12, 1873	141, 691
Hook and ladder apparatus	J. S. Hunt	Richmond, Ind	July 9, 1872	128, 882
Hook and terret	W. R. Ferguson	Marseilles, Ill	Aug. 20, 1867	67, 861
Hook bending and punching	P. L. Weimer	Lebanon, Pa	Oct. 31, 1865	50, 757
Hook-bending machine	W. Malick	Erie, Pa	Sept. 10, 1872	131, 289
Hook-bending machine	R. B. Sears	Providence, R. I	Dec. 24, 1867	72, 547
Hook-eye for wearing-apparel and other purposes	J. C. Howells	Washington, D. C	May 12, 1863	38, 483
Hook-pin for fastening wearing-apparel	A. Lindsay and M. Moses	Malone, N. Y	Oct. 23, 1866	59, 040
Hooks and eyes, Machine for separating	M. Fowler	Northford, Conn	Nov. 29, 1864	45, 238
Hooks and eyes to cards, Attaching	C. Atwood	Birmingham, Conn	Sept. 25, 1849	6, 745
Hooks and eyes to cards, Attaching	C. Atwood	Birmingham, Conn	Dec. 20, 1853	10, 330
Hooks and eyes to cards, Attaching	T. B. De Forest	Birmingham, England	Sept. 15, 1863	39, 893
Hooks and eyes to cards, Machine for attaching	A. Capron and J. S. Dennis	Attleborough and Somerville, Mass.	Aug. 7, 1855	13, 409
Hooks and eyes to cards, Securing	J. D. Franklin	Attleborough, Mass	Feb. 4, 1868	74, 072
Hooks and eyes to paper cards, Attaching	P. Kirkham	Waterbury, Conn	July 30, 1850	7, 526
Hooks and eyes to paper cards, Fastening	C. J. Darrington	Waterbury, Conn	Sept. 9, 1851	8, 354
Hooks and eyes to tapes and dresses, Securing	C. Atwood	Derby, Conn	Aug. 7, 1849	6, 628
Hooks and eyes upon cards, Fastening	E. J. Warner	Waterbury, Conn	Nov. 5, 1850	7, 762
Hooks and staples, Machinery for bending	S. Brooks	Chester, Conn	Nov. 19, 1861	33, 733
Hooks, Elastic mousing for	E. E. Stone	United States Navy	Aug. 8, 1865	49, 316
Hooks, staples, &c., Machine for pointing	W. Malick	Tidioute, Pa	Dec. 1, 1868	84, 562
Hooker-up, Mechanical	D. J. Happersett	Downingtown, Pa	July 8, 1851	8, 204
Hoop:				
See Automatic hoop. Bale hoop. Barrel-hoop. Bucket-hoop. Cheese-hoop. Dowel-hoop. Driving-hoop. Flanged and beaded hoop. Grace-hoop. Guide-rolling hoop. Iron hoop. Jumping-hoop. Mast-hoop. Pail-hoop. Propelling-hoop. Skirt-hoop. Steam coiled hoop. Toy-hoop. Trundling-hoop. Truss-hoop.				
Hoop and barrel dressing machine	Kimbell, Perry, and Spalding	Peterborough, N. H	Sept. 18, 1835	....
Hoop and lath sawing machine	I. Price, jr	Lockport, N. Y	Aug. 19, 1828	....
Hoop and train skirt	L. Guelle	Hoboken, N. J	Mar. 4, 1873	136, 434
Hoop-bending machine	E. Coapman	Rochester, N. Y	Dec. 9, 1873	145, 398
Hoop-bending machine	G. E. Smith	Middleport, N. Y	Oct. 12, 1869	95, 845
Hoop-bolts, Machine for pointing and checking	A. A. Wilder	Detroit, Mich	Mar. 31, 1863	38, 085
Hoop-box, Wooden	S. S. Barrie	Green Point, N. Y	Nov. 22, 1870	109, 373
Hoop cutting and bending	J. Dobbins	Litchfield, Mich	July 4, 1865	48, 535
Hoop cutting and bending machine	J. Dobbins	Litchfield, Mich	Feb. 4, 1873	135, 413
Hoop cutting and dressing machine	F. Ellis	Sylvania, Ohio	Dec. 28, 1869	98, 240
Hoop-cutting machine	W. H. Davis	Lexington, Ind	Aug. 23, 1870	106, 671
Hoop-cutting machine	A. G. Parkhurst	Appleton, Wis	Apr. 1, 1873	137, 384
Hoop-cutting machine	P. Welch	Saint Louis, Mo	July 29, 1873	141, 246
Hoop-dressing machine	S. L. Heywood	Minneapolis, Minn	May 27, 1873	139, 387
Hoop-dressing machine	J. Penney	Rochester, N. Y	Nov. 1, 1870	108, 820
Hoop-dressing machine	A. Prenatt	Buffalo, N. Y	Jan. 11, 1859	22, 581
Hoop-driving and barrel-crozing machine	E. Holmes	Buffalo, N. Y	Feb. 17, 1863	37, 719
Hoop-irons, Machinery for rolling	B. Norton	Boonton, N. J	Sept. 26, 1846	4, 772
Hoop-lock	G. N. Beard	Saint Louis, Mo	Oct. 16, 1866	58, 760
Hoop-lock	C. Hughes	New Orleans, La	Feb. 5, 1861	31, 319
Hoop-lock	D. Hughes	Rochester, N. Y	Dec. 13, 1859	26, 435
Hoop-lock	E. A. Jeffery	Corning, N. Y	Dec. 21, 1858	22, 369
Hoop-lock	T. E. Lucas	Chesterfield, S. C	Oct. 28, 1873	144, 117
Hoop-lock	A. P. Morrill, jr	Natchez, Miss	Jan. 3, 1860	26, 694
Hoop-lock clasp	T. Hanvey	Elma, N. Y	July 12, 1864	43, 495
Hoop-lock cutter	J. G. Logginsand P. P. Wilkins	Williston, Vt	Dec. 30, 1873	145, 956
Hoop-lock cutter	W. and H. A. Tripp	Williamson, N. Y	Oct. 1, 1872	131, 835
Hoop-lock-cutting machine	T. Conklin	Fond du Lac, Wis	Apr. 5, 1870	101, 436
Hoop-lock-cutting machine	A. Cutter	Boston, Mass	Oct. 11, 1870	108, 114
Hoop-lock-cutting machine	J. McGren	Ravenswood, W. Va	June 3, 1873	139, 594
Hoop-lock for metal bale-hoops	C. A. Dubs	Natchez, Miss	May 8, 1860	28, 164
Hoop-machine	J. J. Alvord	Tecumseh, Wis	Apr. 30, 1867	64, 270
Hoop-machine	S. F. Atherton	Fitchburgh, Mass	Nov. 13, 1860	30, 609
Hoop-machine	J. B. Dougherty	Rochester, N. Y	Nov. 17, 1863	40, 613
Hoop-machine	G. W. Holmes	Buckfield, Me	Apr. 22, 1856	14, 753
Hoop-machine	H. C. Peirson	Philadelphia, Pa	Mar. 29, 1859	23, 389
Hoop-machine	J. and S. Sawyer	Fitchburgh, Mass	May 6, 1856	14, 833
Hoop-machine	J. and S. Sawyer	Fitchburgh, Mass	Sept. 23, 1856	15, 780
Hoop-machine	J. Thompson	Rochester, N. Y	Nov. 17, 1863	40, 664
Hoop-machine	W. B. Wood	Fitchburgh, Mass	Oct. 7, 1856	15, 865
Hoop-making	M. Granger	Utica, N. Y	Dec. 31, 1833	....
Hoop-making machine	W. Lawyer	Macomb, N. Y	Nov. 2, 1869	96, 448
Hoop-making machine	A. A. Wilder	Detroit, Mich	Mar. 31, 1863	38, 084
Hoop manufacturing machine	J. B. Dougherty	Rochester, N. Y	Aug. 9, 1870	106, 260
Hoop-manufacturing machine	J. Peirson	Alexandria, Va	Feb. 20, 1855	12, 429
Hoop-notching machine	D. Lamson	East Weymouth, Mass	Sept. 23, 1856	15, 768
Hoop notching and trimming machine	S. Littlefield	West Troy, N. Y	Sept. 14, 1858	21, 507
Hoop-planing machine	S. Sawyer	Fitchburgh, Mass	Jan. 27, 1857	16, 513
Hoop-planing machine	T. S. Scoville	Elmira, N. Y	Apr. 28, 1857	17, 171
Hoop-poles, Machine for splitting	J. Penney	Grand Rapids, Mich	May 21, 1872	126, 983
Hoop-poles, Machine for splitting	J. Penney	Grand Rapids, Mich	July 2, 1872	128, 561
Hoop-poles, Machine for splitting	J. and S. Sawyer	Fitchburgh, Mass	Apr. 7, 1857	17, 014
Hoop-racking machine	R. M. Shaner	Genoa, Ohio	Dec. 20, 1870	110, 295
Hoop-riving machine	G. J. Bentley	Michigan City, Ind	Feb. 21, 1865	46, 439
Hoop-sawing machine	N. Bratt	Lockport, N. Y	July 31, 1828	....
Hoop-sawing machine	A. Lutz	Orangeville, Ill	Jan. 21, 1868	73, 615
Hoop, &c., sawing machine	P. Slayton	Lockport, N. Y	May 1, 1828	....
Hoop-sawing machine	E. Strange and T. B. Smith	Taunton, Mass	Sept. 4, 1855	13, 531
Hoop-sawing machine	E. C. Strange	Taunton, Mass	Jan. 13, 1857	16, 407
Hoop sawing machine, Cooper's	J. O. Woodward	Taunton, Mass	July 22, 1856	15, 399
Hoop-shaving machine	J. A. Douglass and J. Wagner	Chest Springs, Pa	Feb. 11, 1873	135, 637
Hoop-shaving machine	G. Heatwoal	Harrisburgh, Pa	July 27, 1831	....
Hoop-shaving machine	S. Parker	Altoona, Pa	Sept. 20, 1870	107, 622

Index of patents issued from the United States Patent Office from 1790 *to* 1873, *inclusive*—Continued.

Invention.	Inventor.	Residence.	Date.	No.
Hoop-skirt	E. Adams, jr	Attleborough, Mass	June 18, 1867	65, 854
Hoop-skirt	H. B. Ames	Brooklyn, N. Y	Jan. 21, 1862	34, 225
Hoop-skirt	H. B. Aurel	Brooklyn, N. Y	Mar. 1, 1864	41, 809
Hoop-skirt	J. E. Atwood	New York, N. Y	July 24, 1866	56, 512
Hoop-skirt	A. M. Bardwell	Amherst, Mass	May 22, 1866	54, 998
Hoop-skirt	M. Berliner	New York, N. Y	Oct. 1, 1872	131, 840
Hoop-skirt	G. Brering	New York, N. Y	Dec. 7, 1869	97, 750
Hoop-skirt	M. P. Bray	Birmingham, Conn	Mar. 14, 1871	112, 681
Hoop-skirt	F. A. Brewster	Springfield, Mass	July 30, 1867	67, 158
Hoop-skirt	O. R. Burnham	New York, N. Y	Mar. 25, 1862	34, 737
Hoop-skirt	J. P. Buzzell	Clinton, Mass	Feb. 13, 1866	52, 637
Hoop-skirt	C. E. Carpenter	New York, N. Y	June 27, 1871	116, 411
Hoop-skirt	A. Carter	New York, N. Y	Apr. 18, 1871	113, 854
Hoop-skirt	J. F. Chase	Augusta, Me	June 15, 1869	91, 418
Hoop-skirt	W. Coe	New Haven, Conn	Dec. 23, 1873	145, 845
Hoop-skirt	S. Collins	New Haven, Conn	Dec. 20, 1864	45, 557
Hoop-skirt	H. Cook	Manchester, England	Oct. 22, 1861	33, 517
Hoop-skirt	C. Daniels	Birmingham, Conn	Apr. 2, 1867	63, 366
Hoop-skirt	T. D. Day	New York, N. Y	Dec. 13, 1864	45, 392
Hoop-skirt	T. D. Day	New York, N. Y	July 3, 1866	56, 016
Hoop-skirt	T. D. Day	New York, N. Y	Feb. 16, 1869	86, 910
Hoop-skirt	T. B. De Forest	Birmingham, Conn	Jan. 15, 1867	61, 169
Hoop-skirt	T. B. De Forest	Birmingham, Conn	Jan. 15, 1867	61, 170
Hoop-skirt	T. B. De Forest	Birmingham, Conn	Mar. 26, 1867	63, 145
Hoop-skirt	D. H. Fanning	Worcester, Mass	Mar. 31, 1868	76, 064
Hoop-skirt	A. Fellheimer	New York, N. Y	Mar. 26, 1867	63, 233
Hoop-skirt	L. Fellheimer	New York, N. Y	Mar. 26, 1867	63, 234
Hoop-skirt	M. Fishel	New York, N. Y	June 5, 1866	55, 266
Hoop-skirt	E. Fleischer	Cincinnati, Ohio	July 20, 1869	92, 811
Hoop-skirt	L. H. Foy	Worcester, Mass	Aug. 21, 1866	57, 309
Hoop-skirt	H. Freudenburg	New York, N. Y	July 25, 1865	48, 924
Hoop-skirt	T. S. Gilbert	Derby, Conn	Jan. 5, 1864	41, 148
Hoop-skirt	T. S. Gilbert	Derby, Conn	Jan. 5, 1864	41, 149
Hoop-skirt	T. S. Gilbert	New Haven, Conn	June 19, 1866	55, 768
Hoop-skirt	D. Hawkins	Birmingham, Conn	Aug. 4, 1863	39, 396
Hoop-skirt	C. F. Hendee	Waterbury, Conn	Feb. 25, 1862	34, 496
Hoop-skirt	H. A. Horn	New York, N. Y	Jan. 26, 1869	86, 340
Hoop-skirt	W. E. Houston	Birmingham, Conn	May 21, 1867	64, 870
Hoop-skirt	T. Hull	Birmingham, Conn	Nov. 26, 1867	71, 487
Hoop-skirt	J. Levy	Philadelphia, Pa	May 16, 1865	47, 738
Hoop-skirt	L. Lobenstein	New York, N. Y	Feb. 6, 1866	52, 427
Hoop-skirt	H. Loewenberg	New York, N. Y	Nov. 25, 1862	37, 003
Hoop-skirt	G. Mallory	Watertown, Conn	Oct. 19, 1858	21, 839
Hoop-skirt	J. Mann and A. McDonald	Waltham and Cambridge, Mass.	June 9, 1863	38, 832
Hoop-skirt	J. Mayer	Brooklyn, N. Y	June 1, 1869	90, 673
Hoop-skirt	J. McKeever	New York, N. Y	Nov. 19, 1867	71, 033
Hoop-skirt	K. McRae	New York, N. Y	Oct. 5, 1869	95, 625
Hoop-skirt	I. T. Meyer and J. F. J. Gunning	New York, N. Y	Nov. 10, 1868	83, 986
Hoop-skirt	S. A. Moody	New York, N. Y	May 10, 1864	42, 677
Hoop-skirt	E. L. Morris	Boston, Mass	Aug. 14, 1866	57, 169
Hoop skirt	C. Neumann	New York, N. Y	Oct. 20, 1863	40, 355
Hoop-skirt	C. Neumann	New York, N. Y	July 31, 1866	56, 787
Hoop-skirt	C. Neumann	New York, N. Y	Jan. 8, 1867	61, 088
Hoop-skirt	F. S. Otis	Brooklyn, N. Y	Nov. 15, 1864	45, 070
Hoop-skirt	C. E. Pratt	Rahway, N. J	July 5, 1870	105, 124
Hoop-skirt	W. S. Ryerson	Philadelphia, Pa	Aug. 27, 1867	68, 238
Hoop-skirt	L. Sanders	New York, N. Y	Oct. 20, 1863	40, 363
Hoop-skirt	L. Sanders	New York, N. Y	May 10, 1864	42, 736
Hoop-skirt	L. Sanders	New York, N. Y	Aug. 2, 1864	43, 710
Hoop-skirt	J. Schleisinger	New York, N. Y	July 31, 1866	56, 803
Hoop-skirt	L. S. Scofield	Belmont, Mass	Oct. 13, 1863	40, 285
Hoop-skirt	P. E. Sheffield	Pontiac, Mich	May 15, 1866	54, 785
Hoop-skirt	S. S. Sherwood	Acquackamack, N. J	Dec. 15, 1863	40, 985
Hoop-skirt	W. D. Sloan	New York, N. Y	Oct. 14, 1862	36, 677
Hoop-skirt	T. S. Sperry	Chicago, Ill	June 27, 1871	116, 507
Hoop-skirt	A. R. Stanley	Boston, Mass	Mar. 31, 1868	76, 291
Hoop-skirt	W. H. Towers	Boston, Mass	Nov. 17, 1868	84, 149
Hoop-skirt	A. Trager	New York, N. Y	June 18, 1867	65, 986
Hoop-skirt	H. Vincent	New York, N. Y	Jan. 1, 1867	60, 969
Hoop-skirt	E. C. Walker	Newark, N. J	Feb. 26, 1867	62, 381
Hoop-skirt	J. Waterman	New York, N. Y	Oct. 25, 1864	44, 841
Hoop-skirt	J. Waterman	New York, N. Y	Mar. 22, 1870	101, 195
Hoop-skirt	J. O. West	New York, N. Y	Feb. 21, 1871	111, 998
Hoop-skirt	J. Whitehead and J. McKeever	New York, N. Y	Aug. 10, 1869	93, 573
Hoop-skirt	S. R. Wilmot	Brooklyn, N. Y	June 11, 1861	32, 547
Hoop-skirt and bustle combined	A. R. Young	Boston, Mass	Aug. 18, 1868	81, 319
Hoop-skirt bands, Buckle or clasp for	H. P. Brooks	Waterbury, Conn	Aug. 10, 1869	93, 592
Hoop-skirt clasp	A. J. Cross	Greenport, N. Y	June 23, 1868	79, 210
Hoop-skirt clasp	T. B. De Forest	Birmingham, Conn	Oct. 2, 1866	58, 540
Hoop-skirt clasp	A. Delkescamp	Brooklyn, N. Y	Apr. 16, 1867	63, 867
Hoop-skirt clasp	G. P. Evans	Malden, Mass	Dec. 11, 1860	30, 872
Hoop-skirt clasp	L. Fellheimer	New York, N. Y	July 28, 1868	80, 343
Hoop-skirt clasp	J. F. J. Gunning	New York, N. Y	Feb. 11, 1868	74, 349
Hoop-skirt clasp	W. E. Houston	Birmingham, Conn	May 21, 1867	64, 871
Hoop-skirt clasp	J. Ingraham	New York, N. Y	Dec. 1, 1868	84, 492
Hoop-skirt clasp	M. Samuels	New York, N. Y	Dec. 29, 1868	85, 480
Hoop-skirt clasp	A. Smart	New York, N. Y	Aug. 31, 1858	21, 373
Hoop-skirt clasp	S. R. Wilmot	Bridgeport, Conn	Oct. 2, 1866	58, 527
Hoop-skirt clasp	S. R. Wilmot	Bridgeport, Conn	Sept. 10, 1867	68, 677
Hoop-skirt clasps, Forceps for fastening	G. D., S. A., and C. L. Russell	Birmingham, Conn	Dec. 14, 1858	22, 308
Hoop-skirt clasps, Machine for making	A. Delkescamp	Brooklyn, N. Y	June 11, 1867	65, 652
Hoop-skirt clasps, Machinery for forming	F. J. Terrell	Ansonia, Conn	May 3, 1864	42, 608
Hoop-skirt clasps, Making	J. H. Doolittle	Ansonia, Conn	Nov. 15, 1859	26, 144
Hoop-skirt clasps, Making	J. H. Doolittle	Ansonia, Conn	Sept. 26, 1865	50, 102
Hoop-skirt fastening	F. E. Day	New York, N. Y	Sept. 28, 1869	95, 203
Hoop-skirt fastening	A. W. Hale	New York, N. Y	Aug. 30, 1859	25, 260
Hoop-skirt former	A. Raffel	New York, N. Y	Feb. 13, 1866	52, 602

Index of patents issued from the United States Patent Office from 1790 to 1873, inclusive—Continued.

Invention.	Inventor.	Residence.	Date.	No.
Hoop-skirt former	A. R. Stanley	Boston, Mass	Mar. 31, 1868	76, 263
Hoop-skirt former	C. C. Wilson	Baltimore, Md	May 3, 1870	102, 638
Hoop-skirt joint	S. J. Sherman	Brooklyn, N. Y	Aug. 15, 1865	49, 447
Hoop-skirt, Lady's	S. Beberdy	Philadelphia, Pa	Sept. 14, 1858	21, 479
Hoop-skirt, Lady's	J. Holmes	Boston, Mass	May 3, 1859	23, 841
Hoop-skirt, Lady's	F. Hull	Derby, Conn	Feb. 8, 1859	22, 875
Hoop-skirt, Lady's	S. Peberdy	Philadelphia, Pa	Nov. 30, 1858	22, 197
Hoop-skirt-making machine	E. R. Hopkins	New York, N. Y	Nov. 27, 1866	60, 010
Hoop-skirt-making machine	C. Neumann	New York, N. Y	June 26, 1860	28, 893
Hoop-skirt-making machine	S. H. Perkins and T. S. Gilbert.	New Haven, Conn	Feb. 26, 1867	62, 359
Hoop-skirt-making machine	S. H. Perkins and T. S. Gilbert.	New Haven, Conn	Apr. 16, 1867	63, 936
Hoop-skirt, Skeleton	R. J. Mann	Brooklyn, N. Y	Nov. 9, 1858	22, 051
Hoop-skirt, Skeleton	C. Neumann	New York, N. Y	Nov. 1, 1859	25, 976
Hoop-skirt, Spiral	D. G. Rollin	New York, N. Y	Dec. 4, 1860	30, 838
Hoop-skirt spring	T. B. DeForest and T. S. Gilbert	Birmingham, Conn	Feb. 18, 1868	74, 672
Hoop-skirt supporter	E. Blakeslee	Plymouth, Conn	Aug. 14, 1866	57, 077
Hoop-skirt tape	C. C. Carpenter	New York, N. Y	July 22, 1873	141, 112
Hoop-skirt tape	W. E. Houston	Birmingham, Conn	May 21, 1867	64, 872
Hoop-skirt tape	W. E. Houston	Birmingham, Conn	May 21, 1867	64, 873
Hoop-skirt wire	T. B. De Forest	Derby, Conn	Nov. 13, 1866	59, 711
Hoop-skirt wire	J. W. Stiles	New York, N. Y	Dec. 15, 1863	40, 960
Hoop-skirt wires, Attaching	J. E. Earle	New Haven, Conn	June 5, 1866	55, 258
Hoop-skirt wires, Coating	W. H. Towers	Boston, Mass	July 6, 1869	92, 403
Hoop-skirts, Adjustable frame for forming	H. S. Loper	New Haven, Conn	Oct. 16, 1866	58, 949
Hoop-skirts, Apparatus for clasping	C. L. Olmstead	Brooklyn, N. Y	Oct. 31, 1865	50, 728
Hoop-skirts, Apparatus for clinching clasp on	G. F. Wright	Clinton, Mass	Sept. 12, 1865	49, 957
Hoop-skirts, Clasp for bottom hoop of	L. W. McFarland	Brooklyn, N. Y	July 3, 1866	56, 146
Hoop-skirts, Covered clasp for	F. E. Day	New York, N. Y	July 27, 1869	93, 068
Hoop-skirts, Frame for forming	J. F. J. Gunning	New York, N. Y	June 9, 1863	38, 821
Hoop-skirts, Punch for making slides for	H. Scheuerle	New York, N. Y	Jan. 1, 1861	31, 038
Hoop-splitting machine	S. F. Atherton	Fitchburgh, Mass	May 1, 1860	28, 048
Hoop-splitting machine	J. W. Chittenden and W. C. Mead.	Vevay, Ind	Feb. 28, 1854	10, 563
Hoop-splitting machine	W. Rose	Philadelphia, Pa	Sept. 3, 1844	3, 727
Hoops from edges of boards, Machine for cutting	S. C. Blinn, J. J. Alvord, and H. Brewer.	Tecumseh, Mich	Mar. 10, 1868	75, 352
Hoops, Machine for cutting and distancing locks on	A. H. Crozier	Oswego, N. Y	Feb. 21, 1860	27, 206
Hoops, Machine for cutting the ends of	L. N. Hewes	Swanzey, N. H	Dec. 27, 1870	110, 462
Hoops on casks, Machine for driving	H. C. Sherman	Buffalo, N. Y	July 18, 1865	48, 843
Hoops out of split timber, Machine for shaving	D. Shriver	Petersburgh, Pa	Mar. 17, 1810	....
Hoops, Tool for riving	W. Baker	East Templeton, Mass	Aug. 2, 1859	24, 917
Hooped skirt	C. H. De Forest	Birmingham, Conn	Jan. 6, 1863	37, 336
Hooped skirt	C. S. Hutchinson	Burlington, N. J	Nov. 18, 1862	36, 979
Hooped skirt	S. S. Sherwood	Acquackanonck, N. J	Jan. 6, 1863	37, 374
Hooped skirt, Lady's	J. Holmes	Boston, Mass	Dec. 28, 1858	22, 426
Hooped skirts, Apparatus for attaching clasps to	T. B. De Forest	Birmingham, Conn	Oct. 28, 1862	36, 877
Hooped skirts, Machine for making	C. Neumann	New York, N. Y	Aug. 16, 1859	25, 163
Hop and hay press	J. W. Shipman	Springfield Centre, N. Y	Nov. 26, 1861	33, 798
Hop and hay press	J. Wilson	Otsego, N. Y	Feb. 17, 1826	....
Hop-box	W. R. Crandall	Deansville, N. Y	June 2, 1868	78, 433
Hop-curing apparatus	N. E. Hinds	Cooperstown, N. Y	Nov. 19, 1872	133, 152
Hop-curing apparatus	J. S. Sandt	Saint Joseph, Mo	June 2, 1868	78, 485
Hop-curing process	C. C. Mason	Brookfield, N. Y	Nov. 27, 1866	60, 027
Hop-dryer	W. Loofbourow	Fayette, Wis	Mar. 30, 1869	88, 395
Hop-dryer	W. F. Waterhouse	Weyauwega, Wis	Aug. 4, 1868	80, 578
Hop-dryer	J. Whitney	Fort Winnebago, Wis	Mar. 10, 1868	75, 503
Hop-frame	T. D. Aylsworth	Frankfort, N. Y	Feb. 13, 1855	12, 375
Hop-frame	T. D. Aylsworth	Illion, N. Y	Mar. 29, 1859	23, 338
Hop-frame	L. A. Beardsley	South Edmeston, N. Y	Jan. 10, 1860	26, 743
Hop-frame	L. A. Beardsley	South Edmeston, N. Y	Sept. 3, 1861	33, 177
Hop-frame	L. A. Beardsley	South Edmeston, N. Y	May 20, 1862	35, 291
Hop-frame	J. C. and D. P. Leonard	Union City, Mich	May 14, 1867	64, 679
Hop-frame	L. S. Mason	Middlefield Centre, N. Y	Sept. 6, 1864	44, 104
Hop-frame	J. B. Van Dewerker	Cobleskill, N. Y	Mar. 19, 1867	63, 119
Hop-hook	E. and E. C. Denio	Baldwinsville and New Hartford, N. Y.	July 28, 1868	80, 277
Hop-liquor for distillers and brewers, Preparing	A. S. Rollins	Albany, N. Y	Mar. 15, 1859	23, 266
Hop-picker	H. Farncrook, F. J. Shepperd, and A. Garton.	Watertown, Wis	June 30, 1868	79, 336
Hop-picker	M. Mases	Malone, N. Y	Apr. 5, 1870	101, 494
Hop-picking tool	J. Dean	Baraboo, Wis	Mar. 10, 1868	75, 385
Hop-pole	L. B. Clark	Bainbridge, N. Y	July 7, 1868	79, 636
Hop-pole	W. Conner	Rensselaerville, N. Y	June 9, 1868	78, 787
Hop-pole	J. G. Wilber	Kilburn City, Wis	July 21, 1868	80, 104
Hop-pole	D. F. Wilcox	Greenville, N. Y	May 19, 1868	78, 037
Hop-pole cleaner	H. Forncrook	Watertown, Wis	July 28, 1868	80, 468
Hop-pole puller	O. B. Hale	Malone, N. Y	Mar. 16, 1869	87, 843
Hop-pole puller	A. L. and W. A. Hatch	Loyd, Wis	Jan. 5, 1869	85, 587
Hop-pole puller	I. W. Legg	Long Eddy, N. Y	May 19, 1868	78, 103
Hop-pole puller	J. R. Woodworth	Munda, N. Y	Jan. 21, 1878	73, 482
Hop-pole, Pulling-machine for	W. Smith	Munda, N. Y	July 30, 1867	67, 365
Hop-pole sharpener	S. V. Barns	Triangle, N. Y	Aug. 3, 1869	93, 269
Hop-pole-sharpening machine	T. S. Angel	Watertown, N. Y	July 21, 1868	80, 050
Hop-pole-sharpening machine	A. H. West	Hamilton, N. Y	July 19, 1864	43, 614
Hop-poles, Jack for pulling	O. S. Foster	Durhamville, N. Y	Dec. 17, 1867	72, 277
Hop-poles, Machine for dressing	C. D. Brown	Bainbridge, N. Y	Nov. 10, 1868	83, 918
Hop-press	J. Brockway	Richland Centre, Wis	Aug. 9, 1870	106, 252
Hop-press	N. Carpenter and J. Hutchinson.	White Creek, Wis	July 16, 1867	66, 793
Hop-press	M. Coryell	New York, N. Y	Jan. 26, 1869	86, 215
Hop-press	J. Hutchinson	Fond du Lac, Wis	Mar. 3, 1868	75, 166
Hop-press	F. Momburg	Burlington, Wis	June 18, 1872	128, 059
Hop-press	H. Taylor	Middletown, Wis	June 30, 1868	79, 413
Hop-press power	N. Carpenter	White Creek, Wis	Jan. 21, 1868	73, 500
Hop-stripper	S. Holt	Baraboo, Wis	June 23, 1868	79, 072
Hop-trellis	L. D. Snook	Barrington, N. Y	May 12, 1868	77, 775
Hop-vine frame	A. Shoemaker and W. Phelps	Conesville, N. Y	Mar. 10, 1868	75, 474
Hop-vine support	C. T. Bush	Rensselaerville, N. Y	Nov. 19, 1867	71, 128

Index of patents issued from the United States Patent Office from 1790 to 1873, inclusive—Continued.

Invention.	Inventor.	Residence.	Date.	No.
Hop-vine support	P. J. Fuller	Clarksville, N. Y	July 7, 1868	79, 648
Hop-vine support	N. C. Roberts and E. W. Badger	Burlington and Otsego, N. Y	Aug. 21, 1866	57, 381
Hop-vine support	N. C. Roberts and E. W. Badger	Fly Creek, N. Y	July 2, 1867	66, 393
Hop-vine trellis	Y. W. Smith	Bristol, N. Y	Aug. 25, 1868	81, 426
Hop-vines by charring the stems, Process of preserving roots of.	S. Cummings	Middlefield, N. Y	Aug. 8, 1865	49, 238
Hop-vines, Training	B. C. Rogers	Stockbridge, Vt	May 29, 1866	55, 163
Hops, Apparatus for extracting	A. Hammer	Philadelphia, Pa	Jan. 9, 1855	12, 204
Hops, Apparatus for producing extract of.	J. Schneider	Williamsburgh, N. Y	June 26, 1866	55, 915
Hops, Cultivating	N. Baker	Algansee, Mich	June 23, 1868	79, 097
Hops, Cultivating	E. A. Wightman and W. C. Williams.	Livingstonville, N. Y	Jan. 1, 1867	60, 976
Hops, Extract of	H. Burgess	Royer's Ford, Pa	Sept. 28, 1869	95, 317
Hops, Extract of	C. A. Seely	New York, N. Y	Aug. 22, 1871	118, 396
Hops in brewing, Preserving and using	J. Seeger and J. Boyd	Baltimore, Md	Dec. 12, 1871	121, 902
Hops, Kiln for drying and curing	W. M. Haynie	Sacramento County, Cal	Apr. 30, 1867	64, 220
Hops, Manufacture of extracts of	C. A. Seely	New York, N. Y	May 17, 1870	103, 091
Hops, Poling	A. Ingalls	Hartwick, N. Y	Dec. 31, 1867	72, 858
Hops, Poling	G. J. Olendorf	Middlefield, N. Y	Nov. 24, 1868	84, 299
Hops, Poling	G. J. Olendorf and A. O. Parshall.	Middlefield, N. Y	Jan. 28, 1868	73, 750
Hops, Preserving	S. F. Schoonmaker	New York, N. Y	Mar. 3, 1868	75, 203
Hops, Preserving the aromatic principle of	H. Bartholomay	Rochester, N. Y	Mar. 1, 1870	100, 352
Hops, Preserving the aromatic principle of	E. D. Brainard	Albany, N. Y	July 27, 1869	92, 934
Hops, Process for obtaining a condensed extract of.	S. R. Percy and W. S. Wells	New York, N. Y	Mar. 21, 1865	46, 973
Hops, Training	F. W. Collins	Morris, N. Y	Dec. 1, 1863	40, 792
Hops, Training	W. C. Dunn	Greene County, N. Y	Oct. 30, 1866	59, 197
Hops, Training	L. H. Whitney	Vallejo, Cal	Dec. 4, 1866	60, 306
Hopper-boy rake	A. C. Pry	Keedysville, Md	Dec. 28, 1869	98, 412
Hoppers, Feed-regulating mechanism for	J. S. Bodge	Bath, N. Y	July 11, 1865	48, 646
Hopple	G. W. Hyatt	Auburn, N. Y	July 16, 1867	66, 846
Hopple, Animal	G. L. Brent	Gordonsville, Va	June 6, 1871	115, 691
Hopple for horses	R. N. Eagle	Washington, D. C	Mar. 19, 1867	62, 946
Hopple for horses and other animals	R. N. Eagle	Washington, D. C	Apr. 7, 1863	38, 100
Hopple, Training	W. M. Greenwood	Cincinnati, Ohio	July 23, 1867	67, 042
Horizon for quadrants, Artificial	J. Rahill	Ramsgate, England	Aug. 20, 1861	33, 104
Horizontal engines, Expansion-gear for	S. H. Gilman	Cincinnati, Ohio	Dec. 17, 1850	7, 838
Horizontal engines, Frame for	W. Wright	Newburgh, N. Y	Nov. 18, 1873	144, 818
Horizontal mill	J. Kelsey	Hartford, Conn	Aug. 25, 1809	....
Horizontal mill	J. Reynolds	America, N. Y	Mar. 15, 1827	....
Horizontal revolving jointed screen	P. Broughton	Charleston, S. C	Aug. 5, 1831	....
Horn and shell, Machine for splitting	J. Robins	Boston, Mass	June 24, 1851	8, 174
Horn and tortoise-shell combs, Machine for twining	D. E. Noyes	Philadelphia, Pa	Sept. 6, 1817	....
Horn and tortoise-shell combs, Manufacturing	M. J. Littleboy	Philadelphia, Pa	Nov. 19, 1830	....
Horn comb	M. H. Fairchild	Newtown, Conn	May 25, 1869	90, 516
Horn combs with tortoise-shell, Plating	J. P. Spies	Baltimore, Md	Jan. 8, 1810	....
Horn, Composition resembling	W. M. Welling	New York, N. Y	Jan. 11, 1870	98, 727
Horn for lantern-lights, combs, &c., Splitting and shearing.	J. Pulsifer	Newbury, Mass	May 24, 1814	....
Horn handles to knives, Attaching	M. Bradley	Westport, Conn	Aug. 8, 1865	49, 222
Horn, hard rubber, &c., Composition substitute for	R. Haering	New York, N. Y	Sept. 9, 1862	36, 406
Horn, Machine for shaving	W. Noyes, jr	Newburyport, Mass	July 2, 1867	66, 247
Horn, Machine for splitting	E. Prescott	Leominster, Mass	Oct. 10, 1854	11, 793
Horn, Manufacture of articles from pulverized	W. F. Niles and S. G. Pitts	Leominster, Mass	June 20, 1871	116, 213
Horn, Molding and clarifying	E. F. Coffin	Newburyport, Mass	Apr. 5, 1870	101, 587
Horn, Process of whitening	A. Scheller	New York, N. Y	Apr. 16, 1867	63, 949
Horn-splitting and comb-stamp-twining machine	S. Hills and M. Emery, jr	Newburyport, Mass	Oct. 21, 1814	....
Horn to resemble whalebone, Treating	W. Birch	Walworth, Great Britain	Dec. 2, 1873	145, 045
Horns, hoof, &c., Treatment of	E. P. and D. Baugh	Philadelphia, Pa	July 16, 1872	129, 517
Horometer	A. Abbott	Manchester, N. H	Sept. 11, 1855	13, 560
Horoscope	M. Eble	Ellwangen, Würtemberg	Sept. 8, 1863	39, 860
Horse and cattle tie, Self-loosening	J. J. Eshleman	Lancaster, Pa	Mar. 30, 1858	19, 761
Horse and mule shoe	J. Wonderlin	Louisville, Ky	Nov. 29, 1870	109, 701
Horse and ox shoe	N. E. Hinds	Cooperstown, N. Y	Feb. 22, 1859	23, 027
Horse and ox shoe	P. Sternberg	Danube, N. Y	Apr. 10, 1824	....
Horse and vehicle, Auto-propelling	P. W. Mackenzie	Jersey City, N. J	Jan. 19, 1864	41, 310
Horse and wagon brake, Combined	G. Haberland	Pontiac, Ill	Dec. 24, 1867	72, 484
Horse and water wheel, Inclined	L. Wheeler and T. Powell	Coxsackie, N. Y	Dec. 6, 1822	....
Horse-blanket adjuster	A. Z. Neff	Amsterdam, N. Y	Mar. 12, 1872	124, 506
Horse-blinder	J. Dunlap	Madison Township, Pa	Dec. 22, 1868	85, 075
Horse-blinder	S. Ferris	New York, N. Y	Sept. 24, 1867	69, 199
Horse-blinder	R. Lehmicke	Stillwater, Minn	Apr. 30, 1872	126, 218
Horse-blinder	S. B. Rumery	Plainwell, Mich	Sept. 3, 1872	131, 029
Horse-block and hitching-post	G. W. Preston	Corning, N. Y	Nov. 19, 1867	71, 059
Horse-boat	M. Isaacs	Philadelphia, Pa	Mar. 17, 1818	....
Horse-boats, Machinery for	D. Dunham	New York, N. Y	June 28, 1814	....
Horse-boot	P. Murray and F. Koch	Morrisania, N. Y	May 7, 1872	126, 479
Horse-boot	R. Williams	Philadelphia, Pa	Oct. 10, 1871	119, 906
Horse-bracket	H. B. Davis	Lexington, Mass	June 14, 1859	24, 380
Horse bristle-boot	J. J. Davy	Newark, N. J	Jan. 29, 1867	61, 655
Horse-cage	W. G. Hughes	Memano, Ind	Sept. 6, 1864	44, 095
Horse cant-hook	E. W. Gale	Monroeton, Pa	Oct. 12, 1869	95, 675
Horse-checking apparatus	S. S. Ingalls	Chicago, Ill	Mar. 26, 1872	125, 052
Horse-checking apparatus	J. Jackson	Hartford, Conn	Sept. 30, 1873	143, 244
Horse clipping or shearing instrument	H. Knight	Ryde, Isle of Wight, England.	Oct. 18, 1870	108, 489
Horse-collar fastener	A. Steinbach	Evansville, Ind	July 11, 1865	48, 735
Horse-collar, &c., Lining for	I. Lindsley and J. Mackintosh	Pawtucket, R. I	Oct. 28, 1873	143, 987
Horse controller, Carriage	F. Marlow	Cleveland, Ohio	Feb. 5, 1867	61, 748
Horse-controlling device	N. P. Slade	Franklin Grove, Ill	Jan. 11, 1870	98, 715
Horse-cover	E. D. French	Camden, N. J	Sept. 17, 1872	131, 433
Horse-cover	A. H. Morris	Boston, Mass	Jan. 14, 1873	134, 921
Horse-cover	E. L. Perry	New York, N. Y	Oct. 11, 1864	44, 654
Horse-cover, Ventilated	C. P. Eager	Boston, Mass	Apr. 26, 1870	102, 381
Horse-cover, Ventilated	S. Sibley	Chelsea, Mass	July 18, 1871	117, 118
Horse-covers, Ventilating	C. P. Eager	Boston, Mass	Dec. 14, 1869	97, 896
Horse, Extension	R. Hammill	Mineral Point, Wis	June 11, 1867	65, 666
Horse-fastener	F. Vogeli	Newburgh, N. Y	June 13, 1865	48, 225

Index of patents issued from the United States Patent Office from 1790 to 1873, inclusive—Continued.

Invention.	Inventor.	Residence.	Date.	No.
Horse-fastening	J. Bolton	Richmond, Va	Dec. 23, 1856	16, 312
Horse-fetter	A. P. Mason	Gowanda, N. Y	June 15, 1869	91, 462
Horse-fetter	J. Watson	New Canaan, Conn	Jan. 15, 1818	....
Horse foot-cushion	L. Hale	Milford, Mass	May 29, 1860	28, 473
Horse foot-pad	J. Johnson	Lowell, Mass	Feb. 11, 1873	135, 814
Horse foot-pad	J. Johnson	Lowell, Mass	July 5, 1870	104, 960
Horse-frame for facilitating the shoeing of horses, Swinging.	J. Kirk	Weston, Vt	Nov. 8, 1820	....
Horse-grooming apparatus	R. Beaumont and W. Clark, jr.	Albany, N. Y	Dec. 1, 1868	84, 528
Horse-hair woven garments	P. L. Slayton	New York, N. Y	Nov. 22, 1864	45, 209
Horse head-pad	D. K. Humphrey	Buffalo, N. Y	Mar. 16, 1869	87, 938
Horse head-protector	W. H. Wilson	Cynthiana, Ky	June 11, 1872	127, 820
Horse-holder	W. B. Chapman	La Salle, Ill	Nov. 13, 1866	59, 710
Horse-holder	S. Hague	Utica, N. Y	Aug. 30, 1864	43, 986
Horse-holder	J. P. Reynolds	Mirabile, Mo	Dec. 3, 1867	71, 788
Horse-holder	T. S. W. Russell	Kansas, Ky	July 9, 1872	128, 814
Horse-holder	J. M. Whiting	Providence, R. I	Mar. 26, 1867	63, 344
Horse hoof-armor	H. Splitdorf	Brighton, Mass	Oct. 8, 1867	69, 720
Horse hoof-pad	J. Haseltine	Methuen, Mass	May 19, 1868	77, 979
Horse hoof-pad	J. Johnson	Lowell, Mass	June 22, 1869	91, 541
Horse hoof-pad	W. H. Southworth	Mittineague, Mass	Aug. 11, 1868	81, 025
Horse hoof-pad	J. L. Wetherell	Attleborough, Mass	July 28, 1868	80, 319
Horse interfering-boot	A. D. Westbrook	Astoria, N. Y	July 2, 1872	128, 687
Horse interfering-pad	W. H. Hall	Boston, Mass	Feb. 14, 1871	111, 740
Horse interfering-pad	J. Smith	Boston, Mass	Aug. 2, 1870	105, 990
Horse interfering-pad	A. D. Westbrook	Buffalo, N. Y	Apr. 16, 1867	63, 821
Horse leg-fender	S. Rossman	Hudson, N. Y	June 6, 1865	48, 097
Horse machine, Flying	G. L. Newhall and J. F. Cummings.	Chelmsford, Mass	Aug. 31, 1869	93, 234
Horse-mill	E. Brown	....	Apr. 10, 1810	....
Horse-mill	W. Cornwell	Accomack, Va	Dec. 14, 1816	....
Horse-mill	J. Grannis	Cheshire, Conn	May 12, 1812	....
Horse-mill	A. Warren	Saugerties, N. Y	Apr. 15, 1826	....
Horse-mills, Lever for working	J. Galbraith	Maury County, Tenn	Mar. 6, 1828	....
Horse-nail machine	E. W. Kelly	Hamilton, Scotland	Oct. 10, 1871	119, 854
Horse neck-pad	C. J. Fisher	Waukon, Iowa	May 4, 1869	89, 646
Horse neck-protector	J. P. Myer	Waukesha, Wis	Jan. 8, 1867	61, 016
Horse-nicking apparatus	H. Dodge	Buffalo, N. Y	Mar. 17, 1863	37, 904
Horse-power	J. Abbott	South Reading, Mass	June 30, 1836	....
Horse-power	A. Adams	Augusta, Me	May 6, 1836	....
Horse-power	A. Adams	Sandwich, Ill	July 2, 1861	32, 673
Horse-power	J. M. Albertson	New London, Conn	May 4, 1869	89, 825
Horse-power	J. M. Albertson	New London, Conn	Sept. 12, 1871	118, 893
Horse-power	H. Aldridge and W. Bedford	Goshen, Ind., and Chicago, Ill.	May 19, 1868	77, 945
Horse-power	S. S. Allen	Saratoga Springs, N. Y	June 29, 1833	....
Horse-power	S. S. Ammons	Winona, Miss	Oct. 17, 1871	120, 018
Horse-power	D. Anthony	Sharon, N. Y	Aug. 7, 1847	5, 215
Horse-power	W. E. Arnold	Chatham, Conn	Mar. 13, 1835	....
Horse-power	J. E. Atwood	Willimantic, Conn	Mar. 1, 1870	100, 350
Horse-power	C. Aultman and J. Allonas	Mansfield, Ohio	July 13, 1869	92, 560
Horse-power	C. Avery	Tunkhannock, Pa	June 3, 1851	8, 136
Horse-power	J. F. Barrett	North Granville, N. Y	Oct. 30, 1847	5, 348
Horse-power	S. Basket	Crittenden County, Ark	Oct. 24, 1871	120, 232
Horse-power	B. D. Beecher	Cheshire, Conn	Dec. 23, 1833	....
Horse-power	W. N. Berkeley	Cedar Rapids, Iowa	Oct. 26, 1869	96, 192
Horse-power	J. Bish	Dayton, Ohio	Sept. 10, 1867	68, 691
Horse-power	J. M. Blake	Madison, Wis	Mar. 4, 1862	34, 559
Horse-power	B. Bogue	Trenton, Iowa	Sept. 4, 1860	29, 855
Horse-power	T. J. Bottoms	Thomasville, Ga	Feb. 5, 1861	31, 294
Horse-power	G. M. Branch	Winona, Miss	Mar. 18, 1873	136, 899
Horse-power	L. H. Bratt	South Bend, Ind	Jan. 7, 1873	134, 510
Horse-power	G. Brodie	Jefferson County, Ark	Jan. 30, 1872	123, 229
Horse-power	L. Bronson	Buffalo, N. Y	Jan. 7, 1873	134, 511
Horse-power	N. B. Brown	Antwerp, N. Y	Jan. 23, 1872	122, 989
Horse-power	H. L. and J. A. Buckwalter	Kimberton, Pa	July 17, 1866	56, 362
Horse-power	G. E. Burt	Harvard, Mass	Aug. 13, 1861	33, 028
Horse-power	G. E. Burt	Harvard, Mass	Mar. 17, 1863	37, 892
Horse-power	G. E. Burt and A. and G. F. Wright.	Harvard, Mass	Sept. 22, 1857	18, 232
Horse-power	D. Cary	Clarkson, N. Y	June 27, 1846	4, 595
Horse-power	J. L. Cathcart	Washington, D. C	July 16, 1850	7, 504
Horse-power	R. J. Cheney	Petaluma, Cal	Sept. 26, 1871	119, 221
Horse-power	T. C. Churchman	Sacramento, Cal	Feb. 13, 1872	123, 614
Horse-power	P. W. Clark	Oblong, N. Y	May 3, 1870	102, 491
Horse-power	S. Coin	Cazenovia, N. Y	Sept. 17, 1867	68, 957
Horse-power	J. F. Collins	Lodi, Miss	Aug. 1, 1871	117, 516
Horse-power	M. H. Cornell	Feasterville, Pa	Feb. 10, 1852	8, 717
Horse-power	C. Coster and D. Pennypacker	Upper Providence, Pa	May 23, 1836	....
Horse-power	J. C. Cox	Greenville, N. C	July 30, 1867	67, 273
Horse-power	D. D. Craig	Macon, Ga	June 6, 1871	115, 711
Horse-power	A. D. Crane	Newark, N. J	Apr. 8, 1851	8, 028
Horse-power	A. D. Crane	Newark, N. J	June 22, 1852	9, 049
Horse-power	E. Crockett	Macon, Ga	Apr. 1, 1873	137, 294
Horse-power	P. C. Curtis and L. Yale	Otsego, N. Y	Sept. 17, 1833	....
Horse-power	C. Custer and D. Ponnepacker	Providence Township, Pa	Mar. 30, 1835	....
Horse-power	J. Darling	Cincinnati, Ohio	May 18, 1858	20, 257
Horse-power	S. W. Davis	Plattsburgh, N. Y	Mar. 6, 1866	52, 975
Horse-power	S. W. Davis	Brasher Falls, N. Y	Jan. 19, 1869	86, 061
Horse-power	P. K. Dederick	Albany, N. Y	June 25, 1872	128, 214
Horse-power	W. Deering	Louisville, Ky	Dec. 19, 1871	121, 932
Horse-power	A. J. Detrick	Dryden, N. Y	Feb. 9, 1864	41, 493
Horse-power	J. Diffendal and S. Hughes	Westminster, Md	Oct. 5, 1869	95, 440
Horse-power	J. G. Dillaha	Waco, Tex	May 18, 1869	90, 245
Horse-power	W. W. Dingee	Racine, Wis	Apr. 18, 1871	113, 750
Horse-power	N. S. Dodge	Indianapolis, Ind	Feb. 28, 1860	27, 278
Horse-power	C. L. Drury	Rockingham, Vt	Sept. 21, 1869	95, 008

Index of patents issued from the United States Patent Office from 1790 to 1873, inclusive—Continued.

Invention.	Inventor.	Residence.	Date.	No.
Horse-power	J. A. Eberly, J. Lutz, and H. Becker.	East Cocalico Township, Pa.	Feb. 9, 1869	86, 656
Horse-power	R. S. Eddy	La Crosse, Wis	Apr. 21, 1868	77, 015
Horse power	G. Eichenseer	Waterloo, Ill	Dec. 4, 1866	60, 159
Horse-power	W. P. Emerson	Pleasantville, Ky	May 13, 1873	138, 873
Horse-power	W. Emmons	New York, N. Y	Dec. 24, 1834	....
Horse-power	C. M. Erwin	Winona, Miss	Oct. 3, 1871	119, 587
Horse-power	T. A. E. Evans	Albany, Ga	Apr. 4, 1871	113, 410
Horse-power	A. B. Farquhar	York, Pa	May 7, 1872	126, 386
Horse-power	J. B. Fasset	Irasburgh, Vt	Aug. 17, 1869	93, 813
Horse-power	L. R. Faught	Philadelphia, Pa	Nov. 21, 1871	121, 048
Horse-power	L. R. Faught	Philadelphia, Pa	Feb. 20, 1872	123, 774
Horse-power	L. R. Faught	Philadelphia, Pa	Oct. 28, 1873	143, 974
Horse-power	J. A. Fay	Baltimore, Md	Oct. 10, 1838	978
Horse-power	S. M. Feezler	Seneca Falls, N. Y	July 22, 1862	35, 973
Horse-power	M. Fiske	Sparta, Tenn	Nov. 24, 1868	84, 348
Horse-power	D. Fitzgerald	New York, N. Y	Oct. 19, 1836	....
Horse-power	J. Fitzgerald	New York, N. Y	June 16, 1846	4, 579
Horse-power	J. Fraser and W. Thomas	New York, N. Y	Jan. 18, 1870	98, 947
Horse-power	J. Frezer	Newberry, Pa	Jan. 25, 1859	22, 721
Horse-power	H. H. Fultz	Lexington, Mass	July 3, 1855	13, 167
Horse-power	C. F. Gay	Albany, Oreg	Jan. 5, 1869	85, 578
Horse-power	P. Geiser	Waynesborough, Pa	Sept. 29, 1868	82, 617
Horse-power	C. H. Gifford	Potsdam Junction, N. Y	Feb. 8, 1870	99, 555
Horse-power	C. G. Gilbert	Leeds, Me	July 2, 1836	....
Horse-power	W. Gilfillan	Speier, Minn	July 8, 1873	140, 622
Horse-power	M. Gillam	Troy, Pa	June 1, 1858	20, 421
Horse-power	H. Glass	Racine, Wis	May 9, 1871	114, 549
Horse-power	S. Gleason	Westville, N. Y	Sept. 18, 1834	....
Horse-power	W. A. Glidden and A. Starkweather.	Alvaretta, Wis	May 29, 1860	28, 468
Horse-power	W. F. Goodwin	Metuchen, N. J	Feb. 7, 1871	111, 634
Horse-power	C. S. Graves	Elyria, Ohio	Apr. 10, 1860	27, 795
Horse-power	J. A. Green	North Waterford, Me	June 16, 1868	78, 957
Horse-power	M. G. Groff	Vogansville, Pa	July 21, 1868	80, 164
Horse-power	J. A. Hafner	Commerce, Mo	Oct. 13, 1868	83, 059
Horse-power	A. G. Hagerstrom	Red Wing, Minn	Jan. 28, 1873	135, 331
Horse-power	S. B. Haines	Lancaster, Pa	Jan. 17, 1865	45, 921
Horse-power	W. G. Hulbert	Columbus, Miss	June 11, 1872	127, 763
Horse-power	J. and L. Hale	Hollis, N. H	Dec. 24, 1833	....
Horse-power	A. H. Halton	Tyro, Miss	Apr. 9, 1872	125, 455
Horse-power	G. B. Hamlin	Willimantic, Conn	Aug. 16, 1870	106, 483
Horse-power	T. Harrison	Belleville, Ill	Feb. 12, 1867	61, 030
Horse-power	T. Harrison and W. C. Buchanan	Belleville, Ill	May 9, 1871	114, 555
Horse-power	T. Harrison and W. C. Buchanan	Belleville, Ill	July 16, 1872	129, 552
Horse-power	J. Haw	Hanover County, Va	Mar. 15, 1845	3, 953
Horse-power	J. R. Hedges	Glenwood, N. Y	July 26, 1870	105, 680
Horse-power	D. S. and J. D. Heebner	Norristonville, Pa	Oct. 17, 1871	120, 066
Horse-power	G. Hely	Rochester, Wis	Sept. 14, 1858	21, 495
Horse-power	W. Heston	Bedford, Ohio	Apr. 22, 1862	35, 060
Horse-power	J. Heuermann	Davenport, Iowa	Aug. 10, 1869	93, 438
Horse-power	D. W. Hunt	San Francisco, Cal	May 23, 1865	47, 830
Horse-power	R. Hunt	Freeport, Ill	Mar. 25, 1856	14, 547
Horse-power	J. W. Huntoon	Montgomery, Ala	Sept. 13, 1870	107, 379
Horse-power	J. W. Huntoon	Saint Louis, Mo	Feb. 28, 1871	112, 251
Horse-power	H. M. Irwin	Davidson College, N. C	Apr. 8, 1873	137, 684
Horse-power	T. S. Johnson	Winona, Miss	Oct. 8, 1872	132, 011
Horse-power	I. Keller	Randolph, Ohio	June 30, 1868	79, 476
Horse-power	W. King	Richmond, Mo	June 9, 1868	78, 808
Horse-power	J. H. Kleppinger	Cherryville, Pa	June 23, 1868	79, 074
Horse-power	P. Kline	Johnsville, Ohio	June 22, 1869	91, 546
Horse-power	J. W. Knox	Winona, Miss	Oct. 10, 1871	119, 774
Horse-power	R. W. Krouse	Westminster, Md	Nov. 8, 1870	109, 114
Horse-power	G. Kuenne, J. N. Cole, and D. F. Rath.	Fond du Lac, Wis	Oct. 13, 1863	40, 267
Horse-power	D. L. Lamon	Boston, Ga	Feb. 21, 1871	111, 949
Horse-power	J. S. Lamphier	Dresden, N. Y	Jan. 23, 1872	122, 896
Horse-power	Z. P. Landrum	Columbus, Miss	Feb. 18, 1873	136, 075
Horse-power	Z. P. Landrum	Columbus, Miss	July 29, 1873	141, 363
Horse-power	C. Lane	Hamilton, Ohio	Apr. 26, 1859	23, 809
Horse-power	H. B. Larzelere	Doylestown, Pa	May 21, 1872	127, 072
Horse-power	W. Lanver	Peru Mills, Pa	June 15, 1869	91, 456
Horse-power	J. A. Leibey	Davenport, Iowa	Nov. 26, 1867	71, 522
Horse power	S. Leonard	Hampden, Me	Sept. 23, 1834	....
Horse-power	G. Lewis	Panama, N. Y	Aug. 2, 1864	43, 738
Horse-power	W. J. F. Liddell	Madison, Wis	Mar. 5, 1872	124, 212
Horse-power	J. H. Lighter	Freedom Township, Ill	Jan. 6, 1863	37, 350
Horse-power	F. M. C. Liles	Roanoke, Ala	July 11, 1871	116, 843
Horse-power	B. F. Love and J. H. Frazee	Shelbyville, Ind	May 18, 1858	20, 279
Horse-power	J. E. Lutz	East Cocalico Township, Pa	June 1, 1869	90, 671
Horse-power	D. G. Marden	Memphis, Tenn	Apr. 25, 1871	114, 168
Horse-power	M. H. Marmaduke and B. F. Stewart.	Santa Fé, Mo	Dec. 12, 1871	121, 882
Horse-power	H. J. Marnette	Chicago, Ill	June 10, 1873	139, 682
Horse-power	J. Marshall	New Orleans, La	Sept. 26, 1871	119, 279
Horse-power	D. Marvin	New York, N. Y	Feb. 5, 1836	....
Horse-power	W. McCord	Sing Sing, N. Y	July 11, 1854	11, 258
Horse-power	E. G. McMillan	Berlin, Ohio	Aug. 26, 1873	142, 115
Horse-power	C. L. Merrill	Watertown, N. Y	May 4, 1869	89, 678
Horse-power	D. Michaels and J. H. Croskey	Hopedale, Ohio	May 18, 1869	90, 182
Horse-power	J. Milbourn	Millport, Ohio	June 10, 1873	139, 725
Horse-power	S. B. Minnech	Landsville, Pa	Nov. 24, 1868	84, 434
Horse-power	T. Mitchel	Newburgh, N. Y	July 7, 1835	....
Horse-power	L. B. Morris	Hopefield, Ark	Oct. 15, 1872	132, 313
Horse-power	G. W. Moyers	Gordonsville, Va	Oct. 12, 1869	95, 826
Horse-power	J. Müller	Nacogdoches County, Tex	Feb. 4, 1873	135, 439
Horse-power	J. W. Murrell	El Dorado, Ark	Aug. 9, 1870	106, 194
Horse-power	J. Musten	Franklin County, Vt	Feb. 12, 1836	....

Index of patents issued from the United States Patent Office from 1790 *to* 1873, *inclusive*—Continued.

Invention.	Inventor.	Residence.	Date.	No.
Horse-power	S. Newton	Dayton, Ohio	Jan. 20, 1836	....
Horse-power	J. Nichols and E. D. Brown	Battle Creek, Mich	Mar. 11, 1873	136, 754
Horse-power	F. I. Norton	Lower Sandusky, Ohio	Apr. 25, 1848	5, 530
Horse-power	G. Oerllein	Utica, Minn	July 14, 1868	79, 852
Horse-power	G. Oerllein	Utica, Minn	Aug. 17, 1869	93, 737
Horse-power	G. Oerllein	Utica, Minn	Aug. 17, 1869	93, 738
Horse-power	G. Oerllein	Utica, Minn	Apr. 5, 1870	101, 498
Horse-power	W. R. Palmer	Elizabeth City, N. C	July 18, 1854	11, 328
Horse-power	G. Partridge and D. M. Johnson	Saint Louis, Mo., and Coshocton, Ohio.	Apr. 3, 1866	53, 664
Horse-power	J. N. Pease	Harmony, N. Y	Sept. 16, 1862	36, 477
Horse-power	J. N. Pease	Panama, N. Y	Nov. 3, 1863	40, 499
Horse-power	W. L. Peet	Maple Rapids, Mich	Jan. 1, 1867	60, 933
Horse-power	S. Pelton	New Windsor, Md	Dec. 18, 1855	13, 955
Horse-power	S. Pelton	Marysville, Cal	Feb. 6, 1872	123, 416
Horse-power	W. P. Penn	Belleville, Ill	Feb. 21, 1860	27, 235
Horse-power	T. D. Pennington	Forsyth, Ga	Mar. 28, 1871	113, 088
Horse-power	S. Perry	Newport, N. Y	June 10, 1862	35, 567
Horse-power	S. Perry	Newport, N. Y	July 21, 1863	39, 324
Horse-power	S. Perry	Newport, N. Y	July 21, 1863	39, 325
Horse-power	L. A. Peter	Neffs, Pa	Oct. 11, 1870	108, 178
Horse-power	W. Pierpont	Salem, N. J	Dec. 16, 1862	37, 178
Horse-power	J. A. Pitts	Buffalo, N. Y	July 4, 1854	11, 232
Horse-power	R. Porter	Billerica, Mass	June 11, 1836	....
Horse-power	D. J. Powers and H. B. Stevens	Madison, Wis	May 14, 1867	64, 791
Horse-power	T. B. Pyron	Springfield, Mo	Aug. 16, 1870	106, 505
Horse-power	R. Quinn	Whitefield, Miss	July 26, 1870	105, 724
Horse-power	J. M. Rand	Chicago, Ill	Feb. 8, 1870	99, 591
Horse-power	J. M. Randle	Brighton, Ill	Mar. 20, 1866	53, 378
Horse-power	J. W. Reid	New York, N. Y	Apr. 25, 1865	47, 456
Horse-power	J. R. Remington	Lowndes County, Ala	June 20, 1846	4, 590
Horse-power	E. B. Regna	Jersey City, N. J	Apr. 17, 1860	27, 928
Horse-power	L. Price	Attleborough, Pa	July 17, 1837	276
Horse-power	L. Price and D. Congdon	West Chester, Pa	July 17, 1837	277
Horse-power	D. Richardson	Onowa, Iowa	Dec. 24, 1872	134, 167
Horse-power	M. A. Richardson	Sherman, N. Y	Apr. 14, 1868	76, 816
Horse-power	M. A. Richardson	Sherman, N. Y	July 28, 1868	80, 308
Horse-power	W. Rider	Almont, Mich	July 20, 1858	20, 978
Horse-power	W. Rider	Almont, Mich	July 1, 1862	35, 780
Horse-power	C. Roberts and J. A. Throp	Three Rivers, Mich	July 13, 1869	92, 473
Horse-power	C. Roberts and J. A. Throp	Three Rivers, Mich	July 13, 1869	92, 474
Horse-power	C. Roberts and J. A. Throp	Three Rivers, Mich	July 13, 1869	92, 475
Horse-power	C. Roberts and J. A. Throp	Three Rivers, Mich	Feb. 15, 1870	99, 951
Horse-power	F. W. Robinson	Richmond, Ind	Jan. 10, 1860	26, 786
Horse-power	F. W. Robinson	Richmond, Ind	Mar. 22, 1870	101, 045
Horse-power	J. S. Rowell and W. F. Lowth	Beaver Dam, Wis	Sept. 30, 1862	36, 575
Horse-power	C. Russell	Massillon, Ohio	May 1, 1855	12, 782
Horse-power	D. Russell	Saint Louis, Mo	Aug. 24, 1852	9, 221
Horse-power	E. P. Russell	Manlius, N. Y	May 9, 1865	47, 662
Horse-power	W. J. Sage	Steubenville, Ohio	May 5, 1863	38, 421
Horse-power	G. Sanford	New York, N. Y	Jan. 8, 1861	31, 090
Horse-power	G. Sanford	New York, N. Y	June 13, 1865	48, 212
Horse-power	G. W. Sanor and J. Stoffer	Hanoverston and New Chambersburgh, Ohio.	Feb. 20, 1866	52, 753
Horse-power	J. Schley	Savannah, Ga	July 2, 1867	66, 401
Horse-power	W. Schuyler	Orangeville, Pa	Sept. 17, 1872	131, 413
Horse-power	W. Schuyler	Orangeville, Pa	Dec. 3, 1872	133, 546
Horse-power	C. J. Shuttleworth	Springville, N. Y	July 13, 1869	92, 664
Horse-power	J. Simpson	Atlanta, Ga	Apr. 17, 1855	12, 731
Horse-power	R. Skinner	Williamson, N. Y	Mar. 31, 1836	....
Horse-power	A. Smith	Dayton, Oreg	July 21, 1868	80, 228
Horse-power	H. Smith	Shelby Station, Tenn	Nov. 7, 1871	120, 678
Horse-power	F. M. Sofge	Columbus, Ga	Jan. 18, 1859	22, 680
Horse-power	O. W. Stanford	Cincinnati, Ohio	Apr. 24, 1860	28, 018
Horse-power	I. Starr	Wooster, Ohio	Aug. 22, 1871	118, 292
Horse-power	N. Starr, jr	Homer, N. Y	May 1, 1866	54, 433
Horse-power	J. G. Stephenson	Bucyrus, Ohio	Sept. 14, 1869	94, 921
Horse-power	H. B. Stevens	Buffalo, N. Y	May 16, 1871	114, 874
Horse-power	W. S. Stone	Pitt's Point, Ky	Oct. 8, 1872	132, 033
Horse-power	S. Stone	Canton, Ohio	July 12, 1870	105, 276
Horse-power	O. O. Storle	Milwaukee, Wis	May 21, 1872	126, 912
Horse-power	I. Straub	Lewiston, Pa	May 14, 1836	....
Horse-power	J. B. Streeter	Middlesex, N. Y	Aug. 26, 1834	....
Horse-power	J. B. Sweetland	Pontiac, Mich	July 4, 1865	48, 600
Horse-power	J. B. Sweetland	Pontiac, Mich	Sept. 11, 1866	57, 995
Horse-power	J. S. Tadlock	Belmont, Tex	Oct. 28, 1873	144, 040
Horse-power	H. Tarpley	Wesley, Ky	Mar. 17, 1868	75, 597
Horse-power	R. B. Tatum	Helena, Ark	Feb. 13, 1872	123, 745
Horse-power	L. P. Teed	Mechanicsburgh, Pa	Dec. 6, 1870	109, 967
Horse-power	D. J. Tittle	Albany, N. Y	June 23, 1868	79, 282
Horse-power	P. A. Tobey	Caton, N. Y	Mar. 15, 1870	100, 818
Horse-power	F. Tobias	Covington, Ohio	Oct. 15, 1867	69, 865
Horse-power	A. W. Tooker	Chemung, Ill	Dec. 20, 1864	45, 540
Horse-power	S. E. Tooly	Delphi, N. Y	July 16, 1872	129, 627
Horse-power	A. Trahern, H. and J. Heberling, and W. E. Lukens.	Harrison County, Ohio	Oct. 28, 1835	....
Horse-power	J. S. Upton	Battle Creek, Mich	Mar. 15, 1859	23, 280
Horse-power	J. S. Upton	Battle Creek, Mich	Feb. 5, 1861	31, 342
Horse-power	J. S. Upton	Battle Creek, Mich	Sept. 21, 1869	95, 168
Horse-power	J. Urmy	Wilmington, Del	Mar. 6, 1847	5, 001
Horse-power	J. Valentine	Buffalo, N. Y	Oct. 17, 1871	119, 954
Horse power	D. Van Houten	Fuller's Corners, Ind	Feb. 17, 1863	37, 713
Horse power	B. Wales	Hallowell, Me	Aug. 17, 1835	....
Horse-power	W. Ward	Zanesville, Ohio	Sept. 11, 1849	6, 713
Horse-power	A. G. Waterhouse	San Francisco, Cal	Apr. 22, 1873	138, 108
Horse-power	S. Wheeler	Albany, N. Y	Mar. 2, 1869	87, 450
Horse-power	S. Wheeler	Albany, N. Y	June 14, 1870	104, 240
Horse-power	S. Wheeler and E. Jerome	Albany, N. Y	Sept. 27, 1864	44, 479

Index of patents issued from the United States Patent Office from 1790 to 1873, inclusive—Continued.

Invention.	Inventor.	Residence.	Date.	No.
Horse-power	W. Whitman	Haverhill, N. H	June 20, 1836	....
Horse-power	F. Wicks	Kansas, Ill	Nov. 29, 1864	45, 291
Horse-power	B. H. Wilcox	Petroleum Centre, Pa	Oct. 6, 1868	82, 777
Horse-power	J. W. Wilcox	Macon, Ga	Dec. 20, 1870	110, 320
Horse-power	W. H. Wiley	Fredonia, N. Y	Nov. 12, 1867	70, 768
Horse-power	T. H., J. E., J. F., and R. J. Wilson.	Athens, Ga	June 1, 1858	20, 461
Horse-power	T. Wiltse, jr	Panama, N. Y	Oct. 8, 1867	69, 736
Horse-power	A. Wissler and J. Gamber	Brunersville and Petersburgh, Pa.	Dec. 12, 1871	121, 740
Horsé-power	D. Woodbury	Weathersfield, Vt	Aug. 26, 1846	4, 718
Horse-power	D. Woodbury	Rochester, N. Y	Apr. 27, 1869	89, 368
Horse-power	D. Woodbury	Rochester, N. Y	Oct. 5, 1869	95, 549
Horse-power	D. Woodbury	Rochester, N. Y	June 4, 1872	127, 537
Horse-power	D. Woodbury	Rochester, N. Y	Dec. 9, 1873	145, 473
Horse-power	W. R. Wright and D. R. Warnock.	Barnwell County and Beaufort County, S. C.	Oct. 10, 1871	119, 909
Horse-power, Adjustment of gearing for	J. S. Everitt	Oshkosh, Wis	June 2, 1868	78, 445
Horse-power and accelerating machine	R. S. Schevenell	Orangeburgh, S. C	Nov. 13, 1832	....
Horse-power and baling-press, Combined	C. A. Wright	Rodney, Miss	Oct. 4, 1870	107, 990
Horse-power and truck	C. Roberts and J. A. Throp	Three Rivers, Mich	Dec. 8, 1868	84, 766
Horse-power apparatus	A. Gaar	Richmond, Ind	Aug. 30, 1870	106, 979
Horse-power apparatus	E. O. and C. B. Thompson	Thomasville, Ga	Sept. 13, 1870	107, 309
Horse-power, Apparatus for confining	R. Knott	Suisun, Cal	Feb. 9, 1869	86, 677
Horse-power apparatus for elevating hay, &c	F. Wicks	Upper Sandusky, Ohio	Aug. 28, 1866	57, 608
Horse-power, Attaching the arms of	C. Roberts	Belleville, Ill	Feb. 10, 1857	16, 612
Horse-power brake	T. Harvey and N. J. Becky	Amsterdam, N. Y	July 24, 1860	29, 280
Horse-power brake	J. Hull and W. P. Anderson	Hackettstown, N. J	Oct. 2, 1866	58, 547
Horse-power brake	T. G. Palmer	Shultville, N. Y	Nov. 10, 1868	83, 995
Horse-power brake	A. Pursell	New Village, N. J	Sept. 27, 1864	44, 454
Horse-power brake	W. F. Rundell	Geneva, N. Y	Oct. 3, 1865	50, 277
Horse-power brake	A. D. Tingley	Adrian, Mich	Sept. 10, 1867	68, 806
Horse-power, Chain-band for	J. S. and H. S. Pitts	Winthrop and Livermore, Me.	Aug. 15, 1834	....
Horse-power, Chain-wheel for chain	S. Perry	Newport, N. Y	Sept. 24, 1867	69, 243
Horse-power, Circuit	S. Perry	Newport, N. Y	Apr. 22, 1862	35, 063
Horse-power, Circuit	S. Perry	Newport, N. Y	Aug. 26, 1862	36, 327
Horse-power, Connecting the parts of endless floors for.	S. Hale	Hollis, N. H	Jan. 16, 1845	3, 887
Horse-power connection	J. A. Hafner	Commerce, Mo	Feb. 2, 1869	86, 533
Horse-power connection	J. A. Hafner	Commerce, Mo	Oct. 26, 1869	96, 225
Horse-power, Coupling of endless-chain	W. E. Arnold	Rochester, N. Y	May 2, 1854	10, 852
Horse-power, Device for securing	F. W. Randall	Burlington, Mich	May 16, 1871	114, 853
Horse-power draft	J. H. Jones	Rockton, Ill	Nov. 16, 1858	22, 079
Horse-power elevator	W. H. Hawley	Utica, N. Y	Nov. 26, 1867	71, 303
Horse-power elevator and excavator	S. T. Bishop and A. Steveley	Fond du Lac, Wis	Jan. 24, 1865	45, 965
Horse-power elevator and excavator	S. T. Bishop and A. Steveley	Fond du Lac, Wis	Jan. 24, 1865	45, 966
Horse-power elevator and excavator	S. T. Bishop and A. Steveley	Fond du Lac, Wis	Jan. 24, 1865	45, 967
Horse-power elevator and excavator	S. T. Bishop and A. Steveley	Fond du Lac, Wis	Jan. 24, 1865	45, 968
Horse-power, Endless band for	J. G. Stanley and J. C. Howard	Winthrop, Me	Jan. 16, 1835	....
Horse-power, Endless-chain	H. C. Dodge	North Danville, Vt	Mar. 5, 1872	124, 208
Horse-power, Endless-chain	H. L. Emery	Albany, N. Y	Feb. 24, 1852	8, 754
Horse-power, Endless-chain	W. Furbush	Hallowell, Me	Apr. 14, 1835	....
Horse-power, Endless-chain	A. W. Gray	Middletown, Vt	Oct. 26, 1842	2, 833
Horse-power, Endless-chain	J. Kelly	Lewistown, Pa	Mar. 4, 1842	2, 480
Horse-power, Endless-chain	I. R. Lawrence	Green Island, N. Y	Sept. 2, 1862	36, 353
Horse-power, Endless-chain	A. Parker	Windham, Me	Aug. 30, 1838	898
Horse-power, Endless-chain	S. Richards	East Poultney, Vt	Oct. 28, 1843	3, 322
Horse-power, Endless-chain	T. Sharp	Albany, N. Y	Mar. 2, 1852	8, 779
Horse-power, Endless-chain	G. Westinghouse	Central Bridge, N. Y	June 13, 1854	11, 104
Horse-power, Endless chain and tread of	G. E. Burt	Harvard, Mass	Aug. 6, 1861	33, 024
Horse-power, Endless chain for	A. and G. F. Wright	Clinton, Mass	Jan. 26, 1869	86, 197
Horse-power, Endless-chain inclined-plane	O. Badger	Cooperstown, N. Y	Oct. 15, 1836	....
Horse-power engine, Portable endless-screw	I. Van Doren	Hopewell Township, N. J	June 13, 1831	....
Horse-power equalizer	G. Hely	La Porte, Ind	Feb. 15, 1859	22, 950
Horse-power, Equalizing draft in	J. Wilkinson	Prophetstown, Ill	Aug. 11, 1863	39, 516
Horse-power, Equalizing the action of gearing in	C. Caples	Savannah, Mo	July 31, 1849	6, 614
Horse-power, Equalizing the draft of	W. P. Dunlap	Maquoketa, Iowa	July 16, 1867	66, 688
Horse-power fastener	R. Knott	Suisun, Cal	Nov. 10, 1868	83, 862
Horse-power fastening	H. B. Hassler	New Berlin, Ohio	Dec. 12, 1871	121, 874
Horse-power fastening	C. Ross	Hartland, Mich	Aug. 4, 1863	39, 424
Horse-power for driving machinery	S. S. Allen	Miamisburgh, Ohio	Jan. 10, 1840	1, 466
Horse-power for driving machinery	W. E. Arnold	Rochester, N. Y	Apr. 26, 1839	1, 135
Horse-power for driving machinery	A. D. Childs	Rochester, N. Y	Apr. 6, 1844	3, 572
Horse-power for driving machinery	M. Davenport	Pittsburgh, Pa	Sept. 4, 1841	2, 239
Horse-power for driving machinery	S. B. Haines	Greensburgh, Pa	Dec. 31, 1844	3, 871
Horse-power for driving machinery	C. Hibbard	Gilford, N. H	Sept. 14, 1840	1, 788
Horse-power for driving machinery	B. Hinkley	Fayette, Me	Nov. 2[illegible], 1837	487
Horse-power for driving machinery	S. H. Little	Gettysburgh, Pa	June 11, 1841	2, 124
Horse-power for driving machinery	M. C. Mix	Danby, N. Y	June 23, 1838	802
Horse-power for driving machinery	R. Montgomery	Waterville, N. Y	Apr. 10, 1845	3, 988
Horse-power for driving machinery	E. Piper	Camden, Me	June 10, 1840	1, 629
Horse-power for driving machinery	L. S. Rand	Townshend, Vt	Mar. 4, 1843	2, 992
Horse-power for driving machinery	J. Secor	New York, N. Y	Apr. 28, 1838	714
Horse-power for driving machinery	O. Straight	Lycoming County, Pa	Aug. 3, 1838	874
Horse-power for driving machinery	G. Strenge and J. Rohrer	Lancaster, Pa	May 8, 1840	1, 586
Horse-power for driving machinery	E. Warren	New York	Jan. 5, 1841	1, 924
Horse-power for driving machinery	T. J. Wells	New York	July 1, 1841	2, 152
Horse-power for driving machinery, Endless-chain	M. Davenport	Phillips, Me	July 10, 1839	1, 229
Horse-power for driving machinery, Endless-chain	J. G. Hall	Zanesville, Ohio	Sept. 21, 1837	395
Horse-power for driving machinery, Endless-chain	H. G. Hall	Putnam, Ohio	Sept. 28, 1837	417
Horse-power for driving machinery, Endless-chain	I. R. Lawrence	Chatham, N. Y	Oct. 7, 1842	2, 799
Horse-power for driving machinery, Endless-chain	A. Palmer	Akron, N. Y	Sept. 22, 1837	398
Horse-power for driving machinery, Endless-chain	A. and W. C. Wheeler	Chatham, N. Y	July 8, 1841	2, 157
Horse-power for driving machinery, Endless-floor	J. M. Reed	Middlefield, N. Y	Jan. 30, 1841	1, 963
Horse-power for driving machinery, Endless-floor	O. Badger	Cooperstown, N. Y	Dec. 14, 1840	1, 898
Horse-power for machinery	H. Smith	Bethel, Ohio	Oct. 6, 1837	418
Horse-power for propelling machinery	M. Davenport	Phillips, Me	Oct. 10, 1835	....

Index of patents issued from the United States Patent Office from 1790 to 1873, inclusive—Continued.

Invention.	Inventor.	Residence.	Date.	No.
Horse-power for propelling machinery	C. Emmons	New York	Aug. 2, 1831	....
Horse-power for propelling machinery	B. Langdon	Troy, N. Y	July 19, 1837	286
Horse-power for propelling saw-mills, &c	I. Jones	Natchez, Miss	Jan. 13, 1835	....
Horse-power, Frame-work of portable	J. A. Nelson and J. P. Ross	Lewisburgh, Pa	June 30, 1837	252
Horse-power, Gearing for	C. Avery	Tunkhannock, Pa	Apr. 20, 1858	19, 976
Horse-power, &c., Gearing for	C. Avery	Tunkhannock, Pa	June 5, 1860	28, 550
Horse-power, Gearing for	S. Gardiner	Auburn, N. Y	May 14, 1833	....
Horse-power governor	T. B. McConaughey	Newark, Del	Sept. 11, 1866	57, 941
Horse-power governor	L. Pusey	Wilmington, Del	May 25, 1858	20, 368
Horse-power, Hoisting-attachment for portable	P. Cary	Coeymans, N. Y	May 25, 1869	90, 423
Horse-power, Inclined plane	W. G. Johnston	Bridgeton, N. J	Oct. 4, 1834	....
Horse-power jack	S. Perry	Newport, N. Y	Jan. 2, 1872	122, 488
Horse-power link	S. Wheeler	Albany, N. Y	Sept. 2, 1862	36, 378
Horse-power, Link for	A. W. Gray	Middletown, Conn	Sept. 9, 1856	15, 693
Horse-power, Link for endless-chain	G. E. Burt	Harvard, Mass	Oct. 25, 1870	108, 564
Horse-power, Link for endless-chain	J. Casho	Newark, Del	Aug. 18, 1868	81, 251
Horse-power, Link for endless-chain	A. W. Gray	Middletown, Vt	July 29, 1873	141, 268
Horse-power, Link for endless-chain	G. L. Sheldon	Hartsville, Mass	July 12, 1870	105, 263
Horse-power, Link for railway	S. Wheeler	Albany, N. Y	June 2, 1863	38, 778
Horse-power, Link-gearing for	T. D. Burk	Chicago, Ill	Apr. 22, 1856	14, 750
Horse-power machine	J. F. Axtell	Geneva, N. Y	Mar. 13, 1834	....
Horse-power machine	E. Briggs	Fort Covington, N. Y	July 12, 1834	....
Horse-power machine	A. B. Colton	Athens, Ga	May 24, 1859	24, 101
Horse-power machine	J. Eastman	Elbridge, N. Y	May 31, 1834	....
Horse-power machine	W. Field	Providence, R. I	June 7, 1859	24, 291
Horse-power machine	L. R. Fraught	Atlanta, Ga	Sept. 20, 1859	25, 497
Horse-power machine	J. Grant	Rochester, N. Y	Mar. 30, 1858	19, 769
Horse-power machine	E. J. Keep and W. H. Briggs	Stockton, Cal	Sept. 10, 1861	33, 252
Horse-power machine	H. B. Middaugh and A. Clarke	Mansfield, Pa	Sept. 24, 1861	33, 358
Horse-power machine	W. Phelps and W. H. Hanford	Sycamore, Ill	Nov. 22, 1859	26, 203
Horse-power machine	G. Sandford	Poughkeepsie, N. Y	Sept. 20, 1859	25, 528
Horse-power machine	J. A. Stone	Rochester, N. Y	July 27, 1858	21, 032
Horse-power machine	F. B. Williams	Freeport, Ill	Aug. 16, 1859	25, 155
Horse-power-machine brake	D. G. Terrell	Wakefield, Pa	June 23, 1868	79, 281
Horse-power machine, Endless chain for	I. R. Lawrence and G. E. Gould	Green Island, N. Y	Oct. 11, 1859	25, 743
Horse-power machine, Mode of applying and constructing.	W. Zeller	Lebanon County, Pa	Apr. 12, 1859	23, 638
Horse-power machine, Mode of connecting the draft-lever for.	H. Shaw	New Orleans, La	Aug. 25, 1868	81, 544
Horse-power, Master-wheel of	J. A. Taplin	Fishkill, N. Y	June 12, 1849	6, 525
Horse-power, Master-wheel of portable	J. A. Taplin	Hammond, N. Y	Dec. 30, 1841	2, 407
Horse-power, Mode of constructing	T. Schankwiler	Fayette, N. Y	Nov. 19, 1861	33, 755
Horse-power, Mode of fastening	H. C. Drew	Jamestown, Mich	Nov. 28, 1871	121, 281
Horse-power, Mounted	M. B. Erskine	Racine, Wis	Sept. 13, 1870	107, 237
Horse-power, Platform	F. J. Culver	Hartford, Vt	Jan. 31, 1871	111, 326
Horse-power, Portable	H. Aldridge	Goshen, Ind	Sept. 17, 1867	68, 829
Horse-power, Portable	S. S. Allen	Saratoga Springs, N. Y	Apr. 3, 1835	....
Horse-power, Portable	J. Brandon	Williamsport, Pa	Apr. 8, 1835	....
Horse-power, Portable	W. Burk	Whitemarsh Township, Pa	June 25, 1834	....
Horse-power, Portable	J. H. Corey	Bethel Township, Ohio	Feb. 27, 1832	....
Horse-power, Portable	D. Flagg, jr	New York	Dec. 28, 1832	....
Horse-power, Portable	S. Gardner	Rochester, N. Y	Nov. 25, 1834	....
Horse-power, Portable	E. H. Jaques	Springfield, Vt	May 16, 1845	4, 051
Horse-power, Portable	C. M. Keller	Washington, D. C	June 3, 1833	....
Horse-power, Portable	I. J. Richardson	New York	Aug. 26, 1843	3, 229
Horse-power, Portable	I. J. Richardson	New York, N. Y	Feb. 10, 1846	4, 374
Horse-power, Portable	G. W. Swift	Oxford, Miss	May 3, 1859	23, 871
Horse-power, Portable	D. Woodbury	Rochester, N. Y	Aug. 18, 1857	18, 028
Horse-power, Propelling	C. Griffen	Greece, N. Y	May 31, 1853	....
Horse-power, Propelling machinery by	T. Showerman	Covington, N. Y	June 27, 1832	....
Horse-power, Propelling machinery by	I. Stoddard	Great Bend, Pa	Oct. 9, 1860	30, 362
Horse-power, Reversible	P. H. Kells	Hudson, N. Y	July 8, 1856	15, 296
Horse-power, Safety-brake for	D. M. Reynolds	Rising Sun, Md	Dec. 22, 1863	41, 020
Horse-power, &c., Speed-regulator for	S. Perry	Newport, N. Y	Apr. 22, 1862	35, 064
Horse-power, Staking	W. Gregg	Dansville, Mich	Aug. 22, 1871	118, 232
Horse-power staking-apparatus	B. F. Osgood	Coloma, Mich	July 12, 1870	105, 238
Horse-power, Sun and planet	J. Bogardus	New York, N. Y	Aug. 29, 1848	5, 735
Horse-power, Suspending band-wheel in	I. Straub	Lewistown, Pa	July 11, 1837	267
Horse-power, Sweep	R. J. M. King	Ypsilanti, Mich	Apr. 19, 1870	102, 015
Horse-power to machinery, Applying	B. Maltby	New York	Apr. 11, 1825	....
Horse-power to mills, Applying	B. E. Orton	Lyndon, Ill	Mar. 13, 1860	27, 468
Horse-power to the ground, Fastening	W. Buchanan, jr	Maine Prairie, Cal	Nov. 10, 1868	83, 919
Horse-power to the ground, Securing	W. H. Buell	Union City, Mich	Oct. 6, 1868	82, 688
Horse-power to the ground, Securing	F. W. Randall	Burlington, Mich	Apr. 5, 1870	101, 507
Horse-power, Tread	S. Perry	Newport, N. Y	Apr. 22, 1862	35, 062
Horse-power, Water-supported	A. James	Vinton, Ohio	July 23, 1872	129, 734
Horse-power, &c., Wheel and axle attachment of	G. E. Burt and G. F. Wright	Harvard, Mass	Feb. 23, 1858	19, 408
Horse-propelling	J. H. Brown	New York, N. Y	Dec. 11, 1866	60, 335
Horse-pulleying apparatus	A. H. Trego	Lambertville, N. J	July 9, 1861	32, 801
Horse-radish for use, Preparing	J. D. Husbands, jr	Saint Louis, Mo	Apr. 30, 1872	126, 212
Horse-rake	J. W. Acker	Copenhagen, N. Y	Sept. 29, 1868	82, 474
Horse-rake	J. Q. Adams	High Spire, Pa	June 3, 1862	35, 428
Horse-rake	D. G. Adelsberger	Emmetsburgh, Md	Jan. 23, 1866	52, 122
Horse-rake	H. Albright	Lewisburgh, Pa	Feb. 21, 1865	46, 435
Horse-rake	M. Alden	Auburn, N. Y	Apr. 7, 1868	76, 480
Horse-rake	F. M. Allerton	Alliance, Ohio	July 6, 1869	92, 140
Horse-rake	S. E. Ament	Oswego, Ill	Feb. 19, 1867	62, 107
Horse-rake	D. W. Amos	Bedford, Pa	Jan. 12, 1864	41, 187
Horse-rake	H. A. Bailey and A. R. Burdick	Racine, Wis	June 26, 1866	55, 804
Horse-rake	M. W. Baldwin and A. S. Lyman.	Philadelphia, Pa	May 16, 1848	5, 587
Horse-rake	L. B. Ball	Dayton, Ohio	Aug. 13, 1867	67, 705
Horse-rake	L. B. Ball	Dayton, Ohio	Oct. 8, 1867	69, 610
Horse-rake	A. T. Barnes	Tiffin, Ohio	Mar. 24, 1868	75, 836
Horse-rake	W. H. Barton	Olney, Ill	July 28, 1868	80, 383
Horse-rake	J. N. Baxter	Greensburgh, Ind	July 24, 1866	56, 517
Horse-rake	H. L. Beach	Montrose, Pa	Dec. 10, 1867	71, 841
Horse-rake	H. L. Beach	Montrose, Pa	Mar. 17, 1868	75, 514

Index of patents issued from the United States Patent Office from 1790 *to* 1873, *inclusive*—Continued.

Invention.	Inventor.	Residence.	Date.	No.
Horse-rake	H. L. Beach	New York, N. Y	Mar. 31, 1868	76, 040
Horse-rake	L. Beach	Montrose, Pa	Apr. 10, 1860	27, 771
Horse-rake	L. Beach	Montrose, Pa	Jan. 19, 1864	41, 270
Horse-rake	E. L. Bergstresser	Berrysburgh, Pa	Aug. 26, 1862	36, 268
Horse-rake	S. Bingham	Troy, N. Y	Dec. 5, 1865	51, 287
Horse-rake	R. H. Blair and A. W. Beatty	Saltsburgh, Pa	Mar. 4, 1862	34, 558
Horse-rake	I. W. Boatman	Seven Mile, Ohio	Sept. 10, 1867	68, 693
Horse-rake	J. Bohner	Alden, N. Y	May 4, 1869	89, 555
Horse-rake	O. Bonney, jr	San Francisco, Cal	Nov. 23, 1869	97, 157
Horse-rake	L. S. Bortree	Grand Rapids, Mich	Nov. 24, 1868	84, 254
Horse-rake	W. L. Bostwick	Ithaca, N. Y	Mar. 5, 1867	62, 524
Horse-rake	M. Bradley	Dundee, Ill	Aug. 16, 1859	25, 088
Horse-rake	B. Bridendolph	Clear Spring, Md	Jan. 4, 1859	22, 526
Horse-rake	N. Briggs	New Hartford, N. Y	Feb. 11, 1835	......
Horse-rake	S. C. Brinser	Middletown, Pa	Feb. 17, 1863	37, 672
Horse-rake	S. C. Brinser	Middletown, Pa	Apr. 28, 1863	38, 281
Horse-rake	S. C. Brinser	Middletown, Pa	Oct. 27, 1868	83, 452
Horse-rake	H. L. Brown	Adrian, Mich	Nov. 17, 1868	84, 085
Horse-rake	N. H. Brown	Derby, N. H	Apr. 14, 1868	76, 597
Horse-rake	N. T. Brown	Ononwa, Iowa	June 10, 1862	35, 502
Horse-rake	W. H. Brown	Middletown, N. Y	Jan. 4, 1859	22, 482
Horse-rake	E. Buckman	East Greenbush, N. Y	Sept. 3, 1861	33, 183
Horse-rake	F. M. Buckmaster	Galesburgh, Ill	Sept. 28, 1869	95, 315
Horse-rake	D. Bull	Amboy, Ill	July 6, 1869	92, 156
Horse-rake	E. W. Bullard	Barre, Mass	Aug. 27, 1867	68, 283
Horse-rake	E. W. Bullard	Barre, Mass	Feb. 25, 1868	74, 794
Horse-rake	I. L. Bullock	Marcy, Ind	May 28, 1867	65, 164
Horse-rake	I. C. Burgot	Davenport Centre, N. Y	Mar. 8, 1859	23, 155
Horse-rake	G. E. Burt	Harvard, Mass	Mar. 14, 1865	46, 776
Horse-rake	G. E. Burt	Harvard, Mass	Oct. 24, 1865	50, 557
Horse-rake	G. E. Burt	Harvard, Mass	Sept. 17, 1867	68, 950
Horse-rake	G. E. Burt	Harvard, Mass	Dec. 29, 1868	85, 275
Horse-rake	C. Carlisle	Norwich, Vt	May 15, 1847	5, 119
Horse-rake	I. Carman	Sandwich, Ill	May 11, 1869	89, 972
Horse-rake	S. Carpenter	Brookfield, Ill	Dec. 3, 1867	71, 579
Horse-rake	P. S. Carver	Honeoye Falls, N. Y	May 2, 1865	47, 519
Horse-rake	A. H. Chaplin	Adrian, Mich	Nov. 25, 1862	36, 992
Horse-rake	A. Chapman	Delta, N. Y	Feb. 4, 1868	73, 952
Horse-rake	J. Chappel	Greene, N. Y	Nov. 20, 1860	30, 666
Horse-rake	L. Clinton	North Haven, Conn	Nov. 17, 1868	84, 090
Horse-rake	L. Clinton	North Haven, Conn	Feb. 2, 1869	86, 366
Horse-rake	L. Clinton and E. S. Munson	North Haven, Conn	May 22, 1866	54, 862
Horse-rake	A. W. Coates	Alliance, Ohio	Aug. 27, 1867	68, 288
Horse-rake	C. B. Cogswell	Essex, Mass	June 17, 1862	35, 586
Horse-rake	W. H. Cook	Bridgehampton, N. Y	Sept. 22, 1868	82, 292
Horse-rake	E. Crandal	Northville, Mich	Apr. 20, 1869	89, 029
Horse-rake	J. Crites	Orrville, Ohio	Apr. 2, 1861	31, 875
Horse-rake	A. J. Curtis, D. J. Roberts, and W. Curtis.	Swanville and Monroe, Me.	Aug. 28, 1866	57, 482
Horse-rake	G. H. Dailey and R. M. Treat	Morris, Conn	Nov. 11, 1862	36, 897
Horse-rake	M. Davenport	Minerva, Ohio	Sept. 6, 1864	44, 079
Horse-rake	B. W. Davis	Fort Madison, Iowa	Mar. 17, 1868	75, 532
Horse-rake	G. Deal	Wilmot, Ohio	Apr. 7, 1868	76, 307
Horse-rake	G. Deal	Wayne Township, Ohio	July 10, 1866	56, 189
Horse-rake	N. C. Decker	Saint Louis, Mo	Sept. 18, 1866	58, 073
Horse-rake	C. Delano	East Livermore, Me	Feb. 27, 1849	6, 151
Horse-rake	L. S. Deming	Newington, Conn	Jan. 18, 1859	22, 632
Horse-rake	S. L. Denney	Christiana, Pa	Aug. 4, 1863	39, 383
Horse-rake	S. L Denney and J. L. Chalfant	Christiana and Chester County, Pa.	Nov. 10, 1868	83, 937
Horse-rake	D. Dewey	Poultney, Vt	Nov. 23, 1837	472
Horse-rake	J. Dillier	Greensburgh, Ind	Sept. 18, 1866	58, 075
Horse-rake	T. H. Dodge	Worcester, Mass	Dec. 20, 1864	45, 481
Horse-rake	H. S. Doolittle	Kortright, N. Y	Aug. 22, 1846	4, 708
Horse-rake	E. Dorr	Rockford, Ill	July 13, 1869	92, 521
Horse-rake	J. B. Drake	Picture Rocks, Pa	Feb. 23, 1864	41, 740
Horse-rake	D. M. Dunham, and J. and A. Webb.	Bangor, Me	Aug. 28, 1866	57, 488
Horse-rake	O. D. Dunham	Plainville, Mich	Mar. 29, 1864	42, 079
Horse-rake	S. Eberly	Mechanicsburgh, Pa	Aug. 30, 1864	43, 980
Horse-rake	S. Eberly and G. Hauck	Mechanicsburgh, Pa	July 2, 1867	66, 314
Horse-rake	S. Eberly and S. Hauck	Mechanicsburgh, Pa	Oct. 29, 1867	70, 327
Horse-rake	C. Edgar	Dayton, Ohio	Dec. 23, 1873	145, 852
Horse-rake	J. Eliot	Vermillion, Ill	Mar. 3, 1868	75, 006
Horse-rake	W. Emmons	Sandwich, Ill	Apr. 13, 1869	88, 858
Horse-rake	R. M. Ewing	Clinton, Ill	Jan. 30, 1866	52, 360
Horse-rake	J. Farmwalt	German Township, Ohio	Dec. 1, 1863	40, 793
Horse-rake	H. V. Farris	Richmond, Ind	July 16, 1867	66, 696
Horse-rake	E. A. Field	Sidney, Me	Dec. 10, 1867	72, 006
Horse-rake	C. H. Finson	Bangor, Me	Jan. 23, 1866	52, 152
Horse-rake	L. W. Frederick	Ray, Ind	Feb. 16, 1864	41, 613
Horse-rake	L. W. Frederick	Gosport, Ind	Aug. 6, 1867	67, 523
Horse-rake	S. Freet	Upper Strasburgh, Pa	Oct. 1, 1867	69, 332
Horse-rake	C. Furst, D. Bradley, and J. Lacey.	Chicago, Ill	Apr. 15, 1862	34, 953
Horse-rake	C. Garver	Londonderry, Pa	Dec. 7, 1858	22, 232
Horse-rake	E. Geiger	Lancaster, Pa	May 24, 1859	24, 114
Horse-rake	J. Ginther	Mier, Ill	June 23, 1868	79, 065
Horse-rake	D. D. Gitt	Arendtsville, Pa	Jan. 17, 1865	45, 920
Horse-rake	C. N. Goss	Claremont, N. H	June 9, 1868	78, 801
Horse-rake	R. A. Graham	Greensburgh, Ind	Dec. 12, 1865	51, 450
Horse-rake	J. S. Grant	Sidney Centre, Me	S pt. 3, 1867	68, 500
Horse-rake	S. H. Grinnell	Charlestown, N. H	Feb. 20, 1849	6, 123
Horse-rake	J. W. Hadcock and P. Wilcox.	Norway, N. Y	Dec. 7, 1858	22, 235
Horse-rake	E. R. Hall	Ilion, N. Y	Sept. 4, 1866	57, 706
Horse-rake	S. J. Halstead	Margarettville, N. Y	May 25, 1869	90, 446
Horse-rake	O. J. Hardgrove	Massillon, Ohio	Mar. 22, 1864	41, 990
Horse-rake	E. Harris	Princeton, Ill	July 5, 1859	24, 631
Horse-rake	A. L. Haskell	Amity, Pa	Oct. 15, 1867	69, 913

Index of patents issued from the United States Patent Office from 1790 *to* 1873, *inclusive*—Continued.

Invention.	Inventor.	Residence.	Date.	No.
Horse-rake	J. V. Hawkey	Greensburgh, Pa	Dec. 10, 1867	71, 875
Horse-rake	H. R. Hawkins	Akron, Ohio	Dec. 27, 1864	45, 505
Horse-rake	H. R. Hawkins	Akron, Ohio	Apr. 24, 1866	54, 151
Horse-rake	H. R. Hawkins	Akron, Ohio	Mar. 24, 1868	75, 905
Horse-rake	J. D. Heebner	Norrittonville, Pa	Aug. 27, 1867	68, 075
Horse-rake	G. L. Heidler	York, Pa	July 10, 1866	56, 222
Horse-rake	H. Hersh	Lancaster, Pa	June 14, 1859	24, 389
Horse-rake	J. N. Hicks	Barre Centre, N. Y	Mar. 17, 1868	75, 548
Horse-rake	T. Himmelberger	Heidelburgh Township, Pa	Feb. 11, 1868	74, 358
Horse-rake	J. B. Hoag	Oxford, Ill	July 23, 1867	67, 054
Horse-rake	H. W. Holcomb	Northville, Mass	Nov. 16, 1869	96, 917
Horse-rake	C. B. Holden	Worcester, Mass	Apr. 14, 1868	76, 760
Horse-rake	F. Holden	Clyde, Ill	Aug. 8, 1865	49, 269
Horse-rake	F. Holden	Litchfield, Ill	Oct. 8, 1867	69, 667
Horse-rake	J. Hollingsworth	Chicago, Ill	Feb. 2, 1864	41, 433
Horse-rake	J. Hollingsworth	Chicago, Ill	July 25, 1865	48, 944
Horse-rake	J. Hollingsworth	Chicago, Ill	June 11, 1867	65, 573
Horse-rake	J. Hollingsworth	Chicago, Ill	May 25, 1869	90, 362
Horse-rake	W. Horning	New Lebanon, Ohio	Feb. 23, 1858	19, 420
Horse-rake	B. F. Horton	Ithaca, N. Y	Jan. 14, 1868	73, 248
Horse-rake	A. Hovey	Brookfield, Vt	Feb. 12, 1850	7, 084
Horse-rake	C. Howard	Bearsville, N. Y	July 30, 1867	67, 305
Horse-rake	E. Huber	Kelso, Ind	Jan. 24, 1865	46, 001
Horse-rake	J. Hudson	Charleston, Ill	Sept. 14, 1869	94, 889
Horse-rake	J. Hunsberger	Worcester Township, Pa	June 16, 1868	78, 969
Horse-rake	J. P. Hunter	Williamsport, Ind	July 16, 1867	66, 679
Horse-rake	C. S. Huntington	Black River, N. Y	Oct. 16, 1866	58, 832
Horse-rake	E. Huson	Ithaca, N. Y	Feb. 12, 1867	61, 034
Horse-rake	D. G. Hussey	Nantucket, Mass	June 23, 1863	38, 965
Horse-rake	D. G. Hussey	Nantucket, Mass	Oct. 18, 1864	44, 728
Horse-rake	D. G. Hussey	Nantucket, Mass	June 13, 1865	48, 179
Horse-rake	D. G. Hussey	Nantucket, Mass	Nov. 28, 1865	51, 187
Horse-rake	Y. Hyatt	Westfield, Mass	Dec. 15, 1857	18, 850
Horse-rake	J. M. Jay	Canton, Ohio	Apr. 18, 1865	47, 371
Horse-rake	C. Jennings	Easton, Conn	May 10, 1864	42, 664
Horse-rake	R. B. and A. C. Jennings	Livermore, Me	Oct. 3, 1848	5, 833
Horse-rake	A. B. Johnson	Washington, D. C	Dec. 1, 1868	84, 551
Horse-rake	S. Johnson	New Harmony, Ind	July 2, 1867	66, 352
Horse-rake	W. H. Johnson	Northborough, Mass	Oct. 10, 1865	50, 363
Horse-rake	E. O. Jones	Oakwood, Mich	July 10, 1866	56, 227
Horse-rake	J. D. Jones	Pittsburgh, Pa	May 9, 1865	47, 644
Horse-rake	S. A. and L. M. Kays	Independence, Iowa	Apr. 30, 1867	64, 329
Horse-rake	C. P. Kelly	Phelps, N. Y	Apr. 6, 1869	88, 719
Horse-rake	A. S. Kendall	Guilford, Me	Jan. 28, 1868	73, 904
Horse-rake	J. E. Kendall	Plymouth, Ind	Dec. 17, 1867	72, 404
Horse-rake	G. Kimball	Springfield, Vt	Dec. 20, 1864	45, 502
Horse-rake	G. W. King	Schoharie, N. Y	Oct. 22, 1867	70, 004
Horse-rake	J. King	Omaha, Nebr	June 19, 1866	55, 672
Horse-rake	W. King	Springfield, Ill	Dec. 5, 1865	51, 322
Horse-rake	W. King	Springfield, Ill	Apr. 9, 1867	63, 729
Horse-rake	W. King	Springfield, Ill	June 22, 1869	91, 640
Horse-rake	G. S. Kinsey	Reading, Pa	Aug. 28, 1860	29, 795
Horse-rake	F. C. Kneeland	Hartford, Wis	Mar. 1, 1859	23, 091
Horse-rake	S. D. Knight, J. W. Smith, and G. W. Bercaw.	Bryan, Ohio	Feb. 23, 1869	87, 267
Horse-rake	J. B. Koon	Aurelius, N. Y	July 20, 1869	92, 732
Horse-rake	A. Krause	West Liberty, Ohio	Jan. 29, 1867	61, 670
Horse-rake	C. Kugler	Cadiz, Ohio	Aug. 28, 1866	57, 520
Horse-rake	J. La Croix	Chicopee, Mass	Mar. 16, 1869	87, 782
Horse-rake	J. Lacy	Chicago, Ill	May 16, 1865	47, 736
Horse-rake	S. Ladd	Danville, Vt	Mar. 14, 1846	4, 421
Horse-rake	E. E. Sauer and H. W. Eisenhart.	York, Pa	Mar. 2, 1869	87, 345
Horse-rake	A. S. Lazier	North Parma, N. Y	Sept. 28, 1869	95, 361
Horse-rake	P. Lebzelter	Lancaster, Pa	July 5, 1859	24, 645
Horse-rake	W. A. Lewis	Joliet, Ill	June 22, 1869	91, 548
Horse-rake	J. M. Long	Hamilton, Ohio	Oct. 22, 1867	70, 008
Horse-rake	W. H. Long	Lancaster, Pa	Mar. 1, 1859	23, 098
Horse-rake	R. Lownsbury and F. G. Willson.	Fulton and Ontario, N. Y	Jan. 31, 1860	27, 000
Horse-rake	C. O. Luce	Brandon, Vt	Jan. 28, 1868	73, 820
Horse-rake	J. B. Luce	Earlville, Ill	Aug. 13, 1867	67, 778
Horse-rake	P. Lugenbell and T. Barns	Greensburgh, Ind	Aug. 7, 1866	56, 960
Horse-rake	E. Luther	West Troy, N. Y	Feb. 25, 1868	74, 922
Horse-rake	E. Luther	West Troy, N. Y	Oct. 6, 1868	82, 855
Horse-rake	S. P. Macay	Killbourne, Ohio	Apr. 21, 1868	76, 932
Horse-rake	J. Maltby	North Branford, Conn	Dec. 5, 1843	3, 369
Horse-rake	E. C. Martin	West Liberty, Iowa	Jan. 24, 1865	46, 009
Horse-rake	N. Martz	Briar Creek Township, Pa	Feb. 26, 1856	14, 321
Horse-rake	R. W. McClelland	Springfield, Ill	June 29, 1869	91, 858
Horse-rake	W. McCord	Sing Sing, N. Y	Aug. 7, 1866	56, 968
Horse-rake	W. McCord	Sing Sing, N. Y	Feb. 25, 1868	74 924
Horse-rake	G. M. L. McMillen	Dayton, Ohio	Oct. 13, 1868	82, 972
Horse-rake	G. M. L. McMillen	Dayton, Ohio	Mar. 2, 1869	87, 504
Horse-rake	W. H. McPherson	Danby, N. Y	May 7, 1867	64, 550
Horse-rake	W. H. McPherson	Danby, N. Y	July 16, 1867	66, 865
Horse-rake	B., S., jr., and J. Mellinger	Mount Pleasant, Pa	May 13, 1862	35, 245
Horse-rake	B., S., jr., and J. Mellinger	Mount Pleasant, Pa	Nov. 18, 1862	36, 961
Horse-rake	G. W. Middlecoff	Atlanta, Ill	Apr. 20, 1869	89, 160
Horse-rake	S. D. Milam	Leoti, Ind	May 18, 1869	90, 286
Horse-rake	M. Miles	Middlesex, N. Y	Dec. 17, 1867	72, 315
Horse-rake	A. Miller	Hagerstown, Md	Feb. 9, 1869	86, 855
Horse-rake	J. Miller	Paris, Ohio	Feb. 18, 1862	34, 441
Horse-rake	M. Morgan	Lancaster, Pa	Aug. 24, 1858	21, 268
Horse-rake	L. H. Morrill	West Cumberland, Me	Nov. 10, 1868	83, 875
Horse-rake	C. H. Mosey	Mansfield, Ohio	July 6, 1869	92, 339
Horse-rake	W. Mosher	Pittstown, N. Y	May 8, 1866	54, 582
Horse-rake	S. Mowry	Womelsdorf, Pa	May 14, 1861	32, 307

Index of patents issued from the United States Patent Office from 1790 *to* 1873, *inclusive*—Continued.

Invention.	Inventor.	Residence.	Date.	No.
Horse-rake	C. E. Murray	Sugar Valley, Pa	June 30, 1868	79, 378
Horse-rake	G. D. Neal	Mount Vernon, Ohio	Oct. 6, 1868	82, 864
Horse-rake	O. Nivison	Hector, N. Y	Jan. 21, 1868	73, 457
Horse-rake	H. B. Noble	South Windsor, Conn	Sept. 29, 1868	82, 631
Horse-rake	S. R. Nye	Barre, Mass	Mar. 13, 1866	53, 172
Horse-rake	S. R. Nye	Barre, Mass	Jan. 12, 1869	85, 757
Horse-rake	G. Palmer	Littlestown, Pa	Oct. 3, 1865	50, 269
Horse-rake	L. H. Parson and G. Houston	Middletown, N. Y	Aug. 31, 1858	21, 358
Horse-rake	J. Partridge	Pittsfield, Mass	Aug. 2, 1864	43, 702
Horse-rake	G. Peirce	Ercildoun, Pa	Nov. 29, 1859	26, 293
Horse-rake	L. D. Pennington and J. G. Woodfill.	Vernon, Ind	June 9, 1868	78, 824
Horse-rake	J. Pennypacker	Charlestown, Pa	Feb. 28, 1865	46, 583
Horse-rake	C. B. Perkins	Kenduskeag, Me	Apr. 7, 1868	76, 518
Horse-rake	J. Perry	Vernon, Ind	Mar. 2, 1869	87, 363
Horse-rake	W. Phillips	New Brunswick, N. Y	Aug. 15, 1831	......
Horse-rake	W. M. Piatt	West Liberty, Ohio	May 22, 1866	54, 947
Horse-rake	O. Pier	Ludlow, Vt	Sept. 13, 1859	25, 441
Horse-rake	O. Pier	Winhall, Vt	Feb. 5, 1867	61, 862
Horse-rake	C. H. Poage	Perry, Mo	Sept. 22, 1868	82, 345
Horse-rake	L. L. Pollard	Worcester, Mass	Dec. 27, 1864	45, 626
Horse-rake	D. Prest	Marlborough, N. J	May 30, 1865	47, 980
Horse-rake	D. Prest	Marlborough, N. J	Aug. 13, 1867	67, 673
Horse-rake	J. Pudney	Stanford, N. Y	Nov. 7, 1835	......
Horse-rake	J. S. Randall	Grand Rapids, Mich	May 28, 1867	65, 271
Horse-rake	O. E. Randall	Lewiston, Me	May 23, 1865	47, 857
Horse-rake	O. E. Randall	Lewiston, Me	Feb. 4, 1868	74, 136
Horse-rake	W. Read	Vernon, Ind	Aug. 11, 1868	81, 003
Horse-rake	P. Reading	Trenton, N. J	Aug. 1, 1818	......
Horse-rake	S. B. Reed	Stuyvesant, N. Y	Jan. 9, 1866	51, 969
Horse-rake	H. Reese	Petersburgh, Ind	Apr. 27, 1869	89, 343
Horse-rake	A. R. Reese	Phillipsburgh, N. J	Dec. 8, 1868	84, 760
Horse-rake	G. M. and C. C. Richardson	Dana, Mass	Apr. 20, 1869	89, 169
Horse-rake	A. H. Robbins	Copenhagen, N. Y	Aug. 25, 1868	81, 409
Horse-rake	I. Robbins and J. Old	Pittsburgh, Pa	Dec. 6, 1864	45, 346
Horse-rake	R. I. Robeson	Chicago, Ill	July 4, 1865	48, 629
Horse-rake	J. Robinson	Lawrence, Pa	Apr. 12, 1864	42, 309
Horse-rake	E. S. Root	Mount Morris, N. Y	May 17, 1836	......
Horse-rake	A. V. Ryder	Germano, Ohio	Oct. 10, 1865	50, 391
Horse-rake	A. V. Ryder	Germano, Ohio	Jan. 8, 1867	61, 023
Horse-rake	H. W. Sabin	Canandaigua, N. Y	Dec. 3, 1850	7, 813
Horse-rake	C. W. Sanborn	Morrill, Me	Sept. 21, 1869	95, 147
Horse-rake	C. Satterlee	Paris, Ill	Nov. 13, 1866	59, 663
Horse-rake	E. Saunders	Wethersfield, Vt	Oct. 24, 1848	5, 883
Horse-rake	W. and T. Schnebly	Hackensack, N. J	July 10, 1860	29, 105
Horse-rake	S. C. Schofield and A. J. Wise	Chicago, Ill	Oct. 26, 1869	96, 272
Horse-rake	J. Seely	North Java, N. Y	May 26, 1868	78, 239
Horse-rake	F. Seidle	Mechanicsburgh, Pa	Sept. 20, 1864	44, 344
Horse-rake	F. Seidle	Mechanicsburgh, Pa	Sept. 12, 1865	49, 926
Horse-rake	F. Seidle and S. Eberley	Mechanicsburgh, Pa	July 3, 1860	29, 012
Horse-rake	N. Selby	Flora, Ill	Oct. 20, 1868	83, 326
Horse-rake	G. C. Shaler and H. Barlow	Gilboa and Hobart, N. Y	Oct. 13, 1868	83, 000
Horse-rake	D. P. Sharp	Ithaca, N. Y	Jan. 19, 1864	41, 326
Horse-rake	D. P. Sharp	Ithaca, N. Y	Dec. 12, 1865	51, 486
Horse-rake	D. P. Sharp	Ithaca, N. Y	July 16, 1867	66, 894
Horse-rake	S. M. Sherman	Fort Dodge, Iowa	Aug. 1, 1865	49, 164
Horse-rake	J. H. Shireman	East Berlin, Pa	Sept. 9, 1862	36, 426
Horse-rake	J. H. Shireman	York, Pa	Oct. 8, 1867	69, 713
Horse-rake	J. H. Shireman	York, Pa	Mar. 16, 1869	87, 977
Horse-rake	T. J. Shreves	Greenbush, Ill	Aug. 4, 1868	80, 772
Horse-rake	A. J. Shunk	Shanesville, Ohio	Aug. 30, 1864	44, 020
Horse-rake	A. J. Shunk	Millersburgh, Ohio	May 25, 1869	90, 592
Horse-rake	E. Smith and S. Cowles	Northford, Conn	Sept. 10, 1861	33, 267
Horse-rake	F. Smith	Highgate, Vt	Sept. 8, 1868	82, 040
Horse-rake	F. M. Smith and E. Brumfield	Albion, N. Y	Dec 4, 1866	60, 270
Horse-rake	J. B. Smith	Newton, Ill	Oct. 13, 1868	83, 002
Horse-rake	M. Smith	Worcester, Mass	May 23, 1865	47, 872
Horse-rake	M. Smith	Worcester, Mass	May 19, 1868	78, 147
Horse-rake	S. P. Smith	Waterford, N. Y	May 4, 1869	89, 697
Horse-rake	G. W. Snyder	Kalamazoo, Mich	Sept. 17, 1867	69, 036
Horse-rake	E. R. and W. P. Spear	Orland, Ind	Apr. 24, 1866	54, 228
Horse-rake	E. R. and W. P. Spear	Orland, Ind	May 11, 1869	89, 950
Horse-rake	J. A. Spear	Braintree, Vt	Oct. 9, 1866	58, 689
Horse-rake	A. B. Sprout	Hughesville, Pa	Nov. 25, 1862	37, 012
Horse-rake	A. B. Sprout	Hughesville, Pa	Jan. 17, 1865	45, 942
Horse-rake	A. B. Sprout	Hughesville, Pa	June 6, 1865	48, 109
Horse-rake	W. Squire	Edon, Ohio	Feb. 15, 1870	99, 966
Horse-rake	J. M. Stafford	Pike, N. Y	Sept. 11, 1847	5, 291
Horse-rake	C. Starrett	Chicago, Ill	July 21, 1868	80, 097
Horse-rake	T. J. Steffe	Lancaster, Pa	Sept. 20, 1859	25, 535
Horse-rake	T. Stewart	Pittsburgh, Pa	Feb. 6, 1866	52, 458
Horse-rake	H. Stimmel	Canton, Ohio	June 26, 1866	55, 930
Horse-rake	W. Stinson	Cool Spring Township, Pa	Jan. 12, 1869	85, 869
Horse-rake	J. C. Stoddard	Worcester, Mass	Sept. 11, 1860	30, 007
Horse-rake	J. C. Stoddard	Worcester, Mass	Jan. 3, 1865	45, 769
Horse-rake	J. C. Stoddard	Worcester, Mass	Jan. 5, 1869	85, 620
Horse-rake	J. C. Stoddard	Worcester, Mass	Mar. 15, 1870	100, 950
Horse-rake	A. C. Stone	Steeleville, Pa	Nov. 7, 1865	50, 853
Horse-rake	A. C. Stone	Steeleville, Pa	Nov. 21, 1865	51, 097
Horse-rake	H. K. Stoner	Lancaster, Pa	June 2, 1863	38, 769
Horse-rake	H. K. Stoner	Lancaster, Pa	Nov. 5, 1867	70, 643
Horse-rake	O. O. Storlo and L. Swenson	Norway, Wis	Nov. 24, 1868	84, 315
Horse-rake	S. Stoughton	Windsor, Ohio	May 4, 1869	89, 605
Horse-rake	H. A. Streeter	Worcester, Mass	May 4, 1869	89, 808
Horse-rake	D. Strock	Chambersburgh, Pa	Sept. 11, 1860	30, 010
Horse-rake	G. E. Sutphen	Louisiana, Mo	Dec. 8, 1868	84, 777
Horse-rake	E. Sweet	Triangle, N. Y	Jan. 12, 1869	85, 872
Horse-rake	L. Swift	Clarkson, N. Y	Jan. 19, 1847	4, 933

Index of patents issued from the United States Patent Office from 1790 *to* 1873, *inclusive*—Continued.

Invention.	Inventor.	Residence.	Date.	No.
Horse-rake	J. F. Swinnerton	Marion, Ohio	June 25, 1867	66, 186
Horse-rake	B. C. Taylor	Dayton, Ohio	May 22, 1866	54, 977
Horse-rake	B. C. Taylor	Dayton, Ohio	Aug. 6, 1867	67, 609
Horse-rake	P. A. Thayer	Theresa, N. Y	Sept. 21, 1869	95, 164
Horse-rake	J. Thompson	Bridgeport, Ill	Oct. 12, 1869	95, 852
Horse-rake	C. M. Titus and L. Mood	Ithaca, N. Y	Nov. 5, 1872	132, 741
Horse-rake	E. J. Toof	Fort Madison, Iowa	Dec. 15, 1868	84, 917
Horse-rake	R. M. Treat	Morris, Conn	June 10, 1862	35, 572
Horse-rake	A. Tschop and J. Hartman	East Berlin, Pa	May 11, 1869	89, 956
Horse-rake	H. Tunison	White Hall Grove, Ill	May 2, 1865	47, 587
Horse-rake	T. J. Turner	Richmond County, Ill	Aug. 13, 1867	67, 688
Horse-rake	J. A. Varney	Alton, N. H	May 1, 1866	54, 448
Horse-rake	J. E. Voiles	Madison, Ind	Nov. 10, 1868	84, 028
Horse-rake	C. F. Walker	Benford's Store, Pa	July 29, 1862	36, 043
Horse-rake	S. W. Walker	Anson, Me	June 11, 1867	65, 778
Horse-rake	M. N. Ward	Bangor, Me	Feb. 9, 1869	86, 717
Horse-rake	M. N. Ward	Linneus, Me	Apr. 20, 1869	89, 257
Horse-rake	C. W. Warner	Williston, Vt	Nov. 15, 1864	45, 099
Horse-rake	C. W. Warner and H. N. Tracy	Williston and Essex, Vt	Oct. 17, 1865	50, 525
Horse-rake	B. Webb	Unadilla Forks, N. Y	Mar. 28, 1871	113, 121
Horse-rake	J. W. Webb	Mount Morris, N. Y	Feb. 5, 1836	....
Horse-rake	A. Wells	Morgantown, W. Va	Jan. 16, 1866	52, 098
Horse-rake	M. D. Wells	Morgantown, Va	Aug. 15, 1854	11, 538
Horse-rake	M. D. Wells	Morgantown, W. Va	Dec. 12, 1865	51, 503
Horse-rake	T. J. West	Alfred Centre, N. Y	Aug. 17, 1869	93, 777
Horse-rake	W. H. White	Garrattsville, N. Y	Mar. 1, 1859	23, 133
Horse-rake	H. C. Whitney	Coxsackie, N. Y	Sept. 19, 1865	50, 060
Horse-rake	J. R. Whittemore	Chicopee Falls, Mass	Feb. 4, 1868	74, 023
Horse-rake	F. Wicks	Upper Sandusky, Ohio	June 23, 1868	79, 284
Horse-rake	T. Witmer	Williamsville, N. Y	Aug. 29, 1865	49, 673
Horse-rake	H. Wood	East Henrietta, N. Y	Feb. 23, 1869	87, 315
Horse-rake	J. Wood	North Bloomfield, N. Y	Oct. 23, 1866	59, 113
Horse-rake	J. M. Woodcock	Bridgeport, Ohio	Nov. 11, 1862	36, 928
Horse-rake	M. Woodman and L. Atwood	Farmington, Me., and Norwich, Conn.	Jan. 29, 1867	61, 646
Horse-rake	D. B. Woodward	Ercildoun, Pa	Feb. 19, 1861	31, 507
Horse-rake	J. Zimmerman	Powhatan, Md	Aug. 27, 1867	68, 140
Horse-rake and hay-spreader	G. N. Palmer	Greene, N. Y	May 28, 1867	65, 113
Horse-rake and hay-spreader, Combined	J. M. Low	Portlandville, N. Y	Nov. 12, 1867	70, 870
Horse-rake and hay-spreader, Combined	F. E. Nearing	Brookfield, Conn	Jan. 7, 1868	73, 192
Horse-rake and hay-spreader, Combined	F. E. Nearing	Brookfield, Conn	Jan. 8, 1869	91, 037
Horse-rake and hay-spreader, Combined	G. N. Palmer	Greene, N. Y	Dec. 12, 1865	51, 473
Horse-rake and hay-spreader, Combined	C. Rogers	Barker, N. Y	Oct. 1, 1867	69, 487
Horse-rake and hay-spreader, Combined	E. E. Seymour and S. J. Taylor	Rome, N. Y	July 30, 1867	67, 221
Horse-rake and hay-spreader, Combined	J. M. Spangler	Canton, Ohio	Mar. 16, 1869	87, 982
Horse-rake and hay-spreader, Combined	H. C. Varnum	Hartford, Vt	Aug. 24, 1869	94, 048
Horse-rake and tedder combined	G. L. Ives	Rome, N. Y	Jan. 28, 1873	135, 276
Horse-rake clearer-bar	D. E. Bristol	Albany, N. Y	Dec. 10, 1872	133, 745
Horse-rake, harrow, and roller combined	H. Cousins	Whiting, Me	Dec. 24, 1872	134, 127
Horse-rake, Hay, &c	J. Baily	Philadelphia, Pa	Mar. 30, 1827	....
Horse-rake, Hay and grain	M. and S. Pennock	East Marlborough, Pa	Feb. 17, 1827	....
Horse-rake, Revolving	L. S. Edleblute	Tullahoma, Tenn	Oct. 22, 1872	132, 457
Horse-rake, Revolving	M. K. Flory	Viola, Ill	Oct. 14, 1873	143, 684
Horse-rake, Revolving	H. Hunt	Bridgewater, N. Y	Dec. 10, 1836	104
Horse-rake, Revolving	G. Penisten, sr	Buchanan, Ohio	Sept. 9, 1873	142, 581
Horse-rake, Revolving	W. Wells	Salem, Mass	Apr. 15, 1873	137, 987
Horse-rake, Revolving hay and grain	M. Pennock	Kennett's Square, Pa	Nov. 23, 1824	....
Horse-rake, Revolving hay and grain	M. Pennock	East Marlborough, Pa	Feb. 8, 1825	....
Horse-rake, Spring-tooth	L. M. Whitman	Pike, N. Y	Apr. 4, 1846	4, 444
Horse-rake teeth	I. L. and J. B. L. Bartlett	North Jay, Me	Sept. 21, 1869	95, 070
Horse-rake teeth	J. B. L. Bartlett	North Jay, Me	July 25, 1871	117, 243
Horse-rake teeth	C. Coleman	Allegheny City, Pa	Feb. 20, 1866	52, 680
Horse-rake teeth	A. B. Sprout	Hughesville, Pa	Sept. 5, 1865	49, 800
Horse-rake teeth	A. B. Sprout	Hughesville, Pa	May 1, 1866	54, 430
Horse-rake teeth, Machine for bending and hardening.	G. F. Simmonds	Fitchburgh, Mass	June 6, 1871	115, 775
Horse-rake teeth, &c., Tempering	J. A. Ferson	Fitchburgh, Mass	Dec. 24, 1872	134, 264
Horse-rake-tooth fastening	B. C. Taylor	Dayton, Ohio	Apr. 9, 1872	125, 630
Horse-rakes, Bending teeth for	H. Brandt	Columbia, Pa	June 5, 1860	28, 555
Horse-releasing apparatus	J. Harrison	New York, N. Y	Oct. 24, 1871	120, 272
Horseshoe	D. Anthony, jr	Adams, Mass	Apr. 8, 1831	....
Horseshoe	H. Armstrong	Sparta, Wis	June 5, 1866	55, 220
Horseshoe	J. Austin	Rockford, Ill	Jan. 2, 1866	51, 786
Horseshoe	J. Austin	Rockford, Ill	Aug. 21, 1866	57, 433
Horseshoe	J. Austin	Rockford, Ill	Feb. 12, 1867	61, 982
Horseshoe	H. H. Baker	New Market, N. J	July 10, 1866	56, 161
Horseshoe	H. H. Baker	New Market, N. J	Oct. 9, 1866	58, 573
Horseshoe	S. J. Baker	Madison Centre, Me	June 14, 1870	104, 097
Horseshoe	J. Barker	Champlain, N. Y	Dec. 22, 1868	85, 052
Horseshoe	J. Behel and J. M. Buell	Rockport, Ill	May 29, 1866	55, 045
Horseshoe	J. Behel, J. Perrine, and J. M. Buell.	Rockford and Ogle County, Ill.	Nov. 12, 1867	70, 783
Horseshoe	T. B. Bishop	Baltimore, Md	Feb. 12, 1867	61, 990
Horseshoe	T. B. Bishop	Baltimore, Md	Dec. 24, 1867	72, 594
Horseshoe	G. Bonnet	New York, N. Y	Aug. 7, 1866	56, 888
Horseshoe	J. Bracket	Lynn, Mass	Mar. 21, 1871	112, 773
Horseshoe	J. B. Brown and J. Farmer	Boston, Mass	Mar. 11, 1822	....
Horseshoe	J. H. Brown	Watertown, Minn	Jan. 9, 1866	51, 916
Horseshoe	J. Carlin	Cumminsville, Ohio	Jan. 17, 1860	26, 832
Horseshoe	I. Carman	Schoolcraft, Mich	Nov. 19, 1867	70, 954
Horseshoe	N. B. Carpenter	New York, N. Y	May 13, 1856	14, 852
Horseshoe	E. Cate	Franklin, N. H	Feb. 5, 1861	31, 296
Horseshoe	E. Cate	Watertown, Mass	Apr. 5, 1870	101, 581
Horseshoe	G. T. Chapman	New York, N. Y	Feb. 25, 1868	74, 892
Horseshoe	M. Chittenden	Danbury, Conn	Oct. 11, 1864	44, 603
Horseshoe	J. K. Christopher	Dayton, Ohio	Oct. 25, 1870	108, 684
Horseshoe	M. C. Clark	Manchester, N. H	Apr. 15, 1873	137, 891
Horseshoe	J. N. Clarke	Cincinnati, Ohio	July 10, 1866	56, 181

Index of patents issued from the United States Patent Office from 1790 *to* 1873, *inclusive*—Continued.

Invention.	Inventor.	Residence.	Date.	No.
Horseshoe	J. N. Clarke	Cincinnati, Ohio	Aug. 18, 1868	81, 142
Horseshoe	N. Clouse	Valley Grove, W. Va	July 30, 1872	130, 019
Horseshoe	W. Coes	Worcester, Mass	Apr. 25, 1865	47, 397
Horseshoe	T. M. Coleman	Philadelphia, Pa	June 12, 1860	28, 656
Horseshoe	R. F. Cooke	Brooklyn, N. Y	Nov. 25, 1873	144, 833
Horseshoe	W. Cooper	Brooklyn, N. Y	June 30, 1857	17, 672
Horseshoe	G. Copeland	Denver, Colo	May 3, 1870	102, 504
Horseshoe	D. Cumming	Sorrel Horse, Pa	May 12, 1857	17, 265
Horseshoe	G. Custer	Monroe, Mich	June 28, 1864	43, 293
Horseshoe	G. Custer	Monroe, Mich	July 4, 1865	48, 618
Horseshoe	J. M. Cuykendal	Metomen, Wis	June 2, 1868	78, 436
Horseshoe	J. Deitz	New York	Apr. 22, 1831	....
Horseshoe	W. Disbrow	San Francisco, Cal	Jan. 31, 1865	46, 087
Horseshoe	H. Dawnie and I. B. Harris	Corstophine and Edinburgh, Scotland.	Jan. 25, 1870	99, 072
Horseshoe	F. W. Edison	Port Huron, Mich	Dec. 8, 1868	84, 684
Horseshoe	W. L. Edwards	Ellison, Ill	Jan. 23, 1872	123, 007
Horseshoe	J. W. Faulkner	Rockford, Ill	Apr. 10, 1866	53, 801
Horseshoe	H. B. Ferren	Batavia, N. Y	Aug. 20, 1872	130, 629
Horseshoe	W. H. Freleigh	Troy, N. Y	Mar. 5, 1872	124, 203
Horseshoe	E. C. Gers	Galesburgh, Mich	July 17, 1866	56, 401
Horseshoe	C. Goodenough	New York, N. Y	Aug. 26, 1873	142, 097
Horseshoe	R. A. Goodenough	Brooklyn, N. Y	May 29, 1860	28, 469
Horseshoe	D. Grim	Pittsburgh, Pa	Nov. 14, 1871	120, 813
Horseshee	L. M. Guiteau	Batavia, N. Y	May 9, 1865	47, 635
Horseshoe	P. Hanly	New York, N. Y	Dec. 24, 1867	72, 485
Horseshoe	W. T. Harmer	New York, N. Y	Feb. 4, 1868	73, 971
Horseshoe	J. Haseltine	Warren, N. H	July 25, 1865	49, 028
Horseshoe	J. Henderson	Elmira, N. Y	May 20, 1856	14, 915
Horseshoe	J. Henderson	Albion, N. Y	Apr. 19, 1870	102, 121
Horseshoe	J. A. Heyl	Boston, Mass	July 21, 1868	80, 074
Horseshoe	J. A. Heyl	Boston, Mass	Mar. 9, 1869	87, 565
Horseshoe	N. E. Hinds	Cooperstown, N. Y	Nov. 8, 1859	26, 036
Horseshoe	W. Hinds	Worcester, Mass	Oct. 23, 1866	59, 016
Horseshoe	W. Hinds	Worcester, Mass	Apr. 7, 1868	76, 449
Horseshoe	H. S. Hitner	Marble Hall, Pa	Jan. 19, 1869	86, 014
Horseshoe	J. W. Hodges	Baltimore, Md	July 4, 1865	48, 623
Horseshoe	J. W. Hodges	Baltimore, Md	Oct. 22, 1867	70, 091
Horseshoe	A. S. Hopson	Plain View, Minn	Mar. 31, 1868	76, 078
Horseshoe	O. A. Howe	Fort Plain, N. Y	Sept. 15, 1863	39, 926
Horseshoe	N. W. Hubbard	New York, N. Y	Feb. 18, 1868	74, 539
Horseshoe	W. E. Hubbard	Randolph, N. Y	June 29, 1858	20, 713
Horseshoe	B. P. Hutchinson	Cady, Mich	Nov. 19, 1872	133, 105
Horseshoe	H. Ingraham	Armada, Mich	Mar. 15, 1870	100, 768
Horseshoe	R. G. Jameson and W. H. Chamberlain.	Bristol, N. H	Dec. 8, 1868	84, 745
Horseshoe	J. Johnson	Lowell, Mass	May 4, 1869	89, 774
Horseshoe	J. Johnson	Lowell, Mass	May 4, 1869	89, 775
Horseshoe	J. M. Johnson	Washington, D. C	May 24, 1864	42, 857
Horseshoe	P. C. Johnson	Central City, Colo	Aug. 10, 1869	93, 447
Horseshoe	P. C. Johnson and E. Froggott	Central City, Colo	Sept. 29, 1868	82, 528
Horseshoe	J. B. Johnston	New York, N. Y	Sept. 30, 1873	143, 353
Horseshoe	R. K. Jordan	Oakland, Cal	Dec. 2, 1873	145, 108
Horseshoe	J. Jorey	Rocky Hill, Conn	July 5, 1859	24, 643
Horseshoe	J. Jorey	Westville, Conn	Feb. 25, 1868	74, 829
Horseshoe	J. Jorey	Westville, Conn	Mar. 30, 1869	88, 304
Horseshoe	L. H. Kellogg	Monroe, Ohio	Nov. 24, 1868	84, 360
Horseshoe	J. B. Kendall	Boston, Mass	Nov. 12, 1861	33, 709
Horseshoe	T. and R. Kinghorn	Morgan, Ohio	Feb. 6, 1872	123, 404
Horseshoe	A. E. Kroger	Norwalk, Conn	Aug. 25, 1868	81, 512
Horseshoe	B. Ladd	Ottumwa, Iowa	Sept. 1, 1868	81, 796
Horseshoe	P. A. La France	Elmira, N. Y	July 3, 1866	56, 065
Horseshoe	R. Láportà	New York, N. Y	Dec. 8, 1868	84, 834
Horseshoe	A. Leach	Lynn, Mass	Aug. 5, 1873	141, 509
Horseshoe	G. W. Lewis	Providence, R. I	Oct. 29, 1867	70, 231
Horseshoe	W. Litzenberg	Macomb, Ill	Aug. 21, 1866	57, 346
Horseshoe	H. D. Lyman	Kalamazoo, Mich	Sept. 8, 1868	82, 013
Horseshoe	O. P. Macgill	Brooklandville, Md	Apr. 11, 1865	47, 242
Horseshoe	J. Maddock	Bloomington, Ill	Sept. 21, 1858	21, 571
Horseshoe	J. F. Mallett	New York, N. Y	Aug. 30, 1864	44, 009
Horseshoe	S. Mason	Newark, N. J	Nov. 24, 1868	84, 294
Horseshoe	D. L. McDonell	Detroit, Mich	Mar. 12, 1867	62, 867
Horseshoe	J. McPherson	Rockford, Ill	Dec. 26, 1865	51, 738
Horseshoe	J. J. Mervesp	New York, N. Y	Dec. 15, 1868	84, 958
Horseshoe	G. B. Milligan	Baltimore, Md	July 21, 1868	80, 199
Horseshoe	S. A. Moore	Bloomfield, Iowa	July 18, 1865	48, 827
Horseshoe	I. D. Mott	Cannonsville, N. Y	Mar. 25, 1873	137, 128
Horseshoe	A. L. Murphy	Philadelphia, Pa	Feb. 22, 1870	100, 180
Horseshoe	P. Murray	East Morrisania, N. Y	Apr. 7, 1868	76, 501
Horseshoe	H. H. Palmer	Rockford, Ill	Nov. 21, 1865	51, 078
Horseshoe	C. Parish	Elkhart, Ind	July 16, 1872	129, 586
Horseshoe	G. A. Parker	Westford, Mass	Jan. 5, 1869	85, 689
Horseshoe	R. H. Parks	Columbus, Ohio	Apr. 20, 1869	89, 165
Horseshoe	I. Peacock	Shortsville, N. Y	May 5, 1863	38, 451
Horseshoe	C. Peillard	France	Sept. 21, 1869	95, 133
Horseshoe	J. I. Peyton	Washington, D. C	Nov. 13, 1866	59, 725
Horseshoe	E. J. Pleyel	Dallas County, Iowa	Oct. 6, 1863	40, 215
Horseshoe	I. R. Potter	Dartmouth, Mass	July 21, 1868	80, 213
Horseshoe	Z. V. Purdy	Washington, D. C	June 30, 1868	79, 389
Horseshoe	D. Roberge	Mooers, N. Y	May 25, 1869	90, 394
Horseshoe	D. Roberge	Moore's Forks, N. Y	Mar. 1, 1870	100, 328
Horseshoe	D. Roberge	Moore's Forks, N. Y	Mar. 1, 1870	100, 329
Horseshoe	J. S. Robertson	Cathcart, Scotland	Nov. 8, 1870	109, 055
Horseshoe	J. Rowe, jr., and F. B. Brown	Wilmington and Boston, Mass.	Aug. 26, 1873	142, 282
Horseshoe	H. Schreiner	Philadelphia, Pa	May 22, 1866	55, 019
Horseshoe	R. Seiffert	Chicago, Ill	Sept. 26, 1871	119, 415
Horseshoe	G. Sewell	Poughkeepsie, N. Y	Feb. 12, 1867	62, 076

Index of patents issued from the United States Patent Office from 1790 to 1873, inclusive—Continued.

Invention.	Inventor.	Residence.	Date.	No.
Horseshoe	F. Shinn	Rockford, Ill	June 5, 1866	55,374
Horseshoe	S. Short	New London, Conn	July 8, 1856	15,306
Horseshoe	W. H. Shurtleff	Providence, R. I	Sept. 10, 1867	68,800
Horseshoe	T. Skelton	Rockford, Ill	July 18, 1865	48,845
Horseshoe	G. W. Skinner	Rockford, Ill	Feb. 18, 1868	74,620
Horseshoe	S. Sloat	Morgan, Ohio	Oct. 22, 1867	70,035
Horseshoe	S. Sloat	Morgan, Ohio	Jan. 23, 1872	122,972
Horseshoe	A. W. Smith	Manchester, N. H	June 11, 1872	127,932
Horseshoe	A. W. Smith	Manchester, N. H	Sept. 9, 1873	142,590
Horseshoe	H. Smith and J. H. Evans	Sandyville and Bolivar, Ohio.	June 22, 1869	91,571
Horseshoe	L. A. Smith	Pekin, Ill	Aug. 4, 1868	80,778
Horseshoe	L. T. C. Smith	Scott County, Ky	Dec. 15, 1831	....
Horseshoe	E. Sneider	Baltimore, Md	Oct. 2, 1866	58,495
Horseshoe	U. Snyder	Conshohocken, Pa	Nov. 11, 1873	144,483
Horseshoe	A. Soles	Fonda, N. Y	June 4, 1872	127,657
Horseshoe	C. O. Stevens	Auburn, Me	Aug. 18, 1868	81,307
Horseshoe	J. Stickney	Manchester, N. H	May 7, 1872	126,497
Horseshoe	J. Stickney	Manchester, N. H	Dec. 3, 1872	133,549
Horseshoe	P. Thiry	Paris, France	July 22, 1862	35,958
Horseshoe	J. S. Toan	King's Ferry, N. Y	July 14, 1868	80,033
Horseshoe	G. H. Todd	Montgomery, Ala	Dec. 9, 1873	145,463
Horseshoe	W. H. Towers	Philadelphia, Pa	Dec. 20, 1853	10,345
Horseshoe	W. H. Towers	Philadelphia, Pa	July 25, 1854	11,392
Horseshoe	J. H. Tyler	Martin, N. C	Apr. 20, 1869	89,096
Horseshoe	A. Tyrrell	Batavia, N. Y	July 26, 1864	43,650
Horseshoe	A. Tyrrell	Batavia, N. Y	Apr. 25, 1865	47,502
Horseshoe	S. Ward and L. J. Munger	Cambridge and Charlestown, Mass.	May 24, 1864	42,889
Horseshoe	T. Waterhouse and C. F. McKenney.	West Gorham and Saco, Me.	Dec. 28, 1869	98,323
Horseshoe	B. R. Watson	New Bedford, Mass	Feb. 19, 1867	62,170
Horseshoe	W. R. Watson	Stockton, Cal	Mar. 2, 1869	87,384
Horseshoe	A. Weitman	West Union, Iowa	Aug. 8, 1865	49,324
Horseshoe	A. Weitman	West Union, Iowa	Sept. 5, 1865	49,812
Horseshoe	C. Weitman	Hazelton, Iowa	May 7, 1867	64,604
Horseshoe	C. M. Werner	Rockford, Ill	Apr. 10, 1866	53,930
Horseshoe	C. M. Werner	Rockford, Ill	Sept. 25, 1866	58,355
Horseshoe	C. M. Werner	Rockford, Ill	Feb. 19, 1867	62,242
Horseshoe	E. Wheeler	Marlborough, Mass	Mar. 2, 1858	19,526
Horseshoe	J. Wheeler	Huntington, Ind	Nov. 5, 1867	70,660
Horseshoe	W. C. Whitmore	Macon City, Mo	Nov. 24, 1868	84,462
Horseshoe	A. S. Wilkinson	Pawtucket, R. I	July 3, 1866	56,134
Horseshoe	A. S. Wilkinson	Pawtucket, R. I	July 10, 1866	56,307
Horseshoe	A. S. Wilkinson	Pawtucket, R. I	July 10, 1866	56,308
Horseshoe	A. S. Wilkinson	Pawtucket, R. I	July 10, 1866	56,309
Horseshoe	A. S. Wilkinson	Pawtucket, R. I	July 10, 1866	56,310
Horseshoe	A. S. Wilkinson	Pawtucket, R. I	July 17, 1866	56,475
Horseshoe	A. S. Wilkinson	Pawtucket, R. I	July 17, 1866	56,476
Horseshoe	A. S. Wilkinson	Pawtucket, R. I	July 17, 1866	56,477
Horseshoe	A. S. Wilkinson	Pawtucket, R. I	Aug. 7, 1866	57,029
Horseshoe	A. S. Wilkinson	Pawtucket, R. I	Aug. 7, 1866	57,030
Horseshoe	A. S. Wilkinson	Pawtucket, R. I	Aug. 14, 1866	57,235
Horseshoe	A. S. Wilkinson	Pawtucket, R. I	Aug. 21, 1866	57,420
Horseshoe	A. S. Wilkinson	Pawtucket, R. I	Feb. 12, 1867	61,971
Horseshoe	A. S. Wilkinson	Pawtucket, R. I	Feb. 12, 1867	61,972
Horseshoe	A. S. Wilkinson	Pawtucket, R. I	Feb. 12, 1867	61,973
Horseshoe	A. S. Wilkinson	Pawtucket, R. I	Feb. 12, 1867	61,974
Horseshoe	A. S. Wilkinson	Pawtucket, R. I	Feb. 12, 1867	61,975
Horseshoe	A. S. Wilkinson	Pawtucket, R. I	May 28, 1867	65,144
Horseshoe	A. S. Wilkinson	Pawtucket, R. I	May 28, 1867	65,145
Horseshoe	A. S. Wilkinson	Pawtucket, R. I	Sept. 24, 1867	69,146
Horseshoe	A. A. York	De Lancey, N. Y	June 29, 1869	91,996
Horseshoe, Adjustable	H. and P. Moran	New York, N. Y	Mar. 5, 1872	124,215
Horse shoe and boot, Combined	H. G. Haedrich	Philadelphia, Pa	Oct. 26, 1869	96,104
Horse shoe and boot, Combined	H. G. and E. M. Haedrich	Philadelphia, Pa	July 19, 1870	105,566
Horseshoe and calk	C. H. Johnson	Boston, Mass	Apr. 25, 1865	47,495
Horseshoe and other nails, Making	B. W. Pierce	New Bedford, Mass	Feb. 10, 1863	37,640
Horseshoe-attachment	P. H. O'Neill	Saint Louis, Mo	Apr. 15, 1873	137,794
Horseshoe-bar	E. Cate	Watertown, Mass	Nov. 29, 1870	109,586
Harseshoe-bars, Machine for rolling	A. Reese	Pittsburgh, Pa	July 30, 1867	67,348
Horseshoe-bars, Roll for rolling	E. Cato	East Woburn, Mass	Apr. 4, 1871	113,255
Horseshoe bending and channeling machine	A. Oehme	Rock Island, Ill	Oct. 17, 1871	120,087
Horseshoe-bending machine	H. J. Batchelder	Boston, Mass	May 29, 1866	55,042
Horseshoe-bending machine	H. F. Goulding	Providence, R. I	Aug. 28, 1866	57,631
Horseshoe-bending machine	E. Wheeler	Marlborough, Mass	Dec. 22, 1857	18,940
Horseshoe-beveler	E. Quinby	Comstock, Mich	Dec. 14, 1869	97,811
Horseshoe-blank	J. Day	Buffalo, N. Y	Apr. 19, 1870	102,095
Horseshoe-blank	E. B. Turner	Providence, R. I	Nov. 30, 1869	97,327
Horseshoe-blank bar	E. Cate	Watertown, Mass	May 10, 1870	102,775
Horseshoe-blank bar	J. Montgomery	Sing Sing, N. Y	Dec. 28, 1869	98,290
Horseshoe-blanks, Forming	C. A. Perkins	Providence, R. I	Mar. 10, 1868	75,456
Horseshoe-blanks, Machine for making	H. and J. Reese	Pittsburgh, Pa	July 19, 1870	105,490
Horseshoe-blanks, Machine for swaging	C. H. Perkins and R. W. Comstock.	Providence, R. I	May 21, 1867	64,903
Horseshoe-blanks, Manufacture of	H. J. Batchelder	Boston, Mass	Nov. 20, 1866	59,755
Horseshoe-blanks, Roll for	J. W. Kingsbury	New Bedford, Mass	Sept. 1, 1863	39,773
Horseshoe-blanks, Roll for forming	W. Acheson	Pittsburgh, Pa	Mar. 14, 1871	112,668
Horseshoe-blanks, Roll for forming	C. L. Fitzhugh	Pittsburgh, Pa	Dec. 20, 1870	110,349
Horseshoe-blanks, Rolling	A. Reese	Pittsburgh, Pa	Oct. 20, 1868	83,207
Horseshoe-calk	T. M. Clarke	Winsted, Conn	Jan. 21, 1873	135,081
Horseshoe-calk	G. Custer	Monroe, Mich	May 6, 1873	138,618
Horseshoe-calk	J. H. Jennings	New Bedford, Mass	May 21, 1861	32,371
Horseshoe-calk	J. Jorey	North Manchester, Conn	June 15, 1869	91,237
Horseshoe-calk	F. Judson	Jersey City, N. J	May 22, 1866	54,918
Horseshoe-calk	E. Maynard	Williamsburgh, N. Y	Feb. 24, 1857	16,691
Horseshoe-calk	J. J. Mervesp	New York, N. Y	Aug. 29, 1871	118,470
Horseshoe-calk	J. J. Mervesp	New York, N. Y	Mar. 4, 1873	136,379

Index of patents issued from the United States Patent Office from 1790 *to* 1873, *inclusive*—Continued.

Invention.	Inventor.	Residence.	Date.	No.
Horseshoe-calk	C. H. Perkins	Providence, R. I	June 16, 1868	78, 892
Horseshoe-calk	J. L. Pike	Lynn, Mass	Feb. 7, 1865	46, 262
Horseshoe-calk	I. R. Potter	Dartmouth, Mass	Nov. 14, 1865	50, 952
Horseshoe-calk	A. S. Wilkinson	Pawtucket, R. I	Aug. 14, 1866	57, 236
Horseshoe-calk, Adjustable	W. J. Berne	Cincinnati, Ohio	Nov. 5, 1867	70, 507
Horseshoe calk and toe, Movable	S. Lloyd	Washington, D. C	Jan. 31, 1865	46, 192
Horseshoe-calk blank	R. B. Caswell	Palmer, Mass	June 11, 1867	65, 723
Horseshoe-calk, Detachable	W. J. Berne	Cincinnati, Ohio	May 18, 1869	90, 224
Horseshoe-calk, Detachable	J. E. Byers	Butler, Pa	June 11, 1872	127, 843
Horseshoe-calk, Detachable	C. H. Johnson	Boston, Mass	Apr. 18, 1865	47, 360
Horseshoe-calk, Detachable	P. Laflin	Warren, Mass	July 27, 1869	93, 098
Horseshoe-calk die	S. Stone	North Manchester, Conn	Oct. 12, 1869	95, 745
Horseshoe-calk-grinding machine	J. Little	Newburgh, N. Y	Apr. 15, 1873	137, 852
Horseshoe-calk, Removable	J. D. Barnum	Amenia Union, N. Y	June 7, 1870	103, 827
Horseshoe-calk sharpener	R. Crocker	Marshalltown, Iowa	June 30, 1868	79, 319
Horseshoe-calk sharpener	W. Duncan	Vinton, Iowa	Oct. 6, 1868	82, 813
Horseshoe-calk sharpening	A. W. Payne	Morris, N. Y	Apr. 26, 1859	23, 780
Horseshoe-calk-sharpening device	J. Johnson	Barrington, N. Y	Jan. 28, 1868	73, 901
Horseshoe-calk-sharpening device	W. M. Jones	Horicon, Wis	Jan. 7, 1868	73, 102
Horseshoe-calk-sharpening device	P. Winegard	Cold Water, Mich	July 14, 1868	80, 044
Horseshoe-calk-sharpening implement	H. M. Close	Chariton, Iowa	June 30, 1868	79, 316
Horseshoe-calk-sharpening machine	E. A. Bushnell	Dodge County, Wis	June 19, 1866	55, 617
Horseshoe-calk-sharpening machine	E. A. Bushnell	Horicon, Wis	Aug. 11, 1868	80, 805
Horseshoe-calk-sharpening machine	W. M. Butler	Waukegan, Ill	Nov. 5, 1867	70, 519
Horseshoe-calk-sharpening machine	H. Howell	Salem, Ohio	Apr. 4, 1871	113, 670
Horseshoe-calk-sharpening machine	J. B. McClanathan	Horicon, Wis	Feb. 12, 1867	62, 048
Horseshoe-calk-sharpening tool	O. E. Butler, S. P. Dunham, and G. K. Wann.	Marshalltown, Iowa	Jan. 19, 1869	85, 997
Horseshoe-calk-sharpening tool	N. Hays, W. Duncan, and E. H. Bowen.	Winton, Iowa	Dec. 24, 1867	72, 491
Horseshoe-calk-turning machine	A. T. Boon, D. M. Osborn, and G. Geer.	Galesburgh, Ill	Aug. 1, 1871	117, 508
Horseshoe-calks, Attachable and removable	G. S. Norris	Baltimore, Md	Dec. 14, 1869	97, 798
Horseshoe-calks, Attaching	W. J. Berne	Cincinnati, Ohio	July 16, 1867	66, 780
Horseshoe-calks, Clamp for forming	J. Rhoads	Niles, Mich	Feb. 1, 1870	99, 476
Horseshoe-calks, Die for swaging	J. Allen	Palmer, Mass	Apr. 16, 1867	63, 823
Horseshoe-calks, Die for swaging	P. A. Page	Palmer, Mass	May 7, 1867	64, 561
Horseshoe-calks, Swage for	E. D. Withers	Parkton, Md	May 30, 1871	115, 405
Horseshoe-cork, Removable	G. W. Griswold	New York, N. Y	Aug. 2, 1864	43, 747
Horseshoe-cushion	G. E. Rust	Boston, Mass	Jan. 7, 1868	73, 050
Horseshoe, Elastic	J. O. Jones	Newton, Mass	Aug. 3, 1852	9, 173
Horseshoe for cure of hoof-bound horses	D. Homer	Alton, Ill	Apr. 16, 1842	2, 564
Horseshoe interfering-attachment	H. S. Davis	Camden, N. J	Apr. 8, 1873	137, 660
Horseshoe-irons, Formation of	E. Cate	Franklin, N. H	Mar. 26, 1861	31, 781
Horseshoe-irons, Machine for rolling	C. Richardson	Auburn, N. Y	Oct. 8, 1861	33, 452
Horseshoe-jack	J. Shimer	Scranton Station, Iowa	Mar. 12, 1872	124, 452
Horseshoe-machine	D. N. Allard	McConnellsville, Ohio	Apr. 9, 1861	31, 941
Horseshoe-machine	F. D. Althans and J. F. Allen	Morrisania and Tremont, N. Y.	Nov. 16, 1869	96, 762
Horseshoe-machine	W. Anderson	Pittsburgh, Pa	Nov. 23, 1869	97, 021
Horseshoe-machine	R. Y. Babcock	New London, Conn	Apr. 29, 1851	8, 067
Horseshoe-machine	H. J. Bachelder and G. E. Woods.	Marlborough, Mass	Apr. 21, 1868	76, 975
Horseshoe-machine	H. J. Batchelder	Boston, Mass	June 19, 1866	55, 602
Horseshoe-machine	L. H. Bigelow	Philadelphia, Pa	Oct. 17, 1865	50, 440
Horseshoe-machine	U. Billings	New Bedford, Mass	Nov. 12, 1861	33, 692
Horseshoe-machine	U. Billings	New Bedford, Mass	Sept. 9, 1862	36, 390
Horseshoe-machine	U. Billings	New Bedford, Mass	June 9, 1863	38, 804
Horseshoe-machine	U. Billings	Cambridgeport, Mass	Mar. 21, 1871	112, 769
Horseshoe-machine	H. Burden	Troy, N. Y	Nov. 23, 1835	....
Horseshoe-machine	H. Burden	Troy, N. Y	Sept. 14, 1843	3, 261
Horseshoe machine	H. Burden	Troy, N. Y	June 30, 1857	17, 665
Horseshoe-machine	H. Burden	Troy, N. Y	July 1, 1862	35, 746
Horseshoe-machine	J. Christie	Pittsburgh, Pa	May 4, 1869	89, 737
Horseshoe-machine	J. B. Collen	Philadelphia, Pa	May 10, 1859	23, 976
Horseshoe-machine	E. Davis and S. Field	Barre and Oakum, Mass	Oct. 8, 1834	....
Horseshoe-machine	S. W. Davis	Wilmington, Del	Nov. 17, 1863	40, 611
Horseshoe-machine	S. W. Davis	Wilmington, Del	Aug. 30, 1864	43, 977
Horseshoe-machine	S. Decatur and W. Tatham		Oct. 11, 1809	....
Horseshoe-machine	J. Desnos	Troy, N. Y	July 31, 1860	29, 365
Horseshoe-machine	M. Hardaway	Saint Louis, Mo	Nov. 24, 1863	40, 669
Horseshoe-machine	R. E. Hobart	Pottstown, Pa	May 24, 1828	....
Horseshoe-machine	O. A. Howe	Jersey City, N. J	Sept. 22, 1868	82, 409
Horseshoe-machine	W. R. Justus	Pittsburgh, Pa	Mar. 28, 1871	113, 173
Horseshoe-machine	W. R. Justus	Pittsburgh, Pa	Jan. 23, 1872	122, 893
Horseshoe-machine	J. W. Kingsbury	New Bedford, Mass	May 21, 1867	64, 883
Horseshoe-machine	W. W. Lewis	Cincinnati, Ohio	June 22, 1858	20, 646
Horseshoe-machine	W. W. Lewis	Cincinnati, Ohio	Feb. 1, 1859	22, 812
Horseshoe-machine	W. W. Lewis	Cincinnati, Ohio	Aug. 31, 1869	94, 225
Horseshoe-machine	J. McCarty	Philadelphia, Pa	Apr. 27, 1858	20, 079
Horseshoe-machine	J. McCarty	Philadelphia, Pa	Oct. 16, 1860	30, 448
Horseshoe-machine	J. McCarty	Philadelphia, Pa	Nov. 13, 1860	30, 656
Horseshoe-machine	B. Mee	Troy, N. Y	June 9, 1863	38, 873
Horseshoe-machine	J. Neff	Prattsburgh, N. Y	Sept. 16, 1862	36, 475
Horseshoe-machine	C. H. Perkins	Putnam, Conn	June 1, 1858	20, 441
Horseshoe-machine	C. H. Perkins	Providence, R. I	June 25, 1861	32, 645
Horseshoe-machine	C. H. Perkins	Providence, R. I	Oct. 25, 1864	44 838
Horseshoe-machine	C. H. Perkins	Providence, R. I	Oct. 25, 1864	44, 839
Horseshoe-machine	C. H. Perkins	Providence, R. I	Oct. 25, 1864	44, 840
Horseshoe-machine	C. H. Perkins	Providence, R. I	Oct. 21, 1873	143, 781
Horseshoe-machine	C. H. Perkins	Providence, R. I	Oct. 21, 1873	143, 782
Horseshoe-machine	C. H. Perkins and R. W. Comstock.	Providence, R. I	May 28, 1867	65, 265
Horseshoe-machine	D. I. Pruner	Bellefonte, Pa	May 30, 1871	115, 520
Horseshoe-machine	P. P. Read	Durham, Me	Nov. 13, 1847	5, 364
Horseshoe-machine	C. B. Reed	West Bridgewater, Mass	Oct. 20, 1834	....
Horseshoe-machine	A. Reese	Pittsburgh, Pa	July 23, 1867	66, 991
Horseshoe-machine	J. and A. Reese	Pittsburgh, Pa	Nov. 23, 1869	97, 118

Index of patents issued from the United States Patent Office from 1790 *to* 1873, *inclusive*—Continued.

Invention.	Inventor.	Residence.	Date.	No.
Horseshoe-machine	W. D. Rinehart	Pittsburgh, Pa	June 8, 1869	91, 165
Horseshoe-machine	A. J. Roberts	Boston, Mass	Sept. 1, 1863	39, 778
Horseshoe-machine	A. J. Roberts	Boston, Mass	Mar. 28, 1865	47, 071
Horseshoe-machine	A. J. Roberts	Boston, Mass	Oct. 19, 1869	95, 935
Horseshoe-machine	L. D. Roberts	Cleveland, Ohio	Sept. 2, 1862	36, 370
Horseshoe-machine	L. D. Roberts	Cleveland, Ohio	Dec. 22, 1863	41, 022
Horseshoe-machine	L. D. Roberts	Cleveland, Ohio	Apr. 5, 1870	101, 513
Horseshoe-machine	W. Roberts	Cleveland, Ohio	Sept. 19, 1871	119, 051
Horseshoe-machine	D. T. Robinson	Boston, Mass	Apr. 30, 1867	64, 254
Horseshoe-machine	T. H. Russell and A. Morrill	Northfield and Strafford, Vt	Oct. 12, 1858	21, 779
Horseshoe-machine	E. Shaw and C. Carpenter, jr	Providence, R. I., and Pawtucket, Mass.	Apr. 20, 1858	20, 023
Horseshoe-machine	S. Shetter	Allegheny, Pa	Jan. 18, 1859	22, 677
Horseshoe-machine	J. Shinneller and J. Brislin	Temperanceville, Pa	July 10, 1866	56, 280
Horseshoe-machine	J. H. Snyder	Troy, N. Y	Oct. 18, 1870	108, 530
Horseshoe-machine	J. H. Snyder	Richmond, Va	Aug. 22, 1871	118, 399
Horseshoe-machine	G. Stiles, jr., and S. Kneass	Philadelphia, Pa	Apr. 13, 1858	19, 957
Horseshoe-machine	J. G. Stowe	Providence, R. I	May 31, 1870	103, 681
Horseshoe-machine	T. R. Taylor	Cleveland, Ohio	Apr. 3, 1860	27, 750
Horseshoe-machine	T. R. Taylor	Cleveland, Ohio	July 29, 1862	36, 041
Horseshoe-machine	W. H. Thompson	Cleveland, Ohio	Dec. 2, 1862	37, 062
Horseshoe machine	E. Tolles	Winchester, Conn	Oct. 24, 1834	....
Horseshoe-machine	E. B. Turner	Providence, R. I	Oct. 20, 1868	83, 342
Horseshoe-machine	J. T. Walker	Albany, N. Y	Feb. 22, 1870	100, 222
Horseshoe-machine	W. Wallick	Philadelphia, Pa	May 1, 1866	54, 452
Horseshoe-machine	E. Wassell	Pittsburgh, Pa	Dec. 29, 1868	85, 414
Horseshoe-machine	E. Wassell	Wood's Run, Pa	Mar. 8, 1870	100, 570
Horseshoe-machine	H. L. Watts	Chester, Mass	June 28, 1859	24, 596
Horseshoe-machine	T. J. West	Alfred, N. Y	Dec. 27, 1864	45, 661
Horseshoe-machine	C. W. Wettengel	Pittsburgh, Pa	Mar. 8, 1870	100, 574
Horseshoe-machine	C. P. Williamson	Louisville, Ky	Aug. 17, 1869	93, 778
Horseshoe-machine	H. A. Wills	Keeseville, N. Y	Mar. 2, 1858	19, 528
Horseshoe-machine	H. A. Wills	Keeseville, N. Y	July 3, 1860	29, 042
Horseshoe-machine	B. Young and S. Titus	Killingly and Brooklyn, Conn.	July 29, 1837	301
Horseshoe-machine	J. Zepf	Troy, N. Y	Nov. 10, 1868	84, 043
Horseshoe-machinery	S. S. Greene	Lowell, Mass	Nov. 12, 1850	7, 772
Horseshoe-machinery	S. Shetter	Allegheny, Pa	Nov. 9, 1852	9, 390
Horseshoe-nail	T. Fowler	Seymour, Conn	June 5, 1866	55, 270
Horseshoe-nail	J. Jorey	Norwich, Conn	May 14, 1872	126, 712
Horseshoe-nail	A. Reese	McClure Township, Pa	Feb. 2, 1869	86, 450
Horseshoe-nail	A. S. Wilkinson	Pawtucket, R. I	Feb. 12, 1867	61, 970
Horseshoe-nail blank	M. D. Whipple	Cambridge, Mass	Mar. 8, 1864	41, 881
Horseshoe-nail blanks from plate-metal, Machine for punching.	G. L. Hall	Boston, Mass	Mar. 26, 1872	125, 045
Horseshoe-nail clincher	E. E. Fisher and W. H. Mack	Indianola, Ill	Mar. 30, 1869	88, 378
Horseshoe-nail clincher	W. H. Lyman	Springfield, Mo	July 2, 1872	128, 642
Horseshoe-nail clincher	D. Mater, jr	Bellmore, Ind	Aug. 8, 1871	117, 793
Horseshoe-nail clincher	A. Morley	Addison, Mich	Sept. 28, 1869	95, 254
Horseshoe-nail clincher	N. Repp	Waterloo, Iowa	June 8, 1869	91, 044
Horseshoe, &c., nail cutter	E. Rolph	North Providence, R. I	Sept. 16, 1833	....
Horseshoe-nail machine	A. D. Bingham	Vergennes, Vt	Jan. 14, 1873	134, 841
Horseshoe-nail machine	E. L. Brundage	Middletown, N. Y	Aug. 31 1869	94, 391
Horseshoe-nail machine	M. Burnett	Boston, Mass	Apr. 1 1851	8, 006
Horseshoe-nail machine	C. Carpenter, jr	Providence, R. I	June 9, 1857	17, 491
Horseshoe-nail machine	T. Carpenter	Providence, R. I	May 4, 1858	20, 141
Horseshoe-nail machine	A. H. Caryl	Groton, Mass	Sept. 2, 1873	142, 437
Horseshoe-nail machine	H. D. Cowles and G. Stacy	Montreal, Canada	May 2, 1871	114, 413
Horseshoe-nail machine	D. Dodge	Keeseville, N. Y	Jan. 5, 1864	41, 141
Horseshoe-nail machine	L. H. Dwelley	Dorchester, Mass	July 18, 1865	48, 798
Horseshoe-nail machine	L. H. Dwelley	Dorchester, Mass	Feb. 26, 1867	62, 322
Horseshoe-nail machine	D. J. Farmer	Wheeling, W. Va	Aug. 24, 1869	94, 095
Horseshoe-nail machine	T. Fowler	Seymour, Conn	May 21, 1867	64, 964
Horseshoe-nail machine	P. N. Gallaer and I. C. Tate	New London, Conn	July 12, 1870	105, 189
Horseshoe-nail machine	W. C. Grimes	Philadelphia, Pa	July 21, 1863	39, 287
Horseshoe-nail machine	E. W. Kelley	Hamilton, Scotland	May 23, 1871	115, 064
Horseshoe-nail machine	E. W. Kelley	Lowell, Mass	May 27, 1873	139, 250
Horseshoe-nail machine	J. M. Laughlin	Boston, Mass	Feb. 18, 1873	135, 993
Horseshoe-nail machine	B. A. Mason	Newport, R. I	Jan. 18, 1859	22, 663
Horseshoe-nail machine	J. Mills	Keeseville, N. Y	Jan. 21, 1873	135, 140
Horseshoe-nail machine	S. S. Putnam	Dorchester, Mass	June 11, 1861	32, 531
Horseshoe-nail machine	S. S. Putnam	Dorchester, Mass	Dec. 9, 1862	37, 107
Horseshoe-nail machine	S. S. Putnam and L. H. Dwelley	Dorchester, Mass	Mar. 28, 1865	47, 070
Horseshoe-nail machine	S. S. Putnam and L. H. Dwelley	Dorchester, Mass	Apr. 24, 1866	54, 256
Horseshoe-nail machine	S. S. Putnam and L. H. Dwelley	Dorchester, Mass	Mar. 5, 1867	62, 684
Horseshoe-nail machine	S. S. Putnam and L. H. Dwelley	Dorchester, Mass	Mar. 5, 1867	62, 685
Horseshoe-nail machine	G. W. Sargent and B. P. Rider	Chelsea, Mass	Jan. 16, 1866	52, 081
Horseshoe-nail machine	G. W. Sargent and B. P. Rider	Fairhaven and Chelsea, Mass.	June 26, 1866	55, 912
Horseshoe-nail machine	H. F. Sehnders	Buffalo, N. Y	Aug. 29, 1865	49, 657
Horseshoe-nail machine	W. Sharts	Hudson, N. Y	July 9, 1867	66, 642
Horseshoe-nail machine	A. Shaw	Westford, Mass	Dec. 24, 1867	72, 550
Horseshoe-nail machine	A. St. Louis	Keeseville, N. Y	Oct. 29, 1867	70, 374
Horseshoe-nail machine	J. Stone	Batavia, Ill	Mar. 19, 1872	124, 771
Horseshoe-nail machine	W. Tallman	Providence, R. I	Mar. 27, 1860	27, 656
Horseshoe-nail machine	E. Thomas	Middleborough, Mass	Mar. 19, 1867	62, 980
Horseshoe-nail machine	G. D. Walcott	Jackson, Mich	June 30, 1868	79, 417
Horseshoe-nail machine	M. D. Whipple	Cambridge, Mass	Mar. 15, 1864	41, 955
Horseshoe-nail machine	M. D. and L. W. Whipple	Brighton and Boston, Mass	Apr. 9, 1872	125, 423
Horseshoe-nail machine	J. White and J. Madden	Cleveland and Youngstown, Ohio.	May 10, 1864	42, 709
Horseshoe-nail machine	A. Whittemore	Cambridgeport, Mass	Aug. 14, 1860	29, 643
Horseshoe-nail machine	A. Whittemore	Cambridgeport, Mass	Jan. 12, 1864	41, 251
Horseshoe-nail machine	J. A. Whitney	Maryland, N. Y	Jan. 9, 1866	51, 986
Horseshoe-nail machine	W. Wickersham	Boston, Mass	Oct. 8, 1867	69, 734
Horseshoe-nail machine	W. Wickersham	Boston, Mass	Nov. 23, 1869	97, 141
Horseshoe-nail machine	D. Wilson, jr	North Chelmsford, Mass	Jan. 21, 1851	7, 913
Horseshoe-nail machine	C. W. Woodford	Montreal, Canada	Mar. 5, 1872	124, 411

Index of patents issued from the United States Patent Office from 1790 to 1873, inclusive—Continued.

Invention.	Inventor.	Residence.	Date.	No.
Horseshoe-nail machine	H. E. and C. W. Woodford	Keeseville, N. Y	Oct. 30, 1866	59, 327
Horseshoe-nail machine	S. Wootton	Boonton, N. J	Nov. 10, 1857	18, 617
Horseshoe-nail machine	C. D. Wrightington and B. P. Rider.	Fairhaven and Chelsea, Mass.	Sept. 10, 1867	68, 824
Horseshoe-nail manufacture	A. H. Caryl	Groton, Mass	July 18, 1871	117, 148
Horseshoe-nail pointer	W. T. Vann	Macomb, Ill	June 22, 1869	91, 797
Horseshoe-nail-pointing device	C. H. Perkins	Providence, R. I	Aug. 1, 1871	117, 560
Horseshoe-nail-pointing device	A. H. Wills	Vergennes, Vt	Aug. 1, 1871	117, 584
Horseshoe-nail, Steel-headed	J. H. Smith	Allegheny City, Pa	June 28, 1870	104, 785
Horseshoe-nails, Apparatus for pointing	C. H. Perkins	Providence, R. I	Sept. 26, 1871	119, 402
Horseshoe-nails, Clinching	J. Houck	Greencastle, Ind	Apr. 20, 1858	19, 993
Horseshoe-nails from plates, Die for punching	A. H. Caryl	Forge Village, Mass	Sept. 9, 1873	142, 678
Horseshoe-nails, Hammer-roll device for forming	T. Harris and B. Dunn	Coté St. Paul, Canada	Jan. 7, 1873	134, 664
Horseshoe-nails, Machine for cutting off	C. F. Johnson, jr	Oswego, N. Y	May 30, 1871	115, 480
Horseshoe-nails, Machine for feeding	S. S. Putnam	Boston, Mass	Nov. 4, 1873	144, 226
Horseshoe-nails, Machine for finishing	J. Huggett and J. A. Huggett	Eastbourne and Clapham, England.	Oct. 8, 1872	132, 082
Horseshoe-nails, Machine for finishing	C. H. Perkins and C. E. Sheridan.	Providence, R. I	Oct. 8, 1872	131, 967
Horseshoe-nails, Machine for finishing	S. S. Putnam	Boston, Mass	Nov. 25, 1873	144, 922
Horseshoe-nails, Machine for finishing	R. Ross	Vergennes, Vt	May 27, 1873	139, 331
Horseshoe-nails, Machine for finishing	R. Ross	Vergennes, Vt	Sept. 2, 1873	142, 352
Horseshoe-nails, Machine for finishing	H. A. Wills	Vergennes, Vt	Jan. 16, 1872	122, 876
Horseshoe-nails, Machine for forging	R. Cook	South Abington, Mass	June 9, 1857	17, 542
Horseshoe-nails, Machine for forging	J. and J. A. Huggett	Eastbourne, England	Apr. 20, 1869	89, 224
Horseshoe-nails, Machine for forging	C. Parkhurst and C. Weed	Boston, Mass	Aug. 19, 1856	15, 571
Herreshoe-nails, Machine for forging	S. S. Putnam	Dorchester, Mass	Jan. 11, 1870	98, 707
Horseshoe-nails, Machine for forging	F. H. Richards	New Britain, Conn	Aug. 19, 1873	142, 044
Horseshoe-nails, Machine for forging	J. Roy	Boston, Mass	Oct. 2, 1866	58, 485
Horseshoe-nails, Machine for manufacturing	G. D. Hall	Boston, Mass	Dec. 5, 1871	121, 511
Horseshoe-nails, Machine for pointing	D. Armstrong	Chicago, Ill	Sept. 29, 1868	82, 476
Horseshoe-nails, Machine for pointing	A. H. Caryl	Groton, Mass	May 16, 1871	114, 920
Horseshoe-nails, Machine for pointing extremities of.	H. A. Wills	Vergennes, Vt	July 16, 1872	129, 077
Horseshoe-nails, Machine for shaping heads of	N. Dexter	Bower Hill, Pa	Sept. 27, 1870	107, 670
Horseshoe-nails, Machine for stamping, pressing, and pointing.	F. Sandham	Montreal, Canada	Dec. 6, 1870	109, 844
Horseshoe-nails, Machine for swaging and trimming points of.	R. Ross	Vergennes, Vt	May 27, 1873	139, 332
Horseshoe-nails, Machinery for clinching	D. Kirk	Orleans, N. Y	Jan. 26, 1869	86, 166
Horseshoe-nails, Machinery for making	A. Whittemore	Cambridgeport, Mass	Oct. 2, 1866	58, 520
Horseshoe-nails, Manufacture	S. E. Chase	Boston, Mass	June 9, 1868	78, 644
Horseshoe-nails, Making	A. M. Polsey	Boston, Mass	July 6, 1869	92, 355
Horseshoe-nails, Making	A. H. Caryl	Groton, Mass	Sept. 24, 1872	131, 664
Horseshoe-nails, Manufacture of	H. D. Cowles	Bridgeport, Conn	Dec. 3, 1867	71, 710
Horseshoe-nails, Manufacture of	H. D. Cowles	Montreal, Canada	Apr. 15, 1873	137, 762
Horseshoe-nails, Manufacture of	H. D. Cowles	Montreal, Canada	Dec. 9, 1873	145, 336
Horseshoe-nails, Manufacture of	J. C. Paige	Stoneham, Mass	Sept. 2, 1873	142, 506
Horseshoe-nails, Manufacture of	D. Turbayne	Boston, Mass	Sept. 9, 1873	142, 597
Horseshoe-nails, Manufacture of	D. Turbayne and G. M. Wyman	Boston, Mass	Jan. 14, 1873	134, 948
Horseshoe-nails, Manufacture of	H. R. Underhill	Derry, N. H	Dec. 31, 1872	134, 448
Horseshoe-nails, Roll for rolling iron for	J. Huggett and J. A. Huggett	Eastbourne and Clapham, England.	May 6, 1873	138, 499
Horseshoe-nails, Tool for clinching	J. E. Draper	Northville, Mich	Aug. 6, 1861	32, 989
Horseshoe-nails, Tool for clipping and clinching	J. E. Giles	Marshall, Mich	Feb. 23, 1864	41, 695
Horseshoe-rubber pad	D. L. Corbin	Friendship, N. Y	Jan. 10, 1871	110, 828
Horseshoe, Sectional	T. P. Randall	Adrian, Mich	Nov. 26, 1872	133, 337
Horseshoe-sharpener	H. Thompson	Philadelphia, Pa	Feb. 6, 1872	123, 523
Horseshoe-stretcher	J. D. Abbott	Manchester, N. H	Sept. 9, 1873	142, 547
Horseshoe, Supplemental	T. P. Handy and C. B. Kleibacker.	Baltimore, Md	May 11, 1869	89, 991
Horseshoe, Temporary	W. H. Halsey	Philadelphia, Pa	May 30, 1871	115, 311
Horseshoe toe-calk	G. Custer	Monroe, Mich	Feb. 12, 1867	62, 016
Horseshoe toe-calk	G. Custer	Monroe, Mich	Apr. 12, 1870	101, 831
Horseshoe toe-calk die	P. F. Burke	Worcester, Mass	July 6, 1869	92, 259
Horseshoe toe-calk die	E. R. Cheney	South Boston, Mass	Aug. 11, 1868	80, 807
Horseshoe toe-calks, Die for welding and forming	J. Flynn	Portland, Conn	Jan. 30, 1872	123, 256
Horseshoe-tool	R. Stout	Matteawan, N. Y	Apr. 5, 1870	101, 538
Horseshoes, Apparatus for swaging and welding toe-calks to.	A. Cook	Hillsdale, Mich	Apr. 27, 1869	89, 384
Horseshoes, Attaching	H. Schreiner	Philadelphia, Pa	Oct. 29, 1867	70, 274
Horseshoes, Attaching	J. Wagner	Washington, D. C	Mar. 10, 1868	75, 494
Horseshoes, Attaching calks to	J. Forbes	Plainwell, Mich	Sept. 24, 1867	69, 201
Horseshoes, Attaching calks to	P. Laflin	Warren, Mass	Dec. 7, 1869	97, 525
Horseshoes, Attaching soles to	A. H. Knapp	Coxsackie, N. Y	May 29, 1866	55, 119
Horseshoes, Attaching elastic soles to	W. Somerville	Buffalo, N. Y	Sept. 29, 1857	18, 306
Horseshoes, chain-links, &c., Machine for forming	J. F. Winslow and T. Bigwood	Troy, N. Y	Aug. 11, 1843	3, 217
Horseshoes, Detachable calk for	K. Goddard	Richmond, N. Y	May 25, 1869	90, 442
Horseshoes, Device for attaching	H. B. Ferren	Batavia, N. Y	Sept. 15, 1868	82, 215
Horseshoes, Device for forming	E. Cate	Franklin, N. H	Mar. 26, 1861	31, 780
Horseshoes, Device for forming	U. Stewart	Berlin, Wis	Jan. 9, 1866	51, 979
Horseshoes, Expanding	B. P. Sargent	Sutton, N. H	Oct. 25, 1853	10, 161
Horseshoes, Machine for forging	R. Griffiths and G. Shield	Allegheny City, Pa., and Cincinnati, Ohio.	Dec. 19, 1854	12, 112
Horseshoes, Machine for forming	D. F. Hofstatter and W. H. Shaw.	New London, Ohio	Apr. 15, 1873	137, 923
Horseshoes, Machine for punching	C. H. Perkins	Providence, R. I	Nov. 20, 1860	30, 689
Horseshoes, Machine for punching	S. D. Turner	Providence, R. I	Sept. 5, 1865	49, 838
Horseshoes, Machine for punching holes in, and bending the calks on.	A. Oehme	Rock Island, Ill	Oct. 17, 1871	120, 088
Horseshoes, Machine for rolling bars for	A. Reese	McClure Township, Pa	Nov. 23, 1869	97, 117
Horseshoes, Machine for the manufacture of	A. J. Roberts	Boston, Mass	Aug. 20, 1867	67, 910
Horseshoes, Machinery for manufacturing	W. Gibson	Troy, N. Y	Dec. 27, 1843	3, 392
Horseshoes, Making	W. Morehouse	Buffalo, N. Y	Dec. 22, 1868	85, 119
Horseshoes, Making steel	I. Peacock	Shortsville, N. Y	Jan. 20, 1863	37, 458
Horseshoes, Manufacturing	S. King	Sullivan, N. Y	Apr. 1, 1830	....
Horseshoes, &c., Metal for	J. Kennelly	Hartford, Conn	Mar. 31, 1863	38, 046
Horseshoes, Mold for casting	J. Kennelly	Hartford, Conn	Mar. 31, 1863	38, 047

Index of patents issued from the United States Patent Office from 1790 *to* 1873, *inclusive*—Continued.

Invention.	Inventor.	Residence.	Date.	No.
Horseshoes, &c., Process of treating cast iron for the manufacture of.	J. R. Speer	Pittsburgh, Pa	July 6, 1869	92, 220
Horseshoes, Removable calk for	E. Whitehead	Cincinnati, Ohio	Aug. 10, 1869	93, 654
Horseshoes, Securing	L. Carpenter	Lake City, Minn	June 21, 1864	43, 179
Horseshoes, Securing	W. H. Forker	Meadville, Pa	Oct. 4, 1864	44, 522
Horseshoes, Securing	W. C. Stickney and H. B. Taylor.	Putnam, Ohio	Oct. 18, 1864	44, 754
Horseshoes, Securing	C. Weitman	Hazleton, Iowa	Oct. 4, 1864	44, 569
Horseshoes, Toe-calk for	C. H. Perkins	Providence, R. I	Apr. 9, 1861	31, 995
Horseshoes, Tool for making	T. G. Thompson	Oswego, N. Y	Nov. 6, 1866	59, 479
Horseshoeing	P. Charlier	Paris, France	Apr. 23, 1867	64, 073
Horseshoeing	G. R. Stevens	Clarksville, Mo	July 3, 1860	29, 016
Horseshoeing-apparatus	J. B. Brusoe	Lula, Ill	Jan. 30, 1872	123, 231
Horseshoeing-apparatus	N. Warlick	La Fayette, Ala	Aug. 29, 1854	11, 624
Horseshoeing rest	S. S. Blackburn	Fredericktown, Ohio	Apr. 27, 1869	89, 379
Horseshoeing-rest	G. Stansel	Saint Johnsville, N. Y	Oct. 17, 1871	120, 119
Horseshoeing, Tool for clinching nails in	D. W. Bush	Clarence, Mo	Aug. 27, 1867	68, 040
Horseshoer's rest	J. Legg	Rushville, N. Y	Aug. 12, 1873	141, 651
Horse, Spring	H. F. Metzler	New York, N. Y	Feb. 21, 1865	46, 529
Horse, Spring	H. F. Metzler	New York, N. Y	May 5, 1868	77, 513
Horse, Spring rocking	J. A. Crandell	New York, N. Y	Feb. 9, 1864	41, 570
Horse, Spring rocking	C. B. Northrup	New York, N. Y	May 31, 1864	42, 960
Horse-stocking	W. Lewis	Astoria, N. Y	Dec. 12, 1871	121, 880
Horse-tail holder	G. Gray	West Chester, Pa	July 31, 1846	4, 680
Horse-tail holder	F. A. Roberts	North Vassalborough, Me	Mar. 22, 1870	101, 043
Horse-tail protector	C. A. Warren	Watertown, Conn	May 21, 1872	127, 002
Horse-taming apparatus	J. M. Lanier	Eufaula, Ala	Nov. 29, 1859	26, 272
Horse-taming apparatus	D. Springer	Pontiac, Mich	Oct. 8, 1872	131, 978
Horse-taming bit	A. L. Weymouth	Boston, Mass	July 1, 1862	35, 796
Horse-tie, Stable	E. D. Cramer	Hackettstown, N. J	Oct. 19, 1869	95, 883
Horse-troughs, &c., Float-valve for	J. Johnson	New York, N. Y	May 6, 1873	138, 659
Horse-wheel, Hollow inclined	J. Keller and S. B. Burt	Cincinnati, Ohio	Apr. 23, 1824	....
Horse-wheel, Inclined	J. Bradford	Lexington, Ky	May 14, 1824	....
Horse-wheel, Inclined	E. Holliday	Schoharie, N. Y	May 5, 1826	....
Horse-wheel, Inclined	J. B. Robinson	Cincinnati, Ohio	Oct. 14, 1816	....
Horse-wheel, Vertical	J. A. Morton	Baltimore, Md	July 21, 1815	....
Horses, &c., across rivers, Floating	S. P. Heintzelman	Newport Barracks, Ky	Nov. 24, 1857	18, 691
Horses, Administering balls to	C. P. Moyer	Womelsdorf, Pa	Nov. 5, 1867	70, 600
Horses and cattle, Breeding-instrument for	C. Addle	Winthrop, Me	May 9, 1835	....
Horses and cattle, Breaking	D. Cooley	Orange County, N. Y	June 23, 1810	....
Horses and cattle, Condition-powder for	J. H. Woolrich	Woburn, Mass	June 7, 1870	104, 094
Horses and other animals, Apparatus for clipping	J. Tidmarsh	Twickenham, England	Nov. 9, 1869	96, 742
Horses and shafts from carriages, Apparatus for discharging.	G. Hubbard	Sandisfield, Mass	May 26, 1857	17, 376
Horses, Anti-kicking attachment for	O. H. P. Fancher	New York, N. Y	Oct. 29, 1867	70, 185
Horses, Apparatus for breaking	S. Tomlinson	Washington Hollow, N. Y	Sept. 3, 1850	7, 625
Horses, Apparatus for cleaning	O. W. Allison and P. Homeliers	Buffalo, N. Y	Mar. 25, 1873	137, 045
Horses, Apparatus for clipping hair from	S. H. Salom and J. Field	London and Westminster, Great Britain.	June 1, 1869	90, 877
Horses, Apparatus for detaching	J. L. Hamilton	Saint Joseph, Mo	May 9, 1871	114, 553
Horses, Apparatus for detaching	J. C. Hancock and E. P. Richardson.	Charlestown and Somerville, Mass.	July 12, 1870	105, 198
Horses, Apparatus for detaching	H. Latshaw	McKnightstown, Pa	May 20, 1873	139, 068
Horses, Apparatus for grooming	J. J. Greenough	Syracuse, N. Y	Aug. 29, 1871	118, 449
Horses, Apparatus for handling vicious	O. Turner	Clinton, Wis	Oct. 11, 1864	44, 680
Horses, Apparatus for shearing	W. Harrison	Chicago, Ill	Mar. 18, 1873	136, 993
Horses, Apparatus for stopping runaway	S. M. Cooper	Fairfax County, Va	Sept. 10, 1867	68, 606
Horses, Apparatus for throwing	S. Shaw	Russellville, Ohio	Oct. 11, 1864	44, 665
Horses, Apparatus for treating	D. Sullivan	Washington, D. C	May 20, 1873	139, 207
Horses attached to carriages, Device for holding	A. Moore and J. Aylwerd	Mission of San José, Cal	June 1, 1869	90, 865
Horses attached to vehicles, Apparatus for checking.	M. O'Connell	Boston, Mass	July 21, 1868	80, 085
Horses, &c., Attaching and detaching	J. K. Harris	Madison, Ind	May 14, 1867	64, 666
Horses, Attaching shoes to	T. H. Ince	Westminster, England	May 29, 1866	55, 215
Horses, Awning for	W. McCormick	Philadelphia, Pa	Nov. 22, 1870	109, 532
Horses, Boot for	J. Fennell	Cynthiana, Ky	Dec. 10, 1872	133, 768
Horses, Breeching-attachment for	W. F. Schatz	Columbus, Ohio	Aug. 23, 1870	106, 732
Horses, Bridle for stopping	P. Laporte	Louisa County, Va	Mar. 20, 1821	....
Horses, Clipping the hair of	G. F. Evans	Norway, Me	Sept. 13, 1864	44, 171
Horses, Combined boot and shoe for	H. G. and E. M. Haedrich	Philadelphia, Pa	July 5, 1870	105, 068
Horses, Crib for	H. Eddy	North Bridgewater, Mass	Aug. 26, 1862	36, 287
Horses, Curing blindness in	J. Sater	Huntsville, N. C	Aug. 14, 1821	....
Horses, Detaching	P. T. Share	Baltimore, Md	Mar. 18, 1836	....
Horses, Device for attaching and detaching	G. L. Du Laney	Mechanicsburgh, Pa	May 25, 1869	90, 352
Horses, Device for checking	T. Itzstein	New York, N. Y	Mar. 2, 1869	87, 497
Horses, Device for detaching	E. P. Jones	Shell Mound, Miss	Dec. 9, 1873	145, 426
Horses, Device for detaching	R. R. Jones	Pillar Point, N. Y	Sept. 30, 1873	143, 286
Horses, Device for detaching	I. L. Landis	Lancaster, Pa	Dec. 16, 1863	145, 662
Horses, Device for detaching	A. Schmitt	Cincinnati, Ohio	Sept. 27, 1870	107, 727
Horses, Device for detaching	S. Sykes	Portland, Oreg	Oct. 7, 1873	143, 428
Horses, Device for detaching runaway	E. P. Connor	Jeffersonville, Ohio	Jan. 1, 1867	60, 836
Horses, Device for fettering	A. H. Lewis	North Greenbush, N. Y	Apr. 14, 1868	76, 640
Horses, Device for forming the curve in tails of	I. B. Phillips	Philadelphia, Pa	Aug. 27, 1872	130, 939
Horses, Device for moistening the legs of	G. J. Harris	New York, N. Y	June 11, 1872	127, 875
Horses, Device for preventing and curing cribbiting in.	G. H. White	Huntington, N. Y	May 5, 1868	77, 689
Horses, Device for singeing	J. Alexander	Nashua, N. H	Dec. 15, 1868	84, 983
Horses, Device for spurring or driving	J. Davis	Northampton, Ill	Nov. 7, 1865	50, 803
Horses, Device for training	O. S. Pratt	Batavia, N. Y	Oct. 29, 1872	132, 689
Horses, Dock-holder for	E. H. Penfield	Middletown, Conn	Dec. 12, 1854	12, 068
Horses, Dock-holder for	S. Tomlinson	Pleasant Valley, N. Y	July 18, 1854	11, 341
Horses' ears, Improving or setting	S. Janes	....	Dec. 31, 1804	....
Horses, Elastic heel-guard for	W. H. Hall	New Gloucester, Me	June 8, 1869	91, 122
Horses, Fastening for	P. Ricord and J. R. Scattergood	Newark, N. J	May 18, 1869	90, 196
Horses, Feeding	A. Carman	Hyde Park, N. Y	Nov. 27, 1828	....
Horses' feet, Device for measuring	M. S. Woodward	Marshallton, Pa	Oct. 29, 1867	70, 307
Horses' feet, Elastic cushion and guard for	W. H. Hall	New Gloucester, Me	May 1, 1866	54, 335

Index of patents issued from the United States Patent Office from 1790 *to* 1873, *inclusive*—Continued.

Invention.	Inventor.	Residence.	Date.	No.
Horses' feet, Elastic protector for	W. H. Hall	Boston, Mass	Mar. 8, 1870	100, 618
Horses' feet, Stopping for	A. S. Wilkinson	Pawtucket, R. I	Aug. 28, 1866	57, 611
Horses, Foot-rest for	J. E. Tucker	Montfort, Wis	Feb. 5, 1867	61, 897
Horses from biting and crib-biting, Device for preventing.	B. D. Howe	Hanover, N. H	July 16, 1867	66, 840
Horses from carriages, and drag for arresting the motion thereof, Manner of combining an apparatus for disengaging.	J. Harlacher	Lancaster, Pa	Nov. 4, 1842	2, 839
Horses from carriages, Apparatus for detaching	T., jr., and J. Byrd	Williamsville, Mich	Mar. 19, 1872	124, 790
Horses from carriages, Apparatus for detaching	J. Davis	Elmira, N. Y	Nov. 6, 1860	30, 566
Horses from carriages, Apparatus for detaching	G. Gabriel	Pittsburgh, Pa	Sept. 15, 1868	82, 217
Horses from carriages, Apparatus for detaching	S. Kepner	Pottstown, Pa	June 15, 1869	91, 346
Horses from carriages, Apparatus for releasing	T. B. Pyron	Salina, Tenn	Oct. 22, 1850	7, 733
Horses from carriages, Danger-escape to disengage	W. Start	Greensborough, Md	May 3, 1816	....
Horses from carriages, Detaching	O. R. Broyles	Anderson Court-House, S. C	Mar. 17, 1834	....
Horses from carriages, Detaching	J. W. Harrison	Logansport, Ind	Aug. 27, 1850	7, 600
Horses from carriages, Detaching	E. P. Jones	Shell Mound, Miss	Nov. 9, 1869	96, 703
Horses from carriages, Detaching	S. Kepner	North Coventry, Pa	Dec. 18, 1860	30, 972
Horses from carriages, Detaching	J. Madden	Warren, Ohio	July 9, 1844	3, 652
Horses from carriages, Detaching	E. Myers	New Holland, Pa	Apr. 10, 1847	5, 055
Horses from carriages, Detaching	J. Summers	Kokomo, Ind	Mar. 19, 1872	124, 703
Horses from carriages, Detaching	H. C. Wiatt	Weldon, N. C	Oct. 29, 1846	4, 828
Horses from carriages, Device for detaching	C. McElroy	New Baltimore, Mich	Apr. 20, 1869	89, 159
Horses from carriages, Device for detaching	C. McElroy	New Baltimore, Mich	July 6, 1869	92, 333
Horses from carriages, Disengaging	D. Little	Gettysburgh, Pa	Mar. 30, 1843	3, 020
Horses from carriages, Disengaging	J. S. Shnell	Shiremantown, Pa	Apr. 13, 1844	3, 536
Horses from carriages, Releasing	J. T. Guthrie	Leesburgh, Ohio	Jan. 9, 1866	51, 941
Horses from carriages, Releasing	G. Kintzi	Exeter, Pa	May 4, 1842	2, 601
Horses from carriages suddenly, Splinter-bar and swingle-tree to disengage.	H. Dunlap	Georgetown, D. C	Apr. 23, 1807	....
Horses from carts, Device for releasing	G. H. Meier	Cincinnati, Ohio	July 1, 1873	140, 521
Horses from cribbing, Apparatus to prevent	S. S. Bent	Port Chester, N. Y	June 30, 1868	79, 437
Horses from cribbing, Device for preventing	W. H. Bishop and A. H. Low	Warren, Mass	July 31, 1860	29, 354
Horses from cribbing, Device for preventing	G. G. Hickman	Coatesville, Pa	Oct. 1, 1867	69, 341
Horses from cribbing, Device for preventing	R. Jennings and J. A. Marshall	Bordentown, N. J	Jan. 29, 1867	61, 666
Horses from cribbing, Device for preventing	A. Stillwell	Dwaar's Kill, N. Y	Aug. 19, 1873	141, 962
Horses from crowding, Device for preventing	O. C. Ross	Penfield, N. Y	July 29, 1873	141, 386
Horses from dragging weights, Device for preventing.	A. Pope, jr	Dorchester, Mass	Apr. 28, 1868	77, 317
Horses from fire, Apparatus for rescuing	E. French	Braintree, Mass	July 31, 1860	29, 370
Horses from flies, Device for protecting the necks of	E. L. Kurtz	New York, N. Y	Aug. 4, 1857	17, 934
Horses from interfering, Device to prevent	W. Somerville	New York, N. Y	Apr. 24, 1860	28, 017
Horses from kicking, Apparatus to prevent	E. H. Gammon	Batavia, Ill	May 22, 1866	54, 887
Horses from kicking in the stable, Apparatus for preventing.	W. Itschner	Philadelphia, Pa	Nov. 16, 1869	96, 812
Horses from running away, Apparatus for preventing.	W. Hall	Indianapolis, Ind	Oct. 18, 1859	25, 822
Horses from running away, Apparatus for preventing.	C. C. Moore and R. Blair	Philadelphia, Pa	July 16, 1872	129, 358
Horses from running away, Preventing	E. E. Burnham and G. Brown	Gloucester, Mass	Sept. 10, 1867	68, 600
Horses from stables in case of fire, Releasing	E. N. Moore	Boston, Mass	May 16, 1848	5, 583
Horses from sun-stroke, Shield for protecting	J. Anderson	Brooklyn, N. Y	July 6, 1869	92, 142
Horses from vehicles, Apparatus for detaching	T. Y. Vancleve	Cornersville, Tenn	Jan. 21, 1873	135, 177
Horses from vehicles, Apparatus for unhitching	J. K. Harris	Madison, Ind	Feb. 19, 1867	62, 132
Horses from vehicles, Attaching and detaching	G. Switzer	Washington, D. C	Aug. 9, 1870	106, 293
Horses from vehicles, Detaching	F. M. English	Hopkinsville, Ky	June 24, 1856	15, 179
Horses from vehicles, Detaching	A. B. Johnson, M. H. Vaughan, and J. Stinnett.	Clarksville, Ark., and Shelby County, Tenn.	July 3, 1860	28, 988
Horses from vehicles, Detaching	W. D. Mayfield	Bloomington, Ill	Oct. 13, 1857	18, 426
Horses from vehicles, Detaching	J. M. Roberds	Plaisance, La	July 24, 1860	29, 313
Horses from vehicles, Device for detaching	G. B. Lumpkin	Lexington, Ga	July 25, 1871	117, 306
Horses from vehicles, Device for detaching	B. A. McConnaughey	New Market, Ohio	Dec. 18, 1866	60, 533
Horses from vehicles, Device for detaching	J. McNeill	Wells, Minn	June 10, 1873	139, 803
Horses from vehicles, Disconnecting	J. Rancevan	Carthage, Ohio	Nov. 12, 1867	70, 897
Horses from vehicles, Unhitching	O. N. Weaver	Dover, Ky	Feb. 5, 1867	61, 782
Horses from whiffletrees, Mechanism for disengaging.	J. B. Goldsmith	Rockport, Mass	June 11, 1872	127, 760
Horses, Harnessing	J. Murray, jr	Worcester, Mass	Apr. 16, 1872	125, 752
Horses, Head-gear for stopping runaway	J. Koehler	New York, N. Y	Apr. 17, 1860	27, 952
Horses, Heel-boot for	W. Mathis	New York, N. Y	Nov. 18, 1873	144, 782
Horses in carriages from falling, Apparatus to prevent.	R. D. Dwyer	Richmond, Va	Mar. 11, 1856	14, 395
Horses in navigating canals, Disengaging	G. Shultz	Philadelphia, Pa	Sept. 9, 1835	....
Horses, India-rubber fender for interfering	C. Brinckerhoff	Fishkill, N. Y	Dec. 22, 1868	85, 205
Horses, Insect-guard for	B. Treadwell	Reading, Conn	Jan. 28, 1868	73, 851
Horses, Knee-boot for	J. Finlay	New York, N. Y	Dec. 8, 1868	84, 813
Horses, &c., Knee-clasp for	J. A. Warden	Minnesota Station, Wis	Aug. 10, 1869	93, 570
Horses, Making pads, caps, blinds, &c., for	S. Boyden	Newark, N. J	Apr. 10, 1824	....
Horses, Means for stopping	N. Fountain	New York, N. Y	Aug. 18, 1868	81, 157
Horses, Nose-bag for	H. D. McGovern	Brooklyn, N. Y	June 17, 1873	140, 058
Horses of cribbing, Apparatus to cure	G. W. Metcalfe	Hummelstown, Pa	July 2, 1867	66, 375
Horses or mules to rack, Apparatus for training	C. Daniels	Barnwell Court-House, S. C	Apr. 30, 1861	32, 185
Horses, Preventing interfering in	E. M. Gardner	Nantucket, Mass	May 31, 1864	42, 937
Horses, Quarter-boot for	W. H. Jones	Boston, Mass	June 23, 1868	79, 128
Horses, Remedy for spavin in	S. E. Thayer	Manchester, Vt	Aug. 27, 1867	68, 260
Horses, Rubber boot for	J. H. Jennings	Cambridgeport, Mass	May 28, 1867	65, 231
Horses, Safety-check for	J. W. Domet	Boston, Mass	Jan. 28, 1873	135, 208
Horses, Securing and releasing	C. Wills	New York, N. Y	Nov. 19, 1867	71, 107
Horses, Shaving	F. J. Eldred	Webster, N. Y	Oct. 3, 1871	119, 512
Horses, Sun-bonnet for	J. Anderson	Brooklyn, N. Y	Feb. 22, 1870	100, 000
Horses, Sun-bonnet for	J. Anderson	Brooklyn, N. Y	June 3, 1873	139, 491
Horses, Sun-shade for	E. T. Balch	Philadelphia, Pa	Feb. 1, 1870	99, 282
Horses, Sun-shade for	J. L. McIntosh	Brooklyn, N. Y	Aug. 3, 1869	93, 325
Horses, Sun-shade for	S. Ruth	Philadelphia, Pa	Aug. 25, 1868	81, 412
Horses, Sun-shade for	S. Ruth	Philadelphia, Pa	Jan. 7, 1873	134, 564
Horses, Sun-shield for	H. D. McGovern	Brooklyn, N. Y	Oct. 24, 1871	120, 208
Horses, Sun-shield for	H. D. McGovern	Brooklyn, N. Y	July 2, 1872	128, 554
Horses, Suspender for fly-guards upon	G. Shelton	Normal, Ill	Apr. 22, 1873	138, 205

Index of patents issued from the United States Patent Office from 1790 *to* 1873, *inclusive*—Continued.

Invention.	Inventor.	Residence.	Date.	No.
Horses, Throat-check for	S. French	Boston, Mass	Dec. 26, 1871	122, 114
Horses, Throwing	O. S. Pratt	Batavia, N. Y	Feb. 13, 1872	123, 726
Horses to and from carriages, Apparatus for attaching and detaching.	P. W. Hardwick	Williamsburgh, Ind	Jan. 28, 1862	34, 253
Horses to and from vehicles, Attaching and detaching.	A. Colburn	Leominster, Mass	Mar. 18, 1862	34, 671
Horses to carriages, Attaching	L. W. Boynton	Hartford, Conn	July 5, 1864	43, 388
Horses to carriages, Attaching	W. H. Townsend	Camden, Ohio	Sept. 24, 1867	69, 273
Horses to machinery, Applying draft	W. Hart	Accomack, Va	Jan. 22, 1817	....
Horses to one-wheeled carriages, Attaching	J. R. Remington	Lowndes County, Ala	Apr. 4, 1846	4, 436
Horses to plows, Attaching	J.D.Filkins and W.H.De Puy	Lima, Ind	Oct. 25, 1853	10, 153
Horses to shafts of vehicles, Attaching	G. B. Kaign	Lumberton, N. J	June 10, 1856	15, 103
Horses to shoe them, Apparatus for raising and securing the legs of.	J. P. Champion	Phelps, N. Y	Nov. 5, 1867	70, 523
Horses to two-wheeled vehicles, Attaching	J. W. Barnes	Murfreesborough, N. C	Oct. 23, 1860	30, 455
Horses to vehicles, Apparatus for attaching	C. Leroy	Mexico, N. Y	Oct. 6, 1868	82, 852
Horses to vehicles, Attaching	J. K. Andrews	East Lampeter Township, Pa.	Feb. 1, 1870	99, 278
Horses to vehicles, Attaching	A. Avery	South Windham, Conn	Sept. 18, 1860	30, 037
Horses to vehicles, Attaching	G. H. Gray	Clinton, Miss	Sept. 9, 1856	15, 692
Horses to vehicles, Attaching	E. D. Lockwood	Penfield, N. Y	Apr. 5, 1859	23, 476
Horses to vehicles, Device for connecting	A. Shaler	New York, N. Y	Feb. 1, 1870	99, 357
Horses to vehicles, Mechanism for connecting	D. Belcher	Easton, Mass	Mar. 30, 1869	88, 263
Horses to wagons, Attaching	M. Knapp	Springwater, N. Y	Sept. 5, 1846	4, 743
Horses, Training and breaking	T. Cock	New York, N. Y	Mar. 1, 1811	....
Horses, Weight and hitch-strap for fastening	D. S. Bartlett	Roxbury, Mass	Jan. 15, 1861	31, 104
Horses while in harness, Refreshing	W. Compton	Newark, N. J	July 13, 1869	92, 425
Hose	A. Carney	New York, N. Y	Mar. 31, 1868	76, 049
Hose	H. H. Day	Jersey City, N. J	Aug. 16, 1845	4, 150
Hose	B. F. Lee	New York, N. Y	Jan. 4, 1870	98, 603
Hose	A. S. Libby	Lawrence, Mass	Aug. 6, 1872	130, 303
Hose	A. S. Libby	Lawrence, Mass	Nov. 12, 1872	133, 044
Hose	E. L. Perry	New York, N. Y	Dec. 29, 1868	85, 395
Hose	J. Sharp	Cincinnati, Ohio	Feb. 6, 1872	123, 365
Hose	G. C. Smith	Matteawan, N. Y	May 11, 1869	90, 031
Hose and belting, Machine for cutting India-rubber sheets into bands for.	T. J. Mayall	Boston, Mass	Apr. 9, 1872	125, 597
Hose and flexible tubes, Manufacture of	H. A. Alden	Fishkill, N. Y	Dec. 16, 1862	37, 192
Hose and machine for making	G. Coles, J. A. Jaques, and J. A. Fanshawe.	London and Tottenham, England.	Oct. 20, 1868	83, 132
Hose and pipe connection	G. E. Nutting	New York, N. Y	Dec. 24, 1872	134, 158
Hose and pipe coupling	T. E. Button	Waterford, N. Y	Apr. 15, 1873	137, 823
Hose and pipe coupling	M. S. Curtis and W. D. Tewksbury.	New York, N. Y	Mar. 31, 1868	76, 057
Hose and pipe coupling	F. Shaller	Hudson, N. Y	Apr. 12, 1870	101, 927
Hose and pipe coupling	J. C. Thompson	Charlestown, Mass	Oct. 26, 1869	96, 286
Hose and pipe coupling	C. H. Weston	Lowell, Mass	Mar. 16, 1869	87, 993
Hose and tubing	T. J. Mayall	Roxbury, Mass	July 14, 1863	39, 237
Hose and tubing for water, steam, &c., Manufacture of.	S. P. Cook	San Francisco, Cal	Apr. 2, 1872	125, 175
Hose and tubing, Machinery for making India-rubber.	T. J. Mayall	Boston, Mass	Apr. 9, 1872	125, 598
Hose and tubing, Manufacture of elastic	T. J. Mayall	Roxbury, Mass	July 14, 1863	39, 238
Hose and tubing, Vulcanized rubber and combination.	J. Murphy	New York, N. Y	July 25, 1871	117, 447
Hose and tubes with India rubber, &c., Lining flexible and other.	J. B. Forsyth	Boston, Mass	June 23, 1868	79, 220
Hose, Apparatus for oiling, testing, and washing fire-engine.	P. Noyes	Lowell, Mass	Aug. 19, 1873	141, 885
Hose, belting, &c., Machine for making India-rubber.	J. B. Forsyth	Roxbury, Mass	Sept. 13, 1864	44, 254
Hose-bridge	E. Batzel	Philadelphia, Pa	Oct. 7, 1873	143, 400
Hose-bridge	W. E. Cameron	Taunton, Mass	Oct. 4, 1870	107, 870
Hose-bridge	P. Daily	New York, N. Y	Mar. 7, 1871	112, 427
Hose-bridge	W. Donoghue and F. L. Charlton.	Philadelphia, Pa	Nov. 9, 1869	96, 559
Hose-bridge	J. S. Hagerty	Baltimore, Md	Oct. 29, 1872	132, 660
Hose-bridge	L. W. Hodges	Wilmington, N. C	May 14, 1872	126, 635
Hose-bridge	J. P. Maxwell	Baltimore, Md	Nov. 12, 1872	133, 045
Hose-bridge	J. E. McCaullay	Philadelphia, Pa	Mar. 11, 1873	136, 745
Hose-bridge	G. J. Orr and C.E.Gildersleeve	New York, N. Y	Jan. 28, 1873	135, 285
Hose-bridge, Portable	A.L.Wilkinson and E.Y.Beggs	Huntsville, Ala., and Nashville, Tenn.	Dec. 3, 1867	71, 671
Hose-bridge, Railway	J. Burson	Yates City, Ill	Mar. 30, 1869	88, 364
Hose-carriage	N. S. Bean	Manchester, N. H	June 11, 1867	65, 634
Hose-carriage	W. Boate	Philadelphia, Pa	June 22, 1869	91, 705
Hose-carriage	A. A. Justin	Buffalo, N. Y	Aug. 29, 1871	118, 615
Hose-carriage	K. Perrine and S. M. Stewart	Rochester, N. Y	Apr. 18, 1865	47, 330
Hose carriage	J. W. Wilor, S. B. Sturges, and G. McFall.	Mansfield, Ohio	Oct. 13, 1857	18, 421
Hose-cart	W. E. Shaw and C. A. Ashley	Stockton, Cal	May 28, 1872	127, 277
Hose, Clamp for closing ruptures in fire	P. H. Collins	Philadelphia, Pa	Apr. 30, 1867	64, 285
Hose-clasp	T. J. Pettit	Brooklyn, N. Y	Aug. 1, 1871	117, 563
Hose, Compound for lining textile	J. Dolman and F. W. Claessens	Boston, Mass	Dec. 7, 1869	97, 486
Hose connection	P. Schemmals	Chicago, Ill	Sept. 27, 1864	44, 487
Hose-coupling	A. F. Allen	Providence, R. I	Apr. 27, 1869	89, 455
Hose-coupling	A. F. Allen	Providence, R. I	Jan. 30, 1872	123, 070
Hose-coupling	A. E. Barnard	Paterson, N. J	Jan. 15, 1861	31, 103
Hose-coupling	W. G. Bedford	Philadelphia, Pa	Nov. 26, 1867	71, 266
Hose-coupling	W. H. Bliss	Newport, R. I	Feb. 25, 1862	34, 476
Hose-coupling	J. R. Buchanan	Chicago, Ill	Sept. 3, 1867	68, 348
Hose-coupling	H. C. Bull and S. T. Shelley	Louisville, Ky	Oct. 27, 1868	83, 455
Hose-coupling	L. Button and R. Blake	Waterford, N. Y	Jan. 10, 1860	26, 749
Hose-coupling	A. W. Cary	Brockport, N. Y	Feb. 10, 1852	8, 716
Hose coupling	J. T. Condon and F. Jeffers	New Orleans, La., and Pawtucket, R. I.	Dec. 24, 1872	134, 195
Hose-coupling	J. C. Cooke	Middletown, Conn	Nov. 30, 1858	22, 166
Hose-coupling	M. S. Curtis	New York, N. Y	Oct. 15, 1867	69, 778

Index of patents issued from the United States Patent Office from 1790 to 1873, inclusive—**Continued.**

Invention.	Inventor.	Residence.	Date.	No.
Hose, &c., coupling	M. S. Curtis and W. D. Tewksbury.	New York, N. Y	July 16, 1867	66, 804
Hose-coupling	C. H. Cushman	Alexandria, Va	Mar. 29, 1870	101, 403
Hose-coupling	C. H. Cushman	Alexandria, Va	June 28, 1870	104, 837
Hose-coupling	J. W. Douglas	Middletown, Conn	Apr. 30, 1867	64, 292
Hose-coupling	J. Edson	Boston, Mass	Mar. 9, 1869	87, 554
Hose-coupling	J. Edson	Boston, Mass	Apr. 12, 1870	101, 841
Hose-coupling	J. Edson	Boston, Mass	Dec. 23, 1873	145, 731
Hose-coupling	R. J. Falconer	Washington, D. C	June 7, 1853	9, 768
Hose-coupling	L. M. Ferry	Chicopee, Mass	Oct. 7, 1856	15, 846
Hose-coupling	H. Feyh	Columbus, Ohio	Oct. 8, 1861	32, 430
Hose-coupling	J. E. Ford	Memphis, Tenn	Feb. 27, 1866	52, 839
Hose-coupling	L. B. Forester	Clyde, Mich	Aug. 11, 1868	80, 820
Hose-coupling	R. J. Gaines	Portland, Conn	Apr. 14, 1868	76, 622
Hose-coupling	A. M. George	Nashua, N. H	Oct. 25, 1864	44, 794
Hose-coupling	J. H. George	Newark, N. J	July 27, 1869	92, 954
Hose-coupling	G. S. Goble	Saint Louis, Mo	Sept. 2, 1873	142, 388
Hose-coupling	S. Groom	Troy, N. Y	July 12, 1853	9, 841
Hose-coupling	S. Groom	Troy, N. Y	Apr. 3, 1855	12, 626
Hose-coupling	S. Groom	Troy, N. Y	Mar. 29, 1859	23, 368
Hose-coupling	W. Hamilton	Chicopee, Mass	June 30, 1868	79, 343
Hose-coupling	G. Hancock	Providence, R. I	Aug. 7, 1860	29, 548
Hose-coupling	L. W. Hanson and S. Bush	Springfield, Mass	July 28, 1868	80, 407
Hose-coupling	C. F. Hartwig	West Haven, Conn	Feb. 11, 1868	74, 219
Hose-coupling	R. Heneage	Buffalo, N. Y	Aug. 16, 1859	25, 117
Hose-coupling	R. Heneage	Buffalo, N. Y	July 10, 1866	56, 223
Hose coupling	L. E. Hicks	Boston, Mass	May 23, 1855	12, 937
Hose-coupling	B. Holly	Lockport, N. Y	Nov. 20, 1860	30, 678
Hose-coupling	R. Hoskin	Dutch Flat, Cal	Apr. 8, 1862	34, 888
Hose-coupling	S. Ingersoll	Stamford, Conn	July 15, 1873	140, 830
Hose coupling	T. L. Johnson and J. Fitzgerald	Rochester, N. Y	Jan. 11, 1870	98, 774
Hose-coupling	E. B. Jucket	New Haven, Conn	Sept. 9, 1862	36, 410
Hose-coupling	E. B. Jucket	New Haven, Conn	Feb. 17, 1863	37, 721
Hose-coupling	E. S. Kennedy, J. Putman, and H. Smith.	Birmingham, Pa	Jan. 2, 1872	122, 388
Hose-coupling	J. W. Kennedy	Saint Louis, Mo	July 1, 1873	140, 374
Hose-coupling	J. W. Kennedy	Saint Louis, Mo	Aug. 12, 1873	141, 718
Hose-coupling	W. A. Kenyon, jr., and F. S. Wagenhals.	Lancaster, Ohio	Feb. 20, 1872	123, 914
Hose-coupling	J. Kerns	New York, N.Y	Sept. 17, 1867	68, 993
Hose-coupling	W. Kilburn	Napa, Cal	Oct. 7, 1873	143, 514
Hose-coupling	W. Knowles	Boston, Mass	Aug. 18, 1863	39, 580
Hose-coupling	R. B. Lawton and W. H. Bliss	Newport, R. I	Feb. 22, 1859	23, 033
Hose-coupling	H. Lewis	Bennington, Vt	Feb. 25, 1873	136, 328
Hose-coupling	A. S. Libby and L. H. Downing	Lawrence and North Andover, Mass.	Oct. 29, 1872	132, 589
Hose-coupling	C. Locker	Orville, Cal	Dec. 20, 1870	110, 253
Hose-coupling	S. H. Loring	Lawrence, Mass	Oct. 16, 1866	58, 850
Hose-coupling	S. H. Loring	Lawrence, Mass	June 11, 1867	65, 758
Hose-coupling	J. J. Lovell	New York, N. Y	May 9, 1871	114, 575
Hose coupling	A. H. Lowell	Manchester, N. H	July 19, 1859	24, 811
Hose-coupling	J. W. Magill and M. D. Herbert	Little Falls, N. Y	Aug. 13, 1872	130, 436
Hose-coupling	J. Mahony	Newport, R. I	Sept. 3, 1872	131, 013
Hose-coupling	T. J. Mayall	Boston, Mass	Nov. 25, 1873	144, 997
Hose-coupling	N. N. McLeod	Saint Louis, Mo	May 24, 1859	24, 179
Hose-coupling	B. Mee	Troy, N. Y	May 7, 1867	64, 437
Hose-coupling	J. B. Mitchell	Portland, Me	July 10, 1866	56, 324
Hose-coupling	A. J. Morse	Boston, Mass	Nov. 5, 1867	70, 456
Hose-coupling	P. H. Niles	Boston, Mass	Sept. 3, 1867	68, 380
Hose-coupling	P. H. Niles	Boston, Mass	Apr. 7, 1868	76, 503
Hose-coupling	W. J. Osbourne	New York, N. Y	Dec. 21, 1869	98, 184
Hose-coupling	W. J. Osbourne and G. B. Massey.	New York, N. Y	Mar. 5, 1867	62, 675
Hose-coupling	J. C. Osgood	Troy, N. Y	June 5, 1866	55, 354
Hose-coupling	D. T. Perkins and C. F. Hovey	Springfield, Mass	Sept. 10, 1867	68, 650
Hose-coupling	D. T. Perkins and C. F. Hovey	Springfield, Mass	Dec. 31, 1867	72, 892
Hose-coupling	C. Perley	New York, N. Y	Feb. 24, 1863	37, 767
Hose-coupling	L. D. Phillips	Chicago, Ill	Jan. 20, 1857	16, 450
Hose-coupling	C. Powell	Birmingham, England	Apr. 19, 1870	102, 038
Hose-coupling	L. J. Roberts	Corry, Pa	Oct. 8, 1867	69, 706
Hose-coupling	O. Salgee	New York, N. Y	Sept. 10, 1867	68, 656
Hose-coupling	G. Sewell	Brooklyn, N. Y	Feb. 8, 1870	99, 715
Hose-coupling	G. Sewell	Brooklyn, N. Y	May 31, 1870	103, 785
Hose-coupling	J. Singer	Cleveland, Ohio	Jan. 3, 1860	26, 713
Hose-coupling	W. H. Smith	Newport, R. I	Aug. 30, 1859	25, 283
Hose-coupling	C. F. Spencer	Rochester, N. Y	Oct. 2, 1860	30, 261
Hose-coupling	J. Steger	New York, N. Y	Feb. 25, 1868	74, 950
Hose-coupling	N. Thompson	Brooklyn, N. Y	Dec. 24, 1867	72, 565
Hose-coupling	T. J. Trapp	Williamsport, Pa	June 3, 1873	139, 632
Hose coupling	C. Vander Woerd	Cambridge, Mass	June 8, 1858	20, 532
Hose-coupling	G. H. Van Vleck and H. Tupper	Buffalo, N. Y	Aug. 9, 1859	25, 065
Hose-coupling	S. W. Warren	Brooklyn, N. Y	May 8, 1860	28, 221
Hose-coupling	A. M. Waterhouse	New York, N. Y	June 19, 1855	13, 112
Hose-coupling	T. W. Welsh	Pittsburgh, Pa	Jan. 16, 1872	122, 873
Hose-coupling	G. Westinghouse, jr	Pittsburgh, Pa	Mar. 4, 1873	136, 397
Hose-coupling	L. Wharton	Salem, Ohio	June 13, 1871	115, 917
Hose-coupling	J. Williston	Vallejo, Cal	Nov. 25, 1862	37, 021
Hose-coupling	T. Wilson and L. B. Kendall	Kalamazoo, Mich	May 20, 1873	139, 223
Hose-coupling	M. A. Winham	North San Juan, Cal	Sept. 9, 1862	36, 433
Hose-coupling, Escape-valve	A. F. Allen	Providence, R. I	Sept. 3, 1867	68, 478
Hose-coupling, Metallic	E. S. Elmer	Hudson, N. Y	June 3, 1873	139, 459
Hose-coupling spanner	L. Pond	Foxborough, Mass	Nov. 25, 1873	144, 862
Hose couplings, Device for connecting dissimilar	W. H. Paige	Springfield, Mass	July 7, 1868	79, 592
Hose-couplings, Ring for securing	A. Schrader	New York, N. Y	July 5, 1870	105, 003
Hose-covering	T. McAuley and M. L. Cheney	San Francisco and Illinoistown, Cal.	Mar. 13, 1866	53, 161
Hose, Covering the ends of rubber	J. P. Rider and J. R. Bird	Brooklyn, N. Y	Mar. 1, 1870	100, 448
Hose, Device for carrying	P. Frentz	New Albany, Ind	Sept. 11, 1866	57, 886

Index of patents issued from the United States Patent Office from 1790 to 1873, inclusive—Continued.

Invention.	Inventor.	Residence.	Date.	No.
Hose, Device for handling fire-engine	J. Lowe	Washington, D. C	June 18, 1872	128, 155
Hose, Elbow-support for flexible	A. O. Bourn	Cranston, R. I	Aug. 11, 1868	80, 901
Hose, Engine	E. W. French	South Scituate, Mass	Sept. 3, 1867	68, 432
Hose, Engine	C. Lenzmann	Brooklyn, N. Y	Dec. 14, 1858	22, 296
Hose, Fire-engine	J. Boyd	Boston, Mass	May 30, 1821	....
Hose, Flexible	J. Davis	Pawtucket, R. I	Feb. 9, 1869	86, 652
Hose, Flexible play-pipe for	J. Greacen, jr	New York, N.Y	Dec. 9, 1873	145, 415
Hose for water, &c., Wooden	C. H. Proessdorf and E. Bauch	Boston Highland, Mass	Aug. 10, 1869	93, 476
Hose from cloth and gum-elastic, Manufacturing	S. D. Reed	Philadelphia, Pa	June 29, 1833	....
Hose-guard	D. P. Lewis	Huntsville, Ala	Nov. 19, 1867	71, 026
Hose holder, Garden	H. C. Smith	Cleveland, Ohio	Apr. 15, 1873	137, 802
Hose, Hydraulic	L. R. Blake	Brooklyn, N. Y	May 13, 1873	138, 845
Hose, Hydraulic	L. R. Blake	Brooklyn, N. Y	Oct. 14, 1873	143, 661
Hose, Hydraulic	L. H. Downing	North Andover, Mass	May 13, 1873	138, 867
Hose, Hydraulic	A. S. Libby	Lawrence, Mass	Dec. 10, 1872	133, 785
Hose, India-rubber	E. M. Chaffee	Roxbury, Mass	Jan. 9, 1834	....
Hose, India-rubber	H. H. Day and R. McMullin	Jersey City, N. J., and Great Barrington, Mass.	Sept. 17, 1850	7, 650
Hose. India-rubber and combination	W. A. Torry	Mont Clair, N. J	Dec. 13, 1870	110, 172
Hose-ladder strap	A. F. Allen	Providence, R. I	Nov. 16, 1869	96, 863
Hose-leak stopper	W. C. Davol, jr	Fall River, Mass	June 6, 1871	115, 715
Hose, Leather	C. F. W. Meyer	Oconomowoc, Wis	Sept. 10, 1867	68, 641
Hose, &c., Machine for cleaning	J. B. Alden, jr., and E. L. Gates	Worcester, Mass	Aug. 31, 1858	21, 307
Hose, Machine for making rubber	T. J. Mayall	Roxbury, Mass	May 15, 1860	28, 288
Hose, Machine for making rubber	J. Murphy	New York, N. Y	Apr. 9, 1872	125, 479
Hose, &c., Machine for making rubber	J. Murphy and A. H. Hook	New York, N. Y	June 15, 1869	91, 476
Hose, Machine for manufacture of rubber	J. Quinn	Leyland, Great Britain	Jan. 9, 1872	122, 649
Hose, Machine for manufacturing	T. J. Mayall	Boston, Mass	July 15, 1873	140, 934
Hose, &c., Machine for repairing	A. Shedlock	New York, N. Y	Mar. 28, 1871	113, 213
Hose, Machinery for making India-rubber	T. J. Mayall	Roxbury, Mass	Nov, 13, 1860	30, 637
Hose, Machinery for manufacturing India-rubber	T. B. De Forest	Birmingham, Conn	Apr. 14, 1857	17, 029
Hose, Making India-rubber	J. H. Howell	Ansonia, Conn	Oct. 21, 1856	15, 947
Hose, Manufacture of	G. H. Bonnaffon	Allegheny City, Pa	Sept. 3, 1867	68, 344
Hose, Manufacture of flexible rubber	C. McBurney	Boston, Mass	Apr. 23, 1872	126, 069
Hose, Manufacture of hydraulic	E. A. Street	South Orange, N. J	Aug. 19, 1873	142, 054
Hose, Manufacture of India-rubber	I. B. Harris	Newtown, Conn	Oct. 8, 1872	132, 006
Hose, Manufacture of leather	J. and J. A. Punderford	New Haven, Conn	Dec. 18, 1866	60, 555
Hose, Manufacture of rubber	T. J. Mayall	Roxbury, Mass	May 22, 1860	28, 388
Hose, Manufacture of rubber	E. L. Perry and W. A. Torrey	New York, N. Y., and Mont Clair, N. J.	July 9, 1867	66, 518
Hose, Manufacture of rubber or gutta-percha	E. L. Perry and C. Manheim	New York, N. Y	July 6, 1869	92, 353
Hose, Manufacture of textile	L. B. Cooley and J. C. Cooke	Middletown, Conn	Mar. 16, 1858	19, 625
Hose. Manufacture of water-proof	E. M. Chaffee	Providence, R. I	Oct. 2, 1866	58, 377
Hose, Manufacture of water-proof	T. J. Mayall	Roxbury, Mass	Nov. 29, 1859	26, 276
Hose, Manufacture of water-proof	T. S. Reed	Providence, R. I	Apr. 2, 1872	125, 331
Hose, Mending fire-engine	J. S. Mackay	Brooklyn, N. Y	Jan. 15, 1861	31, 143
Hose, Method of lining	H. Hartley	Pittsburgh, Pa	Dec. 24, 1867	72, 488
Hose, Method of stopping accidental breaches in fire.	R. Bulkley	New York, N. Y	May 8, 1840	1, 598
Hose, Mode of protecting the ends of vulcanized-rubber.	H. A. Alden	Matteawan, N. Y	Dec. 14, 1869	97, 853
Hose or pipe coupling, Compound	J. W. Osgood	Columbus, Ohio	May 20, 1851	8, 096
Hose or tubing, Manufacture of rubber	T. J. Mayall	Roxbury, Mass	Apr. 13, 1869	88, 888
Hose or tubing, Rubber	T. J. Mayall	Roxbury, Mass	Apr. 13, 1869	88, 887
Hose-pipe	J. Edson	Boston, Mass	Nov. 9, 1869	96, 682
Hose-pipe	R. Hollings	Boston, Mass	Jan. 4, 1853	9, 520
Hose-pipe	A. R. Reese	Phillipsburgh, N. J	June 13, 1871	115, 895
Hose-pipe	G. Smith	Macon, Ga	June 12, 1860	28, 694
Hose-pipe and sprinkler combined	W. W. Ransom	Buffalo, N. Y	Mar. 25, 1873	137, 095
Hose-pipe, Compound	M. Cronin	Washington, D. C	June 3, 1873	139, 550
Hose-pipe coupling	L. Button and R. Blake	Waterford, N. Y	Dec. 18, 1860	30, 909
Hose-pipe, India-rubber	C. McBurney	Roxbury, Mass	Jan. 11, 1859	22, 566
Hose, pipes, and tubes, Closing leaks in	R. Street	Albany, N. Y	Jan. 18, 1870	98, 893
Hose-pipes, Clamp for stopping leaks in	C. Rubsam	Newark, N. J	May 10, 1864	42, 735
Hose-pipes, Coating	G. H. Hinckley	Stonington, Conn	Sept. 29, 1857	18, 281
Hose-pipes, Manufacture of rubber	J. H. Cheever	Boston, Mass	Feb. 8, 1859	22, 854
Hose pipes, Self-packing patch for	J. B. Button	Milwaukee, Wis	Jan. 2, 1872	122, 435
Hose, Process of making suction rubber	G. C. Smith	Matteawan, N. Y	July 6, 1869	92, 381
Hose, Process of vulcanizing rubber	J. H. Cheever	New York, N. Y	Feb. 6, 1872	123, 454
Hose, Protecting the ends of vulcanized India-rubber or combination.	J. Schenck	Brooklyn, N. Y	Feb. 1, 1870	99, 485
Hose-protector	W. C. Bridges and D. P. Dieterich.	Philadelphia, Pa	Oct. 2, 1860	30, 199
Hose-protector	C. C. Converse	Brooklyn, N. Y	Feb. 2, 1869	86, 508
Hose-protector	D. Demarest	New York, N. Y	Nov. 1, 1853	10, 183
Hose-protector	I. H. Stone	Saint Louis, Mo	June 12, 1866	55, 550
Hose-protector, Railway	J. H. Bellamy	Charlestown, Mass	May 25, 1869	90, 336
Hose-reel	B. F. Lee and H. A. Alden	New York and Fishkill, N. Y.	Sept. 9, 1862	36, 438
Hose-reel	T. J. Mayall	Boston, Mass	July 15, 1873	140, 784
Hose-rest, Garden	C. Ryder	Cambridgeport, Mass	Nov. 11, 1873	144, 415
Hose-ring	G. Sewell	Brooklyn, N. Y	Feb. 21, 1871	111, 981
Hose, Rubber	E. L. Perry	New York, N. Y	Nov. 16, 1869	96, 932
Hose, Rubber	G. C. Smith	Fishkill, N. Y	Jan. 18, 1870	99, 018
Hose, Rubber and gutta-percha	E. M. Chaffee	Providence, R. I	May 26, 1868	78, 260
Hose, Rubber play-pipe for	E. L. Perry	New York, N. Y	May 2, 1871	114, 337
Hose-shield	J. A. Haase	Philadelphia, Pa	Dec. 10, 1867	72, 028
Hose, Sleeve or patch for stopping leaks in	J. F. Bellemere and F. R. Fleer	Reading, Pa	Aug. 5, 1873	141, 536
Hose-sprinkler	W. Anderson	San Francisco, Cal	Mar. 7, 1871	112, 309
Hose, Suction	A. F. Allen	Providence, R. I	Mar. 1, 1870	100, 244
Hose, Suction	W. C. Downs	Providence, R. I	Jan. 7, 1873	134, 654
Hose, Suction	C. McBurney	Roxbury, Mass	May 31, 1859	24, 222
Hose, Suction	J. Riley	Boston, Mass	Feb. 6, 1833	....
Hose, Suction	T. D. Stetson and W. Brandon	New York, N. Y	Mar. 19, 1872	124, 770
Hose-supporter	A. Johnson	Cairo, N. Y	Apr. 27, 1858	20, 069
Hose, syringe, &c., Distributing-apparatus for garden.	S. Barton, jr	London, England	Aug. 12, 1873	141, 620
Hose-tender	H. A. Gilbertson	New York, N. Y	Aug. 11, 1868	80, 941

Index of patents issued from the United States Patent Office from 1790 *to* 1873, *inclusive*—Continued.

Invention.	Inventor.	Residence.	Date.	No.
Hose to couplings, Mode of fastening	C. McBurney	Roxbury, Mass	Feb. 16, 1864	41, 630
Hose-tongs, clamp, wrench, and pick, Combined	H. Bitter	Philadelphia, Pa	Oct. 15, 1867	69, 896
Hose-tubing	H. A. Alden	Matteawan, N. Y	May 8, 1860	28, 231
Hose-tubing	T. J. Mayall	Roxbury, Mass	Mar. 20, 1860	27, 553
Hose-tubing	C. McBurney	Roxbury, Mass	Apr. 10, 1860	27, 819
Hose, tubing, and other rubber fabrics, Manufacture of.	J. Schenck	Brooklyn, N. Y	May 25, 1869	90, 397
Hose-tubing, Elastic	J. C. Boyd	Boston, Mass	Aug. 30, 1859	25, 239
Hose-tubing, Flexible	H. A. Alden	Matteawan, N. Y	May 22, 1860	28, 432
Hose-tubing, Rubber	T. J. Mayall	Roxbury, Mass	May 22, 1860	28, 389
Hose tubes or pipes for conveying fluids under pressure, Manufacture of flexible.	J. P. Ryder	Brooklyn, N. Y	May 25, 1869	90, 393
Hose-tunnel for railways	L. Myers	Philadelphia, Pa	Mar. 14, 1871	112, 731
Hose, Vulcanized India-rubber	D. C. Gately	Newton, Conn	Nov. 19, 1872	133, 219
Hose, Vulcanized rubber	H. A. Alden	Matteawan, N. Y	Feb. 22, 1870	100, 096
Hose, Water-proof	H. A. Alden	Matteawan, N. Y	Sept. 18, 1860	30, 032
Hose, Water-proof	T. J. Mayall	Roxbury, Mass	Mar. 5, 1861	31, 614
Hose, Water-proof	T. L. Reed	Providence, R. I	Mar. 26, 1872	124, 914
Hose, Water-proof leather	J. Penderford	New Haven, Conn	Apr. 26, 1859	23, 781
Hose-wrench	E. Bucklin, jr	Pawtucket, R. I	Nov. 25, 1873	144, 827
Hosiery goods, Uniting edges of	E. E. Killbourn	New Brunswick, N. J	Aug. 27, 1867	68, 087
Hosiery, Manufacture of	W. H. McNary	Brooklyn, N. Y	Dec. 23, 1856	16, 285
Hosiery, Manufacturing seamless	W. Goddard	New York, N. Y	Aug. 26, 1856	15, 607
Hosiery, Pressing	O. Osborne	Philadelphia, Pa	June 17, 1873	140, 068
Hosiery, Stretcher for	S. J. Wright	Ellsworth, N. Y	Sept. 24, 1867	69, 293
Hospital-table	S. J. A. Hussey	Cornwall, N. Y	May 23, 1865	47, 831
Hot-air and cupola furnace	L. V. Badger	Portsmouth, N. H	Dec. 2, 1835	....
Hot-air conductor, Portable	J. B. Oldershaw	Baltimore, Md	Dec. 10, 1867	72, 074
Hot-air engine	C. W. Baldwin	Charlestown, Mass	Feb. 14, 1865	46, 320
Hot-air engine	C. W. Baldwin	Boston, Mass	July 11, 1865	48, 639
Hot-air engine	C. W. Baldwin	Boston, Mass	June 19, 1866	55, 601
Hot-air engine	B. F. Craig	Washington, D. C	Jan. 17, 1860	26, 837
Hot-air engine	M. G. Crane	Chelsea, Mass	Jan. 31, 1865	46, 084
Hot-air engine	W. D. Grimshaw	Newark, N. J	June 19, 1860	28, 748
Hot-air engine	H. Kilbourne	Waterloo, Iowa	Nov. 7, 1865	50, 875
Hot-air engine	H. J. Kritzer	Albion, Mich	July 29, 1862	36, 008
Hot-air engine	C. P. Leavitt	New York, N. Y	Oct. 4, 1870	107, 928
Hot-air engine	C. P. Leavitt	New York, N. Y	Dec. 3, 1872	133, 649
Hot-air engine	W. Lehmann	Nuremberg, Germany	June 15, 1869	91, 239
Hot-air engine	T. McDonough	Middletown, Conn	Jan. 14, 1862	34, 155
Hot-air engine	T. McDonough	Middletown, Conn	Nov. 8, 1864	44, 966
Hot-air engine	T. McDonough	Middletown, Conn	Jan. 30, 1866	52, 305
Hot-air engine	T. McDonough	Newburgh, N. Y	Nov. 2, 1869	96, 335
Hot-air engine	H. Messer	Roxbury, Mass	Apr. 21, 1863	38, 237
Hot-air engine	H. Messer	Roxbury, Mass	July 21, 1863	39, 321
Hot-air engine	H. Messer	Roxbury, Mass	Jan. 5, 1864	41, 084
Hot-air engine	H. Messer	Roxbury, Mass	Mar. 7, 1865	46, 689
Hot-air engine	W. Reinlein	Barcelona, Spain	June 26, 1866	55, 971
Hot-air engine	A. K. Rider	New York, N. Y	Jan. 17, 1871	111, 087
Hot-air engine	A. K. Rider	New York, N. Y	July 16, 1872	128, 979
Hot-air engine	S. H. Roper	Boston, Mass	Feb. 4, 1862	34, 333
Hot-air engine	S. H. Roper	Roxbury, Mass	Mar. 18, 1862	34, 723
Hot-air engine	S. H. Roper	Boston, Mass	June 9, 1863	38, 866
Hot-air engine	S. H. Roper	Roxbury, Mass	June 5, 1866	55, 428
Hot-air engine	P. Shaw	Boston, Mass	May 28, 1861	32, 455
Hot-air engine	P. Shaw	Boston, Mass	Nov. 26, 1861	33, 799
Hot-air engine	P. Shaw	Boston, Mass	May 18, 1869	90, 128
Hot-air engine	L. A. L. Soderstrom	Paris, France	June 29, 1869	92, 117
Hot-air engine	C. Stevens	Boston, Mass	Oct. 17, 1865	50, 506
Hot-air engine	F. H. Wenham	London, England	Sept. 1, 1868	81, 853
Hot-air engine	M. D. Whipple	Cambridge, Mass	Sept. 30, 1862	36, 581
Hot-air engine	S. Wilcox, jr	Westerly, R. I	Feb. 14, 1860	27, 180
Hot-air engine	S. Wilcox, jr	Westerly, R. I	Apr. 2, 1861	31, 924
Hot-air engine	S. Wilcox, jr	Westerly, R. I	May 16, 1865	47, 759
Hot-air engine	S. Wilcox, jr	Westerly, R. I	Sept. 19, 1865	50, 062
Hot-air engine	A. O. Willcox	Philadelphia, Pa	Aug. 2, 1853	9, 909
Hot-air furnace	W. N. Abbott	New York, N. Y	Oct. 29, 1872	132, 554
Hot-air furnace	J. Albee	Boston, Mass	Sept. 10, 1867	68, 589
Hot-air furnace	J. Albee	Boston, Mass	June 30, 1868	79, 294
Hot-air furnace	C. Allen	Hartford, Conn	Apr. 27, 1869	89, 458
Hot-air furnace	C. Allen	Hartford, Conn	Jan. 17, 1871	111, 027
Hot-air furnace	J. Amory	West Roxbury, Mass	Mar. 31, 1868	76, 136
Hot-air furnace	H. Arden	Cincinnati, Ohio	Feb. 11, 1868	74, 270
Hot-air furnace	B. Arthurs	Pittsburgh, Pa	Nov. 12, 1867	70, 679
Hot-air furnace	J. R. Barker	Chicago, Ill	May 28, 1872	127, 140
Hot-air furnace	J. C. Barnes	Albany, N. Y	Oct. 3, 1871	119, 554
Hot-air furnace	A. H. Bartlett	King's Bridge, N. Y	Jan. 30, 1855	12, 305
Hot-air furnace	A. H. Bartlett	Spuyten Duyvil, N. Y	Sept. 18, 1860	30, 039
Hot-air furnace	W. D. Bartlett	Amesbury, Mass	Jan. 10, 1871	110, 889
Hot-air furnace	P. D. Beckwith	Dowagiac, Mich	June 1, 1869	90, 630
Hot-air furnace	O. Bellman	Hagerstown, Md	Nov. 26, 1872	133, 293
Hot-air furnace	O. Bellman and J. W. Garver	Hagerstown, Md	Feb. 22, 1870	100, 107
Hot-air furnace	J. M. Blackman	Decorah, Iowa	May 30, 1871	115, 270
Hot-air furnace	V. W. Blanchard	Bridport, Vt	Jan. 8, 1867	61, 043
Hot-air furnace	J. Bolton	Richmond, Va	Apr. 26, 1853	9, 686
Hot-air furnace	J. Bolton	Richmond, Va	Dec. 20, 1853	10, 333
Hot-air furnace	J. A. Bolton	Leicester, England	Dec. 3, 1861	33, 819
Hot-air furnace	W. H. Bond	Syracuse, N. Y	Apr. 2, 1872	125, 259
Hot-air furnace	L. Bonnell	Milwaukee, Wis	Apr. 5, 1870	101, 422
Hot-air furnace	L. Bonnell	Milwaukee, Wis	Jan. 3, 1871	110, 731
Hot-air furnace	J. Bouis	Baltimore, Md	Mar. 29, 1834	
Hot-air furnace	R. Boyd and J. C. Hart	Rochester, N. Y	Nov. 30, 1869	97, 347
Hot-air furnace	N. A. Boynton	Boston, Mass	Feb. 8, 1853	9, 573
Hot-air furnace	N. A. Boynton	New York, N. Y	Aug. 22, 1854	11, 545
Hot-air furnace	N. A. Boynton	New York, N. Y	May 4, 1869	89, 549
Hot-air furnace	E. Brady and J. Sloan	Philadelphia, Pa	Apr. 14, 1868	76, 590
Hot-air furnace	S. A. Briggs	Providence, R. I	July 31, 1855	13, 374
Hot-air furnace	L. W. Brown	Cleveland, Ohio	Apr. 19, 1870	102, 086

Index of patents issued from the United States Patent Office from 1790 *to* 1873, *inclusive*—Continued.

Invention.	Inventor.	Residence.	Date.	No.
Hot-air furnace	L. W. Brown and I. L. Frankem	Indianapolis, Ind	Feb. 20, 1866	52, 671
Hot-air furnace	R. F. Brown	Chicago, Ill	Nov. 1, 1870	108, 754
Hot-air furnace	B. Brownell	Chicago, Ill	Sept. 18, 1866	58, 057
Hot-air furnace	W. Bryent	Boston, Mass	Sept. 25, 1847	5, 309
Hot-air furnace	J. T. Budd	New York, N. Y	Oct. 2, 1866	58, 373
Hot-air furnace	H. G. Burr	Minneapolis, Minn	Apr. 7, 1868	76, 397
Hot-air furnace	J. H. Burtis	Brooklyn, N. Y	June 6, 1871	115, 699
Hot-air furnace	J. H. Cahill	Philadelphia, Pa	Apr. 14, 1857	17, 022
Hot-air furnace	E. H. Camp	Jackson, Mich	Aug. 6, 1861	32, 974
Hot-air furnace	E. H. Camp	Jackson, Mich	May 7, 1867	64, 485
Hot-air furnace	B. F. Campbell	Boston, Mass	Feb. 28, 1871	112, 122
Hot-air furnace	J. Carton and J. Briggs	Utica, N. Y	Aug. 1, 1854	11, 414
Hot-air furnace	J. E. Chapman	Cannon Falls, Minn	Aug. 17, 1869	93, 675
Hot-air furnace	F. E. Chatard, jr	Baltimore, Md	Aug. 22, 1871	118, 195
Hot-air furnace	T. W. Chatfield	Utica, N. Y	Mar. 28, 1854	10, 700
Hot-air furnace	T. W. Chatfield	Utica, N. Y	Mar. 22, 1870	100, 979
Hot-air furnace	J. Chilcott	Brooklyn, N. Y	Jan. 9, 1866	51, 921
Hot-air furnace	J. Child	Elyria, Ohio	Mar. 23, 1858	19, 683
Hot-air furnace	G. Chilson	Boston, Mass	Aug. 4, 1845	4, 133
Hot-air furnace	G. Chilson	Boston, Mass	May 2, 1848	5, 550
Hot-air furnace	E. Clark	Lancaster, Pa	May 9, 1871	114, 528
Hot-air furnace	T. E. Coles	Troy, Ohio	Jan. 5, 1869	85, 564
Hot-air furnace	M. A. Cushing	Aurora, Ill	July 6, 1869	92, 282
Hot-air furnace	M. A. Cushing	Aurora, Ill	May 30, 1871	115, 444
Hot-air furnace	G. Darby	Augusta, Me	Feb. 2, 1858	19, 239
Hot-air furnace	G. B. Davis	Chicago, Ill	Aug. 13, 1867	67, 732
Hot-air furnace	G. W. Day	Haverhill, Mass	Dec. 16, 1873	145, 489
Hot-air furnace	H. G. Dayton	Maysville, Ky	July 31, 1866	56, 727
Hot-air furnace	W. W. Dodge	Boston, Mass	Aug. 27, 1872	130, 847
Hot-air furnace	C. M. Drennan	Boston, Mass	May 28, 1872	127, 321
Hot-air furnace	M. B. Dyott	Philadelphia, Pa	Aug. 30, 1853	9, 966
Hot-air furnace	H. A. and C. H. Engels and J. Wieland.	San Francisco, Cal	Apr. 2, 1867	63, 490
Hot-air furnace	W. Ennis	New York, N. Y	Mar. 29, 1853	9, 633
Hot-air furnace	A. Ernst and C. Shepard	Milwaukee, Wis	July 17, 1860	29, 150
Hot-air furnace	M. G. Fagan	Troy, N. Y	June 28, 1870	104, 842
Hot-air furnace	J. R. Ferguson	Brooklyn, N. Y	Nov. 30, 1858	22, 173
Hot-air furnace	R. R. Finch	Peekskill, N. Y	Oct. 24, 1871	120, 257
Hot-air furnace	J. Fridley, jr	Carlisle, Pa	Feb. 4, 1873	135, 422
Hot-air furnace	J. R. Gaston	Normal, Ill	Aug. 27, 1872	130, 913
Hot-air furnace	S. Gates	Albion, N. Y	Dec. 7, 1852	9, 444
Hot-air furnace	W. B. Geddes	Rochester, N. Y	June 13, 1871	115, 841
Hot-air furnace	W. B. Geddes	Rochester, N. Y	Jan. 14, 1873	134, 879
Hot-air furnace	S. F. Gold	Englewood, N. J	Apr. 8, 1873	137, 546
Hot-air furnace	B. Gommenginger	Rochester, N. Y	Mar. 1, 1870	100, 281
Hot-air furnace	B. Gommenginger	Rochester, N. Y	July 4, 1871	116, 701
Hot-air furnace	B. Gommenginger	Rochester, N. Y	June 4, 1872	127, 476
Hot-air furnace	B. Gommenginger and C. W. Trotter.	Rochester, N. Y	Oct. 5, 1869	95, 461
Hot-air furnace	J. E. Grant	Charlestown, Mass	Oct. 3, 1854	11, 770
Hot-air furnace	M. Greenebaum	Chicago, Ill	Jan. 23, 1855	12, 277
Hot-air furnace	C. B. Gregory	Beverly, N. J	Aug. 25, 1868	81, 492
Hot-air furnace	W. O. Grover	Boston, Mass	June 3, 1873	139, 505
Hot-air furnace	D. Gusweiler	Cincinnati, Ohio	June 15, 1869	91, 436
Hot-air furnace	J. Gwynn	Tiffin, Ohio	May 18, 1869	90, 095
Hot-air furnace	J. D. Hall	Brooklyn, N. Y	Apr. 15, 1873	137, 838
Hot-air furnace	S. J. Hare	Louisville, Ky	Nov. 24, 1868	84, 356
Hot-air furnace	W. H. Harris	Buffalo, N. Y	Nov. 19, 1872	133, 096
Hot-air furnace	O. N. Hart	Winona, Minn	July 20, 1869	[illegible]
Hot-air furnace	J. B. Haupt	Philadelphia, Pa	July 9, 1872	128, 799
Hot-air furnace	J. P. Hayes	Philadelphia, Pa	May 23, 1854	10, 962
Hot-air furnace	J. I. Hess	Philadelphia, Pa	Oct. 13, 1868	82, 948
Hot-air furnace	F. C. Hesse	Cincinnati, Ohio	June 25, 1867	66, 083
Hot-air furnace	S. E. Hewes	Albany, N. Y	June 13, 1871	115, 956
Hot-air furnace	R. Hillson	Albany, N. Y	Feb. 29, 1848	5, 459
Hot-air furnace	I. H. Hobbs, A. W. Rand, and G. H. Sellers.	Philadelphia, Pa	Oct. 11, 1859	25, 735
Hot-air furnace	H. Holcomb	Painesville, Ohio	Dec. 22, 1863	41, 007
Hot-air furnace	B. Holly	Lockport, N. Y	Jan. 31, 1865	46, 107
Hot-air furnace	W. and W. James, jr	Montreal, Canada	Nov. 1, 1870	108, 790
Hot-air furnace	O. S. Kelsey	Hartford, Conn	Oct. 7, 1873	143, 453
Hot-air furnace	J. L. Kite	Philadelphia, Pa	June 19, 1855	13, 092
Hot-air furnace	A. Knobel	Monroe, Wis	Mar. 31, 1868	76, 203
Hot-air furnace	A. Kohler	Boston, Mass	Oct. 15, 1867	69, 820
Hot-air furnace	B. S. Koll	Pittsburgh, Pa	July 5, 1870	104, 962
Hot-air furnace	T. Kruse	La Fayette, Ind	July 16, 1872	129, 145
Hot-air furnace	J. A. Lawson	Troy, N. Y	Nov. 5, 1867	70, 445
Hot-air furnace	J. A. Lawson	Troy, N. Y	July 4, 1871	116, 723
Hot-air furnace	J. A. Lawson	Troy, N. Y	July 4, 1871	116, 724
Hot-air furnace	W. H. Lee and C. M. Hardenbergh.	Minneapolis, Minn	June 23, 1868	79, 075
Hot-air furnace	W. H. Lee and C. M. Hardenbergh.	Minneapolis, Minn	Nov. 2, 1869	96, 328
Hot-air furnace	J. Leeds	Philadelphia, Pa	Nov. 27, 1860	30, 738
Hot-air furnace	M. W. Lester	Chicago, Ill	Aug. 31, 1869	94, 224
Hot-air furnace	R. Z. Liddle	Brooklyn, N. Y	Nov. 26, 1867	71, 517
Hot-air furnace	H. N. Longfellow	Lawrence, Mass	Jan. 28, 1873	135, 227
Hot-air furnace	S. Macferran	Philadelphia, Pa	Feb. 5, 1856	14, 196
Hot-air furnace	J. Magee	Chelsea, Mass	Dec. 16, 1873	145, 512
Hot-air furnace	J. Magee	Chelsea, Mass	Dec. 16, 1873	145, 513
Hot-air furnace	W. R. Manning	Chicago, Ill	Nov. 26, 1872	133, 375
Hot-air furnace	J. Martin	Florence, Ala	Jan. 26, 1869	86, 239
Hot-air furnace	P. Martin	Cincinnati, Ohio	Jan. 15, 1867	61, 223
Hot-air furnace	P. Martin	Cincinnati, Ohio	Jan. 3, 1871	110, 664
Hot-air furnace	J. McCoy	Burlington, N. J	Aug. 23, 1870	106, 708
Hot-air furnace	S. Mead	New Haven, Conn	Dec. 24, 1861	34, 003
Hot-air furnace	J. Mealey	Ogdensburgh, N. Y	July 2, 1872	128, 555
Hot-air furnace	J. H. Mearns	Philadelphia, Pa	Mar. 19, 1872	124, 842

Index of patents issued from the United States Patent Office from 1790 *to* 1873, *inclusive*—Continued.

Invention.	Inventor.	Residence.	Date.	No.
Hot-air furnace	G. F. Merklee	New York, N. Y	Dec. 24, 1867	72, 520
Hot-air furnace	M. Metcalf	Grand Rapids, Mich	Sept. 1, 1868	81, 809
Hot-air furnace	G. H. Miller	Leavenworth, Kans	May 10, 1870	102, 956
Hot-air furnace	G. H. Miller	Leavenworth, Kans	July 1, 1873	140, 523
Hot-air furnace	J. A. Miller	New York, N. Y	Apr. 25, 1865	47, 443
Hot-air furnace	D. B. Montague	Springfield, Mass	Jan. 21, 1873	135, 003
Hot-air furnace	D. B. Morris	Pittsburgh, Pa	Dec. 30, 1873	146, 089
Hot-air furnace	J. R. Morris	New Haven, Conn	Aug. 12, 1846	4, 689
Hot-air furnace	H. Movers	Chicago, Ill	June 27, 1871	116, 339
Hot-air furnace	J. R. Nichols	Boston, Mass	Mar. 28, 1871	113, 197
Hot-air furnace	E. D. Norcross	Augusta, Me	Feb. 8, 1870	99, 585
Hot-air furnace	E. D. Norcross	Boston, Mass	June 11, 1872	127, 913
Hot-air furnace	S. W. Norton	Lemont, Ill	Feb. 20, 1866	52, 738
Hot-air furnace	A. L. Otis	Normal, Ill	Sept. 28, 1869	95, 256
Hot-air furnace	O. Paddock	Watertown, N. Y	Nov. 27, 1860	30, 752
Hot-air furnace	H. L. Palmer	New York, N. Y	Feb. 11, 1873	135, 838
Hot-air furnace	J. H. H. Perkins	Utica, N. Y	Dec. 23, 1856	16, 287
Hot-air furnace	J. S. Perry	Albany, N. Y	June 22, 1869	91, 560
Hot-air furnace	J. L. Pfau, jr	Quincy, Ill	Apr. 25, 1871	114, 035
Hot-air furnace	A. Pfund	New York, N. Y	Nov. 4, 1873	144, 355
Hot-air furnace	W. M. Phelps	Marshall, Mich	Aug. 20, 1872	130, 740
Hot-air furnace	C. L. Pierce	Natick, Mass	Apr. 22, 1873	138, 188
Hot-air furnace	C. L. Pierce	Natick, Mass	Apr. 22, 1873	138, 189
Hot-air furnace	S. Pierce	Troy, N. Y	May 20, 1851	8, 104
Hot-air furnace	W. T. Powell and R. F. Brown	Chicago, Ill	Aug. 20, 1872	130, 595
Hot-air furnace	H. Randall	Quincy, Ill	Aug. 8, 1871	117, 814
Hot-air furnace	A. M. Rice	Boston, Mass	Oct 26, 1852	9, 358
Hot-air furnace	A. Richmond	Providence, R. I	Oct. 26, 1852	9, 360
Hot-air furnace	S. I. Russell	Chicago, Ill	May 15, 1855	12, 875
Hot-air furnace	S. C. Salisbury	New York, N. Y	Aug. 3, 1869	93, 351
Hot-air furnace	W. Sanford	Brooklyn, N. Y	May 6, 1862	35, 180
Hot-air furnace	W. Sanford	Brooklyn, N. Y	June 22, 1869	91, 569
Hot-air furnace	C. B. Sawyer	Fitchburgh, Mass	Mar. 18, 1862	34, 699
Hot-air furnace	B. C. Sayre	Montrose, Pa	Apr. 15, 1873	137, 800
Hot-air furnace	P. I. Schopp	Louisville, Ky	Oct. 12, 1869	95, 734
Hot-air furnace	J. H. Schwein	Cincinnati, Ohio	Feb. 13, 1866	52, 648
Hot-air furnace	M. D. Seward	Normal, Ill	Apr. 8, 1873	137, 727
Hot-air furnace	S. B. Sexton	Baltimore, Md	Feb. 21, 1871	112, 081
Hot-air furnace	S. B. Sexton and G. W. Beard	Baltimore, Md	Oct. 25, 1870	108, 733
Hot-air furnace	J. H. Shedd and B. Worcester	Waltham, Mass	Oct. 31, 1865	50, 739
Hot-air furnace	C. J. Shepard	Brooklyn, N. Y	Oct. 21, 1873	143, 851
Hot-air furnace	J. Siddons	Rochester, N. Y	Dec. 29, 1868	85, 335
Hot-air furnace	E. Slater	Philadelphia, Pa	Jan. 18, 1870	99, 015
Hot-air furnace	E. Slater and A. H. Platt	Philadelphia, Pa	Oct. 27, 1868	83, 415
Hot-air furnace	J. B. Smith and O. L. Giddings	Manchester and Exeter, N. H.	July 16, 1872	129, 180
Hot-air furnace	S. Smith	Worcester, Mass	June 16, 1868	79, 020
Hot-air furnace	W. Smith	Cicero, Ind	Feb. 15, 1870	99, 963
Hot-air furnace	G. S. G. Spence	Boston, Mass	Aug. 17, 1852	9, 204
Hot-air furnace	G. S. G. Spence	Boston, Mass	Feb. 20, 1855	12, 448
Hot-air furnace	J. Stuber	Utica, N. Y	May 1, 1860	28, 112
Hot-air furnace	J. Stuber and F. Frank	Utica, N. Y	June 1, 1858	20, 454
Hot-air furnace	B. F. Sturtevant	Jamaica Plains, Mass	Sept. 28, 1869	95, 281
Hot-air furnace	P. Sweeney	Buffalo, N. Y	July 11, 1854	11, 278
Hot-air furnace	J. M. Thatcher	Bergen, N. J	July 9, 1872	128, 926
Hot-air furnace	M. A. Thayer	Chicago, Ill	May 24, 1870	103, 390
Hot-air furnace	G. G. Thomas	Saint Louis, Mo	July 26, 1870	105, 864
Hot-air furnace	G. G. Thomas	Saint Louis, Mo	Aug. 22, 1871	118, 299
Hot-air furnace	W. D. Titus	Brooklyn, N. Y	May 4, 1869	89, 610
Hot-air furnace	J. C. Treat	East Hartford, Conn	Aug. 5, 1851	8, 276
Hot-air furnace	C. W. Trotter	Rochester, N. Y	Mar. 5, 1867	62, 577
Hot-air furnace	L. B. Tupper	New York, N. Y	Jan. 9, 1872	122, 684
Hot-air furnace	W. H. Turner	Indianapolis, Ind	Sept. 12, 1871	118, 988
Hot-air furnace	W. Twitchell	Syracuse, N. Y	Sept. 24, 1872	131, 723
Hot-air furnace	J. Van	Cincinnati, Ohio	Aug. 23, 1870	106, 746
Hot-air furnace	J. S. Van Buren	Green Island, N. Y	Apr. 5, 1870	101, 683
Hot-air furnace	J. J. Vogelgesang	Columbus, Ohio	July 18, 1871	117, 225
Hot-air furnace	G. W. Walker	Boston, Mass	Oct. 31, 1871	120, 406
Hot-air furnace	T. Wallace	Chicago, Ill	May 15, 1866	54, 795
Hot-air furnace	G. E. Waring	Stamford, Conn	Dec. 24, 1850	7, 863
Hot-air furnace	E. Webster	Hartford, Conn	June 12, 1866	55, 564
Hot-air furnace	E. Webster	Hartford, Conn	July 5, 1870	105, 017
Hot-air furnace	D. P. Weeks	Malden, Mass	Dec. 12, 1854	12, 080
Hot-air furnace	D. P. Weeks	Boston, Mass	Aug. 11, 1857	18, 002
Hot-air furnace	J. G. Weldon	Pittsburgh, Pa	Oct. 17, 1871	119, 955
Hot-air furnace	J. W. Wentworth	Minneapolis, Minn	Jan. 12, 1869	85, 774
Hot-air furnace	S. Wethered	Baltimore, Md	Jan. 3, 1860	26, 724
Hot-air furnace	T. Whitaker and J. Constantine.	Bolton and Manchester, England.	Nov. 26, 1867	71, 348
Hot-air furnace	C. White	West Roxbury, Mass	Dec. 17, 1872	134, 118
Hot-air furnace	J. Whitehill	Frederick, Md	Nov. 8, 1859	26, 064
Hot-air furnace	J. Whitehill	Frederick, Md	Mar. 24, 1868	76, 018
Hot-air furnace	H. Whittingham	New York, N. Y	Aug. 7, 1866	57, 026
Hot-air furnace	H. Whittingham	New York, N. Y	Aug. 7, 1866	57, 027
Hot-air furnace	H. Whittingham	New York, N. Y	Aug. 7, 1866	57, 028
Hot-air furnace	O. H. Whorf	Saint Louis, Mo	Mar. 3, 1868	75, 097
Hot-air furnace	I. T. Winchester	Boston, Mass	Mar. 9, 1869	87, 610
Hot-air furnace	H. G. Wing	New Bedford, Mass	May 23, 1846	4, 532
Hot-air furnace	C. Wood	Worcester, Mass	July 11, 1871	116, 908
Hot-air furnace	T. Yates	Milwaukee, Wis	Mar. 25, 1873	137, 277
Hot-air furnace and fire-grate for heating apartments, &c.	W. H. Whiteley	Charlestown, Mass	May 11, 1841	2, 085
Hot-air furnace, bake-oven, and heating room	J. Stahl	Baltimore, Md	May 29, 1832	....
Hot-air furnace, Base-burning	J. Gray	Albany, N. Y	Jan. 7, 1868	73, 091
Hot-air furnace, Base-burning	J. G. Porter	New York, N. Y	Apr. 12, 1870	101, 911
Hot-air furnace for heaters	J. B. Driscole	New York, N. Y	Oct. 6, 1868	82, 811
Hot-air furnace, Portable	J. P. Hayes	Boston, Mass	Mar. 20, 1849	6, 201
Hot-air furnace, Portable	H. L. B. Lewis	New York, N. Y	Sept. 19, 1845	4, 198

Index of patents issued from the United States Patent Office from 1790 *to* 1873, *inclusive*—Continued.

Invention.	Inventor.	Residence.	Date.	No.
Hot-air-furnace register	D. Culver	New York, N. Y	Aug. 10, 1848	5, 698
Hot-air-furnace register	J. W. Geddes	Baltimore, Md	Mar. 2, 1858	1[illegible], 562
Hot-air-furnace register	C. F. Tuttle	Williamsburgh, N. Y	Jan. 23, 1849	6, [illegible]60
Hot-air-furnace register	C. F. Tuttle	Williamsburgh, N. Y	Sept. 11, 1849	6, 70[illegible]
Hot-air furnaces and for vapor-draft of grate-bars, Water-vessel for.	W. Moultrie	New York, N. Y	Aug. 25, 1857	1[illegible], 0[illegible]4
Hot-air furnaces, Apparatus to prevent an over-supply of coal in fire-boxes of.	L. W. Leeds	Germantown, Pa	Sept. 4, 1855	13, 525
Hot-air furnaces, Dome for	B. Gommingingér	Rochester, N. Y	Mar. 19, 1872	124, 807
Hot-air furnaces, Heat-regulator for	I. Hayes	PhiladelphiaPa.	July 2, 1872	128, 485
Hot-air furnaces, Heat-regulator for	A. H. Tingley	Providence, R. I	Oct. 30, 1866	59, 294
Hot-air furnaces, Radiating-attachment for	J. H. Keyser	New York, N. Y	Mar. 19, 1867	62, 962
Hot-air furnaces, Setting for	C. J. Shepard	Brooklyn, N. Y	June 18, 1872	128, 179
Hot-air furnaces, Setting for	C. J. Shepard	Brooklyn, N. Y	June 18, 1872	128, 180
Hot-air furnaces, Shield for feed-air pipes for	B. C. Bibb	Baltimore, Md	May 7, 1872	126, 373
Hot-air-pipe elbow	J. M. Thacher	Jersey City, N. J	June 6, 1871	115, 785
Hot-air-pipe evaporator	G. F. J. Colburn	Newark, N. J	Jan. 22, 1861	31, 152
Hot-air-pipe evaporator	J. L. Hutchinson	Baltimore, Md	Feb. 13, 1866	52, 572
Hot-air register	W. S. Bronson	Hartford, Conn	Apr. 26, 1870	102, 217
Hot-air register	T. W. Brown	Reading, Pa	Oct. 27, 1868	83, 359
Hot-air register	B. Bunce and R. Salt	New York, N. Y	Mar. 19, 1872	124, 663
Hot-air register	W. G. Creamer	Brooklyn, N. Y	Sept. 14, 1869	94, 872
Hot-air register	T. Dowling	Salem, Mass	Jan. 6, 1863	37, 338
Hot-air register	O. A. Ebert	Baltimore, Md	Feb. 15, 1870	99, 864
Hot-air register	D. T. Gale	Fort Wayne, Ind	June 27, 1871	116, 294
Hot-air register	W. Highton	Malden, Mass	June 23, 1868	79, 226
Hot-air register	W. Highton	Malden, Mass	Nov. 18, 1873	144, 616
Hot-air register	W. B. Kehew and C. H. Fifield	Salem, Mass	July 22, 1862	35, 938
Hot-air register	J. W. McGlashan	Montreal, Canada	June 27, 1871	116, 337
Hot-air register	T. E. McNeill	Philadelphia, Pa	Mar. 13, 1860	27, 461
Hot-air register	H. M. Phinney	Cambridge, Mass	Oct. 5, 1869	95, 510
Hot-air register	H. M. Phinney	Boston, Mass	Apr. 4, 1871	113, 447
Hot-air register	H. M. Phinney	Cambridge, Mass	Sept. 17, 1872	131, 459
Hot-air register	G. Pollock	Roxbury, Mass	Aug. 20, 1850	7, 582
Hot-air register	C. B. Sawyer	Fitchburgh, Mass	Mar. 25, 1862	34, 783
Hot-air register	S. B. Sexton	Baltimore, Md	July 23, 1861	32, 897
Hot-air register	S. J. Sherman	New York, N. Y	Oct. 6, 1857	18, 356
Hot-air register	J. H. Simonds	New York, N. Y	Feb. 19, 1861	31, 485
Hot-air register	J. H. Simonds	New York, N. Y	Mar. 19, 1861	31, 748
Hot-air register	J. V. Tibbets	New York, N. Y	Aug. 18, 1857	18, 026
Hot-air register	W. H. Towers	Philadelphia, Pa	Nov. 15, 1853	10, 240
Hot-air register	W. Turton	Bushwick, N. Y	Mar. 16, 1852	8, 811
Hot-air register	W. Turton	Brooklyn, N. Y	Oct. 13, 1863	40, 294
Hot-air register	E. A. Tuttle	Williamsburgh, N. Y	Apr. 12, 1853	9, 664
Hot-air register	E. A. Tuttle	Williamsburgh, N. Y	Jan. 3, 1854	10, 371
Hot-air register	E. A. Tuttle	Brooklyn, N. Y	Sept. 6, 1859	25, 355
Hot-air register	E. A. Tuttle	Brooklyn, N. Y	Mar. 11, 1873	136, 682
Hot-air register	A. Watson	Jersey City, N. J	June 28, 1870	104, 802
Hot-air register	W. Young	Easton, Pa	Apr. 25, 1871	114, 247
Hot-air-register attachment	J. D. McBride	Mansfield, Ohio	Feb. 25, 1868	74, 968
Hot-air-register attachment	H. Sinclair	New York, N. Y	Sept. 1, 1868	81, 695
Hot-air-register frame, &c	C. R. Harvy	New York, N. Y	Jan. 29, 1867	61, 538
Hot-air registers, Composition-stone frame for	J. S. Elliott and J. F. Wood	Chelsea and Everett, Mass	Apr. 23, 1872	126, 042
Hot-air registers, Deflector for	S. H. Caughey	Baltimore, Md	July 30, 1867	67, 265
Hot-air, steam, and water gage	J. C. C. Walker	Waco Village, Tex	Oct. 8, 1867	69, 729
Hot-bed box or case	W. Wells	Salem, Mass	Aug. 9, 1870	106, 298
Hot-bed shutter	J. Weed	Muscatine, Iowa	Oct. 10, 1871	119, 805
Hot-blast apparatus	L. Blair	Painesville, Ohio	Mar. 11, 1856	14, 386
Hot-blast apparatus	R. Denholm	Newburgh, Ohio	Sept. 12, 1865	49, 862
Hot-blast apparatus, Furnace	J. Froggett	Youngstown, Ohio	Mar. 26, 1867	63, 237
Hot-blast-furnace lamp	J. H. Wilhelm	Chicago, Ill	Nov. 7, 1865	50, 865
Hot-blast furnaces, Collar for pipes in	H. McCullaugh	Marietta, Pa	Dec. 29, 1868	85, 392
Hot-blast oven	R. Long	Pittsburgh, Pa	Apr. 2, 1872	125, 138
Hot-blast pipe	C. Glidden	Milwaukee, Wis	Feb. 28, 1865	46, 559
Hot-blast pipe	W. B. Pollock	Youngstown, Ohio	Mar. 7, 1865	46, 698
Hot-house, Parlor	P. Griffith	Brooklyn, N. Y	Mar. 1, 1870	100, 287
Hot-water and hot-air heaters, Apparatus for automatic control of combustion in.	J. C. Wrightman and J. H. Mills.	Newton and Boston, Mass	May 6, 1873	138, 547
Hot-water apparatus	J. Brown	New York, N. Y	May 30, 1854	10, 982
Hot-water apparatus	H. Humphreville, jr	Strasburgh, Pa	July 24, 1860	29, 267
Hot-water apparatus	J. F. Hunter	New York, N. Y	Mar. 27, 1860	27, 670
Hot-water apparatus, Boiler for	E. B. Cherevoy	New York, N. Y	Apr. 10, 1860	27, 858
Hot-water apparatus, Boiler for	E. Horeick	New York, N. Y	Apr. 9, 1861	31, 976
Hot-water apparatus for warming buildings	E. L. Miller	Brooklyn, N. Y	July 7, 1846	4, 625
Hot water boiler	P. Lesson	Newark, N. J	Jan. 30, 1872	123, 269
Hot-water boiler	A. Müller	Brooklyn, N. Y	June 15, 1869	91, 475
Hot-water boiler	H. Steeger	New York, N. Y	Dec. 31, 1867	72, 760
Hot-water boiler	J. Tregeser	New York, N. Y	July 27, 1869	93, 139
Hot-water-furnace boiler	C. R. Ellis	Brooklyn, N. Y	Sept. 6, 1864	44, 083
Hot-water heater	E. C. Clay	Malden, Mass	Jan. 30, 1872	123, 153
Hot-water heater	H. Howard	Springfield, Mass	Jan. 1, 1867	60, 732
Hot-water heater	A. Janos	New York, N. Y	Jan. 29, 1850	7, 054
Hot-water heater	G. Nixon	Philadelphia, Pa	May 28, 1872	127, 262
Hot-water heater	A. C. Pentland	Philadelphia, Pa	May 9, 1871	114, 595
Hot-water heater	G. H. Sellers	Phœnixville, Pa	Oct. 27, 1868	83, 553
Hot-water heater for apartments	J. F. Hunter and F. W. Geissenhainer.	New York, N. Y	Sept. 18, 1860	30, 065
Hot-water heater for buildings	G. M. Dexter	Boston, Mass	Dec. 1, 1840	1, 875
Hot-water heater for buildings	A. E. Hitchings	New York, N. Y	Jan. 25, 1848	5, 418
Hot-water heater for buildings	A. Marriott	Saint Louis, Mo	Sept. 12, 1871	118, 953
Hot-water heaters, Automatic air-damper for	W. C. Baker	New York, N. Y	Aug. 30, 1864	44, [illegible]38
Hot-water heaters, Boiler for	C. R. Ellis	Brooklyn, N. Y	July 16, 1872	129, 542
Hot-water or steam heater	S. Williams	Cambridge, Mass	June 11, 1872	127, 817
Hotel and burglar alarm	C. S. Noé	Bergen Point, N. J	Mar. 15, 1870	100, 792
Hotel communicating-apparatus	N. A. Patterson	Nashville, Tenn	Jan. 11, 1870	98, 794
Hotel register	C. L. Hawes	Titusville, Pa	Apr. 16, 1867	63, 889
Hotel-register	J. L. Mitchell	Buffalo, N. Y	Apr. 16, 1867	63, 924
Hotels, &c., Elevator or hoisting-apparatus for	O. Tufts	Boston, Mass	Aug. 9, 1859	25, 06[illegible]

Index of patents issued from the United States Patent Office from 1790 *to* 1873, *inclusive*—Continued.

Invention.	Inventor.	Residence.	Date.	No.
Hound and fifth-wheel combined	G. Archer	Massillon, Ohio	June 24, 1862	35, 662
House: *See* Bath-house. Bee-house. Bird-house. Brick-drying house. Corn-house. Dairy-house. Dry-house. Drying-house. Fire-engine house. Fire-proof house. Floating house. Grain store-house. Hot-house. Ice-house. Ice-preserving house. Iron house. Light-house. Locomotive-engine house. Malt-house. Metallic house. Milk-house. Navigable mercantile house. Portable house. Preserving-house. Refrigerating-house. Sheep-house. Signal-house. Slaughter-house. Smoke-house. Sugar-house. Switch-tender's house. Tobacco-drying house.				
House	O. C. Campbell	Omaha, Nebr	Mar. 11, 1873	136, 583
House	G. W. Clayton	Cleveland, Ohio	July 2, 1872	128, 464
House	D. L. Emerson	Oakland, Cal	Apr. 15, 1873	137, 833
House	H. M. Irwin	Charlotte, N. C	Aug. 24, 1869	94, 116
House	T. W. H. Moseley	Hyde Park, Mass	July 15, 1873	140, 941
House	T. W. H. Moseley	Hyde Park, Mass	July 15, 1873	140, 942
House	W. Ortwine	Baltimore, Md	Feb. 2, 1864	41, 445
House	J. Park	Joliet, Ill	Mar. 5, 1867	62, 676
House	A. Sidle	Minneapolis, Minn	Nov. 20, 1866	59, 929
House	R. B. Varden	Uniontown, Md	Mar. 5, 1872	124, 396
House	F. Walton	Staines, England	June 15, 1869	91, 289
House	W. Ward	New York, N. Y	Oct. 21, 1873	143, 946
House-alarm, Electro-magnetic	W. Whiting	Roxbury, Mass	July 20, 1858	20, 970
House, bridge, boat, and wagon-body, Combined	J. C. Adams	Baltimore, Md	Feb. 18, 1862	34, 399
House-furnaces, Regulating the draft of	S. L. Hay and H. B. Osgood	Reading and Dorchester, Mass.	Nov. 11, 1856	16, 055
House-hole cover	R. Liston	Albany, N. Y	May 31, 1870	103, 630
House-trimmings	J. R. Webber	Morris, Ill	Feb. 15 1870	99, 798
House-ventilation	J. H. Griscom	New York, N. Y	Jan. 18, 1859	22, 646
House-ventilator	R. Boyd	Evansville, Iowa	Nov. 12, 1867	70, 792
House-ventilator	B. J. Burnett	Mount Vernon, N. Y	Aug. 15, 1865	49, 373
House-ventilator	S. W. Williams	Centreville, N. Y	Sept. 17, 1861	33, 325
House-warmer and smoke-driver, Atmospherical	R. Annesly	Philadelphia, Pa	Apr. 27, 1814	....
House-warming furnace	P. Martin	Cincinnati, Ohio	Sept. 20, 1864	44, 329
Houses, &c., Composition for covering	A. Rogers	Hudson, N. Y	Mar. 14, 1817	....
Houses, Composition for covering and flooring	L. De Niroth	Baltimore, Md	Feb. 3, 1807	....
Houses, halls, &c., Ventilating	W. C. Grimes	Philadelphia, Pa	Feb. 9, 1869	86, 663
Houses, Ventilating	R. Mayo and R. Mills	Washington, D. C	Oct. 24, 1836	67
Houses with tin-plate, sheet-iron, or zinc, Covering	R. S. Tilden	Lynchburgh, Va	Sept. 10, 1829	....
Household and culinary operations, Machine for facilitating.	H. S. Shepardson	Shelburne Falls, Mass	Oct. 31, 1865	50, 779
Household-implement	T. Garrick	Providence, R. I	June 29, 1869	91, 839
Household-implement	T. Garrick	Providence, R. I	May 9, 1871	114, 665
Household-implement	A. Guinzburg	Boston, Mass	June 15, 1869	91, 227
Household-implements	R. A. McCauley	Baltimore, Md	July 11, 1871	116, 849
Household-utensil	H. R. Halsey	La Fayette, Ill	Aug. 15, 1871	118, 125
Household-utensil	W. C. McGill	Cincinnati, Ohio	Sept. 24, 1867	69, 111
Howel and croze	J. B. Siegfried	Pittsburgh, Pa	Sept. 15, 1868	82, 163
Hub	A. Huston	Cincinnati, Ohio	Oct. 9, 1866	58, 647
Hub and axle	J. W. and A. W. Beer	Myers' Dale and Rural Valley, Pa.	Aug. 6, 1872	130, 102
Hub and axle	C. R. Donner	Sonora, Cal	Feb. 14, 1871	111, 822
Hub and axle	N. Maxham	Hancock, Vt	Feb. 25, 1868	74, 841
Hub and axle, Carriage	L. D. Cook	West Liberty, Ohio	Jan. 2, 1872	122, 363
Hub and axle fastening	J. Henderson	Horseheads, N. Y	May 29, 1855	12, 946
Hub and axle lubricator	T. Wilson	Garton, England	Aug. 25, 1868	81, 449
Hub and axle, Vehicle	D. M. Buckhout	Mount Kisco, N. Y	Oct. 15, 1867	69, 760
Hub and axle, Vehicle	M. Chapin	Erie, Pa	Feb. 11, 1873	135, 689
Hub and axle-tree for wheels and pulleys	P. Slayton	Lockport, N. Y	Oct. 27, 1828	....
Hub and bearing, Car-wheel	D. Worstell and H. Pirrung	Lawrence County, Ohio	June 25, 1872	128, 348
Hub and box, Wheel	S. Mosher	Winchester, Ill	Oct. 13, 1868	83, 083
Hub and felly boring machine	D. Sperry	Colchester, Conn	May 12, 1812	....
Hub and journal, Carriage-wheel	R. W. McClelland	Springfield, Ill	June 10, 1862	35, 531
Hub and spoke, Carriage-wheel	J. Maris	Marietta, Ohio	Apr. 27, 1869	89, 322
Hub, axle, and box, Wagon	J. W. Pollock	Cross Bridges, Tenn	Mar. 17, 1868	75, 574
Hub band, Carriage	J. Jenkins and J. R. Cooke	Winsted, Conn	Dec. 12, 1854	12, 062
Hub band, Carriage	S. C. Talcott	Ashtabula, Ohio	June 21, 1864	43, 238
Hub-band, Metallic	S. D. Forbes	Wilmington, Del	Mar. 5, 1872	124, 202
Hub band, Vehicle-wheel	J. Ives	Mount Carmel, Conn	July 16, 1872	128, 961
Hub band, Vehicle-wheel	G. H. Johnson	Bridgeport, Conn	May 27, 1873	139, 317
Hub band, Wheel	S. H. Miller	Elizabethtown, N. J	Oct. 7, 1830	....
Hub-bands, Fastening	J. Ives	Mount Carmel, Conn	Feb. 7, 1871	111, 547
Hub-bands fo carriage-wheels, Manufacture of	S. E. Farrand	Newark, N. J	June 11, 1839	1, 171
Hub-bands for wagon-wheels, Making	G. W. Beers	Bridgeport, Conn	Nov. 29, 1859	26, 242
Hub-bands for wagon-wheels, Making	J. A. Boughton	Poughkeepsie, N. Y	Nov. 29, 1859	26, 246
Hub-bands	E. B. Butler	New Britain, Conn	Dec. 31, 1867	72, 719
Hub bands, Roll for rolling wagon	J. Glass	Pittsburgh, Pa	Aug. 12, 1873	141, 641
Hub-blocks for the lathe, Machine for preparing	L. Eames	Kalamazoo, Mich	June 9, 1857	17, 500
Hub-borer	W. I. Casselman	Vernon, N. Y	May 9, 1854	10, 882
Hub-borer	J. Shaerer	Reading, Pa	Jan. 13, 1857	16, 402
Hub-borer	C. B. Wiley	Adrian, Mich	Apr. 5, 1859	23, 535
Hub boring and mortising machine	G. M. Atherton	Friendsville, Ill	Dec. 6, 1859	26, 330
Hub boring and mortising machine	C. H. Guard	Brownville, N. Y	July 24, 1855	13, 307
Hub boring and mortising machine	J. Tompkins	Conesville, N. Y	Oct. 6, 1837	419
Hub boring and mortising machine, Carriage	R. I. R. Stone	Berlin, Ohio	Mar. 7, 1854	10, 603
Hub boring and mortising machinery	J. J. Greenough	Washington, D. C	June 13, 1846	4, 565
Hub-boring and spoke-tenoning machine	J. Gardner	Bigler, Pa	Apr. 16, 1872	125, 803

Index of patents issued from the United States Patent Office from 1790 to 1873, inclusive—Continued.

Invention.	Inventor.	Residence.	Date.	No.
Hub-boring apparatus, Wheel	J. Hinds	Troy, N. Y	Nov. 25, 1837	488
Hub-boring machine	A. Bascomb	Seneca, Mich	May 28, 1872	127, 296
Hub-boring machine	S. L. Bond	Greenwood, S. C	Nov. 29, 1859	26, 245
Hub-boring machine	D. C. Breed	Lyndonville, N. Y	Mar. 27, 1866	53, 400
Hub-boring machine	E. Caswell	Lyons, N. Y	Aug. 5, 1873	141, 425
Hub-boring machine	F. W. Dexter	Randolph, N. Y	Jan. 19, 1869	86, 006
Hub-boring machine	L. A. Dole	Salem, Ohio	Mar. 13, 1866	53, 216
Hub-boring machine	Z. Doolittle	Perry, Ga	Dec. 8, 1857	18, 808
Hub-boring machine	J. Duncan and W. H. Arnold	Buchanan, Mich	July 29, 1873	141, 264
Hub-boring machine	J. Eisenmann	Chicago, Ill	May 13, 1873	138, 742
Hub-boring machine	N. Johnson	Berlin, Wis	Mar. 3, 1868	75, 167
Hub-boring machine	F. Jonas	Freeport, Ill	Mar. 16, 1869	87, 852
Hub-boring machine	F. Jonas	Burlington, Iowa	Nov. 18, 1873	144, 618
Hub-boring machine	D. Murphy	Dubuque, Iowa	Apr. 6, 1869	88, 579
Hub-boring machine	J. A. Newell	Kalamazoo, Mich	Dec. 16, 1873	145, 521
Hub-boring machine	D. J. Owen	Springville, Pa	Dec. 31, 1867	72, 887
Hub-boring machine	W. S. Owen	Oskaloosa, Iowa	Dec. 23, 1873	145, 894
Hub-boring machine	G. T. Pearsall and S. A. Garrison	Apalachin and Union, N. Y.	Sept. 30, 1862	36, 573
Hub-boring machine	D. Quimby	Littleton, N. H	July 12, 1859	24, 756
Hub-boring machine	J. L. Roberts and R. K. Daily	Waverly, Iowa	Apr. 9, 1872	125, 616
Hub-boring machine	P. Schuttler	Chicago, Ill	May 2, 1865	47, 576
Hub-boring machine	A. R. Silver	Salem, Ohio	Aug. 11, 1868	80, 837
Hub-boring machine	A. R. Silver	Salem, Ohio	July 12, 1870	105, 265
Hub-boring machine	A. Troup	Lewisburgh, Pa	May 3, 1870	102, 625
Hub-boring machine	J. W. Walters	Riceville, Iowa	Apr. 21, 1868	77, 137
Hub-boring machine	D. B. Wright	South Amesbury, Mass	Dec. 2, 1873	145, 142
Hub-boring-machine clamp	G. V. Brecht	Saint Louis, Mo	June 16, 1868	78, 922
Hub-boring machine, Wagon	F. Bremerman	Indianapolis, Ind	Dec. 26, 1865	51, 688
Hub-boring machine, Wagon	E. Caswell	Newport, Mich	Mar. 29, 1864	42, 075
Hub-boring machine, Wagon	J. W. Emerson	Rochester, Minn	Feb. 4, 1868	74, 062
Hub-boring machine, Wagon	T. Harper	West Manchester, Pa	Aug. 21, 1866	57, 319
Hub-boring machine, Wagon	J. F. Kendall	Concord, N. H	Apr. 7, 1868	76, 467
Hub-boring machine, Wagon	J. Kirtch	Rochester, N. Y	Oct. 25, 1865	44, 807
Hub-boring machine, Wagon	S. Lavenue	Alton, Ill	Nov. 6, 1866	59, 417
Hub-boring machine, Wagon	L. Mason	Turin, N. Y	May 15, 1866	54, 750
Hub-boring machine, Wagon	J. R. McAlister	Richville, N. Y	Oct. 30, 1866	59, 245
Hub-boring machine, Wagon	W. Snodgrass	Macomb, Ill	July 2, 1867	66, 405
Hub-boring machine, Wagon	J. Thompson	Clifton, N. Y	Aug. 6, 1861	33, 012
Hub-boring machine, Wagon-wheel	J. Wharff	Bangor, Me	May 26, 1868	78, 246
Hub-boring tool	H. C. Garvin and J. H. King	Hagerstown, Md	Apr. 10, 1855	12, 677
Hub-boring tool	H. L. Mooney and W. B. Carter	Astoria, Ill	May 27, 1856	14, 968
Hub box, Carriage	J. A. Cramer	Brooklyn, N. Y	Apr. 2, 1861	31, 873
Hub-box, Cast-iron	J. Witt	Chambersburgh, Pa	Dec. 28, 1825	....
Hub-box molds, Preparing	J. G. Holt	Chicago, Ill	June 5, 1866	55, 296
Hub box, Wheel	I. Cooper	Baltimore, Md	Feb. 7, 1831	....
Hub box, Wheel	E. G. Woodside	San Francisco, Cal	Nov. 5, 1867	70, 670
Hub, Carriage	S. Allaire	New York, N. Y	Mar. 3, 1868	75, 105
Hub, Carriage	S. S. Barry	Brownhelm, Ohio	Jan. 6, 1852	8, 623
Hub, Carriage	S. W. Beach	Chicago, Ill	May 26, 1857	17, 360
Hub, Carriage	H. V. Belding	Oppenheim, N. Y	June 15, 1869	91, 302
Hub, Carriage	N. Bryan	Thomaston, Ga	Oct. 18, 1870	108, 441
Hub, Carriage	J. P. Chandler	Wilton, Me	Feb. 1, 1870	99, 405
Hub, Carriage	W. T. Dole	Peabody, Mass	Dec. 12, 1871	121, 857
Hub, Carriage	H. R. Fry	Wabash, Ind	Feb. 6, 1872	123, 472
Hub, Carriage	G. Kenny	Nashua, N. H	Apr. 6, 1869	88, 720
Hub, Carriage	C. Leavitt	Cleveland, Ohio	Jan. 20, 1863	37, 451
Hub, Carriage	J. A. Maynard	Newtonville, Mass	Oct. 25, 1870	108, 612
Hub, Carriage	J. W. Minor and D. P. Ward	New Bedford, Mass	Mar. 24, 1868	75, 948
Hub, Carriage	H. Nycum	Uniontown, Pa	Mar. 11, 1856	14, 407
Hub, Carriage	J. O'Connor	Jackson, Mo	May 3, 1870	102, 580
Hub, Carriage	J. Pruette	Aurora, Ill	Aug. 2, 1859	24, 955
Hub, Carriage	S. W. Reed	Berkshire, N. Y	Dec. 11, 1855	13, 919
Hub, Carriage	J. Smith	Sunbury, Ohio	Feb. 19, 1856	14, 294
Hub, Carriage	H. W. Stow	New Haven, Conn	May 2, 1871	114, 364
Hub, Carriage	J. W. Weston	New York, N. Y	July 6, 1869	92, 238
Hub, Carriage	M. Whelan	New Haven, Conn	Apr. 21, 1868	77, 147
Hub, Carriage	J. M. Whiting	Providence, R. I	Jan. 15, 1867	61, 293
Hub, Carriage	J. M. Whiting	Providence, R. I	Aug. 3, 1869	93, 259
Hub, Carriage	A. S. Woodward	Pepperell, Mass	Aug. 31, 1869	94, 372
Hub, Carriage-wheel	A. Allcott	Haverhill, Mass	June 11, 1872	127, 823
Hub, Carriage-wheel	E. A. Archibald	Methuen, Mass	Nov. 14, 1871	120, 845
Hub, Carriage-wheel	J. D. Arnold	Murray, N. Y	Oct. 1, 1872	131, 838
Hub, Carriage-wheel	W. J. Arrington	Jefferson County, Ga	Mar. 29, 1870	101, 208
Hub, Carriage-wheel	S. Atha	West Liberty, Ohio	Feb. 15, 1870	99, 809
Hub, Carriage-wheel	S. Atha	West Liberty, Ohio	Nov. 29, 1870	109, 707
Hub, Carriage-wheel	J. B. Bauman	Shepherdstown, Pa	Sept. 24, 1872	131, 650
Hub, Carriage-wheel	J. F. Beckwith	South Alabama, N. Y	Mar. 27, 1860	27, 611
Hub, Carriage-wheel	J. E. Bowers	Bainbridge, Ga	Apr. 2, 1872	125, 164
Hub, Carriage-wheel	R. Brown and W. Aldrich	Dayton, Ohio	Oct. 15, 1872	132, 241
Hub, Carriage-wheel	J. Y. Burwell	Worthington, Pa	Oct. 24, 1871	120, 150
Hub, Carriage-wheel	B. Burr	Batavia, Ill	Mar. 22, 1870	100, 973
Hub, Carriage-wheel	A. F. Cooper	San Francisco, Cal	May 3, 1870	102, 657
Hub, Carriage-wheel	H. Delano	Skaneateles, N. Y	Oct. 28, 1837	448
Hub, Carriage-wheel	L. Dorman	Worcester, Mass	May 21, 1867	64, 955
Hub, Carriage-wheel	J. Dump	Kingston, Ohio	Mar. 15, 1870	100, 873
Hub, Carriage-wheel	J. Dump and A. Moore	Kingston, Ohio	Apr. 16, 1872	125, 727
Hub, Carriage-wheel	J. M. Emmerich	New Haven, Conn	Aug. 23, 1870	106, 678
Hub, Carriage-wheel	J. F. Fowler	Alliance, Ohio	Mar. 19, 1872	124, 731
Hub, Carriage-wheel	J. C. Garretson	Marshland, N. Y	May 7, 1872	126, 389
Hub, Carriage-wheel	H. D. Haraden	Hartford, Vt	May 19, 1868	77, 077
Hub, Carriage-wheel	L. T. Hazen	Coventry, N. Y	May 24, 1859	24, 120
Hub, Carriage-wheel	C. C. Holt	Lawrence, Mass	May 14, 1872	126, 810
Hub, Carriage-wheel	J. W. Jackson and L. W. Burchinal.	Smithfield, Pa	Sept. 1, 1857	18, 097
Hub, Carriage-wheel	J. Kritch	Cleveland, Ohio	Jan. 2, 1872	122, 322
Hub, Carriage-wheel	J. Kritch	Cleveland, Ohio	Jan. 2, 1872	122, 323
Hub, Carriage-wheel	J. Locke and J. Clark	Lewisburgh and Milroy, Pa.	Oct. 1, 1872	131, 887
Hub, Carriage-wheel	C. Merry	New York, N. Y	Dec. 15, 1857	18, 855

Index of patents issued from the United States Patent Office from 1790 *to* 1873, *inclusive*—Continued.

Invention.	Inventor.	Residence.	Date.	No.
Hub, Carriage-wheel	P. Murphy	South Amesbury, Mass	Aug. 12, 1873	141, 659
Hub, Carriage-wheel	W. C. Pearsall	McMinnville, Tenn	Feb. 15, 1870	99, 942
Hub, Carriage-wheel	N. Platt	Jackson, Miss	Aug. 3, 1858	21, 083
Hub, Carriage-wheel	J. Ridge	Richmond, Ind	Nov. 12, 1872	132, 924
Hub, Carriage-wheel	S. T. F. Sterrick	Washington, D. C	June 29, 1869	92, 121
Hub, Carriage-wheel	H. Thomas	Medway, Mass	Feb. 8, 1828	....
Hub, Carriage-wheel	H. Thomas	Middleborough, Mass	Sept. 11, 1829	....
Hub, Carriage-wheel	H. E. Vick	Alliance, Ohio	Nov. 14, 1871	121, 025
Hub, Carriage-wheel	H. E. Vick	Alliance, Ohio	Dec. 5, 1871	121, 694
Hub, Carriage-wheel	A. Warner	Hamden, Conn	Feb. 5, 1867	61, 900
Hub, Carriage-wheel	J. M. Whiting	New Bedford, Mass	Mar. 30, 1858	19, 820
Hub, Carriage-wheel	M. Young	Frederick, Md	Dec. 26, 1865	51, 773
Hub, Cast-iron	C. Washburn	Bridgewater, Mass	Jan. 18, 1833	....
Hub-centering machine and spoke-guide	K. P. Allen	Homer, Mich	July 11, 1871	116, 790
Hub-clamp	J. McClelland	Geneva, N. Y	Sept. 12, 1871	118, 868
Hub dividing and boring machine	J. B. Cleveland	Bergen County, N. J	June 7, 1809	....
Hub, Expanding wheel	A. I. Judge	Baltimore, Md	Jan. 14, 1868	73, 342
Hub-fastening for eccentrics	G. W. Miller and J. D. Stevens	Scranton, Pa	May 12, 1868	77, 747
Hub for carriage-wheels, Metallic	E. A. Archibald	Methuen, Mass	Apr. 9, 1872	125, 514
Hub for carriage-wheels, Metallic	N. T. Edson	New Orleans, La	July 13, 1858	20, 869
Hub for carriage-wheels, Metallic	N. T. Edson	New Orleans, La	Oct. 16, 1860	30, 395
Hub for carriage-wheels, Metallic	J. Johnson	Garysburgh, N. C	Nov. 20, 1860	30, 681
Hub for carriage-wheels, Metallic	S. I. Russell	Chicago, Ill	June 15, 1858	20, 586
Hub for coaches, wagons, &c., Cast	W. Dickinson	Batavia, N. Y	Oct. 11, 1828	....
Hub for heavy-wheeled vehicles	J. J. Anderson	Indianapolis, Ind	Apr. 15, 1873	137, 874
Hub for pulleys, &c., Self-oiling	D. M. Weston	Boston, Mass	Mar. 10, 1868	75, 323
Hub for vehicle-wheels, Metallic	S. T. F. Sterick	Georgetown, D. C	Mar. 26, 1872	125, 095
Hub for vehicles, Metallic	J. Abbott	Washington, Ind	Jan. 30, 1866	52, 255
Hub for wheels, Cast-iron	B. Lyman	Manchester, Conn	Oct. 29, 1825	....
Hub for wheels, Cast-iron	B. Lyman	Manchester, Conn	Nov. 6, 1827	....
Hub for wheels of vehicles	E. D. Ives	Philadelphia, Pa	Feb. 6, 1872	123, 480
Hub for wheels of vehicles	J. Monk	Norwich, Conn	Apr. 1, 1873	137, 314
Hub for wheels of vehicles, Metallic	J. B. Stuart	Bunker Hill, Ill	Feb. 19, 1867	62, 161
Hub-machine	L. Eames	Kalamazoo, Mich	May 11, 1858	20, 197
Hub-machine	I. N. Felch	Hollis, Me	Jan. 22, 1861	31, 161
Hub-machine	J. B. Ripsom	East Kendall, N. Y	Aug. 25, 1863	39, 675
Hub, Metallic	W. W. Ball	Edinburgh, Ind	May 5, 1868	77, 568
Hub, Metallic	H. B. Buch	Litiz, Pa	Sept. 11, 1866	57, 855
Hub, Metallic	J. H. Harper	Pittsburgh, Pa	Mar. 7, 1871	112, 452
Hub, Metallic	J. B. Hayden	Easton, N. Y	Feb. 7, 1854	10, 506
Hub, Metallic	J. Oliphant	Springhill Furnace, Pa	June 30, 1868	79, 496
Hub, Metallic carriage	H. Boardman	Lancaster, Pa	Feb. 14, 1860	27, 097
Hub, Metallic wheel	J. Monk	Norwich, Conn	Dec. 5, 1871	121, 462
Hub-mortising machine	E. Joslin and D. L. Gibbs	Keene, N. H., and Norwich, Conn.	May 27, 1862	35, 420
Hub-mortising machine	P. Snyder	Grand Rapids, Mich	Oct. 17, 1871	120, 117
Hub-mortising machinery	A. Thompson	Ridgeville, Md	Apr. 25, 1843	3, 054
Hub or gig band, Carriage	C. Cornelius	Philadelphia, Pa	June 18, 1824	....
Hub-reamer	S. A. Garrison	Union, N. Y	Aug. 23, 1859	25, 190
Hub-reamer	J. G. Robinson	Springfield, Ill	Mar. 22, 1870	101, 164
Hub, Road-carriage	J. Raddin	Lynn, Mass	May 3, 1870	102, 709
Hub-shaping machine	A. Goodyear	Albion, Mich	Aug. 31, 1869	94, 201
Hub shell, Carriage	J. O'Connor	Jackson, Mo	Oct. 19, 1869	96, 028
Hub turning and mortising machine	E. M. Scott	Auburn, N. Y	Mar. 4, 1862	34, 595
Hub-turning machine	W. G. Beach	New Haven, Conn	Nov. 14, 1871	120, 931
Hub-turning machine	A. Rickart	Schoharie, N. Y	Sept. 7, 1858	21, 443
Hub-turning machine	A. Rickurt	Schoharie, N. Y	June 28, 1859	24, 579
Hub-turning machine	J. J. Zufelt and R. Craig	Sheboygan Falls, Wis	Jan. 5, 1869	85, 629
Hub, Vehicle	H. M. Du Bois	Philadelphia, Pa	Feb. 4, 1873	135, 470
Hub, Vehicle	C. W. Fillmore	Marengo, Ill	Apr. 30, 1872	126, 278
Hub, Vehicle	C. W. Fillmore	Chicago, Ill	Aug. 13, 1872	130, 416
Hub, Vehicle	W. I. Lyman	East Hampton, Mass	Jan. 17, 1871	111, 070
Hub, Vehicle	S. McGee	Madison, N. J	Feb. 4, 1873	135, 570
Hub, Vehicle	A. Moffitt	Brownsville, Pa	Sept. 2, 1862	36, 355
Hub, Vehicle	G. L. Rouse	Cincinnati, Ohio	June 10, 1873	139, 691
Hub, Vehicle	T. Royer and G. L. Rouse	Cincinnati, Ohio	Mar. 4, 1873	136, 459
Hub, Vehicle	S. T. F. Sterick	Georgetown, D. C	Jan. 14, 1873	134, 945
Hub, Vehicle	J. P. Zeller	South Bend, Ind	Sept. 7, 1869	94, 685
Hub, Vehicle-wheel	J. S. Ball	Charleston, Ill	June 25, 1872	128, 354
Hub, Vehicle-wheel	D. Davis	New York, N. Y	Aug. 29, 1871	118, 589
Hub, Vehicle-wheel	D. Davis	New York, N. Y	July 2, 1872	128, 600
Hub, Vehicle-wheel	C. G. Gray	Ansonia, Conn	July 9, 1872	128, 794
Hub, Vehicle-wheel	C. J. Harris	Louisville, Ky	Oct. 8, 1872	132, 071
Hub, Vehicle-wheel	S. B. Hitt and R. W. Chapman	Waterloo, Iowa	Apr. 23, 1872	125, 955
Hub, Vehicle-wheel	J. B. Hubbell	Naugatuck, Conn	July 23, 1872	129, 666
Hub, Vehicle-wheel	E. D. Ives	Philadelphia, Pa	Aug. 22, 1871	118, 369
Hub, Vehicle-wheel	P. Jones	Newark, N. J	July 2, 1872	128, 546
Hub, Vehicle-wheel	P. Jones	Newark, N. J	July 2, 1872	128, 547
Hub, Vehicle-wheel	W. A. Lewis	Chicago, Ill	Dec. 10, 1872	133, 867
Hub, Vehicle-wheel	E. B. Lowe	Bellefontaine, Ohio	Oct. 8, 1872	132, 091
Hub, Vehicle-wheel	W. F. Morton	New Haven, Conn	Dec. 24, 1872	134, 155
Hub, Vehicle-wheel	A. M. Ocobock	Toledo, Ohio	Oct. 29, 1872	132, 596
Hub, Vehicle-wheel	J. Ridge	Richmond, Ind	July 30, 1872	130, 075
Hub, Vehicle-wheel	H. Taylor, A. P. Taylor, and C. Palm.	Milton, Mineral Ridge, and Jackson, Ohio.	Jan. 16, 1872	122, 742
Hub, Vehicle-wheel	O. Vanorman	Fond du Lac, Wis	Oct. 10, 1871	119, 803
Hub, Wagon	E. R. Baker	Fairhaven, Mass	Aug. 18, 1868	81, 127
Hub, Wagon	J. H. Gaines	Durhamville, Tenn	Aug. 8, 1871	117, 878
Hub, Wagon	J. D. Ham	Bethany, Ga	July 26, 1870	105, 674
Hub, Wagon	J. B. Hubbell	Naugatuck, Conn	Sept. 20, 1870	104, 497
Hub, Wagon	T. M. Jones and C. W. Fillmore.	Chicago, Ill	Jan. 16, 1872	122, 835
Hub, Wagon	R. W. McClelland	Springfield, Ill	July 24, 1866	56, 586
Hub, Wagon	N. Rixford	Mansfield Centre, Conn	Feb. 2, 1869	86, 451
Hub, Wagon	S. W. Slocumb	Albany, Ill	Dec. 3, 1867	71, 801
Hub, Wagon	A. E. Smith	Bronxville, N. Y	July 7, 1863	39, 177
Hub, Wagon	W. C. Tucker	Richmond Switch, R. I	Sept. 10, 1867	68, 803
Hub, Wagon	A. S. Woodward	Pepperell, Mass	Dec. 1, 1868	84, 601

Index of patents issued from the United States Patent Office from 1790 to 1873, inclusive—Continued.

Invention.	Inventor.	Residence.	Date.	No.
Hub, Wagon-wheel	S. Altha	West Liberty, Ohio	Sept. 26, 1871	119, 295
Hub, Wagon-wheel	F. Nichols	Newport, Ky	Sept. 29, 1868	82, 630
Hub, Wagon-wheel	W. H. Rodeheaver	Miamisburgh, Ohio	Apr. 14, 1868	76, 818
Hub, Wheel	J. Atherton	Philadelphia, Pa	June 2, 1836	....
Hub, Wheel	W. F. Ehlers	Pottsville, Pa	Aug. 2, 1870	105, 926
Hub, Wheel	J. B. Hards	Chicago, Ill	Aug. 16, 1870	106, 361
Hub, Wheel	W. C. Johnson	Philadelphia, Pa	Mar. 28, 1871	113, 171
Hub, Wheel	O. B. Little	Wheeling, W. Va	Jan. 23, 1872	123, 030
Hub, Wheel	J. Summers	Raleigh Court-House, Va	Apr. 15, 1856	14, 678
Hub, Wheel	B. F. Taft	Groton Junction, Mass	Sept. 3, 1867	68, 397
Hub, Wheel	J. L. Van West	Tolland, Mass	Feb. 7, 1871	111, 703
Hub, Wheel	C. P. Whitman	Charlemont, Mass	May 10, 1870	102, 988
Hub, Wheel	G. E. Whitmore	Housatonic, Mass	June 4, 1867	65, 521
Hub, Wheel	J. V. Woolsey	Sandusky, Ohio	Aug. 8, 1871	117, 849
Hubs and axles, Applying friction-roller to	J. B. and S. Wilson	Townsend's Inlet, N. J., and Kensington, Pa.	Feb. 25, 1851	7, 948
Hubs and axles, Attaching and detaching	R. D. Munson	Williston, Vt	Jan. 2, 1849	5, 995
Hubs and axles, Connecting	C. Chinnock	New York, N. Y	Jan. 9, 1849	6, 004
Hubs and axles, Connecting and disconnecting	A. M. Billings	Claremont, N. H	Jan. 14, 1851	7, 899
Hubs and axles, Connecting and disconnecting	A. M. Billings and T. A. Ambrose.	Claremont, N. H	June 25, 1850	7, 450
Hubs and axles, Manufacture of	S. R. Hunter and M. Merrill	Cortlandville, N. Y	Feb. 27, 1849	6, 147
Hubs and axles of vehicles, Connecting	D. Beard	Shippensburgh, Pa	Mar. 8, 1859	23, 148
Hubs and ships' blocks, Box for wheel	I. Cooper	Baltimore, Md	Jan. 27, 1830	....
Hubs and spokes of wrought-iron wheels, Process for forming the.	T. Ryan	Scott Bar, Cal	Aug. 10, 1869	93, 483
Hubs, Apparatus for setting box in carriage	C. Weidig	New Haven, Conn	Dec. 26, 1871	122, 295
Hubs, Arrangement of cutter for turning	G. Cooper	Berlin, Wis	Nov. 30, 1858	22, 167
Hubs, Attaching caps to	P. Heoter and R. Victor	Grand Rapids, Mich	May 30, 1871	115, 316
Hubs, Boring and mortising	H. Hayes	Quincy, Ill	Sept. 2, 1856	15, 652
Hubs, Box for carriage	A. C. Garratt	Roxbury, Mass	Feb. 26, 1856	14, 310
Hubs, Box-setter for boring in	E. Badlam, jr	Chester, Vt	Sept. 18, 1835	....
Hubs, Box-setter for wheel	F. W. Dexter	Randolph, N. Y	Apr. 22, 1862	35, 015
Hubs. Casting box for wheel	T. Ellis	Philadelphia, Pa	Dec. 6, 1859	26, 394
Hubs, Cutter for boring wheel	L. S. Maring	Westport, Mass	Oct. 4, 1853	10, 087
Hubs, Device for securing the boxing in	M. Turley	Council Bluffs, Iowa	Aug. 22, 1871	118, 406
Hubs for boring, Clamp for centering	J. Thrasher	Avon, N. Y	Nov. 10, 1857	18, 612
Hubs for boxes, Apparatus for boring	H. Sidle	Dillsburgh, Pa	Feb. 24, 1852	8, 761
Hubs for boxes, Boring	S. H. Yocum	Shelbyville, Ind	Sept. 2, 1856	15, 678
Hubs for boxes, Machine for preparing	W. R. Jones	Granville, N. Y	July 29, 1851	8, 267
Hubs for boxes, Machinery for preparing	I. Munden	Allegheny City, Pa	Dec. 11, 1849	6, 941
Hubs for boxes, Tool for preparing	S. Fahrney	Near Boonsborough, Md	Mar. 19, 1850	7, 185
Hubs for containing oil, Constructing carriage and wagon wheel.	A. Randel	Vernon, N. Y	Sept. 8, 1837	374
Hubs for linch-cups, Boring out the inside and the boxes of.	J. R. Morrison	Springfield, Ohio	May 13, 1834	....
Hubs for reception of boxes, Machine for preparing	H. Moore	Seneca County, Ohio	July 29, 1851	8, 265
Hubs for setting boxes, Machine for boring wheel	J. C. Hendry	Manchester, N. H	July 18, 1871	117, 170
Hubs, Greasing and keeping dust, &c., out of the boxes of carriage-wheel.	N. C. Day	Lunenburgh, Mass	July 10, 1840	1, 680
Hubs, Inserting India rubber into carriage	G. F. Wilson	East Providence, R. I	Nov. 24, 1868	84, 329
Hubs, Lathe for turning	J. Z. Wagner and H. R. Fry	Wabash, Ind	Apr. 29, 1873	138, 303
Hubs, Machine for forming	W. Patterson	Constantine, Mich	Dec. 20, 1859	26, 520
Hubs, Machine for hewing out	G. W. Miles and P. P. Lane	Michigan City, Ind., and Cincinnati, Ohio.	Mar. 8, 1859	23, 220
Hubs, Machine for turning the band portion of carriage.	Z. Doolittle	Perry, Ga	Dec. 22, 1857	18, 893
Hubs, Mortising wagon	P. Paullin	Dantown, Ohio	Apr. 17, 1833	....
Hubs of carriage-wheels, Lining metallic boxes for	M. Palmer	Baltimore, Md	Mar. 9, 1844	3, 463
Hubs of carriage-wheels, &c., Making metallic cases for chilling cast-iron-pipe boxes for.	J. Huntington	Zanesville, Ohio	Feb. 24, 1845	3, 929
Hubs of cars and other wheels, Hardening or chilling.	H. Thomas	Beaver Meadow, Pa	Oct. 13, 1838	980
Hubs of propellers, Securing the arms to the	H. O. Perry	Buffalo, N. Y	Dec. 7, 1858	22, 266
Hubs of wagon-wheels and the tenons of spokes to fit in the hubs, Mortising.	D. B. Goewey	Birmingham, Pa	Apr. 9, 1867	63, 628
Hubs of wheels, Boring and mortising	R. McCarty	Martinsburgh, N. Y	Mar. 31, 1834	....
Hubs of wheels, Machine for making	G. Lyon	New York	Dec. 28, 1826	....
Hubs on axles, Securing	A. E. Smith	Bronxville, N. Y	Feb. 17, 1857	16, 661
Hubs on axles, Securing carriage-wheel	J. Scheeper	New York, N. Y	Dec. 24, 1861	34, 015
Hubs, &c., Pattern for metal	J. Johnson	Geneseo, N. Y	July 6, 1852	9, 095
Hubs, Sand-cap for vehicle	G. H. Nevens	Livermore, Cal	May 2, 1871	114, 324
Hubs, Seasoning wagon	A. Bennighofen	Hamilton, Ohio	May 23, 1871	115, 017
Hubs, Securing box-metal in carriage	A. H. Ahlborn	Lawrenceville, Pa	Feb. 4, 1868	74, 029
Hubs, Securing boxes in	W. Greenleaf	Terre Haute, Ind	May 8, 1866	54, 529
Hubs, Securing boxes in metallic	J. B. Stuart	Bunker Hill, Ill	Feb. 5, 1867	61, 890
Hubs, &c., Securing boxes to wheel	J. Kritsch	Binghamton, N. Y	Sept. 9, 1862	36, 412
Hubs to axle-boxes, Attaching	T. C. Maris	Athens, Ohio	Mar. 5, 1867	62, 551
Hubs to axles, Attaching	P. Bargion	Stockton, Cal	Sept. 3, 1872	131, 077
Hubs to axles, Attaching	C. H. Denison	Brattleborough, Vt	Oct. 29, 1861	33, 577
Hubs to axles, Attaching	C. H. Guard	New York, N. Y	Sept. 25, 1866	58, 247
Hubs to axles, Attaching	J. Gunn	Salem Township, Ill	June 15, 1869	91, 228
Hubs to axles, Attaching	J. M. Perkins	New York, N. Y	May 15, 1855	12, 871
Hubs to axles, Attaching	J. M. Riley	Newark, N. J	Sept. 30, 1856	15, 818
Hubs to axles, Attaching	E. Sampson	Claremont, N. H	Feb. 21, 1854	10, 552
Hubs to axles, Attaching	E. S. Scripture	Green Point, N. Y	Dec. 18, 1855	13, 962
Hubs to axles, Attaching	H. B. Simonds	West Hartford, Vt	Mar. 4, 1856	14, 371
Hubs to axles, Attaching	J. Weathers	Greensburgh, Ind	May 12, 1868	77, 785
Hubs to axles, Attaching	J. M. White	Xenia, Ohio	Mar. 24, 1857	16, 891
Hubs to axles, Attaching	W. L. Williams	New York, N. Y	Jan. 2, 1872	122, 506
Hubs to axles, Attaching	L. Winslow	Rochester, N. Y	Apr. 14, 1857	17, 063
Hubs to axles, Attaching carriage	L. Adams	Amherst, Mass	Mar. 16, 1869	87, 815
Hubs to axles, Connecting	J. Kellogg	Madison, Ohio	Nov. 13, 1849	6, 870
Hubs to axles, Device for attaching	O. F. Shepard	Cincinnati, Ohio	July 1, 1873	140, 389
Hubs to axles of carriage-wheels, Attaching	J. London, jr	Auburn, N. Y	Oct. 12, 1839	1, 365
Hubs to axles, Securing	G. E. Clow	Jeffersonville, Ind	Sept. 20, 1870	107, 452
Hubs to axles, Securing	C. Darling	Utica, N. Y	Aug. 15, 1854	11, 521
Hubs to axles, Securing	L. J. Worden	Utica, N. Y	May 5, 1857	17, 247

Index of patents issued from the United States Patent Office from 1790 *to* 1873, *inclusive*—Continued.

Invention.	Inventor.	Residence.	Date.	No.
Hubs to axles, Securing the boxes of carriage-wheel.	L. Smith and G. R. Waring	Derby, Conn	June 17, 1840	1, 640
Hubs to axles, Securing wheel	I. Osgood	Utica, N. Y	Nov. 6, 1866	59, 445
Hubs to let in boxes, Cutting and boring the ends of	J. B. Francis	Loudoun County, Va	Mar. 2, 1832	....
Hubs to receive boxes, Tool for boring	U. Kimble	Penfield, N. Y	Mar. 27, 1855	12, 598
Hubs to shafts, Securing wheel	E. Sanford	Hartford, Conn	Oct. 21, 1873	143, 849
Hubs to the arms of axle-trees, Attaching carriage-wheel.	G. Hunt	Prattsville, N. Y	July 8, 1839	1, 223
Hubs, Tool for fitting bands on	C. E. Stone and A. Herbert	Amesbury and Salisbury, Mass.	July 14, 1868	79, 925
Hubs, tool-handles, &c., Machine for turning	S. Carpenter	Flushing, N. Y	Oct. 10, 1854	11, 777
Hubs, &c., Turning	S. Beers	Naugatuck, Conn	June 6, 1854	11, 016
Hubs, Turning carriage	A. Rickart	Schoharie, N. Y	July 21, 1857	17, 846
Hubs, Turning wagon	W. W. Cleaveland	Cold Water, Mich	Sept. 29, 1868	82, 600
Hubs with axles, Connecting	J. Foster	Bridgeport, Conn	July 24, 1849	6, 608
Hubs with axles, Connecting	J. Foster	Bridgeport, Conn	Feb. 26, 1850	7, 117
Hubs with axles, Connecting	E. Sampson and A. M. Billings	Claremont, N. H	Nov. 20, 1849	6, 887
Hubs with axles, Connecting	E. S. Scripture	Green Point, N. Y	July 30, 1850	7, 535
Hubs while being bored, Centering and holding	A. Moore	Honeoye Falls, N. Y	Mar. 31, 1857	16, 932
Hubs while being bored, Machine for holding	P. Schuttler	Chicago, Ill	Feb. 21, 1865	46, 501
Hull for steamboats	W. Hagerty	Monongahela, Pa	June 23, 1868	79, 114
Hull of vessels	D. De Haven	New Orleans, La	July 19, 1870	105, 437
Hull of vessels	L. P. Rider	Pittsburgh, Pa	Oct. 4, 1870	107, 961
Hull of vessels	J. Van Pelt	Perry, Ill	Oct. 29, 1867	70, 296
Huller: *See* Barley-huller. Cotton-seed huller. Buckwheat-huller. Grain-huller. Clover-huller. Grass-seed huller. Clover-seed huller. Rice-huller. Corn-huller. Strawberry-huller.				
Huller and screen	G. Stevenson and J. J. Crider	Zionsville and Greenfield, Ind.	Mar. 22, 1864	42, 027
Hulling-machine	A. Angell	Newburgh, N. Y	Mar. 2, 1869	87, 388
Hulling-machine	S. Bentz	Carroll County, Md	Feb. 11, 1862	34, 346
Hulling-machine	A. Bertelson	West Salem, Wis	June 10, 1873	139, 758
Hulling-machine	G. A. Buchholz	Shepherd's Bush, England	Nov. 23, 1869	97, 039
Hulling-machine	G. A. Buchholz	Regent's Park, England	May 10, 1870	102, 764
Hulling-machine	J. E. Carver	Bridgewater, Mass	July 26, 1870	105, 643
Hulling-machine	J. M. Hendricks	Philadelphia, Pa	Sept. 2, 1862	36, 350
Hulling-machine	S. R. Hockman	Urbana, Ohio	June 1, 1869	90, 752
Hulling-machine	C. Jordan	East Bridgewater, Mass	Nov. 2, 1869	96, 439
Hulling-machine	D. Kahnweiler	New York, N. Y	Mar. 29, 1870	101, 271
Hulling-machine	C. Keniston and J. H. Sawyer	Somerville and Boston, Mass.	Oct. 5, 1869	95, 485
Hulling-machine	J. McGregor	Saratoga, N. Y	May 12, 1840	1, 600
Hulling-machine	C. Miller	Carroll Township, Pa	Dec. 25, 1855	13, 992
Hulling-machine	D. Pease, jr	Floyd, N. Y	Apr. 3, 1849	6, 271
Hulling-machine	D. Pease, jr	Floyd, N. Y	Apr. 10, 1849	6, 289
Hulling-machine	W. Porter	Brooklyn, N. Y	Apr. 19, 1870	102, 152
Hulling-machine	W. Seck	Frankfort, Prussia	May 7, 1872	126, 416
Hulling machine, Barley, &c	J. J. Johnston and J. E. Weaver	Allegheny City and Temperanceville, Pa.	June 30, 1863	39, 051
Hulling-machine, Cylindrical	S. Houpt	Philadelphia, Pa	Nov. 16, 1812	....
Hulling-machine feed-device	D. Kahnweiler	New York, N. Y	Dec. 19, 1871	122, 028
Hulling machine, Grain	W. Zimmermann	Quincy, Ill	June 14, 1859	24, 423
Hulling-mill	C. S. Bailey	New York, N. Y	Sept. 7, 1869	94, 540
Hulling-mill, Centrifugal	C. S. Bailey	New York, N. Y	July 20, 1869	92, 690
Hulling-mills, Dress of stones for	D. Collins	Jersey City, N. J	Mar. 9, 1858	19, 605
Human power, Economizing	W. C. Moores	Bloomfield, Wis	May 10, 1864	42, 678
Husk-hackling machines, Separating-attachment for.	G. B. Stacy	Richmond, Va	Oct. 18, 1870	108, 532
Husk-splitter	B. Hollis	Randolph, Ill	Dec. 19, 1871	122, 017
Husking-machine	E. Farnum and G. W. Scott	Blackstone, Mass	Feb. 5, 1867	61, 823
Husking-machine	J. Russell	Brooklyn, N. Y	Nov. 26, 1867	71, 539
Husking-palm	D. E. Shaw	Ross County, Ohio	Nov. 10, 1857	18, 607
Husking-pin	F. Brown	Russell, Ohio	June 7, 1864	43, 004
Husking-pin	W. L. Corson	Hennepin, Ill	Feb. 13, 1872	123, 615
Husking-pin	W. Sherwin	Shelburne Falls, Mass	Sept. 24, 1861	33, 369
Husking pin, Corn	E. Blair	Bucyrus, Ohio	Oct. 13, 1868	82, 915
Husking-thimble	J. H. Gould	Smith, Ohio	May 13, 1856	14, 864
Husking-thimble	H. J. Kintz	Greece, N. Y	Mar. 16, 1869	87, 856
Hut, Portable	A. Derrom	Paterson, N. J	July 9, 1861	32, 757
Hydrant	J. Allison	Cincinnati, Ohio	Nov. 10, 1868	83, 899
Hydrant	G. C. Bailey	Pittsburgh, Pa	Dec. 21, 1869	98, 011
Hydrant	G. C. Bailey	Pittsburgh, Pa	Mar. 19, 1872	124, 777
Hydrant	H. J. Bailey	Pittsburgh, Pa	June 16, 1868	78, 917
Hydrant	H. J. Bailey	Pittsburgh, Pa	Dec. 29, 1868	85, 267
Hydrant	T. R. Bailey, jr	Lockport, N. Y	Mar. 10, 1868	75, 344
Hydrant	W. Bailey	Troy, N. Y	July 4, 1865	48, 504
Hydrant	W. Bailey	Troy, N. Y	Sept. 20, 1870	107, 434
Hydrant	J. W. Baker	Parkersburgh, W. Va	Sept. 24, 1867	69, 155
Hydrant	F. H. Bartholomew	New York, N. Y	Mar. 7, 1846	4, 410
Hydrant	F. H. Bartholomew	New York	Aug. 12, 1846	4, 692
Hydrant	F. H. Bartholomew	New York, N. Y	Feb. 15, 1859	22, 927
Hydrant	F. Bauschtliker	Washington, D. C	Apr. 20, 1869	89, 117
Hydrant	H. C. Biggs and A. C. Flinn	Lancaster, Pa	Apr. 17, 1866	53, 944
Hydrant	W. W. Birny	Seneca Falls, N. Y	June 2, 1857	17, 415
Hydrant	B. P. Bower	Cleveland, Ohio	May 22, 1866	54, 845
Hydrant	G. N. Bowman	Pottsville, Pa	July 20, 1869	92, 696
Hydrant	T. C. Bride	Quincy, Ill	Feb. 1, 1870	99, 396
Hydrant	S. H. Brown	Troy, N. Y	May 13, 1862	35, 213
Hydrant	T. Brown	Allegheny City, Pa	June 23, 1868	79, 200
Hydrant	J. G. Bryan	Philadelphia, Pa	Apr. 20, 1869	89, 123
Hydrant	J. Bryant	Brooklyn, N. Y	June 16, 1857	17, 552
Hydrant	J. Bryant	Brooklyn, N. Y	Apr. 5, 1859	23, 444
Hydrant	J. H. Buckley	New Haven, Conn	Mar. 19, 1867	62, 933
Hydrant	S. G. Cabell	Quincy, Ill	Apr. 6, 1869	88, 542

Index of patents issued from the United States Patent Office from 1790 to 1873, inclusive—Continued.

Invention.	Inventor.	Residence.	Date.	No.
Hydrant	S. G. Cabell and A. Q. Ross	Quincy, Ill., and Cincinnati, Ohio.	Mar. 1, 1870	100, 255
Hydrant	J. L. Chapman	Baltimore, Md	Oct. 12, 1842	2, 812
Hydrant	E. Clampitt	Baltimore, Md	Nov. 29, 1870	109, 587
Hydrant	S. P. Clark	Baltimore, Md	Sept. 3, 1861	33, 185
Hydrant	Z. E. Coffin	Boston, Mass	July 21, 1868	80, 143
Hydrant	R. H. Colvin	Columbia, Pa	July 20, 1843	3, 185
Hydrant	C. J. Cowperthwaite	Philadelphia, Pa	Jan. 15, 1856	14, 090
Hydrant	C. J. Cowperthwaite	Philadelphia, Pa	May 6, 1856	14, 805
Hydrant	D. C. Cregier	Chicago, Ill	Sept. 10, 1861	33, 239
Hydrant	T. A. Davies	New York	May 12, 1843	3, 082
Hydrant	R. De Charms	Philadelphia, Pa	Jan. 5, 1858	19, 022
Hydrant	W. P. Dickinson, D. S. Witman, and G. W. Rabold.	Reading, Pa	July 2, 1867	66, 225
Hydrant	W. H. Duffett	Rochester, N. Y	Aug. 15, 1871	118, 116
Hydrant	H. English	Baltimore, Md	Apr. 1, 1856	14, 557
Hydrant	J. Ewing	Philadelphia, Pa	Dec. 19, 1823	....
Hydrant	J. Farman	Cleveland, Ohio	Apr. 4, 1871	113, 507
Hydrant	W. Fields and S. Gerhard	Wilmington, Del	Feb. 3, 1857	16, 536
Hydrant	G. W. Fisher	Saint Louis, Mo	June 4, 1872	127, 587
Hydrant	L. Fitton	Wheeling, W. Va	Feb. 2, 1869	86, 384
Hydrant	A. S. Fort	Cincinnati, Ohio	Mar. 1, 1870	100, 389
Hydrant	S. P. Francisco and W. P. Dickinson.	Reading, Pa	Dec. 21, 1858	22, 357
Hydrant	C. E. Frazier	Baltimore, Md	Nov. 17, 1868	84, 104
Hydrant	J. Fricker	Cincinnati, Ohio	Dec. 12, 1871	121, 771
Hydrant	J. Fricker and A. Warden	Cincinnati, Ohio	June 7, 1870	104, 012
Hydrant	A. Fuller	Cincinnati, Ohio	Sept. 4, 1860	29, 872
Hydrant	B. G. Fuller	Baltimore, Md	Apr. 21, 1868	77, 027
Hydrant	J. P. Gallagher	Saint Louis, Mo	Jan. 5, 1864	41, 144
Hydrant	J. P. Gallagher	Saint Louis, Mo	May 2, 1871	114, 283
Hydrant	J. Gibson and M. Heberger	Cincinnati, Ohio	June 10, 1862	35, 515
Hydrant	T. Gibson	Philadelphia, Pa	Dec. 13, 1823	....
Hydrant	K. Goddard	Philadelphia, Pa	Feb. 9, 1858	19, 330
Hydrant	D. K. Gould	Chestertown, N. Y	Apr. 30, 1867	64, 307
Hydrant	N. B. Gousha	Baltimore, Md	Aug. 7, 1866	56, 927
Hydrant	W. H. Graham	Saint Louis, Mo	Dec. 9, 1873	145, 294
Hydrant	P. H. Griffin	Albany, N. Y	Dec. 12, 1871	121, 716
Hydrant	J. V. Hayes	Buffalo, N. Y	May 6, 1873	138, 562
Hydrant	N. Hayman	New York, N. Y	Sept. 8, 1863	39, 812
Hydrant	S. C. Higbie	Oppenheim, N. Y	Sept. 1, 1843	3, 240
Hydrant	J. R. Higgs	Utica, N. Y	Aug. 31, 1858	21, 338
Hydrant	A. Hoagland	Jersey City, N. J	Apr. 21, 1857	17, 093
Hydrant	B. Holly	Lockport, N. Y	Sept. 14, 1869	94, 749
Hydrant	D. Horne	Baltimore, Md	Mar. 31, 1836	....
Hydrant	J. Hyde	New York, N. Y	Jan. 5, 1858	19, 029
Hydrant	J. P. Hyde	New York, N. Y	Apr. 19, 1870	102, 126
Hydrant	W. James	Baltimore, Md	Nov. 1, 1859	25, 969
Hydrant	E. T. Jenkins	Ravenswood, N. Y	Jan. 19, 1869	86, 076
Hydrant	A. Johnson	Philadelphia, Pa	June 12, 1860	28, 674
Hydrant	W. Kearney	Union Township, N. J	Sept. 22, 1868	82, 326
Hydrant	J. P. Kenyon	Brooklyn, N. Y	Feb. 19, 1861	31, 464
Hydrant	J. R. Manny	Chicago, Ill	Jan. 25, 1870	99, 097
Hydrant	N. B. Marsh	Cincinnati, Ohio	Oct. 4, 1859	25, 660
Hydrant	J. W. Marshall	Williamsburgh, N. Y	Nov. 17, 1868	84, 202
Hydrant	J. McCann	Albany, N. Y	Mar. 21, 1871	112, 825
Hydrant	R. A. McCauley	Baltimore, Md	May 29, 1866	55, 136
Hydrant	S. McElroy	Brooklyn, N. Y	Oct. 14, 1862	36, 663
Hydrant	A. J. and T. McKenna	Pittsburgh, Pa	July 5, 1870	105, 106
Hydrant	J. McLelland	Washington, D. C	Mar. 3, 1863	37, 820
Hydrant	B. B. Moore	Detroit, Mich	Aug. 20, 1861	33, 095
Hydrant	J. G. Morgan	Brooklyn, N. Y	Jan. 20, 1857	16, 448
Hydrant	J. G. Murdock	Cincinnati, Ohio	May 26, 1863	38, 694
Hydrant	J. G. Murdock	Cincinnati, Ohio	Dec. 10, 1867	72, 071
Hydrant	J. W. Murphy	Baltimore, Md	Aug. 6, 1872	130, 233
Hydrant	J. Myers	Cincinnati, Ohio	May 3, 1870	102, 698
Hydrant	J. Neumann	Philadelphia, Pa	May 14, 1861	32, 338
Hydrant	W. P. Newhall	New York, N. Y	Dec. 4, 1866	60, 230
Hydrant	T. W. Newton and J. H. Laning	Philadelphia, Pa	July 30, 1833	....
Hydrant	J. Old	Pittsburgh, Pa	June 26, 1866	55, 893
Hydrant	J. and S. P. Parham	Philadelphia, Pa., and Trenton, N. J.	Mar. 2, 1858	19, 511
Hydrant	R. D. Patterson	Cincinnati, Ohio	Sept. 26, 1865	50, 155
Hydrant	J. Pandler, jr., and F. Bauschtliker.	Washington, D. C	June 5, 1866	55, 357
Hydrant	F. J. Paul	Cincinnati, Ohio	Apr. 1, 1873	137, 476
Hydrant	G. P. Perrine and J. E. Boyle	Richmond, Va	June 23, 1857	17, 632
Hydrant	J. L. Pillsbury	Columbus, Ohio	June 6, 1871	115, 637
Hydrant	A. Pixley and J. Robertson	Brooklyn, N. Y	June 19, 1866	55, 782
Hydrant	J. Powell	Cincinnati, Ohio	Mar. 2, 1858	19, 513
Hydrant	J. Pringle	Jersey City, N. J	Dec. 8, 1863	40, 852
Hydrant	D. J. Pruner	Bellefonte, Pa	Dec. 28, 1869	98, 411
Hydrant	W. Race and S. R. C. Mathews	Seneca Falls, N. Y	Jan. 26, 1858	19, 206
Hydrant	W. Race and S. R. C. Mathews	Seneca Falls, N. Y	Apr. 19, 1859	23, 706
Hydrant	W. Race and S. R. C. Mathews	Lockport, N. Y	Nov. 16, 1869	96, 959
Hydrant	T. Ragan	Philadelphia, Pa	Mar. 4, 1873	136, 545
Hydrant	J. Regester	Baltimore, Md	June 9, 1863	38, 841
Hydrant	J. Regester	Baltimore, Md	June 19, 1866	55, 711
Hydrant	J. Regester	Baltimore, Md	June 19, 1866	55, 712
Hydrant	J. Regester	Baltimore, Md	Oct. 22, 1867	70, 119
Hydrant	R. Reilly	Baltimore, Md	Nov. 10, 1868	83, 882
Hydrant	A. Richmond, jr	Dayton, Ohio	Nov. 22, 1870	109, 547
Hydrant	J. P. Riley	Philadelphia, Pa	Dec. 11, 1866	60, 423
Hydrant	G. W. Robertson	Philadelphia, Pa	Jan. 24, 1860	26, 931
Hydrant	G. H. Rodgers	Baltimore, Md	Sept. 10, 1861	33, 282
Hydrant	G. J. and H. W. Ross	New York, N. Y	June 3, 1862	35, 473
Hydrant	N. Schlesser	Chicago, Ill	Aug. 15, 1871	118, 059
Hydrant	G. M. Selden	Troy, N. Y	Aug. 13, 1861	33, 054

Index of patents issued from the United States Patent Office from 1790 *to* 1873, *inclusive*—Continued.

Invention.	Inventor.	Residence.	Date.	No.
Hydrant	J. Small	Washington, D. C	Nov. 28, 1871	121, 429
Hydrant	J. J. Smith and S. Wood	Cleveland, Ohio	Feb. 2, 1869	86, 466
Hydrant	J. N. Smith	Jersey City, N. J	Nov. 13, 1866	59, 670
Hydrant	J. N. Smith	Jersey City, N. J	Nov. 13, 1866	59, 671
Hydrant	J. N. Smith	Jersey City, N. J	Jan. 22, 1867	61, 366
Hydrant	J. N. Smith	Jersey City, N. J	July 9, 1872	128, 763
Hydrant	C. L. Stacy	Cincinnati, Ohio	Oct. 4, 1859	25, 683
Hydrant	C. L. Stacy	Cincinnati, Ohio	Sept. 10, 1872	131, 191
Hydrant	A. Stephenson	Saint Louis, Mo	Sept. 5, 1865	49, 801
Hydrant	R. Stileman	Philadelphia, Pa	Jan. 20, 1863	37, 466
Hydrant	E. Stocker	Lancaster, Pa	Feb. 20, 1866	52, 797
Hydrant	E. Stocker and A. Henpel	Lancaster, Pa	Oct. 1, 1872	131, 796
Hydrant	J. Stroop	Pittsburgh, Pa	Dec. 16, 1833	....
Hydrant	J. Stroop	Pittsburgh, Pa	Nov. 11, 1834	....
Hydrant	J. Swan	Brooklyn, N. Y	Oct. 19, 1858	21, 858
Hydrant	T. T. Tasker	Philadelphia, Pa	July 11, 1842	2, 714
Hydrant	S. Tice	Cincinnati, Ohio	Dec. 8, 1868	84, 848
Hydrant	T. Van Kannel	Cincinnati, Ohio	Dec. 14, 1869	97, 835
Hydrant	T. Van Kannel and G. Elvenhoefer.	Cincinnati, Ohio	Dec. 23, 1873	145, 768
Hydrant	S. T. Walker	Baltimore, Md	Nov. 26, 1835	....
Hydrant	S. T. Walker	Baltimore, Md	July 1, 1836	....
Hydrant	S. T. Walker	Baltimore, Md	July 2, 1836	....
Hydrant	J. Walsh	Philadelphia, Pa	Jan. 18, 1870	99, 037
Hydrant	J. Walsh	Philadelphia, Pa	May 16, 1871	114, 996
Hydrant	J. Walsh	Philadelphia Pa	June 18, 1872	128, 192
Hydrant	J. M. Ward	New York, N. Y	Aug. 21, 1866	57, 449
Hydrant	L. W. Werner	Saint Louis, Mo	Jan. 12, 1869	85, 879
Hydrant	M. Zwiebel	Pottsville, Pa	Aug. 10, 1869	93, 658
Hydrant and fire-plug	S. H. Davies	Cincinnati, Ohio	July 18, 1840	1, 702
Hydrant and fire-plug	J. I. Healey	Brooklyn, N. Y	June 17, 1873	139, 890
Hydrant and fire-plug, Combined	F. Bauschtliker	Washington, D. C	Nov. 1, 1870	108, 751
Hydrant and gas-pipe stop	F. Shickle	Saint Louis, Mo	Nov. 17, 1868	84, 222
Hydrant and street-washer	J. Farnan	Cleveland, Ohio	Sept. 24, 1872	131, 512
Hydrant, Anti-freezing	J. W. Slocum	Philadelphia, Pa	Apr. 6, 1869	88, 589
Hydrant-cap	N. W. Speers	Cincinnati, Ohio	June 6, 1854	11, 019
Hydrant-curb, Self-adjusting	J. A. Finnegan	Charlestown, Mass	Aug. 18, 1868	81, 266
Hydrant for aqueducts	D. Burt	Schenectady, N. Y	June 22, 1808	....
Hydrant for filtration	J. H. Carter	Cincinnati, Ohio	Oct. 18, 1859	25, 805
Hydrant-fountain	L. Gouley	Philadelphia, Pa	Aug. 1, 1828	....
Hydrant, Non-freezing	J. E. Carter	Portland, Me	Nov. 3, 1868	83, 602
Hydrant, &c., Non-freezing	J. H. Hall	Peoria, Ill	Apr. 30, 1872	126, 202
Hydrant, Non-freezing	H. Merrie	Cincinnati, Ohio	June 13, 1871	115, 882
Hydrant, Non-freezing	J. H. Printz	Zanesville, Ohio	July 30, 1872	130, 069
Hydrant, Non-freezing	J. Walsh	Philadelphia, Pa	Aug. 6, 1872	130, 166
Hydrant stop-cock rod	H. Rausch	Brooklyn, N. Y	Oct. 12, 1869	95, 724
Hydrant waste-device	J. Culver	Baltimore, Md	Apr. 22, 1856	14, 712
Hydrants and fire-plugs, Thawing-device for	J. C. Moore	Philadelphia, Pa	July 26, 1870	105, 830
Hydrants, Incasing	W. Bramwell	New York, N. Y	June 9, 1857	17, 538
Hydrants, Stop-plate for	J. C. Eastman	Nashua, N. H	Aug. 2, 1870	105, 925
Hydrants, Waste-attachment to	E. J. Baker	Baltimore, Md	Apr. 8, 1856	14, 592
Hydraulic and pneumatic elevator for railways	H. H. Day	Bloomingdale, N. J	Sept. 5, 1871	118, 696
Hydraulic and steam engine, Cylinder and piston of	J. Underwood	Lowell, Mass	Dec. 9, 1856	16, 202
Hydraulic apparatus	A. Desgoffe and A. Ollivier	Paris, France	Feb. 7, 1865	46, 315
Hydraulic apparatus	L. Eames	Kalamazoo, Mich	Sept. 9, 1862	36, 397
Hydraulic apparatus	G. H. Herring	Durand, Ill	Feb. 7, 1871	111, 536
Hydraulic apparatus	N. Nolan	New York, N. Y	Nov. 8, 1870	109, 042
Hydraulic apparatus	P. F. Powers	Genoa, Nev	May 2, 1871	114, 339
Hydraulic apparatus	J. F. Thompson	Greensborough, Pa	Dec. 8, 1868	84, 719
Hydraulic apparatus for producing blast	R. Cook	Saratoga Springs, N. Y	June 11, 1850	7, 423
Hydraulic apparatus for transmitting power	D. Hinman	Brunswick, Ohio	Oct. 31, 1848	5, 899
Hydraulic boat and mode of propelling the same	W. Shultz	Philadelphia, Pa	May 28, 1812	....
Hydraulic cement, Manufacture of artificial	W. B. Tucker	Baltimore, Md	May 13, 1873	138, 770
Hydraulic disk	T. Shaw	Philadelphia, Pa	May 30, 1871	115, 365
Hydraulic elevator	C. W. Baldwin	Boston, Mass	May 28, 1872	127, 139
Hydraulic elevator	D. Bickford	Boston, Mass	Nov. 28, 1865	51, 129
Hydraulic elevator	T. Chambers	Saint Louis, Mo	Dec. 17, 1867	72, 165
Hydraulic elevator	D. Corey	New York, N. Y	Aug. 31, 1827	....
Hydraulic elevator	H. L. Ensign	Chicago, Ill	Jan. 30, 1872	123, 161
Hydraulic elevator	H. Flad	Saint Louis, Mo	Sept. 14, 1869	94, 732
Hydraulic elevator	H. Flad and G. P. Herthel, jr	Saint Louis, Mo	May 28, 1867	65, 200
Hydraulic elevator	T. Stebins	San Francisco, Cal	Oct. 8, 1872	132, 111
Hydraulic elevator	T. Stebins	San Francisco, Cal	June 3, 1873	139, 624
Hydraulic engine	W. E. B. Ainsworth and G. W. Scott.	Blackstone, Mass	Apr. 5, 1870	101, 445
Hydraulic engine	J. F. Burgin and A. Koch	Williamsport, Pa	Feb. 14, 1860	27, 104
Hydraulic engine	A. C. Carey and J. Smith	Ipswich, Mass	Mar. 21, 1854	10, 665
Hydraulic engine	R. F. Dobson	Belmont, Wis	Jan. 7, 1873	134, 651
Hydraulic engine	J. Dreisörner	New York, N. Y	Aug. 21, 1866	57, 298
Hydraulic engine	M. Everett	Kalamazoo, Mich	Sept. 17, 1867	68, 863
Hydraulic engine	J. S. Groynne	New York, N. Y	Jan. 26, 1858	19, 224
Hydraulic engine	J. D. Heaton	Dixon, Ill	Sept. 29, 1857	18, 2[illegible]0
Hydraulic engine	J. C. Hodges	Morristown, Tenn	Jan. 14, 1873	13[illegible], 806
Hydraulic engine	K. Horde	Philadelphia, Pa	Aug. 13, 1861	33, 039
Hydraulic engine	H. Hubbell	New Haven, Conn	Aug. 28, 1866	57, 572
Hydraulic engine	H. J. King and B. L. Beebe	Middletown, N. Y	Dec. 14, 1869	97, 779
Hydraulic engine	A. D. Laws	Bridgeport, Conn	May 10, 1870	102, 94[illegible]
Hydraulic engine	T. G. McLaughlin	Philadelphia, Pa	Aug. 7, 1847	5, 214
Hydraulic engine	A. Miller	Grafton, Va	May 20, 1855	14, 922
Hydraulic engine	F. Ransom	Buffalo, N. Y	Apr. 28, 1868	77, 214
Hydraulic engine	L. M. Sabin	Saint Louis, Mo	Oct. 8, 1867	69, 591
Hydraulic engine	W. Snyder	Bullskin Township, Pa	Jan. 3, 1871	11[illegible], 796
Hydraulic engine	J. Strong	....	Mar. 24, 1801	....
Hydraulic engine	H. H. Stuart	Jamaica, N. Y	July 3, 1866	56, 120
Hydraulic engine	W. J. Tate	Philadelphia, Pa	Oct. 31, 1871	120, 548
Hydraulic engine	E. Thomas and W. K. Hall	Beverly and Philippi, Va	Dec. 4, 1860	30, 844
Hydraulic engine	J. H. and R. W. Welch	Georgetown, D. C	Dec. 17, 1872	134, 115

Index of patents issued from the United States Patent Office from 1790 *to* 1873, *inclusive*—Continued.

Invention.	Inventor.	Residence.	Date.	No.
Hydraulic engine and meter	J. S. Barden	Providence, R. I	Oct. 24, 1865	50, 651
Hydraulic engine, Atmospheric and condensing	A. J. Reynolds	Chicago, Ill	Dec. 7, 1869	97, 698
Hydraulic engine, Revolving	A. Hubbard	Windsor, Vt	Apr. 22, 1828	....
Hydraulic heater	W. H. Churchman	Philadelphia, Pa	May 2, 1854	10, 855
Hydraulic heater	L. W. Leeds and R. M. Smith	Philadelphia, Pa	May 16, 1854	10, 942
Hydraulic indicator	R. D. Bradley	Preston, Md	Nov. 8, 1864	44, 931
Hydraulic jack	R. Blackwood	Philadelphia, Pa	June 25, 1861	32, 612
Hydraulic jack	R. Dudgeon	New York, N. Y	Aug. 1, 1865	49, 097
Hydraulic jack	R. Dudgeon	New York, N. Y	Apr. 15, 1873	137, 765
Hydraulic jack	G. Lindsay	New York, N. Y	Mar. 10, 1857	16, 801
Hydraulic jack	A. V. Ojeda	San Francisco, Cal	July 29, 1873	141, 232
Hydraulic jack	J. Ryan	Saint Louis, Mo	Jan. 23, 1866	52, 210
Hydraulic jack	J. Ryan	Saint Louis, Mo	Jan. 30, 1866	52, 324
Hydraulic jack	E. J. Shaw and G. F. Eisenhardt.	Philadelphia, Pa	Jan. 9, 1872	122, 539
Hydraulic jack	T. H. Watson	New York, N. Y	Sept. 20, 1864	44, 358
Hydraulic lifting-jack	J. Tangye	Birmingham, England	Apr. 14, 1863	38, 187
Hydraulic machine	L. Brunier	New York, N. Y	July 8, 1842	2, 711
Hydraulic machine	C. Darmond	Georgetown, D. C	Aug. 7, 1817	....
Hydraulic machine	T. Hutchins	Reading, Pa	May 7, 1834	....
Hydraulic machine	J. James	Lynganore, Md	Nov. 18, 1818	....
Hydraulic machine, Balance-lever	D. Fisk	Springfield, N. Y	July 18, 1820	....
Hydraulic machine for raising water	B. Langdon	....	Feb. 20, 1801	....
Hydraulic machine for raising water	W. Shultz	Philadelphia, Pa	May 26, 1812	....
Hydraulic machinery	R. McCormick	Rockbridge County, Va	Oct. 1, 1830	....
Hydraulic meter	J. S. Barden	New Haven, Conn	Feb. 26, 1856	14, 335
Hydraulic motor	A. Barbarin and J. Albrecht	New Orleans, La	Aug. 1, 1871	117, 504
Hydraulic motor	A. Barbarin and J. Albrecht	New Orleans, La	Aug. 1, 1871	117, 505
Hydraulic motor	B. A. Bloch	Paris, France	Mar. 25, 1873	137, 171
Hydraulic motor	U. A. Boyden	Boston, Mass	Sept. 20, 1853	10, 027
Hydraulic motor	J. Harris	Boston, Mass	July 19, 1870	105, 450
Hydraulic motor	J. H. Jennings	New Bedford, Mass	Nov. 14, 1871	120, 975
Hydraulic motor	M. Keely and G. W. Cressman	Barren Hill, Pa	Oct. 11, 1859	25, 741
Hydraulic motor	W. Keunish	London, England	May 15, 1860	28, 282
Hydraulic motor	V. Kromer	Grand Rapids, Wis	June 13, 1871	115, 869
Hydraulic motor	C. Mesler	Almond, N. Y	Aug. 30, 1864	44, 010
Hydraulic motor	M. Millard	Franklin, Ohio	Mar. 19, 1872	124, 843
Hydraulic motor	M. A. Shepard	Orio, Ill	July 19, 1859	24, 826
Hydraulic motor	W. Walter	Arkada, Wash	Aug. 26, 1873	142, 183
Hydraulic motor	T. Welham	Brownsville, Nebr	May 31, 1864	42, 980
Hydraulic motor, Organ-lever coupling for	H. F. Wheeler	Boston, Mass	May 6, 1873	138, 546
Hydraulic power	Z. and A. Parker	Coshocton County, Ohio	Oct. 19, 1829	....
Hydraulic-power elevator	P. J. Borger	Cincinnati, Ohio	Aug. 13, 1872	130, 468
Hydraulic press	T. Baxter	Petersburgh, Va	Apr. 19, 1859	23, 741
Hydraulic press	J. T. Burr	Brooklyn, N. Y	Aug. 13, 1867	67, 719
Hydraulic press	M. H. Clark	Danville, Va	Sept. 20, 1859	25, 488
Hydraulic press	J. A. Desgoffe	Paris, France	Dec. 20, 1870	110, 346
Hydraulic press	R. Dudgeon	New York, N. Y	Jan. 25, 1859	22, 713
Hydraulic press	A. H. Emery	New York, N. Y	Apr. 16, 1867	63, 875
Hydraulic press	W. Ettenger and H. P. Edmund.	Richmond, Va	Apr. 14, 1868	76, 732
Hydraulic press	C. W. Flippen	Laurel Grove, Va	May 8, 1860	28, 167
Hydraulic press	R. Grant	New York, N. Y	Aug. 1, 1854	11, 455
Hydraulic press	T. Harbottle	Brooklyn, N. Y	May 4, 1869	89, 656
Hydraulic press	M. Hittinger	Charlestown, Mass	Aug. 23, 1864	43, 912
Hydraulic press	C. W. Holbrook	New York, N. Y	Mar. 3, 1868	75, 015
Hydraulic press	A. Judson and S. A. Farrington	Newark, N. J	June 29, 1869	92, 057
Hydraulic press	G. W. Rawson	Cambridgeport, Mass	Aug. 11, 1868	80, 877
Hydraulic press	C. E. Rymes	Charlestown, Mass	May 19, 1863	38, 606
Hydraulic press	J. B. Tunstall	Boydton, Va	Oct. 27, 1868	83, 424
Hydraulic press, Copper-lined cylinder for	C. Sellers	Philadelphia, Pa	May 28, 1872	127, 191
Hydraulic press for cotton, &c	C. Wilson	Williamsburgh, N. Y	Oct. 9, 1849	6, 776
Hydraulic press for peat, brick, &c	A. H. Emery	New York, N. Y	Aug. 21, 1866	57, 303
Hydraulic press, Portable	R. Dudgeon	New York, N. Y	July 8, 1851	8, 203
Hydraulic press, Pressure-accumulator for	W. D. Grimshaw	Ansonia, Conn	Jan. 30, 1872	123, 169
Hydraulic press, Retainer for	M. Madden	Saint Louis, Mo	July 15, 1862	35, 881
Hydraulic press, Retainer for	C. E. Rymes	Charlestown, Mass	Dec. 27, 1859	26, 622
Hydraulic press to prevent corrosion	T. P. Shaffner	Louisville, Ky	Dec. 18, 1866	60, 570
Hydraulic-pressure regulator	I. Judson	New Haven, Conn	June 25, 1867	66, 028
Hydraulic-pressure regulator	J. S. McDonald	New Orleans, La	June 25, 1872	128, 235
Hydraulic ram	B. S. Benson	Harford County, Md	Dec. 26, 1845	4, 328
Hydraulic ram	J. Cerneau and S. S. Hallet	New York, N. Y	Aug. 3, 1809	....
Hydraulic ram	W. Fields, jr	Providence, R. I	Jan. 28, 1851	7, 918
Hydraulic ram	J. L. Gatchel	Chester, Pa	Apr. 10, 1847	5, 052
Hydraulic ram	D. A. Leighton	Middleburgh, N. Y	Aug. 22, 1848	5, 725
Hydraulic ram	J. C. Strode	East Bradford, Pa	Mar. 27, 1847	5, 037
Hydraulic ram	J. C. Strode	East Bradford, Pa	Oct. 4, 1853	10, 079
Hydraulic ram	J. C. Strode	West Chester, Pa	May 23, 1854	10, 969
Hydraulic ram	J. F. Warner	Philadelphia, Pa	June 8, 1858	20, 523
Hydraulic ram	E. Webb	Parkersville, Pa	Dec. 5, 1854	12, 042
Hydraulic ram	J. D. West	New York, N. Y	Aug. 29, 1854	11, 626
Hydraulic ram, Tidal or current	J. W. Middleton	Philadelphia, Pa	Sept. 19, 1854	11, 697
Hydraulic ram, Apparatus for directing water to and maintaining continuous pressure upon.	T. Harbottle and M. F. Fowler	New York, N. Y	Oct. 29, 1861	33, 585
Hydraulic rams, Operating	C. Polly	May's Landing, N. J	Mar. 7, 1854	10, 610
Hydraulic rams, Operating the waste-gate in	J. Osborn	Hamden, Conn	Feb. 11, 1851	7, 932
Hydraulic regulator for machinery	L. B. Pitcher	Syracuse, N. Y	Apr. 9, 1850	7, 272
Hydraulic seal	J. H. Sutton	Honesdale, Pa	July 12, 1870	105, 277
Hydraulic steam-engine, Planetary	J. Black	Philadelphia, Pa	Sept. 20, 1853	10, 024
Hydraulic structures, Obtaining permanent foundations for.	T. W. P. Lewis	Boston, Mass	May 1, 1845	4, 030
Hydraulic wheel for raising water	P. D. Henry	New Orleans, La	Jan. 9, 1841	1, 930
Hydraulic-wheel press	G. A. Gray, jr	Hamilton, Ohio	Apr. 29, 1873	138, 393
Hydraulics	J. Hatheway	....	Oct. 29, 1794	....
Hydraulics	J. Roup	Kanawha County, Va	Oct. 6, 1827	....
Hydraulics	T. Somerby	Nantucket, Mass	Oct. 1, 1830	....
Hydro-atmospheric elevator	C. W. Baldwin	Boston, Mass	Jan. 17, 1871	111, 030
Hydrocarbon and other liquids, Package for	M. W. Brown	New York, N. Y	Apr. 4, 1871	113, 394

Index of patents issued from the United States Patent Office from 1790 to 1873, inclusive—Continued.

Invention.	Inventor.	Residence.	Date.	No.
Hydrocarbon and other liquids, Vessel and tank for holding.	J. G. Moody	New York, N. Y	Feb. 5, 1867	61, 753
Hydrocarbon-apparatus, Self-ventilating safety-can for filling and discharging.	L. D. Towsley	Newark, N. J	Dec. 21, 1869	98, 208
Hydrocarbon as fuel, Apparatus for burning	W. H. Russell	Brooklyn, N. Y	July 9, 1872	128, 914
Hydrocarbon-heater	O. F. Morrill	Chelsea, Mass	Dec. 4, 1866	60, 224
Hydrocarbon-liquid can	I. W. Shaler	Brooklyn, N. Y	Oct. 3, 1871	119, 657
Hydrocarbon liquids, Apparatus for heating	P. B. Kitchen	Philadelphia, Pa	Nov. 8, 1859	26, 070
Hydrocarbon liquids, Apparatus for making light from.	E. H. Covel	New York, N. Y	June 15, 1869	91, 423
Hydrocarbon liquids, Apparatus for vaporizing	W. H. Burr	Philadelphia, Pa	Nov. 5, 1872	132, 800
Hydrocarbon liquids as fuel, Burning	A. J. Works	Fair Haven, Conn	Jan. 8, 1867	61, 131
Hydrocarbon liquids, Burning	C. Fisher	Trenton, N. J	July 23, 1867	67, 108
Hydrocarbon liquids for illumination, Apparatus for vaporizing.	J. Griffin	Meriden, Conn	Apr. 26, 1864	42, 469
Hydrocarbon liquids for light, Burning	T. J. Barrow	Brooklyn, N. Y	Feb. 28, 1871	112, 206
Hydrocarbon liquids for transmitting heat, Use of	W. C. Baker	New York, N. Y	Oct. 16, 1866	58, 755
Hydrocarbon liquids in the presence of steam, Process of vaporizing and decomposing.	F. Cook and J. A. Bassett	New York, N. Y., and Salem, Mass.	Sept. 10, 1867	68, 708
Hydrocarbon liquids, Process of vaporizing	J. A. Bassett	Salem, Mass	Sept. 10, 1867	68, 686
Hydrocarbon liquids, &c., Storing	P. C. P. L. Prefontaine	Paris, France	July 17, 1866	56, 508
Hydrocarbon oils as fuel, Burning	F. Cook	New York, N. Y	Sept. 10, 1867	68, 707
Hydrocarbon oils, Generating and burning vapor of.	H. K. Foote	Oil City, Pa	Sept. 18, 1866	58, 087
Hydrocarbon on locomotives, Burning	A. C. Rand	Chicago, Ill	Oct. 10, 1871	119, 720
Hydrocarbon-vapor apparatus	G. H. Bronson	Cincinnati, Ohio	Dec. 13, 1859	26, 458
Hydrocarbon-vapor apparatus	E. H. Covel	New York, N. Y	May 31, 1859	24, 200
Hydrocarbon-vapor apparatus	A. and A. Davis and C. Cunningham.	Lowell, Mass., and Nashua, N. H.	Jan. 15, 1856	14, 129
Hydrocarbon-vapor apparatus	L. L. Hill	Greenport, N. Y	Dec. 20, 1859	26, 497
Hydrocarbon-vapor apparatus	A. A. Moss	Philadelphia, Pa	Aug. 9, 1859	25, 032
Hydrocarbon-vapor apparatus	F. S. Pease	Buffalo, N. Y	Mar. 13, 1860	27, 470
Hydrocarbon vapor for illumination	W. H. Laubach	Philadelphia, Pa	Feb. 14, 1860	27, 190
Hydrocarbon-vapor machine	J. F. Spence	Brooklyn, N. Y	Sept. 17, 1867	69, 037
Hydrocarbon vapors, Apparatus for burning	F. C. Ambler	New York, N. Y	June 4, 1872	127, 402
Hydrocarbon vapors, Apparatus for generating	W. R. Roberts	Philadelphia, Pa	May 9, 1871	114, 603
Hydrocarbon vapors, Combustion-chamber for burning.	C. J. Eames	New York, N. Y	Mar. 25, 1873	137, 132
Hydrocarbon vapors, Device for generating and burning.	C. Carpenter, jr	Seekonk, Mass	Sept. 23, 1873	143, 061
Hydrocarbons, Apparatus for burning	M. Carbonel	Havana, Cuba	June 17, 1873	140, 009
Hydrocarbons, Apparatus for burning	C. H. Cushing	Tidioute, Pa	Nov. 18, 1873	144, 660
Hydrocarbons, Apparatus for burning	I. Kendrick	Philadelphia, Pa	Jan. 30, 1872	123, 109
Hydrocarbons, Apparatus for burning	I. Kendrick	Philadelphia, Pa	Nov. 5, 1872	132, 839
Hydrocarbons, Apparatus for burning	A. C. Rand	Aurora, Ill	Oct. 22, 1872	132, 491
Hydrocarbons, Apparatus for burning	E. F. Rogers	Chelsea, Mass	Jan. 21, 1873	135, 011
Hydrocarbons, Apparatus for burning	A. Schpakofsky and N. Stange.	St. Petersburg, Russia	Apr. 3, 1866	53, 763
Hydrocarbons, Apparatus for burning	W. Sim and A. Barff	Glasgow, North Britain	Aug. 8, 1865	49, 357
Hydrocarbons, Apparatus for burning	F. T. Suhr	Titusville, Pa	Apr. 21, 1868	77, 124
Hydrocarbons, Apparatus for burning liquid	A. J. Works	Fair Haven, Conn	Aug. 21, 1866	57, 429
Hydrocarbons, Apparatus for generating and burning vapor from.	J. Kidd	New York, N. Y	Jan. 10, 1871	110, 857
Hydrocarbons, Apparatus for separating the products of distillation of.	L. N. Wilcox	Pittsburgh, Pa	July 25, 1865	49, 020
Hydrocarbons, Apparatus for vaporizing	J. K. Caldwell	Philadelphia, Pa	Sept. 12, 1871	118, 903
Hydrocarbons, Apparatus for vaporizing and aërating volatile.	L. Stevens	Fitchburgh, Mass	Dec. 20, 1864	45, 568
Hydrocarbons, Apparatus for vaporizing and burning.	S. J. Whiting	Philadelphia, Pa	Mar. 26, 1872	124, 993
Hydrocarbons, Apparatus for vaporizing and burning.	S. J. Whiting	Philadelphia, Pa	June 11, 1872	127, 723
Hydrocarbons, Apparatus for vaporizing and burning liquid.	F. Cook	New York, N. Y	Sept. 10, 1867	68, 702
Hydrocarbons, Apparatus for vaporizing and burning liquid.	O. F. Morrill	Chelsea, Mass	May 27, 1862	35, 383
Hydrocarbons, Apparatus for vaporizing volatile	J. Butler	New York, N. Y	Apr. 25, 1871	113, 978
Hydrocarbons as fuel, Using liquid	F. Cook	New York, N. Y	Sept. 10, 1867	68, 705
Hydrocarbons, Burning	J. Chandler and D. A. Wray	Pioneer, Pa	Oct. 22, 1872	132, 440
Hydrocarbons, Burning	I. Kendrick	Philadelphia, Pa	Nov. 8, 1870	109, 131
Hydrocarbons for heating, &c., Vaporizing	H. H. Eames and C. J. Eames	Philadelphia, Pa., and New York, N. Y.	Oct. 15, 1872	132, 266
Hydrocarbons, Producing light by the combination of solid and liquid.	J. Phillips	Cologne, Germany	Jan. 18, 1870	98, 883
Hydrocarbons, Production of light from heavy	R. S. Merrill	Cambridge, Mass	Mar. 15, 1870	100, 915
Hydrocarbons, Separating	R. A. Chesebrough	New York, N. Y	June 5, 1866	55, 240
Hydrocarbons, Treating solid and liquid	F. Lambe	London, England	June 13, 1871	115, 871
Hydrocarbons, Vaporizing and burning liquid	F. Cook	New York, N. Y	Sept. 10, 1866	68, 703
Hydrocarbons with oxygen, Apparatus for supplying.	A. H. Webster	Hudson, N. Y	Feb. 8, 1859	22, 907
Hydrodynamic engine	L. C. St. John	Buffalo, N. Y	May 2, 1854	10, 869
Hydrodynamic machine for testing strength of materials.	F. C. Lowthorp	Trenton, N. J	May 26, 1857	17, 383
Hydro-mechanical press	J. Beverly		Dec. 26, 1803	....
Hydro-oxygen light, Apparatus for the production of Hare's.	G. H. Smith	Rochester, N. Y	Sept. 27, 1859	25, 611
Hydropneumatic apparatus	J. Gregg	Rochester, N. Y	Jan. 24, 1842	2, 433
Hydropneumatic engine	R. Crosbie	Newark, N. J	Jan. 6, 1812	....
Hydropneumatic engine	J. A. Huss	Saint Louis, Mo	Aug. 15, 1865	49, 410
Hydropneumatic engine for extinguishing fires, Platuary.	W. Loughbridge	Weverton, Md	Apr. 10, 1855	12, 686
Hydropneumatic machine for exhausting and sealing vessels.	W. H. Elliot	Plattsburgh, N. Y	July 24, 1855	13, 303
Hydropneumatic motor	C. L. Stevens	Galesburgh, Ill	Oct. 31, 1871	120, 545
Hydrofuze fabric	J. Wansbrough	Southwark, England	Jan. 11, 1859	22, 596
Hydrogen and air, Apparatus for mixing	D. Ashworth	Wappinger's Falls, N. Y	Apr. 26, 1870	102, 203
Hydrogen generator and carbureter	B. Sloper	Saint Louis, Mo	July 20, 1869	92, 892
Hydrogen-lighter, Döbereiner	J. Taylor and R. W. Brown	Westerly, R. I	Nov. 18, 1862	36, 971
Hydrolator	Z. Swope	Lancaster, Pa	Feb. 26, 1850	7, 128
Hydrometer	J. Adams	New York, N. Y	Jan. 15, 1861	31, 100

Index of patents issued from the United States Patent Office from 1790 *to* 1873, *inclusive*—Continued.

Invention.	Inventor.	Residence.	Date.	No.
Hydrometer	P. Hogg	Brooklyn, N. Y	May 6, 1862	35, 152
Hydrometer	H. Petrie	Chicago, Ill	June 13, 1865	48, 203
Hydrometer	J. Saxton	Washington, D. C	Sept. 27, 1864	44, 460
Hydrometer	E. Southworth	New York, N. Y	Feb. 28, 1821	....
Hydrometer	G. Tagliabue	New York, N. Y	Oct. 16, 1866	58, 913
Hydrometer	J. C. Tucker and E. Dwelle	Boston, Mass	Dec. 23, 1822	....
Hydrometer	T. Welham	Brownsville, Nebr	May 31, 1864	42, 979
Hydrometric apparatus	L. Brauer	Memphis, Tenn	Sept. 26, 1865	50, 092
Hydrostatic and aërostatic machinery	E. C. Genet	Albany, N. Y	Oct. 31, 1825	....
Hydrostatic and hydraulic elevator	J. McCreary	Nobletown, Pa	Jan. 5, 1828	....
Hydrostatic and steam engine	S. H. Long and G. F. A. Hauto	Germantown, Pa	Sept. 3, 1813	....
Hydrostatic engine	T. Hansbron and B. B. Redding	Sacramento, Cal	Mar. 14, 1865	46, 795
Hydrostatic paradox to govern fluid-pressure, Application of.	C. D. Colden	New York, N. Y	June 2, 1815	....
Hydrostatic paradox to move machinery, Application of.	C. D. Colden	New York, N. Y	May 19, 1815	....
Hydrostatic press	T. Baxter	Petersburgh, Va	Apr. 13, 1836	....
Hydrostatic press	J. T. Burr	Brooklyn, N. Y	July 18, 1871	117, 043
Hydrostatic press	C. Graham	Kingston, Pa	Aug. 6, 1867	67, 531
Hydrostatic press	D. H. Mason and M. W. Baldwin.	Philadelphia, Pa	Dec. 2, 1829	....
Hydrostatic press	E. Merrill	New Bedford, Mass	Mar. 28, 1838	662
Hydrostatic press for cotton, &c	J. Houpt	Forkland, Ala	Aug. 21, 1841	2, 221
Hygienic basin and steam-bath	E. O. Murden	Hightstown, N. J	Jan. 21, 1873	135, 004
Hygieometer	B. M. Lawrence	Galesburgh, Ill	Mar. 23, 1869	88, 047
Hygrometer	R. Baeklen and W. Staehlen	Brooklyn, N. Y	Aug. 8, 1865	49, 221
Hygrometer	A. H. and C. R. Black	Indianapolis, Ind	Mar. 26, 1861	31, 773
Hygrometer	W. Edson	Boston, Mass	July 4, 1865	48, 620
Hygrometer	A. E. Lazell	West Meriden, Conn	Mar. 10, 1868	75, 432
Hygrometer	L. S. Ullman	Nashville, Tenn	Sept. 13, 1859	25, 457
Hygrometers, Device for actuating the index of	C. L. Clark	Rochester, N. Y	May 25, 1858	20, 326
Hygrometric regulator for hot-water apparatus	J. H. Ross	New York, N. Y	Dec. 12, 1854	12, 071
I.				
Ice, Accumulating	J. Dutton	Aston, Pa	May 8, 1843	3, 080
Ice and coal box	R. Hagen	Saint Louis, Mo	Aug. 21, 1866	57, 317
Ice and cooling air, liquids, &c., Manufacture of	C. Tellier	Paris, France	Mar. 8, 1870	100, 689
Ice and cooling, Apparatus for making	F. Windhausen	Brunswick, Germany	Jan. 24, 1871	111, 292
Ice and for other purposes, Manufacture of liquids for.	P. H. Vander Weyde	Philadelphia, Pa	Dec. 17, 1867	72, 431
Ice and in cooling air, &c., Manufacture of	D. E. Somes	Washington, D. C	Sept. 17, 1867	68, 908
Ice and refrigerating, Apparatus for making	A. B. Tripler	New Orleans, La	Mar. 14, 1871	112, 654
Ice and refrigerating machine, Manufacture of	S. B. Martin and J. M. Beath	San Francisco, Cal	May 28, 1872	127, 180
Ice and snow fender for roofs of buildings	L. Howe	Worcester, Mass	May 31, 1864	42, 992
Ice and the refrigerating of air and liquids, Manufacture of.	C. Tellier	Passy, near Paris, France	Jan. 5, 1869	85, 719
Ice, Apparatus for and manner of cutting	N. J. Wyeth	Cambridge, Mass	Mar. 18, 1829	....
Ice, Apparatus for loading	P. F. Whitney	Saugerties, N. Y	Mar. 9, 1860	87, 459
Ice, Apparatus for making	J. H. Bunnell	Massillon, Ohio	Nov. 10, 1863	40, 551
Ice, Apparatus for making	C. P. Leavitt	New York, N. Y	July 18, 1871	117, 087
Ice, Apparatus for making	A. C. Twining	New Haven, Conn	Apr. 15, 1862	34, 993
Ice, Apparatus for making	E. C. Weld	New York, N. Y	Aug. 22, 1871	118, 411
Ice, Apparatus for making	G. Willard	New York, N. Y	Apr. 1, 1873	137, 523
Ice, Apparatus for purifying water for the manufacture of.	S. F. Peterson	New Orleans, La	Mar. 22, 1870	101, 035
Ice, Apparatus for the manufacture of	S. Bennett	Jefferson Parish, La	Nov. 1, 1870	108, 868
Ice, Apparatus for the manufacture of	D. L. Holden	New Orleans, La	Apr. 12, 1870	101, 876
Ice, Apparatus for the manufacture of	T. S. C. H. Lowe	New York, N. Y	Apr. 9, 1867	60, 404
Ice, Apparatus for the manufacture of	A. Mühl	San Antonio, Tex	Nov. 28, 1871	121, 402
Ice, Apparatus for the manufacture of	J. E. Sears	Waco, Tex	Aug. 29, 1871	118, 649
Ice-bag	S. Stroock	New York, N. Y	July 11, 1871	117, 013
Ice, Bag for preserving	W. B. Coates	Philadelphia, Pa	Apr. 30, 1867	64, 283
Ice-boat	T. B. Kelley	Dundee, Ill	Sept. 8, 1868	81, 909
Ice-boat	J. B. Stoddard	Baltimore, Md	Apr. 2, 1867	63, 577
Ice-boat, Arrangement of means attached to	D. Large	Philadelphia, Pa	June 24, 1856	15, 187
Ice-boat for breaking ice	W. Hunt and J. Townsend	New York, N. Y	Oct. 3, 1838	958
Ice, Boat for traveling on	T. Chapin	Canandaigua, N. Y	Sept. 14, 1838	919
Ice, Boat for traveling on	J. R. Halsey	Newark, N. J	Oct. 16, 1866	58, 815
Ice-boat, Velocipede	F. G. Johnson	Brooklyn, N. Y	Feb. 11, 1862	34, 369
Ice-box or cooler	J. W. Campbell	New York, N. Y	Feb. 26, 1867	62, 392
Ice-breaker	C. W. Chapman	Hartford, Conn	July 3, 1860	28, 966
Ice-breaker	C. W. Dunlap	Brooklyn, N. Y	Sept. 18, 1866	58, 078
Ice-breaker	A. Flannigan	Trappe, Ind	Apr. 28, 1868	77, 366
Ice-breaker	M. Freytag	Philadelphia, Pa	July 2, 1836	....
Ice breaker and fender	F. Ford	Point Pleasant, W. Va	Sept. 23, 1873	143, 003
Ice-breaker and propeller in boats, Paddle to be used as an.	W. Van Dusen	Kensington, Pa	July 29, 1837	314
Ice-breaker for boats, &c	S. Nicolson	Boston, Mass	July 16, 1844	3, 668
Ice-breaker for navigating frozen waters	B. S. Gillespie	New York, N. Y	Mar. 8, 1837	138
Ice-breaking boat	H. and W. Brown	Philadelphia, Pa	Aug. 5, 1856	15, 472
Ice-breaking boat	D. Compton	Newport, N. J	Sept. 28, 1869	95, 326
Ice-breaking boat	J. D. Foster, H. C. Foster, and J. Q. Miller.	Cincinnati and Springfield, Ohio.	Dec. 8, 1857	18, 832
Ice-breaking boat	Z. Oram	Camden, N. J	Apr. 14, 1857	17, 052
Ice-breaking instrument	I. H. Griffing	New York, N. Y	Aug. 5, 1856	15, 483
Ice-breaking machine	J. I. Williams	Georgetown, D. C	Mar. 2, 1813	....
Ice-calk	G. W. Farley	Manchester, N. H	May 26, 1868	78, 270
Ice-calk	D. Krauser	Pottsville, Pa	Feb. 24, 1863	37, 788
Ice, Car and receiving-platform for receiving blocks of.	N. J. Wyeth	Cambridge, Mass	Oct. 28, 1843	3, 323
Ice-carriage	H. C. Moore	Springfield, Mass	Mar. 24, 1868	75, 953
Ice-carrier	L. Townsend	Terre Haute, Ind	Apr. 23, 1872	126, 108
Ice-chair	F. Ashley	New York, N. Y	Feb. 5, 1861	31, 290
Ice-chest for soda-apparatus and refrigerator	E. Bigelow	Springfield, Mass	Apr. 20, 1869	89, 010
Ice-chute, Automatic	J. A. Wolfer	Rondout, N. Y	Dec. 22, 1868	85, 156
Ice-clog	J. J. Starr	Cincinnati, Ohio	Oct. 28, 1862	36, 866
Ice, Cooling air and liquids and making	D. E. Somes	Washington, D. C	May 5, 1868	77, 669

Index of patents issued from the United States Patent Office from 1790 *to* 1873, *inclusive*—Continued.

Invention.	Inventor.	Residence.	Date.	No.
Ice, Cooling air in the manufacture of	O. Lugo and J. B. McPherson	Baltimore, Md	Mar. 14, 1871	112, 726
Ice, Cooling and producing	C. E. Haynes	Boston, Mass	June 21, 1870	104, 588
Ice, cooling buildings, &c., Apparatus for the manufacture of.	A. H. Tait	Jersey City, N. J	Nov. 11, 1873	144, 577
Ice-cream freezer	A. H. Austin	Baltimore, Md	Sept. 19, 1848	5, 775
Ice-cream freezer	A. S. Ballard	Mount Pleasant, Iowa	Dec. 29, 1868	85, 420
Ice-cream freezer	R. P. Beil	Peoria, Ill	Oct. 28, 1873	144, 053
Ice-cream freezer	G. W. Brown	New York, N. Y	May 29, 1860	28, 449
Ice-cream freezer	G. W. Brown	New York, N. Y	May 17, 1864	42, 744
Ice-cream freezer	H. B. Campbell	Racine, Wis	June 3, 1873	139, 453
Ice-cream freezer	A. Cavileer, W. McCuddy, and P. N. Woliston.	Springfield, Ohio	Jan. 1, 1867	60, 689
Ice-cream freezer	J. R. Champlin	Laconia, N. H	June 26, 1866	55, 820
Ice-cream freezer	J. R. Champlin	Laconia, N. H	Mar. 16, 1869	87, 909
Ice-cream freezer	G. Coffeen, jr	Warren County, Ohio	Nov. 13, 1849	6, 865
Ice-cream freezer	J. W. Condon	Logansport, Ind	July 9, 1872	128, 856
Ice-cream freezer	M. G. Crane	Chelsea, Mass	Nov. 6, 1866	59, 366
Ice-cream freezer	G. W. Davis	New Orleans, La	May 22, 1860	28, 354
Ice-cream freezer	J. Decker	Bel Air, Md	Aug. 21, 1849	6, 661
Ice-cream freezer	J. Dooling	Boston, Mass	Oct. 20, 1868	83, 265
Ice-cream freezer	J. Dooling	Boston, Mass	July 16, 1872	129, 326
Ice-cream freezer	J. Dooling	Boston, Mass	July 29, 1873	141, 209
Ice-cream freezer	J. W. Dougherty and F. W. Gerecke.	Newburgh, N. Y	July 16, 1867	66, 685
Ice-cream freezer	F. H. Duc	Charleston, S. C	Sept. 17, 1867	66, 968
Ice-cream freezer	A. W. Edwards	Mendota, Ill	Nov. 28, 1865	51, 160
Ice-cream freezer	S. S. Fitch	New York, N. Y	Mar. 4, 1873	136, 496
Ice-cream freezer	W. A. Garloch and W. D. Richards.	Belpre, Ohio	Sept. 29, 1868	82, 509
Ice-cream freezer	C. Gooch	Cincinnati, Ohio	June 11, 1867	65, 559
Ice-cream freezer	C. Gooch	Cincinnati, Ohio	Jan. 21, 1868	73, 596
Ice-cream freezer	C. Gooch	Cincinnati, Ohio	July 4, 1871	116, 584
Ice-cream freezer	M. F. Graves	Sunbury, Pa	Oct. 21, 1873	143, 822
Ice-cream freezer	J. Gray	Milwaukee, Wis	Sept. 10, 1867	68, 734
Ice-cream freezer	G. Guinot	Milton, Fla	Feb. 4, 1873	135, 474
Ice-cream freezer	E. Halloway	Belvidere, Ill	Sept. 24, 1872	131, 611
Ice-cream freezer	S. H. Hamilton and C. A. Ashton	Jacksonville, Ill	Nov. 11, 1862	36, 904
Ice-cream freezer	W. Hawkins	Oregon, Mo	Aug. 31, 1869	94, 313
Ice-cream freezer	B. H. Hilliar	New London, Conn	May 17, 1870	103, 182
Ice-cream freezer	L. A. Lipp	Coatesville, Pa	Feb. 26, 1867	62, 429
Ice-cream freezer	A. Lucetti	New York, N. Y	July 22, 1873	141, 060
Ice-cream freezer	B. G. Martin	Williamsburgh, N. Y	Mar. 1, 1870	100, 305
Ice-cream freezer	H. B. Masser	Sunbury, Pa	Dec. 12, 1848	5, 960
Ice-cream freezer	H. B. Masser	Sunbury, Pa	Jan. 19, 1858	19, 147
Ice-cream freezer	H. B. Masser	Sunbury, Pa	Apr. 23, 1861	32, 138
Ice-cream freezer	H. B. Masser	Sunbury, Pa	Mar. 19, 1867	63, 068
Ice-cream freezer	T. and G. M. Mills	Philadelphia, Pa	Mar. 15, 1870	100, 918
Ice-cream freezer	T. and G. M. Mills	Philadelphia, Pa	June 25, 1872	128, 414
Ice-cream freezer	C. W. Packer	Philadelphia, Pa	May 22, 1860	28, 402
Ice-cream freezer	C. W. Packer	Philadelphia, Pa	July 23, 1867	67, 133
Ice-cream freezer	O. Paddock	Watertown, N. Y	Dec. 3, 1861	33, 850
Ice-cream freezer	O. Paddock	Watertown, N. Y	Oct. 21, 1862	36, 725
Ice-cream freezer	O. Paddock	Watertown, N. Y	Oct. 15, 1867	69, 833
Ice-cream freezer	J. Parisette	Indianapolis, Ind	June 10, 1856	15, 083
Ice-cream freezer	A. L. Platt	Bloomington, Ill	Nov. 19, 1872	133, 251
Ice-cream freezer	A. L. Platt and J. W. Wilmeth	Bloomington, Ill	Aug. 15, 1871	118, 149
Ice-cream freezer	T. M. Powell	Baltimore, Md	Sept. 5, 1854	11, 651
Ice-cream freezer	J. H. Rae	Syracuse, N. Y	July 2, 1867	66, 389
Ice-cream freezer	J. E. Robinson	Boston, Mass	Jan. 22, 1867	61, 360
Ice-cream freezer	M. Rosenstein	Boston, Mass	May 30, 1871	115, 361
Ice-cream freezer	J. F. Rote	Reading, Pa	Apr. 4, 1871	113, 450
Ice-cream freezer	T. Sands	Laconia, N. H	Feb. 27, 1872	124, 015
Ice-cream freezer	P. Schumacher	New York, N. Y	Nov. 2, 1869	96, 489
Ice-cream freezer	J. S. Shattuck	Medford, Mass	Sept. 5, 1865	49, 797
Ice-cream freezer	F. G. Siemers	Winona, Minn	Jan. 21, 1868	73, 658
Ice-cream freezer	J. S. Silva	Savannah, Ga	Oct. 2, 1860	30, 256
Ice-cream freezer	C. E. Simmons	Monroe, Wis	Sept. 16, 1873	142, 951
Ice-cream freezer	J. Sissons	Horncastle, England	May 25, 1869	90, 594
Ice-cream freezer	W. H. Skerrett	Cincinnati, Ohio	Aug. 6, 1867	67, 599
Ice-cream freezer	E. B. Tilden	Prairie City, Iowa	Aug. 12, 1873	141, 738
Ice-cream freezer	J. Tingley	Philadelphia, Pa	Feb. 11, 1868	74, 259
Ice-cream freezer	J. Tingley	Philadelphia, Pa	June 6, 1871	115, 657
Ice-cream freezer	E. P. Torrey	Jersey City, N. J	Jan. 17, 1860	26, 867
Ice-cream freezer	T. Weaver	Harrisburgh, Pa	Mar. 26, 1872	124, 992
Ice-cream freezer	D. Wiggins	Greenport, N. Y	Apr. 20, 1869	89, 259
Ice-cream freezer	W. G. Young	Baltimore, Md	May 30, 1848	5, 601
Ice-cream freezer, Bottom for	E. M. Manigle	Philadelphia, Pa	Dec. 12, 1865	51, 468
Ice-cream-freezer scraper	F. G. Siemers	Winona, Minn	Mar. 3, 1868	75, 209
Ice-cream holder and filler	H. R. Robbins	Baltimore, Md	Apr. 2, 1872	125, 333
Ice-cream, jelly, &c., Mold for	E. M. May	Fond du Lac, Wis	Aug. 19, 1873	141, 882
Ice-cream machine	J. R. Champlin	Laconia, N. H	Nov. 1, 1864	44, 851
Ice-cream, Machine for manufacturing	C. Gooch	Cincinnati, Ohio	May 14, 1872	126, 691
Ice-cream mold	P. Field	Boston, Mass	Feb. 2, 1869	86, 519
Ice-cream pail	G. A. Nash	Niles, Mich	Oct. 4, 1870	107, 949
Ice-cream receptacle	E. A. G. Roulstone	Boston, Mass	Mar. 9, 1869	87, 710
Ice-cream refrigerator	C. Vignal	New York, N. Y	July 19, 1870	105, 521
Ice-cream server	J. Oyarzabal	Malaga, Spain	Oct. 19, 1869	95, 929
Ice-cream, water-ice, &c., Manner of packing and conveying.	I. Allegretti	Philadelphia, Pa	Apr. 4, 1871	113, 239
Ice-creeper	G. L. Bailey	Portland, Me	Apr. 8, 1862	34, 869
Ice-creeper	W. B. Coates	Philadelphia, Pa	Jan. 11, 1870	98, 668
Ice-creeper	E. B. Colby	Franklin, N. H	Feb. 11, 1873	135, 629
Ice-creeper	D. Cumming	Philadelphia, Pa	Feb. 9, 1869	86, 894
Ice-creeper	E. F. Dieterichs	Philadelphia, Pa	June 25, 1872	128, 370
Ice-creeper	R. H. Earle	Saint John's, Newfoundland.	Dec. 9, 1873	145, 340
Ice-creeper	E. S. Ellis	Trenton, N. J	July 4, 1871	116, 575
Ice-creeper	W. Field	Providence, R. I	Jan. 27, 1863	37, 558

Index of patents issued from the United States Patent Office from 1790 *to* 1873, *inclusive*—Continued.

Invention.	Inventor.	Residence.	Date.	No.
Ice-creeper	E. M. Gardner	Nantucket, Mass	Jan. 23, 1866	52, 157
Ice-creeper	D. Greene	Troy, N. Y	Oct. 10, 1865	50, 353
Ice-creeper	A. S. Hadaway	Plymouth, Mass	June 12, 1866	55, 584
Ice-creeper	A. Houlings	Philadelphia, Pa	June 3, 1873	139, 506
Ice-creeper	C. Hoëller	Cincinnati, Ohio	Sept. 14, 1869	94, 888
Ice-creeper	L. G. Hoffman	Albany, N. Y	July 7, 1868	79, 760
Ice-creeper	I. S. and J. W. Hyatt, jr	Chicago, Ill	Aug. 18, 1863	39, 575
Ice-creeper	C. H. Johnson and W. Axford	White Hall, Ill	Dec. 16, 1873	145, 656
Ice-creeper	W. P. Patton	Harrisburgh, Pa	Dec. 24, 1867	72, 534
Ice-creeper	H. L. Schell	Philadelphia, Pa	Jan. 24, 1860	26, 961
Ice-creeper	T. Symonds	Portland, Me	Oct. 17, 1865	50, 512
Ice-creeper	W. A. Towers	Philadelphia, Pa	Mar. 25, 1856	14, 527
Ice-creeper	E. M. Turner	Pittsburgh, Pa	Aug. 8, 1871	117, 945
Ice-creeper	M. Warne	Philadelphia, Pa	May 16, 1871	114, 998
Ice-creeper	W. C. Wells	Philadelphia, Pa	June 2, 1868	78, 560
Ice-creeper	H. H. White	Leominster, Mass	Oct. 1, 1872	131, 836
Ice-creeper	A. Wilke	Brunswick, Germany	Dec. 7, 1869	97, 738
Ice-crusher	W. W. Armington	Lowell, Mass	Sept. 26, 1865	50, 198
Ice-crusher	M. J. Dixon and W. Tunstill	New York, and Brooklyn, N. Y.	June 17, 1873	140, 019
Ice-crusher	F. F. Fowler	Upper Sandusky, Ohio	Mar. 5, 1867	62, 539
Ice-crusher	J. W. Hollingsworth and H. D. Weaver.	Mount Vernon, Ind	Mar. 24, 1868	75, 913
Ice-crusher	F. Hudner and J. R. Little	New York, N. Y	Feb. 4, 1873	135, 556
Ice-crusher	O. B. Kinsey	Newark, N. J	Feb. 7, 1871	111, 652
Ice-crusher	J. Middleton	New York, N. Y	Jan. 1, 1861	31, 027
Ice-crusher	L. H. Moseley	Poughkeepsie, N. Y	Jan. 26, 1864	41, 388
Ice-crusher	E. F. Pryor	Dayton, Ohio	June 4, 1867	65, 504
Ice-crusher	N. Richardson	Gloucester, Mass	June 9, 1868	78, 830
Ice-crusher	J. L. Rowe	New York, N. Y	Mar. 19, 1861	31, 766
Ice-crushing machine	A. C. Hobbs and J. Brown	New York, N. Y	Sept. 4, 1849	6, 690
Ice-crushing table	S. H. and D. W. Davis	Detroit, Mich	Jan. 18, 1870	98, 852
Ice-cutter	E. Bacher	Findley, Ohio	Jan. 30, 1872	123, 076
Ice-cutter	G. L. Cummings	New York, N. Y	Dec. 22, 1868	85, 217
Ice-cutter	C. W. Flint	Washington, D. C	Dec. 22, 1868	85, 080
Ice-cutter	V. H. Hallock	Queens, N. Y	Apr. 5, 1870	101, 613
Ice-cutter	G. R. Marvin	Keokuk, Iowa	Feb. 11, 1868	74, 236
Ice-cutter	F. G. Siemers	Winona, Minn	Aug. 18, 1868	81, 303
Ice-cutter	L. Townsend	Terre Haute, Ind	Oct. 8, 1872	132, 035
Ice-cutter	J. B. Wilson	Malden, Mass	Nov. 6, 1847	5, 362
Ice cutter and marker	N. J. Wyeth	Cambridge, Mass	Oct. 25, 1843	3, 317
Ice, Cutting and removing	W. Blount and J. Cammett	Osterville, Mass	Aug. 26, 1846	4, 713
Ice-cutting apparatus	C. and G. Beatty	Portsmouth, Va	July 21, 1868	80, 119
Ice-cutting apparatus	J. Fielemeyer	Philadelphia, Pa	Feb. 5, 1861	31, 307
Ice-cutting attachment to vessels	T. Estlack	Philadelphia, Pa	May 5, 1857	17, 209
Ice-cutting machine	J. Baker	Philadelphia, Pa	Feb. 11, 1868	74, 278
Ice-cutting machine	L. Barnum	Bridgeport, Conn	Mar. 7, 1871	112, 408
Ice-cutting machine	A. Deutschel	Cleveland, Ohio	July 12, 1870	105, 183
Ice-cutting machine	W. F. Gordon	Detroit, Mich	Nov. 19, 1867	70, 994
Ice-cutting machine	S. Trask	Hallowell, Me	Apr. 21, 1836	....
Ice-cutting machine	N. J. Wyeth	Cambridge, Mass	Dec. 1, 1840	1, 877
Ice, Device for cutting and shaving	S. E. Blake	Worcester, Mass	Feb. 14, 1865	46, 331
Ice, Device to prevent slipping on	W. R. Landfear	Hartford, Conn	Nov. 8, 1864	44, 961
Ice-elevating device	H. Little	Middletown, N. Y	Aug. 27, 1867	68, 218
Ice-elevating machine	N. J. Wyeth	Cambridge, Mass	Nov. 6, 1843	3, 327
Ice-elevating track	A. Pfund	New York, N. Y	Oct. 28, 1873	144, 133
Ice-elevator	E. E. Conklin	New York, N. Y	June 24, 1873	140, 187
Ice-elevator	W. Eighmie	Beekman, N. Y	Nov. 5, 1872	132, 758
Ice-elevator	H. Little	Middletown, N. Y	May 21, 1867	64, 886
Ice-elevator	J. J. Neuman	Middletown, Ohio	Aug. 8, 1871	117, 804
Ice-elevator	W. F. B. Read	Chicago, Ill	Aug. 11, 1868	81, 004
Ice-elevator	L. Townsend	Terre Haute, Ind	Apr. 23, 1872	126, 106
Ice-elevator	T. C. Wolking	Covington, Ky	May 21, 1872	127, 131
Ice-elevator	N. J. Wyeth	Cambridge, Mass	Dec. 10, 1840	1, 886
Ice-elevator	R. S. Zilar	Cincinnati, Ohio	Nov. 3, 1868	83, 752
Ice-float	L. Townsend	Terre Haute, Ind	Apr. 23, 1872	126, 107
Ice for shipping and storing, Machine for preparing	N. J. Wyeth	Cambridge, Mass	Dec. 1, 1837	497
Ice for storage, Cutting	V. H. Hallock	Milton, N. Y	Oct. 4, 1864	44, 529
Ice for transportation, Machinery and car for elevating and depositing.	N. J. Wyeth	Cambridge, Mass	Dec. 10, 1841	2, 381
Ice from ponds, Machine for raising	W. W. Cowling	Richmond, Va	June 5, 1847	5, 145
Ice from rivers, Machinery for raising	C. R. Wortendyke	New York, N. Y	Nov. 13, 1855	13, 809
Ice from the water and depositing on sleds, Machinery for raising blocks of.	N. J. Wyeth	Cambridge, Mass	Dec. 1, 1840	1, 878
Ice, Heel-spur to prevent slipping on	H. Pollard	Boston, Mass	Jan. 26, 1858	19, 205
Ice-house	A. Baierle	Chicago, Ill	Dec. 22, 1868	85, 161
Ice-house	C. Liebmann	Brooklyn, E. D., N. Y	Nov. 7, 1871	120, 754
Ice-house	C. Liebmann	Brooklyn, N. Y	Apr. 16, 1872	125, 820
Ice-house	W. D. Parker	New York, N. Y	June 19, 1855	13, 098
Ice-house	W. Velte and J. Fagan	Pittsburgh, Pa	Mar. 29, 1870	101, 405
Ice house	A. Wilbur	Cedar City, Mo	Nov. 18, 1873	144, 720
Ice-house, Filling for walls of	C. Stoll	Brooklyn, N. Y	Dec. 3, 1872	133, 680
Ice-house floor	J. Barbian	Chicago, Ill	Jan. 7, 1868	72, 960
Ice-house for brewers and butchers	A. Baierle, F. Hartmann, and F. Reese.	Chicago, Ill	June 2, 1868	78, 507
Ice-house for packing and curing meat	H. A. Roberts	Boston, Mass	Aug. 22, 1871	118, 277
Ice-house, Upper floor of	P. Kephart	Baltimore, Md	Sept. 26, 1848	5, 798
Ice, Housing and shipping	B. Wise	Cincinnati, Ohio	Aug. 5, 1862	36, 121
Ice in blocks, Sleds for transportation of	N. J. Wyeth	Cambridge, Mass	Dec. 1, 1840	1, 876
Ice in rivers and harbors, Method of obstructing	P. Voorhis	New York, N. Y	Apr. 16, 1867	63, 968
Ice in rivers, Machine for planing away	R. W. Heywood	Baltimore, Md	Jan. 26, 1858	19, 195
Ice-increaser	L. Townsend	Terre Haute, Ind	Apr. 23, 1872	126, 109
Ice into blocks for storing, Machine for cutting	G. B. Gruman	Ridgefield, Conn	Oct. 1, 1867	69, 338
Ice leveling and smoothing machine	W. Wharton, jr	Philadelphia, Pa	Apr. 11, 1865	47, 237
Ice-lowering trap	C. Egnor	Catskill, N. Y	Aug. 30, 1870	106, 798
Ice-machine	D. Boyle	San Francisco, Cal	June 25, 1872	128, 488
Ice-machine	W. Bray	Lambertsville, N. J	Dec. 23, 1873	145, 838
Ice-machine	D. L. Holden	New Orleans, La	Sept. 28, 1869	95, 347

Index of patents issued from the United States Patent Office from 1790 *to* 1873, *inclusive*—Continued.

Invention.	Inventor.	Residence.	Date.	No.
Ice-machine	W. R. Johnston and W. White-law.	Sedalia, Mo., and Memphis, Tenn.	Apr. 30, 1872	126, 305
Ice-machine	A. Mühl	Waco, Tex	Dec. 12, 1871	121, 888
Ice-machine	H. Norman and C. F. Dietrich	New Orleans, La	Feb. 4, 1873	135, 576
Ice-machine	C. Plagge	New York, N. Y	Aug. 23, 1870	106, 722
Ice-machine	R. Reece	Llandilo, Wales	Oct. 1, 1872	131, 783
Ice-machine	A. Rock	Boston, Mass	Feb. 28, 1871	112, 284
Ice-machine	C. A. Seely	New York, N. Y	Oct. 10, 1871	119, 795
Ice-machine	D. K. Tuttle and O. Lugo	Baltimore, Md	Apr. 5, 1870	101, 682
Ice-machine	D. K. Tuttle and O. Lugo	Baltimore, Md	Jan. 24, 1871	111, 280
Ice-machine	F. Windhausen	Brunswick, Germany	Mar. 22, 1870	101, 198
Ice-machine	F. Windhausen	Brunswick, Germany	Jan. 24, 1871	111, 293
Ice-machine, Carré	A. Vaass and F. Littmann	Halle, Prussia	May 2, 1871	114, 495
Ice-machine using ammonia	A. C. Twining	New Haven, Conn	Dec. 19, 1871	121, 975
Ice, Machinery for raising and depositing	N. J. Wyeth	Cambridge, Mass	Dec. 10, 1841	2, 380
Ice, Machinery for the manufacture of	W. T. Duval	Georgetown, D. C	Aug. 2, 1870	105, 924
Ice-making and air and liquid cooling apparatus	D. E. Somes	Washington, D. C	Jan. 28, 1868	73, 936
Ice, Manufacture of	A. Albertson	Jersey City, N. J	May 30, 1871	115, 409
Ice, Manufacture of	T. G. Bigger	Kansas City, Mo	Mar. 7, 1871	112, 409
Ice, Manufacture of	J. Dutton	Aston, Pa	Aug. 18, 1846	4, 697
Ice, Manufacture of	R. S. Egbert	Colfax, Cal	July 21, 1868	80, 063
Ice, Manufacture of	J. F. Gesner	West Farms, N. Y	Oct 4, 1870	107, 898
Ice, Manufacture of	C. M. Keller and J. Henderson	New York and Brooklyn, N. Y.	Sept. 12, 1865	49, 887
Ice, Manufacture of	S. J. Newsham, W. H. Haines, and W. S. Henson.	Mont Clair and Newark, N. J.	Nov. 1, 1870	108, 816
Ice, Manufacture of	M. Rosenstein	Boston, Mass	Oct. 4, 1870	107, 962
Ice, Manufacture of	P. H. Vander Weyde	New York, N. Y	Nov. 1, 1870	108, 851
Ice, Manufacturing	A. C. Twining	Hudson, Ohio	Nov. 8, 1853	10, 221
Ice-manufacturing machine	P. H. Vander Weyde	Philadelphia, Pa	Feb. 16, 1869	87, 084
Ice, Manufacturing or forming	T. B. Smith	Saint Louis, Mo	Jan. 23, 1841	1, 941
Ice, Mode of manufacturing	T. S. C. Lowe	New York, N. Y	Apr. 2, 1867	63, 413
Ice, Mode of obtaining and securing	J. E. Manuel	Philadelphia, Pa	July 12, 1839	1, 240
Ice, Packing and stowing	F. Tredor	Boston, Mass	May 4, 1838	726
Ice-pick	S. S. Boynton and D. Keefe	Peoria, Ill	Oct. 28, 1873	143, 957
Ice-pick	J. H. Bridgins	Astoria, N. Y	July 9, 1867	66, 557
Ice-pick	E. Brown	Green Point, N. Y	Apr. 5, 1870	101, 424
Ice-pick	E. G. Burnham	Bridgeport, Conn	May 22, 1866	54, 852
Ice-pick	W. T. Eames	New York, N. Y	Aug. 8, 1871	117, 754
Ice-pick	S. G. Hoyt	New York, N. Y	Jan. 28, 1868	73, 809
Ice-pick	J. L. Rowe	New York, N. Y	Dec. 21, 1858	22, 403
Ice-pick	T. A. Taylor	Bangor, Me	Aug. 15, 1871	118, 070
Ice-pick	J. T. Van Kirk	Philadelphia, Pa	July 17, 1860	29, 220
Ice-pick	M., H., H. T., and J. White	Philadelphia, Pa	May 10, 1859	23, 969
Ice pick and hook	E. H. Schmultz and J. Baker	New York, N. Y	Jan. 24, 1871	111, 260
Ice-pick and meat-maul combined	T. A. Conklin	New Britain, Conn	June 18, 1872	128, 020
Ice-pick, nut-cracker, &c	G. L. Witsil	Philadelphia, Pa	May 8, 1866	54, 653
Ice-planer	S. Lewis	Williamsburgh, N. Y	July 6, 1869	92, 325
Ice-planer	W. F. Pough	Esopus, N. Y	July 13, 1869	92, 542
Ice planer or cutter	J. Serrill	Philadelphia, Pa	Feb. 12, 1867	62, 075
Ice-planing machine	S. Lewis	Brooklyn, N. Y	Mar. 3, 1868	75, 029
Ice planing or shaving machine	H. D. J. Pratt	Washington, D. C	May 17, 1859	24, 052
Ice-preserver	J. Dunning	Bangor, Me	Sept. 19, 1871	119, 016
Ice-preserver	J. W. D. Patten	New York, N. Y	Oct. 6, 1868	82, 869
Ice-preserver	J. E. Pilkington	Baltimore, Md	Sept. 13, 1870	107, 405
Ice-preserver and water-cooler	R. Heneage	Buffalo, N. Y	May 25, 1869	90, 360
Ice-preserver, Portable	E. Powers	Hartford, Conn	Aug. 28, 1821	....
Ice-preserving jar	W. L. Faxon	Quincy, Mass	Jan. 16, 1872	122, 821
Ice, Process for the artificial production of	J. Gorrie	New Orleans, La	May 6, 1851	8, 080
Ice-run	G. Burhans	Rondout, N. Y	July 11, 1871	116, 802
Ice-sandal	T. J. Linton	Providence, R. I	Mar. 7, 1865	46, 682
Ice-shaver	R. Gilliland	Hudson, Mich	Feb. 2, 1869	86, 527
Ice-shaver, Rotary	C. C. Clawson	Raleigh, N. C	Sept. 5, 1871	118, 689
Ice-shaving machine	W. H. Collins	Boston, Mass	Sept. 26, 1871	119, 324
Ice-shaving machine	J. D. Freeman	Abbeville, Ala	Aug. 5, 1873	141, 553
Ice shoe or calk	D. Bennet	Stratford, Conn	Apr. 1, 1862	34, 809
Ice-spur	C. Monnin	Buffalo, N. Y	Feb. 16, 1858	19, 376
Ice-spur	M. V. B. White	Ballston, N. Y	Jan. 1, 1867	60, 811
Ice-spur	C. F. Wieland	Darmstadt, Ill	Nov. 24, 1868	84, 400
Ice-stand	H. A. Roberts	Hartford, Conn	Aug. 24, 1858	21, 275
Ice therefrom to platforms, Railway-car for discharging blocks of.	N. J. Wyeth	Cambridge, Mass	Dec. 10, 1841	2, 382
Ice to a uniform thickness, Machine for reducing blocks of.	N. J. Wyeth	Cambridge, Mass	Dec. 1, 1840	1, 879
Ice, Vessel for the formation of	B. T. Babbitt	New York, N. Y	Jan. 12, 1869	85, 781
Ice-water receptacle	W. H. Hart	Medfield, Mass	Feb. 19, 1867	62, 133
Iceland moss and carrageen, Manufacture from	W. J. Rand	Brooklyn, N. Y	Aug. 10, 1869	93, 478
Ichthyocolla, Manufacture of	E. Rowe	Rockport, Mass	Dec. 19, 1848	5, 978
Ichthyocolla or fish-glue, Extr cting	D. Waldron	New York, N. Y	Mar. 4, 1812	....
Identifying-box	L. T. McNeiley	Danville, Mo	June 18, 1867	65, 928
Illuminating-apparatus	G. Cuppers	Brooklyn, E. D., N. Y	Nov. 28, 1871	121, 234
Illuminating-apparatus	C. Geisse	Taycheedah, Wis	July 10, 1866	56, 203
Illuminating-apparatus	J. H. Irwin	Chicago, Ill	May 8, 1866	54, 553
Illuminating-apparatus	J. A. Thompson	Brooklyn, N. Y	Dec. 30, 1873	146, 109
Illuminating-compound	J. Scott	Washington, D. C	Feb. 5, 1867	61, 768
Illuminating-plate, Flexible sheet-metal	T. Hyatt	New York, N. Y	Dec. 2, 1873	145, 185
Illuminating-signals, Igniting	J. J. Detwiller	Greenville, N. J	Dec. 18, 1866	60, 491
Illumination	G. H. Smith	Rochester, N. Y	Oct. 14, 1862	36, 689
Illusory decapitation	J. C. Withington	Brookline, Mass	Apr. 23, 1867	64, 179
Illustrating waves, Apparatus for	C. S. Lyman	New Haven, Conn	Nov. 19, 1867	71, 190
Image, Sectional	B. Day	West Hoboken, N. J	Dec. 20, 1870	110, 213
Immersion and steam bath combined	J. W. Caldwell	Cincinnati, Ohio	Dec. 27, 1870	110, 545
Implement	B. F. Bean	Schuylkill, Pa	June 23, 1868	79, 100
Implement	E. S. Bennett	New York, N. Y	Mar. 16, 1869	87, 754
Implement	J. W. Calep	Boston, Mass	Dec. 20, 1870	110, 199
Implement	T. C. Comstock	Harrodsburgh, Ky	Mar. 3, 1868	74, 992
Implement	E. R. Mecker	Elizabeth, N. J	Feb. 18, 1868	74, 568
Implement	W. K. Rairigh	Rural Valley, Pa	Nov. 10, 1868	83, 997

Index of patents issued from the United States Patent Office from 1790 *to* 1873, *inclusive*—Continued.

Invention.	Inventor.	Residence.	Date.	No.
Implement	W. A. Sharp	Tama City, Iowa	Apr. 12, 1870	101, 773
Implement	A. Thayer	Albany, N. Y	June 16, 1868	79, 027
Implement	C. Wetzler	Chicago, Ill	Jan. 19, 1869	86, 048
Implement	H. B. Whitehead	Holly Springs, Miss	June 10, 1873	139, 748
Implement	E. F. Wilder	Lowell, Mass	May 26, 1868	78, 248
Implement	W. Worle	Newark, W. Va	July 20, 1869	92, 768
Implement, Compound	B. G. Devoe, T. Rodgers, and J. C. Beals.	Fredericktown, Ohio	Jan. 9, 1872	122, 577
Implement, Compound	E. A. Edwards	San Buenaventura, Cal	July 16, 1872	129, 116
Implement, Compound	A. Iske	Lancaster, Pa	Mar. 11, 1873	136, 601
Implement, Compound	J. G. Powell	Philadelphia, Pa	Oct. 1, 1872	131, 904
Implement, Compound	C. W. Russell	Milford, Mass	Apr. 23, 1872	126, 094
Implement, Compound	J. C. Schlarbaum	San José, Cal	Apr. 2, 1872	125, 336
Incendiary compound	H. W. Libbey	Cleveland, Ohio	June 13, 1865	48, 187
Inclined-plane elevator	T. B. Simonton	Williamsburgh, N. Y	Mar. 15, 1870	100, 811
Inclined plane for drawing up vessels	E. Clark	New York, N. Y	June 27, 1825	....
Inclined-plane statical wheels, Machine to facilitate passage of boats in canals, removing earth, &c., by.	J. Williams	....	Mar. 23, 1804	....
Inclined planes and hoists, Brake for	L. Klee	Pittsburgh, Pa	Mar. 12, 1872	124, 592
Inclined planes, Device for assisting railway-car to ascend.	T. Agudio	Turin, Italy	June 9, 1863	38, 800
Inclined planes, Machinery for ascending	R. F. Stevens and S. B. Pitcher	Syracuse, N. Y	Dec. 28, 1846	4, 906
Inclined planes, Machinery for ascending and descending.	G. E. Sellers	Cincinnati, Ohio	Nov. 13, 1847	5, 367
Inclined planes, Method of ascending	J. S. French	Old Point Comfort, Va	June 22, 1852	9, 052
Inclined wheel	G. Lyman	Charles Parish, La	Mar. 4, 1822	....
Inclined wheel	J. Palmer	Cincinnati, Ohio	Dec. 29, 1818	....
Inclined wheel moved by animal weight, Horizontal	M. Isaacs and J. Wilbank	Philadelphia, Pa	Oct. 14, 1817	....
Inclined wheels, Method of propelling	N. H. Shaw	Charlotte Hall, Md	Oct. 9, 1824	....
Inclinometer	A. Chase, jr	Somerville, Mass	Aug. 29, 1865	49, 675
Incombustible, Rendering articles	A. G. Fell	Brooklyn, N. Y	Dec. 31, 1867	72, 830
Incubation, Artificial	W. J. Canteto	England	July 24, 1847	5, 204
Incubation, Artificial	L. G. Hoffman	Albany, N. Y	Feb. 20, 1847	4, 978
Incubator	J. and H. Graves	Boston, Mass	Dec. 27, 1870	110, 454
Incubator	H. J. Haight	New York, N. Y	Dec. 26, 1871	122, 249
Incubator	E. Woodward and N. J. Millet	Charlestown, Mass	Dec. 19, 1871	121, 977
Incubators, Electro-magnetic regulator for	J. Graves	Reading, Mass	June 18, 1872	127, 972
Index	O. F. Bullard	Media, Pa	Mar. 28, 1871	113, 015
Index	H. H. Hall	Boston, Mass	Oct. 27, 1868	83, 378
Index, Alphabetical	W. C. Huston	Eaton, Ohio	Apr. 26, 1870	102, 404
Index gage and calipers	D. F. Elmer	Springfield, Mass	Oct. 1, 1867	69, 418
Index, Ledger	C. A. Fitch and W. O. St. John	San Francisco and Oakland, Cal.	Mar. 22, 1870	100, 994
Index-lettering, Tool for	E. Crawley	Cincinnati, Ohio	July 8, 1856	15, 283
Index or book-marker	J. Johnson	New York, N. Y	Oct. 12, 1858	21, 759
Index, Universal	M. L. Powell	Newcastle, Ind	May 8, 1866	54, 593
Indexes, Case for containing	C. E. Fellows	Philadelphia, Pa	June 3, 1873	139, 503
Indexing books and ledgers	J. H. Swindell	Camden, N. J	Apr. 21, 1868	77, 126
Indexing for records, System of	A. Campbell	Frederick, Md	Feb 11, 1868	74, 299
India-rubber: *See* Gum-elastic.				
India-rubber and gutta-percha compound, Colored.	J. Malcolm	New York, N. Y	Jan. 2, 1866	51, 846
India-rubber and gutta-percha compounds, Manufacture of colored.	J. Malcolm	New York, N. Y	Jan. 2, 1866	51, 847
India-rubber and gutta-percha compounds, Manufacture of colored.	J. Malcolm	New York, N. Y	Jan. 2, 1866	51, 848
India-rubber and gutta-percha compounds, Manufacture of colored vulcanized.	J. Malcolm	New York, N. Y	Jan. 2, 1866	51, 849
India-rubber and gutta-percha goods, Manufacture of.	J. Murphy	New York, N. Y	Feb. 15, 1870	99, 934
India-rubber and pneumatic spring	F. M. Ray	New York, N. Y	Aug. 1, 1848	5, 696
India-rubber and steel springs, Combined	E. T. Bussell	Shelbyville, Ind	Nov. 29, 1853	10, 280
India rubber, Apparatus for curing	J. B. Forsyth	Roxbury, Mass	Nov. 21, 1865	51, 036
India rubber, Apparatus for pressing and vulcanizing.	J. Banigan	Woonsocket, R. I	Aug. 31, 1869	94, 273
India-rubber bat-cloth, Mode of making	C. Goodyear	New Haven, Conn	Oct. 12, 1852	9, 319
India-rubber boat, Portable	H. H. Day	New York, N. Y	Jan. 15, 1846	4, 356
India rubber, Cleaning	A. G. Day	Seymour, Conn	June 10, 1856	15, 007
India-rubber-cloth goods, Manufacture of	W. Cable	Boston, Mass	Mar. 12, 1872	124, 542
India-rubber cloth, Preparing	G. B. Millerd	Colchester, Conn	Feb. 10, 1857	16, 601
India-rubber cloth, Preparing elastic	C. Winslow	Lynn, Mass	June 23, 1857	17, 649
India-rubber cloth, Process for making	H. G. Tyer and J. Helm	New Brunswick, N. J	Jan. 2, 1855	12, 144
India-rubber cloth, Process for making	H. G. Tyer and J. Helm	New Brunswick, N. J	Jan. 30, 1855	12, 334
India-rubber cloth, Process of preparing elastic	N. Hayward	Colchester, Conn	May 6, 1856	14, 811
India-rubber cloth, Process of rolling	F. D. Hayward and J. C. Bickford.	Colchester, Conn	Mar. 19, 1850	7, 189
India rubber, Coating metal articles with	C. Hingher	New Brunswick, N. J	Aug. 23, 1870	106, 585
India rubber, Covering metal articles with	C. Hingher	New Brunswick, N. J	Aug. 30, 1870	106, 819
India-rubber fabric	C. J. Gilbert and G. Gay	New York, N. Y	July 17, 1847	5, 196
India-rubber fabric	C. Goodyear	New York, N. Y	Mar. 9, 1844	3, 461
India-rubber fabric	C. Goodyear	New York, N. Y	Mar. 9, 1844	3, 462
India-rubber fabric	C. Goodyear	New York, N. Y	June 15, 1844	3, 633
India-rubber fabric	C. Goodyear	New Haven, Conn	July 5, 1845	4, 099
India-rubber fabric	C. Goodyear	New Haven, Conn	Aug. 16, 1859	25, 111
India-rubber fabric	E. E. Marcy	New York, N. Y	Dec. 6, 1859	26, 358
India-rubber fabric	E. E. Marcy	New York, N. Y	Dec. 6, 1859	26, 359
India-rubber fabric	E. E. Marcy	New York, N. Y	Dec. 6, 1859	26, 360
India-rubber fabrics, Manufacture of	N. Goodyear	Newtown, Conn	May 13, 1845	4, 047
India-rubber fabrics, Mode of making binding for.	C. A. Ensign	Naugatuck, Conn	Nov. 28, 1865	51, 161
India rubber for the manufacture of hose, belting, packing, &c., Preparation of.	E. L. Simpson	Bridgeport, Conn	Feb. 28, 1865	46, 608
India-rubber goods, Finishing the surface of	J. T. Trotter and I. F. Williams	New York, N. Y	May 22, 1860	28, 420
India-rubber goods, Flocking	F. R. Taylor	New York, N. Y	Feb. 2, 1864	41, 453
India-rubber goods, Machine for manufacturing corrugated or shirred.	H. H. Day	Jersey City, N. J	Oct. 12, 1844	3, 788
India-rubber goods, Manufacture of	H. Hutchinson	Newark, N. J	Feb. 12, 1861	31, 391
India-rubber goods, Manufacture of	J. W. Sutton	Akron, Ohio	Oct. 15, 1872	132, 185

Index of patents issued from the United States Patent Office from 1790 *to* 1873, *inclusive*—Continued.

Invention.	Inventor.	Residence.	Date.	No.
India-rubber goods, Manufacture of	J. W. Sutton	Akron, Ohio	Oct. 15, 1872	132, 186
India rubber, gutta-percha, &c., Manufacture of	E. L. Simpson	Bridgeport, Conn	Oct. 16, 1866	58, 902
India rubber, gutta-percha, &c., Process of manufacturing.	E. L. Simpson	Bridgeport, Conn	Feb. 28, 1865	46, 610
India-rubber hollow-molded articles, Manufacture of.	D. D. Parmelee	Salem, Mass	Dec. 20, 1859	26, 551
India rubber in the liquid state, Preserving	H. L. Norris	New York, N. Y	July 26, 1853	9, 891
India rubber into shreds, Machine for cutting	J. Bogardus	New York, N. Y	Nov. 21, 1845	4, 280
India rubber into threads, Machine for cutting	J. W. Cox	Malden, Mass	June 14, 1859	24, 426
India rubber into threads, Machine for cutting	H. H. Day, H. G. Tyer, and J. Helen.	Jersey City and New Brunswick, N. J.	June 7, 1845	4, 073
India rubber into threads, Machine for cutting	H. Messer	Roxbury, Mass	May 31, 1859	24, 265
India rubber, Machine for cutting	H. G. Tyer and J. Helm	New Brunswick, N. J	Oct. 9, 1844	3, 782
India rubber, Machine for cutting sheets of	C. A. Ensign	Naugatuck, Conn	Nov. 29, 1870	109, 725
India rubber, Machine for joining irregular seams in	C. A. Ensign	Naugatuck, Conn	Nov. 29, 1870	109, 726
India rubber, Making hollow articles of	C. Goodyear	New Haven, Conn	Apr. 25, 1848	5, 536
India rubber, Manufacture of	N. Goodyear	New York, N. Y	May 6, 1851	8, 075
India rubber, Manufacture of	D. Hayward	Providence, R. I	Aug. 29, 1854	11, 608
India rubber, Manufacture of	N. Hayward	Colchester, Conn	Apr. 15, 1856	14, 657
India rubber, Manufacture of	D. McCurdy	Newark, N. J	Apr. 1, 1851	8, 016
India rubber, Manufacture of	R. Solis	New Brunswick, N. J	Feb. 1, 1853	9, 570
India rubber, Manufacture of	J. T. Trotter	New York, N. Y	Jan. 1, 1851	7, 880
India rubber, Manufacture of	H. G. Tyer and J. Helme	New Brunswick, N. J	Jan. 30, 1849	6, 066
India rubber, Mode of treating	S. Bourne	Headstone, Drive Harrow, England.	Feb. 12, 1867	61, 992
India-rubber-packing former	W. Webster	Middletown, Ohio	Apr. 25, 1865	47, 477
India rubber, Preparing	W. F. Ely	New York, N. Y	Apr. 17, 1847	5, 969
India rubber, Preparing	J. Thomas	New York, N. Y	Sept. 4, 1847	5, 271
India rubber, Preserving	F. Bronner	Vera Cruz, Mexico	Sept. 7, 1852	9, 246
India rubber, Printing and ornamenting	G. H. Lewis	Providence, R. I	May 26, 1863	38, 686
India rubber, Process for cleaning	T. Sault	Seymour, Conn	Nov. 11, 1856	16, 069
India rubber, Process for treating	E. D. S. Goodyear	Stapleton, N. Y	Mar. 28, 1854	10, 689
India rubber, Process of obtaining gum resembling	E. H. Rogers	Tuscaloosa, Ala	Mar. 11, 1873	136, 765
India rubber, Process of preparing	W. Mullee	Franklin, Pa	Feb. 12, 1867	62, 055
India-rubber rings, Manufacture of	W. H. Collins and C. E. Longden.	Prospect, Conn	July 1, 1873	140, 472
India-rubber roller, Manufacture of	J. B. Forsyth	Roxbury, Mass	Nov. 13, 1866	59, 580
India-rubber spring	W. F. Converse	Harrison, Ohio	Apr. 17, 1855	12, 738
India-rubber spring	A. Magowan	Trenton, N. J	Feb. 18, 1873	136, 079
India-rubber spring for cars, Manufacture of	F. M. Ray	New York, N. Y	Apr. 2, 1850	7, 251
India-rubber spring for upholstery purposes	F. Colton	New York, N. Y	Nov. 17, 1857	18, 630
India-rubber spring, Vulcanized	F. M. Ray	New York, N. Y	Oct. 8, 1850	7, 706
India-rubber springs, Operating	J. W. Wilder	New York, N. Y	Aug. 28, 1866	57, 647
India rubber, Treating	W. F. Shaw	Boston, Mass	Aug. 12, 1856	15, 531
India rubber, Treatment of	A. K. Eaton	New York, N. Y	June 19, 1860	28, 744
India rubber, Treatment of	H. W. Joslin	Trenton, N. J	Jan. 12, 1859	22, 560
India rubber, Treatment of	H. W. Joslin and A. K. Eaton	Trenton, N. J., and New York, N. Y.	Nov. 22, 1859	26, 233
India rubber, Treatment of	A. Shannon	New York, N. Y	Apr. 19, 1859	23, 717
India-rubber tube, Manufacture of	C. McBurney and J. B. Forsyth.	Roxbury, Mass	Apr. 17, 1866	53, 999
India-rubber tube or strip, Machine for making	B. M. Hotchkiss	Naugatuck, Conn	July 12, 1870	105, 335
India-rubber tubing	F. R. Taylor	New York, N. Y	May 31, 1864	42, 972
India rubber, Use of the oxide of tin in the manufacture of.	J. Pridham	New Brunswick, N. J	Mar. 19, 1850	7, 196
India rubber with leather, Mode of uniting	A. C. Andrews	New Haven, Conn	Apr. 16, 1867	63, 779
Indicating-tubes for ascertaining draft of boats	J. E. Vansant	Louisville, Ky	July 25, 1854	11, 393
Indicator: *See* Billiard-indicator. Canal-lock indicator. Car-indicator. Car-seat indicator. Check-valve indicator. Crank-indicator. Door-indicator. Electrical indicator. Electro-magnetic indicator. Fire-indicator. Fluid-indicator. Fuel-indicator. Gas light and pressure indicator. Gas-meter indicator. Hand-indicator. Heat-indicator. High and low water indicator. High-water indicator. Knitting-machine indicator. Knitting-machine-knot indicator. Laundry-indicator. Leeway-indicator. Loom-indicator. Low-water indicator. Magnetic indicator. Navigator's bearing-indicator. Office-indicator. Padlock-indicator. Plumb and level indicator. Power-indicator. Pressure-indicator. Privy-indicator. Railway-train indicator. Speed-indicator. Station-indicator. Station or street-registering indicator. Steam-boiler indicator. Steam-engine indicator. Steam-generator indicator. Steam-pressure indicator. Street and station indicator. Telegraph-indicator. Telegraph-circuit indicator. Telephonic indicator. Time-indicator. Time-table indicator. Valve-indicator. Wages-indicator. Water-indicator. Water-closet indicator. Water-gage indicator. Water-level indicator.				
Indicator and low-water alarm for boilers	M. Ellwood	Girard, Ohio	Mar. 11, 1873	136, 646
Indicator-lock	E. A. Cooper	Lancaster, N. Y	Mar. 25, 1873	137, 181
Indicator-lock	T. Lalor	Toronto, Canada	Aug. 4, 1868	80, 637
Indicator-lock	E. M. and J. E. Mix	Westfield, N. Y	Jan. 4, 1870	98, 513
Indigo-blue vat for coloring wool and cotton	G. Molt	Millbury, Mass	Jan. 7, 1873	134, 694

Index of patents issued from the United States Patent Office from 1790 *to* 1873, *inclusive*—Continued.

Invention.	Inventor.	Residence.	Date.	No.
Indigo, Compounding a blue color as a substitute for	S. De Riemer	Poughkeepsie, N. Y	May 15, 1809	....
Indigo-dye	T. Weber	Alton, Ill	Jan. 19, 1869	86, 047
Indigo for dyeing and printing, Preparation of	P. Schutzenberger and F. De Lalande.	Paris, France	Mar. 11, 1873	136, 770
Indigo, Manufacture of	C. W. Smith	Highfield, near Stroud, England.	Aug. 20, 1872	130, 666
Indigo, Manufacture of	T. T. Woodruff	Philadelphia, Pa	May 14, 1872	126, 665
Indigo, Process and apparatus for the manufacture of.	T. T. Woodruff	Philadelphia, Pa	May 14, 1872	126, 663
Indigo, Process and apparatus for the manufacture of.	T. T. Woodruff	Philadelphia, Pa	May 14, 1872	126, 664
Indigo, Process of restoring waste	F. A. Sawyer, jr	Boston, Mass	Oct. 24, 1871	120, 215
Induction-coil	D. M. Cook	Mansfield, Ohio	Oct. 10, 1871	119, 825
Induction-coil apparatus and circuit-breaker	C. G. Page	Washington, D. C	Apr. 14, 1868	76, 654
Infant's chair	J. Hayes	Philadelphia, Pa	Aug. 22, 1871	118, 240
Inflammable liquid so as to prevent accidents, Mode of preparing.	T. J. Barron	Brooklyn, N. Y	Mar. 28, 1865	46, 987
Ingot-casting mold	H. Dickenson	Jersey City, N. J	July 30, 1867	67, 171
Ingot-mold	G. Abel and J. Pedder	Temperanceville, Pa	Aug. 31, 1869	94, 170
Ingot-mold	N. Churchman	Pioche, Nev	Dec. 9, 1873	145, 278
Ingot-mold	Z. S. Durfee	Troy, N. Y	Sept. 27, 1870	107, 766
Ingot-mold	Z. S. Durfee	New York, N. Y	Sept. 17, 1872	131, 332
Ingot-mold	A. L. Holley	Harrisburgh, Pa	July 30, 1872	129, 957
Ingot-mold	M. T. Mooney	Johnstown, Pa	Dec. 16, 1873	145, 677
Ingot-mold, Adjustable	M. S. Drake	Newark, N. J	Mar. 18, 1873	136, 821
Ingot mold, Combined steel and iron	P. Doyle	Newark, N. J	Feb. 1, 1870	99, 299
Ingot-mold for gold or silver	J. Feix	San Francisco, Cal	Nov. 18, 1873	144, 605
Ingot molds, Stopper for steel	S. T. Wellman	Nashua, N. H	July 23, 1872	129, 699
Ingots, Casting	Z. S. Durfee	New York, N. Y	Jan. 2, 1872	122, 310
Ingots, Casting brass	L. T. Wooster	Ansonia, Conn	Aug. 17, 1871	118, 174
Ingots, Casting steel	H. Disston and J. Marsden	Philadelphia, Pa	Aug. 17, 1869	93, 863
Ingots, Casting steel	J. A. Holmes	Philadelphia, Pa	Aug. 5, 1873	141, 506
Ingots for alloying gold and silver, Manufacture of copper.	J. Feix	San Francisco, Cal	Nov. 25, 1873	144, 896
Ingots for corrugated metal beams, Form of	R. Montgomery	New York, N. Y	Aug. 23, 1870	106, 713
Ingots in groups, Casting	F. J. Slade	Trenton, N. J	Mar. 29, 1870	101, 389
Ingots, Manufacture of steel	J. B. Tarr	Fairhaven, Mass	Oct. 20, 1868	83, 222
Ingots, Mold for casting steel	J. E. Fry	Johnstown, Pa	Nov. 12, 1867	70, 710
Ingots of iron and steel, Apparatus for casting	A. L. Holley	Brooklyn, N. Y	Nov. 12, 1872	133, 030
Ingots of iron and steel, Manufacture of refined	C. Shunk	Armstrong County, Pa	Nov. 5, 1867	70, 476
Ingots, Working scrap-iron or other metals into	D. D. Parmelee	New York, N. Y	Dec. 30, 1873	146, 092
Inhalation, Apparatus for generating and washing gas for.	A. W. Sprague	Boston, Mass	June 12, 1866	55, 548
Inhalation, Oxygenized compound for	J. L. Martin	Baltimore, Md	Feb. 13, 1872	123, 714
Inhalator	M. Vergnes	New York, N. Y	July 12, 1864	43, 536
Inhaler	G. Bastian and B. Segnitz	New York, N. Y	Sept. 1, 1863	39, 716
Inhaler	G. L. Chapin	Chicago, Ill	Oct. 15, 1872	132, 197
Inhaler	W. R. Crumb	Buffalo, N. Y	Jan. 14, 1873	134, 858
Inhaler	W. H. Cutler	Buffalo, N. Y	Feb. 4, 1873	135, 411
Inhaler	D. H. Goodwillie	New York, N. Y	Nov. 6, 1866	59, 386
Inhaler	R. B. Heintzelman	New York, N. Y	Nov. 14, 1871	120, 816
Inhaler	I. Holmes	Moscow, N. Y	Jan. 8, 1867	61, 008
Inhaler	C. D. Hunter and E. S. Woods	Marlborough, Mass	Apr. 29, 1873	138, 253
Inhaler	C. D. Hunter and E. S. Woods	Marlborough, Mass	July 15, 1873	140, 828
Inhaler	A. Prentiss	Prentiss Vale, Pa	May 14, 1867	64, 793
Inhaler	S. J. Shaw	Marlborough, Mass	July 22, 1873	141, 175
Inhaler, Air	Z. Rogers	Chicago, Ill	May 11, 1869	90, 051
Inhaler and fumigator	E. C. Kirkwood	Washington, D. C	June 4, 1872	127, 420
Inhaler and nasal douche combined	E. Schofield	Worcester, Mass	Jan. 28, 1873	135, 372
Inhaler and remedy for throat disease	G. H. Tichenor	Canton, Miss	Mar. 9, 1869	87, 603
Inhaler and vaporizer for administering anæsthetics.	F. E. Duncanson	Chicago, Ill	Oct. 10, 1871	119, 748
Inhaler, Chloroform	C. B. Porter	Ann Arbor, Mich	May 13, 1862	35, 257
Inhaler, Ether	W. F. G. Morton and A. A. Gould.	Boston, Mass	Nov. 13, 1847	5, 365
Inhaler for medical purposes	J. P. Brower	Syracuse, N. Y	Feb. 9, 1869	86, 638
Inhaler, Gas	S. W. Albee	Charlestown, N. H	May 10, 1864	42, 721
Inhaler, Gas	T. F. Frank	Buffalo, N. Y	Oct. 23, 1866	59, 000
Inhaler, Gas	C. L. Munns	Philadelphia, Pa	June 12, 1866	55, 527
Inhaler, Medical	N. I. Donaldson	Worcester, Mass	Jan. 14, 1873	134, 861
Inhaler, Medical	J. C. Parkinson	Vineland, N. J	Dec. 16, 1873	145, 679
Inhaler or lung-protector	L. P. Haslett	Louisville, Ky	June 12, 1849	6, 529
Inhaler, Vapor	M. M. Mills	Aurora, Ill	Jan. 28, 1868	73, 913
Inhaler, Vapor	D. Russell	Milford, Mass	Oct. 31, 1865	50, 735
Inhaling-apparatus	J. Montgomery	New York, N. Y	Nov. 23, 1869	97, 214
Inhaling-apparatus	E. W. Owen	Brooklyn, N. Y	May 4, 1869	89, 594
Inhaling-apparatus	A. E. Pursell	Indianapolis, Ind	Oct. 1, 1872	131, 780
Inhaling-apparatus	E. Schofield	Worcester, Mass	Dec. 23, 1873	145, 758
Inhaling-apparatus	J. C. Schooley	Cincinnati, Ohio	Aug. 18, 1857	18, 020
Inhaling-apparatus	S. H. T. Tilghman	Snow Hill, Md	Nov. 21, 1854	11, 976
Inhaling-apparatus	C. Warren	Milford, Mass	Sept. 13, 1864	44, 244
Inhaling apparatus, Vapor	A. P. Lighthill	Boston, Mass	Apr. 25, 1865	47, 434
Inhaling-fluid for cure of consumption and other diseases.	N. W. Abbott	Centralia, Ill	July 16, 1867	66, 667
Inhaling gases	W. Z. W. Chapman	New York, N. Y	Apr. 28, 1868	77, 168
Inhaling gases, Apparatus for	W. Z. W. Chapman	New York, N. Y	Sept. 6, 1864	44, 075
Inhaling gases, Apparatus for	P. H. Vander Weyde	New York, N. Y	June 5, 1866	55, 398
Inhaling medicinal agents, Means for	A. G. Hull	New York, N. Y	Apr. 21, 1857	17, 095
Inhaling powders, Instrument for	I. Warren	Boston, Mass	Mar. 16, 1852	8, 813
Inhaling pure air, Apparatus for	B. I. Lane	South Framingham, Mass	Feb. 21, 1865	46, 477
Inhaling-tube	C. Bullock	Cambridge, Mass	July 18, 1865	48, 789
Inhaling-tube	A. P. Messer	Boston, Mass	May 4, 1869	89, 591
Inhaling-tube	D. Minthorn	New York, N. Y	June 27, 1854	11, 171
Inhaling-tube	S. W. Sine	Easton, Pa	Dec. 17, 1867	72, 234
Injecting and douching instrument	J. Singer	Chicago, Ill	Oct. 6, 1863	40, 192
Injections, Apparatus for	C. A. Jozansi	Saint Romain, France	July 28, 1863	39, 348
Injector	A. J. Blakslee and G. C. Williams.	Du Quoin, Ill	June 15, 1869	91, 205

Index of patents issued from the United States Patent Office from 1790 *to* 1873, *inclusive*—Continued.

Invention.	Inventor.	Residence.	Date.	No.
Injector	P. C. Heinz	Funkville, Pa	Nov. 13, 1866	59, 602
Injector	C. Hughes	Niles, Ohio	July 6, 1869	92, 313
Injector	A. Morton	Glasgow, Scotland	Jan. 19, 1869	86, 030
Injector	C. F. Root	West Springfield, Mass	Dec. 20, 1870	110, 289
Injector	S. Rue	Philadelphia, Pa	Dec. 30, 1873	145, 969
Injector, Air	J. A. Bassett and O. C. Smith	Salem, Mass	May 30, 1865	48, 041
Injector and ejector	W. B. Mack	Boston, Mass	Dec. 17, 1872	133, 997
Injector, Boiler	J. Gresham	Manchester, Great Britain	Aug. 7, 1866	57, 057
Injector, Boiler	J. Gresham	Manchester, Great Britain	Feb. 11, 1868	74, 345
Injector, Boiler	J. Gresham	Manchester, Great Britain	Feb. 11, 1868	74, 346
Injector, Boiler	J. T. Hancock	Jamaica Plain, Mass	Jan. 26, 1869	86, 152
Injector, Feed-water	J. Robinson and J. Gresham	Manchester, Great Britain	May 29, 1866	55, 218
Injector, feed-water heater, and condenser, Combined.	E. Korting	Vienna, Austria	Jan. 30, 1872	123, 264
Injector for feeding boilers	W. Sellers	Philadelphia, Pa	Mar. 3, 1868	75, 059
Injector for feeding boilers	W. and C. Sellers	Philadelphia, Pa	July 22, 1873	141, 174
Injector for feeding boilers, Self-adjusting	W. Sellers	Philadelphia, Pa	July 22, 1873	141, 173
Injector for heating liquids	P. Hogg	Brooklyn, N. Y	Aug. 2, 1864	43, 691
Injector for insect-powder	C. Chinnock	Brooklyn, N. Y	Aug. 10, 1869	93, 597
Injector for insect-powder	M. Koppe	New York, N. Y	Sept. 17, 1867	68, 995
Injector for steam and other enginery, Oil	R. Brayton	Fremont, Ohio	Dec. 15, 1868	84, 853
Injector for steam-engines, Oil	G. B. Brayton	Boston, Mass	Aug. 14, 1866	57, 079
Injector for steam-generators	A. Barclay	Kilmarnock, North Britain	Mar. 19, 1867	62, 993
Injector for steam-generators	E. Ferguson	New Berne, N. C	Feb. 4, 1868	74, 067
Injector for steam-generators	S. Rue, jr	Chester County, Pa	Jan. 28, 1868	73, 757
Injector for steam-generators	S. Rue, jr	Paoli, Pa	Sept. 1, 1868	81, 822
Injector for steam-generators	S. Rue, jr	Philadelphia, Pa	Apr. 22, 1873	138, 198
Injector for steam-generators	S. Rue, jr	Philadelphia, Pa	Apr. 22, 1873	138, 199
Injector for steam-generators	E. Ware	Southford, Conn	Aug. 24, 1869	94, 152
Injector for the hair	A. A. Smith	Seneca Falls, N. Y	Oct. 25, 1864	44, 843
Injector, Gaseous liquid	C. Schultz and T. Warker	New York, N. Y	Oct. 17, 1865	50, 500
Injector, Giffard	N. Cope	Cincinnati, Ohio	Jan. 27, 1863	37, 542
Injector, Giffard	J. Gresham	Manchester, England	Apr. 12, 1870	101, 858
Injector, Giffard	J. Millholland	Reading, Pa	June 10, 1862	35, 575
Injector, Giffard	W. Sellers	Philadelphia, Pa	July 21, 1863	39, 312
Injector, Giffard	W. Sellers	Philadelphia, Pa	July 21, 1863	39, 313
Injector, Giffard	W. Sellers	Philadelphia, Pa	Aug. 15, 1865	49, 445
Injector, Metal	J. W. Hollingsworth	Mount Vernon, Ind	May 17, 1870	103, 188
Injector, Mode of operating Giffard's	J. Millholland	Reading, Pa	Dec. 17, 1861	33, 957
Injector, Pile-ointment	L. Heins	Brunswick, Ga	May 14, 1872	126, 700
Injector, Powder	D. Russell	Milford, Mass	Aug. 25, 1863	39, 678
Injector, Steam	S. and J. Benson	Centralia, Ill	Apr. 8, 1873	137, 651
Injector, Steam and water	S. S. Jamison, jr	Saltsburgh, Pa	Nov. 28, 1871	121, 376
Injector, Steam-boiler	H. C. Crowell	Morgan, Ohio	May 24, 1870	103, 304
Injector, Steam-boiler	J. P. F. Datichy	Brooklyn, N. Y	July 12, 1870	105, 181
Injector, Steam-boiler	W. B. Mack	Manchester, England	Sept. 5, 1871	118, 734
Injector, Steam-boiler	S. Maltby and C. Osborn	Dayton, Ohio	Sept. 26, 1865	50, 205
Injector, Steam-boiler	T. O'Rorke	Pittsburgh, Pa	July 16, 1872	129, 491
Injector, Vagina	G. W. King	Saratoga Springs, N. Y	Aug. 18, 1868	81, 278
Injector, Veterinary narcotic	M. Crohn	Saint Louis, Mo	Nov. 5, 1867	70, 418
Injector, Water	J. B. Collen	Philadelphia, Pa	Mar. 22, 1864	42, 044
Injectors, Feed-regulator for	J. A. Marden and T. S. Smith	Boston and Somerville, Mass.	Oct. 28, 1873	143, 988
Ink	C. F. Panknin	Charleston, S. C	Aug. 9, 1870	106, 198
Ink and preparation for indelible marking	E. W. Briggs	Brooklyn, N. Y	Nov. 19, 1872	133, 197
Ink, Black	D. A. Dougherty	Kittaning, Pa	Feb. 23, 1869	87, 094
Ink, &c., Black pigment or mineral black for printing.	J. Cist	Wilkesbarre, Pa	Oct. 28, 1808	....
Ink box, Indelible	C. L. Lochman	Carlisle, Pa	July 16, 1867	66, 858
Ink, coloring-matter, mordant, or other liquids, Supplying or steeping, &c.	J. Rennie	Lodi, N. J	June 9, 1834	....
Ink-composition	B. Owens	Saint Louis, Mo	May 21, 1867	65, 002
Ink, Composition for	J. Shaw	Bridgeport, Conn	Oct. 15, 1867	69, 846
Ink, Composition for bank-note and other	T. S. Hunt	Montreal, Canada	Dec. 8, 1863	40, 839
Ink, Composition for printing and copying	J. Underwood and F. V. Burt	London, England	Mar. 24, 1863	38, 008
Ink, Compound for making writing	A. D. Bowman	New York, N. Y	Nov. 13, 1866	59, 549
Ink, Compound for printers'	H. Loewenberg	New York, N. Y	Apr. 9, 1867	63, 733
Ink, Compound for printers'	H. Lowenberg	New York, N. Y	June 28, 1870	104, 862
Ink-compound for telegraphic and other purposes	G. Little	Rutherford Park, N. J	Oct. 1, 1872	131, 886
Ink, Copying	A. G. Buzby	Philadelphia, Pa	May 26, 1868	78, 258
Ink-cup	P. K. Holbrook	Malden, Mass	Oct. 30, 1866	59, 223
Ink-distributer	J. Prince	Philadelphia, Pa	Nov. 19, 1833	....
Ink-distributer, Self-moving	J. Prince	New York, N. Y	Apr. 23, 1830	....
Ink-distributer, Self-operating	J. Maxson	Schenectady, N. Y	Jan. 9, 1835	....
Ink for hand-stamp, &c	R. H. Rogers	New York, N. Y	Apr. 19, 1864	42, 405
Ink for paper-ruling	L. Francis	New York, N. Y	Dec. 24, 1867	72, 621
Ink for printing, &c	L. Francis	New York, N. Y	Apr. 26, 1870	102, 243
Ink for printing bank-notes, &c	A. K. Eaton	New York, N. Y	Apr. 28, 1863	38, 298
Ink for printing, Copying	C. McIlvaine	Philadelphia, Pa	Oct. 25, 1870	108, 615
Ink for printing postage-stamps, Manufacture of	J. Macdonough	New York, N. Y	Feb. 27, 1866	52, 869
Ink for printing stamps, drafts, and checks	J. P. Simonds	New York, N. Y	Mar. 22, 1870	101, 170
Ink for ruling and printing, Copying	C. McIlvaine	Philadelphia, Pa	July 25, 1871	117, 314
Ink for stamping-purposes	J. B. F. Jud	New York, N. Y	July 15, 1873	140, 782
Ink-fountain	E. Jordan	West Cummington, Mass	Nov. 20, 1849	6, 883
Ink-fountain	H. H. Shorp	Cleveland, Ohio	Feb. 13, 1872	123, 651
Ink from grahamite, Mode of compounding printers'	H. Wurtz	New York, N. Y	Feb. 4, 1868	74, 188
Ink from type, Composition for removing	A. M. Bonton	Newark, N. J	Oct. 2, 1866	58, 370
Ink, Green	G. Smillie	New York, N. Y	Mar. 24, 1863	37, 984
Ink-holder	D. Harrington	Philadelphia, Pa	Apr. 22, 1831	....
Ink, Indelible	J. M. and G. W. Caldwell	Burlington, N. J	May 10, 1870	102, 915
Ink, Making black	P. Ferris	Greenwich, Conn	Dec. 5, 1842	2, 870
Ink, Making printers'	B. Chase	New London, Conn	June 28, 1810	....
Ink, Making printers'	E. Clark	Brooklyn, N. Y	July 25, 1845	4, 102
Ink, Making printers'	C. A. Thompson	Adrian, Mich	Apr. 17, 1855	12, 733
Ink, Manufacture of	J. A. W. and H. H. Sangster	Buffalo, N. Y	Mar. 20, 1866	53, 379
Ink, Manufacture of indelible writing	T. J. Spear	New Orleans, La	July 16, 1841	2, 176
Ink, Manufacture of printing	G. Duryee	New York, N. Y	June 27, 1865	48, 385
Ink, Manufacture of writing	J. W. Carter	Boston, Mass	Mar. 12, 1872	124, 544

Index of patents issued from the United States Patent Office from 1790 *to* 1873, *inclusive*—Continued.

Invention.	Inventor.	Residence.	Date.	No.
Ink, &c., Mixing and grinding apparatus for	J. Martin	New York, N. Y	Dec. 31, 1872	134, 434
Ink or writing-fluid	C. Hebel	Louisville, Ky	Aug. 23, 1870	106, 582
Ink-pad	H. R. Towne and W. H. Taylor	Stamford, Conn	Feb. 18, 1873	135, 949
Ink-powder and dye from aniline colors	J. Zengeler	Chicago, Ill	May 25, 1869	90, 417
Ink-press, Composition for	H. G. Horton	Troy, N. Y	Sept. 17, 1872	131, 444
Ink, Printers'	C. Kréci	Scranton, Pa	Aug. 31, 1869	94, 220
Ink, Printers'	C. McIlvaine	Philadelphia, Pa	July 25, 1871	117, 313
Ink, Printers'	M. Turly	Council Bluffs, Iowa	Apr. 18, 1871	113, 947
Ink, Printers'	M. Turly and B. F. Thomas	Council Bluffs, Iowa	July 5, 1870	105, 014
Ink, Printers'	S. H. Turner	Brooklyn, N. Y	Sept. 6, 1853	10, 006
Ink, Printers'	M. Weissberger	Saint Paul, Minn	June 18, 1867	65, 973
Ink, Printers'	C. Wulsten	La Fayette, Ind	June 16, 1868	79, 045
Ink, Printing	G. W. Casilear	Washington, D. C	June 21, 1870	104, 554
Ink, Printing	A. A. Hulot	Paris, France	May 23, 1865	47, 909
Ink, Printing	J. Kircher	Cannstadt, near Stuttgart, Würtemberg.	Dec. 13, 1870	110, 048
Ink, Printing	J. Kircher	New York, N. Y	Sept. 19, 1871	119, 154
Ink, Printing	J. Kircher	New York, N. Y	Nov. 5, 1872	132, 840
Ink, Printing	G. Matthews	Montreal, Canada	June 30, 1857	17, 688
Ink, Printing	J. C. White	Quincy, Ill	May 7, 1872	126, 601
Ink, Red	T. J. Summus	Lynn, Mass	Mar. 7, 1865	46, 684
Ink-roller	A. A. Hanscom	Saco, Me	June 29, 1858	20, 710
Ink-roller	A. Schimmelfennig and J. Ende.	Washington, D. C	June 8, 1858	20, 512
Ink-roller, Printer's	L. Francis and F. W. Letmate.	New York, N. Y	June 21, 1864	43, 492
Ink slab, India	I. Speyer	Terre Haute, Ind	Nov. 19, 1872	133, 178
Ink-stains, &c., Compound for removing	V. G. Bloede	Brooklyn, N. Y	Feb. 23, 1869	87, 088
Ink, Sympathetic	D. C. McNeil	Osceola, Mo	June 30, 1868	79, 374
Ink, Use of resin oil in printer's	M. M. Mathews	Rochester, N. Y	Oct. 1, 1850	7, 686
Ink, Vessel for making	A. Harrison	Philadelphia, Pa	Feb. 24, 1852	8, 755
Ink-well	T. Bell	New York, N. Y	Dec. 3, 1867	71, 686
Ink-well	F. C. Brownell	Brooklyn, N. Y	Jan. 5, 1864	41, 136
Ink-well	F. C. Brownell	East Orange, N. J	May 9, 1865	47, 616
Ink-well	J. H. Kidder	Lawrence, Mass	Feb. 18, 1873	136, 069
Ink-well	J. W. Ross	Boston, Mass	Sept. 1, 1863	39, 754
Ink-well cover	J. A. Blake	New Haven, Conn	Mar. 19, 1867	62, 925
Ink-well cover	G. Munger	New York, N. Y	Oct. 16, 1866	58, 952
Ink-wells, Fastening the covers of	G. and H. M. Sherwood	Chicago, Ill	Dec. 9, 1862	37, 113
Ink-wells, Fastening the covers of	H. M. Sherwood	Chicago, Ill	Oct. 18, 1864	44, 753
Ink, Writing	R. G. Loftus	Chelsea, Mass	Nov. 15, 1870	109, 327
Ink, Writing	J. D. Myers	New York, N. Y	Sept. 26, 1835	....
Ink, Writing	J. Popper	Phillipsburgh, N. J	Oct. 8, 1872	132, 102
Ink, Writing	E. F. Walsh and S. Duryea	Brooklyn, N. Y	June 13, 1871	115, 915
Inking-apparatus	F. L. Bailey	Boston, Mass	Oct. 7, 1873	143, 489
Inking-apparatus	H. Barth	Cincinnati, Ohio	Apr. 12, 1870	101, 703
Inking-apparatus	G. P. Gordon	Rahway, N. J	Sept. 2, 1873	142, 457
Inking-apparatus	I. Hart	Cincinnati, Ohio	Nov. 18, 1873	144, 673
Inking-apparatus	G. W. Wood	Richmond, Ind	Apr. 16, 1867	63, 975
Inking-apparatus for printing in colors	T. L. Baylies and G. W. Wood	Richmond, Ind	Aug. 6, 1867	67, 400
Inking-pad	J. A. Miles	Charleston, Ill	Jan. 21, 1873	135, 643
Inking-pad	W. A. Whitney and C. G. Seaver.	Adrian, Mich	July 9, 1872	128, 834
Inking-roller	R. M. Hoe	New York, N. Y	Apr. 17, 1844	3, 550
Inking-roller	W. J. Stone	Washington, D. C	Oct. 20, 1829	....
Inking-roller, Composition for printer's	A. Van Bibber	Cincinnati, Ohio	Mar. 28, 1871	113, 224
Inking-roller, Lithographic	S. D. Tucker	New York, N. Y	June 1, 1869	90, 702
Inking-rollers, Regulating temperature of	E. W. Arnold	Boston, Mass	Sept. 21, 1837	394
Inkstand	J. L. Adam	London, England	July 9, 1872	128, 695
Inkstand	H. L. Andrews	Chicago, Ill	Mar. 31, 1868	76, 138
Inkstand	H. P. Andrews and M. E. Rawson.	Cleveland, Ohio	Oct. 20, 1868	83, 126
Inkstand	J. Axtman	East Cambridge, Mass	Aug. 8, 1865	49, 211
Inkstand	S. W. Baldwin	Yonkers, N. Y	Jan. 14, 1873	134, 836
Inkstand	O. Barber	Hartford, Conn	May 13, 1814	....
Inkstand	N. G. Bartlett	Keokuk, Iowa	Jan. 14, 1868	73, 284
Inkstand	J. Barwick	North Woolwich, England	Feb. 4, 1868	73, 946
Inkstand	J. M. Batchelder	Cambridge, Mass	Apr. 27, 1858	20, 028
Inkstand	E. O. Bennett	Mount Pleasant, Iowa	Nov. 26, 1867	71, 443
Inkstand	A. Bingham	Boston, Mass	Aug. 28, 1855	13, 515
Inkstand	J. A. Bowen	Boston, Mass	June 6, 1871	115, 686
Inkstand	H. Bradford	New York, N. Y	Apr. 3, 1866	53, 729
Inkstand	M. Braun	Brooklyn, N. Y	Feb. 21, 1860	27, 200
Inkstand	A. W. Brinkerhoff	Upper Sandusky, Ohio	May 7, 1872	126, 514
Inkstand	W. M. Brooke	Eaton, Ohio	Jan. 7, 1873	134, 587
Inkstand	B. Brower	New York, N. Y	Nov. 25, 1873	144, 825
Inkstand	W. Burnet	New York, N. Y	Aug. 23, 1859	25, 175
Inkstand	G. Burnham	Philadelphia, Pa	Dec. 30, 1841	2, 409
Inkstand	W. E. Carlile	New York, N. Y	Nov. 9, 1869	96, 673
Inkstand	S. C. Catlin	Cleveland, Ohio	May 2, 1871	114, 262
Inkstand	C. T. Chase	Albany, N. Y	June 30, 1868	79, 444
Inkstand	S. D. Clark	Minneapolis, Minn	Aug. 8, 1871	117, 743
Inkstand	D. Cumming, jr	New York, N. Y	Dec. 13, 1864	45, 391
Inkstand	S. Darling	Bangor, Me	Sept. 14, 1858	21, 554
Inkstand	S. Darling	Bangor, Me	Aug. 1, 1865	49, 093
Inkstand	S. Darling	Bangor, Me	Jan. 9, 1866	51, 931
Inkstand	S. Darling	Bangor, Me	Sept. 3, 1867	68, 588
Inkstand	S. Darling	Bangor, Me	Apr. 7, 1868	76, 409
Inkstand	S. Darling	Providence, R. I	Jan. 31, 1871	111, 435
Inkstand	S. Darling and J. E. Hall	Bangor, Me	Apr. 14, 1868	76, 719
Inkstand	O. Dean	Richmond, Va	Feb. 18, 1868	74, 671
Inkstand	T. Duncan	Brookville, Md	May 23, 1871	115, 041
Inkstand	J. R. Ender	Trenton, La	Jan. 3, 1860	26, 661
Inkstand	H. Evans	Philadelphia, Pa	Jan. 10, 1860	26, 755
Inkstand	A. Fife	Philadelphia, Pa	Aug. 21, 1849	6, 664
Inkstand	B. S. Fletcher	Cornish, N. H	Aug. 29, 1865	49, 617
Inkstand	V. Fogerty	Cambridgeport, Mass	Aug. 31, 1858	21, 395
Inkstand	R. Gleason, jr	Dorchester, Mass	Mar. 18, 1856	14, 451
Inkstand	K. Goddard	Philadelphia, Pa	Apr. 28, 1857	17, 147

Index of patents issued from the United States Patent Office from 1790 to 1873, inclusive—Continued.

Invention.	Inventor.	Residence.	Date.	No.
Inkstand	A. P. Griffing	East Cambridge, Mass	Feb. 4, 1862	34, 299
Inkstand	F. T. Grimes	Liberty, Mo	June 14, 1870	104, 300
Inkstand	W. O. Haskell	Boston, Mass	June 6, 1871	115, 733
Inkstand	S. Hawson and T. Sweeney	Jersey City, N. J., and Brooklyn, N. Y.	Apr. 8, 1873	137, 548
Inkstand	A. Haywood	Easton, Mass	Nov. 23, 1818	....
Inkstand	B. J. Heywood	London, England	May 26, 1857	17, 373
Inkstand	G. G. Hickman	Coatesville, Pa	Feb. 9, 1869	86, 670
Inkstand	F. L. Hicks	New York, N. Y	Jan. 3, 1865	45, 794
Inkstand	L. E. Hicks	Boston, Mass	Mar. 9, 1858	19, 613
Inkstand	L. E. Hicks	New York, N. Y	June 9, 1863	38, 863
Inkstand	P. K. Holbrook	Malden, Mass	Dec. 20, 1864	45, 498
Inkstand	T. P. How	New York, N. Y	Jan. 31, 1860	26, 992
Inkstand	T. S. Hudson	East Cambridge, Mass	Dec. 28, 1858	22, 429
Inkstand	T. S. Hudson	East Cambridge, Mass	June 3, 1862	35, 449
Inkstand	W. Hunt	New York, N. Y	May 29, 1845	4, 062
Inkstand	W. Hunt	New York, N. Y	Oct. 7, 1845	4, 221
Inkstand	W. Hunt	New York, N. Y	Dec. 11, 1845	4, 306
Inkstand	O. H. Jadwin	Carbondale, Pa	Nov. 23, 1858	22, 123
Inkstand	L. P. Jenks	Boston, Mass	Dec. 8, 1863	40, 841
Inkstand	G. R. G. Jones	Memphis, Tenn	Dec. 24, 1872	134, 144
Inkstand	W. F. Jones	Circleville, Ohio	July 2, 1872	128, 630
Inkstand	A. D. Judd	New Haven, Conn	Dec. 9, 1873	145, 352
Inkstand	J. H. Kidder and G. E. Hood	Lawrence, Mass	Feb. 18, 1873	136, 070
Inkstand	R. Lapham	New York, N. Y	Dec. 3, 1867	71, 766
Inkstand	J. G. Lucas	Newark, N. J	Nov. 26, 1872	133, 465
Inkstand	T. J. Mayall	Boston, Mass	Apr. 23, 1872	125, 973
Inkstand	G. Merritt	Brooklyn, N. Y	Apr. 14, 1868	76, 792
Inkstand	E. Morgan	Springfield, Mass	June 18, 1872	128, 163
Inkstand	F. E. Oliver	New York, N. Y	June 9, 1863	38, 836
Inkstand	J. Peard	New York, N. Y	Feb. 27, 1872	124, 152
Inkstand	S. and J. J. Perry	London, England	Nov. 12, 1867	70, 893
Inkstand	G. M. Prentiss	Worcester, Mass	Jan. 11, 1859	22, 582
Inkstand	J. S. Rankin	Minneapolis, Minn	Jan. 23, 1872	122, 909
Inkstand	W. Read, jr	Cambridge, Mass	May 27, 1873	139, 261
Inkstand	T. Robjohn	New York, N. Y	Aug. 25, 1857	18, 060
Inkstand	T. Robjohn	New York, N. Y	Aug. 16, 1859	25, 217
Inkstand	J. W. Ross	Boston, Mass	Apr. 30, 1861	32, 207
Inkstand	W. G. Shattuck	Boston, Mass	Oct. 20, 1868	83, 328
Inkstand	T. S. Shenston	Brantford, Canada	Nov. 25, 1873	144, 929
Inkstand	S. Slocomb	Cambridge, Mass	Oct. 2, 1860	30, 257
Inkstand	S. Slocomb	East Cambridge, Mass	June 3, 1862	35, 478
Inkstand	E. Southworth	Saybrook, Conn	June 29, 1833	....
Inkstand	J. C. Sparr	Irondequoit, N. Y	Jan. 30, 1872	123, 205
Inkstand	J. G. Thompson	Carbondale, Pa	Nov. 3, 1863	40, 517
Inkstand	J. B. Thurston and F. M. West	New York, N. Y	Jan. 28, 1873	135, 296
Inkstand	W. C. Tilden	Washington, D. C	Apr. 30, 1872	126, 348
Inkstand	W. H. Towers	New York, N. Y	Oct. 2, 1860	30, 266
Inkstand	H. Whitney, jr	Cambridge, Mass	July 1, 1851	8, 188
Inkstand	H. Whitney, jr	Cambridge, Mass	May 8, 1855	12, 841
Inkstand	C. H. Wright	Baltimore, Md	Apr. 12, 1870	101, 953
Inkstand and bell combined	A. J. Lyon	Mount Vernon, N. Y	Sept. 24, 1872	131, 692
Inkstand and calendar combined	G. G. Percival	Brooklyn, N. Y	July 24, 1866	56, 603
Inkstand and mucilage-holder combined	W. W. Beach	New York, N. Y	July 9, 1867	66, 448
Inkstand and pen-rack	S. Walker	Boston, Mass	Jan. 6, 1863	37, 313
Inkstand, Attaching pen-rack to	C. A. Roberts	West Meriden, Conn	June 8, 1869	91, 166
Inkstand, Barometer	T. S. Hudson	East Cambridge, Mass	June 4, 1861	32, 498
Inkstand-filler	M. C. Stebbins	Springfield, Mass	Feb. 25, 1873	136, 186
Inkstand, Fountain	C. T. Close	New York, N. Y	Dec. 11, 1855	13, 902
Inkstand, Fountain	F. Draper	East Cambridge, Mass	Jan. 7, 1851	7, 885
Inkstand from composition stone	J. S. Elliott and J. F. Wood	Chelsea and Everett, Mass	June 25, 1872	128, 218
Inkstand, pen-cleaner, &c., Arrangement of	F. Bailey	New York, N. Y	Sept. 27, 1864	44, 381
Inkstand, Revolving	J. M. Kennedy	Vicksburgh, Miss	Dec. 15, 1868	84, 950
Inkstand, wafer or sand box, calendar, letter and envelope holder, and pen-rack combined	G. Schmidt	New York, N. Y	July 3, 1866	56, 104
Inkstand with air-tight stopper	J. Farley	Washington, D. C	Jan. 30, 1841	1, 957
Inkstands, Calendar-attachment to	S. R. Dummer	New York, N. Y	Dec. 10, 1867	71, 861
Inkstands, Construction of	R. T. Fry	Spring Garden, Pa	July 11, 1854	11, 270
Inkstands, Construction of	D. Harrington	Philadelphia, Pa	Nov. 8, 1845	4, 258
Inkstands, Manufacture of	T. S. Hudson	East Cambridge, Mass	Aug. 24, 1869	94, 113
Inkstands, Manufacture of	D. J. Mandell	Springfield, Mass	Feb. 21, 1842	2, 460
Inkstands, &c., Method of making	H. Whitney	East Cambridge, Mass	Mar. 30, 1869	88, 354
Inkstands of stone, Manufacturing	A. H. Quincy	Boston, Mass	Sept. 3, 1811	....
Inkstands, Operating-parts of fountain	W. A. Wheeler	New York, N. Y	Mar. 28, 1865	47, 060
Inkstands to desks, Mode of attaching	L. R. Satterlee	Rochester, N. Y	Aug. 12, 1856	15, 527
Inkstands, Tool for molding	A. W. Brinkerhoff	Sandusky, Ohio	Dec. 16, 1873	145, 619
Inlaying	J. W. Hyatt, jr	Albany, N. Y	Oct. 10, 1871	119, 710
Inlaying gold in tortoise-shell	U. Bailey	West Newbury, Mass	Feb. 22, 1827	....
Inlaying, Machine for preparing wood for	C. F. Ritchel	Newark, N. J	Mar. 15, 1870	100, 927
Inlaying metallic surfaces	E. G. Wright	Boston, Mass	Mar. 8, 1870	100, 580
Inlaying wood, Method of	T. W. and H. K. Porter	Boston, Mass	Sept. 6, 1870	107, 097
Inoculating-apparatus	A. Stauch	Philadelphia, Pa	June 12, 1860	28, 697
Insect and reptile destroying apparatus	C. Russell	Pittsburgh, Pa	July 2, 1861	32, 720
Insect-bar	J. V. Hirley	Cincinnati, Ohio	July 16, 1872	129, 476
Insect-destroyer	W. Armistead	Prince William County, Va	Nov. 13, 1810	....
Insect-destroyer	J. M. Bennett	Janesville, Iowa	Nov. 28, 1871	121, 272
Insect-destroyer	T. Byrne and D. Strunk	New York, N. Y., and Lavaca County, Tex.	Dec. 6, 1870	109, 869
Insect-destroyer	S. Creighton	Lithopolis, Ohio	Aug. 29, 1871	118, 517
Insect-destroyer	C. R. Dudley	Canton, Miss	Dec. 24, 1872	134, 130
Insect-destroyer	J. A. Finney	Nashville, Ohio	July 15, 1873	140, 818
Insect-destroyer	J. G. G. Garrett	Port Gibson, Miss	Nov. 12, 1872	133, 023
Insect-destroyer	C. J. Hauck	Brooklyn, N. Y	Nov. 26, 1872	133, 444
Insect-destroyer	W. H. Lewis	New York, N. Y	Jan. 25, 1870	99, 214
Insect-destroyer	A. McKenzie	Henry, Ill	Mar. 17, 1868	75, 560
Insect-destroyer	D. G. Mosher	Mosherville, Mich	Sept. 26, 1871	119, 389
Insect-destroyer	D. G. Mosher	Mosherville, Mich	Apr. 1, 1873	137, 469
Insect-destroyer	J. Orin	Dayton, Ohio	Sept. 3, 1872	131, 116

Index of patents issued from the United States Patent Office from 1790 *to* 1873, *inclusive*—Continued.

Invention.	Inventor.	Residence.	Date.	No.
Insect-destroyer	E. D. Pugh	Fort Plain, Iowa	Aug. 13, 1872	130, 390
Insect-destroyer	P. Reynard	Saint Louis, Mo	July 16, 1872	129, 167
Insect-destroyer	R. V. Shockey	Mount Pleasant, Iowa	Jan. 9, 1872	122, 541
Insect-destroyer	W. B. Stewart	Brooklyn, N. Y	Oct. 8, 1872	132, 116
Insect-destroyer	S. W. Thomas	North Royalton, Ohio	May 27, 1873	139, 277
Insect-destroyer	P. S. Von Wagner	Saltfleet Township, Canada	Oct. 3, 1871	119, 547
Insect-destroying apparatus	H. S. Danziger	New York, N. Y	Dec. 31, 1872	134, 468
Insect-destroying composition	J. Ahearn	Baltimore, Md	Nov. 16, 1869	96, 861
Insect-destroying composition	G. M. Jaques	Boston, Mass	Oct. 2, 1866	58, 423
Insect-destroying composition	D. R. Prindle	East Bethany, N. Y	Mar. 26, 1867	63, 298
Insect-destroying composition	A. M. Turney	Butternuts, N. Y	Jan. 30, 1866	52, 345
Insect-destroying composition	W. Weaver	Phœnixville, Pa	Apr. 2, 1867	63, 589
Insect-destroying composition for trees	B. Best	Dayton, Ohio	June 2, 1868	78, 569
Insect-destroying compound	J. G. Barker	Watertown, Mass	May 28, 1872	127, 141
Insect-destroying compound	B. Best	Dayton, Ohio	July 27, 1869	93, 044
Insect-destroying compound	W. C. Bibb	Madison, Ga	Sept. 23, 1873	142, 985
Insect-destroying compound	W. R. Fairbairn	Ridotte Township, Ill	Dec. 15, 1868	84, 867
Insect-destroying compound	S. Galbraith	Pine Grove Plantation, La	Sept. 17, 1867	68, 867
Insect-destroying compound	M. Haas	New York, N. Y	Oct. 20, 1868	83, 280
Insect-destroying compound	J. Hinds	Hindsburgh, N. Y	Apr. 13, 1869	88, 873
Insect-destroying compound	J. B. Lunbeck	Leon, Iowa	Aug. 5, 1873	141, 512
Insect-destroying compound	A. McDougall	Manchester, England	July 16, 1867	66, 725
Insect-destroying compound	N. T. P. Robertson and T. Niles	Fairbury, Ill	Dec. 7, 1869	97, 555
Insect-destroying compound	W. B. Royal	Brenham, Tex	June 17, 1873	140, 079
Insect-destroying compound	P. B. Sheldon	Bath, N. Y	Feb. 9, 1869	86, 873
Insect-destroying compound for trees	D. Daniels	Fitchburgh, Mass	Feb. 11, 1868	74, 317
Insect-destroying compound for trees, &c	Z. F. De Moss	Pleasanton, Kans	Sept. 24, 1872	131, 603
Insect-destroying varnish	T. J. Elliott	New York, N. Y	Nov. 14, 1871	120, 813
Insect-exterminator	A. B. Ewing	Lewisburgh, Tenn	Apr. 9, 1872	125, 557
Insect-exterminator	E. Gilliam	Allegheny City, Pa	May 8, 1866	54, 528
Insect-guard	J. W. Brook	Lynchburgh, Pa	July 25, 1871	117, 379
Insect-powder	H. S. Danziger	New York, N. Y	Dec. 3, 1872	133, 634
Insect powder blower	P. Reynard and V. Varin	New York and Brooklyn, N. Y.	Sept. 20, 1859	25, 526
Insect-powder blower	P. Reynard and V. Varin	New York and Brooklyn, N. Y.	Nov. 8, 1859	26, 055
Insect-powder gun	W. H. Ball	Brooklyn, N. Y	Aug. 5, 1873	141, 481
Insect-shield for the head	J. Haven	Boston, Mass	Dec. 4, 1860	30, 818
Insect-trap	J. W. Anderson	Portland, Ind	Apr. 8, 1873	137, 588
Insect trap	J. J. Armstrong	Kings County, N. Y	Oct. 17, 1871	120, 020
Insect-trap	A. D. Chesebro	Grand Rapids, Mich	Dec. 31, 1872	134, 419
Insect-trap	C. W. Curtis	Osborn, Mo	Dec. 10, 1872	133, 763
Insect-trap	L. Endslow	Blair, Pa	Sept. 26, 1871	119, 337
Insect-trap	F. G. Fowler	Bridgeport, Conn	Oct. 8, 1872	132, 067
Insect-trap	P. Funk and J. N. Baader	Buffalo, N. Y	May 17, 1870	103, 037
Insect-trap	W. Ogden	Philadelphia, Pa	Jan. 16, 1866	52, 067
Insect-trap	C. E. Penny	Fort Wayne, Ind	Nov. 7, 1871	120, 595
Insect-trap	M. Rigell	Newton, Ala	Jan. 28, 1873	135, 366
Insect-trap	J. M. Smith	Fort Wayne, Ind	Mar. 25, 1873	137, 955
Insect-trap	J. W. Stell	Gonzales, Tex	Dec. 31, 1872	134, 444
Insect-trap	L. I. Way	Annawan, Ill	July 4, 1871	116, 652
Insect-trap	W. Weaver	Phœnixville, Pa	Nov. 26, 1867	71, 346
Insect-trap	J. H. Welch and J. Baker	Fort Wayne, Ind	Feb. 27, 1872	124, 099
Insect-trap	T. Wier	Lacon, Ill	Oct. 10, 1871	119, 905
Insect-trap for protecting fruit while growing	B. M. Quint	Saint Joseph, Mich	May 18, 1869	90, 194
Insects and vermin, Device for destroying	P. Reynard and V. Varin	New York, N. Y	Sept. 20, 1864	44, 341
Insects, Apparatus for killing	H. Hill and L. E. P. Bush	Lexington, Ky	Nov. 26, 1867	71, 486
Insects, Cleansing hair and feathers from	W. Wisdom	Cleveland, Ohio	Dec. 20, 1853	10, 347
Insects, Composition to prevent depredations of	F. G. Johnson	Bellwood, Sag Harbor, N. Y	Mar. 27, 1860	27, 637
Insects, &c., Compound for destroying	E. B. Sears	San Francisco, Cal	Feb. 25, 1873	136, 185
Insects, Construction of apparatus for destroying	A. Isaacsen	New York, N. Y	Aug. 21, 1860	29, 699
Insects injurious to fruit-trees, Composition for destroying.	P. B. Sheldon	Prattsburgh, N. Y	Aug. 30, 1859	25, 281
Insects in trees and plants, Method of destroying	H. A. Graef	Brooklyn, N. Y	Oct. 20, 1868	83, 279
Insects in wheat, Composition for destroying	J. Newcomer	Baltimore, Md	Aug. 4, 1868	80, 655
Insects, Lotion for the destruction of	D. Leibert	Washington, D. C	Aug. 31, 1869	94, 221
Insects on hop-vines and other plants, Process of fumigation for destroying.	J. Deane	Conneaut, Ohio	Dec. 24, 1867	72, 613
Insects on plants, Compound for destroying	W. A. Phillips	Perry Centre, N. Y	Aug. 4, 1868	80, 660
Insects on plants, Mode of destroying	J. Nicolas	La Fourche, La	June 13, 1831	....
Insects on potato-plants, Composition for destroying	J. P. Wilson	Elmwood, Ill	Sept. 22, 1868	82, 468
Insects on trees and plants, Composition for destroying.	J. A. Elias	Le Roy, Ohio	Nov. 27, 1866	59, 986
Insects on trees, Apparatus for destroying	C. Baudouin and A. Fteley	New York, N. Y	Mar. 2, 1869	87, 324
Insects on trees, Apparatus for destroying	J. Hatch	Lynn, Mass	May 14, 1867	64, 667
Insects on trees, Compound for killing	H. D. Flower	Chicago, Ill	Nov. 3, 1868	83, 615
Insects on trees, plants, &c., Compound for destroying.	G. W. Spots	Jacksonville, Ill	Nov. 3, 1868	83, 737
Insects, Trap for destroying	T. C. Silliman	Chester, Conn	Nov. 3, 1868	83, 666
Insole	B. S. Bryant	Hanson, Mass	Feb. 13, 1872	123, 675
Instep-stretcher	C. C. Pease	Jamestown, N. Y	Jan. 21, 1868	73, 546
Instructor, Self	C. Varlé	Baltimore, Md	May 20, 1830	....
Insubmersible boat	A. Du Buc Marentille	....	Dec. 23, 1802	....
Insulating and finishing compound for conducting-wires.	T. L. Reed and E. F. Phillips	Providence, R. I	Nov. 18, 1873	144, 794
Insulating-compound	A. H. Castle	Ann Arbor, Mich	Nov. 23, 1869	97, 045
Insulating-wires for helices	J. J. Clark and H. Splitdorf	East Chester and New York, N. Y.	Sept. 18, 1866	58, 217
Insulation of telegraph line and apparatus, Total	G. W. Nichols	Chicago, Ill	Sept. 10, 1867	68, 779
Insulator	A. H. Castle	Ann Arbor, Mich	Mar. 10, 1868	75, 365
Insulator	A. H. Castle	Ann Arbor, Mich	July 14, 1868	79, 951
Insulator	G. Floyd	Cincinnati, Ohio	May 14, 1867	64, 654
Insulator	F. Scott	Brooklyn, N. Y	Aug. 24, 1869	94, 037
Insulator and bracket, Telegraphic	J. Robertson	Carbondale, Pa	Jan. 30, 1872	123, 198
Insulator, Battery	A. G. Davis	Baltimore, Md	July 16, 1872	129, 465
Insulator, Battery	O. W. Robertson	Milwaukee, Wis	July 12, 1870	105, 252
Insulator, Cramp-hook for telegraphic-wire	W. H. Dechaut	Philadelphia, Pa	July 26, 1870	105, 656
Insulator for battery-cups	J. H. Thomas	Baltimore, Md	Feb. 25, 1873	136, 191

Index of patents issued from the United States Patent Office from 1790 *to* 1873, *inclusive*—Continued.

Index of patents issued from the United States Patent Office from 1790 to 1873, inclusive—Continued.

Invention.	Inventor.	Residence.	Date.	No.
Invalid-chair	A. P. Blunt and J. S. Smith	Washington, D. C	Feb. 16, 1869	86, 899
Invalid-chair	G. T. Fowler	East Somerville, Mass	May 2, 1871	114, 283
Invalid-chair	J. G. Holmes	Charleston, S. C	Sept. 24, 1844	3, 761
Invalid-chair	J. G. Holmes	Charleston, S. C	June 16, 1857	17, 567
Invalid-chair	D. S. James	New Market, Va	May 13, 1856	14, 872
Invalid-chair	G. A. Mansfield	Melrose, Mass	Aug. 25, 1863	39, 663
Invalid-chair	J. N. McMullen	West Liberty, Ohio	Mar. 5, 1867	62, 663
Invalid-chair	C. Messenger	Warren, Ohio	Sept. 25, 1860	30, 149
Invalid-chair	H. F. Siebold	New York, N. Y	Sept. 10, 1872	131, 230
Invalid-chair	C. L. Taillant	New York, N. Y	Sept. 2, 1856	15, 673
Invalid-chair	J. B. Wallace	Franklin, Ohio	Jan. 8, 1867	61, 126
Invalid-chair	L. M. Whitman	Sterling, Ill	Nov. 3, 1868	83, 746
Invalid-chair	R. Witherell	Huntington, Mass	Apr. 7, 1857	17, 008
Invalid chair and lounge	W. Y. Eastes	Summitville, Ind	Aug. 16, 1870	106, 343
Invalid-chair, Conduit for	G. Wells	Bethel, Conn	Mar. 23, 1869	88, 104
Invalid-chair, Traveling	C. L. Bander	Cleveland, Ohio	Nov. 10, 1863	40, 547
Invalid-chair, Traveling	C. L. Bander	Cleveland, Ohio	Dec. 18, 1866	60, 464
Invalid locomotive-chair	T. S. Minniss	Meadville, Pa	May 10, 1853	9, 708
Invalid-rest	T. S. Kennard	Exeter, N. H	Sept. 15, 1868	82, 125
Invalid-supporter	J. T. Alston	Raleigh, N. C	Aug. 12, 1856	15, 504
Invalid-table	J. M. Allen	Worcester, Mass	Mar. 1, 1859	23, 068
Invalid-table	S. Ustick	Philadelphia, Pa	July 18, 1865	48, 854
Invalid-table	T. N. Webb	Baltimore, Md	June 27, 1871	116, 516
Invalid table and apparatus	S. Ustick	Philadelphia, Pa	Sept. 5, 1865	49, 807
Invalid-table and book-holder	P. J. Probeck and J. B. Corlett	Newburgh, Ohio	May 18, 1869	90, 124
Invalids, Apparatus for removing	J. Ruth	Philadelphia, Pa	Oct. 22, 1861	33, 545
Iodine from sea-water, Extracting	R. Cupper	New York, N. Y	Jan. 22, 1867	61, 404
Iodine, Manufacture of	J. Fongerat	New York, N. Y	May 19, 1868	78, 078
Iodine, Process for the manufacture of	J. Fongerat and L. A. Tartierè	Quogue, N. Y	Oct. 27, 1868	83, 372
Iron:				
See Andiron. Axle-iron. Boot and shoe iron. Branding-iron. Carriage-iron. Chafe-iron. Clinching-iron. Clothes-iron. Curling-iron. Draw-bar iron. Edge-iron. Flat-iron. Fluting-iron. Fluting and puffing iron. Fluting and sad iron. Fore-iron. Gas and fluting iron. Glossing and fluting iron. Grappling-iron. Hat-shell iron. Hod-iron. Holdback-iron. Lasting-iron. Laundry-iron. Mop-iron. Plane-iron. Plow-iron. Polishing-iron. Puffing-iron. Rub-iron. Sad-iron. Sad and embossing iron. Sad and fluting iron. Shaft-iron. Sheet-iron. Smoothing-iron. Soldering-iron. Tew-iron. Tuyere-iron. Wafer-iron. Waffle-iron.				
Iron against corrosion, Protecting	C. Godfrey and R. Lighthall	New York, N. Y	Nov. 22, 1870	109, 509
Iron and apparatus therefor, Manufacture of	J. D. Whelpley and J. J. Storer	Boston, Mass	Mar. 22, 1870	101, 067
Iron and castings, Manufacturing sheet, boiler, slit, and bar.	N. Callender	Cumberland County, Pa	Dec. 14, 1832	....
Iron and copper, Process of purifying	K. W. Zenger	Prague, Austria	June 18, 1872	128, 088
Iron and copper with tin, &c., Coating	E. P. Morewood	New York, N. Y	Sept. 17, 1844	3, 746
Iron and cutting nails, Machine for rolling and plating.	S. Couch and A. Towne	Boston, Mass	Dec. 10, 1808	....
Iron and granulating the same, Manufacture of	W. H. Perry	Sharon, Pa	Feb. 8, 1870	99, 588
Iron and making castings, Method of refining cast	Z. S. Durfee	New York, N. Y	Aug. 29, 1871	118, 597
Iron and other material to give greater strength for building, &c., Treating.	W. Haggett	Watford, England	Sept. 3, 1872	130, 992
Iron and other metals, Composition of	W. M. Arnold	New York, N. Y	June 12, 1866	55, 452
Iron and other metals, Construction of converter and furnace for treating.	J. Absterdam	New York, N. Y	Feb. 9, 1869	86, 796
Iron and other metals from potter's clay, Process of separating.	W. J. Lynd	Golden City, Colo	May 25, 1869	90, 565
Iron and other metals with protecting alloys, Coating.	C. Marshall	Philadelphia, Pa	Nov. 11, 1873	144, 403
Iron and other ores, Furnace and process for treating.	J. Y. Smith	Pittsburgh, Pa	Nov. 15, 1870	109, 355
Iron and other ores, Reduction of	H. S. Lucas	Chester, Mass	Nov. 1, 1864	44, 872
Iron and other oxides from clay, porcelain-earth, &c., Method of extracting.	W. J. Lynd	Golden City, Colo	July 27, 1869	92, 981
Iron and removing scale from iron for coating with other metals, Manufacture of sheet.	C. Marshall	Philadelphia, Pa	May 16, 1871	114, 956
Iron and steel and in the deoxidizing of iron-ores, Process of manufacturing malleable cast.	A. K. Eaton	New York, N. Y	June 25, 1861	32, 621
Iron and steel, Apparatus for refining	T. S. C. Lowe	Morristown, Pa	Aug. 13, 1872	130, 380
Iron and steel, Apparatus for the manufacture of	C. Adams	Philadelphia, Pa	Dec. 21, 1869	98, 139
Iron and steel, Apparatus for the manufacture of	W. Gerhardt	Pittsburgh, Pa	Feb. 13, 1866	52, 559
Iron and steel, Apparatus for the manufacture of	A. L. Holley and J. B. Pease	Swatara Township, Pa	Jan. 26, 1869	86, 304
Iron and steel bar, Combined	E. Wheeler	Hudson, Mass	June 14, 1870	104, 237
Iron and steel by applying anthracite coal, Manufacturing.	F. W. Geissenhainer	New York	Dec. 19, 1833	....
Iron and steel by means of blasts of air, Purifying	G. W. Swett	Troy, N. Y	June 30, 1863	39, 078
Iron and steel by rolling, Manufacturing	R. W. Bangs and S. W. Walbridge.	Bennington, Vt	Dec. 31, 1833	....
Iron and steel by the Bessemer or pneumatic process, Manufacture of.	A. L. Holley	Swatara Township, Pa	Jan. 26, 1869	86, 303
Iron and steel, Carbonizing and hardening	F. E. Sessions	Worcester, Mass	Mar. 24, 1868	75, 986
Iron and steel castings, Manufacture of	R. Yeilding	New York, N. Y	Oct. 3, 1871	119, 682
Iron and steel combined, Manufacture of bars and articles of.	W. M. Pickslay	Philadelphia, Pa	Nov. 13, 1866	59, 644
Iron and steel, Composition-flux for manufacture of	J. Jameson	Philadelphia, Pa	Sept. 7, 1869	94, 605
Iron and steel, Compound for hardening	J. McDonald	Kankakee, Ill	Dec. 5, 1871	121, 645

Index of patents issued from the United States Patent Office from 1790 *to* 1873, *inclusive*—Continued.

Invention.	Inventor.	Residence.	Date.	No.
Iron and steel, Compound for hardening	C. Pauvert	Targé, France	Mar. 23, 1858	19,710
Iron and steel direct from the ore, Manufacture of wrought.	W. Henderson	Glasgow, Scotland	Nov. 27, 1866	60,002
Iron and steel direct from the ore, Manufacture of wrought.	G. H. Smith	New York, N. Y	July 20, 1869	92,894
Iron and steel directly from the ore, Manufacture of	C. M. Dupuy	New York, N. Y	Feb. 28, 1865	46,549
Iron and steel, Electro-deposition of copper and brass in.	W. H. Walenn	London, England	June 7, 1870	103,947
Iron and steel, Flux for refining	E. T. Atwood	Minerva, Ohio	Apr. 25, 1871	114,090
Iron and steel, Flux for the manufacture of	L. Sibert	Staunton, Va	Aug. 1, 1871	117,693
Iron and steel, Flux for welding, puddling, and brazing.	A. J. Hindermeyer	Rohrerstown, Pa	Dec. 18, 1866	60,516
Iron and steel from granulated iron, Manufacture of.	C. Wood	Middlesborough-on-Tees, England.	Oct. 28, 1873	144,009
Iron and steel from iron sponge, Manufacture of wrought.	T. S. Blair	Pittsburgh, Pa	May 21, 1872	126,923
Iron and steel from rust, Preserving	M. Sorel	Paris, France	Dec. 7, 1837	510
Iron and steel furnace	T. J. Chubb	Williamsburgh, N. Y	June 8, 1869	90,924
Iron and steel furnace	G. E. Harding	New York, N. Y	Sept. 23, 1873	143,145
Iron and steel furnace	J. W. Nystrom	Philadelphia, Pa	Oct. 8, 1861	33,446
Iron and steel furnace	A. Pousard	Paris, France	Oct. 27, 1868	83,542
Iron and steel furnace	J. M. Roberts	Burlington, N. J	Nov. 28, 1871	121,422
Iron and steel furnace	H. Ross and D. F. Agnew	Pittsburgh, Pa	July 13, 1869	92,478
Iron and steel furnace	H. Ross and J. H. Clemens	Pittsburgh, Pa	Aug. 22, 1871	118,279
Iron and steel furnace	J. G. Trotter	Newark, N. J	Mar. 16, 1869	87,889
Iron and steel in furnace, and apparatus therefor, Manufacture of.	G. J. and T. C. Hinde	Wolverhampton, England, and Ynispenllweh, Wales.	Mar. 30, 1869	88,480
Iron and steel furnace and manufacture	T. C. Hinde	Fownhope, England	Dec. 12, 1871	121,872
Iron and steel, Furnace and process for the manufacture of.	F. Ellershausen	Montreal, Canada	May 12, 1868	77,722
Iron and steel, Furnace and tool for treating	W. Yates	Westminster, England	Feb. 23, 1869	87,231
Iron and steel, Furnace for converting cast-iron into.	W. Fields	Wilmington, Del	May 24, 1870	103,437
Iron and steel, Hardening	G. J. Farmer	Birmingham, England	Apr. 6, 1858	19,836
Iron and steel in run-ways, Refining	S. M. Wickersham	Pittsburgh, Pa	Dec. 3, 1872	133,690
Iron and steel, &c., Increasing the strength of wrought.	W. R. Johnson	Philadelphia, Pa	June 30, 1838	814
Iron and steel, Increasing the strength of wrought	W. R. Johnson	Philadelphia, Pa	July 9, 1838	826
Iron and steel, Machinery and buildings for manufacture of.	H. Bessemer	London, England	Aug. 15, 1871	117,968
Iron and steel, Machinery for the manufacture of	H. Bessemer	London, England	July 25, 1865	49,055
Iron and steel, Machinery to aid in puddling	H. Bennett	Wombridge, England	Feb. 23, 1864	41,671
Iron and steel, Making	W. P. Boyden	New York, N. Y	June 11, 1836	....
Iron and steel, Making	H. G. Spafford	Albany, N. Y	Oct. 30, 1822	
Iron and steel, Manufacture of	G. F. Ansell	London, England	June 28, 1870	104,686
Iron and steel, Manufacture of	H. Bessemer	London, England	Nov. 11, 1856	16,082
Iron and steel, Manufacture of	H. Bessemer	London, England	July 25, 1865	49,051
Iron and steel, Manufacture of	H. Bessemer	London, England	July 25, 1865	49,052
Iron and steel, Manufacture of	H. Bessemer	London, England	July 25, 1865	49,053
Iron and steel, Manufacture of	H. Bessemer	London, England	Dec. 5, 1865	51,397
Iron and steel, Manufacture of	H. Bessemer	London, England	Dec. 5, 1865	51,398
Iron and steel, Manufacture of	H. Bessemer	London, England	Sept. 21, 1869	94,994
Iron and steel, Manufacture of	H. Bessemer	London, England	Sept. 21, 1869	94,995
Iron and steel, Manufacture of	H. Bessemer	London, England	Sept. 21, 1869	94,996
Iron and steel, Manufacture of	H. Bessemer	London, England	Sept. 21, 1869	94,997
Iron and steel, Manufacture of	O. Bolton, jr., and J. Pedder	Pittsburgh, Pa	July 15, 1873	140,761
Iron and steel, Manufacture of	J. B. Bradley and E. De Camp	Morristown, N. J	Aug. 27, 1872	130,787
Iron and steel, Manufacture of	J. P. Budd	Ystalyfera, near Swasea, Wales.	May 10, 1870	102,912
Iron and steel, Manufacture of	C. Carpenter, jr	Seekonk, Mass	Dec. 23, 1873	145,843
Iron and steel, Manufacture of	N. Cutter and E. Savage	Cincinnati, Ohio, and West Meriden, Conn.	Oct. 5, 1869	95,568
Iron and steel, Manufacture of	Z. S. Durfee	New York, N. Y	May 2, 1871	114,277
Iron and steel, Manufacture of	Z. S. Durfee	New York, N. Y	Jan. 2, 1872	122,312
Iron and steel, Manufacture of	F. Ellershausen, A. E. Stayner, and A. Guzman.	Ellershouse, Halifax, Nova Scotia, and New York, N. Y.	Nov. 17, 1868	84,053
Iron and steel, Manufacture of	W. Ennis	Philadelphia, Pa	Dec. 14, 1869	97,897
Iron and steel, Manufacture of	F. P. Fletcher and V. W. Blanchard.	Bridport, Vt	Nov. 16, 1869	96,905
Iron and steel, Manufacture of	A. L. Fleury and C. Adams	Philadelphia and Pittsburgh, Pa.	May 13, 1862	35,276
Iron and steel, Manufacture of	J. L. Floyd	Philadelphia, Pa	June 15, 1869	91,324
Iron and steel, Manufacture of	W. Gerhardt	Pittsburgh, Pa	Feb. 13, 1866	52,560
Iron and steel, Manufacture of	J. Henderson	Brooklyn, N. Y	Oct. 17, 1865	50,474
Iron and steel, Manufacture of	J. Henderson	New York, N. Y	Aug. 17, 1869	93,713
Iron and steel, Manufacture of	J. Henderson	New York, N. Y	Mar. 29, 1870	101,263
Iron and steel, Manufacture of	J. Henderson	New York, N. Y	Aug. 16, 1870	106,365
Iron and steel, Manufacture of	J. Jameson	Philadelphia, Pa	Mar. 23, 1869	88,173
Iron and steel, Manufacture of	J. J. Johnston	Allegheny City, Pa	July 27, 1869	93,155
Iron and steel, Manufacture of	J. A. Jones	Middlesborough, England	June 9, 1868	78,806
Iron and steel, Manufacture of	J. A. Jones	Middlesborough, England	Sept. 8, 1868	81,908
Iron and steel, Manufacture of	C. Low	England	May 28, 1846	4,540
Iron and steel, Manufacture of	J. G. Martien	Newark, N. J	Feb. 24, 1857	16,690
Iron and steel, Manufacture of	J. W. Middleton	Philadelphia, Pa	Mar. 21, 1871	112,828
Iron and steel, Manufacture of	R. Mushet	Coleford, England	May 26, 1857	17,389
Iron and steel, Manufacture of	C. M. Nes	York, Pa	Nov. 29, 1870	109,752
Iron and steel, Manufacture of	C. M. Nes	York, Pa	Jan. 30, 1872	123,191
Iron and steel, Manufacture of	G. Parry	Ebro Vale Iron-Works, England.	Apr. 25, 1865	47,506
Iron and steel, Manufacture of	E. Peckham	Antwerp, N. Y	Oct. 14, 1873	143,637
Iron and steel, Manufacture of	J. Pedder and G. Abel	Pittsburgh, Pa	Apr. 8, 1873	137,621
Iron and steel, Manufacture of	O. M. Phillips	New York, N. Y	Feb. 9, 1869	86,859
Iron and steel, Manufacture of	J. Player	Philadelphia, Pa	Oct. 19, 1869	95,933
Iron and steel, Manufacture of	J. Ralston, A. L. Thomas, and W. Parkenson.	Tamaqua, Pa	Mar. 23, 1869	88,208
Iron and steel, Manufacture of	L. Sibert	Mount Solon, Va	May 21, 1867	64,916
Iron and steel, Manufacture of	L. Sibert	Mount Solon, Va	June 23, 1868	79,152

Index of patents issued from the United States Patent Office from 1790 to 1873, inclusive—Continued.

Invention.	Inventor.	Residence.	Date.	No.
Iron and steel, Manufacture of	L. Sibert	Staunton, Va	Oct. 4, 1870	107, 058
Iron and steel, Manufacture of	C. W. Siemens	Westminster, England	Oct. 12, 1869	95, 843
Iron and steel, Manufacture of	G. H. Smith	New York, N. Y	July 12, 1870	105, 267
Iron and steel, Manufacture of	T. Southall and C. Crudgington	Kidderminster, England	Sept. 14, 1844	3, 742
Iron and steel, Manufacture of	H. Spencer and L. K. Saylor	Philadelphia, Pa	Dec. 7, 1869	97, 718
Iron and steel, Manufacture of	C. M. Tessié du Motay	Paris, France	Mar. 2, 1869	87, 479
Iron and steel, Manufacture of	C. Usher	Iowa Falls, Iowa	Mar. 5, 1867	62, 706
Iron and steel, Manufacture of	R. Yeilding	Detroit, Mich	Oct. 13, 1868	83, 119
Iron and steel, Manufacture of	H. K. York	Cardiff, Great Britain	Dec. 24, 1867	72, 579
Iron and steel, Manufacture of malleable	H. Bessemer	London, England	Dec. 5, 1865	51, 399
Iron and steel, Manufacture of malleable	H. Bessemer	London, England	Dec. 5, 1865	51, 400
Iron and steel, Manufacture of malleable	H. Bessemer	London, England	Dec. 5, 1865	51, 401
Iron and steel, Manufacture of malleable	J. C. Ridley	Newcastle-upon-Tyne, England.	Aug. 1, 1871	117, 684
Iron and steel, Manufacture of malleable	E. B. Wilson	Westminster, England	July 28, 1863	39, 364
Iron and steel, Manufacture of pure	H. Larkin, A. Leighton, and W. White.	Theydon Germon, Liverpool, and Hampstead, England.	Apr. 9, 1872	125, 464
Iron and steel, Manufacturing	O. Bolton, jr., and J. Pedder	Pittsburgh, Pa	Feb. 4, 1873	135, 512
Iron and steel, Metal plates of	J. Myers, jr	Williamsburgh, N. Y	Oct. 4, 1870	108, 042
Iron and steel, Metallic solution for coating	A. A. Lothrop	Neponset, Mass	Nov. 30, 1869	97, 417
Iron and steel, Method of uniting	W. and W. H. Terwilliger and J. S. Lockwood.	New York, N. Y	July 24, 1866	56, 680
Iron and steel, Mode of making bars, shafts, &c., composed of.	C. Sanderson	Sheffield, England	Sept. 19, 1865	50, 084
Iron and steel, Mode of plating, coating, and ornamenting articles of.	K. Frazer	Syracuse, N. Y	Oct. 26, 1869	96, 217
Iron and steel, Mode of utilizing tin-plate cuttings in the manufacture of.	D. D. Parmelee	New York, N. Y	June 29, 1869	91, 962
Iron and steel, Preventing corrosion of	R. A. Fisher	San Francisco, Cal	Dec. 16, 1873	145, 496
Iron and steel, Process and apparatus for making	J. Jameson	Philadelphia, Pa	Mar. 30, 1869	88, 299
Iron and steel, Process and apparatus for the manufacture of.	H. Bessemer	London, Great Britain	Feb. 22, 1870	100, 003
Iron and steel, Process and apparatus for the manufacture of.	J. Jameson	Philadelphia, Pa	Mar. 31, 1868	76, 196
Iron and steel, Process and apparatus for the manufacture of.	J. W. Middleton	Philadelphia, Pa	July 16, 1872	129, 243
Iron and steel, Process for refining	J. Absterdam	New York, N. Y	Jan. 23, 1866	52, 121
Iron and steel, Process of making	M. Lane	Washington, D. C	Dec. 17, 1861	33, 949
Iron and steel, Process of manufacturing axles, &c., from.	H. Bessemer	London, England	July 25, 1865	49, 054
Iron and steel, Process of mixing	J. Cartwright	Youngstown, Ohio	Jan. 7, 1868	73, 163
Iron and steel, Process of purifying	J. F. Bennett	Pittsburgh, Pa	Mar. 10, 1868	75, 240
Iron and steel, Process of treating	E. Savage	West Meriden, Conn	Feb. 16, 1869	86, 944
Iron and steel puddling machinery	T. Harrison	Tudhoe, England	June 7, 1864	43, 067
Iron and steel, &c., Purifying	E. Brady	Philadelphia, Pa	Oct. 5, 1869	95, 419
Iron and steel, Purifying	W. Gerhardt	New York, N. Y	Nov. 3, 1863	40, 472
Iron and steel, Refining	J. E. Atwood	Pittsburgh, Pa	Sept. 13, 1870	107, 322
Iron and steel, Refining	N. B. Hatch	Pittsburgh, Pa	Dec. 17, 1872	133, 937
Iron and steel Refining	W. H. Kimball	Boston, Mass	June 20, 1871	116, 065
Iron and steel, Refining	I. M. Phelps	Philadelphia, Pa	Dec. 16, 1873	145, 680
Iron and steel, Refining	A. H. Siegfried	South Bend, Ind	Apr. 4, 1871	113, 583
Iron and steel shavings, turnings, &c., Melting	G. Whitney	Philadelphia, Pa	Nov. 5, 1872	132, 743
Iron and steel to anneal and toughen them, Treating	A. F. Andrews	New Haven, Conn	Sept. 23, 1873	142, 977
Iron and steel, Welding	J. Popping	New York, N. Y	Sept. 16, 1873	142, 939
Iron and steel wire, Annealing	G. I. Washburn	Worcester, Mass	Oct. 7, 1862	36, 628
Iron and steel with copper, brass, &c., Process of uniting.	R. Savary	Pittsburgh, Pa	Aug. 11, 1863	39, 531
Iron and steel with gold, silver, &c., Coating	J. B. Thompson	Middlesex County, England	Sept. 11, 1866	58, 037
Iron and steel with molten iron, Coating	R. B. Fowler and D. F. Brandon	Chicago, Ill	Mar. 28, 1871	113, 040
Iron and the separation of copper, silver, and other metals from their solutions, Preparing finely divided.	G. Bishop, jr	Swansea, Great Britain	Feb. 25, 1868	74, 791
Iron, Annealing and polishing sheet	J. W. Ells	Pittsburgh, Pa	July 25, 1865	48, 918
Iron, Annealing sheet	W. D. Wood	Borough of McKeesport, Pa	July 9, 1867	66, 546
Iron, Apparatus and machinery for puddling	P. A. Dormoy	Vienna, Austria	May 18, 1869	90, 158
Iron, Apparatus for heating the converters and other vessels used in the Bessemer process of treating.	C. G. Larson	Stockholm, Sweden	June 20, 1871	116, 201
Iron, Apparatus for making pig-blooms in the manufacture of.	J. A. Burden	Troy, N. Y	Aug. 3, 1869	93, 170
Iron, Apparatus for puddling	J. Davies	Knoxville, Tenn	Oct. 21, 1873	143, 811
Iron, Apparatus for puddling	J. Griffiths	Litchurch, England	June 27, 1865	48, 485
Iron, Apparatus for puddling	J. McCarty	Reading, Pa	Oct. 5, 1852	9, 303
Iron, Apparatus for puddling and melting	H. A. V. Post	New York, N. Y	Aug. 6, 1872	130, 241
Iron, Apparatus for purifying	S. M. Wrickersham	Allegheny, Pa	Dec. 14, 1869	97, 844
Iron, Apparatus for refining	J. W. Niptrom	Philadelphia, Pa	Apr. 3, 1866	53, 658
Iron, Apparatus for the manufacture of	J. Yates	Mott Haven, N. Y	Mar. 1, 1864	41, 806
Iron, Apparatus for the manufacture of	J. Yates	Mott Haven, N. Y	Mar. 1, 1864	41, 808
Iron, Apparatus for the manufacture of malleable	A. Dickerson	Newark, N. J	Mar. 13, 1847	5, 013
Iron, Apparatus for the manufacture of pig-bloom from cast.	J. A. Burden	Troy, N. Y	Aug. 17, 1869	93, 857
Iron articles into steel, Process for converting	R. A. Jackson	Pittsburgh, Pa	Feb. 11, 1873	135, 646
Iron at the rolls, Machine for moving	C. Hewitt	Trenton, N. J	June 7, 1859	24, 304
Iron axles, Gage for	W. C. Bamberger	Washington, D. C	Nov. 15, 1859	26, 079
Iron, Balanced "heave-up" for rolling	J. Copley, jr	Allegheny City, Pa	Sept. 24, 1867	69, 077
Iron balls, blooms, and slabs, Manufacture of malleable.	J. Player	Norton, England	Apr. 28 1868	77, 210
Iron bands, Machine for contracting the circumference of wrought.	W. Massey	Greene County, Ill	July 3, 1849	6, 573
Iron bars and rods, Polishing	B. Lauth	Pittsburgh, Pa	Feb. 26, 1861	31, 546
Iron bars, Constructing cast or wrought	C. McCammon	Albany, N. Y	Feb. 14, 1860	27, 1[illegible]
Iron bars, &c., Furnace for heating	J. Pardoe	Worcester, Mass	Mar. 11, 1873	136, 667
Iron bars, Hollowing out	T. Rowand	Gloucester County, Pa	Dec. 24, 1819	......
Iron bars, Machine for straightening	G. H. Sellers	Phœnixville, Pa	Mar. 22, 1864	42, 051
Iron bars, Rolling	B. Lauth	Pittsburgh, Pa	Aug. 21, 1860	29, 702
Iron beam, Wrought	L. Kirkup	East New York, N. Y	Apr. 22, 1873	•138, 029

Index of patents issued from the United States Patent Office from 1790 to 1873, inclusive—**Continued.**

Invention.	Inventor.	Residence.	Date.	No.
Iron beams, Machine for forming flanges on wrought.	J. H. Kroehl	New York, N. Y	Jan. 2, 1855	12, 133
Iron boat for navigating rivers, &c	T. J. Bond	Baltimore, Md	Dec. 21, 1820	....
Iron boiler, Cast	J. A. Miller	New York, N. Y	Oct. 24, 1865	50, 611
Iron boilers, Machine for bending sheet	C. Post	Springport, N. Y	Mar. 11, 1828	....
Iron boiling and puddling furnace	J. Westerman	Sharon, Pa	May 24, 1870	103, 401
Iron bolt, Malleable cast	P. Frost	Springfield, Vt	Jan. 26, 1869	86, 220
Iron borings, drillings, and filings rendered available in furnaces.	M. Maison	Georgetown, D. C	Nov. 8, 1831	....
Iron, brass, &c., Mold for casting	J. Leonard, jr., J. S. Brainard, and A. Sizer.	Wallingford, Conn	Jan. 28, 1830	....
Iron by means of fluxes introduced by suction, Refining and purifying.	J. Absterdam	New York, N. Y	Jan. 19, 1869	86, 050
Iron by the "Ellershausen process," Manufacture of.	H. Davies	Newport, Ky	July 5, 1870	104, 939
Iron, Carbonating and smelting	A. Whitney	Southfield, N. Y	July 17, 1837	269
Iron, Casting	W. Hancock		Feb. 26, 1799	......
Iron chills, molds, pig-beds, &c., Cold-fix for lining	H. A. Laughlin	Pittsburgh, Pa	Jan. 5, 1869	85, 529
Iron chips for remelting, Process for preparing	A. Pevey	Portland, Me	July 1, 1873	140, 383
Iron chips, shavings, &c., for melting, Preparing	E. C. Haserick	Lake Village, N. H	Nov. 28, 1871	121, 245
Iron chips, turnings, &c., Mode of melting and aggregating.	E. C. Haserick	Lake Village, N. H	Mar. 13, 1866	53, 142
Iron-cleaning machine, Sheet	E. A. Harvey	Wilmington, Del	Apr. 4, 1865	47, 103
Iron, Coating	G. A. Mariner and F. Dorsett	Chicago, Ill	Aug. 15, 1871	118, 032
Iron, Coating and bronzing	L. Dockstader	West Meriden, Conn	Apr. 26, 1870	102, 378
Iron, Composition	W. M. Arnold	New York, N. Y	Aug. 23, 1864	43, 891
Iron, Composition and manufacture of	A. H. Everitt	New York, N. Y	Apr. 11, 1865	47, 198
Iron, Composition for carbonizing	C. V. Wilson	Newark, N. J	Aug. 2, 1864	43, 732
Iron, Composition for cementing	J. Johnson	East Brooklyn, N. Y	June 14, 1859	24, 429
Iron, Composition for coating	D. Read and J. F. Galley	New York, N. Y	Mar. 20, 1866	53, 336
Iron, Composition for covering and protecting	Y. Bey	New York, N. Y	Mar. 31, 1863	38, 022
Iron, Composition for refining and toughening	J. H. Wolf	Philadelphia, Pa	May 12, 1868	77, 904
Iron, Composition for treating	S. Weaver	Pottstown, Pa	Mar. 28, 1871	113, 228
Iron, Composition to be used in puddling	I. E. Richards	Columbia, Pa	Oct. 29, 1867	70, 361
Iron, Compound for coating	J. A. Sewall	Normal, Ill	Aug. 22, 1871	118, 397
Iron, Compound for hardening bar	F. E. Blake	Mattoon, Ill	Mar. 26, 1872	125, 009
Iron, Compound for hardening cast	B. W. Nichols	Canton, Ohio	May 11, 1869	90, 014
Iron, Compound for improving the quality of	T. Jones and W. Morgan	Pittsburgh, Pa	Apr. 30, 1867	64, 338
Iron, Corrugating	J. L. Vernam	Rochester, N. Y	Jan. 1, 1861	31, 068
Iron cutting and punching machine	A. J. Peavey	South Montville, Me	Jan. 4, 1859	22, 510
Iron cutting and punching machine	A. Shogren	Mission, Ill	Sept. 1, 1863	39, 757
Iron, Cutting, punching, and slitting	J. V. Buskark	Norwalk, Ohio	Apr. 6, 1833	....
Iron-cutting shears	J. Nichol	New York, N. Y	May 18, 1869	90, 184
Iron, Decarbonizing	J. F. Allen	Tremont, N. Y	Apr. 14, 1868	76, 581
Iron direct from the ore, Apparatus for making wrought.	J. Renton	Newark, N. J	Dec. 23, 1851	8, 613
Iron direct from the ore, Making wrought	J. Renton	Cleveland, Ohio	Oct. 24, 1854	11, 838
Iron direct from the ore, Process for making malleable.	M. S. Salter	Newark, N. J	Nov. 20, 1849	6, 886
Iron directly from the ore, Combination of furnaces for manufacturing wrought.	C. S. Quilliard	Rondout, N. Y	Dec. 23, 1841	2, 394
Iron directly from the ore, Manufacture of wrought	C. M. Dupuy	New York, N. Y	Jan. 23, 1866	52, 149
Iron directly from the ore, Manufacturing malleable	G. A. Whipple	Newark, N. J	May 10, 1853	9, 715
Iron directly from the ore, Method of making wrought.	A. Dickerson	Newark, N. J	July 22, 1850	7, 519
Iron directly from the ore, Mode of manufacturing malleable.	J. F. Winslow	Troy, N. Y	May 16, 1846	4, 526
Iron directly from the ore, Process of obtaining wrought.	J. D. Whelpley and J. J. Storer	Boston, Mass	Sept. 28, 1869	95, 295
Iron-drilling machine	J. H. Currier and W. H. Taber	Fairhaven, Mass	July 8, 1839	1, 224
Iron-drilling machine	I. S. Lauback	New York, N. Y	Feb. 9, 1864	41, 516
Iron, Finery-fire for the manufacture of	T. Ash and Z. Walker	Trenton, N. J	Nov. 25, 1873	144, 821
Iron-finishing furnace, Sheet	W. D. Wood	McKeesport, Pa	Mar. 14, 1865	46, 841
Iron, Fire-plating	W. H. Thoss	San Francisco, Cal	Aug. 30, 1859	25, 291
Iron, Flux for puddling	J. Burnish, J. Talbot, and T. W. Yardley.	Pottsville, Pa	Feb. 1, 1859	22, 779
Iron for car-wheels, &c., Manufacture of cast	C. Burgess	Portsmouth, Ohio	Oct. 21, 1873	143, 874
Iron for fronts of buildings, &c., Coating	J. Alexander	Green Point, N. Y	Sept. 21, 1869	95, 064
Iron for making steel and malleable cast-iron, Mode of treating pig.	J. R. Speer	Pittsburgh, Pa	July 6, 1869	92, 221
Iron for malleable-iron castings, Process of melting and refining.	T. J. Chubb	Williamsburgh, N. Y	June 8, 1869	90, 927
Iron for strap-joint	W. D. Rinehart	Pittsburgh, Pa	Aug. 29, 1869	49, 683
Iron for the manufacture of stoves, pipes, &c., Machine for punching sheet.	S. Davis	Mifflin, Pa	Feb. 21, 1840	1, 495
Iron forging and crushing apparatus	R. Morrison	Newcastle-upon-Tyne, Great Britain.	Mar. 4, 1862	34, 587
Iron, &c., Forming oblique catches, &c., on plates and other pieces of cast.	J. S. Mott	New York, N. Y	Dec. 1, 1840	1, 872
Iron from blast-furnace slag, Manufacture of	J. J. Vinton	Sharon, Pa	Oct. 14, 1873	143, 600
Iron from brass turnings and filings, Magnetic machine for separating.	J. Jouson	Baltimore, Md	Aug. 27, 1867	68, 205
Iron from corrosion, Preserving	C. De Bussey	Paris, France	July 19, 1864	43, 630
Iron from corrosion, Preserving	G. W. Holley	Niagara, N. Y	Dec. 1, 1863	40, 752
Iron from furnace-cinders, Machinery for separating.	O. Walroth and L. Evans	Chittenango and Manlius, N. Y.	Dec. 21, 1852	9, 493
Iron from gas-purifiers, Treating and utilizing oxides of.	W. Cleland	Liverpool, England	May 24, 1864	42, 898
Iron from mill-cinders, Manufacture of	C. M. Nes	York, Pa	Apr. 2, 1872	125, 212
Iron from ore-cinders or slag, Manufacture of wrought.	J. D. Whelpley and J. J. Storer	Boston, Mass	May 3, 1870	102, 740
Iron from oxidation, Method of protecting	E. G. Pomroy	New York, N. Y	June 28, 1859	24, 604
Iron from oxidation, Process of protecting	P. Sumner and P. Naylor	New York, N. Y	Oct. 18, 1839	1, 374
Iron from pig-metal, Manufacturing malleable	T. C. Lewis	Pittsburgh, Pa	Oct. 1, 1830	....
Iron from sand, Machine for separating	G. H. Sanborn	Boston, Mass	Jan. 22, 1867	61, 471
Iron from the fire, Apparatus for drawing	D. N. Williams	Chicago, Ill	Mar. 19, 1867	63, 125
Iron from the ore, Furnace for making	M. Bell and E. B. Isett	Sabbath Rest and Tyrone City, Pa.	Nov. 14, 1854	11, 927

Index of patents issued from the United States Patent Office from 1790 to 1873, inclusive—Continued.

Invention.	Inventor.	Residence.	Date.	No.
Iron from the ore, Furnace for manufacturing wrought.	T. W. Harvey	New York, N. Y	Aug. 22, 1854	11, 584
Iron from the ore, Manufacture of cast	P. E. Jay	Saint Jean Baptiste, Canada.	Sept. 30, 1873	143, 350
Iron from the ore, Manufacture of purified	J. W. Middleton	Philadelphia, Pa	Oct. 4, 1870	107, 942
Iron from the ore, Manufacture of wrought	H. Boardman	New York, N. Y	June 27, 1865	48, 478
Iron from the ore, Manufacturing malleable	W. N. Clay	Flimby, England	July 25, 1845	4, 103
Iron from the ore, Mode of obtaining wrought	S. Broadmeadow	New York, N. Y	May 30, 1844	3, 605
Iron from the ore, Process of making cast	P. E. Jay	Saint Jean Baptiste, Canada.	Dec. 24, 1872	134, 289
Iron from the slag of blast-furnaces, Extracting	E. J. Bird	Frostburgh, Md	Oct. 11, 1870	108, 235
Iron from the water of salt wells and springs, Mode of separating compounds of.	G. H. Cook	New Brunswick, N. J	Oct. 29, 1861	33, 574
Iron from titaniferous iron-ore, Manufacture of	J. Player	New York, N. Y	July 7, 1868	79, 681
Iron-furnace	A. Hamar	New York, N. Y	May 12, 1868	77, 730
Iron-furnace	B. B. Howell	Philadelphia, Pa	Oct. 11, 1828	....
Iron, Furnace and oven for the manufacture of	J. Yates	Mott Haven, N. Y	Mar. 1, 1864	41, 807
Iron, Furnace for boiling	C. D. Baker	Wheeling, W. Va	June 27, 1865	48, 355
Iron, Furnace for making	D. W. Hendrickson and J. P. McLean.	New York, N. Y	Sept. 3, 1867	68, 566
Iron, Furnace for making malleable	M. S. Salter	Saltersville, N. J	Nov. 24, 1863	40, 710
Iron, Furnace for manufacture of	A. Reese	Pittsburgh, Pa	Jan. 9, 1872	122, 651
Iron, Furnace for manufacturing	I. C. Bryant	Philadelphia, Pa	Dec. 31, 1838	1, 057
Iron, Furnace for producing malleable	D. R. Nash	Brooklyn, N. Y	Sept. 9, 1873	142, 716
Iron-furnaces, Machine for charging	E. G. Scovil	Saint John, Canada	May 2, 1871	114, 346
Iron-galvanizing furnace	W. Blake	Boston, Mass	Mar. 19, 1861	31, 696
Iron gate or fence post	T. E., A., and E. King	Cherry Valley, Ohio	May 25, 1858	20, 352
Iron girder	P. H. Jackson	New York, N. Y	Nov. 28, 1871	121, 374
Iron, Hardening	W. C. Dunn	La Porte, Ind	July 24, 1866	56, 539
Iron, Hardening	H. Paddock	Saint Johnsbury, Vt	Feb. 12, 1867	61, 949
Iron, Hardening cast	T. Allin	New York, N. Y	Apr. 26, 1864	42, 453
Iron-hardening process	T. H. Jenkins	New York, N. Y	Dec. 26, 1865	51, 723
Iron, Hardening the surface of	R. T. King	Pana, Ill	Dec. 2, 1873	145, 110
Iron, Hearth for working and refining	R. S. Stewart, J. Christopher, and R. Forward.	Ligonier and Somerset, Pa	Mar. 1, 1859	23, 123
Iron-heating furnace, Bar	H. Burden	Troy, N. Y	Oct. 14, 1834	....
Iron-heating furnaces, Charging-apparatus for	W. F. Maharg	Saint Louis, Mo	Dec. 24, 1872	134, 151
Iron-holder	W. B. Coates	Philadelphia, Pa	Jan. 23, 1866	52, 140
Iron hoops, Straightening	B. Seymour		June 26, 1797	
Iron house	D. D. Badger	New York, N. Y	Aug. 7, 1855	13, 379
Iron houses, Construction of	C. Mettam	New York, N. Y	July 11, 1854	11, 290
Iron in casting chilled rolls, Method of giving a rotary motion to the melted.	J. C. Parry	Pittsburgh, Pa	Oct. 16, 1849	6, 805
Iron in sea-water, Mode of preventing corrosion in pipes, bolts, and similar articles of.	R. Lighthall	Brooklyn, N. Y	Dec. 7, 1869	97, 657
Iron in the hearth of a blast-furnace, Refining	C. Shunk	Canton, Ohio	May 17, 1859	24, 060
Iron into bar iron and steel, Process of converting cast.	J. Heaton	Langley Mill, England	Aug. 13, 1867	67, 762
Iron into malleable iron, Converting cast	A. Hamer and G. H. Sellers	Philadelphia, Pa., and Wilmington, Del.	Feb. 2, 1869	86, 537
Iron into steel and malleable iron, Process of converting cast.	F. P. Fletcher and V. W. Blanchard.	Bridgeport, Vt	Aug. 6, 1867	67, 426
Iron into steel, Apparatus for converting	J. Webster	Birmingham, England	Apr. 8, 1873	137, 742
Iron into steel by cementation, Process of converting.	E. R. Weston	East Corinth, Mo	June 11, 1861	32, 546
Iron into steel by means of hydrocarbon vapor, Apparatus for converting.	T. R. Scowden	Cincinnati, Ohio	Aug. 13, 1872	130, 540
Iron into steel, Composition for converting	T. Sheehan	Dunkirk, N. Y	June 4, 1867	65, 512
Iron into steel, Converting	J. F. Boynton	Syracuse, N. Y	July 16, 1867	66, 785
Iron into steel, Converting	E. F. Houghton	Philadelphia, Pa	June 18, 1872	128, 042
Iron into steel, Converting	S. C. Salisbury	New York, N. Y	Aug. 27, 1867	68, 118
Iron into steel, Converting articles made of wrought	L. La Breche-Viger	Montreal, Canada	Nov. 28, 1871	121, 263
Iron into steel, Converting articles of cast	B. W. Nichols	Canton, Ohio	July 20, 1869	92, 876
Iron into steel, Converting cast	W. Harris and A. Woolever	Allentown, Pa	Apr. 12, 1870	101, 963
Iron into steel, Converting pig	E. Leonard	Canton, Mass	Jan. 6, 1812	....
Iron into steel, Method of converting	E. G. Pomeroy	New York, N. Y	Sept. 4, 1860	29, 909
Iron into steel, Method of converting	T. Sheehan	Dunkirk, N. Y	Sept. 4, 1860	29, 919
Iron into steel, Method of converting cast	A. Hamar	Philadelphia, Pa	Feb. 2, 1869	86, 536
Iron into steel, Mode of converting articles of	R. A. Jackson	Lawrenceville, Pa	Dec. 3, 1867	71, 620
Iron into steel, Mode of converting articles of	B. W. Nichols	Canton, Ohio	Apr. 27, 1869	89, 329
Iron into steel, Mode of converting cast	T. Clark	Louisville, Ky	June 25 1872	128, 362
Iron into steel partially, Converting	E. Jenks	Colebrook, Conn	Jan. 28, 1818	....
Iron into steel, Process and apparatus for converting cast.	A. Thoma	New York, N. Y	Nov. 9, 1869	96, 633
Iron into steel, Process for converting	T. J. Barron	Brooklyn, N. Y	Jan. 1, 1867	60, 823
Iron into wrought iron and steel, Converting cast	J. R. Bradley	Ironton, Ohio	Apr. 15, 1862	34, 937
Iron into wrought iron and steel, Converting cast	A. C. Rand	Aurora, Ill	Mar. 18, 1873	137, 025
Iron, Machine for bending sheet or plate	J. Watchman	Baltimore, Md	June 1, 1843	3, 116
Iron, Machine for cutting rings from	C. H. Bassett	Birmingham, Conn	Aug. 1, 1865	49, 191
Iron, Machine for dressing	J. G. Tibbets	New York, N. Y	June 21, 1839	1, 183
Iron, &c., Machine for drilling	A. Morgan	Wooster, Ohio	July 20, 1843	3, 187
Iron, Machine for molding and rolling	J. White	Philadelphia, Pa	Mar. 14, 1810	....
Iron, Machine for rolling	H. B. Comer	Temperanceville, Pa	Mar. 29, 1859	23, 425
Iron, Machine for rolling bar	J. S. Hartupee and A. Alexander.	Pittsburgh, Pa	Apr. 19, 1853	9, 673
Iron, Machine for scraping	C. Zug	Pittsburgh, Pa	Apr. 4, 1871	113, 611
Iron, Machine for squeezing puddled balls of	S. Gissinger	Lawrenceville, Pa	Dec. 17, 1867	72, 387
Iron, Machine for squeezing puddled balls of	S. Gissinger	Lawrenceville, Pa	Dec. 17, 1867	72, 388
Iron, Machine for working	J. Reese	Pittsburgh, Pa	Oct. 6, 1868	82, 876
Iron, Machinery for bending sheet	H. A. Roe	Erie, Pa	Sept. 11, 1845	4, 187
Iron, Machinery for drawing out and compressing heated.	H. Burden	Troy, N. Y	Oct. 16, 1849	6, 792
Iron, &c., Machinery for puddling	V. S. Bloomhall	Conshohocken, Pa	Feb. 20, 1872	123, 860
Iron, Machinery for twisting and rolling	H. Ames	Falls Village, Conn	Oct. 9, 1847	5, 323
Iron, Machinery for working sheet	A. W. Whitney	Woodstock, Vt	Dec. 11, 1847	5, 389
Iron, &c., Machinery or apparatus operated by steam for forging, stamping, and cutting.	J. Nasmyth	Great Britain	Apr. 10, 1843	3, 042
Iron, Making	A. Hamar	New York, N. Y	Sept. 1, 1868	81, 775

Index of patents issued from the United States Patent Office from 1790 *to* 1873, *inclusive*—Continued.

Invention.	Inventor.	Residence.	Date.	No.
Iron malleable, Making cast	A. K. Eaton	New York, N. Y	June 16, 1857	17, 561
Iron malleable, Process of rendering cast	A. H. Siegfried	Lewisburgh, Pa	Sept. 24, 1872	131, 634
Iron, Manufacture of	J. Absterdam	New York, N. Y	Apr. 2, 1872	125, 245
Iron, Manufacture of	E. Brady	Philadelphia, Pa	Aug. 10, 1869	93, 406
Iron, Manufacture of	W. G. Brown and F. McKee	Birmingham, Pa	Apr. 3, 1860	27, 688
Iron, Manufacture of	J. Burt	Detroit, Mich	Jan. 19, 1869	86, 056
Iron, Manufacture of	J. Burt	Detroit, Mich	May 25, 1869	90, 494
Iron, Manufacture of	W. Bushnell	New York, N. Y	Feb. 18, 1873	135, 886
Iron, Manufacture of	A. G. Cook	Burlington, Vt	Mar. 12, 1867	62, 819
Iron, Manufacture of	G. Crane	London, England	Nov. 29, 1838	1, 024
Iron, Manufacture of	W. H. Dawes	West Bromwich, England	May 24, 1864	42, 899
Iron, Manufacture of	A. H. Everett	New York, N. Y	June 27, 1865	48, 483
Iron, Manufacture of	W. Fields	Wilmington, Del	May 10, 1870	102, 797
Iron, Manufacture of	A. L. Fleury	Philadelphia, Pa	July 16, 1861	32, 826
Iron, Manufacture of	A. L. Fleury	Troy, N. Y	Sept. 15, 1863	39, 991
Iron, Manufacture of	L. S. Goodrich	Waverly, Tenn	Nov. 14, 1871	120, 871
Iron, Manufacture of	D. W. Hendrickson	New York, N. Y	Dec. 3, 1867	71, 754
Iron, Manufacture of	J. Jameson	Philadelphia, Pa	May 7, 1867	64, 425
Iron, Manufacture of	J. J. Johnston	Allegheny City, Pa	Feb. 5, 1867	61, 743
Iron, Manufacture of	J. J. Johnston	Allegheny, Pa	Jan. 9, 1872	122, 524
Iron, Manufacture of	J. J. Johnston	Columbiana, Ohio	May 14, 1872	126, 709
Iron, Manufacture of	S. T. Jones	New York, N. Y	Sept. 16, 1851	8, 357
Iron, Manufacture of	W. Kelly	Lyon County, Ky	June 23, 1857	17, 628
Iron, Manufacture of	A. K. Kerpely	Oravitza, Austria	Feb. 13, 1866	52, 656
Iron, Manufacture of	B. Lauth	Pittsburgh, Pa	Aug. 23, 1859	25, 235
Iron, Manufacture of	W. M. Lyon	Pittsburgh, Pa	May 25, 1869	90, 372
Iron, Manufacture of	W. H. Perry	Saint Louis, Mo	Aug. 6, 1861	33, 006
Iron, Manufacture of	W. H. Richardson	Glasgow, North Britain	July 30, 1867	67, 350
Iron, Manufacture of	D. Stewart	Kittaning, Pa	Dec. 17, 1867	72, 335
Iron, Manufacture of	D. Stewart	Kittaning, Pa	July 13, 1869	92, 667
Iron, Manufacture of	D. Stewart	Kittaning, Pa	Feb. 1, 1870	99, 367
Iron, Manufacture of	D. Stewart	Kittaning, Pa	Feb. 1, 1870	99, 368
Iron, Manufacture of	F. D. Taylor	Brady's Bend Township, Pa.	Aug. 1, 1871	117, 576
Iron, Manufacture of	W. J. Taylor	High Bridge, N. J	Dec. 9, 1873	145, 461
Iron, Manufacture of	W. J. Taylor	High Bridge, N. J	Dec. 9, 1873	145, 462
Iron, Manufacture of	A. Thomas	Howard, Pa	Apr. 5, 1859	23, 512
Iron, Manufacture of	J. D. Williams	Allegheny City, Pa	Aug. 8, 1865	49, 331
Iron, Manufacture of	J. D. Williams	Allegheny, Pa	Dec. 26, 1865	51, 771
Iron, Manufacture of	J. D. Williams	Philadelphia, Pa	June 7, 1870	104, 091
Iron, Manufacture of	J. Yates	Mott Haven, N. Y	Mar. 1, 1864	41, 805
Iron, Manufacture of galvanized	J. D. Grey	Pittsburgh, Pa	Sept. 21, 1869	95, 017
Iron, Manufacture of malleable	J. J. G. Collins	Philadelphia, Pa	May 3, 1870	102, 501
Iron, Manufacture of malleable	A. Manvel	Elizabethport, N. J	July 19, 1864	43, 595
Iron, Manufacture of metal-coated sheet	B. Morison	Philadelphia, Pa	Feb. 20, 1872	123, 787
Iron, Manufacture of pig	J. Burt	Detroit, Mich	Sept. 7, 1869	94, 471
Iron, Manufacture of pig	C. M. Nes	York, Pa	Nov. 16, 1869	96, 927
Iron, Manufacture of refined cast	C. Burgess	Portsmouth, Ohio	July 29, 1873	141, 319
Iron, Manufacture of sheet	R. N. Allen and C. C. Hinsdale	Cleveland, Ohio	Mar. 20, 1866	53, 253
Iron, Manufacture of sheet	G. Atkins	Sharon, Pa	Mar. 23, 1869	88, 002
Iron, Manufacture of sheet	S. Barker	Hartford, Conn	Oct. 5, 1869	95, 554
Iron, Manufacture of sheet	J. Chandler	Attica, Ohio	Oct. 19, 1858	21, 817
Iron, Manufacture of sheet	I. E. Craig	Camden, Ohio	Feb. 11, 1868	74, 314
Iron, Manufacture of sheet	I. E. Craig	Camden, Ohio	May 31, 1870	103, 577
Iron, Manufacture of sheet	J. Dickson	Newcastle, Pa	Feb. 4, 1862	34, 294
Iron, Manufacture of sheet	W. Fields	Wilmington, Del	Dec. 28, 1869	98, 364
Iron, Manufacture of sheet	I. and T. Gray	Pittsburgh, Pa	Sept. 26, 1865	50, 203
Iron, Manufacture of sheet	J. D. Grey	Pittsburgh, Pa	May 24, 1870	103, 323
Iron, Manufacture of sheet	C. C. Hinsdale	Cleveland, Ohio	Apr. 27, 1869	89, 312
Iron, Manufacture of sheet	H. McCarty	Pittsburgh, Pa	Sept. 27, 1863	10, 047
Iron, Manufacture of sheet	D. A. Morris	Pittsburgh, Pa	Oct. 12, 1858	21, 772
Iron Manufacture of sheet	D. A. Morris	Pittsburgh, Pa	Jan. 22, 1861	31, 184
Iron, Manufacture of sheet	E. B. Nock	Cleveland, Ohio	Aug. 25, 1868	81, 528
Iron, Manufacture of sheet	C. H. Perkins	Providence, R. I	Mar. 27, 1866	53, 476
Iron, Manufacture of sheet	E. G. Pomeroy	Pittsburgh, Pa	Jan. 31, 1854	10, 482
Iron, Manufacture of sheet	D. L. Pratt	Bridgeport, Ohio	Feb. 14, 1865	46, 384
Iron, Manufacture of sheet	D. L. Pratt	Bethesda, Ohio	Mar. 26, 1872	125, 079
Iron, Manufacture of sheet	W. Rogers	Apollo, Pa	Jan. 23, 1872	122, 912
Iron, Manufacture of sheet	W. D. Wood	Wilmington, Del	May 14, 1861	32, 341
Iron, Manufacture of sheet and bar	J. M. Jones, B. Spaulding, and S. Parkins.	East Taunton, Mass., Port Richmond, N. Y., and Providence, R. I.	July 31, 1866	56, 759
Iron, Manufacture of sheet and plate	N. C. Gridley	Milwaukee, Wis	July 27, 1869	93, 083
Iron, Manufacture of sheet and plate	C. C. Hinsdale	Cleveland, Ohio	Sept. 8, 1868	81, 903
Iron, Manufacture of wrought	T. C. Coleman	Louisville, Ky	Jan. 19, 1869	86, 004
Iron, Manufacture of wrought	W. M. Lyon	Pittsburgh, Pa	May 25, 1869	90, 373
Iron, Manufacture of wrought	C. Sacre, S. Perkins, and W. Smellie.	Manchester and Gorton, Great Britain.	Oct. 18, 1870	108, 521
Iron, Manufacture of wrought	R. Thomas and G. Edwards	Columbiana, Ala	Aug. 1, 1865	49, 175
Iron, Manufacture of wrought and puddled	W. Ennis	Philadelphia, Pa	Aug. 16, 1870	106, 347
Iron, Manufacturing	J. Butland	Philadelphia, Pa	Nov. 12, 1807	....
Iron, Manufacturing and purifying	J. Webster	Birmingham, England	May 17, 1870	103, 109
Iron-manufacturing furnace	E. Peckham	Antwerp, N. Y	Oct. 24, 1871	120, 318
Iron-manufacturing furnace	H. Ross	Pittsburgh, Pa	Oct. 27, 1868	83, 551
Iron, Manufacturing hoop and sheet	J. H. Pierson	Ramapo Works, Rockland, N. Y.	Dec. 24, 1807	....
Iron, Manufacturing malleable cast	S. Boyden	Newark, N. J	Mar. 9, 1831	....
Iron, Manufacturing malleable cast	S. Boyden	Newark, N. J	Apr. 6, 1831	....
Iron, Manufacturing, puddling, and heating	W. Jones	Haversham, N. Y	Dec. 16, 1833	....
Iron, Manufacturing sheet	W. D. Wood	McKeesport, Pa	Apr. 8, 1873	137, 585
Iron, Manufacturing sheet or boiler	J. McNary	Brooklyn, N. Y	Dec. 17, 1834	....
Iron measuring and cutting machine	L. B. Griffith	Honeybrook, Pa	Nov. 4, 1851	8, 485
Iron, Melting and decarbonizing	C. Peters	Trenton, N. J	Nov. 2, 1869	96, 479
Iron, Melting and refining	G. P. Miller and H. Dougherty	Lancaster, Pa	Oct. 6, 1857	18, 347
Iron-melting furnace	W. McFarland	Saint Louis, Mo	Dec. 7, 1858	22, 257
Iron-melting furnace	G. H. Sellers	Wilmington, Del	Apr. 2, 1872	125, 147
Iron melting and refining furnace	A. F. Du Faur	New York, N. Y	Feb. 1, 1870	99, 415
Iron-mill	J. W. and C. Post	Washington, D. C	Apr. 11, 1829	....

Index of patents issued from the United States Patent Office from 1790 *to* 1873, *inclusive* -Continued.

Invention.	Inventor.	Residence.	Date.	No.
Iron, Mode of purifying	E. Brady	Philadelphia, Pa	Apr. 13, 1869	88, 941
Iron, Mode of treating conglomerates of cast	T. S. Blair	Pittsburgh, Pa	Dec. 14, 1869	97, 754
Iron, Mode of treating slag and cinders for the manufacture of.	R. Keck	Clintonville, N. Y	Oct. 1, 1867	69, 348
Iron or ores to the metallic state by coating them with certain fluxes, Process of reducing.	J. Tower	Madison, Ohio	Dec. 7, 1844	3, 850
Iron or steel bars or rods, Manufacture of	B. Lauth	Reading, Pa	Mar. 12, 1867	62, 758
Iron or steel with copper or brass or other alloys of copper, Coating.	G. J. Hinde	Wolverhampton, England	Sept. 7, 1869	94, 492
Iron-ore for smelting, Preparations of	C. Cochrane	Upper Gornal, England	July 9, 1872	128, 712
Iron-ore from other substances, Machine for separating magnetic.	J. Y. Smith	Pittsburgh, Pa	Nov. 15, 1870	109, 354
Iron-ore, Furnace for producing sponge from	G. Nock	New Monmouth, N. J	May 24, 1870	103, 489
Iron-ore in the manufacture of steel, Preparing and treating.	E. L. Seymour	New York, N. Y	May 23, 1871	115, 243
Iron-ore, Mode of desulphurizing	J. Little	Newburgh, N. Y	May 21, 1867	64, 887
Iron-ore, Mode of smelting	J. Lyon	Pottsville, Pa	July 31, 1837	332
Iron-ore, Mode of smelting	J. Richard	Philadelphia, Pa	Dec. 10, 1838	1, 032
Iron-ore, Reduction of	W. S. Cooke	East Fairfield, Ohio	May 29, 1847	5, 131
Iron-ore, Roasting and treating	H. Aitken	Falkirk, Scotland	Aug. 31, 1869	94, 374
Iron-ore, Smelting	H. Bessemer	London, England	Nov. 18, 1856	16, 083
Iron-ore, Smelting	T. S. Ridgway	Pottsville, Pa	Dec. 17, 1834	......
Iron-ore, &c., Treating	H. Aitken	Falkirk, Scotland	Sept. 29, 1868	82, 576
Iron-ore, Treatment and reduction of titaniferous	C. Martin, W. Barrett, and T. S. Webb.	Chancery Lane and Norton, England.	Sept. 29, 1868	82, 539
Iron pig-bloom, Apparatus for the manufacture of	T. S. Blair	Pittsburgh, Pa	June 29, 1869	91, 901
Iron pig-bloom, Apparatus for the manufacture of	J. Coyne	Allegheny City, Pa	Aug. 31, 1869	94, 286
Iron pins, Making coated	D. and T. Fowler	North Branford, Mass	Jan. 17, 1860	26, 874
Iron pipe. Cast	C. Pomroy	Pottsville, Pa	May 4, 1858	20, 171
Iron-pipe joint	E. G. Blakeslee	Sing Sing, N. Y	July 16, 1867	66, 782
Iron pipe, Tin-lined	C. M. Platt	Waterbury, Conn	Dec. 26, 1871	122, 131
Iron-pipe-welding machinery	J. C. Vaughn and J. F. Winslow.	Greenbush and Troy, N. Y	Aug. 1, 1848	5, 695
Iron pipes, Cleansing galvanized	W. Blake	Boston, Mass	Aug. 21, 1860	29, 663
Iron pipes, Flask for cast	J. Firth and J. Ingham	Phillipsburgh, N. J	Dec. 2, 1862	37, 037
Iron pipes for welding, Machine for bending	E. J. Woodworth, J. Brandy, and F. Merithew.	Birmingham, Pa	Nov. 8, 1870	109, 094
Iron pipes, Machine for cleansing cast	W. Smith	Allegheny City, Pa	June 7, 1870	104, 072
Iron pipes, Method of employing centrifugal force in casting.	T. J. Lovegrove	Baltimore, Md	Dec. 26, 1848	5, 988
Iron pipes, &c., Mold for casting	J. D. Morris	Kensington, Pa	Dec. 2, 1835	....
Iron pipes, Molding	W. Doyle	Auburn, N. Y	May 15, 1860	28, 261
Iron plate, Method of treating sheet	C. H. Perkins	Providence, R. I	Feb. 13, 1866	52, 647
Iron-plating furnace	C. Wray	San Francisco, Cal	Feb. 21, 1860	27, 254
Iron ponton	H. Grundt	Berlin, Prussia	Feb. 4, 1862	34, 329
Iron post for wire fences	D. Kaufman	Boiling Springs, Pa	June 3, 1873	139, 469
Iron, Preserving the surface of cast or wrought	C. F. L. Oudry	Paris, France	Nov. 23, 1858	22, 132
Iron, Process and apparatus for making sheet	G. W. Francis	Hartford, Conn	June 22, 1869	91, 619
Iron, Process and apparatus for treating crude or pig.	H. Bessemer	London, England	Aug. 1, 1871	117, 507
Iron, Process for smelting	S. Macferran	Philadelphia, Pa	July 24, 1855	13, 316
Iron, Process for steelifying	T. Sheehan	Dunkirk, N. Y	Mar. 31, 1868	76, 256
Iron, Process for the manufacture of	C. Adams	Philadelphia, Pa	Nov. 28, 1871	121, 226
Iron, Process of coating	E. G. Pomeroy	Philadelphia, Pa	Oct. 13, 1857	18, 409
Iron, Process of hardening	T. Chappell and W. A. Jones	Winona, Minn	Aug. 9, 1864	43, 813
Iron, Process of manufacturing rolled	G. W. Billings	Chicago, Ill	May 27, 1873	139, 359
Iron, Process of manufacturing sheet	S. H. Kimball	Cleveland, Ohio	Mar. 23, 1869	88, 182
Iron, Process of manufacturing wrought	P. E. Jay	Saint Jean Baptiste, Canada	Dec. 24, 1872	134, 288
Iron, Process of preserving	W. H. Sterling	New York, N. Y	May 21, 1872	126, 909
Iron, Process of refining cast	H. M. Baker	Harlem, N. Y	Sept. 15, 1868	82, 194
Iron, Process of treating cleaned or scaled	W. D. Wood	Keysport, Pa	Jan. 8, 1867	61, 034
Iron, Protecting surfaces of articles of	T. Sellick	Greenwich, Conn	July 5, 1859	24, 665
Iron, Puddling	J. J. Johnston	Columbiana, Ohio	May 14, 1872	126, 710
Iron, Puddling	R. Savary	Steubenville, Ohio	Mar. 11, 1856	14, 412
Iron, Puddling	W. Sellers	Philadelphia, Pa	July 16, 1872	129, 430
Iron, Puddling	J. G. Willans	Bayswater, England	Mar. 12, 1867	62, 795
Iron-puddling apparatus	W. Sellers	Philadelphia, Pa	Mar. 5, 1872	124, 224
Iron, Puddling process for making wrought	P. Merryman and R. McCombs	West Fairview, Pa	Mar. 15, 1870	100, 916
Iron, Puddling process for the manufacture of wrought.	C. Hewitt	Hamilton Township, N. J	Mar. 30, 1869	88, 479
Iron, Purification of cast	J. W. Middleton	Philadelphia, Pa	Jan. 3, 1871	110, 666
Iron, Purifying cast	S. W. Kirk	Coatesville, Pa	Mar. 24, 1863	38, 003
Iron, reducing ore, &c., Refining	J. F. Allen	Tremont, N. Y	Feb. 21, 1871	112, 003
Iron, Refining	W. Kelly	Lyon, Ky	Dec. 22, 1857	18, 910
Iron, Refining	C. Shunk	Canton, Ohio	July 12, 1859	24, 766
Iron refining and carbonizing composition	J. E. Atwood	Trenton, N. J	Feb. 9, 1869	86, 801
Iron, Refining cast	H. S. Osborn	Easton, Pa	Sept. 22, 1868	82, 435
Iron, &c., remelting furnace	W. J. Keep	Troy, N. Y	Oct. 29, 1872	132, 667
Iron, Removing acid from the surface of	D. McDaniel and E. A. Harney	New Castle County and Wilmington, Del.	Dec. 3, 1861	33, 844
Iron, Restoring burned	G. W. Morris and W. Quann	Philadelphia, Pa	May 1, 1860	28, 103
Iron, Restoring tinned sheet	W. E. Brockway	New York, N. Y	July 22, 1873	141, 109
Iron roller for metals, Cast	J. Wood	Philadelphia, Pa	July 20, 1831	....
Iron, Roller for rolling	A. B. Clemons	Derby, Conn	Apr. 19, 1864	42, 433
Iron roller for rolling iron, &c., Mode of casting large.	C. Alger	Boston, Mass	Mar. 30, 1811	....
Iron, Roller for slitting and other mills for rolling	B. Seymour		June 26, 1797	....
Iron, Rolling	A. S. Valentine	Bellefonte, Pa	Jan. 3, 1827	....
Iron, Rolling angle	J. L. Lewis	Pittsburgh, Pa	Apr. 26, 1864	42, 495
Iron-rolling machine	H. Burden	Troy, N. Y	Dec. 10, 1840	1, 890
Iron-rolling machinery	J. F. Lauth	Reading, Pa	Mar. 6, 1866	53, 012
Iron scraps, Remelting	A. Pevey	Lowell, Mass	Jan. 15, 1856	14, 114
Iron-shearing machine	A. Hardy	Boston, Mass	Sept. 22, 1863	40, 034
Iron-shearing machine, Angle	T. Reaney	Chester, Pa	Nov. 16, 1869	96, 960
Iron, Sheet	J. Miller	Cuba, N. Y	Apr. 16, 1867	63, 805
Iron-slitting machine	J. Reed	Marshfield, Mass	Apr. 21, 1825	....
Iron, Smelting and refining	W. W. Hubbell	Philadelphia, Pa	Feb. 8, 1870	99, 677
Iron, Smelting and refining	E. G. Pomeroy	New York, N. Y	Jan. 24, 1860	26, 923

Index of patents issued from the United States Patent Office from 1790 *to* 1873, *inclusive*—Continued.

Invention.	Inventor.	Residence.	Date.	No.
Iron-smelting furnace	S. M. Fales	Baltimore, Md	Feb. 8, 1859	22, 861
Iron-smelting furnace, Scrap	A. Ott	New York, N. Y	Sept. 27, 1870	107, 712
Iron-smelting process	F. Lang and C. A. Frey	Vienna, Austria, and Storré, Styria.	Dec. 12, 1865	51, 531
Iron sockets, Machine for making	L. Morse and H. Pettee	Foxborough, Mass	Dec. 28, 1838	1, 046
Iron, Softening cast or crude	A. Scott, jr., and A. Selden	Winchester, N. H	Nov. 8, 1814	....
Iron spindle and cast-iron driver, Cast or wrought.	J. Barton	Milford, Pa	Apr. 23, 1831	....
Iron splaying or flaring machinery	N. Sutton	New York, N. Y	Apr. 28, 1825	....
Iron sponge	T. S. Blair	Pittsburgh, Pa	May 21, 1872	126, 924
Iron, steel, and malleable iron, Producing refined cast.	J. W. Middleton	Philadelphia, Pa	Jan. 17, 1871	110, 990
Iron, steel, and other material, Tool and wheel for cutting and polishing.	A. K. Eaton	Piermont, N. Y	May 24, 1870	103, 312
Iron, steel, and other metals, Processes for purifying.	J. F. Bennett	Pittsburgh, Pa	June 18, 1872	127, 953
Iron, steel, &c., Compound for treating	W. A. Skinner and J. E. Goodson.	Callao, Mo	May 27, 1873	139, 335
Iron, steel, &c., Manufacture of	A. T. Hay	Burlington, Iowa	Nov. 19, 1872	133, 098
Iron, steel, &c., Process of refining	J. Reese	Pittsburgh, Pa	June 18, 1867	65, 830
Iron structures, Clamp for	J. M. Moorehead	Brooklyn, N. Y	Oct. 6, 1868	82, 738
Iron, Surfacing cast	H. Tucker	Newton, Mass	May 24, 1870	103, 527
Iron-swaging machine	J. Foster	Brooklyn, N. Y	Mar. 3, 1857	16, 775
Iron-swaging machine	J. T. Willmarth	Worcester, Mass	Aug. 19, 1856	15, 588
Iron, Tool for cutting, punching, and upsetting	J. J. Rose	Elmwood, Ill	Nov. 12, 1867	70, 902
Iron, Tool for turning or planing	T. Cooper	Warwick, R. I	July 3, 1866	56, 141
Iron, Tools used in the manufacture of	A. L. Fleury	Philadelphia, Pa	Feb. 19, 1861	31, 450
Iron-tooth rake	N. C. Sanford and E. P. Parmlee	Meriden, Conn	June 8, 1830	....
Iron tubes, Heating skelps for the manufacture of wrought.	J. McCarty	Reading, Pa	Apr. 4, 1854	10, 747
Iron tubes, Welding	T. H. Russell	Wednesbury, England	Mar. 6, 1847	4, 992
Iron tubing, Manufacture of	J. H. Cotton	Boston, Mass	Aug. 12, 1862	36, 141
Iron tubings, &c., Mode of utilizing	C. S. and J. A. Lynch and C. E. Coffin.	Boston, Mass., and Muirkirk, Md.	Apr. 20, 1869	89, 228
Iron vessels, Coppering	W. B. Barnard	Waterbury, Conn	Mar. 24, 1863	37, 998
Iron vessels, Preventing the corrosion of cast	C. O. Greene	Troy, N. Y	Apr. 28, 1868	77, 186
Iron-ware, Coating articles of	H. Tucker	Newton, Mass	June 1, 1869	90, 891
Iron, Welding wrought	J. C. Cooke	Middletown, Conn	Apr. 3, 1860	27, 697
Iron wheels, Molding cast	G. S. Bosworth	Troy, N. Y	Oct. 30, 1860	30, 525
Iron with a harder metal, Mode of coating wrought or cast.	J. Rigg	Iowa Falls, Iowa	July 9, 1867	66, 630
Iron with brass or copper, Method of coating	H. Burgess	Kentish Town, England	July 18, 1854	11, 319
Iron with cast iron, Method of combining wrought	W. M. Arnold	New York, N. Y	May 22, 1866	54, 838
Iron with cast steel, Mode of coating wrought	J. W. Ells	Pittsburgh, Pa	Oct. 15, 1867	69, 904
Iron with copper, Coating	T. G. Bucklin	Troy, N. Y	Sept. 21, 1852	9, 270
Iron with copper or its alloy, Coating	E. G. Pomeroy	Saint Louis, Mo	Jan. 8, 1850	7, 005
Iron with other metals, Coating sheet	J. H. Whitling	Philadelphia, Pa	Nov. 20, 1866	59, 787
Iron with steel surface, Process for the manufacture of.	J. Reese	Pittsburgh, Pa	Apr. 30, 1867	64, 253
Iron with tin and other metals, Coating	E. Morewood	London, England	Sept. 4, 1866	57, 832
Iron with tin, Apparatus for coating sheet	W. O. Davies	Pittsburgh, Pa	June 4, 1872	127, 580
Iron with zinc, Process for coating	J. D. Grey and J. Lippincott	Baltimore, Md	June 4, 1872	127, 412
Iron, wood, and other material, Compound for coating.	H. E. F. De Brion	London, England	Aug. 13, 1867	67, 733
Iron-working furnace	J. Heatley	Etna, Pa	Sept. 22, 1868	82, 313
Iron, Working scrap and other	G. W. Jones	Ormsby, Pa	Feb. 8, 1870	99, 574
Iron, zinc, and copper, Composition metal of	A. Birkholz	Hartford, Conn	Mar. 11, 1862	34, 621
Ironing and kitchen table combined	J. N. Valley	North East, Pa	Sept. 12, 1871	118, 990
Ironing and pressing garments, Apparatus for	R. B. Sanson	London, England	Sept. 2, 1873	142, 516
Ironing and stiffening machine for linen and other fabrics.	J. Decoudun	Paris, France	Dec. 7, 1869	97, 614
Ironing and stretching board	J. W. Davis	Reno, Nev	June 27, 1871	116, 277
Ironing-apparatus	E. Sumption	South Bend, Ind	Feb. 4, 1873	135, 606
Ironing-apparatus	R. A. Tyler	Haverhill, Mass	Mar. 4, 1873	136, 471
Ironing-board	E. Bassett	Topeka, Kans	July 15, 1873	140, 758
Ironing-board	A. G. Beek	Philadelphia, Pa	May 9, 1871	114, 515
Ironing-board	J. H. Beidler	Adrian, Mich	Nov. 29, 1870	109, 712
Ironing-board	C. H. Bennett and W. H. Daggett.	South Vineland, N. J	Nov. 2, 1869	96, 384
Ironing-board	G. I. Birch	New York, N. Y	Apr. 14, 1868	76, 701
Ironing-board	B. D. Brown and D. Moyer	Philadelphia, Pa	Oct. 17, 1871	120, 030
Ironing-board	J. M. Childers	Corry, Pa	July 9, 1872	128, 709
Ironing-board	D. E. Crosby	South Vineland, N. J	Sept. 7, 1869	94, 572
Ironing-board	J. Fischer	Pittsburgh, Pa	Sept. 20, 1870	107, 469
Ironing-board	H. Hamill	New York, N. Y	Oct. 1, 1872	131, 876
Ironing-board	R. N. Herring	Chew's Landing, N. J	Apr. 29, 1873	138, 331
Ironing-board	G. M. Lane	Battle Creek, Mich	Mar. 5, 1872	124, 275
Ironing-board	A. Watson	Elizabeth, N. J	June 8, 1869	91, 033
Ironing-board	J. C. Miller	Lancaster, Ohio	Apr. 20, 1869	89, 162
Ironing-board	S. A. Mort	Dayton, Ohio	Aug. 14, 1866	57, 170
Ironing-board	L. N. Vallett	Providence, R. I	July 2, 1872	128, 569
Ironing-board and closet	J. N. Brewster	Brooklyn, N. Y	May 7, 1867	64, 480
Ironing clothes, Device for	C. S. Whipple	Waterford, Conn	May 21, 1872	126, 925
Ironing-machine	W. H. Bovell and C. Partinsky	San Francisco, Cal	Oct. 18, 1864	44, 765
Ironing-machine	G. Boxley	Troy, N. Y	Apr. 5, 1870	101, 573
Ironing-machine	L. Berenger	Paris, France	Mar. 30, 1869	88, 437
Ironing-machine	G. W. H. Calver	Burlington, N. J	Sept. 3, 1872	131, 052
Ironing-machine	G. W. Cottingham	Saint Mary's, Tex	Nov. 18, 1873	144, 743
Ironing-machine	P. J. Flanedy	San Francisco, Cal	Sept. 10, 1867	68, 617
Ironing-machine	A. G. Gardner	Troy, N. Y	Feb. 20, 1872	123, 819
Ironing-machine	G. Gilbert and A. N. Allen	New Haven, Conn	Aug. 13, 1867	67, 644
Ironing-machine	W. Jones	Oshkosh, Wis	June 28, 1870	104, 740
Ironing-machine	W. T. Nichols	Rutland, Vt	Apr. 22, 1862	35, 034
Ironing-machine	P. O'Thayne	New York, N. Y	Mar. 26, 1867	63, 301
Ironing-machine	C. F. Parker	Greenfield, Ohio	Sept. 27, 1870	107, 714
Ironing-machine	G. B. Perkins	Bridgeport, Conn	Mar. 12, 1867	62, 773
Ironing-machine	L. Ringler	New York, N. Y	July 13, 1869	92, 544
Ironing-machine	A. C. Sawyer	Canton, N. Y	June 28, 1870	104, 776
Ironing-machine	G. F. Taylor	New York, N. Y	Dec. 12, 1871	121, 908

Index of patents issued from the United States Patent Office from 1790 *to* 1873, *inclusive*—Continued.

Invention.	Inventor.	Residence.	Date.	No.
Ironing-machine	C. C. Thomas	Natchez, Miss	Jan. 9, 1872	122, 680
Ironing-machine	J. W. Thorp	Sanbornton Bridge, N. H.	Mar. 12, 1867	62, 906
Ironing-machine	W. M. Tobey	New London, Conn	Mar. 26, 1867	63, 186
Ironing-machine	J. T. Walker	Albany, N. Y	Nov. 25, 1873	145, 034
Ironing-machine	A. B. Walters	Philadelphia, Pa	Aug. 6, 1872	130, 257
Ironing machine, Clothes	J. Shaefer	Lancaster, Pa	Sept. 7, 1858	21, 450
Ironing machine, Clothes	S. Swett, jr	Readfield, Me	Dec. 30, 1835	....
Ironing machine, Clothes	S. W. Woodward	Buffalo, N. Y	Mar. 24, 1863	37, 994
Ironing machine, Hosiery	W. Aiken	Franklin, N. H	Aug. 28, 1866	57, 455
Ironing or fluting machine	M. P. Carpenter	Buffalo, N. Y	May 6, 1862	35, 138
Ironing-stand	J. P. and M. L. Bever	Springfield, Ill	Mar. 3, 1863	37, 804
Ironing-stand and clothes-drier	B. S. Boydston	Richmond, Ind	Mar. 3, 1868	75, 120
Ironing-table	C. Blom, jr	Grand Haven, Mich	Dec. 16, 1873	145, 614
Ironing-table	P. Bostrom	Galesburgh, Ill	Jan. 9, 1872	122, 554
Ironing-table	F. A. Byram	Germantown, Pa	Aug. 8, 1871	117, 737
Ironing-table	P. P. Carroll	Washington, D. C	Oct. 27, 1868	83, 460
Ironing-table	J. R. Carzier	North East, Pa	Apr. 23, 1872	126, 016
Ironing-table	A. A. Chittenden	Boston, Mass	Nov. 19, 1867	70, 958
Ironing-table	J. M. Couthamal	Philadelphia, Pa	Nov. 26, 1872	133, 304
Ironing-table	H. T. De Montigny	West Troy, N. Y	Mar. 30, 1869	88, 281
Ironing-table	E. S. Farson	Philadelphia, Pa	Dec. 2, 1873	145, 162
Ironing-table	J. H. Frank	Ilion, N. Y	Aug. 12, 1873	141, 639
Ironing-table	J. Frankenfield and W. E. Kichline.	Bethlehem, Pa	July 1, 1873	140, 356
Ironing-table	J. F. Galley	New York, N. Y	Aug. 2, 1859	24, 929
Ironing-table	W. B. Grosh and S. H. Foreman	Reading, Pa	Sept. 23, 1873	143, 071
Ironing-table	J. R. Groves	Hightstown, N. J	Feb. 18, 1873	136, 056
Ironing-table	C. C. Hardy	Rutland, N. Y	Oct. 10, 1871	119, 703
Ironing-table	L. Harrington	Saugatuck, Mich	Oct. 27, 1868	83, 379
Ironing-table	O. C. Hotchkiss	Cortlandville, N. Y	Dec. 1, 1868	84, 549
Ironing-table	J. Johnson	New York, N. Y	June 24, 1862	35, 735
Ironing-table	F. Liller	Baltimore, Md	July 16, 1872	129, 042
Ironing-table	J. H. Mallory	La Porte, Ind	Dec. 6, 1870	109, 918
Ironing-table	J. H. Mallory	La Porte, Ind	Oct. 24, 1871	120, 296
Ironing-table	J. J. Märki	Chicago, Ill	Dec. 27, 1870	110, 577
Ironing-table	D. W. Marshall	Pawtucket, R. I	Jan. 3, 1871	110, 769
Ironing-table	J. F. Martin and W. A. Schaffner.	Harrisburgh, Pa	Sept. 13, 1870	107, 271
Ironing-table	H. McChesney	Buffalo, N. Y	Oct. 4, 1870	107, 940
Ironing-table	S. Merritt	Erie, Pa	Jan. 25, 1870	99, 219
Ironing-table	P. O'Thayne	New York, N. Y	Sept. 12, 1871	118, 878
Ironing-table	A. J. Palmberg	Boston, Mass	Sept. 9, 1873	142, 645
Ironing-table	H. Park	Columbus, Ohio	Mar. 31, 1868	76, 098
Ironing-table	W. P. Patten	Harrisburgh, Pa	July 6, 1869	92, 207
Ironing-table	J. T. Piercy	Martinsburgh, Ohio	Oct. 20, 1868	83, 310
Ironing-table	J. T. Plowman, sr	Baltimore, Md	June 11, 1872	127, 792
Ironing-table	T. Reed	Plainwell, Mich	Aug. 29, 1871	118, 643
Ironing-table	T. Reed	Plainwell, Mich	Oct. 17, 1871	120, 101
Ironing-table	T. M. Richards	Philadelphia, Pa	May 11, 1869	89, 943
Ironing-table	J. H. Ruff	Thomas' Run Post-Office, Md	Aug. 16, 1870	106, 509
Ironing-table	M. D. Safford	Boston, Mass	Jan. 29, 1867	61, 569
Ironing-table	B. Schoonmaker	Plainwell, Mich	Nov. 19, 1872	133, 122
Ironing-table	L. Scofield	Grand Haven, Mich	Nov. 25, 1873	144, 869
Ironing-table	H. Soggs	Columbus, Pa	June 8, 1869	90, 966
Ironing-table	W. H. Sparks	Absecom, N. J	Nov. 11, 1873	144, 418
Ironing-table	W. Vandenburg	New York, N. Y	Apr. 6, 1858	19, 883
Ironing-table	W. Vandenburg	New York, N. Y	May 11, 1858	20, 231
Ironing-table	W. Vandenburg and J. Harvey	New York, N. Y	Feb. 16, 1858	19, 390
Ironing-table	B. Van Gaasbeek	Mount Vernon, N. Y	Apr. 23, 1867	64, 170
Ironing-table	A. Wechsler	New York, N. Y	Jan. 28, 1873	135, 303
Ironing-table	A. T. Woolsey	Sandusky, Ohio	Oct. 22, 1867	70, 146
Ironing-table	G. F. Zimmann	Philadelphia, Pa	Jan. 3, 1860	26, 728
Ironing-table and bureau	M. White	Saratoga Springs, N. Y	June 14, 1870	104, 385
Ironing-table and clothes-drier	W. P Adams	Brooklyn, N. Y	Dec. 14, 1869	97, 851
Ironing-table and clothes-drier	J. L. Devol	Parkersburgh, W. Va	June 29, 1869	91, 921
Ironing-table and skirt-board, Combined	A. S. Robinson and R. A. Harris	Albany, N. Y	Aug. 20, 1872	130, 749
Ironing-table bureau	M. White	Saratoga Springs, N. Y	Dec. 20, 1870	110, 410
Ironing-table, Folding	W. Edwards	Cleveland, Ohio	Apr. 8, 1873	137, 602
Ironing-table, lap-board, and step-ladder, Combined	H. E. Smith	Cleveland, Ohio	Mar. 4, 1873	136, 390
Ironing table, Skirt	A. S. Phillips	Boston, Mass	Dec. 17, 1867	72, 228
Irrigating-machine	W. W. Hull	Ashland, N. Y	Sept. 13, 1870	107, 375
Irrigation, Subterranean	W. H. Pugh	Chicago, Ill	Sept. 2, 1873	142, 413
Isinglass	W. Hall	Boston, Mass	Mar. 23, 1822	....
Isinglass, Apparatus for cooling the rollers used in the manufacture of.	E. Rowe	Rockport, Mass	Aug. 9, 1870	106, 212
Isinglass, Manufacture of	J. Manning	Rockport, Mass	Jan. 7, 1873	134, 690
Isinglass, Manufacturing	J. Hastings	Cambridge, Mass	Aug. 14, 1822	....
Isinglass, or ichthyocolla	W. Norwood, J. Burns, J. Rowe, and J. Haskell.	Gloucester, Mass	Jan. 21, 1834	....
Isinglass, Purifying	J. Lewis and I. Stanwood	Gloucester, Mass	Apr. 3, 1866	53, 639
Ivory and similar material, Elastic composition to imitate.	W. M. Welling	New York, N. Y	Apr. 20, 1869	89, 100
Ivory, Artificial	C. F. Dupper	Bridgeport, Conn	Nov. 21, 1865	51, 109
Ivory, Artificial	J. Hackert	New York, N. Y	Feb. 19, 1867	62, 128
Ivory, Artificial	A. Starr and W. M. Welling	New York, N. Y	June 9, 1868	78, 842
Ivory, Artificial	W. M. Welling	New York, N. Y	May 12, 1868	77, 938
Ivory-bleaching apparatus	J. Phyfe	New York, N. Y	Oct. 28, 1856	15, 983
Ivory-bleaching device	W. M. Welling	Brooklyn, N. Y	Dec. 11, 1855	13, 928
Ivory, bone, and wood combs, Method of making	J. Pratt	Meriden, Conn	Dec. 28, 1830	....
Ivory by steam, Machine for junking	L. Pratt and F. Bush	Meriden, Conn	Feb. 9, 1828	....
Ivory combs, Machine for cutting teeth in	P. Pratt	New York	Sept. 9, 1825	....
Ivory combs, Machine for manufacturing	D. Williams, 3d	Saybrook, Conn	Mar. 28, 1810	....
Ivory, Composition for artificial	A. Starr	New York, N. Y	Mar. 3, 1868	75, 067
Ivory, Composition for artificial	W. M. Welling	New York, N. Y	Apr. 27, 1869	89, 531
Ivory, Composition of artificial	J. Hackett	Bridgeport, Conn	May 31, 1864	42, 942
Ivory, Composition resembling	W. M. Welling	New York, N. Y	Apr. 27, 1869	89, 532
Ivory-cutting machine	C. B. Rogers	Deep River, Conn	May 2, 1865	47, 572

Index of patents issued from the United States Patent Office from 1790 *to* 1873, *inclusive*—Continued.

Invention.	Inventor.	Residence.	Date.	No.
Ivory-dust and other materials, Compound of	J. W. Hyatt, jr., and D. Blake	Albany and Spencertown, N. Y.	May 4, 1869	89, 582
Ivory, Factitious	L. Held	Brooklyn, N. Y	Aug. 4, 1857	17, 931
Ivory, Factitious	W. M. Welling	New York, N. Y	Aug. 4, 1857	17, 949
Ivory for piano-forte keys, Machine for gaging and toothing.	E. Savage	West Meriden, Conn	July 30, 1861	32, 967
Ivory-frame composition	I. M. M. Legaré	Aiken, S. C	June 15, 1858	20, 569
Ivory, Frame for bleaching	A. C. Breckenridge	Meriden, Conn	Aug. 19, 1856	15, 590
Ivory key-board, Machine for smoothing	M. Pratt	Meriden, Conn	July 31, 1866	56, 862
Ivory, Molding composition for	J. W. Hyatt, jr	Albany, N. Y	Apr. 6, 1869	88, 633
Ivory or other comb plates, Machine for planing and forming.	W. M. Fowler	North Branford, Conn	Oct. 28, 1840	1, 837
Ivory planing and dressing machine	U. Pratt	Deep River, Conn	Mar. 24, 1863	37, 976
Ivory, Process of bleaching	U. Pratt	Deep River, Conn	Jan. 6, 1852	8, 639
Ivory-slitting machine	P. Pratt	Meriden, Conn	Feb. 12, 1828	....
Ivory, wood, horn, &c., Composition for making imitation.	S. Gradenwitz	New York, N. Y	May 25, 1869	90, 443

J.

Jack:

See Aërostatic jack. / Mechanical jack.
Billiard-table-leveling jack. / Painter's jack.
Boom-jack. / Pegging-jack.
Boot-jack. / Railway-jack.
Builder's jack. / Railway pushing-jack.
Car-pushing jack. / Replacing car-jack.
Car-replacing jack. / Saw-jack.
Carriage-jack. / Screw-jack.
Coupling-jack. / Self-acting jack.
Cross-head jack. / Shackle-jack.
Fence-jack. / Shaft-turning jack.
Fishing-jack. / Shoe-jack.
Hoisting-jack. / Shoe-finishing jack.
Hoop-jack. / Shoe-holding jack.
Hop-pole jack. / Shoemaker's jack.
Horse-power jack. / Spinning-jack.
Horseshoeing-jack. / Supporting-jack.
Hydraulic jack. / Thill-shifting jack.
Knitting-needle jack. / Trimming-jack.
Lasting-jack. / Wagon-jack.
Lever-jack. / Whirling-jack.
Lifting-jack.

Invention.	Inventor.	Residence.	Date.	No.
Jack	A. N. Breneman	Lancaster, Pa	Jan. 29, 1867	61, 603
Jack	T. Wiles	Indianapolis, Ind	Dec. 24, 1867	72, 577
Jack	R. Wood	Grand Ledge, Mich	Mar. 13, 1860	27, 494
Jack-head	N. Thelen	Schenectady, N. Y	Apr. 30, 1872	126, 346
Jacket, Hunting	J. Garand	New York, N. Y	June 17, 1873	139, 950
Jack-screw	S. Vail	Morris, N. J	May 7, 1839	1, 146
Jacquard	W. K. Greene, jr	Schenectady, N. Y	Nov. 28, 1848	5, 939
Jacquard	H. Kelly	Manayunk, Pa	Nov. 28, 1848	5, 937
Jacquard-machine	A. Babbett	Auburn, N. Y	Oct. 18, 1859	25, 796
Jacquard-machine	J. Scott and J. Tannahell	Philadelphia, Pa	Mar. 18, 1851	7, 990
Jail, Iron-plate	E. Jacobs	Cincinnati, Ohio	Dec. 20, 1859	26, 500
Japan applied to leather	W. Gates	Hanover, N. Y	Nov. 14, 1835	....
Japan, &c., over fabrics, Machine for spreading	F. Sautermeister	Newark, N. J	Feb. 3, 1863	37, 592
Japanned metal surfaces, Decorating in gilt and upon	H. Petrie	Chicago, Ill	Nov. 28, 1871	121, 410
Japanning-apparatus	G. W. Hubbell	Derby, Conn	May 23, 1865	47, 826
Japanning eyelets, buttons, &c	S. N. Smith	Providence, R. I	Feb. 14, 1871	111, 882
Japanning metals	G. J. Sturdy and S. W. Young	Providence, R. I	July 21, 1868	80, 236
Japanning pins	J. I. Howe and T. Piper	Derby, Conn	June 10, 1856	15, 111

Jar:

See Candy-jar. / Ice-preserving jar.
Confectionery-jar. / Jelly-jar.
Drill-jar. / Leech-jar.
Drilling-jar. / Oil-tool jar.
Fish-jar. / Preserve-jar.
Flower-jar. / Slop-jar.
Fruit-jar. / Soda-fountain-sirup jar.
Ice-jar.

Invention.	Inventor.	Residence.	Date.	No.
Jar and bottle cover	S. S. Butler	Petaluma, Cal	July 9, 1872	128, 849
Jar and bottle stopper	H. W. Ketcham	New York, N. Y	Apr. 19, 1870	102, 129
Jar and bottle stopper	A. Kline	Philadelphia, Pa	Oct. 27, 1863	40, 415
Jar and bottle stopper	T. O. Oliver	New York, N. Y	Apr. 3, 1866	53, 659
Jar, bottle, &c., stopper	N. Thompson	Saint Johnswood, England	May 16, 1865	47, 779
Jar, can, &c., cap	J. K. Chase	New York, N. Y	June 9, 1863	38, 810
Jar-cover fastening	J. F. Johnson	New York, N. Y	Jan. 7, 1868	73, 014
Jar-mold	G. Scott	Cincinnati, Ohio	June 5, 1860	28, 611
Jar-stopper	G. F. J. Colburn	Newark, N. J	Nov. 1, 1864	44, 852
Jars and bottles, Apparatus for stopping	N. Thompson	Saint Johnswood, England	Aug. 30, 1864	44, 054
Jars and cans, Instrument for lifting	J. E. Higby	West Meriden, Conn	Oct. 17, 1865	50, 476
Jars, cans, &c., Device for closing the mouths of	E. T. Jenkins	Ravenswood, N. Y	Apr. 14, 1868	76, 772
Jars for transportation, Device for securing cover to	M. H. Nichols	Hancock, N. Y	July 6, 1869	92, 343
Jars, Metallic screw-cap for	J. K. Chase	New York, N. Y	Oct. 27, 1857	18, 498
Jars, Sealing-ring for preserve	J. Adams	Pittsburgh, Pa	Jan. 2, 1866	51, 785
Jars, &c., Stopping	W. D. Ludlow	New York, N. Y	Aug. 6, 1861	33, 002
Jaw wrench, Revolving	N. Colver	Boston, Mass	May 14, 1850	7, 362
Jelly-glass	J. and T. B. Atterbury	Pittsburgh, Pa	May 31, 1870	103, 702
Jelly-glass	W. M. Kirchner	Pittsburgh, Pa	Dec. 6, 1870	109, 824
Jelly-glass	W. M. Kirchner	Pittsburgh, Pa	July 16, 1872	129, 037
Jelly-glass	D. C. Ripley	Pittsburgh, Pa	May 20, 1873	139, 081
Jelly-glass	T. A. Zellers	East Birmingham, Pa	Apr. 15, 1873	137, 995
Jelly-glass lid	J. Adams	Pittsburgh, Pa	May 13, 1873	138, 833
Jelly-glass lid	D. Challinor	Pittsburgh, Pa	May 13, 1873	138, 735
Jelly, Hand-press for	J. M. Newton	Dayton, Ky	Aug. 12, 1873	141, 809
Jelly-jar	W. C. King	Pittsburgh, Pa	May 6, 1873	138, 502
Jelly-tumbler	W. Doyle	Birmingham, Pa	Mar. 4, 1873	136, 492
Jet, Adjustable spray	A. Nickerson	East Somerville, Mass	Sept. 9, 1873	142, 719
Jewel-case	C. Beck	Providence, R. I	Nov. 11, 1873	144, 431

Index of patents issued from the United States Patent Office from 1790 *to* 1873, *inclusive*—Continued.

Invention.	Inventor.	Residence.	Date.	No.
Jewel-case	G. F. Kolb	Philadelphia, Pa	Apr. 25, 1865	47, 430
Jewel-case	G. F. Kolb	Philadelphia, Pa	Dec. 17, 1867	72, 206
Jewel-setting instrument	A. A. F. and A. Thoma	Piqua, Ohio	Aug. 6, 1867	67, 462
Jeweler's drilling-device	J. K. Laudermilch	Lebanon, Pa	Sept. 23, 1873	143, 084
Jeweler's scraps, Process for refining	S. B. Darling	Providence, R. I	Mar. 27, 1855	12, 585
Jeweler's stock, Manufacture of	T. Diebold, jr	New York, N. Y	Dec. 10, 1872	133, 835
Jeweler's tool	D. M. Bissell	Shelburne Falls, Mass	Sept. 6, 1870	107, 153
Jeweler's use, Manufacture of catches for	G. H. Fuller	Pawtucket, R. I	Dec. 30, 1873	145, 938
Jewelry	J. J. Huber	Geneva, Switzerland	Jan. 3, 1860	26, 678
Jewelry and silver-ware case	S. Bodwell and C. Beck	Providence, R. I	Mar. 19, 1872	124, 783
Jewelry, Apparatus for ornamenting gum	N. Lanphear	Monmouth, Ill	Sept. 22, 1863	40, 053
Jewelry-bases, Device for manufacturing	S. Cottle	New York, N. Y	July 8, 1873	140, 616
Jewelry-bases, Manufacture of	S. Cottle	New York, N. Y	Jan. 21, 1873	135, 087
Jewelry-bases, Manufacture of	S. Cottle	New York, N. Y	Jan. 21, 1873	135, 088
Jewelry-box	L. L. Burdon	Providence, R. I	Nov. 9, 1869	96, 669
Jewelry-boxes, Fastening for	H. Hoefer	Brooklyn, N. Y	May 28, 1872	127, 238
Jewelry-case	J. Smith	Philadelphia, Pa	June 27, 1871	116, 506
Jewelry, diamonds, &c., Chemical compound for cleaning.	A. H. Longley and E. R. Stickney.	Springfield, Mass	Dec. 2, 1873	145, 065
Jewelry-fastening	R. J. Pond	New York, N. Y	Dec. 26, 1871	122, 192
Jewelry, Fastening for	J. T. Folwell	Philadelphia, Pa	Sept. 29, 1857	18, 277
Jewelry, Loop-chain for	C. W. Dickinson	Newark, N. J	Mar. 2, 1858	19, 497
Jewelry, Making joint-wire or stock for	A. J. Wiley	South Attleborough, Mass	Nov. 26, 1861	33, 807
Jewelry, Manufacture of	S. Cottle	New York, N. Y	Dec. 10, 1872	133, 762
Jewelry, Manufacture of	J. S. Palmer	Providence, R. I	Dec. 17, 1867	72, 224
Jewelry, Method of fastening	J. B. Coppinger	New York, N. Y	Oct. 28, 1856	15, 969
Jewelry, Mode of connecting strung pearl	H. Dubosq	Philadelphia, Pa	Apr. 26, 1859	23, 760
Jewelry, Mode of imitating cluster	W. O. Draper, A. C. Sweetland, and G. H. Draper.	North Attleborough, Mass	Mar. 10, 1868	75, 393
Jewelry, &c., Ornament for	C. E. Richards	North Attleborough, Mass	Feb. 9, 1869	86, 864
Jewelry-ornaments, Pin for	J. P. Courtney	Brooklyn, N. Y	Dec. 30, 1873	145, 990
Jewelry-pin	I. Farjeon and W. H. Horton	New York, N. Y., and Jersey City, N. J.	Oct. 25, 1870	108, 578
Jewelry, plate, &c., Machine for ornamenting	O. S. Parmenter	Providence, R. I	May 23, 1865	47, 853
Jewelry, &c., Production of ornamental surfaces for	W. B. Woodbury	London, England	Apr. 28, 1868	77, 831
Jewelry, Rolling metal for	J. S. Palmer	Providence, R. I	June 14, 1859	24, 432
Jewelry, Setting for	S. J. Smith	New York, N. Y	Apr. 14, 1863	38, 184
Jewelry, Setting stones in	F. Stefani	New York, N. Y	Oct. 2, 1866	58, 501
Jib and stay connection	J. E. Seavey	Kennebunkport, Me	Jan. 20, 1863	37, 476
Jib-boom for vessels	C. L. Linnell	Truro, Mass	May 10, 1859	23, 926
Jib, &c., Hank for	W. A. Pratt	Deep River, Conn	Dec. 24, 1872	134, 164
Jib-hank for vessels	H. K. Eldridge	Cambridge, Mass	Feb. 22, 1870	100, 019
Jib-stays, Backer for	T. Lynch	Peacedale, R. I	Nov. 12, 1872	132, 971
Joiner, General	D. R. Williams, sr	Paris, Ky	Aug. 20, 1872	130, 681
Joiner, Universal	E. Passé	Fort Wayne, Ind	Sept. 2, 1873	142, 348
Joiner, Universal	J. H. Rayner and B. W. and W. P. Hatch.	Boston, Mass	Apr. 29, 1873	138, 284
Joiner's bench	J. E. Cryer	Peoria, Ill	May 24, 1859	24, 102
Joiner's bench	J. W. Mahan	Lexington, Ill	Oct. 6, 1857	18, 345
Joiner's bench-strip	C. T. Pearson	Chelsea, Mass	May 12, 1857	17, 292
Joiner's clamp	J. Clackson	Milford, Pa	May 31, 1859	24, 194
Joiner's clamp	W. H. Goodchild	Centreville, N. J	May 16, 1871	114, 800
Joiner's clamp	W. H. Goodchild	Centreville, N. J	May 16, 1871	114, 801
Joiner's clamp	W. H. Goodchild and S. F. Hay	Bayonne, N. J., and Brooklyn, N. Y.	Feb. 25, 1873	136, 152
Joiner's clamp	J. Kuneman	Canton, Ohio	Aug. 25, 1868	81, 381
Joiner's clamp	W. S. Todd	Mechanicsville, Iowa	Sept. 6, 1859	25, 353
Joiner's clamp	E. A. Walker	New Haven, Conn	July 8, 1873	140, 747
Joiner's gage	M. C. Ames	Hartford, Conn	Nov. 19, 1867	70, 937
Joiner's gage	S. H. Jennings and W. F. Arnold.	Deep River and Winthrop, Conn.	Dec. 30, 1873	146, 075
Joiner's gage	C. Sholl	Mount Joy, Pa	Mar. 8, 1864	41, 867
Joining and planing machine	J. S. Mott	Alburgh, Saint Albans, Vt	Mar. 8, 1806	....
Joint:				

See Air-tight joint.
Ankle-joint.
Ankle-brace joint.
Ball-joint.
Ball-and-socket joint.
Box-joint.
Brace-joint.
Bracelet-joint.
Buggy-top joint.
Car-seat link-joint.
Carriage-prop joint.
Carriage-top joint.
Cement-pipe joint.
Chimney-joint.
Clapboard-joint.
Compass-joint.
Core-spindle joint.
Desk-joint.
Doll-joint.
Dovetailing-joint.
Felly-joint.
Fishing-rod joint.
Flue-joint.
Frame-joint.
Framing-joint.
Friction-joint.
Gas and water pipe joint.
Gas-pipe joint.
Globe-joint.
Iron-pipe joint.
Knee-joint.
Knuckle-joint.
Ladder-joint.
Lap-joint.
Lightning-rod joint.
Miter-joint.
Movable joint.
Pipe-joint.
Pivoted stump-joint.
Rail-joint.
Railway-rail joint.
Railway-track joint.
Rule-joint.
Seam-joint.
Seat and desk joint.
Shackle-joint.
Sheathing-joint.
Sheet-metal joint.
Sink-joint.
Skirt-hook joint.
Spectacle-bow joint.
Stationary joint.
Stave-joint.
Stove-pipe joint.
Stump-joint.
Sucker-rod joint.
Toggle-joint.
Toilet-glass joint.
Tongs-joint.
Tongue and groove joint.
Top-joint.
Tube-joint.
Universal joint.
Universal globe-joint.
Water-wheel joint.
Yielding joint.

Index of patents issued from the United States Patent Office from 1790 *to* 1873, *inclusive*—Continued.

Invention.	Inventor.	Residence.	Date.	No.
Joint of railway-rail	G. Palmer	Littlestown, Pa	Sept. 24, 1867	69, 241
Joint bit and check	R. D. Sterling	New York, N. Y	Oct. 22, 1867	70, 129
Joint-bolt	S. E. Jewett	Haverhill, Mass	Oct. 29, 1867	70, 223
Joint-clamp	D. E. Hall	Detroit, Mich	June 23, 1868	79, 116
Joint mold or flask	W. Culliss	Philadelphia, Pa	Jan. 4, 1870	98, 567
Joint tube, Free	R. Prosser	Birmingham, England	Sept. 21, 1852	9, 278
Jointed molds, Construction of	M. B. Stafford	New York, N. Y	Sept. 4, 1866	57, 791
Jointed-pipe connection	H. Smith	Norwalk, Ohio	Nov. 16, 1869	96, 982
Jointed pipe for steam, &c	L. Kirk	Reading, Pa	Oct. 9, 1847	5, 325
Jointer and gage-machine	J. Wadsworth	Milton, Mass	Feb. 10, 1810	....
Joist-hook	J. S. Miller	Terre Haute, Ind	Aug. 8, 1871	117, 798
Joist-protector	H. Galmann and C. Ruhe	Buchanan, Pa	Feb. 1, 1870	99, 305
Joists, Machine for dressing	J. H. Story	Cincinnati, Ohio	Aug. 21, 1860	29, 749
Joss-stick, Composition to be used instead of	F. Eckstein and J. McIlhenney	Philadelphia, Pa	Oct. 12, 1814	....
Journal	D. R. Quick	New York, N. Y	Aug. 31, 1869	94, 439
Journal, Adjustable	M. Jackson	Bavington, Pa	Feb. 4, 1873	135, 477
Journal and axle box	P. S. Devlan	Jersey City, N. J	July 9, 1867	66, 471
Journal and bearing	J. Whitaker	Woonsocket, R. I	Dec. 9, 1873	145, 467
Journal and bearing composition	M. J. Marcotte	Saint Louis, Mo	Aug. 1, 1871	117, 652
Journal and bearing material	E. D. Murfey	New York, N. Y	Oct. 25, 1870	108, 722
Journal and box	T. Hopper and T. Garrison	New Brunswick, N. J	Jan. 2, 1849	5, 996
Journal and box	W. S. Mead	New York, N. Y	Mar. 6, 1866	53, 023
Journal and box for railway-cars	A. Jackman	Elizabethport, N. J	May 1, 1860	28, 129
Journal and journal-box	J. Stimpson	Baldwinsville, Mass	Sept. 16, 1862	36, 487
Journal and shaft bearing, Frictionless	J. Eccles	Philadelphia, Pa	July 22, 1873	141, 129
Journal, Anti-friction	M. E. Dayton	Syracuse, N. Y	Apr. 22, 1873	138, 011
Journal, Anti-friction	A. W. Hall and M. E. Dayton	Adrian, Ohio, and Syracuse, N. Y.	Mar. 12, 1872	124, 573
Journal, Anti-friction	P. W. Yarrell	Littleton, N. C	Mar. 29, 1870	101, 401
Journal, Anti-friction	P. W. Yarrell	Littleton, N. C	Dec. 20, 1870	110, 323
Journal axle, Carriage	H. McIlroy	Poplar Ridge, N. Y	Sept. 8, 1868	81, 924
Journal, Axle or shaft	C. D. Smith	Baldwinsville, Mass	Jan. 7, 1868	73, 055
Journal bearing	S. P. and H. B. Cook	San Francisco, Cal	Aug. 8, 1871	117, 868
Journal-bearing	D. A. Hopkins	Jersey City, N. J	Nov. 15, 1870	109, 319
Journal-bearing	A. F. Jones	New York, N. Y	June 27, 1871	116, 317
Journal bearing	J. H. Lindsay	Allegheny, Pa	Jan. 23, 1872	122, 898
Journal-bearing	E. D. Murfey	New York, N. Y	Oct. 4, 1870	108, 080
Journal-bearing	E. D. Murfey	New York, N. Y	Dec. 12, 1871	121, 804
Journal-bearing	I. P. Wendell	Philadelphia, Pa	Jan. 11, 1870	98, 825
Journal-bearing	R. Yeilding	Ypsilanti, Mich	Oct. 6, 1863	40, 220
Journal-bearing and lubricating-device	D. N. M. Peregoy	Johnson City, Tenn	Dec. 26, 1871	122, 277
Journal-bearing, Anti-friction	R. G. Hatfield	New York, N. Y	Feb. 8, 1870	99, 566
Journal-bearing, Anti-friction-roller	E. Turner	Chicago, Ill	Nov. 3, 1863	40, 471
Journal-bearing for hoisting-machines, &c., Anti-friction.	R. G. Hatfield	New York, N. Y	Dec. 28, 1869	98, 260
Journal-bearing, Railway-car	I. P. Wendell	Philadelphia, Pa	Nov. 24, 1868	84, 326
Journal-bearings, Composition for	S. Croll	Philadelphia, Pa	Mar. 12, 1872	124, 485
Journal-bearings, &c., Graphitized materials for	A. Hitchcock	New York, N. Y	Nov. 19, 1872	133, 101
Journal-bearings, Lining for	H. F. Boody and E. P. Merrill	Deering, Me	Sept. 24, 1872	131, 653
Journal-bearings, Lubricating	I. P. Wendell	Philadelphia, Pa	June 28, 1870	104, 805
Journal-bearings, &c., Metalline for	S. Gwynn	New York, N. Y	Apr. 12, 1870	101, 861
Journal-bearings, &c., Metalline for	S. Gwynn	New York, N. Y	Apr. 12, 1870	101, 862
Journal-bearings, &c., Metalline for	S. Gwynn	New York, N. Y	Apr. 12, 1870	101, 863
Journal-bearings, &c., Metalline for	S. Gwynn	New York, N. Y	Apr. 12, 1870	101, 864
Journal-bearings, &c., Metalline for	S. Gwynn	New York, N. Y	Apr. 12, 1870	101, 865
Journal-bearings, &c., Metalline for	S. Gwynn	New York, N. Y	Apr. 12, 1870	101, 866
Journal-bearings, &c., Metalline for	S. Gwynn	New York, N. Y	Apr. 12, 1870	101, 867
Journal-bearings, &c., Metalline for	S. Gwynn	New York, N. Y	Apr. 12, 1870	101, 868
Journal-bearings, steps, and other articles liable to friction, Process of forming metalline for.	S. Gwynn	New York, N. Y	Apr. 12, 1870	101, 869
Journal-box	S. Aland	Rome, N. Y	Feb. 25, 1873	136, 198
Journal-box	J. A. Althouse	New Harmony, Ind	May 27, 1873	139, 287
Journal-box	J. E. Atwood	Mansfield, Conn	Sept. 15, 1868	82, 066
Journal-box	J. D. Beers	Philadelphia, Pa	Nov. 21, 1871	121, 040
Journal-box	J. M. Brower	Syracuse, N. Y	June 9, 1863	38, 808
Journal-box	A. G. Brown	Wareham, Mass	Apr. 4, 1871	113, 490
Journal-box	J. T. Bruen	New York, N. Y	Oct. 17, 1865	50, 445
Journal-box	J. Bryant	Brooklyn, N. Y	Jan. 10, 1860	26, 747
Journal-box	T. F. Burgess	Lowell, Mass	Dec. 10, 1867	71, 849
Journal-box	A. B. Caldwell	Syracuse, N. Y	Sept. 1, 1868	81, 748
Journal-box	W. T. Carroll	Medway, Mass	Apr. 4, 1871	113, 494
Journal-box	N. W. Clark	Clarkston, Mich	Feb. 5, 1861	31, 297
Journal-box	R. Colburn and G. W. Gould	Norwich, Conn	July 14, 1868	79, 900
Journal-box	J. Connolly	Boston, Mass	Jan. 1, 1867	60, 696
Journal-box	G. H. Corliss	Providence, R. I	July 25, 1871	117, 386
Journal-box	R. Daniels	Almena, Mich	Apr. 12, 1859	23, 664
Journal-box	P. S. Devlan	Elizabethport, N. J	Sept. 25, 1860	30, 128
Journal-box	E. S. Dickinson and W. D. Nelson.	New York, N. Y	Sept. 27, 1864	44, 407
Journal-box	W. H. Doane	Cincinnati, Ohio	Aug. 22, 1865	49, 584
Journal-box	D. H. Dotterer	Memphis, Tenn	May 7, 1861	32, 243
Journal-box	S. C. Ellis	Jersey City, N. J	Feb. 1, 1870	99, 422
Journal-box	S. C. Ellis	Jersey City, N. J	May 9, 1871	114, 542
Journal-box	J. H. Hayward	New York, N. Y	Dec. 10, 1872	133, 776
Journal-box	O. T. L. Heine and E. Prussing	Chicago, Ill	Dec. 24, 1861	33, 992
Journal-box	G. H. Henfield	San Francisco, Cal	Nov. 10, 1868	83, 961
Journal-box	D. R. Hickok	Morrisville, Vt	Apr. 24, 1866	54, 156
Journal-box	S. B. Higgins	South Boston, Mass	Dec. 19, 1865	51, 654
Journal-box	T. Hill	Saint Louis, Mo	Oct. 17, 1865	50, 477
Journal-box	D. A. Hopkins	Paterson, N. J	Feb. 23, 1858	19, 424
Journal-box	J. Hughes	New Berne, N. C	Aug. 16, 1870	106, 486
Journal-box	G. G. Hunt	Bridgeport, Conn	Jan. 12, 1864	41, 221
Journal-box	J. P. Kenyon	Brooklyn, N. Y	Sept. 15, 1863	39, 932
Journal-box	H. A. Lee	Worcester, Mass	Aug. 1, 1865	49, 124
Journal-box	E. F. Light	Worcester, Mass	Feb. 20, 1866	52, 720
Journal-box	E. F. Light	Worcester, Mass	Aug. 21, 1866	57, 343
Journal-box	E. F. Light	Worcester, Mass	Oct. 15, 1867	69, 822
Journal-box	I. D. Mathews	Worcester, Mass	Mar. 31, 1868	76, 092

Index of patents issued from the United States Patent Office from 1790 *to* 1873, *inclusive*—Continued.

Invention.	Inventor.	Residence.	Date.	No.
Journal-box	J. S. McClure	Gold Hill, Nev	Oct. 1, 1872	131, 889
Journal-box	J. McIlvain	Churchville, Md	Jan. 18, 1870	98, 987
Journal-box	G. R. Meneely	West Troy, N. Y	June 11, 1872	127, 784
Journal-box	G. R. Meneely	West Troy, N. Y	Sept. 17, 1872	131, 454
Journal-box	G. R. Meneely	West Troy, N. Y	Nov. 26, 1872	133, 472
Journal-box	J. Millholland and J. J. Lahaye	Reading, Pa	Mar. 18, 1862	34, 690
Journal-box	J. A. Montgomery	Williamsport, Pa	Aug. 2, 1859	24, 947
Journal-box	J. R. Morris	Houston, Tex	May 13, 1873	138, 916
Journal-box	W. T. Morrow	Chicago, Ill	Feb. 3, 1863	37, 586
Journal-box	E. D. Murfey	New York, N. Y	Oct. 4, 1870	108, 079
Journal-box	E. D. Murfey	New York, N. Y	Oct. 4, 1870	108, 081
Journal-box	J. A. Norris	Philadelphia, Pa	May 25, 1858	20, 363
Journal-box	J. Old	Pittsburgh, Pa	Mar. 20, 1847	5, 027
Journal-box	W. S. Pratt	Brooklyn, N. Y	Apr. 19, 1859	23, 704
Journal-box	H. M. Preston	Unionville, Conn	Aug. 11, 1868	81, 002
Journal-box	M. Queen	Brooklyn, N. Y	Aug. 28, 1860	29, 817
Journal-box	M. J. Rice and W. H. Millen	Boston, Mass	Mar. 14, 1865	46, 823
Journal-box	M. J. Rice and W. H. Millen	Boston, Mass	Aug. 22, 1865	49, 591
Journal-box	J. T. Robinett	Petersburgh, Va	June 9, 1868	78, 832
Journal-box	J. T. Robinett	Petersburgh, Va	Feb. 8, 1870	99, 711
Journal-box	J. B. Sargent	New Haven, Conn	Aug. 29, 1871	118, 647
Journal-box	A. H. Sassaman	Scranton, Pa	Dec. 14, 1869	97, 968
Journal-box	J. Scheider	New York, N. Y	Jan. 31, 1871	111, 479
Journal-box	L. Schmid	Cleveland, Ohio	Sept. 17, 1872	131, 376
Journal-box	W. Sherburne	Charlestown, Mass	July 14, 1868	79, 867
Journal-box	R. Sibley	Greenville, Conn	May 4, 1869	89, 839
Journal-box	E. W. Skinner	Madison, Wis	Mar. 30, 1869	88, 333
Journal-box	E. H. Stearns	Erie, Pa	Dec. 13, 1870	110, 086
Journal-box	L. M. Taylor and W. D. Fowler	Washington, D. C	Oct. 1, 1867	69, 373
Journal-box	H. H. Thayer	Sandwich, Mass	Sept. 7, 1858	21, 474
Journal-box	W. O. Thoms and A. M. Miller	Fond du Lac, Wis	Oct. 14, 1862	36, 682
Journal-box	H. Turner	Charlestown, N. H	Apr. 27, 1852	8, 915
Journal-box	G. L. Weaver	Hartford, Conn	Sept. 29, 1868	82, 665
Journal-box	J. Webster	Brooklyn, N. Y	June 18, 1861	32, 604
Journal-box	I. P. Wendell	Philadelphia, Pa	Apr. 17, 1860	27, 940
Journal-box	D. Whitlock	Newark, N. J	June 18, 1872	128, 002
Journal-box and axle-bearing	S. T. F. Sterick	Georgetown, D. C	July 18, 1871	117, 217
Journal box and bearing	J. B. Caryl	Candor, N. Y	Apr. 9, 1867	63, 609
Journal box and bearing, Self-lubricating	C. Purdy	Bedford, Ohio	Feb. 26, 1867	62, 500
Journal box and bearing, Self-oiling	B. D. Stevens	Lawrence, Mass	June 23, 1863	39, 018
Journal-box, Anti-friction	B. C. Baker	Toledo, Ohio	July 11, 1871	116, 795
Journal-box, Anti-friction	H. M. Le Duc	Washington, D. C	Sept. 18, 1866	58, 109
Journal-box, Anti-friction	W. S. Pratt	Williamsburgh, N. Y	May 8, 1860	28, 195
Journal-box, Anti-friction	R. W. Smith	Toledo, Ohio	July 16, 1872	129, 256
Journal-box, Anti-friction	H. F. Stith	Stanton, Kans	Sept. 10, 1867	68, 667
Journal-box, Anti-friction	H. P. Westcott	Seneca Falls, N. Y	Mar. 14, 1871	112, 660
Journal-box, Car	G. H. Clemens	Baltimore, Md	Nov. 12, 1867	70, 697
Journal-box, Car	J. B. Collin	Altoona, Pa	Sept. 28, 1869	95, 196
Journal-box, Car	E. Von Jeinsen	Omaha, Nebr	Nov. 9, 1870	109, 159
Journal-box composition	F. W. Armstrong	New York, N. Y	May 27, 1862	35, 409
Journal-box for car-wheels	R. J. Hamilton	Chicago, Ill	Mar. 3, 1863	37, 814
Journal-box for carriages	S. B. Batchelor	Lowville, N. Y	Jan. 2, 1855	12, 154
Journal-box for land-carriages	E. P. Palmer	Milton, Del	Nov. 14, 1865	50, 948
Journal-box for lubricating axles	J. Schinneller	Temperanceville, Pa	Jan. 9, 1872	122, 602
Journal-box for railway-car axles	W. B. Gage	Saratoga Springs, N. Y	Apr. 1, 1856	14, 560
Journal-box for saw-mill carriages	W. M. Ferry, jr	Ferrysburgh, Mich	Nov. 29, 1859	26, 262
Journal-box for saw-mill carriages	C. R. Fox	Chicago, Ill	May 9, 1854	10, 882
Journal box, Glass	E. Campbell	Columbus, Ohio	Aug. 21, 1855	13, 454
Journal-box, Grindstone	T. W. Brown	New York, N. Y	Oct. 16, 1866	58, 768
Journal-box lining	J. Corduan	Brooklyn, N. Y	Mar. 12, 1861	31, 655
Journal-box lining	P. S. Devlan	Jersey City, N. J	Nov. 13, 1866	59, 564
Journal-box lining	S. Gwynn	New York, N. Y	May 6, 1873	138, 641
Journal-box lining	S. Gwynn	New York, N. Y	May 6, 1873	138, 642
Journal-box lining	S. Gwynn	New York, N. Y	May 6, 1873	138, 643
Journal-box lining	S. Gwynn	New York, N. Y	May 6, 1873	138, 644
Journal-box lining	S. Gwynn	New York, N. Y	May 6, 1873	138, 645
Journal-box lining	S. Gwynn	New York, N. Y	May 6, 1873	138, 646
Journal-box lining	S. Gwynn	New York, N. Y	July 15, 1873	140, 774
Journal-box lining	S. Gwynn	New York, N. Y	July 15, 1873	140, 775
Journal-box lubricator	A. and F. Brown	New York, N. Y	Sept. 28, 1869	95, 313
Journal-box lubricator, Railway-car	J. C. Geisendorff	Cincinnati, Ohio	Feb. 9, 1858	19, 291
Journal-box-oiling device	G. M. Morris	Cohoes, N. Y	Apr. 23, 1867	64, 130
Journal-box, Railway	O. Beecher and R. E. Rogers	Philadelphia, Pa	Dec. 8, 1863	40, 810
Journal-box, Railway	I. Dripps	Altoona, Pa	Aug. 8, 1871	117, 753
Journal-box, Railway	W. Groat	Green Island, N. Y	Feb. 16, 1864	41, 616
Journal-box, Railway	A. B. Nimbs	Buffalo, N. Y	Oct. 31, 1865	50, 771
Journal-box, Railway	R. C. Wright	Meadville, Pa	July 25, 1865	49, 022
Journal-box, Railway	R. C. Wright	Mead Township, Pa	Nov. 14, 1865	50, 982
Journal-box, Railway-car	W. B. Aitken	Philadelphia, Pa	Dec. 1, 1863	40, 729
Journal-box, Railway-car	R. N. Allen	Pittsford, Vt	July 8, 1873	140, 568
Journal-box, Railway-car	J. Ashenfelder	Philadelphia, Pa	July 31, 1860	29, 349
Journal-box, Railway-car	R. Brewer	Pontiac, Ill	Dec. 26, 1871	122, 219
Journal-box, Railway-car	D. A. Hopkins	Brooklyn, N. Y	Mar. 11, 1862	34, 636
Journal-box, Railway-car	C. Ihrig	Jersey City, N. J	July 26, 1870	105, 689
Journal-box, Railway-car	R. McWilliams	Philadelphia, Pa	Dec. 15, 1857	18, 884
Journal-box, Railway-car	R. McWilliams	Philadelphia, Pa	July 19, 1859	24, 846
Journal-box, Railway-car, &c	J. A. Norris	Philadelphia, Pa	Dec. 29, 1857	18, 984
Journal-box, Railway-car	A. L. Raplee	Edwin, N. Y	Sept. 13, 1870	107, 408
Journal-box, Railway-car	H. Rice	Concord, N. H	Aug. 7, 1860	29, 551
Journal-box, Railway-car	H. B. Rowley	Buffalo, N. Y	Apr. 27, 1869	89, 349
Journal-box, Railway-car	J. O. Scott	New York, N. Y	May 24, 1864	42, 881
Journal-box, Railway-car	M. Thornton	Macon, Ga	May 26, 1868	78, 244
Journal-box, Railway-car	P. Umholtz	Tremont, Pa	Mar. 15, 1859	23, 279
Journal-box, Roller	E. Wadhams	Yorkville, N. Y	Dec. 31, 1867	72, [illegible]42
Journal-box, Self-lubricating	S. F. Bond	Worcester, Mass	Apr. 23, 1867	64, 0[illegible]4
Journal-box, Self-lubricating	P. P. Lane	Cincinnati, Ohio	Nov. 26, 1867	71, 313
Journal-box, Self-oiling	J. Sault	South Manchester, Conn	Sept. 26, 1871	119, 412
Journal-box, Self-oiling	J. Whitlock	Birmingham, Conn	Apr. 23, 1867	64, 054

Index of patents issued from the United States Patent Office from 1790 *to* 1873, *inclusive*—Continued.

Invention.	Inventor.	Residence.	Date.	No.
Journal-box, Bearing for	H. Grogan	Flatbush, N. Y	July 11, 1871	116, 952
Journal-boxes, bearings, &c., Composition for forming.	J. Absterdam	New York, N. Y	May 6, 1862	35, 132
Journal-boxes, Composition for	G. W. Disman	Upper Sandusky, Ohio	Feb. 27, 1866	52, 829
Journal-boxes, &c., Composition for lining	P. S. Devlan	Jersey City, N. J	Aug. 22, 1865	49, 509
Journal-boxes, Composition for lining	P. S. Devlan	Jersey City, N. J	Dec. 26, 1865	51, 700
Journal-boxes, Composition for lining	P. S. Devlan	Jersey City, N. J	Dec. 26, 1865	51, 701
Journal-boxes, Composition for lining	P. S. Devlan	Jersey City, N. J	Dec. 26, 1865	51, 702
Journal-boxes for shafting, &c., Manufacture of	D. Taylor	Carbondale, Pa	June 30, 1857	17, 706
Journal-boxes, Lubricating	E. A. Atwood and H. H. Bodwell.	San Francisco, Cal	Feb. 2, 1869	86, 489
Journal-boxes, Lubricating	J. T. Light	Worcester, Mass	Sept. 26, 1865	50, 143
Journal-boxes, Lubricating	J. Morin	New York, N. Y	Dec. 30, 1873	146, 013
Journal-boxes, Lubricating	A. R. Sherman	Natick, R. I	July 24, 1866	56, 624
Journal-boxes, &c., Lubricating-compound for	H. Colby	New York, N. Y	Sept. 19, 1865	49, 983
Journal-boxes, Machine for dressing brasses of	G. Sanford	Stroudsburgh, Pa	Feb. 25, 1873	136, 338
Journal-boxes, Machine for forming, shaping, and dressing.	J. W. Cole	Brooklyn, N. Y	June 10, 1873	139, 773
Journal-boxes, Material for	P. S. Devlan	Hudson City, N. J	Apr. 21, 1868	77, 012
Journal-boxes of connecting-rod or pitman, Mode of tightening and securing the keys of the.	L. Dederick	Albany, N. Y	Mar. 9, 1858	19, 548
Journal-boxes, Oil-hole cover for	W. A. Wood	Hoosick Falls, N. Y	Nov. 7, 1871	120, 693
Journal boxes or bearings, Process of grinding	J. W. Cole	Brooklyn, N. Y	June 10, 1873	139, 772
Journal-boxes, Packing	W. Groat	Troy, N. Y	Mar. 27, 1855	12, 591
Journal, Car-axle	J. Whitaker	Woonsocket, R. I	Nov. 25, 1873	145, 035
Journal, Compensation	J. M. Vernon	Mount Vernon, Ohio	July 30, 1872	129, 911
Journal for car-wheels, &c., Anti-friction	C. H. Parshall	Detroit, Mich	Nov. 26, 1867	71, 527
Journal-lubricating composition	B. Battle	Pittsburgh, Pa	Jan. 8, 1867	60, 989
Journal-lubricating composition	R. C. Klein	Saint Louis, Mo	Mar. 28, 1871	113, 176
Journal-lubricator	E. Andrews and J. H. Carr	Palo Alto, Pa	June 19, 1860	28, 725
Journal-lubricator	J. A. Cowles	Chicago, Ill	Dec. 6, 1870	109, 870
Journal-lubricator	P. S. Devlan	Jersey City, N. J	Apr. 18, 1871	113, 860
Journal-lubricator	F. Keifel	Cincinnati, Ohio	Jan. 23, 1872	123, 024
Journal-lubricator	A. Overbagh	Scranton, Pa	Dec. 28, 1869	98, 402
Journal-lubricator	C. M. Reid	Greensborough, Ala	Dec. 8, 1868	84, 765
Journal-lubricator	M. Sellers	Keokuk, Iowa	Nov. 26, 1867	71, 332
Journal-lubricator	N. S. Snedeker	Philadelphia, Pa	Jan. 8, 1867	61, 109
Journal-lubricator	I. P. Wendell	Philadelphia, Pa	Jan. 1, 1867	60, 809
Journal lubricator, Car	T. Sayles	Chicago, Ill	Feb. 1, 1870	99, 354
Journal lubricator, Car	S. Ustick	Philadelphia, Pa	Dec. 5, 1871	121, 557
Journal lubricator, Car-axle	E. Collins	New York, N. Y	Sept. 5, 1871	118, 691
Journal lubricator, Car-axle	S. Ustick	Philadelphia, Pa	July 16, 1872	129, 500
Journal lubricator, Car-axle and other	P. E. Proust	Orleans, France	Nov. 18, 1856	16, 099
Journal lubricator, Car-shaft	I. P. Wendell	Philadelphia, Pa	June 9, 1868	78, 704
Journal lubricator, Railway	A. P. Allen	Niagara, N. Y	Aug. 12, 1862	36, 133
Journal lubricator, Railway	C. W. Harvey	Buffalo, N. Y	Sept. 19, 1871	119, 087
Journal of axles with friction-rollers	G. A. Prentiss	Cambridge, Mass	Sept. 22, 1857	18, 248
Journal or shaft hanger	J. Whitaker	Woonsocket, R. I	Dec. 2, 1873	145, 264
Journal-packing rack	W. H. Humphrey	Lansingburgh, N. Y	Sept. 9, 1873	142, 698
Journal, Railway-car	W. G. Smith	Chicago, Ill	Nov. 15, 1864	45, 124
Journal, Rolling or frictionless	D. Holmes	Chelsea, Mass	Apr. 15, 1862	34, 962
Journal, Shaft	L. Patric	Shortsville, N. Y	Sept. 8, 1868	81, 938
Journals and axle-boxes, Lining for	P. S. Devlan	Jersey City, N. J	July 9, 1867	66, 472
Journals and axles, Lubricating	J. B. G. M. F. Pirot	Paris, France	June 2, 1863	38, 760
Journals and bearings and for lubricating, Material for.	A. B. Jones	Wilmington, N. C	Apr. 20, 1869	89, 049
Journals and bearings, Lubricating	I. P. Wendell	Philadelphia, Pa	Apr. 5, 1870	101, 551
Journals and boxes, Coupling	D. H. Dotterer	Philadelphia, Pa	July 9, 1867	66, 474
Journals and other parts of machinery, Packing	C. A. Stevens	New York, N. Y	Mar. 29, 1870	101, 394
Journals, Anti-friction bearing for	A. W. Hall	New York, N. Y	June 6, 1871	115, 602
Journals, axles, &c., Adjustable coupling for	F. Burghardt	Curtisville, Mass	May 24, 1870	103, 295
Journals, axles, &c., Composition for lubricating	C. L. Morehouse	Jackson, Tenn	Jan. 29, 1861	31, 254
Journals, bearings, and packings, Manufacture of material for.	E. D. Murfey	New York, N. Y	Dec. 27, 1870	110, 584
Journals, &c., by a pendulum-valve arrangement, Method of lubricating.	J. B. Tom and S. D. Tucker	New York, N. Y	Jan. 12, 1858	19, 108
Journals, Device for cooling car	H. G. Thompson	Milford, Conn	Mar. 7, 1871	112, 513
Journals from heating, Method of preventing	E. Reid	Columbus, Ga	May 25, 1844	3, 595
Journals in soft metal, Gage and box for casting	C. W. Griffith	Dayton, Ohio	Sept. 20, 1859	25, 504
Journals, Lubricating	C. Andrew	Providence, R. I	Nov. 7, 1865	50, 786
Journals, Lubricating	E. Andrews and J. H. Carr	Palo Alto, Pa	Mar. 13, 1860	27, 416
Journals, Lubricating	C. Bean	Providence, R. I	June 28, 1870	104, 821
Journals, Lubricating	G. A. Chapman	Woonsocket, R. I	Aug. 26, 1873	142, 206
Journals, Lubricating	G. E. Clarke and E. P. Dickey	Racine, Wis	June 22, 1869	91, 603
Journals, Lubricating	S. Nash	Boston, Mass	Oct. 12, 1869	95, 716
Journals, Lubricating	J. J. Stevenson	Auburn, N. Y	June 5, 1866	55, 386
Journals, Lubricating	S. Ustick	Philadelphia, Pa	Sept. 26, 1871	119, 288
Journals, Lubricating	W. Van Auden	Poughkeepsie, N. Y	Oct. 17, 1865	50, 520
Journals, Lubricating-cushion for railway-car	P. S. Devlan	Jersey City, N. J	June 8, 1869	91, 000
Journals, Lubricating railway	C. R. Laman	Painted Post, N. Y	Apr. 25, 1871	114, 157
Journals of axles on railway, Reducing the friction of.	L. J. Pomme de Mirimonde	Paris, France	Feb. 2, 1858	19, 237
Journals of car-axles, Cooling	H. G. Thompson	Milford, Conn	Mar. 7, 1871	112, 511
Journals of car-axles, Device for cooling	H. G. Thompson	Milford, Conn	Mar. 7, 1871	112, 510
Journals of car-wheel axles, Cooling	H. G. Thompson	Milford, Conn	Mar. 7, 1871	112, 512
Journals of railway-cars, locomotives, &c., Mode of constructing the bearings and oil-boxes for.	J. H. Tims	Newark, N. J	Oct. 31, 1839	1, 390
Journals of rolling-mills, Application of hot water to	J. Bryan	Covington, Ky	Nov. 24, 1857	18, 674
Journals of shafts, axles, &c., Friction-roller for	W. H. Main	Litchfield, Ohio	May 19, 1857	17, 333
Journals, Oil-hole cover for	S. A. Skinner	Hoosick Falls, N. Y	Feb. 27, 1872	124, 032
Journals, Oiling	D. B. Jordan	Woonsocket, R. I	Mar. 15, 1859	23, 251
Journals, Oiling	W. G. Wynne	Albany, N. Y	Aug. 8, 1871	117, 845
Journals, Oiling	J. Wood	Red Bank, N. J	Dec. 3, 1861	33, 871
Journals to feed-rollers, Mode of attaching	G. M. Amsden	South Boston, Mass	Sept. 2, 1873	142, 364
Journals, Tool for turning	J. Hall	New Haven, Conn	Dec. 29, 1857	18, 970
Jug and pitcher, Molasses	W. Bennett	Baldwin Township, Pa	June 18, 1872	128, 100
Jug, cruet, &c., Means of attaching top to	H. Wright	Pittsburgh, Pa	Oct. 26, 1869	96, 178
Jug or pitcher, Molasses	M. J. Bennett	Braddock's Field, Pa	Jan 28, 1873	135, 260

Index of patents issued from the United States Patent Office from 1790 to 1873, inclusive—Continued.

Invention.	Inventor.	Residence.	Date.	No.
Jug-top	P. Lauster	Allegheny, Pa	Sept. 8, 1868	82, 010
Jug-top	H. Wright	Pittsburgh, Pa	May 10, 1864	42, 712
Jug-top	H. Wright	Pittsburgh, Pa	Mar. 6, 1866	53, 076
Jug-top	H. Wright	Pittsburgh, Pa	Sept. 24, 1867	69, 151
Jug-top	H. Wright	Pittsburgh, Pa	Apr. 13, 1869	88, 829
Jug-top	H. Wright	Pittsburgh, Pa	July 16, 1872	129, 303
Jug-top, Casting	F. B. Richardson	Boston, Mass	June 22, 1869	91, 669
Jug-tops, Making	G. P. Lang and P. Lauster	Allegheny, Pa	Jan. 1, 1867	60, 750
Jump-seat	N. Warren and T. Underwood	Wilmington, Del	Dec. 5, 1871	121, 563
Jump-seat, Stump-joint	J. D. McAuliff	Saint Louis, Mo	July 2, 1872	128, 645
Jumping-hoop	C. L. Browne	Brooklyn, N. Y	July 23, 1867	67, 101
Jute fiber, Treating	J. Monach	Philadelphia, Pa	Feb. 4, 1868	74, 113
Jute twine, Preparing	J. B. Travers	New York, N. Y	Nov. 3, 1863	40, 520
Jute yarn, Preparing	J. Monach	Philadelphia, Pa	Apr. 21, 1868	77, 073
K.				
Kaleidoscope	J. Earnshaw	East Greenwich, R. I	June 25, 1867	66, 134
Kaleidoscope	M. Keller	Boston, Mass	July 12, 1870	105, 218
Kaleidoscope	A. C. McNulty and D. Lyman, jr.	New York, N. Y	Oct. 2, 1860	30, 238
Kaleidoscope-object	C. G. Bush	Boston, Mass	Sept. 30, 1873	143, 271
Kaleidoscope, Telescopic	J. Pool	Newark, N. J	Nov. 12, 1872	132, 978
Kalsomining wall, &c., Composition for	N. A. Frank	Chicago, Ill	June 30, 1868	79, 337
Kaolin, Apparatus for washing	D. Hull	Newburgh, N. Y	May 1, 1866	54, 355
Keel and bilge block	J. T. Parlour	Brooklyn, N. Y	Jan. 14, 1868	73, 376
Keel block or rest	J. T. Parlour	Brooklyn, N. Y	June 11, 1867	65, 691
Keel for boats, Ballast	H. A. Dirkes	New York, N. Y	Apr. 21, 1868	76, 896
Keel for ships and other navigable vessels	J. B. Tarr	Chicago, Ill	May 23, 1865	47, 879
Keel or bottom of vessels, Apparatus for ascertaining the curvature of.	H. E. Towle	Exeter, N. H	Apr. 23, 1861	32, 157
Keel, Swinging	J. Beale	Alexandria, Va	Jan. 15, 1812	......
Keel, Swinging false	P. Hall	......	Apr. 2, 1816	......
Keels, Apparatus for examining vessels	J. E. Simpson	East Boston, Mass	May 26, 1857	17, 396
Keels for vessels, Combination of center and bilge	S. P. Willeby	Philadelphia, Pa	May 20, 1873	139, 222
Keels of vessels, Construction of	G. Livingston	......	Oct. 22, 1803	......
Keels to vessels, Attaching	T. F. Griffith	New Market, Md	Apr. 26, 1845	4, 016
Keg and barrel for paint and other material	A. L. Freeman	Manchester, England	June 4, 1867	65, 368
Keg, Gunpowder	J. Wilson, C. Green, and W. Wilson, jr.	Brandywine, Del	Mar. 31, 1857	16, 944
Keg, Metallic	W. Hill	Pottsville, Pa	July 27, 1869	93, 085
Kegs, &c., Attaching the heads of metallic powder	J. Wilson, C. Green, and W. Wilson, jr.	Wilmington, Del	July 12, 1859	24, 772
Kegs, Machine for securing heads to metallic	W. Hill	Pottsville, Pa	Oct. 25, 1870	10[illegible], 594
Kegs, Manufacture of oyster	G. Frauenberger	Rochester, N. Y	Aug. 17, 1869	93, 870
Kegs or casks, Machine for chamfering and crozing	J. A. Seaman	Saint Louis, Mo	Nov. 22, 1859	26, 206
Kerosene burning-fluid	A. Gesner	Williamsburgh, N. Y	June 27, 1854	11, 203
Kerosene burning-fluid	A. Gesner	Williamsburgh, N. Y	June 27, 1854	11, 204
Kerosene burning-fluid	A. Gesner	Williamsburgh, N. Y	June 27, 1854	11, 205
Kerosene-heater	Z. B. and C. E. Grandy	Stafford Springs, Conn	July 22, 1873	141, 046
Kerosene, Process for making	A. Gesner	Williamsburgh, N. Y	Mar. 27, 1855	12, 612
Kettle	W. J. Evans and W. Perry	New York, N. Y	June 3, 1873	139, 502
Kettle	H. Matthews	South Yarmouth, Mass	Dec. 31, 1867	72, 746
Kettle	A. N. Merrill	Batavia, Ill	May 14, 1867	64, 685
Kettle	A. Sower	New York, N. Y	May 28, 1867	65, 129
Kettle	D. Stuart	Philadelphia, Pa	Oct. 3, 1871	119, 666
Kettle and boiler for heating water	P. N. Griffith	Orange, N. J	Sept. 26, 1825	......
Kettle and vessel cover	E. K. Dean	Bangor, Me	May 3, 1870	102, 507
Kettle-bail	J. Britton	Williamsburgh, N. Y	Aug. 9, 1870	106, 115
Kettle, &c., bail	A. D. Crocker	Boston, Mass	Sept. 18, 1866	58, 224
Kettle-bail	J. T. Darnell	Florence, N. J	Oct. 22, 1872	132, 450
Kettle-bail	T. H. Dodge	Nashua, N. H	May 24, 1853	9, 744
Kettle bail	W. Hailes	Albany, N. Y	Dec. 17, 1867	72, 195
Kettle, Boiling	H. H. Pember	New York, N. Y	June 9, 1868	78, 823
Kettle, Boiling	S. Spoor	Phelps, N. Y	Feb. 23, 1869	87, 306
Kettle-bottom	C. A. Moore	West Brook, Conn	May 29, 1866	55, 145
Kettle, Camp	C. Avery	Ashtabula, Ohio	May 3, 1864	42, 547
Kettle, Camp	J. C. Mulligan	Elizabeth City, N. J	Jan. 20, 1863	37, 456
Kettle, Cooking	B. W. Dunning	Brooklyn, N. Y	Apr. 16, 1867	63, 871
Kettle-cover	V. R. Holmes	Grand Rapids, Mich	Jan. 12, 1869	85, 827
Kettle, Dinner	J. Wagner	Cumberland, Md	Mar. 31, 1868	76, 121
Kettle-ears, Making	M. Wells	Brooklyn, N. Y	May 8, 1860	28, 222
Kettle for boiling by steam	J. O. Morse and G. D. Hiscox	Englewood, N. J., and Brooklyn, N. Y.	Sept. 14, 1869	94, 907
Kettle for boiling milk	H. Friedländer	New York, N. Y	Feb. 2, 1869	86, 385
Kettle for boiling salt, &c	J. Kendall	Fayette County, Pa	Mar. 18, 1818	......
Kettle for calcining plaster of Paris	J. B. King	New York, N. Y	Nov. 21, 1854	11, 966
Kettle for culinary purposes	J. Burns	New York, N. Y	Dec. 29, 1868	85, 429
Kettle for culinary purposes	D. H. Tuxworth, sr	Baltimore, Md	Jan. 27, 1863	37, 533
Kettle for manufacturing comfits	W. H. Holt	Hartford, Conn	Oct. 8, 1850	7, 702
Kettle for melting, mixing, and casting metals	W. S. Deeds	Baltimore, Md	July 19, 1870	105, 553
Kettle for trying oil	J. L. Alberger	Buffalo, N. Y	Nov. 30, 1858	22, 152
Kettle-furnace, Portable	J. Murdock	South Carver, Mass	June 20, 1865	48, 302
Kettle-handle	H. T. Robbins	Hyde Park, Mass	May 16, 1871	114, 860
Kettle-handle	J. Warner	New Britain, Conn	Jan. 22, 1861	31, 198
Kettle, Hatter's	R. Wildman	Hartford, Conn	Nov. 8, 1845	4, 256
Kettle, Hatter's planking	E. Joyce	Brooklyn, N. Y	June 6, 1825	......
Kettle, Lard-rendering	J. J. Bate	Brooklyn, N. Y	Oct. 21, 1856	15, 942
Kettle, Lard-rendering	J. J. Bate	Brooklyn, N. Y	Sept. 29, 1857	18, 271
Kettle, Lard-rendering	J. J. Bate	Brooklyn, N. Y	July 13, 1858	20, 856
Kettle, Lard-rendering	A. Lapham	Brooklyn, N. Y	Nov. 10, 1857	18, 622
Kettle-lifter	E. A. Thissell	Lowell, Mass	July 22, 1873	141, 020
Kettle machine, Brass	J. Cannon	Warren, R. I	Feb. 3, 1857	16, 564
Kettle machine, Brass	O. W. Minard	Waterbury, Conn	July 1, 1856	15, 247
Kettle-molding flask	G. Walworth	Peekskill, N. Y	Mar. 30, 1869	88, 425
Kettle or boiler used in salt manufacture, Separating and removing bitterings from.	D. Dear	Salina, N. Y	Oct. 18, 1837	428
Kettle, pail, &c., lid	S. B. Cox	Buffalo, N. Y	Aug. 20, 1867	67, 847

Index of patents issued from the United States Patent Office from 1790 *to* 1873, *inclusive*—Continued.

Invention.	Inventor.	Residence.	Date.	No.
Kettle, Shielding-arch for evaporating	J. English	Syracuse, N. Y	Nov. 17, 1868	84, 101
Kettle-spout attachment	M. Saulson	Troy, N. Y	Aug. 3, 1869	93, 354
Kettle, Stove or range	L. D. Lothrop	Dover, N. H	Feb. 13, 1872	123, 569
Kettle, Wash	J. Reist	Philadelphia, Pa	Nov. 29, 1864	45, 272
Kettles and other hollow ware, Machine for making brass.	F. J. Seymour	Wolcottville, Conn	Jan. 14, 1873	134, 938
Kettles and other vessels, Apparatus for discharging liquids from.	A. Brear	Saugatuck, Conn	Apr. 15, 1862	35, 007
Kettles and other vessels, Mode of discharging the contents of sugar.	A. Brear	Saugatuck, Conn	Apr. 1, 1862	34, 811
Kettles, boilers, &c., Bail for	W. M. Stratton	West Troy, N. Y	Feb. 13, 1872	123, 648
Kettles, boilers, stoves, &c., Attaching covers to	W. Hailes	Albany, N. Y	Dec. 18, 1866	60, 508
Kettles, Casting iron	C. McGinnes	Pittsburgh, Pa	July 20, 1858	20, 951
Kettles, Constructing and setting salt	C. C. P. Crosby	Brooklyn, N. Y	Jan. 11, 1840	1, 471
Kettles, Constructing hatters'	W. Porter and A. Sanger	Waltham, Mass	July 8, 1834	....
Kettles, Construction of salt	O. W. Seely	Albany, N. Y	Apr. 9, 1861	32, 005
Kettles, Construction of stamped sheet-metal	F. G. and W. F. Niedringhaus	Saint Louis, Mo	Dec. 10, 1867	71, 900
Kettles, Cover and lifting-device for	A. Selkirk	Albany, N. Y	Jan. 29, 1867	61, 636
Kettles, Die for manufacturing brass	O. Newton	Pittsburgh, Pa	Feb. 11, 1862	34, 374
Kettles, &c., for boilers, Setting	G. W. Robinson	Attleborough, Mass	July 17, 1818	....
Kettles, &c., from disks of metal, Machinery for making.	H. W. Hayden	Waterbury, Conn	Dec. 16, 1851	8, 589
Kettles from metal disks, Machine for forming	L. C. Camp	Berlin, Conn	Jan. 9, 1855	12, 227
Kettles, Grinding the inner surfaces of cast-iron	C. C. Bradley, jr	Syracuse, N. Y	Feb. 24, 1857	16, 679
Kettles, Machine for making brass	E. C. Blakesley, E. Platt, jr., and E. Jordan.	Waterbury, Conn	Oct. 28, 1856	15, 961
Kettles, Making brass	O. W. Minard	Waterbury, Conn	Apr. 15, 1856	14, 696
Kettles, Making brass	O. W. Minard	Waterbury, Conn	Sept. 23, 1856	15, 772
Kettles, Making brass	F. J. Seymour	Waterbury, Conn	May 13, 1856	14, 887
Kettles, Metallic flask for casting large	W. Kelly	Eddyville, Ky	Dec. 3, 1850	7, 810
Kettles, Setting and arranging sugar	M. White	New Orleans, La	Sept. 17, 1839	1, 326
Kettles, Setting evaporating	W. S. Worthington	Newtown, N. Y	Sept. 8, 1863	39, 859
Kettles, Setting potash	D. B. Turner	Florence, Ohio	Sept. 18, 1841	2, 263
Kettles, Setting sugar	J. Malory	New Orleans, La	Sept. 20, 1838	936
Kettles, Setting sugar	H. Roth	Iberville Parish, La	Mar. 2, 1858	19, 515
Kettles with spouts, Molding	W. H. Pease	Dayton, Ohio	Oct. 28, 1851	8, 471
Key: *See* Bed-key. Bedstead-key. Clock-key. Cock-key. Coffin-key. Door-key. Edge-key. Fence-key. Lever-key. Lock-key. Organ-key. Piano-key. Sewing-machine-lock key. Sheet-metal key. Telegram-key. Telegraph-key. Telegraph transmitting-key. Tuning-key. Watch-key. Watch-pendant key. Wire-fence key.				
Key	T. S. Bowman	Saint Louis, Mo	May 15, 1866	54, 678
Key	J. Deally	Louisville, Ky	May 29, 1860	28, 457
Key	H. H. Elwell	South Norwalk, Conn	July 12, 1870	105, 322
Key	J. Linden	Seneca Falls, N. Y	Feb. 8, 1870	99, 577
Key	C. E. Lombard	Springfield, Mass	Mar. 26, 1867	63, 271
Key	E. Parker	New Britain, Conn	May 11, 1869	89, 886
Key	E. Parker	New Britain, Conn	May 18, 1869	90, 121
Key	S. Perry	Newport, N. Y	Aug. 4, 1857	17, 939
Key	A. Vang	Chicago, Ill	Jan. 19, 1869	86, 114
Key-blanks, Die for making	E. Parker	New Britain, Conn	July 2, 1872	128, 650
Key-board	R. Humphreys	Jonesborough, Tenn	June 14, 1859	24, 392
Key-board	J. P. Perry	Yarmouth Port, Mass	May 23, 1871	115, 096
Key-board, Chromatic	H. Downes	New York, N. Y	Jan. 9, 1872	122, 584
Key-board instruments, Key-cap for	J. P. Lord	Manchester, N. H	Nov. 11, 1873	144, 399
Key-board instruments, Melody-attachment to	C. Fogelberg	New York, N. Y	Jan. 26, 1869	86, 219
Key-board, Transposing	A. D. B. Wolff	Paris, France	Sept. 16, 1873	142, 974
Key-body-milling machine	F. Caffrey and J. L. Nettleton	West Cheshire, Conn	June 8, 1869	91, 082
Key bush, Lock	A. F. Whiting	Crestline, Ohio	June 25, 1872	128, 266
Key-check	J. J. De Barry	Chicago, Ill	Jan. 21, 1873	135, 092
Key-fastener	E. H. Bailey	Philadelphia, Pa	Apr. 9, 1861	31, 939
Key-fastener	R. S. Dunning	Fall River, Mass	May 4, 1869	89, 641
Key-fastener	H. E. Fowler	Northford, Conn	Aug. 2, 1864	43, 676
Key-fastener	M. P. Hall	Manchester, N. H	Aug. 7, 1866	56, 933
Key-fastener	J. E. Hills	Orange, Mass	June 8, 1869	90, 945
Key-fastener	H. Hungerford	Brooklyn, N. Y	Feb. 7, 1865	46, 239
Key-fastener	W. Neracker	Cleveland, Ohio	Apr. 29, 1873	138, 272
Key-fastener	L. Tobey	Naples, N. Y	May 16, 1871	114, 992
Key-fastening, Metallic	C. L. Rehn	Philadelphia, Pa	Oct. 9, 1860	30, 350
Key-guard	M. E. Berolzheimer	New York, N. Y	Nov. 23, 1869	97, 024
Key-guard	M. E. Berolzheimer	New York, N. Y	Jan. 11, 1870	98, 736
Key-guard	E. P. Furlong	Portland, Me	Nov. 6, 1866	59, 383
Key-guard	B. R. Hathaway	Mormon Island, Cal	Oct. 12, 1869	95, 682
Key-guard	P. G. Smith	Brooklyn, N. Y	Dec. 7, 1869	97, 713
Key-guard for door-locks	J. Wiard	New Britain, Conn	Nov. 26, 1867	71, 350
Key-guard for locks	J. M. Rix	Boston, Mass	Dec. 19, 1865	51, 621
Key-hole escutcheon	A. L. Crooker	Charlestown, Mass	Oct. 15, 1872	132, 202
Key-hole escutcheon	S. W. Drowne	Norwich, Conn	June 28, 1870	104, 838
Key-hole escutcheon	E. Field	Greenwich, Conn	Mar. 10, 1857	16, 792
Key-hole guard	D. P. Bird	Richwood, Ohio	Sept. 14, 1869	94, 862
Key-hole guard	J. F. Bodtker	Chicago, Ill	May 24, 1870	103, 288
Key-hole guard	G. C. Bovey	Cincinnati, Ohio	Apr. 4, 1871	113, 621
Key-hole guard	C. B. Davies	Dayton, Ohio	Nov. 10, 1868	83, 837
Key-hole guard	W. N. Hall	Mexia, Tex	Dec. 23, 1873	145, 865
Key-hole guard	T. G. Harold	Brooklyn, N. Y	May 29, 1860	28, 535
Key-hole guard	A. Huffnagle	Philadelphia, Pa	Aug. 11, 1868	80, 959
Key-hole guard	F. Jenny	Parkersburgh, W. Va	Oct. 10, 1871	119, 851
Key-hole guard	J. H. Jones	Binghamton, N. Y	Aug. 16, 1870	106, 368
Key-hole guard	C. Kurz	Newark, N. J	Aug. 20, 1872	130, 583
Key-hole guard	M. MacKenzie	Malden, Mass	May 18, 1869	90, 283

Index of patents issued from the United States Patent Office from 1790 *to* 1873, *inclusive*—Continued.

Invention.	Inventor.	Residence.	Date.	No.
Key-hole guard	C. Read	Jersey City, N. J	Sept. 29, 1868	82, 551
Key-hole guard	J. L. Russell	Prairie City, Iowa	Jan. 18, 1870	99, 005
Key-hole guard	E. E. Shepardson	Providence, R. I	Mar. 9, 1869	87, 714
Key-hole guard	W. H. Taylor	Stamford, Conn	Aug. 12, 1873	141, 833
Key-hole guard	G. Wheeler	New York, N. Y	Oct. 9, 1860	30, 308
Key-hole guard and bolt-fastener for locks, Combined.	C. H. Townsend and A. F. Potter.	Oakland, Cal	Apr. 23, 1872	126, 104
Key-hole guard and door-securer, Combined	G. Wheeler	New Springville, N. Y	Apr. 30, 1872	126, 360
Key-hole guard and protector	H. M. Flanagin	Penn's Grove, N. J	Jan. 25, 1870	99, 178
Key-hole guard for door-locks	A. T. Brooks	New Britain, Conn	Apr. 28, 1868	77, 165
Key-hole guard for door-locks	W. Johnson, 2d	Haverhill, Mass	May 7, 1867	64, 426
Key-hole guard for door-locks	D. A. Pratt	Sing Sing, N. Y	Oct. 8, 1867	69, 700
Key-hole guard for door-locks	S. Shloss, F. Veerkamp, and C. F. Leopold.	Philadelphia, Pa	Nov. 19, 1867	71, 235
Key-hole guard for locks	J. M. Alden	New Rochelle, N. Y	Mar. 26, 1872	124, 874
Key-hole guard for locks	J. B. Whitney	New York, N. Y	June 18, 1872	128, 082
Key-hole-guard lock	H. R. Towne	Stamford, Conn	May 17, 1870	103, 256
Key-hole guard, Portable	W. E. Dante	Washington, D. C	Mar. 9, 1869	87, 645
Key-hole guide	W. R. Pomeroy	Millersburgh, Ohio	July 25, 1865	49, 000
Key-hole protector	E. Kershaw	Boston, Mass	May 22, 1849	6, 467
Key-hole protector	J. H. Parker	Waltham, Mass	May 23, 1871	115, 093
Key-hole stop	J. Moulson	Philadelphia, Pa	Sept. 28, 1858	21, 616
Key-ring	W. T. Jenks	Toledo, Ohio	Feb. 11, 1873	135, 712
Key-ring	G. A. Libbey	Milwaukee, Wis	Feb. 2, 1869	86, 424
Key-ring and check	C. A. Wentworth	Boston, Mass	Aug. 10, 1869	93, 506
Key-ring and door-fastener, Combined	B. H. Melendy	Manchester, N. H	July 19, 1870	105, 474
Key-ring and door-fastener, Combined	B. H. Melendy	Manchester, N. H	Feb. 14, 1871	111, 762
Key-seat cutter	L. M. Mooney	Tecumseh, Mich	Sept. 10, 1872	131, 293
Key-seat cutter for wheels and pulleys	J. Barton	Cleveland, Ohio	Nov. 30, 1858	22, 155
Key-seat-cutting machine	T. R. Bailey	Lockport, N. Y	Oct. 15, 1872	132, 239
Key-seat, &c., cutting machine	W. B. Bement	Philadelphia, Pa	June 5, 1860	28, 627
Key-seat-cutting machine	J. C. Morgan	Alliance, Ohio	Mar. 7, 1865	46, 758
Key-seat-cutting machine	W. H. Smith and R. S. Eddy	La Crosse, Wis	Apr. 28, 1868	77, 225
Key-seat-cutting machinery	J. K. Dirner	Honesdale, Pa	Sept. 18, 1866	58, 076
Key-turning machine	J. Terry, jr., and S. M. Hunter	Terrysville, Conn	July 25, 1871	117, 348
Key-way gage	J. Donaldson	Carondelet, Mo	Nov. 19, 1872	133, 086
Keys, Attaching bows to	E. L. Gaylord	Terrysville, Conn	May 29, 1860	28, 466
Keys for locks, Manufacture of	W. Wilcox	Middletown, Conn	Jan. 7, 1873	134, 577
Keys in locks, Fastening	J. H. Desalusse	Philadelphia, Pa	Aug. 8, 1865	49, 338
Keys, Safety-drop for	R. K. Lee	Brooklyn, N. Y	May 18, 1858	20, 280
Keyed instruments, Producing the swell in	C. Pommer	Philadelphia, Pa	Aug. 30, 1819	....
Kiln: *See* Bone-black kiln. Grain-kiln. Brick-kiln. Hop-kiln. Cement-kiln. Lime-kiln. Charcoal-kiln. Malt-kiln. Circular kiln. Ore-roasting kiln. Desiccating-kiln. Pipe-kiln. Drying-kiln. Pottery-kiln. Drying and burning kiln. Progressive kiln. Stone-kiln. Earthenware-kiln. Tile-kiln. Glass-kiln. Wood-charring kiln.				
Kiln	H. Aiken	Philadelphia, Pa	May 21, 1872	127, 007
Kiln	A. Batchelar	Brockham, Great Britain	Apr. 25, 1871	114, 095
Kiln	R. Connable	Jackson, Mich	Mar. 25, 1873	137, 180
Kiln	B. R. Hawley	Normal, Ill	Sept. 20, 1870	107, 612
Kiln	H. Howard	Brazil, Ind	June 25, 1872	128, 228
Kiln	P. C. Taylor	San Antonio, Tex	Sept. 27, 1870	107, 735
Kiln	G. A. Wedekind and H. Dueberg.	Baltimore, Md	May 2, 1871	114, 499
Kiln-drying machine	A. B. Winegar	San Francisco, Cal	June 30, 1868	79, 425
Kiln-furnace	O. Bennett	Boston, Mass	June 11, 1854	127, 836
Kindling-block	W. Loft	Bergen, N. J	Feb. 2, 1869	86, 427
Kindling-material	T. Schwartz	New York, N. Y	Apr. 4, 1871	113, 578
King-bolt	E. A. Keasey	Ligonier, Ind	Oct. 1, 1867	69, 446
King-bolt	F. B. Morse	New Haven, Conn	Feb. 23, 1869	87, 280
King-bolt	F. B. Morse	Plantsville, Conn	Oct. 11, 1870	108, 282
King-bolt	W. Vandersaar	Indianapolis, Ind	Mar. 7, 1871	112, 517
King-bolt and plate, Clip	H. M. Beecher	Plantsville, Conn	June 17, 1873	139, 855
King-bolt and whiffletree-plate for vehicles	L. Adams	Amherst, Mass	Oct. 20, 1868	83, 234
King-bolt blanks, Die for making	F. Van Patten	Auburn, N. Y	Aug. 12, 1873	141, 679
King-bolt blanks, Die for swaging clip	F. Van Patten	Auburn, N. Y	Aug. 12, 1873	141, 681
King-bolt clips, Die for making	A. Dillenbeck	Fort Plain, N. Y	Feb. 25, 1873	136, 143
King-bolt dies, Manufacturing clip	F. B. Prindle	Southington, Conn	Oct. 3, 1871	119, 643
King-bolt dies, Series of	F. B. Morse	Plantsville, Conn	Oct. 26, 1869	96, 251
King-bolt for car-trucks	S. W. Murray and B. P. Lamason.	Milton, Pa	Apr. 25, 1871	114, 028
King-bolt for carriages	W. Clark	Johnsonville, N. Y	Mar. 29, 1870	101, 227
King-bolt for carriages	D. Hyde and E. H. Andrews	Bridgeport, N. Y	Apr. 4, 1871	113, 521
King-bolt for carriages	J. L. H. Mosier	New York, N. Y	Feb. 25, 1873	136, 255
King-bolt for carriages	J. Phelps	Red Creek, N. Y	Nov. 13, 1866	59, 642
King-bolt for carriages	J. Reiber and J. Schrader	Bridgeport, Ill	May 5, 1868	77, 528
King-bolt for perch-carriages	J. L. H. Mosier	New York, N. Y	Apr. 29, 1873	138, 342
King-bolt for vehicles	J. Deeble	Plantsville, Conn	Nov. 4, 1873	144, 323
King-bolt for wagons	L. T. Swartwout	Locke, N. Y	Apr. 4, 1871	113, 595
King-bolt for wagons	J. J. Waldron	East Durham, N. Y	May 12, 1868	77, 782
King-bolt-forging die	F. B. Morse	Plantsville, Conn	June 13, 1871	115, 976
King-bolt-forming die	F. B. Prindle	Southington, Conn	May 3, 1870	102, 707
King-bolt-head die	F. B. Morse	Plantsville, Conn	Feb. 22, 1870	100, 178
King-bolt-head dies, Series of	R. R. Miller	Plantsville, Conn	Feb. 22, 1870	100, 053
King-bolt protector and anti-rattling attachment to wagon.	H. L. Taylor	Fredonia, N. Y	Jan. 2, 1872	122, 415
King-bolt socket, Carriage	F. B. Morse	Plantsville, Conn	May 17, 1870	103, 274
King-bolt-trimming die	F. B. Prindle	Southington, Conn	May 3, 1870	102, 708
King-bolts, Die for manufacture of clip	F. Van Patten	Auburn, N. Y	Aug. 12, 1873	141, 680
Kisses, &c., Machine for wrapping	C. C. Wilson	Baltimore, Md	Jan. 3, 1871	110, 812
Kit, Mess	E. Blakeslee	New Haven, Conn	Feb. 7, 1865	46, 210

Index of patents issued from the United States Patent Office from 1790 *to* 1873, *inclusive*—Continued.

Invention.	Inventor.	Residence.	Date.	No.
Kitchen and ironing table	J. Good	Lancaster, Pa	Oct. 28, 1873	143, 977
Kitchen-boiler	J. and W. C. Bailey	Farmington, Me	May 22, 1835	....
Kitchen-boiler	W. B. Scaife	Pittsburgh, Pa	Sept. 27, 1870	107, 817
Kitchen-boiler	W. B. Scaife	Pittsburgh, Pa	Oct. 18, 1870	108, 524
Kitchen-boiler	W. B. Scaife	Pittsburgh, Pa	Dec. 3, 1872	133, 668
Kitchen-boiler	W. B. Scaife	Pittsburgh, Pa	Dec. 3, 1872	133, 669
Kitchen-cabinet	N. B. Smith	Waukegan, Ill	Sept. 10, 1872	131, 186
Kitchen-implement	J. Frisch	Albany, N. Y	Sept. 29, 1868	82, 615
Kitchen-implement	C. S. Westland and J. B. Allen	Providence, R. I	Sept. 15, 1868	82, 187
Kitchen-knife	J. W. Androvall and T. W. Joline.	Tottenville, N. Y	May 2, 1871	114, 388
Kitchen, Movable	G. Youle	New York, N. Y	July 1, 1814	....
Kitchen or stove-caboose, Portable	J. Truman	Philadelphia, Pa	Dec. 3, 1811	....
Kitchen-table	H. Closterman, jr	Cincinnati, Ohio	Dec. 26, 1871	122, 225
Kitchen-table	F. Ehrhardt	Washington, D. C	Jan. 11, 1870	98, 751
Kitchen-table	J. Shaw	Whitby, Canada	Feb. 21, 1871	112, 083
Kitchen-table	B. Welteck	New York, N. Y	Sept. 19, 1871	119, 208
Kitchen, Tin	G. Richardson	South Reading, Mass	June 14, 1832	....
Kitchen, Traveling	M. Pinner	New York, N. Y	July 7, 1863	39, 170
Kitchen-utensil	J. J. Diehl	York, Pa	Aug. 17, 1869	93, 862
Kite	S. Clark	New York, N. Y	Nov. 9, 1869	96, 550
Kite	I. Ferris	Cincinnati, Ohio	July 25, 1871	117, 270
Kite	C. J. Hardekopf	Philadelphia, Pa	Feb. 18, 1873	135, 987
Kite	E. J. Hughes	Pittsburgh, Pa	Dec. 31, 1867	72, 855
Kite	O. Maddans	Brooklyn, N. Y	Nov. 21, 1871	121, 056
Kite	T. Perrins	Philadelphia, Pa	Jan. 2, 1866	51, 860
Kite	A. Horns and E. J. Smith	New York, N. Y	Mar. 24, 1868	76, 028
Kite-frame	J. Shepard	New Britain, Conn	Nov. 19, 1867	71, 232
Kite-string holder	G. G. Sheldon	Chicago, Ill	May 31, 1870	103, 668
Knapsack	G. D. Arrington	Charlestown, Mass	Sept. 24, 1861	33, 329
Knapsack	J. Bondy	New York, N. Y	Mar. 4, 1862	34, 560
Knapsack	R. C. Buchanan	United States Army	Oct. 4, 1859	25, 625
Knapsack	A. N. Clark	Boston, Mass	Mar. 17, 1863	37, 936
Knapsack	J. W. Frazier	Newark, N. J	Feb. 20, 1872	123, 885
Knapsack	T. Garrick	Providence, R. I	Sept. 24, 1861	33, 343
Knapsack	W. Griffiths	Philadelphia, Pa	May 10, 1859	23, 979
Knapsack	W. B. Johns	United States Army	Apr. 12, 1859	23, 582
Knapsack	J. K. Mizner	Detroit, Mich	Nov. 27, 1866	60, 034
Knapsack	G. H. Palmer	Monmouth, Ill	June 10, 1873	139, 731
Knapsack	A. Perrin	Paris, France	June 13, 1865	48, 246
Knapsack	J. Rush	Philadelphia, Pa	Mar. 25, 1862	34, 778
Knapsack	C. W. Schaefer	Philadelphia, Pa	July 11, 1871	116, 874
Knapsack	J. Short	New York, N. Y	Nov. 12, 1861	33, 726
Knapsack	J. Short	Boston, Mass	Jan. 28, 1862	34, 272
Knapsack	J. Short	Boston, Mass	Dec. 16, 1862	37, 203
Knapsack	E. F. Southward	Boston, Mass	May 26, 1863	38, 701
Knapsack	A. W. Süs	New York, N. Y	May 17, 1864	42, 805
Knapsack	C. E. Sweney	Charlestown, Mass	Feb. 4, 1862	34, 336
Knapsack	J. Weber	New York, N. Y	Jan. 31, 1865	46, 195
Knapsack	O. E. Woods	Philadelphia, Pa	May 24, 1864	42, 895
Knapsack	O. E. Woods	Philadelphia, Pa	Nov. 8, 1864	44, 993
Knapsack and bed combined	B. Frodsham and M. Levett	New York, N. Y	Oct. 1, 1861	33, 385
Knapsack-attachment	J. F. Fog	Philadelphia, Pa	Apr. 2, 1872	125, 188
Knapsack-attachment	J. Sherlock	New York, N. Y	July 27, 1869	93, 130
Knapsack-collar	J. E. Atwood	Washington, D. C	Dec. 16, 1862	37, 146
Knapsack-engine	J. W. Douglas	Middletown, Conn	Aug. 24, 1869	93, 974
Knapsack, Folio extension	J. Boyd	Boston, Mass	Nov. 19, 1833	....
Knapsack, Hospital	J. McEvoy	New York, N. Y	Jan. 7, 1862	34, 117
Knapsack, Military	J. P. Lherbette	Washington, D. C	Oct. 7, 1808	....
Knapsack, overcoat, and tent	J. Short, 2d	New York, N. Y	Oct. 8, 1861	33, 468
Knapsack-sling	D. C. Baxter	Philadelphia, Pa	Mar. 17, 1863	37, 893
Knapsack-sling	J. T. Warren	Stafford, N. Y	Feb. 14, 1865	46, 410
Knapsack-supporter	A. Dickey	Cincinnati, Ohio	Mar. 21, 1865	46, 886
Knapsack, tent, and litter combined	L. Joubert	Paris, France	July 7, 1863	39, 150
Knapsacks, Mode of carrying	W. Clifford	Holly Springs, Miss	Nov. 3, 1868	83, 762
Knapsacks, Mode of carrying	T. S. Truss	London, England	Sept. 20, 1864	44, 354
Knapsacks, Slinging	W. oftHman	United States Army	Jan. 9, 1872	122, 522
Knapsacks, Slinging	T. Miles	Philadelphia, Pa	Jan. 28, 1862	34, 260
Kneading-board	H. P. Jones	Davenport, Iowa	Nov. 18, 1873	144, 768
Kneading-machine	E. McAllister	Rochester, N. Y	Dec. 26, 1871	122, 184
Kneading-machine	W. S. Reinert	Philadelphia, Pa	Sept. 7, 1858	21, 442
Knee-brace lacer	J. Monroe	New York, N. Y	July 16, 1867	66, 728
Knee-joint, Artificial	E. Cotly	Washington, D. C	Dec. 9, 1862	37, 087
Kneeling-case for churches	L. Mooney	Baltimore, Md	Oct. 15, 1867	69, 831

Knife:
See Apple and vegetable knife.
Band-knife.
Band-cutting knife.
Belt-knife.
Blubber-knife.
Budding-knife.
Buffing-knife.
Butcher's knife.
Butter-knife.
Can-opening knife.
Carving-knife.
Chopping-knife.
Conical-edged knife.
Corn-knife.
Corn-stalk knife.
Counter-knife.
Crozing-knife.
Currier's knife.
Currying-knife.
Cylindrical cutting knife.
Dirk-knife.
Drawing-knife.
Folding-knife.
Fruit-knife.
Fruit-coring knife.
Gage-knife.
Harness round-knife.
Harvester-knife.
Hay-knife.
Hay and pruning knife.
Hide-knife.
Honey-comb knife.
Kitchen-knife.
Leather-splitting knife.
Mincing-knife.
Mowing-machine knife.
Nurseryman's knife.
Orange-knife.
Paper-sack knife.
Paring-knife.
Penknife.

Index of patents issued from the United States Patent Office from 1790 *to* 1873, *inclusive*—Continued.

Invention.	Inventor.	Residence.	Date.	No.
Knife—Continued. *See* Planer-knife. Pocket-knife Polishing-knife. Pruning-knife. Putty-knife. Reaper-knife. Reaper and mower knife. Reaping-knife. Round knife. Saw-knife. Sheaf-band knife. Shoe-knife. Skinning-knife. Sole-trimming knife. Spiral knife. Splint-knife. Splitting-knife. Table-knife. Tanner's and currier's knife. Tobacco-knife. Trimming guard-knife. Vegetable-knife. Welt-knife.				
Knife	N. E. Babcock and G. D. Goodsell.	Rockford, Ill	Mar. 29, 1870	101, 209
Knife	N. W. Caughy	Baltimore, Md	Dec. 17, 1867	72, 365
Knife	L. Eddy	Taunton, Mass	July 22, 1873	140, 997
Knife	J. O. Ely	Philadelphia, Pa	Feb. 9, 1869	86, 657
Knife	R. H. Fisher	Beaver Falls, Pa	Dec. 29, 1868	85, 296
Knife	N. P. Mulloy	Waltham, Mass	Jan. 29, 1867	61, 556
Knife	W. Sausser	Hannibal, Mo	Aug. 7, 1866	56, 998
Knife and cane-stripper, Combined	J. Leffel	Springfield, Ohio	Dec. 5, 1865	51, 326
Knife and fork	J. Ball	Brooklyn, N. Y	Nov. 23, 1869	97, 152
Knife and fork	F. C. Beach and A. A. C. Klaucke.	Stratford, Conn., and Washington, D. C.	Sept. 4, 1866	57, 662
Knife and fork	W. D. Smith	Homerville, Ga	Dec. 17, 1872	134, 011
Knife and fork cleaner	S. Brackett	Fall River, Mass	Apr. 10, 1860	27, 773
Knife and fork cleaner	J. L. Hannum	Berea, Ohio	Nov. 26, 1872	133, 367
Knife and fork cleaner	B. F. Hughson	Cold Spring, N. Y	Aug. 27, 1867	68, 079
Knife and fork cleaner	E. Marwedel	Washington, D. C	July 15, 1873	140, 932
Knife and fork cleaner	J. Merritt	New York, N. Y	Dec. 10, 1867	72, 067
Knife and fork cleaner	J. Protz	Easton, Pa	Mar. 13, 1860	27, 472
Knife and fork cleaner	H. and F. Wegner	West Troy, N. Y	Aug. 30, 1870	106, 897
Knife and fork cleaner, knife-sharpener, and can-opener combined.	F. W. Echternach and M. J. Welch.	Philadelphia, Pa	Feb. 6, 1872	123, 386
Knife and fork cleaning and polishing machine	M. N. Armstrong	New York, N. Y	Sept. 14, 1843	3, 268
Knife and fork, Combined	A. W. Cox	Malden, Mass	Mar. 30, 1869	87, 370
Knife and fork, Combined	J. S. Jennings	Buffalo, N. Y	Sept. 11, 1866	57, 918
Knife and fork, Combined	J. McMorries	Columbus, Miss	Oct. 26, 1869	96, 134
Knife and fork, Combined	T. B. Thorpe	New York, N. Y	Mar. 14, 1865	46, 832
Knife and fork handle	C. A. Moore	Westbrook, Conn	Aug. 25, 1868	81, 399
Knife and fork handle	W. Sanderson and J. J. De Barry.	New York, N. Y	Jan. 25, 1870	99, 242
Knife and fork handle, Detachable	L. Boynton and W. F. Sweet	Jackson Township, Pa	Mar. 15, 1870	100, 717
Knife and fork holder	J. W. Patten	North Greenbush, N. Y	Oct. 27, 1868	83, 538
Knife and fork rack	J. B. Hawley	New Haven, Conn	May 8, 1866	54, 541
Knife and fork rest	L. J. Cherrington	Boston, Mass	July 11, 1871	116, 807
Knife and fork scourer	E. J. Marsh	Leominster, Mass	May 9, 1871	114, 578
Knife and fork scourer	D. S. Marvin	Watertown, N. Y	May 5, 1868	77, 632
Knife and fork scourer	M. Merrill	Marengo, Ill	Feb. 25, 1868	74, 763
Knife and fork scourer	R. R. Pattison	Springfield, Ill	Apr. 10, 1866	53, 863
Knife and fork scouring machine	C. Aumock	Columbus, Ohio	Jan. 13, 1852	8, 646
Knife and pencil-case, Combined	R. Cross	Attleborough, Mass	Feb. 26, 1856	14, 306
Knife and pencil-sharpener, Combined	J. McClure	Nashua, N. H	July 4, 1871	116, 732
Knife and scissors combined	C. W. Sykes	Suffield, Conn	Mar. 5, 1867	62, 702
Knife and scissors sharpener	W. H. Alcorn	New York, N. Y	Aug. 9, 1864	43, 752
Knife and scissors sharpener	M. T. Chapman	Galesburgh, Ill	Feb. 26, 1867	62, 472
Knife and scissors sharpener	A. Herthal	Bridgeport, Conn	July 16, 1867	66, 837
Knife and scissors sharpener	M. V. B. Howe	Ashburnham, Mass	June 26, 1866	55, 867
Knife and scissors sharpener	I. S. and J. W. Hyatt, jr	Chicago, Ill	June 17, 1862	35, 652
Knife and scissors sharpener	J. N. Jacobs	Worcester, Mass	Nov. 26, 1867	71, 491
Knife and scissors sharpener	T. T. Prosser and J. Lawson	Chicago, Ill	Aug. 28, 1866	57, 564
Knife and scissors sharpener	J. J. Russ	Worcester, Mass	July 24, 1866	56, 615
Knife and spoon cleaner	J. Macnish	Berlin, Wis	June 29, 1858	20, 724
Knife and spoon polisher	L. J. Williams	Rolla, Mo	Oct. 15, 1872	132, 343
Knife and watch-key combined	E. D. Norton	Bangor, Me	Mar. 26, 1867	63, 171
Knife-blade holder	S. A. Cummings	Middleton, Mass	Dec. 5, 1865	51, 297
Knife-blades, Machine for rolling	W. P. Lathrop	Winchester, Conn	Mar. 22, 1864	42, 005
Knife-blades, Machine for straightening	H. Pierce	Claremont, N. H	June 23, 1857	17, 634
Knife-carrier	P. E. Cummings	Sanford, Me	Oct. 23, 1866	58, 992
Knife-cleaner	J. J. Banta	Jersey City, N. J	July 20, 1858	20, 929
Knife-cleaner	J. T. Barnstead	Peabody, Mass	Apr. 30, 1872	126, 172
Knife-cleaner	W. S. Beebe, J. T. Baynes, and A. A. King.	West Troy, N. Y	Dec. 12, 1871	121, 747
Knife-cleaner	W. J. Burge	Atchison, Kans	June 25, 1867	66, 126
Knife-cleaner	W. Christian and J. H. Morrow	New York, N. Y	Sept. 11, 1866	57, 865
Knife-cleaner	A. B. Clarke	Wyandotte, Mich	Jan. 30, 1872	123, 238
Knife-cleaner	C. F. Dean	Saint Johnsbury, Vt	July 30, 1867	67, 169
Knife-cleaner	I. Detheridge, jr	New York, N. Y	Oct. 23, 1860	30, 468
Knife-cleaner	J. A. Eivins	South Boston, Mass	Oct. 15, 1867	69, 905
Knife-cleaner	T. M. Fell	Brooklyn, N. Y	June 2, 1863	38, 786
Knife-cleaner	H. C. Gilbert	New London, Conn	Mar. 18, 1873	136, 990
Knife-cleaner	T. and L. Gringras	Buffalo, N. Y	May 20, 1873	139, 057
Knife-cleaner	L. Goulding	Medfield, Mass	Dec. 10, 1867	72, 021
Knife-cleaner	J. F. Hammond	Providence, R. I	Aug. 6, 1867	67, 534
Knife-cleaner	C. H. Hardy	Bath, Me	Apr. 20, 1869	89, 218
Knife-cleaner	W. W. Hopkins	Chesterfield Factory, N. H.	Aug. 19, 1856	15, 562
Knife-cleaner	I. Jewett	Boston, Mass	Aug. 6, 1867	67, 434
Knife-cleaner	A. C. Kaiser	Vienna, Mo	Dec. 1, 1868	84, 552
Knife-cleaner	W. T. Kosinski	New York, N. Y	Sept. 12, 1865	49, 892
Knife-cleaner	J. McNamee	Easton, Pa	Nov. 8, 1859	26, 040
Knife-cleaner	W. Miller	Waltham, Mass	May 25, 1858	20, 391
Knife-cleaner	W. Miller	Chicopee, Mass	Dec. 15, 1868	84, 960
Knife-cleaner	C.C. Morgan	Auburn, N. Y	July 30, 1867	67, 336
Knife-cleaner	R. R. Pattison	Chicago, Ill	Jan. 15, 1867	61, 245
Knife-cleaner	J. Seeberger	West Troy, N. Y	July 4, 1871	116, 635
Knife-cleaner	A. J. Sellon	Boston, Mass	May 8, 1866	54, 612

Index of patents issued from the United States Patent Office from 1790 *to* 1873, *inclusive*—Continued.

Invention.	Inventor.	Residence.	Date.	No.
Knife-cleaner	C. B. Sheldon	New York, N. Y	Dec. 23, 1873	145, 909
Knife-cleaner	G. Smith	New York, N. Y	July 23, 1861	32, 902
Knife-cleaner	O. Sweeney	Norwich, Conn	Jan. 22, 1861	31, 193
Knife-cleaner	I. F. Thompson	Providence, R. I	Jan. 21, 1868	73, 673
Knife-cleaner	J. H. Van Riper	New York, N. Y	Mar. 29, 1864	42, 133
Knife-cleaner	W. Vine	Norwalk, Conn	Nov. 9, 1869	96, 642
Knife-cleaner	H. Woodward	London, England	Oct. 8, 1867	69, 601
Knife cleaner and sharpener	S. A. W. Houmann and H. Nielson.	Saint Louis, Mo	Jan. 26, 1872	125, 050
Knife cleaner and sharpener	L. J. Johnson	Montville, Conn	Nov. 5, 1867	70, 574
Knife-cleaning box	C. D. Copeland	Fall River, Mass	Mar. 20, 1866	53, 275
Knife-cleaning box	F. A. Salisbury	Greene, N. Y	May 20, 1862	35, 328
Knife-cleaning machine	A. C. Ketchum	New York, N. Y	Apr. 29, 1856	14, 788
Knife-cleaning machine	T. Roberts	Lynn, Mass	Feb. 1, 1870	99, 480
Knife-cleaning machine	G. Weedon	New York, N. Y	Jan. 27, 1863	37, 537
Knife-die for cutting leather-straps for whips	C. Baeder	New York, N. Y	Oct. 31, 1854	11, 854
Knife, fork, and pie crimper	G. D. Bayley	Boston, Mass	Oct. 18, 1864	44, 699
Knife, fork, and spoon cleaning machine	E. and A. Buckman	East Greenbush, N. Y	Mar. 25, 1862	34, 736
Knife, fork, and spoon, Combination of	J. H. Cables	Plymouth Hollow, Conn	Mar. 18, 1862	34, 712
Knife, fork, and spoon, Combination of	J. W. Hardie and A. S. Hayward.	New York, N. Y., and Boston, Mass.	Jan. 7, 1862	34, 098
Knife, fork, and spoon, Combined	N. Ames	Saugus Centre, Mass	Sept. 17, 1861	33, 285
Knife, fork, and spoon Combined	A. Neill	Boston, Mass	Jan. 7, 1862	34, 069
Knife, fork, and spoon Combined	W. H. Richards	Newton, Mass	July 23, 1861	32, 916
Knife, fork, and spoon holder	G. L. Morse and L. M. Herrick	Harrison, N. J	Mar. 7, 1865	46, 692
Knife-grinding machine	E. S. M. Fernald	Saco, Me	Oct. 4, 1870	107, 894
Knife-grinding machine	A. Hankey and F. Stiles, jr	Leicester, Mass	Aug. 4, 1857	17, 952
Knife-grinding machinery	W. Fosket	Meriden, Conn	Dec. 26, 1865	51, 779
Knife-guard	G. K. Dearborn	Pawtucket, R. I	Mar. 29, 1870	101, 235
Knife-guard	E. A. Goodes	Philadelphia, Pa	Oct. 5, 1869	95, 463
Knife-guard	D. Perry	North Providence, R. I	Feb. 8, 1870	99, 701
Knife-guard	T. T. Woodward	Ansonia, Conn	Nov. 9, 1869	96, 758
Knife-handle	J. B. Bailey	New York, N. Y	May 24, 1870	103, 278
Knife-handle	L. Carrier	East Douglass, Mass	Mar. 6, 1860	27, 349
Knife, &c., handle	M. Chapman	Greenfield, Mass	Aug. 9, 1870	106, 254
Knife-handle	S. B. Everett	Plymouth Hollow, Conn	Aug. 19, 1862	36, 255
Knife-handle	J. L. Frary	New Britain, Conn	Dec. 7, 1869	97, 624
Knife-handle	D. H. Merriam	Fitchburgh, Mass	Apr. 24, 1866	54, 189
Knife-handle	A. L. Taylor	Springfield, Vt	Dec. 14, 1869	97, 991
Knife-handle	W. D. Woods	Bennington, N. H	Mar. 22, 1870	101, 201
Knife-handle and other table-cutlery	Z. K. Murdock	Meriden, Conn	Apr. 16, 1841	2, 052
Knife-handle bolster	W. P. Lathrop	West Winsted, Conn	July 6, 1869	92, 196
Knife, nut cracker and picker	G. A. Fairfield	Hartford, Conn	Apr. 5, 1870	101, 448
Knife-polisher	J. W. Battelle	Worcester, Mass	Dec. 19, 1865	51, 542
Knife-polisher	W. H. Cummings	Oxford, Mass	Nov. 28, 1871	121, 339
Knife-polisher	H. T. Field	New Braintree, Mass	May 25, 1858	20, 340
Knife-polisher	C. H. Lithgow	Chicago, Ill	May 17, 1870	103, 210
Knife-polisher	J. Palmer	Cleveland, Ohio	June 20, 1865	48, 305
Knife-polisher	E. H. Rockwood and C. D. Shepard.	Boston, Mass	July 16, 1872	129, 594
Knife-polisher	R. Shaler	Madison, Conn	Nov. 28, 1848	5, 941
Knife-polisher	F. W. Tarble and H. W. Sawyer.	Charlestown and Chelsea, Mass.	Feb. 25, 1873	136, 188
Knife-polisher	H. J. Wattles	Rockford, Ill	Oct. 15, 1872	131, 339
Knife-polisher and knife and scissors sharpener	J. Wheeler	Athol, Mass	June 5, 1866	55, 404
Knife polisher and sharpener	T. Bootsmann	Tompkinsville, N. Y	Oct. 21, 1873	143, 871
Knife-polishing machine	A. Munger and R. C. Taylor	Auburn, N. Y	Sept. 11, 1849	6, 707
Knife-rest	D. Sherwood	Lowell, Mass	Dec. 28, 1869	98, 431
Knife-ring	C. B. and W. A. N. Long	Worcester, Mass	Aug. 4, 1868	80, 553
Knife-ring	J. C. Wilmarth and A. Fobes	Saint Louis, Mo	Oct. 6, 1868	82, 912
Knife, saw, and spring-scale combined, Butcher's	J. Baggs	Easton, England	Apr. 29, 1873	138, 362
Knife-scourer	J. Q. Adams and S. R. Goodsell	Brooklyn, N. Y	Apr. 12, 1870	101, 698
Knife-scourer	H. E. French	Unity, N. H	Sept. 27, 1870	107, 674
Knife-scourer	S. R. Goodsell and J. Q. Adams	Brooklyn, N. Y	June 8, 1869	91, 011
Knife-scourer	F. O. Harvey	Kansas City, Mo	Dec. 3, 1872	133, 645
Knife-scourer	E. W. Haven	Brandon, Vt	Mar. 11, 1873	136, 598
Knife-scourer	H. B. Hutchins and W. Horter	Philadelphia, Pa	July 24, 1866	56, 564
Knife-scourer	D. S. Neal	Lynn, Mass	June 13, 1871	115, 885
Knife scourer and sharpener	D. Cromwell, jr	Yarmouth Port, Mass	June 20, 1871	116, 029
Knife scourer and sharpener	D. Hodgkins	Newburyport, Mass	Mar. 23, 1869	88, 167
Knife-scouring board	E. Dening	Seneca Falls, N. Y	Apr. 2, 1872	125, 124
Knife-scouring machine	G. M. Morris and J. Newton	Watertown, Conn	Dec. 4, 1855	13, 883
Knife, scythe, &c., sharpener and renovator	I. Jennings	New York	Mar. 17, 1829	......
Knife-sharpener	A. Anan	New York, N. Y	Nov. 16, 1858	22, 055
Knife-sharpener	J. J. and A. T. Armstrong	Brooklyn, N. Y	Aug. 3, 1858	21, 058
Knife-sharpener	A. B. Bean	New Haven, Conn	Nov. 26, 1867	71, 442
Knife-sharpener	C. A. Bogert	Bay City, Mich	Apr. 13, 1869	88, 839
Knife-sharpener	W. J. Dunn	New York	Aug. 18, 1829	......
Knife-sharpener	S. Gissinger	Pittsburgh, Pa	Jan. 9, 1872	122, 597
Knife-sharpener	T. Haynes	Saint Louis, Mo	Oct. 22, 1867	69, 993
Knife-sharpener	G. Hinman	New Haven, Conn	Feb. 8, 1859	22, 919
Knife-sharpener	W. H. Howland	San Francisco, Cal	Nov. 23, 1869	97, 090
Knife-sharpener	W. Hunt	New York	Feb. 19, 1829	......
Knife-sharpener	J. W., jr., and I. S. Hyatt	Chicago, Ill	Feb. 19, 1861	31, 461
Knife-sharpener	T. K. Knapp	Worcester, Mass	July 30, 1867	67, 200
Knife-sharpener	T. K. Knapp	Dansville, N. Y	May 28, 1872	127, 170
Knife-sharpener	G. B. Markham	Plymouth, Mich	Apr. 30, 1867	64, 342
Knife-sharpener	J. Meyer, jr	New York, N. Y	May 14, 1867	64, 687
Knife-sharpener	W. Nash	New Britain, Conn	Feb. 13, 1866	52, 587
Knife-sharpener	D. Porter	Cleveland, Ohio	Nov. 14, 1865	50, 951
Knife-sharpener	J. B. Pratt	Corning, N. Y	July 10, 1866	56, 264
Knife-sharpener	J. Quipp	Buffalo, N. Y	Aug. 20, 1872	130, 660
Knife-sharpener	C. H. Reynolds	New York, N. Y	May 19, 1868	78, 070
Knife-sharpener	J. C. Reynolds	Taunton, Mass	Oct. 29, 1867	70, 360
Knife-sharpener	Z. C. and H. A. Robbins	Washington, D. C	Dec. 7, 1869	97, 701
Knife-sharpener	J. J. Russ	Worcester, Mass	June 27, 1871	116, 495
Knife-sharpener	S. C. Stokes	Manchester, N. H	Mar. 6, 1860	27, 413
Knife-sharpener	S. C. Stokes	Manchester, N. H	May 3, 1870	102, 727

Index of patents issued from the United States Patent Office from 1790 to 1873, inclusive—Continued.

Invention.	Inventor.	Residence.	Date.	No.
Knife-sharpener	T. Vickery	Providence, R. I	Mar. 1, 1870	100, 470
Knife-sharpener	T. H. White	Orange, Mass	Sept. 4, 1866	57, 804
Knife-sharpener	P. M. Withington	Stoughton, Mass	May 17, 1870	103, 113
Knife-sharpener	M. Young	Frederick, Md	July 23, 1867	67, 011
Knife sharpener and cleaner	J. M. Farnam	Hartford, Conn	Apr. 10, 1860	27, 787
Knife sharpener and cleaner	J. J. Mason	Drummondville, Canada	Dec. 10, 1872	133, 711
Knife-sharpener and glass-cutter combined	S. C. Stokes	Manchester, N. H	Oct. 15, 1872	132, 219
Knife sharpener and polisher	E. Beeman	Owego, N. Y	Jan. 2, 1866	51, 789
Knife sharpener and scourer	A. T. Boon	Galesburgh, Ill	Sept. 15, 1863	39, 877
Knife-sharpener, Domestic	P. Cornell	Brutus, N. Y	Jan. 15, 1830	......
Knife to handle, Attaching	W. Clayton	Bristol, Conn	Sept. 4, 1866	57, 677
Knife, tweezers, and ear-spoon, Combined	B. C. English	Springfield, Mass	Sept. 26, 1865	50, 106
Knives and forks, Construction of	J. W. Hardie	New York, N. Y	Nov. 12, 1861	33, 703
Knives and forks, Construction of	P. Ulmer	Philadelphia, Pa	Feb. 4, 1862	34, 337
Knives and forks, Manufacture of	J. D. Frary	New Britain, Conn	Sept. 25, 1866	58, 242
Knives and forks, Manufacture of	J. B. H. Leonard	Meriden, Conn	Sept. 25, 1866	58, 269
Knives and forks, Manufacture of	C. A. Moore	Westbrook, Conn	Dec. 15, 1868	85, 020
Knives and forks, Manufacturing	J. H. Nichols	New Britain, Conn	Oct. 29, 1867	70, 352
Knives, Attaching bolsters to	H. Barber	Greenfield, Mass	Nov. 12, 1867	70, 778
Knives, Attaching handles to	M. Chapman	Greenfield, Mass	Dec. 16, 1862	37, 150
Knives, spoons, &c., Impressed roller for making	J. Owings	Baltimore, Md	Jan. 25, 1814	......
Knit fabric	C. J. Appleton	Cohoes, N. Y	Sept. 27, 1870	107, 749
Knit fabric	H. Boot	Philadelphia, Pa	Aug. 13, 1872	130, 467
Knit fabric	B. J. McGee	Watertown, Mass	Nov. 25, 1873	144, 915
Knit fabric	S. Thacker	Snenton, England	Oct. 17, 1871	120, 126
Knit fabric and method of knitting	D. Bickford	New York, N. Y	Sept. 17, 1872	131, 386
Knit fabric and method of knitting	D. Bickford	New York, N. Y	Sept. 17, 1872	131, 387
Knit fabric and method of knitting	D. Bickford	New York, N. Y	Mar. 11, 1873	136, 639
Knit fabrics, Machine for uniting the selvages of	T. Beven	Loughborough, England	Apr. 22, 1873	137, 997
Knit fabrics, Machinery for drying and finishing tubular	W. K. Greene, jr., and W. M. Pawling	Amsterdam and Hagaman's Mills, N. Y.	Jan. 6, 1863	37, 288
Knit fringe or border	J. Phipps	Philadelphia, Pa	June 10, 1873	139, 810
Knit mitten	O. F. Tripp	Battle Creek, Mich	Aug. 26, 1873	142, 301
Knit or woven fabric, Producing designs for	I. Rehn	Washington, D. C	Nov. 21, 1871	121, 128
Knitted fabric	H. Boot	Philadelphia, Pa	May 2, 1871	114, 397
Knitted fabric	G. S. Harwood	Newton, Mass	Jan. 12, 1864	41, 218
Knitted fabric	G. Jelley	Roxbury, Mass	July 21, 1868	80, 183
Knitted fabric	M. Landenberger, jr	Philadelphia, Pa	Aug. 3, 1869	93, 314
Knitted fabric	J. Phipps	Philadelphia, Pa	Aug. 6, 1872	130, 312
Knitted fabric	J. Vickerstaff	Philadelphia, Pa	June 16, 1857	17, 612
Knitted fabric, Manufacturing	J. K. Kilbourn and E. E. Kilbourn	Pittsfield, Mass., and Litchfield, Conn.	June 7, 1859	24, 308
Knitted fabrics, Machine for drying and finishing tubular	N. P. Akin	Philmont, N. Y	Mar. 2, 1869	87, 317
Knitted fabrics, Machine for napping, brushing, &c., tubular	C. and I. Tompkins	Troy, N. Y	June 24, 1873	140, 320
Knitted goods, Mode of uniting edges of	S. Arnold	Claverack, N. Y	June 21, 1870	104, 532
Knitted hosiery, Machine	B. F. Shaw	South Danvers, Mass	Apr. 23, 1867	64, 154
Knitted mitten, Machine-made	A. C. Carey	Malden, Mass	Apr. 24, 1868	54, 108
Knitted stocking, Machine	A. C. Carey	Malden, Mass	July 30, 1867	67, 264
Knitted stocking, Machine-made	A. C. Carey	Malden, Mass	Apr. 24, 1866	54, 109
Knitted stockings, Apparatus for cutting out	W. Martin	Philadelphia, Pa	Aug. 9, 1864	43, 783
Knitting	C. Tompkins and J. Johnson	Troy, N. Y	Sept. 18, 1855	13, 586
Knitting-burrs, Tool for manufacturing	H. Fisher	Waterford, N. Y	Nov. 10, 1863	40, 587
Knitting-holder	H. P. Joy	South Berwick Junction, Me.	Aug. 12, 1873	141, 796
Knitting-machine	W. H. Abel	Greenville, R. I	Feb. 11, 1868	74, 266
Knitting-machine	W. H. Abel	Greenville, R. I	Nov. 3, 1868	83, 583
Knitting-machine	W. H. Abel	Greenville, R. I	Nov. 3, 1868	83, 584
Knitting-machine	W. H. Abel	Bennington, Vt	July 19, 1870	105, 537
Knitting-machine	W. H. Abel	Lowell, Mass	July 19, 1870	105, 538
Knitting-machine	W. H. Abel	Lowell, Mass	July 19, 1870	105, 539
Knitting-machine	W. H. Abel	Bennington, Vt	Aug. 23, 1870	106, 531
Knitting-machine	J. B. Aiken	Franklin, N. H	May 22, 1855	12, 933
Knitting-machine	J. B. Aiken	Manchester, N. H	Aug. 2, 1859	24, 916
Knitting-machine	J. B. Aiken	New York, N. Y	Feb. 5, 1861	31, 287
Knitting-machine	J. B. and W. Aiken	Franklin, N. H	July 8, 1856	15, 314
Knitting-machine	N. P. Aikin	Troy, N. Y	July 13, 1858	20, 854
Knitting-machine	W. Aiken	Franklin, N. H	Dec. 1, 1857	18, 725
Knitting-machine	W. Aiken	Franklin, N. H	Nov. 9, 1858	22, 004
Knitting-machine	W. Aiken	Franklin, N. H	Aug. 7, 1860	29, 450
Knitting-machine	W. Aiken	Franklin, N. H	July 2, 1861	32, 674
Knitting-machine	W. Aiken	Franklin, N. H	June 21, 1864	43, 170
Knitting-machine	W. Aiken	Franklin, N. H	Dec. 31, 1867	72, 771
Knitting-machine	W. and J. B. Aiken	Franklin, N. H	Sept. 11, 1855	13, 561
Knitting-machine	C. J. Appleton	Philadelphia, Pa	Feb. 14, 1860	27, 183
Knitting-machine	C. J. Appleton	Cohoes, N. Y	Sept. 27, 1870	107, 750
Knitting-machine	C. J. Appleton	Hamilton, Canada	Feb. 13, 1872	123, 545
Knitting-machine	J. M. Armour	Craftsbury, Vt	May 15, 1866	54, 812
Knitting-machine	J. M. Armour	Syracuse, N. Y	Apr. 25, 1871	113, 965
Knitting-machine	J. M. Armour	Syracuse, N. Y	Mar. 4, 1873	136, 480
Knitting-machine	H. L. Arnold	Ottawa, Ill	Sept. 10, 1872	131, 144
Knitting-machine	V. G. Arnold	Providence, R. I	Aug. 2, 1870	106, 025
Knitting-machine	T. Bailey	Ballston Spa, N. Y	Feb. 24, 1852	8, 750
Knitting-machine	T. Bailey	Ballston Spa, N. Y	Nov. 20, 1855	13, 811
Knitting-machine	J. H. Barsantee	Portsmouth, N. H	July 29, 1851	8, 262
Knitting-machine	J. H. Barsantee	Philadelphia, Pa	May 30, 1854	10, 980
Knitting-machine	C. E. Bean	Oswego, N. Y	Aug. 20, 1872	130, 558
Knitting-machine	D. Bickford	Boston, Mass	Sept. 10, 1867	68, 595
Knitting-machine	D. Bickford	Boston, Mass	July 21, 1868	80, 121
Knitting-machine	D. Bickford	Boston, Mass	July 6, 1869	92, 146
Knitting-machine	W. Binkley	Manchester, N. H	Oct. 4, 1859	25, 698
Knitting-machine	C. W. Blakeslee	Northfield, Conn	Sept. 20, 1864	44, 365
Knitting-machine	C. W. Blakeslee, E. B. Beecher, and A. G. Davis.	Watertown and Westville, Conn.	July 14, 1868	79, 897
Knitting-machine	C. W. Blakeslee, A. G. Davis, and E. B. Beecher.	Watertown and Westville, Conn.	Aug. 11, 1868	80, 835
Knitting-machine	H. Bogel	Watertown, Wis	Mar. 10, 1868	75, 353

Index of patents issued from the United States Patent Office from 1790 *to* 1873, *inclusive*—Continued.

Invention.	Inventor.	Residence.	Date.	No.
Knitting-machine	B. Bollinger and G. G. Nodle	New Berlin, Ohio	Sept. 22, 1868	82, 281
Knitting-machine	J. Bradley	Lowell, Mass	May 14, 1872	126, 621
Knitting-machine	J. Bradley	Lowell, Mass	Sept. 24, 1872	131, 505
Knitting-machine	J. L. Branson	Chicago, Ill	June 18, 1872	127, 954
Knitting-machine	F. L. Buell	Manchester, Conn	Aug. 23, 1859	25, 230
Knitting-machine	J. Bullock	Cohoes, N. Y	Nov. 15, 1859	26, 085
Knitting-machine	F. Burns	Upper Gilmanton, N. H	Aug. 24, 1869	94, 071
Knitting-machine	W. W. Burson and J. Nelson	Rockford, Ill	June 23, 1868	79, 202
Knitting-machine	W. W. Burson and J. Nelson	Rockford, Ill	Oct. 4, 1870	108, 003
Knitting-machine	H. Burt	Newark, N. J	May 16, 1854	10, 915
Knitting-machine	H. Burt	Newark, N. J	July 4, 1854	11, 238
Knitting-machine	C. Callahan	Chelsea, Mass	July 15, 1873	140, 809
Knitting-machine	W. and W. Campion	Nottingham and Sneinton, England.	May 20, 1873	139, 114
Knitting-machine	A. C. Carey	Malden, Mass	Apr. 18, 1865	47, 354
Knitting-machine	A. C. Carey	Malden, Mass	July 30, 1867	67, 263
Knitting-machine	J. Chantrell	Bristol, Conn	Mar. 13, 1860	27, 430
Knitting-machine	J. Chantrell	Bristol, Conn	May 15, 1860	28, 255
Knitting-machine	J. Chantrell	Bristol, Conn	Jan. 7, 1868	73, 164
Knitting-machine	M. Christoffers	Hanover, Prussia	July 16, 1872	129, 529
Knitting-machine	W. W. Clay	Nottingham, England	Dec 22, 1863	40, 993
Knitting-machine	W. W. Clay	Philadelphia, Pa	Sept. 19, 1865	49, 980
Knitting-machine	W. W. Clay	Philadelphia, Pa	June 5, 1866	55, 241
Knitting-machine	E. Colvin	Poultney, Vt	Jan. 13, 1857	16, 375
Knitting-machine	E. Colvin	Poultney, Vt	June 7, 1859	24, 284
Knitting-machine	F. M. Comstock	Cleveland, Ohio	Sept. 20, 1870	107, 454
Knitting-machine	F. M. Comstock	Cleveland, Ohio	Apr. 9, 1872	125, 543
Knitting-machine	J. A. Corwin	Newark, N. J	Sept. 26, 1854	11, 720
Knitting-machine	W. Cotton	Loughborough, England	Nov. 20, 1866	59, 892
Knitting-machine	W. Cotton	Loughborough, England	Aug. 13, 1867	67, 727
Knitting-machine	T. Crane	Fort Atkinson, Wis	Oct. 15, 1867	69, 775
Knitting-machine	T. Crane	Fort Atkinson, Wis	Oct. 15, 1867	69, 776
Knitting-machine	T. Crane	Fort Atkinson, Wis	Jan. 28, 1868	73, 697
Knitting-machine	T. Crane	Fort Atkinson, Wis	June 15, 1869	91, 214
Knitting-machine	J. Dalton	Brooklyn, N. Y	May 7, 1861	32, 241
Knitting-machine	J. Dalton	Williamsburgh, N. Y	Oct. 9, 1866	58, 610
Knitting-machine	J. Dalton	Brooklyn, N. Y	Apr. 28, 1868	77, 363
Knitting-machine	A. G. Davis, C. W. Blakeslee, and A. N. Allen.	Westville, Conn	Mar. 22, 1870	101, 106
Knitting-machine	O. Davis	New Lebanon, Ind	Sept. 13, 1870	107, 230
Knitting-machine	J. P. Delahunty	Cohoes, N. Y	July 27, 1858	21, 045
Knitting-machine	J. H. Doolittle	Waterbury, Conn	Oct. 16, 1855	13, 693
Knitting-machine	R. Ellis	Northampton, Mass	June 17, 1851	8, 163
Knitting-machine	E. S. Ells	Troy, N. Y	Aug. 23, 1859	25, 185
Knitting-machine	S. V. Essick	Mansfield, Ohio	Sept. 3, 1867	68, 425
Knitting-machine	S. V. Essick	Alliance, Ohio	July 8, 1873	140, 691
Knitting-machine	W. B. Evans	Holderness, N. H	Dec. 16, 1862	37, 161
Knitting-machine	S. D. Fairbanks	Cohoes, N. Y	Dec. 1, 1857	18, 792
Knitting-machine	G. W. Folts and J. L. Branson	Boston, Mass., and Chicago, Ill.	Feb. 21, 1871	112, 027
Knitting-machine	T. Fowler	Cohoes, N. Y	July 3, 1855	13, 165
Knitting-machine	W. Franz and W. Pope	Crestline, Ohio	Mar. 23, 1869	88, 027
Knitting-machine	W. Franz and W. Pope	Bucyrus and Crestline, Ohio.	July 12, 1870	105, 187
Knitting-machine	A. French	Waterbury, Conn	Nov. 6, 1855	13, 750
Knitting-machine	F. Gardner	Hamilton, Canada	July 6, 1869	92, 300
Knitting-machine	A. J. and D. Goffe	Cohoes, N. Y	Aug. 5, 1856	15, 484
Knitting-machine	A. J. and D. Goffe	Cohoes, N. Y	Nov. 22, 1859	26, 231
Knitting-machine	J. Goodman	Pawtucket, R. I	Mar. 27, 1866	53, 530
Knitting-machine	H. Guenther	New York, N. Y	Mar. 26, 1872	124, 950
Knitting-machine	H. Günther	New York, N. Y	Jan. 14, 1873	134, 881
Knitting-machine	H. V. Hartz	Cleveland, Ohio	Jan. 30, 1872	123, 100
Knitting machine	H. V. Hartz and J. Feiss	Cleveland, Ohio	Feb. 8, 1870	99, 670
Knitting-machine	W. S. Hill	Manchester, N. H	May 24, 1870	103, 332
Knitting-machine	J. Hiller and J. Bullock	Cohoes, N. Y	Jan. 25, 1859	22, 769
Knitting-machine	J. Hinkley	Norwalk, Ohio	May 29, 1866	55, 103
Knitting-machine	J. Hinkley	Norwalk, Ohio	Apr. 12, 1870	101, 875
Knitting-machine	J. Hollen	White Township, Pa	July 16, 1850	7, 509
Knitting-machine	J. Hollen	White Township, Pa	Nov. 28, 1854	11, 995
Knitting-machine	J. Hollen	Fostoria, Pa	Oct. 18, 1859	25, 827
Knitting-machine	J. Hollen	Fostoria, Pa	Jan. 1, 1861	31, 020
Knitting-machine	J. Hollen	Fostoria, Pa	Sept. 2, 1862	36, 351
Knitting-machine	J. Hollen	Fostoria, Pa	July 19, 1870	105, 454
Knitting-machine	W. H. H. Hollen	Fostoria, Pa	Dec. 17, 1867	72, 296
Knitting-machine	W. H. H. Hollen	Fostoria, Pa	Jan. 3, 1871	110, 656
Knitting-machine	W. H. H. Hollen	Fostoria, Pa	Mar. 5, 1872	124, 357
Knitting-machine	I. M. Hopkins	Pascoag, R. I	May 16, 1854	10, 910
Knitting-machine	J. A. and H. A. House	Bridgeport, Conn	Mar. 13, 1866	53, 224
Knitting-machine	H. A. House	Bridgeport, Conn	Jan. 4, 1870	98, 500
Knitting-machine	H. A. House	Bridgeport, Conn	Apr. 12, 1870	101, 878
Knitting-machine	H. A. House	Bridgeport, Conn	Nov. 19, 1872	133, 103
Knitting-machine	J. M. Howe	Rochester, N. Y	Mar. 15, 1870	100, 765
Knitting-machine	J. M. Howe	Rochester, N. Y	Aug. 8, 1871	117, 886
Knitting-machine	S. W. Howland	Adams, Mass	May 14, 1861	32, 295
Knitting-machine	L. B. Hunt	Hyde Park, Mass	July 25, 1871	117, 293
Knitting-machine	L. B. Hunt	Hyde Park, Mass	Oct. 8, 1872	131, 955
Knitting-machine	G. Jacksen	Cohoes, N. Y	Aug. 22, 1854	11, 554
Knitting-machine	G. Jenson	Brooklyn, N. Y	Dec. 20, 1864	45, 501
Knitting-machine	G. Johnstone	Philadelphia, Pa	Aug. 11, 1868	80, 965
Knitting-machine	G. Johnstone	Philadelphia, Pa	Sept. 13, 1870	107, 381
Knitting-machine	L. Kavanaugh	Waterford, N. Y	Jan. 26, 1864	41, 381
Knitting-machine	C. G. Keeny	Manchester, Conn	Jan. 25, 1859	22, 735
Knitting-machine	J. Kent	New York, N. Y	Dec. 28, 1869	98, 272
Knitting-machine	D. Kidder	Franklin, N. H	Oct. 27, 1868	83, 390
Knitting-machine	E. E. Kilbourn	New Brunswick, N. J	May 23, 1865	47, 829
Knitting-machine	J. K. and E. E. Kilbourn	Norfolk, Conn	Feb. 16, 1858	19, 370
Knitting-machine	J. K. and E. E. Kilbourn	Norfolk, Conn	Oct. 12, 1858	21, 762
Knitting-machine	J. K. and E. E. Kilbourn	Norfolk, Conn	Apr. 9, 1861	31, 984
Knitt'ng-machine	I. W. Lamb	Detroit, Mich	Sept. 15, 1863	39, 934

Index of patents issued from the United States Patent Office from 1790 *to* 1873, *inclusive*—Continued.

Invention.	Inventor.	Residence.	Date.	No.
Knitting-machine	I. W. Lamb	Rochester, N. Y	Oct. 10, 1865	50, 369
Knitting-machine	M. T. Lamb	Valparaiso, Ind	June 5, 1866	55, 420
Knitting-machine	T. Langham	Philadelphia, Pa	Jan. 21, 1862	34, 210
Knitting-machine	T. Langham	Philadelphia, Pa	Mar. 6, 1866	53, 013
Knitting-machine	S. Larkin	Bridgeport, Conn	Aug. 11, 1868	80, 866
Knitting-machine	J. and M. Lee and W. Carter	Needham, Mass	Dec. 27, 1870	110, 479
Knitting-machine	J. Leonard	Attica, Ohio	Feb. 11, 1873	135, 823
Knitting-machine	J. Y. Leslie	Cincinnati, Ohio	Oct. 3, 1854	11, 751
Knitting-machine	T. Lovelidge	Germantown, Pa	Aug. 31, 1858	21, 396
Knitting-machine	W. Mansfield	Dracut, Mass	Mar. 22, 1853	9, 626
Knitting-machine	M. Marshall	Lowell, Mass	Mar. 15, 1853	9, 621
Knitting-machine	J. Maxwell	Galesville, N. Y	Mar. 29, 1853	9, 637
Knitting-machine	J. McCune	Auburn, Ind	Sept. 14, 1869	94, 904
Knitting-machine	J. McMullen and J. Hollen, jr	Sinking Valley and Logan's Valley, Pa.	Feb. 11, 1837	126½
Knitting-machine	W. H. McNary	Brooklyn, N. Y	May 15, 1860	28, 290
Knitting-machine	W. H. McNary	Brooklyn, N. Y	Apr. 30, 1867	64, 241
Knitting-machine	G. Merrill	East Orange, N. J	July 8, 1873	140, 635
Knitting-machine	E. Morse	Winchendon, Mass	Aug. 16, 1870	106, 499
Knitting-machine	J. Nesmith	Lowell, Mass	July 29, 1856	15, 435
Knitting-machine	A. Paget	Loughborough, Great Britain.	May 21, 1867	64, 900
Knitting-machine	O. Parkhurst	Cohoes, N. Y	Mar. 5, 1861	31, 647
Knitting-machine	G. M. Patten	Bath, Me	Jan. 19, 1869	86, 033
Knitting-machine	H. Pease	Brockport, N. Y	June 25, 1872	128, 421
Knitting-machine	R. Peberdy	Philadelphia, Pa	Mar. 2, 1869	87, 508
Knitting-machine	J. Pepper	Portsmouth, N. H	Feb. 25, 1851	7, 945
Knitting-machine	J. Pepper	Portsmouth, N. H	June 24, 1851	8, 172
Knitting-machine	J. Pepper	Franklin, N. H	July 17, 1855	13, 289
Knitting-machine	J. Pepper	Lowell, Mass	Dec. 19, 1865	51, 618
Knitting-machine	J. Pepper	Lake Village, N. H	Aug. 27, 1867	68, 107
Knitting-machine	J. Pepper	Lake Village, N. H	Apr. 26, 1870	102, 313
Knitting-machine	J. Pepper, jr	Portsmouth, N. H	Dec. 5, 1854	12, 046
Knitting-machine	D. C. Philips and C. and I. Tompkins.	Philmont and Troy, N. Y	June 14, 1870	104, 346
Knitting-machine	F. Philip	Stockport, N. Y	Mar. 15, 1870	100, 798
Knitting-machine	C. H. Platt and G. A. Stanbery	Norwalk, Ohio	Apr. 8, 1873	137, 568
Knitting-machine	J. Powell	Waterbury, Conn	Oct. 2, 1855	13, 621
Knitting-machine	E. R. Pray	Holderness, N. H	Sept. 1, 1868	81, 683
Knitting-machine	A. B. Prouty	Worcester, Mass	Dec. 28, 1869	98, 410
Knitting-machine	H. Pudder and T. W. Hurlbut	Dayton, Ky., and Cincinnati, Ohio.	June 13, 1871	115, 983
Knitting-machine	W. H. Ramsdell	Lowell, Mass	Aug. 27, 1872	130, 866
Knitting-machine	E. M. Ray	Providence, R. I	May 30, 1854	10, 993
Knitting-machine	J. D. Reiff	Skippackville, Pa	Jan. 12, 1869	85, 765
Knitting-machine	J. W. Rist	Rochester, N. Y	Sept. 22, 1868	82, 348
Knitting-machine	M. L. Roberts	Mount Union, Ohio	Feb. 12, 1861	31, 404
Knitting-machine	M. L. Roberts	New Brunswick, N. J	July 27, 1869	93, 008
Knitting-machine	S. H. Roper	Boston, Mass	Aug. 24, 1869	94, 135
Knitting-machine	S. R. Roper	Roxbury, Mass	Aug. 8, 1871	117, 931
Knitting-machine	J. Rose	Nottingham, England	Jan. 9, 1872	122, 535
Knitting-machine	L. D. Sanborn	Cohoes, N. Y	Oct. 7, 1873	143, 468
Knitting-machine	F. Schott	Brooklyn, N. Y	Nov. 23, 1858	22, 135
Knitting-machine	H. B. Scudder	Needham, Mass	Mar. 15, 1870	100, 810
Knitting-machine	A. Sessions, jr	Springfield, Mass	Sept. 5, 1865	49, 835
Knitting-machine	E. Shore	Conshohocken, Pa	Sept. 2, 1862	36, 373
Knitting-machine	E. Shore	Conshohocken, Pa	Feb. 9, 1864	41, 540
Knitting-machine	J. and J. Smith	Philadelphia, Pa	June 17, 1873	140, 086
Knitting-machine	A. B. Stillings	Springfield, Mass	June 19, 1866	55, 734
Knitting-machine	E. Taélbouis and A. Renevey	St. Just-en-Chaussee, France.	Feb. 18, 1873	136, 012
Knitting-machine	W. A. Tangeman	Lockland, Ohio	Jan. 5, 1869	85, 546
Knitting-machine	J. Teachout	Waterford, N. Y	Dec. 10, 1867	72, 116
Knitting-machine	J. Teachout	Waterford, N. Y	Dec. 10, 1867	72, 117
Knitting-machine	J. Terrell	Philadelphia, Pa	Jan. 1, 1861	31, 042
Knitting-machine	S. T. Thomas	Laconia, N. H	Apr. 1, 1862	34, 852
Knitting-machine	A. Thornton	Nottingham, England	Sept. 27, 1864	44, 494
Knitting-machine	H. Thornton	Nottingham, England	Mar. 20, 1866	53, 368
Knitting-machine	J. and W. Thornton	Pease Hill Rise, Nottingham, England.	June 5, 1866	55, 444
Knitting-machine	E. Tiffany	Thompsonville, Conn	May 1, 1860	28, 133
Knitting-machine	C. Tompkins	Troy, N. Y	Dec. 23, 1856	16, 297
Knitting-machine	O. F. Tripp	Battle Creek, Mich	July 15, 1873	140, 800
Knitting-machine	O. Twombly	Ashland, N. H	Aug. 12, 1873	141, 836
Knitting-machine	O. Twombly and W. Noyes, jr	Holderness, N.H., and Newburyport, Mass.	July 7, 1868	79, 789
Knitting-machine	J. Vickerstaff	Philadelphia, Pa	Mar. 23, 1858	19, 740
Knitting-machine	J. Waldie	Ipswich, Mass	Aug. 11, 1868	80, 887
Knitting-machine	A. J. Wale	Philadelphia, Pa	July 21, 1868	80, 099
Knitting-machine	S. Wallis	Townsend, Mass	Oct. 4, 1864	44, 568
Knitting-machine	S. Wallis	Lowell, Mass	Dec. 19, 1865	51, 666
Knitting-machine	B. Ward	Troy, N.Y	Apr. 14, 1868	76, 854
Knitting-machine	J. F. Waterhouse	Germantown, Pa	Aug. 9, 1859	25, 067
Knitting-machine	J. C. Welsch	Edgerton, Ohio	Feb. 11, 1868	74, 460
Knitting-machine	J. Whittle	Philadelphia, Pa	Apr. 11, 1865	47, 239
Knitting-machine	H. L. Williams	Seneca Falls, N. Y	June 26, 1866	55, 968
Knitting-machine	J. G. Wilson	New York, N. Y	Aug. 12, 1862	36, 199
Knitting-machine	J. G. Wilson	New York, N. Y	May 22, 1866	55, 027
Knitting-machine	J. G. Wilson	New York, N. Y	June 5, 1866	55, 434
Knitting-machine	B. S. Wood	Burrillville, R. I	June 6, 1854	10, 998
Knitting-machine	L. Woodward	Nottingham, England	Dec. 2, 1873	145, 079
Knitting-machine	H. C. Work	Philadelphia, Pa	Nov. 2, 1869	96, 531
Knitting-machine	C. H. Young	Lake Village, N. H	Sept. 23, 1873	143, 051
Knitting-machine and knitted fabric	T. Crane	Fort Atkinson, Wis	June 15, 1869	91, 215
Knitting machine and needle	A. W. Allen	Indianapolis, Ind	Dec. 6, 1870	109, 793
Knitting-machine and process of knitting	N. Clark	Malden, Mass	Dec. 12, 1871	121, 924
Knitting-machine burr	C. Allardice	Cohoes, N. Y	May 3, 1864	42, 545
Knitting machine burr	W. H. Carr	Troy, N. Y	July 26, 1864	43, 636
Knitting-machine burr	J. Clute	Cohoes, N. Y	May 9, 1865	47, 620

Index of patents issued from the United States Patent Office from 1790 *to* 1873, *inclusive*—Continued.

Invention.	Inventor.	Residence.	Date.	No.
Knitting-machine burr	H. Fisher	Waterford, N. Y	Nov. 24, 1863	40, 679
Knitting-machine burr	G. Jackson and G. Campbell	Cohoes, N. Y	July 28, 1863	39, 347
Knitting-machine burr	L. Kavanaugh	Waterford, N. Y	June 10, 1862	35, 565
Knitting-machine burr	L. Kavanaugh	Waterford, N. Y	July 25, 1871	117, 299
Knitting-machine burr	M. Van Auken	Amsterdam, N. Y	June 23, 1863	38, 997
Knitting-machine burr, Self-lubricating	C. Allardice	Cohoes, N. Y	May 29, 1866	55, 034
Knitting-machine burrs, Means for oiling	J. Maxwell	Amsterdam, N. Y	Sept. 27, 1864	44, 437
Knitting machine, Circular	G. Campbell	Waterford, N. Y	July 22, 1862	35, 968
Knitting machine, Circular	C. G. Cole	Bennington, Vt	July 6, 1869	92, 167
Knitting machine, Circular	J. Dalton	Brooklyn, N. Y	Sept. 27, 1864	44, 403
Knitting machine, Circular	W. Franz and W. Pope	Crestline, Ohio	Feb. 1, 1870	99, 425
Knitting machine, Circular	W. Franz and W. Pope	Crestline, Ohio	Feb. 1, 1870	99, 426
Knitting machine, Circular	T. Hawthorne	Philadelphia, Pa	Mar. 14, 1865	46, 848
Knitting machine, Circular	S. L. Otis	Manchester, Conn	Nov. 1, 1864	44, 885
Knitting machine, Circular	D. Scattergood	Nottingham, England	Aug. 11, 1863	39, 499
Knitting machine, Circular	C. Shirtcliff	Philadelphia, Pa	May 2, 1865	47, 579
Knitting-machine indicator	G. P. Fuller	Adrian, Mich	Mar. 10, 1868	75, 260
Knitting-machine indicator	J. C. Potter	Alfred, N. Y	Mar. 3, 1868	75, 194
Knitting-machine indicator	J. W. Rist	Rochester, N. Y	Sept. 29, 1868	82, 554
Knitting-machine knot-indicator	J. F. Steward	Plano, Ill	Apr. 23, 1867	64, 046
Knitting machine, Lace	H. Williamson	Williamsburgh, N. Y	Oct. 31, 1871	120, 474
Knitting-machine needle	W. Aiken	Franklin, N. H	May 24, 1864	42, 824
Knitting-machine needle	W. Aiken	Franklin, N. H	June 2, 1868	78, 506
Knitting-machine needle	R. Allen	Salem, Mich	July 28, 1868	80, 264
Knitting-machine needle	J. M. Armour	Craftsbury, Vt	Jan. 30, 1866	52, 256
Knitting-machine needle	N. H. Baldwin	Laconia, N. H	June 25, 1872	128, 274
Knitting-machine needle	D. Bickford	Boston, Mass	Dec. 1, 1868	84, 472
Knitting-machine needle	D. Bickford	New York, N. Y	Oct. 22, 1872	132, 382
Knitting-machine needle	A. C. Carey	Lynn, Mass	Apr. 25, 1865	47, 488
Knitting-machine needle	J. F. Daniels	Lake Village, N. H	Feb. 8, 1870	99, 647
Knitting-machine needle	R. Ellis	Boston, Mass	June 3, 1856	15, 006
Knitting-machine needle	L. W. Fifield	New Hampton, N. H	Dec. 19, 1865	51, 577
Knitting-machine needle	L. W. Fifield	Holderness, N. H	July 24, 1866	56, 544
Knitting-machine needle	L. W. Fifield	Melrose, Mass	Jan. 22, 1867	61, 413
Knitting-machine needle	J. K. and E. E. Kilbourn	Pittsfield, Mass., and Norfolk, Conn.	Sept. 21, 1858	21, 566
Knitting-machine needle	J. W. Lamb	Rochester, N. Y	Nov. 21, 1865	51, 115
Knitting-machine needle	J. H. Lane and C. D. House	Lake Village, N. H	Dec. 7, 1869	97, 526
Knitting-machine needle	J. Miller	Warren, R. I	Nov. 19, 1867	71, 039
Knitting-machine needle	J. Miller and J. A. Bidwell	Warren, R. I., and East Boston, Mass.	Jan. 29, 1867	61, 630
Knitting-machine needle	J. L. and S. L. Otis	Florence, Mass., and Manchester, Conn.	May 23, 1865	47, 852
Knitting-machine needle	B. F. Peaslee	Lake Village, N. H	Apr. 9, 1867	63, 746
Knitting-machine needle	S. Peberdy	Philadelphia, Pa	May 13, 1862	35, 254
Knitting-machine needle	C. H. Platt and G. A. Stanberg	Norwalk, Ohio	Feb. 25, 1873	136, 262
Knitting-machine needle	S. H. Roper	Boston, Mass	Oct. 3, 1871	119, 651
Knitting-machine needle	C. P. S. Wardwell	Lake Village, N. H	Feb. 26, 1867	62, 382
Knitting-machine-needle shanks, Machine for making.	J. S. Perkins	Lake Village, N. H	July 28, 1868	80, 304
Knitting-machine needles, Manufacture of	J. H. Bullard	Chicopee Falls, Mass	July 8, 1873	140, 611
Knitting-machine register	B. B. Bollinger	Louisville, Ohio	Sept. 8, 1868	81, 870
Knitting-machine register	W. V. Dubois	Covington, Ind	May 25, 1869	90, 435
Knitting-machine register	J. W. Rist	Rochester, N. Y	Sept. 29, 1868	82, 555
Knitting machine register	B. F. Wyman and B. H. Hartshorn.	Lancaster and Ashland, Mass.	Sept. 17, 1867	68, 928
Knitting-machine, Rotary	J. Hollen	Blair County, Pa	Mar. 6, 1866	53, 000
Knitting-machine, Rotary	S. W. Park and E. S. Ells	Troy, N. Y	Aug. 5, 1856	15, 492
Knitting-machine, Rotary	M. L. Roberts	Chatsworth, Ill	May 7, 1867	64, 572
Knitting-machine, Rotary	H. G. Sanford	Worcester, Mass	Nov. 30, 1852	9, 434
Knitting-machine, Rotary	D. Tainter	Worcester, Mass	Nov. 30, 1852	9, 435
Knitting-machine, Rotary	C. Tompkins and J. Johnson	Troy, N. Y	May 27, 1856	14, 975
Knitting-machine sinkers, Blade for	L. Kavanaugh	Waterford, N. Y	Jan. 6, 1863	37, 347
Knitting-machine, Weft-thread	C. Callahan	Chelsea, Mass	Feb. 11, 1873	135, 625
Knitting-machine-weight hook	C. H. Koegel	Buffalo, N. Y	Apr. 15, 1873	137, 778
Knitting-machines, Construction of cylinders of	C. H. Young	Lake Village, N. H	Sept. 13, 1870	107, 318
Knitting-machines, Dividing-wheel of weft-thread	H. Woodman	Saco, Me	Apr. 25, 1871	114, 077
Knitting-machines, looms, &c., Electro-magnetic stop-motion, &c., for.	H. E. Wells and D. H. Moran	Van Wert, Ohio	Jan. 30, 1872	123, 315
Knitting-machines, &c., Machine for producing reciprocating motion in.	S. Haslam, jr	New Britain, Conn	July 28, 1868	80, 283
Knitting-machines, Manufacture of latch-needles for.	S. Woodward	Manchester, N. H	June 27, 1871	116, 386
Knitting machines, Register for tubular	D. A. Lehman	Wakarusa, Ind	Dec. 16, 1873	145, 509
Knitting-machines, Rotary take-up for	J. Teachout	Waterford, N. Y	Dec. 10, 1867	72, 115
Knitting-machines, Set-up device for	H. L. Arnold	Ottawa, Ill	Sept. 2, 1873	142, 429
Knitting-machines, Setting-up and tension comb for.	W. Gaskill	Cincinnati, Ohio	June 7, 1870	104, 114
Knitting machines, Setting-up device for circular	W. Franz and W. Pope	Crestline, Ohio	May 3, 1870	102, 529
Knitting-machines, Setting-up work in	I. W. Lamb	Rochester, N. Y	Sept. 12, 1865	49, 895
Knitting-machines, Sinker-wheel for	H. C. Bradford	Providence, R. I	May 30, 1871	115, 426
Knitting-machines, Stop-motion for	R. Cushman	Pawtucket, R. I	May 22, 1855	12, 896
Knitting-machines, Stop-motion for	J. Dalton	Brooklyn, N. Y	June 28, 1864	43, 294
Knitting-machines, Stop-motion for	E. Kay	Philadelphia, Pa	June 5, 1866	55, 308
Knitting-machines, Stop-motion for	M. Lee	Needham, Mass	Sept. 19, 1865	50, 012
Knitting-machines, Stop-motion for	B. L. Mack	Essex, Conn	May 10, 1864	42, 672
Knitting-machines, Stop-motion for	T. F. Wynn	Atlanta, Ga	June 4, 1872	127, 539
Knitting-machines, Stop-motion and creel-stand for	H. C. Bradford	Providence, R. I	July 4, 1871	116, 677
Knitting-machines, Stop-motion and indicator for	E. Wilder	Chicopee Falls, Mass	Oct. 13, 1868	83, 116
Knitting machines, Stop-motion for circular	P. W. Hart	Stamford, N. Y	Jan. 31, 1865	46, 186
Knitting-machines, Stopping-mechanism for	R. Cook	New Hartford, N. Y	Oct. 3, 1871	119, 448
Knitting-machines, Stopping-mechanism for	J. Kennedy	Claverack, N. Y	Aug 31, 1869	94, 218
Knitting-machines, Take-up device for	N. H. Baldwin	Laconia, N. H	July 16, 1872	129, 387
Knitting machines, Take-up for circular	S. Ward	Amsterdam, N. Y	Jan. 26, 1864	41, 415
Knitting-machines, Take-up mechanism for	T. Crane	Fort Atkinson, Wis	Jan. 29, 1867	61, 608
Knitting-machines, Take-up mechanism for	S. Ward	Amsterdam, N. Y	Feb. 19, 1867	62, 239
Knitting machines, Take-up of circular	H. Brockway	Cohoes, N. Y	Nov. 8, 1864	44, 932
Knitting machines, Taking-up mechanism for circular.	J. Lessels	Troy, N. Y	June 5, 1866	55, 422

Index of patents issued from the United States Patent Office from 1790 *to* 1873, *inclusive*—Continued.

Invention.	Inventor.	Residence.	Date.	No.
Knitting-machines, Transferring and loop-retaining comb for.	J. N. Kendall	Nashua, N. H	May 24, 1870	103, 471
Knitting-machines, Yarn-take-up and tension-device for.	H. L. Arnold	Ottawa, Ill	Aug. 5, 1873	141, 534
Knitting-machines, Yarn-tension device for	J. Crandell	Chicopee Falls, Mass	Sept. 28, 1869	95, 200
Knitting-machines, Yarn-tension device for	W. Franz and W. Pope	Bucyrus and Crestline, Ohio	Feb. 13, 1872	123, 687
Knitting machinery	J. Whitworth	Manchester, England	Feb. 1, 1848	5, 432
Knitting-needle	J. Hibbert	Providence, R. I	Jan. 9, 1840	6, 025
Knitting-needle jack	H. J. Wickham	Manchester, Conn	Aug. 18, 1868	81, 235
Knitting-needles, Machine	T. Sands	Gilford, N. H	June 23, 1863	38, 988
Knitting-needles, Machine for making	C. P. S. Wardwell	Lake Village, N. H	Sept. 26, 1865	50, 188
Knitting-needles, Machine for making the tongues of machine.	W. Aiken	Franklin, N. H	Nov. 12, 1867	70, 678
Knitting-needles, Machine for notching	W. Aiken	Franklin, N. H	Dec. 10, 1867	71, 836
Knitting-needles, Making	T. Sands	Gilford, N. H	June 23, 1863	38, 987
Knitting needles, Method of riveting latch	C. P. S. Wardwell	Lake Village, N. H	Dec. 5, 1871	121, 562
Knitting pile fabric	D. Bickford	Boston, Mass	Dec. 1, 1868	84, 473
Knitting lacings, &c., Machine for	W. W. Westcott and H. L. Walcott.	Providence, R. I., and Charles River Village, Mass.	July 18, 1865	48, 878
Knitting stockings and joining selvages, Method of	D. Bickford	New York, N. Y	Sept. 17, 1872	131, 388
Knitting stockings and stockinets, Machine for	A. Porter, J. Mead, and J. Stedwall.	Queensbury, N. Y	July 31, 1815	....
Knitting stockings, Machine for	H. Burt	Boston, Mass	Sept. 23, 1843	3, 275
Knitting stockings, &c., Machine for	A. French	Springfield, Mass	Mar. 18, 1842	2, 493
Knitting stockings, Machine for	B. Hutchinson	Springfield, Mass	Oct. 22, 1840	1, 834
Knitting stockings, &c., Machine for	J. McMullen and J. Hollen, jr.	Huntingdon County, Pa	Mar. 5, 1831	....
Knitting stockings, Method of	H. A. House	Bridgeport, Conn	Mar. 7, 1871	112, 346
Knob and door fastener, Combined base	H. H. Heskett and M. E. Ferguson.	McLean County, Ill	Jan. 25, 1870	99, 086
Knob, Atmospheric	O. H. Needham	New York, N. Y	Sept. 29, 1868	82, 629
Knob bolt, Door	L. Lillie	Troy, N. Y	July 5, 1859	24, 647
Knob-bolts, Operating and locking	O. Ellsworth	Hartford, Conn	June 7, 1853	9, 767
Knob, Carriage	J. Barclay	Attleborough, Mass	Oct. 22, 1867	69, 959
Knob, Carriage	R. D. Case and J. Barclay	New York, N. Y., and Attleborough Falls, Mass.	Oct. 22, 1867	69, 965
Knob, Commode	E. Skinner	Sandwich, N. H	June 11, 1829	....
Knob, Curtain	C. Z. Kroh	Tiffin, Ohio	Aug. 29, 1865	49, 634
Knob, Door	J. W. Bliss	Hartford, Conn	July 22, 1856	15, 367
Knob, Door	G. W. Cady	Providence, R. I	June 21, 1870	104, 420
Knob, Door	E. Day	Chicago, Ill	Jan. 12, 1869	85, 799
Knob, Door	H. H. Elwell	Meriden, Conn	July 15, 1856	15, 332
Knob, Door	H. H. Elwell	South Norwalk, Conn	Sept. 2, 1873	142, 451
Knob, Door	J. W. Grogan	Brooklyn, N. Y	Feb. 22, 1870	100, 026
Knob, Door	G. Jones	Peekskill, N. Y	May 31, 1870	103, 749
Knob, Door	J. J. King	New York, N. Y	Sept. 21, 1869	95, 024
Knob, Door	W. Leighton	Cambridge, Mass	Jan. 16, 1855	12, 265
Knob, Door	A. Longstreet	Chicago, Ill	Nov. 26, 1867	71, 315
Knob, Door	G. B. Lothrop	Boston, Mass	July 20, 1869	92, 733
Knob, Door	E. Parker	Meriden, Conn	May 5, 1863	38, 406
Knob, Door	T. J. Sloan	Bronxville, N. Y	Nov. 9, 1869	96, 627
Knob, Door	L. L. Smith	New York, N. Y	July 9, 1872	128, 764
Knob, Door	A. E. Young	Dorchester, Mass	Aug. 21, 1855	13, 473
Knob, Door, commode, &c	E. Robinson, F. Draper, and J. H. Lord.	Cambridge and Boston, Mass.	Dec. 2, 1836	98
Knob, Drawer	N. Thompson	Brooklyn, N. Y	Sept. 26, 1871	119, 429
Knob, Drawer, commode, &c	D. Hottman	Baltimore, Md	July 31, 1837	325
Knob, Extension door	W. T. Munger	Branford, Conn	July 18, 1865	48, 874
Knob for doors, &c., Ferrule	E. Robinson, F. Draper, and J. H. Lord.	Cambridge and Boston, Mass.	Oct. 20, 1836	65
Knob for doors, &c., Glass	F. Draper	East Cambridge, Mass	Sept. 10, 1840	1, 784
Knob for doors, Glass	C. D. Kellogg and W. L. Coan.	Boston, Mass	Dec. 22, 1857	18, 911
Knob for fastening curtains, &c	W. Z. M. and J. W. Chapman.	New York, N. Y	Mar. 20, 1855	12, 540
Knob for permutation-locks	H. Isham	New Britain, Conn	Jan. 18, 1870	98, 971
Knob for safe-doors	J. G. Kittredge	San Francisco, Cal	Apr. 19, 1870	102, 132
Knob, Furniture	J. H. Shelton	Waterbury, Conn	Nov. 22, 1870	109, 459
Knob, Glass	T. T. Abbot	Canton, Mass	Nov. 19, 1833	....
Knob, Glass	S. Richards	Cambridge, Mass	Oct. 31, 1831	....
Knob, Glass	W. S. Thompson and A. C. Hobbs.	Cambridge and Boston, Mass.	May 20, 1842	2, 628
Knob, Glass door	A. E. Young	Boston, Mass	Apr. 23, 1872	126, 004
Knob guard, Door	H. Ahrend	Newark, N. J	Apr. 1, 1873	137, 400
Knob guard, Door	J. W. Mellish	Walpole, N. H	June 8, 1869	91, 116
Knob label, Drawer	F. Hale and W. Manley	Philadelphia, Pa	Aug. 3, 1869	93, 2[illegible]9
Knob latch and lock, Combined	J. H. Vickers	Norwich, Conn	Aug. 10, 1869	93, 504
Knob-latch, Reversible	W. H. Andrews	New Haven, Conn	Jan. 17, 1871	111, 028
Knob-lock	J. Adt	Waterbury, Conn	Dec. 11, 1860	30, 902
Knob-lock	F. G. Johnson	Brooklyn, N. Y	June 12, 1866	55, 500
Knob-lock	J. Patterson	Pittsburgh, Pa	Oct. 19, 1829	....
Knob machine, Porcelain	T. J. Sloan	New York, N. Y	Sept. 14, 1869	94, 849
Knob machine, Porcelain	T. J. Sloan	Bronxville, N. Y	Apr. 25, 1871	114, 210
Knob-molding machine	C. H. Palmer	Brooklyn, E. D., N. Y	Jan. 31, 1871	111, 470
Knob or handle, Glass furniture	I. and J. P. Bakewell	Pittsburgh, Pa	May 14, 1828	....
Knob, Picture	E. D. Ives	New Haven, Conn	Feb. 18, 1868	74, 692
Knob, Picture	H. C. Luther and C. E. Richards.	Providence, R. I., and North Attleborough, Mass	June 21, 1870	104, 609
Knob rose, Door	W. H. Andrews	New Haven, Conn	July 14, 1868	79, 936
Knob rose, Door	C. L. Bates	New York, N. Y	May 25, 1860	00, 130
Knob rose, Door	E. A. Crew	Waterbury, Conn	Mar. 26, 1872	124, 936
Knob rose, Door	S. S. Day	New York, N. Y	Nov. 3, 1857	18, 537
Knob rose, Door	H. H. Elwell	South Norwalk, Conn	Dec. 10, 1872	133, 842
Knob rose, Door	W. Hall	Boston, Mass	Oct. 19, 1869	95, 899
Knob rose, Door	J. J. Henderson	New York, N. Y	Nov. 22, 1870	109, 409
Knob rose, Door	W. H. Mattson	Philadelphia, Pa	May 14, 1872	126, 642
Knob rose, Door	W. T. Munger	New Britain, Conn	Mar. 1, 1870	100, 314
Knob rose, Door	M. V. Nobles	Rochester, N. Y	May 30, 1867	48, 023
Knob roses, Bush-nut for door	C. Hood	Hartford, Conn	Aug. 6, 1872	130, 131

Index of patents issued from the United States Patent Office from 1790 *to* 1873, *inclusive*—Continued.

Invention.	Inventor.	Residence.	Date.	No.
Knob-roses, Fastening for door	J. M. Adolphus	Philadelphia, Pa	Jan. 31, 1871	111, 297
Knob-roses to doors, Attaching	L. P. Waterman and C. H. Porter.	Bridgeport, Conn	May 18, 1869	90, 209
Knob shank, Door	L. R. Livingston, J. J. Roggen, and C Adams.	Pittsburgh, Pa	July 7, 1846	4, 620
Knob, Silvered-glass door	J. W. Haines	Cambridgeport, Mass	Aug. 5, 1873	141, 504
Knob spindle	W. Varrah	New Haven, Conn	Mar. 11, 1873	136, 795
Knob spindle	O. Newt n	Pittsburgh, Pa	July 28, 1857	17, 887
Knob-spindles, Fastening door	A. Cooley	Hartford, Conn	Nov. 11, 1856	16, 047
Knob with glass shanks or screw, Glass	D. Jarves	Boston, Mass	June 13, 1829	....
Knobs and other molded articles, Plastic composition for the manufacture of.	H. E. Shepard	New Haven, Conn	July 8, 1873	140, 735
Knobs, Attaching door	D. B. Cobb	Jersey City, N. J	Sept. 17, 1867	68, 956
Knobs, Attaching screws to	W. E. Doolittle	West Haven, Conn	July 20, 1869	92, 801
Knobs, &c., Composition for door	J. Harrison	Stillwater, N. Y	Oct. 2, 1847	5, 315
Knobs, Die for making	J. Wise	Branford. Conn	Feb. 12, 1867	61, 977
Knobs, &c., Fastening commode	W. Price	Pitt, Pa	Nov. 14, 1826	....
Knobs, Fastening door	N. Benham	Hartford, Conn	Apr. 8, 1856	14, 595
Knobs for doors, &c., Making glass	H. Whitney and E. Robinson	Cambridge, Mass	Nov. 4, 1826	....
Knobs for furniture, Fastening for	L. B. Myers	Elmore, Ohio	Oct. 30, 1866	59, 250
Knobs for locks, &c., Attaching necks, shanks, screws, &c., to glass.	J. G. Hotchkiss, J. A. Davenport, and J. W. Quincy.	New Haven, Conn., and New York.	Nov. 16, 1841	2, 361
Knobs, Glass screw for	O. Newton	Pittsburgh, Pa	Oct. 17, 1835	....
Knobs, Grinding off the shanks of glass	J. W. Haines	Cambridge, Mass	Apr. 9, 1872	125, 565
Knobs, Instrument for manufacturing door	A. Rogers	Painesville, Ohio	Aug. 22, 1854	11, 562
Knobs, Machine for making porcelain	T. J. Sloan	Bronxville, N. Y	Dec. 19, 1871	122, 072
Knobs, Machine for stamping clay door	G. Lawton	Trenton, N. J	May 28, 1867	65, 244
Knobs, Making pressed glass	J. Robinson	Pittsburgh, Pa	Oct. 6, 1827	....
Knobs, Manufacture of door	O. Newton	Pittsburgh, Pa	Nov. 25, 1851	8, 547
Knobs, Manufacture of door	B. Nott	Bethlehem, N. Y	Mar. 9, 1852	8, 795
Knobs, Manufacture of door	G. O. Russell	Middletown, Conn	Oct. 7, 1845	4, 220
Knobs, Manufacture of glass	E. D. Ives	New Haven, Conn	Apr. 14, 1868	76, 766
Knobs, Manufacture of iron door	C. Carpenter	Hamilton, Canada	July 16, 1872	129, 458
Knobs, Manufacture of mineral	J. Jones	Brooklyn, N. Y	Sept. 17, 1867	68, 885
Knobs, Manufacturing glass	D. Jarves	Boston, Mass	Oct. 19, 1830	....
Knobs, Mold for casting metal	C. Rebstock and J. Ottner	Bridgeport, Conn	Jan. 14, 1873	134, 931
Knobs, Mold for casting metal door	C. Rebstock and A. Hart	Bridgeport and Meriden, Conn.	July 16, 1872	129, 056
Knobs of clay used in pottery and porcelain, Making door.	J. G. Hotchkiss, J. A. Davenport, and J. W. Quincy.	New Haven, Conn., and New York.	July 29, 1841	2, 197
Knobs of plate-metal, Die for making door and other	L. E. Hicks	Middletown, Conn	June 27, 1840	1, 654
Knobs, Securing the necks to door	T. Kennedy	Branford, Conn	Jan. 10, 1865	45, 836
Knobs to collars, Attaching door	L. R. Livingston, J. J. Roggen, C. Adams, and W. and R. Phillips.	Pittsburgh, Pa	Dec. 10, 1846	4, 883
Knobs to door-latches, Attaching	W. H. Andrews	New Haven, Conn	Apr. 23, 1867	63, 980
Knobs to doors, Attaching	J. A. Crever	Pittsburgh, Pa	Oct. 16, 1849	6, 800
Knobs to doors, Attaching	C. F. Langford	Brooklyn, N. Y	Dec. 7, 1869	97, 654
Knobs to doors, &c., Attaching roses for	N. Matthews	Pittsburgh, Pa	Apr. 6, 1852	8, 857
Knobs to doors, drawers, &c., Securing	C. Ward	Salem, Mass	Nov. 19, 1861	33, 759
Knobs to drawers, &c., Attaching	S. A. Brackett	Boston, Mass	Nov. 18, 1873	144, 597
Knobs to latches, Attaching	L. Michaels	Covington, Ky	Sept. 25, 1873	143, 174
Knobs to metallic sockets, Attaching glass	E. and G. W. Robinson	Boston, Mass	Oct. 20, 1837	434
Knobs to roses, Attaching	C. F. Rebstock	Meriden, Conn	Dec. 10, 1872	133, 889
Knobs to screws, Mode of attaching	C. H. Thurston	Marlborough, N. H	May 5, 1868	77, 550
Knobs to shanks, Attaching	J. Hoeflinger	Saint Joseph, Mo	Jan. 30, 1872	123, 102
Knobs to shanks, Attaching door	W. T. Munger	Branford, Conn	Mar. 19, 1867	62, 967
Knobs to shanks, Clasp for securing	J. W. Haines	Cambridge, Mass	May 14, 1872	126, 805
Knobs to shanks, Fastening door	G. Jones and B. E. Mead	Peekskill, N. Y	Nov. 13, 1866	59, 605
Knobs to shanks, Fastening door	M. V. Nobles	Rochester, N. Y	May 30, 1867	48, 024
Knobs to shanks, Fastening door	M. V. Nobles	Rochester, N. Y	May 30, 1867	48, 025
Knobs to shanks, Securing	T. J. Sloan	New York, N. Y	Apr. 13, 1869	88, 918
Knobs to shanks, Securing	T. J. Sloan	New York, N. Y	Apr. 13, 1869	88, 919
Knobs to spindles, Adjusting	A. Dawes	Hudson, Mass	Dec. 31, 1867	72, 815
Knobs to spindles, Adjusting	T. Kennedy	Branford, Conn	May 29, 1866	55, 115
Knobs to spindles, Attaching	M. Andrew	Melbourne, Colony of Victoria.	Apr. 12, 1870	101, 808
Knobs to spindles, Attaching	M. Andrew	Melbourne, Victoria	June 6, 1871	115, 675
Knobs to spindles, Attaching	W. H. Andrews	New Haven, Conn	Apr. 28, 1868	77, 344
Knobs to spindles, Attaching	M. W. Barse	Olean, N. Y	Oct. 31, 1871	120, 362
Knobs to spindles, Attaching	G. N. Cummings	Meriden, Conn	July 22, 1862	35, 924
Knobs to spindles, Attaching	W. A. Fenn	Wolcott, N. Y	Dec. 15, 1868	84, 868
Knobs to spindles, Attaching	W. A. Fenn	Rochester, N. Y	Feb. 28, 1871	112, 134
Knobs to spindles, Attaching	W. A. Fenn	Rochester, N. Y	July 11, 1871	116, 941
Knobs to spindles, Attaching	J. R. Gill and W. R. Baker	Hamilton and Wellington Square, Canada.	May 21, 1872	127, 046
Knobs to spindles, Attaching	A. G. Gray	Saint John, Canada	May 14, 1872	126, 694
Knobs to spindles, Attaching	A. M. Hill	Pittsburgh, Pa	July 19, 1864	43, 582
Knobs to spindles, Attaching	J. F. Keller and N. Schner	Hagerstown, Md	Mar. 7, 1871	112, 354
Knobs to spindles, Attaching	F. M. and J. B. Merriam	West Meriden, Conn	July 29, 1873	141, 285
Knobs to spindles, Attaching	E. M. and J. E. Mix	Westfield, N. Y	May 21, 1872	127, 089
Knobs to spindles, Attaching	C. Morrill	New York, N. Y	Jan. 17, 1871	110, 992
Knobs to spindles, Attaching	W. T. Munger	Branford, Conn	Feb. 9, 1869	86, 683
Knobs to spindles, Attaching	W. T. Munger	Branford, Conn	Feb. 16, 1869	86, 998
Knobs to spindles, Attaching	S. Selden	Erie, Pa	Jan. 7, 1873	134, 708
Knobs to spindles, Attaching	O. L. Smith	Providence, R. I	June 6, 1871	115, 650
Knobs to spindles, Attaching	H. B. Tuthill	New York, N. Y	May 4, 1869	89, 708
Knobs to spindles, Attaching	H. J. B. Whipple	Meriden, Conn	July 29, 1873	141, 408
Knobs to spindles, Attaching door	W. Boch	Newtown, N. Y	Mar. 24, 1868	75, 847
Knobs to spindles, Attaching door	C. B. Bristol	New Haven, Conn	Mar. 5, 1867	62, 599
Knobs to spindles, Attaching door	C. B. Bristol	New Haven, Conn	Dec. 24, 1867	72, 597
Knobs to spindles, Attaching door	J. Evans	New Haven, Conn	May 31, 1870	103, 580
Knobs to spindles, Attaching door	J. N. Karr	Buffalo, N. Y	Jan. 17, 1871	110, 977
Knobs to spindles, Attaching door	T. Marvin	Cambridge, Mass	May 12, 1868	77, 745
Knobs to spindles Attaching door	E. Parker	West Meriden, Conn	June 28, 1864	43, 332
Knobs to spindles, Attaching door	S. S. Putnam	Dorchester, Mass	July 7, 1868	79, 597
Knobs to spindles, Attaching door	E. Robinson	Boston, Mass	Jan. 10, 1843	2, 904
Knobs to spindles, Fastening	D. Skidmore	Seneca Falls, N. Y	July 15, 1862	35, 899

Index of patents issued from the United States Patent Office from 1790 *to* 1873, *inclusive*—Continued.

Invention.	Inventor.	Residence.	Date.	No.
Knobs to spindles, &c., Fastening door locks and latch.	A. O. Downer	Utica, N. Y	Dec. 21, 1842	2, 891
Knobs to spindles of door-locks, Attaching	W. T. Munger	Branford, Conn	Nov. 26, 1867	71, 502
Knobs, umbrella-handles, and other molded articles, Manufacture of door.	W. Sanderson and J. J. De Barry.	New York, N. Y	Jan. 25, 1870	99, 241
Knobs upon collars, Casting door	W. Heiggs	Utica, N. Y	Dec. 10, 1846	4, 884
Knock-down chair	C. R. Long	Louisville, Ky	Apr. 8, 1873	137, 613
Knock-down chair	V. Stockton	Williamsburgh, Ohio	July 11, 1871	117, 012
Knuckle-joint	J. K. Collins	Huntsville, Ala	Mar. 6, 1866	52, 969
Knuckle-joint	G. B. Garlinghouse	North Madison, Ind	May 14, 1867	64, 752
Knuckle-joint	J. H. Mears and C. W. Yale	Oshkosh, Wis	Dec. 11, 1866	60, 403
Knuckle-joint	S. N. Taylor	Horicon, Wis	Feb. 27, 1866	52, 908
Krout-cutting machine	W. K. Baylor and C. Rapp	Batesville, Ind	Apr. 13, 1869	88, 769
L.				
Label, Apothecary's	G. G. Percival	Philadelphia, Pa	Sept. 8, 1868	82, 028
Label-attaching machine	J. F. Zacharias	Leesburgh, Va	Aug. 13, 1867	67, 833
Label, Baggage	J. M. March	Washington, D. C	Sept. 10, 1867	68, 632
Label, Bottle	H. Frank	Pittsburgh, Pa	Jan. 21, 1873	135, 034
Label-card	J. Sharp	Roxbury, Mass	June 15, 1852	9, 039
Label, Cask	E. A. Locke	Boston, Mass	Mar. 24, 1868	75, 934
Label, Commercial	C. N. Morris	Cincinnati, Ohio	May 26, 1863	38, 692
Label, Direction	G. E. Pevey	Lowell, Mass	Mar. 5, 1867	62, 561
Label, Direction	M. J. Procter	Lowell, Mass	Mar. 13, 1866	53, 233
Label-fastener	C. R. Doane	Brooklyn, N. Y	Aug. 24, 1869	94, 088
Label for hat-lining, Ornamental	T. W Bracher	New York, N. Y	Dec. 9, 1873	145, 391
Label for trees, &c	W. W. Wade and F. T. Cordis	Long Meadow, Mass	Dec. 21, 1850	22, 390
Label, Fruit-can	J. Dunlap	Pittsburgh, Pa	June 29, 1869	92, 023
Label-gumming apparatus	J. H. Hudson	Winona, Minn	Feb. 20, 1872	123, 906
Label-holder	F. J. Collier	Philadelphia, Pa	Oct. 2, 1860	30, 205
Label-holder	C. A. Dickerman	New Haven, Conn	Dec. 8, 1868	84, 804
Label-holder	G. S. True	Leavenworth, Kans	Aug. 4, 1860	80, 683
Label holder and handle combined	H. Manbeck	New York, N. Y	Aug. 16, 1870	106, 494
Label holder, Drawer-pull	J. A. Evarts and P. Cinquin	West Meriden, Conn	Apr. 4, 1871	113, 506
Label-holder for railway-car	J. H. Parsons	Quincy, Mich	Nov. 19, 1867	71, 051
Label holder, Lock	J. S. Ramsay	Baltimore, Md	July 21, 1868	80, 218
Label intended for second use	H. Loewenberg	New York, N. Y	Apr. 5, 1864	42, 208
Label, Mail-bag	S. Andrews	Perth Amboy, N. J	July 25, 1854	11, 348
Label, Mail-bag	O. S. North	New Britain, Conn	Mar. 13, 1844	3, 486
Label, Mail-bag, &c	T. P. Trott	Washington, D. C	Apr. 24, 1860	28, 023
Label or tag	T. B. and L. De Forest	Birmingham, Conn	Nov. 15, 1864	45, 029
Label or tag	E. W. Dennison	Boston, Mass	June 9, 1863	38, 871
Label-pasting machine	W. T. Slocum	Philadelphia, Pa	July 16, 1872	129, 606
Label, Sheep	C. H. Dana	West Lebanon, N. H	June 6, 1865	48, 055
Label, Tag	D. D. Foley	Washington, D. C	Apr. 26, 1870	102, 461
Label, Trunk	S. W. Downey	Piedmont, W. Va	Feb. 11, 1868	74, 322
Labels, Apparatus for cutting and attaching	C. M. Spencer	Manchester, Conn	Oct. 11, 1859	25, 770
Labels, Apparatus for damping and gumming	J. Benn and G. O. Luckman	Goat and Manchester, England.	Oct. 15, 1867	69, 895
Labels, Apparatus for pasting	W. E. Booraem	New York, N. Y	Nov. 10, 1868	83, 825
Labels, Apparatus for softening the gum of adhesive	B. Wilder	North Scituate, Mass	Feb. 21, 1865	46, 514
Labels for periodicals, &c., Accountant	R. Dick	Toronto, Canada	Oct. 4, 1859	25, 635
Labels, hat-linings, &c., Ornamenting	T. W. Bracker	New York, N. Y	Oct. 3, 1871	119, 563
Labels, Machine for making sheep's	C. H. Dana	West Lebanon, N. H	Aug. 29, 1865	49, 609
Labels, Metallic hook for	S. B. Fay	New York, N. Y	July 1, 1856	15, 226
Labels on glass ware, Mode of securing	E. M. Davis	Pittsburgh, Pa	Feb. 25, 1868	74, 804
Labels or tags, Machine for cutting and punching	C. S. Moseley	Boston, Mass	July 5, 1864	43, 458
Labels, tags, &c., Process for preparing wood for the manufacture of.	J. Melling	Rochester, N. Y	July 9, 1867	66, 512
Labels to bottles, Attaching	W. N. Walton	New York, N. Y	Sept. 23, 1862	36, 542
Labels to bottles, Mode of applying	J. S. and T. B. Atterbury	Pittsburgh, Pa	May 15, 1866	54, 665
Labels to newspapers, Machine for attaching	J. F. Zacharias	Leesburgh, Va	Apr. 28, 1868	77, 430
Labeling-machine	C. W. Greene	Jackson, Tenn	Feb. 25, 1873	136, 154
Labeling machine, Spool	Q. S. Backus	Winchendon, Mass	Feb. 9, 1869	86, 627
Labeling machine, Spool	H. Conant	Willimantic, Conn	June 10, 1862	35, 562
Lace and ribbon case	M. Deitzler	Ashland, Pa	Apr. 1, 1873	137, 297
Lace-drying frame	A. Chatain	Washington, D. C	Sept. 26, 1871	119, 318
Lace, &c., Fabric for making imitation	H. Loewenberg	New York, N. Y	Nov. 17, 1863	40, 633
Lace-making machine	G. Osborn	Brooklyn, N. Y	July 27, 1869	92, 995
Lacing and shoe-fastener, Slide	R. Adams	Cincinnati, Ohio	Oct. 2, 1866	58, 364
Lacing-device	E. C. C. Kellogg	Hartford, Conn	Feb. 13, 1866	52, 645
Lacing-device	J. Nealey, jr	Bangor, Me	Mar. 31, 1868	76, 234
Lacing-device	A. Young	Philadelphia, Pa	Dec. 31, 1867	72, 955
Lacing-eye	A. G. Mead	Boston, Mass	Oct. 12, 1869	95, 708
Lacings, Apparatus for tagging	F. J. Seymour	Wolcottville, Conn	May 17, 1864	42, 798
Lacings, Machine for cutting	H. B. Harvey	Northbridge, Mass	Sept. 13, 1864	44, 187
Lacteal instrument	C. H. Davidson	Charlestown, Mass	Nov. 9, 1858	22, 018
Ladder	P. M. Ackerman	Webster, N. Y	June 11, 1867	65, 525
Ladder	F. Aldrich, jr	Augusta, Mich	Sept. 15, 1863	39, 871
Ladder	G. Aldrich	Armada, Mich	Apr. 30, 1861	32, 170
Ladder	H. O. Baker	New York, N. Y	Feb. 26, 1867	62, 305
Ladder	M. D. Boyd	Buffalo, N. Y	Sept. 17, 1867	68, 836
Ladder	W. L. Burlingame	Leslie, Mich	July 7, 1868	79, 727
Ladder	M. Burt	Norton, Mass	Feb. 26, 1867	62, 314
Ladder	E. P. H. Capron	Springfield, Ohio	Sept. 29, 1868	82, 594
Ladder	G. Ckertizza	New York, N. Y	Oct. 23, 1866	58, 984
Ladder	J. H. Conley	Philadelphia, Pa	Mar. 14, 1871	112, 552
Ladder	C. Croley	Dayton, Ohio	Oct. 16, 1866	58, 781
Ladder	C. Croley	Dayton, Ohio	Aug. 13, 1867	67, 636
Ladder	R. L. Dodge	Portland, Me	Apr. 6, 1869	88, 616
Ladder	J. Eagon	Batesville, Ohio	July 22, 1873	141, 126
Ladder	F. Ellis	Walton, N. Y	Dec. 26, 1871	122, 238
Ladder	G. H. Ellis	London, England	Mar. 14, 1871	112, 697
Ladder	D. Fitzgerald	New York, N. Y	Apr. 12, 1859	23, 669
Ladder	F. W. Hovey	Boston, Mass	Jan. 1, 1867	60, 731
Ladder	J. Hughes	New Berne, N. C	Oct. 18, 1870	108, 483

Index of patents issued from the United States Patent Office from 1790 *to* 1873, *inclusive*—Continued.

Invention.	Inventor.	Residence.	Date.	No.
Ladder	E. J. Knowlton	Lyon, Mich	May 19, 1863	38, 586
Ladder	T. B. Luzier and G. A. Haas	Philadelphia, Pa	Feb. 4, 1868	74, 102
Ladder	J. Maxwell and T. Gray	Philadelphia, Pa	July 18, 1871	117, 186
Ladder	W. H. McHench	Cobleskill, N. Y	May 10, 1870	102, 845
Ladder	A. C. McKendree	Conneaut, Ohio	June 16, 1868	78, 988
Ladder	E. T. Olds	Lyon, Mich	Apr. 5, 1864	42, 219
Ladder	P. M. Papin	Saint Louis, Mo	Oct. 13, 1868	83, 084
Ladder	W. G. Philips	Newport, Del	Nov. 3, 1863	40, 500
Ladder	B. Pickering	Dayton, Ohio	May 15, 1866	54, 767
Ladder	B. Schoonmaker	Plainwell, Mich	July 20, 1869	92, 755
Ladder	A. Simmerman and J. S. Simmerman.	Glassborough and Millville, N. J.	Jan. 21, 1868	73, 468
Ladder	A. P. Smith	Sacramento, Cal	Dec. 30, 1873	146, 029
Ladder	M. L. Smith	Battle Creek, Mich	Nov. 2, 1869	96, 495
Ladder	J. Stiles	Salem, Mich	Sept. 15, 1863	39, 970
Ladder	D. B. Taylor	Avon, Mich	Feb. 19, 1867	62, 234
Ladder	C. C. Thomas and F. A. S. Raymond.	Beverly, Mass	June 30, 1868	79, 520
Ladder	J. A. Thompson	Auburn, N. Y	Apr. 12, 1870	101, 943
Ladder	B. F. Turner	Bridgeton, N. J	July 9, 1867	66, 655
Ladder	G. W. Willis	Atchinson, Kans	Mar. 12, 1872	124, 526
Ladder, Adjustable fruit	S. Wright	Hillsborough, Mo	Aug. 23, 1870	106, 645
Ladder, Adjustable lifting	G. Claflin	Miller's Corners, N. Y	Sept. 13, 1870	107, 225
Ladder, Adjustable step	R. R. Croasdale and P. Rink	Reaville, N. J	July 20, 1869	92, 708
Ladder, Adjustable step	J. J. Hardenbrook and S. Belford.	Columbus Grove, Ohio	May 20, 1873	139, 004
Ladder and chair	E. Kohn and J. L. Natcher	Sidney, Ohio	Oct. 1, 1867	69, 448
Ladder and ironing-board, Combined step	H. H. Kendrick	Fulton, N. Y	Dec. 23, 1873	145, 738
Ladder and scaffold combined	R. L. Upchurch	Pana, Ill	Nov. 26, 1872	133, 505
Ladder and staging, Artisan's	C. Monson	New Haven, Conn	Feb. 4, 1862	34, 313
Ladder apparatus, Fire	D. Fobes and H. M. Hartshorn	Boston and Malden, Mass	June 17, 1862	35, 601
Ladder basket-rest	W. E. Ludlow	Cincinnati, Ohio	Aug. 31, 1869	94, 422
Ladder, bench, and clothes-frame, Combined step	H. H. Barker	Hartford, Conn	Dec. 23, 1873	145, 716
Ladder, Convertible	H. B. Malbone	Geneva, N. Y	Aug. 3, 1869	93, 323
Ladder, Elevating	A. Miller	New York, N. Y	May 20, 1873	139, 175
Ladder, Elevating or scaling	S. D. Wollison	Pittsfield, Mass	Sept. 8, 1863	39, 857
Ladder, Extension	H. Barns	Somers, Wis	Dec. 24, 1867	72, 444
Ladder, Extension	C. R. Bryant	Frankfort, N. Y	Mar. 19, 1867	62, 932
Ladder, Extension	M. T. Burbank	Lawrence, Mass	Apr. 13, 1869	88, 842
Ladder, Extension	R. F. Delmot	Flemington, Pa	Apr. 4, 1871	113, 273
Ladder, Extension	C. Eaton	Webster, N. Y	Mar. 7, 1865	46, 648
Ladder, Extension	J. C. Hearne and D. Adams	Pleasant Hill, Mo	Apr. 8, 1873	137, 679
Ladder, Extension	G. L. Johnson	Fairfield, N. Y	Mar. 26, 1867	63, 163
Ladder, Extension	F. and C. Kavemann and B. Hoerstmann.	Cincinnati, Ohio	Apr. 24, 1860	28, 039
Ladder, Extension	J. Kearns	Louisville, Ohio	Feb. 18, 1868	74, 548
Ladder, Extension	M. M. Knowles	Elmira, N. Y	Aug. 18, 1868	81, 279
Ladder, Extension	T. F. Mantey	New Orleans, La	May 26, 1868	78, 302
Ladder, Extension	L. N. Millener	Adams' Basin, N. Y	Feb. 20, 1872	123, 834
Ladder, Extension	W. Morehead	Parkersburgh, W. Va	Mar. 14, 1865	46, 812
Ladder, Extension	W. Morehead	Parkersburgh, W. Va	Dec. 8, 1868	84, 839
Ladder, Extension	J. F. Morse	Montgomery, Ill	May 5, 1868	77, 640
Ladder, Extension	J. Moulton	Boston, Mass	May 17, 1859	24, 044
Ladder, Extension	M. D. Myers	Ilion, N. Y	Feb. 2, 1864	41, 443
Ladder, Extension	G. B. Nickle and J. M. Carville	New York, N. Y	Apr. 10, 1860	27, 821
Ladder, Extension	W. O'Connor	Berville, Mich	Apr. 15, 1873	137, 948
Ladder, Extension	B. Pine	New York, N. Y	Mar. 3, 1868	75, 052
Ladder, Extension	J. Pine	Brooklyn, N. Y	June 10, 1873	139, 689
Ladder, Extension	N. Pullman	New Oregon, Iowa	July 5, 1870	105, 126
Ladder, Extension	H. M. Quackenbush	Herkimer, N. Y	Oct. 22, 1867	70, 016
Ladder, Extension	J. L. Ripley	Fremont, Ohio	June 13, 1865	48, 210
Ladder, Extension	A. Rogers	Painesville, Ill	Aug. 10, 1869	93, 642
Ladder, Extension	J. W. Scott	Philadelphia, Pa	Sept. 27, 1870	107, 822
Ladder, Extension	E. O. Shepardson	Saint Louis, Mo	July 18, 1871	117, 117
Ladder, Extension	J. A. Smith	Lacon, Ill	July 28, 1868	80, 367
Ladder, Extension	W. F. Trautman	Llewellyn, Pa	July 13, 1869	92, 492
Ladder, Extension	C. G. Udell	Chicago, Ill	Mar. 26, 1867	63, 188
Ladder, Extension	C. G. Udell	Chicago, Ill	Aug. 10, 1869	93, 502
Ladder, Extension	L. F. Ward	Marathon, N. Y	Apr. 28, 1863	38, 345
Ladder, Extension	T. Watson and C. Perry	Brooklyn, N. Y	Oct. 30, 1866	59, 297
Ladder, Extension	T. Watson and C. Perry	Brooklyn, N. Y	Oct. 15, 1867	69, 875
Ladder, Extension	T. Watson and C. Perry	Brooklyn, N. Y	Dec. 13, 1870	110, 094
Ladder, Extension	F. Willis	Marathon, N. Y	Jan. 20, 1863	37, 470
Ladder, Extension	T. C. Wood	Augusta, Mich	Jan. 24, 1865	46, 042
Ladder, Extension and step	H. J. Hancock	New York, N. Y	Mar. 9, 1869	87, 666
Ladder, Extension and step	S. E. Howes	Albany, N. Y	Mar. 24, 1868	75, 907
Ladder, Extension fruit	A. Rogers	Painesville, Ohio	May 15, 1866	54, 778
Ladder, Extension fruit	J. E. Treat	Oxford, Mich	July 2, 1867	66, 420
Ladder, Extension platform	G. A. Schachtel	Newark, N. J	Nov. 15, 1870	109, 256
Ladder, Extension step	H. D. Chance	Llewellyn, Pa	Aug. 23, 1870	106, 659
Ladder, Extension step	L. B. Covert	New York, N. Y	May 5, 1868	77, 463
Ladder, Extension step	L. B. Covert	Detroit, Mich	Apr. 29, 1873	138, 234
Ladder, Extension step	G. W. Packer	Toulon, Ill	June 2, 1868	78, 476
Ladder, Extension step	A. W. and J. E. Walker	Detroit, Mich	Mar. 18, 1873	137, 040
Ladder, Fire	J. Blomgren	Galesburgh, Ill	Dec. 10, 1867	71, 962
Ladder, Fire	G. K. Foster	San Francisco, Cal	Apr. 30, 1867	64, 212
Ladder, Fire	D. F. Haasz	Philadelphia, Pa	Aug. 17, 1869	93, 877
Ladder, Fire	J. L. Hannah	New York	Aug. 19, 1834	......
Ladder, Fire	M. B. Hill	Fairfield, Ohio	Sept. 1, 1843	3, 245
Ladder, Fire	S. Lehman	Philadelphia, Pa	Aug. 9, 1826	......
Ladder, Fire	B. Lies	New York	June 20, 1815	......
Ladder, Fire	A. Lotz	Franklin, Tenn	Nov. 9, 1869	96, 710
Ladder, Fire	J. Welte	Buffalo, N. Y	June 29, 1858	20, 752
Ladder, Fire and escape	J. Van Amringe	Cincinnati, Ohio	May 13, 1856	14, 891
Ladder, Fire-escape	C. G. Buttkereit	Toledo, Iowa	Mar. 19, 1872	124, 541
Ladder, Fire-escape	J. H. Grimsley	New Lexington, Ohio	July 13, 1858	20, 875
Ladder, Fire-escape	H. T. Hartman	Norwood, Va	July 13, 1869	92, 527
Ladder, Fire-escape	D. Hayes and W. Free	San Francisco, Cal	Feb. 25, 1868	74, 821

Index of patents issued from the United States Patent Office from 1790 *to* 1873, *inclusive*—Continued.

Invention.	Inventor.	Residence.	Date.	No.
Ladder, Fire-escape	I. Henderson	Philadelphia, Pa	Feb. 26, 1867	62, 418
Ladder, Fire-escape	A. Iske	Lancaster, Pa	Nov. 26, 1861	33, 785
Ladder, Fire-escape	H. Johnston and W. I. Matthews.	Collinsville, Ill	July 26, 1859	24, 909
Ladder, Fire-escape	H. Loewenberg	New York, N. Y	Sept. 22, 1857	18, 262
Ladder, Fire-escape	R. Loomis	Saratoga Springs, N. Y	Apr. 9, 1867	63, 735
Ladder, Fire-escape	H. and G. Luckenbach	Philadelphia, Pa	Sept. 19, 1871	119, 161
Ladder, Fire-escape	H. Lyles	Washington, D. C	Mar. 18, 1873	136, 844
Ladder, Fire-escape	W. H. Paige	Springfield, Mass	July 5, 1864	43, 425
Ladder, Fire-escape	W. W. Parsons	Stanstead, Canada	Sept. 2, 1873	142, 409
Ladder, Fire-escape	W. B. Peregoy	Baltimore, Md	Feb. 28, 1871	112, 276
Ladder, Fire-escape	G. Pfluger	Deerfield, N. Y	Oct. 7, 1873	143, 462
Ladder, Fire-escape	J. S. Pierson	Brooklyn, N. Y	Sept. 23, 1873	143, 182
Ladder, Fire-escape	R. G. Pike	Middletown, Conn	June 21, 1864	43, 261
Ladder, Fire-escape	W. P. Prewitt	Elkton, Ky	Mar. 5, 1867	62, 564
Ladder, Fire-escape	S. R. Roscoe	Carlisle, N. Y	Apr. 3, 1855	12, 643
Ladder, Fire-escape	J. C. Salomon, jr	Georgetown, D. C	June 1, 1852	8, 987
Ladder, Fire-escape	W. A. Shannon	Washington, D. C	Dec. 17, 1861	33, 963
Ladder, Fire-escape	G. Skinner	Brooklyn, N. Y	Sept. 22, 1868	82, 358
Ladder, Fire-escape	C. Weidling	New York, N. Y	June 8, 1869	90, 975
Ladder, Fire-escape	C. H. White	White's Station, Mich	Apr. 29, 1873	138, 304
Ladder, Fire-escape	J. Withers	Collinsville, Ill	Dec. 14, 1858	22, 324
Ladder, Fire-escape	T. Witmer	Buffalo, N. Y	Apr. 18, 1871	113, 958
Ladder, Fireman's	M. Cronin	Washington, D. C	Nov. 25, 1873	144, 958
Ladder, Fireman's	D. Fitzgerald	New York, N. Y	July 12, 1859	24, 728
Ladder, Fireman's	D. Giambastiani	Washington, D. C	Oct. 14, 1856	15, 883
Ladder, Fireman's	P. Porta	Milan, Italy	Sept. 2, 1873	142, 349
Ladder, Fireman's extension	G. W. Harris	Lancaster, N. Y	Dec. 23, 1873	145, 867
Ladder, Fireman's extension	R. H. Jones	San Francisco, Cal	July 7, 1868	79, 763
Ladder, Folding	J. M. White and A. J. Kee	Hannibal, Mo	Apr. 8, 1873	137, 582
Ladder, Folding step	M. Mattern	Poughkeepsie, N. Y	Aug. 27, 1872	130, 929
Ladder, Fruit	C. S. and C. D. Cannon	Chicago, Ill	Sept. 6, 1870	106, 996
Ladder, Fruit	J. Hannan	Lyon, Mich	Dec. 15, 1863	40, 926
Ladder, Fruit	J. Hannan	South Lyon, Mich	Oct. 25, 1864	44, 799
Ladder, Fruit	S. Hudson	Milford, Mich	Jan. 23, 1866	52, 172
Ladder, Fruit	A. and J. B. Longcor	Paw Paw, Mich	June 24, 1873	140, 286
Ladder, Fruit	D. McMaster	Bath, N. Y	Nov. 28, 1865	51, 205
Ladder, Fruit	A. W. Olds	Green Oak, Mich	Apr. 26, 1864	42, 500
Ladder, Fruit	A. W. Olds	Green Oak, Mich	Mar. 7, 1865	46, 694
Ladder, Fruit	E. F. Olds	South Lyon, Mich	Dec. 13, 1864	45, 430
Ladder, Fruit	D. C. Smith	Adrian, Mich	June 18, 1867	65, 959
Ladder, Fruit	B. F. Turner	Bridgeton, N. J	Sept. 17, 1867	69, 049
Ladder, Fruit	J. H. Verity	Lansing, Mich	Feb. 22, 1870	100, 219
Ladder, Fruit	S. S. Whaley	Tidioute, Pa	Aug. 20, 1867	67, 933
Ladder, Fruit and extension	M. Tuell	Penn Yan, N. Y	Apr. 12, 1870	101, 789
Ladder, Fruit extension	E. Slater	Girard, Pa	Nov. 13, 1866	59, 666
Ladder, Fruit or step	W. E. Bond	Cleveland, Ohio	Nov. 14, 1865	50, 894
Ladder, Fruit step	J. F. Winchell	Springfield, Ohio	Jan. 8, 1867	61, 129
Ladder-hook	S. D. Fish	Schuyler Falls, N. Y	Sept. 26, 1871	119, 226
Ladder-hook	J. G. Rockwell	Cortland, N. Y	Oct. 9, 1866	58, 734
Ladder-hook, Adjustable	W. T. Farrar	Concord, Mass	Apr. 9, 1861	31, 963
Ladder joint, Step	S. E. Hewes	Albany, N. Y	Sept. 15, 1868	82, 221
Ladder, Library step	C. C. Schmitt	New York, N. Y	Jan. 23, 1866	52, 211
Ladder, Life, escape, and fire	J. Johnson	Baltimore, Md	Apr. 18, 1831	... ...
Ladder or fire-escape	T. Armitage	Philadelphia, Pa	Apr. 18, 1854	10, 797
Ladder or step for street-lamp lighter	M. M. Smith	Nashville, Tenn	Sept. 8, 1868	81, 954
Ladder, Orchard	C. Hays	Madison, Wis	Jan. 31, 1865	46, 105
Ladder, Orchard	W. H. Heffley	Rochester, Ind	Sept. 17, 1872	131, 442
Ladder, Platform	J. Sheetz	Sunbury, Pa	Nov. 16, 1869	96, 974
Ladder-platform	C. Wolf	Wilmore, Pa	Sept. 3, 1872	131, 042
Ladder-splice and safety-hook combined	J. Edmunds	South Adams, Mass	Mar. 26, 1872	124, 940
Ladder-stand	D. R. Burkholder	Plainfield, Pa	June 6, 1871	115, 696
Ladder, Step	M. E. Abbott	Bethlehem, Pa	Nov. 19, 1867	70, 933
Ladder, Step	E. R. Austin	Elmira, N. Y	Apr. 13, 1869	88, 831
Ladder, Step	E. R. Austin	Elmira, N. Y	Nov. 23, 1869	97, 148
Ladder, Step	J. H. Balsley	Dayton, Ohio	Jan. 7, 1862	34, 100
Ladder, Step	J. H. Balsley	Dayton, Ohio	Feb. 8, 1870	99, 621
Ladder, Step	J. Barnett	Dayton, Ohio	Sept. 26, 1865	50, 091
Ladder, Step	W. W. Berntheisel	West Hempfield Township, Pa.	Aug. 13, 1867	67, 710
Ladder, Step	C. E. Boman	San Francisco, Cal	June 22, 1869	91, 706
Ladder, Step	W. E. Bond	Cleveland, Ohio	Apr. 5, 1864	42, 162
Ladder, Step	J. Booher	Dayton, Ohio	Sept. 27, 1864	44, 391
Ladder, Step	C. W. Brown	Newark, N. J	Apr. 3, 1866	53, 566
Ladder, Step	E. P. H. Capron	Springfield, Ohio	June 26, 1866	55, 817
Ladder, Step	J. Charleville	Saint Louis County, Mo	Nov. 24, 1868	84, 260
Ladder, Step	T. Curtis	New Hudson, Mich	May 3, 1864	42, 565
Ladder, Step	W. J. Emens	Saint Louis, Mo	Apr. 16, 1872	125, 728
Ladder, Step	V. Fountain, jr	Factoryville, N. Y	Sept. 1, 1863	39, 726
Ladder, Step	C. Frizell	Boston, Mass	July 1, 1873	140, 359
Ladder, Step	C. Frizell	Boston, Mass	July 8, 1873	140, 578
Ladder, Step	M. B. Geary	New York, N. Y	Mar. 7, 1871	112, 442
Ladder, Step	H. P. Hammond and W. A. Hathaway.	North Kingston, R. I	June 5, 1866	55, 288
Ladder, Step	D. B. Hedden	Newark, N. J	July 30, 1867	67, 193
Ladder, Step	W. Huey	Galena, Md	Oct. 24, 1871	120, 158
Ladder, Step	J. S. Lash	Philadelphia, Pa	Aug. 7, 1866	56, 957
Ladder, Step	M. C. Longacre	Cleveland, Ohio	Oct. 20, 1868	83, 291
Ladder, Step	M. N. Lovell	Erie, Pa	Apr. 16, 1872	125, 683
Ladder, Step	M. N. Lovell	Erie, Pa	Mar. 11, 1873	136, 659
Ladder, Step	P. Martin	Newark, N. J	Jan. 23, 1872	123, 032
Ladder, Step	A. Moore, jr	Hillsborough, Ohio	May 1, 1866	54, 387
Ladder, Step	E. M. Norton	Bridgeport, Conn	Jan. 30, 1872	123, 287
Ladder, Step	J. S. Oakley	Passaic, N. J	Nov. 21, 1871	121, 190
Ladder, Step	J. S. Oakley and B. M. Post	Passaic, N. J	May 20, 1873	139, 183
Ladder, Step	A. W. O'Blenus	New York, N. Y	Dec. 16, 1873	145, 522
Ladder, Step	W. G. Philips	Newport, Del	June 8, 1869	91, 039
Ladder, Step	W. G. Philips	Newport, Del	Jan. 3, 1871	110, 785

Index of patents issued from the United States Patent Office from 1790 *to* 1873, *inclusive*—Continued.

Invention.	Inventor.	Residence.	Date.	No.
Ladder, Step	C. S. Rouse	Dowagiac, Mich	Nov. 3, 1868	83, 795
Ladder, Step	A. F. Saunders	Boston, Mass	Dec 5, 1865	51, 394
Ladder, Step	H. T. Smith	Brooklyn, N. Y	June 25, 1867	66, 179
Ladder, Step	D. I. Stagg	New York, N. Y	Feb. 3, 1863	37, 594
Ladder, Step	O. M. Sweet	Forestville, N. Y	Mar. 12, 1872	124, 518
Ladder, Step	C. G. Udell	Chicago, Ill	June 8, 1869	90, 973
Ladder, Step	T. Vogelmann	Hamilton, Ohio	July 30, 1867	67, 235
Ladder, Step	S. Wright	Hillsborough, Mo	May 13, 1873	138, 977
Ladder, Step and extension	H. W. Covert	Rochester, N. Y	Oct 1, 1867	69, 321
Ladder, Step and extension	C. J. Komar	Willoughby, Ohio	Aug. 27, 1867	68, 090
Ladder, Step and extension	W. E. Ludlow	Cincinnati, Ohio	Aug. 31, 1869	94, 423
Ladder, Step and extension	G. S. Walker	Erie, Pa	Feb. 4, 1868	74, 177
Ladders, Construction of step	J. P. McCandless	Cincinnati, Ohio	Oct. 15, 1872	132, 212
Ladders, Construction of step	C. G. Udell	Chicago, Ill	Nov. 12, 1872	132, 938
Ladies' dress-spring	S. J. Sherman	Brooklyn, N. Y	Feb. 25, 1862	34, 531
Ladies' figure-stand	J. R. Palmenberg	New York, N. Y	Aug. 1, 1865	49, 145
Ladies' supporting-brace	M. Harris	New York, N. Y	Feb. 28, 1871	112, 244
Ladies' work-stand	J. B. Atwater	Chicago, Ill	July 25, 1865	48, 888
Ladies' work-table	C. R. P. Foster	Canandaigua, N. Y	Apr. 8, 1851	8, 030
Ladle and fork	W. B. Dunbar	Waterbury, Conn	Dec. 6, 1859	26, 393
Ladle, Bent wood handle and socket for	S. Stedman	Hartford, Conn	July 25, 1818	....
Ladle, Bullet	M. Babcock	Charlestown, Mass	June 14, 1864	43, 151
Ladle, Bullet	G. Rugg	Potsdam, N. Y	Apr. 24, 1860	28, 013
Ladle, Culinary	J. C. Haines	Dublin, Ind	Feb. 23, 1858	19, 419
Ladle for pouring metal	H. W. Benton	Lebanon, N. H	July 2, 1867	66, 285
Ladle-stopper	F. A. Ostrander	Troy, N. Y	Aug. 8, 1871	117, 918
Ladle with fork attached	N. S. Warner and H. S. Benedict.	Bridgeport, Conn	Oct. 16, 1860	30, 437
See, also, Culinary ladle. Fruit-canning ladle. Pouring-metal ladle. Wax-ladle.				
Lagging, Machine for cutting	B. B. Slade	Smithfield, R. I	Jan. 27, 1838	586
Lambrequins	H. M. Johnston	New York, N. Y	Jan. 18, 1870	98, 974
Lambrequins	H. M. Johnston	New York, N. Y	July 26, 1870	105, 692
Lambrequins	H. M. Johnston	New York, N. Y	July 26, 1870	105, 693
Lambrequins, Wooden	H. Weber	Detroit, Mich	Apr. 2, 1872	125, 356
Lamp	J. Adair	Pittsburgh, Pa	July 31, 1860	29, 347
Lamp	J. Adair	Pittsburgh, Pa	Oct. 13, 1863	40, 222
Lamp	J. Adair	Pittsburgh, Pa	Jan. 10, 1865	45, 805
Lamp	S. Adam, jr., and J. R. Fogg	Portland, Me	Oct. 14, 1862	36, 692
Lamp	H. W. Adams	Brooklyn, N. Y	Sept. 6, 1859	25, 310
Lamp	F. Adt	Wolcottville, Conn	July 14, 1868	79, 887
Lamp	A. Albertson	Jersey City, N. J	Jan. 10, 1871	110, 816
Lamp	J. B. Alexander	Washington, D. C	Dec. 3, 1867	71, 566
Lamp	J. B. Alexander	Washington, D. C	Apr. 12, 1870	101, 961
Lamp	J. Allen	New York, N. Y	Jan. 14, 1868	73, 278
Lamp	J. E. Ambrose	Batavia, Ill	Oct. 16, 1860	30, 381
Lamp	J. E. Ambrose	Lena, Ill	Apr. 23, 1861	32, 110
Lamp	J. K. Andrews	Antrim, Ohio	July 11, 1865	48, 760
Lamp	J. K. Andrews	Antrim, Ohio	July 2, 1867	66, 278
Lamp	A. C. Arnold and E. Blackman	Norwalk, Conn	Feb. 11, 1868	74, 271
Lamp	J. S. and T. B. Atterbury	Pittsburgh, Pa	Apr. 7, 1863	38, 087
Lamp	J. S. and T. B. Atterbury	Pittsburgh, Pa	Sept. 29, 1868	82, 579
Lamp	J. S. and T. B. Atterbury	Pittsburgh, Pa	Apr. 26, 1870	102, 204
Lamp	J. S. and T. B. Atterbury	Pittsburgh, Pa	Nov. 22, 1870	109, 370
Lamp	T. B. Atterbury	Pittsburgh, Pa	July 22, 1873	140, 988
Lamp	L. J. Atwood	Waterbury, Conn	Sept. 16, 1862	36, 493
Lamp	L. J. Atwood	Waterbury, Conn	Oct. 13, 1863	40, 226
Lamp	L. J. Atwood	Waterbury, Conn	Oct. 13, 1863	40, 227
Lamp	L. J. Atwood	Waterbury, Conn	Mar. 1, 1864	41, 751
Lamp	L. J. Atwood	Waterbury, Conn	Aug. 1, 1865	49, 064
Lamp	L. J. Atwood	Waterbury, Conn	Jan. 21, 1868	73, 488
Lamp	L. J. Atwood	Waterbury, Conn	July 21, 1868	80, 111
Lamp	L. J. Atwood	Waterbury, Conn	June 22, 1869	91, 590
Lamp	L. J. Atwood	Waterbury, Conn	Aug. 29, 1871	118, 421
Lamp	L. J. Atwood	Waterbury, Conn	July 16, 1872	129, 509
Lamp	L. J. Atwood	Waterbury, Conn	Feb. 11, 1873	135, 749
Lamp	S. K. Ayers	Delton, Wis	Mar. 3, 1868	74, 972
Lamp	C. H. Bagley	Elgin, Ill	Aug. 14, 1866	57, 064
Lamp	L. Bailey and R. Thayer	Charlestown and Boston, Mass.	May 4, 1858	20, 134
Lamp	J. R. Baker	Kendallville, Ind	Apr. 12, 1864	42, 266
Lamp	T. J. Barron	Brooklyn, N. Y	Feb. 11, 1862	34, 390
Lamp	J. Barson, E. Daniels, and J. Farrell.	New York, N. Y	Aug. 11, 1868	80, 851
Lamp	H. J. Batchelder	Marlborough, Mass	Sept. 18, 1860	30, 104
Lamp	W. W. Batchelder	New York, N. Y	Dec. 28, 1858	22, 409
Lamp	W. W. Batchelder	New York, N. Y	Feb. 28, 1865	46, 538
Lamp	J. M. Batchelor	Foxcroft, Me	July 12, 1859	24, 711
Lamp	J. T. Beale	London, England	Nov. 29, 1838	1, 023
Lamp	G. A. Beidler	Chicago, Ill	Nov. 26, 1867	71, 268
Lamp	H. M. Beidler	Chicago, Ill	May 7, 1867	64, 474
Lamp	J. H. Beidler	Lincoln, Ill	June 26, 1866	55, 806
Lamp	J. Bellerjean	Philadelphia, Pa	Sept. 15, 1868	82, 196
Lamp	N. Benedict	Washington, D. C	Nov. 9, 1869	96, 664
Lamp	L. Berns	Middletown, N. Y	Sept. 9, 1873	142, 609
Lamp	A. Black	Barnstable, Mass	Oct. 3, 1817	....
Lamp	S. G. Blackman	Waterbury, Conn	Jan. 7, 1862	34, 048
Lamp	W. S. Blaisdell and C. K. Young	Factory Point, Vt	Sept. 2, 1873	142, 432
Lamp	E. C. Blakeslee	Waterbury, Conn	Dec. 2, 1862	37, 068
Lamp	G. Blanchard	New York, N. Y	Aug. 7, 1860	29, 456
Lamp	A. Bliss	New York, N. Y	Jan. 14, 1862	34, 125
Lamp	E. Boesch	San Francisco, Cal	Dec. 26, 1871	122, 217
Lamp	W. G. A. Boneville	Dover, Del	Apr. 12, 1864	42, 269
Lamp	C. Boschan, J. Bindtner, and W. Caffon.	Vienna, Austria	July 4, 1865	48, 635
Lamp	H. H. Boucher	Doylestown, Pa	Aug. 4, 1868	80, 705

Index of patents issued from the United States Patent Office from 1790 *to* 1873, *inclusive*—Continued.

Invention.	Inventor.	Residence.	Date.	No.
Lamp	B. S. Boydston	Richmond, Ind	Sept. 29, 1868	82, 590
Lamp	O. Boynton	Hinesburgh, Vt	Jan. 21, 1868	73, 495
Lamp	S. C. Brockington	Groton, Conn	Aug. 4, 1868	80, 590
Lamp	E. Brown	Burlington, Vt	Apr. 17, 1866	53, 945
Lamp	W. Brown	Newburyport, Mass	Mar. 5, 1872	124, 249
Lamp	A. W. Browne	Brooklyn, N. Y	Jan. 28, 1868	73, 870
Lamp	G. Brownlee	Princeton, Ind	Apr. 22, 1873	138, 070
Lamp	A. Burbank	Rochester, N. Y	May 17, 1870	103, 011
Lamp	M. Burnett	South Boston, Mass	Mar. 22, 1864	41, 972
Lamp	F. Burrows	Peoria, Ill	Jan. 15, 1867	61, 152
Lamp	H. E. Burton	Boston, Mass	Nov. 26, 1867	71, 449
Lamp	C. W. Cahoon	Portland, Me	Feb. 19, 1861	31, 511
Lamp	C. W. Cahoon	Portland, Me	Dec. 3, 1861	33, 824
Lamp	C. W. Cahoon	Portland, Me	Dec. 3, 1861	33, 825
Lamp	C. W. Cahoon	Portland, Me	Sept. 2, 1862	36, 386
Lamp	C. W. Cahoon	Portland, Me	Sept. 16, 1862	36, 451
Lamp	C. W. Cahoon	Portland, Me	Oct. 13, 1863	40, 240
Lamp	C. W. Cahoon	Portland, Me	Oct. 13, 1863	40, 241
Lamp	C. W. Cahoon	Portland, Me	July 21, 1868	80, 133
Lamp	J. Calkins	New York, N. Y	Sept. 1, 1868	81, 749
Lamp	M. L. Callender	New York, N. Y	Apr. 23, 1861	32, 118
Lamp	M. L. Callender	New York, N. Y	Apr. 28, 1863	38, 284
Lamp	M. L. Callender	New York, N. Y	June 6, 1865	48, 128
Lamp	D. Challinor	Birmingham, Pa	July 26, 1870	105, 644
Lamp	O. M. Chamberlain	New York, N. Y	Oct. 11, 1870	108, 106
Lamp	L. Chandor	St. Petersburg, Russia	July 21, 1868	80, 137
Lamp	R. S. Chapin	New York, N. Y	Mar. 4, 1862	34, 562
Lamp	S. Cheney	Cleveland, Ohio	Jan. 25, 1859	22, 703
Lamp	R. Christy	Mansfield Valley, Pa	Aug. 12, 1873	141, 763
Lamp	E. Clark	Philadelphia, Pa	Apr. 27, 1814	....
Lamp	H. M. Clark	Cleveland, Ohio	Mar. 24, 1868	75, 731
Lamp	H. M. Clark	Meriden, Conn	Apr. 5, 1870	101, 584
Lamp	I. Clark	Brooklyn, N. Y	Jan. 16, 1866	52, 031
Lamp	J. M. Clark	Lancaster, Pa	July 15, 1873	140, 764
Lamp	P. J. Clark	Meriden, Conn	June 7, 1839	1, 164
Lamp	J. F. Clegg	Philadelphia, Pa	Apr. 2, 1861	31, 870
Lamp	T. Clough	Dobb's Ferry, N. Y	Nov. 26, 1867	71, 280
Lamp	T. Clough	Dobb's Ferry, N. Y	Nov. 26, 1867	71, 281
Lamp	T. Clough	Dobb's Ferry, N. Y	Oct. 27, 1868	83, 462
Lamp	G. F. J. Colburn	Newark, N. J	Apr. 14, 1863	38, 149
Lamp	M. H. Collins	Chelsea, Mass	Sept. 19, 1865	49, 984
Lamp	M. H. Collins	Chelsea, Mass	Feb. 4, 1868	74, 049
Lamp	A. Combs	Burlingame, Kans	Apr. 23, 1872	126, 025
Lamp	C. J. Conway	New York, N. Y	July 19, 1853	9, 857
Lamp	R. Cornelius	Philadelphia, Pa	Dec. 9, 1862	37, 086
Lamp	E. T. Covell	New Bedford, Mass	Feb. 18, 1862	34, 408
Lamp	N. Cradit	Ripley, Ohio	Dec. 14, 1858	22, 327
Lamp	A. Crook	New York, N. Y	May 25, 1869	90, 428
Lamp	E. E. Dailey, W. H. Johnson, and C. C. Du Bois.	Brooklyn, E. D., N. Y	Feb. 18, 1868	74, 670
Lamp	J. Davidson	Baltimore, Md	July 2, 1846	4, 617
Lamp	J. Davis, jr	Templeton, Mass	Apr. 23, 1867	63, 999
Lamp	C. T. Day	Newark, N. J	Oct. 20, 1863	40, 334
Lamp	F. B. De Keravenan	New York, N. Y	Oct. 23, 1860	30, 466
Lamp	J. M. A. Dew	Chicago, Ill	Aug. 31, 1869	94, 295
Lamp	M. A. Dietz	Brooklyn, N. Y	Mar. 8, 1859	23, 160
Lamp	M. A. Dietz	Brooklyn, N. Y	May 3, 1859	23, 832
Lamp	R. E. ietz	New York, N. Y	Oct. 28, 1873	144, 070
Lamp	J. Dodin	New York, N. Y	Feb. 18, 1862	34, 415
Lamp	C. H. Dolbeare	Boston, Mass	Jan. 1, 1861	31, 012
Lamp	H. W. Dopp	Buffalo, N. Y	Oct. 16, 1860	30, 444
Lamp	J. L. Drake	Cincinnati, Ohio	May 17, 1859	24, 015
Lamp	J. L. Drake	Cincinnati, Ohio	Sept. 20, 1859	25, 493
Lamp	L. F. Drake and E. Egginton	Portland, Me	Dec. 17, 1867	72, 376
Lamp	J. P. Driver	Marengo, Ind	Aug. 29, 1865	49, 613
Lamp	J. Duke	New York, N. Y	Aug. 15, 1821	....
Lamp	J. Dunn	New York, N. Y	Oct. 18, 1870	108, 339
Lamp	M. B. Dyott	Philadelphia, Pa	Sept. 29, 1863	40, 094
Lamp	M. B. Dyott	Philadelphia, Pa	May 29, 1866	55, 075
Lamp	M. B. Dyott	Philadelphia, Pa	May 7, 1867	64, 508
Lamp	R. N. Engle	Washington, D. C	Sept. 6, 1864	44, 081
Lamp	O. C. Evans	New York, N. Y	Jan. 22, 1861	31, 160
Lamp	H. B. Fernald	Boston, Mass	May 17, 1844	3, 591
Lamp	J. S. Fish	Cleveland, Ohio	Apr. 26, 1870	102, 240
Lamp	F. A. Flanegin	Fagundus City, Pa	Mar. 4, 1873	136, 427
Lamp	J. C. Fletcher	Springfield, Ohio	July 12, 1838	834
Lamp	A. L. Fleury	Baltimore, Md	July 5, 1859	24, 622
Lamp	W. Foster, jr., and L. Foster	Boston, Mass	Aug. 20, 1813	....
Lamp	S. W. Fowler	Brooklyn, N. Y	Jan. 21, 1868	73, 443
Lamp	S. W. Fowler	Brooklyn, N. Y	Apr. 12, 1870	101, 723
Lamp	W. Freeman	Mount Carmel, Conn	May 21, 1861	32, 358
Lamp	J. A. Frey	Washington, D. C	Jan. 28, 1868	73, 710
Lamp	G. P. Fuller	Humphrey, N. Y	July 5, 1870	105, 063
Lamp	J. B. Fuller	Norwich, Conn	June 15, 1869	91, 325
Lamp	J. B. Fuller	Norwich, Conn	May 24, 1870	103, 442
Lamp	W. Fulton	Cranberry, N. J	Aug. 3, 1858	21, 069
Lamp	W. Fulton	Cranberry, N. J	July 24, 1860	29, 260
Lamp	J. Funck	Tompkinsville, N. Y	Aug. 17, 1869	93, 871
Lamp	B. Garvey	New York, N. Y	Jan. 20, 1863	37, 442
Lamp	J. Gibbs	Brooklyn, N. Y	Dec. 1, 1868	84, 623
Lamp	D. Gilbert	Brattleborough, Vt	Mar. 29, 1834	....
Lamp	H. J. Goff	Dubuque, Iowa	Mar. 14, 1871	112, 586
Lamp	S. J. Gold	Cornwall, Conn	May 8, 1843	3, 073
Lamp	C. Goodwin	Chicago, Ill	Apr. 26, 1864	42, 534
Lamp	R. A. Goodyear	New Haven, Conn	June 5, 1866	55, 280
Lamp	W. J. Gordon and M. W. House	Cleveland, Ohio	July 23, 1872	129, 728
Lamp	W. J. Gordon and M. W. House	Cleveland, Ohio	Oct. 29, 1872	132, 655
Lamp	E. H. Green	Baltimore, Md	Sept. 5, 1865	49, 752

Index of patents issued from the United States Patent Office from 1790 *to* 1873, *inclusive*—Continued.

Invention.	Inventor.	Residence.	Date.	No.
Lamp	B. F. Greenough	Boston, Mass	Apr. 10, 1841	2, 039
Lamp	F. T. Grimes	Liberty, Mo	Jan. 3, 1871	110, 648
Lamp	F. T. Grimes	Liberty, Mo	Feb. 7, 1871	111, 531
Lamp	F. T. Grimes	Liberty, Mo	Oct. 10, 1871	119, 842
Lamp	P. C. Guion and P. K. Wombaugh.	Cincinnati, Ohio	Aug. 26, 1856	15, 636
Lamp	S. Guthrie	New Orleans, La	Feb. 14, 1860	27, 124
Lamp	J. S. Gwynn	Plainfield, N. J	Aug. 27, 1861	33, 140
Lamp	M. P. Hadley	Bluffton, Wis	Oct. 15, 1872	132, 283
Lamp	E. J. Hale	Foxcroft, Me	June 12, 1860	28, 666
Lamp	E. J. Hale	Foxcroft, Me	June 19, 1860	28, 749
Lamp	E. J. Hale	Foxcroft, Me	June 26, 1860	28, 855
Lamp	E. J. Hale and C. H. Chandler	Foxcroft, Me	Mar. 1, 1859	23, 085
Lamp	H. Halvorson	Cambridge, Mass	Sept. 20, 1859	25, 506
Lamp	H. L. Hanson	Portland, Me	Sept. 24, 1867	69, 091
Lamp	E. Harris	Boston, Mass	Mar. 20, 1855	12, 550
Lamp	J. Harris, jr	Boston, Mass	Aug. 15, 1854	11, 524
Lamp	A. Harroun, jr	Onondaga, N. Y	Feb. 20, 1866	52, 711
Lamp	H. Hassenpflug	Huntingdon, Pa	Aug. 20, 1861	33, 084
Lamp	H. W. Hayden	Waterbury, Conn	Apr. 14, 1863	38, 162
Lamp	H. W. Hayden	Waterbury, Conn	May 9, 1865	47, 680
Lamp	H. W. Hayden	Waterbury, Conn	Jan. 21, 1868	73, 599
Lamp	H. W. Hayden	Waterbury, Conn	Jan. 21, 1868	73, 600
Lamp	H. W. Hayden	Waterbury, Conn	Dec. 14, 1869	97, 773
Lamp	H. W. Hayden	Waterbury, Conn	July 23, 1872	129, 821
Lamp	P. Hayden	Pittsburgh, Pa	Oct. 14, 1862	36, 649
Lamp	E. K. Haynes	Boston, Mass	June 29, 1869	91, 933
Lamp	B. F. Hebard	Dorchester, Mass	Aug. 23, 1864	43, 911
Lamp	F. Heidrick	Philadelphia, Pa	Apr. 2, 1861	31, 887
Lamp	F. Heidrick	Philadelphia, Pa	May 7, 1861	32, 248
Lamp	A. A. Henderson	Buffalo, N. Y	Feb. 17, 1863	37, 690
Lamp	J. E. Hendricks	Waterbury, Conn	Aug. 25, 1868	81, 370
Lamp	A. Hicks	Factoryville, N. Y	Apr. 25, 1865	47, 493
Lamp	J. Higgins	Cambridge, Mass	Feb. 23, 1864	41, 702
Lamp	S. A. Hill and D. Alter	Freeport, Pa	Apr. 12, 1859	23, 579
Lamp	C. F. A. Hinrichs	New York, N. Y	Dec. 27, 1870	110, 464
Lamp	R. Hitchcock	Watertown, N. Y	Jan. 7, 1873	134, 547
Lamp	R. Hitchcock	Watertown, N. Y	Aug. 26, 1873	142, 103
Lamp	R. Hitchcock	Watertown, N. Y	Dec. 2, 1873	145, 176
Lamp	J. H. Hobbs	Wheeling, W. Va	May 24, 1870	103, 460
Lamp	B. J. Hoffacker	New York, N. Y	July 17, 1866	56, 497
Lamp	P. Hoffmann	Constableville, N. Y	Oct. 1, 1867	69, 436
Lamp	E. B. Horn	Boston, Mass	Jan. 8, 1842	2, 413
Lamp	J. Horton	New York, N. Y	Dec. 28, 1869	98, 264
Lamp	T. Houghton	Philadelphia, Pa	Jan. 10, 1860	26, 767
Lamp	M. W. House	Cleveland, Ohio	Apr. 14, 1868	76, 764
Lamp	M. W. House	Cleveland, Ohio	Dec. 5, 1871	121, 521
Lamp	D. Howarth	Portland, Me	June 19, 1866	55, 771
Lamp	J. J. Hoyt	Chelmsford, Mass	July 23, 1872	129, 828
Lamp	D. Hughes	Rochester, N. Y	Dec. 17, 1861	33, 945
Lamp	J. S. Hull	Cincinnati, Ohio	May 14, 1867	64, 768
Lamp	J. S. Hull	Cincinnati, Ohio	Mar. 19, 1872	124, 823
Lamp	W. Hunt	New York, N. Y	May 21, 1861	32, 402
Lamp	T. A. Hunter and J. Blewitt	New York, N. Y	Nov. 3, 1868	83, 711
Lamp	H. C. Hutchinson	Cayuga, N. Y	Apr. 15, 1862	34, 964
Lamp	H. C. Hutchinson	Cayuga, N. Y	Jan. 3, 1865	45, 719
Lamp	J. Ingersoll	Cleveland, Ohio	Feb. 18, 1868	74, 542
Lamp	J. H. Irwin	Beardstown, Ill	Oct. 29, 1861	33, 588
Lamp	J. H. Irwin	Chicago, Ill	May 28, 1867	65, 230
Lamp	J. H. Irwin	Philadelphia, Pa	May 6, 1873	138, 653
Lamp	J. H. Irwin	Philadelphia, Pa	May 6, 1873	138, 654
Lamp	J. Ives	Mount Carmel, Ill	Feb. 21, 1865	46, 471
Lamp	J. Ives	Mount Carmel, Conn	July 18, 1865	48, 816
Lamp	W. W. Jacobs	Hagerstown, Md	Dec. 22, 1868	85, 099
Lamp	R. Jenkins	Covington, Ky	Jan. 25, 1859	22, 729
Lamp	I. Jennings	New York, N. Y	Sept. 22, 1836	31
Lamp	I. Jennings	New York, N. Y	Jan. 19, 1847	4, 935
Lamp	J. C. Jennison and A. Hale	Foxcroft, Me	June 12, 1860	28, 717
Lamp	M. Jincks	Dansville, N. Y	Nov. 12, 1867	70, 856
Lamp	E. F. Jones	Boston, Mass	May 4, 1858	20, 159
Lamp	G. A. Jones	New York, N. Y	Nov. 10, 1863	40, 566
Lamp	A. Judson	Brooklyn, N. Y	Apr. 9, 1861	31, 983
Lamp	A. Judson	Brooklyn, N. Y	Oct. 15, 1867	69, 815
Lamp	A. Judson	Brooklyn, N. Y	Sept. 1, 1868	81, 645
Lamp	A. Kaestner	New York, N. Y	Nov. 5, 1861	33, 649
Lamp	F. Kampfe	New York, N. Y	Oct. 7, 1873	143, 452
Lamp	J. Keim	Philadelphia, Pa	Mar. 29, 1834	....
Lamp	E. G. Kelley	New York, N. Y	Aug. 17, 1869	93, 719
Lamp	S. J. Kelly	Pemberton, N. J	Feb. 3, 1863	37, 581
Lamp	J. P. and E. Kenyon	Brooklyn, N. Y	Aug. 31, 1858	21, 344
Lamp	M. R. Kenyon	Providence, R. I	Apr. 26, 1864	42, 492
Lamp	W. M. Kimball	Rochester, N. Y	Feb. 26, 1856	14, 317
Lamp	J. Kirby, jr	Adrian, Mich	Aug. 19, 1873	141, 933
Lamp	C. A. Kleeman	Erfurt, Prussia	Mar. 10, 1863	37, 867
Lamp	A. Kleinsteiber	Milwaukee, Wis	Oct. 9, 1860	30, 377
Lamp	A. H. Knapp	Newton Centre, Mass	Nov. 8, 1859	26, 071
Lamp	H. Knowles	New London, Conn	Jan. 25, 1859	22, 771
Lamp	W. Kuebler and H. Beierlein	Philadelphia, Pa	Sept. 9, 1862	86, 413
Lamp	E. L. Lambie	Washington, D. C	Mar. 26, 1872	125, 058
Lamp	E. M. Lang and J. Gilman	Westbrook and Portland, Me.	May 9, 1865	47, 681
Lamp	C. B. Lashar	New York, N. Y	Apr. 14, 1863	38, 170
Lamp	W. Lassell	Boston, Mass	Sept. 5, 1865	49, 769
Lamp	A. B. Latta	Cincinnati, Ohio	July 7, 1863	39, 154
Lamp	H. H. Laughlin	Philadelphia, Pa	May 30, 1871	115, 329
Lamp	F. Leclair	New York, N. Y	Dec. 8, 1857	18, 818
Lamp	J. Lee	New York, N. Y	Nov. 12, 1867	70, 867
Lamp	J. K. Leedy	Woodstock, Va	Jan. 24, 1860	26, 910

Index of patents issued from the United States Patent Office from 1790 *to* 1873, *inclusive*—Continued.

Invention.	Inventor.	Residence.	Date.	No.
Lamp	H. Leibert	Norristown, Pa	Jan. 1, 1861	31, 024
Lamp	J. E. Leighton	Lowell, Mass	May 6, 1873	138, 509
Lamp	W. Lewis	Boston, Mass	Jan. 23, 1818	......
Lamp	J. Lindsley	Pawtucket, R. I	Aug. 27, 1872	130, 924
Lamp	E. A. Locke and W. N. Weeden	Boston, Mass	Jan. 21, 1868	73, 538
Lamp	L. Loeffler	East Cambridge, Mass	Aug. 11, 1863	39, 488
Lamp	G. H. Lomax	Somerville, Mass	Sept. 20, 1870	107, 514
Lamp	J. R. Loomis	Winsted, Conn	Sept. 6, 1859	25, 342
Lamp	H. Long	Kittaning, Pa	Sept. 21, 1869	95, 027
Lamp	D. Lubin	New York, N. Y	June 4, 1872	127, 621
Lamp	C. E. Lyon	Worcester, Mass	Apr. 12, 1870	101, 748
Lamp	A. M. Mace	Needham, Mass	July 30, 1872	129, 973
Lamp	C. D. Macqueen	Philadelphia, Pa	Jan. 31, 1871	111, 360
Lamp	J. Mallory	New York, N. Y	Nov. 20, 1812	......
Lamp	L. Mangeon	New York, N. Y	Apr. 13, 1869	88, 968
Lamp	C. B. and S. S. Mann	Baltimore, Md	May 16, 1871	114, 954
Lamp	G. Marlow and M. Ralphe	Cincinnati, Ohio	Aug. 30, 1859	25, 304
Lamp	E. Marsh	Newark, N. J	Jan. 24, 1865	46, 053
Lamp	R. Marsh	Flushing, N. Y	June 17, 1873	139, 964
Lamp	S. Marshall	Wilmington, Del	Aug. 19, 1862	36, 226
Lamp	W. M. Marshall	Philadelphia, Pa	Nov. 19, 1872	133, 107
Lamp	A. J. Martin	Catskill, N. Y	Nov. 12, 1872	132, 972
Lamp	C. F. Martine	Dorchester, Mass	May 20, 1862	35, 349
Lamp	C. F. Martine	Dorchester, Mass	Sept. 19, 1865	50, 019
Lamp	C. Mayer	Baltimore, Md	May 13, 1873	138, 812
Lamp	T. Mayhew	Poughkeepsie, N. Y	Jan. 7, 1862	34, 065
Lamp	F. McDaniels	Philadelphia, Pa	Sept. 13, 1870	107, 276
Lamp	S. T. McDougall	Brooklyn, N. Y	Feb. 18, 1868	74, 563
Lamp	J. K. Mentzer	New Holland, Pa	Dec. 21, 1869	98, 083
Lamp	W. Mentzel and A. and J. W. Geddes.	Baltimore, Md	Mar. 11, 1862	34, 638
Lamp	R. S. Merrill	Lynn, Mass	June 14, 1859	24, 397
Lamp	R. S. Merrill	Boston, Mass	May 22, 1866	54, 935
Lamp	R. S. Merrill	Boston, Mass	Aug. 7, 1866	56, 974
Lamp	R. S. Merrill	Hyde Park, Mass	June 21, 1870	104, 481
Lamp	R. S. Merrill	Hyde Park, Mass	June 21, 1870	104, 623
Lamp	R. S. Merrill	Boston, Mass	Sept. 13, 1870	107, 277
Lamp	R. S. Merrill	Boston, Mass	Sept. 13, 1870	107, 278
Lamp	R. S. Merrill	Boston, Mass	Jan. 17, 1871	111, 072
Lamp	R. S. Merrill and W. Carleton	Boston, Mass	Jan. 28, 1868	73, 912
Lamp	L. Michaels	Covington, Ky	Mar. 18, 1873	136, 849
Lamp	C. Miller	Saint Louis, Mo	June 12, 1860	28, 683
Lamp	J. J. Miller	Chicago, Ill	Mar. 10, 1863	37, 887
Lamp	J. J. Miller	Chicago, Ill	June 30, 1863	39, 101
Lamp	G. W. Mitchell	Saint Louis, Mo	Jan. 2, 1866	51, 852
Lamp	D. A. Moore	Syracuse, N. Y	Dec. 10, 1861	33, 898
Lamp	S. C. Moore	Boston, Mass	Nov. 10, 1868	83, 874
Lamp	L. E. C. Moore and J. S. Hamilton.	Pittston, Pa	Apr. 12, 1870	101, 902
Lamp	W. Morehouse	Buffalo, N. Y	Nov. 19, 1861	33, 749
Lamp	S. G. Morrison	Williamsport, Pa	Jan. 6, 1863	37, 301
Lamp	T. H. Mott	New York, N. Y	Apr. 23, 1872	126, 075
Lamp	C. Mulchahey	Springfield, Mass	Oct. 2, 1866	58, 457
Lamp	W. Mulholland	Brooklyn, N. Y	Sept. 28, 1858	21, 617
Lamp	W. Mullally	Boston, Mass	Oct. 29, 1867	70, 247
Lamp	W. Mullally	Boston, Mass	Feb. 11, 1868	74, 403
Lamp	J. Mulvany	New York, N. Y	Nov. 26, 1861	33, 791
Lamp	G. Neilson	Boston, Mass	Jan. 24, 1860	26, 952
Lamp	G. Neilson	Boston, Mass	Sept. 18, 1860	30, 082
Lamp	G. Neilson	Boston, Mass	Dec. 17, 1867	72, 219
Lamp	G. Neilson	Boston, Mass	Jan. 14, 1868	73, 372
Lamp	C. Newman	Pittsburgh, Pa	Aug. 13, 1861	33, 047
Lamp	A. H. North	Hartford, Conn	Apr. 5, 1859	23, 483
Lamp	E. D. Norton	Bradford, Pa	June 18, 1867	65, 935
Lamp	J. E. Noyes	New Albany, Ind	Sept. 15, 1868	82, 150
Lamp	P. Noyes	Lowell, Mass	July 28, 1868	80, 302
Lamp	R. Nutting	Randolph, Vt	Jan 24, 1871	111, 237
Lamp	A. Odel and W. A. Burrows	New York, N. Y	Oct. 8, 1861	33, 447
Lamp	W. J. Palmer	Flushing, N. Y	Feb. 25, 1862	34, 521
Lamp	J. M. Parker	La Grange, Mo	Feb. 6, 1872	123, 415
Lamp	G. T. Parkhurst	Baltimore, Md	Sept. 13, 1859	25, 438
Lamp	W. P. Patton	Harrisburgh, Pa	Aug. 20, 1861	33, 102
Lamp	J. A. Pease	Boston, Mass	Apr. 22, 1873	138, 185
Lamp	J. A. Pease	Boston, Mass	Oct. 28, 1873	144, 130
Lamp	O. N. Perkins	Meriden, Conn	Aug. 13, 1872	130, 530
Lamp	O. N. Perkins	Meriden, Conn	Oct. 28, 1873	144, 132
Lamp	J. M. Perkins and M. W. House	Cleveland, Ohio	Mar. 14, 1865	46, 819
Lamp	J. M. Perkins and M. W. House	Cleveland, Ohio	Dec. 11, 1866	60, 416
Lamp	J. Phelps	Owego, N. Y	Aug. 3, 1869	93, 224
Lamp	D. L. Pickard	Rochester, N. Y	Jan. 30, 1866	52, 316
Lamp	W. H. Pierce	Somerville, Mass	May 5, 1863	38, 453
Lamp	W. Pitt	Ithaca, N. Y	July 23, 1861	32, 883
Lamp	P. Plant	Washington, D. C	Apr. 6, 1858	19, 896
Lamp	A. H. Platt and W. S. Rosecrans	Cincinnati, Ohio	Oct. 1, 1861	33, 402
Lamp	W. Porter	Belleville Township, N. J	July 23, 1867	67, 070
Lamp	S. S. and A. J. Post	Jersey City, N. J	July 16, 1861	32, 851
Lamp	G. Pugh	Cleveland, Ohio	June 15, 1869	91, 263
Lamp	R. B. Pullam	Cincinnati, Ohio	Dec. 24, 1861	34, 029
Lamp	W. H. Racey	Saint Augustine, Fla	Sept. 28, 1858	21, 627
Lamp	W. H. Racey	Saint Augustine, Fla	Oct. 30, 1860	30, 546
Lamp	M. Rae	Manchester, England	Aug. 12, 1862	36, 167
Lamp	T. Raymond	Franklinville, N. Y	Apr. 8, 1862	34, 904
Lamp	O. Redmond	Rochester, N. Y	July 26, 1853	9, 878
Lamp	C. Reichmand	Philadelphia, Pa	Sept. 21, 1858	21, 576
Lamp	E. B. Requa	Jersey City, N. J	May 6, 1862	35, 175
Lamp	F. Rhind	Brooklyn, N. Y	Sept. 9, 1873	142, 730
Lamp	F. C. Richer	Gilmer, Tex	Aug. 3, 1869	93, 228
Lamp	C. and C. Richman	Philadelphia, Pa	June 11, 1841	2, 129

Index of patents issued from the United States Patent Office from 1790 *to* 1873, *inclusive*—Continued.

Invention.	Inventor.	Residence.	Date.	No.
Lamp	C. W. Richter, sr	Madison, Ga	Feb. 22, 1859	23, 044
Lamp	C. W. Richter	Madison, Ga	Nov. 29, 1859	26, 295
Lamp	F. C. Rider	Providence, R. I	Mar. 13, 1855	12, 520
Lamp	K. Riedel	Guttenberg, N. J	Oct. 2, 1866	58, 481
Lamp	D. C. Ripley	Pittsburgh, Pa	Mar. 2, 1869	87, 367
Lamp	A. J. Ritter	Rahway, N. J	May 31, 1864	42, 998
Lamp	C. H. Robinson	Boston, Mass	July 22, 1862	35, 955
Lamp	S. Rodman	New Bedford, Mass	Oct. 17, 1848	5, 870
Lamp	E. F. Rogers	Boston, Mass	Jan. 30, 1866	52, 369
Lamp	E. D. Rosencrantz	Cincinnati, Ohio	Dec. 24, 1861	34, 030
Lamp	J. Russell	Troy, N. Y	July 2, 1861	32, 721
Lamp	J. F. Russell	Washington, D. C	Feb. 1, 1870	99, 4[illegible]2
Lamp	S. Russell	Waterbury, Conn	July 16, 1872	129, 596
Lamp	S. Russell	Waterbury, Conn	July 16, 1872	129, 597
Lamp	C. W. Russell and N. Clifford	New York, N. Y	Dec. 24, 1867	72, 683
Lamp	S. Rust	New York, N. Y	Apr. 20, 1837	176
Lamp	S. Rust	New York, N. Y	June 30, 1837	246
Lamp	S. Rust	New York, N. Y	Sept. 25, 1837	406
Lamp	S. Rust	New York, N. Y	Jan. 9, 1838	564
Lamp	S. Rust	New York, N. Y	June 7, 1838	772
Lamp	S. Rust	New York, N. Y	Sept. 14, 1843	3, 260
Lamp	L. L. Sagendorph	Boston, Mass	Mar. 10, 1868	75, 467
Lamp	M. Samuels	San Francisco, Cal	May 30, 1871	115, 528
Lamp	J. F. Sanford	Keokuk, Iowa	Feb. 9, 1869	86, 867
Lamp	J. F. Sanford	Keokuk, Iowa	May 4, 1869	89, 600
Lamp	H. Sangster	Buffalo, N. Y	Aug. 30, 1864	44, 017
Lamp	J. Sangster	Buffalo, N. Y	Mar. 25, 1862	34, 782
Lamp	W. H. Sangster	Chicago, Ill	Nov. 27, 1866	60, 068
Lamp	S. Sargent	Watertown, Mass	July 24, 1860	29, 321
Lamp	O. Sanlay	New Orleans, La	May 8, 1860	28, 199
Lamp	W. Scarlett	Aurora, Ill	May 9, 1871	114, 609
Lamp	W. Scarlett	Aurora, Ill	Nov. 14, 1871	121, 008
Lamp	J. Schelly	Savannah, Ga	July 24, 1860	29, 324
Lamp	B. B. Schneider	New York, N. Y	Apr. 19, 1870	102, 163
Lamp	H. A. Schottky and T. Simendinger.	New York, N. Y	Jan. 31, 1871	111, 481
Lamp	C. Seidhof	Lancaster, Mass	July 27, 1852	9, 157
Lamp	I. W. Shaler	Brooklyn, N. Y	Oct. 10, 1871	119, 887
Lamp	I. W. Shaler	Brooklyn, N. Y	Feb. 20, 1872	123, 940
Lamp	W. F. Shaw	Boston, Mass	Jan. 4, 1859	22, 516
Lamp	S. Shea and E. W. Gillman	Long Island City, N. Y	Mar. 17, 1868	75, 587
Lamp	J. W. Shehan	San Francisco, Cal	July 13, 1869	92, 661
Lamp	W. F. Shaw	Boston, Mass	Dec. 31, 1842	2, 893
Lamp	J. W. Shulz and J. Truil	Medford, Mass	Mar. 19, 1831	....
Lamp	A. M. Silber and F. White	London, England	May 21, 1872	127, 108
Lamp	G. H. Simmons	Bennington, Vt	Mar. 4, 1873	136, 553
Lamp	J. B. Slawson	New Orleans, La	Nov. 5, 1867	70, 635
Lamp	F. V. Sleeth	Keokuk, Iowa	Apr. 19, 1870	102, 168
Lamp	E. F. Slocum	Chicago, Ill	June 11, 1861	32, 537
Lamp	A. G. Smith	Jersey City, N. J	Aug. 28, 1866	57, 587
Lamp	J. H. Smith	Brewster Station, N. Y	Apr. 14, 1868	76, 836
Lamp	R. H. Smith	Pittsburgh, Pa	Nov. 16, 1869	96, 9[illegible]6
Lamp	S. P. Smith	Waterford, N. Y	May 4, 1869	89, 696
Lamp	T. B. Smith	Marietta, Ohio	Aug. 28, 1860	29, 831
Lamp	W. H. Smith	New York, N. Y	Jan. 19, 1869	8, 118
Lamp	W. H. Smith	New York, N. Y	Mar. 10, 1868	75, 479
Lamp	D. M. Smyth	Orange, N. J	July 14, 1868	79, 924
Lamp	O. and H. S. Snow	West Meriden, Conn	June 12, 1860	28, 695
Lamp	T. E. Sparks	Norwich, Conn	Aug. 28, 1866	57, 589
Lamp	C. F. Spencer	Rochester, N. Y	Jan. 21, 1868	73, 663
Lamp	S. Spoor	Phelps, N. Y	Jan. 21, 1868	73, 664
Lamp	W. Staehlen	Williamsburgh, N. Y	Jan. 16, 1862	122, 864
Lamp	W. Staehlen	Brooklyn, N. Y	Apr. 16, 1872	125, 852
Lamp	C. C. Stansell	Middleborough, Mass	June 17, 1862	35, 655
Lamp	C. C. Stansell	Middleborough, Mass	July 22, 1862	35, 979
Lamp	W. H. Starr	New York, N. Y	June 6, 1846	4, 561
Lamp	R. Steinmann	Boston, Mass	Apr. 6, 1858	19, 898
Lamp	W. G. Sterling	Bridgeport, Conn	Mar. 17, 1863	37, 928
Lamp	W. G. Sterling	Bridgeport, Conn	Oct. 20, 1863	40, 368
Lamp	L. Sterling and T. W. Willson	New York, N. Y	June 14, 1870	104, 372
Lamp	C. St. John	Charlestown, Mass	Nov. 19, 1867	71, 242
Lamp	G. Stobwasser	Berlin, Prussia	Sept. 25, 1866	58, 361
Lamp	J. Street	Philadelphia, Pa	May 2, 1846	4, 489
Lamp	T. F. Strong	London, England	July 7, 1846	4, 621
Lamp	T. F. Strong	New York, N. Y	July 20, 1846	4, 639
Lamp	J. Stuber	Utica, N. Y	May 1, 1860	28, 132
Lamp	J. Stuber and F. Frank	Utica, N. Y	Apr. 23, 1861	32, 155
Lamp	J. Stuber and T. Harden	Utica, N. Y	Oct. 30, 1855	13, 723
Lamp	J. Stuber and R. Hughes	Utica, N. Y	Feb. 2, 1858	19, 266
Lamp	J. Sutton	New York, N. Y	Apr. 11, 1848	5, 506
Lamp	Z. Swope	Lancaster, Pa	Mar. 13, 1860	27, 500
Lamp	D. Symonds	Lowell, Mass	Mar. 7, 1865	46, 730
Lamp	I. W. Taber	New Bedford, Mass	July 17, 1860	29, 203
Lamp	A. Taplin	Providence, R. I	Oct. 14, 1862	36, 680
Lamp	A. Taplin	Providence, R. I	Mar. 1, 1864	41, 794
Lamp	A. Taplin	Somerville, Mass	Feb. 25, 1868	74, 863
Lamp	J. Thomas	New York, N. Y	Apr. 23, 1861	32, 156
Lamp	J. Thomas	New York, N. Y	July 23, 1861	32, 906
Lamp	W. S. Thompson	Rochester, N. Y	Jan. 7, 1862	34, 080
Lamp	G. W. Thomson	Buffalo, N. Y	July 26, 1870	105, 867
Lamp	G. W. Thompson	New York, N. Y	Oct. 24, 1871	120, 172
Lamp	H. Tilden	Boston, Mass	Jan. 14, 1868	73, 409
Lamp	W. H. Topham	New Bedford, Mass	Feb. 19, 1861	31, 496
Lamp	W. Totheroh	Reading, Pa	July 28, 1868	80, 519
Lamp	G. A. Tremeschini	Vicenza, Austria	Jan. 24, 1865	46, 059
Lamp	C. Tribby	Washington, D. C	Feb. 13, 1872	123, 749
Lamp	E. Tritten	Philadelphia, Pa	Jan. 29, 1861	31, 271
Lamp	E. Tritten	Philadelphia, Pa	Dec. 3, 1861	33, 859

Index of patents issued from the United States Patent Office from 1790 to 1873, inclusive—Continued.

Invention.	Inventor.	Residence.	Date.	No.
Lamp	A. Turnbull	West Meriden, Conn	Jan. 28, 1862	34, 277
Lamp	A. P. Tyler	Cleveland, Ohio	Dec. 31, 1867	72, 764
Lamp	C. N. Tyler	Buffalo, N. Y	Mar. 27, 1866	53, 506
Lamp	C. N. Tyler	Buffalo, N. Y	Nov. 19, 1867	71, 248
Lamp	I. Van Bunschoten	New York, N. Y	Nov. 21, 1854	11, 979
Lamp	J. T. Van Kirk	Philadelphia, Pa	July 17, 1860	29, 221
Lamp	J. T. Vankirk and W. M. Fulton.	Frankfort, Pa., and Cranberry, N. J.	Nov. 29, 1859	26, 310
Lamp	U. B. Vidal	Philadelphia, Pa	July 30, 1861	32, 930
Lamp	C. Von Bonhorst	Hancock, Md	Feb. 21, 1860	27, 248
Lamp	A. V. H. Webb	New York, N. Y	Feb. 19, 1839	1, 083
Lamp	W. Webb	Waterbury, Conn	Apr. 5, 1864	42, 262
Lamp	C. Weber and H. Rimann	West Meriden, Conn	Sept. 22, 1868	82, 458
Lamp	H. B. Wellman	Allegheny City, Pa	Jan. 7, 1868	73, 145
Lamp	H. Weston	Towanda, Pa	Aug. 27, 1867	68, 136
Lamp	J. C. Wharton	Nashville, Tenn	Sept. 2, 1873	142, 425
Lamp	L. White	Hartford, Conn	Sept. 13, 1859	25, 475
Lamp	T. H. White and E. Knight	Cleveland, Ohio	Mar. 4, 1873	136, 573
Lamp	A. Whitlock	Danbury, Conn	Sept. 8, 1868	82, 051
Lamp	C. Wilhelm	Philadelphia, Pa	Feb. 6, 1866	52, 480
Lamp	J. T. Williams	York Borough, Pa	Aug. 28, 1860	29, 838
Lamp	T. S. Williams	Boston, Mass	Dec. 27, 1870	110, 526
Lamp	J. S. Wood	Plainfield, N. J	Apr. 27, 1869	89, 536
Lamp	O. D. Woodbury	Derby, Conn	Mar. 6, 1866	53, 089
Lamp	L. J. Worden and A. Leach	Utica, N. Y	July 23, 1861	32, 918
Lamp	J. C. Wright	Minersville, Pa	July 16, 1861	32, 847
Lamp	F. Yeiser and S. W. Hedger	Lancaster, Ky	May 23, 1871	115, 262
Lamp	H. Young	Cincinnati, Ohio	Nov. 26, 1867	71, 354
Lamp	C. Yaiser	Newark, N. J	June 8, 1869	91, 195
Lamp, Advertising	F. R. Warner	Paris, France	Aug. 19, 1873	141, 971
Lamp, Alcohol and turpentine	I. Jennings	New York, N. Y	Sept. 22, 1836	29
Lamp and blow-pipe, Combined soldering	A. C. Frieseke	Owasso, Mich	Aug. 22, 1871	118, 229
Lamp and burner	I. W. Shaler	Brooklyn, N. Y	Feb. 27, 1872	124, 017
Lamp and burning ingredients for	I. Jennings	New York, N. Y	Mar. 3, 1829	....
Lamp and candle stand and holder	G. Uttley	Chapel Hill, N. C	Aug. 7, 1866	57, 017
Lamp and candle wick	S. R. Weeden	Providence, R. I	Jan. 1, 1861	31, 045
Lamp and gas shade holder	C. W. Emerson	Hartford, Conn	Nov. 19, 1867	71, 150
Lamp and gas stove, Coal-oil	J. E. Ambrose	Middletown, N. Y	Jan. 24, 1865	46, 045
Lamp and lantern	W. Harvie	Glasgow, North Britain	Sept. 13, 1870	107, 366
Lamp and lantern, Combination of chamber	C. H. Peters	Cincinnati, Ohio	May 12, 1863	38, 500
Lamp and lantern, Street	J. N. Aronson	New York, N. Y	May 9, 1871	114, 513
Lamp and liquid to be burned therein	J. Hawkins	Baltimore, Md	Apr. 21, 1810	....
Lamp and producing light	I. Jennings	New York, N. Y	Sept. 10, 1829	....
Lamp and reflector combined	H. J. Goff	Dubuque, Iowa	Apr. 4, 1871	113, 650
Lamp and reflector, Light-house	A. Folger	Nantucket, Mass	Mar. 28, 1842	2, 520
Lamp and stove, Combined	C. B. Guy	Lybrand, Iowa	July 11, 1865	48, 678
Lamp and torch	J. Martin	Boston, Mass	June 30, 1837	248
Lamp, Argand	L. F. Gallup	Woodstock, Vt	Mar. 9, 1831	....
Lamp, Argand	H. W. Hayden	Waterbury, Conn	Nov. 5, 1872	132, 831
Lamp, Argand	B. Hemmenway	Roxbury, Mass	Jan. 20, 1841	1, 934
Lamp, Argand	C. F. A. Hinrichs and R. Knopp	Brooklyn and New York, N. Y.	Dec. 24, 1872	134, 281
Lamp, Argand	A. B. Howland	Titusville, Pa	Sept. 21, 1869	94, 961
Lamp, Argand	D. Melville	Newport, R. I	Nov. 19, 1818	....
Lamp, Argand	J. L. Tough	Baltimore, Md	May 11, 1841	2, 091
Lamp, Argand	J. G. Webb	New York, N. Y	Oct. 9, 1855	13, 674
Lamp-attachment	W. W. Wade and C. Burnham	Long Meadow and Springfield, Mass.	Apr. 6, 1858	19, 885
Lamp-attachment for preventing smoke, &c	R. Thomas	Hoboken, N. J	May 4, 1858	20, 179
Lamp baker, Coal-oil	W. B. Billings	New York, N. Y	Oct. 18, 1864	44, 700
Lamp-base	J. Kintz	Meriden, Conn	June 18, 1872	128, 048
Lamp-basket	P. J. Clark	West Meriden, Conn	Feb. 6, 1872	123, 379
Lamp, Billiard-table	R. Kleeman	Chicago, Ill	Apr. 12, 1870	101, 888
Lamp, Binnacle	W. Lawrence	Meriden, Conn	Nov. 14, 1832	....
Lamp-black and colophane, Manufacture of	E. Clark	Brooklyn, N. Y	Jan. 2, 1849	6, 001
Lamp-black, Apparatus for the manufacture of	M. Matlack	Philadelphia, Pa	Mar. 24, 1868	75, 943
Lamp-black, Making	J. Hastings	Cambridge, Mass	Aug. 15, 1826	....
Lamp-black, Making	G. Mini	Philadelphia, Pa	Nov. 13, 1844	3, 824
Lamp-black, Making	J. A. Roth	Philadelphia, Pa	June 9, 1857	17, 519
Lamp-black, Manufacture of	W. G. W. Iaeger	Baltimore, Md	July 18, 1854	11, 331
Lamp-black, Manufacture of	A. Millochau	New York, N. Y	Dec. 10, 1867	72, 068
Lamp-black, Manufacture of	A. Millochau	New York, N. Y	Nov. 17, 1868	84, 131
Lamp-black, Manufacture of	R. N. Perlee	Jersey City, N. J	Dec. 10, 1867	72, 078
Lamp-black, Manufacture of	A. Prenate	Elizabeth, N. J	Oct. 17, 1865	50, 493
Lamp-boiler	P. H. Sims	Waterloo, Canada	Sept. 24, 1872	131, 712
Lamp boiler, Kerosene	W. T. Rossman	Hudson, N. Y	July 28, 1868	80, 508
Lamp-bowls, &c., to their stands, Means of attaching.	S. S. Barrie	New York, N. Y	Mar. 26, 1872	125, 002
Lamp-boxes, Draft-attachment for	C. M. Bromwick	Boston, Mass	Sept. 2, 1862	36, 337
Lamp-bracket	W. Ascough	Buffalo, N. Y	Aug. 25, 1868	81, 324
Lamp-bracket	T. W. Brown	New York, N. Y	July 17, 1866	56, 361
Lamp-bracket	R. Devereux	Buffalo, N. Y	Nov. 12, 1867	70, 701
Lamp-bracket	E. L. Ferguson	Buffalo, N. Y	Mar. 10, 1868	75, 256
Lamp-bracket	L. Hull	Charlestown, Mass	Aug. 26, 1873	142, 106
Lamp-bracket	T. A. Hunter and J. Blewitt	New York, N. Y	June 29, 1869	92, 053
Lamp-bracket	G. Jones	Peekskill, N. Y	June 3, 1873	139, 582
Lamp-bracket	R. Marsh	New York, N. Y	Sept. 27, 1870	107, 699
Lamp-bracket	W. W. McDonald	Bucyrus, Ohio	Feb. 6, 1872	123, 498
Lamp-bracket	R. S. Merrill	Boston, Mass	Sept. 13, 1870	107, 279
Lamp-bracket	C. H. Miller	Buffalo, N. Y	Mar. 16, 1869	87, 957
Lamp-bracket	R. S. Tay and F. T. Fracker	Milford and Cambridge, Mass.	Nov. 27, 1866	60, 086
Lamp-bracket	A. F. Young and J. F. Foote	Mystic Bridge, Conn	Nov. 5, 1872	132, 785
Lamp, Bracket and chandelier	J. Ives	Mount Carmel, Conn	Oct. 25, 1864	44, 803
Lamp bracket and reflector	E. Boesch	San Francisco, Cal	Apr. 16, 1872	125, 785
Lamp bracket, Coach	S. Boudren	Bridgeport, Conn	Jan. 4, 1870	98, 468
Lamp bracket or support	H. Campbell	San Francisco, Cal	Dec. 5, 1871	121, 583

Index of patents issued from the United States Patent Office from 1790 *to* 1873, *inclusive*—Continued.

Index of patents issued from the United States Patent Office from 1790 *to* 1873, *inclusive*—Continued.

Invention.	Inventor.	Residence.	Date.	No.
Lamp-chimney and dish-washer	C. S. and S. A. Moore	Worcester, Mass	Jan. 31, 1871	111, 368
Lamp chimney and shade combined	J. H. Connelly	Wheeling, W. Va	July 3, 1866	56, 010
Lamp-chimney attachment	W. L. Fish	Newark, N. J	June 17, 1862	35, 598
Lamp-chimney attachment	F. N. Gisborne and H. Allman	London, England	July 14, 1868	79, 823
Lamp-chimney attachment	J. Hughes	New Berne, N. C	Feb. 15, 1870	99, 901
Lamp-chimney attachment	D. M. Shea	Chicopee, Mass	July 21, 1868	80, 093
Lamp-chimney attachment	F. J. Tinker	Cincinnati, Ohio	Oct. 16, 1866	58, 915
Lamp-chimney attachment	J. Zengeler	Chicago, Ill	Jan. 6, 1863	37, 316
Lamp-chimney, bottle, or can cleaner	C. N. Pond	Oberlin, Ohio	July 3, 1866	56, 092
Lamp-chimney cap	J. G. Dreyfus	New York, N. Y	May 4, 1869	89, 743
Lamp-chimney cleaner	F. Ashley	New York, N. Y	Nov. 20, 1866	59, 808
Lamp-chimney cleaner	L. J. Baker	Machias, Me	May 5, 1868	77, 567
Lamp-chimney cleaner	W. B. Barnard	Waterbury, Conn	Aug. 28, 1866	57, 460
Lamp-chimney cleaner	J. S. Black	Oakland, Ill	Mar. 10, 1868	75, 351
Lamp-chimney cleaner	C. Boeckh	Strasburg, France	Oct. 7, 1862	36, 600
Lamp-chimney cleaner	O. B. Brown	Malden, Mass	Nov. 5, 1867	70, 403
Lamp-chimney cleaner	B. E. Chollar and J. G. Johannes.	Leavenworth, Kans	Mar. 11, 1873	136, 704
Lamp-chimney cleaner	C. B. Clark	Armada, Mich	Aug. 8, 1871	117, 741
Lamp-chimney cleaner	H. Fellows	Bloomington, Ind	May 28, 1872	127, 230
Lamp-chimney cleaner	A. A. Ford	North Abington, Mass	June 11, 1872	127, 754
Lamp-chimney cleaner	R. G. Forsyth	Oakland, Ill	Apr. 8, 1873	137, 670
Lamp-chimney cleaner	L. Granger	Armada, Mich	May 14, 1872	126, 693
Lamp-chimney cleaner	J. R. Hamilton	Portland, Oreg	Jan. 14, 1868	73, 324
Lamp-chimney cleaner	E. T. Hazeltine	Warren, Pa	Mar. 22, 1864	41, 994
Lamp-chimney cleaner	E. T. Hazeltine	Warren, Pa	June 5, 1866	55, [illegible]
Lamp-chimney cleaner	J. P. Howell	New York, N. Y	Oct. 1, 1867	69, 439
Lamp-chimney cleaner	R. R. Howell	Fostoria, Ohio	July 9, 1872	128, 879
Lamp-chimney cleaner	F. Imhorst	New York, N. Y	Dec. 20, 1864	45, 500
Lamp-chimney cleaner	F. Imhorst	New York, N. Y	Jan. 2, 1866	51, 835
Lamp-chimney cleaner	W. J. Johnson	Newton, Mass	Aug. 14, 1866	57, 149
Lamp-chimney cleaner	L. Keiler	Catawissa, Pa	Sept. 25, 1866	58, 265
Lamp-chimney cleaner	M. R. Kenyon	Providence, R. I	July 22, 1862	35, 939
Lamp-chimney cleaner	G. Leas	Shirleysburgh, Pa	Dec. 17, 1867	72, 306
Lamp-chimney cleaner	J. Lee	New York, N. Y	June 2, 1868	78, 528
Lamp-chimney cleaner	J. H. Lightner	Shirleysburgh, Pa	Nov. 26, 1867	71, 396
Lamp-chimney cleaner	M. J. Lourrentz	Leavenworth, Kans	Oct. 27, 1868	83, 391
Lamp-chimney cleaner	M. N. Lovell	Erie, Pa	Sept. 8, 1868	81, 917
Lamp-chimney cleaner	E. P. Morse	Batavia, N. Y	Aug. 6, 1872	130, 307
Lamp-chimney cleaner	R. B. Musson	Champaign, Ill	Dec. 10, 1867	72, 072
Lamp-chimney cleaner	A. G. Newkirk	Warren, Pa	Jan. 28, 1868	73, 745
Lamp-chimney cleaner	R. Pattin	Marietta, Ohio	Jan. 8, 1867	61, 092
Lamp-chimney cleaner	R. Pearson	Allegheny City, Pa	July 25, 1871	117, 452
Lamp-chimney cleaner	H. W. Prouty	Boston, Mass	Apr. 23, 1872	126, 084
Lamp-chimney cleaner	H. W. Prouty	Boston, Mass	Feb. 11, 1873	135, 665
Lamp-chimney cleaner	J. Ryan	Brooklyn, N. Y	July 12, 1870	105, 256
Lamp-chimney cleaner	A. D. Smith	Grafton, Ohio	Oct. 18, 1870	108, 527
Lamp-chimney cleaner	I. Smith	New York, N. Y	Oct. 10, 1871	119, 890
Lamp-chimney cleaner	N. A. Vurgason	Brooklyn, N. Y	Sept. 1, 1868	81, 710
Lamp-chimney cleaner	J. J. Wait	Oreana, Nev	May 19, 1868	78, 160
Lamp-chimney cleaner	L. Ward	Poughkeepsie, N. Y	June 20, 1871	116, 121
Lamp chimney, Coal-oil	W. Howard	Flushing, N. Y	June 3, 1862	35, 447
Lamp-chimney fastening	J. A. Allen and E. F. Lewis	Washington, D. C	Jan. 7, 1868	73, 068
Lamp-chimney fastening	W. Fulton	Elizabeth City, N. J	Apr. 8, 1862	34, 884
Lamp-chimney fastening	E. Russell	Naugatuck, Conn	Nov. 26, 1867	71, 538
Lamp-chimney fastening	J. and A. W. Sangster	Buffalo, N. Y	Oct. 28, 1862	36, 802
Lamp-chimney fastening	W. S. Thompson	Rochester, N. Y	Dec. 16, 1862	37, 183
Lamp-chimney fastening	L. C. White	Waterbury, Conn	Dec. 9, 1862	37, 119
Lamp-chimney, Glass	G. F. J. Colburn	Newark, N. J	June 23, 1863	38, 947
Lamp-chimney handle	M. W. Brown	New York, N. Y	Feb. 7, 1865	46, 212
Lamp-chimney heater	J. B. Greene	Providence, R. I	Sept. 30, 1862	36, 584
Lamp-chimney holder	J. F. Hechtle	Waterbury, Conn	Dec. 21, 1869	98, 060
Lamp-chimney holder	J. H. Irwin	Philadelphia, Pa	Dec. 17, 1872	134, 063
Lamp-chimney holder	O. Newton	Pittsburgh, Pa	Apr. 8, 1862	34, 900
Lamp-chimney holder	T. Rider	Providence, R. I	Apr. 23, 1867	64, 039
Lamp-chimney holder, Sectional	C. T. Haughey	Wabash, Ind	June 4, 1872	127, 603
Lamp-chimney, Lens	S. Constant	Brooklyn, N. Y	June 13, 1854	11, 060
Lamp-chimney lifter	J. H. Wilhelm and J. W. Larimore.	Chicago, Ill.	Apr. 24, 1866	54, 239
Lamp-chimney, Metal-top	F. I. Seymour	Wolcottville, Conn	Mar. 10, 1868	75, 302
Lamp-chimney, Mica	J. Baird and W. Fisher	Hamilton, Ohio	Aug. 9, 1870	106, 247
Lamp-chimney, Mica	G. M. Bull	New Baltimore, N. Y	Jan. 9, 1872	122, 560
Lamp-chimney, Mica	E. C. Gaylord	Hartford, Conn	June 24, 1862	35, 675
Lamp-chimney, Mica	B. J. C. Howe	Syracuse, N. Y	Jan. 27, 1863	37, 506
Lamp-chimney, Mica	J. Y. Humphrey	Philadelphia, Pa	July 17, 1860	29, 172
Lamp-chimney, Mica	J. W. Schreiber	New York, N. Y	Aug. 19, 1862	36, 237
Lamp-chimney, Mica	L. Van Order	Ithaca, N. Y	Apr. 8, 1862	34, 917
Lamp-chimney, Mica	W. P. Ware	New York, N. Y	Mar. 3, 1863	37, 837
Lamp-chimney protector	E. Stern and S. Blair	New York, N. Y	Oct. 29, 1872	132, 606
Lamp-chimney, Reflecting	H. L. Hervey	Philadelphia, Pa	Aug. 17, 1869	93, 882
Lamp-chimney, Reflecting	A. Kunkle	Birmingham, Pa	Dec. 31, 1872	134, 432
Lamp-chimney spring	O. Snow	West Meriden, Conn	Apr. 8, 1862	34, 908
Lamp-chimney spring-catch	M. Fowler	Meriden, Conn	June 17, 1862	35, 649
Lamp-chimney supporter	W. Mears and H. Davies	Newport, Ky	June 24, 1873	140, 214
Lamp-chimneys, Apparatus for preventing the breakage of.	B. Vitalis	Athens, Greece	Apr. 23, 1872	125, 915
Lamp-chimeys, bottles, &c., Device for cleaning	J. T. Walker	Palmyra, N. Y	Aug. 15, 1865	49, 458
Lamp-chimneys, Combined holding and supporting spring for.	W. N. Weeden	Waterbury, Conn	Oct. 1, 1872	131, 918
Lamp-chimneys, Construction of glass	C. L. Daboll	New London, Conn	May 26, 1863	38, 657
Lamp-chimneys, Device for attaching	J. B. Alexander	Washington, D. C	Aug. 27, 1867	68, 142
Lamp-chimneys, Device for removing	G. Asmus	Houghton, Mich	June 27, 1865	48, 354
Lamp-chimneys, Device for securing	J. J. Marcy	Meriden, Conn	Dec. 4, 1860	30, 856
Lamp-chimneys, Forming glass	M. H. Collins	Chelsea, Mass	Sept. 3, 1867	68, 416
Lamp-chimneys, Glass mold for forming	M. H. Collins	Chelsea, Mass	Mar. 18, 1873	136, 971
Lamp-chimneys, Implement for handling	A. H. Merrill	Boston, Mass	Aug. 27, 1861	33, 170
Lamp-chimneys, Instrument for cleaning	T. B. De Forest	Birmingham, Conn	Jan. 15, 1861	31, 111
Lamp-chimneys, Machine for forming glass	M. H. Collins	Chelea, Mass	Sept. 26, 1871	119, 222

Index of patents issued from the United States Patent Office from 1790 *to* 1873, *inclusive*—Continued.

Invention.	Inventor.	Residence.	Date.	No.
Lamp chimneys, Mode of affixing argand	D. Jarves	Boston, Mass	July 20, 1829	....
Lamp-chimneys, Mode of cleaning	C. D. Blinn	Port Huron, Mich	Sept. 8, 1863	39, 792
Lamp-chimneys, Mode of elevating	J. G. Leffingwell	Newark, N. J	Mar. 17, 1863	37, 917
Lamp-chimneys, Mode of securing	E. Bowen	Meriden, Conn	Dec. 10, 1861	33, 875
Lamp-chimneys, Mode of securing	T. T. Jacobs	Mount Carroll, Ill	May 13, 1862	35, 239
Lamp-chimneys, Mold for making	E. Dithridge	Pittsburgh, Pa	Mar. 5, 1872	124, 343
Lamp-chimneys, Molding	E. Eltinge	Kingston, N. Y	Sept. 26, 1865	50, 105
Lamp-chimneys, Tool for sizing	L. J. Atwood	Waterbury, Conn	Dec. 10, 1867	71, 838
Lamp-cleaner	B. F. Horn	Boston, Mass	June 26, 1866	55, 862
Lamp-cleaner	E. F. Sutton	New York, N. Y	Apr. 10, 1866	53, 898
Lamp-cleaner	R. White	Princeton, Canada	Oct. 3, 1865	50, 310
Lamp, Coach	C. B. Brown and E. Andrews	Placerville, Cal	Oct. 4, 1864	44, 511
Lamp, Coach	W. Lawrence	Wallingford, Conn	Apr. 7, 1838	683
Lamp, Coal-oil	J. E. Ambrose	Jersey City, N. J	Apr. 26, 1864	42, 455
Lamp, Coal-oil	L. Bader	Philadelphia, Pa	July 7, 1863	39, 195
Lamp, Coal-oil	H. Behn	New York, N. Y	June 17, 1862	35, 579
Lamp, Coal-oil	W. B. Billings	New York, N. Y	June 16, 1863	38, 879
Lamp, Coal-oil	C. E. Corbett	Corbettsville, N. Y	Aug. 18, 1863	39, 551
Lamp, Coal-oil	G. Y. Custer	Morristown, Pa	Aug. 19, 1862	36, 206
Lamp, Coal-oil	B. Garvey	New York, N. Y	Apr. 28, 1863	38, 305
Lamp, Coal-oil	D. E. Hall	Brooklyn, N. Y	Dec. 16, 1862	37, 197
Lamp, Coal-oil	J. M. Hancock	Philadelphia, Pa	Dec. 2, 1862	37, 080
Lamp, Coal-oil	A. J. Harwood	Utica, N. Y	Oct. 25, 1864	44, 800
Lamp, Coal-oil	J. G. Hunt	Cincinnati, Ohio	Jan. 19, 1864	41, 300
Lamp, Coal-oil	J. H. Irwin	Beardstown, Ill	May 6, 1862	35, 158
Lamp, Coal-oil	A. Judson	Brooklyn, N. Y	Apr. 21, 1863	38, 263
Lamp, Coal-oil	A. C. Ketchum	New York, N. Y	Sept. 22, 1863	40, 080
Lamp, Coal-oil	A. Leach	Utica, N. Y	Jan. 6, 1863	37, 349
Lamp, Coal-oil	R. S. Merrill	Lynn, Mass	June 19, 1860	28, 762
Lamp, Coal-oil, &c	R. S. Merrill	Lynn, Mass	June 3, 1862	35, 460
Lamp, Coal-oil	W. Morehouse	Buffalo, N. Y	Apr. 1, 1862	34, 841
Lamp, Coal-oil	J. Mowery	Philadelphia, Pa	Aug. 23, 1864	43, 953
Lamp, Coal-oil	A. H. Platt	Yellow Springs, Ohio	Apr. 25, 1865	47, 451
Lamp, Coal-oil	J. Ridge	Richmond, Ind	Apr. 15, 1862	34, 984
Lamp, Coal-oil	J. Ridge	Richmond, Ind	Jan. 5, 1864	41, 093
Lamp, Coal-oil	J. Ridge	Richmond, Ind	Jan. 5, 1864	41, 094
Lamp, Coal-oil, &c	G. Rimmington	South Brooklyn, N. Y	May 25, 1858	20, 373
Lamp, Coal-oil	J. N. Smith	New York, N. Y	June 10, 1862	35, 549
Lamp, Coal-oil	C. F. Spencer	Rochester, N. Y	Mar. 24, 1863	37, 986
Lamp, Coal-oil	W. D. Taylor	Fort Madison, Iowa	Mar. 22, 1864	42, 028
Lamp, Coal-oil	J. White and A. Agnew	Philadelphia, Pa	Sept. 2, 1862	36, 380
Lamp, Coal-oil	G. L. Witsil	Philadelphia, Pa	Oct. 21, 1862	36, 755
Lamp, Coal-oil hand	E. Roberts	Moorestown, N. J	Sept. 12, 1865	49, 924
Lamp, Coal-oil-stove	J. Bowles	Augusta, Ga	Feb. 7, 1871	111, 512
Lamp, coffee-pot, and boiler combined	L. A. Plumb	Biddeford, Me	Jan. 22, 1867	61, 454
Lamp-collar	J. C. Beers	Brooklyn, N. Y	Jan. 12, 1864	41, 255
Lamp-collar	H. Kelly and W. H. Locke	Boston, Mass	May 2, 1871	114, 302
Lamp-collar	A. Taplin	Providence, R. I	Mar. 1, 1864	41, 795
Lamp-collar and safety-tube	C. B. Mann	Baltimore, Md	Apr. 9, 1872	125, 588
Lamp-cone	C. H. Buckelew	Jersey City, N. J	Apr. 4, 1865	47, 087
Lamp-cone	M. B. Dyott	Philadelphia, Pa	Mar. 24, 1863	37, 956
Lamp-cone	E. B. Requa	Jersey City, N. J	Jan. 5, 1864	41, 092
Lamp cone, Coal-oil	W. Fulton	Elizabeth City, N. J	May 6, 1862	35, 151
Lamp-crane	A. C. Black	Kaukauna, Wis	July 13, 1869	92, 570
Lamp cup, Kerosene	N. L. Bradley	Meriden, Conn	Jan. 31, 1871	111, 430
Lamp, Electric	H. M. Collier and H. N. Baker	Binghamton and New York, N. Y.	May 18, 1858	20, 255
Lamp, Electric	G. G. Percival	Brooklyn, N. Y	Apr. 3, 1866	53, 669
Lamp, Essential-oil	M. B. Dyott	Philadelphia, Pa	May 30, 1842	2, 658
Lamp extension-supporter	L. Hull	Charleston, Mass	Jan. 16, 1872	122, 767
Lamp-extinguisher	C. E. Abbott	Boston, Mass	July 3, 1866	55, 973
Lamp-extinguisher	C. E. Abbott	Malden, Mass	Nov. 13, 1866	59, 534
Lamp-extinguisher	C. E. Abbott	Malden, Mass	Nov. 19, 1867	71, 112
Lamp-extinguisher	C. E. Abbott	Malden, Mass	Aug. 11, 1868	80, 796
Lamp-extinguisher	J. M. Aubrey	Portage City, Wis	June 29, 1869	91, 812
Lamp-extinguisher	S. J. Batchelder	Manchester, N. J	June 23, 1868	79, 187
Lamp-extinguisher	W. P. Bennett	East Pepperell, Mass	Nov. 5, 1867	70, 392
Lamp-extinguisher	G. V. Bunker	Yankton, Dak	Nov. 30, 1869	97, 271
Lamp-extinguisher	W. I. Bunker	Yankton, Dak	Dec. 14, 1869	97, 757
Lamp-extinguisher	S. C. Catlin and A. Corbin	Cleveland, Ohio, and Ellenville, N. Y.	Feb. 1, 1870	99, 289
Lamp-extinguisher	W. C. Ebert	Hannibal, Mo	May 10, 1870	102, 926
Lamp-extinguisher	J. W. Fowle	Boston, Mass	Apr. 7, 1868	76, 319
Lamp-extinguisher	J. M. Goodridge	Norfolk, Va	Mar. 12, 1872	124, 567
Lamp-extinguisher	W. Grayson and D. C. Hyndman	Odell, Ill	Aug. 24, 1869	93, 986
Lamp-extinguisher	M. C. Heptinstall	Enfield, N. C	June 29, 1869	91, 934
Lamp-extinguisher	F. B. Hill and W. H. McCoy	Cleveland, Ohio, and Des Moines, Iowa.	Nov. 5, 1867	70, 567
Lamp-extinguisher	F. Hille	Lyons, Iowa	Dec. 23, 1873	145, 870
Lamp-extinguisher	J. N. Howe	Franklin, N. H	Nov. 13, 1866	59, 607
Lamp-extinguisher	J. Hughes	New Berne, N. C	May 17, 1870	103, 048
Lamp-extinguisher	R. Lange	Baltimore, Md	May 27, 1873	139, 398
Lamp-extinguisher	C. Lanzendorfer	Chicago, Ill	Jan. 19, 1869	86, 084
Lamp-extinguisher	W. D. Lindsley	Wathena, Kans	Aug. 12, 1873	141, 801
Lamp-extinguisher	C. Newman	San Francisco, Cal	May 25, 1869	90, 572
Lamp-extinguisher	J. H. Noyes	Abington, Mass	Feb. 13, 1855	12, 394
Lamp-extinguisher	S. W. Perkins	Geneseo, Ill	Dec. 28, 1869	98, 405
Lamp-extinguisher	D. Pike and O. B. Graham	Bridgeport, Conn	Sept. 2, 1873	142, 508
Lamp-extinguisher	J. Pons	Baltimore, Md	Mar. 2, 1869	87, 511
Lamp-extinguisher	D. J. Powers	Chicago, Ill	May 31, 1870	103, 652
Lamp-extinguisher	J. H. Prentice	Ashtabula, Ohio	July 21, 1868	80, 088
Lamp-extinguisher	W. A. Richardson and H. D. Ward.	Worcester, Mass	Apr. 3, 1866	53, 679
Lamp-extinguisher	W. A. Richardson and H. D. Ward.	Worcester, Mass	Sept. 4, 1866	57, 768
Lamp-extinguisher	E. Richmond	Abington, Mass	Dec. 18, 1855	13, 972
Lamp-extinguisher	F. Rohrer	San Francisco, Cal	Nov. 19, 1867	71, 221

Index of patents issued from the United States Patent Office from 1790 *to* 1873, *inclusive*—Continued.

Invention.	Inventor.	Residence.	Date.	No.
Lamp-extinguisher	W. G. Rugo	Holstein, Mo	Aug. 22, 1871	118, 280
Lamp-extinguisher	M. Safford	Boston, Mass	May 7, 1867	64, 578
Lamp-extinguisher	L. M. Sargent	Worcester, Mass	Oct. 22, 1867	70, 125
Lamp-extinguisher	E. Shaw	Portland, Me	June 11, 1867	65, 696
Lamp-extinguisher	G. Simpson and W. H. Edmunds.	Waterbury, Vt	July 9, 1867	66, 526
Lamp-extinguisher	W. H. and C. W. Terpening	Geneseo, Ill	Aug 10, 1869	93, 565
Lamp-extinguisher	E. J. Toof	Fort Madison, Iowa	Mar. 10, 1868	75, 317
Lamp-extinguisher	F. C. Wireman	New York, N. Y	Aug. 19, 1873	142, 067
Lamp extinguisher and regulator	C. E. Lyon	Worcester, Mass	July 31, 1866	56, 770
Lamp, Extinguisher for fluid	W. B. Carpenter	Brooklyn, N. Y	Sept. 9, 1856	15, 686
Lamp extinguisher, Street	G. S. Dunbar	Pittsfield, Mass	Jan. 16, 1872	122, 762
Lamp-deflectors, Buffer for inserting coiled wire around the edges of.	M. H. Mosman	Waterbury, Conn	Nov. 23, 1869	97, 215
Lamp, Döbereiner self-lighting	G. Müller	Newark, N. J	June 9, 1868	78, 755
Lamp-fastening	L. B. Carpenter	Buffalo, N. Y	July 25, 1854	11, 354
Lamp fastening, Coach	M. De Voursney	Newark, N. J	Nov. 5, 1867	70, 422
Lamp-feed regulator, Automatic	H. S. Hudson	Selma, Ala	May 4, 1869	89, 580
Lamp-feed trap	M. W. House	Cleveland, Ohio	May 24, 1870	103, 334
Lamp-feeder	T. P. Gibbons	Baltimore, Md	Sept. 15, 1868	82, 219
Lamp-feeder	T. Mayhew	Poughkeepsie, N. Y	Jan. 13, 1863	37, 404
Lamp-filler	C. R. Landmann	New York, N. Y	Aug. 29, 1854	11, 612
Lamp-filler	H. H. V. Lilley	Milford, Mass	Apr. 29, 1873	138, 415
Lamp-filler	H. W. Stables	Saco, Me	Oct. 19, 1869	95, 948
Lamp, Fire-kindling	S. J. Lowe	Quincy, Ill	Nov. 26, 1867	71, 515
Lamp, Fluid	S. F Allen	New York, N. Y	Nov. 29, 1853	10, 270
Lamp, Fluid	W. Bennett	Brooklyn, N. Y	Nov. 27, 1855	13, 860
Lamp, Fluid	S. Bidwell	Rochester, N. Y	Sept. 9, 1856	15, 724
Lamp, Fluid	D. H. Chamberlain	West Roxbury, Mass	May 8, 1855	12, 814
Lamp, Fluid	A. Von Schuttenbach	St. Petersburg, Russia	June 21, 1859	24, 509
Lamp, Fluid	T. Ward	Columbus, Ohio	Mar. 22, 1870	101, 192
Lamp for boiling water, &c	T. G. Fessenden	Boston, Mass	Jan. 31, 1827	....
Lamp for burning compound, spirits, or oil	W. Magee	New York, N. Y	June 27, 1832	....
Lamp for burning compound, tallow, lard, beeswax, &c.	N. Rublee	Montpelier, Vt	Dec. 4, 1832	....
Lamp for burning evaporable ingredients	I. Jennings	New York, N. Y	Aug. 1, 1831	....
Lamp for burning fat, &c	J. Warren	Cincinnati, Ohio	Feb. 20, 1821	....
Lamp for burning heavy oil	R. Hitchcock	Watertown, N. Y	Apr. 23, 1872	125, 954
Lamp for burning lard and other concrete substances.	E. T. Williams and L. T. Tew	Newport, R. I	June 26, 1841	2, 140
Lamp for burning off paint	W. W. Wakeman, jr	New York, N. Y	June 11, 1867	65, 621
Lamp for burning tallow	M. S. Underwood	Marshallton, Pa	Sept. 18, 1841	2, 257
Lamp for burning tallow and other fatty substances.	I. Jennings	New York, N. Y	May 20, 1830	....
Lamp for burning turpentine, camphene, &c., Argand.	C. Carr	Philadelphia, Pa	Mar. 28, 1842	2, 514
Lamp for burning vapor of benzole, &c	C. Warner	Washington, D. C	Oct. 14, 1851	8, 433
Lamp for burning volatile ingredients	I. Jennings	New York, N. Y	Sept. 11, 1841	2, 254
Lamp for burning volatile ingredients	I. Jennings	New York, N. Y	Oct. 12, 1844	3, 793
Lamp for burning volatile liquids	E. N. Horsford and J. R. Nichols.	Cambridge and Haverhill, Mass.	Oct. 30, 1855	13, 729
Lamp for burning volatile materials	G. Egles	Boston, Mass	May 22, 1835	....
Lamp for burning volatile materials, Argand	D. Pettibone	Philadelphia, Pa	Mar. 28, 1842	2, 516
Lamp for cooking-purposes	M. E. Hatch	Beloit, Wis	Apr. 13, 1869	88, 784
Lamp for cooking-purposes, Coal-oil	W. B. Billings	Brooklyn, N. Y	Nov. 14, 1865	50, 892
Lamp for heating curling-irons	D. T. Burrell	Bridgewater, Mass	Mar. 14, 1865	46, 775
Lamp for lighting gas	C. McIntosh	Jersey City, N. J	June 15, 1858	20, 573
Lamp for medicated-vapor bath, Alcohol	L. E. Hicks	Middletown, Conn	Mar. 16, 1844	3, 488
Lamp for vessels, Head	H. Langster	Buffalo, N. Y	June 2, 1863	38, 764
Lamp for volatile materials	A. V. H. Webb	New York, N. Y	Nov. 4, 1842	2, 843
Lamp-fount to bracket, chandelier, &c	G. Bohner	Chicago, Ill	Dec. 6, 1870	109, 864
Lamp, Fountain	H. W. Adams	New York, N. Y	June 30, 1857	17, 658
Lamp, Fountain	J. P. Driver	Marengo, Iowa	Aug. 21, 1866	57, 299
Lamp, Fountain	G. W. Harrold	Rochester, N. Y	July 15, 1862	35, 875
Lamp, Fountain	N. Linden	Jersey City, N. J	June 24, 1856	15, 198
Lamp, Fountain	M. Samuels	New York, N. Y	Dec. 15, 1868	85, 029
Lamp, Fountain	J. H. Seaman	Brooklyn, N. Y	June 9, 1863	38, 846
Lamp-fountains, Mold for glass	J. Wing	Boston, Mass	Oct. 21, 1873	143, 864
Lamp, &c., frame, Street	J. Bradston	San Francisco, Cal	May 17, 1870	103, 238
Lamp, Gas	S. Andrews	Perth Amboy, N. J	May 10, 1832	....
Lamp, Gas	J. H. Brown	New York, N. Y	Dec. 13, 1870	110, 005
Lamp, Gas	J. Gray	New York, N. Y	Oct. 17, 1871	120, 058
Lamp, Gas	D. Melville	Newport, R. I	Mar. 18, 1813	....
Lamp, Gas	D. Melville	Newport, R. I	Mar. 24, 1810	....
Lamp, Gas	H. G. Sickel	Philadelphia, Pa	Aug. 7, 1849	6, 624
Lamp, Gas-burning	S. Andrews	Perth Amboy, N. J	June 3, 1856	14, 994
Lamp, Gas-conducting	R. Cornelius	Philadelphia, Pa	Mar. 18, 1841	2, 000
Lamp, Gas hand	H. J. Rice	Cincinnati, Ohio	July 25, 1871	117, 331
Lamp generator, Vapor	S. D. Baldwin	Milwaukee, Wis	May 8, 1860	28, 143
Lamp generator, Vapor	J. S. Hull	Cincinnati, Ohio	May 14, 1867	64, 769
Lamp, Glass	J. Adams	Birmingham, Pa	Jan. 10, 1871	110, 815
Lamp, Glass	J. S. and T. B. Atterbury	Pittsburgh, Pa	Nov. 22, 1870	109, 309
Lamp, Glass	J. Reighard	Birmingham, Pa	July 2, 1861	32, 739
Lamp, Glass	D. C. Ripley	Pittsburgh, Pa	Jan. 7, 1868	73, 122
Lamp, Glass	D. C. Ripley, jr	Birmingham, Pa	Apr. 5, 1870	101, 512
Lamp, Glass	D. C. Ripley	Pittsburgh, Pa	Sept. 20, 1870	107, 544
Lamp globe-shade	H. Whitney	East Cambridge, Mass	Feb. 23, 1869	87, 128
Lamp, Hall, garden, &c	A. W. Wehrhan	Columbia, S. C	May 10, 1870	102, 985
Lamp-hanger	T. S. Hudson	East Cambridge, Mass	Feb. 7, 1865	46, 236
Lamp, Hanging	G. Bohner	Chicago, Ill	Apr. 30, 1872	126, 254
Lamp, Hanging	A. Platt	Middletown, Conn	Oct. 8, 1836	42
Lamp. Hanging	A. Platt	Middletown, Conn	Oct. 20, 1836	....
Lamp-heater	W. T. Eddy	West Hoboken, N. J	Apr. 7, 1863	38, 102
Lamp-heater	W. Friel	San Francisco, Cal	May 20, 1873	139, 140
Lamp-heater	C. J. Hauck	Brooklyn, N. Y	July 16, 1872	129, 555
Lamp-heater	L. E. Truesdell	Warren, Mass	Sept. 24, 1872	131, 578
Lamp-heater for nursery-cups	M. Fox and F. Reinhard	West Eau Claire, Wis	July 16, 1872	129, 275
Lamp-heater for nursery-flasks	S. Hughes	Jersey City, N. J	Aug. 13, 1872	130, 429

Index of patents issued from the United States Patent Office from 1790 to 1873, inclusive—Continued.

Invention.	Inventor.	Residence.	Date.	No.
Lamp heater, Kerosene	W. L. Brown	Shelburne Falls, Mass	Aug. 24, 1869	93, 961
Lamp heater, Vehicle	E. H. Reynolds	Rising Sun, Md	May 21, 1867	65, 010
Lamp heating and shaving apparatus	C. S. Bourne	Springfield, Mass	Sept. 13, 1864	44, 252
Lamp-heating apparatus	A. Rittenhouse	Philadelphia, Pa	Nov. 8, 1870	109, 054
Lamp-heating apparatus	W. N. White	Winchendon, Mass	Nov. 1, 1870	108, 863
Lamp-holder	C. Monson	New Haven, Conn	Jan. 18, 1859	22, 664
Lamp holder and shade	W. H. Hawkins	Cleveland, Ohio	Jan. 23, 1872	122, 886
Lamp, Hunting	J. T. Staples	Bladen Springs, Ala	Apr. 2, 1872	125, 150
Lamp, Hydrocarbon-vapor	J. Cook	New York, N. Y	May 14, 1872	126, 625
Lamp, Hydrocarbon-vapor	R. R. Crosby	Boston, Mass	Feb. 9, 1858	19, 287
Lamp, Hydrocarbon-vapor	A. M. Mace	Springfield, Mass	Apr. 22, 1856	14, 727
Lamp, Hydrocarbon-vapor	I. Suggitt	Providence, R. I	Sept. 29, 1857	18, 307
Lamp, Hydrocarbon-vapor	T. Varney	San Francisco, Cal	Sept. 30, 1856	15, 829
Lamp, Hydrogen	C. Hagen and F. Aurnhammer	New York, N. Y	Mar. 28, 1865	47, 012
Lamp, Hydrogen	J. Pusey	Philadelphia, Pa	July 23, 1872	129, 681
Lamp, Hydropneumatic	S. S. Lee	Providence, R. I	Apr. 21, 1842	2, 570
Lamp, Hydrostatic	E. D. Kendall	New York, N. Y	May 21, 1872	127, 067
Lamp, Hydrostatic safety	H. S. Whitfield	Tuscaloosa, Ala	May 21, 1872	127, 004
Lamp, Insect-destroying	F. M. Briggs	Livonia, Mich	July 28, 1868	80, 446
Lamp, Insect-destroying	G. C. Cranston	South Bend, Ind	Mar. 23, 1869	88, 140
Lamp, Insect-destroying	J. Zimmerman	Royalton Centre, N. Y	July 30, 1867	67, 244
Lamp, Kerosene	S. Adlaur, jr., and J. R. Fogg	Portland, Me	Apr. 21, 1863	28, 208
Lamp, Kerosene	J. S. and T. B. Atterbury	Pittsburgh, Pa	Jan. 6, 1863	37, 267
Lamp, Kerosene	J. H. Beidler and A. R. Crihfield.	Lincoln, Ill	Mar. 13, 1866	53, 103
Lamp, Kerosene	H. H. Dickinson	Hartford, Conn	Dec. 10, 1861	33, 879
Lamp, Kerosene	A. Meucci	Clifton, N. Y	Aug. 12, 1862	36, 192
Lamp, Kerosene or coal-oil	J. Ridge	Richmond, Ind	Dec. 9, 1862	37, 144
Lamp, Lantern	R. M. Merrill	Chicago, Ill	Sept. 30, 1862	36, 570
Lamp, Lantern	J. Silvins and W. T. Hain	Sunbury, Pa	June 11, 1867	65, 698
Lamp, lantern, &c., Street	J. Binney	Boston, Mass	Mar. 7, 1865	46, 631
Lamp, Lard	E. S. Archer	Philadelphia, Pa	June 18, 1842	2, 671
Lamp, Lard	A. H. Baird	New York, N. Y	May 26, 1842	2, 644
Lamp, Lard	H. P. Benton	Norwalk, Ohio	Apr. 10, 1843	3, 040
Lamp, Lard	J. S. Brown	Washington, D. C	Oct. 9, 1855	13, 675
Lamp, Lard	J. S. Brown	Washington, D. C	Feb. 3, 1857	16, 524
Lamp, Lard	D. H. Chamberlain	Boston, Mass	Aug. 29, 1854	11, 633
Lamp, Lard	P. J. Clark	Meriden, Conn	Mar. 21, 1843	3, 009
Lamp, Lard	I. N. Coffin	Washington, D. C	Mar. 17, 1857	16, 825
Lamp, Lard	R. Cornelius	Philadelphia, Pa	Apr. 6, 1843	3, 030
Lamp, Lard	J. T. Creighton	Alexandria, D. C	Aug. 11, 1842	2, 751
Lamp, Lard	S. Davis	New Holland, Pa	May 6, 1856	14, 806
Lamp, Lard	J. Grannis	Oberlin, Ohio	Aug. 25, 1842	2, 763
Lamp, Lard	J. D. Hays	Mount Morris, Ill	July 3, 1855	13, 170
Lamp, Lard	B. H. Horn	Boston, Mass	May 26, 1842	2, 641
Lamp, Lard	T. Houghton and J. F. Wallace	Philadelphia, Pa	Nov. 15, 1843	3, 333
Lamp, Lard	H. Howard	Wooster, Ohio	Feb. 10, 1843	2, 946
Lamp, Lard	L. Jones	Utica, N. Y	May 26, 1843	3, 101
Lamp, Lard	A. Kayser	Fulton, Mo	Aug. 20, 1845	4, 154
Lamp, Lard	D. Kinnear	Circleville, Ohio	Feb. 4, 1851	7, 921
Lamp, Lard	J. Lee	Wellsville, Ohio	Nov. 21, 1842	2, 854
Lamp, Lard	B. K. Maltby and J. Neal	Rootstown and Middlebury, Ohio.	May 4, 1842	2, 604
Lamp, Lard	W. S. Moore	Springfield, Ill	Feb. 10, 1843	2, 950
Lamp, Lard	S. P. Moorehouse	Ludlowville, N. Y	Apr. 26, 1830	....
Lamp, Lard	C. F. Reese	Millersville, Pa	June 12, 1866	55, 535
Lamp, Lard	P. Robinson	Chillicothe, Ohio	Dec. 5, 1842	2, 873
Lamp, Lard	H. Rose	Rushville, Ill	Aug. 4, 1843	3, 206
Lamp, Lard	J. S. Senseny	Chambersburgh, Pa	July 15, 1856	15, 364
Lamp, Lard	T. Sewell	New York, N. Y	Oct. 2, 1847	5, 311
Lamp, Lard	I. Smith and J. Stonesifer	Boonsborough, Md	Aug. 8, 1854	11, 497
Lamp, Lard	L. A. Stockwell	Batavia, N. Y	Sept. 13, 1853	10, 022
Lamp, Lard	F. H. Southworth	Washington, D. C	July 2, 1842	2, 703
Lamp, Lard	F. H. Southworth	Washington, D. C	Oct. 22, 1842	2, 827
Lamp, Lard	T. Terrel	Spring Hill, Ohio	Aug. 14, 1866	57, 215
Lamp, Lard	S. B. Terry	Plymouth, Conn	Feb. 24, 1843	2, 976
Lamp, Lard	J. Tobin	Bloomfield, N. J	Mar. 26, 1844	3, 513
Lamp, Lard	H. Tomlinson	Geneva, N. Y	Sept. 1, 1843	3, 243
Lamp, Lard	W. Thompson	Cincinnati, Ohio	Apr. 10, 1843	3, 044
Lamp, Lard	C. Wilhelm	Philadelphia, Pa	Apr. 6, 1843	3, 032
Lamp, Lard	Z. Worrell	Chester Hill, Ohio	Feb. 7, 1842	2, 448
Lamp, Lard or oil solar	J. G. Webb	Williamsburgh, N. Y	Oct. 14, 1851	8, 436
Lamp, Lard, tallow, &c	G. Carr	Buffalo, N. Y	Sept. 4, 1841	2, 238
Lamp, Lard, tallow, &c	N. S. Cate and J. H. Putnam	Charlestown and Malden, Mass.	Nov. 16, 1841	2, 358
Lamp-light	I. Jennings	New York, N. Y	Oct. 16, 1830	....
Lamp, Light-house, &c	C. Cornelius	Philadelphia, Pa	Dec. 28, 1822	....
Lamp, Light-house	S. K. Jackson	Norfolk, Va	Dec. 3, 1872	133, 647
Lamp, Light-house	W. Lewis, sr., and B. Hemmenway.	Boston and Roxbury, Mass	Aug. 7, 1844	3, 692
Lamp, Light-house	C. Wheeler	Rockport, Mass	Apr. 11, 1846	4, 454
Lamp-lighter	A. Assmann	Rahway, N. J	Feb. 4, 1868	74, 033
Lamp-lighter	C. M. Clinton	Ithaca, N. Y	Dec. 1, 1863	40, 734
Lamp-lighter	J. Rigby	Fort Howard, Wis	Aug. 24, 1869	94, 031
Lamp-lighter	S. and J. Thomas	Brooklyn, N. Y	Mar. 8, 1859	23, 203
Lamp-lighter and burglar and fire alarm	J. B. Irwin	Newark, Ohio	May 24, 1870	103, 338
Lamp lighter, Street	M. McGill and J. E. Tynan	Paterson, N. J	May 14, 1867	64, 684
Lamp-lighting apparatus	M. A. Lynch	Boston, Mass	May 27, 1873	139, 255
Lamp-lighting apparatus	W. Wuerz	New York, N. Y	July 30, 1872	130, 000
Lamp-lighting device	N. Allen	West Meriden, Conn	Jan. 13, 1863	37, 378
Lamp-lighting device	N. L. Cole	Norwich, Conn	Mar. 17, 1868	75, 529
Lamp-lighting device	G. K. Proctor	Beverly, Mass	Jan. 24, 1860	26, 953
Lamp, Locomotive	S. E. and H. B. Cleveland	Buffalo, N. Y	June 24, 1856	15, 172
Lamp, Locomotive	L. A. Hamblen	Chicago, Ill	Jan. 13, 1857	16, 384
Lamp, Locomotive	N. J. Knapp	Chicago, Ill	Sept. 13, 1859	25, 421
Lamp, Locomotive	J. Radley	New York, N. Y	July 31, 1860	29, 402
Lamp, Locomotive	J. Radley	New York, N. Y	June 26, 1866	55, 903
Lamp, Locomotive	T. Snook and S. Hill	Rochester, N. Y	Dec. 21, 1852	9, 490

Index of patents issued from the United States Patent Office from 1790 *to* 1873, *inclusive*—Continued.

Invention.	Inventor.	Residence.	Date.	No.
Lamp, Locomotive	S. H. Trimmons	Memphis, Tenn	Apr. 8, 1862	34, 914
Lamp, Locomotive	I. A. Williams	Utica, N. Y	Oct. 10, 1854	11, 799
Lamp, Locomotive	I. A. Williams	Utica, N. Y	Apr. 29, 1862	35, 122
Lamp, Locomotive and railway	J. Stuber	Utica, N. Y	May 20, 1856	14, 942
Lamp, Locomotive coal-oil	P. Budenbach	New York, N. Y	Mar. 24, 1863	38, 001
Lamp, Locomotive head	P. Budenback	New York, N. Y	Nov. 28, 1865	51, 135
Lamp, Locomotive reflector	F. J. Seymour	Waterbury, Conn	July 8, 1856	15, 305
Lamp, Magnesium	I. S. Russell and H. R. Russell	New Market, Md., and Woodbury, N. J.	Apr. 29, 1873	138, 346
Lamp, Magnesium	R. H. Thurston	Providence, R. I	May 1, 1866	54, 442
Lamp, Magnesium	R. H. Thurston	Providence, R. I	June 5, 1866	55, 393
Lamp, Miner's	W. G. Dowd	Scranton, Pa	May 25, 1869	90, 434
Lamp, Miner's	W. McClare	Hyde Park, Pa	Aug. 15, 1865	49, 477
Lamp, Miner's	W. Seybold	McKeesport, Pa	May 13, 1862	35, 264
Lamp, Miner's	W. Seybold	McKeesport, Pa	June 8, 1869	90, 963
Lamp, Miner's	J. C. Sommerville	Snow Shoe, Pa	June 8, 1869	90, 967
Lamp, Miner's	G. W. Trimble	Bloomsburgh, Pa	Mar. 17, 1868	75, 603
Lamp, Miner's	W. C. Winfield	Hubbard, Ohio	May 23, 1871	115, 143
Lamp, Miner's	W. C. Winfield	Hubbard, Ohio	May 7, 1872	126, 606
Lamp mold, Glass	J. Bridges	Allentown, Pa	July 23, 1872	129, 781
Lamp mold, Glass	H. Dillaway	Sandwich, Mass	July 18, 1871	117, 157
Lamp mold, Glass	J. Jenkins	Boston, Mass	Sept. 18, 1860	30, 066
Lamp, Naval store-room or coach	J. J. Walton	New York, N. Y	July 1, 1873	140, 448
Lamp, Night	T. W. Howchin	Morrisania, N. Y	May 6, 1862	35, 154
Lamp, Nurse	W. Howe	Boston, Mass	Dec. 31, 1812	......
Lamp, Oil	J. Calkins	Long Branch, N. J	July 23, 1872	129, 786
Lamp, Oil	T. S. Speakman	Camden, N. J	Apr. 4, 1865	47, 173
Lamp, Oil or spirit gas	J. M. and G. Truman	Penn Township, Pa	Nov. 19, 1833	......
Lamp or candle stand	F. A. Marshall	Marlborough, Mass	Jan. 29, 1861	31, 251
Lamp or candlestick and match-box combined	T. Shanks	Baltimore, Md	Jan. 19, 1858	19, 158
Lamp or gas-light bracket	C. Monson	New Haven, Conn	Sept. 15, 1863	39, 945
Lamp or gas shade	M. J. Wellman	New York, N. Y	Aug. 4, 1863	39, 438
Lamp or holder, Candle	M. Weis	Boston, Mass	Feb. 20, 1872	123, 803
Lamp, Paraffine	H. Ryder	Somerville, Mass	Apr. 29, 1873	138, 441
Lamp, Petroleum	W. B. Billings	Brooklyn, N. Y	Oct. 11, 1864	44, 592
Lamp, Petroleum	J. and J. Hinks	Birmingham, England	Dec. 10, 1867	72, 040
Lamp, Petroleum argand	A. K. Murray and A. B. Howland.	Titusville, Pa	Jan. 21, 1868	73, 455
Lamp, Plumbers' and painters'	G. Wanier	New York, N. Y	Dec. 3, 1867	71, 826
Lamp, Pocket	J. Espelding and W. E. Huttmann.	Chicago, Ill	June 29, 1869	91, 833
Lamp, Pocket	A. P. Foster and D. B. H. Bartlett.	Lowell, Mass	Aug. 31, 1869	94, 300
Lamp, Pocket	H. L. Hervey	Quincy, Ill	Dec. 9, 1856	16, 180
Lamp, Portable condensing	T. Philipps	New York, N. Y	June 25, 1823	......
Lamp, Portable cooking-apparatus by	C. A. Harper	Little Rock, Ark	May 1, 1866	54, 337
Lamp-post	L. A. Gouch	Yonkers, N. Y	Nov. 21, 1871	121, 165
Lamp-post	L. A. Gouch	Yonkers, N. Y	June 3, 1873	139, 504
Lamp-post	J. W. Graham	Chillicothe, Ohio	May 25, 1869	90, 356
Lamp-post	E. A. Heath	New York, N. Y	Jan. 23, 1872	122, 887
Lamp-post attachment for supporting electrical wires.	J. P. Tirrell	Charlestown, Mass	June 3, 1873	139, 630
Lamp post, Street	P. H. Branson	Saint Louis, Mo	Feb. 21, 1865	46, 444
Lamp post, Street gas	S. W. France	Brooklyn, N. Y	Dec. 5, 1871	121, 456
Lamp post, Street gas	J. T. P. Hunt	Manchester, N. H	Jan. 3, 1865	45, 717
Lamp-post, Sediment-chamber for	G. C. Bovey	Cincinnati, Ohio	Feb. 8, 1870	99, 527
Lamp-post, Sediment-chamber for	J. W. Graham	Chillicothe, Ohio	Sept. 26, 1871	119, 347
Lamp, Railway	J. Carton	Utica, N. Y	Aug. 26, 1862	36, 274
Lamp, Railway	J. Carton	Utica, N. Y	Oct. 7, 1862	36, 602
Lamp, Railway	J. M. A. Dew	Chicago, Ill	Aug. 10, 1869	93, 602
Lamp, Railway	L. Hover	Chicago, Ill	May 24, 1864	42, 853
Lamp, Railway	T. J. Newland	Utica, N. Y	Dec. 23, 1862	37, 239
Lamp, Railway	L. S. White	Chicopee, Mass	Sept. 5, 1854	11, 657
Lamp, Railway and marine signal	J. F. Veronee	Charleston, S. C	July 16, 1872	129, 502
Lamp, Railway-car	J. L. Howard	Hartford, Conn	Oct. 7, 1873	143, 411
Lamp, Railway-car	W. C. Marshall	Hartford, Conn	Aug. 20, 1872	130, 586
Lamp, Railway-car	R. S. Merrill	Boston, Mass	Nov. 22, 1870	109, 537
Lamp, Railway-car	W. H. Paige	Springfield, Mass	Jan. 16, 1872	122, 734
Lamp, Railway-car	T. S. Williams and P. S. Page	Boston, Mass	May 19, 1863	38, 620
Lamp, Railway-car coal-oil	W. O. B. Merrill	Philadelphia, Pa	June 10, 1862	35, 534
Lamp, Reflecting	J. C. Fletcher	Springfield, Ohio	Mar. 30, 1836	......
Lamp-reflector	J. W. Haines	Cambridge, Mass	July 11, 1871	116, 953
Lamp-reflector	F. Judson	Castleton, N. Y	Feb. 15, 1870	99, 908
Lamp, Reflector	J. H. Pease	Reading, Pa	June 8, 1852	9, 001
Lamp-reflector	J. T. Williams	Philadelphia, Pa	June 17, 1862	35, 644
Lamp-reflector, Adjustable	A. Haye	New York, N. Y	July 25, 1871	117, 282
Lamp reflector and chimney protector	G. F. J. Colburn	Newark, N. J	June 10, 1862	35, 507
Lamp-reservoirs, Placing	C. N. Orpen	New York, N. Y	July 29, 1862	36, 060
Lamp, Resin-oil	F. Blake	Needham, Mass	July 17, 1855	13, 259
Lamp, Resin-oil	S. Constant	Brooklyn, N. Y	Jan. 24, 1854	10, 443
Lamp, Resin-oil	S. Constant	Brooklyn, N. Y	Aug. 8, 1854	11, 474
Lamp, Resin-oil	I. Pitman	Reading, Mass	Sept. 19, 1854	11, 701
Lamp, Resin-oil	P. Sargent	Newburyport, Mass	Mar. 4, 1856	14, 369
Lamp, Resin-oil argand	I. Van Bunschoten	New York, N. Y	Mar. 18, 1856	14, 478
Lamp, Safety	J. Defossez	Paris, France	Sept. 2, 1862	36, 341
Lamp, Safety	A. H. Emery	New York, N. Y	Dec. 29, 1868	85, 373
Lamp, Safety	J. W. Hoffman	Southwark District, Pa	Apr. 23, 1850	7, 303
Lamp, Safety	C. R. Laudmann	New York, N. Y	Sept. 26, 1854	11, 730
Lamp, Safety	W. Pratt	Baltimore, Md	Nov. 24, 1857	18, 704
Lamp, Safety	F. Saunders	Aberdeen, Miss	May 4, 1869	89, 795
Lamp, Safety	C. Smith	Hermon, Me	Dec. 7, 1869	97, 560
Lamp safety-apparatus	C. J. Brown	Plymouth, N. H	Jan. 21, 1873	135, 074
Lamp safety-attachment	J. F. Sanford	Keokuk, Iowa	Apr. 27, 1869	89, 508
Lamp, Safety compass	J. Feinour, jr	Philadelphia, Pa	Apr. 5, 1828	......
Lamp, Safety derrick	J. Dillen	Petroleum Centre, Pa	May 3, 1870	102, 663
Lamp, Safety derrick	H. Freeman	Petroleum Centre, Pa	May 14, 1872	126, 688
Lamp safety-tube	G. M. Hopkins and J. A. Straight	Albion, N. Y	Nov. 29, 1870	109, 735
Lamp safety-tube	F. Stewart	Philadelphia, Pa	July 2, 1850	7, 484

Index of patents issued from the United States Patent Office from 1790 *to* 1873, *inclusive*—Continued.

Invention.	Inventor.	Residence.	Date.	No.
Lamp screw-collar	W. H. Russell	Brooklyn, N. Y	June 7, 1870	104, 066
Lamp, Self-acting blow-pipe	D. W. C. McCloskey	New York, N. Y	Aug. 26, 1851	8, 324
Lamp, Self-extinguishing	J. J. Cuthbert	Duck Port, La	June 18, 1872	127, 963
Lamp, Self-generating gas	S. Clayton and Y. Bailey	West Chester, Pa	Aug. 27, 1850	7, 594
Lamp, Self-lighting	W. W. Batchelder	Boston, Mass	May 2, 1871	114, 389
Lamp, Self-lighting	W. W. Batchelder	Boston, Mass	Oct. 3, 1871	119, 497
Lamp, Self-lighting	A. Bennett	New York, N. Y	Mar. 27, 1849	6, 244
Lamp, Self-lighting	T. W. Carroll	Baltimore, Md	Jan. 24, 1860	26, 884
Lamp, Self-lighting	W. A. Leonard	Boston, Mass	May 14, 1872	126, 640
Lamp, Self-supplying	E. B. Horn	Boston, Mass	Sept. 11, 1844	3, 735
Lamp-shade	J. B. Alexander	Washington, D. C	Mar. 30, 1869	88, 430
Lamp-shade	B. Arnold	East Greenwich, R. I	July 9, 1861	32, 740
Lamp-shade	D. W. Bashore	Palmyra, Pa	Aug. 27, 1867	68, 278
Lamp-shade	D. W. Bashore	Palmyra, Pa	June 2, 1868	78, 413
Lamp-shade	R. W. Churchill	Bridgeport, Conn	Sept. 21, 1869	95, 000
Lamp-shade	M. H. Collins	Chelsea, Mass	Apr. 25, 1871	113, 982
Lamp-shade	M. H. Collins	Chelsea, Mass	May 16, 1871	114, 766
Lamp-shade	M. H. Collins	Chelsea, Mass	July 25, 1871	117, 260
Lamp-shade	A. Combs	Burlingame, Kans	Aug. 27, 1872	130, 899
Lamp-shade	G. H. Dimond and G. Doolittle	Bridgeport, Conn	Feb. 9, 1869	86, 825
Lamp-shade	T. B. Doolittle	Bridgeport, Conn	Mar. 10, 1868	75, 390
Lamp-shade	J. V. Dunlap	Hartford, Conn	Nov. 13, 1866	59, 566
Lamp-shade	M. B. Dyott	Philadelphia, Pa	Jan. 16, 1855	12, 239
Lamp-shade	J. Emery	Bucksport, Me	July 9, 1867	66, 576
Lamp-shade	J. Emery	Bucksport, Me	Mar. 3, 1868	75, 136
Lamp-shade	F. E. Foster	Cressona, Pa	Jan. 19, 1869	85, 921
Lamp-shade	H. M. Hartshorn	Malden, Pa	Mar. 10, 1868	75, 418
Lamp-shade	A. W. Hasskell	Boston, Mass	Aug. 2, 1870	105, 940
Lamp-shade	E. K. Haynes	Hanover, N. H	Nov. 19, 1867	71, 168
Lamp-shade	E. K. Haynes	Hanover, N. H	May 19, 1868	78, 084
Lamp-shade	E. K. Haynes	Boston, Mass	Nov. 30, 1869	97, 294
Lamp-shade	A. B. Hendryx	Ansonia, Conn	June 8, 1869	91, 130
Lamp-shade	A. M. Hills	Hockanum, Conn	May 9, 1865	47, 638
Lamp-shade	C. E. L. Holmes	Waterbury, Conn	Apr. 1, 1862	34, 860
Lamp-shade	W. H. Horton	Chelsea, Mass	Feb. 18, 1868	74, 537
Lamp-shade	S. W. Huntington	Augusta, Me	Apr. 7, 1868	76, 462
Lamp-shade	S. W. Huntington	Augusta, Me	Apr. 21, 1868	76, 921
Lamp-shade	W. H. Johnson	Springfield, Mass	Mar. 8, 1870	100, 533
Lamp-shade	W. Kemble and W. H. C. Bartlett.	New York and West Point, N. Y.	Oct. 20, 1857	18, 456
Lamp-shade	A. D. Laws	Bridgeport, Conn	Feb. 16, 1869	86, 987
Lamp-shade	N. F. Leete	Middletown, Conn	Mar. 24, 1868	75, 932
Lamp-shade	E. A. Locke and W. N. Needen	Boston, Mass	Sept. 8, 1868	81, 915
Lamp-shade	G. Macbridge	Carthage, Ill	Sept. 10, 1872	131, 173
Lamp-shade	J. McFarland	Clinton, Ill	Mar. 20, 1866	53, 320
Lamp-shade	J. F. Phelps	Havana, N. Y	Mar. 2, 1869	87, 428
Lamp-shade	W. Reed jr	Boston, Mass	June 7, 1870	103, 928
Lamp-shade	W. Robinson	Spring Valley, N. Y	Mar. 1, 1870	100, 450
Lamp-shade	S. Rust	New York, N. Y	Jan. 9, 1838	565
Lamp-shade	W. F. Shaw	Boston, Mass	Nov. 26, 1867	71, 333
Lamp-shade	A. Shepard	Ashland, Mass	Dec. 18, 1866	60, 577
Lamp-shade	W. Simons	Charleston, S. C	Mar. 18, 1873	137, 034
Lamp-shade	A. A. Smith	Seneca Falls, N. Y	Apr. 21, 1868	77, 115
Lamp-shade	M. Stewart	Philadelphia, Pa	Feb. 20, 1866	52, 768
Lamp-shade	C. St. John	Boston, Mass	July 4, 1865	48, 632
Lamp-shade	C. St. John	Boston, Mass	Oct. 17, 1865	50, 509
Lamp-shade	J. F. Travis	New York, N. Y	May 18, 1869	90, 135
Lamp-shade	G. W. Tucker	Waterbury, Conn	May 4, 1869	89, 707
Lamp-shade	J. H. Webber	Charlestown, Mass	July 24, 1866	56, 647
Lamp-shade	G. Wedekind	Philadelphia, Pa	May 12, 1868	77, 936
Lamp-shade	G. Wedekind	Philadelphia, Pa	June 30, 1868	79, 526
Lamp-shade	G. Wedekind	Philadelphia, Pa	Nov. 30, 1869	97, 463
Lamp-shade	A. M. Weeks	New York, N. Y	June 16, 1868	79, 039
Lamp-shade	M. J. Williams and J. J. Greenough.	New York, N. Y	Feb. 28, 1865	46, 600
Lamp-shade	C. and A. C. Wilhelm	Philadelphia, Pa	May 3, 1859	23, 875
Lamp-shade	H. Zahn	New York, N. Y	Mar. 7, 1865	46, 748
Lamp shade and drip-cup, Combined	J. S. and T. B. Atterbury	Pittsburgh, Pa	Dec. 16, 1873	145, 548
Lamp shade and reflector	J. M. Kenerson	Newport, N. H	Aug. 9, 1870	106, 173
Lamp-shade clasp	J. J. Marcy	West Meriden, Conn	July 2, 1867	66, 366
Lamp-shade clasp	C. Reichman	Philadelphia, Pa	Feb. 4, 1862	34, 332
Lamp-shade clasp	C. Reichman	Philadelphia, Pa	Sept. 12, 1865	49, 921
Lamp-shade, Folding	H. M. Hartshorn	Malden, Mass	Jan. 3, 1871	110, 651
Lamp-shade holder	L. J. Atwood	Waterbury, Conn	May 30, 1865	47, 913
Lamp-shade holder	E. E. Conrad	Philadelphia, Pa	Apr. 8, 1862	34, 921
Lamp shade holder	E. E. Conrad	Philadelphia, Pa	Feb. 11, 1873	135, 778
Lamp-shade holder	J. Fallows	Philadelphia, Pa	July 5, 1864	43, 397
Lamp-shade holder	J. Foller	Washington, D. C	Sept. 6, 1870	107, 026
Lamp-shade holder	E. P. Gleason	New York, N. Y	Nov. 16, 1869	96, 909
Lamp-shade holder	J. Hanley	New York, N. Y	Apr. 25, 1865	47, 418
Lamp-shade holder	M. W. House	Cleveland, Ohio	Mar. 7, 1871	112, 459
Lamp-shade holder	M. W. House	Cleveland, Ohio	July 29, 1873	141, 223
Lamp-shade holder	J. W. Lyon	Brooklyn, N. Y	May 3, 1870	102, 691
Lamp-shade holder	J. W. Lyon	Brooklyn, N. Y	Mar. 28, 1871	113, 181
Lamp-shade holder	W. C. Matthews	Chelsea, Mass	Mar. 25, 1862	34, 805
Lamp-shade holder	C. M. Mitchell	Waterbury, Conn	May 18, 1869	90, 287
Lamp-shade holder	T. B. Peacock	Dresden, Ohio	Aug. 30, 1870	106, 866
Lamp-shade holder	O. N. Perkins	Meriden, Conn	Dec. 10, 1872	133, 885
Lamp-shade holder	J. T. Pope	New York, N. Y	Mar. 23, 1869	88, 206
Lamp-shade holder	J. T. Pope	New York, N. Y	May 11, 1869	90, 021
Lamp-shade holder	C. H. Reichmann	New York, N. Y	Sept. 30, 1862	36, 574
Lamp-shade holder	P. A. Stecher	New York, N. Y	Aug. 26, 1862	36, 308
Lamp-shade holder	C. St. John	Boston, Mass	Feb. 21, 1865	46, 534
Lamp-shade holder	C. St. John	New York, N. Y	Jan. 28, 1870	104, 898
Lamp shade or globe holder	L. J. Atwood	Waterbury, Conn	July 29, 1873	141, 251
Lamp-shade panel	G. W. Bruiker	Philadelphia, Pa	Feb. 12, 1867	61, 994
Lamp-shade protector	C. and A. C. Wilhelm	Philadelphia, Pa	Apr. 3, 1855	12, 646
Lamp-shade support	J. B. Alexander	Washington, D. C	Apr. 5, 1870	101, 409

Index of patents issued from the United States Patent Office from 1790 *to* 1873, *inclusive*—Continued.

Invention.	Inventor.	Residence.	Date.	No.
Lamp-shade support	J. H. Connelly	Wheeling, W. Va	July 3, 1866	56, 011
Lamp-shade support	E. E. Conrad	Philadelphia, Pa	June 21, 1870	104, 558
Lamp-shade support	C. W. Emerson	Hartford, Conn	Oct. 5, 1869	95, 575
Lamp-shade support	G. W. Hatch	Brooklyn, N. Y	June 25, 1872	128, 308
Lamp-shade support	E. W. Holt	Corinna, Me	June 7, 1870	103, 885
Lamp-shade support	J. W. Lyon	Brooklyn, N. Y	Apr. 21, 1868	77, 059
Lamp-shade support	G. C. James	Cincinnati, Ohio	Oct. 22, 1867	70, 094
Lamp-shade support	C. Reichmann	Philadelphia, Pa	Oct. 20, 1863	40, 361
Lamp-shade support	E. Russell	Waterbury, Conn	May 30, 1871	115, 527
Lamp-shade support	W. F. Shaw	Boston, Mass	Dec. 14, 1858	22, 311
Lamp-shade supports, Process of making	W. N. Weeden	Waterbury, Conn	Dec. 16, 1873	145, 702
Lamp-shades, Incombustible paper	G. Wedekind	Philadelphia, Pa	May 12, 1863	38, 523
Lamp-shades, Manufacture of	J. H. Hobbs, C. W. Brockemier, and W. Leighton, jr.	Wheeling, W. Va	Apr. 25, 1871	114, 140
Lamp-shades, Mode of packing	E. A. Locke and W. N. Weeden	Boston, Mass	Nov. 24, 1868	84, 289
Lamp, Shoemaker's	J. M. Sparer	Philadelphia, Pa	Dec. 3, 1867	71, 658
Lamp sign, Street	H. D. Coleman	New Orleans, La	Oct. 1, 1872	131, 741
Lamp, Signal	R. P. Bailey	Niagara, N. Y	May 12, 1857	17, 253
Lamp, Signal	J. and J. Feinour, jr	Philadelphia, Pa	Aug. 1, 1828	....
Lamp, Signal	J. Wall	New York, N. Y	June 25, 1867	66, 191
Lamp, Spirit	A. J. Walker	New York, N. Y	May 24, 1853	9, 751
Lamp-snuffer	R. N. Eagle	New York, N. Y	Sept. 30, 1862	36, 5[illegible]0
Lamp-snuffer	K. Schow	La Fayette, Ind	Apr. 10, 1866	53, 890
Lamp snuffer and extinguisher	M. L. Battle	Bainbridge, Ga	May 7, 1872	126, 437
Lamp, Solar	J. Hassell	Brooklyn, N. Y	June 9, 1857	17, 507
Lamp-stand and clothes-dryer, Combined	J. Donaldson	Rockford, Ill	Aug. 8, 1865	49, 244
Lamp-stands, &c., Mode of ornamenting	H. W. Hayden	Waterbury, Conn	June 9, 1863	38, 823
Lamp, Street	T. P. Austin	New York, N. Y	Aug. 30, 1870	106, 764
Lamp, Street	J. Benson	Yonkers, N. Y	May 9, 1871	114, 518
Lamp, Street	E. Boesch	San Francisco, Cal	Sept. 6, 1870	106, 990
Lamp, Street	C. J. Bowell	Cleveland, Ohio	June 25, 1872	128, 355
Lamp, Street	G. Brandon	New York, N. Y	Mar. 21, 1871	112, 774
Lamp, Street	A. Burger	New York, N. Y	May 14, 1872	126, 783
Lamp, Street	O. Case and B. D. Evans	Columbus, Ohio	Dec. 1, 1868	84, 535
Lamp, Street	C. Cuppers	New York, N. Y	Oct. 4, 1870	107, 884
Lamp, Street	M. B. Dyott	Philadelphia, Pa	Mar. 19, 1872	124, 673
Lamp, Street	M. B. Dyott	Philadelphia, Pa	Apr. 8, 1873	137, 538
Lamp, Street	J. H. Eyrse	Pekin, Ill	Sept. 23, 1873	143, 000
Lamp, Street	J. S. Fish	Cleveland, Ohio	June 11, 1872	127, 685
Lamp, Street	B. Giroux	Chicago, Ill	July 16, 1872	129, 221
Lamp, Street	J. S. Hagerty	Baltimore, Md	Sept. 9, 1873	142, 693
Lamp, Street	J. F. Harly	Kipton Station, Ohio	Feb. 1, 1870	99, 314
Lamp, Street	F. Hartman	Chicago, Ill	Mar. 4, 1873	136, 373
Lamp, Street	C. F. Jacobsen and A. Burger	New York, N. Y	May 27, 1873	139, 394
Lamp, Street	J. F. Kerno	Baltimore, Md	Jan. 10, 1871	110, 922
Lamp, Street	J. H. Kramer and A. Burger	New York, N. Y	Nov. 9, 1869	96, 707
Lamp, Street	F. Lange	Chicago, Ill	Dec. 15, 1868	84, 952
Lamp, Street	T. T. Markland	Philadelphia, Pa	June 19, 1866	55, 686
Lamp, Street	T. T. Markland, jr	Philadelphia, Pa	Apr. 7, 1868	76, 335
Lamp, Street	J. F. Marsh	Dubuque, Iowa	June 24, 18[illegible]3	140, 149
Lamp, Street	A. J. McDowell and D. T. Bates	Richmond, Ind	July 23, 1872	129, 846
Lamp, Street	J. G. Miner	Morrisania, N. Y	May 29, 1866	55, 143
Lamp, Street	J. G. Miner	Morrisania, N. Y	June 17, 1873	140, 062
Lamp, Street	A. Nahe	Brooklyn, N. Y	Sept. 2, 1873	142, 500
Lamp, Street	T. North	Cincinnati, Ohio	June 27, 1871	116, 476
Lamp, Street	J. H. Robinson	Washington, D. C	May 7, 1872	126, 577
Lamp, Street	W. G. Schmidlen and J. W. Driscoll.	New York, N. Y	Apr. 12, 1870	101, 769
Lamp, Street	W. G. Schmidlin and J. W. Driscoll.	Hoboken, N. J., and New York, N. Y.	July 25, 1871	117, 470
Lamp, Street	W. G. Schmidlin and J. W. Driscoll.	New York, N. Y	Feb. 6, 1872	123, 516
Lamp, Street	F. Schumann	Washington, D. C	Dec. 3, 1872	133, 672
Lamp, Street	T. A. Skelton	Middlesex County, England.	Jan. 2, 1872	122, 409
Lamp, Street	R. H. Smith	Pittsburgh, Pa	June 20, 1871	116, 234
Lamp, Street	J. Stratton	Brooklyn, N. Y	June 27, 1865	48, 461
Lamp, Street	W. O. Wheeler	Deposit, N. Y	June 23, 1868	79, 1[illegible]9
Lamp, Street gas	W. A. Shaw and G. Parker	Boston, Mass	Apr. 11, 1854	10, 760
Lamp, Submarine	L. H. Hasse	New York, N. Y	Oct. 9, 1860	30, 314
Lamp-supporter	E. Russell and F. W. Platt	Waterbury, Conn	June 2[illegible], 1871	116, 224
Lamp, Suspended bunker	H. J. Van Thiel	Stapleton, N. Y	Jan. 5, 1864	41, 108
Lamp, Suspension barrel	I. Couch and J. A. Frary	Meriden, Conn	Mar. 13, 1833	....
Lamp, Tallow	J. Love	....	June 11, 1798	....
Lamp, Torch	H. Brewer	East Boston, Mass	Jan. 10, 1854	10, 419
Lamp-top	C. Dorflinger	Brooklyn, N. Y	Aug. 5, 1862	36, 079
Lamp-top	F. Hewitt	Newark, N. J	June 9, 1863	38, 826
Lamp-top	S. E. Winslow	Kensington, Pa	Oct. 24, 1848	5, 880
Lamp tops, rivets, &c., Method of making	L. C. White	Meriden, Conn	Sept. 7, 1852	9, 256
Lamp-train gearing	R. Hitchcock and G. A. Jones	New York, N. Y	Feb. 25, 1868	74, 914
Lamp trimmer and extinguisher	C. P. Snow	Freeport, Ill	Nov. 30, 1869	97, 453
Lamp-trimmer's shears	W. B. Barnard	Waterbury, Conn	Dec. 27, 1864	45, 574
Lamp-tube	I. Jennings	New York, N. Y	July 9, 1850	7, 492
Lamp-tube-forming machine	W. Wallace	Ansonia, Conn	June 16, 1868	79, 036
Lamp-tubes, Device for adjusting wicks in	H. A. De Graw	Newburgh, N. Y	Jan. 27, 1863	37, 495
Lamp-tubes, Method of forming	T. B. Doolittle	Bridgeport, Conn	July 28, 1868	80, 464
Lamp, Turpentine	L. Jones	New York, N. Y	Nov 25, 1838	1, 022
Lamp, Vapor	C. E. Atherton	Paterson, N. J	Nov. 6, 1860	30, 556
Lamp, Vapor	H. Bateman	Boston, Mass	May 25, 1858	20, 386
Lamp, Vapor	W. B. Billings	New York, N. Y	Oct. 2, 1860	30, 194
Lamp, Vapor	M. V. B. Buell	Buffalo, N. Y	May 1, 1860	28, 003
Lamp, Vapor	M. L. Callender	New York, N. Y	Nov. 26, 1861	33, 773
Lamp, Vapor	J. Clarke	Syracuse, N. Y	June 12, 1860	28, 652
Lamp, Vapor	T. G. Clayton	Washington, D. C	May 15, 1860	28, 258
Lamp, Vapor	N. Clifford	Brooklyn, N. Y	July 9, 1861	32, 755
Lamp, Vapor	C. C. Coe	Rome, N. Y	July 9, 1861	32, 808
Lamp, Vapor	L. F. Conover	Philadelphia, Pa	Feb. 28, 1860	27, 272
Lamp, Vapor	E. Crooker	Buffalo, N. Y	July 10, 1860	29, 124

Index of patents issued from the United States Patent Office from 1790 *to* 1873, *inclusive*—Continued.

Invention.	Inventor.	Residence.	Date.	No.
Lamp, Vapor	M. W. Dillingham	Charlestown, Mass	Nov. 6, 1860	30, 568
Lamp, Vapor	H. W. Dopp	Buffalo, N. Y	Nov. 13, 1860	30, 621
Lamp, Vapor	T. Drake	Windsor, Conn	Mar. 3, 1863	37, 811
Lamp, Vapor	D. A. Fiske	Milwaukee, Wis	Jan. 7, 1862	34, 053
Lamp, Vapor	A. Geiger	Dayton, Ohio	May 29, 1860	28, 467
Lamp, Vapor	J. G. Gilbert	New York, N. Y	Sept. 8, 1859	18, 142
Lamp, Vapor	J. S. Gray	New York, N. Y	Dec. 18, 1860	30, 965
Lamp, Vapor	H. Johnson	Washington, D. C	Feb. 28, 1860	27, 293
Lamp, Vapor	H. Johnson	Washington, D. C	Apr. 17, 1860	27, 912
Lamp, Vapor	C. B. Loveless	Tom's River, N. J	Jan. 22, 1861	31, 179
Lamp, Vapor	S. W. Lowe	Philadelphia, Pa	May 29, 1860	28, 536
Lamp, Vapor	A. M. Mace	Springfield, Mass	June 22, 1858	20, 649
Lamp, Vapor	H. N. Macomber	Lynn, Mass	May 18, 1858	20, 283
Lamp, Vapor	J. K. O'Neil	Kingston, N. Y	Nov. 22, 1859	26, 200
Lamp, Vapor	I. W. Pettibone	Norfolk, Conn	May 15, 1860	28, 299
Lamp, Vapor	W. H. Racey	Saint Augustine, Fla	June 29, 1858	20, 729
Lamp, Vapor	T. S. Ray and A. C. Rand	Buffalo, N. Y	May 29, 1860	28, 503
Lamp, Vapor	C. W. Richter	Madison, Ga	June 12, 1860	28, 691
Lamp, Vapor	J. J. Riddle	Cincinnati, Ohio	Feb. 7, 1865	46, 266
Lamp, Vapor	J. H. Rollins	Wapello, Iowa	May 15, 1860	28, 306
Lamp, Vapor	L. Short	Buffalo, N. Y	Jan. 29, 1861	31, 282
Lamp, Vapor	I. Van Bunschoten	New York, N. Y	July 31, 1860	29, 421
Lamp, Vapor	G. Walker	Philadelphia, Pa	June 26, 1860	28, 924
Lamp, Vapor	F. W. Willard	New York, N. Y	Dec. 11, 1860	30, 899
Lamp, Vapor-burning	D. H. Chamberlain	West Roxbury, Mass	Nov. 24, 1857	18, 719
Lamp, Vapor-burning	N. Mason	Chelsea, Mass	July 20, 1858	20, 952
Lamp, Vapor-burning	S. Whitmarsh	Northampton, Mass	Aug. 12, 1856	15, 547
Lamp, Vapor-burning street	C. E. Smith and H. J. Rice	Columbus, Ohio	Aug. 23, 1870	106, 630
Lamp, Vehicle	S. P. Dodge	Boston, Mass	June 5, 1866	55, 255
Lamp, Vehicle	C. F. Waldron	New York, N. Y	Apr. 14, 1868	76, 652
Lamp ventilator and shade	C. Recht	New York, N. Y	July 31, 1866	56, 799
Lamp ventilator, Petroleum-oil	S. D. Pinkham	Fond du Lac, Wis	Apr. 22, 1862	35, 039
Lamp, Vulcanizing	B. W. Franklin	New York, N. Y	Feb. 3, 1863	37, 575
Lamp-wick	P. Backer	Chicago, Ill	Dec. 20, 1870	110, 188
Lamp-wick	T. Bingham	Newburgh, N. Y	Dec. 13, 1864	45, 384
Lamp-wick	E. D. Boyd	Helena, Ark	Sept. 14, 1869	94, 707
Lamp-wick	J. H. Connelly	Wheeling, W. Va	Jan. 6, 1863	37, 276
Lamp-wick	J. H. Connelly and J. Cook	Wheeling, Va	Aug. 26, 1862	36, 277
Lamp-wick	J. Farrell	Brooklyn, N. Y	May 23, 1871	115, 045
Lamp-wick	W. A. Gensch	New York, N. Y	Aug. 25, 1868	81, 359
Lamp-wick	I. L. Hoard	Bristol, R. I	Nov. 20, 1866	59, 839
Lamp-wick	C. P. Ladd	Bloomfield, N. J	Feb. 20, 1872	123, 917
Lamp-wick	C. B. Larchar	New York, N. Y	July 7, 1863	39, 153
Lamp-wick	C. W. Le Count	Norwalk, Conn	Oct. 30, 1866	59, 237
Lamp-wick	C. W. Le Count and S. Chard	Norwalk, Conn	Feb. 27, 1866	52, 862
Lamp-wick	J. Y. Leslie	Brooklyn, N. Y	Oct. 26, 1858	21, 890
Lamp-wick	P. Martin	Medford, Mass	Dec. 16, 1873	145, 667
Lamp-wick	C. F. Martine	Boston, Mass	Mar. 5, 1867	62, 657
Lamp-wick	F. McKee	Pittsburgh, Pa	Dec. 23, 1862	37, 237
Lamp-wick	P. Noyes	Lowell, Mass	Dec. 19, 1865	51, 611
Lamp-wick	H. T. Robbins	Hyde Park, Mass	May 30, 1871	115, 524
Lamp-wick	D. T. Robinson	Boston, Mass	Dec. 27, 1870	110, 591
Lamp-wick	S. Rust	New York, N. Y	May 10, 1845	4, 040
Lamp-wick	E. O. Schartau	Philadelphia, Pa	May 12, 1868	77, 767
Lamp-wick	P. J. Skinner	Oswego, N. Y	Sept. 13, 1870	107, 299
Lamp-wick	C. L. Topliff	New York, N. Y	June 19, 1866	55, 745
Lamp-wick	B. F. Walton	Philadelphia, Pa	Jan. 23, 1866	52, 231
Lamp-wick	A. J. White	Brooklyn, N. Y	Apr. 28, 1863	38, 349
Lamp-wick	J. B. Wortendyke	Godwinsville, N. J	Apr. 26, 1859	23, 801
Lamp-wick, Adjustable	G. Finley	Collins Township, Pa	Apr. 28, 1863	38, 302
Lamp-wick adjuster	L. B. Lathrop	San José, Cal	Sept. 26, 1871	119, 275
Lamp-wick adjuster	C. C. Stansell	Middleborough, Mass	Feb. 13, 1866	52, 618
Lamp-wick adjuster	H. Woodbury	Huntsville, Mo	Apr. 23, 1872	126, 002
Lamp-wick, Automatic lighting	W. H. Weeks	New York, N. Y	May 17, 1870	103, 110
Lamp-wicks, Feed-roller for	C. H. Bagley	Elgin, Ill	June 18, 1867	65, 861
Lamp-wick feeder	G. Cade	Long Branch, N. J	Apr. 5, 1870	101, 427
Lamp-wick gage	J. A. Paddock	Chandlerville, Ill	Dec. 24, 1872	134, 161
Lamp wick-holder, Argand	C. Moeller	Newark, N. J	Feb. 12, 1856	14, 248
Lamp-wick inserter	W. Y. A. Boardman	New Haven, Conn	Dec. 4, 1866	60, 130
Lamp-wick raiser	A. Combs	Philadelphia, Pa	Feb. 18, 1873	136, 034
Lamp-wick regulator	H. Beebe	Jersey City Heights, N. J	May 16, 1871	114, 906
Lamp-wick regulator	L. W. Leary	Norfolk, Va	Sept. 20, 1870	107, 617
Lamp-wick regulator	H. Mund and E. Hoffmann	Chicago, Ill	July 10, 1866	56, 250
Lamp-wick regulator	J. Pomeroy	Derby, Conn	June 16, 1863	38, 933
Lamp-wick trimmer	H. Halvorsen	Cambridge, Mass	July 19, 1859	24, 804
Lamp-wick trimmer	E. C. Jenkins, jr	Worcester, Mass	Dec. 14, 1869	97, 776
Lamp-wick trimmer	S. Naylor and A. Fairchild	Independence, Iowa	Aug. 26, 1873	142, 265
Lamp-wick trimmer	J. F. Sanford	Keokuk, Iowa	Feb. 9, 1869	86, 868
Lamp-wick trimmer	D. Warner	Boston, Mass	Aug. 4, 1868	80, 577
Lamp-wick trimmer, chimney cleaner and lifter, Combined.	C. M. Tyler	Indianapolis, Ind	July 20, 1869	92, 908
Lamp-wick tube	F. H. Fuller and O. S. Severance	Boston, Mass	Dec. 1, 1868	84, 542
Lamp-wick tube	J. H. Gray	Boston, Mass	Mar. 1, 1870	100, 285
Lamp-wick tube	E. K. Haynes	Boston, Mass	June 6, 1871	115, 607
Lamp-wick tube	A. D. Laws	Bridgeport, Conn	Mar. 10, 1868	75, 281
Lamp-wick tube	E. W. Perry	Boston, Mass	Oct. 12, 1842	2, 811
Lamp-wick tube	S. Rust	New York, N. Y	Mar. 21, 1846	4, 428
Lamp wick-tubes, Camphene	J. Maclean	Philadelphia, Pa	Dec. 26, 1845	4, 326
Lamp-wicks, Chemically-prepared	A. M. Daniels	Hartford, Conn	Apr. 4, 1871	113, 269
Lamp-wicks, Device for raising and adjusting	J. B. Alexander	Washington, D. C	Nov. 17, 1868	84, 045
Lamp-wicks, Device for trimming	H. F. Bond	Waltham, Mass	Jan. 3, 1865	45, 692
Lamp-wicks, Elevator-tube for	R. Cornelius and C. Wilhelm	Philadelphia, Pa	July 24, 1849	6, 6[illegible]3
Lamp-wicks, Machine for making	J. H. Connelly	Wheeling, W. Va	Jan. 14, 1864	43, 097
Lamp-wicks, Process for treating	S. F. Sutherland	Baltimore, Md	Jan. 28, 1873	135, 252
Lamp-wicks, Raising	S. Rust	New York, N. Y	Mar. 9, 1844	3, 467
Lamp-wicks, Raising and lowering	S. Rust	New York, N. Y	Dec. 16, 1835	....
Lamp-wicks, Tube for raising	R. C. Overton	New York, N. Y	Nov. 7, 1848	5, 91[illegible]
Lamp with argand burner, Lard	R. Cornelius	Philadelphia, Pa	Apr. 6, 1843	3, 028

Index of patents issued from the United States Patent Office from 1790 *to* 1873, *inclusive*—Continued.

Invention.	Inventor.	Residence.	Date.	No.
Lamp without chimney, Coal-oil	J. W. Schreiber	New York, N. Y	Mar. 24, 1863	37, 983
Lamp, Wooden	J. H. Mather	Saybrook, Conn	July 20, 1831	....
Lamps and economy of light, Construction of	I. Jennings	New York, N. Y	June 11, 1829	....
Lamps and other lights, Chimney and shade for	M. J. Wellman and J. J. Greenough.	New York, N. Y	Dec. 1, 1863	40, 785
Lamps, Apparatus for extinguishing street gas	G. S. Dunbar	Pittsfield, Mass	Oct. 3, 1871	119, 455
Lamps, Apparatus for supplying naphtha to vapor-burning street.	F. M. Randell	Greenburgh, N. Y	Jan. 10, 1871	110, 860
Lamps, Attachment for converting other fluid into coal-oil.	A. G. Tisdel and W. Nash	Watertown, N. Y	Mar. 24, 1863	37, 987
Lamps, binnacles of ships, &c., Lighting street	J. E. Smith	Baltimore, Md	June 1, 1825	....
Lamps by electricity, Method of lighting street	C. W. Smith	Evans, N. Y	Oct. 12, 1858	21, 781
Lamps, Cap of glass	F. Draper	East Cambridge, Mass	Jan. 17, 1842	2, 424
Lamps, Case for ratchet-wheel for	L. Hover	Chicago, Ill	Aug. 18, 1863	39, 574
Lamps, Coating	G. W. Thomson	Buffalo, N. Y	Apr. 4, 1871	113, 365
Lamps, Combination of globe and chimney for	E. B. Regna	Jersey City, N. J	Feb. 24, 1863	37, 773
Lamps, Composition for burning in	I. Jennings	New York, N. Y	Dec. 31, 1839	1, 453
Lamps, Composition for supplying	H. Porter	Bangor, Me	Apr. 8, 1835	....
Lamps, Construction of	J. J. Hoyt and J. E. Crane	Chelmsford and Lowell, Mass.	Mar. 14, 1871	112, 598
Lamps, Construction of	C. West	Baltimore, Md	Oct. 7, 1844	3, 781
Lamps, Construction of burning-fluid	D. F. Randall	Chicopee, Mass	Jan. 13, 1857	16, 398
Lamps, Detachable fluid-reservoir for street	L. A. Gouch	Yonkers, N. Y	May 9, 1871	114, 550
Lamps, Device for securing collars to	E. S. Kennedy	Birmingham, (Buchanan Post-Office,) Pa.	May 9, 1871	114, 680
Lamps, Device for snuffing	J. Ives	Brooklyn, N. Y	July 30, 1867	67, 309
Lamps, Device for suspending	D. F. Hubbell	Bethel, Conn	Apr. 30, 1867	64, 319
Lamps, Electric fan for	C. T. Mason	Sumter, S. C	June 9, 1868	78, 674
Lamps, Fixture for ratchet-wheel for	H. Topping and M. Gally	Marion and Auburn, N. Y.	July 3, 1866	56, 126
Lamps, Flame-contracting cap for night	J. Christison	New York, N. Y	Apr. 12, 1864	42, 328
Lamps, Flame regulator and extinguisher for	W. H. H. Hinds	Groton, Mass	May 14, 1867	64, 671
Lamps, gas-burners, &c., Apparatus for lighting	P. B. Tyler, W. M. Chandler, and L. F. Standish.	Chicopee and Springfield, Mass.	Jan. 15, 1867	61, 284
Lamps, Generator for vapor	W. S. Mead	Buffalo, N. Y	Feb. 7, 1860	27, 061
Lamps, Glass deflector for	W. H. Matthews	Chelsea, Mass	Apr. 22, 1862	35, 061
Lamps, Glass globe for	G. M. Irwin	Birmingham, Pa	Nov. 7, 1871	120, 747
Lamps, Glass globe or reservoir for	A. Otto	Braunfels, Tex	Apr. 16, 1872	125, 754
Lamps, Heating-attachment for oil	E. O. Schartau	Philadelphia, Pa	May 28, 1867	65, 286
Lamps, Hinged collar for	G. F. J. Colburn	Newark, N. J	July 14, 1863	39, 253
Lamps, &c., Indicating filler for	J. A. Ryan	Cleveland, Ohio	Jan. 14, 1873	134, 819
Lamps, lanterns, &c., Glass lights or panes for	B. B. Schneider	New York, N. Y	Feb. 27, 1872	124, 016
Lamps, Lantern globe or shade for street	J. N. Aronson	New York, N. Y	Sept. 24, 1872	131, 649
Lamps, Lenticular globe for	G. W. Morrison	New Albany, Ind	May 9, 1871	114, 701
Lamps, Lighting	E. Hubbard and W. L. Cheney	Chester, Mass	Mar. 14, 1834	....
Lamps, Lighting and trimming	J. Gallagher	Brooklyn, N. Y	June 24, 1862	35, 733
Lamps, Lighting street	H. Elliot	New York, N. Y	Aug. 11, 1863	39, 471
Lamps, Lock-fastener for	J. Harding	Warrington, Great Britain.	Dec. 17, 1867	73, 196
Lamps, Making glass	P. F. Slave and J. Golding	East Cambridge, Mass	Jan. 23, 1845	3, 892
Lamps, Making reservoir of metallic	P. J. Clark	Meriden, Conn	May 28, 1850	7, 393
Lamps, Manufacture of	C. T. Close	New York, N. Y	May 27, 1[illegible]	[illegible]
Lamps, Mechanical movement for	F. B. De Keravenan	New York, N. Y	Feb. 10, 1863	37, 659
Lamps, Mechanical movement for	F. B. De Keravenan	New York, N. Y	June 9, 1863	38, 859
Lamps, Mode of attaching chimneys to	W. Morehouse	Buffalo, N. Y	Apr. 1, 1862	34, 842
Lamps, Mode of attaching chimneys to	H. H. Swift	Hart's Village, N. Y	Jan. 6, 1863	37, 366
Lamps, Mode of attaching extinguishers to	F. A. Jewett	Abington, Mass	Dec. 11, 1855	13, 910
Lamps, Mode of hanging torch	C. H. Cooper	New York, N. Y	Sept. 18, 1860	30, 048
Lamps, Mode of lighting street	H. Lenox	Trenton, N. J	Apr. 26, 1870	102, 284
Lamps, Mode of manufacturing	H. Whitney and T. Leighton	Cambridge, Mass	Jan. 11, 1839	1, 068
Lamps, Mode of removing chimney and filling	J. R. Baker	Kendallville, Ind	June 24, 1862	35, 664
Lamps, Mold for pressing glass fountain	H. W. Adams	New York, N. Y	Aug. 19, 1856	15, 548
Lamps, Mode of preventing explosion of	C. P. Grosvenor	McGrawville, N. Y	Dec. 10, 1867	71, 872
Lamps, Mode of preventing the explosion of	E. Howard	Redhill, England	Nov. 12, 1867	70, 721
Lamps, Name-plate for street	T. T. Markland, jr	Philadelphia, Pa	Apr. 7, 1868	76, 334
Lamps, Name-plate for street	C. J. O'Hara	New Orleans, La	Sept 29, 1868	82, 544
Lamps or torches, Frame for swinging	L. T. Pitkin	Hartford, Conn	Sept. 18, 1860	30, 089
Lamps, Packing-tube for vapor	W. E. Jervey	New Orleans, La	Apr. 4, 1871	113, 524
Lamps, Raising and lowering signal	T. G. Crosby	Buffalo, N. Y	June 20, 1865	48, 334
Lamps, Reflector and elastic tube for	J. and J. Flagg, jr	Boston, Mass	May 17, 1810	....
Lamps, Regulating the flame of	J. S. Trough	Baltimore, Md	July 17, 1839	1, 246
Lamps, Regulating the flow of oil to wick in carcel	A. Coates	New York, N. Y	Mar. 25, 1856	14, 492
Lamps, Regulating the interior draft of	E. Whelan	Philadelphia, Pa	Mar. 26, 1845	3, 966
Lamps, Safety-apparatus for	C. Applebee	Lyndon, Vt	Nov. 17, 1868	84, 078
Lamps, Safety-device for	H. W. M. Washington	Green Plains, Va	Oct. 18, 1870	108, 416
Lamps, Sign for street	W. H. Bell and J. G. Jory	Baltimore, Md	Feb. 6, 1872	123, 324
Lamps, Sign for street	J. T. Foley	New York, N. Y	Dec. 19, 1871	121, 938
Lamps, Sign for street	W. Graham, W. Snyder, and P. O'Brien.	Pittsburgh, Pa	Dec. 5, 1871	121, 614
Lamps, Sign for street	W. P. Ware	New York, N. Y	Mar. 3, 1868	75, 086
Lamps, Slide-pendant for hanging	J. A. Evarts	Meriden, Conn	Dec. 16, 1873	145, 637
Lamps, Spring-catch for	D. A. Draper	Cambridge, Mass	July 7, 1863	39, 131
Lamps, Supplying oil to street	R. F. De Guinon	Jersey City, N. J	July 15, 1873	140, 815
Lamps, &c., Suspension-device for	H. Weed	New Haven, Conn	Nov. 13, 1866	59, 689
Lamps, Suspension-spring for	J. A. Evarts	West Meriden, Conn	Apr. 7, 1868	76, 317
Lamps, Thick tube for	H. W. Hayden	Waterbury, Conn	Mar. 31, 1863	38, 079
Lamps to lanterns, Fastening	E. Sirret	Buffalo, N. Y	Oct. 24, 1854	11, 843
Lamps to lanterns, Method of attaching	J. Fleming	Pittsburgh, Pa	July 6, 1858	20, 785
Lamps to lanterns, Method of fastening	E. Sirret and W. H. Scott	Buffalo, N. Y	Apr. 22, 1856	14, 741
Lamps to lanterns, Securing	W. Porter	Williamsburgh, N. Y	Oct. 24, 1854	11, 849
Lamps, Train for	G. A. Jones	New York, N. Y	Feb. 18, 1868	74, 695
Lamps, Tube for glass	D. Jarves	Cambridge, Mass	Feb. 2, 1822	....
Lamps, Wind-guard and air-heater for	J. B. Capewell	Gloucester, N. J	Oct. 24, 1865	50, 560
Lance, Bomb	O. Allen	San Francisco, Cal	Oct. 27, 1863	40, 387
Lance, Bomb	A. F. and J. H. Andrews	Avon, Conn	Nov. 16, 1858	22, 054
Lance, Bomb	C. C. Brand	Ledyard, Conn	June 22, 1852	9, 047
Lance, Bomb	I. Goodspeed	Norwich, Conn	Aug. 9, 1859	25, 080
Lance, Bomb	J. Grudchos and S. Eggers	New Bedford, Mass	May 26, 1857	17, 370
Lance, Bomb	Z. Kelley	New Bedford, Mass	June 9, 1868	78, 673
Lance, Bomb	E. Pierce	Hallowell, Me	Aug. 22, 1865	49, 548

Index of patents issued from the United States Patent Office from 1790 *to* 1873, *inclusive*—Continued.

Invention.	Inventor.	Residence.	Date.	No.
Lance, Bomb	E. Pierce	Hallowell, Me	June 1, 1869	90, 868
Lance, Bomb	R. Sibley	Greenville, Conn	Apr. 28, 1857	17, 173
Lance, Bomb	R. Sibley	Greenville, Conn	Aug. 17, 1858	21, 219
Lance, Harpoon	N. Scholfield	Norwich, Conn	Aug. 24, 1858	21, 278
Lance, Whaling	O. Allen	Norwich, Conn	Sept. 19, 1846	4, 764
Lances, Cushion for wings of	N. Scholfield	Norwich, Conn	Dec. 8, 1857	18, 824
Lancet	H. Mellish	Walpole, N. H	Apr. 3, 1855	12, 636
Lancet	S. Wilmot	Bridgeport, Conn	Oct. 25, 1832	....
Lancet, Retreating-spring	T. R. Williams	Newport, R. I	July 16, 1824	....
Lancet, Revolving	T. C. Harrison	New Egypt, N. J	June 20, 1836	....
Lancet, Spring	G. J. Capewell	West Cheshire, Conn	June 19, 1866	55, 620
Lancet, Spring	J. Dewey	Chelsea, Vt	Apr. 2, 1824	....
Lancet, Spring	J. H. Genrig	Philadelphia, Pa	Apr. 11, 1846	4, 450
Lancet, Spring	J. W. W. Gordon	Catonsville, Md	Jan. 27, 1857	16, 479
Lancet, Spring	J. Ives	Bristol, Conn	Mar. 27, 1849	6, 240
Lancet, Spring	J. H. Johnson	Saint Louis, Mo	Apr. 10, 1849	6, 288
Lancet, Spring	W. Parkinson	Marshall County, Va	Aug. 11, 1857	17, 994
Lancet, Spring	J. M. Van Osdel	Chicago, Ill	Apr. 24, 1841	2, 061
Land and sea, Ascertaining position and direction on.	B. Garvey	Ashland, N. Y	Feb. 4, 1862	34, 298
Land-clearing machine	T. Oxley	Norfolk, Va	Apr. 18, 1821	....
Land-conveyance	L. F. Hake	Salem, Ohio	Feb. 19, 1867	62, 264
Land-conveyance	R. J. Nunn	Savannah, Ga	Apr. 23, 1867	64, 134
Land-conveyance brake	W. T. Kosinski	Philadelphia, Pa	Oct. 18, 1870	108, 366
Land-conveyance brake	W. T. Koskinski	Philadelphia, Pa	Dec. 3, 1872	133, 587
Land-conveyance, Propeller for	P. Dickson	Utica, Minn	Nov. 25, 1862	36, 995
Land, Device for cutting marshy	A. McFarlane	South Genesee, Wis	Apr. 15, 1862	34, 974
Land-leveler	F. Monroe	Bruce, Mich	Apr. 17, 1866	54, 003
Land-leveler	R. Shepard	Shaker Village, N. H	Apr. 1, 1862	34, 851
Land, Machine for fallowing	R. J. Gatling	Indianapolis, Ind	Jan. 27, 1857	16, 476
Land, Machine for marking and furrowing	J. R. Dikeman and J. J. Hewlett	Hempstead, N. Y	Apr. 29, 1862	35, 087
Land-roller	W. W. Andrew	La Porte, Ind	Sept. 26, 1871	119, 294
Land-roller	E. P. H. Capron	Springfield, Ohio	Jan. 5, 1869	85, 641
Land-roller	J. Cole	Fredericksburgh, Iowa	Nov. 21, 1871	121, 157
Land-roller	H. L. Currier	Oregon, Ill	Sept. 10, 1867	68, 610
Land-roller	W. W. Ballard	Davisburgh, Mich	Oct. 8, 1867	69, 611
Land-roller	G. R. Burt	Perry, N. Y	Sept. 24, 1867	69, 175
Land-roller	J. W. Dilley	Macomb, Ill	Dec. 27, 1870	110, 554
Land-roller	G. C. Dolph	West Andover, Ohio	Dec. 13, 1870	110, 019
Land-roller	D. Fuller and D. Swain	Oakwood, Mich	Dec. 1, 1868	84, 541
Land-roller	J. F. Glidden	De Kalb, Ill	June 11, 1872	127, 687
Land-roller	W. H. Crow and C. M. Sloan	Darien, Kans	Nov. 4, 1873	144, 272
Land-roller	J. T. Hudnetand H. W. Mathews	Reaville and Frenchtown, N. J.	Mar. 28, 1871	113, 169
Land-roller	E. F. Hutchinson	Auburn, Me	Apr. 21, 1868	77, 044
Land-roller	E. O. Jones	Brandon, Mich	Apr. 21, 1868	77, 048
Land roller	E. O. Jones	Brandon, Mich	Mar. 30, 1869	88, 391
Land-roller	E. J. Knowlton	South Lyon, Mich	May 1, 1866	54, 368
Land-roller	J. Lanyon	Mineral Point, Wis	Sept. 10, 1872	131, 222
Land-roller	H. P. Manley	Ellsworth, N. Y	Jan. 19, 1869	86, 087
Land-roller	S. F. Mann	Indianapolis, Ind	Dec. 24, 1867	72, 653
Land-roller	H. W. Mathews	Frenchtown, N. J	May 7, 1872	126, 471
Land-roller	H. W. Mathews	Frenchtown, N. J	Sept. 17, 1872	131, 364
Land-roller	H. W. Mathews	Frenchtown, N. J	Oct. 29, 1872	132, 618
Land-roller	J. D. McAnally	Waterloo, Ind	Nov. 26, 1872	133, 468
Land-roller	N. S. McLay	Olathe, Kans	Jan. 12, 1869	85, 841
Land-roller	M. Miller	East Gaines, N. Y	Apr. 30, 1867	64, 242
Land-roller	E. F. Olds	Lyon, Mich	Nov. 12, 1867	70, 739
Land-roller	R. Provost	Raritan, N. J	Dec. 24, 1872	134, 311
Land-roller	H. Retzlaff	Saint Louis, Mo	Dec. 7, 1869	97, 697
Land-roller	R. Sandiford	Joliet, Ill	Dec. 29, 1868	85, 333
Land-roller	I. W. Searles	Tiffin, Ohio	Oct. 4, 1870	107, 057
Land-roller	J. S. Shafer	Plymouth, Mich	Apr. 9, 1867	63, 657
Land-roller	A. S. Skiff	Trenton Falls, N. J	May 1, 1866	54, 425
Land-roller	W. H. Staats, A. C. Schwanke, and L. Stadler.	La Prairie and Bowen, Ill	May 18, 1869	90, 131
Land-roller	E. A. Uehling	Richwood, Wis	Nov. 2, 1869	96, 513
Land-roller	E. Whitcomb and D. A. Gunn	Waterville, Ohio	July 26, 1870	105, 749
Land-roller	W. Williams	New Berlin, Ill	Nov. 11, 1873	144, 489
Land-roller	J. Winans	Plymouth, Mich	Apr. 2, 1867	63, 597
Land-roller	W. Winer	Freeport, Ill	July 25, 1871	117, 497
Land-roller	J. Woolridge	Dean's Corners, Ill	Jan. 21, 1873	135, 191
Land-roller	W. S. Worley	Tuscola, Ill	Dec. 17, 1867	72, 438
Land-roller and clod-pulverizer, Combined	J. Breiver	New Vienna, Ohio	Oct. 6, 1868	82, 793
Land-roller and harrow	L. French	Crawfordsville, Iowa	Aug. 20, 1872	130, 710
Land roller and marker	A. Mains	Olena, Ill	July 16, 1867	66, 861
Land roller and marker, Combined	H. F. Baker	Centreville, Ind	Dec. 20, 1870	110, 333
Land-roller and plaster-sower, Combined	G. F. Brock and E. Brondige	Davisburgh, Mich	Oct. 22, 1867	70, 070
Land-roller, fertilizer, and seed-sower combined	L. D. Taylor	Granville Centre, Pa	June 23, 1868	79, 280
Land roller, marker, and harrow, Combined	R. Sandiford	Joliet, Ill	June 8, 1869	90, 961
Land-roller, Pulverizing	F. Post	Plano, Ill	Oct. 20, 1868	83, 311
Land-roller, Sulky	P. Schmitt	Stewartsville, Mo	July 20, 1869	92, 888
Land-spading machine	W. E. Ward	Port Chester, N. Y	Oct. 20, 1857	18, 479
Land-tiller	G. W. Zeigler	Maumee, Ohio	Apr. 14, 1868	76, 687
Lantern	J. E. Ambrose	Lombard, Ill	Aug. 10, 1869	93, 577
Lantern	E. S. Archer	New York, N. Y	Mar. 17, 1868	75, 510
Lantern	J. S. and T. B. Atterbury	Pittsburgh, Pa	May 12, 1863	38, 457
Lantern	J. S. and T. B. Atterbury	Pittsburgh, Pa	Nov. 17, 1863	40, 594
Lantern	J. S. and T. B. Atterbury	Pittsburgh, Pa	Apr. 18, 1865	47, 267
Lantern	F. A. Balch	Hingham, Wis	Aug. 29, 1871	118, 576
Lantern	C. S. S. Baron	Bellaire, Ohio	Jan. 23, 1872	122, 931
Lantern	C. S. S. and A. L. Baron	Bellaire, Ohio	Oct. 18, 1870	108, 430
Lantern	H. Beebe	Hudson, N. J	Dec. 24, 1867	72, 589
Lantern	H. Beebe	Jersey City Heights, N. J	Jan. 7, 1873	134, 631
Lantern	J. Bellerjean	Philadelphia, Pa	Dec. 4, 1866	60, 123
Lantern	W. J. Berry	Brooklyn, N. Y	Mar. 15, 1870	100, 845
Lantern	W. J. Berry	Brooklyn, N. Y	Sept. 23, 1873	142, 984
Lantern	L. F. Betts	Saint Louis, Mo	Nov. 21, 1865	51, 005

Index of patents issued from the United States Patent Office from 1790 to 1873, inclusive—Continued.

Invention.	Inventor.	Residence.	Date.	No.
Lantern	L. F. Betts	New York, N. Y	Nov. 13, 1866	59,705
Lantern	L. F. Betts	New York, N. Y	Nov. 26, 1867	71,444
Lantern	L. F. Betts	Chicago, Ill	Sept. 1, 1868	81,584
Lantern	L. F. Betts	Philadelphia, Pa	July 29, 1873	141,311
Lantern	P. Blake	New Haven, Conn	Jan. 13, 1852	8,650
Lantern	W. H. Bonnell	Buffalo, N. Y	Mar. 10, 1868	75,354
Lantern	W. H. Bonnell	Buffalo, N. Y	Nov. 16, 1869	96,772
Lantern	E. Boorse	Philadelphia, Pa	Apr. 2, 1867	63,359
Lantern	G. R. Boynton	Chicago, Ill	July 1, 1862	35,800
Lantern	T. H. Brady	New Britain, Conn	May 26, 1868	78,180
Lantern	J. H. Breckenridge	West Meriden, Conn	Mar. 17, 1863	37,899
Lantern	J. D. Brown	Cincinnati, Ohio	May 29, 1860	28,450
Lantern	T. B. Burgert	Crestline, Ohio	Oct. 26, 1869	96,082
Lantern	W. Burns	Chicago, Ill	Jan. 2, 1866	51,798
Lantern	W. Burns	Chicago, Ill	Mar. 19, 1867	63,010
Lantern	C. H. Butterfield	South Lancaster, Mass	July 17, 1855	13,250
Lantern	J. Caldwell	Providence, R. I	Mar. 3, 1868	74,988
Lantern	D. Challinor	Birmingham, Pa	July 12, 1870	105,171
Lantern	R. Chester	Chicago, Ill	Dec. 4, 1866	60,139
Lantern	R. Chester	Chicago, Ill	Dec. 4, 1866	60,140
Lantern	P. J. Clark	West Meriden, Conn	Feb. 17, 1863	37,718
Lantern	P. J. Clark	West Meriden, Conn	May 19, 1868	78,058
Lantern	P. J. Clark	West Meriden, Conn	July 13, 1869	92,515
Lantern	P. J. Clark and J. Kintz	Meriden, Conn	Apr. 23, 1867	64,075
Lantern	E. B. Coffin	Johnstown, R. I	Apr. 23, 1861	32,161
Lantern	J. P. Connor and M. T. Hynes	Boston, Mass	Apr. 29, 1873	138,232
Lantern	E. Conrad	Philadelphia, Pa	Apr. 17, 1866	54,059
Lantern	J. A. Cowles	Chicago, Ill	May 31, 1864	42,928
Lantern	A. R. Cribfield	Lincoln, Ill	Apr. 2, 1867	63,480
Lantern	J. E. Cross	Chicago, Ill	Aug. 6, 1867	67,510
Lantern	T. A. Davies	New York, N. Y	Feb. 25, 1873	136,218
Lantern	C. Dean	Buffalo, N. Y	Nov. 16, 1869	96,897
Lantern	C. Davies	New York, N. Y	Apr. 28, 1863	38,355
Lantern	C. Deaves and E. S. Archer	New York, N. Y	June 21, 1864	43,257
Lantern	T. B. De Forest	New York, N. Y	Feb. 14, 1860	27,186
Lantern	T. B. De Forest	New York, N. Y	Mar. 27, 1860	27,666
Lantern	T. B. De Forest	New York, N. Y	Apr. 17, 1860	27,892
Lantern	T. B. De Forest	New York, N. Y	Aug. 7, 1860	29,472
Lantern	A. M. Duburn	Chicago, Ill	June 26, 1866	55,839
Lantern	A. M. Duburn	Chicago, Ill	Oct. 2, 1866	58,392
Lantern	A. M. Duburn	Chicago, Ill	Sept. 22, 1868	82,297
Lantern	R. Dunham	Portland, Me	Oct. 24, 1865	50,567
Lantern	M. B. Dyott	Philadelphia, Pa	Oct. 17, 1871	119,920
Lantern	C. Engelskirken	Buffalo, N. Y	Feb. 7, 1865	46,227
Lantern	H. Evans, jr	Newark, N. J	Sept. 26, 1848	5,806
Lantern	A. French	Philadelphia, Pa	Sept. 25, 1866	58,244
Lantern	A. French	Philadelphia, Pa	Nov. 28, 1871	121,241
Lantern	A. French	Philadelphia, Pa	Nov. 12, 1872	132,906
Lantern	F. M. Ford	Sebec, Me	July 25, 1871	117,399
Lantern	M. H. Fowler	New York, N. Y	Oct. 27, 1863	40,401
Lantern	H. A. Fox	Cincinnati, Ohio	June 28, 1864	43,303
Lantern	N. Gear	Newark, Ohio	May 2, 1871	114,432
Lantern	C. Gersten	Brooklyn, N. Y	Jan. 25, 1859	22,723
Lantern	A. H. Golden	La Fayette, Ind	Apr. 6, 1858	19,845
Lantern	O. L. Gridley and C. Engelskerken.	Buffalo, N. Y	June 20, 1871	116,177
Lantern	E. J. Hale	Foxcroft, Me	Mar. 19, 1861	31,713
Lantern	E. J. Hale	Foxcroft, Me	July 7, 1868	79,569
Lantern	J. Hale and J. S. Atterbury	Pittsburgh, Pa	Mar. 19, 1861	31,714
Lantern	H. W. Harkness	New Britain, Conn	Nov. 30, 1869	97,396
Lantern	J. O. Harris	Reading, Pa	Apr. 10, 1866	53,819
Lantern	J. O. Harris	Reading, Pa	Oct. 16, 1866	58,942
Lantern	J. O. Harris	Reading, Pa	Nov. 13, 1866	59,717
Lantern	C. Hart	Wakefield, Mass	Mar. 1, 1870	100,289
Lantern	E. F. Haskell	Sherman, Me	Jan. 18, 1870	98,962
Lantern	E. K. Haynes	Boston, Mass	Mar. 18, 1873	136,994
Lantern	E. K. Haynes	Boston, Mass	Nov. 4, 1873	144,200
Lantern	W. H. H. Hinds	Groton, Mass	May 10, 1864	42,658
Lantern	T. Houghton	Philadelphia, Pa	Apr. 6, 1869	88,713
Lantern	L. Hover	Jersey City, N. J	Jan. 23, 1855	12,291
Lantern	J. Hughes	Buchanan, Pa	Aug. 10, 1869	93,535
Lantern	J. Hughes	Buchanan, Pa	Dec. 28, 1869	98,383
Lantern	S. Hughes	Jersey City Heights, N. J	Feb. 11, 1873	135,711
Lantern	J. J. Hull and J. Haufman	Brooklyn, N. Y	Oct. 3, 1871	119,518
Lantern	J. H. Irwin	Chicago, Ill	Oct. 28, 1862	36,841
Lantern	J. H. Irwin	Chicago, Ill	Nov. 17, 1863	40,624
Lantern	J. H. Irwin	Chicago, Ill	May 24, 1864	42,856
Lantern	J. H. Irwin	Chicago, Ill	May 2, 1865	47,551
Lantern	J. H. Irwin	Chicago, Ill	Oct. 24, 1865	50,591
Lantern	J. H. Irwin	Chicago, Ill	Sept. 11, 1866	57,914
Lantern	J. H. Irwin	Chicago, Ill	May 28, 1867	65,229
Lantern	J. H. Irwin	Chicago, Ill	Jan. 7, 1868	73,012
Lantern	J. H. Irwin	Chicago, Ill	Feb. 2, 1869	86,548
Lantern	J. H. Irwin	Chicago, Ill	Feb. 2, 1869	86,549
Lantern	J. H. Irwin	Chicago, Ill	May 4, 1869	89,770
Lantern	J. H. Irwin	Philadelphia, Pa	Feb. 1, 1870	99,442
Lantern	J. H. Irwin	New York, N. Y	June 14, 1870	104,318
Lantern	J. H. Irwin	New York, N. Y	July 5, 1870	105,083
Lantern	J. H. Irwin	New York, N. Y	July 12, 1870	105,339
Lantern	G. B. Isham	Burlington, Vt	May 13, 1873	138,805
Lantern	J. Ives	Mount Carmel, Conn	Aug. 8, 1865	49,274
Lantern	A. S. Jackson	Kokomo, Ind	Apr. 14, 1868	76,767
Lantern	E. N. Jenkins	Chicago, Ill	July 24, 1866	56,567
Lantern	J. M. Jenness	Boston, Mass	May 3, 1870	102,551
Lantern	C. Jones	Brooklyn, N. Y	June 28, 1864	43,312
Lantern	J. B. Jones	Williamsburgh, N. Y	Feb. 28, 1860	27,332
Lantern	W. M. Kimball and K. Hartmann.	Cleveland, Ohio	Mar. 15, 1859	23,253

Index of patents issued from the United States Patent Office from 1790 to 1873, inclusive—Continued.

Invention.	Inventor.	Residence.	Date.	No.
Lantern	J. Kintz	West Meriden, Conn	Aug. 15, 1865	49, 417
Lantern	J. Kintz	West Meriden, Conn	Dec. 30, 1873	146, 079
Lantern	A. Lanergan	Boston, Mass	July 11, 1854	11, 289
Lantern	E. M. Lang	Portland, Me	May 11, 1869	89, 871
Lantern	T. Langston	Meriden, Conn	May 26, 1868	78, 378
Lantern	T. Langston	Brooklyn, N. Y	Dec. 1, 1868	84, 556
Lantern	T. Langston	Brooklyn, N. Y	Mar. 16, 1869	87, 857
Lantern	T. Langston	Brooklyn, N. Y	June 29, 1869	92, 063
Lantern	T. Langston	Brooklyn, N. Y	June 29, 1869	92, 064
Lantern	T. Langston	Meriden, Conn	Jan. 4, 1870	98, 602
Lantern	L. W. Leary	Norfolk, Va	May 31, 1870	103, 628
Lantern	L. W. Leary	Norfolk, Va	Jan. 28, 1870	104, 745
Lantern	F. Leclére	Watertown, N. Y	Mar. 12, 1867	62, 861
Lantern	J. G. Leffingwell	Newark, N. J	Apr. 11, 1865	47, 212
Lantern	A. Loeffelholz and A. Prior	Milwaukee, Wis	Mar. 25, 1873	137, 223
Lantern	G. H. Magersuppe	New York, N. Y	Apr. 2, 1861	31, 896
Lantern	I. C. Mayo	Gloucester, Mass	Dec. 21, 1869	98, 080
Lantern	W. McKay	Newburyport, Mass	July 22, 1873	141, 008
Lantern	G. C. Merrill	Chicago, Ill	Jan. 24, 1865	46, 010
Lantern	R. M. Merrill	Chicago, Ill	Oct. 31, 1865	50, 725
Lantern	F. Meyrose	Saint Louis, Mo	Sept. 18, 1866	58, 122
Lantern	J. H. Miltimore	Milwaukee, Wis	Aug. 8, 1865	49, 290
Lantern	C. F. Moller	Newark, N. J	Aug. 14, 1866	57, 167
Lantern	R. Mood and H. M. Britton	Cincinnati, Ohio	Feb. 23, 1869	87, 189
Lantern	J. C. Moore	Philadelphia, Pa	Mar. 6, 1866	53, 027
Lantern	F. Morandi	Boston, Mass	Feb. 5, 1856	14, 201
Lantern	P. A. Morley	Brooklyn, N. Y	Apr. 17, 1860	27, 924
Lantern	G. Mortimer	Jersey City Heights, N. J	Mar. 5, 1872	124, 372
Lantern	W. Mullin	Steubenville, Ohio	Sept. 22, 1863	40, 057
Lantern	S. Naylor	Independence, Iowa	July 22, 1873	141, 162
Lantern	L. C. Ober	Boston, Mass	July 8, 1862	35, 835
Lantern	J. Orphy	Buffalo, N. Y	May 2, 1871	114, 471
Lantern	S. Peters	Crescent, N. Y	May 10, 1870	102, 858
Lantern	G. Peugeot	Buffalo, N. Y	Jan. 9, 1867	51, 966
Lantern	E. J. Pohl and J. A. Higgins	Philadelphia, Pa	Nov. 26, 1872	133, 479
Lantern	W. Porter	New York, N. Y	Aug. 18, 1863	39, 586
Lantern	W. Porter, jr	New York, N. Y	June 13, 1871	115, 891
Lantern	W. Porter, sr., and W. Porter, jr.	New York, N. Y	Apr. 23, 1867	64, 141
Lantern	W. Porter, sr., and W. Porter, jr.	New York, N. Y	Aug. 3, 1869	93, 226
Lantern	W. Porter and E. A. Tuttle	Williamsburgh, N. Y	July 5, 1853	9, 833
Lantern	E. T. Prindle and J. Wellfare	Aurora, Ill	Dec. 31, 1867	72, 900
Lantern	G. W. Putnam	Boston, Mass	Oct. 6, 1868	82, 750
Lantern	J. H. Reighard	Birmingham, Pa	Jan. 26, 1858	19, 207
Lantern	J. H. Reighard	Birmingham, Pa	Apr. 6, 1858	19, 897
Lantern	J. H. Richardson	Philadelphia, Pa	Dec. 24, 1867	72, 542
Lantern	S. Roebuck	New York, N. Y	Sept. 12, 1865	49, 953
Lantern	J. H. Rohrman	Philadelphia, Pa	Sept. 1, 1857	18, 105
Lantern	H. Sangster	Buffalo, N. Y	May 28, 1867	65, 285
Lantern	H. and J. Sangster	Buffalo, N. Y	June 10, 1851	8, 154
Lantern	S. Shannon	Buffalo, N. Y	Sept. 23, 1856	15, 782
Lantern	A. G. Smith	Jersey City, N. J	Nov. 20, 1866	59, 865
Lantern	A. G. Smith	Jersey City, N. J	Jan. 28, 1868	73, 934
Lantern	J. O. Smith	New York, N. Y	Dec. 31, 1867	72, 759
Lantern	T. Smith	Cleveland, Ohio	May 20, 1873	139, 030
Lantern	W. P. Smith	Louisville, Ky	Sept. 11, 1866	57, 988
Lantern	S. C. Spaulding	Rutland, Vt	Sept. 14, 1858	21, 521
Lantern	C. F. Spencer	Rochester, N. Y	Mar. 26, 1867	63, 321
Lantern	C. F. Spencer	Rochester, N. Y	July 16, 1867	66, 902
Lantern	J. Staniford, jr., and A. D. Allen.	....	May 10, 1804	....
Lantern	W. G. and C. Sterling	New York, N. Y	Apr. 30, 1867	64, 377
Lantern	J. Straszer	Saint Louis, Mo	Jan. 27, 1863	37, 529
Lantern	P. Sweeny	New York, N. Y	Oct. 4, 1870	108, 064
Lantern	C. J. Sykes	Chicago, Ill	Jan. 30, 1872	123, 208
Lantern	C. J. Sykes	Chicago, Ill	Apr. 30, 1872	126, 345
Lantern	C. Tabor	Craftsbury, Vt	Sept. 27, 1870	107, 835
Lantern	A. E. Taylor	New Britain, Conn	May 30, 1871	115, 385
Lantern	N. Thompson	Brooklyn, N. Y	Nov. 3, 1868	83, 799
Lantern	N. Thompson	Brooklyn, N. Y	Apr. 27, 1869	89, 520
Lantern	N. Thompson	Brooklyn, N. Y	Nov. 23, 1869	97, 246
Lantern	W. S. Thompson	Rochester, N. Y	Apr. 2, 1867	63, 441
Lantern	W. D. Titus	Brooklyn, N. Y	Dec. 5, 1854	12, 041
Lantern	F. G. Tucker and C. Crawford	Albany, N. Y	Apr. 14, 1863	38, 189
Lantern	A. Tufts	Malden, Mass	Mar. 13, 1860	27, 485
Lantern	A. Tufts	Malden, Mass	Aug. 29, 1865	49, 665
Lantern	G. Wallingford	Charlestown, Mass	Nov. 12, 1872	132, 995
Lantern	C. Waters	Brooklyn, N. Y	July 17, 1855	13, 286
Lantern	H. Wentworth and V. B. Clark	Ripon, Wis	Jan. 21, 1868	73, 481
Lantern	W. Westlake	Milwaukee, Wis	Aug. 2, 1864	43, 730
Lantern	W. Westlake	Chicago, Ill	July 18, 1865	48, 858
Lantern	W. Westlake	Chicago, Ill	Sept. 26, 1865	50, 192
Lantern	W. Westlake	Chicago, Ill	Dec. 12, 1865	51, 526
Lantern	W. Westlake	Brooklyn, N. Y	Jan. 1, 1867	60, 810
Lantern	W. Westlake	Brooklyn, N. Y	Apr. 7, 1868	76, 366
Lantern	W. Westlake	Chicago, Ill	Sept. 7, 1869	94, 535
Lantern	W. Westlake	Chicago, Ill	Sept. 7, 1869	94, 536
Lantern	W. Westlake	Chicago, Ill	May 10, 1870	102, 895
Lantern	W. Westlake	Chicago, Ill	May 10, 1870	102, 896
Lantern	W. Westlake	Chicago, Ill	Sept. 6, 1870	107, 140
Lantern	W. Westlake	Chicago, Ill	Oct. 11, 1870	108, 221
Lantern	W. Westlake	Chicago, Ill	Oct. 11, 1870	108, 222
Lantern	W. Westlake	Chicago, Ill	Oct. 3, 1871	119, 549
Lantern	W. Westlake	Chicago, Ill	Aug. 26, 1873	142, 134
Lantern	G. Wheeler	Chicago, Ill	Jan. 14, 1868	73, 415
Lantern	A. Whelden	South Dennis, Mass	July 14, 1868	80, 040
Lantern	H. J. White	Boston, Mass	Dec. 6, 1870	109, 856

Index of patents issued from the United States Patent Office from 1790 *to* 1873, *inclusive*—Continued.

Invention.	Inventor.	Residence.	Date.	No.
Lantern	S. R. Wilmot	Bridgeport, Conn	Dec. 11, 1866	60, 452
Lantern	A. Withmar	Saint Louis, Mo	Mar. 8, 1870	100, 578
Lantern	G. W. Woodward	New York, N. Y	Apr. 5, 1864	42, 249
Lantern	M. B. Wright	Meriden, Conn	Nov. 12, 1867	70, 931
Lantern	H. C. Yerby	Leslie, Mich	June 8, 1869	91, 192
Lantern	M. Young	Frederick, Md	July 26, 1870	105, 762
Lantern, Advertising	S. Kuh	Jefferson, Iowa	Nov. 19, 1872	133, 158
Lantern, Advertising	S. Kuh	Jefferson, Iowa	Aug. 12, 1873	141, 719
Lantern and axle lubricator, Combined carriage	J. Scheeper	New York, N. Y	Dec. 3, 1861	33, 853
Lantern and candlestick, Combined	M. A. Shepard	Fort Branch, Ind	Sept. 20, 1870	107, 551
Lantern and foot-warmer, Combined	W. F. Bartlett	Hillsdale, Mich	Nov. 20, 1866	59, 810
Lantern and foot-warmer, Combined	S. M. Wirts and F. Swift	Hudson, Mich	Aug. 6, 1867	67, 622
Lantern and lamp, Combining	J. S. Grimes	Lansingburgh, N. Y	Apr. 18, 1848	5, 516
Lantern and lamp frame	G. F. J. Coburn	Newark, N. J	May 24, 1864	42, 839
Lantern and oil-can combined	W. G. Russell	New York, N. Y	May 19, 1857	17, 351
Lantern and reflector, Attachment of	W. C. Owen	Brooklyn, N. Y	May 12, 1863	38, 498
Lantern, Astronomical	J. F. Clarke	West Roxbury, Mass	Dec. 27, 1870	110, 435
Lantern-attachment to caps	J. C. Cary	New York, N. Y	Sept. 14, 1858	21, 485
Lantern, Carriage	T. Wigley	Bridgeport, Conn	Dec. 10, 1872	133, 735
Lantern-carrier	S. J. Shaw and H. J. Batchelder.	Marlborough, Mass	Feb. 28, 1860	27, 315
Lantern, Coal-oil	P. J. Clark	West Meriden, Conn	Apr. 14, 1863	38, 199
Lantern, Coal-oil	M. H. Fowler	New York, N. Y	Jan. 13, 1863	37, 394
Lantern, Coal-oil	M. Miller	Brooklyn, N. Y	Aug. 17, 1858	21, 209
Lantern, Coal-oil	S. Sargent	Watertown, Mass	Sept. 17, 1861	33, 312
Lantern, Coal-oil	J. W. Schreiber	New York, N. Y	Sept. 2, 1862	36, 371
Lantern, Dock	T. Langston	Meriden, Conn	June 18, 1872	128, 153
Lantern, Decorative	C. C. E. Schwartz	Philadelphia, Pa	Sept. 2, 1873	142, 520
Lantern-fastening	C. Mounin and W. M. Booth	Buffalo, N. Y	Aug. 1, 1854	11, 437
Lantern, Fishing	J. Goodrich	Muscoda, Wis	Feb. 10, 1863	37, 621
Lantern, foot-warmer, and water-heater, Combined	G. A. Wells	Oskaloosa, Iowa	Jan. 29, 1867	61, 587
Lantern for lighting street-gas	J. Reese and C. N. Tyler	Washington, D. C	June 23, 1857	17, 637
Lantern for marine-telegraph	J. W. Moore and W. H. Elliot	Plattsburgh, N. Y	May 20, 1862	35, 322
Lantern for railway-car, Coal-oil	S. B. H. Vance	New York, N. Y	Jan. 27, 1863	37, 553
Lantern for railway-switch, Signal	S. N. Lennon	Deposit, N. Y	July 27, 1858	21, 006
Lantern-frame	R. S. Laird	Sandwich, Ill	Jan. 31, 1865	46, 114
Lantern-frame	E. F. Parker	Proctorsville, Vt	Feb. 1, 1853	9, 566
Lantern-frame	E. F. Parker	Proctorsville, Vt	Jan. 30, 1855	12, 324
Lantern, Gas-lighting	A. Wilson	Philadelphia, Pa	Dec. 1, 1857	18, 784
Lantern, Glass	P. A. Morley	Brooklyn, N. Y	Aug. 29, 1854	11, 632
Lantern, Globe	J. S. and T. B. Atterbury	Pittsburgh, Pa	Apr. 18, 1865	47, 268
Lantern, Globe	C. P. Lindley	Waterbury, Conn	June 2, 1863	38, 754
Lantern-globe	I. C. Mayo	Gloucester, Mass	May 3, 1870	102, 692
Lantern-globe	M. Sweeney	Wheeling, W. Va	Oct. 19, 1869	96, 053
Lantern-globe	W. Westlake	Brooklyn, N. Y	Feb. 4, 1868	74, 020
Lantern-globe holder	J. S. Dennis and W. F. Kistler	Chicago, Ill	May 23, 1871	115, 035
Lantern-globe holder	J. H. Irwin	Philadelphia, Pa	May 23, 1871	115, 059
Lantern-guard	T. Brown, jr., and J. L. Lowry	Pittsburgh, Pa	Oct. 24, 1865	50, 653
Lantern-guard	C. H. Butterfield	South Lancaster, Mass	Sept. 11, 1855	13, 539
Lantern-guard	C. H. Butterfield	Nashua, N. H	Dec. 25, 1855	14, 006
Lantern-guard	W. R. P. Cross	Portland, Me	Oct. 4, 1864	44, 577
Lantern-guard	J. Kintz	West Meriden, Conn	June 28, 1870	104, 857
Lantern-guard	W. H. Pierce	East Cambridge, Mass	Feb. 18, 1862	34, 447
Lantern-guard	T. Smith	Cleveland, Ohio	Mar. 11, 1873	136, 782
Lantern-guard	W. Westlake	Milwaukee, Wis	Apr. 26, 1864	42, 520
Lantern-guard attachment	T. Brown, jr	Allegheny County, Pa	May 24, 1864	42, 906
Lantern, Hand	C. Engelskircher	Buffalo, N. Y	July 24, 1866	56, 543
Lantern, Hand	G. Peugeot	Buffalo, N. Y	July 24, 1866	56, 604
Lantern handle, Signal	A. N. Towne	Chicago, Ill	Apr. 16, 1867	63, 820
Lantern holder, Body	S. E. Fessenden	Stamford, Conn	Mar. 7, 1871	112, 436
Lantern, Insect-destroying	S. C. Wilt	Hartleton, Pa	Oct. 7, 1846	4, 808
Lantern, Insect-trap	R. W. Pitman	West Point, Iowa	Mar. 5, 1867	62, 563
Lantern, Kerosene	E. G. Tobey	Portland, Me	Aug. 12, 1862	36, 198
Lantern-lamp	R. M. Merrill	Chicago, Ill	May 27, 1862	35, 382
Lantern, Locomotive	J. H. Kelley	Rochester, N. Y	Sept. 18, 1855	13, 577
Lantern, Magic	A. G. Buzby	Philadelphia, Pa	June 17, 1873	139, 865
Lantern, Magic	A. G. Buzby	Philadelphia, Pa	Nov. 4, 1873	144, 314
Lantern, Magic	P. Diehl	Chicago, Ill	Feb. 25, 1873	136, 142
Lantern, Magic	A. Krüss	Hamburg, Germany	Jan. 25, 1870	99, 211
Lantern, Magic	L. J. Marcy	Newport, R. I	Apr. 28, 1868	77, 300
Lantern, Magic	L. J. Marcy	Newport, R. I	July 6, 1869	92, 330
Lantern, Magic	G. Sibbald	Philadelphia, Pa	Apr. 19, 1864	42, 412
Lantern, Miners' safety	M. L. Beaufils and J. Rexroth	Paris, France	Sept. 28, 1869	95, 184
Lantern, Omnibus	F. O. Dechamps	Philadelphia, Pa	July 26, 1853	9, 885
Lantern, Pocket	C. Mackh	Chicago, Ill	May 11, 1869	90, 007
Lantern, Pocket	J. A. Minor	Middletown, Conn	Jan. 24, 1865	46, 011
Lantern, Pocket	G. W. Putnam	Peterborough, N. Y	Nov. 17, 1868	84, 213
Lantern, Pocket	A. Ralston	West Middletown, Pa	Mar. 24, 1857	16, 880
Lantern, Portable	C. Deavs	New York, N. Y	Jan. 31, 1865	46, 184
Lantern, Portable	H. L. Kassebaum	New York, N. Y	May 16, 1865	47, 733
Lantern, Portable	N. Waterman	Boston, Mass	Dec. 25, 1849	6, 978
Lantern, Portable and stationary	L. W. Leary	Norfolk, Va	Dec. 15, 1868	85, 014
Lantern, Railway-signal	S. N. Lennon	Deposit, N. Y	Feb. 15, 1859	22, 960
Lantern, Railway-signal	N. A. Menaar	Buffalo, N. Y	May 21, 1861	32, 377
Lantern, Railway-signal	J. L. Wager	Deposit, N. Y	Feb. 15, 1859	22, 992
Lantern, Railway-switch	C. Byrne	Kingsville, Ohio	Jan. 27, 1863	37, 493
Lantern, Railway-switch	E. E. and A. B. Dickerson	Oshkosh, Wis	Aug. 21, 1866	57, 295
Lantern, Reflecting	E. Boesch	San Francisco, Cal	Oct. 22, 1872	132, 433
Lantern, Reflecting	A. E. Young	Dorchester, Mass	May 6, 1862	35, 194
Lantern, Reflecting and magnifying	W. Lewis	Boston, Mass	June 8, 1810	....
Lantern, Self lighting and extinguishing	A. Roesler and C. Frey	Warsaw, Ill	May 18, 1858	20, 302
Lantern, Ship's	A. Saunders	South Kingston, R. I	Jan. 28, 1873	135, 371
Lantern, Ship's	H. Saunders	South Kingston, R. I	May 6, 1873	138, 532
Lantern, Signal	D. Ammen	United States Navy	Feb. 16, 1858	19, 332
Lantern, Signal	L. V. Badger	Chicago, Ill	Dec. 17, 1867	72, 262
Lantern, Signal	G. Callard	Buffalo, N. Y	July 31, 1849	6, 617
Lantern, Signal	R. Chester	Chicago, Ill	Dec. 2, 1862	37, 032

Index of patents issued from the United States Patent Office from 1790 *to* 1873, *inclusive*—Continued.

Invention.	Inventor.	Residence.	Date.	No.
Lantern, Signal	S. W. Clark and W. R. Sykes	London, England	June 17, 1873	140, 015
Lantern, Signal	T. A. Davis	New York, N. Y	Feb. 11, 1873	135, 783
Lantern, Signal	H. C. Felthousen	Buffalo, N. Y	Dec. 3, 1861	33, 831
Lantern, Signal	H. B. Fernald	Washington, D. C	Aug. 26, 1873	142, 223
Lantern, Signal	J. Graham	Grafton, W. Va	Dec. 15, 1868	85, 004
Lantern, Signal	L. Hover	Flushing, N. Y	Oct. 4, 1859	25, 645
Lantern, Signal	W. Howard	Flushing, N. Y	June 1, 1858	20, 431
Lantern, Signal	S. H. Miller	Brooklyn, N. Y	Aug. 26, 1873	142, 257
Lantern, Signal	S. H. Miller	Brooklyn, N. Y	Nov. 11, 1873	144, 554
Lantern, Signal	C. G. Moeller	Newark, N. J	Sept. 17, 1872	131, 368
Lantern, Signal	J. W. Moffitt	Harrisburgh, Pa	June 18, 1872	128, 058
Lantern, Signal	J. R. Pierce and L. B. Austin	Oswego, N. Y	Nov. 10, 1857	18, 602
Lantern, Signal	W. S. Roberts and E. H. Fiske	East Greenwich, R. I	Aug. 22, 1871	118, 278
Lantern, Signal	H. Sangster	Buffalo, N. Y	Dec. 18, 1849	6, 959
Lantern, Signal	I. W. Shaler	Brooklyn, N. Y	Oct. 7, 1873	143, 425
Lantern, Signal	D. Todd	Detroit, Mich	Nov. 9, 1869	96, 637
Lantern, Signal	E. G. Turner	New Bedford, Mass	July 13, 1869	92, 493
Lantern, Signal	T. F. Woodward	South Reading, Mass	July 10, 1860	29, 118
Lantern-spring	J. A. Cowles	Chicago, Ill	May 17, 1864	42, 751
Lantern, Steam-boat	J. M. Read	Louisville, Ky	Dec. 28, 1838	1, 049
Lantern, Street	J. P. Avery and W. L. Nichols	Norwich, Conn	June 18, 1867	65, 860
Lantern, Street	J. W. Bartlett	New York, N. Y	June 7, 1870	103, 828
Lantern, Street	J. W. Bartlett	New York, N. Y	June 27, 1871	116, 399
Lantern, Street	A. Burger	New York, N. Y	Nov. 28, 1871	121, 327
Lantern, Street	J. Cook	New York, N. Y	Mar. 19, 1872	124, 665
Lantern, Street	J. F. Harley	Kipton Station, Ohio	Sept. 26, 1871	119, 355
Lantern, Street	M. A. Heath	Providence, R. I	Oct. 10, 1871	119, 762
Lantern, Street	C. F. Hollis	Boston, Mass	Mar. 19, 1872	124, 819
Lantern, Street	A. R. and E. A. Hunt	Newark, N. J	Dec. 11, 1866	60, 377
Lantern, Street	B. A. Johnson	Jeffersonville, Ind	Aug. 14, 1866	57, 148
Lantern, Street	J. Neumann	Philadelphia, Pa	Apr. 26, 1870	102, 305
Lantern, Street	J. Reese and C. N. Tyler	Washington, D. C	June 30, 1857	17, 696
Lantern, Street	A. Tufts	Malden, Mass	Mar. 7, 1871	112, 396
Lantern, Street-car	A. A. Young	Boston, Mass	Aug. 18, 1868	81, 124
Lantern, Stove	D. L. Jacques	Hudson, Mich	Mar. 14, 1865	46, 802
Lantern, Submarine	G. W. Fuller	Cambridgeport, Mass	Mar. 24, 1863	37, 959
Lantern, Submarine	C. M. Gould and C. B. Lamb	Worcester, Mass	Apr. 8, 1856	14, 608
Lantern, Submarine telescopic	H. Thompson	Mobile, Ala	Feb. 16, 1869	87, 012
Lantern, Tubular	J. S. Dennis	Chicago, Ill	Feb. 27, 1872	123, 980
Lantern, Tubular	J. S. Dennis	Chicago, Ill	Feb. 27, 1872	123, 981
Lantern, Tubular	J. S. Dennis	Chicago, Ill	Feb. 27, 1872	123, 982
Lantern, Wax	J. Headford	New York, N. Y	July 16, 1872	129, 280
Lanterns and lamps, Glass for	M. V. Shaver	Adrian, Mich	Apr. 16, 1872	125, 849
Lanterns, Attachment for lighting	A. C. Richard	Newtown, Conn	Jan. 5, 1858	19, 044
Lanterns, Lenses for	E. Barrett	New York, N. Y	Nov. 26, 1867	71, 441
Lanterns, Machine for making	W. Westlake	Chicago, Ill	Apr. 25, 1865	47, 478
Lanterns, Machinery for manufacturing	W. Westlake	Chicago, Ill	Oct. 17, 1865	50, 538
Lanterns, Regulator for the wicks of	H. W. Bleyer	Buffalo, N. Y	June 20, 1865	48, 254
Lanterns, Removable flanch-bar for securing glass of.	H. Crout	Baltimore, Md	Jan. 15, 1856	14, 087
Lanterns, Securing glass in	H. Sangster	Buffalo, N. Y	Aug. 22, 1854	11, 570
Lanterns, Transparent slide for magic	S. Solomons	London, England	Apr. 2, 1867	63, 437
Lanterns, Wind-brake for	A. Davis	Chicago, Ill	Aug. 5, 1862	36, 077
Lanyard, Elastic	J. E. Jones	Waretown, N. J	Dec. 29, 1868	85, 388
Lanyard, Elastic	J. E. Jones	Waretown, N. J	Feb. 1, 1870	99, 320
Lanyard, Elastic	J. E. Jones	Watertown, N. J	Jan. 30, 1872	123, 178
Lanyard-scores in dead-eyes, Machine for cutting and boring holes in.	T. Blanchard	New York, N. Y	Aug. 10, 1836	....
Lap-board	M. Church	Chicago, Ill	Nov. 26, 1872	133, 302
Lap-board	D. P. Cook	Hartford, Conn	Feb. 28, 1871	112, 126
Lap-board	W. F. Gammel	Elizabeth, N. J	May 3, 1870	102, 530
Lap-board	S. Mahan	Cleveland, Ohio	July 9, 1872	128, 739
Lap-board	S. Mahan	Cleveland, Ohio	Jan. 21, 1873	135, 040
Lap-board	G. K. Proctor	Salem, Mass	June 17, 1873	140, 074
Lap-board	E. J. Sprague	Youngstown, Ohio	June 24, 1873	140, 313
Lap-board	C. Trefethen	Manchester, N. H	Aug. 22, 1871	118, 303
Lap-board	W. B. White and B. S. Bryant	Abington and Hanson, Mass.	July 30, 1872	130, 094
Lap-board, Dress-maker's	R. L. Woodbury	Lexington, Mass	Aug. 2, 1870	106, 016
Lap-joint for belting	H. Underwood	New York, N. Y	Sept. 11, 1866	58, 604
Lap-shaver and leather-splitter	J. Harvey and F. Herkstroeter	Saint Louis, Mo	Apr. 18, 1865	47, 300
Lap-welded tubes, Apparatus for manufacturing	J. Nicholson	Philadelphia, Pa	Apr. 30, 1867	64, 246
Lap-welded tubes, Machine for finishing	P. L. Weimer	Lebanon, Pa	Oct. 31, 1865	50, 781
Lapping-machine	S. Campbell	New York Mills, N. Y	Oct. 9, 1849	6, 785
Lapping-machine feed-regulater, Cotton	R. Kitson	Lowell, Mass	July 11, 1871	116, 840
Lard and butter box	W. Pratt	New York, N. Y	Feb. 21, 1871	111, 970
Lard and butter box	C. L. Tucker	Chicago, Ill	Dec. 1, 1868	84, 657
Lard and butter cutter	W. M. Bleakley	Verplanck, N. Y	July 30, 1872	129, 921
Lard and tallow, Drying	A. Smith	Cincinnati, Ohio	Sept. 9, 1873	142, 743
Lard, Apparatus for refining	G. C. Napheys	Philadelphia, Pa	Nov. 21, 1865	51, 075
Lard, Apparatus for stirring and cooling	E. F. Ring	Saint Louis Mo	Nov. 22, 1870	109, 548
Lard, Apparatus for stirring and cooling	G. B. Williams	New York, N. Y	Feb. 5, 1867	61, 907
Lard boiling, &c., Apparatus for condensation of vapors in.	C. C. and G. F. Peirson	Philadelphia, Pa	July 14, 1868	80, 006
Lard, butter, &c., Package for	G. M. Huntley	Grand Rapids, Mich	Nov. 29, 1870	109, 739
Lard-boiler	W. Branagan	Burlington, Iowa	Nov. 6, 1866	59, 350
Lard, Can or bucket for transporting	J. A. Curtis	Baltimore, Md	Oct. 22, 1872	132, 356
Lard-cooler	A. E. Camp and C. L. Reids	Louisville, Ky	Oct. 10, 1871	119, 818
Lard-cooler	G. C. Cassard	Baltimore, Md	June 16, 1868	78, 925
Lard-cooler	G. C. Cassard	Baltimore, Md	Oct. 1, 1872	131, 847
Lard-cooler	J. Ring	Saint Louis, Mo	Aug. 13, 1872	130, 534
Lard-cooler	V. E. Rusco	Chicago, Ill	Nov. 17, 1863	40, 644
Lard-cooler	W. J. Wilcox	New York, N. Y	Dec. 9, 1862	37, 120
Lard, Cooling	W. J. Wilcox	New York, N. Y	Sept. 4, 1866	57, 805
Lard-expresser	C. Bixler	Rogersville, Ohio	Jan. 24, 1860	26, 879
Lard, Machine for clarifying, mixing, and bleaching.	O. J. Backus	San Francisco, Cal	Apr. 2, 1872	125, 162
Lard or tallow, Apparatus for drying and cooling	A. Smith	Cincinnati, Ohio	Sept. 9, 1873	142, 744

Index of patents issued from the United States Patent Office from 1790 *to* 1873, *inclusive*—Continued.

Invention.	Inventor.	Residence.	Date.	No.
Lard or tallow into separate substances, Process of converting.	B. Lapham	Lexington, Ky	Mar. 1, 1833	....
Lard-package	J. King	Saint Louis, Mo	May 16, 1871	114, 971
Lard, Package for holding and shipping	C. L. Tucker	Chicago, Ill	July 2, 1867	66, 268
Lard-packing boxes, Cementing and strengthening	C. L. Tucker	Chicago, Ill	Aug. 18, 1868	81, 229
Lard, Preparing	H. A. Amelung	Alton, Ill	Nov. 13, 1844	3, 827
Lard-press	S. S. Avis	Penn's Grove, N. J	Apr. 28, 1868	77, 239
Lard-press	W. F. Pagett and C. F. Rohrer.	Fremont, Ohio	Oct. 22, 1872	132, 486
Lard-press	J. Rayner	Piqua, Ohio	Oct. 17, 1865	50, 494
Lard-press and sausage-stuffer	J. B. Cassell	Worcester Township, Pa	Sept. 1, 1868	81, 751
Lard-press and sausage-stuffer	A. J. Truxell	Salem, Va	July 14, 1868	80, 035
Lard-press, Rotary	B. Hubbe	New York, N. Y	Jan. 29, 1867	61, 541
Lard, Process for refining	H. S. Lewis	Chicago, Ill	June 25, 1861	32, 633
Lard-rendering, &c., Mode of condensing noxious vapors from.	S. Davis	New York, N. Y	June 18, 1867	65, 884
Lard-scraps drying and pressing apparatus	A. K. Howe	New York, N. Y	Nov. 1, 1870	108, 788
Lard stirring and cooling apparatus	A. R. Judson	New York, N. Y	Apr. 18, 1865	47, 361
Lard stirring and cooling apparatus	J. N. Meriam	Cambridgeport, Mass	Dec. 22, 1868	85, 117
Lard-stirring machine	W. J. Wilcox	New York, N. Y	Jan. 20, 1863	37, 469
Lard, tallow, and grease from the refuse of rendering-tanks, Machine for separating.	P. Andrew	Cincinnati, Ohio	Feb. 7, 1865	46, 204
Lard, tallow, &c., Mode of treating	G. B. Turrell	New York, N. Y	June 28, 1864	43, 352
Larder, Ventilated	J. Sloan	Philadelphia, Pa	Aug. 28, 1866	57, 642
Last	J. Auzer	Ashtabula, Ohio	Oct. 29, 1872	132, 556
Last	W. L. Beardsley	Binghamton, N. Y	Aug. 7, 1866	56, 882
Last	T. Bullivant	Newark, N. J	Mar. 30, 1869	88, 269
Last	A. W. Cheever	Lynn, Mass	Aug. 6, 1867	67, 414
Last	P. C. Clapp	Dorchester, Mass	Nov. 6, 1866	59, 359
Last	C. O. Crosby	New Haven, Conn	June 22, 1869	91, 720
Last	C. S. Dunbrack	Swampscott, Mass	Nov. 22, 1870	109, 399
Last	E. T. Green	Stoneham, Mass	Oct. 30, 1866	59, 209
Last	B. Hitchings	Lynn, Mass	Oct. 13, 1868	82, 951
Last	S. T. Hutchins	North Anson, Me	July 24, 1866	56, 565
Last	P. Jackson	Saugus, Mass	Aug. 8, 1865	49, 275
Last	N. Jones	Syracuse, N. Y	Aug. 23, 1870	106, 59
Last	G. Marshall	Brooklyn, N. Y	Oct. 3, 1865	50, 252
Last	S. Mawhinney	Worcester, Mass	Apr. 20, 1869	89, 229
Last	A. W. Merritt	Scituate, Mass	Mar. 16, 1869	87, 788
Last	W. J. B. Mills	Philadelphia, Pa	Feb. 20, 1872	123, 835
Last	W. J. B. Mills	Philadelphia, Pa	Feb. 27, 1872	124, 075
Last	W. J. B. Mills	Philadelphia, Pa	Apr. 8, 1873	137, 616
Last	M. H. Pool	East Abington, Mass	Mar. 9, 1869	87, 590
Last	N. M. Rosinsky	New York, N. Y	July 16, 1872	129, 174
Last	I. N. C. Sands	Stoughton, Mass	May 17, 1864	42, 796
Last	I. N. C. Saville	Worcester, Mass	June 22, 1869	91, 774
Last	I. N. C. Saville	Worcester, Mass	May 20, 1873	139, 196
Last	W. C. Shipherd	Saratoga Springs, N. Y	Dec. 9, 1862	37, 114
Last	O. S. Squire	North Haven, Conn	July 19, 1859	24, 830
Last	G. H. Taylor	Shelburne, Mass	Jan. 4, 1859	22, 534
Last	A. Taylor	Osawatomie, Kans	Nov. 5, 1867	70, 648
Last	J. C. Toot	Clearfield, Pa	Jan. 2, 1872	122, 338
Last	D. M. True	Rockland, Me	Feb. 8, 1859	22, 904
Last	S. S. Turner	Westborough, Mass	May 11, 1858	20, 228
Last	G. M. Wells	London, England	Feb. 4, 1868	74, 180
Last	G. M. Wells	London, England	Feb. 4, 1868	74, 181
Last	W. Wells	South Danvers, Mass	Mar. 27, 1866	53, 510
Last	H. M. Whitmarsh	Abington, Mass	Sept. 13, 1864	44, 266
Last	G. B. Whitney	Natick, Mass	July 2, 1872	128, 688
Last	H. Wight	Malden, Mass	Aug. 29, 1871	118, 663
Last and shoe holder	A. J. Rock	Union Village, Va	May 12, 1868	77, 764
Last and shoe holder	B. J. Tayman	Philadelphia, Pa	May 27, 1873	139, 343
Last-block elevator and instep-stretcher	S. Richmond	Annapolis, Md	Dec. 1, 1868	84, 647
Last-block fastener	D. Huard	Ashland, Wis	July 15, 1873	140, 827
Last-block fastener	N. R. Streeter	Groton, N. Y	Nov. 25, 1873	145, 028
Last-block fastener	L. S. Wright	Groton, N. Y	Sept. 9, 1873	142, 603
Last-block fastener	N. R. Streeter	Groton, N. Y	Nov. 4, 1873	144, 368
Last-block fastening	J. Bullivant	Newark, N. J	Nov. 22, 1870	109, 492
Last-block fastening	J. A. Heckenbach and A. Haertle.	Mayville, Wis	June 17, 1873	140, 038
Last-block fastening	D. Lynahan and H. H. Koch	Buffalo, N. Y	Apr. 18, 1865	47, 314
Last-block fastening	L. R. Lockwood	Upton, Mass	Sept. 16, 1851	8, 370
Last-block fastening	A. J. Tewksbury	Haverhill, Mass	Sept. 20, 1864	44, 352
Last-block holder	D. Daniels	Park's Corners, Ill	July 22, 1873	141, 123
Last, Boot and shoe	C. F. Carr and G. F. Holbrook	Norwich, N. Y	Feb. 25, 1868	74, 890
Last, Boot and shoe	J. H. Noyes	Abington, Mass	Mar. 19, 1861	31, 733
Last, Boot and shoe	J. C. Plumer	Portland, Me	Jan. 3, 1865	45, 748
Last, Boot and shoe	W. H. Rounds	Campello, Mass	Apr. 15, 1873	137, 799
Last, Boot and shoe	S. M. Shorey	Chicago, Ill	Dec. 6, 1870	109, 845
Last, Boot and shoe	J. H. Swain	San Francisco, Cal	May 12, 1868	77, 931
Last, Boot and shoe	P. Ware, jr	Boston, Mass	Sept. 2, 1873	142, 539
Last, Boot and shoe	B. L. White	Westford, Mass	June 30, 1863	39, 104
Last, Boot and shoe	W. Young	Philadelphia, Pa	May 20, 1807	....
Last, Boot and shoe	W. Young	Baltimore, Md	June 10, 1817	....
Last, Boot and shoe nailing	J. A. Safford	Winchester, Mass	Sept. 24, 1872	131, 565
Last, Darning	G. H. Babcock	Providence, R. I	Sept. 25, 1866	58, 199
Last, Darning	D. E. Holden	Cleveland, Ohio	Nov. 21, 1865	51, 047
Last-fastener	H. Brown	Burton, Ohio	Dec. 8, 1868	84, 677
Last-fastening	W. S. Huntington	New York, N. Y	July 30, 1867	67, 307
Last-finishing machine	J. McComber	Herkimer, N. Y	Jan. 5, 1864	41, 121
Last-finishing machine	J. D. Spiller	Malden, Mass	Aug. 6, 1872	130, 251
Last for machine-sewed turned shoe	L. R. Blake	Boston, Mass	Aug. 16, 1870	106, 459
Last-head block	C. F. Pollard	Lynn, Mass	Dec. 16, 1862	37, 179
Last-holder	J. Dickinson	Painesville, Ohio	Nov. 13, 1860	30, 620
Last-holder	A. G. Mack	Rochester, N. Y	Feb. 21, 1860	27, 229
Last-holder	G. H. Smith	Lowell, Mass	Sept. 8, 1863	39, 839
Last-holder	A. J. Tewksbury	Haverhill, Mass	Nov. 10, 1857	18, 614
Last holder or jack	G. M. Wells	Chicago, Ill	Apr. 28, 1868	77, 425
Last-holder, Revolving	H. G. De Witt	Napanock, N. Y	June 29, 1852	9, 067

Index of patents issued from the United States Patent Office from 1790 to 1873, inclusive—Continued.

Invention.	Inventor.	Residence.	Date.	No.
Last-holder, Revolving	B. Marshall	Philadelphia, Pa	Apr. 28, 1857	17, 160
Last-holder, Revolving	J. Mumford	Clarksburgh, Ohio	June 3, 1856	15, 045
Last-holder, Rotary	D. Philbrick	Manchester, N. H	May 25, 1858	20, 393
Last-holder, Shoemaker's	H. Calder and G. Burgess	Richmond, Va	Oct. 26, 1869	96, 297
Last-holder, Shoemaker's	M. J. Stein	New York, N. Y	Mar. 23, 1869	88, 227
Last, Hollow metallic	S. H. Whorf	Malden, Mass	Sept. 29, 1857	18, 310
Last, Iron	J. Godfrey	New York, N. Y	May 18, 1869	90, 165
Last-lock	D. Goodyear	Ithaca, N. Y	June 11, 1867	65, 560
Last-machine	R. R. Frohock	Boston, Mass	Mar. 23, 1869	88, 029
Last-machine	J. W. Town	South Woodbury, Vt	Sept. 8, 1863	39, 847
Last, Metal	G. G. Townsend	Rochester, N. Y	Dec. 8, 1868	84, 720
Last-shaping machine	E. R. Stilwell	Dayton, Ohio	Sept. 17, 1861	33, 313
Last, Shoe	S. K. Abbott	Salem, N. H	Jan. 31, 1865	46, 174
Last, Shoe	H. R. Bean	Marlborough, Mass	Apr. 20, 1869	89, 009
Last, Shoe	T. Daugherty	Erie, Pa	Mar. 21, 1854	10, 679
Last, Shoe	J. C. F. Deecken	New York, N. Y	Feb. 15, 1859	22, 942
Last, Shoe	N. Jones	Homer, N. Y	Sept. 16, 1862	36, 495
Last, Shoe	F. Maynard	Cambridge, Mass	Sept. 25, 1860	30, 148
Last, Shoe	W. J. B. Mills	Philadelphia, Pa	Jan. 30, 1872	123, 116
Last, Shoe	W. J. B. Mills	Philadelphia, Pa	Mar. 5, 1872	124, 280
Last, Shoe	W. J. B. Mills	Philadelphia, Pa	May 14, 1872	126, 824
Last, Shoe	W. J. B. Mills	Philadelphia, Pa	May 21, 1872	126, 896
Last, Shoe	P. P. Paul	Brooklyn, N. Y	Dec. 2, 1873	145, 124
Last, Shoe	J. Schmitt	New Albany, Ind	Aug. 11, 1868	81, 014
Last, Shoe	W. C. Shipherd	New York, N. Y	May 17, 1864	42, 800
Last, Shoe	W. C. Shipherd	Willoughby, Ohio	Oct. 26, 1869	96, 152
Last, Shoe	J. Whistler	Carlisle, Pa	Apr. 24, 1849	6, 381
Last, Shoemaker's	S. Carsley, 2d	Harrison, Me	Apr. 2, 1830	....
Last, Shoemaker's	J. C. Plumer	Portland, Me	July 17, 1860	29, 225
Last, Slide-block	P. A. Newhall	Lynn, Mass	Feb. 18, 1868	74, 580
Last, yarn and needle holder combined, Stocking	D. B. Keith	East Boston, Mass	Nov 22, 1870	109, 519
Lasts, Form-plate for bottoms of shoe	J. G. Ross	Philadelphia, Pa	June 17, 1873	140, 078
Lasts, &c., from the same pattern, Machinery for turning right and left.	S. Huntington	Middlefield, N. Y	Feb. 20, 1849	6, 131
Lasts, Machine for manufacturing shoe	A. and W. P. Haskell	North Brookfield, Mass	Mar. 31, 1857	16, 917
Lasts, Machine for mounting the uppers of boots and shoes on.	J. P. Molliere	Lyons, France	Dec. 18, 1855	13, 951
Lasts, &c., Machinery for turning	E. Webber and C. Hartshorn	Gardiner, Me	Apr. 3, 1849	6, 253
Lasts, Machinery for turning right and left	C. Hartshorn and W. B. Shaw	Gardiner, Me	Nov. 13, 1849	6, 869
Lasts, Securing and releasing blocks of	A. J. Bramhart	Hartford, N. Y	Apr. 15, 1856	14, 647
Laster and binder	J. M. Hirlinger	Red Rock, Pa	Dec. 10, 1867	72, 041
Laster and punch, Shank	F. Henderson	Marietta, Ohio	June 28, 1870	104, 733
Laster, Boot	L. Barnett	Leechburgh, Pa	Mar. 8, 1870	100, 490
Laster, Boot and shoe shank	O. R. Clark	La Fayette, Ind	Oct. 10, 1871	119, 741
Laster, Boot-shank	C. W. Bliss and O. M. Adams	Milford, Mass	Sept. 24, 1867	69, 066
Laster, Shank	J. H. Bean	Cincinnati, Ohio	Mar. 18, 1873	136, 896
Laster, Shank	J. and A. B. Cain	Dubuque, Iowa	June 6, 1865	48, 052
Laster, Shank	D. G. Chase	Boston, Mass	Nov. 1, 1859	25, 995
Laster, Shank	O. R. Clark and F. H. Slyter	Marengo, Ill	Feb. 6, 1866	52, 389
Laster, Shank	W. N. Flagg	Boylston, Mass	Jan. 30, 1866	52, 280
Laster, Shank	W. H. Hanna	Chico, Cal	Nov. 25, 1873	144, 903
Laster, Shank	F. Henderson	Marietta, Ohio	Mar. 16, 1869	87, 773
Laster, Shank	R. B. Perkins	Almond, N. Y	May 13, 1873	138, 926
Laster, Shank	L. Rastetter and A. Simcox	Fort Wayne, Ind	May 28, 1867	65, 117
Laster, Shank	W. Steele and F. Henderson	Sistersville, W. Va	Oct. 22, 1867	70, 040
Laster, Shank	D. Witt	Hubbardstown, Mass	July 16, 1872	129, 301
Laster, Shoe	P. Thompson	Sardis, Ohio	June 23, 1868	79, 087
Lasting and nailing boots and shoes, Device for	L. R. Blake	Fort Wayne, Ind	Nov. 5, 1872	132, 794
Lasting boots and shoes	I. N. Beals	North Bridgewater, Mass	Apr. 20, 1869	89, 007
Lasting boots and shoes	J. C. Wightman	Boston, Mass	Mar. 4, 1873	136, 476
Lasting boots and shoes	J. Brackett	Lynn, Mass	Aug. 5, 1873	141, 420
Lasting boots and shoes	R. C. Lambart	Quincy, Mass	Jan. 21, 1873	135, 133
Lasting boots and shoes, Device for	J. G. Rust	Xenia, Ohio	Mar. 17, 1868	75, 580
Lasting boots and shoes, Machine for	C. W. Glidden	Lynn, Mass	Feb. 4, 1873	135, 539
Lasting boots and shoes, Machine for	C. W. Glidden	Lynn, Mass	Feb. 4, 1873	135, 540
Lasting boots and shoes, Machine for	C. W. Glidden	Lynn, Mass	Feb. 25, 1873	136, 317
Lasting boots and shoes, Machine for	S. Hart	New Haven, N. Y	June 20, 1846	4, 587
Lasting boots and shoes, Machine for	J. Kimball	Boston, Mass	Sept. 8, 1857	18, 152
Lasting boots and shoes, Machine for	J. Purinton, jr	Lynn, Mass	Oct. 4, 1859	25, 673
Lasting boots and shoes, Machinery for	C. H. Trask and H. F. Wheeler	Lynn and Boston, Mass	Sept. 9, 1873	142, 657
Lasting boots and shoes, Method of	W. Chambers	Lynn, Mass	June 28, 1870	104, 828
Lasting boots and shoes, Mode of	I. R. Rogers	Lynn, Mass	Jan. 7, 1868	73, 048
Lasting boots and shoes, Nailer for	G. McKay	Cambridge, Mass	Dec. 17, 1872	134, 083
Lasting boots and shoes, Nailing machine for	H. F. Wheeler	Boston, Mass	Dec. 24, 1872	134, 231
Lasting boots, Implement for	H. Conant	Worcester, Mass	Aug. 24, 1852	9, 213
Lasting boots, Instrument for	B. Livermore	Hartland, Vt	Aug. 24, 1852	9, 216
Lasting boots, &c., Machine for	J. Eells, jr	Unadilla, N. Y	Mar. 25, 1825	....
Lasting implement, Boot and shoe	F. O. Claflin	Brooklyn, N. Y	Sept. 29, 1868	82, 599
Lasting-instrument	T. Daugherty	Erie, Pa	Aug. 1, 1854	11, 453
Lasting-iron	A. J. Smith	Canal Dover, Ohio	June 22, 1869	91, 781
Lasting-jack	G. F. Seaver	Haverhill, Mass	Oct. 8, 1872	131, 976
Lasting-jack	J. C. Wightman	Newton, Mass	Oct. 22, 1872	132, 510
Lasting-jack for boots, &c	J. C. Drew	Auburn, Me	Dec. 9, 1873	145, 407
Lasting-knee	C. H. Haskell	Lynn, Mass	May 28, 1872	127, 343
Lasting-machine	W. E. Fischer	Boston, Mass	Nov. 1, 1864	44, 916
Lasting-machine	C. W. Glidden	Lynn, Mass	Jan. 21, 1873	135, 111
Lasting-machine	C. W. Glidden	Lynn, Mass	June 24, 1873	140, 265
Lasting-machine	K. Grassau	Virginia City, Nev	Apr. 5, 1870	101, 456
Lasting-machine	H. N. Moyon and J. E. Demercier.	Paris, France	Mar. 22, 1870	101, 151
Lasting-machine	W. Wells	Boston, Mass	May 1, 1860	28, 120
Lasting-machine	W. Wells	Middleton, Mass	Mar. 15, 1864	41, 967
Lasting-machine	T. Wolcott	Stowe, Mass	Sept. 8, 1863	39, 870
Lasting machine, Boot	D. Harrington	German Flats, N. Y	Aug. 26, 1845	4, 163
Lasting machine, Boot	B. Livermore and N. F. English.	Hartland, Vt	Apr. 17, 1847	5, 063
Lasting machine, Shoe	C. H. Trask	Lynn, Mass	Oct. 4, 1870	107, 981
Lasting-machines, Toe-piece for	W. E. Fischer	Boston, Mass	June 1, 1869	90, 651

Index of patents issued from the United States Patent Office from 1790 *to* 1873, *inclusive*—Continued.

Invention.	Inventor.	Residence.	Date.	No.
Lasting-machines, Toe-piece for	A. S. McIntire and N. S. Thompson.	Stoneham, Mass	Feb. 14, 1865	46,375
Lasting-mechanism	L. R. Blake	Brooklyn, N. Y	Sept. 30, 1873	143,322
Lasting shoes	W. H. Lovejoy	Lowell, Mass	Feb. 21, 1871	112,157
Lasting-tool	J. H. Bean	Cincinnati, Ohio	July 29, 1873	141,255
Lasting-tool	L. R. Blake	Brooklyn, N. Y	Feb. 25, 1873	136,300
Lasting-tool	C. W. Glidden	Lynn, Mass	Feb. 25, 1873	136,318
Lasting-tool	F. Henderson	Marietta, Ohio	Apr. 13, 1869	88,958
Latch	H. and S. W. Budd	Philadelphia, Pa	Feb. 9, 1869	86,641
Latch	J. A. Clarke	New York, N. Y	Feb. 15, 1870	99,845
Latch	T. Dolan	Albany, N. Y	May 31, 1870	103,586
Latch	S. W. Drowne	Norwich, Conn	Mar. 15, 1870	100,872
Latch	H. H. Elwell	South Norwalk, Conn	Mar. 7, 1865	46,650
Latch	B. Erbe	Snowden, Pa	Dec. 8, 1863	40,829
Latch	M. P. Favor	East Northwood, N. H	May 24, 1870	103,314
Latch	J. H. Fisher	Chicago, Ill	Feb. 20, 1872	123,817
Latch	F. W. Gammell	Spring Valley, Iowa	Jan. 11, 1870	98,684
Latch	R. Geselbracht and F. Frey	Galena, Ill	Dec. 7, 1869	97,498
Latch	H. Isbell	Meriden, Conn	Nov. 8, 1834	....
Latch	R. Kinsly	Springfield, Mass	Aug. 7, 1847	5,225
Latch	G. W. Large	Yellow Springs, Ohio	Dec. 14, 1869	97,934
Latch	N. Petré	New York, N. Y	Oct. 5, 1869	95,508
Latch	S. M. Richardson	New York, N. Y	Nov. 8, 1864	44,976
Latch	J. C. Robie	Binghamton, N. Y	Nov. 15, 1864	45,078
Latch	G. A. Seaver	New York, N. Y	Nov. 9, 1869	96,624
Latch	J. Shepard and J. Sigourney	Bristol, Conn	Aug. 13, 1867	67,811
Latch	A. Spangler	Philadelphia, Pa	Nov. 17, 1868	84,227
Latch	A. A. Stuart	Plainfield, Iowa	Nov. 8, 1870	109,074
Latch	E. A. Tuttle	Brooklyn, N. Y	May 16, 1865	47,758
Latch	M. J. Woodruff	New Britain, Conn	Nov. 30, 1869	97,332
Latch	G. C. Worth	Upper Sandusky, Ohio	June 7, 1864	43,063
Latch and bolt, Combined	P. P. Child	Saint Louis, Mo	Jan. 7, 1868	73,078
Latch and bolt, Combined	M. J. Meyer	Washington, D. C	Oct. 3, 1865	50,260
Latch and catch	J. B. Farmer	Indianapolis, Ind	Aug. 6, 1867	67,518
Latch and door-check	R. L. Omensetter	Philadelphia, Pa	Dec. 16, 1873	145,523
Latch and door-lock, Combined knob	H. W. Busse	Chicago, Ill	Mar. 10, 1868	75,360
Latch and lock, Case for reversible	J. Kinzer	Pittsburgh, Pa	Feb. 13, 1872	123,567
Latch and lock combined	J. Adair	Pittsburgh, Pa	Jan. 16, 1872	122,794
Latch and lock, Combined	A. M. Adams	Sacramento, Cal	July 29, 1873	141,193
Latch and lock, Combined	J. H. Allen and J. Schwab	Louisville, Ky	Aug. 31, 1869	94,172
Latch and lock, Combined	W. N. Bailey	Duplain, Mich	Jan. 18, 1870	98,907
Latch and lock, Combined	J. Brady	Norwich, Conn	Jan. 26, 1869	86,276
Latch and lock, Combined	G. W. Cilley	Norwich, Conn	May 11, 1869	89,853
Latch and lock, Combined	W. H. Cloud	Fremont, Ohio	Jan. 5, 1869	85,644
Latch and lock, Combined	C. L. Dean	Newark, N. J	Mar. 15, 1870	100,734
Latch and lock, Combined	N. Edwards	Newark, Ohio	Sept. 8, 1868	81,990
Latch and lock, Combined	V. Frazee	San Francisco, Cal	Dec. 12, 1871	121,769
Latch and lock, Combined	V. Frazee	San Francisco, Cal	Sept. 3, 1872	131,092
Latch and lock, Combined	S. A. Green	Lexington, Ind	Sept. 1, 1868	81,625
Latch and lock, Combined	T. Hahn	New York, N. Y	July 12, 1870	105,197
Latch and lock, Combined	A. J. Hollenback	North Bangor, N. Y	Mar. 11, 1873	136,723
Latch and lock, Combined	F. Hoppe	Stuttgart, Germany	May 10, 1870	102,820
Latch and lock, Combined	F. L. Johnson	Wallingford, Conn	Sept. 22, 1868	82,418
Latch and lock, Combined	G. Müllar	New York, N. Y	Mar. 4, 1873	136,448
Latch and lock, Combined	J. H. Morse	Peoria, Ill	Nov. 28, 1871	121,253
Latch and lock, Combined	A. Ochsner	New Haven, Conn	June 15, 1869	91,479
Latch and lock, Combined	H. G. Pein	Peoria, Ill	June 29, 1869	91,865
Latch and lock, Combined	N. Petré	New York, N. Y	Oct. 5, 1869	95,506
Latch and lock, Combined	N. Petré	New York, N. Y	Oct. 5, 1869	95,507
Latch and lock, Combined	F. P. Pfleghar	New Haven, Conn	Mar. 8, 1870	100,661
Latch and lock, Combined	A. Sprague	Poland, N. Y	Nov. 24, 1868	84,391
Latch and lock, Combined	M. P. Warner and E. W. Payne	Morrison, Ill	Oct. 3, 1871	119,673
Latch and lock, Combined	T. Weaver	Harrisburgh, Pa	Oct. 26, 1869	96,175
Latch and lock, Combined knob	C. Brown	Goshen, N. Y	Jan. 21, 1868	73,498
Latch and lock, Combined knob	M. B. Foote	New England Village, Mass	Oct. 22, 1867	70,081
Latch and lock, Combined knob	J. Imray	London, England	Apr. 23, 1872	125,957
Latch and lock, Combined knob	J. B. Kelley	Brandon, Vt	Mar. 16, 1869	88,791
Latch and lock, Combined knob	J. McLeod	San Francisco, Cal	Apr. 13, 1869	88,974
Latch and lock, Combined knob	J. Peck	New Haven, Conn	July 16, 1872	129,422
Latch and lock, Combined knob	F. Raymond	Woodhaven, N. Y	Apr. 6, 1869	88,737
Latch and lock, Combined knob	G. Schumacher	New York, N. Y	Mar. 3, 1868	75,204
Latch and lock, Combined knob	B. Seegmuller	New York, N. Y	July 23, 1867	67,140
Latch and lock, Combined knob	W. H. Sullenberger	Harrisburgh, Pa	Sept. 7, 1869	94,524
Latch and lock, Combined knob	J. H. Vickers	Norwich, Conn	May 5, 1868	77,552
Latch and lock, Door	A. Bingham	Unity, Me	July 7, 1835	....
Latch and lock, Door	A. Iske and J. Teufel	Lancaster, Pa	Oct. 25, 1859	25,900
Latch and lock for doors, Gravitating combined	G. E. Sellers	Cincinnati, Ohio	May 12, 1842	2,621
Latch and lock for gates, Combined	H. R. Van Eps	Peoria, Ill	Aug. 16, 1870	106,519
Latch and lock for sliding doors, Combined	C. W. Chappell	Watertown, Wis	Apr. 2, 1872	125,168
Latch and lock for sliding doors, Combined	A. F. Whiting	Bath, Me	Feb. 6, 1872	123,533
Latch and lock, Knob	S. H. Wheeler	Dowagiac, Mich	July 11, 1871	117,023
Latch and lock, Spring	W. A. Ives	New Haven, Conn	Nov. 18, 1856	16,089
Latch, Automatic	E. Voigt	Philadelphia, Pa	Feb. 22, 1870	100,220
Latch-bolt	J. Adt	Waterbury, Conn	May 14, 1861	32,270
Latch-bolt, Constructing	S. Oppenheimer	Peru, Ind	Jan. 8, 1867	61,090
Latch, bolt, and lock, Spiral-spring	A. Blauvelt	New Brunswick, N. J	Feb. 9, 1825	....
Latch-bolt, Reversible	B. G. Hosmer	Nashua, N. H	Jan. 26, 1864	41,379
Latch-bolt, Spring	E. M. Ray	Norfolk County, Mass	Oct. 23, 1849	6,808
Latch, Carriage-door	N. Belvallette	Paris, France	Nov. 19, 1867	71,123
Latch, Coach-door	J. Thielemann	Newark, N. J	June 3, 1873	139,485
Latch, Cupboard	E. Doen	New Britain, Conn	June 14, 1864	43,105
Latch, Cupboard	B. P. Grover	Holyoke, Mass	Aug. 15, 1865	49,398
Latch, Cupboard	A. D. Judd	New Haven, Conn	July 9, 1867	66,595
Latch, Cupboard	A. D. Judd	New Haven, Conn	Apr. 25, 1871	114,152
Latch, Cupboard	W. E. Sparks	New Haven, Conn	June 14, 1870	104,370
Latch, Cupboard	W. E. Sparks	New Haven, Conn	Sept. 13, 1870	107,418
Latch, Cupboard	W. E. Sparks and P. Bradford	New Haven, Conn	Mar. 26, 1872	124,981
Latch, Cupboard	W. B. Wadsworth	Cleveland, Ohio	June 9, 1863	38,852

Index of patents issued from the United States Patent Office from 1790 *to* 1873, *inclusive*—Continued.

Invention.	Inventor.	Residence.	Date.	No.
Latch, Door	J. Alrichs	Wilmington, Del	May 24, 1845	4, 060
Latch, Door	F. and F. Babcock	Middletown, Conn	Sept. 29, 1868	82, 581
Latch, Door	T. C. Ball	Keene, N. H	Mar. 16, 1858	19, 614
Latch, Door	E. V. Bitner	Lock Haven, Pa	Apr. 18, 1871	113, 729
Latch, Door	B. M. Bosworth	Warren, R. I	Mar. 6, 1840	1, 507
Latch, Door	P. Bradford	New Haven, Conn	Dec. 10, 1867	71, 965
Latch, Door	E. K. Breckenridge	West Meriden, Conn	May 16, 1871	114, 910
Latch, Door	E. W. Brettell	Elizabeth, N. J	Apr. 20, 1869	89, 197
Latch, Door	E. W. Brettell	Elizabeth, N. J	Oct. 19, 1869	95, 978
Latch, Door	F. Clymer	Galion, Ohio	Dec. 10, 1867	71, 982
Latch, Door	J. H. Cooper	Philadelphia, Pa	Feb. 3, 1863	37, 567
Latch, Door	F. M. Crossett	Piermont, N. Y	Jan. 13, 1863	37, 386
Latch, Door	J. C. Curryer and W. C. Young	Thorntown, Ind	July 6, 1869	92, 280
Latch, Door	F. W. Dean	Tremont, Ill	Aug. 17, 1869	93, 685
Latch, Door	J. Dearborn	Seabrook, N. H	Aug. 19, 1862	36, 207
Latch, Door	H. L. De Zeng	Geneva, N. Y	Apr. 3, 1866	53, 585
Latch, Door	H. Duntze	New Haven, Conn	July 6, 1839	1, 220
Latch, Door	B. Erbe	Pittsburgh, Pa	Nov. 6, 1866	59, 373
Latch, Door	J. B. Evans	Millville, N. J	Nov. 26, 1867	71, 474
Latch, Door	S. W. Fosdick and A. C. Dakin	Clinton, Mass	Mar. 7, 1865	46, 653
Latch, Door	H. Hackman, jr	Paque, Pa	Dec. 7, 1858	22, 234
Latch, Door	C. F. Herrick	Independence, Iowa	June 29, 1869	92, 047
Latch, Door	J. M. Hoggan	New Haven, Conn	Nov. 25, 1841	2, 376
Latch, Door	M. Howland	Waterbury, Conn	May 31, 1859	24, 214
Latch, Door	O. Judd	Cherry Valley, N. Y	Dec. 10, 1840	1, 888
Latch, Door	E. King	Taunton, Mass	Nov. 12, 1867	70, 860
Latch, Door	E. Manley	Marion, N. Y	Jan. 9, 1866	51, 954
Latch, Door	E. Manley	Marion, N. Y	Aug. 3, 1869	93, 324
Latch, Door	J. Moser	Mendota, Ill	July 27, 1869	93, 112
Latch, Door	O. B. Rand	Kalamazoo, Mich	June 24, 1873	140, 307
Latch, Door	O. B. Rand	Kalamazoo, Mich	Nov. 4, 1873	144, 288
Latch, Door	A. Rix	San Francisco, Cal	Oct. 21, 1862	36, 732
Latch, Door	A. H. Rowe	Hartford, Conn	Feb. 20, 1866	52, 750
Latch, Door	H. Sanders	Utica, N. Y	May 9, 1871	114, 608
Latch, Door	S. W. Skinner	Lyons, N. Y	May 21, 1872	126, 907
Latch, Door	T. Slaight	Newark, N. J	Nov. 6, 1860	30, 594
Latch, Door	F. Stowe	Huron, Ohio	Nov. 25, 1873	144, 931
Latch, Door	R. S. Underhill	Bath, N. Y	May 22, 1860	28, 425
Latch, Door	H. Underwood	Tolland, Conn	Sept. 27, 1864	44, 476
Latch, Door	J. Walton	Brooklyn, N. Y	Sept. 11, 1860	30, 016
Latch, Door and gate	W. Broomhall	Circleville, Ohio	Jan. 29, 1867	61, 604
Latch, Door and gate	L. Brumbach	Reading, Pa	Mar. 16, 1869	87, 907
Latch, Door and gate	J. A. Park	Lansing, Mich	Feb. 12, 1867	62, 061
Latch, Door and window	S. W. Gear	Whitestone, N. Y	June 2, 1868	78, 521
Latch fastening, Door	E. Morris	Burlington, N. J	May 30, 1848	5, 606
Latch, Door knob	A. Bingham	Newtonville, Mass	Feb. 25, 1868	74, 746
Latch, Door knob	L. and W. Hotchkiss	Derby, Conn	Mar. 6, 1840	1, 508
Latch, Door knob	A. W. Upton	Lowell, Mass	Mar. 12, 1867	62, 908
Latch, door-lock, &c	E. Robinson and W. Hall	Boston, Mass	Mar. 3, 1841	1, 995
Latch, Door-lock	J. P. Sherwood	Sandy Hill, N. Y	May 6, 1841	2, 082
Latch, Door locking	E. Field	Greenwich, Conn	July 3, 1855	13, 163
Latch, Door locking	E. Halsey	San José, Cal	Oct. 14, 1873	143, 692
Latch, Door thumb	E. Parker	Meriden, Conn	Mar. 28, 1842	2, 518
Latch, Double-door	J. W. Still	San Francisco, Cal	Dec. 14, 1869	97, 985
Latch-fastening	H. N. Avery	New York, N. Y	Jan. 2, 1866	51, 787
Latch-fastening	H. M. Jones	West Meriden, Conn	Aug. 14, 1866	57, 150
Latch-fastening	L. R. Livingston, J. J. Roggen, and C. Adams.	Pittsburgh, Pa	Nov. 24, 1846	4, 862
Latch fastening, Door	J. F. Tozer	Binghamton, N. Y	Feb. 3, 1863	37, 598
Latch for blinds or shutters	B. S. Huntington	New York, N. Y	Nov. 28, 1865	51, 186
Latch for doors, Locking knob	J. Wertsbaugher	La Grange, Ind	Dec. 17, 1867	72, 347
Latch for farm-gates, Catch	J. Summers	Raleigh, Va	Apr. 20, 1858	20, 008
Latch, Gate	H. D. Alderfer	Grater's Ford, Pa	Oct. 3, 1871	119, 491
Latch, Gate and door	J. H. Allison	Charlottesville, Ind	Oct. 3, 1871	119, 552
Latch, Gate	E. Barks and L. H. Emmons	Noblesville, Ind	Sept. 20, 1870	107, 589
Latch, Gate	C. Bartlow	Richmond, Ind	Dec. 31, 1872	134, 454
Latch, Gate	M. J. Briar	Oxford, Ind	Nov. 12, 1867	70, 794
Latch, Gate	J. T. Bryan	Lebanon, Ind	Apr. 2, 1867	63, 466
Latch, Gate	C. B. Clark	Buffalo, N. Y	Sept. 26, 1871	119, 263
Latch, Gate	H. Clomo	Galena, Ill	May 17, 1870	103, 143
Latch, Gate	C. Cole	Ithaca, N. Y	Nov. 8, 1870	108, 971
Latch, Gate	C. Cole	Dayton, Ohio	Jan. 17, 1871	110, 957
Latch, Gate	J. P. Curry	Vincennes, Ind	July 16, 1872	129, 324
Latch, Gate	A. K. Davis	Carey, Ohio	Nov. 5, 1867	70, 537
Latch, Gate	B. F. Dickey	Marshall, Mich	Apr. 30, 1872	126, 191
Latch, Gate	E. E. Earll and J. B. Hunter	Brooklyn and New York, N. Y.	June 14, 1870	104, 287
Latch, Gate	J. Ellis	Detroit, Mich	Aug. 6, 1861	32, 990
Latch, Gate	J. Ellis	Detroit, Mich	Nov. 25, 1862	37, 028
Latch, Gate	J. R. Finley	Delphi, Ind	Jan. 14, 1873	134, 877
Latch, Gate	W. Fosket	Meriden, Conn	Feb. 7, 1871	111, 626
Latch, Gate	C. W. Fox	Saint Louis, Mo	June 20, 1871	116, 042
Latch, Gate	C. H. Frietog	Springfield, Ill	July 25, 1871	117, 401
Latch, Gate	E. O. Frink	Indianapolis, Ind	Dec. 26, 1865	51, 710
Latch, Gate	W. R. Goodrich	Whitestone, N. Y	June 21, 1870	104, 447
Latch, Gate	B. Hendrickson	Huntington, N. Y	Dec. 15, 1868	84, 945
Latch, Gate	G. B. Howland	Gardner, Ill	Apr. 21, 1868	76, 918
Latch, Gate	S. Ingersoll	Mianus, Conn	Feb. 25, 1868	74, 915
Latch, Gate	J. Johnson and S. Ingersoll	Brooklyn, N. Y	June 14, 1870	104, 320
Latch, Gate	W. H. Kellogg	Duquoin, Ill	Mar. 5, 1867	62, 637
Latch, Gate	F. Ketcham	Elizabeth Township, Pa	Aug. 14, 1866	57, 152
Latch, Gate	J. Kindel	Wilmington, Ohio	Feb. 22, 1870	100, 044
Latch, Gate	J. E. Klein	Oskaloosa, Iowa	Feb. 5, 1867	61, 839
Latch, Gate	G. W. Large	Yellow Springs, Ohio	Feb. 5, 1867	61, 744
Latch, Gate	G. W. Large	Yellow Springs, Ohio	Jan. 30, 1872	123, 267
Latch, Gate	J. H. Lee	Marshall, Tex	Sept. 26, 1871	119, 374
Latch, Gate	J. Leonard	Wilmington, Ohio	June 20, 1865	48, 292
Latch, Gate	E. Manross	Bristol, Conn	Jan. 10, 1860	26, 774

Index of patents issued from the United States Patent Office from 1790 *to* 1873, *inclusive*—Continued.

Invention.	Inventor.	Residence.	Date.	No.
Latch, Gate	J. A. Martin	Strasburgh, Pa	Jan. 12, 1869	85, 839
Latch, Gate	J. W. McLean	Indianapolis, Ind	June 5, 1866	55, 326
Latch, Gate	C. E. Merrifield	Indianapolis, Ind	Jan. 21, 1868	73, 624
Latch, Gate	A. E. Morgan	Poughkeepsie, N. Y	July 28, 1857	17, 908
Latch, Gate	E. Nicholson	Rockport, Ohio	July 9, 1867	66, 515
Latch, Gate, &c	W. Patton	Towanda, Pa	Aug. 30, 1870	106, 955
Latch, Gate	G. M. D. Pomeroy	Ithaca, Ind	Apr. 23, 1867	64, 140
Latch, Gate	F. M. Ranons	Little Shasta Valley, Cal	Oct. 26, 1869	96, 147
Latch, Gate	P. Rasar and D. J. Mayes	Illiopolis, Ill	Aug. 18, 1868	81, 209
Latch, Gate	W. Redhead	Des Moines, Iowa	Feb. 16, 1869	87, 065
Latch, Gate	C. W. Rhoads	Indianapolis, Ind	Apr. 2, 1867	63, 562
Latch, Gate	C. Roberts	Winchester, Ill	Feb. 5, 1867	61, 763
Latch, Gate, &c	C. Roberts	Winchester, Ill	Feb. 25, 1873	136, 183
Latch, Gate	H. W. Robie	Binghamton, N. Y	Dec. 19, 1871	122, 061
Latch, Gate	J. C. Rogers	Alden, N. Y	Dec. 8, 1868	84, 711
Latch, Gate, &c	G. N. Sharp	La Plata, Mo	June 18, 1872	128, 075
Latch, Gate, &c	G. N. Sharp	La Plata, Mo	Dec. 9, 1873	145, 368
Latch, Gate	D. Sheets	Suisun, Cal	Oct. 19, 1869	96, 041
Latch, Gate	W. Shumard	Richmond, Ind	Mar. 22, 1870	101, 055
Latch, Gate	J. W. Still	San Francisco, Cal	Feb. 2, 1869	86, 601
Latch, Gate	W. T. Wells	Decatur, Ill	Feb. 12, 1867	62, 102
Latch, Gate	W. T. Wells	Decatur, Ill	Apr. 9, 1867	63, 769
Latch, Gate	S. B. Williams	Leavenworth, Kans	Dec. 8, 1863	40, 875
Latch, Gate	L. Woodruff	Ann Arbor, Mich	Jan. 29, 1867	61, 594
Latch, Gate and door	C. C. Holcomb	Madison, Wis	Apr. 21, 1868	77, 040
Latch, Gate and door	B. D. Shaw	Beverly, Ohio	Dec. 18, 1866	60, 644
Latch, Gate locking	J. L. Devol	Parkersburgh, W. Va	May 30, 1871	115, 448
Latch, Guard-bolt	E. H. Kent	New York, N. Y	Oct. 18, 1870	108, 361
Latch handle, Door	J. Ambrun and E. Olfermann	Leavenworth, Kans	Apr. 22, 1873	138, 112
Latch keeper, Door	W. Harvey	Albany, N. Y	Dec. 19, 1871	122, 011
Latch, Knob	T. C. Ball	Bellows Falls, Vt	Sept. 15, 1868	82, 070
Latch, Knob	J. H. Barnes	Brooklyn, N. Y	Nov. 22, 1864	45, 129
Latch, Knob	A. T. Brooks	New Britain, Conn	July 6, 1869	92, 151
Latch, Knob	A. T. Brooks	New Britain, Conn	July 6, 1869	92, 152
Latch, Knob	A. T. Brooks	New Britain, Conn	July 6, 1869	92, 153
Latch, Knob	A. B. Clemons	Ansonia, Conn	Mar. 23, 1869	88, 133
Latch, Knob	C. J. Colby	Waterbury, Vt	July 25, 1865	48, 909
Latch, Knob	J. Davis	Terre Haute, Ind	June 28, 1870	104, 712
Latch, Knob	J. M. A. Dew	Chicago, Ill	Feb. 25, 1873	136, 141
Latch, Knob	H. H. Elwell	Norwalk, Conn	Mar. 27, 1866	63, 526
Latch, Knob	A. Heyse	Terre Haute, Ind	Aug. 12, 1873	141, 714
Latch, Knob	C. M. Jordan	Stillwater, Minn	June 11, 1872	127, 773
Latch, Knob	R. Lee	Cincinnati, Ohio	Sept. 11, 1866	57, 930
Latch, Knob	W. T. Munger	Branford, Conn	Apr. 3, 1866	53, 745
Latch, Knob	W. T. Munger	Branford, Conn	Nov. 3, 1868	83, 652
Latch, Knob	E. Parker	New Britain, Conn	Jan. 21, 1873	135, 147
Latch, Knob	F. P. Pfleghar	New Haven, Conn	Jan. 11, 1870	98, 795
Latch, Knob	F. B. Pye	Trenton, N. J	Jan. 23, 1866	52, 201
Latch, Knob	O. B. Rand	Kalamazoo, Mich	Feb. 2, 1869	86, 440
Latch, Knob	S. M. Richardson	New York, N. Y	Nov. 8, 1864	44, 977
Latch, Knob	H. Richmond and A. Cloude	West Meriden, Conn	Oct. 3, 1865	50, 273
Latch, Knob	T. Slaight	Newark, N. J	June 8, 1869	91, 174
Latch, Knob	A. W. Upton	Lowell, Mass	Mar. 20, 1866	53, 364
Latch, Knob	R. L. Webb	West Meriden, Conn	Nov. 15, 1864	45, 100
Latch, Knob	A. F. Whiting	Greenville, Conn	June 15, 1869	91, 394
Latch, Knob	A. Williams	Norwich, Conn	June 6, 1865	48, 123
Latch, Knob	J. T. Williams	Chicago, Ill	Sept. 20, 1870	107, 578
Latch-lever	C. C. Lewis	Gainesville, Ala	Dec. 3, 1872	133, 588
Latch, Lever spring	J. N. Houston	Decatur, Ind	July 5, 1864	43, 410
Latch, Lock	N. Petre	New York, N. Y	July 30, 1867	67, 213
Latch lock	N. Petré	New York, N. Y	Feb. 7, 1871	111, 567
Latch lock, Door	H. Bosh	Mount Vernon, N. Y	Oct. 8, 1867	69, 619
Latch lock, Door	J. C. Kline	Pittsburgh, Pa	Jan. 16, 1855	12, 240
Latch, Locking	J. Hardy, 2d, and B. B. Floyd	Andover and Lawrence, Mass.	Sept. 1, 1868	81, 779
Latch, Locking knob	S. W. Drowne	Norwich, Conn	Mar. 24, 1868	75, 879
Latch, Locking knob	C. H. Eiffler	New York, N. Y	Dec. 3, 1867	71, 721
Latch, Locking knob	F. Hegner	Boston, Mass	Apr. 16, 1872	125, 734
Latch, Locking knob	W. L. Imlay	Philadelphia, Pa	Jan. 14, 1868	73, 334
Latch, Locking knob	H. C. Storrs	New York, N. Y	Jan. 21, 1868	73, 666
Latch, Locking-spindle door	W. H. McNamee	Philadelphia, Pa	Jan. 30, 1855	12, 320
Latch-mortise	W. Coover	Erie, Pa	Feb. 17, 1836	......
Latch, Mortise door	L. Foster	Boston, Mass	June 27, 1838	811
Latch, Mortise door	C. S. Gay	Nashua, N. H	Sept. 28, 1837	413
Latch, Mortise door	R. Kinsley	Springfield, Mass	Mar. 7, 1846	4, 402
Latch, Mortise door	W. T. Munger	New Britain, Conn	May 6, 1873	138, 680
Latch, Mortise door	D. N. Ropes	Portland, Me	Sept. 28, 1837	414
Latch, Mortise door	W. Wilson	Northampton, Mass	Nov. 26, 1844	3, 839
Latch, Mortise knob	G. H. Palmer	New Bedford, Mass	Aug. 18, 1867	67, 795
Latch-needles, Manufacture of	T. T. Mayne	Philadelphia, Pa	June 24, 1873	140, 150
Latch, Night	E. W. Brettell	Newark, N. J	July 23, 1867	66, 940
Latch, Night	C. C. Dickerman	Boston, Mass	Feb. 14, 1871	111, 732
Latch, Night	R. S. Foster	Madison, N. J	Jan. 25, 1870	99, 076
Latch, Oblique door	R. T. Fry	Spring Garden, Pa	Dec. 26, 1848	5, 984
Latch or fastener for doors	J. R. Webber	Morris, Ill	June 21, 1870	104, 672
Latch or fastener for window shutters and blinds	A. Hill	Steubenville, Ohio	Feb. 24, 1843	2, 977
Latch, Railway-car-door	J. Stephenson	New York, N. Y	Dec. 13, 1864	45, 442
Latch, Reversible	W. H. Andrews	New Haven, Conn	July 5, 1870	105, 026
Latch, Reversible	W. E. and H. H. Clark	New Haven, Conn	Jan. 26, 1869	86, 326
Latch, Reversible	H. Dotzenroth	Pittsburgh, Pa	Jan. 3, 1871	110, 637
Latch, Reversible	C. R. Fisher	Chelsea, Mass	Aug. 4, 1868	80, 539
Latch, Reversible	E. Halley	Branford, Conn	Nov. 14, 1865	50, 931
Latch, Reversible	J. Hamill	Allegheny, Pa	June 11, 1872	127, 870
Latch, Reversible	C. F. Herrick	Independence, Iowa	July 5, 1870	105, 075
Latch, Reversible	A. D. Judd	New Haven, Conn	Mar. 26, 1872	124, 960
Latch, Reversible	J. Kinzer	Pittsburgh, Pa	Aug. 11, 1868	80, 974
Latch, Reversible	J. Kinzer	Pittsburgh, Pa	May 17, 1870	103, 053
Latch, Reversible	J. Kinzer	Pittsburgh, Pa	May 17, 1870	103, 054

Index of patents issued from the United States Patent Office from 1790 *to* 1873, *inclusive*—Continued.

Invention.	Inventor.	Residence.	Date.	No.
Latch, Reversible	W. T. Munger	Branford, Conn	Sept. 8, 1868	81, 931
Latch, Reversible	W. T. Munger	Branford, Conn	Dec. 8, 1868	84, 704
Latch, Reversible	W. T. Munger	Branford, Conn	Jan. 12, 1869	85, 753
Latch, Reversible	W. T. Munger	Branford, Conn	Jan. 12, 1869	85, 754
Latch, Reversible	W. T. Munger	New Britain, Conn	May 18, 1869	90, 288
Latch, Reversible	W. T. Munger	New Britain, Conn	Mar. 1, 1870	100, 313
Latch, Reversible	W. T. Munger	New Britain, Conn	May 13, 1873	138, 919
Latch, Reversible	G. A. Seaver	New York, N. Y	Nov. 24, 1868	84, 382
Latch, Reversible	W. R. Smith	Branford, Conn	Apr. 5, 1870	101, 673
Latch, Reversible	W. E. Sparks	New Haven, Conn	Oct. 19, 1869	96, 049
Latch, Reversible	W. E. Sparks	New Haven, Conn	May 3, 1870	102, 723
Latch, Reversible	R. L. Webb	New Britain, Conn	Dec. 31, 1867	72, 946
Latch, Reversible	E. Whitney	New Haven, Conn	Dec. 29, 1868	85, 497
Latch, Reversible door	T. Lyons	Hartford, Conn	Mar. 16, 1869	87, 784
Latch, Reversible knob	W. H. Andrews	New Haven, Conn	Mar. 25, 1873	137, 166
Latch, Reversible knob	H. P. Appleton	Norwich, Conn	Sept. 20, 1870	107, 387
Latch, Reversible knob	A. Aston	New Britain, Conn	Mar. 30, 1869	88, 261
Latch, Reversible knob	H. H. Elwell	South Norwalk, Conn	Apr. 28, 1868	77, 264
Latch, Reversible knob	H. H. Elwell	South Norwalk, Conn	July 4, 1861	116, 697
Latch, Reversible knob	B. Erbe	Birmingham, Pa	May 19, 1868	77, 966
Latch, Reversible knob	B. Erbe	East Birmingham, Pa	Feb. 23, 1869	87, 096
Latch, Reversible knob	B. Erbe	Pittsburgh, Pa	Oct. 1, 1872	131, 860
Latch, Reversible knob	B. Erbe	Pittsburgh, Pa	Aug. 5, 1873	141, 430
Latch, Reversible knob	A. M. Hill	New Haven, Conn	Oct. 11, 1870	108, 261
Latch, Reversible knob	W. A. Hopkins	New York, N. Y	July 12, 1870	105, 334
Latch, Reversible knob	C. S. Jennings	Meriden, Conn	Apr. 8, 1873	137, 685
Latch, Reversible knob	C. Moody	Toronto, Canada	Aug. 19, 1873	141, 883
Latch, Reversible knob	W. T. Munger	Branford, Conn	Mar. 3, 1868	75, 184
Latch, Reversible knob	W. T. Munger	New Britain, Conn	Apr. 18, 1871	113, 909
Latch, Reversible knob	W. A. C. Oaks	Reading, Pa	Aug. 25, 1868	81, 529
Latch, Reversible knob	J. E. Parker	Meriden, Conn	May 5, 1868	77, 650
Latch, Reversible knob	J. E. Parker	Meriden, Conn	May 12, 1868	77, 909
Latch, Reversible knob	F. P. Pfleghar	New Haven, Conn	Mar. 2, 1869	87, 510
Latch, Reversible knob	F. P. Pfleghar	New Haven, Conn	July 6, 1869	92, 354
Latch, Reversible knob	H. M. Ritter	Cincinnati, Ohio	May 7, 1867	64, 570
Latch, Reversible knob	H. M. Ritter	Cincinnati, Ohio	May 7, 1867	64, 571
Latch, Reversible knob	H. M. Ritter	Cincinnati, Ohio	Mar. 28, 1871	113, 096
Latch, Reversible knob	H. M. Ritter	Cincinnati, Ohio	Mar. 28, 1871	113, 097
Laich, Reversible knob	R. S. Webb	New Britain, Conn	Mar. 25, 1873	137, 159
Latch, Reversible knob	A. F. Whiting	Crestline, Ohio	Mar. 25, 1873	137, 271
Latch, Reversible knob	A. F. Whiting	Crestline, Ohio	Mar. 25, 1873	137, 272
Latch, Reversible knob	J. Whittingham	Pittsburgh, Pa	Mar. 15, 1870	100, 830
Latch, Right and left door	G. E. Sellers	Cincinnati, Ohio	May 7, 1842	2, 610
Latch, Safety	J. W. Fifield	Franklin, N. H	Oct. 4, 1870	107, 895
Latch, Sliding-door	H. Belfield	Philadelphia, Pa	Jan. 10, 1860	26, 809
Latch, Stop mortise	F. A. Richardson	Poultney, Vt	Oct. 23, 1866	59, 076
Latch, Thumb	P. Eli and W. and J. A. Blake	New Haven, Conn	July 21, 1840	1, 704
Latch, Thumb	J. Range	Meriden, Conn	May 14, 1861	32, 339
Latch, Window	A. Bingham	Newtonville, Mass	Apr. 23, 1867	63, 988
Latch, Window	E. F. Hofmann	Pougkeepsie, N. Y	Jan. 1, 1867	60, 727
Latches, Adjustable escutcheon for night	W. T. Munger	Branford, Conn	Mar. 19, 1867	62, 968
Latches, Catch for gate and door	G. C. Lawton	Algona, Iowa	May 30, 1871	115, 330
Latches, Guard-attachment for door	W. Miller	Boston, Mass	Nov. 5, 1861	33, 682
Latches, Handle-plate for door	F. Corbin	New Britain, Conn	May 13, 1873	138, 862
Latches, Strike for door	J. E. Terry	Philadelphia, Pa	Apr. 14, 1863	38, 203
Lath and billets for rolling, Metallic	J. Reese	Pittsburgh, Pa	July 5, 1870	104, 997
Lath and clapboard sawing-machine	E. C. Gilman	Canaan, N. H	Aug. 23, 1844	3, 715
Lath-bolts, Machine for cutting	T. Crispin	Bay City, Mich	Aug. 22, 1871	118, 205
Lath-bolting machine	J. C. McIntyre	Fort Edward, N. Y	Sept. 2, 1873	142, 489
Lath-bolting machine	J. M. Stowell	Milwaukee, Wis	Apr. 30, 1872	126, 162
Lath-cutting machine	W. Bullock	Philadelphia, Pa	July 9, 1850	7, 487
Lath-cutting machine	I. Carpenter	Cincinnati, Ohio	May 2, 1832	....
Lath-cutting machine	D. M. Cradit	Ithaca, N. Y	May 22, 1835	....
Lath-cutting machine	O. H. Dibble	Evans, N. Y	Jan. 22, 1833	....
Lath-cutting machine	S. F. Finch and J. Wheeler	Rootstown, Ohio	Aug. 16, 1845	4, 147
Lath-cutting machine	R. Haynes	Oberlin, Ohio	Oct. 5, 1858	21, 675
Lath-cutting machine	D. Hubba	Wheeling, Va	July 18, 1834	....
Lath-cutting machine	C. Learned	Indianapolis, Ind	Apr. 24, 1866	54, 180
Lath-cutting machine	W. W. Lesner	Urbana, Ohio	Oct. 8, 1834	....
Lath-cutting machine	J. N. Lynch	Dillsburgh, Pa	Feb. 16, 1830	....
Lath-cutting machine	C. F. Packard	Greenwich, Conn	Feb. 21, 1854	10, 556
Lath-cutting machine	W. Pierson	Lebanon, Ohio	Dec. 31, 1833	....
Lath-cutting machine	S. Willard	Cincinnati, Ohio	Mar. 16, 1832	....
Lath-cutting machine	J. M. Winslow	Campbell's Port, Ohio	Oct. 3, 1846	4, 787
Lath cutting and splitting machine	B. Langdon	Troy, N. Y	Dec. 15, 1835	....
Lath for building	J. L. Brabyn	New York, N. Y	June 16, 1857	17, 550
Lath for building	D. Phillips	Shaftsbury, Vt	Jan. 17, 1865	45, 936
Lath for fire-proof ceilings of houses, Metallic	P. Sumner	New York, N. Y	Apr. 25, 1844	3, 566
Lath for plastering	E. D. Averell	New York, N. Y	May 28, 1867	65, 153
Lath-frame	A. Reed	Mankato, Minn.	July 23, 1867	67, 073
Lath from boards, Cutting	J. Jameson	Springfield, Ohio	Aug. 24, 1831	....
Lath-machine	J. Black	Memphis, Tenn	Nov. 16, 1858	22, 058
Lath-machine	A. Blaikie and W. Clark	Saint Clair, Mich	Nov. 6, 1855	13, 746
Lath-machine	S. P. Brower and W. H. Knox	Glen's Falls, N. Y	Sept. 30, 1873	143, 325
Lath-machine	J. L. Brown	Indianapolis, Ind	Jan. 15, 1856	14, 126
Lath-machine	T. Bruno	Saginaw, Mich	Dec. 16, 1873	145, 486
Lath-machine	J. H. Butler	Hampden, Me	Feb. 27, 1872	124, 034
Lath-machine	J. Dobbins	Litchfield, Mich	Sept. 28, 1869	95, 207
Lath-machine	H. Frisbee	Olmstead, Ohio	May 16, 1854	10, 905
Lath-machine	R. J. Gatling	Indianapolis, Ind	July 10, 1860	29, 072
Lath-machine	J. Gilman	Nashua, N. H	Mar. 25, 1856	14, 499
Lath-machine	A. Gummer	Indianapolis, Ind	Apr. 16, 1861	32, 103
Lath-machine	J. T. Hall	Alma, Mich	Nov. 8, 1870	109, 005
Lath-machine	S. and C. Howard	Easton and Bristol, Mass	Dec. 17, 1834	....
Lath-machine	J. L. Knowlton	Philadelphia, Pa	Aug. 16, 1870	106, 374
Lath-machine	J. C. Mackay	Ionia, Mich	Feb. 28, 1871	112, 261
Lath-machine	O. C. Macklett	Saint Paul, Minn	June 30, 1868	79, 366
Lath-machine	J. W. McLean and A. Gummer	Indianapolis, Ind	June 26, 1860	28, 883

Index of patents issued from the United States Patent Office from 1790 to 1873, inclusive—Continued.

Invention.	Inventor.	Residence.	Date.	No.
Lath-machine	O. C. Meigs	Dubuque, Iowa	Aug. 20, 1872	130, 650
Lath-machine	W. Merrell	Randolph, Ohio	Sept. 23, 1851	8, 376
Lath-machine	W. B. Noyes	Manchester, N. H	Jan. 18, 1870	98, 996
Lath-machine	J. B. Okey	Indianapolis, Ind	Jan. 27, 1857	16, 493
Lath-machine	J. Oothoudt	Minneapolis, Minn	Apr. 5, 1870	101, 653
Lath-machine	S. M. Palmer	Glen's Falls, N. Y	July 16, 1872	129, 492
Lath-machine	S. M. and G. E. Palmer	Sandy Hill, N. Y	Jan. 21, 1870	104, 637
Lath-machine	J. Pefley	Bainbridge, Ind	Dec. 28, 1858	22, 449
Lath-machine	J. H. and A. E. Redstone	Indianapolis, Ind	Nov. 15, 1859	26, 125
Lath-machine	S. B. Rittenhouse	Larwill, Ind	Feb. 25, 1873	136, 182
Lath-machine	C. Schleicher	Louisville, Ky	Jan. 24, 1871	111, 259
Lath-machine	I. R. Shank	Buffalo, Va	June 21, 1853	9, 804
Lath-machine	I. R. Shank	Buffalo, Va	June 6, 1854	11, 035
Lath-machine	E. Smith	Ithaca, N. Y	Apr. 28, 1836	....
Lath-machine	H. C. Smith	Cleveland, Ohio	Sept. 28, 1852	9, 286
Lath-machine	G. W. Tolhurst	Cleveland, Ohio	Dec. 9, 1851	8, 584
Lath-machine	T. Wright and A. P. Howell	Cincinnati, Ohio	Oct. 9, 1827	....
Lath, Machine for making siding	L. Rice	Lockport, N. Y	July 23, 1828	....
Lath machine, Sheet-metal	E. May	Indianapolis, Ind	Apr. 13, 1869	88, 972
Lath-machines, Method of feeding the bolts in	J. Nevison	Morgan, Ohio	May 18, 1858	20, 292
Lath, Metallic	J. B. Cornell	New York, N. Y	June 22, 1858	20, 629
Lath, Metallic	I. V. Holmes	New York, N. Y	Dec. 15, 1868	84, 881
Lath, Metallic	I. V. Holmes	New York, N. Y	Jan. 26, 1869	86, 305
Lath, Metallic	I. V. Holmes	New York, N. Y	Dec. 26, 1871	122, 171
Lath, Metallic	J. M. Manterstock	New York, N. Y	Apr. 12, 1859	23, 597
Lath, Metallic	K. Nirison	Chicago, Ill	Mar. 12, 1872	124, 611
Lath, Metallic	T. O'Callahan	Boston, Mass	Aug. 19, 1873	141, 945
Lath, Metallic	J. F. Walter, jr	New York, N. Y	Jan. 28, 1868	73, 765
Lath, Metallic	C. Welding	New York, N. Y	Apr. 8, 1873	137, 743
Lath, Metallic	W. E. Worthen	New York, N. Y	Feb. 24, 1857	16, 702
Lath, Metallic	W. E. Worthen	New York, N. Y	May 3, 1859	23, 877
Lath-mill	J. S. Hyde	Pent Water, Mich	Nov. 16, 1869	96, 811
Lath-sawing machine	G. P. Appleton	Suncook, N. H	July 8, 1873	140, 570
Lath-sawing machine	S. P. Brower	Glen's Falls, N. Y	Jan. 14, 1873	134, 846
Lath-sawing machine	T. Bruno	Saginaw, Mich	May 5, 1868	77, 449
Lath-sawing machine	W. G. Bulgin	Vienna, N. J	Aug. 25, 1868	81, 475
Lath-sawing machine	J. H. Butler	Hampden, Me	Mar. 15, 1870	100, 722
Lath-sawing machine	E. H. Ewell	Saint Louis, Mich	Oct. 19, 1869	95, 888
Lath-sawing machine	W. P. Hale	Ionia, Mich	Oct. 5, 1869	95, 582
Lath-sawing machine	E. H. Hancock	Augusta, Ga	Jan. 4, 1859	22, 499
Lath-sawing machine	I. A. Hedges and J. M. Story	Cincinnati, Ohio	May 5, 1868	77, 484
Lath-sawing machine	A. Rodgers	Muskegon, Mich	July 8, 1873	140, 647
Lath-sawing machine	W. Tuxworth	Sheridan, Mich	Jan. 18, 1870	98, 898
Lath-sawing machine	E. T. Wheeler and W. H. Vaughn.	Cannelton, Ind	Jan. 7, 1868	73, 214
Lath-sawing machine, Bed for	T. R. Markillie	Winchester, Ill	Jan. 1, 1856	14, 027
Lath, &c., slitting gage	B. K. Crandell	Lockport, N. Y	Aug. 25, 1828	....
Lath surface, Continuous sheet-metal	J. B. Cornell	New York, N. Y	May 13, 1856	14, 854
Lath surface, Metallic	J. B. Cornell	New York, N. Y	Aug. 10, 1858	21, 118
Lath with circular saw, Sawing	C. Madison	Miamisburgh, Ohio	Apr. 4, 1831	....
Laths from the block, Machine for riving	J. L. Brown	Indianapolis, Ind	Feb. 8, 1859	22, 853
Laths, Machine for bundling	J. S. Anderson	Flintville, Wis	Aug. 20, 1872	130, 687
Laths, Machine for sawing bevels on	G. W. Corson	Corson's Post-Office, Pa	July 10, 1860	29, 062
Laths, Machinery for cutting	S. Cheney	Cleveland, Ohio	Sept. 27, 1845	4, 206
Laths, palings, &c., Machine for sawing	S. Head	Millersburgh, Pa	Nov. 24, 1863	40, 690
Lathing and plastering, Mode of	J. G. Vaughn	Middleborough, Mass	Jan. 13, 1857	16, 425
Lathing-apparatus	W. L. Beardsley	Binghamton, N. Y	May 28, 1867	65, 048
Lathing buildings	B. F. Gold	New Haven, Conn	Aug. 1, 1854	11, 423
Lathing, Continuous metallic	B. Cornell	New York, N. Y	Mar. 2, 1858	19, 487
Lathing-machine	G. N. Creamer	Trenton, N. J	Oct. 5, 1869	95, 435
Lathing-machine	G. N. Creamer	Trenton, N. J	Nov. 22, 1870	109, 391
Lathing-machine	T. Hill	New Centreville, Wis	Jan. 21, 1868	73, 604
Lathing-machine	W. S. Wright	Ithaca, N. Y	Apr. 20, 1869	89, 112
Lathe	W. Aldrich	Lowell, Mass	Apr. 10, 1855	12, 662
Lathe	J. P. Allen	Springfield, Ohio	May 9, 1871	114, 512
Lathe	E. and N. Avord	Westfield, Mass	Oct. 12, 1836	....
Lathe	T. R. Bailey	Lockport, N. Y	July 1, 1851	8, 196
Lathe	W. B. Bement	Philadelphia, Pa	May 21, 1867	64, 938
Lathe	A. H. Brown	Albany, N. Y	Apr. 29, 1856	14, 787
Lathe	J. T. Bunce	East Haddam, Conn	May 25, 1858	20, 323
Lathe	H. Chavons	Union City, Ind	Sept. 19, 1871	119, 012
Lathe	C. S. Clark	Providence, R. I	June 29, 1869	91, 910
Lathe	H. O. Clark	Worcester, Mass	June 6, 1854	11, 037
Lathe	T. Collins and A. Castiel	Saint Louis, Mo	Mar. 30, 1869	88, 452
Lathe	J. Cook	Buffalo, N. Y	Apr. 3, 1860	27, 696
Lathe	L. L. Crane	Cleveland, Ohio	June 5, 1866	55, 412
Lathe	M. Gore	Shelby, Mich	June 29, 1869	91, 930
Lathe	O. M. Grimes	Newark, N. J	June 29, 1869	92, 041
Lathe	L. Griscom	Mahanoy Plane, Pa	May 12, 1868	77, 812
Lathe	A. Hatch	New Haven, Conn	Mar. 20, 1866	53, 374
Lathe	D. Heer	Philadelphia, Pa	July 29, 1873	141, 347
Lathe	J. Hess	Niagara Falls, N. Y	July 19, 1859	24, 805
Lathe	H. R. Hill and N. W. Twiss	New Haven, Conn	Sept. 12, 1871	118, 938
Lathe	J. Kievlan	Chicago, Ill	Oct. 12, 1869	95, 694
Lathe	J. Kievlan	Chicago, Ill	Feb. 21, 1871	112, 051
Lathe	S. U. King	Windsor, Vt	Dec. 6, 1870	109, 912
Lathe	B. Lawrence	Lowell, Mass	Dec. 12, 1871	121, 878
Lathe	B. M. Lewy	Montgomery, Ala	Oct. 8, 1867	69, 685
Lathe	F. B. Mattson	Rockford, Ill	June 28, 1870	104, 751
Lathe	D. W. Maurice	Springfield, Ohio	Apr. 24, 1866	54, 187
Lathe	L. F. Munger	Rochester, N. Y	Jan. 7, 1868	73, 027
Lathe	H. E. Nickerson	Atchison, Kans	Dec. 3, 1872	133, 593
Lathe	L. Pastlewait	Russellville, Ohio	Oct. 16, 1860	30, 420
Lathe	P. H. Pitts	Waverly, Mo	Aug. 16, 1870	106, 404
Lathe	E. Putman	Gardner, Mass	Dec. 9, 1829	....
Lathe	S. W. Putnam and J. Q. Wright	Fitchburgh, Mass	July 21, 1863	39, 326
Lathe	E. Rhodes	Kingsbury, N. Y	Nov. 11, 1830	....

Index of patents issued from the United States Patent Office from 1790 *to* 1873, *inclusive*—Continued.

Invention.	Inventor.	Residence.	Date.	No.
Lathe	J. F. C. Rider and E. P. Brownell.	South New Market, N. H., and Providence, R. I.	Apr. 18, 1871	113, 798
Lathe	M. Roberts	Worcester, Mass	July 10, 1860	29, 103
Lathe	J. M. Scribner	Middleburgh, N. Y	May 15, 1860	28, 308
Lathe	W. Sellers	Philadelphia, Pa	Dec. 28, 1869	98, 423
Lathe	H. C. Spalding	New York, N. Y	Apr. 1, 1856	14, 578
Lathe	J. Stark	Waltham, Mass	May 30, 1865	47, 997
Lathe	T. Sullivan	Belleville, Canada	Jan. 23, 1872	123, 058
Lathe	J. Watson	Philadelphia, Pa	Mar. 1, 1870	100, 473
Lathe	B. D. Whitney	Winchendon, Mass	Aug. 7, 1860	29, 534
Lathe and chuck, Watchmaker's	C. Hopkins	Philadelphia, Pa	Nov. 5, 1872	132, 763
Lathe and gear cutter combined	U. Opperman	Washington, D. C	Aug. 22, 1871	118, 262
Lathe-arbors, Machine for grinding	C. Coester, jr., and A. B. Lawther.	Bridgeport, Conn	Sept. 10, 1867	68, 698
Lathe-attachment	C. Kilburn	Burlington, Vt	Nov. 22, 1859	26, 192
Lathe-attachment for chasing designs on molds	J. E. Miller	Birmingham, Pa	Nov. 7, 1871	120, 593
Lathe-attachment for cutting veneers	B. F. Sturtevant	Boston, Mass	Dec. 27, 1859	26, 627
Lathe-attachment for turning irregular forms	M. Roberts	Belfast, Me	May 20, 1856	14, 941
Lathe-attachment for turning tapers on bars	C. Jillson	Worcester, Mass	June 28, 1864	43, 311
Lathe, Automatic	A. Edmonds	Mount Pulaski, Ill	July 14, 1857	17, 782
Lathe, Automatic	J. McNary	Brooklyn, N. Y	May 4, 1858	20, 166
Lathe-back center	A. Lefever and G. C. Barnes	Battle Creek, Mich	Jan. 31, 1860	26, 998
Lathe-back rest	W. H. Hendrick and J. Jacobs	Mount Vernon, Ohio	Jan. 31, 1860	26, 988
Lathe, Block-making gang	E. Whiting		Apr. 1, 1802	
Lathe boring-attachment	C. E. McBeth	Hamilton, Ohio	Sept. 25, 1866	58, 273
Lathe boring-attachment	F. B. Williams	Sterling, Ill	Apr. 29, 1873	138, 460
Lathe box and journal	A. Wood	Worcester, Mass	Dec. 10, 1867	72, 147
Lathe burnishing-attachment	J. F. Ray	East Haddam, Conn	Dec. 28, 1858	22, 452
Lathe-carriage	G. W. Lewin	Worcester, Mass	Mar. 24, 1868	75, 933
Lathe-carrier	F. Ferrin	Camden, N. J	Apr. 1, 1873	137, 298
Lathe-carrier	J. Zook	Harrisburgh, Pa	Feb. 28, 1854	10, 585
Lathe-carrier, Extension	W. A. Lorenz	Newark, N. J	Apr. 2, 1872	125, 203
Lathe carrier or dog	L. P. Wilcox	Brooklyn, N. Y	Dec. 11, 1866	60, 451
Lathe-center	H. K. Porter	Boston, Mass	July 5, 1870	104, 995
Lathe-centers, Apparatus for grinding	W. S. Jarboe	New York, N. Y	Dec. 24, 1872	134, 287
Lathe-centers, Tool for grinding	D. Slate	Hartford, Conn	Apr. 30, 1872	126, 160
Lathe, Centering	B. F. Bee	Harwich, Mass	July 28, 1868	80, 327
Lathe centering-device	H. Gray	Bristol, Conn	Aug. 19, 1873	142, 014
Lathe centering-device	J. A. Talpey	Somerville, Mass	June 16, 1863	38, 922
Lathe centering-machine	D. Kelly	Philadelphia, Pa	Mar. 15, 1870	100, 772
Lathe-chuck	J. C. Miller	Danville, Ky	May 9, 1871	114, 700
Lathe, Cooper's	I. Hoover	Miamisburgh, Ohio	July 2, 1835	
Lathe cutter-head	M. Roberts	Belfast, Me	May 13, 1856	14, 899
Lathe cutter, Wood-turning	J. Shannon	Cohoes, N. Y	Aug. 1, 1865	49, 161
Lathe-dog	C. Buss	Marlborough, N. H	Oct. 18, 1870	108, 325
Lathe-dog	E. Cliff and O. A. Haynes	South Saint Louis, Mo	Nov. 5, 1872	132, 804
Lathe-dog	W. Emmett	Paterson, N. J	Aug. 18, 1868	81, 264
Lathe-dog	R. H. Lecky	Allegheny City, Pa	June 23, 1868	79, 134
Lathe-dog	C. W. Le Count	Norwalk, Conn	Oct. 24, 1865	50, 604
Lathe-dog	C. W. Le Count	Norwalk, Conn	Jan. 1, 1867	60, 751
Lathe-dog	N. M. Phillips	New York, N. Y	May 18, 1858	20, 298
Lathe-dog	W. Pimbott	Syracuse, N. Y	Dec. 4, 1866	60, 238
Lathe-dog	J. W. Russell	Springfield, Mass	Aug. 4, 1868	80, 770
Lathe-dog	J. S. Skinner	Lebanon, N. H	Aug. 9, 1870	106, 224
Lathe-dog	D. M. Smith	Springfield, Vt	May 16, 1854	10, 939
Lathe-dog	W. Vine	Norwalk, Conn	Aug. 1, 1865	49, 180
Lathe-dog	J. B. West	Geneseo, N. Y	Jan. 25, 1870	99, 269
Lathe-dog	H. K. White	Chelsea, Mass	Mar. 14, 1871	112, 759
Lathe-dog	L. P. Whiting	Poughkeepsie, N. Y	Nov. 5, 1872	132, 882
Lathe-dog	C. L. Wickham	Utica, N. Y	Apr. 29, 1873	138, 305
Lathe-dog and bench-vise jaw	N. Wilton	Groton, N. H	Oct. 15, 1867	69, 883
Lathe-dog, Automatic	A. E. Wenzel	New York, N. Y	Dec. 18, 1860	30, 997
Lathe, Engine	A. B. Couch	Worcester, Mass	Apr. 27, 1869	89, 468
Lathe, Engine	J. L. Johnson	Ashburnham, Mass	Mar. 7, 1865	46, 677
Lathe, Engine	S. Teal	Rochester, N. Y	Aug. 18, 1868	81, 309
Lathe, Engine	A. E. Whitmore	Boston, Mass	Jan. 5, 1869	85, 552
Lathe, Engine-turning	C. W. Dickinson	Newark, N. J	Dec. 13, 1864	45, 455
Lathe-fastening	J. M. Stone	North Andover, Mass	May 23, 1865	47, 876
Lathe-fastening	J. G. Stowe	Providence, R. I	Dec. 14, 1869	97, 986
Lathe for chasing and backing down taps	W. X. Stevens	Worcester, Mass	Mar. 19, 1867	62, 977
Lathe for cutting fluted moldings	J. Anderson, J. McLaren, and J. Bryant.	New York, N. Y	Nov. 25, 1856	16, 108
Lathe for cutting grooves in metal rollers	H. Waters	Boston, Mass	May 16, 1871	115, 000
Lathe for cutting irregular forms	P. Barker	Battle Creek, Mich	Apr. 28, 1868	77, 158
Lathe for cutting screws from wire	G. W. Daniels	Waltham, Mass	Oct. 19, 1858	21, 864
Lathe for cutting tenons for clock-movement	R. Peck	Bristol, Conn	Feb. 23, 1858	19, 472
Lathe for finishing the driving-wheels of locomotives.	H. D. Stover	New York, N. Y	June 22, 1869	91, 788
Lathe for forming shafting	A. Wood	Worcester, Mass	Dec. 10, 1867	72, 148
Lathe for grinding and polishing pivots of watch-work.	C. V. Woerd	Waltham, Mass	May 25, 1869	90, 620
Lathe for making awl-handles	F. Reed	Canton, Mass	June 21, 1870	104, 645
Lathe for manufacture of clothes-pins, &c	J. Humphrey	Keene, N. H	Sept. 22, 1857	18, 268
Lathe for setting jewels in time-pieces	H. H. Heskett	McLean County, Ill	Mar. 15, 1870	100, 891
Lathe for squaring nuts	J. Flower	Detroit, Mich	Mar. 29, 1870	101, 249
Lathe for turning a peculiar species of curves	H. G. Thompson	New York, N. Y	Apr. 23, 1850	7, 311
Lathe for turning and polishing soft-metal and composition buttons.	L. Meriam	New Haven, Conn	Jan. 4, 1815	
Lathe for turning axles	A. J. Bower	Albion, Ill	May 30, 1871	115, 424
Lathe for turning balls	J. B. West	Geneseo, N. Y	Aug. 18, 1868	81, 315
Lathe for turning beaded work	F. Baldwin	South Wardsborough, Vt	Aug. 24, 1858	21, 240
Lathe for turning bent sticks	T. Ott	Green Township, Pa	Aug. 6, 1872	130, 237
Lathe for turning billiard-balls	L. A. Johnson	San Francisco, Cal	Sept. 8, 1863	39, 817
Lathe for turning broom-handles	P. Prescott	Boonville, N. Y	Oct. 8, 1861	33, 466
Lathe for turning buttons	D. C. Robie	Springfield, Mass	Mar. 10, 1868	75, 462
Lathe for turning eccentrics	J. B. Gayle	Portsmouth, Va	July 30, 1867	67, 288
Lathe for turning fancy handles	L. Wentworth	Burlington, Iowa	Jan. 9, 1855	12, 224
Lathe for turning gun-barrels	D. Dana and A. Olney	Canton, Mass	Aug. 24, 1818	

Index of patents issued from the United States Patent Office from 1790 *to* 1873, *inclusive*—Continued.

Invention.	Inventor.	Residence.	Date.	No.
Lathe for turning heads of nails, tacks, &c	W. H. Nichols and H. H. Abbe.	Chatham, Conn	Aug. 1, 1865	49, 130
Lathe for turning hoe-handles, &c	R. D. Bartlett	Harmony, Me	Dec. 26, 1848	5, 983
Lathe for turning hubs	A. Goodyear	Albion, Mich	Aug. 31, 1869	94, 200
Lathe for turning interior and exterior surfaces	N. Chapin	New York, N. Y	Jan. 11, 1853	9, 529
Lathe for turning interior surfaces of hollow-ware	P. Teal and C. Tyler	Philadelphia, Pa	May 23, 1854	10, 946
Lathe for turning iron	W. Sellers	Philadelphia, Pa	Feb. 18, 1868	74, 609
Lathe for turning irregular curved forms	W. Payton	London, England	Jan. 28, 1868	73, 830
Lathe for turning irregular forms	F. Baker	Pepperell, Mass	May 16, 1854	10, 922
Lathe for turning irregular forms	F. T. Grant	Augusta, Me	Mar. 24, 1863	37, 960
Lathe for turning irregular forms	B. F. Jenkins and L. L. Knight.	Barre, Mass	Jan. 4, 1853	9, 521
Lathe for turning irregular forms	W. P. Lathrop	West Winsted, Conn	Aug. 20, 1872	130, 725
Lathe for turning irregular forms	E. Lumley	Elizabeth City, N. J	May 5, 1863	38, 399
Lathe for turning irregular forms	L. Smith	Plymouth, Conn	Dec. 9, 1856	16, 192
Lathe for turning irregular forms	C. and A. Spring	Boston, Mass	May 10, 1859	23, 957
Lathe for turning irregular forms	D. Werst and A. Puderbaugh	Waltz Township, Ind	Feb. 14, 1860	27, 178
Lathe for turning irregular forms	C. and M. T. Whipple and J. Sprague.	Douglass, Mass	Apr. 3, 1835	....
Lathe for turning irregular forms	E. K. Wisell	Warren, Ohio	May 28, 1872	127, 286
Lathe for turning irregular forms, Automatic	S. N. Baker	New Haven, Conn	July 28, 1857	17, 866
Lathe for turning irregular forms, Automatic	W. D. Sloan	New York, N. Y	Mar. 31, 1857	16, 936
Lathe for turning irregular forms, Wood	A. R. Stewart	Douglass Harbor, Canning, New Brunswick.	Apr. 2, 1867	63, 574
Lathe for turning knobs, Wood	J. Stever and J. A. Way	Bristol, Conn	Jan. 8, 1867	61, 111
Lathe for turning locomotive-crank pins	S. S. and D. Cheney	Galesburgh, Ill	Aug. 11, 1863	39, 465
Lathe for turning locomotive-drivers	J. M. Stone	Manchester, N. H	Apr. 10, 1855	12, 708
Lathe for turning masts, &c	P. H. Niles	Boston, Mass	Dec. 28, 1858	22, 447
Lathe for turning metal shafting	W. Sellers	Philadelphia, Pa	June 1, 1858	20, 446
Lathe for turning oval frames	J., W., and G. Gardner	New York, N. Y	June 29, 1858	20, 705
Lathe for turning ovals	R. Eickemeyer	Yonkers, N. Y	Apr. 5, 1870	101, 447
Lathe for turning ovals	R. Lawson	Shelburne Falls, Mass	Nov. 23, 1869	97, 096
Lathe for turning ovals	F. Smith	Allegheny City, Pa	Feb. 27, 1866	52, 900
Lathe for turning pen-handles	J. Keith	Worcester, Mass	Nov. 6, 1866	59, 407
Lathe for turning rake-handles, &c	J. Haven	Newport, N. H	Oct. 14, 1835	....
Lathe for turning regular forms	J. Moessinger	New York, N. Y	Nov. 19, 1872	133, 241
Lathe for turning scythe-snaths	S. Hinton	Jackson, Mich	Sept. 4, 1866	57, 716
Lathe for turning shafting	J. J. Conley	New York, N. Y	Oct. 9, 1866	58, 605
Lathe for turning shafting, Engine	A. Wood	Worcester, Mass	Mar. 31, 1868	76, 130
Lathe for turning spherical shapes	R. B. Carsley	New Bedford, Mass	Jan. 19, 1864	41, 277
Lathe for turning spherical shot and shell	C. Foster	Pittsburgh, Pa	Jan. 23, 1866	52, 244
Lathe for turning spokes	T. Derington	Duquoin, Ill	Jan. 26, 1864	41, 368
Lathe for turning spools	L. W. Boynton	Hartford, Conn	Nov. 26, 1872	133, 403
Lathe for turning spools	J. T. Hawkins	Annapolis, Md	Mar. 16, 1869	87, 929
Lathe for turning spools	A. T. Wing	Old Town, Me	Aug. 26, 1873	142, 314
Lathe for turning tobacco-pipes	C. Kapf	Philadelphia, Pa	July 29, 1873	141, 279
Lathe for turning tool-handles	H. K. Jones	Kensington, Conn	June 27, 1865	48, 409
Lathe for turning wagon-hubs	H. Inman	Hagaman's Mills, N. Y	Aug. 14, 1866	57, 145
Lathe for turning wagon-hubs	H. C. Sisco	Indianapolis, Ind	Mar. 20, 1866	53, 382
Lathe for turning whip-stocks	G. Dayton and A. Mallory	Westfield, Mass	Apr. 27, 1832	....
Lathe for turning whip-stocks	L. Hull	Charlestown, Mass	July 17, 1866	56, 419
Lathe for turning wire-eyed buttons, Old pin	W. Lawrence	Meriden, Conn	Apr. 12, 1815	....
Lathe-gearing, Adjustable	J. Humphreys	Chicopee, Mass	Nov. 3, 1868	83, 774
Lathe, Gun-stock	A. Town	Woodbury, Vt	Feb. 28, 1836	....
Lathe-head	A. F. Shaw	West Roxbury, Mass	Sept. 8, 1868	81, 952
Lathe, Metal-planing	W. W. Hubbard	Boston, Mass	Nov. 25, 1856	16, 118
Lathe, Metal-turning	E. F. and G. A. Barnes	New Haven, Conn	Jan. 21, 1873	135, 064
Lathe, Metal-turning	G. Henderson and J. Steelle	Allegheny, Pa	July 6, 1858	20, 794
Lathe, Metal-turning	J. W. Hyatt, jr	Albany, N. Y	Apr. 9, 1872	125, 571
Lathe, Metal-turning	W. Kitson	Lowell, Mass	Jan. 16, 1872	122, 769
Lathe, Metal-turning	F. W. Pomroy	Sandusky, Ohio	Aug. 13, 1872	130, 387
Lathe, Metal-turning	S. W. Putnam	Fitchburgh, Mass	Nov. 26, 1872	133, 483
Lathe, Metal-turning	W. Sellers	Philadelphia, Pa	May 28, 1872	127, 195
Lathe, Metal-turning	M. G. Stolp	Ithaca, N. Y	May 14, 1872	126, 758
Lathe-mill, Vertical	E. Skinner and J. Webster	Sandwich, N. H	Aug. 26, 1831	....
Lathe-pattern for turning lasts, &c	H. Mellish	Walpole, N. H	Dec. 23, 1834	....
Lathe, Pivot	R. Lissenbee	Greenville, Tenn	Apr. 17, 1866	53, 994
Lathe polishing-apparatus, Watch-maker's	J. M. Bottum	New York, N. Y	Mar. 13, 1855	12, 502
Lathe, Potters'	W. Meek	Anderson, Ind	July 22, 1873	141, 157
Lathe, Potters'	H. B. Morris	Burlington, N. J	Apr. 30, 1872	126, 318
Lathe, Prism	A. Kelsey	Charlestown, Mass	Dec. 5, 1865	51, 388
Lathe, Railway-car-axle	G. A. Gray, jr	Cincinnati, Ohio	May 9, 1871	114, 670
Lathe-rest	C. A. Bauer	Hamilton, Ohio	Mar. 11, 1873	136, 579
Lathe-rest	H. K. Smith	Norwich, Conn	Jan. 1, 1867	60, 950
Lathe-rest	H. K. Smith	Norwich, Conn	Feb. 18, 1868	74, 621
Lathe-rest	E. Stahlbrodt	Rochester, N. Y	May 5, 1868	77, 672
Lathe-rest	W. Thompson	Worcester, Mass	July 14, 1868	79, 877
Lathe-rest, Adjustable	J. E. Burdge	Cincinnati, Ohio	Aug. 13, 1867	67, 630
Lathe-rest attachment	D. White, jr	Lowell, Mass	Feb. 16, 1858	19, 395
Lathe-rest engine	H. Morrison	Paterson, N. J	Oct. 1, 1861	33, 399
Lathe-rest for turning balls	W. F. Vose	Newton, Mass	June 19, 1866	55, 749
Lathe-rest for turning spheres	P. Wenzel	Mayence-on-the-Rhine, Germany.	Sept. 14, 1869	94, 930
Lathe-rest, Sliding	E. S. Gardner	Philadelphia, Pa	Sept. 1, 1857	18, 120
Lathe rest, Turning	W. H. Barney	Columbus, Ohio	July 5, 1864	43, 386
Lathe, Rose-engine	T. Lippiatt	New York, N. Y	Aug. 14, 1866	57, 158
Lathe, Rose-engine	T. Lippiatt	New York, N. Y	May 28, 1867	65, 245
Lathe, Rose-engine	T. Lippiatt	New York, N. Y	July 28, 1868	80, 292
Lathe, Rotary	G. Tugnot	New York, N. Y	Nov. 28, 1854	12, 008
Lathe, Screw-cutting	T. D. Bassett	Charlestown, Mass	July 7, 1868	79, 721
Lathe, Screw-cutting	C. Ellis	Canton, Mass	Apr. 14, 1868	76, 612
Lathe, Screw-cutting	J. P. T. Lang	Washington, D. C	July 7, 1868	79, 767
Lathe, Screw-cutting	H. F. Shaw	West Roxbury, Mass	Oct. 27, 1868	83, 554
Lathe, Screw-cutting	J. Tangye	Birmingham, England	July 10, 1866	56, 339
Lathe-slide	E. K. Root	Hartford, Conn	May 15, 1855	12, 874
Lathe slide-rest	C. A. Noyes	Pittsfield, Mass	Apr. 17, 1855	12, 742
Lathe slide-rest	J. G. Rominger	Philadelphia, Pa	Sept. 10, 1867	68, 654
Lathe slide-rest	W. Stephens	Richmond, Ind	Feb. 20, 1855	12, 423
Lathe slide-rest	C. Van Haagen	Philadelphia, Pa	Mar. 11, 1873	136, 794
Lathe slide-rest	C. Van Horn	Springfield, Mass	Apr. 17, 1855	12, 747

Index of patents issued from the United States Patent Office from 1790 *to* 1873, *inclusive*—Continued.

Invention.	Inventor.	Residence.	Date.	No.
Lathe-spindle	J. E. Boutelle	Fisherville, N. H	Apr. 19, 1870	101, 975
Lathe spindles, Hold-fast for	L. R. Faught	Philadelphia, Pa	May 30, 1871	115, 290
Lathe, Spoke	G. F. Almy	Toledo, Ohio	Mar. 15, 1870	100, 836
Lathe, Spoke	I. S. Roland	Reading, Pa	June 1, 1869	90, 689
Lathe, Spoke	C. B. Conant and H. Thompson.	Worcester, Mass	Mar. 1, 1870	100, 261
Lathe, Spoke	J. Pierce and A. B. Curtis	New Haven, Conn	Aug. 15, 1871	118, 148
Lathe-stand	H. D. Stover	New York, N. Y	Oct. 17, 1865	50, 510
Lathe taper-turning attachment	M. Neckermann	Pittsburgh, Pa	Sept. 30, 1873	143, 296
Lathe to prevent counterfeiting, Figure	D. H. Mason	Philadelphia, Pa	June 7, 1832	....
Lathe-tool	J. C. Shackleton	Lawrence, Mass	July 9, 1867	66, 641
Lathe-tool	T. Shaw	Philadelphia, Pa	Mar. 3, 1868	75, 060
Lathe tool and rest holder	J. Walfenden	Jersey City, N. J	July 31, 1866	56, 840
Lathe tool-elevator	W. Hamilton	Chicopee, Mass	Nov. 29, 1864	45, 297
Lathe tool-holder	C. E. Albro	Syracuse, N. Y	Dec. 31, 1872	134, 343
Lathe tool-holder	J. Baillie	Salem, Ohio	Feb. 25, 1868	74, 877
Lathe tool holder	W. Coulter	Birmingham, Conn	Dec. 10, 1872	133, 833
Lathe tool-holder	J. Edson	Boston, Mass	Dec. 24, 1867	72, 468
Lathe tool-holder	C. H. Fowler	Roxbury, Mass	June 2, 1868	78, 520
Lathe tool-holder	W. O. Hickok and G. W. Reisinger.	Harrisburgh, Pa	Nov. 10, 1868	83, 963
Lathe tool-holder	C. Peck	New Haven, Conn	June 28, 1859	24, 574
Lathe tool-holder	L. Reder	Wilmington, Del	Dec. 16, 1873	145, 684
Lathe tool-holder	I. P. Richards	Providence, R. I	Dec. 24, 1872	134, 219
Lathe tool-rest	T. J. Currier and A. M. Black	Worcester, Mass	Jan. 15, 1867	61, 166
Lathe tool-rest	C. F. Hadley	Chicopee, Mass	Oct. 29, 1872	132, 573
Lathe tool-rest	J. Service	Greenville, Conn	Dec. 4, 1866	60, 266
Lathe tool-rest	J. G. Stowe	Providence, R. I	Dec. 7, 1869	97, 723
Lathe tool-rest, Engine	C. Newhall	Hinsdale, N. H	June 30, 1868	79, 381
Lathe tool-supporter	I. F. Brown	New London, Conn	July 4, 1871	116, 545
Lathe, Turning	W. Aldrich	Lowell, Mass	Mar. 15, 1853	9, 616
Lathe, Turning	E. and N. Alvord	Westfield, Mass	Oct. 11, 1836	43
Lathe, Turning	F. A. Armbruster	New York, N. Y	Jan. 15, 1867	61, 135
Lathe, Turning	J. Ball	New York, N. Y	Oct. 4, 1864	44, 503
Lathe, Turning	J. G. Baker	Philadelphia, Pa	Nov. 1, 1864	44, 915
Lathe, Turning	E. Bancroft and W. Sellers	Philadelphia, Pa	Feb. 7, 1854	10, 491
Lathe, Turning	J. Bisbee	Plainfield, Mass	Nov. 8, 1832	....
Lathe, Turning	C. G. Bloomer	Wickford, R. I	Dec. 20, 1864	45, 555
Lathe, Turning	J. Burt	Sturgis, Mich	Nov. 13, 1866	59, 552
Lathe, Turning	R. J. Cole	Poultney, Vt	Dec. 22, 1863	40, 994
Lathe, Turning	F. W. Coy	Boston, Mass	Oct. 8, 1867	69, 544
Lathe, Turning	H. Doane and A. Goodman	Dana, Mass	Nov. 6, 1849	6, 843
Lathe, Turning	R. W. Drew	Lowell, Mass	Dec. 4, 1866	60, 155
Lathe, Turning	L. R. Faught	Philadelphia, Pa	Dec. 4, 1866	60, 163
Lathe, Turning	M. J. Gardner	Yorkborough, Pa	Dec. 14, 1830	....
Lathe, Turning	A. Goodyear	Springport, Mich	Aug. 23, 1864	43, 908
Lathe, Turning	N. Green	Bedford County, Tenn	Apr. 27, 1832	....
Lathe, Turning	P. J. Hardy	New York, N. Y	Mar. 28, 1871	113, 048
Lathe, Turning	N. Harper	Newark, N. J	Feb. 14, 1865	46, 352
Lathe, Turning	J. Hunt	Shutesbury, Mass	June 8, 1826	....
Lathe, Turning	W. Johnson	Lambertville, N. J	May 28, 1867	65, 089
Lathe, Turning	F. Kirst	Westfield, Mass	July 12, 1864	43, 512
Lathe, Turning	S. R. Lewis	Rockford, Ill	Oct. 20, 1868	83, 181
Lathe, Turning	H. Locke	Grand Rapids, Mich	Sept. 20, 1870	107, 513
Lathe, Turning	J. P. Luther	Berlin, Wis	Aug. 19, 1873	141, 940
Lathe, Turning	G. Meldrum	West Philadelphia, Pa	Apr. 18, 1854	10, 783
Lathe, Turning	F. B. Miles	Philadelphia, Pa	Feb. 14, 1871	111, 859
Lathe, Turning	J. Moore	Leverett, Mass	Jan. 19, 1828	....
Lathe, Turning	H. L. Morse	New Bedford, Mass	Apr. 16, 1867	63, 928
Lathe, Turning	W. Patrick	Leverett, Mass	Apr. 24, 1827	....
Lathe, Turning	J. Richards	Cincinnati, Ohio	Aug. 13, 1867	67, 676
Lathe, Turning	W. Ross	Chester County, Pa	July 10, 1809	....
Lathe, Turning	W. Sellers	Philadelphia, Pa	Mar. 13, 1860	27, 478
Lathe, Turning	W. Sellers	Philadelphia, Pa	June 11, 1872	127, 929
Lathe, Turning	F. Shaller	Hudson, N. Y	Aug. 13, 1867	67, 810
Lathe, Turning	F. G. Sheldon	Hudson, Mich	Mar. 19, 1872	124, 861
Lathe, Turning	S. D. Sheldon	Fitchburgh, Mass	Mar. 21, 1871	112, 971
Lathe, Turning	W. Simpson, Z. Oman, and S. Foreman.	Nottawa, Mich	Sept. 11, 1866	57, 984
Lathe, Turning	D. Slate	Hartford, Conn	Jan. 31, 1865	46, 152
Lathe, Turning	W. D. Sloan	New York, N. Y	Jan. 5, 1858	19, 051
Lathe, Turning	Q. H. Spencer	North Providence, R. I	Aug. 10, 1869	93, 492
Lathe, Turning	A. Warth	New York, N. Y	Oct. 10, 1854	11, 798
Lathe, Turning	J. D. White	Hartford, Conn	May 21, 1850	7, 390
Lathe, Turning	C. White and A. Hill	Hadley, Mass	Mar. 30, 1824	....
Lathe, Turning	J. C. Wybell	West Meriden, Conn	Aug. 7, 1866	57, 054
Lathe, Turning	L. D. Wynhoop	Owassa, Mich	Apr. 5, 1864	42, 250
Lathe, Turning	F. Zibell	Saint Louis, Mo	Mar. 24, 1868	75, 822
Lathe, Uniform-turning	A. Haggett	Middlesex County, Mass	June 10, 1809	....
Lathe, Variety-frame	A. C. Wicker and L. W. Williams.	Fair Haven, Vt	Aug. 6, 1867	67, 473
Lathe, Watch-maker's	E. J. Chapin	Ottawa, Ill	Nov. 18, 1862	36, 942
Lathe, Watch-maker's	R. Cowles	Cleveland, Ohio	Jan. 14, 1873	134, 787
Lathe, Watch-maker's	E. Harris	Princeton, Ill	Mar. 29, 1859	23, 370
Lathe, Watch-maker's	G. Hunziker	Summit, Miss	Jan. 16, 1872	122, 768
Lathe, Watch-maker's	E. H. Kelly	Jackson, Tenn	Nov. 11, 1873	144, 547
Lathe, Watch-maker's	G. Jackson	Cincinnati, Ohio	Oct. 28, 1862	36, 842
Lathe, Watch-maker's	J. M. Oram	Gosport, Ind	Oct. 7, 1873	143, 458
Lathe, Watch-maker's	C. C. Powell and J. Belknap	Saint Johnsbury, Vt	July 13, 1869	92, 545
Lathe, Watch-maker's	F. Shaller	Hudson, N. Y	Sept. 4, 1866	57, 775
Lathe, Watch-maker's	R. H. St. John	Bellefontaine, Ohio	Sept. 15, 1857	18, 213
Lathe, Watch-maker's	R. H. St. John	Bellefontaine, Ohio	Mar. 29, 1859	23, 406
Lathe, Watch-maker's	C. Tribby	Winchester, Va	Feb. 5, 1861	31, 340
Lathe, Watch-maker's	A. Wild	New York, N. Y	July 17, 1860	29, 227
Lathe, Watch-maker's	C. V. Woerd	Waltham, Mass	Oct. 28, 1873	144, 178
Lathe way-smoother	A. Thomas	Worcester, Mass	Nov. 5, 1867	70, 483
Lathe-wheel	B. Hoadley	New Haven, Conn	Dec. 16, 1814	....
Lathe, Wheelwright's felly	J. Sitton and J. A. Black	Columbia, S. C	May 5, 1828	....
Lathe, Wood	H. T. Clay	Philadelphia, Pa	Oct. 8, 1867	69, 543
Lathe, Wood	L. H. Dwelley	Dorchester, Mass	May 17, 1870	103, 033

Index of patents issued from the United States Patent Office from 1790 *to* 1873, *inclusive*—Continued.

Invention.	Inventor.	Residence.	Date.	No.
Lathe, Wood	G. W. Feltes	Carbondale, Ill	Apr. 14, 1868	76, 733
Lathe, Wood	A. Pries and H. Arnd	Saint Louis, Mo	Nov. 28, 1871	121, 417
Lathe, Wood	W. H. Wussow	Aurora, Ill	Dec. 4, 1866	60, 317
Lathe, Wood-turning	L. Atwood and E. Ripley	Rutland, Vt	Feb. 15, 1870	99, 810
Lathe, Wood-turning	F. Baldwin	Brattleborough, Vt	June 18, 1867	65, 864
Lathe, Wood-turning	F. Baldwin	Janesville, Wis	Aug. 30, 1870	106, 767
Lathe, Wood-turning	A. J. and N. M. Barnes	Tiffin, Ohio	Nov. 9, 1869	96, 659
Lathe, Wood-turning	A. Basse	Quincy, Ill	Oct. 16, 1866	58, 758
Lathe, Wood-turning	H. C. Berry	Wauseon, Ohio	Apr. 16, 1867	63, 836
Lathe, Wood-turning	C. A. Blessing	Philadelphia, Pa	Apr. 22, 1873	137, 999
Lathe, Wood-turning	A. P. C. Bonte	Cincinnati, Ohio	June 16, 1868	78, 919
Lathe, Wood-turning	J. C. Brown	Washington, D. C	July 12, 1870	105, 166
Lathe, Wood-turning	W. W. Carey	Lowell, Mass	Dec. 19, 1865	51, 554
Lathe, Wood-turning	J. Chase	Rochester, N. Y	Mar. 26, 1867	63, 213
Lathe, Wood-turning	J. Chase	Rochester, N. Y	Mar. 15, 1870	100, 855
Lathe, Wood-turning	J. Chase	Rochester, N. Y	Sept. 30, 1873	143, 221
Lathe, Wood-turning	J. Coleman	Argyle, N. Y	July 12, 1864	43, 477
Lathe, Wood-turning	M. F. Connett	Ladoga, Ind	Mar. 24, 1868	75, 736
Lathe, Wood-turning	D. Dick	Corning, N. Y	Dec. 17, 1867	72, 273
Lathe, Wood-turning	D. Dick	Corning, N. Y	May 5, 1868	77, 596
Lathe, Wood-turning	L. H. Dwelley	Dorchester, Mass	Mar. 5, 1867	62, 618
Lathe, Wood-turning	R. W. George	Boston, Mass	Aug. 22, 1865	49, 516
Lathe, Wood-turning	A. Goodyear	Springport, Mich	June 25, 1867	66, 146
Lathe, Wood-turning	T. Gray	Lowell, Mass	July 12, 1864	43, 492
Lathe, Wood-turning	S. L. Hart	Milwaukee, Wis	Aug. 13, 1867	67, 757
Lathe, Wood-turning	H. R. Hawkins	Akron, Ohio	June 11, 1867	65, 568
Lathe, Wood-turning	L. Hull	Charlestown, Mass	Mar. 6, 1866	53, 003
Lathe, Wood-turning	S. U. King	Windsor, Vt	May 5, 1868	77, 495
Lathe, Wood-turning	J. E. F. Leland	New York, N. Y	Oct. 16, 1866	58, 947
Lathe, Wood-turning	G. Lewis	Westfield, Ohio	Feb. 4, 1868	73, 985
Lathe, Wood-turning	H. Locke	Boston, Mass	July 25, 1865	49, 035
Lathe, Wood-turning	J. McMicheal	Philadelphia, Pa	Jan. 22, 1867	61, 442
Lathe, Wood-turning	N. Norris and C. S. and H. S. Black.	Buchanan, Mich	Feb. 8, 1870	99, 696
Lathe, Wood-turning	G. H. Ober	Newburgh, Ohio	June 27, 1865	48, 428
Lathe, Wood-turning	A. R. Park	Columbia, Tex	May 20, 1873	139, 076
Lathe, Wood-turning	O. H. Perry	Golconda, Ill	Apr. 5, 1870	101, 502
Lathe, Wood-turning	J. Phillips, jr	Chicago, Ill	Mar. 19, 1867	63, 089
Lathe, Wood-turning	J. Richards	Cincinnati, Ohio	July 16, 1867	66, 885
Lathe, Wood-turning	I. Rood	Elyria, Ohio	Jan. 10, 1871	110, 871
Lathe, Wood turning	O. S. Smiley and C. O. Small	Augusta, Me	May 20, 1873	139, 028
Lathe, Wood-turning	M. Spenli	Detroit, Mich	Dec. 18, 1866	60, 647
Lathe, Wood-turning	M. Spenli	Detroit, Mich	Apr. 2, 1867	63, 571
Lathe, Wood-turning	J. J. Urmston	Rahway, N. J	Apr. 5, 1870	101, 548
Lathe, Wood-turning	A. J. Van Ornum	Hartford, Vt	May 26, 1868	78, 341
Lathe, Wood-turning	A. Warth	Stapleton, N. Y	Nov. 9, 1858	22, 043
Lathe, Wood-turning	W. J. Watson	Benton Centre, N. Y	Apr. 7, 1868	76, 363
Lathe, Wood-turning	J. B. West	Lakeville, N. Y	Apr. 17, 1866	54, 047
Lathe, Wood-turning	A. N. Wilcox	Watervliet, N. Y	Jan. 5, 1858	19, 056
Lathe, Wood-turning	E. Williams	Rowlesburgh, W. Va	Mar. 24, 1868	76, 022
Lathe, Wood-turning	J. Wilson and R. Hughes	Boston, Mass	Aug. 20, 1867	68, 020
Lathe, Wood-turning	E. K. Wisell	Warren, Ohio	Jan. 14, 1868	73, 423
Lathes, Adjustable tool-rest for	J. S., A. N., and O. Wheeler	Worcester, Mass	June 27, 1871	116, 517
Lathes and other machinery, Mechanism for transmitting power to.	J. F. Stewart, F. Klinkerman, and J. Lamb.	Aurora, Ind	Apr. 26, 1870	102, 445
Lathes, Centering heavy articles in	J. S. French	Boston, Mass	Jan. 19, 1864	41, 289
Lathes, Cutting-tool for turning	C. Rahsskopff	San Francisco, Cal	Dec. 24, 1872	134, 313
Lathes, Device for securing the tail-stocks of	A. Thomas	Worcester, Mass	Mar. 6, 1866	53, 058
Lathes, Drill-rest for chucking	G. E. Johnson and W. Boston	Biddeford, Me	Aug. 29, 1871	118, 614
Lathes, Drill-rest for turning	H. H. Folsom	Lowell, Mass	Feb. 14, 1871	111, 736
Lathes, Drilling-attachment for turning	J. McCrum	Locust Grove, Ohio	Aug. 28, 1866	57, 536
Lathes, Expanding and contracting or universal chuck for.	S. Fairman	Waterford, N. Y	July 18, 1840	1, 692
Lathes, Eye-protector or chip-arrester for	C. T. Lamphere	Greenfield, Mass	July 7, 1868	79, 766
Lathes, Feed-stop for	A. P. Brown	Worcester, Mass	July 10, 1866	56, 172
Lathes for beaded work, Mode of operating radial cutters in.	G. W. Walton and H. Edgarton	Wilmington, Del	July 7, 1857	17, 762
Lathes, Gear-cutting attachment to	W. P. Hopkins	Lawrence, Mass	Sept. 30, 1873	143, 242
Lathes, Gear-cutting attachment to	T. O. Mills	Rockford, Ill	Nov. 4, 1873	144, 214
Lathes, Gearing for	W. Fitzgerald	Boston, Mass	May 8, 1866	54, 519
Lathes, Method of centering in watch-maker's	P. Leffel and J. H. Mulholland	Springfield, Ohio	Sept. 6, 1859	25, 359
Lathes, Method of feeding the tool-carriage in turning.	A. Rennie	Binghamton, N. Y	July 20, 1858	20, 956
Lathes, Method of operating feed-nuts in	W. A. Patrick	Ludlow, Vt	Feb. 14, 1860	27, 149
Lathes or planing-machines, Tool-post for	W. H. Leach	Uxbridge, Mass	May 12, 1868	77, 742
Lathes, Revolving center-rest for wood	H. J. Dargin	Rochester, N. Y	Apr. 5, 1870	101, 596
Lathes, Screw-cutting attachment for	W. Gleason	Rochester, N. Y	Oct. 21, 1873	143, 898
Lathes, Securing buttons in	C. W. Robinson	Providence, R. I	June 27, 1809	......
Lathes, Securing pinions, &c., of watches in	J. M. Bottum	New York, N. Y	July 15, 1851	8, 216
Lathes, Socket-coupling for	G. N. Trowbridge	Lowell, Mass	Sept. 1, 1857	18, 112
Lathes, Steady-rest for	J. Brodie	San Francisco, Cal	Aug. 17, 1869	93, 801
Lathes, Tail-stock for turning	L. B. Tyng	Lowell, Mass	Mar. 7, 1854	10, 602
Lathes to turn tapering shafts, Tool for adjusting.	W. N. Berkeley	Cedar Rapids, Iowa	July 10, 1866	56, 164
Lathes, Tool-adjuster for	L. J. Parsons	New Haven, Conn	Apr. 27, 1869	89, 335
Lathes, Tool-elevator for	C. Knox	Chicopee, Mass	Apr. 26, 1870	102, 278
Lathes, Tool for turning	A. A. Burr	Rockdale, N. Y	Feb. 9, 1864	41, 478
Lathes, Tool-grinding attachment to	J. Schaughnessy	Cincinnati, Ohio	May 17, 1870	103, 242
Lathes, Tool-holder for turning	S. Gissinger	Lawrenceville, Pa	Sept. 17, 1867	68, 870
Lathes, Tool-rest for turning	A. Hathaway	Chicopee, Mass	Mar. 8, 1864	41, 889
Lathes, Tool-rest for turning	M. H. Merriam and W. W. Nichols.	Chelsea and Boston, Mass	Aug. 22, 1854	11, 586
Lathes, Tool-rest for turning	F. A. Pratt	Hartford, Conn	Sept. 16, 1862	36, 479
Lathes, Tool-rest for turning	G. A. Rollins	Nashua, N. H	Feb. 28, 1854	10, 582
Lathes, Tool-holder for turning	W. Sellers	Philadelphia, Pa	Aug. 5, 1862	36, 111
Lathes, Universal slide-rest for	E. G. Chormann	Philadelphia, Pa	Aug. 1, 1871	117, 513
Lathes, Varying the speed of the mandrel in	W. A. Chapin, jr	Saint Johnsbury, Vt	Oct. 2, 1849	6, 755
Lathes, Work-holder for	W. P. Hopkins	Lawrence, Mass	Oct. 22, 1872	132, 400
Lathes, Work-holder for	W. P. Hopkins	Lawrence, Mass	Dec. 31, 1872	134, 381
Latitude and longitude, Instrument for determining	J. Stinson	Danville, N. J	Sept. 18, 1855	13, 584

Index of patents issued from the United States Patent Office from 1790 *to* 1873, *inclusive*—Continued.

Invention.	Inventor.	Residence.	Date.	No.
Latitude at sea, Method of determining approximate.	E. Cavendy	New York, N. Y	Nov. 10, 1857	18, 572
Latitude, longitude, and difference of time, Apparatus for showing.	S. D. Tillman	Jersey City, N. J	July 30, 1872	129, 909
Launching vessels, Apparatus for	T. F. Rowland	Green Point, N. Y	May 5, 1863	38, 418
Launching vessels, Greasing ways for	W. Mead	Greenwich, Conn	Oct. 25, 1832	....
Laundry and tailor's press	G. W. Jennings	Boston, Mass	Feb. 14, 1860	27, 131
Laundry-bench	A. H. Spencer	Providence, R. I	Feb. 8, 1870	99, 608
Laundry-box	M. A. H. Saurman	Philadelphia, Pa	May 23, 1871	115, 109
Laundry, Family	H. E. Smith	Cincinnati, Ohio	Nov. 14, 1865	50, 964
Laundry-heater	C. H. Goss	Troy, N. Y	June 15, 1869	91, 433
Laundry-indicator	R. Clarke	Macon, Ga	Mar. 8, 1870	100, 601
Laundry-iron heater	R. Diven	New York, N. Y	June 7, 1870	103, 997
Laundry, Portable	J. Morrison	Indianapolis, Ind	Feb. 14, 1871	111, 864
Laundry, Portable	J. Van	Cincinnati, Ohio	May 4, 1869	89, 815
Lavatory-apparatus	D. Wellington	Boston, Mass	May 3, 1870	102, 737
Law-blank case	C. I. Walker	Charleston, S. C	May 20, 1873	139, 094
Lawn or garden sprinkler	H. E. Cooke	Cleveland, Ohio	Aug. 12, 1873	141, 632
Lawn-seat	H. H. Gratz	Lexington, Ky	Nov. 11, 1873	144, 533
Lawn-seat	W. C. Medcalf	Rochester, N. Y	Sept. 16, 1873	142, 927
Lawn-sprinkler	J. Lessler	Buffalo, N. Y	Dec. 19, 1871	121, 949
Lawns, Machinery for dressing	R. L. Hawes	Worcester, Mass	Apr. 24, 1855	12, 771
Lay figures, dolls, &c	C. L. Parent	Havre, France	July 30, 1872	130, 068
Leach, Tanner's	S. Snyder	Cincinnati, Ohio	Feb. 23, 1869	87, 119
Leach-tub	W. Bauzelt	Brooklyn, N. Y	Oct. 13, 1868	83, 020
Leaching and tanning apparatus	C. Korn	Brooklyn, N. Y	Mar. 18, 1873	137, 004
Leaching-vat	T. H. Baker	Springfield, Ohio	Feb. 23, 1869	87, 235
Lead, acetate of copper, and acetate of iron, Manufacture of carbonate of.	O. Jacobi	Philadelphia, Pa	Aug. 31, 1869	94, 214
Lead and acetic acid, Manufacture of sugar of	L. D. Gale and I. M. Gattman	Washington, D. C., and New York, N. Y.	Aug. 17, 1869	93, 817
Lead and flake-white, Manufacturing white	A. J. Hamilton	New York	Nov. 8, 1813	....
Lead, and in the purification of the products of combustion for the same, Manufacture of white.	H. Hannen	Philadelphia, Pa	Feb. 9, 1869	86, 835
Lead and packing can, White	C. J. Fortin and D. H. Drake	Cincinnati, Ohio	Aug. 3, 1869	93, 191
Lead and Paris-white, Grinding white	S. Watrous	Groton, Conn	Jan. 26, 1833	....
Lead and pulpy materials, Apparatus for drying white.	J. B. Pollock	Port Richmond, N. Y	Jan. 2, 1872	122, 404
Lead and other metallic pipe, Manufacture of	R. M. Seydle and L. Ward	Milton, Pa	Aug. 1, 1838	862
Lead and saltpeter, Manufacture of white	C. Delafield	Factoryville, N. Y	June 5, 1866	55, 249
Lead and zinc for the manufacture of pigments, Treating the ores of.	E. O. Bartlett	Birmingham, Pa	Oct. 18, 1870	108, 433
Lead and zinc, Refining and whitening the products from ores of.	G. S. Lewis	Philadelphia, Pa	Mar. 14, 1871	112, 607
Lead and zinc, Refining and whitening the product from sulphurets of.	G. S. Lewis	Philadelphia, Pa	Mar. 14, 1871	112, 606
Lead and zinc, Sheet-metal from	C. C. C. Morgan	Auburn, N. Y	May 11, 1869	89, 881
Lead, Apparatus for making white	T. J. Coggeshall	New York, N. Y	Dec. 27, 1864	45, 587
Lead, Apparatus for manufacture of white	C. L. Wheeler	Pittsburgh, Pa	May 28, 1872	127, 395
Lead, Apparatus for manufacturing white	H. Hannen	Dubuque, Iowa	Mar. 30, 1858	19, 771
Lead, Apparatus for manufacturing white	R. Rowland	New York, N. Y	June 29, 1858	20, 731
Lead, Apparatus for refining and desilvering	G. Luce	Marseilles, France	Nov. 25, 1873	144, 993
Lead apparatus, White	D. R. Erdmann	Philadelphia, Pa	Aug. 16, 1859	25, 106
Lead by precipitation, Manufacture of white	R. Baker	Newark, N. Y	Apr. 3, 1855	12, 616
Lead-cutting machine	C. Koehler	Evansville, Ind	Apr. 9, 1872	125, 462
Lead, Die for making sheet	N. York	Boston, Mass	July 1, 1873	140, 567
Lead-drier	A. D. Armstrong	Allegheny City, Pa	Dec. 23, 1873	145, 713
Lead-foil coated with tin, Manufacture of	D. S. Hines	Brooklyn, N. Y	Mar. 26, 1867	63, 249
Lead for pigments, Manufacture of white	E. Milner	Springfield, England	July 8, 1873	140, 721
Lead from dross, Method of obtaining	C. Pickering	Saint Louis, Mo	Oct. 23, 1866	59, 063
Lead from precious metals, Separating	S. W. Kirk	Philadelphia, Pa	Nov. 29, 1870	109, 742
Lead, &c., from mixing-tubs to millstones, Apparatus for feeding.	J. B. Pollock	Port Richmond, N. Y	Nov. 28, 1871	121, 416
Lead-holder for pencil	P. Gabriel	Seymour, Conn	Aug. 6, 1867	67, 528
Lead, Machine for cutting sheet	S. E. Chubbuck	Roxbury, Mass	May 28, 1867	65, 173
Lead, Machine for making sheet	D. G. Batchelder	Salem, Mass	Feb. 25, 1873	136, 124
Lead, Machine for making sheet	H. Hannen	Philadelphia, Pa	June 6, 1871	115, 730
Lead, Machine for making sheet	J. Millingar	Philadelphia, Pa	July 5, 1870	105, 112
Lead, Machine for manufacturing sheet	J. Robertson	Brooklyn, N. Y	Sept. 19, 1865	50, 036
Lead, Machine for preparing bar	E. J. Bailey	Pittsburgh, Pa	June 29, 1869	91, 814
Lead machines, Combination of dies for sheet	J. Robertson	Brooklyn, N. Y	May 27, 1851	8, 112
Lead, &c., Machinery for making pipes continuously from.	G. E. Sellers	Upper Darby, Pa	Dec. 17, 1840	1, 908
Lead, Machinery for manufacture of carbonate of	C. Ripley	Saugerties, N. Y	July 11, 1837	264
Lead, Making sulphate, carbonate, and other pigments of.	H. Holland	Westfield, Mass	Nov. 3, 1838	994
Lead, Making tinned sheet	T. Ervbank	New York	May 30, 1823	....
Lead, Making white	E. Clark	Philadelphia, Pa	Mar. 28, 1818	....
Lead, Making white	E. De Butts	Baltimore, Md	May 5, 1814	....
Lead, Manufacture of acetate of	J. E. Fournier	Courville, France	July 17, 1866	56, 505
Lead, Manufacture of carbonate of	H. Hannen	Philadelphia, Pa	May 14, 1867	64, 763
Lead, Manufacture of carbonate of	G. S. Lewis	Philadelphia, Pa	Mar. 14, 1871	112, 608
Lead, Manufacture of carbonate of	H. Hannen	Philadelphia, Pa	Feb. 19, 1867	62, 130
Lead, Manufacture of chloride of	F. F. Meyer	New York, N. Y	Oct. 26, 1860	30, 521
Lead, Manufacture of dry white	E. O. Bartlett	Birmingham, Pa	Mar. 1, 1870	100, 353
Lead, Manufacture of oxychloride of	L. Brumlen	Hoboken, N. J	Jan. 29, 1861	31, 224
Lead, Manufacture of red	S. Witherell, jr	Philadelphia, Pa	Nov. 1, 1811	....
Lead, Manufacture of the subacetate of	L. Brumlen	Hoboken, N. J	Aug. 13, 1872	130, 410
Lead, Manufacture of white	F. Albert	Brooklyn, N. Y	May 3, 1859	23, 815
Lead, Manufacture of white	W. Archer and C. Rice	New York, N. Y	Mar. 7, 1865	46, 706
Lead, Manufacture of white	W. Baker	Sheffield, England	June 13, 1865	48, 243
Lead, Manufacture of white	S. R. Bradley	New York, N. Y	Sept. 21, 1869	95, 075
Lead, Manufacture of white	L. Brumlen	Hoboken, N. J	Mar. 28, 1871	113, 014
Lead, Manufacture of white	T. H. Burridge	Saint Louis, Mo	May 2, 1871	114, 405
Lead, Manufacture of white	T. Cobley	London, England	Apr. 28, 1863	38, 283
Lead, Manufacture of white	H. Cory	England	Oct. 8, 1840	1, 804
Lead, Manufacture of white	J. Cuddy	Pittsburgh, Pa	Sept. 28, 1869	95, 201

Index of patents issued from the United States Patent Office from 1790 *to* 1873, *inclusive*—Continued.

Invention.	Inventor.	Residence.	Date.	No.
Lead, Manufacture of white	J. Cuddy	Pittsburgh, Pa	July 19, 1870	105, 431
Lead, Manufacture of white	J. Cuddy and G. S. Selden	Pittsburgh and Philadelphia, Pa.	June 21, 1870	104, 434
Lead, Manufacture of white	C. Delafield	Staten Island, N. Y	Jan. 23, 1866	52, 144
Lead, Manufacture of white	C. Delafield	Factoryville, N. Y	Apr. 3, 1866	53, 583
Lead, Manufacture of white	C. W. Dwelle	Saint Louis, Mo	Sept. 21, 1869	95, 097
Lead, Manufacture of white	C. W. Dwelle	Saint Louis, Mo	Oct. 25, 1870	108, 571
Lead, Manufacture of white	T. M. and A. G. Fell	New York, N. Y	July 24, 1866	56, 685
Lead, Manufacture of white	T. M. and A. G. Fell	Brooklyn, N. Y	Nov. 20, 1866	59, 901
Lead, Manufacture of white	T. M. and A. G. Fell	Brooklyn, N. Y	Nov. 20, 1866	59, 902
Lead, Manufacture of white	T. M. and A. G. Fell	Brooklyn, N. Y	June 25, 1867	66, 137
Lead, Manufacture of white	T. M. and A. G. Fell	Brooklyn, N. Y	June 25, 1867	66, 138
Lead, Manufacture of white	T. M. and A. G. Fell	Brooklyn, N. Y	June 25, 1867	66, 139
Lead, Manufacture of white	S. Gardner	New York	Aug. 26, 1843	3, 232
Lead, Manufacture of white	I. M. Gattman	New York, N. Y	Nov. 19, 1867	70, 990
Lead, Manufacture of white	I. M. Gattman	New York, N. Y	July 20, 1869	92, 816
Lead, Manufacture of white	H. S. Hannen	Philadelphia, Pa	July 21, 1868	80, 168
Lead, Manufacture of white	R. F. Hatfield	New York, N. Y	Nov. 8, 1870	109, 125
Lead, Manufacture of white	O. Jacobi	Philadelphia, Pa	May 12, 1868	77, 818
Lead, Manufacture of white	G. T. Lewis and E. O. Bartlett	Philadelphia and Birmingham, Pa.	Dec. 14, 1869	97, 936
Lead, Manufacture of white	F. F. Mayer	New York, N. Y	June 15, 1869	91, 466
Lead, Manufacture of white	A. P. Meylert	New Britain, Conn	Mar. 4, 1873	136, 446
Lead, Manufacture of white	H. J. Overman	New York, N. Y	Oct. 23, 1866	59, 135
Lead, Manufacture of white	T. Reakish	Philadelphia, Pa	Apr. 20, 1869	89, 074
Lead, Manufacture of white	J. Richards	Philadelphia, Pa	Dec. 2, 1836	95
Lead, Manufacture of white	R. Rowland	New York, N. Y	Apr. 19, 1864	42, 407
Lead, Manufacture of white	B. F. Smith	Few York, N. Y	Nov. 9, 1858	22, 036
Lead, Manufacture of white	B. F. Smith	New York, N. Y	Jan. 18, 1859	22, 679
Lead, Manufacture of white	P. Spence	Newton Heath, Great Britain.	Mar. 6, 1866	53, 093
Lead, Manufacture of white	P. H. Vander Weyde	Philadelphia, Pa	Feb. 12, 1867	62, 097
Lead, Manufacture of white	D. Wadsworth	Brooklyn, N. Y	Nov. 14, 1871	120, 916
Lead, Manufacture of white	J. Welsh and T. Evans	Philadelphia, Pa	Aug. 15, 1814	....
Lead, Manufacturing a white pigment to be used as a substitute for white.	W. Cumberland	New York	June 7, 1838	767
Lead, Manufacturing white	E. A. Boehne	Saint Louis, Mo	Aug. 26, 1873	142, 199
Lead, Manufacturing white	G. F. Hagner	Philadelphia, Pa	Oct. 13, 1817	....
Lead, Manufacturing white	G. A. Harrison	New York, N. Y	July 20, 1831	....
Lead, Manufacturing white	P. Phillips	Campbell County, Ky	Apr. 17, 1837	160
Lead, Manufacturing white	J. Richards	Philadelphia, Pa	May 28, 1818	....
Lead, Method of manufacturing sheet	J. Robertson	Brooklyn, N. Y	Oct. 3, 1848	5, 820
Lead, Method of producing white pigments from	J. G. Dale and E. Milner	Warrington, England	Jan. 12, 1869	85, 796
Lead mixing and feeding machine, White	J. Haslett, jr	Allegheny City, Pa	Dec. 3, 1867	71, 612
Lead, Mode of making oxychloride of	L. Brumlen	Hoboken, N. J	Aug. 21, 1860	29, 665
Lead, Mode of producing white	J. G. Dale and E. Milner	Warrington, England	Nov. 30, 1869	97, 365
Lead, Mode of separating the oxide from the metal in the manufacture of white.	E. Clark	Saugerties, N. Y	July 11, 1839	1, 231
Lead, Mode of separating the oxide from the metal during the manufacture of white.	E. Clark	Saugerties, N. Y	Dec. 5, 1839	1, 424
Lead, Mode of treating lead salts for the manufacture of white.	T. M. and A. G. Fell	Brooklyn, N. Y	June 25, 1867	66, 140
Lead, Mode of washing white	S. Witherlee, jr	Philadelphia, Pa	Oct. 29, 1811	....
Lead-ore, Treating galena, or	J. E. Schwabe	New York, N. Y	Dec. 18, 1855	13, 961
Lead or other pipes from frost, &c., Method of securing.	D. Lownes	....	July 21, 1803	....
Lead-ores and collecting precious metals, Reducing.	J. V. Z. Blaney and R. F. Adams	Chicago, Ill	Dec. 31, 1872	134, 457
Lead, Packing bar	Z. Kinsey	Dubuque, Iowa	Feb. 22, 1859	23, C31
Lead-pipe against the action of water, Mode of protecting.	L. Brandeis	Brooklyn, N. Y	Feb. 23, 1864	41, 675
Lead pipe against the action of water, Protecting	L. Brandeis	Brooklyn, N. Y	Dec. 15, 1863	40, 904
Lead pipe, Apparatus for making	W. Spillman	Columbus, Miss	Feb. 26, 1867	62, 450
Lead pipe, Coating	R. P. Berry	Newport, R. I	Nov. 5, 1872	132, 746
Lead pipe, &c., Composition for lining	T. Hodgson	Brooklyn, N. Y	Aug. 18, 1863	39, 615
Lead-pipe connection	W. D. Richardson	Springfield, Ill	Aug. 4, 1868	80, 667
Lead pipe, Die for manufacturing	B. Tatham	New York, N. Y	Oct. 12, 1869	95, 850
Lead-pipe die, Tin-lined	W. A. Shaw	New York, N. Y	July 24, 1866	56, 622
Lead pipe for conveying water, Making	A. Todd	Oxford, N. H	Apr. 18, 1814	....
Lead pipe for stills, &c., Tin-plated	T. Ewbank	New York	May 9, 1823	....
Lead-pipe machine	S. Parks, jr	Brooklyn, N. Y	Oct. 17, 1848	5, 852
Lead-pipe machine	C. E. Rockwell	New York, N. Y	Feb. 9, 1858	19, 313
Lead pipe, Machine for forming tin-lined	A. Hamon	Paris, France	Nov. 17, 1868	84, 114
Lead pipe, Machine for making	J. Laing	Bordentown, N. J	Mar. 30, 1843	3, 023
Lead pipe, Machine for making tin-lined	H. J. Bailey	Pittsburgh, Pa	June 29, 1869	91, 815
Lead pipe, Machine for making tin-lined	S. E. Chubbuck	Boston, Mass	July 7, 1868	79, 548
Lead pipe, Machine for making tin-lined	S. E. Chubbuck and J. H. Chadwick.	Boston, Mass	July 7, 1868	79, 549
Lead pipe, Machine for making tin-lined	J. E. Granniss	New York, N. Y	Mar. 12, 1872	124, 568
Lead pipe, Machine for making tin-lined	H. Merric	Cincinnati, Ohio	Apr. 27, 1868	89, 326
Lead pipe, Machine for making tin-lined	J. Robertson	Brooklyn, N. Y	June 1, 1869	90, 872
Lead pipe, Machine for pressing	R. Layng	New York, N. Y	June 5, 1866	55, 315
Lead-pipe machinery	I. Adams	Boston, Mass	Mar. 8, 1848	5, 466
Lead-pipe machinery	N. Buttrick, jr	Chelmsford, Mass	Nov. 8, 1845	4, 259
Lead-pipe machinery	S. G. Cornell	Greenwich, Conn	Aug. 21, 1847	5, 253
Lead-pipe machinery	C. and G. E. Sellers	Cincinnati, Ohio	Mar. 9, 1844	3, 475
Lead-pipe machinery	B. Tatham	New York, N. Y	May 11, 1852	8, 943
Lead pipe, Making	J. Robertson	Brooklyn, N. Y	Aug. 26, 1856	15, 620
Lead pipe, Manufacture and coating of	F. Bennett	Watford, England	Jan. 1, 1867	60, 851
Lead pipe, Manufacture of	W. P. Tatham	Philadelphia, Pa	Sept. 3, 1850	7, 624
Lead pipe, Manufacture of tin-coated	D. Turner	Cambridgeport, Mass	Sept. 3, 1872	131, 072
Lead pipe, Manufacture of tin-lined	H. J. Bailey	Pittsburgh, Pa	Feb. 2, 1869	86, 347
Lead pipe, Manufacture of tin-lined	J. Farrell	Pittsburgh, Pa	Feb. 25, 1868	74, 755
Lead pipe, Manufacture of tin-lined	E. W. Newton	Franklin Grove, Ill	Mar. 26, 1872	124, 911
Lead pipe, Manufacture of tin-lined	W. A. Shaw	New York, N. Y	Feb. 18, 1868	74, 611
Lead pipe, Manufacture of tin-lined	W. A. Shaw	New York, N. Y	Feb. 18, 1868	74, 612
Lead pipe, Manufacture of tin-lined	W. A. Shaw	New York, N. Y	Feb. 18, 1868	74, 613
Lead pipe, Manufacture of tinned	W. A. Shaw and G. Willard	New York, N. Y	Jan. 26, 1864	41, 401

Index of patents issued from the United States Patent Office from 1790 *to* 1873, *inclusive*—Continued.

Invention.	Inventor.	Residence.	Date.	No.
Lead pipe, Method of tinning	H. M. Ward, S. L. Selden, and E. Y. Kneeland.	Rochester, N. Y	Mar. 12, 1845	3, 944
Lead pipe, Manufacturing	G. W. Potter	Otsego, N. Y	Mar. 23, 1833	....
Lead pipe, Mode of making	R. Ward	Waterbury, Conn	July 5, 1822	....
Lead pipe, Mold for casting and machine for drawing.	J. C. Vaughn and F. Leach	Tioga, N. Y	Dec. 31, 1839	1, 463
Lead pipe with tin, Lining	W. A. Shaw	New York, N. Y	Mar. 10, 1863	37, 877
Lead pipes, Uniting the ends of	N. F. Weston	Boston, Mass	July 9, 1867	66, 658
Lead plates, Apparatus for molding	S. E. Chubbuck	Roxbury, Mass	Aug. 13, 1867	67, 721
Lead, Pot for the manufacture of white	I. H. Chadwick	Boston, Mass	Nov. 21, 1865	51, 018
Lead, Process and apparatus for the manufacture of sulphate of.	H. A. Whiting	San Francisco, Cal	Apr. 2, 1872	125, 153
Lead, Process of making white	H. Hannen	Dubuque, Iowa	Sept. 22, 1857	18, 244
Lead, Process of making white	C. L. Wheeler	Pittsburgh, Pa	Oct. 31, 1871	120, 556
Lead, Process of manufacturing carbonate or white	C. Button and H. G. Dyar	London, England	Apr. 10, 1839	1, 115
Lead, Process of manufacturing white	G. Cary	Cleveland, Ohio	Sept. 24, 1861	33, 337
Lead, Process of manufacturing white	S. Gardner	New York	Aug. 28, 1840	1, 744
Lead, Process of refining	O. Wassermann	Call, Prussia	Aug. 27, 1867	68, 135
Lead, Process of refining and softening	A. H. Everett	New York, N. Y	May 24, 1864	42, 844
Lead, Reducing oxide of	T. Taylor	Washington, D. C	July 24, 1866	56, 635
Lead, Refining	J. J. Crooke	New York, N. Y	Nov. 7, 1865	50, 800
Lead, Screening and separating white	S. Witherill, jr	Philadelphia, Pa	Nov. 1, 1811	....
Lead, Setting the beds or stacks in making white	S. Witherill, jr	Philadelphia, Pa	Oct. 29, 1811	....
Lead shavings, Machine for making	J. Repetti	Philadelphia, Pa	June 15, 1869	91, 267
Lead-smelting furnace	J. V. Woodhouse	Mine La Motte, Mo	June 13, 1871	115, 921
Lead to collect the dust, Manufacture of white	M. Tolle	Saint Louis, Mo	Sept. 2, 1873	142, 419
Lead to destroy its crystalline character, Mode of treating precipitated.	G. T. Lewis	Philadelphia, Pa	Aug. 20, 1867	67, 992
Lead-washing machine, White	W. Davison	Baltimore, Md	Sept. 12, 1871	118, 794
Lead, White, &c	E. Clark	Saugerties, N. Y	June 20, 1836	....
Lead, White	H. Holland	Westfield, Mass	Mar. 18, 1836	....
Lead without grinding, Mode of preparing white	J. Burney	New Haven Township, Ohio	Jan. 2, 1829	....
Leads, Method of fastening	J. Merlett	Boundbrook, N. J	June 21, 1870	104, 622
Leader-bracket	S. J. Dwyer	Albany, N. Y	Feb. 20, 1872	123, 877
Leader-hook	C. D. Woodruff	Toledo, Ohio	Apr. 9, 1872	125, 652
Leaf-holder	J. Nock	Washington, D. C	Aug. 11, 1843	3, 218
Leaf-turner, Atmospheric	C. Phelps and A. K. Tuttle	Clayton and Cape Vincent, N. Y.	Nov. 22, 1870	109, 448
Leak alarm and indicator	T. P. Akers	New York, N. Y	Mar. 10, 1868	75, 235
Leak-indicator for ships	G. B. Massey	Mobile, Ala	Aug. 11, 1857	17, 975
Leak-signal for vessel	N. D. Lamb	Norwich, Conn	Jan. 1, 1867	60, 749
Leakage-alarm for vessel	G. B. Massey	New York, N. Y	Feb. 19, 1867	62, 144
Leather and cloth water-proof, Making	J. L. Comstock	Hartford, Conn	Jan. 21, 1828	....
Leather and harness polishing machine	W. Crane	Brooklyn, N. Y	Nov. 25, 1856	16, 114
Leather and hides, Process of splitting	I. Lippmann	Paris, France	June 2, 1857	17, 471
Leather and hides, Machine for softening	J. Tidd	Woburn, Mass	Jan. 14, 1868	73, 408
Leather and morocco, Machine for polishing	N. Ames	Saugus, Mass	May 1, 1855	12, 806
Leather and process of manufacturing the same	G. Bottero	Boston, Mass	Feb. 21, 1865	46, 443
Leather and similar fabrics, Manufacture of artificial.	T. J. Mayall	Boston, Mass	June 24, 1873	140, 209
Leather and skins, Machine for splitting	W. Bent	Boston, Mass	Feb. 24, 1808	....
Leather and skins, Machine for splitting	J. Woodcock	Worcester, Mass	May 8, 1809	....
Leather and skins, Splitting, shaving, and cutting	A. S. Dawley	Boston, Mass	Mar. 22, 1831	....
Leather, &c., Apparatus for molding	D. C. Taylor	New York, N. Y	Oct. 8, 1872	132, 117
Leather, Apparatus for stoning, glassing, and pebbling.	H. Cunningham	Albany, N. Y	Nov. 12, 1872	132, 901
Leather, Apparatus for stretching	S. D. Castle	Bridgeport, Conn	Feb. 27, 1872	123, 979
Leather, Artificial	H. A. Clark	Boston, Mass	July 22, 1873	140, 117
Leather, Artificial	W. Elmer	New York, N. Y	June 2, 1863	38, 785
Leather, Artificial	W. K. Hall	West Hoboken, N. J	Oct. 5, 1858	21, 721
Leather, Artificial	L. Montier	New York, N. Y	Feb. 4, 1868	73, 991
Leather, Attaching dicing-stone to wheel for polishing.	H. C. Havemeyer and P. D. Burdon.	New York, N. Y	Nov. 15, 1870	109, 205
Leather banding for machinery, Manufacturing	G. Miller	Providence, R. I	Nov. 7, 1854	11, 902
Leather bands, Stretching	W. Kumble	New York, N. Y	Apr. 11, 1846	4, 455
Leather blacking and polishing composition	A. Bona	Philadelphia, Pa	Sept. 15, 1863	39, 986
Leather-blacking composition	P. W. Keating	Norwich, Conn	May 30, 1865	47, 957
Leather, Blacking for harness	J. N. Knapp	Syracuse, N. Y	Sept. 14, 1869	94, 897
Leather, Bleaching and dyeing	A. P. Viol and C. P. Duflo, jr	Paris, France	Sept. 1, 1868	81, 709
Leather, Bleaching buffed sole	H. W. Cattle	Lynn, Mass	Feb. 11, 1873	135, 632
Leather board, Manufacture of	S. Moore	Sudbury, Mass	June 3, 1873	139, 598
Leather, Boarding	M. B. Bishop	Whitingham, Vt	Oct. 10, 1870	108, 319
Leather boarding and graining machine	L. Townsend	Terre Haute, Ind	Apr. 23, 1872	126, 105
Leather-boarding machine	O. Coogan	Pittsfield, Mass	Apr. 22, 1873	138, 133
Leather-boarding machine	J. W. Hildreth	Boston, Mass	June 21, 1870	104, 454
Leather-boarding machine	W. H. Moore	Salem, Mass	July 25, 1865	48, 971
Leather-boarding machine	W. R. Williams and W. P. Martin.	Salem, Mass	Jan. 10, 1871	110, 944
Leather boarding and graining machine	L. A. Gignac	Troy, N. Y	Aug. 31, 1869	94, 196
Leather boarding and graining machine	J. Parker	Woburn, Mass	May 15, 1866	54, 821
Leather boarding, pebbling, and glossing machine.	O. Coogan	Pittsfield, Mass	Oct. 10, 1871	119, 743
Leather, Bronze dressing for	M. S. Cahill	Boston, Mass	Nov. 10, 1868	83, 925
Leather brushing and finishing machine	G. H. Parker	Detroit, Mich	Aug. 15, 1871	118, 146
Leather-buffing machine	S. D. Tripp	Lynn, Mass	Oct. 11, 1870	108, 216
Leather buffing and reducing machine	J. Turner	Cambridgeport, Mass	June 5, 1860	28, 641
Leather buffing, whitening, glassing, polishing, and stoning machine.	A. W. Pratt	Salem, Mass	May 4, 1869	89, 789
Leather by rollers, Compressing	W. Edwards	Northampton, Mass	Oct. 19, 1812	....
Leather-cement	B. F. Pettingell	Newburyport, Mass	Jan. 30, 1866	52, 315
Leather-chamfering machine	C. H. Helms	Poughkeepsie, N. Y	Aug. 7, 1866	56, 938
Leather channeling and folding tool	A. H. Bailey and W. G. Bratton	Marseilles, Ill	Jan. 18, 1870	98, 906
Leather-channeling, Pattern for	S. Golcher	Newark, N. J	Nov. 19, 1872	133, 149
Leather-channeling tool	E. H. Crane	Jonesville, Mich	Jan. 24, 1865	45, 978
Leather clamping and stretching machine	E. Shaw	Milwaukee, Wis	Aug. 21, 1866	57, 392
Leather cleaning and renovating machine	D. Budd	Valatie, N. Y	June 2, 1868	78, 421
Leather cloth	M. Altmann	Baltimore, Md	Feb. 14, 1871	111, 716

Index of patents issued from the United States Patent Office from 1790 *to* 1873, *inclusive*—Continued.

Invention.	Inventor.	Residence.	Date.	No.
Leather cloth, imitation leather, &c., Manufacture of.	N. C. Szerelmey	Clapham Common, England	Nov. 1, 1864	44, 910
Leather, cloth, &c., Machine for cutting objects from	W. H. Johnson	Springfield, Mass	Nov. 17, 1863	40, 626
Leather, cloth, &c., Manufacture of enameled	F. Longhurst and A. L. Murdock.	Boston, Mass	May 3, 1864	42, 584
Leather cloth. Ornamenting	A. Pellet	Paris, France	Oct. 2, 1860	30, 242
Leather, cloth, &c., to render water and fire proof, Mode of treating.	R. O. Lowrey	Salem, N. Y	May 19, 1868	77, 990
Leather, Coloring	J. Koppitz and F. B. Mayer	Boston and Cambridgeport, Mass.	Sept. 16, 1873	142, 797
Leather, Coloring and finishing	H. Hibbard	Attica, N. Y	Dec. 15, 1837	511
Leather-coloring composition	L. C. May	Cochituate, Mass	Mar. 26, 1872	124, 965
Leather-coloring compound	G. Jäger	Indianapolis, Ind	Nov. 28, 1871	121, 375
Leather, Coloring tanned	B. H. Lightfoot	Philadelphia, Pa	Dec. 22, 1863	41, 011
Leather, Composition for artificial	S. Whitmarsh	Northampton, Mass	May 25, 1858	20, 383
Leather, Composition for blacking	J. Hayward	Cleveland, Ohio	May 9, 1846	4, 498
Leather, Composition for bleaching and stuffing	L. W. Fiske	Louisville, Ky	Feb. 6, 1855	12, 368
Leather, Composition for coating	E. Brown	Indianapolis, Ind	Apr. 16, 1867	63, 847
Leather, Composition for currying	I. S. Hershey	Hagerstown, Md	June 12, 1847	5, 151
Leather, Composition for dressing and coloring	Z. A. Taylor	Haverbill, Mass	July 11, 1871	116, 886
Leather, Composition for finishing wax	P. Maguire	New Boston, Mass	May 3, 1870	102, 565
Leather, Composition for preserving and coloring	O. F. Battey	Worcester, Mass	July 18, 1871	117, 144
Leather, Composition for preserving and waterproofing.	R. K. Wright	Philadelphia, Pa	Apr. 28, 1863	38, 353
Leather, Composition for stuffing	R. Andrews	Milwaukee, Wis	May 25, 1869	90, 333
Leather, Composition for stuffing	J. Haseltine	Warren, N. H	Mar. 10, 1868	75, 264
Leather, Composition for stuffing	J. Rose	Newark, N. J	May 6, 1856	14, 832
Leather, Composition for stuffing and finishing	T. D. Williams	Chicago, Ill	Sept. 20, 1870	107, 579
Leather, Composition for the manufacture and preserving of.	R. Andrews	Milwaukee, Wis	Jan. 22, 1867	61, 379
Leather, Composition for treating	A. Taw	Philadelphia, Pa	Aug. 30, 1864	44, 025
Leather, Composition for water-proofing	R. Goldenblum and F. Steiner.	East Hampton, Mass	Feb. 19, 1861	31, 453
Leather, Composition to be applied to	W. Lehman	Newville, Pa	May 26, 1868	78, 295
Leather, Compound for oiling, polishing, and blacking.	G. F. Whitney	Boston, Mass	May 24, 1870	103, 402
Leather, Compound for stuffing	J. Merril	Boston, Mass	Mar. 8, 1870	100, 652
Leather, Compound for stuffing	A. Doepp	Newark, N. J	Nov. 17, 1868	84, 096
Leather, Compound for treating	S. S. Walbank	Superior City, Wis	Dec. 27, 1870	110, 610
Leather-compressing machine, Sole	J. Elliot	Sharon, Conn	Jan. 18, 1810	....
Leather cord or rope	O. L. Harrington and E. Weaver	Wood's Run, Pa	Sept. 7, 1869	94, 690
Leather-creasing machine	W. S. Bullen	Indianapolis, Ind	Nov. 27, 1860	30, 785
Leather-creasing machine	B. R. Hamilton and S. Swan	South Deerfield and Conway, Mass.	May 30, 1871	115, 312
Leather creasing or ornamenting	J. M. Bent	Wayland, Mass	Oct. 16, 1866	58, 761
Leather-creasing roller	S. B. Randall	Cincinnati, Ohio	Dec. 2, 1873	145, 240
Leather creasing, slicking, and skiving	C. C. Bellows	New Ipswich, N. H	Oct. 23, 1866	58, 978
Leather crimping and folding machine	G. Platts and J. Walden	Newark, N. J	Oct. 21, 1873	143, 783
Leather-crimping machine	G. and M. Algar	Greenport, N. Y	Nov. 25, 1838	1, 016
Leather-crimping machine	L. Upham	Putney, Vt	Jan. 9, 1838	541
Leather-crimping machine	H. Wing	Buffalo, N. Y	July 31, 1860	29, 429
Leather-crimping machine for boots, &c	J. Adams	Fair Haven, Vt	Mar. 26, 1838	656
Leather-crimping machine for boots, &c	C. H. Jaquith	Keene, N. H	Mar. 21, 1838	642
Leather-crimping machine, Mode of regulating the jaws of.	J. Goodwin, jr	Sterling, Mass	May 30, 1839	1, 162
Leather cushioning and ventilating machine	E. Thomas	Philadelphia, Pa	Aug. 2, 1870	105, 998
Leather-cutter	C. S. Stearns	Marlborough, Mass	June 9, 1868	78, 696
Leather, Cutter for cutting washers from	E. R. Brown and J. Long	Mauch Chunk and Packerton, Pa.	May 23, 1871	115, 157
Leather cutting and embossing machine	S. D. Tripp	Lynn, Mass	June 7, 1864	43, 050
Leather cutting and punching machine	I. Genung	Newark, N. J	July 30, 1831	....
Leather cutting and splitting machine	J. Edson	Boston, Mass	Jan. 31, 1860	26, 977
Leather-cutting machine	H. F. Baker	Pawtucket, R. I	Feb. 21, 1871	111, 900
Leather-cutting machine	H. H. Bigelow	Worcester, Mass	Apr. 4, 1871	113, 618
Leather-cutting machine	J. S. Bishop	Wayne, Me	June 29, 1833	....
Leather-cutting machine	R. Crocker	Marshalltown, Iowa	Jan. 28, 1873	135, 266
Leather-cutting machine	G. Domett	Boston, Mass	July 20, 1831	....
Leather-cutting machine	H. Eichler	New Lisbon, Wis	Mar. 2, 1869	87, 481
Leather-cutting machine	E. A. Holbrook	East Randolph, Mass	Apr. 9, 1872	125, 457
Leather-cutting machine	A. Keith	North Bridgewater, Mass	Mar. 7, 1871	112, 353
Leather-cutting machine	S. H. King	Tunbridge, Vt	Dec. 22, 1868	85, 102
Leather-cutting machine	L. N. Leland	Grafton, Mass	Sept. 28, 1837	416
Leather-cutting machine	N. Paquette	Summerfield, Mich	Apr. 15, 1873	137, 952
Leather-cutting machine	J. W. Richardson	South Braintree, Mass	Nov. 27, 1860	30, 761
Leather-cutting machine	J. F. Severance	East Bridgewater, Mass	May 23, 1865	47, 865
Leather-cutting machine	C. S. Stearns	Marlborough, Mass	Sept. 25, 1866	58, 315
Leather-cutting machine	A. L. Zent	Roanoke, Ind	Oct. 11, 1870	108, 229
Leather, Cutting out	A. B. Keith and T. K. Reed	North Bridgewater and East Bridgewater, Mass.	Apr. 30, 1867	64, 228
Leather-cutting press	N. J. Simonds	Woburn, Mass	Apr. 12, 1870	101, 931
Leather-cutting tool	J. Demarest	Belleville, N. J	May 27, 1873	139, 240
Leather-cutting tool	J. Sweesy	Elizabethville, Pa	Oct. 29, 1872	132, 609
Leather-cutting tool for soles	L. Marble	Henrietta, N. Y	Mar. 28, 1828	....
Leather, Cutting up and punching sole	L. D. Hawkins	Stoneham, Mass	Nov. 7, 1871	120, 642
Leather, Currying and finishing	S. Parker	Bellerica, Mass	July 9, 1808	....
Leather, Currying and finishing	S. Parker	....	Apr. 26, 1809	....
Leather, Device for stretching	W. Strevell	Jersey City, N. J	Oct. 30, 1866	59, 292
Leather, Device for trimming	W. F. Foley	Albany, N. Y	Dec. 30, 1873	145, 937
Leather, Dicing and polishing	R. and H. Brackett	Boston and Woburn, Mass.	Mar. 15, 1845	3, 957
Leather dicing and polishing machine	L. Hedge	Windsor, Vt	July 22, 1819	....
Leather-dicing machine	A. Shidloch	New York, N. Y	Dec. 26, 1871	122, 136
Leather-dresser's table	C. T. Woodman	Boston, Mass	Feb. 2, 1864	41, 438
Leather-dressing	J. N. Baratta	Ayer, Mass	Nov. 26, 1872	133, 400
Leather dressing and scouring machine	E. Fitzhenry	Boston, Mass	Apr. 14, 1868	76, 619
Leather-dressing composition	M. Shaw	Abington, Mass	Feb. 25, 1862	34, 530
Leather-dressing composition	C. Werner	Philadelphia, Pa	Jan. 2, 1855	12, 171
Leather-dressing compound	T. G. Bell	New York, N. Y	Jan. 28, 1873	135, 310
Leather-dressing machine	J. A. Enos	Peabody, Mass	May 13, 1873	138, 874
Leather-dressing machine	E. Fitzhenry	Boston, Mass	Feb. 26, 1867	62, 524

Index of patents issued from the United States Patent Office from 1790 *to* 1873, *inclusive*—Continued.

Invention.	Inventor.	Residence.	Date.	No.
Leather-dressing machine	E. Fitzhenry	Boston, Mass	Mar. 1, 1870	100, 387
Leather-dressing machine	E. Fitzhenry	Boston, Mass	Aug. 15, 1871	118, 003
Leather-dressing machine	C. Korn	Meriden, Conn	June 25, 1861	32, 629
Leather-dressing machine	C. Korn	Wurtsborough, N. Y	June 18, 1867	65, 919
Leather-dressing machine	R. Lee	Newark, N. J	June 13, 1865	48, 186
Leather-dressing machine	C. A. McDonald	Woburn, Mass	Jan. 2, 1872	122, 395
Leather-dressing machine	H. P. Reed	Peabody, Mass	July 2, 1872	128, 658
Leather-dressing machine	C. Slawson	Norwich, N. Y	Nov. 13, 1849	6, 875
Leather, Drop for hammering sole	J. H. Walker	Worcester, Mass	May 10, 1864	42, 703
Leather drying and pressing machine	G. Harvey	New York, N. Y	Aug. 15, 1865	49, 402
Leather drying and polishing machine	S. Couillard, jr	Boston, Mass	June 27, 1827	....
Leather-edge-trimming machine	W. F. Foley	Albany, N. Y	Apr. 23, 1872	125, 946
Leather, Enameling	J. Rose	Newark, N. J	Mar. 9, 1858	19, 583
Leather, Fabric from waste	E. Waite	Franklin, Mass	May 2, 1871	114, 373
Leather, Fabrics prepared to imitate	H. A. Clark	Boston, Mass	Sept. 23, 1873	143, 122
Leather, Factitious enameled	J. W. Munroe	Fall River, Mass	May 10, 1859	23, 987
Leather fastener, Metallic	H. Beals	East Stoughton, Mass	Mar. 26, 1872	125, 007
Leather fastening	H. Beals	East Stoughton, Mass	Aug. 6, 1872	130, 181
Leather filling, &c., Machine for cutting	R. C. Turner	Mendon, Mich	July 21, 1868	80, 098
Leather-finisher	C. Bassett	Boston, Mass	Mar. 25, 1835	....
Leather-finishing	A. Bertram	New Albany, Ind	May 12, 1868	77, 863
Leather, Finishing and dressing sheep-skins for	R. Andrews	Milwaukee, Wis	July 13, 1869	92, 416
Leather-finishing machine	J. Burr	Johnstown, N. Y	Nov. 24, 1820	....
Leather-finishing machine	S. P. Cobb	South Danvers, Mass	June 5, 1860	28, 562
Leather-finishing machine	S. P. Cobb	South Danvers, Mass	Dec. 1, 1863	40, 735
Leather-finishing machine	W. Ellard	Woburn, Mass	Apr. 2, 1861	31, 879
Leather-finishing machine	E. Fitzhenry and J. Ball	Portland, Oreg	Jan. 15, 1867	61, 182
Leather-finishing machine	J. P. Friend	Peabody, Mass	Aug. 8, 1871	117, 877
Leather-finishing machine	W. P. Martin	Salem, Mass	Feb. 28, 1860	27, 300
Leather-finishing machine	W. P. Martin	Salem, Mass	May 1, 1860	28, 108
Leather-finishing machine	J. Pyle	Wilmington, Del	Sept. 30, 1856	15, 816
Leather-finishing machine	R. L. and C. Smith	Stockport, N. Y.	Jan. 10, 1860	26, 792
Leather-finishing machine	T. F. Weston	Salem, Mass	June 7, 1859	24, 344
Leather-finishing machine	C., T. F., and J. W. Weston	Salem, Mass	Sept. 25, 1855	13, 605
Leather-finishing, Method of	L. B. Fox	Williamsport, Pa	Apr. 21, 1868	77, 025
Leather, Finishing split	J. Putnam	Danvers, Mass	July 27, 1869	93, 002
Leather flinting and glazing machine	G. Crossley	Philadelphia, Pa	Feb. 13, 1872	123, 681
Leather-folding machine	J. Lombard	Springfield, Me	Oct. 1, 1867	69, 454
Leather-folding machine	E. B. Stimpson	New York, N. Y	July 29, 1873	141, 398
Leather for boots and shoes, Method of preparing sole.	D. M. Ayer	Lewiston, Me	Jan. 1, 1867	60, 810
Leather for boots, &c., Machine for crimping	N. Woodbury	Calais, Me	Aug. 16, 1838	885
Leather for cards, Twilling-machine for pricking	P. Earle		Dec. 6, 1803	....
Leather for cotton or woolen cards, Machine for finishing.	G. S. Quick	Auburn, N. Y	Apr. 19, 1864	42, 399
Leather for enameling, Treating	T. P. Howell and N. F. Blanchard.	Newark, N. J.	Nov. 20, 1855	13, 819
Leather for floor-covering, Artificial	S. M. Allen	Woburn, Mass	Oct. 15, 1867	69, 742
Leather for gloves, wafers, &c., Machine for cutting.	E. M. Ely	Paris, France	Jan. 28, 1829	....
Leather for harness, Machine for creasing and blacking.	A. Stempel	Oquawka, Ill	Nov. 2, 1858	21, 989
Leather for shoe-binding, Mode of striping alum-dressed.	S. Frothingham and G. Harriss	Middletown, Conn	Apr. 20, 1808	....
Leather for soling, Process of preparing the flanking of.	D. Kern	York, Pa	June 25, 1861	32, 628
Leather for suspender-straps, Mode of preparing	O. Webb	New York, N. Y	Oct. 14, 1807	....
Leather for the manufacture of boots and shoes, Machine for preparing.	C. Rice and S. H. Whorf	Boston and Roxbury, Mass	Nov. 20, 1855	13, 827
Leather for the soles and heels of boots and shoes, Machine for hammering.	J. P. Molliere	Lyons, France	Jan. 22, 1856	14, 140
Leather for wear, Preparing	G. V. Sheffield and J. F. Coburn	Hopkinton, Mass	May 7, 1867	64, 587
Leather from curriers' shavings or buffings, Making sheets of.	C. F. Crockett	Newark, N. J	June 17, 1856	15, 121
Leather from scrap-leather, &c., Preparation of artificial.	S. Sorenson	Ebeltoft, Denmark	May 24, 1870	103, 517
Leather from vats, Apparatus for raising	C. E. Robinson and L. D. Sanborn.	Concord, N. H	Apr. 27, 1858	20, 093
Leather-gage	L. W. Beecher	Avon, N. Y	May 4, 1852	8, 917
Leather, Gilding and silvering	A. Allen	New York, N. Y	Dec. 26, 1817	....
Leather, Glazed	E. G. Adams	Decatur, Ga	June 30, 1836	....
Leather goods, Water-proof	S. La Forge	Cleveland, Ohio	Feb. 28, 1860	27, 337
Leather-graining composition	O. P. Whitman	Lynn, Mass	Feb. 27, 1866	52, 920
Leather, Gripe for holding	B. Rowe	Albany, N. Y	July 19, 1853	9, 862
Leather-handling machine	W. G. Waterman	Sullivan, N. Y	Sept. 16, 1834	....
Leather, harness, &c., Blacking for	T. James	Medford, Mass	May 29, 1866	55, 203
Leather, hides, and skins, Composition for dyeing and coloring.	C. Bond	New York, N. Y	Jan. 4, 1870	98, 466
Leather-holder	C. Bates	Conestoga, Pa	Sept. 14, 1869	94, 697
Leather-holding clamp	R. Emans	Mansfield, N. J	Oct. 6, 1837	420
Leather-holding clamp	A. Rittenhouse	Philadelphia, Pa	Aug. 18, 1868	81, 293
Leather impermeable to water, Rendering	J. Board	New York, N. Y	Nov. 4, 1813	....
Leather, &c., impervious to gas, Process of rendering.	E. Osgood	Boston, Mass	July 30, 1872	129, 901
Leather impervious to hydrocarbon liquids, Rendering.	A. Warth	Stapleton, N. Y	Oct. 4, 1870	108, 072
Leather, &c., Infusing oil into	H. Guest		July 16, 1801	....
Leather into bales, Machine for rolling	N. Burk	Fulton, N. Y	Aug. 10, 1858	21, 114
Leather into counters, Machine for cutting	A. Keith	North Bridgewater, Mass.	Sept. 20, 1864	44, 318
Leather into hollow-ware forms, Machine for cutting.	F. Durand and O. Recqueur	Paris, France	June 11, 1850	7, 424
Leather into round bands, Machine for cutting	C. G. Burnham	East Hartford, Conn	Aug. 17, 1869	93, 803
Leather into strips for boot and shoe soles and heels, Machine for cutting.	J. P. Molliere	Lyons, France	June 19, 1855	13, 095
Leather, Japanned	F. S. Merritt	Boston, Mass	May 9, 1871	114, 586
Leather lacings, Instrument for cutting	C. A. Woodbury	Woodstock, Vt	July 2, 1872	128, 519
Leather, Lap cutting and leveling	H. Underwood	Tolland, Conn	Oct. 16, 1847	5, 329
Leather, Machine for creasing straps of	D. H. Hovey	Kilborn, Ohio	Nov. 15, 1853	10, 231

Index of patents issued from the United States Patent Office from 1790 *to* 1873, *inclusive*—Continued.

Invention.	Inventor.	Residence.	Date.	No.
Leather, Machine for cutting articles from	C. Rice and S H. Whorf	Boston and Roxbury, Mass	Dec. 11, 1855	13, 920
Leather, Machine for forming staple seams in	S. W. Shorey	Boston, Mass	Feb. 25, 1873	136, 340
Leather, Machine for removing grease from	W. A. Perkins	Salem, Mass	Feb. 13, 1872	123, 643
Leather, Machine for removing grease from	J. Perkins and G. L. Newcomb	Peabody and Salem, Mass	Dec. 26, 1871	122, 130
Leather, Machine for removing grease from	J. Starratt	Salem, Mass	July 19, 1870	105, 506
Leather, Machinery for striking out	S. Hutchinson	Leeds, England	Apr. 26, 1870	102, 270
Leather, Making	R. Downey	New Albany, Ind	June 15, 1844	3, 632
Leather, Making scabbards, sheaths, &c., of	F. Durand and O. Pecquer	France	Feb. 15, 1847	4, 986
Leather, Making sod-oil for	G. Poyzer	Philadelphia, Pa	May 10, 1805	....
Leather, Manufacture of	W. Adamson	Philadelphia, Pa	June 23, 1868	79, 177
Leather, Manufacture of	F. A. Holcomb	Grand Rapids, Mich	Dec. 27, 1870	110, 465
Leather, Manufacture of	S. B. Jenks and F. A. Holcomb	Grand Rapids, Mich	Jan. 24, 1871	111, 214
Leather, Manufacture of	W. Pyle	Wilmington, Del	July 12, 1870	105, 247
Leather, Manufacture of	G. Rawle and W. N. Evans	Bristol and Bedminster, England.	Aug. 5, 1873	141, 459
Leather, Manufacture of	C. J. Tinnerholm	Quincy, Ill	Feb. 13, 1872	123, 748
Leather, Manufacture of	A. H. Van Gieson	Newark, N. J	Nov. 24, 1863	40, 718
Leather, Manufacture of artificial	H. B. Meech	Fort Edward, N. Y	Mar. 17, 1868	75, 690
Leather, Manufacture of artificial	C. Saffray	New York, N. Y	Feb. 26, 1867	62, 503
Leather, Manufacture of artificial	W. W. Waite	South Natick, Mass	Dec. 26, 1865	51, 781
Leather, Manufacture of artificial	F. Walton	Staines, England	Jan. 17, 1871	111, 100
Leather, Manufacture of enameled and japanned	I. W. Dawson	Newark, N. J	Oct. 29, 1867	70, 176
Leather, Manufacture of japanned	F. S. Merritt	Boston, Mass	May 23, 1871	115, 083
Leather, Manufacture of japanned	J. L. Newton	Boston, Mass	Oct. 9, 1866	58, 733
Leather, Manufacture of water-proof	R. Sponouse	Jersey Shore, Pa	May 7, 1867	64, 589
Leather, Manufacturing artificial	E. and J. K. Cushman	Amherst, Mass	Apr. 5, 1859	23, 454
Leather, Method of laying off patterns for stitching on.	W. P. Wolfington	Louisville, Ky	Mar. 8, 1870	100, 579
Leather, Method of oiling	L. Holcomb	Granby, Conn	Oct. 9, 1860	30, 320
Leather, Method of stuffing	F. A. White	Roxbury, Mass	Aug. 5, 1856	15, 499
Leather, Mode of covering rounded articles with	J. H. Osgood	Boston, Mass	May 21, 1872	127, 098
Leather, Mode of economizing the manufacture of articles of.	W. Adamson	Philadelphia, Pa	Jan. 31, 1865	46, 061
Leather, Mode of embossing	G. W. Pratt	Salem, Mass	July 18, 1865	48, 876
Leather, Mode of manufacturing	A. Hickman and E. S. Davenport.	Abingdon, Va	Aug. 1, 1838	866
Leather, Mode of manufacturing fair	J. L. Turner	Philadelphia, Pa	July 15, 1840	1, 688
Leather, Mode of polishing	A. Bayrd	South Reading, Mass	Dec. 29, 1828	....
Leather, Mode of softening	H. Cunningham	Albany, N. Y	Oct. 12, 1869	95, 779
Leather, Mode of treating tanned	B. H. Lightfoot	Philadelphia, Pa	June 14, 1864	43, 120
Leather, Mode of whitening	J. C. Booth	Philadelphia, Pa	Dec. 5, 1840	1, 881
Leather more durable and flexible, Mode of rendering.	L. G. Sourzar and L. Bombail	Bordeaux, France	Nov. 13, 1866	59, 736
Leather, Oil-blacking for	J. L. Baumer	Columbus, Ohio	Dec. 14, 1869	97, 857
Leather oil or polish	S. M. Farnham	Tully, N. Y	Apr. 21, 1868	77, 021
Leather-oiling machine, Tanner's	G. Huttelmaier	Allegheny, Pa	Sept. 22, 1863	40, 079
Leather, Ornamenting	L. Wolfson	Boston, Mass	Sept. 26, 1871	119, 291
Leather or skins, Machine for softening or dressing.	F. J. Burcham	Racine, Wis	June 25, 1867	66, 125
Leather-ornamenting machine	C. T. Woodman	Boston, Mass	Mar. 29, 1864	42, 136
Leather, paper, &c., Drying	S. M. Allen	Woburn, Mass	Jan. 19, 1864	41, 348
Leather, paper, &c., finishing machine	P. Farrell	Albany, N. Y	June 15, 1869	91, 219
Leather, paste-board, and paper, Manufacture of	A. N. Mathiew	Paris, France	Apr. 20, 1858	20, 020
Leather pebbling and embossing machine	B. Merritt, jr	Chelsea, Mass	Aug. 19, 1862	36, 228
Leather-perforating machinery for harness-makers	L. D. Woodmansee	Hamilton, Ohio	Dec. 16, 1873	145, 543
Leather pieces, Mode of connecting	B. F. Sturtevant	Boston, Mass	June 21, 1864	43, 236
Leather-piercing machine	A. Eggleston	Fall River, Mass	Apr. 18, 1871	113, 863
Leather-piercing machine	G. V. Sheffield and J. F. Coburn	Hopkinton, Mass	Sept. 4, 1866	57, 780
Leather-polish	H. A. and R. G. Sawyer	Milwaukee, Wis	May 2, 1871	114, 344
Leather, Polish for	J. Herold and M. Brown	Frederick, Md	May 26, 1868	78, 372
Leather polishing and graining machine, Morocco	J. Perkins	Boston, Mass	June 26, 1809	....
Leather-polishing machine	G. T. Adler	Philadelphia, Pa	Feb. 7, 1860	27, 028
Leather-polishing machine	W. P. Gamble	Philadelphia, Pa	Apr. 8, 1856	14, 606
Leather-polishing machine	J. W. Hildreth	Boston, Mass	Apr. 21, 1868	76, 914
Leather-polishing machine	J. M. Poole	Wilmington, Del	Sept. 28, 1852	9, 292
Leather-polishing machine	F. Seibert	Williamsburgh, N. Y	Nov. 29, 1853	10, 287
Leather-polishing machine	R. A. Stratton	Philadelphia, Pa	Jan. 24, 1860	26, 932
Leather-polishing machine	P. P. Tapley	Lynn, Mass	Jan. 3, 1854	10, 379
Leather, Preparation of	M. W. Page	Franklin, N. H	May 8, 1866	54, 587
Leather, Prepared	S. Dyar	Charlestown, Mass	Feb. 19, 1867	62, 120
Leather, Preparing cloth and paper, &c., to imitate	H. A. Clark	Boston, Mass	July 22, 1873	141, 118
Leather, Preservative blacking for	W. B. Moore	Winchester, Mo	Nov. 12, 1867	70, 737
Leather-preserving composition	J. Lamplugh	Philadelphia, Pa	Dec. 30, 1873	146, 080
Leather-preserving composition	T. P. Merriam	New Bedford, Mass	Nov. 4, 1842	2, 844
Leather-pressing machine	G. F. Dressing	Buffalo, N. Y	Oct. 11, 1864	44, 616
Leather-pricking machine	G. C. Converse	Spencer, Mass	Feb. 18, 1873	135, 969
Leather-pricking machine	R. U. and M. Richards	Norfolk, Conn	Apr. 24, 1810	....
Leather-pricking machine	S. Sheldon	Cincinnati, Ohio	Oct. 10, 1840	1, 817
Leather-pricking machine	S. Sheldon	Cincinnati, Ohio	Oct. 10, 1840	1, 818
Leather-pricking machine	J. H. Walker	Worcester, Mass	Feb. 25, 1862	34, 537
Leather-pricking machine for harness and coach makers.	J. W. Briggs, and L. C. and J. S. Carmer.	Painesville, Ohio	Mar. 26, 1838	658
Leather, Process and apparatus for stuffing	W. A. Perkins and J. A. Enos	Salem and Peabody, Mass	Oct. 1, 1872	131, 777
Leather, Process for making japanned	H. L. Hall	Beverly, Mass	Jan. 9, 1855	12, 226
Leather, Process for the manufacture of artificial	H. A. Clark	Boston, Mass	July 22, 1873	141, 116
Leather, Process of and machinery for stamping and staining.	R. Robinson	Newburyport, Mass	Mar. 19, 1810	....
Leather, Process of finishing	H. C. Williams	Lancaster, Pa	May 12, 1863	38, 525
Leather, Process of oiling, preserving, and strengthening.	L. S. Robbins	New York, N. Y	Mar. 24, 1868	75, 795
Leather, Punching	J. M. Bent	Wayland, Mass	Oct. 16, 1866	58, 762
Leather punching and cutting machine	E. F. Hardy and N. Dubrul	Joliet, Ill	Feb. 7, 1871	111, 533
Leather punching and feeding machine	W. Barry	Carthage, N. Y	Feb. 14, 1871	111, 719
Leather-punching machine	T. Burkitt and J. K. Krieg	Brooklyn and New York, N. Y.	Sept. 17, 1872	131, 330
Leather-punching machine	J. Matheis	Ottawa, Ill	Feb. 21, 1871	111, 957
Leather-punching machine	L. H. Wood	Marlborough, Mass	June 27, 1865	48, 474

Index of patents issued from the United States Patent Office from 1790 *to* 1873, *inclusive*—Continued.

Invention.	Inventor.	Residence.	Date.	No.
Leather-quilting machine	A. G. Brewer	Washington, D. C	Oct. 29, 1867	70, 158
Leather raising, creasing, and slicking machine	C. W. Guest	Dexter, Mich	Nov. 10, 1863	40, 561
Leather-reducing machine	W. C. Joslin	Putnam, Conn	Dec. 1, 1868	84, 631
Leather remnants, Mode of treating	B. C. Young	Boston, Mass	Feb. 4, 1873	135, 615
Leather, Removing color from	E. S. Frye	Salem, Mass	Mar. 22, 1870	100, 997
Leather, Removing grease from waste	J. P. Rush	Peabody, Mass	Oct. 11, 1870	108, 191
Leather, Restoring old	B. Schmidt	Hoboken, N. J	Apr. 14, 1868	76, 824
Leather, &c., Rigid hoop-knife for splitting	D. H. Chamberlian	West Roxbury, Mass	June 16, 1857	17, 554
Leather-roller	J. Gardiner	Petersburgh, Pa	Apr. 25, 1843	3, 055
Leather-roller	J. T. Harris	Swampscott, Mass	June 23, 1868	79, 070
Leather-roller	D. H. Priest	Watertown, Mass	Sept. 19, 1865	50, 079
Leather, Rolling	J. Whitney	Manchester, Mass	Sept. 22, 1863	40, 069
Leather-rolling machine	J. G. Busfield	Feltonville, Mass	Jan. 26, 1864	41, 363
Leather-rolling machine	W. Coburn	Lancaster, Pa	Oct. 31, 1839	1, 388
Leather-rolling machine	J. G. Curtis	Emporium, Pa	Mar. 29, 1870	101, 234
Leather-rolling machine	J. G. Curtis	Warren, Pa	May 30, 1871	115, 443
Leather-rolling machine	W. H. Leach	Uxbridge, Mass	June 30, 1868	79, 360
Leather-rolling machine	N. Linsley	Lena, Ill	Nov. 28, 1871	121, 389
Leather-rolling machine	W. P. Martin	Salem, Mass	Feb. 20, 1866	52, 728
Leather-rolling machine	J. McLaughlin and H. Hill	Sunderland, Vt	Apr. 28, 1836	....
Leather-rolling machine	J. E. Merritt	Winn, Me	Feb. 27, 1866	52, 871
Leather-rolling machine	C. W. Monson	Upton, Iowa	Aug. 10, 1869	93, 465
Leather-rolling machine	H. Paign	Alexander, N. Y	Dec. 23, 1828	....
Leather-rolling machine	J. A. Safford	Winchester, Mass	Jan. 18, 1870	98, 889
Leather-rolling machine	Q. Stoddard	Jackson, Mich	Dec. 4, 1866	60, 277
Leather-rolling machine	J. H. Walker	Worcester, Mass	Dec. 10, 1867	71, 929
Leather-rolling machine	H. J. Weston	Buffalo, N. Y	Mar. 22, 1870	101, 19[illegible]
Leather-rolling machine	J. Whitney	Winchester, Mass	Mar. 19, 1872	124, 709
Leather rolling, shaving, softening, and finishing machine.	T. Chace	New York, N. Y	Sept. 12, 1838	914
Leather-rounding machine	P. Beckman	Napierville, Ill	July 28, 1863	39, 335
Leather rolling machine, Green or wet	J. Whitney	Winchester, Mass	Mar. 24, 1863	37, 991
Leather-rounding machine	J. H. Tizzard	Eaton, Ohio	Apr. 4, 1871	113, 707
Leather-rounding machine	J. Yeager	Berrysburgh, Pa	Jan. 8, 1867	61, 132
Leather-rounding tool	L. A. Sweatt	San Francisco, Cal	Aug. 1, 1871	117, 575
Leather, Rubber-coated	H. W. Joslin	Brooklyn, N. Y	Nov. 6, 1866	59, 402
Leather rubbing and polishing machine	J. T. Flanders	Newburyport, Mass	Oct. 4, 1853	10, 067
Leather-sammier	D. C. Guttridge	Pittsburgh, Pa	Feb. 1, 1870	99, 431
Leather-scalloping machine	A. Goodyear	Albion, Mich	Jan. 14, 1868	73, 243
Leather-scarfing machine	C. H. Helms	Poughkeepsie, N. Y	Aug. 11, 1868	80, 952
Leather-scarfing machine	C. H. Helms	Poughkeepsie, N. Y	Mar. 15, 1870	100, 890
Leather scouring and setting machine	P. E. Hummel	Pulaski, N. Y	June 16, 1857	17, 576
Leather scouring and setting-out machine	C. Holmes	Boston, Mass	June 1, 1869	90, 664
Leather scouring and setting-out machine	A. W. Reid	Schenectady, N. Y	Apr. 5, 1870	101, 508
Leather scouring and stretching machine	J. C. Parmerlee	Bean Blossom, Ind	Aug. 8, 1871	117, 921
Leather scouring, blacking, and finishing machine	F. W. Rust	Umatilla, Oreg	Nov. 10, 1868	84, 001
Leather, Scouring, cleansing, &c	R. Emes	Boston, Mass	Nov. 21, 1831	....
Leather scouring, glassing, or setting machine	F. A. Lockwood	Winthrop, Mass	Oct. 21, 1873	143, 829
Leather-scouring machine	D. P. Burdon	New York, N. Y	July 19, 1870	105, 419
Leather-scouring machine	W. M. Clark	Butternuts, N. Y	Aug. 29, 1865	49, 606
Leather-scouring machine	F. Davis	Lawrence, Kans	Dec 4, 1866	60, 149
Leather-scouring machine	H. C. Havemeyer and D. P. Burdon.	New York, N. Y	Nov. 1, 1870	108, 782
Leather-scouring machine	H. C. Havemeyer and D. P. Burdon.	New York, N. Y	Nov. 1, 1870	108, 783
Leather-scouring machine	I. W. Pray and E. Fitzhenry	Portland, Oreg	Jan. 15, 1867	61, 250
Leather-scouring machine	A. W. Roberts	Hartford, Conn	Mar. 26, 1867	63, 307
Leather-scouring machine	A. W. Reid	Schenectady, N. Y	July 16, 1872	129, 251
Leather-scouring machine	N. F. Snow	Salem, Mass	Oct. 1, 1872	131, 831
Leather scouring, polishing, and glassing machine	D. Harrington	Boston, Mass	May 16, 1871	114, 809
Leather-scraping machine	J. T. Barnstead	Peabody, Mass	Mar. 22, 1870	101, 081
Leather scraps, Method of treating	H. B. Farwell	Boston, Mass	Apr. 10, 1866	53, 800
Leather, Seam for	E. Shaw	Milwaukee, Wis	Dec. 10, 1872	133, 897
Leather, Shaving	H. Johnson	Brooklyn, Conn	June 16, 1836	....
Leather, Shaving and splitting	S. Parker	Billerica, Mass	Apr 5, 1813	....
Leather, Shaving and splitting	M. Tullar	Royalton, Vt	July 29, 1808	....
Leather-shaving knife	J. B. Wentworth	Lynn, Mass	July 13, 1858	20, 911
Leather-shaving machine	D. Y. Haas	Hepler, Pa	Nov. 4, 1873	144, 273
Leather-shaving machine	J. P. Houston and O. Batchelder.	Bedford, N. H	Apr. 11, 1825	....
Leather-shaving machine	J. Reilly	Waynesborough, Pa	Feb. 28, 1833	....
Leather, Silvering and gilding	L. Kenton	Philadelphia, Pa	June 21, 1830	....
Leather, skins, &c., Machine for pressing and finishing.	N. D. Morey	Saratoga Springs, N. Y	Dec. 12, 1871	121, 727
Leather, Skiving	B. S. Mathews	Stamford, Conn	Apr. 10, 1849	6, 290
Leather skiving and splitting machine	P. D. Cummings	Portland, Me	Mar. 25, 1873	137, 183
Leather skiving and whitening machine	S. Graham	Roxbury, Mass	May 10, 1838	732
Leather-skiving machine	G. Andrews	Oxford, Me	Oct. 7, 1873	143, 398
Leather-skiving machine	J. W. Chandler	East Corinth, Me	June 3, 1873	139, 495
Leather-skiving machine	J. P. Crooks	Hopkinton, Mass	Oct. 25, 1870	108, 687
Leather-skiving machine	E. T. Ingalls	Haverhill, Mass	June 5, 1860	28, 580
Leather-skiving machine	S. Keen	East Bridgewater, Mass	Nov. 27, 1860	30, 736
Leather-skiving machine	E. B. Stimpson	New York, N. Y	May 13, 1873	138, 951
Leather-skiving machine	W. S. Williams	Lynn, Mass	July 31, 1860	29, 428
Leather-slicker	G. B. Fowle	Boston, Mass	June 11, 1872	127, 756
Leather-slicker	H. L. Sultzbach	Marietta, Pa	Apr. 27, 1858	20, 098
Leather so as to render it suitable for the manufacture of gloves, &c., Process of treating.	N. C. Russell	Gloversville, N. Y	Aug. 17, 1869	93, 910
Leather-softening machine	H. Cunningham	Albany, N. Y	Nov. 12, 1872	132, 902
Leather-softening machine	J. Greenleaf	Lowell, Mass	Sept. 30, 1856	15, 807
Leather-softening machine	J. B. Wentworth	Lynn, Mass	Feb. 5, 1856	14, 211
Leather, Splitting	E. Howard and J. Butters	Boston, Mass	May 3, 1820	....
Leather, Splitting	C. S. Stearns	Marlborough, Mass	June 14, 1864	43, 159
Leather splitting and fleshing machine	J. A. Safford	Winchester, Mass	Jan. 18, 1870	98, 888
Leather splitting and paring machine	H. Aiken	Middlesex, Mass	June 12, 1835	....
Leather splitting and rolling machine	C. S. Stearns	Marlborough, Mass	June 9, 1868	78, 697
Leather splitting and skiving machine	A. Cochran	Athens, Ohio	Apr. 4, 1871	113, 631
Leather splitting and skiving machine	S. H. Schenck	Zionsville, Ind	Oct. 9, 1866	58, 681

Index of patents issued from the United States Patent Office from 1790 to 1873, inclusive—Continued.

Invention.	Inventor.	Residence.	Date.	No.
Leather splitting and stretching apparatus	B. Rowe	Albany, N. Y	Apr. 30, 1850	7, 327
Leather splitting and stripping machine	C. S. Stearns	Marlborough, Mass	Feb. 9, 1864	41, 583
Leather-splitting gage	J. N. Baird	Bloomfield, Iowa	May 14, 1872	126, 668
Leather-splitting knife	H. Aiken	Brighton, Mass	Mar. 12, 1823	....
Leather-splitting knife	A. H. Van Gieson	Newark, N. J	June 9, 1868	78, 701
Leather-splitting machine	J. H. Abbott and J. A. Marden	Malden and Boston, Mass.	Mar. 24, 1868	75, 823
Leather-splitting machine	C. W. Baldwin and L. D. Hawkins.	Charlestown and Stoneham, Mass.	Apr. 10, 1866	53, 771
Leather-splitting machine	J. Butters	New York, N. Y	May 31, 1822	....
Leather-splitting machine	D. H. Chamberlain	West Roxbury, Mass	May 10, 1859	23, 900
Leather-splitting machine	H. E. Chapman	Albany, N. Y	Nov. 23, 1858	22, 108
Leather-splitting machine	H. Cunningham	Albany, N. Y	Oct. 12, 1869	95, 780
Leather-splitting machine	A. Dawes	Hudson Mass	Oct. 29, 1867	70, 175
Leather-splitting machine	J. P. Fairlamb	Wilmington, Del	Feb. 22, 1848	5, 456
Leather-splitting machine	J. F. Flanders	Boston, Mass	Aug. 14, 1860	29, 649
Leather-splitting machine	J. F. Flanders and J. A. Marden.	Newburyport, Mass	Aug. 29, 1854	11, 604
Leather-splitting machine	J. Goebel and J. Preis	Caledonia, Wis	Nov. 25, 1873	144, 89[illegible]
Leather-splitting machine	C. Keniston	Somerville, Mass	Dec. 21, 1869	98, 068
Leather-splitting machine	J. A. Marden and H. A. Butters	Newburyport and Haverhill, Mass.	Nov. 6, 1855	13, 756
Leather-splitting machine	J. A. Marden	Newburyport, Mass	Apr. 3, 1866	53, 741
Leather-splitting machine	J. A. Marden	Newburyport, Mass	May 8, 1866	54, 571
Leather-splitting machine	M. H. Merriam and J. B. Crosby.	Chelsea and Stoneham, Mass.	Feb. 13, 1855	12, 392
Leather-splitting machine	W. Panton	Milton, Mass	July 15, 1851	8, 227
Leather-splitting machine	E. Pratt	Salem, Mass	Dec. 19, 1854	12, 114
Leather-splitting machine	E. Pratt	Salem, Mass	Mar. 11, 1856	14, 430
Leather-splitting machine	A. Richardson	Boston, Mass	Apr. 23, 1831	....
Leather-splitting machine	A. Richardson	Boston, Mass	Feb. 9, 1841	1, 967
Leather-splitting machine	A. Richardson	Boston, Mass	Apr. 17, 1844	3, 541
Leather-splitting machine	A. Richardson	North Enfield, N. H	Sept. 16, 1851	8, 369
Leather-splitting machine	B. Rowe	Albany, N. Y	June 2, 1863	38, 763
Leather-splitting machine	J. A. Safford	Boston, Mass	Mar. 19, 1861	31, 746
Leather-splitting machine	J. A. Safford	Boston, Mass	Feb. 2, 1864	41, 448
Leather-splitting machine	J. P. Shaw and J. C. Briggs	Boston, Mass	Dec. 31, 1833	....
Leather-splitting machine	C. S. Stearns	Marlborough, Mass	Feb. 22, 1870	100, 082
Leather-splitting machine	A. F. Stowe	Worcester, Mass	June 10, 1873	139, 744
Leather-splitting machine	J. Taggart	Boston, Mass	Nov. 10, 1868	83, 888
Leather-splitting machine	S. S. Turner	Westborough, Mass	Oct. 30, 1860	30, 553
Leather-splitting machine	A. H. Van Gieson	Newark, N. J	July 8, 1862	35, 850
Leather-splitting machine	A. H. Van Gieson	Newark, N. J	Apr. 17, 1866	54, 043
Leather-splitting machine	F. J. Vittum	Newburyport, Mass	Feb. 18, 1868	74, 734
Leather-splitting machine	C. Weston	Salem, Mass	Aug. 30, 1853	9, 980
Leather-splitting machine	H. White	Binghamton, N. Y	Aug. 2, 1839	1, 272
Leather-splitting machine	H. Wing	Buffalo, N. Y	Aug. 25, 1863	39, 695
Leather-splitting machine, Bed-spring of	J. B. Tay	North Woburn, Mass	Aug. 7, 1855	13, 407
Leather-splitting machine, Feed of	D. H. Chamberlain	West Roxbury, Mass	June 5, 1860	28, 559
Leather splitting machine, Roll for	C. S. Stearns	Marlborough, Mass	Feb. 13, 1872	123, 580
Leather splitting machine, Sole, &c	E. Putnam	Danvers, Mass	Nov. 20, 1838	1, 010
Leather-splitting mill, Cylindrical	P. Dow	Boston, Mass	July 12, 1810	....
Leather stamping and shaping machine	B. B. Harris	Lockport, Ill	Nov. 5, 1867	70, 558
Leather-stitching machine	S. Sheldon	Cincinnati, Ohio	Aug. 3, 1839	1, 277
Leather stock, &c., Machine for cutting up	F. L. Walker	Boston, Mass	Feb. 2, 1869	86, 611
Leather strap and tube, Seamless	A. Escheulohr	Munich, Bavaria	May 12, 1868	77, 724
Leather-strap clasp	A. B. Conde	Albany, N. Y	Jan. 9, 1866	51, 926
Leather straps, Clasp for securing end of	O. O. Storle	Milwaukee, Wis	June 3, 1873	139, 626
Leather straps, Creasing-roll for	J. F. Moloney	Cincinnati, Ohio	Dec. 3, 1872	133, 660
Leather straps for harness, &c., Process for forming.	G. Earle	Dover, Ohio	Sept. 30, 1862	36, 562
Leather straps, Machine for creasing	S. Cronce	Flemington, N. J	Oct. 24, 1848	5, 882
Leather straps, Machine for creasing the edges of	W. M. Thornton	Pottsville, Pa	May 15, 1855	12, 878
Leather straps, Machine for creasing and finishing	W. M. K. Thornton	Niles, Mich	Jan. 8, 1861	31, 094
Leather straps, &c., Machine for cutting	J. Haberbush, J. Bentz, and V. S. Vogel.	Lancaster, Pa	Feb. 23, 1869	87, 167
Leather straps, Machine for cutting and punching	J. F. Hollister	Plano, Ill	May 27, 1873	139, 388
Leather straps, Machine for raising and creasing	G. W. Pruyne	Mexico, N. Y	Apr. 15, 1856	14, 698
Leather straps, Machine for shecoring	C. Morris	New Haven, Conn	Feb. 20, 1855	12, 444
Leather straps, Machine for waving or embossing	D. H. Unger	Philadelphia, Pa	Oct. 7, 1873	143, 477
Leather straps, Tool for chamfering	J. Bridger	Richland, Iowa	Nov. 2, 1858	21, 937
Leather-stretching frame	C. P. Oliver	Newark, N. J	Feb. 11, 1873	135, 836
Leather-stretching frame, Wet	R. W. Dawson	Newark, N. J	June 25, 1867	66, 131
Leather-stretching machine	W. R. Andrews and R. Dingwell.	Newark, N. J	Sept. 15, 1868	82, 063
Leather-stretching machine	J. F. Connelly and W. B. Hughes	Newark, N. J	Oct. 8, 1867	69, 633
Leather-stretching machine	J. H. Haskell	Baltimore, Md	May 15, 1860	28, 271
Leather-stretching machine	A. W. Roberts	Hartford, Conn	Feb. 8, 1859	22, 893
Leather-stretching machine	B. Rowe	Albany, N. Y	Apr. 22, 1851	8, 052
Leather-stretching machine	W. Strevel and D. Brown	Albany, N. Y	Apr. 22, 1851	8, 054
Leather-stretching pole	H. Dawson	Baltimore, Md	Apr 6, 1869	88, 697
Leather stripping and cutting machine	J. E. Coffin, C. E. Morrill, and G. F. Hall.	Portland and Westbrook, Me.	Mar. 28, 1871	113, 189
Leather-stripping machine	A. R. E. Falck and P. Stoerger.	Newark, N. J	Oct. 20, 1857	18, 441
Leather-stripping machine	D. Ridgway	....	Oct. 26, 1809	....
Leather, Stuffing	J. Armstrong	Woburn Centre, Mass	Feb. 10, 1857	16, 573
Leather, Stuffing and currying	T. McDonald	Roxbury, Mass	Aug. 14, 1866	57, 165
Leather stuffing and currying machine	F. Carl	Charlestown, Mass	Apr. 16, 1867	63, 856
Leather-stuffing machine	F. Fischer	Cambridge, Mass	Mar. 24, 1868	75, 896
Leather-stuffing machine	H. Muller	North Cambridge, Mass	June 9, 1868	78, 815
Leather-stuffing machine	J. W. Schayer	Boston, Mass	June 9, 1868	78, 835
Leather, Substitute for	P. Wenzel	Mayence, Hesse-Darmstadt	Mar. 27, 1866	53, 549
Leather-tapering machine	W. Mannheim	New York, N. Y	Sept. 24, 1867	69, 229
Leather-tearing machine	E. S. Hidden	Millburn, N. Y	June 20, 1871	116, 055
Leather to a uniform thickness, Machine for reducing roller.	W. C. Joslin	West Thompson, Conn	Sept. 24, 1867	69, 219
Leather-trimmer	L. P. Curtis	Marlborough, Mass	Apr. 14, 1868	76, 608
Leather trimmer, Cue	L. A. Johnson, H. W. Collinder, and J. E. Boyle.	New York, N. Y	Aug. 31, 1869	94, 317

Index of patents issued from the United States Patent Office from 1790 to 1873, inclusive—Continued.

Invention.	Inventor.	Residence.	Date.	No.
Leather trimming, chamfering, and skiving machine.	W. H. Rounds	Campello, Mass	Feb. 7, 1860	27, 073
Leather-trimming machine	S. H. Jones	East Albany, N. Y	July 8, 1873	140, 631
Leather-trimming machine	S. P. Pope	Burlington, N. Y	Sept. 10, 1861	33, 264
Leather-trimming machine, Cue	J. E. Boyle	New York, N. Y	May 4, 1869	89, 624
Leather, &c., Treating	J. Burrill	Lynn, Mass	May 3, 1864	42, 619
Leather, Treating tanned	B. K. Lightfoot	Philadelphia, Pa	Nov. 17, 1863	40, 669
Leather, Treatment of tanned	A. Dietz	New York, N. Y	Oct. 16, 1860	30, 393
Leather-varnish	C. Brumby	Rochester, N. Y	Sept. 12, 1871	118, 842
Leather-varnishing composition	O. S. Boyden and M. C. Fredericks.	Newark, N. J	May 25, 1858	28, 320
Leather-varnishing machine	L. L. Allen	Hallowell, Me	Sept. 3, 1872	130, 969
Leather-washer	G. Miller and C. M. Andrews	Providence, R. I	Dec. 20, 1859	26, 510
Leather-washer, Annular	P. L. Gibbs	Chicago, Ill	Dec. 30, 1873	146, 062
Leather-washing machine	A. Howard and G. F. Howard	Wellsville, Md., and Chicago, Ill	May 28, 1867	65, 224
Leather-washing machine	A. Howard and G. F. Howard	Wellsville, N. Y., and Chicago, Ill.	July 14, 1868	79, 832
Leather-washing machine	I. Sipes	Boston, Mass	Dec. 12, 1865	51, 487
Leather water-proof and mode of applying same, Composition for rendering.	C. F. Miller	Baltimore, Md	Aug. 28, 1840	1, 749
Leather, Water-proof composition for	W. J. Roome	New York, N. Y	Jan. 6, 1844	3, 401
Leather water-proof, Composition for rendering	W. A. Ronald and H. Miller	Rowan County, N. Y	Jan. 24, 1842	2, 431
Leather water-proof, Making	A. Straub	Milton, Pa	June 30, 1828	....
Leather water-proof, Making	D. Kizer	New York, N. Y	Nov. 19, 1827	....
Leather water-proof, Method of dressing and making.	O. A. Goold	Portland, Me	July 9, 1872	128, 873
Leather water-proof, Process for rendering	G. Conkling	Conklingsville, N. Y	Sept. 5, 1865	49, 826
Leather-work, Metal fastening material for	C. Keniston	Somerville, Mass	Dec. 16, 1873	145, 658
Leather-work, Nail for	S. W. Baldwin	Baldwinsville, N. Y	Sept. 27, 1864	44, 382
Leather-work ornament	L. E. Sallee	Peoria, Ill	June 23, 1868	79, 148
Leather-work, Seam for	C. Keniston	Somerville, Mass	Aug. 19, 1873	142, 026
Leather-work, Seam for	S. W. Shorey	Boston, Mass	Sept. 10, 1872	131, 308
Leather-work, Seam for	S. W. Shorey	Boston, Mass	May 27, 1873	139, 429
Leather-working tool	J. S. Alexander	Pittsburgh, Pa	July 1, 1873	140, 336
Leathern tubes, Machine for making	N. Wyllys	South Glastonbury, Conn.	Dec. 23, 1851	8, 604
Leathern tubes or hose, &c., Making	A. L. Pennock and J. Sellers	Philadelphia, Pa	July 6, 1818	....
Ledger account, Keeping	A. J. Folger	Nantucket, Mass	Oct. 30, 1849	6, 825
Lee-board	J. H. and J. Swain	Cape May Court-House, N. J	Apr. 10, 1811	....
Lee-board for vessels	A. Jouan	San Francisco, Cal	Nov. 18, 1856	16, 090
Lee-way indicator	A. A. Wilder	Detroit, Mich	Jan. 21, 1851	7, 912
Lee-way indicator for vessels	T. Byrne	New York, N. Y	July 3, 1866	56, 003
Leech, Artificial	F. Wolff	New York, N. Y	Oct. 30, 1866	59, 306
Leech, Mechanical	M. Delluc	New York, N. Y	May 7, 1850	7, 341
Leech, Mechanical	F. A. Stohlmann and A. H. Smith.	Brooklyn and New York, N. Y.	Feb. 22, 1870	100, 210
Leeches, Bleeding with and breeding	J. Kunitz	Philadelphia, Pa	May 7, 1805	....
Leg and foot, Artificial	M. H. Hawking	New Haven, Conn	May 18, 1869	90, 099
Leg and foot rest	E. Withall	Rochester, N. Y	Oct. 24, 1871	120, 176
Leg, Artificial	J. J. Austin	New York, N. Y	June 20, 1865	48, 251
Leg, Artificial	D. Bly	Rochester, N. Y	Apr. 19, 1859	23, 656
Leg, Artificial	D. Bly	Rochester, N. Y	May 17, 1859	24, 002
Leg, Artificial	D. Bly	Rochester, N. Y	Aug. 30, 1859	25, 238
Leg, Artificial	D. Bly	Rochester, N. Y	Feb. 19, 1861	31, 438
Leg, Artificial	D. Bly	Rochester, N. Y	May 19, 1863	38, 549
Leg, Artificial	D. Bly	Rochester, N. Y	May 19, 1863	38, 550
Leg, Artificial	D. Bly	New York, N. Y	Sept. 4, 1866	57, 666
Leg, Artificial	D. Bly	Macon, Ga	Mar. 9, 1869	87, 264
Leg, Artificial	J. Bringhurst	Philadelphia, Pa	June 12, 1866	55, 459
Leg, Artificial	T. Burr	Battle Creek, Mich	Apr. 18, 1865	47, 353
Leg, Artificial	H. L. Byrd	Augusta, Ga	Mar. 6, 1866	52, 964
Leg, Artificial	T. J. Cain	Cleveland, Ohio	Oct. 18, 1864	44, 766
Leg, Artificial	E. Carlton and E. Goss	Portland, Me	Mar. 12, 1867	62, 731
Leg, Artificial	R. Clement	Philadelphia, Pa	Apr. 18, 1865	47, 281
Leg, Artificial	J. Condell	Morristown, N. Y	July 11, 1865	48, 660
Leg, Artificial	J. Condell	Morristown, N. Y	July 18, 1865	48, 792
Leg, Artificial	P. Daniels	Le Roy, N. Y	Jan. 19, 1864	41, 282
Leg, Artificial	D. D. Douglass	Springfield, Mass	Jan. 10, 1860	26, 753
Leg, Artificial	J. S. Drake	New York, N. Y	Aug, 31, 1852	9, 232
Leg, Artificial	J. S. Drake	New York, N. Y	Sept. 4, 1866	57, 691
Leg, Artificial	E. Elliott and C. E. Newton	Lowell, Mass	Jan. 28, 1868	73, 787
Leg, Artificial	J. Emery	Cedar Falls, Iowa	May 28, 1867	65, 187
Leg, Artificial	T. F. Englebrecht, R. Baeklen, and W. Staehlen.	New York and Brooklyn, N. Y.	Jan. 6, 1863	37, 282
Leg, Artificial	J. A. Foster	West Stockholm, N. Y	Aug. 8, 1865	49, 253
Leg, Artificial	J. A. Foster	Detroit, Mich	June 29, 1869	92, 031
Leg, Artificial	J. Travel	Saint Louis, Mo	Sept. 15, 1863	39, 912
Leg, Artificial	D. Gilson	Nashua, N. H	June 19, 1866	55, 645
Leg, Artificial	S. G. Gregory	Albany, N. Y	Aug. 17, 1869	93, 876
Leg, Artificial	G. W. Hall	Lyndonville, N. Y	June 2, 1863	38, 739
Leg, Artificial	J. E. Hanger	Staunton, Va	Feb. 14, 1871	111, 741
Leg, Artificial	W. M. Hawkins	Elba, Ala	July 16, 1872	129, 340
Leg, Artificial	G. B. Head	Albany, N. Y	Oct. 27, 1868	83, 496
Leg, Artificial	W. Hinds	Little Falls, N. Y	Oct. 18, 1864	44, 726
Leg, Artificial	B. W. Jewett	Gilford, N. H	Jan. 6, 1857	16, 360
Leg, Artificial	B. W. Jewett	Gilford, N. H	Aug. 7, 1860	29, 494
Leg, Artificial	G. B. Jewett	Salem, Mass	June 24, 1862	35, 686
Leg, Artificial	G. B. Jewett	Salem, Mass	July 22, 1862	35, 937
Leg, Artificial	G. B. Jewett	Salem, Mass	Oct. 4, 1864	44, 534
Leg, Artificial	G. B. Jewett	Salem, Mass	Aug. 22, 1865	49, 528
Leg, Artificial	G. B. Jewett	Salem, Mass	Aug. 22, 1865	49, 529
Leg, Artificial	G. B. Jewett	Salem, Mass	Dec. 19, 1865	51, 593
Leg, Artificial	S. B. Jewett	Laconia, N. H	Dec. 7, 1869	97, 647
Leg, Artificial	H. A. Kimball and A. J. Lawrence.	Philadelphia, Pa	May 1, 1866	54, 364
Leg, Artificial	L. Legran	Allegheny City, Pa	Jan. 16, 1866	52, 057
Leg, Artificial	L. Legran	Allegheny City, Pa	Sept. 3, 1867	68, 758

Index of patents issued from the United States Patent Office from 1790 to 1873, inclusive—Continued.

Invention.	Inventor.	Residence.	Date.	No.
Leg, Artificial	P. Lockie	Rochester, N. Y	Oct. 27, 1863	40, 417
Leg, Artificial	R. G. Lockwood	Battle Creek, Mich	Oct. 31, 1865	50, 770
Leg, Artificial	J. Madden	Cleveland, Ohio	Jan. 9, 1866	51, 953
Leg, Artificial	A. A. Marks	New York, N. Y	Mar. 7, 1865	46, 687
Leg, Artificial	D. B. Marks	New York, N. Y	Mar. 7, 1854	10, 611
Leg, Artificial	A. McOmber	Schenectady, N. Y	Nov. 19, 1867	71, 197
Leg, Artificial	A. Mennel	New York, N. Y	Aug. 29, 1865	49, 645
Leg, Artificial	H. L. Mills	Saint Paul, Minn	Oct. 15, 1867	69, 829
Leg, Artificial	J. Monroe	New York, N. Y	Mar. 15, 1864	41, 934
Leg, Artificial	J. Monroe	New York, N. Y	Oct. 11, 1864	44, 644
Leg, Artificial	J. Monroe	New York, N. Y	July 25, 1865	49, 038
Leg, Artificial	K. Moore	Oswego, N. Y	July 20, 1869	92, 739
Leg, Artificial	F. W. Neubert	Pittsburgh, Pa	Nov. 22, 1864	45, 169
Leg, Artificial	R. H. Nicholas	Chicago, Ill	Aug. 7, 1866	56, 983
Leg, Artificial	R. H. Nicholas and D. Bly	Rochester, N. Y	July 28, 1857	17, 888
Leg, Artificial	B. F. Palmer	Meredith, N. H	Nov. 4, 1846	4, 834
Leg, Artificial	B. F. Palmer	Meredith, N. H	Feb. 20, 1849	6, 122
Leg, Artificial	B. F. Balmer	Philadelphia, Pa	Aug. 17, 1852	9, 200
Leg, Artificial	B. F. Palmer	Philadelphia, Pa	Apr. 8, 1873	137, 711
Leg, Artificial	D. D. Parmelee	New York, N. Y	Feb. 10, 1863	37, 637
Leg, Artificial	L. F. Pingree	Portland, Me	July 5, 1870	104, 994
Leg, Artificial	H. D. Reinhardt	Baltimore, Md	Feb. 9, 1864	41, 535
Leg, Artificial	J. Reichenbach	Pittsburgh, Pa	Jan. 12, 1864	41, 237
Leg, Artificial	J. Reichenbach	Pittsburgh, Pa	Jan. 12, 1864	41, 238
Leg, Artificial	W. H. Rhodes	Berlin, N. Y	July 31, 1855	13, 360
Leg, Artificial	J. Russell	Philadelphia, Pa	Aug. 17, 1852	9, 202
Leg, Artificial	F. Schmitt	Springfield, Ill	July 16, 1867	66, 744
Leg, Artificial	J. Schneider	Cincinnatti, Ohio	Aug. 15, 1865	49, 443
Leg, Artificial	W. Selpho	New York, N. Y	May 6, 1856	14, 836
Leg, Artificial	G. L. Shephard	New York, N. Y	May 17, 1864	42, 799
Leg, Artificial	S. P. Sleppy	Wilkesbarre, Pa	May 22, 1866	54, 970
Leg, Artificial	I. D. Small	North Fairfield, Ohio	Dec. 15, 1863	40, 956
Leg, Artificial	U. Smith	Battle Creek, Mich	July 28, 1863	39, 361
Leg, Artificial	A. Spaulding	Lowell, Mass	Aug. 7, 1855	13, 404
Leg, Artificial	W. C. Stone	Boston, Mass	Dec. 17, 1850	7, 847
Leg, Artificial	A. Strasser	Montgomery, Ala	Mar. 31, 1868	76, 267
Leg, Artificial	C. Sweet	Vicksburgh, Miss	Nov. 26, 1867	71, 424
Leg, Artificial	J. Taggart	Roxbury, Mass	Sept. 25, 1855	13, 611
Leg, Artificial	L. Tassius	Norwalk, Ohio	Aug. 11, 1868	81, 033
Leg, Artificial	T. Uren	San Francisco, Cal	Aug. 18, 1863	39, 599
Leg, Artificial	R. M. Vaughan	Glasgow, Mo	Dec. 22, 1863	41, 033
Leg, Artificial	J. Walber	New York, N. Y	Sept. 12, 1865	49, 936
Leg, Artificial	A. T. Watson	New York, N. Y	Feb. 5, 1867	61, 780
Leg, Artificial	J. W. Weston, F. Buchner, and R. Bocklen.	New York and Buffalo, N. Y.	Apr. 10, 1866	53, 931
Leg Artificial	J. W. Weston and T. B. Stanley	New York, N. Y	June 6, 1865	48, 138
Leg, Artificial	T. E. M. White	New Bedford, Mass	June 24, 1862	35, 723
Leg Artificial	O. D. Wilcox	Easton, Pa	Jan. 13, 1857	16, 490
Leg, Artificial	O. D. Wilcox	Easton, Pa	Sept. 30, 1856	15, 831
Legs Artificial	G. W. Yerger	Philadelphia, Pa	Mar. 19, 1850	7, 204
Legs, Apparatus for relief of debility or weakness in.	S. P. W. Douglass	Palmyra, N. Y	May 15, 1841	2, 092
Legs, Attachment to artificial	O. D. Wilcox	Elmira, N. Y	Aug. 24, 1858	21, 289
Legs, Boring-tool for making wooden	E. Osborne	Philadelphia, Pa	Sept. 4, 1866	57, 756
Legs, Support for artificial	P. Daniels	Le Roy, N. Y	Feb. 24, 1863	37, 738
Legs, Support for artificial	J. W. Weston	New York, N. Y	Dec. 27, 1864	45, 662
Legs, Tool for making wooden	E. Osborne	Philadelphia, Pa	Sept. 4, 1866	57, 755
Legging	A. L. Munson	New Haven, Conn	Mar. 13, 1866	53, 931
Legging	W. G. Rule	New York, N. Y	Jan 15, 1867	61, 265
Legging	A. P. Smith	Rock Falls, Ill	June 4, 1872	127, 652
Legging, Ladies'	E. F. and S. S. Bedford	Johnstown, N. Y	Apr. 5, 1864	42, 157
Legging, Saddle	W. B. Johns	Georgetown, D. C	Dec. 24, 1861	33, 996
Leggotyping	W. A. Leggo	Montreal, Canada	Mar. 21, 1871	112, 933
Leghorn, straw, &c., Mode of whitening	H. Cooper	Franklin County, Pa	Nov. 12, 1828	......
Lemonade, Composition for	A. Warner	Cleveland, Ohio	Jan. 6, 1863	37, 314
Lemonade, Preparing materials for	J. Warren, jr	New York, N. Y	Sept. 11, 1847	5, 292
Lemon squeezer	A. Barbarin	New Orleans, La	Oct. 8, 1867	69, 531
Lemon-squeezer	A. Bassford	Fort Wayne, Ind	Jan. 21, 1873	135, 067
Lemon-squeezer	A. Bigelow	Killingworth, Conn	Feb. 18, 1873	136, 023
Lemon-squeezer	L. S. Chichester	New York, N. Y	July 3, 1860	28, 967
Lemon-squeezer	V. Fogerty	Boston, Mass	Feb. 27, 1866	52, 928
Lemon-squeezer	O. Hesselbacher and H. Moesta	Detroit, Mich	June 18, 1867	65, 809
Lemon-squeezer	V. Himmer	New York, N. Y	Mar. 22, 1870	101, 128
Lemon-squeezer	J. L. Jensen	Brooklyn, N. Y	June 14, 1870	104, 159
Lemon-squeezer	J. Klepzig	San Francisco, Cal	Nov. 17, 1868	84, 125
Lemon-squeezer	J. H. Mead	New York, N. Y	July 15, 1873	140, 785
Lemon-squeezer	T. Reece and A. Clarke	Philadelphia, Pa	Jan. 15, 1867	61, 251
Lemon-squeezer	A. J. T. Reuter	Boston, Mass	Mar. 26, 1867	63, 304
Lemon squeezer	E. M. Sammis	Babylon, N. Y	Sept. 2, 1873	142, 414
Lemon-squeezer	T. C. Smith	New York, N. Y	Apr. 7, 1868	76, 539
Lemon-squeezer	G. M. Thomas	New York, N. Y	June 10, 1862	35, 554
Lemon-squeezer	W. A. and J. W. Whitney	Brooklyn, N. Y	Nov. 26, 1872	133, 511
Lenitive, Pain	E. Andler	New York, N. Y	Oct. 26, 1818	......
Lens	C. B. Boyle	New York, N. Y	Feb. 20, 1866	52, 672
Lens, Achromatic	H. Fitz	New York, N. Y	Sept. 27, 1864	44, 483
Lens, Double-globe	J. Schnitzer	New York, N. Y	May 24, 1864	42, 880
Lens, Fluid	S. Kakeles	New York, N. Y	Apr. 24, 1860	27, 988
Lens, Fluid	O. Warden	Cincinnati, Ohio	Aug. 28, 1866	57, 602
Lens, Fluid	D. A. Woodward	Baltimore, Md	Nov. 27, 1866	60, 109
Lens, Glass	H. M. Paine	North Oxford, Mass	Oct. 3, 1846	4, 786
Lenses, Adjusting	W. and W. H. Lewis	New York, N. Y	Dec. 16, 1851	8, 590
Lenses, &c., Machine for grinding and polishing	H. D. Knight	Elgin, Ill	Nov. 15, 1870	109, 325
Lenses, Machine for grinding and polishing	M. Witmer	South Pekin, N. Y	Oct. 13, 1863	40, 303
Lenses, Manufacture of glass	J. L. Gilliland	New York, N. Y	Aug. 10, 1852	9, 184
Lenses, Mounting fluid	A. M. Cole	Windham, Me	Mar. 16, 1858	19, 624
Let-off mechanism	G. Bailey	Putnam, Conn	June 13, 1871	115, 806
Let-off mechanism	A. J. Woodman	Indian Orchard, Mass	July 30, 1872	129, 915
Letter and envelope combined	E. B. Gleason	Boston, Mass	Mar. 15, 1859	23, 242

Index of patents issued from the United States Patent Office from 1790 *to* 1873, *inclusive*—Continued.

Invention.	Inventor.	Residence.	Date.	No.
Letter and figure, Metallic projecting	L. Child	New York	Oct. 4, 1817	....
Letter-balance	B. Chambers, jr	Washington, D. C	Mar. 10, 1868	75, 244
Letter-balance and pen-holder, Combined	R. B. Kepner	Philadelphia, Pa	Aug. 17, 1869	93, 720
Letter-board for object-teaching	J. H. Palm	Mansfield, Ohio	May 30, 1871	115, 349
Letter-box	C. R. Bancroft	Boston, Mass	June 7, 1870	103, 826
Letter-box	W. H. Bramble	Decatur, Ill	Apr. 3, 1866	53, 562
Letter-box	J. W. Brown	New York, N. Y	Apr. 15, 1862	34, 940
Letter-box	L. De Mets	Elizabeth, N. J	Nov. 12, 1872	132, 954
Letter-box	J. A. Farrington	Brooklyn, N. Y	Nov. 7, 1871	120, 578
Letter-box	C. P. Gorely	Boston, Mass	Dec. 17, 1867	72, 190
Letter-box	C. P. Gorely	Boston, Mass	June 23, 1868	79, 222
Letter-box	B. B. Gurley	Minneapolis, Minn	Nov. 25, 1873	144, 976
Letter-box	G. C. Jenks	New York, N. Y	Oct. 9, 1860	30, 324
Letter-box	D. P. Jordan	Chicago, Ill	June 30, 1868	79, 355
Letter-box	C. Lewis	Philadelphia, Pa	Jan. 31, 1871	111, 461
Letter-box	S. Maltby	Washington, D. C	July 19, 1864	43, 594
Letter-box	E. T. Marsh	Rochester, N. Y	June 7, 1870	104, 043
Letter-box	E. Mill and S. L. Chubbuck	Cleveland, Ohio	Apr. 30, 1872	126, 151
Letter-box	A. Potts	Philadelphia, Pa	Dec. 6, 1870	109, 939
Letter-box	A. Rosenstjerna	Chicago, Ill	July 29, 1873	141, 237
Letter-box	J. A. Ryan	Cleveland, Ohio	Aug. 15, 1871	118, 057
Letter-box	A. T. Sinclaire	New York, N. Y	Nov. 5, 1872	132, 865
Letter-box	J. W. and J. D. Smith	Washington, D. C	Feb. 4, 1868	74, 015
Letter-box	J. H. Springer	Philadelphia, Pa	Mar. 22, 1864	42, 052
Letter-box	S. Strong	Washington, D. C	Aug. 31, 1869	94, 449
Letter-box	E. C. Weld	New York, N. Y	Oct. 4, 1870	108, 073
Letter-box	S. S. Williams	Chicago, Ill	Dec. 3, 1872	133, 691
Letter-box	F. Wittram	San Francisco, Cal	June 22, 1869	91, 583
Letter-box	J. E. Woodruff	Buffalo, N. Y	Mar. 24, 1868	76, 026
Letter-box, Alarm	S. H. Morris	Charlestown, Mass	May 7, 1872	126, 477
Letter-box alarm	E. H. Ripley	Boston Highlands, Mass	Nov. 16, 1869	96, 965
Letter-box alarm	A. Taylor	Brooklyn, N. Y	Nov. 21, 1871	121, 218
Letter-box alarm	A. Taylor	Brooklyn, N. Y	June 10, 1873	139, 833
Letter-box alarm and door-plate, Combined	Z. E. Fobes	Troy, N. Y	Feb. 20, 1872	123, 880
Letter-box attachment for door	E. K. Mull	Reading, Pa	Nov. 26, 1872	133, 474
Letter box, Drop	J. North	Middletown, Conn	Mar. 13, 1860	27, 466
Letter-box, Postal	D. D. Foley	Washington, D. C	June 19, 1866	55, 636
Letter-box, Post-office	W. H. Taylor	Stamford, Conn	Aug. 12, 1873	141, 834
Letter-box, Post-office	L. Yale, jr	Shelburne Falls, Mass	Sept. 19, 1871	119, 212
Letter-box, Street	C. L. Evert and A. Johnson	Washington, D. C., and New York, N. Y.	Feb. 25, 1873	136, 313
Letter-box, Street	S. Strong	Washington, D. C	Mar. 30 1869	88, 528
Letter-box, Street	D. and G. H. White	Chicago, Ill	Apr. 14, 1868	76, 864
Letter-box to lamp-posts, Mode of attaching metallic.	A. Potts	Philadelphia, Pa	Mar. 9, 1858	19, 578
Letter-clip	L. E. Osborne	New Haven, Conn	May 10, 1864	42, 683
Letter-clip and paper-binder, Combined	G. W. McGill	New York, N. Y	Dec. 23, 1873	145, 809
Letter-copying apparatus	A. L. Adams	Philadelphia, Pa	Oct. 9, 1860	30, 283
Letter-lock	H. L. Arnold	Chicago, Ill	May 4, 1869	89, 826
Letter-opener	H. Gross	Tiffin, Ohio	Jan. 26, 1869	86, 300
Letter-opener	R. Johnson	Urbana, N. Y	May 17, 1864	42, 777
Letter-package	J. W. Burns	Medway, Ohio	Nov. 3, 1868	83, 690
Letter, parcel, &c., Method of receiving and delivering.	A. E. Beach	Stratford, Conn	Nov. 13, 1866	59, 739
Letter-pouch	P. Davis	Newport News, Va	June 30, 1868	79, 324
Letter-press for proof, &c	A. Ramage	Philadelphia, Pa	May 19, 1823	....
Letter sheet and envelope combined	T. Orton	Chicago, Ill	Mar. 11, 1873	136, 666
Letter-sheet blank	A. C. Fletcher	New York, N. Y	May 28, 1872	127, 330
Letter writer, Manifold	J. K. Park	New York, N. Y	Nov. 13, 1844	3, 825
Letters and parcels, Method of transporting	A. E. Beach	Stratford, Conn	Oct. 26, 1869	96, 187
Letters, &c., Device for folding	G. W. R. Lewin	Rochester, N. Y	Aug. 24, 1869	94, 012
Letters for signs, &c., Composition	A. M. and J. P. Todd	Des Moines, Iowa	Aug. 17, 1869	93, 924
Letters for signs, &c., Mode of constructing	T. Motley	Brooklyn, N. Y	Apr. 28, 1857	17, 167
Letters, Machinery for post-marking	E. N. Moore	Boston, Mass	Jan. 16, 1849	6, 036
Letters, Making block	L. Katen	New York, N. Y	Sept. 20, 1844	3, 750
Letters, Making shaded	G. Bruce	New York	Apr. 4, 1820	....
Letters, &c., Mode of sealing	J. Saxton	Washington, D. C	Feb. 15, 1859	22, 982
Letters, &c., Mold for casting	G. F. Sack	New York, N. Y	Aug. 4, 1868	80, 567
Letters on metal plates, Device for raising	R. R. Foote	Chicago, Ill	Aug. 15, 1871	118, 005
Letters on the circumference of metal disks, Device for forming.	S. M. Ott	Newark, N. J	Mar. 26, 1872	124, 970
Letters, packages, &c., Mode of transporting and delivering.	A. E. Beach	Stratford, Conn	Sept. 5, 1865	49, 699
Letters without discovery, Device to prevent opening.	C. S. Westcott	New York, N. Y	May 20, 1862	35, 342
Lettering-block, Adjustable	E. J. Griffin	Hart's Falls, N. Y	Jan. 11, 1870	98, 762
Levee	A. C. Brawner	Frankfort, Ky	July 13, 1869	92, 575
Levee	W. P. Craig	Milton, Ky	Sept. 21, 1869	95, 690
Levees, Inclined way for building	J. Bradford	Wilmington, Del	July 1, 1873	140, 458
Levees, Machine for making	E. J. Brown	Carroll Parish, La	July 13, 1869	92, 421
Levees, Machine for making	E. Comeaux	Bayou Goula, La	Dec. 10, 1867	71, 854
Level	W. P. Cutter	Chelsea, Mass	July 28, 1868	80, 337
Level	H. G. Loomis	Hartford, Conn	July 7, 1868	79, 582
Level	D. Masten	Bingham, N. Y	Sept. 25, 1866	58, 271
Level	W. L. Richardson	Reading, Mass	Jan. 1, 1867	60, 788
Level, Adjustable spirit	P. Clifford	Holyoke, Mass	Nov. 26, 1867	71, 279
Level, Adjustable spirit	L. L. Davis	Springfield, Mass	Sept. 17, 1867	68, 961
Level, Adjustable spirit	L. L. Davis	Springfield Mass	Mar. 17, 1868	75, 534
Level, Adjustable spirit	J. Morss and F. B. Abel	Philadelphia, Pa	Apr. 25, 1871	114, 027
Level, Adjustable spirit	E. A. and C. M. Stratton	Greenfield, Mass	July 16, 1872	129, 183
Level, Adjustable spirit	W. J. Tate	New Haven, Conn	Jan. 21, 1868	73, 670
Level, Adjusting spirit	S. N. Chapin and A. Stanley	New Britain, Conn	Sept. 10, 1867	68, 603
Level, Adjusting spirit	J. A. Traut	New Britain, Conn	Oct. 6, 1868	82, 769
Level and clinometer, Pendulum	W. Johnson	Edisto Island, S. C	Apr. 25, 1871	114, 014
Level and clinometer, Plumb	N. Barnum	La Porte, Ind	May 7, 1872	126, 372
Level and measure, Land	F. A. Archibald	Concord, N. H	July 20, 1869	92, 773
Level and plumb, Balance and pendulum	L. Lewis	Newfield, N. Y	Oct. 3, 1838	956
Level and plumb, Combined	P. Clifford	Holyoke, Mass	June 11, 1867	65, 726

Index of patents issued from the United States Patent Office from 1790 to 1873, inclusive—Continued.

Invention.	Inventor.	Residence.	Date.	No.
Level and plumb instrument	J. E. Eldred	Rochester, N. Y	Aug. 25, 1831	......
Level and square, Plumb	D. G. Davison, E. Pullen, and S. Davison.	Prospect Plain and Cranberry, N. J.	July 7, 1863	39, 124
Level and square, Reflecting spirit	F. Wilbar	Roxbury, Mass	Apr. 20, 1852	8, 807
Level, Collimating	J. Locke	Cincinnati, Ohio	July 2, 1850	7, 477
Level, Combined spirit	A. Cahoon, jr	Harwich, Mass	May 20, 1862	35, 298
Level, Combined square and	O. R. Chaplin	Salem, Mass	May 8, 1866	54, 503
Level, Fluid	W. G. Ladd, jr	Cambridge, Mass	Apr. 9, 1850	7, 263
Level, Grading	J. Thornley	Charlottesville, Va	Oct. 21, 1873	143, 942
Level, Graduating	N. Hollingsworth	Rosetta, Ill	July 2, 1867	66, 342
Level or inclinometer	T. A. Chandler	Rockford, Ill	Apr. 14, 1857	17, 023
Level, Pendulum	T. A. Chandler	Rockford, Ill	May 17, 1853	9, 722
Level, Pendulum	A. C. L. Delsarte	Paris, France	Aug. 19, 1873	142, 005
Level, Pendulum	A. Munger	Oberlin, Ohio	Aug. 17, 1835	......
Level, Pendulum	B. F. St. John	Shelbyville, Ind	Sept. 5, 1865	49, 802
Level, Plumb	G. Cuppers	New York, N. Y	Dec. 19, 1865	51, 564
Level, Plumb	M. W. E. Doran	Indianapolis, Ind	Apr. 10, 1866	53, 795
Level, Plumb	C. Ensminger and A. W. Elmer	Springfield, Mass	July 16, 1867	66, 695
Level, Plumb	A. Girard	Mobile, Ala	July 2, 1846	4, 611
Level, Plumb	W. H. and D. D. Polleys	La Crosse, Wis	Oct. 27, 1863	40, 428
Level, Plumb	J. T. Zimmerman and H. Baker	Lancaster, Pa	Oct. 8, 1867	69, 602
Level, Spirit	L. Brooks	Great Falls, N. H	Aug. 29, 1854	11, 596
Level, Spirit	R. T. Burnett	New York, N. Y	Sept. 5, 1865	49, 710
Level, Spirit	L. L. Davis	Springfield, Mass	Mar. 17, 1868	75, 533
Level, Spirit	L. L. Davis	Springfield, Mass	Nov. 21, 1871	121, 088
Level, Spirit	H. W. Evans	Philadelphia, Pa	Feb. 13, 1855	12, 384
Level, Spirit	C. F. Hill	Hamilton, Ohio	Dec. 15, 1868	84, 880
Level, Spirit	T. N. Hosmer	Todd's Valley, Cal	Nov. 11, 1862	36, 906
Level, Spirit	R. Ledig	Newark, N. J	Feb. 13, 1866	52, 579
Level, Spirit	H. Lewis	Bennington, Vt	June 30, 1868	79, 363
Level, Spirit	W. T. Nicholson	Providence, R. I	May 1, 1860	28, 104
Level, Spirit	F. W. Marston	Philadelphia, Pa	June 17, 1873	140, 055
Level, Spirit	T. S. Scoville	Rochester, N. Y	Sept. 13, 1859	25, 446
Level, Spirit	G. A. Shelley	Madison, Conn	Nov. 7, 1871	120, 675
Level, Spirit	H. S. Shepardson	Shelburne Falls, Mass	Sept. 13, 1864	44, 225
Level, Spirit	J. D. Sibley	Middletown, Conn	June 25, 1868	79, 266
Level, Spirit	E. A. and C. M. Stratton	Greenfield, Mass	Mar. 1, 1870	100, 463
Level, Spirit	J. A. Traut	New Britain, Conn	July 2, 1872	128, 513
Level, Spirit	A. J. Vandegrift	Cincinnati, Ohio	May 8, 1866	54, 626
Level, square, compass, and plumb-staff, Combined	J. R. Abbott	Midway, Ind	Dec. 26, 1865	51, 675
Level, Surveying	C. Becker	Brooklyn, N. Y	Dec. 1, 1857	18, 728
Level, Surveyor's	S. D. Hailey	Jackson, Tenn	Jan. 10, 1860	26, 762
Levels, Adjustments applied to pendulum	C. Cole	Tarrytown, N. Y	July 28, 1857	17, 870
Levels, Mounting spirit	S. I. Sherman	New York, N. Y	July 19, 1853	9, 867
Levels to squares, Attaching spirit	J. Steger	Matteawan, N. Y	Jan. 24, 1860	26, 956
Leveler, Rut	C. Marshall	North Easton, Mass	Jan. 3, 1871	110, 663
Leveling and grading instrument	H. E. Towle	New York, N. Y	May 10, 1870	102, 882
Leveling-instrument	J. Rohrer	South Bend, Ind	Mar. 5, 1872	124, 389
Leveling-instrument for ditching, &c	J. Gray	Raymond, Miss	Nov. 24, 1857	18, 689
Leveling-instrumegt, Self-adjustable	J. Redhead	Woodville, Miss	Dec. 21, 1858	22, 378
Leveling-instrument, Surveyor's	L. Lea	Jackson, Tenn	Aug. 21, 1860	29, 703
Lever, Alternate propelling	R. Smith	Port Gibson, Miss	June 9, 1825	......
Lever and pendulum power engine	S. Woods	Freeport, Me	Jan. 7, 1835	......
Lever and pulley power applied to a standing press	G. W. Grater	Boston, Mass	Oct. 1, 1830	......
Lever and sling, Truss	S. Wells	Boston, Mass	July 24, 1822	......
Lever and stump-puller	J. G. Fox	Oregon, Wis	Jan. 25, 1870	99, 179
Lever, Angle	J. Barron	Norfolk, Va	May 14, 1822	......
Lever balance, Bent	R. G. Kimball	Albany, N. Y	Sept. 14, 1869	94, 755
Lever, Balance spring-catch	L. Kendall	New York, N. Y	May 8, 1809	......
Lever-bottom press, Combined	E. Venear	Halifax, N. C	May 10, 1826	......
Lever, Circular propelling	J. Turner	Marcellus, N. Y	Mar. 12, 1825	......
Lever-clasp, Spring	G. D. Pike	Chatham, Conn	Apr. 3, 1866	53, 671
Lever, Compound	J. Simpson	Marietta, Ga	Jan. 12, 1869	85, 864
Lever, Compound pendulum	C. Berry	Poughkeepsie, N. Y	Nov. 1, 1825	......
Lever-crane	E. Smith	Paris, N. Y	Dec. 31, 1818	......
Lever, crank, weight, and balance wheel, Combination of.	E. T. Merrel	Parkman, Me	Apr. 22, 1835	......
Lever, Differential	G. I. Washburn	Worcester, Mass	Mar. 14, 1865	46, 838
Lever, Engine	T. W. Goodwin	Portsmouth, Va	Apr. 28, 1863	38, 364
Lever, Engine	T. W. Goodwin	Portsmouth, Va	Aug. 18, 1863	39, 564
Lever escapement	P. Humbert	Boston, Mass	Jan. 1, 1861	31, 022
Lever escapement	C. E. Jacob	New York, N. Y	July 28, 1846	4, 664
Lever escapement, Detached	S. B. Terry	Waterbury, Conn	July 1, 1873	140, 560
Lever for drawing framed timber together	J. N. Eames	Wilmington, Mass	Sept. 9, 1873	142, 618
Lever for presses	J. P. Gates	Red River County, Tex	Mar. 4, 1873	136, 431
Lever for supporting ships, Bilge	J. Thomas	New York, N. Y	Nov. 6, 1826	......
Lever for turning wheels, rowing boats, &c	R. Hopkins	Hopkinton, N. Y	Mar. 25, 1808	......
Lever-grapnel	E. B. Dewey	Pontiac, Mich	Dec. 8, 1868	84, 683
Lever, Hand	E. J. Durant	Lebanon, N. H	May 31, 1859	24, 202
Lever-jack	N. Badgley	New York, N. Y	June 23, 1863	38, 936
Lever-jack	J. Bernheisel, sr	Green Park, Pa	Aug. 25, 1868	81, 467
Lever-jack	L. H. Davis	Kennett's Square, Pa	May 4, 1852	8, 919
Lever-jack	T. M. Kane	Goshen, N. Y	Dec. 15, 1863	40, 981
Lever-jack	T. J. Kindleberger	Eaton, Ohio	Nov. 19, 1867	71, 183
Lever-jack	J. Leffel	Springfield, Ohio	Dec. 10, 1850	7, 820
Lever-jack	J. Leffel	Springfield, Ohio	Nov. 15, 1864	45, 056
Lever-jack	F. Stamm	Lampter, Pa	Mar. 15, 1859	23, 273
Lever-jack, Extension	W. H. Hartman	Fostoria, Ohio	Sept. 19, 1865	49, 999
Lever key and lock	A. Prutyman	Philadelphia, Pa	Mar. 4, 1836	......
Lever, Lifting	R. Sealy	New York, N. Y	July 1, 1822	......
Lever-paddle	J. Harris	London, England	June 21, 1864	43, 271
Lever-power	T. W. Lafetra	New York, N. Y	Jan. 23, 1834	......
Lever-power	H. A. Woodman and M. Person	Charlotte, Mich	Sept. 3, 1872	131, 142
Lever-power, Accumulating	J. Cochran	South Hadley, Mass	Mar. 26, 1834	......
Lever-power and inclined-wheel	P. Wykoff	Westport, Ky	July 20, 1831	......
Lever-power, Crossing or collapsing	W. Scarbrough	Darien, Ga	May 19, 1824	......
Lever-power, Double-ratchet	J. S. Williams	Chicago, Ill	Sept. 8, 1868	81, 967
Lever-power for pressing	R. Sanderson	Athens, Ohio	Feb. 20, 1844	3, 446

Index of patents issued from the United States Patent Office from 1790 to 1873, inclusive—Continued.

Invention.	Inventor.	Residence.	Date.	No.
Lever-power for sewing and knitting machines	G. Stackpole	New York, N. Y	Nov. 17, 1868	84, 144
Lever, Power gained	E. G. Fitch	Blakely, Ala	Oct. 5, 1827	
Lever-power, Mode of applying	G. E. Clay	Stillwater, Minn	June 22, 1858	20, 624
Lever-power, Mode of applying	E. Harris	Princeton, Ill	May 17, 1859	24, 025
Lever-power, Projectile	G. Wood	Vernon, Ind	June 9, 1830	
Lever-power, Vibrating	N. Tripp	Niagara Falls, N. Y	Feb. 18, 1868	74, 732
Lever-press	T. Blake	Turner, Mo	Feb. 20, 1821	
Lever-press	H. G. Guyon	New York, N. Y	July 2, 1836	
Lever-press	W. Linn	Danville, Va	Apr. 25, 1828	
Lever-press	B. Morris	Binghamton, N. Y	Feb. 2, 1830	
Lever-press	J. Outland	Barnesville, Ohio	Aug. 8, 1871	117, 919
Lever-press	J. Payne	Russellville, Ky	Sept. 9, 1835	
Lever-press	H. Sharman	Scriba, N. Y	Nov. 1, 1830	
Lever-press, Compound	J. Rodgers	Maury County, Tenn	Jan. 21, 1829	
Lever-press, Progressive	J. Perkins	Philadelphia, Pa	Feb. 27, 1819	
Lever-press, Spiral	A. O. Stansbury	New York, N. Y	Mar. 21, 1821	
Lever-presses, Windlass tackle for	J. B. Carpenter	Henderson, N. Y	Jan. 30, 1830	
Lever, Pumping	J. S. Appel	Kulpsville, Pa	Jan. 1, 1869	90, 718
Lever-purchase	J. B. Case	Fletcher, Vt	Mar. 31, 1868	76, 050
Lever-purchase, rivet, and bolt-cutter	H. Seely	Unadilla, N. Y	May 12, 1825	
Lever, Ratchet	H. Baldwin	Nashua, N. H	Mar. 23, 1854	10, 955
Lever, Ratchet	H. W. Millar	Utica, N. Y	May 31, 1870	103, 763
Lever shears	J. J. Sandgren	Lyons, Iowa	Apr. 16, 1867	63, 945
Lever-spring, Compensating	W. S. Pratt	Williamsburgh, N. Y	July 24, 1860	29, 304
Lever-truss and sling for yards of vessels	S. A. Wells	Boston, Mass	Dec. 23, 1824	
Levers, &c., Adjusting attachment to reversing	G. W. Jordan	Passaic, N. J	Apr. 1, 1873	137, 374
Levers, Combination of	C. Tomkins	Montgomery, Ala	Nov. 14, 1834	
Levers, Hanging the griping-jaw of spike machines in weighted.	J. H. Swett	Pittsburgh, Pa	Mar. 14, 1854	10, 645
Leverage	W. W. Wills	Janesville, Wis	Oct. 10, 1865	50, 409
Lewis-lever	T. Lidgerwood	Brooklyn, N. Y	Mar. 12, 1850	7, 172
Library-register	W. T. Ray	Philadelphia, Pa	Sept. 5, 1865	49, 788
Library-shears	L. Prang	Boston, Mass	July 5, 1870	104, 996
Library-system	J. M. W. Geist	Lancaster, Pa	Jan. 2, 1866	51, 823
Lid and plate lifter and pot-hook	T. Simpson	Newark, Ohio	May 17, 1870	103, 247
Lid-lifter	W. Van Gaasbeck	Hudson, N. Y	Dec. 30, 1873	145, 035
Lid-supporter	J. C. Barlow	Brimfield, Mass	Oct. 16, 1866	58, 936
Life and anchor boat	J. Francis	New York, N. Y	Jan. 11, 1839	1, 067
Life and property saving vessels, Arrangement of means for balancing and propelling.	J. Minifie	Baltimore, Md	July 8, 1856	15, 298
Life and treasure buoy	F. D. Lee	Charleston, S. C	Apr. 27, 1858	20, 072
Life and treasure buoy for vessel, Arrangement of	F. D. Lee	Charleston, S. C	Dec. 8, 1857	18, 819
Life-berth for vessel	J. P. McLean	New York, N. Y	Dec. 7, 1858	22, 258
Life-boat	J. Adams	Gloucester, N. J	Sept. 16, 1873	142, 883
Life-boat	J. Allen	New York, N. Y	Mar. 6, 1855	12, 473
Life-boat	L. Ball	Auburn, N. Y	Apr. 20, 1858	19, 977
Life-boat	E. A. Barrett	New York, N. Y	Apr. 29, 1873	138, 225
Life-boat	P. K. Beaupré	Metropolis City, Ill	Aug. 29, 1871	118, 580
Life-boat	H. Berdan	New York, N. Y	Feb. 13, 1855	12, 375
Life-boat	M. E. D. Brown	Utica, N. Y	Apr. 5, 1859	23, 442
Life-boat	M. M. Camp	New Haven, Conn	Sept. 22, 1857	18, 233
Life-boat	D. Dodge and P. Burgess	New York, N. Y., and East Boston, Mass.	Aug. 9, 1853	9, 915
Life-boat	C. D. Flynt	Philadelphia, Pa	June 1, 1869	90, 833
Life-boat	Y. Foreman	New York, N. Y	Oct. 11, 1853	10, 120
Life-boat, &c	J. Francis	New York, N. Y	Mar. 26, 1841	2, 018
Life-boat	L. F. Frazee	New Brunswick, N. J	Nov. 22, 1853	10, 266
Life-boat	L. F. Frazee	Jersey City, N. J	June 23, 1868	79, 111
Life-boat	J. R. Grace	Brooklyn, N. Y	Sept. 1, 1868	81, 623
Life-boat	P. Henrich	Allegheny City, Pa	Apr. 26, 1870	102, 261
Life-boat	H. Hensel	Carver, Minn	Aug. 28, 1866	57, 633
Life-boat	W. Hughes	Waupun, Wis	July 31, 1866	56, 855
Life-boat	R. Humble	Milwaukee, Wis	Oct. 4, 1870	107, 912
Life-boat	O. R. Ingersoll	Brooklyn, N. Y	Feb. 20, 1872	123, 908
Life-boat	G. W. La Baw	Jersey City, N. J	Apr. 1, 1856	14, 565
Life-boat	G. W. La Baw	Jersey City, N. J	Nov. 1, 1859	25, 971
Life-boat	M. Ludlum	Essex, N. Y	Jan. 27, 1857	16, 491
Life-boat	M. Ludlum	Fair Haven, Vt	Mar. 29, 1859	23, 380
Life-boat	T. Masac	Good Hope Plantation, La	May 31, 1870	103, 637
Life-boat	S. Mills	New York, N. Y	Aug. 7, 1860	29, 508
Life-boat	M. V. Nobles	Elmira, N. Y	June 18, 1867	65, 824
Life-boat	L. Raymond	New York, N. Y	Feb. 4, 1843	2, 938
Life-boat	R. J. Robeson	Oskaloosa, Iowa	Aug. 8, 1871	117, 817
Life-boat	J. T. Scholl	Port Washington, Wis	Feb. 26, 1861	31, 562
Life-boat	A. L. Shears	Omro, Wis	May 25, 1858	20, 374
Life-boat	A. L. Shears	Flint, Mich	Aug. 1, 1865	49, 197
Life-boat	J. M. Starr	Fond du Lac, Wis	Dec. 27, 1870	110, 600
Life-boat	H. Thompson	Mobile, Ala	Oct. 5, 1869	95, 538
Life-boat	G. H. Tier	Fremont, Ohio	Aug. 7, 1866	57, 012
Life-boat	W. A. Vail	Clinton, Conn	Aug. 5, 1873	141, 609
Life-boat	F. Vié	Havre, France	May 4, 1869	89, 816
Life-boat	W. H. Wylly	Savannah, Ga	Jan. 22, 1867	61, 500
Life-boat and bed-bottom, Combined	C. Bütgenbach	Louisville, Ky	Aug. 15, 1871	117, 977
Life-boat constructed of mattresses	J. M. Woodward	New York, N. Y	Sept. 7, 1858	21, 402
Life-boat, Folding	E. S. Berthon	Fareham, England	Mar. 20, 1855	12, 537
Life-boat, Folding	C. Locher	New York, N. Y	Jan. 2, 1855	12, 184
Life-boat, Folding	H. Martin	Louisville, Ky	Apr. 12, 1859	23, 595
Life-boat, Reversible	G. P. Tewksbury	Boston, Mass	Aug. 7, 1849	6, 640
Life-boat, Reversible	N. Thompson, jr	Williamsburgh, N. Y	Apr. 11, 1854	10, 766
Life-boat, Sectional	C. Pond	Philadelphia, Pa	July 9, 1872	128, 750
Life-boat, Self-inflating and folding	W. and T. Schnebly	New York, N. Y	Jan. 23, 1849	6, 053
Life-boat, Sofa	P. Van Zile, S. M. Griffin, and W. S. Dey.	New York, N. Y	Nov. 6, 1855	13, 771
Life-boat, Sporting	J. M. Cayce	Franklin, Tenn	Aug. 20, 1867	67, 846
Life-boat, Surf	W. M. Camp	New Haven, Conn	Oct. 25, 1859	25, 882
Life-boat, Surf	J. R. Grace	Brooklyn, N. Y	Mar. 6, 1860	27, 362
Life-boat, Water-cask	W. N. Clark	Chester, Conn	May 3, 1859	23, 824
Life-boats, Apparatus for supplying air to	P. F. Schenck	Riceville, N. J	June 18, 1867	65, 953

Index of patents issued from the United States Patent Office from 1790 *to* 1873, *inclusive*—Continued.

Invention.	Inventor.	Residence.	Date.	No.
Life-boats, Expansible float for	C. Segros	New York, N. Y	Sept. 21, 1858	21, 570
Life-boats, Form of air-chambers of	J. D. Greene	Cambridge, Mass	Sept. 25, 1849	6, 737
Life-boats, Mode of constructing	G. James	Philadelphia, Pa	Jan. 21, 1839	1, 071
Life-buoy	W. Flower	Philadelphia, Pa	May 14, 1808	....
Life-pole and fire-engine assistant	W. W. Van Loan	Catskill, N. Y	Feb. 2, 1822	....
Life-preserver	S. Albro	Buffalo, N. Y	Mar. 2, 1852	8, 767
Life-preserver	G. M. Allerton	New York, N. Y	July 14, 1868	79, 934
Life-preserver	E. S. Bennett	New York, N. Y	Jan. 26, 1869	86, 127
Life-preserver	J. M. Billhofer	Irvington, N. Y	May 21, 1867	64, 939
Life-preserver	J. Bond	Norfolk, Va	Apr. 6, 1869	88, 692
Life-preserver	H. B. Bourne	New York, N. Y	May 8, 1840	1, 596
Life-preserver	T. B. Boyd	Saint Louis, Mo	June 21, 1870	104, 546
Life-preserver	R. Bulkley	New York, N. Y	May 8, 1840	1, 595
Life-preserver	C. J. Bunker	New York, N. Y	Sept. 29, 1857	18, 274
Life-preserver	A. Chandler	New York, N. Y	Oct. 3, 1844	3, 774
Life-preserver	E. M. Crandal	Marshalltown, Iowa	Oct. 4, 1870	108, 006
Life-preserver	E. G. Fitch	New Orleans, La	Aug. 18, 1846	4, 699
Life-preserver	A. J. Gibson	Worcester, Mass	Dec. 8, 1857	18, 809
Life-preserver	A. Gregory	New York, N. Y	Apr. 27, 1869	89, 402
Life-preserver	N. E. Guérin	New York	Nov. 16, 1841	2, 359
Life-preserver	D. H. Heyen	New York, N. Y	Dec. 3, 1867	71, 755
Life-preserver	L. Jacobs	Albion, Mich	May 10, 1870	102, 823
Life-preserver	J. Knight	New York, N. Y	June 2, 1857	17, 434
Life-preserver	C. Krejci	Scranton, Pa	Mar. 15, 1870	100, 906
Life-preserver	M. Ormsbee	Brooklyn, N. Y	Oct. 17, 1871	120, 089
Life-preserver	H. Palmer	Augusta, Mich	Oct. 12, 1858	21, 776
Life-preserver	N. Platt	Jackson, Miss	Feb. 7, 1860	27, 066
Life-preserver	R. Robinson	San Francisco, Cal	Oct. 22, 1867	70, 121
Life-preserver	E. Rousseil	West Newark, N. J	Jan. 9, 1866	51, 971
Life-preserver	S. Scholfield	Norwich, Conn	July 7, 1863	39, 175
Life-preserver	T. R. Scott	New York, N. Y	Nov. 21, 1871	121, 132
Life-preserver	J. E. Serrell and W. Davies	New York, N. Y	Dec. 15, 1857	18, 869
Life-preserver	J. Shaw	Bridgeport, Conn	May 28, 1867	65, 287
Life-preserver	W. A. Simonds	Boston, Mass	Feb. 3, 1857	16, 555
Life-preserver	B. W. Taylor	Henderson, Ky	May 30, 1871	115, 386
Life-preserver	J. E. Thomson and E. Clark	Buffalo, N. Y., and Louisville, Ky.	Aug. 15, 1871	118, 168
Life preserver, buoy, &c	C. A. De Liancourt	Paris, France	May 10, 1844	3, 584
Life-preserver, Carriage	H. G. De Grandval	Portsmouth, N. H	Oct. 3, 1817	....
Life-preserver for railway-cars, Brakesman's	J. Worsley	Providence, R. I	Sept. 6, 1864	44, 134
Life-preserver for steamboats, &c	M. Pearson	Newburyport, Mass	Jan. 27, 1843	2, 937
Life-preserver, Marine	J. B. Verdier	Paris, France	Feb. 16, 1864	41, 654
Life-preserver or buoyant dress	R. Porter	Billerica, Mass	May 25, 1840	1, 619
Life-preserver, Safety	J. J. White	Philadelphia, Pa	Apr. 7, 1838	679
Life-preservers, &c., Float for	J. W. Weston	New York, N. Y	Jan. 26, 1869	86, 193
Life-preservers to vests, Mode of attaching	R. L. Nelson	Ocala, Fla	Nov. 21, 1854	11, 969
Life-preserving and swimming apparatus	H. Olsen	San Francisco, Cal	Feb. 25, 1868	74, 931
Life-preserving apparatus	J. B. Stoner	New York, N. Y	Feb. 4, 1868	74, 168
Life preserving bed for ship	G. K. Hooper	Boston, Mass	Sept 11, 1855	13, 546
Life-preserving bedstead and sofa	J. T. Garlick	New York, N. Y	Mar. 31, 1857	16, 946
Life-preserving berth	J. J. Clyde	Williamsburgh, N. Y	Mar. 5, 1867	62, 609
Life-preserving berth for steam and other vessels	E. Foster	Hartford, Conn	Sept. 1, 1857	18, 090
Life-preserving boat, Automatic	A. Carson	Memphis, Tenn	July 16, 1867	66, 933
Life preserving bucket	N. Thompson, jr	Williamsburgh, N. Y	Oct. 18, 1853	10, 140
Life-preserving bucket-raft	C. French	Jersey City, N. J	Apr. 20, 1858	19, 989
Life-preserving buoy	B. Burling	Buffalo, N. Y	Mar. 16, 1858	19, 618
Life-preserving buoy	O. E. Woods	Philadelphia, Pa	Oct. 11, 1859	25, 781
Life-preserving cape	G. and C. Palmer	Morris Run, Pa	Nov. 11, 1873	144, 561
Life preserving chair, stool, &c	W. H. Shecut	New York	Mar. [illegible]	[illegible]
Life-preserving door	J. T. Pheatt	Toledo, Ohio	Apr. 17, 1855	12, 743
Life-preserving dress	C. S. Merriman	Vallisca, Iowa	July 16, 1872	128, 971
Life-preserving float	G. W. Hamilton	Watkins, N. Y	Mar. 16, 1858	19, 632
Life-preserving float	T. Hosmer	Sandusky, Ohio	Mar. 4, 1873	136, 436
Life-preserving garment	E. R. Cogswell	New York, N. Y	Aug. 12, 1873	141, 631
Life-preserving garment	W. Morris	Philadelphia, Pa	Apr. 26, 1870	102, 419
Life-preserving hammock, Arrangement of the sections in a.	S. I. Seely	New York, N. Y	July 10, 1849	6, 581
Life-preserving hat	S. W. White	England	Oct. 14, 1840	1, 826
Life-preserving mattress	J. Adams	Gloucester, N. J	Apr. 16, 1872	125, 779
Life-preserving mattress	L. Bauhoefer	Philadelphia, Pa	Dec. 4, 1860	30, 794
Life-preserving mattress	C. P. Crossman and E. M. Quimby.	Warren, Mass	Feb. 16, 1858	19, 350
Life-preserving mattress	J. Golding	New York, N. Y	June 18, 1867	65, 901
Life-preserving mattress	J. D. Greene	Cambridge, Mass	Aug. 17, 1869	93, 819
Life-preserving mattress	J. D. Greene	New York, N. Y	Apr. 4, 1871	113, 652
Life-preserving mattress	D. E. Hall	Detroit, Mich	Aug. 3, 1869	93, 195
Life-preserving mattress	J. Hunt	Providence, R. I	Sept. 20, 1870	104, 499
Life-preserving mattress	H. B. Mountain	New York, N. Y	Mar. 11, 1873	136, 749
Life-preserving mattress	W. H. Pack and J. S. Vanhorn	Jersey City, N. J	Oct. 29, 1872	132, 686
Life-preserving mattress	J. F. Woodfine	Hoboken, N. J	Sept. 30, 1873	143, 313
Life-preserving mattress and raft	J. Golding	New York, N. Y	July 23, 1867	67, 039
Life-preserving raft	A. Baker	Appleton, Wis	Jan. 17, 1860	26, 825
Life-preserving raft	A. G. Mack	Rochester, N. Y	July 12, 1859	24, 747
Life-preserving raft	F. Z. Tucker	Brooklyn, N. Y	Jan. 2, 1855	12, 143
Life-preserving raft	C. W. Wailey	New Orleans, La	Mar. 12, 1867	62, 909
Life-preserving raft of bouyant mattresses	W. Urquhart	New York, N. Y	Mar. 9, 1858	19, 593
Life-preserving rafts, Canvas sheet connected with.	L. Taggart	Philadelphia, Pa	Jan. 26, 1858	19, 216
Life-preserving seat	H. Matthews	Brooklyn, N. Y	Feb. 26, 1867	62, 433
Life-preserving seat	G. B. Towksbury	Boston, Mass	Oct. 19, 1852	9, 349
Life-preserving seat	N. Thompson, jr	Williamsburgh, N. Y	Oct. 18, 1853	10, 141
Life-preserving seat	N. Thompson, jr	Williamsburgh, N. Y	Dec. 19, 1854	12, 108
Life-preserving seat	N. Thompson, jr	Williamsburgh, N. Y	Jan. 9, 1855	12, 220
Life-preserving skirt	S. E. Saul	Brooklyn, N. Y	July 26, 1870	105, 730
Life.preserving state-room for navigable vessel, Bouyant.	H. Hallock	Brookhaven, N. Y	June 1, 1858	20, 426
Life-preserving stool	H. T. Pratt	New York, N. Y	Mar. 15, 1870	100, 801
Life-preserving trunk	L. Rebstock	Hollidaysburgh, Pa	May 2, 1871	114, 475
Life-preserving trunk	O. E. Woods	Philadelphia, Pa	Dec. 28, 1858	22, 467

Index of patents issued from the United States Patent Office from 1790 *to* 1873, *inclusive*—Continued.

Invention.	Inventor.	Residence.	Date.	No.
Life-preserving vessel	J. Macintosh	New York, N. Y	Nov. 11, 1837	462
Life-preserving vest	T. A. Delano	New York, N. Y	Nov. 9, 1858	22, 021
Life-raft	H. C. Calkins	New York, N. Y	Nov. 28, 1871	121, 275
Life-raft	W. B. Davies	New York, N. Y	Feb. 11, 1873	135, 784
Life-raft	L. F. Frazee	South Amboy, N. J	July 10, 1866	56, 201
Life-raft	D. McFarland	New York, N. Y	Dec. 13, 1870	110, 059
Life-raft	J. Murtaugh	New York, N. Y	Dec. 18, 1866	60, 640
Life-raft	E. L. Perry	New York, N. Y	Nov. 15, 1864	45, 073
Life-raft	J. Rider	New York, N. Y	Feb. 22, 1870	100, 191
Life-raft, Extensible	C. Furbush	Kittery, Me	Nov. 30, 1858	22, 175
Life-rafts, Float for	J. Mason and G. W. Rogers	Detroit, Mich., and Brooklyn, N. Y.	Sept. 16, 1873	142, 801
Life-saving apparatus	J. B. Stoner	New York, N. Y	Sept. 24, 1872	131, 719
Life-saving raft	G. Blanchard	Washington, D. C	Feb. 20, 1855	12, 405
Life-saving raft	S. B. Broad	New York, N. Y	May 8, 1860	28, 150
Lifter: *See* Barrel-lifter. Box-lifter. Exercising-lifter. Griddle-lifter. Kettle-lifter. Lamp-chimney lifter. Lid lifter. Log-lifter. Pan-lifter. Pattern-lifter. Pitcher-lid lifter. Plate-lifter. Shoe-lifter. Skirt-lifter. Stone-lifter. Stove-lifter. Stove-cover lifter. Stove-lid lifter. Tie-lifter. Track-lifter. Transom-lifter. Wagon-lifter. Wagon-body lifter.				
Lifter and tongs, Combined	J. Hyslop, jr., and C. E. Phillips.	Abington, Mass	Mar. 10, 1868	75, 427
Lifter for kitchen use	T. S. Coffin	Harrington, Me	Jan. 25, 1870	99, 061
Lifting-apparatus	D. P. Butler	Boston, Mass	Mar. 2, 1869	87, 465
Lifting-apparatus	C. H. Douglas	Hartford, Conn	June 13, 1871	115, 942
Lifting-apparatus	W. L. Jones	Baldwin City, Kans	Apr. 2, 1867	63, 526
Lifting-apparatus	F. W. Reilly	Chicago, Ill	Apr. 21, 1868	76, 944
Lifting-apparatus	G. B. Windship	Boston, Mass	Sept. 12, 1865	49, 945
Lifting-apparatus	G. B. Windship	Boston, Mass	Feb. 20, 1872	123, 804
Lifting-apparatus, Spring	D. P. Butler	Boston, Mass	July 20, 1869	92, 793
Lifting-bar	D. P. Butler	Boston, Mass	June 19, 1866	55, 618
Lifting-bar	D. P. Butler	Boston, Mass	July 20, 1869	92, 792
Lifting-handle	P. Bradford	New Haven, Conn	Mar. 12, 1867	62, 729
Lifting-handle	J. Ottner	New Britain, Conn	June 26, 1860	28, 940
Lifting-handle	J. B. Sargent	New Britain, Conn	Nov. 2, 1858	21, 981
Lifting-jack	G. H. Alger	Ames, N. Y	Apr. 21, 1868	76, 967
Lifting-jack	W. C. Anderson	Saint Louis, Mo	May 15, 1860	28, 245
Lifting-jack	O. A. Anthony	Mayfield, N. Y	Jan. 10, 1871	110, 818
Lifting-jack	W. F. Arnold	New Britain, Conn	Dec. 29, 1868	85, 264
Lifting-jack	W. F. Arnold	New Britain, Conn	Jan. 7, 1873	134, 625
Lifting-jack	J. B. Ausbourne	Milwaukee, Wis	Aug. 10, 1869	93, 512
Lifting-jack	H. Ballentine	Philadelphia, Pa	July 4, 1871	116, 534
Lifting-jack	J. H. Bean	Marietta, Ohio	May 21, 1867	64, 937
Lifting-jack	J. W. Bemis	Fall River, Mass	Feb. 4, 1868	74, 040
Lifting-jack	G. Benjamin	Avoca, N. Y	Sept. 18, 1860	30, 043
Lifting-jack	L. J. Blades and J. Mahoney	Harrington and Wilmington, Del.	Mar. 15, 1870	100, 846
Lifting-jack	J. W. Bliss	Hartford, Conn	Mar. 21, 1854	10, 663
Lifting-jack	J. S. Bodge	La Porte, Ind	Apr. 16, 1872	125, 784
Lifting-jack	W. A. Bowyer	Helen Furnace, Pa	Aug. 10, 1869	93, 405
Lifting-jack	W. and C. H. Brady	Mount Joy, Pa	July 6, 1869	92, 255
Lifting-jack	A. C. Brinfer	Middletown, Pa	Sept. 17, 1867	68, 837
Lifting-jack	W. A. Brunker	Farmersburgh, Ind	May 20, 1873	139, 041
Lifting-jack	W. S. Burgin	Washington, Vt	Nov. 21, 1871	121, 081
Lifting-jack	C. Butterworth	Miamisburgh, Ohio	May 21, 1867	64, 946
Lifting-jack	J. Callahan	Stroud Glades, Va	Dec. 29, 1857	18, 955
Lifting-jack	J. Camp	Olney, Ill	Sept. 22, 1868	82, 381
Lifting-jack	J. S. Chesnut	Philadelphia, Pa	Apr. 14, 1857	17, 024
Lifting-jack	S. J. Clark	Detroit, Mich	July 23, 1867	67, 103
Lifting-jack	W. Clark	Decatur, Ill	May 3, 1870	102, 493
Lifting-jack	H. Clement	Rising Sun, Ind	Mar. 19, 1872	124, 721
Lifting-jack	C. T. Close	Brooklyn, N. Y	Feb. 12, 1867	62, 009
Lifting-jack	C. B. Conant	Hardwick, Mass	Feb. 25, 1862	34, 482
Lifting-jack	J. Cook	Palmersville, Pa	May 26, 1863	38, 654
Lifting-jack	J. Coulter, jr	Xenia, Ohio	Feb. 4, 1868	73, 954
Lifting-jack	I. A. Crippen	Philadelphia, Pa	Sept. 9, 1873	142, 680
Lifting-jack	J. Q. Crosby	Northborough, Mass	Dec. 1, 1868	84, 536
Lifting-jack	C. Crow	Onargo, Ill	May 21, 1867	64, 949
Lifting-jack	A. M. Culver	Bedford, Ohio	Feb. 19, 1867	62, 117
Lifting-jack	A. M. Culver	Bedford, Ohio	Feb. 18, 1868	74, 512
Lifting-jack	H. Culver	Richfield, Ohio	Jan. 2, 1866	51, 807
Lifting-jack	J. Dampman	Lebanon, Pa	Mar. 23, 1869	88, 015
Lifting-jack	A. A. Davis	Clark's Green, Pa	Nov. 14, 1871	120, 864
Lifting-jack	R. W. and D. Davis	Yellow Springs, Ohio	Apr. 14, 1857	17, 027
Lifting-jack	D. Diver	Boone, Iowa	Mar. 5, 1867	62, 616
Lifting-jack	A. Dom	Mount Healthy, Ohio	Nov. 5, 1872	132, 753
Lifting-jack	W. M. Doty	New York, N. Y	Apr. 22, 1873	138, 078
Lifting-jack	F. Drew	South Boston, Mass	Nov. 20, 1855	13, 835
Lifting-jack	S. S. Eccleston	South New Berlin, N. Y	Jan. 14, 1873	134, 865
Lifting-jack	W. C. Edis	Patriot, Ind	June 18, 1872	127, 968
Lifting-jack	J. H. Elward	Polo, Ill	Nov. 2, 1869	96, 408
Lifting-jack	D. Fasig	Rowsburgh, Ohio	Aug. 12, 1862	36, 144
Lifting-jack	D. Fasig	Rowsburgh, Ohio	May 15, 1866	54, 706
Lifting-jack	L. W. Fifield	Worcester, Mass	Apr. 5, 1870	101, 450
Lifting-jack	L. P. Garcin	San Francisco, Cal	Apr. 5, 1870	101, 454
Lifting-jack	E. R. Gard	Chicago, Ill	June 27, 1871	116, 296
Lifting-jack	E. R. Gard	Chicago, Ill	Jan. 23, 1872	123, 010
Lifting-jack	R. W. Genung	Blooming Grove, N. Y	July 11, 1854	11, 298
Lifting-jack	A. M. Gilmore	Greensburgh, Ind	May 14, 1872	126, 800
Lifting-jack	A. E. Goddard	Essex, Conn	Dec. 12, 1871	121, 774

Index of patents issued from the United States Patent Office from 1790 to 1873, inclusive—Continued.

Invention.	Inventor.	Residence.	Date.	No.
Lifting-jack	G. F. Graves	Mount Upton, N. Y	Jan. 1, 1867	60, 881
Lifting-jack	H. Gray	Bristol, Conn	Dec. 16, 1856	16, 233
Lifting-jack	W. Green	Holly, Mich	Oct. 1, 1867	69, 400
Lifting-jack	W. Green	Holly, Mich	Sept. 15, 1868	82, 111
Lifting-jack	W. H. Greenwalt	Strickersville, Pa	Oct. 31, 1871	120, 511
Lifting-jack	W. H. Greenwalt	Strickersville, Pa	Apr. 30, 1872	126, 286
Lifting-jack	S. Gulick	Kline's Grove, Pa	Feb. 19, 1867	62, 193
Lifting-jack	J. S. Haldeman	Kansas City, Mo	Apr. 1, 1873	137, 363
Lifting-jack	J. T. Hamilton	Greensburgh, Ind	Mar. 19, 1872	124, 814
Lifting-jack	J. T. Hamilton and E. F. Conner.	Boone County and Decatur, Ind.	Nov. 25, 1873	144, 977
Lifting-jack	J. F. Hammond	Providence, R. I	Aug. 21, 1866	57, 439
Lifting-jack	A. Healy	Kalamazoo, Mich	June 11, 1867	65, 569
Lifting-jack	W. G. Hermance	Albany, N. Y	Feb. 19, 1867	62, 268
Lifting-jack	A. Higley	South Bend, Ind	May 29, 1866	55, 102
Lifting-jack	J. H. Hinton	Wayne County, Ohio	Feb. 4, 1873	135, 551
Lifting-jack	C. Holmes	Washington, Ohio	Oct. 29, 1867	70, 214
Lifting-jack	E. Hoyt	Stamford, Conn	Feb. 6, 1872	123, 347
Lifting-jack	A. Jackson	Cliffon Springs, N. Y	Sept. 10, 1867	68, 750
Lifting-jack	J. C. Jackson	Rochester, N. Y	Aug. 31, 1858	21, 342
Lifting-jack	B. F. Johnson	Glasgow, Mo	Feb. 21, 1871	111, 942
Lifting-jack	I. D. Johnson	Kennett's Square, Pa	Dec. 12, 1871	121, 721
Lifting-jack	S. G. Jones	Fitzwatertown, Pa	Jan. 23, 1855	12, 280
Lifting-jack	S. D. Jones	Blue Island, Ill	Aug. 15, 1871	118, 020
Lifting-jack	L. J. Knowles	Warren, Mass	Nov. 10, 1857	18, 592
Lifting-jack	J. Kohler	New York, N. Y	Apr. 21, 1868	76, 927
Lifting-jack	F. Krick	Fidelity, Ohio	Nov. 24, 1868	84, 285
Lifting-jack	I. L. Landis	Manheim Township, Lancaster County, Pa.	Jan. 7, 1862	34, 063
Lifting-jack	W. J. Lane	Chappaqua, N. Y	Oct. 23, 1860	30, 484
Lifting-jack	S. C. Leonard	Oberlin, Ohio	Nov. 29, 1870	109, 635
Lifting-jack	S. M. Lusk	Centreton, Ohio	Jan. 30, 1872	123, 270
Lifting-jack	E. L. Marsh	Greenwich, Ohio	Sept. 17, 1867	68, 891
Lifting-jack	C. A. Masterson	Decatur, Ill	Feb. 20, 1872	123, 784
Lifting-jack	T. Maxon	Springfield, Ohio	Sept. 12, 1871	118, 955
Lifting-jack	T. Maxon	Springfield, Ohio	Aug. 20, 1872	130, 733
Lifting-jack	S. McGuffin	Rising Sun, Ind	Mar. 26, 1872	125, 065
Lifting-jack	W. McMille	Milan, Ohio	Nov. 5, 1867	70, 591
Lifting-jack	W. A. Middleton	Newville, Pa	Dec. 17, 1872	134, 086
Lifting-jack	D. L. Miller	Madison, N. J	Dec. 1, 1857	18, 760
Lifting-jack	D. L. H. Mitchell	Forest, Miss	June 17, 1873	140, 063
Lifting-jack	H. T. Morrison	Lawrenceville, Va	May 23, 1871	115, 086
Lifting-jack	T. W. H. Moseley	Hyde Park, Mass	Mar. 4, 1873	136, 533
Lifting-jack	C. W. Mosher	East Leon, N. Y	Jan. 18, 1870	98, 992
Lifting-jack	D. Mulligan and J. C. Imlay	Greensburgh, Ind	Apr. 23, 1872	126, 076
Lifting-jack	B. Newbury	Catskill, N. Y	May 27, 1851	8, 110
Lifting-jack	J. D. Otstot	Springfield, Ohio	Dec. 2, 1862	37, 051
Lifting-jack	J. D. Otstot	Springfield, Ohio	Dec. 17, 1872	133, 949
Lifting-jack	J. N. Parker	Darlington, Wis	Aug. 27, 1867	68, 230
Lifting-jack	S. J. Parmele	Killingworth, Conn	Jan. 9, 1867	51, 963
Lifting-jack	S. Perry	Seaford, Del	May 12, 1868	77, 756
Lifting-jack	J. W. Pettingill	Rockford, Ill	Feb. 13, 1866	52, 596
Lifting-jack	O. L. Pinney	Brunswick, Ohio	Oct. 22, 1867	70, 115
Lifting-jack	L. B. Prindle	Litchfield, Conn	Jan. 28, 1868	73, 833
Lifting-jack	D. Putnam	Stockton, N. Y	Sept. 2, 1873	142, 510
Lifting-jack	G. Race	Norwich, N. Y	Feb. 26, 1867	62, 442
Lifting-jack	C. E. and F. A. Randall	Boston, Mass	Nov. 26, 1872	133, 484
Lifting-jack	W. S. Rayment	Union City, Mich	Nov. 14, 1871	120, 002
Lifting-jack	R. M. Reynolds	Oakville, Mich	Nov. 30, 1869	97, 444
Lifting-jack	A. C. Richard	Newtown, Conn	May 25, 1858	20, 372
Lifting-jack	J. Riddlesberger	Waynesborough, Pa	Feb. 23, 1869	87, 292
Lifting-jack	E. S. Robertson and A. B. Collins.	Mount Liberty, Ohio	Apr. 28, 1868	77 406
Lifting-jack	J. Rosecrans	Berkshire, Ohio	Apr. 12, 1870	101, 920
Lifting-jack	W. F. Rundell	East Genoa, N. Y	Feb. 17, 1863	37, 707
Lifting-jack	B. and C. Schopf	Pittsburgh, Ind	Jan. 25, 1870	99, 244
Lifting-jack	H. G. Seekins and C. H. Goss	Elyria, Ohio	July 7, 1857	17, 757
Lifting-jack	J. W. Shaukland	Summerfield, Ohio	Jan. 7, 1873	134, 709
Lifting-jack	A. Shaver	Warnerville, N. Y	Feb. 8, 1870	99, 717
Lifting-jack	T. Sheldon	Enfield, Conn	June 5, 1866	55, 430
Lifting-jack	H. S. Shepardson	Shelburne Falls, Mass	Oct. 3, 1865	50, 306
Lifting-jack	T. Shiver	Newburgh, Ind	June 4, 1867	65, 514
Lifting-jack	S. Shoemaker	Smithville, Ohio	Oct. 29, 1861	33, 608
Lifting-jack	E. Shopbell	Ashland, Ohio	Oct. 22, 1867	70, 034
Lifting-jack	E. Shopbell	Ashland, Ohio	Jan. 21, 1868	73, 657
Lifting-jack	D. Shumusk	Castile, N. Y	Aug. 13, 1872	130, 542
Lifting-jack	A. S. Skinner	Windsor, Mich	Oct. 29, 1872	132, 697
Lifting-jack	J. Slonecker	Jefferson, Pa	May 7, 1872	126, 496
Lifting-jack	R. T. and R. T. Smart, jr	Troy, N. Y	Sept. 17, 1872	131, 472
Lifting-jack	C. A. Smith	New York, N. Y	Aug. 25, 1868	81, 424
Lifting-jack	F. B. Smith	Craigsville, N. Y	July 11, 1854	11, 303
Lifting-jack	F. S. Smith	Geneva, Ohio	Dec. 10, 1872	133, 728
Lifting-jack	J. U. Smith	Orion, Mich	June 20, 1871	116, 232
Lifting-jack	L. Smith	Rochester, Minn	July 3, 1866	56, 111
Lifting-jack	L. P. Smith	Middletown, Pa	Sept. 21, 1869	95, 157
Lifting-jack	L. P. Smith	Middletown, Pa	Oct. 31, 1871	120, 402
Lifting-jack	T. Stebins	San Francisco, Cal	Apr. 13, 1869	88, 923
Lifting-jack	J. St. John	New York, N. Y	July 8, 1851	8, 214
Lifting-jack	J. Stoody	Ripley, Ohio	Nov. 5, 1867	70, 644
Lifting-jack	J. J. Stewart	New York, N. Y	Aug. 13, 1872	130, 449
Lifting-jack	G. P. Sweezy	Riverhead, N. Y	Nov. 26, 1872	133, 499
Lifting-jack	S. Taber	Saint Joseph, Mo	June 4, 1867	65, 447
Lifting-jack	S. Taylor	Cold Water, Mich	Feb. 25, 1873	136, 281
Lifting-jack	J. Terrell	Royal Centre, Ind	Feb. 13, 1872	123, 593
Lifting-jack	W. Thomas	Hingham, Mass	May 19, 1857	17, 344
Lifting-jack	J. M. and R. M. Thompson	Coshocton, Ohio	Sept. 23, 1862	36, 540
Lifting-jack	R. M. Thompson	Coshocton, Ohio	May 5, 1868	77, 678
Lifting-jack	W. M. K. Thornton	Clinton Junction, Wis	Feb. 28, 1865	46, 599

Index of patents issued from the United States Patent Office from 1790 *to* 1873, *inclusive*—Continued.

Invention.	Inventor.	Residence.	Date.	No.
Lifting-jack	W. Thurber	Oleon, N. Y	Dec. 8, 1863	40, 867
Lifting-jack	B. B. Tomlinson	Mount Carroll, Ill	Mar. 12, 1872	124, 519
Lifting-jack	I. Vaughan	Harrisburgh, Pa	Oct. 15, 1872	132, 189
Lifting-jack	A. F. Wagner	Ilion, N. Y	Jan. 16, 1866	52, 096
Lifting-jack	R. Walker	Batavia, N. Y	July 30, 1867	67, 386
Lifting-jack	H. H. Warren	Buffalo, N. Y	Apr. 22, 1873	138, 059
Lifting-jack	L. D. Warren	Havana, Ill	Oct. 18, 1870	108, 414
Lifting-jack	A. M. Waters	Cuyahoga Falls, Ohio	Aug. 23, 1870	106, 639
Lifting-jack	F. C. White	Euclid, Ohio	Mar. 15, 1870	100, 828
Lifting-jack	J. Wilkinson	Bowling Green, Mo	May 19, 1868	78, 167
Lifting-jack	H. J. Wilson	Mason, Mich	Apr. 4, 1871	113, 381
Lifting-jack	G. W. Windsor	Allegheny City, Pa	July 19, 1870	105, 534
Lifting-jack	E. Young	Camden Centre, Mich	Feb. 23, 1864	47, 738
Lifting-jack	E. Zimerman	Pamelia Four Corners, N.Y	Feb. 19, 1867	62, 175
Lifting-jack	J. Zimmerman	Bloomfield, Pa	June 17, 1862	35, 646
Lifting-jack and cant-hook	D. Fasig	Rowsburgh, Ohio	Nov. 10, 1868	83, 947
Lifting-jack and derrick, Combined	H. Senseman and W. F. Pagett	Tremont, Ohio	July 30, 1872	130, 079
Lifting-jack and spike-extractor	J. Douglass	McConnellstown, Pa	Mar. 3, 1868	75, 034
Lifting-jack and wrench, Combined	R. A. York	Reading, Mich	May 4, 1869	89, 723
Lifting jack, Carriage	J. Jenkins	Monroe, N. Y	Nov. 7, 1854	11, 920
Lifting-jack for dumping-cart	R. B. Little	Providence, R. I	Dec. 19, 1871	122, 036
Lifting-jack, Hydraulic	J. Ryan	Saint Louis, Mo	Mar. 2, 1869	87, 435
Lifting-jack, Hydrostatic	J. Robertson	Brooklyn, N. Y	July 12, 1859	24, 759
Lifting-jack, Suspended	H. C. Havemeyer	New York, N. Y	Oct. 24, 1871	120, 275
Lifting-jacks, Trammel for	E. H. Hako	Canfield, Ohio	Feb. 25, 1873	136, 156
Lifting-machine	T. S. Crane	Newark, N. J	Mar. 26, 1872	124, 885
Lifting-machine	A. Kriebel	Hereford, Pa	Sept. 15, 1868	82, 128
Lifting-machine	C. H. Mann	Orange, N. J	July 5, 1870	104, 973
Lifting-machine	F. B. O'Conner	New York, N. Y	Sept. 5, 1871	118, 740
Lifting-machine	C. A. Simmons	Waldo, Fla	July 25, 1871	117, 339
Lifting-shovel	W. E. Davis	Brooklyn, N. Y	Oct. 27, 1863	40, 398
Light: *See* Artificial light. Billiard-light. Buoy-light. Calcium-light. Deck-light. Drop-light. Electric-light. Entry-light. Fire-light. Foot-light. Gas-light. Head-light. Hydro-oxygen light. Locomotive signal-light. Port-light. Ship-light. Signal-light. Sky-light. Torch-light. Vault-light. Vessel-light.				
Light and air through steps, &c., Admitting	J. B. Cornell	New York, N. Y	June 8, 1858	20, 484
Light and heat, and applying the same, Producing	S. S. Hill	New York, N. Y	Apr. 7, 1863	38, 137
Light and heat, Generating	H. L. Barnum	New York, N. Y	June 2, 1836	....
Light and heat, Method of using explosive liquids for the production of.	W. Beschke	Philadelphia, Pa	Aug. 14, 1866	57, 245
Light from fuel, Mode of obtaining	B. Henfrey	....	Apr. 16, 1802	....
Light-house, Floating	J. B. Stoner	New York, N. Y	Apr. 29, 1873	138, 292
Light-house, Floating	J. B. Stoner	New York, N. Y	Apr. 29, 1873	138, 293
Light-houses, piers, &c., Construction of foundation for.	L. Bail	New Haven, Conn	Feb. 25, 1862	34, 474
Light, Method of producing	P. Carlevaris	Turin, Italy	July 10, 1866	56, 336
Light produced by combination of liquids	I. Jennings	New York	June 13, 1831	....
Light the same as daylight, Means for rendering artificial.	N. H. Gillet	New York, N. Y	July 19, 1864	43, 581
Lighter and alarm	T. N. Howell	Circleville, Ohio	Dec. 31, 1867	72, 854
Lighters, Tool for making	A. Kleinschmidt and F. Schlater.	Philadelphia, Pa	Jan. 23, 1866	52, 173
Lighting and heating apparatus	A. M. Silber and F. White	London, England	Sept. 12, 1871	118, 979
Lighting-apparatus, Catoptric	W. Hodgins	Albany, N. Y	Feb. 2, 1864	41, 432
Lighting-apparatus, Electric	A. P. Berlioz	Paris, France	May 23, 1871	115, 154
Lighting-apparatus, &c., System of flambeau-like	E. A. L. d'Argy	Paris, France	Feb. 18, 1868	74, 513
Lighting factories and other buildings, Mode of	A. J. White	Ballston Spa, N. Y	Sept. 17, 1867	68, 923
Lighting, heating, vaporizing, and drying apparatus.	J. Kidd	New York, N. Y	Oct. 22, 1872	132, 403
Lighting rooms, Method of	D. G. Haskins and J. Winlock	Cambridge, Mass	Nov. 5, 1867	70, 433
Lighting up picture-galleries	E. M. Smith	New York, N. Y	Aug. 4, 1868	80, 570
Lightning-arrester	A. Barbarin	New Orleans, La	Sept. 3, 1867	68, 407
Lightning-conductor	R. L'Anglois	Assumption, La	Oct. 7, 1846	4, 807
Lightning-conductor	N. Brittan	Chicago, Ill	July 19, 1864	43, 565
Lightning-conductor	I. Johnson	Lodi Station, Ill.	Apr. 18, 1865	47, 310
Lightning-conductor	J. A. Kissell and N. Blickensdufer.	Chicago, Ill	May 14, 1867	64, 774
Lightning-conductor	D. Munson	Indianapolis, Ind	Nov. 1, 1864	44, 880
Lightning-conductor	W. G. Pike	Philadelphia, Pa	Nov. 12, 1867	70, 741
Lightning-conductor	O. Preston	South Dansville, N. Y	Oct. 31, 1871	120, 457
Lightning-conductor	O. White	Racine, Wis	Mar. 30, 1858	19, 819
Lightning-conductors, Attachment for	J. Spratt	Cincinnati, Ohio	Feb. 5, 1850	7, 076
Lightning-conductors, Machine for making tubular	B. F. Housel and S. O. Thayer	Winona, Minn	Sept. 6, 1870	107, 051
Lightning-conductors, Making	C. Stearns	Lowell, Mass	Sept. 20, 1859	25, 554
Lightning-conductors, Mode of attaching the receiving and discharging points of.	W. A. Orcutt	Boston, Mass	Oct. 9, 1841	2, 284
Lightning-rod	J. S. Barber	Gloucester, Mass	Mar. 5, 1839	1, 036
Lightning-rod	N. Brittan	Lockport, N. Y	Nov. 6, 1860	30, 560
Lightning-rod	E. Calender	Boston, Mass	Oct. 3, 1808	....
Lightning-rod	H. B. Conner and J. Denton	Pittsburgh, Pa	Oct. 2, 1866	58, 382
Lightning-rod	S. D. Cushman	New Lisbon, Ohio	Feb. 13, 1866	52, 541
Lightning-rod	S. D. Cushman	New Lisbon, Ohio	Mar. 7, 1871	112, 426
Lightning-rod	J. Drew	Wenona, Ill	Dec. 23, 1873	145, 851
Lightning-rod	H. W. Farley	Oswego, Ill	June 22, 1869	91, 530
Lightning-rod	D. A. Foot and A. Chadwick	Winona, Minn	Aug. 10, 1869	93, 609
Lightning-rod	D. A. Foot and G. S. Knapp	Winona, Minn	Mar. 28, 1870	113, 154
Lightning-rod	J. R. Fricke	Pittsburgh, Pa	Jan. 17, 1871	111, 052
Lightning-rod	J. R. Fricke	Pittsburgh, Pa	Feb. 28, 1871	112, 137
Lightning-rod	A. A. Gaylord	East Cleveland, Ohio	Apr. 18, 1871	113, 869
Lightning-rod	F. O. Goodwin	Philadelphia, Pa	Mar. 5, 1872	124, 264

Index of patents issued from the United States Patent Office from 1790 *to* 1873, *inclusive*—Continued.

Invention.	Inventor.	Residence.	Date.	No.
Lightning-rod	W. Hall	Indianapolis, Ind	Oct. 18, 1859	25, 823
Lightning-rod	W. Hall	Dubuque, Iowa	May 12, 1868	77, 729
Lightning-rod	J. W. Hankenson	Minneapolis, Minn	July 2, 1872	128, 617
Lightning-rod	L. J. Hawley	Baltimore, Md	Feb. 6, 1866	52, 411
Lightning-rod	H. H. Homan	Cincinnati, Ohio	Sept. 14, 1852	9, 260
Lightning-rod	C. Jillson	Worcester, Mass	Feb. 5, 1867	61, 741
Lightning-rod	L. King	East Cleveland, Ohio	Apr. 4, 1871	113, 530
Lightning-rod	T. T. Kinsey	Philadelphia, Pa	June 8, 1869	90, 949
Lightning-rod	G. Kirtland	New Haven, Conn	June 2, 1868	78, 459
Lightning-rod	A. Lyon	Worcester, Mass	July 11, 1854	11, 261
Lightning-rod	S. H. Miner	Winona, Minn	Dec. 12, 1871	121, 884
Lightning-rod	S. J. Mitchell	Saint Louis, Mo	May 23, 1865	47, 846
Lightning-rod	J. M. Mott	Chicago, Ill	May 21, 1872	127, 094
Lightning-rod	W. B. Munn	Cleveland, Ohio	Feb. 4, 1873	135, 574
Lightning-rod	D. Munson	Indianapolis, Ind	Aug. 5, 1856	15, 491
Lightning-rod	D. Munson	Indianapolis, Ind	Feb. 11, 1868	74, 406
Lightning-rod	D. Munson	Indianapolis, Ind	Nov. 17, 1868	84, 210
Lightning-rod	D. Munson	Indianapolis, Ind	Feb. 1, 1870	99, 461
Lightning-rod	D. Munson	Indianapolis, Ind	Mar. 8, 1870	100, 549
Lightning-rod	D. Munson	Indianapolis, Ind	Jan. 3, 1871	110, 778
Lightning-rod	D. Munson	Indianapolis, Ind	Sept. 19, 1871	119, 043
Lightning-rod	D. Munson	Indianapolis, Ind	July 23, 1872	129, 675
Lightning-rod	D. Munson	Indianapolis, Ind	July 23, 1872	129, 676
Lightning-rod	D. Munson	Indianapolis, Ind	July 23, 1872	129, 677
Lightning-rod	G. W. Otis	Lynn, Mass	July 21, 1868	80, 205
Lightning-rod	N. Otis	Charles City, Iowa	July 13, 1869	92, 638
Lightning-rod	M. D. Phelps	Bristolville, Ohio	Dec. 19, 1871	122, 055
Lightning-rod	J. Pratt	Chicago, Ill	Dec. 27, 1864	45, 632
Lightning-rod	W. S. Reyburn and F. J. Martin.	Philadelphia, Pa	Sept. 14, 1869	94, 773
Lightning-rod	W. S. Reyburn and F. J. Martin.	Philadelphia, Pa	Oct. 26, 1869	96, 268
Lightning-rod	J. Robertson	New York, N. Y	July 23, 1872	129, 683
Lightning-rod	G. Row	Indiana, Pa	Nov. 22, 1870	109, 455
Lightning-rod	A. S. Sherwood	Detroit, Mich	Jan. 30, 1866	52, 329
Lightning-rod	J. C. Schoonmaker	Winona, Minn	July 9, 1872	128, 818
Lightning-rod	C. P. Snow	Freeport, Ill	Dec. 26, 1871	122, 290
Lightning-rod	J. Spratt	Cincinnati, Ohio	May 4, 1852	8, 930
Lightning-rod	C. Stearns	Lowell, Mass	June 11, 1867	65, 775
Lightning-rod	G. A. Stephenson and O. L. Sutliff.	Wooster, Ohio	Aug. 15, 1871	118, 166
Lightning-rod	L. D. Vermilye, W. S. Reyburn, and E. A. W. Hunter.	Dayton, Ohio, and Philadelphia, Pa.	July 13, 1869	92, 551
Lightning-rod	W. A. Wells and G. House	Saint Paul, Minn., and Gallipolis, Ohio.	Feb. 13, 1872	123, 600
Lightning-rod	D. F. Welsh	Nevada, Ohio	Feb. 20, 1872	123, 958
Lightning-rod	J. D. West	New York, N. Y	Feb. 1, 1870	99, 381
Lightning-rod	J. J. White	Juliustown, N. J	Oct. 21, 1873	143, 862
Lightning-rod	D. Wooster	Seymour, Conn	Sept. 4, 1860	29, 933
Lightning-rod	E. L. Yancey	Batavia, N. Y	Oct. 29, 1872	132, 615
Lightning rod and conductor	T. Garlick	Cleveland, Ohio	Dec. 7, 1869	97, 490
Lightning rod and conductor	J. W. Hankenson	Minneapolis, Minn	Jan. 25, 1870	99, 184
Lightning-rod attachment	R. D. Dwyer	Richmond, Va	July 3, 1855	13, 162
Lightning rod bracket, Pointed	M. Wells	New York, N. Y	July 7, 1868	79, 618
Lightning-rod clasp	R. Street	Albany, N. Y	Aug. 16, 1870	106, 426
Lightning-rod connection	A. Codington	Bound Brook, N. J	Sept. 26, 1871	119, 323
Lightning-rod, Corrugated	J. A. Kissell and N. Blickensderfer.	Chicago, Ill	July 16, 1867	66, 854
Lightning-rod coupling	D. W. Demorest	Newark, N. J	Oct. 24, 1871	120, 251
Lightning-rod coupling	W. S. Reyburn and E. A. W. Hunter.	Philadelphia, Pa	May 25, 1869	90, 578
Lightning-rod coupling, Tubular	J. W. Fritsch	Cincinnati, Ohio	Dec. 16, 1873	145, 641
Lightning-rod coupling, Tubular	J. H. Weston	Cincinnati, Ohio	June 10, 1873	139, 841
Lightning-rod for vessels	R. B. Forbes	Boston, Mass	July 4, 1854	11, 217
Lightning-rod joint	L. King	East Cleveland, Ohio	Aug. 29, 1865	49, 633
Lightning-rod joint	J. B. Lyon	Cleveland, Ohio	Oct. 10, 1865	50, 372
Lightning-rod joint	N. E. Smith	Cleveland, Ohio	Oct. 10, 1865	50, 398
Lightning-rod point	J. F. Boynton	Syracuse, N. Y	Oct. 26, 1869	96, 194
Lightning-rods, Connecting	J. E. Strong	Boston, Mass	Apr. 19, 1841	2, 056
Lightning-rods, Construction of	L. S. Baldwin and L. Parks	Le Roy, N. Y	Aug. 9, 1859	25, 077
Lightning-rods, Construction of	J. M. Patterson	Woodbury, N. J	July 31, 1860	29, 398
Lightning-rods, Device for attaching	J. B. Elliott	Philadelphia, Pa	Dec. 29, 1857	18, 963
Lightning-rods, Device for securing	J. A. Enggren	Brooklyn, N. Y	May 24, 1859	24, 110
Lightning-rods, Device for securing	V. Schrage	Cincinnati, Ohio	July 13, 1858	20, 916
Lightning-rods, Machine for joining tubular	G. S. Knapp	Winona, Minn	June 27, 1871	116, 453
Lightning-rods, Machine for making	G. S. Knapp	Winona, Minn	Oct. 17, 1871	119, 987
Lightning-rods, Method of insulating and supporting.	M. Fox	Stamford, Conn	Apr. 24, 1860	27, 977
Lightning-rods, Method of insulating and supporting.	E. C. Rogers	Boston, Mass	Oct. 26, 1858	21, 905
Lightning-rods, Mode of constructing	G. B. Hamilton and J. S. Stevenes.	Wellington and Cleveland, Ohio.	Jan. 26, 1869	86, 151
Lightning-rods to houses, Mode of securing	J. Brown and G. W. Robinson	Providence, R. I	Sept. 24, 1825	....
Lightning-rods with sheet metal, Machine for covering.	W. S. Reyburn and F. J. Martin.	Philadelphia, Pa	Aug. 17, 1869	93, 834
Limb, Artificial	R. Briody	Detroit, Mich	May 19, 1868	78, 048
Limb, Artificial	J. Coombs	Greenfield, Mass	Aug. 8, 1865	49, 234
Limb, Artificial	R. H. Dutton	Philadelphia, Pa	Apr. 12, 1859	23, 559
Limb, Artificial	S. G. Gregory	Albany, N. Y	Feb. 2, 1869	86, 530
Limb, Artificial	S. B. Kepperling and T. B. Kreiter.	Neffsville, Pa	May 23, 1871	115, 065
Limb, Artificial	H. A. Kimball	Philadelphia, Pa	Aug. 18, 1863	39, 578
Limb, Artificial	G. C. Kirschmann	New York, N. Y	June 7, 1864	43, 031
Limb, Artificial	A. A. Marks	New York, N. Y	Dec. 1, 1863	40, 763
Limb, Artificial	A. McOmber	Schenectady, N. Y	July 10, 1866	56, 243
Limb, Artificial	A. Mennel	New York, N. Y	Dec. 20, 1864	45, 511
Limb, Artificial	J. Monroe	New York, N. Y	Sept. 25, 1866	58, 351
Limb, Artificial	J. Reichenbach	Pittsburgh, Pa	July 9, 1872	128, 907
Limb, Artificial	J. W. Weston	New York, N. Y	Feb. 2, 1864	41, 456

Index of patents issued from the United States Patent Office from 1790 *to* 1873, *inclusive*—Continued.

Invention.	Inventor.	Residence.	Date.	No.
Limb-extension, Counter	S. H. Whittlesey	Central Mine, Mich	Oct. 22, 1872	132, 509
Limb-supporter	D. H. B. Allen	Chelsea, Vt	Oct. 4, 1864	44, 499
Limbs, Apparatus for fractured	J. M. Pitts	Sumpter, S. C	Nov. 6, 1860	30, 588
Limbs, Apparatus for supporting and ventilating wounded.	G. S. Towles	New Castle, Me	Oct. 6, 1863	40, 164
Limbs, Clasp for artificial	S. F. Burd and J. F. Denniston	Mercer and Pittsburgh, Pa	Mar. 14, 1871	112, 683
Limbs, Lining for artificial	J. W. Weston	New York, N. Y	Mar. 13, 1866	53, 206
Limbs, Screw for extending crooked or setting fractured.	J. Merrill	Glasgow, Ky	June 22, 1832	....
Limbs, Ventilating apparatus for wounded	T. C. Ball	Springfield, Vt	Aug. 4, 1863	39, 453
Lime, Acetate of	C. J. T. Burcey	Black Rock, Conn	Sept. 12, 1871	118, 787
Lime, and apparatus therefor, Protecting unslacked	H. Bisbing	Bridgeton, N. J	Sept. 5, 1871	118, 680
Lime and brick and boiling-kettles, Burning	S. Hill	New Milford, Conn	Feb. 12, 1827	....
Lime and guano spreader	W. Croasdale	Hartsville, Pa	June 24, 1856	15, 171
Lime and manure, Spreading	L. Cooper	Coopersville, Pa	Oct. 19, 1852	9, 339
Lime and marl ashes on land and exchanging soils, Machine for spreading.	D. T. Hill	Plainfield, N. J	Feb. 10, 1838	598
Lime and mortar, Preparation of	L. B. Pitcher	Salina, N. Y	Aug. 9, 1870	106, 282
Lime and other fertilizers, Machine for spreading	P. Seymour	East Bloomfield, N. Y	Dec. 1, 1857	18, 774
Lime and sand for mortar, Machine for mixing	H. W. Hunt and J. Sands	Peekskill, N. Y., and Greenwich, Conn.	Apr. 8, 1856	14, 613
Lime and yeast-powder, Manufacture of phosphate of.	E. N. Horsford	Cambridge, Mass	Aug. 6, 1872	130, 298
Lime, Apparatus for making acid sulphite of	L. Gamotis and S. Martin	New Orleans, La	July 21, 1857	17, 830
Lime, Apparatus for preparing the oxymuriate of	J. Kendall	Lincoln, Mass	Jan. 28, 1820	....
Lime, ashes, &c., Machine for spreading	F. H. Smith	Baltimore, Md	July 5, 1837	258
Lime-bin	J. Smith	Fairton, N. J	Dec. 10, 1872	133, 729
Lime, Burning	S. Garber and H. Swartzengrover.	Norristown, Pa	Mar. 25, 1837	152
Lime, Burning	P. Lossing	....	Aug. 4, 1800	....
Lime, Burning	E. A. Smith	Saint Albans, Vt	Apr. 19, 1864	42, 413
Lime-burning and ore-smelting furnace	J. Owings	Adams County, Pa	May 29, 1835	....
Lime-burning furnace	T. R. Hartell	Philadelphia, Pa	Dec. 28, 1858	22, 424
Lime, dirt, and sediment from boiling liquid, Device for removing.	A. J. Simmons	Indianapolis, Ind	May 30, 1871	115, 367
Lime from pyroligneous acid, Manufacture of acetate of.	J. Bell	Dover, N. H	May 9, 1871	114, 517
Lime from water, Apparatus for heating and separating.	S. F. Jackson and J. A. Davis	Eureka, Ill	Mar. 30, 1869	88, 484
Lime-kiln	A. G. Anderson	Quincy, Ill	Nov. 3, 1857	18, 531
Lime-kiln	G. Atkins	Sharon, Pa	May 21, 1867	64, 827
Lime-kiln	G. Atkins	Sharon, Pa	Feb. 23, 1869	87, 131
Lime-kiln	L. Averill	Elmira, N. Y	Aug. 19, 1856	15, 549
Lime-kiln	D. T. Barrett	Port Byron, Ill	Jan. 9, 1872	122, 550
Lime-kiln	L. Bomeisler	Philadelphia, Pa	Sept. 10, 1829	....
Lime-kiln	S. Brown	Berwick, Pa	Aug. 26, 1851	8, 319
Lime-kiln	G. W. Calkins and H. White	Cleveland, Ohio	June 15, 1858	20, 549
Lime-kiln	J. Clarkson and G. W. Decker	Washington, D. C	May 3, 1870	102, 496
Lime-kiln	C. B. Corey	Cleveland, Ohio	Dec. 17, 1872	133, 925
Lime-kiln	J. Coryeon	Bay City, Mich	Dec. 31, 1872	134, 421
Lime-kiln	U. Cummings	Buffalo, N. Y	Nov. 20, 1866	59, 760
Lime-kiln	N. Davis	Ida, Mich	July 16, 1872	129, 539
Lime-kiln	R. Donaldson	Mount Nebo, Pa	Feb. 19, 1861	31, 445
Lime-kiln	J. M. Downing	Bristol, Pa	Aug. 25, 1831	....
Lime-kiln	E. B. English	Philadelphia, Pa	Feb. 23, 1864	41, 691
Lime-kiln	T. Ennett	Rockford, Ill	June 14, 1870	104, 291
Lime-kiln	H. R. Fell	Texas, Md	Jan. 5, 1858	19, 023
Lime-kiln	P. J. Gerbault Guichard	Saint Berthevin-le-Laval, France.	Aug. 2, 1870	106, 054
Lime-kiln	L. Gibbs	Fremont, Ohio	Aug. 27, 1867	68, 063
Lime-kiln	P. Griscom and C. S. Denn	Baltimore, Md	Nov. 17, 1857	18, 635
Lime-kiln	P. Griscom and J. H. Miller	Baltimore, Md	May 15, 1866	54, 717
Lime-kiln	M. Groh and J. V. Weitz	Cleveland, Ohio	Oct. 31, 1871	120, 378
Lime-kiln	G. Harper	Jenkerstown, Pa	Jan. 22, 1817	....
Lime kiln	G. Hensler	Kankakee, Ill	Oct. 26, 1869	96, 109
Lime-kiln	D. Hills	Richville, N. Y	Aug. 30, 1870	106, 818
Lime-kiln	C. Hinkley	Williamsville, N. Y	June 25, 1867	66, 024
Lime-kiln	S. C. Hotchkiss	Sylvania, Ohio	July 23, 1867	66, 962
Lime-kiln	A. Jeffries	Allegheny County, Pa	Apr. 21, 1857	17, 098
Lime-kiln	D. Keeports	Logansport, Ind	Dec. 24, 1872	134, 292
Lime-kiln	T. A. Kirk	Kansas City, Mo	Mar. 7, 1871	112, 468
Lime-kiln	J. L. Livingston	Mount Carroll, Ill	Jan. 16, 1866	52, 061
Lime-kiln	J. L. Livingston	Mount Carroll, Ill	Oct. 29, 1867	70, 234
Lime-kiln	J. McAdams	Mount Pleasant, Pa	Sept. 10, 1872	131, 290
Lime-kiln	W. McCoy	Fannettsburgh, Pa	Nov. 5, 1850	7, 757
Lime-kiln	J. McGregor	Selma, Ala	July 28, 1857	17, 885
Lime-kiln	J. Q. Merriam and A. J. Dietrick.	Fort Scott, Kans	Oct. 24, 1871	120, 307
Lime-kiln	J. L. Messer	Troy, N. Y	Sept. 16, 1873	142, 805
Lime-kiln	L. Montgomery	Newstead, N. Y	Aug. 13, 1867	67, 667
Lime-kiln	L. Montgomery	Akron, N. Y	Nov. 11, 1873	144, 555
Lime-kiln	R. Neisch	New York, N. Y	Aug. 15, 1854	11, 529
Lime-kiln	J. Newkirk	Factoryville, N. Y	Dec. 1, 1857	18, 764
Lime-kiln	E. Orcutt	Bennington, Vt	Sept. 26, 1848	5, 808
Lime-kiln	C. D. Page	Rochester, N. Y	Feb. 7, 1854	10, 510
Lime-kiln	C. D. Page	Rochester, N. Y	July 28, 1857	17, 889
Lime-kiln	C. D. Page	Rochester, N. Y	Dec. 7, 1858	22, 239
Lime-kiln	C. D. Page	Grand Rapids, Mich	Apr. 26, 1864	42, 501
Lime-kiln	C. D. Page	Rochester, N. Y	June 19, 1866	55, 699
Lime-kiln	C. D. Page	Rochester, N. Y	Sept. 8, 1868	82, 023
Lime-kiln	W. C. Pettijohn	Saint Louis, Mo	Sept. 15, 1868	82, 242
Lime-kiln	H. Petton	Erie, Pa	May 22, 1866	54, 943
Lime-kiln	D. H. Pequea	Lancaster, Pa	Aug. 28, 1855	13, 494
Lime-kiln	L. Phleger	Philadelphia, Pa	July 14, 1857	17, 807
Lime-kiln	W. W. Potts	Bridgeport, Pa	Feb. 4, 1868	73, 998
Lime-kiln	T. Power	....	July 12, 1804	....
Lime-kiln	T. Power	Hudson, N. Y	Jan. 7, 1817	....
Lime-kiln	T. H. Powers	Sodus, N. Y	Apr. 17, 1833	....

Index of patents issued from the United States Patent Office from 1790 *to* 1873, *inclusive*—Continued.

Invention.	Inventor.	Residence.	Date.	No.
Lime-kiln	F. W. Raeder	Saint Louis, Mo	Jan. 21, 1873	135, 045
Lime-kiln	E. Randall	Mason City, Iowa	July 30, 1872	129, 985
Lime-kiln	S. H. Robinson	Baltimore, Md	Jan. 16, 1855	12, 242
Lime-kiln	W. Robinson	Baltimore, Md	Apr. 14, 1857	17, 056
Lime-kiln	J. Sands	Sands' Mills, N. Y	Mar. 11, 1856	14, 411
Lime-kiln	R. E. Schroeder	Rochester, N. Y	May 6, 1851	8, 079
Lime-kiln	S. I. Seely	New York, N. Y	May 17, 1853	9, 736
Lime-kiln	F. Shelly	Alton, Ill	Jan. 18, 1870	99, 012
Lime-kiln	R. A. Smith	Atchison, Kans	Mar. 22, 1870	101, 173
Lime-kiln	J. B. Speed	Louisville, Ky	Nov. 23, 1869	97, 243
Lime-kiln	D. Stephens	Elmira, N. Y	Aug. 11, 1857	17, 986
Lime-kiln	W. P. Stuart	Hannibal, Mo	July 16, 1872	129, 617
Lime-kiln	J. Swope	Lancaster, Pa	Mar. 14, 1817	....
Lime-kiln	D. H. Turbett	New Bloomfield, Pa	July 18, 1871	117, 222
Lime-kiln	A. H. Tyson	Baltimore, Md	Sept. 28, 1839	1, 343
Lime-kiln	M. Verhoeven	Rochester, N. Y	Apr. 7, 1868	76, 563
Lime-kiln	A. B. Weeks	Rockland, Me	Mar. 2, 1858	19, 525
Lime-kiln	G. W. White	Greensburgh, Ind	Jan. 22, 1867	61, 495
Lime-kiln	B. Zwart	Keokuk, Iowa	Apr. 20, 1858	20, 015
Lime-kiln arch	A. McBride	Lowell, Mich	Nov. 21, 1871	121, 183
Lime-kiln blast-attachment	W. Hollenbaugh	New Germantown, Pa	Oct. 24, 1871	120, 278
Lime-kiln chimney	W. Rennyson	Morristown, Pa	Feb. 14, 1871	111, 874
Lime-kiln dome	D. Wentzel	New Bloomfield, Pa	Dec. 31, 1872	134, 501
Lime-kiln for anthracite coal	S. Griscom	Reading, Pa	June 26, 1832	....
Lime-kiln, Perpetual	W. Gorsuch	Baltimore, Md	Aug. 17, 1811	....
Lime-kiln, Perpetual	A. Jeans	Mill Creek Hundred, Del	Feb. 15, 1827	....
Lime-kilns, Construction of	J. H. Bower	Walnut, Pa	Sept. 4, 1874	5, 270
Lime-kilns, Mode of regulating draft of	I. Richardson	Paoli, Pa	Feb. 21, 1840	1, 496
Lime, Manufacture and application of bisulphate of.	W. Marr	New York, N. Y	Nov. 5, 1867	70, 588
Lime, Manufacture and use of neutral sulphate of.	E. N. Horsford	Cambridge, Mass	Sept. 15, 1863	39, 922
Lime, Manufacture of acid sulphite of	P. L. Bernard and J. Albrecht	New Orleans, La	Oct. 9, 1855	13, 632
Lime, Manufacture of superphosphate of	A. De Figaniére	Philadelphia, Pa	Nov. 27, 1866	59, 978
Lime, Manufacture of superphosphate of	C. Morfit	Sudbrook Park, England	Jan. 31, 1871	111, 370
Lime, Manufacture of superphosphate of	B. Tanner	New Brighton, England	Feb. 13, 1872	123, 744
Lime, Manufacture of superphosphate of	B. Tanner	New Brighton, England	July 1, 1873	140, 559
Lime, Manufacturing marl into	J. Peasely	....	Feb. 1, 1803	....
Lime, Mode of manufacturing superphosphate of	E. P. Baugh	Philadelphia, Pa	May 9, 1865	47, 610
Lime on land, Spreading	M. L. Wilson	Quakertown, Pa	June 22, 1832	....
Lime-oven	A. Califf	Danville, Ill	Apr. 25, 1871	114, 104
Lime, plaster, &c., spreading and sowing machine	L. Rice	West Chester, Pa	Aug. 31, 1837	372
Lime, Preparing tannate of	O. Rich	Cambridge, Mass	Dec. 9, 1856	16, 189
Lime, Reducing calcareous matter into	C. Varle	....	May 2, 1806	....
Lime-spreader	W. C. Burnett	Burns' Mills, Pa	June 28, 1870	104, 694
Lime-spreader	H. Ludington and S. R. Supton	Addison, Pa	Mar. 13, 1855	12, 508
Limestone, Hardening magnesian	J. McMurty	Mount Sterling, Ky	June 21, 1870	104, 619
Limestone or asphalt, Machine for cutting or screening bituminous.	Q. A. Gillmore	New York, N. Y	Aug. 30, 1859	25, 254
Linch-pin	T. Laly	Philadelphia, Pa	Sept. 20, 1864	44, 321
Linch-pin	G. Wright	Washington, D. C	May 9, 1865	47, 674
Linch-pin and washer	W. Mansfield and H. S. Thistle	South Braintree, Mass., and Washington, D. C.	May 8, 1847	5, 109
Linch-pin, Securing	C. M. Risley	Woodbury, N. J	Aug. 29, 1865	49, 651
Line and strap holder	F. W. Jay	Canton, Ohio	Apr. 2, 1872	125, 136
Line and wire tightener	T. Crosby	Manchester, Iowa	Nov. 1, 1870	108, 766
Line-clamp	W. A. Ford	Greensburgh, Ind	Mar. 26, 1872	125, 038
Line-connector	C. Jarvis	New York, N. Y	Dec. 19, 1871	122, 021
Line-fastener	W. P. W. Dana	Newport, R. I	Feb. 11, 1868	74, 203
Line-holder	J. Bryant	Akron, Ind	Apr. 6, 1869	88, 606
Line-holder	S. J. Clark	Detroit, Mich	Dec. 3, 1867	71, 702
Line-holder	C. S. Hutchins	Canton, Conn	June 15, 1869	91, 340
Line-holder	D. W. C. McMaster	Southborough, Mass	Oct. 13, 1868	83, 075
Line-holder	W. Morse	Boston, Mass	Aug. 27, 1867	68, 304
Line-holder	D. Peters and R. F. Williams	Keokuk, Iowa	Oct. 8, 1867	69, 696
Line-holder	J. D. Sater	Greensburgh, Ind	Nov. 14, 1871	121, 006
Line-holder	J. H. Zinn	Harrisburgh, Pa	June 23, 1868	79, 288
Line-holder and wardrobe-hook, Combined	C. Goddard	Alliance, Ohio	Aug. 22, 1871	118, 361
Line holder for mason-work, &c	L. A. Green	Rocky Hill, Conn	Mar. 11, 1862	34, 621
Line holder, Mackerel	E. L. Decker	Southport, Me	Mar. 7, 1871	112, 326
Line-loop	D. McMillan	Osage, Iowa	Aug. 12, 1873	141, 804
Line-machine, Self-winding	W. A. Coventry	Paterson, N. J	Sept. 21, 1869	95, 002
Line-reel	L. Martin	Indianapolis, Ind	Jan. 7, 1868	73, 019
Lines, Device for rounding	L. B. Gates	Bane Centre, N. Y	Sept. 17, 1867	68, 972
Linen bosoms, Apparatus for folding	H. Beebe	New Haven, Conn	Feb. 27, 1866	52, 817
Linen, &c., Compound for removing mildew from	M. Faurot	Scranton, Pa	Feb. 16, 1869	87, 036
Linen, &c., Glazing and polishing	W. Smith	New York, N. Y	May 9, 1805	....
Linen, Machine for smoothing	W. P. Smith and J. Odell	Durham, N. H	July 17, 1810	....
Linen-smother	H. Rodd	Chestnut Hill, Pa	Apr. 9, 1861	32, 001
Liniment	M. P. Munder	Baltimore, Md	Sept. 2, 1873	142, 499
Liniment	W. H. Burr	Monterey, Ky	May 24, 1870	103, 411
Liniment	J. M. Barrett	Plymouth, N. C	Aug. 31, 1869	94, 379
Liniment	P. Baumann, jr	New Athens, Ill	Aug. 13, 1867	67, 706
Liniment	J. Benda	Yatton, Iowa	Apr. 29, 1873	138, 313
Liniment	J. C. Branch and H. P. Quinn	Washington, Ga	Nov. 8, 1870	108, 963
Liniment	J. W. Burnham	Winterport, Me	Feb. 19, 1867	62, 111
Liniment	E. N. Carpenter	Elkhart, Ind	May 28, 1867	65, 170
Liniment	E. Conway	Dayton, Ohio	Mar. 17, 1863	37, 901
Liniment	L. B. Cram	Weathersfield, Vt	Apr. 23, 1867	63, 995
Liniment	A. J. Creel	Hopkinton, Iowa	May 19, 1868	78, 068
Liniment	J. C. Dustan	Mount Vernon, N. J	May 14, 1867	64, 649
Liniment	D. Edwards	Saint Anthony, Minn	Feb. 20, 1866	52, 697
Liniment	E. C. Evens	Forest Hill, Ind	Nov. 23, 1869	97, 180
Liniment	H. Fedder	Lancaster, N. Y	Sept. 15, 1868	82, 212
Liniment	J. Gifford	Smithport, Pa	Feb. 26, 1867	62, 403
Liniment	D. J. Gish	Hopkinsville, Ky	July 8, 1873	140, 698
Liniment	J. E. Gould	Jamestown, N. Y	Aug. 16, 1864	43, 847
Liniment	J. R. Grows	Brunswick, Me	July 5, 1864	43, 408
Liniment	W. P. Hamlin	Exira, Iowa	May 12, 1868	77, 814

Index of patents issued from the United States Patent Office from 1790 *to* 1873, *inclusive*—Continued.

Invention.	Inventor.	Residence.	Date.	No.
Liniment	O. P. Hare	Petersburgh, Va	June 14, 1870	104, 146
Liniment	H. J. Kintz	Greece, N. Y	Jan. 14, 1873	134, 808
Liniment	J. Knights	Biddeford, Me	Aug. 20, 1872	130, 644
Liniment	J. S. Lightner	Westford, Wis	Mar. 26, 1867	63, 269
Liniment	J. D. Love	Harrisburgh, Oreg	Oct. 18, 1870	108, 371
Liniment	B. Marstella	Wolf Creek, Va	May 9, 1865	47, 653
Liniment	J. H. M. Morris	Reading, Ill	June 13, 1865	48, 198
Liniment	R. M. Moyle	South Manchester, Conn	June 5, 1866	55, 342
Liniment	P. R. Myers	Oxford, Ohio	Mar. 26, 1872	124, 910
Liniment	R. Newcomb	South Brooklyn, N. Y	June 11, 1867	65, 688
Liniment	T. B. Randell	New York, N. Y	Sept. 19, 1871	119, 179
Liniment	W. E. Rose	Hornellsville, N. Y	Apr. 8, 1873	137, 624
Liniment	G. W. Smith	North Whitehall Township, Pa.	Mar. 7, 1865	46, 715
Liniment	H. Stonebraker	Baltimore, Md	Feb. 19, 1867	62, 298
Liniment	W. H. Wagoner	Hurd Post-Office, Pa	May 28, 1872	127, 283
Liniment	W. H. Hallack	Corunna, Ind	July 27, 1869	93, 143
Liniment	W. W. Wells	Freehold, N. J	Apr. 13, 1869	88, 826
Liniment	G. S. Wood	Vassalborough, Me	Feb. 28, 1871	112, 201
Liniment	J. W. Woodring	Greensburgh, Ky	Feb. 20, 1866	52, 778
Liniment and medicine	J. King	Warren County, Ohio	Oct. 12, 1869	95, 695
Liniment, Anti-rheumatic	J. Galette	New York, N. Y	July 9, 1867	66, 482
Liniment, Compound for	M. C. Ross	New York, N. Y	Nov. 17, 1863	40, 642
Liniment for burns, scalds, &c	L. Wolff	Chicago, Ill	Mar. 29, 1870	101, 340
Liniment for cure of foot-rot in sheep	W. H. Lawes	Someville, N. J	Mar. 5, 1867	62, 644
Liniment for horses, &c	D. O'Halloran	New York, N. Y	June 22, 1869	91, 659
Liniment for horses and cattle	H. H. Hammer	Nashville, Tenn	Nov. 15, 1870	109, 203
Liniment for horses and other animals	W. Christman	Philadelphia, Pa	Mar. 24, 1868	75, 729
Liniment for neuralgia, rheumatism, &c	B. Bissell	New London, N. Y	Nov. 30, 1869	97, 345
Liniment for treating neuralgia, &c	G. L. Fearis	Connersville, Ind	June 28, 1870	104, 721
Liniment for treating nervous diseases	J. F. Smith	Forest Hill, Cal	Jan. 11, 1870	98, 809
Liniment, Mammalial	S. Rueger	Kansas City, Mo	Dec. 1, 1868	84, 580
Liniment, Medical vegetable	T. L. Upton	Farmington, W. Va	Apr. 16, 1867	63, 965
Liniment or medical compound	G. H. Fairchild	Saint Louis, Mo	Sept. 5, 1871	118, 706
Liniment or medical compound	D. C. Randall	Niles City, Mich	Dec. 22, 1868	85, 242
Liniment, Rheumatic	L. E. Anderson	Saint Louis, Mo	Nov. 17, 1863	40, 668
Liniment, Rheumatic	A. Brown		Oct. 17, 1804	......
Liniment for rheumatism	J. M. Cantrell	Polk County, Oreg	Mar. 19, 1872	124, 718
Liniment for rheumatism	A. M. Dennen	Folsom City, Cal	Oct. 6, 1868	82, 696
Liniment for rheumatism	S. Galland	Jefferson City, Mo	Nov. 24, 1863	40, 686
Liniment for rheumatism	J. W. Helmes	Bainbridge, Ga	Sept. 19, 1871	119, 144
Liniment for rheumatism	A. A. Riddick	Franklin, Va	Dec. 20, 1870	110, 288
Liniment for rheumatism, &c	G. Shepherd	Philadelphia, Pa	Mar. 29, 1864	42, 115
Liniment for rheumatism	J. A. Schuetz	Saint Louis, Mo	Mar. 6, 1860	27, 389
Liniment, Surgical and flesh-wound	J. S. Stratton	Newfane, Vt	Apr. 9, 1872	125, 628
Link	G. W. Dyer	Crestline, Ohio	Mar. 12, 1872	124, 556
Link	A. Goodhart	Newville, Pa	May 26, 1868	78, 275
Link	J. H. McIntire	Crestline, Ohio	Dec. 12, 1871	121, 725
Link	R. C. Schenck, jr	Dayton, Ohio	Mar. 4, 1873	136, 550
Link, Adjustable	A. J. Dexter	North Foster, R. I	Aug. 2, 1870	105, 921
Link, Adjustable	J. Hinkley	Norwalk, Ohio	Dec. 23, 1862	37, 229
Link, Adjustable	F. Pfeiffer	Giesborough, D. C	Dec. 19, 1865	51, 661
Link, Adjustable	C. J. E. Thompson	Providence, R. I	Aug. 5, 1862	36, 118
Link and hook, Connecting	R. Creuzbaur	New York, N. Y	Apr. 23, 1867	63, 996
Link-boring apparatus	C. Kellogg	Detroit, Mich	Feb. 11, 1868	74, 376
Link, Connecting	R. Creuzbaur	Newark, N. J	July 2, 1867	66, 303
Link, Connecting	R. Creuzbaur	New York, N. Y	Oct. 29, 1867	70, 172
Link, Connecting	W. N. Pelton	New London, Conn	Apr. 7, 1868	76, 517
Link-motion engines, Reversing	T. C. Crowley	Parker Township, Pa	Feb. 4, 1873	135, 410
Link-motion for operating valves	W. Knowles	Boston, Mass	Apr. 12, 1864	42, 297
Link-motion for steam-engines	J. F. Allen	New York, N. Y	Apr. 29, 1862	35, 070
Link-motion for steam-engines	N. V. Stevens	Lowell, Mass	Feb. 20, 1866	52, 766
Link, Snap	C. W. Saladee	Newark, Ohio	Sept. 26, 1865	50, 167
Link, Snap	C. W. Saladee	Newark, Ohio	Nov. 21, 1865	51, 088
Link-welding die	F. Van Patten and O. A. Anthony.	Ilion, N. Y	June 12, 1866	55, 570
Linseed, &c., Apparatus for triturating and heating	T. Rowe	Brooklyn, N. Y	Mar. 21, 1865	56, 945
Linseed, &c., Machine for crushing	T. Rowe	New York, N. Y	Oct. 28, 1862	36, 799
Lint and basket holder	W. Richardson	Baltimore, Md	Mar. 31, 1868	76, 248
Lint for manufacture, Preparing back	A. K. Smede	Lexington, Ky	Oct. 11, 1828	....
Lint, Manufacture of	R. D. Dwyer	New York, N. Y	Aug. 2, 1859	24, 970
Lint-room and cotton-press, Combined	I. H. Albertson	Bedias, Tex	Dec. 20, 1870	110, 327
Lip-shield	E. C. Philbrick	Bath, Me	Nov. 16, 1869	96, 953
Liquid and fluid meter, and in the method of operating valves.	J. C. Horton and S. K. Hawkins	New York and Lansingburgh, N. Y.	Sept. 10, 1867	68, 746
Liquid and gas meter	J. Mason	Paterson, N. J	Sept. 17, 1867	69, 007
Liquid and gas meter	D. B. Spooner	Syracuse, N. Y	Apr. 4, 1871	113, 591
Liquid and gas meter	D. B. Spooner	Syracuse, N. Y	Oct. 22, 1872	132, 497
Liquid and spirit meter	E. S. Hutchinson	Baltimore, Md	Aug. 27, 1867	68, 081
Liquid and spirit meter	W. Murphy	Cork, Ireland	July 3, 1866	56, 156
Liquid-can	A. V. Smith	San Francisco, Cal	July 2, 1872	128, 668
Liquid-concentrating apparatus	C. A. Wood	Dorchester, Mass	Apr. 4, 1865	47, 168
Liquid-cooler	J. L. Baudelot	Harancourt, France	Nov. 1, 1859	25, 992
Liquid-cooler	J. W. Campbell, sr	New York, N. Y	May 16, 1871	114, 916
Liquid-cooler	J. W. Campbell, sr	New York, N. Y	Apr. 2, 1872	125, 268
Liquid-cooler	W. A. Colston	Great Bend, Pa	Dec. 15, 1868	84, 994
Liquid-cooler	A. Fischer	New York, N. Y	May 15, 1866	54, 708
Liquid-cooler	H. C. Fowler	Great Bend, Pa	Mar. 31, 1868	76, 181
Liquid-cooler	M. Gould	New York, N. Y	Oct. 15, 1867	69, 797
Liquid-cooler	C. Greenlee and W. H. Redfield	Belvidere, Ill	May 18, 1869	90, 094
Liquid-cooler	E. and M. A. F. Haas	Mendota, Ill	Aug. 11, 1868	80, 945
Liquid-cooler	V. Hall	New York, N. Y	Sept. 20, 1859	25, 505
Liquid-cooler	P. and F. Kinkel	New York, N. Y	July 25, 1865	48, 941
Liquid-cooler	B. G. Martin	Philadelphia, Pa	Mar. 6, 1866	53, 082
Liquid-cooler	H. Meidinger	Carlsruhe, Germany	Mar. 4, 1873	136, 445
Liquid-cooler	R. Morton	Stockton-on-Tees, England	Mar. 3, 1868	75, 044
Liquid-cooler	J. R. Neil	Walcottville, Conn	July 1, 1873	140, 529
Liquid-cooler	H. Pietsch	New York, N. Y	Dec. 22, 1868	85, 125

Index of patents issued from the United States Patent Office from 1790 to 1873, inclusive—Continued.

Invention.	Inventor.	Residence.	Date.	No.
Liquid-cooler	P. Reutlinger and A. Herbert	Williamsburgh, N. Y	Apr. 17, 1866	54, 016
Liquid-cooler	J. R. B. Schwarze	Washington, D. C	June 26, 1866	55, 962
Liquid-cooler	A. Smith	New York, N. Y	Dec. 22, 1863	41, 024
Liquid-cooler	E. L. Strohecker and A. Iverson, jr.	Macon, Ga	Aug. 6, 1872	130, 331
Liquid-cooler	C. P. Zimmerman	Newark, N. J	Nov. 28, 1865	51, 275
Liquid cooler and filter	D. E. Somes	Washington, D. C	Sept. 29, 1868	82, 651
Liquid-cooler and ice-water stand	G. Giebrich	Burlington, Iowa	Dec. 17, 1872	133, 979
Liquid-cooler for liquids under pressure	E. Bigelow	Springfield, Mass	Oct. 18, 1870	108, 318
Liquid cooling and aerating apparatus	F. Haeck	Brussels, Belgium	Nov. 8, 1864	45, 004
Liquid-cooling coil	W. Gee	New York, N. Y	Oct. 14, 1873	143, 689
Liquid-cooling vessel	H. Waldstein and M. Fausky	New York, N. Y	Nov. 13, 1866	59, 687
Liquid-decanting apparatus	T. Molinier	New Orleans, La	Oct. 19, 1869	95, 924
Liquid-defecating apparatus	J. E. Wilson	Baltimore, Md	June 17, 1862	35, 645
Liquid-diffusing apparatus	C. H. Hudson	Roxbury, Mass	Apr. 28, 1868	77, 286
Liquid drawing and measuring apparatus	R. F. Stevens	Syracuse, N. Y	Mar. 25, 1851	7, 999
Liquid drawing and weighing apparatus	E. S. Harris and S. S. Robinson	Morrison, Ill	June 25, 1867	66, 149
Liquid-drawing apparatus	A. Bain	New York, N. Y	Nov. 15, 1864	45, 013
Liquid-drawing apparatus, Effervescent	T. Warker	New York, N. Y	Nov. 18, 1873	144, 809
Liquid-effervescing bottle	H. Codd	Camberwell, England	Apr. 29, 1873	138, 230
Liquid-elevator	R. Porter	Melrose, Mass	Sept. 3, 1861	33, 226
Liquid-elevator, Pneumatic	J. P. Gruber	New York, N. Y	June 14, 1870	104, 301
Liquid, Evaporating	C. Sholes	Phelps, N. J	July 26, 1826	....
Liquid evaporating and decomposing apparatus	H. Eckstein	Vienna, Austria	Oct. 5, 1869	95, 445
Liquid-evaporating apparatus and process	J. J. Sherman	Albany, N. Y	Feb. 16, 1869	86, 948
Liquid-evaporator	E. Constantine	New York, N. Y	May 8, 1860	28, 158
Liquid-evaporator	J. P. Dake	Salem, Ohio	July 21, 1868	80, 149
Liquid-evaporator	L. C. England	Philadelphia, Pa	May 28, 1867	65, 065
Liquid-evaporator	G. F. Gray	Brooklyn, N. Y	May 31, 1870	103, 602
Liquid-evaporator	H. Hathaway and B. Lathrop	Detroit, Mich., and Tolland, Conn.	Jan. 1, 1861	31, 018
Liquid-evaporator	J. Howarth	Salem, Mass	May 2, 1871	114, 295
Liquid-evaporator	J. J. Johnston	Allegheny City, Pa	Nov. 14, 1865	50, 935
Liquid-evaporator	T. Oxnard	Marseilles, France	May 17, 1864	42, 789
Liquid-evaporator	A. Peck	Manchester, England	Jan. 5, 1864	41, 089
Liquid-evaporator	G. Stephenson	Zionsville, Ind	May 8, 1860	28, 211
Liquid-evaporator	J. Trageser	New York, N. Y	June 11, 1861	32, 542
Liquid-evaporator, Air-blast	G. Clark	Buffalo, N. Y	June 6, 1871	115, 572
Liquid evaporator and concentrator	J. Howarth	Salem, Mass	Mar. 7, 1871	112, 347
Liquid evaporator and concentrator	J. Howarth	Salem, Mass	Mar. 7, 1871	112, 348
Liquid-evaporator for boiling food for stock	A. J. Andrews	Linden, Mich	July 18, 1871	117, 030
Liquid-flask	H. W. Dee	London, England	May 30, 1871	115, 283
Liquid, Forcing	J. Van Norman	Easton, Pa	Mar. 28, 1871	113, 115
Liquid-forcing apparatus	M. E. Ogden	New York, N. Y	Nov. 14, 1871	120, 895
Liquid-freezer	F. P. E. Carré	Paris, France	Oct. 2, 1860	30, 201
Liquid-freezer	S. S. Fitch	New York, N. Y	Oct. 7, 1873	143, 446
Liquid-freezer	F. Sagno	Milan, Italy	May 13, 1873	138, 941
Liquid-gage	J. J. Greenough	Cincinnati, Ohio	Oct. 19, 1858	21, 814
Liquid gage and hydrometer, Combined	E. T. Bussell and J. Smith	New York, N. Y	June 7, 1870	103, 976
Liquid, gas, &c., meter	S. R. P. Camp	Syracuse, N. Y	Sept. 9, 1873	142, 336
Liquids heater and evaporator, Steam	B. T. Babbitt	New York, N. Y	Mar. 12, 1872	124, 530
Liquid heater, Portable	W. N. Hancock	Salem, N. J	Mar. 7, 1865	46, 665
Liquid-heating apparatus	D. H. Burrell	Little Falls, N. Y	Feb. 28, 1871	112, 120
Liquid-heating boiler	W. Nims	Fort Ann, N. Y	Oct. 24, 1846	4, 826
Liquid-heating machine	S. Bolton	Philadelphia, Pa	Jan. 16, 1809	....
Liquid-heating machine	R. Volentine	Canaan, N. Y	June 13, 1812	....
Liquid-measuring apparatus	E. K. Dutton	Manchester, England	Nov. 20, 1866	59, 761
Liquid-meter	R. H. Atwell	Baltimore, Md	Sept. 17, 1867	68, 830
Liquid-meter	R. H. Atwell	Baltimore, Md	Mar. 24, 1868	75, 721
Liquid-meter	P. Ball and B. Fitts	Worcester, Mass	Nov. 22, 1870	109, 372
Liquid-meter	J. S. Barden	Providence, R. I	Apr. 2, 1872	125, 114
Liquid-meter	N. L. Blanchard	Spuyten Duyvil, N. Y	May 3, 1870	102, 650
Liquid-meter	J. A. Bradshaw and W. H. Brown.	Lowell, Mass	Sept. 28, 1869	95, 189
Liquid-meter	F. Buttner	New York, N. Y	Mar. 28, 1871	113, 017
Liquid-meter	H. Chandler	Buffalo, N. Y	Sept. 19, 1871	119, 076
Liquid-meter	H. Chandler	Buffalo, N. Y	May 14, 1872	126, 623
Liquid-meter	W. F. Class and W. Napp	Cleveland, Ohio	Jan. 11, 1870	98, 667
Liquid-meter	I. Cook	Saint Louis, Mo	June 4, 1872	127, 575
Liquid-meter	T. Cook	New York, N. Y	May 2, 1871	114, 411
Liquid-meter	J. C. Cooke	Middletown, Conn	Nov. 5, 1861	33, 675
Liquid-meter	J. W. Cremin	New York, N. Y	Dec. 26, 1871	122, 231
Liquid-meter	R. Creuzbaur	Brooklyn, E. D., N. Y	June 7, 1870	103, 989
Liquid-meter	R. Creuzbaur	Brooklyn, E. D., N. Y	June 7, 1870	103, 990
Liquid-meter	T. A. Curtis	Springfield, Mass	Jan. 16, 1872	122, 759
Liquid-meter	T. A. Curtis and J. S. Curtis	Springfield, Mass., and Hartford, Conn.	Dec. 2, 1873	145, 096
Liquid-meter	J. F. De Navarro and H. C. Sergeant.	New York, N. Y., and Newark, N. J.	May 2, 1871	114, 420
Liquid-meter	J. F. De Navarro and H. C. Sergeant.	New York, N. Y	Apr. 16, 1872	125, 794
Liquid-meter	G. W. Devoe	New York, N. Y	Feb. 9, 1869	86, 823
Liquid-meter	E. M. Du Boys	Paris, France	Oct. 6, 1868	82, 812
Liquid-meter	N. Finck	Elizabeth, N. J	Oct. 10, 1871	119, 699
Liquid-meter	W. Fischer	Essen, Prussia	Sept. 27, 1870	107, 769
Liquid-meter	W. H. Fruen	Boston, Mass	Sept. 20, 1870	107, 607
Liquid-meter	C. A. and G. W. Geissenhainer	Pittsburgh, Pa	Sept. 15, 1868	82, 105
Liquid-meter	O. Gilmore	Raynham, Mass	Mar. 16, 1869	87, 837
Liquid-meter	T. C. Hargrave	Boston, Mass	Mar. 22, 1870	101, 125
Liquid-meter	D. W. Huntington and W. A. Hempstead.	South Coventry, Conn	Dec. 26, 1871	122, 174
Liquid-meter	O. G. Leopold	Cincinnati, Ohio	Apr. 7, 1857	16, 983
Liquid-meter	J. Mason	Paterson, N. J	Sept. 15, 1868	82, 138
Liquid-meter	G. B. Massey	New York, N. Y	Feb. 23, 1869	87, 183
Liquid-meter	G. B. Massey	New York, N. Y	Nov. 1, 1870	108, 804
Liquid-meter	H. W. Mather	New York, N. Y	Sept. 12, 1871	118, 867

Index of patents issued from the United States Patent Office from 1790 to 1873, inclusive—Continued.

Invention.	Inventor.	Residence.	Date.	No.
Liquid-meter	H. S. Maxim	Brooklyn, N. Y	June 11, 1872	127, 907
Liquid-meter	J. Mead.	Charlestown, Mass	July 18, 1871	117, 094
Liquid-meter	J. Minor and N. W. and G. W. Nesmith.	Peoria and Metamora, Ill	Mar. 16, 1869	87, 959
Liquid-meter	C. Moore	New York, N. Y	Dec. 13, 1870	110, 062
Liquid-meter	C. Moore	New York, N. Y	Jan. 24, 1871	111, 234
Liquid-meter	F. A. Norley	Syracuse, N. Y	Apr. 8, 1873	137, 559
Liquid-meter	H. B. Nickerson	Boston, Mass	July 11, 1871	116, 859
Liquid-meter	C. Nida	Greenville, N. J	Aug. 31, 1869	93, 235
Liquid-meter	A. K. Rider	Elizabeth City, N. J	Jan. 4, 1870	98, 634
Liquid-meter	C. H. Riggs	Warwick, N. Y	Aug. 25, 1868	81, 408
Liquid-meter	W. Robjohn	New York, N. Y	July 2, 1867	66, 395
Liquid-meter	T. A. Searle	Providence, R. I	Mar. 2, 1869	87, 516
Liquid-meter	H. C. Sergeant	Newark, N. J	Aug. 30, 1870	106, 878
Liquid-meter	H. C. Sergeant	Newark, N. J	Jan. 24, 1871	111, 262
Liquid-meter	H. C. Sergeant	Newark, N. J	May 2, 1871	114, 480
Liquid-meter	H. C. Sergeant	New York, N. Y	Apr. 16, 1872	125, 848
Liquid-meter	J. H. Shedd	Waltham, Mass	Mar. 23, 1869	88, 259
Liquid-meter	G. Sickels	Boston, Mass	Sept. 24, 1872	131, 633
Liquid-meter	J. P. Smith	Cleveland, Ohio	June 15, 1869	91, 374
Liquid-meter	J. P. Smith	Buffalo, N. Y	July 12, 1870	105, 379
Liquid-meter	W. E. Snediker	New York, N. Y	Nov. 1, 1870	108, 842
Liquid-meter	W. E. Snediker	New York, N. Y	Jan. 24, 1871	111, 268
Liquid-meter	W. E. Snediker	New York, N. Y	Feb. 6, 1872	123, 521
Liquid-meter	D. B. Spooner	Syracuse, N. Y	Nov. 21, 1871	121, 211
Liquid-meter	W. M. Storm	New York, N. Y	May 7, 1867	64, 457
Liquid-meter	W. G. Stuart	Chicopee, Mass	Feb. 21, 1871	112, 091
Liquid-meter	J. Sutherland	Brooklyn, N. Y	June 2, 1868	78, 551
Liquid-meter	I. P. Tice	New York, N. Y	Apr. 2, 1872	125, 152
Liquid-meter	S. C. Treat	Tabor, Iowa	May 10, 1870	102, 883
Liquid-meter	F. Wagner	New York, N. Y	Jan. 25, 1870	99, 266
Liquid-meter	A. Werckmeister	Berlin, Prussia	Nov. 22, 1870	109, 564
Liquid-meter	A. M. White	Bridgeport, Conn	Mar. 11, 1873	136, 799
Liquid-meter	D. Williamson	New York, N. Y	July 18, 1871	117, 229
Liquid-meter	J. Winsborrow	Dalston, England	Nov. 16, 1869	96, 858
Liquid-meter	R. W. Young	Richmond, Va	Aug. 20, 1867	68, 025
Liquid-meter, Piston	R. Creuzbaur	Brooklyn, N. Y	Dec. 28, 1869	98, 353
Liquid-meters, Valve arrangement for	J. Johnson	New York, N. Y	Sept. 26, 1871	119, 364
Liquid purifying and rectifying apparatus	T. H. Sinclaire	New York, N. Y	Apr. 27, 1869	89, 353
Liquid receptacle and funnel	W. H. Mumber	Boston, Mass	Apr. 18, 1871	113, 786
Liquid-removing press	J. Waters	West Sutton, Mass	Oct. 25, 1870	108, 657
Liquid-sampler	A. Barbarin	New Orleans, La	Oct. 13, 1868	83, 024
Liquid-separator	J. J. Miller	Chicago, Ill	Aug. 23, 1864	43, 951
Liquid-still separator	G. W. Shaw	Buffalo, N. Y	Feb. 22, 1870	100, 196
Liquids and other substances, Mode of cooling and freezing.	A. H. Tait	New York, N. Y	Aug. 31, 1869	94, 450
Liquids, Boiler for evaporating	J. Moffat		Feb. 1, 1803	....
Liquids by compressed air, Device for drawing	A. L. Webster	New York, N. Y	Aug. 2, 1870	106, 008
Liquids by steam, Heating	J. B. Armstrong	Wilkes County, Ga	May 29, 1833	....
Liquids by the expansion and condensation of gas, Apparatus for forcing.	J. S. Baldwin	Newark, N. J	Dec. 5, 1871	121, 482
Liquids, Can for holding	L. Jennings	Erving, Mass	Oct. 10, 1854	11, 789
Liquids, Conveying	T. Donelly and J. H. and A. Anchors.	Pittsburgh and Bear Creek Station, Pa.	June 14, 1870	104, 284
Liquids, Decalcifying	F. Garcia	New Orleans, La	Mar. 5, 1861	31, 650
Liquids, Device for mixing	G. Watkins	Brooklyn, N. Y	Sept. 17, 1867	69, 054
Liquids, Evaporating	W. Henry	La Porte, Ind	Jan. 8, 1839	1, 061
Liquids for aiding digestion, Process for preparing	J. J. Sherman	Albany, N. Y	May 8, 1855	12, 834
Liquids for heating by means of pipes, &c	L. Grimm and J. Corvin	Magdeburg, Prussia	Apr. 22, 1873	138, 082
Liquids from boiling over the sides of vessels, Device for preventing.	J. Liblong	Waterbury, Conn	Aug. 26, 1856	15, 633
Liquids from paints and other solid substances, Machine for separating.	D. M. Weston	Boston, Mass	Feb. 4, 1868	74, 021
Liquids from solids, Method of expressing	D. A. James	Cincinnati, Ohio	Apr. 28, 1868	77, 194
Liquids on draft, Apparatus for cooling	J. Link	United States Army	Dec. 15, 1868	84, 888
Liquids on draft, Apparatus for cooling	E. Whyte	Philadelphia, Pa	May 28, 1867	65, 320
Liquids on draft, Apparatus for cooling and dispensing.	B. M. Bastian	New York, N. Y	Oct. 22, 1872	132, 430
Liquids, Pneumatic device for forcing	C. F. Bowman and S. Slyker	Wilkesbarre, Pa	June 22, 1869	91, 707
Liquids under pressure, Regulator and controller for the flow of.	J. S. Baldwin	Newark, N. J	May 4, 1869	89, 617
Liquids, Vessel for holding	J. Stimpson	Baltimore, Md	Oct. 17, 1854	11, 819
Liquids with carbonic acid, Apparatus for charging	J. Matthews	New York, N. Y	Aug. 5, 1873	141, 571
Liquids with nitric acid, Apparatus for treatment of	C. W. Volney	Boston, Mass	Apr. 9, 1872	125, 635
Liquor, Apparatus for improving	C. Walters and I. Rudigier	Pittsburgh, Pa	May 13, 1873	138, 776
Liquor-cooler	J. W. Collier	New York, N. Y	May 6, 1873	138, 478
Liquor-cooler	T. H. Miller	Allentown, Pa	Aug. 7, 1866	56, 978
Liquor-cooler	D. Lownes		July 21, 1803	....
Liquor-cooler	E. Wright	Philadelphia, Pa	Aug. 1, 1871	117, 712
Liquor-discharge, Automatic	A. C. Rowe	Chicago, Ill	May 9, 1871	114, 607
Liquor drawing and measuring apparatus	J. H. Hecker	Hinklotown, Pa	Dec. 19, 1848	5, 970
Liquor-drawing apparatus, Compressed-air	J. Johnson	Genesee, N. Y	May 30, 1837	217
Liquor-flask	R. Heneage	Buffalo, N. Y	Jan. 14, 1864	43, 154
Liquor-forcing apparatus	J. F. Bennett	Pittsburgh, Pa	Apr. 23, 1872	126, 010
Liquor-fount	J. Douglass	New York	June 25, 1821	....
Liquor-gaging device	A. and A. S. Wolcott	East Bloomfield, N. Y	Mar. 3, 1857	16, 766
Liquor-gate	O. H. and C. H. Bush	Fall River, Mass	Mar. 30, 1843	3, 002
Liquor on tap, Device to prevent injury to	J. G. Cullmann	Cincinnati, Ohio	June 8, 1869	91, 092
Liquor purifying and ageing apparatus	A. J. Gibson	Cincinnati, Ohio	Dec. 22, 1868	85, 225
Liquor-ripening process	W. M. Storm	New York, N. Y	Aug. 7, 1866	57, 009
Liquor-thief	J. F. Collins	New York, N. Y	June 22, 1869	91, 522
Liquor thief and measure	S. C. Catlin	Cleveland, Ohio	Dec. 20, 1870	110, 202
Liquor thief and test	N. Stafford	Brooklyn, N. Y	Jan. 2, 1866	51, 879
Liquors and for distillation, Boiling	J. Morris	New Haven, Conn	Mar. 3, 1813	....
Liquors, Apparatus for treating and ageing spirituous.	G. Goewey	Philadelphia, Pa	June 29, 1869	91, 840
Liquors, Compound for fining, purifying, and mellowing spirituous.	W. Thompson	Dublin, Ireland	Feb. 27, 1866	52, 943

Index of patents issued from the United States Patent Office from 1790 *to* 1873, *inclusive*—Continued.

Invention.	Inventor.	Residence.	Date.	No.
Liquors, Cooling and discharging fermented	F. Brunon	Philadelphia, Pa	Dec. 15, 1863	40,977
Liquors, Drawing and corking sparkling	S. A. Bille	New York	Nov. 8, 1828	....
Liquors, Elevating	J. A. Smith	New London, Conn	Aug. 7, 1826	....
Liquors, Extracting specimen of	J. M. Na,lee	Philadelphia, Pa	July 24, 1866	56,596
Liquors, Method of determining the strength of	W. Cornell	Brooklyn, N. Y	Aug. 20, 1827	....
Liquors on draft, Cooler for	J. Gateley	Philadelphia, Pa	Aug. 20, 1867	67,869
Liquors, Refining spirituous	G. W. Ferris	Quincy, Ill	July 10, 1866	56,197
Litharge, paint, &c., Machine for separating light from heavy particles of.	W. Atwood	Cape Elizabeth, Me	July 14, 1868	79,802
Lithographic and other impressions, Mode of preparing cloth for receiving.	C. Volkmar, jr	New York, N. Y	Nov. 12, 1867	70,920
Lithographic copy from photographic negatives, Mode of producing.	E. Rye	Copenhagen, Denmark	Nov. 22, 1870	109,551
Lithographic power-press	B. Ackermann	New York, N. Y	July 2, 1861	32,672
Lithographic power-press	E. Reynolds	Mansfield, Conn	Aug. 9, 1864	43,796
Lithographic press	G. Gooper	New York, N. Y	June 16, 1868	78,930
Lithographic press	J. Crawley	Brooklyn, N. Y	July 6, 1869	92,276
Lithographic press	J. Donlevy	New York, N. Y	Sept. 11, 1847	5,283
Lithographic press	G. Dunlop	Brooklyn, N. Y	Apr. 2, 1867	63,489
Lithographic press	A. Hoen	Baltimore, Md	May 11, 1869	89,997
Lithographic press	B. Huber	Boston, Mass	Dec. 9, 1873	145,420
Lithographic press	T. Hunter	Philadelphia, Pa	Feb. 14, 1871	111,749
Lithographic press	C. C. Maurice	New York, N. Y	Feb. 25, 1868	74,840
Lithographic press	C. C. Maurice	New York, N. Y	June 27, 1871	116,335
Lithographic press	C. C. Maurice	New York, N. Y	May 20, 1873	139,171
Lithographic press	R. McNie	New York, N. Y	Mar. 26, 1861	31,818
Lithographic press	J. Taggart	Roxbury, Mass	Sept. 8, 1863	39,867
Lithographic press	J. Walton	Brooklyn, N. Y	July 20, 1869	92,765
Lithographic press	J. Koehler	New York, N. Y	Mar. 20, 1866	53,309
Lithographic printing	O. Dubois	Fall River, Mass	Apr. 4, 1871	113,276
Lithographic printing	R. C. Manners	Boston, Mass	Feb. 15, 1833	....
Lithographic printing-form	I. Reynolds	Dayton, Ohio	Nov. 18, 1873	144,796
Lithographic printing-press	E. S. Boynton	Brooklyn, N. Y	June 27, 1871	116,406
Lithographic printing-press	A. H. Marinoni	Paris, France	Mar. 16, 1869	87,950
Lithographic printing-press	E. Reynolds	Mansfield, Conn	Mar. 8, 1864	41,862
Lithographic printing-press	E. Reynolds	Mansfield, Conn	Feb. 14, 1865	46,390
Lithographic printing-press	G. H. Reynolds	New York, N. Y	Feb. 17, 1863	37,727
Lithographic printing-press	J. W. H. Stübee	Boston, Mass	Mar. 29, 1864	42,125
Lithographic printing-press	W. H. Stubbe	Boston, Mass	Jan. 4, 1859	22,519
Lithographic prints to glass, Process of transferring.	O. P. Wolff	Cincinnati, Ohio	Dec. 9, 1873	145,472
Lithographic stone	G. H. Moore and L. Bagger	Norwich, Conn., and Washington, D. C.	Aug. 19, 1873	141,943
Lithographic stone, Mounting	G. H. Reynolds	New York, N. Y	Jan. 1, 1861	31,033
Lithographic transferred print	O. P. Wolff	Cincinnati, Ohio	Dec. 16, 1873	145,606
Lithography	D. Henderson	Jersey City, N. J	Sept. 17, 1824	....
Lithontriptor	I. Lukens	Philadelphia, Pa	Dec. 30, 1826	....
Live spindle and flier	W. MacLardy and J. Lewis	Manchester, England	July 3, 1849	6,572
Live spindles, Mode of steadying	A. Anderson	Paterson, N. J	Nov. 21, 1845	4,281
Loader, Car	D. F. Halliburton	Rutherford Station, Tenn	July 16, 1872	129,549
Loader for locomotive-tenders	A. D. Smith	Grafton, Ohio	May 19, 1868	78,022
Loader for locomotive-tenders	A. H. Spencer	Providence, R. I	Feb. 8, 1870	99,728
Loading and dumping machine	M. Phillips	Springfield, Ill	July 25, 1871	117,322
Loading and dumping machine, Portable	W. D. Dorsey	Decatur, Ill	Dec. 20, 1870	110,216
Loading and unloading vessels, Method of	G. A. Des Corats	Paris, France	July 23, 1867	67,031
Loading and unloading vessels, Portable machine for.	J. E. Walsh	New York, N. Y	Oct. 8, 1872	131,982
Loading and unloading wagons	R. Thelfall	Centreville, Cal	Dec. 24, 1872	134,330
Loading apparatus, Cart	J. Dyerly	Frederick, Md	July 15, 1873	140,884
Loading apparatus, Cart	B. G. Fitzhugh	Frederick, Md	Aug. 6, 1872	130,121
Loading apparatus for railway-cars, Wood	J. Nicholson	Monticello, Ind	Aug. 9, 1870	106,274
Loading apparatus, Stone	H. S. Perkins	Rutland, Vt	Mar. 19, 1872	124,852
Loading cars, Adjusting-rolls for	W. Dame	Woonsocket, R. I	Oct. 22, 1872	132,449
Loading carts, wagons, &c	N. Osborn	Fairfield, Conn	May 23, 1826	....
Loading coal, Apparatus for	S. S. Roberts	Elizabeth, Pa	July 13, 1869	92,476
Loading coal, Device for	J. F. Logan	Coal Bluff, Pa	Jan. 11, 1870	98,782
Loading logs, stone, and hay, Wagon for	J. Sutherland	Morris, Ill	Feb. 11, 1868	74,253
Loading lumber, Apparatus for	J. S. Preston	Winona, Minn	Mar. 5, 1872	124,385
Loading of boats, &c., Ascertaining the weight of goods in.	T. Cohoon	Troy, N. Y	June 18, 1829	....
Loading vessels, Swinging railway for	J. L. Cheeseman	Gardiner, Me	Nov. 28, 1871	121,333
Loading vessels, Winch and gin for	E. Allen	New York	Feb. 29, 1832	....
Loam-luting, Mode of revivifying	J. Chilcott	Brooklyn, N. Y	Oct. 3, 1865	50,220
Lock:				

See Alarm-lock.
Bag-lock.
Bank-lock.
Barrel-hoop lock.
Berth-lock.
Bottle-lock.
Brake-lock.
Burglar-alarm lock.
Can-cover lock.
Canal-lock.
Car-lock.
Car-seat lock.
Cell-lock.
Chain-lock.
Chronometric lock.
Combination lock.
Cross-bar lock.
Door-lock.
Drawer-lock.
Egyptian lock.
Faucet-lock.
Fire-arm lock.
Freight-car lock.
Fruit-box lock.
Furniture-lock.
Fuse-lock.
Gun-lock.
Gun and pistol lock.
Hand-cuff lock.
Hasp-lock.
Hoop-lock.
Indicator-lock.
Knob-lock.
Last-lock.
Magnetic lock.
Mail-bag lock.
Master-key lock.
Maze-lock.
Mortise-lock.
Mortise and latch lock.
Nut-lock.
Oar-lock.
Padlock.
Percussion-lock.
Permutation-lock.
Perplexing unpickable lock.

Index of patents issued from the United States Patent Office from 1790 to 1873, inclusive—Continued.

Invention.	Inventor.	Residence.	Date.	No.
Lock—Continued. *See* Piano-lock. Pin-lock. Pipe-coupling lock. Powder-proof lock. Prison-lock. Pump-lock. Railway-switch lock. Reversible lock. Ring-lock. Row-lock. Safe-lock. Sash-lock. Seal-lock. Sewing-machine lock. Ship-lock. Ship's pump lock. Sleigh-lock. Spool-holder lock. Spring-lock. Switch-lock. Till-lock. Till alarm-lock. Time-lock. Trace-lock. Trick-lock. Trunk-lock. Tumbler-lock. Umbrella-lock. Umbrella-stand lock. Valise-lock. Vault-cover lock. Vehicle-seat lock. Vehicle-wheel lock. Vest-chain lock. Vine-lock. Wagon-lock. Wagon-brake lock. Wagon-seat lock. Watch-chain lock.				
Lock	C. Ackerman	Newark, N. J	Sept. 21, 1858	21, 543
Lock	J. Adt	Waterbury, Conn	June 25, 1861	32, 605
Lock	J. Adt	Waterbury, Conn	Dec. 23, 1862	37, 209
Lock	W. H. Aikins	Berkshire, N. Y	May 13, 1856	14, 848
Lock	H. Allman	London, England	Jan. 30, 1866	52, 372
Lock	E. G. F. Arndt and A. Hübne	Rondout, N. Y	Nov. 11, 1862	36, 886
Lock	T. A. Auberlin	San Francisco, Cal	Apr. 17, 1866	53, 941
Lock	S. T. Bacon	Boston, Mass	July 12, 1859	24, 709
Lock	L. Baier	Pittsburgh, Pa	June 30, 1857	17, 714
Lock	L. Baier	Cincinnati, Ohio	Apr. 27, 1858	20, 027
Lock	H. H. Baker	New Market, N. J	Nov. 15, 1864	45, 014
Lock	M. O. Baker	New York, N. Y	May 31, 1870	103, 542
Lock	G. Bayer	New York, N. Y	Feb. 6, 1872	123, 444
Lock	L. Beer	New York, N. Y	Jan. 4, 1870	98, 545
Lock	E. S. Bennett	Brooklyn, N. Y	Feb. 18, 1862	34, 460
Lock	A. Betteley	Boston, Mass	Apr. 6, 1852	8, 851
Lock	O. Billings	La Grange, Ohio	May 17, 1859	24, 075
Lock	I. H. A. Bleckman	Ronsdor, Prussia	Nov. 20, 1855	13, 814
Lock	C. Bodmer	Fayetteville, Ill	Feb. 11, 1873	135, 702
Lock	W. Bohannan	New York, N. Y	May 29, 1866	55, 047
Lock	W. Bohannan and F. G. Johnson.	New York and Brooklyn, N. Y.	May 29, 1866	55, 048
Lock	T. Bowles	New York, N. Y	Dec. 18, 1855	13, 968
Lock	J. A. Braden	La Grange, Ga	June 8, 1858	20, 476
Lock	J. Brady	Branford, Conn	June 7, 1870	103, 838
Lock	J. Brady	Branford, Conn	Oct. 11, 1870	108, 098
Lock	J. Brady	Branford, Conn	Oct. 11, 1870	108, 099
Lock	J. B. Brennan	Mount Vernon, N. Y	Nov. 7, 1854	11, 885
Lock	E. W. Brettell	Newark, N. J	June 26, 1860	28, 828
Lock	E. W. Brettell	Newark, N. J	Mar. 8, 1864	41, 827
Lock	E. W. Brettell	Newark, N. J	Aug. 8, 1865	49, 223
Lock	B. V. M. Bronse	Kokomo, Ind	June 12, 1866	55, 460
Lock	C. F. Brown	Bridgeport, Conn	July 10, 1860	29, 120
Lock	H. and S. W. Budd	Philadelphia, Pa	June 26, 1866	55, 814
Lock	R. B. Burchell	Brooklyn, N. Y	Sept. 11, 1866	57, 857
Lock	I. D. Bush	Detroit, Mich	June 19, 1866	55, 616
Lock	W. C. Bussey	Jackson, Cal	Sept. 5, 1865	49, 712
Lock	W. H. Butler	New York, N. Y	Oct. 28, 1856	15, 962
Lock	A. Clabaugh	Altoona, Pa	July 7, 1863	39, 117
Lock	C. Claude	Newark, N. J	Oct. 15, 1861	33, 476
Lock	C. Claude	New York, N. Y	May 1, 1866	54, 297
Lock	G. Clay	New York, N. Y	Apr. 12, 1859	23, 662
Lock	J. M. Cook	Hinsdale, N. Y	June 2, 1857	17, 424
Lock	G. W. Coppernoll	Ohio, N. Y	Sept. 30, 1856	15, 800
Lock	W. Corbett and W. Burns	Saint Louis, Mo	June 5, 1866	55, 244
Lock	H. W. Covert	Rochester, N. Y	Sept. 15, 1857	18, 228
Lock	G. W. Dana	Durand, Ill	Dec. 27, 1859	26, 569
Lock	V. R. David	Newark, Ill	Mar. 31, 1857	16, 908
Lock	H. and J. Day	New York, N. Y	Apr. 4, 1820	....
Lock	L. P. Decker	Williamsburgh, N. Y	Jan. 8, 1867	61, 055
Lock	A. E. Deitz	Brooklyn, N. Y	Nov. 28, 1865	51, 152
Lock	W. Denney	Philadelphia, Pa	Mar. 16, 1858	19, 628
Lock	L. Derby	New York, N. Y	May 8, 1860	28, 162
Lock	C. C. Dickerman	Boston, Mass	Mar. 29, 1870	101, 355
Lock	C. C. Dickerman	Boston, Mass	Nov. 1, 1870	108, 770
Lock	W. Dickson	Albany, N. Y	Mar. 26, 1867	63, 230
Lock	L. Diss	Ilion, N. Y	Apr. 28, 1857	17, 139
Lock	L. Diss	Oriskany, N. Y	Jan. 3, 1860	26, 659
Lock	W. B. Dodds	Cincinnati, Ohio	Mar. 14, 1865	46, 858
Lock	W. B. Dodds	Cincinnati, Ohio	Apr. 17, 1866	53, 959
Lock	T. Dougherty	Macon, Ga	June 7, 1859	24, 287
Lock	M. Ducharme	Cohoes, N. Y	May 14, 1861	32, 334
Lock	C. Duckworth	Hartford, Conn	Nov. 1, 1859	25, 959
Lock	J. Ebling	New York, N. Y	May 4, 1869	89, 746
Lock	R. S. Eddy and A. O. Miles	Nashua, N. H	Dec. 17, 1861	33, 931
Lock	G. and G. F. Elliott	Manchester, Conn	Feb. 21, 1860	27, 262
Lock	H. H. Elwell	South Norwalk, Conn	Oct. 14, 1862	36, 645
Lock	H. H. Elwell	South Norwalk, Conn	Jan. 13, 1863	37, 390
Lock	H. H. Elwell	South Norwalk, Conn	July 21, 1863	39, 289
Lock	H. H. Elwell	South Norwalk, Conn	Feb. 9, 1864	41, 573
Lock	J. Euteneuer	Peoria, Ill	Sept. 19, 1865	49, 991
Lock	M. Erb and F. C. Goffin	Newark, N. J	June 17, 1856	15, 124
Lock	J. Farrel	New York, N. Y	Nov. 18, 1862	36, 946
Lock	E. Fay	Washington, D. C	Jan. 4, 1870	98, 577
Lock	P. S. Felter	Cincinnatus, N. Y	Jan. 12, 1864	41, 211
Lock	P. S. Felter	Cincinnatus, N. Y	Nov. 1, 1864	44, 859
Lock	P. S. Felter	Cincinnatus, N. Y	July 17, 1866	56, 391

Index of patents issued from the United States Patent Office from 1790 *to* 1873, *inclusive*—Continued.

Invention.	Inventor.	Residence.	Date.	No.
Lock	R. S. Foster	Sing Sing, N. Y	June 23, 1863	39,002
Lock	R. S. Foster	Sing Sing, N. Y	June 23, 1863	39,003
Lock	R. S. Foster	Sing Sing, N. Y	June 23, 1863	39,004
Lock	R. S. Foster	Sing Sing, N. Y	June 23, 1863	39,005
Lock	R. S. Foster	Sing Sing, N. Y	June 23, 1863	39,006
Lock	R. S. Foster	Sing Sing, N. Y	June 23, 1863	39,007
Lock	R. S. Foster	Sing Sing, N. Y	June 23, 1863	39,008
Lock	R. S. Foster	Sing Sing, N. Y	June 23, 1863	39,009
Lock	R. S. Foster	Sing Sing, N. Y	June 23, 1863	39,010
Lock	R. S. Foster	Sing Sing, N. Y	Sept. 22, 1863	40,077
Lock	J. B. Frost	Jackson, Miss	Feb. 27, 1872	124,125
Lock	F. Garachon	New York, N. Y	June 29, 1852	9,070
Lock	J. Gerard	Trenton, N. J	June 7, 1870	103,868
Lock	C. T. Gibson	Baltimore, Md	Aug. 1, 1865	49,101
Lock	F. C. Goffin	New York, N. Y	Oct. 26, 1852	9,361
Lock	F. Gorris	Cleveland, Ohio	Sept. 2, 1873	142,335
Lock	F. Gould	Huntington, N. Y	Aug. 17, 1858	21,193
Lock	W. H. Greenwood	Galva, Ill	Aug. 14, 1860	29,588
Lock	J. T. Guthrie	Leesburgh, Ohio	Sept. 4, 1866	57,704
Lock	J. L. Hall	Cincinnati, Ohio	Sept. 22, 1857	18,243
Lock	J. L. Hall	Cincinnati, Ohio	Sept. 25, 1860	30,140
Lock	J. L. Hall	Cincinnati, Ohio	May 5, 1863	38,384
Lock	W. Hall	Brookline, Mass	Jan. 6, 1863	37,290
Lock	W. Hall	Brookline, Mass	May 23, 1865	47,817
Lock	G. Hopson	Bridgeport, Conn	July 12, 1864	43,501
Lock	W. W. Harder	New York, N. Y	Aug. 28, 1866	57,503
Lock	A. Hardy and F. L. and G. A. Walker.	Boston, Mass	Mar. 6, 1866	52,995
Lock	T. G. Harold	Brooklyn, N. Y	Aug. 5, 1862	36,124
Lock	T. G. Harold	Brooklyn, N. Y	Nov. 24, 1863	40,724
Lock	T. G. Harold	Brooklyn, N. Y	Dec. 8, 1863	40,882
Lock	T. G. Harold	Brooklyn, N. Y	Jan. 5, 1864	41,177
Lock	T. G. Harold	Brooklyn, N. Y	Jan. 5, 1864	41,178
Lock	H. Hartwig	New York, N. Y	Apr. 30, 1861	32,225
Lock	H. Hartwig	New York, N. Y	Apr. 30, 1861	32,226
Lock	S. Hiatt	Indianapolis, Ind	Dec. 28, 1858	22,425
Lock	A. M. Hill	Branford, Conn	July 29, 1862	36,053
Lock	A. M. Hill	Philadelphia, Pa	Mar. 31, 1863	38,080
Lock	J. J. Hirshbuhl	Louisville, Ky	May 14, 1861	32,289
Lock	A. Hoagland	Jersey City, N. J	Mar. 9, 1858	19,564
Lock	E. D. Hodgson	London, England	Nov. 7, 1871	120,743
Lock	L. G. Hoffman	Albany, N. Y	Apr. 24, 1866	54,162
Lock	R. G. Holmes and W. H. Butler	New York, N. Y	Oct. 30, 1855	13,722
Lock	E. R. Hopkins	New York, N. Y	Aug. 28, 1866	57,508
Lock	D. W. G. Humphrey	Gray, Me	Nov. 6, 1855	13,753
Lock	J. Hutson	Janesville, Wis	Dec. 27, 1864	45,611
Lock	A. Inglis	Indianapolis, Ind	Nov. 23, 1869	97,196
Lock	H. Isham	New Britain, Conn	July 1, 1856	15,239
Lock	H. Isham	New Britain, Conn	July 7, 1857	17,740
Lock	H. Isham	New Britain, Conn	Feb. 28, 1860	27,294
Lock	H. Jackson	Brooklyn, N. Y	Feb. 7, 1865	46,241
Lock	C. Jagy	New York, N. Y	May 27, 1862	35,376
Lock	C. Jagy and F. Denzler	New York, N. Y	Mar. 1, 1864	41,817
Lock	C. F. and C. F., jr., and W. W. Johnson.	Owego, N. Y	Sept. 11, 1860	29,975
Lock	F. G. Johnson	Brooklyn, N. Y	Feb. 5, 1861	31,320
Lock	W. Jones	Brooklyn, N. Y	Nov. 1, 1864	44,869
Lock	H. W. Kahlke	Brooklyn, N. Y	May 17, 1864	42,778
Lock	J. Keath	Ash Grove, Ill	Dec. 17, 1872	133,991
Lock	W. M. Keefer	Chicago, Ill	June 21, 1870	104,598
Lock	E. Kershaw	Boston, Mass	Dec. 4, 1855	13,860
Lock	R. Ketchum	Seneca Castle, N. Y	Dec. 7, 1852	9,451
Lock	W. S. Kirkham	Brantford, Conn	Nov. 6, 1860	30,580
Lock	E. Lawshé	Atlanta, Ga	Nov. 13, 1866	59,615
Lock	L. Layman	Westfield, N. Y	Nov. 6, 1860	30,606
Lock	W. C. Leach and M. J. Knox	Knox Corners, N. Y	Aug. 19, 1862	36,221
Lock	E. Leman, jr	Boston, Mass	Nov. 28, 1822	......
Lock	L. Lillie	Troy, N. Y	Aug. 8, 1865	49,282
Lock	M. T. Lincoln	Washington, D. C	May 31, 1864	42,954
Lock	J. M. Lippincott	Pittsburgh, Pa	May 13, 1856	14,886
Lock	J. M. Lippincott	Pittsburgh, Pa	Aug. 5, 1856	15,489
Lock	J. P. Lipps	Newark, N. J	July 6, 1858	20,850
Lock	J. Loch and G. Bayer	New York, N. Y	Feb. 27, 1866	52,864
Lock	S. N. Long	South Chatham, Mass	Dec. 9, 1862	37,135
Lock	S. N. Long and M. E. Hathaway.	Chatham and Wareham, Mass.	June 30, 1863	39,056
Lock	J. P. Lord	Manchester, N. H	Aug. 31, 1858	21,346
Lock	W. Lorenz	Lebanon, Pa	Aug. 30, 1864	44,007
Lock	J. W. Lyon	Brooklyn, N. Y	Aug. 19, 1862	36,224
Lock	N. Macneale	Cincinnati, Ohio	July 5, 1864	43,457
Lock	N. Macneale and W. B. Dodds	Cincinnati, Ohio	Nov. 11, 1862	36,90[illegible]
Lock	L. H. Magott	Springfield, Mass	Dec. 12, 1865	51,467
Lock	W. K. Marvin	New York, N. Y	Sept. 13, 1864	44,207
Lock	W. K. Marvin	New York, N. Y	May 23, 1865	47,842
Lock	W. Maurer	New York, N. Y	Apr. 8, 1856	14,616
Lock	F. McCollum	Rockville, Conn	Feb. 9, 1864	41,518
Lock	W. C. McGill	Cincinnati, Ohio	May 1, 1866	54,383
Lock	M. McGonnigle	Allegheny City, Pa	Jan. 23, 1866	52,183
Lock	A. O. Miles	Nashua, N. H	Oct. 23, 1866	59,048
Lock	J. W. Miller	Waynesborough, Pa	June 7, 1864	43,078
Lock	J. C. Mix	Terryville, Conn	Jan. 27, 1863	37,515
Lock	W. Moore	Brooklyn, N. Y	Nov. 23, 1858	22,146
Lock	A. B. Mullett	Washington, D. C	Nov. 13, 1866	59,628
Lock	L. F. Munger	Le Roy, N. Y	July 14, 1857	17,804
Lock	L. F. Munger	Rochester, N. Y	Feb. 22, 1859	23,040
Lock	L. F. Munger	Rochester, N. Y	Feb. 21, 1865	46,531
Lock	J. T. Mygatt	Binghamton, N. Y	Feb. 17, 1863	37,700
Lock	R. Newell	New York	June 14, 1843	3,135

Index of patents issued from the United States Patent Office from 1790 *to* 1873, *inclusive*—Continued.

Invention.	Inventor.	Residence.	Date.	No.
Lock	H. Oaks	Waynesborough, Pa	Apr. 18, 1865	47, 325
Lock	J. M. Perkins	New York, N. Y	Mar. 2, 1858	19, 533
Lock	J. M. Perkins	Cleveland, Ohio	Aug. 5, 1862	36, 106
Lock	J. M. Perkins	Cleveland, Ohio	Dec. 23, 1862	37, 240
Lock	S. Perry	Newport, N. Y	May 12, 1857	17, 293
Lock	N. Petré	New York, N. Y	Oct. 5, 1869	95, 509
Lock	N. Petré	New York, N. Y	Dec. 28, 1869	98, 406
Lock	N. Petré	New York, N. Y	Feb. 1, 1870	99, 344
Lock	S. Pettes	Bellows Falls, Vt	July 12, 1834	....
Lock	W. Pio	New York	Aug. 5, 1813	....
Lock	J. A. Pimentel and W. H. Shute	New York, N. Y	July 29, 1862	36, 624
Lock	J. H. Pomeroy	Bloomington, Ill	Dec. 4, 1855	13, 884
Lock	J. Post	Newark, N. J	Aug. 8, 1865	49, 297
Lock	J. Post	Newark, N. J	Aug. 8, 1865	49, 298
Lock	D. Powers	Philadelphia, Pa	Mar. 1, 1859	23, 113
Lock	T. Powers	Philadelphia, Pa	Jan. 29, 1861	31, 258
Lock	T. Powers	New York, N. Y	Sept. 6, 1870	107, 100
Lock	D. A. Pratt	New York, N. Y	May 22, 1866	54, 952
Lock	F. Prockert	New York, N. Y	Nov. 23, 1869	97, 227
Lock	T. L. Pye	New York, N. Y	Sept. 28, 1858	21, 636
Lock	P. B. Quimby	Belfast, Me	May 23, 1836	....
Lock	A. Rankin	Philadelphia, Pa	May 22, 1860	28, 406
Lock	D. B. Read and J. H. Clapp	Providence, R. I	Mar. 8, 1870	100, 556
Lock	E. S. Renwick	New York, N. Y	May 17, 1864	42, 793
Lock	A. A. Richards	Urbana, Ohio	Feb. 15, 1859	22, 980
Lock	H. D. Richardson	Florence, Mass	June 13, 1865	48, 209
Lock	H. D. Richardson	Northampton, Mass	May 7, 1867	64, 568
Lock	J. C. Riethmüller	Pittsburgh, Pa	Mar. 3, 1857	16, 749
Lock	G. A. Rogers	Augusta, Me	Jan. 20, 1830	....
Lock	G. Rossner	Rochester, N. Y	Sept. 18, 1860	30, 092
Lock	E. A. G. Roulstone	Roxbury, Mass	Dec. 4, 1866	60, 257
Lock	F. Rudolph	New York, N. Y	July 25, 1865	48, 980
Lock	H. D. Russell	Naugatuck, Conn	Sept. 9, 1856	15, 708
Lock	J. Sargent	Rochester, N. Y	Aug 28, 1866	57, 574
Lock	J. Sargent and H. W. Covert	Rochester, N. Y	May 2, 1865	47, 575
Lock	J. Sargent and H. W. Covert	Rochester, N. Y	Jan. 9, 1866	51, 973
Lock	Z. Sargent	Rochester, N. H	Feb. 22, 1870	100, 195
Lock	W. Sellers	New York, N. Y	Oct. 16, 1866	58, 898
Lock	F. G. Shalling	Somerset, Mass	Nov. 11, 1862	36, 919
Lock	E. M. Shaw	Baltimore, Md	Apr. 6, 1858	19, 879
Lock	H. L. Shepardson	Shelburne Falls, Mass	Jan. 18, 1870	99, 013
Lock	G. A. Sherlock	Boston, Mass	July 31, 1866	56, 808
Lock	J. P. Sherwood	Fort Edward, N. Y	Sept. 8, 1857	18, 162
Lock	A. W. Snow	Norwich, Conn	Oct. 9, 1866	58, 688
Lock	M. Stachelin and H. Young	Port Chester, N. Y	June 12, 1866	55, 549
Lock	N. Stafford	Brooklyn, N. Y	Mar. 7, 1865	46, 721
Lock	H. Stein	New York, N. Y	Dec. 9, 1873	145, 370
Lock	T. Stewart	Philadelphia, Pa	Dec. 5, 1871	121, 680
Lock	S. C. St. John	Edmeston, N. Y	Jan. 29, 1861	31, 267
Lock	O. B. Thompson	Hudson, Ohio	Nov. 2, 1858	21, 994
Lock	C. F. Toll	Boston, Mass	Aug. 9, 1864	43, 806
Lock	C. F. Toll	Boston, Mass	Jan. 23, 1866	52, 226
Lock	H. R. Towne	Stamford, Conn	Sept. 12, 1871	118, 986
Lock	H. B. Tyler	Norwich, Conn	Aug. 15, 1865	49, 481
Lock	B. M. Van Der Veer	Clyde, N. Y	Apr. 25, 1865	47, 470
Lock	B. M. Van Der Veer	Clyde, N. Y	Dec. 26, 1865	51, 763
Lock	E. Voigh and P. P. Mathes	Philadelphia, Pa	Apr. 5, 1870	101, 684
Lock	R. Vollschwitz	New York, N. Y	Dec. 12, 1865	51, 498
Lock	H. Underwood	Tolland, Conn	Sept. 27, 1864	44, 475
Lock	C. P. Wagner	New York, N. Y	May 2, 1865	47, 588
Lock	H. B. Weaver	Hartford, Conn	Nov. 2, 1869	96, 369
Lock	R. L. Webb	West Meriden, Conn	Jan. 12, 1864	41, 247
Lock	T. K. Webster	Lawrence, Mass	May 10, 1859	23, 967
Lock	C. S. Westcott	New York, N. Y	Dec. 14, 1858	22, 319
Lock	W. Whiting and H. Pickford	Roxbury and Boston, Mass	Mar. 24, 1857	16, 892
Lock	W. Whiting and H. Pickford	Roxbury and Boston, Mass.	July 14, 1857	17, 815
Lock	E. P. Whitney	Westfield, N. Y	Sept. 4, 1860	29, 929
Lock	T. Wilbur	Woodbury, Conn	Dec. 23, 1834	....
Lock	H. Willard	Evansville, Ind	Dec. 20, 1870	110, 412
Lock	H. Willard and P. Trunz	Evansville, Ind	Sept. 6, 1870	107, 145
Lock	A. Williams and E. P. Cummings.	Philadelphia, Pa	May 5, 1857	17, 245
Lock	C. Wilson	Springfield, Mass	Nov. 7, 1854	11, 916
Lock	G. M. Wood	Decatur, Ill	Aug. 21, 1866	57, 423
Lock	L. Woodruff	New Britain, Conn	July 15, 1862	35, 912
Lock	T. B. Worrell and T. Walker	Philadelphia, Pa	Dec. 7, 1869	97, 581
Lock	G. C. Worth	Upper Sandusky, Ohio	Sept. 23, 1862	36, 547
Lock	W. E. Worthen	New York, N. Y	Apr. 10, 1860	27, 854
Lock	H. Wynblad	New York, N. Y	Aug. 19, 1856	15, 589
Lock	H. Wynblad	West Hoboken, N. J	Aug. 24, 1858	21, 293
Lock	L. Yale, jr	Newport, N. Y	June 3, 1856	15, 031
Lock	L. Yale, jr	Philadelphia, Pa	Nov. 9, 1858	22, 048
Lock	L. Yale, jr	Philadelphia, Pa	June 12, 1860	28, 710
Lock	L. Yale, jr	Philadelphia, Pa	Jan. 29, 1861	31, 278
Lock	L. Yale, jr	Philadelphia, Pa	May 14, 1861	32, 330
Lock	L. Yale, jr	Philadelphia, Pa	May 14, 1861	32, 331
Lock	L. Yale, jr	Shelburne Falls, Mass	June 27, 1865	48, 475
Lock	L. Yale, jr	Bernardston, Mass	Oct. 12, 1869	95, 865
Lock and bolt	E. Fink and C. E. McDonald	Indianapolis, Ind	July 14, 1863	39, 223
Lock and bolt	H. C. Howells	Putnam, Ohio	Feb. 10, 1834	....
Lock and detector	J. H. Lyons	New York, N. Y	Sept. 13, 1859	25, 428
Lock and escutcheon	R. Kinsley	Springfield, Mass	Sept. 19, 1848	5, 791
Lock and key	J. R. Bugbee	Boston, Mass	Apr. 22, 1851	8, 060
Lock and key	F. A. H. Gaebel	New York, N. Y	May 12, 1863	38, 476
Lock and key	E. M. Hendrickson	Brooklyn, N. Y	May 27, 1856	14, 958
Lock and key	J. Hoffacker	New York, N. Y	Nov. 2, 1858	21, 962
Lock and key	H. B. Weaver	Hartford, Conn	Jan. 1, 1867	60, 974
Lock and key	W. M. Williams	England	Dec. 1, 1840	1, 874

Index of patents issued from the United States Patent Office from 1790 to 1873, inclusive—Continued.

Invention.	Inventor.	Residence.	Date.	No.
Lock and key	W. Wilson	Northampton, Mass	Jan. 27, 1843	2, 931
Lock and key	L. Yale, jr	Newport, N. Y	May 6, 1851	8, 071
Lock and knob-latch	A. M. Hill	Branford, Conn	Jan. 11, 1861	32, 551
Lock and knob-latch	W. S. Kirkham	Branford, Conn	Jan. 11, 1861	32, 521
Lock and knob-latch combined	E. G. F. Arndt and C. E. L. Molbius.	New York, N. Y	Aug. 20, 1867	67, 938
Lock and label-sheath, Compound	S. W. Marsh	Washington, D. C	Apr. 3, 1860	27, 728
Lock and latch	J. Chittenden	Burlington, Vt	Aug. 12, 1820	....
Lock and latch	H. H. Elwell	South Norwalk, Conn	Feb. 23, 1864	41, 741
Lock and latch	W. S. Kirkham	Branford, Conn	Mar. 15, 1859	23, 254
Lock and latch	B. Mallory	New Haven, Conn	May 5, 1863	38, 400
Lock and latch	B. Mallory	New Haven, Conn	June 7, 1864	43, 035
Lock and latch	J. F. Thayer	Boston, Mass	May 18, 1822	....
Lock and latch, Combined	G. B. Allen	Norwalk, Conn	Mar. 15, 1870	100, 708
Lock and latch, Combined	G. W. Cilley	Norwich, Conn	May 11, 1869	89, 852
Lock and latch, Combined	R. Kinsley	Springfield, Mass	Apr. 16, 1845	4, 060
Lock and latch, Combined	F. M. Ranous	Yreka City, Cal	Dec. 13, 1870	110, 071
Lock and latch, Combined	S. C. Weddington	Jonesborough, Ind	Sept. 6, 1870	107, 207
Lock and latch, Combined	S. H. Wheeler	Dowagiac, Mich	Apr. 4, 1871	113, 377
Lock-attachment	P. P. Stephan	Newark, N. J	Feb. 5, 1861	31, 337
Lock-attachment	J. M. Wilson	Philadelphia, Pa	June 7, 1859	24, 346
Lock-bolt	T. Kromer	Neustadt, Baden	Sept. 19, 1871	119, 036
Lock-bolt	G. Washburn	New York, N. Y	July 14, 1868	79, 879
Lock-bolt spindle	J. Sargent	Rochester, N. Y	Jan. 4, 1870	98, 623
Lock-bolts, Cam for throwing	H. W. Covert	Rochester, N. Y	Oct. 5, 1858	21, 655
Lock-bolts, Method of operating	T. Slaight	Newark, N. J	Mar. 12, 1850	7, 176
Lock-case, Machine for riveting	T. W. Baxter and E. Brown	Chicago, Ill., and Green Point, N. Y.	June 25, 1872	128, 205
Lock catch or stopper	E. P. McCeney	Washington County, D. C	Jan. 14, 1868	73, 358
Lock, Combination	W. C. McGill	Cincinnati, Ohio	Mar. 11, 1873	136, 606
Lock-cutting machine	I. L. Penney	Minneapolis, Minn	May 13, 1873	138, 815
Lock-fastener	H. S. Wilcox	West Meriden, Conn	June 14, 1864	43, 147
Lock-gate wicket	J. Jack	Nunda, N. Y	May 21, 1850	7, 379
Lock-gates by water-power, Method of working	J. D. Willoughby	Susquehanna Post-Office, Pa.	Oct. 24, 1848	5, 885
Lock-gates, Manner of suspending, opening, and closing.	H. McCarty	Pittsburgh, Pa	Mar. 16, 1844	3, 493
Lock-guard	L. Schroder	Cincinnati, Ohio	June 21, 1859	24, 523
Lock-guard attachment	P. S. Felter	Cincinnatus, N. Y	Dec. 17, 1861	33, 933
Lock guard or tumbler	L. Kellogg	Charleston, Ohio	Aug. 15, 1848	5, 706
Lock-hasp	M. Newman, 2d	Oak Hill, N. Y	Apr. 15, 1856	14, 675
Lock-joints in sheet-metal, Machine for forming	J. Shaw	Hinckley, Ohio	Sept. 11, 1866	57, 980
Lock-keeper, Adjustable	T. Strobridge	Pittsburgh, Pa	July 31, 1866	56, 828
Lock-key	A. G. Batchelder	Lowell, Mass	Sept. 3, 1867	68, 338
Lock-key	P. A. Gladwin	Boston, Mass	June 12, 1866	55, 582
Lock-key	W. E. Hawkins	Wallingford, Conn	Mar. 5, 1872	124, 266
Lock-key	J. Koenig	Erie, Pa	Feb. 27, 1872	124, 066
Lock-key	W. Park	Norwich, Conn	Dec. 27, 1870	110, 491
Lock-key	E. Reynolds	Mansfield, Conn	Mar. 7, 1865	46, 704
Lock-key	F. W. Smith, jr	Bridgeport, Conn	Mar. 2, 1869	87, 518
Lock-key	A. A. Stuart	Cedar Rapids, Iowa	Dec. 24, 1872	134, 325
Lock-key	P. Van Antwerp	New York, N. Y	Oct. 25, 1859	25, 923
Lock-key	A. F. Whiting	Bath, Me	Feb. 6, 1872	123, 532
Lock-key fastener	J. B. Ayer	Malden, Mass	Sept. 12, 1865	51, 410
Lock-key, Device for fastening	J. R. Tempest	Philadelphia, Pa	July 4, 1865	48, 602
Lock-key guard	C. Leavitt	Cleveland, Ohio	Mar. 21, 1865	46, 914
Lock-key, Sheet-metal	T. Powers	New York, N. Y	July 18, 1871	117, 110
Lock label and seal, Combined	E. J. Brooks	New York, N. Y	June 10, 1873	139, 764
Lock-nut	L. Arnold and J. B. Atwood	Belchertown and Palmer, Mass.	July 23, 1872	129, 707
Lock-nut	A. G. Binns	Pittsburgh, Pa	Apr. 5, 1870	101, 421
Lock-nut	C. F. Brush	Cleveland, Ohio	Jan. 30, 1872	123, 148
Lock-nut	A. W. Bunnell	Linesville, Pa	Sept. 5, 1871	118, 684
Lock-nut	A. W. Bunnell	Linesville, Pa	Sept. 5, 1871	118, 685
Lock-nut	J. Dinsmore	Dinsmore, Pa	May 28, 1872	127, 320
Lock-nut	P. Dyer, jr., A. Parker, and W. B. Way.	Pontiac, Mich	Nov. 8, 1870	108, 967
Lock-nut	W. P. Ewing and I. S. De Ford	Elkton, Md	Sept. 7, 1869	94, 585
Lock-nut	P. L. Gibbs	Dunleith, Ill	Oct. 26, 1869	96, 218
Lock-nut	S. Barber	Beaver, Pa	July 7, 1868	79, 649
Lock-nut	M. W. Griswold	New York, N. Y	Oct. 19, 1869	95, 897
Lock-nut	T. Hagan	Rochester, Pa	July 20, 1869	92, 721
Lock-nut	R. W. Hamilton	New York, N. Y	Aug. 8, 1871	117, 881
Lock-nut	W. Hamilton	Toronto, Canada	Dec. 29, 1868	85, 381
Lock-nut	J. Harvey	Painesville, Ohio	Aug. 22, 1871	118, 238
Lock-nut	G. G. Hermance	Claverack, N. Y	Feb. 1, 1870	99, 315
Lock-nut	J. W. Hilton	Bradford, Pa	Mar. 30, 1869	88, 385
Lock-nut	S. W. Kirk	Coatesville, Pa	Mar 31, 1868	76, 201
Lock-nut	J. S. Kirkpatrick	Punta Arenas, Cal	Aug. 13, 1872	130, 377
Lock-nut	S. T. Lamb	New Albany, Ind	Nov. 21, 1871	121, 175
Lock-nut	S. T. Lamb	New Albany, Ind	Nov. 21, 1871	121, 176
Lock-nut	M. Langhorne	Washington, D. C	Feb. 15, 1870	99, 919
Lock-nut	E. T. Ligon	Demopolis, Ala	May 24, 1870	103, 348
Lock-nut	W. C. Mason	Beaver Falls, Pa	Oct. 12, 1869	95, 704
Lock-nut	J. Miller, jr	Marshalltown, Iowa	Nov. 1, 1870	103, 927
Lock-nut	A. T. Morris	Nevada, Ohio	Jan. 16, 1872	122, 845
Lock-nut	J. A. Morrison	Parker's Landing, Pa	Apr. 30, 1872	126, 223
Lock-nut	W. H. Nichols	East Hampton, Conn	Jan. 28, 1873	135, 242
Lock-nut	F. Oakley	Toronto, Canada	Aug. 17, 1869	93, 943
Lock-nut	G. Palmer	Littlestown, Pa	Feb. 2, 1869	83, 582
Lock-nut	J. A. Peabody and A. F. Champlin.	Westerly, R. I	May 7, 1872	126, 482
Lock-nut	J. L. Randolph	Berkeley Springs, W. Va	Dec. 27, 1870	110, 499
Lock-nut	A. Roff	Southport, Conn	Apr. 13, 1869	88, 908
Lock-nut	A. Roff	Southport, Conn	Aug. 3, 1869	93, 529
Lock-nut	J. Rogers	Sterling, Ill	June 30, 1868	79, 397
Lock-nut	H. Rosamyer	Rochester, Pa	Aug. 13, 1869	93, 230
Lock-nut	G. P. Rose	Milwaukee, Wis	Mar. 5, 1872	124, 390

Index of patents issued from the United States Patent Office from 1790 *to* 1873, *inclusive*—Continued.

Invention.	Inventor.	Residence.	Date.	No.
Lock-nut	E. R. Shepard	Scranton, Pa	May 3, 1870	102, 600
Lock-nut	A. D. Smith	Grafton, Ohio	June 16, 1868	78, 899
Lock-nut	L. J. Smith	New York, N. Y	May 23, 1871	115, 249
Lock-nut	H. C. Stouffer	Columbiana, Ohio	Jan. 9, 1872	122, 676
Lock-nut	W. J. Stowell	Baltimore, Md	Aug. 11, 1868	84, 050
Lock-nut	A. N. Towne	Chicago, Ill	July 14, 1868	80, 034
Lock-nut	U. B. Vidal	Philadelphia, Pa	May 31, 1870	103, 689
Lock-nut	R. White	Mechanicsburgh, Pa	June 8, 1869	91, 188
Lock-nut and axle, Vehicle	T. F. Van Keuren	Saugerties, N. Y	May 14, 1872	126, 762
Lock-nut and splice-bar, Combined	R. L. McGowan	New Brighton, Pa	Aug. 30, 1870	106, 849
Lock-nut and tightener	H. W. Olney, R. R. Logan, and J. H. Fisher.	Allegheny City, Pa	Sept. 22, 1868	82, 473
Lock-nut, Axle and skein box	J. R. Smith	Bethel, Conn	Apr. 28, 1868	77, 411
Lock-nut, Railway-joint	W. P. Horton	Milwaukee, Wis	Oct. 10, 1871	119, 767
Lock-nut, Spring-clip	E. M. Turner	Allegheny, Pa	Apr. 23, 1872	126, 110
Lock-nut washer	D. Elliot and E. Seely	New York, N. Y	Feb. 4, 1868	74, 060
Lock safety-guard	O. Lund	Nashua, N. H	Aug. 30, 1864	44, 008
Lock-seal	F. W. Brooks	New York, N. Y	July 22, 1873	141, 110
Lock, Seal	G. Ulman	Ivey-sur-Seine, near Paris, France.	Mar. 8, 1870	100, 568
Lock-spindle	J. L. Hall	Cincinnati, Ohio	July 26, 1870	105, 800
Lock-spindle socket	O. Gallagher	Boston, Mass	Oct. 18, 1870	108, 343
Lock-washer	W. H. Van Cleve	Ypsilanti, Mich	Mar. 12, 1872	124, 644
Lock with double catch-bolt	C. Leibrich	Philadelphia, Pa	Oct. 5, 1839	1, 355
Locks and bolts, Alarm-attachment for	C. E. Pierce	New York, N. Y	June 14, 1870	104, 195
Locks and catches, Keeper for	A. Patterson	Birmingham, Pa	June 16, 1857	17, 611
Locks, Face-plate and strike for	J. M. Spring	New Britain, Conn	May 14, 1872	126, 840
Locks, Face-plate for	T. B. Atterbury and W. Warwich.	Pittsburgh, Pa	July 29, 1856	15, 456
Locks, Guard-attachment for	W. Miller	Boston, Mass	Sept. 29, 1863	40, 115
Locks, Manufacturing	J. Meneely	Watervleit, N. Y	Jan. 12, 1832	....
Locks, Mode of fastening keys in	S. B. Loughborough	Canandaigua, N. Y	May 29, 1866	55, 126
Locks, Nosing for	J. L. Rowe	New York, N. Y	Feb. 5, 1861	31, 356
Locks on sheet-metal, Machine for forming	J. Walker	East Bloomfield, N. Y	Apr. 1, 1851	8, 012
Locks on sheet-metal, Machine for turning	J. Wright	Rochester, N. Y	Mar. 20, 1849	6, 216
Locks, Safety-device for	G. W. Robinson	Attleborough, Mass	July 17, 1818	....
Locks, Safety-escutcheon for	H. Hungerford	Brooklyn, N. Y	July 12, 1864	43, 504
Locks, Safety-guard for	W. C. Sinclair	New York, N. Y	Sept. 22, 1868	82, 444
Locks, Strike for	W. Weisner	Elizabethtown, N. Mex	Jan. 18, 1870	99, 039
Locks, Tumbler of	H. Blakely	New York, N. Y	May 25, 1852	8, 964
Locking-bolt	H. Feyh	New York, N. Y	Sept. 13, 1870	107, 350
Locking-bolt device	S. Vanstone	North Providence, R. I	Aug. 15, 1871	118, 075
Locking-nut	C. F. Keller	Nevada, Ohio	May 4, 1869	89, 586
Locking-nut	W. Morehouse	Buffalo, N. Y	May 25, 1869	90, 377
Locking-washer	H. C. Stouffer	Lewistown, Ohio	Apr. 2, 1872	125, 226
Locket	S. J. Smith	New York, N. Y	Oct. 28, 1873	144, 153
Locket for photographs, &c., Tablet	G. W. Pitcher	Brooklyn, N. Y	July 26, 1870	105, 721
Locket, Miniature	E. N. Foote	New England Village, Mass.	Mar. 14, 1865	46, 788
Lockets, Die for forming basil of	H. Henrich	New York, N. Y	Dec. 17, 1872	133, 986
Lockets, &c., Mode of constructing	C. G. Bloomer	Wickford, R. I	Apr. 28, 1857	17, 137
Locomotion, Apparatus for	E. Matteson	Brooklyn, N. Y	May 28, 1861	32, 447
Locomotive	W. D. Arnett	Denver, Colo	July 18, 1871	117, 031
Locomotive	W. D. Arnett	Denver City, Colo	Oct. 24, 1871	120, 180
Locomotive	J. Cooke	Paterson, N. J	Sept. 19, 1871	119, 013
Locomotive	P. H. Corlett	West Manchester, Pa	Nov. 10, 1863	40, 554
Locomotive	M. N. Forney	Baltimore, Md	Feb. 6, 1866	52, 406
Locomotive	W. A. Foster	Fitchburgh, Mass	Sept. 20, 1870	107, 472
Locomotive	J. S. French	Alexandria, Va	Nov. 4, 1873	144, 271
Locomotive	R. S. Gillespie	New York, N. Y	Nov. 28, 1871	121, 283
Locomotive	H. N. Hamilton	Armonk, N. Y	Apr. 23, 1872	125, 891
Locomotive	J. Harrison, jr	Philadelphia, Pa	July 16, 1872	129, 133
Locomotive	W. S. Hudson	Paterson, N. J	July 16, 1872	129, 229
Locomotive	W. S. Hudson	Paterson, N. J	July 16, 1872	129, 230
Locomotive	W. S. Hudson	Paterson, N. J	July 16, 1872	129, 231
Locomotive	W. S. Hudson	Paterson, N. J	July 16, 1872	129, 232
Locomotive	W. S. Hudson	Paterson, N. J	July 16, 1872	129, 233
Locomotive	W. S. Hudson	Paterson, N. J	July 16, 1872	129, 234
Locomotive	W. S. Hudson	Paterson, N. J	Mar. 11, 1873	136, 729
Locomotive	J. L. Lay	Buffalo, N. Y	Oct. 29, 1867	70, 341
Locomotive	L. P. Magoon	Saint Johnsbury, Vt	July 25, 1871	117, 308
Locomotive	G. A. Nicolls	Reading, Pa	Apr. 25, 1848	5, 552
Locomotive	G. A. Nicolls	Reading, Pa	Sept. 19, 1848	5, 787
Locomotive	J. C. Parker	Chicago, Ill	Sept. 18, 1866	58, 127
Locomotive	J. B. Root	New York, N. Y	Dec. 26, 1865	51, 753
Locomotive	A. S. Smith	Boston, Mass	Mar. 28, 1871	113, 216
Locomotive	G. Thomas	Frankfort-on-the-Main, Germany.	Aug. 30, 1864	44, 055
Locomotive	J. M. Ure	Glasgow, Great Britain	Oct. 4, 1870	107, 983
Locomotive	J. L. Whetstone	Cincinnati, Ohio	Nov. 19, 1861	33, 760
Locomotive	H. Whitaker	New York, N. Y	July 23, 1872	129, 701
Locomotive and car brake	J. E. Wilson	Winnecorne, Wis	Apr. 19, 1870	102, 072
Locomotive and other engines for railways	S. H. Long	Philadelphia, Pa	Jan. 17, 1833	....
Locomotive and other wheels	G. Tefft	Salem, N. Y	May 7, 1867	64, 595
Locomotive and railway engine	I. W. Edgar	Wayne County, Ohio	July 2, 1856	....
Locomotive and railway, Traction	W. B. Hyde	Oakland, Cal	Sept. 24, 1872	131, 685
Locomotive and steam-boiler furnace	W. P. Parrott	Boston, Mass	Sept. 16, 1856	15, 742
Locomotive and steam-carriage for common roads.	J. K. Fisher	New York, N. Y	Aug. 6, 1861	32, 991
Locomotive and steam-engine boiler	J. Perkins	England	Dec. 15, 1838	1, 034
Locomotive and tender coupling	T. D. Simpson	Mount Vernon, Ohio	May 10, 1870	102, 978
Locomotive and traction engine	T. Aveling	Rochester, England	Nov. 7, 1871	120, 611
Locomotive, Animal-power	C. Masserano	Turin, Sardinia	Oct. 7, 1851	8, 417
Locomotive-apparatus	E. R. Morrison	South Bergen, N. J	Aug. 6, 1861	33, 019
Locomotive ash-pan	A. Ohlenslager	Jersey City, N. J	July 9, 1867	66, 618
Locomotive ash-pan and fire-grate	D. Upton and C. H. Nichols	Rochester and Buffalo, N. Y	Mar. 12, 1867	62, 787
Locomotive-attachment	C. Lewis	Cassville, N. Y	Sept. 20, 1870	107, 511
Locomotive-attachment	J. E. Wootten and H. Hazel	Cressona, Pa	Apr. 2, 1867	63, 600
Locomotive-axle	D. Williams	Syracuse, N. Y	Jan. 14, 1862	34, 175

Index of patents issued from the United States Patent Office from 1790 *to* 1873, *inclusive*—Continued.

Invention.	Inventor.	Residence.	Date.	No.
Locomotive-axle bearings	J. P. Laird	Altoona, Pa	Mar. 6, 1866	53, 009
Locomotive-axle bearings	D. Matthew	Philadelphia, Pa	Dec. 28, 1858	22, 439
Locomotive-boiler	J. Briggs	Louisville, Ky	Oct. 27, 1863	40, 389
Locomotive-boiler	B. Crawford	Pittsburgh, Pa	July 7, 1863	39, 123
Locomotive-boiler	S. Crawford	New York, N. Y	Aug. 8, 1865	49, 236
Locomotive-boiler	J. J. Dutcher	New Haven, Conn	Apr. 10, 1855	12, 673
Locomotive-boiler	B. L. Griffith	Hazleton, Pa	July 26, 1859	24, 862
Locomotive-boiler	S. J. Hayes and E. S. Jeffrey	Chicago, Ill	July 11, 1871	116, 836
Locomotive-boiler	G. F. Johnson	Philadelphia, Pa	May 19, 1863	38, 584
Locomotive-boiler	J. E. McConnell	Wolverton, England	June 2, 1857	17, 436
Locomotive-boiler	J. Millholland	Reading, Pa	Jan. 19, 1864	41, 316
Locomotive-boiler	J. Thompson	East Boston, Mass	May 29, 1860	28, 520
Locomotive-boiler furnace	P. W. Mackenzie	Blauveltville, N. Y	Jan. 16, 1872	122, 841
Locomotive-boiler furnace	A. J. Stevens	San Francisco, Cal	Feb. 14, 1871	111, 884
Locomotive-boiler furnace	R. Winans	Baltimore, Md	Apr. 27, 1859	20, 117
Locomotive-boiler furnace	J. Wood, jr	Conshohocken, Pa	Jan. 23, 1872	123, 069
Locomotive-boilers, Attaching steam-gage to	J. L. Eastman	Boston, Mass	Nov. 10, 1857	18, 578
Locomotive-boilers, Fire-box of	R. Winans	Baltimore, Md	Apr. 27, 1859	20, 115
Locomotive-boilers, Process for using anthracite-coal dust or slack as fuel in.	A. Berney	Jersey City, N. J	Mar. 18, 1873	136, 960
Locomotive-brake	A. De Bergue	Paris, France	Mar. 24, 1868	75, 873
Locomotive, Carriage	M. W. Baldwin	Philadelphia, Pa	May 16, 1846	4, 530
Locomotive, Carriage	W. Heston	Philadelphia, Pa	Oct. 11, 1830	......
Locomotive, Carriage	W. Howard	Baltimore, Md	Dec. 10, 1828	......
Locomotive-carriage and rail therefor	E. Kimber	Kimberton, Pa	Mar. 5, 1831	......
Locomotive-carriage, Machinery of a	S. Olcott	Harsimus, N. J	Sept. 11, 1829	......
Locomotive carrying its own railway, Traction	C. F. Mann	Troy, N. Y	Nov. 22, 1859	26, 195
Locomotive center-bearing	B. W. Healey	Providence, R. I	June 8, 1869	91, 015
Locomotive cow-catcher	J. Mitchell	Osceola, Iowa	Oct. 6, 1857	18, 348
Locomotive cow-catcher	J. Noble	Pittsburgh, Pa	Apr. 21, 1868	76, 938
Locomotive cow-catcher	J. Wisner	Aurora, N. Y	Feb. 10, 1857	16, 623
Locomotive crank-pins from true center, Apparatus for determining deviation of.	C. J. Clifford	New Hampton, N. J	Jan. 14, 1868	73, 233
Locomotive-cross-head jack	J. S. Funk	Marysville, Pa	Mar. 1, 1870	100, 390
Locomotive cross-heads, Rest for	S. A. Alexander and E. Dunn	Sunbury, Pa	Aug. 31, 1869	94, 171
Locomotive draft-pipe	J. M. Foss	Concord, N. H	June 12, 1866	55, 479
Locomotive draft-pipe	A. Pearsall	Atlanta, Ga	Mar. 5, 1867	62, 680
Locomotive draw-bar	D. C. Cannell	La Fayette, Ind	Apr. 23, 1867	64, 071
Locomotive driving-box	J. E. Wootton	Philadelphia, Pa	Dec. 1, 1857	18, 786
Locomotive driving-power	W. D. Arnett	Denver, Colo	Feb. 7, 1871	111, 603
Locomotive driving-wheel	J. Birkenhead	Canton, Mass	July 28, 1868	80, 384
Locomotive driving-wheel	W. J. Johnson	New Orleans, La	Mar. 23, 1869	88, 178
Locomotive driving-wheels, Apparatus for distributing sand to produce adhesion of.	E. Tolles	New York	Oct. 9, 1841	2, 283
Locomotive driving wheels, Apparatus for turning crank-pins on.	C. J. Clifford	New Hampton, N. J	Sept. 3, 1867	68, 415
Locomotive driving-wheels, Increasing adhesion of.	E. L. Miller	Charleston, S. C	June 19, 1834	......
Locomotive driving-wheels, Mode of increasing adhesion of.	J. L. Mott	New York	Aug. 28, 1841	2, 228
Locomotive-engine	H. W. Adams	Philadelphia, Pa	Aug. 20, 1872	130, 685
Locomotive-engine	W. S. G. Baker	Baltimore, Md	Nov. 6, 1866	59, 340
Locomotive-engine	M. W. Baldwin	Philadelphia, Pa	Dec. 31, 1840	1, 921
Locomotive-engine	F. Barker	Baltimore, Md	Sept. —, 1836	......
Locomotive-engine	C. W. Cahoon	Portland, Me	Aug. 31, 1869	94, 396
Locomotive-engine	H. R. Campbell	Philadelphia, Pa	Feb. 5, 1836	......
Locomotive-engine	J. Clark	London, England	Nov. 7, 1871	120, 713
Locomotive-engine	J. M. Coale	Baltimore, Md	May 21, 1867	64, 842
Locomotive engine	I. H. Congdon	Springfield, Ill	Aug. 23, 1864	43, 898
Locomotive-engine	E. Crane	Dorchester, Mass	Nov. 22, 1859	26, 165
Locomotive-engine	R. F. Fairlie	London, England	Jan. 16, 1866	52, 117
Locomotive-engine	R. F. Fairlie	Westminster, England	Oct. 3, 1871	119, 591
Locomotive-engine	A. J. Graham	Portland, Oreg	Sept. 4, 1860	29, 879
Locomotive-engine	H. Gray	Boston, Mass	May 5, 1857	17, 213
Locomotive-engine	W. S. Hudson	Paterson, N. J	Dec. 10, 1872	133, 859
Locomotive-engine	S. L. Langdon	New Orleans, La	Mar. 10, 1868	75, 280
Locomotive-engine	J. Lewis	Manchester, England	Nov. 26, 1872	133, 321
Locomotive-engine	Z. H. Mann and L. B. Thyng	Lowell, Mass	Mar. 10, 1838	628
Locomotive-engine	L. Rarchaert	Paris, France	May 21, 1867	65, 008
Locomotive-engine	H. R. Remsen and P. M. Hutton	Troy, N. Y	June 29, 1852	9, 078
Locomotive-engine	S. Samuels and W. J. Brassington.	Mott Haven and Brooklyn, N. Y.	June 18, 1867	65, 951
Locomotive-engine	C. and G. Sellers	Philadelphia, Pa	May 22, 1835	......
Locomotive-engine	S. Skillman	Jersey City, N. J	Feb. 6, 1872	123, 367
Locomotive-engine	S. Skillman	Jersey City, N. J	Aug. 27, 1872	130, 875
Locomotive-engine	R. Winans	Baltimore, Md	Apr. 6, 1858	19, 889
Locomotive-engine	R. Winans	Baltimore, Md	Apr. 13, 1858	19, 962
Locomotive-engine	R. Winans	Baltimore, Md	Aug. 24, 1858	21, 290
Locomotive-engine	R. Winans	Baltimore, Md	Jan. 11, 1859	22, 597
Locomotive-engine	D. Winder	Xenia, Ohio	May 10, 1853	9, 716
Locomotive-engine	S. Wright	Philadelphia, Pa	Dec. 26, 1837	540
Locomotive-engine alarm	R. A. Filkins	North Adams, Mass	Jan. 3, 1871	110, 613
Locomotive-engine and car, Combining	M. McDowell and N. W. Wheeler.	New York and Brooklyn, N. Y.	Jan. 31, 1860	27, 002
Locomotive-engine and railway-car	M. W. Baldwin	Philadelphia, Pa	Sept. 10, 1834	......
Locomotive-engine boiler	J. Penniman	Baltimore, Md	Apr. 24, 1840	1, 568
Locomotive-engine boiler	R. Winans	Baltimore, Md	Apr. 27, 1859	20, 116
Locomotive-engine boilers, Fire-box of	R. Winans	Baltimore, Md	Apr. 27, 1859	20, 114
Locomotive-engine, car, and carriage	A. C. Jones	Philadelphia, Pa	June 14, 1834	......
Locomotive-engine, Condensing double-end	J. P. Woodbury	West Roxbury, Mass	Feb. 21, 1871	112, 001
Locomotive-engine driving-wheel	A. F. Cooper	San Francisco, Cal	Sept. 19, 1871	119, 014
Locomotive-engine for ascending inclined planes	L. Burnell	Elyria, Ohio	June 26, 1834	......
Locomotive-engine for ascending inclined planes	D. H. Dotterer and T. Jackson	Reading, Pa	Mar. 25, 1839	1, 108
Locomotive-engine for ascending inclined planes	S. Marsh	West Roxbury, Mass	Sept. 10, 1861	33, 255
Locomotive-engine for inclined planes	J. Ruggles	Thomaston, Me	July 13, 1836	......
Locomotive-engine for inclined planes, Railway	E. F. Aldrich	New York	Oct. 12, 1837	426
Locomotive-engine for producing increased adhesion to the rails.	F. Windhausen	Duderstadt, Hanover	Dec. 29, 1857	19, 000
Locomotive-engine for railways, Constructing	H. Waterman	Hudson, N. Y	Feb. 10, 1841	1, 969

110 P

Index of patents issued from the United States Patent Office from 1790 *to* 1873, *inclusive*—Continued.

Invention.	Inventor.	Residence.	Date.	No.
Locomotive-engine furnace	R. Gill and G. W. Grier	Altoona, Pa	July 26, 1859	24, 867
Locomotive-engine furnace	D. W. Wyman	New York, N. Y	Apr. 20, 1869	89, 261
Locomotive-engine furnaces, Apparatus for increasing draft of.	J. Sweeney	Chicago, Ill	Aug. 6, 1861	33, 011
Locomotive-engine houses, Smoke-stack for	H. Clayton	Tamaqua, Pa	May 3, 1859	23, 825
Locomotive-engine lubricator	E. C. Hamlin	Pavilion, N. Y	June 23, 1863	38, 962
Locomotive-engine running-gear	S. Norris	Philadelphia, Pa	Sept. 26, 1854	11, 733
Locomotive-engine running-gear	J. L. Whetstone	Cincinnati, Ohio	Apr. 10, 1860	27, 850
Locomotive-engine smoke-stack	R. Frazer	Camden County, N. J	Sept. 30, 1873	143, 339
Locomotive-engine smoke-stack	L. Fulton	Cincinnati, Ohio	Apr. 13, 1869	88, 864
Locomotive-engine smoke-stack	C. P. Noble	Chicago, Ill	Aug. 11, 1863	39, 493
Locomotive-engine wheel	G. S. Griggs	Roxbury, Mass	Dec. 29, 1857	18, 966
Locomotive-engine wheel and boiler-tube	M. W. Baldwin	Philadelphia, Pa	Apr. 3, 1835	....
Locomotive-engine wheel and boiler-tube for railways.	M. W. Baldwin	Philadelphia, Pa	Aug. 17, 1835	....
Locomotive-engine wheel, axle, and railway	R. Berrian	New York	Nov. 13, 1832	....
Locomotive-engines and railway-cars, Making springs for.	I. Oberhausser	Charleston, S. C	Feb. 15, 1838	605
Locomotive-engines, Apparatus for reversing or stopping.	J. Cunningham	Reading, Pa	Oct. 22, 1850	7, 730
Locomotive-engines, Apparatus for supplying pure water to.	H. V. P. Draper and N. Covington.	Hannibal, Mo	Dec. 24, 1872	134, 263
Locomotive-engines, Arrangement for disposing of the sparks from.	L. P. Teed	White Deer Mills, Pa	Aug. 7, 1860	29, 531
Locomotive-engines, Arrangement of cylinders and their connection for.	A. Smethurst	Philadelphia, Pa	Nov. 24, 1857	18, 712
Locomotive-engines, Arrangement of exhaust-pipes in.	J. Williams	Dunkirk, N. Y	Mar. 6, 1855	12, 495
Locomotive-engines, Ash-pan for	G. Davis	Roxbury, Mass	Oct. 11, 1853	10, 104
Locomotive-engines, Boiler and water-heater of	T. Perkins	Baltimore, Md	June 26, 1849	6, 561
Locomotive-engines by hand-power, Moving	J. E. Wootten	Philadelphia, Pa	Nov. 29, 1859	26, 318
Locomotive-engines, Device for discharging the smoke, cinders, &c., from.	J. Greacen, jr	New York, N. Y	Sept. 4, 1860	29, 880
Locomotive-engines, Exhaust-device for	J. Densmore, sr	Erie, Pa	Jan. 5, 1869	85, 571
Locomotive-engines, Exhaust-pipe of	G. W. Lathrop	Weedsport, N. Y	Dec. 18, 1860	30, 976
Locomotive-engines, Fire-box for	L. Crossman and S. Atkinson	Elizabeth City, N. J	July 20, 1858	20, 937
Locomotive-engines, Fire-box of	R. Greenwood	Altoona, Pa	Aug. 9, 1859	25, 009
Locomotive-engines for removing objects from the track, Attachment to.	C. H. Eisenbrandt	Baltimore, Md	June 14, 1859	24, 383
Locomotive-engines for working heavy grades, Boiler and gearing of.	G. E. Sellers	Cincinnati, Ohio	July 9, 1850	7, 498
Locomotive-engines, Managing and supplying fire for generating steam in.	M. W. Baldwin	Philadelphia, Pa	Oct. 15, 1836	54
Locomotive-engines on railways, Braking	M. W. Verdin	Baltimore, Md	July 5, 1859	24, 680
Locomotive-engines on railways, Propelling	M. W. Verdin	Baltimore, Md	July 5, 1859	24, 679
Locomotive-engines, Sand-box for	T. Price and J. H. Wilson	Covington, Ky	Mar. 13, 1866	53, 182
Locomotive-engines, Signal-mechanism for	J. M. Cook	Taunton, Mass	Apr. 29, 1862	35, 082
Locomotive-engines, Spark-arrester and chimney of.	W. A. Peaslee and J. O. D. Lilly.	Indianapolis, Ind	Dec. 6, 1859	26, 373
Locomotive-engines, Superheating steam for	J. Martin	Toronto, Canada	Dec. 4, 1860	30, 857
Locomotive-engines with water, Supplying	A. D. Merritt and C. Kemplen	Woodstock, Ill	Dec. 4, 1860	30, 829
Locomotive-engines, Traction for	T. Selleck	Greenwich, Conn	Sept. 16, 1862	36, 486
Locomotive exhaust-pipe	A. S. Smith	Boston, Mass	Apr. 9, 1872	125, 414
Locomotive, Farm	D. F. Leach	Forsyth, Ill	Aug. 3, 1869	93, 316
Locomotive-fender with fuel, Apparatus for supplying.	A. C. Land	Garlandville, Miss	May 7, 1872	126, 466
Locomotive fire-box	C. Dean	Saint Catharine's, Canada	Dec. 1, 1863	40, 743
Locomotive fire-box	C. H. Farley	Portland, Me	Jan. 14, 1873	134, 741
Locomotive fire-box	J. P. Laird	Altoona, Pa	Nov. 17, 1863	40, 630
Locomotive fire-box	W. R. Thomas	Catasauqua, Pa	Mar. 15, 1859	23, 278
Locomotive fire-box	J. L. Vauclain	La Fayette, Ind	Aug. 13, 1861	33, 057
Locomotive fire-box	R. and T. Winans	Baltimore, Md	May 9, 1854	10, 901
Locomotive-fire-box crown	H. C. Darby	Wyandotte, Kans	June 24, 1873	140, 251
Locomotive fire-box, Removable	J. I. De Haven	Reading, Pa	Apr. 24, 1849	6, 376
Locomotive foot-board	G. F. Morse	Portland, Me	Apr. 26, 1870	102, 299
Locomotive for ascending and descending inclined planes.	E. Coleman	Philadelphia, Pa	Dec. 31, 1845	4, 342
Locomotive for ascending inclined planes	A. Cathcart	Madison, Iowa	Oct. 23, 1849	6, 818
Locomotive for ascending inclined planes	N. Riggenbach	Olten, Switzerland	Feb. 13, 1872	123, 729
Locomotive for consuming sparks	F. B. Longmire and H. J. Brooke.	Philadelphia, Pa	July 10, 1840	1, 681
Locomotive for plowing, &c	T. S. Minniss	Meadville, Pa	Jan. 15, 1867	61, 231
Locomotive, Friction	T. S. Minniss	Meadville, Pa	Sept. 27, 1870	107, 702
Locomotive-furnace	O. W. Bayley	Manchester, N. H	Apr. 17, 1855	12, 740
Locomotive-furnace	O. W. Bayley	Boston, Mass	Feb. 2, 1858	19, 277
Locomotive-furnace	T. Davis	Cleveland, Ohio	Dec. 31, 1872	134, 422
Locomotive-furnace	G. S. Griggs	Roxbury, Mass	Dec. 15, 1857	18, 882
Locomotive-furnace	A. J. Stevens	Sacramento, Cal	Dec. 23, 1873	145, 819
Locomotive guide-wheel	S. Norris	Philadelphia, Pa	Apr. 24, 1860	28, 007
Locomotive, Horse-power	J. C. Miller	Union Township, Ohio	Oct. 11, 1859	25, 752
Locomotive Horse-power	J. R. Sleeper	Philadelphia, Pa	Oct. 30, 1834	....
Locomotive-links, Machine for slotting	W. H. Denny	Philadelphia, Pa	June 24, 1873	140, 121
Locomotive, Marine	S. Lee	New York, N. Y	May 16, 1871	114, 832
Locomotive-pilot	B. F. Partridge, jr	Columbus, Ky	May 28, 1867	65, 264
Locomotive, portable, and other steam-engines	L. Perkins	London, England	Dec. 16, 1873	145, 525
Locomotive, Portable traction	J. Barrans	Peckham, England	Nov. 8, 1859	26, 074
Locomotive-power machine for removing houses	S. Compton, jr	Elmira, N. Y	July 31, 1837	331
Locomotive recording-instrument	J. D. Richardson	Houston, Tex	Apr. 29, 1873	138, 437
Locomotive, Road	E. C. James	Baltimore, Md	Dec. 17, 1867	72, 397
Locomotive, Road, &c	J. Robingson	New Brighton, Pa	Sept. 30, 1856	15, 820
Locomotive running-gear	G. S. Griggs	Roxbury, Mass	June 17, 1851	8, 166
Locomotive running-gear	J. H. Murrill	Manchester, Va	Oct. 7, 1851	8, 410
Locomotive running-gear	T. H. Neal	Pittsburgh, Pa	June 30, 1863	39, 060
Locomotive running-gear	J. C. Story	Cincinnati, Ohio	Nov. 21, 1865	51, 098
Locomotive running-gear	R. Winans	Baltimore, Md	Dec. 2, 1851	8, 571
Locomotive sanding-device	M. V. Nobles	Elmira, N. Y	Apr. 1, 1873	137, 472
Locomotive-signal	A. E. Turnbull	Springfield, Ohio	June 15, 1858	20, 590
Locomotive signal-light	A. Dick	Hamilton, Canada	July 23, 1872	129, 797

Index of patents issued from the United States Patent Office from 1790 *to* 1873, *inclusive*—Continued.

Invention.	Inventor.	Residence.	Date.	No.
Locomotive smoke and cinder conveyer	A. M. Rodgers	Brooklyn, N. Y	May 9, 1871	114, 605
Locomotive smoke and spark conveyer	E. A. C. Fox	Frederick City, Md	Sept. 16, 1873	142, 906
Locomotive smoke-conductor	A. Storm	Matteawan, N. Y	Sept. 24, 1872	131, 639
Locomotive smoke-stack	J. Atkins	Washington, D. C	July 28, 1868	80, 437
Locomotive smoke-stack	L. Bell	Washington, D. C	May 23, 1871	115, 153
Locomotive smoke-stack	J. V. Bishop	Atlanta, Ga	Nov. 25, 1873	144, 884
Locomotive smoke-stack	H. Brooks and J. Ball	Zanesville, Ohio	Oct. 30, 1866	59, 175
Locomotive smoke-stack	E. A. Castellaw	Savannah, Ga	Apr. 25, 1871	113, 980
Locomotive smoke-stack	E. Fontaine	Fort Wayne, Ind	Nov. 1, 1870	108, 899
Locomotive smoke-stack	S. Ham	Philadelphia, Pa	Nov. 22, 1864	45, 204
Locomotive smoke-stack	J. Hughes	Scranton, Pa	Sept. 16, 1873	142, 851
Locomotive smoke-stack	J. A. W. Justi	Savannah, Ga	Oct. 27, 1868	83, 506
Locomotive smoke-stack	W. J. Mehary	Philadelphia, Pa	July 5, 1870	105, 107
Locomotive smoke-stack	A. S. Sweet, jr	Detroit, Mich	June 23, 1863	38, 992
Locomotive smoke-stack	J. L. Vauclain	La Fayette, Ind	Aug. 20, 1861	33, 114
Locomotive smoke-stack	E. H. Winchell	New York, N. Y	Oct. 10, 1871	119, 803
Locomotive smoke-stack, Self-cleaning	S. M. Cummings and H. Isreal	Allegheny, Pa	Dec. 27, 1870	110, 441
Locomotive smoke-stack, Arrangement of vertical-tube feed-water heaters in.	M. W. Baldwin and D. Clark	Philadelphia, Pa	Feb. 14, 1854	10, 514
Locomotive smoke-stack, Base-piece of	P. S. Ebbert	Chicago, Ill	July 1, 1856	15, 225
Locomotive smoke-stack, Cone for	D. B. Strope	Fort Wayne, Ind	Nov. 14, 1871	120, 835
Locomotive smoke-stack, Exhaust-device for	R. H. Lecky	Allegheny City, Pa	July 13, 1869	92, 617
Locomotive spark and smoke conductor	J. Viall	Somerville, Mass	June 1, 1869	90, 899
Locomotive spark-arrester	M. Brassill	Hartford, Conn	Oct. 14, 1873	143, 664
Locomotive spark-extinguisher	P. H. Corlett	Manchester, Pa	Aug. 18, 1863	39, 552
Locomotive steam-boiler	E. Ayer	Norwich, Conn	May 2, 1848	5, 546
Locomotive steam-engine	M. W. Baldwin	Philadelphia, Pa	Aug. 25, 1842	2, 759
Locomotive steam-engine	A. Haenpel and J. Reinhardt	Philadelphia, Pa	June 30, 1868	79, 341
Locomotive steam-engine	R. F. Fairlie	London, England	Dec. 22, 1868	85, 076
Locomotive steam-engine	E. F. Johnson	Middletown, Conn	Dec. 31, 1844	3, 866
Locomotive steam-engine	G. G. Jones	Rushsylvania, Ohio	July 6, 1869	92, 193
Locomotive steam-engine, &c	S. H. Long	Philadelphia, Pa	Dec. 28, 1832	......
Locomotive steam-engine	J. S. Stuart	Philadelphia, Pa	May 4, 1869	89, 809
Locomotive steam-engine	A. Whitney	Rotterdam, N. Y	June 27, 1840	1, 653
Locomotive steam-engine	R. Winans	Baltimore, Md	July 29, 1837	308
Locomotive steam-engine	R. Winans	Baltimore, Md	July 29, 1837	311
Locomotive steam-engine	R. Winans	Baltimore, Md	July 28, 1843	3, 201
Locomotive steam-engine carriage	S. Norris and W. Knight	Philadelphia, Pa	Feb. 10, 1843	2, 951
Locomotive steam-engine for rail and other roads	J. Ruggles	United States	July 28, 1836	1
Locomotive steam-engines, Framing of	R. Winans	Baltimore, Md	July 29, 1837	305
Locomotive steam-engines, Method of connecting driving-wheels of.	R. S. Stevens	New York, N. Y	Sept. 3, 1842	2, 773
Locomotive-tender	W. Ross and W. E. Rutter	Providence, R. I	May 16, 1848	5, 584
Locomotive-tender	A. Sturrock	Doncaster, England	Nov. 24, 1863	40, 715
Locomotive-tender	R. and T. Winans	Baltimore, Md	May 23, 1854	10, 971
Locomotive-tender with water, Device for supplying.	W. E. Prall	Washington, D. C	July 4, 1871	116, 752
Locomotive-tender with water, Method of supplying.	W. J. Brassington and W. B. Burtnett.	Brooklyn and New York, N. Y.	June 18, 1867	65, 869
Locomotive-tenders, Machine for loading	J. N. Jackson	Brookhaven, Miss	June 15, 1869	91, 236
Locomotive, Traction	T. S. Winniss	Meadville, Pa	Jan. 28, 1873	135, 240
Locomotive vehicle for running on ice or in water.	N. Wiard	Janesville, Wis	Jan. 24, 1860	26, 960
Locomotive, Water	G. A. Fall	Hoboken, N. J	Oct. 12, 1869	95, 785
Locomotive water-supply pipe	F. Gerrard	Kansas City, Mo	Jan. 11, 1870	98, 685
Locomotive-wheel	M. J. Montgomery	New York, N. Y	May 31, 1864	42, 958
Locomotive-wheel	J. R. Richardson	Max Meadows, Va	July 19, 1870	105, 598
Locomotive-wheel for ascending inclined planes on railways.	E. Town	Montpelier, Vt	July 31, 1837	339
Locomotive-wheels, Apparatus for applying chalk to.	N. Sehner	Hagerstown, Md	Feb. 27, 1872	124, 165
Locomotive-wheels, Counterpoise to cast	H. A. Chase	Boston, Mass	May 23, 1854	10, 958
Locomotive-wheels, Mode of chilling rims for	H. W. Moore	Jersey City, N. J	Apr. 26, 1859	23, 775
Locomotive-window	J. H. Dinsmore	Boston, Mass	Dec. 23, 1873	145, 790
Locomotive-window	H. Skinner	Fulton, N. Y	Sept. 29, 1857	18, 304
Locomotive with driving-axle above the boiler	R. H. Emerson	Portland, Me	May 1, 1849	6, 401
Locomotive with spading apparatus, Agricultural	C. Locher	Oroville, Cal	Oct. 13, 1868	82, 963
Locomotives, Adjusting cross-heads of	J. D. Akley and R. English	Oil City, Pa	May 24, 1870	103, 407
Locomotives, Apparatus for burning soft coal in	A. C. Rand	Aurora, Ill	Dec. 24, 1872	134, 314
Locomotives, Arrangement of deflecting-plates and spark-receivers in.	W. H. Bullock	Boston, Mass	Dec. 29, 1857	18, 953
Locomotives, Arrangement of means for regulating the discharge of exhaust steam in.	J. E. Wootten	Philadelphia, Pa	May 1, 1855	12, 805
Locomotives, Arrangement of means for regulating the fire of coal-burning.	J. M. Hartnett	Waukegan, Ill	July 21, 1857	17, 834
Locomotives, Automatic apparatus for operating valves of exhaust-pipes of.	R. C. Huse, jr	Georgetown, Mass	Apr. 4, 1871	113, 299
Locomotives by stationary power, Propelling	J. A. Etzler	Philadelphia, Pa	Dec. 23, 1841	2, 396
Locomotives, Conveyer of smoke and cinders for	A. M. Rodgers	Brooklyn, N. Y	Oct. 10, 1871	119, 790
Locomotives, Device for preventing collision of	H. Payne, sr	Mount Vernon, Ohio	Jan. 22, 1867	61, 354
Locomotives for ascending inclined planes, Cog-gearing of.	W. Hoyt	Dupont, Ind	Apr. 17, 1849	6, 321
Locomotives for railways, Constructing and operating.	R. P. Morgan	Chicago, Ill	Nov. 8, 1864	44, 997
Locomotives, Heating feed-water apparatus of	P. S. Ebbert	Chicago, Ill	May 5, 1857	17, 208
Locomotives in engine-houses, Arrangement for carrying off smoke from.	J. O. D. Lilly, J. S. Vanclain, and J. W. Lilly.	La Fayette, Ind	Feb. 23, 1858	19, 469
Locomotives, Machinery for oiling the journals of	S. Scotton	Richmond, Ind	Jan. 11, 1859	22, 586
Locomotives, Manner of attaching legs to walking	S. G. Hoge	Bellefontaine, Ohio	Feb. 23, 1858	19, 468
Locomotives, Means for increasing draft in	C. F. Thomas	Taunton, Mass	July 10, 1855	13, 238
Locomotives, Method of increasing traction in	C. W. T. Krausch	Philadelphia, Pa	Jan. 29, 1867	61, 546
Locomotives, &c., Mode of applying steam-power to.	J. Ericsson	Sweden	Nov. 5, 1840	1, 847
Locomotives, Mode of supplying water to	S. Vail	Morristown, N. J	July 12, 1839	1, 234
Locomotives, Mode of turning	G. M. Cole	Folsom City, Cal	Sept. 17, 1861	33, 294
Locomotives, Petticoat-pipe for	W. G. Freeman	Richmond, Va	June 9, 1868	78, 658
Locomotives, Storing power in pneumatic	H. F. C. Krumme	Ridgway, Pa	Apr. 5, 1870	101, 631
Locomotives with compressed air from fixed stations, Mode of charging the receivers of.	S. Carson	New York, N. Y	Dec. 9, 1856	16, 220

Index of patents issued from the United States Patent Office from 1790 to 1873, inclusive—Continued.

Invention.	Inventor.	Residence.	Date.	No.
Log and leeway indicator, Marine	A. E. Lozier	New York, N. Y	Mar. 15, 1864	41, 932
Log and sounding line, Combined	A. Pécoul	Marseilles, France	Dec. 25, 1855	13, 994
Log-binder	W. L. Dean	Dayton, N. Y	Sept. 5, 1871	118, 697
Log-boat	A. Olmsted	Windsor, Mich	Apr. 27, 1869	89, 332
Log-canting apparatus	B. R. Stevens	Grand Rapids, Mich	Jan. 5, 1869	85, 543
Log-casting machine	A. Rodgers	Muskegon, Mich	Aug. 26, 1873	142, 280
Log, Electro-magnetic	W. I. Reid	Brooklyn, N. Y	Apr. 7, 1868	76, 521
Log, Hand	J. R. St. John	New York, N. Y	May 6, 1851	8, 074
Log-lifter	G. B. Sims	Elizabeth, Ind	Mar. 5, 1872	124, 294
Log-loader	J. H. Harvey	Chanticleer, Ohio	May 31, 1870	103, 608
Log loader and piler	J. G. Brady	Forest Hill, Mich	June 21, 1870	104, 548
Log-moving machine	G. M. Hinckley and A. Rodgers	Milwaukee, Wis., and Muskegon, Mich.	June 3, 1873	139, 578
Log, Nautical	T. Hotchkiss	Stratford, Conn	Nov. 15, 1864	45, 042
Log, Registering marine	A. Gordan	New York, N. Y	Dec. 8, 1863	40, 834
Log-rolling device	T. Emery	Peshtigo, Wis	Feb. 25, 1873	136, 145
Log-rolling machine	J. Torrent	Muskegon, Mich	Oct. 24, 1871	120, 220
Log-rolling machine	J. Torrent	Muskegon, Mich	July 11, 1871	117, 017
Log-rossing machine	G. W. Nichols	Clinton, Iowa	June 18, 1872	128, 060
Log-turner	H. T. Hunter	Spring Lake, Mich	Oct. 8, 1872	131, 956
Log-turner	J. C. Moore	Ypsilanti, Mich	Apr. 29, 1873	138, 270
Log-turner	I. W. Pool	Eau Claire, Wis	July 1, 1873	140, 386
Log-turner	E. Tarrant	Muskegon, Mich	Aug. 12, 1873	141, 736
Log-turner	J. Torrent	Muskegon, Mich	Aug. 12, 1873	141, 739
Log-turner	C. Van Vleck	Sidney, Mich	Aug. 19, 1873	141, 969
Log turning and canting device	D. S. Griffes	Flint, Mich	May 23, 1871	115, 194
Log turning and loading device	S. Snyder	Delaware, Ohio	July 6, 1869	92, 388
Log-turning apparatus	L. J. Stannard and M. L. Perry	Newark, N. Y	Mar. 23, 1869	88, 093
Log-turning device for mills	I. H. Newton	Grand Rapids, Mich	Mar. 23, 1869	88, 063
Log-turning machine	G. W. Baker	Elizabeth City, N. C	Nov. 19, 1872	133, 185
Log-turning machine	E. C. Dicey	Whitehall, Mich	Aug. 27, 1872	130, 904
Log-turning machine	W. E. Hill	Erie, Pa	Aug. 9, 1870	106, 160
Log-turning machine	E. H. Stearns	Erie, Pa	Dec. 20, 1870	110, 398
Log-turning machine	E. H. Stearns	Erie, Pa	Aug. 8, 1871	117, 828
Logs, Building flumes for floating	J. Du Bois	Williamsport, Pa	Jan. 19, 1864	41, 285
Logs, Hauling up	S. H. Richardson	Bangor, Me	May 24, 1870	103, 372
Logs, Implement for rolling and piling	W. Todd	Cherryfield, Me	Aug. 31, 1858	21, 386
Logs on wagons, Apparatus for loading	P. Gilbert	Alexandria, Ohio	Sept. 22, 1857	18, 242
Logging-skid	G. W. Nichols	River Falls, Wis	Jan. 1, 1867	60, 926
Logotrope	C. Richardson and J. Graeme, jr	New York, N. Y	Oct. 1, 1867	69, 482
Logotype	W. H. Wilkinson	Southwick, Mass	Aug. 23, 1870	106, 641
Longimeter	A. Bayrd	South Reading, Mass	July 16, 1828	....
Longitude, Mode of finding	J. Henry	Maysville, Ky	Sept. 13, 1820	....
Looking-glass	G. H. Chinnock	Brooklyn, N. Y	Sept. 24, 1872	131, 665
Looking-glass	H. Willard	Vergennes, Vt	May 20, 1862	35, 344
Looking-glass	U. A. Woodbury	Morrisville, Vt	Apr. 28, 1868	77, 229
Looking-glass and match-holder	D. Cumming, jr	New York, N. Y	July 21, 1863	39, 277
Looking-glass attachment	C. J. Hartmann	London, England	June 1, 1869	90, 746
Looking-glass bracket	F. C. Brock	Oriskany Falls, N. Y	Aug. 20, 1872	130, 563
Looking-glass frames to bureaus, Attaching	C. Kilburn	Philadelphia, Pa	Aug. 2, 1870	106, 065
Looking-glass or mirror	S. F. Brooks	Weston, Mass	Apr. 24, 1860	27, 965
Looking-glass support	W. H. Gray	New York, N. Y	Dec. 15, 1868	84, 944
Looking-glasses, Silvering	T. Drayton	Brighton, England	Aug. 12, 1844	3, 702
Looking-glasses, Silvering	J. Webster	Brooklyn, N. Y	May 8, 1855	12, 840
Looking-glasses and pictures, Device for hanging	D. Crowell, jr	Yarmouth Port, Mass	June 6, 1871	115, 713
Looking-glasses in dressing-cases, Device for the Adjustment of.	A. A. Gray and W. C. Hyde	Detroit, Mich	Nov. 12, 1867	70, 713
Looking-glasses, Process for making	L. P. Augenard	New York, N. Y	Jan. 31, 1865	46, 062
Loom	R. W. Andrews	Staffordville, Conn	Oct. 3, 1865	50, 215
Loom	H. Bachofner	Springfield, Mass	Oct. 30, 1849	6, 823
Loom	C. F. Bartling	Greenville, Ohio	Apr. 17, 1866	53, 943
Loom	C. Bergen	New York	June 14, 1826	....
Loom	A. Bigelow and J. Butler	Granville, Ohio	Jan. 16, 1849	6, 035
Loom	E. B. Bigelow	Boston, Mass	Apr. 8, 1856	14, 590
Loom	E. B. Bigelow	Boston, Mass	Apr. 20, 1869	89, 011
Loom	E. B. Bigelow	Boston, Mass	Mar. 22, 1870	101, 088
Loom	S. Boorn	Lowell, Mass	July 22, 1862	35, 918
Loom	S. Boorn	Lowell, Mass	May 16, 1871	114, 756
Loom	W. H. Boozer	Potter's Mills, Pa	Mar. 14, 1871	112, 679
Loom	A. H. Boyd	Saco, Me	Oct. 15, 1850	7, 717
Loom	J. L. Branson	Pittsburgh, Pa	Nov. 23, 1869	97, 158
Loom	J. Braun	Philadelphia, Pa	Aug. 15, 1865	49, 369
Loom	J. Braun	Philadelphia, Pa	Mar. 20, 1866	53, 265
Loom	D. Briggs	Canaan, Conn	Feb. 19, 1814	....
Loom	S. Briggs	Canaan, Conn	May 27, 1813	....
Loom	A. Brigham	Manchester, N. H	Jan. 2, 1855	12, 120
Loom	W. Breitenstein	New York, N. Y	Apr. 14, 1863	38, 195
Loom	W. Breitenstein	New York, N. Y	Jan. 24, 1865	45, 969
Loom	J. Broadbent	Oak Grove, Ky	Aug. 7, 1855	13, 382
Loom	S. Bronson	Kent, Conn	Nov. 25, 1814	....
Loom	E. W. Brown	Fall River Mass	July 25, 1854	11, 352
Loom	O. C. Burr	Millbury, Mass	July 17, 1835	....
Loom	A. Campbell	Kingston, Pa	Mar. 28, 1812	....
Loom	A. Canis and F. Higgins	Manchester, N. H	Feb. 25, 1873	136, 133
Loom	N. B. Carney	New York, N. Y	May 19, 1857	17, 353
Loom	S. Chapman	Otsego, N. Y	Aug. 30, 1816	....
Loom	S. Chidester	Windham, N. Y	Sept. 2, 1826	....
Loom	T. Clarkson	South Adams, Mass	May 1, 1860	28, 125
Loom	G. Cliff	Memphis, Mich	Apr. 28, 1868	77, 169
Loom	J. H. Clifton	Newcastle, Pa	May 29, 1860	28, 453
Loom	R. Collins	Cabotsville, Mass	Oct. 2, 1847	5, 318
Loom	A. Conant	Willimantic, Conn	Apr. 17, 1860	27, 889
Loom	J. C. Cooke	Middletown, Conn	Apr. 10, 1860	27, 860
Loom	C. Cooper	Lebanon, Pa	Nov. 4, 1808	....
Loom	C. Cooper and G. Shanke	Lebanon, Pa	Aug. 28, 1812	....
Loom	C. Cooper and G. Shalk	Lebanon, Pa	May 31, 1815	....
Loom	H. M. Cooper	Lindley, Mo	Nov. 19, 1867	70, 964
Loom	G. Copeland	Lewiston, Me	Jan. 23, 1855	12, 293

Index of patents issued from the United States Patent Office from 1790 *to* 1873, *inclusive*—Continued.

Invention.	Inventor.	Residence.	Date.	No.
Loom	G. Copeland	North Gray, Me	Apr. 9, 1861	31, 956
Loom	J. D. Cottrell	Milford, Mass	Oct. 27, 1863	40, 442
Loom	J. D. Cottrell and G. Draper	Milford, Mass	Oct. 1, 1867	69, 320
Loom	J. Coulter and S. Gano	Berkeley County, Va	Apr. 18, 1814	....
Loom	S. Craige	Philadelphia, Pa	Oct. 18, 1814	....
Loom	G. Crompton	Worcester, Mass	Apr. 27, 1858	20, 044
Loom	G. Crompton	Worcester, Mass	Apr. 23, 1861	32, 123
Loom	G. Crompton	Worcester, Mass	Oct. 22, 1861	33, 520
Loom	G. Crompton	Worcester, Mass	July 16, 1867	66, 682
Loom	G. Crompton	Worcester, Mass	Mar. 17, 1868	75, 530
Loom	G. Crompton	Worcester, Mass	Apr. 28, 1868	77, 361
Loom	G. Crompton	Worcester, Mass	Aug. 4, 1868	80, 608
Loom	G. Crompton	Worcester, Mass	Aug. 11, 1868	80, 810
Loom	G. Crompton	Worcester, Mass	Aug. 18, 1868	81, 070
Loom	G. Crompton	Worcester, Mass	Aug. 25, 1868	81, 347
Loom	G. Crompton	Worcester, Mass	Dec. 29, 1868	85, 432
Loom	G. Crompton	Worcester, Mass	Aug. 31, 1869	94, 401
Loom	G. Crompton	Worcester, Mass	Sept. 7, 1869	94, 571
Loom	G. Crompton	Worcester, Mass	Sept. 14, 1869	94, 873
Loom	G. Crompton	Worcester, Mass	Sept. 21, 1869	95, 092
Loom	G. Crompton	Worcester, Mass	July 15, 1873	140, 894
Loom	R. Crosbie	Newark, N. J	Jan. 6, 1812	....
Loom	C. Grossley	Ellington, Conn	June 14, 1859	24, 378
Loom	J. F. Crowley	Philadelphia, Pa	May 16, 1871	114, 929
Loom	B. Cummings	Townsend, Mass	Oct. 3, 1817	....
Loom	P. C. Curtis	Paris, N. Y	Nov. 17, 1810	....
Loom	H. Dale	Philadelphia, Pa	Oct. 9, 1866	58, 609
Loom, Circular	W. Darker, jr	Philadelphia, Pa	July 7, 1863	39, 197
Loom	J. F. Davies and W. E. Yates	Manchester, Great Britain	June 13, 1871	115, 938
Loom	H. D. Davis	North Andover, Mass	Aug. 16, 1870	106, 334
Loom	H. D. Davis	North Andover, Mass	Jan. 10, 1871	110, 904
Loom	H. G. Davis	Worcester, Mass	Nov. 21, 1871	121, 161
Loom	H. D. Davis	North Andover, Mass	July 29, 1873	141, 262
Loom	B. G. Dawley	North Providence, R. I	Dec. 23, 1856	16, 306
Loom	W. Day	Newark, N. J	July 16, 1872	129, 216
Loom	J. Deakin	Gloucester, N. J	Jan. 14, 1868	73, 237
Loom	J. Detweiler	West Liberty, Ohio	Aug. 15, 1871	118, 114
Loom	R. M. Dill	Holyoke, Mass	July 10, 1855	13, 217
Loom	W. F. Draper	Hopedale, Mass	June 16, 1868	78, 941
Loom	J. W. Drummond	New York, N. Y	Nov. 24, 1863	40, 685
Loom	G. W., J., and J. C. Duckworth	Pittsfield, Mass	Nov. 23, 1869	97, 175
Loom	J. C. Duckworth	Mount Carmel, Conn	Apr. 15, 1873	137, 898
Loom	F. Durand	Paris, France	Jan. 6, 1857	16, 354
Loom	W. Dutcher	Bennington, Vt	June 27, 1846	4, 606
Loom	J. Earnshaw	East Greenwich, R. I	July 9, 1867	66, 574
Loom	J. Eccles	Philadelphia, Pa	Jan. 23, 1855	12, 274
Loom	R. Elliott	Chester, Pa	Jan. 3, 1871	110, 640
Loom	H. A. Ellis	Mystic River, Conn	Nov. 27, 1866	59, 987
Loom	D. Farrior and D. P. J. Murphy	Aberfoil, Ala	Dec. 4, 1843	3, 364
Loom	A. Faulkner	Walpole, N. H	Oct. 23, 1849	6, 813
Loom	G. S. Faulkner	Indianapolis, Ind	Jan. 25, 1870	99, 177
Loom	A. R. Field	Centre Falls, R. I	Apr. 19, 1870	101, 989
Loom	W. T. Flinn	Bridesburgh, Pa	Dec. 14, 1868	84, 872
Loom	J. F. Fosdick	Lowell, Mass	Dec. 23, 1863	37, 223
Loom	D. K. Fretz	Cono, Iowa	Oct. 1, 1867	69, 422
Loom	A. Frey	New York, N.Y	May 7, 1861	32, 245
Loom	J. G. Frick	Pottsville, Pa	Aug. 1, 1871	117, 616
Loom	A. L. Fuller	Clinton, Mass	Dec. 23, 1856	16, 271
Loom	M. A. Furbush	Philadelphia, Pa	Nov. 24, 1863	40, 683
Loom	W. Gadd and J. Moore	Manchester, Great Britain	July 13, 1869	92, 523
Loom	J. G. Garretson	Cincinnati, Ohio	Nov. 3, 1868	83, 621
Loom	J. C. and A. P. Garretson	Jackson Township, Iowa	May 13, 1862	35, 229
Loom	W. V. Gee	New Haven, Conn	Feb. 27, 1855	12, 457
Loom	W. V. Gee	Philadelphia, Pa	Oct. 31, 1871	120, 510
Loom	W. R., W. R., jr., and J. A. Giffard.	Milo, Me	Apr. 4, 1871	113, 288
Loom	C. W. Gilbert	Worcester, Mass	May 31, 1870	103, 600
Loom	L. M. Gilbert	Cow Run, Ohio	Aug. 30, 1870	106, 809
Loom	R. B. Goodyear	Wilmington, Del	Aug. 1, 1871	117, 533
Loom	O. W. Gordon and N. T. Frame	Salem, Iowa	Jan. 23, 1866	52, 161
Loom	J. Graham	New York, N. Y	May 7, 1867	64, 525
Loom	W. Graichen and C. H. Hoffman	Clinton, Mass	May 7, 1861	32, 269
Loom	D. Grieve	....	June 8, 1797	....
Loom	J. M. M. Guiramond	Baltimore, Md	Jan. 7, 1814	....
Loom	F. Haigh	Methuen, Mass	Nov. 26, 1867	71, 480
Loom	W. Hainsworth	Philadelphia, Pa	Nov. 26, 1867	71, 299
Loom	H. Halvorson	Northampton, Mass	Oct. 1, 1850	7, 682
Loom	G. Hancock	Holyoke, Mass	Aug. 16, 1864	43, 848
Loom	M. M. Hankins	Vandalia, Ill	Nov. 12, 1867	70, 837
Loom	D. S. Harris	Coventry, R. I	Mar. 27, 1855	12, 593
Loom	E. B. Hastings	Palmer, Mass	Jan. 31, 1871	111, 343
Loom	J. Haworth	Frankford, Pa	Mar. 20, 1847	5, 021
Loom	J. and J. Haworth	Frankford, Pa	Oct. 31, 1848	5, 891
Loom	J. Heavin	Montgomery County, Va	Oct. 16, 1812	....
Loom	J. Heavin	Montgomery County, Va	Feb. 18, 1814	....
Loom	J. G. Henderson	Keokuk, Iowa	Nov. 2, 1869	96, 320
Loom	B. W. and H. Hendrick	Woonsocket Falls, R. I	May 12, 1842	2, 616
Loom	J. J. Herbert	Philadelphia, Pa	June 7, 1870	103, 880
Loom	J. Hillsley	West Manayunk, Pa	May 21, 1872	127, 055
Loom	I. N. Hodson	Mount Pleasant, Iowa	Sept. 4, 1866	57, 717
Loom	S. Holdsworth	Durham, England	June 3, 1862	35, 445
Loom	W. J. Horstman	Philadelphia, Pa	July 8, 1856	15, 295
Loom	W. H. Howard	Philadelphia, Pa	May 26, 1857	17, 375
Loom	J. Hubbard	Jamestown, N. C	Nov. 6, 1826	....
Loom	O. B. Hubbard	Lowell, Mass	Mar. 21, 1865	46, 971
Loom	J. P. Humaston	New York, N. Y	July 14, 1868	79, 908
Loom	F. W. Huppelsberg	New York, N. Y	July 16, 1867	66, 844
Loom	W. S. Irish	Middlebury, N. Y	May 27, 1855	12, 596

Index of patents issued from the United States Patent Office from 1790 *to* 1873, *inclusive*—Continued.

Invention.	Inventor.	Residence.	Date.	No.
Loom	R. Jackson	Attleborough, Pa	July 23, 1816	....
Loom	W. H. Jackson and G. Merrill	Brooklyn, N. Y., and Newburyport, Mass.	May 5, 1868	77, 619
Loom	W. Janes	Ashford, Conn	Sept. 1, 1810	....
Loom	J. Jelleffs	Butternuts, N. Y	May 24, 1815	....
Loom	B. H. Jenks	Bridesburgh, Pa	Oct. 24, 1854	11, 833
Loom	B. H. Jenks	Bridesburgh, Pa	Apr. 3, 1855	12, 630
Loom	B. H. Jenks	Bridesburgh, Pa	May 12, 1863	38, 489
Loom	B. H. Jenks	Bridesburgh, Pa	Nov. 24, 1868	84, 423
Loom	B. H. Jenks	Bridesburgh, Pa	Mar. 2, 1869	87, 341
Loom	B. H. Jenks and R. B. Goodyear	Philadelphia, Pa	May 22, 1866	55, 011
Loom	C. J. Kane	Milford, Conn	Mar. 25, 1873	137, 077
Loom	J. J. Kendall	Corinth, Miss	Feb. 28, 1860	27, 296
Loom	O. A. Kelly	Woonsocket, R. I	Nov. 22, 1853	10, 252
Loom	J. B. Kershaw	Indianapolis, Ind	Sept. 11, 1866	58, 024
Loom	J. Knowles	Cohoes, N. Y	Nov. 8, 1853	10, 209
Loom	L. J. Knowles	Warren, Mass	June 24, 1856	15, 186
Loom	L. K. Knowles	Warren, Mass	Nov. 4, 1856	16, 015
Loom	L. J. Knowles	Worcester, Mass	Jan. 21, 1873	134, 992
Loom	W. S. Laycock	Sheffield, England	Dec. 13, 1870	110, 050
Loom	J. O. Leach	Ballston, N. Y	Oct. 30, 1855	13, 724
Loom	T. Loveridge	Philadelphia, Pa	Feb. 14, 1860	27, 136
Loom	F. C. Lowell and P. T. Jackson	Boston, Mass	Feb. 23, 1815	....
Loom	J. Lyall	New York, N. Y	Aug. 11, 1868	80, 982
Loom	J. Lyall	New York, N. Y	Dec. 10, 1872	133, 868
Loom	J. A. Marden	Newburyport, Mass	Oct. 1, 1861	33, 397
Loom	B. A. Mann	West Meriden, Conn	Nov. 24, 1863	40, 698
Loom	J. G. Melville and W. Brayshaw	Wetheredville, Md	Apr. 24, 1855	12, 762
Loom	S. C. Mendenhall	Richmond, Ind	Oct. 10, 1854	11, 790
Loom	S. C. Menbenhall	Richmond, Ind	Jan. 4, 1859	22, 533
Loom	S. C. Mendenhall	Richmond, Ind	Aug. 29, 1865	49, 644
Loom	C. Miller	Saint Louis, Mo	Apr. 23, 1867	64, 127
Loom	J. Miller	Eldridge, Ill	Oct. 11, 1870	108, 281
Loom	R. Mueller	New York, N. Y	Apr. 30, 1872	126, 319
Loom	W. Murkland	Lowell, Mass	Feb. 5, 1861	31, 324
Loom	W. and J. W. Murkland	Lowell, Mass	Jan. 23, 1872	123, 037
Loom	J. Nields	Taunton, Mass	Nov. 27, 1847	5, 379
Loom	A. Nimmo	Philadelphia, Pa	May 30, 1871	115, 508
Loom	J. R. Norfolk	Salem, Mass	Feb. 8, 1870	99, 586
Loom	E. L. Norfolk, S. S. Stanley, and J. A. Marden.	Salem and Newburyport, Mass.	Feb. 22, 1848	5, 450
Loom	B. Oldfield	Williamsburgh, N. Y	Jan. 23, 1866	52, 192
Loom	B. and E. Oldfield	Newark, N. J	Jan. 14, 1868	73, 374
Loom	B. and E. Oldfield	Norwich, Conn	Mar. 30, 1869	88, 503
Loom	E. Oldfield	Norwich, Conn	Aug. 1, 1871	117, 672
Loom	I. Orndorff	Russellville, Ky	Apr. 10, 1866	53, 857
Loom	I. Orndorff	Russellville, Ky	Apr. 10, 1866	53, 858
Loom	F. Painter	East Hampton, Mass	May 26, 1857	17, 404
Loom	A. Parkinson	Norwich, Conn	Dec. 18, 1860	30, 948
Loom	F. Peabody	Salem, Mass	Mar. 5, 1861	31, 617
Loom	N. Perry	Boston, Mass	Sept. 25, 1811	....
Loom	J. Phelps	Manlius, N. Y	Aug. 27, 1812	....
Loom	R. Pilson and S. P. Heath	Laurel, Md	May 27, 1856	14, 971
Loom	W. J. Porter and W. Cross	New York, N. Y., and Jersey City, N. J.	Oct. 11, 1870	108, 292
Loom	W. J. Quinn	Philadelphia, Pa	Feb. 25, 1868	74, 938
Loom	P. F. Ramel and J. Dogat	Lyons, France	Apr. 19, 1870	102, 039
Loom	R. Reynolds	Stockport, N. Y	Sept. 23, 1862	36, 529
Loom	W. Reynolds	Manchester, N. H	July 22, 1862	35, 981
Loom	E. Robbins	Milford, Mass	May 27, 1862	35, 394
Loom	E. Robinson	Augusta, Mass	July 24, 1813	....
Loom	T. Robjohn	New York, N. Y	May 7, 1867	64, 573
Loom	R. C. Rogers	Maine, Mass	Nov. 29, 1811	....
Loom	R. C. Rogers	District of Maine, Mass	June 23, 1812	....
Loom	W. Rossetter	Accrington, England	May 4, 1869	89, 692
Loom	G. Roth	New York, N. Y	Dec. 12, 1854	12, 072
Loom	J. Salsbury	Central Falls, R. I	Aug. 18, 1868	81, 113
Loom	J. Schofield	Worcester, Mass	Feb. 5, 1867	61, 767
Loom	E. A. Scholfield	Westerly, R. I	Aug. 25, 1857	18, 061
Loom	J. Scholfield	Stonington, Conn	Sept. 9, 1825	....
Loom	C. W. Schonherr	Saxony	May 8, 1837	187
Loom	L. Scofield	Farmington, Wis	Oct. 8, 1867	69, 711
Loom	L. Schofield	Farmington, Wis	Aug. 21, 1866	57, 444
Loom	E. M. Scott	Auburn, N. Y	Sept. 7, 1858	21, 448
Loom	J. Shaw	Ballardvale, Mass	Mar. 10, 1868	75, 305
Loom	S. Shepard	Taunton, Mass	July 29, 1862	36, 029
Loom	J. Shinn	Leverington, Pa	Apr. 30, 1861	30, 236
Loom	J. Short	New Brunswick, N. J	July 16, 1872	129, 602
Loom	T. Siddall	Bristol Township, Pa	July 9, 1814	....
Loom	J. Silbermann and G. Unger	New York, N. Y	Mar. 3, 1868	75, 062
Loom	A. W. Silvis	Birmingham, Iowa	Oct. 27, 1868	83, 557
Loom	J. Smith and P. McMahon	Somerville and Charlestown, Mass.	Feb. 13, 1872	123, 647
Loom	O. C. Smith	Salem, Mass	Apr. 21, 1863	38, 265
Loom	D. W. Snell	Woonsocket, R. I	May 29, 1855	12, 970
Loom	D. W. Snell and S. S. Bartlett	Woonsocket, R. I	Jan. 13, 1857	16, 405
Loom	J. Sprinkle	Wythe County, Va	Apr. 27, 1814	....
Loom	J. Standish	Providence, R. I	Nov. 11, 1830	....
Loom	T. Stibbs	Wooster, Ohio	Aug. 12, 1862	36, 178
Loom	A. Stockwell	Providence, R. I	June 1, 1869	90, 888
Loom	A. Stockwell and B. D. Humes	Millbury, Mass	Feb. 18, 1862	34, 451
Loom	L. Stone	Nelson, N. H	Dec. 14, 1869	97, 826
Loom	R. Sugden	Boston, Mass	Dec. 2, 1813	....
Loom	J. S. Templeton	Glasgow, Scotland	July 2, 1872	128, 675
Loom	S. T. Thomas	Lawrence, Mass	July 3, 1855	13, 186
Loom	S. T. Thomas	Lawrence, Mass	July 3, 1855	13, 187
Loom	S. T. Thomas	Gilford, N. H	Jan. 7, 1873	134, 57[illegible]
Loom	S. T. Thomas	Gilford, N. H	Dec. 9, 1873	145, 316

Index of patents issued from the United States Patent Office from 1790 *to* 1873, *inclusive*—Continued.

Invention.	Inventor.	Residence.	Date.	No.
Loom	S. T. Thomas and J. H. Dolley	Gilford, N. H	Dec. 10, 1867	72, 119
Loom	T. G. Thompson and B. Ballard	Richmond, Ind	Sept. 3, 1867	68, 583
Loom	W. Tongue	Philadelphia, Pa	Jan. 9, 1855	12, 229
Loom	H. E. Towle	Newark, N. J	Jan. 24, 1871	111, 278
Loom	W. Townsend	Seneca Falls, N. Y	Aug. 10, 1869	93, 500
Loom	W. Townshend	Hinsdale, Mass	Mar. 1, 1853	9, 603
Loom	W. Townshend	Hinsdale, Mass	Nov. 15, 1853	10, 241
Loom	W. Tunstill	New York, N. Y	May 9, 1865	47, 687
Loom	A. Urbahu	Paterson, N. J	Jan. 30, 1872	123, 210
Loom	L. Van Rifer	Spring Valley, N. H	Mar. 20, 1855	12, 565
Loom	A. O. Very	Andover, N. Y	Oct. 1, 1867	69, 515
Loom	J. Wade	Palmer, Mass	Jan. 23, 1872	122, 976
Loom	W. B. Walker	Salem, Iowa	Oct. 1, 1867	69, 516
Loom	A. Ware, jr	Franklin, Mass	June 18, 1813	....
Loom	E. Warren	New York	Aug. 27, 1818	....
Loom	E. Warren	New York	Dec. 11, 1821	....
Loom	E. Warren	New York	May 1, 1822	....
Loom	L. R. Wattles	Newton, Mass	July 31, 1860	29, 427
Loom	J. R. Weber	Bourbon, Ind	Mar. 12, 1867	62, 791
Loom	J. Welsh	Philadelphia, Pa	Oct. 3, 1854	11, 768
Loom	J. Welsh	Philadelphia, Pa	Jan. 9, 1855	12, 225
Loom	J. Welsh	Philadelphia, Pa	May 29, 1855	12, 981
Loom	J. Welsh	Philadelphia, Pa	July 20, 1858	20, 969
Loom	J. Welsh	Philadelphia, Pa	Aug. 3, 1858	21, 098
Loom	J. Welsh	Philadelphia, Pa	Sept. 5, 1865	49, 814
Loom	J. Welsh	Philadelphia, Pa	July 17, 1866	56, 474
Loom	G. L. White	Woonsocket, R. I	Nov. 27, 1866	60, 101
Loom	R. Whitehill	New York, N. Y	Dec. 27, 1870	110, 524
Loom	W. Whiteside and J. Shinn	Philadelphia, Pa	June 5, 1855	13, 022
Loom	B. Wilcox	Franklin County, Ohio	Apr. 21, 1826	....
Loom	T. R. Williams	Newport, R. I	July 3, 1813	....
Loom	J. Wilson	Abbeville District, S. C	May 29, 1849	6, 487
Loom	F. M. Wolf	Glanchan, Saxony	Sept. 13, 1864	44, 271
Loom	E. Wood	Philadelphia, Pa	July 17, 1855	13, 284
Loom	J. C. Wood	Conshohocken, Pa	Jan. 19, 1869	85, 981
Loom	E. Wright and B. Fitts	Worcester, Mass	May 17, 1864	42, 812
Loom	W. Wright	Brookline, N. H	Oct. 12, 1814	....
Loom	H. Wyman	Worcester, Mass	Oct. 29, 1867	70, 369
Loom	H. Wyman	Worcester, Mass	July 2, 1872	128, 694
Loom	G. Yates and E. Clayton	Lancaster, Pa	June 13, 1854	11, 100
Loom	J. Zürcher	Lancaster, Pa	Aug. 19, 1873	141, 907
Loom	C. Zwicki	Pittsburgh, Pa	June 4, 1861	32, 501
Loom	C. Zwicki	Pittsburgh, Pa	Jan. 12, 1864	41, 254
Loom	C. Zwicki	Chicago, Ill	Mar. 26, 1867	63, 353
Loom, Bag	L. B. Jillson and G. Sparhawk	Lewiston, Me	Nov. 27, 1855	13, 848
Loom-beam	S. T. Thomas and E. Everett	Lawrence, Mass	Nov. 7, 1854	11, 919
Loom beam and roller	G. L. Garsed	Wilmington, Del	Apr. 8, 1873	137, 672
Loom-box-operating mechanism	H. Wyman	Worcester, Mass	Jan. 31, 1871	111, 417
Loom, Broad power	J. Leland	Milbury, Mass	Mar. 9, 1833	....
Loom, Brussels	E. B. Bigelow	Boston, Mass	Mar. 20, 1847	5, 020
Loom-cam	G. S. Faulkner	Staffordsville, Conn	July 16, 1867	66, 818
Loom-cam	F. E. Howe and L. Washburn	Stafford, Conn	Oct. 1, 1867	69, 438
Loom-cam	G. O. Wickers and T. J. McClary.	North Andover, Mass	Mar. 8, 1870	100, 575
Loom, Carpet	J. R. Clark	South Coventry, Conn	May 7, 1830	....
Loom, Carpet	J. Haight	Harsimus, N. J	Mar. 17, 1834	....
Loom, Carpet	W. Sherwood	Ridgefield, Conn	July 20, 1846	4, 644
Loom, Carpet	J. A. Van Riper	New York, N. Y	Nov. 16, 1852	9, 401
Loom, Carpet and rug	W. Bacon	Philadelphia, Pa	Apr. 7, 1830	....
Loom, Cassinet	J. Hammond and J. McClelland	Williamsport, Pa	July 3, 1839	....
Loom, Carpet, &c., power	T. Flint	Boston, Mass	June 27, 1842	2, 696
Loom, Check and plaid power	H. Burt and O. D. and A. H. Boyd	Manchester, Conn	Aug. 19, 1828	....
Loom, Circular	M. C. Bryant	Lowell, Mass	Jan. 11, 1870	98, 738
Loom, Circular	J. J. Greenough	New York, N. Y	Nov. 21, 1865	51, 040
Loom, Circular	J. A. Grunwald	New York, N. Y	Dec. 27, 1859	26, 585
Loom, Circular	W. H. Walton and A. Nandam	New York and West Farms, N. Y.	Nov. 1, 1864	44, 902
Loom, Circular hat-weaving	J. V. Reed	New York, N. Y	Mar. 5, 1872	124, 288
Loom, Circular-weaving	J. Buser	New York, N. Y	Dec. 17, 1867	72, 362
Loom, Circular-weaving	A. Wagner	New York, N. Y	Aug. 25, 1868	81, 438
Loom cloth-beam	J. P. Hillard	Fall River, Mass	Aug. 1, 1871	117, 630
Loom-combs, Machine for shaping the elastic dents of.	M. Robinson	Accrington, England	Feb. 4, 1873	135, 444
Loom, Common and power	B. Lapham	Waterford, N. Y	Jan. 20, 1838	575
Loom, Cross-weaving	C. Roder	Philadelphia, Pa	Nov. 14, 1865	50, 990
Loom, Crank	E. Cutter	Walpole, N. H	Sept. 27, 1811	....
Loom, Crank-power	E. Burt	Manchester, Conn	Aug. 8, 1837	348
Loom, Damask	Tompkins and Gilroy	North Providence, R. I	May 9, 1835	....
Loom, Domestic	B. Maltby	New York, N. Y	Apr. 13, 1822	....
Loom, Domestic and factory	W. Janes	Ashford, Conn	Mar. 20, 1813	....
Loom, Family	C. Hathaway	Walton, N. Y	Oct. 11, 1814	....
Loom, Fancy	G. Crompton	Worcester, Mass	Jan. 9, 1866	51, 928
Loom, Fancy	L. J. Knowles	Warren, Mass	Feb. 24, 1863	37, 760
Loom, Fancy	C. Roder	Ceralvo, Ky	Oct. 18, 1859	25, 869
Loom, Fancy	S. T. Thomas	Laconia, N. H	Feb. 11, 1862	34, 381
Loom, Fancy check	E. Burt	Manchester, Conn	Jan. 24, 1854	10, 442
Loom, Fancy check power	E. Burt	Manchester, Conn	Feb. 4, 1851	7, 925
Loom, Figure power	W. Crompton	Taunton, Mass	Nov. 25, 1837	491
Loom, Figure power	J. M. Hoggan	New Haven, Conn	Oct. 14, 1834	....
Loom filling-fork	W. G. Duce and A. C. Eddy	Baltic, Conn., and Providence, R. I.	Oct. 20, 1868	83, 267
Loom filling-fork	J. H. Knowles	Lawrence, Mass	May 27, 1873	139, 251
Loom for figured fabrics	S. Eccles	Kensington, Pa	Mar. 5, 1850	7, 137
Loom for figured fabrics	J. Reynolds	Providence, R. I	Oct. 16, 1849	6, 797
Loom for making fishing and other nets	B. F. Jouannin and F. M. and A. Baudouin.	Paris, France	Aug. 16, 1864	43, 888
Loom for making fringe	G. Roth	New York, N. Y	Sept. 21, 1869	95, 145
Loom for making weavers' harness	K. Vogel	Chelsea, Mass	Oct. 25, 1853	10, 173

Index of patents issued from the United States Patent Office from 1790 *to* 1873, *inclusive*—Continued.

Invention.	Inventor.	Residence.	Date.	No.
Loom for operating shuttle-boxes	J. Ashworth	North Andover, Mass	Mar. 9, 1869	87,616
Loom for piled fabric	J., jr., and J. Turnbull	Simsbury, Conn	Jan. 29, 1850	7,061
Loom for regulating the delivery of the warp from the warp-beam.	W. H. Brayton	Warren, R. I	Jan. 6, 1844	3,397
Loom for warping	C. Whitney	Dedham, Mass	Feb. 1, 1814	....
Loom for weaving	A. Faulkner	Walpole, N. H	Apr. 17, 1849	6,316
Loom for weaving	C. Whiting, J. Goulding, and W. Janes.	Dedham, Mass	June 25, 1824	....
Loom for weaving bags	C. Baldwin	Manchester, N. H	Dec. 2, 1851	8,553
Loom for weaving bags	W. Talbot	Sanford, Me	Nov. 28, 1854	12,005
Loom for weaving bags	S. T. Thomas	Lawrence, Mass	Apr. 22, 1856	14,746
Loom for weaving blinds, &c	B. R. Murphy	Parkersburgh, W. Va	Nov. 26, 1872	133,332
Loom for weaving Brussels carpets, &c	E. B. Bigelow	Boston, Mass	Mar. 10, 1849	6,153
Loom for weaving Brussels carpets, &c	E. B. Bigelow	Boston, Mass	Mar. 13, 1849	6,186
Loom for weaving carpets	H. Baker	North Salem, N. Y	Apr. 14, 1826	....
Loom for weaving carpets, &c	E. B. Bigelow	Lancaster, Mass	May 16, 1842	2,625
Loom for weaving carpets, &c	E. B. Bigelow	Lancaster, Mass	May 26, 1842	2,639
Loom for weaving carpets	J. Marsden	Halifax, England	July 20, 1869	92,856
Loom for weaving carpets, &c., Power	H. Skinner	Yonkers, N. Y	May 4, 1869	88,694
Loom for weaving chip	P. Bennet	Rochester, Mass	Feb. 8, 1806	....
Loom for weaving cloth	A. Whittemore	....	Nov. 17, 1796	....
Loom for weaving cloth for stocks	C. Kile	Philadelphia, Pa	Oct. 11, 1836	47
Loom for weaving cloth with swells or gores	M. Opper	New York, N. Y	Aug. 14, 1866	57,177
Loom for weaving coach-lace	J. H. Murrill	Richmond, Va	Oct. 4, 1853	10,096
Loom for weaving coach-lace, &c., Power	E. B. Bigelow	West Boylston, Mass	Apr. 20, 1837	169
Loom for weaving concave and convex surfaces, Power.	W. Breitenstein	New York, N. Y	Oct. 9, 1866	58,589
Loom for weaving corsets	W. Breitenstein	New York, N. Y	Jan. 27, 1863	37,556
Loom for weaving corsets, &c	W. P. Brown	New York, N. Y	May 31, 1864	42,986
Loom for weaving corsets	B. J. Goullioud	Paris, France	Jan. 27, 1863	37,547
Loom for weaving counterpanes, &c., Power	E. B. Bigelow	Lancaster, Mass	July 28, 1842	2,741
Loom for weaving counterpanes, &c., Power	E. B. Bigelow	Lancaster, Mass	Aug. 2, 1842	2,744
Loom for weaving the covering of cord, &c., Circular	J. E. Palmer	Middletown, Conn	Dec. 27, 1864	45,629
Loom for weaving coverlet	J. Meily, jr., and S. Mellinger	Lebanon, Pa	June 26, 1835	....
Loom for weaving cut-pile fabric	E. B. Bigelow	Boston, Mass	Oct. 17, 1854	11,803
Loom for weaving cut-pile fabric	M. C. Bryant	Lowell, Mass	June 25, 1850	7,452
Loom for weaving cut-pile fabric	M. C. Bryant	Lowell, Mass	Aug. 5, 1851	8,283
Loom for weaving cut-pile fabrics, Jacquard	E. B. Bigelow	Clinton, Mass	Mar. 18, 1851	7,983
Loom for weaving double-faced pile-fabrics	P. Joyot, jr	Paris, France	Feb. 14, 1865	46,433
Loom for weaving embroidered fabrics	J. G. Spitzli	Millville, Mass	Nov. 21, 1865	51,095
Loom for weaving fancy goods	B. F. Rice	Clinton, Mass	Oct. 18, 1853	10,138
Loom for weaving figured cloth	W. Levally	Canterbury, Conn	Mar. 20, 1829	....
Loom for weaving figured fabrics	C. W. Blanchard	Clinton, Mass	Aug. 3, 1852	9,162
Loom for weaving figured fabrics	G. Crompton	Worcester, Mass	Nov. 14, 1854	11,933
Loom for weaving figured fabrics	S. and J. Eccles	Kensington, Pa	Aug. 3, 1852	9,168
Loom for weaving figured fabric	R. Garsed	Frankford, Pa	Nov. 6, 1849	6,845
Loom for weaving figured fabrics, &c	C. G. Gilroy	Great Britain	Mar. 12, 1842	2,486
Loom for weaving figured fabric	C. G. Gilroy	New York	Apr. 15, 1843	3,047
Loom for weaving figured fabrics	B. H. Jenks and R. B. Goodyer	Bridesburgh, and Philadelphia, Pa.	Apr. 13, 1852	8,874
Loom for weaving figured fabrics	M. Marshall	Lowell, Mass	Dec. 11, 1849	6,939
Loom for weaving figured fabrics	S. T. Thomas and E. Everett	Lowell and Lawrence, Mass	Dec. 24, 1850	7,861
Loom for weaving figured fabrics, Hand	W. Townshend	Rochester, N. H	Apr. 26, 1845	4,015
Loom for weaving figured goods	A. Babbett	Auburn, N. Y	Oct. 8, 1850	7,714
Loom for weaving figured goods	H. Baker	North Salem, N. Y	Aug. 30, 1827	....
Loom for weaving figured work	J. Smith	Shafferstown, Pa	Apr. 22, 1835	....
Loom for weaving fringe	L. D. Valletton	Philadelphia, Pa	Aug. 4, 1868	80,786
Loom for weaving fringe, Hand	E. A. B. Judkins	Portland, Me	Feb. 2, 1839	1,075
Loom for weaving fringed counterpanes, Power	E. B. Bigelow	Lancaster, Mass	Apr. 24, 1840	1,561
Loom for weaving hair-cloth	I. Angell	Pawtucket, R. I	Nov. 13, 1860	30,632
Loom for weaving hair-cloth	S. B. and S. M. Chaffee	Providence, R. I	Oct. 12, 1858	21,793
Loom for weaving hair-cloth	H. Halvorson	Hartford, Conn	Sept. 27, 1853	10,044
Loom for weaving hair-cloth	W. S. Laycock	Sheffield, England	Oct. 8, 1872	132,014
Loom for weaving hair-cloth	I. Lindsley	Providence, R. I	June 25, 1861	32,634
Loom for weaving hair-cloth	I. Lindsley	North Providence, R. I	Nov. 15, 1864	45,107
Loom for weaving hair-cloth	I. Lindsley	Pawtucket, R. I	Sept. 26, 1871	119,277
Loom for weaving hair-cloth	I. Lindsley	Pawtucket, R. I	Sept. 26, 1871	119,278
Loom for weaving hair-cloth	J. Turpie	New York, N. Y	June 17, 1873	139,981
Loom for weaving hair-cloth	J. S. Winsor	Providence, R. I	May 31, 1864	42,982
Loom for weaving hair-cloth and fabrics produced thereon.	I. Lindsley	Pawtucket, R. I	Sept. 26, 1871	119,276
Loom for weaving hair-cloth, Power	I. Lindsley	Pawtucket, R. I	Oct. 25, 1864	44,808
Loom for weaving hats	P. Brooks	New Haven, Conn	June 15, 1869	91,305
Loom for weaving hats, &c	P. L. Slayton	New York, N. Y	Nov. 22, 1864	45,208
Loom for weaving hats, Circular	L. Bonard	New York, N. Y	June 4, 1861	32,461
Loom for weaving hats, &c., Circular	P. L. Slayton	New York, N. Y	Feb. 2, 1864	41,466
Loom for weaving hats, Circular	P. L. Slayton and C. I. Kane	New York, N. Y	Oct. 23, 1866	59,138
Loom for weaving ingrain carpets	E. B. Bigelow	Boston, Mass	Feb. 9, 1869	86,806
Loom for weaving ingrain carpets, Power	W. and J. W. Murkland	Lowell, Mass	Nov. 23, 1869	97,106
Loom for weaving ingrain carpets, Power	A. Murray	Lowell, Mass	Apr. 20, 1869	89,065
Loom for weaving irregular fabrics	A. Goullioud	Barcelona, Spain	Mar. 22, 1870	101,006
Loom for weaving irregular fabrics	C. Heptonstall	Providence, R. I	July 20, 1869	92,722
Loom for weaving knotted counterpanes, &c	E. B. Bigelow	West Boylston, Mass	Jan. 6, 1838	546
Loom for weaving palm-leaf, &c	G. W. Chandler	Fitchburgh, Mass	Dec. 10, 1867	71,852
Loom for weaving palm-leaf	J. C. Smith	Chicopee, Mass	July 14, 1868	79,923
Loom for weaving palm-leaf, straw, &c	J. M. Baker	Providence, R. I	Nov. 15, 1864	45,115
Loom for weaving pile-fabrics	E. B. Bigelow	Clinton, Mass	Jan. 14, 1851	7,898
Loom for weaving pile-fabrics	E. B. Bigelow	Boston, Mass	Nov. 15, 1853	10,222
Loom for weaving pile-fabrics	E. B. Bigelow	Boston, Mass	Dec. 18, 1855	13,936
Loom for weaving pile-fabrics	E. B. Bigelow	Boston, Mass	May 5, 1857	17,189
Loom for weaving pile-fabrics	M. C. Bryant	Lowell, Mass	Mar. 19, 1850	7,180
Loom for weaving pile-fabrics	E. K. Davis	New York, N. Y	Feb. 9, 1869	86,651
Loom for weaving pile-fabrics	J. C. Ellison	Philadelphia, Pa	Oct. 28, 1873	144,077
Loom for weaving pile-fabrics	L. Ferguson	Lowell, Mass	Aug. 25, 1868	81,487
Loom for weaving pile-fabrics	L. Ferguson	Lowell, Mass	Nov. 9, 1869	96,570
Loom for weaving pile-fabrics	W. G. Hartley	Saxonville, Mass	Apr. 9, 1867	63,631
Loom for weaving pile-fabrics	J. Johnson	Troy, N. Y	Mar. 12, 1850	7,168
Loom for weaving pile-fabrics	J. Johnson	Troy, N. Y	Aug. 5, 1851	8,281

Index of patents issued from the United States Patent Office from 1790 to 1873, inclusive—Continued.

Invention.	Inventor.	Residence.	Date.	No.
Loom for weaving pile-fabrics	C. A. Maxfield	Troy, N. Y	Jan. 13, 1852	8, 656
Loom for weaving pile-fabrics	E. Pickford	New Brunswick, N. J	May 27, 1873	139, 328
Loom for weaving pile-fabrics	S. Richardson	Claremont, N. H	Aug. 10, 1852	9, 188
Loom for weaving pile-fabrics	H. Skinner	Yonkers, N. Y	Dec. 30, 1873	146, 101
Loom for weaving pile-fabrics	W. Webster	Morrisania, N. Y	Oct. 6, 1868	82, 773
Loom for weaving pile-fabrics	W. Webster	Morrisania, N. Y	Aug. 27, 1872	130, 960
Loom for weaving pile-fabrics	W. Webster	Morrisania, N. Y	Aug. 27, 1872	130, 961
Loom for weaving pile-fabrics	W. Webster	Morrisania, N. Y	Sept. 10, 1872	131, 197
Loom for weaving pile-fabrics double	E. B. Bigelow	Boston, Mass	Jan. 13, 1857	16, 370
Loom for weaving pile-fabrics, Power	E. B. Bigelow	Boston, Mass	Aug. 17, 1869	93, 800
Loom for weaving pile-fabrics without the figuring-wires.	R. W. Sievier	Cavendish Square, England.	June 1, 1852	8, 988
Loom for weaving plaids, &c	M. A. Furbush and G. Crompton.	Worcester, Mass	May 31, 1859	24, 206
Loom for weaving plaids, &c., Power	E. B. Bigelow	Boston, Mass	Apr. 10, 1846	3, 987
Loom for weaving plain cloth, Power	F. Downing	Enfield, Mass	Jan. 27, 1843	2, 935
Loom for weaving plush or pile fabric	S. Holt	Newark, N. J	Mar. 7, 1865	46, 754
Loom for weaving narrow goods, Silk	C. Bergen	Brooklyn, N. Y	Dec. 18, 1839	1, 432
Loom for weaving narrow ware	B. Oldfield	Newark, N. J	Oct. 2, 1866	58, 464
Loom for weaving ribbon, &c	J. Rushworth	New York, N. Y	Oct. 8, 1867	69, 708
Loom for weaving satinet, kerseymere, and other cloth.	J. D. Seagrave	Uxbridge, Mass	May 17, 1838	740
Loom for weaving seamless bags	S. Northrop	New Milford, Conn	Jan. 1, 1851	7, 876
Loom for weaving skirt-fringe	J. Beck	New York, N. Y	Aug. 31, 1858	21, 312
Loom for weaving skirts, Warp-beam for	F. X. Loughery	Kelleysville, Pa	May 11, 1869	89, 935
Loom for weaving slat-blinds	J. L. Devol	Parkersburgh, W. Va	May 28, 1872	127, 318
Loom for weaving slat-blinds	M. Free	Philadelphia, Pa	July 17, 1866	56, 493
Loom for weaving slats, window-shades, &c	E. B. Hastings, W. L. Merritt, and W. E. Nichols.	Palmer and Templeton, Mass.	Feb. 4, 1873	135, 427
Loom for weaving slatted window-shades	G. Hasecoster	Richmond, Va	Sept. 11, 1866	57, 898
Loom for weaving stock-frames, Power	F. Goodell and F. W. Harvey	Ramapo, N. Y	Dec. 2, 1835	....
Loom for weaving stocks	C. Kile	Erie, Pa	Sept. 18, 1835	....
Loom for weaving suspender-webbing	W. V. Gee	New Haven, Conn	Sept. 18, 1855	13, 571
Loom for weaving tape, &c	J. Duckworth	Pittsfield, Mass	Nov. 9, 1869	96, 564
Loom for weaving tape, ribbon, &c	L. J. Knowles	Warren, Mass	May 15, 1866	54, 742
Loom for weaving tapestry and Brussels carpets	E. B. Bigelow	Clinton, Mass	Sept. 24, 1850	7, 660
Loom for weaving tapestry carpets with parti-colored warp.	E. B. Bigelow	Clintonville, Mass	Jan. 7, 1851	7, 884
Loom for weaving trimmings	L. D. Valetton	Philadelphia, Pa	Apr. 12, 1864	42, 335
Loom for weaving tufted pile-fabrics, Power	H. Skinner	West Farms, N. Y	Sept. 1, 1863	39, 759
Loom for weaving, Water, steam, or other power	S. Blydenburgh and H. Healy	Worcester, Mass	Feb. 20, 1815	....
Loom for weaving webbing, tape, &c	A. G. Bill and G. Spalding	Middletown, Conn	Mar. 28, 1831	....
Loom for weaving wire	S. Holdsworth	Maspeth, N. Y	Apr. 22, 1873	138, 090
Loom for weaving wire	G. W. Smith	Mauch Chunk, Pa	Dec. 25, 1855	14, 000
Loom for weaving wire-cloth	L. Kittinger	Massillon, Ohio	Jan. 26, 1869	86, 233
Loom for weaving wire-cloth	C. H. Waters	Groton, Mass	Sept. 2, 1862	36, 377
Loom for weaving wire-cloth	G. F. Wright and C. K. Sawyer	Clinton, Mass	Sept. 24, 1872	131, 584
Loom for weaving wire-cloth, Power	E. B. Bigelow	Boston, Mass	Oct. 6, 1857	18, 320
Loom for weaving wire-gauze	M. Bretzger	Pittsburgh, Pa	Dec. 5, 1846	4, 873
Loom for weaving wire, Power	J. S. Gustin	New York	Feb. 23, 1806	....
Loom for working any number of heddles, Weaver's	G. McCrae	Baltimore, Md	Oct. 18, 1843	3, 309
Loom, Fringe	S. Walker	Roxbury, Mass	Nov. 9, 1858	22, 042
Loom, Garment-weaving	S. Leather	Dalton, England	May 5, 1868	77, 498
Loom, Hair-cloth	J. Noblit	Philadelphia, Pa	Apr. 9, 1861	31, 992
Loom, Hand	J. Albertson and S. C. Byers	Richmond, Ind	July 21, 1868	80, 047
Loom, Hand	J. L. Branson	Cincinnati, Ohio	Mar. 27, 1866	53, 398
Loom, Hand	J. L. Branson	Cincinnati, Ohio	July 14, 1868	79, 946
Loom, Hand	J. D. Browne	Cincinnati, Ohio	June 25, 1867	65, 997
Loom, Hand	A. Carter and R. Spake	Salem, Iowa	May 8, 1866	54, 501
Loom, Hand	T. A. Dugdale	Richmond, Ind	Sept. 17, 1867	69, 909
Loom, Hand	G. W. Firestone	Fredericksburgh, Ohio	Oct. 29, 1867	70, 186
Loom, Hand	W. S. Freeman	West Union, Ohio	Aug. 4, 1868	80, 717
Loom, Hand	I. H. Garretson	Clay, Iowa	Feb. 18, 1851	7, 936
Loom, Hand	J. G. Garretson	Salem, Iowa	May 21, 1850	7, 378
Loom, Hand	J. G. Garretson	Salem, Iowa	Jan. 19, 1864	41, 290
Loom, Hand	G. Harsin and T. M. Kirkpatrick.	Kirkville, Iowa	Feb. 11, 1868	74, 351
Loom, Hand	J. W. Hayse	Salem, Iowa	Aug. 22, 1865	49, 589
Loom, Hand	J. G. Henderson	Palmyra, Mo	Jan. 1, 1861	31, 019
Loom, Hand	H. D. Hunt	Danville, Ill	Oct. 8, 1867	69, 673
Loom, Hand	J. G. and H. T. Henderson	Salem, Iowa	Mar. 14, 1865	46, 798
Loom, Hand	A. Jones	Clinton, Ill	Oct. 26, 1869	96, 238
Loom, Hand	R. Lloyd	Philadelphia, Pa	Feb. 8, 1810	....
Loom, Hand	E. Lowe	Burrows, Ind	Sept. 1, 1868	81, 656
Loom, Hand	C. L. McDowell	Wassonville, Iowa	Jan. 30, 1866	52, 306
Loom, Hand	D. Mendenhall	Fairfield, Iowa	Oct. 16, 1866	58, 865
Loom, Hand	S. C. Mendenhall	Richmond, Ind	Jan. 13, 1857	16, 392
Loom, Hand	S. C. Mendenhall	Richmond, Ind	Jan. 19, 1864	41, 313
Loom, Hand	S. C. Mendenhall	Richmond, Ind	Nov. 12, 1867	70, 877
Loom, Hand	S. C. Mendenhall and O. and E. King.	Richmond, Ind., and Salem, Iowa.	Nov. 9, 1852	9, 388
Loom, Hand	S. C. Mendenhall and S. Sparks	Richmond, Ind	Oct. 23, 1866	59, 132
Loom, Hand	A. and P. P. Meredith	Maxintuckee, Ind	Nov. 19, 1867	71, 035
Loom, Hand	H. H. Mitchell	Mineral Point, Wis	Dec. 31, 1867	72, 879
Loom, Hand	.. A. Mitchell	Ringgold, Ga	Dec. 20, 1853	10, 340
Loom, Hand	A. R. Nixon	Rhea Springs, Tenn	Oct. 11, 1859	25, 756
Loom, Hand	J. E. Nute	Lincoln, Me	Sept. 6, 1870	107, 094
Loom, Hand	J. E. Nute and G. H. Hathorn	Lincoln, Me	Jan. 14, 1868	73, 256
Loom, Hand	J. Pelsor	Brooklyn, Ill	Mar. 20, 1866	53, 330
Loom, Hand	C. Roder	Ceralvo, Ky	Mar. 8, 1864	41, 863
Loom, Hand	A. Rosenberger	Brandonville, W. Va	Apr. 9, 1868	63, 752
Loom, Hand	J. Seaman and W. Y. Henderson.	Andover, N. Y	Sept 19, 1865	50, 041
Loom, Hand	O. Strong	Green Centre, Ind	Jan. 14, 1868	73, 266
Loom, Hand	T. G. Thompson and A. F. Fox.	Richmond and Greensborough, Ind.	Sept. 17, 1867	69, 047
Loom, Hand	T. H. Tibbles	Kansas City, Mo	Nov. 19, 1867	71, 087
Loom, Hand	C. Unverzagt	Richmond, Ind	Feb. 13, 1866	52, 626

Index of patents issued from the United States Patent Office from 1790 to 1873, inclusive—Continued.

Invention.	Inventor.	Residence.	Date.	No.
Loom, Hand	C. Unverzagt	Terre Haute, Ind	Jan. 28, 1868	73, 852
Loom, Hand	W. B. Walker and N. D. Hartley	Salem, Iowa	Aug. 14, 1866	57, 226
Loom, Hand	C. Wandel	Milton, Iowa	Sept. 17, 1867	68, 918
Loom, Hand	J. Whitehead	Oskaloosa, Iowa	Dec. 4, 1863	60, 305
Loom, Hand and power	J. Thorp	Providence, R. I	Mar. 28, 1812	....
Loom, Hand-power	G. A. Brown and J. A. Hunt	Oakalla, Ill	Jan. 19, 1869	86, 055
Loom, Hand-power	J. M. Deen, W. B. Bolding, and H. Perry.	Dayton, Iowa	Mar. 26, 1867	63, 143
Loom-harness	J. Blackmar	Brooklyn, Conn	Oct. 20, 1836	64
Loom, Harness	D. C. Brown	Lowell, Mass	Jan. 2, 1866	51, 794
Loom-harness	D. C. Brown	Lowell, Mass	Jan. 7, 1868	72, 967
Loom-harness	A. B. Corey	Providence, R. I	Sept. 28, 1869	95, 198
Loom-harness	G. Matoon	Chicopee Falls, Mass	Sept. 15, 1857	18, 208
Loom, Harness	H. Parsons	Waterloo, N. Y	Aug. 12, 1862	36, 166
Loom-harness-actuating mechanism	J. F. Foss	Lowell, Mass	July 2, 1872	128, 480
Loom-harness board, Jacquard	E. Everett and S. T. Thomas	Lawrence and Lowell, Mass	Jan. 18, 1853	9, 545
Loom-harness clasp	G. Copeland	Danville, Me	Sept. 26, 1854	11, 719
Loom-harness connection	J. T. Holden	Elmwood, R. I	Mar. 16, 1869	87, 848
Loom-harness frame	D. C. Brown	Lowell, Mass	May 5, 1868	77, 446
Loom-harness frame	J. Greenhalgh, sr	Pascoag, R. I	May 15, 1860	28, 270
Loom-harness, Heddle-eye for	J. L. Cheney	Lowell, Mass	Jan 9, 1866	51, 920
Loom-harness, Machine for making	J. Blackmar	Brooklyn, Conn	Oct. 20, 1836	61
Loom-harness, Machine for making	J. Senneff	Philadelphia, Pa	June 26, 1855	13, 152
Loom-harness, Machine for making	J. Sladdin	Lawrence, Mass	Aug. 4, 1868	80, 774
Loom-harness, Machine for making	J. Sladdin	Lawrence, Mass	Nov. 28, 1871	121, 258
Loom-harness, Machine for making wire heddles for.	D. C. Brown and J. Ashworth	Lowell and North Andover, Mass.	May 12, 1868	77, 713
Loom-harness, Machine for making wire heddles for.	A. Howe and S. S. Granniss	Morrisville, N. Y	Sept. 30, 1841	2, 273
Loom-harness, Machine for manufacturing	D. C. Brown	Lowell, Mass	Feb. 15, 1853	9, 584
Loom-harness, Machine for manufacturing	J. H. Crowell	Providence, R. I	July 25, 1871	117, 389
Loom-harness, Machine for varnishing and dressing	E. J. Ellis	Lewiston, Me	Aug. 23, 1870	106, 676
Loom-harness, Machine for weaving	G. W. and W. R. Harris	Lowell, Mass., and Manchester, England.	Sept. 13, 1864	44, 186
Loom-harness, Machine for weaving	S. Holton, jr., and W. R. Harris	Middlebury, Vt	Sept. 4, 1849	6, 691
Loom-harness, Manufacture of wire heddles for	A. Howe and S. S. Grannis	Morrisville, N. Y	Oct. 11, 1841	2, 299
Loom-harness mechanism	J. Booth	Pottstown, Pa	Dec. 12, 1871	121, 842
Loom-harness mechanism	A. F. Gibboney	Belleville, Pa	Sept. 28, 1869	95, 221
Loom-harness mechanism	J. Greenhalgh	Woonsocket, R. I	Feb. 25, 1868	74, 819
Loom-harness, Mechanism for operating	O. Plummer and J. Schofield	Worcester, Mass	Aug. 11, 1868	80, 876
Loom-harness, Mode of counterbalancing	J. Greenhalgh	Waterford, Mass	Nov. 2, 1852	9, 377
Loom-harness, Mode of hanging and guiding	L. S. Fisher	Broadhead, Wis	Apr. 16, 1867	63, 877
Loom-harness needle, Machinery for filling	L. S. Reynolds	Manchester, N. H	Mar. 29, 1859	23, 393
Loom-harness, Operating	F. Traub	Philadelphia, Pa	Apr. 28, 1868	77, 335
Loom-harness-operating mechanism	W. R. Andrews	Mystic River, Conn	Oct. 23, 1866	58, 969
Loom-harness-operating mechanism	J. Ashworth	North Andover, Mass	Aug. 16, 1870	106, 306
Loom-harness-operating mechanism	J. Bachelder and W. H. Bliss, 2d.	Norwich, Conn	Mar. 31, 1868	76, 036
Loom-harness-operating mechanism	E. B. Bigelow	Boston, Mass	Aug. 17, 1869	93, 799
Loom-harness-operating mechanism	G. Crempton	Worcester, Mass	Apr. 7, 1868	76, 406
Loom-harness-operating mechanism	G. Crompton	Worcester, Mass	Jan. 31, 1871	111, 324
Loom-harness-operating mechanism	J. C. Duckworth	Pittsfield, Mass	Dec. 6, 1870	109, 884
Loom-harness-operating mechanism	G. S. Faulkner	Indianapolis, Ind	Jan. 25, 1870	99, 176
Loom-harness-operating mechanism	A. R. Field	Central Falls, R. I	Aug. 23, 1870	106, 571
Loom-harness-operating mechanism	J. F. Gebhart	New Albany, Ind	June 15, 1869	91, 431
Loom-harness-operating mechanism	R. B. Goodyear	Wilmington, Del	Oct. 3, 1871	119, 459
Loom-harness-operating mechanism	O. Plummer	Worcester, Mass	Sept. 12, 1871	118, 969
Loom-harness, Shuttles for machines for knitting	D. C. Brown and J. Ashworth	Lowell, Mass	June 21, 1864	43, 252
Loom-harness, Wire	E. Brown	Cazenovia, N. Y	Oct. 30, 1828	....
Loom-harness, Wire heddle for	M. Finkle	New York, N. Y	Aug. 8, 1865	49, 251
Loom harness, Warp-eyes of wire heddle for	D. C. Brown	Lowell, Mass	May 21, 1867	64, 944
Loom-heddle	J. Ashworth	Andover, Mass	Jan. 16, 1872	122, 795
Loom-heddle	E. G. Jelley	Pawtucket, R. I	May 25, 1869	90, 364
Loom-heddle	C. Kennedy	Philadelphia, Pa	Sept. 3, 1867	68, 444
Loom-heddle	J. Senneff	Philadelphia, Pa	Dec. 31, 1867	72, 917
Loom-heddle	C. Whipple	Providence, R. I	Feb. 13, 1872	123, 601
Loom heddle and harness	B. Hartford and W. B. Tilton	Enfield, N. H	Dec. 29, 1837	544
Loom-heddle frame	M. Finkle	New York, N. Y	June 6, 1865	48, 057
Loom-heddle, Metallic	W. Janes and D. Bolles	Ashford, Conn	Apr. 26, 1815	....
Loom-heddle, Metallic	C. Strong	Hartford, Vt	Apr. 24, 1840	1, 563
Loom-heddle motion	B. H. Jenks and R. B. Goodyear	Philadelphia, Pa	May 22, 1866	55, 010
Loom-heddles, Apparatus for varnishing	J. L. Lairdieson	Troy, N. Y	Dec. 17, 1861	33, 948
Loom index-chain	B. H. Jenks	Bridesburgh, Pa	Feb. 19, 1867	62, 204
Loom-indicator	A. Fellows, T. B. Harrison, and H. Dyer.	Marquoketa, Iowa	Feb. 4, 1868	73, 962
Loom, Ingrain carpet	W. Sherwood	Worcester, Mass	Jan. 28, 1830	....
Loom-jack-operating mechanism	C. H. Knowlton	Camden, N. J	Nov. 16, 1869	96, 925
Loom, Jacquard	C. B. Bigelow	Clintonville, Mass	Oct. 23, 1849	6, 806
Loom, Jacquard	A. Calderhead	Philadelphia, Pa	Feb. 3, 1841	1, 964
Loom, Jacquard	J. C. Cooke	Waterbury, Conn	Sept. 9, 1856	15, 717
Loom, Jacquard	J. Goulding	Worcester, Mass	Aug. 3, 1852	9, 171
Loom, Jacquard	J. Goulding	Worcester, Mass	July 8, 1856	15, 291
Loom, Jacquard	H. W. Hensel and L. D. Valetton.	Philadelphia, Pa	June 16, 1863	38, 929
Loom, Jacquard	L. Holms	Andover, Mass	Mar. 27, 1847	5, 033
Loom, Jacquard	J. Perrins	Philadelphia, Pa	May 28, 1846	4, 537
Loom, Jacquard	M. Wolf	Philadelphia, Pa	June 9, 1868	78, 779
Loom jaw-temple	E. B. Bigelow	Boston, Mass	Feb. 24, 1845	3, 925
Loom jaw-temple	G. Draper	Ware, Mass	Feb. 27, 1849	6, 144
Loom jaw-temple	B. Peck	Rehoboth, Mass	Sept. 26, 1848	5, 797
Loom jaw-temples, Opening and closing	E. Williams and D. L. Huntington.	Norwich, Conn	Nov. 16, 1841	2, 363
Loom, Knitting	W. Henson	Newark, N. J	Mar. 2, 1852	8, 773
Loom, Knitting	P. E. Ladrange	Vignory, France	Oct. 16, 1844	3, 798
Loom, Knitting	J. Mee	Lowell, Mass	May 10, 1853	9, 718
Loom, Knitting	R. Walker and J. McIntire	Portsmouth, N. H	Feb. 12, 1844	3, 436
Loom, Lappet	W. Aspinall	Philadelphia, Pa	June 28, 1870	104, 687
Loom, Lappet	J. Clough and J. Crompton	Chicopee, Mass	Mar. 21, 1871	112, 785

Index of patents issued from the United States Patent Office from 1790 to 1873, inclusive—Continued.

Invention.	Inventor.	Residence.	Date.	No.
Loom, Lappet or embroidering	F. W. Newton	South Orange, N. J	Dec. 8, 1868	84, 753
Loom, Lappet-weaving	W. Aspinall	Manayunk, Pa	Oct. 31, 1865	50, 764
Loom let-off	L. C. Briggs	Boston, Mass	Aug. 18, 1868	81, 133
Loom let-off	G. Crompton	Worcester, Mass	Apr. 26, 1864	42, 461
Loom let-off	G. Draper	Hopedale, Mass	Aug. 4, 1868	80, 534
Loom let-off	S. Estes	Newburyport, Mass	Apr. 17, 1866	54, 066
Loom let-off	H. Fiske	Farnumsville, Mass	Aug. 15, 1865	49, 472
Loom let-off	W. H. Gray	Dover, N. H	Apr. 9, 1861	31, 969
Loom let-off	D. Hussey	Nashua, N. H	Oct. 23, 1866	59, 024
Loom let-off	J. A. Marden	Chelsea, Mass	Jan. 14, 1868	73, 353
Loom let-off	J. Phillips	Pawtucket, R. I	Sept. 26, 1865	50, 156
Loom let-off	W. W. Pomeroy	East Hampton, Mass	Aug. 15, 1865	49, 479
Loom let-off	G. Richardson	Lowell, Mass	Apr. 23, 1867	64, 147
Loom let-off	G. Richardson	Lowell, Mass	June 11, 1867	65, 606
Loom let-off	J. Remick	Newburyport, Mass	Oct. 29, 1867	70, 265
Loom let-off	T. S. Smith	Boston, Mass	Jan. 21, 1868	73, 470
Loom let-off	E. Wright	Worcester, Mass	Feb. 20, 1866	52, 780
Loom let-off and take-up	J. Pender	Worcester, Mass	Feb. 4, 1868	73, 993
Loom let-off and take-up	J. F. Kirkwood	Thistle, Md	Oct. 2, 1866	58, 433
Loom let-off and take-up mechanism	J. A. Marden	Newburyport, Mass	Apr. 30, 1867	64, 235
Loom let-off and take-up motion	D. Bassett	Killingly, Conn	Nov. 21, 1865	51, 003
Loom let-off and tension mechanism, Power	E. B. Bigelow	Boston, Mass	Dec. 13, 1870	110, 000
Loom let-off mechanism	E. B. Bigelow	Boston, Mass	May 24, 1870	103, 415
Loom let-off mechanism	E. B. Bigelow	Boston, Mass	Apr. 4, 1871	113, 483
Loom let-off mechanism	T. Booth and C. C. Sanderson	Norway, Me	Nov. 24, 1868	84, 253
Loom let-off mechanism	L. C. Briggs and A. Howard	Boston, Mass	June 15, 1869	91, 207
Loom let-off mechanism	M. Brookfield	Brookfield, Iowa	Apr. 9, 1867	63, 699
Loom let-off mechanism	M. C. Burleigh	Somersworth, N. H	Aug. 23, 1870	106, 543
Loom let-off mechanism	M. C. Burleigh	Somersworth, N. H	Aug. 6, 1872	130, 107
Loom let-off mechanism	B. F. Carter	Manville, R. I	Aug. 25, 1868	81, 342
Loom let-off mechanism	B. F. Carter	Woonsocket, R. I	Jan. 19, 1869	85, 905
Loom let-off mechanism	W. R. Clark	North Adams, Mass	June 8, 1869	90, 996
Loom let-off mechanism	J. Clegg	Warwick, R. I	Feb. 14, 1871	111, 816
Loom let-off mechanism	D. M. Collins	Lowell, Mass	Sept. 28, 1869	95, 197
Loom let-off mechanism	J. Day	Paterson, N. J	May 3, 1870	102, 506
Loom let-off mechanism	G. Draper	Milford, Mass	Aug. 11, 1863	39, 469
Loom let-off mechanism	G. Draper	Milford, Mass	Jan. 7, 1868	72, 987
Loom let-off mechanism	L. N. Fletcher and J. M. Page	Lowell, Mass	July 5, 1870	104, 949
Loom let-off mechanism	W. T. Flinn	Philadelphia, Pa	Apr 4, 1871	113, 415
Loom let-off mechanism	C. Gahren and L. Langlotz	New York, N. Y	Apr. 29, 1873	138, 391
Loom let-off mechanism	T. Gorrell	Warren, R. I	May 9, 1871	114, 667
Loom let-off mechanism	F. W. Graichen	Olneyville, R. I	Aug. 8, 1871	117, 768
Loom let-off mechanism	W. Hall	North Adams, Mass	Aug. 4, 1868	80, 625
Loom let-off mechanism	W. Hall	North Adams, Mass	July 6, 1869	92, 305
Loom let-off mechanism	W. H. Howard	Philadelphia, Pa	Aug. 8, 1871	117, 779
Loom let-off mechanism	D. Hussey	Nashua, N. H	Apr. 14, 1868	76, 636
Loom let-off mechanism	S. M. Linscott	Buxton, Me	Aug. 12, 1873	141, 720
Loom let-off mechanism	D. Long and J. Preston	Fairview, Pa	Jan. 18, 1870	99, 044
Loom let-off mechanism	J. Magee	Usquepaugh, R. I	June 14, 1870	104, 328
Loom let-off mechanism	E. McDaniel	Lowell, Mass	Dec. 21, 1869	98, 082
Loom let-off mechanism	P. McGee	North Providence, R. I	Sept. 29, 1868	82, 625
Loom let-off mechanism	W. Potter and L. J. Labounty	Lowell, Mass	Sept. 14, 1869	94, 769
Loom let-off mechanism	H. A. Remington	Anthony, R. I	Jan. 17, 1871	111, 002
Loom let-off mechanism	R. Reynolds	Stockport, N. Y	Apr. 13, 1869	88, 906
Loom let-off mechanism	C. R. Saatweber	New York, N. Y	Mar. 12, 1872	124, 632
Loom let-off mechanism	C. Schilling	Auburn, N. Y	Nov. 21, 1871	121, 131
Loom let-off mechanism	T. S. Smith	Charlestown, Mass	Dec. 1, 1868	84, 515
Loom let-off mechanism	T. S. Smith and A. B. Ely	Boston and Newton, Mass	Jan. 14, 1868	73, 399
Loom let-off mechanism	B. Tweedle	Philadelphia, Pa	July 16, 1872	129, 628
Loom let-off mechanism	A. J. Woodman	Indian Orchard, Mass	Nov. 14, 1871	120, 843
Loom let-off mechanism	E. Wright	Worcester, Mass	Dec. 31, 1867	72, 769
Loom let-off mechanism	C. H. Young	River Point, R. I	Aug. 25, 1868	81, 453
Loom let-off mechanism, Narrow-ware	J. N. Leavenworth	Hamden, Conn	Jan. 15, 1867	61, 218
Loom let-off mechanism, Power	P. Boylan	Gloucester, N. J	May 10, 1864	42, 638
Loom let-off motion	S. Estes	Newburyport, Mass	Sept. 12, 1865	49, 950
Loom let-off motion	F. W. Graichen	Providence, R. I	June 27, 1871	116, 437
Loom let-off motion	W. H. Gray	Dover, N. H	June 28, 1859	24, 553
Loom let-off motion	W. H. Gray	Dover, N. H	June 28, 1859	24, 602
Loom let-off motion	J. Knowles	Buffalo, N. Y	Apr. 30, 1850	7, 324
Loom let-off motion	J. Myers	Biddeford, Me	Mar. 10, 1849	6, 159
Loom let-off motion	R. Reynolds	Stockport, N. H	May 21, 1861	32, 382
Loom let-off motion	G. Smith	Cumberland, R. I	Dec. 17, 1867	72, 235
Loom let-off motion	A. B. Taylor and S. Wilcox, jr	Mystic, Conn., and Westerly, R. I.	Feb. 22, 1853	9, 596
Loom let-off motion	R. Walker	Milford, Mass	Feb. 19, 1867	62, 168
Loom let-off motion	R. Walker	Milford, Mass	Oct. 29, 1867	70, 297
Loom let-off motion, Power	S. O. Colvin	Coventry, R. I	Jan. 12, 1858	19, 073
Loom let-off motion, Power	W. H. Gray	Dover, N. H	Aug. 4, 1857	17, 926
Loom let-off motion, Power	R. Reynolds	Stockport, N. Y	June 28, 1864	43, 338
Loom let-off motion, Power	N. Wyllys	South Glastenbury, Conn	Mar. 16, 1858	19, 664
Loom, Lint	F. Hall	Charlestown, Mass	Nov. 28, 1817	....
Loom measuring-apparatus	J. A. Bradshaw and S. L. Crockett.	Lowell, Mass	Nov. 1, 1870	108, 871
Loom, Narrow-ware	R. B. Fowler	Worcester, Mass	Nov. 18, 1873	144, 610
Loom, Narrow-ware	L. J. Knowles	Warren, Mass	Nov. 13, 1866	59, 613
Loom, Narrow-ware	L. J. Knowles	Warren, Mass	Dec. 11, 1866	60, 388
Loom, Narrow-ware	S. Walker	Boston, Mass	Jan. 29, 1867	61, 583
Loom, Narrow-ware	A. Williamson	New York, N. Y	May 1, 1860	28, 135
Loom or lathe for weaving	W. Harris	....	Mar. 15, 1800	....
Loom-picker	J. Bardsley and A. W. Gordon	Willimantic, Conn	Apr. 26, 1864	42, 456
Loom-picker	S. Boorn	Lowell, Mass	Mar. 26, 1861	31, 776
Loom-picker	S. Boorn	Lowell, Mass	Apr. 9, 1861	31, 948
Loom-picker	S. Boorn	Lowell, Mass	Sept. 30, 1873	143, 218
Loom-picker	J. Cady	Staffordville, Conn	Sept. 1, 1863	39, 714
Loom-picker	W. E. Card and P. Andrews	Phœnix, R. I	Oct. 29, 1867	70, 164
Loom-picker	A. H. Carroll	Baltimore, Md	Apr. 13, 1869	88, 845
Loom-picker	I. M. Colburn	Derby, Conn	Sept. 26, 1848	5, 796
Loom-picker	B. F. Day and C. H. Nelson	Biddeford, Me	Oct. 3, 1865	50, 227

Index of patents issued from the United States Patent Office from 1790 *to* 1873, *inclusive*—Continued.

Invention.	Inventor.	Residence.	Date.	No.
Loom-picker	J. Hunt and A. Stockwell	Providence, R. I	Apr. 12, 1870	101, 734
Loom-picker	R. Joslin	Pawtucket, R. I	Feb. 18, 1868	74, 545
Loom-picker	R. Leach	Linwood Station, Pa	Jan. 28, 1868	73, 818
Loom-picker	Z. Lyford	Lowell, Mass	Feb. 23, 1858	19, 428
Loom-picker	T. J. Mayall	Roxbury, Mass	June 2, 1857	17, 468
Loom-picker	G. P. Medbery	Sprague, Conn	Jan. 5, 1864	41, 165
Loom-picker	B. W. Nichols	Phœnix Village, R. I	Oct. 1, 1867	69, 472
Loom-picker	J. M. Parker	Leicester, Mass	Jan. 31, 1871	111, 378
Loom-picker	O. A. Sawyer	Lowell, Mass	Dec. 23, 1873	145, 905
Loom-picker	O. B. Smith	Palmer, Mass	July 23, 1867	66, 996
Loom-picker	W. Taylor	Maynard, Mass	Jan. 28, 1873	135, 253
Loom-picker	E. Wright	Worcester, Mass	Oct. 1, 1867	69, 380
Loom-picker arm	G. W. Paterson	Newburyport, Mass	June 26, 1866	55, 959
Loom-picker check	L. J. Labounty	Lowell, Mass	Sept. 5, 1865	49, 830
Loom-picker check	W. Montgomery	Northampton, Mass	Apr. 19, 1870	102, 145
Loom-picker cushion	G. L. Richardson and J. C. Moody.	Brunswick, Me	July 13, 1869	92, 471
Loom-picker cushion	W. J. Thorn	Westbrook, Me	Feb. 12, 1867	62, 087
Loom-picker motion	S. Boorn	Lowell, Mass	May 5, 1857	17, 193
Loom-picker motion	J. Cady	Staffordville, Conn	May 2, 1865	47, 517
Loom-picker motion	J. C. Fisher	Providence, R. I	June 20, 1871	116, 251
Loom-picker motion	W. Nugent	North Providence, R. I	Apr. 30, 1867	64, 354
Loom-picker motion	J. Pilkington	Frankford, Pa	Jan. 7, 1868	73, 037
Loom-picker motion	J. Robinson	North Andover, Mass	June 18, 1861	32, 586
Loom-picker motion	E. Robbins	Milford, Mass	Mar. 11, 1862	34, 647
Loom-picker motion	J. Stover	Bristol, Conn	Sept. 24, 1872	131, 575
Loom-picker motion, Power	H. Elliott	Globe Village, Mass	Aug. 7, 1866	56, 913
Loom-picker motion, Power	W. Stearns	Manchester, N. H	July 5, 1859	24, 668
Loom-picker shoe	E. J. Hall	Cambridge, Mass	June 1, 1869	90, 838
Loom-picker staff	R. Collins	Chicopee, Mass	Oct. 1, 1867	69, 319
Loom-picker staff	S. Estes	Newburyport, Mass	Nov. 23, 1858	22, 114
Loom-picker staff	E. Robbins	Hopedale, Mass	May 27, 1862	35, 395
Loom-picker-staff check	C. A. Hooper	Lewiston, Me	Oct. 17, 1871	120, 068
Loom-picker-staff check	E. and H. C. Phillips	Blackstone, Mass	Feb. 4, 1868	74, 124
Loom-picker-staff connection	W. Weiland	Dedham, Mass	Aug. 1, 1865	49, 182
Loom-picker-operating mechanism	E. A. Wareham and W. W. Waggoner.	Kirkville, Iowa	Sept. 13, 1870	107, 313
Loom-picker-staff-actuating mechanism	M. C. Burleigh	Somersworth, N. H	Aug. 9, 1870	106, 116
Loom-picker-staff arrester	E. and H. C. Phillips	Blackstone, Mass	Jan. 30, 1866	52, 307
Loom-picker-staff-operating mechanism	S. Mortimer	Leicester, Mass	Feb. 5, 1867	61, 754
Loom-picker-staff-operating mechanism	W. W. and V. J. Phillips	Wellsville, N. Y	Apr. 2, 1867	63, 421
Loom-picker-staff-operating mechanism	E. P. Terrel	West Liberty, Ohio	July 20, 1869	92, 900
Loom-picker staves, Operating	W. Rouse	Taunton, Mass	Sept. 13, 1870	107, 412
Loom-picker staves, Operating	P. A. Sherman and F. Baxter	Pawtucket, R I	Apr. 10, 1866	53, 892
Loom-picker staves, Operating	E. Wright	Worcester, Mass	Oct. 29, 1867	70, 308
Loom-picker staves, Operating	J. F. Wicks	Millville, Mass	Apr. 27, 1869	89, 365
Loom picking-mechanism	M. C. Burleigh	Somersworth, N. H	Aug. 6, 1872	130, 108
Loom picking mechanism	J. C. Fisher	Providence, R. I	Jan. 30, 1872	123, 094
Loom picking-mechanism	H. R. Fry	Wabash, Ind	Aug. 29, 1871	118, 446
Loom picking-mechanism	E. D. Gove	Holyoke, Mass	Nov. 28, 1871	121, 359
Loom picking-mechanism	H. T. Henderson	Keokuk, Iowa	Aug. 20, 1852	130, 636
Loom picking-mechanism	O. D. Lombard	Lowell, Mass	Sept. 14, 1869	94, 760
Loom picking-mechanism	E. W. Moffatt	Hyde Park, Mass	Dec. 24, 1872	134, 153
Loom picking-mechanism	L. E. Ross	Providence, R. I	Apr. 15, 1873	137, 798
Loom picking-mechanism	L. E. Ross	Providence, R. I	Sept. 30, 1873	143, 257
Loom picking-mechanism	C. E. Smith	Lowell, Mass	Nov. 26, 1872	133, 493
Loom picking-mechanism	W. Stearns	Manchester, N. Y	Sept. 9, 1873	142, 592
Loom picking-mechanism	J. Stever	Bristol, Conn	June 17, 1873	140, 091
Loom picking-mechanism	A. Stockwell	Woonsocket, R. I	Mar. 11, 1873	136, 786
Loom picking-mechanism	E. P. Terrel	West Liberty, Ohio	Feb. 20, 1872	123, 952
Loom picking-mechanism	W. Townsend	Auburn, N. Y	July 22, 1873	141, 187
Loom picking-mechanism	H. A. Whitten	New Bedford, Mass	Mar. 26, 1872	124, 995
Loom picking-motion	H. A. Whitten and E. D. Gove	Holyoke, Mass	Feb. 7, 1871	111, 595
Loom picking-stick check	R. Davis	Lawrence, Mass	Aug. 5, 1873	151, 496
Loom, Pneumatic	C. Richardson	London, England	Nov. 26, 1872	133, 386
Loom, Power	A. Allen	Wilmington, Del	Apr. 15, 1856	14, 644
Loom, Power	A. Allen	Wilmington, Del	Aug. 4, 1857	17, 912
Loom, Power	W. Baird	Philadelphia, Pa	Nov. 29, 1853	10, 290
Loom, Power	J. T. Barnes	Manayunk, Pa	Oct. 17, 1854	11, 804
Loom, Power	D. Barnum	Bridgeport, Conn	Mar. 26, 1845	3, 968
Loom, Power	E. B. Bigelow	Boston, Mass	Aug. 18, 1846	4, 696
Loom, Power	E. B. Bigelow	Boston, Mass	Feb. 12, 1856	14, 222
Loom, Power	E. B. Bigelow	Boston, Mass	Feb. 9, 1869	86, 805
Loom, Power	J. Bullough	Accrington, England	Jan. 18, 1870	98, 919
Loom, Power	W. H. Cheetham, jr	New York, N. Y	Apr. 5, 1859	23, 446
Loom, Power	J. L. Cheney	Lowell, Mass	Mar. 17, 1857	16, 824
Loom, Power	J. Crawshaw	Rochester, N. Y	Nov. 16, 1858	22, 065
Loom, Power	J. C. Duckworth	Pittsfield, Mass	May 18, 1869	90, 085
Loom, Power	C. Duxbury and J. Nield	Taunton, Mass	Apr. 21, 1842	2, 574
Loom, Power	J. Earnshaw	Providence, R. I	Apr. 9, 1867	63, 622
Loom, Power	E. Fairman	Stafford, Conn	Feb. 6, 1838	595
Loom, Power	M. A. Furbush and G. Crompton.	Worcester, Mass	Sept. 4, 1860	29, 873
Loom, Power	W. G. Gavit	Washington County, R. I	Apr. 8, 1835	....
Loom, Power	W. Gilmore	Smithfield, R. I	Oct. 28, 1820	....
Loom, Power	J. Gledhill	New York, N. Y	Nov. 15, 1853	10, 223
Loom, Power	W. Graichen and C. Hoffman	Clinton, Mass	Jan. 14, 1862	34, 141
Loom, Power	W. H. Gray	Dover, N. H	Jan. 17, 1860	26, 875
Loom, Power	J. Greenhalgh, jr	Waterford, Mass	Nov. 8, 1853	10, 205
Loom, Power	J. Greenhalgh, sr	Waterford, Mass	Mar. 4, 1856	14, 358
Loom, Power	E. Hall	Rochester, N. Y	Feb. 12, 1856	14, 237
Loom, Power	P. W. Hart	Stamford, Conn	Sept. 4, 1866	57, 712
Loom, Power	A. Holbrook and B. F. French	Dunstable, N. H	June 17, 1834	....
Loom, Power	H. Holcroft and C. S. Smith	Media and Chester Valley, Pa.	July 19, 1864	43, 583
Loom, Power	T. H. and H. James	Stockport, N. Y	Jan. 28, 1862	34, 255
Loom, Power	B. H. Jenks and J. Shinn	Bridesburgh, Pa	May 12, 1863	38, 534
Loom, Power	J. Johnson	Troy, N. Y	Feb. 19, 1856	14, 285

Index of patents issued from the United States Patent Office from 1790 *to* 1873, *inclusive*—Continued.

Invention.	Inventor.	Residence.	Date.	No.
Loom, Power	J. C. Kempton	Manayunk, Pa	June 14, 1834	....
Loom, Power	T. King	West Farms, N. Y	Apr. 16, 1861	32, 068
Loom, Power	B. Lapham	Waterford, N. Y	June 28, 1836	....
Loom, Power	W. B. Leonard	Fishkill, N. Y	May 23, 1827	....
Loom, Power	W. B. Leonard	Fishkill, N. Y	May 23, 1827	....
Loom, Power	F. C. Lewis	Grafton, Mass	Mar. 30, 1836	....
Loom, Power	R. Lightbown	Eaton, N. Y	Oct. 30, 1849	6, 832
Loom, Power	W. Mason	Taunton, Mass	Oct. 18, 1853	10, 135
Loom, Power	T. H. Müller	New York, N. Y	Jan. 25, 1870	99, 223
Loom, Power	J. Nield	Taunton, Mass	May 25, 1844	3, 599
Loom, Power	J. Nield	Taunton, Mass	Mar. 15, 1845	3, 954
Loom, Power	J. Nield	Taunton Mass	Mar. 15, 1845	3, 955
Loom, Power	J. Pender	Worcester, Mass	Oct. 18, 1853	10, 137
Loom, Power	W. B. Pender and N. C. Horn	Wolfsborough, N. H	Aug. 15, 1838	884
Loom, Power	R. Reynolds	Valatia Village, N. Y	June 1, 1852	8, 984
Loom, Power	C. Roder	Ceralvo, Ky	June 21, 1864	43, 263
Loom, Power	G. B. Sanford	Putney, Vt	Mar. 17, 1843	3, 005
Loom, Power	J. Shinn	Philadelphia, Pa	Mar. 25, 1873	137, 252
Loom, Power	J. Shuttleworth	Frankford, Pa	May 21, 1850	7, 387
Loom, Power	J. Shuttleworth	Frankford, Pa	Jan. 3, 1854	10, 382
Loom, Power	A. Smith and H. Skinner	West Farms, N. Y	Nov. 14, 1856	16, 037
Loom, Power	A. Stone	Providence, R. I	Apr. 30, 1829	....
Loom, Power	W. Weild	Manchester, Great Britain	Jan. 13, 1857	16, 415
Loom-power	D. Whitman	Windham, Conn	Mar. 20, 1835	....
Loom power, Silk	G. Gay	Poughkeepsie, N. Y	Sept. 26, 1835	....
Loom, Rattan-weaving	S. L. Fitts	Ashburnham, Mass	Aug. 16, 1870	106, 477
Loom-reed	H. Holt	Rutland Township, Ohio	Oct. 17, 1831	....
Loom-reed	J. Senereff	Philadelphia, Pa	Oct. 1, 1830	....
Loom-reed	B. F. Whitcomb	Claremont, N. H	June 29, 1869	91, 888
Loom-reed	J. H. Williams	Pleasant Hill, Ohio	June 28, 1870	104, 808
Loom, Revolving weaving	J. Hess	Hempfield, Pa	Oct. 2, 1813	....
Loom, Rib-knitting	B. Wainwright	East Boston, Mass	Mar. 5, 1876	62, 579
Loom, Ribbon	C. V. Ganahl	Innspruck, Austria	Oct. 3, 1846	4, 782
Loom, Ribbon	W. J. Horstmann	Philadelphia, Pa	Mar. 23, 1858	19, 698
Loom, Rotary-power stocking	R. Walker	Portsmouth, N. H	Dec. 5, 1839	1, 421
Loom, Satinet power	R. Mitchell and N. Butterworth.	Troy, Mass	Mar. 22, 1828	....
Loom shedding-mechanism	G. Crompton	Worcester, Mass	July 1, 1873	140, 476
Loom shedding-mechanism	G. Crompton	Worcester, Mass	July 8, 1873	140, 682
Loom shedding-mechanism	G. Crompton	Worcester, Mass	Aug. 12, 1873	141, 768
Loom shedding-mechanism	C. H. Landenberger and H. Atkinson.	Philadelphia, Pa	Dec. 2, 1873	145, 216
Loom shedding-mechanism	J. Lomas	Kellyville, Pa	Apr. 15, 1873	137, 937
Loom shedding-mechanism	E. Redfearn	Philadelphia, Pa	Aug. 5, 1873	141, 589
Loom shedding-mechanism	I. L. Wilber	Taunton, Mass	Oct. 28, 1873	144, 176
Loom-shuttle	E. Baggett	Fall River, Mass	Sept. 1, 1868	81, 578
Loom-shuttle	E. P. Ball	Chicopee, Mass	Dec. 23, 1873	145, 831
Loom-shuttle	F. and A. J. Blanding	Lowell, Mass	Jan. 21, 1873	134, 971
Loom-shuttle	J. Brown	Brooklyn, N. Y	Sept. 2, 1873	142, 070
Loom-shuttle	H. H. Bryant	Boston, Mass	Jan. 14, 1873	134, 848
Loom-shuttle	W. H. Burns	Grafton, Mass	Jan. 25, 1870	99, 152
Loom-shuttle	E. F. Burrows	Mystic River, Conn	Apr. 26, 1870	102, 368
Loom-shuttle	H. Carstaedt	New York, N. Y	Sept. 13, 1870	107, 220
Loom-shuttle	H. Carstaedt	New York, N. Y	Feb. 27, 1872	123, 978
Loom-shuttle	N. D. Chapman	Rollingsford, N. H	Nov. 18, 1873	144, 657
Loom-shuttle	D. G. Chase	Boston, Mass	Mar. 17, 1868	75, 522
Loom-shuttle	S. E. Chase	Boston, Mass	Apr. 14, 1868	76, 604
Loom-shuttle	A. D. Clark	Wilkinsonville, Mass	June 21, 1864	43, 182
Loom-shuttle	A. D. Clark	Wilkinsonville, Mass	Aug. 25, 1868	81, 344
Loom-shuttle	T. Clark	Ware, Mass	Oct. 11, 1864	44, 604
Loom-shuttle	J. H. Coburn	Lowell, Mass	Mar. 12, 1842	2, 489
Loom-shuttle	J. H. Coburn	Lowell, Mass	June 13, 1865	48, 154
Loom-shuttle	J. H. Coburn	Lowell, Mass	Dec. 14, 1869	97, 882
Loom-shuttle	G. Crompton	Worcester, Mass	Nov. 9, 1869	96, 676
Loom-shuttle	G. Crompton	Worcester, Mass	Nov. 9, 1869	96, 677
Loom-shuttle	E. Cross	Southbridge, Mass	Feb. 4, 1868	73, 955
Loom-shuttle	A. M. Damon and J. Whitaker	Lowell, Mass	July 7, 1868	79, 557
Loom-shuttle	D. C. G. Field	Lowell, Mass	Mar. 11, 1873	136, 715
Loom-shuttle	M. F. Fields	Lawrence, Mass	Oct. 28, 1873	143, 976
Loom-shuttle	C. H. Fiske	Lowell, Mass	Aug. 10, 1869	93, 608
Loom-shuttle	E. H. Graham	Biddeford, Me	June 27, 1871	116, 435
Loom-shuttle	T. Isherwood	Stonington, Conn	June 6, 1871	115, 614
Loom-shuttle	J. Kuttner	New York, N. Y	Apr. 25, 1871	114, 155
Loom-shuttle	J. E. Ladd and H. T. Henderson.	Bristol, Conn	Dec. 17, 1872	134, 074
Loom-shuttle	P. A. Lambert	Ansonia, Conn	Dec. 17, 1872	134, 075
Loom-shuttle	L. Litchfield	Southbridge, Mass	May 1, 1855	12, 780
Loom-shuttle	J. Lofvendahl	Woonsocket, R. I	May 31, 1870	103, 757
Loom-shuttle	J. Lofvendahl	Woonsocket, R. I	July 18, 1871	117, 183
Loom-shuttle	J. Lyall	New York, N. Y	July 4, 1871	116, 609
Loom-shuttle	E. P. Marble	New Worcester, Mass	May 1, 1855	12, 781
Loom-shuttle	E. W. Marble	Sutton, Mass	Apr. 25, 1871	114, 167
Loom-shuttle	J. Martin	Lowell, Mass	Jan. 30, 1872	123, 184
Loom-shuttle	T. Martin	Chelsea, Mass	Feb. 11, 1873	135, 652
Loom-shuttle	T. K. McIntyre	Salem, Mass	Aug. 12, 1873	141, 803
Loom-shuttle	F. Miller	New York, N. Y	Mar. 26, 1872	124, 967
Loom-shuttle	L. L. Northup	Olneyville, R. I	May 7, 1872	126, 569
Loom-shuttle	E. A. Paine	Grafton, Mass	Mar. 1, 1870	100, 317
Loom-shuttle	C. L. F. Pfefferkorn and W. Aust.	Manchester, N. H	Oct. 21, 1873	143, 927
Loom-shuttle	C. R. Sawyer and G. F. Wright	Clinton, Mass	Feb. 4, 1873	135, 446
Loom-shuttle	F. W. Sawyer	Grafton, Mass	Oct. 22, 1867	70, 026
Loom-shuttle	L. Scofield	Farmington, Wis	Feb. 4, 1868	74, 013
Loom-shuttle	A. Simpson	Cumberland, R. I	Jan. 19, 1869	85, 969
Loom-shuttle	C. E. Smith	Lowell, Mass	Mar. 14, 1871	112, 643
Loom-shuttle	C. E. Smith and F. T. Jaques	Lowell, Mass	Oct. 27, 1868	83, 559
Loom-shuttle	E. A. Thissell	Lowell, Mass	July 18, 1871	117, 220

Index of patents issued from the United States Patent Office from 1790 *to* 1873, *inclusive*—Continued.

Invention.	Inventor.	Residence.	Date.	No.
Loom-shuttle	E. A. Thissell	Lowell, Mass	July 23, 1872	129, 692
Loom-shuttle	F. O. Tucker	Stonington, Conn	Aug. 22, 1871	118, 405
Loom-shuttle	F. O. Tucker	Stonington, Conn	Sept. 19, 1871	119, 201
Loom-shuttle	F. O. Tucker	Stonington, Conn	Nov. 28, 1871	121, 262
Loom-shuttle	W. Tucker	Blackstone, Mass	Dec. 28, 1852	9, 507
Loom-shuttle	W. Tucker	Fiskedale, Mass	June 17, 1873	140, 098
Loom-shuttle	W. Tunstill	New York, N. Y	Dec. 27, 1864	45, 682
Loom-shuttle	C. H. Waters and W. Orr, jr	Groton and Clinton, Mass	Nov. 12, 1871	121, 830
Loom-shuttle	R. Webendörfer	New York, N. Y	Jan. 11, 1870	98, 822
Loom-shuttle	H. Wells	Hopkinton, R. I	May 24, 1870	103, 400
Loom-shuttle	R. Whitehill	New York, N. Y	Apr. 23, 1872	126, 119
Loom-shuttle	W. Wilder	Wilkinsonville, Mass	Aug. 16, 1864	43, 877
Loom-shuttle	W. Wilder	Wilkinsonville, Mass	Aug. 23, 1864	43, 940
Loom-shuttle	W. Wilder	Wilkinsonville, Mass	Jan. 24, 1865	46, 040
Loom-shuttle	N. A. Williams	Utica, N. Y	Feb. 2, 1869	86, 483
Loom-shuttle	D. Wright	Waltham, Mass	May 16, 1871	115, 006
Loom-shuttle-acting mechanism	T. Martin	Chelsea, Mass	Mar. 12, 1872	124, 501
Loom-shuttle-actuating mechanism	J. R. Norfolk	Salem, Mass	Nov. 4, 1873	144, 218
Loom-shuttle-actuating mechanism	G. V. Sheffield and W. S. Horton.	Providence, R. I	Apr. 9, 1872	125, 493
Loom-shuttle and bobbin	C. Unverzagt	Richmond, Ind	May 1, 1866	54, 445
Loom-shuttle binder	C. Duckworth	Mount Carmel, Conn	Aug. 21, 1866	57, 300
Loom-shuttle binder	G. B. Hammond and J. A. Tucker.	Putnam, Conn., and Attleborough, Mass.	Apr. 22, 1873	138, 020
Loom-shuttle binder	M. E. Haskell	Lowell, Mass	Apr. 20, 1869	89, 043
Loom-shuttle binder	J. C. Poland, jr., and B. R. Cotton.	Auburn and Lewiston, Me	Jan. 22, 1867	61, 357
Loom-shuttle binder	E. Prentice	Wauregan, Conn	Feb. 9, 1869	86, 688
Loom-shuttle box	A. F. Gibboney	Union Township, Pa	Dec. 28, 1858	22, 420
Loom-shuttle-box mechanism	J. Ashworth	North Andover, Mass	Dec. 12, 1871	121, 838
Loom-shuttle-box mechanism	W. Atkinson and C. C. Klein	Philadelphia, Pa	Feb. 11, 1873	135, 748
Loom-shuttle-box mechanism	R. B. Goodyear	Wilmington, Del	May 27, 1873	139, 384
Loom-shuttle-box mechanism	J. M. Stone	North Andover, Mass	Dec. 12, 1871	121, 907
Loom-shuttle-box motion	C. Duckworth	Thompsonville, Conn	June 28, 1853	9, 815
Loom-shuttle-box-operating mechanism	R. B. Goodyear	Elkton, Md	Mar. 2, 1869	87, 337
Loom-shuttle-box-operating mechanism	L. J. Knowles	Warren, Mass	Dec. 13, 1870	110, 146
Loom-shuttle-box-operating mechanism	A. Nimmo	Philadelphia, Pa	Aug. 10, 1869	93, 637
Loom-shuttle-box-operating mechanism	A. Nimmo	Philadelphia, Pa	Dec. 7, 1869	97, 679
Loom-shuttle-box-operating mechanism	W. Whiteside	Philadelphia, Pa	May 18, 1869	90, 212
Loom-shuttle box, Actuating	M. Rice	Upland, Pa	Aug. 11, 1868	81, 005
Loom-shuttle boxes, Apparatus for operating	A. Allen	Wilmington, Del	Sept. 4, 1849	6, 693
Loom-shuttle boxes, Apparatus for operating	R. B. Goodyear	Philadelphia, Pa	Mar. 13, 1849	6, 170
Loom-shuttle boxes, Operating	J. Ashworth	North Andover, Mass	Sept. 8, 1868	81, 865
Loom-shuttle boxes, Operating	J. Brierley	Worcester, Mass	July 28, 1868	80, 385
Loom-shuttle boxes, Operating	T. Cleary	Millbury, Mass	Feb. 2, 1869	86, 365
Loom-shuttle boxes, Operating	R. B. Goodyear and B. Hirst	Philadelphia and Manayunk, Pa.	Jan. 1, 1850	6, 986
Loom-shuttle check	E. W. Bray	Mooresville, Ind	Dec. 3, 1872	133, 624
Loom-shuttle check	J. N. Kingston	Bristol County, Mass	Oct. 1, 1872	131, 761
Loom-shuttle check	D. Pickman	Lowell, Mass	Dec. 21, 1869	98, 188
Loom-shuttle driver, Power	A. S. Lyon	Lawrence, Mass	Jan. 27, 1863	37, 513
Loom-shuttle-driving mechanism	J. Stever	Bristol, Conn	Jan. 18, 1870	99, 024
Loom-shuttle guard	C. H. Hudson	Clinton, Mass	Apr. 8, 1873	137, 551
Loom-shuttle guard	J. Rycroft and L. A. White	Millbury, Mass	June 27, 1871	116, 499
Loom-shuttle guard	E. M. Stevens	Boston, Mass	Apr. 30, 1867	64, 261
Loom-shuttle guard	E. M. Stevens	Boston, Mass	Dec. 23, 1823	145, 763
Loom-shuttle guard, Power	P. Miggett	Hoosick Falls, N. Y	Aug. 8, 1854	11, 505
Loom-shuttle guide	J. B. Bancroft	Hopedale, Mass	Jan. 14, 1873	134, 837
Loom-shuttle guide	J. B. Bancroft	Hopedale, Mass	Dec. 30, 1873	145, 925
Loom-shuttle guide	W. A. Hastings	Palmer, Mass	Feb. 9, 1869	86, 668
Loom-shuttle guide	H. T. Robbins	Lowell, Mass	Sept. 14, 1852	9, 263
Loom-shuttle guide	H. T. Robbins	Lowell, Mass	July 24, 1855	13, 322
Loom-shuttle guide or guard	S. R. Bixby and J. B. Bancroft	Worcester and Milford, Mass.	Aug. 16, 1864	43, 828
Loom-shuttle motion	J. Avery	Lowell, Mass	May 29, 1855	12, 941
Loom-shuttle motion	L. Ferguson	Lowell, Mass	May 19, 1857	17, 323
Loom-shuttle motion	A. Gartenmann	New York, N. Y	Mar. 4, 1873	136, 501
Loom-shuttle motion	O. A. Kelly	Woonsocket, R. I	June 4, 1850	7, 417
Loom-shuttle motion	G. W. Perry	Thompson, Conn	Nov. 11, 1851	8, 506
Loom-shuttle motion	G. J. Wardwell	Hanover, Me	Aug. 5, 1851	8, 278
Loom-shuttle motion	T. T. Willcox	Norwich, Conn	Nov 26, 1850	7, 803
Loom-shuttle motion, Power	W. Crighton	Fall River, Mass	Oct. 18, 1853	10, 126
Loom-shuttle-operating device	J. J. and E. Harrison	Manchester, England	Nov. 20, 1866	59, 767
Loom-shuttle-operating machinery	R. P. Cunningham	Abington, Conn	Dec. 16, 1845	4, 308
Loom-shuttle-operating mechanism	D. Brickford	Boston, Mass	Nov. 21, 1865	51, 006
Loom-shuttle-operating mechanism	E. P. Terrel	West Liberty, Ohio	Dec. 13, 1870	110, 171
Loom-shuttle, Narrow-ware	J. Y. Hamilton	Clinton, Mass	Jan. 29, 1867	61, 535
Loom-shuttle relief-mechanism	W. C. Macomber	Mystic River, Conn	Dec. 23, 1873	145, 882
Loom-shuttle spindle	N. I. Allen and J. C. Moody	Brunswick, Me	Nov. 23, 1869	97, 147
Loom-shuttle spindle	A. Morton	Salmon Falls, N. H	Aug. 1, 1871	117, 556
Loom-shuttle stopping-mechanism	S. Boorn	Lowell, Mass	Apr. 12, 1870	101, 817
Loom-shuttle stopping-device, Power	D. S. Esten	Hinsdale, Mass	Sept. 8, 1863	39, 803
Loom-shuttle tongue	C. B. Thorp	Smithfield, R. I	Apr. 17, 1837	162
Loom-shuttles, Operating	J. Schottenfels	New York, N. Y	Mar. 20, 1866	53, 350
Loom, Stocking	J. Bazier, jr	Canton, Mass	Oct. 28, 1814	....
Loom, Stocking	E. Herrick	Stockbridge, Mass	Oct. 22, 1813	....
Loom, Stocking	A. Porter	Queensbury, N. Y	Aug. 22, 1816	....
Loom, Stocking	J. Vickerstaff	Philadelphia, Pa	Mar. 20, 1847	5, 023
Loom stop-apparatus	E. Burt	Manchester, Conn	June 20, 1845	4, 088
Loom-stop-actuating mechanism	A. J. Woodman	Springfield, Mass	July 9, 1872	128, 772
Loom-stop, Electrical	G. F. Luibery	Montargis, France	Sept. 16, 1873	142, 925
Loom stop-mechanism	C. E. Barnes	Boston, Mass	May 20, 1873	138, 987
Loom stop-mechanism	J. Cowgill	Philadelphia, Pa	Dec. 20, 1870	110 208
Loom stop-mechanism	G. Crompton	Worcester, Mass	Feb. 9, 1869	86, 735
Loom stop-mechanism	G. K. Dearborn	Smithfield, R. I	Aug. 24, 1869	93, 971
Loom stop-mechanism	W. Edson	Boston, Mass	June 3, 1873	139, 560
Loom stop-mechanism	T. Isherwood and W. Nuttall	Westerly, R. I	May 13, 1873	138, 893
Loom stop-mechanism	L. J. Knowles	Worcester, Mass	Sept. 2, 1873	142, 401

Index of patents issued from the United States Patent Office from 1790 to 1873, inclusive—Continued.

Invention.	Inventor.	Residence.	Date.	No.
Loom stop-mechanism	J. J. Switzer	Chelsea, Mass	June 8, 1869	91,050
Loom stop-mechanism	J. J. Switzer	Boston, Mass	Mar. 18, 1873	136,880
Loom stop-mechanism	H. F. Wheeler and W. O. Wakefield.	Boston, Mass	Oct. 7, 1873	143,552
Loom stop-motion	A. A. Barker	Lewiston, Me	June 1, 1869	90,627
Loom stop-mechanism, Power	E. B. Bigelow	Boston, Mass	Apr. 20, 1869	89,012
Loom stop-motion	A. Carmichel and W. Barney	Westerly, R. I	Nov. 27, 1866	59,962
Loom stop-motion	G. Draper	Milford, Mass	Sept. 1, 1863	39,723
Loom stop-motion	C. Duckworth	Mount Carmel, Conn	Sept. 6, 1864	44,080
Loom stop-motion	E. Hall	Cabotville, Mass	Sept. 24, 1850	7,665
Loom stop-motion	L. B. Hoit	Millbury, Mass	Feb. 17, 1852	8,740
Loom stop-motion	A. J. Loisean	New York, N. Y	Jan. 22, 1867	61,439
Loom stop-motion	A. Marchington	Upland, Pa	Aug. 21, 1866	57,350
Loom stop-motion	G. E. Milroy	Lowell, Mass	Mar. 30, 1869	88,322
Loom stop-motion	W. and D. Pilkington	Frankford and Chester, Pa	July 16, 1867	66,734
Loom stop-motion	J. J. Switzer	Roxbury, Mass	Aug. 11, 1868	80,885
Loom stop-motion	J. Taylor and J. Woodhead	Middletown, Pa	Apr. 21, 1831	
Loom stop-motion and shuttle	W. W. Tucker	Hartford, Conn	Jan. 16, 1872	122,745
Loom stop-motion, Hair-cloth	R. J. Stafford	Smithfield, R. I	Mar. 23, 1858	19,719
Loom stop-motion, Power	T. Foden	Holyoke, Mass	Oct. 1, 1861	33,383
Loom take-up	C. Gahren	New York, N. Y	Dec. 10, 1872	133,845
Loom take-up	G. H. Holmes	New Brunswick, N. J	Aug. 10, 1869	93,441
Loom take-up	W. Rouse	Taunton, Mass	May 9, 1871	114,606
Loom take-up, Corset	J. Kuttner	New York, N. Y	Mar. 7, 1871	112,356
Loom-take-up mechanism	S. Estes	Newburyport, Mass	Mar. 25, 1873	137,188
Loom-take-up mechanism	C. Gahren	New York, N. Y	Nov. 18, 1873	144,756
Loom-take-up mechanism	J. Michna and G. Fischer	New York, N. Y	Mar. 7, 1871	112,478
Loom-take-up mechanism	A. Mitchell and H. S. Combs	Providence, R. I., and Warren, Mass.	Sept. 17, 1872	131,455
Loom-take-up mechanism	J. Sparks	Concord, Ky	Apr. 27, 1869	89,445
Loom-take-up motion	L. L. Shaw	Lewiston, Me	Aug. 16, 1864	43,868
Loom-take-up motion	A. Stone	Johnston, R. I	Aug. 17, 1835	
Loom-take-up motion, Power	J. P. Comins	Killingly, Conn	Oct. 20, 1836	
Loom-take-up motion, Power	J. Whitaker	Philadelphia, Pa	Oct. 28, 1862	36,875
Loom take-up, Narrow-ware	M. Wolf	New York, N. Y	Oct. 9, 1866	58,708
Loom take-up, Trimming	S. Walker	Roxbury, Mass	Sept. 20, 1859	25,538
Loom tappet-chain	J. Nuttall	Walmersly, England	Mar. 15, 1870	100,794
Loom-temple	J. H. Allen	Biddeford, Me	Jan 16, 1855	12,236
Loom-temple	N. J. Allen and J. C. Moody	Auburn and Brunswick, Me.	Sept. 24, 1872	131,491
Loom-temple	N. J. Allen and E. S. Stimpson	Milford, Mass	Nov. 18, 1873	144,647
Loom-temple	E. A. Angell	Killingly, Conn	Oct. 19, 1838	987
Loom-temple	J. Beard and A. Whitney	Waltham, Mass	Apr. 6, 1842	2,536
Loom-temple	L. Briggs	Grosvenor Dale, Conn	Oct. 6, 1868	82,684
Loom-temple	N. Chapman	Hopedale, Mass	Apr. 22, 1873	138,004
Loom-temple	N. Chapman	Hopedale, Mass	June 17, 1873	140,013
Loom-temple	N. Chapman	Hopedale, Mass	Nov. 11, 1873	144,507
Loom-temple	L. K. and P. Day	Sacarappa, Me	Feb. 27, 1849	6,143
Loom-temple	E. and W. W. Dutcher	North Bennington, Vt	Dec. 16, 1851	8,586
Loom-temple	E. and W. W. Dutcher	North Bennington, Vt	Dec. 28, 1852	9,502
Loom-temple	F. J. Dutcher	Hopedale, Mass	Mar. 11, 1873	136,644
Loom-temple	W. W. Dutcher	Milford, Mass	Apr. 2, 1867	63,372
Loom-temple	W. W. Dutcher	Hopedale, Mass	June 2, 1868	78,441
Loom-temple	W. W. Dutcher	Hopedale, Mass	Dec. 14, 1869	97,895
Loom-temple	W. W. Dutcher	Hopedale, Mass	Aug. 2, 1870	106,038
Loom-temple	W. W. Dutcher	Hopedale, Mass	Sept. 6, 1870	107,168
Loom-temple	W. W. Dutcher	Hopedale, Mass	June 17, 1873	139,884
Loom-temple	W. W. Dutcher and G. Draper	Milford, Mass	Apr. 16, 1867	63,872
Loom temple	A. H. Gilman	Boston, Mass	June 14, 1870	104,137
Loom-temple	J. B. Greene	Worcester, Mass	Nov. 8, 1863	10,206
Loom-temple	W. H. Howard	Media, Pa	May 31, 1870	103,616
Loom-temple	A. Jillson	Woonsocket, R. I	Sept. 16, 1851	8,368
Loom-temple	P. Newell	Waterford, N. Y	Jan. 5, 1833	
Loom-temple	P. Newell	Waterford, N. Y	Dec. 2, 1836	96
Loom-temple	W. and J. W. Markland	Lowell, Mass	Feb. 27, 1872	124,004
Loom-temple	A. Palmer	Lebanon, N. Y	Feb. 10, 1846	4,379
Loom-temple	R. Pilson	Laurel, Md	Sept. 14, 1858	21,515
Loom-temple	R. Reynolds	Stockport, N. Y	Feb. 19, 1856	14,292
Loom-temple	B. Rogers	Beaver Creek, Ohio	Nov. 24, 1814	
Loom-temple	C. T. K. Rollins	Lowell, Mass	Aug. 30, 1833	
Loom-temple	J. A. Scholfield	Westerly, R. I	Aug. 2, 1853	9,900
Loom-temple	E. F. Shaw	Boston, Mass	Nov. 26, 1867	71,541
Loom-temple	J. Simpson	Millbury, Mass	July 20, 1869	92,890
Loom-temple	J. Smith	Laurel, Md	Aug. 7, 1855	13,413
Loom-temple	E. S. Stimpson	Hopedale, Mass	June 17, 1873	139,929
Loom-temple	J. C. Tilton	Sanbornton Bridge, N. H	May 15, 1855	12,879
Loom-temple	J. C. Tilton	Sanbornton Bridge, N. H	Jan. 11, 1859	22,594
Loom-temple	G. L. White	Cumberland, R. I	Aug. 28, 1866	57,646
Loom temple, Cloth	J. Bazin, jr	Canton, Mass	Aug. 8, 1823	
Loom temple, Double-cloth	S. Everett	Biddeford, Me	Apr. 2, 1850	7,237
Loom temple, Perpetual	D. A. Smith	Townsend, Mass	Feb. 23, 1810	
Loom temple, Power	J. Standish	Providence, R. I	Nov. 11, 1830	
Loom temple, Revolving-burr	E. B. Harris and A. R. Arnold	Woodstock, Conn	Apr. 13, 1830	
Loom-temple, Rotary	I. Draper	Saugus, Mass	Apr. 1, 1829	
Loom-temple, Rotary	I. C. Lane	Waltham, Mass	Mar. 26, 1844	3,511
Loom-temple, Self-acting rotary	W. Craig and J. Cochrane	England	Nov. 25, 1841	2,377
Loom-temple, Self-adjusting	S. P. Mason	Newport, R. I	July 22, 1837	291
Loom-temple, Self-operating	E. R. Otis	River Head, Conn	Apr. 2, 1830	
Loom-temples, Making	O. M. Stillman	Brookfield, N. Y	Jan. 30, 1826	
Loom-tension-regulating device	M. Sigler	Paterson, N. J	Feb. 2, 1869	86,465
Loom-tension regulator	E. B. Bigelow	Boston, Mass	Mar. 12, 1845	3,948
Loom-tension regulator	F. Painter	East Hampton, Mass	Oct. 29, 1867	70,253
Loom-treadles and harness-shafts, Operating	O. W. Gordon	Mount Pleasant, Iowa	Apr. 30, 1867	64,216
Loom-treadles, Operating	R. W. Andrews	Stafford, Conn	Jan. 18, 1853	9,540
Loom, Upright power	S. Shepard and J. Thorp	Taunton, Mass	July 25, 1816	
Loom, Vibrating-cam	W. H. Howard	Worcester, Mass	Feb. 21, 1830	

Index of patents issued from the United States Patent Office from 1790 *to* 1873, *inclusive*—Continued.

Invention.	Inventor.	Residence.	Date.	No.
Loom warp-feeding mechanism	A. B. Ely	Newton, Mass	Dec. 17, 1867	72, 379
Loom warp-shedding mechanism	J. Holding and J. Eccles	Manchester, Great Britain	Oct. 10, 1871	119, 848
Loom warp-stand	P. D. Tifft	Eagleville, Ohio	May 10, 1870	102, 989
Loom warp-suspender	W. Squire	New York, N. Y	Jan. 17, 1815	....
Loom warp-tension mechanism	J. F. Randall	Warren, Ohio	Nov. 1, 1870	108, 826
Loom, Water	B. Brown	Albemarle, Va	Mar. 21, 1814	....
Loom, Water	P. C. Curtis	Paris, N. Y	May 28, 1816	....
Loom, Water	T. Mussy	Philadelphia, Pa	Mar. 4, 1811	....
Loom, Water	T. Mussey	New London, Conn	June 10, 1817	....
Loom, Water	S. Shepard	Philadelphia, Pa	Apr. 27, 1812	....
Loom, Weaving	B. Slingerland	Paterson, N. J	Apr. 1, 1845	3, 977
Loom, Weaving	R. Wilcox	Columbus, Ohio	Mar. 21, 1825	....
Loom yarn-beam	B. A. Bailey	Lewiston, Me	Aug. 4, 1868	80, 529
Loom yarn-beam	A. W. Cole	West Killingly, Conn	Sept. 12, 1871	118, 910
Loom yarn-beam	S. Pettee	Foxborough, Mass	June 15, 1813	....
Loom yarn-beams, Friction-device for	A. Beebe and F. L. Bestor	Enfield, Mass	Aug. 23, 1864	43, 895
Looms, Belt-shifting mechanism for	L. J. Knowles	Warren, Mass	Aug. 9, 1870	106, 267
Looms, Belt-shipper for	L. J. Knowles	Warren, Mass	July 14, 1868	79, 986
Looms, Belt-shipping mechanism for	L. J. Knowles	Warren, Mass	Feb. 2, 1869	86, 417
Looms, Brake for power	R. and G. B. Reynolds	Stockport, N. Y	June 21, 1859	24, 488
Looms, Brake for yarn-beam of	J. J. and E. Harrison	Broughton, England	Sept. 15, 1868	82, 220
Looms by electricity, Operating	G. Bonelli	Turin, Sardinia	Dec. 12, 1854	12, 050
Looms, Cloth-registering attachment for	C. C. Temple	Saco, Me	June 20, 1865	48, 325
Looms, Cloth-winding mechanism for	W. A. Arnold	Rockport, Mass	Dec. 6, 1870	109, 794
Looms, Cylinder for figuring	E. M. Hastings and J. Shepherdson.	Jamestown, N. Y	Mar. 18, 1851	7, 988
Looms, Delivery and take-up motion of	A. H. Boyd	Saco, Me	Mar. 10, 1849	6, 157
Looms, Device for supplying weft to the shuttles in	T. Ingram	Brantford, England	July 12, 1864	43, 556
Looms, Eye of wire heddles for	J. Ashworth	North Andover, Mass	Jan. 29, 1867	61, 501
Looms, Finger for shuttle-stop rod in	E. S. Laney	Waterloo, N. Y	May 12, 1868	77, 820
Looms for twilling, Operating treadle-cams in	R. Garsed	Frankford, Pa	July 20, 1846	4, 645
Looms for weaving bolting-cloth, Mounting and using harness in.	R. White	Williamstown, Vt	Apr. 1, 1842	2, 528
Looms for weaving counterpanes, &c., Mounting	E. B. Bigelow	Lancaster, Mass	May 30, 1842	2, 653
Looms for weaving figured fabrics, Shuttle-box motion in.	C. Duckworth	Mount Carmel, Conn	Jan. 9, 1866	51, 932
Looms for weaving goods with elastic strands, Tension-mechanism for.	L. Hull	Charlestown, Mass	Aug. 8, 1865	49, 271
Looms for weaving irregular fabrics, Take-up mechanism for.	H. Carstaedt	New York, N. Y	Mar. 30, 1869	88, 365
Looms, Friction-mechanism for warp-beams of	O. Kenison and A. J. McClary	Lawrence, Mass	Feb. 21, 1865	46, 475
Looms, Graduating take-up motion for power	H. Hendrick	Killingly, Conn	Sept. 22, 1836	30
Looms, Harness-motion for	A. L. Anderson	Ware, Mass	May 15, 1866	54, 661
Looms, Harness-motion for	R. Collins	Chicopee, Mass	Apr. 17, 1866	54, 058
Looms, Harness-motion for	G. Crompton	Worcester, Mass	Nov. 27, 1866	59, 972
Looms, Harness-motion for	J. F. Gebhart	New Albany, Ind	July 16, 1867	66, 827
Looms, Harness-motion for	J. F. Gebhart	New Albany, Ind	July 16, 1867	66, 828
Looms, Harness-motion for	B. F. Knowles	Providence, R. I	Jan. 22, 1861	31, 176
Looms, Harness-motion for	L. J. Knowles	Warren, Mass	Aug. 27, 1867	68, 303
Looms, Harness-motion for	W. W. McGregor	Dedham, Mass	July 3, 1866	56, 077
Looms, Harness-motion for	C. Miller	Saint Louis, Mo	Aug. 7, 1866	56, 976
Looms, Harness-motion for	C. Schilling	Auburn, N. Y	Sept. 10, 1867	68, 795
Looms, Harness-motion for	H. Yount	Dayton, Ohio	Aug. 13, 1867	67, 699
Looms, Harness-motion for power	D. M. Ayer	Lewiston, Me	Oct. 21, 1862	36, 697
Looms, Harness-motion for power	C. Duckworth	Mount Carmel, Conn	Aug. 1, 1865	49, 096
Looms, Harness-motion for power	J. Greenhalgh	Pascoag, R. I	May 3, 1864	42, 575
Looms, Heddle-connection for	T. K. McIntyre	Warner, N. H	July 9, 1872	128, 807
Looms, Heddle-eye for	T. Clegg	North Andover, Mass	Mar. 7, 1854	10, 623
Looms, Hook-temple for	W. W. Dutcher and G. Draper	Milford, Mass	May 12, 1857	17, 267
Looms, Jacquard apparatus for	J. A. Elder	Westbrook, Me	June 21, 1853	9, 795
Looms, Jacquard apparatus for	L. D. Valetton	Philadelphia, Pa	Nov. 14, 1865	50, 993
Looms, Jacquard mechanism for	E. K. Davis	New York, N. Y	Nov. 9, 1869	96, 553
Looms, Lay for power	B. Brooks	New Market, N. H	Dec. 17, 1834	....
Looms, Lubricating temple for	E. S. Stimpson	Milford, Mass	May 18, 1869	90, 133
Looms, Lubricator for picker-spindles in	T. Parker	Shelby, N. C	Sept. 19, 1871	119, 171
Looms, Machine for casting metallic eyes or mail of heddles for.	J. Senneff	Philadelphia, Pa	Aug. 22, 1854	11, 589
Looms, Making and using	J. Goulding	Dedham, Mass	July 7, 1830	....
Looms, Measuring cloth on	J. G. Webster	Middlesex County, Mass	Feb. 19, 1850	7, 112
Looms, Mechanism for operating pile-wire	W. Weild	Manchester, England	Aug. 25, 1868	81, 443
Looms, Mechanism for presenting palm-leaf to	I. Angell	Malden, Mass	Dec. 10, 1867	71, 945
Looms, Mode of applying friction to the yarn-beam of power.	S. Kimball	Putney, Vt	May 30, 1838	758
Looms, Mode of delivering warp in	J. Coulter	Xenia, Ohio	Apr. 25, 1843	3, 057
Looms, Mode of operating the treadle in power	E. Horton	Stafford, Conn	Feb. 22, 1838	617
Looms, Mode of throwing shuttle in	S. C. Mendenhall	Richmond, Ind	Nov. 9, 1852	9, 387
Looms, Mode of throwing shuttle in power	R. P. Cunningham	Pomfret, Conn	Apr. 25, 1843	3, 062
Looms, Motion of the lay in	J. Goulding	Worcester, Mass	June 15, 1852	9, 024
Looms, Mounting lease-rod for	W. A. Hastings	Thorndike, Mass	Sept. 21, 1869	94, 957
Looms, Needle of knitting	W. Bamford	Ipswich, Mass	Sept. 12, 1848	5, 761
Looms, Pattern-card for Jacquard	S. T. Thomas and E. Everett	Lowell and Lawrence, Mass	Mar. 16, 1852	8, 810
Looms, Pattern-chain for	H. D. Davis	North Andover, Mass	Apr. 19, 1870	102, 094
Looms, Pattern-chain for	B. H. Jenks	Philadelphia, Pa	July 24, 1860	29, 283
Looms, Reeds, heddles, or harness for	J. A. Wilkinson	Providence, R. I	July 17, 1835	....
Looms, Regulating motion of yarn-beam in power	W. A. Potter	Cranston, R. I	Nov. 23, 1837	475
Looms, Roller for pattern-chain for	G. Crompton and M. A. Furbush.	Worcester, Mass	Nov. 14, 1854	11, 956
Looms, Roller jaw-temple for	W. H. Burns	Grafton, Mass	Sept. 28, 1869	95, 314
Looms, Roller jaw-temple for	W. H. Burns	Grafton, Mass	Feb. 1, 1870	99, 287
Looms, Roller-temple for	W. H. Burns	Grafton, Mass	Nov. 29, 1870	109, 583
Looms, Roller-temple for	N. Chapman	Hopedale, Mass	Jan. 8, 1867	61, 050
Looms, Roller-temple for	W. W. Dutcher	Milford, Mass	June 16, 1857	17, 559
Looms, Roller-temple for	W. W. Dutcher	Milford, Mass	Mar. 24, 1863	37, 954
Looms, Roller-temple for	W. W. Dutcher	Milford, Mass	Mar. 27, 1866	53, 423
Looms, Roller-temple for	W. W. Dutcher	Hopedale, Mass	Aug. 30, 1870	106, 797
Looms, Roller-temple for	J. Mathis	Dornbein, Austria	Apr. 24, 1866	54, 269
Looms, Rotary cam for	R. Sargent	Norwich, Conn	Nov. 5, 1867	70, 623
Looms, Rotary temple for power	G. Draper	Saugus, Mass	Feb. 21, 1842	2, 464

Index of patents issued from the United States Patent Office from 1790 *to* 1873, *inclusive*—Continued.

Invention.	Inventor.	Residence.	Date.	No.
Looms, Rotary temple for weaving	G. Draper	Palmer, Mass	Oct. 28, 1840	1, 838
Looms, Sectional take-up for corset	S. Ottenheimer	New York, N. Y	May 28, 1867	65, 112
Looms, Self-acting temple for	K. Gibbs	South Berwick, Me	Feb. 13, 1839	1, 080
Looms, Shipper-lever for	G. W. Hathaway	Hinsdale, Mass	Feb. 3, 1863	37, 577
Looms, Shuttle-box-operating lever for	B. H. Jenks	Bridesburgh, Pa	July 13, 1869	92, 612
Looms, Spooling-machine for tape	I. Gibbs	Warren, Mass	Nov. 21, 1865	51, 038
Looms, Stopping power	O. M. Stillman	Stonington, Conn	Nov. 10, 1841	2, 333
Looms, Treadle-cam for	B. H. Jenks	Bridesburgh, Pa	Mar. 19, 1867	63, 055
Looms, Weft-feeding device for hair-cloth	J. Blanchard	Pawtucket, R. I	Feb. 21, 1865	46, 442
Looms, Weft-fork for	J. H. Knowles	Lawrence, Mass	May 2, 1871	114, 307
Looms, Weft-stop mechanism for	W. Comey and S. S. Turner	Westborough, Mass	July 16, 1872	129, 212
Looms, Weft-stop mechanism for	J. D. Cottrell	Hopedale, Mass	May 4, 1869	89, 561
Looms, Weft-stop mechanism for	T. Naylor	Manchester, Iowa	May 7, 1872	126, 410
Looms, Weft-stop mechanism for	J. J. Switzer	Boston, Mass	Oct. 10, 1871	119, 798
Looms, Weft-stop motion for	G. Crompton	Worcester, Mass	Oct. 22, 1867	69, 974
Looms, Weft-stop motion for	J. C. Fifield	Lowell, Mass	Mar. 24, 1868	75, 888
Looms, Wire heddle for	D. C. Brown	Lowell, Mass	May 3, 1864	42, 553
Looms, Yarn-delivering mechanism for	G. Draper	Milford, Mass	Mar. 8, 1864	41, 834
Loop-banding machine	G. W. Weeks	Clinton, Mass	Nov. 14, 1871	121, 031
Loop, Body	R. R. Miller	Plantsville, Conn	June 27, 1871	116, 472
Loop-making machine	A. Babbett	Auburn, N. Y	Feb. 9, 1864	41, 470
Loop, Metallic	C. H. Littlefield	Turner, Me	July 30, 1867	67, 322
Lotteries, and mode of drawing, System of	F. W. Dana	Boston, Mass	Dec. 28, 1825	....
Lotteries, Calculating and drawing	W. Seger	Bergen, N. J	May 25, 1825	....
Lotteries, Constructing and drawing	J. J. Cohen	Baltimore, Md	Aug. 28, 1828	....
Lotteries, Constructing and drawing	M. Davis	Cumberland City, Mo	May 9, 1825	....
Lotteries, Drawing	J. I. Cohen	Baltimore, Md	Mar. 3, 1825	....
Lotteries, Drawing	J. C. Rives	Washington, D. C	Dec. 22, 1826	....
Lotteries, Instituting and drawing	J. K. Casey	New York, N. Y	Mar. 28, 1831	....
Lotteries, Method of drawing	W. C. Conine	Baltimore, Md	Nov. 14, 1826	....
Lotteries, Method of drawing	J. Howard and A. Hayward	Providence, R. I	Nov. 8, 1825	....
Lotteries, Mode of drawing	D. Malcomb	Orleans, La	July 30, 1829	....
Lotteries, Mode of drawing	W. E. Spalding	Brooklyn, Conn	Feb. 15, 1828	....
Lottery	J. Vannini	Washington, D. C	Dec. 4, 1815	....
Lottery, American	E. Nott	Union College, N. Y	Sept. 9, 1825	....
Lottery-scheme	E. Gratton	Providence, R. I	May 16, 1826	....
Lottery-scheme	J. Vannini	Philadelphia, Pa	Oct. 18, 1823	....
Lottery-schemes, Construction of	J. W. Taguo	Philadelphia, Pa	May 16, 1825	....
Lottery-schemes, Mathematical operation of drawing.	J. Vannini	New York, N. Y	July 18, 1840	1, 700
Lounge	L. H. Baker	Tarrytown, N. Y	June 23, 1868	79, 050
Lounge	J. G. Broemser	Saint Louis, Mo	Jan. 25, 1859	22, 699
Lounge	W. H. Colley	Leavenworth, Kans	May 24, 1870	103, 301
Lounge	A. Eliaers	Boston, Mass	Oct. 25, 1853	10, 150
Lounge	W. C. Hart	Nantucket, Mass	May 26, 1868	78, 204
Lounge	W. H. Hart	Lowell, Mass	Sept. 30, 1873	143, 283
Lounge	F. Hayek	New York, N. Y	Mar. 2, 1869	87, 411
Lounge	A. Hayer	Philadelphia, Pa	Dec. 9, 1873	145, 418
Lounge	F. Krater	Pittsburgh, Pa	Jan. 14, 1873	134, 752
Lounge	O. Kubitschky	Chicago, Ill	Jan. 10, 1871	110, 093
Lounge	C. H. Levy and S. T. Herschmann.	New York, N. Y	Oct. 15, 1872	132, 162
Lounge	A. and D. Short	West Liberty, Ohio	July 31, 1866	56, 809
Lounge and bath-tub, Combined	C. Wendel	New York, N. Y	Apr. 29, 1873	138, 457
Lounge and bed	J. C. Hall	Cincinnati, Ohio	Feb. 21, 1871	112, 031
Lounge and bed	A. Hansen	San Francisco, Cal	Jan. 4, 1870	98, 589
Lounge and bed, Combined	P. Frank	New York, N. Y	Aug. 6, 1872	130, 290
Lounge and bedstead	A. R. Harper and C. B. Dake	Hobart, Ind	Dec. 21, 1869	98, 056
Lounge and bedstead, Combined	J. C. Hall	Cincinnati, Ohio	Oct. 24, 1871	120, 270
Lounge and bedstead, Combined	S. Higgins	Portland, Me	June 17, 1873	139, 957
Lounge and camp-stool, Combined	O. P. Pell	Flushing, N. Y	Sept. 24, 1861	33, 362
Lounge and chair, Combined	S. Herschmann	New York, N. Y	May 9, 1871	114, 679
Lounge and chair, Combined	A. T. Linikin	Roxbury, Mass	July 17, 1849	6, 594
Lounge and chair, Combined	M. P. Robinson	Fayette County, Ky	Dec. 9, 1873	145, 452
Lounge and chair seat	M. Le Page and F. Raymond	Woodhaven, N. Y	Apr. 23, 1867	64, 115
Lounge and sofa-bed	R. M. Reeby	Worcester, Mass	Oct. 26, 1869	96, 266
Lounge, bureau, and table	A. Faller	New York, N. Y	July 21, 1868	80, 159
Lounge, Extension	V. O. Hash and J. Benjamin	Naples, N. Y	Aug. 8, 1871	117, 775
Lounge, Extension	N. H. Hill	Cincinnati, Ohio	Sept. 20, 1870	107, 494
Lounge, Extension	J. S. Paine	Cambridge, Mass	Apr. 4, 1871	113, 333
Lounge, Extension	C. Streit	Indianapolis, Ind	Nov. 17, 1868	84, 147
Lounge, Extension	C. Streit	Cincinnati, Ohio	Feb. 13, 1872	123, 649
Lounge, Extension	G. H. Thomas	Quincy Point, Mass	July 23, 1872	129, 766
Lounge, Extension bed	C. F. Vollmer	Harrisburgh, Pa	Dec. 24, 1867	72, 704
Lounge, Folding	J. Beiersdorf	Chicago, Ill	Apr. 26, 1870	102, 360
Lounge, Folding	W. B. Coats	Philadelphia, Pa	Apr. 30, 1872	1[illegible]6, 266
Lounge, Folding	D. Forbes	Chicago, Ill	Nov. 10, 1868	83, 949
Lounge, Folding	W. H. Lotz	Chicago, Ill	Dec. 29, 1868	85, 391
Lounge, Folding	A. Lundberg	Chicago, Ill	Mar. 18, 1873	136, 843
Lounge, Folding	M. Marso	Chicago, Ill	Dec. 27, 1870	110, 578
Lounge, Folding	J. W. McDonough	Chicago, Ill	Dec. 11, 1866	60, 400
Lounge, Folding	W. Prufrock	Saint Louis, Mo	June 27, 1871	116, 350
Lounge, Hammock	J. C. Craft	Baltimore, Md	Sept. 2, 1873	142, 327
Lounge, Invalid	R. Leaman	Hillsborough, Ohio	Jan. 1, 1867	60, 911
Lounge, Nursery	S. Buttenheim	New York, N. Y	Aug. 27, 1867	68, 164
Lounge, Ottoman	G. H. C. Williams	Chicago, Ill	July 28, 1868	80, 322
Lounge, Reversible	S. Lloyd	Washington, D. C	Feb. 26, 1867	62, 349
Lounges, Leg for folding	W. Seng	Chicago, Ill	Sept. 3, 1872	131, 031
Lounges, Leg for folding	J. F. Tobey	Medford, Mass	May 6, 1873	138, 716
Lowering and detaching boats from davits, Device for.	C. H. Ramsten	Carlskrona, Sweden	Oct. 23, 1866	59, 153
Low land, Device for filling	G. Howell	Philadelphia, Pa	Dec. 27, 1870	110, 468
Low-water alarm	J. C. Leistner and A. Kayser	Cincinnati, Ohio	Sept. 19, 1871	119, 037
Low-water alarm	J. Stanton	Philadelphia, Pa	Apr. 25, 1871	114, 216
Low-water alarm and fire-extinguisher for steam-boiler.	B. R. Singleton	Saint Louis, Mo	Feb. 20, 1872	123, 795
Low-water alarm and indicator	J. Ross	Greenville, Mich	Mar. 25, 1873	136, 097
Low-water-alarm apparatus for steam-boilers	C. H. Brown	Fitchburgh, Mass	Aug. 7, 1860	29, 455

Index of patents issued from the United States Patent-Office from 1790 *to* 1873, *inclusive*—Continued.

Invention.	Inventor.	Residence.	Date.	No.
Low-water alarm for boilers	J. Atkins	Washington, D. C	June 30, 1868	79, 431
Low-water alarm for boilers	F. S. Davenport	Jerseyville, Ill	Jan. 7, 1868	73, 168
Low-water alarm for boilers	E. W. Mandeville	Ithaca, N. Y	Nov. 17, 1868	84, 201
Low-water alarm in boilers	A. S. Lyman	Upper Alton, Ill	Dec. 18, 1849	6, 955
Low-water alarm for steam-boilers	J. R. Brown and W. A. Fosket	New Haven, Conn	Feb. 4, 1873	135, 519
Low-water alarm for steam-boilers	S. Dustin	Detroit, Mich	Apr. 26, 1859	23, 761
Low-water alarm for steam-boilers	A. Carr	Paterson, N. J	Apr. 16, 1861	32, 045
Low-water alarm for steam-boilers	J. C. Leistner and A. Kayser	Cincinnati, Ohio	June 25, 1872	128, 318
Low-water alarm for steam-boilers	W. H. Miller	Philadelphia, Pa	May 14, 1861	32, 305
Low-water alarm for steam-boilers	L. Savage	Ashtabula, Ohio	Jan. 23, 1872	122, 970
Low-water alarm for steam-boilers, Electro-magnetic.	W. Duryea	Glen Cove, N. Y	Aug. 9, 1870	106, 141
Low-water alarm for steam-generators	R. F. Crane	Chicago, Ill	Aug. 6, 1867	67, 506
Low-water alarm for steam-generators	S. B. Palmer	Syracuse, N. Y	Feb. 19, 1867	62, 150
Low-water alarm for steam-generators	S. B. Palmer	Syracuse, N. Y	Sept. 8, 1868	82, 025
Low-water alarm for steam-generators	J. H. Springer and W. M. Bartram.	Philadelphia, Pa	Feb. 19, 1867	62, 230
			Aug. 23, 1870	106, 598
Low-water and high-pressure alarm	G. B. Massey	New York, N. Y		
Low-water and high-pressure indicator	G. M. Hopkins and J. A. Straight	Albion, N. Y	Aug. 30, 1870	106, 821
Low-water and high-steam indicator	L. F. Smith	Philadelphia, Pa	Nov. 24, 1868	84, 444
Low-water and high-steam indicator	L. F. Smith	Philadelphia, Pa	Sept. 6, 1870	107, 114
Low-water detector	J. Ashcroft	Brooklyn, N. Y	June 5, 1866	55, 221
Low-water detector	W. C. Baker	New York, N. Y	Apr. 21, 1863	28, 212
Low-water detector	B. H. Bartol	Philadelphia, Pa	Dec. 18, 1866	60, 463
Low-water detector	C. F. Cosfeldt, jr	Philadelphia, Pa	Nov. 12, 1867	70, 807
Low-water detector	J. Cosfeldt	Philadelphia, Pa	June 13, 1865	48, 158
Low-water detector	J. Delery	Saint Bernard Parish, La	July 3, 1866	56, 017
Low-water detector	T. Firth	Cincinati, Ohio	Nov. 13, 1866	59, 574
Low-water detector	J. E. Gillespie	Boston, Mass	Mar. 12, 1867	62, 837
Low-water detector	J. Harrison, jr	Philadelphia, Pa	Mar. 24, 1868	75, 754
Low-water detector	G. W. Hewitt and J. B. Haley	Cincinnati, Ohio	May 8, 1866	54, 544
Low-water detector	L. L. Lee	Milwaukee, Wis	June 27, 1871	116, 460
Low-water detector	G. B. Massey	New York, N. Y	Sept. 5, 1871	118, 737
Low-water detector	M. Peck	New Haven, Conn	Jan. 23, 1866	52, 194
Low-water detector	J. R. Shupplee and B. K. Wright	Philadelphia, Pa	Dec. 26, 1865	51, 759
Low-water detector	S. W. Warren	Brooklyn, N. Y	Oct. 21, 1862	36, 741
Low-water detector	J. Yates	Mott Haven, N. Y	Jan. 31, 1865	46, 173
Low-water detector and safety-valve, Combined	D. Burns	Bay City, Mich	May 5, 1868	77, 451
Low-water detector and steam-pressure alarm	D. McFarland	New York, N. Y	July 7, 1868	79, 673
Low-water detector for boilers	J. Ashcroft	New York, N. Y	Aug. 4, 1868	80, 700
Low-water detector for boilers	C. F. Cosfeldt, jr	Philadelphia, Pa	Jan. 14, 1868	73, 301
Low-water detector for boilers	T. Dutton	Port Jervis, N. Y	Sept. 8, 1868	81, 989
Low-water detector for boilers	G. B. Massey	New York, N. Y	Sept. 28, 1869	95, 245
Low-water detector for boiler-feeds	J. N. B. Bond	New York, N. Y	Feb. 25, 1868	74, 883
Low-water detector for steam-boilers	E. S. Bennett	New York, N. Y	Jan. 26, 1869	86, 125
Low-water detector for steam-boilers	G. W. Blake	New York, N. Y	Aug. 20, 1861	33, 067
Low-water detector for steam-boilers	J. W. Hopper	New York, N. Y	July 24, 1860	29, 277
Low-water detector for steam-boilers	B. Schaffer	Buckau, Magdeburg, Prussia.	Nov. 22, 1864	45, 207
Low-water detector for steam-boilers	M. S. Vosburgh	Attica, N. Y	Dec. 31, 1872	134, 499
Low-water detector for steam-boilers	L. Youmans	Fulton, N. Y	Sept. 3, 1861	33, 224
Low-water detector for steam-boilers, Electro-magnetic.	W. Duryea	Glen Cove, N. Y	Aug. 9, 1870	106, 142
Low-water detector for steam-generators	W. W. Bailey	New York, N. Y	Mar. 17, 1868	75, 618
Low-water detector for steam-generators	G. B. Massey	New York, N. Y	Feb. 18, 1868	74, 706
Low-water detector for steam-generators	G. B. Massey	New York, N. Y	Feb. 23, 1869	87, 181
Low-water detector for steam-generators	D. C. Mead and C. Maggi	Pittsburgh, Pa	Oct. 24, 1865	50, 660
Low-water detector for steam-generators	T. Savill	Philadelphia, Pa	Sept. 18, 1866	58, 179
Low-water detector for steam-generators	J. V. Weitz	Cleveland, Ohio	Sept. 3, 1867	68, 473
Low-water indicator	J. O. Alter	Saint Louis, Mo	Aug. 13, 1867	67, 702
Low-water indicator	F. S. Barns	New York, N. Y	July 13, 1869	92, 564
Low-water indicator	R. Berryman	Philadelphia, Pa	Feb. 15, 1870	99, 996
Low-water indicator	J. W. Bishop	New Haven, Conn	Apr. 11, 1865	47, 183
Low-water indicator	W. A. Bradford	Cincinnati, Ohio	Dec. 21, 1869	98, 019
Low-water indicator	W. A. Bradford	Cincinnati, Ohio	Apr. 18, 1871	113, 843
Low-water indicator	J. Brahn	Jersey City, N. J	June 22, 1869	91, 599
Low-water indicator	R. R. Carpenter	Tippecanoe, Ohio	July 19, 1870	105, 422
Low-water indicator	T. G. Eiswald	Providence, R. I	July 28, 1868	80, 399
Low-water indicator	T. G. Eiswald	Providence, R. I	July 28, 1868	80, 400
Low-water indicator	T. G. Eiswald and J. Barbour	Providence, R. I	Aug. 25, 1868	81, 483
Low-water indicator	R. A. Filkins	North Adams, Mass	Dec. 3, 1867	71, 729
Low-water indicator	L. Fulton	Edinburgh, Ind	Apr. 23, 1867	64, 007
Low-water indicator	T. C. Hargrave and W. B. Charlton.	Boston, Mass	May 19, 1868	77, 978
Low-water indicator	G. M. Hopkins	Albion, N. Y	Sept. 15, 1868	82, 222
Low-water indicator	G. M. Hopkins	Albion, N. Y	Feb. 23, 1869	87, 102
Low-water indicator	G. M. Hopkins and J. A. Straight.	Albion, N. Y	Aug. 31, 1869	94, 209
Low-water indicator	G. M. Hopkins and J. A. Straight.	Albion, N. Y	Aug. 30, 1870	106, 820
Low-water indicator	E. Joyce	New York, N. Y	Nov. 3, 1868	83, 638
Low-water indicator	H. Kimball	Randolph, Vt	Feb. 28, 1871	112, 256
Low-water indicator	W. H. Laubach	Philadelphia, Pa	Mar. 19, 1867	63, 061
Low-water indicator	L. L. Lee	Milwaukee, Wis	Oct. 5, 1869	95, 595
Low-water indicator	G. McAllister	San Francisco, Cal	May 14, 1867	64, 781
Low-water indicator	W. Moore	Kokomo, Ind	Oct. 13, 1868	82, 974
Low-water indicator	A. F. W. Neynaber	Philadelphia, Pa	Oct. 25, 1870	108, 618
Low-water indicator	E. D. Pritchard	New York, N. Y	June 22, 1869	91, 769
Low-water indicator	J. C. Raymond and F. T. Allyn	Brooklyn, N. Y	June 2, 1868	78, 481
Low-water indicator	F. Reynsford	Grand Rapids, Mich	Nov. 1, 1870	108, 828
Low-water indicator	P. W. Reinshagen	Cincinnati, Ohio	Apr. 26, 1870	102, 433
Low-water indicator	H. Reynolds	New Haven, Conn	Nov. 23, 1869	97, 228
Low-water indicator	J. W. Richards	Newark, N. J	Dec. 1, 1868	84, 646
Low-water indicator	L. F. Smith	Philadelphia, Pa	Nov. 12, 1867	70, 758
Low-water indicator	L. F. Smith	Philadelphia, Pa	Dec. 3, 1867	71, 803
Low-water indicator	F. Steele	Brooklyn, N. Y	Dec. 9, 1873	145, 369
Low-water indicator	W. W. Virdin	Baltimore, Md	May 4, 1869	89, 714
Low-water indicator	C. S. Watson	Philadelphia, Pa	July 7, 1868	79, 708

Index of patents issued from the United States Patent Office from 1790 *to* 1873, *inclusive*—Continued.

Invention.	Inventor.	Residence.	Date.	No.
Low-water indicator	D. Wiehl	Cincinnati, Ohio	July 5, 1870	105, 155
Low-water indicator and alarm	W. Moore	Kokomo, Ind	July 16, 1872	129, 159
Low-water indicator and alarm	F. Strange	Detroit, Mich	Aug. 26, 1873	142, 181
Low-water indicator and safety-valve, Combined	C. Burley	Cincinnati, Ohio	Dec. 17, 1867	72, 361
Low-water indicator and steam-pressure alarm	D. McFarland	New York, N. Y	July 7, 1868	79, 672
Low-water indicator and feed-water regulator	R. Berryman	Philadelphia, Pa	Jan. 25, 1870	99, 234
Low-water indicator for boilers	J. S. Griffith	Saint Louis, Mo	Feb. 2, 1869	86, 531
Low-water indicator for boilers	H. Kimball	Randolph, Vt	June 1, 1869	90, 853
Low-water indicator for boilers	C. S. Watson	Philadelphia, Pa	May 31, 1870	103, 690
Low-water indicator for boilers	P. D. Wesson	Providence, R. I	July 6, 1869	92, 237
Low-water indicator for steam-boilers	C. Brooke	Norfolk, Va	Aug. 6, 1872	130, 185
Low-water indicator for steam-boilers	G. A. Riedel	Philadelphia, Pa	Apr. 18, 1865	47, 365
Low-water indicator for steam-boilers	L. Fulton	Edinburgh, Ind	July 21, 1868	80, 162
Low-water indicator for steam-generators	L. H. Colborn	Chicago, Ill	Apr. 21, 1868	76, 893
Low-water register, Steam-boiler	W. S. Belt	Cincinnati, Ohio	Aug. 27, 1872	130, 786
Low-water safety-apparatus for steam-boilers	J. Ashcroft	Lynn, Mass	Mar. 20, 1860	27, 508
Low-water signal	T. Shaw	Philadelphia, Pa	July 11, 1865	48, 730
Low-water signal for steam-boilers	T. Shaw	Philadelphia, Pa	Dec. 27, 1864	45, 680
Lozenge	F. Ernst	New York, N. Y	Nov. 27, 1866	59, 988
Lozenge and cracker machine	C. A. Oehl	Portsmouth, N. H	Jan. 19, 1869	86, 096
Lozenge and other paste, Machine for rolling	C. A. Oehl	New York, N. Y	May 2, 1871	114, 468
Lozenge, Cathartic	W. M. Du Bois	Poughkeepsie, N. Y	July 7, 1868	79, 641
Lozenge-cutting machine	G. H. Copping	Toronto, Canada	Nov. 4, 1873	144, 258
Lozenge, &c., cutting machine	W. E. Damant	West Hoboken, N. J	Apr. 4, 1871	113, 635
Lozenge-cutting machine	W. P. Eayrs	Nashua, N. H	Feb. 28, 1871	112, 231
Lozenge-cutting machine	W. J. McClelland	New York, N. Y	May 8, 1860	28, 239
Lozenge-cutting machine	J. W. Pepper	Salem, Mass	July 2, 1850	7, 482
Lozenge-machine	E. Belling	New York, N. Y	Aug. 30, 1859	25, 236
Lozenge-machine	O. B. and S. E. Chase	Boston and Charlestown, Mass.	May 12, 1857	17, 262
Lozenge-machine	O. R. Chase	Birmingham, England	July 7, 1863	39, 196
Lozenge-machine	E. Greenfield	New York, N. Y	July 16, 1872	129, 277
Lozenge-machine	E. Günther	New York, N. Y	Jan. 9, 1872	122, 602
Lozenge-machine	J. S. Howell and C. W. Carter	Portsmouth, N. H	June 11, 1867	65, 575
Lozenge-machine	G. Kober	New York, N. Y	May 22, 1860	28, 381
Lozenge-machine	C. A. Oehl	Portsmouth, N. H	Feb. 15, 1870	99, 939
Lozenge-machine	D. Sowle	Fall River, Mass	June 19, 1860	28, 815
Lozenge-package	H. W. Booth	Toronto, Canada	June 18, 1872	128, 011
Lozenges and confectionery, Manufacture of	E. Greenfield	New York, N. Y	Feb. 15, 1870	99, 379
Lubricating-apparatus	J. C. Eggleston	Waterbury, Conn	Sept. 25, 1866	58, 339
Lubricating-axle	M. F. Kellogg	Pittsfield, Ohio	Apr. 12, 1870	101, 885
Lubricating-box	C. Andrews	Providence, R. I	Aug. 30, 1870	106, 762
Lubricating-box	T. J. Rowley and W. Roland	Chillicothe, Ohio	Mar. 10, 1868	75, 465
Lubricating-box and bearing for upright shafts	J. P. Grosvenor	Lowell, Mass	Feb. 13, 1872	123, 562
Lubricating-composition	A. Bauer	Philadelphia, Pa	May 12, 1863	38, 459
Lubricating-composition	C. Chitterling	Dunkirk, N. Y	July 24, 1860	29, 243
Lubricating-composition	J. Cordnan	Brooklyn, N. Y	Mar. 31, 1868	76, 165
Lubricating-composition	G. W. Goodhue	Cincinnati, Ohio	Jan. 5, 1864	41, 151
Lubricating-composition	C. A. Grandey	Rutland, Vt	Nov. 26, 1867	71, 296
Lubricating-composition	C. Grath	Saint Louis, Mo	Dec. 8, 1863	40, 905
Lubricating-composition	A. Hamill	Baltimore, Md	June 9, 1863	38, 822
Lubricating-composition	E. E. Hendrick	New York, N. Y	July 1, 1862	35, 753
Lubricating-composition	W. Hilton	Belvidere, N. J	Jan. 5, 1864	41, 068
Lubricating-composition	J. Turner	New York, N. Y	July 7, 1863	39, 185
Lubricating-composition for machinery	E. E. Hendrick	New York, N. Y	July 1, 1862	35, 754
Lubricating-composition, Metallic	C. B. Lawrence	Munda, N. Y	Oct. 8, 1861	33, 438
Lubricating-compound	W. N. Abbott	Boston, Mass	Nov. 30, 1869	97, 335
Lubricating-compound	F. T. Allyn	New York, N. Y	Nov. 26, 1867	71, 356
Lubricating-compound	E. Brown, jr	Binghamton, N. Y	July 10, 1855	13, 211
Lubricating-compound	R. R. Brown	Buffalo, N. Y	Jan. 18, 1859	22, 021
Lubricating-compound	G. A. Colehamer	Reading, Pa	July 25, 1854	11, 359
Lubricating-compound	J. and W. W. Cumberland	Mobile, Ala., and New Albany, Ind.	Apr. 3, 1849	6, 263
Lubricating-compound	T. E. Curtiss	Titusville, Pa	May 26, 1868	78, 189
Lubricating-compound	P. S. Devlan	Reading, Pa	Jan. 16, 1849	6, 028
Lubricating-compound	N. Dresser	Rochester, N. Y	Apr. 17, 1855	12, 716
Lubricating-compound	E. Dyer, jr	Providence, R. I	Jan. 31, 1871	111, 441
Lubricating-compound	B. French	Rochester, N. Y	Apr. 30, 1872	126, 282
Lubricating-compound	B. French	Rochester, N. Y	Sept. 10, 1872	131, 264
Lubricating-compound	E. F. Gerdom and C. W. Schindler.	Albany, N. Y	Jan. 7, 1868	73, 089
Lubricating-compound	E. F. Gerdom and C. W. Schindler.	Albany, N. Y	July 28, 1868	80, 471
Lubricating-compound	G. W. Gladden	Cincinnati, Ohio	Mar. 8, 1864	41, 840
Lubricating-compound	A. S. Grenville	Westborough, Mass	Jan. 30, 1849	6, 071
Lubricating-compound	H. Grogan	Flatbush, N. Y	Aug. 2, 1870	106, 053
Lubricating-compound	H. Grogan	Flatbush, N. Y	July 4, 1871	116, 704
Lubricating-compound	T. Hull and A. H. Vail	Poughkeepsie, N. Y	Feb. 11, 1868	74, 366
Lubricating-compound	H. R. Hutchinson	Empire Iron Works, Ky	Jan. 14, 1873	134, 890
Lubricating-compound	L. Kirk and J. Dodsworth	Reading, Pa	Oct. 16, 1847	5, 333
Lubricating-compound	A. Lebkücher	Belleville, Ill	June 25, 1861	32, 631
Lubricating-compound	I. Lossiel	Philadelphia, Pa	Sept. 19, 1865	50, 015
Lubricating-compound	S. Y. Love	Philadelphia, Pa	Sept. 30, 1873	143, 362
Lubricating-compound	J. Marshall	Reading, Pa	May 22, 1855	12, 918
Lubricating-compound	J. B. McNum	Port Jervis, N. Y	Jan. 3, 1860	26, 693
Lubricating-compound	A. Millochau	New York, N. Y	May 30, 1871	115, 342
Lubricating-compound	C. S. Moore	Erie, Pa	Nov. 23, 1869	97, 262
Lubricating-compound	J. B. Norris	Richmond, Va	Dec. 23, 1873	145, 812
Lubricating-compound	V. Radspinner and W. H. Moss	New Richmond, Ohio	July 3, 1860	29, 005
Lubricating-compound	C. Keck	Floraville, Ill	June 4, 1872	127, 429
Lubricating-compound	E. Roat	Buffalo, N. Y	Mar. 18, 1873	136, 868
Lubricating-compound	E. F. Prentiss	Philadelphia, Pa	May 29, 1855	12, 964
Lubricating-compound	V. Purdy	Poughkeepsie, N. Y	Mar. 7, 1871	112, 489
Lubricating-compound	J. Selgrath	Pottsville, Pa	June 24, 1851	8, 176
Lubricating-compound	E. B. S. Shoemaker	Towsontown, Md	July 19, 1870	105, 499
Lubricating-compound	N. D. Smith	Prairie City, Iowa	Nov. 25, 1873	144, 875
Lubricating-compound	J. H. Smyser	Pittsburgh, Pa	Mar. 29, 1870	101, 325
Lubricating-compound	J. H. Smyser	Pittsburgh, Pa	Jan. 3, 1871	110, 690

Index of patents issued from the United States Patent Office from 1790 *to* 1873, *inclusive*—Continued.

Invention.	Inventor.	Residence.	Date.	No.
Lubricating-compound	D. C. Taylor	Goshen, N. Y	Sept. 19, 1865	50,049
Lubricating-compound	E. A. Tonner	Saint Louis, Mo	Mar. 18, 1873	136,945
Lubricating-compound	W. Turner	Phœnixville, Pa	Feb. 19, 1861	31,519
Lubricating-compound	H. Vaughn and W. Hutton	Providence, R. I., and Baltimore, Md.	Aug. 2, 1859	24,965
Lubricating-compound	J. Williams	Sharpsburgh, Pa	Oct. 22, 1872	132,379
Lubricating-compounds, Solution for thinning	R. Patterson	Philadelphia, Pa	Oct. 11, 1859	25,792
Lubricating-cup	M. F. Carson	Cleveland, Ohio	Dec. 22, 1868	85,164
Lubricating-cup	J. B. Collin	Altoona, Pa	Dec. 17, 1867	72,168
Lubricating-cup	T. Cooper	Warwick, R. I	May 6, 1873	138,558
Lubricating-cup	J. Songster	Buffalo, N. Y	July 11, 1865	48,776
Lubricating-cup	C. S. Westlandt	Providence, R. I	Oct. 22, 1861	33,562
Lubricating-cup, Automatic	J. A. Bryan and W. Stamfield	Kent, Ohio	Jan. 24, 1871	111,105
Lubricating-cup for steam and other enginery	I. Dreyfus	New York, N. Y	Jan. 19, 1869	86,009
Lubricating-device	T. S. Brown	Poughkeepsie, N. Y	Mar. 5, 1867	62,601
Lubricating-device	F. Colligon	Buffalo, N. Y	July 28, 1868	80,457
Lubricating-device	C. Hirsch	Buffalo, N. Y	Mar. 26, 1872	124,954
Lubricating-device	G. M. Morris	Cohoes, N. Y	Jan. 1, 1867	60,925
Lubricating-device	B. H. Reynolds and J. Bachelder.	Canterbury and Norwich, Conn.	June 16, 1868	78,895
Lubricating-device	M. Senior	Frankford, Pa	Mar. 17, 1868	75,584
Lubricating machinery, Composition for	S. Balsden	Brooklyn, N. Y	Oct. 13, 1863	40,229
Lubricating machinery, Composition for	E. Q. Henderson	Charlotte, N. C	Mar. 7, 1871	112,453
Lubricating machinery, Composition for	P. Zieber, P. S. Devlan, and J. Hancock.	Reading and Philadelphia, Pa.	Sept. 11, 1847	5,288
Lubricating machinery, Method of	P. E. Proust	Paris, France	Mar. 14, 1865	46,856
Lubricating machinery, Mode of	A. P. Gross	Vallejo, Cal	Oct. 22, 1872	132,395
Lubricating-material	C. Carpenter, jr	Astoria, N. Y	Sept. 15, 1868	82,083
Lubricating-material	W. P. Ewing	Rochester, N. Y	Sept. 11, 1866	58,020
Lubricating-material	W. Little	Middlesex County, England	Aug. 1, 1854	11,406
Lubricating-sleeve	G. A. Lloyd	San Francisco, Cal	Dec. 21, 1869	98,172
Lubricating-spindle	A. H. Gilman	Hopedale, Mass	Dec. 17, 1867	72,185
Lubricating substances, Mode of applying	J. Dougall	Sterlingshire, Scotland	May 10, 1864	42,714
Lubricating under pressure, Method of	J. D. Custer	Norristown, Pa	Mar. 24, 1857	16,871
Lubricator: *See* Atmospheric lubricator. Automatic lubricator. Axle-lubricator. Axle-box lubricator. Car-wheel lubricator. Cylinder-lubricator. Hammer-lubricator. Hub and axle lubricator. Journal-lubricator. Machinery-lubricator. Nail-machine lubricator. Piston-lubricator. Portable lubricator. Pulley-lubricator. Shaft-lubricator. Shaft-bearing lubricator. Ship's-block lubricator. Spindle-lubricator. Spindle-step lubricator. Spinning-machine lubricator. Steam lubricator. Steam-engine lubricator. Steam-valve lubricator. Valve lubricator. Veloci-trot lubricator. Ventilator-lubricator. Wheel-lubricator. Wood lubricator.				
Lubricator	D. Adamson	New York, N. Y	Dec. 7, 1869	97,470
Lubricator	D. Adamson	Bremen, Germany	June 21, 1870	104,400
Lubricator	J. F. A. Aerts	Antwerp, Belgium	Nov. 28, 1865	51,276
Lubricator	J. F. A. and P. F. Aerts	Antwerp and Brussels, Belgium.	Jan. 31, 1865	46,196
Lubricator	H. V. Aiken	Gibsonburgh, Pa	May 27, 1873	139,286
Lubricator	C. D. Austin	Newcastle-upon-Tyne, England.	Dec. 6, 1870	109,795
Lubricator	E. J. Baker	Baltimore, Md	Apr. 1, 1856	14,549
Lubricator	I. Barnett and P. Eckford	Cincinnati, Ohio	Jan. 2, 1872	122,427
Lubricator	C. A. Baumgart	Allegheny City, Pa	Oct. 12, 1869	95,635
Lubricator	M. Bergner and O. Netzow	Baltimore, Md	May 18, 1869	90,223
Lubricator	A. Breaer	Saugatuck, Conn	May 6, 1856	14,797
Lubricator	F. Bresson	Paris, France	Apr. 18, 1865	47,367
Lubricator	J. Broughton	New York, N. Y	Jan. 3, 1865	45,691
Lubricator	J. Broughton	New York, N. Y	Dec. 12, 1865	51,419
Lubricator	J. Broughton	New York, N. Y	Nov. 2, 1869	96,389
Lubricator	P. G. Brown	Schenectady, N. Y	Apr. 12, 1859	23,661
Lubricator	H. A. Burr and L. E. Rockwell	New York, N. Y	Dec. 9, 1862	37,084
Lubricator	A. S. Cameron	New York, N. Y	Nov. 30, 1869	97,354
Lubricator	H. Campbell	Newton, Conn	Nov. 24, 1863	40,673
Lubricator	E. Clampitt	Baltimore, Md	Oct. 19, 1858	21,816
Lubricator	W. A. Clarke	Westville, Conn	Jan. 2, 1872	122,361
Lubricator	B. F. Clemenshaw	Troy, N. Y	Feb. 11, 1873	135,777
Lubricator	J. L. Courcier	Paris, France	Feb. 12, 1867	62,013
Lubricator	H. Crossley	Brooklyn, N. Y	Aug. 11, 1868	80,923
Lubricator	D. Currie	Saint Louis, Mo	June 14, 1870	104,279
Lubricator	H. A. Daniels	Thomaston, Conn	July 6, 1869	92,283
Lubricator	A. C. Dewios	Crefeld, Prussia	Oct. 21, 1862	36,703
Lubricator	J. L. Dickinson	Dubuque, Iowa	Dec. 20, 1870	110,215
Lubricator	I. Dreyfus	New York, N. Y	Aug. 9, 1870	106,261
Lubricator	I. Dreyfus	New York, N. Y	Nov. 8, 1870	109,118
Lubricator	E. Ehlin	San Francisco, Cal	Oct. 10, 1871	119,751
Lubricator	W. E. Everett	New York, N. Y	Mar. 4, 1856	14,352
Lubricator	J. Fanyon	Bridgeport, Conn	June 11, 1867	65,734
Lubricator	E. Faull	Maldon, Australia	Aug. 13, 1867	67,743
Lubricator	J. H. Ferguson	Springfield, Mass	Oct. 3, 1865	50,233
Lubricator	F. Ficht	Marinette, Wis	May 28, 1872	127,161
Lubricator	W. T. Garratt	San Francisco, Cal	Dec. 5, 1871	121,611
Lubricator	W. T. Garratt	San Francisco, Cal	May 14, 1872	126,799
Lubricator	W. T. Garratt	San Francisco, Cal	Nov. 19, 1872	133,217
Lubricator	J. Gates	Portland, Ohio	Sept. 20, 1870	107,478
Lubricator	J. Gates	Portland, Ohio	Apr. 29, 1873	138,243

Index of patents issued from the United States Patent Office from 1790 *to* 1873, *inclusive*—Continued.

Invention.	Inventor.	Residence.	Date.	No.
Lubricator	S. F. Gates	Cambridge, Mass	Feb. 21, 1871	112, 029
Lubricator	W. Gee	New York, N. Y	Feb. 12, 1856	14, 236
Lubricator	W. Gee	New York, N. Y	Aug. 9, 1870	106, 150
Lubricator	G. Gerdom	New York, N. Y	Apr. 26, 1870	102, 391
Lubricator	G. Gerdom	Albany, N. Y	May 17, 1870	103, 170
Lubricator	E. F. Gerdom and C. W. Schindler.	Albany, N. Y	May 19, 1868	77, 974
Lubricator	F. R. Glascock	Hillsborough, Ohio	Dec. 3, 1872	133, 581
Lubricator	T. W. Godwin	Portsmouth, Va	Nov. 3, 1863	40, 475
Lubricator	T. W. Godwin	Portsmouth, Va	Nov. 10, 1863	40, 560
Lubricator	T. W. Godwin	Portsmouth, Va	June 28, 1864	43, 305
Lubricator	T. W. Godwin	Portsmouth, Va	Feb. 7, 1865	46, 231
Lubricator	D. A. Greene	New York, N. Y	June 14, 1870	104, 139
Lubricator	W. Hamilton	Chicopee, Mass	Oct. 2, 1866	58, 409
Lubricator	J. Harlin	New York, N. Y	Sept. 29, 1868	82, 522
Lubricator	A. W. Harris	Providence, R. I	July 6, 1869	92, 307
Lubricator	H. J. Hawkins	Mobile, Ala	July 14, 1857	17, 822
Lubricator	J. E. Hendrick	Waterbury, Conn	Sept. 6, 1870	107, 044
Lubricator	J. W. and R. Hewitt	Allegheny City, Pa	May 11, 1869	89, 865
Lubricator	R. S. Hildreth	South Adams, Mass	Jan. 4, 1870	98, 655
Lubricator	M. Hinmann	West Stockbridge, Mass	Dec. 13, 1870	110, 040
Lubricator	J. Hodge	Harrison, N. J	June 27, 1871	116, 442
Lubricator	T. Holland	New York, N. Y	Nov. 10, 1868	83, 965
Lubricator	T. Holland	New York, N. Y	Feb. 15, 1870	99, 894
Lubricator	T. Holland and J. T. Cody	Cincinnati, Ohio	June 2, 1868	78, 594
Lubricator	J. J. Hoyt	Chelmsford, Mass	Nov. 23, 1869	97, 091
Lubricator	S. Hutchinson, jr	Salem, Mass	Nov. 26, 1872	133, 316
Lubricator	E. Johnson	Wilkins, Pa	Apr. 14, 1868	76, 774
Lubricator	A. J. Judge	Baltimore, Md	Feb. 10, 1863	37, 627
Lubricator	A. J. Judge	Baltimore, Md	Jan. 19, 1864	41, 302
Lubricator	S. E. Kleinschmidt	Cleveland, Ohio	Oct. 31, 1865	50, 719
Lubricator	F. A. Lane	Swanzey, N. H	Feb. 2, 1869	86, 421
Lubricator	L. S. Lapham	Providence, R. I	Mar. 20, 1860	27, 548
Lubricator	L. Leigh	Seymour, Conn	Aug. 30, 1864	44, 005
Lubricator	S. Lemon, jr	Hoboken, N. J	Aug. 6, 1867	67, 438
Lubricator	J. E. Lonergan	Sacramento, Cal	Oct. 22, 1872	132, 477
Lubricator	F. Lunkenheimer and J. Meehan.	Cincinnati, Ohio, and Covington, Ky.	Apr. 22, 1873	138, 169
Lubricator	C. Mather	Steubenville, Ohio	May 2, 1871	114, 317
Lubricator	E. McCoy	Ypsilanti, Mich	May 27, 1873	139, 407
Lubricator	F. P. McCullon and W. Woodcock.	Philadelphia and Scranton, Pa.	Jan. 26, 1869	86, 174
Lubricator	W. McCully	Paterson, N. J	Mar. 2, 1869	87, 423
Lubricator	H. McGraw	Detroit, Mich	Dec. 31, 1872	134, 436
Lubricator	J. Meehan	Covington, Ky	Aug. 23, 1870	106, 712
Lubricator	J. F. Monroe	Fitchburgh, Mass	Nov. 4, 1856	16, 018
Lubricator	T. J. Moors	Blossburgh, Pa	Sept. 6, 1870	107, 192
Lubricator	W. Morris	Dayton, Ohio	Apr. 29, 1873	1[illegible]8, 341
Lubricator	D. F. Mosman	Chelsea, Mass	Nov. 7, 1871	120, 660
Lubricator	C. Nelson	Brooklyn, N. Y	Aug. 30, 1870	106, 857
Lubricator	H. T. Neuss	Williamsburgh, N. Y	Aug. 16, 1864	43, 859
Lubricator	T. J. Nottingham	Cincinnati, Ohio	Mar. 1, 1870	100, 437
Lubricator	T. J. Nottingham	Cincinnati, Ohio	Dec. 13, 1870	110, 156
Lubricator	T. J. Nottingham	Cincinnati, Ohio	Mar. 28, 1871	113, 200
Lubricator	L. H. Olmsted	Newark, N. J	May 2, 1865	47, 561
Lubricator	J. A. Osenbrück	Hemelingen, Germany	Apr. 29, 1873	138, 274
Lubricator	E. Painter	East Hampton, Mass	Feb. 9, 1864	41, 531
Lubricator	E., B., and T. S. Parker	Schenectady, N. Y	Aug. 7, 1860	29, 515
Lubricator	C. H. Parshall	Detroit, Mich	June 18, 1872	128, 167
Lubricator	T. G. Pelton and J. Brewer	Lyons, Iowa, and Albany, Ill.	Aug. 1, 1865	[illegible]0, 148
Lubricator	W. E. Phillips	Silver City, Idaho	Apr. 25, 1871	114, 194
Lubricator	N. W. Pomeroy	Meriden, Conn	Sept. 23, 1856	15, 775
Lubricator	N. W. Pomeroy	Meriden, Conn	Feb. 3, 1857	16, 553
Lubricator	J. M. Power	Port Discovery, Wash. Ty	Jan. 7, 1873	134, 611
Lubricator	J. M. Power	Port Discovery, Wash. Ty	Oct. 14, 1873	143, 639
Lubricator	W. A. Pratt	Baltimore, Md	Oct. 7, 1873	143, 530
Lubricator	W. Pratt and N. B. Williams	Providence, R. I., and New York, N. Y.	Jan. 3, 1871	110, 677
Lubricator	D. T. Pray	Boston, Mass	July 22, 1873	141, 168
Lubricator	H. Pringle	Green Point, N. Y	Aug. 30, 1870	106, 959
Lubricator	T. Reeves	Little River, Cal	Mar. 21, 1871	112, 847
Lubricator	H. W. Regan	Renovo, Pa	Apr. 29, 1873	138, 436
Lubricator	J. Regester	Baltimore, Md	Dec. 5, 1854	12, 047
Lubricator	W. M. Rennyson	Pottsville, Pa	Nov. 17, 1868	84, 135
Lubricator	D. M. Reynolds	Port Deposit, Md	Oct. 6, 1868	82, 878
Lubricator	J. Richey	Cincinnati, Ohio	June 30, 1868	79, 501
Lubricator	J. Richter	Cincinnati, Ohio	Jan. 10, 1871	110, 868
Lubricator	T. R. Robinson and R. E. Jones	Providence, R. I	Dec. 4, 1866	60, 251
Lubricator	J. Ross	Charlestown, Mass	Oct. 7, 1873	143, 422
Lubricator	R. Ross	Bethlehem, Pa	Mar. 1, 1864	41, 785
Lubricator	R. Ross and B. E. Lehman	Bethlehem, Pa	June 11, 1867	65, 609
Lubricator	M. G. Ryan	Frostburgh, Md	May 9, 1871	114, 716
Lubricator	C. Schott	Nashville, Tenn	Aug. 14, 1866	57, 196
Lubricator	G. Scott	Kensington, Pa	Mar. 13, 1866	53, 190
Lubricator	J. R. Sees	Philadelphia, Pa	Oct. 27, 1863	40, 433
Lubricator	N. Seibert	Nevada, Cal	Sept. 14, 1869	94, 780
Lubricator	N. Seibert	San Francisco, Cal	Feb. 14, 1871	111, 881
Lubricator	A. Shafer	Wellsville, N. Y	Apr. 20, 1869	89, 175
Lubricator	W. R. Shaw	Buffalo, N. Y	Sept. 24, 1872	131, 570
Lubricator	L. F. Smith	Philadelphia, Pa	Sept. 27, 1870	107, 826
Lubricator	W. K. Stevens	Alexandria, La	Feb. 16, 1858	19, 385
Lubricator	H. Strait	Covington, Ky	Apr. 21, 1857	17, 118
Lubricator	H. Taylor	Cincinnati, Ohio	May 12, 1868	[illegible]7, 851
Lubricator	H. Taylor	Cincinnati, Ohio	June 23, 1868	79, 279
Lubricator	J. Fenwick	Grantham, England	Jan. 31, 1871	111, 401
Lubricator	H. Thomas	New York, N. Y	Feb. 16, 1869	87, 079
Lubricator	H. Thomas	New York, N. Y	July 6, 1869	92, 399

Index of patents issued from the United States Patent Office from 1790 *to* 1873, *inclusive*—Continued.

Invention.	Inventor.	Residence.	Date.	No.
Lubricator	R. H. Tradenick	Pittsburgh, Pa	Sept. 15, 1868	82, 182
Lubricator	J. E. Uhl	Renovo, Pa	Mar. 7, 1871	112, 515
Lubricator	W. Van Anden	Poughkeepsie, N. Y	Dec. 6, 1864	45, 365
Lubricator	E. Von Jeinsen	San Francisco, Cal	Sept. 27, 1870	107, 739
Lubricator	R. M. Wade	Wadesville, Va	June 6, 1854	11, 024
Lubricator	R. M. Wade	Wadesville, Va	May 1, 1855	12, 803
Lubricator	G. Waters	Cincinnati, Ohio	Aug. 6, 1867	67, 469
Lubricator	G. Waters	Cincinnati, Ohio	June 23, 1868	79, 167
Lubricator	G. Waters	Cincinnati, Ohio	Sept. 22, 1868	82, 368
Lubricator	G. Waters	Cincinnati, Ohio	June 29, 1869	92, 131
Lubricator	J. Webster	New York, N. Y	Apr. 11, 1854	10, 767
Lubricator	D. M. Weston	Boston, Mass	Jan. 19, 1864	41, 335
Lubricator	J. B. Wickersham	Philadelphia, Pa	Oct. 22, 1867	70, 058
Lubricator	J. B. Wickersham	Philadelphia, Pa	Sept. 29, 1868	82, 667
Lubricator	J. B. Wickersham	Philadelphia, Pa	July 26, 1870	105, 750
Lubricator	J. L. Wiggin and E. Folsom	South New Market, N. H	July 10, 1866	56, 306
Lubricator	J. H. Wilkinson	South New Market, N. H	Apr. 12, 1870	101, 954
Lubricator	W. W. W. Wood	Philadelphia, Pa	May 12, 1863	38, 526
Lubricator and anti-friction bearing	T. Roddick and J. Lockhead	Stranraer and Glasgow, Scotland.	Oct. 24, 1871	120, 213
Lubricator and water-conductor, Combined	T. A. Shock	Boston, Mass	Jan. 14, 1868	73, 262
Lubricators, Oil-hole cover for	S. A. Skinner	Hoosick Falls, N. Y	Nov. 7, 1871	120, 677
Lumber, Apparatus for washing and elevating	S. E. Worrell	Quincy, Ill	Mar. 15, 1870	100, 831
Lumber, Arrangement of devices for dressing pieces of.	H. Brown	New York, N. Y	May 19, 1857	17, 313
Lumber by superheated steam, Apparatus for drying and seasoning.	C. F. Allen and L. W. Campbell	Aurora, Ill	May 7, 1867	64, 398
Lumber, Device for canting or turning logs during the process of sawing them into.	C. Foster	Oshkosh, Wis	Jan. 6, 1863	37, 341
Lumber, Device for tonguing and grooving	N. G. Norcross	Lowell, Mass	May 2, 1854	10, 844
Lumber-dryer	J. Allonas	Mansfield, Ohio	Jan. 4, 1870	98, 541
Lumber-dryer	J. Brakeley	Bordentown, N. J	Apr. 14, 1868	76, 591
Lumber-dryer	J. Du Bois	Williamsport, Pa	Aug. 18, 1868	81, 074
Lumber-dryer	A. Edwards	New Haven, Conn	Jan. 7, 1873	134, 529
Lumber-dryer	R. E. Ferguson	Chicago, Ill	Oct. 4, 1870	107, 892
Lumber-dryer	J. W. Hanna	Wabash, Ind	Nov. 27, 1866	59, 998
Lumber-dryer	R. P. Johnson	Wabash, Ind	July 9, 1867	66, 594
Lumber dryer	J. B. and T. E. Johnson	Indianapolis, Ind	Dec. 14, 1869	97, 777
Lumber-dryer	R. P. Johnson and E. J. Sumner	Wabash, Ind	July 7, 1868	79, 661
Lumber-dryer	F. J. Norton	Fremont, Ohio	Nov. 2, 1869	96, 471
Lumber-dryer	J. H. Osgood, jr	Boston, Mass	Nov. 27, 1866	60, 043
Lumber-dryer	J. H. Rae	Syracuse, N. Y	June 9, 1868	78, 692
Lumber-dryer	W. C. Scott	Richmond, Ind	May 11, 1869	90, 027
Lumber-dryer	E. J. Sumner	Chicago, Ill	Mar. 26, 1872	125, 098
Lumber, Drying and seasoning	E. C. Bender and W. Steffe	York and Philadelphia, Pa	Dec. 17, 1867	72, 157
Lumber-drying kiln	S. R. Kirby	Detroit, Mich	Oct. 21, 1873	143, 912
Lumber-drying kiln	H. W. Oliver	New Haven, Conn	June 24, 1862	35, 700
Lumber-edging machine	W. G. Caldwell	Three Rivers, Mich	Aug. 19, 1873	141, 852
Lumber-edging machine	L. A. Ensworth and B. Barker	Williamsport, Pa	Jan. 27, 1863	37, 496
Lumber, Feed-roll for	W. P. Hale	Lock Haven, Pa	Dec. 10, 1872	133, 775
Lumber-feeding rollers, Device for governing the parallel yielding of.	J. B. Pomroy	Chicago, Ill	Nov. 18, 1856	16, 105
Lumber for building purposes, Machine for preparing.	J. F. Gyles	Chicago, Ill	June 25, 1872	128, 388
Lumber for joiner's use, Prepared	J. Mullay	Somerville, Mass	Feb. 25, 1873	136, 332
Lumber from logs, Machine for dressing	A. Wolcott	Detroit, Mich	July 24, 1855	13, 342
Lumber-grooving machine	M. E. Carter	Syracuse, N. Y	Aug. 28, 1866	57, 474
Lumber-grooving machine	B. Holly and J. W. Wheeler	Seneca Falls, N. Y	July 8, 1851	8, 206
Lumber into irregular forms, Machinery for working.	R. Powers	Prescott, Mass	May 8, 1849	6, 436
Lumber-jointing machines, Clamp and mouth-piece for.	C. F. Bauersfeld	Cincinnati, Ohio	Mar. 27, 1855	12, 580
Lumber, Machine for cutting thin	B. F. Betts	Tonawanda, N. Y	July 7, 1863	39, 113
Lumber, Machine for resawing	E. H. Titus and J. Sharp	Wilkesbarre and Phillipsburgh, Pa.	June 29, 1858	20, 745
Lumber, &c., Machine for tallying	G. R. Lewis	Ashtabula, Ohio	June 6, 1865	48, 078
Lumber, Manufacturing	E. B. Rowe	Chicago, Ill	Apr. 20, 1869	89, 172
Lumber-marker	W. Merritt	Weymouth, Mass	Oct. 21, 1873	143, 773
Lumber, Process of seasoning	H. H. Beach	Rome, N. Y	Oct. 4, 1870	107, 854
Lumber-rack	W. H. Powers	Grand Rapids, Mich	June 11, 1872	127, 921
Lumber-rack	C. Sach	Grand Rapids, Mich	Jan. 18, 1870	99, 006
Lumber-rack	R. B. Woodcock	Grand Rapids, Mich	May 16, 1871	114, 897
Lumber-register	F. Nehauus	Ellenburgh Centre, N. Y	June 5, 1866	55, 327
Lumber preserving and drying	J. Oliver	Toronto, Canada	Sept. 2, 1873	142, 347
Lumber-scribing machine	J. Shellenberger	Indianapolis, Ind	Sept. 10, 1850	7, 636
Lumber-seasoning apparatus	M. R. Moore	Philadelphia, Pa	May 31, 1859	24, 230
Lumber-slitting gage	F. P. Hart	Chandlersville, Pa	May 15, 1855	12, 861
Lumber to rotary planers, Feeding tapering	P. H. Woolsey	Andes, N. Y	Feb. 19, 1861	31, 508
Lumber tonguing and grooving cutter	D. Perrin	McGregor, Iowa	Apr. 15, 1873	137, 861
Lumber-trimming machine	C. Lamb and T. J. Frazier	Clinton, Iowa	May 6, 1873	138, 505
Lunch-bag, Traveling	J. H. Noyes	Oneida, N. Y	Apr. 18, 1865	47, 324
Lunch-box	R. Cook	Saratoga Springs, N. Y	Jan. 28, 1862	34, 243
Lunch-box	J. Elson	Northampton, Mass	Nov. 8, 1870	108, 989
Lunch-box	J. Erpelding	Chicago, Ill	Jan. 12, 1869	85, 807
Lunch-box	J. Erpelding	Chicago, Ill	Mar. 22, 1870	101, 113
Lunch-box	M. G. Fagan	Troy, N. Y	Sept. 14, 1869	94, 729
Lunch-box	E. C. Lockwood	Bridgeport, Conn	Feb. 11, 1873	135, 651
Lunch-box	D. Miller	Allegheny City, Pa	Jan. 17, 1871	110, 991
Lunch-box	J. F. Morgan	Boston, Mass	May 2, 1865	47, 559
Lunch-box	J. F. Morgan	Boston, Mass	Nov. 5, 1867	70, 598
Lunch-box	P. H. Niles	Boston, Mass	Mar. 10, 1868	75, 451
Lunch-box	A. Norman	Rochester, N. Y	Aug. 3, 1869	93, 334
Lunch-box	J. S. Tibbals	Milford, Conn	Sept. 24, 1867	69, 272
Lunch-box	D. Troxell	Newark, N. J	May 2, 1871	114, 370
Lunch-box	W. H. Van Allen	Albany, N. Y	Nov. 19, 1872	133, 129
Lunch-box and shopping-bag, Combined	C. C. Cobleigh	Brighton, Mass	Nov. 11, 1873	144, 385
Lunch-box, Folding	C. S. Hurlburt	Springfield, Mass	Jan. 30, 1866	52, 291
Lunch-box, Folding	J. A. Minor	Middletown, Conn	Jan. 9, 1867	51, 957

Index of patents issued from the United States Patent Office from 1790 *to* 1873, *inclusive*—Continued.

Invention.	Inventor.	Residence.	Date.	No.
Lunch-box, Folding	F. B. Parks	Cambridgeport, Mass	July 16, 1867	66, 881
Lunch-heater	H. M. Kinsley	Chicago, Ill	Jan. 1, 1867	60, 743
Lye and melting potash, Cleansing	J. Jones	Fabius, N. Y	Feb. 11, 1809	....
Linch-pin	L. B. Gusman	Philadelphia, Pa	Mar. 13, 1866	53, 220
M.				
Macaroni-server	A. L. Lincoln	Boston, Mass	July 1, 1856	15, 266
Macaroni, vermicelli, Italian pastes, &c., Making	J. B. Sartori	Philadelphia, Pa	July 13, 1807	....
Mace, Policeman's	M. Warne	Philadelphia, Pa	Oct. 20, 1868	83, 228
Machine-drill	N. P. Eddy	Fall River, Mass	Apr. 24, 1866	54, 248
Machine for propelling, Hydrostatic and pneumatic.	R. Mills and H. B. Fernald	Washington, D. C., and Portland, Me.	Oct. 10, 1835	....
Machine, Horizontal double-rotary	W. C. Sullivan	Jefferson County, Ind	June 5, 1824	....
Machine-stand	W. B. Bement	Philadelphia, Pa	June 3, 1862	35, 433
Machinery, Apparatus for propelling	W. Z. W. Chapman	New York, N. Y	Sept. 14, 1869	94, 809
Machinery, Apparatus for propelling	E. J. Leyburn	Lexington, Va	Sept. 7, 1869	94, 618
Machinery, Apparatus to prevent rapping noises in	G. Ames	Rochester, N. Y	May 27, 1873	139, 227
Machinery, Application of the screw in propelling	W. H. Godfrey	Rochester, N. Y	Nov. 27, 1826	....
Machinery, Applying water in propelling	G. M. Gibbs	Prince William's Park, S. C	Mar. 10, 1828	....
Machinery-brake	D. S. Baker	West Bloomfield, N. Y	Sept. 1, 1868	81, 729
Machinery-brake	R. D. Napier	Birkenhead, England	Aug. 10, 1869	93, 469
Machinery-brake	R. D. Napier	Limehouse, England	Oct. 31, 1871	120, 527
Machinery by balance-lever, Propelling	S. H. Emmons and G. Upham	Massillon, Ohio	May 16, 1833	....
Machinery by combining air, water, and animal power, Propelling.	T. P. Codman	Boston, Mass	Apr. 20, 1833	....
Machinery by helping-power, Propelling	W. Loomis	Ashford, N. Y	Apr. 29, 1833	....
Machinery by means of cams and inclined planes, Propelling.	P. C. Curtis	Utica, N. Y	May 29, 1835	....
Machinery by weights, Propelling	C. Broyles	Tellico, Tenn	Oct. 19, 1827	....
Machinery by weights, Propelling	E. Turner	North Pownal, Me	June 12, 1835	....
Machinery, Device for connecting parts of	E. S. Pierce	Hartford, Conn	Apr. 13, 1869	88, 902
Machinery, Driving	W. S. Wells and S. B. Wells	New York and Middleburgh, N. Y.	June 27, 1865	48, 467
Machinery-lubricator	B. Hilbert	Cincinnati, Ohio	Apr. 23, 1867	64, 105
Machinery-lubricator	G. E. Smith	Fitchburgh, Mass	May 23, 1871	115, 247
Machinery-register	S. F. Payne	Chicopee, Mass	Nov. 25, 1873	144, 861
Machinery, Regulating	N. Schoolfield	Norwich, Conn	May 17, 1836	....
Machinery-regulator	T. C. Van Wyck and W. Kent	Poughkeepsie, N. Y	Feb. 1, 1870	99, 376
Machinery, Propelling	L. M. Edwards	Trenton, Tenn	Mar. 2, 1835	....
Machinery, Propelling	T. D. Newsom	Nashville, Tenn	Jan. 22, 1831	....
Machinery, Propelling	T. D. Newsom and J. C. Shule	Nashville, Tenn	Dec. 14, 1830	....
Machinery, Propelling	D. Russel	Tuscumbia, Ala	Oct. 27, 1835	....
Machinery, Propelling	W. Stanton	Centre Township, Ind	Apr. 23, 1827	....
Machinery with the foot, Driving light	A. Clarke	Greenwich, Conn	Jan. 20, 1841	1, 936
Machinists' gage	S. H. Bellows	Middletown, Conn	Sept. 16, 1873	142, 761
Mackerel-latch	C. S. H. Foster	Deer Isle, Me	Mar. 23, 1869	88, 026
Mackerel, Machine for splitting	S. S. Turner	Lewiston, Me	Oct. 21, 1856	15, 941
Madder and other plants, Process of separating coloring matter from.	A. Paraf	Boston, Mass	Feb. 25, 1868	74, 935
Madder colors, Dyeing and printing	A. Paraf	New York, N. Y	Oct. 31, 1871	120, 393
Madder-extractor	J. Hunter	Philadelphia, Pa	Apr. 12, 1870	101, 735
Madder extracts, Obtaining	J. Hunter	Philadelphia, Pa	Feb. 15, 1870	99, 904
Madder for dyeing and printing, Extract of	A. Paraf	New York, N. Y	Sept. 21, 1869	95, 039
Madder for dyeing, Composition of	A. Paraf	New York, N. Y	Oct. 31, 1871	120, 392
Madder for dyeing, Preparation of coloring-matter from.	T. Briston	Cranston, R. I	June 14, 1870	104, 259
Madder for printing cloth, &c., Composition of	A. Paraf	New York, N. Y	Apr. 18, 1871	113, 919
Madder, Material for dyeing and printing obtained from.	A. Paraf	New York, N. Y	Aug. 17, 1869	93, 900
Madder, Obtaining extract of	A. Paraf	New York, N. Y	Jan. 24, 1871	111, 142
Madder, Process of extracting the coloring-matter of.	A. Paraf	New York, N. Y	Feb. 16, 1869	86, 939
Madder, Process of extracting the coloring-matter of.	A. Paraf	New York, N. Y	Jan. 17, 1871	110, 995
Madder, Products from	A. Paraf	New York, N. Y	Apr. 18, 1871	113, 918
Magazines, Flooding and entering powder	C. W. Copeland	Brooklyn, N. Y	Nov. 6, 1849	6, 842
Magic lantern: *See* Lantern.				
Magnesia compounds in chemical manufacture, Employment of.	C. G. Clemm	Dresden, Germany	Jan. 19, 1864	41, 349
Magnesia, Making hydrate or milk of	C. H. Phillips and L. Reid	New York, N. Y	July 22, 1873	141, 167
Magnesia, Manufacture of hydrate of	C. Wandel	Waldau, Germany	July 6, 1869	92, 405
Magnesia, Preparing	W. Dunn	Boston, Mass	Apr. 1, 1812	....
Magnesia, Preparing	W. Dunn	Boston, Mass	Apr. 8, 1813	....
Magnesium for burning, Preparing	C. H. Wing	Newton Corner, Mass	Apr. 24, 1866	54, 266
Magnesium, Manufacture and purification of	E. Sonstadt	Loughborough, Great Britain.	Dec. 27, 1864	45, 684
Magnesium, Manufacture of	J. O. Christian and J. and H. Charlton.	Manchester and Strangeways, England.	Sept. 17, 1867	68, 955
Magnesium Manufacture of	E. Sonstadt	Loughborough, Great Britain.	Dec. 6, 1864	45, 370
Magnet, Electro	T. A. Edison	Newark, N. J	Aug. 27, 1872	130, 795
Magnet, Receiving	N. Parks	Rome, N. Y	Feb. 16, 1858	19, 379
Magnet, Receiving	S. F. Van Choate	New York, N. Y	Sept. 29, 1863	40, 133
Magnet, Relay	W. G. Brownson	Wellsville, Ohio	July 30, 1867	67, 160
Magnet, Relay	L. G. D'Arlincourt	Paris, France	July 26, 1870	105, 653
Magnet, Relay	C. Durant	Jersey City, N. J	May 19, 1868	78, 076
Magnet, Relay	C. Durant	Jersey City, N. J	June 30, 1868	79, 330
Magnet, Relay	C. Durant	Jersey City, N. J	June 30, 1868	79, 331
Magnet, Relay	T. A. Edison	Newark, N. J	Aug. 12, 1873	141, 771
Magnet, Relay	S. Marcus	Vienna, Austria	Sept. 10, 1861	33, 254
Magnet, Reversible	J. L. Churchill	Nokomis, Ill	Feb. 11, 1873	135, 690
Magnet, Self-adjusting relay	J. M. Fairchild	New Haven, Conn	Dec. 10, 1867	71, 863
Magnet, Sounder	J. J. Clark and H. Splitdorf	New York, N. Y	Sept. 12, 1865	49, 857
Magnet, Sounder	S. F. Day	Ballston, N. Y	June 19, 1866	55, 627
Magnet spring, Relay	W. N. McInnis	Northumberland, Pa	Feb. 25, 1868	74, 925
Magnet, Telegraph	S. F. Van Choate	New York, N. Y	June 2, 1863	38, 774

Index of patents issued from the United States Patent Office from 1790 *to* 1873, *inclusive*—Continued.

Invention.	Inventor.	Residence.	Date.	No.
Magnets, Helix for	L. Bradley	New York, N. Y	Aug. 1, 1865	49, 074
Magnetic cylinder	L. Browning	Franconia, N. H	Oct. 13, 1812	....
Magnetic electric machine	J. Kidder	New York, N. Y	Mar. 15, 1864	41, 927
Magnetic engine	H. M. Paine	Newark, N. J	May 17, 1870	103, 229
Magnetic from non-magnetic substances, Process and apparatus for separating.	F. A. H. La Rue and O. Audet	Quebec, Canada	Jan. 26, 1869	86, 167
Magnetic lock	L. C. Springer	Chicago, Ill	Jan. 1, 1867	60, 953
Magnetic needle on ships, Mode of compensating the local attraction of the.	C. Kline	New York, N. Y	Mar. 17, 1857	16, 845
Magnetic needles, Correcting	C. Guiteau	Syracuse, N. Y	Mar. 26, 1850	7, 216
Magnetic signals on railway, Transmitting	H. Maule	Philadelphia, Pa	Oct. 5, 1858	21, 688
Magnetism and electro-magnetism, Propelling machinery by.	T. Davenport	Brandon, Vt	Feb. 25, 1837	132
Magneto-electric apparatus	J. Kidder	New York, N. Y	Sept. 18, 1866	58, 165
Magneto-electric battery for firing fuses, &c	B. G. Noble	Brooklyn, N. Y	Mar. 5, 1872	124, 216
Magneto-electric machine	A. N. Allen	Pittsfield, Mass	Feb. 6, 1872	123, 438
Magneto-electric machine	G. W. Beardslee	Flushing, N. Y	Dec. 27, 1859	26, 557
Magneto-electric machine	G. W. Beardslee	Flushing, N. Y	Dec. 27, 1859	26, 558
Magneto-electric machine	C. Carpenter, jr	Pawtucket, Mass	Nov. 1, 1853	10, 175
Magneto-electric machine	C. Carpenter, jr	Providence, R. I	Apr. 8, 1856	14, 598
Magneto-electric machine	A. Davis	New York, N. Y	Aug. 1, 1854	11, 415
Magneto-electric machine	L. T. Gramme and E. L. C. D'Ivernois.	Paris, France	Oct. 17, 1871	120, 057
Magneto-electric machine	D. F. J. Lontin and E. L. C. de Ivernois.	Paris, France	Aug. 24, 1869	94, 014
Magneto-electric machine	E. C. Shepard	New York, N. Y	Aug. 19, 1856	15, 596
Magneto-electric machine	H. Wilde	Manchester, England	Nov. 13, 1866	59, 738
Magneto-electric machine for giving shocks	G. H. and B. H. Horn	Boston, Mass., and New York, N. Y.	Apr. 11, 1848	5, 507
Mail and packages from railway-cars while in motion, Method of delivering.	C. D. Everett	Cleveland, Ohio	Nov. 28, 1865	51, 162
Mail and packages on railway-cars, Mode of receiving and delivering.	W. J. Ketcham	Washington, D. C	July 25, 1865	48, 954
Mail and traveling bag, India-rubber	E. M. Chaffee	Boston, Mass	Dec. 31, 1833	....
Mail-axle	W. H. Saunders	Hastings-upon-Hudson, N. Y	Mar. 5, 1850	7, 150
Mail axle and hub	A. E. Smith	Bronxville, N. Y	Jan. 13, 1857	16, 404
Mail-bag	F. C. Brest and P. Wonsey	Spencerport, N. Y	June 8, 1869	91, 075
Mail-bag	D. F. Dodge	Lowville, N. Y	June 23, 1868	79, 214
Mail-bag	J. Fye	Hamilton, Ohio	June 5, 1866	55, 416
Mail-bag	R. Gornall	Baltimore, Md	May 10, 1859	23, 916
Mail-bag	T. J. Hardaway	Macon, Ga	May 20, 1873	139, 061
Mail-bag	T. J. Lamdin	Baltimore, Md	May 10, 1859	23, 924
Mail-bag	T. J. Lamdin	Baltimore, Md	Jan. 30, 1872	123, 113
Mail-bag	P. Laporte	Richmond, Va	May 5, 1824	....
Mail-bag	R. O. Lowrey	Salem, N. Y	Oct. 12, 1869	95, 817
Mail-bag	W. Ruddach	Baltimore, Md	May 3, 1859	23, 863
Mail-bag	L. M. Sims	Lincoln, Ill	Nov. 12, 1867	70, 907
Mail-bag	E. Stevens and J. A. Knight	Saint Louis, Mo	Nov. 5, 1867	70, 480
Mail-bag	Z. T. Sweet	Davisville, Cal	Sept. 7, 1869	94, 526
Mail-bag, Air-tight	C. A. Robbins and H. Allen	Iowa City, Iowa, and Allen Grove, Wis.	Sept. 7, 1852	9, 253
Mail-bag and carpet-sack lock	J. B. Logan	Thorntown, Ind	July 3, 1866	56, 069
Mail-bag bolt	J. Atkins	Hanover, N. H	Feb. 17, 1836	....
Mail-bag catcher for railway-cars	L. F. Ward	Elyria, Ohio	Jan. 29, 1867	61, 584
Mail-bag clasp	L. Fox	New York, N. Y	Sept. 18, 1866	58, 091
Mail-bag clasp and lock	H. C. Jones	Newark, N. Y	Oct. 23, 1837	440
Mail-bag delivering and receiving device	G. W. Hildreth	Lockport, N. Y	Sept. 12, 1871	118, 937
Mail-bag fastener	S. Denison	Portlandville, N. Y	Dec. 10, 1867	71, 997
Mail-bag fastener	D. G. Gay	Eugene City, Oreg	Oct. 11, 1870	108, 251
Mail-bag fastener	R. O. Lowrey	Salem, N. Y	Oct. 12, 1869	95, 818
Mail-bag fastener	H. M. Stephenson and J. B. Tyer.	Wabash, Ind	Apr. 4, 1871	113, 359
Mail-bag fastening	B. X. Blair	Huntingdon, Pa	June 3, 1873	139, 536
Mail-bag fastening	D. F. Dodge	Lowville, N. Y	Feb. 7, 1871	111, 521
Mail-bag fastening	J. C. Garland	Chicago, Ill	Feb. 8, 1859	22, 866
Mail-bag fastening	T. McGrane	New York, N. Y	Dec. 5, 1871	121, 646
Mail-bag fastening	J. A. Paul	Huntingdon, Pa	July 8, 1873	140, 639
Mail-bag fastening	A. D. Perry	New York	May 23, 1848	5, 596
Mail-bag fastening	W. J. Stowell	Baltimore, Md	Jan. 7, 1873	134, 571
Mail-bag fastening	J. A. Truitt	Oakland, Pa	Apr. 27, 1869	89, 447
Mail-bag fastening, &c	M. V. B. White	Fort Edwards, N. Y	Apr. 25, 1871	114, 239
Mail-bag lock	W. W. Gingrich and L. S. Coates.	Mexico, Pa	Dec. 16, 1862	37, 164
Mail-bag lock	R. O. Lowrey	Salem, N. Y	Oct. 12, 1869	95, 816
Mail-bag receiver, Railway	C. D. Everett	Cleveland, Ohio	Oct. 31, 1865	50, 698
Mail-bag tag	C. A. Snyder	Richmond, Va	Dec. 28, 1869	98, 441
Mail-bag, Water-proof	J. M. Jarrett	Brooklyn, N. Y	Oct. 30, 1866	59, 228
Mail bag, Way	H. Johnson	Washington, D. C	Apr. 25, 1846	4, 475
Mail-bags, Adjustable label-holder for	G. A. Lamb	Jeffersonville, N. Y	Oct. 29, 1867	70, 228
Mail-bags, Apparatus for receiving and delivering	J. J. Cooper	Olmsted, Ohio	Mar. 27, 1866	53, 417
Mail-bags, &c., by nailing, Making	A. L. Pennock and J. Sellers	Philadelphia, Pa	July 6, 1818	....
Mail-bags, Device for closing	G. M. Rhoades	East Hamilton, N. Y	Feb. 10, 1863	37, 641
Mail-bags, &c., Mouth-piece of	J. Sellers and A. L. Pennock	Philadelphia, Pa	June 12, 1840	1, 633
Mail-bags on railway-cars, Receiving and delivering	P. N. Maine	Olmsted Falls, Ohio	Sept. 11, 1866	57, 936
Mail-bags to and from railway trains and stations, Apparatus for receiving and delivering.	A. Chavanne	Paris, France	July 25, 1865	49, 056
Mail-bags to cars, Device for delivering	J. B. McLain	Newark, Ohio	Nov. 15, 1870	109, 233
Mail-bags upon railway-cars, Receiving and delivering.	J. Fredenburgh and G. A. Davidson.	Green, N. Y	May 15, 1866	54, 710
Mail-box	A. W. Allen and C. Reitz	Indianapolis, Ind	Sept. 20, 1870	107, 586
Mail-lock	T. Ascherfeld	Elkton, Md	Mar. 13, 1866	53, 096
Mail-pouch	M. Smith	Saint Louis, Mo	Apr. 28, 1863	38, 362
Mail-pouch	M. Smith	Saint Louis, Mo	May 17, 1864	42, 802
Mail-pouch holder and catcher, Railway	B. F. Bean	Pawling, Pa	Oct. 14, 1873	143, 657
Mail pouch or box	M. Smith	Saint Louis, Mo	May 15, 1866	54, 787
Mails, Device for receiving and delivering	F. K. Sibley and L. C. Wade	Auburndale and Newton Upper Falls, Mass.	Dec. 22, 1868	85, 183
Main: *See* Gas-main. Hydraulic main. Water-main.				

Index of patents issued from the United States Patent Office from 1790 to 1873, inclusive—Continued.

Invention.	Inventor.	Residence.	Date.	No.
Maize-husking machine	T. C. Hargreaves	Schenectady, N. Y	Oct. 4, 1853	10,070
Maize, Process of mashing	F. Seitz	Easton, Pa	Jan. 20, 1852	8,665
Maize suitable for grinding, Preparation to render	S. P. McCroskey	Monroe, Iowa	May 28, 1861	32,448
Malleable-iron and steel furnace	H. Bessemer	London, England	July 25, 1871	117,247
Malleable metal pipes, Connection for	W. H. Harrison	Philadelphia, Pa	Jan. 18, 1870	98,961
Mallet	L. W. Blanchard	Whitingham, Vt	Mar. 29, 1859	23,346
Mallet	J. B. Davids	New York, N Y	Aug. 1, 1871	117,609
Mallet	A. Holbrook	Providence, R. I	Aug. 19, 1873	141,930
Mallet	W. Lance	Olney, Ill	July 12, 1859	24,744
Mallet	A. Partridge	Medway, Mass	Mar. 21, 1865	46,972
Mallet	A. Partridge	Medway, Mass	May 28, 1872	127,363
Mallet	S. M. Willis	Athol, Mass	July 1, 1873	140,395
Mallet, Calker's	S. C. Searles	Wilmington, Del	Mar. 8, 1870	100,562
Mallet, Elastic	A. C. Eddy	Providence, R. I	Feb. 20, 1866	52,696
Mallet for driving tubes, &c	C. Stewart	Worcester, Mass	Dec. 16, 1873	145,537
Mallet, Serving	T. Batty	New York, N. Y	Aug. 20, 1850	7,571
Mallet, Serving	D. H. Southworth	New York, N. Y	Nov. 16, 1852	9,412
Mallet, Serving and worming	J. B. Petitval	Charleston, S. C	July 11, 1837	266
Malt and other liquors, Apparatus for fermenting	G. Wallace	Cincinnati, Ohio	Mar. 5, 1867	62,581
Malt and other liquors, Treating	G. Storey	Wheeling, W. Va	Mar. 31, 1868	76,266
Malt, Apparatus for sprouting	J. Geemen	Chicago, Ill	Mar. 13, 1866	53,135
Malt, Apparatus for sprouting	J. Geemen	Chicago, Ill	Apr. 14, 1868	76,624
Malt, Condensed extract of	T. Hawks	Rochester, N. Y	Dec. 11, 1866	60,370
Malt-degerminating machine	C. Santer	New York, N. Y	Feb. 7, 1871	111,686
Malt-dryer	H. A. and C. H. Engels and J. Wieland.	San Francisco, Cal	Apr. 2, 1867	63,374
Malt-dryer	W. L. Phillips	Normal, Ill	June 14, 1870	104,347
Malt-dryer	J. G. Schiffer	San Francisco, Cal	Dec. 30, 1873	146,025
Malt-dryer	P. Schuff	New York, N. Y	Sept. 16, 1873	142,946
Malt, Drying	W. W. Hughes	Philadelphia, Pa	May 25, 1869	90,545
Malt-drying kiln	J. Geemen	New York, N. Y	Mar. 25, 1873	137,194
Malt-drying kiln	C. Oefinger and S. Grupp	San Francisco, Cal	Apr. 7, 1868	76,508
Malt-drying kiln	J. A. Remer	New York, N. Y	Feb. 7, 1871	111,569
Malt-extract	L. Hoff	New York, N. Y	Dec. 18, 1866	60,631
Malt-extract, Concentrated	T. Hawks	Rochester, N. Y	June 16, 1868	78,875
Malt, &c., for making beer, ale, and porter, Extract of	T. Hawks	Rochester, N. Y	Feb. 3, 1863	37,578
Malt-house	T. Krausch	New York, N. Y	Nov. 28, 1871	121,387
Malt house or kiln	W. Appleton	Albany, N. Y	Nov. 19, 1867	71,117
Malt in mash-tubs, Self-acting sparger for distributing water upon.	H. J. Brooks and F. B. Longmire.	Philadelphia, Pa	May 8, 1840	1,597
Malt in the kiln, Apparatus for stirring	W. Toepfer	Milwaukee, Wis	May 30, 1871	115,390
Malt-kiln	C. P. Berckhemer	Cincinnati, Ohio	Nov. 8, 1870	109,105
Malt-kiln	W. Blakey	Baltimore, Md	Oct. 8, 1867	69,534
Malt-kiln	J. Geemen	Chicago, Ill	Nov. 28, 1865	51,169
Malt-kiln	J. Geemen	Chicago, Ill	Nov. 13, 1866	59,583
Malt-kiln	J. Geemen	Chicago, Ill	Sept. 17, 1867	68,869
Malt-kiln	J. G. White	Albany, N. Y	Feb. 22, 1870	100,091
Malt-kiln and malt-house	W. W. Hughes	Philadelphia, Pa	Aug. 24, 1869	94,114
Malt-kiln floor	H. Franz	Philadelphia, Pa	Sept. 25, 1860	30,137
Malt-kiln floor	W. Gerhard, jr	Florence, Mass	Aug. 17, 1869	93,701
Malt-kiln floor	T. S. Smith	Cincinnati, Ohio	Feb. 21, 1860	27,237
Malt-kiln floor	G. Taylor	Camden, N. J	Sept. 11, 1866	57,997
Malt-kiln-floor plate	M. Stewart	Philadelphia, Pa	Jan. 3, 1854	10,370
Malt-kiln flooring	W. W. Hughes and J. C. Adams	Philadelphia, Pa	Jan. 15, 1867	61,202
Malt-kiln, Steam-drying	D. S. Davis	Hibernia, Tenn	Aug. 30, 1822	......
Malt-kilns, Device to operate the tilting doors of	J. C. Rhodes	Chicago, Ill	June 29, 1869	92,096
Malt-kilns, Material for floor of	R. Speidel	New York, N. Y	Sept. 10, 1867	68,664
Malt-kilns, Stirrer for	E. Schmidt	Pittsburgh, Pa	Apr. 29, 1873	138,288
Malt liquor cooler	O. Hœpfner and C. Schnepf	Philadelphia, Pa	Oct. 13, 1863	40,262
Malt-liquor cooler	W. Partington	Philadelphia, Pa	Oct. 15, 1867	69,930
Malt-liquor preserving and discharging apparatus	M. Reeder	Philadelphia, Pa	Jan. 24, 1860	26,930
Malt-liquors and cider, Ripening and keeping	R. Hare	Philadelphia, Pa	Aug. 22, 1811	....
Malt-liquors, Apparatus for boiling, cooling, and fermenting.	T. Haigh and R. A. Robertson	Liverpool, England, and Philadelphia, Pa.	Apr. 23, 1867	64,011
Malt-liquors, Apparatus for cooling	W. Allenderff	Philadelphia, Pa	Nov. 6, 1866	59,336
Malt-liquors, Apparatus for cooling	F. L. Wissmann	Philadelphia, Pa	Oct. 3, 1865	50,293
Malt-liquors, Apparatus for drawing and preserving	H. Mittendorf	York, Pa	Sept. 7, 1867	69,013
Malt-liquors, &c., Apparatus for fermenting	A. Hammer	New York, N. Y	Dec. 13, 1864	45,406
Malt-liquors, Apparatus for preserving and discharging.	L. Dungan	Philadelphia, Pa	Nov. 5, 1861	33,637
Malt-liquors, Compound for treating	W. Zinsser	New York, N. Y	Dec. 6, 1870	109,990
Malt-liquors, Decolorizing	C. R. M. Wall	Brooklyn, N. Y	Oct. 17, 1865	50,523
Malt-liquors from becoming flat, Apparatus for preventing.	F. M. Ruschhaupt	New York, N. Y	Mar. 18, 1862	34,695
Malt, Machine for bruising	J. F. Tibble	Detroit, Mich	May 16, 1871	114,882
Malt-polishing apparatus	W. W. Stoll	Brooklyn, N. Y	Nov. 25, 1873	145,027
Malt-polishing machine	C. Stoll	Brooklyn, N. Y	June 24, 1873	140,316
Malt-reservoir	F. C. Spiess and A. Dobler	New York, and Brooklyn, N. Y.	June 6, 1871	115,652
Malt-reservoir	C. Stoll	Brooklyn, N. Y	Dec. 27, 1870	110,691
Malt-shovel	W. Beach	Albany, N. Y	Aug. 14, 1866	57,069
Malt-steep	W. W. Stoll	Brooklyn, N. Y	Oct. 21, 1873	143,938
Malt, Steeping, growing, and drying	A. Kreusler	New Lebanon, N. Y	Sept. 5, 1865	49,768
Malt-sirup, Manufacture of	T. Hawks	Rochester, N. Y	June 27, 1865	48,396
Malting and brewing vessel	C. Berckhemer	Cincinnati, Ohio	May 14, 1867	64,622
Malting and drying kiln	H. Parsons	Colebrook, N. H	July 9, 1812	....
Malting-apparatus	M. Riely	Morrow, Ohio	Nov. 14, 1865	50,958
Man-hole covers, Device for securing	J. J. Craven	Jersey City, N. J	July 15, 1873	140,893
Man-power, Machine for application of	T. H. Hoskings	Detroit, Mich	May 21, 1861	32,369
Man's assistant, One-armed	G. W. Dalbey	Carrollton, Miss	May 17, 1870	103,023
Manacle	A. Delestatins	Philadelphia, Pa	Dec. 10, 1861	33,917
Manacle	A. Rankin	Philadelphia, Pa	Feb. 20, 1866	52,745
Mandrel, Adjustable	C. W. Le Count	Norwalk, Conn	June 19, 1866	55,681
Mandrel, Buffing	G. B. Dunham	Lynn, Mass	Mar. 25, 1873	137,187
Mandrel, Buffing	N. S. Thompson	Augusta, Me	Aug. 5, 1873	141,523
Mandrel, Circular-saw	T. H. Keeney	Newport, Ky	July 31, 1855	13,357
Mandrel-cutters in turning tapering sticks, Device to operate.	P. H. Niles	Boston, Mass	Feb. 24, 1857	16,705

Index of patents issued from the United States Patent Office from 1790 *to* 1873, *inclusive*—Continued.

Invention.	Inventor.	Residence.	Date.	No.
Mandrel, Expanding	D. L. Allen	Williamsport, Pa	Apr. 7, 1868	76, 372
Mandrel, Expanding	D. L. Allen	Williamsport, Pa	Sept. 21, 1869	95, 065
Mandrel, Expanding	I. Beetison	New Britain, Conn	Nov. 27, 1866	59, 948
Mandrel, Expanding	J. Brewer	Philadelphia, Pa	June 23, 1868	79, 198
Mandrel, Expanding	M. Gardner, sr	Carlisle, Pa	Apr. 28, 1868	77, 369
Mandrel, Expanding	J. B. Mallalieu	Chicago, Ill	Apr. 30, 1867	64, 234
Mandrel, Expanding	A. F. Nagle	Providence, R. I	Aug. 4, 1868	80, 560
Mandrel, Expanding	W. Sherrod	Providence, R. I	Dec. 2, 1851	8, 562
Mandrel, Expanding	J. P. Simons	Houston, Tex	Jan. 23, 1872	122, 920
Mandrel, Expanding	J. D. Smith	New York, N. Y	May 13, 1873	138, 947
Mandrel for cutting tapering sticks	G. Turner	Edinborough, Pa	May 22, 1855	12, 926
Mandrel for holding carriage-hubs, &c	N. Hawley	Rome, N. Y	Feb. 27, 1855	12, 455
Mandrel for holding tapered cast rings	G. E. Brettell	Rochester, N. Y	June 24, 1873	140, 113
Mandrel for lining cylinders with metal, Revolving	G. Potts	Cincinnati, Ohio	Sept. 13, 1853	10, 013
Mandrel for loading case-shot, &c	S. Sawyer	Fitchburgh, Mass	Dec. 24, 1861	34, 041
Mandrel for making gutta-percha tubing	J. Reynolds	New York, N. Y	June 10, 1856	15, 086
Mandrel for turning-machinery, Expanding	W. Sherrod	Providence, R. I	June 21, 1853	9, 805
Mandrel or boring-tool	J. C. Millerd	River Point, R. I	Sept. 1, 1868	81, 664
Mandrel, Riveting	J. Berry	Buffalo, N. Y	Nov. 29, 1870	109, 575
Mandrel, Universal	J. Gallatin, jr	New York, N. Y	Mar. 14, 1871	112, 580
Mandrel, Wheelwright's guide	J. Sykes	Mercer, Pa	June 26, 1855	13, 142
Mane-turner, Elastic	C. Phelps	Clayton, N. Y	Dec. 13, 1870	110, 068
Magnates and permagnates, Preparations of	B. Schmidt	Hoboken, N. J	Apr. 14, 1868	76, 825
Manganese, ferro-manganese, and spie gelcisen, Manufacfacture of.	O. Bolton, jr	Pittsburgh, Pa	Aug. 5, 1873	141, 419
Manganese, Manufacture of oxide of	C. T. Dunlop	Glasgow, North Britain	Aug. 7, 1860	29, 474
Manganese ores, Reducing	C. Adams	Philadelphia, Pa	Nov. 26, 1867	71, 355
Manganese, Process for preparing blue peroxide of	J. W. Hobbs	Boston, Mass	Apr. 6, 1869	88, 570
Manger	C. E. Albright	Muncy, Pa	June 7, 1870	103, 956
Manger	E. Carlin	United States Army	June 4, 1872	127, 563
Manger	A. Chambers	Unionville, N. Y	Aug. 17, 1869	93, 673
Manger	F. Denzler and J. Miller	Brooklyn, N. Y	Sept. 17, 1867	68, 965
Manger	J. Johnson	Kent, Ind	Sept. 25, 1866	58, 263
Manger	J. E. Kelly	New York, N. Y	Apr. 10, 1860	27, 812
Manger	D. Sager	Albany, N. Y	Nov. 15, 1864	45, 081
Manger	W. F. Stanley	Cazenovia, N. Y	Nov. 3, 1868	83, 738
Manger, Automatic	D. A. Townsend	Unionville, S. C	Feb. 25, 1873	136, 192
Manger for horses	J. C. Higgins	Millstone, N. J	Jan. 7, 1873	134, 601
Manger, Hay	J. Packer	Philadelphia, Pa	Mar. 29, 1859	23, 386
Manger, Portable	T. Fisler	Camden, N. J	Sept. 10, 1861	33, 242
Mangle	J. Beaumont	Chambersburgh, Pa	June 23, 1868	79, 188
Mangle	C. C. Converse	Elmira, N. Y	May 20, 1862	35, 301
Mangle	C. C. Converse	Dubuque, Iowa	Sept. 22, 1863	40, 022
Mangle	G. Coombs	West Falls, N. Y	Apr. 17, 1860	27, 890
Mangle	J. T. Coxell	Brooklyn, N. Y	Nov. 22, 1859	26, 227
Mangle	D. Cumming, jr	Mobile, Ala	July 27, 1858	21, 044
Mangle	R. A. Duncan	Brooklyn, N. Y	Sept. 23, 1873	142, 997
Mangle	T. Farnsworth	Cleveland, Ohio	May 16, 1865	47, 708
Mangle	R. Gage	Kingston, Canada	June 18, 1872	127, 971
Mangle	H. Gransden	Dubuque, Iowa	July 30, 1867	67, 293
Mangle	W. D. Grimshaw	New York, N. Y	Dec. 4, 1860	30, 812
Mangle	E. Gundlach	Hackensack, N. J	Nov. 18, 1873	144, 671
Mangle	P. H. Hink and H. Kaack	Moline, Ill	Aug. 31, 1869	94, 206
Mangle	W. W. Hollman	Eddyville, Ky	Sept. 13, 1859	25, 413
Mangle	J. Johansen	Springfield, Ill	Oct. 15, 1867	69, 813
Mangle	J. Johnson	New York, N. Y	Apr. 4, 1865	47, 152
Mangle	W. T. Littlejohn	Kalamazoo, Mich	Oct. 4, 1859	25, 656
Mangle	S. W., J. F., and N. Palmer	Auburn, N. Y	Jan. 8, 1867	61, 018
Mangle	W. Price	Cincinnati, Ohio	Jan. 17, 1865	45, 938
Mangle	H. W. Putnam	Cleveland, Ohio	Aug. 26, 1862	36, 303
Mangle	W. Radbourne	Rahway, N. J	Feb. 7, 1865	46, 263
Mangle	N. Soderstrom	Chicago, Ill	Aug. 10, 1869	93, 491
Mangle	R. A. Stratton	Philadelphia, Pa	Mar. 24, 1857	16, 887
Mangle	R. A. Stratton	Philadelphia, Pa	May 28, 1867	65, 134
Mangle	H. Tregellas	Calumet, Mich	Aug. 29, 1871	118, 654
Mangle	H. Wehdeking	Edgerton, Ohio	Nov. 12, 1867	70, 766
Mangle	J. B. Westwick	Galena, Ill	Nov. 9, 1869	96, 646
Mangle and ironing-machine	J. Seaman	Chicago, Ill	July 14, 1868	80, 019
Mangle and wringer	D. Cumming, jr	New York, N. Y	Apr. 3, 1866	53, 580
Mangle, Clothes	J. B. Greenhut	Chicago, Ill	Dec. 12, 1865	51, 452
Mangle, Clothes	H. E. Smith	New York, N. Y	Nov. 2, 1869	96, 361
Mangle, Domestic	S. Nowlan	New York, N. Y	Apr. 20, 1858	20, 002
Mangle, Domestic-calender	I. Doolittle	Bennington, Vt	Feb. 1, 1833	......
Mangle or rolling-press	A. Hermann	New Haven, Conn	May 8, 1866	54, 543
Mangle, Steam	P. Rundquist	New York, N. Y	May 16, 1871	114, 863
Mangles, washing-machines, &c., Composition bowl for.	T. Hardcastle	Bolton, England	May 25, 1869	90, 447
Mangling and ironing machine	S. Williams	Philadelphia, Pa	Nov. 29, 1870	109, 699
Mangling and wringing machine	A. Hollings	Providence, R. I	Sept. 3, 1872	131, 001
Mangling or wringing machine	T. Hall and J. Newton	Lawrence, Mass	June 3, 1873	139, 574
Manila and other fibrous substances, Machine for braiding.	D., J., and E. Fitzgerald	New York, N. Y	Apr. 24, 1840	1, 566
Mantel-bar	W. P. Chadwick	Edgartown, Mass	June 1, 1858	20, 404
Mantel, False	J. T. Fleehearty	Baltimore, Md	May 20, 1873	139, 137
Mantel for fire-place	E. Skinner	Sandwich, N. Y	July 20, 1831	
Mantel-piece	H. Tucker	Cambridge, Mass	Apr. 2, 1850	7, 253
Mantel-pieces of cast iron or other soft metals, Making.	I. Deaves	Philadelphia, Pa	Feb. 21, 1821	......
Manual-power	I. E. Overpeck	Overpeck's Station, Ohio	Apr. 25, 1865	47, 447
Manual-power	J. H. Yager	Trenton, Ohio	Oct. 23, 1866	59, 117
Manual-power, Driving machinery by	A. West	Green, Me	Aug. 16, 1839	1, 289
Manual-power machine	T. L. Kenworthy and A. Silvers	Collinsville, Ohio	July 30, 1867	67, 315
Manure	W. D. Hall	Hamden, Conn	Mar. 14, 1865	46, 847
Manure	P. G. Renny	Rahway, N. J	July 2, 1867	66, 357
Manure and fertilizer, Composting	D. Ruggles	Fredericksburgh, Va	May 17, 1870	103, 085
Manure and hay fork	G. B. Flint	Sing Sing, N. Y	May 18, 1869	90, 087
Manure and lime spreader	J. W. Fawkes	Christiana, Pa	Aug. 29, 1854	11, 602
Manure and oil, Treating fish for	R. C. Demolon and G. A. C. Thurneyssen.	Paris, France	Mar. 6, 1855	12, 480

Index of patents issued from the United States Patent Office from 1790 *to* 1873, *inclusive*—Continued.

Invention.	Inventor.	Residence.	Date.	No.
Manure and sand loader	H. G. Marchant	Annisquam, Mass	May 9, 1854	10, 892
Manure, Apparatus for distributing liquid	J. W. Clark	Iowa City, Iowa	Feb. 12, 1867	62, 006
Manure, Artificial	W. S. Amies	Guernsey, Great Britain	July 2, 1872	128, 578
Manure, Artificial	D. Bruce	Paspebiac, Canada	Jan. 11, 1859	22, 544
Manure, Artificial	P. S. and W. H. Chappell	Baltimore, Md	Mar. 27, 1849	6, 234
Manure, Artificial	E. Von Nordhausen	New York, N. Y	Jan. 19, 1864	41, 331
Manure-beds, Preparing	C. F. Spicker	New York, N. Y	Apr. 13, 1858	19, 974
Manure crusher and sower	T. T. Nelson	Clarke County, Va	Dec. 20, 1853	10, 325
Manure, Device for spreading	J. H. Stevens	East Durham, N. Y	Jan. 3, 1865	45, 767
Manure-distributer	J. W. Barnes	Murfreesborough, N. C	Oct. 14, 1856	15, 876
Manure-distributer	J. Cadwell	Dexter, Mich	Nov. 17, 1863	40, 605
Manure-distributer	D. E. Cripo	Pyrmont, Ind	Apr. 30, 1872	126, 187
Manure-distributer	J. B. Crowell	Greencastle, Pa	June 23, 1863	38, 949
Manure-distributer	H. S. Palmer	Norvell, Mich	May 14, 1867	64, 697
Manure-distributer and seed-sower	S. L. Fraser	West Town, N. Y	Oct. 18, 1870	108, 469
Manure-drag	W. Geahr	East Earl Township, Pa	Jan. 4, 1870	98, 487
Manure-drag	C. H. and J. H. Harnley	Penn Township, Pa	May 21, 1867	64, 861
Manure-drag	J. D. Heebner	Norrittonville, Pa	Aug. 9, 1870	106, 158
Manure-drag	U. Hummer	White Oak, Pa	Jan. 28, 1868	73, 894
Manure-drag	A. H. Shock	Piqua, Pa	July 9, 1867	66, 525
Manure-drag	A. H. Shock and H. R. Shirk	Lancaster, Pa	Oct. 5, 1869	95, 523
Manure-drag	J. S. and H. D. Spangler and D. Madlem.	Ephratah, Pa	Feb. 16, 1869	86, 951
Manure-drag	D. Wingenroth	Ephratah, Pa	Mar. 16, 1869	87, 812
Manure-drag, Stop-jointed	S. B. Minnich	Landisville, Pa	Apr. 9, 1867	63, 650
Manure-drill	G. B. Singeltary	Greenville, N. C	Dec. 13, 1859	26, 448
Manure-excavator	A. R. Hurst	Harrisburgh, Pa	Aug. 29, 1854	11, 610
Manure for transportation, Preparing	S. W. Boynton	Hartford, Conn	Nov. 26, 1872	133, 404
Manure for transportation, storage, or market, Preparing.	H. C. Babcock	Hartford, Conn	Aug. 20, 1872	130, 616
Manure-fork	A. S. Brinser and H. Bricker	Falmouth, Pa	Jan. 3, 1871	110, 734
Manure-fork	B. H. Franklin	Worcester, Mass	Dec. 20, 1853	10, 322
Manure-fork	J. W. Horst	Annville, Pa	July 14, 1868	79, 831
Manure-fork	W. Truby	Brush Valley, Pa	Oct. 22, 1867	70, 135
Manure-fork	P. Yengst	Union Deposit, Pa	Nov. 26, 1867	71, 353
Manure fork and hook, Combined	C. C. Cole	Phelps, N. Y	Feb. 25, 1868	74, 895
Manure fork and hook, Combined	R. Dean	Cohoctah, Mich	Feb. 23, 1869	87, 249
Manure fork and scraper, Combined	J. W. Clark	Kingston, Wis	Apr. 19, 1870	102, 092
Manure from fish, Manufacture of	J. B. Hyde	Newark, N. J	Nov. 12, 1861	33, 706
Manure-hook	S. B. Minnich	Landisville, Pa	Mar. 1, 1870	100, 311
Manure-hook	M. Stoll and H. Gross	Middletown, Pa	Dec. 8, 1868	84, 691
Manure hook or drag	H. Gross	Middletown, Pa	May 25, 1869	90, 526
Manure hook or drag	H. Gross	Middletown, Pa	May 25, 1869	90, 527
Manure hook or drag	H. Gross	Middletown, Pa	Nov. 9, 1869	96, 583
Manure, Machine for sowing pulverulent	J. G. Stegall	Thomasville, Ga	Jan. 5, 1869	85, 619
Manure, Machine for sowing pulverulent	N. G. Swift	Hart's Village, N. Y	July 9, 1861	32, 800
Manure, Making	C. Baer and J. Gouliart	Baltimore, Md	June 24, 1843	3, 139
Manure, Manufacture of	P. Eley	Philadelphia, Pa	Aug. 11, 1863	39, 525
Manure, Manufacture of	A. F. Mosselman	Paris, France	Sept. 6, 1861	44, 147
Manure, Preparation of animal and other	R. Hare	Philadelphia, Pa	Jan. 29, 1850	7, 053
Manure, Preparing night-soil for	R. B. Fitts	Philadelphia, Pa	Feb. 17, 1863	37, 685
Manure, Process of treating feldspar for	C. Bickell	Baltimore, Md	Nov. 25, 1856	16, 111
Manure-spreader	P. Eley	New York, N. Y	May 17, 1864	42, 758
Manure-spreader	S. A. Hedges	Lancaster, Ohio	Jan. 11, 1853	9, 535
Manure-spreader	D. Hill	New Vienna, Ohio	Mar. 15, 1870	100, 762
Manure-spreader	E. Marshall	Clinton, N. J	Aug. 22, 1854	11, 557
Manure-spreading machine	J. H. Stevens	East Durham, N. Y	Mar. 25, 1862	34, 790
Manure, Treating	E. P. Baugh	Philadelphia, Pa	May 9, 1865	47, 611
Map and chart holder	E. A. and A. C. Apgar	Trenton, N. J	Mar. 10, 1868	75, 343
Map and chart rack	F. G. Johnson	Brooklyn, N. Y	Apr. 23, 1872	125, 960
Map, Dissected	C. J. Higgins	Indianapolis, Ind	Sept. 2, 1873	142, 338
Map-drawing apparatus	G. S. Ormsby	Xenia, Ohio	Nov. 30, 1869	97, 432
Map exhibiter and cabinet	W. A. and G. Rice	Framingham, Mass	Oct. 14, 1873	143, 717
Map, Geographical	E. A. and A. C. Apgar	Philadelphia, Pa	Oct. 16, 1866	58, 748
Map-holder	J. Highfield	Centreville, Ind	July 30, 1872	130, 050
Map-holder, Portable	G. Rice	Framingham, Mass	Mar. 9, 1869	87, 591
Map, Ribbon	M. Coloney and S. B. Fairchild	Saint Louis, Mo	July 9, 1867	66, 463
Map-stand	W. F. Phelps	Winona, Minn	Oct. 25, 1870	108, 625
Map-suspending frame	W. F. Phelps	Winona, Minn	Nov. 5, 1867	70, 467
Map to teach geography, &c., Outline	E. F. Anderson	Mansfield, Conn	Oct. 20, 1868	83, 236
Maps and charts, Roller-case for	J. S. Ostrander	Albany, N.Y	June 22, 1869	91, 766
Maps, charts, &c., Apparatus for exhibiting	N. K. Lombard	Boston, Mass	Oct. 27, 1836	....
Maps, charts, &c., Apparatus for preserving and exhibiting.	N. K. Lombard, jr	Boston, Mass	Oct. 27, 1836	70
Maps, charts, &c., Instrument for delineating	A. Bailey	Newberry, S. C	Sept. 4, 1840	1, 770
Maps, Coloring	L. Stebbins	Hartford, Conn	Mar. 12, 1840	1, 510
Maps, Constructing block or plate for printing	G. W. Bacon	London, Great Britain	Aug. 7, 1866	57, 056
Maps, Making dissected	S. McCleary and J. Pierce	Hoosick, N. Y	Sept. 25, 1849	6, 747
Maps, Process and apparatus for coloring	L. Stebbins	Hartford, Conn	July 11, 1837	263
Maps, songs, curtains, &c., Apparatus for winding	J. S. Ostrander	Albany, N. Y	June 22, 1869	91, 660
Mapping lands, &c., Instrument for	J. Deneale	Dumfries, Va	Aug. 3, 1820	....
Marble and freestone scouring machine	R. P. Henry	Akron, Ohio	Oct. 14, 1862	36, 650
Marble and other stone or wood, Sawing	C. B. Austin	Kensington, Pa	Dec. 24, 1833	....
Marble and other stone, Sawing, molding, and polishing.	I. D. Kirk	Philadelphia, Pa	Feb. 26, 1833	....
Marble and other substances, Composition for grinding and polishing.	J. C. McAfee	West Alexander, Pa	Aug. 25, 1868	81, 520
Marble and other substances, Process for imitating	W. Bonney	New York, N. Y	Aug. 8, 1854	11, 468
Marble and stone sawing machine	G. J. Wardwell	Hatley, Canada	Nov. 4, 1856	16, 033
Marble and veining the same, Process of manufacturing artificial.	F. H. Hall	Hackensack, N. J	Dec. 9, 1873	145, 345
Marble and wood polishing machine	J. C. Mateer	Kankakee, Ill	Oct. 22, 1872	132, 404
Marble, Artificial	B. Auguste	Chicago, Ill	July 8, 1873	140, 668
Marble, Artificial	G. H. Mellen	New York, N. Y	Dec. 24, 1872	134, 300
Marble, Artificial	C. Vlockmann	Columbus, Ohio	Nov. 16, 1869	97, 004
Marble, Artificial variegated	A. and I. Straub	Milton, Pa., and Cincinnati, Ohio.	Nov. 17, 1863	40, 650
Marble, &c., Carving machine	W. A. Pease	Goshen, Ind	Apr. 10, 1860	27, 827
Marble, Coloring	S. Gardner	New York, N. Y	Feb. 1, 1870	99, 306

Index of patents issued from the United States Patent Office from 1790 *to* 1873, *inclusive*—Continued.

Invention.	Inventor.	Residence.	Date.	No.
Marble, Coloring and staining	J. Zengeler	Chicago, Ill	Mar. 28, 1871	113, 127
Marble, &c., Composition for cleaning	W. Ginnaugh	Niles, Mich	Apr. 18, 1865	47, 295
Marble, Compound for polishing and cleaning	P. Doyle and F. P. Colton	Hartford, Conn	June 14, 1870	104, 285
Marble, Compound for the manufacture of artificial	W. Meyer	Canton, Ohio	Feb. 16, 1869	86, 962
Marble, &c., Compounding China clay to imitate	D. Kaempfe and C. List	Nevohaus, Schwarzburg-Rudolstadt.	July 5, 1870	105, 089
Marble cutting and finishing machine	J. W. Maloy	Boston, Mass	Apr. 10, 1866	53, 845
Marble, Cutting and polishing	J. F. Mangin		Feb. 16, 1797	
Marble cutting and polishing machine	S. Mulliken		Mar. 11, 1791	
Marble cutting, grooving, and beading machine	J. D. Buzzell	Cape Elizabeth, Me	Aug. 13, 1838	880
Marble-cutting machine	G. W. Wheeler and H. I. Stevens	New Fairfield and Bethel, Conn.	Dec. 17, 1867	72, 250
Marble-cutting machine	W. Wickersham	Boston, Mass	Mar. 3, 1868	75, 230
Marble dressing and carving machine	G. V. Black	Jacksonville, Ill	Jan. 16, 1872	122, 754
Marble-dressing machine	R. P. Bailey	Niagara Falls, N. Y	June 25, 1867	65, 990
Marble-finishing machine	H. W. Kent	Battle Creek, Mich	Nov. 7, 1865	50, 829
Marble for painting on, Preparing	A. Robertson		Mar. 5, 1804	
Marble-graining tool	S. Wiggins	Bridgeport, Conn	June 28, 1864	43, 359
Marble, granite, &c., Composition for making	L. Matthey	Brooklyn, N. Y	Mar. 7, 1827	
Marble, granite, &c., Process for cleaning	J. Sawyer	Northfield, Vt	Dec. 30, 1873	145, 971
Marble grinding and polishing machine	J. H. Volk	Chicago, Ill	Apr. 19, 1870	102, 184
Marble, Grinding, smoothing, and polishing	S. Risley	Philadelphia, Pa	June 14, 1834	
Marble, Grooving and beading	A. J. Torrey	New York	Nov. 1, 1825	
Marble, Imitating	B. Trembley	New York	Nov. 13, 1827	
Marble, Imitating	H. Tucker	Cambridge, Mass	Oct. 14, 1851	8, 420
Marble, Imitation	S. W. Davis	Cincinnati, Ohio	May 22, 1849	6, 471
Marble, Imitation	A. and W. Fisher	New York, N. Y	Mar. 25, 1862	34, 740
Marble in kerfs of varying angles, Machine for sawing.	S. Nickelson	Pulaski, Tenn	May 6, 1856	14, 823
Marble in obelisk forms, Machine for sawing	J. A. Bailey	Detroit, Mich	Mar. 25, 1856	14, 532
Marble in obelisk forms, Machine for sawing	J. E. Haviland	Galveston, Tex	Apr. 15, 1856	14, 658
Marble in obelisk forms, Machine for sawing	I. A. Heald	Springfield, Mass	Mar. 25, 1856	14, 536
Marble in obelisk forms, Machine for sawing	J. Miller	Buffalo, N. Y	Apr. 22, 1856	14, 729
Marble in obelisk forms, Machine for sawing	P. Schrag and W. J. Von Kammerhueber.	Washington, D. C	Feb. 19, 1856	14, 296
Marble in taper forms, Machine for sawing	C. Amazeen	New Castle, N. H	Mar. 4, 1856	14, 342
Marble in taper forms, Machine for sawing	M. M. Manly	South Dorset, Vt	Sept. 30, 1856	15, 814
Marble in taper forms, Machine for sawing	G. T. Pearsall	Apalachin, N. Y	Dec. 11, 1855	13, 916
Marble in taper forms, Machine for sawing	C. A. Schultz	Chicago, Ill	Mar. 18, 1856	14, 471
Marble, Lettering	J. Tasel	Kendallville, Ind	May 12, 1868	77, 781
Marble, Machine for cutting, chiseling and polishing.	A. Clark	West Stockbridge, Mass	Apr. 10, 1810	
Marble, Machine for sawing out tapering blocks of	J. A. Cole	Washington, D. C	Dec. 4, 1855	13, 866
Marble, Machine for scouring	H. C. Lull	Montpelier, Vt	Feb. 19, 1867	62, 142
Marble, Making artificial	C. F. Barker	Boston, Mass	Aug. 12, 1873	141, 689
Marble, Manufacture of artificial	R. Lamb	New York, N. Y	Jan. 31, 1860	27, 022
Marble, Manufacture of artificial	W. Myer	Canton, Ohio	Feb. 16, 1869	86, 963
Marble, Manufacture of artificial	G. A. Trear	Chicago, Ill	Aug. 9, 1870	106, 263
Marble, Manufacture of imitation	G. Davey	London, England	Apr. 2, 1872	125, 122
Marble, Manufacture of imitation	J. S. Elliott and J. F. Wood	Chelsea and Everett, Mass	Mar. 12, 1872	124, 557
Marble, Manufacture of imitation	A. Fischer	New York, N. Y	Mar. 12, 1872	124, 562
Marble-molding machine	J. Finn	Boston, Mass	Jan. 7, 1873	134, 532
Marble-molding machine	J. Finn	Boston, Mass	July 29, 1873	141, 213
Marble obelisks, Adjusting angles in machines for sawing.	L. Brooks	Great Falls, N. H	Apr. 15, 1856	14, 688
Marble obelisks, Machine for sawing	A. Straub	Milton, Pa	Jan. 8, 1856	14, 072
Marble or pipe-stone turner	D. Clark	West Stockbridge, Mass	Apr. 9, 1810	
Marble or plastic material, Artificial	H. A. Garvey	Memphis, Tenn	Sept. 14, 1869	94, 736
Marble or stone, Machine for grinding the surface of	J. A. Bachman	Lambertville, N. J	Apr. 23, 1867	63, 982
Marble, Ornamenting	S. Gardner	New York, N. Y	Oct. 22, 1867	69, 985
Marble, Ornamenting	A. Hill	Norwalk, Conn	Nov. 27, 1866	60, 007
Marble, Ornamenting	H. Hoffman and C. F. Hill	New York, N. Y	Apr. 15, 1851	8, 046
Marble planing and carving machine	G. A. Haley	Chicago, Ill	Dec. 10, 1872	133, 704
Marble-polishing machine	W. Emmet	Galveston, Tex	Mar. 20, 1860	27, 538
Marble-polishing machine	M. Mallon	Rahway, N. J	Jan. 23, 1872	122, 951
Marble-polishing machine	J. W. Maloy	Boston, Mass	Oct. 16, 1866	58, 852
Marble-polishing machine	J. W. Maloy	Boston, Mass	Nov. 20, 1866	59, 916
Marble-polishing machine	E. T. Nichols	Joliet, Ill	July 18, 1865	48, 832
Marble-polishing machine	E. and E. B. Price	Norwalk, Conn	Feb. 1, 1870	99, 348
Marble-polishing machine	A. M. Tomb	Lyons, N. Y	Feb. 14, 1865	46, 407
Marble-polishing machine	J. Ziegler	Philadelphia, Pa	Mar. 12, 1845	3, 937
Marble, Producing mosaics upon	C. C. Fitzgerald	New York, N. Y	June 7, 1870	104, 008
Marble-running machinery	E. Ferris	West Chester, N. Y	Mar. 7, 1826	
Marble, Sawing	I. D. Kirk	Philadelphia, Pa	Dec. 28, 1832	
Marble, Sawing and polishing	L. Bissel	Otsego, N. Y	June 7, 1811	
Marble, Sawing and polishing	J. Dickerson		Jan. 9, 1796	
Marble, Sawing and polishing	J. F. Mangin		July 2, 1796	
Marble sawing and polishing machine	A. McAllister	Salem, N. Y	Feb. 18, 1828	
Marble-sawing machine	J. Ashenfelder	Philadelphia, Pa	June 3, 1856	14, 995
Marble-sawing machine	C. Avery	Tunkhannock, Pa	June 17, 1856	15, 115
Marble-sawing machine	G. W. Bishup	Brooklyn, N. Y	Nov. 13, 1855	13, 773
Marble-sawing machine	W. and G. Bull	Tonawanda, Pa	Feb. 19, 1856	14, 277
Marble-sawing machine	H. Burt	Newark, N. J	Nov. 6, 1855	13, 742
Marble-sawing machine	D. M. Campbell and J. Stevens	Oswego, N. Y	Dec. 29, 1868	85, 280
Marble-sawing machine	I. Carter	Malone, N. Y	July 15, 1856	15, 328
Marble-sawing machine	W. C. Chipman	Sandwich, Mass	Nov. 13, 1855	13, 777
Marble-sawing machine	J. Cochrane	Baltimore, Md	Sept. 11, 1855	13 540
Marble-sawing machine	R. S. Craig and A. H. Woodward.	Dover, N. Y	Mar. 1, 1870	100, 378
Marble-sawing machine	L. B. Fisher	Branch County, Mich	Nov. 13, 1855	13, 784
Marble-sawing machine	L. S. Fisher	Waynesborough, Pa	July 29, 1856	15, 419
Marble-sawing machine	J. E. French and J. M. Stephenson.	Pendleton, Ind	Sept. 27, 1870	107, 675
Marble-sawing machine	W. D. Gallaher	Bensalem Township, Pa	Nov. 18, 1856	16, 086
Marble-sawing machine	G. W. Hubbard	Middletown, Conn	Nov. 13, 1855	13, 788
Marble-sawing machine	E. W. Judd	Middlebury, Vt	Aug. 14, 1822	
Marble-sawing machine	W. B. Kimball	Peterborough, N. H	Nov. 13, 1855	13, 791
Marble-sawing machine	H. Lawrence	New York, N. Y	July 1, 1856	15, 242

Index of patents issued from the United States Patent Office from 1790 to 1873, inclusive—Continued.

Invention.	Inventor.	Residence.	Date.	No.
Marble-sawing machine	J. Lyon	New York, N. Y	Jan. 11, 1859	22, 605
Marble-sawing machine	J. M. Mott, jr	Lansingburgh, N. Y	July 22, 1856	15, 363
Marble-sawing machine	R. Myers	Factory Point, Vt	June 3, 1856	15, 015
Marble-sawing machine	F. Noette and A. Schmidt	Brooklyn, N. Y	Nov. 20, 1855	13, 829
Marble-sawing machine	C. H. G. Pease	Danbury, Conn	Mar. 9, 1869	87, 701
Marble-sawing machine	R. G. Pine	Newark, N. J	Nov. 6, 1855	13, 762
Marble-sawing machine	A. H. Tingley	Providence, R. I	Mar. 19, 1850	7, 201
Marble-sawing machine	J. Toll	Locust Grove, Ohio	Sept. 9, 1856	15, 713
Marble-sawing machine	J. A. Toll	Sugar Ridge, Ohio	June 3, 1856	15, 024
Marble-sawing machine	P. J. Torney	Washington, D. C	Dec. 1, 1868	84, 518
Marble-sawing machine	A. F. Ward	Louisville, Ky	May 6, 1856	14, 839
Marble-sawing machine	C. T. Warren	Melden, Mass	Nov. 13, 1855	13, 805
Marble-sawing machine	A. Webster and D. K. Bennett	Montpelier, Vt	Sept. 23, 1856	15, 792
Marble-scouring device	W. Weaver	Nashua, N. H	Nov. 5, 1867	70, 656
Marble shooter	W. F. Falls	Boston, Mass	Nov. 12, 1867	70, 822
Marble slabs, Machine for molding the edges of	J. Finn	Boston, Mass	Apr. 22, 1873	138, 079
Marble, Smoothing and polishing	H. A. Torrey	New York	Nov. 1, 1825	....
Marble, stone, &c., Composition for cleaning	A. C. Ford	Lynn, Mass	May 25, 1869	90, 438
Marble, stone, &c., Machine for carving	W. I. Casselman	Vernon Village, N. Y	July 11, 1854	11, 286
Marble, stone, &c., Machine for sawing	J. Norman and A. R. McLean	West Dresden, N. Y	Sept. 28, 1858	21, 622
Marble to preserve it, Treating	C. C. Fitzgerald	New York, N. Y	June 7, 1870	104, 007
Marble with a revolving saw, Sawing	I. D. Kirk	Philadelphia, Pa	July 3, 1832	....
Marble-working machine	A. Saffer	New York, N. Y	Apr. 23, 1872	125, 990
Marble-working machine	C. Warner	Washington, D. C	June 1, 1858	20, 458
Marine alarm and fog-signal	L. Lewenberg	New York, N. Y	May 25, 1858	20, 354
Marine boilers, Arrangement of flues in	R. F. Soper	Philadelphia, Pa	Apr. 17, 1849	6, 367
Marine camel	M. Toulmin	New Orleans, La	Apr. 2, 1872	125, 352
Marine camel	W. P. Walker	Memphis, Tenn	Oct. 22, 1872	132, 506
Marine camel	S. Woolston	Vincent Town, N. J	Oct. 21, 1862	36, 745
Marine camel	S. Woolston	Vincent Town, N. J	May 12, 1863	38, 527
Marine drag	J. C. Beals	Searsport, Me	Sept. 30, 1873	143, 318
Marine drag	O. R. Ingersoll	New York, N. Y	Aug. 13, 1872	130, 503
Marine drag or floating anchor	T. Wilson and A. Crawford	Detroit, Mich	July 23, 1872	129, 878
Marine engine	J. Ericsson	New York, N. Y	Aug. 30, 1870	106, 800
Marine engine	W. B. Reaney	Chester, Pa	Dec. 6, 1870	109, 944
Marine-engine indicator and register	A. R. Harris	Saint Louis, Mo	June 13, 1871	115, 846
Marine engines, Apparatus for signaling the directions of motions of.	A. R. Harris	Saint Louis, Mo	July 5, 1870	105, 071
Marine engines, Refrigerator for	J. Merrill and G. Patten	Boston, Mass	Aug. 8, 1854	11, 485
Marine foundation	E. Manico	London, Great Britain	Jan. 21, 1868	73, 617
Marine leak-signal	W. Huston	Wilmington, Del	Mar. 1, 1864	41, 777
Marine motor	W. P. Kirkland	San Francisco, Cal	Jan. 15, 1867	61, 211
Marine signal	T. H. Dodge	Nashua, N. H	Mar. 23, 1852	8, 822
Marine signal	T. H. Dodge	Nashua, N. H	Nov. 9, 1852	9, 384
Marine steam-engine	W. Kennish, jr	New York, N. Y	Jan. 25, 1859	22, 736
Marine steam-engine	W. A. Lighthall	Albany, N. Y	Apr. 11, 1842	2, 545
Marine steam-engine	J. Maudslay and J. Field	Lambeth, England	June 11, 1842	2, 668
Marker and furrower, Ground	G. W. Martin, W. G. Parish, and J. A. Petrie.	Elizabeth, N. J	Oct. 3, 1871	119, 628
Marker and planter, Combined	R. A. Green	Martinsville, Ohio	Mar. 30, 1869	88, 382
Marker, Carpenter's	G. N. Nickason	Ellenville, N. Y	Jan. 4, 1870	98, 617
Marker, Corn	J. H. Beam	Woodside, Ill	Apr. 20, 1869	89, 008
Marker, Corn	J. Bearden	Bath, Ill	Sept. 14, 1869	94, 698
Marker, Corn	A. G. Brassfield	Henry, Ill	Aug. 29, 1871	118, 584
Marker, Corn	J. T. Corbitt	Des Moines, Iowa	Dec. 28, 1869	98, 350
Marker, Corn	J. W. Eardly	Grand Rapids, Mich	Nov. 3, 1868	83, 612
Marker, Corn	J. P. Olin	Westfield, Ohio	Mar. 3, 1868	75, 190
Marker, Corn	W. E. Phelps	Elmwood, Ill	June 16, 1868	78, 997
Marker, Corn	J. B. Thomas	Centreville, Ind	July 26, 1870	105, 865
Marker, Corn	J. W. Tucker	Eugene, Ill	Mar. 23, 1869	88, 234
Marker, Corn	S. J. Woland	Lincoln, Ill	June 1, 1869	90, 908
Marker, Corn-ground	O. H. Catey	Williamsburgh, Ind	Aug. 31, 1869	94, 185
Marker, Corn-ground	G. Johnson	Decorah, Iowa	Dec. 26, 1865	51, 725
Marker, Corn-ground	F. B. Kendall	Monmouth, Ill	July 11, 1871	116, 838
Marker, Corn-ground	J. H. Rynerson	Clayton, Ind	Oct. 28, 1873	144, 146
Marker, Corn-row	A. Cooke and D. W. Cross	Brooklandville, Md	June 3, 1873	139, 456
Marker, Corn-row	J. Roberts	Greenfield, Ind	Aug. 24, 1869	94, 032
Marker, Field	J. M. Canterbury	Mexico, Mo	Feb. 11, 1868	74, 302
Marker for corn-planting, Land	A. C. Smith	Joyner's Depot, N. C	Oct. 25, 1870	108, 644
Marker for planting corn	J. Burnham	La Salle, Ill	Nov. 19, 1867	70, 953
Marker for planting corn	D. A. Freeman	Belleville, Mich	Aug. 27, 1867	68, 297
Marker for planting, Field	W. Goltry	La Grange, Iowa	July 4, 1865	48, 551
Marker for seeding-machines	G. Armstrong	Elmira, Ill	Sept. 21, 1869	95, 067
Marker, Ground	S. P. Lionberger	Saint Mary's Township, Ill	May 18, 1869	90, 277
Marker, Land	J. H. Anderson	Hillsborough, Ohio	Sept. 3, 1872	131, 045
Marker, Land	G. W. Betts	Shadeville, Ohio	Aug. 19, 1873	141, 913
Marker, Land	W. L. Bower	Joliet, Ill	Sept. 22, 1868	82, 282
Marker, Land	J. Cuff	Emery, Ohio	Feb. 11, 1873	135, 695
Marker, Land	J. V. Gray	Washington, Ind	Feb. 6, 1872	123, 474
Marker, planter, and cultivator, Combined corn	W. Stirk	Fort Wayne, Ind	Mar. 15, 1870	100, 815
Marker, planter, and cultivator, Corn	E. Barto	Tiffin, Ohio	May 11, 1869	89, 843
Marker, Ready	T. W. Wisner	Oceola, Mich	July 22, 1862	35, 964
Marker, Seed	S. D. Fisher	Normal, Ill	Aug. 2, 1870	106, 042
Marker, seed-planter, and cultivator	C. H. and F. Logan	Pawling Post-Office, Pa	Dec. 2, 1873	145, 220
Market-box	F. Gearing	Pittsburgh, Pa	Sept. 24, 1867	69, 204
Marking and cutting gage, Dressmaker's	M. Blauvelt	Ithica, N. Y	Aug. 19, 1873	141, 849
Marking-brand	B. B. Hill	Chicopee, Mass	Dec. 3, 1861	33, 868
Marking-brand	G. H. Strong	Buffalo, N. Y	July 14, 1863	39, 261
Marking-can	F. W. Marvin	Sacramento, Cal	July 7, 1868	79, 666
Marking-compound	H. Loewenberg	New York, N. Y	Apr. 9, 1867	63, 734
Marking corn-furrows, Device for	G. G. Crose	Schoolcraft, Mich	Oct. 15, 1861	33, 478
Marking corn-ground, Machine for	G. Sprague	Spring Hill, Kans	June 4, 1867	65, 443
Marking corn-ground, with rake-attachment, Machine for.	M. Dickle and E. P. Cowan	Ottumwa, Iowa	Mar. 9, 1869	87, 552
Marking ground for planting, Device for	W. W. Potts	Rushville, Ill	Nov. 22, 1864	45, 174
Marking ground for planting, Machine for	W. R. McKinley	Lucas County, Iowa	Mar. 10, 1868	75, 288
Marking-machine	E. Spencer	Saint Louis, Mo	Aug. 19, 1862	36, 243
Marking-pot	J. F. W. Dorman	Baltimore, Md	Aug. 6, 1872	130, 111

Index of patents issued from the United States Patent Office from 1790 *to* 1873, *inclusive*—Continued.

Invention.	Inventor.	Residence.	Date.	No.
Marking-pot	W. H. Green	New York, N. Y	Sept. 13, 1870	107, 250
Marking-pot	J. L. Tarbox	New York, N. Y	May 7, 1872	126, 501
Marking-roller, Elastic	L. R. Witherell	Galesburgh, Ill	Feb. 12, 1867	62, 104
Marking-roller for planting	F. Roth and B. Fürst	Lacon, Ill	June 13, 1871	115, 900
Marking stock	G. W. Devin	Ottumwa, Iowa	Apr. 24, 1866	54, 127
Marking-wheel	H. Holt	Brooklyn, N. Y	Jan. 23, 1866	52, 169
Marl, dirt, &c., Machine for raising	T. F. Christman	Wilson, N. C	June 22, 1858	20, 623
Marl, Process for treating and compounding	R. B. Fitts	Philadelphia, Pa	May 30, 1865	47, 941
Marquetry	J. Dill	Grand Rapids, Mich	Aug. 29, 1871	118, 595
Marquetry	L. F. Groebl	Philadelphia, Pa	Jan. 23, 1855	12, 276
Marsh and swamp lands, Mode of reclaiming	S. B. Driggs	New York, N. Y	June 27, 1865	48, 382
Marsh-drainer	J. Blanc	New Orleans, La	July 2, 1836	....
Marsh-filling device	G. Howell and W. Smith	Philadelphia, Pa	Sept. 15, 1868	82, 224
Marsh-filling machine	G. Howell	Philadelphia, Pa	Nov. 12, 1867	70, 849
Marshes, Filling	G. Howell	Philadelphia, Pa	Mar. 31, 1868	76, 194
Marshes, Filling	S. B. Woods and J. T. Chapman	Jersey City, N. J., and Brooklyn, N. Y.	Nov. 10, 1868	83, 895
Martingale	P. J. McGuiness	New York, N. Y	Dec. 1, 1868	84, 643
Martingale	R. R. Miller	Southington, Conn	Feb. 11, 1873	135, 657
Martingale	F. G. Moody	North Monmouth, Me	Aug. 26, 1873	142, 261
Martingale	W. B. Perrie	Horse Head, Md	Sept. 8, 1868	82, 029
Martingale for preventing horses or mules from throwing or breaking fences, Pricking.	D. and J. Pfouts	Holmes County, Ohio	Sept. 2, 1862	36, 363
Martingale-ring	G. T. Bushnell	Birmingham, Conn	Nov. 15, 1859	26, 086
Martingale-ring	W. F. Gilbert	Birmingham, Conn	Nov. 9, 1869	96, 690
Martingale-ring	D. C. Lockwood	Derby, Conn	Oct. 2, 1860	30, 233
Martingale-ring	W. M. Welling	New York, N. Y	Mar. 17, 1863	37, 941
Martingale-ring, Composition	C. B. Bristol	New Haven, Conn	Feb. 23, 1869	87, 240
Mash and beer cooler	C. Schenck	Manheim, Baden	Aug. 27, 1867	68, 242
Mash, Apparatus for ascertaining the proof-spirits in fermented.	G. Tagliabue	New York, N. Y	Feb. 14, 1871	111, 885
Mash-cooler	F. P. Bush	Cincinnati, Ohio	Nov. 5, 1867	70, 518
Mash-cooling apparatus	A. Naegeli	Wegeleben, Germany	Nov. 5, 1872	132, 730
Mash-cooling machine	J. Shilling	Troy, Ohio	Aug. 11, 1857	17, 984
Mash for distillation, Preparing	G. Seitz	Easton, Pa	Jan. 26, 1858	19, 210
Mash from grain, Saccharifying	C. H. Frings	Centreton, Mo	Dec. 6, 1870	109, 887
Mash-machine	M. Brand and C. P. Hoffman	Chicago, Ill	Dec. 11, 1866	60, 330
Mash-machine	J. T. Forrer	Peoria, Ill	Apr. 3, 1866	53, 597
Mash-machine	A. Hammer	Philadelphia, Pa	Jan. 9, 1855	12, 205
Mash of beer, &c., Mode of cooling	A. Smith	New York, N. Y	Dec. 22, 1863	41, 025
Mash, Process and apparatus for forming	F. W. Wolf	Chicago, Ill	Apr. 9, 1872	125, 645
Mash-tub	L. Klee	Pittsburgh, Pa	June 16, 1868	78, 973
Mash-tub	A. Miller and J. R. Stauffer	Pennsville, Pa	Sept. 1, 1863	39, 739
Mash-tub	B. Roop	Pekin, Ohio	May 6, 1844	3, 575
Mash-tub and vapor-cooler	A. W. Cram	Saint Louis, Mo	Mar. 9, 1869	87, 640
Mash-tub, Distiller's	D. P. Patterson	Fayette County, Pa	Aug. 21, 1860	29, 713
Mash-tubs, Apparatus for heating	A. Hammer	Reading, Pa	Aug. 17, 1858	21, 195
Mash-tubs, Heating and cooling coil for	M. J. Allen	New York, N. Y	Apr. 27, 1869	89, 374
Mash-tun	J. Chilcott	Brooklyn, N. Y	May 30, 1865	47, 931
Mash-tun	C. Stoll	Brooklyn, N. Y	Dec. 3, 1872	133, 679
Mash-tun	R. Wicks and J. Faulkner, jr.	Williamsburgh, N. Y	May 11, 1852	8, 946
Mash-tun pipes, Preventing coating of	A. Adams	Springfield, Mass	Aug. 28, 1866	57, 454
Mashing	N. G. Thorn	Dayton, Ohio	Mar. 1, 1859	23, 125
Mashing-apparatus	E. Haeckel	Cincinnati, Ohio	Sept. 6, 1859	25, 329
Mashing-machine	L. C. Field	Galesburgh, Ill	Sept. 14, 1869	94, 730
Mashing-machine	A. Harris	New York, N. Y	July 12, 1870	105, 331
Mashing, Treating cereal grains for	C. H. Frings	Central, Mo	Apr. 30, 1872	126, 283
Mashing-tub	J. C. Birket	Peoria, Ill	Sept. 19, 1871	119, 072
Mashing-tub	J. Wright	Waterloo, N. Y	June 24, 1851	8, 183
Mask and respirator, Fireman's	I. P. Nelson	Cambridge, Mass	Mar. 17, 1857	16, 863
Mask, Safety	D. W. Flora	Newaygo, Mich	May 28, 1872	127, 331
Masonic emblems, Apparatus for display of	R. H. Lyon and G. B. Rand	Dubuque, Iowa	Nov. 19, 1872	133, 236
Mast and rigging	G. T. May	Tompkinsville, N. Y	July 15, 1862	35, 882
Mast and spar for vessels	W. T. Griffenberg	Wilmington, Del	May 21, 1872	127, 047
Mast and truss hoops, Bending	J. Mulford	Northern Liberties, Pa	Jan. 16, 1835	....
Mast coat	A. J. Gore	San Francisco, Cal	July 11, 1865	48, 767
Mast-coat	G. Perkins and J. A. Stetson	Lynnfield and Gloucester, Mass.	Nov. 5, 1872	132, 774
Mast for navigable vessels	C. P. Coles	Southsea, England	Apr. 14, 1863	38, 151
Mast for vessels	W. T. Griffenberg	Wilmington, Del	July 30, 1872	130, 035
Mast for vessels	E. E. Holley	New York, N. Y	Feb. 22, 1870	100, 151
Mast for vessels, Hinged	J. E. Hammond	New Bedford, Mass	Dec. 16, 1873	145, 648
Mast-head	S. T. Nickerson	South Orrington, Me	Feb. 27, 1872	124, 006
Mast-hoop	E. F. Burrows	Mystic River, Conn	July 13, 1869	92, 577
Mast-hoop	F. B. Dunton	Centre Lincolnville, Me	Apr. 20, 1869	89, 135
Mast-hoop	W. T. Maddocks	Northport, Me	Feb. 23, 1869	87, 105
Mast-hoop	W. McKay	Newburyport, Mass	Dec. 22, 1868	85, 234
Mast-hoop	D. R. Proctor	Gloucester, Mass	Mar. 10, 1863	37, 872
Mast-hoop, Anti-friction	D. H. Hussey	Portsmouth, N. H	Aug. 3, 1869	93, 303
Mast-hoop balance-line	J. Conway	Harrison, Md	May 4, 1869	89, 632
Mast-hoops, Machine for manufacturing	E. W. Scott	Lowell, Mass	Oct. 31, 1854	11, 875
Mast-hoops, Sail-link to	J. E. Seavey	Kennebunkport, Me	June 17, 1862	35, 654
Mast, steeple, &c., Iron	E. S. Boynton	Alexandria, Va	Mar. 26, 1861	31, 777
Masts and spars, Connection of telescopic	C. F. Brown	Warren, R. I	June 17, 1851	8, 164
Masts, Apparatus for lowering ship's	G. Dowling	Fair Haven, Conn	Mar. 17, 1868	75, 671
Masts, Attaching topmasts and top-gallant	L. Brown	New York, N. Y	Apr. 18, 1871	113, 738
Masts, Ball for	E. C. Seely	Port Medway, Canada	Aug. 27, 1872	130, 825
Masts, Cheek-piece for ships'	J. Baymore	Philadelphia, Pa	Apr. 4, 1871	113, 386
Masts in the decks of vessels, Supporting	T. J. Woodworth	Salem, N. J	Jan. 27, 1857	16, 510
Masts of navigable vessels, Connection of the gaff to the	C. R. Fisher	Chelsea, Mass	Mar. 28, 1865	47, 004
Masts of ships, Fid for the	B. Rotch	Bristol, Great Britain	Dec. 10, 1823	....
Masts of vessels, Cap or withe for	J. W. Sikes	Plymouth, N. C	Aug. 1, 1854	11, 461
Masts of vessels, Purchase for hoisting and lowering top-	W. Winchester	Portland, Me	Sept. 14, 1869	94, 804
Masts of vessels, Securing	D. S. Stevens and L. Snedecor	Red Bank, N. J	Aug. 11, 1868	80, 884
Masts, Supporting ship's top	T. Batty	Brooklyn, N. Y	Sept. 4, 1855	13, 518
Master-key lock	A. Rankin	Philadelphia, Pa	July 29, 1873	141, 381

Index of patents issued from the United States Patent Office from 1790 to 1873, inclusive—Continued.

Invention.	Inventor.	Residence.	Date.	No.
Master-wheel, Double-lever	D. E. Mitchell	Rural Dale, Ohio	June 21, 1870	104, 484
Mastic, Composition	A. Thiele	Yickfaw, La	Sept. 9, 1873	142, 595
Mastic composition	J. Thompson	North Wrentham, Mass	Mar. 30, 1858	19, 802
Masting and rigging vessels	W. Webster	Jefferson County, Wash. T.	June 22, 1858	20, 673
Mat: *See* Bird-cage mat. Rubber mat. Cage-mat. Straw mat. Door-mat. Table-mat. Metallic mat. Toilet-mat. Oil-press mat. Wooden mat.				
Mat	J. M. Groh	Benevola, Md	Sept. 22, 1868	82, 308
Mat	D. M. Harman	Baltimore, Md	Sept. 17, 1872	131, 440
Mat	S. Lewis	Williamsburgh, N. Y	Oct. 11, 1870	108, 160
Mat	P. W. Neefus	New York, N. Y	Feb. 7, 1871	111, 559
Mat and scraper	W. Farrington	Morrisania, N. Y	June 29, 1869	92, 026
Mat, Husk	J. P. Smucker	Ashland, Ohio	Aug. 31, 1869	94, 252
Mats, &c., Machinery for making	D. Hodgman and A. D. Wyckoff	New York, N. Y	May 14, 1849	6, 412
Mats, Manufacture of Manila	S. S. Williams	Roxbury, Mass	Aug. 22, 1828	....
Match and cigar box	G. Graetz	Alexandria, Va	Dec. 22, 1868	85, 170
Match-blocks, Machine for making	E. Andrews and W. Tucker	Portland, Me., and Philadelphia, Pa.	Nov. 16, 1869	96, 764
Match-blocks, Machine for splitting	V. R. Powell	Troy, N. Y	June 16, 1863	38, 911
Match-box	E. B. and L. W. Beecher	New Haven, Conn	Feb. 23, 1869	87, 236
Match-box	L. W. Beecher	Westville, Conn	June 22, 1869	91, 703
Match-box	J. Bonsfield	Cleveland, Ohio	Nov. 2, 1869	96, 303
Match-box	D. Burhans	Burlington, Iowa	May 30, 1871	115, 430
Match-box	T. Crommelin	Brooklyn, N. Y	Mar. 26, 1872	124, 886
Match-box	G. Dowler	Birmingham, England	Mar. 22, 1864	41, 976
Match-box	S. W. Francis	New York, N. Y	Apr. 30, 1861	32, 187
Match-box	J. H. Goodfellow and M. Russell, jr.	Troy, N. Y	May 10, 1870	102, 807
Match-box	J. Holtz	Baltimore, Md	Aug. 12, 1873	141, 793
Match-box	H. Howson	Philadelphia, Pa	Jan. 21, 1862	34, 230
Match-box	H. Howson	Philadelphia, Pa	Sept. 15, 1863	39, 994
Match-box	A. D. Judd	Hew Haven, Conn	Dec. 12, 1871	121, 788
Match-box	A. D. Judd	Hew Haven, Conn	Jan. 16, 1872	122, 836
Match-box	J. Kirchfeld and F. Heyl	Riegelsville, Pa	Mar. 31, 1868	76, 084
Match-box	A. Martin	Basel, Switzerland	Apr. 14, 1868	76, 786
Match-box	J. Matthias	New York, N. Y	Nov. 11, 1873	144, 404
Match-box	L. O. P. Meyer	Newtown, Conn	Aug. 29, 1871	118, 627
Match-box	L. O. P. Meyer	Newtown, Conn	Nov. 14, 1871	120, 889
Match-box	A Meyersberg	New York, N. Y	Aug. 5, 1873	141, 581
Match-box	J. Monaghan	Tuckahoe, N. Y	Nov. 15, 1870	109, 235
Match-box	M. L Orum	Philadelphia, Pa	Nov. 4, 1873	144, 283
Match-box	O. Paddock	Watertown, N. Y	Jan. 14, 1873	134, 764
Match-box	C. Prahl	Stapleton, N. Y	July 15, 1873	140, 949
Match-box	L. Schoerken	Kohn, Prussia	Mar. 19, 1867	63, 104
Match-box	J. H. Wheeler	Addison, Vt	Jan. 28, 1866	73, 855
Match-box	H. J. Wickham	Manchester, Conn	May 4, 1869	89, 719
Match box and candle stick combined	C. R. Stickney	Hartford, Conn	June 4, 1867	65, 445
Match-box and taper-holder combined	J. A. Whipple	Cambridge, Mass	May 28, 1867	65, 142
Match-box, Pocket	S. W. Francis	New York, N. Y	Jan. 28, 1862	34, 249
Match-box, Pocket	A. Roesler and C. Frey	Warsaw, Ill	Feb. 15, 1859	22, 979
Match-boxing machine	N. B. Forrest	Auburn, N. Y	June 15, 1869	91, 429
Match-card	M. G. Crane	Boston, Mass	Sept. 5, 1865	49, 727
Match cards, Rack for holding comb	E. G. Byam and B. E. Parkhurst.	Boston, Mass., and Brunswick, Me.	Oct. 12, 1858	21, 794
Match-case	F. B. Coleman	Southampton, Mass	May 3, 1870	102, 499
Match-case	H. Ellig	Bridgeport, Conn	Feb. 19, 1867	62, 260
Match-case, Pocket	L. O. P. Meyer	Newtown, Conn	Nov. 14, 1871	120, 892
Match-case, Portable	D. Ellis and P. Hine	Waterbury, Conn	Apr. 24, 1860	27, 974
Match-composition	J. F. Babcock, W. A. Leonard, and E. B. Crane.	Boston, Mass	Mar. 18, 1873	136, 953
Match-composition	W. H. Rogers	New York, N. Y	Oct. 27, 1868	83, 412
Match composition, Friction	N. B. and D. Shaw	Sanbornton, N. H	Aug. 29, 1865	49, 659
Match-compound	L. Lanszweert	San Francisco, Cal	Mar. 27, 1866	53, 454
Match cord, Pocket-safe for friction	W. H. Rogers	New York, N. Y	Oct. 13, 1868	83, 097
Match-cutting machine	P. D. Cummings, M. Smith, and J. C. Jordan.	Portland, Me	June 3, 1873	139, 457
Match-cutting machine	T. G. Murphy	Akron, Ohio	Jan. 7, 1873	134, 565
Match for lighting cigars, &c	G. Graetz	Alexandria, Va	Dec. 8, 1868	84, 821
Match for lighting cigars, &c	J. Howe	Allegheny, Pa	July 2, 1872	128, 626
Match for retaining fire, Friction	J. H. Stevens	New York, N. Y	Nov. 16, 1839	1, 414
Match-frames, Machine for filling	W. Gates, jr	Frankfort, N. Y	Oct. 31, 1854	11, 860
Match-frames, Machine for filling	A. Sohn	Monroeville, Ohio	Apr. 18, 1854	10, 809
Match, Friction	W. K. Ashard	New York	June 24, 1843	3, 146
Match, Friction	G. W. Carleton	Bath, Me	May 20, 1842	2, 635
Match, Friction	J. R. Dey	Hudson City, N. J	Apr. 7, 1863	38, 096
Match, Friction	S. C. Moore	Boston, Mass	May 23, 1865	47, 848
Match, Friction	A. D. Phillips	Springfield, Mass	Oct. 24, 1836	68
Match, Friction	W. H. Rogers	New York, N. Y	Oct. 12, 1869	95, 730
Match, Friction	C. W. Smith	New York, N. Y	Mar. 11, 1862	34, 649
Match, Friction	E. Smith	Erving, Mass	Oct. 3, 1844	3, 773
Match, Friction	B. F. Woodside	McDonald, Tenn	Jan. 14, 1868	73, 275
Match-holder	L. J. Johnson	Philadelphia, Pa	July 2, 1861	32, 735
Match-holder	L. O. P. Meyer	Newtown, Conn	Aug. 29, 1871	118, 628
Match-holder	L. O. P. Meyer	Newtown, Conn	Nov. 14, 1871	120, 890
Match-holder	L. O. P. Meyer	Newtown, Conn	Nov. 14, 1871	120, 891
Match-hook	J. D. Leach	Penobscot, Me	May 2, 1871	114, 451
Match-igniter	J. H. Merrill	Norwalk, Conn	Aug. 22, 1865	49, 542
Match light and box	R. and N. S. Allison	Burlington, N. J	Mar. 26, 1814	......
Match-lighter	W. Adamson	Philadelphia, Pa	July 16, 1867	66, 770
Match-machine	E. Andrews and W. Tucker	Portland, Me	Dec. 1, 1868	84, 464
Match-machine	E. Andrews and W. Tucker	Portland, Me	July 13, 1869	92, 415
Match-machine	W. B. Eltonhead	Philadelphia, Pa	Aug. 10, 1869	93, 525
Match-machine	L. Holms	Paterson, N. J	July 1, 1856	15, 238
Match-machine	L. T. Luther	Oak Grove, Pa	Feb. 15, 1870	99, 780
Match-machine	S. Miller and W. Gates, jr	Hammond and Frankfort, N. Y.	Mar. 9, 1858	19, 608

Index of patents issued from the United States Patent Office from 1790 to 1873, inclusive—Continued.

Invention.	Inventor.	Residence.	Date.	No.
Match-machine	L. and J. Thomas	Allegheny City, Pa	Jan. 23, 1855	12, 297
Match machine, Friction	C. D. Smith and H. Patterson	Baldwinville, Mass	July 22, 1856	15, 393
Match-making machine	M. Young	Frederick, Md	Aug. 29, 1871	118, 502
Match, Parlor	F. Zaiss	Philadelphia, Pa	Apr. 16, 1872	125, 874
Match, Percussion	W. Servant	Providence, R. I	Apr. 18, 1871	113, 937
Match-plates, Method and apparatus for molding	J. R. Davies	Racine, Wis	July 25, 1871	117, 391
Match-safe	W. H. Andrews	New Haven, Conn	Dec. 20, 1864	45, 554
Match-safe	J. E. Auld	Buffalo, N. Y	Oct. 2, 1866	58, 536
Match-safe	C. A. Babcock	Frankfort, N. Y	May 28, 1867	65, 041
Match-safe	C. Bünger	New Haven, Conn	June 11, 1867	65, 722
Match-safe	L. Burnell	Milwaukee, Wis	Jan. 3, 1860	26, 648
Match-safe	P. D. M. Carmichael	Le Roy, N. Y	Jan. 7, 1868	72, 973
Match-safe	L. Darozir	Worcester, Mass	Jan. 7, 1873	134, 523
Match-safe	S. De Mackiewiez	New York, N. Y	June 21, 1864	43, 186
Match-safe	J. W. Durham	Ripley, Tenn	July 6, 1869	92, 290
Match-safe	J. A. Evarts	West Meriden, Conn	Jan. 28, 1868	73, 706
Match-safe	G. Geer	Meriden, Conn	May 3, 1870	102, 676
Match-safe	J. Gibbs	Brooklyn, N. Y	July 27, 1869	92, 955
Match-safe	C. Goldthwait	South Weymouth, Mass	Oct. 18, 1870	108, 346
Match-safe	T. S. Hathaway	Decatur, Ill	Sept. 25, 1866	58, 253
Match safe	J. and A. Helm	Washington, D. C	Nov. 18, 1873	144, 615
Match-safe	A. Hoyt	New York, N. Y	June 30, 1868	79, 353
Match-safe	R. W. Jenks	Providence, R. I	Sept. 5, 1865	49, 760
Match-safe	M. Jincks	Dansville, N. Y	May 19, 1868	77, 984
Match-safe	A. D. Judd	New Haven, Conn	May 10, 1870	102, 828
Match-safe	P. Killin and H. C. Yates	Decatur, Ill	July 30, 1867	67, 317
Match-safe	C. H. Leggett	Elizabeth, N. J	Apr. 29, 1873	138, 261
Match-safe	H. Richmond	West Meriden, Conn	Oct. 6, 1868	82, 752
Match-safe	J. Roebuck	New York, N. Y	Apr. 16, 1867	63, 943
Match-safe	A. Roesler	Warsaw, Ill	Oct. 29, 1867	70, 362
Match-safe	G. H. Snow	New Haven, Conn	Apr. 19, 1864	42, 414
Match-safe	A. Steinbök	New York, N. Y	Oct. 26, 1869	96, 163
Match-safe	W. Stine	Elmore, Ohio	Apr. 4, 1871	113, 360
Match-safe	H. M. Woodford	Kensington, Conn	June 30, 1868	79, 426
Match-safe, Pocket	C. E. Fowler	Carmel, N. Y	Nov. 20, 1866	59, 830
Match-safe, Pocket	A. M. Smith	New York, N. Y	Jan. 17, 1860	26, 865
Match safe, Portable and water-proof friction	P. Merrill	Port Sanilac, Mich	Oct. 12, 1858	21, 770
Match-safe, Sheet-metal	J. Musgrove	Newark, N. J	May 2, 1871	114, 323
Match-safes, Manufacture of metallic	J. Fallows	Philadelphia, Pa	Feb. 27, 1872	124, 043
Match, Safety	H. House	London, Great Britain	Feb. 20, 1872	123, 905
Match-slitting machine	M. Smith and J. C. Jordan	Portland, Me	Apr. 2, 1872	125, 149
Match-splint card	B. Hotchkiss	New Haven, Conn	June 13, 1865	48, 176
Match-splint-cutting machine	T. Cook	New York, N. Y	May 12, 1857	17, 263
Match-splint-cutting machine	E. Fitzgerald	New York	July 18, 1840	1, 695
Match-splint-cutting machine	H. Law	Wilmington, N. C	Aug. 28, 1844	3, 719
Match-splint-cutting machine	C. E. Warner	New York	July 2, 1842	2, 700
Match-splint-framing machine	J. Bentz	Brooklyn, N. Y	Sept. 3, 1867	68, 342
Match-splint machine	R. F. Gustine	Chicago, Ill	Apr. 12, 1853	9, 660
Match-splint-splitting machinery	L. Smith	New York, N. Y	Mar. 28, 1848	5, 496
Match splints and arranging them in the dipping-frame, Machine for making.	A. Fessenden and L. L. Knight	Templeton and Barre, Mass	Apr. 26, 1845	4, 013
Match splints for dipping, Machine for framing lucifer.	A. and E. B. Beecher	New Haven, Conn	Feb. 3, 1863	37, 562
Match-splints, Machine for making	E. Andrews and W. Tucker	Springfield, Mass	Dec. 20, 1864	45, 465
Match-splints, Machine for making	E. Andrews and W. Tucker	Portland, Me	Aug. 27, 1867	68, 144
Match-splints, Machine for making	D. Burhans	Burlington, Iowa	Aug. 16, 1870	106, 319
Match splints, Machine for making	C. L. Zeidler	Cincinnati, Ohio	June 18, 1867	65, 853
Match splints or sticks, Manufacture of friction and other.	N. T. Winans and T. Hyatt	New York	Nov. 26, 1840	1, 867
Match-splints, Splitting	M. D. Murphy and O. C. Barber	Middlebury, Ohio	Feb. 2, 1869	86, 576
Match-stand	G. G. Taylor	Charlestown, Mass	Apr. 7, 1868	76, 549
Match stand, Friction	N. Waterman	Boston, Mass	June 16, 1863	38, 923
Match-stick	J. K. Robinson	Middlebury, Ohio	July 26, 1870	105, 727
Match-sticks, Machine for making	F. D. Bowens	Philadelphia, Pa	Apr. 27, 1869	89, 472
Match-sticks, Machine for making	F. Zaiss	Philadelphia, Pa	Mar. 18, 1873	136, 893
Match-sticks, Machinery for cutting	F. De Bowens	Philadelphia, Pa	Aug. 31, 1869	94, 189
Match-sticks, Manufacture of	S. C. Ellis	Jersey City, N. J	Sept. 22, 1863	40, 027
Match, Water-proof friction	L. T. Luther	Oak Grove, Pa	Nov. 19, 1867	71, 189
Matches, Apparatus for dipping lucifer	S. A. Bell and T. Higgins	Stratford and Bow, England.	June 16, 1863	38, 878
Matches, Composition for	E. K. Smith	Philadelphia, Pa	June 18, 1867	65, 838
Matches, Composition for friction	W. P. Allen	Dubuque, Iowa	Jan. 11, 1859	22, 538
Matches, Composition for friction	L. Langszweert	San Francisco, Cal	Oct. 10, 1865	50, 370
Matches, Composition for the manufacture of friction.	J. J. Karlen	Eilenbach, Switzerland	Apr. 5, 1870	101, 625
Matches, Composition for the manufacture of friction.	N. T. Winans and T. and T. Hyatt.	New York	Dec. 23, 1841	2, 402
Matches, Composition for manufacture of friction	N. T. Winans and T. and T. Hyatt.	New York	Dec. 23, 1841	2, 403
Matches, Composition for the manufacture of safety and other friction.	W. Austin	London, England	Nov. 3, 1868	83, 683
Matches, Composition-matter of friction	J. H. Stevens	New York, N. Y	Nov. 16, 1839	1, 413
Matches, Compound for making friction	O. C. Green	Copenhagen, Denmark	June 8, 1869	91, 114
Matches, Dipping-frame for the manufacture of	D. Helmer	Mohawk, N. Y	Mar. 15, 1864	41, 918
Matches, Dipping loco-foco	J. Hatfield	Stillwater, N. Y	June 3, 1837	219
Matches, Drying	M. Young	Frederick, Md	Aug. 27, 1872	130, 834
Matches for preserving from accidental ignition, Manufacture of friction.	J. H. Stevens	New York, N. Y	Nov. 16, 1839	1, 412
Matches, Friction-mat for lighting	J. B. Colt	Brooklyn, N. Y	July 23, 1872	129, 716
Matches, Ignitible compound for friction	S. Blaisdell	Brunswick, Me	Mar. 18, 1842	2, 494
Matches in frames for dipping, Machine for placing friction.	G. Sebold	Durlach, Germany	June 9, 1868	78, 837
Matches, Machine for cutting	A. Welsch	Chicago, Ill	June 28, 1864	43, 358
Matches, Machine for framing	J. Bentz	Brooklyn, N. Y	June 26, 1866	55, 808
Matches, Machine for gathering and depositing dipped.	T. Cook	New York, N. Y	May 19, 1857	17, 316
Matches, Machine for making and dipping	J. F. Cranston	Springfield, Mass	Jan. 21, 1868	73, 507
Matches, Machine for making friction	W. Gates, jr., and H. J. Harwood.	Franklin and Utica, N. Y	Apr. 4, 1854	10, 737

Index of patents issued from the United States Patent Office from 1790 *to* 1873, *inclusive*—Continued.

Invention.	Inventor.	Residence.	Date.	No.
Matches, Machine for manufacturing friction	A. Underwood	German Flats, N. Y	Apr. 29, 1856	14, 782
Matches, Machine for matting the ends of blocks in making.	H. E. Pierce	Charlemont, Mass	Jan. 10, 1854	10, 413
Matches, Machinery for making	E. Adler	New York, N. Y	Aug. 1, 1854	11, 405
Matches, Machinery for making	I. H. Smith	Wolcott, Conn	Apr. 29, 1851	8, 066
Matches, Manufacture and packing of friction	E. Precht and V. Joepken	New York, N. Y	Nov. 18, 1862	36, 967
Matches, Manufacture of	E. Andrews	Portland, Me	Oct. 15, 1867	69, 891
Matches, Manufacture of	E. Andrews and W. Tucker	Fiskedale, Mass	Mar. 26, 1867	63, 197
Matches, Manufacture of	G. Graham	New Haven, Conn	Jan. 26, 1869	86, 299
Matches, Manufacture of	E. J. Hill	Milwaukee, Wis	Dec. 24, 1867	72, 637
Matches, Manufacture of friction	W. Baustian	Davenport, Iowa	Mar. 19, 1867	62, 922
Matches, Manufacture of friction	C. F. Bonback	New York, N. Y	Aug. 5, 1873	141, 483
Matches, Manufacture of friction	G. G. Dennis	Dover, N. H	July 25, 1865	48, 913
Matches, Manufacture of friction	J. W. Hjerpe	Stockholm, Sweden	Oct. 13, 1863	40, 259
Matches, Manufacture of friction	S. Krackowizer	New York, N. Y	Apr. 18, 1865	47, 311
Matches, Manufacture of friction	V. Powell	Troy, N. Y	July 25, 1865	49, 002
Matches, Manufacture of friction	V. Powell	Troy, N. Y	Aug. 22, 1865	49, 549
Matches, Manufacture of friction	P. B. Tyler, W. M. Chandler, and L. F. Standish.	Springfield and Chicopee Falls, Mass.	Nov. 7, 1865	50, 869
Matches, Manufacture of safety	L. O. P. Meyer	Bethlehem, Pa	June 25, 1867	66, 101
Matches, Manufacture of safety	L. O. P. Meyer	Newtown, Conn	Jan. 17, 1871	111, 075
Matches, Manufacturing	O. H. Hicks	Baltimore, Md	Apr. 15, 1873	137, 840
Matches, Manufacturing splints for friction	S. Dole	Concord, N. H	May 30, 1837	214
Matches, Pocket-safe for friction	J. S. Morton	Hartford, Conn	Jan. 25, 1870	99, 224
Matches, Putting up	E. J. Hill	Milwaukee, Wis	July 23, 1867	67, 052
Matches, tapers, &c., Machine for filling dipping-clamps in the manufacture of.	T. Higgins	Bon, England	Sept. 16, 1862	36, 463
Matches water-proof, Rendering friction	L. J. Henry	New York, N. Y	Apr. 5, 1859	23, 465
Matrices, &c., Mode of constructing	J. J. C. Smith	Covington, Ky	Sept. 20, 1859	25, 533
Matter, Composition of	G. F. J. Colburn	Newark, N. J	Jan. 28, 1868	73, 694
Matter, Composition of	L. Francis	New York, N. Y	June 26, 1866	55, 951
Matter, Composition of	W. K. Wyckoff	Ripon, Wis	Oct. 2, 1866	58, 532
Matting-apparatus	J. H. Reilly	Brooklyn, N. Y	Jan. 17, 1871	111, 086
Matting for carpet-linings, &c	H. C. Heermance	Claverack, N. Y	Feb. 5, 1867	61, 831
Matting for cars, Slat	W. Barton	Troy, N. Y	Oct. 6, 1868	82, 678
Matting for floors	J. Michell	West Farms, N. Y	Apr. 16, 1867	63, 921
Matting, India-rubber binding for	F. S. Beardsley	Bridgeport, Conn	Jan. 23, 1872	122, 933
Matting, Slat	W. Barton	Troy, N. Y	Mar. 3, 1868	74, 975
Mattock, hoe, &c., swaging die	C. Konald	Snowden, Pa	Sept. 3, 1867	68, 446
Mattocks, picks, &c., Machine for forming the eyes of.	J. Donald	Cohoes, N. Y	Nov. 26, 1872	133, 360
Mattress	E. F. Bassett	Seymour, Conn	Apr. 15, 1862	34, 935
Mattress	T. G. Clinton	Cincinnati, Ohio	Apr. 20, 1852	8, 883
Mattress	C. Enscoe	New York, N. Y	Oct. 15, 1872	132, 268
Mattress	W. Foehlich	Harburg, Hanover	June 24, 1862	35, 732
Mattress	W. P. Ford	Cheneyville, La	July 14, 1857	17, 785
Mattress	J. J. Haley	Newton, Mass	Feb. 13, 1872	123, 628
Mattress	H. W. Henley	New York, N. Y	June 7, 1859	24, 302
Mattress	R. Krause	New York, N. Y	Apr. 4, 1865	47, 112
Mattress	J. Maas	Westfield, Mass	Dec. 12, 1871	121, 729
Mattress	A. G. Morey	Chicago, Ill	Oct. 13, 1868	82, 975
Mattress	W. H. Pack and J. S. Vanhorn	Jersey City, N. J	Feb. 28, 1871	112, 175
Mattress	W. H. Robertson	New London, Conn	Dec. 17, 1846	4, 896
Mattress	T. Roller	Allegheny City, Pa	Mar. 18, 1873	137, 030
Mattress	G. Schott	New York, N. Y	Jan. 21, 1868	73, 466
Mattress	H. E. Smith	Fitchburgh, Mass	May 20, 1873	139, 202
Mattress	B. F. Walton	Philadelphia, Pa	Oct. 1, 1867	69, 376
Mattress	W. F. Wehrmann	Cincinnati, Ohio	Sept. 24, 1872	131, 727
Mattress and bed	F. Elder	Winnsborough, S. C	July 10, 1860	29, 066
Mattress and chair spring	W. Hersee	Buffalo, N. Y	Nov. 10, 1857	18, 583
Mattress and cushion tacker	T. A. Watson and A. H. Phillips.	Brenham, Tex	July 2, 1872	128, 571
Mattress and life-preserving float	L. Bauhoefer	Philadelphia, Pa	Apr. 16, 1867	63, 831
Mattress, bolster, &c	A. Salisbury and J. Uram	Troy, N. Y	July 1, 1836	....
Mattress, Columnar	H. E. Smith	New York, N. Y	May 17, 1870	103, 096
Mattress, Earth	T. W. Phinney	Newport, R. I	Apr. 18, 1871	113, 931
Mattress, Floating	W. Williams	Saint Louis, Mo	Jan. 1, 1861	31, 048
Mattress, Folding	M. Sulzbacher	New York, N. Y	July 6, 1869	92, 396
Mattress, Folding	W. Wells	Harrisburgh, Pa	Apr. 27, 1859	20, 112
Mattress, Folding-box spring	S. P. Kittle	Brooklyn, N. Y	Dec. 7, 1869	97, 522
Mattress for invalids, Spring	D. Baird	New York, N. Y	Jan. 7, 1851	7, 882
Mattress frame, Elastic-fabric	J. E. Whittlesey and J. W. C. Peters.	Chicago, Ill	Sept. 16, 1873	142, 827
Mattress frame, Wire	G. V. and W. I. Bunker	Yankton, Dak	Dec. 16, 1873	145, 553
Mattress frame, Wire	G. C. Perkins	Hartford, Conn	Apr. 4, 1871	113, 559
Mattress frame, Wire	J. G. Smith	Saint Louis, Mo	Feb. 25, 1873	136, 271
Mattress frame, Woven-wire	G. C. Perkins	Hartford, Conn	Oct. 8, 1872	131, 968
Mattress frame, Woven-wire	D. J. Powers	Chicago, Ill	Feb. 25, 1873	136, 180
Mattress, Hair	W. F. Phyfe	New York, N. Y	Oct. 1, 1830	....
Mattress or seaman's bed	I. Martin	Boston, Mass	Sept. 17, 1833	....
Mattress, Palm-leaf	E. Howe	Cambridge, Mass	Sept. 13, 1834	....
Mattress, sofa, &c	R. Cromelon	New York, N. Y	Nov. 7, 1831	....
Mattress, Spring	E. L. Bushnell	Poughkeepsie, N. Y	Apr. 12, 1853	9, 658
Mattress, Spring	E. L. Bushnell	Poughkeepsie, N. Y	Oct. 15, 1872	132, 248
Mattress, Spring	C. and G. H. Chinnock	New York, N. Y	Aug. 20, 1872	130, 698
Mattress, Spring	R. T. Cornel	Poughkeepsie, N. Y	May 9, 1871	114, 648
Mattress, Spring	F. Forst	Philadelphia, Pa	June 16, 1846	4, 578
Mattress, Spring	C. Fulton	New York, N. Y	Oct. 11, 1870	108, 128
Mattress, Spring	H. Gardiner, R. Lowe, J. and J. Wood, and J. Pickering.	Manchester, Great Britain.	Oct. 24, 1871	120, 154
Mattress, Spring	H. A. Gaston	San José, Cal	Apr. 8, 1873	137, 673
Mattress, Spring	A. Gebhard	Indianapolis, Ind	Feb. 18, 1863	74, 528
Mattress, Spring	J. G. F. Grote	Cincinnati, Ohio	Apr. 19, 1864	42, 368
Mattress, Spring	J. G. F. Grote	Cincinnati, Ohio	Apr. 19, 1864	42, 369
Mattress, Spring	W. B. Judson	Poughkeepsie, N. Y	Sept. 13, 1870	107, 383
Mattress, Spring	H. Jury	New York, N. Y	Aug. 14, 1860	29, 600
Mattress, Spring	S. P. Kittle	Brooklyn, N. Y	Nov. 8, 1864	44, 9[illegible]0
Mattress, Spring	J. V. McElwee	Philadelphia, Pa	Apr. 2, 1850	7, 248

Index of patents issued from the United States Patent Office from 1790 *to* 1873, *inclusive*—Continued.

Invention.	Inventor.	Residence.	Date.	No.
Mattress, Spring	G. W. Mitchell	Saint Louis, Mo	Feb. 28, 1865	46, 573
Mattress, Spring	P. O'Neil	Philadelphia, Pa	Sept. 4, 1849	6, 686
Mattress, Spring	W. F. Ressegine	Cincinnati, Ohio	June 11, 1850	7, 429
Mattress, Spring	J. W. H. Scott	Rochester, N. Y	Apr. 14, 1868	76, 663
Mattress, Spring	D. N. Sellig	Newburgh, N. Y	Dec. 16, 1873	145, 597
Mattress, Spring	R. Stillwell	New York, N. Y	Sept. 20, 1864	44, 375
Mattress, Spring	C. Van Dyeck	Nashville, Tenn	Aug. 14, 1866	57, 223
Mattress, Spring	H. H. Vere	New York, N. Y	Aug. 13, 1867	67, 822
Mattress, Spring	J. Waters	Southwark, Pa	Jan. 20, 1852	8, 669
Mattress, Spring	L. Whitehead and S. P. Kittle	New York, N. Y	Aug. 28, 1860	29, 846
Mattress-spring	E. L. Bushnell	Poughkeepsie, N. Y	Oct. 19, 1869	95, 984
Mattress-spring	W. B. Judson	Poughkeepsie, N. Y	Sept. 19, 1871	119, 153
Mattress-stuffer	T. A. Watson	Brenham, Tex	Apr. 2, 1872	125, 233
Mattress-stuffing	F. A. Lane	Swanzey, N. H	May 25, 1869	90, 368
Mattress-stuffing machine	E. L. Wright	Sterling, Ill	Sept. 12, 1871	118, 892
Mattress-stuffing, Maching for cutting wood into shreds and crimping them for.	E. K. Browning	Utica, N. Y	July 15, 1851	8, 217
Mattress-stuffing, &c., Process of preparing moss for.	A. Rock	New Orleans, La	Jan. 7, 1873	134, 562
Mattress, Ventilating	T. Tolman	West Townsend, Mass	Dec. 15, 1857	18, 886
Mattress, Ventilating spring	R. W. Geraghty	Newark, N. J	Mar. 27, 1860	27, 667
Mattress, Wire	E. Parker	New Britain, Conn	May 20, 1873	139, 077
Mattress, Wire	D. J. Powers	Chicago, Ill	Nov. 11, 1873	144, 564
Mattress, Wire	J. G. Smith	Saint Louis, Mo	Dec. 2, 1873	145, 249
Mattress, Wire spring	F. W. Hoffmann	Morrisania, N. Y	Dec. 3, 1872	133, 533
Mattrêss, Wire spring	F. R. Wegman	Saxony	June 16, 1868	79, 040
Mattress, Woven-wire	G. C. Perkins	Hartford, Conn	Nov. 22, 1870	109, 446
Mattress, Woven-wire	J. W. C. Peters	Chicago, Ill	Oct. 15, 1872	132, 175
Mattresses and cusions, Elastic material for	T. B. Smith	Marietta, Ohio	Nov. 9, 1858	22, 037
Mattresses and making paper, Material for filling	F. C. Cone	San Francisco, Cal	Feb. 20, 1872	123, 810
Mattresses, Bracing and fastening spiral springs for.	O. Fuller	San Francisco, Cal	Oct. 25, 1864	44, 793
Mattresses, cushions, and other articles of upholstery, Material for filling.	W. H. Sowers	Boston, Mass	July 11, 1871	117, 018
Mattresses, cushions, &c., Making	C. L. Fleischmann	Washington, D. C	Sept. 4, 1847	5, 275
Mattresses, cushions, &c., Material for	B. Frodsham	New York, N. Y	Nov. 26, 1861	33, 777
Mattresses, Frame for wire	G. C. Perkins	Hartford, Conn	July 2, 1872	128, 652
Mattresses, Machine for cutting shavings for	F. Skinner	New Haven, Conn	Jan. 10, 1860	26, 791
Mattresses, Machine for hackling shucks for	D. A. Cole	Nashville, Tenn	Feb. 9, 1869	86, 734
Mattresses, Machine for preparing husks for	W. Nash and M. Kenfield	Malden, Mich	Oct. 27, 1868	83, 402
Mattresses, Machinery for preparing husks for	A. Olcott	Newark, N. J	Nov. 20, 1847	5, 373
Mattresses, &c., Material for filling	W. J. Woodley	San Francisco, Cal	Aug. 6, 1872	130, 171
Mattresses, Material for filling	B. Harrington	China, Me	Jan. 23, 1872	123, 016
Mattresses, &c., Material for stuffing	H. R. Hildreth and W. H. Smith.	Dutch Flat, Cal	Jan. 15, 1867	61, 194
Mattresses, &c., Material for stuffing	R. J. Kellett	San Francisco, Cal	Aug. 6, 1872	130, 141
Mattresses, pillows, &c., Stuffing for	A. C. Crondal	New York, N. Y	Sept. 22, 1863	40, 024
Mattresses, Process for treating moss for	S. Barker	New York, N. Y	Apr. 7, 1857	16, 961
Mattresses, seats, backs of sofas, &c., from paper pulp, sponge, &c., Manufacture of.	J. L. Kendall	Foxborough, Mass	Aug. 6, 1872	130, 223
Mattresses, seats, &c., Flexible support for	C. Pullinger	Philadelphia, Pa	Apr. 8, 1873	137, 718
Mattresses, seats, &c., Spring-coupling for	W. B. Judson	New York, N. Y	Jan. 14, 1873	134, 893
Mattresses, sofas, &c., from paper pulp, sponge, &c., Manufacture of.	J. L. Kendall	Foxborough, Mass	Aug. 6, 1872	130, 222
Mattresses, sofas, seats, &c., Stuffing for	J. M. Gilbert	Troy, N. Y	Feb. 11, 1868	74, 340
Mattresses, Stuffing for	H. A. Alden	Matteawan, N. Y	Feb. 16, 1864	41, 589
Mattresses, Tightening-device for woven-wire	D. J. Powers	Chicago, Ill	Sept. 16, 1873	142, 940
Mattresses, &c., Treating cork for	L. Bauhoefer	Philadelphia, Pa	Nov. 6, 1866	59, 342
Mattresses, &c., Treating moss for	C. G. Augeroth, jr	Philadelphia, Pa	Aug. 30, 1864	43, 962
Mattresses, &c., Treating sponge for stuffing	R. O. Doremus	New York, N. Y	Nov. 13, 1866	59, 714
Maxiliary compress	C. E. Davis	Saint Helena, Cal	Sept. 6, 1870	107, 011
Maze-lock	T. Nicholson	Falmouth, Va	Sept. 30, 1851	8, 396
Mead, Carbonated	J. Heaton	New York	July 28, 1819	....
Mead, Composition of	T. T. Kimball and A. H. White	Dedham, Mass	Sept. 26, 1835	....
Mead, Making carbonated sarsaparilla	J. S. Schieffelin	New York	July 19, 1824	....
Mead, Portable sarsaparilla	J. C. Brigham	Methuen, Mass	July 25, 1833	....
Meal and flour dryer	H. Cutter	Ashland, Mass	Sept. 7, 1869	94, 574
Meal and flour making machine	G. W. Holman	Beloit, Wis	Feb. 1, 1859	22, 806
Meal and flour mill	J. C. Smith	Lacon, Ill	Sept. 29, 1837	396
Meal, Cooling and drying	J. Denchfield	Oswego, N. Y	Apr. 20, 1858	19, 984
Meal-separator	W. Hawkins	Brooklyn, N. Y	June 8, 1869	91, 128
Measure, Adjustable	H. Kraut	Saint Louis, Mo	Oct. 16, 1866	58, 838
Measure and funnel combined	H. L. Dickenson	East Berlin, Conn	May 14, 1867	64, 752
Measure and funnel, Combined	S. R. Dummer	New York, N. Y	Apr. 5, 1864	42, 174
Measure and funnel, Combined	J. Fanyon	Providence, R. I	May 4, 1869	89, 569
Measure and funnel, Combined	E. Grattem	Williamstown, Mich	Nov. 13, 1866	59, 590
Measure and funnel, Liquid	J. Huff	Ironton, Mo	Aug. 23, 1870	106, 588
Measure and weigher, Combined	A. B. Hurd	Watkins, N. Y	Oct. 6, 1868	82, 842
Measure, Beer	E. Bagot	New York, N. Y	Jan. 10, 1860	26, 807
Measure, Beer	C. Chinnock	Brooklyn, N. Y	Apr. 15, 1862	34, 943
Measure, Board	A. D. Hoffman	Belleville, Mich	Feb. 24, 1863	37, 755
Measure, Board	G. S. Tiffany	Palmyra, Mich	July 25, 1865	49, 011
Measure, Cloth	C. V. Hemenway	New London, Ohio	Feb. 11, 1873	135, 804
Measure, Dry	J. H. Teabl	Eberly's Mills, Pa	Jan. 5, 1869	85, 706
Measure, Foot	J. C. F. Bremser	Saint Louis, Mo	Aug. 16, 1870	106, 314
Measure for cutting dresses	S. R. Windle	Chillicothe, Ohio	Mar. 31, 1868	76, 128
Measure for cutting garments	S. Severson	Baltimore, Md	Feb. 15, 1826	....
Measure for gaging the liquid contents of vessels, Tape.	H. E. Wrigley	Titusville, Pa	July 16, 1872	129, 639
Measure for laying out garments, Tailor's	F. A. Fairchild	Columbus, Ga	Oct. 31, 1835	....
Measure for stock, Feed	C. E. Spaulding	Theresa, N. Y	Sept. 28, 1869	95, 280
Measure for the human body	G. Beard	Salineville, Ohio	July 11, 1865	48, 644
Measure, Funnel	J. J. Hillman	Boston, Mass	July 28, 1863	39, 346
Measure, funnel, and faucet, Combined	G. W. McCann	Springfield, Ohio	July 2, 1867	66, 371
Measure, funnel, and faucet, Combined	H. Mitchell	Richmond, Ind	Mar. 7, 1865	46, 690
Measure, Grain	J. Brown	Lawn Ridge, Ill	Dec. 28, 1858	22, 411
Measure, Grain	N. M. Lurton	Newbern, Ill	Sept. 3, 1867	68, 761
Measure, Grain	E. O. Melvin	Brooklyn, Wis	Jan. 7, 1868	73, 112
Measure, Grain	J. A. Rosbeck	Hermon, N. Y	Aug. 3, 1869	93, 231

Index of patents issued from the United States Patent Office from 1790 *to* 1873, *inclusive*—Continued.

Invention.	Inventor.	Residence.	Date.	No.
Measure, Grain	H. C. Smith	Chicago, Ill	June 25, 1867	66, 050
Measure, Hat	J. Wehle	New York, N. Y	Aug. 16, 1859	25, 154
Measure, Hatter's head	H. Bonham	Philadelphia, Pa	Aug. 2, 1870	105, 895
Measure, Hatter's head	F. A. Fenton	Camden, N. J	Nov. 18, 1873	144, 606
Measure, Heel	J. T. Siegert	Washington, D. C	Aug. 6, 1867	67, 598
Measure-holder	G. W. Burwell	Zanesville, Ohio	July 20, 1869	92, 791
Measure, Liquid	J. L. Abbott	North Providence, R. I	Sept. 17, 1867	68, 931
Measure, Liquid	P. Brinkerhoff	Chillicothe, Mo	June 4, 1867	65, 335
Measure, Liquid	P. Brinkerhoff	Chillicothe, Mo	Aug. 20, 1867	67, 948
Measure, Liquid	J. Dailey	Albany, N. Y	Aug. 14, 1866	57, 097
Measure, Liquid	T. W. Ellis	Macon, Ga	June 14, 1870	104, 129
Measure, Liquid	A. Hicks	Flushing, N. Y	Feb. 16, 1864	41, 620
Measure, Liquid	S. Mainster and J. F. Kirkwood	Thistle, Md	Mar. 12, 1867	62, 762
Measure, Liquid	M. McDevitt	Hampton, Va	Mar. 1, 1870	100, 367
Measure, Liquid	T. Rogers	Montgomery Square, Pa	Feb. 25, 1862	34, 526
Measure, Liquid	W. Sprague	Sandy Creek, N. Y	July 28, 1868	80, 370
Measure, Liquid	J. S. Tough	Baltimore, Md	July 23, 1841	2, 187
Measure, Liquid	J. Trent	Millerton, N. Y	Jan. 1, 1867	60, 806
Measure, Liquid	W. L. Wever and A. W. Johnson	Plattsburgh, N. Y	Dec. 9, 1873	145, 377
Measure, Liquid	W. M. Wright	Chambersburgh, Pa	Mar. 28, 1871	113, 237
Measure, Lumber	F. S. Baldwin	Saint Louis, Mo	Apr. 29, 1873	138, 310
Measure, Lumber	A. M. Olds	Chicago, Ill	Aug. 9, 1864	43, 789
Measure, Lumber	A. M. Olds	Chicago, Ill	Jan. 3, 1865	45, 741
Measure, Lumber	W. S. Poulson	Cadiz, Ohio	Feb. 20, 1872	123, 931
Measure, Lumber	W. E. Walson	Chester, Pa	Nov. 7, 1871	120, 685
Measure, Pantaloon	E. Grimston	Danvers, Mass	Sept. 21, 1837	388
Measure, Rotary disk	E. V. Lawrence	Brooklyn, N. Y	Nov. 25, 1862	37, 002
Measure, Self-counting	E. Dexter	Holmes' Hole, Mass	May 13, 1856	14, 857
Measure, Self-registering lumber	D. E. Pease and C. Richards	Richland Centre, Wis	Mar. 23, 1869	88, 203
Measure, Shoemaker's	M. Lane	Washington, D. C	Sept. 17, 1861	33, 307
Measure, Shoemaker's foot	C. Cross	Louisville, Ky	May 11, 1869	89, 919
Measure, Shoemaker's foot	J. McNichol	Pontiac, Ill	June 6, 1871	115, 761
Measure, Shoemaker's foot	W. Wilson	Boston, Mass	Aug. 22, 1871	118, 414
Measure, Spring tape	W. H. Bangs, jr	West Meriden, Conn	Dec. 6, 1864	45, 372
Measure, Surveyor's	W. H. Paine	Sheboygan, Wis	Aug. 7, 1860	29, 514
Measure, Tailor's	W. R. Acton	Richmond, Va	Sept. 5, 1846	4, 742
Measure, Tailor's	W. W. Allen	Bordentown, N. J	Sept. 10, 1850	7, 641
Measure, Tailor's	G. Beard	Salineville, Ohio	Mar. 6, 1866	52, 950
Measure, Tailor's	G. Beard	Philadelphia, Pa	Mar. 6, 1866	52, 951
Measure, Tailor's	J. Beaudry	Montreal, Canada	Oct. 14, 1873	143, 556
Measure, Tailor's	A. A. Bogardus	Newburgh, N. Y	Nov. 29, 1845	4, 294
Measure, Tailor's	J. Carpenter	Uniontown, Pa	Apr. 10, 1849	6, 286
Measure, Tailor's	J. P. Combs	Trenton, N. J	Nov. 9, 1844	3, 820
Measure, Tailor's	R. Dame	Hanover, N. H	May 8, 1840	1, 584
Measure, Tailor's	L. Derby	New York, N. Y	Jan. 27, 1857	16, 472
Measure, Tailor's	H. Donges	Newport, Pa	Apr. 25, 1846	4, 477
Measure, Tailor's	I. Du Bois	Brooklyn, N. Y	May 23, 1871	115, 180
Measure, Tailor's	G. Ecklee and S. X. Ball	Flint Creek, N. Y	Sept. 28, 1843	3, 286
Measure, Tailor's	G. W. T. Harley	Frederick, Md	Feb. 13, 1866	52, 566
Measure, Tailor's	G. W. T. Harley	Frederick, Md	Jan. 30, 1872	123, 170
Measure, Tailor's	H. Isham	Montpelier, Vt	May 30, 1844	3, 603
Measure, Tailor's	J. A. Johnston	Montreal, Canada	July 1, 1873	140, 507
Measure, Tailor's	C. Kile	Nashville, Ohio	June 27, 1846	4, 596
Measure, Tailor's	L. P. Lemont	Bath, Me	Jan. 9, 1829	......
Measure, Tailor's	W. and C. F. Lillibridge	Zanesville, Ohio	Oct. 31, 1854	11, 868
Measure, Tailor's	J. Maginnis	Lockport, N. Y	Dec. 2, 1851	8, 566
Measure, Tailor's	B. G. Martin	Richmond, Va	Oct. 29, 1846	4, 831
Measure, Tailor's	L. B. and E. Miller	Wallkill, N. Y	May 29, 1841	2, 106
Measure, Tailor's	C. Morey and D. Hummer	McArthurstown, Ohio	July 8, 1843	3, 161
Measure, Tailor's	I. Moses	New York, N. Y	May 24, 1870	103, 487
Measure, Tailor's	T. Oliver	New York, N. Y	Mar. 30, 1843	3, 024
Measure, Tailor's	M. T. Rowlands	Pittston Ferry, Pa	Apr. 18, 1854	10, 779
Measure, Tailor's	M. G. Simril	Chesterville, S. C	Jan. 7, 1847	4, 923
Measure, Tailor's	W. Sinnott and J. McNaughton	Brooklyn, N. Y	June 23, 1868	79, 083
Measure, Tailor's	W. R. Stace	Rochester, N. Y	July 6, 1858	20, 826
Measure, Tailor's	A. Stocker	Ogdensburgh, N. Y	May 28, 1850	7, 402
Measure, Tailor's	A. Stocker	Rome, N. Y	Sept. 30, 1856	15, 824
Measure, Tailor's	E. Virtue	Philadelphia, Pa	Dec. 16, 1851	8, 600
Measure, Tailor's	A. Ward	Camden, N. J	Dec. 26, 1845	4, 327
Measure, Tailor's	T. Watt	Hubbard Township, Ohio	Feb. 20, 1847	4, 975
Measure, Tailor's	W. T. Wells	Shelbyville, Tenn	Apr. 20, 1852	8, 895
Measure, Tailor's	J. W. Whitham	Washington, D. C	Nov. 6, 1849	6, 856
Measure, Tape	L. P. Bradley	New Haven, Conn	July 13, 1869	92, 573
Measure, Tape	J. A. Evarts	West Meriden, Conn	Feb. 13, 1872	123, 624
Measure, Tape	A. J. Fellows	New Haven, Conn	July 14, 1868	79, 965
Measure, Tape	H. B. Tyler	West Meriden, Conn	July 8, 1873	140, 745
Measure, Variable	L. Coates	Collamer, Pa	Oct. 30, 1866	59, 186
Measure, Yard	J. Douglass	McConnellstown, Pa	Dec. 3, 1867	71, 719
Measures, Apparatus for tapering	A. Hatch	New Haven, Conn	Feb. 19, 1867	62, 267
Measures, Construction of funnel-attachments for liquid.	C. C. Judwin	Honesdale, Pa	Mar. 12, 1872	124, 492
Measures, Forming sheet-metal	J. Coover	Chambersburgh, Pa	Oct. 22, 1872	132, 386
Measures, Machine for forming shell for dry	J. Gage	Henniker, N. H	Sept. 7, 1869	94, 591
Measures, Machine for graduating linear	S. C. Hubbard	Middletown, Conn	June 16, 1857	17, 606
Measures, Manufacture of graduated glass	W. Hodgson, jr	Philadelphia, Pa	Feb. 18, 1862	34, 424
Measurements and laying out garments, Pattern for applying	S. C. Ewing	Indianapolis, Ind	Feb. 21, 1872	112, 024
Measurements, Apparatus for electric	L. Bradley	Jersey City, N. J	Jan. 7, 1873	134, 636
Measurements, Device for taking tailor's	J. Smith	Burton, Ohio	Aug. 6, 1872	130, 161
Measurer and register, Grain	R. B. Clark	Lyle, Minn	Mar. 11, 1873	136, 643
Measurer and register, Grain	J. A. Cleuxton	Bentonville, Ohio	Feb. 26, 1861	31, 530
Measurer, Board	A. Lafever	Battle Creek, Mich	Sept. 13, 1859	25, 423
Measurer, Fluid	J. L. Perry and M. Burt	Mansfield and Morton, Mass.	Apr 19, 1859	23, 701
Measurer, Grain	J. W. Smith and H. Griesa	North Cohocton, N. Y	Sept. 7, 1869	94, 659
Measurer, Self-regulating grain	G. W. Atkins	Milton, Del	May 11, 1858	20, 186
Measurer, Ship's model	A. S. Hosley	New York, N. Y	Aug. 19, 1851	8, 307
Measurer, Standard coat	C. Barber	Boston, Mass	Dec. 26, 1837	539

Index of patents issued from the United States Patent Office from 1790 *to* 1873, *inclusive*—Continued.

Invention.	Inventor.	Residence.	Date.	No.
Measurer, Tailor's	J. S. Rockafellow	Flemington, N. J	Sept. 18, 1835	....
Measuring altitudes, &c., Instrument for	G. C. Ayling	Boston, Mass	Dec. 21, 1858	22, 396
Measuring and cutting garments	J. Henderson and C. K. Watson	Zanesville, Ohio	Oct. 22, 1831	....
Measuring and cutting garments	W. and C. Kahler	Bloomsburgh, Pa	Jan. 20, 1838	574
Measuring and cutting garments	D. L. Pendill	Gilboa, N. Y	June 14, 1843	3, 130
Measuring and cutting garments	J. Pudney	Waterford, N. Y	Sept. 3, 1831	....
Measuring and cutting garments	H. Seger	Macon, Ga	Apr. 29, 1842	2, 590
Measuring and cutting garments	J. G. Wilson	New York	May 3, 1833	....
Measuring and cutting out dress-waists, Pattern for.	J. M. Lent	Schuyler's Lake, N. Y	June 22, 1869	91, 642
Measuring and cutting out dresses, System of	J. Malnight	Grass Lake, Mich	Nov. 19, 1867	71, 192
Measuring and drafting garments	I. J. Hendryx	Troy, N. Y	Apr. 18, 1840	1, 557
Measuring and folding cloth	L. Hillman	Newton, N. J	Aug. 10, 1869	93, 619
Measuring and folding machine, Cloth	L. Harvey	Newcastle, Ind	July 16, 1872	129, 554
Measuring and laying out garments	J. Lenley, jr	Newtown, Va	Nov. 26, 1867	71, 520
Measuring and laying out garments	H. Wengel	Philadelphia, Pa	May 5, 1868	77, 704
Measuring and laying out garments, Device for	I. J. Ordway	West Edmeston, N. Y	Jan. 4, 1870	98, 618
Measuring and laying out garments, Pattern for	F. Wetmore	Chicago, Ill	Dec. 13, 1870	110, 097
Measuring and marking, Combination-tool for	C. M. Lane	Cincinnati, Ohio	June 16, 1868	78, 974
Measuring and marking out garments	A. Ward	Philadelphia, Pa	Jan. 7, 1835	....
Measuring and recording by the tape	E. A. Preston	Battle Creek, Mich	Dec. 7, 1858	22, 241
Measuring and registering device, Liquid	A. Werrkmeister	Charlottenburg, Prussia	Mar. 1, 1870	100, 475
Measuring and registering grain	W. H. Pfrimmer	Lanesville, Ind	May 5, 1868	77, 521
Measuring and sacking grain	D. M. Cochran	Richmond, Ind	Sept. 3, 1861	33, 186
Measuring and transferring liquid	T. W. Whitley	Paterson, N. J	Feb. 1, 1833	....
Measuring and weighing apparatus	N. Smith	Lansing, Iowa	May 12, 1863	38, 511
Measuring and weighing grain, Machine for	C. A. Postley	Philadelphia, Pa	Aug. 7, 1855	13, 397
Measuring and winding machine, Cloth	J. E. Race and H. Whitney	Chicago, Ill	June 15, 1869	91, 264
Measuring apparatus, Silk and thread	L. Dimock	Leeds, Mass	Apr. 25, 1871	114, 114
Measuring-apparatus, Tailor's	F. Mueller and H. Koeller	New York, N. Y	July 20, 1869	92, 873
Measuring apparatus, Thread	L. F. Dunn	Oneida, N. Y	Dec. 26, 1871	122, 236
Measuring-attachment for packaged fabrics	E. Morgan	Washington, D. C	Jan. 24, 1871	111, 235
Measuring boards, Apparatus for	I. J. W. Adams	Salisbury, Md	Apr. 19, 1870	101, 967
Measuring boards, Implement for	M. Tuells	Penn Yan, N. Y	May 10, 1870	102, 885
Measuring boards, Instrument for	J. Jones	Rochester, N. Y	Jan. 6, 1857	16, 359
Measuring boards, Machine for	W. S. Finley	Fond-du-Lac, Wis	Aug. 6, 1872	130, 120
Measuring-can	D. H. Sumner	South Boston, Mass	Sept. 17, 1867	68, 913
Measuring-can	A. W. Tice and N. F. Smalley	Boston, Mass	Nov. 18, 1873	144, 713
Measuring-can	A. G. Wilson	Chicago, Ill	Nov. 15, 1870	109, 283
Measuring-can, Automatic	T. D. Arkle and H. C. Greer	Bridgeport, Ohio	Sept. 3, 1867	68, 480
Measuring-can for liquids	W. Barry	Carthage, N. Y	June 8, 1869	90, 916
Measuring-can for liquids	C. M. Bridge	Leon, Iowa	Sept. 19, 1871	119, 114
Measuring-can for liquids	J. S. Gold	Springfield, Ill	May 11, 1869	90, 044
Measuring-can for oils, &c	W. B. Sherman	Clinton, N. Y	Feb. 18, 1873	136, 103
Measuring cloth, &c	A. Smith	Perrysburgh, Ohio	June 19, 1847	5, 167
Measuring cloth, Apparatus for	T. Bisbing	Buckstown, Pa	Feb. 23, 1869	87, 134
Measuring cloth, Apparatus for	T. M. Brintnall	Medina, Ohio	Apr. 19, 1870	101, 979
Measuring cloth, Apparatus for	R. H. Kent	Middlebury, Ohio	Nov. 24, 1868	84, 281
Measuring cloth, Apparatus for	S. B. Luckett	Corydon, Ind	Nov. 24, 1868	84, 290
Measuring cloth, Apparatus for	S. B. Luckett	Corydon, Ind	Mar. 1, 1870	100, 425
Measuring cloth, Apparatus for	G. R. McIntire	Houghton, Mich	Apr. 28, 1868	77, 305
Measuring cloth, Apparatus for	J. E. Race and A. Smith	Chicago, Ill	Dec. 15, 1868	84, 603
Measuring cloth, Apparatus for	C. L. Shotnell	Allamuchy, N. J	May 26, 1868	78, 399
Measuring cloth, Device for	O. R. Holbrook	Staffordville, Conn	Sept. 16, 1873	142, 787
Measuring cloth in the piece or roll, Device for	W. Beaton	Grinnell, Iowa	Nov. 22, 1864	45, 131
Measuring cloth, Machine for	J. I. Benham	New London, Conn	Feb. 6, 1866	52, 378
Measuring cloth, Machine for	W. F. Boggs	Petersburgh, Ill	Mar. 24, 1868	75, 848
Measuring cloth, Machine for	J. M. Brintnall	Medina, Ohio	Sept. 26, 1871	119, 303
Measuring cloth, Machine for	J. W. Drummond	New York, N. Y	May 31, 1859	24, 260
Measuring cloth, Machine for	W. Hebdon	New York, N. Y	May 18, 1869	90, 262
Measuring cloth, Machine for	D. Max	Newton, Ill	Sept. 8, 1868	81, 921
Measuring cloth, Machine for	C. M. Swany	Philadelphia, Pa	Dec. 15, 1863	40, 962
Measuring cloth, Machine for	T. Weeden and T. Tribe	Hillsdale, Mich	Sept. 6, 1870	107, 205
Measuring cloth, Machine for	W. W. Wythes	Saint Clair, Pa	Sept. 29, 1857	18, 313
Measuring cloth on looms, Mechanism for	H. Halvorson	Boston, Mass	Aug. 1, 1854	11, 435
Measuring cloth in the cloth-beam, Method of	W. H. Woodworth	Salmon Falls, N. H	Dec. 21, 1852	9, 496
Measuring coats	T. E. Tilden	Baltimore, M. D	Dec. 5, 1840	1, 880
Measuring-device for horses' feet	H. B. Ferren	Batavia, N. Y	Sept. 15, 1868	82, 213
Measuring-device for horses' feet	H. B. Ferren	Batavia, N. Y	Sept. 15, 1868	82, 214
Measuring device, Skirt	L. G. Rice	Montague, Mass	Dec. 1, 1868	84, 579
Measuring distances	R. Porter	Billerica, Mass	June 11, 1836	....
Measuring distances from a single station, Instrument for.	E. A. Crandall	Friendship, N. Y	Aug. 26, 1856	15, 602
Measuring distances, Instrument for	G. Achelis and H. Poppenhusen	Westchester, Pa., and New York, N. Y.	Mar. 24, 1868	75, 824
Measuring distances, Instrument for	W. H. Chaffee	Flint, Mich	Apr. 8, 1862	34, 879
Measuring distances, Instrument for	J. J. Daly	New Orleans, La	Mar. 27, 1866	53, 420
Measuring distances, Instrument for	J. Frankle	Amesbury, Mass	Sept. 25, 1866	58, 241
Measuring distances, Parallactic instrument for	H. L. Hervey	Quincy, Ill	June 3, 1856	15, 040
Measuring distances, Parallactic instrument for	W. Würdemann	Washington, D. C	Sept. 11, 1849	6, 709
Measuring, drafting, and cutting garments	W. W. Allen	Philadelphia, Pa	Oct. 23, 1837	435
Measuring dry goods, Instrument for	T. A. Shinn	Baden, Pa	Aug. 27, 1867	68, 245
Measuring fluids, Apparatus for	J. C. Horton and S. K. Hawkins	New York and Lansingburgh, N. Y.	Apr. 2, 1867	63, 388
Measuring fluids, Apparatus for	E. G. Shortt	Carthage, N. Y	June 29, 1869	92, 110
Measuring fluids, Machine for	J. Bogardus	New York, N. Y	Oct. 12, 1837	425
Measuring fluids while drawing, Method of	S. Kranser	Reading, Pa	Oct. 7, 1856	15, 853
Measuring for garments	G. W. M. Bacon	New York, N. Y	Apr. 20, 1833	....
Measuring for garments	J. Mendenhall	Westchester, Pa	Apr. 16, 1833	....
Measuring for garments	D. Williams	New York, N. Y	May 22, 1833	....
Measuring garments	W. C. Bishop	Ovid, N. Y	Apr. 25, 1837	179
Measuring garments	J. B. West	New York, N. Y	Jan. 3, 1865	45, 780
Measuring garments, Instrument for	H. C. Brundage	Middletown, N. Y	Feb. 7, 1842	2, 450
Measuring garments, Instrument for	J. and J. F. Knowland	Brownsborough, Ky	July 20, 1842	2, 730
Measuring garments, Instrument for	P. F. L. Verot	Warrenton, Ga	May 26, 1842	2, 640
Measuring garments, Instrument for	D. Williams	New York, N. Y	Apr. 26, 1839	1, 136
Measuring grain	D. Murray	Fairfield, N. C	Aug. 23, 1859	25, 212
Measuring grain and seed	P. H. Schenck	New York, N. Y	Feb. 15, 1810	....

Index of patents issued from the United States Patent Office from 1790 to 1873, inclusive—Continued.

Invention.	Inventor.	Residence.	Date.	No.
Measuring grain, Machine for	C. Ross	Hartland, Mich	Sept. 15, 1863	39,959
Measuring grain, Machine for	T. Wilson	Matamora, Ill	Sept. 15, 1863	39,980
Measuring grain, Tally-box for	A. J. Burke	Grundy Centre, Iowa	Oct. 9, 1866	58,593
Measuring heights and distances	B. Wilmer	Baltimore, Md	Aug. 20, 1819	....
Measuring heights and distances	C. Wilson	Clinton, Pa	Sept. 29, 1868	82,669
Measuring-instrument, Tailor's	J. B. Barnett and F. Story	Catskill, N. Y	Nov. 12, 1839	1,406
Measuring-instrument, Tailor's	J. M. Krider	Newtown Stephensburgh, Va.	Oct. 31, 1854	11,866
Measuring-instrument, Tailor's	J. M. Krider	Madison Court-House, Va	Aug. 13, 1867	67,774
Measuring-instrument, Tailor's	W. J. Lemmond	Lancaster, S. C	Apr. 18, 1840	1,556
Measuring, laying out, and cutting garments, Device for.	G. P. Sweezy	Riverhead, N. Y	Apr. 20, 1869	89,091
Measuring liquids	G. W. Devoe	New York, N. Y	Dec. 11, 1866	60,344
Measuring liquids, Apparatus for	F. H. Houghton	Brooklyn, N. Y	May 8, 1866	54,548
Measuring liquids, Apparatus for	H. James	Barclay, Ill	June 5, 1860	28,581
Measuring liquids, Apparatus for	P. Noyes	Lowell, Mass	Nov. 2, 1869	96,340
Measuring liquids, Apparatus for	J. C. Rankin	Mount Vernon, N. Y	June 12, 1860	28,690
Measuring liquids, Apparatus for	J. Schmidt	Vienna, Austria	Dec. 26, 1871	122,195
Measuring liquids, Apparatus for	E. F. Wilder	Lowell, Mass	Mar. 22, 1870	101,069
Measuring liquids, Apparatus for	M. H. Wiley and T. Miller	Boston, Mass	May 24, 1870	103,404
Measuring liquids, Device for	A. Fickett and J. C. Ware	Titusville, Pa	June 18, 1867	65,897
Measuring liquids, Device for	R. F. Fisher and S. L. Bell	Boston and Wellfleet, Mass	June 4, 1872	127,410
Measuring lumber, Instrument for	C. Fleming	Ypsilanti, Mich	Feb. 5, 1861	31,308
Measuring lumber, Instrument for	C. Littlefield	Kennebunk, Me	July 21, 1868	80,077
Measuring lumber, Instrument for	T. B. Stevenson	Dayton, Ohio	Sept. 17, 1867	68,911
Measuring packages, Rod for	D. Stanbury	Laurel Hill, N. J	Oct. 26, 1818	....
Measuring register, Cloth	S. Crocker	Port Allen, Iowa	Apr. 15, 1873	137,829
Measuring register, Land	W. Frasius	Chatsworth, Ill	May 23, 1871	115,186
Measuring register, Liquid	M. Springwater	Evansville, Ind	Mar. 25, 1873	137,155
Measuring-rod	A. D. King	Granville Corners, Mass	Mar. 12, 1867	62,756
Measuring-rod and dividers, Combined extension	G. H. Disher	Mobile, Ala	Nov. 4, 1873	144,264
Measuring sails, Instrument for	J. Dominis	Sandwich Islands	Sept. 30, 1842	2,790
Measuring-scale	J. J. Hendryx	Bennington, Vt	May 29, 1822	....
Measuring scale, Log	L. Smith	Lowell, Mass	June 14, 1870	104,368
Measuring-tape	O. W. Minard	Waterbury, Conn	May 1, 1860	28,101
Measuring the body of garments, Pattern for	M. Palmer, jr., and E. W. Anderson.	Boston, Mass., and Washington, D. C.	July 19, 1870	105,486
Measuring the foot and leg and cutting leather for boots.	T. Howe	Worcester, Mass	Apr. 18, 1820	....
Measuring the length of braces in carpentry, Instrument for.	H. Whipple	South Shaftsbury, Vt	Feb. 26, 1856	14,329
Measuring the person and cutting out ladies' dresses, Guide for.	L. L. Jackson	Richmond, Ind	May 25, 1869	90,363
Measuring the superfices of boards, Machine for	S. C. Kennard	South Newmarket, N. H	Jan 1, 1858	19,031
Measuring-wheel	T. R. Way	Springfield, Ohio	Jan. 21, 1873	135,185
Measuring-wheel for wagon-tires	T. Whiting	Niles, Mich	May 13, 1873	138,829
Measuring-wheel, Revolving	L. Young	New York, N. Y	Nov. 20, 1855	13,833
Meat and brine, Device for keeping	J. Burgum	Concord, N. H	June 12, 1866	55,463
Meat, and for other purposes, Portable apparatus for curing.	H. J. Noyes and S. C. Talcott	Ashtabula, Ohio	Jan. 21, 1870	104,632
Meat and other articles, Machine for cutting	P. Vande Sande	Rochester, N. Y	Nov. 3, 1868	83,676
Meat and other biscuit, Manufacture of	J. Carr and C. Lucop	Clapham and Drummond Road, England.	Apr. 21, 1868	76,889
Meat and vegetable chopper	L. S. Chichester	New York, N. Y	Aug. 28, 1860	29,766
Meat and vegetable chopper	S. F. Emerson	Seville, Ohio	Feb. 1, 1870	99,301
Meat and vegetable chopper	W. H. Peirce	Bangor, Me	May 9, 1871	114,594
Meat and vegetable chopper	E. Smith	Claremont, N. H	Apr. 2, 1867	63,566
Meat and vegetable chopping machine	H. W. Russell	Stoughton, Mass	Feb. 6, 1866	52,449
Meat and vegetable cutter	P. Bonfils and J. Grossman	West Hoboken, N. J	Sept. 20, 1870	107,592
Meat and vegetable cutter	G. E. Bringman	Philadelphia, Pa	Dec. 13, 1870	110,111
Meat and vegetable cutter	E. Covert	Farmers' Village, N. Y	June 6, 1871	115,579
Meat and vegetable cutter	M. L. Edwards	Salem, Ohio	June 3, 1873	139,501
Meat and vegetable cutter	J. R. Hall and C. H. Ellis	Salem, Ohio	July 15, 1873	140,824
Meat and vegetable cutter	J. Shepard	Bristol, Conn	Apr. 24, 1866	54,223
Meat and vegetable cutter	A. R. Silver	Salem, Ohio	July 11, 1871	117,003
Meat and vegetable cutter	J. Stepp	Somerville, Mass	Aug. 13, 1872	130,544
Meat and vegetable cutter	E. P. Whitney	Stamford, Conn	July 10, 1866	56,305
Meat and vegetable cutter and grater	E. Culver	Shelburne, Mass	Aug. 16, 1864	43,837
Meat and vegetable cutting machine	S. Blim	Salem, Ohio	Dec. 31, 1872	134,458
Meat and vegetable masher	G. A. Anderson and C. J. Baker	Albany, N. Y	Oct. 11, 1870	108,084
Meat, Apparatus for slaughtering and curing	G. W. Fulton	Fulton, Tex	June 29, 1869	92,035
Meat-baker	W. K. Wyckoff	Ripon, Wis	Sept. 17, 1872	131,383
Meat-biscuit	A. Kirkwood	Toronto, Canada	May 28, 1871	113,063
Meat-block	L. H. Ives	Syracuse, N. Y	May 18, 1869	90,104
Meat-broiler	G. B. Ransom	Chester, Conn	Jan. 20, 1863	37,460
Meat, Building for packing	D. E. Somes	Biddeford, Me	Nov. 27, 1860	30,770
Meat by steam, Apparatus for cooking	G. H. Munroe	New York, N. Y	Apr. 26, 1870	102,302
Meat chest or bin	W. B. Smith	Oskaloosa, Iowa	Oct. 18, 1870	108,403
Meat-chopper	F. Ashley	New York, N. Y	May 1, 1866	54,274
Meat-chopper	D. W. Baker	Harwich, Mass	May 15, 1866	54,667
Meat-chopper	J. Battey	Manchester, N. J	Jan. 28, 1873	135,258
Meat-chopper	C. B. Becker	Lancaster County, Pa	May 8, 1860	28,145
Meat-chopper	J. H. Blake	Westbrook, Me	Sept 28, 1869	95,309
Meat-chopper	C. N. Brumm	Minersville, Pa	Mar. 23, 1869	88,126
Meat-chopper	P. H. Chesley	Lynn, Mass	Mar. 23, 1858	19,681
Meat-chopper	P. Clareton	New York, N. Y	June 8, 1869	90,994
Meat-chopper	A. J. Curtis	Monroe, Me	Aug. 11, 1868	80,811
Meat-chopper	J. H. De Poe	Boonton, N. J	July 5, 1870	104,940
Meat-chopper	A. F. Doebert	Lancaster, N. Y	Aug. 20, 1867	67,854
Meat-chopper	L. A. Dole	Salem, Ohio	May 18, 1858	20,253
Meat-chopper	J. A. Eberly	Reamstown, Pa	Sept. 20, 1870	107,466
Meat-chopper	M. L. Edwards	Salem, Ohio	Aug. 15, 1871	117,999
Meat-chopper	C. H. Finson	Bangor, Me	June 1, 1869	90,737
Meat-chopper	C. A. Foster	Winchendon, Mass	June 5, 1866	55,414
Meat-chopper	O. Gardner	New York Mills, N. Y	July 8, 1873	140,696
Meat-chopper	C. L. Gilpatric	Boston, Mass	June 30, 1868	79,465
Meat-chopper	J. L. Good	Elizabethtown, Pa	May 16, 1871	114,799
Meat-chopper	J. A. Hard	Lawrence, Kans	Mar. 25, 1873	137,074

Index of patents issued from the United States Patent Office from 1790 to 1873, inclusive—Continued.

Invention.	Inventor.	Residence.	Date.	No.
Meat-chopper	N. L. Hatch	Cape Elizabeth, Me	Mar. 31, 1868	76, 187
Meat-chopper	F. C. Heberhart	Madison, Ind	Sept. 23, 1862	36, 518
Meat-chopper	J. G. Hirzel	Wilmington, Del	Aug. 11, 1868	80, 955
Meat-chopper	J. H. Landis	Eden, Pa	Mar. 12, 1861	31, 669
Meat-chopper	J. Massey	New York, N. Y	July 25, 1865	48, 965
Meat-chopper	A. McCarter	Salem, Ohio	Dec. 14, 1869	97, 946
Meat-chopper	A. McCarter	Salem, Ohio	June 7, 1870	104, 049
Meat-chopper	G. H. Melleu	New York, N. Y	Nov. 1, 1870	108, 809
Meat-chopper	H. D. Musselman	Lancaster, Pa	Feb. 28, 1860	27, 334
Meat-chopper	H. Obrecht	Mahanoy City, Pa	July 23, 1867	66, 989
Meat-chopper	W. P. Patton	Harrisburgh, Pa	Dec. 19, 1871	122, 051
Meat-chopper	T. Payne	Grand Rapids, Mich	Sept. 8, 1868	81, 939
Meat-chopper	D. Peters	Fort Madison, Iowa	Feb. 13, 1872	123, 725
Meat-chopper	H. P. Rankin	Allegheny, Pa	July 15, 1873	140, 950
Meat-chopper	W. Reitz	Cleveland, Ohio	Aug. 20, 1872	130, 745
Meat-chopper	N. E. Russell	China, Me	June 15, 1869	91, 273
Meat-chopper	S. Rutschman	Philadelphia, Pa	Sept. 1, 1868	81, 824
Meat-chopper	F. G. Siemers	Winona, Minn	Feb. 9, 1869	86, 783
Meat-chopper	A. R. Silver	Salem, Ohio	Aug. 30, 1870	106, 879
Meat-chopper	A. R. Silver	Salem, Ohio	Aug. 12, 1873	141, 731
Meat-chopper	A. F. Spaulding and S. M. Scott	Winchester, Mass	July 11, 1865	48, 734
Meat-chopper	L. Steigert	Cincinnati, Ohio	Jan. 11, 1870	98, 812
Meat-chopper	W. H. H. Walker	Bangor, Me	July 16, 1867	66, 916
Meat-chopper	N. T. Worthley	Brunswick, Me	Oct. 24, 1871	120, 359
Meat-chopper and vegetable-slicer	J. M. Taft	Arcadia, Wis	Jan. 3, 1871	110, 801
Meat-chopper, Rotating	G. W. Dechant	Berrysburgh, Pa	Apr. 22, 1862	35, 014
Meat, Chopping	J. Masser and S. Smith	Mcgentown, Pa	June 2, 1836	......
Meat, &c., chopping machine	W. H. Allen	Lowell, Mass	Feb. 13, 1855	12, 373
Meat-chopping machine	H. Chamberlain	Calais, Me	Sept. 24, 1867	69, 073
Meat-chopping machine	J. L. Good	Elizabethtown, Pa	May 14, 1872	126, 632
Meat-chopping machine	G. W. Sargent and P. H. Chesley	Chelsea, Mass	June 27, 1865	48, 451
Meat-chopping machine	H. Smith	Philadelphia, Pa	July 19, 1811	......
Meat-chopping machine	A. F. Spaulding and S. M. Scott	Winchendon, Mass	Jan. 31, 1865	46, 153
Meat-chopping machine	G. Spiehlman	Strasburgh, Pa	Mar. 11, 1873	136, 624
Meat-cleaver	E. Pollard	Albany, N. Y	Feb. 1, 1859	22, 839
Meat, Composition for covering	J. J. Bate and F. S. Lowe	Brooklyn, N. Y., and Jersey City, N. J.	Sept. 29, 1857	18, 270
Meat-compound	B. M. Fowler	Hackensack, N. J	Nov. 28, 1865	51, 259
Meat, Cooling and packing	D. E. Somes	Washington, D. C	Jan. 29, 1867	61, 638
Meat-crusher	A. J. Crane	Waterbury, Vt	Feb. 11, 1873	135, 781
Meat-crusher	O. P. Dennison	Mulberry, Ohio	June 20, 1871	116, 166
Meat-crusher	R. V. Jones	Canton, Ohio	June 6, 1865	48, 072
Meat, &c., Curing	A. C. Allen	Houston, Tex	Sept. 12, 1846	4, 754
Meat, Curing	G. A. Scherpf	New York, N. Y	May 7, 1845	4, 035
Meat, Curing	J. C. Schooley	Cincinnati, Ohio	Mar. 13, 1855	12, 530
Meat, Curing	W. H. Silberhorn	New York, N. Y	June 22, 1869	91, 678
Meat, Curing	G. Starkweather	Hartford, Conn	Nov. 5, 1850	7, 766
Meat, Curing and packing	I. Atkinson	Hamilton, Canada	Jan. 21, 1873	134, 965
Meat, &c., Curing and preserving	A. S. Lyman	New York, N. Y	June 22, 1869	91, 552
Meat, &c., Curing and preserving	A. Rock	New Orleans, La	July 26, 1870	105, 851
Meat, Curing and preserving	M. B. Sherwood	Buffalo, N. Y	Dec. 29, 1868	85, 485
Meat-curing apparatus	W. G. Bell	Boston, Mass	Dec. 19, 1871	121, 925
Meat-curing apparatus	M. B. Sherwood, jr	Buffalo, N. Y	Nov. 16, 1869	96, 978
Meat, Curing ham, beef, and other	O. M. Martin	Ann Arbor, Mich	Dec. 15, 1868	84, 893
Meat-cutter	C. Adams	Pittsburgh, Pa	July 23, 1861	32, 852
Meat-cutter	J. S. Artley	Danville, Pa	Jan. 3, 1871	110, 721
Meat-cutter	D. Bearly	Newcastle, Ind	Nov. 14, 1865	50, 984
Meat-cutter	W. G. Bell	Boston, Mass	Jan. 3, 1865	45, 788
Meat-cutter	L. Bonnet	New York, N. Y	July 24, 1860	29, 239
Meat-cutter	A. Burdick	Saratoga County, N. Y	Aug. 21, 1849	6, 652
Meat-cutter	W. Burns	Rome, Ohio	May 11, 1852	8, 936
Meat-cutter	A. Clark	Doe Run Post-Office, Pa	May 9, 1846	4, 507
Meat-cutter	A. B. Davenport	Petersham, Mass	Mar. 9, 1858	19, 547
Meat-cutter	P. Demeure	Brooklyn, N. Y	Apr. 20, 1858	20, 019
Meat-cutter	J. Derr	Oley, Pa	June 5, 1866	55, 251
Meat-cutter	D. Deshon, 2d	New London, Conn	June 13, 1848	5, 629
Meat-cutter	B. Dewalt and C. E. Schrador	Reading, Pa	Feb. 22, 1859	23, 016
Meat-cutter	J. G. Divoll	Sonora, Cal	June 22, 1869	91, 526
Meat-cutter	D. S. Early	Hummelstown, Pa	June 11, 1867	65, 551
Meat-cutter	A. J. Eddy	New Britain, Conn	Apr. 21, 1868	77, 014
Meat-cutter	H. H. Everts	Mount Morris, N. Y	Oct. 12, 1843	3, 303
Meat-cutter	C. A. Foster	Fitchburgh, Mass	July 14, 1868	79, 902
Meat-cutter	C. L. Gilpatric	Lewiston, Me	Oct. 4, 1864	44, 527
Meat-cutter	J. C. Hasfele	New York, N. Y	Aug. 27, 1867	68, 068
Meat-cutter	S. Hartley	Bridgeport, Ohio	Sept. 13, 1870	107, 364
Meat-cutter	H. Havell	Newark, N. J	May 31, 1859	24, 210
Meat-cutter	J. L. Haven	Cincinnati, Ohio	Jan. 22, 1867	61, 424
Meat-cutter	W. B. Heintze	Williamsburgh, N. Y	Feb. 22, 1870	100, 035
Meat-cutter	J. Hetzel and S. H. Hager	Miamisburgh, Ohio	Feb. 9, 1869	86, 669
Meat-cutter	J. C. Howe	Worcester, Mass	Aug. 18, 1868	81, 088
Meat-cutter	J. K. Hoyer	Reading, Pa	Sept. 7, 1858	21, 421
Meat-cutter	R. V. Jones	Canton, Ohio	Sept. 5, 1865	49, 762
Meat-cutter	D. R. Kenyon	Raritan, N. J	Mar. 19, 1872	124, 748
Meat-cutter	A. Klein	New York, N. Y	July 27, 1869	93, 094
Meat-cutter	P. Miles	New Haven, Conn	May 22, 1860	28, 393
Meat-cutter	P. Miles	New Haven, Conn	Jan. 8, 1861	31, 097
Meat-cutter	W. M. Miller	Tulpehocken, Pa	June 25, 1867	66, 033
Meat-cutter	S. Millet	New York, N. Y	May 24, 1853	9, 748
Meat-cutter	J. Morris	Derby, Conn	Feb. 25, 1835	......
Meat cutter	G. Mosman	Boston, Mass	Mar. 16, 1869	87, 961
Meat-cutter	D. H. Mundy and H. W. Hoffman	Camden, N. J	May 4, 1869	89, 787
Meat-cutter	H. B. Murphy	Allegheny City, Pa	Feb. 7, 1871	111, 670
Meat-cutter	J. Nacher	La Crosse, Wis	Feb. 11, 1868	74, 409
Meat-cutter	M. Newman	Oak Hill, N. Y	Sept. 14, 1858	21, 514
Meat-cutter	A. Nittinger, jr	Philadelphia, Pa	Aug. 11, 1863	39, 492
Meat-cutter	A. Nittinger, jr	Philadelphia, Pa	Sept. 21, 1869	95, 038
Meat-cutter	A. Nittinger, jr	Philadelphia, Pa	May 24, 1870	103, 360

Index of patents issued from the United States Patent Office from 1790 *to* 1873, *inclusive*—Continued.

Invention.	Inventor.	Residence.	Date.	No.
Meat-cutter	H. Obrecht	Mahanoy City, Pa	July 7, 1868	79, 678
Meat-cutter	J. G. Perry	Kingston, R. I	Mar. 15, 1859	23, 262
Meat-cutter	J. G. Perry	Kingston, R. I	Aug. 2, 1859	24, 953
Meat-cutter	J. G. Perry	South Kingston, R. I	July 31, 1860	29, 400
Meat-cutter	J. G. Perry	South Kingston, R. I	Sept. 18, 1860	30, 088
Meat-cutter	J. G. Perry	South Kingston, R. I	Dec. 2, 1862	37, 053
Meat-cutter	J. G. Perry	South Kingston, R. I	July 14, 1863	39, 243
Meat-cutter	J. G. Perry	South Kingston, R. I	July 5, 1864	43, 427
Meat-cutter	J. G. Perry	South Kingston, R. I	July 5, 1864	43, 428
Meat-cutter	J. G. Perry	South Kingston, R. I	Jan. 3, 1865	45, 744
Meat-cutter	J. G. Perry	Kingston, R. I	Oct. 20, 1868	83, 308
Meat-cutter	J. G. Perry	Kingston, R. I	Apr. 6, 1869	88, 666
Meat-cutter	J. G. Perry	Kingston, R. I	Apr. 6, 1869	88, 667
Meat-cutter	J. Potts	Yocumtown, Pa	June 15, 1852	9, 035
Meat-cutter	F. S. Rutschman	Philadelphia, Pa	May 25, 1869	90, 588
Meat-cutter	C. Schiller	Baltimore, Md	Jan. 5, 1869	85, 615
Meat-cutter	H. Seib	New York, N. Y	June 14, 1870	104, 363
Meat-cutter	I. Siegrist	Steinsburgh, Pa	Apr. 16, 1872	125, 699
Meat-cutter	F. G. Siemers	Winona, Minn	Aug. 25, 1868	81, 421
Meat-cutter	F. G. Siemers	Winona, Minn	June 13, 1871	115, 987
Meat-cutter	D. Slaughter	West Hempfield Township, Pa.	Aug. 4, 1868	80, 675
Meat-cutter	J. E. Smith	Buffalo, N. Y	Feb. 25, 1868	74, 947
Meat-cutter	J. E. Smith	Buffalo, N. Y	June 21, 1870	104, 506
Meat-cutter	L. S. Starrett	Newburyport, Mass	May 23, 1865	47, 875
Meat-cutter	S. L. Stockstill and H. H. Dille	Medway, Ohio	June 23, 1868	79, 085
Meat-cutter	E. D. Tippett	Georgetown, D. C	Dec. 28, 1846	4, 909
Meat-cutter	G. P. Trenlieb	Baltimore, Md	Dec. 28, 1869	98, 318
Meat-cutter	C Welte	Frankford, Pa	Apr. 21, 1868	77, 143
Meat-cutter	J. M. Wilder	Peterborough, N. H	May 10, 1845	4, 042
Meat-cutter	O. D. Woodruff	Southington, Conn	Jan. 10, 1860	26, 806
Meat-cutter	O. D. Woodruff	Southington, Conn	Mar. 9, 1869	87, 611
Meat-cutter	J. Zeitler and J. Zintl	Dayton, Ohio	Aug. 13, 1861	33, 060
Meat-cutters, Casting cylinders for	S. R. Plumb	Southington, Conn	July 31, 1860	29, 440
Meat-cutting apparatus	J. G. Perry	South Kingston, R. I	Feb. 26, 1850	7, 123
Meat-cutting machine	W. Biiesner	Saint Louis, Mo	July 30, 1867	67, 254
Meat-cutting machine	G. V. Brecht	Saint Louis, Mo	May 20, 1856	14, 901
Meat-cutting machine	A. Kirn	Buffalo, N. Y	Nov. 2, 1869	96, 326
Meat-cutting machine	J. Morris	Derby, Conn	Nov. 20, 1837	459
Meat-cutting machine	J. Morris	Derby, Conn	Feb. 28, 1842	2, 476
Meat-cutting machine	W. E. Richardson	Chicago, Ill	Mar. 21, 1865	46, 942
Meat-cutting machine	T. Vanderslice	Valley Forge, Pa	May 6, 1851	8, 072
Meat-cutting machine	F. Wobfersberger	Salem Station, Ohio	Mar. 23, 1858	19, 728
Meat-holder	S. Beissel	Shamokin, Pa	Dec. 9, 1873	145, 329
Meat-hook	S. Weaver	Pottstown, Pa	Sept. 27, 1870	107, 839
Meat in market-places, Apparatus for preserving and holding butcher's.	A. Seltzer	Baltimore, Md	Sept. 5, 1840	1, 774
Meat-mangle	G. A. Cover	Macomb, Ill	June 18, 1867	65, 878
Meat-mangle	S. D. Ingham	Ripley, Ohio	Sept. 20, 1870	110, 150
Meat-mangler	A. T. Adams	Indianapolis, Ind	Nov. 12, 1867	70, 675
Meat-mangler	J. T. Harvey and W. Dixon	Marysville, Pa	Oct. 4, 1870	108, 021
Meat-mangler	G. H. Tift	Morrisville, Vt	Sept. 11, 1866	58, 000
Meat, Manufacture of fluid	S. Darby	London, England	Jan. 9, 1872	122, 574
Meat-masher	J. Lefeber	Cambridge City, Ind	Aug. 20, 1867	67, 991
Meat-masher	G. W. Putnam	Peterborough, township of Smithfield, N. Y.	Dec. 27, 1864	45, 633
Meat-masher	G. Storer	New Britain, Conn	July 19, 1859	24, 831
Meat-mill	J. Gardiner	Philadelphia, Pa	Sept. 8, 1868	81, 904
Meat-mincer	J. M. Dronyer	Carondelet, Mo	June 16, 1868	78, 942
Meat-mincer	A. W. Hale	New Britain, Conn	Mar. 15, 1859	23, 246
Meat-mincer	L. Holzwarth, H. Froelsch, and T. Gerhards.	Philadelphia, Pa	Feb. 20, 1872	123, 899
Meat-mincing machine	A. Lightheiser	Reading, Pa	Dec. 25, 1855	13, 990
Meat-mincing machine	A. McInturff	Liberty, Va	Jan. 17, 1854	10, 430
Meat-mincing machine	P. Miles	New Haven, Conn	Aug. 19, 1862	36, 229
Meat-mincing machine	P. Miles	New York, N. Y	July 12, 1864	43, 520
Meat-mincing machine	O. Moses	Malone, N. Y	July 1, 1856	15, 248
Meat or fruit press	L. S. Starrett	Athol Depot, Mass	Apr. 15, 1873	137, 803
Meat, paint, &c., Metallic can for	E. M. Wood	Needham, Mass	Mar. 18, 1873	136, 890
Meat-pounder and ice-pick	G. Murray, jr	Cambridgeport, Mass	May 12, 1868	77, 752
Meat-pounder and potato-masher	J. A. McNeil	Grand Rapids, Mich	Oct. 31, 1865	50, 724
Meat pounder, block, and chopping bowl	G. B. Mill	Buffalo, N. Y	Mar. 8, 1870	100, 545
Meat, Preparing and preserving	H. Endemann	New York, N. Y	May 31, 1870	103, 728
Meat, Preparing extract of	M. S. Valentine	Richmond, Va	Apr. 25, 1871	114, 234
Meat-press	J. I. Danforth	Newburyport, Mass	Nov. 4, 1873	144, 322
Meat, &c., Process and apparatus for curing and packing.	D. E. Somes	Washington, D. C	Feb. 26, 1867	62, 449
Meat, pulverizing spice, &c., Chopping	B. P. Coston	Philadelphia, Pa	June 13, 1831	....
Meat-roaster	S. Pierce	New York, N. Y	July 12, 1838	835
Meat-roasting apparatus	J. G. Brown and J. P. Derby	South Reading, Mass	May 12, 1857	17, 259
Meat-roasting spit	E. W. Bigelow	Worcester, Mass	Apr. 28, 1868	77, 349
Meat, Salting	T. Davison	New York, N. Y	Aug. 7, 1849	6, 623
Meat, Salting	T. Spencer	Syracuse, N. Y	Oct. 21, 1862	36, 735
Meat, &c., Salting and preserving	E. Darnell	Fox, Ill	Apr 2, 1867	63, 484
Meat, Salting and seasoning	W. Sadler	Lockport, N. Y	Apr. 2, 1867	63, 432
Meat-smoking apparatus	E. Bickford	Ogden, N. Y	June 17, 1862	35, 580
Meat-smoking apparatus	R. T. Burnett	New York, N. Y	Jan. 19, 1869	85, 998
Meat-smoking apparatus	J. Wright	Wilmington, Del	July 29, 1856	15, 452
Meat-spit	P. Fisher	Williamsburgh, N. Y	Nov. 19, 1867	70, 984
Meat-stuffer	A. R. Silver	Salem, Ohio	Nov. 26, 1872	133, 312
Meat-tenderer	W. Beach	Philadelphia, Pa	Aug. 16, 1853	9, 932
Meat-tenderer	M. M. Pettes	Worcester, Mass	Mar. 5, 1872	124, 383
Meat-tenderer	L. B. Tarbox	Collinsville, N. Y	Feb. 21, 1871	112, 092
Meat tendering or chopping device	F. W. Codding	West Rutland, Mass	Jan. 28, 1870	104, 706
Meat, Utensil for compressing	E. Mingay	Boston, Mass	Dec. 10, 1872	133, 791
Meat with brine, Apparatus for injecting	J. C. Adams	Baltimore, Md	Sept. 10, 1861	33, 228
Mechanical adjustment	W. F. Parker	Meriden, Conn	Jan. 19, 1869	86, 098
Mechanical adjustment	R. B. Perkins	Meriden, Conn	Apr. 6, 1869	88, 733

Index of patents issued from the United States Patent Office from 1790 *to* 1873, *inclusive*—Continued.

Invention.	Inventor.	Residence.	Date.	No.
Mechanical adjustment	R. B. Perkins	Meriden, Conn	Apr. 6, 1869	88, 734
Mechanical device	J. M. Folsom	Louisville, Ky	Feb. 13, 1872	123, 686
Mechanical escapement	E. C. Hopping	Madison, N. J	Nov. 5, 1872	132, 833
Mechanical jack	A. Jones	Lebanon, N. H	Aug. 3, 1858	21, 107
Mechanical power	N. S. Baker	Covington, Ky	Mar. 12, 1872	124, 415
Mechanical power	F. S. Bremer	Montgomery, Ill	Aug. 20, 1872	130, 696
Mechanical power	C. P. Carter	Poughkeepsie, N. Y	Apr. 28, 1868	77, 167
Mechanical power	E. Mattesou	Troy, N. Y	Sept. 28, 1858	21, 611
Mechanical power	E. W. Morton	Brooklyn, N. Y	Feb. 18, 1873	136, 088
Mechanical power	C. Van Derzee	Albany, N. Y	Aug. 12, 1873	141, 741
Mechanical power	A. B. Wood	Hamburg, Ark	June 18, 1867	65, 981
Mechanical power, Mode of applying	L. Butts	Canterbury, Conn	July 9, 1823	....
Mechanical power, Rotary	A. G. Waterhouse	San Francisco, Cal	Sept. 21, 1869	95, 057
Medallion case	J. Powell	Cincinnati, Ohio	Nov. 13, 1866	59, 648
Medals, breast-pins, &c., Pin-fastening for	G. O. Monroe	New York, N. Y	Nov. 15, 1864	45, 121
Medals, &c., Machine for producing reduced copies of.	C. J. Hill	London, England	Apr. 14, 1868	76, 631
Medical apparatus, Electro	L. Drescher	New York, N. Y	July 4, 1871	116, 695
Medical apparatus, Electro-magnetic	H. Soltmann	New York, N. Y	Mar. 15, 1859	23, 272
Medical apparatus, Electro-magnetic	A. J. Steel	Brooklyn, N. Y	June 20, 1871	116, 110
Medical apparatus for treating diseases by vacuum.	G. Hadfield	Cincinnati, Ohio	May 8, 1866	54, 530
Medical apparatus, Sponge-holder for electro	J. Kidder	New York, N. Y	June 20, 1871	116, 197
[Medical.] Burns and scalds, Method of curing	L. Maxwell	Mitchell, Ind	Feb. 28, 1865	46, 570
[Medical] Catarrhal douche	R. M. Moyle	New York, N. Y	Mar. 28, 1871	113, 080
Medical extract	H. S. Draper	Rochester, N. Y	Sept. 28, 1869	95, 209
Medical extract from hemlock-bark	S. H. Kennedy	Johnstown, N. Y	Dec. 5, 1871	121, 631
Medical extracts, Mode of evaporating solutions, decoctions, &c., and preparing.	J. W. W. Gordon	Baltimore, Md	Oct. 8, 1840	1, 805
[Medical.] Means of applying heat and cold in the treatment of diseases.	J. Chapman	Portman Square, England	Feb. 21, 1865	46, 535
Medical or aromatic substances, Apparatus for diffusing the vapors of.	A. L. Fleury	New York, N. Y	Nov. 13, 1866	59, 576
Medical purposes, Apparatus for forming vapor for.	J. Gardette and H. Rance	New Orleans, La	Jan. 3, 1860	26, 667
Medical purposes, Citrates of iron and manganese for.	A. K. Clapp	Boston, Mass	May 2, 1871	114, 264
Medical purposes, Electrical instrument for	A. Page	Springfield, Mass	Aug. 19, 1862	36, 231
Medical purposes, Electro-magnetic machine for	L. Drescher	New York, N. Y	Feb. 1, 1870	99, 414
Medical purposes, Voltaic pile for	A. C. Garratt	Boston, Mass	Dec. 29, 1868	85, 300
Medical rubbing-apparatus	G. H. Taylor	New York, N. Y	Jan. 2, 1872	122, 500
Medical steam-bath	J. Deverin and J. B. Isabelle	Lexington, Ky	July 6, 1818	....
Medical topical application, Apparatus for	G. E. B. French	Washington, D. C	Jan. 3, 1860	26, 663
Medical use, Electro-magnetic apparatus for	C. Reitz	Indianapolis, Ind	Sept. 20, 1870	107, 626
Medical use, Oscillating rubbing-machine for	G. H. Taylor	New York, N. Y	May 12, 1868	77, 933
Medical vacuum-apparatus	G. Hadfield	Cincinnati, Ohio	Jan. 1, 1867	60, 883
Medical vacuum-apparatus	J. G. Hadfield	Cincinnati, Ohio	Dec. 24, 1867	72, 631
Medical vibrating and kneading machine	G. H. Taylor	New York, N. Y	Feb. 2, 1869	86, 604
Medicated balsam	L. F. Griffin	New York, N. Y	Dec. 3, 1867	71, 743
Medicated bath	B. Swett	Mount Morris, N. Y	Dec. 31, 1845	4, 347
Medicated-bath apparatus	J. Davenport	Philadelphia, Pa	May 21, 1872	126, 878
Medicated bath, Compound for saline	C. Lennig	Philadelphia, Pa	Jan. 1, 1867	60, 753
Medicated candy	B. H. Bener and M. H. Burgess	Erie, Pa	July 11, 1865	48, 645
Medicated candy	S. B Colton	Bloomfield, Conn	Jan. 21, 1873	134, 979
Medicated candy	L. Morgenthau	Manheim, Baden	Apr. 25, 1865	47, 504
Medicated confections	N. Saltabassi	New York, N. Y	Nov. 5, 1872	132, 861
Medicated cracker	J. L. Halliman	Grand Rapids, Mich	May 4, 1869	89, 654
Medicated fabric	H. Glynn	Baltimore, Md	Dec. 21, 1858	22, 362
Medicated liquid magnesia	J. Cullen	Philadelphia, Pa	May 4, 1818	....
Medicated liquid magnesia	R. Jordan and M. Anderson	Philadelphia, Pa	June 25, 1834	....
Medicated lotion	J. G. Popp	Vienna, Austria	Dec. 4, 1860	30, 834
Medicated malt-liquor, Manufacture of	H. H. Kessler	Detroit, Mich	May 10, 1870	102, 833
Medicated or pure steam to brutes, Apparatus for administering.	L. Quetsch	Urbana, Ill	Oct. 19, 1869	96, 036
Medicated pad	W. D. Titus	Brooklyn, N. Y	Jan. 3, 1860	26, 719
Medicated pad or belt	A. F. Cooper	Cambridge, Mass	Nov. 4, 1873	144, 315
Medicated paper	C. J. Eames	New York, N. Y	Jan. 21, 1873	134, 983
Medicated plaster	A. D. Richards	Somerville, Mass	Feb. 7, 1871	111, 682
Medicated powders, Instrument for depositing	A. M. A. Laforgue	France	Mar. 16, 1869	87, 946
Medicated shampoo-bath	R. D. Mott	Boston, Mass	Dec. 16, 1833	....
Medicated-steam bath, Machine used for	B. Marshall	New York, N. Y	Dec. 7, 1821	....
Medicated towel	L. Stewart	San Francisco, Cal	Mar. 4, 1873	136, 466
Medicated troches	C. H. Needles	Philadelphia, Pa	Nov. 27, 1866	60, 039
Medicated-vapor apparatus	A. F. Rose	Brooklyn, N. Y	July 13, 1858	20, 896
Medicated vapors, Apparatus for administering	J. C. Cook	New Haven, Conn	Feb. 1, 1870	99, 408
Medicated vapors, Chair for disseminating	J. W. Smith	Iowa Point, Kans	May 21, 1861	32, 384
Medicated voltaic plaster	W. C. Collins	Bucksport, Me	July 4, 1871	116, 562
Medicator, Dermatic	I. R. Weisiger	Danville, Ky	May 11, 1869	90, 058
Medicator, Mechanical	P. A. Royce	Buffalo, N. Y	Sept. 25, 1866	58, 301
Medicinal beverage	L. F. Lastreto	San Francisco, Cal	Mar. 22, 1870	101, 027
Medicinal beverage	A. S. Taylor	San Francisco, Cal	Aug. 8, 1871	117, 942
Medicinal beverage, Aerated	A. F. Drickwitz	New York, N. Y	Oct. 26, 1869	96, 089
Medicinal extracts, Preparation of	A. J. Despinoy	Lille, France	Sept. 8, 1860	30, 050
Medicinal preparation of iron	W. A. Aspinoll	Brooklyn, N. Y	June 13, 1871	115, 805
Medicinal purposes, &c., Solution of balsamic gum for.	O. Oldberg	Washington, D. C	Sept. 12, 1871	118, 813
Medicine	H. Adolph	Clinton, Kans	June 1, 1869	90, 715
Medicine	A. Alexander	Eugene City, Oreg	Apr. 28, 1868	77, 343
Medicine	J. S. Anders	North Wales, Pa	June 12, 1866	55, 449
Medicine	J. Baker	Jefferson Township, Ohio.	May 5, 1831	....
Medicine	J. P. Barnett	Navasota, Tex	Oct. 7, 1873	143, 399
Medicine	T. A. Barry and B. A. Patten	San Francisco, Cal	Mar. 24, 1868	75, 837
Medicine	S. Bass	Savannah, Ga	Nov. 3, 1868	83, 755
Medicine	J. Bates	Salineville, Ohio	Feb. 26, 1867	62, 389
Medicine	P. Becker	South Bethlehem, Pa	May 31, 1870	103, 545
Medicine	D. D. T. Benedict	Havanna, N. Y	Nov. 20, 1866	59, 756
Medicine	R. S. Bernard	Norfolk, Va	Aug. 17, 1835	....
Medicine	E. M. and L. M. Berry	Saltillo, Ind.	Nov. 10, 1868	83, 823
Medicine	G. B. Bieler	Cincinnati, Ohio	Jan. 9, 1866	51, 914

Index of patents issued from the United States Patent Office from 1790 to 1873, inclusive—Continued.

Invention.	Inventor.	Residence.	Date.	No.
Medicine	F. Bird	Hancock, Ga	Apr. 16, 1828	....
Medicine	O. W. Blanchard	Delavan, Wis	Aug. 27, 1867	68, 159
Medicine	G. Bode	Milwaukee, Wis	June 5, 1866	55, 234
Medicine	A. Brown	Fond du Lac, Wis	May 6, 1873	138, 473
Medicine	A. R. Brown	Litchfield, Mich	Feb. 5, 1867	61, 802
Medicine	E. D. Burrill	Providence, R. I	Mar. 22, 1870	101, 094
Medicine	T. J. Butcher	Wenona Station, Ill	July 20, 1869	92, 937
Medicine	L. J. Buttrill	Jackson, Ga	July 12, 1870	105, 168
Medicine	J. H. Butts	Stroudsburgh, Pa	Sept. 8, 1868	81, 876
Medicine	J. Cahn	Selma, Ala	June 18, 1872	128, 198
Medicine	W. R. Call and T. F. Griffin	Gloucester, Mass	Jan. 4, 1870	98, 554
Medicine	M. H. Campbell	Syracuse, N. Y	June 18, 1872	128, 110
Medicine	J. Carnrick	New York, N. Y	Mar. 2, 1869	87, 393
Medicine	F. J. Carrall	Millville, N. J	Apr. 17, 1866	53, 946
Medicine	M. Cary	Racine, Wis	Dec. 8, 1868	84, 796
Medicine	W. Cawein	Louisville, Ky	Feb. 13, 1866	52, 530
Medicine	G. W. Chambers	Talladega, Ala	Sept. 5, 1871	118, 686
Medicine	H. Clark	Northampton, Mass	Mar. 24, 1868	75, 732
Medicine	S. P. Clayton	South Amboy, N. J	Jan. 16, 1866	52, 032
Medicine	H. W. Cloud	Evansville, Ind	Oct. 4, 1870	107, 877
Medicine	J. H. Conway	Richmond, Va	May 29, 1866	55, 214
Medicine	J. Conzelman	Saint Louis, Mo	May 20, 1873	139, 045
Medicine	T. Copland	Hamilton, Canada	Dec. 17, 1872	133, 924
Medicine	A. P. Coryell	Janesville, Wis	Feb. 21, 1865	46, 449
Medicine	A. M. Cox	Elizabeth, N. J	July 2, 1867	66, 301
Medicine	J. L. A. Creuse	Brooklyn, N. Y	Nov. 28, 1871	121, 338
Medicine	J. D. Curl and J. G. Bartlett	Mokena, Ill	May 4, 1869	89, 562
Medicine	W. Davidson	Binghamton, N. Y	Dec. 3, 1867	71, 712
Medicine	D. Davis	Sheakleyville, Pa	Dec. 24, 1872	134, 129
Medicine	J. Dean	Freeport, Ill	Aug. 27, 1867	68, 051
Medicine	G. Déclat	Paris, France	Dec. 23, 1873	145, 850
Medicine	H. De Lapaturelle	New York, N. Y	May 1, 1866	54, 304
Medicine	J. Dent	Augusta, Ga	July 2, 1828	....
Medicine	P. M. Devos	New York, N. Y	July 24, 1866	56, 533
Medicine	C. A. F. Dietz	New York, N. Y	Feb. 11, 1868	74, 205
Medicine	O. G. Ditmars	New York, N. Y	Mar. 17, 1868	75, 536
Medicine	O. G. Ditmars	New York, N. Y	Mar. 17, 1868	75, 537
Medicine	B. W. Donaldson	Dixfield, Me	Apr. 7, 1868	76, 308
Medicine	L. Dow	Hebron, Conn	Nov. 24, 1820	....
Medicine	J. D. Doyle	Rochester, N. Y	June 25, 1872	128, 373
Medicine	N. T. Drake	High Point, N. C	Sept. 28, 1869	95, 330
Medicine	C. J. Eames	New York, N. Y	May 10, 1870	102, 788
Medicine	M. C. Edey	New York, N. Y	Mar. 31, 1868	76, 176
Medicine	J. B. Edwards	Knightstown, Ind	Sept. 11, 1866	57, 884
Medicine	J. P. Eisenhut	Monroe, Mich	June 12, 1866	55, 474
Medicine	W. H. Farnham	Sparta, Wis	Apr. 3, 1866	53, 595
Medicine	J. Fay	Lacon, Ill	May 9, 1871	114, 544
Medicine	S. Field	Galva, Ill	Dec. 16, 1873	145, 560
Medicine	J. Fischer	Saint Louis, Mo	July 2, 1867	66, 315
Medicine	W. B. Foster	Ridgeville, Ohio	Mar. 5, 1867	62, 622
Medicine	E. Frese	San Francisco, Cal	Aug. 4, 1868	80, 718
Medicine	A. Fuqua	Woodville, Ky	July 1, 1873	140, 408
Medicine	G. C. Furber	Yreka, Cal	June 28, 1870	104, 727
Medicine	H. Gahn	Upsala, Sweden	Apr. 8, 1873	137, 542
Medicine	G. W. C. Gamble	Millersburgh, Iowa	Feb. 18, 1868	74, 680
Medicine	O. V. Garnett	Versailles, Ky	Apr. 9, 1867	63, 627
Medicine	L. L. Gebhart	New Providence, Ind	Aug. 23, 1870	106, 573
Medicine	R. A. Gettings	Marion, Ky	Oct. 15, 1872	132, 275
Medicine	F. Goetsch	Bloomington, Ill	Apr. 2, 1872	125, 286
Medicine	C. Goffinet	Leopold, Ind	June 25, 1872	128, 385
Medicine	G. H. Goltry and F. W. Hogarth	Port Allegheny, Pa	May 30, 1871	115, 460
Medicine	D. C. Gould	Sterling, Ill	Oct. 15, 1867	69, 796
Medicine	J. W. Gray	Riceville, Pa	Aug. 13, 1872	130, 368
Medicine	J. Greenwald	Cincinnati, Ohio	Dec. 17, 1867	72, 193
Medicine	J. Gregory	Marion, Ohio	Jan. 8, 1867	61, 063
Medicine	H. H. Grigg	Philadelphia, Pa	Jan. 11, 1870	98, 690
Medicine	B. Gutierrez	Santa Barbara, Cal	Sept. 5, 1871	118, 714
Medicine	J. M. F. Hall	Davenport, Iowa	Sept. 25, 1866	58, 249
Medicine	C. L. Hammond	Java, N. Y	Sept. 14, 1869	94, 886
Medicine	M. J. Hanson	Mauston, Wis	Jan. 18, 1870	98, 959
Medicine	J. Harrigan	East Boston, Mass	Nov. 12, 1867	70, 838
Medicine	O. F. Harris	Norwich, Conn	Dec. 6, 1870	109, 895
Medicine	J. B. Haskins	Brooklyn, N. Y	July 5, 1870	105, 073
Medicine	G. M. Hay	Americus, Ga	Jan. 5, 1869	85, 659
Medicine	W. Habbert	Manchester, England	Apr. 27, 1869	89, 481
Medicine	M. A. Hilt	Syracuse, N. Y	Oct. 22, 1867	69, 995
Medicine	A. J. Hobbs	Van Wirt, Ga	June 30, 1868	79, 352
Medicine	W. Horner	Washington, D. C	Dec. 13, 1864	45, 410
Medicine	J. M. Hughes	Menomonee, Wis	Dec. 29, 1868	85, 385
Medicine	J. P. Hulse	Acra, N. Y	May 8, 1866	54, 550
Medicine	J. P. Humes	Winnebago City, Minn	May 19, 1868	78, 096
Medicine	S. W. Ingraham	Wooster, Ohio	Feb. 2, 1869	86, 404
Medicine	J. G. Jeffrey	South New Berlin, N. Y	Apr. 16, 1867	63, 902
Medicine	H. D. and I. D. Jewett	Saint Omer, Ind	May 21, 1872	127, 060
Medicine	A. Johnson	Brockport, N. Y	Mar. 2, 1869	87, 343
Medicine	N. Joly	Paris, France	Aug. 6, 1867	67, 555
Medicine	W. N. Jordan	Cambridge, Mass	June 22, 1869	91, 751
Medicine	C. Judson	Omro, Wis	Aug. 13, 1867	67, 657
Medicine	A. W. Kidder	South Norridgewock, Me	Apr. 6, 1869	88, 644
Medicine	N. Kieffer	New Orleans, La	Sept. 18, 1866	58, 106
Medicine	J. W. M. Kirkpatrick	Hamburgh, Ark	Feb. 16, 1869	86, 986
Medicine	J. B. Knoebel	Shoal Creek, Ill	June 14, 1864	43, 118
Medicine	M. Knotts	Carondelet, Mo	May 28, 1867	65, 235
Medicine	J. Kornitzer	New York, N. Y	May 14, 1872	126, 637
Medicine	D. Laugell	Apple Creek, Ohio	Sept. 7, 1869	94, 616
Medicine	P. Lee and L. Matthews	Lebanon, Oreg	Mar. 29, 1870	101, 279
Medicine	C. L. Lege	San Antonio, Tex	June 11, 1867	65, 580
Medicine	J. Leich	Brooklyn, N. Y	Sept. 20, 1864	44, 323

Index of patents issued from the United States Patent Office from 1790 *to* 1873, *inclusive*—Continued.

Invention.	Inventor.	Residence.	Date.	No.
Medicine	G. H. Leithead	East Birmingham, Pa	June 4, 1867	65, 401
Medicine	J. Levy	Chicago, Ill	Aug. 14, 1866	57, 155
Medicine	N. C. Lincoln	Brunswick, Me	Oct. 9, 1866	58, 652
Medicine	H. Lister	Houston, Tex	Feb. 7, 1871	111, 655
Medicine	H. H. Lockwood	Madison, Wis	May 29, 1866	55, 123
Medicine	A. Loosley	Philadelphia, Pa	Mar. 5, 1867	62, 650
Medicine	E. A. Lowdermill	Grand Junction, Tenn	Mar. 24, 1868	75, 776
Medicine	W. S. Lyon	Tranquility, Ohio	Sept. 10, 1867	68, 632
Medicine	L. Marmaduke	Shelbyville, Mo	July 3, 1866	56, 072
Medicine	A. A. Marsh	Frankfort, Mich	Aug. 30, 1870	106, 846
Medicine	Z. Marshall	Andersonville, Ind	Nov. 20, 1866	59, 776
Medicine	P. and P. Mays	Clearfield, Pa	Jan. 18, 1870	98, 875
Medicine	W. L. McCord	Abbeville, S. C	June 29, 1869	91, 954
Medicine	G. Mershon	Brookville, Iowa	Apr. 10, 1866	53, 852
Medicine	A. Messex	Waynesborough, Ga	Feb. 19, 1867	62, 282
Medicine	L. B. Millard	San José, Cal	Nov. 26, 1872	133, 328
Medicine	E. Miller	New York, N. Y	Oct. 18, 1870	108, 504
Medicine	W. J. Miller	Linesville, Pa	Jan. 29, 1867	61, 631
Medicine	C. H. Mitchell	Bristow Station, Ky	May 14, 1867	64, 787
Medicine	G. Mohler	Yates City, Ill	Sept. 29, 1868	82, 541
Medicine	G. Montgomery	Canton, Ill	Aug. 21, 1866	57, 357
Medicine	L. O. Moreno	New York, N. Y	Oct. 23, 1866	59, 052
Medicine	T. Marley	Cambridge, Mass	Jan. 14, 1873	134, 762
Medicine	C. A. Morse	Williamsport, Ohio	Sept. 6, 1864	44, 106
Medicine	G. F. Munro	Leon, Iowa	Nov. 25, 1873	145, 002
Medicine	E. Myers	Davis, Ill	June 28, 1870	104, 758
Medicine	W. B. Myres	Frenchtown, N. J	June 29, 1869	92, 086
Medicine	F. H. Norton	New York, N. Y	June 14, 1864	43, 128
Medicine	W. W. Oglesby	Benton County, Oreg	Oct. 18, 1870	108, 508
Medicine	H. C. S. Otto	New York, N. Y	Mar. 29, 1870	101, 305
Medicine	S. E. Paddock	Delaware, Ohio	Sept. 2, 1873	142, 408
Medicine	B. Paine	McMinnville, Tenn	Nov. 5, 1867	70, 602
Medicine	S. Payne	Louisville, Ky	Nov. 26, 1867	71, 407
Medicine	E. Penfield	Oberlin, Ohio	Mar. 28, 1865	47, 035
Medicine	J. Penoyer	Massillon, Ohio	Jan. 12, 1869	85, 761
Medicine	M. Perl	New Orleans, La	Dec. 10, 1867	71, 907
Medicine	E. G. Perry	Knoxville, Ill	May 22, 1866	54, 944
Medicine	W. H. H. Peters	Tuskegee, Ala	June 1, 1869	90, 779
Medicine	N. Petre	New York, N. Y	Aug. 28, 1866	57, 561
Medicine	W. W. Pettengill	Georgia, Vt	Oct. 1, 1872	131, 778
Medicine	S. Pitcher	Barnstable, Mass	May 12, 1868	77, 758
Medicine	H. Pool	Montgomery County, Tenn	June 22, 1869	91, 768
Medicine	H. G. Pope and H. F. Herrick	New Berlin, N. Y	June 11, 1867	65, 693
Medicine	P. Poucin	Minneapolis, Minn	June 12, 1866	55, 534
Medicine	J. Prentiss	New London, Conn	Mar. 27, 1822	....
Medicine	J. L. Putegnat	Brownsville, Tex	Mar. 18, 1873	136, 937
Medicine	M. H. Ramsaur	Lincolnton, N. C	July 6, 1869	92, 209
Medicine	J. Ramsburgh, sr	New Madrid, Mo	Oct. 6, 1868	82, 873
Medicine	W. Ranson	Portage, Wis	Oct. 23, 1866	59, 070
Medicine	J. J. Reeves	Sulphur Springs, Tex	June 26, 1860	28, 904
Medicine	M. Richardson	Norristown, Pa	May 29, 1866	55, 160
Medicine	J. A. Robbins	Medford, Mass	Apr. 4, 1871	113, 569
Medicine	E. Robinson	Fairport, N. Y	May 18, 1869	90, 308
Medicine	L. Rogers	Morehouse Parish, La	May 19, 1868	78, 017
Medicine	L. Rogers	Morehouse Parish, La	May 19, 1868	78, 018
Medicine	T. Rosbrugh	Bellefontaine, Ohio	Feb. 4, 1868	74, 147
Medicine	P. G. Rosenblatt	Greenville, Tenn	June 12, 1866	55, 540
Medicine	A. Rullmann	New York, N. Y	Sept. 24, 1867	69, 252
Medicine	A. J. Runyan	Ashland, Ind	Mar. 20, 1866	53, 348
Medicine	W. T. Salie	Bowdoinham, Me	June 26, 1866	55, 911
Medicine	B. Samson	Cortlandville, N. Y	Nov. 20, 1866	59, 750
Medicine	L. Schultz	Buffalo, N. Y	May 16, 1865	47, 750
Medicine	I. W. Scranton	West Liberty, Iowa	Sept. 8, 1868	82, 038
Medicine	S. P. Sedgwick	Wheaton, Ill	Apr. 14, 1868	76, 832
Medicine	A. F. Shannon	Quincy, Ill	July 11, 1871	116, 875
Medicine	G. V. and J. A. Sheffield	Northbridge Centre, Mass	Feb. 15, 1870	99, 959
Medicine	O. Skidmore	Albany, N. Y	June 8, 1869	90, 962
Medicine	G. H. Smith	New Orleans, La	Apr. 2, 1867	63, 567
Medicine	G. H. Smith	New Orleans, La	May 10, 1870	102, 980
Medicine	J. Smith	Schodack Centre, N. Y	Feb. 18, 1868	74, 622
Medicine	W. C. Smith	Lafargeville, N. Y	Feb. 20, 1866	52, 760
Medicine	F. W. Speissegger	Charleston, S. C	June 19, 1866	55, 730
Medicine	J. H. Sperling	Peru, Ind	June 18, 1867	65, 840
Medicine	J. T. Stewart	Peoria, Ill	Jan. 21, 1868	73, 552
Medicine	J. T. Stewart	Peoria, Ill	May 31, 1870	103, 675
Medicine	L. Stroever	Philadelphia, Pa	Aug. 20, 1867	67, 926
Medicine	P. O. R. Stromski	Boston, Mass	Mar. 16, 1869	87, 802
Medicine	D. W. Stutsman	Upshur, Ohio	Nov. 20, 1866	59, 803
Medicine	W. C. Tait	Alexandria, La	June 28, 1870	104, 791
Medicine	A. S. Talbert	Lexington, Ky	Jan. 16, 1866	52, 091
Medicine	W. H. Tate	Orleans, Ind	Sept. 3, 1867	68, 582
Medicine	D. W. Taylor	Bell County, Tex	Jan. 7, 1868	73, 135
Medicine	T. H. Taylor	Saratoga Springs, N. Y	Nov. 26, 1867	71, 549
Medicine	C. K. Tayntor	Cuyler, N. Y	Mar. 31, 1868	76, 272
Medicine	H. Themel	Esconawba, Mich	Nov. 5, 1872	132, 876
Medicine	H. Themel	Esconawba, Mich	Mar. 11, 1873	136, 680
Medicine	J. M. Thompson	Saltilloville, Ind	Jan. 28, 1868	73, 939
Medicine	S. Thompson	Boston, Mass	Jan. 28, 1823	....
Medicine	A. A. Thomson	Newburgh, Pa	Nov. 30, 1869	97, 459
Medicine	W. P. Thurber	Chicago, Ill	Apr. 7, 1868	76, 555
Medicine	J. Thurmon	Pike County, Mo	Apr. 4, 1865	47, 140
Medicine	J. P. and L. Thurmon	Warrenton, Mo	May 28, 1867	65, 302
Medicine	H. A. Tilden	New Lebanon, N. Y	July 16, 1872	129, 626
Medicine	T. S. Tuggle	Columbus, Ga	Apr. 6, 1869	88, 680
Medicine	T. S. Tuggle	Columbus, Ga	Apr. 6, 1869	88, 681
Medicine	T. H. W. Upshur	Norfolk, Va	Sept. 8, 1868	82, 046
Medicine	J. B. Vann	Elyton, Ala	Jan. 26, 1869	86, 331
Medicine	P. W. Vaughn	Columbia, Ky	Mar. 9, 1869	87, 605

Index of patents issued from the United States Patent Office from 1790 to 1873, inclusive—Continued.

Invention.	Inventor.	Residence.	Date.	No.
Medicine	J. Ward	East Hardwick, Vt	May 10, 1864	42, 704
Medicine	J. Ward	Evansville, Ind	Sept. 21, 1869	95, 173
Medicine	J. Weaver	Knightstown, Ind	Jan. 5, 1864	41, 111
Medicine	W. Weber	Cincinnati, Ohio	June 29, 1869	91, 886
Medicine	R. B. Weese	Charlottesville, Ind	May 24, 1864	42, 890
Medicine	C. H. Whittemore	Lewiston, Me	Aug. 18, 1868	81, 122
Medicine	A. D. Willis	Crawfordsville, Ind	Jan. 4, 1870	98, 535
Medicine	J. A. Willis	Cherry Valley, N. Y	May 7, 1867	64, 611
Medicine	E. Wills	Winslow, N. J	Sept. 28, 1869	95, 298
Medicine	J. T. Wilson	Brooklyn, N. Y	Dec. 10, 1867	72, 143
Medicine	W. M. Wilson and D. S. Denton.	Polk County, Mo	Aug. 30, 1870	106, 905
Medicine	L. W. Wollenweber	Jeffersonville, Ind	Oct. 22, 1872	132, 424
Medicine	T. T. Wright	Memphis, Tenn	July 15, 1873	140, 866
Medicine	J. Yount	Gettysburgh, Pa	July 23, 1867	67, 012
Medicine	H. Zoeger	New York, N. Y	Feb. 26, 1867	62, 516
Medicine, Ague	T. M. Daniel	Athens, Ga	Oct. 1, 1867	69, 414
Medicine, Ague	J. Monfort	Jessamine County, Ky	Dec. 18, 1866	60, 538
Medicine and apparatus for making the same	P. Fahrney	Chicago, Ill	Dec. 13, 1870	110, 025
Medicine and apparatus for the cure of asthma	J. Pinchard	San Francisco, Cal	Mar. 25, 1873	137, 093
Medicine and disinfectant	J. Walton	Newark, Ohio	June 11, 1872	127, 814
Medicine and liniment	W. S. Crooker	Shamburgh, Pa	Aug. 9, 1870	106, 127
Medicine and medicated food	J. M. O. Tamin	New York, N. Y	Mar. 11, 1873	136, 678
Medicine and remedial agents, and apparatus therefor, Mode of applying.	J. Allen	New York, N. Y	Mar. 12, 1867	62, 916
Medicine and surgery, Application of carbonated water in.	A. D. Puffer	Somerville, Mass	Aug. 24, 1869	94, 029
[Medicine,] Anodyne and alterative sirup	R. Thompson	Rome, N. Y	Nov. 14, 1835	....
Medicine, Anti-bilious	J. J. Giraud	Baltimore, Md	Sept. 9, 1817	....
Medicine, Anti-rheumatic	J. Schmoll	New York, N. Y	Mar. 5, 1867	62, 692
Medicine, Anti-spasmodic tincture	H. Howard	Columbus, Ohio	Aug. 25, 1832	....
Medicine, Anti-typhus	J. P. Fertig	Saint Louis, Mo	Sept 8, 1863	39, 861
Medicine, Apparatus for administering	S. Curtiss, jr	Williamsburgh, N. Y	Apr. 10, 1866	53, 790
Medicine, Apparatus for administering pulverulent	J. Moore and D. P. Adams	Marietta, Ohio	July 24, 1855	13, 318
Medicine, Astringent	E. W. Ferris	Macon, Miss	Oct. 16, 1860	30, 396
Medicine, Astringent tonic	H. Howard	Columbus, Ohio	Aug. 25, 1832	....
Medicine, Bitter tonic	H. Howard	Columbus, Ohio	Aug. 25, 1832	....
[Medicine.] Bitters	T. Beck	Omaha, Nebr	Dec. 27, 1870	110, 423
[Medicine.] Bitters	J. Bender	Lonaconing, Md	Apr. 28, 1868	77, 245
[Medicine.] Bitters	G. W. Brown	Portland, Oreg	Aug. 13, 1872	130, 409
[Medicine.] Bitters	E. E. Crady	Sioux City Iowa	Oct. 17, 1871	120, 042
[Medicine.] Bitters	A. Desaulniers	Oswego, N. Y	Mar. 14, 1871	112, 558
[Medicine.] Bitters	J. Frechette	Chicago, Ill	May 21, 1872	126, 948
[Medicine.] Bitters	W. P. and J. Haubert	Canton, Ohio	Feb. 23, 1869	87, 169
[Medicine.] Bitters	M. Holst	Memphis, Tenn	June 27, 1871	116, 310
[Medicine.] Bitters	J. D. Holtzermann	Piqua, Ohio	May 7, 1867	64, 421
[Medicine.] Bitters	A. T. Hyde	Rochester, Minn	Aug. 25, 1868	81, 508
[Medicine.] Bitters	L. Kappler	Woodstock, Ill	Dec. 2, 1873	145, 062
[Medicine.] Bitters	J. Lladó	New Orleans, La	Aug. 3, 1869	93, 209
[Medicine.] Bitters	A. M. Loryea	East Portland, Oreg	July 11, 1871	116, 846
[Medicine.] Bitters	E. Phillips	Peoria, Ill	June 18, 1872	128, 169
[Medicine.] Bitters	G. V. Rambaut	Petersburgh, Va	Feb. 25, 1868	74, 940
[Medicine.] Bitters	W. T. Sherwood	Ripon, Wis	Apr. 18, 1871	113, 938
[Medicine.] Bitters	R. G. Turner	Columbia, Tex	Jan. 23, 1872	122, 975
[Medicine.] Bitters	S. R. Whitlow	Limestone Township, Ill	Apr. 4, 1871	113, 713
[Medicine.] Bitters, Anti-dyspeptic	S. M. Barnes	La Fayette, Va	Oct. 15, 1872	132, 233
[Medicine.] Bitters, Anti-dyspeptic	C. T. Provost	New York, N. Y	Aug. 20, 1867	68, 004
[Medicine.] Bitters for chills and fevers	T. B. Owens	Gatesville, Tex	May 20, 1873	139, 185
[Medicine.] Bitters, Jaundice	J. Wheaton	....	Jan. 17, 1801	....
[Medicine.] Bitters, Stomach	L. W. Harris	Chicago, Ill	Jan. 5, 1869	85, 657
[Medicine.] Bitters, Stomach	D. Rinkle	Indianapolis, Ind	Mar. 22, 1870	101, 049
[Medicine.] Bitters, Stomach	F. Stutzel	Wabashaw, Minn	Oct. 7, 1873	143, 427
[Medicine.] Bitters, Tonic	R. Blacklidge	Bridgeport, Conn	Apr. 6, 1869	88, 537
[Medicine.] Bitters, Tonic	F. Fullerton	Williamsport, Pa	Oct. 20, 1868	83, 273
[Medicine.] Bitters, Tonic	I. Hellman	Saint Louis, Mo	Apr. 11, 1865	47, 204
[Medicine.] Bitters, Tonic	R. P. Jenkins	Nashville, Tenn	Feb. 2, 1869	86, 551
[Medicine.] Bitters, Tonic	J. H. McCartney	Dansville, N. Y	May 11, 1869	90, 048
Medicine, Botanic	S. Thompson	Boston, Mass	May 6, 1836	....
[Medicine.] Bronchial troche	S. C. Henszey, jr	Westchester, Pa	June 5, 1866	55, 292
Medicine by means of steam, Administering	C. Whitlaw	New York	Feb. 16, 1825	....
Medicine called bilious cordial	S. Chamberlaine	....	Dec. 31, 1804	....
Medicine called canker-drops	D. Albrook	Onondaga County, N. Y	Mar. 28, 1814	....
Medicine called the essence of tansy	I. Newton	Norwich, Vt	Feb. 28, 1806	....
[Medicine.] Carbolated cod-liver oil	J. H. Willson	Brooklyn, N. Y	May 21, 1872	127, 005
Medicine-case	M. O. Baldwin	Wamego, Kans	Sept. 16, 1873	142, 887
Medicine-case	A. Button	Dunkirk, N. Y	May 17, 1870	103, 012
Medicine-case	W. H. Cutler	Buffalo, N. Y	Mar. 18, 1873	136, 973
Medicine-case	E. H. Hance	Philadelphia, Pa	May 31, 1864	42, 943
Medicine-case	A. Harrington	Otsego County, N. Y	July 16, 1811	....
Medicine-case	J. R. Haynes and A. F. Worthington.	Newport, Ky., and Cincinnati, Ohio.	Aug. 27, 1867	68, 073
Medicine-case	T. M. Perot	Philadelphia, Pa	Sept. 15, 1863	39, 952
Medicine-case	H. M. Smith	New York, N. Y	May 23, 1871	115, 121
Medicine, Chemical anti-dysenteric	J. G. Vought	Rochester, N. Y	Dec. 4, 1821	....
Medicine-chest	R. B. Parkinson and J. M. Maris	Philadelphia, Pa	Nov. 29, 1864	45, 266
Medicine-chest	E. Sander	Saint Louis, Mo	Apr. 20, 1869	89, 173
Medicine-chest	E. M. Skinner	Boston, Mass	Oct. 23, 1866	59, 086
Medicine-chest for veterinary surgeon	J. W. Gladsden	Philadelphia, Pa	Mar. 12, 1872	124, 432
[Medicine.] Chlorine, Cosmetic	D. West	Hudson, N. Y	Jan. 11, 1833	....
[Medicine.] Compound tincture of myrrh	H. Howard	Columbus, Ohio	Aug. 25, 1832	....
[Medicine.] Cough-lozenges	E. Gauvreau	Quebec, Canada	July 30, 1872	129, 942
[Medicine.] Cough-mixture	M. Connell	Jersey City, N. Y	Apr. 16, 1872	125, 723
[Medicine.] Cough-mixture	P. M. Huffman	Harvard, Ill	June 18, 1867	65, 813
[Medicine.] Cough-mixture	S. R. Whitlow	Rosefield, Ill	Aug. 25, 1868	81, 446
[Medicine.] Cough-mixture for candies	L. Violet	New Lebanon, N. Y	June 6, 1871	115, 790
[Medicine.] Cough-sirup	C. Benedict, jr	Omro, Wis	Apr. 24, 1866	54, 098
[Medicine.] Cough-sirup	B. F. Burroughs	West Perry Township, Pa	June 6, 1871	115, 698
[Medicine.] Cough-sirup	G. W. Chatfield	New Haven, Conn	Mar. 22, 1870	101, 096
[Medicine.] Cough-sirup	J. Conzelman	Saint Louis, Mo	Aug. 2, 1870	105, 916

Index of patents issued from the United States Patent Office from 1790 *to* 1873, *inclusive*—Continued.

Invention.	Inventor.	Residence.	Date.	No.
[Medicine.] Curing sores	R. Rood	Centre Lisle, N. Y	Feb. 20, 1836	....
Medicine-dropper	P. J. McElroy	Cambridge, Mass	June 30, 1868	79, 487
[Medicine.] Elixir for the cure of bronchitis, &c., Vegetable.	P. Faulkner	Rockville, Pa	Sept. 23, 1843	3, 282
Medicine-evaporator	E. O. Schartau	Philadelphia, Pa	Mar. 23, 1869	88, 218
Medicine, Eye	D. R. Morgan	San Francisco, Cal	Jan. 14, 1868	73, 369
Medicine, Eye	H. M. Sanderson	Noble, Ill	Feb. 4, 1868	74, 151
Medicine, Eye	A. Schuster	Saint Louis, Mo	Jan. 2, 1866	51, 868
[Medicine.] Eye-balsam	W. Klingbeil	Champaign City, Ill	June 20, 1871	116, 199
[Medicine.] Eye-wash	P. Paul	Black Earth, Wis	Dec. 5, 1871	121, 467
[Medicine.] Eye-wash	F. Pawlewski and J. Schulz	Milwaukee, Wis	Dec. 23, 1873	145, 749
Medicine, Febrifuge	A. Johnson	Northumberland County, Pa.	Jan. 11, 1812	....
[Medicine.] Felon, &c., curing compound	R. Feibelman	Columbus, Ind	Aug. 18, 1868	81, 152
Medicine, Fever	C. F. Holing	Hudson City, N. J	Aug. 13, 1867	67, 763
Medicine, Fever	S. Thompson	Surry, N. H	Mar. 2, 1813	....
Medicine, Fever and ague	W. H. Barker	West Farms, N. Y	May 21, 1872	127, 014
Medicine, Fever and ague	G. Behrens	Brunswick, Mo	Sept. 3, 1872	131, 078
Medicine, Fever and ague	L. Bodenheimer	Paducah, Ky	Dec. 19, 1871	121, 984
Medicine, Fever and ague	W. Campbell	Camden, N. J	Oct. 29, 1872	132, 629
Medicine, Fever and ague	W. Campbell	Camden, N. J	Jan. 7, 1873	134, 639
Medicine, Fever and ague	M. Cannon	New Orleans, La	Oct. 1, 1830	....
Medicine, Fever and ague	J. M. Ferguson	Summit, Miss	Oct. 26, 1869	96, 215
Medicine, Fever and ague	M. Ferro	Iuka, Miss	Sept. 6, 1870	107, 024
Medicine, Fever and ague	G. W. Gay	Bath, Me	May 21, 1872	126, 882
Medicine, Fever and ague	S. Laighton	Newark, N. J	July 12, 1870	105, 224
Medicine, Fever and ague	A. V. Lee	Clayton, Ala	Aug. 18, 1868	81, 181
Medicine, Fever and ague	J. Mabrey	Jefferson City, Mo	June 23, 1868	79, 241
Medicine, Fever and ague	J. H. Morris	Eaton, Ohio	Feb. 8, 1870	99, 583
Medicine, Fever and ague	O. Scidmore	Albany, N. Y	Sept. 19, 1871	119, 053
Medicine, Fever and ague	O. M. Spiller	Akron, Ohio	July 5, 1870	105, 011
Medicine, Fever and ague	G. E. Swan	Mount Vernon, Ohio	Mar. 7, 1871	112, 509
Medicine for asthma	C. B. Hurst	Rochester, Pa	Sept. 10, 1872	131, 167
Medicine for asthma and other diseases	J. C. Lewis	Belfast, Me	May 18, 1869	90, 179
Medicine for bronchial and lung diseases	S. W. Helm	Bachelor Valley, Cal	Mar. 14, 1871	112, 595
Medicine for cancer	J. Andrus	Hillsborough, N. H	July 30, 1816	....
Medicine for cancer, &c	F. Baker	New York, N. Y	June 27, 1871	116, 530
Medicine for cancer	J. Harrigan	Boston, Mass	Nov. 26, 1872	133, 443
Medicine for cancer	J. Harrison	Castalia, N. C	May 16, 1871	114, 810
Medicine for cancer	N. McKelfresh	Elizabeth, Ind	Dec. 5, 1871	121, 534
Medicine for cancer	L. Roy	Plattsburgh, Mo	Oct. 26, 1869	96, 269
Medicine for cancer	M. Williams	Middletown, Ohio	Nov. 2, 1869	96, 371
Medicine for cancer and other diseases	P. A. Cobb	Lynchburgh, Va	Apr. 19, 1870	102, 093
Medicine for catarrh	J. A. Carpenter	Waverly, N. Y	Jan. 21, 1873	135, 077
Medicine for catarrh	F. P. Gardiner	New Haven, Conn	Feb. 5, 1867	61, 730
Medicine for catarrh	W. V. Green and J. M. Kintzel	Wellsville, Ohio	June 17, 1873	139, 951
Medicine for catarrh	J. Snow	Grand Rapids, Mich	Nov. 2, 1869	96, 496
Medicine for catarrh and asthma	E. Field	Ostrander, Ohio	Mar. 7, 1871	112, 329
Medicine for cattle and other animals	O. E. Hornidy	Chauncey, Ill	Apr. 27, 1869	89, 314
Medicine for cattle and other animals	D. P. Mathews	Winthrop, Mass	Sept. 8, 1868	81, 920
Medicine for chills and fever	R. Bevil	Bowie County, Tex	Oct. 10, 1871	119, 812
Medicine for chills and fever	I. Brown	Philadelphia, Pa	Dec. 19, 1871	121, 989
Medicine for chills and fever	W. P. Fennell	Louisville, Ky	May 16, 1871	114, 784
Medicine for cholera	J. Houck	Baltimore, Md	Oct. 25, 1832	....
Medicine for cholera	A. Hunn, sr	Lancaster, Ky	Aug. 12, 1833	....
Medicine for cholera	A. E. Parrott	Norfolk, Va	June 26, 1866	55, 898
Medicine for cholera	A. Racicot	Troy, N. Y	July 8, 1873	140, 642
Medicine for cholera-infantum, &c	A. Haws	Battle Mountain Station, Nev.	July 16, 1872	129, 343
Medicine for cholera-morbus	J. Pool	Hermon, N. Y	Sept. 28, 1869	95, 376
Medicine for colds	W. E. Chilson	Troy, Pa	Jan. 17, 1871	111, 044
Medicine for colic	W. W. Gardner	Russellville, Ky	July 30, 1872	129, 886
Medicine for colic	J. A. McKinnon	Selma, Ala	Mar. 22, 1870	101, 145
Medicine for constipation	C. A. Simmons	Waldo, Fla	Dec. 12, 1871	121, 904
Medicine for consumption	T. Bell	New York, N. Y	Apr. 24, 1866	54, 097
Medicine for consumption	T. W. Bethel	Brooklyn, N. Y	Sept. 26, 1871	119, 301
Medicine for consumption	N. C. Jarrell	High Point, N. C	May 7, 1872	126, 397
Medicine for consumption	J. E. Larkin	Newark, N. J	Oct. 3, 1871	119, 620
Medicine for consumption, &c	C. S. Long	Philadelphia, Pa	May 27, 1812	....
Medicine for consumption, &c	W. H. H. White	San Francisco, Cal	June 24, 1873	140, 320
Medicine for corns and bunions	J. L. S. Hall	Wheeling, W. Va	Sept. 14, 1869	94, 819
Medicine for coughs, &c	J. Cushings	Wellington, Ohio	Aug. 16, 1870	106, 333
Medicine for coughs and colds	D. E. Smith	Cornwall, Conn	June 30, 1828	....
Medicine for coughs and colds	A. Young	Pittsborough, N. C	Apr. 19, 1870	102, 194
Medicine for coughs, colds, &c	C. P. Crossman	West Warren, Mass	May 23, 1871	115, 032
Medicine for coughs, colds, &c	W. H. H. Irvin	Philadelphia, Pa	Feb. 22, 1870	100, 039
Medicine for coughs, colds, &c	J. D. Richmond, jr	New Liberty, Ky	Aug. 8, 1871	117, 930
Medicine for coughs, colds, &c	J. Willey	Oakland, Cal	July 2, 1872	128, 692
Medicine for croup	C. Sanborn	South Berwick, Me	Apr. 8, 1862	34, 905
Medicine for cure of diseases in cattle	E. A. Wilder	Dennysville, Me	June 21, 1870	104, 677
Medicine for cure of gravel	B. W. Deal	New Market, Md	Apr. 6, 1869	88, 554
Medicine for cure of gravel	T. G. Eiswald	Providence, R. I	Aug. 16, 1870	106, 474
Medicine for cure of foot-rot in sheep	G. Eckert	Charlestown, Ohio	Jan. 5, 1869	85, 650
Medicine for cure of foot-rot in sheep	W. T. Sherman	Marengo, Ohio	Apr. 4, 1871	113, 459
Medicine for curing foot-rot in horses, &c	E. W. Wakefield	San Francisco, Cal	Oct. 13, 1863	40, 297
Medicine for cutaneous diseases	E. A. Shewell	Boston, Mass	Nov. 21, 1871	121, 205
Medicine for diarrhœa, &c	A. B. Dorman	Cape Girardeau, Mo	Nov. 19, 1872	133, 213
Medicine for diarrhœa	R. A. Walton	Shawneetown, Ill	May 30, 1871	115, 547
Medicine for diptheria	A. J. Denison	Paris, Mich	May 25, 1869	90, 347
Medicine for diptheria, &c	T. J. Glines	Hebron, N. Y	Sept. 20, 1864	44, 299
Medicine for diptheria	H. Pruden	Erie, Pa	June 24, 1873	140, 163
Medicine for diptheria	H. L. Sheffer	Marston, Wis	May 24, 1870	103, 512
Medicine for dropsy	C. Brown	New Albany, Ind	Oct. 20, 1868	83, 249
Medicine for dropsy	J. Newton and I. Barfield	Richland Parish, La	Feb. 1, 1870	99, 464
Medicine for dropsy	J. R. Strickland	Sayville, N. Y	July 6, 1869	92, 394
Medicine for dropsy and epilepsy	J. S. Fall	Rattlesnake Springs, Ga	Nov. 25, 1831	....
Medicine for dysentery and dyspepsia	T. Powell	Burlington, Vt	Feb. 2, 1828	....
Medicine for dyspepsia	C. Brightman	Crittenden County, Ky	Sept. 3, 1872	131, 048

Index of patents issued from the United States Patent Office from 1790 to 1873, inclusive—Continued.

Invention.	Inventor.	Residence.	Date.	No.
Medicine for dyspepsia	N. H. Chesebrough	New York, N. Y	Apr. 17, 1866	54, 056
Medicine for dyspepsia, &c	D. Mayon and E. Champlain	Cloverdale, Cal	Aug. 13, 1872	130, 518
Medicine for erysipelas	H. A. Lamb	Portland, Me	July 4, 1865	48, 567
Medicine for fowls	S. H. Kipp	Millerstown, Pa	Oct. 25, 1870	108, 603
Medicine for gonorrhœa, &c	S. F. Alford	Goshen Springs, Miss	Oct. 22, 1872	132, 425
Medicine for gout	E. Smith	New York	Dec. 15, 1828	......
Medicine for gout and rheumatism	W. A. Parker	Accomac County, Va	Nov. 4, 1831	......
Medicine for healing wounds, &c	N. W. Gaddy	Nichols, S. C	July 16, 1872	129, 469
Medicine for heart-disease, &c	M. D. Britten	Eaton, Mich	July 2, 1872	128, 528
Medicine for kidney-diseases	R. Hawkins and A. A. Hill	Beallsville, Pa	Aug. 1, 1871	117, 537
Medicine for kidney-diseases	W. A. Ruth	Wyoming, Del	Apr. 1, 1873	137, 488
Medicine for liver and other diseases	C. A. Simmons	Waldo, Fla	July 25, 1871	117, 338
Medicine for liver-disease	J. M. Cunningham	Mount Morris, Ill	June 10, 1873	139, 707
Medicine for lung and other diseases	W. Trinder	Philadelphia, Pa	Oct. 8, 1872	132, 119
Medicine for lung-diseases	W. A. Smith	Concord, N. C	Aug. 13, 1872	130, 392
Medicine for miasmatic diseases	J. C. Hascall	Corunna, Mich	Apr. 22, 1862	35, 022
Medicine for neutralizing the taste of cod-liver oil	C. D. Bradley	Taunton, Mass	July 22, 1873	141, 030
Medicine for organs of the voice	F. Frisiani	New York, N. Y	Aug. 24, 1869	93, 982
Medicine for piles	L. E. Brady	Gatesville, N. C	July 9, 1872	128, 784
Medicine for piles	W. Carr	Bath, Me	May 12, 1863	38, 466
Medicine for piles	S. S. Davis	Edgerton, Wis	July 13, 1869	92, 587
Medicine for piles	L. Heins	New York, N. Y	Jan. 9, 1872	122, 608
Medicine for piles	L. L. Moore and R. B. S. Whayre	Calhoun, Ky	June 27, 1871	116, 474
Medicine for piles	M. Rinehart	Lockport, Ill	Nov. 24, 1868	84, 441
Medicine for piles	P. Roskopf	Brooklyn, E. D., N. Y	Feb. 20, 1872	123, 934
Medicine for piles	J. W. Ward	Lowell, Ohio	June 11, 1872	127, 719
Medicine for purifying the blood	R. Guinn	Baltimore, Md	Apr. 20, 1869	89, 039
Medicine for purifying the blood	P. H. Schmid	New York, N. Y	Oct. 24, 1871	120, 333
Medicine for rheumatism, &c	H. A. Chase	Tully, N. Y	July 19, 1870	105, 425
Medicine for rheumatism	G. Conroy	Mendocino County, Cal	Oct. 12, 1869	95, 771
Medicine for rheumatism, &c	W. Curless	Trucker, Cal	Aug. 8, 1871	117, 870
Medicine for rheumatism, &c	W. H. Farrar	Richmond, Va	Apr. 25, 1871	113, 995
Medicine for rheumatism	R. Gilkinson	New York, N. Y	Jan. 24, 1871	111, 195
Medicine for rheumatism, &c	T. B. Hick	New Haven, Conn	Aug. 13, 1872	130, 497
Medicine for rheumatism	A. J. Jenkins	Virginia City, Nev	Nov. 5, 1870	109, 216
Medicine for rheumatism	N. Jenkins	New Orleans, La	Dec. 27, 1870	110, 470
Medicine for rheumatism	J. Keiser	York, Pa	Oct. 17, 1871	120, 074
Medicine for rheumatism	W. Landert and J. Deggeller	Chicago, Ill	May 3, 1870	102, 686
Medicine for rheumatism	R. Lighthall	Brooklyn, N. Y	June 29, 1869	92, 065
Medicine for rheumatism, &c	G. Lucy	Mobile, Ala	Apr. 29, 1873	138, 335
Medicine for rheumatism	H. H. Munroe	Louisville, Ky	Dec. 14, 1869	97, 794
Medicine for rheumatism	H. Sawyer	Wilson, N. C	May 9, 1871	114, 717
Medicine for rheumatism	C. F. Washburn	San Francisco, Cal	Nov. 7, 1871	120, 802
Medicine for rheumatism	A. Whitney	Boston, Mass	Sept. 5, 1871	118, 766
Medicine for rheumatism and other diseases	J. B. Rodgers	Saint Louis, Mo	Apr. 19, 1870	102, 161
Medicine for ringbone, spavin, &c	W. A. Cleveland	Waterville, N. Y	Mar. 23, 1869	88, 134
Medicine for ringbone, spavin, &c	A. M. French	Burton, Ohio	May 28, 1867	65, 203
Medicine for ringbone, spavin, &c	H. H. Vance	New Corydon, Ind	Nov. 24, 1868	84, 322
Medicine for ringbone, spavin, splint, &c	W. A. Cleveland	Waterville, N. Y	Oct. 22, 1867	69, 970
Medicine for scarlatina, sore-throat, &c	M. A. Ragsdale	Grass Valley, Cal	Aug. 16, 1864	43, 866
Medicine for scrofula and cancer	E. Willard	Albany, N. Y	Oct. 8, 1810	......
Medicine for scrofula, liver-complaint, &c	P. Fay, sr	Lacon, Ill	Mar. 28, 1870	113, 151
Medicine for scrofula, rheumatism, &c	G. Jaques	Wilmington, Del	July 16, 1824	......
Medicine for skin-diseases	F. W. A. Bergengren	Stockholm, Sweden	Mar. 19, 1872	124, 660
Medicine for skin-diseases	A. B. Simpson	Camden, Ohio	Mar. 4, 1873	136, 554
Medicine for small-pox	A. Haws	Battle Mountain Station, Nev.	July 16, 1872	129, 342
Medicine for sore-throat	F. M. Moore	Chico, Cal	Oct. 10, 1871	119, 782
Medicine for spavin	W. J. Andrews	East Machias, Me	June 20, 1871	116, 135
Medicine for the teeth	J. C. Hassell	Nevada City, Cal	July 2, 1872	128, 620
Medicine for toothache	F. J. Oswald	New York, N. Y	Dec. 24, 1872	134, 217
Medicine for toothache	S. Pennington	Mount Pleasant, Ohio	July 30, 1829	......
Medicine for toothache	W. P. Sigsby	Delta, Ohio	Oct. 5, 1869	95, 610
Medicine for toothache	J. Utley	Hartford, Conn	Nov. 5, 1817	......
Medicine for toothache	T. White	Saint Clairsville, Ohio	July 29, 1829	......
Medicine for treating catarrh, &c., by inhalation	D. Slade	Chicago, Ill	Mar. 21, 1871	112, 858
Medicine for treating diseases of the lungs, chest, &c.	M. E. Perrin	Montreal, Canada	July 11, 1871	116, 863
Medicine for treating horses' hoofs, &c	J. C. Burroughs	Clinton, Mich	Dec. 2, 1873	145, 047
Medicine for treating ringworm, &c	P. Roskopf	Brooklyn, N. Y	June 11, 1872	127, 925
Medicine for wounds, inflammation, &c	O. Troemel	Manitowoc, Wis	Mar. 8, 1864	41, 873
Medicine for yellow-fever, &c	T. H. Thompson	Grenada, Miss	Mar. 12, 1872	124, 458
Medicine from chloral	T. H. Hazard	Richmond, Va	Mar. 15, 1870	100, 760
Medicine from globe-flowers	J. S. Pemberton	Atlanta, Ga	Mar. 1, 1870	100, 439
[Medicine.] Glanders in horses, Compound for cure of.	J. Althouse	East Cocalico Township, Pa.	Mar. 26, 1867	63, 195
Medicine-glass	T. G. Boggs	Philadelphia, Pa	Dec. 22, 1868	85, 203
Medicine, Hog	J. Shannon	Palmyra, Mo	Apr. 9, 1872	125, 492
Medicine, Hog	D. W. Stow	Thorntown, Ind	Dec. 18, 1866	60, 591
Medicine, Hog and chicken cholera	M. N. George	Evansville, Ind	July 18, 1871	117, 065
Medicine, Hog and chicken cholera	A. C. McMahan	Lincoln, Ill	Aug. 31, 1869	94, 428
Medicine, Hog-cholera	J. B. Ball	Lebanon, Ind	June 23, 1868	79, 098
Medicine, Hog-cholera	S. S. Barger	Golconda, Ill	July 6, 1869	92, 248
Medicine, Hog-cholera	G. H. Baugh	Oskaloosa, Iowa	Nov. 27, 1866	59, 946
Medicine, Hog-cholera	M. Boyes	Pocahontas, Ill	Oct. 2, 1866	58, 371
Medicine, Hog-cholera	A. J. Carver and E. P. Horn	Greenhill, Tenn	Sept. 15, 1868	82, 205
Medicine, Hog-cholera	N. H. Cass	Henryville, Ind	Aug. 4, 1868	80, 597
Medicine, Hog-cholera	T. L. Cotten	Madison County, Miss	Oct. 18, 1870	108, 453
Medicine, Hog-cholera	E. Fabra	Covington, Ky	Oct. 15, 1872	132, 204
Medicine, Hog-cholera	H. C. Fahlbush	Marshall, Ill	June 25, 1872	128, 380
Medicine, Hog-cholera	G. W. Gish and W. H. Ferguson.	Daviess County, Ky	Jan. 7, 1868	72, 996
Medicine, Hog-cholera	J. and M. J. Holton	Mahaska County, Iowa	May 21, 1872	126, 961
Medicine, Hog-cholera	A. M. Johnston and H. H. Avrit.	Clarksville, Tenn	Aug. 3, 1869	93, 206
Medicine, Hog-cholera	C. L. Jones	Pedler Township, Va	Oct. 5, 1871	121, 523
Medicine, Hog-cholera	F. La Rew	Hamilton, Ohio	Apr. 14, 1863	38, 169
Medicine, Hog-cholera	J. Lighter	Clay Village, Ky	Mar. 19, 1861	31, 723

Index of patents issued from the United States Patent Office from 1790 *to* 1873, *inclusive*—Continued.

Invention.	Inventor.	Residence.	Date.	No.
Medicine, Hog-cholera	J. S. Mason	Jefferson County, Ky	July 3, 1866	56, 075
Medicine, Hog-cholera	J. H. Mesler	Symmes' Corners, Ohio	May 28, 1867	65, 256
Medicine, Hog-cholera	W. B. Robuck	Oxford, Miss	Jan. 12, 1869	85, 855
Medicine, Hog-cholera	W. M. Runyon, R. H. Hailer, and D. B. Morris.	Oskaloosa, Iowa	Oct. 16, 1866	58, 890
Medicine, Horse	J. Andrews	Hardin County, Ohio	Jan. 14, 1873	134, 721
Medicine, Horse	T. Burns	Jefferson County, Ohio	Apr. 23, 1872	125, 933
Medicine, Horse	P. Cope	Perryopolis, Pa	Apr. 9, 1872	125, 441
Medicine, Horse	C. L. Hammond	North Java, N. Y	Oct. 30, 1866	59, 216
Medicine, Horse	V. Vance	Havana, N. Y	July 25, 1865	49, 015
Medicine, Horse and cattle powder	A. Rudisill and M. Sell	Sell Station, Pa	Feb. 14, 1871	111, 876
Medicine, Horse, cattle, &c	G. Van Wagenen	Racine, Wis	Sept. 1, 1868	81, 711
[Medicine.] Horse-powder	T. H. Russell	Lebanon, N. H	Aug. 15, 1871	118, 056
[Medicine.] Lotion	J. Hopkins	Philadelphia, Pa	Aug. 24, 1809	....
Medicine, Nutritive	S. C. Upham	Philadelphia, Pa	Feb. 12, 1867	62, 091
Medicine of vegetable alkaloids	G. W. Scollay	New York, N. Y	Dec. 19, 1871	122, 065
Medicine or apple ginger	D. Nixon	Philadelphia, Pa	Sept. 12, 1871	118, 812
Medicine or blood-purifier	D. Parlow	Philadelphia, Pa	May 13, 1873	138, 759
Medicine or capsicum plaster	J. and I. Coddington	New York, N. Y	Apr. 2, 1872	125, 173
Medicine or coffee-antidote	E. D. D. St. Cyr	Lowell, Mass	Dec. 5, 1871	121, 665
Medicine or cordial	J. Ambrose	Nashville, Tenn	Oct. 5, 1869	95, 407
Medicine or cordial	J. Shapley and A. D. Hutchinson.	Rosefield, Ill	June 29, 1869	92, 107
Medicine or ethereal oil for treating rheumatism, &c	L. J. Bell	College Corner, Ind	Feb. 13, 1872	123, 666
Medicine or infant's powders	J. Fehr	Hoboken, N. J	Sept. 23, 1873	143, 133
Medicine or liver-invigorator	W. S. Simmons	Weatherford, Tex	Jan. 9, 1872	122, 672
Medicine or liver-lozenge	S. Gilbert	Portsmouth, N. H	Dec. 16, 1873	145, 498
Medicine or lotion for the skin	A. Field	Springfield, Mass	June 24, 1873	140, 126
Medicine or milk of magnesia	C. H. Phillips and L. Reid	New York and Brooklyn, N. Y.	Apr. 29, 1873	138, 282
Medicine or mosquito-lotion	A. D. Breazeale	Selma, Ala	Sept. 24, 1872	131, 655
Medicine from oak bark	S. J. Miller	Economy, Ind	Nov. 16, 1869	96, 943
Medicine or opium-cure	U. H. Kellogg	Jamestown, N. Y	Sept. 23, 1873	143, 164
Medicine or pain-remedy	J. J. Mills and W. Perry	Frederick's Hall, Va	Nov. 1, 1870	108, 811
Medicine or pectoral sirup	G. La Montagne	Muskegon, Mich	Sept. 6, 1870	107, 066
Medicine or piles-liniment	P. H. Steenbergen	Chico, Cal	Aug. 19, 1873	141, 961
Medicine or pill	J. P. Waddell	Brookhaven, Miss	Aug. 9, 1870	106, 239
Medicine or powders for horses, &c	J. Norcross	West Newton, Pa	Mar. 28, 1871	113, 198
Medicine or powders for treating cattle and hogs	R. H. Pullin	McDowell, Va	Aug. 15, 1871	118, 052
Medicine or restorative balsam	G. H. Brecht	Burton, Ill	Nov. 7, 1871	120, 705
Medicine or worm-candy	J. C. Wills	Owingsville, Ky	Jan. 2, 1872	122, 507
[Medicine.] Pain-destroying compound	S. T. Bond	Edenton, N. C	July 21, 1868	80, 123
[Medicine.] Pain-killer	H. Stonebraker	Baltimore, Md	Feb. 19, 1867	62, 297
Medicine, Preparation of aerated drink for	P. H. Vander Weyde and J. Matthews, jr.	New York, N. Y	Mar. 30, 1869	88, 348
Medicine, Preparation of cod-liver and other oils for	T. Hyatt	Atchison, Kans	Jan. 5, 1869	85, 668
Medicine, Preparation of iron for	J. E. Siebel	Chicago, Ill	July 27, 1869	93, 131
Medicine, Process of preparing homeopathic	B. Fincke	Brooklyn, N. Y	Aug. 24, 1869	93, 980
[Medicine.] Prophylactic remedy	C. H. Thomas	New Orleans, La	Oct. 16, 1860	30, 434
Medicine, Putting up	F. Kraus	Cincinnati, Ohio	Oct. 4, 1870	108, 034
[Medicine.] Salt-rheums, Remedy for	W. B. Trufant	Bath, Me	Feb. 10, 1838	597
Medicine-spoon and bottle-stopper combined	S. C. Currie	New York, N. Y	July 6, 1869	92, 278
Medicine, Sugar tablet for containing	H. F. Wiesecke	New York, N. Y	Oct. 28, 1862	36, 816
[Medicine.] Sweating-powders	H. Howard	Columbus, Ohio	Aug. 25, 1832	....
[Medicine.] Syphilis, Cure of	N. W. Badeau	New York	Mar. 8, 1832	....
[Medicine.] Syphilis, Cure of	J. Moser	Philadelphia, Pa	Jan. 26, 1816	....
Medicine to animals, Instrument for giving fluid	H. A. Brandes	Newark, N. J	June 21, 1870	104, 416
Medicine to diseased parts, Instrument for conveying.	P. J. Frank	Ashford, N. Y	Nov. 2, 1869	96, 415
Medicine to prevent or cure the effects of the bite of a mad-dog.	W. Stoy	Lebanon, Pa	June 9, 1809	....
Medicine to prevent scurvy	J. W. Armour	Fredericktown, Md	Sept. 28, 1827	....
Medicine to the ear, Apparatus for applying	S. Van Etten	Corning, N. Y	May 21, 1872	127, 123
[Medicine.] Tonic or bitters	J. Heisler	Schuylkill Haven, Pa	Mar. 3, 1868	75, 161
Medicine, Worm	A. E. Wright	Hamlin, N. Y	May 30, 1871	115, 407
Medicine, Worm-destroying	J. J. Oellig	Waynesborough, Pa	Oct. 28, 1837	445
Medicine: *See* Liniment, Ointment, Salve.				
Melodeon	C. Austin	Concord, N. H	June 19, 1849	6, 543
Melodeon	C. Austin	Concord, N. H	Jan. 22, 1867	61, 305
Melodeon	J. Berger	Knoxville, Ill	May 14, 1867	64, 741
Melodeon	W. H. Bigelow	South Framingham, Mass	Sept. 3, 1861	33, 180
Melodeon	T. Brett	Geneva, Ohio	Oct. 25, 1870	108, 560
Melodeon	J. C. Briggs	Woodbury, Conn	Feb. 10, 1857	16, 574
Melodeon	J. C. Briggs	Ansonia, Conn	Sept. 27, 1870	107, 655
Melodeon	R. Burditt	Brattleborough, Vt	Sept. 10, 1861	33, 278
Melodeon	R. Burditt and H. P. Green	Brattleborough, Vt	Mar. 10, 1857	16, 786
Melodeon	C. G. Burke	Utica, N. Y	Dec. 6, 1859	26, 344
Melodeon	P. Corbach	Cleveland, Ohio	Aug. 28, 1866	57, 472
Melodeon	J. Carhart	New York, N. Y	Apr. 17, 1855	12, 713
Melodeon	J. Carhart	New York, N. Y	July 1, 1856	15, 218
Melodeon	J. R. Cluxton	Russellville, Ohio	Sept. 14, 1869	94, 713
Melodeon	W. Cooper	Deposit, N. Y	Jan. 7, 1868	73, 080
Melodeon	K. Ebermeyer	Ellwangen, Germany	Dec. 3, 1867	71, 593
Melodeon	P. Engers	Jefferson Furnace, Pa	Apr. 25, 1871	113, 994
Melodeon	W. Evans	Lockport, Ill	June 9, 1857	17, 501
Melodeon	O. N. Frary	Ansonia, Conn	Aug. 8, 1854	11, 479
Melodeon	H. N. Goodman	New Haven, Conn	June 28, 1853	9, 816
Melodeon	R. Goodrich	Pittsfield, Mass	Mar. 15, 1864	41, 915
Melodeon	E. Hamlin	Winchester, Mass	July 21, 1868	80, 167
Melodeon	F. Hoddick	Buffalo, N. Y	Oct. 23, 1866	59, 128
Melodeon	G. G. Hunt	Wolcottville, Conn	Oct. 23, 1855	13, 704
Melodeon	S. A. Jewett	Cleveland, Ohio	Oct. 13, 1857	18, 399
Melodeon	S. A. Jewett	Cleveland, Ohio	May 28, 1867	65, 088
Melodeon	S. H. Jones	Jamaica Plains, Mass	July 3, 1860	29, 037
Melodeon	W. J. Kinnard and B. Dreher	Cleveland, Ohio	Apr. 4, 1865	47, 110

Index of patents issued from the United States Patent Office from 1790 *to* 1873, *inclusive*—Continued.

Index of patents issued from the United States Patent Office from 1790 *to* 1873, *inclusive*—Continued.

Invention.	Inventor.	Residence.	Date.	No.
Metal can	J. Pollock and T. J. Diederick	Philadelphia, Pa	Nov. 3, 1868	83, 660
Metal can and box for paints, &c	F. W. Devoe	New York, N. Y	May 28, 1867	65, 181
Metal can, case, box, &c., for preserving food, gunpowder, liquids, paints, &c.	J. Bouvet	La Rochelle, France	June 28, 1864	43, 378
Metal can or box	H. Everett	Philadelphia, Pa	May 30, 1865	47, 939
Metal can or case for putting up alkalies	J. McCoy	Philadelphia, Pa	May 12, 1868	77, 899
Metal cans, Machine for cutting and pressing the heads of	R. Brady	New York, N. Y	Jan. 4, 1870	98, 550
Metal cans, Machine for manufacturing	J. Mays and E. W. Bliss	Brooklyn, N. Y	June 15, 1869	91, 248
Metal cans, Securing caps to	D. W. Pepper	Philadelphia, Pa	Jan. 29, 1867	61, 563
Metal caps for boxes, Machine for making	C. Diederick and W. T. Slocum	Philadelphia, Pa	Apr. 3, 1860	27, 756
Metal-casting mold	C. B. Cotter	Milford, Pa	Nov. 18, 1862	36, 943
Metal-casting mold	H. Davies	Covington, Ky	Jan. 11, 1870	98, 673
Metal castings, Machine for grinding	G. H. Spencer	York, Pa	Mar. 25, 1873	137, 104
Metal cleaning and silvering composition	C. P. Brockett, E. Todd, and J. Brockett.	New Haven, Conn	Mar. 13, 1860	27, 425
Metal-cleaning powder	J. H. Musgrave and J. M. Biedel.	Chambersburgh, Pa	May 7, 1872	126, 409
Metal-coated ware	G. Booth	Philadelphia, Pa	Feb. 18, 1873	136, 025
Metal-coating composition	J. Weisman	Philadelphia, Pa	July 5, 1864	43, 444
Metals, Coloring	J. Kintz	West Meriden, Conn	May 27, 1873	139, 319
Metal comb	H. Durell	New York, N. Y	June 20, 1836	....
Metal comb	R. A. Ives	Bristol, Conn	July 1, 1836	....
Metal combs, Making	R. A. Ives	Bristol, Conn	Sept. 25, 1838	945
Metal, Composition	J. Hackert	Bridgeport, Conn	May 17, 1864	42, 762
Metal-compressing and wire-pointing machine	A. B. Hendryx and A. W. Webster.	Ansonia, Conn	Nov. 19, 1872	133, 224
Metal-corrugating machine	J. B. Williams	Glastenbury, Conn	June 29, 1869	91, 889
Metal cupping and raising die	S. R. Wilmot	Bridgeport, Conn	June 26, 1866	55, 944
Metal cutter and shearer	I. Dubois	Boonesborough, Iowa	Nov. 15, 1870	109, 192
Metal cutter and shearer	T. C. Robinson	Boston, Mass	July 6, 1869	92, 368
Metal-cutter, Rotary	A. P. Stephens	Brooklyn, N. Y	Mar. 16, 1869	87, 985
Metal cutting and planing machine	D. Saunders	New York, N. Y	Oct. 28, 1862	36, 803
Metal cutting and punching machine	N. De Telescheff	St. Petersburg, Russia	Apl. 26, 1864	42, 531
Metal cutting and punching machine	A. L. Hastings	Horton, Iowa	May 7, 1872	126, 543
Metal cutting and shaping machine	C. Van Haagen	Philadelphia, Pa	Dec. 27, 1870	110, 609
Metal-cutting machine	A. S. Bunker	Lawrence, Mass	Apl. 30, 1872	126, 179
Metal-cutting machine	J. Tetlon	Salem, Mass	July 14, 1857	17, 813
Metal cutting, punching, and upsetting machine	H. B. Sevey	Vienna, Me	July 9, 1872	128, 758
Metal-cutting shears	H. Barth	Cincinnati, Ohio	Oct. 10, 1865	50, 325
Metal-cutting shears	G. Christian	Mentz, N. Y	Sept. 20, 1834	....
Metal-cutting shears	J. Hornig	Oswego, N. Y	Feb. 7, 1865	46, 237
Metal-cutting shears	G. M. Marshall	New Haven, Wis	Dec. 26, 1871	122, 270
Metal-cutting shears	T. F. Taft	Worcester, Mass	Aug. 18, 1857	18, 025
Metal-cutting tool	L. F. Goodyear	New Haven, Conn	Mar. 29, 1859	23, 364
Metal-cutting tool	J. Mooney	Providence, R. I	June 24, 1856	15, 190
Metal-die	B. Mead and G. Richards	Boston, Mass	July 2, 1814	....
Metal, Die for forming articles of	E. J. Manville	Waterbury, Conn	June 5, 1866	55, 425
Metal-drill	W. C. Burch	Gloucester, N. J	Dec. 31, 1872	134, 355
Metal-drill	W. R. Jones	Granville, N. Y	Jan. 11, 1836	....
Metal-drill	J. A. Kirkpatrick and G. W. Hornby.	Evansville, Ind	Aug. 10, 1869	93, 451
Metal-drill	W. Lyon	New York, N. Y	Sept. 20, 1853	10, 035
Metal-drill	J. Murphy	Half Moon, Pa	May 17, 1859	24, 045
Metal-drill	A. Pease	Enfield, Conn	Apr. 4, 1854	10, 744
Metal-drill	W. Wakeley	Homer, N. Y	Nov. 28, 1854	12, 010
Metal drill or borer, Governing the feed of a	J. R. Grout	Birmingham, Mich	Apr. 25, 1844	3, 564
Metal drilling and boring machine	C. Sellers	Philadelphia, Pa	Mar. 1, 1864	41, 789
Metal-drilling machine	R. Wilson	Milton, Pa	Dec. 14, 1858	22, 323
Metal-drilling machine	A. Woodworth	Cambridge Village, N. Y	Dec. 23, 1873	145, 923
Metal-drilling tool	E. Davies and R. H. Taunton	Birmingham, England	Nov. 26, 1867	71, 463
Metal expansion and contraction to generate power	H. Knowles	Hartford, Conn	Nov. 8, 1836	....
Metal-folding machine	D. Newton	Southington, Conn	Jan. 23, 1855	12, 282
Metal-forging machine	E. Wheeler	Feltonville, Mass	Jan. 27, 1857	16, 505
Metal, glass, and pottery, Manufacturing	E. W. Nobl	Ripon, Wis	Apr. 30, 1867	64, 353
Metal, glass, &c., Furnace for melting	W. P. Prickett	Philadelphia, Pa	May 19, 1868	78, 007
Metal-heating apparatus	J. Roy	Boston, Mass	Feb. 2, 1869	86, 456
Metal-heating furnace	I. J. Conklin	Philadelphia, Pa	Oct. 15, 1872	132, 139
Metal hoop	S. Lewis	Rochester, N. Y	Aug. 22, 1871	118, 252
Metal hoop, Flaring	H. D. Barnes	New Haven, Conn	Nov. 15, 1864	45, 015
Metal, Joining plates of	E. Jacobs	Cincinnati, Ohio	Feb. 21, 1860	27, 222
Metal melting and refining furnace	T. and J. Barnhurst and W. Walker.	Philadelphia, Pa	May 4, 1814	....
Metal-melting furnace	J. Harrison	East Hampton, Conn	July 22, 1873	141, 139
Metal-melting furnace	W. Shea and L. D. Harvey	Harvey, Mich	July 14, 1868	79, 866
Metal-melting furnace	W. A. Sweet	Syracuse, N. Y	July 11, 1865	48, 739
Metal-melting furnace	J. Thomson and T. Widdowfield.	New York and Brooklyn, N. Y.	June 7, 1864	43, 048
Metal-melting furnace and pot	S. Grilley	Waterbury, Conn	Mar. 17, 1838	644
Metal-mold	J. W. Hollingsworth	Mount Vernon, Ind	Apr. 12, 1870	101, 877
Metal or wood into the ground, Apparatus for driving.	W. W. Winter	Cortlandville, N. Y	Feb. 20, 1866	52, 799
Metal pipe and tube for conveying water, gas, &c	D. D. Parmelee	New York, N. Y	June 18, 1872	128, 166
Metal-pipe-bending machine	J. Perkins and W. H. Burnet	Newark, N. J	Oct. 14, 1856	15, 899
Metal pipe or tube, Composition	B. Titus	Ulysses, N. Y	Apr. 19, 1831	....
Metal pipes, Casting	B. S. Benson	Baltimore, Md	Dec. 3, 1867	71, 687
Metal pipes, Casting	J. J. C. Smith	Philadelphia, Pa	May 18, 1869	90, 201
Metal pipes, Composition for lining	W. Johnston and H. Forbes	Brooklyn, N. Y	Jan. 18, 1859	22, 634
Metal pipes, Connection for soft	I. Davis	Brooklyn, N. Y	June 23, 1868	79, 055
Metal pipes, Determining the thickness of	H. Allen	New York, N. Y	July 8, 1843	3, 145
Metal pipes, Machine for coiling	P. L. Weimer	Lebanon, Pa	Nov. 9, 1858	22, 044
Metal pipes, Machine for coiling	P. L. Weimer	Lebanon, Pa	Aug. 30, 1859	25, 294
Metal pipes, Molding	H. Parmelee	Philadelphia, Pa	Sept. 4, 1860	29, 908
Metal pipes, roofing, &c., Composition for coating	J. Specht	Memphis, Tenn	Jan. 14, 1873	134, 822
Metal pipes with gutta-percha lining	A. D. Puffer	Somerville, Mass	May 20, 1856	14, 929
Metal-planing gage	H. H. Jennings	New Haven, Conn	Aug. 4, 1863	39, 404
Metal-planing machine	F. A. Pratt	Hartford, Conn	Aug. 17, 1869	93, 903
Metal-planing machine	A. Whitcomb	Worcester, Mass	Aug. 17, 1869	93, 847

Index of patents issued from the United States Patent Office from 1790 *to* 1873, *inclusive*—Continued.

Invention.	Inventor.	Residence.	Date.	No.
Metal-plate-bending machine	J. W. Easby	Washington, D. C	July 4, 1865	48, 538
Metal-plate-bending machine	R. Tippett	Harrisburgh, Pa	Feb. 23, 1869	87, 224
Metal-plate-heating and ore-calcining furnace	H. Chess	Pittsburgh, Pa	July 25, 1871	117, 382
Metal-plate-polishing machine	E. C. Atkins	Indianapolis, Ind	June 16, 1868	78, 915
Metal-plates, Apparatus for cutting tined implement from.	A. Clark	Saint Johnsville, N. Y	Nov. 25, 1862	36, 993
Metal plates, Beveling the edges of	W. H. Singer	Pittsburgh, Pa	Dec. 6, 1870	109, 953
Metal plates, Device for grinding	T. Handy	Decatur, Ill	Dec. 4, 1866	60, 181
Metal plates, Machine for manufacturing waved and corrugated.	R. Montgomery	New York, N. Y	July 26, 1859	24, 882
Metal plates, Machine for manufacturing waved and corrugated.	R. Montgomery	New York, N. Y	July 26, 1859	24, 883
Metal plates, Machine for straightening rolled	R. Crooker	Boston, Mass	June 25, 1872	128, 287
Metal plates of unequal thickness, Machine for uniting.	J. Carhart	New York, N. Y	Aug. 1, 1854	11, 413
Metal plates or sheets, Machine for cutting	M. Gregg	Wilmington, Del	Apr. 1, 1842	2, 532
Metal-pointing machine	A. Moltz	New York, N. Y	Dec. 22, 1863	41, 014
Metal-polishing machine	R. Cave	Louisville, Ky	Feb. 1, 1859	22, 781
Metal-pouring ladle	A. S. Wells	New Britain, Conn	Mar. 15, 1870	100, 826
Metal-press	C. H. Schubens	Newark, N. J	May 7, 1867	64, 584
Metal punching and shearing machine	E. A. Sloat	Stone Mill, N. Y	Jan. 16, 1872	122, 860
Metal-punching machine	M. Burnett and C. Vander Woerd.	Boston, Mass	Apr. 17, 1855	12, 712
Metal-punching machine	D. and G. Fowler	Wallingford, Conn	Apr. 17, 1855	12, 723
Metal raising and trimming die	S. Simpson	Wallingford, Conn	May 25, 1869	90, 596
Metal-rod-bending machine	G. J. Neveil	Philadelphia, Pa	Aug. 8, 1865	49, 294
Metal rods and tubes, Making	J. Newman	Birmingham, England	Nov. 28, 1854	12, 000
Metal rods, Machine for forming eyes on	C. Kellogg	Detroit, Mich	Feb. 11, 1868	74, 377
Metal, Rolling: *See* Rolling metal.				
Metal scouring and burnishing machine	J. B. Lyons	Milton, Conn	Apr. 5, 1870	101, 479
Metal-scraping machine	J. Stever	Bristol, Conn	Dec. 12, 1854	12, 076
Metal screen	J. Hopkins	New York, N. Y	Nov. 18, 1862	36, 957
Metal screens, Machine for punching	J. W. Nesmith	Black Hawk, Colo	Aug. 31, 1869	94, 333
Metal-separator	E. Borlase	Bristol, Conn	July 7, 1857	17, 721
Metal-separator	T. J. Chubb	New York, N. Y	Dec. 19, 1854	12, 090
Metal-separator	H. W. Lathrop	Baltimore, Md	June 25, 1872	128, 404
Metal-separator	S. J. Peet	Boston, Mass	Apr. 29, 1873	138, 276
Metal-separator, Magnetic	P. Righter	Newark, N. J	Sept. 24, 1867	69, 128
Metal-shaping die	L. Dodge	Waterford, N. Y	July 19, 1859	24, 842
Metal-shaping machine	W. E. Cass	Newark, N. J	July 30, 1872	130, 016
Metal-shaping machine	M. Wells	Williamsburgh, N. Y	Apr. 23, 1867	64, 176
Metal-shearing apparatus	I. Lamplugh	Springfield, Ill	Mar. 22, 1864	42, 003
Metal-shearing device	J. E. Heath	Niles, Mich	June 11, 1867	65, 745
Metal-shearing machine	W. J. Adams	Grand Rapids, Mich	May 2, 1871	114, 385
Metal-shearing machine	R. Briggs	Philadelphia, Pa	June 22, 1869	91, 512
Metal-shearing machine	E. Doty and H. Richardson	Janesville, Wis	Mar. 29, 1870	101, 356
Metal-shearing machine	A. A. Kent	Lyons, Iowa	Sept. 6, 1870	107, 179
Metal-shearing machine	N. B. Reynolds	Auburn, N. Y	Feb. 19, 1867	62, 154
Metal shearing, punching, and bending apparatus	I. Lamplugh	Springfield, Ill	Mar. 22, 1864	42, 004
Metal-sheathing composition	G. F. Muntz	Birmingham, England	Aug. 1, 1848	5, 694
Metal springs, Machine for polishing	A. B. Doolittle	Hartford, Conn	May 7, 1867	64, 505
Metal-striking-up press	J. North	Brooklyn, N. Y	Jan. 9, 1867	51, 960
Metal strips, Machine for connecting	H. A. Bartlett	Philadelphia, Pa	Sept. 3, 1867	68, 547
Metal surfaces, Apparatus for stippling	R. Dimes	New York, N. Y	June 25, 1872	128, 290
Metal surfaces, Mode of uniting	W. B. Barnard	Waterbury, Conn	Jan. 27, 1863	37, 555
Metal surfaces, Tool for matting	D. Mosman	West Meriden, Conn	May 7, 1872	126, 408
Metal tools, Setting diamond or carbon points in the faces of	W. McKenna	Boston, Mass	Sept. 23, 1873	143, 021
Metal-trimming machine	J. H. Ferguson and H. W. Lovejoy.	New York, N. Y	Apr. 2, 1867	63, 491
Metal tubes and columns, Manufacture of	F. H. Smith	Baltimore, Md	Feb. 21, 1871	112, 884
Metal tubes, Apparatus for brazing	T. G. Gorman and N. W. Peregoy.	Springfield, Ill	Mar. 25, 1873	137, 068
Metal tubes, Arrangement of dies and stocks for ornamenting.	S. M. Cate and E. Jordan	Waterbury, Conn	Apr. 10, 1855	12, 669
Metal tubes, Arrangement of rollers for making	M. R. Griswold	Watertown, Conn	Apr. 17, 1855	12, 739
Metal tubes, Die for making seamless	T. D. Jackson	New York, N. Y	Feb. 28, 1854	10, 569
Metal tubes from forming-mandrels, Liberating	J. Cutler	Birmingham, England	July 29, 1851	8, 250
Metal tubes, Implement for cutting	T. J. Lloyd	Pottsville, Pa	Dec. 22, 1857	18, 918
Metal tubes, Machine for forming	E. Ells	Ansonia, Conn	July 17, 1855	13, 255
Metal tubes, Machine for making	H. Hotchkiss	Plainfield, N. J	July 24, 1866	56, 561
Metal tubes, Machine for making	C. G. Smith	Chelsea, Mass	Sept. 4, 1866	57, 783
Metal tubes, Machine for making	P. L. Weimer	Lebanon, Pa	July 3, 1866	56, 152
Metal tubes, Machine for polishing	J. A. Thomas	Buffalo, N. Y	Feb. 5, 1867	61, 894
Metal tubes, Making	S. Vanstone	Providence, R. I	May 18, 1869	90, 322
Metal tubes, Making seamless	W. F. Brooks	New York, N. Y	Apr. 1, 1856	14, 551
Metal tubes, Making seamless	J. Pratt	Taunton, Mass	June 6, 1854	11, 009
Metal tubes, Making seamless	J. J. Speed, jr., and J. A. Bailey	Detroit, Mich	July 15, 1856	15, 348
Metal tubes, Manufacture of	G. F. Muntz, jr	Birmingham, England	June 14, 1853	9, 782
Metal tubes, Manufacturing	W. Beasley	Smithwick, England	Jan. 9, 1855	12, 228
Metal tubes to glass cups or vessels, Fastening	G. Gordon	Albany, N. Y	Aug. 31, 1869	94, 408
Metal-twisting machine	T. Smith	Green Island, N. Y	June 4, 1872	127, 436
Metal vessels, Pressing and polishing	J. Newmann	New York, N. Y	Mar. 3, 1863	37, 822
Metal wheel	A. Haines and J. Kirkman	Pekin and Kickapoo, Ill	July 3, 1866	56, 042
Metal with metal, Process of coating	J. Poleux	New York, N. Y	Oct. 21, 1856	15, 953
Metal with rubber, Inlaying	B. L. Rowley	New Britain, Conn	Nov. 19, 1872	133, 259
Metal-working machine, Compound	G. L. Jones	Vanville, Wis	Sept. 2, 1873	142, 398
Metal-working machine, Compound	H. B. Sevey	Vienna, Me	Sept. 9, 1873	142, 588
Metal-working machine, Compound	H. B. Sevey	Vienna, Me	Oct. 14, 1873	143, 722
Metals and minerals, Treating	Z. A. Willard and W. G. Adams	Franklin, Mass	May 19, 1868	78, 168
Metals, Apparatus for amalgamating	O. G. Warren	New York, N. Y	Jan. 17, 1865	45, 944
Metals, Apparatus for collecting precious	J. T. McDougall	San Francisco, Cal	Jan. 7, 1868	73, 021
Metals, Apparatus for collecting precious	J. T. McDougall	San Francisco, Cal	Nov. 10, 1868	83, 868
Metals, Apparatus for collecting precious	J. T. McDougall	San Francisco, Cal	Nov. 22, 1870	109, 533
Metals, Apparatus for melting	J. A. Thayer	Cambridgeport, Mass	Aug. 20, 1872	130, 601
Metals, Apparatus for refining	A. Millochau	New York, N. Y	Mar. 21, 1871	112, 831
Metals, Apparatus for saving precious	W. C. Knight	Yankee Jim's, Cal	Dec. 17, 1867	72, 205

Index of patents issued from the United States Patent Office from 1790 *to* 1873, *inclusive*—Continued.

Invention.	Inventor.	Residence.	Date.	No.
Metals, Apparatus for testing the tensile strength of	A. B. Davis	Philadelphia, Pa	Aug. 6, 1872	130, 284
Metals, Apparatus for testing the tensile strength of.	T. Olsen	Philadelphia, Pa	Sept. 16, 1873	142, 937
Metals, Bath for coating metals with other	G. Rogers	Enfield, England	May 16, 1854	10, 914
Metals, Casting molten	J. D. Rosthorn	Vienna, Austria	Jan. 17, 1865	45, 947
Metals, Coating	J. D. Güneberg	Camden, N. J	Apr. 26, 1870	102, 254
Metals, Coating	S. Hiler	Haverstraw, N. Y	Oct. 12, 1858	21, 797
Metals, Coating and bronzing	F. Weil	Paris, France	July 12, 1864	43, 557
Metals, Combining wrought and cast	E. L. Brown	Philadelphia, Pa	June 9, 1868	78, 786
Metals, &c., Composition for coating	B. Oertly and H. Fendrick	Washington, D. C	July 21, 1868	80, 086
Metals, Composition for covering	J. H. Green	Christiansburgh, Iowa	Nov. 29, 1859	26, 267
Metals, &c., Compound for cleaning	W. Z. Moore	Vallejo, Cal	Nov. 4, 1873	144, 215
Metals, Compound for cleaning polished	M. McGlenn	Aurora, Ill	July 22, 1873	141, 156
Metals, Compressing, condensing, and extending	J. F. Shearman	Brooklyn, N. Y	July 31, 1866	56, 807
Metals, Core for casting	O. D. Hunter	Plymouth, Conn	Sept. 16, 1873	142, 790
Metals, Cutting	R. Anderson and A. Vancleve	United States Army, and Trenton, N. J.	Nov. 4, 1856	16, 001
Metals from other substances, Apparatus for separating the precious.	V. B. Ryerson	New York, N. Y	Apr. 10, 1866	53, 885
Metals from the beds of rivers, Process of obtaining.	J. Johnson	Saco, Me	Jan. 1, 1867	60, 898
Metals, &c., Furnace for reheating	H. Chisholm	Cleveland, Ohio	Mar. 4, 1873	136, 413
Metals, Granulating	J. Feix	San Francisco, Cal	Sept. 16, 1856	15, 733
Metals, Hardening	A. Wheeler	Warwick, Mass	July 31, 1849	6, 615
Metals in liquid state, Apparatus for compressing cast.	H. W. Barnum	Omaha, Nebr	Dec. 9, 1873	145, 325
Metals, ivory, &c., Method of forming designs upon.	T. Skinner	Pittsburgh, Pa	Dec. 24, 1867	72, 553
Metals, Machine for cleaning and brightening particles of precious.	J. D. Whelpley and J. J. Storer	Boston, Mass	Sept. 11, 1866	58, 011
Metals, Machine for forging	E. Wheeler	Feltonville, Mass	Sept. 1, 1857	18, 115
Metals, Machine for routing	H. Cottrell	Newark, N. J	July 2, 1872	128, 596
Metals, Machinery for hammering	J. Ramsbottom	Crewe, England	July 25, 1865	49, 058
Metals, Manufacture and preservation of	J. Corson	Washington, D. C	Oct. 19, 1869	95, 882
Metals, Manufacture of	A. Wall	Poplar Blackwell, England	Aug. 10, 1844	3, 698
Metals, &c., Melting	J. Walworth	New York	Oct. 17, 1809	....
Metals or ores, Composition to be used in the fusion, refining, or working of	H. B. Chew and E. V. Freeman	Baltimore, Md	Apr. 17, 1829	....
Metals, Quality and ornamentation of	W. Rose	Halesowen, England	July 7, 1863	39, 174
Metals, Refining	J. Ramdohr	Virginia City, Nev	June 27, 1865	48, 438
Metals, Separating	W. Elmer	New York, N. Y	Mar. 19, 1867	63, 026
Metals, Separating and refining	C. S. Eyster	Denver, Colo	June 14, 1870	104, 130
Metals, Shaping and punching	J. C. Dickey	Saratoga Springs, N. Y	Mar. 2, 1858	19, 498
Metals, Treating	M. Lane	Washington, D. C	Aug. 20, 1861	33, 090
Metals while in the molten state, Treating	H. W. Woodruff	Watertown, N. Y	Oct. 11, 1853	10, 115
Metals with anthracite coal, Manufacturing	R. Willcox	Paterson, N. J	Apr. 5, 1831	....
Metals with bituminous coal, charcoal, &c., Manufacturing.	R. Willcox	Paterson, N. J	Apr. 5, 1831	....
Metals with metal, Coating	I. Adams, jr	Boston, Mass	Aug. 21, 1866	57, 271
Metals, wood, &c., Composition for cleaning	C. S. Toms	Utica, N. Y	May 28, 1867	65, 139
Metallic and alkaline sulphates, Separating metal from a mixture of.	A. Monnier	Philadelphia, Pa	Dec. 12, 1871	121, 798
Metallic and alkaline sulphates to separate copper, &c., Treating.	A. Monnier	Philadelphia, Pa	Dec. 12, 1871	121, 799
Metallic and non-metallic surfaces, Plastic compound for protecting.	T. L. Roux	Toulon, France	Apr. 9, 1867	63, 654
Metallic and other hard substances, Machine for cutting.	J. H. Hall	Harper's Ferry, Va	Mar. 7, 1827	....
Metallic axles and other hollow metallic articles, Process of making hollow.	W. A. Lewis	Chicago, Ill	Oct. 10, 1871	119, 864
Metallic axles, Process of making hollow	W. A. Lewis	Chicago, Ill	Mar. 19, 1872	124, 833
Metallic bar	S. Stone	North Manchester, Conn	Oct. 12, 1869	95, 746
Metallic-bar cutter	T. Snell and W. Tucker	Philadelphia, Pa	Oct. 6, 1868	82, 886
Metallic bar, Triple-ribbed	J. Y. Smith	Alexandria, Va	May 15, 1866	54, 786
Metallic bars and rods, Cutter for	J. Gallagher	New York, N. Y	May 2, 1854	10, 858
Metallic bars, Machine for cutting	S. Hall	New York, N. Y	Dec. 8, 1857	18, 811
Metallic-bath furnace	J. S. Brown	Pawtucket, R. I.	June 7, 1864	43, 066
Metallic boat	J. Francis	New York, N. Y	Mar. 23, 1858	19, 693
Metallic boat	L. Raymond	New York, N. Y	June 23, 1863	38, 984
Metallic bodies, Machine for squeezing and compressing.	E. A. Lester	Boston, Mass	Jan. 10, 1854	10, 410
Metallic box	I. C. Mayo	Gloucester, Mass	June 11, 1872	127, 782
Metallic box for presses, &c	J. Foster, jr., and P. Evens, jr.	Cincinnati, Ohio	June 28, 1853	9, 823
Metallic bracket	A. D. Judd	New Haven, Conn	June 21, 1870	104, 463
Metallic bracket	A. D. Judd	New Haven, Conn	Dec. 9, 1873	145, 351
Metallic bracket	A. D. and E. M. Judd	New Haven, Conn	June 21, 1870	104, 464
Metallic-can bottom	H. W. Shepard	Mannsville, N. Y	Jan. 4, 1870	98, 526
Metallic-can cap	G. H. Perkins	Brooklyn, N. Y	Feb. 22, 1870	100, 186
Metallic-can cover	H. W. Shepard	Mannsville, N. Y	Apr. 12, 1870	101, 929
Metallic case or stove for kettles, boilers, stills, &c.	C. Gaston	Williamsport, Pa	Jan. 31, 1812	....
Metallic chair	A. W. Hopkins	New York, N. Y	Oct. 15, 1867	69, 808
Metallic chair-bottom	V. Stockton	Williamsburgh, Ohio	Aug. 7, 1860	29, 553
Metallic clasp	C. Marshall	Lockport, N. Y	June 24, 1873	140, 207
Metallic combs, Making	R. A. Ives	Bristol, Conn	Mar. 28, 1838	661
Metallic compound or alloy	J. Hackert	New York, N. Y	June 11, 1867	65, 563
Metallic cup and stand	S. J. Ladd	Providence, R. I	May 8, 1860	28, 238
Metallic fabric for grain scouring and percolating purposes.	A. J. Vandegrift	Covington, Ky	Feb. 4, 1873	135, 608
Metallic fabric for scouring and percolating purposes.	A. J. Vandegrift	Covington, Ky	Jan. 7, 1873	134, 619
Metallic handle	J. Hatton	New York, N. Y	Jan. 28, 1868	73, 804
Metallic hoop	J. L. Alberger	Buffalo, N. Y	Apr. 14, 1863	38, 139
Metallic hoop	E. C. Hamlin	Pavilion, N. Y	Oct. 10, 1871	119, 844
Metallic-hoop clasp	J. R. Speer	Pittsburgh, Pa	Dec. 1, 1857	18, 779
Metallic house	S. J. Seely	Brooklyn, N. Y	Dec. 23, 1862	37, 248
Metallic mat	F. G. Johnson	Brooklyn, N. Y	Aug. 27, 1872	130, 808
Metallic mill	J. C. Gentry	Dayton, Ohio	Mar. 18, 1836	....
Metallic mills, Dressing	S. G. Reynolds	Providence, R. I	Mar. 18, 1835	....
Metallic molds for casting metals, Preparing	J. Revere	Boston, Mass	Feb. 25, 1862	34, 524
Metallic molds, Process in casting in	J. Williams	Philadelphia, Pa	May 13, 1873	138, 775

Index of patents issued from the United States Patent Office from 1790 *to* 1873, *inclusive*—Continued.

Invention.	Inventor.	Residence.	Date.	No.
Metallic molds, Wash or coating for	J. Kinniburgh	Motherwell, Scotland	Mar. 18, 1862	34, 719
Metallic moldings, Machine for making	F. M. Campbell	Cleveland, Ohio	July 16, 1872	129, 001
Metallic oxides and in refining the metal resulting therefrom, Reducing.	J. Reese	Pittsburgh, Pa	Sept. 11, 1866	57, 969
Metallic pipe	W. S. Mayo	New York, N. Y	May 24, 1859	24, 132
Metallic pipe or tubing	P. Naylor	New York, N. Y	Jan. 2, 1872	122, 482
Metallic pipes and tubes, Machine for making	J. and C. Hanson	Huddersfield, England	Mar. 29, 1841	2, 021
Metallic pipes, Machinery and process of manufacturing.	J. E. Serrell	New York	Jan. 20, 1843	2, 918
Metallic pipes, Machinery for making	G. N. and B. Tathem, jr	Philadelphia, Pa	Oct. 11, 1841	2, 296
Metallic plate, Corrugated	R. Montgomery	New York, N. Y	July 31, 1866	56, 779
Metallic plates, Apparatus for bending	J. W. Bean	Henry, Ill	Feb. 25, 1868	74, 880
Metallic plates, Joining and riveting	W. Beschke	Alexandria, Va	Nov. 22, 1853	10, 247
Metallic plates, Machine for flanging and wiring	O. W. Stow	Plantsville, Conn	Nov. 12, 1867	70, 917
Metallic plates, Press for bending	J. Cochrane	New York, N. Y	Sept. 15, 1863	39, 886
Metallic plates to each other, Method of uniting	S. Pratt	Cohasset, Mass	Aug. 14, 1849	6, 647
Metallic points, Removing pains, &c., by	E. Perkins	......	Feb. 19, 1796	......
Metallic rods, Apparatus for rolling tapered	W. Clay	Clifton Lodge, England	Apr. 22, 1851	8, 055
Metallic rods, &c., Device for cutting tenons on	C. Gillson	Worcester, Mass	Aug. 9, 1864	43, 776
Metallic rods or bars, Machine for upsetting and forming articles from.	M. Seward	New Haven, Conn	Nov. 6, 1866	59, 465
Metallic spring	L. Bissell	New York, N. Y	Feb. 20, 1855	12, 406
Metallic spring	J. Harrison, jr	New York, N. Y	Aug. 24, 1858	21, 255
Metallic spring	G. W. McMinn	Covington, Ky	Feb. 19, 1851	31, 517
Metallic substances, Apparatus for concentrating and condensing volatile.	J. C. Coult	San Francisco, Cal	Apr. 2, 1867	63, 365
Metallic surfaces, Coating	W. and W. A. Butcher	Philadelphia, Pa	June 29, 1858	20, 697
Metallic surfaces, Composition for coating	G. Brown, E. E. Burnham, and J. Morrise.	Gloucester, Mass	June 11, 1867	65, 639
Metallic surfaces, Device for producing pearl finish on.	B. D. Beiderhase and C. Witteck.	New York, N. Y	Jan. 7, 1873	134, 581
Metallic surfaces, Device for producing satin finish on.	B. D. Beiderhase and C. Witteck.	New York, N. Y	Jan. 7, 1873	134, 582
Metallic surfaces, Machine for finishing	G. Cowing	Seneca Falls, N. Y	May 19, 1863	38, 565
Metallic surfaces, Mode of uniting	W. B. Barnard	Waterbury, Conn	Nov. 8, 1864	44, 928
Metallic surfaces to prevent oxidation, Compound for coating.	A. Wall	Shadwell, England	June 22, 1841	2, 136
Metallic tip	B. F. Sparrow	Boston, Mass	Aug. 20, 1867	67, 920
Metallic tubes, Casting	J. Smith, jr	Norton, Mass	Dec. 2, 1856	16, 166
Metallic tubes, Device for cutting off	S. P. M. Tasker	Philadelphia, Pa	May 31, 1870	103, 685
Metallic tubes, Galvano-plastic coating for the interior of.	J. Matthews, jr	New York, N. Y	June 5, 1860	28, 590
Metallic tubes, Machine for making	E. Valentine and M. T. Ridout.	Milwaukee, Wis	Feb. 7, 1865	46, 311
Metallic tubes, Making	J. B. Root	New York, N. Y	Oct. 19, 1869	96, 037
Metallic tubes or spouts, Machine for making	E. Valentine and M. T. Ridout.	Milwaukee, Wis	July 25, 1865	49, 045
Metallic tubes, Punching	B. Mackerley	New Petersburgh, Ohio	May 4, 1858	20, 165
Metallic washer	M. G. Hubbard	Syracuse, N. Y	Feb. 11, 1873	135, 809
Metallurgic and other furnaces	T. Weiss	Dresden, Saxony	Nov. 1, 1870	108, 862
Metallurgic and other furnaces, Draft-apparatus for.	G. Wingate	Boston, Mass	Nov. 11, 1873	144, 585
Metallurgic and other furnaces, Feeding fuel to	J. D. Whelpley and J. J. Storer	Boston, Mass	May 31, 1870	103, 695
Metallurgic and other furnaces, Feeding pulverized fuel to.	J. D. Whelpley and J. J. Storer	Boston, Mass	May 31, 1870	103, 804
Metallurgic and other furnaces, Heating	J. D. Whelpley and J. J. Storer.	Boston, Mass	Nov. 29, 1870	109, 795
Metallurgic and other furnaces, Lining for	J. C. Hill	Alexandria, Va	Nov. 26, 1872	133, 446
Metallurgic furnace	A. G. Bevin	East Hampton, Conn	May 4, 1869	89, 620
Metallurgic furnace	H. Boetius	Hanover, Prussia	Oct. 22, 1867	69, 963
Metallurgic furnace	Z. S. Durfee	New York, N. Y	Aug. 6, 1872	130, 200
Metallurgic furnace	J. L. Hamilton and H. McDonald.	Pittsburgh, Pa	Jan. 2, 1872	122, 380
Metallurgic furnace	J. Montgomery	Sing Sing, N. Y	Feb. 1, 1870	99, 456
Metallurgic furnace	G. Perry and J. Webb	Chicago, Ill	May 6, 1873	138, 523
Metallurgic furnace	J. Y. Smith	Pittsburgh, Pa	Oct. 17, 1871	120, 005
Metallurgic furnace for iron and steel	J. Y. Smith	Pittsburgh, Pa	June 14, 1870	104, 219
Metallurgic-furnace lining	S. Lansdowne	Sharon, Pa	Apr. 8, 1873	137, 554
Metallurgic-furnace lining	J. Y. Smith	Pittsburgh, Pa	Oct. 17, 1871	120, 006
Metallurgic-furnace regenerator	H. Frank	Pittsburgh, Pa	Dec. 31, 1872	134, 373
Metallurgic-furnace regenerator	A. Ponsard	Paris, France	Aug. 6, 1872	130, 313
Metallurgic furnaces, Apparatus for condensing and purifying the smoke of.	W. J. Johnson	Allendale, England	Feb. 18, 1873	136, 066
Metallurgic furnaces, Apparatus for feeding the charge to.	G. Edwards	Tannehill, Ala	Aug. 27, 1872	130, 849
Metallurgic furnaces, Apparatus for generating and burning gas in.	J. W. Ells	Pittsburgh, Pa	July 19, 1870	105, 557
Metallurgic furnaces, Constructing and heating	C. E. Detmold	New York, N. Y	July 15, 1843	3, 176
Metallurgic furnace, Construction and operation of.	H. Bessemer	London, England	July 25, 1871	117, 250
Metallurgic furnaces, Gas and air heating apparatus for.	J. W. Ells	Pittsburgh, Pa	July 19, 1870	105, 558
Metallurgic furnaces so as to use the waste heat under steam-boiler, Arrangement of.	G. Nimmo	Jersey City, N. J	Nov. 19, 1872	133, 114
Metallurgic operations by the introduction of chemical agents.	R. Willcox	Paterson, N. J	Apr. 5, 1831	......
Metallurgical process and furnace	C. W. Siemens	Westminster, England	Apr. 27, 1869	89, 441
Metamorphoscope	E. A. Goodes	Philadelphia, Pa	Nov. 12, 1867	70, 831
Meteorological changes, Electro-magnetic apparatus for noting.	S. Chester	Elizabeth, N. J	Jan. 2, 1872	122, 437

Meter:

See Cable-meter.
Dynamometer.
Electro-magnetic meter.
Fluid meter.
Gas-meter.
Gas and water meter.
Grain-meter.
Heliometer.
Horometer.
Hydraulic meter.
Hydrometer.
Hygieometer.
Hygrometer.
Liquid-meter.
Power meter.
Proof-meter.
Pyrometer.
Rotary meter.
Sounding-meter.
Spirit-meter.
Water-meter.
Way-meter.

Index of patents issued from the United States Patent Office from 1790 *to* 1873, *inclusive*—Continued.

Invention.	Inventor.	Residence.	Date.	No.
Meter	W. Ball	New Vienna, Ohio	July 25, 1871	117, 366
Meter	T. Cook and J. Watson	Old Kent Road and Westminster, England	May 2, 1871	114, 267
Meter	D. P. Davis	Jersey City, N. J	Aug. 15, 1871	118, 112
Meter	V. Fogerty	Boston Highlands, Mass	Dec. 19, 1871	121, 937
Meter	J. Harris	Boston, Mass	Mar. 30, 1869	88, 475
Meter	A. D. Laws	Bridgeport, Conn	Mar. 14, 1871	112, 722
Meter	C. Moore	New York, N. Y	Jan. 24, 1871	111, 134
Meter	S. P. Ruggles	Boston, Mass	Mar. 23, 1869	88, 215
Meter	G. Sewell	Brooklyn, N. Y	Aug. 2, 1870	106, 103
Meter	F. Wagner	New York, N. Y	Mar. 21, 1871	112, 992
Meter-safes, Self-locking bolt for	A. W. Adams	New York, N. Y	July 14, 1868	79, 801
Metronome	H. S. Blunt	New York, N. Y	Mar. 10, 1868	75, 241
Metronome	H. C. Carden	Paris, France	June 11, 1867	65, 542
Metronome	U. Emmons	New York, N. Y	Apr. 7, 1831	....
Metronome	J. A. Hackenback	Mayville, Wis	Mar. 4, 1873	136, 435
Mica for tablets, &c., Preparing	J. Stevens and J. Johnson	New York, N. Y., and Saco, Me.	Feb. 26, 1867	62, 377
Mica, Preparation of	H. M. Johnson and F. Beck	New York, N. Y	Nov. 22, 1870	109, 416
Micrometer-gage	A. Bonnaz	Paris, France	Aug, 19, 1873	141, 915
Microscope	J. J. Bausch	Rochester, N. Y	Apr. 25, 1865	47, 382
Microscope	O. N. Chase	Boston, Mass	July 10, 1866	56, 178
Microscope	H. Craig	Cleveland, Ohio	Feb. 18, 1862	34, 409
Microscope	J. Ellis	New York, N. Y	May 24, 1864	42, 843
Microscope	H. L. Smith	Gambier, Ohio	Feb. 27, 1866	52, 901
Microscope	W. Wales	Fort Lee, N. J	Feb. 21, 1865	46, 511
Microscope, Portable	R. P. P. Dagron	Paris, France	Aug. 13, 1861	33, 031
Microscope, Single	J. H. Logan	Allegheny, Pa	Aug. 17, 1869	93, 895
Middlings-bolt	A. G. and H. W. Mowbray	Stockton, Minn	Nov. 18, 1873	144, 783
Middlings dresser or purifier	G. T. Smith	Minneapolis, Minn	Dec. 10, 1872	133, 898
Middlings-purifier	J. Bandon and J. Wiggim, jr	Dayton, Ohio	Oct. 7, 1873	143, 490
Middlings-purifier	L. G. Binkley	Baughman, Ohio	Dec. 27, 1870	110, 425
Middlings-purifier	F. Bleckman	Washington, Mo	Sept. 23, 1873	143, 118
Middlings-purifier	H. J. Burdick and C. S. Fuller	Oswego, N. Y	Oct. 28, 1873	144, 056
Middlings-purifier	E. Clark	Lancaster, Pa	Dec. 31, 1872	134, 462
Middlings-purifier	W. A. Clarke	Milwaukee, Wis	Oct. 28, 1873	144, 017
Middlings-purifier	J. H. Dedrick	Milwaukee, Wis	Aug. 26, 1873	142, 086
Middlings-purifier	C. S. Fuller	Oswego, N. Y	Nov. 4, 1873	144, 329
Middlings-purifier	A. R. Guilder	Minneapolis, Minn	Dec. 2, 1873	145, 170
Middlings-purifier	C. M. Hardenburgh, A. L. Miner, and W. J. Fender.	Minneapolis, Minn	May 27, 1873	139, 386
Middlings-purifier	A. Herr	Georgetown, D. C	Aug. 5, 1873	141, 557
Middlings-purifier	J. Hollingworth	New York, N. Y	Sept. 30, 1873	143, 346
Middlings-purifier	A. Hunter and C. E. Whitmore	Quincy, Ill	Aug. 19, 1873	142, 022
Middlings-purifier	A. Hunter and C. E. Whitmore	Quincy, Ill	Aug. 19, 1873	142, 023
Middlings-purifier	A. Hunter and C. E. Whitmore	Quincy, Ill	Mar. 25, 1873	137, 207
Middlings-purifier	W. W. Huntly and A. P. Holcomb.	Silver Creek, N. Y	Feb. 11, 1873	135, 810
Middlings-purifier	C. Janney	Minneapolis, Minn	Jan. 14, 1873	134, 750
Middlings-purifier	E. N. Lacroix	Minneapolis, Minn	June 3, 1873	139, 585
Middlings-purifier	E. N. Lacroix	Minneapolis, Minn	Nov. 25, 1873	144, 988
Middlings-purifier	N. Lacroix	Milwaukee, Wis	May 27, 1873	139, 397
Middlings-purifier	C. E. McNeal	Silver Creek, N. Y	Feb. 4, 1873	135, 571
Middlings-purifier	L. Mowry	Minneapolis, Minn	Dec. 3, 1872	133, 662
Middlings-purifier	G. Parker	Poughkeepsie, N. Y	Nov. 18, 1873	144, 697
Middlings-purifier	L. S. Reynolds	Minneapolis, Minn	Sept. 9, 1873	142, 729
Middlings-purifier	I. Scholfield	Dunlap, Iowa	Dec. 23, 1873	145, 906
Middlings-purifier	E. P. Welch	Georgetown, D. C	Nov. 26, 1872	133, 509
Middlings-purifier	E. P. Welch	Georgetown, D. C	Feb. 18, 1873	135, 952
Middlings-purifier	E. P. Welch	Georgetown, D. C	Feb. 18, 1873	135, 953
Middlings-purifier	J. W. Wilson	Pawtucket, R. I	June 10, 1873	139, 843
Middlings-purifier screen	W. W. Huntley	Silver Creek, N. Y	Sept. 23, 1873	143, 014
Middlings-purifiers, Feeding-device for	R. Craik	Money Creek, Minn	Oct. 21, 1873	143, 887
Middlings-purifying process	W. W. Huntly and A. P. Holcomb.	Silver Creek, N. Y	Feb. 11, 1873	135, 811
Middlings receiver, cooler, and duster	A. Hunter and C. E. Whitmore	Quincy, Ill	Apr. 1, 1873	137, 449
Middlings-separator	J. Barker	Chicago, Ill	Nov. 30, 1869	97, 341
Middlings-separator	L. G. Binkly	Baughman, Ohio	Sept. 3, 1872	131, 079
Middlings-separator	J. G. Birely	Danville, Ill	Nov. 19, 1872	133, 193
Middlings-separator	J. B. Brennan and W. Tucker	Paris, Ill	Nov. 7, 1871	120, 706
Middlings-separator	R. L. Downton	Collinsville, Ill	Oct. 7, 1873	143, 442
Middlings-separator	A. R. Guilder	Minneapolis, Minn	June 4, 1872	127, 413
Middlings-separator	A. R. Guilder	Minneapolis, Minn	Apr. 22, 1873	138, 019
Middlings-separator	H. P. Jones	Springfield, Ohio	Apr. 29, 1873	138, 408
Middlings-separator	E. N. Lacroix	Minneapolis, Minn	May 14, 1872	126, 719
Middlings separator	W. R. Middleton	Cleveland, Ohio	Nov. 1, 1870	108, 926
Middlings-separator	E. Yeagly	West Earl, Pa	June 11, 1872	127, 727
Middlings-separators, Feed-regulator for	A. G. Mowbray	Stockton, Minn	Sept. 24, 1872	131, 625
Mildew in canvas, cloth, &c., Mode of preventing	W. Stacey	Kittery, Me	June 21, 1864	43, 233
Mile-mark, Cast-iron	W. M. Summers	New York, N. Y	Jan. 13, 1825	....
Mileage-register	B. Horn	Sergeantville, N. J	May 5, 1868	77, 613
Military equipments	W. H. Penrose	Fort Lyon, Colo	Sept. 26, 1871	119, 460
Military insignia woven in cloth	A. M. Dorman	Philadelphia, Pa	Aug. 8, 1865	49, 339
Military tactics, Apparatus for teaching	W. S. Engle	Brooklyn, N. Y	Dec. 16, 1862	37, 160
Milk and beer cooler	O. L. Mayhew	Sanborn, N. Y	Feb. 2, 1869	86, 430
Milk and cheese rack	J. G. Cross	Brattleborough, Vt	July 31, 1866	56, 722
Milk and cream cooler	H. C. Baldwin	North Wolcott, Vt	Aug. 12, 1873	141, 688
Milk and cream cooler	J. Prince and N. D. Martin	North Craftsbury, Vt	Mar. 4, 1873	136, 542
Milk and liquid cooler	A. P. Bussey	Westernville, N. Y	Oct. 18, 1870	108, 448
Milk and oyster can	J. Buckley	Baltimore, Md	Sept. 10, 1867	68, 696
Milk and provision rack	E. Osborn	Rome Centre, Mich	Feb. 25, 1868	74, 766
Milk and unchrystallizable sugar, Compound of condensed.	G. R. Percy	New York, N. Y	Jan. 24, 1865	46, 022
Milk, &c., Apparatus for concentrating	J. R. Pond	New Hartford, Conn	Apr. 19, 1864	42, 398
Milk, Apparatus for cooling and transporting	E. D. Gird	Syracuse, N. Y	Feb. 25, 1873	136, 150
Milk, Apparatus for obtaining cream from	A. Pope	Randolph, N. Y	Feb. 9, 1869	86, 860
Milk, Apparatus for purifying	J. M. Schermerhorn and S. Perry.	North Gage and Newport, N. Y.	June 11, 1872	127, 802

Index of patents issued from the United States Patent Office from 1790 *to* 1873, *inclusive*—Continued.

Invention.	Inventor.	Residence.	Date.	No.
Milk, Apparatus for testing	C. S. Brown	New York, N. Y	June 20, 1865	48, 256
Milk, Apparatus for transporting	J. H. Whitney	Brooklyn, N. Y	Dec. 17, 1872	134, 020
Milk, Apparatus for treating	G. D. Greenleaf and D. C. Larkins.	Depauville, N. Y	Apr. 14, 1868	76, 749
Milk, Apparatus for treating	J. A. Otis and T. Barber	Watertown, N. Y	Oct. 6, 1868	82, 866
Milk-boiler	O. Zwicker	Philadelphia, Pa	Nov. 18, 1873	144, 819
Milk-boiler alarm	S. Mangold	New York, N. Y	May 20, 1873	139, 072
Milk-box	C. W. Eastwood	New York, N. Y	Feb. 1, 1870	99, 417
Milk-can	E. Abbott	Chagrin Falls, Ohio	Sept. 25, 1866	58, 194
Milk-can	E. Abbott	Willoughby, Ohio	Jan. 28, 1868	73, 687
Milk-can	T. W. Akin	Patterson, N. Y	Nov. 10, 1868	83, 898
Milk-can	S. O. Avery	Brewster's Station, N. Y	Oct. 29, 1867	70, 150
Milk-can	N. P. Barnes	Carmel, N. Y	Mar. 20, 1866	53, 256
Milk-can	T. M. Bell	New York, N. Y	Jan. 31, 1871	111, 302
Milk-can	T. M. Bell	New York, N. Y	Oct. 3, 1871	119, 559
Milk-can	J. A. Bennett	Millerton, N. Y	Aug. 11, 1868	80, 800
Milk-can	N. C. Burnap	rgusville, N. Y	Nov. 5, 1867	70, 516
Milk-can	D. Burnett	Bedford Station, N. Y	Nov. 21, 1871	121, 154
Milk-can	C. L. Camp and J. H. Chappel	Brooklyn, N. Y	May 17, 1870	103, 137
Milk-can	J. F. Cass	L'Original, Canada	Dec. 9, 1873	145, 324
Milk-can	J. Cochran	Purdy's Station, N. Y	Sept. 28, 1869	95, 324
Milk-can	J. Cochran	Purdy's Station, N. Y	Mar. 1, 1870	100, 376
Milk-can	A. P. Cook	Collins Centre, N. Y	Aug. 25, 1868	81, 345
Milk-can	A. Crittenden	West Turin, N. Y	July 14, 1832	....
Milk-can	A. P. Curry	Chagrin Falls, Ohio	June 11, 1867	65, 548
Milk-can	J. L. Davis	Salem, N. J	Feb. 20, 1866	52, 691
Milk-can	J. E. Dean	Canaan, Conn	May 19, 1868	78, 072
Milk-can	E. R. Denniston	Middletown, N. Y	June 14, 1859	24, 381
Milk-can	S. J. Dwyer	Albany, N. Y	Apr. 27, 1869	89, 392
Milk-can	J. H. Farley	Lowell, Mass	Mar. 19, 1867	62, 949
Milk-can	J. L. Finch	Warwick, N. Y	Mar. 19, 1867	63, 029
Milk-can	W. Frost	Amenia, N. Y	Nov. 15, 1859	26, 098
Milk-can	E. A. Grant	Louisville, Ky	Sept. 17, 1872	131, 436
Milk-can	G. A. Huggins	Mannsville, N. Y	July 14, 1868	79, 907
Milk-can	H. L. McAvoy and E. Mills	Baltimore, Md	Nov. 5, 1867	70, 450
Milk-can	J. C. Milligan	Brooklyn, N. Y	Sept. 20, 1870	107, 521
Milk-can	O. J. Nutting	Warwick, N. Y	Nov. 3, 1868	83, 724
Milk-can	H. Preston and J. Mahood	Philadelphia, Pa	Mar. 24, 1863	37, 977
Milk-can	W. Ralph	Utica, N. Y	June 18, 1867	65, 828
Milk-can	C. A. Reight	Middleton, N. Y	Apr. 21, 1868	77, 092
Milk-can	H. Langster	Buffalo, N. Y	Mar. 1, 1870	100, 454
Milk-can	S. P. Sauer	Wheaton, Ill	Aug. 20, 1872	130, 752
Milk-can	D. W. Shaw	Baltimore, Md	July 7, 1868	79, 693
Milk-can	A. C. Shorte and H. B. Todd	Plymouth, Conn	Jan. 19, 1869	86, 039
Milk-can	S. Stroock	New York, N. Y	Sept. 12, 1871	118, 887
Milk-can	L. A. Sunderland	Chagrin Falls, Ohio	Dec. 15, 1868	84, 973
Milk-can	L. A. Sunderland	Madison, Ohio	Apr. 12, 1870	101, 784
Milk-can	P. Teets	New York, N. Y	May 28, 1861	32, 439
Milk-can	I. Vanderslice	Philadelphia, Pa	Aug. 18, 1868	81, 119
Milk can	I. F. Van Duzen	Middletown, N. Y	June 19, 1866	55, 791
Milk-can	H. M. Viet	Carlisle, Ohio	Dec. 15, 1868	84, 921
Milk-can	R. C. Wickham	Pawlet, Vt	Dec. 18, 1866	60, 603
Milk can and cooler	S. Shattuc	Kipton, Ohio	Sept. 6, 1870	107, 107
Milk-can bottom, &c	T. M. Bell	New York, N. Y	July 5, 1870	104, 924
Milk-can bottom	A. Burnham	Arkwright, N. Y	Sept. 5, 1865	49, 711
Milk-can bottom	M. Wiles and J. C. Wock	Fort Plain, N. Y	Aug. 6, 1867	67, 474
Milk-can cover	A. Brightman	New Bedford, Mass	Nov. 1, 1864	44, 848
Milk-can cover	J. Fandel	Boston, Mass	Apr. 21, 1868	77, 020
Milk-can covers, Locking	D. Burnett	Bedford Station, N. Y	June 6, 1871	115, 697
Milk-can covers, Securing	N. C. Burnap	Argusville, N. Y	Mar. 12, 1872	124, 539
Milk-can fastener	D. D. Edgerton	Ava, N. Y	Nov. 22, 1870	109, 490
Milk, Can for preserving and transporting	M. M. Clark	Monroe, N. Y	Mar. 7, 1865	46, 640
Milk-can handle	A. C. Beckwith and G. H. Graham.	Oriskany, N. Y	Oct. 4, 1870	107, 855
Milk-can handle	J. W. Hannan	Elyria, Ohio	Nov. 21, 1871	121, 168
Milk-can, &c., handle	C. Millar	Utica, N. Y	Apr. 3, 1866	53, 649
Milk-can handle	H. W. Shepard	Mannsville, N. Y	June 7, 1870	103, 933
Milk-can holder, Condensed	R. J. Tanner	Lake View, N. J	Dec. 17, 1872	134, 015
Milk-can, Refrigerating	W. W. White and M. King	Lowville, N. Y	June 9, 1868	78, 777
Milk-can stopper	J. M. Burghartt	Great Barrington, Mass	Aug. 3, 1869	93, 171
Milk-can, Ventilating	L. B. Arnold	Lansing, N. Y	Mar. 14, 1871	112, 674
Milk-cans, Apparatus for stamping	W. M. Storm	New York, N. Y	Sept. 7, 1858	21, 473
Milk-cans, &c., Clasp for securing covers to	G. D. C. Ransom and C. Smith	Brooklyn, N. Y	June 11, 1872	127, 794
Milk-cans, Metallic bottom for	T. M. Bell	New York, N. Y	Jan. 14, 1873	134, 723
Milk-carrier	L. Morris	Havre de Grace, Md	Dec. 13, 1870	110, 152
Milk-concentrating apparatus	A. S. Lyman	New York, N. Y	Sept. 2, 1862	36, 354
Milk, Concentration of	G. Borden, jr	Brooklyn, N. Y	Aug. 19, 1856	15, 553
Milk, Condensed	G. Borden and J. G. Borden	White Plains and South East, N. Y.	Nov. 4, 1873	144, 311
Milk, Condensing	J. G. Borden	South East, N. Y	Oct. 29, 1872	132, 621
Milk, Condensing	J. R. Pond	New Hartford, Conn	Jan. 5, 1864	41, 090
Milk, Condensing	J. R. Pond	New Hartford, Conn	Nov. 14, 1865	50, 950
Milk, Condensing	J. R. Pond	New Hartford, Conn	Dec. 26, 1865	51, 749
Milk-cooler	L. B. Arnold	Lansing, N. Y	Dec. 22, 1868	85, 160
Milk-cooler	A. E. Baldwin	Newark, N. J	Aug. 10, 1869	93, 583
Milk-cooler	L. Bauer	McHenry County, Ill	Sept 17, 1872	131, 325
Milk-cooler	A. Beeman	Potter's Corners, Pa	June 18, 1872	127, 952
Milk-cooler	H. Blake	Panama, N. Y	Oct. 8, 1872	131, 993
Milk-cooler	A. M. Blanchard	Ellington, N. Y	Apr. 9, 1872	125, 531
Milk-cooler	B. C. Bort and T. Bryant	Chateaugay, N. Y	June 18, 1872	128, 102
Milk cooler	B. C. Bort and T. Bryant	Chateaugay, N. Y	Nov. 5, 1872	132, 735
Milk-cooler	N. C. Burnap	Argusville, N. Y	Aug. 13, 1867	67, 717
Milk-cooler	A. P. Bussey	Westernville, N. Y	Nov. 2, 1869	96, 390
Milk-cooler	W. O. Campbell	Montgomery Centre, Vt	Mar. 11, 1873	136, 584
Milk-cooler	S. F. Cowles	Coventry, Vt	Feb. 25, 1873	136, 215
Milk-cooler	A. J. Crane and D. J. Hadley	Cambridge, Vt	Nov. 25, 1873	144, 835
Milk-cooler	J. Dingee	Downington, Pa	Aug. 10, 1869	93, 424
Milk-cooler	C. A. Douglas	Franklin, N. Y	Feb. 20, 1872	123, 813

Index of patents issued from the United States Patent Office from 1790 *to* 1873, *inclusive*—Continued.

Index of patents issued from the United States Patent Office from 1790 *to* 1873, *inclusive*—Continued.

Invention.	Inventor.	Residence.	Date.	No.
Milk-vat	L. C. and J. M. Schermerhorn	North Gage, N. Y	Apr. 9, 1867	63, 755
Milk-vat	O. H. Willard and A. Pope	Randolph and Conewago, N. Y.	Jan. 30, 1872	123, 318
Milk-vessel	M. L. Shade	Fair Grove, Mo	Sept. 3, 1872	131, 069
Milker, Cow	L. O. Colvin	Cincinnatus, N. Y	May 22, 1860	28, 351
Milker, Cow	L. O. Colvin	Cincinnatus, N. Y	May 29, 1860	28, 455
Milker, Cow	L. O. Colvin	Philadelphia, Pa	Feb. 17, 1863	37, 676
Milker, Cow	L. O. Colvin	Philadelphia, Pa	Jan. 12, 1864	41, 196
Milker, Cow	L. O. Colvin	Philadelphia, Pa	Mar. 28, 1865	46, 994
Milker, Cow	L. O. Colvin	New York, N. Y	May 22, 1866	54, 865
Milker, Cow	L. O. Colvin	New York, N. Y	Apr. 5, 1870	101, 589
Milker, Cow	L. C. Colvin	Humphreyville, Pa	Nov. 1, 1870	108, 881
Milker, Cow	G. H. Gardner	Philadelphia, Pa	Dec. 20, 1864	45, 486
Milker, Cow	G. H. Gardner	Philadelphia, Pa	Sept. 3, 1867	68, 358
Milker, Cow	E. A. Hewitt	Groton, Conn	Feb. 21, 1871	111, 932
Milker, Cow	C. Knapp	New York, N. Y	Nov. 27, 1849	6, 904
Milker, Cow	W. Reading	Offutt's Cross-Roads, Md	Oct. 17, 1871	120, 100
Milker's protector	J. M. Weare	Seabrook, N. H	May 2, 1854	10, 870
Milkers, Tail-clasp for	H. H. Dickinson	West Northfield, Mass	June 16, 1868	78, 936
Milking-apparatus	A. C. Black	Kaukauna, Wis	Jan. 25, 1870	99, 053
Milking-apparatus	E. Spedden	Astoria, Oreg	Oct. 19, 1869	95, 947
Milking cows	L. T. Blake	New Haven, Conn	Jan. 29, 1867	61, 539
Milking cows, Apparatus for	M. L. Baker	Milford, N. Y	May 21, 1861	32, 343
Milking cows, Device for	W. D. Nichols	Davenport, Iowa	May 21, 1861	32, 379
Milking implement, Cow	W. H. Whitman	Bailey Hollow, Pa	Aug. 26, 1856	15, 629
Milking-machine	H. V. Belding	Oppenheimer, N. Y	Aug. 9, 1864	43, 754
Milking-machine	B. F. Graves	Groton, Mass	Mar. 10, 1868	75, 261
Milking machine, Cow	L. O. Colvin	New York, N. Y	June 30, 1868	79, 317
Milking machine, Cow	I. Cook	Philadelphia, Pa	Dec. 3, 1867	71, 582
Milking machine, Cow	J. W. Kingman	Dover, N. H	Aug. 9, 1859	25, 019
Milking machine, Cow	T. H. Lindley	Taunton, Mass	Nov. 3, 1868	83, 777
Milking machine, Cow	S. W. Lowe	Philadelphia, Pa	Aug. 9, 1859	25, 022
Milking machinery, Cow	L. O. Colvin	New York, N. Y	Feb. 18, 1868	74, 507
Milking-seat, Folding	S. Snow	Somers, Conn	Apr. 13, 1869	88, 921
Milking-shield	O. H. Needham	New York, N. Y	Jan. 6, 1857	16, 361
Milking, Teat-cup for	B. F. Graves	Groton, Mass	July 7, 1868	79, 568

Mill:

See Apple-mill.
Apple-grinding mill.
Bark-mill.
Bone-grinding mill.
Boring-mill.
Cane-mill.
Cider-mill.
Clamp-mill.
Clay-mill.
Coffee-mill.
Corn-mill.
Corn and cob mill.
Cotton-seed mill.
Cracking-mill.
Crushing-mill.
Curd-mill.
Cylinder-mill.
Family mill.
Fanning-mill.
Feed-mill.
Feed-grinding mill.
Felly-mill.
Fish-bait mill.
Flax and hemp mill.
Floating-mill.
Flour-mill.
Flouring-mill.
Flouring and grist mill.
Foot-mill.
Founder's cleansing-mill.
Fulling-mill.
Gig-mill.
Grain-mill.
Grape-mill.
Grinding-mill.
Grinding and bolting mill.
Grinding and chopping mill.
Grinding and crushing mill.
Grinding and hulling mill.
Grinding and pounding mill.
Grist-mill.
Grist and coffee mill.
Grist and flour mill.
Grist and saw mill.
Gunpowder-mill.
Hide and fulling mill.
Hominy-mill.
Hominy and pearling mill.
Hominy and smut mill.
Horizontal mill.
Horse-mill.
Hulling-mill.
Iron mill.
Meat-mill.
Metallic mill.
Mixing-mill.
Mortar-mill.
Mortar-tub mill.
Mud and ore mill.
Ore-crushing mill.
Ore-stamp mill.
Paint-mill.
Paper-mill.
Pendulum-mill.
Percussion-mill.
Platting-mill.
Pug-mill.
Quartz-mill.
Rolling-mill.
Saw-mill.
Screw-mill.
Shingle-mill.
Slitting-mill.
Smut-mill.
Smut-fanning mill.
Spice-mill.
Stamp-mill.
Stamping-mill.
Starch-mill.
Steam rice-mill.
Steel mill.
Sugar-mill.
Sugar-cane mill.
Tide-mill.
Tide or current mill.
Turning and boring mill.
Universal mill.
Warping-mill.
Water-mill.
Windmill.
Winnowing-mill.

Invention.	Inventor.	Residence.	Date.	No.
Mill	S. Brewster	Middlesex, N. J	Mar. 15, 1816	....
Mill	E. Bryant	New London, Conn	Feb. 8, 1814	....
Mill	H. Chase		Dec. 16, 1799	....
Mill	E. D. Clark	Earlville, N. Y	June 26, 1860	28, 834
Mill	S. Coleman	New Orleans, La	June 12, 1860	28, 655
Mill	E. D. Curtis	Boston, Mass	Nov. 23, 1837	477
Mill	J. W. Curtiss		Aug. 24, 1797	....
Mill	J. Frisbie	Pennsylvania	June 30, 1810	....
Mill	J. Goulding	Leicester, Mass	Mar. 1, 1811	....
Mill	C. Hamaker	Lancaster	May 21, 1807	....
Mill	J. Harman, jr	New York	Mar. 30, 1836	....

Index of patents issued from the United States Patent Office from 1790 to 1873, inclusive—Continued.

Invention.	Inventor.	Residence.	Date.	No.
Mill	F. B. Hunt	Cincinnati, Ohio	Jan. 10, 1860	26, 818
Mill	G. D. Jones	New York, N. Y	Feb. 21, 1860	27, 224
Mill	W. Joslin	Cleveland, Ohio	July 3, 1860	28, 989
Mill	D. L. Latourette	Saint Louis, Mo	June 13, 1854	11, 075
Mill	A. E. Pirkey	Bradford, Ill	May 31, 1859	24, 235
Mill	J. Quinby	Wilmington, N. C	Feb. 3, 1812	......
Mill	W. Stewart	Philadelphia, Pa	Feb. 19, 1861	31, 492
Mill	H. C. Velie	Poughkeepsie, N. Y	June 26, 1860	28, 923
Mill	A. H. Wagner	Staunton, Va	July 31, 1860	29, 422
Mill	W. Warwick	Birmingham, Pa	Oct. 24, 1854	11, 845
Mill	J. Woodward	Philadelphia, Pa	Mar. 30, 1858	19, 807
Mill and grain-drill hopper	R. P. Johnson	Griffin, Ga	July 26, 1870	105, 807
Mill and machinery	L. Valcourt	Lewisville, Ky	June 25, 1805	......
Mill, and means of communicating a rotary motion by an auger, spindle, &c.	S. Mans and J. Black	Pennsylvania	Feb. 18, 1809	......
Mill-attachment	N. Bryan	Thomaston, Ga	Oct. 18, 1870	108, 442
Mill-bolt	W. H. Berdan	Mooreville, Mich	June 7, 1870	103, 833
Mill-bolt	B. C. White	Des Moines, Iowa	Oct. 17, 1871	120, 137
Mill-buhr dresser	W. P. Stalcup	Brookville, Ind	Oct. 29, 1867	70, 373
Mill-buhr feeder	A. P. Lawshaw	Harper's Ferry, W. Va	Apr. 26, 1870	102, 279
Mill-buhrs, Device for cooling	E. Embry and T. J. Blackburn	West Liberty, Ohio	June 18, 1872	128, 132
Mill-buhrs, Preventing the heating of	M. Bradley	Peru, Ind	Dec. 3, 1867	71, 690
Mill-buhrs, Stopping and starting	W. Crawford and S. Clarke	Attica, Ind	Mar. 22, 1870	101, 105
Mill-bush	J. Aulabaugh	East Berlin, Pa	July 18, 1840	1, 701
Mill-bush	E. Casner	Penn Yan, N. Y	Oct. 9, 1860	30, 296
Mill-bush	G. L. Dulaney	Long Meadow, Va	Oct. 24, 1854	11, 828
Mill-bush	G. L. Dulaney	Mount Jackson, Va	June 12, 1855	13, 029
Mill-bush	J. M. Evril	Centre, Pa	Apr. 28, 1868	77, 265
Mill-bush	H. F. Frisbie	Danville, Ill	Jan. 10, 1871	110, 843
Mill-bush	J. Heck	Boonsborough, Md	Mar. 26, 1844	3, 505
Mill-bush	H. Knowles	Washington, D. C	Jan. 16, 1849	6, 039
Mill-bush	J. G. Shafer	Fulton County, Pa	July 3, 1860	29, 013
Mill-bush	G. Strause	Boonsborough, Md	Feb. 16, 1858	19, 386
Mill-bush	R. M. Wade	Summit Point, Va	May 25, 1844	3, 601
Mill-bush	J. Wells	Baltimore, Md	Mar. 23, 1858	19, 727
Mill-bush and other spindles	C. T. Weston	Scranton, Pa	Nov. 22, 1870	110, 565
Mill-bush and spindle	J. Williams	Sullivan, Ill	May 4, 1869	89, 721
Mill-bush, Self-regulating and self-oiling	S. Moore	Chambersburgh, Pa	Apr. 8, 1840	1, 541
Mill-can shoe	J. Dennis	Providence, R. I	Sept. 8, 1834	......
Mill-dress	G. L. Dulaney	Mount Jackson, Va	June 19, 1855	13, 115
Mill-dress	J. W. Kane	New Carlisle, Ohio	Aug. 10, 1852	9, 192
Mill-driver	F. Walters	Covington, Ky	May 31, 1859	24, 268
Mill-feed, Adjustable	W. T. Duvall	Georgetown, D. C	Mar. 21, 1871	112, 909
Mill-feed apparatus	J. Sherlock and W. Brackbill	Portugal, Pa	June 18, 1850	7, 445
Mill-feed regulator	J. Ross	Brooklyn, N. Y	June 2, 1868	78, 541
Mill-feeder	J. M. Clark	Lancaster, Pa	Jan. 28, 1862	34, 239
Mill for mashing vegetables and mixing clay	C. Alvord	Geddes, N. Y	Aug. 31, 1852	9, 226
Mill for rolling irregular shapes by means of a cam-pattern.	J. S. Hall	Columbus, Ohio	Jan. 30, 1849	6, 079
Mill for tempering oleagenous seeds	W. Wilber	New Orleans, La	Jan. 27, 1857	16, 508
Mill-gudgeon	M. Withers	Strasburgh, Pa	Aug. 24, 1813	......
Mill or machine wheel, and applying water, wind, or steam thereto.	L. Copley	New Lebanon, N. Y	May 15, 1834	......
Mill-pick	A. Babcock and J. B. Martin	Silver Creek, N. Y	Aug. 6, 1872	130, 178
Mill-pick	C. Crossley	Philadelphia, Pa	Feb. 20, 1866	52, 688
Mill-pick	J. Cummings	West Charleston, Vt	July 11, 1871	116, 817
Mill-pick	J. Cummings	West Charleston, Vt	June 24, 1873	140, 118
Mill-pick	I. Faver	New York, N. Y	Apr. 9, 1861	31, 964
Mill-pick	E. S. Forgy	Dayton, Ohio	Apr. 22, 1873	138, 080
Mill-pick	H. H. Gillett	Warsaw, Mo	Sept. 15, 1868	82, 107
Mill-pick	B. Hostler	Brookfield, N. Y	Apr. 30, 1861	32, 190
Mill-pick	F. Kortick	Mendota, Ill	June 11, 1872	127, 774
Mill-pick	C. McNeal	Silver Creek, N. Y	Apr. 23, 1872	126, 072
Mill-pick	C. Metzger and G. R. Roraback	De Soto, Mo	Sept. 28, 1869	95, 252
Mill-pick	L. M. Osborn	Hamilton, N. Y	Oct. 6, 1863	40, 183
Mill-pick	C. Pennoyer	Lockport, N. Y	Oct. 29, 1872	132, 598
Mill-pick	A. Rasner	Dayton, Ohio	Oct. 31, 1871	120, 536
Mill-pick	H. N. Relyea	Warsaw, N. Y	Jan. 15, 1867	61, 252
Mill-pick	G. M. Rhoades	East Hamilton, N. Y	May 19, 1863	38, 603
Mill-pick	T. Sheehan	Dunkirk, N. Y	May 24, 1864	42, 882
Mill-pick	W. B. Stephens	Stephens' Mills, N. Y	Sept. 3, 1867	68, 394
Mill-pick	U. Stewart	Berlin, Wis	Dec. 3, 1867	71, 811
Mill-pick	A. Van Doren	Fredericksburgh, Va	Oct. 8, 1872	132, 121
Mill-pick	T. R. Way	Springfield, Ohio	Mar. 25, 1873	137, 116
Mill-pick	J. A. Wheeler	Cleveland, Ohio	Oct. 4, 1864	44, 570
Mill-pick	S. G. Williams	Trumansburgh, N. Y	Sept. 27, 1870	107, 745
Mill-pick	M. N. Young	Kelseyville, Cal	Aug. 26, 1873	142, 321
Mill-pick, Combination	J. W. Nance and D. Lillard	Cornersville, Tenn	June 10, 1873	139, 806
Mill pick handle	N. Rose	Belmont, N. Y	July 6, 1869	92, 369
Mill-pick holder	J. P. Brady	Mount Joy, Pa	Aug. 17, 1858	21, 184
Mill-pick holder	J. P. Sinclair	Mottville, N. Y	Sept. 5, 1871	118, 753
Mill power, Floating	A. G. Heittmann	Brooklyn, N. Y	Mar. 15, 1870	100, 761
Mill power, Increasing	R. Burns	Mifflin County, Pa	June 20, 1822	......
Mill-races, Floating sluice for	M. A. Shepard	Parkersburgh, Ill	May 8, 1860	28, 204
Mill-roller	W. S. Seymour	Ravenna, Ohio	Oct. 3, 1865	50, 281
Mill-shaft coupling	W. Kean	Chicago, Ill	Feb. 7, 1871	111, 550
Mill-spindle	E. T. Butler	Buffalo, N. Y	Jan. 27, 1852	8, 682
Mill-spindle	J. C. Gentry	Dayton, Ohio	Sept. 25, 1839	1, 340
Mill-spindle	J. A. Hafner	Pittsburgh, Pa	Dec. 30, 1873	146, 063
Mill-spindle	S. Hoyt	Wilmington, Del	Jan. 31, 1860	26, 993
Mill-spindle	J. H. McMinn	Logansport, Ind	Dec. 24, 1867	72, 661
Mill-spindle, &c	E. G. Patter	Lebanon, Ill	July 18, 1840	1, 703
Mill-spindle	S. P. Ruff	Weaver's Old Stand, Pa	Mar. 6, 1860	27, 386
Mill-spindle, &c., bush	R. E. and F. A. Howe	Somonauk, Ill	Aug. 13, 1872	130, 373
Mill-spindle bush	H. W. Vett	Union, Mo	Aug. 15, 1871	118, 077
Mill-spindle driver	J. M. Laogan	Springfield, Ill	May 17, 1870	103, 211
Mill-spindle driver	J. McConnell	Beaver, Pa	Apr. 17, 1847	5, 074
Mill-spindle spring	T. Alsop	Elkhart, Ill	July 7, 1868	79, 537

Index of patents issued from the United States Patent Office from 1790 *to* 1873, *inclusive*—Continued.

Invention.	Inventor.	Residence.	Date.	No.
Mill-spindle springs, Attaching	T. Alsop	Elkbart City, Ill	Aug. 25, 1868	81, 456
Mill-spindle step	J. N. Parker	Lewiston, Me	Dec. 18, 1855	13, 954
Mill-spindle step	G. S. Young	Clearfield, Pa	Jan. 3, 1871	110, 814
Mill-spindle step and tram-box	W. G. Norment	Dyersburgh, Tenn	Feb. 14, 1871	111, 768
Mill-spindle-step support	J. Russell	Prairie, Mo	Feb. 1, 1870	99, 353
Mill-spindles from dirt, Oiling and protecting	J. Hubbard	Watertown, Conn	Apr. 8, 1840	1, 538
Mill-spindles, &c., Gudgeon or pivot and step of	J. Staub	Georgetown, D. C	May 4, 1841	2, 075
Mill-spindles, Hanging	W. H. Naracon	Auburn, N. Y	June 15, 1852	9, 033
Mill-spindles, Hanging step of	G. Hotchkiss	Windsor, N. Y	June 29, 1852	9, 072
Mill-spindles, Securing and adjusting the steps of	G. Hotchkiss	Windsor, N. Y	Aug. 17, 1858	21, 199
Mill-spindles, Self-tightening bush for	H. Flinchbaugh	Lampeter Township, Pa	Oct. 19, 1838	984
Mill-spindles, Spring equalizer for	T. L. Clark	Mount Vernon, Ohio	Mar. 3, 1868	75, 124
Mill-spindles, Step and bearing of	T. S. Minniss	Meadville, Pa	June 22, 1852	9, 057
Mill-spindles, Tightening followers to	G. W. Wilson	Galesburgh, Ill	Nov. 18, 1862	36, 975
Mill-staff	A. Eshelman	Earlville, Pa	Feb. 4, 1869	86, 518
Mill-step	C. Corbit	Christiana, Pa	Mar. 24, 1868	75, 865
Mili-step	J. C. Dickey	Saratoga Springs, N. Y	July 3, 1855	13, 161
Mill-wheel	J. Cowen		Dec. 14, 1802	
Mill-wheel	S. H. Freeman	Cecilton, Md	May 17, 1836	
Mill-wheel	R. Lathrop	West Springfield, Mass	Nov. 12, 1812	
Mill-wheel	E. Rew	Canandaigua, N. Y	Feb. 22, 1809	
Mill wheel dresser	J. Yurk	Columbus, Pa	July 1, 1836	
Mill-wheel, Tide	D. Wheeler	Bristol, Mass	Mar. 2, 1811	
Mill-wheel, Trailing	J. Read	Lime, Conn	Mar. 6, 1818	
Mill-wheels, Applying water to	W. Kendall	Fairfield, Me	Jan. 12, 1831	
Mill-wheels, Applying water to	J. Mi·henor	Clinton, Ohio	Mar. 9, 1831	
Mills and machinery, Impelling-power for	B. S. Ridgway	Charleston, S. C	Feb. 6, 1829	
Mills, Application of the pendulum to	I. Cobb	Walpole, N. H	Mar. 1, 1809	
Mills, Application of water-power to	Dart and Wood	Fabius, N. Y	Sept. 9, 1835	
Mills, boats, &c., by supporting-chains, Propelling	B. Todd	Marietta, Ohio	Aug. 24, 1831	
Mills, boats, cars, &c., Propelling machinery for	A. Parkhurst and S. Bacon	Scriba, N. Y	Apr. 11, 1831	
Mills, boats, land-vehicles, &c., by wind, Propelling	B. Dugdale	Trenton, N. J	May 3, 1832	
Mills, boats, &c., Lever-power machinery for propelling.	W. Rhodes	Trenton, Tenn	Apr. 27, 1832	
Mills, Bushing and regulating	E. W. Welsh	Paris, Va	Jan. 28, 1840	1, 481
Mills by concentric machinery, Propelling	H. Buell	Randolph, Tenn	May 3, 1833	
Mills, &c., by rolling and balance wheel power, Propelling.	J. Estes	Brownsville, Tenn	Aug. 15, 1831	
Mills, &c., by weights, Propelling	O. R. Marston	Java, N. Y	Jan. 9, 1835	
Mills, Chain-band, with buckets or float-boards, to give motion to the machinery of.	S. Belknap	New York	Dec. 28, 1809	
Mills, Constructing and securing wing-gudgeons for	D. Philips	Georgetown, Pa	Dec. 28, 1840	1, 917
Mills, Cooling and feeding material to	B. F. Harrington and U. B. Burriss.	Missouri City, Mo	Apr. 19, 1859	23, 682
Mills, cotton-gins, &c., Construction of driving-shaft for.	J. Massey	Thomasville, Ga	June 21, 1859	24, 475
Mills, Feed-regulator for	M. Weaver	East Earl Township, Pa	May 11, 1869	89, 904
Mills, Feeding device and tell-tale for	J. D. Mines	Moffatt's Creek, Va	July 16, 1872	129, 577
Mills, Flume for	A. Messer	Providence, R. I	Nov. 19, [illegible]	
Mills, Frame-work of	W. S. Parham	Sussex County, Va	June 14, 1834	
Mills, Grinding-surface of	O. N. Stanford	Cincinnati, Ohio	Jan. 11, 1859	22, 588
Mills, Hanging runner-stone in	N. Oakley	Babylon, N. Y	Oct. 31, 1848	5, 893
Mills, Machinery for	M. B. Poiteaux	Richmond, Va	May 20, 1830	
Mills, Metallic grinding-rings for	G. and H. O'Connor	Mishawaka, Ind	Nov. 26, 1872	133, 333
Mills, Plate to	S. Hull	Poughkeepsie, N. Y	May 22, 1860	28, 373
Mills, Raising water for	A. Brookfield		Oct. 24, 1800	
Mills, Spindle, bush, ring, and ball for	W. P. Wing	Greenwich, Mass	Feb. 20, 1835	
Mills, thrashing-machines, &c., Machinery for	A. Sawyer	Hopewell, N. Y	Oct. 1, 1830	
Mills, Tram and level for	J. M. Seldomridge	Spring Valley, Ohio	May 12, 1863	38, 508
Mills, vessels, &c., by manual or other power, Propelling.	J. G. Aldebert	Opelousas, La	Dec. 29, 1831	
Mills, Working	J. Rumsay		Aug. 26, 1791	
Miller's stone-staff	S. Barnes	Rochester, Mich	Sept. 16, 1862	36, 444
Milling	J. Reeder	Lebanon, Ohio	Oct. 31, 1829	
Milling and cutting metals, Machine for	M. G. Wilder	West Meriden, Conn	Oct. 21, 1862	36, 744
Milling-machine	A. H. Brainard	Hyde Park, Mass	Oct. 1, 1872	131, 733
Milling-machine	J. R. Brown	Providence, R. I	Feb. 21, 1865	46, 521
Milling-machine	L. Chapman	Collinsville, Conn	Jan. 25, 1870	99, 059
Milling-machine	W. Hawkins	San Francisco, Cal	Oct. 31, 1871	120, 433
Milling-machine	B. Lawrence	Lowell, Mass	Apr. 29, 1873	138, 260
Milling-machine	W. A. N. Long	Worcester, Mass	Sept. 5, 1871	118, 732
Milling-machine	G. T. Pillings	Philadelphia, Pa	July 22, 1873	141, 012
Milling-machine	W. H. Robertson	Hartford, Conn	Oct. 5, 1852	9, 307
Milling-machine	E. S. Stiles	Portland, Me	Feb. 14, 1865	46, 401
Milling-machine	J. Watson	Philadelphia, Pa	Jan. 11, 1870	98, 821
Milling-machine	J. Wheelock	Worcester, Mass	July 23, 1872	129, 876
Milling-machine head-stock	A. H. Brainard	Hyde Park, Mass	Jan. 31, 1871	111, 311
Milling-machines, Back-center for	A. H. Brainard	Hyde Park, Mass	Jan. 17, 1871	110, 951
Milling-tool	J. Buckingham and J. H. Baird	Watertown, Conn	Apr. 15, 1851	8, 036
Milling-tool	A. W. Gifford	Worcester, Mass	Aug. 27, 1867	68, 064
Milling-tool	L. Griswold	Branford, Conn	Feb. 21, 1871	111, 927
Milling-tool	A. J. Lutz and H. Reiss	New York, N. Y	Aug. 3, 1869	93, 212
Milling-tool	J. T. Smith	Middletown, Conn	Mar. 27, 1866	53, 496
Milling-tools, Machine for cutting	B. F. Bee	Harwick, Mass	July 13, 1869	92, 565
Millstone	W. Bahme	New Media, Pa	Dec. 24, 1867	72, 587
Millstone	T. Barnett	Beverly, England	Oct. 12, 1852	9, 313
Millstone	J. H. Burrows	Cincinnati, Ohio	Apr. 23, 1842	2, 581
Millstone	D. Drawbaugh	Eberly's Mills, Pa	Apr. 28, 1863	38, 296
Millstone	E. T. Hanon-Valcke	Paris, France	Apr. 15, 1851	8, 037
Millstone	F. Kelsey	New York, N. Y	Aug. 29, 1848	5, 738
Millstone	J. W. Masury	Brooklyn, N. Y	Feb. 21, 1871	111, 956
Millstone	W. Miller	Lampeter, Pa	Aug. 15, 1811	
Millstone	D. Philips	Jefferson County, Miss	Apr. 3, 1829	
Millstone	J. Preslow	Auburn, N. Y	Feb. 12, 1833	
Millstone-adjusting device	A. W. Winall	Cincinnati, Ohio	Apr. 8, 1873	137, 748
Millstone-adjustment	J. G. Siemers	Saint Louis, Mo	July 15, 1856	15, 346
Millstone, Artificial	E. Cutter and S. Blanchard	Cincinnati, Ohio	Sept. 10, 1846	4, 753
Millstone, Artificial	C. Rands	Englewood, N. J	Nov. 8, 1864	44, 973

Index of patents issued from the United States Patent Office from 1790 *to* 1873, *inclusive*—Continued.

Invention.	Inventor.	Residence.	Date.	No.
Millstone bail and driver	A. G. Waldo	Milwaukee, Wis	Feb. 7, 1871	111, 704
Millstone-balance	J. A. Althouse	New Harmony, Ind	May 2, 1871	114, 249
Millstone-balance	W. C. Benn	San Francisco, Cal	Sept. 22, 1868	82, 278
Millstone-balance	D. Collins	Zanesfield, Ohio	Feb. 13, 1872	123, 678
Millstone-balance	Z. Dawson	Cole Creek, Ind	Aug. 15, 1871	118, 113
Millstone-balance	A. Frederick	Toledo, Ohio	Mar. 9, 1869	87, 656
Millstone-balance	J. M. Schramm	Pontoosuc, Ill	Dec. 31, 1872	134, 491
Millstone-balance	W. A. Vance	Kalamazoo, Mich	May 16, 1871	114, 890
Millstone-balance	J. Walsh	Galena, Ill	Jan. 24, 1871	111, 282
Millstone-balance	G. W. Wilson	Tolono, Ill	Aug. 31, 1869	94, 370
Millstone-balance	G. W. Wilson	Chebanse, Ill	Feb. 11, 1873	135, 740
Millstone balance and cooler	W. B. Dolsen	Waterloo, Iowa	June 14, 1870	104, 283
Millstone-bearing	P. Plamondon	Atchison, Kans	Aug. 9, 1870	106, 283
Millstone, Buhr	O. Evans	....	May 28, 1796	....
Millstone-bush	J. Brown	Fonda, N. Y	Aug. 15, 1871	118, 101
Millstone-bush	C. Custer	Philadelphia, Pa	Dec. 31, 1867	72, 812
Millstone-bush	M. De Camp	South Bend, Ind	May 24, 1859	24, 106
Millstone-bush	M. French	Leesville, Ind	Apr. 3, 1860	27, 709
Millstone-bush	L. S. Ives	Brooklyn, N. Y	Aug. 16, 1859	25, 120
Millstone-bush	G. W. Landon	Graham, Ind	Mar. 15, 1864	41, 928
Millstone-bush	J. F. McKray	Harmonsburgh, Pa	June 23, 1863	38, 971
Millstone-bush	N. Taylor	Urbana, N. Y	July 23, 1827	....
Millstone bush and bearing	A. Ortlip	East Vincent, Pa	Oct. 4, 1870	108, 045
Millstone bush and spindle	R. S. Cathcart	Cincinnati, Ohio	Apr. 1, 1872	137, 418
Millstone-cleaning composition	G. M. Pettee	Detroit, Mich	Nov. 30, 1869	97, 439
Millstone, Concave and convex	J. Sawyer and E. Clark	Royalston, Mass	Feb. 18, 1826	....
Millstone condenser and ventilator	D. Heffner	Independence, Iowa	May 9, 1871	114, 678
Millstone-cooling device	J. J. Roth	Philadelphia, Pa	Sept. 30, 1873	143, 258
Millstone-cooling device	E. D. Shanton and S. Shaver	Minneapolis, Minn	Apr. 22, 1873	138, 050
Millstone-curb	G. W. Eggleston	Monroe County, N. Y	July 18, 1871	117, 160
Millstone-curb	O. L. Richardson	Athens, Ga	Aug. 14, 1860	29, 617
Millstone-curb	A. Van Vleck and T. Phillips	Jordan, N. Y	July 19, 1870	105, 610
Millstone-dress	W. Ager	Rohrsburgh, Pa	June 29, 1852	9, 063
Millstone-dress	F. Bellinger	Lockport, N. Y	Apr. 27, 1858	20, 029
Millstone-dress	D. Bowman	Knoxville, Tenn	Apr. 2, 1867	63, 360
Millstone-dress	D. Bowman	Johnson City, Tenn	June 14, 1870	104, 107
Millstone-dress	J. Broughton	New York, N. Y	May 22, 1860	28, 344
Millstone-dress	W. P. Coleman	New Orleans, La	Oct. 7, 1856	15, 868
Millstone-dress	E. Deer	Annapolis, Ind	Jan. 2, 1872	122, 369
Millstone-dress	P. Dickson	Woodcock Township, Pa	Sept. 12, 1854	11, 665
Millstone-dress	W. G. Dunnivay and H. Osborn	New Cumberland, Ind	July 12, 1870	105, 186
Millstone-dress	J. Fairclough	Saint Joseph, Mo	Nov. 30, 1869	97, 373
Millstone-dress	W. Finkle	Cole Creek, Ind	July 4, 1854	11, 216
Millstone-dress	M. Fries	Philadelphia, Pa	Sept. 19, 1871	119, 134
Millstone-dress	E. P. Gaines	Melrose, Tex	July 25, 1854	11, 372
Millstone-dress	J. W. Gaines	Melrose, Tex	June 26, 1860	28, 850
Millstone-dress	J. W. Gaines	Clarksville, Tex	Nov. 24, 1868	84, 349
Millstone-dress	A. N. Garland	West Charleston, Vt	Dec. 3, 1867	71, 733
Millstone-dress	J. P. Harris	Greensborough, Ga	Apr. 22, 1873	138, 021
Millstone-dress	A. C. Hartsock	Douglas, Ill	July 21, 1868	80, 171
Millstone-dress	N. Haywood	Cleveland, Ohio	Dec. 15, 1857	18, 848
Millstone-dress	C. V. Littlepage	Austin, Tex	Mar. 20, 1860	27, 551
Millstone-dress	G. W. Loy	Jefferson, Tex	July 20, 1858	20, 950
Millstone-dress	G. W. Loy	Nacogdoches, Tex	Mar. 8, 1870	100, 537
Millstone-dress	G. Natcher	Indianapolis, Ind	Apr. 27, 1858	20, 083
Millstone-dress	G. Natcher	Indianapolis, Ind	Apr. 27, 1858	20, 084
Millstone-dress	G. Natcher	Sidney, Ohio	July 18, 1865	48, 831
Millstone-dress	G. and H. O'Conner	Mishawaka, Ind	Nov. 22, 1870	109, 542
Millstone-dress	H. O'Conner	Mishawaka, Ind	Oct. 22, 1872	132, 514
Millstone-dress	D. N. M. Peregoy	Elizabethtown, Tenn	May 16, 1871	114, 967
Millstone-dress	J. Sedgebeer	Cincinnati, Ohio	July 8, 1862	35, 858
Millstone-dress	B. C. Stephens	Honston, Mo	Dec. 15, 1868	84, 971
Millstone-dress	T. B. Stout	Keyport, N. J	Nov. 11, 1856	16, 074
Millstone-dress	H. L. Spencer	Social Circle, Ga	June 9, 1868	78, 768
Millstone-dress	D. W. Thompson	Somerset, Pa	Sept. 27, 1864	44, 471
Millstone-dress	J. Wilson	Davis' Mills, Va	May 16, 1871	114, 896
Millstone-dress	J. P. H. Wohlenberg	Lyons, Iowa	Oct. 18, 1870	108, 422
Millstone-dress	S. Wolff	Vicksburgh, Miss	June 1, 1858	20, 462
Millstone-dress	N. W. Wortham	Union Point, Ga	Dec. 17, 1867	72, 256
Millstone-dress	P. W. Yarrell	Garysburgh, N. C	Apr. 2, 1872	125, 157
Millstone-dress for cleaning grain	W. Ager	Rohrsburgh, Pa	Sept. 5, 1854	11, 636
Millstone-dresser	Z. G. Hurd	El Dorado, Iowa	Dec. 9, 1862	37, 097
Millstone-dresser	J. E. Karelsen	New York, N. Y	Oct. 21, 1862	36, 716
Millstone-dressing	J. Black	Helena, Ark	Aug. 10, 1844	3, 699
Millstone-dressing	C. Carlisle and E. Estabrook	Norwich, Vt	Aug. 28, 1846	4, 724
Millstone-dressing	H. Dolmetch	Canton, Pa	Oct. 31, 1871	120, 505
Millstone-dressing	I. W. Elmore	Lyons, N. Y	Sept. 25, 1834	....
Millstone-dressing	E. P. Gaines	Nacogdoches County, Tex	Mar. 4, 1851	7, 980
Millstone-dressing	M. P. Gardner	Huntington County, Ind	Jan. 7, 1862	34, 054
Millstone-dressing	S. Golay	Noyon, Switzerland	Jan. 21, 1868	73, 524
Millstone-dressing	M. Holden	Lawrenceburgh, Ind	Aug. 5, 1851	8, 272
Millstone-dressing	N. Jacobs	Newark, Ohio	Mar. 11, 1837	145
Millstone-dressing	J. Pennabecker	Eagle Mills, Durlach, Pa	Aug. 4, 1868	80, 762
Millstone-dressing	T. Rogers	Leesville, Ohio	Apr. 16, 1872	125, 760
Millstone-dressing	A. Sheek	Smith Grove, N. C	Feb. 1, 1859	22, 831
Millstone-dressing	R. M. Smith	Rutherford County, Pa	Feb. 8, 1840	1, 488
Millstone-dressing	T. W. Trussell	Winchester, Va	Nov. 21, 1854	11, 978
Millstone-dressing	I. Whissen	Mount Jackson, Va	Feb. 2, 1858	19, 273
Millstone-dressing	A. Wing	Mayville, N. Y	Mar. 21, 1865	46, 963
Millstone-dressing	J. Yorbrough	Milton, N. C	Jan. 24, 1860	26, 943
Millstone dressing and cracking machine	J. J. Faulkner	McMinnville, Tenn	Oct. 29, 1872	132, 651
Millstone dressing and furrowing apparatus	J. L. Norton	London, England	Dec. 6, 1870	109, 928
Millstone dressing and furrowing apparatus	J. L. Norton	London, England	Dec. 13, 1870	110, 064
Millstone-dressing apparatus	J. Dickinson	New York, N. Y	Oct. 28, 1862	36, 770
Millstone-dressing apparatus	H. Robinson and J. Smith	Lewishan, Carshalton, England.	Nov. 29, 1870	109, 760
Millstone-dressing gage	R. Ruston	Bockville, Ind	June 22, 1869	91, 773
Millstone-dressing guide	J. North	New York, N. Y	May 2, 1871	114, 326

Index of patents issued from the United States Patent Office from 1790 *to* 1873, *inclusive*—Continued.

Invention.	Inventor.	Residence.	Date.	No.
Millstone-dressing machine	M. H. Bacon	Mystic, Conn	July 10, 1860	29, 047
Millstone-dressing machine	M. H. Bacon	Mystic, Conn	July 31, 1860	29, 350
Millstone-dressing machine	E. C. Badger	Warner, N. H	Sept. 19, 1854	11, 686
Millstone-dressing machine	W. Bold	Sheboygan Falls, Wis	Oct. 20, 1868	83, 243
Millstone-dressing machine	J. Bowman	Somerset, Ohio	June 5, 1860	28, 553
Millstone-dressing machine	C. D. Brewer	Lewisburgh, Pa	Sept. 4, 1860	29, 857
Millstone-dressing machine	J. S. Carr	Alliance, Ohio	Dec. 14, 1869	97, 878
Millstone-dressing machine	Z. Chesebrough	Alden, N. Y	Dec. 10, 1838	1, 029
Millstone-dressing machine	W. B. Cummings and N. P. Dadman.	Tyngsborough and Chelmsford, Mass.	Nov. 22, 1853	10, 269
Millstone-dressing machine	S. W. Draper and R. M. Draper	South Dedham and Roxborough, Mass.	May 13, 1856	14, 859
Millstone-dressing machine	W. A. Dryden and J. H. Montgomery.	Monmouth, Ill	Oct. 24, 1865	50, 565
Millstone-dressing machine	J. East	Romeo, Mich	Aug. 17, 1869	93, 867
Millstone-dressing machine	S. East	Memphis, Mich	May 2, 1871	114, 425
Millstone-dressing machine	W. Farrow	Alexandria, Ill	Aug. 19, 1873	142, 009
Millstone-dressing machine	H. B. Gill	Ogden, N. Y	June 14, 1859	24, 385
Millstone-dressing machine	J. T. Gilmore and L. Anderson	Painesville, Ohio	Jan. 26, 1869	86, 297
Millstone-dressing machine	J. T. Gilmore and J. S. Crane	Lake Village, N. H	June 6, 1871	115, 600
Millstone-dressing machine	J. B. Harris	Ottawa, Ill	Sept. 21, 1869	94, 956
Millstone-dressing machine	J. B. Harris	Ottawa, Ill	Apr. 1, 1873	137, 364
Millstone-dressing machine	E. W. Hazard and C. H. Jenner	Binghamton and Rochester, N. Y.	Sept. 16, 1851	8, 364
Millstone-dressing machine	J. Hine	Cockermouth, England	Oct. 19, 1869	95, 901
Millstone-dressing machine	G. Z. Hockenburry	Pittsburgh, Pa	July 24, 1860	29, 275
Millstone-dressing machine	F. A. Hoyt	La Crosse, Wis	Mar. 4, 1873	136, 517
Millstone-dressing machine	J. W. Kennedy and J. T. Plummer.	Plainfield, Conn	July 12, 1859	24, 742
Millstone-dressing machine	S. K. Landes	West Cocalico, Pa	Aug. 21, 1860	29, 701
Millstone-dressing machine	A. Lane	Addison, N. Y	June 30, 1868	79, 359
Millstone-dressing machine	W. A. McGlaughlin	Greenland, Pa	Jan. 19, 1869	85, 948
Millstone-dressing machine	R. D. Nesmith	Lake Village, N. Y	Jan. 15, 1856	14, 109
Millstone-dressing machine	L. Norcross	Livermore, Me	May 6, 1823	....
Millstone-dressing machine	J. Norman	Glasgow, Scotland	Apr. 18, 1871	113, 914
Millstone-dressing machine	F. Ogden and F. A. Gilbert	Mansfield, Ohio	June 25, 1872	128, 419
Millstone-dressing machine	J. W. Parish	McFarland's, Va	July 29, 1873	141, 288
Millstone-dressing machine	J. Pepler	Bath, England	Nov. 8, 1870	109, 046
Millstone-dressing machine	W. Pickens and P. Dalrymple	Chicago, Ill	May 10, 1870	102, 968
Millstone-dressing machine	S. Prettyman and J. Wallis	Mumford and Greenwich, England.	May 25, 1869	90, 461
Millstone-dressing machine	L. Randolph	Jerseyville, Ill	Aug. 19, 1873	141, 949
Millstone-dressing machine	J. Richmond	Lockport, N. Y	July 15, 1862	35, 894
Millstone-dressing machine	J. B. Sacket	Lawton, Mich	Mar. 25, 1862	34, 780
Millstone-dressing machine	I. G. Shands	Saint Louis, Mo	Mar. 21, 1854	10, 674
Millstone-dressing machine	S. Shine and D. Savage	Forks and Orangeville, Pa	July 29, 1873	141, 392
Millstone-dressing machine	M. Shirk	Lancaster County, Pa	July 24, 1860	29, 345
Millstone-dressing machine	F. Simmons	New Orleans, La	Nov. 30, 1869	97, 320
Millstone-dressing machine	S. Teague	Newton, Ohio	Apr. 5, 1859	23, 511
Millstone-dressing machine	H. B. Weaver	South Windham, Conn	Jan. 15, 1861	31, 142
Millstone-dressing machine	C. Woolverton and B. Ridgeway, jr.	Morris Villa, Pa	May 24, 1811	....
Millstone-dressing machine	J. I. Yount	Tippecanoe, Ohio	Oct. 3, 1871	119, 551
Millstone-dressing machine, Diamond	F. S. Davenport	Jerseyville, Ill	Aug. 26, 1873	142, 153
Millstone-dressing machine, Diamond	D. Larer	Pottsville, Pa	Nov. 25, 1873	144, 851
Millstone-dressing machinery	A. Fisk, jr., L. D. Ramsay, and O. S. Gregory.	Sullivan, Pa	Feb. 20, 1847	4, 973
Millstone-dressing machinery	A. Woodworth	Cambridge, N. Y	Aug. 6, 1872	130, 265
Millstone dressing or pecking machine	S. Trumbull	Suffield, Conn	Aug. 2, 1839	1, 269
Millstone-dressing, Staff or gage for	L. B. Woolever	Mechanicsville, Ohio	Feb. 18, 1873	135, 956
Millstone-dressing, Tool for laying off furrows for	J. C. Hunt and J. Temple	Terre Haute, Ind	Sept. 1, 1868	81, 639
Millstone-driver	S. Darkess	Deerfield, Ind	Nov. 2, 1869	96, 312
Millstone-driver	W. A. Gustine	Ipava, Ill	Apr. 30, 1872	126, 143
Millstone-driver	A. Miller	New Berne, N. C	Aug. 9, 1859	25, 028
Millstone-driver	J. Plance	Constantine, Mich	Mar. 18, 1873	136, [illegible]63
Millstone driver	D. B. Ritter	Glasgow, Ky	Sept. 7, 1869	94, 650
Millstone-driver	J. J. Tomlinson	Bazeman City, Mont	Jan. 24, 1871	111, 277
Millstone-drivers, Bearing for	E. Clark	Lancaster, Pa	Sept. 22, 1857	18, 234
Millstone-elevator	J. B Hall	Lebanon, N. H	June 13, 1871	115, 844
Millstone-exhaust	D. Baird	Bloody Run, Pa	Dec. 3, 1867	71, 677
Millstone-exhaust	D. Baird	Bloody Run, Pa	Apr. 21, 1868	76, 972
Millstone-exhaust	J. Lingenfelter	Bloody Run, Pa	Apr. 4, 1871	113, 676
Millstone-exhaust	J. Lingenfelter	Bloody Run, Pa	June 6, 1871	115, 747
Millstone-face tester	J. Kuhn	Mount Pleasant, Pa	Sept. 8, 1868	81, 912
Millstone-feed	J. W. Anderson	Portland, Ind	Nov. 26, 1872	133, 288
Millstone-feed	M. De Camp	South Bend, Ind	Apr. 16, 1867	63, 786
Millstone-feed apparatus	A. Dewey	Princeton, Ky	July 25, 1871	117, 267
Millstone-feed regulator	G. W. Clapper	Martinsville, Ind	May 18, 1869	90, 079
Millstone feeder and grain scourer	P. H. Massey	South Bend, Ind	Feb. 14, 1871	111, 760
Millstone for dressing and cooling flour	J. B. Mirick	Manlius, N. Y	Aug. 10, 1833	....
Millstone for flour-mills	J. Clyde	Crawford County, Pa	July 27, 1831	....
Millstone, Grist	D. Stern	Vanderburgh County, Ind	Jan. 5, 1832	....
Millstone leveling	D. A. Balmer	Lexington, Ind	Oct. 23, 1860	30, 454
Millstone-leveling apparatus	B. H. Meitzler and P. L. Hoos	Fort Seneca, Ohio	Apr. 9, 1872	125, 472
Millstone-leveling device	A. Dray	Portland, Oreg	Jan. 18, 1861	32, 596
Millstone-leveling device	W. Ring	Owen County, Ind	June 3, 1873	139, 479
Millstone-machine	E. C. and R. A. Henderson	Albia, Iowa	Aug. 11, 1868	80, 954
Millstone-pick	H. J. Brunner	Nazareth, Pa	Feb. 14, 1871	111, 810
Millstone-pick	C. R. Elmer	Bridgeton, N. J	Nov. 21, 1865	51, 030
Millstone-pick	S. Etheridge	Tecumseh, Mich	Mar. 2, 1863	[illegible]
Millstone-pick	A. S. Jones	Joliet, Ill	Apr. 15, 1862	34, 966
Millstone-pick	D. C. Stone	Kingston, N. Y	Jan. 24, 1865	46, 035
Millstone-pick	H. P. Straub	Cincinnati, Ohio	June 15, 1869	91, 284
Millstone-pick	T. R. Way	Springfield, Ohio	Aug. 5, 1873	141, 613
Millstone-pick	C. Whitehouse	Bridgetown, England	Mar. 2, 1869	87, 530
Millstone-pick, Handle for	A. Rowe	Atalissa, Iowa	Aug. 12, 1862	36, 170
Millstone-pick, Stock for	F. Trump	Springfield, Ohio	Apr. 15, 1873	137, [illegible]79
Millstone-picking machine	S. A. Bell	Newtown, Ohio	Oct. 24, 1871	120, 181

Index of patents issued from the United States Patent Office from 1790 *to* 1873, *inclusive*—Continued.

Invention.	Inventor.	Residence.	Date.	No.
Millstone-picking machine	E. W. Daniels	Springfield, Mass	Jan. 24, 1860	26, 897
Millstone-picking machine	J. H. Gray	Boston, Mass	Dec. 4, 1866	60, 176
Millstone-picking machine	J. Kelly	West Milton, Ohio	Aug. 21, 1860	29, 748
Millstone-picking machine	R. D. Nesmith	Franklin, N. H	Aug. 30, 1859	25, 274
Millstone-propelling machinery	J. Case	Sodus, N. Y	June 13, 1831	....
Millstone-redressing machine	W. Y. Gill	Henderson, Ky	May 19, 1857	17, 326
Millstone-regulating device	S. Benson	Centralia, Ill	June 30, 1868	79, 528
Millstone-spindles, Oiling	J. J. Chubb	Decatur, Ind	Apr. 22, 1873	138, 074
Millstone to grind grain and hull clover-seed	B. Myers	Chambersburgh, Pa	Sept. 28, 1831	....
Millstone tram and level combined	J. Hutchins	Portsmouth, Ohio	Feb. 11, 1873	135, 812
Millstone tramming, staffing, and fine-dressing machine.	J. T. Gilmore	Burton, Ohio	Oct. 28, 1862	36, 776
Millstone trimming and dressing machine	J. I. Gilmore	Painesville, Ohio	Jan. 12, 1869	85, 737
Millstone-ventilating device	R. Symes	Saint Charles, Mo	June 30, 1868	79, 517
Millstone-ventilator	W. K. Fuller	Modena, Ill	Dec. 24, 1867	72, 474
Millstone-ventilator	A. Grube	West Hempfield Township, Pa.	Feb. 8, 1870	99, 564
Millstone-ventilator	J. Metherell	Rockford, Ill	Sept. 8, 1868	81, 927
Millstone, with ventilator for cooling the flour, Dressing.	P. Cheek	Flat Rock, Ga	July 26, 1845	2, 222
Millstones and curbs, Cooling	J. Mallin	Bedford, Ohio	Dec. 31, 1867	72, 744
Millstones, Balance-iron for	J. H. Glover	Skreggs Creek, Ky	Dec. 1, 1857	18, 741
Millstones, Balancing	S. Culver	Sandwich, Ill	Jan. 2, 1872	122, 367
Millstones, Balancing	E. and Z. Dawson and B. Hilton.	Brumersburgh, Ohio	Sept. 17, 1867	68, 963
Millstones, Balancing	J. Fairclough	Louisville, Ky	Dec. 21, 1858	22, 356
Millstones, Balancing	D. Fellenbaum	Lancaster, Pa	Sept. 4, 1860	29, 869
Millstones, Balancing	J. Foley	Cleveland, Ohio	Sept. 17, 1867	68, 865
Millstones, Balancing	C. V. Foreman	Mechanicstown, Md	Sept. 6, 1870	107, 029
Millstones, Balancing	W. J. Lemuth	Greencastle, Ind	July 29, 1862	36, 009
Millstones, Balancing	G. S. Thompson	Philadelphia, Pa	Nov. 23, 1869	97, 134
Millstones, Balancing	R. Van Ormer and W. J. Bell	McAllisterville, Pa	May 6, 1862	35, 187
Millstones, Balancing and ventilating	S. N. Page	Salona, Pa	July 7, 1863	39, 167
Millstones, Balancing or adjusting	M. L. Chase	Baltimore, Md	June 13, 1831	....
Millstones, Blacking-staff for facing	A. Heartsill	Louisville, Tenn	Mar. 9, 1869	87, 669
Millstones, Casting the face of	G. Mauter	Londonderry, N. H	Dec. 8, 1818	....
Millstones, Cementing	S. Hoyt	Wilmington, Del	May 1, 1860	28, 083
Millstones, Composition for cleaning	D. Kindig	Newville, Pa	Feb. 25, 1868	74, 833
Millstones, Composition for dressing	W. H. Hayes	Rock Island, Ill	Sept. 13, 1864	44, 189
Millstones, Composition for taking off the glaze from.	J. Noble	Derby, Conn	June 10, 1813	....
Millstones, Cooling	F. Freleigh	Stowe, Ohio	Mar. 26, 1845	3, 969
Millstones, Cooling	A. Taylor	Littleton, N. H	Mar. 31, 1836	....
Millstones, Crane, screw, and bale for raising	T. Reynolds	West Fallowfield, Pa	Nov. 8, 1814	....
Millstones, Device for adjusting	A. W. Winall	Cincinnati, Ohio	Jan. 28, 1873	135, 393
Millstones, Device for guiding diamonds for dressing.	B. D. Tripp	Moravia, N. Y	Dec. 18, 1860	30, 995
Millstones, Diamond-protector for dressing	J. Dickinson	Brooklyn, N. Y	Sept. 3, 1861	33, 187
Millstones, Elevating	A. M. Bruckart	Brunnersville, Pa	Feb. 24, 1863	37, 732
Millstones, Eye for	E. Munson	Utica, N. Y	July 19, 1853	9, 859
Millstones, Feeding	M. Smith	Princeton, Iowa	Jan. 19, 1858	19, 156
Millstones, Finishing and balancing	G. Todd	Saint Louis, Mo	Nov. 8, 1851	8, 538
Millstones for scouring and hulling buckwheat, Dressing.	B. I. Harris	Auburn, Pa	Dec. 25, 1855	13, 985
Millstones, Forming and balancing	E. Munson	Utica, N. Y	Aug. 7, 1849	6, 639
Millstones, Grain and middlings feeder for	B. B. Broyles	Freedom, Tenn	July 9, 1872	128, 702
Millstones, Hand-mill for picking	S. Hodge and J. Dorr	....	Oct. 23, 1792	....
Millstones, Hanging	E. Clark	Lancaster, Pa	Sept. 15, 1857	18, 191
Millstones, Hanging	W. A. Clark, S. D. Porter, and W. D. Simpson.	Saint Louis, Mo	June 2, 1857	17, 421
Millstones, Hanging	R. Cochran	Cincinnati, Ohio	Aug. 7, 1855	13, 384
Millstones, Hanging	G. P. Dance	Columbia, Tex	May 29, 1860	28, 534
Millstones, Hanging	S. Fagin	Cincinnati, Ohio	Oct. 30, 1866	59, 201
Millstones, Hanging	J. A. Forsman	Cincinnati, Ohio	Aug. 31, 1858	21, 330
Millstones, Hanging	J. H. Glover	Glasgow, Ky	May 22, 1860	28, 362
Millstones, Hanging	D. Marsh	Bridgeport, Conn	Jan. 15, 1856	14, 128
Millstones, Hanging	Z. McDaniel	Bowling Green, Ky	Sept. 11, 1860	30, 025
Millstones, Hanging	C. Schneider	Galion, Ohio	July 19, 1870	105, 496
Millstones, Hanging	H. P. Straub	Cincinnati, Ohio	Oct. 18, 1870	108, 405
Millstones, Hanging	A. W. Winall	Cincinnati, Ohio	June 15, 1869	91, 295
Millstones, Hanging	A. W. Winall	Cincinnati, Ohio	June 3, 1873	139, 642
Millstones, Hanging and using	J. McDuffee	Bradford, Vt	Feb. 21, 1833	....
Millstones, Hanging-device for	P. Zimmerman	Delaware Water Gap, Pa	July 7, 1868	79, 717
Millstones, Instrument for facilitating the facing of	B. D. Sanders	Holliday's Cove, Va	Feb. 10, 1857	16, 615
Millstones, Laying off and dressing the runners of	C. Vest	Stokes County, N. C	Oct. 2, 1838	954
Millstones, Machine for facing and polishing	E. Munson	Utica, N. Y	Jan. 29, 1861	31, 255
Millstones, Machine for furrowing	J. J. Zinn	Albion, Pa	July 2, 1867	66, 439
Millstones, Machine for leveling the faces of	D. Drawbaugh	Eberle's Mills, Pa	May 12, 1863	38, 472
Millstones, Manufacture of	J. S. Elliott and J. F. Wood	Chelsea and Everett, Mass	Mar. 12, 1872	124, 558
Millstones, Method of putting face-dress on	J. S. Braley and J. A. and P. L. Schmitt	Utica, Mo	Aug. 23, 1870	106, 540
Millstones, Process of repairing	J. S. Elliott and J. F. Wood	Chelsea and Everett, Mass	Mar. 12, 1872	124, 559
Millstones, Steadying the motion of	J. J. Doughty	Lake City, Minn	July 13, 1869	92, 433
Millstones, Stopping	A. Reeder	Buckingham, Pa	June 2, 1863	38, 761
Millstones, Swinging spout for feeding	M. and C. Painter	Owing's Mills, Md	July 1, 1856	15, 250
Millstones, Tram for gaging	T. R. James	Saint Louis, Mo	Oct. 27, 1868	83, 503
Millstones, Ventilating	J. Campbell	Peoria, Ill	Jan. 28, 1868	73, 776
Millstones, Ventilating	C. Cerisier	Mung sur Loire, France	Jan. 4, 1870	98, 555
Millstones, Ventilating	J. Gray	Dubuque, Iowa	Oct. 29, 1867	70, 198
Millstones, Ventilating	A. D. Lagoguey	Paris, France	Apr. 24, 1866	54, 268
Millstones, Ventilating	H. McEldowney	Dixon, Ill	Oct. 1, 1867	69, 459
Millstones, Ventilating	S. G. Morrison	Williamsport, Pa	June 18, 1861	32, 579
Millstones, Ventilating	G. P. Plant and J. Raith	Saint Louis, Mo	Oct. 23, 1860	30, 498
Millstones, Ventilating	L. Racine	Joliet, Ill	July 6, 1858	20, 818
Mincing-cleaver	S. J. Tongue	Philadelphia, Pa	July 21, 1868	80, 242
Mincing-knife	S. C. Ketchum	Winchendon, Mass	May 28, 1867	65, 233
Mincing-knife	H. M. Remington	Springfield, Mass	Feb. 12, 1867	61, 954
Mincing-machine	W. H. Peirce	Bangor, Me	Aug. 31, 1869	93, 237

Index of patents issued from the United States Patent Office from 1790 *to* 1873, *inclusive*—Continued.

Invention.	Inventor.	Residence.	Date.	No.
Mine-ventilator	J. B. Jones	Laurel Run, Pa	May 3, 1870	102, 681
Mine-water, Compound for purifying	F. J. Delker	Ashland, Pa	Dec. 20, 1870	110, 345
Mines, Device for raising tailings from	W. A. Rogers	Folsom, Cal	Jan. 9, 1872	122, 657
Mines, Door for	N. Jensen	Carbon Station, Wyo	June 25, 1872	128, 312
Mines, Purifying water from	H. Burgess	Royer's Ford, Pa	Nov. 18, 1873	144, 737
Mines, Safety-cage for	J. Evans	Virginia City, Nev	Aug. 7, 1866	56, 915
Mines, Safety dumping-cage for	W. Z. Hatcher	Plymouth, Pa	June 28, 1870	104, 848
Mines, Ventilating	J. S. Fisk and J. Westerman	Meadville and Sharon, Pa	Jan. 17, 1865	45, 918
Mines, Ventilation of	J. L. Beadle	Ashland, Pa	May 16, 1865	47, 694
Mines, Ventilation of	H. Haupt	Cambridge, Mass	June 6, 1865	48, 065
Mines, Winding-machinery for	E. M. Ivens	Tamaqua, Pa	Apr. 14, 1857	17, 065
Mineral and fossil substances, Machine for separating.	R. George	Denver City, Colo	Aug. 2, 1870	106, 047
Mineral and fossil substances with bitumenized clay to form a plastic material, Mixing.	P. Lea	Goliad, Tex	Nov. 11, 1873	144, 397
Mineral and metallic substances, Apparatus for collecting and separating.	W. T. Duvall	Georgetown, D. C	July 28, 1868	80, 398
Mineral and other hard substances, Machine for crushing.	D. C. Ebaugh	Charleston, S. C	July 26, 1870	105, 659
Mineral bath to imitate mineral-water, Preparation of.	O. Gavron	New York, N. Y	June 21, 1870	104, 446
Mineral composition resembling jasper	J. P. Pepper	New Britain, Conn	Dec. 16, 1851	8, 592
Minerals, &c., crushing and grinding machine	T. O. Cutler	New York, N. Y	Dec. 12, 1854	12, 054
Mineral-fountain, Family	G. E. Pendergast	Philadelphia, Pa	Aug. 18, 1815	......
Mineral-washing machine	P. G. Gardiner	New York, N. Y	Oct. 4, 1864	44, 526
Mineral-water	C. E. Michel	Saint Louis, Mo	Dec. 3, 1867	71, 777
Mineral-water, Apparatus for making	J. S. Ewing	Philadelphia, Pa	Feb. 10, 1817	......
Mineral-water, Apparatus for making	S. Fahnestock	Lancaster, Pa	Apr. 23, 1819	......
Mineral-water, Giving temperature to artificial	J. Hawkins	Philadelphia, Pa	Apr. 5, 1823	......
Mineral-water, Imitating natural	J. Hawkins	Baltimore, Md	Sept. 7, 1809	......
Mineral-water, Manufacture of	J. Matthews	New York, N. Y	July 16, 1872	129, 044
Mineral-water, Manufacture of bottled	T. Maloney	Des Moines, Iowa	Mar. 14, 1871	112, 610
Mineral-water, Manufacturing	B. Marshall	Albany, N. Y	Dec. 5, 1815	......
Mineral-water, Manufacturing	W. Mead	Newburgh, N. Y	Apr. 12, 1831	......
Miners' bar	R. B. Platt	Hazleton, Pa	Apr. 22, 1873	138, 098
Miners' pick	S. H. Wallace	Parnassus, Pa	Oct. 29, 1867	70, 298
Miners' pick	R. K. Walton	Clarington, Ohio	Nov. 18, 1873	144, 808
Miners' squib	J. Holmes	Saint Clair, Pa	Dec. 12, 1871	121, 783
Miniature-case	H. Halvorson	Boston, Mass	Mar. 25, 1856	14, 501
Miniature-cases, &c., Composition for	M. Tomlinson	Birmingham, Conn	Aug. 24, 1858	21, 285
Miniature-cases, Press for making	H. T. Anthony	New York, N. Y	Jan. 31, 1854	10, 465
Minie-balls, Apparatus for casting	C. B. Tatham	Brooklyn, N. Y	May 20, 1862	35, 334
Minie-balls, Device for making	P. Naylor	New York, N. Y	May 17, 1864	42, 785
Minie-balls, Grooving and sizing	P. Naylor	New York, N. Y	May 17, 1864	42, 786
Mining and tunneling	H. Haupt and J. Y. Smith	Cambridge, Mass., and Alexandria, Va.	Apr. 4, 1865	47, 168
Mining and tunneling machine	R. C. M. Lovell	Covington, Ky	July 30, 1867	67, 323
Mining-apparatus, Coal	G. E. Donisthorpe, W. Firth, and R. Ridley.	Leeds, England	May 10, 1864	42, 724
Mining apparatus, Coal	R. H. Lamborn	Altoona, Pa	May 10, 1864	42, 669
Mining apparatus, Hydraulic	F. H. Fisher	Nevada City, Cal	Dec. 20, 1870	110, 222
Mining apparatus, Hydraulic	T. Watson	Nevada City, Cal	Oct. 25, 1870	108, 658
Mining-drill	H. Burk	Mineral Point, Ohio	Apr. 15, 1862	34, 941
Mining-machine	D. Morris	Bartlett, Ohio	July 20, 1869	92, 871
Mining-machine	D. Williamson	Elizabeth, Pa	Mar. 2, 1869	87, 453
Mining-pan	J. A. Brock	Chicago, Ill	Apr. 23, 1861	32, 115
Mining-pick	G. Hofman	Scott Bar, Cal	May 12, 1863	38, 481
Mining-pick	H. L. Lowman	Virginia City, Nev	June 6, 1865	48, 080
Mining, Placer	C. H. Smith	Rock Island, Ill	Mar. 28, 1865	47, 046
Mining-shaft and building ventilator	T. Boyd	Cambridgeport, Mass	Jan. 1, 1867	60, 677
Mining-shaft safety-guard	E. O. Leermo	Gold Hill, Nev	June 23, 1868	79, 076
Mining sluice and rifle	C. J. Garland	Gwin Mine, Cal	Oct. 17, 1871	119, 978
Mining-sluices, Apparatus for collecting the precious metals in.	J. B. Beers	San Francisco, Cal	July 23, 1872	129, 644
Mirror	P. A. Dailey	New York, N. Y	July 31, 1866	56, 723
Mirror	S. R. Scottron	Springfield, Mass	Mar. 31, 1866	76, 253
Mirror, Adjustable	L. F. Neaglo	Philadelphia, Pa	Nov. 26, 1867	71, 320
Mirror, Adjustable	G. S. Roberts	Meredith Village, N. H	June 13, 1871	115, 898
Mirror, Adjustable duplex	G. S. Roberts	Meredith Village, N. H	Apr. 23, 1872	126, 091
Mirror and frame hanger	J. S. Thomson	Camptown, Pa	Apr. 23, 1872	125, 913
Mirror and picture-frame	W. McConnell	Clarksville, N. J	June 21, 1870	104, 617
Mirror and picture-frame support	D. Hartmann	Mansfield, Ohio	July 4, 1871	116, 589
Mirror and stand, Adjustable	H. W. Southworth	Mittineague, Mass	Feb. 2, 1869	86, 599
Mirror, Back-reflecting	R. Mason	Newark, N. J	Mar. 7, 1871	112, 474
Mirror for attachment to windows	G. P. Bertrand	Easton, Pa	July 22, 1862	35, 917
Mirror-frame	O. L. Gardner	New York, N. Y	Sept. 10, 1867	68, 730
Mirror-frame, Hand-toilet	E. P. Williams and G. H. Chinnock.	Elizabeth, N. J., and New York, N. Y.	May 31, 1870	103, 807
Mirror frame, Hand toilet	E. P. Williams and G. H. Chinnock.	Elizabeth, N. J., and New York, N. Y.	May 31, 1870	103, 808
Mirror frame, Hand toilet	E. P. Williams and G. H. Chinnock.	Elizabeth, N. J., and New York, N. Y.	May 31, 1870	103, 809
Mirror-frame, Swivel	C. W. Hafermaly	Providence, R. I	Jan. 12, 1869	85, 815
Mirror, Hand	J. F. Dohan	Binghampton, N. Y	Dec. 23, 1873	145, 729
Mirror, Hand	W. U. Dudley	New York, N. Y	July 20, 1869	92, 942
Mirror, Hand	A. C. Estabrook	Northamton, Mass	Jan. 23, 1872	123, 008
Mirror, Hand	J. E. Faris	Baltimore, Md	Feb. 13, 1866	52, 551
Mirror, Hand	W. M. Welling	New York, N. Y	Apr. 16, 1872	125, 867
Mirror-holder	W. Simpson	Berlin, Canada	July 15, 1873	140, 851
Mirror, &c., Ornamental	W. M. Davis	Brook Haven, N. Y	Apr. 16, 1872	125, 726
Mirror, Suspension toilet	R. T. Sargent	Norwich, Vt	Mar. 28, 1871	113, 100
Mirror, Toilet	G. H. Chinnock	New York, N. Y	Sept. 14, 1869	94, 711
Mirror, Toilet	G. H. Chinnock	New York, N. Y	Sept. 14, 1869	94, 712
Mirror, Toilet	G. H. Chinnock	New York, N. Y	Mar. 22, 1870	100, 980
Mirror, Toilet	G. H. Chinnock and E. P. Williams.	Elizabeth N. J	June 7, 1870	103, 980
Mirror, Toilet	J. Johnson	Saco, Me	Oct. 24, 1865	50, 659
Mirror, Toilet	W. H. King	Newark, N. J	Mar. 22, 1870	101, 022

Index of patents issued from the United States Patent Office from 1790 to 1873, inclusive—Continued.

Invention.	Inventor.	Residence.	Date.	No.
Mirror, Toilet	W. F. Patterson	Boston, Mass	July 2, 1872	128, 503
Mirror, Toilet	W. F. Patterson	Boston, Mass	July 2, 1872	128, 651
Mirror, Toilet	L. H. Rogers	Boston, Mass	Mar. 23, 1869	88, 077
Mirror, Toilet	G. Wattis	New York, N. Y	Sept. 14, 1869	94, 798
Mirror, Window	A. and A. Olander	Glen Gardner, N. J., and New York, N. Y.	Apr. 1, 1873	137, 383
Mirror, Window	H. Voss and H. P. Brandt	Davenport, Iowa	Aug. 5, 1873	141, 476
Mirrors, &c., Adjustable hanger for	J. Wright	New York, N. Y	Dec. 23, 1873	145, 825
Mirrors, Coating and protecting the silvering of	D. Briansky	St. Petersburg, Russia	Apr. 12, 1864	42, 276
Mirrors, Composition for silvering	H. Poissonnier	New York, N. Y	June 19, 1860	28, 773
Mirrors, Device for adjusting	O. L. and W. Gardner	Glen Gardners, N. J	Apr. 18, 1871	113, 868
Mirrors, Hanging	F. Brown	Detroit, Mich	July 9, 1867	66, 458
Mirrors, Hanging	W. C. Cumming	Peekskill, N. Y	Jan. 1, 1867	60, 699
Mirrors, Hanging	J. S. and H. F. Gray	Chelsea, Mass	June 19, 1866	55, 647
Mirrors in traps, Arrangement of	J. Stevens	Middleton, Md	June 11, 1850	7, 431
Mirrors, Manufacture of	R. Kleck	New York, N. Y	Nov. 7, 1865	50, 874
Mirrors, Manufacture of silvered	D. Durand	New York, N. Y	Jan. 30, 1872	123, 247
Mirrors, Metallizing	E. Dodé	Paris, France	Nov. 13, 1866	59, 734
Mirrors, Ornamenting	W. M. Davis	Mount Sinai, N. Y	Feb. 14, 1871	111, 731
Mirrors, Ornamenting	T. C. March	London, England	Mar. 5, 1867	62, 654
Mirrors, Silvering	R. Keck	Chicago, Ill	July 17, 1866	56, 507
Mirrors, Silvering	T. Petitjean	Tottenham Court Road, England.	Oct. 21, 1856	15, 950
Mirrors to furniture, Suspending	L. Pierce	Charleston, Mass	Mar. 8, 1870	100, 663
Missiles, Means of checking and resisting	A. C. Twining	New Haven, Conn	July 28, 1863	39, 363
Miter and bevel square, Carpenters'	J. S. Halsted and C. J. Ackerman.	New York, N. Y	June 26, 1855	13, 127
Miter and beveling machine	L. Holliday	Rogersville, N. Y	May 15, 1855	12, 862
Miter and other joints, Machine for cutting	F. A. Gleason	Rome, N. Y	Mar. 6, 1855	12, 482
Miter-box	J. W. Bower and A. D. Mohler	Greencastle and Huntington, Ind.	May 14, 1872	126, 669
Miter-box	S. D. Bowker	Kansas City, Mo	Aug. 10, 1869	93, 589
Miter-box	D. Bull	Amboy, Ill	June 12, 1866	55, 576
Miter-box	J. Bullard	Hyde Park, Vt	Feb. 1, 1870	99, 398
Miter-box	R. and R. T. Burchell	Trenton, N. J	Apr. 20, 1869	89, 124
Miter-box	H. N. Burr	Mansfield, Ohio	July 19, 1870	105, 421
Miter-box	C. Caton	Coshocton, Ohio	May 21, 1872	126, 873
Miter-box	C. Caton	Coshocton, Ohio	June 17, 1873	139, 867
Miter-box	G. L. Chapin	Perrysburgh, N. Y	June 9, 1857	17, 494
Miter-box	E. C. Cheek	Placerville, Cal	July 2, 1867	66, 217
Miter-box	J. Devoge	Meadville, Pa	Nov. 15, 1870	109, 190
Miter-box	T. B. Dooley	Boston, Mass	Mar. 19, 1872	124, 797
Miter-box	A. C. Hall	Carbondale, Pa	July 9, 1872	128, 795
Miter-box	J. and T. Hanks	New Brunswick and Jersey City, N. J.	July 16, 1872	129, 557
Miter-box	G. E. Hedges	Ashland, Nebr	Aug. 1, 1871	117, 538
Miter-box	W. H. Herbert	Blissfield, Mich	Sept. 1, 1868	81, 782
Miter-box	D. Howell, jr., and M. K. Kellam.	New York, N. Y	May 3, 1864	42, 582
Miter-box	G. Keating	Thomaston, Me	Aug. 21, 1866	57, 334
Miter-box	J. King	Northampton, Mass	July 16, 1872	128, 963
Miter-box	L. W. Langdon	Northampton, Mass	Nov. 15, 1864	45, 055
Miter-box	C. F. Linscott	Chicago, Ill	Dec. 6, 1870	109, 915
Miter-box	H. Markle	Spencer, Ind	June 6, 1871	115, 752
Miter-box	J. A. McKinstry	Monson, Mass	Oct. 23, 1866	59, 046
Miter-box	J. A. McKinstry and W. Walden.	Springfield, Mass	Feb. 11, 1873	135, 832
Miter-box	H. B. Nash	Sandy Hill, N. Y	Sept. 11, 1860	30, 027
Miter-box	J. Pons	Baltimore, Md	Mar. 9, 1869	87, 703
Miter-box	J. W. Richard son	Boston, Mass	Dec. 31, 1867	72, 906
Miter-box	C. Robinson	Fox Lake, Wis	Apr. 28, 1868	77, 217
Miter-box	C. Robinson	Fox Lake, Wis	Sept. 22, 1868	82, 351
Miter-box	E. Root	Parma, Mich	June 8, 1869	90, 960
Miter-box	M. O. Royce	Boston, Mass	July 20, 1869	92, 887
Miter-box	M. O. Royce	East Cambridge, Mass	Sept. 5, 1871	118, 748
Miter-box	S. D. Skiff	Columbia, Ohio	Aug. 20, 1872	130, 665
Miter-box	M. Spear	Bowdoinham, Me	Sept. 30, 1851	8, 401
Miter-bok	M. Spear	Bowdoinham, Me	May 16, 1854	10, 936
Miter-box	M. Spear	Bowdoinham, Me	May 1, 1855	12, 796
Miter-box	G. M. Stevens	Portland, Me	July 15, 1873	140, 853
Miter-box	A. F. Tarr	Rockport, Mass	Dec. 21, 1858	22, 386
Miter-box	J. P. Tierney	Sacramento, Cal	Sept. 16, 1873	142, 878
Miter-box	W. Tinsley	Glen's Falls, N. Y	Nov. 28, 1854	12, 006
Miter-box	C. H. Underwood	Dorchester, Mass	Mar. 23, 1869	88, 236
Miter-box	W. S. Wheeler and S. E. Brickford.	Franklin, N. H	Mar. 12, 1867	62, 910
Miter-box	E. M. Wilcox	Bloomer, Wis	July 26, 1870	105, 752
Miter-box	W. P. Wood	Washington, D. C	June 3, 1856	15, 053
Miter-box and miter-saw	D. McAllaster	Malden, Mass	July 5, 1870	104, 975
Miter-box for cutting printers' rules	T. H. Mead	New York, N. Y	Nov. 11, 1873	144, 406
Miter-cutting machine	F. D. Green	Williamsport, Pa	Oct. 8, 1872	131, 949
Miter-cutting machine	S. W. Hall	Williamsport, Pa	Aug. 17, 1858	21, 194
Miter-cutting machine	Z. Kelley	Easton, Md	Apr. 22, 1812	....
Miter-gage	G. W. La Baw, A. T. Hutchinson, and I. W. Fleming.	Jersey City, N. J., Harlem, N. Y., and Jersey City, N. J.	Apr. 21, 1868	77, 053
Miter-joints, Machine for trimming	A. Johnson	Putnam, Conn	June 12, 1866	55, 499
Miter-machine	J. H. Brown	Brockport, N. Y	May 4, 1869	89, 627
Miter-machine	D. Bull	Amboy, Ill	July 15, 1873	140, 763
Miter-machine	J. H. Estes	Boston, Mass	Sept. 29, 1868	82, 504
Miter-machine	L. Henkle	Rochester, N. Y	July 2, 1872	128, 486
Miter-machine	J. R. Howell	Buffalo, N. Y	June 21, 1870	104, 458
Miter-machine	F. C. Jones	Ouachita Parish, La	May 24, 1870	103, 341
Miter-machine	T. E. King	Boston, Mass	Mar. 7, 1871	112, 355
Miter-machine	G. W. La Baw	Jersey City, N. J	June 27, 1854	11, 164
Miter-machine	G. W. La Baw	Jersey City, N. J	May 29, 1855	12, 956
Miter-machine	J. Nonnenbacher	New York, N. Y	Apr. 5, 1870	101, 497
Miter-machine	T. Pooley	Blackberry Station, Ill	Dec. 23, 1873	145, 751

Index of patents issued from the United States Patent Office from 1790 *to* 1873, *inclusive*—Continued.

Invention.	Inventor.	Residence.	Date.	No.
Miter-machine	L. W. Rosecrans	Marshalltown, Iowa	Oct. 11, 1870	108, 296
Miter-machine	D. B. Ware	Athens, Mich	Aug. 30, 1870	106, 894
Miter-planing machine	J. Mannebach	New York, N. Y	Dec. 3, 1872	133, 653
Miter-point borer	B. Hall	Canaan, N. Y	Mar. 8, 1810	....
Miters, Templet for cutting	A. H. Bartlett	New Haven, Conn	July 16, 1872	129, 516
Mitering-bench	J. W. Mahan	Lexington, Ill	Apr. 1, 1856	14, 569
Mitering-machine	F. D. Green	Williamsport, Pa	June 24, 1873	140, 267
Mitering-machine	J. Holzberger	Newark, N. J	July 26, 1870	105, 686
Mitering-machine	F. A. Howard	Belfast, Me	Aug. 21, 1866	57, 325
Mitering-machine	F. A. Howard	Belfast, Me	Aug. 25, 1868	81, 373
Mitering-machine	C. Loetscher	Dubuque, Iowa	Dec. 9, 1873	145, 354
Mitering-machine	H., G. W., and A. D. Malin	Pleasantville, Pa	Apr. 16, 1872	125, 745
Mitering-machine	J. J. Landers, jr	New York, N. Y	May 26, 1868	78, 330
Mitering-machine	J. M. Seymour	Newark, N. J	July 7, 1868	79, 783
Mitering-machine	E. Shaw	Tarr Farm, Pa	Aug. 9, 1870	106, 218
Mitering-machine	G. M. Stevens	Portland, Me	May 7, 1872	126, 588
Mitering-machine	R. F. Tompkins and H. T. Williams.	New York, N. Y	Nov. 10, 1868	83, 891
Mitering-machine	G. Trimble	Philadelphia, Pa	Dec. 19, 1865	51, 633
Mitering-machine	J. H. Van Ness	Charlotte, N. C	Sept. 30, 1873	143, 265
Mitering machine, Wood	R. F. Tompkins	New York, N. Y	Oct. 1, 1867	69, 510
Mitten	C. Neumann	Boston, Mass	July 8, 1862	35, 833
Mitten	F. L. Oakley	New York, N. Y	Apr. 29, 1873	138, 273
Mitten	A. P. Smith	Sterling, Ill	Jan. 1, 1867	60, 949
Mitten	G. Topping	Chicago, Ill	Nov. 3, 1868	83, 800
Mitten	E. V. Whitaker	Gloversville, N. Y	Dec. 3, 1872	133, 689
Mitten	J. L. Whitten	Essex Junction, Vt	Nov. 19, 1872	133, 183
Mixer:				
See Batter-mixer. Mortar-mixer.				
Clay-mixer. Paint-mixer.				
Mixing, heating, and cooling substances, Apparatus for.	M. Nolden	Frankfort-on-the-Main, Prussia.	Sept. 20, 1870	107, 621
Mixing-machine	J. W. Stockwell	Portland, Me	June 24, 1873	140, 171
Mixing-mill	W. Martien	Baltimore, Md	Jan. 2, 1872	122, 393
Moccasin	C. Elliott	New York, N. Y	Jan. 26, 1825	....
Moccasin	T. Hersey	Bangor, Me	Feb. 4, 1873	135, 548
Moccasin	A. Margesson	Bangor, Me	Aug. 26, 1873	142, 252
Moccasin	A. Tillotson	New York, N. Y	Nov. 22, 1823	....
Moccasin and sock	W. Bolton	Northampton, Mass	Oct. 21, 1814	....
Moccasin, Boot	J. J. Drown	Plattsburgh, N. Y	Dec. 17, 1872	134, 041
Moccasin, sock, &c	D. Williams	New York, N. Y	Oct. 25, 1832	....
Moccasin-sock or overshoe	W. Brower	Rensselaer County, N. Y	Mar. 6, 1822	....
Moccasin, Water-proof	J. Syms	New York, N. Y	Nov. 10, 1827	....
Moccasins, Application of hat-trimmings in making	J. Jackson	Burlington, N. J	Apr. 1, 1815	....
Moccasins, &c., in molds, Making	J. J. Staples and W. H. Cornell	New York, N. Y	Dec. 15, 1824	....
Model-stand, Artist's	G. A. Gilbert	New York, N. Y	May 14, 1872	126, 690
Molasses-cup	E. R. Cook	Trenton, N. J	May 17, 1864	42, 750
Molasses-cup	D. W. Messer	Boston, Mass	Apr. 14, 1857	17, 048
Molasses-cup	W. S. Rooney	Albany, N. Y	Sept. 3, 1867	68, 532
Molasses-cup, Sheet-metal	G. B. Halsted	New York, N. Y	Feb. 4, 1868	74, 081
Molasses-gate	J. Hegarty	Jersey City, N. J	Jan. 5, 1869	85, 660
Molasses-gate	G. W. Hubbard	Meriden, Conn	Jan. 10, 1860	26, 769
Molasses-gate	J. D. Kellogg and J. Wright	Northampton, Mass	Aug. 16, 1839	1, 290
Molasses-gate	T. Mace	New York, N. Y	Mar. 5, 1867	62, 652
Molasses-gate	E. Stebbins	Chicopee, Mass	Oct. 15, 1850	7, 724
Molasses-gate	A. West	Greene, Me	Aug. 28, 1840	1, 750
Molasses-gate	J. Whittemore	Boston, Mass	Oct. 9, 1839	1, 359
Molasses gate and cock for drawing liquors	L. Lincoln	Hartford, Conn	Nov. 10, 1841	2, 342
Molasses-gate, Automatic	P. H. Kimball	Prophetstown, Ill	July 30, 1867	67, 319
Molasses gate or faucet	E. Stebbins	Chicopee, Mass	July 15, 1851	8, 228
Molasses-gates, Casting	M. R. Chace	Fall River, Mass	Apr. 24, 1860	27, 969
Molasses, Manufacture of	G. W. Sayre	Pisgah, Ohio	Apr. 10, 1866	53, 887
Molasses, Process for improving the color of	J. Von Bohm	Melbourne, Australia	Feb. 14, 1865	46, 431
Molasses-vessel	W. Briggs	Bristol, R. I	May 28, 1867	65, 159
Mold:				
See Baking-mold. Electrotype-mold.				
Bonnet-front mold. Gig-mold.				
Boot and shoe sole mold. Glass-mold.				
Glass-blowers' mold.				
Bottle-mold. Glass-jar mold.				
Brick-mold. Glass-lamp mold.				
Building-block mold. Glass-makers' mold.				
Bullet-mold. Glass-pressing mold.				
Burial-case mold. Glass-ware mold.				
Butter-mold. Hard-rubber mold.				
Button-mold. Hat-forming mold.				
Candle-mold. Hub-box mold.				
Car-wheel mold. Ice-cream mold.				
Casting-mold. Ingot-mold.				
Cement-mold. Ingot-casting mold.				
Chill-mold. Jar-mold.				
Cigar-mold. Joint-mold.				
Cigar-bunch mold. Jointed mold.				
Cigar-drying mold. Lamp-mold.				
Clay-mold. Lamp-chimney mold.				
Clay and cement mold. Melon-shaped mold.				
Clay and cement pipe mold. Metal-mold.				
Metallic mold.				
Collar-mold. Needle-threader mold.				
Concrete-block mold. Oil-cake mold.				
Confectioners' mold. Pencil-tip mold.				
Corn-cake mold. Pipe-mold.				
Cornice-mold. Pipe-casting mold.				
Cylindrical mold. Plaster-mold.				
Die-mold. Plow-point mold.				
Doughnut-mold. Post-mold.				
Drain-pipe mold. Printing-type mold.				
Elastic-mold. Projectile-mold.				

Index of patents issued from the United States Patent Office from 1790 *to* 1873, *inclusive*—Continued.

Invention.	Inventor.	Residence.	Date.	No.
Mold—Continued. *See* Pudding-mold. Rifle-ball mold. Saddle-tree mold. Sealing-wax mold. Sectional mold. Sewer-pipe mold. Sheaf-mold. Soap-mold. Sole-mold. Spoon-mold. Stereotype-mold. Sugar-mold. Suppository-mold. Syringe-mold. Tile-mold. Tooth-mold. Type-mold. Vulcanizable-gum mold. Water-trap mold. Wrought-iron mold.				
Mold	T. J. Barron	Brooklyn, N. Y	May 16, 1871	114, 905
Mold	D. N. Spogle	Annapolis, Md	July 25, 1871	11[illegible], 343
Mold and core for metallic castings	W. Hainsworth	Allegheny, Pa	Dec. 6, 1870	109, 894
Mold-board and colter, Adjustable	G. D. Rowell	Menomonee Falls, Wis	July 27, 1869	93, 011
Mold-board blank	W. M. Watson	Tonica, Ill	Nov. 5, 1861	33, 672
Mold-board press	A. Thompson	Ottawa, Ill	June 7, 1870	103, 942
Mold-board, Revolving	J. S. Godfrey	Leslie, Mich	Apr. 28, 1868	77, 184
Mold-boards, Casting	J. Oliver	South Bend, Ind	Apr. 21, 1868	76, 939
Mold-boards, Chill for casting	J. Oliver	South Bend, Ind	Feb. 2, 1869	86, 579
Mold-facing material	J. B. Hyde	Newark, N. J	June 5, 1860	28, 579
Mold for casting molds	J. S. Elliott	Chelsea, Mass	Apr. 15, 1873	137, 901
Mold for composition articles	I. Smith	New York, N. Y	July 18, 1871	117, 119
Mold-forming machine	S. J. Peet and L. Sawyer	Boston, Mass	Mar. 22, 1870	101, 033
Molds, Composition for facing	R. B. Smith	Mount Pleasant, Ohio	July 31, 1866	56, 817
Molds, Device for blackwashing	J. B. Aston	Pittsburgh, Pa	Sept. 9, 1873	142, 661
Molds, Facing	J. and A. Hursh	Philadelphia, Pa	Aug. 8, 1865	49, 272
Molds for casting off alto-relievo figures, &c., Process of making.	N. Heintzelman	New York, N. Y	June 7, 1870	103, 879
Molds, Heating	H. Mooers	Toledo, Ohio	Dec. 3, 1861	33, 846
Molds, Machine for producing	S. J. Peet	New York, N. Y	Mar. 1, 1870	100, 319
Molds, Powder for facing	W. Batty	Cincinnati, Ohio	Jan. 1, 1867	60, 824
Molds, Powder for facing	J. Nichols and W. Batty	Cincinnati, Ohio	Feb. 28, 1865	46, 578
Molds, &c., with metallic surface, Covering	J. A. Adams	Brooklyn, N. Y	Feb. 15, 1870	99, 806
Molded and coated articles, Composition for forming	S. Whitmarsh	Northampton, Mass	July 7, 1868	79, 794
Molded articles from plaster materials, Manufacture of.	A. C. Estabrook	Northampton, Mass	Jan. 30, 1872	123, 254
Molded articles, Manufacture of	W. B. Gleason	Boston, Mass	Mar. 3, 1868	75, 149
Molder's bench	C. L. Bishop	Meriden, Conn	Feb. 7, 1865	46, 209
Molder's clamp	J. B. Crowley	Cincinnati, Ohio	Jan. 9, 1866	51, 995
Molder's clamp	W. W. Tice	California, Ohio	Nov. 27, 1866	60, 091
Molder's clamp	C. Trusdale	Cincinnati, Ohio	Jan. 2, 1867	51, 903
Molder's flask	J. A. Blake	Providence, R. I	Apr. 23, 1872	126, 012
Molder's flask	M. M. Donnelly	Cincinnati, Ohio	Aug. 20, 1867	67, 855
Molder's flask	M. M. Donnelly	Cincinnati, Ohio	Aug. 20, 1867	67, 856
Molder's flask	J. G. Holliday	Wheeling, W. Va	Sept. 11, 1866	57, 903
Molder's flask	E. F. Jones	Foxborough, Mass	May 23, 1871	115, 214
Molder's flask	E. C. Little	Saint Louis, Mo	Sept. 18, 1866	58, 173
Molder's flask	L. K. Livingston, J. J. Roggen, and C. Adams.	Pittsburgh, Pa	Oct. 10, 1848	5, 841
Molder's flask	A. Merriman	Plantsville, Conn	June 13, 1871	115, 973
Molder's flask	C. Truesdale	Cincinnati, Ohio	May 13, 1873	138, 960
Molder's flask	A. Wilcox	Green Island, N. Y	Jan. 11, 1870	98, 728
Molder's flask	S. Williamson	Cincinnati, Ohio	May 28, 1872	127, 398
Molder's flasks, Packer for packing sand in	B. S. Benson	Baltimore, Md	Oct. 13, 1868	83, 029
Molder's gate or sprue	J. M. Killin	Pittsburgh, Pa	May 28, 1872	127, 351
Molder's match-board	W. H. Jefts	Lowell, Mass	July 16, 1872	129, 033
Molder's match-plates, Composition for	G. P. Darrow	Cincinnati, Ohio	July 10, 1866	56, 187
Molder's match-plates, Manufacture of	C. Truesdale and A. J. Sennett	Cincinnati, Ohio	Dec. 5, 1865	51, 366
Molder's pot or ladle	J. McCleary	Martin's Ferry, Ohio	Mar. 18, 1873	136, 927
Molder's slicker	W. J. Fryer, jr	New York, N. Y	May 23, 1871	115, 049
Molder's sprue	M. R. Howell	Elizabethport, N. J	Jan. 3, 1865	45, 716
Molder's sprue	P. W. Lamb	Albany, N. Y	Jan. 1, 1867	60, 909
Molder's table	J. Lebeau	Cincinnati, Ohio	Mar. 14, 1865	46, 859
Molding	P. Conver	Farmington, Ill	Aug. 27, 1872	130, 900
Molding	J. Dougherty	Cold Spring, N. Y	May 15, 1860	28, 260
Molding	A. King	Utica, N. Y	Sept. 4, 1866	57, 822
Molding and carving machine	A. S. Gear	Boston, Mass	Apr. 30, 1872	126, 198
Molding and casting apparatus	G. Ross	Newport, Ky	Apr. 10, 1866	53, 883
Molding and glazing forms made of plastic material	S. Hart	Marietta, Ohio	Jan. 25, 1870	99, 083
Molding and pressing bricks by hydraulic pressure, Machine for.	E. Rogers	Cleveland, Ohio	Sept. 23, 1873	143, 188
Molding and venting machine	W. Newsham	Philadelphia, Pa	Aug. 9, 1870	106, 273
Molding-apparatus	W. C. Amish	Hainesport, N. J	Dec. 23, 1873	145, 828
Molding-apparatus	S. Ashford	Philadelphia, Pa	Sept. 13, 1864	44, 149
Molding beads or hollow ware	C. Neal	Philadelphia, Pa	July 5, 1859	24, 701
Molding-box	T. L. Luders	Olney, Ill	July 9, 1867	66, 568
Molding circular and undercut work	W. Sellers and J. Walker	Cincinnati, Ohio	Oct. 2, 1855	13, 625
Molding-cutter	N. Jenkins	New York, N. Y	Oct. 4, 1870	108, 026
Molding-cutter	T. J. Shannon	Lawrenceburgh, Ind	Feb. 8, 1870	99, 716
Molding-cutter head	E. H. Hinners	New York, N. Y	July 8, 1873	140, 626
Molding-cutter head	D. Strauley	Cincinnati, Ohio	Apr. 16, 1872	125, 765
Molding-cutter head	J. K. Wood	Allegheny, Pa	Aug. 20, 1872	130, 684
Molding earthenware, Machine for	T. G. Green	Church Gresley Pottery, England.	Nov. 1, 1870	108, 779
Molding-facing machine	R. Howdon	Cincinnati, Ohio	Nov. 12, 1867	70, 847
Molding-flask	J. J. Johnston	Allegheny, Pa	Apr. 22, 1856	14, 724
Molding-flask	T. G. Lucas	Middletown, Conn	Nov. 16, 1869	96, 819
Molding-flask	J. McLaughlin	Philadelphia, Pa	Aug. 30, 1870	106, 850
Molding-flask	J. Yocom, jr	Philadelphia Pa	Jan. 15, 1867	61, 303
Molding for cast-iron plates with dovetailed recesses.	T. A. Smith	Albany, N. Y	Mar. 8, 1853	9, 613
Molding from plaster of Paris, Composition for	G. Schlueter	Brooklyn, N. Y	Feb. 1, 1870	99, 355
Molding hollow ware	J. J. Johnston and J. V. Cunningham.	Allegheny and Pittsburgh, Pa.	June 6, 1854	11, 044
Molding hollow ware	W. Rogers	Philadelphia, Pa	Feb. 8, 1848	5, 442
Molding in flasks, Apparatus for	E. Satterlee	Albany, N. Y	Jan. 13, 1852	8, 661

Index of patents issued from the United States Patent Office from 1790 *to* 1873, *inclusive*—Continued.

Invention.	Inventor.	Residence.	Date.	No.
Molding in flasks, Machine for	L. A. Orcutt	Albany, N. Y	Mar. 8, 1853	9, 612
Molding-machine	J. Albright	Midway, Tenn	May 16, 1871	114, 900
Molding-machine	W. J. Boda	Dayton, Ohio	Sept. 16, 1873	142, 890
Molding-machine	M. W. Clark	Worcester, Mass	Sept. 24, 1872	131, 666
Molding-machine	J. H. Culver	San Francisco, Cal	Mar. 14, 1871	112, 691
Molding-machine	J. Demarest	Mott Haven, N. Y	July 25, 1871	117, 266
Molding-machine	J. S. Dewing	Orange, Mass	June 25, 1872	128, 289
Molding-machine	A. S. Gear	New Haven, Conn	Nov. 23, 1869	97, 188
Molding-machine	J. W. Gheen	Wilmington, Del	Apr. 22, 1873	138, 015
Molding-machine	T. Glover	Philadelphia, Pa	Aug. 9, 1870	106, 264
Molding-machine	J. P. Grosvenor	Lowell, Mass	Sept. 8, 1868	81, 997
Molding-machine	G. S. Hudson	Ellisburgh, N. Y	May 28, 1867	65, 226
Molding-machine	N. Jenkins	Madison, Conn	Dec. 23, 1873	145, 804
Molding-machine	I. L. Lamb	Darien, Wis	July 18, 1871	117, 083
Molding-machine	H. A. Lee	Worcester, Mass	Oct. 13, 1863	40, 269
Molding-machine	H. A. Lee	Worcester, Mass	July 7, 1868	79, 663
Molding-machine	W. McConnell	Clarksville, N. J	Nov. 23, 1869	97, 210
Molding-machine	C. H. Mellor	Philadelphia, Pa	Aug. 4, 1868	80, 648
Molding-machine	C. B. Morse	Rhinebeck, N. Y	Jan. 16, 1855	12, 248
Molding-machine	O. H. Perry	Cincinnati, Ohio	Nov. 26, 1867	71, 531
Molding-machine	G. M. Ramsay	New York, N. Y	Jan. 16, 1855	12, 243
Molding-machine	C. Schilling	Pekin, Ill	June 27, 1871	116, 503
Molding-machine	H. I. Seymour	Troy, N. Y	Oct. 13, 1863	40, 286
Molding-machine	E. M. Smith	Indianapolis, Ind	Feb. 21, 1860	27, 261
Molding-machine	I. P. Tice	New York, N. Y	Feb. 2, 1869	86, 605
Molding-machine	C. Warner	New York, N. Y	May 24, 1859	24, 168
Molding-machine	J. A. Woodbury	Boston, Mass	Aug. 1, 1871	117, 586
Molding-machine cutter	E. Benjamin	Chicago, Ill	Dec. 5, 1871	121, 447
Molding-machine-cutter guide	D. Jordan	Charlestown, Mass	Apr. 30, 1867	64, 227
Molding-machine-cutter head	J. Alloways	Decatur, Ill	Sept. 9, 1873	142, 548
Molding-machine-cutter head	M. W. Clark	Worcester, Mass	Sept. 12, 1871	118, 907
Molding-machine-cutter head	J. M. Smith	New York, N. Y	Sept. 13, 1853	10, 017
Molding-machine cutter, Wood	D. Jordan	Charlestown, Mass	June 7, 1870	103, 890
Molding-machine cutters, Machine for shaping	W. H. Brown	Worcester, Mass	Dec. 3, 1872	133, 627
Molding-machine feed	L. Gould	Norwich Conn	July 14, 1863	39, 226
Molding machine, Power	H. A. Billings	Providence, R. I	Sept. 17, 1861	33, 291
Molding machine, Variety	J. Hatch	San Francisco, Cal	July 13, 1869	92, 609
Molding machine, Variety	N. Jenkins	New York, N. Y	Jan. 25, 1870	99, 202
Molding machine, Wood	N. Jenkins	New Haven, Conn	Jan. 2, 1872	122, 321
Molding-machines, Adjusting the planes in	W. D. Jones	Dayton, Ohio	Apr. 24, 1860	27, 986
Molding-machinery	W. T. Horrobin	Bennington, Vt	Nov. 24, 1868	84, 279
Molding materials admitting of cohesion, Machine for.	J. Greacen, jr., A. Foster, and J. H. Cooper.	New York, N. Y., and Philadelphia, Pa.	July 17, 1866	56, 496
Molding-pit for casting cylinders and pipes	R. Cartwright	Chicago, Ill	May 31, 1870	103, 560
Molding plaster cornices, Device for	S. Ferris	New York, N. Y	Sept. 13, 1870	107, 246
Molding pots and crucibles, Apparatus for	R. Taylor and F. Strow	Philadelphia, Pa	Dec. 1, 1868	84, 593
Molding-sand, &c., Apparatus for crushing, grinding, and mixing.	J. O. Hanctin	Paris, France	Mar. 11, 1873	136, 721
Molding-tool	A. P. Odholm	Bridgeport, Conn	Feb. 23, 1869	87, 282
Molding, Wood	L. Bushnell	New Bedford, Mass	Nov. 11, 1873	144, 438
Moldings and beads, Machine for turning	E. Coddington	Thompson, N. Y	July 31, 1837	321
Moldings and frames, Machine for making	G. Henze	New York, N. Y	Sept. 5, 1865	49, 829
Moldings and other purposes, Composition for manufacture of.	H. Lowenberg	Boston, Mass	Mar. 25, 1862	34, 757
Moldings, &c., Apparatus for applying metal leaf to.	R. J. Marcher	New York, N. Y	Sept. 22, 1863	40, 055
Moldings, Apparatus for filling, polishing, and varnishing.	M. Hamburger, I. J. Siskind, and A. Klein.	New York, N. Y	Jan. 7, 1873	134, 599
Moldings, &c., Apparatus for laying metal leaf on	R. Marcher	New York, N. Y	May 10, 1859	23, 930
Moldings, &c., Apparatus for laying metal leaf on.	R. Marcher	New York, N. Y	July 30, 1861	32, 950
Moldings, Arrangement of device for planing	H. B. Smith	Lowell, Mass	July 6, 1858	20, 824
Moldings, Burnishing	R. Marcher	New York, N. Y	June 7, 1859	24, 319
Moldings, Composition for	J. and W. Thiem	Lawrenceburgh, Ind	Feb. 9, 1869	86, 710
Moldings, Composition for ornamental	C. E. Bonnet	Philadelphia, Pa	Dec. 22, 1868	85, 055
Moldings, Cutter for planing	R. M. Evans	Gilford, N. H	Dec. 6, 1853	10, 304
Moldings, Cutting curved	J. J. Westerfield	New Brunswick, N. J	Dec. 9, 1856	16, 197
Moldings, Forming	J. Dill and E. Rice	Grand Rapids, Mich	Oct. 5, 1869	95, 441
Moldings, Machine for cutting	S. Kennedy	New York, N. Y	Apr. 10, 1830	....
Moldings, Machine for cutting	C. B. Rogers	Norwich, Conn	Jan. 10, 1860	26, 787
Moldings, Machine for cutting	J. B. Winslow	Charlestown, Mass	May 29, 1860	28, 527
Moldings, Machine for cutting ornamental	H. and R. S. Schevenell	Athens, Ga	Sept. 25, 1855	13, 602
Moldings, Machine for cutting twist	H. Newhouse	Chicago, Ill	Feb. 18, 1862	34, 467
Moldings, Machine for cutting wood	W. A. McDonald	Morrisania, N. Y	May 11, 1869	89, 937
Moldings, Machine for cutting wooden	A. H. Brown	Albany, N. Y	Jan. 22, 1861	31, 207
Moldings, Machine for cutting wooden curved	I. P. Tice	Baltimore, Md	May 3, 1859	23, 872
Moldings, Machine for fluting	R. Van Riper	Brooklyn, N. Y	July 5, 1870	105, 145
Moldings, Machine for forming sheet-metal	V. Fischer	New York, N. Y	Feb. 4, 1868	74, 068
Moldings, Machine for forming sheet-metal	H. G. Fiske	San Francisco, Cal	Aug. 30, 1870	106, 893
Moldings, Machine for making	T. J. Close	Philadelphia, Pa	Oct. 9, 1866	58, 599
Moldings, Machine for making	N. Jenkins	New York, N. Y	Aug. 6, 1867	67, 553
Moldings, Machine for making metallic	H. Adler and C. Rogers	Pittsburgh, Pa	Sept. 10, 1872	131, 239
Moldings, Machine for making metallic	J. Parkin and J. H. Smith	Cleveland, Ohio	Sept. 10, 1872	131, 179
Moldings, Machine for making metallic	B. K. Price	Cleveland, Ohio	Apr. 16, 1872	125, 691
Moldings, Machine for manufacture of	J. C. Cooke and H. A. Whiteley	Preston, Conn	Feb. 14, 1871	111, 817
Moldings, Machine for ornamenting	J. W. Campbell	New York, N. Y	Nov. 20, 1866	59, 817
Moldings, Machine for ornamenting	G. F. Goetze	New York, N. Y	Dec. 4, 1866	60, 173
Moldings, Machine for planing	J. B. H. Chatain	New York, N. Y	Mar. 19, 1850	7, 181
Moldings, Machine for planing oval	F. Brandon	Albany, N. Y	May 12, 1863	38, 461
Moldings, Machine for planing oval	G. Hammer	Cincinnati, Ohio	Mar. 11, 1862	34, 663
Moldings, Machine for polishing and varnishing	C. and J. Gschwind	Union Hill, N. J	Jan. 23, 1872	122, 945
Moldings, Machine for preparing	G. Henze	New York, N. Y	Nov. 8, 1864	44, 952
Moldings, Machine for sand-papering	J. Barker	Chicago, Ill	Aug. 23, 1870	106, 536
Moldings, Machine for sand-papering	F. G. Chapman	Chicago, Ill	May 31, 1870	103, 717
Moldings, Machine for smoothing	J. S. Loomis	Brooklyn, N. Y	Aug. 29, 1871	118, 466
Moldings, Machine for turning spiral	P. P. Ruger	New York, N. Y	Oct. 4, 1853	10, 089
Moldings, Machine for turning spiral	E. A. Stockton	San Francisco, Cal	July 25, 1871	117, 477
Moldings, Machinery for making	A. T. Serrell	New York	May 16, 1848	5, 575

Index of patents issued from the United States Patent Office from 1790 *to* 1873, *inclusive*—Continued.

Invention.	Inventor.	Residence.	Date.	No.
Moldings, Manufacture of imitation gold	H. W. Ladd	New York, N. Y	May 19, 1863	38, 588
Moldings, Metal	A. Destouy	New York, N. Y	Aug. 30, 1864	43, 979
Moldings on marble, Machine for cutting	H. L. Houghton	Springfield, Vt	Jan. 29, 1856	14, 177
Moldings on planks, piles, &c., Sticking	J. Judge	Washington, D. C	June 13, 1831	....
Moldings on wood, Machine for planing	A. Church, jr	Canandaigua, N. Y	July 29, 1837	296
Moldings, Pattern for	J. Alexander	Brooklyn, N. Y	Aug. 16, 1859	25, 157
Moldings, Planing-machine for	E. H. Ripley	North Chelmsford, Mass	June 16, 1868	79, 006
Moldings, Rest for finishing	D. A. Stevens	Toledo, Ohio	Oct. 8, 1872	132, 115
Moldings, Rolled-iron	M. M. Manly and S. Sellers	Philadelphia, Pa	Jan. 7, 1873	134, 553
Moldings, Tool for cutting	E. C. Austin	Monroe Village, Wis	Jan. 22, 1867	61, 306
Moldings, Tool for cutting	D. W. Perry	Wilkesbarre, Pa	Dec. 15, 1868	84, 964
Moldings, Tool for cutting wood	C. E. Boynton	San Francisco, Cal	Sept. 27, 1870	107, 653
Moldings, Tool for grooving	R. I. Marcher	Salisbury Mills, N. Y	May 22, 1855	12, 916
Moldings, Tool for grooving	R. I. Marcher	Salisbury Mills, N. Y	May 22, 1855	12, 917
Moldings, Tool for turning	W. W. Carey	Lowell, Mass	Aug. 25, 1868	81, 338
Moldings, Wood	W. J. Miller and J. W. Campbell.	New York, N. Y	Oct. 5, 1869	95, 598
Mole-killer	J. Wilson	Little Falls, N. Y	Dec. 7, 1869	97, 742
Mole-trap	J. Adams	Greencastle, Ind	Jan. 3, 1871	110, 618
Mole-trap	W. C. Akers	Petersburgh, Va	Jan. 4, 1870	98, 539
Mole-trap	L. Andrews	Plantsville, Conn	Nov. 4, 1862	36, 827
Mole-trap	D. Boswell	Greencastle, Ind	Aug. 22, 1871	118, 189
Mole-trap	T. G. Brown	Tarrytown, N. Y	Nov. 26, 1872	133, 407
Mole-trap	G. W. Hardwick	Wyandotte, Ind	Sept. 26, 1871	119, 354
Mole-trap	R. I. Huggins	Bethel, Ohio	Dec. 30, 1873	146, 003
Mole-trap	J. A. Myers	Lovely Dale, Ind	Aug. 20, 1872	130, 737
Mole-trap	C. Polley	Sinking Spring, Ohio	May 19, 1868	78, 128
Mole-trap	C. Polley	McMinnville, Tenn	May 7, 1872	126, 573
Money-box	G. W. McDaniel	Georgetown, D. C	Mar. 22, 1870	101, 144
Money-box for stages, &c	T. W. Gibbons	Franklin, N. J	June 7, 1859	24, 295
Money-counting apparatus	J. W. Meaker	Chicago, Ill	Jan. 28, 1868	73, 911
Money-drawer	C. E. Hays	Lancaster, Pa	Feb. 19, 1867	62, 195
Money-drawer	C. L. Page	Cambridge, Mass	Mar. 23, 1869	88, 068
Money-drawer	A. G. and P. M. Shults	Avoca, N. Y	Oct. 2, 1866	58, 490
Money-drawer	H. Unna	San Francisco, Cal	June 25, 1872	128, 439
Money-drawer alarm	R. M. Campbell	East Cambridge, Mass	July 12, 1859	24, 775
Money-drawer alarm	J. M. Case	Lansing, Mich	May 14, 1872	126, 785
Money-drawer alarm	I. Robbins	Hughesville, Pa	Sept. 3, 1867	68, 531
Money-drawer alarm	J. F. Winchell	Springfield, Ohio	Jan. 15, 1867	61, 297
Money-drawer catch	S. Hubbell, jr	Urbana, Ohio	Apr. 10, 1866	53, 827
Money-drawer lock and alarm attachment	J. H. Weaver	Columbus, Ohio	Sept. 10, 1867	68, 816
Money-table	W. Painter	Wilmington, Del	Aug. 3, 1858	21, 082
Monkey-wrench	L. Coes	Worcester, Mass	Aug. 10, 1869	93, 521
Monkey-wrench	S. Zarley	Niantic, Ill	Dec. 14, 1869	97, 849
Monument	F. M. Jones	Independence, Miss	Nov. 28, 1871	121, 380
Monument	J. M. Martin	Cleveland, Ohio	May 13, 1862	35, 276
Monument	J. C. Taylor	Roanoke, Ala	Dec. 2, 1873	145, 254
Monuments, &c., Casting	H. Siebenau	New York, N. Y	Nov. 17, 1830	....
Monuments, Fixing likenesses in	W. Boyd	Garrettsville, Ohio	Jan. 31, 1854	10, 468
Monuments, Protecting likenesses in	I. H. Wells	Pagetown, Ohio	Dec. 17, 1867	72, 345
Mop	C. L. W. Baker	Hartford, Conn	Sept. 6, 1870	106, 986
Mop	A. T. Boon and W. W. Spaulding	Galesburgh, Ill	Mar. 21, 1865	46, 872
Mop	P. Cook, jr	Sioux City, Iowa	Dec. 14, 1869	97, 885
Mop	R. H. Ewing	Elizabethtown, Ohio	June 2, 1863	38, 734
Mop	A. Hemmenway	Townsend, Ohio	Oct. 1, 1861	33, 390
Mop	M. Jincks	Danville, N. Y	May 16, 1865	47, 730
Mop	M. W. Kirkwood and S. H. Riley	Iowa City, Iowa	Apr. 18, 1871	113, 777
Mop	J. Law	New York, N. Y	May 25, 1869	90, 562
Mop	J. and H. R. Lee	Galesburgh, Ill	Mar. 28, 1865	47, 066
Mop	W. Mallerd	Bridgeport, Conn	July 13, 1869	92, 624
Mop	T. F. Rooney	Chicago, Ill	June 13, 1871	115, 899
Mop	J. Sangster and M. Boyd	Buffalo, N. Y	June 5, 1866	55, 429
Mop	M. H. Wiley	Bucksport, Me	May 1, 1866	54, 473
Mop	G. W. Tolhurst	Circleville, Ohio	Feb. 14, 1865	46, 406
Mop	J. A. Wright	Keene, N. H	Sept. 8, 1868	82, 055
Mop and brush	A. C. Bacon	Cleveland, Ohio	June 11, 1867	65, 528
Mop and brush combined	H. McClay	Niles, Mich	Oct. 19, 1858	21, 842
Mop and brush holder	D. Edward	Montreal, Canada	July 4, 1871	116, 574
Mop and clothes wringer	E. Youngs	Tuscarora, N. Y	Aug. 4, 1868	80, 794
Mop and mop-wringer combined	J. L. Boyden	Bridgeport, Conn	Feb. 4, 1873	135, 400
Mop and rubber scrubber	B. C. Davis	Binghamton, N. Y	Feb. 8, 1870	99, 648
Mop and scourer	W. T. Grant	Jacksonville, Ill	Feb. 20, 1866	52, 705
Mop and scrubber	W. T. Grant	Jacksonville, Ill	Sept. 4, 1866	57, 703
Mop and scrubber, Combination of	R. Price	New York, N. Y	Mar. 6, 1860	27, 383
Mop and scrubber combined	E. S. Bennett	New York, N. Y	Feb. 2, 1869	86, 351
Mop and scrubber head	W. H. Kline	Jersey Shore, Pa	Oct. 18, 1870	108, 364
Mop and scrubber head	E. M. Shoemaker and S. McCleery.	Franklin County, Ohio	Feb. 8, 1870	99, 603
Mop and scrubbing-brush holder	P. O'Brian	Philadelphia, Pa	May 19, 1868	78, 121
Mop and wringer	J. M. Enos	Saint Joseph, Mich	June 5, 1866	55, 413
Mop and wringer combined	J. A. Wright	Keene, N. H	June 30, 1868	79, 428
Mop-cloth	W. B. Link	Taberg, N. Y	Feb. 26, 1867	62, 347
Mop for crockery and lamp-chimnies	C. S. Moore and H. P. Boyd	Worcester, Mass	Dec. 19, 1871	121, 955
Mop-handle	H. B. Malbone	Middletown, Conn	May 29, 1866	55, 135
Mop-handle	H. and J. S. B. Norton	Farmington, Me	May 17, 1859	24, 049
Mop-head	F. Allen	Worcester, Mass	Dec. 2, 1856	16, 137
Mop-head	J. W. Barbour	Winooski Falls, Vt	Feb. 11, 1873	135, 753
Mop-head	A. Barns	Ashtabula, Ohio	Nov. 20, 1855	13, 812
Mop-head	W. H. Beatty	Columbus, Ohio	May 24, 1870	103, 413
Mop-head	G. E. Brettell	Rochester, N. Y	May 18, 1869	90, 153
Mop-head	J. Brizee	Alvarado, Cal	June 13, 1871	115, 815
Mop-head	J. D. Browne	Cincinnati, Ohio	Sept. 22, 1868	82, 286
Mop-head	C. B. Clark	Buffalo, N. Y	Aug. 13, 1867	67, 722
Mop-head	C. B. Clark	Buffalo, N. Y	July 9, 1872	128 711
Mop-head	C. B. Clark	Buffalo, N. Y	Dec. 9, 1873	145, 397
Mop-head	C. B. Clark and E. L. Ferguson	Buffalo, N. Y	Sept. 22, 1868	82, 383
Mop-head	C. B. Clark and O. S. Garretson	Buffalo, N. Y	May 22, 1866	54, 869

Index of patents issued from the United States Patent Office from 1790 *to* 1873, *inclusive*—Continued.

Invention.	Inventor.	Residence.	Date.	No.
Mop-head	S. Cooper	Antwerp, N. Y	Apr. 25, 1865	47, 398
Mop-head	W. S. Crinklaw	Lanark, Ill	Mar. 22, 1870	100, 987
Mop-head	J. Davis	Chicago, Ill	Mar. 14, 1871	112, 556
Mop-head	J. Davis	Chicago, Ill	Sept. 2, 1873	142, 328
Mop head	H. Dodge	Albany, N. Y	Apr. 12, 1870	101, 716
Mop-head	J. Doty, jr	West Falls, N. Y	July 17, 1860	29, 148
Mop-head	W. Duncan	Lebanon, N. H	Jan. 14, 1873	134, 739
Mop head	L. H. Emmons	Noblesville, Ind	June 2, 1868	78, 444
Mop-head	R. W. English	Buffalo, N. Y	Mar. 10, 1868	75, 253
Mop-head	J. Fahrney	Boonsborough, Md	Mar. 23, 1869	88, 150
Mop-head	J. Fazig	West Salem, Ohio	Sept. 13, 1859	25, 400
Mop-head	A. Field	Jericho, Vt	May 21, 1872	127, 037
Mop-head	B. French	Rochester, N. Y	June 20, 1871	116, 174
Mop head	S. Gantz	Beaver Creek, Md	May 25, 1869	90, 520
Mop-head	O. S. Garretson	Buffalo, N. Y	June 11, 1867	65, 662
Mop-head	O. S. Garretson	Cincinnati, Ohio	Aug. 13, 1867	67, 643
Mop-head	O. S. and J. G. Garretson	Buffalo, N. Y	Oct. 4, 1870	107, 897
Mop-head	H. M. Guild	Springfield, Mass	Aug. 20, 1867	67, 873
Mop-head	G. H. Hammer	Newville, Pa	Apr. 24, 1866	54, 146
Mop-head	F. M. Hardison and J. A. Hooper.	South Berwick, Me	Oct. 16, 1866	58, 817
Mop-head	J. S. Harris	East Poultney, Vt	July 28, 1857	17, 877
Mop-head	J. S. Harris	Poultney, Vt	Sept. 1, 1863	39, 730
Mop-head	P. R. Higley	Buffalo, N. Y	Dec. 12, 1865	51, 514
Mop-head	N. Homes	Laona, N. Y	May 9, 1865	47, 642
Mop-head	J. H. Hood	Dansville, N. Y	Oct. 29, 1872	132, 580
Mop-head	M. L. Horton	Claremont, N. H	Mar. 18, 1862	34, 687
Mop-head	C. Karr	Buffalo, N. Y	Sept. 26, 1865	50, 137
Mop-head	F. C. Kendall	Hartford, Wis	Nov. 22, 1870	109, 421
Mop-head	S. C. Ketchum	Winchendon, Mass	Apr. 23, 1872	125, 895
Mop-head	W. A. Lewis	Springfield, Vt	Apr. 9, 1867	63, 731
Mop-head	D. W. Marston and J. S. Skinner.	Lebanon, N. H	Dec. 30, 1873	145, 958
Mop-head	H. H. Mason and J. Messinger.	Springfield, Vt	Feb. 19, 1867	62, 214
Mop-head	H. H. Mason and J. Messinger.	Springfield, Vt	May 14, 1867	64, 683
Mop-head	H. H. Mason and J. Messinger.	Springfield, Vt	July 16, 1867	66, 863
Mop-head	H. H. Mason and S. W. Shelden	Springfield, Vt	Jan. 5, 1864	41, 081
Mop-head	G. Meader	Prairie Centre, Ill	June 11, 1867	65, 584
Mop-head	J. Messinger	Springfield, Vt	Mar. 14, 1871	112, 615
Mop-head	H. Murch	Lebanon, N. H	June 14, 1853	9, 781
Mop-head	E. D. Murphy	Claremont, N. H	July 21, 1863	39, 305
Mop-head	H. C. Parsons	Dexter, Me	Sept. 28, 1869	95, 259
Mop-head	T. Randlett	Endfield, N. H	Nov. 15, 1853	10, 237
Mop-head	E. R. and N. F. Reed	Hyde Park, Vt	July 22, 1862	35, 951
Mop-head	W. A. Robinson	Grand Rapids, Mich	Mar. 24, 1868	75, 796
Mop-head	L. C. Rodier	Springfield, Mass	June 26, 1866	55, 961
Mop-head	O. Root	Wendell, Mass	Nov. 10, 1868	84, 000
Mop-head	G. W. Sanders	Springfield, Vt	Oct. 1, 1867	69, 368
Mop-head	S. Seldon and M. Griswold, jr	Erie, Pa	Apr. 4, 1871	113, 457
Mop head	J. W. Shaw	Concord, N. H	Feb. 26, 1867	62, 373
Mop head	E. and E. G. Sirret	Buffalo, N. Y	Apr. 13, 1869	88, 917
Mop-head	W. W. Spaulding	Galesburgh, Ill	Aug. 22, 1865	49, 566
Mop-head	J. A. Taylor	Alden, N. Y	Feb. 6, 1855	12, 365
Mop-head	L. Taylor	North Springfield, Vt	Feb. 15, 1859	22, 990
Mop-head	L. W. Taylor	Upper Falls Post-Office, Vt	Aug. 8, 1871	117, 943
Mop-head	L. H. Thomas	Waterbury, Vt	Apr. 3, 1866	53, 704
Mop-head	E. P. Thompson	Worcester, Mass	June 9, 1857	17, 527
Mop-head	H. Thompson and R. Q. Tuson	Lebanon, N. H	July 15, 1856	15, 352
Mop-head	J. Todd	Webster, N. Y	June 27, 1871	116, 512
Mop-head	J. Troxel	Reedsburgh, Ohio	Oct. 1, 1867	69, 512
Mop-head	W. P. Valentine	Buffalo, N. Y	Oct. 11, 1870	108, 304
Mop-head	I. E. Weston	Winchendon, Mass	Oct. 30, 1866	59, 299
Mop-head	H. H. Wetmore	Barre, Vt	Jan. 24, 1871	111, 161
Mop-head	G. W. Wheat	Philipsburgh, Pa	Feb. 11, 1873	135, 679
Mop-head	R. W. Whitney and A. C. Stockin.	South Berwick, Me	Oct. 9, 1866	58, 703
Mop-head	L. Williams	Arlington, Vt	Dec. 13, 1870	110, 180
Mop-head	J. A. Wilson	Spencer, Mass	May 7, 1867	64, 612
Mop-head and scrubbing-brush, Combined	W. S. Van Hoesen	Saugerties, N. Y	Nov. 3, 1868	83, 806
Mop-head, and securing same to handles	J. Howe	Worcester, Mass	June 15, 1837	241
Mop head and wringer	G. Gullman	Poughkeepsie, N. Y	Aug. 18, 1868	81, 272
Mop head and wringer, Combined	W. Gage	Buffalo, N. Y	Aug. 23, 1870	106, 684
Mop-holder	W. R. Axe	Beloit, Wis	Mar. 5, 1861	31, 586
Mop-holder	T. C. Ball	Springfield, Vt	Feb. 7, 1865	46, 290
Mop-holder	J. Brizee	Alvarado, Cal	Feb. 21, 1871	112, 014
Mop-holder	B. W. Bruel	Beloit, Wis	Aug. 14. 1860	29, 566
Mop-holder	G. Fliedner	Portland, Oreg	July 16, 1872	129, 019
Mop-holder	C. L. Haines	Carmel, Me	Mar. 18, 1873	136, 912
Mop-holder	W. Morehouse	Buffalo, N. Y	Oct. 16, 1866	58, 872
Mop-holder	E. M Naramore	Underhill, Vt	Oct. 14, 1873	143, 633
Mop-holder	G. Stackpole	New York, N. Y	May 4, 1869	89, 803
Mop-holder	S. S. Whitman	Little Falls, N. Y	Jan. 24, 1842	2, 436
Mop-iron	S. Johnson	Lee, N. Y	Oct. 18, 1843	3, 306
Mop, mat, and wiping-cushion	R. E. Smith	Newark, N. J	Apr. 12, 1870	101, 777
Mop-press	L. S. Covey and J. Duffy	Saint Croix County, Wis	Mar. 10, 1868	75, 247
Mop-squeezer	E. N. Porter	Morrisville, Vt	Mar. 26, 1867	63, 294
Mop-squeezer	E. S. Wilkins and J. Straw	Stowe, Vt	Apr. 16, 1867	63, 973
Mop, Tar	J. W. Midwinter	Port Washington, N. J	Oct. 27, 1863	40, 420
Mop-wringer	J. Adams	Janesville, Wis	Oct. 29, 1867	70, 310
Mop-wringer	M. J. Barcalo	Mount Morris, N. Y	Sept. 1, 1868	81, 583
Mop-wringer	O. C. Barnes	Stowe, Vt	June 4, 1867	65, 330
Mop-wringer	O. D. Barrett	Fulton, N. Y	July 31, 1855	13, 347
Mop-wringer	C. Bradway	Maquoketa, Iowa	May 10, 1870	102, 759
Mop-wringer	E. E. Brewster	Holly, Mich	Jan. 16, 1872	122, 804
Mop-wringer	O. M. Brooks and E. J. Matteson	Janesville, Wis	Nov. 12, 1867	70, 692
Mop-wringer	A. W. Bunnell	Linesville, Pa	Nov. 25, 1873	144, 828
Mop-wringer	M. P. Carpenter	Buffalo, N. Y	Jan. 23, 1866	52, 136

Index of patents issued from the United States Patent Office from 1790 *to* 1873, *inclusive*—Continued.

Invention.	Inventor.	Residence.	Date.	No.
Mop-wringer	B. B. Choate	Springfield, Vt	Apr. 28, 1868	77, 255
Mop-wringer	O. Choate	Morrisville, Vt	May 15, 1860	28, 227
Mop-wringer	A. J. Davis	Hartford, Mich	May 19, 1868	78, 071
Mop-wringer	H. L. Ennes	Birmingham, Ohio	Nov. 15, 1870	109, 307
Mop-wringer	J. Filkins	Sandwich, Ill	June 9, 1868	78, 657
Mop-wringer	W. W. Finch	Mishawaka, Ind	June 11, 1867	65, 737
Mop-wringer	M. M. Follett	Westborough, Mass	Apr. 7, 1868	76, 430
Mop-wringer	C. W. Gage and J. Northrup	Homer, N. Y	June 11, 1867	65, 558
Mop-wringer	W. and W. S. Gillett	Stowe, Vt	Nov. 13, 1866	59, 586
Mop-wringer	J. H. Gridley	Washington, D. C	Mar. 17, 1868	75, 679
Mop-wringer	W. Hall	North Adams, Mass	Apr. 20, 1869	89, 041
Mop-wringer	Z. Rowe	Lowell, Mass	Dec. 10, 1867	72, 043
Mop-wringer	A. S. Lesner	Boston, Mass	July 16, 1867	66, 720
Mop wringer	J. M. May and W. M. Colton	Jonesville and Stoughton, Wis.	Apr. 28, 1868	77, 391
Mop-wringer	J. H. Mears	Oshkosh, Wis	Feb 11, 1868	74, 400
Mop-wringer	W. R. Mills	Hartford, Mich	May 5, 1868	77, 639
Mop-wringer	J. H. Newton	Paxton, Ill	June 24, 1873	140, 216
Mop-wringer	D. Peck	Rochelle, Ill	Oct. 8, 1867	69, 694
Mop-wringer	R. T. Reed	Binghamton, N. Y	Oct. 22, 1867	70, 017
Mop-wringer	F. Rhines	Watertown, N. Y	Mar. 27, 1866	53, 489
Mop-wringer	A. J. Robinson	Troy, N. Y	Jan. 1, 1867	60, 790
Mop-wringer	L. F. Rollins	Bangor, Me	Mar. 5, 1867	62, 567
Mop-wringer	H. B. Rorke	California, Mo	May 12, 1868	77, 919
Mop-wringer	H. Russell	New Richmond, Wis	July 9, 1867	66, 639
Mop-wringer	C. S. Schmidt	New York, N. Y	Sept. 25, 1860	30, 159
Mop-wringer	R. W. Soper	Janesville, Wis	Oct. 15, 1867	69, 942
Mop-wringer	A. G. Starkweather	Burlington, Vt	July 9, 1867	66, 647
Mop-wringer	D. J. Stone	Norwich, R. I	June 2, 1868	78, 618
Mop-wringer	A. D. Tate	Peekskill, N. Y	Jan. 26, 1869	86, 328
Mop-wringer	B. Van Alstine	Clannahan Township, Ill	Feb. 19, 1867	62, 237
Mop-wringer	C. E. Wareham	Sedalia, Mo	July 30, 1867	67, 388
Mop-wringer	G. Wells and S. A. Haynes	Island Pond, Vt	Nov. 10, 1868	84, 034
Mop-wringer	J. F. White	Brattleborough, Vt	Nov. 27, 1866	60, 102
Mop-wringer	H. B. Willoughby	Ottawa, Ill	July 7, 1868	79, 795
Mop-wringer and brush	W. S. Simpson	Berea, Ohio	May 5, 1868	77, 543
Mop-wringer and scrubber, Combined	W. R. Winter	Cedar Falls, Iowa	July 1, 1873	140, 566
Mop-wringer and scrubbing-brush	L. L. Gordon	Detroit, Mich	Feb. 15, 1870	99, 772
Mops, Wringing	R. Holgate and C. Hart	Wyoming and Farmington, Ill	June 4, 1872	127, 416
Mordant	E. F. Prentiss	Philadelphia, Pa	Apr. 24, 1866	54, 203
Mordant for fixing aniline colors	A. Schultz	Lyons, France	Jan. 13, 1863	37, 426
Morocco and leather, Polishing and graining	R. Emes	Boston, Mass	Apr. 20, 1831	....
Morocco, Machine for figuring and polishing	E. L. Norton	Charlestown, Mass	May 6, 1856	14, 821
Morocco, Machine for finishing	E. Bookhout and H. Cochen, jr	Williamsburgh, N. Y.	June 18, 1850	7, 433
Morocco, Machine for finishing and dicing	M. Hall	Charlestown, Mass	Dec. 31, 1817	....
Morocco, Machine for graining	G. R. Johnson	Wilmington, Del	May 1, 1866	54, 360
Morocco, Polishing, graining, and dicing	R. Emes	Boston, Mass	Nov. 21, 1831	....
Morocco, Tool for figuring	S. Green	Lynn, Mass	Mar. 25, 1856	14, 534
Mortar	J. A. Fremon	Montgomery, Ala	Dec. 22, 1868	85, 223
Mortar and brick elevator	P. Anderson and J. A. Wait	Chicago, Ill	Aug. 27, 1872	130, 889
Mortar and cement, and material therefor, Manufacture of.	E. A. Ellsworth	Washington, D. C	Jan. 28, 1868	73, 882
Mortar and making bricks, Tempering	A. Kinsley	....	Dec. 20, 1794	....
Mortar, cement, paint, &c., Composition for	C. B. Hutchins	Ann Arbor, Mich	Feb. 2, 1869	86, 546
Mortar, cement, &c., Preparation of	H. Y. D. Scott	Ealing, England	Nov. 7, 1871	120, 672
Mortar for bricks, Preparing	O. W. Seely	Williamson, N. Y	July 20, 1831	....
Mortar for making building-blocks, &c., Apparatus for mixing.	J. Shelley	Harlem, N. Y	Nov. 19, 1867	71, 231
Mortar, Hominy	J. Keezer	Chillicothe, Ohio	Mar. 2, 1858	19, 507
Mortar-machine	S. H. Hinsdell	Camillus, N. Y	Aug. 1, 1871	117, 540
Mortar-machine and mason's tender	S. Whitman	Danville, Ill	Mar. 11, 1835	....
Mortar-making machine	D. Hunt	Stamford, Conn	Mar. 22, 1810	....
Mortar-mill	A. A. Anderson	Galesburgh, Ill	Dec. 24, 1867	72, 440
Mortar-mill	H. Hill	Chicago, Ill	Apr. 2, 1872	125, 295
Mortar-mill	J. McIntyre	Syracuse, N. Y	June 28, 1870	104, 870
Mortar-mill	L. B. Pitcher	Salina, N. Y	Nov. 27, 1866	60, 055
Mortar-mill	L. B. Pitcher	Salina, N. Y	July 26, 1870	105, 840
Mortar-mixer	A. A. Hoagland and W. Nickel.	Oneonta, N. Y	Oct. 14, 1873	143, 575
Mortar-mixer	C. Pierce	Portland, Mich	Aug. 19, 1873	141, 947
Mortar-mixer	C. W. Wasson	Friendship, N. Y	Mar. 4, 1873	136, 571
Mortar-mixer	S. Wetmore	Wellsborough, Pa	June 21, 1870	104, 519
Mortar mixer and grinder	E. Spaulding	Keene, N. H	May 27, 1873	139, 338
Mortar-mixer, Portable	H. Rosen	Elkhart, Ind	Aug. 13, 1872	130, 536
Mortar, Mixing	J. Peck	Buffalo, N. Y	June 29, 1852	9, 077
Mortar-mixing machine	A. C. Ellithorpe	Chicago, Ill	June 20, 1871	116, 172
Mortar-mixing machine, &c	Everhart and Swimley	Washington, D. C	June 20, 1835	....
Mortar-mixing machine	B. F. Field	Beloit, Wis	Oct. 7, 1856	15, 847
Mortar-mixing machine	A. Kirkpatrick	Urbana, Ohio	Nov. 12, 1831	....
Mortar and molding and pressing brick, Machine for mixing.	J. M. Brookings	Wiscasset, Me	Jan. 23, 1826	....
Mortar or grinding apples, Preparing	T. Streeter and J. S. Wibert	Chili, N. Y	June 12, 1827	....
Mortar-tub mill	J. Parker	Gardiner, Me	Mar. 28, 1826	....
Mortars, Mounting hand	W. F. Goodwin	New York, N. Y	Jan. 31, 1865	46, 101
Mortars, Process and machinery for making	L. B. Pitcher	Salina, N. Y	Sept. 20, 1870	187, 535
Mortise and common door-lock	T. Whitehouse	Boston, Mass	Sept. 29, 1825	....
Mortise and draw-bore delineator	R. Bloss	Rochester, N. Y	May 3, 1864	42, 550
Mortise and latch lock	J. G. Hotchkiss	New Haven, Conn	June 19, 1835	....
Mortise-lock	P. and E. W. Blake	New Haven, Conn	Feb. 5, 1836	....
Mortises, Cutting straight or curved	E. Q. Smith	Cincinnati, Ohio	July 3, 1855	13, 184
Mortises, Machine for boring dovetailed	H. J. Betjemann	Cincinnati, Ohio	Dec. 3, 1850	7, 807
Mortises, Machine for cutting square	C. H. Thompson	Cleveland, Ohio	Mar. 5, 1872	124, 230
Mortises, Machine for laying out	J. M. Rodkey	Indiana, Pa	July 25, 1871	117, 334
Mortises, Tool for dressing	C. E. Littlefield	Carver's Harbor, Me	June 18, 1872	128, 050
Mortising	J. B. Stark	Fisherville, N. H	July 31, 1866	56, 820
Mortising and boring	G. Page	Keene, N. H	Mar. 13, 1836	....
Mortising and boring machine	J. M. Jay	Canton, Ohio	July 28, 1857	17, 880
Mortising and boring machine	M. Sands	Franklin, N. Y	Oct. 1, 1830	....
Mortising and boring machine	G. N. Stearns	Syracuse, N. Y	Mar. 11, 1856	14, 416

Index of patents issued from the United States Patent Office from 1790 *to* 1873, *inclusive*—Continued.

Invention.	Inventor.	Residence.	Date.	No.
Mortising and dovetailing machine	I. Brainerd	Aurora, Ohio	Jan. 9, 1838	570
Mortising and making tenons on the ends of spokes of wheels, panel-work, carpetery, &c.	L. Davis	Northampton, Mass	Apr. 14, 1829	....
Mortising and tenoning	J. H. Darby	Leominster, Mass	June 15, 1835	....
Mortising and tenoning machine	A. Bailey	Jefferson, Ohio	June 20, 1840	1, 649
Mortising and tenoning machine	S. Baldwin	Montpelier, Vt	Mar. 17, 1810	....
Mortising and tenoning machine	T. A. Chandler	Rockford. Ill	Dec. 9, 1846	4, 876
Mortising and tenoning machine	E. Mudge	Genesee County, N. Y	Oct. 1, 1830	....
Mortising and tenoning machine	W. L. Peters	Frankfort, Pa	Dec. 12, 1842	2, 878
Mortising and tenoning machine	E. M. Shaw	Brooklyn, N. Y	Oct. 20, 1834	....
Mortising and tenoning machine	E. M. Shaw	Brooklyn, N. Y	Apr. 8, 1835	....
Mortising and tenoning machine	W. H. Sible	Harrisburgh, N. Y	Aug. 24, 1869	94, 140
Mortising and tenoning machine	L. Sprague, M. Noyes, T. Lund, jr., and J. Ayres.	Dunstable, N. H	Nov. 19, 1833	....
Mortising and tenoning machine	H. D. Stover	New York, N. Y	Nov. 3, 1868	83, 671
Mortising and tenoning machine	E. Street	Montville, Conn	Mar. 20, 1855	12, 563
Mortising and tenoning timber	J. McClintic	Chambersburgh, Pa	Oct. 22, 1835	....
Mortising and tenoning timber, Machine for	H. Barnes	Munson, Ohio	Apr. 28, 1838	716
Mortising and tenoning timber, Machine for	W. Jackson and J. J. Speed, jr	Speedville, N. Y	Sept. 10, 1828	....
Mortising and tenoning timber, Machine for	J. McClintic	Chambersburgh, Pa	Oct. 8, 1827	....
Mortising in wood	G. Page	Keene, N. H	Mar. 13, 1833	....
Mortising-machine	W. R. Axe	Beloit, Wis	Sept. 20, 1859	25, 479
Mortising-machine	S. E. Babcock	Olstead, N. H	June 16, 1836	....
Mortising-machine	C. Baggerman and J. Green	Saint Louis, Mo	Jan. 21, 1868	73, 428
Mortising-machine	T. R. Bailey	Lockport, N. Y	Aug. 5, 1856	15, 467
Mortising-machine	S. S. Bartlett	Woonsocket, R. I	Sept. 24, 1861	33, 330
Mortising-machine	C. Bennett	Pepperell, Mass	Sept. 17, 1845	4, 195
Mortising-machine	N. Bickford	Cincinnati, Ohio	Sept. 19, 1871	119, 007
Mortising-machine	G. Bradley	Auburn, N. Y	Feb. 25, 1835	....
Mortising-machine	J. E. Brown and S. S. Bartlett	Woonsocket, R. I	Jan. 24, 1854	17, 464
Mortising-machine	S. C. Brown	Richmond, Ind	June 25, 1867	65, 996
Mortising-machine	M. E. Campfield	Newark, N. J	Jan. 25, 1870	99, 155
Mortising-machine	C. Carter	Auburn, N. Y	Oct. 6, 1868	82, 692
Mortising-machine	J. B. Chambers	Pittsburgh, Pa	May 25, 1852	8, 966
Mortising-machine	J. H. Chandler	Worcester, Mass	Dec. 17, 1834	....
Mortising-machine	J. C. Channell	Dunstable, N. H	June 16, 1836	....
Mortising-machine	F. G. Chapman	Chicago, Ill	Mar. 8, 1870	100, 502
Mortising-machine	D. Clark	Brooklyn, Conn	Sept. 14, 1836	24
Mortising-machine	J. G. Clifton	Middletown, Ohio	Feb. 26, 1867	62, 394
Mortising-machine	E. Coffin	Indianapolis, Ind	July 17, 1860	29, 141
Mortising-machine	J. Cox	Portland, Oreg	Jan. 25, 1870	99, 064
Mortising-machine	E. Dayton	Meriden, Conn	Apr. 2, 1872	125, 123
Mortising-machine	F. A. Deland and L. Phillips	Memphis, Mich	Oct. 13, 1868	82, 926
Mortising-machine	J. Dennis and I. J. Richardson	Palmyra, N. Y	Nov. 19, 1833	....
Mortising-machine	W. Downing and W. H. Soley	Philadelphia, Pa	Feb. 28, 1871	112, 228
Mortising-machine	J. Driver	Marysville, Cal	June 24, 1873	140, 257
Mortising-machine	L. Eames	Kalamazoo, Mich	Apr. 17, 1860	27, 895
Mortising-machine	W. L. Epperson	Louisville, Ky	Dec. 1, 1868	84, 620
Mortising machine	H. K. Forbis	Danville, Ky	Jan. 26, 1869	86, 200
Mortising-machine	A. Foster	Phillipstown, Mass	Apr. 29, 1830	....
Mortising-machine	J. W. Fowle	Boston, Mass	Feb. 14, 1871	111, 834
Mortising-machine	C. Gates	Centre Antrim, N. H	Sept. 26, 1835	....
Mortising-machine	D. L. Gibbs	Worcester, Mass	Aug. 27, 1867	68, 298
Mortising machine	D. L. Gibbs	Worcester, Mass	May 12, 1868	77, 877
Mortising-machine	D. L. Gibbs	Worcester, Mass	Dec. 8, 1868	84, 817
Mortising-machine	D. L. Gibbs	Worcester, Mass	Feb. 2, 1869	86, 389
Mortising-machine	D. L. Gibbs	Worcester, Mass	Nov. 7, 1871	120, 737
Mortising-machine	S. Glover and D. Parmelee	Newtown, Conn	June 8, 1807	....
Mortising-machine	C. Green	Bethel, Ohio	Jan. 6, 1857	16, 332
Mortising-machine	A. Greenleaf and H. Amidons	Mexico, N. Y	Dec. 28, 1827	....
Mortising-machine	J. Guild	Cincinnati, Ohio	Nov. 30, 1852	9, 431
Mortising-machine	E. Hammond	Cumberland, Md	May 6, 1873	138, 649
Mortising-machine	F. H. Harwood	Rushville, N. Y	Jan. 31, 1860	26, 986
Mortising-machine	J. Hawkins	Stockbridge, Mass	June 20, 1836	....
Mortising-machine	B. Holly	Seneca Falls, N. Y	July 25, 1854	11, 403
Mortising-machine	E. J. Houston	Nacoochee, Ga	Feb. 18, 1873	135, 917
Mortising-machine	L. Houston	Smithville, N. J.	Jan. 28, 1873	135, 337
Mortising-machine	J. W. Ingle	Upperville, Va	Jan. 21, 1847	4, 936
Mortising-machine	J. M. Johnson and J. Herig	Cleveland, Ohio	Sept. 8, 1868	82, 064
Mortising-machine	E. Joslyn	Keene, N. H	Apr. 1, 1856	14, 564
Mortising-machine	W. Kegg	Lasellsville, N. Y	Apr. 12, 1859	23, 586
Mortising machine	J. J. Kellogg	Richmond, N. Y	Jan. 24, 1828	....
Mortising-machine	J. King	Morristown, N. J	Mar. 18, 1841	2, 010
Mortising-machine	J. Lemman	Cincinnati, Ohio	May 20, 1862	35, 318
Mortising-machine	S. Leroy	Mexico, N. Y	July 10, 1827	....
Mortising-machine	H. Marsh	Morristown, N. J	June 13, 1831	....
Mortising-machine	W. E. Marsh	Westfield Township, Essex County, N. J.	Apr. 18, 1828	....
Mortising-machine	J. McBride	Richmond, Ind	Nov. 26, 1835	....
Mortising-machine	N. Mead and J. Cutel	Walpole, N. H	Nov. 3, 1817	....
Mortising-machine	H. Mellish	Walpole, N. H	Nov. 19, 1833	....
Mortising-machine	L. G. Merrill	Angels, Cal	Oct. 18, 1870	108, 503
Mortising-machine	J. A. Merriman	Hinsdale, Mass	Jan. 15, 1856	14, 106
Mortising-machine	J. Moreland	Adrian, Mich	Feb. 22, 1853	9, 595
Mortising-machine	W. H. Morrison and M. W. E. Doran.	Indianapolis, Ind	Aug. 1, 1854	11, 430
Mortising-machine	P. A. Mowers	Cleversburgh, Pa	Jan. 12, 1869	85, 752
Mortising-machine	J. Munsell	Painted Post, N. Y	Aug. 12, 1846	4, 695
Mortising-machine	C. Muschar	Pittsburgh, Pa	Jan. 14, 1873	134, 815
Mortising-machine	E. Myers and S. R. Smith	Cincinnati, Ohio	Nov. 14, 1871	120, 003
Mortising-machine	W. Nangel	Philadelphia, Pa	Feb. 14, 1860	27, 145
Mortising-machine	W. Nangel	Philadelphia, Pa	Oct. 16, 1866	58, 878
Mortising-machine	A. O. Neal	Hyde Park, Mass	Sept. 21, 1869	95, 129
Mortising-machine	G. V. Orton and J. Richards	Cincinnati, Ohio	June 23, 1868	79, 249
Mortising-machine	B. H. Otis	Dedham, Mass	Feb. 20, 1846	4, 387
Mortising-machine	G. F. Outten	Norfolk, Va	Apr. 15, 1878	137, 950
Mortising-machine	G. Page	Keene, N. H	Mar. 18, 1836	....
Mortising-machine	J. Page	Henniker, N. H	June 12, 1835	....

Index of patents issued from the United States Patent Office from 1790 *to* 1873, *inclusive*—Continued.

Invention.	Inventor.	Residence.	Date.	No.
Mortising-machine	F. Parker	Petaluma, Cal	Oct. 20, 1868	83, 306
Mortising-machine	H. Parmele	Killingworth, Conn	Feb. 15, 1810	....
Mortising-machine	L. E. Payne and O. Pier	Stowe, Vt	Nov. 6, 1855	13, 759
Mortising-machine	J. A. Peabody	Philadelphia, Pa	Aug. 18, 1868	81, 286
Mortising-machine	G. B. Phillips	Poughkeepsie, N. Y	Apr. 14, 1868	76, 808
Mortising-machine	H. and S. H. Plumb	Honesdale, Pa	June 27, 1854	11, 177
Mortising-machine	F. Purden	Baltimore, Md	June 14, 1853	9, 784
Mortising-machine	J. Richards	Cincinnati, Ohio	Aug. 25, 1868	81, 407
Mortising-machine	J. Richards and W. H. Doane	Cincinnati, Ohio	Sept. 10, 1867	68, 791
Mortising-machine	I. J. Richardson	Palmyra, N. Y	July 7, 1835	....
Mortising-machine	E. Ripley	New York, N. Y	Dec. 22, 1827	....
Mortising-machine	E. J. Rowe	Eureka, Cal	Dec. 5, 1871	121, 551
Mortising-machine	G. T. Savary	Newburyport, Mass	Aug. 16, 1870	106, 511
Mortising-machine	A. Schmackers	Cincinnati, Ohio	Mar. 8, 1870	100, 672
Mortising-machine	H. Selick	Lewistown, Pa	Mar. 16, 1869	87, 879
Mortising-machine	E. M. Shaw	Wilbraham, Mass	Mar. 23, 1836	....
Mortising-machine	W. C. Shaw	Madison, Ind	July 27, 1852	9, 156
Mortising-machine	D. L. Shearer	Waupun, Wis	Sept. 13, 1870	107, 413
Mortising-machine	W. H. Sible	Harrisburgh, Pa	Dec. 13, 1870	110, 079
Mortising-machine	J. Skipp	Newark, N. J	Mar. 15, 1870	100, 813
Mortising-machine	H. B. Smith	Manchester, N. H	Apr. 17, 1849	6, 343
Mortising-machine	H. B. Smith	Lowell, Mass	Jan. 10, 1854	10, 422
Mortising-machine	H. B. Smith	Lowell, Mass	Oct. 9, 1855	13, 663
Mortising-machine	H. B. Smith	Lowell, Mass	June 30, 1857	17, 701
Mortising-machine	H. B. Smith	Lowell, Mass	Aug. 16, 1859	25, 221
Mortising-machine	H. C. Smith	Clarksville, Ohio	Sept. 4, 1860	29, 920
Mortising-machine	H. C. Smith	Clarksville, Ohio	Aug. 11, 1863	39, 502
Mortising-machine	H. C. Smith	Lawrence, Kans	July 19, 1870	105, 603
Mortising-machine	A. Spencer, jr	Southport, N. Y	Aug. 9, 1859	25, 055
Mortising-machine	S. Spencer	Angelica, N. Y	Sept. 10, 1850	7, 635
Mortising-machine	J. H. Stahl	Stanton's Mills, Pa	Feb. 4, 1873	135, 604
Mortising-machine	W. Stoddard	Lowell, Mass	Jan. 8, 1856	14, 071
Mortising-machine	J. Stufflebern	Milwaukee, Wis	Nov. 21, 1865	51, 100
Mortising-machine	J. Swan	Seymour, Conn	Oct. 29, 1872	132, 608
Mortising-machine	C. R. Tompkins	Rochester, N. Y	Oct. 22, 1867	70, 051
Mortising-machine	A. C. Vaughn	Rainsburgh, Pa	Mar. 6, 1860	27, 397
Mortising-machine	E. Wallace	Huntsville, Pa	Sept. 10, 1872	131, 236
Mortising-machine	J. J. Weeks	Buckram, N. Y	Apr. 17, 1849	6, 350
Mortising machine	L. W. Wolfe	Jacksonville, Ill	Feb. 18, 1868	74, 739
Mortising-machine	I. Wright	Centre Antrim, N. H	July 17, 1835	....
Mortising-machine	C. L. Zeidler	Cincinnati, Ohio	July 18, 1865	48, 863
Mortising-machine	C. L. Zeidler	Cincinnati, Ohio	Nov. 12, 1867	70, 771
Mortising-machine, Bearded-chisel	S. Metcalf	Wilmington, Vt	Jan. 17, 1827	....
Mortising-machine clamp	A. Schmackers	Cincinnati, Ohio	Sept. 26, 1871	119, 414
Mortising-machine gage	T. Beach	Freeport, Pa	July 4, 1871	116, 538
Mortising-machine, Lever-power	T. Green	Maulius, N. Y	Apr. 12, 1826	....
Mortising-machine, Power	G. W. Gould	Norwich, Conn	June 16, 1863	38, 898
Mortising machine, Timber	J. Andrews	Sudbury, Mass	May 30, 1838	752
Mortising machine, Timber	F. and T. Burdick	Brooklyn, N. Y	Apr. 7, 1838	684
Mortising machine, Timber	J. A. Fay	Keene, N. H	Jan. 17, 1842	2, 425
Mortising machine, Timber	T. H. Haskings	Springfield, Ohio	June 30, 1837	247
Mortising machine, Timber	I. McLaughlin	Sunderland, Vt	Apr. 7, 1838	685
Mortising machine, Timber	E. M. Shaw	Baltimore, Md	Sept. 22, 1838	937
Mortising-machines, Device for throwing into and out of gear the tools of.	L. Kittinger	East Greenville, Ohio	Dec. 29, 1857	18, 977
Mortising-machines, Feeding	R. P. Benton	Rochester, N. Y.	Jan. 23, 1855	12, 287
Mortising-machines, Regulating length of stroke in.	E. Gould	Newark, N. J	Sept. 25, 1855	13, 594
Mortising-machines, Reversing the chisels in	M. Marshall	Lowell, Mass	Apr. 28, 1857	17, 183
Mortising-machines, Reversing the chisels in	F. Stamm	Lancaster, Pa	Oct. 12, 1858	21, 783
Mortising, sawing, and boring machine	R. Medley	Bloomfield, Ky	July 7, 1829	....
Mortising, slotting, and dovetailing machine	J. Felber	Saint Louis, Mo	Oct. 6, 1868	82, 816
Mortising, tenoning, and boring machine	B. Richtmyre and J. H. Martin	Conesville, N. Y	July 16, 1839	1, 241
Mortising timber	J. W. Bliss	Somers, Conn	Mar. 20, 1834	....
Mortising timber	G. Page	Keene, N. H	Mar. 21, 1836	....
Mortising-tool	T. Board and C. N. Austin	Jackson Court-House, Va	Jan. 10, 1860	26, 745
Mortising-tool	T. and J. H. Burdick	Albany, N. Y	Jan. 26, 1869	86, 209
Mortising-tool	L. Eames	Kalamazoo, Mich	Apr. 10, 1860	27, 782
Mortising-tool	H. K. Forbis	Danville, Ky	Dec. 30, 1873	146, 058
Mortising-tool	A. C. Hitchcock and C. H. Amidon.	Greenfield, Mass	Mar. 18, 1856	14, 454
Mortising-tool	H. Knowles	New York, N. Y	Jan. 29, 1856	14, 160
Mortising-tool, Revolving	W. Zimmerman	Quincy, Ill	July 11, 1865	48, 759
Mortising wood	C. Thompson	Poughkeepsie, N. Y	Jan. 30, 1834	....
Mosaics, Manufacture of wooden	L. De Forest	Derby, Conn	Jan. 24, 1860	26, 898
Mosquito and fly net	E. O. Carrington	Philadelphia, Pa	Dec. 24, 1867	72, 452
Mosquito and fly net	J. W. Cran and A. S. Randolph.	Norwalk, Conn., and Plainfield, N. J.	Aug. 13, 1867	67, 729
Mosquito and fly net	A. M. Rodgers	Brooklyn, N. Y	May 11, 1869	90, 050
Mosquito and fly net	W. Weston, jr	West Newton, Mass	Aug. 20, 1867	68, 018
Mosquito and fly screen	J. W. Boughton	Philadelphia, Pa	July 2, 1872	128, 582
Mosquito-bar	A. L. Carrier	Washington, D. C	July 7, 1863	39, 116
Mosquito-bar	W. Field	Providence, R. I	Nov. 15, 1864	45, 032
Mosquito-bar	R. F. S. Heath	Philadelphia, Pa	May 9, 1871	114, 556
Mosquito-bar	J. S. Hunter	Hartford, Conn	May 28, 1867	65, 085
Mosquito-bar	S. Kemper	Berger, Mo	May 10, 1870	102, 830
Mosquito-bar	T. Masac	Good Hope Plantation, La	Oct. 25, 1870	108, 611
Mosquito-bar	S. Roebuck	Brooklyn, N. Y	Sept. 24, 1861	33, 365
Mosquito-bar	E. A. G. Roulstone	Roxbury, Mass	Aug. 7, 1866	56, 994
Mosquito-bar	L. Sawyer	Springfield, Vt	Apr. 24, 1866	54, 259
Mosquito-bar	T. S. Scoville	New York, N. Y	Feb. 21, 1860	27, 236
Mosquito-bar	E. Steinel	Amsterdam, N. Y	May 19, 1868	78, 024
Mosquito-bar	C. L. S. Walker	Newark, N. J	July 3, 1866	56, 128
Mosquito-bar	T. S. Williams	Enterprise, Miss	May 31, 1859	24, 255
Mosquito-bar and window-screen	A. J. Whittier	Roxbury, Mass	Aug. 27, 1867	68, 326
Mosquito-bar, Canopy or	J. B. Platt	Augusta, Ga	Mar. 9, 1869	87, 589
Mosquito-bar for windows, &c	V. Barker	Otisfield, Me	Feb. 12, 1867	61, 986
Mosquito-bar for windows	A. C. Flint	Boston, Mass	June 16, 1868	78, 948

Index of patents issued from the United States Patent Office from 1790 *to* 1873, *inclusive*—Continued.

Invention.	Inventor.	Residence.	Date.	No.
Mosquito-bar for windows	C. T. Warren	Linden, N. J	June 30, 1868	79, 419
Mosquito-bar frame	J. Higgins	Chicago, Ill	July 30, 1861	32, 943
Mosquito-bar frame	L. A. P. Jacques	Cincinnati, Ohio	Apr. 24, 1849	6, 390
Mosquito-bar frame	D. McHugh	Mainville, Ohio	Aug. 13, 1867	67, 784
Mosquito-bar frame	H. Searle	Washington, D. C	May 11, 1869	90, 054
Mosquito-bar frame, Adjustable	L. J. Parsons	New Bedford, Mass	Oct. 20, 1868	83, 202
Mosquito-bar netting, Elastic frame for	R. M. Holland and A. J. Hibbs	Philadelphia, Pa	June 11, 1867	65, 673
Mosquito bar or tent	A. W. Price	Adrian, Mich	June 20, 1865	48, 308
Mosquito-bars, Adjusting	F. C. Payne	Hebron, Conn	July 27, 1858	21, 019
Mosquito-bars to window-blinds, doors, &c., Attaching.	J. Hebron	Buffalo, N. Y	Dec. 14, 1869	97, 916
Mosquito-bars to windows, Fastening	A. F. Tarr	Rockport, Mass	Feb. 16, 1864	41, 650
Mosquito-canopy	M. Bliss	Grinnell, Iowa	May 15, 1866	54, 674
Mosquito-canopy	W. Bourquignon	Providence, R. I	Sept. 23, 1873	143, 119
Mosquito-canopy	L. J. Henry	New York, N. Y	Aug. 19, 1856	15, 592
Mosquito-canopy	I. E. Palmer	Hackensack, N. J	Feb. 25, 1868	74, 933
Mosquito-canopy	S. Roebuck	Brooklyn, N. Y	Apr. 28, 1863	38, 339
Mosquito-canopy	S. and J. Roebuck	New York, N. Y	Mar. 30, 1869	88, 514
Mosquito-canopy frame	I. E. Palmer	Hackensack, N. J	Sept. 25, 1866	58, 283
Mosquito-curtain	J. S. Martin	Boston, Mass	Dec. 11, 1855	13, 931
Mosquito-curtain	B. B. Webster	Boston, Mass	Oct. 3, 1854	11, 764
Mosquito-frame	L. S. Thompson	Brooklyn, N. Y	Feb. 24, 1863	37, 779
Mosquito-frame, Folding	S. C. Maine	Boston, Mass	Aug. 11, 1868	80, 830
Mosquito-frame for windows	A. E. Horton	North Leominster, Mass	Oct. 1, 1867	69, 343
Mosquito-guard	H. Harris	Newark, N. J	Sept. 10, 1867	68, 739
Mosquito-guard	A. D. Puffer	Somerville, Mass	Mar. 20, 1866	53, 334
Mosquito-guard	R. Themar	Sheboygan, Wis	June 22, 1869	91, 793
Mosquito-guard for bedsteads	A. D. Puffer	Somerville, Mass	May 8, 1866	54, 595
Mosquito-killer	H. D. Forbes	Cambridge, Mass	June 16, 1868	78, 950
Mosquito-net	J. D. De Coursey	Philadelphia, Pa	June 12, 1866	55, 473
Mosquito-net	J. J. A. Ergenzinger	Atlanta, Ga	Feb. 21, 1871	112, 023
Mosquito-net	J. B. Holmes	Philadelphia, Pa	Mar. 16, 1869	87, 849
Mosquito-net	T. G. Voorhis and W. B. Whiteman.	New York, N. Y	Apr. 9, 1861	32, 018
Mosquito-net	T. S. Winslow	New York, N. Y	Mar. 5, 1872	124, 239
Mosquito-net	J. Zengler	Chicago, Ill	Mar. 15, 1864	41, 957
Mosquito-net, Adjustable frame for	E. Howard	New York, N. Y	July 23, 1872	129, 665
Mosquito-net and shades for windows	R. B. Burchell	Brooklyn, N. Y	Feb. 14, 1860	27, 103
Mosquito-net frame	U. W. Armstrong	Evansville, Ind	Feb. 16, 1869	86, 966
Mosquito-net frame	W. A. Griffith	Boston, Mass	Aug. 6, 1867	67, 532
Mosquito-net frame	S. E. Hartwell	New York, N. Y	Apr. 15, 1856	14, 655
Mosquito-net frame	S. Hughes	Jersey City, N. J	July 16, 1872	129, 030
Mosquito-net frame	G. T. Palmer	Brooklyn, N. Y	Apr. 28, 1868	77, 207
Mosquito-net frame	N. Peterson and R. Roecher	Memphis, Tenn	Mar. 4, 1873	136, 456
Mosquito-net frame	M. L. Treadwell	New York, N. Y	Aug. 6, 1867	67, 612
Mosquito-net frame	W. M. White	Canton, Conn	Nov. 19, 1872	133, 279
Mosquito-net frames, Metallic connection for	U. W. Armstrong and I. Keeney	Evansville, Ind	June 14, 1870	104, 245
Mosquito-net frames, Rubber connection for	U. W. Armstrong and I. Keeney	Evansville, Ind	Mar. 22, 1870	100, 966
Mosquito-net holder	C. Messenger	Cleveland, Ohio	Aug. 16, 1870	106, 386
Mosquito-net in window-blinds	G. W. Miles	Philadelphia, Pa	Dec. 3, 1867	71, 778
Mosquito-net stand	A. Strasser and B. M. Lowy	Montgomery, Ala	Aug. 4, 1868	80, 782
Mosquito-net support	B. M. Lewy and A. Strasser	Montgomery, Ala	June 8, 1869	91, 030
Mosquito-netting, Machine for stretching and folding.	J. A. and L. Van Riper	Spring Valley, N. Y	Sept. 8, 1863	39, 868
Mosquito-screen	J. M. Griswold	Auburndale, Mass	July 13, 1869	92, 441
Mosquito-screen	E. Spaulding	Brooklyn, N. Y	June 9, 1868	78, 767
Mosquito-screen for windows	G. A. Keene	Newburyport, Mass	June 4, 1872	127, 609
Mosquito-screen for windows	T. Stover	Cambridgeport, Mass	July 9, 1867	66, 534
Mosquito screen frame	W. W. Wooley	New York, N. Y	June 28, 1870	101, 800
Mosquito tent or frame	J. Stewart	New York, N. Y	Mar. 15, 1864	41, 946
Moss, Cleaning and carding	L. Boudreaux	Thibodeaux, La	Aug. 4, 1857	17, 954
Moss-cleaning machine	H. Hull	Petersonville, La	June 25, 1867	66, 026
Moth-destroying apparatus	C. F. Worch	New York, N. Y	June 4, 1867	65, 462
Moth-killer	W. A. Flanders	Sharon, Vt	June 6, 1854	11, 013
Moth-proof box	R. M. Seldis	New York, N. Y	Nov. 14, 1871	120, 904
Moth-proof case	J. W. Aiken and J. H. Stone	Philadelphia, Pa	Nov. 12, 1867	70, 676
Moth-proof clothing-chest	R. S. Jennings	Philadelphia, Pa	Nov. 14, 1871	120, 976
Moth-proof lining	J. R. Smith	Chicago, Ill	Aug. 31, 1869	94, 357
Moth-protector	F. F. Voigt	New Orleans, La	Oct. 3, 1871	119, 481
Moth-trap	J. Frew	Meadville, Pa	Feb. 10, 1863	37, 618
Moth-trap	J. M. Heard	Aberdeen, Miss	Feb. 7, 1860	27, 048
Motion and multiplying power, Extending	W. Kendall, jr	Waterville, Me	Jan. 16, 1826	....
Motion and preventing reaction, Transmitting rotary.	C. Sellers	Philadelphia, Pa	May 28, 1872	127, 192
Motion, Apparatus for converting	J. J. Chenal	Génissiat, France	Nov. 7, 1871	120, 622
Motion, Apparatus for converting	L. S. Fithian	Brooklyn, N. Y	June 6, 1871	115, 589
Motion, Apparatus for converting	C. L. Spencer	Providence, R. I	Feb. 23, 1869	87, 218
Motion, Apparatus for converting	K. J. Winslow	Twickenham, England	June 23, 1868	79, 092
Motion, Apparatus for converting reciprocating into rotary.	R. M. Fryer	Nashville, Tenn	Mar. 19, 1872	124, 805
Motion, Apparatus for converting reciprocating into rotary.	C. L. Spencer	Providence, R. I	Feb. 23, 1869	87, 217
Motion, Apparatus for converting reciprocating into rotary.	F. Wagner	New York, N. Y	Mar. 21, 1871	112, 991
Motion, Apparatus for converting rotary into reciprocating.	D. Morrison	Portland, Me	Nov. 26, 1867	71, 506
Motion, Apparatus for preventing back	A. H. and J. H. Race	Brooklyn, N. Y	Sept. 30, 1873	143, 252
Motion, Apparatus for producing reciprocating	J. J. P. Lyon	Ypsilanti, Mich	Oct. 13, 1868	82, 964
Motion, Apparatus for reproducing	C. T. Chester	Englewood, N. J	May 27, 1873	139, 294
Motion, Apparatus for transmitting	E. H. Bancroft	Syracuse, N. Y	Aug. 1, 1871	117, 503
Motion, Apparatus for transmitting	F. M. La Boiteaux	Cincinnati, Ohio	Oct. 29, 1872	132, 671
Motion, Apparatus for transmitting	P. Merhof and F. Healy	Croton, N. Y., and Barrington, R. I.	Jan. 2, 1872	122, 398
Motion, Apparatus for transmitting	S. S. Rembert	Memphis, Tenn	Aug. 29, 1871	118, 554
Motion, Apparatus for transmitting rotary	M. Clemens	Springfield, Mass	Nov. 2, 1869	96, 395
Motion, Apparatus for transmitting rotary	S. R. Morgan	Philadelphia, Pa	Mar. 23, 1869	88, 061
Motion, Application of steam to produce rotary	J. Mount	Nashville, Tenn	July 16, 1829	....
Motion by belts, Transmitting	J. Rand	Roxbury, Mass	Aug. 30, 1864	44, 014

Index of patents issued from the United States Patent Office from 1790 *to* 1873, *inclusive*—Continued.

Invention.	Inventor.	Residence.	Date.	No.
Motion by means of friction, Apparatus for converting rotary into reciprocating.	R. Sammer	Vineland, N. J	Apr. 18, 1871	113, 801
Motion by means of friction, Device for obtaining	S. Marden	Newton, Mass	Feb. 11, 1868	74, 234
Motion by varying and irregular power, Machinery to produce uniform.	J. Kello	New York, N. Y	Apr. 1, 1815	....
Motion, Changing	R. G. Turner and H. Stone	Dedham, Mass	Dec. 15, 1863	40, 965
Motion, Changing reciprocating, eccentric, or irregular to rotary.	J. Woodhull	Rochester, N. Y	Apr. 6, 1829	....
Motion, Changing reciprocating into rotary	A. M. Bouton and A. Perry	Newark, N. J	Aug. 18, 1842	2, 753
Motion, Changing reciprocating into rotary	J. Harris, jr	Boston, Mass	Jan. 14, 1851	7, 902
Motion, Changing reciprocating into rotary	T. J. Morton	Racine, Wis	June 25, 1872	128, 240
Motion, Changing reciprocating into rotary	J. V. Strait	Litchfield, Ohio	July 22, 1851	8, 233
Motion, Changing reciprocating into rotary	P. Yates	Milwaukee, Wis	Apr. 23, 1850	7, 316
Motion, Changing rotary into reciprocating	A. Dean	Penn Yan, N. Y	Nov. 29, 1864	45, 230
Motion, Changing rotary into reciprocating	I. D. Garlick	Lyons, N. Y	Aug. 20, 1850	7, 577
Motion, Changing rotary into reciprocating	J. H. and A. E. Redstone	Indianapolis, Ind	Apr. 29, 1862	35, 131
Motion, Changing rotary into reciprocating	S. L. Wiegand	Philadelphia, Pa	May 25, 1858	20, 384
Motion, Combined lever and crank	E. Nolt	Earl Township, Pa	Apr 14, 1863	38, 176
Motion, Communicating	W. Bicknell	Hartford, Me	Dec. 4, 1866	60, 126
Motion, Compensation for loss of	T. Shaw	Philadelphia, Pa	Feb. 7, 1865	46, 309
Motion, Converting	W. H. Akins	Dryden, N. Y	July 19, 1864	43, 558
Motion, Converting	H. Burk	Mineral Point, Ohio	May 28, 1867	65, 166
Motion, Converting	P. Dickson	Utica, Minn	Mar. 24, 1863	37, 953
Motion, Converting	J. W. Drummond	New York, N. Y	Mar. 8, 1864	41, 835
Motion, Converting	A. Eckert	Trenton, Ohio	Jan. 15, 1867	61, 179
Motion, Converting	W. H. Hurlbut	Elgin, Ill	Mar. 26, 1867	63, 254
Motion, Converting	S. C. Ketchum	Winchendon, Mass	Apr. 28, 1863	38, 311
Motion, Converting	A. E. Kline	Woodville, Pa	Nov. 29, 1864	45, 252
Motion, Converting	T. D. Lakin	Greenfield, N. H	Dec. 17, 1872	133, 994
Motion, Converting	T. A. Macaulay	New York, N. Y	Mar. 1, 1864	41, 780
Motion, Converting	W. H. McNary	Brooklyn, N. Y	July 15, 1862	35, 883
Motion, Converting	L. Planer	New York, N. Y	May 22, 1860	28, 403
Motion, Converting	C. L. Spencer	Providence, R. I	Mar. 4, 1863	31, 597
Motion, Converting alternate rectilinear into rotary	P. Cooper	New York, N. Y	Apr. 28, 1828	....
Motion, Converting circular into reciprocating	T. and C. Plumleigh	Dundee, Ill	Oct. 8, 1867	69, 584
Motion-converting mechanism	E. S. Pierce	Hartford, Conn	June 29, 1869	92, 091
Motion, Converting reciprocal into rotary	D. Morrison	Portland, Me	Jan. 21, 1868	73, 631
Motion, Converting reciprocating into rotary	J. C. Butterworth and B. Arnold.	Providence and East Greenwich, R. I.	July 9, 1861	32, 749
Motion, Converting reciprocating into rotary	I. Chapman	New York, N. Y	Aug. 3, 1858	21, 065
Motion, Converting reciprocating into rotary	J. Cupps and A. R. Harper	Chicago, Ill., and Grandville, Mich.	Jan. 1, 1867	60, 837
Motion, Converting reciprocating into rotary	J. F. Foss	Lowell, Mass	Aug. 29, 1865	49, 618
Motion, Converting reciprocating into rotary	C. P. Gallagher	San Francisco, Cal	Nov. 28, 1854	11, 993
Motion, Converting reciprocating into rotary	S. Gissinger and J. W. Kellberg	Allegheny, Pa	Dec. 23, 1856	16, 303
Motion, Converting reciprocating into rotary	C. A. Harper	Fort Worth, Tex	Nov. 1, 1859	25, 966
Motion, Converting reciprocating into rotary	J. F. Hartmann	Richmond, Ind	Apr. 26, 1870	102, 260
Motion, Converting reciprocating into rotary	F. H. Harwood	Rushville, N. Y	Apr. 6, 1858	19, 818
Motion, Converting reciprocating into rotary	C. Howard	Alton, Ill	Aug. 10, 1852	9, 185
Motion, Converting reciprocating into rotary	R. Maffett	Bradford, Pa	Apr. 1, 1856	14, 568
Motion, Converting reciprocating into rotary	E. Matteson	Brooklyn, N. Y	Aug. 28, 1860	29, 802
Motion, Converting reciprocating into rotary	C. B. Parsons	Burr Oak, Mich	Feb. 28, 1860	27, 307
Motion, Converting reciprocating into rotary	E. A. Smead	Tioga, Pa	Feb. 15, 1859	22, 984
Motion, Converting reciprocating into rotary	A. T. Underhill	New York, N. Y	Sept. 20, 1859	25, 550
Motion, Converting reciprocating into rotary	C. A. Watson	Green River, Ky	Aug. 1, 1838	864
Motion, Converting reciprocating into rotary	T. Williams	Providence, R. I	Mar. 5, 1861	31, 649
Motion, Converting reciprocating rectilinear into continuous circular.	B. Babbett	Bangor, Me	Oct. 11, 1836	....
Motion, Converting reciprocating rotary into reciprocating rectilinear.	A Carson	New York, N. Y	July 20, 1852	9, 131
Motion, Converting rectilinear into circular	R. Dennis	Rahway, N. J	Nov. 3, 1825	....
Motion, Converting rectilinear into rotary	J. B. Eads	Saint Louis, Mo	Jan. 2, 1866	51, 815
Motion, Converting rectilinear into rotary	J. A. Ehle	Greenbush, Wis	May 21, 1867	64, 852
Motion, Converting rectilinear into rotary	J. G. Harroun	Sag Harbor, N. Y	July 25, 1871	117, 280
Motion, Converting rectilinear into rotary	J. and T. G. McLaughlin	Pennsylvania	Aug. 29, 1848	5, 741
Motion, Converting rectilinear into rotary	G. W. Richardson and R. Glover	Grayville, Ill	Feb. 15, 1859	23, 002
Motion, Converting rectilinear into rotary	J. T. Savage	Raleigh, N. C	Aug. 9, 1823	....
Motion, Converting rotary into double-acting reciprocating.	C. Willson	Bedford, N. Y	Oct. 5, 1839	1, 356
Motion, Converting rotary into reciprocating	S. F. Ames	Stamford, Conn	July 4, 1865	48, 502
Motion, Converting rotary into reciprocating	H. Baker	Catskill, N. Y	June 7, 1853	9, 761
Motion, Converting rotary into reciprocating	A. E. Kline	Goodville, Pa	Sept. 20, 1864	44, 319
Motion, Converting rotary into reciprocating	P. C. Van Brocklin	Buffalo, N. Y	July 15, 1862	35, 903
Motion, Converting rotary into reciprocating	A. Warth	New York, N. Y	Dec. 2, 1856	16, 155
Motion, Converting rotary into reciprocating	J. J. Weeks	Locust Valley, N. Y	July 20, 1858	20, 980
Motion, Converting rotary into reciprocating	P. Werni	Manchester, Mich	Jan. 3, 1855	45, 779
Motion, Converting rotary into reciprocating rectilinear.	A. Broughton	Malone, N. Y	Sept. 13, 1859	25, 465
Motion, Converting rotary into reciprocating rectilinear.	W. S. Lazelle	New York, N. Y	Feb. 7, 1860	27, 057
Motion, Crank	E. P. Brownell	East Haddam, Conn	Mar. 25, 1862	34, 735
Motion, Crank	M. A. Rowe	Martinsville, Ill	Apr. 26, 1870	102, 323
Motion, Crank	T. Shaw	Philadelphia, Pa	May 19, 1863	38, 603
Motion, Device for arresting	J. Everding	Philadelphia, Pa	Nov. 7, 1871	120, 730
Motion, Device for changing rotary into reciprocal	D. Shively	Bayard, Ohio	May 28, 1872	127, 196
Motion, Device for converting	W. H. Abel	Greenville, R. I	Feb. 11, 1868	74, 265
Motion, Device for converting	H. H. Bishop	Bristol, Conn	Mar. 3, 1863	37, 806
Motion, Device for converting	F. Brewer	Collinsville, Ill	June 27, 1865	48, 369
Motion, Device for converting	W. M. Cox	Blanket Hill, Pa	Dec. 30, 1873	145, 931
Motion, Device for converting	B. Eybel	New York, N. Y	May 7, 1867	64, 511
Motion, Device for converting	M. M. Follett	Westborough, Mass	Apr. 7, 1868	76, 429
Motion, Device for converting	J. Hawthorn	Coshocton, Ohio	Feb. 5, 1867	61, 829
Motion, Device for converting	T. A. Mitchell	Washington, D. C	June 22, 1869	91, 555
Motion, Device for converting	J. P. Taylor	Hudson City, N. J	Dec. 16, 1873	145, 599
Motion, Device for converting	P. Werni	Elizabeth, N. J	July 23, 1872	129, 771
Motion, Device for converting	H. Wheeler	Racine, Wis	Dec. 16, 1873	145, 540
Motion, Device for converting alternate circular motion into direct circular.	A. Bartholf	New York, N. Y	Apr. 19, 1859	23, 651

Index of patents issued from the United States Patent Office from 1790 *to* 1873, *inclusive*—Continued.

Invention.	Inventor.	Residence.	Date.	No.
Motion, Device for convertng circular into reciprocating.	L. Dame	Newburyport, Mass	Feb. 11, 1873	135, 697
Motion, Device for converting reciprocating into alternate circular.	H. Ehrenfeld	New York, N. Y	June 21, 1859	24, 448
Motion, Device for converting reciprocating into rotary.	G. L. Gavett	Sandstone, Mich	Nov. 28, 1871	131, 354
Motion, Device for converting reciprocating into rotary.	J. Hathaway	Marietta, Ga	Apr. 3, 1860	27, 715
Motion, Device for converting reciprocating into rotary.	G. S. Nutter	Bunker Hill, Ill	July 16, 1872	129, 582
Motion, Device for converting rotary into reciprocating.	E. Walker	Dundee, Ill	Mar. 10, 1868	75, 495
Motion, Device for convert ingrotary into reciprocating rectilinear.	C. F. Hadley	Chicopee, Mass	Mar. 23, 1869	88, 163
Motion, Device for reversing	C. F. Hadley	Chicopee, Mass	Mar. 1, 1870	100, 395
Motion, Device for stopping and changing	F. A. Pratt	Hartford, Conn	Sept. 4, 1860	29, 939
Motion, Device for transferring	W. H. Benson	Waynesborough, Va	July 15, 1873	140, 878
Motion, Device for transmitting	J. Brizee	Alvarado, Cal	July 10, 1866	56, 171
Motion, Device for transmitting	J. W. Browning	Mattoon, Ill	Oct. 11, 1864	44, 594
Motion, Device for transmitting	T. C. Entwistle	New York, N. Y	Sept. 4, 1866	57, 693
Motion, Device for transmitting	E. Gallagher	Brooklyn, N. Y	Feb. 11, 1868	74, 337
Motion, Device for transmitting	P. Palmlund	Brooklyn, N. Y	July 4, 1871	116, 745
Motion, Device for transmitting	N. Thompson	Farmington, Mich	Dec. 3, 1867	71, 820
Motion, Device for transmitting	L. Tilton	Brooklyn, E. D., N. Y	May 28, 1867	65, 304
Motion, Device for transmitting and reversing rotary.	W. H. Merrick	Philadelphia, Pa	Feb. 28, 1871	112, 269
Motion, Device for transmitting rotary	L. B. Flanders	Philadelphia, Pa	Oct. 15, 1867	69, 787
Motion, Device for transmitting rotary	H. J. Hancock	New York, N. Y	Aug. 2, 1870	105, 937
Motion, Device for transmitting rotary	H. Morris	West Philadelphia, Pa	Dec. 28, 1858	22, 445
Motion, Directing	A. Buchanan	Jersey City, N. J	Sept. 22, 1363	40, 075
Motion, Feed	A. J. Shipley	Waterbury, Conn	Dec. 22, 1868	85, 249
Motion for machinery, Electrical stop	E. Maertens	Philadelphia, Pa	June 24, 1873	140, 287
Motion for machinery, Pendulum	J. Jacob	Montreal, Canada	Dec. 29, 1826	......
Motion for preserving rolling contact, &c	G. P. Gordon and F. O. Degener	New York, N. Y	June 16, 1857	17, 565
Motion for propelling machinery, Alternate	B. Elliott	Danby, N. Y	Oct. 4, 1834	......
Motion for stamping and other machines, Progressive reciprocating.	R. L. Barclay	Brooklyn, N. Y	Nov. 23, 1869	97, 153
Motion from reciprocating rectilinear motion, Mechanism for obtaining rotary.	A. R. Gilmore	Bath, Me	Apr. 5, 1859	23, 461
Motion in car-brakes and other machinery, Transmitting.	J. Steger	New York, N. Y	Apr. 13, 1869	88, 990
Motion, Instrument for registering reciprocating and rotary.	F. B. Hall	Hartford, Conn	Apr. 23, 1861	32, 133
Motion, Machine for converting oscillating motion into direct circular.	L. Planer	New York, N. Y	June 7, 1859	24, 359
Motion, Machine for converting reciprocating into intermittent rotary.	H. Ehrenfeld	New York, N. Y	Nov. 29, 1859	26, 261
Motion, Machine for converting reciprocating into rotary.	J. G. Deshler	Allentown, Pa	Sept. 22, 1868	82, 296
Motion, Machine for converting reciprocating into rotary.	J. B. Page	Chicago, Ill	Aug. 14, 1866	57, 178
Motion, Machine for converting rotary into reciprocating.	C. C. Welsh	Pleasant Valley, Pa	Oct. 6, 1868	82, 907
Motion, Machine for producing rotary	D. F. Mosman	Chelsea, Mass	Dec. 17, 1872	133, 999
Motion, Machine for transmitting	C. B. Grout	New York, N. Y	Apr. 16, 1872	125, 809
Motion, Machinery for changing	C. L. Pyron and R. Bruce	Manchester, Tenn	Jan. 3, 1860	26, 709
Motion, Machinery for transmitting	M. Kaefer	New York, N. Y	May 31, 1859	24, 218
Motion, Mechanical	A. Kloman	Pittsburgh, Pa	Aug. 2, 1870	105, 951
Motion, Mechanical	D. Lynahan	Buffalo, N. Y	Dec. 6, 1864	45, 330
Motion, Mechanism for converting reciprocating into rotary.	W. Simpson and A. Gardner	Ilford, England	Nov. 23, 1869	97, 240
Motion, Mechanism for converting rotary into reciprocating.	W. N. Brown	New York, N. Y	Oct. 25, 1859	25, 880
Motion, Mechanism for obtaining intermitted rotary.	L. J. Knowles	Warren, Mass	Mar. 20, 1866	53, 308
Motion, Mechanism for producing rotary	C. E. De Lorierè	London, England	June 11, 1872	127, 682
Motion, Mechanism for transmitting	C. C. Hull	Williamsburgh, N. Y	Apr. 13, 1869	88, 876
Motion, Mechanism for transmitting rotary	G. Bancker and A. Campbell	New York, N. Y	Nov. 2, 1858	21, 934
Motion, Metallic band for communicating	J. Eve	Augusta, Ga	May 1, 1828	......
Motion obtained from rectilinear, Rotary	A. Smith	Rhinebeck, N. Y	Aug. 30, 1823	......
Motion of machinery, Changing the	O. Pettee	Newton, Mass	Mar. 15, 1825	......
Motion of machinery, Registering	S. L. Wiegand	Philadelphia, Pa	Aug. 3, 1858	21, 101
Motion of shafts into reciprocating motion, Converting rotary.	J. B. Siner	Lawrence, Mass	Mar. 11, 1873	136, 675
Motion, Producing alternate lateral	N. Moody	Hallowell, Mass	Feb. 19, 1819	......
Motion, Producing continuous circular from reciprocating rectilinear.	C. S. Harris	Holyoke, Mass	June 20, 1854	11, 123
Motion, Producing reciprocating and lateral	M. Ingalls	Burlington, Pa	Mar. 13, 1847	5, 015
Motion, Producing vertical and horizontal reciprocating.	M. Steigers	Saint Louis, Mo	Mar. 9, 1858	19, 586
Motion, Reciprocating and rotary	I. Van Doren	Somerville, N. J	Mar. 23, 1858	19, 726
Motion, Reciprocating crank	B. K. Dorwart	Lancaster, Pa	Oct. 31, 1865	50, 695
Motion, Stop	B. Stott	Westerly, R. I	Aug. 30, 1864	44, 024
Motion, Stopping and changing	F. A. Pratt	Hartford, Conn	Nov. 26, 1861	33, 816
Motion to machinery, Apparatus for communicating	S. Kilburn	Sterling, Mass	Apr. 16, 1831	......
Motion to machinery, Communicating	C. H. Miller	Buffalo, N. Y	Aug. 2, 1870	105, 964
Motion to machinery, Transmitting	W. Phelps	Sycamore, Ill	July 3, 1860	29, 000
Motion to machines, Device for communicating	L. Cheetham	Lewiston, Me	July 16, 1872	129, 316
Motion, Transmitting	M. M. Ammidown	Boston, Mass	Oct. 6, 1868	82, 674
Motion, Transmitting	S. Andress	Chesaning, Mich	Feb. 26, 1861	31, 522
Motion, Transmitting	J. W. Howlett	Greensborough, N. C	Apr. 9, 1861	31, 978
Motion, Transmitting	M. Kaefer	Alexandria, Pa	May 5, 1857	17, 222
Motion, Transmitting	M. Kaefer	Factoryville, N. Y	Mar. 5, 1861	31, 607
Motion, Transmitting	N. Read	Jewett City, Conn	Nov. 9, 1869	96, 615
Motion, Transmitting	W. Rowell	New York, N. Y	May 7, 1867	64, 430
Motion, Transmitting	W. Rowell	New York, N. Y	May 7, 1867	64, 431
Motion, Transmitting	C. Sellers	Philadelphia, Pa	Sept. 10, 1861	33, 2[illegible]3
Motion, Transmitting	H. F. Shaw	West Roxbury, Mass	Oct. 3, 1865	50, 283

Index of patents issued from the United States Patent Office from 1790 *to* 1873, *inclusive*—Continued.

Invention.	Inventor.	Residence.	Date.	No.
Motion, Transmitting	E. Wadhams	Hamilton, Canada	July 11, 1865	48, 780
Motion, Transmitting	H. F. Wheeler	Boston, Mass	Sept. 26, 1871	119, 437
Motion, Transmitting	F. Yeiser	Danville, Ky	Jan. 30, 1866	52, 354
Motion, Transmitting and arresting	W. Sellers	Philadelphia, Pa	Jan. 21, 1862	34, 217
Motion, Traverse	D. Walker	East Hampton, Mass	Jan. 16, 1872	122, 790
Motion, Weft-stop	F. O. Tucker	Westerly, R. I	Aug. 1, 1871	117, 698
Motive-power	J. B. Atwater	Chicago, Ill	Apr. 13, 1869	88, 767
Motive-power	A. M. Bacon	Boston, Mass	Nov. 9, 1869	96, 657
Motive-power	C. Batcheller	Polk County, Iowa	Aug. 23, 1870	106, 537
Motive-power	J. N. Bethune	Warrenton, Va	Sept. 26, 1871	119, 260
Motive-power	A. Bouchard	New Orleans, La	Jan. 16, 1872	122, 802
Motive-power	I. W. Brown	New Haven, Conn	Feb. 11, 1868	74, 295
Motive-power	J. Bruce	Brooklyn, N. Y	Sept. 23, 1862	36, 506
Motive-power	J. M. Cayce	Franklin, Tenn	Sept. 13, 1870	107, 221
Motive-power	A. Giraudat	New York, N. Y	Aug. 27, 1861	33, 139
Motive-power	J. K. Glenn	New York, N. Y	Oct. 22, 1867	69, 987
Motive-power	J. Green	Providence, R. I	Sept. 13, 1864	44, 181
Motive-power	W. H. Hartman	Fostoria, Ohio	May 16, 1865	47, 717
Motive-power	J. H. Haven	Lewiston, N. Y	Jan. 30, 1866	52, 287
Motive-power	J. B. Hunter	Ashley, Ill	Sept. 26, 1871	119, 362
Motive-power	A. Johnson and W. H. Elliott	Bloomington, Ill	May 14, 1867	64, 770
Motive-power	W. B. Jones	Franklin, Ky	May 21, 1867	64, 881
Motive-power	F. Kettler	Milwaukee, Wis	Feb. 3, 1863	37, 582
Motive-power	C. Mans	Danville, Pa	Mar. 22, 1859	23, 305
Motive-power	J. McKnight and W. S. Deisher	Reading, Pa	June 5, 1866	55, 325
Motive-power	G. Meyer	New Richmond, Mich	Apr. 15, 1873	137, 856
Motive-power	J. G. Mitchell	Collington, Md	Mar. 1, 1859	23, 104
Motive-power	W. T. Nichols	Rutland, Vt	Dec. 2, 1862	37, 050
Motive-power	T. J. Polson	Kilmichael, Miss	Aug. 29, 1871	118, 641
Motive-power	W. S. Reeder	Saint Louis, Mo	May 28, 1872	127, 368
Motive-power	B. J. Sage	New Orleans, La	May 28, 1872	127, 189
Motive-power	J. A. Schulo	New York, N. Y	Apr. 24, 1866	54, 220
Motive-power	J. S. Swan	Kanawha County, W. Va	Sept. 12, 1871	118, 982
Motive-power	L. Tilliers	New York, N. Y	Aug. 20, 1861	33, 112
Motive-power	J. Toll	Locust Grove, Ohio	Jan. 26, 1864	41, 406
Motive-power	L. D. Towsley and E. Matteson	New York and Brooklyn, N. Y.	July 24, 1860	29, 329
Motive-power	W. M. Watson	Tonica, Ill	Aug. 16, 1870	106, 523
Motive-power and balance-car	J. Bayma	San Francisco, Cal	Nov. 28, 1871	121, 316
Motive-power apparatus	T. Davis	Detroit, Mich	Dec. 20, 1870	110, 212
Motive-power apparatus	J. T. Gilbert	Asbury, Ill	Apr. 4, 1871	113, 417
Motive-power apparatus	A. Jackson	Lebanon, Tenn	June 21, 1870	104, 595
Motive-power apparatus	L. W. Lathrop	Poughkeepsie, N. Y	Nov. 1, 1870	108, 915
Motive-power apparatus	J. B. Martel	Philadelphia, Pa	June 25, 1872	128, 408
Motive-power apparatus	C. A. Mills	Bridgeport, Conn	Jan. 3, 1871	110, 667
Motive-power apparatus	S. G. Monce	Marathon, Ohio	June 23, 1868	79, 289
Motive-power apparatus	H. Wilkins and W. H. Sangster	Paris, Ky	Feb. 15, 1870	99, 992
Motive-power, Apparatus for obtaining	R. N. Thomson	Edinburgh, Scotland	May 22, 1866	55, 030
Motive-power, Apparatus for obtaining and applying.	W. Huston	Wilmington, Del	Jan. 22, 1867	61, 339
Motive-power, Apparatus for producing	E. Stockton and W. O. St. John	Folsom, Cal	May 5, 1868	77, 674
Motive-power, Apparatus for the application of electro-magnetism as a.	S. Stimpson	New York	Sept. 12, 1838	910
Motive-power, Applying	J. Schley	Savannah, Ga	May 16, 1871	114, 866
Motive-power for the vertical rise and fall of the tide, Device for producing.	A. W. Scharit	Saint Louis, Mo	Jan. 10, 1865	45, 867
Motive-power, Combining springs for	G. Terry	New York, N. Y	Dec. 8, 1863	40, 866
Motive-power, Controlling	F. E. Sickels	New York, N. Y	Sept. 12, 1848	5, 765
Motive-power, Device for transmitting	R. T. Smith	Nashua, N. H	Jan. 28, 1873	135, 293
Motive-power engine	J. S. Johnstone	Ayr, North Britain	June 13, 1871	115, 963
Motive-power engine	J. Robertson	Glasgow, Scotland	Aug. 2, 1870	106, 078
Motive-power, Engine for employing steam, &c., under pressure to obtain.	J. W. Durham	Durhamville, Tenn	July 17, 1860	29, 149
Motive-power engines, Apparatus for forming an explosive mixture of air and hydrocarbon-vapor for use in.	J. Kidd	London, England	Mar. 12, 1867	62, 856
Motive-power engines, Recovering and regenerating steam in.	F. M. A. Motard	Paris, France	Aug. 19, 1873	142, 038
Motive-power for driving street-cars	P. E. McDonnell	Lyons, Ill	Aug. 23, 1870	106, 600
Motive-power for locomotion, &c	D. A. Pratt	Tremont, N. Y	Oct. 8, 1867	69, 699
Motive-power from petroleum and other oils, Obtaining.	R. D. McCreary	Oil City, Pa	Nov. 12, 1867	70, 732
Motive-power, Liquid used as a	J. C. Salomon, jr	Baltimore, Md	July 22, 1856	15, 391
Motive-power machine	H. Wickham, jr	Chicago, Ill	July 26, 1870	105, 870
Motive-power, Machine for obtaining	R. Lide	Union Street Borough, England.	Aug. 18, 1868	81, 302
Motive-power, Machinery for transmitting and distributing.	J. Richmond	Lockport, N. Y	Jan. 24, 1871	111, 251
Motive-power, Obtaining	P. Daniel	Franklin County, Ky	June 29, 1858	20, 701
Motive-power, Obtaining	R. M. Marchant	London, England	June 13, 1871	115, 877
Motive-power, Obtaining	J. Pringle	Summerhill, Pa	July 24, 1860	29, 303
Motive-power, Obtaining	W. M. Storm	Troy, N. Y	Feb. 4, 1851	7, 922
Motive-power, Pipe for transmitting pneumatic currents for.	R. Spear	New Haven, Conn	May 17, 1870	103, 252
Motive-power regulator	J. E. Gillespie	Boston, Mass	Nov. 19, 1867	71, 160
Motive power, Self	J. J. Girand	Baltimore, Md	Mar. 31, 1836	....
Motive-power, Spring	T. B. Fogarty	New York, N. Y	Jan. 7, 1873	134, 533
Motive-power, Spring	J. B. Howell	Wilkesbarre, Pa	Sept. 24, 1872	131, 614
Motive-power, Steam-enginery	A. C. Pilliner and J. C. Hill	Newport, England	Oct. 26, 1869	96, 262
Motive-power, Tidal	W. R. Close	Bangor, Me	Dec. 27, 1870	110, 430
Motive-power to machinery, Applying	R. R. Moffatt	Brooklyn, N. Y	Dec. 20, 1870	110, 268
Motive-power, Transmitting	T. Hanson	New York, N. Y	June 5, 1866	55, 289
Motive-power, Transmitting	J. Richmond	Lockport, N. Y	June 20, 1871	116, 221
Motive-power, Transmitting	R. S. Smith	Nashua, N. H	Oct. 23, 1866	59, 089
Motive-powers, Adaptation of substances as	F. M. Ruschhaupt	New York, N. Y	Mar. 20, 1860	27, 569

Index of patents issued from the United States Patent Office from 1790 *to* 1873, *inclusive*—Continued.

Invention.	Inventor.	Residence.	Date.	No.
Motor: *See* Air-motor. Atmospheric motor. Ball-motor. Churn-motor. Compressed-air motor. Electro-motor. Electro-magnetic motor. Fluid-motor. Grinding-motor. Hand-motor. Hydraulic motor. Hydro-pneumatic motor. Marine motor. Rotary motor. Rotary fluid-motor. Sewing-machine motor. Signal-alarm motor. Spring-motor. Steam-motor. Thermal motor. Tremolo-motor. Tydal motor. Water-motor.				
Motor	M. Comstock	Chicago, Ill	May 13, 1873	138, 737
Motor	L. H. Dean	Tecumseh, Mich	Dec. 16, 1873	145, 490
Motor	P. Guzman	Paris, France	Oct. 31, 1871	120, 379
Motor	C. L. Langdon	New Orleans, La	Mar. 19, 1872	124, 829
Motor	C. J. Schumacher	Portland, Me	Sept. 24, 1872	131, 631
Motor	L. E. Truesdell	Warren, Mass	Feb. 25, 1873	136, 286
Motor-engine	T. D. Richardson	North Providence, R. I	Apr. 15, 1873	137, 865
Motor for driving machinery	A. N. Proctor	Boston, Mass	Aug. 19, 1873	142, 041
Motor-machine	I. G. Hubbard	Nokomis, Ill	May 23, 1871	115, 058
Mouse and animal trap	G. L. Hart	New Britain, Conn	Mar. 21, 1871	112, 807
Mouse-trap	W. K. Bachman	Columbia, S. C	Sept. 27, 1870	107, 647
Mouse-trap	E. and A. Buckman	Brooklyn, N. Y	Oct. 11, 1870	108, 104
Mouse-trap	J. N. Bunnell	Unionville, Conn	May 4, 1869	89, 551
Mouse-trap	J. N. Bunnell	Unionville, Conn	Nov. 1, 1870	108, 876
Mouse-trap	A. G. Davis	Watertown, Conn	Mar. 30, 1869	88, 456
Mouse-trap	A. G. Davis	Watertown, Conn	June 22, 1869	91, 722
Mouse-trap	A. G. Davis and H. S. Frost	Watertown, Conn	July 5, 1870	105, 048
Mouse-trap	G. L. Hart	New Britain, Conn	Jan. 2, 1872	122, 318
Mouse-trap	H. C. Hart	Unionville, Conn	Dec. 20, 1870	110, 357
Mouse-trap	H. C. Hart	Unionville, Conn	Mar. 28, 1871	113, 051
Mouse-trap	A. A. Hotchkiss	Sharon, Conn	Apr. 5, 1870	101, 620
Mouse-trap	C. A. Hotchkiss	Bridgeport, Conn	Sept. 5, 1871	118, 721
Mouse-trap	A. Ovaitt	Unionville, Conn	July 4, 1871	116, 743
Mouse-trap	H. S. Weller	Watertown, Conn	May 9, 1871	114, 629
Mousing-hook	S. G. Coleman	Providence, R. I	May 1, 1860	28, 126
Mousing-hook	J. North	Middletown, Conn	Nov. 20, 1860	30, 687
Mousing-hook	H. Wynblad	West Hoboken, N. J	July 3, 1860	29, 029
Movable axle	G. N. Reynolds	Charleston, S. C	Apr. 29, 1819	....
Movable joint for tables, &c	H. N. House	Washington, D. C	May 16, 1865	47, 726
Movement and gas-burner combined, Mechanical	E. Omensetter	Philadelphia, Pa	Jan. 23, 1872	123, 042
Movement, Apparatus for producing intermittent rotative.	R. C. Reynolds	Lawrence, Mass	July 15, 1873	140, 951
Movement, Apparatus for producing plane or parallel.	A. G. Barrett	Barrett, Kans	Dec. 26, 1871	122, 145
Movement, Apparatus for treating diseases by mechanical.	A. L. Wood	New York, N. Y	Dec. 7, 1869	97, 744
Movement, Application of wind-power to produce a reciprocating.	C. C. Moore	New York, N. Y	Aug. 5, 1862	36, 101
Movement, Cam	J. E. Goodwin	Brewerton, N. Y	Sept. 30, 1873	143, 235
Movement-cure apparatus	D. Wark	Montreal, Canada	Aug. 1, 1871	117, 580
Movement, Electro-magnetic	F. L. Pope	Elizabeth, N. J	May 17, 1870	103, 077
Movement for actuating presses, Mechanical	P. C. Ingersoll	Green Point, N. Y	Dec. 28, 1869	98, 385
Movement for converting motion, Mechanical	A. B. Hendryx	Ansonia, Conn	Aug. 13, 1872	130, 371
Movement for converting oscillating into rotary motion, Mechanical.	J. Peabody	Bangor, Me	Jan. 28, 1873	135, 288
Movement for converting power into speed, Mechanical.	W. F. Goodwin	East New York, N. Y	Dec. 31, 1867	72, 842
Movement, Mechanical	E. Allen	Norwich, Conn	Oct. 22, 1867	70, 063
Movement, Mechanical	J. Armstrong	Toledo, Ohio	Sept. 2, 1873	142, 365
Movement, Mechanical	H. H. Baker	New Market, N. J	May 28, 1872	127, 294
Movement, Mechanical	W. H. Baker	Tamaqua, Pa	Nov. 15, 1859	26, 078
Movement, Mechanical	J. S. Barden	Providence, R. I	Aug. 25, 1868	81, 329
Movement, Mechanical	J. and W. F. Barnes	Rockford, Ill	May 9, 1871	114, 514
Movement, Mechanical	W. B. Bartram	Danbury, Conn	Aug. 27, 1872	130, 784
Movement, Mechanical	A. Benneckendorf	Hoboken, N. J	Oct. 10, 1871	119, 735
Movement, Mechanical	P. Bloomsburg, jr., and J. Molyneux.	Bordentown, N. J	May 28, 1867	65, 051
Movement, Mechanical	M. Bockman	Brooklyn, N. Y	Mar. 3, 1868	74, 979
Movement, Mechanical	N. Bradford	Addison, Me	June 7, 1870	103, 971
Movement, Mechanical	W. Brant	Paris, Ill	July 17, 1866	56, 354
Movement, Mechanical	A. W. Browne	Brooklyn, N. Y	Dec. 31, 1867	72, 792
Movement, Mechanical	A. W. Browne	Brooklyn, N. Y	Aug. 24, 1869	94, 068
Movement, Mechanical	A. W. Browne and W. F. Goodwin.	Brooklyn and New York, N. Y.	Aug. 18, 1868	81, 248
Movement, Mechanical	A. R. Buffington	United States Army	Dec. 8, 1868	84, 678
Movement, Mechanical	H. C. Burk	Mineral Point, Ohio	Jan. 7, 1868	72, 972
Movement, Mechanical	A. C. Burner	Green Bank, W. Va	May 28, 1872	127, 219
Movement, Mechanical	C. W. Carr	Paola, Kans	Jan. 28, 1873	135, 263
Movement, Mechanical	H. J. Case	Auburn, N. Y	Apr. 20, 1869	89, 026
Movement, Mechanical	E. Chapman	Rochester, Minn	Jan. 30, 1872	123, 237
Movement, Mechanical	A. Clark	Albany, Ill	Oct. 10, 1871	119, 821
Movement, Mechanical	T. J. and G. M. Clark	Higganum, Conn	Aug. 10, 1869	93, 598
Movement, Mechanical	I. A. Clippinger	Newton, Iowa	Oct. 31, 1865	50, 687
Movement, Mechanical	R. F. Cochran	Barry, Md	Feb. 27, 1872	124, 036
Movement, Mechanical	J. H. Cooper	Philadelphia, Pa	Oct. 3, 1871	110, 576
Movement, Mechanical	J. Corley	Near Coalfield, Kans	Apr. 25, 1871	114, 111
Movement, Mechanical	E. Courtright	Bartello, Mich	Oct. 8, 1872	131, 940
Movement, Mechanical	J. H. Cranston	Norwich, Conn	Dec. 9, 1873	145, 402
Movement, Mechanical	F. Creamer	Brooklyn, N. Y	Jan. 19, 1869	86, 060
Movement, Mechanical	C. M. Currey	Pontiac, Mich	Dec. 24, 1867	72, 608
Movement, Mechanical	J. Daguier	Cincinnati, Ohio	July 8, 1873	140, 683
Movement, Mechanical	R. B. Davidson	Greenville, Ky	Sept. 15, 1863	39, 891
Movement, Mechanical	J. H. Davis	Woburn, Mass	Nov. 2, 1858	22, 003

Index of patents issued from the United States Patent Office from 1790 *to* 1873, *inclusive*—Continued.

Invention.	Inventor.	Residence.	Date.	No.
Movement, Mechanical	J. P. Davis	Stiles, Wis	May 5, 1868	77, 464
Movement, Mechanical	M. S. Davis	Vincennes, Ind	June 11, 1872	127, 855
Movement, Mechanical	J. B. Doolittle	Wallingford, Conn	June 3, 1873	139, 499
Movement, Mechanical	A. Duvall	Baltimore, Md	Feb. 9, 1869	86, 826
Movement, Mechanical	A. Eckert	Trenton, Ohio	Oct. 18, 1870	108, 340
Movement, Mechanical	R. Eickemeyer	Yonkers, N. Y	May 27, 1873	139, 379
Movement, Mechanical	W. H. Fenn	Rochester, N. Y	Mar. 14, 1871	112, 699
Movement, Mechanical	P. Ferguson and F. G. Bates	New Haven, Conn., and Springfield, Mass.	Nov. 14, 1871	120, 955
Movement, Mechanical	E. E. Furney	Chicopee, Mass	May 18, 1869	90, 254
Movement, Mechanical	D. T. Gale	Poughkeepsie, N. Y	Nov. 2, 1869	96, 419
Movement, Mechanical	W. Galladay	Sheboygan Falls, Wis	May 21, 1867	64, 858
Movement, Mechanical	W. Garrison	Clarkstown, N. Y.	Mar. 15, 1870	100, 747
Movement, Mechanical	L. Goodall	Deering, N. H	Apr. 30, 1872	126, 284
Movement, Mechanical	W. F. Goodwin	East New York, N. Y	Mar. 17, 1868	75, 676
Movement, Mechanical	W. F. Goodwin	New York, N. Y	Aug. 18, 1868	81, 271
Movement, Mechanical	R. B. Hamel and J. B. Holden	Jersey City, N. J	May 17, 1870	103, 175
Movement, Mechanical	W. Hammill	Parma, Mich	June 21, 1870	104, 451
Movement, Mechanical	M. A. Hardy	Cambridge, Mass	Nov. 2, 1869	96, 316
Movement, Mechanical	W. M. Henderson	Philadelphia, Pa	Oct. 5, 1869	95, 586
Movement, Mechanical	A. B Hendryx	Ansonia, Conn	Aug. 13, 1872	130, 372
Movement, Mechanical	E. C. Hopping	Madison, N. J	Apr. 15, 1873	137, 924
Movement, Mechanical	H. R. Huling	Boston, Mass	Mar. 23, 1869	88, 040
Movement, Mechanical	C. W. Hurd	Comstock's Landing, N. Y.	Aug. 15, 1871	118, 016
Movement, Mechanical	J. James	Ogdensburgh, N. Y.	Mar. 5, 1861	31, 606
Movement, Mechanical	J. M. Johnson	Northcutt's Store, N. Y.	July 6, 1869	92, 192
Movement, Mechanical	W. F. Jones	Easton, Kans	Nov. 14, 1871	120, 883
Movement, Mechanical	D. E. Keating	Oswego Falls, N. Y.	Oct. 3, 1871	119, 617
Movement, Mechanical	J. J. Kimball	Naperville, Ill	Sept. 5, 1871	118, 723
Movement, Mechanical	M. Laemmel	Bay Ridge, N. Y	Dec. 1, 1868	84, 632
Movement, Mechanical	S. H. Lancaster	Le Claire, Iowa	Dec. 27, 1864	45, 616
Movement, Mechanical	C. B. Lewis	Clifton, Ohio	Mar. 20, 1866	53, 312
Movement, Mechanical	L. V. Lewis	Sun Prairie, Wis	Aug. 12, 1873	141, 800
Movement, Mechanical	E. Matteson	South Brooklyn, N. Y	June 20, 1365	48, 294
Movement, Mechanical	S. C. Matteson	Osceola, Wis	Nov. 14, 1865	50, 943
Movement, Mechanical	J. H. McCanny	Wytheville, Va	July 4, 1871	116, 617
Movement, Mechanical	J. H. McCanny	Wytheville, Va	Jan. 9, 1872	122, 530
Movement, Mechanical	W. C. McGill	Cincinnati, Ohio	Apr. 18, 1865	47, 315
Movement, Mechanical	W. C. McGill	Cincinnati, Ohio	Sept. 12, 1865	49, 905
Movement, Mechanical	W. S. Mead	New York, N. Y	Sept. 17, 1867	69, 009
Movement, Mechanical	C. Meiners	Indianapolis, Ind	Apr. 29, 1873	138, 424
Movement, Mechanical	E. Melton	Flemingsburgh, Ky	July 5, 1870	114, 977
Movement, Mechanical	H. Merriman	Bloomington, Ill	June 15, 1869	91, 251
Movement, Mechanical	D. S. Merritt	Mount Morris, Mich	Aug. 25, 1868	81, 393
Movement, Mechanical	G. R. Metten	Cleveland, Ohio	Oct. 6, 1868	82, 860
Movement, Mechanical	G. R. Metten	Cleveland, Ohio	Mar. 2, 1869	87, 353
Movement, Mechanical	J. Niebergall	New York, N. Y	Oct. 15, 1867	69, 929
Movement, Mechanical	A. Nimmo	Philadelphia, Pa	Apr. 4, 1871	113, 687
Movement, Mechanical	W. Oliver	Haynesville, La	Sept. 9, 1873	142, 578
Movement, Mechanical	I. E. Palmer	Hackensack, N. J	Sept. 22, 1868	82, 436
Movement, Mechanical	W. Park	Norwich, Conn	Jan. 14, 1873	134, 766
Movement, Mechanical	J. H. Pelton	Cleveland, Tenn	July 9, 1867	66, 623
Movement, Mechanical	T. H. Percival	Harper's Ferry, W. Va	Dec. 12, 1871	121, 893
Movement, Mechanical	O. Plummer	Worcester, Mass	Dec. 14, 1869	97, 959
Movement, Mechanical	N. Read	Winchendon, Mass	July 27, 1869	93, 004
Movement, Mechanical	G. Rich	Smithville, Mo	Apr. 28, 1868	77, 322
Movement, Mechanical	G. N. Rhodes	Hamilton, N. Y.	Jan. 4, 1870	98, 632
Movement, Mechanical	E. O. Rood	Lodi, Ill	Sept. 1, 1868	81, 819
Movement, Mechanical	E. O. Rood	Lodi, Ill	Feb. 9, 1869	86, 692
Movement, Mechanical	E. G. Russell	Ravenna, Ohio	Mar. 8, 1871	112, 498
Movement, Mechanical	C. W. Saladee and W. Veach	Newark, Ohio	Sept. 25, 1866	58, 304
Movement, Mechanical	E. W. Sargent	Lowell, Mass	June 16, 1868	79, 013
Movement, Mechanical	C. B. Sawyer	Fitchburgh, Mass	Sept. 24, 1872	131, 566
Movement, Mechanical	L. Scofield	Watertown, Wis	Aug. 6, 1872	130, 158
Movement, Mechanical	J. M. Scott	Kinsman, Ohio	Dec. 31, 1867	72, 915
Movement, Mechanical	J. See	Mitchell, Ind	June 30, 1868	79, 503
Movement, Mechanical	J. A. Shanner	Plain View, Ill	Nov. 10, 1868	84, 007
Movement, Mechanical	H. F. Shaw	West Roxbury, Mass	Aug. 10, 1869	93, 485
Movement, Mechanical	A. Shear	Plymouth, Mich	May 16, 1871	114, 870
Movement, Mechanical	H. Shutts	Oregon, Mo	June 8, 1869	91, 048
Movement, Mechanical	H. Smith	Salem, Mass	Jan. 30, 1866	52, 334
Movement, Mechanical	P. J. Smith	Philadelphia, Pa	May 22, 1866	54, 972
Movement, Mechanical	C. D. Snell and J. W. Renney	Mechanic Falls, Me	Oct. 8, 1867	69, 719
Movement, Mechanical	E. Soper	New York, N. Y	Jan. 14, 1868	73, 401
Movement, Mechanical	M. F. Spore	Preble, N. Y	Mar. 17, 1868	75, 591
Movement, Mechanical	C. R. Squier	Cleveland, Ohio	May 7, 1872	126, 421
Movement, Mechanical	W. R. Swinnerton	Peoria, Ill	Feb. 25, 1868	74, 952
Movement, Mechanical	H. Thomas and R. Wallace	New York, N. Y	Sept. 21, 1869	95, 165
Movement, Mechanical	J. C. Vancleave	Hamburgh, Ark	Feb. 8, 1870	99, 611
Movement, Mechanical	A. Van Guysling	Albany, N. Y	Nov. 9, 1869	96, 746
Movement, Mechanical	J. H. Wait	Portsmouth, Ohio	May 15, 1860	28, 318
Movement, Mechanical	W. Walker	Odin, Ill	Dec. 13, 1870	110, 175
Movement, Mechanical	A. Wallace	Jacksonville, Fla	Jan. 2, 1872	122, 502
Movement, Mechanical	M. Wappich	Sacramento, Cal	Oct. 11, 1864	44, 681
Movement, Mechanical	A. G. Waterhouse	Portage, Wis	Mar. 25, 1873	137, 115
Movement, Mechanical	W. Weaver	Greenwich, N. Y	Oct. 3, 1871	119, 674
Movement, Mechanical	M. Webb	Chenango Forks, N. Y	Nov. 24, 1868	84, 451
Movement, Mechanical	J. H. Whitney	Rochester, Minn	Nov. 8, 1870	109, 163
Movement, Mechanical	J. Woolf	Burr Oak, Mich	Oct. 18, 1870	108, 547
Movement, Mechanical	J. Woolf	Burr Oak, Mich	Nov. 29, 1870	109, 790
Movement, Mechanical	J. Woolf	Burr Oak, Mich	Dec. 13, 1870	110, 185
Movement, Mechanical	J. Woolf	Burr Oak, Mich	Jan. 31, 1871	111, 503
Movement, Mechanical	J. Woolf	Burr Oak, Mich	May 30, 1871	115, 406
Movement, Mechanical	E. E. Young	York, Me	Aug. 29, 1871	118, 572
Movement, Mechanical	D. Zeigler	Lewistown, Pa	Sept. 13, 1870	107, 432
Movement, Mechanical	C. Zeitler	Benton's Port, Iowa	Mar. 19, 1872	124, 712
Movement or substitute for cog-wheels, Mechanical	G. H. Knight	Cincinnati, Ohio	Sept. 10, 1867	68, 754
Movement, Parallel	A. J. Vandegrift	Cincinnati, Ohio	Nov. 26, 1867	71, 342

Index of patents issued from the United States Patent Office from 1790 to 1873, inclusive—Continued.

Invention.	Inventor.	Residence.	Date.	No.
Movement, Pneumatic reciprocating	D. V. Wood	Ann Arbor, Mich	Jan. 18, 1870	98, 901
Movement, Rotary	W. H. Mitchell	San Francisco, Cal	Aug. 16, 1859	25, 132
Movement, Rotary mechanical	A. G. Waterhouse	San Francisco, Cal	Apr. 26, 1870	102, 454
Moving heavy bodies, Apparatus for	C. Whittaker	Milwaukee, Wis	July 14, 1868	79, 883
Moving heavy bodies, Device for	J. A. Woodworth	Hickory Corners, Mich	Apr. 20, 1869	89, 111
Moving-machine for the sick	J. C. Jenckes	Providence, R. I	May 29, 1823	....
Mower and hedge-trimmer	J. O. Taber	Salem, Ohio	Mar. 22, 1870	101, 060
Mower and reaper knife sharpener	W. B. Deuel	Ithaca, N. Y	Nov. 3, 1868	83, 767
Mower and reaper knives, Machine for grinding	H. Millard	New York, N. Y	Sept. 21, 1869	95, 033
Mower and reaper sickle sharpener	E. Dowd	Oshkosh, Wis	July 12, 1870	105, 184
Mower-cutters, Machine for grinding	H. C. Fisk	Wellsville, N. Y	Mar. 7, 1871	112, 330
Mower, Lawn	W. Allen	Worcester, Mass	July 11, 1871	116, 791
Mower, Lawn	W. Allen	Worcester, Mass	Nov. 18, 1873	144, 725
Mower, Lawn	J. Arbeiter	East Hartford, Conn	Apr. 20, 1869	89, 004
Mower, Lawn	J. Arbeiter	East Hartford, Conn	Mar. 15, 1870	100, 810
Mower, Lawn	D. H. Chamberlain	West Roxbury, Mass	Dec. 12, 1871	121, 705
Mower, Lawn	L. Chapman	Collinsville, Conn	Dec. 28, 1869	98, 348
Mower, Lawn	L. Chapman	Collinsville, Conn	July 16, 1872	129, 099
Mower, Lawn	C. M. Clinton	Ithaca, N. Y	Apr. 5, 1870	101, 585
Mower, Lawn	T. Coldwell and G. L. Chadborn	Newburgh, N. Y	July 26, 1870	105, 781
Mower, Lawn	T. Coldwell and G. L. Chadborn	Newburgh, N. Y	Mar. 18, 1873	136, 969
Mower, Lawn	W. H. Drake	Musconetcong, N. J	Nov. 22, 1870	109, 398
Mower, Lawn	N. Eaton	Woburn, Mass	Apr. 19, 1870	102, 101
Mower, Lawn	J. C. Field	Chicago, Ill	Mar. 15, 1870	100, 742
Mower, Lawn	G. W. Fischer	Dayton, Ohio	Apr. 23, 1872	126, 102
Mower, Lawn	E. R. Gard	Chicago, Ill	June 27, 1871	116, 295
Mower, Lawn	T. Garrick	Providence, R. I	May 11, 1869	89, 981
Mower, Lawn	H. W. Harkness	New Britain, Conn	Jan. 4, 1870	98, 590
Mower, Lawn	B. and D. H. Harnish	Lancaster and Pequea, Pa	Mar. 7, 1871	112, 338
Mower, Lawn	H. C. Hart	Unionville, Conn	Dec. 27, 1870	110, 457
Mower, Lawn	A. M. Hills	Hockanum, Conn	Jan. 28, 1868	73, 807
Mower, Lawn	A. M. Hills	Hockanum, Conn	June 20, 1871	116, 188
Mower, Lawn	A. Ingrham	Philadelphia, Pa	Feb. 28, 1871	112, 146
Mower, Lawn	S. D. King	Middletown, N. Y	Feb 4, 1873	135, 563
Mower, Lawn	S. D. King	Middletown, N. Y	Apr. 8, 1873	137, 691
Mower, Lawn	W. J. Lane	Millbrook, N. Y	Oct. 15, 1872	132, 160
Mower, Lawn	B. Merritt, jr	Newton, Mass	Apr. 20, 1869	89, 230
Mower, Lawn	J. S. Oakley and O. D. Wood	Passaic, N. J., and Newburgh, N. Y.	Nov. 14, 1871	120, 994
Mower, Lawn	A. J. Ohmer	Hamilton, Ohio	Dec. 20, 1870	110, 388
Mower, Lawn	E. G. Passmore	Philadelphia, Pa	Feb. 23, 1869	87, 286
Mower, Lawn	E. G. Passmore	Philadelphia, Pa	Apr. 4, 1871	113, 336
Mower, Lawn	L. Ross	Worcester, Mass	Dec. 27, 1870	110, 562
Mower, Lawn	S. W. Sears	New York, N. Y	Apr. 14, 1868	76, 831
Mower, Lawn	W. Sellers	Haverhill, Mass	Oct. 21, 1873	143, 787
Mower, Lawn	J. Shaw	Brooklyn, N. Y	Oct. 13, 1868	83, 101
Mower, Lawn	T. Loetbeer	Irvington, N. Y	June 10, 1873	139, 741
Mower, Lawn	B. L. Walker	Sing Sing, N. Y	Sept. 24, 1872	131, 726
Mower, Lawn	M. E. C. Wilde	Somerville, Mass	Feb. 16, 1860	96, 960
Mower, Lawn	A. W. C. Williams	London, England	May 2, 1871	114, 501
Mowers and reapers, Frame for combined	J. A. Moore and A. H. Patch	Louisville, Ky	June 30, 1857	17, 693
Mowers and reapers, Grinding the teeth of	D. Hinman	Berea, Ohio	Sept. 13, 1859	25, 412
Mowing and reaping machine	A. Amsden	Rochester, N. Y	May 26, 1857	17, 357
Mowing and reaping machine	W. Burgess	London, England	Sept. 18, 1855	13, 565
Mowing and reaping machine	J. Butter	Buffalo, N. Y	Feb. 7, 1860	27, 034
Mowing and reaping machine	V. M. Chafee	Xenia, Ohio	Mar. 20, 1860	27, 523
Mowing and reaping machine	R. Ketcham	South Dansville, N. Y	May 15, 1860	28, 283
Mowing and reaping machine	J. W. Mulley	Amsterdam, N. Y	Dec. 16, 1856	16, 247
Mowing and reaping machine	A. M. Wilson	Rhinebeck, N. Y	June 10, 1837	197
Mowing and reaping machine frame	M. G. Hubbard	Penn Yan, N. Y	July 15, 1856	15, 338
Mowing and reaping machine, Grain, grass, &c	A. M. Wilson	Rhinebeck, N. Y	May 15, 1837	196
Mowing and reaping machines, Cutting-apparatus for.	C. H. McCormick	Chicago, Ill	Nov. 5, 1861	33, 656
Mowing and reaping machines, Driver's seat for	F. Getz	Amherst, N. Y	Nov. 6, 1860	30, 572
Mowing and thrashing	E. Briggs and G. G. Carpenter	Fort Covington, N. Y	Feb. 5, 1836	....
Mowing grass and cutting grain, Machine for	A. S. Hathaway	Columbia, Me	July 1, 1856	15, 265
Mowing-machine	H. Adkins	Round Prairie, Ill	Jan. 15, 1850	7, 012
Mowing-machine	A. B. Allen	New York, N. Y	July 24, 1860	29, 228
Mowing-machine	H. Allen	Fayetteville, Tenn	June 2, 1836	....
Mowing-machine	W. Allen	Worcester, Mass	Mar. 24, 1868	75, 720
Mowing-machine	T. A. Anderson	McMinn County, Tenn	June 29, 1833	....
Mowing-machine	W. Anderson	Newton, N. J	Apr. 9, 1872	125, 429
Mowing-machine	B. Atwood	Stanstead, Canada	June 11, 1872	127, 730
Mowing-machine	C. Aultman and L. Miller	Canton, Ohio	June 17, 1856	15, 160
Mowing-machine	W. Bacheller	West Newbury, Mass	Aug. 11, 1857	17, 956
Mowing-machine	J. Baily	Chester County, Pa	Feb. 13, 1822	...
Mowing-machine	J. C. Baker	Mechanicsburgh, Ohio	Sept. 5, 1871	118, 673
Mowing-machine	E. Ball	Canton, Ohio	Aug. 12, 1856	15, 507
Mowing-machine	E. Ball	Canton, Ohio	Dec. 1, 1857	18, 788
Mowing-machine	A. Bank	Salem, Ohio	June 26, 1866	55, 905
Mowing-machine	J. F. Barrett	North Granville, N. Y	May 13, 1856	14, 898
Mowing-machine	S. S. Bartlett	Providence, R. I	May 16, 1865	47, 692
Mowing-machine	L. M. Batty	Canton, Ohio	Sept. 19, 1865	49, 962
Mowing-machine	L. D. Bidwell	Birmingham, Conn	Dec. 8, 1868	84, 790
Mowing-machine	A. Bolander	Akron, Ohio	Mar. 21, 1871	112, 892
Mowing-machine	S. P. Briggs	Saratoga Springs, N. Y	June 2, 1857	17, 417
Mowing-machine	A. Brown, L. G. Kniffen, and T. H. Dodge.	Worcester, Mass	Nov. 8, 1864	44, 935
Mowing-machine	C. B. Brown	Alton, Ill	Sept. 4, 1855	13, 517
Mowing-machine	J. J. and F. Bulfinch	Freeport and Waldoborough, Me.	Feb. 20, 1872	123, 864
Mowing-machine	C. Bullock	Jamestown, N. Y	June 5, 1860	28, 628
Mowing-machine	H. M. Burdick and T. V. Le Roy	Ilion, N. Y	Nov. 12, 1872	133, 013
Mowing-machine	O. H. Burdick and O. F. Dagget.	Auburn, N. Y	Apr. 2, 1872	125, 264
Mowing-machine	T. D. Burrall	Geneva, N. Y	Apr. 27, 1858	20, 035
Mowing-machine	G. E. Burt	Harvard, Mass	Mar. 3, 1868	74, 986
Mowing-machine	G. E. Burt	Harvard, Mass	Mar. 9, 1869	87, 539

Index of patents issued from the United States Patent Office from 1790 *to* 1873, *inclusive*—Continued.

Invention.	Inventor.	Residence.	Date.	No.
Mowing-machine	J. Butler	Buffalo, N. Y	June 8, 1858	20,479
Mowing-machine	A. H. Caryl	Sandusky, Ohio	Sept. 15, 1857	18,187
Mowing-machine	O. R. Chaplin	Saint Johnsbury, Vt	Jan. 10, 1860	26,814
Mowing-machine	L. S. Clark	Bethel, Conn	Aug. 8, 1871	117,864
Mowing-machine	T. J. and G. M. Clark	Higganum, Conn	Oct. 12, 1869	95,769
Mowing-machine	S. Clarridge	Mount Sterling, Ohio	May 28, 1872	127,229
Mowing-machine	C. Colahan	Cleveland, Ohio	July 15, 1873	140,890
Mowing-machine	S. Comfort, jr	Morrisville, Pa	Mar. 18, 1856	14,445
Mowing-machine	A. J. Cook	Delphi, Ind	Nov. 20, 1846	4,861
Mowing-machine	E. Cope and T. Hoopes, jr	Chester County, Pa	May 18, 1825	......
Mowing-machine	W. Crook	New Hope, Pa	Apr. 13, 1858	19,913
Mowing-machine	M. V. Cumming	Winthrop, Me	May 12, 1868	77,719
Mowing-machine	A. D. Darby	White Creek, N. Y	Sept. 17, 1861	33,296
Mowing-machine	T. H. Dodge	Washington, D. C	Nov. 15, 1859	26,095
Mowing-machine	T. H. Dodge	Washington, D. C	Feb. 19, 1861	31,520
Mowing-machine	G. C. Dolph	West Andover, Ohio	Sept. 8, 1859	18,141
Mowing-machine	J. Drummond	Waterford, N. Y	June 30, 1836	......
Mowing-machine	J. G. Dunham	Raritan, N. J	Dec. 18, 1860	30,961
Mowing-machine	R. Dutton	Dayton, Ohio	Mar. 19, 1861	31,705
Mowing-machine	R. Dutton	Brooklyn, N. Y	Feb. 2, 1864	41,460
Mowing-machine	R. Dutton	Brooklyn, N. Y	Mar. 15, 1864	41,958
Mowing-machine	R. Emerson, jr., and F. Graham	Rockford, Ill	Jan. 14, 1862	34,180
Mowing-machine	H. Fisher	Canton, Ohio	Jan. 12, 1858	19,083
Mowing-machine	H. Fisher	Alliance, Ohio	May 7, 1861	32,244
Mowing-machine	H. L. Frailey	Lancaster, Pa	Dec. 5, 1865	51,301
Mowing-machine	A. Franklin, W. J. Hastings, and A. Gates.	Florence, Ind	Feb. 2, 1869	86,523
Mowing-machine	S. Fuller	Somers, Conn	Mar. 14, 1871	112,579
Mowing-machine	P. Gaillard	Lancaster, Pa	Dec. 4, 1812	......
Mowing-machine	A. Gale	Poughkeepsie, N. Y	July 25, 1854	11,367
Mowing-machine	A. Gale	Poughkeepsie, N. Y	Aug. 11, 1857	17,964
Mowing-machine	F. Gardiner	Gardiner, Me	July 31, 1860	29,371
Mowing-machine	F. A. Geisler	Bristol, R. I	Dec. 29, 1868	85,378
Mowing-machine	A. M. George	Nashua, N. H	May 1, 1860	28,072
Mowing-machine	L. Gorden	Vanlue, Ohio	Mar. 19, 1872	124,678
Mowing-machine	J. P. Greeley and L. W. Buxton	Nashua, N. H	July 21, 1863	39,286
Mowing-machine	W. C. Greenleaf	Andover, Me	July 1, 1836	......
Mowing-machine	A. M. Hall	West Falmouth, Me	Dec. 23, 1856	16,274
Mowing-machine	T. Harding	Springfield, Ohio	May 19, 1857	17,350
Mowing-machine	D. K. and J. K. Harris	Allensville, Ind	Nov. 6, 1849	6,846
Mowing-machine	R. Heath	West Newbury, Mass	Apr. 26, 1833	......
Mowing-machine	E. F. and J. Hennington	West Hoosick, N. Y	Apr. 9, 1861	31,973
Mowing-machine	W. Heston	Bedford, Ohio	Feb. 6, 1872	123,399
Mowing-machine	E. and D. Hinchley	Worcester, Mass	June 19, 1866	55,654
Mowing-machine	E. C. Hopping	Madison, N. J	Dec. 23, 1873	145,872
Mowing-machine	C. Howell	Cleveland, Ohio	Mar. 2, 1858	19,504
Mowing-machine	M. G. Hubbard	Syracuse, N. Y	Oct. 11, 1864	44,628
Mowing-machine	M. G. Hubbard	Syracuse, N. Y	Oct. 11, 1864	44,629
Mowing-machine	M. G. Hubbard	Syracuse, N. Y	Nov. 22, 1864	45,158
Mowing-machine	M. G. Hubbard	Syracuse, N. Y	Nov. 29, 1864	45,246
Mowing-machine	O. Hussey	Baltimore, Md	July 5, 1859	24,641
Mowing-machine	B. Illingworth	Le Roy, Minn	Nov. 15, 1870	109,322
Mowing-machine	E. Ingersoll	Farmington, Mich	May 7, 1830	......
Mowing-machine	S. E. and M. P. Jackson	Booneville, N. Y	Dec. 29, 1857	18,975
Mowing-machine	S. E. and M. P. Jackson	Booneville, N. Y	Dec. 29, 1857	18,976
Mowing-machine	J. Jann	New Windsor, Md	Feb. 28, 1865	46,566
Mowing-machine	G. W. Jennings	Boston, Mass	Mar. 12, 1861	31,668
Mowing-machine	G. T. and M. Jerome	Mineola, N. Y	Sept. 28, 1858	21,607
Mowing-machine	S. Johnston	Brockport, N. Y	June 27, 1871	116,315
Mowing-machine	W. G. Kenyon	Wakefield, R. I	Oct. 12, 1869	95,807
Mowing-machine	W. G. Kenyon	Wakefield, R. I	Mar. 12, 1872	124,440
Mowing-machine	W. F. Ketchum	Buffalo, N. Y	May 27, 1856	14,961
Mowing-machine	W. A. Kirby	Auburn, N. Y	Aug. 26, 1873	142,111
Mowing-machine	L. G. Kniffen	Worcester, Mass	Dec. 24, 1861	33,998
Mowing-machine	L. G. Kniffen	Worcester, Mass	Dec. 13, 1864	45,416
Mowing-machine	W. H. Knight	East Machias, Me	Apr. 27, 1869	89,412
Mowing-machine	S. Lamb	New York, N. Y	June 20, 1840	1,645
Mowing-machine	I. Lard	Ashley, Mo	Nov. 20, 1846	4,859
Mowing-machine	H. F. Lovejoy	Nineveh, N. Y	Nov. 7, 1865	50,830
Mowing-machine	C. M. Lufkin	Ackworth, N. H	May 13, 1856	14,874
Mowing-machine	J. J. Mann	Westville, Ind	Mar. 11, 1856	14,404
Mowing-machine	J. S. Manning	Philadelphia, Pa	Jan. 22, 1856	14,138
Mowing-machine	J. P. Manny	Rockford, Ill	Oct. 27, 1857	18,510
Mowing-machine	J. P. Manny	Rockford, Ill	Mar. 25, 1862	34,763
Mowing-machine	C. D. Mansfield	Lynn, Mass	Apr. 6, 1869	88,650
Mowing-machine	H. Marcellus	Amsterdam, N. Y	May 4, 1858	20,164
Mowing-machine	H. Marcellus	Amsterdam, N. Y	Sept. 4, 1860	29,895
Mowing-machine	H. C. and D. C. Markham	Collinsville, N. Y	Mar. 7, 1871	112,365
Mowing-machine	N. F. Mathewson	Barrington, R. I	May 5, 1868	77,634
Mowing-machine	N. F. Mathewson	Barrington, R. I	Apr. 6, 1869	88,653
Mowing-machine	H. Michaux	Hockanum, Conn	Mar. 4, 1873	136,380
Mowing-machine	H. Michaux	Higganum, Conn	Sept. 23, 1873	143,025
Mowing-machine	S. M. Moore	Beloit, Wis	Oct. 28, 1862	36,850
Mowing-machine	A. Palmer	Brockport, N. Y	Dec. 26, 1865	51,746
Mowing-machine	E. R. Pease	Poughkeepsie, N. Y	Mar. 12, 1861	31,692
Mowing-machine	H. Pease	Brockport, N. Y	Jan. 8, 1856	14,078
Mowing-machine	J. G. Perry	South Kingston, R. I	Mar. 21, 1865	46,933
Mowing-machine	J. G. Perry	South Kingston, R. I	Mar. 6, 1866	53,039
Mowing-machine	J. W. Pierce	Milbury, Mass	Mar. 17, 1868	75,573
Mowing-machine	R. G. Pike	Middletown, Conn	Jan. 30, 1866	52,317
Mowing-machine	J. Pine	Troy, N. Y	Oct. 11, 1864	44,655
Mowing-machine	S. Ray and M. R. Shalters	Alliance, Ohio	Apr. 10, 1860	27,869
Mowing-machine	C. Reed	Springfield, Ohio	July 3, 1866	56,093
Mowing-machine	J. H. Rible	Dayton, Ohio	Apr. 2, 1861	31,907
Mowing-machine	E. S. Ritchie	Brookline, Mass	May 13, 1862	35,263
Mowing-machine	S. Rockafellow	Coatsville, Pa	July 3, 1855	13,181
Mowing-machine	H. J. Ruggles	Poultney, Vt	Apr. 28, 1868	77,407
Mowing-machine	F. Russell	Boston, Mass	Feb. 27, 1855	12,463

Index of patents issued from the United States Patent Office from 1790 *to* 1873, *inclusive*—Continued.

Invention.	Inventor.	Residence.	Date.	No.
Mowing-machine	F. Russell	Boston, Mass	Mar. 20, 1855	12, 559
Mowing-machine	F. Russell	South Boston, Mass	Oct. 12, 1858	21, 777
Mowing-machine	G. T. Savery	Newburyport, Mass	Jan. 24, 1871	111, 258
Mowing-machine	W. and T. Schnebly	Hagerstown, Md	Aug. 22, 1833	....
Mowing-machine	W. H. Seymour	Brockport, N. Y	Nov. 23, 1869	97, 126
Mowing-machine	M. R. Shalters and S. Ray	Alliance, Ohio	Aug. 22, 1871	118, 284
Mowing-machine	G. F. Shaw	West Roxbury, Mass	Sept. 6, 1870	107, 108
Mowing-machine	H. F. Shaw	West Roxbury, Mass	Sept. 13, 1870	107, 296
Mowing-machine	J. W. Shipman	Springfield Centre, N. Y	July 24, 1860	29, 325
Mowing-machine	A. Soule	Yarmouth, Me	June 1, 1869	90, 697
Mowing-machine	W. S. Stone	New Philadelphia, Ohio	Aug. 19, 1873	142, 053
Mowing-machine	Z. Swope	Lancaster, Pa	Jan. 3, 1871	110, 799
Mowing-machine	J. Taggart	Roxbury, Mass	Mar. 17, 1857	16, 855
Mowing-machine	J. W. Thompson	Greenfield, Mass	July 15, 1856	15, 354
Mowing-machine	J. B. Tinker	Buffalo, N. Y	Dec. 5, 1865	51, 364
Mowing-machine	A. P. Trask and D. Aldrich	Ellington, N. Y	Oct. 16, 1839	1, 370
Mowing-machine	W. Van Auden	Poughkeepsie, N. Y	Sept. 27, 1864	44, 477
Mowing-machine	W. V. Walker	Dresserville, N. Y	Aug. 5, 1873	141, 611
Mowing-machine	J. B. Wardwell	Methuen, Mass	June 16, 1857	17, 597
Mowing-machine	F. W. Warner	East Haddam, Conn	July 31, 1860	29, 425
Mowing-machine	J. D. L. Watkins and R. Bryson.	Schenectady, N. Y	July 23, 1861	32, 908
Mowing-machine	A. Wemple	Chicago, Ill	Jan. 29, 1867	61, 644
Mowing-machine	R. W. Whitney and F. M. Hardison.	South Berwick, Me	Dec. 5, 1865	51, 396
Mowing-machine	J. D. Wilber	Poughkeepsie, N. Y	Feb. 10, 1863	37, 656
Mowing-machine	J. D. Wilber	Poughkeepsie, N. Y	Apr. 5, 1870	101, 554
Mowing-machine	A. M. Wilson	New York, N. Y	Sept. 3, 1846	4, 734
Mowing-machine	C. Wilson and W. Moore, jr	Yardleyville, Pa	May 29, 1855	12, 980
Mowing-machine	T. Windell	New Albany, Ind	Jan. 11, 1859	22, 608
Mowing-machine	T. H. Witherby	Worcester, Mass	Mar. 29, 1870	101, 338
Mowing-machine	J. E. Wood	Worcester, Mass	Mar. 17, 1863	37, 932
Mowing-machine	W. A. Wood	Hoosick Falls, N. Y	Feb. 22, 1859	23, 057
Mowing-machine	G. W. N. Yost	Corry, Pa	June 9, 1868	78, 712
Mowing-machine	G. W. N. Yost	Corry, Pa	Apr. 15, 1873	137, 817
Mowing-machine and harvester	J. H. Manny	Waddam's Grove, Ill	Sept. 23, 1851	8, 385
Mowing-machine and hay-spreader, Combined	J. G. Perry	Kingston, R. I	Apr. 13, 1869	88, 899
Mowing-machine and hay-tedder, Combined	J. H. Dater	Eagle Mills, N. Y	Feb. 22, 1870	100, 124
Mowing-machine blade	G. Stone	Beloit, Wis	Jan. 8, 1856	14, 070
Mowing-machine cutter	F. Russell	Manchester, N. H	Feb. 14, 1860	27, 159
Mowing-machine-cutter grinder	C. B. Curtis	Jordan, N. Y	Oct. 6, 1868	82, 808
Mowing-machine-cutter grinder	A. French	Philadelphia, Pa	June 2, 1868	78, 448
Mowing-machine-cutter grinder	D. W. Jameson	Warren, Ohio	Oct. 13, 1868	83, 067
Mowing-machine-cutter grinder	D. W. Jameson	Warren, Ohio	Aug. 31, 1869	94, 415
Mowing-machine-cutter grinder	S. W. and J. C. Palmer	Auburn, N. Y	Aug. 4, 1868	80, 761
Mowing-machine-cutter grinder	B. B. Snow and T. J. Dickerson.	Auburn, N. Y	Aug. 4, 1868	80, 571
Mowing-machine-cutter grinder	A. P. Thayer	Syracuse, N. Y	May 26, 1868	78, 401
Mowing-machine-cutter grinder	G. P. Yorke and W. H. Wilson	Westfield, N. Y	June 16, 1868	78, 909
Mowing-machine cutters, Clamp-bar for holding	M. C. Cronk	Auburn, N. Y	Mar. 30, 1869	88, 279
Mowing-machine cutters, Machine for sharpening	G. Sanford	Bergen Point, N. J	Oct. 20, 1868	83, 322
Mowing-machine cutters while being ground, Machine for holding.	D. W. Jameson	Warren, Ohio	Nov. 17, 1868	84, 191
Mowing machine draft-pole	C. W. Cardot	Fredonia, N. Y	May 14, 1867	64, 746
Mowing-machine gearing	J. V. Strait	Litchfield, Ohio	Oct. 19, 1869	95, 950
Mowing machine, Grass, &c	P. Baker	Long Island, N. Y	Feb. 19, 1814	....
Mowing machine, Grass and grain	D. Lewis, jr	Bern, N. Y	Apr. 14, 1838	692
Mowing machine, Grass, grain, &c	I. Wheeler	Salem, N. H	May 30, 1838	754
Mowing-machine guards, Manufacture of	A. Simonds	Fitchburgh, Mass	May 8, 1866	54, 652
Mowing machine, Hand	L. M. Dondwa	Amherst, N. H	Oct. 29, 1861	33, 578
Mowing machine, Hand	H. Fisher	Alliance, Ohio	May 5, 1863	38, 381
Mowing machine, Hand	G. W. Jennings	Boston, Mass	Mar. 28, 1865	47, 022
Mowing machine, Hand	J. M. Spencer	Enfield, Me	Sept. 4, 1860	29, 922
Mowing-machine-knife grinder	M. C. Cronk	Auburn, N. Y	June 16, 1868	78, 863
Mowing-machine-knife grinder	M. W. Knox	Sheridan, N. Y	Jan. 14, 1873	134, 809
Mowing-machine-knife grinder	D. F. Welsh	Nevada, Ohio	Nov. 8, 1870	109, 162
Mowing-machine knives, Machine for grinding	J. A. Thompson	Auburn, N. Y	June 16, 1868	79, 030
Mowing-machine knives while being ground, Clamp for.	H. J. Case	Auburn, N. Y	Aug. 4, 1868	80, 596
Mowing machine, Lawn	J. A. and H. A. House	Bridgeport, Conn	July 25, 1865	48, 947
Mowing-machine thill	O. A. Hillman	Providence, R. I	Dec. 10, 1872	133, 706
Mowing-machine track-cleaner	W. J. Andrews	Columbia, Tenn	July 2, 1872	128, 453
Mowing-machine track-cleaner	B. F. Power	McConnellsville, Ohio	Jan. 17, 1871	111, 083
Mowing-machine track-cleaner	J. M. Davis	Saint Louis, Mo	Apr. 24, 1866	54, 124
Mowing-machine track-clearer	A. Marcellus	Amsterdam, N. Y	Dec. 29, 1857	18, 983
Mowing-machine track-clearer	A. Marcellus	Amsterdam, N. Y	Apr. 13, 1858	19, 936
Mowing-machine track-clearer	A. W. Morse	Eaton, N. Y	Jan. 21, 1862	34, 213
Mowing-machine track-clearer	P. A. Spicer	Marshall, Mich	Dec. 7, 1869	97, 563
Mowing-machine track-clearer	J. Timms	Malta, Ohio	Nov. 28, 1871	121, 300
Mowing-machines, Attaching draft-pole to	C. W. Cardot	Fredonia, N. Y	May 14, 1867	64, 745
Mowing-machines, Cutting-apparatus for	C. Bullock	Jamestown, N. Y	Dec. 8, 1857	18, 800
Mowing-machines, Cutting-apparatus for	M. G. Hubbard	Syracuse, N. Y	Sept. 13, 1864	44, 193
Mowing-machines, Cutting-apparatus for	H. F. Shaw	West Roxbury, Mass	June 14, 1870	104, 364
Mowing-machines, Device for sharpening the cutters of.	W. Tanner	Chicago, Ill	May 25, 1869	90, 407
Mowing-machines, Dividing-shoe for	W. A. Wood	Hoosick Falls, N. Y	June 24, 1856	15, 203
Mowing-machines, Friction-clutch for	G. S. Reynolds	Lebanon, N. H	Apr. 9, 1872	125, 409
Mowing-machines, Machine for grinding cutters of	H. Dodge	Washington Mills, N. Y	July 13, 1869	92, 431
Mowing-machines, Machine for grinding cutters of	M. Fowks	Leeds, N. Y	Jan. 19, 1869	86, 010
Mowing-machines, &c., Machine for grinding cutters of.	W. H. Laubach	Philadelphia, Pa	Feb. 18, 1868	74, 701
Mowing-machines, Machine for grinding cutters of	H. T. Phillips and H. W Leonard.	Auburn, N. Y	Aug. 18, 1868	81, 105
Mowing-machines, Machine for grinding cutters of	A. P. Thayer	Syracuse, N. Y	Nov. 16, 1869	97, 000
Mowing-machines, Machine for grinding cutters of	S. D. Wackman	Auburn, N. Y	June 2, 1868	78, 623
Mowing-machines, Mounting the cutters of	G. Hart	Dillsborough, Ind	Oct. 8, 1850	7, 700
Mucilage	A. Klozcowskie and D. Mindeleff.	Washington, D. C	Mar. 25, 1873	137, 213

Index of patents issued from the United States Patent Office from 1790 *to* 1873, *inclusive*—Continued.

Index of patents issued from the United States Patent Office from 1790 *to* 1873, *inclusive*—Continued.

Invention.	Inventor.	Residence.	Date.	No.
Music, Keyed instrument of	F. Peabody	Salem, Mass	Nov. 10, 1863	40, 573
Music-leaf holder	A. Extein and J. C. Mills	Springfield, Mass	July 4, 1871	116, 577
Music-leaf turner	C. I. Adkins	Pataskala, Ohio	Apr. 2, 1872	125, 113
Music-leaf turner	A. Altenburg and G. J. Lambrix	Buffalo, N. Y	Oct. 10, 1871	119, 810
Music-leaf turner	C. P. Brown	Spring Lake, Mich	May 7, 1872	126, 516
Music-leaf turner	A. W. Bush and M. McComb	Saint Cloud, Minn	Aug. 1, 1871	117, 511
Music-leaf turner	A. B. Clark	Richmond, Ind	Jan. 1, 1867	60, 833
Music-leaf turner	G. C. A. Class	Chicago, Ill	Oct. 17, 1871	120, 038
Music-leaf turner	G. Downing	Weymouth, England	Jan. 28, 1873	135, 322
Music-leaf turner	A. M. Ernsberger	Indianapolis, Ind	Dec. 31, 1872	134, 367
Music-leaf turner	T. Folks	Melrose, N. Y	Dec. 31, 1872	134, 369
Music-leaf turner	J. H. Gerry	Boston, Mass	May 20, 1873	139, 141
Music-leaf turner	J. B. Geyser	Pittsburgh, Pa	May 6, 1873	138, 631
Music-leaf turner	T. Goodman	Providence, R. I	Feb. 4, 1868	73, 970
Music-leaf turner	N. P. Hall	Janesville, Wis	Aug. 12, 1873	141, 644
Music-leaf turner	C. Heyer	Racine, Wis	Sept. 12, 1871	118, 935
Music-leaf turner	G. N. Hopkins	Albion, N. Y	Jan. 9, 1866	51, 999
Music-leaf turner	J. W. Mellor	Philadelphia, Pa	Sept. 20, 1870	107, 519
Music-leaf turner	L. F. Morawetz	Baltimore, Md	Jan. 7, 1873	134, 557
Music-leaf turner	S. Nickelson	Gallatin, Tenn	Aug. 20, 1872	130, 588
Music-leaf turner	A. K. Noyes	Lynn, Mass	Dec. 15, 1868	84, 962
Music-leaf turner	E. B. Robinson	Portland, Me	July 26, 1870	105, 726
Music-leaf turner	G. Robb	McDonald, Pa	Dec. 24, 1872	134, 221
Music-leaf turner	A. Rosenfield	San Francisco, Cal	Apr. 15, 1873	137, 797
Music-leaf turner	W. Weaver	Phœnixville, Pa	Mar. 26, 1872	124, 924
Music-leaf turner	G. W. White	Brooklyn, N. Y	July 8, 1873	140, 664
Music, Machine for registering	H. F. Bond	Hudson, Wis	Nov. 29, 1859	26, 244
Music-notator	S. R. Brooks	Saint Louis, Mo	June 13, 1871	115, 930
Music-note block	C. J. and J. Costello	Kingston and New York, N. Y.	June 28, 1870	104, 833
Music, Notes in	W. C. Phillips	Lunenburgh Court-House, Va.	June 18, 1829	....
Music, Printing	A. Law		May 12, 1802	....
Music-portfolio	J. C. Koch	Brooklyn, N. Y	Jan. 21, 1873	135, 130
Music-portfolio	F. C. Schumann	New York, N. Y	Jan. 7, 1873	134, 613
Music-rack	C. S. Ely	Branford, Conn	Jan. 7, 1873	134, 530
Music-rack	J. R. Hill	Hyde Park, Mass	May 17, 1870	103, 046
Music-rack	A. Schäffer	Dubuque, Iowa	Apr. 20, 1869	89, 084
Music-rack	G. Tefft	Salem, N. Y	Apr. 7, 1868	76, 551
Music-rack	T. Ward	Birmingham, Pa	May 20, 1856	14, 935
Music rack or desk	F. X. Rizy	Monroe, Mich	Apr. 23, 1867	64, 148
Music-registering machine	H. B. Horton	Akron, Ohio	Dec. 18, 1855	13, 946
Music, Scale for instrumental	A. S. Smith	Rochester, N. Y	Oct. 21, 1856	15, 937
Music-sheets, Medium for turning	A. Krafft	Berlin, Prussia	Nov. 8, 1870	109, 023
Music-staff	V. C. Taylor	Des Moines, Iowa	Nov. 26, 1867	71, 550
Music-stand	O. Behr	Saint Joseph, Mo	July 16, 1872	129, 388
Music-stand	L. V. Brown	Salisbury, N. C	Sept. 13, 1870	107, 219
Music-stand	L. V. Brown	Salisbury, N. C	Aug. 6, 1872	130, 186
Music-stand	E. Conley	Cincinnati, Ohio	Nov. 30, 1869	97, 360
Music-stand	J. David	Brooklyn, N. Y	Oct. 17, 1865	50, 160
Music-stand	H. W. Holly	Stamford, Conn	Feb. 6, 1849	6, 094
Music-stand	W. C. James	Fishersville, N. H	Nov. 14, 1871	120, 880
Music-stand	C. Lehnert	West Roxbury, Mass	Sept. 27, 1870	107, 791
Music-stand	A. R. Nettleton	Unionville, Conn	Oct. 19, 1869	96, 026
Music-stand	J. Newton	Hanover, N. H	Dec. 20, 1870	110, 272
Music-stand	O. Schulze	Saint Louis, Mo	July 1, 1873	140, 547
Music-stand	A. W. White	Boston, Mass	Oct. 29, 1872	132, 704
Music-stand clamp	I. H. K. Downes	Charlestown, Mass	Nov. 19, 1872	133, 214
Music-stand, Portable	A. Iske	Lancaster, Pa	June 18, 1872	127, 978
Music stand, Portable	D. M. White	Malden, Mass	May 12, 1868	77, 940
Music-stand standard	J. Thornton	Oswego, N. Y	Feb. 13, 1872	123, 594
Music-stands, Metallic frame for	W. M. Greenwood and B. Roux	Cincinnati, Ohio	Nov. 2, 1869	96, 424
Music, Teaching	J. C. Clime	Philadelphia, Pa	Sept. 24, 1867	69, 179
Music, &c., Temporary binder for	J. C. Clarke	Jersey City, N. J	Dec. 3, 1867	71, 703
Music to printed books, Arranging	M. Elliot	Boscawen, N. H	June 9, 1818	....
Music, Transposing scales for teaching	P. G. Bryan	Louisville, Ky	Nov. 8, 1870	109, 109
Music turner and holder, Sheet	F. J. Herpers and M. M. Sommer	Newark, N. J	Feb. 8, 1870	99, 568
Music, Turning the leaves of	I. M. Ross	Petersburgh, Ind	Nov. 2, 1869	96, 483
Music-type	E. L. Balch	Boston, Mass	June 16, 1868	78, 855
Music-type	U. K. Hill	New York, N. Y	Feb. 7, 1812	....
Music-type by combining printer's type	G. Bruce	New York	Nov. 27, 1830	....
Music, Writing	T. Harrison	Springfield, Ohio	Oct. 26, 1839	1, 383
Musical blackboard	J. M. Thomson and J. Cordley	Adrian, Mich	May 6, 1873	138, 714
Musical box	A. Paillard	St. Croix, Switzerland	Aug. 2, 1870	105, 972
Musical cage for animals	A. Suppernio	Millburn, N. J	Dec. 26, 1871	122, 138
[Musical.] Cornet and other wind instruments	W. C. Fritz	Boston, Mass	July 5, 1870	105, 059
Musical demonstrating-board	W. H. and G. W. Bowlsby	Monroe, Mich	July 12, 1864	43, 473
Musical dial	N. Petersen	Columbus, Miss	Apr. 30, 1867	64, 360
Musical instrument	J. D. Akin	Spartansburgh, Pa	June 1, 1858	20, 397
Musical instrument	T. Atkins and H. Drewer	Cincinnati, Ohio	July 11, 1871	116, 793
Musical instrument	M. Cannon	Salt Lake City, Utah	May 27, 1873	139, 293
Musical instrument	R. W. Carpenter	Brattleborough, Vt	Sept. 11, 1866	57, 861
Musical instrument	M. H. Collins	Chelsea, Mass	July 23, 1872	129, 653
Musical instrument	W. David	Tamaqua, Pa	Oct. 20, 1863	40, 333
Musical instrument	J. and E. Foster	Keene, N. H	Apr. 10, 1866	53, 803
Musical instrument	J. J. Hawkins		Oct. 24, 1800	....
Musical instrument	G. Herrick	Waverly, N. Y	Aug. 24, 1869	94, 108
Musical instrument	C. F. Hill	New York, N. Y	Dec. 21, 1869	98, 163
Musical instrument	U. C. and C. F. Hill	New York, N. Y	June 19, 1847	5, 164
Musical instrument	U. C. and C. F. Hill	Jersey City, N. J., and New York, N. Y	Feb. 9, 1858	19, 296
Musical instrument	U. C. Hill and H. J. Newton	Jersey City, N. J., and New York, N. Y.	Feb. 28, 1860	27, 289
Musical instrument	A. F. Hunt and J. S. Bradish	Warren, Ohio	Jan. 9, 1849	6, 006
Musical instrument	W. Leigh	Bridgeport, Conn	Mar. 25, 1873	137, 219
Musical instrument	L. Marques	New York, N. Y	Mar. 20, 1866	53, 375
Musical instrument	H. T. Merrill	Galena, Ill	June 14, 1859	24, 396
Musical instrument	A. L. Mora	New York, N. Y	July 23, 1872	129, 746

Index of patents issued from the United States Patent Office from 1790 *to* 1873, *inclusive*—Continued.

Invention.	Inventor.	Residence.	Date.	No.
Musical instrument	J. F. Munson	Washington, D. C	May 23, 1865	47, 851
Musical instrument	F. Peabody	Salem, Mass	Oct. 10, 1865	50, 381
Musical instrument	J. W. Prescott	Concord, N. H	Apr. 17, 1849	6, 356
Musical instrument	A. Schoenhut	Philadelphia, Pa	Mar. 25, 1873	137, 243
Musical instrument	D. Schuyler	Buffalo, N. Y	July 3, 1866	56, 106
Musical instrument	K. Smith	Lowell, Mass	July 11, 1871	116, 879
Musical instrument	W. Standing, sr	Duquoin, Ill	July 2, 1872	128, 566
Musical instrument	F. Suter	Williamsburgh, N. Y	June 1, 1869	90, 793
Musical instrument	G. W. Van Dusen	Williamsburgh, N. Y	Dec. 10, 1867	72, 129
Musical instrument	W. Vogel	Norwich, Conn	Nov. 10, 1868	84, 027
Musical instrument	A. White	New Haven, Conn	May 8, 1860	28, 224
Musical instrument	J. and H. W. Whitney	Boston, Mass	Apr. 8, 1873	137, 643
Musical instrument	G. Woods	Cambridge, Mass	Jan. 3, 1865	45, 800
Musical instrument	G. Woods	Cambridge, Mass	Aug. 15, 1865	49, 483
Musical instrument	G. Woods	Cambridge, Mass	Sept. 11, 1866	58, 032
Musical instrument	A. Young	Norfolk, Va	Feb. 20, 1872	123, 969
Musical instrument, Automatic	H. Groves	New York, N. Y	Nov. 4, 1856	16, 009
Musical instrument, Electro-magnetic	L. Wesson	Chillicothe, Ohio	Apr. 4, 1865	47, 144
Musical instrument, Keyed	J. Alley and H. W. Poole	Newburyport and Worcester, Mass.	July 3, 1849	6, 565
Musical instrument, Keyed	H. C. Baudet	Paris, France	Oct. 16, 1866	58, 959
Musical instrument, Keyed	E. Hamlin	Winchester, Mass	July 11, 1871	116, 834
Musical-instrument keys	E. C. Riley	New York	Aug. 22, 1833	....
Musical instrument, Military-band	R. H. Gates	Lancaster, Ohio	June 4, 1872	127, 591
Musical instrument, Military brass	B. F. Quinby	Boston, Mass	Apr. 9, 1872	125, 614
Musical instrument, Reed	C. Austin	Concord, N. H	Apr. 30, 1850	7, 317
Musical instrument, Reed	J. A. Bazin	Canton, Mass	Aug. 2, 1853	9, 892
Musical instrument, Reed	B. T. Blodget and H. B. Horton	Akron, Ohio	June 19, 1849	6, 531
Musical instrument, Reed	J. C. Briggs	Ansonia, Conn	Mar. 19, 1867	63, 606
Musical instrument, Reed	R. Burdett	Chicago, Ill	Nov. 6, 1866	59, 354
Musical instrument, Reed	R. Burditt	Brattleborough, Vt	Jan. 9, 1866	51, 994
Musical instrument, Reed	J. Foster	Keene, N. H	Aug. 29, 1871	118, 445
Musical instrument, Reed	L. K. Fuller and H. K. White	Brattleborough, Vt	Dec. 1, 1868	84, 486
Musical instrument, Reed	W. H. Gerrish	Boston, Mass	Aug. 2, 1870	105, 936
Musical instrument, Reed	F. A. Gleason	Rome, N. Y	June 20, 1854	11, 121
Musical instrument, Reed	E. Hamlin	Boston, Mass	May 27, 1856	14, 955
Musical instrument, Reed	E. P. Needham	New York, N. Y	Apr. 12, 1859	23, 601
Musical instrument, Reed	E. P. Needham	New York, N. Y	June 2, 1868	78, 536
Musical instrument, Reed	I. T. Packard	Chicago, Ill	May 1, 1866	54, 395
Musical instrument, Reed	I. T. Packard	Chicago, Ill	July 26, 1870	105, 718
Musical instrument, Reed	G. S. Shepard	Canaan, N. H	July 31, 1855	13, 365
Musical instrument, Reed	J. P. Sleeper	Worcester, Mass	Oct. 29, 1850	7, 747
Musical instrument, Reed	G. Woods	Boston, Mass	May 8, 1860	28, 230
Musical instrument, Reed and pipe	I. T. Packard	Chicago, Ill	Jan. 15, 1867	61, 241
Musical instrument, Rotary-valve wind	T. D. Paine	Smithfield, R. I	Nov. 14, 1848	5, 919
Musical-instrument sounding-board	C. Bogart	Charlestown, Mass	Dec. 9, 1851	8, 575
Musical-instrument sounding-board	J. H. Schucht	London, England	Dec. 26, 1871	122, 196
Musical instrument, Stringed	G. Schleicher	Mount Vernon, N. Y	Aug. 11, 1868	81, 012
Musical instrument, Stringed	G. L. Wild	Baltimore, Md	Sept. 5, 1854	11, 655
Musical instrument, Wind	C. H. Eisenbrandt	Baltimore, Md	Jan. 26, 1858	19, 187
Musical instrument, Wind	W. Fischer	New York, N. Y	Dec. 22, 1868	85, 222
Musical instrument, Wind	I. Fiske	Worcester, Mass	Apr. 29, 1873	138, 389
Musical instrument, Wind	I. Fiske	Worcester, Mass	Sept. 23, 1873	143, 134
Musical instrument, Wind	G. E. Parker	Newark, N. J	June 17, 1873	140, 069
Musical instrument, Wind	C. J. Van Oeckelen	New York, N. Y	Nov. 23, 1858	22, 139
Musical instruments, Automatic attachment for keyed.	E. D. Bootman	Janesville, Wis	Dec. 18, 1860	30, 955
Musical instruments, Automatic attachment for keyed.	H. Downes	New York, N. Y	May 30, 1871	115, 286
Musical instruments, Bellows for	J. Carhart	Buffalo, N. Y	Dec. 28, 1846	4, 912
Musical instruments, Bellows for	O. Follett	Pittsfield, Mass	Jan. 21, 1868	73, 592
Musical instruments, Bellows for reed	A. W. Wilcox	Providence, R. I	Apr. 28, 1868	77, 428
Musical instruments by means of bellows, Applying wind to.	P. L. and G. Grosh	Lancaster, Pa	Apr. 8, 1833	....
Musical instruments, Crook for	I. Fiske	Worcester, Mass	Nov. 12, 1867	70, 824
Musical instruments, Device for carrying off water from.	G. F. Dallon	Flatbush, N. Y	Apr. 21, 1863	38, 218
Musical instruments, Electro-pneumatic action for.	W. F. and H. Schmoele, jr	Philadelphia, Pa	Dec. 16, 1873	145, 532
Musical instruments, Key-board attachment for	E. Tourjee	Providence, R. I	Nov. 26, 1867	71, 427
Musical instruments, Key-board for	W. D. Edgar	Ottawa, Kans	Sept. 26, 1871	119, 335
Musical instruments, Key-board for	E. Hamlin	Winchester, Mass	Dec. 12, 1871	121, 778
Musical instruments, Key-board for	G. B. Kirkham	New York, N. Y	Jan. 1, 1867	60, 744
Musical instruments, Key-coupling for	J. Berger	Knoxville, Ill	Jan. 7, 1868	73, 072
Musical instruments, Machine for cutting key-boards, &c., for.	D. E. Butler	Chesterfield, Ohio	Feb. 16, 1858	19, 345
Musical instruments, Making pivot-holes in keys of.	S. H. Wilder	Deep River, Conn	Apr. 2, 1872	125, 239
Musical instruments, Material for	H. Huebscher	Klingenthal, Germany	Aug. 20, 1872	130, 641
Musical instruments, Metallic reed for	C. N. Cutter	Worcester, Mass	June 30, 1868	79, 321
Musical instruments, Metallic reed for	C. N. Cutter	Worcester, Mass	June 30, 1868	79, 322
Musical instruments, Metallic reed for	A. H. Hammond	Worcester, Mass	Aug. 27, 1861	33, 143
Musical instruments, Metallic reed for	A. H. Hammond	Worcester, Mass	Feb. 2, 1869	86, 394
Musical instruments, Mute for	J. F. Stratton	New York, N. Y	Dec. 5, 1865	51, 363
Musical instruments, Operating swell in	G. Woods	Boston, Mass	Oct. 21, 1862	36, 756
Musical instruments, Pedal-attachment for	S. Schoenbrum	New York, N. Y	Feb. 27, 1872	124, 163
Musical instruments, Pedal for	H. Haas	New York, N. Y	Oct. 7, 1873	143, 509
Musical instruments, Preparing wood for	W. H. May	Bridgeport, Conn	Dec. 31, 1867	72, 877
Musical instruments, Reed-box for	M. F. Landfeur	Manchester, Conn	May 2, 1854	10, 848
Musical instruments, Reed for	J. C. Briggs	Woodbury, Conn	Oct. 21, 1856	15, 921
Musical instruments, Reed for	R. R. Martino	New York, N. Y	Jan. 19, 1869	86, 022
Musical instruments, Reed-stop for	A. P. Hughes	Philadelphia, Pa	Sept. 29, 1857	18, 284
Musical instruments, Sounding-board bridge for	G. Woods	Cambridgeport, Mass	May 6, 1873	138, 725
Musical instruments, Sounding-plate for	A. Schoenhut	Philadelphia, Pa	Oct. 28, 1873	144, 148
Musical instruments, String for	W. Randle	Florida, N. Y	Apr. 7, 1857	16, 995
Musical instruments, Swell for	T. Loud	Philadelphia, Pa	Aug. 1, 1865	49, 127
Musical instruments, Swell for	H. Woodbury	Buffalo, N. Y	Sept. 17, 1861	33, 322
Musical instruments, Tremolo-motor for	J. R. Lomas	New Haven, Conn	Nov. 28, 1871	121, 390
Musical instruments, Tremolo-regulator for reed	I. T. Packard	Chicago, Ill	Mar. 30, 1869	88, 328

Index of patents issued from the United States Patent Office from 1790 *to* 1873, *inclusive*—Continued.

Invention.	Inventor.	Residence.	Date.	No.
Musical instruments, Tuning-pin for	S. Clarke	New York, N. Y	Mar. 5, 1861	31, 589
Musical instruments, Valve for wind	C. H. Eisenbrandt	Baltimore, Md	July 4, 1854	11, 215
Musical instruments, Valve for wind	G. Hammer	Cincinnati, Ohio	Apr. 3, 1855	12, 628
Musical notation	F. W. Acee	Columbus, Ga	Dec. 26, 1871	122, 096
Musical notation	P. Phillips	Middletown Point, N. J	May 27, 1856	14, 970
Musical notation	P. Phillips	Matawan, N. J	Mar. 17, 1868	75, 572
Musical notation	J. P. Powell	Cincinnati, Ohio	July 22, 1873	141, 013
Musical notation	E. P. Stewart	Cotton Plant, Miss	Apr. 22, 1873	138, 104
Musical notation	E. Van Heeringen	Pickensville, Ala	June 12, 1849	6, 528
Musical picture-case	F. F. Kullrich	Berlin, Prussia	Oct. 19, 1869	95, 915
Musical pitch-pipe	W. H. Clark	Dayton, Ohio	Jan. 23, 1872	122, 879
Musical reed	J. C. Briggs	Woodbury, Conn	May 1, 1860	28, 060
Musical reed	N. B. Jewett	Worcester, Mass	Dec. 31, 1845	4, 339
Musical return-ball	J. Burke	Brooklyn, N. Y	Apr. 7, 1868	76, 396
Musical scale	W. E. Catlin	Wayne Township, Pa	Aug. 27, 1867	68, 287
Musical scale, Revolving	S. D. Tillman	Seneca Falls, N. Y	Nov. 8, 1853	10, 217
Musical scales, Representing	W. B. Billings	Eastport, Me	Sept. 10, 1850	7, 628
Musical staff	J. C. Cline	Philadelphia, Pa	Feb. 13, 1866	52, 534
Musical staff	H. Wright	Akron, Ohio	June 14, 1870	104, 393
Musical tablet	J. Branique	New York, N. Y	Nov. 26, 1867	71, 359
Musical-tone index	W. H. Clarke	Dayton, Ohio	July 2, 1872	128, 591
Musical transposing-board	J. M. Bruner	Roanoke, Ohio	May 30, 1871	115, 428
Musical transposing-index	J. L. Ober	Beverly Farms, Mass	Dec. 17, 1872	134, 001
Musician's hand-exerciser	J. Monestier	Saint Denis, near Paris, France.	Mar. 30, 1858	19, 814
Musket and gun barrels, Altering the caliber of	C. E. Bailey	Springfield, Mass	Dec. 31, 1867	72, 777
Musket-balls, Apparatus for compressing	C. S. Bulkley	New York, N. Y	May 27, 1862	35, 411
Musket-balls, Machine for compressing	P. Naylor	New York, N. Y	Apr. 1, 1862	34, 844
Mustache-guard	M. C. Heptinstall	Enfield, N. C	June 29, 1869	91, 936
Mustache-guard	C. E. Mitchell and M. Moriarty	Bangor, Me	July 23, 1867	66, 985
Mustache-guard	E. J. F. Randolph	New York, N. Y	Feb. 20, 1872	123, 839
Mustache-guard for drinking-vessels	E. W. H. Bass	Quincy, Mass	Mar. 8, 1870	100, 586
Mustache-shield for cups	G. P. Cutler	Lawrence, Mass	June 28, 1870	104, 711
Mustard-box, &c	R. W. Betts	New York, N. Y	May 3, 1864	42, 549
Mustard-pot	T. M. Schleier	Knoxville, Tenn	Feb. 4, 1873	135, 595
Mustard, Preparation of flour of	S. Hopkins	Rahway, N. J	Aug. 19, 1813	....
Mustard, &c., Preparation of flour of	S. Hopkins	Philadelphia, Pa	May 31, 1817	....
Muzzle, Dog	H. Belmer and C. H. S. Schultz	Cincinnati, Ohio	July 23, 1867	67, 016
Muzzle, Dog	C. D. Quillfelat	New York, N. Y	Aug. 27, 1872	130, 846
Muzzle, Dog	H. Kaempf	Newark, N. J	July 6, 1869	92, 315
Muzzle for dogs and other animals	C. F. Schmidt	Williamsburgh, N. Y	Apr. 2, 1861	31, 911
N.				
Nail: *See* Barbed nail. Brass, copper, and composition nail. Button-nail. Capped nail. Carpet-nail. Coffin-nail. Copper sheathing-nail. Cut nail. Galvanized nail. Harness-nail. Horseshoe-nail. Leather-work nail. Picture-nail. Screw-nail. Sheathing-nail. Shoe-nail. Siding-nail. Slating-nail. Tree nail. Trunk-nail. Upholstery-nail. Weather-tiling nail. Weather tiling and siding nail. Wrought nail.				
Nail	W. N. Hall	Mexia, Tex	Jan. 21, 1873	135, 111
Nail	H. A. Harvey	Orange, N. J	July 2, 1867	66, 331
Nail	W. E. Lockwood	Philadelphia, Pa	July 9, 1867	66, 601
Nail	B. T. Nichols	Roselle, N. J	Aug. 12, 1873	141, 810
Nail	M. Otis	East Bridgewater, Mass	Mar. 20, 1834	....
Nail	A. Patterson	Worcester, Mass	Aug. 16, 1870	106, 501
Nail and nail machine	W. Hunt	New York, N. Y	Nov. 12, 1839	1, 407
Nail and peg driver	F. O. Claflin	New York, N. Y	Dec. 21, 1869	98, 030
Nail and screw head	W. H. Nichols and F. D. Strong	East Hampton, Conn	June 11, 1861	32, 527
Nail and spike machine	H. Burden	Troy, N. Y	Dec. 2, 1834	....
Nail and spike, Grooved and flanged	T. W. Harvey	Jamestown, N. Y	Sept. 13, 1833	....
Nail and spike making machine	B. Allison and R. French	....	June 8, 1804	....
Nail and spike making machine	T. Wood	Matteawan, N. Y	Feb. 22, 1870	100, 230
Nail and spike pointing machine	W. Spink	Providence, R. I	July 5, 1859	24, 705
Nail and tack	R. Speer	Passaic, N. J	July 3, 1866	56, 114
Nail and tack cutting and heading machine	J. Reed	Marshfield, Mass	Apr. 22, 1825	....
Nail and tack feeding apparatus	J. Reed	Marshfield, Mass	Apr. 22, 1825	....
Nail and tack heading machine	L. Bolles	New York, N. Y	Sept. 19, 1822	....
Nail and tack machine	D. J. Farmer	Wheeling, W. Va	Aug. 22, 1871	118, 357
Nail and tack making machinery	O. L. Bassett, T. R. Bearse, and W. B. Wilber.	Taunton, Mass	Oct. 9, 1866	58, 577
Nail and tack machine feeder	W. Hunt	New York, N. Y	Oct. 12, 1843	3, 305
Nail and tack plate feeder	D. J. Farmer	Wheeling, W. Va	Apr. 4, 1871	113, 644
Nail-assorting apparatus	J. Coyne	Pittsburgh, Pa	*June 3, 1873*	*139, 548*
Nail-assorting machine	B. Berelander	Boston, Mass	**May 13, 1873**	**138, 783**
Nail-brad-cutting machine, Cylindrical vibrating	E. A. Lester	Baltimore, Md	Dec. 27, 1815	....
Nail-button	E. Kolben	New York, N. Y	Jan. 2, 1872	122, 468
Nail-clincher	J. Koyl	Rockford, Ill	Sept. 3, 1867	68, 567
Nail-clincher	J. Kunz	Napoleon, Ohio	Jan. 7, 1873	134, 680
Nail-cutter, Cylindrical	J. Perkins	Newburyport, Mass	Nov. 1, 1815	....
Nail cutters, Manufacture of barbed	E. Riley	Cleveland, Ohio	Dec. 19, 1871	121, 964
Nail cutting and heading machine	H. Hayes	Philadelphia, Pa	Feb. 24, 1816	....
Nail cutting and heading machine	J. Jelleff	Butternuts, N. Y	June 14, 1838	....
Nail cutting and heading machine	R. C. Lambert	Quincy, Mass	Feb. 27, 1872	124, 067
Nail cutting and heading machine	W. Leslie	....	Nov. 5, 1801	....
Nail cutting and heading machine	W. Miller	New York, N. Y	Sept. 26, 1810	....
Nail cutting and heading machine	T. Morgan	Hempstead, N. Y	Sept. 5, 1821	....
Nail cutting and heading machine	M. Otis	Boston, Mass	Dec. 31, 1817	....
Nail cutting and heading machine	N. Read	....	Jan. 8, 1798	....
Nail cutting and heading machine	J. Reed	Boston, Mass	Feb. 22, 1807	....

Index of patents issued from the United States Patent Office from 1790 *to* 1873, *inclusive*—Continued.

Invention.	Inventor.	Residence.	Date.	No.
Nail cutting and heading machine	J. Reed	Kingston, Mass	Sept. 16, 1810	....
Nail cutting and heading machine	M. and R. Reeve	Philadelphia, Pa	Nov. 14, 1811	....
Nail cutting and heading machine	B. R. Reid	Salem, Mass	Aug. 18, 1810	....
Nail cutting and heading machine	G. Thompson	Philadelphia, Pa	Sept. 15, 1821	....
Nail cutting and heading machine	F. Young	....	Aug. 23, 1800	....
Nail-cutting machine	F. J. Ayers	Roxbury, Mass	July 2, 1842	2, 698
Nail-cutting machine	W. H. Battelle	Youngstown, Ohio	Oct. 6, 1868	82, 679
Nail-cutting machine	E. Bigelow	Baltimore, Md	Apr. 1, 1807	....
Nail-cutting machine	L. Bissel	Hartwick, N. Y	June 27, 1809	....
Nail-cutting machine	E. Bless	Indianapolis, Ind	Jan. 3, 1871	110, 624
Nail-cutting machine	N. Boureau	....	Aug. 21, 1804	....
Nail-cutting machine	J. Coyne	Allegheny City, Pa	Aug. 31, 1869	94, 285
Nail-cutting machine	D. J. Farmer	Wheeling, W. Va	Nov. 1, 1870	108, 895
Nail-cutting machine	F. W. Geyssenhayner	....	Dec. 22, 1804	....
Nail-cutting machine	C. W. Glidden	Lynn, Mass	Feb. 11, 1873	135, 797
Nail-cutting machine	J. C. Gould	Oxford, N. J	Mar. 16, 1869	87, 924
Nail-cutting machine	M. and D. Hain	Hermann, Mo	Dec. 17, 1872	133, 935
Nail-cutting machine	J. Hicks	Hadley, Mass	Jan. 28, 1808	....
Nail-cutting machine	S. P. Hollis and H. B. Chest	Pittsburgh, Pa	Feb. 18, 1873	135, 915
Nail-cutting machine	C. D. Hunt	Fairhaven, Mass	Nov. 17, 1868	84, 189
Nail-cutting machine	C. D. Hunt	Fairhaven, Mass	Jan. 5, 1869	85, 665
Nail-cutting machine	C. D. Hunt	Fairhaven, Mass	Oct. 12, 1869	95, 802
Nail-cutting machine	T. M. Lawrence	Piqua, Ohio	Jan. 2, 1872	122, 324
Nail-cutting machine	H. Lutgen	Marietta, Pa	Sept. 8, 1813	....
Nail-cutting machine	F. B. Miles and J. Lawrence	Philadelphia, Pa	July 16, 1872	129, 156
Nail-cutting machine	R. Moore	Lycoming County, Pa	Jan. 13, 1814	....
Nail-cutting machine	J. G. Peerson	....	Mar. 23, 1794	....
Nail-cutting machine	J. Perkins	....	Jan. 16, 1795	....
Nail-cutting machine	B. R. Reid	Danvers, Mass	Mar. 8, 1810	....
Nail-cutting machine	E. Reynolds	Mansfield, Conn	May 2, 1865	47, 605
Nail-cutting machine	M. J. Rice	Birmingham, England	Feb. 16, 1869	87, 067
Nail-cutting machine	P. Richards	Philadelphia, Pa	Mar. 1, 1870	100, 447
Nail-cutting machine	W. A. Sweet	Syracuse, N. Y	Nov. 3, 1863	40, 514
Nail-cutting machine	M. Thibault	Hull, Canada	Dec. 16, 1873	145, 696
Nail-cutting machine	E. West	....	July 6, 1802	....
Nail-cutting machine	A. Whittemore	....	Nov. 19, 1796	....
Nail-cutting machine	W. Wickersham	Boston, Mass	June 26, 1860	28, 928
Nail-cutting machine, Vertical balance-wheel	E. A. Lester	Herkimer, N. Y	May 8, 1811	....
Nail-cutting machine, Vibrating	E. A. Lester	Herkimer, N. Y	Apr. 2, 1814	....
Nail-cutting machines, Feeder for	C. Isbister	Allegheny, Pa	Dec. 31, 1844	3, 868
Nail-delivering machine	S. S. Putnam	Boston, Mass	Dec. 23, 1873	145, 755
Nail-distributer	A. Morrison	Harrisburgh, Pa	Nov. 11, 1873	144, 407
Nail-distributing machine	C. M. Chase	Boston, Mass	Jan. 14, 1873	134, 852
Nail-distributing machine	A. Knowlton	Boston, Mass	Aug. 22, 1871	118, 250
Nail-drawer	S. Marden	Newton, Mass	Dec. 24, 1867	72, 655
Nail-drawing implement	H. W. Holly	Norwich, Conn	Jan. 14, 1868	73, 330
Nail-drawing instrument	E. P. Whitney	Stamford, Conn	Apr. 28, 1868	77, 427
Nail-drawing tool	D. T. Robinson	Boston, Mass	Nov. 13, 1866	59, 657
Nail-driving instrument	S. P. Carpenter	Milford, Mass	July 13, 1852	9, 112
Nail-driving machine	D. L. Falardo	New York, N. Y	June 4, 1867	65, 360
Nail-driving machine	S. W. Phelps	Sandusky, Ohio	Oct. 25, 1870	108, 624
Nail-driving machine	A. Smith	Perrysburgh, Ohio	Dec. 23, 1873	145, 818
Nail-extracting implement	D. Morris	Bartlett, Ohio	Feb. 11, 1868	74, 401
Nail-extractor	J. B. Breathitt	Cooper County, Mo	June 30, 1868	79, 307
Nail-extractor	G. J. Capewell	Cheshire, Conn	July 16, 1872	129, 210
Nail-extractor	G. J. Capewell	Cheshire, Conn	Oct. 7, 1873	143, 496
Nail-extractor	I. F. Davis	Rockford, Ill	Aug. 10, 1869	93, 522
Nail-extractor	J. Edson	Boston, Mass	Jan. 21, 1868	73, 587
Nail-extractor	E. A. Franklin	Brenham, Tex	May 14, 1872	126, 686
Nail-extractor	J. H. Hogan	Saint Louis, Mo	May 29, 1866	55, 105
Nail-extractor	H. Jeffrey	Saint Charles, Mo	Aug. 6, 1867	67, 552
Nail-extractor	W. Knaus	Otterville, Mo	June 15, 1869	91, 452
Nail-extractor	A. Marden and A. H. Burgess	Philadelphia, Pa	June 25, 1867	66, 032
Nail-extractor	R. McConnell	Lawrenceville, Pa	Nov. 7, 1865	50, 834
Nail-extractor	D. R. Morland and C. Blocher	Broad Ford, Pa	Apr. 15, 1873	137, 945
Nail-extractor	I. A. Pinnell	Boonville, Mo	June 2, 1868	78, 479
Nail-extractor	G. C. Taft	Worcester, Mass	Sept. 6, 1870	107, 121
Nail-extractor	G. C. Taft	Worcester, Mass	Dec. 30, 1873	145, 975
Nail-extractor	J. Tyzick	Saint John, New Brunswick	June 9, 1868	78, 699
Nail-extractor	J. Tyzick and H. W. Eskildson	New York, N. Y	Sept. 29, 1868	82, 563
Nail-extractor	W. G. Ward	Fayette, N. Y	Dec. 13, 1870	110, 176
Nail-forging machine	S. S. Putnam	Boston, Mass	Aug. 17, 1858	21, 213
Nail-forging machine	S. J. Seely	New York, N. Y	Aug. 4, 1857	17, 941
Nail-forging machine, Wrought	H. W. Mather	Deep River, Conn	Apr. 6, 1869	88, 652
Nail-head-plating machine	W. H. Van Gilson	Newark, N. J	Nov. 30, 1858	22, 211
Nail heads, Machine for cutting metal caps for	Z. Walsh	Newark, N. J	Oct. 6, 1857	18, 372
Nail heading and cutting machine	I. Garrettson	....	Nov. 16, 1796	....
Nail-heading machine	E. Bigelow	Massachusetts	Jan. 29, 1817	....
Nail-heading machine	A. Chandler and S. Shepard	Taunton, Mass	Sept. 16, 1808	....
Nail-heading machine	S. Rogers	Plymouth, Mass	Dec. 15, 1807	....
Nail-heading machine	S. Rogers	Bridgewater, Mass	Jan. 26, 1814	....
Nail-heading machine	H. Sedam	Troy, N. Y	Sept. 6, 1815	....
Nail-heading machine	B. S. Walcott	....	Sept. 4, 1802	....
Nail-machine	J. M. Allen	Cambridge, Mass	Nov. 5, 1867	70, 496
Nail-machine	E. Bartholomew and W. Church	Boston, Mass	Aug. 18, 1815	....
Nail-machine	S., sr., and S. Briggs, jr	....	Aug. 2, 1791	....
Nail-machine	W. Caruthers	....	Dec. 13, 1802	....
Nail-machine	F. Davison	Richmond, Va	Apr. 28, 1868	77, 262
Nail-machine	F. Davison	Richmond, Va	Aug. 31, 1869	94, 290
Nail-machine	F. Davison	Richmond, Va	July 21, 1868	80, 150
Nail-machine	D. Dodge	Keesville, N. Y	June 3, 1856	15, 054
Nail-machine	D. Dodge	Keesville, N. Y	Aug. 23, 1859	25, 183
Nail-machine	D. Dodge	Keesville, N. Y	Aug. 30, 1859	25, 309
Nail-machine	J. Elgar	York, Pa	June 21, 1815	....
Nail-machine	N. Elgar	York, Pa	June 19, 1815	....
Nail-machine	J., sr., and J. Elgar	Yorktown, Pa	Sept. 1, 1809	....
Nail-machine	D. J. Farmer	Wheeling, W. Va	May 31, 1870	103, 730
Nail-machine	W. H. Field	Taunton, Mass	May 26, 1863	38, 665

Index of patents issued from the United States Patent Office from 1790 to 1873, inclusive—Continued.

Invention.	Inventor.	Residence.	Date.	No.
Nail-machine	N. Fobes		Aug. 2, 1802	....
Nail-machine	F. A. Gleason	Brooklyn, N. Y	Jan. 5, 1869	85, 522
Nail-machine	S. L. Gould	Skowhegan, Me	Oct. 9, 1866	58, 637
Nail-machine	R. C. Grant	Middleport, Ohio	May 5, 1868	77, 604
Nail-machine	H. Greene and W. J. Jordon	Philadelphia, Pa	Apr. 27, 1858	20, 126
Nail-machine	E. G. Hall	New York, N. Y	May 7, 1861	32, 263
Nail-machine	S. Hart		Jan. 4, 1799	....
Nail-machine	J. Kello and J. Carter	Philadelphia, Pa	June 10, 1817	....
Nail-machine	J. S. King	Raynham, Mass	Oct. 20, 1857	18, 457
Nail-machine	J. B. Kingham	East Bridgewater, Mass	Oct. 11, 1864	44, 637
Nail-machine	J. Krauser	Philadelphia, Pa	Sept. 23, 1826	....
Nail-machine	J. L. Krauser	Reading, Pa	May 18, 1858	20, 312
Nail-machine	J. L. Krauser	Philadelphia, Pa	Aug. 6, 1861	33, 018
Nail-machine	R. Lario	Reading, Pa	May 31, 1864	42, 993
Nail-machine	H. McCluer	Homer, N. Y	Apr. 18, 1817	....
Nail-machine	M. McManus	Brooklyn, N. Y	June 28, 1864	43, 324
Nail-machine	C. H. Merrick	Pittsburgh, Pa	Sept. 12, 1865	49, 906
Nail-machine	A. V. B. Orr	Lancaster, Pa	Oct. 25, 1859	25, 910
Nail-machine	G. Osborn	Lakeville, Mass	July 27, 1869	92, 994
Nail-machine	A. Owen	Buffalo, N. Y	Feb. 12, 1861	31, 397
Nail-machine	A. W. Paull and J. Morgan, jr	Wheeling, W. Va	July 4, 1871	116, 747
Nail-machine	C. Philips, jr	Sackett's Harbor, N. Y	Oct. 21, 1815	....
Nail-machine	A. M. Polsey	Boston, Mass	Nov. 27, 1866	60, 056
Nail-machine	A. M. Polsey	Boston, Mass	Mar. 5, 1867	62, 682
Nail-machine	H. Reese	Baltimore, Md	Feb. 28, 1871	112, 280
Nail-machine	H. Reese	Baltimore, Md	Sept. 5, 1871	118, 744
Nail-machine	M. Reeve	Philadelphia, Pa	June 27, 1816	....
Nail-machine	M. and R. Reeve	Philadelphia, Pa	Sept. 28, 1813	....
Nail-machine	S. G. Reynolds	Worcester, Mass	Jan. 20, 1852	8, 677
Nail-machine	S. G. Reynolds	Bristol, R. I	Sept. 18, 1866	58, 137
Nail-machine	S. G. Reynolds	Bristol, R. I	Apr. 23, 1867	64, 035
Nail-machine	B. Robinson	Boston, Mass	Dec. 17, 1867	72, 230
Nail-machine	M. T. Rounds	Taunton, Mass	Oct. 11, 1864	44, 661
Nail-machine	J. Russell	Brooklyn, N. Y	Aug. 29, 1865	49, 653
Nail-machine	J. Russell	Brooklyn, N. Y	Apr. 9, 1867	63, 655
Nail-machine	G. W. Sargent	New York, N. Y	Mar. 31, 1868	76, 104
Nail-machine	H. Scheuerle	New York, N. Y	May 24, 1870	103, 376
Nail-machine	E. W. Scott and A. M. George	Lowell, Mass., and Nashua, N. H.	June 9, 1857	17, 523
Nail-machine	H. W. Taylor	Birmingham, Pa	July 6, 1858	20, 829
Nail-machine	R. Ward	Alder Creek, N. Y	May 22, 1866	54, 981
Nail-machine	W. Waring and C. Philips, jr	Jefferson, N. Y	Oct. 21, 1815	....
Nail-machine	W. Wickersham	Boston, Mass	Feb. 6, 1866	52, 479
Nail-machine	G. B. Wiggin and J. W. Hoard	Providence, R. I	Oct. 7, 1862	36, 629
Nail-machine	P. A. Wilbur	Newcastle, Pa	Oct. 14, 1856	15, 910
Nail-machine	A. P. Winslow	Cleveland, Ohio	Oct. 11, 1870	108, 225
Nail-machine	J. Wooton	Boonton, N. J	Aug. 29, 1854	11, 628
Nail-machine, Compound-lever and parallel-shaft	R. Turner	Boston, Mass	Nov. 25, 1811	....
Nail machine, Cut	T. H. Barlow	Lexington, Ky	June 20, 1854	11, 114
Nail machine, Cut	J. Berry	New York, N. Y	Feb. 14, 1860	27, 096
Nail machine, Cut	F. Davison	Liberty, Va	Oct. 31, 1871	120, 501
Nail machine, Cut	D. J. Farmer	Wheeling, W. Va	Aug. 22, 1871	118, 355
Nail machine, Cut	D. J. Farmer	Wheeling, W. Va	Aug. 22, 1871	118, 356
Nail machine, Cut	S. K. Jones and G. H. Snow	New Haven, Conn	May 11, 1869	90, 003
Nail machine, Cut	J. B. Kingham	Dorchester, Mass	May 18, 1869	90, 108
Nail machine, Cut	G. C. Goodhaus	Jamestown, Ohio	Mar. 16, 1858	19, 631
Nail machine, Cut	A. W. Paull and J. Morgan, jr	Wheeling, W. Va	June 20, 1871	116, 091
Nail machine, Cut	L. Richards	Providence, R. I	Dec. 21, 1869	98, 109
Nail machine, Cut	J. P. Sherwood	Fort Edward, N. Y	Dec. 19, 1854	12, 103
Nail machine, Cut	J. E. Sweet and J. B. Elliott	Syracuse and New York, N. Y.	Dec. 22, 1868	85, 145
Nail machine, Cylindrical wrought	C. I. Richards	New York, N. Y	Oct. 17, 1848	5, 853
Nail-machine, Eccentric-roller	J. Fairbanks	Boston, Mass	Aug. 14, 1810	....
Nail-machine feeder	H. B. Landers	Williamsburgh, N. Y	Apr. 4, 1871	113, 310
Nail-machine-feeding apparatus	J. Nolan	Oxford, N. J	Apr. 5, 1870	101, 496
Nail-machine-feeding apparatus	U. S. Wolff	Burrel Township, Pa	Apr. 19, 1870	102, 073
Nail-machine-feeding machinery	T. Odiorne	Malden, Mass	Apr. 3, 1829	....
Nail-machine-grinding die	G. B. Wiggin and J. W. Hoard	Providence, R. I	Feb. 10, 1863	37, 657
Nail-machine lubricator	L. A. Dodge	Keeseville, N. Y	June 30, 1868	79, 326
Nail-machine, Self-feeding	T. Morgan	Paterson, N. J	June 17, 1824	....
Nail-machine toggle-pin	W. H. Ridley	Pittsburgh, Pa	May 16, 1871	114, 857
Nail machine, Wrought	D. Armstrong	Chicago, Ill	May 30, 1871	115, 267
Nail machine, Wrought	C. Clareni	New York, N. Y	July 5, 1859	24, 616
Nail machine, Wrought	J. Nichols	Rhode Island	July 11, 1807	....
Nail machine, Wrought	L. Norcross	Livermore, Me	May 18, 1824	....
Nail machine, Wrought	A. V. B. Orr and G. Bautz	Frederick, Md	Dec. 7, 1858	22, 238
Nail machine, Wrought	H. Reese	Baltimore, Md	Nov. 26, 1872	133, 486
Nail machine, Wrought	J. P. Sherwood	Fort Edward, N. Y	May 10, 1853	9, 712
Nail machine, Wrought	T. Somerby	Wells, Me	Jan. 22, 1840	1, 477
Nail machine, Wrought	J. Taggart	Boston, Mass	Apr. 6, 1869	88, 753
Nail-machines, Grinding the upper cutters of	G. B. Wiggin and J. W. Hoard	Providence, R. I	May 5, 1863	38, 439
Nail-machines, Nippers for	S. Chubbuck	Wareham, Mass	Jan. 13, 1835	....
Nail-machines, Swedge for forming cutters for	E. Bless	Indianapolis, Ind	June 6, 1871	115, 563
Nail-machinery	T. Fowler	Seymour, Conn	May 21, 1867	64, 963
Nail-machinery	J. Holcomb	Brandon, Vt	July 14, 1846	4, 634
Nail machinery, Wrought	D. Dodge	Keeseville, N. Y	June 22, 1852	9, 051
Nail machinery, Wrought	W. M. Gooding	Newark, N. J	Mar. 28, 1848	5, 489
Nail machinery, Wrought	A. W. Gray	Middletown, Vt	May 16, 1846	4, 525
Nail-plate-cutting machine	T. Searls	Pottstown, Pa	Dec. 3, 1872	133, 673
Nail-plate feeder	P. S. Bradford	Bridgewater, Mass	May 17, 1864	42, 742
Nail-plate feeder	W. Diehl	Norristown, Pa	Apr. 10, 1849	6, 291
Nail-plate feeder	D. Drawbaugh	Eberle's Mills, Pa	Dec. 12, 1865	51, 435
Nail-plate feeder	D. Drawbaugh	Eberle's Mill, Pa	Nov. 19, 1867	71, 148
Nail-plate feeder	J. Ferguson and J. Turner	Bridgewater, Mass	Aug. 22, 1871	118, 224
Nail-plate feeder	J. S. Fisk	Youngstown, Ohio	June 30, 1863	39, 105
Nail-plate feeder	J. C. Gould	Boonton, N. J	May 12, 1857	17, 273
Nail-plate feeder	A. Heddaeus	Pittsburgh, Pa	Aug. 12, 1856	15, 515
Nail-plate feeder	J. W. Hoard and T. A. Searle	Providence, R. I	Aug. 17, 1858	21, 198

Index of patents issued from the United States Patent Office from 1790 *to* 1873, *inclusive*—Continued.

Invention.	Inventor.	Residence.	Date.	No.
Nail-plate feeder	J. W. Hoard and T. A. Searle	Providence, R. I	Feb. 28, 1860	27, 289
Nail-plate feeder	C. D. Hunt	Fairhaven, Mass	Apr. 10, 1866	53, 924
Nail-plate feeder	C. D. Hunt	Fairhaven, Mass	Feb. 4, 1868	74, 094
Nail-plate feeder	J. Iler and W. Fitzpatrick	Troy, N. Y	June 13, 1854	11, 053
Nail-plate feeder	C. Isbister	Allegheny, Pa	Jan. 27, 1852	8, 687
Nail-plate feeder	E. B. Lake	Bridgeport, N. J	Nov. 26, 1867	71, 521
Nail-plate feeder	J. McNeely	Pittsburgh, Pa	Sept. 23, 1873	143, 022
Nail-plate feeder	S. K. Paden	Pulaski, Pa	Apr. 8, 1873	137, 710
Nail-plate feeder	H. B. and E. C. Perkins	Oxford, N. J	Jan. 30, 1872	123, 290
Nail-plate feeder	J. Poulson	Frenchtown, N. J	Dec. 16, 1873	145, 681
Nail-plate feeder	J. Newell	Lowell, Mass	Dec. 20, 1859	26, 513
Nail-plate feeder	D. Reed	Birmingham, Conn	Dec. 29, 1868	85, 478
Nail-plate feeder	J. R. Richardson	Newcastle, Pa	Apr. 13, 1869	88, 809
Nail-plate feeder	W. Riley, jr	Reading, Pa	Aug. 14, 1860	29, 613
Nail-plate feeder	T. A. Searle	Providence, R. I	June 19, 1866	55, 722
Nail-plate feeder	W. Sherman and J. Russell	New York and Brooklyn, N. Y.	Sept. 1, 1868	81, 830
Nail-plate feeder	J. P. Sherwood	Fort Edward, N. Y	Aug. 9, 1859	25, 051
Nail-plate feeder	J. H. Swett	Pittsburgh, Pa	Aug. 17, 1858	21, 222
Nail-plate feeder	J. L. Wiggin	South New Market, N. H	June 19, 1866	55, 755
Nail-plate feeder and turner	M. Otis	East Bridgewater, Mass	Dec. 3, 1850	7, 812
Nail-plate-feeding machine	F. J. Ayers	St. John, New Brunswick	Nov. 26, 1850	7, 792
Nail-plate-feeding machine	J. C. Gould	Oxford, N. J	Sept. 26, 1871	119, 228
Nail-plate-feeding machine	J. C. Gould	Oxford, N. J	Oct. 24, 1871	120, 190
Nail-plate-feeding machine	F. Hain	Gasconade County, Mo	Dec. 17, 1867	72, 392
Nail-plate-feeding machine	L. Richards	Providence, R. I	Sept. 10, 1872	131, 181
Nail-plate-feeding machine	J. P. Sherwood	Fort Edward, N. Y	Mar. 18, 1856	14, 474
Nail-plate-feeding machine	W. Wallick	Philadelphia, Pa	June 25, 1867	66, 192
Nail-plate holder	W. H. Battelle	Newcastle, Pa	July 14, 1857	17, 778
Nail-plate holder	J. C. Rhodes	South Abington, Mass	Nov. 19, 1867	71, 064
Nail-plate nippers	G. B. Johnson	Wheeling, W. Va	Apr. 2, 1872	125, 196
Nail-plates and sheet-strips, Process of manufacturing.	J. Trowen, E. Hemmings, and J. Sheldon.	Niles, Ohio	June 20, 1871	116, 046
Nail-plates, Feeding	F. E. Boyd	Boston, Mass	Mar. 17, 1868	75, 620
Nail-plates, Feeding	A. L. Reed	Pittsburgh, Pa	July 24, 1846	4, 657
Nail-plates, Feeding	P. A. Wilbur	Newcastle, Pa	Oct. 21, 1856	15, 938
Nail-plates, Mechanism for feeding	R. C. Grant	Middleport, Ohio	Feb. 25, 1873	136, 235
Nail-plates, Rolling	O. H. Dewey, J. Dudley, and F. Lindeman.	Wheeling, W. Va	Mar. 28, 1871	113, 029
Nail-plates while cutting, Vibrating	J. Bernard	Troy, N. Y	Apr. 16, 1811	....
Nail-separating apparatus	A. Morrison	Harrisburgh, Pa	Sept. 30, 1873	143, 246
Nail-separating device	H. B. Chess	Pittsburgh, Pa	Dec. 23, 1873	145, 720
Nail-separating machine	B. Bevelander	Boston, Mass	Jan. 28, 1873	135, 312
Nail slitting and heading machine	M. Garber	....	Apr. 17, 1804	....
Nail, spike, &c	W. Osgood and E. Hunt	Troy, N. Y	Mar. 14, 1834	....
Nail, spike, tack, horseshoe and nail, screw-bolt, &c., Wrought-iron.	W. and T. Schnebly	Hagerstown, Md	Mar. 3, 1834	....
Nails and brads, Machine for cutting	W. Hunt	New York, N. Y	Nov. 13, 1840	1, 853
Nails and brads, Machine for cutting	I. Kimball	New Hampshire	May 1, 1805	....
Nails and brads, Machine for cutting	P. Zacharie	....	May 4, 1796	....
Nails and making wrought nails, Machine for cutting and heading.	M. and R. Reeve	Burlington, N. J	Feb. 3, 1812	....
Nails and making wrought nails, Machine for cutting and heading.	M. and R. Reeve	Philadelphia, Pa	Apr. 16, 1813	....
Nails by rolling, Method of making	E. H. Collier	Scituate, Mass	June 18, 1850	7, 437
Nails, Casting iron, brass, and composition	J. Walworth	New York	Nov. 22, 1809	....
Nails, Claw for drawing	J. S. Allen and O. Gillmore	Norwich, Conn	May 14, 1867	64, 733
Nails, Covering heads of	W. H. Van Gieson	Waterbury, Conn	Dec. 29, 1857	18, 007
Nails, Cutting and heading	G. Chandlee	....	Dec. 12, 1796	....
Nails, Cutting and heading	N. and D. Chickering	Dover, Mass	Mar. 9, 1812	....
Nails, Cutting and heading	N. Elgar	Yorktown, Pa	July 25, 1815	....
Nails, Cutting and heading	J. Kersey	....	Feb. 24, 1797	....
Nails, Cutting and heading	E. Mason	Milton, Vt	June 6, 1832	....
Nails, Cutting and heading	J. Nevill	....	Aug. 12, 1797	....
Nails, Cutting and heading	T. Newhall, jr	Essex County, Mass	Sept. 7, 1816	....
Nails, Cutting and heading	I. Northrop	Oneida County, N. Y	Jan. 7, 1817	....
Nails, Cutting and heading	J. Reed	Kingston, Mass	Aug. 14, 1811	....
Nails, Cutting and heading	J. Reed	Hanover, Mass	Dec. 16, 1814	....
Nails, Cutting and heading	M. Reeve	Philadelphia, Pa	June 26, 1816	....
Nails, Cutting and heading	M. and R. Reeve	Burlington, N. J	Oct. 31, 1812	....
Nails, Cutting and heading	S. Rogers	Bridgewater, Mass	Oct. 21, 1814	....
Nails, Cutting and heading	F. A. Russel	Georgetown, D. C	Mar. 14, 1816	....
Nails, Cutting and heading	J. Spence	....	Feb. 16, 1797	....
Nails, Cutting and heading	J. White	Philadelphia, Pa	Jan. 11, 1812	....
Nails, Cutting and heading	S. Willard	Hudson, N. Y	Nov. 30, 1807	....
Nails, Cutting, griping, and heading	J. Reed	Hanover, Mass	Oct. 22, 1814	....
Nails from felt, Process of separating	W. Jones	New York, N. Y	Aug. 9, 1870	106, 171
Nails, Heading	J. Byington	....	Dec. 23, 1796	....
Nails, Heading	J. Frost	....	Dec. 23, 1796	....
Nails, Heading and cutting	E. West	....	July 6, 1802	....
Nails, Horizontal wheel for cutting and heading	J. Reed	Kingston, Mass	Apr. 15, 1809	....
Nails, Machine for cutting off	M. Gormley	Wilna, N. Y	July 21, 1868	80, 068
Nails, Machine for making and heading	M. Garber	....	Feb. 20, 1801	....
Nails, Machine for selecting and delivering	A. Knowlton	Boston, Mass	Jan. 28, 1873	135, 345
Nails, Machine for selecting and delivering	A. Knowlton	Boston, Mass	Feb. 18, 1873	136, 074
Nails, Making	J. Byington	....	Jan. 15, 1796	....
Nails, Making	J. M. Cooper	Philadelphia, Pa	Mar. 30, 1869	88, 454
Nails, Making	L. Fling	....	Dec. 19, 1797	....
Nails, Making	J. Perkins	....	Feb. 14, 1799	....
Nails, Manufacturing	J. Bigelow	....	Nov. 19, 1796	....
Nails, Manufacturing	J. Brest and E. Gilmore	Roxbury, Mass	June 1, 1830	....
Nails, Manufacturing	J. Hearsey	Wareham, Mass	May 13, 1830	....
Nails, Manufacturing	J. L. Krauser	Reading, Pa	July 27, 1858	21, 005
Nails, Manufacturing	F. Palmer	Buffalo, N. Y	Oct. 25, 1832	....
Nails, Manufacturing	F. Palmer	Buffalo, N. Y	Aug. 7, 1834	....
Nails, Manufacturing	T. Perkins	....	Feb. 7, 1794	....
Nails, Method of rolling iron for	J. Reed	....	July 15, 1802	....
Nails, Mode of pointing	H. A. Wills	Vergennes, Vt	Dec. 6, 1870	109, 987

Index of patents issued from the United States Patent Office from 1790 *to* 1873, *inclusive*—Continued.

Invention.	Inventor.	Residence.	Date.	No.
Nails milled out of heated rods	J. Reed		June 9, 1801	
Nails or brads, Machine for cutting	J. P. Sawin and J. Skinner	Roxbury, Mass	Feb. 8, 1811	
Nails, screws, &c., Attaching ornamental head to	T. C. Richards	New York, N. Y	Dec. 31, 1867	72, 905
Nails, Tool for clinching	D. J. Hendrickson	Otego, N. Y	Feb. 16, 1858	19, 364
Nailing and pegging machine	J. F. Sargent	Boston, Mass	Apr. 21, 1868	77, 104
Nailing-machine	L. R. Blake and A. S. Libby	Boston and Lawrence, Mass.	Mar. 31, 1868	76, 150
Nailing or pegging machine	J. B. Crosby	Boston, Mass	Mar. 2, 1869	87, 473
Name-plate and letter-chute, Combination of	A. M. Damon and S. G. Lyford	Lowell, Mass	Dec. 22, 1868	85, 218
Name-plate and letter-slide, Combined	J. M. Coombs	Boston, Mass	Mar. 26, 1867	63, 141
Naphtha storing and discharging apparatus	T. Forstall	New Orleans, La	June 27, 1871	116, 423
Napkin-holder	W. L. Dewey	Bridgeport, Conn	June 11, 1867	65, 653
Napkin holder and clasp, Child's table	D. Essex	New York, N. Y	Feb. 25, 1873	136, 146
Napkin-ring	C. E. Green and J. H. Bodwell	Newark, N. J	Aug. 20, 1872	130, 576
Napkin-ring and salt-cup combined	G. Pine	Trenton, N. J	Mar. 24, 1868	75, 968
Napkin, Table	A. L. Munson	New Haven, Conn	Mar. 27, 1866	53, 468
Napkins to the person, Device for attaching	P. H. Raiford	Houston, Tex	May 23, 1871	115, 238
Napping-machine	C. P. Ladd	Bloomfield, N. J	July 30, 1872	129, 895
Napping-machine	S. Marsh	Jericho, Vt	Jan. 11, 1836	
Napping-machine, Metallic	R. Daniels	Woodstock, Vt	June 26, 1835	
Nasal irrigator	M. Mattson	New York, N. Y	Mar. 10, 1868	75, 285
Nasal plug, Elastic	J. J. Essex	Newport, R. I	Sept. 13, 1870	107, 238
Nautical alarm	S. G. Cabell	Quincy, Ill	June 20, 1871	116, 152
Nautical alarm	J. F. Haskins	Fitchburgh, Mass	Dec. 13, 1870	110, 036
Nautical alarm	E. L. Seymour	New York, N. Y	Mar. 3, 1857	16, 776
Nautical ventilator	B. Wynkoop		June 19, 1795	
Naval architecture	J. Thomas	Washington, D. C	Apr. 14, 1825	
Naval architecture	B. F. Wells	Georgetown, D. C	Oct. 18, 1859	25, 857
Navigable mercantile house	R. Haskell	Geneva, N. Y	July 7, 1821	
Navigating and propelling vessels by action of wind and waves.	J. A. Etzler	Philadelphia, Pa	Apr. 1, 1842	2, 533
Navigating-machine	J. J. Giraud	Baltimore, Md	Jan. 31, 1827	
Navigating rivers and shoal water	M. Williams	Albany, N. Y	Feb. 24, 1809	
Navigation	W. W. Van Loan	New York	Jan. 15, 1831	
Navigation, Chute for river	I. A. Putman	Mexico, Me	Apr. 28, 1868	77, 212
Navigation, Internal	J. Renwick	New York	Nov. 7, 1823	
Navigation, Machine for the improvement of	A. Hunn		Mar. 24, 1804	
Navigation, Machine to be used in	A. Heilbroun	New York	Mar. 16, 1829	
Navigator's bearing-indicator	J. D. Leach	Penobscot, Me	July 6, 1869	92, 197
Navigator's bearing-indicator	J. D. Leach	Penobscot, Me	Apr. 26, 1870	102, 281
Neck-scarf and collar supporter	E. A. Bushee	East Boston, Mass	Feb. 9, 1864	41, 568
Neck-scarf, Reversible	A. S. Saroni	New York, N. Y	Dec. 23, 1862	37, 247
Neck-stock	C. Fanshaw	New York	Oct. 3, 1846	4, 788
Neck-stock	W. Watson	Lowell, Ill	June 12, 1860	28, 705
Neck-stock and cravat-stiffener, Metallic frame for	J. Johnson	New York	Dec. 31, 1838	1, 058
Neck-tie	M. Altman	Baltimore, Md	Aug. 30, 1870	106, 761
Neck-tie	J. Bachelder	Norwich, Conn	June 15, 1869	91, 405
Neck-tie	J. Bachelder	Norwich, Conn	Feb. 8, 1870	99, 619
Neck-tie	H. Bendix and J. H. Fleisch	New York, N. Y	Mar. 13, 1866	53, 102
Neck-tie	J. R. Cluxton	Russellville, Ohio	Mar. 20, 1866	53, 272
Neck-tie	A. M. Dexter	Philadelphia, Pa	Dec. 5, 1865	51, 290
Neck-tie	F. Field	Troy, N. Y	Sept. 21, 1869	94, 951
Neck-tie	H. G. Fisk and T. J. Flagg	New York, N. Y	Jan. 28, 1868	73, 790
Neck-tie	F. D. Ford	New Bedford, Mass	Oct. 29, 1867	70, 331
Neck-tie	J. B. Gardiner	Springfield, Mass	Feb. 13, 1866	52, 557
Neck-tie	L. A. Grill	New York, N. Y	Oct. 11, 1870	108, 133
Neck-tie	I. W. Hamlet	Nashua, N. H	Sept. 18, 1866	58, 168
Neck-tie	W. H. Hart, jr	Philadelphia, Pa	May 8, 1866	54, 536
Neck-tie	W. H. Hart, jr	Philadelphia, Pa	Dec. 10, 1867	72, 034
Neck-tie	J. G. Hitchcock	New York, N. Y	Oct. 4, 1870	107, 907
Neck-tie	A. Hoffstadt	New York, N. Y	Oct. 31, 1871	120, 437
Neck-tie	A. Johnson	Brooklyn, N. Y	Sept. 22, 1868	82, 323
Neck-tie	G. A. Keene	Boston, Mass	Mar. 26, 1867	63, 164
Neck-tie	G. Kennedy	Philadelphia, Pa	Aug. 10, 1869	93, 624
Neck-tie	O. Kueppers	Union Hill, N. J	Nov. 26, 1872	133, 457
Neck-tie	G. Lane	New York, N. Y	Feb. 15, 1870	99, 918
Neck-tie	W. A. Laverty	Philadelphia, Pa	Feb. 19, 1867	62, 210
Neck-tie	W. A. Laverty	Philadelphia, Pa	Oct. 1, 1872	131, 884
Neck-tie	J. Silberman and G. Unger	New York, N. Y	Feb. 11, 1868	74, 436
Neck-tie	J. H. Newman	Boston, Mass	July 10, 1866	56, 253
Neck-tie	R. R. Parker	Indianapolis, Ind	Apr. 22, 1873	138, 097
Neck-tie	J. K. P. Pine	Troy, N. Y	Sept. 4, 1866	57, 762
Neck-tie	P. F. Smith	New York, N. Y	Jan. 29, 1861	31, 265
Neck-tie	W. S. Smoot	Washington, D. C	Dec. 15, 1868	85, 036
Neck-tie	G. B. Taylor	New York, N. Y	Mar. 13, 1866	53, 197
Neck-tie	E. Thomas	Philadelphia, Pa	May 19, 1868	78, 027
Neck-tie	E. C. Townsend	New York, N. Y	Nov. 19, 1861	33, 766
Neck-tie	H. P. Wetmore and J. G. Hitchcock.	Elizabeth, N. J., and New York, N. Y.	Oct. 11, 1870	108, 309
Neck-tie	J. R. Wilber and O. W. Peirce	Providence, R. I	Apr. 18, 1871	113, 957
Neck-tie	T. S. Wiles	Troy, N. Y	Mar. 21, 1865	46, 962
Neck-tie and bow	E. Woodward	Brooklyn, N. Y	July 20, 1869	92, 921
Neck-tie and collar combined	J. Varley	Hudson, N. J	Dec. 21, 1869	98, 129
Neck-tie and watch-guard combined	T. J. Flagg	New York, N. Y	June 23, 1868	79, 063
Neck-tie clasp	W. S. Barnes	Watertown, N. Y	June 27, 1865	48, 356
Neck-tie clasp	C. M. Hyatt	Albany, N. Y	Sept. 4, 1866	57, 821
Neck-tie, collar, and bosom combined	G. V. Woods	Belchertown, Mass	Oct. 2, 1866	58, 561
Neck-tie fastener	W. B. Dean	New York, N. Y	Oct. 8, 1872	131, 999
Neck-tie fastener	J. A. Hard	Lawrence, Kans	July 1, 1873	140, 412
Neck-tie fastener	E. Slde	Brooklyn, N. Y	Nov. 7, 1871	120, 600
Neck-tie fastener	P. S. White	Providence, R. I	Nov. 30, 1869	97, 331
Neck-tie fastening	J. Bachelder	Norwich, Conn	June 15, 1869	91, 406
Neck-tie fastening	J. Bachelder	Norwich, Conn	Mar. 1, 1870	100, 250
Neck-tie fastening	G. W. Bishop	Baltimore, Md	Nov. 8, 1870	109, 106
Neck-tie fastening	W. J. Cowing	Washington, D. C	Dec. 27, 1870	110, 439
Neck-tie fastening	R. A. Goodyear	New Haven, Conn	Jan. 14, 1868	73, 244
Neck-tie fastening	H. M. Heineman	Williamsburgh, N. Y	Oct. 2, 1866	58, 414
Neck-tie fastening	H. M. Heineman	San Francisco, Cal	Dec. 7, 1869	97, 637

Index of patents issued from the United States Patent Office from 1790 to 1873, inclusive—Continued.

Invention.	Inventor.	Residence.	Date.	No.
Neck-tie fastening	J. G. Hitchcock	New York, N. Y	June 26, 1866	55, 859
Neck-tie fastening	A. Komp	New York, N. Y	Mar. 28, 1871	113, 177
Neck-tie fastening	P. G. Shirts	Highland Falls, N. Y	Sept. 20, 1870	107, 634
Neck-tie fastening	W. J. Terry	Walla Walla, Wash	June 11, 1867	65, 776
Neck-tie fastening	D. H. Tierney	Waterbury, Conn	Dec. 15, 1868	84, 974
Neck-tie fastening	W. A. Wicks	Baltimore, Md	Sept. 27, 1870	107, 744
Neck-tie fastening	C. Winterbottom	Philadelphia, Pa	May 14, 1872	126, 766
Neck-tie fastening	E. Woodward	Brooklyn, N. Y	July 7, 1868	79, 796
Neck-tie frame	S. W. Roof	New York, N. Y	May 12, 1868	77, 765
Neck-tie frame	P. Welch	New York, N. Y	May 12, 1868	77, 788
Neck-tie holder	A. F. Bixby	Decatur, Ill	Apr. 2, 1872	125, 116
Neck-tie holder	F. H. Brown	Stamford, Conn	Feb. 11, 1868	74, 294
Neck-tie holder	J. A. Eshleman	Philadelphia, Pa	Jan. 31, 1865	46, 096
Neck-tie holder	J. A. Eshleman	Philadelphia, Pa	July 10, 1866	56, 196
Neck-tie holder	W. H. Hart, jr	Philadelphia, Pa	Apr. 26, 1870	102, 259
Neck-tie holder	W. H. Hart, jr., and H. H. Thayer.	Philadelphia, Pa	July 5, 1870	105, 072
Neck-tie holder	J. Hayden and W. H. Hart, jr.	Philadelphia, Pa	Oct. 3, 1871	119, 608
Neck-tie holder	J. O. Schoener	Saint Louis, Mo	Feb. 27, 1872	124, 164
Neck-tie holder	H. H. Thayer	Philadelphia, Pa	Sept. 19, 1871	119, 197
Neck-tie, India-rubber	W. H. Halsey	Hoboken, N. J	Sept. 25, 1866	58, 344
Neck-tie or scarf, Reversible	L. Kahn	New York, N. Y	Apr. 2, 1872	125, 305
Neck-tie retainer	J. B. Bishop	New Haven, Conn	Jan. 11, 1870	98, 660
Neck-tie retainer	S. A. Fite	Philadelphia, Pa	July 13, 1869	92, 522
Neck-tie retainer	E. B. Gibbud	Waterbury, Conn	Feb. 28, 1871	112, 237
Neck-tie retainer	W. J. Ketcham	Washington, D. C	Apr. 7, 1868	76, 468
Neck-tie retainer	H. Laurence	New Orleans, La	Nov. 8, 1870	109, 133
Neck-tie retainer	H. Laurence	New Orleans, La	Oct. 17, 1871	120, 077
Neck-tie retainer	C. P. Lawton	Webster, Mass	Apr. 18, 1871	113, 780
Neck-tie retainer	J. McGee	Lancaster, N. H	July 18, 1871	117, 189
Neck-tie retainer	J. C. Merritt	New York, N. Y	Mar. 5, 1867	62, 665
Neck-tie retainer	P. C. Moulton	New Haven, Conn	June 8, 1869	91, 153
Neck-tie retainer	P. C. Moulton	New Haven, Conn	Oct. 12, 1869	95, 713
Neck-tie retainer	J. Varley	Hudson City, N. J	Mar. 23, 1869	88, 239
Neck-tie supporter	B. F. Bean	Lynn, Mass	June 7, 1864	43, 068
Neck-tie supporter	S. Bruhl	New York, N. Y	Apr. 27, 1869	89, 382
Neck-tie supporter	G. R. Eager	East Boston, Mass	Apr. 7, 1868	76, 312
Neck-tie supporter	J. A. Eshleman	Philadelphia, Pa	Sept. 4, 1866	57, 694
Neck-tie supporter	W. H. Hart, jr	Philadelphia, Pa	Aug. 14, 1866	57, 129
Neck-tie supporter	A. Komp	New York, N. Y	June 17, 1873	140, 049
Neck-tie supporter	E. Peltz	New York, N. Y	Mar. 4, 1873	136, 538
Neck-tie supporter	T. Rosenthal	New York, N. Y	Aug. 7, 1866	57, 048
Neck-tie supporter	G. K. Snow	Watertown, Mass	Feb. 14, 1865	46, 426
Neck-tie supporter	W. A. Wicks	Baltimore, Md	Dec. 13, 1870	110, 178
Neck-tie supporter, Collar	C. L. Lockwood	New York, N. Y	June 5, 1866	55, 317
Neck-ties, Device for adjusting and buttoning	O. Wilson	Sandusky City, Ohio	Aug. 10, 1869	93, 656
Neck-ties, Device for securing fastening-loop to	B. F. Weishampel	Baltimore, Md	Jan. 12, 1869	85, 773
Neck-ties, Mode of packing	W. H. Hart, jr	Philadelphia, Pa	Mar. 31, 1868	76, 186
Necklace-fastening	E. S. Dodge	Providence, R. I	July 16, 1872	129, 111
Needle	A. I. Ambler	Milwaukee, Wis	Mar. 24, 1863	37, 996
Needle	G. Cooper	Thompsonville, Conn	Jan. 22, 1861	31, 153
Needle	J. W. Donaldson and D. Sheets.	Suisun, Cal	Apr. 26, 1870	102, 380
Needle	S. Ivers	New Bedford, Mass	Oct. 27, 1868	83, 501
Needle	W. H. Marriott	Baltimore, Md	Feb. 9, 1869	86, 769
Needle	H. G. Suplee	San Francisco, Cal	Oct. 25, 1870	108, 737
Needle and tool holder for work-table	H. H. Wescott	Portland, Me	July 25, 1871	117, 354
Needle-arranging machine	C. O. Crosby	New Haven, Conn	Apr. 12, 1870	101, 829
Needle-blanks, Machine for reducing wire for	T. Fowler	Seymour, Conn	Sept. 3, 1867	68, 430
Needle, Belt-lacing	J. W. Smith	Bryan, Ohio	May 6, 1873	138, 540
Needle-book	J. F. Umpleby	Albany, N. Y	May 5, 1868	77, 551
Needle-case	W. Avery and A. Fenton	Redditch, England	Jan. 18, 1870	98, 904
Needle-case	W. Avery and A. Fenton	Redditch, England	May 3, 1870	102, 471
Needle-case	W. B. Field and B. Beach	Rockville and Meriden, Conn.	Mar. 15, 1870	100, 879
Needle-case	T. G. Harold	Brooklyn, N. Y	Apr. 24, 1860	28, 037
Needle-case	W. J. Marchant and H. B. Hart	Philadelphia, Pa	Sept. 9, 1873	142, 639
Needle-case	J. F. Newhall	Waltham, Mass	June 25, 1867	66, 166
Needle-case	G. L. Turney	London, England	Nov. 13, 1866	59, 737
Needle-case	G. Wempe	San Francisco, Cal	Jan. 14, 1873	134, 829
Needle case and index	C. D. Wheeler	New York, N. Y	May 17, 1859	24, 070
Needle case, Sewing	J. O. Whitcomb	Brooklyn, N. Y	Apr. 15, 1862	35, 004
Needle, Crochet	J. M. Hoadley	Derby, Conn	Aug. 18, 1863	39, 572
Needle, Crochet	J. M. Hoadley	Birmingham, Conn	Apr. 21, 1868	76, 916
Needle-cushion	W. H. Bixler	Easton, Pa	July 8, 1873	140, 609
Needle-electrodes for electro-surgical uses	J. Kidder	New York, N. Y	Dec. 19, 1871	122, 031
Needle-envelope	A. James	Redditch, England	Oct. 20, 1868	83, 284
Needle for caning chairs	M. E. Hurley	Baltimore, Md	July 17, 1866	56, 421
Needle-grinding machine	C. Marsh	Bridgeport, Conn	Dec. 7, 1869	97, 664
Needle-guard	J. C. Howells	Madison, Wis	Apr. 10, 1860	27, 805
Needle-machine	W. Aiken	Franklin, N. H	Nov. 12, 1867	70, 677
Needle-making machine	R. T. Barton	New Haven, Conn	Sept. 10, 1867	68, 593
Needle-making machine	C. O. Crosby	New Haven, Conn	Nov. 28, 1865	51, 150
Needle-making machine	C. O. Crosby	New Haven, Conn	June 14, 1870	104, 278
Needle-making machinery	C. Kaiser	New York, N. Y	Nov. 12, 1861	33, 707
Needle-making machine	F. W. Mallett	New Haven, Conn	Apr. 29, 1873	138, 419
Needle-making machine	F. Plant	New York, N. Y	June 19, 1860	28, 772
Needle-making machine	C. P. S. Wardwell	Lake Village, N. H	Sept. 24, 1867	69, 280
Needle-making machinery	C. P. S. Wardwell	Lake Village, N. H	Sept. 24, 1867	69, 281
Needle-manufacturing machine	J. B. Blanchard	Boston, Mass	Aug. 2, 1870	106, 026
Needle, Neutralizing local attraction of the	C. Kline	Brooklyn, N. Y	Nov. 23, 1858	22, 125
Needle-paper wrapper	M. E. Baylis	Redditch, England	Aug. 15, 1871	117, 967
Needle-papering machine	C. O. Crosby	New Haven, Conn	July 12, 1870	105, 313
Needle-papering machine	G. P. Farmer	Philadelphia, Pa	July 1, 1862	35, 748
Needle-papering machine	G. P. Farmer	Philadelphia, Pa	Apr. 10, 1866	53, 799
Needle, Plated	R. J. Roberts	New York, N. Y	Feb. 15, 1870	99, 952
Needle-points, Machine for grinding	C. Kaiser	New York, N. Y	Aug. 13, 1861	33, 042
Needle polisher, sharpener, and lubricator	H. W. Little	Muncie, Ind	Nov. 16, 1869	96, 931
Needle-polishing machine	C. O. Crosby	New Haven, Conn	Mar. 16, 1869	87, 826

Index of patents issued from the United States Patent Office from 1790 *to* 1873, *inclusive*—Continued.

Invention.	Inventor.	Residence.	Date.	No.
Needle-polishing machine	C. O. Crosby	New Haven, Conn	July 12, 1870	105, 312
Needle-polishing machine	F. W. Mallett	New Haven, Conn	Oct. 3, 1871	119, 525
Needle scouring machine	C. O. Crosby	New Haven, Conn	Feb. 9, 1869	86, 819
Needle-scouring machine	F. W. Mallett	New Haven, Conn	Mar. 28, 1871	113, 182
Needle-setter and thread-pinchers	C. T. Barber and B. T. Loomis	New York, N. Y	May 4, 1869	89, 618
Needle setter and threader	R. W. Brown	Providence, R. I	June 14, 1870	104, 260
Needle setter and threader	C. F. Martine	Boston, Mass	Apr. 4, 1871	113, 542
Needle-setter, needle-sharpener, needle-case, and ripper.	E. E. Hendrick	Carbondale, Pa	May 16, 1871	114, 815
Needle setter, threader, and cutter	A. Johnston	Ottumwa, Iowa	Feb. 4, 1873	135, 479
Needle, Sewing	J. Cottrill	Studley, England	June 1, 1858	20, 409
Needle, Sewing	B. Garvey	New York, N. Y	May 12, 1857	17, 272
Needle, Sewing	G. A. Lloyd and S. Fetlow	San Francisco, Cal	Oct. 1, 1867	69, 453
Needle, Sewing	A. Morrall	Studley, England	May 25, 1869	90, 460
Needle, Sewing	J. Wilcox and S. H. Whitridge	Philadelphia, Pa	Oct. 3, 1854	11, 769
Needle-sharpener	A. S. Dinsmore	New York, N. Y	May 25, 1869	90, 433
Needle-sharpener	T. Harris	Cote Saint Paul, Canada	Dec. 19, 1871	121, 940
Needle-sharpener	E. K. Haynes	Boston, Mass	June 28, 1870	104, 732
Needle-sharpener	J. M. Mason	Chelsea, Mass	July 23, 1872	129, 673
Needle-sharpening device	R. S. Barnum and H. C. Goodrich.	Chicago, Ill	Sept. 9, 1873	142, 552
Needle shuttle, threader, and knife	J. Slack	Brooklyn, N. Y	Oct. 14, 1873	143, 726
Needle, Spiral fissure	S. Hodgins	Saint Louis, Mo	Aug. 6, 1867	67, 545
Needle, Tag	O. Low	Chelsea, Mass	Apr. 3, 1866	53, 638
Needle-threader	N. Barnum	La Porte, Ind	July 26, 1870	105, 628
Needle-threader	O. Cox	Alexandria, Va	Feb. 9, 1869	86, 647
Needle-threader	F. R. Gaspary	Berlin, Prussia	Nov. 25, 1873	144, 974
Needle-threader	A. Hunter	Buffalo, N. Y	Apr. 6, 1869	88, 717
Needle-threader	A. Huston	Bristol, Me	Oct. 22, 1867	69, 997
Needle-threader	S. L. Mercer	Washington, D. C	Jan. 30, 1872	123, 277
Needle-threader	J. O'Kane	New York, N. Y	Apr. 19, 1864	42, 394
Needle-threader	C. W. Simpson	Bangor, Me	Sept. 21, 1869	94, 980
Needle-threader	C. L. Spencer	Providence, R. I	Sept. 25, 1866	58, 314
Needle-threader	E. J. Stanley	Montezuma, Iowa	Oct. 8, 1872	132, 110
Needle-threader and spool combined	G. P. Farmer	Brooklyn, N. Y	Dec. 26, 1871	122, 241
Needle-threader, Magnet	O. Cox	Washington, D. C	June 14, 1864	43, 100
Needle-threaders, Mold for casting	S. S. Burlingame	Warwick, R. I	July 24, 1860	29, 230
Needle-threading apparatus	J. O'Kane	Philadelphia, Pa	Feb. 3, 1863	37, 587
Needle-threading hook	H. Wells	Woburn, Mass	Dec. 30, 1873	145, 978
Needle-threading instrument	S. S. Burlingame	Warwick, R. I	Feb. 15, 1859	22, 934
Needle-wrapper	R. Bennett	Redditch, England	Apr. 19, 1859	23, 742
Needle-wrapper	O. H. Blood and F. C. Treadwell, jr.	New York, N. Y	Oct. 13, 1863	40, 305
Needle-wrapper	W. Byfield	New York, N. Y	May 22, 1866	54, 856
Needle-wrapper	J. Clark	Redditch, England	May 19, 1868	78, 059
Needle-wrapper	R. Crowley	New York, N. Y	Dec. 13, 1859	26, 418
Needle-wrapper	D. Evans	Studley, England	Mar. 19, 1867	63, 028
Needle-wrapper	G. Ewart	New York, N. Y	May 11, 1869	89, 921
Needle-wrapper	J. Hill	Redditch, England	Nov. 12, 1867	70, 716
Needle-wrapper	C. B. James	Redditch, England	Feb. 9, 1869	86, 757
Needle-wrapper	V. Milward	Ipsley, England	July 5, 1870	105, 113
Needle-wrapper	J. Pitzler	New York, N. Y	July 25, 1871	117, 323
Needle-wrapper	J. Pitzler	New York, N. Y	July 25, 1871	117, 324
Needle-wrapper	C. Schleicher	Schoenthal, Rhenish Prussia.	May 30, 1871	115, 529
Needle-wrapper	A. Shrimpton	Redditch, England	Feb. 8, 1870	99, 605
Needle-wrapper	S. E. Tolten	Brooklyn, N. Y	Oct. 23, 1866	59, 095
Needle-wrapper	J. W. Whitfield	New York, N. Y	Apr. 23, 1867	64, 053
Needles, Machine for forming the hooks of machine	N. Paine	Milford, Mass	Sept. 16, 1873	142, 863
Needles, Machine for polishing the eyes of	W. H. Dayton and J. Aldis	Wolcottville, Conn	June 4, 1872	127, 582
Needles, Machine for polishing the eyes of	C. Kaiser	New York, N. Y	Dec. 17, 1861	33, 947
Needles, Machine for polishing the eyes of	R. Thompson	Waterbury, Conn	June 4, 1872	127, 529
Needles, Machine for scouring the eyes of	H. A. Nettleton and E. R. Lawton.	West Cheshire, Conn	Nov. 8, 1870	109, 040
Needles, Manufacture of	A. Morral	Studley, Great Britain	Dec. 21, 1839	1, 437
Needles, Manufacture of sewing	H. Walker	London, England	Nov. 23, 1858	22, 140
Needles, Manufacture of split	C. H. Palmer	New York, N. Y	May 2, 1871	114, 332
Needles, Manufacture of split	C. H. Palmer	New York, N. Y	Oct. 10, 1871	119, 876
Needles, Papering or putting up	D. Evans	Studley, England	Dec. 12, 1871	121, 860
Needles, Polishing	C. O. Crosby	New Haven, Conn	Feb. 9, 1869	86, 817
Needles, Polishing	C. O. Crosby	New Haven, Conn	Feb. 9, 1869	86, 818
Negative plate for "Smee" battery	L. L. Smith	Brooklyn, N. Y	Feb. 22, 1870	100, 202
Ne plus ultra, Machine called the	J. Baily	Philadelphia, Pa	Mar. 21, 1825	
Nest, Hen's	S. S. Bent	Port Chester, N. Y	June 16, 1868	78, 857
Nest, Hen's	C. W. Blackman	Bridgeport, Conn	Oct. 1, 1867	69, 392
Nest, Hen's	C. Campbell	Yellow Head, Ill	Feb. 20, 1866	52, 675
Nest, Hen's	B. F. Hayward	Nebraska City, Nebr	Sept. 22, 1868	82, 312
Nest, Hen's	D. P. Leach	Franklin, Ind	Oct. 5, 1869	95, 594
Nest, Hen's	N. Stevens	New Haven, Conn	May 22, 1866	54, 974
Nest, Wooden bird's	J. A. Deknatel	New York, N. Y	July 2, 1872	128, 538
Nests, Device for preserving eggs in hens'	C. V. Ament	Dansville, N. Y	Mar. 21, 1854	10, 675
Net, Adjustment of fishing	W. Randolph	Bloomington, Ill	Aug. 25, 1863	39, 676
Net and sash, Combined window	I. Wiswell	Springfield, Vt	Oct. 9, 1860	30, 370
Net, Fish	W. and J. Carr and J. Shannon	Sunbury, Pa	Sept. 14, 1844	3, 741
Net, Fish	T. Cartwright	Davenport, Iowa	Sept. 29, 1868	82, 490
Net, Fish	C. Livandais	New Orleans, La	Oct. 22, 1872	132, 476
Net, Fish	P. G. Sabins	Westport, Mass	Feb. 8, 1870	99, 713
Net, Fishing	L. H. Alexander	Gloucester, Mass	Aug. 15, 1871	117, 957
Net, Fishing, &c	B. Arnold	East Greenwich, R. I	July 16, 1867	66, 669
Net, Fishing	B. Arnold	East Greenwich, R. I	June 9, 1868	78, 716
Net, Fishing	E. A. Field	Sidney, Me	June 19, 1866	55, 635
Net, Fishing	T. Hall	Gloucester, Mass	Apr. 27, 1858	20, 125
Net, Fishing	S. Harper	Leipersville, Pa	Oct. 27, 1868	83, 493
Net, Fishing	W. Maxwell	Washington, N. J	Nov. 6, 1866	59, 429
Net, Fishing	F. A. Werdmüller	New York, N. Y	Mar. 9, 1869	87, 740
Net, Fly	C. K. Burkholder	York Springs, Pa	Feb. 18, 1868	74, 664
Net, Fly	J. Frymire	Orangeville, Pa	Nov. 9, 1869	96, 574

Index of patents issued from the United States Patent Office from 1790 *to* 1873, *inclusive*—Continued.

Invention.	Inventor.	Residence.	Date.	No.
Net, Fly	L. B. and G. W. Lee	Jerusalem, N. Y	Sept. 3, 1867	68, 518
Net, Fly	W. Sohier	Boston, Mass	May 30, 1871	115, 374
Net, Fly	R. Wilson	Milton, Pa	May 11, 1858	20, 235
Net for anglers, Landing-net	C. De Saxe	New York, N. Y	Apr. 18, 1854	10, 794
Net for catching fish at sea	B. Merritt, jr	Charlestown, Mass	June 29, 1858	20, 725
Net for catching mackerel, &c., in deep water	B. W. Hale	Newbury, Mass	June 4, 1838	763
Net for fishing, Pound	P. E. Tierman	Waukegan, Ill	Apr. 18, 1871	113, 817
Net for horses	H. Korn	Philadelphia, Pa	June 22, 1832	....
Net for horses	H Korn	Philadelphia, Pa	Apr. 19, 1834	....
Net for horses	H. Korn	Philadelphia, Pa	Apr. 19, 1834	....
Net for horses, Fly	J. Cantner	Millheim, Pa	Mar. 9, 1869	87, 631
Net for horses, Fly	J. Cantner and D. W. Zeigler	Millheim, Pa	Apr. 16, 1872	125, 659
Net for horses, Fly	J. Graham	New York, N. Y	Apr. 19, 1870	102, 113
Net for horses, Fly	J. M. Harman	Orangeville, Pa	Feb. 1, 1870	99, 434
Net for horses, Fly	W. H. Herstman	Philadelphia, Pa	July 28, 1831	....
Net for horses, Fly	D. W. Hurst	Petersburgh, Pa	July 21, 1868	80, 182
Net for horses, Fly	J. S. Huston	Mechanicsburgh, Pa	May 24, 1870	103, 337
Net for horses, Fly	S. B. Kline	Mechanicsburgh, Pa	Mar. 2, 1869	87, 499
Net for horses, Fly	H. Korn	Philadelphia, Pa	Sept. 12, 1829	....
Net for horses, Fly	H. Korn	Philadelphia, Pa	Dec. 8, 1831	....
Net for horses, Fly	H. Korn	Philadelphia, Pa	July 2, 1836	....
Net for horses, Fly	P. Mintzer	Philadelphia, Pa	July 20, 1831	....
Net for horses, Tail	F. Parson and E. W. Parker	Saint Louis, Mo., and Penn Yan, N. Y.	June 11, 1872	127, 918
Net for window, Fly	W. C. McGowan and J. M. Hale.	Georgia Plains, Vt	Aug. 18, 1868	81, 099
Net frame, Insect	H. Stoltz	Evansville, Ind	Aug. 6, 1872	130, 163
Net gear, Fishing	C. C. Crosman	Portland, Me	Feb. 26, 1867	62, 481
Net, Gill	H. Cook	Haddam, Mass	Mar. 17, 1843	3, 004
Net, Gill	J. Drummond	New York	Aug. 27, 1818	....
Net, Gill	D. Wills	Camden, N. J	Mar. 31, 1868	76, 284
Net, Insect	C. B. Seaman	Honesdale, Pa	Nov. 24, 1868	84, 381
Net, Lady's hair	J. Dalton	New York, N. Y	Mar. 5, 1872	124, 340
Net, Lady's head	D. M. Smyth	New York, N. Y	Feb. 2, 1864	41, 465
Net-making machine	B. Arnold	East Greenwich, R. I	Jan. 4, 1870	98, 461
Net, Metallic fish	W. Beck	New York, N. Y	Oct. 25, 1870	108, 553
Net or seine, Fish	S. May	Columbia, Pa	Jan. 11, 1812	....
Net, Portable fly and mosquito	T. M. Prentiss	Boston, Mass	Nov. 18, 1873	144, 792
Net straps, Machine for cutting and punching fly	J. Yeager	Berrysburgh, Pa	Nov. 19, 1867	71, 259
Net straps, Machine for rounding fly	C. K. Burkholder and H. Lerew	York Springs, Pa	Sept. 17, 1867	68, 949
Net supporter, Fish	B. Ryder, jr	South Orrington, Me	Apr. 4, 1871	113, 572
Net, Trap	W. S. Wilcox	Wellington, Ohio	Oct. 27, 1868	83, 429
Nets, Machine for cutting fly	A. D. Hoffman	Detroit, Mich	June 26, 1866	55, 860
Nets, Machine for cutting fly	H. D. Martin	Ypsilanti, Mich	Jan. 1, 1867	60, 764
Nets, Machine for cutting leather fly	A. Worden	Ypsilanti, Mich	Feb. 2, 1861	41, 459
Nets, Machine for making fly	A. Prutzmann	Canton, Ohio	Dec. 14, 1869	97, 809
Nets, Machine for making seine	B. Arnold	East Greenwich, R. I	Sept. 23, 1862	36, 499
Nets, Machine for punching leather straps for fly	J. Matheis	Ottawa, Ill	July 21, 1868	80, 081
Nets, Machine for punching leather straps for fly	J. Matheis	Ottawa, Ill	May 18, 1869	90, 281
Nets, Machine for weaving fish	L. Van Hoesen	New Haven, Conn	Oct. 7, 1842	2, 798
Nets, Securing lash in fly	M. Heilman	Lebanon, Pa	Jan. 4, 1870	98, 493
Nets, Securing lash in fly	J. S. Huston	Mechanicsburgh, Pa	Dec. 21, 1869	98, 164
Netted or laced fabric, Machine for making	H. A. Oesterle	Philadelphia, Pa	Dec. 26, 1865	51, 744
Netting, Horse	J. Cantner and M. Ulrich	Millheim, Pa	Aug. 7, 1866	56, 900
Netting-machine	J. McMullen	Baltimore, Md	June 27, 1846	4, 608
Netting-machine	J. McMullen	Baltimore, Md	July 1, 1856	15, 245
Netting-machine	P. Moulton	New Rochelle, N. Y	Aug. 22, 1846	4, 707
Netting, Machine for cutting leather fly	T. Tulley	Springfield, Ill	June 11, 1872	127, 716
Netting, Machine for making	A. Vivarttas	New York, N. Y	June 14, 1870	104, 381
Netting, Machine for making wire	J. Nesmith	Lowell, Mass	Apr. 4, 1854	10, 743
Netting, Window	J. R. Wharry	Moundsville, W. Va	Mar. 17, 1868	75, 612
News-distributer	J. H. Pratt	New York, N. Y	Feb. 3, 1863	37, 590
Newspaper-clamp	J. D. Hall	Philadelphia, Pa	Aug. 2, 1864	43, 685
Newspaper-wrapper	L. P. Mara	New York, N. Y	Feb. 19, 1861	31, 516
Newspapers, &c., Machine for feeding up, cutting, and pasting directions on.	R. W. Wright	New Haven, Conn	Dec. 20, 1859	26, 543
Newspapers, music, &c., Ready-binder for	E. Ripley	Troy, N. Y	Dec. 26, 1837	536
Nickel, Electro-deposition of	I. Adams, jr	Boston, Mass	Aug. 3, 1869	93, 157
Nickel, Electro-deposition of	I. Adams, jr	Boston, Mass	May 10, 1870	102, 748
Nickel, Melting, casting, and hardening	I. Adams, jr	Boston, Mass	May 25, 1869	90, 476
Nickel-plating	N. S. Keith	New York, N. Y	Nov. 28, 1871	121, 383
Nickel-plating	W. J. Kuhns	Brooklyn, N. Y	May 17, 1870	103, 201
Night-chair	J. Macferran	Philadelphia, Pa	Mar. 24, 1868	75, 778
Night-lock, Traveler's	A. V. Thomas	Frederick, Md	Oct. 31, 1865	50, 751
Night-signals, Mode of firing	I. Edge and C. C. Hyde	Jersey City, N. J., and Stonington, Conn.	Apr. 29, 1862	35, 089
Night-soil, Boat for transporting	J. Paterson	Hoboken, N. J	July 30, 1872	129, 982
Night-soil for agricultural purposes, Treating	R. B. Fitts	Philadelphia, Pa	Aug. 11, 1863	39, 472
Night-soil, Treating	R. B. Fitts	Philadelphia, Pa	Sept. 9, 1862	36, 399
Night-soil, Vessel for reception and transportation of.	R. A. Smith	Philadelphia, Pa	July 18, 1865	48, 847
Nine-pin ball	J. Taggart	Roxbury, Mass	July 26, 1859	24, 893
Nipper-block, Self-acting	W. Waley	New London, Conn	May 30, 1854	10, 997
Nipper-block, Self-acting	J. Whipple, jr	Milford, Mass	May 22, 1855	12, 928
Nippers	T. G. Hall	New York, N. Y	May 14, 1867	64, 664
Nippers	E. F. Stacy	Gloucester, Mass	June 2, 1868	78, 546
Nippers and pinchers, Combined	D. Sweetman	Homer, N. Y	Oct. 13, 1863	40, 291
Nippers, Clinching	J. B. Wilder	Mannsville, N. Y	July 14, 1868	79, 928
Nippers, Cutting	F. A. Adams	Shelburne, Mass	Oct. 15, 1872	132, 191
Nippers, &c., Cutting	P. Broadbooks	Batavia, N. Y	Nov. 18, 1873	144, 734
Nippers, Cutting	W. A. Stevens	East Brookfield, Mass	May 7, 1872	126, 423
Nippers, Cutting	N. Thompson	Brooklyn, N. Y	July 14, 1868	80, 031
Nippers, Friction	D. Thomas	Hingham, Mass	Aug. 4, 1868	80, 575
Nippers, Police	J. B. Craig and M. Haughey	Saint Louis, Mo	Mar. 4, 1873	136, 419
Nippers, Police	W. G. Phillips	Brooklyn, N. Y	Aug. 10, 1869	93, 474
Nippers, Policeman's	G. Lutz, D. H. Royce, M. Trenor, and R. Chadwick.	Columbus, Ohio	May 3, 1870	102, 563
Nipple, Artificial	H. D. Lockwood	Charlestown, Mass	Dec. 7, 1869	97, 659

Index of patents issued from the United States Patent Office from 1790 *to* 1873, *inclusive*—Continued.

Invention.	Inventor.	Residence.	Date.	No.
Nipple, Artificial	E. Pratt	New York, N. Y	Aug. 4, 1855	4, 131
Nipple, Nursing	F. H. Holton	Brooklyn, N. Y	June 9, 1868	78, 741
Nipple, Rubber	G. Stevenson	Zionsville, Ind	Sept. 3, 1872	131, 130
Nipple-shield	W. Baxton	Woburn, Mass	Apr. 2, 1835	....
Nipple-shield	S. C. Foster	New York, N. Y	May 2, 1871	114, 281
Nipple-shield	T. McLaughlin	Millville, N. J	Apr. 28, 1868	77, 393
Nipple-shield	J. Parker	Boston, Mass	Jan. 13, 1857	16, 396
Nipple-shield	C. H. and J. M. Wilder	New York, N. Y	June 18, 1867	65, 978
Nipple-shield, Artificial metallic	J. J. Heintzleman	Philadelphia, Pa	Mar. 16, 1833	....
Nitric acid, Method of purifying	G. M. Mowbray	Titusville, Pa	Sept. 21, 1869	94, 969
Nitro-glycerine	A. Nobel	Hamburg, Germany	Oct. 24, 1865	50, 617
Nitro-glycerine	A. Nobel	New York, N. Y	Aug. 14, 1866	57, 175
Nitro-glycerine	A. Nobel	Hamburg, Germany	May 26, 1868	78, 317
Nitro-glycerine	A. Nobel	Hamburg, Germany	Aug. 5, 1873	141, 455
Nitro-glycerine, Apparatus for the manufacture of	G. M. Mowbray	North Adams, Mass	Aug. 23, 1870	106, 606
Nitro-glycerine can	J. Taylor	Petroleum Centre, Pa	Aug. 6, 1871	117, 577
Nitro-glycerine, &c., Can for transporting	A. Hamar	Philadelphia, Pa	Apr. 1, 1873	137, 439
Nitro-glycerine, Manufacture of	A. Hamar	Philadelphia, Pa	Apr. 1, 1873	137, 440
Nitro-glycerine, Manufacture of	G. M. Mowbray	Titusville, Pa	Apr. 7, 1868	76, 499
Nitro-glycerine, Manufacture of	E. A. L. Roberts	Titusville, Pa	Mar. 21, 1871	112, 848
Nitro-glycerine, Manufacture of	E. A. L. Roberts	Titusville, Pa	Mar. 21, 1871	112, 849
Nitro-glycerine, Manufacture of	E. A. L. Roberts	Titusville, Pa	Dec. 12, 1871	121, 898
Nitro-glycerine, Manufacture of	T. P. Shaffner	Louisville, Ky	Aug. 17, 1869	93, 756
Nitro-glycerine, Method of exploding	G. M. Mowbray	Titusville, Pa	July 27, 1869	93, 113
Nitro-glycerine, Method of packing	T. P. Shaffner	Louisville, Ky	July 24, 1866	56, 620
Nitro-glycerine, Method of preparing	S. Chester and O. Bürstenbinder	New York, N. Y	Jan. 19, 1869	85, 906
Nitro-gylcerine, nitro-benzole, &c., Apparatus for the manufacture of	G. M. Mowbray	North Adams, Mass	Aug. 23, 1870	106, 607
Nitro-glycerine, Preserving	T. P. Shaffner	Louisville, Ky	Feb. 9, 1869	86, 701
Nitroleum and other explosive liquids, Process of preserving	T. P. Shaffner	Louisville, Ky	Dec. 28, 1869	98, 426
Nitroleum, Method of blasting with	T. P. Shaffner	Louisville, Ky	Dec. 18, 1866	60, 573
Nitrous oxide and other gases, Apparatus for washing	E. E. Pursell	Indianapolis, Ind	July 9, 1872	128, 753
Nitrous oxide, &c., Apparatus for making	M. B. Renslow	Springfield, Mass	Aug. 20, 1867	67, 907
Nitrous oxide as an anaesthetic agent, Use of	W. P. Barker	Grand Rapids, Mich	Mar. 2, 1869	87, 319
Non-freezing battery	E. H. Ashcroft	Boston, Mass	Dec. 2, 1873	145, 143
Noodle-machine	D. Obergfell	Wheeling, W. Va	May 10, 1870	102, 962
Nose-bag, Automatic	G. A. Hankinson	Manahocking, N. J	Mar. 17, 1863	37, 914
Notations, Phonetic and diacritical	J. W. Shearer	Madison, N. C	July 16, 1872	129, 600
Notes, checks, &c., Device for canceling	E. M. Scott	Auburn, N. Y	Apr. 1, 1862	34, 850
Notes, checks, &c., from alteration, Protecting	E. Watson	Albany, N. Y	Aug. 3, 1838	871
Notes, checks, &c., To prevent alteration in	C. Folsom	New York, N. Y	Dec. 13, 1870	110, 129
Nozzle	R. R. and J. Craig	Nevada City, Cal	Apr. 23, 1872	125, 884
Nozzle	C. Oyston	Little Falls, N. Y	Aug. 25, 1863	39, 674
Nozzle	T. Watson	Nevada, Cal	Nov. 28, 1871	121, 265
Nozzle and cap, Oil-can	E. T. Covell	Brooklyn, N. Y	June 21, 1870	104, 430
Nozzle and cap, Oil-can	J. A. Bostwick	New York, N. Y	Nov. 7, 1871	120, 702
Nozzle and sealing-cap for sheet-metal cans	J. A. Bostwick	New York, N. Y	June 21, 1870	104, 413
Nozzle and stopper for cans, &c	M. Bray	Boston, Mass	May 24, 1870	103, 420
Nozzle and turn-pipe, Hydraulic	D. L. Gorman	Michigan Bluff, Cal	Aug. 13, 1872	130, 366
Nozzle and ventilating oil-can, Extension	W. Bonner	Saint Louis, Mo	Sept. 24, 1867	69, 067
Nozzle-bending die, Coffee-pot	W. H. Miller	Brandenburgh, Ky	Jan. 29, 1867	61, 678
Nozzle, Can	F. W. Devol	New York, N. Y	Oct. 20, 1868	83, 140
Nozzle, Can	C. Pratt	New York, N. Y	Apr. 20, 1869	89, 167
Nozzle, Elastic hose	J. Greacen, jr	New York, N. Y	Jan. 17, 1872	122, 719
Nozzle, Exhaust	A. F. Du Faur	New York, N. Y	Sept. 20, 1870	107, 603
Nozzle, Exhaust	C. H. Frisbie	Chicago, Ill	May 31, 1870	103, 596
Nozzle, Exhaust	J. Lewis	Chicago, Ill	July 4, 1871	116, 605
Nozzle, Expansible hose	C. Crook	Yonkers, N. Y	Sept. 10, 1867	68, 712
Nozzle, Fire-engine	C. Burchardt	New York, N. Y	Mar. 13, 1866	53, 109
Nozzle, Fire-engine	L. Button and R. Blake	Waterford, N. Y	Nov. 15, 1859	26, 088
Nozzle, Fire-engine, &c	J. J. Hofer	New Orleans, La	Dec. 3, 1867	71, 756
Nozzle, Fire-engine	J. C. Howells	Madison, Wis	June 26, 1860	28, 865
Nozzle, Fire-engine	C. Oyston	Little Falls, N. Y	Mar. 13, 1866	53, 175
Nozzle for cans and casks	A. L. Webster	Cleveland, Ohio	July 5, 1870	105, 151
Nozzle for cans, Cut-off	J. M. Murphy	New York, N. Y	Oct. 19, 1869	95, 927
Nozzle for cans, Screw	F. W. Devoe	New York, N. Y	Mar. 16, 1869	87, 915
Nozzle for cans, Tap	H. Miller	New York, N. Y	July 26, 1870	105, 713
Nozzle for drawing liquor, Cut-off	F. C. Edwards	Chicago, Ill	Sept. 2, 1873	142, 383
Nozzle for exhaust-pipes of locomotives	W. E. Cooper	Dunkirk, N. Y	Dec. 18, 1855	13, 939
Nozzle for exhaust-pipes of locomotives, Movable tapering	F. Espenschade	Mifflintown, Pa	Mar. 14, 1854	10, 634
Nozzle for fire-engines, Water-spreading	W. Gurley	Troy, N. Y	June 5, 1866	55, 285
Nozzle for hose-pipes, hydrants, &c	I. O. Endicott	Manchester, N. H	Nov. 19, 1872	133, 215
Nozzle for hose-pipes, Stop	J. N. Allen	Providence, R. I	Nov. 25, 1873	144, 942
Nozzle for lead-pipe machine	J. B. Collan	Reading, Pa	Jan. 1, 1851	7, 867
Nozzle for liquid-jars, Measuring	J. H. Pein	New York, N. Y	Aug. 2, 1864	43, 703
Nozzle for liquid-package	H. H. Rogers	Brooklyn, N. Y	Aug. 15, 1871	118, 154
Nozzle for liquid-package, Screw	H. H. Rogers and G. F. Walter	Brooklyn, N. Y	Aug. 15, 1871	118, 155
Nozzle for locomotives, Exhaust	G. W. Richardson	Troy, N. Y	Mar. 11, 1873	136, 619
Nozzle for steam-engines, Exhaust	W. A. Foster	Fitchburgh, Mass	Sept. 20, 1870	107, 605
Nozzle for steam-engines, Exhaust	C. H. Frisbie	Chicago, Ill	Aug. 3, 1869	93, 294
Nozzle, Hose	M. S. Curtis and G. W. Harris	New York, N. Y	Dec. 17, 1867	72, 372
Nozzle, Hose	E. A. Day	Oberlin, Ohio	Feb. 25, 1873	136, 309
Nozzle, Hose	W. Jeffers	Pawtucket, R. I	Sept. 3, 1861	33, 200
Nozzle, Hose	C. F. Macy and S. Martin	Little York, Cal	Dec. 8, 1863	40, 847
Nozzle, Hose	J. B. Mitchell	Portland, Me	Aug. 25, 1868	81, 396
Nozzle, Hose	J. Trees	Greenburgh, Pa	Aug. 6, 1867	67, 613
Nozzle, Hose	G. O. Wickers	Lawrence, Mass	Apr. 9, 1872	125, 510
Nozzle, Hose and pipe	H. B. Morrison	Mount Morris, N. Y	Oct. 14, 1862	36, 600
Nozzle, Hose-pipe	A. F. Allen	Providence, R. I	Apr. 27, 1869	89, 456
Nozzle, Hose-pipe	A. F. Allen	Providence, R. I	Nov. 16, 1869	96, 862
Nozzle, Hose-pipe	A. F. Allen	Providence, R. I	Oct. 29, 1872	132, 617
Nozzle, Hose-pipe	O. J. Backus	San Francisco, Cal	Oct. 6, 1868	82, 676
Nozzle, Hose-pipe	B. A. Berryman and A. B. Wakefield	Saint Louis, Mo	Apr. 15, 1873	137, 881
Nozzle, Hose-pipe	J. A. Cushman	Seneca Falls, N. Y	Aug. 18, 1868	81, 257

Index of patents issued from the United States Patent Office from 1790 *to* 1873, *inclusive*—Continued.

Invention.	Inventor.	Residence.	Date.	No.
Nozzle, Hose-pipe	F. E. Hall	Boston, Mass	Aug. 12, 1873	141, 787
Nozzle, Hose-pipe	N. Hotz	Trenton, N. J	Jan. 31, 1860	26, 994
Nozzle, Hose-pipe	R. F. Moore	Manchester, N. H	Nov. 27, 1866	60, 036
Nozzle, Hose-pipe	A. W. Roberts	Hartford, Conn	May 8, 1855	12, 831
Nozzle, Hose-pipe	A. M. White	New York, N. Y	Apr. 9, 1867	63, 680
Nozzle, Hose-pipe	A. Williscroft	Wilmington, Del	Aug. 31, 1869	94, 369
Nozzle, Hose-pipe	E. M. Wright	Marysville, Cal	Oct. 15, 1861	33, 504
Nozzle, Hydraulic	A. Harris	La Porte, Cal	Dec. 28, 1869	98, 257
Nozzle, Hydraulic	H. Shaw	Nevada City, Cal	Mar. 21, 1871	112, 969
Nozzle of oil-cans, Sealing	J. A. Bostwick	New York, N. Y	Jan. 9, 1872	122, 512
Nozzle, Oil-can	J. A. Boswick	New York, N. Y	Mar. 28, 1871	113, 137
Nozzle, Oil-can	S. P. Doane	San Francisco, Cal	Sept. 19, 1871	119, 081
Nozzle, Oil-can	G. H. Perkins	Philadelphia, Pa	Dec. 23, 1873	145, 898
Nozzle, Oil-can	C. Pratt	New York, N. Y	Dec. 28, 1869	98, 408
Nozzle, Pipe	F. S. Babbitt	Taunton, Mass	Aug. 18, 1868	81, 242
Nozzle, Pump	C. Wilson	Columbia, Ohio	Mar. 4, 1873	136, 478
Nozzle, Revolving hose	H. B. Morrison	Le Roy, N. Y	June 30, 1868	79, 493
Nozzle, Revolving hose	I. Smith and H. D. Tewksbury	New York, N. Y	July 10, 1866	56, 284
Nozzle, Steam	W. Ebbitt	New York, N. Y	July 15, 1873	140, 902
Nozzle-stopper for oil-cans	J. H. Noyes	Abington, Mass	May 2, 1871	114, 467
Nozzle, Valved oil-can	E. J. Durant	Lebanon, N. H	July 2, 1872	128, 473
Nozzle, Variable exhaust	M. E. Brown	Buffalo, N. Y	Feb. 21, 1865	46, 445
Nozzles and screw-caps for oil-can, Manufacture of.	E. T. Covell	Brooklyn, N. Y	Nov. 16, 1869	96, 890
Nozzles, Attaching revolving tip to hose	H. B. Morrison	Le Roy, N. Y	Aug. 25, 1863	39, 700
Nozzles, Blast-protector for exhaust	J. H. Setchel	Cincinnati, Ohio	July 11, 1871	116, 873
Nozzles, Elastic clasp for elastic	E. A. Day	Oberlin, Ohio	Dec. 31, 1872	134, 469
Nubia	T. Dolan	Philadelphia, Pa	July 18, 1871	117, 158
Nubia and veil, Combined	J. W. Tuttle	Watertown, Mass	Dec. 30, 1873	145, 977
Numbering and paging machine	G. J. Hill	Buffalo, N. Y	June 27, 1865	48, 488
Numbering and paging machine	A. C. Sine	Cincinnati, Ohio	Feb. 28, 1871	112, 292
Numbering-head	C. W. Dickinson	Belleville, N. J	July 15, 1873	140, 899
Numbering-machine	J. C. Clapp	South Boston, Mass	Apr. 12, 1864	42, 274
Numbering-machine	W. H. Forbush	Buffalo, N. Y	Feb. 4, 1868	73, 963
Numbering-machine	C. L. Sholes	Milwaukee, Wis	Sept. 12, 1871	118, 978
Numbering-machine	J. D. Smith	Washington, D. C	Sept. 6, 1870	107, 202
Numbering-machine	J. D. Smith	Washington, D. C	Dec. 16, 1873	145, 598
Numbering-machine	S. W. Soulé	Milwaukee, Wis	June 18, 1867	65, 839
Numbering-machine	S. W. Soulé and C. L. Sholes	Milwaukee, Wis	Nov. 13, 1866	59, 675
Numbering-machine	H. Sutcliffe	Brooklyn, N. Y	Sept. 28, 1869	95, 282
Numeral-frame	J. H. R. Reffelt	Hoboken, N. J	Mar. 3, 1863	37, 825
Nursery-chair	A. B. Anderson, jr	Brooklyn, N. Y	June 30, 1863	39, 026
Nursery-chair	C. M. Loring and E. Averell	Charlestown, Mass	Nov. 26, 1867	71, 397
Nursery-chair	S. S. May	Sterling, Mass	June 4, 1850	7, 418
Nursery-chair	S. Rainey	New Orleans, La	Nov. 28, 1865	51, 215
Nursery-cup	J. F. Leslie and E. A. Tibbetts	Woburn, Mass	July 28, 1868	80, 355
Nursery-gate	J. W. Boughton	Philadelphia, Pa	May 27, 1873	139, 232
Nursery-gate	E. A. Tuttle	Brooklyn, N. Y	Aug. 12, 1873	141, 677
Nursery gate, Child's	E. Howard	Milford, Mass	July 11, 1871	116, 960
Nursery gate, Child's	E. Howard	Milford, Mass	Feb. 20, 1872	123, 904
Nurseryman's knife	S. S. Jackson	Cincinnati, Ohio	Aug. 30, 1864	43, 996
Nursing and other chairs	A. Woods	Charlestown, Va	Mar. 13, 1822	....
Nursing-bottle	L. P. Dodge	New York, N. Y	Nov. 7, 1871	120, 575
Nursing-cup	S. W. Talbot	Dedham, Mass	Apr. 19, 1833	....
Nursing-table	J. Larkin	Unionville, S. C	Apr. 13, 1869	88, 881
Nut	F. L. Delfer	Burlington, Iowa	Jan. 7, 1873	134, 649
Nut	W. Harris and C. Browning	Rush Run, Ohio	Jan. 29, 1867	61, 537
Nut	B. D. Sanders	Wellsburgh, W. Va	Oct 20, 1868	83, 213
Nut	B. D. Sanders	Wellsburgh, W. Va	Oct. 20, 1868	83, 214
Nut	L. Till	Sandusky, Ohio	July 3, 1866	56, 124
Nut	W. Van Anden	Poughkeepsie, N. Y	Apr. 1, 1873	137, 395
Nut, Adjustable	J. Russell	Springfield, Mass	Feb. 25, 1873	136, 184
Nut and bolt dressing machine	J. Kohler	Cincinnati, Ohio	Jan. 21, 1873	135, 131
Nut and bolt fastening	J. T. Antill and W. J. Sloan	Smith's Ferry, Pa., and Liverpool Township, Ohio.	Aug. 19, 1873	141, 980
Nut and bolt fastening	L. L. Dunlap	Pontiac, Mich	Oct. 8, 1872	131, 945
Nut and bolt fastening	P. Hayden	Allegheny, Pa	Oct. 1, 1872	131, 755
Nut and bolt fastening	P. F. King	Saint Louis, Mo	Aug. 19, 1873	141, 877
Nut and bolt fastening	M. F. McIntyer	Girard, Pa	Dec. 10, 1872	133, 873
Nut and bolt fastening	A. C. Smith	Fort Madison, Iowa	Nov. 5, 1872	132, 867
Nut and bolt fastening	J. L. Williams	Jackson, Tenn	Oct. 1, 1872	131, 922
Nut and bolt fastening	E. B. Wingate	Friendship, N. Y	Jan. 14, 1873	134, 963
Nut and coffee roaster	D. A. T. Gale	Poughkeepsie, N. Y	June 8, 1869	91, 009
Nut and coffee roaster	D. T. Gale	Fort Wayne, Ind	May 20, 1873	139, 054
Nut and washer	D. B. Hart	Mentor, Ohio	Aug. 6, 1867	67, 539
Nut and washer machine	H. Carter and J. Rees	Pittsburgh, Pa	Aug. 26, 1851	8, 322
Nut and washer machine	W. R. Wilbur	Cleveland, Ohio	June 10, 1873	139, 696
Nut and washer machine	J. T. Wood and E. C. Smith	Pittsburgh, Pa	Jan. 12, 1864	41, 252
Nut, Axle	E. A. Stanley	Brewer, Me	May 10, 1870	102, 876
Nut-bar	S. Vanstone	Providence, R. I	June 1, 1869	90, 897
Nut blank, Pressed	F. Washbourne	Brooklyn, N. Y	Feb. 21, 1871	111, 996
Nut-blanks, Making	R. H. Cole	Saint Louis, Mo	Sept. 28, 1858	21, 599
Nut-blanks, Rolled bar for	H. W. Oliver, jr	Pittsburgh, Pa	Feb. 14, 1871	111, 772
Nut-blanks, tubes, &c., Piles for	J. Ostrander	Manchester, Va	Apr. 25, 1871	114, 187
Nut-blanks, tubes, &c., Piles for	J. Ostrander	Manchester, Va	Apr. 25, 1871	114, 188
Nut-box	R. H. Cole	Saint Louis, Mo	Jan. 1, 1856	14, 011
Nut-box, Adjustable	J. Turner	Richmond, Va	Nov. 13, 1866	59, 731
Nut, Cast	G. P. Darrow	Cincinnati, Ohio	Aug. 25, 1868	81, 480
Nut, Clamp	W. Pearson	Windsor Locks, Conn	Aug. 11, 1868	80, 997
Nut-cracker	P., E. W., and J. A. Blake	New Haven, Conn	Sept. 6, 1853	9, 985
Nut-cracker	P. Ceredo	Dusseldorf, Prussia	Nov. 22, 1870	109, 495
Nut-cracker	L. A. Clark	Bridgeport, Conn	Jan. 24, 1860	26, 885
Nut-cracker	T. Earle	Providence, R. I	Dec. 8, 1863	40, 825
Nut-cracker	R. Frisbie	Middletown, Conn	May 17, 1859	24, 018
Nut-cracker	C. Hayden	Collinsville, Conn	Nov. 10, 1868	83, 959
Nut-cracker	E. L. Pratt	Boston, Mass	Mar. 31, 1868	76, 247
Nut-cracker	J. Pusey	Philadelphia, Pa	Nov. 8, 1870	119, 144
Nut-cracker	E. Ripley	Troy, N. Y	May 31, 1859	24, 238

Index of patents issued from the United States Patent Office from 1790 *to* 1873, *inclusive*—Continued.

Invention.	Inventor.	Residence.	Date.	No.
Nut-cracker	S. J. Smith	New York, N. Y	May 15, 1860	28, 311
Nut-fastener	C. Buckley, jr	Rochester, N. Y	Sept. 24, 1867	69, 173
Nut-fastener	J. Christy	Philadelphia, Pa	Apr. 21, 1868	76, 995
Nut-fastener	J. Davis	New Bedford, Mass	Apr. 20, 1869	89, 206
Nut-fastener	M. Grover	Clyde, Ohio	Sept. 14, 1869	94, 885
Nut-fastener	W. Harris and C. Browning	Rush Run, Ohio	Jan. 29, 1867	61, 620
Nut-fastening	W. Harris	Rush Run, Ohio	Dec. 24, 1867	72, 486
Nut-fastening	F. Myers	New York, N. Y	Oct. 18, 1870	108, 381
Nut-fastening	F. Myers	New York, N. Y	Oct. 18, 1870	108, 382
Nut-fastening	A. M. Rouse	Saint Louis, Mo	Dec. 31, 1872	134, 489
Nut-fastening	S. A. Todd	Wellsville, Ohio	Dec. 10, 1872	133, 812
Nut-fastening device	D. Sawyer	Topeka, Kans	Sept. 24, 1872	131, 630
Nut-fastenings, Washer for	H. P. Hood	Indianapolis, Ind	Jan. 21, 1873	134, 990
Nut-finishing machine	F. P. Pleghar and W. Schöllhorn	New Haven, Conn	Oct. 25, 1864	44, 818
Nut for joint-bolts	J. Baldwin	Manchester, N. H	Apr. 7, 1868	76, 380
Nut for screw-bolts	N. Thompson	Brooklyn, N. Y	Feb. 8, 1870	99, 729
Nut for securing piston-heads to piston-rods	J. Wheelock	Worcester, Mass	June 24, 1873	140, 328
Nut-forging machine	E. Paye and S. Hall	New York, N. Y	Sept. 8, 1859	18, 156
Nut-forging machine	A. P. Plant and A. Shepard	Plantsville, Conn	June 23, 1868	79, 255
Nut, Jam	B. W. Nichols	Canton, Ohio	Apr. 20, 1869	89, 066
Nut, Jam	W. H. Van Cleve	Ypsilanti, Mich	Nov. 7, 1871	120, 798
Nut, Jam	J. M. Winslow	Rochester, N. Y	June 4, 1872	127, 445
Nut-lock	S. C. Adams	Buffalo, N. Y	June 15, 1869	91, 197
Nut-lock	J. M. Allison	Salina, Pa	Apr. 8, 1873	137, 646
Nut-lock	W. E. Ball	Belmont, Ohio	Apr. 27, 1869	89, 460
Nut-lock	W. E. Ball	Bethesda, Ohio	Aug. 20, 1872	130, 689
Nut-lock	M. L. Ballard	Canton, Ohio	Apr. 1, 1873	137, 401
Nut-lock	H. Beagle, jr	Philadelphia, Pa	May 31, 1870	103, 544
Nut-lock	J. Bell	New York, N. Y	Oct. 4, 1870	107, 996
Nut-lock	F. A. Bishop	Shingle Springs, Cal	Dec. 10, 1872	133, 694
Nut-lock	W. H. Bowman	London, Ohio	Nov. 25, 1873	144, 886
Nut-lock	F. H. Bradley	Mystic River, Conn	July 23, 1872	129, 710
Nut-lock	K. Brown	Gardner, Ill	Apr. 27, 1869	89, 283
Nut-lock	J. M. Connel	Newark, Ohio	Aug. 1, 1871	117, 517
Nut-lock	E. A. Cooper	Lancaster, N. Y	Nov. 11, 1873	144, 511
Nut-lock	J. R. Cribbs	Gardner, Ill	Mar. 16, 1869	87, 761
Nut-lock	D. Cumming, jr	Brooklyn, N. Y	Aug. 23, 1870	106, 668
Nut-lock	M. A. Cushing and O. R. Glover	Ottawa, Ill	Dec. 6, 1870	109, 809
Nut-lock	E. Czarniecki	Allegheny, Pa	June 24, 1873	140, 119
Nut-lock	B. W. Davis	Fort Madison, Iowa	Mar. 18, 1873	136, 906
Nut-lock	J. Dennis	Churchville, N. Y	Aug. 9, 1870	106, 138
Nut-lock	L. L. Deweese	Canton, Ohio	June 22, 1869	91, 724
Nut-lock	C. Dittman	Leacock, Pa	May 24, 1870	103, 308
Nut-lock	C. Dittman	Leacock Post-Office, Pa	Nov. 25, 1873	144, 964
Nut-lock	E. H. Dooley	New York, N. Y	Jan. 16, 1872	122, 761
Nut-lock	H. W. Dopp	Buffalo, N. Y	Aug. 26, 1873	142, 215
Nut-lock	J. R. Eichenberger and C. G. Binkly.	Burton City, Ohio	Apr. 1, 1873	137, 426
Nut-lock	L. Fay	Worcester, Mass	Mar. 14, 1871	112, 574
Nut-lock	J. H. Fisher	Chicago, Ill	Oct. 19, 1869	95, 890
Nut-lock	O. S. Freeland	Newport, R. I	Feb. 22, 1870	100, 023
Nut-lock	B. Furst and P. Oettinger	Lacon, Ill	Dec. 31, 1872	134, 426
Nut-lock	R. Gilliland	Hudson, Mich	Sept. 20, 1870	107, 609
Nut-lock	G. Hart	Columbiana, Ohio	June 25, 1872	128, 391
Nut-lock	F. C. Hays	Uhricksville, Ohio	July 23, 1872	129, 822
Nut-lock	M. Hays	Binghamton, N. Y	Apr. 15, 1873	137, 919
Nut-lock	D. Hoffman and W. Johnston	Noblestown and Havelock, Pa.	June 7, 1870	104, 033
Nut-lock	H. P. Hood and W. Coombs	Indianapolis, Ind., and Bangor, Me.	Feb. 4, 1873	135, 553
Nut-lock	W. P. Horton	Milwaukee, Wis	Dec. 24, 1872	134, 204
Nut-lock	C. A. Howard	Pontiac, Mich	Nov. 25, 1873	144, 905
Nut-lock	A. C. Hull and J. S. Brown	Saint Louis, Mo., and Mount Pleasant, D. C.	June 3, 1873	139, 467
Nut-lock	E. Kaylor	Pittsburgh, Pa	Dec. 9, 1873	145, 427
Nut-lock	G. W. Keen	Zanesville, Ohio	Apr. 1, 1873	137, 453
Nut-lock	L. Leeds	New London, Conn	Dec. 30, 1873	146, 081
Nut-lock	K. H. Loomis	New York, N. Y	Nov. 26, 1872	133, 324
Nut-lock	K. H. Loomis	Cleveland, Ohio	Sept. 9, 1873	142, 709
Nut-lock	H. C. Lowe	North East, Md	Sept. 16, 1873	142, 858
Nut-lock	H. W. McAuley	Sterling, Wis	Oct. 25, 1870	108, 613
Nut-lock	T. B. McConaughey and J. Adams.	Newark, Del	May 17, 1870	103, 220
Nut-lock	H. McCown	Enon Valley, Pa	May 25, 1869	90, 567
Nut-lock	J. Miller, jr	Richmond, Ind	Aug. 20, 1872	130, 736
Nut-lock	L. J. Miller	Orville, Ohio	Apr. 29, 1873	138, 268
Nut-lock	J. Moorcroft	Newport, R. I	Sept. 27, 1870	107, 704
Nut-lock	A. Morley	New Troy, Mich	May 28, 1872	127, 359
Nut-lock	I. E. Nagle	New Orleans, La	Apr. 1, 1873	137, 382
Nut-lock	G. Palmer	Littletown, Pa	June 1, 1869	90, 867
Nut-lock	M. Payne	Cardington, Ohio	Dec. 21, 1869	98, 187
Nut-lock	S. Peatfield	Ipswich, Mass	Oct. 28, 1873	144, 131
Nut-lock	R. G. Peterson and J. Coulter	Perryville, Ohio	June 4, 1872	127, 509
Nut-lock	W. H. Phipps	Southborough, Mass	June 13, 1871	115, 889
Nut-lock	A. Porter	Chicago, Ill	May 27, 1873	139, 269
Nut-lock	W. P. Porter	Pittsburgh, Pa	June 1, 1869	90, 685
Nut-lock	P. F. Randolph	Jerseyville, Ill	May 2, 1871	114, 474
Nut-lock	J. A. Reed	Dunellen, N. J	July 8, 1873	140, 644
Nut-lock	T. E. Rhine	Quaker City, Ohio	Nov. 19, 1872	133, 253
Nut-lock	J. F. Saiger	Shelby, Ohio	July 9, 1872	128, 815
Nut-lock	D. Sawyer	Washington, Ind	Sept. 23, 1873	143, 097
Nut-lock	A. C. Smith	Fort Madison, Iowa	Dec. 30, 1873	146, 102
Nut-lock	C. S. Southwick and D. H. Baker.	Newport, R. I	Apr. 5, 1870	101, 674
Nut-lock	G. C. Stamper	Osceola, Iowa	May 18, 1869	90, 132
Nut-lock	J. B. Sweetland	Pontiac, Mich	Dec. 30, 1873	146, 106
Nut-lock	C. H. Taylor	New York, N. Y	Sept. 16, 1873	142, 877
Nut-lock	D. O. Thompson and B. A. Rice	Pontiac, Mich	Feb. 11, 1873	135, 864

Index of patents issued from the United States Patent Office from 1790 *to* 1873, *inclusive*—Continued.

Invention.	Inventor.	Residence.	Date.	No.
Nut-lock	F. P. Thompson	Frederickton, Canada	July 29, 1873	141, 403
Nut-lock	S. I. Thompson	New Waterford, Ohio	May 16, 1871	114, 991
Nut-lock	E. Turner	Greensburgh, Pa	Apr. 1, 1873	137, 513
Nut-lock	W. B. Wait	New York, N. Y	Apr. 22, 1873	138, 214
Nut-lock	C. R. Watrous	Mystic, Conn	Dec. 30, 1873	146, 113
Nut-lock	S. H. Wheeler	Dowagiac, Mich	Oct. 11, 1870	108, 223
Nut-lock	G. S. Willard	Salesville, Ohio	Oct. 8, 1872	132, 079
Nut-lock	A. Williams	Burrell, Pa	Jan. 14, 1873	134, 781
Nut-lock and washer	E. A. Ellsworth	Washington, D. C	Dec. 3, 1867	71, 722
Nut-lock bolt	T. T. Prosser	Chicago, Ill	May 31, 1870	103, 775
Nut-lock for fish-plates	J. W. Hazelton, A. A. Southard, and O. Merwin.	Drayton Plains and Elba, Mich.	May 4, 1869	89, 659
Nut-lock plate	E. D. Taylor	Hornellsville, N. Y	Apr. 5, 1870	101, 541
Nut-lock washer	M. F. McIntyer	Girard, Pa	Mar. 18, 1873	136, 848
Nut-locking device	I. Allen	Taunton, Mass	July 30, 1872	130, 003
Nut-locking device	J. L. Estel	Salem, Ohio	Nov. 28, 1871	121, 348
Nut-locking device	A. B. Davis	Philadelphia, Pa	Jan. 23, 1872	122, 999
Nut-locking device	J. M. Horton	Milwaukee, Wis	Sept. 26, 1871	119, 360
Nut-locking device	P. L. Gibbs	Dunleith, Ill	May 28, 1872	127, 336
Nut-locking device	T. W. Kirkwood	McKeesport, Pa	Apr. 2, 1872	125, 309
Nut-locking device	K. H. Loomis	New York, N. Y	Jan. 23, 1872	122, 900
Nut-locking device	S. B. Lowe	Chattanooga, Tenn	Feb. 20, 1872	123, 831
Nut-locking device	J. Maitland	Newburgh, Ohio	Dec. 12, 1871	121, 723
Nut-locking device	H. McCown	Enon Valley, Pa	Mar. 16, 1869	87, 953
Nut-locking device	A. McKenney	Maumee, Ohio	Aug. 22, 1871	118, 256
Nut-locking device	W. Moorehouse	Buffalo, N. Y	Oct. 13, 1868	83, 081
Nut-locking device	P. Philippi	Beardstown, Ill	Apr. 6, 1869	88, 668
Nut-locking device	H. L. Purdie	Buffalo, N. Y	Dec. 5, 1871	121, 546
Nut-locking device	H. L. Purdie	Buffalo, N. Y	Dec. 5, 1871	121, 547
Nut-locking device	H. L. Purdie	Buffalo, N. Y	Dec. 5, 1871	121, 548
Nut-locking device	H. L. Purdie	Buffalo, N. Y	Dec. 5, 1871	121, 549
Nut-locking device	J. Wetmore	Salem, Ohio	Jan. 21, 1873	135, 052
Nut-locking washer	J. H. Gridley	Washington, D. C	Feb. 26, 1867	62, 483
Nut-locking washer	R. Rutter	Vallejo, Cal	May 31, 1870	103, 780
Nut-locking washer	W. H. Williams	Canton, Ohio	July 6, 1869	92, 241
Nut-locking washer	L. Winslow	Rochester, N. Y	July 30, 1872	130, 097
Nut-machine	A. B. Bean	New Haven, Conn	Apr. 28, 1868	77, 347
Nut-machine	J. R. Blakeslee	Youngstown, Ohio	Oct. 8, 1872	132, 047
Nut-machine	J. Braun	Philadelphia, Pa	Apr. 22, 1873	138, 066
Nut-machine	R. Brayton	Buffalo, N. Y	Aug. 4, 1857	17, 914
Nut-machine	O. C. Burdict	New Haven, Conn	Apr. 10, 1866	53, 782
Nut-machine	O. C. Burdict	New Haven, Conn	Sept. 3, 1867	68, 556
Nut-machine	H. Carter and J. Rees	Pittsburgh, Pa	Nov. 22, 1853	10, 249
Nut-machine	R. H. Cole	Saint Louis, Mo	July 17, 1855	13, 252
Nut-machine	R. H. Cole	Saint Louis, Mo	June 3, 1856	15, 001
Nut-machine	R. H. Cole	Saint Louis, Mo	May 5, 1857	17, 197
Nut-machine	R. H. Cole	Saint Louis, Mo	Oct 27, 1857	18, 499
Nut-machine	R. H. Cole	Saint Louis, Mo	May 4, 1858	20, 145
Nut-machine	R. H. Cole	Saint Louis, Mo	Sept. 21, 1858	21, 551
Nut-machine	J. C. Day	Jersey City, N. J	Dec. 22, 1857	18, 892
Nut-machine	G. Dunham	Unionville, Conn	June 27, 1865	48, 383
Nut-machine	G. Dunham	Unionville, Conn	Nov. 14, 1865	50, 923
Nut-machine	G. H. Fuller	Unionville, Conn	May 9, 1871	114, 664
Nut-machine	A. B. Glover	Birmingham, Conn	July 7, 1868	79, 749
Nut-machine	R. Griffiths	Allegheny, Pa	Oct. 30, 1855	13, 720
Nut-machine	R. Griffiths	Allegheny City, Pa	Mar. 18, 1856	14, 452
Nut-machine	R. Griffiths	Philadelphia, Pa	Dec. 2, 1856	16, 142
Nut-machine	J. S. Hall	Pittsburgh, Pa	Aug. 11, 1868	80, 946
Nut-machine	J. Harlem	Philadelphia, Pa	June 4, 1867	65, 380
Nut-machine	D. Howell	Louisville, Ky	July 2, 1867	66, 238
Nut-machine	E. Hubner and C. Hall	New York, N. Y	Dec. 5, 1865	51, 316
Nut-machine	P. Koch	Manchester, England	May 9, 1871	114, 691
Nut-machine	P. Miles	New Haven, Conn	Jan 1, 1861	31, 056
Nut-machine	C. Ratcliff	Cincinnati, Ohio	Dec. 9, 1856	16, 188
Nut-machine	J. B. Savage	Southington, Conn	Dec. 14, 1858	22, 310
Nut-machine	A. Schwebel	New Haven, Conn	Sept. 24, 1867	69, 256
Nut-machine	A. J. Shepard	Buffalo, N. Y	Dec. 13, 1859	26, 446
Nut-machine	L. Thomas	Allegheny City, Pa	Aug. 11, 1863	39, 504
Nut-machine	W. E. Ward	Port Chester, N. Y	Oct. 7, 1856	15, 861
Nut-machine	F. Watkins	Birmingham, England	Dec. 8, 1868	84, 781
Nut-machine	S. H. Whitaker	Cincinnati, Ohio	Jan. 27, 1857	16, 507
Nut-machine	S. H. Whitaker	Cincinnati, Ohio	June 9, 1857	17, 534
Nut-machine	S. H. Whitaker	Cincinnati, Ohio	Sept. 22, 1857	18, 259
Nut-machine	S. H. Whitaker	Cincinnati, Ohio	Oct. 19, 1858	21, 860
Nut-machine die	O. C. Burdict	New Haven, Conn	Jan. 5, 1864	41, 137
Nut-machine die	O. C. Burdict	New Haven, Conn	Apr. 10, 1866	53, 780
Nut-machines, Reversible die-box for	W. Anderson	Pittsburgh, Pa	Nov. 16, 1869	96, 763
Nut-making machine	E. R. Addison	Wheeling, W. Va	July 21, 1868	80, 108
Nut-making machine	A. B. Bean	New Haven, Conn	Mar. 19, 1867	62, 923
Nut-making machine	J. R. Blakeslee	Youngstown, Ohio	July 16, 1872	129, 520
Nut-making machine	B. H. Bradley	New York, N. Y	July 5, 1870	105, 033
Nut-making machine	J. R. Bridges	Pittsburgh, Pa	Aug. 6, 1867	67, 403
Nut-making machine	O. C. Burdict	New Haven, Conn	Dec. 8, 1863	40, 815
Nut-making machine	W. Chisholm	Cleveland, Ohio	Nov. 17, 1863	40, 606
Nut-making machine	P. Coleman	Philadelphia, Pa	Feb. 14, 1865	46, 337
Nut-making machine	G. Dunham	Unionville, Conn	Aug. 6, 1867	67, 421
Nut-making machine	A. Emerson	New York, N. Y	May 7, 1867	64, 510
Nut-making machine	M. H. Foster and H. C. Hart	Unionville, Conn	May 5, 1868	77, 476
Nut-making machine	G. H. Fuller	Unionville, Conn	Feb. 16, 1869	86, 914
Nut-making machine	A. B. Glover	Yonkers, N. Y	May 10, 1864	42, 654
Nut-making machine	W. Horsfall	Sing Sing, N. Y	Feb. 16, 1869	87, 046
Nut-making machine	W. W. Hubbard	Philadelphia, Pa	Sept. 18, 1866	58, 101
Nut-making machine	E. Kaylor	ittsburgh, Pa	July 3, 1866	56, 062
Nut-making machine	E. Kaylor	Perrysville, Pa	Dec. 1, 1868	84, 494
Nut-making machine	P. Koch	New Haven, Conn	May 5, 1863	38, 397
Nut-making machine	L. A. Livingood	Womelsdorf, Pa	Sept. 3, 1872	131, 105
Nut-making machine	J. Paton	Newburgh, Ohio	Nov. 29, 1864	45, 268
Nut-making machine	J. Reese	Sharon, Pa	Feb. 7, 1854	10, 493

Index of patents issued from the United States Patent Office from 1790 *to* 1873, *inclusive*—Continued.

Invention.	Inventor.	Residence.	Date.	No.
Nut-making machine	T. R. Taylor	Cleveland, Ohio	Oct. 21, 1862	36, 737
Nut-making machine	S. Vanstone	Providence, R. I	Nov. 3, 1868	83, 745
Nut-making machine	L. Weckesser	New Haven, Conn	Dec. 13, 1864	45, 446
Nut-mandrel	G. A. Gray, jr	Cincinnati, Ohio	June 29, 1869	91, 841
Nut-roaster	D. A. T. Gale	Poughkeepsie, N. Y	Aug. 18, 1868	81, 159
Nut, Rubber jam	W. Todd	Allegheny City, Pa	Jan. 14, 1873	134, 823
Nut, Screw	L. Derby	New York, N. Y	Jan. 20, 1863	37, 435
Nut, Self-locking	J. H. Raymond and W. J. Brassington.	Brooklyn, N. Y	Feb. 14, 1865	46, 389
Nut, Spring	D. R. Pratt	Worcester, Mass	Jan. 25, 1870	99, 232
Nut, Take-up	J. Ross and W. W. Graham	Rutland, Vt	Nov. 15, 1870	109, 348
Nut-tapping and screw-cutting machine	G. Bennett and R. Dalzell	Waddington, N. Y	Jan. 22, 1861	31, 148
Nut-tapping machine	J. W. Durrell	Dunkirk, N. Y	May 30, 1871	115, 451
Nut-tapping machine	J. L. Gill, jr	Columbus, Ohio	Apr. 4, 1871	113, 512
Nut-tapping machine	A. B. Glover	Burmingham, Conn	July 7, 1857	17, 734
Nut-tapping machine	T. Hull and N. Thomas	Chicago, Ill	Aug. 13, 1872	130, 502
Nut-tapping machine	J. Kirkley	Chicago, Ill	Nov. 12, 1867	70, 862
Nut-tapping machine	S. W. Putnam, jr	Fitchburgh, Mass	Sept. 2, 1873	142, 511
Nut-tapping machine	S. W. Putnam, jr	Fitchburgh, Mass	July 8, 1873	147, 727
Nut-tapping machine	W. Scully	Detroit, Mich	Aug. 9, 1870	106, 215
Nut-tapping machine	T. Votteler	Cincinnati, Ohio	June 26, 1866	55, 965
Nut-tapping machine	F. Watkins	Birmingham, England	Nov. 15, 1864	45, 110
Nut-tapping machinery	D. Reese	Newburgh, Ohio	July 3, 1866	56, 095
Nut-threading machine	O. P. and L. W. Briggs	Chicago, Ill	Apr. 15, 1873	137, 757
Nut-threading machine	W. W. Hubbard	Philadelphia, Pa	Nov. 8, 1864	44, 954
Nut-turning device	B. Boardman	Norwich, Conn	Apr. 21, 1868	76, 983
Nut-wrench and drill	T. A. Chandler	Rockford, Ill	Sept. 26, 1871	119, 317
Nuts and bolt-heads, Machine for dressing	J. King	Bordentown, N. J	Feb. 27, 1849	6, 140
Nuts and bolts, Clamp for holding	W. J. Lewis	Pittsburgh, Pa	Apr. 12, 1864	42, 299
Nuts and washers, Machine for making	R. Brayton	Buffalo, N. Y	Jan. 9, 1855	12, 194
Nuts and washers, Machine for making	C. P. Geissenhainer	Pittsburgh, Pa	Aug. 7, 1866	57, 042
Nuts and washers, &c., Machine for making	W. Kenyon	Steubenville, Ohio	Oct. 14, 1851	8, 427
Nuts and washers, Machine for making	I. Scoville	Chicago, Ill	Aug. 18, 1863	39, 590
Nuts, bolts, &c., Manufacturing	J. Marsden	Orrell, England	Apr. 7, 1863	38, 116
Nuts, Device for dressing the sides of	I. Doeg	New Market, N. H	Nov. 4, 1873	144, 265
Nuts, Device for locking screw	W. F. Vernier	Philadelphia, Pa	Nov. 10, 1863	40, 583
Nuts from unscrewing, Mode of preventing	S. Noblet	Halifax, Pa	Sept. 21, 1858	21, 574
Nuts, Locking	D. Cumming, jr	New York, N. Y	June 16, 1868	78, 864
Nuts, Locking	W Mullins	Pittsburgh, Pa	Apr. 7, 1868	76, 500
Nuts, Machine for making hot-pressed	L. Thierry and G. B. Hill	Detroit, Mich	July 2, 1867	66, 414
Nuts, Machine for making metallic	A. S. Upson	Unionville, Conn	June 15, 1869	91, 288
Nuts, Machine for manufacturing iron	B. Hobbs	Newburgh, Ohio	Jan. 26, 1869	86, 302
Nuts, Machine for polishing metallic	R. H. and J. C. Cole	Saint Louis, Mo	June 3, 1856	15, 004
Nuts, Machine for rolling hollow bars for	J. Ostrander	Manchester, Va	Apr. 4, 1871	113, 332
Nuts, Making	R. H. Cole	Saint Louis, Mo	June 3, 1856	15, 003
Nuts, Making	I. H. Steer	Winchester, Va	June 19, 1855	13, 118
Nuts, Making	J. H. Sternbergh	Reading, Pa	Aug. 18, 1868	81, 224
Nuts, Making	O. W. Yale	Hartford, Conn	Sept. 22, 1868	82, 470
Nuts, Making metallic	E. W. Wood	Washington, D. C	Apr. 27, 1858	20, 110
Nuts, Making screw	F. Rheydt	Chicago, Ill	Dec. 29, 1868	85, 399
Nuts, Manufacture of	J. W. Gaskill and J. Christie	Phillipsburgh, N. J	Dec. 14, 1869	97, 901
Nuts, Manufacture of	L. Thomas	Allegheny City, Pa	Apr. 12, 1864	42, 320
Nuts, Manufacture of screw-threaded	C. H. Guard	Toronto, Canada	Feb. 20, 1872	123, 823
Nuts on axles, Securing	T. W. Williams	Philadelphia, Pa	May 26, 1857	17, 410
Nuts on bolts, Locking	G. J. Harris	Brooklyn, N. Y	June 25, 1872	128, 389
Nuts on bolts, Machine for putting	B. S. Walker	Sing Sing, N. Y	Jan. 23, 1872	122, 927
Nuts, Stop-washer for	H. N. Armstrong	Erie, Pa	June 27, 1865	48, 353
Nuts, Substitute for jam	T. B. Young	Louisville, Ky	May 3, 1870	102, 644
Nuts, Tapping	H. C. Hart and J. B. Blakeslee	Unionville, Conn	Dec. 24, 1867	72, 487
Nuts to bolts, Device for applying rubber	J. Minetree	Petersburgh, Va	Dec. 10, 1872	133, 716
Nutritive and curative preparation	S. C. Upham	Philadelphia, Pa	Feb. 26, 1867	62, 512
O.				
Oakum by machinery, Picking	E. Waterhouse	Gardiner, Me	Apr. 24, 1826	....
Oakum, Combination of machinery for picking	J. Stansbury and W. Ridgaway	Baltimore, Md	Jan. 17, 1842	2, 428
Oakum, curled hair, &c., Machine for picking	H. Burnham	Boston, Mass	Oct. 5, 1838	962
Oakum from junk, Manufacturing	E. Cook	Haddam, Conn	July 9, 1832	....
Oakum-machine	J. Tibbals	Haddam, Conn	Feb. 8, 1839	1, 077
Oakum, Machine for balling	S. G. Archibald	Edinburgh, North Britain	Apr. 18, 1871	113, 831
Oakum, Machine for picking	O. Allen	Tewksbury, Mass	July 16, 1842	2, 723
Oakum, Machine for picking	M. Morrison	Newburyport, Mass	Feb. 14, 1806	....
Oakum, Machine for tarring	R. Mansley	Philadelphia, Pa	Jan. 18, 1859	22, 662
Oakum, Machine for twisting	L. and C. Howard	Watkins, N. Y	Jan. 28, 1873	135, 338
Oakum, Machinery for the manufacture of	T. Henry and D. Tibbal	Chatham, Conn	July 8, 1843	3, 163
Oakum, Making	D. Rider	New York	Apr. 26, 1808	....
Oakum, Manufacture of	T. H. Dunham	Boston, Mass	Oct. 1, 1872	131, 859
Oakum, Manufacture of	L. S. Robbins and J. A. Southmayd.	New York, N. Y., and Elizabeth, N. J.	Mar. 2, 1869	87, 434
Oakum, Manufacturing	E. Cook and S. Usher	Haddam, Conn	June 7, 1834	....
Oakum, Manufacturing old junk or old rigging into	D. French	Middletown, Conn	Oct. 9, 1805	....
Oakum, Process for preparing	J. A. and G. Cormack	New York, N. Y	June 8, 1852	8, 995
Oakum with tar, Coating	J. F. Stairs	Halifax, Canada	Dec. 30, 1873	146, 105
Oar	H. W. Connor	Troy, N. Y	Mar. 23, 1869	88, 013
Oar	N. Davenport	Troy, N. Y	June 6, 1871	115, 584
Oar	S. W. Francis	New York, N. Y	Oct. 22, 1867	69, 983
Oar	R. E. Gleason	Libertyville, Ill	June 15, 1869	91, 226
Oar	W. J. Hough	Martinez, Cal	Jan. 18, 1870	98, 868
Oar	A. S. Jacobs	Saint Louis, Mo	July 16, 1867	66, 847
Oar	R. J. Kellam	Tremont, N. Y	Mar. 12, 1867	62, 755
Oar	T. Lindsley	Moline, Ill	Aug. 6, 1861	33, 001
Oar, Boat	R. Rode	Manchester Township, Pa	Sept. 23, 1856	15, 794
Oar, Boat	R. Rode	Manchester, Pa	Feb. 17, 1857	16, 656
Oar-collar	J. Robison	Curwinsville, Pa	Aug. 13, 1867	67, 804
Oar for steering, Head	C. Seaman	Washington County, Md	Aug. 15, 1817	....
Oar, Jointed	C. Dann	La Crosse, Wis	Sept. 27, 1870	107, 669
Oar-lock	W. E. Beman	Portland, Me	June 29, 1869	91, 817
Oar-lock	W. P. Glading	New York, N. Y	Feb. 21, 1854	10, 543

Index of patents issued from the United States Patent Office from 1790 *to* 1873, *inclusive*—Continued.

Invention.	Inventor.	Residence.	Date.	No.
Oar-lock	H. Hempstead	Greenport, N. Y	Aug. 24, 1869	93, 992
Oar-lock	M. A. Lanagan	Brooklyn, N. Y	Dec. 31, 1867	72, 865
Oar or boat-fin, Rigged	R. Smith	Brooklyn, N. Y	Sept. 12, 1865	49, 930
Oar, Sculling	F. T. Angers	Canastota, N. Y	Oct. 12, 1869	95, 754
Oar, Steel-plated	W. H. McMillan	New York, N. Y	Apr. 19, 1864	42, 389
Oar-swivel	M. Fryer	Greenbush, N. Y	Oct. 2, 1866	58, 402
Oars for boats, Machine for sawing boards into	J. Benson, E. Page, and R. T. Hough.	Boston, Mass., and West Leyden, N. Y.	May 26, 1842	2, 643
Oars, Machine for turning boat	B. and A. F. Potter	Hubbardston, Mass	Jan. 20, 1844	3, 408
Oars, &c., Machinery for turning	E. Page	Barcelona, N. Y	Dec. 26, 1845	4, 325
Oat-cleaner	S. Dickens, jr	Milwaukee, Wis	Dec. 13, 1870	110, 122
Oat-cleaning machine	W. D. Freeman	Tomales, Cal	June 8, 1869	91, 110
Oat-dusting machine	R. Exelby and G. W. Marshall	Buffalo, N. Y	Oct. 13, 1868	83, 054
Oat-separator	L. Patrie	Victor, N. Y	May 19, 1863	38, 630
Oats, barley, &c., Machine for hulling	J. Andrews and E. Piper	Camden, Me	Dec. 10, 1840	1, 894
Oats from wheat, Screen for separating	A. C. Ferren	Decorah, Iowa	July 29, 1862	36, 052
Observatory	L. B. Sawyer	Charlestown, Mass	Oct. 9, 1866	58, 680
Observatory and signal-tower, Portable	E. Tanner	Bowmansville, N. Y	Aug. 25, 1863	39, 689
Observatory, Military	T. Welham	Nemaha County, Nev	Dec. 16, 1862	37, 207
Obstetric chair and supporter	A. Blood	Janesville, Wis	Aug. 27, 1850	7, 590
Obstetrics, Manakin, with fœtus, &c., for illustrating the practice of.	B. H. Aylworth	Oxford, N. Y	Mar. 30, 1869	88, 432
Obstetrical bandage	J. O. Hamilton	Jerseyville, Ill	Oct. 8, 1867	69, 659
Obstetrical chair	C. C. Wingo	Newport, Va	Aug. 24, 1858	21, 291
Obstetrical extractor	A. C. Buffum	Chicago, Ill	Aug. 21, 1855	13, 453
Obstetrical supporter	F. H. Chase, A. Weston, and L. Babbit.	Clintonville and Au Sable, N. Y.	Aug. 20, 1850	7, 573
Obstetrical supporter	W. S. Daniels	Panama, N. Y	Mar. 14, 1854	10, 649
Obstetrical supporter	W. W. Finch, J. Blaisdell, and L. Babbit.	Essex County, N. Y	Jan. 15, 1850	7, 019
Obstetrical supporter	S. B. Manly	Corry, Pa	Sept. 3, 1867	68, 521
Obstetrical supporter	A. Pollard and S. Minkler	Clinton County, N. Y	Apr. 24, 1849	6, 385
Octave-coupling for reed-instruments	B. O. Church and H. Smith	Brattleborough, Vt	July 9, 1867	66, 460
Odometer	S. Beers	Naugatuck, Conn	Aug. 12, 1856	15, 509
Odometer	D. L. Branning	Tampa, Fla	Feb. 28, 1871	112, 116
Odometer	A. Carter	Forestville, Conn	Nov. 4, 1856	16, 003
Odometer	J. Clarke	Powhatan County, Va	Nov. 30, 1818	....
Odometer	N. R. Coburn	Lowell, Mass	Oct. 17, 1865	50, 456
Odometer	A. D. Hoffman	Wayne, Mich	Nov. 22, 1864	45, 157
Odometer	A. F. Howard	Hartford, Conn	July 3, 1860	28, 985
Odometer	L. W. Nicholls	North Brookfield, N. Y	Nov. 13, 1860	30, 640
Odometer	W. Oldroyd	Mount Vernon, Ohio	Nov. 14, 1848	5, 914
Odometer	W. H. Prescott and W. Judson	Galesburgh, Ill	Feb. 19, 1867	62, 152
Odometer	W. H. Prescott and W. Judson	Galesburgh, Ill	Mar. 24, 1868	75, 789
Odometer	J. C. Spencer	Phelps, N. Y	Sept. 17, 1867	69, 038
Odometer	M. W. Stevens and E. H. Drake	Stoughton, Mass	Oct. 1, 1867	69, 504
Odometer	J. Thompson	Middleborough, Mass	Oct. 31, 1854	11, 878
Odometer	S. R. Thorp	Batavia, N. Y	Oct. 31, 1854	11, 877
Odometer	H. Walker	Hartford, Vt	Mar. 22, 1859	23, 329
Odometer	J. M. Whitney	Bolton, Mass	Mar. 20, 1860	27, 589
Odometer	T. K. Work	Hartford, Conn	Feb. 8, 1859	22, 912
Odometer and counting-machine	J. L. Martin	Baltimore, Md	June 17, 1856	15, 140
Odometer-case	J. P. Seipel and C. B. Alsover	Easton, Pa	Jan. 17, 1871	111, 008
Odometer for carriage-wheels	S. Beers	Waterbury, Conn	Sept. 14, 1839	1, 325
Odometer for carriages	W. A. Turner	Plymouth, N. C	Nov. 19, 1833	
Odometer, Railway	M. F. Potter	Charlemont, Mass	June 13, 1854	11, 082
Odometer-register	H. F. Hart	New York, N. Y	Dec. 10, 1867	72, 033
Odometer-register	W. Yorke	Portland, Me	May 17, 1870	103, 272
Odor-trap for sinks, &c	W. E. Hatfield	Newark, N. J	June 17, 1862	35, 606
Odors, Compound for cleansing the human body from offensive.	H. D. Bird	Petersburgh, Va	Apr. 30, 1867	64, 189
Offal in the manufacture of fertilizers, Treating	J. J. Storer	Boston, Mass	Mar. 18, 1873	136, 943
Offal and manufacturing gas, Treating	J. Turner	Chicago, Ill	July 1, 1873	140, 391
Offal, Apparatus for treating	J. N. B. Bond	New York, N. Y	June 10, 1873	139, 759
Offal, Apparatus for treating	W. L. Bradley	Boston, Mass	Jan. 14, 1873	134, 844
Offal, Apparatus for treating	J. B. Chenoweth and E. P. Baugh.	Philadelphia, Pa	May 26, 1868	78, 261
Offal, Apparatus for treating	A. Edwards	New Haven, Conn	Jan. 28, 1873	135, 324
Offal, Apparatus for treating	A. W. Louth	Philadelphia, Pa	Mar. 28, 1865	47, 027
Offal, Apparatus for treating	J. A. Slater	New Haven, Conn	Feb. 11, 1873	135, 733
Offal-drier	M. Anderson	Chicago, Ill	June 3, 1873	139, 528
Offal, Treating	W. Adamson	Philadelphia, Pa	Feb. 14, 1865	46, 318
Office-chair	R. Fitts, jr	Fitchburgh, Mass	July 23, 1867	67, 034
Office-indicator	L. Burger	Chicago, Ill	June 20, 1871	116, 018
Office indicator and register	J. M. Keep	New York, N. Y	June 22, 1869	91, 543
Office register and directory	H. Rentchler	Bellville, Ill	May 3, 1870	102, 592
Office table and bed combined	D. Walker	Newark, N. J	Jan. 21, 1873	135, 021
Oil and burning-fluid by pneumatic pressure, Raising	J. H. Smith	Allegheny City, Pa	Dec. 10, 1867	72, 101
Oil and cider press	W. H. Hoeg	Ashtabula County, Ohio	May 20, 1826	....
Oil and naphtha from paraffine wax, &c., Process of separating.	R. B. and W. W. Lucas	Cleveland, Ohio	June 6, 1871	115, 622
Oil and other cans, Packing-case for	T. Scantlin	Evansville, Ind	May 14, 1872	126, 750
Oil and other liquids, Vessel for	G. W. Banker	New York, N. Y	June 27, 1871	116, 393
Oil and other mixed liquids, Apparatus for testing coal.	H. J. Smith and W. Jones	Philadelphia, Pa	May 6, 1862	35, 184
Oil and paint box and can	H. B. Everest and A. P. Ross	Rochester, N. Y	Nov. 12, 1872	132, 955
Oil and paint can	J. T. Williams	Philadelphia, Pa	Mar. 4, 1873	136, 575
Oil and paint, Can for transporting	H. Everett	Philadelphia, Pa	Oct. 1, 1872	131, 863
Oil and spirits of turpentine for paint, Preparing	J. G. Pendergast	Palmyra, N. Y	July 9, 1830	....
Oil and spirits of turpentine from pine wood, Producing.	S. L. Cole	Burlington, Vt	Dec. 1, 1863	40, 737
Oil and tallow cup	R. A. Copeland	Brooklyn, N. Y	July 6, 1869	92, 170
Oil and varnish can	E. T. Woodward	Charlestown, Mass	Aug. 12, 1862	36, 188
Oil, Anti-friction	J. F. Boynton	Syracuse, N. Y	Jan. 1, 1867	60, 829
Oil, Apparatus for coal	R. Shroder	Darlington, Pa	Dec. 16, 1856	16, 255
Oil, Apparatus for collecting floating	A. Ralston	West Middletown, Pa	Apr. 17, 1866	54, 014
Oil, Apparatus for collecting floating	J. J. Serrell	Hudson County, N. J	Feb. 5, 1867	61, 880
Oil, Apparatus for containing and measuring	E. F. Wilder	Lowell, Mass	Aug. 27, 1872	130, 833

Index of patents issued from the United States Patent Office from 1790 *to* 1873, *inclusive*—Continued.

Invention.	Inventor.	Residence.	Date.	No.
Oil, Apparatus for drawing and measuring	P. Noyes	Lowell, Mass	Oct. 4, 1870	107, 951
Oil, &c., Apparatus for extracting	L. Smith	Erie, Pa	Jan. 24, 1865	46, 033
Oil, Apparatus for filtering and refining	M. H. Krüger	New York, N. Y	Jan. 1, 1867	60, 747
Oil, Apparatus for manufacture of coal	H. K. Symmes	Newton, Mass	Nov. 1, 1859	26, 000
Oil, Apparatus for purifying	S. Gwynn and S. M. Clark	New York, N. Y., and Washington, D. C.	Aug. 25, 1868	81, 496
Oil, Apparatus for purifying and filtering	D. M. Lamb	Strathroy, Canada	Nov. 18, 1873	144, 621
Oil, Apparatus for purifying coal	B. Crawford	Allegheny, Pa	Mar. 28, 1871	113, 023
Oil, Apparatus for purifying mineral	W. Adamson	Philadelphia, Pa	Nov. 15, 1864	45, 007
Oil, Apparatus for redistillation of coal	L. Atwood	New York, N. Y	May 15, 1860	28, 246
Oil, Apparatus for testing coal	J. Tagliabue	New York, N. Y	Sept. 16, 1862	36, 488
Oil, Apparatus for testing coal	G. Tagliabue	New York, N. Y	Oct. 28, 1862	36, 826
Oil, Apparatus for treating	T. H. Burridge	Saint Louis, Mo	July 5, 1870	105, 038
Oil, Apparatus for treating linseed	D. D. Templeton	New York, N. Y	Mar. 18, 1873	136, 881
Oil as a substitute for linseed-oil, Process of preparing.	A. Millochau	New York, N. Y	May 19, 1863	38, 641
Oil, Ascertaining the quality of lamp	J. W. Harris	Dorchester, Mass	Sept. 3, 1840	1, 764
Oil-barrels, Process for lining	C. B. Hutchinson	Auburn, N. Y	July 4, 1865	48, 625
Oil, Bleaching and refining	O. Loew	New York, N. Y	Mar. 29, 1870	101, 284
Oil-boring apparatus	L. Atwood	Norwich, Conn	May 9, 1865	47, 609
Oil-box	J. W. Rhoades	Clyde, Ohio	Nov. 17, 1868	84, 215
Oil, Box and axle for saving	B. Kraft	Reading, Pa	July 1, 1851	8, 187
Oil-box for axles with conical journals	W. D. Titus	Brooklyn, N. Y	Feb. 12, 1856	14, 261
Oil, Burning	H. C. Dewitt	Saint Louis, Mo	Mar. 26, 1867	63, 229
Oil-burning apparatus,	A. J. White	Balston Spa, N. Y	Feb. 18, 1868	74, 648
Oil-bush and cap-neck spindle	J. Hinman	Clinton Township, Pa	Mar. 2, 1835	....
Oil-cabinet	T. Miller	East Boston, Mass	Nov. 25, 1873	144, 856
Oil-cabinet	M. H. Wiley	East Boston, Mass	Mar. 22, 1870	101, 070
Oil-cabinet	M. H. Wiley	East Boston, Mass	June 14, 1870	104, 389
Oil-cake mold	R. Macdonald	Hull, England	Nov. 11, 1873	144, 400
Oil-cake-packing apparatus	W. Hawes	Port Richmond, N. Y	Nov. 15, 1870	109, 2[illegible]6
Oil-cake stripper	W. Hawes	Port Richmond, N. Y	Dec. 31, 1872	134, 429
Oil-cake trimmer	W. Hawes	Port Richmond, N. Y	Nov. 12, 1872	132, 962
Oil-cake trimmer	A. Judson	Newark, N. J	June 6, 1871	115, 739
Oil, Camphene	A. V. H. Webb	New York, N. Y	Feb. 19, 1839	1, 082
Oil-can	T. K. Anderson	Addison, N. Y	Oct. 15, 1861	33, 471
Oil-can	W. C. Arthur	Baltimore, Md	Apr. 10, 1860	27, 876
Oil-can	J. Ashton	Fall River, Mass	July 3, 1866	55, 975
Oil-can	N. Aubin	Montreal, Canada	Dec. 28, 1869	98, 332
Oil-can	J. E. Auld	Buffalo, N. Y	Aug. 12, 1873	141, 619
Oil-can	B. F. Barnes	Boston, Mass	Nov. 24, 1868	84, 405
Oil-can	J. F. Bérendorf	Paris, France	Dec. 29, 1857	18, 949
Oil-can	J. Born	New Orleans, La	Sept. 24, 1872	131, 495
Oil-can	J. A. Bostwick	New York, N. Y	July 9, 1872	128, 699
Oil-can	J. Broughton	New York, N. Y	Mar. 7, 1865	46, 635
Oil-can	C. J. Brown	Plymouth, N. H	July 23, 1872	129, 783
Oil-can	J. Burson	Yates City, Ill	Aug. 22, 1871	118, 194
Oil-can	P. Childs	Northampton, Mass	July 11, 1871	116, 808
Oil-can	C. Chinnock	Brooklyn, N. Y	Sept. 7, 1869	94, 561
Oil, &c., can	G. H. Chinnock	Brooklyn, N. Y	Sept. 9, 1873	142, 613
Oil-can	B. Clark	New York, N. Y	Sept. 19, 1865	49, 979
Oil-can	P. C. Clark	Philadelphia, Pa	Feb. 1, 1870	99, 407
Oil-can	J. R. Cole	Morrisania, N. Y	Aug. 19, 1873	141, 920
Oil-can	J. R. Compton and C. M. Toms.	Rahway, N. J	May 24, 1870	103, 428
Oil-can	W. G. Cowell	Wallingford, Conn	Oct. 21, 1873	143, 810
Oil-can	P. J. Dwyer	Elizabethport, N. J	Apr 12, 1870	101, 839
Oil-can	L. S. Enos	Olean, N. Y	Feb. 12, 1856	14, 229
Oil-can	J. G. Evans	Saint Louis, Mo	Aug. 24, 1869	94, 093
Oil-can	W. A. Fenn	Rochester, N. Y	Apr. 27, 1869	89, 299
Oil-can	S. Field and C.W. Heald	Barre, Mass	Aug. 31, 1852	9, 233
Oil-can	J. L. Folsom	East Boston, Mass	Mar. 1, 1870	100, 279
Oil-can	W. A. Foster	Fitchburgh, Mass	Oct. 14, 1873	143, 567
Oil-can	O. H. Gardner	Fulton, N. Y	Jan. 5, 1869	85, 521
Oil-can	J. B. Gale	Raleigh, N. C	Dec. 17, 1867	72, 384
Oil-can	J. D. Gray	Cincinnati, Ohio	Mar. 1, 1870	100, 393
Oil-can	W. R. Hallock	Westfield, N. J	Dec. 30, 1873	145, 999
Oil-can	G. Hatch	Pomeroy, Ohio	Sept. 3, 1867	68, 503
Oil-can	C. J. Hauck	Williamsburgh, N. Y	Mar. 24, 1868	75, 904
Oil-can	L. W. Hemp and J. Z. Skinner	Saint Louis, Mo	June 18, 1872	128, 144
Oil-can	L. W. Hemp and J. Z. Skinner	Saint Louis, Mo	June 25, 1872	128, 394
Oil-can	L. W. Hemp and J. Z. Skinner	Saint Louis, Mo	Sept. 3, 1872	130, 999
Oil-can	G. L. Holt	Springfield, Mass	Apr. 24, 1866	54, 250
Oil, &c., can	M. W. House	Cleveland, Ohio	July 29, 1873	141, 224
Oil-can	G. P. Hunt	New York, N. Y	Oct. 18, 1859	25, 831
Oil-can	E. D. Hurst	Lancaster, Pa	Dec. 27, 1864	45, 610
Oil-can	J. Jackson, jr	Westerly, R. I	July 31, 1860	29, 377
Oil-can	W. E. Jenkins	Auburn, N. Y	July 27, 1869	93, 092
Oil-can	M. S. Kavanagh	Detroit, Mich	Aug. 1, 1871	117, 544
Oil-can	H. Keller	Sauk Centre, Minn	May 27, 1873	139, 318
Oil-can	L. W. Kent	Cleveland, Ohio	Sept. 12, 1871	118, 860
Oil-can	G. A. Knowlton	Natick, Mass	Dec. 10, 1867	71, 887
Oil-can	D. D. Mackay and C. Butler	Whitestone and New York, N. Y.	Nov. 7, 1871	120, 653
Oil-can	J. J. Marcy	West Meriden, Conn	Apr. 7, 1868	76, 483
Oil-can	J. Mayher	East Hampton, Mass	July 7, 1863	39, 157
Oil-can	J. Mayher	East Hampton, Mass	Sept. 5, 1865	49, 776
Oil-can	J. S. McIntire	Chicago, Ill	May 7, 1867	64, 436
Oil-can	M. McNamara	Bridgeport, Conn	Oct. 29, 1872	132, 680
Oil-can	W. C. Newkirk	Piqua, Ohio	Dec. 11, 1866	60, 409
Oil-can	J. Ogdin	Philadelphia, Pa	May 12, 1868	77, 755
Oil-can	L. H. Olmsted	Newark, N. J	May 3, 1864	42, 593
Oil-can	G. C. Overshiser	Binghamton, N. Y	Sept. 27, 1864	44, 485
Oil-can	J. M. Perkins and M.W. House	Cleveland, Ohio	July 4, 1865	48, 585
Oil-can	A. H. Phillippi	Reading, Pa	Feb. 12, 1867	62, 065
Oil-can	W. Polyblank	Cleveland, Ohio	Apr. 13, 1869	88, 903
Oil-can	W. T. Prall	Pomeroy, Ohio	Oct. 23, 1866	59, 066
Oil-can	C. Pratt	New York, N. Y	Nov. 16, 1869	96, 958
Oil-can	T. Priestley	Saxonville, Mass	Apr. 8, 1856	14, 641

Index of patents issued from the United States Patent Office from 1790 *to* 1873, *inclusive*—Continued.

Invention.	Inventor.	Residence.	Date.	No.
Oil-can	G. S. Prior	Boston, Mass	Dec. 5, 1871	121, 545
Oil-can	A. P. Quinby	Newark, N. J	Mar. 4, 1873	136, 544
Oil-can	F. W. Read	Marquette, Mich	Dec. 12, 1871	121, 813
Oil-can	M. Robbins	Cincinnati, Ohio	Oct. 1, 1867	69, 485
Oil-can	S. Sargent	Lowell, Mass	Dec. 11, 1866	60, 429
Oil-can	P. Scanlan	Indianapolis, Ind	Feb. 15, 1870	99, 789
Oil-can	E. S. Scripture	New York, N. Y	Feb. 25, 1862	34, 529
Oil-can	E. S. Scripture	Brooklyn, N. Y	Feb. 9, 1864	41, 537
Oil-can	E. S. Scripture	Brooklyn, N. Y	June 28, 1864	43, 345
Oil-can	S. Short and E. S. Scripture	Brooklyn, N. Y	Oct. 31, 1865	50, 742
Oil-can	G. W. and G. H. Simmons	Bennington, Vt	July 14, 1857	17, 810
Oil-can	F. Skinner	Cleveland, Ohio	Nov. 9, 1869	96, 626
Oil-can	H. M. Smith	Chicago, Ill	June 17, 1873	139, 927
Oil-can	D. G. Starkey	New York, N. Y	Apr. 9, 1850	7, 276
Oil-can	A. C. Stoessiger	Chicago, Ill	Nov. 5, 1872	132, 875
Oil-can	J. M. Thompson	Holyoke, Mass	Oct. 28, 1856	15, 986
Oil-can	B. W. Tuttle	Galena, Ill	Dec. 3, 1872	133, 683
Oil-can	C. O. Twining	Saint Louis, Mo	Sept. 12, 1871	118, 989
Oil-can	R. Wallace	Wallingford, Conn	July 2, 1872	128, 570
Oil-can	H. C. Warfel	Philipsburgh, Pa	Nov. 29, 1870	109, 693
Oil-can	O. H. Warren	Baldwinsville, N. Y	Sept. 23, 1873	143, 107
Oil-can	H. B. Wellman	Indianapolis, Ind	June 16, 1868	78, 906
Oil-can	H. Wells	Florence, Mass	Apr. 21, 1857	17, 124
Oil-can	W. Westlake	Chicago, Ill	Sept. 6, 1870	107, 141
Oil-can	J. A. Whitman	Auburn, Me	Apr. 2, 1867	63, 501
Oil-can	F. C. Wilson	Chicago, Ill	Aug. 5, 1873	141, 615
Oil-can	G. D. Winchell	Cincinnati, Ohio	Sept. 26, 1871	119, 441
Oil-can and lamp, Lubricating	E. Ashcroft	South Boston, Mass	Dec. 4, 1866	60, 117
Oil-can and oiler	J. E. Weaver	Temperanceville, Pa	Sept. 18, 1866	58, 157
Oil-can and powder-canister	E. W. Woodruff	Washington, D. C	July 26, 1864	43, 652
Oil-can and torch combined	W. Kelley	Baltimore, Md	Mar. 11, 1873	136, 738
Oil-can cap	J. H. Breckenridge	Meriden, Conn	Apr. 2, 1861	31, 866
Oil-can cap	J. H. Noyes	Centre Abington, Mass	Apr. 14, 1868	76, 802
Oil-can cap and nozzle	J. R. Hathway	Cortland, N. Y	July 23, 1872	129, 820
Oil-can cover	E. A. More	Saint Louis, Mo	Feb. 26, 1867	62, 357
Oil-can, Dripping	L. R. Boyd	New Brunswick, N. J	Jan. 21, 1873	135, 072
Oil, Can for transporting	J. E. Pimley	Newark, N. J	Dec. 30, 1873	146, 020
Oil-can from paper-stuff	E. G. Kelley	New York, N. Y	July 6, 1869	92, 316
Oil-can holder	W. S. Brick	Fitchburgh, Mass	Mar. 14, 1871	112, 537
Oil-can holder	J. N. Wilkins	Chicago, Ill	July 2, 1872	128, 517
Oil-can, Lubricating	F. Fildes	Media, Pa	Apr. 12, 1859	23, 560
Oil-can, Metallic	G. W. Banker	New York, N. Y	May 20, 1873	138, 986
Oil can or tank, Coal	C. H. Phelps	New York, N. Y	Sept. 16, 1862	36, 478
Oil can, Plumbago	D. D. Mackay and C. Butler	Whitestone and New York, N. Y.	Apr. 18, 1871	113, 783
Oil-can, Portable	C. J. Hauck	Brooklyn, N. Y	Apr. 16, 1872	125, 813
Oil-can, Railway	S. W. Murray and B. P. Lamason.	Milton, Pa	Feb. 22, 1870	100, 058
Oil-can, Safety	J. M. Perkins and M. W. House	Cleveland, Ohio	Dec. 6, 1864	45, 340
Oil-can spout	S. Moyle, jr	Bridgeport, Conn	May 31, 1870	103, 644
Oil-can spout	J. Robinson	Baltimore, Md	Oct. 15, 1872	132, 323
Oil-can stopper	L. R. Boyd	New Brunswick, N. J	Jan. 14, 1873	134, 842
Oil-can stopper	E. C. Godwin	Norfolk, Va	Nov. 12, 1872	133, 024
Oil-can stopper	C. Seimel	Brooklyn, N. Y	Oct. 26, 1869	96, 274
Oil-can top	M. Fowler	Meriden, Conn	Oct. 1, 1872	131, 748
Oil-can, Transportation	J. C. Moore	Philadelphia, Pa	May 21, 1872	127, 092
Oil-can, Transportation	J. C. Moore	Philadelphia, Pa	Nov. 19, 1872	133, 242
Oil-can tube	E. B. Beach	West Meriden, Conn	July 7, 1868	79, 723
Oil-can vent-spout	J. J. Marcy	Meriden, Conn	Jan. 18, 1870	98, 986
Oil-cans, Apparatus for filling	J. A. Bostwick	New York, N. Y	June 21, 1870	104, 414
Oil-cans, Apparatus for filling	E. F. Wilder	Lowell, Mass	Feb. 11, 1873	135, 680
Oil-cans, Boxing	D. Saunderson	Saint Louis, Mo	June 18, 1867	65, 836
Oil-cans, Packing-case for	J. M. Murphy	New York, N. Y	Dec. 14, 1869	97, 954
Oil-cans, Reversible cap and spout for	E. T. Covell	Brooklyn, N. Y	June 14, 1870	104, 275
Oil-cans, Screw-cap for	W. Rigg	London, England	Feb. 25, 1868	74, 850
Oil-cans, Ventilating decanting-apparatus for	W. Cleveland	Orange, N. J	Aug. 15, 1871	117, 984
Oil-case	I. Van Hagen	New York, N. Y	Jan. 27, 1857	16, 501
Oil, &c., case or box	S. Selden	Erie, Pa	June 30, 1863	39, 071
Oil, Cask for transporting	W. Jenkins	Abergavenny, England	Oct. 28, 1873	144, 103
Oil-casks, Sink for	T. Miller	Boston, Mass	Mar. 5, 1872	124, 279
Oil, Clarifying castor	J. Hallock	Burlington County, N. J	Nov. 5, 1818	....
Oil, Cleansing fish	P. Fitz Simmons	West Troy, N. Y	Apr. 4, 1846	4, 439
Oil, Cleansing whale	B. Folger		Jan. 2, 1792	....
Oil-cloth: *See* Floor-cloth.				
Oil-cloth and similar fabrics, Drying	O. S. Washburn	Philadelphia, Pa	Jan. 3, 1871	110, 702
Oil-cloth by machinery, Manufacturing	W. Powers	Lansingburgh, N. Y	Feb. 18, 1832	....
Oil-cloth carpeting, Printing	J. Albro, jr	Elizabethtown, N. J	Sept. 18, 1847	5, 299
Oil-cloth, carpets, &c., Metal binding for	J. Piper	Utica, N. Y	May 4, 1869	89, 599
Oil-cloth, Composition for	R. Hoskin	Brooklyn, N. Y	Oct. 27, 1863	40, 411
Oil-cloth, Composition for making	O. C. Washburn	Philadelphia, Pa	Feb. 18, 1862	34, 453
Oil-cloth, &c., Compound for coating	J. L. Tenney and J. W. Bailey	Skowhegan, Me	May 28, 1867	65, 301
Oil-cloth cutter	J. Rauch	Selin's Grove, Pa	Feb. 1, 1870	99, 474
Oil-cloth-cutting instrument	M. A. Sunderland	Utica, N. Y	Jan. 23, 1866	52, 123
Oil-cloth-drying machine	D. Sampson	Winthrop, Me	May 30, 1837	206
Oil-cloth, Machine for coating	M. Sawyer	Cincinnati, Ohio	Oct. 20, 1863	40, 364
Oil-cloth, Machine for painting floor	C. W. Strout and A. Wilder	Hallowell, Me	Sept. 21, 1869	94, 985
Oil-cloth, Machine for preparing floor	S. W. Herrick and C. G. Gilbert, jr.	Salem, N. J	July 30, 1867	67, 195
Oil-cloth, Machine for varnishing floor	C. W. Strout	Hallowell, Me	Mar. 16, 1869	87, 803
Oil-cloth, Machine for varnishing floor	C. W. Strout and A. Wilder	Hallowell, Me	July 27, 1869	93, 135
Oil-cloth, Making furniture	I. Macauley	Philadelphia, Pa	Apr. 4, 1825	....
Oil-cloth, Manufacture of	W. Berri, jr	Brooklyn, N. Y	Oct. 18, 1870	108, 437
Oil-cloth, Manufacture of	G. Sampson	Manchester, Me	June 20, 1865	48, 314
Oil-cloth, Manufacture of	J. B. Stevenson, jr	Philadelphia, Pa	Dec. 31, 1867	72, 932
Oil-cloth, Manufacture of floor	J. and W. Weems	Johnstone, Great Britain	Aug. 17, 1869	93, 776
Oil-cloth, Marbleized	T. Carson	Brooklyn, N. Y	Aug. 1, 1871	117, 512
Oil-cloth, Metal binding for	D. M. Knowles	New Bedford, Mass	Sept. 19, 1871	119, 035
Oil cloth, Oil-paint for coating	T. Potter	Philadelphia, Pa	Jan. 17, 1871	110, 999

Index of patents issued from the United States Patent Office from 1790 *to* 1873, *inclusive*—Continued.

Invention.	Inventor.	Residence.	Date.	No.
Oil-cloth, Printing	J. Albro	Elizabeth, N. J	June 7, 1859	24, 270
Oil-cloth-printing apparatus	J. Marchbank	Lansingburgh, N. Y	Jan. 2, 1867	51, 898
Oil-cloth-printing block	J. Jenkins	Elizabeth, N. J	May 11, 1852	8, 941
Oil-cloth, Printing floor	L. Moore	Ballston Spa, N. Y	Mar. 26, 1850	7, 221
Oil-cloth, Printing floor	N. B. Powers	Lansingburgh, N. Y	Mar. 26, 1850	7, 228
Oil-cloth-printing machine	C. Rommel	Elizabeth, N. J	Apr. 4, 1871	113, 570
Oil-cloth-printing machine	C. Rommel	Elizabeth, N. J	Apr. 15, 1873	137, 962
Oil-cloth-printing machine	W. E. Worth	San Francisco, Cal	Dec. 9, 1873	145, 379
Oil-cloth-printing machinery	A. F. Buchanan	East Newark, N. J	Oct. 25, 1870	108, 563
Oil-cloth-printing machinery	E. A. Goodes	Philadelphia, Pa	Mar. 7, 1871	112, 443
Oil-cloth-printing machinery	W. H. Halsey	Philadelphia, Pa	Feb. 21, 1871	112, 032
Oil-cloth-printing machinery	J. Kraft	Newark, N. J	Aug. 30, 1870	106, 834
Oil-cloth, Registering-block for printing	J. Albro	Elizabethtown, N. J	June 5, 1855	12, 989
Oil-cloth, Surfacing floor	W. Berry	Bedford, N. Y	Oct. 23, 1849	6, 816
Oil, Coloring lamp	E. Bourne	New York, N. Y	Mar. 4, 1834	....
Oil, Composition for paint	J. D. Baldwin	Columbus, Ga	Oct. 11, 1859	25, 712
Oil, Composition paint	J. McCafferty	New York, N. Y	Apr. 15, 1873	137, 854
Oil, Composition to prevent the absorption of animal and fish.	N. Hathaway	Fairhaven, Mass	May 22, 1835	... .
Oil, Compound	A. Dart	Dartford, Wis	Nov. 15, 1864	45, 028
Oil, Compound for the manufacture of lubricating	E. E. Hendrick	Carbondale, Pa	May 18, 1869	90, 100
Oil, Compound paint	Z. S. Doty	Janesville, Wis	May 17, 1864	42, 757
Oil, Converting the residuum of petroleum into	C. C. Mengel and A. P. Von Pohrnhoff.	Brooklyn, N. Y., and Saint Catharine's, Canada.	July 11, 1871	116, 852
Oil cool in lamps or burners, Keeping	J. Allen	Washington, D. C	Sept. 26, 1865	50, 087
Oil-cup	A. C. Ancona	Evansville, Ind	Apr. 12, 1870	101, 807
Oil-cup	A. T. Ballantine	Titusville, Pa	Sept. 30, 1873	143, 216
Oil-cup	C. W. Bancroft	Columbus, Ohio	June 7, 1864	43, 000
Oil-cup	S. Charnley	Portage City, Wis	July 28, 1868	80, 333
Oil cup	C. F. Davis	New Market, N. H	Nov. 27, 1866	59, 976
Oil-cup	C. Ferger	New York, N. Y	Apr. 14, 1868	76, 616
Oil-cup	J. Fogle	Putnam, Ohio	Sept. 20, 1864	44, 297
Oil-cup	J. H. Gomer	New York, N. Y	Dec. 10, 1867	71, 869
Oil-cup	J. P. Haines	New York, N. Y	July 20, 1869	92, 820
Oil-cup	E. Hurd	Virginia City, Nev	Sept. 22, 1868	82, 321
Oil-cup	T. C. Longton	Trenton, N. J	Mar. 24, 1868	75, 937
Oil-cup	H. McGraw	Detroit, Mich	Apr. 9, 1872	125, 602
Oil-cup	C. H. Parshall	Detroit, Mich	Apr. 9, 1872	125, 405
Oil-cup	W. P. Patton and J. R. Miller	Harrisburgh, Pa	Oct. 8, 1867	69, 582
Oil-cup	R. Poole	Baltimore, Md	Aug. 8, 1865	49, 352
Oil-cup	W. Rider	Almont, Mich	Aug. 22, 1871	118, 276
Oil-cup	J. Ross	North Cambridge, Mass	May 10, 1870	102, 973
Oil-cup	H. K. Sears and S. L. Holt	Hartford, Conn	May 28, 1867	65, 126
Oil cup	H. Stanley	Saint Louis, Mo	Oct. 6, 1868	82, 888
Oil-cup	E. Weissenborn	Hudson City N. J	Jan. 17, 1865	45, 945
Oil-cup	C. Williams	Vineland, N. J	Oct. 15, 1867	69, 882
Oil-cup	N. B. Williams	Providence, R. I	Mar. 3, 1868	75, 100
Oil-cup	N. B. Williams	Providence, R. I	Aug. 11, 1868	81, 051
Oil-cup and bearing for spindles	E. Convers	Oswego, N. Y	June 14, 1870	104, 115
Oil-cup, Automatic	J. H. Wilkinson	South New Market, N. H	May 28, 1872	127, 205
Oil-cup for adjustable boxes	C. C. Klein	Philadelphia, Pa	Feb. 9, 1869	86, 761
Oil-cup for journal-boxes	A. Richardson	Bellows Falls, Vt	July 29, 1851	8, 251
Oil-cup for lubricating engines	S. H. Whitmore	Cincinnati, Ohio	June 22, 1858	20, 674
Oil-cup for lubricating shafts	E. J. Gerdom and C. W. Schindler.	Albany, N. Y	Jan. 28, 1868	73, 795
Oil-cup for machinery	W. Douglas and H. M. Ingler	Bellaire, Ohio	May 14, 1867	64, 646
Oil-cup for machinery	T. W. Godwin	Portsmouth, Va	Sept. 22, 1863	40, 032
Oil-cup for machinery	R. Ross and W. Holland	Philadelphia, Pa	June 22, 1858	20, 665
Oil-cup for movable bearings	D. Harrigan	Somerville, Mass	Nov. 2, 1869	96, 427
Oil-cup for steam-engines	E. H. Ashcroft	Lynn, Mass	Nov. 3, 1868	83, 587
Oil-cup for steam-engines	J. C. Carroll	Litchfield, Ill	Aug. 11, 1868	80, 911
Oil-cup for steam-engines	T. Chatterton	Cleveland, Ohio	Aug. 6, 1867	67, 496
Oil-cup for steam-engines	D. Clark	Philadelphia, Pa	Jan. 10, 1854	10, 399
Oil-cup for steam-engines	F. H. Furniss	Cleveland, Ohio	Aug. 20, 1867	67, 867
Oil-cup for steam-engines	M. Hawkins	Birmingham, Conn	Mar. 19, 1867	63, 045
Oil-cup for steam-engines	F. Lunkenheimer	Cincinnati, Ohio	June 25, 1867	66, 157
Oil-cup for steam-engines	G. Trott	Pittsburgh, Pa	Mar. 28, 1854	10, 712
Oil-cup for steam-engines	N. J. White	Lawrence, Mass	Aug. 21, 1866	57, 418
Oil-cup for steam-engines	J. Wildhack	Pekin, Ill	Feb. 12, 1867	62, 103
Oil-cup for steam-pressure	F. P. McCullon and W. Woodcock.	Philadelphia, Pa	May 12, 1868	77, 900
Oil-cup, Globe	V. Giroud	New York, N. Y	Aug. 29, 1865	49, 624
Oil-cup, Lubricating	A. Higley	Cleveland, Ohio	Apr. 2, 1872	125, 133
Oil-cup, Lubricating	J. H. Palmer	Greenbush, N. Y	Apr. 17, 1866	54, 070
Oil-cup, Lubricating	E. N. Roland	Baltimore, Md	Dec. 15, 1857	18, 863
Oil-cup of connecting-rods or movable bearings of machinery.	D. Harrigan	Somerville, Mass	Feb. 2, 1869	86, 540
Oil-cup or lubricator	R. Ross	Pittsburgh, Pa	Oct. 15, 1861	33, 568
Oil-cups and inkstands, Spring-closing cap for	H. H. Heskett	Le Roy, Ill	June 7, 1870	104, 032
Oil-cups, Tip for	F. P. Pfleghar and W. Shölhorn	New Haven, Conn	Feb. 28, 1865	46, 586
Oil, &c., Device for drawing off and skimming	I. Peck and W. H. H. Glover	Southold and New York, N. Y.	July 7, 1863	39, 169
Oil, Discoloring and drying linseed	N. S. Parmantier	Philadelphia, Pa	Jan. 30, 1809	....
Oil-drill	L. H. Bowman	Morristown, Pa	Sept. 5, 1865	49, 704
Oil-dripper	J. M. Thompson	Holyoke, Mass	Aug. 7, 1855	13, 406
Oil, Factitious	H. W. Adams	New York, N. Y	Apr. 3, 1855	12, 614
Oil, Factitious	J. W. Harmon	Elizabethtown, N. J	Sept. 29, 1857	18, 279
Oil-feeder	J. Benson	Boston, Mass	Feb. 28, 1844	3, 454
Oil-feeder	R. Cornelius	Philadelphia, Pa	Apr. 6, 1843	3, 031
Oil-feeder	E. S. Scripture	Syracuse, N. Y	Mar. 15, 1845	3, 952
Oil-feeder	J. Turner	Cambridgeport, Mass	July 31, 1860	29, 446
Oil-filter	T. Antisell	New York, N. Y	Nov. 12, 1850	7, 767
Oil fire-tester, Carbon	G. E. Shaw	Pittsburgh, Pa	July 3, 1866	56, 167
Oil, Fish	W. D. Hall	Hamden, Conn	July 17, 1860	29, 164
Oil for burning and lubricating, Compound	R. A. Gilman	Woodland, Wis	Dec. 15, 1863	40, 924
Oil for burning and lubricating, Compound	W. G. Hermance	West Sandlake, N. Y	June 9, 1863	38, 825
Oil for coating leather and metal, Compound	W. K. Wyckoff	Ripon, Wis	July 20, 1869	92, 924
Oil for curriers' use	J. B. Kendall	Boston, Mass	June 15, 1869	91, 446

Index of patents issued from the United States Patent Office from 1790 *to* 1873, *inclusive*—Continued.

Invention.	Inventor.	Residence.	Date.	No.
Oil for lubricating, Compound for the treatment of	C. J. Eames and C. A. Seely	New York, N. Y	July 9, 1867	66, 573
Oil for lubricating from petroleum	J. K. Truax	Pittsburgh, Pa	Mar. 25, 1873	137, 157
Oil for lubricating machinery, &c	J. H. Lester	Brooklyn, N. Y	Jan. 30, 1866	52, 301
Oil for lubricating, paint, &c., Compound	W. H. Spooner	Bristol, R. I	Nov. 15, 1864	45, 090
Oil for manufacture of varnish and other purposes, Treating and drying.	F. Walton	Denton, England	Nov. 12, 1861	33, 722
Oil for mixing paint, Componnd	P. M. Wallover	Smith's Ferry, Pa	Dec. 21, 1869	98, 211
Oil for mixing paint, Compound	E. K. Wood and R. W. Henry	De Witt, Iowa	Jan. 29, 1867	61, 696
Oil for paint, Manufacture of drying	F. Huot	New York, N. Y	Sept. 3, 1867	68, 367
Oil for paintings, Preparing linseed or other	D. E. Breinig	New York, N. Y	Jan. 29, 1867	61, 653
Oil for purposes of illumination, Mode of using volatile.	B. F. Greenough	Boston, Mass	Nov. 15, 1843	3, 339
Oil for wool	W. H. Moss	New Richmond, Ohio	Aug. 4, 1868	80, 559
Oil from acid residuum of petroleum, Process of obtaining useful.	A. Farrar	Longwood, Mass	Oct. 26, 1869	96, 097
Oil from animal and vegetable substances, Apparatus for extracting.	C. O. Heyl	Berlin, Prussia	Sept. 3, 1867	68, 506
Oil from animal and vegetable substances, Apparatus for extracting.	H. A. Sterns	Smithfield, R. I.	May 24, 1870	103, 519
Oil from bitumens, Preparing	L. and W. Atwood	Waltham, Mass	Aug. 12, 1856	15, 506
Oil from bone-dust, Apparatus for extracting	W. W. Lucas	Cleveland, Ohio	Feb. 7, 1871	111, 658
Oil from cannel-coal, Production of	L. and W. Atwood	Waltham, Mass	Aug. 12, 1856	15, 505
Oil from cans, Apparatus for drawing	J. G. Evenden	Chicago, Ill	June 17, 1873	140, 022
Oil from castor beans, Making warm-pressed	T. Pharo	Tuckerton, N. J	Feb. 5, 1822	....
Oil from coal, Extracting	E. N. Horner	New Brighton, Pa	Jan. 25, 1859	22, 727
Oil, &c., from coal, Extraction of volatile	L. Atwood	Brooklyn, N. Y	Oct. 19, 1858	21, 805
Oil from cotton-seed, Extracting	C. L. Fleischmann	Washington, D. C	May 4, 1869	89, 720
Oil from cotton-seed, Extracting	A. A. Hayes	Boston, Mass	Apr. 8, 1856	14, 610
Oil from cotton-seed, Extracting	G. Palmer	Montville, Pa	Dec. 14, 1830	......
Oil from cotton-seed, Extracting	C. Whiting		Mar. 2, 1799	......
Oil from cotton-seed, Manufacture of	G. G. Henry	Mobile, Ala	Jan. 17, 1860	26, 845
Oil from cotton-seed, Process of making	W. M. Force	Newark, N. J	Aug. 20, 1872	130, 573
Oil from cotton-waste, Apparatus for separating	A. N. Cole	Brockville, Canada	July 2, 1872	128, 594
Oil from cotton-waste, Separating	A. N. Cole	Brockville, Canada	June 6, 1871	115, 708
Oil from fish, Cylinder-press for extracting	J. G. Stoddard and B. F. Gallup	Groton, Conn	June 18, 1867	65, 962
Oil from fish, Extracting	W. H. H. Glover	Southold, N. Y	Nov. 2, 1869	96, 315
Oil from fish, Extracting	A. M. Millochau	New York, N. Y	Mar. 19, 1861	31, 765
Oil from flax-seed, Extracting	D. Dodge	Hamilton, Mass	May 14, 1827	....
Oil from flax-seed, Pressing	C. Howe, F. Brewster, and J. Johnson.	Burlington, Vt	May 24, 1828	....
Oil from grain and other materials, Apparatus for separating.	E. T. Hutchinson	Baltimore, Md	Jan. 17, 1871	111, 064
Oil from grain and seed, Apparatus and process for removing.	E. T. Hutchinson	Baltimore, Md	Mar. 7, 1871	112, 350
Oil from herbs, and for other purposes, Apparatus for extracting.	L. Danbert	Louisville, Ky	May 7, 1867	64, 499
Oil from leather, &c., Extracting	J. A. Dean	Easton, Mass	May 1, 1866	54, 303
Oil, &c., from minerals, Apparatus for extracting	H. P. Gengembre	Pittsburgh, Pa	Jan. 30, 1866	52, 284
Oil, &c., from minerals, Process for extracting	H. P. Gengembre	Pittsburgh, Pa	Jan. 30, 1866	52, 283
Oil from oils extracted from bituminous minerals, Preparation of drying.	C. Cherry	Pittsburgh, Pa	Sept. 2, 1856	15, 644
Oil from palma christi, Extracting	J. G. Gebhard		Feb. 4, 1800	....
Oil from petroleum, Lubricating	T. E. Merrick	Cleveland, Ohio	June 22, 1869	91, 654
Oil from petroleum, Manufacturing lubricating	S. Gibbons	Freedom, Pa	Mar. 2, 1869	87, 485
Oil from petroleum residuum, Preparing paint	A. Millochau	New York, N. Y	Mar. 17, 1863	37, 918
Oil from pigs' feet, Extracting	C. Wahl	Chicago, Ill	Aug. 12, 1862	36, 184
Oil from poppy-seed, Obtaining	D. Markham		Dec. 2, 1824	....
Oil from resin, Lubricating	L. S. Robbins	New York, N. Y	Nov. 4, 1851	8, 489
Oil from resin, Paint	L. S. Robbins	New York, N. Y	Nov. 4, 1851	8, 491
Oil from resin, Tanners'	L. S. Robbins	New York, N. Y	Nov. 4, 1851	8, 488
Oil from running streams, Apparatus for obtaining	T. S. Scoville	Williamsport, Pa	July 18, 1865	48, 841
Oil from seed, Machinery for expressing	J. Hallock	Little Egg Harbor, N. J	Jan. 22, 1822	....
Oil from seed, Extracting	J. Robertson	Brooklyn, N. Y	Jan. 22, 1867	61, 463
Oils from seed, Extracting fat	J. Preston	Dorchester, Mass	Apr. 13, 1858	19, 948
Oil from seed, meal, &c., Apparatus for removing	E. S. Hutchinson	Baltimore, Md	July 26, 1870	105, 688
Oil from streams, Apparatus for collecting floating	E. C. Summers	Huntingdon, Pa	June 5, 1866	55, 388
Oil from streams, Apparatus for collecting floating	H. E. Towle	New York, N. Y	May 29, 1866	55, 178
Oil from sunflowers, Manufacturing	C. A. Barnitz	Spring Garden, Pa	Oct. 20, 1830	....
Oil from tanks, Device for discharging	W. J. Brundred	Oil City, Pa	Jan. 17, 1871	110, 953
Oil from vegetable and animal substances, Extracting.	T. Richardson, J. J. Lundy, and R. Irvine.	Newcastle-upon-Tyne, Leith, and Midlothian, Great Britain.	May 31, 1864	42, 987
Oil from vegetable and other matter, Apparatus and process for extracting.	T. Sim and E. S. Hutchinson	Baltimore, Md	Mar. 23, 1869	88, 222
Oil from vegetable and other matter, Apparatus for extracting.	E. S. Hutchinson	Baltimore, Md	Apr. 30, 1872	126, 300
Oil from vegetable and other matter, Apparatus for removing.	E. S. Hutchinson	Baltimore, Md	Mar. 7, 1871	112, 349
Oil from vegetable and other matter, Apparatus for removing.	T. Sim	Baltimore, Md	Mar. 7, 1871	112, 389
Oil from vegetable and other matter, Process and apparatus for extracting.	T. Sim	Baltimore, Md	Aug. 10, 1869	93, 645
Oil from water and other impurities, Apparatus for separating fish.	T. L. Robinson	Boston, Mass	Apr. 18, 1865	47, 333
Oil from wells, Apparatus for inducing the flow of.	J. B. Christian	Mount Carroll, Ill	May 22, 1866	54, 859
Oil, &c., from wells, Apparatus for raising	H. W. Faucett	Titusville, Pa	July 8, 1873	140, 692
Oil from wells, Obtaining	L. Phleger and G. G. Lobdell	Philadelphia, Pa., and Wilmington, Del.	Aug. 7, 1866	56, 989
Oil, &c., Furnace for the combustion of coal	J. Calkins, P. A. Hover, and C. W. Grannis.	Hudson and Gowanda, N. Y.	May 8, 1860	28, 154
Oil, Gelatinizing	E. Quern	New York, N. Y	Aug. 30, 1859	25, 277
Oil-globe for steam-chest	C. A. Wilson	Cincinnati, Ohio	Sept. 22, 1868	82, 467
Oil, grease, gums, &c., from cotton and woolen waste, Apparatus for removing and recovering.	J. W. Krepps	Chicago, Ill	Apr. 19, 1870	102, 018
Oil, Grinding cotton and other seed for	W. Wilber	New Orleans, La	Sept. 11, 1855	13, 558
Oil, Illuminating	A. W. Burrows	Cleveland, Ohio	Nov. 5 1867	70, 405
Oil, Illuminating	D. W. Fowler	Goshen, Ind	Mar. 3, 1868	75, 147
Oil, Illuminating	J. E. Noyes	New Albany, Ind	Sept. 15, 1868	82, 151

Index of patents issued from the United States Patent Office from 1790 *to* 1873, *inclusive*—Continued.

Invention.	Inventor.	Residence.	Date.	No.
Oil in making paints, Manufacturing a mucilagenous compound as a substitute for.	L. J. Vankleek	Poughkeepsie, N. Y	Nov. 23, 1829	....
Oil, Instrument for testing the inflammability of illuminating.	H. M. Hartshorn	Malden, Mass	June 29, 1869	91, 843
Oil, Lubricating	E. L. Brady	New Orleans, La	Sept. 17, 1867	68, 942
Oil, Lubricating	R. A. Chesebrough	New York, N. Y	May 19, 1868	77, 959
Oil, Lubricating	H. and C. Fink	Baltimore, Md	May 28, 1872	127, 328
Oil, Lubricating	G. W. Gladden	Cincinnati, Ohio	Nov. 5, 1872	132, 759
Oil, Lubricating	D. R. P. Hill	Fredericktown, Pa	Dec. 20, 1870	110, 361
Oil, Lubricating	J. A. Kestler	Chicago, Ill	Aug. 18, 1868	81, 093
Oil, Lubricating	J. M. Lippincott	Pittsburgh, Pa	Sept. 4, 1866	57, 737
Oil, Lubricating	W. H. Mason	Boston, Mass	May 25, 1852	8, 971
Oil, Lubricating	C. Moore	East Saginaw, Mich	Nov. 5, 1867	70, 454
Oil, Lubricating	G. O. Spence	Titusville, Pa	Apr. 14, 1868	76, 838
Oil, Lubricating	P. H. Vander Weyde	Philadelphia, Pa	Feb. 12, 1867	62, 092
Oil, Making benne	W. Few	New York	Mar. 1, 1809	....
Oil, Making paraffine	J. Young	Manchester, England	Mar. 23, 1852	8, 833
Oil, Manufacture of coal	L. Atwood	New York, N. Y	May 29, 1860	28, 448
Oil, Manufacture of coal	H. P. Gengembre	Allegheny, Pa	Aug. 16, 1859	25, 109
Oil, Manufacture of compound	F. L. De Gerbeth	Dalston, England	Aug. 18, 1868	81, 071
Oil, Manufacture of illuminating	C. C. Menger and A. P. Von Pohrnhoff.	Brooklyn, N. Y., and St. Catharine's, Canada.	Mar. 26, 1872	125, 067
Oil, Manufacture of lubricating	L. M. Mott	Chicago, Ill	Apr. 24, 1866	54, 192
Oil, Manufacture of lubricating	S. M. Mott	Wellsville, N. Y	July 23, 1861	32, 885
Oil, Manufacture of lubricating	H. T. Slemmer	Norristown, Pa	Feb. 27, 1866	52, 897
Oil, Manufacture of pyrogenic	L. Atwood	Brooklyn, N. Y	Dec. 28, 1858	22, 406
Oil, Manufacture of resin	W. T. Clough	Jersey City, N. J	Apr. 22, 1845	4, 008
Oil, Manufacturing resin	S. W. Hawes	Boston, Mass	Apr. 12, 1853	9, 662
Oil, Manufacture of resin	J. Merrill	Boston, Mass	Apr. 27, 1869	89, 493
Oil, Manufacture of resin	J. Treat	New York, N. Y	Mar. 15, 1870	100, 953
Oil, Manufacturing fish, whale, and sperm	D. C. and C. H. Knapp and A. H. Howland.	Oswego, N. Y	May 29, 1832	....
Oil obtained from mineral coal, Apparatus for purifying.	C. Cherry	Pittsburgh, Pa	Sept. 2, 1856	15, 642
Oil, Obtaining flaxseed	E. Horner	East Brook, Pa	Feb. 9, 1847	4, 961
Oil, Odorized kerosene	S. Jarden	Baltimore, Md	Mar. 4, 1862	34, 577
Oil of cotton-seed for all purposes of linseed-oil, Application of.	G. P. Digges	Albemarle County, Va	Dec. 16, 1820	....
Oil of hazze, Preparation of	Preswick and Fisher	New York	Aug. 17, 1835	....
Oil of vitriol, Concentrating	W. H. Belmain	Saint Helen's, Great Britain	Apr. 26, 1870	102, 205
Oil of vitriol, Storing and transporting	W. H. Belmain	Saint Helen's, Great Britain	Apr. 26, 1870	102, 206
Oil on surface of rivers, Mode of collecting	J. Cannon	New Richmond, Pa	Mar. 29, 1864	42, 073
Oil or blubber press	W. P. Chadwick	Edgartown, Mass	July 11, 1854	11, 299
Oil or fluid can	J. R. Nichols	Haverhill, Mass	Sept. 20, 1853	10, 035
Oil or suet cup	R. Ross	Bethlehem, Pa	Aug. 25, 1868	81, 542
Oil, Paint	R. Bartholow	Cincinnati, Ohio	Apr. 4, 1865	47, 083
Oil, Paint	V. Cordier	Paris, France	Sept. 29, 1868	82, 604
Oil, Paint	W. Johnson	Allegheny City, Pa	Sept. 20, 1864	44, 316
Oil, Paint	S. Melsom	Erie, Pa	Apr. 23, 1867	64, 125
Oil, Paint	A. Millochau	New York, N. Y	May 19, 1863	38, 640
Oil, Paint	W. W. Nichols	Lockport, N. Y	Nov. 21, 1865	51, 118
Oil, Paint, &c	D. S. Robinson	Oswego, N. Y	Dec. 18, 1866	60, 557
Oil, Paint, &c	W. Ward	Cleveland, Ohio	Mar. 10, 1868	75, 497
Oil-paintings from one surface to another, Method of transferring.	J. F. Wachsmuth	Highland, Ill	Apr. 20, 1869	89, 188
Oil, &c., Portable tank for	J. F. Keeler	Pittsburgh, Pa	Mar. 27, 1866	53, 451
Oil, Preparation of cod-liver	M. O'Reilly	New York, N. Y	Dec. 20, 1870	110, 275
Oil, Preparation of wool	T. Barrows	Dedham, Mass	Aug. 28, 1855	13, 486
Oil, Preparing	S. Gwynn	New York, N. Y	Aug. 25, 1868	81, 495
Oil, Preparing lubricating	L. Atwood	Boston, Mass	Mar. 29, 1853	9, 630
Oil, Preparing paraffine	W. Brown	Glasgow, Scotland	Sept. 27, 1853	10, 055
Oil-press	W. P. Chadwick	Edgartown, Mass	Nov. 9, 1852	9, 382
Oil-press	I. Clowes	Hampton, Va	Nov. 3, 1825	....
Oil-press	W. M. Force	Newark, N. J	Mar. 11, 1873	136, 716
Oil-press	E. Hills	Cincinnati, Ohio	Mar. 12, 1850	7, 166
Oil-press	A. Judson	Newark, N. J	Aug. 1, 1871	117, 641
Oil-press	D. L. Latourette	Saint Louis, Mo	Oct. 28, 1851	8, 469
Oil-press	W. W. Marsh	Jacksonville, Ill	Jan. 13, 1857	16, 391
Oil-press	J. Marshall	Pentonville Road, England	June 6, 1865	48, 140
Oil-press	W. V. McKenzie	Jersey City, N. J	May 6, 1862	35, 168
Oil-press	W. V. McKenzie	Jersey City, N. J	Apr. 25, 1865	47, 441
Oil-press	I. S. Schuyler	New York, N. Y	Aug. 27, 1861	33, 173
Oil-press	J. White		Nov. 27, 1804	....
Oil-press case and wrapping	W. M. Force	Newark, N. J	July 16, 1872	129, 021
Oil-press, Double hydrostatic	O. Badger and O. Lull	Otsego, N. Y	Aug. 9, 1832	....
Oil-press, Double hydrostatic	O. Badger and O. Lull	Otsego and Waterloo, N. Y	Oct. 5, 1836	40
Oil, Press for packing pulp for linseed	C. Moore	Trenton, N. J	Jan. 19, 1858	19, 149
Oil-press, Hydraulic	W. R. Fee	Cincinnati, Ohio	Sept. 13, 1859	25, 402
Oil-press, Hydraulic	W. Wilber	New Orleans, La	Dec. 25, 1855	14, 005
Oil press, Linseed	C. Moore	Trenton, N. J	Mar. 23, 1858	19, 708
Oil-press mat	H. Nutt	Brooklyn, N. Y	Aug. 25, 1863	39, 671
Oil-presses, Press-board for	J. Shinn	Philadelphia, Pa	Apr. 16, 1867	63, 951
Oil-presses, Securing and guiding the boxes of	W. W. Marsh	Jacksonville, Ill	Apr. 28, 1857	17, 161
Oil-presses, Steam-heating seed for	J. Criswell	Pittsburgh, Pa	July 31, 1837	334
Oil-pressing machinery	W. Wilber	New Orleans, La	Jan. 13, 1857	16, 422
Oil-pressing machinery	W. Wilber	New York, N. Y	Oct. 6, 1857	18, 367
Oil, &c., Process of extracting	C. A. Seely	New York, N. Y	Aug. 31, 1869	94, 246
Oil, Producing Columbian	T. Paul	Baltimore, Md	Feb. 23, 1811	....
Oil, Purifying	T. Drayton	Brooklyn, N. Y	July 4, 1854	11, 239
Oil, Purifying	H. Halvorson	Cambridge, Mass	Apr. 28, 1857	17, 181
Oil, Purifying and deodorizing	D. Symonds	Lowell, Mass	May 28, 1867	65, 137
Oil, Purifying and refining	W. M. Sloane	New York, N. Y	Nov. 29, 1870	109, 772
Oil, Purifying animal	A. Traband	New York	Nov. 21, 1842	2, 855
Oil, &c., Purifying coal	R. A. Chesebrough	New York, N. Y	Aug. 22, 1865	49, 502
Oil, Purifying cotton-seed	J. J. Gerrard	New York	Aug. 14, 1833	....
Oil, Purifying linseed	R. T. Clarke	Cincinnati, Ohio	May 17, 1870	103, 141
Oil, Purifying resin	S. L. Dana	Lowell, Mass	Apr. 19, 1853	9, 680
Oil, Purifying sperm and whale	J. L. Embree	New York	June 13, 1831	....

Index of patents issued from the United States Patent Office from 1790 *to* 1873, *inclusive*—Continued.

Invention.	Inventor.	Residence.	Date.	No.
Oil, Purifying spermaceti	R. Robotham		Aug. 14, 1799	....
Oil, Pyrogenous lubricating	S. Downer and J. Merrill	Boston, Mass	July 29, 1856	15, 418
Oil-refineries, Process of treating acid residuum from.	A. Farrar	Brookline, Mass	Mar. 15, 1870	100, 876
Oil, Solidifying	C. A. Jordery	Paris, France	May 7, 1872	126, 552
Oil, Steam apparatus for extracting vegetable	W. Wilber	New Orleans, La	Sept. 11, 1855	13, 557
Oil-still: *See* Still.				
Oil-stone box or case	G. O. Wichers	Lawrence, Mass	Feb. 9, 1864	41, 561
Oil-stone holder	H. Brown	Hamilton, Ill	Apr. 26, 1870	102, 218
Oil-storing tank	P. Noyes	Lowell, Mass	Apr. 12, 1870	101, 761
Oil, Substitute for linseed	Todd and Peabody	Washington, D. C	July 7, 1835	....
Oil, tallow, &c., Extracting	A. Randel	New York, N. Y	Mar. 18, 1862	34, 722
Oil-tank	G. W. Glass	Pittsburgh, Pa	Oct. 24, 1871	120, 263
Oil-tank	G. W. Howard	Pontiac, Mich	Feb. 18, 1862	34, 426
Oil-tank	J. K. Ingall	New York, N. Y	Apr. 10, 1866	53, 829
Oil-tank	S. H. Ingalls	New Bedford, Mass	Aug. 4, 1863	39, 446
Oil-tank	T. C. Joy	Titusville, Pa	Oct. 9, 1866	58, 648
Oil-tank	J. Lessler	Buffalo, N. Y	Nov. 5, 1872	132, 843
Oil-tank	H. Pierce and J. B. Button	Cleveland, Ohio	Jan. 22, 1867	61, 453
Oil-tank	J. Reese	Pittsburgh, Pa	Feb. 11, 1862	34, 373
Oil-tank	H. F., G. S., and A. Snyder	Williamsport and Blairsville, Pa.	Aug. 17, 1869	93, 918
Oil-tank	H. F., G. S., and A. Snyder	Williamsport and Freeport, Pa.	May 2, 1871	114, 361
Oil-tank	H. F., G. S., and A. Snyder	Williamsport and Freeport, Pa.	May 23, 1871	115, 127
Oil-tank	H. F., G. S., and A. Snyder	Williamsport and Freeport, Pa.	June 27, 1871	116, 366
Oil-tank	W. C. Strickler	Chicago, Ill	Aug. 20, 1872	130, 762
Oil-tank	A. G. Wilson	New York, N. Y	Sept. 8, 1868	81, 968
Oil-tank fitting	H. F. Snyder	Williamsport, Pa	Sept. 10, 1872	131, 190
Oil-tank, Incased	J. Shea and T. Campbell	Boston, Mass	Nov. 19, 1872	133, 123
Oil-tank safety-chamber	E. H. Knight	Washington, D. C	Jan. 15, 1867	61, 213
Oil-tank, Self-measuring	J. Schalk, jr	Guttenbergh, N. J	May 20, 1873	139, 084
Oil-tanks, Apparatus for filling and emptying	D. Birdsall	Brooklyn, N. Y	Feb. 20, 1872	123, 859
Oil-tanks, grain-cars, &c., Discharge-apparatus for	W. J. Brundred	Oil City, Pa	June 13, 1871	115, 817
Oil-tanks, Machine for turning the flanges on the heads of.	G. Selden	Erie, Pa	July 4, 1871	116, 636
Oil-tanks, Man-hole cover for	H. F. Snyder	Williamsport, Pa	Mar. 19, 1872	124, 768
Oil, tobacco, and cotton press	G. Palmer	Montville, Conn	Dec. 14, 1830	....
Oil, tobacco, and other presses	E. Thomas	Craigsville, Va	July 14, 1868	79, 875
Oil to render it drying, Treating cotton-seed	R. Farley	Cincinnati, Ohio	Jan. 31, 1871	111, 444
Oil to render it drying, Treating cotton-seed	H. Goldmann	New York, N. Y	Aug. 5, 1873	141, 503
Oil-tool extractor	S. Buckley and P. H. Lawrence	Lawrenceburgh, Pa	July 1, 1873	140, 345
Oil-tool jar	R. H. Lecky	Allegheny City, Pa	May 30, 1865	47, 961
Oil-tools, Socket-joint for	R. Gracey	Pittsburgh, Pa	May 29, 1866	55, 086
Oil, Transporting	W. W. Horton	Freeport, Ill	Mar. 21, 1865	46, 906
Oil, Treating	P. Marsh	South Adams, Mass	Jan. 1, 1856	14, 042
Oil, &c., Treating	H. L. Smith	Gambier, Ohio	Nov. 27, 1866	60, 076
Oil, Treating	H. K. Taylor and D. M. Graham	Cleveland, Ohio	May 22, 1866	54, 978
Oil, Treating cod-liver and castor	G. W. Fox	Manchester, Great Britain	Sept. 13, 1870	107, 244
Oil, Treating vegetable	R. T. Clark	Covington, Ky	Jan. 2, 1872	122, 439
Oil-trying apparatus	J. M. Hunter	New York, N. Y	Nov. 6, 1860	30, 575
Oil-vessel	G. W. Banker	Saint Louis, Mo	Oct. 13, 1863	40, 230
Oil, Vessel for holding	A. T. Woodward	New York, N. Y	Feb. 7, 1871	111, 599
Oil, Vessel for storing and transporting	T. J. McGarry	Cleveland, Ohio	Apr. 23, 1867	64, 123
Oil-vessels, Coating for	S. Gwynn	New York, N. Y	July 4, 1865	48, 552
Oil which has been used in lubricating machinery, Purifying.	C. Houlker	Frankford, Pa	Dec. 20, 1870	110, 364
Oil which wool contains, Composition to start the animal.	J. Goulding	Dedham, Mass	Aug. 24, 1827	....
Oils and fats by acids, Purification of	R. G. Loftus	Chelsea, Mass	Apr. 18, 1871	113, 782
Oils and fats for burning in lamps, lubricating machinery, &c., Treating.	S. Lewis	Rochester, N. Y	June 10, 1862	35, 527
Oils and fats, Process of chilling	J. E. Richardson	New York, N. Y	May 28, 1867	65, 275
Oils and fats, Purifying	R. C. Barton	Brooklyn, N. Y	Apr. 21, 1868	76, 974
Oils and fats, Treating and refining	J. W. Burton	Leeds, England	Aug. 6, 1872	130, 277
Oils and fats to form composition for illuminating and other purposes, Treating.	S. Lewis	Rochester, N. Y	June 14, 1864	43, 156
Oils, Apparatus for burning hydrocarbon	M. T. Gosnell	Baltimore, Md	Nov. 12, 1867	70, 712
Oils, Apparatus for burning hydrocarbon	J. R. Lee	Grass Valley, Cal	Mar. 7, 1871	112, 358
Oils, Apparatus for distilling heavy	H. Ryder	Somerville, Mass	Sept. 2, 1873	142, 515
Oils, Apparatus for extracting essential	G. G. Percival	Waterville, Me	Jan. 17, 1871	110, 998
Oils, Apparatus for gasifying and burning carbon	O. F. Morrill	Chelsea, Mass	Aug. 5, 1862	36, 102
Oils, Apparatus for testing hydrocarbon	J. B. Blair	Philadelphia, Pa	June 10, 1873	139, 654
Oils, Apparatus for treating hydrocarbon	J. J. Johnston	Allegheny, Pa	July 25, 1871	117, 426
Oils, Apparatus for treating hydrocarbon	L. M. Rice and S. E. Adams	Hartford, Conn., and Charlestown, Mass.	May 25, 1869	90, 392
Oils as fuel, Using hydrocarbon	G. B. Hill	New York, N. Y	Sept. 15, 1863	39, 918
Oils, Bleaching vegetable	T. Leonhard	Paterson, N. J	Dec. 3, 1867	71, 768
Oils, Burning hydrocarbon	D. D. Van Norman, L. B. Brown, and E. R. Morrison.	Petroleum Centre, Pa	Aug. 1, 1865	49, 179
Oils, Burning petroleum and other hydrocarbon	L. E. Truesdale	Warren, Mass	Apr. 12, 1870	101, 964
Oils, Clarifying sperm, whale, cotton-seed, and other.	E. C. Moss	New York	Dec. 28, 1832	....
Oils, Drier for petroleum and heavy	S. F. Rogers	Malden, Mass	Dec. 18, 1866	60, 558
Oils for use, Preparing hydrocarbon	E. Dennis	Sing Sing, N. Y	Feb. 13, 1872	123, 683
Oil in casks, Putting up	P. G. Finn	Erie, Pa	May 21, 1867	64, 855
Oils, Machine for extracting essential	V. Squarza	San Francisco, Cal	Feb. 20, 1866	52, 764
Oils, Machine for testing lubricating	R. H. Thurston	Hoboken, N. J	Dec. 24, 1872	134, 229
Oils, Manufacture of deodorized heavy hydrocarbon	J. Merrill	Boston, Mass	May 18, 1869	90, 284
Oils, Manufacture of hydrocarbon	L. Atwood	New York, N. Y	Mar. 26, 1861	31, 858
Oils, Manufacture of hydrocarbon	J. Merrill	Boston, Mass	July 2, 1861	32, 705
Oils, Manufacture of hydrocarbon	W. Spears	Jamestown, N. Y	Sept. 27, 1870	107, 734
Oils, Manufacture of siccative	J. Roux	New York, N. Y	Jan. 24, 1860	26, 929
Oils, &c., Pressing and straining	J. Cunningham and S. S. Edmonston.	New York, N. Y	June 22, 1812	....
Oils, Process and apparatus for extracting essential.	G. G. Percival	Waterville, Me	Nov. 7, 1871	120, 596

Index of patents issued from the United States Patent Office from 1790 *to* 1873, *inclusive*—Continued.

Invention.	Inventor.	Residence.	Date.	No.
Oils, Process and apparatus for extracting essential.	G. G. Percival	Waterville, Me	Dec. 10, 1872	133, 719
Oils, &c., Process and apparatus for producing	J. D. Stanley	Baltimore, Md	Sept. 10, 1872	131, 312
Oils, Process of preparing linseed for pressing in extracting.	C. Moore	Trenton, N. J	Aug. 19, 1856	15, 594
Oils, Purifying fish and other	L. Lecesne	New York, N. Y	June 23, 1812	....
Oils, Purifying hydrocarbon	J. Fordred	Blackheath, England	Apr. 24, 1866	54, 267
Oils, Purifying hydrocarbon	J. Merrill	Boston, Mass	June 28, 1864	43, 325
Oils, Storing	H. P. Genembre	Tarentum, Pa	Feb. 18, 1862	34, 462
Oils, Treating hydrocarbon	A. M. Burke and S. Wright	Cleveland, Ohio	June 25, 1867	65, 999
Oils, Treating hydrocarbon	H. K. Taylor and D. M. Graham	Cleveland, Ohio	Nov. 20, 1866	59, 751
Oiled fabric	G. Streat	New York, N. Y	Jan. 21, 1868	73, 473
Oiled silk and linen, Manufacturing	R. Hodgson		Feb. 1, 1793	....
Oiler	J. Broughton	New York, N. Y	Mar. 6, 1866	52, 959
Oiler	G. J. Capewell	West Cheshire, Conn	July 3, 1866	56, 004
Oiler	J. H. Godwin	Scotland Neck, N. C	Aug. 20, 1867	67, 974
Oiler	C. E. and Z. B. Grandy	Stafford Springs, Conn	Apr. 1, 1873	137, 361
Oiler	W. H. Hart	Meriden, Conn	May 23, 1865	47, 896
Oiler	C. J. Hauck	Williamsburgh, N. Y	May 8, 1866	54, 540
Oiler	C. J. Hauck	Brooklyn, N. Y	Sept. 17, 1872	131, 441
Oiler	G. L. Holt	Springfield, Mass	Apr. 23, 1867	64, 014
Oiler	J. King	Ansonia, Conn	July 10, 1866	56, 233
Oiler	J. Kips and W. Allmendinger	Melrose, N. Y	Feb. 12, 1867	61, 039
Oiler	P. Laflin	Warren, Mass	Sept. 18, 1866	58, 107
Oiler	A. D. Laws	Bridgeport, Conn	Jan. 17, 1871	111, 068
Oiler	E. McDuff and E. D. Forron	Warwick, R. I	Mar. 7, 1871	112, 367
Oiler	J. A. Metcalf	Lawrence, Mass	Aug. 31, 1869	94, 430
Oiler	L. H. Olmsted	Binghamton, N. Y	Nov. 19, 1861	33, 751
Oiler	L. H. Olmsted	Stamford, Conn	May 1, 1866	54, 394
Oiler	F. P. Pfleghar and W. Shollhorn.	New Haven, Conn	Mar. 5, 1867	62, 562
Oiler	F. Stone	New York, N. Y	July 30, 1867	67, 225
Oiler	C. B. Wood	Mansfield, Mass	Dec. 24, 1872	134, 342
Oiler and filler	A. Millar	Roxbury, Mass	Aug. 20, 1867	67, 896
Oiler for looms, Automatic	J. Crook	Ludlow, Vt	Dec. 26, 1871	122, 159
Oiler for loose pulleys	E. L. Conkey	Boston, Mass	Mar. 5, 1872	124, 333
Oiler for loose pulleys	C. A. King	Springfield, Mass	June 14, 1870	104, 322
Oiler for loose pulleys	S. R. Norris	New York, N. Y	Mar. 28, 1871	113, 199
Oiler for machinery	C. A. Morton	Biddeford, Me	July 20, 1869	92, 872
Oiler for machinery	D. C. Smiley	New York, N. Y	Apr. 25, 1854	10, 836
Oiler for machinery	N. Tallman	West New Brighton, N. Y	Sept. 1, 1868	81, 705
Oiler for steam-engines, Automatic	T. Sims	Philadelphia, Pa	May 14, 1872	126, 651
Oiler for the slide of steam-engines	C. C. Tracy	New York, N. Y	Aug. 17, 1869	93, 772
Oiler, Journal	J. A. Weaver	Temperanceville, Pa	June 26, 1866	55, 938
Oiler, Pneumatic	C. E. Decker	Chicago, Ill	Aug. 12, 1873	141, 703
Oiler, Portable	G. W. Crossley	Cleveland, Ohio	Jan. 23, 1872	122, 995
Oiler, Portable	J. A. Mathewson	Johnston, R. I	Aug. 20, 1872	130, 732
Oiler, Steam-cylinder	S. F. Stanton and O. Ripley	Manchester, N. H., and Charlestown, Mass	Aug. 24, 1869	94, 040
Oiler, Steam-engine	A. Worden	Ypsilanti, Mich	July 29, 1873	141, 304
Oiler-tube	F. J. Seymour	Meriden, Conn	Dec. 29, 1868	85, 484
Oilers, Attaching tubes to	S. V. Beckwith	Hamden, Conn	July 10, 1866	56, 163
Oiling-apparatus for crank-pins	J. F. Allen	Mott Haven, N. Y	Dec. 12, 1871	121, 744
Oiling-device	A. M. and J. M. Schoenfelt	Waterside, Pa	July 6, 1869	92, 378
Oiling spindles, top-rolls, &c., of spinning and other machinery, Device for.	S. H. Barber	East Greenwich, R. I	Feb. 12, 1867	61, 915
Oily matter from water, Apparatus for separating.	J. Naughten	Cincinnati, Ohio	Sept. 29, 1857	18, 293
Oily particles held in suspension by steam, Apparatus for separating the.	R. Hale	Roxbury, Mass	June 30, 1857	17, 675
Ointment	J. Bogan and J. B. McCrary	Clarksville, Ohio	Aug. 11, 1868	80, 854
Ointment	T. Combs	Yonkers, N. Y	Apr. 26, 1870	102, 224
Ointment	M. A. Curtis	San Francisco, Cal	Feb. 20, 1873	136, 308
Ointment	W. Dony	Honesdale, Pa	Nov. 28, 1865	51, 154
Ointment	G. D. Field	New Orleans, La	Mar. 8, 1870	100, 516
Ointment	W. W. Gray	Richmond, Va	Mar. 18, 1835	....
Ointment	J. G. Hucks	San Francisco, Cal	May 27, 1873	139, 315
Ointment	W. H. Jones	Henry County, Ala	May 7, 1872	126, 551
Ointment	L. H. Moseley	Franklin, Tenn	Feb. 4, 1868	74, 118
Ointment	N. Shepherd	Belmont County, Ohio	July 9, 1830	....
Ointment	L. Strober	Jersey City, N. J	Nov. 21, 1865	51, 099
Ointment	D. Wigg	Hyde Park, N. Y	Jan. 5, 1869	85, 714
Ointment	H. F. Wood	San Francisco, Cal	May 26, 1868	78, 409
Ointment, Cancer	E. Gilman	Licking, Ohio	Mar. 31, 1836	....
Ointment for bruises, burns, &c	J. T. Mulkey	Walton's Ford, Ga	Feb. 22, 1870	100, 057
Ointment for curing spavin, splint, &c., in horses.	B. Lehmann	Piqua, Ohio	Mar. 12, 1867	62, 759
Ointment for horses	G. P. Barnum	Marion, Iowa	Aug. 13, 1867	67, 627
Ointment for horses, cattle, &c	R. Jones	New York, N. Y	July 28, 1868	80, 353
Ointment for horses' hoofs	A. McDougall	Goffstown, N. H	Aug. 6, 1872	130, 229
Ointment for piles	A. M. Irwin	Brooklyn, N. Y	Nov. 14, 1871	120, 879
Ointment for piles	W. W. Riley	Mansfield, Ohio	Jan. 31, 1844	3, 417
Ointment for treating diseases in horses and other animals.	J. S. Williams	Warsaw, Ohio	Mar. 5, 1867	62, 717
Ointment for various uses	W. Judkins	Smithfield, Ohio	June 26, 1816	....
Ointment, Machine for making mercurial	J. W. W. Gordon	Baltimore, Md	June 5, 1844	3, 619
Ointment, Nitrated mercurial	C. Larned	Columbus, Ohio	Feb. 17, 1863	37, 697
Ointment or salve, Vegetable	W. C. and J. H. Roney	Gallupville, N. Y	June 28, 1864	43, 341
Ointment, Rheumatic	J. Gecmen	New York, N. Y	Feb. 18, 1873	135, 983
Oleine from fatty matter, Extracting	C. F. A. Simonin	Philadelphia, Pa	Oct. 28, 1873	144, 000
Omnibus	D. C. Macomber	New York, N. Y	Sept. 23, 1856	15, 769
Omnibus	H. M. Stow	San Francisco, Cal	Jan. 19, 1869	86, 042
Omnibus and car register	R. Cromelien and W. R. Crisp.	Washington, D. C	May 26, 1863	38, 655
Omnibus and car register	S. R. Stinard	Canandaigua, N. Y	May 26, 1863	38, 704
Omnibus-brake	H. Bothe	Saint Louis, Mo	Aug. 15, 1871	117, 972
Omnibus-coffer	J. T. Curtis	New York, N. Y	May 5, 1857	17, 206
Omnibus-register	L. Brauer	Washington, D. C	July 27, 1858	20, 986
Omnibus-register	F. Brunon	Philadelphia, Pa	Mar. 26, 1861	31, 778
Omnibus-register	L. Cromwell	Baltimore, Md	Apr. 15, 1856	14, 652
Omnibus-register	E. S. Dawson and A. Weeks	Syracuse, N. Y	Mar. 26, 1861	31, 788

Index of patents issued from the United States Patent Office from 1790 *to* 1873, *inclusive*—Continued.

Invention.	Inventor.	Residence.	Date.	No.
Omnibus-register	F. O. Deschamps	Philadelphia, Pa	Mar. 2, 1852	8, 770
Omnibus-register	F. O. Deschamps	Philadelphia, Pa	Feb. 7, 1854	10, 503
Omnibus-register	R. E. House	Binghamton, N. Y	May 25, 1858	20, 349
Omnibus-register	H. C. Howells and J. C. Howells	New York, N. Y., and Madison, Wis.	June 14, 1859	24, 390
Omnibus-register	W. M. Keague	Brooklyn, N. Y	Oct. 11, 1859	25, 740
Omnibus-register	L. W. Mallory	Philadelphia, Pa	June 20, 1854	11, 148
Omnibus-register	W. Morris	Philadelphia, Pa	Nov. 14, 1854	11, 939
Omnibus-register	I. B. Person and J. L. Brockett	Baltimore, Md	Aug. 19, 1851	8, 304
Omnibus-register	J. Rodgers	New York, N. Y	Mar. 18, 1856	14, 470
Omnibus-register	P. Taltavull	Washington, D. C	Jan. 25, 1853	9, 563
Omnibus-register	I. Z. A. Wagner	Philadelphia, Pa	Sept. 28, 1852	9, 289
Omnibus-register	R. F. White	New York, N. Y	June 7, 1859	24, 345
Omnibus-registers, Self-locking device for	M. Offley	Baltimore, Md	Aug. 7, 1860	29, 549
Omnibus-spring	J. H. Dennis	Louisville, Ky	Jan. 14, 1862	34, 136
Omnibus-step	J. Ashenfelder	Philadelphia, Pa	Mar. 23, 1852	8, 816
Omnibus-step	W. H. Hoyt	New York, N. Y	May 27, 1851	8, 119
Omnibus-step protector	T. Coles	New York, N. Y	July 25, 1854	11, 360
Omnibus-steps, Turning up	S. Burdett	New York, N. Y	Aug. 27, 1850	7, 591
Onion-seed from grain, Machine for separating	L. Ruggles and D. Tomlinson	Brookfield, Conn	Mar. 16, 1810	....
Open or middle ring	B. F. Zinn	Mount Rock, Pa	Nov. 17, 1868	84, 155
Open-ring	A. H. Bixler	Carlisle, Pa	May 5, 1868	77, 441
Opera-chair	A. Abel	New York, N. Y	Jan. 30, 1872	123, 140
Opera-chair	B. Koechling	New York, N. Y	Feb. 8, 1870	99, 576
Opera-chairs, Perfuming	S. Fredrick	New York, N. Y	Aug. 13, 1872	130, 421
Opera-glass, Folding	J. Saxton	Washington, D. C	June 29, 1869	91, 969
Operating-chair	J. B. Morrison	Saint Louis, Mo	Sept. 29, 1868	82, 542
Operating-drill	T. D. Keith	Mayville, Wis	May 11, 1869	89, 932
Operating-drill	E. G. Lamson	Shelburne Falls, Mass	Sept. 17, 1867	68, 888
Opium from the poppy-plant, Pressing	L. Cook	Shrewsbury, N. J	Apr. 20, 1832	....
Ophthalmic vapor-apparatus	T. F. Frank	Ischua, N. Y	Aug. 30, 1859	25, 252
Optical instrument	O. B. Brown	Malden, Mass	Aug. 10, 1869	93, 594
Optical instrument	L. Heath	Boston, Mass	May 8, 1866	54, 542
Optical instrument	R. B. Tolles	Canastota, N. Y	Sept. 25, 1855	13, 603
Optical instruments, Adjustment for	C. B. Richard	Hartford, Conn	May 23, 1865	47, 860
Optical instruments, Binocular eye-piece for	R. B. Tolles	Canastota, N. Y	July 3, 1866	56, 125
Optical instruments, Double eye-piece for	A. Beckers	New York, N. Y	Dec. 13, 1859	26, 407
Oracular wheel or center-table	W. O. George	Richmond, Va	May 20, 1856	14, 910
Orange-knife	E. L. Cooke	Hartford, Conn	Mar. 15, 1870	100, 862
Ordnance	B. T. Babbitt	New York, N. Y	Feb. 5, 1861	31, 291
Ordnance	B. T. Babbitt	New York, N. Y	Feb. 25, 1862	34, 472
Ordnance	E. D. Baker	Claremont, N. H	Feb. 4, 1862	34, 287
Ordnance	T. A. Blakly	London, England	Feb. 16, 1864	41, 662
Ordnance	B. Britten	Red Hill, England	June 27, 1871	116, 408
Ordnance	L. W. Broadwell	New Orleans, La	June 19, 1866	55, 609
Ordnance	J. B. Eads	Saint Louis, Mo	Sept. 4, 1866	57, 692
Ordnance	J. Ericsson	New York, N. Y	Jan. 12, 1864	41, 208
Ordnance	G. H. Ferriss	Utica, N. Y	May 3, 1864	42, 571
Ordnance	T. Fowlds	Trevorton, Pa	May 6, 1862	35, 148
Ordnance	E. Heaton	New Haven, Conn	June 3, 1862	35, 443
Ordnance	J. C. C. Holenshade	Cincinnati, Ohio	Apr. 7, 1863	38, 110
Ordnance	W. W. Hubbell	Philadelphia, Pa	Sept. 13, 1864	44, 194
Ordnance	J. L. Lowry	Pittsburgh, Pa	May 16, 1865	47, 740
Ordnance	T. J. Mayall	Roxbury, Mass	Nov. 27, 1860	30, 742
Ordnance	T. J. Mayall	Roxbury, Mass	May 21, 1861	32, 376
Ordnance	R. Montgomery	New York, N. Y	Mar. 11, 1862	34, 666
Ordnance	W. Ostrander, I. D. Reeder, and J. Cordwan.	Brooklyn, N. Y	Apr. 15, 1862	34, 977
Ordnance	W. Palliser	Pall Mall, England	June 29, 1869	91, 864
Ordnance	J. B. Read	Tuscaloosa, Ala	Oct. 5, 1869	95, 604
Ordnance	C. W. Stafford	New York, N. Y	Sept. 6, 1864	44, 119
Ordnance	C. W. Stafford	Old Saybrook, Conn	Feb. 21, 1865	46, 506
Ordnance	N. Wiard	New York, N. Y	May 26, 1863	38, 709
Ordnance	N. Wiard	New York, N. Y	Aug. 18, 1863	39, 604
Ordnance	N. Wiard	New York, N. Y	Apr. 12, 1864	42, 315
Ordnance	C. Walter	Bridgeport, Conn	Mar. 13, 1855	12, 522
Ordnance	W. E. Woodbridge	Little Falls, N. Y	Jan. 1, 1867	60, 979
Ordnance	L. W. Wright	Thorndike, Mass	May 29, 1866	55, 193
Ordnance	D. F. Yeakel	La Fayette, Ind	Apr. 29, 1862	35, 124
Ordnance, Adjustable front sight for	J. B. Learock	Boston, Mass	Mar. 29, 1864	42, 091
Ordnance, Adjustable sight for	C. B. Long	Worcester, Mass	Dec. 2, 1862	37, 076
Ordnance and fire-arm	R. White	Bridgeport, Conn	Aug. 28, 1866	57, 607
Ordnance and other fire-arms	O. Abbruzzo	New York, N. Y	Dec. 1, 1868	84, 525
Ordnance and other fire-arms, Chamber of	J. P. Schenkl	Worcester, Mass	Apr. 19, 1859	23, 746
Ordnance and other fire-arms, Wad for	E. D. Williams	Philadelphia, Pa	May 13, 1862	35, 273
Ordnance and projectile	B. F. Bates and C. R. Macy	New York, N. Y	Mar. 7, 1865	46, 625
Ordnance and projectile	C. Heaton	New York, N. Y	June 25, 1872	128, 392
Ordnance and projectile therefor	B. C. Pole	Washington, D. C	Sept. 30, 1873	143, 251
Ordnance, Apparatus for casting	J. Millholland	Reading, Pa	July 15, 1862	35, 884
Ordnance, Arranging water-tubes for cooling the breech of.	J. Hursh	Philadelphia, Pa	July 15, 1862	35, 877
Ordnance, Attaching trunnions to	R. Gibbons	Oakland, Cal	Aug. 2, 1864	43, 681
Ordnance, Automatic revolving	T. J. Campbell	Lincoln, Ill	June 10, 1862	35, 504
Ordnance, Breech-loading	C. Alger	Hudson, N. Y	Dec. 24, 1861	33, 976
Ordnance, Breech-loading	W. Bacon	Monticello, Kans	May 11, 1869	89, 965
Ordnance, Breech-loading	P. R. Beaupre	Metropolis, Ill	Apr. 14, 1868	76, 586
Ordnance, Breech-loading	G. W. Bishop	Brooklyn, N. Y	Sept. 9, 1856	15, 682
Ordnance, Breech-loading	R. C. Bristol	Chicago, Ill	Mar. 25, 1862	34, 730
Ordnance, Breech-loading	L. W. Broadwell	St. Petersburg, Russia	Dec. 10, 1861	33, 876
Ordnance, Breech-loading	L. W. Broadwell	London, Great Britain	July 12, 1864	43, 553
Ordnance, Breech-loading	L. W. Broadwell	New Orleans, La	June 19, 1866	55, 762
Ordnance, Breech-loading	E. B. Butterfield	New York, N. Y	Nov. 5, 1861	33, 635
Ordnance, Breech-loading	J. S. Butterfield	Philadelphia, Pa	Apr. 30, 1861	30, 238
Ordnance, Breech-loading	J. A. De Brame	New York, N. Y	Dec. 24, 1861	34, 025
Ordnance, Breech-loading	L. Evans	Morgantown, Va	Oct. 9, 1860	30, 303
Ordnance, Breech-loading	G. H. Ferris	Utica, N. Y	Mar. 22, 1864	41, 984
Ordnance, Breech-loading	T. L. Fortune	Weston, Mo	July 26, 1864	43, 655
Ordnance, Breech-loading	J. A. France	Cobleskill, N. Y	Sept. 10, 1861	33, 244

Index of patents issued from the United States Patent Office from 1790 *to* 1873, *inclusive*—Continued.

Invention.	Inventor.	Residence.	Date.	No.
Ordnance, Breech-loading	G. P. Ganster	New York, N. Y	Sept. 9, 1862	36, 401
Ordnance, Breech-loading	W. F. Goodwin	Powhatan, Ohio	May 20, 1862	35, 311
Ordnance, Breech-loading	W. F. Goodwin	Powhatan, Ohio	June 14, 1864	43, 110
Ordnance, Breech-loading	W. F. Goodwin	Powhatan, Ohio	June 14, 1864	43, 111
Ordnance, Breech-loading	W. F. Goodwin	Powhatan, Ohio	June 28, 1864	43, 306
Ordnance, Breech-loading	C. Gordon	Goswell Road, England	Apr. 22, 1873	138, 146
Ordnance, Breech-loading	C. Grimshaw	Milwaukee, Wis	Oct. 29, 1861	33, 582
Ordnance, Breech-loading	J. S. Hall	Pittsburgh, Pa	Apr. 29, 1862	35, 095
Ordnance, Breech-loading	E. Hamilton	Chicago, Ill	July 9, 1861	32, 768
Ordnance, Breech-loading	W. S. Henson	Newark, N. J	Nov. 5, 1861	33, 646
Ordnance, Breech-loading	W. W. Hubbell	Philadelphia, Pa	May 29, 1860	28, 486
Ordnance, Breech-loading	W. W. Hubbell	Philadelphia, Pa	July 5, 1864	43, 412
Ordnance, Breech-loading	J. Lee	Bolivar, Ohio	May 19, 1863	38, 638
Ordnance, Breech-loading	C. F. Leisen	Philadelphia, Pa	Oct. 22, 1861	33, 530
Ordnance, Breech-loading	L. M. Lull and J. T. Starr	Walnut Grove, Ill	Feb. 18, 1868	74, 557
Ordnance, Breech-loading	H. F. Mann	La Porte, Ind	Nov. 26, 1861	33, 787
Ordnance, Breech-loading	T. J. Mayall	Roxbury, Mass	Oct. 9, 1860	30, 335
Ordnance, Breech-loading	E. R. McCabe	Rochester, Iowa	May 27, 1862	35, 380
Ordnance, Breech-loading	J. A. Miller	Paducah, Ky	Feb. 7, 1865	46, 259
Ordnance, Breech-loading	R. L. Mills	Columbus, Ohio	Feb. 16, 1864	41, 631
Ordnance, Breech-loading	R. R. Moffatt	La Crosse, Wis	July 8, 1862	35, 863
Ordnance, Breech-loading	J. B. Moody	Cincinnati, Ohio	Mar. 6, 1866	53, 026
Ordnance, Breech-loading	C. Morrill	New York, N. Y	June 17, 1862	35, 619
Ordnance, Breech-loading	C. C. W. Müller	New Orleans, La	Aug. 13, 1867	67, 792
Ordnance, Breech-loading	M. J. Murphy	Cork, Ireland	Dec. 18, 1866	60, 540
Ordnance, Breech-loading	A. F. Potter	San Francisco, Cal	May 17, 1870	103, 078
Ordnance, Breech-loading	R. B. Reynolds	United States Navy	Sept. 29, 1863	40, 121
Ordnance, Breech-loading	T. C. Rice	Cambridgeport, Mass	Jan. 28, 1862	34, 266
Ordnance, Breech-loading	W. Schmoele, jr	Philadelphia, Pa	Jan. 19, 1864	41, 325
Ordnance, Breech-loading	W. G. Sherwin, J. McFarland, and C. Thieme.	Cincinnati, Ohio	Nov. 19, 1861	33, 756
Ordnance, Breech-loading	F. Shilling	Fort Delaware	Oct. 13, 1863	40, 295
Ordnance, Breech-loading	A. B. Smith	Clinton, Pa	July 28, 1863	39, 359
Ordnance, Breech-loading	H. Stanhope	Philadelphia, Pa	July 23, 1861	32, 921
Ordnance, Breech-loading	W. Stow	Utica, N. Y	July 29, 1862	36, 036
Ordnance, Breech-loading	E. A. Sutcliffe	New York, N. Y	Aug. 18, 1863	39, 596
Ordnance, Breech-loading	N. Thompson	Brooklyn, N. Y	May 28, 1872	127, 202
Ordnance, Breech-loading	N. Thompson	Brooklyn, N. Y	Aug. 20, 1872	130, 673
Ordnance, Breech-loading	N. Thompson	Brooklyn, N. Y	Oct. 1, 1872	131, 833
Ordnance, Breech-loading	N. Thompson	Brooklyn, N. Y	Oct. 15, 1872	132, 220
Ordnance, Breech-loading	E. H. Tobey	Chicago, Ill	Dec. 28, 1869	98, 316
Ordnance, Breech-loading	W. B. Treadwell	Albany, N. Y	June 17, 1862	35, 637
Ordnance, Breech-loading	T. Tully	Waukegan, Ill	June 28, 1864	43, 351
Ordnance, Breech-loading	E. A. Turpin	New York, N. Y	Aug. 26, 1862	36, 325
Ordnance, Breech-loading	W. Wallace	Ansonia, Conn	Sept. 20, 1864	44, 356
Ordnance, Breech-loading	G. I. Washburn	Worcester, Mass	Oct. 6, 1863	40, 205
Ordnance, Breech-loading	L. C. T. Weber	Rochester, N. Y	Jan. 1, 1861	31, 044
Ordnance, Breech-loading	A. Weeks	Minneapolis, Minn	Oct. 1, 1867	69, 519
Ordnance, Breech-loading	J. Whitworth	Manchester, England	Nov. 14, 1871	120, 842
Ordnance, Breech-loading	W. Williams	Saint Louis, Mo	Nov. 1, 1864	44, 905
Ordnance, Breech-loading	T. Yates	Milwaukee, Wis	Feb. 14, 1865	46, 417
Ordnance, Breech-loading	D. F. Yeakel	La Fayette, Ind	Feb. 11, 1862	34, 388
Ordnance, Bronze	S. B. Dean	Boston, Mass	May 18, 1869	90, 244
Ordnance, Carriage for raising and transporting	C. Bates	Kingston, Mass	Sept. 27, 1864	44, 390
Ordnance, Cast-iron	J. A. Dahlgren	Philadelphia, Pa	Aug. 6, 1861	32, 983
Ordnance, Cast-iron	J. A. Dahlgren	Philadelphia, Pa	Aug. 6, 1861	32, 984
Ordnance, Cast-iron	J. A. Dahlgren	Philadelphia, Pa	Aug. 6, 1861	32, 985
Ordnance, &c., Casting	T. J. Rodman	Pittsburgh, Pa	Aug. 14, 1847	5, 236
Ordnance, Chambered trunnion for disabling	P. B. Lawson and A. Berney	Cold Spring, N. Y., and Jersey City, N. J.	Mar. 10, 1863	37, 870
Ordnance, Counterpoise-platform for	A. T. Brewer	Brighton, Mass	Apr. 26, 1870	102, 213
Ordnance, Device for indicating the elevation of	C. B. Long	Worcester, Mass	July 29, 1862	36, 054
Ordnance, Discharging	W. Johnson	Milwaukee, Wis	May 26, 1863	38, 683
Ordnance-discharging apparatus	H. Arden	Brooklyn, N. Y	Oct. 24, 1871	120, 179
Ordnance, Elastic breech for	J. F. Clew	New York, N. Y	Dec. 27, 1864	45, 586
Ordnance, Elastic breech for	H. H. Day	New York, N. Y	Dec. 2, 1862	37, 035
Ordnance exhausted of air, Tampion-cap for the muzzles of.	J. F. Clew	New York, N. Y	Nov. 15, 1864	45, 020
Ordnance, Firing-mechanism for	N. Thompson	Brooklyn, N. Y	Aug. 13, 1872	130, 452
Ordnance for use under water	J. C. Tilton	Geneseo, Ill	Apr. 22, 1862	35, 048
Ordnance gas-check	J. W. Pearson	Watertown, Mass	Oct. 12, 1869	95, 720
Ordnance, Hooped	R. P. Parrott	Cold Spring, N. Y	May 6, 1862	35, 171
Ordnance, Implement for disabling	E. E. Bean	Boston, Mass	Jan. 5, 1864	41, 115
Ordnance, Implement for disabling	A. Bonzano	Detroit, Mich	Mar. 24, 1863	37, 946
Ordnance in boring-mills, Apparatus for adjusting	S. P. Dean	Boston, Mass	Dec. 8, 1863	40, 824
Ordnance lighting and firing device	G. K. Farrington	Alcatraz Island, Cal	Dec. 27, 1870	110, 448
Ordnance, Loading	J. B. Eads	Saint Louis, Mo	Dec. 3, 1867	71, 592
Ordnance, Loading	W. E. Moore	Crawfordsville, Ind	Mar. 15, 1859	23, 258
Ordnance, Loading	D. E. Rice	Detroit, Mich	June 28, 1864	43, 339
Ordnance-lock	N. Thompson	Brooklyn, N. Y	Sept. 10, 1872	131, 194
Ordnance-lock	J. Vavasseur	Southwark, England	May 18, 1869	90, 323
Ordnance, Machinery for finishing rim-bases of	E. F. Howard	Boston, Mass	Sept. 22, 1863	40, 039
Ordnance, Manufacture of	H. Ames	Falls Village, Conn	Apr. 11, 1865	47, 177
Ordnance, Manufacture of	G. P. Harding	Chiswick, England	Feb. 19, 1867	62, 266
Ordnance, Manufacture of	R. P. Parrott	Cold Spring, N. Y	Oct. 1, 1861	33, 401
Ordnance, Manufacture of	T. E. Vickers	Sheffield, Great Britain	Mar. 21, 1865	46, 978
Ordnance, Mounting	B. H. Bartol	Philadelphia, Pa	Apr. 7, 1863	38, 089
Ordnance, Mounting	C. F. Brown	Warren, R. I	July 17, 1855	13, 249
Ordnance, Mounting	C. F. Brown	Warren, R. I	Aug. 26, 1862	36, 273
Ordnance, Mounting	M. Stoddard	Buffalo, N. Y	Aug. 25, 1863	39, 687
Ordnance, Mounting and operating	J. B. Eads	Saint Louis, Mo	Jan. 3, 1865	45, 704
Ordnance, Mounting and operating	J. Taggert	Roxbury, Mass	Apr. 21, 1863	38, 266
Ordnance, Mounting and working	A. Moncrieff	Culfargie, Scotland	May 30, 1871	115, 502
Ordnance, Mounting field	F. A. De Mey	New York, N. Y	Sept. 16, 1862	36, 457
Ordnance, Mounting field	W. F. Goodwin	Powhatan, Ohio	Sept. 15, 1863	39, 914
Ordnance, Muzzle-loading	F. W. Alexander	Baltimore, Md	Feb. 18, 1868	74, 478

Index of patents issued from the United States Patent Office from 1790 *to* 1873, *inclusive*—Continued.

Invention.	Inventor.	Residence.	Date.	No.
Ordnance on gun-boats, &c., Operating	W. L. Winans	Baltimore, Md., now residing in St. Petersburg, Russia.	Feb. 21, 1865	46, 516
Ordnance on war-vessels, Operating	W. P. Miller	Marysville, Cal	Apr. 7, 1863	38, 118
Ordnance, Operating	M. L. Callender and N. W. Northrup.	New York and Greene, N. Y.	Mar. 17, 1863	37, 935
Ordnance, Operating	J. B. Eads	Saint Louis, Mo	Feb. 16, 1864	41, 611
Ordnance, Operating	J. B. Eads	Saint Louis, Mo	May 3, 1864	42, 566
Ordnance, Operating	J. B. Eads	Saint Louis, Mo	Feb. 14, 1865	46, 342
Ordnance, Operating	J. B. Eads	Saint Louis, Mo	July 3, 1866	56, 021
Ordnance, Operating	J. B. Eads	Saint Louis, Mo	Feb. 28, 1871	112, 230
Ordnance, Operating	J. H. Field	Saugerties, N. Y	July 17, 1866	56, 395
Ordnance, Operating	C. Perley	New York, N. Y	Dec. 12, 1865	51, 475
Ordnance, Operating	J. Ridgway	Boston, Mass	Oct. 21, 1862	36, 730
Ordnance, Operating	E. A. Stevens	Hoboken, N. J	Jan. 6, 1863	37, 364
Ordnance, Operating	O. Tufts	Boston, Mass	Dec. 23, 1862	37, 250
Ordnance, Operating	M. Wappich	Sacramento, Cal	Sept. 22, 1863	40, 067
Ordnance, Operating heavy	J. B. Eads	Saint Louis, Mo	Jan. 12, 1864	41, 206
Ordnance, Pendulum-sight for	W. F. Goodwin	New York, N. Y	Nov. 8, 1864	44, 947
Ordnance, Provision for the recoil of	J. B. Atwater	Chicago, Ill	July 5, 1864	43, 382
Ordnance, Recoil-obviator for	S. F. Hawley	Constableville, N. Y	Sept. 12, 1871	118, 933
Ordnance, Repeating	E. C. C. Kellogg	Hartford, Conn	Nov. 3, 1863	40, 540
Ordnance, Repeating	J. A. Matthews	Saint Louis, Mo	July 31, 1860	29, 437
Ordnance, &c., Repeating	W. McCord and E. Maher	Sing Sing and New York, N. Y.	Nov. 26, 1861	33, 813
Ordnance Repeating	J. P. Taylor	Elizabethton, Tenn	July 4, 1871	116, 775
Ordnance Repeating	A. H. Townsend	Georgetown, Colo	June 6, 1871	115, 659
Ordnance, Repeating	G. Utley	Louisburgh, N. C	May 11, 1858	20, 229
Ordnance, Reversible	J. D. Murphy	Baltimore, Md	June 30, 1868	79, 377
Ordnance, Revolving	J. L. Bonham	Hellen, Pa	Sept. 30, 1862	36, 355
Ordnance, Revolving	J. A. De Brame	New York, N. Y	Dec. 24, 1861	34, 024
Ordnance, Revolving	B. Franke	New York, N. Y	July 29, 1862	35, 998
Ordnance, Revolving	M. F. Hardy	Seward, N. Y	Aug. 12, 1862	36, 148
Ordnance, Revolving	N. L. Milburn	Saint Louis, Mo	Sept. 4, 1866	57, 751
Ordnance, Revolving	N. Taliaferro	Augusta, Ky	Jan. 14, 1862	34, 171
Ordnance, Rifling	H. E. Dimick	Saint Louis, Mo	Jan. 13, 1857	16, 377
Ordnance, Rifling	B. B. Hotchkiss	New York, N. Y	Jan. 21, 1868	73, 447
Ordnance, Rifling	J. Seipel	Washington, D. C	Oct. 17, 1865	50, 502
Ordnance, Rifling	T. Steece	United States Navy	Mar. 17, 1863	37, 924
Ordnance, Serving	J. H. Coon	Deposit, N. Y	Jan. 6, 1863	37, 279
Ordnance, Shield for	P. Andrew	Cincinnati, Ohio	Nov. 3, 1863	40, 453
Ordnance, Sight for	J. Brady	Philadelphia, Pa	Feb. 14, 1865	46, 329
Ordnance, Strengthening	P. M. Parsons	Blackheath, England	July 19, 1864	43, 629
Ordnance, Submarine	D. Fitzgerald	New York, N. Y	Dec. 5, 1871	121, 455
Ordnance to facilitate unspiking, Vent-bushing for	B. G. Martin	New York, N. Y	Dec. 23, 1862	37, 257
Ordnance, Vent-hole of	W. W. Gould	Skowhegan, Me	Apr. 22, 1862	35, 019
Ordnance-vent stopper	J. J. Hirschbuhl	Louisville, Ky	May 28, 1861	32, 422
Ore after being pulverized, Machine for washing and separating.	H. Trumbull	Jersey City, N. J	Aug. 9, 1859	25, 060
Ore and bone crusher	A. Newell	Red Wing, Minn	Sept. 8, 1868	81, 932
Ore and coal, Desulphurizing	W. H. Letterman	Philadelphia, Pa	June 26, 1860	28, 875
Ore and extracting gold and silver, Sulphurizing	W. W. Hubbell	Philadelphia, Pa	June 4, 1867	65, 387
Ore and grain separator	S. Duvall	Philadelphia, Pa	Apr. 16, 1872	125, 797
Ore and metal furnace	W. S. Wood	Newtown, N. Y	June 20, 1871	116, 129
Ore and mineral separator	S. R. Krom	New York, N. Y	Apr. 7, 1868	76, 331
Ore and mineral washer	E. Platt	Charleston, S. C	Oct. 4, 1870	107, 957
Ore and mineral washer	J. B. Platt	Augusta, Ga	Oct. 4, 1870	107, 958
Ore and mineral washing pans, Arrangement of	S. Porter	Hartford, Conn	Dec. 9, 1851	8, 582
Ore and rock, Machine for pulverizing	V. Ryerson	New York, N. Y	Nov. 30, 1869	97, 445
Ore and stone crusher	R. Learmonth	Buffalo, N. Y	Feb. 6, 1872	123, 406
Ore by means of fuel, chemicals, and fluxes, Treatment of.	J. B. Whelpley and J. J. Storer	Boston, Mass	July 11, 1871	116, 903
Ore-calcining apparatus	G. W. White	New York, N. Y	Feb. 7, 1865	46, 287
Ore-calcining furnace	J. B. Britton	Philadelphia, Pa	June 7, 1864	43, 003
Ore-calcining furnace	J. H. Dunstan	Belleville, Canada	Apr. 9, 1872	125, 384
Ore-chloridizing furnace	I. M. Phelps	Chicago, Ill	Feb. 25, 1873	136, 261
Ore, Chlorodizing silver	E. N. Riotte	San Francisco, Cal	Mar. 28, 1871	113, 208
Ore-cleaner	D. Pollock	Lancaster, Pa	Jan. 27, 1857	16, 495
Ore cleaner and separator	J. H. Hillman	Trigg Furnace, Ky	Oct. 29, 1872	132, 577
Ore cleaning and separating machine	N. Carpenter	Ellenville, N. Y	May 3, 1864	42, 558
Ore cleaning machine	B. O. Bryan	Marietta, Pa	June 26, 1855	13, 136
Ore, coal, and similar material, Apparatus for the discharge and preparation of granular.	P. Osterspey	Mechernich, Prussia	Nov. 2, 1869	96, 472
Ore, Concentrating and cleaning	V. Baron	France	May 6, 1862	35, 197
Ore-concentrating apparatus	J. Hepburn	Mokelumne Hill, Cal	Dec. 15, 1863	40, 932
Ore-concentrating machine	T. Varney	San Francisco, Cal	Mar. 19, 1867	62, 983
Ore-concentrating table	G. Kustel	Dayton, Nev	Mar. 14, 1865	46, 806
Ore-concentrating table	H. A. Thompson	Grant, Gipps Land, Victoria.	Nov. 27, 1866	60, 089
Ore-concentrator	H. Donnelly	Virginia City, Nev	Jan. 14, 1868	73, 306
Ore-concentrator	L. Goodwin and S. A. West	San Francisco, Cal	Sept. 17, 1867	68, 978
Ore-concentrator	M. Hungerford	San Francisco, Cal	Mar. 12, 1867	62, 749
Ore-concentrator	M. Hungerford	San Francisco, Cal	July 2, 1872	128, 628
Ore-concentrator	J. A. Peer	Grass Valley, Cal	Sept. 9, 1873	142, 646
Ore-concentrator	J. W. V. Rawlins and S. Stephens.	Houghton, Mich	June 29, 1869	91, 966
Ore-concentrator	E. L. Seymour	New York, N. Y	Aug. 30, 1859	25, 280
Ore-concentrator	W. C. Stiles	Nevada City, Cal	May 2, 1871	114, 486
Ore-concentrator	G. W. and W. L. Strong	San Francisco, Cal	May 18, 1869	90, 205
Ore concentrator and amalgamator	J. S. Bradford	New York, N. Y	Oct. 12, 1869	95, 640
Ore concentrator and amalgamator	S. Fountain	Silver City, Nev	Dec. 17, 1867	72, 182
Ore concentrator and endless sluice-blanket	C. D. Smith	Drytown, Cal	July 20, 1869	92, 893
Ore concentrator and separator	J. Edgar	New York, N. Y	Dec. 7, 1869	97, 488
Ore, Concentrator for dressing	H. Weston and G. C. Langtry	Dayton, Nev	Oct. 20, 1868	83, 231
Ore-concentrator, Self-discharging blanket	J. M. Bryan	Lincoln, Cal	May 4, 1869	89, 734
Ore-crusher	J. Burns	New York, N. Y	June 4, 1872	127, 561
Ore-crusher	J. W. Cummings	Georgetown, Colo	Apr. 23, 1872	126, 034
Ore-crusher	M. B. Dodge	New York, N. Y	Dec. 4, 1866	60, 154

Index of patents issued from the United States Patent Office from 1790 *to* 1873, *inclusive*—Continued.

Invention.	Inventor.	Residence.	Date.	No.
Ore-crusher	M. B. Dodge	Brooklyn, N. Y	May 25, 1869	90, 510
Ore-crusher	J. Fowler	Rahway, N. J	May 22, 1866	54, 884
Ore-crusher	A. W. Hall	New York, N. Y	Oct. 31, 1865	50, 703
Ore-crusher	J. Hamilton, L. E. Hanson, and G. W. and J. Hamilton.	Wheeling, W. Va	May 25, 1869	90, 532
Ore crusher	W. P. Hammond	Napa City, Cal	Sept. 24, 1872	131, 612
Ore-crusher	W. P. Hammond	Napa City, Cal	Dec. 24, 1872	134, 201
Ore crusher	B. Hotchkiss and S. C. Goodsell	New Haven, Conn	Mar. 26, 1867	63, 160
Ore-crusher	S. Ingersoll	Stamford, Conn	July 25, 1865	49, 032
Ore-crusher	S. R. Krom	New York, N. Y	June 22, 1869	91, 641
Ore-crusher	S. R. Krom	New York, N. Y	July 16, 1872	129, 238
Ore-crusher	G. Mitchell	Philadelphia, Pa	May 28, 1872	127, 260
Ore-crusher	G. Mitchell	Philadelphia, Pa	Nov. 19, 1872	133, 111
Ore-crusher	W. R. Parrott and J. J. Boardman.	Boston, Mass	Jan. 8, 1867	61, 019
Ore-crusher	A. F. W. Partz	Wurtsborough, N. Y	Sept. 27, 1864	44, 450
Ore-crusher	J. V. Pomeroy	Utica, N. Y	July 25, 1865	48, 999
Ore-crusher	J. Reese	Pittsburgh, Pa	Mar. 31, 1868	76, 100
Ore-crusher	C. W. Stafford	Saybrook, Conn	Nov. 13, 1866	59, 676
Ore-crusher	E. and W. Thornton	Sharon, Pa	July 30, 1872	130, 084
Ore-crusher	J. Webster and J. G. Morgan	Brooklyn, N. Y	Nov. 21, 1865	51, 104
Ore-crusher	Z. Wheeler	San Francisco, Cal	Mar. 6, 1866	53, 066
Ore-crusher	J. T. Whelpley	Boston, Mass	Nov. 8, 1864	44, 989
Ore crusher and amalgamator	L. Griswold	Denver, Colo	Mar. 21, 1871	112, 804
Ore crusher and amalgamator	J. Hart	Melbourne, Victoria	May 7, 1867	64, 416
Ore crusher and grinder	S. Hughes	San Francisco, Cal	Apr. 7, 1868	76, 461
Ore crusher, grinder, and amalgamator	J. A. Collins	Virginia City, Nev	July 14, 1868	79, 954
Ore-crushers, Chilled plate for	J. L. Agnew and C. E. Wright	Negaunee, Mich	Mar. 16, 1869	87, 750
Ore crushing and pulverizing machine	A. Buffum	Perth Amboy, N. J	Jan. 9, 1855	12, 191
Ore crushing and pulverizing machine	S. Gardiner, jr	New York, N. Y	July 25, 1854	11, 368
Ore-crushing machine	N. Conkling	Brooklyn, N. Y	Mar. 16, 1858	19, 670
Ore, &c., crushing machine	A. W. Hall	New York, N. Y	Apr. 18, 1865	47, 297
Ore-crushing machine	A. W. Hall	New York, N. Y	Oct. 17, 1865	50, 467
Ore-crushing machine	S. F. Hodge	Detroit, Mich	May 26, 1857	17, 374
Ore-crushing machine	S. Hughes	Charleston, S. C	Nov. 14, 1871	120, 972
Ore-crushing machine	J. T. Plass	New York, N. Y	Oct. 25, 1864	44, 819
Ore-crushing machine	W. H. Plumb	New York, N. Y	Dec. 5, 1854	12, 031
Ore, Machine for crushing mineral	W. P. Parrott	Boston, Mass	May 29, 1860	28, 499
Ore-crushing mill	S. F. Mackie	New York, N. Y	Apr. 30, 1867	64, 339
Ore-crushing mill	J. Moore	San Francisco, Cal	May 6, 1862	35, 202
Ore-crushing roller	W. H. Plumb	New York, N. Y	Nov. 14, 1856	16, 039
Ore-crushing stamp	G. D. Crocker	San Francisco, Cal	Aug. 5, 1873	141, 493
Ore-crushing stamp	J. M. McFarland	Golden City, Colo	July 1, 1873	140, 425
Ore, Method of decomposing and desulphurizing	R. Spencer	New York, N. Y	Nov. 8, 1864	44, 986
Ore, Deodorizing	I. Rogers	North Haverstraw, N. Y	Aug. 21, 1860	29, 746
Ore, Deoxidizing and carbonizing iron	C. Adams	Philadelphia, Pa	Feb. 22, 1870	100, 095
Ore-deoxidizing furnace, Iron	A. H. Brainerd	Rome, N. Y	Nov. 15, 1870	109, 173
Ore, Desulphurizing	H. Bradford	New York, N. Y	June 28, 1864	43, 363
Ore, Desulphurizing	E. A. Burns	Meadow Lake, Cal	Mar. 8, 1870	100, 497
Ore, Desulphurizing	M. B. Dodge	New York, N. Y	Apr. 4, 1865	47, 094
Ore, Desulpurizing	J. A. Hitchings	Denver City, Colo	Sept. 18, 1866	58, 100
Ore, Desulphurizing	J. F. Osgood	Boston, Mass	Apr. 5, 1870	101, 654
Ore, Desulphurizing	V. Ryerson	New York, N. Y	Aug. 14, 1866	57, 194
Ore desulphurizing and amalgamating apparatus	E. P. Gardiner	New York, N. Y	Jan. 24, 1865	45, 991
Ore, Desulphurizing and disintegrating	J. C. Ayer	Lowell, Mass	Mar. 7, 1865	46, 619
Ore, Desulphurizing and disintegrating	J. C. Ayer	Lowell, Mass	Mar. 7, 1865	46, 620
Ore, Desulphurizing and disintegrating	D. Minthorn	New York, N. Y	Feb. 27, 1866	52, 875
Ore, Desulphurizing and disintegrating	G. Vining	Boston, Mass	Aug. 8, 1865	49, 322
Ore desulphurizing and oxidizing apparatus	J. M. Rohrer	Pine Grove, Pa	Apr. 6, 1869	88, 741
Ore, Desulphurizing and oxidizing metallic	M. B. Mason	New York, N. Y	Jan. 3, 1865	45, 803
Ore, Desulphurizing and treating	J. W. Bailey	Hamilton, Nev	July 25, 1871	117, 363
Ore-desulphurizing apparatus	J. S. Briggs	Chicago, Ill	Mar. 20, 1866	53, 266
Ore-desulphurizing apparatus	F. W. Crosby and W. Helen	New York, N. Y	Aug. 21, 1866	57, 293
Ore-desulphurizing apparatus	B. G. Noble	New York, N. Y	Jan. 23, 1866	52, 191
Ore-desulphurizing apparatus	I. N. Stanley	Brooklyn, N. Y	Jan. 29, 1867	61, 577
Ore-desulphurizing apparatus	A. H. Tait and J. W. Avis	New York, N. Y	Jan. 16, 1866	52, 090
Ore desulphurizing furnace	E. C. Hegeler and F. W. Matthiesen.	La Salle, Ill	Nov. 25, 1873	144, 904
Ore-desulphurizing furnace	M. B. Mason	New York, N. Y	Jan. 3, 1865	45, 804
Ore-desulphurizing furnace	J. G. Trotter	Newark, N. J	Nov. 22, 1870	109, 559
Ore-desulphurizing furnace, Isomeric diaphragm	D. Menthorn	New York, N. Y	Nov. 7, 1865	50, 836
Ore, Desulphurizing gold and silver	A. L. Fleury	Pittsburgh, Pa	July 3, 1866	56, 027
Ore-desulphurizing machine, Metallic	J. S. Sullivan	New York, N. Y	Aug. 18, 1834	....
Ore-desulphurizing process	J. Absterdam	New York, N. Y	Jan. 23, 1866	52, 120
Ore-desulphurizing process	J. F. Alexander	Shelby, N. C	Aug. 11, 1868	80, 798
Ore-desulphurizing process	M. B. Mason	New York, N. Y	Apr. 3, 1866	53, 742
Ore-desulphurizing process	J. S. Oliver	New York, N. Y	Feb. 27, 1872	124, 077
Ore-disintegrating apparatus	J. B. Cox	San Francisco, Cal	Apr. 28, 1868	77, 259
Ore-disintegrating apparatus	A. B. Paul	San Francisco, Cal	Apr. 5, 1870	101, 500
Ore, Disintegrating, desulphurizing, and oxidizing	J. C. Ayer	Lowell, Mass	Mar. 7, 1865	46, 621
Ore-dressing apparatus	C. F. Collom	Calstock, Great Britain	Aug. 29, 1871	118, 437
Ore-feeder for grinding and crushing mills	C. P. Stanford	San Francisco, Cal	May 20, 1873	139, 204
Ore furnace and apparatus, Pyritous	J. Fretz	Angel, Cal	Sept. 20, 1859	25, 500
Ore, Furnace and process for treating and reducing	V. W. Blanchard	Bridport, Vt	Nov. 16, 1869	96, 872
Ore, Furnace for making iron direct from the	J. M. Quinby, A. H. Brown, G. H. Penton, and J. Criswell.	Newark, N. J	Aug. 9, 1859	25, 044
Ore, Furnace for producing iron direct from the	W. Griffith, jr	Pottsville, Pa	Aug. 31, 1869	94, 306
Ore-grinder	J. W. V. Rawlins and S. Stephens.	Houghton, Mich	Sept. 14, 1869	94, 771
Ore grinding and amalgamating apparatus	M. B. Dodge	New York, N. Y	May 23, 1865	47, 808
Ore grinding and amalgamating machine	P. Hinkle	San Francisco, Cal	Apr. 16, 1867	63, 979
Ore-grinding mill	W. Ball	Chicopee, Mass	Mar. 30, 1852	8, 835
Ore-grinding mill	W. Ball	Chicopee, Mass	Apr. 11, 1854	10, 754
Ore-grinding mill	J. S. Niswander	Oakland, Cal	Mar. 28, 1871	113, 085
Ore-heater	A. C. Vandyke	Greenupsburgh, Ky	Apr. 24, 1860	28, 025
Ore-jigger	J. F. Utsch	Iserlohn, Germany	Dec. 3, 1872	133, 606
Ore, Jigging-machine for dressing	W. W. Spalding	Greenland, Mich	June 11, 1867	65, 703

Index of patents issued from the United States Patent Office from 1790 *to* 1873, *inclusive*—Continued.

Invention.	Inventor.	Residence.	Date.	No.
Ore of copper and other metals to obtain metal and other products therefrom, Treating	W. Henderson	Glasgow, Scotland	Dec. 18, 1866	60, 514
Ore or quartz crusher	J. Fowler	Rahway, N. J	Sept. 18, 1866	58, 089
Ore or quartz crusher	E. P. McCarthy	San Francisco, Cal	Sept. 18, 1866	58, 116
Ore-oxidizing furnace	T. J. Chubb	Brooklyn, N. Y	Aug. 6, 1867	67, 497
Ore-oxidizing furnace	I. M. Phelps	Chicago, Ill	June 6, 1871	115, 769
Ore, Process and apparatus for reducing iron	T. S. Blair	Pittsburgh, Pa	May 21, 1872	126, 922
Ore-pulverizer	S. Gardner	New York, N. Y	Nov. 4, 1873	144, 330
Ore-pulverizer	F. C. Morse	Buckskin, Colo	Oct. 25, 1870	108, 617
Ore-pulverizing machine	A. K. Eaton	New York, N. Y	Feb. 28 1854	10, 567
Ore, Purifying and preparing glass	C. Carter	Newburgh, N. Y	Apr. 9, 1867	63, 702
Ore reducing and smelting furnace	L. G. Marshall	Philadelphia, Pa	June 16, 1863	38, 906
Ore-reducing apparatus	G. B. Simpson	Washington, D. C	Aug. 21, 1866	57, 394
Ore-reducing furnace	J. E. Ware	Saint Louis, Mo	May 9, 1871	114, 734
Ore-reducing furnace for refractory metals	W. Quann	Philadelphia, Pa	Sept. 14, 1869	94, 911
Ore-reducing furnace, Iron	A. Thorna	New York, N. Y	Oct. 22, 1867	70, 047
Ore-reducing furnace, Iron	J. Wilson	Dover, N. J	July 16, 1872	128, 993
Ore-reducing furnace, Iron	J. Wilson	Dover, N. J	Dec. 9, 1873	145, 471
Ore-reducing furnace, Metallic	E. W. Nohl	Ripon, Wis	June 11, 1867	65, 586
Ore-reducing furnaces, Apparatus for feeding fuel to.	J. H. Boyd	Chicago, Ill	Nov. 4, 1873	144, 185
Ore, Reducing sulphur	C. W. Moore	San Francisco, Cal	June 1, 1869	90, 677
Ore, Reduction of iron	W. Quann	Philadelphia, Pa	Aug. 6, 1861	33, 020
Ore, Reduction of refractory iron	C. Martin	Chancery Lane, England	Feb. 11, 1868	74, 397
Ore-roaster	F. W. Crosby	Toledo, Iowa	Jan. 7, 1868	72, 920
Ore-roaster	C. A. Griffin	Chelsea, Mass	Oct. 18, 1864	44, 769
Ore, Roasting	G. Kustel and F. W. Smith	San Francisco, Cal., and Ellsworth, Nev.	Oct. 3, 1872	131, 882
Ore roasting and calcining furnace	E. Westman	Stockholm, Sweden	Jan. 12, 1869	85, 881
Ore roasting and chloridizing furnace	H. Tindall	Chicago, Ill	Oct. 13, 1868	83, 122
Ore roasting and chloridizing, Process of	H. Tindall	Chicago, Ill	Oct. 13, 1868	83, 121
Ore, Roasting and desulphurizing	G. W. Baker	New York, N. Y	Mar. 28, 1865	46, 984
Ore, Roasting and desulphurizing	H. Bradford	New York, N. Y	Feb. 21, 1865	46, 520
Ore, Roasting and desulphurizing	R. P. Wilson	New York, N. Y	Nov. 1, 1864	44, 906
Ore, Roasting and desulphurizing	R. P. Wilson	New York, N. Y	Mar. 7, 1865	46, 745
Ore roasting and reducing apparatus	G. W. White	New York, N. Y	Sept. 6, 1864	44, 145
Ore roasting and reducing furnace	J. Maunton	New York, N. Y	Dec. 3, 1867	71, 776
Ore roasting and smelting furnace	W. Quann	Philadelphia, Pa	Oct. 24, 1871	120, 165
Ore roasting and smelting furnace	L. Stevens	Washington, D. C	July 22, 1873	141, 181
Ore, Roasting and treating	J. W. Bailey	San Francisco, Cal	Nov. 7, 1871	120, 695
Ore roasting and treating furnace	R. George	Mineral Point, Wis	July 21, 1868	80, 065
Ore roasting and treating furnace	C. Stetefeldt	Austin, Nev	Dec. 31, 1867	72, 931
Ore roasting and treating furnace	Z. A. Willard	Boston, Mass	Jan. 14, 1873	134, 782
Ore roasting by waste gas, Iron	A. Hamar	New York, N. Y	Nov. 17, 1868	84, 113
Ore, Roasting, desulphurizing, and smelting	J. D. Whelpley and J. J. Storer	Boston, Mass	Sept. 11, 1866	58, 012
Ore-roasting furnace	J. Agrell and J. Klepzig	San Francisco, Cal	Sept. 3, 1867	68, 406
Ore-roasting furnace	S. F. Ambler	Monitor, Cal	July 25, 1871	117, 360
Ore-roasting furnace	J. P. Arey	Georgetown, Colo	June 6, 1871	115, 559
Ore-roasting furnace	C. N. Atkins and A. Govan	Pottsville, Pa	May 24, 1870	103, 409
Ore-roasting furnace	J. W. Bailey	San Francisco, Cal	Nov. 7, 1871	120, 696
Ore-roasting furnace	N. Bartlett	Centreville, N. J	May 19, 1868	77, 950
Ore-roasting furnace	M. P. Boss	Eureka, Nev	Nov. 12, 1872	132, 892
Ore-roasting furnace	E. Brady and J. Sloan	Philadelphia, Pa	Nov. 26, 1867	71, 271
Ore-roasting furnace	W. Buckner	Central City, Colo	Nov. 26, 1867	71, 448
Ore-roasting furnace	S. W. Bullock	Elizabeth, N. J	Mar. 2, 1869	87, 227
Ore-roasting furnace	D. C. Collier, S. Cushman, and N. E. Farrell.	Central City, Colo	June 16, 1868	78, 928
Ore-roasting furnace	J. Collom	Empire City, Colo	Aug. 23, 1870	106, 553
Ore-roasting furnace	G. B. Field	New York, N. Y	Aug. 20, 1867	67, 862
Ore-roasting furnace	G. B. Field	New York, N. Y	Aug. 27, 1867	68, 058
Ore-roasting furnace	H. Goulding	Silver City, Nev	Nov. 19, 1867	70, 995
Ore-roasting furnace	C. B. Grubb	Lancaster, Pa	Jan. 26, 1864	41, 376
Ore-roasting furnace	E. P. Hudson	New York, N. Y	June 2, 1868	78, 456
Ore-roasting furnace	J. Jenkins and H. Weissenborn	Elizabeth and Newark, N. J.	Nov. 18, 1862	36, 954
Ore-roasting furnace	F. Kesseler	San Francisco, Cal	June 27, 1871	116, 324
Ore-roasting furnace	R. F. Knox and J. Osborn	San Francisco, Cal	July 25, 1871	117, 430
Ore-roasting furnace	R. F. Knox and J. Osborn	San Francisco, Cal	July 2, 1872	128, 637
Ore-roasting furnace	C. Millinger	Cornwall, Pa	May 19, 1868	78, 113
Ore-roasting furnace	R. B. Norman	Sacramento, Cal	Apr. 8, 1862	34, 901
Ore-roasting furnace	D. J. O'Harra and C. B. Thompson.	Empire City, Nev	Aug. 6, 1867	67, 445
Ore-roasting furnace for precious metals	J. S. Akin	Rye Patch, Nev	June 6, 1871	115, 673
Ore-roasting furnace, Metallic	F. W. Crosby	Washington, D. C	May 23, 1871	115, 031
Ore-roasting furnace, Revolving	F. Ernst	San Francisco, Cal	Sept. 1, 1868	81, 762
Ore-roasting furnace, Feeder for	J. P. Arey	Georgetown, Colo	Aug. 8, 1871	117, 718
Ore-roasting furnace, Feeder for	C. Stetefeldt	Austin, Nev	Mar. 1, 1870	100, 337
Ore, Roasting iron	H. Aitken	Falkirk, Scotland	Apr. 13, 1860	88, 939
Ore-roasting kiln	J. M. Rohrer and J. H. Bassler.	Pine Grove, Pa	Aug. 10, 1869	93, 643
Ore-roasting kiln, Iron	A. Thoma	New York, N. Y	Oct. 22, 1867	70, 045
Ore roasting, oxidizing, and chloridizing furnace	J. P. Arey	Georgetown, Colo	May 17, 1870	103, 006
Ore, Separating	T. J. Chubb	New York, N. Y	Aug. 25, 1857	18, 038
Ore, Separating	A. Huet and A. Geyler	Paris, France	Mar. 12, 1872	124, 578
Ore separating and concentrating apparatus	A. Hunter	Solano County, Cal	May 23, 1865	47, 828
Ore separating and concentrating apparatus	J. Jenkins	South Bethlehem, Pa	June 6, 1871	115, 737
Ore separating and concentrating apparatus	E. L. Seymour	New York, N. Y	Dec. 19, 1865	51, 662
Ore separating and dressing machine	H. Trumbull	Jersey City, N. J	June 2, 1863	38, 773
Ore separating and dressing process	N. Troughton	Swansea, England	July 22, 1844	3, 674
Ore, Separating and sorting	E. Lawson	Island Pond, Vt	Feb. 23, 1864	41, 710
Ore separating and treating machine	S. R. Krom	New York, N. Y	Dec. 5, 1871	121, 526
Ore-separating apparatus	H. Bradford and E. Fitzgerald.	New York, N. Y	Feb. 22, 1853	9, 590
Ore separating, concentrating, and amalgamating apparatus.	H. H. Eames	Gold Hill, N. C	June 3, 1873	139, 556
Ore-separating jigger	W. H. Plumb	Paterson, N. J	July 1, 1873	140, 535
Ore, Separating precious metals from	S. W. Kirk	Philadelphia, Pa	Nov. 18, 1873	144, 772
Ore-separator	S. F. Ambler	Brooklyn, N. Y	Dec. 26, 1865	51, 678
Ore-separator	H. Ball	New York, N. Y	June 7, 1870	103, 825
Ore-separator	Y. Bates	Pinos Altos, N. Mex	May 3, 1870	102, 476

Index of patents issued from the United States Patent Office from 1790 to 1873, inclusive—Continued.

Invention.	Inventor.	Residence.	Date.	No.
Ore-separator	W. O. Bourne	New York, N. Y	Nov. 24, 1857	18,672
Ore-separator	W. O. Bourne	New York, N. Y	June 14, 1859	24,367
Ore-separator	W. O. Bourne	New York, N. Y	Oct. 9, 1860	30,290
Ore-separator	H. Bradford	New York, N. Y	June 29, 1858	20,756
Ore-separator	H. Bradford	Reading, Pa	Oct. 7, 1873	143,492
Ore separator	F. Cazin	Frumet, Mo	July 2, 1872	128,536
Ore-separator	T. J. Chubb	New York, N. Y	Sept. 1, 1857	18,085
Ore-separator	T. J. Chubb	New York, N. Y	Oct. 13, 1857	18,388
Ore-separator	T. J. Chubb	Williamsburgh, N. Y	Aug. 13, 1872	130,478
Ore-separator	G. Copeland	Denver, Colo	May 31, 1870	103,574
Ore-separator	A. Goodhart	Newville, Pa	June 10, 1873	139,782
Ore-separator	D. Gross	Maxatawny, Pa	Aug. 13, 1872	130,425
Ore-separator	C. D. Hicks	Denver, Colo	Apr. 19, 1864	42,372
Ore-separator	W. Hooper	Ticonderoga, N. Y	Mar. 22, 1870	101,132
Ore-separator	W. Hooper	Ticonderoga, N. Y	Mar. 21, 1871	112,918
Ore-separator	W. Hooper	Ticonderoga, N. Y	Mar. 21, 1871	112,919
Ore-separator	S. R. Krom	New York, N. Y	Sept. 1, 1868	81,794
Ore-separator	S. Man	Chicago, Ill	July 31, 1860	29,438
Ore-separator	J. J. Müller	New York, N. Y	Aug. 27, 1861	33,171
Ore-separator	H. P. Minot	Chicago, Ill	Dec. 9, 1873	145,441
Ore-separator	F. A. Morley	Sodus Point, N. Y	Apr. 19, 1864	42,392
Ore-separator	R. C. Morton	West Lubec, Me	June 2, 1868	78,468
Ore-separator	D. Nevin	Georgetown, Colo	July 19, 1870	105,480
Ore-separator	D. Nevin	Georgetown, Colo	Sept. 3, 1872	131,064
Ore-separator	P. P. Parkhurst	Princeton, Mass	Nov. 22, 1859	26,236
Ore-separator	H. P. Russ	Russville, Cal	June 22, 1858	20,666
Ore-separator	E. L. Seymour	New York, N. Y	July 17, 1860	20,198
Ore-separator	E. L. Seymour	New York, N. Y	Jan. 3, 1865	45,757
Ore-separator	R. Shaler	Madison, Conn	Mar. 20, 1855	12,576
Ore-separator	L. Stadtmuller	Bristol, Conn	Nov. 23, 1858	22,138
Ore-separator	J. F. Utsch	Iserholn, Germany	Oct. 15, 1872	131,336
Ore-separator	J. Watson	Cliff Mine, Mich	Apr. 25, 1865	47,476
Ore-separator	J. D. Whelpley and J. J. Storer	Boston, Mass	Apr. 3, 1866	53,718
Ore separator and amalgamator	H. Pietsch	New York, N. Y	May 3, 1864	42,596
Ore separator and concentrator	R. George	Denver City, Colo	Aug. 2, 1870	106,049
Ore separator and concentrator	T. N. Paine and S. Stephens	Grass Valley, Cal	Nov. 5, 1867	70,603
Ore separator and concentrator	R. D. Symons, J. T. Harvy, and S. Stephens.	Grass Valley, Cal	Oct. 20, 1868	83,221
Ore separator and washer	J. J. Müller	New York, N. Y	Mar. 18, 1862	34,721
Ore-separator, Centrifugal	S. T. Pearce	New York, N. Y	Mar. 2, 1869	87,360
Ore-separator, Electro-magnetic	R. Cook	Plattsburgh, N. Y	Feb. 20, 1849	6,121
Ore separator, Gold	H. Merrill	Brooklyn, N. Y	Aug. 22, 1871	118,379
Ore separator, Gold	T. Rhoads	Ottawa, Ill	Jan. 7, 1868	73,041
Ore separator, Iron	J. Goulding	Reeseville, N. Y	July 18, 1832	......
Ore-separator or jigging machine	T. Davey	Houghton, Mich	May 15, 1866	54,697
Ores, Smelting and desulphurizing iron	A. Hamar	New York, N. Y	Feb. 11, 1868	74,217
Ore, Smelting copper	W. L. Faber	New York, N. Y	Apr. 25, 1865	47,407
Ore, Smelting copper	S. F. Tracy	New York, N. Y	Dec. 16, 1851	8,599
Ore-smelting flux	S. D. Young	Elizabethtown, Ill	Aug. 26, 1873	142,186
Ore-smelting furnace	E. Balbach, jr	Newark, N. J	Aug. 19, 1873	141,912
Ore-smelting furnace	E. J. Hall	San Francisco, Cal	Mar. 22, 1864	41,989
Ore-smelting furnace	J. Neville	Jersey City, N. J	Apr. 29, 1873	138,428
Ore smelting furnace	G. W. Swett	Troy, N. Y	Apr. 14, 1863	38,186
Ore-smelting furnace	J. Winchester	New York, N. Y	June 7, 1864	43,061
Ore smelting furnace, Gold, silver, &c	E. W. Nohl	Chicago, Ill	Sept. 1, 1868	81,671
Ore-smelting furnace, Iron	P. Kerr	New Bethlehem, Pa	May 28, 1861	32,426
Ore-smelting furnace, Iron	A. Thoma	New York, N. Y	Oct. 22, 1867	70,046
Ore-smelting furnace, Lead	D. P. Webster	New York, N. Y	Apr. 7, 1868	76,364
Ore, Smelting iron	C. Cochrane	The Ellowes, Upper Gornal, England.	July 4, 1871	116,558
Ore, Smelting lead	A. H. Everett	New York, N. Y	May 3, 1864	42,570
Ore, Smelting lead	D. P. Webster	New York, N. Y	Apr. 14, 1868	76,679
Ore-stamp	L. D. Webb	Greenville, Cal	Oct. 7, 1873	143,480
Ore-stamp feeder	J. D. Cusenbary and J. A. Mars.	San Francisco, Cal	June 24, 1873	140,250
Ore-stamp feeder	J. Tullock	Jamestown, Cal	Nov. 18, 1873	144,714
Ore, Stamp-head for crushing	F. Murray	Baltimore, Md	Aug. 26, 1862	36,302
Ore-stamp mill	G. S. Olin	Deer Lodge, Mont	Aug. 26, 1873	142,172
Ore-stamps, Sectional cam for	J. M. Thompson	San Francisco, Cal	June 24, 1873	140,318
Ore-stamper	T. Reaney	Philadelphia, Pa	June 15, 1852	9,036
Ore-stamping machine	W. Ball	Chicopee, Mass	Mar. 23, 1852	8,818
Ore-stamping machine	W. Ball	Chicopee, Mass	Dec. 31, 1867	72,715
Ore-stamping machine	J. F. Laird	Philadelphia, Pa	July 4, 1854	11,229
Ore-stamping machine	V. Woodcock	Swanzey, N. H	June 15, 1852	9,042
Ore, Stamping-machine for crushing	W. Murray	Baltimore, Md	Aug. 23, 1859	25,213
Ore, Sweeping and washing	W. Davis	New York, N. Y	Apr. 28, 1836	......
Ore to obtain precious metals, Process of desulphurizing.	H. H. Eames	Saint Paul, Minn	Feb. 2, 1869	86,513
Ore, Treating	G. W. Baker	New York, N. Y	Mar. 28, 1865	46,983
Ore, Treating	C. F. Carpenter	Louisville, Ky	June 26, 1866	55,818
Ore, Treating	M. Laflin	Chicago, Ill	July 22, 1873	141,147
Ore, Treating	S. F. Mackie	New York, N. Y	Aug. 29, 1865	49,637
Ore, Treating	H. B. Seidel	Wilmington, Del	Aug. 29, 1865	49,658
Ore-treating and iron and steel manufacturing apparatus.	E. Peckham	Antwerp, N. Y	June 24, 1873	140,1[illegible]8
Ore-treating apparatus	J. C. Coult and J. Roach	San Francisco, Cal	July 17, 1866	56,378
Ore-treating apparatus	H. Halvorson	North Cambridge, Mass	Oct. 17, 1865	50,533
Ore-treating apparatus	R. Spencer	New York, N. Y	May 23, 1865	47,873
Ore-treating apparatus	R. Spencer	New York, N. Y	May 23, 1865	47,874
Ore, Treating copper	H. J. M. Priestienne	Paris, France	Mar. 4, 1862	34,591
Ore-treating furnace	W. Ennis	Philadelphia, Pa	Mar. 14, 1871	112,698
Ore-treating furnace	S. H. Folsom	Winchester, Mass	July 28, 1868	80,279
Ore-treating furnace	W. Hendrick	New York, N. Y	Oct. 31, 1865	50,717
Ore-treating furnace	L. Stevens	Washington, D. C	Oct. 15, 1872	132,331
Ore-treating furnace, Auriferous	A. E. Griffiths	Philadelphia, Pa	Mar. 27, 1866	53,440
Ore-treating furnace, Iron	I. Rogers	Haverstraw, N. Y	Feb. 26, 1861	31,580
Ore, Treating gold	H. Halvorson	North Cambridge, Mass	Feb. 6, 1866	52,490
Ore, Treating metalliferous	W. L. Rabt	Baltimore, Md	Aug. 21, 1866	57,376

Index of patents issued from the United States Patent Office from 1790 *to* 1873, *inclusive*—Continued.

Invention.	Inventor.	Residence.	Date.	No.
Ore, Treatment of	C. M. Du Motay	Paris, France	May 24, 1870	103, 434
Ore, Treatment of sulphureted	I. Gattman	Philadelphia, Pa	Apr. 20, 1858	19, 991
Ore, Utilizing the sulphur fumes from copper	A. Bigelow	Newark, N. J	Sept. 28, 1869	95, 307
Ore-washer	W. Ball	Chicopee, Mass	Feb. 5, 1856	14, 182
Ore-washer	H. Barnard	Morristown, N. J	Feb. 16, 1858	19, 337
Ore-washer	K. L. Bean	Chicago, Ill	Nov. 2, 1869	96, 327
Ore-washer	R. Boehm	Chicago, Ill	May 18, 1869	90, 227
Ore-washer	H. Bradford	New York, N. Y	Aug. 12, 1856	15, 554
Ore-washer	H. Bradford	Reading, Pa	Nov. 8, 1870	108, 962
Ore-washer	A. Buffam	Brooklyn, N. Y	Oct. 21, 1851	8, 443
Ore-washer	W. L. Carter	Marietta, Pa	Mar. 11, 1856	14, 388
Ore-washer	W. L. Carter	Marietta, Pa	Nov. 15, 1859	26, 089
Ore-washer	W. L. Carter	Marietta, Pa	Aug. 24, 1869	93, 965
Ore-washer	S. T. Dorland	Beekman, N. Y	Oct. 29, 1872	132, 642
Ore-washer	R. Gidley	Moore's Mills, N. Y	June 3, 1873	139, 570
Ore-washer	L. M. Gochnauer	Marietta, Pa	June 11, 1872	127, 688
Ore-washer	I. T. Halstead	Fredonia, N. Y	Sept. 16, 1873	142, 848
Ore-washer	G. B. Hamilton	Washington, D. C	Mar. 13, 1866	53, 221
Ore-washer	W. Hooper	Ticonderoga, N. Y	May 27, 1873	139, 390
Ore-washer	W. M. Hughes	Howard County, Mo	Jan. 8, 1850	7, 002
Ore-washer	P. P. Martin	Paris, France	May 26, 1857	17, 385
Ore-washer	G. E. Mills	New York, N. Y	Mar. 13, 1860	27, 463
Ore-washer	B. O'Bryan	Lancaster, Pa	Sept. 8, 1860	30, 086
Ore-washer	C. B. Parson and G. W. Fisher	San Francisco and Saint Louis, Mo.	July 16, 1872	128, 973
Ore-washer	J. Paull	Clifton, Mich	Oct. 13, 1857	18, 406
Ore-washer	M. Peckham and L. O. Palmer	Utica, N. Y	Jan. 4, 1853	9, 522
Ore-washer	J. A. Peer	Grass Valley, Cal	Sept. 9, 1873	142, 647
Ore-washer	J. Pritchett	Philadelphia, Pa	Oct. 9, 1849	6, 781
Ore-washer	G. H. Reynolds	New York, N. Y	Sept. 17, 1872	131, 373
Ore-washer	G. H. Reynolds	New York, N. Y	Sept. 17, 1872	131, 374
Ore-washer	W. T. Rickard	Monitor, Cal	May 14, 1872	126, 744
Ore-washer	N. S. Ryder	Greenland, Mich	Sept. 17, 1867	69, 030
Ore-washer	R. Solliday	Allentown, Pa	May 13, 1873	138, 948
Ore-washer	R. Solliday	Allentown, Pa	Dec. 9, 1873	145, 455
Ore-washer	S. Thomas	Allentown, Pa	Sept. 30, 1856	15, 827
Ore-washer	P. Von Schmidt	New York, N. Y	Oct. 16, 1849	6, 791
Ore-washer	S. Wheeler	Albany, N. Y	Dec. 23, 1873	145, 771
Ore-washer	J. Wicks	Greenland, Mich	Apr. 23, 1867	64, 178
Ore-washer	T. Wise	Boston, Mass	Mar. 24, 1863	38, 016
Ore-washer	T. Wren	Hamilton, Nev	Apr. 29, 1873	138, 307
Ore washer and amalgamator	J. N. Wyckoff and T. M. Fell	Brooklyn, N. Y., and Orange Mines, Va.	July 26, 1859	24, 901
Ore washer and concentrator	W. G. Heslep and T. H. Cochrane.	Jamestown, Cal	June 22, 1869	91, 627
Ore washer and separator	J. H. Rae	Syracuse, N. Y	Jan. 26, 1869	86, 249
Ore-washer, Gold	S. L. Beckwith	San Francisco, Cal	Feb. 11, 1868	74, 281
Ore-washer gudgeon or bearing	S. Thomas	Catasauqua, Pa	Sept. 2, 1873	142, 530
Ore-washing apparatus	P. Scheneman	Hancock, Mich	Dec. 15, 1863	40, 950
Ore-washing apparatus	R. Uren	Houghton, Mich	Jan. 8, 1867	61, 124
Ore-washing apparatus	J. Watson	Cliff Mine, Mich	May 23, 1865	47, 884
Ore-washing machine	J. Collom	Houghton, Mich	May 12, 1863	38, 467
Ore-washing machine	R. Edwards	Eagle River, Mich	Nov. 29, 1853	10, 282
Ore-washing machine	F. Fredly	Logan Township, Centre County, Pa.	Apr. 20, 1837	171
Ore-washing machine	N. H. Lebby	Charleston, S. C	Apr. 12, 1870	101, 890
Ore, &c., washing machine	R. R. Osgood	Green Island, N. Y	July 23, 1872	129, 678
Ore-washing machine	M. A. Woodside	Georgetown, Cal	Oct. 16, 1866	58, 931
Ore with chlorine, Apparatus for treating	E. Gaussoin	Baltimore, Md	July 24, 1866	56, 549
Ore with copper amalgam by means of electric currents, Process of treating.	A. L. Nolf and F. L. A. Pioche	San Francisco, Cal	Aug. 17, 1869	93, 899
Ore with spongy iron, Treating metallic	J. J. A. De Bronac and A. J. M. Deherrypon.	Paris, France	Sept. 6, 1859	25, 320
Ore with superheated steam, Treating	L. E. Rivot	Paris, France	Jan. 28, 1868	73, 840
Ore, Working	V. and J. King	Bel Air, Md	May 30, 1811	....
Ore, Working Franklinite	T. Selleck	Greenwich, Conn	Jan. 30, 1855	12, 329
Ore, Working silver	W. L. Faber	New York, N. Y	Apr. 18, 1865	47, 286
Ore, Working silver	H. Janin	Virginia City, Nev	Mar. 2, 1869	87, 340
Ores and amalgamating precious metals, Apparatus for grinding.	C. H. Griffin	Lynn, Mass	July 22, 1862	35, 980
Ores and amalgamating precious metals, Machine for grinding.	J. G. Randall	Canyon City, Colo	Nov. 3, 1863	40, 501
Ores and extracting precious metals, Furnace for reducing.	J. H. Boyd	Chicago, Ill	Nov. 4, 1873	144, 186
Ores and fertilizers, Machine for washing and screening.	A. Duvall	Baltimore, Md	Sept. 6, 1870	107, 169
Ores and metals, Apparatus for assaying and testing.	J. S. Phillips	San Francisco, Cal	Sept. 7, 1869	94, 508
Ores and metals, Flux for treating	P. N. Mackay	San Francisco, Cal	Dec. 16, 1873	145, 580
Ores and metals, Melting	D. Walworth	Middletown, Conn	May 27, 1814	....
Ores and minerals, Apparatus for concentrating	S. K. Krom	New York, N. Y	Aug. 4, 1868	80, 747
Ores and minerals, Calcining	B. Keith, A. Behr, and N. S. Keith.	New York, N. Y	Sept. 9, 1862	36, 437
Ores and minerals, Disintegrating and desulphurizing.	W. F. Goodwin and C. R. Squires.	East New York and New York, N. Y.	Sept. 3, 1867	68, 561
Ores and minerals, Machine for concentrating and separating.	R. George	Denver City, Colo	Apr. 27, 1869	89, 476
Ores and minerals, Process and machinery for obtaining metals and other products from.	J. D. Whelpley and J. J. Storer	Boston, Mass	Nov. 13, 1866	59, 696
Ores and minerals, Roasting and desulphurizing	S. Stevens	New York, N. Y	June 7, 1864	43, 046
Ores and other granular substances, Machine for separating.	S. T. Pearce	New York, N. Y	Aug. 4, 1868	80, 764
Ores and other granular substances, Machine for separating.	S. T. Pearce	New York, N. Y	Mar. 2, 1869	87, 361
Ores and other materials, Apparatus for roasting and drying.	A. Duvall	Baltimore, Md	Sept. 20, 1870	107, 600
Ores and other materials, Machine for separating	S. T. Pearce	New York, N. Y	Aug. 4, 1868	80, 763

Index of patents issued from the United States Patent Office from 1790 *to* 1873, *inclusive*—Continued.

Invention.	Inventor.	Residence.	Date.	No.
Ores a d other materials of different specific gravities, Machine for separating and concentrating.	R. George	Denver City, Colo	Aug. 2, 1870	106, 048
Ores and oxides, Reducing and refining metallic	S. C. Salisbury	New York, N. Y	May 28, 1867	65, 122
Ores and refining metals, Flux for reducing	W. Quann	Philadelphia, Pa	Oct. 17, 1871	120, 099
Ores, Apparatus and process for separating, concentrating, and amalgamating	W. C. Shaw	Philadelphia, Pa	July 4, 1871	116, 763
Ores, Apparatus for chlorinating and leaching	M. H. Stowe	Washington, D. C	May 7, 1872	126, 424
Ores, Apparatus for cleaning sulphurets and other	C. C. Coleman	San Francisco, Cal	Feb. 7, 1871	111, 612
Ores, Apparatus for concentrating auriferous	J. A. Bertola	New York, N. Y	Mar. 29, 1864	42, 065
Ores, Apparatus for extracting precious metals from.	A. Ott	New York, N. Y	Mar. 10, 1868	75, 294
Ores, Apparatus for reducing quicksilver	J. C. Coult	San Francisco, Cal	Oct. 29, 1867	70, 321
Ores, Apparatus for separating metals from	J. H. Elward	Polo, Ill	June 5, 1866	55, 261
Ores, Apparatus for separating metals from	S. R. Krom	New York, N. Y	Oct. 16, 1866	58, 839
Ores, Apparatus for separating metals from	S. R. Krom	New York, N. Y	Nov. 6, 1866	59, 509
Ores, Apparatus for separating metals from	S. R. Krom	New York, N. Y	Nov. 6, 1866	59, 510
Ores, Apparatus for separating metals from	S. R Krom	New York, N. Y	Nov. 20, 1866	59, 773
Ores, Apparatus for separating metals from	J. D. Whelpley and J. J. Storer	Boston, Mass	June 13, 1865	48, 226
Ores, Apparatus for separating volatile metals from.	J. C. Coult and J. Roach	San Francisco, Cal	Apr. 17, 1866	53, 951
Ores, Apparatus for smelting zinc	A. Borgnet	Swansea, England	Feb. 5, 1867	61, 706
Ores, Burning, roasting, and smelting	J. D. Whelpley and J. J. Storer	Boston, Mass	Jan. 12, 1864	41, 250
Ores, "Chilian mill" for pulverizing metallic	J. A. Bertola	New York, N. Y	Apr. 19, 1864	42, 341
Ores, Collecting gold, silver, &c., from	J. Corson	Washington, D. C	May 5, 1868	77, 591
Ores containing precious metals, Furnace for reducing.	J. H. Boyd	Chicago, Ill	Nov. 4, 1873	144, 184
Ores containing sulphur, &c., Stall for roasting	J. T. Reese	Baltimore, Md	Mar. 19, 1867	63, 094
Ores, Desulphurizing and volatilizing lead and silver.	G. T. Lewis	Philadelphia, Pa	July 4, 1871	116, 604
Ores, Desulphurizing auriferous pyrites and other sulphuret.	J. H. Rae and T. T. Davis	Syracuse, N. Y	Aug. 10, 1869	93, 640
Ores, Desulphurizing auropyrites and other	L. Sibert	Staunton, Va	Sept. 20, 1870	107, 553
Ores, Disintegrating and desulphurizing gold, silver, and copper.	F. F. Brower and G. C. Campbell.	Ottawa, Ill	Jan. 23, 1866	52, 132
Ores, Extracting gold and silver from	C. F. Carpenter	Louisville, Ky	Apr. 3, 1866	53, 569
Ores, Extracting metals from	W. P. McConnell	Washington, D. C	Apr. 25, 1871	114, 175
Ores, Extracting precious metals from	G. P. B. Hill	Virginia City, Nev	Sept. 22, 1868	82, 315
Ores, Extracting precious metals from	A. Ott	New York, N. Y	Mar. 10, 1868	75, 293
Ores, Extracting precious metals from	V. Ryerson	New York, N. Y	May 1, 1866	54, 412
Ores for smelting-furnaces, Device for heating	A. Royer	Reed's Mills, Ohio	May 5, 1863	38, 419
Ores for the extraction of precious metal, Desulphurizing and treating.	H. H. Eames	Saint Paul, Minn	Feb. 2, 1869	86, 514
Ores, Flux for extracting precious metals from their.	W. W. Hubbell	Philadelphia, Pa	May 4, 1869	89, 579
Ores, Furnace and condenser for reducing cinnabar and other volatile.	R. F. Knox and J. Osborn	San Francisco, Cal	June 14, 1870	104, 323
Ores, Hydraulic machine for washing	J. M. Allenwood	Timbuctoo, Cal	July 12, 1864	43, 468
Ores into powder and preparing it for various uses, Grinding and reducing.	H. Alexander	Baltimore, Md	Oct. 27, 1813	....
Ores, Lifting-stamp for crushing	D. E. Rice	Detroit, Mich	June 28, 1859	24, 578
Ores, limestones, &c., Kiln for calcining	C. W. Siemens	Westminster, England	Jan. 21, 1870	104, 655
Ores, Machine for separating and concentrating magnetic and other.	E. L. Seymour	New York, N. Y	Jan. 5, 1869	85, 700
Ores, Machinery for separating metal from	J. A. Hitchings	Denver City, Colo	May 15, 1866	54, 726
Ores, metals, and minerals, Treating	C. A. Stevens	New York, N. Y	July 7, 1868	79, 701
Ores, mills, &c., Machine for concentrating the dust from.	S. H. Rounds and W. L. Strong	San Francisco, Cal	July 26, 1864	43, 646
Ores of copper and silver, Reducing	J. T. McDougall	San Francisco, Cal	Mar. 29, 1864	42, 100
Ores of copper, &c., Treating	C. M. T. Du Motay and W. Hillegeirt.	Paris, France, and Clausthal, Hanover.	June 18, 1872	128, 026
Ores of copper, Treating sulphurous	J. D. Whelpley and J. J. Storer	Boston, Mass	Nov. 13, 1866	59, 693
Ores of gold and silver, Treatment of	J. A Bertola	New York, N. Y	Dec. 1, 1857	18, 789
Ores of gold and silver with vapor of mercury, Machine for treating.	W. A. Thompson	New York, N. Y	Jan. 7, 1868	73, 137
Ores of gold or silver, Triturating and amalgamating apparatus for treating.	L. Wray	Ramsgate, England	Mar. 9, 1869	87, 748
Ores of gold, silver, and copper, Treating	J. McCullock	San Francisco, Cal	June 26, 1860	28, 882
Ores of gold, silver, and other metals, Flux for smelting.	C. W. Moore	San Francisco, Cal	June 29, 1869	91, 862
Ores of gold, silver, copper, &c., Smelting	W. Quann	Philadelphia, Pa	Dec. 16, 1862	37, 198
Ores of gold, silver, &c., Roasting, desulphurizing, and disintegrating.	A. K. Johnson	New York, N. Y	June 30, 1865	48, 284
Ores of iron, Magnetic machine for cleaning and separating.	J. Y. Smith	Pittsburgh, Pa	June 14, 1870	104, 221
Ores of iron, Purifying and reducing magnetic	J. Y. Smith	Pittsburgh, Pa	June 14, 1870	104, 220
Ores of precious metal, Reducing	A. I. Frick and J. B. Le Clerc	San Francisco, Cal	Nov. 16, 1869	96, 790
Ores of precious metals, Treating the	J. S. Wethered and S. E. Woodworth.	San Francisco, Cal	Feb. 19, 1861	31, 499
Ores of zinc, lead, iron, &c., Treating certain	R. George	Mineral Point, Wis	Apr. 3, 1860	27, 710
Ores, Preparing	E. N. Kent	New York, N. Y	Nov. 15, 1864	45, 048
Ores, Reducing	J. H. Boyd	Chicago, Ill	Oct. 14, 1873	143, 622
Ores, Reducing	T. J. Chubb	Williamsburgh, N. Y	Dec. 14, 1869	97, 761
Ores, Reducing	C. D. Williams and W. H. Nobles.	Saint Croix, Minn	Nov. 9, 1869	96, 755
Ores, Reducing lead	C. F. Carpenter	Louisville, Ky	Jan. 28, 1868	73, 874
Ores, Reducing refractory gold, silver, and copper	C. W. Harvey	Buffalo, N. Y	Mar. 27, 1866	53, 444
Ores, &c., Reduction of	A. T. Hay	Burlington, Iowa	Nov. 19, 1872	133, 099
Ores, Reduction of	A. Parkes	Birmingham, Great Britain	Jan. 30, 1849	6, 062
Ores, Refining	J. L. Constable	New York, N. Y	July 14, 1863	39, 257
Ores, Refining and smelting	J. P. McLean	New York, N. Y	July 14, 1868	79, 848
Ores, Roasting sulphurets and other	A. F. W. Partz	Wurtsborough, N. Y	June 14, 1864	43, 129
Ores, Separating cobalt and nickel from other	A. Monnier	Philadelphia, Pa	May 19, 1868	78, 001
Ores, Separating gold and silver from	W. Crooks	Wine-Office Court, England	May 15, 1866	54, 829
Ores, Separating gold and silver from	J. H. Elward and J. L. Hayes	Polo, Ill., and Boston, Mass	Feb. 27, 1866	52, 834
Ores, Separating gold and silver from	M. Laflin	Chicago, Ill	Sept. 20, 1864	44, 320
Ores, Separating metals from	S. W. Kirk	Philadelphia, Pa	Feb. 28, 1871	112, 153
Ores, Shaft-furnace for roasting	T. McGlew	Austin, Nev	July 15, 1873	140, 837
Ores, Shaking-machine for separating	A. W. Schell	Clausthal, Hanover	Mar. 10, 1863	37, 888

Index of patents issued from the United States Patent Office from 1790 *to* 1873, *inclusive*—Continued.

Index of patents issued from the United States Patent Office from 1790 *to* 1873, *inclusive*—Continued.

Invention.	Inventor.	Residence.	Date.	No.
Organ, Reed	F. Hoddick	Buffalo, N. Y	Apr. 14, 1868	76,759
Organ, Reed	W. J. Kent	Buffalo, N. Y	Sept. 2, 1873	142,399
Organ, Reed	W. J. Kent	Buffalo, N. Y	Nov. 11, 1873	144,549
Organ, Reed	R. E. Letton	Quincy, Ill	Jan. 14, 1873	134,754
Organ, Reed	R. E. Letton	Quincy, Ill	Aug. 5, 1873	141,510
Organ, Reed	E. Oakes	Danville, Ill	Mar. 11, 1873	136,614
Organ, Reed	C. W. Palmer	Cleveland, Ohio	June 17, 1873	139,913
Organ, Reed	B. E. Riggs	Homer, Mich	Nov. 19, 1872	133,255
Organ, Reed	T. A. Rousseau	Belleville, near Paris, France.	May 3, 1859	23,884
Organ, Reed	G. W. Scribner	Detroit, Mich	June 21, 1870	104,653
Organ, Reed	J. A. Smith	Erie, Pa	Aug. 5, 1873	141,468
Organ, Reed	A. J. Sorensen	Erie, Pa	June 3, 1873	139,483
Organ, Reed	D. Tripp	Owego, N. Y	Aug. 27, 1872	130,956
Organ, Reed	F. H. Wells	Brattleborough, Vt	Mar. 24, 1868	76,013
Organ, Reed	H. K. White and E. A. Foster	Brattleborough, Vt	Aug. 20, 1872	130,677
Organ, Reed	A. W. Wilcox	New Haven, Conn	Jan. 26, 1869	86,335
Organ, Reed	G. W. Woodruff	Hartford, Conn	Apr. 4, 1871	113,714
Organ, Reed	G. W. Woodruff	Hartford, Conn	Dec. 5, 1871	121,700
Organ, Reed	G. Woods	Cambridgeport, Mass	Sept. 13, 1870	107,316
Organ, Reed	G. Woods	Cambridgeport, Mass	Jan. 23, 1872	122,979
Organ, Reed	G. Woods	Cambridgeport, Mass	Jan. 14, 1873	134,830
Organ, Reed	G. Woods	Cambridgeport, Mass	Dec. 2, 1873	145,078
Organ-reed	A. M. Brush	Clayton, N. Y	Sept. 4, 1866	57,671
Organ-reed	M. C. Morgan	Boston, Mass	June 17, 1873	139,912
Organ reed, Reed	M. O. Nichols	Clyde, Ohio	Nov. 11, 1873	144,409
Organ reed, Reed	J. R. Perry	Wilkesbarre, Pa	Feb. 4, 1873	135,581
Organ-reeds, Adjustable tongue for	M. Procopé	Stockholm, Sweden	Sept. 23, 1873	143,093
Organ, Seraphina or harmonicon	L. Zwahlen	New York	May 5, 1832	....
Organ-stop	A. R. Hebard	Boston, Mass	July 19, 1870	105,571
Organ-stop action	J. A. Smith	Erie, Pa	Aug. 5, 1873	141,469
Organ-stop action, Reed	I. T. Packard	Fort Wayne, Ind	July 1, 1873	140,381
Organ-stop action, Reed	J. A. Smith	Erie, Pa	Oct. 22, 1872	132,419
Organ-stop action, Reed	J. A. Smith	Erie, Pa	Mar. 25, 1873	137,102
Organ-stop handle	W. Boyrer	New York, N. Y	Nov. 9, 1869	96,543
Organ-stops, Name-plate for	C. E. Bacon and W. J. Kent	Buffalo, N. Y	May 23, 1871	115,146
Organ swell, Mechanism for reed	M. J. Kerigan	Boston, Mass	Dec. 26, 1871	122,257
Organ swell, Reed	J. R. Lomas	New Haven, Conn	July 29, 1873	141,282
Organ-treadle	J. A. Smith	Erie, Pa	Mar. 25, 1873	137,101
Organ-valve arrangement	M. Baumgarten, jr	New Haven, Conn	Oct. 20, 1868	83,241
Organs and melodeons, Knee-swell for	S. Taylor	Worcester, Mass	Jan. 30, 1866	52,339
Organs, Blowing-apparatus for	J. F. Barker	Springfield, Mass	Nov. 28, 1871	121,269
Organs, &c., Blowing-apparatus for	M. J. Kerigan	Boston, Mass	Feb. 1, 1870	99,322
Organs, &c., Chime for reed and pipe	C. Lehnert	Boston, Mass	Dec. 12, 1871	121,790
Organs, &c., Enharmonic key-board for	H. W. Poole	South Danvers, Mass	Jan. 22, 1868	73,753
Organs, Fall for parlor	W. O. Trowbridge	Newton, Mass	Oct. 28, 1873	144,167
Organs, &c., Key-board for	M. Baumgarten	New Haven, Conn	Apr. 3, 1866	53,728
Organs, Key-board for reed and pipe	C. Fogelberg	Boston, Mass	Dec. 12, 1871	121,768
Organs, Mechanism for operating swell of reed	A. E. Thompson	Boston, Mass	Feb. 20, 1866	52,771
Organs, &c., Melody-attachment for	C. Fogelberg	Boston, Mass	Apr. 5, 1870	101,605
Organs, Motor-regulator and register-attachment for	W. H. Topham	New Haven, Conn	June 4, 1867	65,452
Organs, Operating fan-tremolo for	A. C. Bradley	Washington, D. C	July 9, 1872	128,845
Organs, Operating the stop for	H. R. Moore	Erie, Pa	Mar. 11, 1873	136,664
Organs, Operating tremolo in	R. W. Carpenter	Chicago, Ill	Mar. 2, 1869	87,395
Organs or melodeons, Reverberating reed-cells for	C. W. Small	Worcester, Mass	Apr. 18, 1871	113,940
Organs, &c., Pedal for	T. Robjohn	New York, N. Y	Feb. 9, 1858	19,312
Organs, Pneumatic action for	T. Winans	Baltimore, Md	Oct. 14, 1873	143,602
Organs, Producing a vibrating swell in	A. Hitchcock	New York, N. Y	Jan. 28, 1868	73,719
Organs, Reed-board for	R. Burdett	Chicago, Ill	Sept. 27, 1870	107,755
Organs, Reed-board for	W. Munroe	Worcester, Mass	Oct. 28, 1873	144,121
Organs, Reed-board for	J. R. Perry	Wilkesbarre, Pa	Aug. 6, 1872	130,239
Organs, Reed-board for	A. W. Wilcox	New Haven, Conn	July 8, 1873	140,750
Organs, &c., Reed for	W. Munroe	Cambridge, Mass	Jan. 7, 1868	73,114
Organs, Swell for pipe	F. H. Hastings	Boston, Mass	Dec. 3, 1872	133,531
Organs, Transposing-mechanism for	W. G. Day	Baltimore, Md	Sept. 13, 1870	107,342
Organs, Tremolo-attachment for reed or pipe	M. J. Kerigan	Boston, Mass	Apr. 12, 1870	101,742
Organs, Tremolo for	J. R. Lomas	New Haven, Conn	Feb. 22, 1870	100,048
Organs, Tremolo for reed	L. K. Fuller	Brattleborough, Vt	Dec. 5, 1871	121,609
Organs, Tremolo for reed	L. K. Fuller	Brattleborough, Vt	Dec. 5, 1871	121,610
Organs, Tremolo for reed	J. R. Lomas	New Haven, Conn	Aug. 8, 1871	117,788
Organs, Tremulant for	T. P. Sanborn	Boston, Mass	Sept. 20, 1870	107,549
Organs, &c., Valve for	J. R. Mortimore	New York, N. Y	Jan. 11, 1870	98,790
Organs, Valve for pneumatic draw in	A. Stein	Westfield, Mass	Aug. 30, 1870	106,884
Ornament manufacture, Metal	F. J. Emery	Chicago, Ill	May 4, 1869	89,642
Ornaments, &c., Composition-matter for forming	E. A. Hildreth	Wheeling, W. Va	Feb. 11, 1868	74,225
Ornaments from anthracite or other coal, Making	J. W. Kirk	Schuylkill County, Pa	June 13, 1831	....
Ornaments of sheet-metal, Constructing architectural.	H. Baeuerle	New York, N. Y	June 29, 1869	91,813
Ornaments, Producing metal	W. Henigst	Columbus, Ohio	Nov. 12, 1872	133,029
Ornaments used for picture and mirror frames and for architectural purposes, Device for cutting up composition.	R. J. Marcher	New York, N. Y	Sept. 9, 1862	36,416
Ornamental bars or rods of metal, Manufacture of	S. Tuddenham	London, England	Jan. 17, 1871	111,097
Ornamental compounds, Cementing-material for	C. L. Grau	Hamburg, Germany	Apr. 26, 1853	9,693
Ornamental purposes, Composition of matter for	A. H. Wright	Camden, N. J	Aug. 9, 1859	25,074
Ornamental transfer	P. Kneipp	Philadelphia, Pa	Sept. 28, 1869	95,356
Ornamentation, Composition for articles of	W. E. Houston	New Haven, Conn	Oct. 1, 1861	33,391
Ornamentation of metal, glass, &c	B. G. George	London, England	Apr. 18, 1871	113,758
Ornamenting	H. Harrop	Greenwich, N. Y	Feb. 14, 1865	46,354
Ornamenting and dressing glass and metal surfaces, &c.	G. F. Morse	New York, N. Y	Nov. 21, 1871	121,119
Ornamenting and lettering glass	T. R. Hartell	Philadelphia, Pa	Oct. 2, 1866	58,410
Ornamenting and lettering hard and uneven surfaces	W. H. McElcheran	Hamilton, Canada	Nov. 7, 1871	120,656
Ornamenting and lettering looking-glasses, signs, &c., Machine for.	W. M. Davis	Brook Haven, N. Y	Oct. 18, 1870	108,458
Ornamenting columns, chairs, ivory, &c	R. Thompson	Washington, Ohio	Oct. 1, 1830	....
Ornamenting hollow articles of metal	D. R. Pruden	West Meriden, Conn	July 16, 1861	32,850
Ornamenting metal	N. A. Batchelor	New York, N. Y	Jan. 5, 1864	41,132

Index of patents issued from the United States Patent Office from 1790 *to* 1873, *inclusive*—Continued.

Invention.	Inventor.	Residence.	Date.	No.
Ornamenting metallic surfaces	H. Tucker	Newton, Mass	Apr. 27, 1869	89, 523
Ornamenting textile fabrics, Process and material for.	C. Günther	Berlin, Prussia	Feb. 15, 1870	99, 773
Ornamenting tin, &c	L. Fitzmaier	New York, N. Y	July 30, 1867	67, 184
Ornamenting surface of metals by electro-depositions from solutions.	R. O'Neil	New York, N. Y	Apr. 4, 1871	113, 331
Ornamenting surface of wood, &c., Machine for	T. T. Ponsonby	Nottingham, England	Sept. 27, 1870	107, 718
Ornamenting wood	A. Beckers	New York, N. Y	June 19, 1866	55, 604
Ornamenting wood, Composition for and mode of	T. H. Davis	Philadelphia, Pa	May 3, 1870	102, 660
Ornamenting wood, metals, &c	R. Parke	New York, N. Y	Mar. 12, 1872	124, 616
Orrery	J. G. Moore	Philadelphia, Pa	Nov. 21, 1865	51, 072
Orthography, System of pronouncing	E. Leigh	Saint Louis, Mo	May 26, 1868	78, 296
Oscillating chair	W. T. Doremus	New York, N. Y	Dec. 31, 1872	134, 423
Oscillating chair	W. T. Doremus	New York, N. Y	Feb. 18, 1873	135, 974
Oscillating engine	W. E. Bird	New York, N. Y	May 21, 1867	64, 940
Oscillating engine	M. E. Bollinger	Littlestown, Pa	Mar. 26, 1861	31, 775
Oscillating engine	F. Brown	New York, N. Y	Jan. 16, 1866	52, 024
Oscillating engine	F. Brown	New York, N. Y	June 12, 1866	55, 575
Oscillating engine	H. Chavons	Union City, Ind.	Oct. 15, 1872	132, 251
Oscillating engine	W. Craig	New York, N. Y	Sept. 12, 1854	11, 664
Oscillating engine	M. V. Cummings	Winthrop, Me	Aug. 6, 1867	67, 511
Oscillating engine	T. W. Godwin	Norfolk, Va	Nov. 28, 1871	131, 355
Oscillating engine	A. Goulding	Worcester, Mass	Dec. 5, 1871	121, 457
Oscillating engine	J. Goulding	Worcester, Mass	June 13, 1871	115, 952
Oscillating engine	L. Griscom	Port Carbon, Pa	Sept. 3, 1872	130, 988
Oscillating engine	T. Hill	Vallejo, Cal	July 12, 1870	105, 204
Oscillating engine	M. C. Kilgore and W. Eberhard	Washington, Iowa	Nov. 22, 1864	45, 159
Oscillating engine	W. H. King	Philadelphia, Pa	July 4, 1865	48, 563
Oscillating engine	A. B. Latta	Cincinnati, Ohio	Oct. 11, 1853	10, 119
Oscillating engine	H. B. Martin	San Francisco, Cal	Mar. 2, 1869	87, 419
Oscillating engine	C. R. Otis	Yonkers, N. Y	Oct. 2, 1860	30, 240
Oscillating engine	E. G. Otis	Yonkers, N. Y	Oct. 2, 1860	30, 241
Oscillating engine	C. Reed	Georgetown, D. C	Sept. 17, 1872	131, 461
Oscillating engine	J. A. Reed	New York, N. Y	Jan. 9, 1855	12, 216
Oscillating engine	M. Runkel	New York, N. Y	Apr. 12, 1859	23, 612
Oscillating engine	A. Schmid	Zurich, Switzerland	Feb. 27, 1872	124, 162
Oscillating engine	W. Sellers	Philadelphia, Pa	June 11, 1872	127, 928
Oscillating engine	H. Shlarbaum	New York, N. Y	Sept. 1, 1863	39, 756
Oscillating engine	J. W. Van Sant	Perth Amboy, N. Y	Apr. 16, 1872	125, 773
Oscillating engine	H. Wightman and W. Warden	Allegheny, Pa	Jan. 15, 1856	14, 124
Oscillating engine.	G. F. Wood	Ulysses, N. Y	Jan. 23, 1855	12, 299
Oscillating engines and pumps, Arrangement of	G. Sprenkel and T. W. Basford	Harrisonburgh, Pa	Dec. 22, 1857	18, 933
Oscillating engines, Mechanism for	H. Broomell	Christiana, Pa	June 7, 1870	103, 840
Oscillating engines, Trunnion-box lining for	J. A. Reed	Jersey City, N. Y	July 5, 1859	24, 663
Oscillating piston-engine	W. M. Elrod	Saint Louis, Mo	Aug. 20, 1867	67, 966
Oscillating steam-engine	W. D. Andrews	New York, N. Y	Nov. 11, 1862	36, 885
Oscillating steam-engine	J. S. Barden	New Haven, Conn	Feb. 23, 1858	19, 464
Oscillating steam-engine	M. Cridge and S. Wadsworth	Pittsburgh, Pa	Dec. 12, 1854	12, 052
Oscillating steam-engine	I. M. Forrester	Bridgeport, Conn	Dec. 20, 1870	110, 352
Oscillating steam-engine	M. J. Gardner	York, Pa	Aug. 23, 1853	9, 954
Oscillating steam-engine	R. J. King	Lancaster City, Pa	Sept. 15, 1868	82, 126
Oscillating steam-engine	G. F. Lowry	Lake Mississippi, Miss	Mar. 19, 1872	124, 754
Oscillating steam-engine	F. Millward	Cincinnati, Ohio	Nov. 17, 1868	84, 206
Oscillating steam-engine	T. Murgatroyd	Cleveland, Ohio	Jan. 14, 1862	34, 161
Oscillating steam-engine	D. H. Nation and T. B. Hall	Saint Louis, Mo	Aug. 8, 1865	49, 293
Oscillating steam-engine	A. Nittinger, jr	Philadelphia, Pa	Aug. 30, 1871	106, 869
Oscillating steam-engine	J. Pattison	Brooklyn, N. Y	Feb. 5, 1856	14, 204
Oscillating steam-engine	F. C. Ricker	Gilmer, Tex	July 6, 1869	92, 363
Oscillating steam-engine	J. Robertson	Brooklyn, N. Y	May 7, 1872	126, 576
Oscillating steam-engine	M. Runkel	New York, N. Y	Mar. 6, 1860	27, 387
Oscillating steam-engine	E. Russell	Brooklyn, N. Y	July 9, 1861	32, 787
Oscillating steam-engine	E. P. Ryder	Brooklyn, N. Y	Aug. 6, 1872	130, 157
Oscillating steam-engine	J. B. Sweetland	Pontiac, Mich	July 9, 1872	128, 766
Oscillating steam-engine	J. Wallace	Pittsburgh, Pa	July 28, 1857	17, 903
Ottoman	A. O. and E. G. Ganiard	New York, N. Y	Feb. 9, 1869	86, 744
Ottoman and hassock	G. B. Green	Staffordshire, England	Feb. 20, 1872	123, 821
Ottoman and hassock filler	E. G. Ganiard	New York, N. Y	Aug. 4, 1868	80, 619
Oval-cutting machine	C. W. Packer	Philadelphia, Pa	Aug. 16, 1864	43, 883
Oval frames, Machine for jointing	D. Garrison	Philadelphia, Pa	May 22, 1866	54, 890
Oval frames, Machine for jointing	J. E. Rogers	Chelsea, Mass	Feb. 7, 1865	46, 268
Oval frames, Machine for turning	I. P. Tice	Baltimore, Md	Mar. 26, 1861	31, 837
Oval-turning machine	A. H. Brown	Albany, N. Y	Sept. 24, 1861	33, 334
Oval-turning machine	J. Irving	New York, N. Y	July 5, 1859	24, 642
Oval-turning machine	P. Prybil	New York, N. Y	Aug. 24, 1869	94, 130
Oven	W. H. Akins	Berkshire, N. Y	June 2, 1836	....
Oven	J. Allen and E. Pick	Brooklyn, N. Y	June 24, 1862	35, 661
Oven	H. Ball	New York, N. Y	Sept. 23, 1856	15, 753
Oven	H. Ball	New York, N. Y	July 19, 1870	105, 541
Oven	J. Barnitt	Upper Providence, Pa	Mar. 3, 1809	....
Oven	B. Blaney	Boston, Mass	Dec. 17, 1834	....
Oven	J. S. Brown	Washington, D. C	Apr. 13, 1858	19, 965
Oven	D. P. Burdon	Brooklyn, N. Y	Feb. 1, 1859	22, 778
Oven	E. Bussey	Troy, N. Y	May 5, 1868	77, 581
Oven	J. C. Carlisle	Chesterville, Me	Apr. 19, 1833	....
Oven	D. S. Coburn	Amboy, Ill	Jan. 30, 1872	123, 086
Oven	A. Crumbie	Jersey City, N. J	Feb. 8, 1870	99, 645
Oven	J. S. Dunham and J. Green	Saint Louis, Mo	Oct. 12, 1869	95, 671
Oven	M. Dunn	Bloomington, Ill	Mar. 29, 1870	101, 357
Oven	S. S. Eaton	Mansfield, Conn	Dec. 17, 1834	....
Oven	D. J. George	Arcade, Ind	Aug. 16, 1870	106, 352
Oven	P. J. Gindre and J. Doerler	Cincinnati, Ohio	June 30, 1863	39, 042
Oven	G. Y. Gray	Niles, Mich	Mar. 30, 1869	88, 472
Oven	J. P. Hayes	Philadelphia, Pa	July 31, 1855	13, 375
Oven	J. P. Hayes	Philadelphia, Pa	Dec. 2, 1856	16, 143
Oven	J. F. Hoffmeister	Alton, Ill	May 24, 1859	24, 122
Oven	G. C. Jennison	Ware, Mass	Feb. 1, 1859	22, 809
Oven	D. A. Kennedy	Beloit, Wis	June 15, 1869	91, 451
Oven	D. McKenzie	Brooklyn, N. Y	May 1, 1860	28, 130

Index of patents issued from the United States Patent Office from 1790 *to* 1873, *inclusive*—Continued.

Invention.	Inventor.	Residence.	Date.	No.
Oven	D. McKenzie	Brooklyn, N. Y	June 6, 1871	115, 626
Oven	D. Moore	Davenport, Iowa	Feb. 8, 1870	99, 694
Oven	G. R. Moore	Philadelphia, Pa	Nov. 2, 1869	96, 338
Oven	J. Ohmert	Mount Morris, Ill	May 6, 1856	14, 825
Oven	E. G. Otis	Yonkers, N. Y	Aug. 24, 1858	21, 271
Oven	S. Pollard	Orono, Me	Feb. 3, 1836	....
Oven	C. Preston	Wheeling, W. Va	Sept. 7, 1869	94, 643
Oven	J. Rainey	Brooklyn, N. Y	Oct. 17, 1870	108, 389
Oven	J. M. Read	Boston, Mass	Aug. 28, 1860	29, 818
Oven	T. N. Reid	Baltimore, Md	May 18, 1852	8, 960
Oven	T. Russell	New York, N. Y	Feb. 1, 1859	22, 828
Oven	W. Sellers	Philadelphia, Pa	Jan. 22, 1861	31, 192
Oven	I. H. Shaver	Cedar Rapids, Iowa	July 4, 1871	116, 762
Oven	C. Swain	Laconia, N. H	May 3, 1870	102, 614
Oven	R. R. Tongue	Fryburgh, Me	Aug. 15, 1834	....
Oven	E. Treadwell	New York, N. Y	July 19, 1853	9, 864
Oven	F. C. Treadwell, jr	New York, N. Y	Aug. 22, 1854	11, 578
Oven	J. R. Treadwell	Brooklyn, N. Y	Oct. 12, 1869	95, 750
Oven	J. Vatter	Philipsburgh, Ohio	May 4, 1869	89, 713
Oven	W. C. Wedge	Chicopee, Mass	July 27, 1869	93, 145
Oven	J. West	Stafford County, Va	Oct. 27, 1808	....
Oven	A. Whittock	Danbury, Conn	Oct. 2, 1866	58, 522
Oven	A. Wilson	Buffalo, N. Y	Oct. 3, 1871	119, 680
Oven and bath, Combined	J. Perry	Brooklyn, N. Y	May 26, 1868	78, 231
Oven and drum, Combined	J. J. Stout, S. J. Russell, and N. Mendenhall.	Greensburgh, Ind	Sept. 21, 1869	95, 161
Oven and drum, Detachable	C. A. Harper and I. A. Crane	Rahway, N. J	Nov. 26, 1867	71, 481
Oven and range, Combined baker's	J. Williams	Topeka, Kans	June 24, 1873	140, 330
Oven and stove-pipe drum, Combined	J. W. Clough	Montville, Me	Feb. 16, 1869	86, 907
Oven, Bake	A. A. and J. A. Aull	Bellefontaine, Ohio	Oct. 11, 1870	108, 086
Oven, Bake	H. Ball	Philadelphia, Pa	Nov. 19, 1850	7, 778
Oven, Bake	J. Bonis	Baltimore, Md	Dec. 30, 1812	....
Oven, Bake	P. Grouvell and L. N. and E. Monchat.	Paris, France	Mar. 20, 1847	5, 023
Oven, Bake	J. Hall	Cincinnati, Ohio	Oct. 11, 1870	108, 258
Oven, Bake	I. A. Hammer	Newton, Iowa	Sept. 21, 1869	95, 106
Oven, Bake	J. P. Hayes	Philadelphia, Pa	July 29, 1856	15, 422
Oven, Bake	D. Moore	Davenport, Iowa	June 1, 1869	90, 771
Oven, Bake	W. Pettet	New York, N. Y	Aug. 10, 1858	21, 147
Oven, Bake	J. J. Piggott	Saint Louis, Mo	Apr. 22, 1873	138, 042
Oven, Bake	H. Reip	Baltimore, Md	Sept. 27, 1825	....
Oven, Bake	J. Vale	Beloit, Wis	July 7, 1868	79, 615
Oven, Baker's	G. E. Bailey	Mansfield, Mass	Dec. 5, 1871	121, 573
Oven, Baker's	G. E. Bailey	Mansfield, Mass	July 16, 1872	129, 080
Oven, Baker's	H. Berdan	New York, N. Y	Oct. 20, 1857	18, 429
Oven, Baker's, &c	H. Chatain	Washington, D. C	Apr. 9, 1872	125, 430
Oven, Baker's	J. Chilcott	Brooklyn, N. Y	June 9, 1857	17, 495
Oven, Baker's	A. Crumbie	Jersey City, N. J	Apr. 9, 1872	125, 546
Oven, Baker's	D. Donald	Williamsburgh, N. Y	July 8, 1873	140, 686
Oven, Baker's	C. F. Holmes	Saint Louis, Mo	Apr. 16, 1872	125, 735
Oven, Baker's	W. Johnson and H. Davies	Brooklyn, N. Y	Feb. 25, 1862	34, 502
Oven, Baker's	J. G. Kluge	New York, N. Y	Oct. 12, 1869	95, 811
Oven, Baker's	I. W. Knapp	New York, N. Y	Feb. 18, 1862	34, 432
Oven, Baker's	T. P. Mahon	New York, N. Y	Apr. 26, 1870	102, 415
Oven, Baker's	A. Martin	Allegheny, Pa	Feb. 14, 1871	111, 759
Oven, Baker's	D. McKenzie	Brooklyn, N. Y	Oct. 15, 1872	132, 305
Oven, Baker's	D. McKenzie	Brooklyn, N. Y	June 10, 1873	139, 802
Oven, Baker's	D. McKenzie	Brooklyn, N. Y	Oct. 7, 1873	143, 522
Oven, Baker's	W. R. Nevins and J. J. Yates	New York, N. Y	Sept. 28, 1858	21, 620
Oven, Baker's	J. Rayney	Brooklyn, N. Y	Oct. 18, 1870	108, 517
Oven, Baker's	J. Rayney and W. Cairns	Brooklyn, N. Y., and Jersey City, N. J.	Mar. 4, 1873	136, 546
Oven, Baker's	N. F. Rice	New Orleans, La	Nov. 15, 1859	26, 126
Oven, Baker's	C. Robbins	Charlestown, Mass	Apr. 15, 1873	137, 961
Oven, Baker's	R. Sanderson	Cleveland, Ohio	Nov. 14, 1871	120, 901
Oven, Baker's	A. B. Smith	Geneva, N. Y	July 30, 1872	129, 993
Oven, Baker's	G. H. Smith	Galesburgh, Ill	June 25, 1872	128, 433
Oven, Baker's	E. B. Strong	Buffalo, N. Y	Apr. 11, 1836	....
Oven, Baker's	J. Taylor	Galesburgh, Ill	Oct. 29, 1872	132, 548
Oven, Baker's	J. and C. L. Vale	Chicago, Ill	June 7, 1870	104, 083
Oven, Baker's	J. H. Vance and F. B. Hall	Ann Arbor, Mich	Sept. 10, 1872	131, 315
Oven, Baking	A. A. and J. A. Aull	Bellefontaine, Ohio	Dec. 12, 1871	121, 839
Oven, Baking	G. S. Blodgett and P. T. Sweet	Burlington, Vt	Dec. 5, 1854	12, 018
Oven, Baking	E. Moody	Northfield, Mass	Apr. 5, 1828	....
Oven, Baking and cooking	S. J. Gold	Cornwall, Conn	Mar. 29, 1834	....
Oven, Chimney	E. Smith	Ithaca, N. Y	July 1, 1836	....
Oven, Circular coke	F. J. F. Laumonier	Angers, France	June 18, 1867	65, 820
Oven, Coke	L. F. and A. Beckwith	New York, N. Y	July 19, 1870	105, 413
Oven, Coke	J. Bowers	Connellsville, Pa	Oct. 24, 1865	50, 552
Oven, Coke	J. Erichsen and J. G. Maardt	London, England	July 23, 1872	129, 803
Oven, Coke	T. G. Kenny	Prospect, Pa	June 18, 1872	128, 151
Oven, Coke	G. Lambert	Mons, Belgium	Jan. 9, 1855	12, 239
Oven, Coke	T. G. Meier	Saint Louis, Mo	Sept. 19, 1871	119, 092
Oven, Coke	L. Schantl	Saint Louis, Mo	May 24, 1870	103, 507
Oven, Combined furnace and tempering	R. L. and A. F. Hewett	Worcester and Milbury, Mass.	May 7, 1872	126, 393
Oven, Cooking	J. Chilcott	Brooklyn, N. Y	Aug. 22, 1865	49, 503
Oven, Desulphurizing	J. C. Brewster	New York, N. Y	Jan. 25, 1870	99, 143
Oven, Detachable	A. B. Nott	Fairhaven, Mass	June 13, 1865	48, 200
Oven-door	C. H. Finn	Syracuse, N. Y	Jan. 12, 1869	85, 809
Oven-door flue	B. Simonds, jr	Bedford, Mass	Oct. 15, 1821	....
Oven-door, Hollow	G. W. Eddy	Waterford, N. Y	Feb. 10, 1846	4, 377
Oven-doors, Attachment for	S. Morton	Valparaiso, Ind	June 9, 1868	78, 754
Oven, Dutch	C. Schmeidt	Chicago, Ill	Aug. 13, 1873	130, 444
Oven, Elevated	J. C. Kennedy	Albany, N. Y	July 11, 1854	11, 295
Oven, Enameling	V. Keller	Allegheny, Pa	Jan. 11, 1870	98, 777
Oven, Fire-heating	C. G. Nye	Onondaga, N. Y	June 1, 1869	90, 775
Oven for annealing iron	S. Reynolds	Pittsburgh, Pa	Feb. 13, 1866	52, 605

Index of patents issued from the United States Patent Office from 1790 *to* 1873, *inclusive*—Continued.

Invention.	Inventor.	Residence.	Date.	No.
Oven for bakers, Reel	J. A. Groffet	Saint Louis, Mo	Dec. 17, 1867	72, 194
Oven for baking and heating houses, &c	J. Baldwin	New York, N. Y	Oct. 27, 1835	....
Oven for baking bread and other aliments	J. L. Rolland	Paris, France	Nov. 27, 1855	13, 855
Oven for baking bread, Railway	S. Short	Nantucket, Mass	Oct. 6, 1837	421
Oven for baking fire-brick	L. A. Boisson	Lyons, France	Jan. 22, 1861	31, 159
Oven for baking, &c., Revolving	S. Harris	Norfolk, Va	Dec. 5, 1825	....
Oven for converting iron into steel	W. A. Sweet	Syracuse, N. Y	Jan. 10, 1865	45, 878
Oven for cooling and annealing glass	J. Franck	Millville, N. J	May 12, 1843	3, 081
Oven for cooling castings	P. F. Geisse	Wellsville, Ohio	Apr. 12, 1859	23, 671
Oven for cooling window-glass	D. Bievez	Haine St. Pierre, Belgium	Dec. 3, 1867	71, 571
Oven for drying fruit, Portable	G. Diffenderfer	Lewisburgh, Pa	July 30, 1867	67, 172
Oven for plate-glass, Flattening	J. B. Boulicault	New Albany, Ind	Nov. 25, 1873	144, 946
Oven, Fruit-drying	D. Lippy and S. Linn	Mansfield, Ohio	Sept. 20, 1864	44, 324
Oven, Gas and air-heating	G. F. Wilson	Providence, R. I	June 18, 1872	128, 004
Oven-heating furnace	A. H. Quincy	Boston, Mass	Feb. 17, 1812	....
Oven in chimney or large bakery, Double	R. Bacon and W. E. Marshall	Walpole, N. H	Feb. 9, 1831	....
Oven, Peanut	D. Dill	New Haven, Conn	June 8, 1869	91, 098
Oven, Perpetual	J. Deneale	Dumfries, Va	Dec. 4, 1806	....
Oven, Perpetual	W. T. Hunter	New York	Jan. 8, 1818	....
Oven, Perpetual rotary	J. L. D. Mathies	Rochester, N. Y	Dec. 21, 1831	....
Oven, Portable	G. W. Ayres	Rahway, N. J	Mar. 18, 1861	34, 726
Oven, Portable	T. B. Curtis	New Haven, Conn	June 27, 1839	1, 201
Oven, Portable	C. Doane	Wareham, Mass	Oct. 9, 1849	6, 772
Oven, Portable	J. A. Frey	Washington, D. C	Mar. 15, 1864	41, 914
Oven, Portable	W. Goddard	Portsmouth, N. H	Oct. 12, 1831	....
Oven, Portable	F. L. Hedenberg	New York	Apr. 26, 1828	....
Oven, Portable	M. H. Leland	Millbury, Mass	July 30, 1867	67, 321
Oven, Portable	F. C. Miller	Evans Centre, N. Y	May 26, 1868	78, 226
Oven, Portable	Z. N. Morrell	Cameron, Tex	July 12, 1859	24, 752
Oven, Portable	W. S. Robinson	Oskaloosa, Iowa	July 25, 1871	117, 464
Oven, Portable	E. Wassenich	Cincinnati, Ohio	Feb. 25, 1862	34, 540
Oven, Portable bake	D. Hull	Utica, N. Y	Mar. 28, 1834	....
Oven, Portable elevated	P. Killin	Mount Healthy, Ohio	Oct. 7, 1851	8, 415
Oven, Portable rotary	R. Wilcox	New York, N. Y	Apr. 20, 1820	....
Oven-reel	A. Crumbie	Brooklyn, N. Y	Aug. 19, 1873	142, 003
Oven, Reflecting	B. Ames	Ithaca, N. Y	June 25, 1836	....
Oven, Reflecting	C. D. Van Allen	Penn Yan, N. Y	Mar. 31, 1836	....
Oven, Revolving	J. A. Kinkele	Sacramento City, Cal	Dec. 24, 1867	72, 645
Oven, Rotary	D. A. Kennedy	Darien, Wis	July 20, 1869	92, 840
Oven, Rotary bake	J. Vale	Beloit, Wis	Oct. 4, 1870	107, 984
Oven, Steam	E. Brown	Burlington, Vt	Jan. 30, 1872	123, 230
Oven, Steam	J. G. Whitlock	New York, N. Y	Jan. 23, 1866	52, 235
Oven, Tin bake	W. Lewis	Franklin, N. Y	Jan. 6, 1832	....
Oven, Union	J. S. Gold	Norwich, Conn	Dec. 16, 1853	....
Oven with reflector	J. Hurd, jr	Boston, Mass	Nov. 10, 1829	....
Ovens, Admission of steam to baking	J. Y. Betts	Coventry, England	Mar. 2, 1869	87, 390
Ovens and boilers for generation of steam, Combining cooking.	R. McMillen	Middleburgh, Ohio	Dec. 14, 1841	2, 388
Ovens and other like furnaces, Heating pottery	H. Speeler	Trenton, N. J	Dec. 24, 1867	72, 694
Ovens and rooms, Heating	M. B. Poitiaux	Richmond, Va	Jan. 17, 1827	....
Ovens by anthracite, Heating	F. C. Treadwell	Brooklyn, N. Y	June 16, 1836	....
Ovens by steam, Heating	H. Lyon	Cincinnati, Ohio	Sept. 28, 1858	21, 610
Ovens, Construction and application of	S. Pollard	Bucksport, Me	June 12, 1835	....
Ovens, Heating elevated	C. W. Grannis	Collins, N. Y	Feb. 1, 1847	4, 950
Ovens, Heating elevated	P. A. Palmer	Le Roy, N. Y	Sept. 24, 1850	7, 672
Ovens to be combined with cooking-stoves, Method of constructing flues of elevated.	R. D. Granger	Albany, N. Y	Oct. 11, 1841	2, 308
Ovens, Turn-table for baker's	R. Shaler	Madison, Conn	Oct. 2, 1866	58, 489
Ovens, Valve for elevated	H. Stanley	West Poultney, Vt	June 11, 1842	2, 664
Ovens with cooking-stoves so as to render them movable, Combining elevated.	E. C. Robinson	Troy, N. Y	Dec. 30, 1841	2, 404
Ovens with stoves, Combining elevated	S. B. Spaulding	Brandon, Vt	Aug. 28, 1841	2, 235
Overall	E. F. and J. H. Stacy	Gloucester, Mass	Apr. 23, 1872	125, 994
Overall	E. Weil	New York, N. Y	Oct. 21, 1873	143, 947
Overcoat and tent, Convertible	H. J. Phillips	New York, N. Y	Nov. 19, 1861	33, 752
Overflow-alarm	T. Mayes	Albany, N. Y	Nov. 18, 1873	144, 687
Overshoe	D. H. Bond	Canterbury, Conn	June 30, 1836	....
Overshoe	A. O. Bourn	Cranston, R. I	Feb. 23, 1869	87, 137
Overshoe	A. O. Bourn	Providence, R. I	Nov. 28, 1871	121, 446
Overshoe	P. Dorn	Philadelphia, Pa	May 7, 1850	7, 343
Overshoe	J. Evans	New Brunswick, N. J	Jan. 30, 1872	123, 162
Overshoe	D. Hodgman	New York, N. Y	Mar. 23, 1842	2, 504
Overshoe	H. L. Hotchkiss	New Haven, Conn	June 2, 1868	78, 595
Overshoe	J. J. Mellon	Pana, Ill	Oct. 29, 1872	132, 681
Overshoe	W. W. Swann	Richmond, Va	May 24, 1870	103, 522
Overshoe	H. G. Tyer	Andover, Mass	Sept. 3, 1867	68, 398
Overshoe	A. Vetter	Philadelphia, Pa	May 2, 1846	4, 497
Overshoe	A. G. Waterhouse	San Francisco, Cal	Sept. 7, 1869	94, 531
Overshoe	G. Watkinson	New Haven, Conn	Nov. 18, 1873	144, 810
Overshoe	P. S. Whitman	Providence, R. I	Aug. 29, 1871	118, 568
Overshoe	J. Wild	Woonsocket, R. I	Mar. 12, 1872	124, 525
Overshoe	I. F. Williams	Bristol, R. I	Sept. 10, 1872	131, 199
Overshoe	I. F. Williams	Bristol, R. I	Sept. 10, 1872	131, 200
Overshoe	I. F. Williams	Bristol, R. I	Sept. 10, 1872	131, 201
Overshoe and boot and shoe	W. S. McEwen and N. A. Patterson.	Kingston, Tenn	Apr. 17, 1860	27, 918
Overshoe-fastening	H. E. Starrett	Lawrence, Kans	Feb. 22, 1870	100, 207
Overshoe, India-rubber	E. F. Bickford	Malden, Mass	July 16, 1872	129, 389
Overshoe-spike	G. W. Bradley	Weston, Conn	Mar. 30, 1869	88, 441
Overshoe spur-attachment	H. Schwandt	New Orleans, La	Feb. 14, 1871	111, 880
Overshoe, Straw and wood	F. W. Mitchel, W. C. Wilcox, and H. T. Miller.	Utica, N. Y	June 1, 1858	20, 439
Overshot-bucket wheel, used in hydraulic works	D. S. Howard	Lyonsdale, N. Y	Feb. 16, 1831	....
Ox-bow	A. L. D. Moore	La Grange, Tex	May 24, 1870	103, 355
Ox-bow fastening	E. Kenney	Livermore, Me	May 5, 1863	38, 395
Ox-bow pin	I. Glover	Newtown, Conn	Aug. 1, 1871	117, 618
Ox-bow pin	E. N. Hartwell	Sharon, Conn	July 30, 1872	129, 888

Index of patents issued from the United States Patent Office from 1790 *to* 1873, *inclusive*—Continued.

Invention.	Inventor.	Residence.	Date.	No.
Ox-bow pin	B. B. Hotchkiss	New York, N. Y	Apr. 16, 1867	63, 894
Ox-bow pin	O. D. Hunter	Terrysville, Conn	Apr. 16, 1867	63, 796
Ox-bow pin	H. P. Judson	Bethlehem, Conn	Dec. 8, 1868	84, 831
Ox-bow pin	J. Low	New Britain, Conn	Apr. 9, 1867	63, 736
Ox-bow pin	E. S. Woodford	Winchester, Conn	Feb. 27, 1866	52, 923
Ox-bows, Shaping wood for	H. S. Denison	Coleraine, Mass	Oct. 13, 1863	40, 249
Ox-shoe	H. Lake	Shelburne, Vt	Oct. 29, 1867	70, 227
Ox-shoe	I. Merrill and A. Maxwell	Shelburne, Mass	July 19, 1864	43, 623
Ox-shoe	A. Van Valkenburgh	Griffin's Corners, N. Y	May 8, 1860	28, 218
Ox-shoe-forging die	H. Colburn	Stafford, Conn	Apr. 26, 1870	102, 372
Ox-shoe machine, with movable dies, Roller	P. P. Read	Bowdoin, Me	Jan. 23, 1849	6, 048
Ox-shoes, Die for forming	J. Deeble	Plantsville, Conn	Dec. 16, 1873	145, 633
Ox-shoes, Die for manufacturing	B. S. Parker	Greenfield, Mass	Feb. 20, 1872	123, 927
Ox-shoes, Machine for making	R. French	Seymour, Conn	Dec. 16, 1873	145, 561
Ox-shoes, Machine for making	W. Hamilton	Neversink, N. Y	Mar. 18, 1873	136, 915
Ox-shoes, Machine for making	A. L. Wooding	Bristol, Conn	June 7, 1870	103, 954
Oxides, Purifying metallic	A. Monnier	Philadelphia, Pa	Mar. 21, 1865	46, 924
Oxides, Reducing metallic	J. Reese	Pittsburgh, Pa	June 19, 1866	55, 710
Oxidizing metal, Apparatus for	W. Atwood	Cape Elizabeth, Me	Mar. 7, 1865	46, 618
Oxygen, and applying the same to useful purposes, Mode of preparing.	H. A. Archereau	Paris, France	July 2, 1867	66, 279
Oxygen and chlorine, Process for producing	J. T. A. Mallet	Paris, France	Jan. 21, 1868	73, 540
Oxygenating-furnace	G. Stamm	Pittsburgh, Pa	June 29, 1869	92, 120
Oyster and other dredger	J. Walmer	Ludlow, Ky	Oct. 28, 1873	144, 169
Oyster-basket	C. D. Martin	Port Chester, N. Y	Apr. 9, 1872	125, 470
Oyster-lock	W. Rankin	New York, N. Y	Feb. 6, 1872	123, 509
Oyster-boats, Dredge-roller for	C. T. Belbin	Baltimore, Md	June 5, 1866	55, 228
Oyster-bucket, Shell	E. H. Frazier	Baltimore, Md	Aug. 20, 1872	130, 631
Oyster-can	C. H. Dexter	Baltimore, Md	July 22, 1873	141, 124
Oyster-can	J. A. Tillery	Baltimore, Md	May 28, 1872	127, 388
Oyster-dredge	W. C. Baker	Baltimore, Md	Nov. 28, 1871	121, 227
Oyster-dredge	C. T. Belbin	Baltimore, Md	June 2, 1868	78, 509
Oyster-dredge	W. Belbin	Baltimore, Md	Jan 17, 1865	45, 904
Oyster-dredge	E. Fairbanks	Baltimore, Md	Aug. 8, 1865	49, 340
Oyster-dredge	W. L. Force	Keyport, N. J	Feb. 21, 1860	27, 213
Oyster-dredge	W. C. Hornfager and M. Fitz Gibbons.	New York, N. Y	Aug. 5, 1873	141, 439
Oyster-dredge	I. A. Ketcham	Breslau, N. Y	Oct. 29, 1872	132, 668
Oyster-dredge	I. A. Ketcham	Breslau, N. Y	Apr. 22, 1873	138, 164
Oyster-dredge	E. B. Lake	Mauricetown, N. J	Nov. 28, 1871	121, 249
Oyster-dredge	T. W. Landon	Fairmount, Md	Apr. 23, 1872	125, 964
Oyster-dredge	T. F. Mayhew	Port Morris, N. J	Apr. 27, 1869	89, 323
Oyster-dredge	T. F. Mayhew	Port Morris, N. J	Nov. 30, 1869	97, 420
Oyster-dredge	T. P. Sink	Fairton, N. J	Oct. 4, 1859	25, 680
Oyster-dredge	T. P. Sink	Fairton, N. J	Aug. 18, 1868	81, 304
Oyster-dredge	T. P. Sink	Fairton, N. J	Oct. 31, 1871	120, 463
Oyster-dredge winding-apparatus	T. F. Mayhew	Port Morris, N. J	Jan. 16, 1872	122, 843
Oyster-dredge windlass	W. T. Howard and O. Reeder	Baltimore, Md	Feb. 15, 1870	99, 900
Oyster-dredges, Hoisting	J. Whitecar	Philadelphia, Pa	May 5, 1863	38, 106
Oyster-dredges, Machine for raising	W. T. Howard	Baltimore, Md	Apr. 8, 1873	137, 550
Oyster-dredges, Roller for boarding	T. P. Sink	Fairton, N. J	June 4, 1867	65, 442
Oyster-dredges, Winder for	C. T. Belbin	Baltimore, Md	Nov. 8, 1870	109, 104
Oyster-dredges, Winder for	H. S. Lawson	Baltimore, Md	Mar. 2, 1865	46, 912
Oyster-dredging apparatus	W. Belbin	Baltimore, Md	Nov. 20, 1866	59, 812
Oyster-dredging machinery	T. P. Sink	Fairton, N. J	Feb. 25, 1868	74, 857
Oyster-keg stoppers, Device for removing	S. Swartz	Buffalo, N. Y	Apr. 17, 1866	54, 039
Oyster-knife	P. Blake	New Haven, Conn	Apr. 18, 1854	10, 798
Oyster-knife	G. W. Huffnagle	New Hope, Pa	Mar. 24, 1868	75, 917
Oyster-knife and ice-pick, Combined	W. Pattberger	Philadelphia, Pa	Dec. 21, 1869	98, 102
Oyster-nursery	B. F. Syford	San Francisco, Cal	June 11, 1872	127, 903
Oyster-opener	J. E. Alger	New York, N. Y	July 30, 1867	67, 246
Oyster-opener	J. Seipel and W. Rupp	Washington, D. C	Jan. 5, 1858	19, 050
Oyster-opener	J. H. Starin, jr	New York, N. Y	Nov. 19, 1872	133, 267
Oyster-opening device	M. C. Boyer	Norristown, Pa	Mar. 21, 1871	112, 895
Oyster-opening device	D. M. Cleary	Baltimore, Md	Oct. 28, 1873	144, 063
Oyster-opening machine	W. H. Towers	Philadelphia, Pa	Apr. 18, 1854	10, 810
Oyster-packing can	L. R. Comstock	Baltimore, Md	Jan. 21, 1873	135, 083
Oyster-platform	J. Vreeland	New York	Apr. 29, 1828	....
Oyster-rake	A. Barrett	Baltimore, Md	Apr. 14, 1868	76, 697
Oyster-rake	J. Johnson	Brooklyn, N. Y	Oct. 2, 1866	58, 426
Oyster-shucker	G. Holtzman	Baltimore, Md	May 30, 1871	115, 474
Oyster-tongs	J. Johnson	Brooklyn, N. Y	Mar. 17, 1868	75, 550
Oyster-tongs	J. W. Sands	Annapolis, Md	July 19, 1870	105, 495
Oyster-tongs	J. Smith	Bruceport, Wash	Jan. 21, 1873	135, 167
Oyster-tongs	E. Ward	Smyrna, Del	Sept. 27, 1870	107, 740
Oyster-tongs, Working	G. Jury	Baltimore, Md	Oct. 11, 1864	44, 634
Oyster-winder	W. H. Kanne	Baltimore, Md	Mar. 11, 1873	136, 737
Oyster-winder	S. S. Shaw	Newport, N. J	Apr. 21, 1868	77, 110
Oyster, Apparatus for opening	W. Beach	Baltimore, Md	Sept. 29, 1857	18, 273
Oysters, clams, &c., Package for	M. W. Brown	New York, N. Y	Apr. 4, 1871	113, 395
Oysters, &c., Cooking	L. P. Keach	Baltimore, Md	Apr. 18, 1854	10, 806
Ozone, and treating liquids with, Generation of	T. A. Hoffman	Beardstown, Ill	June 25, 1872	128, 227
Ozone, Apparatus for making	C. H. Johnson	New Brighton, Pa	Jan. 28, 1873	135, 226
Ozone, Apparatus for the manufacture of	C. F. Dunderdale	New York, N. Y	Nov. 29, 1870	109, 601
Ozone, Apparatus for the production of	C. F. Dunderdale	New York, N. Y	Mar. 15, 1870	100, 736
Ozone, Generating	P. A. Royce	Suspension Bridge, N. Y	Sept. 12, 1871	118, 976
Ozone-generator	W. Elmer and H. G. Hubert	New York, N. Y	Nov. 5, 1867	70, 426
Ozone-generator	R. Heneage	Buffalo, N. Y	Mar. 4, 1873	136, 511
Ozone-generator	R. Heneage	Buffalo, N. Y	Mar. 4, 1873	136, 512
Ozone-generator	O. Loew	New York, N. Y	May 28, 1872	127, 175
Ozone or ozonized air, Process of obtaining	O. Loew	New York, N. Y	Sept. 6, 1870	107, 071
P.				
Package band or fastener	F. R. Hunt	Leavenworth, Kans	Mar. 15, 1870	100, 897
Package-clamp	D. Dick	New York, N. Y	May 9, 1871	114, 651
Package-register	G. W. Moore	Philadelphia, Pa	Mar. 25, 1873	137, 231
Package-strap	J. C. Carey	New York, N. Y	Apr. 26, 1870	102, 370

Index of patents issued from the United States Patent Office from 1790 *to* 1873, *inclusive*—Continued.

Invention.	Inventor.	Residence.	Date.	No.
Packages for dry goods	A. Robertson	Upper Holloway, England	Aug. 24, 1858	21, 274
Packages, Marking and lettering	W. Francis and W. Johnston	Wayneville, N. C	Oct. 3, 1844	3, 771
Packages, Sling for	F. Hohorst	New York, N. Y	Oct. 10, 1871	119, 765
Packages, Spring frame for	H. B. Osgood	Dorchester, Mass	Nov. 4, 1856	16, 019
Packer, Flour	S. Taggart	Indianapolis, Ind	Aug. 2, 1859	24, 963
Packing	J. F. Barrett	Granville, N. Y	Jan. 23, 1836	....
Packing	L. Katzenstein	New York, N. Y	July 19, 1870	105, 462
Packing and atomizing can for insect-powder	F. L. Palmer, sr	New York, N. Y	Dec. 21, 1869	98, 101
Packing and bearing	E. D. Murfey	New York, N. Y	July 12, 1870	105, 361
Packing and bearing	E. D. Murfey	New York, N. Y	Oct. 11, 1870	108, 286
Packing and bearing	E. D. Murfey	New York, N. Y	Dec. 12, 1871	121, 805
Packing and tubing, Manufacture of	T. J. Mayall	Roxbury, Mass	Nov. 29, 1859	26, 278
Packing-apparatus	C. E. Haynes	Boston, Mass	Mar. 15, 1870	100, 889
Packing-bag	E. A. Merrill	New York, N. Y	Aug. 29, 1871	118, 626
Packing, belting, and hose, India-rubber	J. Murphy	New York, N. Y	Apr. 12, 1870	101, 905
Packing-box	D. L. Bartlett	Ogdensburgh, N. Y	May 20, 1873	139, 106
Packing-box	J. H. Foster	Chicago, Ill	July 18, 1871	117, 061
Packing-box	L. A. Fullgraff	New York, N. Y	Jan. 21, 1873	135, 108
Packing-box	A. Gregg	Watertown, Mass	May 30, 1871	115, 306
Packing-box and show-case, Combined	J. Nelson	Bridgeport, Conn	Dec. 31, 1872	134, 480
Packing-box for bottles	J. J. Soloman	Philadelphia, Pa	Dec. 6, 1870	109, 961
Packing-box for fruit, provisions, &c	G. M. Huston	Putnam, Ohio	Dec. 6, 1870	109, 821
Packing-box, Metallic	H. L. Hopkins	San Francisco, Cal	July 11, 1865	48, 686
Packing-box, Paper	J. W. Tuttle	Newton, Mass	July 25, 1871	117, 350
Packing-box, Pasteboard	J. W. Tuttle	Newton, Mass	July 25, 1871	117, 349
Packing-box, Rotary-steam-cylinder	S. Deacon and J. Russell	Lawrence, Mass	Nov. 29, 1870	109, 722
Packing-boxes, Construction of	E. L. Perkins	Roxbury, Mass	Sept. 27, 1859	25, 581
Packing-boxes, Metallic strap for	H. C. Dixon	Brooklyn, N. Y	Sept. 24, 1872	131, 605
Packing-can	N. P. Lindergreen	Boston, Mass	Oct. 13, 1868	83, 070
Packing-case	W. W. Cadwell and J. M. Case	Lansing, Mich	July 23, 1872	129, 713
Packing-case	E. A. Pharo	Philadelphia, Pa	Feb. 4, 1873	135, 583
Packing-case	H. D. Stover	New York, N. Y	Feb. 26, 1861	31, 569
Packing case, Bottle and jar	W. Thompson	Dublin, Ireland	June 29, 1869	91, 884
Packing-case, Folding	L. Selling	Detroit, Mich	Dec. 17, 1872	134, 103
Packing-case for perfumery, &c	T. P. Spencer	Jersey City, N. J	July 12, 1870	105, 269
Packing-cases, Safety-strip for	S. Baker	Newark, N. J	Feb. 2, 1869	86, 492
Packing, Composition for steam	W. R. Bunnell	Jersey City, N. J	June 6, 1871	115, 694
Packing, Compound hard and soft metal	A. Fulton	Pittsburgh, Ind	Mar. 26, 1850	7, 214
Packing-device	W. Riddle	10 Larkhall Lane, England	Aug. 10, 1869	93, 555
Packing for artesian wells	P. C. Heinz	Pioneer, Pa	Mar. 9, 1869	87, 670
Packing for artesian wells	S. Swartz	Buffalo, N. Y	July 4, 1865	48, 599
Packing for axle oil-boxes, Adjusting the	W. Groat	Troy, N. Y	June 11, 1850	7, 426
Packing for barometers	J. P. Simmons	Fulton, N. Y	Mar. 12, 1861	31, 679
Packing for car axles and boxes	S. R. Dummer	New York, N. Y	July 28, 1868	80, 465
Packing for car-wheels, Elastic	J. N. Farrar	Trumansburgh, N. Y	Aug. 6, 1872	130, 118
Packing for carriage-shackles	F. B. Morse	New Haven, Conn	Apr. 28, 1868	77, 399
Packing for deep-well tubes	J. J. Parker	Marietta, Ohio	Oct. 2, 1866	58, 468
Packing for deep wells	A. D. Griffin	Titusville, Pa	Aug. 13, 1867	67, 749
Packing for deep wells	F. Martin	New York, N. Y	Sept. 12, 1865	49, 903
Packing for deep wells	J. R. Robinson and D. A. Strong	Washington, D. C	Feb. 6, 1866	52, 448
Packing for deep wells, Piston	C. H. Jackson	Angola, N. Y	Sept. 5, 1865	49, 759
Packing for doors and windows	S. Strook	New York, N. Y	Aug. 8, 1871	117, 939
Packing for engines, &c	H. C. White	New York, N. Y	Feb. 27, 1872	124, 183
Packing for engines, pumps, &c	W. H. Miller	Philadelphia, Pa	Apr. 28, 1868	77, 204
Packing for exterior of pumps in deep oil-wells, &c., Elastic	J. Moulton	Boston, Mass	Mar. 14, 1865	46, 860
Packing for hose-couplings	R. J. Gould	New York, N. Y	Mar. 3, 1868	75, 151
Packing for hydrocarbon-burners	J. Waterman	New York, N. Y	Sept. 6, 1864	44, 144
Packing for joints	J. E. Wooten	Reading, Pa	Mar. 31, 1868	76, 286
Packing for joints of doors, lids, &c., Elastic	D. E. Somes	Washington, D. C	Nov. 12, 1867	70, 910
Packing for joints of steam and water pipes	H. Brocard	Paris, France	May 19, 1868	77, 954
Packing for joints of steam-engines	W. J. Towne	Newtonville, Mass	Nov. 10, 1868	84, 022
Packing for joints of steam-engines, Paper	J. Ross	Somerville, Mass	Nov. 6, 1866	59, 456
Packing for joints, valves, &c	N. Jenkins	Boston, Mass	Oct. 15, 1867	69, 811
Packing for journal-boxes	J. Conk	Red Bank, N. J	Jan. 12, 1864	41, 197
Packing for journal-boxes	E. S. Hanna	Pittsburgh, Pa	Feb. 27, 1872	124, 133
Packing for journal-boxes, joints, &c., Composition for.	J. T. Davis and W. C. Selden	Jersey City, N. J., and Brooklyn, N. Y.	Nov. 5, 1867	70, 421
Packing for journal-boxes of railway-cars	N. Monroe	Boston, Mass	May 17, 1870	103, 071
Packing for journals and bearings, Fibrous mineral.	C. A. Stevens	New York, N. Y	Mar. 14, 1871	112, 648
Packing for journals, Lubricating	J. T. Robinson	New York, N. Y	May 21, 1872	126, 987
Packing for journals, shafting, &c., Lubricating	S. Baxendale	Boston, Mass	Dec. 9, 1873	145, 384
Packing for machinery and linings for journal-boxes and bearing-surfaces.	C. A. Stevens and C. Butler	New York, N. Y	Aug. 22, 1871	118, 295
Packing for man-holes of steam-generators	J. P. McLean	Brooklyn, N. Y	Mar. 19, 1867	63, 073
Packing for oil-well tubes	G. E. Mills	New York, N. Y	Sept. 5, 1865	49, 778
Packing for oil-well tubes	C. L. Noe	Bergen Point, N. J	Aug. 22, 1865	49, 544
Packing for oil-well tubes	J. Parham, jr	Philadelphia, Pa	Sept. 5, 1865	49, 783
Packing for oil-wells	H. S. Cate	Deerfield, Pa	Aug. 15, 1871	117, 980
Packing for oil-wells	J. R. Cross	Chicago, Ill	Feb. 7, 1865	46, 217
Packing for oil-wells	J. R. Cross	Chicago, Ill	Nov. 14, 1865	50, 910
Packing for oil-wells	O. Redmond	Rochester, N. Y	Oct. 30, 1866	59, 319
Packing for oil-wells	W. S. Wilkinson	Baltimore, Md	Mar. 6, 1866	53, 073
Packing for piston and other rods	J. Johnson	Roxbury, Mass	Feb. 10, 1863	37, 663
Packing for piston and valve rods	R. Martin	Brooklyn, N. Y	June 24, 1862	35, 698
Packing for piston-heads	J. W. Adams	Richfield, Mich	June 16, 1868	78, 911
Packing for piston-rods and other enginery, Wooden	C. N. Peterson	Chicago, Ill	Jan. 26, 1869	86, 316
Packing for piston-rods and valve-stems	W. Heston	Akron, Ohio	July 22, 1873	141, 052
Packing for piston stuffing-boxes	C. McBurney	Roxbury, Mass	June 28, 1859	24, 569
Packing for pistons and stuffing-boxes, Tubular	W. C. Moat	London, England	Dec. 25, 1849	6, 974
Packing for pistons, &c., Lubricating	D. J. Browne and C. W. Baldwin	Boston, Mass	Nov. 22, 1864	45, 136
Packing for pistons, &c., Manufacture of	I. B. and W. H. Miller	Philadelphia, Pa	Apr. 4, 1865	47, 119
Packing for printing-presses, Piston	C. B. Cottrell	Westerly, R. I	Apr. 2, 1872	125, 177
Packing for pumps, Adjustable	C. Hood	Seneca Falls, N. Y	Jan. 2, 1866	51, 831
Packing for pumps, hydrants, &c	J. W. Murphy	Baltimore, Md	Aug. 26, 1873	142, 264
Packing for pumps, &c., Self-regulating water	J. Smart	Philadelphia, Pa	July 10, 1855	13, 233
Packing for rifled projectiles	J. P. Schenkl	Boston, Mass	Jan. 17, 1865	45, 951

Index of patents issued from the United States Patent Office from 1790 *to* 1873, *inclusive*—Continued.

Invention.	Inventor.	Residence.	Date.	No.
Packing for stationary joints, Metallic	H. T. Lee	Marysville, Cal	Apr. 26, 1870	102, 282
Packing for stationary joints, Metallic	I. J. Saunders	Davisville, Cal	July 2, 1872	128, 604
Packing for steam-boiler pistons	T. J. Hudson	New Berne, N. C	Oct. 18, 1859	25, 829
Packing for steam-engine-piston rods	D. Neahr	Fort Yuma, Cal	Jan. 5, 1869	85, 606
Packing for steam-engine-piston rods, &c	C. F. Pike	Providence, R. I	Aug. 12, 1840	1, 715
Packing for steam-engine pistons	J. Barwick and S. Tindall	Silvertown, England	Nov. 19, 1867	71, 121
Packing for steam-engine pistons	F. A. Brown	Ithaca, N. Y	Jan. 5, 1869	85, 561
Packing for steam-engine pistons	L. P. Garner	Ashland, Pa	Aug. 17, 1869	93, 697
Packing for steam-engine pistons	J. Gates	Portland, Oreg	June 8, 1869	91, 010
Packing for steam-engine pistons	H. Horton	New York, N. Y	Oct. 5, 1858	21, 678
Packing for steam-engine pistons	J. C. Merriam	Olneyville, R. I	Apr. 4, 1871	113, 547
Packing for steam-engine pistons	T. R. Morgan	Pittsburgh, Pa	Mar. 16, 1869	87, 789
Packing for steam-engine pistons	J. W. Pettis	Hillsdale, Mich	Dec. 11, 1855	13, 917
Packing for steam-engine pistons	E. Sullivan	Pittsburgh, Pa	Oct. 26, 1869	96, 284
Packing for steam-engine pistons	J. Williamson	New York, N. Y	Aug. 31, 1837	368
Packing for steam-engine pistons	J. Young	Franklin Furnace, Ohio	Mar. 26, 1861	31, 847
Packing for steam-engine pistons and stuffing-boxes	P. Clark	Rahway, N. J	June 2, 1857	17, 422
Packing for steam-engine pistons, Metallic	G. H. Corliss	Providence, R. I	Sept. 8, 1859	18, 136
Packing for steam and other enginery, Gasket	B. Densmore	New York, N. Y	Dec. 15, 1868	84, 864
Packing for steam and other engines	W. Peters	Baltimore, Md	Jan. 28, 1862	34, 283
Packing for steam and water joints	H. T. Lee	Chicago, Ill	July 9, 1872	128, 737
Packing for steam-cylinder pistons	W. Johnson	Lambertville, N. J	Aug. 17, 1869	93, 889
Packing for steam-engines	J. L. Bates	Providence, R. I	Nov. 25, 1862	36, 987
Packing for steam-engines	J. H. Gould	Philadelphia, Pa	Feb. 19, 1861	31, 512
Packing for steam-engines	W. V. Grinnel	Middletown, Conn	Apr. 5, 1832	....
Packing for steam-engines	F. Raymond, H. Dunn, and C. Colt.	Nashville, Tenn	July 9, 1872	128, 754
Packing for steam-engines, Fibrous	J. L. Husband	Philadelphia, Pa	July 3, 1866	56, 054
Packing for steam-engines, Fibrous	W. H. Miller	Philadelphia, Pa	Nov. 29, 1864	45, 261
Packing for steam-engines, Fibrous	W. H. Miller	Philadelphia, Pa	May 14, 1867	64, 688
Packing for steam-engines, gas-pipes, water-pipes, &c.	A. J. Simmons	Indianapolis, Ind	Oct. 17, 1871	120, 003
Packing for steam-engines, Manufacture of	J. Glanding	Philadelphia, Pa	Apr. 28, 1868	77, 275
Packing for steam-engines, &c., Manufacture of	W. H. Miller	Philadelphia, Pa	May 12, 1868	77, 902
Packing for steam-engines, Metallic	G. S. Cox	Barbour County, Ala	Oct. 2, 1849	6, 757
Packing for steam-engines, Pasteboard	J. Ross	Somerville, Mass	Oct. 8, 1867	69, 590
Packing for steam-engines, pumps, &c	W. Beschke	Philadelphia, Pa	Feb. 8, 1870	99, 627
Packing for steam-piston and other rods	H. Bertrand and F. Velut	Paris, France	Jan. 19, 1869	85, 988
Packing for steam-pistons	T. Hanson	New York, N. Y	July 18, 1871	117, 071
Packing for steam-pistons	T. Thurber	Auburn, N. Y	July 9, 1867	66, 537
Packing for steam-pistons	J. Wheelock	Worcester, Mass	Feb. 15, 1870	99, 990
Packing for steam-pistons	W. Wilson	Galesburgh, Ill	Aug. 11, 1868	81, 052
Packing for steam-pistons, Metallic	H. D. Dunbar	Springfield, Mass	Oct. 31, 1865	50, 697
Packing for steam-pistons, Metallic	D. Lasher	Brooklyn, N. Y	June 30, 1857	17, 683
Packing for stuffing-boxes	W. W. Allmand	East Boston, Mass	Dec. 7, 1869	97, 585
Packing for stuffing-boxes	F. J. Roth	Schenectady, N. Y	May 5, 1868	77, 534
Packing for stuffing-boxes	C. M. Templeton	Concord, N. H	Dec. 3, 1867	71, 664
Packing for stuffing-boxes, &c	J. H. Tuck	Pall Mall, England	June 26, 1855	13, 148
Packing for stuffing-boxes	R. and T. Winans	Baltimore, Md	Feb. 5, 1861	31, 346
Packing for stuffing-boxes, Metallic	J. F. Chuse	Litchfield, Ill	Mar. 5, 1867	62, 527
Packing for stuffing-boxes, Metallic	A. H. Hall and T. Locher	Sacramento, Cal	May 30, 1871	115, 463
Packing for stuffing-boxes, Metallic	C. F. Jauriet	Aurora, Ill	Feb. 13, 1866	52, 575
Packing for stuffing-box of steam and other engines.	M. Botticher	Washington, D. C	Oct. 4, 1864	44, 575
Packing for stuffing-boxes of steam-engines, pumps, &c., Manufacture of.	W. H. Miller	Philadelphia, Pa	May 21, 1867	64, 995
Packing for tubes of boilers or condensers	J. Newkirk	Factoryville, N. Y	Aug. 8, 1865	49, 295
Packing from asbestus and other fibrous materials, Machinery for forming.	C. A. Stevens and I. Lindsley	New York, N. Y., and Pawtucket, R. I.	Mar. 14, 1871	112, 651
Packing from asbestus and other fibrous minerals	C. A. Stevens	New York, N. Y	Mar. 14, 1871	112, 647
Packing from stuffing-boxes, Instrument for extracting.	H. N. Smade	Manistee, Mich	Sept. 12, 1871	118, 823
Packing, Gasket	G. W. Coffee	San Francisco, Cal	May 14, 1872	126, 624
Packing, Gasket	W. W. Girdwood	Barking Road, Bromley, Great Britain.	Sept. 19, 1871	119, 027
Packing, Impregnating fibrous materials for	E. D. Murfey	New York, N. Y	Oct. 11, 1870	108, 284
Packing in steam-engines, Metallic stuffing-box for	E. Winship	New York, N. Y	Aug. 31, 1852	9, 239
Packing, India-rubber	A. C. Andrews	New Haven, Conn	Nov. 9, 1869	96, 654
Packing journals and bearings, Material for	E. D. Murfey	New York, N. Y	May 10, 1870	102, 958
Packing, Lubricating-compound for steam and other.	W. M. Canfield	Philadelphia, Pa	Dec. 27, 1870	110, 432
Packing-machine	A. Ralph	Philadelphia, Pa	Sept. 9, 1873	142, 651
Packing, Making gaskets for steam	M. C. Gardiner	Rochester, N. Y	Aug. 2, 1870	105, 934
Packing, Manufacture of elastic	W. Jenkins	Boston, Mass	May 8, 1866	54, 554
Packing, Manufacturing Ind a-rubber piston	I. B. Harris	Edinburgh, Scotland	May 21, 1872	126, 953
Packing, Metallic piston	A. G. Bill	Cuyahoga Falls, Ohio	Apr. 5, 1859	23, 439
Packing, Metallic piston	W. A. Boyden	Harrisburgh, Pa	Oct. 28, 1873	143, 956
Packing, Metallic piston	D. H. Fairbanks	South Norwalk, Conn	Oct. 1, 1872	131, 817
Packing, Metallic piston	J. Massey	Chester, Pa	Sept. 9, 1873	142, 640
Packing, Metallic piston	H. L. Russell	Hudson, Mich	Oct. 25, 1853	10, 159
Packing, Metallic piston	W. Wright	Providence, R. I	Feb. 27, 1849	6, 146
Packing, Metallic piston and valve rod	G. M. Cruickshank	Providence, R. I	Mar. 7, 1871	112, 423
Packing, Metallic piston-rod	C. S. Barry	Providence, R. I	June 11, 1872	127, 834
Packing, Metallic piston-rod	D. Devore	Philadelphia, Pa	Jan. 9, 1872	122, 578
Packing, Metallic piston-rod	J. C. Furness	Boston, Mass	Oct. 22, 1872	132, 361
Packing, Metallic piston-rod	J. C. Furness	Boston, Mass	May 6, 1873	138, 627
Packing, Metallic piston-rod	F. W. Pettee	Boston, Mass	Feb. 11, 1873	135, 663
Packing, Metallic piston-rod	P. W. Richards	Boston, Mass	Oct. 8, 1872	132, 024
Packing, Metallic piston-rod	J. H. Webster	Saint Louis, Mo	July 16, 1867	66, 758
Packing of engine and pump pistons, Lightening the.	J. Crabtree and J. Hopkinson	Philadelphia, Pa	Nov. 29, 1853	10, 283
Packing of rotary engines, Adjusting the	H. G. Thompson	New York, N. Y	Feb. 4, 1851	7, 926
Packing of pistons for steam-engines, Adjusting the.	J. Crabtree	Philadelphia, Pa	May 2, 1854	10, 856
Packing of pistons for steam-engines, Setting out the.	G. H. Hoagland	Port Jervis, N. Y	July 14, 1857	17, 788
Packing of pistons in deep wells, Adjusting the	E. D. Brown	Buffalo, N. Y	Nov. 14, 1865	50, 898
Packing of stuffing-boxes, &c., Lubricating the	I. B. and W. H. Miller	Philadelphia, Pa	Apr. 4, 1865	47, 170
Packing, Oil-proof rubbers for steam	J. M. Flagg	Providence, R. I	Aug. 15, 1871	118, 004

Index of patents issued from the United States Patent Office from 1790 *to* 1873, *inclusive*—Continued.

Invention.	Inventor.	Residence.	Date.	No.
Packing or bearing for journals, &c	E. D. Murfey	New York, N. Y	Dec. 12, 1871	121, 806
Packing or show case	G. G. Bates	New York, N. Y	Aug. 26, 1873	142, 073
Packing, Paper	A. L. Jones	New York, N. Y	Dec. 19, 1871	122, 023
Packing, Piston	J. R. Abbe	Providence, R. I	June 7, 1864	42, 999
Packing, Piston	E. B. Allen	Portland, Me	Aug. 6, 1867	67, 476
Packing, Piston	J. Anthony and T. B. Purves	Greenbush, N. Y	Aug. 10, 1869	93, 578
Packing, Piston	J. Askwith	Chicago, Ill	Mar. 5, 1867	62, 590
Packing, Piston	G. R. Babbitt	Providence, R. I	Feb. 27, 1872	123, 971
Packing, Piston	G. R. Babbitt and W. A. Harris	Providence, R. I	Mar. 4, 1873	136, 405
Packing, Piston	M. W. Bailey	West Chester, Pa	Oct. 15, 1867	69, 747
Packing, Piston	C. S. Barry	Providence, R. I	June 11, 1872	127, 833
Packing, Piston	C. S. Barry	Providence, R. I	July 8, 1873	140, 608
Packing, Piston	T. A. Birbec	Dayton, Ohio	Apr. 7, 1868	76, 295
Packing, Piston	G. F. Blake	Boston, Mass	Sept. 12, 1871	118, 838
Packing, Piston	G. F. Blake	Boston, Mass	Nov. 7, 1871	120, 565
Packing, Piston	J. Broughton	Lambertsville, N. J	Oct. 16, 1866	58, 766
Packing, Piston	W. Buchanan	New York, N. Y	Sept. 12, 1865	49, 853
Packing, Piston	J. C. Budd	Union Township, N. J	Sept. 23, 1873	143, 057
Packing, Piston	L. W. Campbell	Aurora, Ill	Nov. 5, 1867	70, 521
Packing, Piston	L. W. Campbell	Aurora, Ill	Nov. 5, 1867	70, 522
Packing, Piston	A. S. Cameron	New York, N. Y	July 31, 1866	56, 699
Packing, Piston	A. S. Cameron	New York, N. Y	Mar. 14, 1871	112, 685
Packing, Piston	J. W. Carey	Baltimore, Md	June 24, 1873	140, 244
Packing, Piston	T. Carpenter	Providence, R. I	Apr. 8, 1862	34, 878
Packing, Piston	C. H. Clark	Wilmington, Del	Apr. 17, 1866	53, 949
Packing, Piston	C. H. Clark	Wilmington, Del	May 7, 1867	64, 491
Packing, Piston	D. Clark	Hazleton, Pa	Aug. 15, 1865	49, 379
Packing, Piston	J. Clark	Harrisburgh, Pa	Feb. 15, 1870	99, 846
Packing, Piston	J. Clark	Harrisburgh, Pa	May 17, 1870	103, 140
Packing, Piston	J. Clark	Harrisburgh, Pa	Sept. 12, 1871	118, 843
Packing, Piston	J. J. Clanse	Cleveland, Ohio	Jan. 9, 1872	122, 570
Packing, Piston	J. M. Clay	Lancaster, Pa	Dec. 3, 1872	133, 519
Packing, Piston	J. H. Cooper	Philadelphia, Pa	July 23, 1872	129, 793
Packing, Piston	G. Dryden	Worcester, Mass	Jan. 29, 1867	61, 614
Packing, Piston	G. Dryden	South Boston, Mass	July 23, 1872	129, 799
Packing, Piston	P. Estes	Leavenworth, Kans	May 2, 1871	114, 427
Packing, Piston	W. J. Ford	Chicago, Ill	Sept. 13, 1870	107, 243
Packing, Piston	D. R. Fraser	Chicago, Ill	Mar. 25, 1862	34, 749
Packing, Piston	D. R. Fraser	Chicago, Ill	June 3, 1862	35, 439
Packing, Piston	A. Fulton	Pittsburgh, Pa	July 4, 1865	48, 547
Packing, Piston	J. Gates	Portland, Oreg	Nov. 8, 1870	108, 999
Packing, Piston	S. M. Goodale	Saint Louis, Mo	Aug. 6, 1872	130, 214
Packing, Piston	S. Goodfellow	Troy, N. Y	Dec. 12, 1865	51, 449
Packing, Piston	G. Gwynn	New York, N. Y	Feb. 20, 1872	123, 891
Packing, Piston	A. W. Harris	Providence, R. I	Apr. 2, 1872	125, 132
Packing, Piston	C. A. Hodge	Knoxville, Tenn	Sept. 16, 1873	142, 914
Packing, Piston	J. W. Holloway	Akron, Ohio	July 25, 1865	48, 945
Packing, Piston	W. S. Hudson	Paterson, N. J	July 23, 1867	66, 964
Packing, Piston	J. Hughes	Saxton, Pa	June 21, 1870	104, 459
Packing, Piston	A. W. Jackson	Centralia, Ill	Sept. 11, 1866	57, 916
Packing, Piston	T. J. Jones	Summit, N. J	Feb. 13, 1866	52, 644
Packing, Piston	J. Keesey	Chester, Pa	Sept. 6, 1870	107, 062
Packing, Piston	O. Kelsey	Worcester, Mass	May 24, 1870	103, 343
Packing, Piston	E. Kendall	New Lebanon, N. Y	July 11, 1865	48, 692
Packing, Piston	J. King	Hoboken, N. J	Mar. 24, 1868	75, 927
Packing, Piston	L. J. Knowles	Worcester, Mass	July 19, 1870	105, 466
Packing, Piston	P. L. Kreuter	Bloomington, Ill	May 27, 1862	35, 378
Packing, Piston	B. Lowe	Fall River, Mass	Dec. 11, 1866	60, 395
Packing, Piston	T. J. Mayall	Boston, Mass	Oct. 14, 1873	143, 705
Packing, Piston	J. McAlonan	New York, N. Y	June 27, 1871	116, 336
Packing, Piston	H. P. McCarroll	Pittsburgh, Pa	Nov. 14, 1871	120, 888
Packing, Piston	F. McConnell	Dowagiac, Mich	Sept. 13, 1870	107, 274
Packing, Piston	J. R. McCormick	Cincinnati, Ohio	June 4, 1872	127, 625
Packing, Piston	J. P. McLean	Brooklyn, N. Y	Mar. 19, 1867	63, 071
Packing, Piston	A. McMullin	Paterson, N. J	June 15, 1869	91, 250
Packing, Piston	J. J. Miller	Chicago, Ill	Aug. 23, 1864	43, 950
Packing, Piston, &c	E. D. Murfey	New York, N. Y	Mar. 19, 1872	124, 846
Packing, Piston	C. A. Murray	Pittsburgh, Pa	July 29, 1862	36, 016
Packing, Piston	J. Myers	Allegheny City, Pa	Oct. 24, 1865	50, 661
Packing, Piston	W. Ord	Brooklyn, Ohio	Nov. 15, 1870	109, 242
Packing, Piston	C. E. Perkins	Malone, N. Y	May 6, 1873	138, 690
Packing, Piston	W. A. and T. F. Powers	Brooklyn, N. Y	Sept. 18, 1866	58, 176
Packing, Piston	E. T. Prindle	Aurora, Ill	Aug. 21, 1866	57, 375
Packing, Piston	G. W. Reisinger	Harrisburgh, Pa	Nov. 12, 1872	133, 054
Packing, Piston	G. Robinson	Detroit, Mich	July 16, 1867	66, 888
Packing, Piston	F. J. Roth	Newark, Ohio	Nov. 27, 1866	60, 062
Packing, Piston	F. J. Roth	Schenectady, N. Y	May 5, 1868	77, 532
Packing, Piston	F. J. Roth	Schenectady, N. Y	May 5, 1868	77, 533
Packing, Piston	O. C. Smith	Salem, Mass	May 13, 1862	35, 267
Packing, Piston	W. G. Snook and O. C. Patchell	Corning, N. Y	Oct. 23, 1866	59, 143
Packing, Piston	A. J. Stevens	San Francisco, Cal	Mar. 7, 1865	46, 723
Packing, Piston	J. H. Strehli	Cincinnati, Ohio	Jan. 23, 1872	122, 923
Packing, Piston	E. Sullivan	Pittsburgh, Pa	Apr. 9, 1867	63, 672
Packing, Piston	E. Sullivan	Pittsburgh, Pa	Dec. 7, 1869	97, 564
Packing, Piston	E. Sullivan	Mount Washington, Pa	Sept. 26, 1871	119, 248
Packing, Piston	A. Tannock	Litchfield, Ill	Nov. 7, 1865	50, 855
Packing, Piston	J. H. Teal	Memphis, Tenn	Apr. 16, 1872	125, 857
Packing, Piston	W. R. Thomas	Catasauqua, Pa	Mar. 7, 1865	46, 733
Packing, Piston	T. Thurber	Auburn, N. Y	Mar. 26, 1867	63, 185
Packing, Piston	C. Tibbetts and D. L. Weaver	Riverton, Ky	July 16, 1872	129, 070
Packing, Piston	S. Tucker	Worcester, Mass	Sept. 17, 1867	68, 914
Packing, Piston	L. Turner	New Orleans, La	Oct. 25, 1870	108, 653
Packing, Piston	E. A. Walker	Hyannis, Mass	Apr. 12, 1870	101, 792
Packing, Piston	S. L. Weigand	Philadelphia, Pa	Sept. 19, 1871	119, 100
Packing, Piston	G. Wells	Providence, R. I	Feb. 21, 1871	111, 997
Packing, Piston	J. Wheelock	Worcester, Mass	Apr. 5, 1864	42, 245
Packing, Piston	J. Wheelock	Worcester, Mass	Nov. 28, 1865	51, 250
Packing, Piston	W. D. Whitmore	Bloomington, Ill	Mar. 1, 1870	100, 477

Index of patents issued from the United States Patent Office from 1790 *to* 1873, *inclusive*—Continued.

Invention.	Inventor.	Residence.	Date.	No.
Packing, Piston	W. F. Williams	Schenley, Pa	Sept. 24, 1872	131, 645
Packing, Piston and stop-valve	P. Giffard	Paris, France	May 6, 1873	138, 633
Packing, Piston and valve	W. M. Stevenson and A. Pearce	Harmony, Pa	Nov. 16, 1869	96, 993
Packing, Piston and valve rod	G. Tetley and C. D. B. Fisk	Providence, R. I	May 21, 1872	126, 997
Packing, Piston, cylinder, valve, &c	J. Schenck	Brooklyn, N. Y	Aug. 22, 1871	118, 394
Packing, Piston-rod	C. P. Benoit	Detroit, Mich	Sept. 25, 1866	58, 204
Packing, Piston-rod	J. H. Blessing	Albany, N. Y	Mar. 21, 1871	112, 770
Packing, Piston-rod	O. Collier	Sacramento, Cal	Oct. 13, 1868	83, 123
Packing, Piston-rod	P. and J. Eckford	Cincinnati, Ohio	Apr. 12, 1870	101, 840
Packing, Piston-rod	J. M. Flagg	Providence, R. I	Aug. 13, 1872	130, 418
Packing, Piston-rod	T. R. Grant	Newark, Ohio	Nov. 20, 1866	59, 907
Packing, Piston-rod	W. Hartley	Rockford, Ill	Dec. 13, 1870	110, 035
Packing, Piston-rod	J. F. Hermance	New Haven, Conn	Apr. 1, 1873	137, 444
Packing, Piston-rod	O. H. Jewell	Chicago, Ill	Jan. 12, 1869	85, 890
Packing, Piston-rod, &c	S. Katzenstein	New York, N. Y	Jan. 21, 1873	135, 127
Packing, Piston-rod, &c	R. C. Koch and T. B. McLaughlin.	Pittston and South Easton, Pa.	Sept. 17, 1872	131, 355
Packing, Piston-rod	H. T. Lee	Marysville, Cal	Apr. 19, 1870	102, 137
Packing, Piston-rod	S. Lockard	La Grange, Ind	Sept. 22, 1868	82, 331
Packing, Piston-rod	T. Lord	Providence, R. I	Apr. 15, 1873	137, 783
Packing, Piston-rod	J. W. Lynch	Richmond, Va	May 28, 1872	127, 254
Packing, Piston-rod	R. Martin	Brooklyn, N. Y	Aug. 14, 1866	57, 163
Packing, Piston-rod	J. P. McLean	Brooklyn, N. Y	Mar. 19, 1867	63, 072
Packing, Piston-rod	J. P. McLean and J. Vandercar	Brooklyn, N. Y	Nov. 20, 1866	59, 918
Packing, Piston-rod	W. H. Miller	Philadelphia, Pa	Mar. 26, 1867	63, 285
Packing, Piston-rod	W. H. Miller	Philadelphia, Pa	Jan. 21, 1868	73, 454
Packing, Piston-rod	E. T. Prindle	Aurora, Ill	Mar. 27, 1866	53, 542
Packing, Piston-rod	G. S. Prindle	Aurora, Ill	Feb. 20, 1866	52, 742
Packing, Piston-rod	E. A. Richmond	San Francisco, Cal	Oct. 11, 1870	108, 188
Packing, Piston-rod	C. Thayer and D. McPherson	Rome, N. Y	Feb. 22, 1870	100, 213
Packing, Piston-rod	S. S. Turner	Westborough, Mass	Dec. 17, 1867	72, 340
Packing, Piston-rod	W. W. Vanderbilt	New York, N. Y	July 15, 1873	140, 973
Packing, Piston-rod	H. C. White	New York, N. Y	Feb. 27, 1872	124, 182
Packing, Piston-rod	W. P. Woodruff	New York, N. Y	Aug. 14, 1866	57, 243
Packing, Piston-rod	F. Wright	Galesburgh, Ill	July 24, 1866	56, 658
Packing, Piston-rod	J. Young	Brooklyn, N. Y	Feb. 5, 1867	61, 789
Packing-press	J. Y. Parce	Fairport, N. Y	Oct. 9, 1860	30, 342
Packing, Pump	W. Race	Seneca Falls, N. Y	Nov. 24, 1857	18, 705
Packing pump-joints	B. Fitts	Newark, N. J	June 18, 1867	65, 803
Packing, Pump-piston	M. N. Allen, J. Rouse, and J. Melcher.	Titusville, Pa., and Waterbury, Conn.	Jan. 9, 1866	51, 990
Packing, Pump-piston	M. J. Althouse	Waupun, Wis	Apr. 2, 1867	63, 454
Packing, Pump-piston	J. Carter	Philadelphia, Pa	May 1, 1866	54, 295
Packing, Pump-piston	E. T. Ford	Stillwater, N. Y	July 6, 1869	92, 180
Packing, Pump-piston	D. B. Fuller	Buffalo, N. Y	Jan. 24, 1865	45, 989
Packing, Pump-piston	E. A. Jeffery	Corning, N. Y	Dec. 11, 1849	6, 937
Packing ring, Piston	J. B. Root	New York, N. Y	Aug. 14, 1866	57, 189
Packing ring, Piston-rod	W. C. Conwell	Scranton, Pa	Sept. 5, 1865	49, 725
Packing-ring for pistons of steam-engines	J. F. Bogardus	Brooklyn, N. Y	Aug. 7, 1866	57, 037
Packing, Rotary-engine	J. Loader	London, England	June 7, 1870	103, 901
Packing, Rotary-pump	A. W. Cary	Rockport, N. Y	May 15, 1849	6, 456
Packing, Rotary-pump	O. Sweet and M. E. Hicks	Providence, R. I	Nov. 25, 1862	37, 015
Packing, Rotary steam-engine	G. Sickels	Brooklyn, N. Y	Aug. 25, 1857	18, 063
Packing, Rubber	C. L. Frink	Rockville, Conn	May 8, 1866	54, 523
Packing, Rubber	J. Johnson	Brooklyn, E. D., N. Y	Oct. 26, 1869	96, 237
Packing, &c., Rubber flanged tubing for	T. J. Mayall	Boston, Mass	Apr. 9, 1872	125, 594
Packing, Spring for piston	D. Maydole	Norwich, N. Y	Aug. 16, 1870	106, 495
Packing, Steam	A. O. Bourn	Providence, R. I	July 4, 1871	116, 671
Packing, Steam	A. O. Bourn	Providence, R. I	June 4, 1872	127, 555
Packing, Steam	W. Brown	Hoboken, N. J	July 5, 1870	105, 036
Packing, Steam	A. H. Hall and H. T. Lee	Marysville, Cal	June 8, 1869	91, 120
Packing, Steam and hydraulic	W. M. Canfield	Philadelphia, Pa	Apr. 4, 1871	113, 252
Packing, Steam and water	W. C. Selden	Brooklyn, N. Y	Mar. 10, 1868	75, 301
Packing to projectiles, Securing soft-metal	H. F. Mann	La Porte, Ind	Jan. 6, 1863	37, 352
Packing, Tube	S. L. Fox	Philadelphia, Pa	Jan. 10, 1865	45, 822
Packing, Valve and piston	J. Marks	Boston, Mass	Feb. 1, 1870	99, 333
Packing, Vulcanized-rubber	D. C. Gately	Newtown, Conn	Jan. 26, 1869	86, 296
Packing, Well	J. Calkins and J. Fraser	Buffalo, N. Y	Oct. 24, 1865	50, 558
Packing, Well	J. H. Luther	Petroleum Centre, Pa	Jan. 21, 1873	134, 995
Packing, Well	J. J. Parker	Marietta, Ohio	Oct. 2, 1866	58, 469
Packing, Well-tube	S. E. Howes	Albany, N. Y	Jan. 9, 1866	51, 945
Packing, Well-tube	P. Sicourt	Saragossa, Spain	Aug. 22, 1865	49, 599
Pad:				
See Blotting-pad. Bosom-pad. Breast-pad. Bronzing-pad. Carpet-pad. Coach-pad. Collar-pad. Corset-pad. Electrotype-backing pad. Foot-pad. Furniture-pad. Harness-pad. Harness and saddle pad. Harness coach-pad. Harness-saddle pad. Harness-tree pad. Hernia-pad. Hernial pad. Hoof-pad. Horse foot-pad. Horse head-pad. Horse hoof-pad. Horse interfering-pad Horse neck-pad. Horseshoe rubber pad. Ink-pad. Inking-pad. Interfering-pad. Medicated pad. Neck-pad. Paper pad. Rouge-pad. Saddle-pad. Scrubbing-pad. Shoulder-pad. Stair-pad. Truss-pad. Ventilating-pad.				
Pad and trunk lock	A. M. Adams	Washington, D. C	Nov. 8, 1870	108, 951
Pad-billet	L. Hays	Ames, Iowa	June 9, 1868	78, 736
Pad-crimp press	H. H. Beers	Toulon, Ill	Aug. 27, 1867	68, 156
Pad-crimp press	G. Kennedy	Clarksville, Iowa	Jan. 7, 1868	73, 186

Index of patents issued from the United States Patent Office from 1790 *to* 1873, *inclusive*—Continued.

Invention.	Inventor.	Residence.	Date.	No.
Pad-hook	G. Roos and M. White	Buffalo, N. Y	Sept. 25, 1866	58,352
Pad-hooks to pads, Attaching	G. D. Gillett	Meridian, N. Y	July 14, 1868	79,822
Pad-tree	A. J. Cronk	Peoria, Ill	Aug. 6, 1867	67,507
Paddle	C. C. Everson	Palmyra, N. Y	July 16, 1872	129,012
Paddle and water wheel, Bracing the arms of	W. F. Julian	Hatsville, Ind	June 7, 1841	2,117
Paddle, Boat	P. E. Barbour	Louisville, Ky	Jan. 27, 1835	....
Paddle, Boat	J. Cochran	Baltimore, Md	Apr. 28, 1836	....
Paddle-floats, Means of adjusting	J. B. Baptista	New York, N. Y	Sept. 16, 1873	142,835
Paddle, Folding-boat	E. Jenks	Colebrook, Conn	June 6, 1817	....
Paddle for propelling boats	R. Claiborne	....	Feb. 23, 1802	....
Paddle for propelling ships, &c	J. L. Sullivan	Boston, Mass	Apr. 20, 1818	....
Paddle for steam-vessels, Elastic-plate	A. Jouan	San Francisco, Cal	Oct. 7, 1856	15,850
Paddle for vessels	A. C. Semple	Cincinnati, Ohio	July 5, 1853	9,834
Paddle-gate	E. Trumbull	Little Falls, N. Y	Apr. 4, 1833	....
Paddle-mechanism for boats	C. Howard	New York, N. Y	Mar. 19, 1872	124,746
Paddle-mechanism for boats	C. Howard	New York, N. Y	Feb. 11, 1873	135,710
Paddle of paddle-wheels for vessels, Arranging	P. G. Gardiner	New York, N. Y	May 4, 1841	2,076
Paddle, Propelling	J. J. Giraud	Baltimore, Md	Sept. 18, 1827	....
Paddle, Pendulum	A. F. W. Neynaber	Philadelphia, Pa	Aug. 12, 1862	36,165
Paddle, Reciprocating	P. C. Clark	Reading, Pa	May 25, 1858	20,328
Paddle, Revolving	J. W. McDermott	New York, N. Y	Feb. 11, 1873	135,653
Paddle-wheel	L. Ames and M. Miles	Waubeck and Pepin, Wis	Oct. 20, 1863	40,377
Paddle-wheel	D. Anderson	Philadelphia, Pa	Feb. 15, 1870	99,807
Paddle-wheel	J. Atkins	Louisville, Ky	Sept. 2, 1873	142,322
Paddle-wheel	E. H. Bailey	Philadelphia, Pa	Dec. 15, 1863	40,896
Paddle-wheel	E. Banks	Millport, N. Y	Jan. 8, 1867	60,988
Paddle-wheel	S. Bateman	Asniere, near Paris, France	Nov. 15, 1870	109,286
Paddle-wheel	T. S. Bigelow	Lake Mills, Wis	Aug. 11, 1863	39,522
Paddle-wheel	A. T. Boon	Galesburgh, Ill	Nov. 3, 1863	40,455
Paddle-wheel	E. T. Bostrom	Newnan, Ga	Nov. 19, 1867	70,947
Paddle-wheel	A. Buchanan	New York, N. Y	Mar. 2, 1858	19,482
Paddle-wheel	J. Burson	Yates, Ill	June 2, 1868	78,574
Paddle-wheel	F. W. Capen	Newton, Mass	Dec. 18, 1855	13,940
Paddle-wheel	J. C. Chaffee	Titusville, Pa	May 13, 1873	138,856
Paddle-wheel	A. Chapman	Fairfax, Vt	Oct. 3, 1854	11,741
Paddle-wheel	A. R. Chase	Cincinnati, Ohio	Nov. 9, 1842	2,845
Paddle-wheel	W. Choate	Newburyport, Mass	Oct. 10, 1865	50,417
Paddle-wheel	E. T. Colburn	Boston, Mass	July 31, 1866	56,713
Paddle-wheel	M. H. Collins and W. H. Holland.	Chelsea, Mass	Oct. 9, 1866	58,603
Paddle-wheel	G. Collyer and A. H. Patterson	Philadelphia, Pa	Mar. 13, 1860	27,618
Paddle-wheel	A. M. Comstock	Old Lyme, Conn	May 10, 1864	42,640
Paddle-wheel	R. H. Connolly	Philadelphia, Pa	Aug. 16, 1870	106,331
Paddle-wheel	R. Cromelien	Washington, D. C	Jan. 28, 1868	73,783
Paddle-wheel	M. A. Crooker	New York, N. Y	Sept. 13, 1870	107,340
Paddle-wheel	G. H. Cushman	Plymouth County, Mass	Apr. 9, 1872	125,381
Paddle-wheel	D. S. Darling	Brooklyn, N. Y	Aug. 15, 1871	117,901
Paddle-wheel	D. De Haven	New Orleans, La	Sept. 6, 1870	107,014
Paddle-wheel	D. Deshon, 2d	New London, Conn	Apr. 4, 1846	4,446
Paddle-wheel	G. W. Dickinson	Charleston, Ill	Oct. 29, 1872	132,641
Paddle-wheel	N. T. Edson	New Orleans, La	Apr. 30, 1867	64,295
Paddle-wheel	N. T. Edson	New Orleans, La	June 25, 1872	128,217
Paddle-wheel	H. Ehrhart	Muscatine, Iowa	Oct. 19, 1858	21,826
Paddle-wheel	P. Emerson	Carondelet, Mo	Sept. 22, 1868	82,395
Paddle-wheel	H. B. Fay	New York, N. Y	Feb. 7, 1860	27,042
Paddle-wheel	A. C. Fletcher	New York, N. Y	Aug. 11, 1863	39,473
Paddle-wheel	A. C. Fletcher	New York, N. Y	June 15, 1869	91,221
Paddle-wheel	A. C. Fletcher	New York, N. Y	Dec. 23, 1873	145,857
Paddle-wheel	C. Fletcher	Cincinnati, Ohio	Mar. 25, 1856	14,497
Paddle-wheel	W. C. Ford	Brooklyn, N. Y	Feb. 3, 1863	37,572
Paddle-wheel	R. Germain	Buffalo, N. Y	Apr. 10, 1860	27,794
Paddle-wheel	R. Germain	Buffalo, N. Y	Sept. 8, 1863	39,807
Paddle-wheel	J. Goodier and J. F. Kilshaw	Chester and New Brighton, England.	June 5, 1866	55,439
Paddle-wheel	W. Goodwin	Boston, Mass	Oct. 22, 1867	70,084
Paddle-wheel	W. Gorman	New York, N. Y	July 5, 1859	24,627
Paddle-wheel	J. Granger	Zanesville, Ohio	Apr. 10, 1866	53,812
Paddle-wheel	E. Haight	Buffalo, N. Y	Jan. 31, 1860	26,984
Paddle-wheel	J. W. Harris	Durhamville, N. Y	June 7, 1859	24,300
Paddle-wheel	A. Heuston	San Francisco, Cal	Sept. 11, 1866	57,901
Paddle-wheel	B. Hill	Rochester, N. Y	Dec. 25, 1855	13,988
Paddle-wheel	J. C. Hite	Mound City, Ill	Aug. 29, 1871	118,532
Paddle-wheel	W. H. Holland	Boston, Mass	June 19, 1866	55,770
Paddle-wheel	W. H. Holland	Chelsea, Mass	Dec. 18, 1866	60,517
Paddle-wheel	W. H. Holland	Chelsea, Mass	Dec. 18, 1866	60,518
Paddle-wheel	W. H. Holland	Boston, Mass	Aug. 2, 1870	105,943
Paddle-wheel	A. Houseworth	New York, N. Y	Aug. 19, 1856	15,564
Paddle-wheel	W. Huffman	Oshkosh, Wis	Nov. 30, 1869	97,404
Paddle-wheel	W. Hunter	Detroit, Mich	Nov. 12, 1867	70,851
Paddle-wheel	B. Irving	Green Point, N. Y	Sept. 6, 1853	10,000
Paddle-wheel	E. Jones	San Francisco, Cal	Mar. 6, 1866	53,008
Paddle-wheel	G. A. Keene	Lynn, Mass	Oct. 8, 1872	132,013
Paddle-wheel	S. Kepner	Pottstown, Pa	Apr. 15, 1862	34,969
Paddle-wheel	C. A. Kirkpatrick	Somerville, Mass	July 25, 1865	48,956
Paddle-wheel	W. F. Knowlton	Saint Cloud, Minn	June 21, 1870	104,603
Paddle-wheel	R. B. Locke	Stapleton, N. Y	Oct. 26, 1858	21,892
Paddle-wheel	J. Mahony	Newport, R. I	Nov. 9, 1869	96,603
Paddle-wheel	W. R. Manley	New York, N. Y	May 26, 1868	78,301
Paddle-wheel	W. R. Manley	New York, N. Y	Mar. 23, 1869	88,052
Paddle-wheel	B. G. Martin	Philadelphia, Pa	July 11, 1865	48,771
Paddle-wheel	E. Mathers	Harrisville, W. Va	Apr. 25, 1871	114,022
Paddle-wheel	E. Matteson	South Brooklyn, N. Y	June 4, 1867	65,495
Paddle-wheel	J. May	Columbus, Ga	Feb. 8, 1859	22,884
Paddle-wheel	E. J. McCarthy	Saugerties, N. Y	Dec. 28, 1847	5,405
Paddle-wheel	A. McKenzie	Newport, Ky	May 28, 1867	65,252
Paddle-wheel	J. Merkel	Mount Pleasant, Iowa	Aug. 7, 1866	56,973
Paddle-wheel	W. H. Muntz	Norton, Mass	Nov. 15, 1853	10,235
Paddle-wheel	W. H. Muntz	Norton, Mass	July 4, 1854	11,231

Index of patents issued from the United States Patent Office from 1790 *to* 1873, *inclusive*—Continued.

Invention.	Inventor.	Residence.	Date.	No.
Paddle-wheel	L. F. Noe	New York, N. Y	Mar. 25, 1862	34, 767
Paddle-wheel	L. A. Norton	Healdsburgh, Cal	June 5, 1864	43, 424
Paddle-wheel	N. Orcutt	Binghamton, N. Y	Jan. 18, 1859	22, 688
Paddle-wheel	J. Perkins	London, England	Nov. 30, 1829	....
Paddle-wheel	J. W. Post	Castile, Pa	Oct. 25, 1870	108, 626
Paddle-wheel	E. Pratt	New York, N. Y	Apr. 4, 1871	113, 561
Paddle-wheel	J. Rees	Pittsburgh, Pa	Aug. 23, 1870	106, 620
Paddle-wheel	W. C. Rice	Oquawka, Ill	May 25, 1869	90, 465
Paddle-wheel	F. B. Scott	Buffalo, N. Y	July 14, 1863	39, 248
Paddle-wheel	T. Shaw	Philadelphia, Pa	Nov. 26, 1872	133, 341
Paddle-wheel	N. P. Sheldon	San Francisco, Cal	Aug. 8, 1871	117, 822
Paddle-wheel	E. C. Smith	Old Ripley, Ill	Nov. 5, 1867	70, 637
Paddle-wheel	N. Smith	Berwick, La	Apr. 27, 1858	20, 096
Paddle-wheel	J. Speers	West Manchester, Pa	Oct. 18, 1859	25, 871
Paddle-wheel	E. Spencer	Ottawa, Canada	May 21, 1867	65, 023
Paddle-wheel	J. S. Swann	Kanawha County, W. Va	May 14, 1872	126, 760
Paddle-wheel	J. Thompson and M. L. Doty	Chariton, Iowa	May 17, 1859	24, 067
Paddle-wheel	N. Thompson	Bridgeport, Conn	Mar. 22, 1859	23, 324
Paddle-wheel	W. Thompson	Madison, Wis	Dec. 5, 1871	121, 689
Paddle-wheel	N. Thorn	Philadelphia, Pa	Jan. 26, 1869	86, 258
Paddle-wheel	G. W. Tinsley	Minneapolis, Minn	Nov. 19, 1872	133, 179
Paddle-wheel	C. A. Todd	New York, N. Y	July 31, 1866	56, 834
Paddle-wheel	B. W. Tucker	Brooklyn, N. Y	Mar. 15, 1870	100, 820
Paddle-wheel	A. Van Antwerp	Albany, N. Y	July 25, 1854	11, 393
Paddle-wheel	T. G. Walker and H. H. Holbrooke.	New York, N. Y	Jan. 7, 1873	134, 620
Paddle-wheel	W. P. Walker	Memphis, Tenn	Dec. 10, 1872	133, 905
Paddle-wheel	J. U. Wallis	Dansville, N. Y	Jan. 23, 1855	12, 298
Paddle-wheel	J. U. Wallis	Dansville, N. Y	July 3, 1855	13, 190
Paddle-wheel	A. Wingard	San Francisco, Cal	Feb. 1, 1870	99, 510
Paddle-wheel	G. Wingate	Philadelphia, Pa	June 22, 1858	20, 676
Paddle-wheel	L. W. Wright	Brooklyn, N. Y	Jan. 21, 1870	104, 528
Paddle-wheel, Adjustable	L. M. Dehart	Reading, Pa	Apr. 3, 1855	12, 621
Paddle-wheel and mode of propelling canal-boat	B. M. Smith	Rochester, N. Y	May 22, 1835	....
Paddle-wheel bucket	A. M. Glover	Waterborough, S. C	June 12, 1855	13, 058
Paddle-wheel bucket	A. C. Loud	San Francisco, Cal	Aug. 23, 1870	106, 704
Paddle-wheel buckets, Arrangement of	M. A. Crooker	New York, N. Y	Oct. 28, 1856	15, 967
Paddle-wheel, Chain	A. Rogers	Middletown, Pa	Apr. 6, 1830	....
Paddle-wheel for and manner of connecting to vessels.	B. Phillips	Philadelphia, Pa	June 13, 1831	....
Paddle-wheel for propelling boats	J. J. Greenough	Boston, Mass	Aug. 18, 1837	360
Paddle-wheel for propelling boats	W. F. Kearsing	New York	Dec. 8, 1826	....
Paddle-wheel for propelling boats	I. McCord	Harrisburgh, Pa	Sept. 22, 1838	938
Paddle-wheel for propelling boats	J. Ong	North Huntingdon, Pa	May 22, 1837	199
Paddle wheel for propelling boats, &c., Inclined float.	F. W. Stevens	Chigwell, England	Sept. 5, 1840	1, 771
Paddle-wheel for propelling steam and other boats.	W. A. Douglas	Albany, N. Y	July 31, 1837	326
Paddle-wheel for propelling steamboats, &c	M. W. King	New York, N. Y	Jan. 15, 1840	1, 473
Paddle-wheel for propelling vessels, Constructing and arranging.	W. W. Van Loan	Catskill, N. Y	Mar. 29, 1841	2, 022
Paddle-wheel for steam-vessels	A. Connison	Newark, N. J	June 18, 1842	2, 672
Paddle-wheel for steamers, Feathering	A. H. Brown	Washington, D. C	July 19, 1853	9, 854
Paddle-wheel, Horizontal	P. Lear and E. Buck	Boston, Mass	Feb. 20, 1844	3, 448
Paddle-wheel, Oblique-bucket	G. S. Weeks	Oswego, N. Y	Apr. 13, 1852	8, 880
Paddle-wheel propulsion, Engine for	I. L. Thompson	Sun Fish, Ohio	Sept. 3, 1872	131, 135
Paddle-wheel, Steam and horse boat	J. P. Espy	Philadelphia, Pa	May 22, 1833	....
Paddle-wheel, Steamboat, &c	R. D. Chatterton	Derby, England	July 24, 1844	3, 679
Paddle-wheel, Steamboat	T. Hunt	Boston, Mass	Feb. 17, 1831	....
Paddle-wheel, Stern	D. K. Peoples	Philadelphia, Pa	Sept. 18, 1860	30, 087
Paddle-wheel, Submerged	S. B. Howd	Arcadia, N. Y	June 14, 1845	4, 081
Paddle-wheel, Submerged	W. F. Ketchum	Buffalo, N. Y	Aug. 1, 1854	11, 429
Paddle-wheels, Application of steam to	W. Thornton	Washington, D. C	Dec. 23, 1814	....
Paddle-wheels, Arrangement of feathering floats with.	J. P. Worthing	Binghamton, N. Y	May 27, 1862	35, 425
Paddle-wheels, Construction and location of	H. Randall	Philadelphia, Pa	Jan. 13, 1863	37, 421
Paddle-wheels, Driving	J. M. Story	Cincinnati, Ohio	July 3, 1866	56, 118
Paddle-wheels, Driving	J. M. Story	Cincinnati, Ohio	May 30, 1871	115, 380
Paddle-wheels, Fastening	E. Reid	Bazetta, Ohio	Aug. 28, 1860	29, 819
Paddle-wheels, Feathering	F. Barbaires	Solano County, Cal	June 21, 1864	43, 173
Paddle-wheels, Feathering	R. Bell	East Saginaw, Mich	June 2, 1868	78, 415
Paddle-wheels, Feathering	J. Burson	Yates City, Ill	Aug. 8, 1865	49, 226
Paddle-wheels, Feathering	A. Buttrick	Chelsea, Vt	Sept. 22, 1863	40, 009
Paddle-wheels, Feathering	S. and T. Champion	Washington, D. C	Jan. 3, 1854	10, 359
Paddle-wheels, Feathering	T. and S. Champion	Washington, D. C	June 6, 1854	11, 031
Paddle-wheels, Feathering	M. G. Collins	Cumberland, Md	Oct. 10, 1865	50, 338
Paddle-wheels, Feathering	T. Cooley	Brockport, N. Y	Oct. 22, 1861	33, 518
Paddle-wheels, Feathering	B. Densmore	Brockport, N. Y	Dec. 4, 1860	30, 804
Paddle-wheels, Feathering	J. V. Dinsmore	Milford, Mass	Dec. 11, 1866	60, 345
Paddle-wheels, Feathering	P. Emerson	Carondelet, Mo	Sept. 28, 1869	95, 333
Paddle-wheels, Feathering	F. Felter	Perth Amboy, N. J	Nov. 28, 1854	11, 992
Paddle-wheels, Feathering	S. F. Gates	Boston, Mass	Oct. 31, 1865	50, 702
Paddle-wheels, Feathering	A. Gilman	Charlestown, Mass	Aug. 1, 1865	49, 102
Paddle-wheels, Feathering	G. K. Glenn	Ashton, Ill	July 22, 1873	141, 135
Paddle-wheels, Feathering	E. Haight	Buffalo, N. Y	Nov. 7, 1865	50, 817
Paddle-wheels, Feathering	L. T. Howard	Smith's Mills, Miss	Feb. 10, 1857	16, 589
Paddle-wheels, Feathering	L. T. Howard	Smith's Mills, Miss	Sept. 15, 1857	18, 202
Paddle-wheels, Feathering	T. L. Jones	Poughkeepsie, N. Y	Jan. 10, 1854	10, 424
Paddle-wheels, Feathering	G. A. Keene	Newburyport, Mass	Jan. 10, 1865	45, 835
Paddle-wheels, Feathering	G. A. Keene	Newburyport, Mass	Aug. 6, 1867	67, 437
Paddle-wheels, Feathering	G. A. Keene	Lynn, Mass	Oct. 8, 1872	132, 012
Paddle-wheels, Feathering	A. Lefebvre	San Francisco, Cal	May 13, 1873	138, 900
Paddle-wheels, Feathering	H. Lull	Hoboken, N. J	May 20, 1856	14, 920
Paddle-wheels, Feathering	W. R. Manley	New York, N. Y	June 24, 1862	35, 697
Paddle-wheels, Feathering	W. R. Manley	New York, N. Y	Mar. 16, 1869	87, 860
Paddle-wheels, Feathering	W. R. Manley	New York, N. Y	Mar. 16, 1869	87, 861
Paddle-wheels, Feathering	W. L. R. Mattason	Rochester, N. Y	Mar. 13, 1860	27, 499
Paddle-wheels, Feathering	M. T. Miles	Chicago, Ill	Sept. 3, 1872	131, 019
Paddle-wheels, Feathering	J. G. Shands	Saint Louis, Mo	June 17, 1856	15, 149
Paddle-wheels, Feathering	J. C. H. Stut	San Francisco, Cal	Dec. 24, 1872	134, 326

Index of patents issued from the United States Patent Office from 1790 *to* 1873, *inclusive*—Continued.

Invention.	Inventor.	Residence.	Date.	No.
Paddle-wheels, Feathering	R. Williams and S. Wilson	Buffalo, N. Y	June 12, 1860	28, 707
Paddle-wheels, Method of arranging and operating submerged horizontal.	P. Lear	Boston, Mass	Feb. 27, 1855	12, 460
Paddle-wheels of steam and other boats, Machine for bending the stirrup for the.	A. B. Latta	Cincinnati, Ohio	Mar. 30, 1843	3, 022
Paddle-wheels of vessels, Machinery for gearing and ungearing.	B. S. Doxey	United States Navy	Feb. 9, 1821	....
Paddle-wheels, Opening and closing bucket for	G. Tingle	New York, N. Y	Aug. 20, 1850	7, 585
Paddle wheels, Valve or gate for oblique float	J. C. Carncross	Philadelphia, Pa	June 15, 1852	9, 012
Padlock	A. M. Adams	Washington, D. C	Apr. 12, 1870	101, 803
Padlock	W. H. Akins and H. E. Abell	Ithaca and Brooklyn, N. Y	May 2, 1871	114, 386
Padlock	J. H. Ames	Stamford, Conn	Jan. 17, 1871	110, 948
Padlock	S. Andrews	Perth Amboy, N. J	July 8, 1856	15, 270
Padlock	W. Ball	Washington, D. C	Apr. 16, 1842	2, 565
Padlock	T. Bernhard	Hartford, Conn	Dec 21, 1869	98, 015
Padlock	A. S. Blake	Waterbury, Conn	Nov. 17, 1868	84, 162
Padlock	H. D. Blake	New Britain, Conn	Aug. 15, 1865	49, 468
Padlock	W. Bohannan	Baltimore, Md	Apr. 17, 1860	27, 883
Padlock	W. Bohannan	New York, N. Y	Feb. 28, 1865	46, 539
Padlock, &c.	W. Bohannan	Brooklyn, N. Y	Aug. 6, 1867	67, 401
Padlock	D. T. Brown	Newton, N. Y	Apr. 18, 1865	47, 352
Padlock	D. T. Brown	Plainfield, N. J	July 4, 1871	116, 544
Padlock	S. W. Budd	Philadelphia, Pa	Jan. 23, 1872	122, 991
Padlock	I. D. Bush	Detroit, Mich	Apr. 2, 1867	63, 467
Padlock	S. G. Cabell	Quincy, Ill	Dec. 14, 1869	97, 875
Padlock	G. L. Chamberlin	Marietta, Ohio	June 11, 1872	127, 678
Padlock	B. Chambers	Washington, D. C	Apr. 10, 1847	5, 057
Padlock	C. J. Clements	New York, N. Y	Dec. 17, 1867	72, 166
Padlock	J. Corbett	Brooklyn, N. Y	Nov. 29, 1870	109, 717
Padlock	E. Coyle	Albany, N. Y	Mar. 7, 1865	46, 643
Padlock	A. Crosby	Westfield, N. Y	June 22, 1869	91, 610
Padlock	G. W. Dana	Racine, Wis	Oct. 6, 1868	82, 694
Padlock	J. M. A. Dew	Chicago, Ill	May 3, 1870	102, 661
Padlock	B. J., S. B., and A. H. Ebert	Frederick, Md	Oct. 8, 1867	69, 645
Padlock	F. Egge	Bridgeport, Conn	Feb. 2, 1869	86, 377
Padlock	F. Egge	Bridgeport, Conn	Dec. 23, 1873	145, 853
Padlock	C. H. Eiffler	New York, N. Y	Jan. 12, 1869	85, 805
Padlock	H. Eiffler	New York, N. Y	Dec. 17, 1867	72, 378
Padlock	V. Enders	New York, N. Y	Dec. 19, 1865	51, 576
Padlock	H. Essex	Meadville, Pa	Aug. 6, 1872	130, 203
Padlock	H. Essex	Meadville, Pa	July 8, 1873	140, 617
Padlock	E. L. Gaylord	Terryville, Conn	June 21, 1870	104, 572
Padlock	C. F. Gerlach	Milwaukee, Wis	Mar. 19, 1872	124, 677
Padlock	C. T. Gibson	Baltimore, Md	Apr. 5, 1870	101, 607
Padlock	J. A. Grewey	Albany, N. Y	Feb. 8, 1859	22, 869
Padlock	F. C. Goffin and C. Liebrick	Philadelphia, Pa	June 12, 1849	6, 522
Padlock	J. B. Green	Darien, Conn	Dec. 31, 1867	72, 845
Padlock	H. F. Haack	New York, N. Y	May 10, 1870	102, 810
Padlock	E. P. Hall	Chicago, Ill	May 17, 1870	103, 043
Padlock	J. Harrison, jr	Milwaukee, Wis	Jan. 8, 1856	14, 059
Padlock	W. Harvey	Albany, N. Y	Feb. 11, 1868	74, 220
Padlock	L. Hillebrand	Philadelphia, Pa	Mar. 1, 1870	100, 402
Padlock	L. Hillebrand	Philadelphia, Pa	Mar. 1, 1870	100, 403
Padlock	J. J. Hirschbuhl	Louisville, Ky	June 18, 1861	32, 563
Padlock	A. Huffer and N. Sehner	Hagerstown, Md	July 4, 1865	48, 558
Padlock	J. Ingels	Milton, Ind	July 26, 1870	105, 691
Padlock	H. Jackson	New York, N. Y	Aug. 1, 1865	49, 113
Padlock	H. Jackson	Brooklyn, N. Y	June 4, 1867	65, 388
Padlock	B. Rensler	Hartford, Conn	June 11, 1872	127, 695
Padlock	R. Kinsley	Springfield, Mass	Dec. 7, 1852	9, 452
Padlock	J. E. Kohout	Brooklyn, N. Y	Apr. 14, 1868	76, 780
Padlock	I. W. Lamb	Salem, Mich	Feb. 2, 1869	86, 420
Padlock	F. F. Landis	Lancaster, Pa	May 12, 1868	77, 740
Padlock	T. Lanston	Washington, D. C	Mar. 8, 1870	100, 644
Padlock	C. Liebrick	Philadelphia, Pa	Aug. 11, 1863	39, 486
Padlock	J. W. Lyon	Brooklyn, N. Y	Apr. 22, 1862	35, 030
Padlock	O. D. Madge	Washington, D. C	July 11, 1871	116, 972
Padlock	G. McGregor, R. Lee, and T. [illegible]. Clinton.	Cincinnati, Ohio	Aug. 26, 1851	8, 314
Padlock	W. McIntyre	New York, N. Y	Dec. 6, 1870	109, 922
Padlock	J. H. McWilliams	New York, N. Y	July 11, 1871	116, 977
Padlock	G. Merkel and C. H. Meyer	New York, N. Y	Apr. 4, 1871	113, 322
Padlock	C. H. Miller	Frederick, Pa	Feb. 26, 1867	62, 353
Padlock	D. K. Miller	Reading, Pa	July 26, 1870	105, 710
Padlock, &c.	D. K. Miller	Philadelphia, Pa	Oct. 28, 1873	143, 990
Padlock	E. M. and J. E. Mix	Ithaca, N. Y	Nov. 2, 1858	22, 000
Padlock	E. M. and J. E. Mix	Ithaca, N. Y	Dec. 10, 1861	33, 920
Padlock	E. M. and J. E. Mix	Westfield, N. Y	Sept. 25, 1866	58, 278
Padlock, &c.	W. H. Murphy	Versailles, Ohio	Sept. 3, 1867	68, 527
Padlock	H. Nelson	Jerome, N. Y	Aug. 29, 1871	118, 473
Padlock, &c.	J. Nock	Philadelphia, Pa	July 16, 1839	1, 243
Padlock	J. North	Middletown, Conn	Dec. 18, 1860	31, 003
Padlock	I. J. Oldis	Wheeler, N. Y	Jan. 1, 1856	14, 030
Padlock	D. F. Randall	Chicopee, Mass	July 14, 1868	79, 859
Padlock	J. S. Rankin	Ann Arbor, Mich	Dec. 14, 1869	97, 812
Padlock	M. F. Ridout	Milwaukee, Wis	Sept. 9, 1862	36, 421
Padlock	H. Ritchie	Newark, N. J	Aug. 23, 1853	9, 963
Padlock	L. C. Rodier	Springfield, Mass	Mar. 8, 1864	41, 864
Padlock	E. H. Roper and W. Ball	Washington, D. C	Nov. 9, 1842	2, 848
Padlock	F. Roos and F. Spoehr	New York, N. Y	July 31, 1860	29, 406
Padlock	C. W. Saladee	Newark, Ohio	May 29, 1866	55, 166
Padlock	C. W. Saladee	Newark, Ohio	May 29, 1866	55, 167
Padlock	C. W. Saladee	Newark, Ohio	July 17, 1866	56, 450
Padlock	C. W. Saladee	Newark, Ohio	July 24, 1866	56, 616
Padlock	C. W. Saladee	Newark, Ohio	July 24, 1866	56, 617
Padlock	C. W. Saladee and W. Armstrong.	Newark, Ohio	July 17, 1866	56, 451
Padlock	J. Schneider	Chicago, Ill	Jan. 26, 1858	19, 208

Index of patents issued from the United States Patent Office from 1790 *to* 1873, *inclusive*—Continued.

Invention.	Inventor.	Residence.	Date.	No.
Padlock	A. Seeger	New York, N. Y	Jan. 9, 1872	122, 666
Padlock	H. S. Shepardson	Shelburne Falls, Mass	Apr. 26, 1870	102, 440
Padlock	T. Slaight	Newark, N. J	Oct. 14, 1851	8, 431
Padlock	T. Slaight	Newark, N. J	Jan. 2, 1855	12, 186
Padlock	T. Slaight	Newark, N. J	Nov. 12, 1861	33, 715
Padlock	T. Slaight	Newark, N. J	Dec. 12, 1865	51, 488
Padlock	T. Slaight	Newark, N. J	Nov. 23, 1869	97, 127
Padlock	A. F. Smith and J. H. Vickers	Norwich, Conn	July 7, 1868	79, 694
Padlock	F. W. Smith, jr	Bridgeport, Conn	Jan. 25, 1870	99, 115
Padlock	F. W. Smith, jr	Bridgeport, Conn	Jan. 25, 1870	99, 116
Padlock	D. Snell	Little Falls, N. Y	Mar. 29, 1870	101, 391
Padlock	M. P. Thatcher	Pontiac, Mich	Jan. 7, 1868	73, 136
Padlock	J. E. Thompson	Buffalo, N. Y	Oct. 18, 1864	44, 757
Padlock	D. Tilton	Stoneham, Mass	Aug. 26, 1851	8, 318
Padlock	C. A. Wallin	Stockholm, Sweden	Sept. 10, 1872	131, 237
Padlock	S. White	Newark, N. J	June 20, 1854	11, 149
Padlock	W. Wilcox	Middletown, Conn	Oct. 31, 1871	120, 557
Padlock	W. Wilcox	Middletown, Conn	July 23, 1872	129, 879
Padlock	J. E. Wootten	Cressona, Pa	Dec. 11, 1866	60, 454
Padlock	L. Yale	Newport, N. Y	Sept. 8, 1857	18, 169
Padlock	D. Zeiler	Sumneytown, Pa	Feb. 27, 1872	124, 189
Padlock-case	S. Andrews	Perth Amboy, N. J	Dec. 16, 1856	16, 224
Padlock, Combination	G. K. Farrington	Alcatraz Island, Cal	Feb. 15, 1870	99, 872
Padlock, Combination	J. L. Hall	Cincinnati, Ohio	Sept. 1, 1868	81, 630
Padlock, Combination	W. N. Hall	Mexia, Tex	Dec. 23, 1873	145, 864
Padlock, Combination	A. S. Harding and N. Reed	Otisville, N. Y	Apr. 9, 1867	63, 722
Padlock, Combination	J. Kittle	East Bend, S. C	Feb. 18, 1873	135, 992
Padlock, Combination	W. C. Langenau	Cleveland, Ohio	Oct. 28, 1873	144, 114
Padlock, Combination	A. Leyden	Atlanta, Ga	Mar. 12, 1867	62, 862
Padlock, Combination	E. E. Stubbs	West Elkton, Ohio	Mar. 3, 1868	75, 216
Padlock, Combination	G. Thompson	Trenton, N. J	May 3, 1870	102, 620
Padlock for mail-bags, &c	P. A. Altmaier	Harrisburgh, Pa	Jan. 5, 1869	85, 630
Padlock for mail-bags, &c	S. Andrews	Perth Amboy, N. J	Dec. 5, 1840	1, 882
Padlock for mail-bags	H. C. Jones	Newark, N. J	Apr. 1, 1842	2, 525
Padlock, Hermetically closed and keyless	S. Bickerstaff	Cincinnati, Ohio	May 19, 1868	77, 953
Padlock, Indicator	H. W. Dopp	Buffalo, N. Y	Feb. 13, 1872	123, 558
Padlock, Indicator	F. J. Hoyt	New York, N. Y	Mar. 5, 1872	124, 270
Padlock-key	E. L. Gaylord	Bridgeport, Conn	July 18, 1871	117, 064
Padlock, Keyless	W. Bohannan	Brooklyn, N. Y	Feb. 28, 1871	112, 211
Padlock, Permutation	S. Beadle and W. P. Yates	Elmira, N. Y	Aug. 31, 1869	94, 274
Padlock, Permutation	L. H. Gano	New York, N. Y	Sept. 12, 1871	118, 798
Padlock, Permutation	S. Loyd	New York, N. Y	Nov. 14, 1871	120, 822
Padlock, Permutation	J. E. Treat	Oxford, Mich	Aug. 10, 1869	93, 501
Padlock, Register	J. L. Chambers	Brooklyn, N. Y	Nov. 5, 1867	70, 409
Padlock, Sealing	Z. Kelly	New Bedford, Mass	Mar. 5, 1867	62, 636
Padlock, Sealing	R. M. Webb and I. Hermann	New York, N. Y	July 23, 1867	67, 089
Padlock-shield	C. C. Gale	Indianapolis, Ind	June 21, 1870	104, 441
Paging and numbering machine	P. Koch and G. Schüle	New York, N. Y	Feb. 9, 1869	86, 763
Paging-machine	F. O. Degener	New York, N. Y	July 24, 1855	13, 302
Paging-machine	G. Hodgkinson	Cincinnati, Ohio	Nov. 14, 1854	11, 936
Paging-machine	C. L. Sholes	Milwaukee, Wis	June 15, 1869	91, 277
Paging-machine	S. W. Soule	Milwaukee, Wis	Nov. 16, 1869	96, 990
Paging-machine	S. W. Soule and C. L. Sholes	Milwaukee, Wis	Sept. 27, 1864	44, 488
Pail	C. W. Bartlett	New York, N. Y	Apr. 16, 1872	125, 715
Pail	J. J. Dutcher	Brooklyn, N. Y	Feb. 15, 1859	22, 999
Pail and can fastening	J. K. Chace	New York, N. Y	Apr. 5, 1870	101, 430
Pail and commode, Combined slop	G. Howland	Brunswick, N. Y	Mar. 18, 1873	136, 999
Pail and commode, Combined slop	D. Paterson	New York, N. Y	July 16, 1872	129, 294
Pail and lantern, Dinner	C. Britain	Saint Joseph, Mich	Jan. 19, 1864	41, 274
Pail and lantern, Dinner	N. C. Burnap	Argusville, N. Y	May 31, 1870	103, 558
Pail and night-stool combined, Chamber	S. Smith	Brooklyn, N. Y	May 28, 1872	127, 378
Pail and stool combined, Milking	M. A. Green	Fitchburgh, Mass	Nov. 5, 1872	132, 720
Pail and strainer, Milk	D. S. Curtiss	Washington, D. C	Oct. 12, 1869	95, 781
Pail and wringer, Mop	E. H. Lord and E. Hinman	Homer and Syracuse, N. Y.	Sept. 3, 1867	68, 447
Pail, Ash	G. W. Walker	Boston, Mass	Aug. 13, 1861	33, 058
Pail bail and handle	E. T. Covell	Brooklyn, N. Y	Nov. 16, 1869	96, 889
Pail-bail ear	W. J. McLea	Leroy, N. Y	Aug. 25, 1868	81, 390
Pail-bail ear	J. Walton	Brooklyn, N. Y	May 12, 1868	77, 853
Pail-bails, &c., Machinery for bending	R. Bunker	Rochester, N. Y	Dec. 7, 1852	9, 438
Pail, Butter	H. P. Westcott	Seneca Falls, N. Y	Sept. 8, 1868	82, 048
Pail, Butter	H. P. Westcott	Seneca Falls, N. Y	Sept. 6, 1870	107, 139
Pail, Chamber	G. A. Giggins	New York, N. Y	Feb. 1, 1870	99, 437
Pail, Chamber	J. S. Jennings	Brooklyn, N. Y	Dec. 7, 1869	97, 516
Pail, Chamber	J. H. Orr	Long Island City, N. Y	July 23, 1867	67, 067
Pail, Dinner	N. C. Burnap	Argusville, N. Y	Dec. 14, 1869	97, 873
Pail, Dinner	S. B. Cox	Buffalo, N. Y	Sept. 10, 1867	68, 608
Pail, Dinner	J. O. Fairbairn	Milwaukee, Wis	Nov. 23, 1869	97, 181
Pail, Dinner	J. H. Foote	Pittsfield, Mass	Apr. 21, 1868	76, 905
Pail, Dinner	D. Howarth	Portland, Me	May 15, 1866	54, 817
Pail, Dinner	D. Howarth	Portland, Me	Nov. 26, 1867	71, 305
Pail, Dinner	H. Joyce and A. Ernst	Troy, N. Y	Apr. 18, 1871	113, 893
Pail, Dinner	H. C. and W. W. Ketcham	Newark, N. J	Oct. 31, 1871	120, 442
Pail, Dinner	E. C. Luks	Massillon, Ohio	May 6, 1873	138, 569
Pail, Dinner	A. Macqueen, jr	Philadelphia, Pa	Mar. 9, 1869	87, 574
Pail, Dinner	P. Molan	New York, N. Y	Mar. 25, 1873	137, 230
Pail, Dinner	A. Nevin	Philadelphia, Pa	Nov. 19, 1872	133, 113
Pail, Dinner	M. Saulson	Troy, N. Y	Oct. 20, 1868	83, 323
Pail, Dinner	M. Saulson	Troy, N. Y	June 8, 1869	91, 168
Pail, Dinner	H. L. and A. R. Simonson	Graniteville, N. Y	Feb. 27, 1872	124, 019
Pail, Dinner	R. S. Squires	Cleveland, Ohio	Feb. 18, 1873	135, 946
Pail, Dinner	J. T. Sturtevant	Cleveland, Ohio	Dec. 24, 1872	134, 175
Pail, Dinner	G. Wetzler	Peoria, Ill	July 4, 1871	116, 780
Pail, Dinner	L. K. Williams	Hudson, Mich	Feb. 20, 1872	123, 963
Pail-ear	G. E. Eastman	Washington Mills, N. Y	July 28, 1868	80, 341
Pail-ear	J. A. Otis	Rutland, N. Y	June 21, 1870	104, 488
Pail car, Water	J. G. Kritchbaum	Youngstown, Ohio	May 4, 1869	89, 666
Pail, Earth-chamber	H. Terry	Waterbury, Conn	July 19, 1870	105, 607

Index of patents issued from the United States Patent Office from 1790 *to* 1873, *inclusive*—Continued.

Invention.	Inventor.	Residence.	Date.	No.
Pail, Folding	A. Du Chateau and J. D. Williams.	Green Bay, Wis	Nov. 25, 1873	144, 893
Pail, Folding	R. Sabin	Benona, Mich	Sept. 10, 1872	131, 303
Pail, Garbage and slop	C. T. Voss	Stapleton, N. Y	Apr. 16, 1872	125, 860
Pail, Heat-retaining	J. C. Brain	Brooklyn, N. Y	Feb. 9, 1869	86, 727
Pail-hoop	E. Hill	Orange, Mass	June 25, 1872	128, 310
Pail lid, Slop	W. B. Sawyer	New York, N. Y	Feb. 9, 1869	86, 698
Pail-making machine	R. Hyde	Winchester, Mass	Apr. 10, 1822	....
Pail, Measuring	J. B. Wightman	Brooklyn, N. Y	May 16, 1871	114, 894
Pail, Milk	J. Carton and W. Ralph	Utica, N. Y	June 19, 1866	55, 621
Pail, Milk	H. Marshall	Atlanta, Ga	Jan. 12, 1869	85, 837
Pail, Milk	S. Oppenheimer	Peru, Ind	Dec. 1, 1868	84, 572
Pail, Milk	S. E. Oviatt	West Richfield, Ohio	Aug. 6, 1872	130, 147
Pail, Milk	C. Tabor	Craftsbury, Vt	Sept. 27, 1870	107, 834
Pail, Milk-straining	H. Buckins	Canton, Ohio	Sept. 18, 1860	30, 046
Pail, Milking	R. A. Fish	Worcester, Mass	Nov. 17, 1868	84, 179
Pail, Milking	A. J. Johnson	Vicksburgh, Mich	Sept. 16, 1873	142, 917
Pail, Milking	G. T. Lincoln	Leominster, Mass	June 10, 1873	139, 679
Pail, Milking	S. Oppenheimer	Peru, Ind	Mar. 16, 1858	19, 648
Pail, Milking	H. Springer	Brady Township, Mich	Apr. 8, 1873	137, 731
Pail, Paint	C. R. Otis	Chicago, Ill	Apr. 26, 1870	102, 308
Pail, Paper	A. and I. Jennings	Fairfield, Conn	Sept. 29, 1868	82, 526
Pail rest, Milk	G. C. Taft	Worcester, Mass	Mar. 26, 1872	124, 919
Pail, Scrubbing	A. P. Hawse and L. J. Adams	Morrisville, Vt	Feb. 1, 1859	22, 799
Pail, Self-sealing	C. A. Marshall	Cleveland, Ohio	Feb. 27, 1872	124, 145
Pail, Slop	J. S. Jennings	Brooklyn, N. Y	Nov. 26, 1872	133, 370
Pail, Slop	J. G. Roth	New York, N. Y	May 14, 1872	126, 648
Pail, Tin	H. Sangster	Buffalo, N. Y	Apr. 5, 1870	101, 516
Pail with strainer attached, Milk	L. M. Doddridge	New Mount Pleasant, Ind	May 17, 1870	103, 158
Pail, Wooden	L. A. Fleming	West Mount Vernon, N. Y	Jan. 4, 1870	98, 578
Pails, Apparatus for graining	J. R. and A. J. Cross	Chicago, Ill	Dec. 27, 1864	45, 590
Pails, Apparatus for painting or graining	J. Carter	Winchendon, Mass	Dec. 10, 1867	71, 978
Pails, Attachment of handle to tin	T. Evans	Watkins, N. Y	June 21, 1859	24, 451
Pails, buckets, &c., Mode of making and affixing the ears and bales of.	J. F. Phelps	Havana, N. Y	May 12, 1840	1, 605
Pails, firkins, &c., Cover for	T. F. Canklin	Fond du Lac, Wis	July 5, 1870	104, 933
Pails, &c., Machine for crozing and dressing the inside of.	R. W. How and C. E. Patterson	Brooklyn, N. Y	Oct. 22, 1872	132, 401
Pails, &c., Machine for driving hoops on	W. Raymond	Marlborough, N. H	June 28, 1859	24, 576
Pails, &c., Making	A. Miner	Camillus, N. Y	Apr. 25, 1823	....
Pails, &c., Manufacturing water	J. W. Jarboe	Green Point, N. Y	Aug. 20, 1867	67, 984
Pails, &c., Non-conducting filling for the walls of dinner.	O. M. Spiller	Cleveland, Ohio	Nov. 4, 1873	144, 232
Pails, Tool for painting	J. Carter	Winchendon, Mass	Jan. 31, 1865	46, 075
Paint	E. Battley	Mont Clair, N. J	Aug. 15, 1865	49, 467
Paint	H. W. Bradley	New Berlin, N. Y	Mar. 5, 1867	62, 598
Paint	E. S. Burns	Portland, Me	Aug. 17, 1869	93, 804
Paint	W. J. Byrne	Russellville, Ky	Nov. 28, 1871	121, 330
Paint	T. H. Currey	Cincinnati, Ohio	Feb. 15, 1870	99, 856
Paint	W. J. Dodge	Syracuse, N. Y	Dec. 24, 1867	72, 614
Paint	J. S. D'Orsey	New York, N. Y	July 27, 1858	20, 993
Paint	J. L. Geroldsek	Livingston, N. Y	July 10, 1866	56, 322
Paint	J. B. Hodgskin	New York, N. Y	Dec. 5, 1865	51, 317
Paint	W. N. Jordon	Cambridge, Mass	Apr. 25, 1871	114, 017
Paint	M. W. Kidder	Lowell, Mass	July 25, 1871	117, 301
Paint	B. F. King	Annapolis Junction, Md	Nov. 7, 1871	120, 751
Paint	T. C. Lamb	Chicago, Ill	Oct. 16, 1866	58, 840
Paint	H. A. and D. E. Longsdorf	Mechanicsburgh, Pa	June 12, 1866	55, 510
Paint	J. Lucas	Gibbsborough, N. J	Aug. 12, 1873	141, 802
Paint	F. C. Semelroth	Logansport, Ind	July 27, 1869	93, 129
Paint	R. B. Smith	Mount Pleasant, Ohio	July 31, 1866	56, 814
Paint	I. Tyson, jr	Baltimore, Md	June 4, 1861	32, 491
Paint	J. P. Vinsonheller	Urbana, Ohio	July 30, 1867	67, 385
Paint and artificial stone, Compound for	T. Hodgson	Brooklyn, N. Y	Jan. 12, 1869	85, 825
Paint and chemical mixing machine	R. Poole	Baltimore, Md	Jan. 12, 1869	85, 762
Paint and drug mill	M. Bishop	La Fayette, Ind	Mar. 6, 1866	52, 953
Paint and drugs, Mill for grinding	G. D. Jones	Jersey City, N. J	Apr. 1, 1851	8, 009
Paint, and in coating wood, stone, &c	W. Coggeshall	Springfield, Ohio	Dec. 18, 1866	60, 478
Paint and other compounds from bituminous slates, &c., Manufacture of.	W. F. Patterson	Vanceburgh, Ky	Mar. 5, 1867	62, 679
Paint and liquid can	H. Miller	New York, N. Y	Feb. 18, 1873	136, 085
Paint and package for	A. T. Stevens	New York, N. Y	Nov. 5, 1872	132, 874
Paint and pigment	J. Brenton	Pittston, Pa	Oct. 19, 1869	95, 977
Paint and pigment	S. W. Isham	Hinckley, Ohio	Apr. 12, 1870	101, 880
Paint and putty, Compounds for	R. M. Breinig	Brooklyn, N. Y	Apr. 2, 1872	125, 261
Paint and varnish, Manufacture of	G. F. Cornelius	Westminster, Great Britain	Aug. 30, 1870	102, 784
Paint, Apparatus for mixing and grinding oil	W. H. Dolson	New York, N. Y	Nov. 3, 1857	18, 538
Paint box or can	H. Tucker	Newton, Mass	May 20, 1873	139, 214
Paint-burner	W. W. Wakeman, jr., and R. Ross.	New York and Brooklyn, N. Y.	Jan. 15, 1867	61, 288
Paint-can	L. U. Bean and H. A. Stevenson	Philadelphia, Pa	May 14, 1872	126, 775
Paint-can	G. W. Bennett	Brooklyn, N. Y	Aug. 27, 1867	68, 158
Paint-can	G. M. Bligh	New York, N. Y	Feb. 28, 1860	27, 265
Paint-can	P. Brown	Brooklyn, N. Y	Nov. 29, 1859	26, 248
Paint-can	C. Burnham	Philadelphia, Pa	Aug. 13, 1867	67, 718
Paint, &c., can	G. H. Chinnock	Brooklyn, N. Y	Sept. 9, 1873	142, 612
Paint, &c., can	J. R. Cole	New York, N. Y	Dec. 9, 1873	145, 400
Paint-can	J. F. Drummond	New York, N. Y	May 21, 1861	32, 397
Paint-can	J. F. Drummond	New York, N. Y	May 21, 1872	127, 034
Paint-can	H. Everett	Philadelphia, Pa	July 30, 1867	67, 278
Paint-can	H. Everett	Philadelphia, Pa	Sept. 3, 1872	131, 089
Paint-can	W. L. Gilroy	Philadelphia, Pa	Oct. 9, 1860	30, 310
Paint-can	E. B. Hamlin	Saint Louis, Mo	Nov. 11, 1868	83, 849
Paint, &c., can	W. A. Hopkins	New York, N. Y	Jan. 22, 1867	61, 337
Paint-can	E. A. Leland	Brooklyn, N. Y	Aug. 14, 1860	29, 651
Paint-can, &c	J. W. Masury	Brooklyn, N. Y	July 12, 1859	24, 748
Paint-can	J. W. Masury	New York, N. Y	July 10, 1860	29, 090
Paint-can	J. W. Masury	Brooklyn, N. Y	Dec. 16, 1873	145, 670

Index of patents issued from the United States Patent Office from 1790 *to* 1873, *inclusive*—Continued.

Invention.	Inventor.	Residence.	Date.	No.
Paint-can	H. Miller	Hoboken, N. J	Mar. 26, 1867	63, 283
Paint-can	D. Miller	Allegheny City, Pa	Mar. 31, 1868	76, 229
Paint, &c., can	H. Miller	New York, N. Y	Dec. 9, 1873	145, 440
Paint-can	O. E. Walker	Cincinnati, Ohio	Sept. 16, 1873	142, 971
Paint can and box	F. W. Devoe	New York, N. Y	June 4, 1867	65, 476
Paint can and box	J. F. Drummond	New York, N. Y	June 4, 1867	65, 477
Paint-can and device for removing the contents	F. L. Miller	Brooklyn, N. Y	Dec. 9, 1873	145, 439
Paint can and pail	F. W. Devoe	New York, N. Y	June 25, 1867	66, 077
Paint-can car	C. F. Brand	Philadelphia, Pa	Jan. 31, 1865	46, 176
Paint, Can for packing and transporting	H. Tucker	Newton, Mass	Jan. 14, 1873	134, 947
Paint-can, Metallic	G. H. Chinnock	Brooklyn, N. Y	Nov. 19, 1872	133, 203
Paint-can, Metallic	G. H. Chinnock	Brooklyn, N. Y	June 17, 1873	139, 868
Paint-can, Metallic	G. H. Chinnock	Brooklyn, N. Y	June 17, 1873	139, 869
Paint-can, Sealing	J. F. Walter, jr	Brooklyn, N. Y	July 8, 1873	140, 748
Paint-cans, Machine for filling	J. Wilcox	Springfield, Mass	Jan. 14, 1868	73, 272
Paint-canister, Metallic	J. W. Masury	Brooklyn, N. Y	Apr. 28, 1857	17, 163
Paint, cement, &c., Composition for	C. W. Westover	Saint Joseph, Mo	Apr. 5, 1870	101, 687
Paint, cement, hard and soft rubber, &c., Composition to be used in the manufacture of.	F. Dickenson, jr	Hartford, Conn	June 15, 1869	91, 216
Paint, chemicals, &c., Machine for rubbing and mixing.	R. Poole	Baltimore, Md	May 28, 1867	65, 268
Paint, Combination of ochers for	D. S. Wood	Tiskilwa, Ill	Aug. 23, 1870	106, 643
Paint-composition	J. Ball and J. P. Ford	Zanesville, Ohio	Aug. 1, 1865	49, 066
Paint-composition	E. F. Barnes	New York, N. Y	June 23, 1863	39, 000
Paint-composition	J. H. Beardsley	New York, N. Y	Oct. 19, 1858	21, 810
Paint-composition	G. Blondin	New York, N. Y	June 20, 1854	11, 112
Paint-composition	R. Brannon	Baltimore, Md	Feb. 5, 1847	4, 958
Paint-composition	A. B. W. Bullard	Worcester, Mass	Jan. 5, 1864	41, 119
Paint-composition	P. Caubet	Paris, France	June 23, 1863	39, 020
Paint-composition	W. F. Cronkhite	Syracuse, N. Y	Aug. 23, 1864	43, 944
Paint-composition	E. P. Emerson	Blairsville, Pa	Feb. 28, 1860	27, 280
Paint-composition	A. Ferris	Benville, Ind	Nov. 13, 1866	59, 573
Paint-composition	W. H. Foran	Boston, Mass	June 18, 1872	128, 136
Paint-composition	J. L. Geroldsex	Livingston, N. Y	Sept. 13, 1864	44, 178
Paint-composition	R. P. Hinds	Chicago, Ill	Dec. 29, 1868	85, 308
Paint-composition	W. C. Hurd	New York, N. Y	Oct. 2, 1866	58, 421
Paint-composition	P. A. Klipstine	New Baltimore, Va	July 20, 1831	....
Paint-composition	R. Love	New York, N. Y	Feb. 1, 1870	99, 332
Paint-composition	C. Metcalf	Hanover, N. H	Feb. 25, 1831	....
Paint-composition	J. Miller	Moore Township, Pa	July 7, 1863	39, 202
Paint-composition	E. Oberlander, H. Weiss, and E. Bagnicki.	New York, N. Y	Mar. 20, 1866	53, 387
Paint-composition	W. Ten Eyck	Pike Township, Pa	Feb. 23, 1864	47, 729
Paint-composition	J. Trippe	Orange, N. J	Dec. 6, 1864	45, 362
Paint-composition	J. C. Wendrem	Albany, N. Y	Aug. 1, 1865	49, 201
Paint, Composition bronze	A. Towne	Auburn, Me	June 21, 1870	104, 665
Paint, Composition for drying	N. Hemenway	West Springfield, Mass	Feb. 12, 1833	....
Paint, Composition for mixing with	U. Lee	Burlington, Mich	Jan. 3, 1860	26, 685
Paint, Composition for mixing with	G. W. Slagle	Washington, D. C	Oct. 25, 1859	25, 935
Paint, Composition in preparing	J. Seabury	Brooklyn, N. Y	July 7, 1863	39, 204
Paint, Composition of	H. Hibbard	Darien, N. Y	Sept. 20, 1836	....
Paint, Composition of	H. Hibbard	United States	Sept. 20, 1836	27
Paint, Composition of metallic oxide for white	F. Sheppard	Fredericksburgh, Va	Mar. 18, 1835	....
Paint, &c., Composition of water-proof	D. M. Cummings	Enfield, N. H	Dec. 22, 1868	85, 216
Paint, Composition stone	E. Skinner and J. Webster	Sandwich, N. H	May 28, 1833	....
Paint-compound	D. R. Averill	Newburgh, Ohio	July 16, 1867	66, 773
Paint-compound	E. Blakeley	Neponset, Ill	Mar. 8, 1870	100, 558
Paint compound	A. H. Bourne	Fort Scott, Kans	July 26, 1870	105, 634
Paint compound	H. W. Bradley	Binghamton, N. Y	June 23, 1868	79, 196
Paint-compound	H. W. Bradley	Binghamton, N. Y	May 18, 1869	90, 073
Paint-compound	W. T. Burrell	Weymouth, Mass	Apr. 9, 1872	125, 436
Paint-compound	C. Campbell	New York, N. Y	Aug. 5, 1873	141, 544
Paint-compound	D. P. Flinn	Geneva, N. Y	Sept. 19, 1865	50, 068
Paint-compound	I. Gattman	Philadelphia, Pa	Sept. 30, 1856	15, 806
Paint-compound	J. Gould	Lexington, Mass	Sept. 28, 1869	95, 403
Paint-compound	W. G. Huyett	Williamsburgh, Pa	May 11, 1858	20, 205
Paint-compound	W. N. Jordan	Cambridge, Mass	Sept. 6, 1870	107, 059
Paint-compound	S. F. Mathews	Harrisburgh, Pa	Apr. 23, 1872	125, 972
Paint-compound	J. M. Merrymon	Logansport, Ind	Dec. 29, 1857	18, 006
Paint-compound	J. D. Mills	Dayton, Ohio	Jan. 21, 1868	73, 627
Paint-compound	H. Nelson	New York, N. Y	June 10, 1873	139, 729
Paint-compound	J. Quarterman	Flushing, N. Y	Oct. 21, 1873	143, 841
Paint-compound	M. Schneider	Cleveland, Ohio	July 2, 1867	66, 257
Paint-compound	C. F. Sibbald	Philadelphia, Pa	May 10, 1853	9, 717
Paint-compound	C. A. Sitzler	Kitzingen, Germany	Dec. 2, 1873	145, 130
Paint-compound	F. Staling	Harrisonburgh, Va	Dec. 24, 1872	134, 323
Paint-compound	F. A. Stall	Boston, Mass	Aug. 19, 1873	141, 897
Paint-compound	J. S. Vosseller	Plainfield, N. J	Jan. 21, 1873	135, 049
Paint-compound	A. F. Willson	Fort Wayne, Ind	Sept. 1, 1868	81, 721
Paint, Compound for mixing	J. L. Newell	Port Huron, Mich	July 29, 1873	141, 231
Paint-compound, Vehicle for	F. Kuhlman	Lille, France	Aug. 12, 1856	15, 520
Paint, Construction of vessel or apparatus for preserving.	J. Rand	United States	Sept. 11, 1841	2, 252
Paint, Drying	A. Jones	New York, N. Y	Nov. 12, 1850	7, 773
Paint, drugs, &c., Machine for grinding	I. Eddy	Boston, Mass	Mar. 19, 1824	....
Paint, dye-stuffs, &c., Mill for grinding	W. G. Johnson	Philadelphia, Pa	May 3, 1832	....
Paint, Fire and water proof	T. Brinkmann	Greeneville, Tenn	Dec. 21, 1869	98, 022
Paint, Fire and water proof	J. Weisman	Philadelphia, Pa	Jan. 31, 1844	3, 420
Paint, Fire-proof	C. P. Crossman	West Warren, Mass	Jan. 7, 1873	134, 522
Paint, Fire-proof	E. Kunzendorf	New York, N. Y	Oct. 12, 1869	95, 696
Paint, Fire-proof	L. Paimbeouf	Washington, D. C	Nov. 11, 1837	449
Paint, Fire-proof	G. W. Powell	Chinese Camp, Cal	Oct. 27, 1863	40, 429
Paint fire-proof, Composition for rendering	J. B. Harris	Germantown, Ky	July 3, 1866	56, 044
Paint for bituminous roofing-fabrics	A. Robinson	New York, N. Y	Nov. 29, 1864	45, 275
Paint for buildings, roofs, &c.	J. H. and G. O. Smith	Chicago, Ill	Nov. 10, 1868	84, 011
Paint for houses, Composition for	W. Cox	Dayton, Ohio	Oct. 23, 1837	438
Paint for liquid compasses	E. S. Ritchie	Brookline, Mass	May 12, 1868	77, 763
Paint for marbleizing wood	F. G. Pokorny	New York, N. Y	July 19, 1870	105, 596

Index of patents issued from the United States Patent Office from 1790 *to* 1873, *inclusive*—Continued.

Invention.	Inventor.	Residence.	Date.	No.
Paint for oil-barrels, &c	W. L. Elkins	Philadelphia, Pa	July 15, 1873	140, 817
Paint for plastered houses, Oil	A. Thompson	Bethany, N. Y	Feb. 2, 1828	....
Paint for protecting iron from rust, &c., Galvanic	L. Knapp	New York	July 1, 1840	1, 664
Paint for roofs	T. C. Rice	Worcester, Mass	June 21, 1870	104, 648
Paint for roofing, wood, metal, &c	J. C. and C. M. Bills	Ottumwa and Albia, Iowa.	July 22, 1873	140, 991
Paint for ships' bottoms	G. B. Bair	Baltimore, Md	July 9, 1872	128, 777
Paint for ships' bottoms, &c	J. Bowker	Baltimore, Md	May 24, 1870	103, 290
Paint for ships' bottoms	J. G. Fuller	Brooklyn, N. Y	Feb. 28, 1860	27, 331
Paint for ships' bottoms	S. A. Gilman	Boston, Mass	Mar. 5, 1872	124, 204
Paint for ships' bottoms	L. A. Messinger	Philadelphia, Pa	Apr. 12, 1870	101, 899
Paint for ships' bottoms	G. W. Morse	New York, N. Y	Oct. 2, 1866	58, 458
Paint for ships' bottoms	D. Parkhurst	Gloucester, Mass	July 4, 1865	48, 583
Paint for ships' bottoms	H. Roundy	San Francisco, Cal	May 25, 1869	90, 395
Paint for ships' bottoms	J. G. Tarr and A. H. Wonson	Gloucester, Mass	Nov. 3, 1863	40, 515
Paint for ships' bottoms	J. G. Tarr and A. H. Wonson	Gloucester, Mass	June 13, 1865	48, 221
Paint for ships' bottoms	J. G. Tarr and A. H. Wonson	Gloucester, Mass	June 20, 1865	48, 323
Paint for ships' bottoms	T. D. Teal	Philadelphia, Pa	Feb. 1, 1870	99, 497
Paint for ships' bottoms	I. J. Wyman	New York, N. Y	Jan. 24, 1871	111, 295
Paint for ships' bottoms and other surfaces	W. Jesty	Gosport, England	Jan. 23, 1872	122, 949
Paint for ships' bottoms, Copper	T. F. Griffin and R. Tarr, jr	Gloucester, Mass	Oct. 18, 1870	108, 474
Paint for wood and other materials	J. K. Palmer	Cambridge, Mass	Oct. 19, 1869	96, 031
Paint for wood, metal, and woven fabrics	C. D. Smith	Chicago, Ill	Sept. 10, 1867	68, 661
Paint from bituminous coal, Process of making	C. Mortimer	Philadelphia, Pa	Apr. 9, 1850	7, 266
Paint from iron-ore, Manufacturing	R. Nicholson	Bairdstown, Ky	June 23, 1812	....
Paint from iron-ore, Manufacturing	B. Wilbor	Whitestown, N. Y	Feb. 10, 1814	....
Paint from magnesite	J. Briggs	Boston, Mass	Jan. 16, 1872	122, 755
Paint from zinc-ores, Manufacture of white	N. Bartlett and G. T. Lewis	Birmingham and Philadelphia, Pa.	Aug. 20, 1867	67, 839
Paint, fruit, &c., can	H. Miller	Hoboken, N. J	June 28, 1864	43, 326
Paint, Grinding	P. Breasted	Greene County, N. Y	Apr. 18, 1814	....
Paint-grinding machine	E. Skinner	Sandwich, N. H	Dec. 20, 1818	....
Paint-grinding mill	S. Cook	Mendon, Mass	Mar. 3, 1829	....
Paint-grinding mill	A. Holcomb	Butternuts, N. Y	May 14, 1827	....
Paint-grinding mill	T. R. Kingdon and L. P. Dodge	New York, N. Y	Mar. 18, 1873	136, 838
Paint-grinding mill	C. Thomas	West Newbury, Mass	Apr. 27, 1858	20, 102
Paint-grinding mill	J. W. Webb	Mount Morris, N. Y	Feb. 15, 1839	1, 081
Paint-guard and brush-holder	W. T. Bailey	Fairfax County, Va	Mar. 26, 1872	124, 928
Paint, India-rubber	W. and W. A. Butcher	Philadelphia, Pa	Sept. 15, 1857	18, 183
Paint, indigo, &c., Machine for grinding	O. C. Harris	Waterville, N. Y	Mar. 15, 1832	....
Paint, Hydraulic	D. R. Prindle	East Bethany, N. Y	Mar. 26, 1867	63, 209
Paint-keg, Metallic	J. C. Adams	Philadelphia, Pa	June 18, 1867	65, 855
Paint, Machine for coating cloth with	D. Cushing	Wheeling, Va	June 10, 1856	15, 065
Paint, Machine for grinding oil	H. W. Gear	New York, N. Y	Apr. 19, 1864	42, 364
Paint, Manufacture of	D. R. Averill	New Centreville, N. Y	Nov. 21, 1871	121, 146
Paint, Manufacture of	D. R. Averill	New Centreville, N. Y	Nov. 21, 1871	121, 147
Paint, Manufacture of	D. R. Averill	Centreville, N. Y	Dec. 5, 1871	121, 478
Paint, Manufacture of	H. Curtis	New York, N. Y	Dec. 3, 1867	71, 585
Paint, Manufacture of	G. F. De Douhet	Paris, France	Sept. 9, 1851	8, 356
Paint, Manufacture of	F. W. Gerdes	Allegheny City, Pa	Dec. 13, 1870	110, 027
Paint, Manufacture of	W. C. Hurd	New York, N. Y	Feb. 26, 1867	62, 490
Paint, Manufacture of	W. C. Hurd	New York, N. Y	Sept. 22, 1868	82, 416
Paint, Manufacture of	T. C. Rice	Worcester, Mass	Nov. 29, 1870	109, 757
Paint, Manufacture of brown metallic	P. Winter	Horicon, Wis	May 21, 1867	65, 032
Paint, Manufacture of pigment for	R. S. Stenton	New York, N. Y	Oct. 5, 1869	95, 534
Paint, Marine	J. J. Currier and L. Cook	Gloucester, Mass	Dec. 26, 1871	122, 110
Paint, Marine	G. Hook	West Harwich, Mass	Mar. 24, 1868	75, 915
Paint, Marine	W. H. Lewis	Boston, Mass	Oct. 5, 1869	95, 494
Paint-material	W. H. Hubbard	West Meriden, Conn	Apr. 6, 1869	88, 715
Paint, Metallic	H. E. Colton	New York, N. Y	Sept. 24, 1867	69, 185
Paint, Metallic	F. G. Holland	Washington, D. C	July 21, 1868	80, 177
Paint, Metallic alloy	C. Wetterstedt	Marseilles, France	Aug. 5, 1851	8, 275
Paint-mill	W. R. Axe	Rockton, Ill	July 13, 1869	92, 561
Paint-mill	W. R. Axe	Rockton, Ill	July 4, 1871	116, 662
Paint-mill	C. Belcher	Newark, N. J	Apr. 7, 1868	76, 386
Paint-mill	J. A. Berrill	Waterville, N. Y	Oct. 9, 1860	30, 288
Paint-mill	J. A. Berrill	Waterville, N. Y	Mar. 30, 1869	88, 360
Paint-mill	C. W. Brown	Boston, Mass	Feb. 20, 1855	12, 431
Paint-mill	R. Byrne	Brooklyn, N. Y	Nov. 18, 1873	144, 656
Paint-mill	C. Clifton	Jersey City, N. J	Jan. 1, 1867	60, 691
Paint-mill	W. W. Draper	Greenfield, Mass	Sept. 10, 1850	7, 630
Paint-mill	S. J. Goodwin	Rockton, Ill	Apr. 2, 1867	63, 502
Paint-mill	J. W. Masury	Brooklyn, N. Y	Oct. 4, 1870	107, 939
Paint-mill	R. J. McGrew	Evansville, Ind	Oct. 15, 1872	132, 304
Paint-mill	D. E. Paynter	Philadelphia, Pa	Apr. 10, 1855	12, 707
Paint-mill	G. P. Zindgraf	Philadelphia, Pa	June 12, 1866	55, 572
Paint-mill, Horizontal cast-iron	O. Packard	Wilmington, Vt	Feb. 12, 1827	....
Paint, Mineral	E. B. Hexel	Vincent, Pa	Apr. 19, 1870	102, 118
Paint, Mineral	J. S. Kenyon and L. Fox	Baldwinsville, N. Y	Nov. 1, 1870	108, 914
Paint, Mineral	J. R. Smith	Keyport, N. J	Oct. 15, 1867	69, 852
Paint, Mineral	W. H. Towers	Boston, Mass	Mar. 19, 1872	124, 772
Paint-mixing machine	C. W. Brown and G. W. Bauker	Boston, Mass	May 15, 1860	28, 324
Paint-mixing machine	S. G. Cheever	Boston, Mass	June 5, 1860	28, 560
Paint, Mode of making brown	T. Miller, jr	Madison County, New York	Jan. 25, 1815	....
Paint, Mode of mixing	J. Linnell, jr	Orleans, Mass	Mar. 26, 1831	....
Paint, Mode of preparing white-lead	J. N. Troville	Christiansburgh, Va	Mar. 31, 1840	1, 535
Paint-oil	D. R. P. Hill	Morgantown, W. Va	Sept. 29, 1868	82, 523
Paint-oil compound	W. E. Tascott	Cleveland, Ohio	June 22, 1869	91, 578
Paint-oils, varnish, &c., Vessel for	R. Mansure, jr	Philadelphia, Pa	June 24, 1873	140, 148
Paint or mastic, Machine for spreading	J. W. Wheeler	Cleveland, Ohio	Nov. 17, 1868	84, 074
Paint-pots to sides of buildings, Device for hanging	J. H. Flagg	Perkinsville, Vt	Jan. 8, 1867	61, 002
Paint, Preparing hydrocarbon-liquids to serve as vehicle for.	A. Meucci	Clifton, N. Y	May 26, 1863	38, 714
Paint, Preparing white	H. C. Otto	Philadelphia, Pa	Feb. 28, 1842	2, 470
Paint-preserving can	E. Clark	New York, N. Y	Oct. 12, 1858	21, 796
Paint, Process for making	W. F. Davis	New York, N. Y	Aug. 17, 1852	9, 208
Paint, Process for preparing	H. S. Lucas	Chester, Mass	Nov. 23, 1852	9, 422
Paint, Process for treating	G. Blondin	New York, N. Y	June 20, 1854	11, 111
Paint, Process of separating impalpable powder for.	G. W. Griswold	Carbondale, Pa	July 18, 1854	11, 326

Index of patents issued from the United States Patent Office from 1790 *to* 1873, *inclusive*—Continued.

Invention.	Inventor.	Residence.	Date.	No.
Paint-restoring compound	W. O. Bradstreet	Gratis, Ohio	Nov. 19, 1872	133, 195
Paint, Roofing	C. W. Langworthy	Bergen, N. J	June 15, 1869	91, 455
Paint, Rubber	S. Truscott	Cleveland, Ohio	Apr. 4, 1871	113, 368
Paint, soap, &c., Machine for mixing	J. Stainthorp and I. Cole	New York, N. Y	Apr. 26, 1870	102, 330
Paint, Stone	B. V. Betterton	Piqua, Ohio	Apr. 21, 1868	76, 882
Paint to stencil-plates, Apparatus for applying	C. Bates	Kingston, Mass	May 30, 1865	47, 917
Paint to surfaces, Applying	J. W. Kingman	North Bridgewater, Mass	May 1, 1866	54, 365
Paint, &c., Treating oil for	D. D. Cattanach	Providence, R. I	Dec. 30, 1873	146, 044
Paint, Varnish	W. B. Finch	Chicago, Ill	Dec. 24, 1867	72, 617
Paint, varnish, &c., Apparatus for removing	T. F. Moody	Toledo, Ohio	Mar. 21, 1871	112, 832
Paint, varnish, &c., Composition for removing	H. C. Kearney and J. W. Harrison.	Philadelphia, Pa	Apr. 12, 1864	42, 295
Paint-vehicle	A. C. Church	Union City, Mich	Jan. 5, 1858	19, 014
Paint-vehicle	I. Gattam	Philadelphia, Pa	Dec. 1, 1857	18, 794
Paint vessel and package	J. W. Masury	Brooklyn, N. Y	July 22, 1873	141, 061
Paint, Water-color	M. J. Green	Boston, Mass	Apr. 30, 1872	126, 142
Paint, Water-proof	E. E. and J. F. Ellery	New York, N. Y	July 5, 1859	24, 620
Paint, Water-proof	J. A. Moffitt	Boston, Mass	Dec. 8, 1868	84, 702
Paint, Water-proof	T. E. Warren	Troy, N. Y	Mar. 15, 1845	3, 956
Paint, White	F. Shepherd	New Haven, Conn	July 19, 1837	284
Paint, White water-color	F. Shepherd	New Haven, Conn	July 17, 1837	285
Painting wire-cloth, Machine for	J. H. De Witt	Orange, N. J	Sept. 30, 1873	143, 227
Paint, Yellow	Baron Carrendeffez	New York	Sept. 2, 1806	....
Painted-cloth cover	W. D. Richardson	Pawtucket, R. I	Sept. 16, 1873	142, 868
Painted or varnished surfaces, Compound for filling or stuffing.	D. R. Averill	New Centreville, N. Y	Nov. 28, 1871	121, 311
Painted surfaces, Apparatus for sanding	E. May	Indianapolis, Ind	Dec. 6, 1859	26, 363
Painted surfaces, Composition for cleaning	B. F. Shaw	Peabody, Mass	Feb. 9, 1869	86, 702
Painted surfaces, Finish for	H. T. Payne and W. Ayres	San Francisco, Cal	Jan. 14, 1868	73, 377
Painters' canvas	C. Volkmar	Baltimore, Md	Jan. 7, 1868	73, 060
Painters' canvas-stretcher	J. Fairman	New York, N. Y	Mar. 7, 1871	112, 435
Painters' canvas-stretcher	M. K. Kellogg	Baltimore, Md	Oct. 1, 1867	69, 447
Painters' colors, Mill for grinding	C. Greene	....	July 23, 1801	....
Painters' colors, Mill for grinding	J. J. Wells	Hartford, Conn	June 13, 1809	....
Painters' composition, Coach	J. W. Tully	Delano, Pa	Feb. 8, 1870	99, 610
Painters' composition, Filling for	R. Sharp	Pittsburgh, Pa	Aug. 5, 1873	141, 518
Painters' holder	S. J. Newell	Dirigo, Me	Mar. 12, 1872	124, 610
Painters' hook	J. W. Pattee	Thornton, N. H	July 14, 1868	79, 853
Painters' jack, Carriage	A. G. Rykert	Attica, N. Y	Feb. 4, 1873	135, 593
Painters' palette	O. L. R. Andrews	Boston, Mass	July 2, 1872	128, 522
Painters' panel	A. G. Collins	Washington, D. C	Aug. 25, 1863	39, 632
Painter's pot	T. E. Costello	Brooklyn, N. Y	Apr. 1, 1872	137, 423
Painters' stand, Carriage	J. Burtch	Hall's Corners, N. Y	Feb. 11, 1873	135, 771
Painter stopper, Shank	C. Perley and J. Terry	New York, N. Y	June 5, 1849	6, 509
Painters' striping-implement	D. Lydon and C. Halsey	San Francisco, Cal	Nov. 19, 1872	133, 163
Painters' striping-instrument	J. J. McCormick and G. Crossingham.	New York and Croton Falls, N. Y.	Nov. 24, 1857	18, 699
Painters' trestle	D. Moritz	New York, N. Y	Dec. 12, 1871	121, 801
Painting and coating wood, Compound for	H. F. Snow and J. H. Davis	Dover, N. H	Oct. 7, 1873	143, 472
Painting and staining furniture	H. K. Wilson	Barbourville, Ky	Nov. 7, 1871	120, 605
Painting and varnishing, Composition for	S. Page	Chelsea, Mass	May 7, 1867	64, 443
Painting and varnishing machine	H. Thayer and L. L. Martin	Warsaw, N. Y	Feb. 9, 1858	19, 316
Painting and varnishing wood and metals	W. Piotrowski	New York, N. Y	Feb. 19, 1867	62, 222
Painting-apparatus	A. P. Merritt	Charlotte, Mich	May 31, 1870	103, 640
Painting barrels, casks, &c., Machine for	W. Burnet and G. B. Fisher	Chicago, Ill	Aug. 5, 1873	141, 542
Painting cloth, Machine for	O. Ferrin	Lansingburgh, N. Y	Nov. 28, 1831	....
Painting coach-bodies, &c., Trestle for	W. M. Knapp	Muncie, Ind	Nov. 4, 1873	144, 339
Painting, Composition for encaustic	W. H. F. Kehrwieder	Philadelphia, Pa	July 26, 1870	105, 809
Painting, Decorative oil	J. M. Lasché	Paris, France	Jan. 16, 1872	122, 771
Painting for ornamenting walls, floors, &c., Elastic-stamp.	T. Boynton	Windsor, Vt	July 12, 1832	....
Painting, graining, &c., Compound for ornamental	B. S. Harrington	Pontiac, Mich	Feb. 9, 1869	86, 836
Painting metallic roofs, Composition for	C. D. Smith	Chicago, Ill	May 1, 1866	54, 426
Painting, Oil	J. C. Walker	Washington, D. C	Jan. 14, 1873	134, 827
Painting on cloth	L. Jarosson	Jersey City, N. J	June 7, 1853	9, 771
Painting on glass, Ornamental	J. W. Bowers	Brookline, Mass	Jan. 13, 1852	8, 651
Painting on translucent surfaces	J. B. Hall	Philadelphia, Pa	Mar. 28, 1848	5, 495
Painting, Pigments for distemper	H. M. Johnston	New York, N. Y	Feb. 15, 1870	99, 907
Painting, Preparing pigments for water-color	E. R. Campbell	New York, N. Y	Oct. 17, 1871	119, 966
Painting, Process for decorating oil	P. Cousin and P. Oury	Paris, France	Dec. 7, 1869	97, 607
Painting rooms	J. Selby		Dec. 30, 1803	
Painting signs on cloth, Process of	A. Stempel	Chicago, Ill	Nov. 11, 1873	144, 574
Palette and brush for graining	F. Cole	Kankakee, Ill	Oct. 26, 1869	96, 085
Palliasses, Apparatus for making	J. Pigot	Brooklyn, N. Y	May 23, 1854	10, 966
Palm-leaf fabrics, Manufacture of	F. Perrin	Cambridge, Mass	Dec. 9, 1862	37, 138
Palm leaf-splitting machine	C. McFarland	Barre, Mass	Mar. 31, 1841	2, 024
Palm-leaf-splitting machine	C. Wadsworth	Barre, Mass	Nov. 17, 1834	....
Palm-leaf-splitting machine	C. Wadsworth	Barre, Mass	Dec. 23, 1834	....
Palm-leaf warp and woof for weaving, Preparation of.	F. Perren	Cambridge, Mass	May 28, 1867	65, 266
Pamphlet-cover, Self-sealing	W. H. Nichols	East Hampton, Conn	Feb. 4, 1868	74, 121
Pamphlet-covering machine	W. H. Clague and R. B. Randall.	Rochester, N. Y	Feb. 13, 1872	123, 555

Pan:
See Amalgamating-pan.
Ash-pan.
Bake pan.
Baking-pan.
Bed-pan.
Biscuit-pan.
Bread-pan.
Cake-pan.
Confection-pan.
Dish-pan.
Drip-pan.
Dripping-pan.
Dust-pan.
Egg-poaching pan.
Evaporating-pan.
Frying-pan.
Gravel-pan.
Grinding-pan.
Locomotive ash-pan.
Milk-pan.
Mining-pan.
Nitro-glycerine pan.
Refrigerator-pan.
Rendering-pan.
Saucepan.
Scale-pan.
Sheet-metal pan.
Stereotype-pan.
Stew-pan.
Urinal-pan.
Vacuum-pan.
Water-closet pan.

Index of patents issued from the United States Patent Office from 1790 *to* 1873, *inclusive*—Continued.

Invention.	Inventor.	Residence.	Date.	No.
Pan-folding machine	W. Hamilton	Philadelphia, Pa	Aug. 25, 1868	81, 498
Pan-former	J. Finn	Decorah, Iowa	May 12, 1868	77, 726
Pan-forming machine	E. Kuessner	Cincinnati, Ohio	Dec. 16, 1873	145, 660
Pan-lifter	W. H. S. Lawrence and C. I. Collamore.	Bangor, Me	June 27, 1871	116, 330
Pan-lifter	M. A. Shepard	Bridgeport, Ill	Feb. 25, 1868	74, 855
Pan-making, Machine for	J. Dane, jr	Newark, N. J	Mar. 26, 1872	124, 938
Pan-wiring machine	J. Shephard	Bristol, Conn	May 19, 1868	78, 021
Pans and other cooking-utensils, Handle for	J. A. Lawson	Troy, N. Y	Oct. 26, 1869	96, 122
Panacea	J. Houck	Baltimore, Md	May 9, 1833	......
Panel-cutting machine	F. D. Green	Williamsport, Pa	Aug. 25, 1868	81, 361
Panel-molding machine	C. H. Williamson and J. H. Allyn.	Whitestown, N. Y	Mar. 11, 1873	136, 800
Panel-raising machine	F. D. Green	Williamsport, Pa	Aug. 19, 1873	142, 015
Panel-raising machine	J. H. Millspaugh	Williamsport, Pa	Jan. 16, 1872	122, 774
Panel-raising machine	D. F. Walker	Minneapolis, Minn	Oct. 31, 1871	120, 405
Panel-table	J. G. Greene	Port Henry, N. Y	Oct. 15, 1867	69, 799
Panel, Wooden	E. Maitre	Veuxaulles, France	Dec. 2, 1873	145, 221
Panels to wooden frames, Constructing and attaching iron.	J. Shaefer	Lancaster, Pa	Sept. 9, 1862	36, 427
Paneling-machine	S. Heyser	Jackson, Mich	Mar. 5, 1872	124, 267
Paneling-machine	N. Jenkins	New York, N. Y	Nov. 23, 1869	97, 092
Paneling-machine	N. Jenkins	New York, N. Y	Oct. 4, 1870	108, 027
Paneling-machine	D. F. Walker	Minneapolis, Minn	July 11, 1871	117, 020
Paneling-machine	H. P. Westcott	Seneca Falls, N. Y	Jan. 7, 1862	34, 083
Panier	J. S. Colby	New York, N. Y	Aug. 19, 1873	141, 854
Panoramas, Electro-magnetic apparatus for moving	L. Finger	Cambridge, Mass	July 26, 1870	105, 664
Pantaloons	G. R. Eager	Boston, Mass	June 13, 1871	115, 832
Pantaloons	G. R. Eager	Boston, Mass	June 11, 1872	127, 750
Pantaloons	J. D. Ehlers and J. P. Herron	Baltimore, Md., and Washington, D. C.	Mar. 11, 1862	34, 635
Pantaloons	W. Franklin	New Haven, Conn	Jan. 17, 1860	26, 842
Pantaloons	B. J. Greely	Springfield, Mass	June 29, 1858	20, 708
Pantaloons	B. J. Greely	New York, N. Y	Oct. 3, 1865	50, 242
Pantaloons	F. T. Hoyt	Brooklyn, N. Y	June 24, 1873	140, 197
Pantaloons	H. Osler	Philadelphia, Pa	Aug. 18, 1863	39, 584
Pantaloons	T. R. Sloan	Brooklyn, N. Y	July 26, 1870	105, 856
Pantaloons	E. T. Taylor	Philadelphia, Pa	Oct. 21, 1873	143, 792
Pantaloons, Attaching strap to	S. Heller	New York, N. Y	June 17, 1862	35, 608
Pantaloons, Device for preserving shape of	W. D. Sinclair	Trenton, N. J	Aug. 2, 1864	43, 714
Pantaloons, Drying and pressing apparatus for	J. Braun	Rochester, N. Y	Dec. 22, 1868	85, 059
Pantaloons, Fastening for	I. Stratton	Keene, N. H	Nov. 5, 1867	70, 645
Pantaloons, Guard for	S. B. Smith and E. Lindsley	Cleveland, Ohio	Apr. 23, 1867	64, 161
Pantaloons, Guide for cutting out	J. Ramsey and A. B. Smith	Clinton, Pa	Apr. 23, 1861	32, 145
Pantaloons, Presser-block for	L. Cohen and J. C. Walker	San Francisco, Cal	Apr. 1, 1873	137, 346
Pantaloons, Protecting-guard for	E. Lindsley	Cleveland, Ohio	Jan. 7, 1868	73, 106
Pantaloons, Protector for	C. W. S. Heaton	Belleville, Ill	July 10, 1866	56, 221
Pantaloons, Steaming and drying apparatus for shaping.	E. B. Viets	Boston, Mass	Nov. 18, 1873	144, 644
Pantaloons, Strap-clasp for	W. H. Miller	New York	July 8, 1842	2, 705
Pantaloons, Strap-fastening for	W. W. Riley	Columbus, Ohio	Oct. 31, 1848	5, 887
Pantaloons, Strapfor	S. Heller	New York, N. Y	Dec. 15, 1863	40, 931
Pantaloons, Strap for	M. H. Simpson	Boston, Mass	May 30, 1839	1, 161
Pantaloons, Strap for	C. F. Walker	Benford's Store Post-Office, Pa.	Aug. 2, 1864	43, 727
Pantaloons, Strap for	E. Wilder	Boston, Mass	Nov. 19, 1833	......
Pantaloons, Stretcher for	L. Kaltenbacher	Louisville, Ky	Aug. 19, 1873	142, 025
Pantaloons, Stretcher for	J. D. Ryan	New York, N. Y	Nov. 18, 1873	144, 703
Pantaloons, Stretching-apparatus for	J. Mottet	Philadelphia, Pa	July 19, 1864	43, 596
Pantaloons, Stretching-board for	E. F. Smith	Salem, Mass	Mar. 26, 1872	124, 916
Pantaloons, Stretching-device for	S. C. Wells	Le Roy, N. Y	Sept. 29, 1868	82, 666
Pantaloons, Stretching-frame for	J. H. Hulse	Baltimore, Md	Apr. 21, 1868	76, 919
Pantaloons, Tree for	M. Taine	South Boston, Mass	June 24, 1873	140, 172
Pantaloons, Tree for	E. B. Viets	Boston, Mass	May 27, 1873	139, 283
Pantaloons, &c., Waistband-fastening for	J. A. Haarvig	Chicago, Ill	Nov. 4, 1873	144, 334
Pantiles, Making	W. Harwood	Richmond, Va	July 3, 1806	......
Pantry	J. Shattuck	Brookline, Mass	July 4, 1865	48, 631
Pantograph	L. F. Bruce and N. J. Wolcott.	Springfield, Mass	Sept. 12, 1871	118, 902
Pantograph	A. Judd	Chicopee, Mass	Aug. 13, 1850	7, 564
Pantograph	H. Neumeyer	Macungie, Pa	Dec. 2, 1856	16, 148
Pantograph and parallel ruler	J. J. Hawkins	......	May 17, 1803	......
Pantographic engraving-machine	B. L. Phillips	Providence, R. I	Dec. 16, 1862	37, 177
Pantographic machine	E. Oldham	Brooklyn, N. Y	May 15, 1866	54, 759
Pantographic machine	R. Whittman	Accrington, England	Oct. 1, 1861	33, 421
Pantographic reversing-instrument	C. L. Getz	Philadelphia, Pa	Aug. 19, 1862	36, 212
Paper	H. E. Wagner	Dresden, Germany	Apr. 29, 1873	138, 302
Paper, Adhesive composition for fly	G. N. Harcy	Lyme, Ohio	June 25, 1872	128, 223
Paper, Adhesive or gum	S. L. Weigand	Philadelphia, Pa	Aug. 2, 1864	43, 731
Paper and card-board to fix marks by metallic pencils. Treating.	H. M. Johnston	New York, N. Y	Nov. 18, 1873	144, 678
Paper and cloth collar	J. Barton	Battle Creek, Mich	Feb. 20, 1866	52, 659
Paper and cloth combined, Machine for forming shirt-bosoms from.	E. P. Furlong	Portland, Me	June 1, 1869	90, 739
Paper and cloth, Machine for uniting	G. K. Snow	Watertown, Mass	Dec. 17, 1872	134, 105
Paper and leather board, Manufacture of	C. B. Hutchins	Ann Arbor, Mich	Feb. 21, 1871	112, 045
Paper and other fabrics, Apparatus for sizing, stretching, and coloring.	P. Hild	New York, N. Y	June 11, 1872	127, 692
Paper and other fabrics, Friction-calender for	W. W. Harding and W. H. Soley	Philadelphia, Pa	Feb. 16, 1869	87, 044
Paper and other fabrics incorrodible, Rendering	T. G. Chase	Philadelphia, Pa	Nov. 9, 1858	22, 015
Paper and other fabrics. Machine for drying	F. Beck	New York, N. Y	Dec. 27, 1870	110, 540
Paper and other fabrics, Producing colored prints on	M. Laemmel	Bay Ridge, N. Y	July 4, 1871	116, 720
Paper and other materials with solutions, Process of coating sheets of.	J. C. Crosman	Boston, Mass	Feb. 19, 1867	62, 256
Paper and paper-boards, Machine for making	J. F. Jones	Rochester, N. Y	Oct. 13, 1863	40, 265
Paper and paper-pulp, Manufacture of	J. M. Allen	Marion, Mass	Aug. 22, 1871	118, 177
Paper and paper-pulp, Manufacture of	J. Brown	London, England	May 10, 1859	23, 897
Paper and paper-pulp, Treatment of	T. Taylor	Grove End Road, England	May 16, 1871	114, 880
Paper and print roller	H. J. Smith	Boston, Mass	June 1, 1869	90, 695

Index of patents issued from the United States Patent Office from 1790 *to* 1873, *inclusive*—Continued.

Invention.	Inventor.	Residence.	Date.	No.
Paper and trimming books, Machine for cutting	F. J. Austin	New York, N. Y	June 16, 1841	2, 141
Paper and vegetable fibrous substances, Treating	A. T. Schmidt	Pittsburgh, Pa	Apr. 4, 1871	113, 454
Paper and woven fabrics to produce water-proof sheets and slabs, Treating.	J. Scoffern	Finsbury, England	Jan. 19, 1869	86, 103
Paper, Apparatus for the manufacture of albumenized.	J. Klein	Hesse-Darmstadt, North German Confederation.	July 19, 1870	105, 464
Paper, Apparatus for pasting lining to	B. F. Fields	Beloit, Wis	July 18, 1871	117, 060
Paper, Apparatus for securing pulverized and other materials to.	W. Adamson	Philadelphia, Pa	May 28, 1867	65, 038
Paper, &c., Apparatus for separating wood fiber for.	A. Fickett	Rochester, N. Y	Apr. 26, 1870	102, 239
Paper articles, Apparatus for making	H. Grambo	Philadelphia, Pa	Dec. 3, 1867	71, 742
Paper, &c., Application of the hibiscus moscheutos to the manufacture of.	W. J. Cantello	Philadelphia, Pa	May 13, 1862	35, 215
Paper articles of apparel, Button-hole for	B. M. Smith	New York, N. Y	Mar. 3, 1868	75, 212
Paper articles of wearing-apparel, Button-hole for	W. C. Reeves and L. L. Solomon	Charlestown, Mass	Apr. 7, 1868	76, 347
Paper bag	A. Adams	Chagrin Falls, Ohio	Aug. 2, 1870	105, 877
Paper bag	J. Arkell and B. and A. Smith	Canajoharie, N. Y	Apr. 25, 1865	47, 376
Paper bag	J. Arkell and B. Smith	Canajoharie, N. Y	June 6, 1865	48, 036
Paper bag	L. D. Benner	Boston, Mass	Dec. 24, 1872	134, 244
Paper bag	L. D. Benner	Boston, Mass	Sept. 30, 1873	143, 321
Paper bag	G. H. Chinnock	Brooklyn, N. Y	June 17, 1873	139, 870
Paper bag	L. C. Crowell	West Dennis, Mass	May 28, 1867	65, 176
Paper bag	L. C. Crowell	Boston, Mass	Feb. 20, 1872	123, 811
Paper bag	L. C. Crowell	Boston, Mass	Apr. 8, 1873	137, 533
Paper bag	G. and D. Dare	Auburn, N. Y	Nov. 19, 1867	71, 143
Paper bag	D. and C. F. Manuel	Boston, Mass	Nov. 26, 1867	71, 398
Paper bag	O. W. Stow	Plantsville, Conn	Nov. 4, 1873	144, 238
Paper bag and box making machine	J. Miller, jr	Baltimore, Md	July 9, 1861	32, 777
Paper-bag and envelope making machine	J. A. Smith and S. E. Pettee	Clinton and Foxborough, Mass.	May 1, 1855	12, 786
Paper-bag and material therefor	A. Jameson	Trenton, N. J	June 22, 1869	91, 748
Paper-bag cutter	L. D. Barrand	New York, N. Y	Apr. 24, 1860	27, 959
Paper-bag engine	I. N. Crehore and F. Stiles, jr	Boston and Leicester, Mass	Jan. 25, 1859	22, 707
Paper-bag-folding machine	E. A. Hollingsworth	South Braintree, Mass	Jan. 24, 1865	45, 999
Paper-bag-folding machine	G. J. Moffatt	Brooklyn, N. Y	Mar. 29, 1870	101, 299
Paper-bag machine	A. Adams	Chagrin Falls, Ohio	May 16, 1871	114, 743
Paper-bag machine	C. Amazeen	New York, N. Y	Nov. 17, 1868	84, 076
Paper-bag machine	C. F. Annan	Boston, Mass	June 1, 1869	90, 624
Paper-bag machine	C. F. Annan	Boston, Mass	Dec. 27, 1870	110, 536
Paper-bag machine	C. F. Annan	Boston, Mass	Feb. 14, 1871	111, 802
Paper-bag machine	C. F. Annan	Boston, Mass	Feb. 14, 1871	111, 803
Paper-bag machine	C. F. Annan	Boston, Mass	Nov. 12, 1872	132, 890
Paper-bag machine	C. F. Annan	Boston, Mass	Jan. 7, 1873	134, 580
Paper-bag machine	H. G. Armstrong	Philadelphia, Pa	Oct. 2, 1860	30, 191
Paper-bag machine	H. G. Armstrong	Trenton, N. J	May 20, 1873	139, 104
Paper-bag machine	P. E. Armstrong	Philadelphia, Pa	Feb. 21, 1871	112, 005
Paper-bag machine	J. Arkell	Canajoharie, N. Y	Dec. 26, 1871	122, 099
Paper-bag machine	L. D. Benner	Boston, Mass	Jan. 9, 1872	1[illegible], 510
Paper-bag machine	L. D. Benner	Boston, Mass	May 13, 1873	138, 844
Paper-bag machine	T. and J. Bilby, J. Baron, and W. Baron.	Burnley and Rochdale, England.	Oct. 1, 1872	141, 841
Paper-bag machine	B. S. Binney	Somerville, Mass	Oct. 17, 1871	119, 915
Paper-bag machine	J. N., J. P., and S. H. Bryant	Temperanceville, Pa	Sept. 26, 1871	119, 307
Paper-bag machine	B. Cole	Brooklyn, N. Y	Dec. 17, 1872	134, 035
Paper-bag machine	L. C. Crowell	Boston, Mass	Feb. 20, 1872	123, 812
Paper-bag machine	L. C. Crowell	Boston, Mass	Oct. 14, 1873	143, 674
Paper-bag machine	G. Dunham	Unionville, Conn	Aug. 19, 1873	141, 862
Paper-bag machine	B. F. Ellis	Dayton, Ohio	Sept. 5, 1865	49, 736
Paper-bag machine	E. W. Goodale	Clinton, Mass	May 29, 1855	12, 945
Paper-bag machine	W. Goodale	Clinton, Mass	July 12, 1859	24, 734
Paper-bag machine	W. Goodale	Clinton, Mass	Aug. 23, 1859	25, 191
Paper-bag machine	E. W. Goodale	Clinton, Mass	Sept. 12, 1865	49, 951
Paper-bag machine	J. J. Greenough	New York, N. Y	Feb. 3, 1863	37, 573
Paper-bag machine	G. Guild	Saint Louis, Mo	Jan. 23, 1872	123, 013
Paper-bag machine	T. Hotchkiss	Stratford, Conn	Jan. 28, 1873	135, 275
Paper-bag machine	G. L. Jaeger	New York, N. Y	May 7, 1867	64, 537
Paper-bag machine	J. Keller	York County, Pa	Mar. 2, 1858	19, 506
Paper-bag machine	M. E. Knight	Boston, Mass	July 11, 1871	116, 842
Paper-bag machine	H. Law	Chatham, N. J	June 14, 1870	104, 169
Paper-bag machine	W. Liddell	Buffalo, N. Y	Sept. 30, 1873	143, 358
Paper-bag machine	H. C. Lockwood	Baltimore, Md	Mar. 9, 1869	87, 689
Paper-bag machine	N. Lorton and J. S. Davison	Cranberry, N. J	July 5, 1870	105, 099
Paper-bag machine	G. F. Lufberg	New York, N. Y	May 8, 1860	28, 188
Paper-bag machine	G. H. Mallary	New York, N. Y	July 28, 1868	80, 298
Paper-bag machine	G. H. Mallary	Poughkeepsie, N. Y	Nov. 3, 1868	83, 648
Paper-bag machine	C. H. Morgan	Philadelphia, Pa	Feb. 17, 1863	37, 726
Paper-bag machine	H. B. Morris	Burlington, N. J	Oct. 15, 1872	132, 312
Paper-bag machine	M. Murphy	Cincinnati, Ohio	Jan 21, 1873	135, 145
Paper-bag machine	E. B. Olmsted	Washington, D. C	Nov. 5, 1867	70, 601
Paper-bag machine	J. S. Ostrander	Dayton, Ohio	Oct. 21, 1873	143, 925
Paper-bag machine	S. E. Pettee	Philadelphia, Pa	May 29, 1860	28, 537
Paper-bag machine	S. E. Pettee	Philadelphia, Pa	May 5, 1863	38, 452
Paper-bag machine	J. P. Pultz	Plantsville, Conn	Sept. 7, 1869	94, 511
Paper-bag machine	J. P. Raymond	Cincinnati, Ohio	Dec. 2, 1873	145, 125
Paper-bag machine	B. F. Rice	Clinton, Mass	Apr. 28, 1857	17, 184
Paper-bag machine	J. Wells	Hoboken, N. J	Apr. 21, 1863	38, 253
Paper-bag machine	J. Wells	Hoboken, N. J	Sept. 15, 1863	40, 001
Paper-bag machine	J. Wells	Hoboken, N. J	Apr. 12, 1864	42, 313
Paper-bag machine	F. Wolle	Bethlehem, Pa	May 29, 1855	12, 982
Paper-bag machine	F. Wolle	Bethlehem, Pa	July 6, 1858	20, 838
Paper-bag machine	F. Wolle	Bethlehem, Pa	Oct. 26, 1852	9, 355
Paper-bag machines, Pasting-apparatus for	J. Arkell	Canajoharie, N. Y	Nov. 26, 1872	133 395
Paper-bag machinery	J. Wells	Brooklyn, N. Y	Mar. 9, 1869	87, 608
Paper-bag-manufacturing tool	S. M. Kirk and E. J. Howlett	Philadelphia, Pa	Feb. 26, 1867	62, 342
Paper bags, Apparatus for making	B. B. Taggart	Watertown, N. Y	Aug. 15, 1865	49, 454
Paper bags, Closing	M. W. Brown	New York, N. Y	Sept. 7, 1869	94, 558
Paper bags, &c., Knife to cut	H. R. David	New York, N. Y	Oct. 5, 1858	21, 657

Index of patents issued from the United States Patent Office from 1790 *to* 1873, *inclusive*—Continued.

Invention.	Inventor.	Residence.	Date.	No.
Paper barrel	W. H. Bailey	New York, N. Y	Mar. 19, 1872	124, 714
Paper-binder, Temporary	F. M. Smith	Syracuse, N. Y	Oct. 4, 1870	107, 968
Paper-binding	W. P. Read	Long Meadow, Mass	Aug. 27, 1867	68, 236
Paper, Bleaching stock for	S. T. Merrill	Beloit, Wis	Nov. 12, 1867	70, 878
Paper, Blotting	N. Floyd	New York, N. Y	Sept. 3, 1872	131, 056
Paper, Bluing compound for the manufacture of	J. Hogben	Cleveland, Ohio	Nov. 2, 1869	96, 321
Paper board, Attachment to roll for calendering	E. A. Seeley	Scotch Plains, N. J	Oct. 19, 1869	96, 039
Paper board for building	C. B. Ayer	Beloit, Wis	Mar. 5, 1872	124, 315
Paper board for building	F. N. Davis	Beloit, Wis	Mar. 19, 1872	124, 795
Paper board for building, &c., Manufacture of	F. N. Davis	Beloit, Wis	July 9, 1872	128, 863
Paper board, Machine for making	J. F. Jones	Rochester, N. Y	Aug. 1, 1865	49, 119
Paper board, Machine for making	J. F. Jones	Rochester, N. Y	Sept. 12, 1865	49, 884
Paper-board manufacture	A. E. Seeley	Scotch Plains, N. J	May 16, 1871	114, 868
Paper board, pipe, &c., Manufacture of	N. C. Szerelmey	Middlesex, Great Britain	Apr. 17, 1866	54, 082
Paper boat	P. S. Tyler	Boston, Mass	Nov. 3, 1868	83, 804
Paper, books, &c., Cutting	J. Woodhouse	Otsego, N. Y	May 30, 1825	......
Paper bosom	J. C. Arms	Northampton, Mass	Feb. 27, 1866	52, 812
Paper bosom	T. A. Curtis	Springfield, Mass	Feb. 20, 1866	52, 784
Paper bosom	E. P. Furlong	Portland, Me	Nov. 9, 1869	96, 687
Paper bosom	J. B. Gardiner	Springfield, Mass	Sept. 5, 1865	49, 747
Paper bosom	C. M. Lee	Springfield, Mass	Mar. 27, 1866	53, 455
Paper bosom	W. E. Lockwood	Philadelphia, Pa	Feb. 28, 1865	46, 567
Paper bosom	C. C. Taylor	Springfield, Mass	Feb. 27, 1866	52, 907
Paper bosom and collar	J. C. Arms	Northampton, Mass	June 12, 1866	55, 451
Paper-bosom machine	S. B. Hill	Chicopee, Mass	Apr. 3, 1866	53, 615
Paper box	W. Armour	Belfast, Ireland	June 9, 1868	78, 715
Paper box	E. B. Beecher and W. H. Swift	Westville, Conn., and Wilmington, Del.	Mar. 5, 1872	124, 319
Paper box	P. Beer	Detroit, Mich	May 16, 1871	114, 752
Paper box	H. H. Corbin	New Britain, Conn	Aug. 17, 1869	93, 809
Paper box	C. O. Crosby	New Haven, Conn	Mar. 15, 1870	100, 865
Paper box	H. A. Devendorf	Port Jackson, N. Y	June 8, 1869	91, 096
Paper box	M. W. Dillingham	Amsterdam, N. Y	Feb. 4, 1873	135, 469
Paper box	F. Field	Troy, N. Y	Jan. 18, 1870	98, 860
Paper box	G. M. Hendrickson	Albany, N. Y	Oct. 26, 1869	96, 108
Paper box	A. B. Hendryx	Ansonia, Conn	Mar. 23, 1869	88, 165
Paper box	G. L. Jaeger	New York, N. Y	July 20, 1869	92, 831
Paper box	G. L. Jaeger	New York, N. Y	July 25, 1871	117, 423
Paper box	B. J. Magee and J. F. Wall	Watertown, Mass	Apr. 4, 1871	113, 319
Paper box	H. Matier	Belfast, Ireland	May 9, 1871	114, 582
Paper box	J. W. Millet	Albany, N. Y	Mar. 28, 1865	47, 069
Paper box	B. Osborn	New York, N. Y	Mar. 8, 1870	100, 658
Paper box	B. Osborn	Newark, N. J	Dec. 31, 1872	134, 394
Paper box	C. L. Palmer	Norwich, Conn	Oct. 22, 1872	132, 368
Paper box	J. L. Reber	Philadelphia, Pa	June 22, 1869	91, 666
Paper box	J. Root	New Haven, Conn	Mar. 15, 1870	100, 806
Paper box	J. Root	New Haven, Conn	June 7, 1870	103, 932
Paper box	H. D. Scudder	Amsterdam, N. Y	Apr. 14, 1868	76, 830
Paper box	G. K. Snow	Watertown, Mass	July 8, 1873	140, 736
Paper box	R. N. Stewart	Philadelphia, Pa	Jan. 8, 1867	61, 028
Paper box	J. W. Tuttle	Newton Corner, Mass	Apr. 5, 1870	101, 546
Paper box	J. W. Wilcox	New York, N. Y	Feb. 28, 1871	112, 306
Paper box cutter, Tubular	J. and E. Spooner	New York, N. Y	Aug. 31, 1869	94, 358
Paper-box fastening	D. Crannell	Amsterdam, N. Y	Nov. 11, 1873	144, 513
Paper-box fastening	G. F. Wright	Clinton, Mass	Feb. 20, 1866	52, 800
Paper-box-finishing machine	R. Smith	Sherbrook, Canada	June 22, 1869	91, 780
Paper-box machine	G. R. Clarke	Pompton, N. J	May 7, 1872	126, 378
Paper-box machine	W. Gates, D. J. Lloyd, and S. Miller	Frankfort, and South Hammond, N. Y.	Dec. 29, 1868	85, 301
Paper-box machine	C. R. Grimm	New York, N. Y	Oct. 1, 1872	131, 751
Paper-box machine	J. Hatfield	Troy, N. Y	Jan. 21, 1862	34, 229
Paper-box machine	C. B. Hatfield	Philadelphia, Pa	July 23, 1867	67, 050
Paper-box machine	C. B. Hatfield	Philadelphia, Pa	July 23, 1867	67, 051
Paper-box machine	C. B. Hatfield	Philadelphia, Pa	Mar. 8, 1870	100, 621
Paper-box machine	R. L. Hawes	Worcester, Mass	Jan. 16, 1855	12, 255
Paper-box machine	H. R. Heyl	Philadelphia, Pa	Oct. 31, 1871	120, 436
Paper-box machine	H. R. Heyl	Philadelphia, Pa	Oct. 8, 1872	132, 076
Paper-box machine	H. R. Heyl	Philadelphia, Pa	Oct. 8, 1872	132, 077
Paper-box machine	H. R. Heyl and A. Brehmer	Philadelphia, Pa	Oct. 8, 1872	132, 078
Paper-box machine	G. L. Jaeger	New York, N. Y	Feb. 25, 1873	136, 246
Paper-box machine	J. M. D. Keating and T. V. Waymouth.	New York, N. Y	June 8, 1869	91, 025
Paper-box machine	L. Koch	New York, N. Y	Mar. 13, 1855	12, 511
Paper-box machine	T. C. Luther	Waterbury, Conn	June 27, 1865	48, 490
Paper-box machine	C. A. Maxfield	New York, N. Y	Feb. 6, 1866	52, 432
Paper-box machine	C. A. Maxfield	New York, N. Y	June 21, 1870	104, 477
Paper-box machine	C. A. Maxfield	New York, N. Y	Oct. 3, 1871	119, 631
Paper-box machine	W. Orr, jr., and G. F. Wright	Clinton, Mass	Aug. 13, 1867	67, 669
Paper-box machine	B. F. Quinby	Boston, Mass	Apr. 27, 1869	89, 433
Paper-box machine	D. Simmons, jr	New York, N. Y	Feb. 22, 1870	100, 198
Paper-box machine	D. Simmons	New York, N. Y	Mar. 1, 1870	100, 459
Paper-box machine	R. Smith	Sherbrooke, Canada	Nov. 17, 1868	84, 246
Paper-box machine	J. and E. Spooner	New York, N. Y	June 29, 1869	91, 983
Paper-box machine	S. B. Terry	Terrysville, Conn	Sept. 6, 1859	25, 373
Paper-box machine	J. Worell	Philadelphia, Pa	Mar. 19, 1872	124, 872
Paper boxes, Apparatus for making	R. T. Barton	New Haven, Conn	Nov. 27, 1866	59, 944
Paper boxes, Apparatus for making	F. Leclére	Boston, Mass	Dec. 8, 1868	84, 835
Paper boxes, Finishing	S. Wheeler and E. Jerome	Albany, N. Y	Mar. 21, 1871	112, 874
Paper boxes from pulp, Manufacture of	S. Wheeler and E. Jerome	Albany, N. Y	Apr. 13, 1869	88, 827
Paper boxes, Machinery for making	C. O. Crosby	New Haven, Conn	Mar. 29, 1870	101, 354
Paper boxes, Machinery for making	F. Gates	Vineland, N. J	Apr. 5, 1870	101, 455
Paper boxes, Making blanks for	S. Wheeler and E. Jerome	Albany, N. Y	July 16, 1867	66, 920
Paper boxes, Manufacture of	S. Wheeler and E. Jerome	Albany, N. Y	Feb. 12, 1867	61, 968
Paper boxes, Manufacturing angular	E. Waters	Troy, N. Y	Feb. 2, 1858	19, 270
Paper braid	C. E. Richards	North Attleborough, Mass	June 18, 1867	65, 833
Paper, Burnisher for enameled	S. Shepard and A. M. George	Nashua, N. H	Oct. 6, 1868	82, 883
Paper by machinery, Making	T. Gilpin	Philadelphia, Pa	Dec. 24, 1816	......
Paper by machinery, Manufacturing	R. Fairchild	Trumbull, Conn	May 4, 1829	......

Index of patents issued from the United States Patent Office from 1790 *to* 1873, *inclusive*—Continued.

Invention.	Inventor.	Residence.	Date.	No.
Paper by single sheets, Separating	J. P. Comly	Dayton, Ohio	Mar. 22, 1853	9, 623
Paper case or envelope	J. W. Wilcox	New York, N. Y	Apr. 5, 1864	42, 247
Paper cask	J. W. Jarboe	Green Point, N. Y	Apr. 14, 1868	76, 771
Paper-calendering machine	S. Moore	Sudbury, Mass	Dec. 12, 1871	121, 800
Paper-calendering machine	T. B. De Forest	Birmingham, Conn	Apr. 9, 1872	125, 548
Paper-calendering roll	J. H. Garfield	Newton Lower Falls, Mass	Feb. 27, 1872	124, 048
Paper cap	M. G. Imbach and I. Weidenman.	Hartford, Conn	July 28, 1868	80, 352
Paper cap	N. Waterman and A. T. Perkins.	Toledo, Ohio	June 30, 1868	79, 420
Paper cap	P. S. White	Providence, R. I	Nov. 22, 1870	109, 566
Paper, Cap and other	D. D. Foley and J. J. Johnson	Washington, D. C	Oct. 24, 1871	120, 152
Paper, Carbonized	L. H. Rogers	New York, N. Y	Oct. 15, 1872	132, 324
Paper, Cigarette	F. X. Hazman and L. L. Arnold	New York, N. Y	July 25, 1865	48, 936
Paper-clamp	A. Faber	Washington, D. C	Feb. 6, 1866	52, 403
Paper-clamp	A. Palmer	Lee, Mass	Oct. 12, 1858	21, 775
Paper-clamp and inkstand combined	E. J. Toof	Fort Madison, Iowa	Apr. 21, 1868	77, 131
Paper-clasp	J. C. Arms	Northampton, Mass	Sept. 1, 1868	81, 727
Paper, Cleaning and preparing rags for the manufacture of.	W. E. Newton	London, England	May 10, 1870	102, 854
Paper-cleaning machine, Rag	W. Debit	Hartford, Conn	July 21, 1835	....
Paper-clip	W. A. Amberg	Chicago, Ill	July 6, 1869	92, 141
Paper-clip	C. E. Candee	Jersey City, N. J	Feb. 11, 1868	74, 301
Paper-clip	J. M. Crull	Harrisburgh, Pa	Mar. 28, 1871	113, 026
Paper-clip	G. W. Emerson	Chicago, Ill	Jan. 4, 1870	98, 479
Paper-clip	A. H. Fatzinger	Washington, N. J	May 26, 1868	78, 271
Paper-clip	E. A. Franklin	Brenham, Tex	June 4, 1872	127, 472
Paper-clip	J. A. Harvey and C. A. Mills	Dubuque, Iowa	July 1, 1856	15, 232
Paper-clip	T. Orton	Chicago, Ill	May 20, 1873	139, 019
Paper-clip	J. H. Parsons	Quincy, Mich	Dec. 3, 1867	71, 635
Paper-clip	M. A. Wheeler	San Francisco, Cal	May 27, 1873	139, 285
Paper-clip, Suspension	W. Stokes	New York, N. Y	July 12, 1870	105, 274
Paper, cloth, and other materials, Coating and decorating.	F. Beck	New York, N. Y	Jan. 3, 1871	110, 727
Paper-cloth bosom	C. Arnold	Chicago, Ill	July 3, 1866	55, 974
Paper, cloth, &c., Coating and ornamenting the surface of.	F. Beck	New York, N. Y	Nov. 22, 1870	109, 486
Paper, cloth, &c., Staining and printing	B. Mesteyer	New York	June 28, 1815	....
Paper collar	W. S. Bell, jr	Boston, Mass	June 13, 1865	48, 148
Paper collar	G. F. Bigelow	Chicago, Ill	June 27, 1865	48, 359
Paper collar	C. A. Brown	Troy, N. Y	May 13, 1873	138, 849
Paper collar	C. K. Brown	Troy, N. Y	Jan. 20, 1863	37, 431
Paper collar	L. M. Crane	Ballston Spa, N. Y	Sept. 18, 1866	58, 223
Paper collar	W. L. Duff	Quincy, Ill	May 30, 1865	48, 016
Paper collar	A. A. Evans	Boston, Mass	May 26, 1863	38, 664
Paper collar	N. Evans, jr	Boston, Mass	Mar. 14, 1871	112, 571
Paper collar	G. A. Goldsmith	New York, N. Y	Apr. 12, 1870	101, 855
Paper collar	S. S. Gray	Boston, Mass	Apr. 14, 1863	38, 160
Paper collar	S. S. Gray	Boston, Mass	July 5, 1864	43, 400
Paper collar	S. S. Gray	Boston, Mass	Jan. 23, 1866	52, 163
Paper collar	S. S. Gray	Boston, Mass	July 11, 1871	116, 948
Paper collar	W. B. Harris	Springfield, Mass	May 15, 1866	54, 720
Paper collar	S. F. Hilton	Providence, R. I	Sept. 12, 1871	118, 803
Paper collar	S. B. Hutchinson	Nashua, N. H	Aug. 8, 1865	49, 347
Paper collar	A. Kaufmann	New York, N. Y	Jan. 29, 1867	61, 669
Paper collar	E. E. Mack	Albany, N. Y	Mar. 29, 1870	101, 286
Paper collar	G. W. Ray	Springfield, Mass	Jan. 8, 1867	61, 100
Paper collar	C. E. Richards	North Attleborough, Mass	Feb. 13, 1866	52, 607
Paper collar	S. J. Shaw	Marlborough, Mass	Oct. 24, 1865	50, 663
Paper collar	D. F. Smith	Sullivan, N. H	Apr. 12, 1870	101, 932
Paper collar	G. K. Snow	Watertown, Mass	Feb. 7, 1865	46, 310
Paper collar	C. Spofford and W. S. Bell, jr	Boston, Mass	Aug. 30, 1864	44, 022
Paper collar	C. Spofford and V. Fogerty	Boston, Mass	Dec. 27, 1864	45, 681
Paper collar	S. S. Stone	Troy, N. Y	July 31, 1866	56, 827
Paper collar and bosom	M. Pinner	New York, N. Y	Oct. 9, 1866	58, 673
Paper collar and cuff	H. Parmenter	New York, N. Y	Apr. 16, 1872	125, 842
Paper-collar band	G. K. Snow	Watertown, Mass	Jan. 30, 1866	52, 371
Paper-collar binding and folding apparatus	W. E. Lockwood	Philadelphia, Pa	Oct. 11, 1864	44, 640
Paper-collar binding and folding apparatus	J. F. Schuyler	Philadelphia, Pa	Feb. 24, 1863	37, 792
Paper-collar box	L. Churchill	Troy, N. Y	Feb. 7, 1871	111, 514
Paper-collar box	S. Kaufmann	New York, N. Y	June 28, 1870	104, 853
Paper-collar box	E. A. Locke and W. N. Weeden	Boston, Mass	July 20, 1869	92, 850
Paper-collar-boxing machine	D. Stoner and J. Ligwalt, jr	Chicago, Ill	Dec. 4, 1866	60, 278
Paper-collar button-hole	J. Merwin	New York, N. Y	Apr. 23, 1867	64, 022
Paper-collar button-hole	B. F. Porter	Nashua, N. H	Feb. 13, 1866	52, 599
Paper-collar button-hole	W. H. Robinson	Rochester, N. Y	Jan. 5, 1869	85, 535
Paper-collar button-hole	D. Stoner	Chicago, Ill	Apr. 10, 1866	53, 897
Paper-collar button-hole	N. F. Whiting	Lawrence, Mass	June 19, 1866	55, 753
Paper-collar button-hole patch, Self-sealing	T. Griffin	Roxbury, Mass	Aug. 21, 1866	57, 314
Paper-collar card, &c., from printing or embossing press, Removing.	T. Tebbetts	New York, N. Y	Oct. 17, 1865	50, 514
Paper collar, Colored	S. W. H. Ward	New York, N. Y	Aug. 2, 1864	43, 728
Paper-collar-creasing apparatus	T. Tebbetts	New York, N. Y	Nov. 13, 1866	59, 680
Paper-collar-cutting die	E. Jeffers	Boston, Mass	Jan. 26, 1869	86, 162
Paper-collar-cutting die	G. K. Snow	Watertown, Mass	Nov. 28, 1865	51, 235
Paper-collar-cutting machine	A. L. Elliott	Boston, Mass	Jan. 24, 1871	111, 113
Paper-collar-cutting machine	D. M. Smythe	Orange, N. J	Nov. 19, 1867	71, 074
Paper-collar die, Adjustable	C. Spofford	Boston, Mass	Nov. 19, 1867	71, 076
Paper-collar fastening	C. R. Everson	Palmyra, N. Y	June 18, 1867	65, 894
Paper-collar fastening	S. Hodgins	Saint Louis, Mo	July 31, 1866	56, 854
Paper-collar finishing and boxing apparatus	J. J. Currier and S. Wells, jr	Boston, Mass	Dec. 18, 1866	60, 484
Paper-collar-folding apparatus	A. H. Hook	New York, N. Y	Mar. 7, 1865	46, 669
Paper-collar-folding machine	J. J. Currier	Boston, Mass	May 22, 1866	55, 005
Paper-collar-folding machine	C. E. Moore	Roxbury, Mass	Oct. 23, 1866	59, 134
Paper-collar-folding machine	E. Vossnack	New York, N. Y	Oct. 10, 1865	50, 426
Paper collar, Ladies'	S. S. Gray	Boston, Mass	Feb. 7, 1865	46, 232
Paper-collar machine	C. K. Brown and W. Wright	Troy, N. Y	Apr. 24, 1866	54, 244

Index of patents issued from the United States Patent Office from 1790 *to* 1873, *inclusive*—Continued.

Invention.	Inventor.	Residence.	Date.	No.
Paper-collar machine	G. W. Cilley and G. H. Spaulding.	Norwich, Conn	Jan. 28, 1868	73, 693
Paper-collar machine	C. H. Dennison	New York, N. Y	Jan. 2, 1872	122, 444
Paper-collar machine	M. P. Dorsch	New York, N. Y	Feb. 14, 1865	46, 340
Paper-collar machine	J. C. Ford	Cambridge, Mass	Aug. 21, 1866	57, 308
Paper-collar machine	J. W. Griswold and J. Sigwalt, jr	Chicago, Ill	Oct. 16, 1866	58, 812
Paper-collar machine	H. F. Knapp	New York, N. Y	May 8, 1866	54, 561
Paper-collar machine	H. F. Knapp	New York, N. Y	Nov. 6, 1866	59, 410
Paper-collar machine	G. W. Livermore	Cambridge, Mass	Nov. 1, 1864	44, 871
Paper-collar machine	T. McSpedon	New York, N. Y	Dec. 27, 1864	45, 621
Paper-collar machine	C. E. Moore and M. L. Wyman	Boston and Melrose, Mass	Sept. 13, 1870	107, 399
Paper-collar machine	S. Shepherd	Nashua, N. H	Nov. 2, 1869	96, 491
Paper-collar machine	S. Shepherd and A. M. George	Nashua, N. H	Sept. 12, 1865	49, 927
Paper-collar machine	D. M. Smyth	New York, N. Y	Dec. 5, 1865	51, 361
Paper-collar machine	G. K. Snow	Watertown, Mass	Nov. 28, 1865	51, 233
Paper-collar machine	G. K. Snow	Watertown, Mass	Nov. 28, 1865	51, 234
Paper-collar machine	C. Spofford and C. H. Montague	Boston, Mass	Dec. 15, 1868	85, 037
Paper-collar machine	W. H. Walton	Brooklyn, N. Y	July 2, 1867	66, 429
Paper-collar machine	O. F. Washburn	Bridgewater, Vt	Dec. 24, 1867	72, 705
Paper-collar machinery, Die for	G. Harrington and G. D. Rollins.	Springfield, Mass	May 20, 1873	139, 146
Paper-collar-making apparatus	D. M. Smyth	New York, N. Y	Aug. 16, 1864	43, 870
Paper-collar-molding machine	S. S. Gray	Boston, Mass	Mar. 16, 1869	87, 841
Paper-collar press	J. H. Darlington	New York, N. Y	Feb. 20, 1866	52, 785
Paper-collar-shaping apparatus	W. E. Lockwood and H. Howson.	Philadelphia, Pa	Sept. 12, 1865	49, 952
Paper-collar-stretching machine	C. Spofford and W. S. Bell, jr	Boston, Mass	Feb. 14, 1865	46, 400
Paper-collar, Turn-down enameled	J. H. Hoffman	New York, N. Y	Jan. 24, 1865	45, 998
Paper collars and cuffs, Manufacture of	D. A. Alden	Roxbury, Mass	Nov. 27, 1866	59, 938
Paper collars, cuffs, &c., Fabric for the manufacture of.	J. Restein	Philadelphia, Pa	Dec. 29, 1868	85, 397
Paper-collars, Device for applying cloth patch to	J. P. Courtney and C. Redmayne.	Brooklyn, N. Y	Aug. 18, 1868	81, 255
Paper collars, Folding	W. S. Bell	Boston, Mass	Feb. 7, 1865	46, 291
Paper collars, Folding	J. H. Darlington	New York, N. Y	Aug. 2, 1864	43, 737
Paper collars, Folding	M. P. Dorsch	New York, N. Y	Oct. 11, 1864	44, 615
Paper collars, Machine for applying cloth patch to	C. H. Denison	Springfield, Mass	June 30, 1868	79, 460
Paper collars, Machine for patching, punching, and embossing button-holes of.	C. Lang	Jersey City, N. J	Feb. 28, 1871	112, 155
Paper collars, Making	A. F. Gray	Boston, Mass	Nov. 8, 1864	44, 948
Paper collars, &c., Manufacture of	G. W. Ray	Springfield, Mass	May 1, 1866	54, 404
Paper collars, Manufacture of sweat-proof	J. H. Hoffman	New York, N. Y	Apr. 4, 1865	47, 107
Paper collars, Pressing	E. E. Ranous	New York, N. Y	Aug. 2, 1864	43, 741
Paper, Colored	G. La Monte, G. G. Saxe, and C. H. Clayton.	New York, N. Y	Dec. 19, 1871	121, 946
Paper, Coloring and drying	C. C. Fitzgerald	New York, N. Y	June 7, 1870	104, 006
Paper, &c., coloring apparatus	C. Williams	Philadelphia, Pa	Sept. 21, 1858	21, 584
Paper-coloring machine	C. K. Brown	Troy, N. Y	Dec. 29, 1868	85, 426
Paper, Composition for chemical-telegraph	G. Little	Rutherford Park, N. J	Aug. 27, 1872	130, 810
Paper, &c, Composition for saturating	A. T. Boon and J. Stafford	Galesburgh, Ill	Oct. 8, 1867	69, 618
Paper, Composition for treating vegetable	C. S. Dikeman	New York, N. Y	Oct. 21, 1862	36, 704
Paper, Compound for making water-proof	C. S. Rauh	Boston, Mass	Oct. 26, 1869	96, 148
Paper cop-tube	A. McCausland	Providence, R. I	July 7, 1857	17, 746
Paper, cordage, twine, &c., Manufacture of	J. H. McConnell	Springfield, Ill	Feb. 7, 1871	111, 661
Paper corset	J. H. Beal, E. J. Sawyer, and G. S. Webster.	Boston, Mass	July 30, 1867	67, 250
Paper cuff	S. A. Bemis, A. E. Foth, and J. W. Goodrich.	Springfield, Mass	Oct. 25, 1870	108, 676
Paper cuff	G. K. Snow	Watertown, Mass	May 25, 1869	90, 599
Paper cuff or wristband	A. A. Evans	Boston, Mass	July 31, 1866	56, 737
Paper-cutter	C. Cropper	Cincinnati, Ohio	Mar. 5, 1867	62, 530
Paper-cutter	St. C. Denny and I. L. G. Rice	Pittsburgh, Pa., and Cambridge, Mass.	Nov. 1, 1870	108, 891
Paper-cutter	J. Hatch	South Windham, Conn	June 22, 1869	91, 742
Paper-cutter	I. L. G. Rice	Cambridge, Mass	Nov. 15, 1870	109, 249
Paper-cutter	S. R. Timby	Saratoga, N. Y	Jan. 21, 1868	73, 674
Paper-cutter	J. Worrell	Philadelphia, Pa	June 13, 1871	115, 922
Paper-cutter and rule, Combined	F. E. Oliver	New York, N. Y	Feb. 26, 1861	31, 556
Paper-cutter and ruler	S. W. Wilcox	Mendon, Mass	Oct. 13, 1868	83, 115
Paper-cutter, Book-binder's	M. Riehl	Philadelphia, Pa	Oct. 2, 1866	58, 482
Paper-cutter, Desk	J. W. Wetmore	Erie, Pa	May 14, 1872	126, 659
Paper-cutter for continuous sheets	G. Keeble	Suffolk, Great Britain	Oct. 2, 1866	58, 563
Paper-cutter gage	C. C. Child	Boston, Mass	Feb. 15, 1870	99, 759
Paper-cutter, Rotary	H. S. Miller	Philadelphia, Pa	May 20, 1873	139, 176
Paper, Cutting	J. P. Farnum	Andover, Mass	Dec. 28, 1852	9, 503
Paper cutting and arranging machinery	J. C. Kneeland and G. M. Phelps	Troy, N. Y	Sept. 12, 1848	5, 767
Paper cutting and folding machine	C. Chambers, jr	Philadelphia, Pa	Apr. 4, 1871	113, 256
Paper cutting and folding machine	C. Chambers, jr., and W. Mendham.	Philadelphia, Pa	Apr. 4, 1871	113, 257
Paper cutting and folding machine	E. H. McArthur	Hillsdale, N. J	Sept. 3, 1867	68, 450
Paper cutting and folding machine	C. Moore	Hartford, Conn	Dec. 16, 1856	16, 256
Paper cutting and punching machine	J. T. S. Smith	New York, N. Y	Aug. 13, 1867	67, 813
Paper cutting and scoring machine	G. Bates	Philadelphia, Pa	Nov. 7, 1871	120, 697
Paper, Cutting and trimming edges of	J. Bateman	Harvard, Mass	Mar. 12, 1834	....
Paper cutting and twisting machine	I. P. Tice	New York, N. Y	Aug. 16, 1864	43, 874
Paper-cutting implement	J. Cook	Grand Manan, Canada	June 18, 1872	128, 021
Paper-cutting machine	J. L. Allen	New York, N. Y	Apr. 7, 1868	76, 373
Paper-cutting machine	J. Arkell	Canajoharie, N. Y	Aug. 22, 1871	118, 327
Paper-cutting machine	G. Bates	Philadelphia, Pa	Jan. 31, 1871	111, 423
Paper-cutting machine	E. P. Beckwith	Windham, Conn	Dec. 19, 1865	51, 669
Paper-cutting machine	M. B. Bigelow	Boston, Mass	July 13, 1858	20, 858
Paper-cutting machine	J. Bole	Grand Rapids, Mich	Aug. 2, 1870	105, 894
Paper-cutting machine	S. Brown	Philadelphia, Pa	Mar. 7, 1871	112, 414
Paper-cutting machine	E. Burroughs	Rochester, N. Y	Aug. 9, 1859	24, 989
Paper-cutting machine	S. M. Clark	Washington, D. C	July 16, 1872	129, 319
Paper-cutting machine	J. M. Clintic	Chambersburgh, Pa	Dec. 17, 1825	....
Paper-cutting machine	J. E. Coffin	Portland, Me	Jan. 3, 1871	110, 631
Paper-cutting machine	E. Cowles	Cleveland, Ohio	Oct. 10, 1871	119, 827

Index of patents issued from the United States Patent Office from 1790 *to* 1873, *inclusive*—Continued.

Invention.	Inventor.	Residence.	Date.	No.
Paper-cutting machine	A. W. Currier	Grand Rapids, Mich	Mar. 8, 1870	100, 506
Paper-cutting machine	A. T. De Puy	New York, N. Y	Oct. 29, 1872	132, 525
Paper-cutting machine	T. B. Dooley	Boston, Mass	May 28, 1872	127, 226
Paper-cutting machine	A. Dougherty	Brooklyn, N. Y	June 17, 1862	35, 592
Paper-cutting machine	S. Ellsworth	Lacon, Ill	Oct. 20, 1868	83, 143
Paper-cutting machine	F. A. Fletcher	Newark, Del	Nov. 28, 1871	121, 239
Paper-cutting machine	H. A. Gage	Manchester, N. H	Mar. 1, 1870	100, 391
Paper-cutting machine	H. Gilman	Troy, N. Y	Sept. 4, 1849	6, 692
Paper-cutting machine	J. L. Gregorie	Chicago, Ill	Mar. 5, 1872	124, 352
Paper-cutting machine	A. Hardy	Boston, Mass	Apr. 12, 1870	101, 873
Paper-cutting machine	F. Hesse	Bethlehem, Pa	July 19, 1853	9, 872
Paper-cutting machine	P. H. Hopkins	Brooklyn, N. Y	Feb. 20, 1872	123, 901
Paper-cutting machine	T. W. Houchin	Worcester, Mass	Dec. 20, 1859	26, 499
Paper-cutting machine	F. B. Howell	Lockport, Ohio	May 31, 1830	....
Paper-cutting machine	H. Law	New York, N. Y	Sept. 16, 1856	15, 738
Paper-cutting machine	H. Law	Chatham, N. J	Apr. 5, 1864	42, 200
Paper-cutting machine	H. Law	Chatham, N. J	Oct. 27, 1868	83, 513
Paper-cutting machine	J. Leviness	Brooklyn, N. Y	Aug. 1, 1871	117, 650
Paper-cutting machine	J. Leviness and P. Van Horn	Brooklyn, N. Y	Mar. 19, 1872	124, 690
Paper-cutting machine	J. E. Mallory	New York, N. Y	Sept. 21, 1852	9, 276
Paper-cutting machine	C. Montague	Hartford, Conn	Mar. 13, 1866	53, 169
Paper-cutting machine	C. Montague	Boston, Mass	Jan. 4, 1870	98, 611
Paper-cutting machine	C. W. L. Montague	Brooklyn, N. Y	Jan. 16, 1872	122, 844
Paper-cutting machine	C. Paine	Boston, Mass	Mar. 14, 1871	112, 623
Paper-cutting machine	E. Pine	Troy, N. Y	July 20, 1831	....
Paper-cutting machine	T. C. Robinson	Boston, Mass	Nov. 23, 1869	97, 120
Paper-cutting machine	T. C. Robinson	Boston, Mass	Dec. 14, 1869	97, 816
Paper-cutting machine	T. C. Robinson	Boston, Mass	May 2, 1871	114, 340
Paper-cutting machine	G. H. Sanborn	Boston, Mass	Dec. 3, 1867	71, 793
Paper-cutting machine	G. H. Sanborn	New York, N. Y	Nov. 24, 1868	84, 306
Paper-cutting machine	G. H. Sanborn	New York, N. Y	Nov. 24, 1868	84, 307
Paper-cutting machine	F. Schoettle	Philadelphia, Pa	July 9, 1872	128, 817
Paper-cutting machine	E. R. and T. W. Sheridan	New York, N. Y	Aug. 27, 1872	130, 873
Paper-cutting machine	E. R. and T. W. Sheridan	New York, N. Y	Oct. 15, 1872	132, 327
Paper-cutting machine	J. Shugert	Quincy, Pa	Dec. 14, 1830	....
Paper-cutting machine	H. Skidmore	Mount Vernon, N. Y	Aug. 11, 1868	81, 021
Paper-cutting machine	F. B. Sweetland	New Haven, Conn	Mar. 1, 1870	100, 465
Paper-cutting machine	J. F. and G. W. Tapley	Springfield, Mass	July 24, 1866	56, 679
Paper-cutting machine	W. H. Topham	New York, N. Y	Feb. 7, 1871	111, 702
Paper-cutting machine	F. L. Walker	Boston, Mass	Nov. 23, 1869	97, 140
Paper-cutting machine	G. A. Walker	Boston, Mass	Aug. 22, 1871	118, 308
Paper-cutting machine	G. A. Walker	Boston, Mass	Mar. 19, 1872	124, 774
Paper-cutting machine	G. A. Walker	Boston, Mass	Dec. 9, 1873	145, 464
Paper-cutting machine	B. Weaver	Springfield, Mass	Mar. 5, 1872	124, 236
Paper-cutting machine	C. Wells and H. Barth	Cincinnati, Ohio	July 25, 1865	49, 018
Paper-cutting machine	J. Worell	Philadelphia, Pa	Mar. 12, 1872	124, 650
Paper-cutting machine, Rotary	W. Bullock	Philadelphia, Pa	Mar. 1, 1870	100, 367
Paper-cutting machines, Clamp for	E. T. Hazeltine and C. Park	Warren, Pa	Apr. 15, 1873	137, 920
Paper-cutting machinery	Z. M. Crane	Dalton, Mass	July 18, 1848	5, 670
Paper-cutting machinery	N. Gavit	Philadelphia, Pa	May 9, 1854	10, 889
Paper-cutting machinery	M. Wilder	Peterborough, N. H	July 18, 1848	5, 672
Paper-cutting machinery	G. L. Wright	Springfield, Mass	July 18, 1848	5, 671
Paper-cutting shears	C. Brombacher	New York, N. Y	July 11, 1865	48, 650
Paper-cutting shears	W. P. Deane	Canton, Mass	May 27, 1873	139, 373
Paper, Cylinder-molds for making	J. F. Jones	Rochester, N. Y	May 26, 1863	38, 684
Paper-dampener	J. A. Lynch	Boston, Mass	Apr. 7, 1868	76, 479
Paper-damping apparatus	C. A. Waterbury	New York, N. Y	Apr. 27, 1859	20, 111
Paper damping and printing machine	A. Overend	Philadelphia, Pa	Mar. 14, 1854	10, 627
Paper-damping machine	A. Dougherty	New York, N. Y	July 9, 1861	32, 759
Paper-damping machine	R. Martin	Philadelphia, Pa	Nov. 27, 1860	30, 788
Paper, Designer's	A. Akeroyd	Manchester, N. H	Dec. 9, 1873	145, 269
Paper, Die for cutting out ornamental forms of	A. Delkescamp	Brooklyn, N. Y	Mar. 25, 1873	137, 063
Paper doll	J. T. Brown	Albion, N. Y	May 14, 1872	126, 622
Paper-dryer	A. E. Harding	Middletown, Ohio	Feb. 9, 1869	86, 666
Paper-dryer	E. A. Seeley	Scotch Plains, N. J	Oct. 28, 1873	143, 999
Paper-dryer	W. H. Soley	Philadelphia, Pa	May 30, 1871	115, 375
Paper dryer, Flock or velvet	T. A. Blanchard	New York, N. Y	Aug. 13, 1872	130, 406
Paper-dryer regulator	D. Crosby	Hampden, Me	Mar. 7, 1871	112, 422
Paper, Drying	J. Hartin	New York, N. Y	Aug. 9, 1853	9, 918
Paper, Drying	H. Howe	Shirley, Mass	Mar. 12, 1836	....
Paper drying and finishing apparatus	H. Dodge	Albany, N. Y	Nov. 14, 1871	120, 810
Paper drying and pressing machine	J. North	Middletown, Conn	Apr. 14, 1857	17, 050
Paper-drying cylinder	H. P. Howe	Shirley, Mass	Sept. 20, 1836	....
Paper-drying machine	J. Hoyt	Cleveland, Ohio	July 23, 1861	32, 870
Paper-drying machine	G. S. Rogers	Thetford Centre, Vt	Jan. 24, 1865	46, 026
Paper-drying machine	N. W. Taylor and J. W. Brightman.	Cleveland, Ohio	June 23, 1863	38, 993
Paper-drying machinery	E. L. Perkins	Roxbury, Mass	June 7, 1859	24, 328
Paper, Drying sized	J. Kingsland, jr., and N. White	Saugerties and New York, N. Y.	Aug. 10, 1852	9, 186
Paper, Drying sized	J. Norton, jr	Boston, Mass	Apr. 25, 1846	4, 480
Paper, Drying thick	E. and J. R. Cushman	Amherst, Mass	Aug. 1, 1854	11, 411
Paper during manufacture, Apparatus for smoothing and pressing.	D. Crosby	Hampden, Me	Feb. 14, 1871	111, 820
Paper-engine	W. Dickinson	Worcester, Mass	Sept. 3, 1840	1, 760
Paper-engine	E. Hawkins	Islip, N. Y	Jan. 11, 1870	98, 691
Paper-engine bed-plate	W. Clarke	Dayton, Ohio	Oct. 9, 1849	6, 784
Paper-engine grinding-cylinder	C. T. Frost	Medfield, Mass	Apr. 3, 1866	53, 598
Paper-engine washer	C. Rice	Watertown, N. Y	Apr. 26, 1834	....
Paper-fastener	C. L. Alexander	Washington, D. C	Oct. 2, 1866	58, 365
Paper-fastener	Z. W. Denham	Washington, D. C	May 2, 1865	47, 526
Paper-fastener	J. C. Jensen	Chicago, Ill	Apr. 16, 1872	125, 682
Paper-fastener	A. J. Kletzker	Saint Louis, Mo	Nov. 3, 1868	83, 640
Paper-fastener	J. W. McGill	Washington, D. C	Aug. 20, 1867	67, 895
Paper-fastener	G. G. W. Morgan	Washington, D. C	Oct. 9, 1866	58, 667
Paper-fastener	A. Smith	Perrysburgh, Ohio	July 5, 1864	43, 435
Paper-fastener	J. F. Tapley	Springfield, Mass	May 11, 1869	89, 898
Paper-fastener	W. M. Tileston	New York, N. Y	Sept. 15, 1868	82, 181

Index of patents issued from the United States Patent Office from 1790 *to* 1873, *inclusive*—Continued.

Invention.	Inventor.	Residence.	Date.	No.
Paper-fastener, Metallic	G. W. McGill	Washington, D. C	July 24, 1866	56, 587
Paper-fasteners, Press for attaching	G. W. McGill	Washington, D. C	Aug. 13, 1867	67, 665
Paper-fastening	J. M. Keep	New York, N. Y	Apr. 16, 1867	63, 905
Paper-fastening	M. H. N. Kendig	Washington, D. C	July 23, 1867	66, 968
Paper-feeder	E. Allen	Newark, N. J	Sept. 15, 1863	39, 872
Paper-feeder	J. T. Ashley	Brooklyn, E. D., N. Y	Sept. 13, 1870	107, 320
Paper-feeder	J. T. Ashley	Brooklyn, E. D., N. Y	Oct. 4, 1870	107, 851
Paper-feeder	J. A. Graves	Washington, D. C	Apr. 4, 1871	113, 291
Paper-feeder	J. H. Gray and W. B. Turner	Saint Anthony, Minn	Dec. 28, 1869	98, 253
Paper-feeder	O. Norelius	Minneapolis, Minn	Feb. 22, 1870	100, 059
Paper-feeding device	A. A. Dunk	Philadelphia, Pa	Nov. 12, 1872	132, 904
Paper-feeding machine	E. R. Andrews, R. B. Randall, and W. H. Clague.	Rochester, N. Y	Apr. 25, 1871	114, 087
Paper-feeding machine	J. T. Ashley	Williamsburgh, N. Y	May 20, 1873	138, 983
Paper-feeding machine	J. T. and F. Ashley	Brooklyn, N. Y	Oct. 21, 1873	143, 740
Paper feeding machine	A. M. Crane	Staunton, Va	Mar. 12, 1872	124, 550
Paper-feeding machine	D. Dick	New York, N. Y	Jan. 30, 1872	123, 157
Paper-feeding machine	M. E. Knight	Boston, Mass	Nov. 15, 1870	109, 224
Paper-feeding machine	W. Mowry	Minneapolis, Minn	Feb. 2, 1869	86, 575
Paper-feeding machine	M. Piedra	Williamsburgh, N. Y	Mar. 11, 1873	136, 668
Paper-feeding machine	S. Scholfield and C. E. Baker	Providence, R. I., and Mont Clair, N. J.	Feb. 4, 1873	135, 597
Paper-feeding machine	S. Scholfield and C. E. Baker	Providence, R. I., and Mont Clair, N. J.	Feb. 4, 1873	135, 598
Paper-feeding machine	S. Scholfield and C. E. Baker	Providence, R. I., and Mont Clair, N. J.	Feb. 4, 1873	135, 599
Paper-feeding machine	C. M. Wielings	San Francisco, Cal	Dec. 16, 1873	145, 542
Paper felt or wadding	W. W. Glentworth and W. H. Gandey.	Philadelphia, Pa., and Lambertville, N. J.	Nov. 23, 1869	97, 189
Paper, Figure-tinted	G. La Monte and G. G. Saxe	New York, N. Y	Feb. 20, 1872	123, 782
Paper, Finishing	J. White and L. Gale	Newburgh, Vt	Feb. 28, 1827	....
Paper-finishing machine	T. Gilpin	Philadelphia, Pa	June 25, 1830	....
Paper flour-sack	J. M. Hurd	Auburn, N. Y	Nov. 12, 1867	70, 852
Paper, Fly	T. E. Peck	Bridgeport, Conn	Apr. 2, 1872	125, 326
Paper-folder	E. Gomez	New York, N. Y	Aug. 9, 1859	25, 007
Paper-folder	E. P. Tiffany	Hartford, Conn	Apr. 26, 1870	102, 334
Paper folding and pasting machine	G. K. Snow	Watertown, Mass	Mar. 6, 1860	27, 392
Paper folding and registering machine	A. F. Endriss	New York, N. Y	Mar. 8, 1859	23, 161
Paper, Folding and ruling	J. P. Herron	Washington, D. C	June 30, 1863	39, 045
Paper folding and stitching machine	J. H. Tanner	Franenfeld, Switzerland	Sept. 9, 1862	36, 428
Paper folding, cutting, and trimming machine	C. Chambers, jr	Philadelphia, Pa	Aug. 5, 1873	141, 487
Paper-folding machine	C. Chambers, jr	Kennett's Square, Pa	Oct. 7, 1856	15, 842
Paper-folding machine	C. Chambers, jr	Philadelphia, Pa	Nov. 3, 1857	18, 533
Paper-folding machine	C. Chambers, jr	Philadelphia, Pa	Apr. 5, 1859	23, 445
Paper-folding machine	C. Chambers, jr	Philadelphia, Pa	Nov. 15, 1859	26, 090
Paper-folding machine	C. Chambers, jr	Philadelphia, Pa	Nov. 27, 1860	30, 719
Paper-folding machine	C. Chambers, jr	Philadelphia, Pa	Dec. 18, 1860	30, 910
Paper-folding machine	C. Chambers, jr	Philadelphia, Pa	July 19, 1870	105, 424
Paper-folding machine	C. Chambers, jr., and W. Mendham.	Philadelphia, Pa	Aug. 5, 1873	141, 489
Paper-folding machine	C. Chambers, jr., and W. Mendham.	Philadelphia, Pa	Aug. 5, 1873	141, 490
Paper-folding machine	J. S. Gallaher, jr	Washington, D. C	Dec. 4, 1860	30, 853
Paper-folding machine	R. R. Gubbins	Troy, N. Y	Nov. 7, 1871	120, 639
Paper-folding machine	R. R. Gubbins	West Troy, N. Y	Dec. 5, 1871	121, 615
Paper-folding machine	J. Macnair	New York, N. Y	June 29, 1869	91, 948
Paper-folding machine	J. McAdams	Brooklyn, N. Y	Feb. 28, 1871	112, 264
Paper-folding machine	W. Mendham	Philadelphia, Pa	Feb. 1, 1870	99, 336
Paper-folding machine	W. Mendham	Philadelphia, Pa	June 21, 1870	104, 621
Paper-folding machine	W. H. and J. Milliken	Manchester, N. H	Feb. 26, 1861	31, 578
Paper-folding machine	O. Norelius	Minneapolis, Minn	Apr. 25, 1871	114, 031
Paper-folding machine	J. North	Middletown, Conn	Apr. 15, 1856	14, 697
Paper-folding machine	J. North	Middletown, Conn	June 24, 1862	35, 738
Paper-folding machine	J. North	Middletown, Conn	June 9, 1863	38, 874
Paper-folding machine	L. E. Osborn	New Haven, Conn	May 13, 1863	35, 249
Paper-folding machine	E. N. Smith	West Brookfield, Mass	Nov. 27, 1849	6, 896
Paper-folding machine	E. N. Smith	Springfield, Mass	May 19, 1857	17, 352
Paper-folding machine	G. K. Snow	Boston, Mass	Oct. 15, 1850	7, 722
Paper-folding machine	R. J. Stuart	Yonkers, N. Y	Oct. 28, 1873	144, 160
Paper-folding machine	T. Thompson	Niversville, N. Y	Feb. 12, 1856	14, 260
Paper-folding machine	G. W. D. Upton	Springfield, Mass	July 29, 1873	141, 405
Paper-folding machine	A. Washburn	Medina, Ohio	Nov. 26, 1872	133, 393
Paper-folding machine	A. Washburn	Medina, Ohio	Aug. 12, 1873	141, 742
Paper-folding machine	J. F. Weeks	Columbus, Ohio	June 9, 1857	17, 535
Paper-folding machine	C. P. Wiggins, A. H. Nordyke, and B. Strawbridge.	Richmond, Ind	Aug. 25, 1857	18, 075
Paper-folding machines, Packing and carrying device for.	W. Farrington	Brooklyn, N. Y	Aug. 5, 1873	141, 501
Paper-folding machines, Packing-box for	C. Chambers, jr	Philadelphia, Pa	Aug. 5, 1873	141, 486
Paper-folding machines, Packing-device for	C. Clark	New York, N. Y	Aug. 5, 1873	141, 491
Paper-folding machinery	C. O. Crosby	New Haven, Conn	Dec. 23, 1856	16, 266
Paper folding, pasting, and cutting machine	C. Chambers, jr	Philadelphia, Pa	Mar. 5, 1861	31, 588
Paper folding, pasting, and insetting machine	C. Chambers, jr	Philadelphia, Pa	Aug. 5, 1873	141, 488
Paper for automatic telegraphy and in recovering chemicals from waste-paper, Preparing.	G. Little	Rutherford Park, N. Y	July 23, 1872	129, 841
Paper for automatic transmitter, Apparatus for perforating.	G. Little	Hudson City, N. J	June 15, 1869	91, 241
Paper for bank-notes, bonds, &c	J. M. Wilcox	Glen Mills, Pa	May 16, 1871	115, 005
Paper for bank-notes, &c., Safety	J. Jameson	Gateshead, England	Apr. 15, 1873	137, 775
Paper for book-binders, Folding	G. K. Snow	Watertown, Mass	Feb. 14, 1860	27, 166
Paper for bonds, drafts, &c	R. Price	New York, N. Y	July 5, 1870	105, 125
Paper for boxes, Machine for cutting, cornering, and sewing.	A. Dennison	Brunswick, Me	Apr. 15, 1851	8, 044
Paper for boxes, Machine for cutting, scoring, and cornering.	J. M. Keer	Philadelphia, Pa	Nov. 29, 1870	109, 630
Paper for buildings, Preparing	C. B. Ayer	Beloit, Wis	Mar. 5, 1872	124, 314
Paper for buildings	F. N. Davis	Beloit, Wis	July 18, 1871	117, 155
Paper for checks, drafts, notes, &c	G. F. Thomas, jr	Brooklyn, N. Y	Apr. 12, 1870	101, 786

Index of patents issued from the United States Patent Office from 1790 *to* 1873, *inclusive*—Continued.

Invention.	Inventor.	Residence.	Date.	No.
Paper for chemical telegraph, &c	T. A. Edison	Newark, N. J	Oct. 22, 1872	132, 455
Paper for collars, &c., Manufacture of	S. M. Allen	Woburn, Mass	Mar. 31, 1863	38, 019
Paper for collars, Manufacture of	J. M. Willcox	Glen Mills, Pa	July 21, 1868	80, 105
Paper for collars, Surface-sized	W. F. Moseley	Brooklyn, N. Y	Nov. 7, 1871	120, 765
Paper for copying-presses, Damping	G. Burnham	Philadelphia, Pa	Dec. 24, 1850	7, 851
Paper for covering walls, Water and damp proof	C. Goessling	Jersey City, N. J	Aug. 4, 1868	80, 620
Paper for envelopes, &c., Apparatus for cutting and creasing.	M. Dimock	Newark, N. J	Oct. 11, 1864	44, 687
Paper for hangings and other purposes, Composite	W. Campbell	Millburn, N. J	May 25, 1869	90, 497
Paper for keeping off mosquitoes, Prepared	J. Bagnol	Baltimore, Md	Sept. 10, 1872	131, 242
Paper for manufacture of floor-covering, belting, window-shades, &c., Preparing.	J. J. Ott	Washington, D. C	Aug. 18, 1868	81, 199
Paper for manufacture of letter and invoice files	J. L. Rile	New York, N. Y	June 4, 1867	65, 436
Paper for manufacture of neck-ties, cravats, &c., Composition for coating.	M. W. Brown	New York, N. Y	Apr. 20, 1869	89, 198
Paper for manufacture of paper bags	W. E. Farrell	Philadelphia, Pa	June 15, 1869	91, 220
Paper for napkins, &c	E. Parrish	Philadelphia, Pa	May 30, 1871	115, 511
Paper for notes, checks, &c., Manufacture of	A. Varnham	London, England	Aug. 9, 1845	4, 143
Paper for postage-stamps	H. Loewenberg	New York, N. Y	Mar. 6, 1866	53, 081
Paper for printing postage and revenue stamps, Mode of preparing.	S. Lenher and H. H. Spencer	Philadelphia, Pa	Dec. 7, 1869	97, 528
Paper for printing, writing, and other purposes	J. Langtree	New York, N. Y	Sept. 6, 1870	107, 067
Paper for protecting goods from moths, &c., Manufacture of.	R. W. Russell	New York, N. Y	Mar. 30, 1869	88, 519
Paper for roofing and similar purposes, Preparing.	C. H. Smith	Baltimore, Md	Mar. 29, 1870	101, 406
Paper for transferring stamps, &c., Composition for preparing.	M. Rosenthal	Philadelphia, Pa	June 2, 1868	78, 610
Paper for twine, &c., Machine for cutting	J. B. Wortendyke	Goodwinville, N. J	Sept. 13, 1864	44, 249
Paper for various purposes, Process of treating	C. F. Crebose	Newton Lower Falls, Mass	Mar. 3, 1868	74, 996
Paper for water-closets, Medicated	G. W. Thompson	Brooklyn, N. Y	Dec. 22, 1868	85, 188
Paper for wrapping tobacco, snuff, soap, &c., Method of preparing.	M. W. Brown	West Farms, N. Y	Feb. 11, 1868	74, 195
Paper, &c., Forming metallic characters on	J. Lanza	New York, N. Y	June 19, 1866	55, 679
Paper fringe, Machine for cutting	P. and G. Pengeot	Buffalo, N. Y	Dec. 19, 1848	5, 972
Paper from alga or sea-weed, Mode of making	S. Green	New London, Conn	Feb. 15, 1809	....
Paper from corn-husks, Manufacturing	B. Allison and J. Hawkins	....	Dec. 30, 1802	....
Paper from curriers' shavings, Making	J. Condit, jr	....	Dec. 28, 1801	....
Paper from cutting-engines, Machinery for taking and laying.	J. M. Hollingsworth	Milton, Mass	Apr. 17, 1849	6, 337
Paper from grain, Process of manufacturing	C. V. and J. Stehlin and H. A. Haan.	New York and Brooklyn, N. Y.	Nov. 4, 1873	144, 294
Paper, &c., from husks of Indian corn, Manufacture of.	A. C. De Welsbach	Vienna, Austria	Apr. 21, 1863	38, 220
Paper from rags and straw and corn-husks, Mode of making white.	J. W. Cooper	Washington Township, Pa	Feb. 7, 1829	....
Paper from reeds	H. Lowe	Belleville, N. Y	Mar. 1, 1859	23, 099
Paper from resinous bark, Preparation of	C. C. Hall	Portland, Me	Feb. 20, 1855	12, 414
Paper from sand-grass, Manufacture of brown	I. Sanderson	Milton, Mass	Feb. 22, 1838	614
Paper from sorghum, Manufacture of	H. Pemberton	East Tarentum, Pa	Mar. 31, 1863	38, 050
Paper from Spanish grass, Manufacture of	W. B. Newbery	Dorchester, Mass	May 24, 1864	42, 866
Paper from straw, hay, &c., Making	W. Magaw	Meadville, Pa	Mar. 8, 1830	....
Paper from straw, Manufacture of	R. T., I. W., and A. I. Smart	Troy, N. Y	Mar. 27, 1860	27, 653
Paper from straw, Manufacture of fine white writing.	L. Bomeisler	Philadelphia, Pa	Sept. 10, 1829	....
Paper from straw, Process for making	W. Clarke	Dayton, Ohio	May 6, 1856	14, 804
Paper from the "ulvamarina," Making	E. H. Collier	Plymouth County, Mass	Apr. 15, 1828	....
Paper from vulcanized gum, Detaching	A. C. Richard	Newtown, Conn	Jan. 11, 1859	22, 584
Paper from wood, Manufacture of	S. M. Allen	Woburn, Mass	Mar. 31, 1863	38, 020
Paper from wood, Manufacture of	C. Watt and H. Burgess	London, England	July 18, 1854	11, 343
Paper from wood, Manufacturing	L. Wooster and J. B. Holmes	Meadville, Pa	Aug. 3, 1830	....
Paper-glazing	F. Beck	New York, N. Y	Jan. 1, 1867	60, 672
Paper goods, Manufacture of corded-edge	A. T. Denison	Poland, Me	Apr. 20, 1869	89, 132
Paper-hanger's apparatus	W. F. Trautman	Llewellyn, Pa	May 24, 1870	103, 393
Paper-hangings, Apparatus for hanging up and carrying off.	T. Van Deventer	New Brunswick, N. J	July 20, 1858	20, 965
Paper-hangings, Apparatus for making stamp-gilt	W. Bailey	New York, N. Y	Nov. 19, 1867	71, 120
Paper-hangings, Apparatus for trimming	E. Boothby	Saco, Me	Oct. 18, 1864	44, 764
Paper-hangings, Machine for coloring	J. Heist	New York, N. Y	Apr. 12, 1870	101, 732
Paper-hangings, Machine for marking	H. Lightbown	Pendleton, Great Britain	July 16, 1872	129, 573
Paper-hangings, Machine for trimming the edges of	J. Waugh	Elmira, N. Y	Oct. 5, 1858	21, 710
Paper-hangings, Paste for	J. Jones	New York, N. Y	Mar. 7, 1871	112, 464
Paper-hangings, Printing	R. L. Hawes	Worcester, Mass	Apr. 18, 1848	5, 525
Paper-hangings, Printing	W. M. Shaw and E. Gould	Newark, N. J	May 1, 1849	6, 404
Paper-hangings, Printing and embossing	E. S. Ormsby	New York, N. Y	Nov. 23, 1869	97, 221
Paper-hangings with light-colored satin ground, Manufacturing.	H. Steel	Hudson, N. Y	Sept. 8, 1813	....
Paper-holder	C. A. Barrows	Willimantic, Conn	Aug. 13, 1872	130, 464
Paper-holder	J. W. Foard	San Francisco, Cal	Sept. 5, 1865	49, 744
Paper-holder	Y. Orton	Chicago, Ill	Dec. 24, 1872	134, 160
Paper-holder	D. M. Smith	Springfield, Vt	Oct. 9, 1866	58, 687
Paper-holder	E. J. Toop	Fort Madison, Iowa	Jan. 30, 1866	52, 343
Paper-holder, Abrasive	G. A. Howe	Niles, Michigan	Dec. 19, 1871	122, 018
Paper-holder shelf	G. V. Hawkins	Clayton, Mich	Nov. 18, 1873	144, 763
Paper, Hot-pressing	F. Bailey	Salisbury, Pa	July 31, 1809	....
Paper, Implements for perforating and severing	A. H. Hook	New York, N. Y	Sept. 13, 1864	44, 190
Paper, &c., in making cards, pasteboard, &c., Machine for applying paste to sheets of.	D. H. Gilbert	Dorchester, Mass	Apr. 11, 1842	2, 553
Paper in paper-making machine, Machinery for drying.	O. Ellsworth	Boston, Mass	Feb. 12, 1867	61, 930
Paper, &c., in rolls, Apparatus for winding	G. D. Harrington	Columbus, Ohio	July 23, 1872	129, 660
Paper in the manufacture of boxes, Machine for cutting the corners of.	D. Whitlock	Newark, N. J	Sept. 18, 1866	58, 183
Paper in the manufacture of paper cop-tubes, Rolling or winding.	S. Burgess	Providence, R. I	Aug. 13, 1867	67, 716
Paper in the ream, Cutting and trimming	J. Ames	Springfield, Mass	Feb. 28, 1834	....
Paper in the ream, Cutting and trimming	J. Ames	Springfield, Mass	Feb. 28, 1834	....
Paper in the sheet, Machine for making	M. Haddock	New York	July 17, 1828	....

Index of patents issued from the United States Patent Office from 1790 *to* 1873, *inclusive*—Continued.

Invention.	Inventor.	Residence.	Date.	No.
Paper in writing and drawing, Clamp for holding..	E. B. Forbush	Buffalo, N. Y	Sept, 3, 1850	7, 617
Paper into sheets, Machine for cutting	J. Hatch	South Windham, Conn	June 27, 1865	48, 395
Paper-knife handle	E. Kelsey	Centre Brook, Conn	July 25, 1865	48, 953
Paper, Leather	S. M. Allen	Woburn, Mass	Aug. 4, 1863	39, 371
Paper, Leather	E. F. and T. Blank	New York	Feb. 16, 1830	....
Paper, leather, and other thin substances, Machine for cutting,	H. A. Snowdon	Brooklyn, N. Y	Sept. 17, 1872	131, 378
Paper, leather, &c , Apparatus for cutting ornaments in.	J. D. Mets	Dubuque, Iowa	May 5, 1863	38, 402
Paper, Letter	T. H. Dodge	Washington, D. C	Apr. 23, 1861	32, 167
Paper, Letter	R. Magee	Philadelphia, Pa	May 10, 1864	42, 673
Paper-machine	J. Ames	Springfield, Mass	Feb. 20, 1835	....
Paper-machine	M. J. Kearney	Philadelphia, Pa	Sept. 3, 1872	131, 103
Paper-machine	W. and A. L. Knight and E. F. Condit.	Whippany, N. J	Sept. 25, 1839	1, 336
Paper-machine	P. Scanlan	Indianapolis, Ind	Nov. 22, 1870	109, 552
Paper-machine	C. Whealen	Dayton, Ohio	Oct. 28, 1873	144, 172
Paper-machine attachment to prevent the straining and breaking of paper during the manufacture.	L. Dean	Fort Edward, N. Y	Aug. 9, 1870	106, 134
Paper-machine-belt guide	R. Hutton	Holyoke, Mass	Ju'y 1, 1873	140, 418
Paper-machine-crimping attachment	W. H. Bleasdale	Chagrin Falls, Ohio	May 10, 1870	102, 754
Paper-machine, Cylinder	I. Sanderson	Milton, Mass	Apr. 18, 1829	....
Paper-machine, Cylinder	J. P. Sherwood	Fort Edward, N. Y	Sept. 21, 1869	95, 153
Paper-machine deckel	C. McBurney and L. Hollingsworth.	Boston, Mass	Aug. 29, 1871	118, 624
Paper-machine-felt guide	P. A. Wait	Sandy Hill, N. Y	Apr. 8, 1856	14, 621
Paper-machine-felt washer	A. Anderson	Milwaukee, Wis	Sept. 6, 1864	44, 059
Paper, Machine for boiling and washing rags for manufacturing.	G. Spafford	Windham, Conn	Sept. 2, 1840	1, 753
Paper, Machine for cutting, printing, and folding..	W. P. Corsa	Catskill, N. Y	May 31, 1870	103, 575
Paper, Machine for cutting rags for	M. Y. Beach	Springfield, Mass	Sept. 11, 1828	....
Paper, Machine for drying sized	N. W. Taylor and J. W. Brightman.	Cleveland, Ohio	Apr. 29, 1862	35, 117
Paper, Machine for folding and cutting the edges of	J. E. Coffin	Portland, Me	Nov. 17, 1868	84, 091
Paper, &c., Machine for grinding wood-pulp for	J. S. Elliott and J. F. Wood	Chelsea and Everett, Mass	Apr. 23, 1872	126, 041
Paper, Machine for making and sizing	G. W. Turner	London, Great Britain	Jan. 27, 1852	8, 698
Paper, Machine for making card	E. L. Perkins	Boston, Mass	July 10, 1840	1, 679
Paper, Machine for making lace	C. Lang	Worcester, Mass	Jan. 31, 1865	46, 115
Paper, Machine for polishing enameled	S. Shepherd and A. M. George	Nashua, N. H	July 17, 1866	56, 457
Paper, Machine for polishing enameled	W. F. Wright	Nashua, N. H	July 24, 1866	56, 682
Paper, Machine for preparing rags for making	B. Cox	Northampton, Mass	Mar. 28, 1818	....
Paper, Machine for preparing top and swingled tow for.	A. Frost	....	Mar. 19, 1804	....
Paper, Machine for printing both sides of a continuous sheet of.	T. French	Ithaca, N. Y	Nov. 20, 1837	468
Paper, Machine for printing house	M. D. Whipple	Lowell, Mass	Sept. 16, 1851	8, 372
Paper, &c., Machine for punching and perforating.	G. C. Howard	Philadelphia, Pa	May 21, 1861	32, 370
Paper, Machine for receiving and piling	J. A. Wilkinson	Brooklyn, N. Y	Aug. 9, 1859	25, 068
Paper, &c., Machine for smoothing and polishing..	W. Coolidge	Boston, Mass	Apr. 18, 1808	....
Paper, Machine for stamping lace	A. Gerandat	New York, N. Y	Mar. 15, 1870	100, 749
Paper, Machine for tearing and dusting rags, &c., used in the manufacture of.	H. Clark and W. Albertson	New London, Conn	Sept. 19, 1838	927
Paper, Machine for trimming edges of	J. McClintic	Chambersburgh, Pa	Mar. 31, 1827	....
Paper, Machine for washing rags in the manufacture of.	R. Carter	Elkton, Md	Feb. 22, 1838	615
Paper-machine, Fourdrinier	C. Hofmann	Elkton, Md	May 4, 1869	89, 766
Paper-machine suction-apparatus	W. McLaughlin	Lee, Mass	Apr. 22, 1873	138, 173
Paper-machine suction-box	F. Curtis	Foxborough, Mass	July 2, 1872	128, 469
Paper-machine vat, Cylinder	A. Howland	Sandy Hill, N. Y	Oct. 20, 1868	83, 165
Paper-machines, &c., Apparatus for controlling the motion of traveling webs in.	F. Thiry	Huy, Belgium	Dec. 21, 1867	72, 564
Paper-machines, Apparatus for removing and replacing the dandy-rolls of.	J. F. Marshall	Louisville, Ky	Feb. 13, 1872	123, 573
Paper-machines, Apparatus for repairing knotters, strainers, or screen-plates of.	J. Robertson	Lasswade, Great Britain	June 24, 1873	140, 166
Paper-machines, Bearings for the table-rolls of Fourdrinier.	G. S. Barton	Worcester, Mass	Mar. 29, 1870	101, 345
Paper-machines, Dandy-roller for	J. Whitehead	Manchester, England	Oct. 21, 1873	143, 801
Paper-machines, Elastic apron for	A. B. Lovell	Pomfret, N. Y	Oct. 6, 1868	82, 854
Paper-machines, Jacket-stretcher for couch-roller of	R. L. Howe	Westbrook, Me	May 24, 1864	42, 854
Paper-machines, Rubber roller for	R. A. Kelty	Lambertsville, N. J	Feb. 14, 1871	111, 751
Paper-machines, Stuff-regulator for	D. Hamel	Holyoke, Mass	Nov. 25, 1873	144, 902
Paper-machinery	H. Chapman	Catawissa, Pa	Dec. 5, 1865	51, 293
Paper, Machinery for cutting rags for making	A. S. Woodward and B. F. Bartlett.	Lowell and Pepperell, Mass	Oct. 31, 1854	11, 882
Paper, Machinery for measuring pulp in the manufacture of.	H. Pohl	Paterson, N. J	July 9, 1850	7, 497
Paper, Machinery for manufacturing	S. Rossman	Stuyvesant, N. Y	Jan. 5, 1858	19, 045
Paper, Machinery for manufacturing composite	W. Campbell	Millburn, N. J	July 6, 1869	92, 161
Paper, Machinery for piling	J. C. Kneeland	Northampton, Mass	July 27, 1858	21, 004
Paper, Machinery for plating or finishing sheets of.	C. T. Bainbridge	Brooklyn, N. Y	Sept. 5, 1865	49, 691
Paper, Machinery for separating sand, &c., from pulp in the manufacture of.	W. Bishop	Coventry, Conn	Dec. 31, 1845	4, 341
Paper-machinery, Screen-plate for	F. Curtis	Auburndale, Mass	Mar. 19, 1867	62, 942
Paper-machinery, Suction-box of	C. J. Bradbury	Charlestown, Mass	Oct. 1, 1872	131, 732
Paper-maker's felt, Washing and cleansing	S. E. Foster	Brattleborough, Vt	May 25, 1832	....
Paper-maker's molds, Twisting or bobbing machine for weaving the wire-work of.	P. Bernard	Whitestown, N. Y	July 7, 1809	....
Paper, Making	C. Forbes	East Hartford, Conn	Feb. 20, 1836	....
Paper, Making	J. M. and L. Hollingsworth	Boston, Mass	Dec. 4, 1843	3, 362
Paper, Making	J. Truman	Bridgeport, Pa	Feb. 28, 1834	....
Paper, Making	L. W. Wright	London, England	Mar. 27, 1847	5, 041
Paper, &c., Making cane-hemp for making	B. C. Smith	Burlington, N. J	Nov. 21, 1843	3, 349
Paper-making, Filtering for	T. Trench	Ithaca, N. Y	May 26, 1832	....
Paper-making machine	J. Ames	Springfield, Mass	May 14, 1822	....
Paper-making machine	E. B. Bingham	Brooklyn, N. Y	June 30, 1857	17, 663
Paper-making machine	J. Burns and J. Campbell	Bloomfield, N. J	Mar. 26, 1872	124, 881
Paper-making machine	P. Clark	Rahway, N. J	Aug. 4, 1857	17, 917

Index of patents issued from the United States Patent Office from 1790 *to* 1873, *inclusive*—Continued.

Invention.	Inventor.	Residence.	Date.	No.
Paper-making machine	E. N. Foote	Saratoga Springs, N. Y	Nov. 22, 1864	45, 149
Paper-making machine	E. T. Ford	Stillwater, N. Y	Nov. 3, 1868	83, 617
Paper-making machine	E. T. Ford	Stillwater, N. Y	July 13, 1869	92, 596
Paper-making machine	R. C. Harris	Harrisville, N. J	June 29, 1869	91, 842
Paper-making machine	I. Jennings	Fairfield, Conn	July 7, 1868	79, 659
Paper-making machine	N. Keely	Philadelphia, Pa	July 29, 1873	141, 358
Paper-making machine	C. Kinsey	Essex, N. J	May 8, 1807	....
Paper-making machine	T. Lindsay and W. Geddes	Westville and Seymour, Conn.	July 27, 1858	21, 008
Paper-making machine	O. Marland	Boston, Mass	Dec. 5, 1854	12, 028
Paper-making machine	O. Marland	Boston, Mass	Dec. 5, 1854	12, 027
Paper-making machine	T. Nugent	Whippany, N. J	Mar. 12, 1872	124, 612
Paper-making machine	J. B. Pignatelli	New York	Dec. 2, 1819	....
Paper-making machine	G. E. Rutledge	Dayton, Ohio	May 26, 1863	38, 698
Paper-making machine	J. Scanlan	Lebanon, Pa	June 20, 1865	48, 347
Paper-making machine	C. B. Van Valkenburgh	Valatie, N. Y	June 1, 1869	90, 898
Paper-making machine	J. Viney	Manchester, N. H	Nov. 17, 1868	84, 235
Paper-making machine	E. Wilmont	Laona, N. Y	Nov. 19, 1867	71, 108
Paper-making machines, Drier-felts for	S. W. Baker	Providence, R. I	Oct. 10, 1865	50, 323
Paper-making machines, Feeding pulp to	I. Kinsey	Hobokus, N. J	Oct. 7, 1856	15, 852
Paper-making machines, Self-acting felt-guide for	T. Baker	Stillwater, N. Y	June 28, 1864	43, 280
Paper-making machines, Vacuum-box of	J. L. Seaverns	Worcester, Mass	Aug. 11, 1863	39, 500
Paper-making machinery	E. B. Bingham	Newark, N. J	Sept. 18, 1866	58, 051
Paper-making machinery	J. S. Blake	Claremont, N. H	Jan. 20, 1857	16, 430
Paper-making machinery	G. Burbank	Worcester, Mass	Apr. 12, 1826	....
Paper-making machinery	W. W. Harding	Philadelphia, Pa	Mar. 15, 1870	100, 755
Paper-making machinery	J. Harper	East Haven, Conn	Mar. 11, 1862	34, 633
Paper-making machinery	S. G. and G. S. Rogers	Thetford, Vt	Nov. 13, 1866	59, 661
Paper-making machinery	G. J. Wheeler, G. W. Dunnell, and W. Sharp.	Bloomfield, N. J	Jan. 22, 1861	31, 215
Paper-making machinery, Rag-engine of	T. Lindsay	Montville, Conn	May 16, 1865	47, 739
Paper-making, Manufacture of suction-boxes for	F. Curtis	Newton, Mass	Nov. 5, 1867	70, 534
Paper-making, Preparing resin size for use in	T. Gray	London, England	Nov. 3, 1868	83, 707
Paper-making, Rag engine for	J. M. Shew	Paper Mills, Md	Dec. 18, 1866	60, 645
Paper-making roll and drier	O. Marland	Boston, Mass	Feb. 27, 1855	12, 442
Paper, Making, ruling, and cutting	J. Ames	Springfield, Mass	July 31, 1840	1, 709
Paper, Making the leaves of books from a continuous sheet of.	T. T. Morrell	Farmington, N. H	Oct. 21, 1862	36, 722
Paper, Making thick	S. G. Levis	Delaware County, Pa	Feb. 14, 1854	10, 519
Paper-making, Washing rags for	J. Ames	Springfield, Mass	Apr. 6, 1831	....
Paper-making, Washing rags for	S. Eckstein	Philadelphia, Pa	June 13, 1831	....
Paper-manufacture and treatment of paper-pulp	A. T. Schmidt	Pittsburgh, Pa	Jan. 15, 1867	61, 267
Paper, Manufacture of	J. M. Allen	Marion, Mass	Aug. 8, 1865	49, 209
Paper, Manufacture of	S. D. Baldwin	Marysville, Cal	Jan. 9, 1872	122, 548
Paper, Manufacture of	E. B. Bingham	Newark, N. J	Feb. 22, 1870	100, 109
Paper, Manufacture of	J. W. Dixon	Philadelphia, Pa	Dec. 19, 1865	51, 568
Paper, Manufacture of	L. Dodge	Waterford, N. Y	June 14, 1870	104, 281
Paper, Manufacture of	L. Dodge	Waterford, N. Y	June 14, 1870	104, 282
Paper, Manufacture of	E. T. Ford	Stillwater, N. Y	Nov. 3, 1868	83, 616
Paper, Manufacture of	I. Hoffman	Oregon, N. Y	Apr. 26, 1870	102, 265
Paper, Manufacture of	G. W. Hurlbut	Fair Haven, Vt	Oct. 16, 1866	58, 944
Paper, Manufacture of	C. B. Hutchins	Ann Arbor, Mich	Nov. 29, 1870	109, 621
Paper, Manufacture of	J. S. Kenyon and L. Fox	Baldwinsville, N. Y	Nov. 1, 1870	108, 913
Paper, Manufacture of	G. E. Marshall	Laurel, N. Y	July 19, 1870	105, 585
Paper, Manufacture of	S. Nowlan	New York, N. Y	Aug. 16, 1864	43, 860
Paper, Manufacture of	C. E. O'Hara	New York, N. Y	Oct. 18, 1870	108, 509
Paper, Manufacture of	H. Pemberton	Allegheny City, Pa	July 19, 1870	105, 594
Paper, Manufacture of	H. Pemberton	Allegheny City, Pa	Oct. 11, 1870	108, 177
Paper, Manufacture of	J. Pickles	Wigan, England	June 15, 1860	01, 180
Paper, Manufacture of	J. B. Read	Tuscaloosa, Ala	Dec. 26, 1865	51, 751
Paper, Manufacture of	J. B. Read	Tuscaloosa, Ala	Aug. 24, 1869	94, 131
Paper, Manufacture of	A. E. Reed	Wookey, England	July 16, 1872	128, 978
Paper, Manufacture of	W. A. Russell	Lawrence, Mass	Aug. 14, 1866	57, 192
Paper, Manufacture of	T. P. Shaffner	Louisville, Ky	Oct. 30, 1866	59, 281
Paper, Manufacture of	A. and N. A. Sprague	Fredonia, N. Y	Oct. 31, 1828	....
Paper, Manufacture of	J. H. Tiemann	New York, N. Y	Feb. 13, 1872	123, 747
Paper, Manufacture of	Z. C. Warren	Brooklyn, N. Y	Mar. 23, 1869	88, 102
Paper, Manufacture of	R. Waterman and G. W. Annis	Providence, R. I	Aug. 30, 1828	....
Paper, Manufacture of lace	C. Lang	Bergen, N. J	Apr. 17, 1866	53, 991
Paper, Manufacture of parchment	A. J. Sheldon	Buffalo, N. Y	Jan. 25, 1870	99, 248
Paper, Manufacture of printing and writing	W. W. Harding	Philadelphia, Pa	Jan. 19, 1869	86, 012
Paper, Manufacture of tarred	H. F. Evans	Beloit, Wis	Feb. 2, 1869	86, 380
Paper, Manufacture of tobacco	R. Antigüedad	New York, N. Y	July 12, 1870	105, 160
Paper, Manufacture of tobacco	P. M. Consuegra and R. Antigüedad.	New York, N. Y	July 13, 1869	92, 427
Paper, Manufacture of water and fire proof	T. Irving, J. McNeil, G. W. Rich, and C. J. Fay.	Elwood, N. J	Dec. 18, 1866	60, 635
Paper, Manufacture of water-proof	S. M. Allen	Woburn, Mass	Mar. 13, 1866	53, 094
Paper, Manufacturing	J. Ames	Springfield, Mass	May 14, 1832	....
Paper, Manufacturing	C. Austin	....	Dec. 14, 1798	....
Paper, Manufacturing	J. Biddis	....	Mar. 31, 1794	....
Paper, Manufacturing	I. Burbank	Worcester, Mass	Sept. 8, 1824	....
Paper, Manufacturing	S. S. Crocker and G. E. Marshall.	Lawrence, Mass	June 14, 1859	24, 377
Paper, Manufacturing	R. R. Livingston	....	Oct. 28, 1799	....
Paper, Manufacturing	S. Stimpson	Newbury, Vt	Mar. 12, 1831	....
Paper, Manufacturing	S. Wheeler	Albany, N. Y	July 16, 1872	128, 992
Paper, Manufacturing copying	W. Mann	Philadelphia, Pa	Jan. 11, 1853	9, 536
Paper, Marbleizing	T. Carson	Brooklyn, N. Y	Mar. 14, 1871	112, 544
Paper, Mark of opacity in	J. Reich and E. Star	Philadelphia, Pa	May 11, 1816	....
Paper, Marking and ornamenting	T. Mackenzie and A. Trochsler	Boston, Mass	Mar. 22, 1859	23, 304
Paper, Material for making	J. J. Gillet-Damitte, H. D. Dubois, and A. Boissonneau.	Paris, France	Feb. 16, 1869	86, 917
Paper, Material for making	W. Magaw	Meadville, Pa	Mar. 8, 1828	....
Paper-mill	T. Langstroth	....	May 1, 1804	....
Paper-mill engines, Bed-plate for	G. A. Corser	Leicester, Mass	June 7, 1864	43, 070
Paper-mill engines, Bed-plate for	O. Morse	Needham Lower Falls, Mass.	May 32, 1865	47, 849

Index of patents issued from the United States Patent Office from 1790 *to* 1873, *inclusive*—Continued.

Invention.	Inventor.	Residence.	Date.	No.
Paper-mills, Machine for cleaning rags for	W. Debit	East Hartford, Conn	Jan. 13, 1829	....
Paper-mills, Self-adjusting guide-roll for	R. L. Howe	Westbrook, Me	Mar. 19, 1867	62, 958
Paper-mold	W. Brewer and J. Smith	Surrey County, England	Mar. 4, 1851	7, 959
Paper-mold	J. Carnes, jr	....	Apr. 11, 1793	....
Paper-molding	W. W. Webster	Chelsea, Mass	Jan. 3, 1871	110, 704
Paper molding, cornice, &c	L. W. Kimball	Pittsford, Vt	Dec. 29, 1868	87, 312
Paper, moth-repellent	S. Crane	Dalton, Mass	Mar. 5, 1872	124, 336
Paper neck-tie	E. Child, jr	Springfield, Mass	May 22, 1866	54, 858
Paper neck-tie	H. Whitney	Watertown, Mass	July 23, 1867	67, 091
Paper neck-tie	J. Sangster and O. W. Seely	Buffalo, N. Y	May 8, 1866	54, 607
Paper on straw-board, &c., Machine for pasting	M. Fitzgibbons	New York, N. Y	June 20, 1871	116, 173
Paper or paper and cloth neck-ties	R. L. Walter	Washington, D. C	June 14, 1870	104, 382
Paper or linen, Machine for preparing and backling tow for.	J. Tatterson	Southampton, Long Island, N. Y.	Dec. 7, 1805	....
Paper, Packing in cylinders for drying	W. B. Fowler	Lawrence, Mass	June 8, 1869	91, 007
Paper pad	S. Van Campen	New York, N. Y	Oct. 1, 1872	131, 915
Paper pail	P. C. Schuyler	New York, N. Y	Apr. 30, 1872	126, 335
Paper pan, bowl, box, and dish	J. W. Jarboe	Green Point, N. Y	Jan. 25, 1870	99, 290
Paper pantalets	E. P. Furlong	Portland, Me	Jan. 15, 1867	61, 187
Paper, pasteboard, &c., Machine for cutting	S. D. Tucker	New York, N. Y	Oct. 29, 1867	70, 292
Paper-perforating apparatus, Telegraph	L. Bradley	Jersey City, N. J	June 27, 1865	48, 479
Paper-perforating apparatus, Telegraph	T. A. Edison	Newark, N. J	Oct. 22, 1872	132, 456
Paper-perforating apparatus, Telegraph	G. Little	Hudson City, N. J	June 15, 1869	91, 240
Paper-perforating apparatus, Telegraph	G. Little	Rutherford Park, N. J	Nov. 2, 1869	96, 330
Paper-perforating apparatus, Telegraph	G. Little	Rutherford Park, N. J	Nov. 2, 1869	96, 331
Paper-perforating machine	G. C. Howard	Philadelphia, Pa	July 2, 1861	32, 693
Paper-perforator	J. C. Gaston	Cincinnati, Ohio	July 13, 1869	92, 600
Paper-planishing machine	J. F. Schuyler	Philadelphia, Pa	Feb. 24, 1863	37, 790
Paper-polisher, Composition	H. T. Cushman	North Bennington, Vt	June 1, 1869	90, 733
Paper polishing and calendering machine	P. Coburn	East Walpole, Mass	May 31, 1870	103, 569
Paper-polishing machine	E. L. Perkins	Roxbury, Mass	Oct. 24, 1854	11, 837
Paper-polishing machine	S Shepard and A. M. George	Nashua, N. H	Oct. 6, 1868	82, 882
Paper-polishing machine	A. and G. F. Wright	Clinton, Mass	Jan. 10, 1871	110, 947
Paper, Preparation of	J. L. Jullion	Aberdeen, Great Britain	May 18, 1860	28, 182
Paper, Preparation of copying	C. Cowan	New York, N. Y	May 4, 1869	89, 738
Paper, Preparation of fiber for the manufacture of	J. E. Mallory	New York, N. Y	Mar. 26, 1861	31, 814
Paper, Preparation of paper-pulp and manufacture of.	J. Denis	London, England	Apr. 4, 1871	113, 502
Paper, Preparation of straw for the manufacture of.	L. Dean	Fort Edward, N. Y	Nov. 29, 1870	109, 596
Paper, Preparing albumenized	H. F. Anthony	New York, N. Y	Mar. 1, 1864	41, 750
Paper, &c., Preparing corn-husks for the manufacture of.	H. Holland	Westfield, Mass	Aug. 13, 1838	878
Paper, &c., Preparing vegetable fiber for	J. B. Fuller	Claremont, N. H	Nov. 17, 1863	40, 659
Paper, Preparing vegetable substances for making	W. Magaw	Meadville, Pa	May 22, 1828	....
Paper-press	W. R. Dingman	Stuyvesant Falls, N. Y	Oct. 20, 1863	40, 336
Paper-press cylinder, Hot and cold	A. H. Jervis and T. Trench	Ithaca, N. Y	Nov. 6, 1832	....
Paper-press, Vertical-cutting	B. Morris	Oxford, N. Y	Feb. 15, 1835	....
Paper, Printing and cutting	J. F. Tapley	Springfield, Mass	Mar. 4, 1862	34, 601
Paper, Printing	J. E. Hover	Philadelphia, Pa	Apr. 16, 1867	63, 896
Paper, Producing designs on	W. B. Woodbury	London, England	Apr. 28, 1868	77, 230
Paper-product	J. L. Kendall	Foxborough, Mass	Nov. 11, 1873	144, 548
Paper-punching machine	S. M. Clark	Washington, D. C	Apr. 30, 1867	64, 197
Paper-puncturing device	T. F. Corry	Madison, Ind	Dec. 3, 1872	133, 567
Paper-pulp	J. W. Dixon	Philadelphia, Pa	Dec. 12, 1865	51, 433
Paper-pulp	J. W. Dixon	Philadelphia, Pa	Dec. 19, 1865	51, 572
Paper-pulp	C. C. Fitzgerald	Phœnix, N. Y	Dec. 22, 1868	85, 079
Paper-pulp	C. B. Sawyer	Fitchburgh, Mass	Nov. 29, 1870	109, 766
Paper-pulp	P. F. Schliecker	Baltimore, Md	July 4, 1871	116, 759
Paper-pulp and bleaching apparatus	J. W. Dixon	Philadelphia, Pa	June 26, 1866	55, 834
Paper pulp and fiber, Treating plants to produce	J. Dupont	Nimes, France	Aug. 6, 1872	130, 114
Paper-pulp, Apparatus for making	J. W. Dixon	Philadelphia, Pa	Feb. 13, 1866	52, 543
Paper-pulp, Apparatus for making	J. W. Dixon	Philadelphia, Pa	Feb. 13, 1866	52, 544
Paper-pulp, Apparatus for making	H. and F. Marx	Baltimore, Md	Oct. 23, 1866	59, 042
Paper-pulp, Apparatus for manufacture of	J. W. Dixon	Philadelphia, Pa	Dec. 12, 1865	51, 430
Paper-pulp, Apparatus for manufacture of	J. W. Dixon	Philadelphia, Pa	Dec. 12, 1865	51, 431
Paper-pulp, Apparatus for manufacture of	J. W. Dixon	Philadelphia, Pa	Feb. 20, 1866	52, 694
Paper-pulp, Apparatus for manufacture of	J. W. Dixon and G. Harding	Philadelphia, Pa	May 8, 1866	54, 510
Paper-pulp, Apparatus for manufacture of	J. Easton, jr., and F. Thiry	Grove Southwark, England, and Huy, Belgium.	Feb. 27, 1866	52, 941
Paper-pulp, Apparatus for manufacture of	M. L. Keen	Jersey City, N. J	May 2, 1871	114, 301
Paper-pulp, Apparatus for manufacture of	H. B. Meech	Troy, N. Y	July 11, 1871	116, 980
Paper-pulp, Apparatus for manufacture of	J. B. Palser and G. Howland	Fort Edward, N. Y	June 21, 1859	24, 484
Paper-pulp, Apparatus for molding hollow articles from.	J. L. Kendall and R. H. Trested	Foxborough, Mass., and New York, N. Y.	Apr. 16, 1872	125, 740
Paper-pulp, Apparatus for reducing wood to	H. Marx	Pikesville, Md	Dec. 1, 1868	84, 640
Paper-pulp, Apparatus for straining	L. Hollingsworth	Boston, Mass	July 2, 1872	128, 625
Paper-pulp-beating engine	G. Ames	New York, N. Y	Aug. 15, 1871	118, 092
Paper-pulp, Bleaching	J. Campbell	Chatham Village, N. Y	Apr. 16, 1872	125, 658
Paper-pulp, Bleaching	J. W. Dixon	Philadelphia, Pa	Dec. 19, 1865	51, 569
Paper-pulp, Bleaching	W. C. Joy and J. Campbell	Penn Yan, N. Y	July 2, 1867	66, 353
Paper-pulp, Bleaching	J. G. and J. H. Kendall	Leominster, Mass	July 2, 1846	4, 616
Paper-pulp-bleaching apparatus	H. L. Jones and D. S. Farquharson.	Rochester, N. Y	Mar. 13, 1866	53, 152
Paper-pulp, Boiler for making	M. L. Keen	Royer's Ford, Pa	June 16, 1863	38, 901
Paper-pulp, Boiler for preparing	L. Dean	Fort Edward, N. Y	Nov. 29, 1870	109, 595
Paper-pulp boiler or digester	W. F. Ladd	New York, N. Y	May 30, 1871	115, 327
Paper-pulp boilers, Device for filling and packing rotary.	A. Frickett	Rochester, N. Y	Dec. 3, 1867	71, 728
Paper-pulp, Disintegrating and bleaching wood, &c., to form.	J. B. Biron	Carpentras, France	Aug. 20, 1867	67, 941
Paper-pulp, Disintegrating fibrous materials for	A. H. F. Diminger	Berlin, Prussia	Oct. 11, 1870	108, 241
Paper-pulp, Disintegrating vegetable fiber for	W. Riddell	London, England	Aug. 1, 1871	117, 683
Paper-pulp, Disintegrating vegetable substances for.	Z. G. A. W. P. Oriolé, A. A. Fredet, and P. A. H. Matussiere.	Paris, France	Apr. 25, 1865	47, 505
Paper-pulp, Disintegrating woody fiber for	T. B. Armitage	New York, N. Y	Sept. 9, 1873	142, 550
Paper-pulp-dressing machine	N. Hebard	Dorchester, Mass	Dec. 27, 1839	1, 441
Paper-pulp-dressing machine	G. E. Sellers	Seller's Landing, Ill	July 2, 1867	66, 258

Index of patents issued from the United States Patent Office from 1790 *to* 1873, *inclusive*—Continued.

Invention.	Inventor.	Residence.	Date.	No.
Paper-pulp engine	S. L. Gould	Skowhegan, Me	Oct. 24, 1871	120, 265
Paper-pulp engine	J. Hatch	South Windham, Conn	Dec. 12, 1871	121, 780
Paper-pulp engine	J. Kingsland, jr	Franklin, N. J	Dec. 23, 1856	16, 278
Paper-pulp engine	H. B. Meech	Fort Edward, N. Y	July 11, 1871	116, 978
Paper-pulp engine	S. Moore and R. H. Hurlbut	Sudbury, Mass	Nov. 11, 1873	144, 557
Paper-pulp engine	T. Nugent	Whippany, N. J	July 30, 1872	130, 067
Paper-pulp engine	W. Parkison	Monongahela City, Pa	Feb. 9, 1869	86, 858
Paper-pulp engine	P. Rose	Norwich, Conn	Sept. 14, 1869	94, 843
Paper-pulp engine	C. Smith	South Windham, Conn	Dec. 19, 1871	121, 970
Paper-pulp engines, Bed-plate for	P. Frost	Medfield, Mass	June 20, 1871	116, 045
Paper-pulp engines, Grinding-plate for	P. Frost	Medfield, Mass	Sept. 14, 1869	94, 816
Paper-pulp for packing and transportation, Preparing	W. H. Merrick	Philadelphia, Pa	July 12, 1870	105, 354
Paper-pulp from beet and other refuse, Preparing	R. H. Collyer	Camden, N. J	Oct. 13, 1857	18, 389
Paper-pulp from coloring-matter, Cleansing	S. W. Wilder	Lawrence, Mass	Feb. 8, 1870	99, 735
Paper-pulp from corn-stalks, Making	J. W. Dixon	Philadelphia, Pa	Dec. 12, 1865	51, 432
Paper-pulp from ivory, Making	W. N. Clark	Chester, Conn	Sept. 15, 1857	18, 190
Paper-pulp from reeds, &c	R. W. Russell and T. Howland	Brooklyn and Stockport, N. Y.	Nov. 5, 1867	70, 474
Paper-pulp from reeds, Preparing	H. Lowe	Baltimore, Md	May 25, 1858	20, 355
Paper-pulp from straw, Manufacture of	C. M. Cresson	Philadelphia, Pa	July 11, 1871	116, 933
Paper-pulp from straw, &c., Manufacture of	A. K. Eaton	New York, N. Y	Mar. 22, 1864	41, 982
Paper-pulp from straw, Manufacture of	A. K. Haxstun	Fort Edward, N. Y	Mar. 6, 1866	52, 994
Paper-pulp from straw, &c., Manufacture of	J. A. Rothe	Philadelphia, Pa	Nov. 21, 1871	121, 130
Paper-pulp from straw, Preparing	J. Priestly and T. C. Bradbury	New York and Poughkeepsie N. Y.	June 19, 1866	55, 706
Paper-pulp from straw, &c., Preparing	J. Tiffany	Albany, N. Y	Aug. 27, 1867	68, 261
Paper-pulp from wood	H. Voelter	Heidenheim, Germany	May 22, 1866	55, 031
Paper-pulp from wood, Apparatus for making	J. Bridge	Augusta, Me	Jan. 2, 1872	122, 353
Paper-pulp from wood, Apparatus for making	H. W. Higley	Curtisville, Mass	Aug. 27, 1872	130, 803
Paper-pulp from wood, Boiler for making	M. L. Keen	Royer's Ford, Pa	Sept. 13, 1859	25, 418
Paper-pulp from wood, Making	A. Meucci	Clifton, N. Y	Mar. 13, 1866	53, 165
Paper-pulp from wood, Manufacture of	J. W. Dixon	Philadelphia, Pa	Dec. 20, 1864	45, 480
Paper-pulp from wood, Manufacture of	C. Marzoni	New York, N. Y	Dec. 21, 1858	22, 401
Paper-pulp from wood, Manufacture of	H. B. Meech	Fort Edward, N. Y	Aug. 23, 1870	106, 710
Paper-pulp from wood, straw, &c., Apparatus for making.	J. W. Dixon	Philadelphia, Pa	June 26, 1866	55, 835
Paper-pulp from wood, straw, &c., Making	J. W. Dixon	Philadelphia, Pa	Feb. 13, 1866	52, 545
Paper-pulp from wood, straw, &c., Making	J. W. Dixon	Philadelphia, Pa	Feb. 13, 1866	52, 546
Paper-pulp from wood, Manufacture of	G. Vining	Pittsfield, Mass	Dec. 21, 1869	98, 210
Paper-pulp grinder, Wood	M. S. and M. E. Otis	Rochester, N. Y	Nov. 4, 1873	144, 354
Paper-pulp, Grinding	J. Kingsland, jr	Franklin, N. J	Dec. 23, 1856	16, 316
Paper-pulp-grinding and sizing machine	J. Jordan, jr., and T. Eustice	East Hartford and Hartford, Conn.	May 18, 1858	20, 277
Paper-pulp-grinding machine	J. F. Jones	Rochester, N. Y	Apr. 25, 1865	47, 425
Paper-pulp-grinding machine, Wood	J. G. Moore	Lisbon, N. H	Nov. 19, 1872	133, 243
Paper-pulp-grinding machinery	J. Kingsland, jr	Franklin, N. J	Dec. 16, 1856	16, 239
Paper-pulp-grinding mill	J. Jordan, jr	East Hartford, Conn	Feb. 5, 1861	31, 321
Paper-pulp-grinding mill	G. Sanford	New York, N. Y	Feb. 12, 1861	31, 408
Paper pulp in vats, Heating	J. Robeson	Montgomery, Pa	May 22, 1810	....
Paper-pulp machine	J. M. Burghardt and F. Burghardt.	Great Barrington and Curtisville, Mass.	July 9, 1872	128, 788
Paper-pulp, Machine for making	W. Miller	Herkimer, N. Y	May 12, 1868	77, 829
Paper-pulp, Machine for making	J. Taggart	Roxbury, Mass	Mar. 31, 1868	76, 270
Paper-pulp, Machine for disintegrating wood for	H. Dodge	Albany, N. Y	Jan. 25, 1870	99, 071
Paper-pulp, Machine for mixing coloring-matter with.	J. Wrinkle	Lee, Mass	Dec. 22, 1868	85, 157
Paper-pulp, Machine for preparing wood for the manufacture of.	F. Burghardt	Curtisville, Mass	Feb. 23, 1869	87, 139
Paper-pulp, Machine for reducing wood for the manufacture of.	J. Bridge	Augusta, Me	Apr. 4, 1871	113, 488
Paper-pulp machines, Water-regulator for	D. Hunter	North Bennington, Vt	Dec. 29, 1868	85, 386
Paper-pulp, Machinery for cleaning	S. S. Crocker	Lawrence, Mass	Apr. 15, 1862	34, 945
Paper-pulp, Machinery for making	L. Koch	New York, N. Y	Aug. 7, 1855	13, 412
Paper-pulp, Machinery for reducing wood to fiber for.	C. D. Negri	Hornsey Road, Great Britain.	Oct. 8, 1872	131, 944
Paper-pulp, Manufacture of	H. E. Bailliere	Hoboken, N. J	Jan. 21, 1868	73, 429
Paper-pulp, Manufacture of	L. Bardoux	Poitiers, France	Apr. 5, 1864	42, 155
Paper-pulp, Manufacture of	C. D. Brown and E. B. Denison	Portland, Me	Dec. 16, 1873	145, 620
Paper-pulp, Manufacture of	E. Clems	Toronto, Canada	July 10, 1860	29, 059
Paper-pulp, Manufacture of	F. De Campoloro	France	Aug. 7, 1860	29, 471
Paper-pulp, Manufacture of	G. Demailly	Argenteuil, France	Mar. 5, 1872	124, 196
Paper-pulp, Manufacture of	J. W. Dixon	Philadelphia, Pa	Dec. 6, 1864	45, 321
Paper-pulp, Manufacture of	J. W. Dixon	Philadelphia, Pa	Dec. 19, 1865	51, 570
Paper-pulp, Manufacture of	J. W. Dixon	Philadelphia, Pa	Dec. 19, 1865	51, 571
Paper-pulp, Manufacture of	J. W. Dixon	Philadelphia, Pa	Jan. 2, 1866	51, 813
Paper-pulp, Manufacture of	J. W. Dixon	Philadelphia, Pa	May 1, 1866	54, 308
Paper-pulp, Manufacture of	A. K. Eaton	Piermont, N. Y	Aug. 9, 1870	106, 143
Paper-pulp, Manufacture of	A. K. Eaton	Brooklyn, N. Y	Sept. 26, 1871	119, 224
Paper-pulp, Manufacture of	D. A. Fyfe	Manchester, England	Oct. 1, 1872	131, 749
Paper-pulp, Manufacture of	H. Glynn	Baltimore, Md	Feb. 6, 1855	12, 361
Paper-pulp, Manufacture of	M. L. Keen	Jersey City, N. J	Oct. 3, 1871	119, 464
Paper-pulp, Manufacture of	H. B. Meech	Fort Edward, N. Y	May 22, 1866	54, 932
Paper-pulp, Manufacture of	H. B. Meech	Troy, N. Y	July 11, 1871	116, 979
Paper-pulp, Manufacture of	M. A. C. Mellier	Paris, France	May 26, 1857	17, 387
Paper-pulp, Manufacture of	J. B. Palser and G. Howland	Fort Edward, N. Y	Nov. 22, 1859	26, 202
Paper-pulp, Manufacture of	C. A. Rose	Columbus, Ga	May 7, 1867	64, 449
Paper-pulp, Manufacture of	J. A. Roth	Philadelphia, Pa	Aug. 15, 1865	49, 480
Paper-pulp, Manufacture of	C. F. Sturgis	Dallas County, Ala	Mar. 31, 1857	16, 949
Paper-pulp, Manufacture of	A. H. Tait and W. H. Holbrook	Jersey City, N. J., and New York, N. Y.	Nov. 24, 1863	40, 728
Paper-pulp, Manufacture of	F. W. Zanders	Erfert, Germany	Feb. 13, 1872	123, 757
Paper-pulp, Manufacture of articles from	F. Curtis	Foxborough, Mass	Jan. 10, 1871	110, 833
Paper-pulp, Material for manufacture of	A. De Gogorza	New York, N. Y	Jan. 2, 1866	51, 810
Paper-pulp, Material for manufacture of	H. V. P. Draper	Hannibal, Mo	Sept. 19, 1871	119, 084
Paper-pulp, Molding articles from	E. H. Knight	Washington, D. C	Apr. 3, 1866	53, 631
Paper pulp or stock	H. Lowe	Baltimore, Md	Aug. 20, 1861	33, 092
Paper pulp or stock, Manufacture of	M. L. Keen	Jersey City, N. J	Oct. 3, 1871	119, 465
Paper-pulp, Preparation of fibers for	M. Nixon	Philadelphia, Pa	May 18, 1858	20, 294

Index of patents issued from the United States Patent Office from 1790 *to* 1873, *inclusive*—Continued.

Invention.	Inventor.	Residence.	Date.	No.
Paper-pulp, Preparation of straw for	J. B. Palser and G. Howland	Fort Edward, N. Y	Mar. 20, 1860	27, 564
Paper-pulp, Preparing and bleaching	J. Campbell	Chatham Village, N. Y	June 20, 1871	116, 020
Paper-pulp, Preparing and bleaching	W. C. Joy and J. Campbell	Penn Yan, N. Y	Mar. 16, 1869	87, 779
Paper-pulp, Preparing	J. Mayrhofer	New York, N. Y	Nov. 1, 1859	25, 975
Paper-pulp, Preparing sawdust for	H. B. Meech	Fort Edward, N. Y	Dec. 20, 1864	45, 510
Paper-pulp, Preparing wood for	M. D. Whipple	Charlestown, Mass	May 29, 1855	12, 978
Paper-pulp, Recovering waste alkali used in treating.	C. M. T. Du Motay	Paris, France	Oct. 22, 1872	132, 452
Paper-pulp, Reducing straw and other fibrous substances for the manufacture of.	R. Sherwood	Fort Edward, N. Y	Dec. 13, 1864	45, 440
Paper-pulp, Reducing wood and woody fiber for	V. E. Keegan	Boston, Mass	Oct. 26, 1869	96, 239
Paper-pulp, Reducing wood fiber to	H. Voelter	Heidenheim, Würtemburg, Germany.	Aug. 10, 1858	21, 161
Paper-pulp, Rotary boiler for the manufacture of	H. B. Meech	Fort Edward, N. Y	Jan. 10, 1865	45, 845
Paper-pulp screen, Metallic	A. St. Clair	Boston, Mass	July 26, 1870	105, 755
Paper-pulp screen, Metallic	A. S. Winchester	Boston, Mass	Mar. 2, 1869	87, 385
Paper-pulp, Separating fibers of palm, palmetto, &c., for the manufacture of.	J. W. Dixon	Philadelphia, Pa	May 1, 1866	54, 309
Paper-pulp, Separating fibers of wood, &c., for the manufacture of.	A. S. Lyman	New York, N. Y	Mar. 4, 1862	34, 581
Paper-pulp, Treating fibrous substances for	A. Ungurer	Pforzheim, Germany	Oct. 7, 1873	143, 546
Paper-pulp, Treating hemp, flax, &c., for the manufacture of.	M. A. Cushing	Glen's Falls, N. Y	Oct. 10, 1865	50, 419
Paper-pulp, Treating straw and other materials for manufacture of.	H. B. Meech	Fort Edward, N. Y	Feb. 5, 1867	61, 848
Paper-pulp, Treating straw for	H. B. Meech	Fort Edward, N. Y	Nov. 7, 1865	50, 835
Paper-pulp, Treating straw for	F. A. Nixon	Philadelphia, Pa	Apr. 11, 1865	47, 217
Paper-pulp, Treating vegetable fiber for manufacture of.	J. W. Dixon	Philadelphia, Pa	Dec. 26, 1865	51, 706
Paper-pulp, &c., Treating vegetable substances for making.	H. B. Meech	Fort Edward, N. Y	Sept. 13, 1864	44, 209
Paper-pulp, Treating vegetable substances for making.	B. C. Tilghman	Philadelphia, Pa	Nov. 5, 1867	70, 485
Paper-pulp, Treating wood for manufacture of	V. E. Keegan	Boston, Mass	Apr. 13, 1869	88, 879
Paper-pulp, Treating wood, straw, &c., for manufacture of.	H. L. Jones and D. S. Farquharson.	Rochester, N. Y	June 5, 1866	55, 418
Paper-pulp washer	J. Piercy	Bloomfield, N. J	Jan. 21, 1862	34, 214
Paper-pulp washer	L. M. Wright	Fort Edward, N. Y	May 22, 1866	54, 993
Paper-pulp-washing machine	S. and J. Deacon	Lawrence, Mass	Dec. 2, 1873	145, 159
Paper, &c., Purifying and cleaning sizing for	N. J. Wells	Huntington, Mass	Jan. 22, 1867	61, 371
Paper-rag engine	N. W. Taylor and J. W. Brightman.	Cleveland, Ohio	Nov. 14, 1871	120, 837
Paper-rag engine	J. Storm	Woonsocket, R. I	Feb. 14, 1860	57, 167
Paper-rag engine, Regulator for	J. M. Hollingsworth	Braintree, Mass	Dec. 31, 1838	1, 059
Paper-stock, Recovering soda used in the manufacture of.	H. Lowe	Baltimore, Md	Dec. 17, 1861	33, 953
Paper, Recovering waste alkali used in the manufacture of.	T. F. Lehmann	Allegheny, Pa	Apr. 10, 1866	53, 839
Paper, Register for sheets of	S. T. Bacon	Boston, Mass	Mar. 8, 1859	23, 146
Paper, Register for sheets of	J. North	Middletown, Conn	Mar. 8, 1859	23, 221
Paper, Removing ink and colors from printed	J. A. Veazie	Boston, Mass	Feb. 11, 1868	74, 260
Paper-ribbon holder	W. Orr, jr., and G. F. Wright	Clinton, Mass	Oct. 2, 1866	58, 466
Paper, Roll for pressing, sizing, and calendering	F. Curtis	Malden, Mass	Jan. 9, 1866	51, 929
Paper-ruler	D. Munson	Indianapolis, Ind	Mar. 22, 1864	42, 012
Paper-ruling machine, Striker for	J. D. Connolly	Lincoln, Nebr	Oct. 14, 1873	143, 672
Paper-sack knife	L. H. Mealey	Alpha, Ohio	July 21, 1868	80, 197
Paper, Safety	L. M. Crane	Ballston, N. Y	Jan. 22, 1867	61, 321
Paper, Safety	H. Hayward	Chicago, Ill	Mar. 11, 1862	34, 634
Paper, Safety	M. A. Howell jr	Ottawa, Ill	May 22, 1860	28, 370
Paper, Safety	J. P. Oliver	Paris, France	June 9, 1863	38, 835
Paper, Safety	J. M. Willcox	Glen Mills, Pa	July 24, 1866	56, 650
Paper-"satining" machine	T. Christy	New York, N. Y	Jan. 19, 1869	86, 058
Paper, Shears for separating	J. A. Wilkinson	Brooklyn, N. Y	Aug. 30, 1859	25, 298
Paper shirt	H. M. Remington	Springfield, Mass	May 1, 1866	54, 408
Paper-sizing	J. Ames	Springfield, Mass	Sept. 1, 1822	....
Paper-sizing	E. Blake	Alstead, N. H	Nov. 19, 1833	....
Paper, &c., Sizing	J. M. Dorlan	East Brandywine Township, Pa.	June 11, 1872	127, 858
Paper-sizing	P. Grady	New York, N. Y	Jan. 14, 1873	134, 797
Paper-sizing	J. E. Hover	Philadelphia, Pa	Mar. 23, 1869	88, 169
Paper-sizing	G. E. Van Derburgh	Mamaroneck, N. Y	Apr. 19, 1864	42, 450
Paper-sizing	Z. C. Warren and H. C. Hulbert	Brooklyn, N. Y	Sept. 14, 1869	94, 797
Paper-sizing	W. W. Wilson and C. Dickerman	Westfield, Mass	Aug. 3, 1839	1, 286
Paper, &c., Sizing and water-proofing	J. N. Sigel	Alexandria, Va	Jan. 12, 1864	41, 241
Paper, Sizing for bank-note	J. M. Sturgeon	New York, N. Y	Dec. 3, 1867	71, 663
Paper, Sizing for colored	C. Williams	Philadelphia, Pa	Feb. 1, 1859	22, 836
Paper-sizing machine	J. Ames	Springfield, Mass	Dec. 1, 1837	495
Paper-sizing machine	L. D. Brown	Lee, Mass	Mar. 4, 1842	2, 479
Paper skirt, Lady's	J. H. Hayward	New York, N. Y	Oct. 9, 1866	58, 641
Paper sock	J. W. B. Covington	New York, N. Y	Feb. 13, 1866	52, 540
Paper-staining machine	C. Bartholomew	New York, N. Y	Dec. 3, 1867	71, 681
Paper stamper and feeder	J. O. Montignani and J. Gibson, jr.	Albany, N. Y	Mar. 26, 1872	125, 070
Paper, Stamping lace	A. Rhorbeck	New York, N. Y	May 11, 1869	89, 945
Paper-stock	M. L. Keen	Jersey City, N. J	Nov. 29, 1870	109, 742
Paper-stock	H. S. Lucas	Chester, Mass	Sept. 3, 1867	68, 370
Paper-stock and other fibers, Recovering waste alkalies from.	C. D. J. Seitz	Bury, England	Nov. 3, 1868	83, 733
Paper-stock, Apparatus for boiling and treating	G. Sinclair	Leith, Scotland	Jan. 10, 1871	110, 873
Paper-stock, Apparatus for manufacturing	M. L. Keen	Jersey City, N. J	July 9, 1872	128, 732
Paper-stock, Apparatus for preparing	J. Tiffany	Albany, N. Y	Jan. 7, 1868	73, 138
Paper-stock bleach	J. W. Rossman	Stockport, N. Y	Jan. 16, 1872	122, 783
Paper-stock, Bleaching	G. E. Marshall	Laurel, Ind	Sept. 28, 1869	95, 365
Paper-stock, Bleaching	S. T. Merrill	Beloit, Wis	Mar. 17, 1868	75, 691
Paper-stock, Bleaching	J. Tiffany and H. B. Meech	Albany and Fort Edward, N. Y.	July 31, 1866	56, 833
Paper-stock, Bleaching	F. Shelden	Fitchburgh, Mass	June 28, 1870	104, 781
Paper-stock, Boiler for treating	M. Nixon	Philadelphia, Pa	Nov. 22, 1859	26, 199

Index of patents issued from the United States Patent Office from 1790 *to* 1873, *inclusive*—Continued.

Invention.	Inventor.	Residence.	Date.	No.
Paper-stock, box-board, roofing-paper, &c	R. W. Russell	New York, N. Y	Mar. 30, 1869	88, 515
Paper-stock-cutting machine	A. L. Knight	Baltimore, Md	Nov. 12, 1867	70, 863
Paper-stock engine	V. O. Balcom and C. H. Hill	Bedford and Billerica, Mass	Dec. 2, 1856	16, 162
Paper-stock engine	J. G. Fuller	Brooklyn, N. Y	Mar. 21, 1865	46, 893
Paper-stock, &c., Forming, drying, and packing	G. S. Sellers	Seller's Landing, Ill	Jan. 5, 1864	41, 102
Paper-stock from reeds	H. Lowe	Baltimore, Md	July 13, 1858	20, 884
Paper-stock from wood, Boiler for preparation of	M. L. Keene	Jersey City, N. J	July 25, 1871	117, 427
Paper-stock from wood, Manufacture of	P. A. Chadbourne	Williamstown, Mass	Mar. 24, 1863	37, 951
Paper-stock from wood, Obtaining	W. Adamson	Philadelphia, Pa	July 18, 1871	117, 136
Paper-stock from wood, Preparing	J. H. Hawes	Stockbridge, Mass	Apr. 20, 1869	89, 220
Paper-stock, Machine for disintegrating wood for	J. Stutt	Fermanagh County, Ireland	Apr. 20, 1869	89, 255
Paper-stock, &c., Machine for pulping wood for	B. F. Barker	Curtisville, Mass	Sept. 19, 1871	119, 107
Paper-stock, Manufacture of	W. Adamson	Philadelphia, Pa	July 18, 1871	117, 134
Paper-stock, Manufacture of	W. P. Arnold	New York, N. Y	Aug. 13, 1872	130, 462
Paper-stock, Manufacture of	H. Betts	Norwalk, Conn	Aug. 1, 1865	49, 069
Paper-stock, Manufacture of	W. Deltour	New York, N. Y	Jan. 3, 1865	45, 791
Paper-stock, Manufacture of	M. L. Keen	Jersey City, N. J	Oct. 18, 1870	108, 487
Paper-stock, Manufacture of	A. T. Sturdevant	Mount Pleasant, N. Y	Oct. 21, 1873	143, 940
Paper-stock, Manufacture of	J. Tiffany	Albany, N. Y	July 31, 1866	56, 832
Paper stock, Manufacture of leather	E. and J. R. Cushman	Amherst, Mass	Aug. 21, 1860	29, 672
Paper stock, Manufacture of wood	G. H. Bliss and M. Rees	Stockbridge, Mass	June 7, 1870	103, 968
Paper-stock, Manufacturing	A. T. Sturdevant	Scipio Centre, N. Y	July 16, 1872	129, 185
Paper-stock, Preparation of	H. M. Baker	Washington, D. C	Aug. 2, 1870	105, 884
Paper-stock, Preparation of	G. E. Marshall	Louisville, Ky	May 25, 1869	90, 566
Paper-stock, Preparing	A. Randel	New York, N. Y	Apr. 30, 1861	32, 203
Paper-stock, Preparing vegetable fiber for	G. E. Sellers	Seller's Landing, Ill	Jan. 5, 1864	41, 101
Paper-stock, Preparing wood fiber for	G. E. Marshall	Laurel, Ind	Aug. 31, 1869	94, 228
Paper-stock, Preparing wood for	A. Fickett	Rochester, N. Y	June 29, 1869	92, 027
Paper-stock, Preparing woody fiber for	G. E. Sellers	Hardin County, Ill	Oct. 6, 1863	40, 217
Paper-stock, Pulp-washer for	G. E. Sellers	Seller's Landing, Ill	Jan. 24, 1865	46, 030
Paper-stock, Pulping and bleaching	J. W. Goodwyn	Petersburgh, Va	Mar. 8, 1870	100, 523
Paper-stock, &c., Rag whipper and duster for treating.	L. Brainard	Hartford, Conn	Mar. 15, 1870	100, 718
Paper-stock, Reducing and bleaching	M. B. Meech	Fort Edward, N. Y	Aug. 23, 1870	106, 711
Paper-stock to make pulp, Treating	J. Tiffany	Albany, N. Y	July 30, 1867	67, 229
Paper-stock, Treating fibrous substances for	T. Routledge	Sunderland, England	Apr. 1, 1873	137, 484
Paper-stock, Treating fibrous substances in the manufacture of.	T. Routledge	Sunderland, England	July 22, 1873	141, 016
Paper-stock, &c., Treating tarred rope, cordage, &c., for the manufacture of.	C. F. A. Simonin	Philadelphia, Pa	Sept. 19, 1871	119, 186
Paper-stock machine, Washing and bleaching	H. Monroe	Baldwinsville, N. Y	June 27, 1871	116, 338
Paper-stock-washing machine	J. E. Andrews	Coeymans Hollow, N. Y	July 14, 1868	79, 935
Paper-stock-washing machine	G. L. Lovett	Fitchburgh, Mass	Apr. 8, 1873	137, 696
Paper-stock-washing machine	H. W. Peaslee	Malden Bridge, N. Y	Jan. 23, 1855	12, 283
Paper-stuff, Apparatus for washing	S. Lenher and H. H. Spencer	Philadelphia, Pa	Mar. 21, 1865	46, 915
Paper-stuff, Boiler for preparing	C. S. Buchanan	Ballston Spa, N. Y	May 1, 1860	28, 062
Paper-stuff, Manufacture of	J. T. Coupier and M. A. C. Mellier.	Paris, France	Aug. 2, 1853	9, 910
Paper-stuff, Treating	J. A. Roth	Philadelphia, Pa	July 28, 1857	17, 895
Paper, Substitute for lining	G. Munger	New Haven, Conn	Oct. 17, 1865	50, 485
Paper-tins	E. H. Boswell	Philadelphia, Pa	May 15, 1866	54, 676
Paper to cutting-machines, Device for feeding	J. F. Schuyler	Philadelphia, Pa	Feb. 24, 1863	37, 791
Paper to printing-press, Feeding	O. Talcott	Chicago, Ill	Sept. 3, 1861	33, 217
Paper to remove ink and recover the pulp, Treatment of printed.	B. Lambert	Dover Road, Surrey, England.	July 23, 1861	32, 879
Paper to wood, Transferring impression from	A. C. Baker and M. Biddle	Albany, N. Y	Feb. 7, 1822	....
Paper, Tobacco	H. J. Hale	New York, N. Y	Feb. 7, 1865	46, 233
Paper together, Mode of fastening sheets of	E. L. Swartwout	Utica, N. Y	Mar. 22, 1859	23, 322
Paper, Top-press roller for making	M. Hunting	Watertown, Mass	Oct. 20, 1828	
Paper, Transfer	W. Grüne	Berlin, Germany	Oct. 29, 1872	132, 659
Paper, Treating straw, wood, &c., for the manufacture of.	A. M. Hastings and S. Pettibone.	Rochester and Niagara Falls, N. Y.	Oct. 8, 1867	69, 663
Paper, Treating tracing	J. Moog	Karlsruhe, Germany	Apr. 19, 1870	102, 030
Paper, &c., Treating vegetables to obtain fiber for	J. A. Rothe	Philadelphia, Pa	Jan. 11, 1870	98, 712
Paper-trimmer	J. Hatch	South Windham, Conn	Feb. 26, 1867	62, 486
Paper-trimmer	J. Kotch	South Adams, Mass	Sept. 4, 1860	29, 891
Paper-trimmer	P. Shee	Philadelphia, Pa	Feb. 7, 1807	....
Paper-trimming machine	W. G. Ayres and S. L. Cole	Brooklyn, N. Y	Aug. 29, 1871	118, 423
Paper-trimming machine	W. G. Ayres and S. L. Cole	Brooklyn, N. Y	Feb. 13, 1872	123, 605
Paper-trimming machine	D. Pirie and A. Croom	Dundee, North Britain	Feb. 27, 1872	124, 155
Paper-trimming machine	J. F. Schuyler	Tiffin, Ohio	Nov. 29, 1870	109, 767
Paper-trimming machine	O. L. Starkey and E. W. Poston	Fort Wayne, Ind	Aug. 15, 1871	118, 066
Paper-trimming machine	H. L. Todd	Corning, N. Y	Sept. 12, 1871	118, 985
Paper-trimming machine	W. P. Yeoman	Waukegan, Ill	Feb. 14, 1871	111, 797
Paper tubes, &c., Machine for making	J. Arkell	Canajoharie, N. Y	Feb. 11, 1868	74, 190
Paper tubes, Machine for making	C. Hotz	Zurich, Switzerland	June 14, 1870	104, 312
Paper tubes, Machine for making	J. A. Olney	Providence, R. I	June 11, 1867	65, 590
Paper twine, Manufacturing	E. B. Bingham	Newark, N. J	Feb. 7, 1865	46, 208
Paper twine, Manufacturing	I. P. Till	New York, N. Y	Feb. 14, 1865	46, 405
Paper twine, Machine for making	R. V. De Guinon	South Bergen, N. J	Feb. 8, 1870	99, 654
Paper twine, Machine for making	J. B. Wortendyke	Goodwinville, N. J	May 24, 1864	42, 896
Paper under-sleeve, Ladies'	H. M. Remington	Springfield, Mass	May 15, 1866	54, 773
Paper, Using pelt for manufacturing	J. M. Thorndyke	New York, N. Y	Mar. 7, 1814	....
Paper, Utilizing the waste chloride of zinc in treating.	D. W. Harma	Pittsburgh, Pa	Oct. 31, 1871	120, 380
Paper veneers, Manufacture of	C. Walker and G. Willson	Windsor County, Vt	Mar. 20, 1849	6, 208
Paper vessel	E. M. Slayton	Conquest, N. Y	Nov. 18, 1873	144, 600
Paper vessels, Manufacture of	A. and I. Jennings	Fairfield, Conn	Nov. 26, 1867	71, 390
Paper, wadding, &c., Machine for drying	E. C. Wilson	Medway, Mass	Feb. 20, 1872	123, 850
Paper, wadding, &c., Machine for drying	E. C. Wilson	Medway, Mass	June 18, 1872	128, 085
Paper, Water-marks on ready-made	B. Westerberg	Saint Paul, Minn	Aug. 20, 1872	130, 676
Paper, Water-proof or damp-proof	S. C. Bishop	New York, N. Y	Jan. 14, 1868	73, 287
Paper washer	E. S. Hanna	Pittsburgh, Pa	Oct. 3, 1871	119, 604
Paper wearing-apparel	G. W. Day	Charlestown, Mass	June 11, 1867	65, 730
Paper wearing-apparel	A. T. Denison and E. P. Furlong.	Poland and Portland, Me	Dec. 18, 1866	60, 488
Paper wearing-apparel	T. W. Mann	Holyoke, Mass	Mar. 5, 1867	62, 550
Paper-weight	J. B. Wilson	Philadelphia, Pa	Dec. 10, 1872	133, 912

Index of patents issued from the United States Patent Office from 1790 *to* 1873, *inclusive*—Continued.

Invention.	Inventor.	Residence.	Date.	No.
Paper-weight and pen-wiper, Combined	D. W. Wright	New York, N. Y	July 9, 1867	66, 547
Paper-weight and sponge-holder, Combined	D. R. Manning	Brooklyn, N. Y	Dec. 23, 1873	145, 883
Paper-wetting apparatus	M. S Beach	Brooklyn, N. Y	Nov. 9, 1858	22, 009
Paper-wetting machine	J. A. Lynch	Boston, Mass	Apr. 27, 1858	20, 077
Paper-wetting machine	W. Overend	Cincinnati, Ohio	Jan. 24, 1854	10, 458
Paper-wetting machine	A. Overend	Philadelphia, Pa	June 26, 1860	28, 895
Paper window-shade	L. A. Colbert	Baltimore, Md	Apr. 10, 1866	53, 786
Paper with mucilage, Machine for coating	H. E. Rile	New York, N. Y	June 18, 1867	65, 948
Paper, Wrapping	B. E. Hale	New York, N. Y	Oct. 10, 1871	119, 843
Paper, Wrapping	S. Wheeler	Albany, N. Y	July 25, 1871	117, 355
Paper, Writing	J. E. Hover	Philadelphia, Pa	Jan. 22, 1867	61, 338
Papers, Adhesive fastening for	G. R. Burdon	Waltham, Mass	Jan. 31, 1865	46, 071
Papers, Appai atus for filing and binding	P. Mac Vicar	Philadelphia, Pa	Jan. 14, 1873	134, 908
Papers for soda-powders, Machine for crimping package.	C. A. Cook	Lowell, Mass	Dec. 2, 1851	8, 556
Papers Machine for filling and folding medical powder.	M. S. Palmer	New Bedford, Mass	June 25, 1861	32, 642
Papering-machine	G. M. Lane	De Graff, Ohio	Sept. 13, 1870	107, 388
Papier-maché compound	G. F. Goetze	New York, N. Y	July 6, 1869	92, 303
Papier-maché, Making stereotype-molds of	A. Chase	Ithaca, N. Y	June 25, 1872	128, 285
Papier-maché, Mode of producing an extra surface on.	E. S. Judge	Baltimore, Md	Mar. 24, 1868	75, 766
Paraffine and obtaining it in crystals, Treating	F. Lambe	London, England	Apr. 19, 1870	102, 135
Paraffine and other hydrocarbon oils, Apparatus for burning.	H. H. Doty	London, England	Nov. 15, 1870	109, 303
Paraffine and paraffine oil, Manufacture of	H. W. C. Tweddle	Pittsburgh, Pa	Feb. 15, 1870	99, 975
Paraffine, &c., from oil, Apparatus for extracting	J. B. Meriam	Cleveland, Ohio	Feb. 12, 1867	61, 946
Paraffine, &c., Obtaining oil from	S. L. Wiegand	Philadelphia, Pa	Mar. 5, 1867	62, 583
Paraffine, Purifying	F. X. Byerley	Cleveland, Ohio	Oct. 22, 1872	132, 353
Paraffine, Purifying	S. H. Crocker	Pittsburgh, Pa	July 16, 1872	129, 463
Paraffine, Purifying	C. C. Parsons	New York, N. Y	Aug. 17, 1869	93, 739
Paraffine, Saturating wood, cloth, paper, &c., with	S. Gwynn	New York, N. Y	Feb. 20, 1866	52, 788
Paraffine, Treatment and purification of	R. M. Letchford and W. B. Nation.	Bethnal Green, England	Nov. 12, 1872	133, 042
Parallel or other rods	C. H. Clark	Wilmington, Del	Dec. 26, 1865	51, 694
Parasol	L. Battles	Gloucester, Mass	May 24, 1870	103, 412
Parasol	A. G. Davis and H. S. Frost	Watertown, Conn	Sept. 25, 1860	30, 127
Parasol	J. L. Jacquin	New York, N. Y	Dec. 5, 1871	121, 625
Parasol	C. St. John	Charlestown, Mass	Dec. 24, 1867	72, 695
Parasol and fan	J. T. Eichberg	New York, N. Y	Oct. 2, 1860	30, 213
Parasol and umbrella	E. Young	Philadelphia, Pa	Nov. 23, 1858	22, 142
Parasol, Feather-covered	G. Anton	Philadelphia, Pa	July 31, 1866	56, 691
Parasol, Reversible	J. Williams	New York, N. Y	June 29, 1869	91, 994
Parasols, Machine for retaining, adjusting, and sewing.	W. J. Tate	Philadelphia, Pa	July 26, 1870	105, 862
Parceling, Machine for manufacturing	J. H. Raynard	Lynn, Mass	Dec. 1, 1868	84, 578
Parchment, Manufacture of vegetable	E. P. Hudson	New York, N. Y	Mar. 16, 1869	87, 937
Parchment, Manufacture of vegetable	X. Karcheski	New York, N. Y	Dec. 18, 1860	30, 945
Parchment or water-proof paper, Apparatus for preparing.	E P. Hudson	New York, N. Y	Sept. 27, 1870	107, 686
Parchment-paper, Manufacture of rubber-coated	E. P. Hudson	New York, N. Y	Sept. 27, 1870	107, 687
Parchment, Solution for treating vegetable fiber for the manufacture of.	S. Gwynn	New York, N. Y	Jan. 14, 1868	73, 322
Parer and corer, Apple	S. Cruttenden	Guilford, Conn	Aug. 25, 1809	....
Parer and corer, Apple	H. S. Leonard	Attica, Ohio	Dec. 17, 1872	134, 078
Parer and corer, Apple	R. W. Mitchell	Springfield, Ohio	Apr. 13, 1838	686
Parer and corer, Apple	H. Selick	Lewiston, Pa	Aug. 21, 1866	57, 391
Parer and cutter, Peach	H. S. Hilsman	Madison, Ga	Dec. 4, 1860	30, 820
Parer and quarterer, Apple	C. S. Pratt	Paris, Me	Dec. 28, 1833	....
Parer and slicer, Apple	G. W. Brokaw	Lodi, N. Y	May 11, 1869	89, 971
Parer and slicer, Apple	C. A. Foster	Fitchburgh, Mass	July 11, 1871	116, 943
Parer and slicer, Apple	D. H. Whittemore	Worcester, Mass	Aug. 10, 1869	93, 574
Parer, Apple	A. G. Batchelder	Lowell, Mass	Mar. 2, 1869	87, 322
Parer, Apple	G. Bergner	Washington, Mo	Feb. 11, 1873	135, 622
Parer, Apple	J. D. Browne	Cincinnati, Ohio	Sept. 9, 1856	15, 683
Parer, Apple	J. Bullock, jr., and S. Benson	New York, N. Y	July 24, 1847	5, 197
Parer, Apple	C. P. Carter	Ware, Mass	Oct. 16, 1849	6, 789
Parer, Apple	C. P. Carter	Ware, Mass	Aug. 26, 1856	15, 603
Parer, Apple	C. P. Carter	Ware, Mass	Nov. 18, 1856	16, 104
Parer, Apple	R. P. Clark	Johnstown, N. Y	Sept. 20, 1859	25, 489
Pare , Apple	M. Coates	....	Feb. 14, 1803	....
Parer, Apple	H. A. Frost	Worcester, Mass	Aug. 6, 1861	33, 016
Parer, Apple	W. M. Griscom	Reading, Pa	May 14, 1872	126, 803
Parer, Apple	W. M. and C. W. Hardy	East Strong, Me	Jan. 29, 1861	31, 238
Parer, Apple	J. W. Hatch	Bedford County, Va	Feb. 3, 1836	....
Parer, Apple	S. S. Hersey	Farmington, Me	June 18, 1861	32, 561
Parer, Apple	S. S. Hersey	Farmington, Me	Aug. 30, 1864	43, 990
Parer, Apple	F. W. Hudson	Leominster, Mass	Dec. 2, 1862	37, 038
Parer, Apple	F. W. Hudson	Leominster, Mass	Mar. 5, 1872	124, 272
Parer, Apple	H. Keyes	Terre Haute, Ind	Nov. 20, 1866	59, 843
Parer, Apple	H. Keyes	Leominster, Mass	Dec. 16, 1856	16, 240
Parer, Apple	W. H. Lazelle	New York, N. Y	Jan. 25, 1853	9, 558
Parer, Apple	E. Mauley	Marion, N. Y	Nov. 17, 1863	40, 640
Parer, Apple	S. N. Maxam	Shelburne Falls, Mass	Apr. 10, 1855	12, 689
Parer, Apple	J. F. and E. P. Monroe	Fitchburgh, Mass	Mar. 24, 1868	75, 951
Parer, Apple	R. and A. Mosher	Galway, N. Y	Dec. 28, 1829	....
Parer, Apple	W. A. C. Oakes	Reading, Pa	Dec. 10, 1872	133, 796
Parer, Apple	E. L. Pratt	Worcester, Mass	Oct. 4, 1853	10, 078
Parer, Apple	E. L. Pratt	Philadelphia, Pa	Apr. 29, 1856	14, 775
Parer, Apple	E. L. Pratt	Boston, Mass	Oct. 6, 1863	40, 185
Parer, Apple	E L. Pratt	Boston, Mass	Jan. 12, 1864	41, 266
Parer, Apple	E. L. Pratt	Boston, Mass	Aug. 23, 1864	43, 955
Parer, Apple	W. Robb, jr	South Stoddard, N. H	Nov. 22, 1870	109, 454
Parer, Apple	M. Smith	New Haven, Conn	Aug. 26, 1856	15, 625
Parer, Apple	J. Voak	Penn Yan, N. Y	July 12, 1864	43, 537
Parer, Apple	J. White	Antrim, N. H	June 3, 1862	35, 482
Parer, Apple	J. White	Antrim, N. H	Feb. 16, 1864	41, 658
Parer, Apple	D. H. Whittemore	Worcester, Mass	Nov. 20, 1866	59, 884

Index of patents issued from the United States Patent Office from 1790 *to* 1873, *inclusive*—Continued.

Invention.	Inventor.	Residence.	Date.	No.
Parer, Apple	C. A. Wiggin	North Sandwich, N. H	Aug. 4, 1868	80, 580
Parer, Apple	G. H. Wilde	Aurora, Ill	July 6, 1869	92, 240
Parer, Apple and potato	J. L. Clewell, jr., and W. F. Schatz.	Nazareth, Pa	Nov. 20, 1860	38, 667
Parer, corer, &c., Apple	G. W. Bennett	Smithville, Ind	Feb. 25, 1873	136, 206
Parer, corer, and cutter, Apple	W. A. Coe	Greensborough, N. C	Feb. 19, 1867	62, 184
Parer, corer, and slicer, Apple	J. I. Armfield	Jamestown, N. C	Dec. 20, 1859	26, 464
Parer, corer, and slicer, Apple	A. Clark	La Fayette, Ind	June 9, 1868	78, 721
Parer, corer, and slicer, Apple	J. J. Parker	Marietta, Ohio	Sept. 11, 1860	29, 988
Parer, corer, and slicer, Apple	J. Shobe	Upper Principio, Md	Nov. 15, 1870	109, 350
Parer, corer, and slicer combined, Apple	G. Bergner	Washington, Mo	Jan. 10, 1872	122, 553
Parer, cutter, and corer, Apple	G. W. Bennett	Harrodsburgh, Ind	May 24, 1870	103, 284
Parer, Cutter-head for apple	E L. Pratt	Boston, Mass	Oct. 6, 1863	40, 216
Parer, Fork for peach	D. H. Goodell	Antrim, N. H	May 10, 1870	102, 934
Parer, Fork for peach	C. D. House	Lake Village, N. H	Aug. 17, 1869	93, 887
Parer, Fruit	A. G. Batchelder	Lowell, Mass	Jan. 3, 1872	122, 305
Parer, Fruit	L. D. Farwell and A. W. Goddard.	Lancaster and Clinton, Mass.	May 5, 1868	77, 600
Parer, Fruit	D. H. Goodell	Antrim, N. H	June 18, 1867	65, 804
Parer, Fruit	R. P. Scott	Cadiz, Ohio	May 16, 1871	114, 867
Parer, Fruit	A. Turnbull and R. L. Webb	New Britain, Conn	June 10, 1873	139, 837
Parer, Fruit	L. S. Woodbury	South Antrim, N. H	Dec. 20, 1870	110, 322
Parer, Fruit and vegetable	A. G. Batchelder	Lowell, Mass	June 7, 1870	103, 830
Parer, Fruit and vegetable	A. G. Batchelder	Lowell, Mass	Mar. 26, 1872	125, 004
Parer, Fruit and vegetable	T. Learing	South Norwalk, Conn	Dec. 5, 1871	121, 669
Parer, Fruit and vegetable	H. Soggs	Columbus, Pa	Dec. 13, 1870	110, 167
Parer, Peach	J. H. Brown	Mitchell, Ind	Oct. 6, 1868	82, 794
Parer, Peach	W. A. Coe	Greensborough, N. C	Oct. 30, 1860	30, 527
Parer, Peach	A. Hermans	Henderson, Tex	Dec. 27, 1859	26, 640
Parer, Peach	E. L. Pratt	Boston, Mass	Aug. 23, 1864	43, 956
Parer, Peach	M. Smith	New Haven, Conn	May 15, 1860	28, 310
Parer, Potato	W. B. Coates	Philadelphia, Pa	Nov. 29, 1859	26, 254
Parer, Potato	C. Disston	Philadelphia, Pa	July 3, 1860	28, 973
Parer, Potato	H. Underwood	Tolland, Conn	Aug. 2, 1864	43, 725
Parer, Potato	W. Zeiger	Elmore, Ohio	Apr. 20, 1869	89, 190
Parer, Potato and fruit	M. B. Atkinson	Georgetown, D. C	Sept. 13, 1870	107, 321
Parer, Vegetable	M. P. Smith	Baltimore, Md	Apr. 2, 1872	125, 225
Parer, Vegetable and fruit	E. D. Averell and J. Malan	Brooklyn, N. Y	Mar. 8, 1870	100, 583
Paring and chopping machine	S. H. Morse	North Jay, Me	June 27, 1871	116, 342
Paring and coring fruit	P. W. Hardwick	Wayne County, Ind	Sept. 25, 1849	6, 735
Paring and cutting machine, Peach	J. O. Ward	Pleasant Valley, N. Y	Oct. 21, 1851	8, 460
Paring and slicing knife	H. P. Brooks	Waterbury, Conn	Oct. 17, 1871	119, 963
Paring and slicing machine, Apple	G. H. Hubbard	Shelburne Falls, Mass	Jan. 27, 1857	16, 517
Paring and slicing machine, Apple	R. W. Thickins	Brasher Iron Works, N. Y	July 28, 1857	17, 901
Paring and slicing machine, Apple	L. Van Hoesen	New Haven, Conn	Dec. 4, 1855	13, 891
Paring and slicing machine, Apple	D. H. Whittemore	Chicopee Falls, Mass	Jan. 13, 1857	16, 417
Paring apples, Machine for	J. D. Browne	Cincinnati, Ohio	May 6, 1856	14, 800
Paring apples, Machine for	J. O. M. Ingersoll	Ithaca, N. Y	Jan. 20, 1857	16, 443
Paring apples, Machine for	B. F. Joslyn	Worcester, Mass	Mar. 17, 1857	16, 640
Paring apples, Machine for	H. Keyes	Leominster, Mass	June 17, 1856	15, 133
Paring apples, Machine for	J. F. and E. P. Monroe	Fitchburgh, Mass	Jan. 27, 1863	37, 516
Paring apples, Machine for	J. J. Parker	Marietta, Ohio	Apr. 7, 1857	16, 993
Paring apples, Machine for	J. D. Seagrave	Milford, Mass	Apr. 18, 1854	10, 785
Paring apples, Machine for	J. D. Seagrave	Worcester, Mass	June 17, 1856	15, 148
Paring apples, Machine for	D. H. Whittemore	Worcester, Mass	Feb. 17, 1857	16, 666
Paring apples, potatoes, &c., Machine for	E. L. Pratt	Philadelphia, Pa	Nov. 11, 1856	16, 080
Paring, coring, and quartering apples, Machine for.	C. F. Bosworth	Petersham, Mass	June 9, 1857	17, 484
Paring, coring, and slicing apples	J. Weed	Painesville, Ohio	July 31, 1849	6, 619
Paring, coring, and slicing apples, Machine for	A. M. Blain	Deerfield, Va	June 11, 1867	65, 534
Paring, coring, and slicing apples, Machine for	J. Stewart and W. Campbell	Fowler, Ill	June 24, 1873	140, 315
Paring, cutting, and coring machine	W. Weaver	Phœnixville, Pa	Apr. 3, 1866	53, 754
Paring-knife	H. P. Brooks	Waterbury, Conn	Sept. 17, 1872	131, 329
Paring-knife	J. H. Bruen	Elmira, N. Y	Jan. 23, 1872	122, 934
Paring-knife	J. Lebeau	Cincinnati, Ohio	July 10, 1866	56, 235
Paring-knife	G. H. Vickroy	Washington, D. C	June 18, 1872	127, 998
Paring knife, Apple	A. Oot	Minetto, N. Y	Oct. 5, 1858	21, 695
Paring knife, Vegetable	W. Vober, jr	Shingle Creek, N. Y	May 25, 1869	90, 474
Paring knife, Vegetable and fruit	H. Soggs	Columbus, Pa	Jan. 25, 1870	99, 119
Paring machine, Fruit	W. M. Howland	Leominster, Mass	Aug. 30, 1864	44, 044
Paring, quartering, and coring apples, Machine for.	W. Badger	Boston, Mass	Feb. 16, 1809	......
Paring, slicing, and coring machine, Apple	J. J. Parker	Marietta, Ohio	July 6, 1858	20, 814
Park and garden seat	S. Macferran	Philadelphia, Pa	May 4, 1869	89, 779
Parlor-bath	J. R. Worster	New York, N. Y	Apr. 27, 1869	89, 370
Parlor-fountain for diffusing liquids	W. Altick	Dayton, Ohio	Mar. 1, 1870	100, 349
Parlor-heater	J. Carton	Utica, N. Y	Oct. 15, 1861	33, 474
Parlors and saloons, Covering for walls of	J. M. Soutrenon	New York, N. Y	Nov. 5, 1867	70, 640
Paris white and whiting	P. Ferris	Greenwich, Conn	Apr. 22, 1835	....
Parral and bow	J. Damo	Portsmouth, N. H	Sept. 26, 1854	11, 721
Partition, Fire-proof	C. F. Brand	Philadelphia, Pa	Nov. 11, 1873	144, 501
Partition, Movable	D. J. Williams	Birmingham, England	Apr. 4, 1871	113, 476
Passe-partout frame	P. Mignot	New York, N. Y	Apr. 26, 1870	102, 296
Passenger and freight elevator	S. Nash	Cambridge, Mass	Aug. 5, 1873	141, 513
Passenger and station register	J. C. Hackett	Sacramento, Cal	Aug. 10, 1869	93, 531
Passenger-boat	N. Cushman and I. Loomis	Whitehall, N. Y	Jan. 8, 1831	......
Passenger-elevator	J. S. Baldwin	Newark, N. J	Apr. 14, 1868	76, 694
Passenger-recorder	G. H. Aldrich	New York, N. Y	Apr. 25, 1871	114, 084
Passenger-recorder, Vehicle	J. T. Aldrich	New York, N. Y	June 13, 1871	115, 801
Passenger-register	S. L. Cooper	Lewistown, Pa	Aug. 5, 1873	141, 492
Passenger-register	J. Enright	Louisville, Ky	June 23, 1868	79, 060
Passenger-register	E. Hackett	New York, N. Y	Mar. 7, 1865	46, 663
Passenger-register	E. Hambujer	New York, N. Y	Jan. 16, 1866	52, 043
Passenger-register	W. Helffricht	Philadelphia, Pa	Jan. 8, 1867	61, 065
Passenger-register	W. Helffricht	Philadelphia, Pa	Mar. 26, 1867	63, 245
Passenger-register	T. S. Huntington and A. Fulton	Bellefontaine, Ohio	July 14, 1868	79, 911
Passenger-register	T. Jacobs	Philadelphia, Pa	Aug. 4, 1868	80, 741
Passenger-register	J. Melling	Rochester, N. Y	Oct. 8, 1867	69, 688
Passenger-register	A. F. Nagle	Providence, R. I	Apr. 7, 1868	76, 502

Index of patents issued from the United States Patent Office from 1790 *to* 1873, *inclusive*—Continued.

Invention.	Inventor.	Residence.	Date.	No.
Passenger-register	M. Springwoter	Louisville, Ky	Apr. 25, 1871	114, 060
Passenger-register, Automatic	H. H. Trenor	New York, N. Y	Oct. 5, 1869	95, 540
Passenger-register, Automatic vehicle	C. Ottinger	Philadelphia, Pa	Apr. 23, 1870	106, 610
Passenger-register, Car	J. Kurz	Philadelphia, Pa	Feb. 21, 1871	111, 948
Passenger-register, Horse-car	E. L. Fitch	West Eau Claire, Wis	June 21, 1870	104, 566
Passenger-register, Street-car	W. Willis	Chicago, Ill	Nov. 2, 1869	96, 526
Passenger-register, Street-car, &c., electro-magnetic.	W. H. Mumler	Boston, Mass	July 2, 1872	128, 500
Passenger-register, Vehicle	T. Ollis	Liverpool, England	Dec. 21, 1869	98, 098
Passenger-register, Vehicle	J. Rhoads	Harrisburgh, Pa	Aug. 2, 1870	105, 980
Passing-box	E. A. G. Roulstone	Roxbury, Mass	Mar. 25, 1862	34, 777
Paste	J. E. Hover	Philadelphia, Pa	Nov. 19, 1867	71, 174
Paste, Broom-corn-seed flour for manufacture of	J. W. Tallmadge	Boston, Mass	Dec. 30, 1873	145, 032
Paste-making apparatus	G. G. Noah	Charlestown, Mass	May 6, 1873	138, 519
Paste-making machine	G. G. Noah	Boston, Mass	Jan. 7, 1868	73, 115
Paste, Manufacture of	G. G. Noah	Boston, Mass	June 14, 1870	104, 190
Paste, Prepared	S. Merwin	Springfield, Mass	Oct. 2, 1866	58, 454
Paste, Preserving and making	G. G. Noah	Charlestown, Mass	May 21, 1872	127, 097
Pasteboard, Adhesive compound for preparing	G. L. Jaeger	New York, N. Y	July 5, 1870	105, 084
Pasteboard box	L. Carrier	Providence, R. I	July 23, 1867	67, 022
Pasteboard box	S. W. Millichamp	Rochester, N. Y	Nov. 26, 1872	133, 330
Pasteboard box	W. Pratt	New York, N. Y	Feb. 21, 1871	112, 075
Pasteboard boxes, Machine for forming	C. A. Maxfield	New York, N. Y	May 5, 1868	77, 502
Pasteboard boxes, Machine for making	J. T. Foster	Jersey City, N. J	Mar. 19, 1867	62, 952
Pasteboard, composition-board, panels, &c., from reed, Manufacture of.	W. P. Arnold	New York, N. Y	Aug. 13, 1872	130, 463
Pasteboard, &c., cutting machine	D. Burhans	Burlington, Iowa	July 7, 1857	17, 723
Pasteboard cutting machine	E. E. Clarke	New Haven, Conn	Feb. 21, 1865	46, 448
Pasteboard-cutting machine	H. A. Gage	Manchester, N. H	Nov. 16, 1869	96, 791
Pasteboard-cutting machine	S. Orth	Philadelphia, Pa	Nov. 19, 1867	71, 048
Pasteboard cutting and scoring machine	S. Orth	Philadelphia, Pa	Sept. 12, 1865	49, 910
Pasteboard, Drying	J. H. Patterson	Schaghticoke, N. Y	Mar. 26, 1861	31, 822
Pasteboard for boxes, Machine for cutting	E. E. Clarke	New Haven, Conn	Mar. 3, 1857	16, 719
Pasteboard for boxes, Machine for cutting	E. E. Clarke	New Haven, Conn	Feb. 21, 1865	46, 522
Pasteboard for boxes, Machine for cutting	E. E. Clarke	New Haven, Conn	Feb. 28, 1865	46, 604
Pasteboard lining and drying machine	G. L. Jaeger	New York, N. Y	May 10, 1870	102, 942
Pasteboard-lining machine	G. H. Dickerman	Somerville, Mass	Mar. 5, 1872	124, 258
Pasteboard-lining machine	G. L. Jaeger	New York, N. Y	Feb. 14, 1871	111, 849
Pasteboard-lining machine	G. L. Jaeger	New York, N. Y	Apr. 30, 1872	126, 301
Pasteboard, Machine for making	L. Koch	New York, N. Y	Mar. 31, 1857	16, 927
Pasteboard, Machinery for making	O. W. Fiske	Dedham, Mass	July 11, 1854	11, 313
Pasteboard. Machinery for pressing water out of	L. Koch	New York, N. Y	Mar. 31, 1857	16, 928
Pasteboard, Manufacture of	W. E. Hale	Chicago, Ill	Mar. 2, 1869	87, 407
Pasteboard, Manufacture of	H. L. Palmer	Stillwater, N. Y	Mar. 2, 1869	87, 359
Pasteboard, Manufacture of stone	J. Smith	Mendon, N. Y	Mar. 13, 1855	12, 531
Pasteboard, &c., Manufacturing	F. A. Taft	Dedham, Mass	July 20, 1831	....
Pasteboard or paper, Machine for drying	P. Clark	Rahway, N. J	June 9, 1857	17, 489
Pasteboard paper for bandbox, &c	I. Sanderson	Milton, Mass	May 15, 1830	....
Pasteboard, sheet-iron, &c., Shears for cutting	A. Hardy	Boston, Mass	Aug. 12, 1831	....
Pastile, Disinfecting	A. Lanergan	Boston, Mass	July 1, 1856	15, 243
Pastry-board	L. E. Higby	Shelburne Falls, Mass	Jan. 31, 1860	26, 989
Pastry cutter and crimper	W. R. Marie	Boston, Mass	Apr. 13, 1869	88, 883
Pastry-cutter	J. Stephon	Womelsdorf, Pa	Aug. 6, 1867	67, 601
Pastry-jigger	J. Redding and J. B. Coe	Charlestown and Boston, Mass.	Apr. 14, 1868	76, 813
Pastry-roller	J. S. Foster	San Francisco, Cal	Nov. 20, 1866	59, 905
Pastry-roller	H. S. Maltby	Cincinnati, Ohio	Aug. 31, 1869	94, 227
Pastry-roller	W. C. Pierce	Philadelphia, Pa	Nov. 5, 1872	132, 736
Pastry-roller	A. L. Taylor	Springfield, Vt	July 16, 1867	66, 909
Patch-bolts, Machine for making	J. Kaylor	Reserve Township, Pa	Oct. 15, 1867	69, 816
Patient-elevator	C. Marx	New York, N. Y	May 19, 1863	38, 595
Pattern:				
See Car-seat-end pattern. Oil-cloth pattern.				
Carpet-pattern. Pipe-pattern.				
Casting-pattern. Pipe-elbow pattern.				
Coat-pattern. Plaster pattern.				
Dress-pattern. Shawl-pattern.				
Embroidery-pattern. Shirt-pattern.				
Garment-pattern. Shirt-bosom pattern.				
Hat-brim pattern. Waving pattern.				
Heel-pattern. Weaving-pattern.				
Pattern chart	C. Freetsha	Paterson, N. J	Jan. 29, 1867	61, 531
Pattern-cutting machine, Segment	J. Bean, jr	Pent Water, Mich	June 27, 1871	116, 258
Pattern-lifter	E. Card	North Providence, R. I	May 28, 1867	65, 169
Pattern-making machine	S. Huffman	Carthage Township, Ill	June 26, 1866	55, 869
Pattern, Transferable embroidery	C. H. Hudson	Asko Terrace, England	Mar. 1, 1864	41, 776
Patterns for flaring vessels, Drawing	O. B. Vandenburg	Findlay, Ohio	July 4, 1871	116, 647
Patterns for molding, Preparing	H. Wright	Milford, N. Y	Jan. 1, 1861	31, 051
Patterns, Measuring and laying out ladies' dress	H. M. Burrows	Norwich, N. Y	May 4, 1869	89, 735
Patterns to supporting-plates, Attaching	J. H. Harper	Pittsburgh, Pa	Jan. 7, 1873	134, 663
Pavement	W. W. Boyington	Chicago, Ill	May 19, 1868	78, 046
Pavement	G. H. Chinnock	New York, N. Y	Nov. 5, 1872	132, 801
Pavement	S. J. Davenport and J. Ward	Albany, N. Y	Oct. 28, 1873	143, 965
Pavement	W. H. De Valin	Sacramento, Cal	Nov. 18, 1873	144, 748
Pavement	A. Dilger	Saint Louis, Mo	July 5, 1870	104, 941
Pavement	L. S. Filbert	Philadelphia, Pa	Oct. 25, 1870	108, 696
Pavement	R. Foley	New York, N. Y	Sept. 15, 1868	82, 102
Pavement	E. Gomez	New York, N. Y	Dec. 17, 1872	134, 049
Pavement	H. A. Gunther	New York, N. Y	Oct. 24, 1871	120, 268
Pavement	I. H. Hobbs	Philadelphia, Pa	July 25, 1871	117, 286
Pavement	B. B. Hotchkiss	New York, N. Y	Mar. 26, 1867	63, 162
Pavement	W. W. Hubbell	Philadelphia, Pa	May 30, 1871	115, 475
Pavement	D. Huestis	Cold Spring, N. Y	July 24, 1866	56, 563
Pavement	J. S. Kelly	New York, N. Y	Aug. 30, 1870	106, 832
Pavement	J. S. Kelly	New York, N. Y	Dec. 5, 1871	121, 524
Pavement	W. Kilburn	Napa, Cal	July 15, 1873	140, 835
Pavement	H. A. Lacey	Detroit, Mich	Dec. 17, 1872	134, 073
Pavement	H. G. McGonegal	New York, N. Y	Feb. 4, 1868	74, 110

Index of patents issued from the United States Patent Office from 1790 *to* 1873, *inclusive*—Continued.

Invention.	Inventor.	Residence.	Date.	No.
Pavement	A. B. McKeon	Rutherford Park, N. J	May 10, 1870	102, 846
Pavement	J. Merlette, sr	Boundbrook, N. J	June 4, 1872	127, 628
Pavement	B. F. Miller	New York, N. Y	Apr. 23, 1867	64, 126
Pavement	B. F. Miller	New York, N. Y	July 7, 1868	79, 770
Pavement	S. C. Prescott	Jersey City, N. J	June 18, 1872	127, 986
Pavement	S. C. Prescott	Jersey City, N. J	July 23, 1872	129, 861
Pavement	W. H. Shurtliff	Providence, R. I	Feb. 12, 1867	61, 958
Pavement	C. W. Stafford	Saybrook, Conn	Mar. 26, 1867	63, 324
Pavement	C. W. Stafford	Saybrook, Conn	June 25, 1867	66, 201
Pavement	C. W. Stafford	Saybrook, Conn	Sept. 24, 1867	69, 297
Pavement	C. W. Stafford	Saybrook, Conn	Mar. 10, 1868	75, 309
Pavement	A. Stevens and L. A. Cauvet	New York, N. Y	July 5, 1870	105, 138
Pavement	A. Van Camp and M. M. Hodgman.	Washington, D. C., and Saint Louis, Mo.	May 17, 1870	103, 105
Pavement	C. G. Von Tagen	Philadelphia, Pa	Feb. 27, 1872	124, 022
Pavement	E. Whitehead	Cincinnati, Ohio	July 10, 1866	56, 303
Pavement	A. Wyckoff	Elmira, N. Y	June 2, 1868	78, 635
Pavement	C. Williams	New York, N. Y	Mar. 10, 1868	75, 504
Pavement	C. Williams	New York, N. Y	Mar. 24, 1868	76, 021
Pavement	C. Williams	New York, N. Y	July 21, 1868	80, 260
Pavement	C. Williams	New York, N. Y	July 21, 1868	80, 261
Pavement	M. E. Worrell	Monmouth, Ill	Jan. 7, 1873	134, 622
Pavement and for other purposes, Composition for street.	J. W. Brown	Washington, D. C	Feb. 2, 1869	86, 355
Pavement and foundation for the same, Composition	G. H. Moore	Norwich, Conn	Apr. 16, 1872	125, 831
Pavement and gutter, Cast-iron	M. Pennock	Kennett's Square, Pa	Jan. 26, 1864	41, 412
Pavement, and machinery for, Manufacture of concrete	W. H. Smith	New York, N. Y	Jan. 2, 1872	122, 498
Pavement and railway combined	B. C. Smith	Burlington, N. J	Jan. 15, 1861	31, 131
Pavement and railway track, Mode of constructing a combined street.	R. Montgomery	New York, N. Y	Jan. 18, 1859	22, 665
Pavement and sidewalk, Composition	N. B. Abbott	Brooklyn, N. Y	Feb. 13, 1872	123, 658
Pavement, Apparatus for laying concrete	G. H. Moore	Norwich, Conn	Feb. 27, 1872	124, 148
Pavement, Apparatus for the manufacture of asphalt.	A. D. Foote	Guilford, Conn	Aug. 6, 1872	130, 289
Pavement, Asphalt	C. P. Alsing	New York, N. Y	May 5, 1868	77, 565
Pavement, Asphalt	C. P. Alsing	New York, N. Y	Jan. 4, 1870	98, 460
Pavement, Asphalt	C. Burlew	Washington, D. C	Nov. 21, 1871	121, 082
Pavement, Asphalt	A. C. Campbell	New York, N. Y	Oct. 25, 1870	108, 566
Pavement, Asphalt	A. J. Crawford	Brooklyn, N. Y	Feb. 6, 1872	123, 458
Pavement, Asphalt	E. J. De Smedt	New York, N. Y	Nov. 29, 1870	109, 597
Pavement, Asphalt	S. R. Scharf	Baltimore, Md	Apr. 6, 1869	88, 746
Pavement, Asphalt	S. R. Scharf	Baltimore, Md	Jan. 24, 1871	111, 151
Pavement, Asphalt and crushed-rock	A. Van Camp	Washington, D. C	Nov. 14, 1871	120, 914
Pavement, Asphalt road and	~~E. J. De Smedt~~	New York, N. Y	May 31, 1870	103, 582
Pavement, Asphaltum or concrete	A. G. Day	Seymour, Conn	June 21, 1870	104, 562
Pavement, Belgian	C. Guidet	New York, N. Y	Oct. 2, 1866	58, 407
Pavement, Bituminous rock	J. L. Fulton and J. Brace	Covington, Ky	May 17, 1870	103, 168
Pavement, Block-connecting device for wood	D. H. Mulford	Saratoga Springs, N. Y	Aug. 8, 1871	117, 802
Pavement-blocks, Forming	G. L. Eagan	San Francisco, Cal	Apr. 16, 1872	125, 798
Pavement-blocks from furnace-slag	T. A. Luckenbach	New York, N. Y	June 11, 1872	127, 699
Pavement, building-blocks, &c., Composition for	J. K. Griffin	Waterdown, Canada	Oct. 15, 1872	132, 153
Pavement, Cast-iron	A. Farran	Longwood, Mass	Jan. 4, 1870	98, 480
Pavement, Cast-iron	P. M. Hutton	Troy, N. Y	Mar. 4, 1856	14, 384
Pavement, Cast-iron	G. Neilson	Boston, Mass	Apr. 17, 1855	12, 755
Pavement, Cast-iron	G. M. Ramsay	New York, N. Y	Sept. 23, 1856	15, 776
Pavement, Cast-iron	A. P. Robinson	New York, N. Y	Apr. 22, 1856	14, 736
Pavement, Cast-iron	S. T. Savage	Albany, N. Y	Feb. 8, 1859	22, 896
Pavement, Cast-iron	C. J. Shepard	Brooklyn, N. Y	Mar. 3, 1857	16, 757
Pavement, Cast iron	C. Warner	New York, N. Y	Jan. 9, 1855	12, 172
Pavement, Cast iron	C. Warner	New York, N. Y	Feb. 1, 1859	22, 835
Pavement, Cellular-iron	S. H. Titus and O. Des Granges	Saint Louis, Mo	Oct. 13, 1857	18, 417
Pavement, Cementing and water-proofing block	B. F. Camp	New York, N. Y	Oct. 29, 1872	132, 561
Pavement, Combination	L. S. Filbert and J. Taylor	Philadelphia, Pa., and New York, N. Y.	Jan. 2, 1872	122, 448
Pavement, Combined concrete and wood	M. Flanigan	Detroit, Mich	May 30, 1871	115, 457
Pavement, Combined wood and concrete	J. W. Brown	Washington, D. C	June 21, 1870	104, 551
Pavement, Combined wood and concrete	S. C. Prescott	Jersey City, N. J	June 20, 1871	116, 217
Pavement, Composition	N. B. Abbott	Brooklyn, N. Y	June 17, 1873	139, 848
Pavement, Composition	A. G. Anderson	Hoboken, N. J	June 13, 1871	115, 924
Pavement, Composition	H. L. Crawford	Brooklyn, N. Y	Aug. 3, 1869	93, 280
Pavement, Composition	J. P. Cranford	Brooklyn, N. Y	Mar. 23, 1869	88, 139
Pavement, Composition	J. R. Hayes	Philadelphia, Pa	Feb. 27, 1872	124, 059
Pavement, Composition	D. C. Heller	Reading, Pa	Dec. 12, 1865	51, 513
Pavement, Composition	D. C. Heller	Reading, Pa	Apr. 22, 1873	138, 023
Pavement, Composition	E. S Martin	Rahway, N. J	July 9, 1872	128, 805
Pavement, Composition	T. Price	Pittsburgh, Pa	Mar. 12, 1872	124, 509
Paving-composition	A. Thiele	Tickfaw, La	Sept. 9, 1873	142, 594
Pavement, Composition	Z. Waters and H. B. Bellamy.	Bloomington, Ill	Dec. 31, 1872	134, 500
Pavement, &c., Composition blocks for	C. W. M. Smith	San Francisco, Cal	Aug. 27, 1872	130, 949
Pavement, Composition cement for	A. B. McKeon	Rutherford Park, N. J	Aug. 10, 1869	93, 659
Pavement, Composition for	N. B. Abbott	Brooklyn, N. Y	Mar. 21, 1871	112, 764
Pavement, Composition for	R. Atkinson	Mount Vernon, N. Y	Oct. 11, 1864	44, 589
Pavement, Composition for	C. P. Burges and J. R. Stevenson.	Rochester, N. Y	Feb. 14, 1871	111, 724
Pavement, Composition for	G. W. Davis	Fitchburgh, Mass	Oct. 9, 1866	58, 718
Pavement, &c., Composition for	W. C. Dodge	Washington, D. C	Oct. 16, 1866	58, 789
Pavement, Composition for	G. Dubellé	Boston, Mass	Nov. 24, 1868	84, 272
Pavement, Composition for	G. L. Eagan	San Francisco, Cal	May 20, 1873	139, 127
Pavement, Composition for	L. S. Filbert	Philadelphia, Pa	Nov. 14, 1871	120, 956
Pavement, Composition for	W. P. Ford and A. A. Moore	Concord, N. H	June 11, 1867	65, 660
Pavement, Composition for	P. Harder	Danville, Pa	May 19, 1863	38, 582
Pavement, Composition for	C. B. Harris	New York, N. Y	Mar. 7, 1871	112, 339
Pavement, &c., Composition for	J. Hartlieb	Reading, Pa	Nov. 6, 1866	59, 391
Pavement, Composition for	J. R. Hayes	New York, N. Y	Sept. 20, 1870	107, 489
Pavement, Composition for	F. E. Matthews	Chicago, Ill	Apr. 25, 1871	114, 172
Pavement, Composition for	G. H. Moore	Norwich, Conn	Nov. 28, 1871	121, 397
Pavement, &c., Composition for	G. Scrimshaw	Milesburgh, Pa	Aug. 21, 1860	29, 722

Index of patents issued from the United States Patent Office from 1790 *to* 1873, *inclusive*—Continued.

Invention.	Inventor.	Residence.	Date.	No.
Pavement, Composition for	H. F. Snow and J. H. Davis	Dover, N. H	Sept. 1, 1868	81, 698
Pavement, Composition for	H. F. Snow and J. H. Davis	Dover, N. H	Nov. 16, 1869	96, 988
Pavement, Composition for	H. F. Snow and J. H. Davis	Dover, N. H	June 18, 1872	127, 992
Pavement, Composition for	A. Wanner	Chicago, Ill	Dec. 17, 1872	134, 112
Pavement, Composition for	E. Wenger	Richmond, Ind	Aug. 25, 1868	81, 564
Pavement, Composition for	S. J. Whiting	Philadelphia, Pa	Sept. 9, 1873	142, 601
Pavement, Composition for concrete	H. Burlew	Lockhaven, Pa	Sept. 2, 1862	36, 338
Pavement, Composition for concrete	R. Fisk	New York, N. Y	Oct. 8, 1867	69, 738
Pavement, Composition for concrete	E. W. Ranney	New York, N. Y	Jan. 4, 1870	98, 522
Pavement, Composition for concrete	E. Seeley	Scranton, Pa	Sept. 6, 1864	44, 117
Pavement, Composition for concrete	A. Van Camp and M. M. Hodgman.	Washington, D. C., and Saint Louis, Mo.	Mar. 15, 1870	100, 954
Pavement, Composition for forming blocks for	J. C. Tucker	New York, N. Y	July 2, 1872	128, 680
Pavement, Compound for	J. C. and M. V. Campbell	Syracuse, N. Y	Jan. 21, 1868	73, 434
Pavement, Compound for	R. Crawford, jr	Brooklyn, N. Y	Apr. 9, 1872	125, 545
Pavement, Compound for making concrete	H. M. Conklin	Carlstadt, N. J	Mar. 15, 1870	100, 730
Pavement, Concrete	D. W. Bailey	Chelsea, Mass	Nov. 23, 1869	97, 149
Pavement, Concrete	G. Bassett	Syracuse, N. Y	Oct. 29, 1872	132, 620
Pavement, Concrete	G. W. Dean	Stamford, Conn	Sept. 2, 1873	142, 444
Pavement, Concrete	N. H. Downs	Birmingham, Conn	Apr. 19, 1870	102, 097
Pavement, Concrete	S. Filbert	Philadelphia, Pa	Jan. 9, 1872	122, 591
Pavement, Concrete	W. Gilbert	Detroit, Mich	Apr. 2, 1872	125, 284
Pavement, Concrete	J. M. Hawes	Covington, Ky	Oct. 3, 1871	119, 607
Pavement, Concrete	R. A. Jackson and S. Gissinger	Pittsburgh, Pa	Oct. 10, 1871	119, 850
Pavement, Concrete	B. N. Lampman	Rutland, Vt	Apr. 16, 1867	63, 908
Pavement, Concrete	G. Leverich and A. H. Emery	New York, N. Y	June 14, 1870	104, 325
Pavement, Concrete	A. B. McKeon	Rutherford Park, N. J	Nov. 15, 1870	109, 333
Pavement, Concrete	H. Meyers	Hyde Park, Pa	May 3, 1864	42, 589
Pavement, Concrete	G. H. Moore	Norwich, Conn	Nov. 28, 1871	121, 294
Pavement, Concrete	G. H. Moore	Norwich, Conn	Jan. 30, 1872	123, 280
Pavement, Concrete	A. H. Perkins	Chicago, Ill	Mar. 12, 1872	124, 620
Pavement, Concrete	J. J. Schillinger	New York, N. Y	July 19, 1870	105, 599
Pavement, Concrete	J. J. Schillinger	New York, N. Y	Feb. 14, 1871	111, 879
Pavement, Concrete	C. Scriber	Washington, D. C	May 28, 1872	127, 190
Pavement, Concrete	R. Skinner and B. Bonnett	San Francisco, Cal	July 19, 1870	105, 502
Pavement, Concrete	T. Smith	Washington, D. C	Feb. 16, 1869	87, 007
Pavement, Concrete	W. H. Smith	New York, N. Y	Jan. 3, 1872	122, 497
Pavement, Concrete	H. B. Steele	Winchester, Conn	July 6, 1869	92, 390
Pavement, Concrete	A. Van Camp	Washington, D. C	July 27, 1869	93, 142
Pavement, Concrete	W. H. White	Troy, Ohio	Feb. 25, 1868	74, 963
Pavement, Concrete and stone	S. B. Brittan	Newark, N. J	Oct. 24, 1871	120, 236
Pavement, Concrete and tile	G. A. Aschback	New York, N. Y	Mar. 5, 1872	124, 312
Pavement, Concrete and wooden	W. B. Coates	Philadelphia, Pa	Mar. 15, 1870	100, 727
Pavement, Concrete asphaltic	W. B. Parisen	Newark, N. J	June 13, 1871	115, 887
Pavement, Concrete compound for	J. W. Smith	Washington, D. C	Nov. 16, 1869	96, 984
Pavement, Concrete for	G. H. S. Duffus	Baltimore, Md	Oct. 25, 1870	108, 693
Pavement, &c., Concrete for	A. Van Camp and M. M. Hodgman.	Washington, D. C., and Saint Louis, Mo.	Nov. 16, 1869	96, 850
Pavement, Connecting and disconnecting the blocks of iron or other.	B. C. Smith	Burlington, N. J	Sept. 22, 1857	18, 251
Pavement, &c., Constructing	J. F. Le Moulnier	New York, N. Y	Jan. 2, 1855	12, 136
Pavement, Construction of metallic side	P. H. Jackson	New York, N. Y	Oct. 19, 1858	21, 834
Pavement, Corrugated-iron	J. Montgomery	New York, N. Y	Feb. 22, 1859	23, 038
Pavement, Corrugated metallic	J. Wilkes	New York, N. Y	Oct. 4, 1870	107, 989
Pavement, Cutting blocks for wood	W. W. Ballard and B. B. Waddell.	Elmira, N. Y., and Memphis, Tenn.	Aug. 24, 1869	94, 068
Pavement-driver	H. Landhop	New York, N. Y	Mar. 22, 1864	42, 022
Pavement, floors, walks, &c., Cement composition for	B. Doud	Cortland, N. Y	Jan. 8, 1867	61, 056
Pavement for streets, Cast-iron	W. D. Ferry	Boston, Mass	July 12, 1843	3, 168
Pavement for streets, Cemented	J. L. Sullivan	New York, N. Y	Oct. 20, 1831	....
Pavement for streets, Cemented	J. L. Sullivan	New York, N. Y	Nov. 9, 1831	....
Pavement for streets, Composition	J. W. Byrnes	Washington, D. C	Mar. 24, 1868	75, 855
Pavement for streets, Iron	G. W. Bishup	Brooklyn, N. Y	June 30, 1857	17, 662
Pavement, Form for laying brick	S. C. Brewer	Water Valley, Miss	Oct. 29, 1872	132, 560
Pavement, &c., Forming blocks for wooden	A. C. De Lisle	Paris, France	July 10, 1840	1, 683
Pavement from rot, Process for preserving wooden	A. B. Tripler	Philadelphia, Pa	May 7, 1872	126, 592
Pavement-foundation	S. Reeve	Chicago, Ill	Dec. 20, 1870	110, 395
Pavement-foundation	J. B. Wickersham	New York, N. Y	July 11, 1854	11, 251
Pavement, Gutter in foot	J. Read	Philadelphia, Pa	Nov. 3, 1868	83, 663
Pavement, Heating and preparing stone for	C. Allen	Albany, N. Y	Oct. 17, 1871	120, 016
Pavement, Implement for laying wood	P. Hinkle	San Francisco, Cal	Oct. 26, 1869	96, 113
Pavement, Inlaying plastic	G. H. Moore	Norwich, Conn	Dec. 5, 1871	121, 651
Pavement, Iron	C. Mettam	New York, N. Y	Feb. 24, 1857	16, 692
Pavement, Iron	R. Montgomery	New York, N. Y	Dec. 28, 1858	22, 444
Pavement, Iron	J. Montgomery	New York, N. Y	Mar. 8, 1859	23, 188
Pavement, Iron	J. Montgomery	New York, N. Y	Mar. 8, 1859	23, 189
Pavement, Iron	B. C. Smith	Philadelphia, Pa	Jan. 8, 1861	31, 093
Pavement, Iron	L. Stebbins	New York, N. Y	Oct. 29, 1861	33, 610
Pavement, Iron	A. R. Tewkesbury	East Boston, Mass	Mar. 9, 1858	19, 592
Pavement, Laying cement	A. F. Shyrma	New York, N. Y	July 15, 1873	140, 956
Pavement, Laying concrete	J. Hopke and J. W. Heesch	Hoboken, N. J	Feb. 27, 1872	124, 135
Pavement, Laying concrete	J. W. and A. W. Krause, jr	Philadelphia, Pa	Apr. 22, 1873	138, 030
Pavement, Laying concrete	J. W. Snyder	Chicago, Ill	May 27, 1873	139, 272
Pavement, Laying street	G. Warren	Boston, Mass	Oct. 19, 1869	96, 062
Pavement, Laying wood	E. P. Morong	Boston, Mass	Oct. 29, 1872	132, 682
Pavement, Laying wooden-block	D. L. De Golyer	Chicago, Ill	Apr. 6, 1869	88, 765
Pavement, Machine for grooving blocks for wood	P. D. Cummings	Portland, Me	Dec. 27, 1870	110, 440
Pavement, Machine for making key-board for wood	P. D. Cummings	Portland, Me	Nov. 29, 1870	109, 591
Pavement, Machine for roughing	E. Hambujer	New York, N. Y	Aug. 2, 1864	43, 687
Pavement, Machinery for preparing wooden blocks for.	W. O. Robbins and C. W. Stafford.	New York, N. Y	Dec. 7, 1869	97, 554
Pavement, Manufacture of slabs for	R. E. Stephens	Owen Sound, Canada	July 2, 1872	128, 670
Pavement, Manufacturing, laying, and finishing artificial-stone.	G. L. Eagan	San Francisco, Cal	July 23, 1872	129, 800
Pavement, Metallic	S. B. Ellithorp	New York, N. Y	Aug. 5, 1856	15, 479
Pavement, Metallic	J. Dean	West Roxbury, Mass	Oct. 22, 1867	70, 076
Pavement, Metallic	J. B. Tarr	Chicago, Ill	Jan. 29, 1867	61, 580
Pavement, Metallic	S. D. Tillman	Jersey City, N. J	June 21, 1870	104, 513

Index of patents issued from the United States Patent Office from 1790 *to* 1873, *inclusive*—Continued.

Invention.	Inventor.	Residence.	Date.	No.
Pavement, Mixer and heater for preparing concrete for.	H. Franko	Brooklyn, N. Y	Nov. 28, 1871	131, 351
Pavement, Mode of making	S. D. Tillman	New York, N. Y	Mar. 27, 1860	27, 658
Pavement or road, Laying asphalt or concrete	E. J. De Smedt	New York, N. Y	May 31, 1870	103, 581
Pavement, Preparation of composition for	J. R. Hayes	New York, N. Y	Sept. 20, 1870	107, 490
Pavement, Preparing blocks for wood	H. M. Stow	San Francisco, Cal	Apr. 9, 1872	125, 418
Pavement, Preserving wooden	E. P. Morong	Boston, Mass	Dec. 31, 1872	134, 479
Pavement, Prismatic blocks or wooden	J. Abbott	Wilton, N. H	Sept. 25, 1841	2, 265
Pavement, Ramming-machine for wood and other	A. L. Lansing	Philadelphia, Pa	Nov. 29, 1870	109, 745
Pavement, Road	M. K. Couzens	Yonkers, N. Y	Mar. 19, 1872	124, 722
Pavement, roads, &c., Manufacture of asphaltic composition for.	A. Ruttkay	New York, N. Y	Sept. 27, 1870	107, 815
Pavement, Roadway	J. T. Parson	Washington, D. C	Apr. 28, 1868	77, 208
Pavement, roadways, &c., Compounds for	R. Fisk	New York, N. Y	Oct. 12, 1869	95, 673
Pavement-roller	E. M. Sealand	Cleveland, Ohio	Apr. 27, 1869	89, 351
Pavement, roofing, &c., Composition for	J. Clarke	Syracuse, N. Y	Nov. 8, 1864	44, 938
Pavement, roofing, &c., Composition for	J. Clarke and D. French	Syracuse, N. Y	Feb. 18, 1862	34, 404
Pavement, roofing, &c., Composition for	J. E. Dotch and E. Duempelman	Washington, D. C	May 18, 1869	90, 084
Pavement, roofing, &c., Composition for	E. Duempelman and J. E. Dotch	Washington, D. C	Feb. 16, 1869	87, 031
Pavement, roofing, &c., Composition for	H. Staples	Nashua, N. H	July 27, 1869	93, 018
Pavement, roofing, drain-pipes, &c., Composition for.	J. L. Kidwell	Washington, D. C	May 18, 1869	90, 106
Pavement, Securing ranges of short plank in	J. E. Ware	Saint Louis, Mo	Feb. 11, 1851	7, 933
Pavement, Sidewalk	J. B. Cornell	New York, N. Y	Apr. 28, 1857	17, 138
Pavement, sidewalk, &c., Composition for	A. G. Anderson	Hoboken, N. J	Feb. 9, 1869	86, 799
Pavement, Stone	J. Bolliger	San Francisco, Cal	Oct. 8, 1872	131, 931
Pavement, Stone	C. Guideh	New York, N. Y	Jan. 12, 1869	85, 814
Pavement, Stone	A. Meckert	Guttenberg, N. J	Mar. 11, 1873	136, 747
Pavement, Stone	T. D. Owens	Pittsburgh, Pa	Oct. 14, 1873	143, 587
Pavement, Stone	C. G. Waterbury	New York, N. Y	May 3, 1870	102, 629
Pavement, Street	J. C. Blake	Elizabeth, N. J	Apr. 26, 1870	102, 361
Pavement, Street	P. Cadac and W. H. DeValin	San Francisco and Sacramento, Cal.	Aug. 11, 1868	80, 856
Pavement, Street	D. C. Colby	Washington, D. C	Mar. 22, 1870	101, 100
Pavement, Street	W. H. De Valin	Sacramento, Cal	Aug. 29, 1871	118, 592
Pavement, Street	A. Dilger	Saint Louis, Mo	July 5, 1870	104, 942
Pavement, Street	E. M. Fowler	New York, N. Y	Feb. 22, 1870	100, 134
Pavement, Street	G. W. Grader and M. H. Baldwin.	Memphis, Tenn	Apr. 23, 1867	64, 094
Pavement, Street	A. Hamar	New York, N. Y	Dec. 3, 1867	71, 746
Pavement, Street	L. F. Noe	New York, N. Y	Apr. 5, 1864	42, 218
Pavement, Street	N. M. Shafer	New York, N. Y	Sept. 27, 1864	44, 464
Pavement, Street	W. E. Shaw	Portland, Me	Jan. 3, 1871	110, 794
Pavement, Street	H. M. Stow	San Francisco, Cal	Dec. 10, 1867	72, 110
Pavement, Street	H. M. Stow	San Francisco, Cal	Dec. 10, 1867	72, 111
Pavement, Street	H. M. Stow	San Francisco, Cal	Feb. 25, 1868	74, 862
Pavement, Street	W. H. Williams	Little Falls, N. Y	Apr. 5, 1870	101, 691
Pavement, Street and walk	A. Hoyt	Chicago, Ill	June 2, 1868	78, 455
Pavement, Substrata for	H. P. Russ	New York, N. Y	Mar. 14, 1848	5, 475
Pavement, &c., Treating bituminous substances for	J. O'Friel	Brooklyn, N. Y	Aug. 23, 1870	106, 717
Pavement, Treating brick for	W. H. DeValin	San Rafael, Cal	Nov. 18, 1873	144, 749
Pavement-washer	A. C. Neall	Philadelphia, Pa	July 25, 1871	117, 316
Pavement-washer, hose-hydrant, and hitching-post	C. M. Alburger	Philadelphia, Pa	Aug. 22, 1854	11, 563
Pavement, Wedge-driving machine for	W. J. Harris	Elizabeth, N. J	Oct. 24, 1871	120, 193
Pavement, Wood	A. M. Adams	Washington, D. C	Aug. 16, 1870	106, 447
Pavement, Wood	M. H. Alexander	Newark, N. J	Feb. 4, 1873	135, 460
Pavement, Wood	H. J. Alvord	Washington, D. C	Aug. 30, 1870	106, 910
Pavement, Wood	W. W. Ballard	Elmira, N. Y	Feb. 1, 1870	99, 391
Pavement, Wood	W. W. Ballard	Elmira, N. Y	July 12, 1870	105, 291
Pavement, Wood	W. W. Ballard	Elmira, N. Y	Mar. 26, 1872	125, 001
Pavement, Wood	W. W. Ballard	Elmira, N. Y	Apr. 30, 1872	126, 171
Pavement, Wood	W. W. Ballard and B. B. Waddell.	Elmira, N. Y., and Memphis, Tenn.	Aug. 24, 1869	94, 062
Pavement, Wood	W. O. Barton	Elizabeth, N. J	Oct. 31, 1871	120, 481
Pavement, Wood	G. A. Beidler	Philadelphia, Pa	Jan. 30, 1872	123, 219
Pavement, Wood	G. A. Beidler	Philadelphia, Pa	Feb. 6, 1872	123, 445
Pavement, Wood	G. A. Beidler	Philadelphia, Pa	May 14, 1872	126, 776
Pavement, Wood	H. M. Beidler	Philadelphia, Pa	Jan. 31, 1871	111, 425
Pavement, Wood	A. Betteley	Boston, Mass	Aug. 24, 1869	94, 066
Pavement, Wood	A. Betteley	Boston, Mass	Nov. 16, 1869	96, 870
Pavement, Wood	A. Betteley	Boston, Mass	Mar. 29, 1870	101, 346
Pavement, Wood	A. Betteley	Boston, Mass	Sept. 6, 1870	106, 989
Pavement, Wood	A. Betteley	Boston, Mass	Sept. 6, 1870	107, 152
Pavement, Wood	*L. H. Boole*	*New York, N. Y*	*Mar. 1, 1870*	*100, 363*
Pavement, Wood	J. W. Brocklebank and W. W. Tubbs.	New York, N. Y., and Elizabeth, N. J.	May 31, 1870	103, 713
Pavement, Wood	J. L. Brown	New York, N. Y	Nov. 5, 1867	70, 514
Pavement, Wood	C. P. Burgess	Rochester, N. Y	Sept. 19, 1871	119, 010
Pavement, Wood	J. J. Burrows	Washington, D. C	May 28, 1872	127, 304
Pavement, Wood	W. Bushnell	Elizabeth, N. J	Jan. 17, 1871	110, 954
Pavement, Wood	W. Bushnell	Elizabeth, N. J	Feb. 21, 1871	111, 904
Pavement, Wood	L. Caldwell	Elmira, N. Y	Jan. 2, 1872	122, 358
Pavement, Wood	L. Caldwell	Elmira, N. Y	Mar. 26, 1872	125, 019
Pavement, Wood	L. Cauvet	New York, N. Y	June 7, 1870	103, 979
Pavement, Wood	G. H. Chinnock	New York, N. Y	May 7, 1872	126, 521
Pavement, Wood	C. C. Converse	Brooklyn, N. Y	Aug. 31, 1869	94, 284
Pavement, Wood	T. Cowing	San Francisco, Cal	Apr. 5, 1870	101, 590
Pavement, Wood	J. Defoe	Detroit, Mich	July 2, 1872	128, 471
Pavement, Wood	D. L. De Golyer	Chicago, Ill	Nov. 30, 1869	97, 278
Pavement, Wood	H. Dowson	Springfield, Ill	July 26, 1870	105, 658
Pavement, Wood	G. W. Dyer	Washington, D. C	Mar. 19, 1872	124, 799
Pavement, Wood	J. Dyer	Rouseville, Pa	July 30, 1872	130, 027
Pavement, Wood	J. Dyer	Rouseville, Pa	July 30, 1872	130, 028
Pavement, Wood	J. Q. Felt	Detroit, Mich	Jan. 21, 1873	134, 985
Pavement, Wood	M. Fitzgibbons	New York	Sept. 13, 1870	107, 352
Pavement, Wood	M. Fitzgibbons	New York, N. Y	Oct. 10, 1871	119, 837
Pavement, Wood	M. Flanigan	Detroit, Mich	May 21, 1872	126, 881
Pavement, Wood	L. T. Follansbee	Washington, D. C	Oct. 25, 1870	108, 697

Index of patents issued from the United States Patent Office from 1790 *to* 1873, *inclusive*—Continued.

Invention.	Inventor.	Residence.	Date.	No.
Pavement, Wood	Z. E. Forbes	Troy, N. Y	May 7, 1872	126, 454
Pavement, Wood	C. Gaytes	Chicago, Ill	Oct. 25, 1870	108, 582
Pavement, Wood	J. F. Gyles	Chicago, Ill	Aug. 29, 1871	118, 528
Pavement, Wood	H. Haupt	Philadelphia, Pa	May 24, 1870	103, 328
Pavement, Wood	L. Hill	Alexandria, Va	June 13, 1871	115, 960
Pavement, Wood	D. W. Hunt	San Francisco, Cal	Sept. 22, 1868	82, 320
Pavement, Wood	S. H. Ingersoll	New York, N. Y	Mar. 29, 1870	101, 270
Pavement, Wood	S. H. Ingersoll	New York, N. Y	Oct. 3, 1871	119, 519
Pavement, Wood	S. H. Ingersoll	Brooklyn, N. Y	Feb. 27, 1872	124, 064
Pavement, Wood	J. Judge	New York, N. Y	July 18, 1871	117, 177
Pavement, Wood	G. A. May	Chicago, Ill	Mar. 21, 1871	112, 945
Pavement, Wood	H. G. McGonegal	New York, N. Y	Jan. 19, 1869	86, 025
Pavement, Wood	H. G. McGonegal	New York, N. Y	Nov. 12, 1872	133, 047
Pavement, Wood	D. McKenzie	Brooklyn, N. Y	July 26, 1870	105, 707
Pavement, Wood	D. McKenzie	Utica, N. Y	Nov. 8, 1870	109, 140
Pavement, Wood	E. McMullen	Montreal, Canada	July 4, 1871	116, 734
Pavement, Wood	A. R. McNair	New York, N. Y	Aug. 24, 1869	94, 019
Pavement, Wood	A. Miller	Chicago, Ill	June 29, 1869	91, 861
Pavement, Wood	A. Miller and C. Mason	Chicago, Ill	Mar. 31, 1868	76, 227
Pavement, Wood	W. S. Morse	Washington, D. C	Dec. 13, 1870	110, 153
Pavement, Wood	D. H. Mulford	Saratoga Springs, N. Y	Aug. 29, 1871	118, 471
Pavement, Wood	H. E. Paine	Milwaukee, Wis	Apr. 25, 1871	114, 190
Pavement, Wood	J. F. Paul	Boston, Mass	June 8, 1869	91, 158
Pavement, Wood	E. W. Perrin	Portland, Oreg	Dec. 30, 1873	146, 019
Pavement, Wood	H. E. Perry	San Francisco, Cal	Feb. 27, 1872	124, 081
Pavement, Wood	J. I. Peyton	Washington, D. C	Apr. 9, 1872	125, 482
Pavement, Wood	J. I. Peyton	Washington, D. C	June 4, 1872	127, 510
Pavement, Wood	R. C. Phillips	Cincinnati, Ohio	Dec. 5, 1871	121, 544
Pavement, Wood	A. W. Platt	New York, N. Y	Mar. 8, 1870	100, 554
Pavement, Wood	A. W. Platt	New York, N. Y	Oct. 11, 1870	108, 181
Pavement, Wood	A. Potts	Philadelphia, Pa	Apr. 25, 1871	114, 038
Pavement, Wood	D. C. Reeves	Washington, D. C	Feb. 13, 1872	123, 728
Pavement, Wood	P. H. Reinhard and E. T. M. Faehtz.	Washington, D. C	Apr. 18, 1871	113, 929
Pavement, Wood	W. O. Robbins	New York, N. Y	Aug. 17, 1869	93, 837
Pavement, Wood	J. J. Schroyer	Springfield, Ill	June 28, 1870	104, 778
Pavement, Wood	W. E. Shaw	Portland, Me	July 18, 1871	117, 212
Pavement, Wood	F. H. Smith	Washington, D. C	Dec. 31, 1872	134, 492
Pavement, Wood	H. M. Stow	San Francisco, Cal	Apr. 9, 1872	125, 416
Pavement, Wood	H. M. Stow	San Francisco, Cal	Apr. 9, 1872	125, 417
Pavement, Wood	H. M. Stowe	San Francisco, Cal	Dec. 31, 1872	134, 494
Pavement, Wood	H. M. Stow	San Francisco, Cal	Apr. 15, 1873	137, 973
Pavement, Wood	H. H. Thayer	Philadelphia, Pa	May 9, 1871	114, 727
Pavement, Wood	A. Van Camp and M. M. Hodgman.	Washington, D. C., and Saint Louis, Mo.	Nov. 16, 1869	96, 849
Pavement, Wood	C. Wheeler, jr	Auburn, N. Y	Mar. 4, 1873	136, 572
Pavement, Wood	J. H. Wilkinson	Concord, N. H	Apr. 12, 1870	101, 955
Pavement, Wood	R. H. Willet	Washington, D. C	May 16, 1871	114, 895
Pavement, Wood	A. Wyckoff	Elmira, N. Y	May 7, 1872	126, 617
Pavement, Wood	G. F. Ziegler	Jersey City, N. J	June 4, 1872	127, 669
Pavement, Wood and concrete	E. A. L. Roberts	Titusville, Pa	Aug. 1, 1871	117, 685
Pavement, Wooden	G. A. Beidler	Philadelphia, Pa	Aug. 23, 1870	106, 651
Pavement, Wooden	G. A. Beidler	Philadelphia, Pa	Apr. 18, 1871	113, 838
Pavement, Wooden	J. W. Brocklebank and C. Trainer.	New York, N. Y	Jan. 12, 1869	85, 786
Pavement, Wooden	W. H. Chappell	Saint Louis, Mo	Apr. 19, 1864	42, 347
Pavement, Wooden	J. R. Cushier	New York, N. Y	Mar. 30, 1869	88, 455
Pavement, Wooden	C. K. Deutsch	New York, N. Y	Aug. 22, 1871	118, 216
Pavement, Wooden	H. Fayette	Port Chester, N. Y	May 21, 1867	64, 959
Pavement, Wooden	H. Fayette	Port Chester, N. Y	May 21, 1867	64, 960
Pavement, Wooden	J. Grant	Milwaukee, Wis	Feb. 28, 1871	112, 239
Pavement, Wooden	L. Holms	Paterson, N. J	Nov. 2, 1869	96, 433
Pavement, Wooden	P. Howard	San Francisco, Cal	Nov. 16, 1869	96, 918
Pavement, Wooden	S. Kneass	Philadelphia, Pa	Apr. 26, 1870	102, 277
Pavement, Wooden	P. Koch	New Haven, Conn	Aug. 27, 1867	68, 089
Pavement, Wooden	H. McDougall	Chicago, Ill	Mar. 24, 1868	75, 780
Pavement, Wooden	D. McKenzie	New York, N. Y	July 7, 1868	79, 674
Pavement, Wooden	S. Nicolson	Boston, Mass	Aug. 8, 1854	11, 491
Pavement, Wooden	R. L. Ream	New York, N. Y	Oct. 8, 1867	69, 703
Pavement, Wooden	W. D. Richardson	Springfield, Ill	Aug. 25, 1868	81, 540
Pavement, Wooden	E. Shaw	Portland, Me	Mar. 7, 1871	112, 503
Pavement, Wooden	J. K. Thompson	Chicago, Ill	Mar. 1, 1870	100, 339
Pavement, Wooden	C. G. Waterbury	New York, N. Y	Oct. 1, 1867	69, 517
Pavement, Wooden-block	A. Trenaunay	Neuilly Sur Seine, near Paris, France.	May 10, 1870	102, 884
Pavement, Wooden street	C. G. Waterbury	New York, N. Y	May 10, 1870	102, 981
Paving and roofing, Composition	A. B. Vandermark	Jersey City, N. J	Aug. 8, 1871	117, 946
Paving and roofing, Concrete for	E. Duempelman	New York, N. Y	Nov. 29, 1870	109, 724
Paving-block	R. Shaler	Madison, Conn	Dec. 21, 1869	98, 197
Paving-block, Concrete	R. Norwood	Baltimore, Md	Jan. 11, 1870	98, 791
Paving-blocks, Device for keying	P. D. Cummings	Portland, Me	Nov. 8, 1870	108, 977
Paving-blocks, Frame for setting	W. Gilbert	Detroit, Mich	Dec. 31, 1872	134, 427
Paving-blocks, Machine for making	S. W. Brooks	Brownsville, Tex	Oct. 24, 1871	120, 237
Paving blocks, Machine for preparing	W. O. Robbins and C. W. Stafford.	New York, N. Y	Dec. 7, 1869	97, 553
Paving-blocks, Machine for sawing	W. W. Ballard	Elmira, N. Y	Aug. 24, 1869	94, 669
Paving-blocks of wood, Machine for sawing	A. Nash	Calais, Me	Nov. 13, 1840	1, 857
Paving-composition	H. Saunders	Chester, Pa	Oct. 3, 1871	119, 476
Paving-composition	J. W. Smith	Washington, D. C	Sept. 14, 1869	94, 785
Paving, &c., Concrete	J. E. Dotch	Washington, D. C	Dec. 14, 1869	97, 893
Paving, Concrete and tile	C. Burlew	Lockhaven, Pa	Aug. 27, 1867	68, 284
Paving, Concrete blocks for	J. J. Schillinger	New York, N. Y	Jan. 31, 1871	111, 480
Paving machine, Street	T. Davidson, jr	Kensington, Pa	Jan. 8, 1856	14, 054
Paving, Rammer for street	E. F. M. Faehtz	Washington, D. C	Feb. 27, 1872	123, 986
Paving-roller	E. W. Kittridge	Cincinnati, Ohio	Sept. 8, 1868	82, 009
Paving, roofing, and other purposes, Composition for.	J. E. Doth and E. Duempelman	Washington, D. C	June 1, 1869	90, 825
Paving, roofing, &c., Conglomerate for	F. N. Hopkins	Baltimore, Md	Nov. 23, 1869	97, 088

Index of patents issued from the United States Patent Office from 1790 *to* 1873, *inclusive*—Continued.

Invention.	Inventor.	Residence.	Date.	No.
Paving, &c., Stone and metal conglomerate for	G. H. Knight	Cincinnati, Ohio	Apr. 1, 1851	8, 020
Paving streets, &c., Forming blocks of wood for	S. Carey	New Orleans, La	Feb. 3, 1841	1, 966
Paving streets, &c., Forming blocks of wood for	J. H. Patterson	New York	Jan. 27, 1841	1, 953
Pawl and ratchet	G. G. Taylor	Worcester, Mass	Sept. 22, 1863	40, 086
Pawl and ratchet device	A. M. Hills	East Hartford, Conn	Apr. 9, 1872	125, 391
Pawl-attachment	N. A. Baker	Covington, Ky	Mar. 12, 1872	124, 416
Pawl-drill	S. Ingersoll	New York, N. Y	Oct. 10, 1854	11, 787
Pawl, Friction	J. Moore	San Francisco, Cal	May 7, 1867	64, 554
Pawl, Jointed	S. S. Walley	Philadelphia, Pa	Sept. 11, 1849	6, 705
Pawl or ratchet	E. P. Russell	Manlius, N. Y	July 24, 1860	29, 310
Pawl, Take-up-detaining	G. Richardson	Lowell, Mass	Nov. 15, 1870	109, 345
Pea and bean sheller	L. Cutting	San Francisco, Cal	Apr. 17, 1866	53, 956
Pea and bean sheller	W. K. Lewis	Boston, Mass	May 22, 1866	54, 927
Pea-dropper	A. J. Williams	Barnesville, Ga	May 4, 1869	89, 722
Pea-picker	A. Quinn	Wilmington, N. C	Mar. 9, 1869	87, 705
Pea-rake	E. W. Rowley, jr	Antwerp, N. Y	July 16, 1867	66, 891
Pea-rake	S. Skinner	Clayton, N. Y	Mar. 19, 1869	87, 718
Pea-sheller	W. H. Bridgens	New York, N. Y	Nov. 26, 1867	71, 273
Pea-sheller	G. B. Price	Watervliet, N. Y	Aug. 16, 1864	43, 864
Pea-sheller	G. T. Savary	Groveland, Mass	Sept. 25, 1866	58, 305
Pea-sheller	S. Ustick	Philadelphia, Pa	May 5, 1868	77, 683
Pea-sheller and cherry-stoner	G. Sanford	New York, N. Y	Oct. 3, 1865	50, 278
Pea-shelling machine	G. Clark, jr	Boston, Mass	July 17, 1866	56, 370
Pea-shelling machine	M. Gray and J. A. Talpey	Boston and Somerville, Mass.	Jan. 30, 1866	52, 356
Pea-shelling machine	W. J. Stevenson	New York, N. Y	Mar. 30, 1858	19, 800
Pea-vines, Machine for gathering	J. B. Stanley	Copiah County, Miss	Jan. 9, 1849	6, 009
Peach-cutter	W. J. Hill	Fayetteville, Tenn	June 10, 1873	139, 714
Peach-cutter	J. T. Vaughn	Griffin, Ga	Jan. 19, 1869	86, 045
Peach cutter and stoner	J. C. Kuhn	Booneville, Ark	Mar. 15, 1859	23, 255
Peach cutting and stoning machine	W. D. Mayfield	Ashley, Ill	Apr. 3, 1866	53, 642
Peach-stoner	W. D. Hatch	Antrim, N. J	Dec. 10, 1872	133, 855
Peach-stoner	J. S. Lester	Atlanta, Ga	May 28, 1872	127, 355
Peach-stone extractor	T. E. Johnson	Indianapolis, Ind	Aug. 8, 1871	117, 892
Peaches, Box for assorting	E. W. Lockwood	Middletown, Del	Feb. 25, 1873	136, 252
Pea-nut and bean sheller	J. H. McNash	Waverly Station, Va	Oct. 1, 1872	131, 768
Pea-nut and coffee polisher	B. F. Walters	Norfolk, Va	July 2, 1872	128, 515
Pea-nut cleaning and polishing machine	J. M. Keating	Norfolk, Va	Mar. 8, 1870	100, 639
Pea-nut digger	C. W. Nicholson	Southampton County, Va	Oct. 1, 1872	131, 891
Pea-nut heater	D. Kellogg	Ypsilanti, Mich	Oct. 28, 1873	144, 108
Pea-nut huller	J. Johnson	New York, N. Y	Apr. 18, 1871	113, 889
Pea-nut picker	W. A. Crocker	Norfolk, Va	Nov. 30, 1869	97, 364
Pea-nut picker	S. C. Newell and G. Baars	Beardstown, Tenn	July 16, 1872	129, 581
Pea-nut picking and cleaning machine	J. C. Underwood	Surry Court-House, Va	Aug. 25, 1868	81, 562
Pea-nut sheller	J. W. Sands and B. F. Walters	Norfolk, Va	Jan. 16, 1872	122, 856
Pea-nut warmer	W. Chrysler	Lockport, N. Y	July 9, 1872	128, 710
Pea-nuts, Bleaching	F. M. Ironmonger	Norfolk, Va	May 7, 1872	126, 550
Pea-nuts, Steaming and polishing	S. Shepherd	Nashua, N. H	Sept. 29, 1857	18, 302
Pearl on solid substances, Imitation of	C. Sticht	Paris, France	Feb. 26, 1867	62, 507
Pearl ornaments in handles of cast metal, Securing	C. Dickinson and W. Bellamy	Newark, N. J	July 8, 1856	15, 286
Pearl-shell, Process for cutting and shaping	E. Carrington	West Meriden, Conn	Jan. 14, 1873	134, 851
Pearl-ash	J. and N. Paree	Lincklean, N. Y	May 14, 1836	....
Pearl-ash in kettles without the use of ovens, Making	R. Ainsworth	....	May 14, 1808	....
Pearl-ash, Manufacture of	J. W. Brown	Washington, D. C	Feb. 26, 1867	62, 312
Pearl-ash, Manufacture of	W. A. Edwards	Clinton, Mich	Feb. 20, 1849	6, 117
Pearl-ash, Manufacture of	W. Wentworth and G. W. Cleveland	Constable, N. Y	Aug. 20, 1872	130, 613
Peat and brick machine	G. B. Fisher	Chicago, Ill	Nov. 19, 1867	70, 983
Peat and brick machine	E. Hall	Dunnamora, N. Y	Dec. 31, 1867	72, 847
Peat and brick machine	C. D. Wrightington and B. P. Rider	Fairhaven and Boston, Mass.	Dec. 10, 1867	71, 939
Peat and feeding it to furnaces, Method of preparing	J. Yates	Mott Haven, N. Y	Apr. 17, 1866	54, 052
Peat-carbonizing apparatus	J. Adams	Rochester, N. Y	Nov. 9, 1869	96, 649
Peat charcoal, Apparatus for making	J. B. Lyons	Litchfield, Conn	May 14, 1867	64, 777
Peat, Cistern for draining	E. H. Ashcroft	Lynn, Mass	Apr. 10, 1866	53, 769
Peat, clay, &c., Machine for pressing	E. M. Hamilton	Minneapolis, Minn	Nov. 19, 1867	71, 163
Peat-cleaning machine	E. H. Ashcroft	Lynn, Mass	Oct. 25, 1864	44, 776
Peat-compressing machine	J. Jones	Baltimore, Md	Oct. 8, 1867	69, 677
Peat-condensing machine	N. C. Sawyer	Boston, Mass	Oct. 4, 1864	44, 557
Peat-crib	M. S. Roberts	Racine, Wis	Feb. 4, 1868	74, 141
Peat-digger	J. B. Lyons	Litchfield, Conn	Dec. 24, 1867	72, 513
Peat-digging machine	A. Fitts	Worcester, Mass	Apr. 22, 1856	14, 716
Peat-disintegrating machine	A. Michelbacher	New York, N. Y	July 2, 1867	66, 376
Peat-dryer	H. Bradford	New York, N. Y	Aug. 28, 1866	57, 469
Peat-dryer	G. Weissenborn	New York, N. Y	Oct. 16, 1866	58, 926
Peat, Drying and charring	F. L. H. Danchell	London, England	Apr. 4, 1865	47, 162
Peat, Drying and preparing	T. G. Walker	New York, N. Y	July 2, 1867	66, 428
Peat-drying floor	J. B. Lyons	Milton, Conn	Mar. 1, 1870	100, 301
Peat-elevator	E. J. Hulbert and A. N. N. Aubin	Portland, Conn	Aug. 1, 1871	117, 634
Peat, &c., Expelling volatile matter from	T. G. Walker	New York, N. Y	Feb. 8, 1870	99, 733
Peat, Expelling volatile matter from	T. G. Walker	New York, N. Y	Nov. 1, 1870	108, 853
Peat for composting, Treatment of	J. B. Hyde	Newark, N. J	Apr. 5, 1859	23, 467
Peat for fuel, Apparatus for preparing	J. Webster	Malden, Mass	Aug. 20, 1867	68, 016
Peat for fuel, Machine for preparing	S. Marden	Newton, Mass	Mar. 27, 1866	53, 538
Peat for fuel, Preparing	W. S. Tisdale	New York, N. Y	Sept. 24, 1872	131, 577
Peat for fuel, Treatment of	E. J. Hulbert and A. N. N. Aubin	Portland, Conn	Dec. 26, 1871	122, 173
Peat fuel, Machine and process for manufacture of	T. H. Leavitt	Boston, Mass	Dec. 26, 1871	122, 181
Peat fuel, Manufacture of	D. Aikman	Montreal, Canada	Apr. 4, 1871	113, 478
Peat fuel, Manufacture of	R. A. Griffin	Montreal, Canada	Apr. 4, 1871	113, 423
Peat-gathering machine	A. Michelbacher	New York, N. Y	May 14, 1867	64, 784
Peat-grinder	J. B. Lyons	Cornwall, Conn	Dec. 11, 1866	60, 298
Peat-grinding machine	S. Mills	Madison, Wis	Aug. 27, 1867	68, 101
Peat into charcoal, Converting	J. B. Hyde	New York, N. Y	June 29, 1858	20, 758
Peat, Kiln for drying and preparing	S. Chapman	Newark, N. J	Aug. 27, 1867	68, 044
Peat-machine	E. Atkinson	Brookline, Mass	Mar. 5, 1867	62, 519

Index of patents issued from the United States Patent-Office from 1790 *to* 1873, *inclusive*—Continued.

Invention.	Inventor.	Residence.	Date.	No.
Peat-machine	A. N. N. Aubin	Portland, Conn	Aug. 2, 1870	105, 881
Peat-machine	N. Aubin	Plattsburgh, N. Y	Oct. 27, 1868	83, 441
Peat-machine	N. Aubin	Montreal, Canada	Dec. 28, 1869	98, 333
Peat-machine	N. A. Barbour	New York, N. Y	Jan. 1, 1867	60, 667
Peat-machine	A. Bridges	Newton, Mass	May 21, 1867	64, 832
Peat-machine	H. and H. Clayton, jr., and F. Howlett.	London, England	Oct. 14, 1873	143, 617
Peat-machine	J. T. Foster	Jersey City, N. J	Mar. 19, 1867	62, 951
Peat-machine	S. D. and S. Gilson	Oswego Falls, N. Y	July 10, 1866	56, 206
Peat-machine	G. D. Goodrich	Chicago, Ill	Dec. 10, 1867	71, 870
Peat-machine	S. B. Greacen	Norwich, Conn	Jan. 22, 1867	61, 418
Peat-machine	J. Hodges	Penny Hill, Bagshot, England.	Apr. 10, 1866	53, 935
Peat-machine	B. Hotchkiss	New Haven, Conn	Jan. 1, 1867	60, 730
Peat-machine	L. P. Jenks	Boston, Mass	Mar. 12, 1867	62, 752
Peat-machine	J. S. Kelly	New York, N. Y	Oct. 12, 1869	95, 806
Peat-machine	T. H. Leavitt	Boston, Mass	Mar. 6, 1866	53, 011
Peat-machine	Z. Ludington	Uniontown, Pa	Dec. 22, 1868	85, 178
Peat-machine	Z. Ludington	Fayette County, Pa	Mar. 28, 1871	113, 180
Peat-machine	C. Luxton	Hudson, N. J	May 14, 1867	64, 681
Peat-machine	C. Luxton	Hudson City, N. J	Apr. 5, 1870	101, 478
Peat-machine	J. B. Lyons	Cornwall, Conn	Dec. 11, 1866	60, 397
Peat-machine	J. B. Lyons	Milton, Conn	Mar. 1, 1870	100, 302
Peat-machine	J. B. Lyons	New Haven, Conn	Apr. 26, 1870	102, 414
Peat-machine	S. Marden	Newton, Mass	Apr. 24, 1866	54, 186
Peat-machine	S. Marden	Newton, Mass	Jan. 1, 1867	60, 918
Peat-machine	H. Mielisch	Racine, Wis	Mar. 21, 1871	112, 830
Peat-machine	A. Moffatt	Washington, D. C	June 5, 1866	55, 339
Peat-machine	H. C. Moore	Springfield, Mass	May 7, 1867	64, 553
Peat-machine	N. F. Potter	Providence, R. I	May 1, 1866	54, 402
Peat-machine	J. H. Rae	Syracuse, N. Y	May 22, 1866	55, 018
Peat-machine	M. S. Roberts	Lewiston, N. Y	Apr. 9, 1867	63, 751
Peat-machine	M. S. Roberts	Racine, Wis	July 28, 1868	80, 309
Peat-machine	A. Robinson	McLean, N. Y	Jan. 22, 1867	61, 464
Peat-machine	A. Root	New Haven, Conn	Feb. 19, 1867	62, 289
Peat-machine	M. B. Stafford	New York, N. Y	July 24, 1866	56, 626
Peat-machine	J. Tisdale	South Dedham, Mass	Feb. 18, 1868	74, 635
Peat-machine	W. S. Tisdale	New York, N. Y	Oct. 10, 1871	119, 801
Peat-machine	D. Wellington	Boston, Mass	Mar. 6, 1866	53, 065
Peat-machine	D. Wellington	Boston, Mass	Feb. 12, 1867	61, 965
Peat-machine	T. J. Wells	New York, N. Y	Feb. 6, 1866	52, 472
Peat-machine	T. J. Wells	New York, N. Y	Nov. 20, 1866	59, 882
Peat-machine	T. J. Wells	Saint Anthony, Minn	Mar. 26, 1867	63, 342
Peat, Machine for slicing and drying	A. H. Emery	New York, N. Y	Sept. 18, 1866	58, 190
Peat, Machine for tempering and preparing	W. F. Potter	Providence, R. I	Apr. 18, 1865	47, 331
Peat-machines, Material for dusting the mold of	T. H. Leavitt	Boston, Mass	Mar. 6, 1866	53, 014
Peat-mold	K. Goddard	Richmond, N. Y	Aug. 31, 1869	94, 302
Peat molding and drying apparatus	E. H. Ashcroft	Lynn, Mass	May 15, 1866	54, 663
Peat-molding apparatus	E. Weissenborn	Hudson City, N. J	June 12, 1866	55, 565
Peat-molding machine	N. Aubin	Montreal, Canada	June 25, 1870	99, 130
Peat-molding machine	J. B. Lyons	New Haven, Conn	Jan. 21, 1868	73, 616
Peat, Preparation of	C. E. L. Holmes	New York, N. Y	Nov. 8, 1870	109, 009
Peat, Preparing	W. H. Buckland	Glamorgan County, Great Britain.	Feb. 11, 1862	34, 391
Peat-preparing apparatus	E. H. Ashcroft	Lynn, Mass	May 22, 1866	54, 830
Peat-preparing apparatus	C. W. Baldwin	Boston, Mass	Dec. 17, 1867	72, 357
Peat-preparing apparatus	L. Elsberg	New York, N. Y	Aug. 28, 1866	57, 489
Peat-preparing apparatus	H. Y. Hind	Fredrickton, New Brunswick.	Mar. 27, 1866	53, 547
Peat-preparing apparatus	S. Marden	Newton, Mass	May 29, 1866	55, 205
Peat-preparing apparatus	M. S. Roberts	Lewiston, N. Y	Aug. 15, 1865	49, 438
Peat-preparing machine	E. H. Ashcroft	Lynn, Mass	Apr. 24, 1866	54, 088
Peat-preparing machine	C. W. Inglis	Paterson, N. J	Mar. 24, 1868	75, 763
Peat-pressing machine	F. Leach	Tioga, N. Y	June 25, 1867	66, 093
Peat, Process for solidifying	L. Elsberg	New York, N. Y	Oct. 4, 1864	44, 520
Peat, Process of treating	B. R. Hawley and A. R. Morgan.	Normal, Ill., and New York, N. Y.	June 4, 1872	127, 415
Peat, Tank for preparing	A. Betteley	Boston, Mass	Jan. 9, 1866	51, 913
Peat, Treating	H. Ball	New York, N. Y	Feb. 12, 1867	61, 985
Peat, Treating	J. H. Smith	New York, N. Y	Oct. 31, 1865	50, 743
Peat, Treating	J. H. Smith	New York, N. Y	Nov. 28, 1865	51, 231
Peat-wringer	A. N. N. Aubin	Portland, Conn	Aug. 1, 1871	117, 591
Pedecycle	G. Brownlee	Princeton, Ind	July 27, 1869	92, 936
Pedestal to column, Fastening	W. and W. H. Lewis	New York, N. Y	June 24, 1851	8, 181
Pedo-motive, Infant's	E. J. Gorham	Bangor, Me	Aug. 4, 1863	39, 392
Pedrometer, Sea	B. Lies	New York	June 20, 1815	....
Peg, Condensed-leather	C. and J. G. Rowland	Quincy, Ill	Aug. 20, 1867	68, 005
Peg-cutter	C. H. Bacon	Boston, Mass	Dec. 23, 1873	145, 714
Peg-cutter	H. M. Buell	Waterbury, Conn	Sept. 3, 1872	130, 974
Peg-cutter	C. Holmes	Lynn, Mass	June 18, 1872	128, 147
Peg-cutter	J. G. Klock	Mansfield, Ohio	Mar. 19, 1872	124, 826
Peg-cutter	E. R. Pease and R. R. Hayman	Poughkeepsie, N. Y	Jan. 11, 1859	22, 577
Peg-cutter	D. B. Sanderson	Lewiston, Me	Apr. 21, 1868	76, 949
Peg-cutter	L. S. Smith	Gorsuch's Mills, Md	Sept. 24, 1867	69, 265
Peg-cutter	W. M. Watt	Como, Miss	Aug. 8, 1871	117, 839
Peg-cutter	A. Whittemore	Cambridgeport, Mass	Dec. 9, 1873	145, 469
Peg cutter, Boot and shoe	H. E. Chapman	Albany, N. Y	Dec. 11, 1855	13, 901
Peg cutter, Boot and shoe	S. R. Jones	Baltimore, Md	Jan. 8, 1856	14, 060
Peg cutter, Boot and shoe	W. H. Sellers	Keokuk, Iowa	Apr. 6, 1869	88, 588
Peg cutter, Boot and shoe	N. S. Wakefield	Camden, N. J	Dec. 31, 1872	134, 449
Peg-cutter, Hand	P. Summiter	Marlborough, Mass	July 8, 1873	140, 599
Peg-cutting instrument, Boot and shoe	F. Osborne	South Hanson, Mass	July 13, 1869	92, 637
Peg-cutting machine	H. Thurber	Painted Post, N. Y	May 22, 1828	....
Peg-cutting machine	J. D. Vinnedge	Indianapolis, Ind	May 27, 1873	139, 441
Peg-cutting machine, Boot and shoe	J. Essex	Killingly, Conn	July 17, 1837	281, 000
Peg-cutting machine, Boot and shoe	A. Whittemore	Cambridgeport, Mass	July 15, 1873	140, 803
Peg-cutting machine, Shoe	S. K. Baldwin	Guilford, N. H	July 16, 1842	2, 275
Peg-cutting tool, Boot and shoe	D. D. Allen	Adams, Mass	Oct. 19, 1852	9, 340

Index of patents issued from the United States Patent Office from 1790 *to* 1873, *inclusive*—Continued.

Invention.	Inventor.	Residence.	Date.	No.
Peg-feeding apparatus	S. D. Tripp	Stoneham, Mass	Sept. 13, 1859	25, 472
Peg-float	W. Miller, J. J. Becker, and A. Simcox.	Fort Wayne, Ind	May 26, 1868	78, 227
Peg-float	C. H. Odell	Poughkeepsie, N. Y	Dec. 22, 1863	41, 015
Peg-float	P. A. Schoellhorn	Buffalo, N. Y	Mar. 4, 1873	136, 551
Peg float, Shoe	G. G. Townsend	Rochester, N. Y	Dec. 18, 1866	60, 596
Peg machine, Shoe	S., S., jr., and E. Beckwith	Westmoreland, N. Y	Jan. 27, 1829	....
Peg machine, Shoe	A. H. Boyd	Saco, Me	Mar. 23, 1858	19, 730
Peg machine, Shoe	A. Brown	West Waterford, Vt	Nov. 16, 1858	22, 061
Peg machine, Shoe	C. Cook	Hillsborough, N. H	Apr. 5, 1859	23, 451
Peg machine, Shoe	C. A. Priest	Winslow, Me	Aug. 7, 1860	29, 519
Peg machine, Shoe	E. T. Weeks	Franconia, N. H	Nov. 22, 1859	26, 238
Peg machine, Shoe	A. Woodward	Keene N. H	Feb. 23, 1858	19, 461
Peg pointing and splitting machine, Shoe	J. Reid	Marshfield, Mass	Apr. 5, 1859	23, 494
Peg-pointing machine	W. A. Greenwood	Palmer, N. Y	Feb. 19, 1831	....
Peg-pointing machine, Shoe	S. G. Brett	Newport, N. H	Aug. 14, 1860	29, 565
Peg-pointing machine, Shoe	J. Ladd	Holderness, N. H	Dec. 23, 1856	16, 280
Peg-rasp	J. W. Foard	San Francisco, Cal	Feb. 4, 1873	135, 419
Peg-rasp	J. Sawyer and L. Clark	South Royalston, Mass	Dec. 13, 1853	10, 316
Peg-rasper	E. Townsend	Boston, Mass	Mar. 13, 1866	53, 200
Peg-sharpener	A. Kehl	Indianapolis, Ind	July 18, 1871	117, 081
Peg, Shoe	O. G. Healy	Abington, Mass	Apr. 30, 1872	126, 206
Peg, Shoe	J. H. Oliver	Baltimore, Md	July 16, 1872	129, 490
Peg, Shoe	R. H. Thompson	Rochester, N. Y	Mar. 23, 1836	....
Peg-splitting machine, Shoe	N. H. Shaw	Tamworth, N. H	Feb. 17, 1857	16, 659
Peg-splitting machine, Shoe	W. D. Shaw	Tamworth, N. H	Apr. 12, 1859	23, 614
Peg-splitting machine, Shoe	M. Wilder	Peterborough, N. H	Aug. 15, 1835	....
Peg strip, Continuous leather	C. Rowland	Quincy, Ill	Feb. 4, 1868	74, 148
Peg tube and driver	J. P. Kemp	Charlestown, Mass	Dec. 6, 1859	26, 398
Pegs and cement into soles, &c., Machine for introducing.	E. Townsend	Boston, Mass	Aug. 30, 1864	44, 029
Pegs and grain, Apparatus for drying shoe	S. Kimball and W. Sawyer	Boxford, Mass	Apr. 19, 1859	23, 689
Pegs for boots and shoes, Machine for making wire for.	W. Wickersham	Boston, Mass	Aug. 22, 1871	118, 318
Pegs for boots and shoes, Screw	J. M. Estabrook	Milford, Mass	Dec. 29, 1868	85, 374
Pegs for boots and shoes, Wire for making	D. H. Campbell and E. Woodward.	Sunderland, Scotland, and Charlestown, Mass.	Dec. 27, 1870	110, 431
Pegs for boots and shoes, Wooden	T. Rowell	Hartford, Vt	Nov. 24, 1820	....
Pegs for shoes, &c., Pointing, splitting, and waxing	T. Rowell	Hartford, Vt	Feb. 11, 1825	....
Pegs in boots and shoes, Machine for inserting metallic.	T. T. Prosser	Chicago, Ill	Dec. 10, 1872	133, 798
Pegs, Machine for arranging	J. Ladd	Holderness, N. H	Sept. 13, 1859	25, 422
Pegs, Machine for making	J. S. L. Babbs	New Albany, Ind	Apr. 28, 1868	77, 241
Pegs, Machine for making condensed-leather	C. and J. G. Rowland	Quincy, Ill	Aug. 20, 1867	68, 006
Pegs, Machine for making metallic shoe	T. T. Prosser	Chicago, Ill	June 20, 1871	116, 219
Pegs, Machine for making wooden	T. A. Robertson	Georgetown, D. C	Aug. 16, 1845	4, 148
Pegs, Machine for making wooden	H. P. Westcott	Seneca Falls, N. Y	Nov. 28, 1848	5, 946
Pegs, Machine for manufacturing shoe	I. G. Worth	Vassalborough, Me	Aug. 3, 1858	21, 104
Pegs, Manufacture of shoe	G. W. Day	Clarlestown, Mass	Mar. 13, 1866	53, 124
Pegs or pins, Pointing	J. Hull	North Bridgewater, Mass	July 7, 1839	....
Pegs, Preparation for shoe	B. F. Sturtevant	Boston, Mass	July 15, 1862	35, 902
Pegs, Preparing shoe	C. Gifford	Braintree, Mass	Sept. 19, 1848	5, 780
Pegs, Putting leather together by wooden or metal	S. B. Browne	Westchester County, N. Y.	Apr. 2, 1814	....
Pegger, Hand	J. H. Brown	Boston, Mass	Oct. 3, 1865	50, 298
Pegging and nailing machine, Shoe	L. R. Blake	Boston, Mass	Jan. 19, 1869	86, 054
Pegging boots and shoes	N. A. Fisher	Westborough, Mass	Apr. 3, 1834	....
Pegging boots and shoes	A. Swingle	Boston, Mass	Feb. 12, 1856	14, 269
Pegging boots and shoes	A. Woodruff	Westford, Mass	Sept. 6, 1833	....
Pegging-jack	A. Bailey	Amesbury, Mass	July 29, 1856	15, 406
Pegging-jack	T. D. Bailey	Lowell, Mass	Dec. 21, 1858	22, 340
Pegging-jack	W. Billings	Brooklyn, N. Y	May 3, 1864	42, 618
Pegging-jack	E. Everson	Haverhill, Mass	July 24, 1860	29, 257
Pegging-jack	E. L. Harlow	Monmouth, Me	July 31, 1860	29, 434
Pegging-jack	J. Jenkins	Andover, Mass	Oct. 15, 1850	7, 721
Pegging-jack	G. A. Knowlton	Natick, Mass	May 1, 1866	54, 369
Pegging-jack	J. K. Krieg	New York, N. Y	Nov. 5, 1867	70, 577
Pegging-jack	W. R. Landfear	Hartford, Conn	Nov. 28, 1865	51, 262
Pegging-jack	D. Mead and E. W. Watson	Danversport, Mass	July 12, 1870	105, 353
Pegging-jack	H. Ricard	Haverhill, Mass	July 5, 1870	104, 999
Pegging-jack	W. H. Sweetland	Marblehead, Mass	Aug. 10, 1869	93, 497
Pegging-jack	A. Swingle	Boston, Mass	July 29, 1856	15, 462
Pegging-jack	C. Varney	East Brimfield, Mass	May 10, 1870	102, 888
Pegging-jack	A. K. Washburn	Bridgewater, Mass	Jan. 16, 1866	52, 095
Pegging-jack	J. G. Ziegler	Salt River, Mich	Oct. 29, 1872	132, 616
Pegging jack, Boot	A. W. Moore	Stafford, Conn	Feb. 21, 1865	46, 484
Pegging-jack or shoemaker's head-block	T. D. Bailey	Lowell, Mass	May 6, 1856	14, 799
Pegging-jack post	D. Bowker	Boston, Mass	June 7, 1870	103, 835
Pegging-machine	J. C. Briggs	Saratoga, N. Y	Oct. 9, 1845	4, 228
Pegging-machine	B. Q. Budding	Milford, Mass	May 7, 1867	64, 481
Pegging-machine	W. Fitzgerald	Salem, Mass	July 1, 1862	35, 749
Pegging-machine	W. Fitzgerald	Boston, Mass	Nov. 20, 1866	59, 903
Pegging-machine	J. J. Greenough	New York, N. Y	June 26, 1860	28, 852
Pegging-machine	L. Hall	Boston, Mass	Mar. 8, 1864	41, 888
Pegging-machine	O. G. Healy	Abington, Mass	Apr. 30, 1872	126, 205
Pegging-machine	H. Kuhlman	Glückstadt, Germany	Aug. 6, 1872	130, 301
Pegging-machine	W. R. Landfear	Hartford, Conn	Aug. 23, 1859	25, 204
Pegging-machine	G. W. Manson	Auburn, Me	Mar. 11, 1873	136, 743
Pegging-machine	M. Marshall	Lowell, Mass	Nov. 5, 1861	33, 679
Pegging-machine	W. B. Randall	Fayette, Me	Aug. 6, 1832	....
Pegging-machine	G. L. Roberts	Boston, Mass	Dec. 23, 1873	145, 900
Pegging-machine	J. and A. W. Sangster	Buffalo, N. Y	Apr. 5, 1859	23, 501
Pegging-machine	J. F. Sargent	Boston, Mass	Aug. 30, 1864	44, 048
Pegging machine	J. F. Sargent	Boston, Mass	Aug. 30, 1864	44, 049
Pegging-machine	J. F. Sargent	Boston, Mass	Oct. 11, 1864	44, 691
Pegging-machine	J. W. Soule	Boston, Mass	Sept. 22, 1868	82, 448
Pegging-machine	E. M. Stevens	Boston, Mass	Aug. 3, 1858	21, 091
Pegging-machine	E. M. Stevens	Boston, Mass	May 28, 1867	65, 294
Pegging-machine	H. C. Stone	Brookfield, Mass	Sept. 17, 1867	69, 040

Index of patents issued from the United States Patent Office from 1790 *to* 1873, *inclusive*—Continued.

Invention.	Inventor.	Residence.	Date.	No.
Pegging-machine	J. Taggart	Roxbury, Mass	Jan. 14, 1862	34, 170
Pegging-machine	E. Townsend	Boston, Mass	Feb. 7, 1860	27, 085
Pegging-machine	C. Varney	East Brimfield, Mass	June 21, 1870	104, 668
Pegging-machine	P. Wells	Middleton, Mass	Dec. 18, 1860	30, 950
Pegging-machine	L. D. Wheeler	Fitchburgh, Mass	Nov. 5, 1872	132, 742
Pegging-machine	L. H. Wood	Marlborough, Mass	Nov. 1, 1859	25, 989
Pegging-machine awl	E. P. Richardson	Lawrence, Mass	Feb. 20, 1872	123, 841
Pegging machine, Boot and shoe	J. A. Bradshaw	Lowell, Mass	Aug. 15, 1854	11, 514
Pegging machine, Boot and shoe	A. C. Gallahue	Metamoras, Ohio	Oct. 28, 1851	8, 465
Pegging machine, Boot and shoe	A. C. Gallahue	Allegheny City, Pa	Aug. 16, 1853	9, 947
Pegging machine, Boot and shoe	A. C. Gallahue	New York, N. Y	July 11, 1854	11, 287
Pegging machine, Boot and shoe	A. C. Gallahue	North East Centre, N. Y	Mar. 29, 1859	23, 361
Pegging machine, Boot and shoe	A. C. Gallahue	Katonah, N. Y	Aug. 26, 1862	36, 292
Pegging machine, Boot and shoe	F. Gray	Rowley, Mass	Apr. 6, 1832	....
Pegging machine, Boot and shoe	J. J. Greenough	New York, N. Y	Jan. 17, 1854	10, 427
Pegging machine, Boot and shoe	H. Halvorson	Hartford, Conn	Jan. 10, 1854	10, 407
Pegging machine, Boot and shoe	W. N. Hawley	Hartford, Conn	Jan. 31, 1860	26, 987
Pegging machine, Boot and shoe	W. B. Johnson	Sandwich, N. H	Jan. 1, 1856	14, 020
Pegging machine, Boot and shoe	W. Kidder	Newburyport, Mass	Aug. 15, 1854	11, 542
Pegging machine, Boot and shoe	L. Lackey	Sutton, Mass	May 16, 1854	10, 917
Pegging machine, Boot and shoe	J. La Dow	Granville, Ohio	July 31, 1849	6, 613
Pegging machine, Boot and shoe	W. R. Landfear	Hartford, Conn	Sept. 13, 1864	44, 203
Pegging machine, Boot and shoe	W. R. Landfear	Hartford, Conn	Nov. 5, 1867	70, 581
Pegging machine, Boot and shoe	N. Leonard	Merrimack, N. H	June 11, 1829	....
Pegging machine, Boot and shoe	M. Marshall	Lowell, Mass	Feb. 11, 1862	34, 370
Pegging machine, Boot and shoe	M. Marshall	Lowell, Mass	Dec. 9, 1862	37, 136
Pegging machine, Boot and shoe	W. Miller	Boston, Mass	Oct. 20, 1863	40, 351
Pegging machine, Boot and shoe	D. D. Palmer	Waltham, Mass	Dec. 31, 1867	72, 753
Pegging machine, Boot and shoe	T. T. Prosser	Chicago, Ill	Dec. 23, 1873	145, 754
Pegging machine, Boot and shoe	J. F. Sargent	Boston, Mass	Feb. 4, 1862	34, 335
Pegging machine, Boot and shoe	G. Schuh and P. L. Slayton	Madison, Ind	Mar. 4, 1856	14, 370
Pegging machine, Boot and shoe	J. Standish	Cuyahoga Falls, Ohio	Feb. 14, 1854	10, 521
Pegging machine, Boot and shoe	A. F. Strong	Northampton, Mass	Feb. 13, 1872	123, 592
Pegging machine, Boot and shoe	B. F. Sturtevant	Boston, Mass	June 9, 1857	17, 544
Pegging machine, Boot and shoe	B. F. Sturtevant	Boston, Mass	Aug. 11, 1857	17, 998
Pegging machine, Boot and shoe	B. F. Sturtevant	Boston, Mass	Sept. 21, 1858	21, 593
Pegging machine, Boot and shoe	S. D. Tripp	Rochester, Mass	Apr. 12, 1853	9, 629
Pegging machine, Boot and shoe	S. D. Tripp	Winchester, Mass	Sept. 8, 1857	18, 170
Pegging machine, Boot and shoe	G. J. Wardwell	Andover, Me	July 18, 1854	11, 346
Pegging machine, Boot and shoe	W. Wells	Boston, Mass	Dec. 15, 1857	18, 879
Pegging machine, Boot and shoe	D. Whittemore	North Bridgewater, Mass	Nov. 5, 1867	70, 662
Pegging machine, Boot and shoe	E. Woodward	Charlestown, Mass	Feb. 11, 1873	135, 681
Pegging machine, Boot and shoe hand	R. H. Thompson	Buffalo, N. Y	June 26, 1855	13, 144
Pegging machine, Boot and shoe hand	F. J. Vittum	Chelsea, Mass	Dec. 4, 1860	40, 846
Pegging-machine feeder	G. M. Cram	East Rindge, N. H	Mar. 5, 1872	124, 335
Pegging-machine, Hand	W. W. Batchelder	New York, N. Y	June 10, 1856	15, 055
Pegging-machine, Hand	C. H. Binger and W. E. Fischer	Boston, Mass	Mar. 8, 1864	41, 884
Pegging-machine, Hand	J. H. Brown	Watertown, Mass	Jan. 1, 1867	60, 681
Pegging-machine, Hand	J. H. Brown	Watertown, Mass	Mar. 5, 1867	62, 525
Pegging-machine, Hand	L. Goddu	Lowell, Mass	Dec. 5, 1865	51, 387
Pegging-machine, Hand	E. T. Ingalls	Haverhill, Mass	Aug. 12, 1862	36, 157
Pegging-machine, Hand	E. M. Stevens	Medford, Mass	Aug. 6, 1861	33, 022
Pegging-machine, Hand	A. Swingle	Boston, Mass	May 29, 1855	12, 985
Pegging-machine, Hand	F. J. Vittum	Newburyport, Mass	Nov. 10, 1868	84, 025
Pegging-machine jack	W. Fitzgerald	Boston, Mass	May 22, 1860	28, 358
Pegging-machine peg-box	A. W. Moore	East Brimfield, Mass	Oct. 25, 1870	108, 719
Pegging machine, Shoe	J. H. Brown	Watertown, Mass	Dec. 22, 1868	85, 206
Pegging machine, Shoe	W. G. Budlong	Providence, R. I	May 12, 1863	38, 463
Pegging machine, Shoe	A. C. Gallahue	Riverdale, N. Y	June 29, 1869	92, 037
Pegging machine, Shoe	C. Keniston	West Cambridge, Mass	Aug. 27, 1861	33, 151
Pegging machine, Shoe	L. Lackey	Sutton, Mass	July 27, 1858	21, 051
Pegging machine, Shoe	J. Robinson	Methuen, Mass	Oct. 31, 1848	5, 896
Pegging machines, Blank for shoe	B. F. Sturtevant	Boston, Mass	Aug. 16, 1859	25, 149
Pegging-machinery, Peg-feed stop for	S. A. Holt and C. H. Williams	Hudson, Mass	Aug. 18, 1868	81, 275
Pen	T. B. Briggs	Brooklyn, N. Y	July 16, 1872	129, 455
Pen	R. M. and D. Cameron	Edinburgh, North Britain	Jan. 15, 1867	61, 156
Pen	R. H. Chinn	Washington, D. C	Dec. 20, 1870	110, 342
Pen	G. Crandell	Washington, D. C	Oct. 18, 1870	108, 455
Pen	L. M. De Spraugh	New York	June 29, 1818	....
Pen	D. D. Foley	Washington, D. C	Jan. 18, 1870	98, 945
Pen	W. R. Glover	Glasgow, Ky	Aug. 1, 1854	11, 421
Pen	H. S. Goodspeed	New York, N. Y	Mar. 9, 1869	87, 661
Pen	R. Hirst	Hudson, N. Y	Mar. 24, 1868	75, 909
Pen	J. H. Holland	Hancock, Mich	Apr. 7, 1868	76, 454
Pen	T. S. Hudson	East Cambridge, Mass	Apr. 3, 1866	53, 622
Pen	W. A. Morse	Philadelphia, Pa	Jan. 14, 1868	73, 255
Pen	W. A. Morse	Philadelphia, Pa	Dec. 7, 1869	97, 676
Pen	E. L. Pratt	Boston, Mass	Feb. 18, 1868	74, 720
Pen	E. L. Pratt	Beverly, Mass	May 11, 1869	89, 940
Pen	J. T. Price	Arrow Rock, Mo	Jan. 19, 1869	85, 959
Pen	E. P. Tiffany	Hartford, Conn	Mar. 1, 1870	100, 340
Pen	J. W. Truman	Macon, Ga	June 8, 1869	91, 056
Pen	M. Wagner	Cincinnati, Ohio	June 20, 1871	116, 242
Pen	A. F. Warren	Brooklyn, N. Y	June 30, 1863	39, 084
Pen	A. G. Waterhouse	San Francisco, Cal	Dec. 7, 1869	97, 735
Pen	C. S. Westcott	Elizabeth, N. J	Aug. 13, 1872	130, 458
Pen	G. W. Woolley	Washington, D. C	Dec. 13, 1870	110, 186
Pen and eraser	W. F. Beaton	Philadelphia Pa	Aug. 27, 1867	68, 153
Pen and eraser combined	J. Schott	Chicago, Ill	July 17, 1866	56, 455
Pen and ink case	A. G. Buzby	Philadelphia, Pa	Jan. 15, 1867	61, 154
Pen and pen-holder	J. T. Foster	Jersey City, N. J	Aug. 7, 1866	56, 920
Pen and pencil calendar	T. B. Briggs	Brooklyn, N. Y	Dec. 10, 1872	133, 695
Pen and pencil case	T. Addison	New York, N. Y	May 10, 1838	736
Pen and pencil case	A. G. Bagley	New York, N. Y	Jan. 1, 1850	6, 981
Pen and pencil case	E. Baptis	Hoboken, N. J	Apr. 29, 1856	14, 760
Pen and pencil case	E. Baptis	Hudson, N. J	July 14, 1857	17, 776
Pen and pencil case	G. S. Clark	New York, N. Y	Nov. 1, 1853	10, 177
Pen and pencil case	J. M. Clark	Jersey City, N. J	Apr. 25, 1871	114, 110
Pen and pencil case	J. M. Clark	Jersey City, N. J	June 6, 1871	115, 704

Index of patents issued from the United States Patent Office from 1790 *to* 1873, *inclusive*—Continued.

Invention.	Inventor.	Residence.	Date.	No.
Pen and pencil case	J. M. Clark	Jersey City, N. J	Aug. 29, 1871	118, 434
Pen and pencil case	J. Cockburn	New York, N. Y	Apr. 6, 1858	19, 831
Pen and pencil case	J. Cockburn	New York, N. Y	Apr. 10, 1860	27, 777
Pen and pencil case	J. Cockburn	New York, N. Y	Jan. 16, 1872	122, 708
Pen and pencil case	F. W. Cox	Brooklyn, N. Y	June 27, 1865	48, 374
Pen and pencil case	F. Edgell	Buffalo, N. Y	Apr. 15, 1873	137, 900
Pen and pencil case	G. E. Frew	Brooklyn, N. Y	June 12, 1860	28, 664
Pen and pencil case	G. E. Frew	Brooklyn, N. Y	July 31, 1860	29, 432
Pen and pencil case	J. J. Hatcher	Philadelphia, Pa	Nov. 6, 1846	4, 839
Pen and pencil case	J. J. Hatcher	Philadelphia, Pa	July 11, 1854	11, 314
Pen and pencil case	W. S. Hicks	New York, N. Y	Sept. 12, 1865	49, 878
Pen and pencil case	W. S. Hicks	New York, N. Y	Oct. 31, 1871	120, 520
Pen and pencil case	W. S. Hicks	New York, N. Y	Dec. 26, 1871	122, 169
Pen and pencil case	J. H. Knapp	New York, N. Y	Sept. 2, 1856	15, 660
Pen and pencil case	J. H. Knapp	New York, N. Y	Aug. 26, 1873	142, 243
Pen and pencil case	C. W. Lord	New York, N. Y	Dec. 10, 1861	33, 894
Pen and pencil case	J. J. Lownds	New York, N. Y	Oct. 3, 1854	11, 752
Pen and pencil case	J. J. Lownds	New York, N. Y	Dec. 8, 1863	40, 846
Pen and pencil case	J. Mabie	English Neighborhood, N. Y	Oct. 3, 1854	11, 762
Pen and pencil case	J. Monaghan and T. Flynn	New York and Brooklyn, N. Y.	Dec. 19, 1871	122, 047
Pen and pencil case	J. H. Rauch	New York, N. Y	Jan. 6, 1852	8, 640
Pen and pencil case	J. H. Rauch	New York, N. Y	Dec. 8, 1863	40, 855
Pen and pencil case	J. Richardson	New York, N. Y	Oct. 17, 1854	11, 820
Pen and pencil case	J. Richardson	New York, N. Y	Oct. 24, 1854	11, 840
Pen and pencil case	J. Richardson	New York, N. Y	Oct. 31, 1854	11, 873
Pen and pencil case	J. Richardson	New York, N. Y	July 12, 1859	24, 758
Pen and pencil case	J. Richardson	New York, N. Y	Feb. 28, 1860	27, 311
Pen and pencil case	R. H. Ryne	New York, N. Y	Dec. 24, 1867	72, 684
Pen and pencil case	L. F. Standish	Springfield, Mass	Aug. 4, 1868	80, 780
Pen and pencil case	C. H. Streightoff	New York, N. Y	Oct. 21, 1873	143, 859
Pen and pencil case	H. Withers	New York, N. Y	Mar. 19, 1836	....
Pen and pencil case, Sliding	J. H. Rauch	New York, N. Y	Oct. 24, 1854	11, 839
Pen and pencil holder	A. G. Bagley	New York, N. Y	June 6, 1846	4, 557
Pen and pencil holder	G. H. Byron	Governor's Island, N. Y	Mar. 3, 1857	16, 711
Pen and pencil holder	J. Jacobs	Columbus, Ohio	Oct. 14, 1862	36, 652
Pen and pencil holder	J. T. Price	Arrow Rock, Mo	Sept. 24, 1867	69, 126
Pen and pencil holder	A. Tower	New York, N. Y	June 9, 1868	78, 698
Pen and pencil holder	T. D. Richardson	New York, N. Y	Oct. 11, 1859	25, 793
Pen and pencil holder	R. Ryne	New York, N. Y	Sept. 13, 1864	44, 261
Pen and pencil holder	R. H. Ryne	New York, N. Y	Oct. 19, 1867	62, 227
Pen and pencil holder and knife, Combined	W. A. Morse	Philadelphia, Pa	Mar. 9, 1869	87, 583
Pen-barrels by drawing, Forming metallic	A. G. Bagley	New York, N. Y	Feb. 27, 1847	4, 991
Pen blanks, Machinery for punching steel	E. A. Marsh and J. P. Kelly	Chicopee, Mass	Jan. 1, 1867	60, 763
Pen, button, and jewelry press	J. M. Riley	Newark, N. J	Apr. 20, 1869	89, 246
Pen-cleaner	S. Darling	Providence, R. I	Jan. 17, 1871	110, 959
Pen-cleaner	J. Warren	Brooklyn, N. Y	Oct. 2, 1860	30, 270
Pen cleaner and holder	T. S. Hudson	Boston, Mass	Apr. 27, 1858	20, 065
Pen-distributer	S. A. Potter	Philadelphia, Pa	July 11, 1865	48, 717
Pen for weaning calves	J. R. Dow	Davenport, Iowa	Apr. 3, 1866	53, 389
Pen, Fountain	N. Bartlett	Canajoharie, N. Y	Sept. 9, 1843	3, 253
Pen, Fountain	G. A. Becker	Seymour, Conn	Jan. 25, 1870	99, 134
Pen, Fountain	G. A. Becker	Seymour, Conn	Sept. 6, 1870	106, 987
Pen, Fountain	H. A. Brown and J. Wiley	Brooklyn, N. Y	Feb. 19, 1856	14, 276
Pen, Fountain	A. S. Carleton	Providence, R. I	June 14, 1870	104, 109
Pen, Fountain	J. S. Charles	Omaha, Nebr	Aug. 13, 1867	67, 720
Pen, Fountain	R. H. Chinn	Washington, D. C	Sept. 29, 1868	82, 598
Pen, Fountain	R. H. Chinn	Washington, D. C	Feb. 8, 1870	99 635
Pen, Fountain	P. C. Clark	Reading, Pa	Feb. 5, 1861	31, 298
Pen, Fountain	W. Cleveland	Orange, N. J	Jan. 31, 1854	10, 469
Pen, Fountain	F. C. Cone	San Francisco, Cal	Oct. 11, 1870	108, 110
Pen, Fountain	J. S. Cutts	Philadelphia, Pa	Nov. 9, 1858	22, 017
Pen, Fountain	T. M. Davis	Philadelphia, Pa	Feb. 6, 1872	123, 460
Pen, Fountain	A. G. Day	Seymour, Conn	July 29, 1856	15, 417
Pen, Fountain	J. Goodyear	Groton, N. Y	July 15, 1873	140, 771
Pen, Fountain	T. W. Grinter	Russellville, Ky	May 31, 1870	103, 737
Pen, Fountain	H. N. Hamilton	White Plains, N. Y	Dec. 2, 1873	145, 102
Pen, Fountain	G. F. Hawkes	New York, N. Y	Oct. 17, 1865	50, 470
Pen, Fountain	G. F. Hawkes	New York, N. Y	Apr. 2, 1872	125, 291
Pen, Fountain	M. F. Hoit	Livingston, Ala	Sept. 11, 1847	5, 286
Pen, Fountain	W. Hunt	New York, N. Y	Jan. 13, 1847	4, 927
Pen, Fountain	D. Hyde	Reading, Pa	May 20, 1830	....
Pen, Fountain	E. H. Hyde and R. Dawson	Haydenville, Mass	Mar. 26, 1850	7, 217
Pen, Fountain	J. Johnson	New York, N. Y	Oct. 26, 1858	21, 881
Pen, Fountain	J. G. Kenyon	Ferndale, Cal	Nov. 9, 1869	96, 598
Pen, Fountain	M. Klein and H. W. Wynne	Keokuk, Iowa	Sept. 3, 1867	68, 445
Pen, Fountain	G. Kneip	New York, N. Y	Oct. 6, 1868	82, 850
Pen, Fountain	L. H. Knisely	New Philadelphia, Ohio	Jan. 30, 1872	123, 263
Pen, Fountain	C. W. Krebs	Baltimore, Md	Nov. 26, 1850	7, 798
Pen, Fountain	B. J. La Mothe	New York, N. Y	Apr. 25, 1865	47, 432
Pen, Fountain	D. A. Macomber	New York, N. Y	Aug. 28, 1849	6, 672
Pen, Fountain	H. Madeheim	Brooklyn, N. Y	Aug. 14, 1866	57, 162
Pen, Fountain	E. Mathers	Fairmount, Va	Dec. 4, 1860	30, 828
Pen, Fountain	H. K. McClelland	Eldersville, Pa	Apr. 17, 1855	12, 727
Pen, Fountain	G. R. Metten	Saint Louis, Mo	Oct. 9, 1866	58, 732
Pen, Fountain	G. R. Metten	Saint Louis, Mo	Nov. 5, 1867	70, 453
Pen, Fountain	G. J. Nolty	New York, N. Y	Apr. 26, 1864	42, 536
Pen, Fountain	F. A. Odermatt and F. Ettlin	San Francisco, Cal	Apr. 6, 1869	88, 656
Pen, Fountain	N. A. Prince	New Gloucester, Me	Sept. 30, 1851	8, 399
Pen, Fountain	N. A. Prince	Brooklyn, N. Y	Jan. 23, 1855	12, 301
Pen, Fountain	N. A. Prince	Brooklyn, N. Y	Dec. 25, 1855	13, 995
Pen, Fountain	J. Reid	Fort Wayne, Ind	Feb. 9, 1864	41, 534
Pen, Fountain	C. A. Rosefield	Columbus, Ga	May 12, 1857	17, 298
Pen, Fountain	L. M. Sandford and J. P. Beebe	Clinton, Iowa, and Morris, Ill.	Nov. 21, 1865	51, 090
Pen, Fountain	F. Schifferle	Saint Louis, Mo	Nov. 15, 1870	109, 257
Pen, Fountain	J. C. Silvy	New Orleans, La	Jan. 27, 1857	16, 514
Pen, Fountain	S. A. Skinner	West Berkshire, Vt	Dec. 18, 1860	30, 935

Index of patents issued from the United States Patent Office from 1790 *to* 1873, *inclusive*—Continued.

Invention.	Inventor.	Residence.	Date.	No.
Pen, Fountain	N. B. Slayton	Madison, Ind	Aug. 26, 1856	15, 622
Pen, Fountain	S. E. Taylor	East Cambridge, Mass	June 29, 1858	20, 741
Pen, Fountain	M. Wagner	Cincinnati, Ohio	Nov. 12, 1867	70, 765
Pen, Fountain	W. R. Walker	Concord, N. H	July 27, 1869	93, 024
Pen, Fountain	A. F. Warren	Brooklyn, N. Y	Dec. 23, 1856	16, 299
Pen, Fountain	A. F. Warren	Brooklyn, N. Y	Oct. 6, 1857	18, 365
Pen, Fountain	A. F. and C. M. M. Warren	Brooklyn, N. Y	Mar. 11, 1856	14, 425
Pen, Fountain	J. Weller	Washington Court-House, Ohio.	Sept. 29, 1863	40, 135
Pen, Fountain	G. W. White	Mount Vernon, N. Y	Sept. 4, 1855	13, 534
Pen, Fountain	G. W. Woolley	Philadelphia, Pa	Dec. 4, 1860	30, 851
Pen, Fountain ruling	C. Ketcham	Penn Yan, N. Y	Sept. 2, 1856	15, 659
Pen, Gold	E. H. Bard and H. H. Wilson	Philadelphia, Pa	Dec. 20, 1853	10, 348
Pen, Gold	A. W. Rapp	Philadelphia, Pa	Jan. 6, 1852	8, 641
Pen-handle	W. A. Morse	Boston, Mass	Feb. 7, 1860	27, 063
Pen-handle tube, Machine for finishing the ends of	S. Wesson	Worcester, Mass	May 1, 1866	54, 455
Pen, Hog	R. M. Abbe	Thompsonville, Conn	Aug. 29, 1854	11, 592
Pen, Hog	B. Gifford	Pedee, Iowa	Jan. 15, 1867	61, 189
Pen-holder	A. Alden	Barre, Mass	Mar. 4, 1843	2, 981
Pen-holder	E. D. Babbitt	New York, N. Y	Oct. 9, 1866	58, 572
Pen-holder	F. Brackett	Calais, Me	Nov. 14, 1865	50, 897
Pen-holder	J. Bryant	Brooklyn, N. Y	Jan. 29, 1861	31, 225
Pen-holder	B. Charles	Akron, Ohio	Mar. 21, 1871	112, 781
Pen-holder	G. H. Christian	New York, N. Y	Sept. 23, 1862	36, 507
Ped-holder	B. Cole	Brooklyn, N. Y	Feb. 15, 1859	22, 938
Pen-holder	A. T. Cross	Providence, R. I	July 1, 1873	140, 478
Pen-holder	W. J. Dewey	New York, N. Y	Apr. 9, 1872	125, 443
Pen-holder	H. G. Eastman	Poughkeepsie, N. Y	June 9, 1868	78, 655
Pen-holder	W. Fife	Philadelphia, Pa	Sept. 28, 1839	1, 345
Pen-holder	D. D. Foley	Washington, D. C	Dec. 17, 1867	72, 382
Pen-holder	P. Gabriel	Seymour, Conn	Sept. 10, 1867	68, 727
Pen-holder	A. M. George	Sand Fly, Tex	Jan. 24, 1871	111, 194
Pen-holder	D. E. Hall	Detroit, Mich	Jan. 19, 1869	85, 928
Pen-holder	E. W. Hanson	Spring Garden, Pa	Dec. 6, 1853	10, 294
Pen-holder	D. Harrington	Philadelphia, Pa	Sept. 2, 1845	4, 174
Pen-holder	G. Harrison	New York, N. Y	May 18, 1869	90, 168
Pen-holder	J. Holland	Cincinnati, Ohio	Aug. 22, 1871	118, 367
Pen-holder	I. Jacobs	New York, N. Y	Feb. 14, 1871	111, 846
Pen-holder	J. Johnson	New York, N. Y	Oct. 12, 1858	21, 758
Pen-holder	R. B. Lawrence	Wheeling, W. Va	Aug. 23, 1870	106, 596
Pen-holder	R. B. Lawrence	Wheeling, W. Va	Nov. 1, 1870	108, 916
Pen-holder	T. K. Lyon	Richmond, Va	Aug. 5, 1856	15, 490
Pen-holder	A. Masson and P. H. Cary	Paris, France	Oct. 17, 1865	50, 543
Pen-holder	J. C. Mullett	Fredonia, N. Y	Dec. 24, 1872	134, 156
Pen-holder	J. S. Orndorff	Virginia City, Nev	July 29, 1873	141, 287
Pen-holder	O. A. Pennoyer	Washington, D. C	Apr. 4, 1871	113, 558
Pen-holder	M. Phineas	New York, N. Y	Oct. 24, 1854	11, 836
Pen-holder	G. G. Ray	Boston, Mass	Apr. 15, 1862	34, 983
Pen-holder	H. Roth	Virginia City, Nev	Feb. 7, 1871	111, 683
Pen-holder	H. Smith	Georgetown, D. C	Jan. 15, 1818	......
Pen-holder	D. M. Somers	Brooklyn, N. Y	Apr. 30, 1872	126, 339
Pen-holder	D. M. Somers	Brooklyn, N. Y	Aug. 20, 1872	130, 670
Pen-holder	H. A. Spencer and R. S. Cutting.	Cleveland, Ohio, and Providence, R. I.	Apr. 27, 1869	89, 354
Pen-holder	W. H. Towers	Philadelphia, Pa	Apr. 17, 1855	12, 734
Pen-holder	W. H. Towers	New York, N. Y	Nov. 18, 1862	36, 972
Pen-holder	A. R. Turner	Malden, Mass	Apr. 26, 1859	23, 800
Pen-holder	S. Walker	Boston, Mass	Jan. 10, 1865	45, 884
Pen-holder	A. F. Warren	Brooklyn, N. Y	June 19, 1860	28, 797
Pen-holder	C. M. H. Warren	Brooklyn, N. Y	Feb. 4, 1868	74, 179
Pen-holder	J. Warren	Brooklyn, N. Y	July 31, 1860	29, 426
Pen-holder	A. Warth	Stapleton, N. Y	Apr. 28, 1868	77, 422
Pen-holder	H. F. Wheeler	Boston, Mass	Dec. 17, 1867	72, 435
Pen-holder	C. T. Widstrand	Washington, D. C	Oct. 1, 1872	131, 802
Pen-holder	C. G. Wilson	Brooklyn, N. Y	Aug. 17, 1869	93, 780
Pen-holder	O. O. Witherell	Plaistow, N. H	Mar. 12, 1867	62, 913
Pen-holder and eraser	W. A. Morse	Philadelphia, Pa	Aug. 20, 1867	67, 899
Pen-holder and ink-eraser	D. E. Holmes	Halifax, Mass	May 17, 1864	42, 769
Pen-holder and letter-balance, Combined	D. C. Lawrence	Cedar Falls, Iowa	Aug. 20, 1861	33, 091
Pen holder and nib, Fountain	A. S. Lyman and M. W. Baldwin.	Philadelphia, Pa	Sept. 19, 1848	5, 789
Pen-holder, Elastic	J. W. Pearson	Winchester, Mass	Aug. 27, 1861	33, 160
Pen holder, Flexible	F. J. Klein	New York, N. Y	Feb. 19, 1856	14, 286
Pen holder, Fountain	C. Cleveland	Middlebury, Vt	June 1, 1852	8, 977
Pen-holder, Metallic	A. Granger	New York, N. Y	Mar. 29, 1859	23, 366
Pen-holder or pencil, Weighing-attachment for	D. A. B. Savy	Paris, France	July 25, 1865	49, 059
Pen-holder, pen-case, and money-safe, Combination of.	W. E. Rose	Waukon, Iowa	May 3, 1864	42, 599
Pen-holder, Reservoir	J. Darling	Stane, North Britain	Sept. 3, 1867	68, 418
Pen-holder, Spring	D. Hall	Brooklyn, N. Y	Apr. 19, 1864	42, 436
Pen-holder springs, Machine for forming	J. Keith	Worcester, Mass	May 1, 1866	54, 362
Pen-holder tip	D. M. Somers	Brooklyn, N. Y	May 21, 1872	127, 113
Pen machine, Microscopic	J. Bennett	Lowell, Mass	Oct. 11, 1828	....
Pen, Marking	D. S. Holman	Philadelphia, Pa	Sept. 30, 1873	143, 347
Pen, Metallic	T. Alden	Barre, Mass	Dec. 12, 1842	2, 877
Pen, Metallic	A. Craytey	Brooklyn, N. Y	July 1, 1856	15, 223
Pen, Metallic	M. S. Fife	Philadelphia, Pa	Apr. 3, 1849	6, 278
Pen, Metallic	W. Fife	Philadelphia, Pa	Aug. 29, 1848	5, 737
Pen, Metallic	R. Griswold	Bainbridge, N. Y	May 20, 1862	35, 312
Pen, Metallic	M. Phineas	New York, N. Y	July 12, 1853	9, 843
Pen, Metallic	M. Phineas	New York, N. Y	Feb. 5, 1856	14, 203
Pen, Metallic	W. H. Towers	Philadelphia, Pa	Nov. 1, 1853	10, 192
Pen, Metallic	F. A. Wait	Philadelphia, Pa	July 7, 1857	17, 761
Pen, Metallic	J. Wilcox	Philadelphia, Pa	Oct. 28, 1856	15, 992
Pen, Metallic	P. Williamson	New York	Mar. 30, 1835	....
Pen, Metallic	T. Woodward	Brooklyn, N. Y	Dec. 1, 1842	2, 874
Pen, Metallic pointed	B. R. Norton	Syracuse, N. Y	June 21, 1853	9, 809
Pen, Metallic writing	P. Williamson	Baltimore, Md	Nov. 22, 1809	......

Index of patents issued from the United States Patent Office from 1790 to 1873, inclusive—Continued.

Invention.	Inventor.	Residence.	Date.	No.
Pen, pencil, knife, and tooth-pick case, Combined..	J. F. Sturdy	Attleborough, Mass	June 21, 1859	24, 502
Pen-rack	C. J. Bouché	Louisville, Ky	Oct. 13, 1868	82, 916
Pen-rack	E. Clarke	Winchester, Conn	Oct. 8, 1872	131, 937
Pen-rack	W. E. Clarke	Attleborough, Mass	July 21, 1868	80, 140
Pen-rack	G. Jedamski	New York, N. Y	July 23, 1867	67, 120
Pen-rack	M. O'Reardon	New York, N. Y	July 12, 1870	105, 237
Pen-rack	O. P. Smith	New York, N. Y	July 1, 1862	35, 786
Pen-rack	J. Young	Boston, Mass	Jan. 3, 1860	26, 727
Pen-rack and bill-holder	C. P. Crossman	West Warren, Mass	Oct. 9, 1866	58, 716
Pen-rack, calendar, and letter-balance, Combination of.	H. N. Taft	Washington, D. C	Jan. 3, 1865	45, 770
Pen rack, cleaner, and pencil-sharpener	H. C. Haskell	Marshall, Mich	Feb. 28, 1860	27, 287
Pen-rack, Spring	J. Adair	Pittsburgh, Pa	Oct. 2, 1866	58, 363
Pen-rack, Spring	J. M. Keep	New York, N. Y	Jan. 21, 1868	73, 448
Pen-rest	R. B. Huguuin	Cleveland, Ohio	Mar. 3, 1868	75, 164
Pen, Ruling	A. Hathaway	Boston, Mass	Jan. 28, 1851	7, 920
Pen, Ruling	A. Hathaway	Charlestown, Mass	Jan. 5, 1869	85, 535
Pen, Ruling	W. O. Hickok	Harrisburgh, Pa	June 24, 1873	140, 272
Pen, Ruling	E. Ingram	Springfield, Mass	Feb. 20, 1872	123, 827
Pen, Ruling	R. McVeen	Cleveland, Ohio	July 6, 1869	92, 336
Pen, Self-supplying	C. Cleveland	Middleburgh, Vt	Dec. 16, 1833	....
Pen, Self-supplying fountain	M. T. C. Gould	Philadelphia, Pa	Oct. 1, 1830	....
Pen, Sheep	B. Gifford	Pedee, Iowa	Jan. 15, 1867	61, 190
Pen-stand	C. J. Rooney and D. Renshaw	New York, N. Y	Apr. 3, 1860	27, 739
Pen, Steel writing	D. Thomas	Hingham, Mass	July 3, 1840	1, 671
Pen-wiper	H. S. Ball	Spartanburgh, S. C	Dec. 23, 1873	145, 778
Pen-wiper	S. Darling	Bangor, Me	Feb. 4, 1868	73, 956
Pen-wiper	J. H. Kidder	Lawrence, Mass	Sept. 2, 1873	142, 480
Pen-wiper and paper-weight	J. L. Rowe	New York, N. Y	Mar. 8, 1859	23, 195
Pen, Writing	R. H. Chinn	Washington, D. C	Jan. 21, 1873	135, 080
Pen, Writing	J. Holland	Cincinnati, Ohio	Jan. 21, 1873	135, 118
Pen, Writing	J. F. Reeve	Richmond, Va	Jan. 27, 1857	16, 496
Pen, Writing	S. Warrington	Philadelphia, Pa	July 24, 1866	56, 645
Pens, Alloy for manufacture of	J. Holland	Cincinnati, Ohio	Oct. 8, 1872	132, 008
Pens, &c., Box, case, and card for	J. Mason	Birmingham, Great Britain.	May 6, 1862	35, 201
Pens for ruling paper, Mode of setting and holding.	S. V. Collins	Charlestown, Mass	June 20, 1854	11, 146
Pens from holders, Device for extracting	F. W. Mattern	Chicago, Ill	Aug. 22, 1871	118, 376
Pens, Ink-reservoir for	R. B. Fitts	Philadelphia, Pa	May 8, 1860	28, 235
Pens, Ink-retaining attachment for	A. G. Brown	Hartford, Conn	Jan. 25, 1870	99, 147
Pens, Machine for rolling gold	A. Morton	New York, N. Y	Aug. 28, 1860	29, 809
Pens, Making metallic	H. C. Windle, J. Gillott, and S. Morris.	Great Britain	Mar. 21, 1838	648
Pens, Manufacture of	G. Stimpson, jr	New York, N. Y	Mar. 20, 1866	53, 356
Pens, Manufacture of	E. Wiley	Brooklyn, N. Y	Jan. 14, 1868	73, 419
Pens, Ruling-guide for fountain	J. H. Hobbs	Philadelphia, Pa	Feb. 19, 1861	31, 515
Pencil	O. Cleveland	Jersey City, N. J	July 5, 1864	43, 391
Pencil	H. T. Cushman	North Bennington, Vt	Aug. 8, 1871	117, 748
Pencil	H. T. Cushman	North Bennington, Vt	Nov. 5, 1872	132, 713
Pencil	J. L. Faber	Stein, Bavaria	Aug. 13, 1861	33, 034
Pencil	J. Reckendorfer	New York, N. Y	Oct. 28, 1862	36, 854
Pencil	F. Rowell and E. C. Loud	Springfield, Mass	Feb. 15, 1870	99, 955
Pencil and eraser	W. R. Evans and L. D. Benner	Thomaston, Me	Apr. 25, 1865	47, 406
Pencil and eraser, Combination of lead	H. L. Lipman	Philadelphia, Pa	Mar. 30, 1858	19, 783
Pencil and pen case, Ever-pointed	J. I. Lownds	Philadelphia, Pa	Sept. 22, 1836	32
Pencil and rubber eraser, Combined lead	J. Illfelder	New York, N. Y	Nov. 4, 1873	144, 337
Pencil and sponge holder for cleaning slates, &c..	J. L. Rowe	New York, N. Y	Aug. 25, 1863	39, 704
Pencil-attachment	E. Weissenborn	Hudson City, N. J	Oct. 12, 1869	95, 861
Pencil blank, Machine for grooving lead	P. Hufeland	New York, N. Y	May 31, 1870	103, 746
Pencil-case	W. N. Bartholomew	Newton Centre, Mass	Mar. 29, 1870	101, 344
Pencil-case	A. T. Cross	Providence, R. I	July 1, 1873	140, 477
Pencil-case	W. H. Davis	Hartsville, Ind	May 7, 1872	126, 448
Pencil-case	C. H. Downes	Hudson City, N. J	Sept. 27, 1870	107, 672
Pencil-case	J. Durant	New York, N. Y	June 27, 1846	4, 602
Pencil-case	F. R. Goulding	Roswell, Ga	June 25, 1871	117, 273
Pencil-case	W. S. Hicks	New York, N. Y	Mar. 21, 1871	112, 917
Pencil-case	J. A. Kemmis	New Orleans, La	Nov. 9, 1869	96, 597
Pencil-case	J. H. Knapp	New York, N. Y	Aug. 6, 1867	67, 558
Pencil-case	J. H. Knapp	New York, N. Y	Feb. 6, 1872	123, 485
Pencil-case	J. H. Knapp	New York, N. Y	Feb. 6, 1872	123, 486
Pencil-case	W. A. Ludden	Brooklyn, N. Y	July 20, 1869	92, 853
Pencil-case	W. A. Ludden	Flushing, N. Y	Oct. 17, 1871	119, 935
Pencil-case	W. Maginn	New York, N. Y	Apr. 21, 1868	77, 062
Pencil-case	D. A. Peirce	East Greenwich, R. I	Apr. 5, 1859	23, 487
Pencil-case	J. H. Rauch	New York, N. Y	June 22, 1869	91, 665
Pencil-case	S. S. Rembert	Memphis, Tenn	July 15, 1873	140, 842
Pencil-case	D. T. Warren	New York, N. Y	June 26, 1866	55, 966
Pencil case, Ever-pointed	J. Hague	New York, N. Y	Aug. 16, 1839	1, 291
Pencil case, Ever-pointed	J. Saxton	Philadelphia, Pa	Apr. 11, 1829	....
Pencil case, Ever-pointed	G. W. Simons	Philadelphia, Pa	Oct. 12, 1839	1, 364
Pencil case, Ever-pointed	T. Woodward	Brooklyn, N. Y	Oct. 14, 1840	1, 823
Pencil case, Lead and slate	T. B. McCaughan	Moscow, Tenn	Apr. 7, 1868	76, 488
Pencil-cases, Ornamenting	E. S. Johnson	Jersey City, N. J	Dec. 5, 1871	121, 627
Pencil, &c., coloring and polishing machine, Lead..	E. Weissenborn	Hudson City, N. J	Apr. 2, 1872	125, 360
Pencil, Composition steatite	R. Langtrom	Cincinnati, Ohio	July 18, 1871	117, 084
Pencil, Eraser and stamp	E. Faber	New York, N. Y	June 16, 1863	38, 892
Pencil, Ever-pointed	A. G. Bagley	New York	Nov. 6, 1846	4, 840
Pencil, Ever-pointed	J. Boss	Philadelphia, Pa	July 20, 1846	4, 651
Pencil, Ever-pointed	W. Jackson	Philadelphia, Pa	July 27, 1829	......
Pencil, Ever-pointed	T. Woodward	Brooklyn, N. Y	June 10, 1840	1, 625
Pencil. Ever-pointed lead	E. Meeds	Philadelphia, Pa	June 26, 1835	....
Pencil-finishing machine	T. H. Müller	Yonkers, N. Y	July 15, 1873	140, 946
Pencil-head mold, India-rubber	A. Neill	Boston, Mass	Jan 22, 1861	31, 187
Pencil-holder	J. M. Clark	Jersey City, N. J	July 3, 1866	56, 007
Pencil-holder	N. B. Cooper	Liberty, Ind	Nov. 26, 1867	71, 456
Pencil-holder	C. A. Eaton	Minneapolis, Minn	Oct. 25, 1870	108, 573
Pencil-holder	A. J. Hall	San Francisco, Cal	Apr. 5, 1870	101, 457
Pencil-holder	H. M. Hally	Norwich, Conn	Sept. 18, 1866	58, 102
Pencil-holder	J. L. Moore	Bridgeport, Conn	Mar. 23, 1869	88, 160

Index of patents issued from the United States Patent Office from 1790 *to* 1873, *inclusive*—Continued.

Invention.	Inventor.	Residence.	Date.	No.
Pencil-holder	L. B. Myers	Elmore, Ohio	July 10, 1866	56, 251
Pencil-holder	T. Orton	Chicago, Ill	Oct. 15, 1872	132, 174
Pencil-holder	G. A. Smith	Boston, Mass	Feb. 4, 1873	135, 603
Pencil-holder	J. Smith	Tiffin, Ohio	Dec. 31, 1867	72, 926
Pencil-holder	E. J. Toof	Fort Madison, Iowa	May 19, 1868	78, 158
Pencil-holder	H. J. Wickham	Manchester, Conn	May 16, 1871	114, 893
Pencil holder and adjuster, Lead	E. Weissenborn	Hudson City, N. J	Apr. 2, 1872	125, 359
Pencil-holder for compass	W. G. Hillegass	Philadelphia, Pa	Nov. 12, 1867	70, 840
Pencil holder, Slate	E. G. Ward	Hoboken, N. J	Oct. 12, 1869	95, 859
Pencil, Indelible	E. P. Clark	Northampton, Mass	July 10, 1866	56, 180
Pencil, Indelible	W. B. Hale	Northampton, Mass	Jan. 1, 1867	60, 885
Pencil, Lead	E. Faber	New York, N. Y	Jan. 28, 1868	73, 883
Pencil, Lead	J. Gray	Medford, Mass	Feb. 25, 1873	136, 319
Pencil, Lead	S. D. Hovey	Brooklyn, N. Y	Dec. 31, 1872	134, 382
Pencil, Lead	P. Hufeland	New York, N. Y	Feb. 18, 1873	136, 122
Pencil, Lead	W. A. Morse	Philadelphia, Pa	Dec. 24, 1872	134, 215
Pencil, Lead	T. H. Müller	Yonkers, N. Y	May 21, 1872	126, 980
Pencil, Lead	T. H. Müller	Yonkers, N. Y	June 4, 1872	127, 033
Pencil, Lead	T. H. Müller	Yonkers, N. Y	Mar. 25, 1873	137, 089
Pencil, Lead	J. Reckendorfer	New York, N. Y	Jan. 6, 1863	37, 360
Pencil, Lead	J. Reckendorfer	New York, N. Y	Jan. 19, 1869	86, 101
Pencil machine, Slate	D. J. Tittle	Albany, N. Y	June 4, 1872	127, 440
Pencil-point protector	G. Merritt	New York, N. Y	Mar. 5, 1867	62, 555
Pencil-point protector and mark-eraser	J. B. Hodgskin	New York, N. Y	Feb. 14, 1865	46, 358
Pencil-points, Composition of matter for, and making.	G. C. Baldwin	Ticonderoga, N. Y	Dec. 2, 1835	....
Pencil-points, Machine for making, and composition for.	G. C. Baldwin	Ticonderoga, N. Y	Dec. 2, 1835	....
Pencil-polishing machine	G. Braun	Hoboken, N. J	Oct. 29, 1872	132, 519
Pencil-scale	B. Worcester	Waltham, Mass	Dec. 4, 1866	60, 815
Pencil-sharpener	H. P. Andrews	Cleveland, Ohio	Aug. 8, 1865	49, 210
Pencil-sharpener	A. G. Batchelder	Lowell, Mass	Nov. 8, 1870	109, 100
Pencil-sharpener	H. Burgess	San Francisco, Cal	May 28, 1867	65, 165
Pencil-sharpener	H. Burgess	Oakland, Cal	May 10, 1870	102, 766
Pencil-sharpener	J. L. and D. H. Coles	New York, N. Y	July 21, 1868	80, 145
Pencil-sharpener	E. M. Crandal	Chicago, Ill	Mar. 11, 1873	136, 590
Pencil-sharpener	S. Darling	Providence, R. I	June 27, 1871	116, 276
Pencil-sharpener	S. W. Davis and C. P. Elliott	Norristown, Pa	Dec. 19, 1871	121, 996
Pencil-sharpener	M. W. Dillingham	Amsterdam, N. Y	Aug. 17, 1869	93, 687
Pencil-sharpener	M. W. Dillingham	Amsterdam, N. Y	July 19, 1870	105, 440
Pencil-sharpener	M. W. Dillingham	Amsterdam, N. Y	Aug. 30, 1870	106, 795
Pencil-sharpener	W. K. Foster	Bangor, Me	Apr. 27, 1858	20, 056
Pencil-sharpener	W. K. Foster	Bangor, Me	May 18, 1858	20, 262
Pencil-sharpener	D. Hoffman	Pottsville, Pa	July 30, 1872	129, 956
Pencil-sharpener	J. MacMullen	New York, N. Y	Jan. 2, 1866	51, 845
Pencil-sharpener	J. McClure	Nashua, N. H	Feb. 1, 1870	99, 335
Pencil-sharpener	L. B. Myers	Elmore, Ohio	May 18, 1869	90, 289
Pencil-sharpener	E. P. Needham	New York, N. Y	Mar. 21, 1871	112, 951
Pencil-sharpener	C. C. Plaisted	Hartford, Conn	Mar. 16, 1869	87, 967
Pencil-sharpener	A. E. Schatz	New York, N. Y	May 13, 1873	138, 943
Pencil-sharpener	E. Spencer	Lambertville, N. J	Aug. 20, 1867	67, 922
Pencil-sharpener	O. C. Squyer	Penn Yan, N. Y	Feb. 18, 1868	74, 728
Pencil-sharpener	J. W. Strange and S. Darling	Bangor, Me	Sept. 22, 1857	18, 265
Pencil-sharpener	W. N. Weeden	Boston, Mass	Nov. 9, 1869	96, 748
Pencil-sharpener	S. S. Woodcock	Somerville, Mass	Apr. 20, 1869	89, 109
Pencil-sharpener	R. Wright	New York, N. Y	Feb. 7, 1865	46, 289
Pencil-sharpener	H. L. D. Zeng	Geneva, N. Y	Feb. 6, 1872	123, 462
Pencil sharpener and holder, Slate	W. H. Alcorn	New York, N. Y	July 30, 1867	67, 245
Pencil sharpener, Slate	F. G. Bottner	Bridgeport, Conn	Aug. 6, 1867	67, 488
Pencil sharpener, Slate	W. Burnet	New York, N. Y	Oct. 5, 1858	21, 649
Pencil sharpener, Slate	A. C. Funston	Philadelphia, Pa	July 30, 1861	32, 940
Pencil sharpener, Slate	J. M. Hicks	Boston, Mass	Sept. 29, 1863	40, 102
Pencil sharpener, Slate and lead	J. Soumeillan	Philadelphia, Pa	Oct. 8, 1872	131, 977
Pencil-sharpeners, Making blade for	W. K. Foster	Bangor, Me	Jan. 26, 1858	19, 191
Pencil-sharpeners, Mold for casting	W. K. Foster	Bangor, Me	Apr. 17, 1855	12, 722
Pencil-sharpening instrument, Slate	G. Sickles	Brooklyn, N. Y	May 11, 1858	20, 219
Pencil-sheath	C. E. Abbott and R. S. Merrill	Malden and Hyde Park, Mass.	Aug. 8, 1871	117, 851
Pencil-sheath	S. Ayres	Danville, Ky	Sept. 22, 1868	82, 272
Pencil-sheath	J. Danner	Canton, Ohio	June 16, 1868	78, 931
Pencil-sleeve and eraser, Combination of	F. E. Oliver	New York, N. Y	June 3, 1862	35, 467
Pencil, Tailor's marking	W. Coover	Erie, Pa	Aug. 26, 1843	3, 236
Pencil-tip molds, Vulcanizing rubber	J. Banigan and G. W. Miller	Smithfield, R. I	Jan. 18, 1870	99, 045
Pencil-tips, Mold for making rubber	W. Weicker	Blackstone, Mass	May 10, 1870	102, 894
Pencil-varnishing machine	T. H. Müller and H. C. Benson	New York, N. Y	Nov. 29, 1870	109, 750
Pencil, &c., varnishing machine	A. Warth, P. Hufeland, and G. Braun.	Stapleton and New York, N. Y.	May 10, 1870	102, 893
Pencil varnishing or coloring machine, Lead	E. Weissenborn	Hudson City, N. J	Apr. 2, 1872	125, 362
Pencils, Attaching eraser to	G. L. Holt	Springfield, Mass	Feb. 9, 1869	86, 753
Pencils, Attaching India-rubber to	W. C. Vosburgh and W. A. Ludden.	Brooklyn, N. Y	May 20, 1862	35, 355
Pencils, Attaching rubber eraser to lead	T. H. Müller	Yonkers, N. Y	Apr. 30, 1872	126, 224
Pencils, Attaching rubber to	J. Reckendorfer	New York, N. Y	Jan. 19, 1869	85, 961
Pencils, Composition for	E. P. Clark	Holyoke, Mass	May 31, 1859	24, 195
Pencils, Composition for	W. Geller	New York, N. Y	Nov. 8, 1870	109, 000
Pencils, Composition for	S. C. Pruden	Athens, Ohio	May 14, 1867	64, 702
Pencils, Composition for making black lead	C. Osgood	Salem, Mass	Mar. 26, 1814	....
Pencils, Composition of slate	L. J. Cohen	New York, N. Y	Nov. 7, 1848	5, 903
Pencils, Cutter-head for wood of lead	F. G. Jenkins	New York, N. Y	May 12, 1863	38, 488
Pencils, Device for polishing	E. Weissenborn	Hudson City, N. J	Apr. 27, 1869	89, 530
Pencils for drawing-machines, Arrangement of	M. Nutting	Portland, Me	Nov. 21, 1854	11, 970
Pencils, Machine for cutting and smoothing ends of	T. H. Müller and H. C. Benson	New York, N. Y	Oct. 25, 1870	108, 721
Pencils, Machine for cutting ends of wooden	T. H. Müller	New York, N. Y	Jan. 3, 1871	110, 777
Pencils, Machine for cutting lead	A. Warth	Stapleton, N. Y	June 21, 1864	43, 267
Pencils, Machine for cutting off lead	P. Schrag and P. Hufeland	New York, N. Y	May 17, 1870	103, 241
Pencils, Machine for forming lead and other	E. Weissenborn	Hudson City, N. J	Apr. 2, 1872	125, 361
Pencils, Machine for making lead	E. Weissenborn	Hudson City, N. J	Sept. 10, 1867	68, 819

Index of patents issued from the United States Patent Office from 1790 *to* 1873, *inclusive*—Continued.

Invention.	Inventor.	Residence.	Date.	No.
Pencils, Machine for making slate	D. R. Satterlee	New Haven, Conn	Aug. 24, 1869	94, 136
Pencils, Machine for making wooden case for lead	A. Weiller	New York, N. Y	June 23, 1863	39, 019
Pencils, Machine for manufacture of slate	D. R. Satterlee	New Haven, Conn	Apr. 21, 1868	77, 105
Pencils, Machine for sand-papering	P. Schrag	New York, N. Y	Dec. 27, 1864	45, 679
Pencils, Manufacture of lead	F. G. Jenkins	Brooklyn, E. D., N. Y	May 18, 1869	90, 269
Pencils, Manufacture of slate	N. C. Harris	Poultney, Vt	Apr. 24, 1855	12, 759
Pencils, Metallic slide and case for ever-pointed	J. Bogardus	New York, N. Y	Sept. 17, 1833	....
Pencils, Packing-board for	O. Cleveland	Jersey City, N. J	Oct. 29, 1872	132, 562
Pencils, Press for stamping lead	T. H. Müller	Yonkers, N. Y	Feb. 20, 1872	123, 925
Pencils, Producing slate	H. O. Brown	Castleton, Vt	July 16, 1872	129, 096
Pencils, Rubber head for	S. D. Hovey	Brooklyn, N. Y	July 28, 1868	80, 485
Pencils, Rubber head for lead	J. B. Blair	Philadelphia, Pa	July 23, 1867	66, 938
Pencils, Rubber head for lead	W. W. Shaw	Troy, N. Y	Apr. 5, 1859	23, 504
Pendulums and levers, Concentrating the power of	A. and C. Berry	Poughkeepsie, N. Y	Mar. 25, 1826	....
Pendulum-balance	E. Sampson	Claremont, N. H	Nov. 6, 1849	6, 852
Pendulum, Balance	H. Twiss	Meriden, Conn	May 13, 1834	....
Pendulum, Compensating	M. Bemis	Ashburnham, Mass	Nov. 8, 1859	26, 008
Pendulum, Compensating	L. Bradley	Hartford, Conn	Feb. 26, 1861	31, 524
Pendulum, Compensating	G. Buchanan	Hickory, Pa	Dec. 4, 1860	30, 798
Pendulum, Compensation	H. B. James	Trenton, N. J	Oct. 31, 1871	120, 385
Pendulum, Compound	D. Bickford	Westerly, R. I	Mar. 2, 1858	19, 479
Pendulum, Compound	C. M. Rice and J. E. Harrington	Worcester and Millbury, Mass.	Jan. 19, 1858	19, 153
Pendulum, Double	R. Leslie		Jan. 30, 1793	....
Pendulum, Electro-magnetic	J. Hamblet, jr., and B. F. Edmands.	Boston, Mass	Jan. 12, 1864	41, 217
Pendulum for time-pieces, Torsion	A. D. Crane	Newark, N. J	Jan. 9, 1855	12, 196
Pendulum in calendar-clocks, Arrangement of	H. B. Horton	Ithaca, N. Y	Jan. 4, 1870	98, 498
Pendulum level and sight combined	C. Morrill	New York, N. Y	Apr. 16, 1867	63, 807
Pendulum-lock, Overbalanced	S. Goodwin	Baltimore, Md	July 7, 1809	....
Pendulum-machine	J. Lefever	Providence, R. I	June 13, 1810	....
Pendulum-machine for mechanical operations	C. L. Allen	Newark, N. J	Oct. 13, 1815	....
Pendulum-mill	W. Grandin	Hector, N. Y	Jan. 6, 1813	....
Pendulum-mill	S. Thayer	Braintree, Mass	Oct. 9, 1811	....
Pendulum-power	J. Tallman	Albany, N. Y	Mar. 31, 1808	....
Pendulum-power, Applying	A. Slevin	Ann Arbor, Mich	Mar. 30, 1858	19, 798
Pendulum-power for propelling machinery	A. Wade	Eagletown, N. Y	Mar. 25, 1835	....
Pendulum-power regulated by a fly-wheel	W. W. McIntosh and W. Barnheart.	Aurelius, Ohio	Apr. 13, 1825	....
Pendulum-screen	L. Dupré	Charleston, S. C	Apr. 1, 1807	....
Pendulum, Self-adjusting	J. S. Greig	Walden, N. Y	Mar. 13, 1847	5, 017
Pendulums upon the earth's surface, Apparatus for illustrating the motion of.	G. M. Dimmock	Springfield, Mass	July 12, 1873	9, 839
Penman's assistant	H. G. Eastman	Poughkeepsie, N. Y	Nov. 5, 1861	33, 638
Penman's assistant	H. G. Eastman	Poughkeepsie, N. Y	Feb. 9, 1864	41, 495
Penman's assistant	H. A. Hutson	Newbury, N. Y	Jan. 1, 1867	60, 896
Penman's assistant	W. King	Hopewell, New Brunswick	Dec. 12, 1865	51, 530
Penman's finger-shield	C. W. Brainerd	Hartford, Conn	Apr. 13, 1869	88, 942
Penmanship, Method of teaching	W. E. MacLawrin	New York, N. Y	Feb. 13, 1855	12, 391
Pepper and ink top	W. Markland	New York, N. Y	Oct. 1, 1830	....
Pepper and salt boxes, Cap for	A. A. Bingham and W. H. Ashfield.	Syracuse, N. Y	Dec. 31, 1872	134, 455
Pepper and spice box	B. Morahan	Brooklyn, N. Y	Nov. 28, 1871	121, 399
Pepper and spice box	A. E. Turnbull, jr	Clifton, Ohio	Dec. 27, 1870	110, 608
Pepper-box	J. Fallows	Philadelphia, Pa	May 8, 1866	54, 517
Pepper-box	A. H. Newton	Worcester, Mass	Oct. 31, 1865	50, 727
Pepper-box, Air-tight	E. Brown	Lynn, Mass	Jan. 26, 1858	19, 182
Pepper-box cover	H. E. Thomas	San Francisco, Cal	June 25, 1872	128, 261
Pepper-box top	J. S. Ewbank	New York, N. Y	May 31, 1864	42, 934
Pepper-box-top fastener	J. Bounds	Bridgeport, Conn	June 16, 1868	78, 858
Pepper-boxes, Manufacture of	J. H. Stone	Philadelphia, Pa	Feb. 26, 1867	62, 451
Pepper-cruet	H. T. Clawson	New Berne, N. C	Dec. 21, 1858	22, 349
Peppermint-dropper	H. H. Snow	New Haven, Conn	Mar. 4, 1851	7, 963
Perambulator	T. Galt	Rock Island, Ill	July 30, 1872	129, 941
Perambulator	A. Clifford	West Point, Mass	Oct. 27, 1829	....
Perambulator	C. Lyne	Padstow, England	Aug. 25, 1868	81, 387
Perambulator	A. W. Richards	Indianola, Iowa	Mar. 18, 1873	136, 867
Perambulator, Folding	A. Christian	New York, N. Y	Oct. 27, 1868	83, 363
Perambulator, Folding	C. Lyne	Padstow, England	Apr. 20, 1869	89, 058
Perambulator or child's carriage	J. A. Crandall	New York, N. Y	Jan. 15, 1861	31, 110
Perch-irons, Die for making	J. W. Sheppard	Plantsville, Conn	Feb. 20, 1872	123, 943
Perch-plate-forming die	R. R. Miller	Plantsville, Conn	Mar. 15, 1870	100, 783
Perch-spring clip	J. Doeble	Plantsville, Conn	Apr. 12, 1870	101, 834
Percolating apparatus	C. A. Smith	Cincinnati, Ohio	Nov. 20, 1847	5, 372
Percolator	P. H. Vander Weyde	Philadelphia, Pa	Apr. 10, 1866	53, 905
Percolator, Adjustable	J. Q. Hill	Worcester, Mass	Sept. 20, 1864	44, 311
Percolator and filtering-machine, Continuous	P. H. Vander Weyde	Philadelphia, Pa	Dec. 11, 1866	60, 445
Percolator for druggists and others	A. Merrell	Cincinnati, Ohio	June 14, 1870	104, 181
Percolator, Reversible	S. Woolston and W. Corfield	Vincentown, N. J., and Philadelphia, Pa.	Feb. 22, 1870	100, 232
Percussion-cap	A. C. Hobbs	Bridgeport, Conn	Sept. 14, 1869	94, 743
Percussion-cap	B. F. Woodside	McDonald Station, Tenn	July 26, 1870	105, 874
Percussion-cap box	T. Harvey	Baltimore, Md	July 12, 1864	43, 497
Percussion-cap, &c., box pocket	G. Gros	Bordeaux, France	Aug. 10, 1869	93, 432
Percussion-cap carrier	R. Paulson	Washington, D. C	Oct. 10, 1871	119, 879
Percussion-cap holder	T. B. Lamb	Hamilton, Mich	Nov. 3, 1863	40, 487
Percussion-cap holder	F. J. Seymour and O. N. Perkins.	Meriden, Conn	June 15, 1869	91, 276
Percussion-cap holder	C. Verniaud and W. Spencer	La Grange, Mo	Sept. 24, 1872	131, 642
Percussion-cap holder	J. F. Warren	Stafford, N. Y	Feb. 16, 1864	41, 633
Percussion-cap primer	C. R. Alsop	Middletown, Conn	Apr. 8, 1862	34, 919
Percussion-cap primer	G. W. Baker	Burlington, Vt	Sept. 1, 1857	18, 117
Percussion-cap primer	E. D. Seely	Brookline, Mass	Oct. 27, 1863	40, 432
Percussion-cap primer for fire-arms	H. R. Jones	Addison, N. Y	June 10, 1862	35, 524
Percussion-cap primer, Portable	J. K. Ely and R. Cook	Franklin, Ohio	Aug. 6, 1867	67, 424
Percussion-cap-trimming machine	J. H. Fowler and A. J. French	Waterbury, Conn	July 17, 1866	56, 492
Percussion-cap-trimming machine	D. N. Goff	Wolcottville, Conn	July 30, 1867	67, 190
Percussion-cap, Water-proof	J. Chattaway	Hampden County, Mass	June 10, 1856	15, 063

Index of patents issued from the United States Patent Office from 1790 *to* 1873, *inclusive*—Continued.

Invention.	Inventor.	Residence.	Date.	No.
Percussion-caps, &c., Composition for	A. Hockstätter	Langen, Hesse-Darmstadt.	Jan. 12, 1864	41, 259
Percussion-caps, Facing ends of	J. Goldmark	New York, N. Y	Nov. 22, 1853	10, 262
Percussion-caps, Feeding	A. J. French	Bridgeport, Conn	Dec. 22, 1868	85, 224
Percussion-caps, Implement for feeding	L. Lelong and J. Decamp	Newark, N. J	Oct. 22, 1861	33, 531
Percussion-caps, Machine for forming and charging	G. Wright	Washington, D. C	Sept. 24, 1850	7, 675
Percussion-caps, Machine for lining	A. S. Blake	Waterbury, Conn	July 30, 1867	67, 253
Percussion-caps, Machine for lining	A. J. French	Waterbury, Conn	Sept. 10, 1867	68, 725
Percussion-caps, Machine for lining	A. J. French	Waterbury, Conn	Apr. 16, 1872	125, 875
Percussion-caps, Machine for lining	D. N. Goff	Wolcottville, Conn	July 30, 1867	67, 189
Percussion-caps, Machine for making	R. M. Bouton	West Troy, N. Y	Mar. 20, 1849	6, 196
Percussion-caps, Machine for ramming	C. Hicks	Haverstraw, N. Y	Feb. 10, 1857	16, 587
Percussion-caps, Machine for varnishing	C. Hicks	Haverstraw, N. Y	Feb. 17, 1857	16, 646
Percussion-caps, Machine for varnishing and lining	W. A. McIntire	Troy, N. Y	Feb. 6, 1866	52, 435
Percussion-caps, &c., Manufacture of water-proof.	B. Burton	Brooklyn, N. Y	Aug. 11, 1868	81, 057
Percussion-caps, Self-feeding machine for charging	M. W. Fisher	Washington, D. C	Nov. 21, 1848	5, 928
Percussion-lock	M. Carleton	Haverhill, N. H	Dec. 23, 1830	....
Percussion-lock	D. G. Colburn	Port Byron, N. Y	Oct. 25, 1832	....
Percussion-lock	F. Dowler	Wayne Township, Ohio	Apr. 9, 1832	....
Percussion-lock	S. Forker	Meadville, Pa	Feb. 13, 1830	....
Percussion-lock	I. B. Richardson	Palmyra, N. Y	Feb. 17, 1832	....
Percussion-lock and walking-cane rifle and pistol	A. D. Cushing	Troy, N. Y	July 20, 1831	....
Percussion-lock, Concealed	J. Newbury	Poughkeepsie, N. Y	Apr. 27, 1830	....
Percussion-lock for discharging mining-blasts	E. Hughes	McCartysville, Cal	Oct. 11, 1864	44, 630
Percussion-mill for reducing or grinding grain	F. McCarthy	Demopolis, Ala	Feb. 24, 1845	3, 924
Percussion-pellet	C. Sharps	Hartford, Conn	June 28, 1853	9, 820
Percussion-powder	M. Kling	Reading, Pa	Aug. 18, 1857	18, 016
Percussion-powder	F. M. Ruschampt and J. Schutte	New York, N. Y	June 3, 1862	35, 477
Percussion-powder for discharging arms	S. Guthrie	Sackett's Harbor, N. Y	Aug. 21, 1834	....
Percussion-primer and gun-lock	E. Maynard	Washington, D. C	Sept. 22, 1845	4, 208
Percussion-tape primer	J. Chattaway	Springfield, Mass	July 22, 1856	15, 370
Perforating-machine	F. Anderson	Peekskill, N. Y	Dec. 26, 1871	122, 098
Perfume and volatile liquids, Device for nebulizing	E. P. Roche	Bath, Me	May 24, 1870	103, 374
Perfumery	G. Smith	Groton Junction, Mass	Nov. 29, 1870	109, 773
Perfuming and disinfecting apparatus	O. Boldeman	New York, N. Y	Apr. 4, 1871	113, 619
Permutation-lock	J. T. Adams	Washington, D. C	May 9, 1871	114, 510
Permutation-lock	G. B. Atwood	Philadelphia, Pa	Mar. 26, 1867	63, 133
Permutation-lock	G. B. Atwood	Philadelphia, Pa	Apr. 16, 1867	63, 824
Permutation-lock	T. A. Auberlin	Detroit, Mich	Sept. 28, 1869	95, 181
Permutation-lock	J. E. and L. P. Barnes	Davers Centre and Fitchburgh, Mass.	Sept. 20, 1870	107, 590
Permutation-lock	E. A. Barrows	Willimantic, Conn	Feb. 26, 1867	62, 388
Permutation-lock	E. W. Brettell	Newark, N. J	Aug. 27, 1867	68, 281
Permutation-lock	E. W. Brettell	Elizabeth, N. J	Oct. 20, 1868	83, 129
Permutation-lock	E. W. Brettell	Elizabeth, N. J	Oct. 25, 1870	108, 561
Permutation-lock	E. W. Brettell	New Haven, Conn	Mar. 18, 1873	136, 963
Permutation-lock	F. H. Brown	Chicago, Ill	Nov. 8, 1870	108, 964
Permutation-lock	W. C. Bussey	San Francisco, Cal	Mar. 12, 1872	124, 540
Permutation-lock	W. A. Carpenter	Elgin, Ill	Feb. 21, 1860	27, 202
Permutation-lock	S. L. Cole and W. G. Ayres	Brooklyn, N. Y	Feb. 16, 1869	86, 908
Permutation-lock	J. Connell	Rochester, N. Y	June 9, 1863	38, 858
Permutation-lock	J. Corbett	Salt Lake, Utah	Feb. 5, 1867	61, 810
Permutation-lock	G. L. Damon	Cambridge, Mass	May 30, 1871	115, 445
Permutation-lock	G. L. Damon	Boston, Mass	Apr. 2, 1872	125, 180
Permutation-lock	G. W. Dana	Racine, Wis	Jan. 19, 1869	86, 085
Permutation-lock	C. Diebold and J. Obernesser	Cincinnati, Ohio	Aug. 16, 1870	106, 472
Permutation-lock	W. F. Ensign	Troy, N. Y	Oct. 20, 1868	83, 144
Permutation-lock	W. F. Ensign	New York, N. Y	Nov. 17, 1868	84, 177
Permutation-lock	J. Farrel	New York, N. Y	Apr. 16, 1872	125, 669
Permutation-lock	C. Flesch	Rochester, N. Y	Feb. 19, 1867	62, 191
Permutation-lock	C. Flesch	Rochester, N. Y	Nov. 26, 1867	71, 373
Permutation-lock	C. Flesch	Rochester, N. Y	Dec. 27, 1870	110, 560
Permutation-lock	C. R. C. French	Berkley, Mass	Sept. 15, 1868	82, 104
Permutation-lock	L. H. Gano	New York, N. Y	Jan. 3, 1872	122, 452
Permutation-lock	J. B. Gray	Fredericksburgh, Va	Sept. 18, 1841	2, 261
Permutation-lock	H. Gross	Tiffin, Ohio	Mar. 29, 1870	101, 362
Permutation-lock	H. Gross	Tiffin, Ohio	May 24, 1870	103, 450
Permutation-lock	H. Gross	Cincinnati, Ohio	Oct. 11, 1870	108, 134
Permutation-lock	H. Gross	Cincinnati, Ohio	Nov. 18, 1873	144, 670
Permutation-lock	H. Gross	Cincinnati, Ohio	Dec. 2, 1873	145, 171
Permutation-lock	H. Gross and J. L. Hall	Cincinnati, Ohio	Sept. 6, 1870	107, 174
Permutation-lock	M. Hainque	San Francisco, Cal	Sept. 13, 1870	107, 251
Permutation-lock	P. W. Hall	Calvert, Tex	Sept. 9, 1873	142, 695
Permutation-lock	W. Hall	Boston, Mass	May 18, 1869	90, 096
Permutation-lock	W. N. Hall	Springfield, Tex	Sept. 27, 1870	107, 775
Permutation-lock	J. C. Hintz, jr	Cincinnati, Ohio	May 17, 1870	103, 183
Permutation-lock	J. C. Hintz, jr	Cincinnati, Ohio	Oct. 18, 1870	108, 481
Permutation-lock	H. Isham	Hartford, Conn	Jan. 18, 1870	98, 972
Permutation-lock	A. W. Johnson and G. Thompson.	New York, N. Y	May 21, 1867	64, 880
Permutation-lock	F. G. Johnson	Brooklyn, N. Y	June 30, 1857	17, 681
Permutation-lock	W. Johnson	Milwaukee, Wis	Nov. 17, 1868	84, 192
Permutation-lock	G. Kaiser	New York, N. Y	May 19, 1868	78, 097
Permutation-lock	W. F. Kistler	Chicago, Ill	July 9, 1867	66, 502
Permutation-lock	W. F. Kistler	Cincinnati, Ohio	Sept. 7, 1869	94, 613
Permutation-lock	W. F. Kistler	Chicago, Ill	July 5, 1870	105, 091
Permutation-lock	J. Klein	Williamsburgh, N. Y	May 5, 1868	77, 623
Permutation-lock	W. Kock	Cincinnati, Ohio	July 5, 1870	104, 961
Permutation-lock	I. W. Lamb	Salem, Mich	Dec. 17, 1867	72, 408
Permutation-lock	L. W. Langdon	Northampton, Mass	June 8, 1869	91, 139
Permutation-lock	J. H. Larry	Weston, Mass	Apr. 5, 1870	101, 632
Permutation-lock	R. A. Lee	Cleveland, Ohio	Oct. 15, 1872	132, 161
Permutation-lock	S. Loyd	New York, N. Y	Mar. 14, 1871	112, 724
Permutation-lock	C. L. Lucas	Plymouth, Mass	Aug. 27, 1867	68, 219
Permutation-lock	N. Macneale	Cincinnati, Ohio	July 5, 1870	105, 102
Permutation-lock	D. K. Miller	Reading, Pa	July 4, 1871	116, 737
Permutation-lock	D. K. Miller	Reading, Pa	May 26, 1868	78, 310
Permutation-lock	L. H. Miller	Baltimore, Md	Jan. 28, 1868	73, 742
Permutation-lock	L. H. Miller	Baltimore, Md	Sept. 6, 1870	107, 190

Index of patents issued from the United States Patent Office from 1790 *to* 1873, *inclusive*—Continued.

Invention.	Inventor.	Residence.	Date.	No.
Permutation-lock	S. Miller	Gratis, Ohio	Apr. 23, 1872	126, 074
Permutation-lock	S. Miller	Gratis, Ohio	Dec. 2, 1873	145, 223
Permutation-lock	J. H. Morse	Peoria, Ill	Mar. 30, 1858	19, 815
Permutation-lock	J. H. Morse	Peoria, Ill	June 15, 1869	91, 359
Permutation-lock	J. Obernesser	Cincinnati, Ohio	May 23, 1871	115, 230
Permutation-lock	J. Obernesser	Cincinnati, Ohio	May 23, 1871	115, 231
Permutation-lock	O. E. Pillard	New Britain, Conn	Jan. 28, 1868	73, 831
Permutation-lock	O. E. Pillard	New Britain, Conn	Mar. 24, 1868	75, 967
Permutation-lock	O. E. Pillard	New Britain, Conn	June 30, 1868	79, 388
Permutation-lock	O. E. Pillard	New Britain, Conn	June 1, 1869	90, 683
Permutation-lock	O. E. Pillard	New Britain, Conn	Dec. 7, 1869	97, 550
Permutation-lock	O. E. Pillard	New Britain, Conn	May 9, 1871	114, 706
Permutation-lock	O. E. Pillard	New Britain, Conn	Oct. 31, 1871	120, 396
Permutation-lock	O. E. Pillard	New Britain, Conn	Apr. 23, 1872	125, 905
Permutation-lock	F. B. Pye	New York, N. Y	Mar. 7, 1846	4, 406
Permutation-lock	J. Sargent	Rochester, N. Y	Jan. 4, 1870	98, 622
Permutation-lock	J. Sargent	Rochester, N. Y	Dec. 22, 1868	85, 245
Permutation-lock	J. P. Schmucker	Ashland, Ohio	Jan. 4, 1870	98, 523
Permutation-lock	D. Snell	Little Falls, N. Y	Oct. 12, 1869	95, 739
Permutation-lock	E. Stockwell	Stamford, Conn	July 25, 1871	117, 478
Permutation-lock	E. Stockwell	Stamford, Conn	Sept. 2, 1873	142, 529
Permutation-lock	W. Streeter	Rochester, N. Y	July 12, 1870	105, 383
Permutation-lock	T. J. Sullivan	Albany, N. Y	Aug. 25, 1868	81, 430
Permutation-lock	T. J. Sullivan	Albany, N. Y	Sept. 14, 1869	94, 789
Permutation-lock	T. J. Sullivan	Albany, N. Y	Nov. 12, 1872	132, 936
Permutation-lock	J. T. Taylor	Newman, Ga	May 21, 1872	127, 117
Permutation-lock	W. Terwilliger	New York, N. Y	Apr. 4, 1871	113, 465
Permutation-lock	H. C. Thrall	Springfield, Mass	Sept. 24, 1872	131, 576
Permutation-lock	D. L. Tower	New York, N. Y	Oct. 17, 1871	120, 128
Permutation-lock	H. R. Towne	Stamford, Conn	Feb. 7, 1871	111, 587
Permutation-lock	J. F. Vinton and G. A. Hines	Brattleborough, Vt	Nov. 14, 1871	120, 915
Permutation-lock	S. C. Weddington	Jonesborough, Ind	Oct. 31, 1871	120, 472
Permutation-lock	J. Weimar	Mount Vernon, Ohio	Aug. 24, 1869	94, 156
Permutation-lock	A. Wetzel	Cincinnati, Ohio	Feb. 25, 1868	74, 781
Permutation-lock	S. Wheeler	Albany, N. Y	May 7, 1867	64, 605
Permutation-lock	S. Wheeler	Albany, N. Y	Sept. 17, 1867	68, 922
Permutation-lock	T. B. Worrell	Philadelphia, Pa	Feb. 18, 1873	135, 957
Permutation-lock	L. Yale, jr	Shelburne Falls, Mass	Sept. 15, 1868	82, 192
Permutation-lock	L. Yale, jr	Bernardstown, Mass	Jan. 4, 1870	98, 536
Permutation-lock, Auricular	E. E. Stubbs	West Elkton, Ohio	July 2, 1872	128, 672
Permutation-lock for doors, &c	J. H. Morse	Peoria, Ill	Nov. 5, 1867	70, 599
Permutation-lock for doors, &c	T. J. Sullivan	Rochester, N. Y	Aug. 20, 1867	67, 927
Permutation-lock for doors, vaults, &c., Manifold	R. Newell	New York, N. Y	Sept. 25, 1838	944
Permutation-lock for vaults, safes, &c	D. W. Maples	Geneva, N. Y	Dec. 4, 1844	3, 842
Permutation-lock indicator	A. Hardy	Boston, Mass	Dec. 22, 1868	85, 227
Permutation-locks, Adjustable tumbler for	O. E. Pillard	New Britain, Conn	Dec. 3, 1867	71, 639
Permutation-locks, Adjustable tumbler for	O. E. Pillard	New Britain, Conn	Sept. 8, 1868	82, 030
Permutation-locks, Adjustable tumbler for	J. H. Porter	New York, N. Y	Dec. 3, 1867	71, 640
Permutation-locks, Tumbler for	E. W. Brettell	New Haven, Conn	Dec. 16, 1873	145, 618
Permutation-locks, Tumbler for	C. Flesch	Rochester, N. Y	Apr. 4, 1871	113, 414
Permutation-locks, Tumbler for	W. Kock	Cincinnati, Ohio	July 25, 1871	117, 302
Permutation plate-lock, Rotating	H. Ritchie	Newark, N. J	June 26, 1849	6, 555
Permutation-wheel for lock	E. W. Brettell	Elizabeth, N. J	June 13, 1871	115, 814
Perpetual register	J. S. and J. H. Hood	Washington, D. C	June 16, 1868	78, 967
Perplexing unpickable lock	J. C. S. Daubreville	Ramapo, N. Y	May 13, 1833	....
Perspective protracter	B. Otis	Philadelphia, Pa	Mar. 14, 1815	....
Pessary	J. B. Andrews	New York, N. Y	Nov. 13, 1849	6, 861
Pessary	B. Atkinson	Davenport, Iowa	Feb. 18, 1868	74, 482
Pessary	E. P. Banning, sr	New York, N. Y	July 29, 1873	141, 253
Pessary	L. Barker	New York, N. Y	Sept. 12, 1822	....
Pessary	E. T. Brigham	Philadelphia, Pa	Dec. 3, 1867	71, 692
Pessary	W. F. Chrisman	Trenton, Tenn	Sept. 8, 1868	81, 877
Pessary	H. H. Christie	Herkimer, N. Y	Nov. 13, 1866	59, 556
Pessary	W. Elmer	New York, N. Y	Aug. 17, 1858	21, 189
Pessary	D. Ewing	Indianapolis, Ind	Apr. 30, 1867	64, 211
Pessary	C. E. Flack	Jacksonville, Ill	Dec. 23, 1873	145, 854
Pessary	W. R. Gardner	Leonardsville, N. Y	July 12, 1870	105, 191
Pessary	L. A. Gescheidt	New York, N. Y	May 9, 1848	5, 556
Pessary	C. R. Gorgas	Brooklyn, N. Y	July 21, 1868	80, 163
Pessary	C. R. Gorgas	Roughsburgh, Ohio	May 18, 1869	90, 166
Pessary	W. G. Grant	Clyde, Ohio	June 18, 1867	65, 903
Pessary	E. B. Harding	Northampton, Mass	Nov. 27, 1866	59, 999
Pessary	J. M. Heard	Aberdeen, Wis	May 29, 1860	28, 480
Pessary	E. F. Hofman	Poughkeepsie, N. Y	June 4, 1867	65, 382
Pessary	E. F. Hofmann	New York, N. Y	Jan. 11, 1870	98, 769
Pessary	B. Joseph	Philadelphia, Pa	June 23, 1868	79, 231
Pessary	J. B. Merriman	Sheffield, Mass	Oct. 24, 1846	4, 825
Pessary	O. M. Muncaster	Washington, D. C	July 22, 1873	141, 069
Pessary	M. J. Rhees	Mount Holly, N. J	Nov. 5, 1867	70, 616
Pessary	J. H. Robinson	Charlestown, Mass	Aug. 7, 1849	6, 629
Pessary	J. H. Robinson	Charlestown, Mass	Nov. 19, 1850	7, 788
Pessary	J. R. Rowand	Philadelphia, Pa	Oct. 24, 1848	5, 877
Pessary	T. C. Sachse	Chicago, Ill	Oct. 22, 1867	70, 124
Pessary	H. V. Scattergood	Albany, N. Y	Dec. 15, 1863	40, 949
Pessary	C. Schmidt	Saint Louis, Mo	Mar. 29, 1870	101, 317
Pessary	I. Stealy	Crestline, Ohio	June 19, 1866	55, 733
Pessary	B. F. Taft	Blackstone, Mass	July 29, 1862	36, 040
Pessary	J. Warrington and W. Scattergood.	Philadelphia, Pa	Sept. 5, 1831	....
Pessary	F. F. Wells	Texana, Tex	Feb. 22, 1859	23, 052
Pessary	F. F. Wells	Texana, Tex	Mar. 13, 1860	27, 491
Pessary	J. P. Willms	Baltimore, Md	Dec. 4, 1866	60, 312
Pessary	D. D. Young	Dayton, Ohio	Aug. 3, 1869	93, 387
Pessary-adjuster	O. M. Muncaster	Washington, D. C	Mar. 4, 1873	136, 449
Pessary, Elastic	J. A. Wadsworth	Providence, R. I	Jan. 24, 1860	26, 941
Pessary for prolapsus uteri	S. K. Jennings	Baltimore, Md	Jan. 20, 1843	2, 921
Pessaries, Construction of	F. Roesler	New York, N. Y	Feb. 19, 1856	14, 293
Pestles with mortars, Arrangement of	P. C. Ingersoll	Elmira, N. Y	May 9, 1854	10, 890

Index of patents issued from the United States Patent Office from 1790 *to* 1873, *inclusive*—Continued.

Invention.	Inventor.	Residence.	Date.	No.
Petroleum and fluids made therefrom, Apparatus for burning.	F. Cook	New York, N. Y	Sept. 10, 1867	68, 704
Petroleum and other hydrocarbons, Apparatus for burning.	H. C. Van Tine	Pittsburgh, Pa	June 18, 1867	65, 846
Petroleum and other hydrocarbon liquids, Apparatus for combustion of.	H. Taylor	Montreal, Canada	May 3, 1870	102, 619
Petroleum and other hydrocarbon oils, Treating	J. Young	Kelly, North Britain	June 4, 1872	127, 446
Petroleum and other inflammable liquids, Apparatus for storing.	F. Bizard and P. Labarre	Marseilles, France	Jan. 15, 1867	61, 148
Petroleum and other liquid fuel, Process of burning.	D. Dick	Meadville, Pa	June 2, 1863	38, 732
Petroleum and other liquids from gas, Apparatus for freeing.	A. H. Hook	New York, N. Y	May 11, 1869	89, 998
Petroleum and other liquids to prevent loss by fire, Mode of storing.	S. Stevens	New York, N. Y	Jan. 1, 1867	60, 956
Petroleum and other liquids, Vessel for holding	E. Waters	Troy, N. Y	Aug. 25, 1868	81, 441
Petroleum and other oils to produce a vehicle for paint and varnish, Treating.	A. Meucci	Clifton, N. Y	Sept. 9, 1862	36, 419
Petroleum, Apparatus and process for treating	S. Van Syckel	Titusville, Pa	Oct. 21, 1873	143, 945
Petroleum, Apparatus for burning	E. McKinney	Middletown, Pa	July 25, 1865	48, 967
Petroleum, Apparatus for burning crude	H. Baldwin	Titusville, Pa	Mar. 10, 1868	75, 238
Petroleum, Apparatus for burning crude	C. Saffray	New York, N. Y	Sept. 24, 1867	69, 253
Petroleum, Apparatus for refining and distilling	J. Perkins and W. H. Burnet	Newark, N. J	Apr. 4, 1865	47, 125
Petroleum, Apparatus for treating	D. H. Burket and J. C. Gray	Half Moon and Putneyville, Pa.	Aug. 21, 1866	57, 285
Petroleum, Apparatus for treating	A. Thirault	Williamsburgh, N. Y	Apr. 16, 1867	63, 963
Petroleum as fuel, Apparatus for burning	F. Cook	New York, N. Y	Sept. 10, 1867	68, 706
Petroleum, benzole, &c., Apparatus for deodorizing	J. Green	Rochester, N. Y	Mar. 14, 1865	46, 794
Petroleum, Burning	A. C. Rand	Aurora, Ill	July 2, 1872	128, 656
Petroleum, &c., Composition for coating vessel for.	W. L. Dooling and D. Christman, jr.	Zanesville, Ohio	Mar. 20, 1866	53, 280
Petroleum, &c., Deodorizing	S. Lewis	Rochester, N. Y	May 10, 1864	42, 671
Petroleum, Deodorizing	O. Lugo	New York, N. Y	Jan. 1, 1867	60, 757
Petroleum, Deodorizing	R. Newall	Marietta, Ohio	Apr. 3, 1866	53, 656
Petroleum, Deodorizing	T. Restieaux	Boston, Mass	Apr. 9, 1867	63, 749
Petroleum, &c., Deodorizing	J. W. W. Tindall	Liverpool, England	Mar. 31, 1863	38, 069
Petroleum, Device for burning	F. A. Hull	Beloit, Wis	Feb. 4, 1873	135, 476
Petroleum, Device for heating and conveying	J. Casey	Washington, D. C	May 16, 1865	47, 701
Petroleum for lubricating, Preparing	W. H. Young	Athens, Ohio	Mar. 12, 1867	62, 798
Petroleum for manufacture of lubricating-oil, Treating.	G. O. Spence	Titusville, Pa	June 2, 1868	78, 545
Petroleum for manufacture of paints, &c., Preparing.	R. Bartholow	Cincinnati, Ohio	Apr. 4, 1865	47, 084
Petroleum-forge or blow-pipe	H. S. Saroni	Cincinnati, Ohio	Apr. 30, 1872	126, 158
Petroleum from the surface of water-courses, Device for collecting.	L. H. Cowley	Silver Creek, N. Y	Sept. 21, 1869	95, 089
Petroleum, &c., in conjunction with steam or heated air, Apparatus for burning.	G. O. Spence	Titusville, Pa	Apr. 30, 1867	64, 260
Petroleum, Manufacture of oils from	C. Toppan	Wakefield, Mass	Feb. 7, 1870	99, 500
Petroleum, Material for purifying and discoloring.	J. Ellis	New York, N. Y	May 5, 1868	77, 470
Petroleum, naphtha, &c., Deodorizing	S. Lewis	Rochester, N. Y	Sept. 13, 1864	44, 258
Petroleum, naphtha, &c., Deodorizing	R. N. Warfield	Rochester, N. Y	Sept. 22, 1863	40, 068
Petroleum, Obtaining useful products from the "tarry residuum" of.	A. Millochaw	New York, N. Y	Jan. 5, 1864	41, 085
Petroleum, Packing-vessels for	J. W. Masury	Brooklyn, N. Y	Sept. 25, 1866	58, 272
Petroleum, Product from	R. A. Chesebrough	New York, N. Y	June 4, 1872	127, 568
Petroleum, &c., Purifying	R. A. Chesebrough	New York, N. Y	July 10, 1866	56, 179
Petroleum, Purifying	C. C. Parsons	New York, N. Y	Apr. 13, 1869	88, 978
Petroleum, Recovering waste acid for refining	O. W. Farrar	Pittsburgh, Pa	June 4, 1867	65, 361
Petroleum, Refined heavy-oil	H. B. Everest	Rochester, N. Y	Feb. 23, 1869	87, 157
Petroleum, Refining	S. Van Syckel	Titusville, Pa	July 15, 1873	140, 801
Petroleum, &c., Storing	E. Quinn	Washington, D. C	Oct. 2, 1866	58, 475
Petroleum-storing apparatus	I. Mathei	Antwerp, Belgium	Oct. 20, 1868	83, 192
Petroleum-tank	M. C. C. Church	Parkersburgh, W. Va	Aug. 28, 1866	57, 479
Petroleum, &c., tank	M. C. C. Church and E. H. Knight.	Parkersburgh, W. Va., and Washington, D. C.	Sept. 11, 1866	57, 866
Petroleum-tank	P. Jacovence	Bucharest, Wallachia	June 26, 1866	55, 970
Petroleum, Tank for storing	P. Andrew	Cincinnati, Ohio	May 28, 1867	65, 045
Petroleum, Tank for storing	J. Fraser and J. Calkins	Buffalo, N. Y	Oct. 10, 1865	50, 348
Petroleum to produce oil and gas, Apparatus for treating.	A. C. Rand and W. M. Sloane	New York, N. Y	Feb. 23, 1869	87, 199
Petroleum to remove the more volatile portions, Treating.	F. Huot	New York, N. Y	Dec. 3, 1867	71, 619
Petroleum to remove the more volatile portions, Treating.	R. G. Loftus	Chelsea, Mass	Sept. 1, 1868	81, 654
Petroleum, Transporting	J. H. Smith	Allegheny City, Pa	Dec. 10, 1867	72, 102
Petroleum, Transportation of	H. J. Lombaert	Philadelphia, Pa	Sept. 12, 1865	49, 901
Petroleum, Treating	J. C. Pedrick	Washington, D. C	Aug. 13, 1867	67, 796
Petroleum, Treating	C. S. Potter	Brooklyn, N. Y	July 5, 1864	43, 429
Petroleum, Treating	E. Schalk	New York, N. Y	Dec. 3, 1872	133, 598
Petroleum, Treating	J. A. Tatro	Hartford, Conn	Feb. 8, 1870	99, 728
Petroleum, Treating heavy	J. J. Looney	Saint Petersburgh, Pa	May 20, 1873	139, 009
Petroleum, &c., Vaporizing	H. R. Foote	Boston, Mass	Dec. 21, 1869	98, 046
Petroleum, &c., vessel	E. S. Allen	Fair Haven, Vt	May 8, 1866	54, 657
Petroleum-vessel	J. Brakeley	Philadelphia, Pa	Jan. 16, 1866	52, 023
Petroleum, Vessel for holding	J. W. Barnum and P. M. McNoah.	Detroit, Mich	July 25, 1865	48, 891
Petroleum, Vessel for holding	J. M. Batchelder	Cambridge, Mass	Feb. 7, 1865	46, 206
Pew-shelf	D. Buchanan	Philadelphia, Pa	Dec. 20, 1870	110, 197
Phaeton, Pony carriage	J. C. Ham	New York, N. Y	May 31, 1870	103, 605
Phaeton, Shifting top for basket	C. P. Kimball	Portland, Me	Apr. 18, 1871	113, 775
Pharmaceutic composition	J. C. Bay	Mount Pleasant, Ohio	Jan. 26, 1825	....
Phial, Tubular essence	J. Staniford	Boston, Mass	Oct. 17, 1831	....
Phosphate and fertilizer, Manufacture of soluble	D. W. Prescott	Edinburgh, Va	Oct. 17, 1871	119, 994
Phosphate and for other purposes, Mixing liquid with dust or powder for the manufacture of.	A. Duvall	Baltimore, Md	Jan. 26, 1869	86, 289
Phosphate and in extracting phosphoric acid from bones, Manufacture of.	G. F. Wilson and E. N. Horsford.	East Providence, R. I., and Cambridge, Mass.	Mar. 10, 1868	75, 336

Index of patents issued from the United States Patent Office from 1790 *to* 1873, *inclusive*—Continued.

Invention.	Inventor.	Residence.	Date.	No.
Phosphate and other acid liquids, Apparatus for conveying acid.	G. F. Wilson	East Providence, R. I	Mar. 10, 1868	75, 335
Phosphate-drying apparatus	F. J. Kimball	Philadelphia, Pa	May 2, 1871	114, 304
Phosphate-drying apparatus	E. Frank and J. B. Adt	Baltimore, Md	Aug. 9, 1870	106, 147
Phosphate, &c., Bag for	B. R. Croasdale	Philadelphia, Pa	Mar. 5, 1872	124, 254
Phosphate, Burning bones for the manufacture of acid.	G. F. Wilson	East Providence, R. I	Mar. 10, 1868	75, 332
Phosphate, Drying acid	G. F. Wilson	East Providence, R. I	Mar. 10, 1868	75, 330
Phosphate for agricultural purposes, Manufacture of.	G. F. Wilson	East Providence, R. I	Mar. 10, 1868	75, 327
Phosphate, Granulating acid	G. F. Wilson	East Providence, R. I	Mar. 10, 1868	75, 334
Phosphate, Grinding and pulverizing acid	G. F. Wilson	East Providence, R. I	Mar. 10, 1868	75, 333
Phosphate in the manufacture of yeast-powder, &c., Treating acid.	G. F. Wilson and E. N. Horsford	East Providence, R. I., and Cambridge, Mass.	Mar. 10, 1868	75, 338
Phosphate, Manufacture of a non-hygroscopic pulverulent acid.	E. N. Horsford	Cambridge, Mass	June 7, 1870	104, 035
Phosphate, Manufacture of acid	H. Storck and F. M. Lyte	Asnieres, near Paris, France.	Apr. 8, 1873	137, 635
Phosphate, Manufacture of acid	G. F. Wilson	East Providence, R. I	Mar. 10, 1868	75, 328
Phosphate, Manufacturing fertilizing	G. A. Liebig and E. K. Cooper	Baltimore, Md	Jan. 17, 1865	45, 961
Phosphate of lime and farinaceous matter in order to granulate them, Treating the mixture of acid.	G. F. Wilson and E. N. Horsford.	East Providence, R. I., and Cambridge, Mass.	Mar. 10, 1868	75, 337
Phosphate of lime and soda for culinary and other purposes, Double.	E. N. Horsford	Cambridge, Mass	Mar. 29, 1864	42, 140
Phosphate of lime, Apparatus for concentrating acid.	G. F. Wilson and E. N. Horsford.	East Providence, R. I., and Cambridge, Mass.	Mar. 10, 1868	75, 339
Phosphate of lime, Apparatus for the manufacture of super.	R. B. Potts and F. Klett	Camden, N. J	Dec. 27, 1864	45, 631
Phosphate of lime for culinary and other purposes, Preparation of.	E. N. Horsford	Cambridge, Mass	Mar. 1, 1864	41, 815
Phosphate of lime, Preparation of acid	E. N. Horsford	Cambridge, Mass	Mar. 10, 1868	75, 271
Phosphate, Preparation of bones for the manufacture of phosphoric acid.	G. F. Wilson	East Providence, R. I	Mar. 10, 1868	75, 326
Phosphate, Prepared	A. A. Moses	Charleston, S. C	Feb. 2, 1869	86, 574
Phosphate, Preparing bones for the manufacture of acid.	G. F. Wilson	East Providence, R. I	Mar. 10, 1868	75, 329
Phosphate, Process and apparatus for the manufacture of acid.	G. F. Wilson	East Providence, R. I	Mar. 10, 1868	75, 331
Phosphate to be used in food, Manufacture of acid.	E. N. Horsford	Cambridge, Mass	Mar. 10, 1868	75, 272
Phosphate, Treating insoluble	H. Bower	Philadelphia, Pa	Jan. 26, 1869	86, 275
Phosphate, Treating vitriolized	A. Duvall	Baltimore, Md	July 12, 1870	105, 319
Phosphatic rock, Treating	N. A. Pratt and G. T. Lewis	Charleston, S. C., and Philadelphia, Pa.	July 9, 1872	128, 752
Phosphatic substances, Grinding	G. T. Lewis	Philadelphia, Pa	May 9, 1871	114, 693
Photo-electrotype	W. A. Leggo and G. E. Desbarats.	Quebec, Canada	May 30, 1865	48, 035
Photo-engraving on metal	W. A. McGill and R. G. Pine	Memphis, Tenn	Mar. 26, 1872	124, 905
Photo-galvanographic process for printing	P. Pretsch	Austria	Aug. 25, 1857	18, 056
Photo-galvanography	P. E. Placet	Paris, France	Aug. 9, 1864	43, 821
Photo-lithographic transfer	J. W. Osborne	Melbourne, Australia	Aug. 27, 1861	33, 172
Photo-lithographic transfers, Producing	I. Rehn	Washington, D. C	Mar. 15, 1870	100, 924
Photo-lithography	J. A. Cutting and L. H. Bradford.	Boston, Mass	Mar. 16, 1858	19, 626
Photo-lithography	J. W. Osborne	Melbourne, Australia	June 25, 1861	32, 668
Photo-sculpture	F. Willéme	Paris, France	Aug. 9, 1864	43, 822
Photograph-camera	J. A. Scott	Lexington, Va	Dec. 22, 1868	85, 247
Photograph-chair, Child's	M. H. Prescott, jr	La Crosse, Wis	May 27, 1873	139, 329
Photograph, Compound	T. Miltenberger	Bellefontaine, Ohio	May 11, 1858	20, 213
Photograph cutter	S. W. Robinson	Champaign, Ill	Nov. 21, 1871	121, 198
Photograph-drier	A. C. Platt	Oberlin, Ohio	Sept. 11, 1866	57, 963
Photograph-exhibiting stand	J. S. Reid	Orange, Ind	Nov. 16, 1869	96, 961
Photograph, Granulated	W. A. Leggo	Montreal, Canada	May 30, 1871	115, 489
Photograph-holder	J. E. Treat	Boston, Mass	June 9, 1863	38, 849
Photograph-mount	A. C. Partridge	Boston, Mass	June 11, 1872	127, 705
Photograph on uneven surfaces, Apparatus to	J. H. Pein	Hoboken, N. J	Aug. 30, 1859	25, 276
Photograph-press	R. Sibley	Greenville, Conn	June 10, 1862	35, 570
Photograph-washing machine	L. V. Moulton	Beaver Dam, Wis	Aug. 12, 1873	141, 658
Photographs and engravings, Mounting	J. L. Duffee	Washington, D. C	May 26, 1868	78, 195
Photographs, Apparatus for cutting	L. A. Foulley	Boston, Mass	Jan. 31, 1865	46, 099
Photographs, Apparatus for exhibiting	M. Dimock	Newark, N. J	Oct. 12, 1869	95, 666
Photographs, Apparatus for exhibiting	W. H. Fay	Chester, Mass	June 28, 1864	43, 299
Photographs, Apparatus for fixing	C. A. Gale	Piqua, Ohio	Oct. 10, 1871	119, 753
Photographs, &c., Apparatus for pasting and mounting.	M. Ormsbee	New York, N. Y	July 7, 1863	39, 166
Photographs, Apparatus for producing vignette	E. R. Scott	Philadelphia, Pa	July 8, 1862	35, 857
Photographs, Apparatus for taking stereoscopic	J. H. Marston	Philadelphia, Pa	June 19, 1855	13, 093
Photographs, Apparatus for washing and drying	A. Pearsch	New Orleans, La	Oct. 11, 1864	44, 648
Photographs, Apparatus to preserve and exhibit	C. Robinson	Springfield, Mass	Apr. 11, 1865	47, 222
Photographs, Applying tints to	H. Vander Weyde	New York, N. Y	Dec. 5, 1871	121, 475
Photographs, Burnisher for	C. B. Conant	Lewiston, Me	Dec. 2, 1873	145, 152
Photographs, Burnisher for	W. G. Entrekin	Philadelphia, Pa	Dec. 2, 1873	145, 161
Photographs, Burnisher for	E. R. Weston	East Corinth, Me	Sept. 10, 1872	131, 319
Photographs, Burnishing-apparatus for	E. R. Weston	East Corinth, Me	Sept. 10, 1872	131, 320
Photographs, Burnishing-press for	W. G. Entrekin and J. Brambles, jr.	Philadelphia, Pa	May 20, 1873	139, 132
Photographs, Coloring	J. F. Bodtker	Madison, Wis	Apr. 14, 1863	38, 144
Photographs, Coloring	R. Winter	San Francisco, Cal	June 14, 1870	104, 241
Photographs, &c., Composition for coloring and water-proofing.	W. F. Spieler	Philadelphia, Pa	June 9, 1863	38, 847
Photographs, Cutting and pasting	J. Franke	Quincy, Ill	Feb. 11, 1868	74, 330
Photographs, &c., Finishing	S. A. L. Hardinge	Brooklyn, N. Y	Jan. 5, 1869	85, 584
Photographs for exhibition, Mounting	I. Rowell and F. E. Mills	San Francisco, Cal	Jan. 22, 1867	61, 469
Photographs from glass to paper, Removing	E. Howell	Ashtabula, Ohio	May 19, 1857	17, 330
Photographs, &c., Glazing	T. J. Denne and A. Hentschel	London, England	Oct. 29, 1872	132, 640
Photographs, Guide for cutting out	J. Schofield	Philadelphia, Pa	Apr. 22, 1873	138, 202
Photographs in the camera, Making positive and negative.	F. B. Gage	Saint Johnsbury, Vt	July 9, 1867	66, 581
Photographs, Instrument for cutting	T. Bergner	Philadelphia, Pa	Jan. 31, 1865	46, 066

Index of patents issued from the United States Patent Office from 1790 *to* 1873, *inclusive*—Continued.

Invention.	Inventor.	Residence.	Date.	No.
Photographs, Instrument for enlarging	D. Shire	Philadelphia, Pa	Mar. 22, 1859	23, 316
Photographs, Machine for pressing and smoothing	D. and J. Rupp	New York, N. Y	Oct. 24, 1865	50, 632
Photographs, Mounting	A. C. Platt	Sandusky, Ohio	June 11, 1872	127, 920
Photographs on cards, Apparatus for pasting	C. S. Lucas	Poughkeepsie, N. Y	Nov. 29, 1864	45, 253
Photographs on glass, Background for	J. W. Wykes	Wheeling, Va	June 23, 1857	17, 651
Photographs on paper, Varnishing	D. W. S. Rawson	Galena, Ill	Apr. 2, 1861	31, 906
Photographs, Printing	J. Albert	Munich, Bavaria	Nov. 30, 1869	97, 336
Photographs, Printing	A. S. Kilby	Huntington, Ind	Aug. 18, 1868	81, 277
Photographs, Printing	C. Meinerth	Newburyport, Mass	July 16, 1867	66, 726
Photographs, Printing	J. W. Swan	Newcastle - upon - Tyne, England.	Jan. 22, 1867	61, 368
Photographs, Printing	W. B. Woodbury	Manchester, England	Feb. 20, 1866	52, 803
Photographs, Producing surfaces for printing from	W. B. Woodbury	London, England	Apr. 28, 1868	77, 232
Photographs, Roller for finishing	W. J. Gordon	Philadelphia, Pa	Nov. 6, 1866	59, 507
Photographs, &c., Roller-press for finishing	D. Marshall and B. Marshall	Pittsburgh, Pa., and Marietta, Ohio.	July 12, 1864	43, 515
Photographs, Roller-press for polishing	N. S. Bowdish	Richfield Springs, N. Y	July 1, 1873	140, 456
Photographs, Solution for toning	J. C. Rutherford	Derby Line, Vt	Feb. 7, 1860	27, 076
Photographs, Tinting	C. Elveena	San Francisco, Cal	Aug. 7, 1866	56, 914
Photographs to tomb-stones, Securing	N. W. Langley, H. Jones, and A. S. Drake.	East Cambridge and Stoughton, Mass.	Dec. 6, 1859	26, 370
Photographs, &c., Treating	E. C. Hawkins	Cincinnati, Ohio	Dec. 22, 1857	18, 901
Photographers and others, Enameled flexible metallic plate for.	W. Hodge and J. R. Peck	Washington, Pa	June 6, 1871	115, 612
Photographer's clip or paper-holder	V. M. Griswold	Peekskill, N. Y	Nov. 23, 1869	97, 081
Photographer's curtain	L. G. Bigelow	Grand Rapids, Mich	Sept. 19, 1871	119, 111
Photographer's decanter	G. W. Doty and E. A. and W. F. Stein.	Ravenna, Ohio	July 11, 1865	48, 664
Photographer's dripping and drying rack	V. M. Griswold	Peekskill, N. Y	Nov. 23, 1869	97, 082
Photographer's movable screen	E. J. Foss	Cambridge, Mass	July 26, 1870	105, 665
Photographer's plate-vise	V. M. Griswold	Peekskill, N. Y	Oct. 19, 1869	95, 999
Photographer's posing-apparatus	G. Cramer and J. Gross	Saint Louis, Mo	Nov. 10, 1868	83, 835
Photographer's refrigerator	M. A. Thornton	Perrysburgh, Ohio	July 18, 1871	117, 124
Photographer's rest	E. L. Wilson	Philadelphia, Pa	Oct. 26, 1869	96, 292
Photographer's steeping-tank	J. W. Osborne	Brooklyn, N. Y	Apr. 25, 1871	114, 186
Photographic album	H. T. Anthony and F. Phoebus	New York, N. Y	May 28, 1861	32, 404
Photographic and other negatives, Producing pictures from.	W. B. Woodbury	Greenhitke, England	May 14, 1872	126, 861
Photographic apparatus	C. G. Anthoni	Paris, France	Sept. 10, 1861	33, 230
Photographic apparatus	N. F. English	Hartland, Vt	May 20, 1862	35, 308
Photographic apparatus	D. H. Houston	Cambria, Wis	Aug. 20, 1867	67, 981
Photographic apparatus	C. A. Leach	Philadelphia, Pa	June 19, 1866	55, 682
Photographic apparatus	W. A. Leggo	Montreal, Canada	May 6, 1873	138, 663
Photographic apparatus	R. H. Mims	Edgefield Court-House, S.C	Mar. 12, 1872	124, 607
Photographic apparatus	E. Müller	New York, N. Y	June 12, 1866	55, 525
Photographic apparatus	O. Nicour	Paris, France	Nov. 19, 1867	71, 205
Photographic apparatus	J. and J. Stock	New York, N. Y	July 4, 1871	116, 771
Photographic apparatus	J. and J. Stock	New York, N. Y	Feb. 6, 1872	123, 522
Photographic apparatus	N. Wright	New York, N. Y	Jan. 23, 1866	52, 239
Photographic background	D. and D. Bendann	Baltimore, Md	Apr. 9, 1872	125, 522
Photographic background	L. G. Bigelow	Grand Rapids, Mich	June 13, 1871	115, 812
Photographic background	J. Buchtel	Portland, Oreg	Dec. 16, 1873	145, 487
Photographic background	A. B. Costello	Jersey City, N. J	Dec. 16, 1873	145, 557
Photographic background	P. F. Finch	Lebanon, Ohio	July 16, 1872	129, 122
Photographic background	A. R. Gould	Carrollton, Ohio	Dec. 12, 1871	121, 715
Photographic background	D. W. Van Riper	Columbus, Ga	Sept. 6, 1870	107, 129
Photographic bank-note	L. Eidlitz	New York, N. Y	Feb. 14, 1860	27, 116
Photographic bath	A. Cobbs	Pittsburgh, Pa	Jan. 1, 1867	60, 834
Photographic bath	B. Hufnagel	New York, N. Y	Oct. 5, 1858	21, 679
Photographic bath	W. and W. H. Lewis	New York, N. Y	Dec. 16, 1856	16, 242
Photographic bath	W. and W. H. Lewis	New York, N. Y	June 26, 1860	28, 876
Photographic bath	J. H. Morrow	Baltimore, Md	Apr. 14, 1857	17, 066
Photographic bath	I. Rehn	Philadelphia, Pa	Dec. 4, 1855	13, 885
Photographic bath	R. E. Sisson	Wickford, R. I	June 12, 1866	55, 590
Photographic bath	N. Wright	New York, N. Y	Nov. 14, 1865	50, 981
Photographic bath and pans, Constructing	G. Mathiot	Washington, D. C	Apr. 28, 1857	17, 162
Photographic-camera embossing-press	E. B. Barker	Brooklyn, N. Y	Oct. 14, 1873	143, 608
Photographic camera	T. Barbour	Boston, Mass	Jan. 15, 1867	61, 139
Photographic camera	B. M. Clinedinst	Staunton, Va	July 16, 1872	129, 104
Photographic camera	E. M. Corbett	New York, N. Y	June 19, 1860	28, 739
Photographic camera	N. Flammang	New York, N. Y	Aug. 27, 1872	130, 800
Photographic camera	F. B. Gage	Saint Johnsbury, Vt	Dec. 24, 1867	72, 627
Photographic camera	W. A. Leggo	Montreal, Canada	Mar. 28, 1871	113, 067
Photographic camera	H. C. and C. L. Moore	Springfield, Mass	Feb. 16, 1864	41, 634
Photographic camera	G. W. Parker	Watkins, N. Y	July 18, 1871	117, 106
Photographic camera	A. Semmendinger	New York, N. Y	Feb. 21, 1860	27, 241
Photographic camera	A. Semmendinger	New York, N. Y	Aug. 7, 1860	29, 523
Photographic camera	E. P. Spahn	Newark, N. J	May 20, 1873	139, 090
Photographic camera	J. Stock	New York, N. Y	July 5, 1859	24, 671
Photographic camera	J. Stock	New York, N. Y	Aug. 4, 1863	39, 432
Photographic camera	J. Stock	New York, N, Y	May 31, 1864	42, 971
Photographic camera	J. Stock	New York, N. Y	Aug. 8, 1865	49, 315
Photographic camera	J. and J. Stock	New York, N. Y	June 6, 1871	115, 655
Photographic camera	J. and J. Stock	New York, N. Y	Dec. 5, 1871	121, 553
Photographic camera	J. and J. Stock	New York. N, Y	Jan. 16, 1872	122, 785
Photographic camera	J. and J. Stock	New York, N. Y	May 7, 1872	126, 498
Photographic camera	I. H. Stoddard	Ansonia, Conn	Jan. 18, 1870	99, 026
Photographic camera	H. W. Vaughan	San Francisco, Cal	Apr. 23, 1872	125, 914
Photographic camera	G. W. Venner	Charlestown, Mass	May 28, 1867	65, 314
Photographic camera	C. A. Waterbury	New York, N. Y	July 9, 1872	128, 832
Photographic camera	C. A. Waterbury	New York, N. Y	Nov. 26, 1872	133, 394
Photographic camera	J. H. and W. K. Wickoff	Ripon, Wis	Aug. 11, 1868	81, 055
Photographic camera	A. B. Wilson	Waterbury, Conn	Mar. 25, 1862	34, 800
Photographic camera	S. Wing	Waterville, Me	Dec. 4, 1860	30, 850
Photographic camera	S. Wing	Boston, Mass	May 26, 1868	78, 408
Photographic camera	T. E. Wood	Boston, Mass	Apr. 26, 1864	42, 524
Photographic-camera box	L. M. Boles and W. G. Smith	Cooperstown, N. Y	Feb. 17, 1857	16, 637
Photographic-camera box	M. Flammang	New York, N. Y	Mar. 19, 1872	124, 674

Index of patents issued from the United States Patent Office from 1790 *to* 1873, *inclusive*—Continued.

Invention.	Inventor.	Residence.	Date.	No.
Photographic camera lens	C. C. Harrison and J. Schnitzer	New York, N. Y	June 17, 1862	35, 605
Photographic-camera plate-holder	A. D. Bollens	Newburgh, N. Y	June 1, 1858	20, 401
Photographic-camera plate-holder	W. and W. H. Lewis	New York, N. Y	Oct. 7, 1856	15, 854
Photographic-camera stand	J. Haworth	Frankford, Pa	Apr. 9, 1867	63, 633
Photographic-camera stand	H. Manger	Philadelphia, Pa	June 13, 1865	48, 193
Photographic-camera stand	G. S. Knapp	Winona, Minn	Aug. 24, 1869	94, 120
Photographic-camera stand	T. M. Schleier	Knoxville, Tenn	July 9, 1872	128, 915
Photographic camera stand	J. Scouler	San Francisco, Cal	Feb. 14, 1865	46, 396
Photographic-camera stand	F. E. Wilke	Brooklyn, N. Y	Aug. 13, 1867	67, 830
Photographic cameras, &c., Achromatic object-glass for.	H. Roettger	Philadelphia, Pa	Apr. 18, 1865	47, 336
Photographic cameras, Cover for the lenses of	O. W. Noble	Darlington, Wis	July 25, 1871	117, 318
Photographic cameras, Diaphragm for	F. Miller and A. Wirsching	New York, N. Y	June 7, 1859	24, 356
Photographic cameras, Diaphragm for	C. C. Harrison and J. Schnitzer	New York, N. Y	Sept. 7, 1858	21, 470
Photographic cameras, Diaphragm for	J. R. Werner	New York, N. Y	Sept. 15, 1857	18, 218
Photographic cameras, Multiplying-reflector for	D. W. S. Rawson	Peru, Ill	Dec. 10, 1867	71, 911
Photographic cameras, Plate-frame for	W. and W. H. Lewis	New York, N. Y	Feb. 2, 1858	19, 252
Photographic-card case	W. Miller	New York, N. Y	Oct. 15, 1861	33, 507
Photographic-card holder	W. E. Eastman	Derby Line, Vt	June 22, 1869	91, 728
Photographic-card mount	T. Mayhew	Poughkeepsie, N. Y	Jan. 24, 1865	46, 008
Photographic-card mount	S. Wing	Boston, Mass	Oct. 13, 1863	40, 302
Photographic-card press	J. B. Blair	Philadelphia, Pa	June 14, 1864	43, 083
Photographic cards, Apparatus for mounting and printing.	G. W. Doty	Lockport, N. Y	Aug. 29, 1865	49, 612
Photographic chair	C. G. Pease	Charlestown, Mass	Mar. 6, 1866	53, 038
Photographic-copying board	W. D. Blackman	Defiance, Ohio	July 2, 1867	66, 288
Photographic dark-chambers, Portable	I. F. Woodward	McMinnville, Tenn	Oct. 31, 1871	120, 558
Photographic-embossing apparatus	H. Holt	Brooklyn, N. Y	May 28, 1872	127, 348
Photographic filtering-apparatus	R. M. Linn	Lookout Mountain, Tenn	Sept. 26, 1871	119, 375
Photographic gallery, Portable	S. Weaver	Gettysburgh, Pa	Dec. 15, 1863	40, 970
Photographic-glass holder	J. Longking	New Windsor Township, N. Y.	Feb. 24, 1857	16, 689
Photographic-glass rack	W. G. Smith	Carlisle, Pa	Dec. 27, 1864	45, 648
Photographic grounds for wood-engravers, Varnish to prepare.	R. Price	Worcester, Mass	May 5, 1857	17, 231
Photographic head-rest	G. Cramer and J. Gross	Saint Louis, Mo	Aug. 3, 1869	93, 279
Photographic head-rest	G. A. Emery	Boston, Mass	May 10, 1864	42, 650
Photographic head-rest	W. Kenyon	Crawfordsville, Ind	Sept. 14, 1869	94, 754
Photographic images, Apparatus for multiplying	D. W. S. Rawson	Peru, Ill	Apr. 27, 1869	89, 342
Photographic impressions, Preparation of oil ground to receive.	J. H. Tatum	Baltimore, Md	Apr. 15, 1856	14, 679
Photographic instrument	D. J. Kellogg	Rochester, N. Y	Sept. 30, 1856	15, 809
Photographic lens	C. B. Boyle	New York, N. Y	Oct. 31, 1865	50, 681
Photographic lens	J. H. Dallmeyer	London, England	Feb. 5, 1867	61, 812
Photographic lens	R. Morrison	Brooklyn, N. Y	May 21, 1872	126, 979
Photographic lens	J. Schnitzer	New York, N. Y	Aug. 1, 1865	49, 160
Photographic lenses, Combining	C. B. Boyle	New York, N. Y	Jan. 23, 1866	52, 129
Photographic medal	H. E. Coply	Waterbury, Conn	Apr. 2, 1861	31, 871
Photographic medal	D. F. Maltby	Waterbury, Conn	Aug. 14, 1860	29, 652
Photographic mercury-bath	J. Moulson	Philadelphia, Pa	Sept. 2, 1851	8, 335
Photographic monocular-glass	A. J. Dallemagne and L. Triboulet.	Paris, France	Dec. 30, 1873	146, 052
Photographic name-plate	T. E. Mackerley	Paint, Ohio	Apr. 25, 1865	47, 438
Photographic negative	J. Kirk	Newark, N. J	Mar. 4, 1873	136, 439
Photographic-negative baths, Heating	H. J. Sunter	Buffalo, N. Y	Mar. 25, 1873	137, 110
Photographic negatives, Apparatus for vignetting	F. Wolf	Philadelphia, Pa	Feb. 18, 1868	74, 654
Photographic negatives, Holder for retouching	N. Sarony	New York, N. Y	Feb. 5, 1867	61, 877
Photographic negatives, Retouching	D. H. Wright	Terre Haute, Ind	Nov. 18, 1873	144, 723
Photographic negatives, Varnish for coating	J. W. Morgeneier	Sheboygan, Wis	Dec. 6, 1870	109, 833
Photographic panoramic views, Apparatus for taking.	J. R. Johnson and J. A. Harrison.	London, England	Nov. 28, 1865	51, 279
Photographic-paper	H. M. Hedden	Worcester, Mass	Dec. 28, 1869	98, 375
Photographic-paper, &c., Apparatus for flowing and sensitizing.	W. S. Lukenbach	Newport, Pa	Apr. 27, 1869	89, 488
Photographic paper, &c., Apparatus for fumigating	J. L. Winner	Elizabeth City, N. C	May 30, 1871	115, 554
Photographic paper, Box for silvering and albumenizing.	E. C. Thompson and M. B. Wheaton.	New York, N. Y	July 31, 1860	29, 418
Photographic paper, Sensitized	J. G. Coffin	Portsmouth, Ohio	Jan. 30, 1872	123, 240
Photographic-paper-sensitizing frame	L. V. Moulton	Muskegon, Mich	July 12, 1870	105, 233
Photographic picture	W. H. F. Talbot	Lacock Abbey, England	June 26, 1847	5, 171
Photographic-picture holder	J. Buckett	Harlem, N. Y	May 2, 1865	47, 515
Photographic picture on glass	J. A. Cutting	Boston, Mass	July 11, 1854	11, 267
Photographic pictures, Apparatus for enlarging	J. A. Whitley	Owego, N. Y	Jan. 1, 1861	31, 060
Photographic pictures, Apparatus for obtaining	J. J. L. R. De Lafargo	Paris, France	June 21, 1864	43, 270
Photographic pictures, Apparatus for preserving and exhibiting.	C. Robinson	Springfield, Mass	Oct. 17, 1865	50, 498
Photographic pictures, Bath for toning	R. Anson	New York, N. Y	Feb. 19, 1861	31, 432
Photographic pictures, Bituminous ground for	V. M. Griswold	Lancaster, Ohio	Oct. 21, 1856	15, 924
Photographic pictures by artificial light, Taking	P. F. and W. S. Dodge	West Cambridge, Mass	Feb. 19, 1861	31, 444
Photographic pictures, Cabinet for exhibiting	J. A. Thompson	Auburn, N. Y	Sept. 6, 1864	44, 125
Photographic pictures, Collodion for	V. M. Griswold	Lancaster, Ohio	July 15, 1856	15, 336
Photographic pictures, Coloring	A. O. Dayton	Washington, D. C	Mar. 12, 1850	7, 160
Photographic pictures, Composition for coating	H. Happel	New York, N. Y	Apr. 4, 1871	113, 426
Photographic pictures, Composition for making	J. A. Cutting	Boston, Mass	July 11, 1854	11, 266
Photographic pictures, Device for exhibiting	A. G. Walton	San Francisco, Cal	Nov. 22, 1870	109, 472
Photographic pictures, engravings, &c., Preparation of.	S. D. Humphrey	New York, N. Y	Apr. 28, 1857	17, 152
Photographic pictures, Frame for printing	T. E. Sexton	Wilmington, Del	May 1, 1866	54, 416
Photographic pictures, Guide for cutting	L. T. Young	Philadelphia, Pa	Dec. 16, 1873	145, 711
Photographic pictures, &c., Machine for rolling	C. Sellers	Philadelphia, Pa	July 29, 1862	36, 028
Photographic pictures, Manufacture of	J. R. Johnson	London, England	July 20, 1869	92, 836
Photographic pictures on ceramic ware, Securing	J. B. Obernetter	Munich, Bavaria	Feb. 6, 1866	52, 503
Photographic pictures on glass	A. Bisbee and Y. Day	Columbus, Ohio, and Nashville, Tenn.	May 27, 1856	14, 946
Photographic pictures on glass, &c	F. Langenheim	Philadelphia, Pa	Nov. 19, 1850	7, 784
Photographic pictures on glass, Coloring	D. B. and H. B. Spooner	Springfield, Mass	Aug. 5, 1856	15, 497
Photographic pictures on japanned surfaces	H. L. Smith	Gambier, Ohio	Feb. 19, 1856	14, 300
Photographic pictures, Preparation of collodion for.	J. A. Cutting	Boston, Mass	July 4, 1854	11, 213
Photographic pictures, Printing	J. Rehn	Philadelphia, Pa	Apr. 14, 1868	76, 660

Index of patents issued from the United States Patent Office from 1790 *to* 1873, *inclusive*—Continued.

Invention.	Inventor.	Residence.	Date.	No.
Photographic pictures, Process and composition for printing.	V. M. Griswold	Peekskill, N. Y	Apr. 10, 1866	53, 815
Photographic pictures, Revolving tablet for multiplying.	J. Lawrence	Ripon, Wis	May 28, 1867	65, 243
Photographic pictures, Tinting	G. Langdell and M. A. Root	Philadelphia, Pa	July 15, 1856	15, 341
Photographic pictures, Treating	J. B. Hall	New York, N. Y	Jan. 20, 1857	16, 438
Photographic pictures upon transparent media, Producing.	J. A. Whipple and W. B. Jones	Boston, Mass	June 25, 1850	7, 458
Photographic-plate dipper	D. H. Cross	Bennington, Vt	Mar. 15, 1870	100, 866
Photographic-plate holder	H. T. Anthony	New York, N. Y	Oct. 1, 1872	131, 837
Photographic-plate holder	H. T. Anthony	New York, N. Y	Apr. 29, 1873	138, 359
Photographic-plate holder	H. K. Averill, jr	Decorah, Iowa	May 13, 1862	35, 206
Photographic-plate holder	J. Buchtel	Portland, Oreg	Aug. 17, 1869	93, 669
Photographic-plate holder	J. Buchtel	Portland, Oreg	Jan. 7, 1873	134, 512
Photographic-plate holder	R. F. Channell and A. Bonzano	Phœnixville, Pa	Oct. 15, 1872	132, 250
Photographic-plate holder	W. L. Gill	Lancaster, Pa	Dec. 3, 1872	133, 526
Photographic-plate holder	W. Lewis	Brooklyn, N. Y	Apr. 29, 1873	138, 414
Photographic-plate holder	W. and W. H. Lewis	New York, N. Y	Mar. 3, 1857	16, 738
Photographic-plate holder	J. W. Moore	Bellefonte, Pa	June 7, 1870	104, 054
Photographic-plate holder	A. Semmendinger	Fort Lee, N. J	Nov. 25, 1873	145, 020
Photographic-plate holder	J. Stock	New York, N. Y	Dec. 1, 1857	18, 780
Photographic-plate holder	A. B. Wilson	Waterbury, Conn	Dec. 23, 1862	37, 252
Photographic-plate holder, Rotary	C. H. Shute	Edgartown, Mass	Feb. 21, 1865	46, 503
Photographic-plate holders, Catch for	H. C. Eddy	Jersey City, N. J	Sept. 13, 1864	44, 169
Photographic-plate holders, Corner for	J. L. Winner	Elizabeth City, N. C	Oct. 17, 1871	120, 141
Photographic-plate shield	H. Bryant and R. D. O. Smith	Washington, D. C	Nov. 30, 1858	22, 158
Photographic-plate shield	W. Campbell	Jersey City, N. J	Feb. 28, 1860	27, 269
Photographic-plate vise	L. Chapman	New York, N. Y	Feb. 5, 1856	14, 184
Photographic-plate vise	J. W. Jarboe	New York, N. Y	Mar. 17, 1857	16, 841
Photographic-plate vise	S. W. and W. L. Pearsall	New York, N. Y	Feb. 1, 1859	22, 822
Photographic-plate vise	M. H. Rison	Paris, Tenn	Jan. 25, 1859	22, 748
Photographic-plate vise	C. W. Stimson	Cleveland, Ohio	Jan. 17, 1854	10, 433
Photographic plates, Apparatus for drying	J. Thompson	Taunton, Mass	Oct. 11, 1864	44, 678
Photographic plates, Box for dry	J. and J. Stock	New York, N. Y	Oct. 15, 1861	33, 500
Photographic portraiture, Lens for	J. H. Dallmeyer	London, England	June 11, 1867	65, 729
Photographic posing-chair	N. S. Bowdish	Richfield Springs, N. Y	Oct. 31, 1871	120, 485
Photographic posing-chair	W. H. Lewis	New York, N. Y	Sept. 19, 1871	119, 090
Photographic posing-chair	C. A. Schindler	West Hoboken, N. J	Oct. 17, 1871	120, 110
Photographic posing-chair	I. O. Swett	Brighton, Mass	June 4, 1872	127, 439
Photographic-print cutter	J. Haworth	Philadelphia, Pa	Nov. 29, 1870	109, 615
Photographic-print cutter	T. M. Saurman	Norristown, Pa	Aug. 2, 1870	106, 080
Photographic-print-cutting apparatus	W. S. Burgess and G. A. Leuzi	Norristown, Pa	Mar. 29, 1870	101, 222
Photographic prints, Apparatus for gumming, cutting, and mounting.	D. H. Cross	Shaftsbury, Vt	June 21, 1864	43, 184
Photographic prints, Coloring	I. O. Beyse	Saint Louis, Mo	Nov. 7, 1865	50, 871
Photographic prints, Washing	M. Ormsbee	New York, N. Y	Mar. 21, 1865	46, 927
Photographic printing	E. Edwards	Firs Willesden, Great Britain.	May 25, 1869	90, 514
Photographic printing	E. G. Fowx	Baltimore, Md	Oct. 19, 1869	95, 892
Photographic-printing apparatus	E. and E. W. Brown	Roxbury and Boston, Mass	Feb. 4, 1868	73, 948
Photographic-printing apparatus	L. J. Marcy	Newport, R. I	Nov. 30, 1869	97, 419
Photographic-printing apparatus	J. Stehman	Lancaster, Pa	Mar. 10, 1868	75, 482
Photographic printing, Contact-pad for	J. Buchtel	Portland, Oreg	Nov. 17, 1868	84, 168
Photographic-printing frame	G. W. and W. Bowlsby	Monroe, Mich	May 10, 1864	42, 637
Photographic-printing frame	F. Burrows	Peoria, Ill	Aug. 16, 1864	43, 831
Photographic-printing frame	J. W. Campbell	New York, N. Y	Aug. 14, 1866	57, 083
Photographic-printing frame	J. G. Coffin	Portsmouth, Ohio	Oct. 24, 1871	120, 245
Photographic-printing frame	G. W. Cook	Saint Paul, Minn	May 12, 1863	38, 468
Photographic-printing frame	N. F. English	Hartland Four Corners, Vt	Mar. 28, 1865	47, 063
Photographic-printing frame	J. Heddon	Elkhart, Ind	Nov. 5, 1867	70, 564
Photographic-printing frame	W. H. Jacoby	Minneapolis, Minn	Oct. 14, 1873	143, 577
Photographic-printing frame	S. K. Jones	New Haven, Conn	Aug. 22, 1865	49, 531
Photographic-printing frame	C. L. Lochman	Carlisle, Pa	Nov. 20, 1866	59, 800
Photographic-printing frame	P. Murphy	New York, N. Y	Feb. 1, 1870	99, 462
Photographic-printing frame	B. Reed	Kokomo, Ind	Oct. 27, 1868	83, 548
Photographic-printing frame	D. Shive	Philadelphia, Pa	Feb. 27, 1866	52, 896
Photographic-printing frame	J. F. Shoemaker	Van Wert County, Ohio	Apr. 30, 1867	64, 374
Photographic-printing frame	M. Witt	Columbus, Ohio	July 7, 1863	39, 191
Photographic-printing frame and slide	M. Ormsbee	New York, N. Y	Apr. 28, 1863	38, 326
Photographic-printing frame for porcelain or glass pictures.	W. J. Kuhns	Brooklyn, N. Y	Dec. 5, 1865	51, 323
Photographic-printing press	C. Fontayne	Cincinnati, Ohio	Sept. 20, 1859	25, 540
Photographic-printing press	J. H. Hamilton	Sioux City, Iowa	Apr. 12, 1870	101, 728
Photographic printing, Pressure-frame for	L. E. Denison	Winthrop, Conn	Dec. 26, 1865	51, 699
Photographic process	M. M. Griswold	Columbus, Ohio	Nov. 5, 1867	70, 551
Photographic process	C. M. Du Motay and C. R. Maréchal.	Metz, France	Nov. 6, 1866	59, 520
Photographic process	J. Wothly	Paris, France	Aug. 15, 1865	49, 488
Photographic process for copying drawings, &c	W. Willis	Birmingham, England	June 12, 1866	55, 592
Photographic purposes, Arrangement of solar mirrors for.	W. Crane and W. H. Pease	Goshen, Ind	Apr. 24, 1866	54, 118
Photographic purposes, Combination of lenses for.	J. Zentmayer	Philadelphia, Pa	May 29, 1866	55, 195
Photographic purposes, Composition for	T. Cummings	Lancaster, Pa	July 5, 1870	104, 937
Photographic purposes, Curvette and tray for	H. T. Anthony	New York, N. Y	Mar. 29, 1864	42, 061
Photographic purposes, Duplicating-deflector for	D. Shive	Philadelphia, Pa	Oct. 3, 1865	50, 284
Photographic purposes, Preparing paper for	P. H. Murray	Portsmouth, Ohio	Apr. 4, 1871	113, 327
Photographic reflector	W. Kurtz	New York, N. Y	July 5, 1870	104, 963
Photographic rest	C. E. Krüger	New York, N. Y	Nov. 3, 1868	83, 714
Photographic rest	N. Sarony	New York, N. Y	Feb. 18, 1868	74, 604
Photographic rest	O. Sarony	Scarborough, England	June 5, 1866	55, 443
Photograhic roller-press	E. B. McCoy	Winsted, Conn	Feb. 25, 1862	34, 515
Photographic room	G. K. Proctor	Salem, Mass	Oct. 27, 1868	83, 545
Photographic screen	J. A. Anderson	Chicago, Ill	Feb. 22, 1870	100, 099
Photographic sensitizing-box	W., jr., and A. L. Hudson	Hingham, Mass	Jan. 2, 1866	51, 834
Photographic shield	E. Gordon	New York, N. Y	Oct. 19, 1858	21, 829
Photographic show-case	J. F. Adams	Buffalo, N. Y	Mar. 22, 1870	100, 962
Photographic sky-light	C. D. Mosher	Cook County, Ill	Apr. 15, 1873	137, 790
Photographic tent	I. F. Woodward	McMinnville, Tenn	Oct. 17, 1871	119, 957

Index of patents issued from the United States Patent Office from 1790 *to* 1873, *inclusive*—Continued.

Invention.	Inventor.	Residence.	Date.	No.
Photographic transfer	O. P. Howe	Augusta, Me	Apr. 7, 1868	76, 458
Photographic transfers, Making	A. G. Mowan	South Bergen, N. J	June 25, 1867	66, 102
Photographic tray	D. J. Kellogg	Rochester, N. Y	Apr. 7, 1857	16, 979
Photographic use, Compound lens for	J. H. Dallmeyer	Middlesex County, England.	June 30, 1868	79, 323
Photographic use, Preparation of paper, &c., for	W. Gibson	New York, N. Y	Dec. 18, 1866	60, 626
Photographic use, Preparing paper for	J. D. Brinckerhoff	New York, N. Y	Nov. 21, 1865	51, 010
Photographic vignettes, Apparatus for printing	J. E. Richard	Columbia, S. C	Aug. 17, 1869	93, 835
Photographing children, Apparatus for	W. W. Sloan	Jefferson, Tex	July 4, 1871	116, 765
Photographing engravers' blocks, Mode of	B. F. Taylor	Philadelphia, Pa	Feb. 19, 1867	62, 299
Photographing, Preparing articles of gold, silver, and glass for.	J. Wilder and A. B. Stow	Middletown, Conn	Mar. 16, 1869	87, 996
Photography	E. L. Bergstresser	Hublersburgh, Pa	June 17, 1873	140, 001
Photography	E. J. Foss	Boston, Mass	Aug. 15, 1871	118, 006
Photography	F. Glessner	Cincinnati, Ohio	Nov. 9, 1869	96, 691
Photography	C. A. Guilmette	Boston, Mass	Apr. 3, 1866	53, 608
Photography	W. H. Hill	Worcester, Mass	July 29, 1873	141, 351
Photography	H. A. Marchant	Philadelphia, Pa	July 21, 1857	17, 858
Photography	W. H. Smith	London, England	Oct. 17, 1865	50, 544
Photography	F. A. Wenderoth	Philadelphia, Pa	Oct. 17, 1871	120, 136
Photography on wood	C. B. Boyle	Albany, N. Y	Feb. 8, 1859	22, 852
Photography to printing, Application of	A. L. Poitevin	Paris, France	Oct. 28, 1862	36, 821
Photometer	S. G. Elliott	San Francisco, Cal	June 19, 1866	55, 797
Photometer	H. Vogel	Berlin, Prussia	Mar. 10, 1868	75, 493
Phrenology, Apparatus for teaching	G. P. Wilcox and W. Butler	Little Falls, N. Y	Apr. 15, 1856	14, 685
Physical culture, Apparatus for	G. B. Windship	Boston, Mass	Jan. 7, 1873	134, 578
Physician's prescription-case	A. S. Hamlin	Saint Louis, Mo	Oct. 21, 1862	36, 710
Physiognotrace	D. Atherton	Richmond, Va	May 4, 1805	....
Physiognotrace	I. Todd, A. Day, and W. Bache		June 15, 1803	....
Physiological battery	A. C. Garratt	Boston, Mass	July 6, 1869	92, 301
Piano	S. P. Brooks	Somerville, Mass	Jan. 9, 1872	122, 557
Piano	O. Bull	New York, N. Y	Nov. 15, 1870	109, 172
Piano	C. F. Chew	George street, Chalk Farm Road, England.	Aug. 23, 1870	106, 661
Piano	A. F. Dessau	Washington, D. C	Feb. 8, 1870	99, 655
Piano	A. Faas	Philadelphia, Pa	Aug. 22, 1871	118, 353
Piano	A. Gunther	Philadelphia, Pa	Aug. 15, 1865	49, 399
Piano	A. H. Hastings	New York, N. Y	Feb. 25, 1862	34, 491
Piano	C. Hattersley	Trenton, N. J	May 24, 1870	103, 456
Piano	C. Kurtzman	Buffalo, N. Y	May 24, 1870	103, 345
Piano	G. C. Manner	New York, N. Y	July 22, 1873	141, 152
Piano	F. Mathushek	New York, N. Y	Mar. 28, 1871	113, 073
Piano	F. B. McGregor	Pontiac, Mich	Sept. 22, 1868	82, 428
Piano	W. Nordhoff	Baltimore, Md	Aug. 14, 1866	57, 257
Piano	C. F. Oliver	Lynn, Mass	June 14, 1870	104, 341
Piano	J. W. Otto	Saint Louis, Mo	Aug. 28, 1866	57, 558
Piano	S. W. Parker	Somerville, Mass	Aug. 23, 1870	106, 612
Piano	R. Raven	New York, N. Y	Aug. 14, 1866	57, 186
Piano	C. Robbins	Chicago, Ill	Feb. 13, 1866	52, 610
Piano	G. Trayser	Indianapolis, Ind	July 9, 1867	66, 650
Piano	W. F. Ulman	Boston, Mass	Jan. 10, 1871	110, 940
Piano	T. Winans	Baltimore, Md	Dec. 24, 1872	134, 340
Piano-action	S. P. Brooks	Somerville, Mass	Nov. 5, 1872	132, 709
Piano-action	F. Koth	North New York, N. Y	Oct. 28, 1873	143, 986
Piano-action	R. Kreter	New York, N. Y	Dec. 23, 1873	145, 879
Piano-action	L. Kühner	New York, N. Y	Feb. 11, 1873	135, 820
Piano-action	L. Matt and B. Grueter	Boston, Mass	May 23, 1871	115, 077
Piano-action	F. B. McGregor	Pontiac, Mich	Dec. 6, 1870	109, 921
Piano-action	E. G. Newman and P. Anderson	New York, N. Y	July 1, 1873	140, 428
Piano-action	C. F. Oliver	Lynn, Mass	Aug. 16, 1870	106, 396
Piano-action	C. F. T. Steinway	New York, N. Y	June 6, 1871	115, 782
Piano-action	F. L. Trayser	Maysville, Ky	Oct. 14, 1873	143, 647
Piano action, Grand	D. F. Haaz	Philadelphia, Pa	Apr. 28, 1857	17, 148
Piano-action, Hammer-flange butt of	D. L. Bollermann	Mount Vernon, N. Y	Nov. 21, 1871	121, 077
Piano action, Upright	T. Atkins and H. Drewer	Cincinnati, Ohio	Oct. 29, 1872	132, 557
Piano action, Upright	R. Kreter	New York, N. Y	Mar. 18, 1873	137, 005
Piano action, Upright	G. C. Manner	Mott Haven, N. Y	Apr. 1, 1873	137, 377
Piano-agraffe	C. F. Chickering	New York, N. Y	Dec. 24, 1872	134, 194
Piano and organ attachment	L. J. Trémaux	New Orleans, La	Dec. 23, 1873	145, 793
Piano and organ key	C. F. L. Goffrie and J. H. Schucht.	London, England	Apr. 8, 1873	137, 544
Piano and organ printing-attachment	G. Heydrich	New Ulm, Minn	Dec. 17, 1872	133, 987
Piano, Bell	C. G. Buttkereit	Toledo, Iowa	Oct. 31, 1871	120, 415
Piano, Bell	C. G. Buttkereit	Toledo, Iowa	Sept. 16, 1873	142, 768
Piano, Bell	C. Williams and E. F. Falconnet.	Nashville, Tenn	Sept. 24, 1861	33, 372
Piano-bridge	C. W. Brewer	Racine, Wis	Apr. 27, 1869	89, 381
Piano-bridge	W. Compton	New York, N. Y	July 10, 1860	29, 061
Piano, couch, and bureau, Combined	C. Hess	Cincinnati, Ohio	July 17, 1866	56, 413
Piano-coupling attachment	J. Clark	Philadelphia, Pa	Oct. 25, 1870	108, 685
Piano-cover Water-proof	H. F. Herkner and J. W. Post	Brooklyn, N. Y	Feb. 7, 1871	111, 535
Piano, Double-action	J. J. Bender	New York, N. Y	Aug. 21, 1866	57, 434
Piano-forte hammers, Composition for paste for attaching the covering to.	J. Munson	Burlington, Vt	June 19, 1860	28, 768
Piano-forte	C. F. L. Albrecht	Philadelphia, Pa	Mar. 18, 1828	....
Piano-forte	A. Babcock	Boston, Mass	Dec. 17, 1825	....
Piano-forte	A. Babcock	Boston, Mass	Oct. 31, 1839	1, 389
Piano-forte	E. Badlam	Potsdam, N. Y	Oct. 25, 1845	4, 241
Piano-forte	W. W. Batchelder	New York, N. Y	June 21, 1864	43, 174
Piano-forte	W. W. Bennett	Jersey City, N. J	Aug. 26, 1873	142, 194
Piano-forte	E. Bloomfield and D. P. Otis	New York, N. Y	May 18, 1860	00, 150
Piano-forte	C. Bossert and J. Shomacker	Philadelphia, Pa	Apr. 29, 1842	2, 595
Piano-forte	W. Bourne	Boston, Mass	Feb. 17, 1863	37, 717
Piano-forte	W. Bourne	Boston, Mass	June 14, 1870	104, 256
Piano-forte	S. P. Brooks	Boston, Mass	Nov. 24, 1857	18, 673
Piano-forte	E. Brown	Boston, Mass	Nov. 20, 1838	1, 014
Piano-forte	E. Brown	Boston, Mass	Jan. 27, 1843	2, 934
Piano-forte	L. H. Browne	Boston, Mass	Sept. 23, 1851	8, 383

Index of patents issued from the United States Patent Office from 1790 *to* 1873, *inclusive*—Continued.

Invention.	Inventor.	Residence.	Date.	No.
Piano-forte	L. Caldera and L. Montu	Turin, Italy	Mar. 10, 1868	75, 362
Piano-forte	G. Charters	New York	July 8, 1816	....
Piano-forte	J. Chickering	Boston, Mass	Oct. 8, 1840	1, 802
Piano-forte	J. Chickering	Boston, Mass	Sept. 1, 1843	3, 238
Piano-forte	P. E. Chollet	New York, N. Y	May 7, 1867	64, 407
Piano-forte	I. Clark	Cincinnati, Ohio	Mar. 2, 1836	....
Piano-forte	O. M. Coleman	Philadelphia, Pa	Apr. 17, 1844	3, 548
Piano-forte	B. E. Colley	Cambridge, Mass	June 21, 1853	9, 793
Piano-forte	W. Compton	New York, N. Y	June 15, 1852	9, 014
Piano-forte	W. Compton	New York, N. Y	Sept. 6, 1853	9, 928
Piano-forte	W. Cumston	Boston, Mass	Aug. 3, 1839	1, 275
Piano-forte	D. Decker	New York, N. Y	June 2, 1863	38, 731
Piano-forte	D. Decker	New York, N. Y	Jan. 10, 1865	45, 818
Piano-forte	J. J. and D. Decker	New York, N. Y	Jan. 29, 1867	61, 612
Piano-forte	S. W. Draper	Boston, Mass	June 25, 1845	4, 082
Piano-forte	S. B. Driggs	Detroit, Mich	Dec. 18, 1855	13, 942
Piano-forte	S. B. Driggs	New York, N. Y	Jan. 12, 1858	19, 081
Piano-forte	S. B. Driggs	New York, N. Y	May 3, 1859	23, 834
Piano-forte	S. B. Driggs	New York, N. Y	Mar. 13, 1860	27, 435
Piano-forte	S. B. Driggs	New York, N. Y	Mar. 22, 1864	41, 977
Piano-forte	S. B. Driggs	New York, N. Y	Apr. 11, 1865	47, 196
Piano-forte	G. Ely	New York, N. Y	May 3, 1870	102, 668
Piano-forte	L. Farnesworth	Nashua, N. H	Aug. 9, 1870	106, 145
Piano-forte	J. U. Fischer	New York, N. Y	July 26, 1859	24, 905
Piano-forte	J. U. Fischer	New York, N. Y	July 10, 1860	29, 068
Piano-forte	L. Fissore	Baltimore, Md	July 22, 1833	....
Piano-forte	J. Geib	New York, N. Y	Oct. 3, 1817	....
Piano-forte	L. Gilbert	Boston, Mass	July 10, 1841	2, 167
Piano-forte	T. Gilbert	Boston, Mass	Sept. 30, 1851	8, 389
Piano-forte	C. Glassborow	Philadelphia, Pa	July 26, 1859	24, 865
Piano-forte	P. Gmehlin	New York, N. Y	July 23, 1872	129, 727
Piano-forte	H. Goldsmith	Philadelphia, Pa	Dec. 8, 1857	18, 810
Piano-forte	O. Gori and P. Ernstz	New York, N. Y	Mar. 26, 1844	3, 504
Piano-forte	J. A. Gray	Albany, N. Y	Mar. 27, 1849	6, 223
Piano-forte	J. A. Gray	Albany, N. Y	May 1, 1860	28, 137
Piano-forte	J. A. Gattwaldt	New York, N. Y	Aug. 27, 1818	....
Piano-forte	A. Hall	Lloydsville, Ohio	Jan. 24, 1854	10, 456
Piano-forte	A. Hall	Lloydsville, Ohio	Jan. 30, 1855	12, 315
Piano-forte	H. Hartye	Baltimore, Md	Mar. 12, 1836	....
Piano-forte	A. H. Hastings	New York, N. Y	Mar. 15, 1870	100, 888
Piano-forte	A. H. Hastings	New York, N. Y	Oct. 10, 1871	119, 760
Piano-forte	J. J. Hawkins	....	Feb. 12, 1800	....
Piano-forte	H. Herrick	New York, N. Y	Oct. 26, 1839	1, 379
Piano-forte	H. Herrick	Boston, Mass	July 21, 1868	80, 073
Piano-forte	G. Hews, R. C. Marsh, and N. M. Tileston.	Boston and Dorchester, Mass.	Apr. 10, 1843	3, 045
Piano-forte	C. Horst	New Orleans, La	Apr. 17, 1849	6, 342
Piano-forte	G. H. Hulskamp	Troy, N. Y	July 14, 1857	17, 789
Piano-forte	G. H. Hulskamp	Troy, N. Y	July 10, 1860	29, 081
Piano-forte	G. H. Hulskamp	New York, N. Y	Dec. 18, 1866	60, 519
Piano-forte	T., H. O., G. T., and W. F. Kearsing.	New York, N. Y	June 13, 1831	....
Piano-forte	J. G. Kunze	New York, N. Y	Dec. 20, 1859	26, 503
Piano-forte	F. C. Lighte	New York, N. Y	Sept. 13, 1859	25, 426
Piano-forte	F. C. Lighte	New York, N. Y	July 1, 1862	35, 766
Piano-forte	H. Lindeman	New York, N. Y	Aug. 7, 1860	29, 502
Piano-forte	G. C. Manner	New York, N. Y	Nov. 13, 1866	59, 619
Piano-forte	T. Marschall	New York, N. Y	Jan. 7, 1862	34, 114
Piano-forte	T. Marschall	New York, N. Y	Apr. 13, 1867	88, 970
Piano-forte	W. H. Mason	Boston, Mass	Oct. 16, 1866	58, 950
Piano-forte	F. Mathushek	New York, N. Y	Oct. 28, 1851	8, 470
Piano-forte	F. Mathushek	New York, N. Y	Oct. 2, 1860	30, 279
Piano-forte	L. Matt	Boston, Mass	Aug. 25, 1863	39, 664
Piano-forte	J. and J. McDonald	New York, N. Y	Oct. 5, 1852	9, 304
Piano-forte	W. H. McDonald	Brooklyn, N. Y	Apr. 2, 1867	63, 547
Piano-forte	J. S. McLean	....	May 27, 1796	....
Piano-forte	M. Miller	Rochester, N. Y	July 1, 1851	8, 194
Piano-forte	G. W. Neill	Boston, Mass	July 7, 1868	79, 591
Piano-forte	D. B. Newhall	Boston, Mass	Nov. 3, 1841	2, 330
Piano-forte	C. F. Oliver and G. M. Jackson	Lynn, Mass	May 1, 1845	4, 019
Piano-forte	S. T. Parmelee	New Haven, Conn	June 24, 1862	35, 703
Piano-forte	S. T. Parmelee	New Haven, Conn	June 24, 1862	35, 704
Piano-forte	S. T. Parmelee	New Haven, Conn	Mar. 7, 1865	46, 759
Piano-forte	D. T. Peck	New York, N. Y	Mar. 2, 1869	87, 509
Piano-forte	C. A. Peterson	New York, N. Y	Aug. 10, 1869	93, 473
Piano-forte	J. Pethick	Mount Morris, N. Y	Feb. 12, 1836	....
Piano-forte	L. Philleo	Utica, N. Y	July 2, 1846	4, 612
Piano-forte	J. Pirsson	New York, N. Y	Aug. 13, 1850	7, 568
Piano-forte	L. Reuckert	Baltimore, Md	Mar. 12, 1845	3, 940
Piano-forte	L. Ricketts	Baltimore, Md	June 24, 1844	3, 643
Piano-forte	T. J. V. Roz	Paris, France	Apr. 30, 1867	64, 371
Piano-forte	J. Ruck	New York, N. Y	Apr. 23, 1850	7, 308
Piano-forte	C. S. Sackmeister	New York	May 17, 1830	....
Piano-forte	J. H. Schomacker	Philadelphia, Pa	June 13, 1848	5, 631
Piano-forte	H. Schonacker	Detroit, Mich	Dec. 18, 1855	13, 960
Piano-forte	J. Schriber	New York	Oct. 29, 1846	4, 832
Piano-forte	P. Schuler	Philadelphia, Pa	Oct. 16, 1866	58, 896
Piano-forte	C. P. Seabury	New York	May 20, 1830	....
Piano-forte, &c	R. Shaw	Boston, Mass	July 2, 1807	....
Piano-forte	M. Stange	New York, N. Y	Jan. 21, 1862	34, 220
Piano-forte	C. F. T. Steinway	New York, N. Y	Dec. 14, 1869	97, 982
Piano-forte	C. F. T. Steinway	New York, N. Y	May 28, 1872	127, 383
Piano-forte	C. F. T. Steinway	New York, N. Y	May 28, 1872	127, 384
Piano-forte	H. Steinway, jr	New York, N. Y	Nov. 29, 1859	26, 300
Piano-forte	T. Steinway	New York, N. Y	Aug. 18, 1868	81, 306
Piano-forte	W. Steinway	New York, N. Y	June 5, 1866	55, 385
Piano-forte	D. Stirn	Milwaukee, Wis	May 4, 1869	89, 806
Piano-forte	W. F. Ulman	Boston, Mass	Aug. 22, 1871	118, 407

Index of patents issued from the United States Patent Office from 1790 *to* 1873, *inclusive*—Continued.

Invention.	Inventor.	Residence.	Date.	No.
Piano-forte	M. Vergnes	New York, N. Y	Oct. 4, 1864	44, 566
Piano-forte	M. Vergnes	New York, N. Y	Jan. 17, 1865	45, 943
Piano-forte	G. Vogt	Philadelphia, Pa	Mar. 1, 1859	23, 130
Piano-forte	E. L. Walker and G. W. Cherry	Carlisle, Pa., and Alexandria, D. C.	Jan. 16, 1845	3, 888
Piano-forte	S. R. Warren	Montreal, Canada	July 10, 1845	4, 109
Piano-forte action	J. Becker	New York, N. Y	June 3, 1856	14, 998
Piano-forte action	J. D. Boedicker	New York, N. Y	Nov. 5, 1861	33, 633
Piano-forte action	S. P. Brooks	Boston, Mass	Apr. 17, 1855	12, 737
Piano-forte action	S. P. Brooks	Somerville, Mass	May 30, 1865	47, 922
Piano-forte action	G. Brown	Boston, Mass	Jan. 27, 1852	8, 680
Piano-forte action	H. S. Calenberg	New York, N. Y	Oct. 8, 1861	32, 427
Piano-forte action	B. E. Colley	Boston, Mass	Oct. 26, 1869	96, 204
Piano-forte action	B. E. Colley	Boston, Mass	Oct. 15, 1872	132, 198
Piano-forte action	W. Compton	New York, N. Y	May 22, 1860	28, 352
Piano-forte action	W. G. Day	Baltimore, Md	Dec. 13, 1870	110, 120
Piano-forte action	D. Decker	New York, N. Y	Sept 13, 1859	25, 393
Piano-forte action	S. B. Driggs	New York, N. Y	May 19, 1857	17, 320
Piano-forte action	T. C. Faulder	Albany, N. Y	Mar. 10, 1863	37, 860
Piano-forte action	F. Frickinger	Shodack, N. Y	July 23, 1861	32, 863
Piano-forte action	T. Gilbert	Boston, Mass	Feb. 10, 1841	1, 970
Piano-forte action	T. Gilbert	Boston, Mass	Aug. 7, 1847	5, 216
Piano-forte action	J. A. Gray	Albany, N. Y	Sept. 9, 1851	8, 352
Piano-forte action	J. A. Gray	Albany, N. Y	Mar. 17, 1857	16, 832
Piano-forte action	N. J. Haines	New York, N. Y	May 24, 1859	24, 119
Piano-forte action	A. Hall	Lloydsville, Ohio	Apr. 11, 1854	10, 776
Piano-forte action	I. I. Harwood	Boston, Mass	Feb. 12, 1861	31, 385
Piano-forte action	A. K. Hebard	Cambridge, Mass	Dec. 9, 1873	145, 417
Piano-forte action	H. Herrick	Boston, Mass	Feb. 19, 1867	62, 134
Piano-forte action	M. Herter	New York, N. Y	Dec. 19, 1865	51, 588
Piano-forte action	G. Howe	Boston, Mass	Sept. 28, 1852	9, 282
Piano-forte action	G. Howe	Roxbury, Mass	Oct. 20, 1857	18, 453
Piano-forte action	G. H. Hulskamp	Troy, N. Y	May 22, 1860	28, 374
Piano-forte action	G. H. Hulskamp	Troy, N. Y	Oct. 21, 1862	36, 712
Piano-forte action	A. B. Irving	Indianapolis, Ind	Dec. 31, 1867	72, 738
Piano-forte action	C. L. and A. B. Irving	Fort Wayne, Ind	Oct. 23, 1866	59, 155
Piano-forte action	R. M. Kerrison	Philadelphia, Pa	Sept. 9, 1851	8, 353
Piano-forte action	R. M. Kerrison	Philadelphia, Pa	June 19, 1855	13, 091
Piano-forte action	J. Kohnle	New York, N. Y	Jan. 24, 1860	26, 909
Piano-forte action	F. Koth	New York, N. Y	July 4, 1865	48, 565
Piano-forte action	R. Kreter	New York, N. Y	Sept. 9, 1851	8, 350
Piano-forte action	H. A. Leaman	New York, N. Y	Apr. 6, 1858	19, 857
Piano-forte action	E. A. Lee	Roxbury, Mass	May 23, 1854	10, 948
Piano-forte action, &c	T. Loud	Philadelphia, Pa	Dec. 7, 1837	504
Piano-forte action	T. Loud	Spring Garden, Pa	Apr. 24, 1847	5, 086
Piano-forte action	J. H. Low	Boston, Mass	Dec. 26, 1848	5, 985
Piano-forte action	N. M. Lowe	Boston, Mass	Mar. 25, 1856	14, 509
Piano-forte action	F. Marschall	New York, N. Y	Aug. 30, 1859	25, 305
Piano-forte action	J. V. Marshall	Albany, N. Y	June 8, 1858	20, 500
Piano-forte action	F. Mathushek	New York, N. Y	Dec. 20, 1859	26, 550
Piano-forte action	F. Mathushek	New York, N. Y	Mar. 28, 1871	113, 074
Piano-forte action	L. Matt	Boston, Mass	Dec. 19, 1871	121, 952
Piano-forte action	F. B. McGregor	Pontiac, Mich	Apr. 18, 1871	113, 785
Piano-forte action	J. S. Morton	New York, N. Y	Sept. 18, 1855	13, 580
Piano-forte action	W. Munroe	Boston, Mass	Apr. 24, 1855	12, 763
Piano-forte action	F. W. Niehaus	Boston, Mass	July 3, 1860	28, 997
Piano-forte action	J. F. Nunns	New York, N. Y	May 5, 1831	....
Piano-forte action	A. W. Perry	Saint Joseph, Mo	Dec. 7, 1869	97, 686
Piano-forte action	F. Pistorias	Chicago, Ill	July 12, 1864	43, 524
Piano-forte action	C. Pratt	West Roxbury, Mass	Dec. 27, 1864	45, 075
Piano-forte action	C. E. Rogers	Boston, Mass	Mar. 5, 1872	124, 289
Piano-forte action	J. Ruck	New York, N. Y	Apr. 16, 1850	7, 292
Piano-forte action	J. Ruck	New York, N. Y	Mar. 11, 1851	7, 976
Piano-forte action	T. S. Seabury	Stony Brook, N. Y	Oct. 25, 1859	25, 919
Piano-forte action	H. Seidel	Roxbury, Mass	May 12, 1868	77, 845
Piano-forte action	J. Shaudelle	Huntsville, Ala	Jan. 14, 1873	134, 940
Piano-forte action	D. H. Shirley	Boston, Mass	Nov. 28, 1854	12, 003
Piano-forte action	D. H. Shirley	Boston, Mass	Mar. 30, 1869	88, 522
Piano-forte action	H. Steinway	New York, N. Y	May 5, 1857	17, 238
Piano-forte action	H. Steinway	New York, N. Y	June 15, 1858	20, 595
Piano-forte action	H. Steinway, jr	New York, N. Y	May 21, 1861	32, 386
Piano-forte action	H. Steinway, jr	New York, N. Y	May 21, 1861	32, 387
Piano-forte action	H. Steinway, jr	New York, N. Y	Apr. 8, 1862	34, 910
Piano-forte action	T. Steinway	New York, N. Y	Aug. 10, 1869	93, 647
Piano forte action	F. Taylor	New York, N. Y	Dec. 11, 1855	13, 924
Piano-forte action	J. Thompson	New York	Aug. 6, 1831	....
Piano-forte action	J. H. Tibbits	Omaha City, Nebr	July 11, 1865	48, 741
Piano-forte action	W. V. Wallace	New York, N. Y	Oct. 30, 1866	59, 295
Piano-forte action	J. J. Wise	Baltimore, Md	June 27, 1839	1, 205
Piano-forte action	J. J. Wise	Baltimore, Md	Dec. 26, 1848	5, 990
Piano-forte action, Upright	A. H. Hastings	Jersey City, N. J	June 20, 1871	116, 182
Piano-forte and cabinet, Combination	E. Cotter	Boston, Mass	Dec. 21, 1869	98, 038
Piano-forte and organ key	W. F. Furgang	Albany, N. Y	Apr. 20, 1852	8, 887
Piano-forte and organ key	E. P. Needham	New York, N. Y	July 1, 1873	140, 528
Piano-forte attachment	S. B. Driggs	Detroit, Mich	Jan. 24, 1854	10, 446
Piano-forte bottoms, Construction of	W. F. Senior	New York, N. Y	Nov. 21, 1845	4, 282
Piano-forte bridge	C. H. De Vine	Buffalo, N. Y	Oct. 13, 1868	82, 928
Piano-forte bridge	W. C. Ellis	Worcester, Mass	Apr. 22, 1873	138, 012
Piano-forte bridge	T. E. Power	Columbia, Mo	May 12, 1857	17, 296
Piano-forte bridge, Metallic	G. H. Hulskamp	Troy, N. Y	July 21, 1857	17, 838
Piano-forte cases, Machinery for making	S. Porter	Leominster, Mass	Dec. 7, 1869	97, 551
Piano-forte covers, &c., Ornamenting	J. F. Burgess	Brooklyn, N. Y	Nov. 27, 1866	59, 959
Piano-forte frame	H. S. Ackerly	New York, N. Y	Feb. 27, 1855	12, 432
Piano-forte frame	A. F. Dessau	Washington, D. C	Mar. 3, 1868	75, 132
Piano-forte frame	G. W. Neill	Boston, Mass	Apr. 6, 1869	88, 729
Piano-forte frame	J. Piffant	New Orleans, La	Jan. 25, 1853	9, 559
Piano-forte, Grand	F. C. Lighte	New York, N. Y	Feb. 21, 1860	27, 226
Piano-forte hammers, Covering	R. Kreter	New York, N. Y	Jan. 4, 1853	9, 526

Index of patents issued from the United States Patent Office from 1790 to 1873, inclusive—Continued.

Invention.	Inventor.	Residence.	Date.	No.
Piano-forte, Horizontal	E. R. Currier	Boston, Mass	Apr. 22, 1831	....
Piano-forte, Horizontal	F. C. Reichenbach	Philadelphia, Pa	May 19, 1841	2, 099
Piano-forte, Horizontal square	G. Bacon and R. Raven	New York, N. Y	Aug. 26, 1851	8, 320
Piano-forte, Instruments for teaching music with the.	E. Von Heeringen	Pickensville, Ala	June 26, 1849	6, 558
Piano-forte key	L. H. Browne	Boston, Mass	Aug. 6, 1861	32, 972
Piano-forte key	H. W. Hindsley	New Haven, Conn	Feb. 7, 1860	27, 050
Piano-forte key	J. Hoffacker and J. Richards	New York, N. Y	June 14, 1859	24, 404
Piano-forte key	C. J. Schoensmann	New York, N. Y	Aug. 7, 1860	29, 522
Piano-forte key-board	M. Philippi	Troy, N. Y	Oct. 11, 1859	25, 760
Piano-forte leg	A. Goodman and L. Hale	North Dana, Mass	Oct. 6, 1863	40, 166
Piano-forte lids, Raising	H. N. Oliver	Rahway, N. J	Feb. 18, 1873	136, 090
Piano-forte lock	P. F. Dodge	West Cambridge, Mass	May 3, 1859	23, 833
Piano-forte lock	S. Walker	New York, N. Y	Apr. 18, 1865	47, 349
Piano-forte pedal-attachment	G. A. Scott and W. B. Frisbee	San Francisco, Cal	July 5, 1870	105, 002
Piano-forte, Portable	M. Vergnes	New York, N. Y	Apr. 26, 1864	42, 518
Piano-forte, Portable	M. Vergnes	New York, N. Y	June 21, 1864	43, 245
Piano-forte soft-pedal attachment	J. Greener	Elmira, N. Y	Feb. 9, 1869	86, 747
Piano-forte sound-board	L. Matt	Boston, Mass	Dec. 31, 1867	72, 745
Piano-forte sound-board	J. Newman	Baltimore, Md	Apr. 7, 1857	16, 990
Piano-forte sound-board	C. F. Th. Steinway	New York, N. Y	Apr. 6, 1869	88, 749
Piano-forte sounding-board	J. A. Gray	Albany, N. Y	Feb. 6, 1855	12, 362
Piano-forte sounding-board	A. Speer and E. Marx	Aquackanonck, N. J	Sept. 28, 1852	9, 287
Piano-forte sounding-board	R. Swan, jr	New Bedford, Mass	Nov. 20, 1849	6, 889
Piano-forte, Square	C. F. Chickering	New York, N. Y	Apr. 23, 1861	32, 119
Piano-forte string	H. J. Newton	New York, N. Y	Oct. 21, 1851	8, 452
Piano-forte tuning-attachment	R. Beebe	West Springfield, Mass	Sept. 22, 1863	40, 015
Piano-forte, Upright	G. H. Davis	Boston, Mass	Oct. 28, 1873	143, 967
Piano-forte, Upright	L. Gilbert	Boston, Mass	June 18, 1850	7, 441
Piano-forte, Upright	H. Klepfer	Cincinnati, Ohio	Mar. 25, 1851	8, 002
Piano-forte, Upright	R. E. Letton	Quincy, Ill	Oct. 5, 1852	9, 301
Piano-forte, Upright	J. Thompson	New York	Oct. 1, 1830	....
Piano-fortes, Apparatus for augmenting sound in	M. Vergnes	New York, N. Y	Aug. 14, 1866	57, 225
Piano-fortes, Arranging the keys in	J. Dwight and D. B. Newhall	Boston, Mass	May 6, 1841	2, 081
Piano-fortes, Combination of reed instrument with	R. W. Carpenter	Brooklyn, N. Y	Dec. 23, 1862	37, 217
Piano-fortes, Combining metallic reeds with	M. Coburn	Savannah, Ga	Feb. 1, 1847	4, 948
Piano-fortes, Compensating tube for	T. Loud	Philadelphia, Pa	July 7, 1835	....
Piano-fortes, Composition finger-key for	L. Wolf	West Meriden, Conn	July 16, 1867	66, 928
Piano-fortes, Constructing hammer-heads in	E. Forbes	Boston, Mass	Feb. 10, 1841	1, 971
Piano-fortes, Constructing rest-pin for	D. Walker	New York, N. Y	June 19, 1838	790
Piano-fortes, Construction and action of	A. Babcock	Philadelphia, Pa	Dec. 31, 1833	....
Piano-fortes, Cross-stringing	A. Babcock	Philadelphia, Pa	May 24, 1830	....
Piano-fortes, Damper-action for upright	G. W. Neill	Boston, Mass	Mar. 8, 1870	100, 550
Piano-fortes, Duplex agraffe-scales for	C. F. T. Steinway	New York, N. Y	May 14, 1872	126, 848
Piano-fortes, Elevating the tops of	C. Meyer	Philadelphia, Pa	Apr. 10, 1849	6, 282
Piano-fortes, Glass string for	R. Bury	Albany, N. Y	Aug. 21, 1819	....
Piano-fortes, Harmonicon-attachment to	C. Peters	New London, Conn	July 16, 1872	128, 976
Piano-fortes, Iron frame for	A. Ludolff	New York, N. Y	Feb. 20, 1866	52, 725
Piano-fortes, Iron frame for upright	S. P. Brooks	Boston, Mass	Apr. 11, 1854	10, 770
Piano-fortes, Longitudinal bar in	J. Dwight	Boston, Mass	July 29, 1824	....
Piano-fortes, Melodeon-attachment for	L. Louis	New York, N. Y	Apr. 13, 1869	88, 967
Piano-fortes, Metallic frame for	T. Gilbert	Boston, Mass	July 24, 1847	5, 202
Piano-fortes, Mover for	D. I. Langworthy	Jamestown, N. Y	Apr. 2, 1867	63, 533
Piano-fortes, Prop-stick for	C. E. McDonald	Brooklyn, N. Y	Apr. 2, 1867	63, 546
Piano-fortes, Repeating-action for	J. Hura	Boston, Mass	Dec. 18, 1866	60, 520
Piano-fortes, Shifting-movement for square or horizontal.	T. Loud	Philadelphia, Pa	Apr. 1, 1842	2, 523
Piano-fortes, String-bearing for	R. A. Tooker	New York, N. Y	Oct. 20, 1863	40, 384
Piano-fortes, Stringing and tuning	W. J. Richards, F. M. Sofge, and J. H. Richards.	La Fayette, Ind	Feb. 8, 1870	99, 708
Piano-fortes, Swell-mute attachment to	A. G. Corliss	Portland, Me	Apr. 11, 1854	10, 773
Piano-fortes, Treble-attachment for	A. V. T. Barberie	Brooklyn, N. Y	Jan. 12, 1869	85, 889
Piano-fortes, &c., Working pedals of	R. H. Hooper	West Roxbury, Mass	July 14, 1868	79, 830
Piano frame, Upright	J. Whitney	Boston, Mass	Jan. 30, 1872	123, 212
Piano-frames, Brace for	D. Gibbons	Rochester, N. Y	Jan. 2, 1855	12, 188
Piano, Grand	G. H. Davis	Boston, Mass	Mar. 1, 1870	100, 266
Piano, Grand	D. F. Haasz	Philadelphia, Pa	Mar. 4, 1856	14, 383
Piano, Grand	G. Steck	New York, N. Y	June 20, 1871	116, 109
Piano, Grand	G. Steck	New York, N. Y	Oct. 21, 1873	143, 789
Piano, Grand	H. Steinway	New York, N. Y	Dec. 20, 1859	26, 532
Piano-hammer	C. W. Brewer	Racine, Wis	Sept. 8, 1868	81, 872
Piano-hammer	J. Percival	Auburn, N. Y	July 5, 1859	24, 659
Piano-hammers, Capping	J. Bullard	Readville, Mass	Apr. 22, 1873	138, 001
Piano, Horizontal	T. Loud, jr	Philadelphia, Pa	May 15, 1827	....
Piano-key	W. A. Reed	Deep River, Conn	Aug. 22, 1871	118, 389
Piano key-board	J. S. Allen and A. P. Wilkins	Allen's Grove, Wis	June 30, 1868	79, 295
Piano key-board arrangement	A. G. and C. Marsh	Seneca Falls, N. Y	May 17, 1859	24, 021
Piano-key-shaping machine	M. Pratt	Deep River, Conn	July 29, 1873	141, 380
Piano-keys, &c., Mode of balancing	F. J. Steinhauser	Lancaster, Pa	Aug. 3, 1869	93, 243
Piano-leg	F. and C. Gelin	New York, N. Y	June 7, 1859	24, 294
Piano-leg attachment	W. Clark	New York, N. Y	May 27, 1856	14, 948
Piano-legs, Making	H. Willgohs	New York, N. Y	Dec. 14, 1869	97, 845
Piano-lid prop-stick	M. Doherty	Cambridgeport, Mass	May 20, 1873	139, 125
Piano, &c., lock	H. Ahrend	Brooklyn, N. Y	June 1, 1869	90, 805
Piano-lock	H. Ahrend	Newark, N. J	Sept. 20, 1870	107, 585
Piano-lock	E. F. French	New York, N. Y	Mar. 16, 1869	87, 834
Piano-lock	E. L. Gaylord	Terryville, Conn	Aug. 1, 1865	49, 100
Piano-lock	E. L. Gaylord	Terryville, Conn	Sept. 3, 1867	68, 496
Piano-lock	E. L. Gaylord	Terryville, Conn	Sept. 3, 1867	68, 497
Piano-lock	A. Helmstaedter	Newark, N. J	Dec. 10, 1867	72, 038
Piano-lock	R. J. Jordan	New York, N. Y	Mar. 30, 1869	88, 488
Piano-lock	J. Murphy	Roslindale, Mass	Oct. 4, 1870	107, 947
Piano-lock	A. F. Pfeifer	Newark, N. J	Aug. 1, 1865	49, 149
Piano-lock	A. F. Pfeifer	Newark, N. J	Feb. 23, 1869	87, 196
Piano-lock	T. Powers	New York, N. Y	Mar. 26, 1872	124, 912
Piano-lock	J. Thielemann	Newark, N. J	Aug. 18, 1868	81, 226
Piano-lock	J. Thielemann and P. Meyer	Newark, N. J	May 24, 1870	103, 525
Piano-lock	E. D. Weyburn	Pittsburgh, Pa	Nov. 7, 1871	120, 804

Index of patents issued from the United States Patent Office from 1790 *to* 1873, *inclusive*—Continued.

Invention.	Inventor.	Residence.	Date.	No.
Piano-lock	N. Wilton	Boston, Mass	Mar. 2, 1858	19, 529
Piano-lock, Eccentric	P. H. Niles	Boston, Mass	Aug. 7, 1849	6, 636
Piano music-desk	M. Doherty	Cambridgeport, Mass	Jan. 14, 1873	134, 860
Piano-orchestra	J. B. Schalkenbach	Triers, Prussia	June 3, 1862	35, 490
Piano, Organ	R. Nutting, 2d	Romeo, Mich	Feb. 8, 1848	5, 438
Piano, &c., pedal	A. A. Mixer	Hamilton, Ohio	Oct. 26, 1869	96, 137
Piano, &c., pedal	E. Zachariae	Loehnberg, Prussia	Feb. 2, 1869	86, 621
Piano-pedal attachment	G. C. A. Class	Philadelphia, Pa	May 13, 1873	138, 857
Piano-pedal attachment	C. W. Held, jr., and A. M. Baugh	Brooklyn, N. Y	Jan. 14, 1873	134, 886
Piano-pedal attachment	S. Schoenbrun	New York, N. Y	Sept. 19, 1871	119, 184
Piano-pedal attachment	W. B. Stetson	Taylor, N. Y	Nov. 2, 1858	21, 9[illegible]0
Piano-pedal attachment	A. A. Stinson	Herkimer, N. Y	Feb. 6, 1872	123, 429
Piano pedal-stool	T. Springer	Hunter's Point, N. Y	Dec. 30, 1873	146, 103
Piano pedal-stool	J. H. Wells	Chicago, Ill	Jan. 2, 1872	122, 343
Piano-pins, Screw-clamp for	C. M. Lindsay	Foreston, Ill	Sept. 13, 1870	107, 391
Piano-seat	L. Postawka	Boston, Mass	Oct. 30, 1866	59, 231
Piano-seat	L. Postawka and A. Krasinski	Boston, Mass	May 1, 1866	54, 401
Piano ound-board	J. Benz	Philadelphia, Pa	Oct. 16, 1866	58, 763
Piano sound-board	P. Schuler	Philadelphia, Pa	Apr. 3, 1866	53, 687
Piano sounding-board	A. De Kuhn	New York, N. Y	Sept. 11, 1866	57, 912
Piano sounding-board	G. M. Guild	Boston, Mass	May 26, 1868	78, 276
Piano sounding-board	T. Kater	Hamilton, Canada	Jan. 21, 1873	135, 126
Piano sounding-board	W. Kells	Albany, N. Y	Oct. 1, 1872	131, 760
Piano sounding-board	C. Meyer	Philadelphia, Pa	July 9, 1850	7, 494
Piano sounding-board	F. E. Ramm	Philadelphia, Pa	Oct. 23, 1866	59, 069
Piano sounding-board	C. F. T. Steinway	New York, N. Y	Feb. 11, 1873	135, 857
Piano sounding-board	J. Stewart	Boston, Mass	Nov. 14, 1822	....
Piano sounding-board	F. Strothmann	Louisville, Ky	Jan. 22, 1867	61, 483
Piano sounding-board	E. L. Taylor	Jersey City Heights, N. J.	Dec. 5, 1871	121, 686
Piano-string bridge	W. C. Ellis	Worcester, Mass	Nov. 25, 1873	144, 842
Piano, Upright	O. Altenburg	New York, N. Y	Mar. 12, 1872	124, 470
Piano, Upright	C. F. Chickering	New York, N. Y	Nov. 28, 1871	121, 334
Piano, Upright	G. C. Manner	New York, N. Y	Dec. 14, 1869	97, 943
Piano, Vertical	E. Fobes	Boston, Mass	May 17, 1853	9, 724
Piano, Vertical	G. Tracyser	Cincinnati, Ohio	Mar. 29, 1853	9, 640
Piano with melodeon-attachment	L. Louis	Buffalo, N. Y	June 10, 1862	35, 528
Pianos, Adjustable extension-pedal for	A. G. Brewer	Hopkinton, Mass	Jan. 19, 1869	85, 901
Pianos, Auxiliary key-board for	T. J. V. Roz	New York, N. Y	Jan. 19, 1869	85, 965
Pianos, Carriage for moving	H. C. Camp and R. C. Munger	Saint Paul, Minn	June 25, 1872	128, 210
Pianos, Child's pedal for	C. A. Class	Saint Louis, Mo	Oct. 13, 1868	83, 039
Pianos, Clamp for moving	D. Benson	Syracuse, N. Y	June 11, 1872	127, 837
Pianos, desks, &c., Lid-support for	C. T. Faber	New York, N. Y	Oct. 1, 1867	69, 329
Pianos, Device for moving	S. D. Reynolds	Rochelle, Ill	Jan. 23, 1872	122, 966
Pianos, Fitting hammer-heads for	J. Mackay	Boston, Mass	Aug. 14, 1828	....
Pianos, Former for bending and gluing cases of grand.	D. Decker	New York, N. Y	July 5, 1870	105, 049
Pianos, Front-fall for	A. H. Hastings	New York, N. Y	May 23, 1871	115, 053
Pianos, Hand-guide for	M. Gether	Saint Louis, Mo	Nov. 23, 1869	97, 074
Pianos, &c., Hand-supporter for	C. Sangalli	New York, N. Y	Dec. 8, 1868	84, 844
Pianos, Instructing-scale for	S. Winner	Philadelphia, Pa	Dec. 22, 1863	41, 043
Pianos, Iron frame for	G. Steck	New York, N. Y	Aug. 9, 1870	106, 227
Pianos, Iron frame for	H. Worcester	New York, N. Y	June 3, 1862	35, 484
Pianos, Iron frame for upright	G. Steck	New York, N. Y	Mar. 15, 1870	100, 948
Pianos, Metal frame for	M. Martins	New York, N. Y	Sept. 4, 1866	57, 743
Pianos, Metal spring for	I. B. Thompson	Philadelphia, Pa	Nov. 29, 1859	26, 304
Pianos, Octave-coupler for	J. Clark	Philadelphia, Pa	July 16, 1872	129, 530
Pianos, Prepared-wood sounding-board for	J. B. Dunham	East Chester, N. Y	July 25, 1871	117, 393
Pianos, Repeating-action for	F. Farholtz	Louisville, Ky	Aug. 15, 1865	49, 394
Pianos, Stringing	A. Choplain and P. E. Chollet	New York, N. Y	Dec. 19, 1865	51, 559
Pianos, tables, &c., Machine for cutting legs for	A. Stoeckel	New York, N. Y	July 10, 1855	13, 236
Pianos, Tension-scale for tuning	C. R. Edwards	Niagara City, N. Y	Aug. 2, 1864	43, 674
Pianos, Wrest-pin for	G. Schilling	Hoboken, N. J	July 14, 1857	17, 812
Pick	N. Rose	Belmont, N. Y	July 6, 1869	92, 370
Pick	J. D. Cochran	Kern River, Cal	July 26, 1864	43, 638
Pick	J. C. Reed	Cincinnati, Ohio	June 5, 1860	28, 638
Pick and ax combined	H. L. Lowman	New York, N. Y	Oct. 2, 1866	58, 438
Pick and pickax	E. P. Dickie	Morristown, N. J	July 14, 1868	79, 815
Pick and rammer socket	J. Pearce	Barton, Md	July 16, 1872	129, 164
Pickax	L. Baker	Pultney, Vt	Jan. 15, 1822	....
Pickax	J. C. Conklin	Peekskill, N. Y	Dec. 20, 1853	10, 356
Pickax	J. C. Conklin	Yorktown, N. Y	July 21, 1868	80, 146
Pickax	M. Gale	San Antonio, Mexico	Oct. 20, 1868	83, 274
Pickax-eye	G. Nimmo	Jersey City, N. J	Nov. 5, 1867	70, 459
Pickax, &c., eyes, Machine for forming	H. L. Lowman	New York, N. Y	June 30, 1868	79, 364
Pick, Digging and tamping	C. Carroll	North Vernon, Ind	May 17, 1870	103, 014
Pick-eyes, Die for forming	H. M. Hamilton	New York, N. Y	Aug. 27, 1867	68, 187
Pick-handle	F. Bellinger	Lockport, N. Y	Mar. 29, 1870	101, 214
Pick-handle	W. Blay	Helena, Mont	June 9, 1868	78, 718
Pick-handle	J. E. Emerson	Sacramento, Cal	Mar. 29, 1859	23, 358
Pick-handles, Attaching	R. C. Grover and C. W. Randall	Newton, Mass	June 1, 1869	90, 742
Pick-handles, Attaching	P. J. Hogan	Cincinnati, Ohio	Nov. 3, 1868	83, 710
Pick-handles, Attaching	H. W. Knight	Columbus, Ohio	Nov. 19, 1867	71, 184
Pick handles, Attaching	W. H. Livingston	New York, N. Y	Oct. 1, 1861	33, 394
Pick-handles, Attaching	T. H. Neal	Allegheny, Pa	June 8, 1869	91, 154
Picks to handles, Attaching	J. P. Walsh	Helena, Mont	Dec. 17, 1867	72, 245
Pick-handles, Fastening	J. H. Fisher	Placerville, Cal	July 17, 1860	29, 155
Pick handles, Fastening mining	J. B. Pressey and D. Sheets	Buffalo, N. Y	Oct. 8, 1861	33, 451
Pick, Miner's	H. M. Hamilton	New York, N. Y	Aug. 20, 1867	67, 875
Pick or ax	A. Moore	San Francisco, Cal	Mar. 24, 1863	37, 970
Pick-pocket alarm	I. T. Dyer	Quincy, Ill	Jan. 5, 1869	85, 649
Pick-pocket detector	S. W. Ruggles	Fitchburgh, Mass	July 29, 1856	15, 440
Picks and rammers, Manufacture of	D. McNally	Mount Savage, Md	Nov. 7, 1871	120, 658
Picks and similar tools, Machine for making	A. Buerkle	Pittsburgh, Pa	Apr. 25, 1871	113, 977
Picks, Apparatus for opening the eyes of	R. Blake	Scranton, Pa	Jan. 31, 1871	111, 305
Picks, &c., Machine for forming the eyes of	H. M. Hamilton	New York, N. Y	Sept. 24, 1867	69, 090
Picks, &c., Making mining	E. Hughes	McCartysville, Cal	May 7, 1861	32, 24[illegible]
Picks, Manufacturing	D. Collins	Newark, N. J	June 3, 1873	139, 455

Index of patents issued from the United States Patent Office from 1790 to 1873, inclusive—Continued.

Invention.	Inventor.	Residence.	Date.	No.
Picker: *See* Clover-picker. Cotton-picker. Cotton-seed picker. Cranberry-picker. Fruit-picker. Grain-picker. Hair-picker. Hop-picker. Loom-picker. Nut-picker. Pea-picker. Pea-nut picker. Waste-picker. Wool-picker.				
Picker	A. Holbrook	Providence, R. I	July 9, 1861	32, 770
Picker	A. Holbrook	Providence, R. I	July 9, 1861	32, 771
Picker-collar	J. M. Gotham	Blackstone, Mass	Sept. 27, 1870	107, 772
Picker-cylinders, Tooth-clothing for	R. Heneage	Lowell, Mass	Oct. 31, 1854	11, 861
Picker-house for opening and cleaning cotton, &c.	R. Kitson	Lowell, Mass	May 29, 1866	55, 118
Picker-motion	W. Nugent	Chicopee, Mass	May 14, 1861	32, 308
Picker-staff check	E. Robbins	Worcester, Mass	Jan. 7, 1868	73, 045
Picker-staff motion	N. S. Bean	Manchester, N. H	Jan. 22, 1861	31, 205
Picker-staff motion	E. H. Graham	Manchester, N. H	Oct. 16, 1860	30, 441
Picker-staff, Mounting	W. Mason	Taunton, Mass	June 2, 1868	78, 463
Picker-staff, Stop for	J. H. Knowles	Lawrence, Mass	Sept. 17, 1872	131, 354
Picker to picker-staff, Attaching	J. D. Barrie	Lawrence, Mass	May 3, 1870	102, 475
Picket-cutter	J. W. Clark	Outville, Ohio	Apr. 4, 1871	113, 629
Picket for fences and walls	A. Dabb	Elizabethport, N. J	Apr. 2, 1867	63, 482
Picket-heading machine	H. D. Heiser, and H. F. and G. F. Snyder.	Williamsport, Pa	May 14, 1872	126, 634
Picket-holding clamp	P. McCarthy	Rochester, N. Y	Apr. 14, 1868	76, 644
Picket-making machine, Wooden	T. J. Jolly	Versailles, Ind	Jan. 1, 1867	60, 900
Picket-pointer	J. W. Minor	Middleborough, Mass	Sept. 12, 1871	118, 964
Picket-pointing machine	H. Balcom	Ann Arbor, Mich	Mar. 18, 1873	136, 807
Picket-pointing machine	A. B. Corby	Binghamton, N. Y	Feb. 4, 1873	135, 467
Picket-pointing machine	W. W. Johnson	Nashville, Tenn	Jan. 14, 1868	73, 339
Picket-shaping machine	A. J. Sutherland	Ann Arbor, Mich	Dec. 23, 1873	145, 766
Picking, carding, &c., machines, Means of feeding wool, &c., to.	S. R. Parkhurst	Bloomfield, N. J	May 30, 1865	47, 976
Pickle and cruet stand	T. Leach	Taunton, Mass	May 21, 1872	127, 073
Pickle-fork	H. Laurence	New Orleans, La	Aug. 27, 1872	130, 923
Picnic or excursion seat	J. M. Perkins	Chicago, Ill	Oct. 9, 1860	30, 344
Picture and card case, Metallic	H., J., T., and H. Hall, jr	Philadelphia, Pa	Feb. 25, 1862	34, 489
Picture and card exhibiter	B. F. Bostian	New Columbia, Pa	Sept. 2, 1873	142, 310
Picture and curtain knob	J. Gardner	New Haven, Conn	Jan. 7, 1868	73, 088
Picture and looking-glass frames, Ornamenting and constructing.	J. Parker and L. P. Clover	New York, N. Y	Mar. 23, 1829	....
Picture and mirror cord retainer	W. Read, jr	Boston, Mass	Feb. 25, 1868	74, 771
Picture and mirror hanging device	A. A. Fielding	Boston, Mass	Aug. 1, 1871	117, 529
Picture and other frames, Device for driving nails, &c., into.	W. Bascom	New York, N. Y	Sept. 15, 1863	39, 985
Picture and other frames, Device for hanging	C. B. Davies	Dayton, Ohio	Nov. 24, 1868	84, 266
Picture-card frame	R. W. Potter	New York, N. Y	Mar. 7, 1865	46, 699
Picture-case	G. Brownlee	Princeton, Ind	Nov. 23, 1869	97, 034
Picture-case	A. R. Stone	Baldwinsville, N. Y	July 23, 1872	129, 764
Picture-case and cane, Combination	H. D. McGeorge	Rochester, N. Y	Feb. 16, 1869	86, 991
Picture cord	T. Kohn	Hartford, Conn	Apr. 5, 1870	101, 629
Picture-cord, Attachment for adjusting	R. D'Heurenso	New York, N. Y	Mar. 30, 1869	88, 373
Picture-frame	E. W. Appleton	Cincinnati, Ohio	July 16, 1872	129, 450
Picture-frame	J. A. Burch	Buffalo, N. Y	Nov. 18, 1873	144, 599
Picture-frame	J. S. Cannon	New Haven, Conn	June 21, 1864	43, 255
Picture-frame	L. S. Chase	New York, N. Y	May 7, 1867	64, 488
Picture-frame	G. E. and W. C. Collins	Bucksport, Me	Dec. 26, 1871	122, 158
Picture-frame	J. D. Crocker	Norwich, Conn	Aug. 16, 1870	106, 332
Picture-frame	L. M. B. Gray	Boston, Mass	May 14, 1872	126, 633
Picture-frame	H. S. Hale	Philadelphia, Pa	May 13, 1873	138, 884
Picture-frame	C. A. Hodgman	Tuckahoe, N. Y	Mar. 30, 1869	88, 482
Picture-frame	A. J. Holmes	Saratoga Springs, N. Y	Nov. 13, 1866	59, 719
Picture-frame	C. H. Hutchinson	Philadelphia, Pa	Dec. 10, 1872	133, 707
Picture-frame	D. W. Marshall	Pawtucket, R. I	Apr. 6, 1869	88, 651
Picture-frame	A. A. Murfey	Montville, Conn	Sept. 7, 1869	94, 632
Picture-frame	M. Ormsbee	New York, N. Y	Feb. 14, 1865	46, 382
Picture-frame	C. P. Poinier	Boston, Mass	Oct. 22, 1867	70, 116
Picture-frame	F. Reifschneider	New York, N. Y	Oct. 7, 1873	143, 464
Picture-frame	J. E. Rice	Boston, Mass	Apr. 24, 1866	54, 208
Picture-frame	J. T. Schmitt	Brooklyn, N. Y	Nov. 3, 1868	83, 797
Picture-frame	G. Schneider	Buffalo, N. Y	June 22, 1869	91, 673
Picture-frame	J. Sperry	New York, N. Y	Jan. 18, 1870	99, 022
Picture-frame	J. L. Warren and J. W. Adderly	Detroit, Mich	Oct. 21, 1873	143, 861
Picture-frame and exhibiter, Combined	B. Anyan	Fitchville, Ohio	Nov. 18, 1873	144, 648
Picture-frame-beveling machine	O. F. Bedell	New York, N. Y	Sept. 19, 1865	49, 965
Picture-frame clamp	J. A. H. Dunne	Boston, Mass	May 1, 1866	54, 403
Picture-frame clamp	L. A. Johnson	Candor, N. Y	Sept. 19, 1871	119, 151
Picture-frame, Easel	R. Wright	Brooklyn, N. Y	July 18, 1871	117, 133
Picture frame fastening	J. D. Crocker	Norwich, Conn	Feb. 15, 1870	99, 854
Picture-frame hanger	C. H. Moulton	Des Moines, Iowa	Dec. 5, 1871	121, 652
Picture-frame-hanging device	T. C. Barnard	West Meriden, Conn	Sept. 23, 1873	142, 981
Picture-frame, &c., hanging device	J. N. Gillespie	Waltham, Mass	May 10, 1870	102, 805
Picture-frame support	L. L. Hodges	Boston, Mass	Jan. 28, 1868	73, 808
Picture-frames and mirrors, Device for suspending.	R. E. Bean	Franklin, N. H	July 27, 1868	93, 040
Picture-frames, Composition for molding for	H. Wurtz	New York, N. Y	Feb. 8, 1870	99, 739
Picture-frames, &c., Corner-piece for	T. Callanan	Brooklyn, N. Y	May 6, 1873	138, 611
Picture-frames, Machine for enameling	J. Sperry and C. W. Sherwood.	Brooklyn, N. Y	Apr. 2, 1861	31, 913
Picture-frames, Machine for enameling oval	F. Smith	Allegheny City, Pa	Aug. 2, 1864	43, 715
Picture-frames, Machine for preparing moldings for.	E. S. Gardner	New York, N. Y	Mar. 29, 1859	23, 426
Picture-frames, Machinery for making oval	J. E. Sweet	Syracuse, N. Y	Feb. 12, 1867	62, 086
Picture-frames, Machinery for preparing oval	W. Gardner	New York, N. Y	Aug. 17, 1858	21, 192
Picture-frames, &c., Manufacture of moldings for.	A. C. Engert	Middlesex County, England.	May 7, 1872	126, 385
Picture-frames, Wooden mat for	H. W. Curtis	Philadelphia, Pa	Oct. 24, 1871	120, 250
Picture-hanger	F. W. Ely	Du Luth, Minn	Aug. 26, 1873	142, 156
Picture-hanger	G. Lamb	Boston, Mass	May 31, 1870	103, 627
Picture-hanging device	C. W. Simpson	Bangor, Me	Apr. 19, 1870	102, 167
Picture-holder	A. Barber	Port Byron, Ill	June 25, 1867	66, 115

Index of patents issued from the United States Patent Office from 1790 *to* 1873, *inclusive*—Continued.

Invention.	Inventor.	Residence.	Date.	No.
Picture-holder	W. Walker	New Haven, Conn	June 12, 1866	55, 561
Picture-hook	W. D. Gridley	New Britain, Conn	Jan. 12, 1869	85, 738
Picture hook and cornice	W. Potts	Handsworth, England	Dec. 1, 1868	84, 645
Picture medal, button, &c	H. C. Griggs	Waterbury, Conn	Jan. 3, 1865	45, 710
Picture-mount	J. H. Fitzgibbon	Saint Louis, Mo	Sept. 10, 1872	131, 260
Picture-nail	J. W. Bishop	New Haven, Conn	May 11, 1869	89, 970
Picture-nail	B. H. Bradley	Waterbury, Conn	Nov. 10, 1868	89, 970
Picture-nail	S. C. Cary	Brooklyn, N. Y	Mar. 26, 1872	124, 933
Picture-nail	H. Hickman	Omaha, Nebr	June 22, 1869	91, 629
Picture-nail	C. B. Jenkins	New York, N. Y	Dec. 10, 1872	133, 778
Picture-nail	A. D. Judd	New Haven, Conn	Nov. 6, 1866	59, 404
Picture-nail	H. L. Judd	Brooklyn, N. Y	June 14, 1870	104, 163
Picture-nail	H. L. Judd	Brooklyn, N. Y	Sept. 26, 1871	119, 273
Picture-nail	T. C. Richards	New York, N. Y	Dec. 27, 1870	110, 500
Picture-nail	T. C. Richards	New York, N. Y	Jan. 10, 1871	110, 867
Picture-nail	T. C. Richards	New York, N. Y	Oct. 3, 1871	119, 648
Picture-nail	F. J. Seymour	Wolcottville, Conn	June 26, 1866	55, 917
Picture-nail	F. W. Smith, jr	Bridgeport, Conn	Dec. 28, 1869	98, 438
Picture-nail	J. H. Squier and E. J. Warner	Newark, N. J	Apr. 12, 1870	101, 781
Picture-nail	J. Uster	Brooklyn, N. Y	Apr. 16, 1872	125, 859
Picture-nail	G. Vintschger	New York, N. Y	Mar. 4, 1873	136, 566
Picture-nail head	H. P. Brooks	Wolcottville, Conn	Feb. 20, 1866	52, 781
Picture-nail head	J. B. Sargent	New Britain, Conn	Aug. 21, 1860	29, 720
Picture-nail knob	W. E. Sparks	New Haven, Conn	Aug. 22, 1871	118, 400
Picture-nail, Ornamental-headed	T. C. Richards	New York, N. Y	Dec. 20, 1870	110, 287
Picture-nails, Attaching ornamental heads to	A. D. Judd	New Haven, Conn	Mar. 28, 1865	47, 023
Picture-nails, Mechanism for manufacturing heads for.	L. Wolf	Meriden, Conn	Mar. 21, 1871	112, 880
Picture-poise	J. J. Glover	Quincy, Mass	May 24, 1870	103, 445
Picture-stand, Revolving	E. B. Sturdevant	Germantown, Ohio	June 4, 1867	65, 446
Picture-stretcher	J. D. Crocker and J. A. Brand	Norwich, Conn	Oct. 3, 1871	119, 507
Picture-type	J. McElheran	Brooklyn, N. Y	Mar. 16, 1858	19, 645
Picture with looking-glass back-ground	J. Jennings	New York	Nov. 11, 1831	....
Pictures, Apparatus for composing and exhibiting group of card.	S. F. Mills	San Francisco, Cal	Mar. 30, 1869	88, 321
Pictures, &c., Apparatus for mounting and exhibiting.	J. W. Hardie	New York, N. Y	Mar. 26, 1872	124, 953
Pictures by metallic leaves, Shading	E. Harmon	Cleveland, Ohio	Mar. 27, 1849	6, 241
Pictures, Coloring	C. J. Mettais	New York, N. Y	Jan. 7, 1873	134, 693
Pictures, looking-glasses, &c., Method of hanging	W. A. Foster	Chester, Conn	July 12, 1859	24, 729
Pictures, Machine for mounting	E. P. Christmas	Mansfield, Ohio	Apr. 8, 1873	137, 656
Pictures, Molding for hanging	H. Hochstrasser	Philadelphia, Pa	May 8, 1860	28, 174
Pictures, Mounting	C. J. Billinghurst	McArthur, Ohio	July 20, 1869	92, 694
Pictures, Mounting	E. L. Courtney	Philadelphia, Pa	June 18, 1872	128, 125
Pictures on glass, Backing up	T. Fry and C. A. Seely	Brooklyn and New York, N. Y.	Jan. 3, 1860	26, 733
Pictures, Preparation of transparent	G. Wedekind	Philadelphia, Pa	Nov. 13, 1860	30, 649
Pictures, Screw-support for hanging	M. Judd	New Britain, Conn	Apr. 1, 1862	34, 837
Pictures to frames, Attaching	S. W. Hanks	Lowell, Mass	Nov. 20, 1866	59, 836
Pictures to translucent surfaces, Transferring	J. T. Foley	New York, N. Y	Aug. 20, 1872	130, 030
Pictures transparent, Compound for making	S. D. McPherson	Normal, Ill	Mar. 31, 1868	76, 225
Pie-case	H. H. Olds	New Haven, Conn	Mar. 21, 1871	112, 953
Pie-crimper	C. A. Shaw	Biddeford, Me	Nov. 6, 1860	30, 592
Pie cutter and crimper	H. Mathews	Cambridge, Mass	Apr. 6, 1869	88, 726
Pie-marker	T. S. Macomber	Hamilton, N. Y	Dec. 23, 1873	145, 808
Pie-rimmer	N. N. Brown	Reading, Pa	Nov. 26, 1867	71, 274
Pie-rimmer	J. Stephen and W. Zeller	Womelsdorf, Pa	Sept. 11, 1866	57, 990
Pie stamper and cutter	A. Hawkes	Providence, R. I	Apr. 19, 1870	102, 117
Pie-tube	N. M. Selden	Chatham, Conn	Oct. 20, 1868	83, 327
Pier and bridge	W. H. Wood	Hudson, N. J	Oct. 21, 1862	36, 747
Pier and bulk-head	S. J. Seely	New York, N. Y	Jan. 31, 1865	46, 144
Pier, dock, and wharf	R. A. Gilpin	Chester County, Pa	June 4, 1867	65, 484
Pier-glass frame and cornice	D. A. Hall and D. Garrison	Philadelphia, Pa	Mar. 28, 1871	113, 046
Pier, Iron	W. B. Porter	Plattsmouth, Nebr	Sept. 1, 1868	81, 682
Pier or breakwater	C. T. Harvey	Marquette, Mich	Apr. 12, 1859	23, 574
Pier, Sectional	O. A. Howe	Jersey City, N. J	Feb. 23, 1869	87, 171
Piers, &c., Construction of	G. A. Parker	Lancaster, Mass	Mar. 15, 1864	41, 935
Piers, Crib for laying foundation for	J. Parish	Chicago, Ill	Mar. 5, 1872	124, 284
Piers for bridges, &c., Building	J. Du Bois	Williamsport, Pa	Sept. 23, 1862	36, 512
Piers into the water, Lowering	C. Van Horn	Springfield, Mass	Oct. 29, 1861	33, 615
Piers, Iron front for stone	J. Abbott	Canton, Ohio	Apr. 29, 1873	138, 222
Pig-singeing apparatus	A. and E. M. Denny	Waterford, Ireland	Oct. 23, 1860	30, 467
Pigeon-hole	C. J. Clements and E. P. Fowler	New York, N. Y	Apr. 15, 1873	137, 892
Pigeon-trap	W. F. Parker	Meriden, Conn	July 1, 1873	140, 532
Pigeon-trap	N. B. Tyler	Warren, Ohio	June 10, 1873	139, 836
Pigment and vehicle for mixing paints	J. M. Merrymon	Indianapolis, Ind	Dec. 13, 1864	45, 426
Pigment from anthracite coal, Black	N. Chater	New York, N. Y	Oct. 5, 1839	1, 358
Pigments from iron-ore, Manufacturing	J. H. Davis	Morristown, N. J	Aug. 8, 1854	11, 476
Pigments from sulphurets of zinc and lead, Manufacture of.	N. Bartlett	Centreville, N. J	Oct. 27, 1868	83, 357
Pigments, Manufacture of	J. Dale	Manchester, England	Aug. 14, 1866	57, 264
Pigments, Manufacture of	H. L. Pattinson	Scott's House, England	Aug. 12, 1851	8, 292
Pigments, Manufacture of opaque	S. Gwynn	New York, N. Y	Dec. 19, 1865	51, 584
Pigments, oils, and gums from cloth used by engravers, Process for recovering.	H. M. Baker	New York, N. Y	Nov. 24, 1868	84, 404
Pigments, Producing carbon	A. Farrar	Boston, Mass	Nov. 2, 1869	96, 409
Pigments to fibrous and textile material, Mixing	A. Paraf	New York, N. Y	Jan. 25, 1870	99, 105
Pike and revolving fire-arm combined	J. C. Campbell	New York, N. Y	June 30, 1863	39, 032
Pile	S. B. Cushing	Providence, R. I	Mar. 23, 1869	88, 141
Pile and post driver	N. Havermale	Canton, Ill	Dec. 20, 1870	110, 359
Pile-driver	S. E. Baker	Chicago, Ill	June 17, 1873	139, 853
Pile-driver	J. T. Baldwin	Petaluma, Cal	July 16, 1872	129, 083
Pile-driver	J. A. Borgort	Hudson City, N. J	Nov. 3, 1868	83, 687
Pile-driver	W. P. Craig	Newport, Ky	Mar. 29, 1859	23, 354
Pile-driver	H. Flad	Saint Louis, Mo	Oct. 12, 1869	95, 786
Pile-driver	C. D. Hall and E. Badger	New Haven, Conn., and Waukegan, Ill.	Dec. 17, 1872	133, 936
Pile-driver	J. W. Hoard	Providence, R. I	Mar. 25, 1856	14, 502

Index of patents issued from the United States Patent Office from 1790 *to* 1873, *inclusive*—Continued.

Invention.	Inventor.	Residence.	Date.	No.
Pile-driver	J. Huy	Whistler, Ala	Nov. 29, 1870	109, 623
Pile-driver	H. Roehrs	Washington, D. C	July 25, 1871	117, 465
Pile-driver	T. Shaw	Philadelphia, Pa	Nov. 24, 1868	84, 383
Pile-driver	T. Shaw	Philadelphia, Pa	June 14, 1870	104, 215
Pile-driver	J. J. Simons	East Saint Louis, Mo	Jan. 5, 1869	85, 703
Pile-driver	A. Smith and J. W. Galbraith	Sedalia, Mo	Feb. 16, 1869	87, 010
Pile-driver	C. H. Smith	Bloomer, Wis	July 22, 1873	141, 086
Pile-driver	J. J. Studer	Richmond, Ind	Nov. 6, 1866	59, 476
Pile-driver	J. Wood	Brooklyn, N. Y	Mar. 22, 1864	42, 038
Pile-driver, Adjustable	T. W. Loveless	Corning, N. Y	July 13, 1858	20, 883
Pile-driver, Adjustable	T. Place	Alfred Centre, N. Y	May 3, 1859	23, 858
Pile-driver, Explosive	H. Vogler	Baltimore, Md	Apr. 1, 1873	137, 514
Pile-driver, Steam	J. Nasmyth	Patricroft, England	June 26, 1847	5, 172
Piles, Driving	J. McClay	Hartford, Conn	June 12, 1866	55, 586
Piles, Driving	J. A. Whipple	Boston, Mass	Nov. 8, 1859	26, 073
Pile-driving, Apparatus for atmospheric	A. Holmstrom	New York, N. Y	Jan. 2, 1855	12, 130
Pile-driving machine	R. Benson	New Orleans, La	Sept. 18, 1841	2, 264
Pile-driving machine	J. Du Bois	Williamsport, Pa	May 19, 1863	38, 570
Pile-driving machine by the aid of shifting-wheels, Mode of progressing with.	E. P. Williams	Utica, N. Y	July 22, 1839	1, 253
Pile-fabric	G. Crompton	Worcester, Mass	Mar. 26, 1872	125, 026
Pile-fabric	G. Crompton	Worcester, Mass	June 25, 1872	128, 286
Pile-fabrics, Apparatus for cutting pile of	J. Johnson	Troy, N. Y	Jan. 13, 1852	8, 649
Pile-fabrics, Cutting	E. B. Bigelow	Boston, Mass	Dec. 4, 1855	13, 862
Pile-fabrics, Machinery for manufacturing	C. Miller	New York, N. Y	Mar. 27, 1860	27, 671
Pile-fabrics, Making	J. W. Crossley	Bridgeport, Conn	May 22, 1866	54, 870
Pile-fabrics, Manufacture of	N. Hill	Caton, N. Y	May 29, 1860	28, 484
Pile-fabrics, Weaving	J. Shinn	Philadelphia, Pa	Dec. 3, 1872	133, 676
Pile-fabrics, Weaving	W. Webster	Morrisania, N. Y	May 19, 1868	78, 163
Pile for beams	J. Stokes	Trenton, N. J	Dec. 6, 1870	109, 850
Pile for corrugated beams	R. Montgomery	New York, N. Y	Oct. 11, 1870	108, 165
Pile for engineering purposes	T. W. H. Moseley	Boston, Mass	Nov. 1, 1870	108, 814
Pile for girder-irons	A. Kloman	Pittsburgh, Pa	Mar. 24, 1868	75, 769
Pile for railway-chairs	D. Eynon	Richmond, Va	Sept. 21, 1869	95, 099
Pile for rolling beams	J. Griffen	Phœnixville, Pa	Dec. 1, 1857	18, 738
Pile for wrought-iron beams or girders	G. Walters and T. Shaffer	Phœnixville, Pa	Aug. 27, 1867	68, 266
Pile for wrought-iron beams or girders	G. Walters and T. Shaffer	Phœnixville, Pa	Aug. 27, 1867	68, 267
Pile-hook, Revolving	J. D. Leach and S. Hutchings	Penobscot, Me	Dec. 15, 1868	84, 886
Pile or fagot	J. Barker	Covington, Ky	Sept. 9, 1873	142, 666
Pile-sawing attachment for boats	H. Vogler	Baltimore, Md	Oct. 15, 1872	131, 337
Pile, Secondary electric	G. G. Percival	Brooklyn, N. Y	Apr. 3, 1866	53, 668
Piles, Apparatus for sinking pneumatic	F. E. Sickels	Kennett's Square, Pa	Feb. 1, 1870	99, 359
Piles, Apparatus for sinking pneumatic	F. E. Sickels	Chicago, Ill	Feb. 1, 1870	99, 360
Piles by atmospheric pressure, Driving	W. S. Smith	Trenton, N. J	Sept. 4, 1860	29, 921
Piles, &c., by exhausting the air from the same, Method of sinking hollow.	L. H. Potts	London, England	Jan. 29, 1850	7, 060
Piles for beams, Method of constructing	C. Hewitt	Trenton, N. J	May 4, 1869	89, 763
Piles for forming axles, &c., Method of constructing.	P. Roberts	Philadelphia, Pa	June 15, 1869	91, 486
Piles for girder-beams, Making	A. Kloman	Pittsburgh, Pa	Mar. 24, 1868	75, 770
Piles for wharves, &c., Construction of	C. Walton	Washington, D. C	June 28, 1864	43, 357
Piles for wharves, piers, &c., Mode of staying	E. H. Angamar	New Orleans, La	July 12, 1859	24, 708
Piles, Instrument for curing	J. P. Gilbert	Long Island City, N. Y	May 2, 1865	47, 540
Piles, Instrument for treating	H. Schevenell and S. S. Rembert.	Memphis, Tenn	May 5, 1868	77, 539
Piles, Machine for sinking hollow	W. S. Smith	Noyesville, Ill	Sept. 4, 1866	57, 784
Piles or caissons, Sinking pneumatic	F. E. Sickels	Marshall, Tex	May 13, 1873	138, 945
Piles or fagots, Mode of securing bars of	G. Walters and T. Shaffer	Phœnixville, Pa	Oct. 8, 1867	69, 597
Piles, Preserving wooden	W. H. Smith	Memphis, Tenn	July 30, 1867	67, 366
Piles, Protecting	W. J. L. Moulton	San Francisco, Cal	June 19, 1866	55, 694
Piles, Protecting the surface of wooden	L. J. Henry	San Francisco, Cal	Jan. 24, 1865	46, 049
Piles, Sinking	L. J. Seely	New York, N. Y	Oct. 18, 1864	44, 751
Piles, Sinking broken	E. C. Boobar	San Francisco, Cal	Dec. 16, 1873	145, 483
Piles, Sinking metallic	J. Du Bois	Williamsport, Pa	Oct. 1, 1872	131, 746
Piles-supporter	C. West	Pittsburgh, Pa	Sept. 16, 1873	142, 972
Piles under water, Machine for sawing off	J. Fleming	Portsmouth, Va	May 1, 1855	12, 789
Piles under water, Mechanism for sawing off	V. Palen	Portsmouth, Va	Oct. 31, 1854	11, 872
Pill-box	N. and E. P. Cary	Albany County, N. Y	Apr. 6, 1830	....
Pill, &c., box	G. H. Hawkins	New York, N. Y	Dec. 3, 1867	71, 751
Pill-box	B. F. Stephens	Brooklyn, N. Y	Dec. 15, 1868	84, 972
Pill-boxes, Machine for making	N. D. White	Winchendon, Mass	Dec. 10, 1850	7, 829
Pill-boxes, Machinery for making	A. Fessenden	Baldwinsville, Mass	Mar. 12, 1850	7, 162
Pill cutting and shaping machine	J. Cooper	Philadelphia, Pa	Oct. 15, 1872	132, 200
Pill-machine	J. C. Ayer	Lowell, Mass	July 28, 1857	17, 865
Pill-machine	P. Cauhape	New York, N. Y	Jan. 3, 1871	110, 630
Pill-machine	H. E. Chapman	Albany, N. Y	May 20, 1856	14, 904
Pill-machine	L. G. Merrell	New Bedford, Pa	Jan. 17, 1854	10, 431
Pill-machine	J. H. Shaw	Saco, Me	May 28, 1867	65, 288
Pill-machine	D. J. Tittle	Albany, N. Y	Oct. 3, 1865	50, 288
Pill-machine	W. V. V. Wilson	Savannah, Ga	June 25, 1867	66, 199
Pill-machine	A. H. Wirz	Philadelphia, Pa	Oct. 1, 1867	69, 379
Pill-making machine	F. Bushby	Manchester, England	Dec. 29, 1868	85, 277
Pill-making machine	N. W. Kumler	Cincinnati, Ohio	Jan. 29, 1856	14, 161
Pill-making machine	W. Lewis	....	July 10, 1807	....
Pill-making machine	C. B. Littlefield	Manchester, N. H	Mar. 18, 1873	137, 008
Pill-making machine	E. A. Pond	Rutland, Vt	Dec. 7, 1852	9, 455
Pill-making machine	E. A. Pond	Rutland, Vt	May 29, 1855	12, 960
Pills	D. Coit	New York	June 17, 1820	....
Pills, Anti-bilious	B. Duval	....	May 3, 1797	....
Pills, Anti-bilious	T. H. Rawson	....	July 24, 1802	....
Pills, Anti-bilious	T. H. Rawson	....	Apr. 4, 1803	....
Pills, Anti-dysenteric	J. C. M. Brockway	Lyme, Conn	Mar. 13, 1822	....
Pills, Anti-dyspeptic	G. Smith	New York	Aug. 7, 1821	....
Pills, Apparatus for sugar-coating	F. Marriott	Detroit, Mich	June 20, 1871	116, 204
Pills, Bilious	S. Lee	New London, Conn	May 24, 1810	....
Pills, Bilious	S. H. P. Lee	....	June 26, 1799	....
Pills, Bilious	S. H. P. Lee	New London, Conn	Feb. 8, 1814	....
Pills, Composition for	S. Cooley	....	June 6, 1798	....

Index of patents issued from the United States Patent Office from 1790 to 1873, inclusive—Continued.

Invention.	Inventor.	Residence.	Date.	No.
Pills, Composition of bilious	S. Lee, jr		Apr. 30, 1796	....
Pills, Device for administering	J. Sullivan	Thornton, Canada	Dec. 5, 1871	121, 684
Pills, Family	D. Coit		Oct. 5, 1803	....
Pills, Hawks's	J. Hawks		Dec. 14, 1799	....
Pills, Mode of counting	J. P. Peters	New York, N. Y	June 27, 1839	1, 202
Pills, Rheumatic	E. Deane	Biddeford, Mass	Jan. 13, 1814	....
Pills, Rheumatic	E. Deane	Biddeford, Me	Feb. 2, 1828	....
Pills, Rheumatic	G. B. Dexter	Boston, Mass	Dec. 18, 1805	....
Pillow, Adjustable	H. Eilers	Cincinnati, Ohio	Oct. 11, 1870	108, 122
Pillow and bolster	T. S. Sperry	Chicago, Ill	Apr. 16, 1872	125, 851
Pillow-block	W. R. Manley	New York, N. Y	Dec. 22, 1868	85, 115
Pillow-spread frame	P. W. Van Est	Rochester, N. Y	Jan. 28, 1873	135, 389
Pillow-support	E. T. Annis	Mount Morris, N. Y	Feb. 25, 1868	74, 786
Pillows and bolsters into cases, Putting	D. B. Tiffany	Xenia, Ohio	Sept. 30, 1856	15, 828
Pillows, Formers for making frames for spring	T. S. Sperry	Chicago, Ill	Feb. 13, 1872	123, 738
Pin:				
See Bench-pin. Linchpin.				
Breast-pin. Nine-pin.				
Car-coupling pin. Ox-bow pin.				
Clothes-pin. Piano-pin.				
Crank-pin. Plow-clevis pin.				
Crimping-pin. Rafting-pin.				
Curling-pin. Rolling-pin.				
Dental pin. Safety-pin.				
Diaper-pin. Shawl-pin.				
Dowel pin. Shawl and dress pin.				
Dress-pin. Shielded pin.				
Flask-pin. Swivel-pin.				
Gage-pin. Ten-pin.				
Hair-pin. Tidy-pin.				
Hair-curling pin. Toggle-pin.				
Hook-pin. Tooth-pin.				
Husking-pin. Tuning-pin.				
Iron pin. Wooden pin.				
Jewelry-pin. Wrist-pin.				
Pin	D. R. Pratt	Worcester, Mass	Feb. 26, 1867	62, 361
Pin	W. H. Towers	New York, N. Y	Aug. 26, 1862	36, 313
Pin and needle making machine	W. F. Hill	New York	Oct. 15, 1814	....
Pin-book	T. Piper	Birmingham, Conn	Apr. 14, 1868	76, 809
Pin-case, Toilet	T. R. Timby	Saratoga Springs, N. Y	May 25, 1869	90, 608
Pin-case, Toilet	T. R. Timby	Saratoga, N. Y	Aug. 10, 1869	93, 499
Pin-cushion	L. J. Atwood	Waterbury, Conn	Nov. 10, 1868	83, 819
Pin-cushion	T. Earle	Valley Falls, R. I	June 9, 1863	38, 817
Pin-cushion	M. Ormsbee	Brooklyn, N. Y	Nov. 8, 1870	109, 044
Pin, Dress	W. Hunt	New York, N. Y	Apr. 10, 1849	6, 281
Pin-drill	T. Prosser	New York, N. Y	Apr. 27, 1869	89, 502
Pin-fastening	D. Brown	Norfolk, Va	Mar. 20, 1860	27, 519
Pin for attaching wearing-apparel	R. J. Nunn	Savannah, Ga	Dec. 11, 1866	60, 411
Pin for attachment of bows and rosettes	L. H. Foy	Worcester, Mass	Apr. 9, 1867	63, 625
Pin lock	A. Herman and W. H. Taylor	Stamford, Conn	Apr. 4, 1871	113, 295
Pin-machine	J. I. Howe	Derby, Conn	Mar. 24, 1841	2, 013
Pin-machine	M. L. Morse	Boston, Mass	Aug. 22, 1814	....
Pin-machine	H. Weston	Philadelphia, Pa	July 23, 1823	....
Pin machine, Wooden	N. W. Brewer	Williamsport, Pa	Mar. 27, 1866	53, 402
Pin-machines, Head-roll for	E. F. Bradley	Birmingham, Conn	Dec. 31, 1867	72, 718
Pin-package	C. O. Crosby	Milford, Conn	Dec. 5, 1871	121, 491
Pin-package	G. Fowler	Seymour, Conn	Dec. 24, 1872	134, 136
Pin-package case	C. O. Crosby	Milford, Conn	Dec. 5, 1871	121, 492
Pin-sticking machine	C. Atwood	New York, N. Y	Dec. 9, 1856	16, 199
Pin-sticking machine	C. O. Crosby	New Haven, Conn	Nov. 1, 1853	10, 180
Pin-sticking machine	C. O. Crosby	New Haven, Conn	Nov. 1, 1853	10, 181
Pin-sticking machine	C. O. Crosby	New Haven, Conn	Nov. 1, 1853	10, 182
Pin-sticking machine	C. O. Crosby	New Haven, Conn	Dec. 5, 1871	121, 493
Pin-sticking machine	T. Fowler	Waterbury, Conn	Aug. 25, 1857	18, 043
Pin-sticking machine	T. Fowler	Waterbury, Conn	Mar. 9, 1858	19, 556
Pin-sticking machine	T. W. Harvey	New York, N. Y	Jan. 3, 1854	10, 385
Pin-sticking machine	J. J. Howe	Derby, Conn	Feb. 24, 1843	2, 970
Pin-sticking machine	J. J. Howe and T. Piper	Derby, Conn	June 10, 1856	15, 112
Pin-sticking machine	J. W. Naramore	Derby, Conn	July 5, 1859	24, 654
Pin-sticking machine	T. Piper	Birmingham, Conn	Jan. 26, 1869	86, 317
Pin-sticking machine	T. Piper and E. F. Bradley	Birmingham, Conn	May 12, 1868	77, 912
Pin-sticking machine	J. B. Terry	Hartford, Conn	Jan. 3, 1854	10, 372
Pin-sticking machine	J. B. Terry	Hartford, Conn	May 29, 1855	12, 975
Pin-sticking machine	J. B. Terry	Hartford, Conn	June 10, 1856	15, 091
Pin-sticking machine	C. W. Van Vliet	Winsted, Conn	Sept. 14, 1858	21, 541
Pins, Device for coating	T. Fowler	Seymour, Conn	Mar. 19, 1861	31, 708
Pins, Device for feeding	C. L. Topliff	New York, N. Y	Mar. 13, 1866	53, 199
Pins in papers, Machine for arranging and sticking	D. Fowler	North Branford, Conn	Sept. 20, 1844	3, 751
Pins in papers, Machine for sticking	C. O. Crosby	New Haven, Conn	Apr. 1, 1851	8, 007
Pins in papers, Machine for sticking	T. Fowler	Waterbury, Conn	Dec. 8, 1857	18, 831
Pins in papers, Machine for sticking	S. Slocum	Poughkeepsie, N. Y	Sept. 30, 1841	2, 275
Pins in papers, Sticking	C. Atwood	New York, N. Y	Oct. 14, 1856	15, 871
Pins in papers, Sticking	W. B. Bartram	Waterbury, Conn	Oct. 14, 1856	15, 877
Pins in papers, Sticking	T. Fowler	Waterbury, Conn	Feb. 12, 1856	14, 234
Pins, Machine for manufacture of	T. and D. Fowler	Northford, Conn	July 31, 1860	29, 431
Pins, Machine for crimping paper for sticking	J. B. Terry	Hartford, Conn	Sept. 11, 1855	13, 553
Pins, &c., Machinery for heading and pointing	S. G. Reynolds	Bristol, R. I	Dec. 31, 1845	4, 346
Pins, Making	L. W. Wright	Manchester, England	Mar. 12, 1825	....
Pins, Manufacturing	J. J. Howe	North Salem, N. Y	June 22, 1832	....
Pins, Papering	C. Atwood	New York, N. Y	Oct. 14, 1856	15, 874
Pins, Papering	C. O. Crosby	New Haven, Conn	July 8, 1851	8, 202
Pins, Papering	C. O. Crosby	New Haven, Conn	Dec. 2, 1851	8, 564
Pins, Papering	C. O. Crosby	New Haven, Conn	Jan. 11, 1870	98, 745
Pins, Papering	G. Fowler	Seymour, Conn	Jan. 17, 1871	111, 051
Pins, Papering	G. L. Turney	London, England	Sept. 10, 1867	68, 809
Pins, Papering-machine for	T. Fowler	Seymour, Conn	Mar. 17, 1868	75, 540
Pins, rivets, &c., Cutting off	A. Stephens	Andover, Mass	Nov. 11, 1828	....
Pins, Separating	T. Fowler	Waterbury, Conn	Nov. 13, 1855	13, 785

Index of patents issued from the United States Patent Office from 1790 *to* 1873, *inclusive*—Continued.

Invention.	Inventor.	Residence.	Date.	No.
Pins upon paper or any other material, Machine for sewing.	E. S. Woodford	Winchester, Conn	July 22, 1856	15, 404
Pinch-bar	A. Reese	Pittsburgh, Pa	May 30, 1871	115, 358
Pinch-bar	P. Spalding	Chicago, Ill	Sept. 21, 1869	95, 159
Pinch-bar for moving heavy weights	W. Siefert	New York, N. Y	Feb. 26, 1867	62, 374
Pinchers	D. Feiler	Marlborough Township, Pa	Oct. 17, 1871	120, 144
Pinchers and grappling-tool, Combined adjustable.	S. B. Dexter	Mason City, Iowa	Mar. 18, 1873	136, 907
Pinchers and nailer, Lasting	F. O. Claflin	Brooklyn, N. Y	Apr. 27, 1869	89, 467
Pinchers, Cutting	P. Broadbooks	Batavia, N. Y	Jan. 25, 1870	99, 057
Pinchers, Die for the manufacture of	W. E. Snediker	Charlotteburgh, N. J	Sept. 16, 1873	142, 955
Pinchers, Double	W. L. Maund	Milan, Tex	Mar. 29, 1870	101, 371
Pinchers for lasting boots and shoes	B. F. Sturtevant	Skowhegan, Me	Oct. 7, 1856	15, 866
Pinchers for operating pile-wires	A. Faulkner	Walpole, N. H	Nov. 23, 1852	9, 417
Pinchers, Hog-ringing	J. C. Schoettle	Collinsville, Ill	Nov. 25, 1873	144, 867
Pinchers, Lasting	G. Kump	Xenia, Ohio	Jan. 16, 1866	52, 055
Pinchers, Lasting	L. B. Richardson	Athol, Mass	Oct. 11, 1859	25, 762
Pinchers, Lasting	B. F. Sturtevant	Skowhegan, Me	Oct. 13, 1857	18, 427
Pinchers, Lasting	A. West	Canton, Ohio	Apr. 19, 1864	42, 451
Pinchers, nail maker and driver for lasting boots	L. R. Blake	Boston, Mass	Feb. 2, 1869	86, 353
Pinchers, Shoemaker's	A. Clarke	Boston, Mass	Oct. 3, 1871	119, 447
Pinchers, Shoemaker's	W. A. Hanna	Chico, Cal	July 22, 1873	141, 047
Pinchers, Shoemaker's	T. Pilote	Marlborough, Mass	Jan. 11, 1870	98, 796
Pinchers, Shoemaker's	T. B. Shelly	Chicago, Ill	Oct. 14, 1873	143, 594
Pinchers, Shoemaker's lasting	A. Dufoult	Spencer, Mass	Dec. 3, 1872	133, 521
Pine leaves for use in the arts, Treating	C. E. Ramus	Lawrence, Kans	Mar. 11, 1873	136, 671
Pine trees, Collecting products of	J. Johnson	Saco, Me	Feb. 2, 1869	86, 553
Pinion	V. W. Blanchard	Bridgeport, Vt	Mar. 9, 1869	87, 623
Pinion	V. W. Blanchard	Bridgeport, Vt	Apr. 26, 1870	102, 363
Pinking-machine	F. L. Hagadom	Baltimore, Md	June 8, 1869	90, 943
Pinking-machine	T. Hagerty	Richmond, Va	Dec. 19, 1871	122, 008
Pinking-machine	W. C. Hooper	Philadelphia, Pa	Oct. 29, 1872	132, 532
Pinking-tool	J. L. McIntosh	Boston, Mass	Sept. 22, 1868	82, 335
Pipe:				

See Aqueduct-pipe. Asphalt pipe. Bit-pipe. Blast-oven pipe. Blow-pipe. Blow-off pipe. Breast-pipe. Building service-pipe. Car-brake pipe. Casting pipe. Cement-pipe. Cement-lined pipe. Chain-locker pipe. Cigar-shaped pipe. Clay pipe. Clay and cement pipe. Composition pipe. Concrete pipe. Conduit-pipe. Confluent-pipe. Copper-coated iron pipe. Coupling-pipe. Dip-pipe. Discharge-pipe. Distributing-pipe. Draft-pipe. Drain-pipe. Drop-pipe. Earthen-pipe. Earthenware-pipe. Escape-pipe. Fire-pipe. Flexible pipe. Fluid-pipe. Furnace air-pipe. Gas-pipe. Gas and water pipe. Gas service-pipe. Glass pipe. Glass-smelting pipe. Hawse pipe. Heating-pipe. Hose-pipe. Hot-air pipe. Hot-blast pipe. Injection-pipe. Iron pipe. Jointed pipe. Lead pipe. Locomotive exhaust-pipe. Locomotive-water-supply pipe. Metallic pipe. Molding pipe. Musical-pitch pipe. Organ-pipe. Oven-pipe. Pump-pipe. Sewer-pipe. Sheet-metal pipe. Smoke-pipe. Smoking-pipe. Stand-pipe. Steam-pipe. Steam and water pipe. Steam-boiler-feed pipe. Steam-conveying pipe. Stove-pipe. Straddle-pipe. Tobacco-pipe. Vapor-bath pipe. Waste-pipe. Waste-water pipe. Water-pipe. Water and air proof pipe. Water-circulating pipe. Wind-pipe. Wooden pipe.

Invention.	Inventor.	Residence.	Date.	No.
Pipe and bolt cutter	G. R. Kirk	Newark, N. J	Jan. 28, 1868	73, 813
Pipe and bolt cutter	C. C. Parsons	Boston, Mass	Oct. 1, 1867	69, 473
Pipe and bolt wrench	C. Frey and G. Macardle	Newark, N. J., and Brooklyn, N. Y.	Apr. 26, 1870	102, 390
Pipe and cigar-holder	V. A. Bond	Cotton Gin, Tex	Dec. 5, 1871	121, 483
Pipe and cylinder flanges, Machine for drilling and boring.	H. S. Fairbanks	Central Falls, R. I	Dec. 29, 1868	85, 293
	J. Elliott	New York, N. Y	Dec. 29, 1868	85, 372
Pipe and faucet clamp				
Pipe and hose coupling	A. H. Brown	Albany, N. Y	Apr. 30, 1850	7, 318
Pipe and other boxes, Machine for fitting	W. Thurber	Olean, N. Y	Dec. 1, 1863	40, 779
Pipe and pump for conveying water	M. Reeve		Dec. 14, 1798	
Pipe and rod cutter	D. F. Hartford	Boston, Mass	Jan. 21, 1868	73, 528
Pipe and stud wrench	J. B. Barnes	Fort Wayne, Ind	June 4, 1867	65, 465
Pipe and tobacco box	L. G. Carr	Philadelphia, Pa	June 22, 1867	91, 713
Pipe, artificial stone, &c., Machine for the manufacture of.	J. W. Stockwell	Portland, Me	Aug. 6, 1872	130, 252
Pipe-attachment in plumbing	A. Barrows	Philadelphia, Pa	May 9, 1871	114, 635
Pipe-bending machine	T. J. Harrison	New York, N. Y	July 21, 1868	80, 069
Pipe-boring machine	T. W. Purcell	Fond du Lac, Wis	Aug. 2, 1870	106, 077
Pipe-butt	L. S. Bunnell	Troy, N. Y	June 4, 1861	32, 464
Pipe, Casting	S. Fulton	Conshohocken, Pa	May 21, 1861	32, 359
Pipe cleaner, Cast	H. Davies	Newport, Ky	May 24, 1870	103, 305
Pipe-cleaner, Pocket	G. E. Matile	Washington, D. C	Nov. 29, 1864	45, 258
Pipe-cleansing apparatus	J. Van Slooten, C. S. Hunt, and W. McCulloch.	New Orleans, La	July 5, 1870	105, 015

Index of patents issued from the United States Patent Office from 1790 *to* 1873, *inclusive*—Continued.

Invention.	Inventor.	Residence.	Date.	No.
Pipe-clearing device	W. Young	Easton, Pa	May 12, 1868	77, 857
Pipe-connection	J. M. Morehead	Brooklyn, N. Y	Mar. 14, 1871	112, 618
Pipe-connection	A. Weyermann	St. Gall, Switzerland	Oct. 5, 1869	95, 621
Pipe-connections, Clamped molds for making lead joints in.	E. Groyn	Tiffin, Ohio	Mar. 1, 1870	100, 394
Pipe-cores, Device for supporting	C. J. C. Petersen	Port Chester, N. Y	Dec. 19, 1871	121, 961
Pipe-coupling	L. Abbott	Lewiston, Me	Apr. 27, 1869	89, 373
Pipe-coupling	W. N. Abbott	Boston, Mass	Feb. 28, 1865	46, 6[illegible]3
Pipe-coupling	W. D. Alford	Cuyahoga Falls, Ohio	May 9, 1871	114, 511
Pipe-coupling	W. D. Alford and J. H. Pitkin	Cuyahoga Falls, Ohio	Feb. 15, 1870	99, 744
Pipe-coupling	W. C. Allison	Philadelphia, Pa	July 12, 1870	105, 290
Pipe-coupling	V. D. Anderson	Kewanee, Ill	May 16, 1871	114, 903
Pipe-coupling	J. F. Andrews	Nashua, N. H	Aug. 15, 1871	117, 960
Pipe-coupling	D. Ashworth	Wappinger's Falls, N. Y	Nov. 25, 1873	144, 943
Pipe-coupling	E. Barbaroux	Louisville, Ky	Jan. 9, 1866	51, 910
Pipe-coupling	A. E. Barnard	Cleveland, Ohio	June 6, 1865	48, 038
Pipe-coupling	J. R. Brown	New Haven, Conn	Aug. 24, 1869	94, 069
Pipe-coupling	J. R. Brown	New Haven, Conn	May 24, 1870	103, 423
Pipe-coupling	C. Burger	Reading, Pa	Dec. 19, 1871	121, 929
Pipe-coupling	J. Chambers	Boston, Mass	July 4, 1865	48, 517
Pipe-coupling	J. Conner	Philadelphia, Pa	Mar. 4, 1873	136, 365
Pipe-coupling	W. Craig	Newark, N. J	Jan. 29, 1867	61, 607
Pipe-coupling	E. Duffee	Haverhill, Mass	Nov. 27, 1866	59, 982
Pipe-coupling	W. B. Dunning	Geneva, N. Y	Sept. 30, 1873	143, 229
Pipe-coupling	W. Dutemple	Malden, Mass	July 18, 1865	48, 797
Pipe-coupling	C. W. Emery	Dorchester, Mass	July 11, 1865	48, 769
Pipe-coupling	L. A. Farjon	Brussels, Belgium	Aug. 2, 1870	106, 041
Pipe-coupling	G. Fetter and J. S. McClintock	Philadelphia, Pa., and Libertyville, Ill.	Aug. 19, 1856	15, 560
Pipe-coupling	J. J. Fifield	East Boston, Mass	June 15, 1869	91, 319
Pipe-coupling	J. B. Fink	Mechanicsville, Pa	Jan. 24, 1871	111, 187
Pipe-coupling	B. Fitts	Newark, N. J	July 16, 1867	66, 820
Pipe-coupling	H. P. Garabedian	Philadelphia, Pa	Dec. 15, 1868	84, 875
Pipe-coupling	A. M. George	Nashua, N. H	July 11, 1865	48, 674
Pipe-coupling	G. C. Germain	Cuyahoga Falls, Ohio	Dec. 13, 1870	110, 028
Pipe-coupling	A. Gwynne	New York, N. Y	Dec. 4, 1866	60, 178
Pipe-coupling	J. M. Hale	Georgia Plains, Vt	Nov. 9, 1869	96, 585
Pipe-coupling	R. Halo	Roxbury, Mass	Sept. 3, 1861	33, 193
Pipe-coupling	R. Hill	East Boston, Mass	Nov. 16, 1869	96, 914
Pipe-coupling	R. Hoskin	Dutch Flat, Cal	July 19, 1870	105, 456
Pipe-coupling	J. V. Jepson	Brooklyn, N. Y	Jan. 4, 1870	98, 597
Pipe-coupling	D. Kahnweiler	Wilmington, N. C	June 29, 1858	20, 717
Pipe-coupling	W. Kearney	Bergen County, N. J	Jan. 9, 1872	122, 614
Pipe-coupling	D. C. Kellam	Pontiac, Mich	May 7, 1872	126, 553
Pipe-coupling	W. A. Lightball	New York, N. Y	July 23, 1870	105, 817
Pipe-coupling	J. Mathews, jr	New York, N. Y	Jan. 29, 1867	61, 626
Pipe-coupling	J. H. McGowan	Cincinnati Ohio	Nov. 45, 1870	109, 332
Pipe-coupling	D. Myers	Chicago, Ill	May 6, 1873	138, 515
Pipe-coupling	J. Old	Pittsburgh, Pa	Aug. 1, 1865	49, 142
Pipe-coupling	J. Old	Pittsburgh, Pa	Oct. 24, 1865	50, 619
Pipe-coupling	C. G. Page and R. J. Falconer	Washington, D. C	Dec. 20, 1859	26, 517
Pipe-coupling	H. D. Parker	Geneseo, N. Y	Feb. 20, 1866	52, 793
Pipe-coupling	J. B. Ramp	Cuyahoga Falls, Ohio	Mar. 21, 1871	112, 958
Pipe-coupling	L. W. Reed	East Cambridge, Mass	Dec. 14, 1869	97, 963
Pipe-coupling	L. T. Scofield	Cleveland, Ohio	Aug. 19, 1873	141, 893
Pipe-coupling	G. Shone	Carondelet, Mo	Jan. 23, 1866	52, 214
Pipe-coupling	E. Smith	Hamburg, Germany	Jan. 11, 1870	98, 717
Pipe-coupling	T. Smith	Baltimore, Md	Oct. 5 1869	95, 528
Pipe-coupling	R. B. Smith and W. H. Whitney.	Cambridge, Mass	Oct. 29, 1872	132, 604
Pipe-coupling	M. Stephens	Brooklyn, N. Y	Oct. 29, 1872	132, 699
Pipe-coupling	A. C. Sweetland	North Attleborough, Mass	Aug. 12, 1873	141, 830
Pipe-coupling	S. P. M. Tasker	Philadelphia, Pa	July 23, 1872	129, 691
Pipe-coupling	N. Thompson	St. John's Wood, England	Sept. 24, 1867	69, 142
Pipe-coupling	N. Thompson	Brooklyn, N. Y	Dec. 24, 1867	72, 566
Pipe-coupling	J. F. Ward	Phillipsburgh, N. J	Aug. 25, 1863	39, 691
Pipe-coupling	J. D. Ware	Savannah, Ga	Dec. 21, 1869	98, 131
Pipe-coupling	C. Warner	Louisville, Ky	Apr. 30, 1850	7, 332
Pipe-coupling	S. R. Warner	New Haven, Conn	Mar. 17, 1863	37, 929
Pipe-coupling	F. R. Wegman	Hartford, Conn	June 15, 1869	91, 291
Pipe-coupling	N. West and N. Thompson	New York	June 27, 1848	5, 651
Pipe-coupling	L. Wilson	Ovid, N. Y	Jan. 14, 1868	73, 422
Pipe-coupling	E. Wright	Boston, Mass	Sept. 1, 1857	18, 116
Pipe-coupling	W. H. Yeaton	Philadelphia, Pa	Oct. 13, 1868	83, 118
Pipe-coupling, Extension	J. McCarthy	New York, N. Y	May 5, 1868	77, 503
Pipe-coupling, Flexible	Q. Rice	Nevada, Cal	Dec. 7, 1869	97, 699
Pipe coupling for car-heaters	T. S. Speakman	Camden, N. J	Apr. 28, 1868	77, 331
Pipe-coupling for connecting inlet-pipes and retorts of gas-machines.	J. Butler	New York, N. Y	Mar. 29, 1870	101, 224
Pipe coupling, Gas and water	J. Holmes	Evanston, Ill	Sept. 30, 1873	143, 348
Pipe-coupling holder	B. Vitalis	Athens, Greece	Apr. 23, 1872	125, 916
Pipe coupling, Lead	J. Hoyt	New York, N. Y	Jan. 4, 1870	98, 501
Pipe coupling, Leader	J. Demarest	Mott Haven, N. Y	Sept. 19, 1871	119, 127
Pipe-coupling lock	G. A. McIlhenny	Washington, D. C	July 5, 1870	105, 105
Pipe-coupling lock	G. A. McIlhenny	Washington, D. C	Jan. 2, 1872	122, 396
Pipe-coupling, Railway-car	M. Henszey, jr	Philadelphia, Pa	May 20, 1873	139, 150
Pipe-coupling, Union	R. M. Potter	Jersey City, N. J	Mar. 8, 1870	100, 665
Pipe couplings, Molding clay	J. Putnam	Salem, Mass	May 16, 1854	10, 911
Pipe-couplings, Wire joint for combined rubber and copper.	F. Kibler	Baltimore, Md	Aug. 1, 1871	117, 546
Pipe-cutter	J. Balmore	New York, N. Y	Feb. 19, 1867	62, 177
Pipe cutter	J. R. Brown	Boston, Mass	Dec. 20, 1859	26, 544
Pipe-cutter	E. Clarkson	Carbondale, Pa	July 13, 1869	92, 516
Pipe-cutter	T. S. Foster	Fitchburgh, Mass	May 28, 1867	65, 0[illegible]6
Pipe-cutter	D. Harrigan	North Somerville, Mass	May 16, 1871	114, 939
Pipe-cutter	J. V. Jepson	Brooklyn, N. Y	Apr. 2, 1867	63, 524
Pipe-cutter	E. S. Moulton	Chelsea, Mass	June 20, 1871	116, 210
Pipe-cutter	J. Peace	Camden, N. J	Aug. 25, 1868	81, 402
Pipe-cutter	R. S. Sanborn	Rockford, Ill	Dec. 17, 1872	134, 007

Index of patents issued from the United States Patent Office from 1790 *to* 1873, *inclusive*—Continued.

Invention.	Inventor.	Residence.	Date.	No.
Pipe-cutter	G. W. Sower	Cambridge, Mass	July 11, 1871	116, 890
Pipe-cutter	A. G. Wilder	Cohoes, N. Y	May 11, 1869	89, 959
Pipe cutter, Gas	W. Kenyon	Steubenville, Ohio	Aug. 14, 1860	29, 602
Pipe-cutter, Self-feeding	M. H. Freeman	Somerville, Mass	Feb. 18, 1868	74, 527
Pipe cutting and threading machine	W. D. Chase	New York, N. Y	Apr. 27, 1869	89, 466
Pipe cutting and threading machine	W. D. Chase	Marion, N. J	Sept. 30, 1873	143, 273
Pipe-cutting device	A. J. Pennock and T. A. Chandler.	Rockford, Ill	Feb. 18, 1873	136, 004
Pipe-cutting machine	M. Bowes	Charlotte, N. C	May 25, 1858	20, 387
Pipe-cutting tool	C. E. Haynes	Boston, Mass	Feb. 11, 1873	135, 644
Pipe-drilling machine	J. Peace and W. P. Cox	Merchantville and Camden, N. J.	Dec. 24, 1872	134, 307
Pipe-driver	L. M. Rumsey and W. P. Smith	Saint Louis, Mo	June 1, 1869	90, 786
Pipe-elbow	W. Austin and W. Opdyke	Philadelphia, Pa	Apr. 4, 1871	113, 614
Pipe-elbow	A. E. Chamberlain and J. B. Crowley.	Cincinnati, Ohio	Sept. 24, 1872	131, 499
Pipe-elbow	I. Leas and W. H. France	Terre Haute, Ind	Oct. 3, 1871	119, 620
Pipe-elbow pattern	J. K. Home	Almonte, Canada	Aug. 12, 1873	141, 645
Pipe elbow, Sheet-metal	F. Dieckmann	Cincinnati, Ohio	June 4, 1872	127, 583
Pipe-elbows, Machine for crimping	C. E. Bell	Greenfield, Ohio	July 9, 1872	128, 781
Pipe-elbows, Machine for forming	C. Hoeller	Cincinnati, Ohio	Mar. 28, 1871	113, 167
Pipe-elbows, Machine for making	L. C. Goodale	Cincinnati, Ohio	May 21, 1872	126, 884
Pipe-elbows, Machine for making	L. C. Goodale	Cincinnati, Ohio	Dec. 24, 1872	134, 267
Pipe elbows, Manufacture of cast	J. G. Weaver, jr	Cincinnati, Ohio	June 28, 1870	104, 908
Pipe-elbows, Molding and casting	G. Ross	Newport, Ky	Feb. 28, 1871	112, 286
Pipe expansion-joint	J. E. Jones	Tidioute, Pa	Apr. 19, 1870	102, 010
Pipe-fittings, Cutting screw-threads on	C. C. Walworth	Boston, Mass	Oct. 11, 1859	25, 779
Pipe-fittings, Machine for tapping	J. L. Pope	Cleveland, Ohio	June 3, 1873	139, 476
Pipe, Flask for casting	J. Demarest	Mott Haven, N. Y	Apr. 25, 1871	113, 986
Pipe, Flask for molding	J. Farrar and L. F. Whiting	Providence, R. I., and Boston, Mass.	Jan. 26, 1869	86, 144
Pipe-joint	P. Ball	Worcester, Mass	Sept. 26, 1865	50, 090
Pipe-joint	P. Clark	Rahway, N. J	Aug. 9, 1870	106, 122
Pipe-joint	R. B. Coar	Jersey City, N. J	Mar. 15, 1870	100, 726
Pipe-joint	R. B. Coar	Jersey City, N. J	May 31, 1870	103, 567
Pipe-joint	J. Demarest	Mott Haven, N. Y	Nov. 18, 1873	144, 663
Pipe-joint	B. Garvin and R. J. Pettibone	Oshkosh, Wis	Mar. 10, 1868	75, 407
Pipe-joint	W. H. Hammond	Tipton, Iowa	June 25, 1872	128, 303
Pipe-joint	A. C. Jones	Philadelphia, Pa	Apr. 23, 1861	32, 135
Pipe-joint	W. R. Maffit	Wilkesbarre, Pa	Mar. 14, 1865	46, 808
Pipe-joint	T. T. Prosser and G. C. Morgan	Chicago, Ill	Apr. 8, 1873	137, 717
Pipe-joint	J. H. Rhodes	Brooklyn, N. Y	Sept. 3, 1867	68, 388
Pipe-joint	H. Smith	Norwalk, Ohio	July 12, 1870	105, 268
Pipe-joint	H. E. Towle	New York, N. Y	June 30, 1863	39, 081
Pipe-joint	S. Trumbore	Easton, Pa	Feb. 20, 1872	123, 954
Pipe-joint	S. R. Warner	New Haven, Conn	Sept. 10, 1861	33, 275
Pipe joint, Adjustable	J. H. Rhodes	Brooklyn, N. Y	Aug. 27, 1867	68, 112
Pipe joint, Cement	D. G. Phipps	New Haven, Conn	June 14, 1870	104, 348
Pipe-joint coupling, Flexible	S. Ainsworth	Pittsburgh, Pa	Sept. 1, 1868	81, 572
Pipe-joint couplings, Sealing-apparatus for	C. Perks	Philadelphia, Pa	Apr. 5, 1870	101, 501
Pipe-joint, Flexible	R. M. Shurtleff	Hartford, Conn	July 9, 1872	128, 760
Pipe joint, Gas	R. C. Robbins	Jersey City, N. J	Jan. 13, 1863	37, 425
Pipe-joint, Metal	W. H. Harrison	Philadelphia, Pa	Jan. 18, 1870	98, 960
Pipe-joint-sealing apparatus	W. Cassidy	New Bedford, Mass	Apr. 12, 1870	101, 708
Pipe joint, Swing	J. R. Worswick and E. Lewis	Cleveland, Ohio	Mar. 4, 1873	136, 577
Pipe joints, Connection for lead	I. Smith	New York, N. Y	Apr. 26, 1870	102, 443
Pipe-joints, &c., Machinery for forging	J. A. Shipton and R. Mitchell	Wolverhampton, England	Oct. 16, 1866	58, 963
Pipe kiln, Clay	J. Dimelow	Philadelphia, Pa	Sept. 10, 1867	68, 716
Pipe-leak stopper	S. Moore	Sudbury, Mass	July 6, 1869	92, 338
Pipe machinery, Socket	F. M. Mattice	Cleveland, Ohio	June 7, 1870	104, 046
Pipe-mold	G. W. Cornell and B. B. Quinn	New York, N. Y	Mar. 6, 1866	52, 971
Pipe-mold	E. Walsh	Rochester, N. Y	Apr. 30, 1872	126, 354
Pipe, Mold for casting	H. M. Bird	Cambridgeport, Mass	Apr. 16, 1867	63, 781
Pipe, Mold for casting	J. Edson	Boston, Mass	Mar. 23, 1869	88, 020
Pipe, Mold for casting	J. Edson	Boston, Mass	Apr. 12, 1870	101, 842
Pipe, Mold for casting	S. Fulton	Conshohocken, Pa	Apr. 16, 1861	32, 078
Pipe, Molding	W. D. Alford	Cuyahoga Falls, Ohio	Sept. 26, 1871	119, 259
Pipe, Molding	J. Aston	Pittsburgh, Pa	Sept. 15, 1868	82, 065
Pipe, Molding	J. Firth and J. Ingham	Phillipsburgh, N. Y	Dec. 20, 1859	26, 486
Pipe, Molding	G. H. Garrett	Saint Louis, Mo	Dec. 29, 1868	85, 440
Pipe, Molding	J. D. Johnson	Hainesport, N. Y	Dec. 3, 1872	133, 584
Pipe, Molding	A. W. A. Logen	Rochdale, England	July 22, 1873	141, 150
Pipe, Molding	F. Shickle	Saint Louis, Mo	Sept. 28, 1869	95, 388
Pipe, Molding	W. Smith	Pittsburgh, Pa	Oct. 25, 1870	108, 647
Pipe, Molding and casting	B. S. Benson	Baltimore, Md	Feb. 17, 1863	37, 670
Pipe, Molding and casting	I. M. Kenworthy	Carbondale, Pa	July 2, 1872	128, 493
Pipe, Molding and casting	L. Martaresche	Pittsburgh, Pa	Oct. 26, 1869	96, 131
Pipe molding and casting device	J. McClelland	Washington, D. C	June 21, 1870	104, 479
Pipe-molding apparatus	F. Fuller	Providence, R. I	Apr. 12, 1870	101, 853
Pipe-molding apparatus	F. Shickle	Saint Louis, Mo	May 25, 1869	90, 591
Pipe-molding machine	B. S. Benson	Baltimore, Md	July 30, 1861	32, 956
Pipe-molding machine	B. S. Benson	Baltimore, Md	Sept. 3, 1861	33, 178
Pipe-molding machine	B. S. Benson	Baltimore, Md	Oct. 13, 1868	83, 028
Pipe-molding machine	B. S. Benson	Baltimore, Md	June 20, 1871	116, 141
Pipe-molding machine	J. T. Glass and L. E. Morrell	Columbus, Ohio	July 16, 1872	129, 547
Pipe-molding machine	D. Long	Louisville, Ky	Dec. 3, 1872	133, 652
Pipe-molding machine	G. Richardson	Milwaukee, Wis	Mar. 8, 1871	112, 495
Pipe molding machine	W. Smith	Pittsburgh, Pa	Oct. 15, 1867	69, 854
Pipe-molding machine	W. Smith	Allegheny City, Pa	Nov. 3, 1868	83, 668
Pipe-molding pattern	G. Ross	Newport, Ky	Mar. 24, 1863	37, 981
Pipe-nippers	G. Smith	New York, N. Y	Dec. 20, 1859	26, 530
Pipe or car coupling	A. G. Story	Waterloo, N. Y	May 13, 1873	138, 822
Pipe or grate boiler and heater	S. Bolton	Philadelphia, Pa	July 7, 1809	....
Pipe or smoking-apparatus	E. S. Franks	Boston, Mass	Sept. 27, 1864	44, 414
Pipe patterns, Mold for plaster	N. P. Bowler	Cleveland, Ohio	Sept. 16, 1873	142, 765
Pipe reaming and squaring tool	W. Barry	Chicago, Ill	June 15, 1869	91, 201
Pipe-stem	J. Davis	Saint Paul, Minn	June 17, 1873	139, 880
Pipe-stem	S. S. Scranton	New Haven, Conn	Apr. 14, 1868	76, 827

Index of patents issued from the United States Patent Office from 1790 *to* 1873, *inclusive*—Continued.

Invention.	Inventor.	Residence.	Date.	No.
Pipe-stem-bending machine	J. Harvey	Philadelphia, Pa	May 20, 1873	139, 148
Pipe-tips, Machine for milling	A. R. Weiss	Brooklyn, N. Y	May 13, 1873	138, 825
Pipe-trap	C. H. Burleigh	Worcester, Mass	Nov. 23, 1869	97, 160
Pipe-trap, Cement	A. A. Lovell	Worcester, Mass	Nov. 4, 1873	144, 277
Pipe-welding furnace	J. Fieldhouse	Taunton, Mass	Aug. 7, 1866	56, 919
Pipe-wrench	W. C. Abbe	Petroleum Centre, Pa	Aug. 20, 1867	67, 937
Pipe-wrench	L. Anderson	Washington, D. C	Mar. 19, 1872	124, 654
Pipe-wrench	R. Bain	Brooklyn, N. Y	Apr. 9, 1867	63, 689
Pipe-wrench	O. G. Barrett	Boston Highlands, Mass	Jan. 31, 1871	111, 422
Pipe-wrench	W. H. Barwick	Montreal, Canada	June 6, 1871	115, 678
Pipe-wrench	W. F. Beecher	Chicago, Ill	May 1, 1860	28, 053
Pipe-wrench	J. R. Brown	Boston, Mass	Oct. 18, 1864	44, 703
Pipe-wrench	J. R. Brown	Boston, Mass	May 28, 1867	65, 162
Pipe-wrench	W. H. Burton	Selma, Ala	Feb. 11, 1873	135, 773
Pipe-wrench	A. Call	Springfield, Mass	Aug. 28, 1866	57, 621
Pipe-wrench	J. W. Close	Buffalo, N. Y	Feb. 5, 1867	61, 715
Pipe-wrench	D. W. Coggeshall	Providence, R. I	June 12, 1866	55, 578
Pipe-wrench	J. Craig	Detroit, Mich	Sept. 10, 1872	131, 282
Pipe-wrench	I. H. Doolittle	Ansonia, Conn	Mar. 13, 1860	27, 622
Pipe-wrench	W. H. Downing	Pioneer, Pa	Aug. 17, 1869	93, 688
Pipe-wrench	D. E. Eaton	Boston, Mass	Nov. 5, 1872	132, 819
Pipe-wrench	H. V. Faries	Indianapolis, Ind	Aug. 26, 1862	36, 289
Pipe wrench	D. Frank and T. Snyder	Allentown, Pa	Oct. 10, 1871	119, 838
Pipe-wrench	M. H. Freeman	Somerville, Mass	May 14, 1867	64, 656
Pipe-wrench	A. Gauntt	Chagrin Falls, Ohio	Oct. 29, 1872	132, 571
Pipe-wrench	I. S. Hamilton	Hamilton, Ohio	Nov. 7, 1871	120, 641
Pipe-wrench	M. Hastings	Brooklyn, N. Y	Aug. 24, 1869	94, 107
Pipe-wrench	H. A. Hyle	Shamburgh, Pa	Nov. 28, 1871	121, 372
Pipe-wrench	D. D. Ingram	Manistee, Mich	Oct. 21, 1873	143, 908
Pipe-wrench	J. C. Johnson	Oil City, Pa	Apr. 26, 1864	42, 489
Pipe wrench	R. H. Lecky	Allegheny City, Pa	June 30, 1868	79, 361
Pipe-wrench	N. F. Loi	New York, N. Y	May 26, 1868	78, 299
Pipe-wrench	G. P. Moloney	Detroit, Mich	May 20, 1873	139, 177
Pipe-wrench	E. S. Moulton	Chelsea, Mass	Feb. 18, 1873	136, 089
Pipe-wrench	C. Neames	New Orleans, La	Jan. 9, 1872	122, 638
Pipe-wrench	J. L. Ordner	Cleveland, Ohio	May 28, 1867	65, 111
Pipe-wrench	W. Pearson	Windsor's Locks, Conn	Aug. 23, 1864	43, 926
Pipe-wrench	C. Pomeroy	Mattoon, Ill	Nov. 16, 1869	96, 957
Pipe-wrench	H. F. Read	Brooklyn, N. Y	July 22, 1862	35, 950
Pipe-wrench	J. Rigg	Cleveland, Ohio	June 3, 1873	139, 520
Pipe-wrench	E. H. Robbins	Pittsfield, Mass	May 2, 1871	114, 318
Pipe-wrench	A. Secor	North Adams, Mass	Dec. 10, 1872	133, 804
Pipe wrench	D. C. Stillson	Charlestown, Mass	Apr. 30, 1872	126, 161
Pipe-wrench	T. Stone	Carbondale, Ill	Dec. 10, 1872	133, 902
Pipe-wrench	G. Warsop	Nottingham, England	July 1, 1873	140, 561
Pipe-wrench	W. C. Westerfield	Fairbury, Ill	May 7, 1872	126, 576
Pipe-wrench	H. B. Wheatcroft	New York, N. Y	Dec. 2, 1873	145, 263
Pipe-wrench	J. A. Wilcox	Rocky Hill, Conn	Sept. 3, 1861	33, 223
Pipe-wrench	J. A. Wilcox	Rocky Hill, Conn	June 28, 1870	104, 913
Pipe-wrench	T. Williams	Providence, R. I	Feb. 25, 1873	136, 350
Pipe-wrench	W. W. Wills	Janesville, Wis	May 29, 1866	55, 190
Pipe-wrench	H. Wilson	Tarr Farm, Pa	Sept. 12, 1871	118, 997
Pipe wrench and cutter	J. L. Brierly	Auburn, Mass	May 19, 1868	78, 047
Pipe wrench and cutter	R. A. Copeland	Baltimore, Md	Jan. 26, 1869	86, 138
Pipe wrench and cutter	A. Robes and J. C. Chapman	Somerville and Cambridgeport, Mass.	Aug. 31, 1869	94, 241
Pipe-wrench attachment for monkey-wrench	A. H. Woodruff	Lansing, Iowa	Oct. 31, 1871	120, 475
Pipe-wrench, Self-adjusting	T. Williams	Providence, R. I	Feb. 25, 1873	136, 351
Pipes across rivers, Mode of laying	J. F. Ward	Jersey City, N. J	Jan. 31, 1871	111, 498
Pipes and hose, Apparatus for transporting, extending, and elevating.	L. Day	Buffalo, N. Y	Oct. 27, 1868	83, 475
Pipes and manifolds, Arrangement of conducting	C. C. Walworth	Boston, Mass	May 5, 1863	38, 433
Pipes and tubes while being screw-threaded, Chuck for holding.	W. T. Cole	New York, N. Y	Aug. 23, 1870	106, 551
Pipes, Construction and joining of	T. L. Truss	Darlington, England	Aug. 21, 1860	29, 731
Pipes, Cutting threads on	J. M. Carpenter	Florence, Mass	Mar. 9, 1869	87, 632
Pipes, Device for cutting off	W. H. Barwick and W. T. Farre	Montreal, Canada	Mar. 19, 1872	124, 659
Pipes, Elastic water-stop for checking the force or momentum of water in.	H. Allen	New York, N. Y	Oct. 25, 1843	3, 315
Pipes from percussion, Regulator to protect	E. A. Chameroy	Paris, France	May 7, 1872	126, 444
Pipes, Hydrostatic apparatus for testing	T. J. McGowan	Cincinnati, Ohio	Sept. 17, 1872	131, 405
Pipes, joints, bottles, casks, &c., Making and coating.	B. Rhodes	Bow, England	Jan. 19, 1864	41, 351
Pipes, Lug and link for connecting	C. Warner	Louisville, Ky	May 8, 1849	6, 428
Pipes, Machine for screwing together	R. T. Crane	Chicago, Ill	Feb. 6, 1872	123, 382
Pipes, Machinery for riveting	J. Ball	New York, N. Y	Dec. 20, 1845	4, 319
Pipes, Manufacture of	G. L. Eagan	San Francisco, Cal	Sept. 2, 1873	142, 448
Pipes, Method and apparatus for casting	A. T. Brodie, R. R. Smith, and J. J. Tyler.	Pittsburgh, Pa	Nov. 21, 1871	121, 151
Pipes of plastic materials, Mold for molding	J. Chilver	Jersey City, N. J	July 9, 1861	32, 754
Pipes or tubes, Casting	W. Badger	Boston, Mass	Dec. 26, 1808	....
Pipes, Stool for casting	C. J. Ellis	Louisville, Ky	Apr. 23, 1872	126, 043
Pipes, tiles, sidewalks, &c., Compound for	W. A. Battersby	Williamsburgh, N. Y	Nov. 8, 1870	108, 957
Pipes, Tool for cutting off	H. Collinson	Boston, Mass	Apr. 9, 1872	125, 380
Pipes, tubes, barrels, &c., Process and apparatus for manufacturing composition.	J. K. Mayo	Williamsburgh, N. Y	July 6, 1869	92, 332
Pipes, tubes, &c., Mode of making	J. Putman	Salem, Mass	Jan. 17, 1827	....
Pipes, tubes, &c., with silver or other metal by the electro-depositing process, Coating the interior of.	D. D. Parmelee	New York, N. Y	Oct. 18, 1870	108, 510
Pipes with tin, Coating	T. Ewbank	New York, N. Y	May 16, 1832	....
Piping, Machine for making	J. Ayers	Chicago, Ill	May 21, 1872	126, 864
Pisciculture	W. H. Furman	Maspeth, N. Y	June 16, 1868	78, 952
Pistol	W. S. Butler	Rockey Hill, Conn	Feb. 3, 1857	16, 571
Pistol	B. and B. M. Darling	Bellingham, Mass	Apr. 23, 1836	....
Pistol, Air	R. Brooks, jr	Rockport, Mass	Feb. 15, 1870	99, 754
Pistol, Air	G. H. Snow and E. H. Hawley	New Haven, Conn., and Kalamazoo, Mich.	Sept. 12, 1871	118, 886
Pistol and dirk, Combined	F. Wesson	Worcester, Mass	July 20, 1869	92, 918

Index of patents issued from the United States Patent Office from 1790 *to* 1873, *inclusive*—Continued.

Invention.	Inventor.	Residence.	Date.	No.
Pistol and other fire-arms	E. Allen	Norwich, Conn	Apr. 16, 1845	3 998
Pistol and pocket-knife, Combined	A. J. Beavey	South Montville, Me	Mar. 27, 1866	53, 473
Pistol and sword, Combined	C. E. Billings	Springfield, Mass	Sept. 22, 1868	82, 279
Pistol and sword, Combined	D. A. Courter	Beloit, Wis	Mar. 11, 1862	34, 625
Pistol, Breech-loading	S. M. Perry	Brooklyn, N. Y	June 21, 1864	43, 259
Pistol, Breech-loading	S. M. Perry	Brooklyn, N. Y	June 21, 1864	43, 260
Pistol, Breech-loading	S. M. Perry and E. Goddard	Plainfield, N. J., and Brooklyn, N. Y.	Apr. 26, 1870	102, 429
Pistol, Breech-loading	D. Werner	Saint Louis, Mo	Oct. 6, 1868	82, 908
Pistol-breech pins, Device for moving and holding	W. W. Marston	New York, N. Y	June 18, 1850	7, 443
Pistol, Burglar-alarm	J. G. Clark	Augusta, Ga	June 7, 1859	24, 349
Pistol-frames, Die for forging and shaping	S. P. Legg	Springfield, Mass	Dec. 5, 1865	51, 327
Pistol-frames, Die for swaging	C. E. Billings	Windsor, Vt	Aug. 7, 1866	56, 885
Pistol-gallery, Portable	F. A. Spofford and M. G. Raffington.	Columbus, Ohio	Feb. 12, 1867	61, 960
Pistol-gallery, Portable	F. A. Spofford and M. G. Raffington.	Columbus, Ohio	Feb. 4, 1868	74, 163
			Apr. 21, 1863	28, 213
Pistol-holster	A. A. Bennett	Cincinnati, Ohio		
Pistol knife or cutlass	G. Elgin	New York, N. Y	July 5, 1837	254
Pistol, Parlor air	B. Haviland and G. P. Gunn	Hudson and Ilion, N. Y	Apr. 18, 1871	113, 766
Pistol, Pocket	V. M. Wallace	West Topham, Vt	Aug. 17, 1835	....
Pistol, Repeating	G. Tigneres	Covington, La	Dec. 20, 1859	29, 538
Pistol, Revolving-breech	J. Stevens	Chicopee, Mass	Oct. 7, 1851	8, 412
Pistol-saber	R. B. Lawton	Newport, R. I	Nov. 23, 1837	481
Pistol safety-guard	R. D. Hay	Crooked Creek, N. C	Nov. 22, 1870	109, 513
Pistols, Base-pin and rammer of revolving	W. H. Elliot	Plattsburgh, N. Y	Dec. 17, 1861	33, 932
Pistols, Mode of attaching knives or dirks to	R. W. Andrews	Stafford, Conn	July 31, 1837	328
Pistols, Tool for manufacturing	A. Rebetey	Norwich, Conn	May 10, 1859	23, 990
Piston	C. B. Allen	Philadelphia, Pa	Sept. 9, 1873	142, 659
Piston	P. A. Brown	Philadelphia, Pa	Oct. 9, 1816	....
Piston	I. D. Buck	Conshohocken, Pa	Oct. 25, 1870	108, 682
Piston	S. E. and I. Y. Chubbuck	Roxbury, Mass	Feb. 25, 1868	74, 750
Piston	W. H. Holland	Boston, Mass	Nov. 21, 1871	121, 103
Piston	H. S. Hopkins	Boston, Mass	May 7, 1872	126, 548
Piston	L. Huntoon	Milford, Mass	July 17, 1866	56, 420
Piston	A. H. Smith	New York, N. Y	Jan. 10, 1871	110, 874
Piston	W. Swayne	United States Army	Mar. 31, 1868	76, 268
Piston	J. Wheelock	Worcester, Mass	Mar. 17, 1868	75, 714
Piston and packing	A. K. Rider	New York, N. Y	Sept. 30, 1873	143, 255
Piston and piston-packing	O. Collier	Sacramento, Cal	Dec. 28, 1869	98, 232
Piston and piston-packing	G. H. Reynolds	New York, N. Y	May 16, 1871	114, 969
Piston and piston-packing	E. Sullivan	Mount Washington, Pa	Sept. 26, 1871	119, 247
Piston and piston-rod connections	A. P. Samuel	New York, N. Y	Mar. 23, 1858	19, 722
Piston and piston-valve, Steam-engine	T. S. Davis	Jersey City, N. J	Feb. 5, 1861	31, 303
Piston and slide-valve, Horizontal steam-engine	B. H. Brown	Philadelphia, Pa	Sept. 1, 1843	3, 241
Piston and valve-cock	J. Lacey	Philadelphia, Pa	May 6, 1823	....
Piston, Balancing	L. Findlay	Saint Louis, Mo	Aug. 22, 1871	118, 227
Piston-blowers	C. W. Isbell	New York, N. Y	Apr. 8, 1873	137, 553
Piston-connection	H. F. Pease	Brooklyn, N. Y	May 3, 1870	102, 703
Piston, Engine	T. T. Dwelley	Charlestown, Mass	May 12, 1863	38, 474
Piston, Engine	M. Hunt	Salem, Ohio	Sept. 17, 1867	68, 989
Piston, Engine	W. Keemle	Philadelphia, Pa	June 11, 1867	65, 753
Piston-engine and stamping-machine, Spring and weight.	E. F. and J. McFarland	Worcester, Mass	Dec. 26, 1865	51, 737
Piston engine, Double	A. W. Morrell	Niles, Mich	Sept. 20, 1870	107, 524
Piston engine, Oscillating	J. B. and S. M. Davis	Harrisonville, Mo	Nov. 29, 1870	109, 721
Piston engine, Vibrating	J. B. Root	New York, N. Y	Apr. 25, 1865	47, 459
Piston-facing machine	W. W. Loweree and G. A. Sanderson.	Albany, N. Y	July 4, 1871	116, 730
Pistons for steam-engines, Arrangement of passages and valves for cushioning.	N. W. Wheeler	New York, N. Y	Jan. 26, 1858	19, 220
Piston head and valve	G. Weinman	Columbia, Ohio	Jan. 28, 1870	104, 909
Piston-head-centering device	O. S. Howard	Bangor, Me	July 29, 1873	141, 221
Piston head, Steam-engine	O. S. Howard	Bangor, Me	Nov. 7, 1871	120, 586
Piston head, Steam-engine	E. Sullivan	Pittsburgh, Pa	June 5, 1866	55, 387
Piston, Hydraulic	T. Critchlow	Baldwin, Pa	Oct. 22, 1872	132, 445
Piston of pneumatic spring for railway-car, &c	A. Connison	Belleville, N. J	Dec. 23, 1841	2, 395
Piston, Printing-press	C. B. Cottrell	Westerly, R. I	Dec. 12, 1871	121, 759
Piston-ring and method of deriving motion therefrom in rotary engines	J. Tremper	Little Britain, N. Y	Apr. 17, 1849	6, 359
Piston-rod adjuster	D. Bly	Macon, Ga	May 19, 1868	78, 045
Piston-rod, Compound	J. R. St. John	Cleveland, Ohio	Apr. 27, 1841	2, 070
Piston-rod guide	E. Dunscomb	Boston, Mass	Apr. 18, 1865	47, 283
Piston rod, Steam-engine	J. F. Carl	Pleasantville, Pa	May 21, 1872	127, 025
Piston-rods, &c., Bushing for	G. W. Welch	Fort Wayne, Ind	Jan. 14, 1873	134, 955
Piston-rods for steam and other powers, Connecting	E. P. Curtis	Madison, Wis	Mar. 12, 1867	62, 825
Piston-rods, Machine for	W. B. Davis	Brooklyn, N. Y	Nov. 21, 1865	51, 027
Piston-rods with pistons of steam-cylinders, Mode of connecting.	M. W. Baldwin	Philadelphia, Pa	Dec. 17, 1840	1, 904
Piston, Rotative	J. M. Cooper	Guildhall, Vt	July 16, 1827	....
Piston-spring	W. R. Brown	Bath, N. Y	Aug. 3, 1869	93, 273
Piston, Steam	E. L. Horton	Hartford, Conn	Sept. 5, 1838	905
Piston, Steam	E. Merriman	Allegheny, Pa	July 11, 1871	116, 853
Piston, Steam	J. Richards	New York	Oct. 3, 1846	4, 781
Piston, Steam-cylinder	M. B. Mason	Aurora, Ind	Sept. 4, 1866	57, 823
Piston, Steam-engine	L. B. Batcheller	Rochester, N. Y	Apr. 30, 1861	32, 174
Piston, Steam-engine	C. Bollinger	Harrisburgh, Pa	May 26, 1868	78, 179
Piston, Steam-engine	G. Bower and J. Qualfer	Ashton-under-Lyne and Barnsley, England.	Aug. 28, 1866	57, 648
Piston, Steam-engine	S. G. Cabell	Quincy, Ill	Aug. 7, 1866	56, 899
Piston, Steam-engine	H. D. Dunbar	Memphis, Tenn	Aug. 14, 1860	29, 576
Piston, Steam-engine	H. D. Dunbar	Hartland, Vt	June 16, 1863	38, 890
Piston, Steam-engine	H. D. Dunbar	Hartland Four Corners, Vt	Jan. 12, 1864	41, 205
Piston, Steam-engine	L. B. Flanders	Philadelphia, Pa	Feb. 5, 1867	61, 728
Piston, Steam-engine	F. H. Furniss and J. Hovey	Cleveland, Ohio	Nov. 10, 1863	40, 559
Piston, Steam-engine	J. O. Haight	Albany, N. Y	Apr. 9, 1861	31, 971
Piston, Steam-engine	J. and H. Hall, sr	Philadelphia, Pa	Mar. 15, 1864	41, 961
Piston, Steam-engine	B. F. Hidden	Norwich, Conn	Aug. 4, 1863	39, 397
Piston, Steam-engine	J. Hinds and G. H. Lodge	Salem and Swampscott, Mass.	Apr. 7, 1868	76, 448

Index of patents issued from the United States Patent Office from 1790 *to* 1873, *inclusive*—Continued.

Invention.	Inventor.	Residence.	Date.	No.
Piston, Steam-engine	T. S. La France	Elmira, N. Y	Oct. 9, 1860	30, 328
Piston, Steam-engine	H. N. J. Mansfield	Malone, N. Y	Oct. 20, 1868	83, 188
Piston, Steam-engine	A. Nadon	Springfield, Mass	Jan. 21, 1868	73, 456
Piston, Steam-engine	W. Ord	Brooklyn, Ohio	Apr. 27, 1869	89, 428
Piston, Steam-engine	J. K. Robinson and J. M. Clark	Bellaire, Ohio	Aug. 14, 1860	29, 621
Piston, Steam-engine	J. K. Robinson and J. M. Clark	Bellaire, Ohio	Oct. 9, 1860	30, 351
Piston, Steam-engine	F. J. Roth and D. R. Gamble	Newark, Ohio	Dec. 19, 1865	51, 622
Piston, Steam-engine	A. J. Scoville and A. H. De Clercq.	Bloomington, Ill	Feb. 25, 1862	34, 552
Piston, Steam-engine	S. M. Seley and P. Hopkins	Peoria, Ill	June 24, 1862	35, 711
Piston, Steam-engine	J. Y. Smith	Pittsburgh, Pa	Apr. 23, 1867	64, 043
Piston, Steam-engine	W. G. Snook	Corning, N. Y	June 1, 1869	90, 884
Piston, Steam-engine	A. M. Sprague	Mobile, Ala	May 9, 1854	10, 898
Piston, Steam-engine	N. P. Stevens	Boston, Mass	Aug. 4, 1863	39, 431
Piston, Steam-engine	N. P. Stevens	Boston, Mass	Oct. 11, 1864	44, 674
Piston, Steam-engine	N. P. Stevens	Boston, Mass	July 18, 1865	48, 851
Piston, Steam-engine	J. Wheelock	Worcester, Mass	Apr. 23, 1867	64, 052
Piston, Steam-engine	W. D. Whitmore	Boston, Mass	Jan. 22, 1867	61, 373
Piston, Steam-engine	R. Winans	Baltimore, Md	Apr. 6, 1858	19, 888
Piston, Steam-engine	R. Witty	Chicago, Ill	Aug. 31, 1869	94, 371
Piston, Steam-engine	Wright and Ketchum	Calhoun County, Mont	June 19, 1835	....
Piston, Steam-engine and pump	J. A. Huss	Bowling Green, Ky	June 13, 1871	115, 858
Piston, Steam-engine metal-packed	G. W. Cotton	Saint Louis, Mo	May 5, 1857	17, 200
Piston, Steam-engine, pump, &c., Metallic	J. Swainson	New York, N. Y	Mar. 30, 1837	159
Pistons and pitmen of steam-engines, Mode of connecting.	R. S. Edwards	Savannah, Mo	Feb. 13, 1872	123, 623
Pistons, Arrangement for lubricating	B. Garvey	New York, N. Y	July 17, 1860	29, 216
Pistons, cylinders, and safety-valves, Steam apparatus substituted for.	A. Patterson	Rush, Pa	Feb. 26, 1839	1, 091
Pistons, Lubricating rotary steam-engine	S. W. Wicks	Seneca Falls, N. Y	Aug. 12, 1873	141, 845
Pistons, Lubricating steam-engine	S. Hall	Basford, England	Oct. 30, 1834	....
Pistons, Method of expanding metallic	J. Touchstone and J. H. Clark	Philadelphia, Pa	Apr. 17, 1849	6, 318
Pistons, Mode of cushioning steam	G. F. Blake	Boston, Mass	Mar. 29, 1870	101, 217
Pistons to piston-rods, Device for fastening	A. T. Norgan	Palo Alto, Pa	Dec. 14, 1869	97, 797
Pitch-board	J. R. Drew	San Francisco, Cal	Nov. 30, 1869	97, 281
Pitch, Composition	H. Ruggles	New York, N. Y	Dec. 2, 1834	....
Pitch, Composition	T. A. Sherman	Scriba, N. Y	Mar. 4, 1836	....
Pitch, Making	H. Ruggles	New York, N. Y	Mar. 19, 1836	....
Pitch, Manufacture of	A. H. Emery	New York, N. Y	Aug. 8, 1865	49, 248
Pitcher	J. H. Hobbs	Wheeling, W. Va	May 11, 1869	89, 996
Pitcher	N. T. Whiting	Lawrence, Mass	Aug. 7, 1866	57, 025
Pitcher and coffee-pot, Revolving ice	J. P. Adams	New York, N. Y	Aug. 10, 1869	93, 575
Pitcher, Beer	W. P. Eayrs	Nashua, N. H	May 29, 1866	55, 076
Pitcher, Beer	L. Hermance	Saratoga Springs, N. Y	Apr. 17, 1860	27, 904
Pitcher, Beer	W. S. Mathews	Meriden, Conn	Nov. 15, 1859	26, 115
Pitcher, Beer	O. F. Pelton	Middletown, Conn	June 19, 1860	28, 771
Pitcher, Beer	D. W. S. Rawson	Peru, Ill	Feb. 2, 1864	41, 447
Pitcher, coffee-pot, &c., Device for tilting	J. Gibson, jr	Albany, N. Y	Dec. 28, 1869	98, 245
Pitcher-cover	W. Bradley	Providence, R. I	Jan. 7, 1873	134, 586
Pitcher, Double-walled	G. Jepson	Chelsea, Mass	Nov. 10, 1868	83, 969
Pitcher, Enameled metallic ice	C. C. Foote	West Meriden, Conn	June 30, 1868	79, 335
Pitcher for cooling liquids	H. Pietsch	New York, N. Y	July 27, 1869	93, 001
Pitcher-holder	B. T. Currier	Boston, Mass	Nov. 5, 1872	132, 808
Pitcher, Ice	J. B. Bailey	New York, N. Y	Apr. 7, 1868	76, 378
Pitcher, Ice	G. H. Bechtel	Philadelphia, Pa	Apr. 23, 1867	63, 986
Pitcher, Ice	W. Bellamy	Newark, N. J	July 16, 1867	66, 779
Pitcher, Ice	W. Bellamy	Newark, N. J	Feb. 11, 1868	74, 285
Pitcher, Ice	W. Bellamy	Newark, N. J	Sept. 7, 1869	94, 550
Pitcher, Ice	C. Conradt	Philadelphia, Pa	Sept. 5, 1865	49, 734
Pitcher, Ice	S. Curtis, jr	New York, N. Y	Jan. 21, 1862	34, 228
Pitcher, Ice	J. Dawson	New York, N. Y	Jan. 9, 1872	122, 576
Pitcher, Ice	C. Dickinson and W. Bellamy	Newark, N. J	Apr. 5, 1859	23, 455
Pitcher, Ice	E. A. Dodge	Washington, D. C	Dec. 10, 1872	133, 837
Pitcher, Ice	S. Eakins	Philadelphia, Pa	June 26, 1855	13, 125
Pitcher, Ice	C. Englebert and J. S. Von Nieda.	Philadelphia, Pa	Dec. 5, 1871	121, 500
Pitcher, Ice	G. A. Eno	Philadelphia, Pa	Feb. 12, 1867	61, 021
Pitcher, Ice	T. B. Fitts and A. D. Cook	New York, N. Y	Apr. 12, 1870	101, 850
Pitcher, Ice	N. F. Griswold	Meriden, Conn	May 1, 1860	28, 076
Pitcher, Ice	R. Holmes	Middletown, Conn	Apr. 14, 1868	76, 761
Pitcher, Ice	E. Kauffman	Philadelphia, Pa	Apr. 6, 1858	19, 855
Pitcher, Ice	G. Lane	New York, N. Y	Apr. 14, 1868	76, 638
Pitcher, Ice	N. Lawrence	Taunton, Mass	Nov. 5, 1867	70, 582
Pitcher, Ice	T. Leach	Taunton, Mass	Aug. 4, 1868	80, 748
Pitcher, Ice	E. B. Manning	Middletown, Conn	Feb. 5, 1867	61, 747
Pitcher, Ice	F. C. Meyer	Philadelphia, Pa	Aug. 8, 1865	49, 351
Pitcher, Ice	F. Miller	Wallingford, Conn	Aug. 29, 1871	118, 630
Pitcher, Ice	F. J. Miller	Brooklyn, N. Y	Apr. 21, 1868	76, 934
Pitcher, Ice	B. Morahan	Brooklyn, N. Y	Aug. 17, 1869	93, 734
Pitcher, Ice	E. A. Parker	West Meriden, Conn	July 9, 1872	128, 810
Pitcher, Ice	E. A. Parker	West Meriden, Conn	July 9, 1872	128, 811
Pitcher, Ice	E. A. Parker	West Meriden, Conn	May 6, 1873	138, 686
Pitcher, Ice	M. Simons	Middletown, Conn	June 18, 1867	65, 958
Pitcher, Ice	G. W. Smith	Hartford, Conn	June 15, 1858	20, 592
Pitcher, Ice	J. Stimpson	Baltimore, Md	Oct. 5, 1858	21, 717
Pitcher, Ice	J. H. Stimpson	Baltimore, Md	Mar. 8, 1859	23, 209
Pitcher, Ice	W. C. Wood	Washington, D. C	Nov. 24, 1868	84, 458
Pitcher-lid lifter	F. W. L. Knuschke	Providence, R. I	June 11, 1867	65, 757
Pitcher, Molasses	T. B. Atterbury	Pittsburgh, Pa	June 24, 1873	130, 236
Pitcher, Molasses	H. W. Goodrich	Boston, Mass	June 17, 1856	15, 128
Pitcher, Molasses	W. C. King	Pittsburgh, Pa	May 6, 1873	138, 501
Pitcher, Molasses	E. B. Manning	West Meriden, Conn	July 1, 1873	140, 515
Pitcher, Molasses, &c	E. Mingay	Boston, Mass	Nov. 17, 1857	18, 645
Pitcher, Molasses	E. Page	Worcester, Mass	Aug. 7, 1855	13, 400
Pitcher, Molasses	J. A. Parise and C. W. MacCord.	New York, N. Y	Oct. 24, 1865	50, 620
Pitcher, Molasses	C. Reistle	Brooklyn, N. Y	Feb. 23, 1869	87, 291
Pitcher, Molasses	D. C. Ripley	Pittsburgh, Pa	July 15, 1873	140, 793

Index of patents issued from the United States Patent Office from 1790 *to* 1873, *inclusive*—Continued.

Invention.	Inventor.	Residence.	Date.	No.
Pitcher, Molasses	M. Robbins	Cincinnati, Ohio	Aug. 7, 1866	56, 991
Pitcher, Molasses	H. Wright	Pittsburgh, Pa	June 10, 1873	139, 698
Pitcher, Refrigerating	F. D. Hall	Philadelphia, Pa	July 1, 1856	15, 231
Pitcher, Refrigerating	W. W. Lyman	West Meriden, Conn	June 8, 1858	20, 499
Pitcher-shield	F. B. Carleton	Cambridge, Vt	Mar. 1, 1870	100, 374
Pitcher, Sirup	H. Bullard	Middletown, Conn	Feb. 18, 1868	74, 663
Pitcher, Sirup	J. H. Hobbs	Wheeling, W. Va	Apr. 19, 1870	102, 124
Pitcher, Sirup	J. Hyslop and C. E. Phillips	Abington, Mass	July 2, 1867	66, 349
Pitcher, Sirup	D. Scanlin	Rochester, N. Y	Oct. 29, 1687	70, 367
Pitcher, Sirup	J. P. Whipple	Woonsocket, R. I	Apr. 13, 1869	88, 999
Pitcher spout and lid	D. Baker	Harwich, Mass	June 19, 1860	28, 726
Pitcher, Water-cooling	A. Hebbard	New York, N. Y	Oct. 3, 1857	18, 546
Pitchers and other vessels, Mode of constructing water	K. Goddard	Richmond, N. Y	Nov. 30, 1869	97, 390
Pitchers, Lining metal ice	L. W. Wilbur and A. Tyler	Fitchburgh, Mass	July 2, 1872	128, 690
Pitchers, Manufacture of double-walled ice	D. C. Wilcox	Meriden, Conn	Nov. 3, 1868	83, 747
Pitchers, Method of securing cover to glass	C. Ballinger	Pittsburgh, Pa	Mar. 2, 1869	87, 318
Pitchers, Silvering glass	J. W. Haines	Somerville, Mass	Apr. 4, 1865	47, 101
Pitchers, Top for molasses	G. M. Irwin	Birmingham, Pa	Nov. 7, 1871	120, 748
Pitchers, Top for molasses	G. M. Irwin	Birmingham, Pa	Nov. 26, 1872	133, 317
Pitchers, Top for molasses	G. P. Lang, jr	Allegheny, Pa	Feb. 18, 1873	136, 076
Pitchers, Top for molasses	G. P. Lang, jr	Allegheny City, Pa	Mar. 25, 1873	137, 139
Pitchers, Top for sirup	G. P. Lang and P. Lauster	Allegheny City, Pa	Oct. 31, 1871	120, 387
Pitchers, Top for sirup	T. Smith, jr	Boston, Mass	Nov. 16, 1869	96, 987
Pitchfork	J. Byington	Hinesborough, Vt	Jan. 12, 1813	....
Pitchfork	J. Herald and C. B. Tompkins	Trumansburgh, N. Y	Dec. 6, 1859	26, 354
Pitchfork	I. D. Peck	South Bristol, N. Y	Oct. 11, 1864	44, 652
Pitchfork	C. Perkins	Waterbury, Conn	Jan. 27, 1829	....
Pitchfork	R. Schuyler and W. Crowninshield	Seneca Falls, N. Y	Jan. 5, 1869	85, 699
Pitchfork	G. Van Sickle	Shorterville, N. Y	Jan. 4, 1870	98, 648
Pitchfork, Horse	W. H. Palmer and W. Crumb	Orleans, N. Y	Feb. 4, 1862	34, 317
Pitchfork, Horse	S. Raymond	Venice, N. Y	Nov. 11, 1862	36, 915
Pitchfork, Horse	S. Raymond	Genoa, N. Y	Nov. 24, 1863	40, 709
Pitman	H. Aufdembrinke	Cincinnati, Ohio	Mar. 11, 1873	136, 696
Pitman	E. C. Bacon	New York, N. Y	Dec. 29, 1868	85, 266
Pitman	A. Blaikie and W. Clark	Saint Clair, Mich	Feb. 26, 1856	14, 302
Pitman	J. Butler	Buffalo, N. Y	Nov. 26, 1867	71, 450
Pitman	J. H. Carothers	Lewisburgh, Pa	Sept. 9, 1873	142, 611
Pitman	M. D. Drake	Scituate, R. I	June 22, 1869	91, 613
Pitman	J. H. Holly and J. F. Robertson	Warwick, N. Y	Dec. 30, 1873	145, 947
Pitman	M. Grover	Clyde, Ohio	Jan. 19, 1869	85, 926
Pitman	F. Graham	Rockford, Ill	Jan. 20, 1863	37, 473
Pitman	G. W. Jason	Lodi, Ohio	Mar. 7, 1871	112, 463
Pitman	T. Kealy	Lewisville, Tex	May 17, 1870	103, 196
Pitman	J. H. McGowan	Cincinnati, Ohio	Nov. 8, 1870	109, 138
Pitman	C. H. Perkins	Providence, R. I	Nov. 8, 1870	109, 047
Pitman	J. Swan	Richmond, Ohio	Oct. 7, 1873	143, 475
Pitman	O. F. Thomas	Barnes's Corners, N. Y	Aug. 19, 1873	141, 902
Pitman	T. Welch	Churchville, N. Y	Dec. 17, 1867	72, 344
Pitman	S. Wheeler	Albany, N. Y	Dec. 17, 1872	134, 117
Pitman and cutter-bar connection	W. Green	Ashland, Ohio	Feb. 8, 1870	99, 563
Pitman and mode of attachment to band-wheel	L. N. Rouse	Covington, Ky	Feb. 6, 1872	123, 303
Pitman-box	A. S. Acker	Albion, N. Y	Aug. 14, 1866	57, 061
Pitman-connection	M. L. Ballard	Canton, Ohio	Oct. 30, 1866	59, 310
Pitman-connection	A. Ketchum	Estherville, Iowa	Dec. 12, 1871	121, 789
Pitman-connection	S. T. Lamb	Albany, Ind	Feb. 22, 1870	100, 157
Pitman-connection	W. K. Miller	Canton, Ohio	July 13, 1869	92, 628
Pitman-connection	S. Mahoney and J. Geller	Terre Haute, Ind	Dec. 10, 1872	133, 869
Pitman-connection	A. Neer	Bellefontaine, Ohio	Jan. 30, 1872	123, 285
Pitman-connection	J. A. Shephard	Portsmouth, Ohio	June 24, 1873	140, 312
Pitman-connection for operating cranks	C. W. Reed	Chagrin Falls, Ohio	Aug. 20, 1872	130, 596
Pitman-connection for steam-engines	S. Van Emon	Covington, Ky	Oct. 24, 1871	120, 222
Pitman-coupling	G. W. Clark	Manchester, Ind	Dec. 10, 1867	71, 853
Pitman-coupling	W. J. Keeney	Florence, Ind	Oct. 29, 1867	70, 224
Pitman-coupling	W. J. Keeney	Florence, Ind	Oct. 29, 1867	70, 225
Pitman for presses, punches, &c., Adjustable	N. C. Stiles	Meriden, Conn	July 24, 1866	56, 631
Pitman-head	J. L. Krester	Tustin, Wis	Sept. 27, 1870	107, 788
Pitman-head and crank-wrist connection	W. N. Whiteley	Springfield, Ohio	Nov. 5, 1867	70, 661
Pitman-head and wrist-pin	W. N. Whiteley	Springfield, Ohio	Oct. 22, 1867	70, 143
Pitman-heads, sickles of harvesters, &c., Joint for	I. Poffenberger	Champaign County, Ohio	Apr. 19, 1870	102, 151
Pitman, Reverse-lever	J. and W. W. Porter	Wauconda, Ill	Sept. 17, 1867	68, 900
Pitman-rod	E. S. Blake	Pittsburgh, Pa	July 4, 1871	116, 542
Pitman-rod	S. N. Wate, jr	Danville, Pa	Oct. 14, 1873	143, 648
Pitman-rod connection	E. G. Fish	Colfax, Iowa	Mar. 28, 1871	113, 039
Pitman-rod connection	E. G. Fish	Colfax, Iowa	Jan. 16, 1872	122, 715
Pitmen, Adjusting length of	N. C. Stiles	Middletown, Conn	Jan. 28, 1873	135, 380
Pitmen, Attaching	A. J. Sweeney	Wheeling, W. Va	Sept. 3, 1872	131, 132
Pitmen, Connecting	T. Hall	Northampton, Mass	Apr. 23, 1872	126, 051
Pitmen, &c., Hanging	E. H. Craige	Brooklyn, N. Y	Aug. 21, 1866	57, 292
Pitmen, Securing close joints in	A. Harroun	Perryville, N. Y	July 16, 1872	128, 959
Pitmen with shafts, Connecting	R. Cleaveland	Covington, Pa	Sept. 9, 1873	142, 616
Pivot-bearings	F. C. Lowthorp	Trenton, N. J	Jan. 3, 1860	26, 689
Pivot for seat	J. J. Wilson	New York, N. Y	Apr. 4, 1871	113, 606
Pivoted stump-joint	A. Searls	New York, N. Y	Feb. 11, 1868	74, 434
Placards, &c., Device for suspending and exhibiting	E. Worrell and B. C. Moore	Philadelphia, Pa	Dec. 16, 1873	145, 709
Plain and figured fabrics, Manufacture of	F. W. Norton	Lasswade, Great Britain	Sept. 13, 1853	10, 019
Plaiting-device	W. Walker	Brooklyn, N. Y	Nov. 1, 1870	108, 854
Plaiting-device	W. Walker	Brooklyn, N. Y	Nov. 1, 1870	108, 855
Plaiting-machine	J. F. McKenney	Baltimore, Md	Feb. 2, 1864	41, 441
Plaiting-machine	G. C. Nelson	Petersham, Mass	Oct. 27, 1863	40, 424
Plaiting-machine	J. E. Wheat	Rochester, N. Y	July 16, 1872	129, 104
Plaiting machine, Box	O. M. Chamberlain	New York, N. Y	Mar 4, 1873	136, 362
Plane	L. A. Alexander	Pittsfield, Mass	July 16, 1872	129, 508
Plane	V. Bitsch	Saint Louis, Mo	Sept. 15, 1868	82, 074
Plane	G. Buckel	Detroit, Mich	Aug. 25, 1868	81, 335
Plane	L. Bundy	Moore's Forks, N. Y	Nov. 15, 1870	109, 174
Plane	T. O. Callahan	Boston, Mass	Sept. 27, 1870	107, 757

Index of patents issued from the United States Patent Office from 1790 *to* 1873, *inclusive*—Continued.

Invention.	Inventor.	Residence.	Date.	No.
Plane	A. H. Comp	Mount Joy, Pa	Sept. 8, 1868	81, 879
Plane	A. S. Cross	Ripon, Wis	Sept. 10, 1861	33, 240
Plane	W. B. Glover	Boston, Mass	Oct. 25, 1870	108, 586
Plane	C. Jensen	Boston, Mass	May 14, 1872	126, 707
Plane	J. T. Jones	Philadelphia, Pa	May 14, 1836	....
Plane	E. H. Morris	Salem, Ohio	Nov. 8, 1870	109, 037
Plane	E. Odell	Winterset, Iowa	Jan. 19, 1864	41, 317
Plane	F. Smith and I. Carpenter	Lancaster, Pa	Aug. 25, 1868	81, 425
Plane	J. K. P. Smith	Jeffersonville, Ind	May 9, 1871	114, 613
Plane	J. C. Spencer	Phelps, N. Y	May 6, 1873	138, 591
Plane	E. G. Storke	Auburn, N. Y	Oct. 19, 1869	96, 052
Plane	J. Vendrand	Paris, France	June 24, 1862	35, 719
Plane, Accelerating	J. W. Newberry	Avon, N. Y	Nov. 15, 1831	....
Plane and manner of fastening the iron, Carpenter's	P. Meigs	Madison, Wis	Feb. 9, 1831	....
Plane, Bench	L. C. Ashley	Troy, N. Y	Mar. 18, 1856	14, 436
Plane, Bench	J. R. Bailey	Woonsocket, R. I	July 26, 1870	105, 767
Plane, Bench	J. R. Bailey	Woonsocket, R. I	Mar. 14, 1871	112, 675
Plane, Bench	L. Bailey	Boston, Mass	Aug. 6, 1867	67, 398
Plane, Bench	C. S. Beardsley and S. Wood	Auburn, N. Y	May 22, 1849	6, 459
Plane, Bench	B. A. Blandin	Charlestown, Mass	May 7, 1867	64, 477
Plane, Bench	L. C. Bliss	Richmond, Ind	Oct. 17, 1865	50, 530
Plane, Bench	J. Brooks	Boston, Mass	June 11, 1872	127, 842
Plane, Bench	W. H. Brown and D. F. Williams.	Worcester, Mass., and Woonsocket, R. I.	Nov. 11, 1873	144, 381
Plane, Bench	M. Chittenden	Danbury, Iowa	May 21, 1872	127, 026
Plane, Bench	P. S. Foster	Richmond, Me	Jan. 26, 1869	86, 295
Plane, Bench	C. H. Hardy	Boston, Mass	Sept. 24, 1872	131, 544
Plane, Bench	C. H. Hardy	Boston, Mass	Sept. 23, 1873	143, 072
Plane, Bench	W. C. Hopper	Pittsburgh, Pa	Jan. 16, 1855	12, 234
Plane, Bench	S. C. Howes	South Chatham, Mass	Feb. 17, 1863	37, 694
Plane, Bench	G. W. Huber and A. E. Flickinger.	Norwalk, Ohio	Dec. 2, 1873	145, 106
Plane, Bench	H. C. Hunt	Ottumwa, Iowa	Apr. 24, 1860	27, 983
Plane, Bench	H. L. Kendall	Baltimore, Md	June 8, 1858	20, 493
Plane, Bench	J. Lehner	Galena, Ill	Nov. 19, 1872	133, 162
Plane, Bench	W. S. Loughborough	Rochester, N. Y	May 10, 1859	23, 928
Plane, Bench	E. Mathew	Morgantown, Va	Mar. 4, 1856	14, 363
Plane, Bench	G. Müllcar	San Francisco, Cal	Oct. 10, 1865	50, 378
Plane, Bench	N. Palmer	Auburn, N. Y	May 14, 1867	64, 790
Plane, Bench	S. W. Palmer and E. G. Storke	Auburn, N. Y	Nov. 28, 1871	121, 406
Plane, Bench	C. H. Sawyer	Hollis, Me	Apr. 16, 1867	63, 948
Plane, Bench	W. H. Taber	Lowell, Mass	Feb. 28, 1865	46, 614
Plane, Bench	C. E. Torrance	Holyoke, Mass	Jan. 2, 1872	122, 339
Plane, Bench	L. D. Tredway	Saint Louis, Mo	Jan. 25, 1870	99, 275
Plane, Bench	T. Vaughan	Boston, Mass	July 23, 1872	129, 695
Plane, Bench	R. Washburn	Ramapo, N. Y	June 7, 1864	43, 053
Plane, Bench	S. Williams	Philadelphia, Pa	June 28, 1864	43, 360
Plane, Bench	T. D. Worrall	Lowell, Mass	June 23, 1857	17, 657
Plane, Beveling	H. W. Lewis	Bath, N. Y	Jan. 13, 1852	8, 655
Plane, Beveling	M. J. Wheeler, G. W. Rogers, H. W. Pierce, and M. B. Tidey.	Dundee, N. Y	July 4, 1854	11, 235
Plane-bit	H. Harris	Gorham, N. Y	Sept. 18, 1855	13, 575
Plane-bits, Device for securing	T. M. Richardson	Stockton, Me	Oct. 2, 1860	30, 248
Plane-bits, Machine for whetting	J. M. Gilstrap	Washington County, Ark	Aug. 11, 1857	17, 965
Plane-bits, Securing	T. D. Worrall	Boston, Mass	May 27, 1856	14, 979
Plane, Boot	A. V. Hill and R. Arnold	Hamburgh, N. Y	Apr. 11, 1848	5, 501
Plane, Carpenter's	L. Bailey	Boston, Mass	Dec. 24, 1867	72, 443
Plane, Carpenter's	J. R. and W. Brown	Boston, Mass	Feb. 28, 1871	112, 218
Plane, Carpenter's	F. M. Chapin and S. Rust	Pine Meadow, Conn	Mar. 31, 1868	76, 051
Plane, Carpenter's	O. R. Chaplin	Boston, Mass	May 7, 1872	126, 519
Plane, Carpenter's	H. N. Frederick	Hancock, N. Y	Sept. 19, 1871	119, 133
Plane, Carpenter's	H. A. Holt	Wilton, N. H	Jan. 9, 1872	122, 609
Plane, Carpenter's	C. G. Miller	Brattleborough, Vt.	June 28, 1870	104, 753
Plane, Carpenter's	W. Miller	Boston, Mass	Feb. 21, 1871	112, 062
Plane, Carpenter's	G. Müller	New York, N. Y	May 29, 1866	55, 207
Plane, Carpenter's	G. Müller	New York, N. Y	Oct. 26, 1869	96, 258
Plane, Carpenter's	O. Nichols	Lowell, Mass	Mar. 10, 1857	16, 805
Plane, Carpenter's	R. Phillips	Gardiner, Me	Aug. 13, 1867	67, 671
Plane, Carpenter's	R. Phillips	Boston, Mass	Aug. 30, 1870	106, 868
Plane, Carpenter's	R. Phillips	Boston, Mass	Oct. 24, 1871	120, 212
Plane, Carpenter's	Z. Phillips	Dixon, Ill	May 10, 1870	102, 966
Plane, Carpenter's	G. D. Spooner and L. N. Johnson	Rutland and Brandon, Vt.	Aug. 6, 1867	67, 458
Plane, Carpenter's	J. B. Tarr	Chicago, Ill	Sept. 22, 1868	82, 450
Plane, Carpenter's	M. B. Tidey	Ithaca, N. Y	Mar. 2, 1857	16, 889
Plane, Carpenter's	J. A. Traut	New Britain, Conn	Mar. 4, 1873	136, 469
Plane, Carpenter's	C. E. Tucker	Boston, Mass	July 26, 1870	105, 869
Plane, Carpenter's	R. Wellford and J. H. Deas	Philadelphia, Pa	May 14, 1821	....
Plane, Carpenter's	J. Woodville	Cincinnati, Ohio	Nov. 6, 1866	59, 498
Plane, Carpenter's	S. W. Woodward	Buffalo, N. Y	Apr. 27, 1869	89, 369
Plane, Carpenter's grooving	T. Duval	Hartford, Conn	Nov. 23, 1869	97, 177
Plane, Combination	A. Johnson	San Francisco, Cal	Dec. 30, 1873	146, 004
Plane, Cooper's crozing	H. and J. C. Taylor	Cincinnati, Ohio	Oct. 2, 1855	13, 626
Plane, Core-box	E. W. Lewis	Philadelphia, Pa	Dec. 4, 1866	60, 206
Plane, Crozing	S. G. Crane	Rochester, N. Y	Jan. 19, 1858	19, 130
Plane, Crozing and chamfering	A. M. Strattan	Ladoga, Ind	Aug. 12, 1873	141, 828
Plane, Dado	R. H. Dorn	Port Henry, N. Y	July 16, 1872	129, 010
Plane, Double-iron trying	W. B. Reynolds	Saint Clairsville, Ohio	July 7, 1832	....
Plane, Floor	C. E. Barlow	Philadelphia, Pa	Mar. 9, 1858	19, 539
Plane for bevel edges	W. W. Blye	De Ruyter, N. Y	Apr. 10, 1849	6, 304
Plane for boards, Bevel wheel	J. Newhall	Washington, Me	Jan. 25, 1831	....
Plane for crossing railways, Revolving	J. P. Fairlamb	New Castle, Del	July 7, 1830	....
Plane for cutting blind-slats	J. L. Bess and A. Hagny	Keokuk, Iowa	July 30, 1867	67, 157
Plane for cutting blind-slats	C. Kupfer	Madison, Wis	Sept. 1, 1868	81, 795
Plane for cutting blind-slats	R. E. Lowe	Upper Alton, Ill	Sept. 22, 1868	82, 421
Plane for cutting blind-slats	J. H. Miller	Oskaloosa, Iowa	Sept. 8, 1868	81, 929
Plane for finishing grooves in patterns	J. P. Robinson	Matteawan, N. Y	Dec. 18, 1855	13, 957
Plane for jointing table-leaves	T. P. Granger	Pocatonica, Ill	Jan. 12, 1864	41, 258
Plane for making blind-slats	E. K. Thomas and H. H. Andresen.	Rock Island, Ill., and Davenport, Iowa.	Apr. 10, 1866	53, 899

Index of patents issued from the United States Patent Office from 1790 *to* 1873, *inclusive*—Continued.

Invention.	Inventor.	Residence.	Date.	No.
Plane for scraping	J. Jones	Newark, N. J	Jan. 28, 1873	135, 341
Plane for shaving leather	F. Woodward	New Lisbon, N. Y	July 27, 1812	....
Plane for shaving whalebone	J. A. Sevey	Boston, Mass	Dec. 28, 1869	98, 305
Plane for tonguing and grooving boards	J. A. Woodbury	Boston, Mass	June 11, 1850	7, 432
Plane, Groove	E. W. Carpenter	Lancaster, Pa	Jan. 30, 1830	....
Plane, Groove	J. Longanecker and C. Myers	Lancaster County, Pa	Feb. 7, 1829	....
Plane, Grooving	J. Herman	Lancaster, Ohio	Aug. 27, 1835	....
Plane-guide	M. Garland	West Eau Claire, Wis	Aug. 30, 1870	106, 808
Plane-guide	J. Woodville	Cincinnati, Ohio	Mar. 23, 1869	88, 109
Plane, Hand	B. F. Bee	Harwich, Mass	Nov. 11, 1851	8, 503
Plane, Hand	S. S. Dodge	Sunapee, N. H	May 10, 1859	23, 978
Plane, Hand	B. Holly	Seneca Falls, N. Y	July 6, 1852	9, 094
Plane, Horizontal knife and roller	U. Emmons	New York	Mar. 6, 1832	....
Plane, Inclined circular	M. Isaacs	New York	Nov. 17, 1819	....
Plane-iron	I. Almy and S. A. Drake	Covert, N. Y	Apr. 25, 1871	114, 085
Plane-iron	W. H. Eckert	Syracuse, N. Y	Apr. 23, 1867	64, 001
Plane-iron	W. Hovey	Boston, Mass	Mar. 10, 1830	....
Plane-iron	S. Markee	Auburn, N. Y	Apr. 30, 1867	64, 341
Plane-iron	E. Quast	Jerseyville, Ill	Dec. 9, 1873	145, 311
Plane-iron	P. Williamson	Baltimore, Md	Sept. 9, 1825	....
Plane iron, Bench	I. H. A. Bleckman	Ronsdorf, Prussia	Nov. 6, 1855	13, 745
Plane iron, Carpenter's	A. M. Cross	Necedah, Wis	Dec. 3, 1872	133, 632
Plane-iron, Double	F. Beals	Pittsfield, Mass	Mar. 16, 1852	8, 796
Plane-iron roller	A. R. Reynolds	Auburn, N. Y	Nov. 9, 1869	96, 728
Plane-iron, scythe, &c	D. Pettibone	Philadelphia, Pa	May 6, 1813	....
Plane-iron sharpener	J. Turner	Cambridgeport, Mass	June 26, 1860	28, 946
Plane-irons and regulating the throats of planes, Adjusting the position of.	E. W. Carpenter	Lancaster, Pa	Mar. 27, 1849	[illegible] 226
Plane-irons, Cap for	C. N. Tuttle	Auburn, N. Y	Nov. 1, 1870	108, [illegible]46
Plane-irons, Device for adjusting	L. Bailey	Winchester, Mass	June 22, 1858	20, [illegible]15
Plane-irons, Fastening and adjusting	L. Bliss	Truxton, N. Y	June 16, 1846	4, 5[illegible]6
Plane-irons, Forming projections on caps of	N. B. Reynolds	Auburn, N. Y	June 11, 1867	65, 6[illegible]4
Plane-irons in their sockets, Securing and adjusting.	W. Stoddard	Lowell, Mass	June 23, 1857	17, 64[illegible]
Plane-irons in their stocks, Holding and adjusting.	W. W. Chipman	Lowell, Mass	June 23, 1857	17, 618
Plane-irons to the stock of bench-planes, Securing.	L. Bailey	Winchester, Mass	Aug. 31, 1858	21, 311
Plane-irons to their stocks, Securing	P. A. Gladwin	Boston, Mass	Feb. 16, 1858	19, 359
Plane, Joiner's	G. C. Beckwith	Boston, Mass	Jan. 25, 1870	99, 137
Plane, Joiner's	W. A. Cole	New York, N. Y	June 6, 1848	5, 629
Plane, Joiner's	A. Gray	Naples, Me	June 11, 1867	65, 562
Plane, Joiner's	B. I. Lane	Newburyport, Mass	May 12, 1857	17, 286
Plane, Joiner's	J. Lashbrooks	Owensborough, Ky	May 19, 1857	17, 332
Plane, Joiner's	J. F. Palmer	Auburn, N. Y	Feb. 3, 1857	16, 569
Plane, Joiner's	B. F. Shellabarger	Mifflintown, Pa	Mar. 28, 1848	5, 486
Plane, Joiner's	G. A. Warren	North Bridgewater, Mass	Feb. 14, 1871	111, 860
Plane, Joiner's	T. D. Worrall	Lowell, Mass	Aug. 4, 1857	17, 951
Plane, Joiner's beveling	T. A. Chandler	Rockport, Ill	Mar. 16, 1858	19, 620
Plane, Match	J. Edwards	New York, N. Y	June 10, 1873	139, 710
Plane, Match	C. E. Marshall	Boston, Mass	Oct. 8, 1872	131, 959
Plane, Match	C. G. Miller	New Britain, Conn	Aug. 19, 1873	142, 037
Plane, Metallic	J. A. Baines	New York, N. Y	Aug. 5, 1873	151, 535
Plane, Metallic	J. F. Baldwin	Boston, Mass	Nov. 25, 1873	144, 823
Plane, Miter	J. Sawyer	Moravia, N. Y	Dec. 4, 1866	60, 265
Plane, Molding	J. E. Donaldson	Montezuma, Ind	May 16, 1871	114, 779
Plane, Molding	C. Fleming	Ypsilanti, Mich	Sept. 11, 1860	29, 962
Plane, Molding	A. W. Maxwell	Milton, Pa	Feb. 9, 1869	86, 851
Plane, Molding	E. H. Morris	Canton, Ohio	Mar. 21, 1871	112, 949
Plane, Molding	A. S. Robertson	Boston, Mass	Jan. 21, 1873	135, 046
Plane, Multiform-molding	T. Worrall	Mount Holly, N. J	Aug. 29, 1854	11, 635
Plane or scraper, Box	A. F. Cushman	Hartford, Conn	Feb. 9, 1864	41, 571
Plane, Plow	I. White	Philadelphia, Pa	Jan. 9, 1834	....
Plane, Rabbet	G. M. Darley	Nebraska City, Nebr	July 2, 1872	128, 470
Plane, Rabbet	F. Smith	Boston, Mass	Sept. 23, 1873	143, 101
Plane, Revolving	L. Hedge	Brattleborough, Vt	June 28, 1836	....
Plane, Revolving timber	D. N. Smith	Warwick, Mass	Dec. 28, 1826	....
Plane, Saw rabbet	D. D. Whitker	Hudson, N. Y	Feb. 6, 1866	52, 478
Plane, Shoemaker's edge	F. Killbrith	Pembroke, Mass	July 13, 1858	20, 882
Plane, Splint	D. E. and A. A. Aiken	Adrian, Mich	May 5, 1868	77, 434
Plane, Splint	J. Dempsey	Richmond, Ind	Nov. 28, 1865	51, 153
Plane, Splint	P. N. Drake and D. Drummond	McGregor, Iowa	Sept. 27, 1870	107, 765
Plane, Splint	H. Ogborn	Richmond, Ind	Nov. 14, 1865	50, 947
Plane, Splint	H. L. Weagant	Morrisburg, Canada	Oct. 14, 1873	143, 737
Plane-stock	S. M. Adams	Fitchburgh, Mass	June 4, 1872	127, 541
Plane-stock	L. Bailey	New Britain, Conn	Mar. 28, 1871	113, 003
Plane-stock	G. F. Evans	Norway, Me	Jan. 28, 1862	34, 248
Plane-stock	G. F. Evans	Norway, Me	Mar. 22, 1864	41, 983
Plane-stock	J. Katz	Cincinnati, Ohio	Apr. 26, 1870	102, 406
Plane-stock	J. B. Thomas	Cincinnati, Ohio	Mar. 11, 1856	14, 423
Plane stock, Bench	J. Bryant	Brooklyn, N. Y	June 16, 1857	17, 553
Plane stock, Bench	G. E. Davis	Lowell, Mass	May 1, 1855	12, 787
Plane stock, Bench	J. Gorham	Bairdstown, Ga	Apr. 19, 1859	23, 678
Plane-stock, Cast-iron	W. Foster	Washington, D. C	Nov. 24, 1843	3, 355
Plane-stock, Cast-iron	H. Knowles	Colchester, Conn	Aug. 24, 1827	....
Plane-stocks and mouth-pieces, Hanging	M. G. Hubbard	New York, N. Y	July 3, 1855	13, 174
Plane-stocks, Hanging	M. G. Hubbard	New York, N. Y	Sept. 5, 1854	11, 648
Plane-stocks, Machine for cutting throats of carpenter's.	H. S. Dewey	Bethel, Vt	Mar. 31, 1857	16, 954
Plane-stocks, Machine for dressing throats in	F. B. Marble	Columbus, Ohio	Feb. 14, 1865	46, 372
Plane-stocks, Machine for mortising	J. Richards	Columbus, Ohio	Feb. 14, 1865	46, 391
Plane-stocks, Machine for mortising	J. Richards	Columbus, Ohio	Feb. 14, 1865	46, 392
Plane, Surfacing	G. E. Franklin	Natick, Mass	May 6, 1873	138, 625
Plane turner, Sliding	J. Sparrow	Portland, Me	Dec. 26, 1827	....
Plane, Wood	P. Viccellio	New Orleans, La	June 29, 1869	91, 990
Planes, Adjusting bits of carpenter's	T. D. Worrall	Lowell, Mass	Dec. 23, 1856	16, 309
Planes, Adjusting size of mouth in	T. J. Tolman	South Scituate, Mass	Jan. 13, 1857	16, 412
Planes, Attaching adjustable handles to joiner's	T. D. Worrall	Lowell, Mass	Sept. 29, 1857	18, 312
Planes, Bevel-attachment for bench	L. O. Fairbanks	Nashua, N. H	Mar. 19, 1861	31, 707
Planes, Constructing screw-arms of	E. W. Carpenter	Lancaster, Pa	Feb. 6, 1838	594

Index of patents issued from the United States Patent Office from 1790 to 1873, inclusive—Continued.

Invention.	Inventor.	Residence.	Date.	No.
Planes or jointers, Making irons for	C. E. West	Colchester, Conn	Jan. 10, 1827	....
Planes, Setting the bit in bench	L. Sanford	East Solon, N. Y	Nov. 26, 1844	3, 838
Planes, Stock for smoothing	J. F. W. Erdmann	Philadelphia, Pa	Aug. 4, 1857	17, 921
Planes, Tonguing and grooving hand	P. A. Gladwin	Boston, Mass	June 9, 1857	17, 541
Planer, Crank	W. H. Warren	Worcester, Mass	Jan. 14, 1868	73, 412
Planer cutter-stock, Metal	J. Mason	Paterson, N. J	July 22, 1856	15, 379
Planer for fellies, Rotary	C. H. Denison	Green River, Vt	Feb. 12, 1856	14, 228
Planer, Iron	C. Carr	Boston, Mass	Jan. 19, 1869	85, 904
Planer-knife grinder	J. Grant	Northampton, Mass	Mar. 10, 1868	75, 411
Planer, Metal	E. H. Cleveland	Worcester, Mass	July 29, 1856	15, 413
Planer, Metal	J. H. Sternbergh	Reading, Pa	Nov. 5, 1867	70, 479
Planer, Metal	W. H. Warren	Worcester, Mass	June 20, 1871	116, 243
Planer, Nut	J. T. Campbell	Altoona, Pa	Nov. 24, 1868	84, 410
Planer, Veneer	G. Williamson	Newark, N. J	Jan. 1, 1861	31, 070
Planers and lathes, Belt-shifting mechanism for	T. Reeve	Brooklyn, N. Y	Sept. 16, 1873	142, 814
Planers, Arrangement of parts in rotary	A. J. Kramer	Marion, Iowa	Dec. 11, 1860	30, 903
Planers, Attachment for	A. S. Hewlett	Sebastopol, Cal	Nov. [illegible]	[illegible], 248
Planers, Cutter-head for	J. More	New York, N. Y	Oct. 4, 1870	107, 943
Planers, Cutter-head for rotary	U. Nixon	Adrian, Mich	Nov. 6, 1855	13, 757
Planers, Feed for iron	H. B. Weaver	Hartford, Conn	Apr. 5, 1870	101, 550
Planers, Feeding-roller for rotary	A. T. Serrell	New York, N. Y	Mar. 12, 1861	31, 678
Planers, Gearing for metal	W. M. Barr	Williamsport, Pa	Nov. 8, 1870	108, 955
Planers, Holder for lathe	J. P. Manton	Providence, R. I	May 26, 1868	78, 387
Planers, Shipping-apparatus for metal	D. Slate	Hartford, Conn	Apr. 20, 1869	89, 177
Planers, Tool-holder for	J. S. Eltenborough	Easton, Pa	Dec. 24, 1872	134, 198
Planetarium	L. Allen	Pekin, Ill	Sept. 20, 1859	25, 476
Planetarium	J. Davis	Allegheny City, Pa	Dec. 24, 1867	72, 612
Planetarium	B. O. Swain	Annisquam, Mass	Aug. 21, 1849	6, 656
Planing and joining, tonguing and grooving, &c	J. McGregor, jr	Wilton, N. Y	Aug. 28, 1833	....
Planing and matching machine	L. B. Grover and D. Lawton	Springwater, N. Y	Dec. 23, 1834	....
Planing and matching machine, Combined	W. H. Doane and W. E. London	Cincinnati, Ohio	Nov. 11, 1862	36, 901
Planing and matching machine, Rotary	C. B. Morse	Rhinebeck, N. Y	Jan. 9, 1855	12, 211
Planing and milling machine, Cutting-tool for	E. T. Prindle	Aurora, Ill	Apr. 28, 1868	77, 211
Planing and molding machine	F. Douglas	Norwich, Conn	July 26, 1870	105, 657
Planing and molding machine	J. P. Grosvenor	Lowell, Mass	Sept. 15, 1868	82, 113
Planing and other machines, Device for changing speed and reversing motion of.	W. Heckert	Newcastle, Pa	Apr. 16, 1872	125, 676
Planing and sawing machines, Dust and shaving conveyer for.	D. R. Miller	Harrisburgh, Pa	Oct. 2, 1866	58, 456
Planing and slotting machine	C. A. Meinhard	Fort Wayne, Ind	Nov. 26, 1867	71, 401
Planing boards	E. Lane	Cincinnati, Ohio	Oct. 26, 1831	....
Planing boards	J. Percival	Philadelphia, Pa	July 20, 1831	....
Planing boards and planks and shaving shingles, Machine for.	W. Badger	Madison County, Miss	Mar. 2, 1813	....
Planing boards, grinding and polishing marble, and sawing tenons with circular saws.	M. Lancaster	Philadelphia, Pa	Apr. 5, 1832	....
Planing boards, Machine for	C. Ritter	Beaver, Pa	Aug. 11, 1809	....
Planing by machinery	J. Bennock	Boston, Mass	June 1, 1805	....
Planing curved surfaces, Machine for	J. P. Grosvenor	Lowell, Mass	Dec. 27, 1859	26, 584
Planing-cutter head	J. M. Patton and W. F. Fergus	Philadelphia, Pa	July 6, 1852	9, 100
Planing-cutter head, Rotary	H. D. Stover	New York, N. Y	Aug. 14, 1860	29, 633
Planing-cutter, Rotary	H. H. Baker	New Market, N. J	Aug. 18, 1857	18, 007
Planing-cutter, Rotary	J. Sperry	New York, N. Y	Oct. 12, 1858	21, 782
Planing-cutter, Rotary	H. D. Stover	Boston, Mass	May 19, 1857	17, 343
Planing-cutter, Rotary	H. D. Stover and J. W. Bicknell	Boston, Mass	Aug. 30, 1859	25, 286
Planing, edging, and grooving boards	J. Tees	Kensington, Pa	June 13, 1831	....
Planing-engine, Circular	U. Emmons	New York	Sept. 10, 1829	....
Planing irregular forms	A. Goodman and J. W. Gibbs	Dana, Mass	Nov. 21, 1848	5, 926
Planing irregular surfaces, Machine for	J. H. Nelson	Oskaloosa, Iowa	Sept. 28, 1858	21, 618
Planing-knife, Rotary	E. Webber	Gardiner, Me	May 2, 1854	10, 871
Planing-knives, Arrangement of rotary	D. N. Hurlbut	Utica, N. Y	Mar. 18, 1856	14, 455
Planing-knives while grinding, Holder for	L. Jennings	Ewing, Mass	Mar. 16, 1858	19, 641
Planing lumber, Fastening the cutter in machine for.	B. Becknell	Cincinnati, Ohio	Mar. 21, 1845	3, 961
Planing lumber, Machine for	F. Walcott and J. H. Hutchinson.	Boston, Mass	July 16, 1839	1, 244
Planing lumber "out of wind," Machine for	S. S. Gray	South Boston, Mass	Aug. 22, 1854	11, 582
Planing-machine	E. G. Allen	Boston, Mass	Apr. 17, 1849	6, 365
Planing-machine	E. G. Allen	Boston, Mass	Oct. 23, 1849	6, 809
Planing-machine	M. L. Andrew	Cincinnati, Ohio	Feb. 15, 1870	99, 808
Planing-machine	J. E. Andrews	Boston, Mass	Nov. 21, 1845	4, 283
Planing-machine	J. Atkins	Augusta, Me	June 24, 1873	140, 235
Planing-machine	N. Barlow	Saint Louis, Mo	May 27, 1851	8, 125
Planing-machine	D. Barnum and T. J. Wells	New York, N. Y	Mar. 13, 1849	6, 185
Planing-machine	C. B. Beall and J. K. Leach	Hamilton, Ohio	June 18, 1872	128, 007
Planing-machine	G. W. Beardslee	Buffalo, N. Y	May 20, 1851	8, 098
Planing-machine	G. W. Beardslee	Albany, N. Y	Nov. 4, 1851	8, 497
Planing-machine	J. D. Beers and I. Winslow	Philadelphia, Pa	Feb. 25, 1851	7, 949
Planing-machine	R. and C. S. Bixby and J. Garst	Dayton, Ohio	May 13, 1851	8, 086
Planing-machine	A. T. Boon and J. Collins	Galesburgh, Ill	June 7, 1864	43, 001
Planing-machine	O. Bridgeman	Addison, N. Y	Dec. 11, 1866	60, 332
Planing-machine	A. L. Brooks	Lowell, Mass	Jan. 7, 1835	....
Planing-machine	B. Brown	Burlington, Vt	Oct. 9, 1845	4, 225
Planing-machine	J. B. Brown	Lowell, Mass	Dec. 7, 1869	97, 476
Planing-machine	W. W. Carey and G. W. Harris	Lowell, Mass	Jan. 25, 1870	99, 157
Planing-machine	T. L. Carley and M. Broughton	Homer, N. Y	Apr. 26, 1870	102, 219
Planing-machine	J. Casson	Sheffield Parish, England	May 4, 1869	89, 556
Planing-machine	H. Climer and J. D. Riley	Cincinnati, Ohio	Nov. 27, 1866	59, 906
Planing-machine	J. Close	Brooklyn, N. Y	Sept. 19, 1865	49, 981
Planing-machine	G. W. Cole	Canton, Ill	Dec. 28, 1869	98, 231
Planing-machine	J. Copplnger	Baltimore, Md	Oct. 31, 1809	....
Planing-machine	L. Curtis	Sherburne, N. Y	June 16, 1836	....
Planing-machine	J. D. Dale	Philadelphia, Pa	Dec. 8, 1857	18, 806
Planing-machine	T. E. Daniels	Worcester, Mass	Dec. 23, 1834	....
Planing-machine	J. Dixon	New York, N. Y	May 8, 1866	54, 511
Planing-machine	W. H. Doane, G. V. Orton, and W. L. London.	Cincinnati, Ohio	May 21, 1867	64, 849
Planing-machine	L. Doolittle	Bushville, N. Y	June 30, 1863	39, 037

Index of patents issued from the United States Patent Office from 1790 *to* 1873, *inclusive*—Continued.

Invention.	Inventor.	Residence.	Date.	No.
Planing-machine	F. Douglas	Norwich, Conn	Oct. 12, 1869	95, 782
Planing-machine	F. Douglas	Norwich, Conn	Aug. 9, 1870	106, 139
Planing-machine	J. H. Draper	Mooresville, Ind	Jan. 5, 1869	85, 648
Planing-machine	G. B. Durkee	Chicago, Ill	May 7, 1872	126, 381
Planing-machine	C. Emmons	New York	June 27, 1848	5, 648
Planing-machine	B. Fitts	Worcester, Mass	July 31, 1860	29, 369
Planing-machine	B. Fitts	Newark, N. J	Dec. 24, 1867	72, 618
Planing-machine	B. Fitts	Newark, N. J	Dec. 24, 1867	72, 619
Planing-machine	M. R. Fletcher	Cambridge, Mass	Feb. 11, 1868	74, 333
Planing-machine	A. H. Frank and G. Spire	Buffalo, N. Y	Sept. 30, 1873	143, 232
Planing-machine	A. Fuller	Milford, N. H	June 5, 1866	55, 272
Planing-machine	T. F. Fuller	Bristol, Pa	Oct. 25, 1832	....
Planing-machine	J. Garfield	Groton, Mass	July 24, 1866	56, 548
Planing-machine	I. Gay	Dunstable, N. H	June 25, 1836	....
Planing-machine	J. Goodrich and H. J. Colburn	Fitchburgh, Mass	Feb. 7, 1871	111, 632
Planing-machine	L. Gould	Norwich, Conn	Apr. 29, 1873	138, 244
Planing-machine	L. Gould	Norwich, Conn	Nov. 4, 1873	144, 199
Planing-machine	S. S. Gray	Boston, Mass	Jan. 24, 1860	26, 902
Planing-machine	W. H. Gray	Boston, Mass	Sept. 2, 1873	142, 460
Planing-machine	J. Griffen	Phœnixville, Pa	Mar. 19, 1872	124, 813
Planing-machine	J. S Graham	Rochester, N. Y	Jan. 9, 1872	122, 599
Planing-machine	E. P. Halsted	Worcester, Mass	May 11, 1869	89, 990
Planing-machine	S. M. Hamilton	Baltimore, Md	Sept. 1, 1868	81, 776
Planing-machine	S. M. Hamilton	Baltimore, Md	Sept. 8, 1868	81, 898
Planing-machine	W. D. Hatch	South Antrim, N. H	Oct. 29, 1867	70, 208
Planing-machine	J. Hawkins	Philadelphia, Pa	Feb. 19, 1806	....
Planing-machine	A. M. Hills	Lowell, Mass	Apr. 5, 1870	101, 462
Planing-machine	J. Howarth	Salem, Mass	Mar. 23, 1852	8, 823
Planing-machine	C. L. and L. P. Hoyt	Aurora, Ill	Oct. 29, 1872	132, 663
Planing-machine	A. G. Hull	Brooklyn, N. Y	Sept. 20, 1837	385
Planing-machine	H. Jeter	Lexington, Ky	Oct. 30, 1849	6, 829
Planing-machine	A. Judson	Brooklyn, N. Y	Sept. 20, 1870	104, 502
Planing-machine	S. U. King	Windsor, Vt	Nov. 1, 1864	44, 870
Planing-machine	H. Knowles	Washington, D. C	Apr. 10, 1849	6, 294
Planing-machine	B. Kugler	Philadelphia, Pa	June 3, 1833	....
Planing-machine	H. Law	Wilmington, N. C	Apr. 10, 1849	6, 309
Planing-machine	H. A. Lee	Worcester, Mass	June 13, 1865	48, 185
Planing-machine	H. A. Lee	Worcester, Mass	Nov. 7, 1871	120, 589
Planing-machine	C. Levey	Toronto, Canada	Mar. 21, 1871	112, 821
Planing-machine	T. Llewellyn	London, England	Nov. 1, 1870	108, 801
Planing-machine	A. Lockwood	Chicago, Ill	July 22, 1856	15, 403
Planing-machine	W. N. Manning	Salem, Mass	July 10, 1860	29, 089
Planing-machine	G. H. Mansfield	Concord, N. H	May 10, 1870	102, 842
Planing-machine	G. H. Mansfield	Concord, N. H	Sept. 19, 1871	119, 163
Planing-machine	P. M. Martz	Marion County, Ind	June 25, 1836	....
Planing-machine	C. E. McBeth, F. Bentel, W. C. Margedant, and H. Climer.	Hamilton, Ohio, and Muscatine, Iowa.	Oct. 31, 1871	120, 448
Planing-machine	H. G. McDuffe	Bradford, Vt	Apr. 21, 1868	77, 067
Planing-machine	J. McGregor, jr	Wilton, N. Y	Jan. 9, 1838	557
Planing-machine	J. McGregor, jr	Savage Factory, Md	July 15, 1840	1, 690
Planing-machine	W. B. McIver	Worcester, Mass	Mar. 27, 1866	53, 466
Planing-machine	McLaughlin and Hill	Sunderland, Vt	Oct. 28, 1835	....
Planing-machine	R. N. Meriam	Worcester, Mass	Nov. 5, 1867	70, 592
Planing-machine	F. B. Miles	Philadelphia, Pa	Sept. 2, 1873	142, 493
Planing-machine	J. Miller	New York, N. Y	Sept. 2, 1873	142, 495
Planing-machine	C. B. Morse	Rhinebeck, N. Y	May 13, 1856	14, 880
Planing-machine	E. Myers	Cincinnati, Ohio	Nov. 26, 1867	71, 403
Planing-machine	S. W. Nelson	Worcester, Mass	Sept. 16, 1873	142, 936
Planing-machine	D. Niles	Fly Creek, N. Y	July 10, 1866	56, 254
Planing-machine	N. G. Norcross	Middlesex County, Mass	Feb. 12, 1850	7, 087
Planing-machine	N. G. Norcross	Lowell, Mass	June 22, 1852	9, 058
Planing-machine	H. Osgood	Waterville, Me	Dec. 9, 1856	16, 185
Planing-machine	J. F. Ostrander	New York, N. Y	Sept. 3, 1850	7, 621
Planing-machine	G. T. Pearsall	Apalachin, N. Y	Aug. 9, 1870	106, 281
Planing-machine	C. H. Peck and C. Hicks	Saint Louis, Mo	Apr. 24, 1849	6, 392
Planing-machine	C. R. Penfield	Lockport, N. Y	Oct. 18, 1864	44, 774
Planing-machine	R. H. Pindell	Fayette County, Ky	Oct. 4, 1853	10, 100
Planing-machine	F. K. Plumbly	Buffalo, N. Y	Oct. 13, 1868	83, 089
Planing-machine	F. J. Plummer	Worcester, Mass	July 9, 1867	66, 519
Planing-machine	J. Rankin	Binghamton, N. Y	Nov. 25, 1873	145, 012
Planing-machine	L. Read	North Brookfield, N. Y	May 14, 1867	64, 706
Planing-machine	J. Reihm	Savage Factory, Md	Nov. 1, 1827	....
Planing-machine	J. Richards	Philadelphia, Pa	May 17, 1870	103, 080
Planing-machine	S. M. Richardson	Worcester, Mass	Jan. 1, 1867	60, 787
Planing-machine	G. J. Rugg	Worcester, Mass	Jan. 15, 1867	61, 264
Planing-machine	J. J. Russ	Worcester, Mass	Aug. 14, 1866	57, 191
Planing-machine	J. J. Russ	Worcester, Mass	June 27, 1871	116, 496
Planing-machine	J. B. Schenck	Matteawan, N. Y	Feb. 2, 1869	86, 460
Planing-machine	J. B. Schenck	Matteawan, N. Y	Feb. 2, 1869	86, 461
Planing-machine	J. B. Schenck	Matteawan, N. Y	Jan. 18, 1870	99, 008
Planing-machine	J. B. Schenck	Matteawan, N. Y	Feb. 8, 1870	99, 599
Planing-machine	W. Sellers	Philadelphia, Pa	June 19, 1866	55, 723
Planing-machine	H. F. Shaw	West Roxbury, Mass	Jan. 29, 1867	61, 573
Planing-machine	T. Shaw	Philadelphia, Pa	Feb. 18, 1873	135, 941
Planing-machine	J. Sheldon	New Haven, Conn	Jan. 26, 1847	4, 941
Planing-machine	J. Sheldon and J. S. Barden	New Haven, Conn	Apr. 17, 1849	6, 339
Planing-machine	A. H. Shipman	Albany, N. Y	May 6, 1873	138, 703
Planing-machine	L. T. Smart	Centre Ossipee, N. H	Sept. 5, 1871	118, 754
Planing-machine	H. B. Smith	Lowell, Mass	Sept. 26, 1865	50, 178
Planing-machine	H. B. Smith	Lowell, Mass	Oct. 24, 1865	50, 637
Planing-machine	D. H. Southworth	New York, N. Y	Dec. 10, 1850	7, 827
Planing-machine	D. H. Southworth	New York, N. Y	Feb. 18, 1851	7, 942
Planing-machine	C. A. Spring and W. H. Derick	Kensington, Pa	Apr. 3, 1849	6, 219
Planing-machine	D. Stearns	Rome, N. Y	Mar. 16, 1852	8, 808
Planing-machine	F. Stedman	Aquackanckin, N. J	Aug. 17, 1835	....
Planing-machine	H. D. Stover	New York, N. Y	Aug. 21, 1860	29, 728
Planing-machine	H. D. Stover	New York, N. Y	Sept. 4, 1860	29, 923
Planing-machine	H. D. Stover	New York, N. Y	July 23, 1861	32, 904

Index of patents issued from the United States Patent Office from 1790 *to* 1873, *inclusive*—Continued.

Invention.	Inventor.	Residence.	Date.	No.
Planing machine	H. D. Stover	New York, N. Y	Dec. 8, 1868	84, 847
Planing-machine	T. J. Talbot	Milford, N. H.	May 1, 1866	54, 441
Planing-machine	J. B. Tarr	Chicago, Ill	June 5, 1866	55, 389
Planing-machine	W. Teal	Rochester, N. Y	Aug. 2, 1870	105, 997
Planing-machine	J. Tesseyman	Dayton, Ohio	Apr. 9, 1867	63, 673
Planing-machine	R. R. Throckmorton	New York	Oct. 6, 1835	....
Planing-machine	R. R. Throckmorton	New York	Oct. 22, 1835	....
Planing-machine	R. R. Throckmorton	Brooklyn, N. Y	Aug. 28, 1849	6, 667
Planing-machine	G. W. Tolhurst	Cleveland, Ohio	Jan. 20, 1852	8, 666
Planing-machine	M. Tuells	Milo, N. Y	Feb. 10, 1836	....
Planing-machine	J. Utter	Central Falls, R. I	Oct. 14, 1873	143, 734
Planing-machine	D. Van Fleet	Sandusky, Ohio	Nov. 28, 1854	12, 009
Planing-machine	A. Van Haagen	Philadelphia, Pa	June 18, 1872	128, 190
Planing-machine	L. B. Walker	Chicago, Ill	Mar. 8, 1870	100, 695
Planing-machine	C. P. S. Wardwell	Lake Village, N. H	Jan. 14, 1862	34, 174
Planing-machine	A. J. Watson	Lyons Township, Mich	Nov. 14, 1871	121, 028
Planing-machine	W. Watson	Chicago, Ill	July 6, 1852	9, 108
Planing-machine	T. J. Wells	New York, N. Y	Mar. 20, 1849	6, 195
Planing-machine	A. Whitcomb	Worcester, Mass	Sept. 1, 1868	81, 854
Planing-machine	S. Whitesides	De Pere, Wis	May 30, 1871	115, 401
Planing-machine	B. D. Whitney	Winchendon, Mass	Jan. 1, 1867	60, 812
Planing-machine	A. A. Wilder	Detroit, Mich	Dec. 21, 1852	9, 496
Planing-machine	A. A. Wilder	Detroit, Mich	May 7, 1867	64, 609
Planing-machine	S. Wilmarth	Malden, Mass	June 1, 1869	90, 907
Planing-machine	G. E. Woodbury	East Cambridge, Mass	Aug. 6, 1867	67, 623
Planing-machine	G. E. Wooobury	Cambridge, Mass	Mar. 22, 1870	101, 072
Planing-machine	J. A. Woodbury	Winchester, Mass	Feb. 7, 1854	10, 512
Planing-machine	J. A. Woodbury	Winchester, Mass	Nov. 13, 1855	13, 808
Planing-machine	J. A. Woodbury	Winchester, Mass	June 8, 1858	20, 527
Planing-machine	J. A. Woodbury	Boston, Mass	July 24, 1866	56, 656
Planing-machine	J. P. Woodbury	Boston, Mass	Mar. 20, 1849	6, 211
Planing-machine	J. P. Woodbury	Boston, Mass	Apr. 20, 1873	138, 462
Planing-machine	S. A. Woods	Boston, Mass	Feb. 14, 1871	111, 894
Planing-machine	S. A. Woods	Boston, Mass	Nov. 11, 1873	144, 588
Planing-machine	S. A. Woods and G. E. Woodbury	Boston and Cambridge, Mass.	July 18, 1871	117, 230
Planing-machine	W. Woodworth	New York, N. Y	Nov. 15, 1836	....
Planing-machine	M. Wright	Montpelier, Vt	July 16, 1872	128, 995
Planing-machine attachment	L. C. Brastow and A. M. Zwicbel.	Wilkesbarre, Pa	Nov. 28, 1871	121, 230
Planing machine, Board	H. Allen	Randolph County, N. C	Dec. 22, 1826	....
Planing machine, Board	J. Lombard	Boston, Mass	Oct. 13, 1838	979
Planing machine, Board	R. Luscomb	Penn Yan, N. Y	June 12, 1838	779
Planing machine, Board	S. Whitney	Dunstable, N. H	June 3, 1837	223
Planing machine, Board	W. Woodworth	Hudson, N. Y	Dec. 27, 1828	....
Planing machine, Board	W. Woodworth	New York, N. Y	Nov. 15, 1836	80
Planing machine, Board and timber	H. Law	Wilmington, N. C	Sept. 30, 1841	2, 272
Planing-machine cutter	E. G. Allen and C. Briggs	Boston and New Bedford, Mass.	Nov. 26, 1850	7, 793
Planing-machine cutter	J. M. Patten and W. F. Fergus	Philadelphia, Pa	Dec. 23, 1851	8, 612
Planing-machine-cutter head	B. J. Barber	Ballston Spa, N. Y	Apr. 12, 1870	101, 701
Planing-machine-cutter head	L. M. Berry	Boston, Mass	July 22, 1856	15, 365
Planing-machine-cutter head	A. Caldwell	Lexington, Ky	Oct. 16, 1847	5, 334
Planing-machine-cutter head	W. H. Christie	Albany, N. Y	Nov. 26, 1867	71, 455
Planing-machine-cutter head	H. Crimer and C. E. McBeth	Hamilton, Ohio	Apr. 4, 1871	113, 262
Planing-machine-cutter head	M. F. Connett	Ladoga, Ind	Aug. 27, 1867	68, 048
Planing-machine-cutter head	J. D. Dale	Philadelphia, Pa	Oct. 10, 1854	11, 778
Planing-machine-cutter head	J. Kuehnle	Cincinnati, Ohio	May 17, 1870	103, 200
Planing machine-cutter head	H. A. Lee	Worcester, Mass	July 16, 1867	66, 719
Planing-machine-cutter head	E. G. Richards	Detroit, Mich	Sept. 9, 1873	142, 731
Planing-machine-cutter head	A. T. Stearns	Dorchester, Mass	June 2, 1868	78, 616
Planing-machine-cutter head	D. Stevens	Danbury, Conn	Aug. 16, 1870	106, 421
Planing-machine-cutter head	L. D. Towne	Worcester, Mass	Feb. 12, 1856	14, 263
Planing-machine-cutter head, Wood	T. Tostevin	Council Bluffs, Iowa	Apr. 7, 1868	76, 556
Planing-machine-cutter holder	W. Walkington	Leeds, England	Jan. 21, 1873	135, 182
Planing-machine-cutter-setting gage	A. Evans	Cincinnati, Ohio	Apr. 27, 1869	89, 394
Planing-machine cutter, Wood	C. Livingston	Redwood City, Cal	July 17, 1866	56, 424
Planing-machine, Cylindrical	U. Emmons	New York	Apr. 25, 1829	
Planing machine feed-motion	S. C. and W. W. Hurlbut	Boonville, N. Y	Oct. 2, 1855	13, 618
Planing-machine feed-roll	S. Inman	Rockford, Ill	Jan. 28, 1873	135, 224
Planing-machine, &c., feed-roller	J. Hall	Worcester, Mass	July 21, 1857	17, 833
Planing-machine feed-roller	F. J. Plummer	Worcester, Mass	Nov. 6, 1866	59, 514
Planing-machine feeder	J. Cumberland	Mobile, Ala	Feb. 3, 1852	8, 702
Planing-machine-feeding device	C. B. Cottrell	Westerly, R. I	May 17, 1859	24, 078
Planing-machine for cutting blind-slats	M. Free	Philadelphia, Pa	June 5, 1866	55, 415
Planing-machine for dressing the edges of boards	W. E. Cornell	Boston, Mass	Jan. 1, 1851	7, 868
Planing-machine for sawing bellows boards, &c	J. Hinman	New Haven, Conn	June 1, 1805	....
Planing-machine guide	F. Keagey	Chambersburgh, Pa	Jan. 4, 1870	98, 600
Planing-machine, Hand	T. P. Butterfield	Indianapolis, Ind	May 17, 1859	24, 076
Planing-machine hood	B. D. Mott	Green Island, N. Y	Aug. 30, 1870	106, 856
Planing machine, Iron	A. B. Couch	Worcester, Mass	Feb. 1, 1870	99, 409
Planing machine, Iron	H. Gastel	Philadelphia, Pa	Mar. 19, 1872	124, 676
Planing machine, Iron	W. Sellers	Philadelphia, Pa	May 17, 1864	42, 797
Planing machine, Metal	M. Allan	Utica, N. Y	Nov. 22, 1859	26, 151
Planing machine, Metal	R. A. Belden and E. H. Cutler	New Haven, Conn	Sept. 14, 1869	94, 860
Planing machine, Metal	J. Carhart	New York, N. Y	July 10, 1859	24, 792
Planing machine, Metal	R. Harper	Glen Mills, Pa	July 22, 1873	141, 049
Planing machine, Metal	F. W. Howe	Windsor, Vt	June 21, 1853	9, 797
Planing machine, Metal	J. T. Kichner and W. H. Odenatt	Philadelphia, Pa	Sept. 23, 1873	143, 080
Planing machine, Metal	M. Kremser	Indianapolis, Ind	Sept. 19, 1871	119, 156
Planing machine, Metal	G. C. Larned	Worcester, Mass	July 22, 1873	141, 006
Planing machine, Metal	W. McPherson	New York, N. Y	Nov. 12, 1867	70, 876
Planing machine, Metal	D. F. Mellen	Wentworth, N. H	Aug. 1, 1854	11, 463
Planing machine, Metal	T. F. Rowland	Green Point, N. Y	Oct. 7, 1862	36, 624
Planing machine, Metal	J. H. Thompson	Paterson, N. J	Apr. 24, 1855	12, 767
Planing-machine presser bar	H. Snow	Dubuque, Iowa	Nov. 21, 1854	11, 984
Planing-machine, Rotary	H. C. and H. S. Ingraham	Guilford and Granger, Ohio	Mar. 15, 1859	23, 2 0
Planing-machine, Rotary	W. H. Smith	Newport, R. I	May 3, 1859	2 , 8 0
Planing-machine, Shell-roller-bed	G. Darby and J. E. Young	Augusta, Me	July 28, 1857	17, 873

Index of patents issued from the United States Patent Office from 1790 to 1873, inclusive—Continued.

Invention.	Inventor.	Residence.	Date.	No.
Planing machine, Stave	J. De Libatcheff	Yardslawl, Russia	Oct. 2, 1860	30, 210
Planing machine suction-tube	G. C. Westover	Vernon, Ind	Dec. 30, 1873	146, 115
Planing-machine-tool head	S. W. Paine	Williamsport, Pa	Sept. 26, 1871	119, 395
Planing-machine-tool holder	C. Hall	New York, N. Y	Jan. 22, 1867	61, 420
Planing machine, Wood	C. B. Cottrell	Westerly, R. I	Oct. 5, 1858	21, 720
Planing machine, Wood	W. H. Deane and W. E. London	Cincinnati, Ohio	June 18, 1867	65, 796
Planing machine, Wood	F. Douglas	Norwich, Conn	Apr. 21, 1868	77, 013
Planing machine, Wood	F. Douglas	Norwich, Conn	June 9, 1868	78, 728
Planing machine, Wood	G. B. Durkee and W. H. Murray	Chicago, Ill	Dec. 10, 1867	71, 862
Planing machine, Wood	N. C. Freck and S. Strock	Millersburgh, Pa	Oct. 31, 1871	120, 426
Planing machine, Wood	O. P. Furman	Addison, N. Y	Apr. 14, 1868	76, 740
Planing machine, Wood	D. A. Harris	Ithaca, N. Y	Feb. 11, 1868	74, 218
Planing machine, Wood	R. N. Meriam	Worcester, Mass	Apr. 14, 1868	76, 791
Planing machine, Wood	G. H. Ober	Newbury, Ohio	July 23, 1867	66, 988
Planing machine, Wood	G. Rinewalt, jr	Pendleton, Ind	July 30, 1861	32, 926
Planing machine, Wood	J. J. Russ	Worcester, Mass	Mar. 24, 1868	75, 984
Planing machine, Wood	F. Schmidt	Cincinnati, Ohio	June 18, 1867	65, 954
Planing machine, Wood	P. T. Smith	Salem, Ohio	May 28, 1867	65, 128
Planing machine, Wood	H. D. Stover	New York, N. Y	Dec. 18, 1860	30, 693
Planing machine, Wood	A. and H. Streit	Cincinnati, Ohio	Dec. 3, 1867	71, 814
Planing machine, Wood	W. Tucker	Paris, Ill	May 28, 1867	65, 308
Planing-machines, Adjusting cutter-heads to	C. R. Tompkins	Rochester, N. Y	Nov. 26, 1867	71, 341
Planing-machines, Adjusting stuff in	W. B. Emery	Albany, N. Y	Feb. 27, 1855	12, 438
Planing-machines, Arrangement of feed-rollers for	B. Fitts	Worcester, Mass	Aug. 11, 1857	17, 963
Planing-machines, Arrangement of feed-rollers for	H. C. Wight	Worcester, Mass	Jan. 1, 1856	14, 038
Planing machines, Arrangement of feed-rollers in wood.	H. H. Baker	New Market, N. J	Feb. 15, 1859	22, 926
Planing-machines, Arrangement of pressure and feed rollers in.	C. A. Spring and P. Boon	Kensington, Pa	July 30, 1850	7, 536
Planing-machines, Certain devices in	V. Houck	Buffalo, N. Y	June 17, 1856	15, 129
Planing-machines, Clamping cutters in cutter-heads for.	J. P. Grosvenor	Lowell, Mass	Dec. 2, 1856	16, 144
Planing-machines, Clamping polygonal pieces in	J. W. Killam	East Wilton, N. H	Mar. 23, 1858	19, 702
Planing-machines, Cutter-arm for	R. N. Meriam	Worcester, Mass	Sept. 20, 1864	44, 333
Planing-machines, Device for dogging lumber in	D. N. B. Coffin and H. D. Stover	Newton and Boston, Mass	May 19, 1857	17, 315
Planing-machines, Device for operating cutter-heads of.	T. F. Taft	Worcester, Mass	June 6, 1874	11, 017
Planing-machines, Device for securing cutters in rotary.	S. F. Forman	New York, N. Y	June 29, 1858	20, 762
Planing-machines, Feed-apparatus of	J. Whitney	Winchester, Mass	Apr. 13, 1852	8, 881
Planing-machines, Feeding planks to	N. Barlow	Newark, N. J	July 31, 1855	13, 345
Planing-machines, Gearing for feed-roller of	C. Burleigh	Fitchburgh, Mass	Feb. 12, 1856	14, 272
Planing-machines, Hanging knives in	M. G. Hubbard	New York, N. Y	Jan. 2, 1855	12, 162
Planing-machines, Mechanism for operating bed of	A. F. Shaw	West Roxbury, Mass	Sept. 8, 1868	81, 953
Planing machines, Molding	T. Dickinson	Newark, N. J	June 26, 1866	55, 833
Planing-machines, Operating feed-roller for	B. Fitts	Worcester, Mass	Mar. 15, 1859	23, 240
Planing-machines, Pressure-block for	J. Lawrence	East Morrisania, N. Y	Sept. 15, 1863	39, 996
Planing-machines, Preventing back-lash in feed-motion of.	T. H. Burridge	Jersey City, N. J	Dec. 24, 1850	7, 852
Planing-machines, Rotary cutter for	D. McKinley	Philadelphia, Pa	May 26, 1863	38, 712
Planing-machines, Rotary cutter-head for	T. Christian	New York, N. Y	Aug. 28, 1860	29, 768
Planing-machines, Shaving-conductor for	W. Weaver	Burlington, Vt	Dec. 9, 1873	145, 375
Planing-machines, Stock for holding cutters in rotary.	I. Gibbs	Worcester, Mass	July 27, 1858	20, 999
Planing-machines, Universal dog for	S. S. Gray	South Boston, Mass	Sept. 4, 1855	13, 536
Planing-machinery, Device for adjusting	L. Tilton	New York, N. Y	May 15, 1855	12, 880
Planing machinery, Metal	W. Sellers	Philadelphia, Pa	Aug. 5, 1862	36, 113
Planing machinery, Metal	W. Sellers	Philadelphia, Pa	June 14, 1864	43, 137
Planing machinery, Metal	W. W. Spofford	Boston, Mass	Sept. 6, 1853	9, 996
Planing machinery, Portable metal	A. C. Jones	New Orleans, La	Sept. 18, 1847	5, 301
Planing-mechanism	C. Van Haagen	Philadelphia, Pa	Oct. 11, 1870	108, 305
Planing metal	C. Van Horn	Springfield, Mass	Aug. 12, 1856	15, 538
Planing metals, &c., Mounting the cutters of machines for.	P. Saulnier	New York, N. Y	Dec. 28, 1852	9, 504
Planing moldings and smooth-planing	C. Thompson	Poughkeepsie, N. Y	Jan. 30, 1834	....
Planing moldings, Cutter for	E. J. Arens	Boston, Mass	July 2, 1867	66, 280
Planing moldings, Machine for	J. D. Dale	Philadelphia, Pa	Jan. 4, 1853	9, 515
Planing moldings, Machine for	J. D. Dale	Philadelphia, Pa	Jan. 4, 1853	9, 516
Planing moldings, Machine for	W. Rankin	Richmond, Va	Sept. 11, 1860	29, 992
Planing or molding machines, Cutter-head for	J. D. Kimball	Reading, Mass	Sept. 16, 1873	142, 794
Planing planks, boards, and clapboards, Machine for.	B. Langdon	Troy, N. Y	Jan. 9, 1838	555
Planing shavings for upholsterers, Machine for	S. A. and W. H. Post	Durham, N. Y	Sept. 1, 1863	39, 747
Planing facets of polygonal bars, nuts, &c., Machine for.	W. F. Batho	Birmingham, England	May 28, 1872	127, 142
Planing, tonguing, and grooving boards	J. Drummond	Whitestown, N. Y	Apr. 27, 1832	....
Planing, tonguing, and grooving machine	J. Powell, N. Barlow, and E. Holden.	Saint Louis, Mo	Feb. 27, 1847	4, 983
Planing, tonguing, and grooving, &c., planks and other lumber.	R. R. Throckmorton	Brooklyn, N. Y	May 1, 1845	4, 017
Planing-tool	N. Russell	Plymouth, Mass	Apr. 18, 1871	113, 800
Planing valve-seats of steam-engines	C. B. Long	Worcester, Mass	Oct. 2, 1860	30, 234
Planing warped surfaces, Machine for	J. Green	Brooklyn, N. Y	June 19, 1860	28, 808
Planing-wheel, Bramah	E. Jones	Greenfield, Mass	Dec. 2, 1856	16, 163
Planing wood, Adjusting knives of rotary cutter-heads for.	B. Fitts	Worcester, Mass	Apr. 26, 1859	23, 763
Planishing and hammering machine, Metal	C. D. Goodrich	Ann Arbor, Mich	Oct. 3, 1848	5, 818
Plank and timber dresser	E. Baily	Cincinnati, Ohio	Nov. 3, 1868	83, 684
Plank, Gang	F. G. Johnson	Brooklyn, N. Y	Apr. 15, 1873	137, 776
Planking-clamp	W. Glenn	Philadelphia, Pa	May 1, 1866	54, 327
Planking-clamp	J. Macotter	Baltimore, Md	Nov. 14, 1865	50, 941
Planking-clamp	J. C. Thomas	Kennebunk, Me	June 12, 1866	55, 554
Planking-clamp, Floor	H. B. Gregg	Camden, Ohio	Sept. 18, 1866	58, 167
Planking-clamp for ships	S. Staples	Bath, Me	Mar. 25, 1856	14, 522
Planking-clamp for ship's sides or floors	J. J. Hill	Sodus Point, N. Y	Nov. 26, 1867	71, 387
Planking-set	E. H. Parker	Bucksport, Me	July 26, 1870	105, 719
Plant-fender	C. Berber	Lostant, Ill	July 28, 1863	39, 333
Plant-grower	B. B. Chadwick	Buffalo, N. Y	May 23, 1871	115, 162

Index of patents issued from the United States Patent Office from 1790 *to* 1873, *inclusive*—Continued.

Invention.	Inventor.	Residence.	Date.	No.
Plant-protector	R. M. Bartlett	Storrs Township, Ohio	Feb. 25, 1868	74, 879
Plant-protector	J. M. Hurt	Blacks and Whites, Va	Oct. 27, 1868	83, 386
Plant-protector	I. Mayfield	Mayfield, Ky	July 11, 1871	116, 974
Plant-protector	E. Mosher	Flushing, Mich	May 10, 1859	23, 935
Plant-protector	H. K. Nelson	Penn Yan, N. Y	Nov. 2, 1869	96, 468
Plant-protector	W. N. Sprague	Keene, N. H	Jan. 18, 1870	98, 892
Plant-protector	T. R. Timby	Tarrytown, N. Y	Jan. 21, 1873	135, 173
Plant-protector	A. Tuttle	Mannsville, N. Y	Jan. 4, 1870	98, 645
Plant-protector	J. M. Watson	Sharon, Mass	June 30, 1868	79, 422
Plant-protector	J. M. Watson	Sharon, Mass	Nov. 23, 1869	97, 259
Plant-protector	J. Weed	Muscatine, Iowa	Dec. 8, 1863	40, 873
Plant-protector	W. B. Wickes	Sharon, Mass	Feb. 23, 1869	87, 313
Plant-setting device	J. C. Fuller	Phœnix, N. Y	Feb. 16, 1869	87, 037
Plant-stand, Folding	B. B. Nourse	Westborough, Mass	Dec. 5, 1871	121, 541
Plants and animals, Accelerating growth of	A. I. Pleasonton	Philadelphia, Pa	Sept. 26, 1871	119, 242
Plants and fruits, Growing	A. C. Chamberlain	Newport, R. I	Nov. 26, 1861	33, 774
Plants and trees, Compound for destroying vermin on.	F. S. Townsend	South Seaville, N. J	Jan. 26, 1869	86, 259
Plants, Apparatus for fumigating	C. C. Preston	Bayland, Tex	Dec. 17, 1867	72, 419
Plants, Box for propagating	E. B. Newman, jr	Watkins, N. Y	Apr. 30, 1867	64, 245
Plants, Box for transplanting	J. Jenkins	Salem, Ohio	Mar. 12, 1867	62, 751
Plants, Box for transporting	M. L. Thompson	Brooklyn, N. Y	June 21, 1864	43, 242
Plants, Box for transporting	T. F. Wardwell	Penn Yan, N. Y	Apr. 11, 1865	47, 234
Plants, Compound for destroying worms in	J. Bingaman	Jersey Shore, Pa	Mar. 16, 1869	87, 820
Plants, Extracting saline matters from marine	C. F. Brackett and G. L. Goodale.	Brunswick and Saco, Me	Oct. 29, 1867	70, 317
Plants for useful purposes, Preparation of the roots of.	J. M. Katzenmayer	New York, N. Y	Sept. 29, 1863	40, 143
Plants, Forcing	C. Reese	Baltimore, Md	Apr. 2, 1872	125, 332
Plants from frost, Preservation of	G. Morris		July 10, 1792	....
Plants from hoes, Implement for shielding	A. Kirkpatrick	Newark, N. J	June 23, 1868	79, 133
Plants from worms, Device for protecting young	J. W. Colburn	Rose, N. Y	Mar. 16, 1869	87, 911
Plants, Hand-lights for protecting	O. R. L. and M. P. A. Crozier	Paris, Mich	May 26, 1863	38, 656
Plants, Implement for setting out	E. W. Packer	Paulsborough, N. J	Apr. 3, 1866	53, 662
Plants, Instrument for transplanting	W. C. S. Ellerbe	Camden, S. C	Oct. 16, 1866	58, 794
Plants, Machine for treating	A. Bouchard	New Orleans, La	May 2, 1871	114, 400
Plants, Machine for treating fibrous	W. B. Shedd	Boston, Mass	Apr. 30, 1872	126, 337
Plants, Machinery for treating fibrous	G. E. Hopkins and W. B. Shedd	Boston, Mass	Jan. 30, 1872	123, 103
Plants, Package-case for	P. L. Suine	Shirleysburgh, Pa	July 10, 1866	56, 290
Plants, Potting and packing	B. L. Ryder	Chambersburgh, Pa	Nov. 10, 1868	84, 002
Plants, Propagation and treatment of house	W. Lilley	Washington, D. C	Mar. 17, 1868	75, 555
Plants, &c., Protecting	J. Shepard	Bristol, Conn	Sept. 1, 1868	81, 693
Plants to prevent the approach of worms, Implement for trenching around.	W. H. Halleck	Ann Arbor, Mich	Dec. 8, 1868	84, 692
Planter	W. Beall	Hainesville, W. Va	May 7, 1872	126, 438
Planter	W. F. Caldwell	Oxford, Me	Sept. 18, 1866	58, 060
Planter and broad-cast seeder, Combined corn	A. J. Edgett	Hornellsville, N. Y	Sept. 18, 1866	58, 166
Planter and chopper, Combined cotton	A. B. Nixon	Polo, Ill	July 4, 1871	116, 741
Planter and cultivator	S. Bacher	Edina, Mo	Dec. 20, 1870	110, 187
Planter and cultivator	N. Earlywine	Centreville, Iowa	Sept. 13, 1870	107, 235
Planter and cultivator	A. Little	Jamestown, Ohio	Apr. 30, 1872	126, 147
Planter and cultivator	A. J. Misenhimer	Oskaloosa, Ill	Sept. 14, 1869	94, 835
Planter and cultivator	C. Rubbles	Lowville, N. Y	May 5, 1868	77, 535
Planter and cultivator	N. Whitehall	Newton, Ind	Aug. 31, 1869	94, 365
Planter and cultivator	A. Q. Withers	Holly Springs, Miss	Jan. 17, 1871	111, 101
Planter and cultivator combined	J. Adams	Clarksville, Tex	Aug. 20, 1867	68, 026
Planter and cultivator, Combined	T. M. and J. Brooks	Cincinnati, Ohio	Apr. 30, 1872	126, 258
Planter and cultivator combined	J. N. Burton	Griffin, Ga	Mar. 26, 1872	125, 018
Planter and cultivator combined	I. H. Chappell	Lawrence, Kans	Mar. 19, 1867	63, 019
Planter and cultivator combined	I. H. Chappell	Decatur, Ill	Apr. 16, 1867	63, 858
Planter and cultivator, Combined	I. H. Chappell	Decatur, Ill	Nov. 19, 1867	71, 134
Planter and cultivator, Combined	J. A. Currie	Xenia, Ohio	Aug. 31, 1869	94, 289
Planter and cultivator, Combined	W. L. Gobby	New Richland, Ohio	May 7, 1867	64, 519
Planter and cultivator combined	G. W. Kinzer	Linden Station, Ohio	Aug. 18, 1868	81, 177
Planter and cultivator combined	E. B. and J. F. McClellan	Alexandria, Ala	Jan. 30, 1872	123, 186
Planter and cultivator, Combined	I. W. McGaffey	Chicago, Ill	Mar. 28, 1865	47, 029
Planter and cultivator, Combined	A. G. Perry	Union County, Miss	Oct. 8, 1872	131, 969
Planter and cultivator, Combined	A. W. Putnam	Suisun, Cal	Apr. 2, 1867	63, 424
Planter and cultivator, Combined	J. A. Rockwood	Kinderhook, Ill	Nov. 23, 1869	97, 121
Planter and cultivator, Combined	W. C. Switzer	Nelsonville, Tex	Mar. 16, 1869	87, 885
Planter and cultivator, Combined	J. Vaughn	College Grove, Tenn	Dec. 24, 1867	72, 703
Planter and cultivator combined	C. L. Wilcox	Wayne, Ohio	July 16, 1872	129, 443
Planter and cultivator, Combined	T. Wilson	Garton, England	Nov. 30, 1869	97, 467
Planter and cultivator, Combined corn	A. G. Aikin	Somerton, Ohio	Aug. 10, 1869	93, 510
Planter and cultivator, Combined corn	E. Braggins	Mount Vernon, Ohio	July 4, 1871	116, 678
Planter and cultivator, Combined corn	J. Campbell	London, Ohio	Aug. 3, 1869	93, 172
Planter and cultivator, Combined corn	G. De Vany, jr	Darbyville, Ohio	June 24, 1873	140, 254
Planter and cultivator, Combined corn	C. Dyer	Coal Run, Ohio	Oct. 27, 1868	83, 478
Planter and cultivator, Combined corn	A. N. Gow	Mount Vernon, Ohio	Feb. 19, 1867	62, 263
Planter and cultivator, Combined corn	J. C. Hazen	West Independence, Ohio	Nov. 2, 1869	96, 319
Planter and cultivator, Combined corn	J. W. Hollaman	Maple Grove, Ill	Jan. 12, 1869	85, 826
Planter and cultivator, Combined corn	M. J. Hunt	Rising Sun, Md	Jan. 8, 1867	61, 071
Planter and cultivator, Combined corn	D. W. Jacoby	Shelbyville, Ill	Aug. 27, 1867	68, 201
Planter and cultivator, Combined corn	M. J. Kavanagh and M. Gregg.	Joliet and Chicago, Ill	July 20, 1869	92, 837
Planter and cultivator, Combined corn	J. S. Mason	Coal Run, Ohio	Sept. 15, 1868	82, 137
Planter and cultivator, Combined corn	J. Palmer	Oriskany, N. Y	July 25, 1865	48, 991
Planter and cultivator, Combined corn	W. Scott	High Banks, Ind	Jan. 19, 1869	86, 105
Planter and cultivator, Combined corn	J. F. Sterett and C. M. J. Reynolds	Ottumwa, Iowa	Oct. 1, 1867	69, 502
Planter and cultivator, Combined corn	S. J. Taylor	Rome, N. Y	July 16, 1867	66, 910
Planter and cultivator, Combined corn	F. Underwood	South Rutland, N. Y	Mar. 7, 1871	112, 516
Planter and cultivator, Combined cotton-seed	J. A. Wright	Marietta, Ga	Apr. 4, 1871	113, 383
Planter and cultivator, Combined potato	C. F. Hoffman	New Orleans, La	Aug. 25, 1868	81, 371
Planter and cultivator, Combined seed	G. W. Carpenter	Butler, Ind	Nov. 16, 1869	96, 774
Planter and cultivator, Combined seed	P. Limhold	Saint Louis, Mo	Aug. 29, 1865	49, 661
Planter and cultivator, Convertible	S. E. Harrington	Greenfield, Mass	Apr. 19, 1864	42, 370
Planter and cultivator, Corn	N. Breed	Jeffersonville, Ind	Jan. 11, 1870	98, 663
Planter and cultivator, Corn	A. Edmister	Westfield, Ohio	Nov. 10, 1868	83, 943

Index of patents issued from the United States Patent Office from 1790 *to* 1873, *inclusive*—Continued.

Invention.	Inventor.	Residence.	Date.	No.
Planter and cultivator, Corn	J. Jenkins	Andrew County, Mo	Oct. 19, 1869	95, 906
Planter and cultivator, Seed	D. W. Shares	Hamden, Conn	Dec. 12, 1854	12, 075
Planter and cultivator, Seed	M. R. Snodgrass	Jamestown, Ohio	Dec. 24, 1867	72, 556
Planter and cultivator, Seed	W. D. Stoud	Oshkosh, Wis	Apr. 12, 1870	101, 939
Planter and digger, Potato	J. C. Walker	Farmington, Mich	Mar. 21, 1871	112, 870
Planter and distributer	D. McKellar	Selma, Ala	Sept. 27, 1870	107, 796
Planter and drill, Cotton-seed	W. J. Arrington	Jefferson County, Ga	July 6, 1869	92, 144
Planter and earth-excavator, Potato	L. Rice	Robbinston, Me	June 29, 1833	....
Planter and fertilizer	M. Barbour	Oxford, Ohio	Apr. 30, 1872	126, 250
Planter and fertilizer, Combined corn	B. F. Grimes	Dawsonville, Md	May 28, 1867	65, 073
Planter and fertilizer, Corn	A. L. Holcomb	Hopewell, N. J	June 21, 1870	104, 591
Planter and fertilizer, Corn	H. K. Roberts	Jefferson County, Ky	Jan. 18, 1870	98, 886
Planter and fertilizer, Corn	M. M. Sprinkle	Rochelle, Va	June 1, 1869	90, 698
Planter and fertilizer-distributer, Combined corn	C. W. Barrick	Walkersville, Md	July 25, 1871	117, 369
Planter and fertilizer-distributer, Combined cotton-seed.	J. Johnson	Northampton County, N. C	July 16, 1867	66, 713
Planter and fertilizer-distributer, Combined cotton-seed.	W. M. McLendon	Greenville, Ga	May 17, 1870	103, 067
Planter and fertilizer-distributer, Cotton	B. H. Washington	Columbia County, Ga	Oct. 18, 1870	108, 415
Planter and fertilizer-distributer, Cotton-seed	H. C. Harris	Fort Valley, Ga	Mar. 8, 1870	100, 526
Planter and fertilizer-distributer Seed	J. Arrington	Livingston, Ala	Feb. 22, 1870	100, 001
Planter and fertilizer-distributer, Seed	E. P. Jones	Greensborough, N. C	June 17, 1873	140, 046
Planter and fertilizer-distributer, Seed	R. Montfort	Butler, Ga	Oct. 21, 1873	143, 832
Planter and fertilizer-distributer, Seed	J. Samplo	Franklin County, Miss	Sept. 26, 1871	119, 411
Planter and fertilizer, Seed	D. P. Webster	Boston, Mass	Aug. 23, 1870	106, 751
Planter and grain-drill, Corn	G. W. Dickinson	Charleston, Ill	Aug. 31, 1869	94, 404
Planter and guano-distributer, Combined cotton	J. H. Nicholes	Sumter, S. C	Feb. 28, 1871	112, 169
Planter and guano-distributer, Combined cotton	J. E. Ross	Greensborough, Ga	Dec. 20, 1870	110, 290
Planter and guano-distributer, Combined cotton and seed.	S. L. Donnell	Humboldt, Tenn	Mar. 22, 1870	100, 990
Planter and guano-distributer, Combined cotton-seed.	L. Garnett	Camilla, Ga	Jan. 3, 1871	110, 644
Planter and guano-distributer, Cotton-seed	W. W. Croom	Montgomery, Ala	Jan. 10, 1871	110, 832
Planter and guano-distributer, Cotton-seed	J. Rafter	Winona, Miss	May 9, 1871	114, 600
Planter and guano-distributer, Cotton-seed	C. G. Wilson	Milledgeville, Ga	Dec 10, 1872	133, 911
Planter and guano-distributer, Cotton-seed	C. G. Wilson	Milledgeville, Ga	Sept. 23, 1873	143, 211
Planter and guano-distributer, Seed	H. L. Tillery	Halifax, N. C	Jan. 3, 1871	110, 694
Planter and guano-distributer, Seed	W. L. Traynham	Warrenton, Ga	Oct. 11, 1870	108, 213
Planter and guano-sower, Corn	J. B. Gemmill	Strawbridge, Pa	July 23, 1867	66, 959
Planter and hoe, Combined corn	H. Soggs	Columbus, Pa	Feb. 26, 1867	62, 376
Planter and lime-spreader, Combined corn	M. M. Mackerly	South Salem, Ohio	Feb. 18, 1862	34, 438
Planter and manure-distributer	B. F. Whitner	Madison, Fla	May 21, 1867	64, 926
Planter and manure-distributer, Combined	B. F. Whitner	Madison, Mich	Apr. 14, 1868	76, 865
Planter and manure-distributer, Cotton	N. Donaldson	Line Creek, S. C	Jan. 3, 1871	110, 636
Planter and marker, Corn	F. Ogden	Fisherville, Ky	Oct. 4, 1870	108, 044
Planter and marker, Corn	T. Ryan	Saint Martin's, Ohio	Feb. 6, 1872	123, 514
Planter and plaster-dropper, Combined corn	H. Rhodes	Clarence Centre, N. Y	Oct. 1, 1867	69, 486
Planter and plow, Combined corn	J. T. McElhiney	Moberly, Mo	Apr. 1, 1873	137, 309
Planter and plow, Combined corn	P. McIsaac	Waterloo, Iowa	Jan. 7, 1868	73, 022
Planter and plow, Combined corn	S. M. Swartz	Millheim, Pa	Sept. 11, 1866	57, 994
Planter and plow, Combined seed	E. Seibel	Witenburgh, Mo	Feb. 2, 1869	86, 457
Planter and plow, Combined sulky corn	R. and A. S. Robins	Dundas, Ill	June 29, 1869	91, 874
Planter and plow, Corn	J. H. Frampton	Hopewell, Ohio	Apr. 14, 1868	76, 738
Planter and roller, Combined	J. A. Comstock	Bowling Green, Mo	Aug. 22, 1871	118, 342
Planter and seed-drill, Corn	J. Weaver, jr	Elizabethville, Pa	June 16, 1868	79, 038
Planter and seeder	H. H. Webster	Claremont, N. H	Feb. 6, 1866	52, 500
Planter and seeder, Combined	W. E. Fricke	Mexico, Mo	Aug. 30, 1870	106, 806
Planter and seeder, Combined	S. Hiestand	Hillsborough, Ohio	Apr. 4, 1871	113, 718
Planter and seeder, Combined corn	R. Cowan	Bloomfield, Ill	Jan. 14, 1873	134, 855
Planter and seeding-machine	S. S. Kimball	Laconia, N. H	Mar. 16, 1869	87, 855
Planter and shovel-plow, Combined corn	A. M. Franklin, W. J. Hastings, and J. A. Holford.	Rising Sun, Ind	Oct. 20, 1868	83, 271
Planter, Automatic check-row corn	J. L. Kreider	Chesnut Level, Pa	July 13, 1869	92, 616
Planter, Automatic seed	D. Lorriaux	Ottawa, Ill	May 14, 1872	126, 721
Planter, Broom-corn-seed	J. Bush	Darien, Wis	May 24, 1870	103, 427
Planter, Combined corn and cotton	W. B. Townes	Grenada, Miss	May 9, 1871	114, 625
Planter, Combined corn and cotton	A. H. Wooton	Bartow, Ga	Sept. 21, 1869	95, 062
Planter, Combined cotton and corn	L. A. Perrault	Natchez, Miss	Mar. 7, 1871	112, 486
Planter, Combined hoe and hand	H. Fessler and I. E. Betz	Canton, Ohio	May 14, 1867	64, 755
Planter, Combined seed and potato	O. N. Chase	Boston, Mass	Jan. 24, 1865	45, 976
Planter, Convertible corn	A. Runstetler	Peoria, Ill	June 27, 1871	116, 493
Planter, Corn	D. B. Abbey	Horseheads, N. Y	Nov. 12, 1861	33, 688
Planter, Corn	G. Abbott	White's Corners, N. Y	Oct. 15, 1867	69, 739
Planter, Corn	H. Ackerman	Pittsburgh, Pa	Aug. 11, 1868	80, 892
Planter, Corn	M. Ackerman	Steamboat Rock, Iowa	Sept. 6, 1870	106, 981
Planter, Corn	J. C. Adams	Greensburgh, Ind	Oct. 4, 1859	25, 613
Planter, Corn	J. W. Adams	Birmingham, Iowa	May 9, 1865	47, 603
Planter, Corn	S. W. Adams	Moultrie County, Ill	Dec. 11, 1860	30, 861
Planter, Corn	W. H. Adle, P. D. Miles, and G. Custer.	Norristown, Pa	Nov. 27, 1860	30, 708
Planter, Corn	J. Agnew	Bath, Pa	Apr. 5, 1864	42, 151
Planter, Corn	J. M. Aitchison	Omar, N. Y	Feb. 21, 1871	112, 006
Planter, Corn	D. R. Alden	Unionville, Ohio	Sept. 8, 1857	18, 127
Planter, Corn	T. K. Alexander	Decatur, Ill	July 19, 1864	43, 559
Planter, Corn	E. C. Allen	Le Roy, N. Y	Nov. 8, 1859	26, 006
Planter, Corn	E. W. Allen	Auburn, N. Y	Apr. 14, 1868	76, 689
Planter, Corn	H. R. Allen	Athens, Ohio	Sept. 8, 1857	18, 128
Planter, Corn	T. Allen	Arrow Rock, Mo	Oct. 1, 1867	69, 301
Planter, Corn	J. M. Allison	Cranberry, Pa	May 19, 1868	78, 039
Planter, Corn	J. M. Allison	Cranberry, Pa	Dec. 1, 1868	84, 604
Planter, Corn	D. Altman	Nashville, Tenn	Nov. 14, 1871	120, 926
Planter, Corn	D. S. Ames	Laclede, Mo	Mar. 24, 1868	75, 825
Planter, Corn	D. W. Amos	Bedford, Pa	Nov. 1, 1864	44, 845
Planter, Corn	A. Anable	Middlesex, N. Y	Jan. 3, 1860	26, 642
Planter, Corn	J. K. Andrews and J. D. Green	Antrim, Ohio	Sept. 3, 1867	68, 514
Planter, Corn	A. Armstrong	Gillespie, Ill	Apr. 21, 1868	76, 971
Planter, Corn	J. Armstrong, jr	Elmira, Ill	July 22, 1862	35, 914
Planter, Corn	J. Armstrong, jr	Elmira, Ill	Sept. 20, 1864	44, 273

Index of patents issued from the United States Patent Office from 1790 *to* 1873, *inclusive*—Continued.

Invention.	Inventor.	Residence.	Date.	No.
Planter, Corn	J. Armstrong, jr	Elmira, Ill	Sept. 21, 1869	95, 068
Planter, Corn	J. N. Arvin and J. M. Whitmore	Valparaiso, Ind	Feb. 12, 1867	61, 981
Planter, Corn	J. N. Arvin and J. M. Whitmore	Valparaiso, Ind	Sept. 29, 1868	82, 578
Planter, Corn	F. J. Ashburn	West Union, W. Va	Jan. 12, 1869	85, 721
Planter, Corn	S. Avery	Pisgah, Mo	July 10, 1860	29, 046
Planter, Corn	A. G. Babcock	Galesburgh, Ill	Sept. 7, 1858	21, 404
Planter, Corn	W. R. Baldwin	Philadelphia, Pa	Nov. 13, 1866	59, 543
Planter, Corn	T. H. Ballard	Elk Grove, Wis	Dec. 3, 1872	133, 618
Planter, Corn	W. C. Banks	Como Depot, Miss	May 1, 1860	28, 051
Planter, Corn	W. C. Banks	Como Depot, Miss	May 1, 1860	28, 052
Planter, Corn	A. Barber	Port Byron, Ill	Feb. 5, 1867	61, 701
Planter, Corn	H. F. and T. R. Bargar	Border Plains, Iowa	Nov. 17, 1863	40, 595
Planter, Corn	A. Barnes	Bloomington, Ill	Nov. 5, 1872	132, 792
Planter, Corn	A. J. Barnhart	Schoolcraft, Mich	Aug. 14, 1855	13, 419
Planter, Corn	H. F. Batcheller	Sterling, Ill	Oct. 13, 1863	40, 233
Planter, Corn	M. Bates	Carlton, N. Y	Sept. 23, 1856	15, 755
Planter, Corn	H. Baughman	Columbus, Ohio	Sept. 7, 1869	94, 546
Planter, Corn	H. Baughman	Sandusky, Ohio	May 2, 1871	114, 390
Planter, Corn	L. Becker	Jackson Township, Pa	Feb. 7, 1871	111, 510
Planter, Corn	T. M. Bedgood	Cleveland, Ind	Aug. 17, 1858	21, 180
Planter, Corn	U. Beebee	Oakland, Mich	June 19, 1855	13, 077
Planter, Corn	W. T. Beekman	Petersburgh, Ill	Feb. 23, 1869	87, 237
Planter, Corn	M. Bemiss	Lyme, Ohio	July 15, 1856	15, 322
Planter, Corn	G. I. Bergen	Galesburgh, Ill	Dec. 1, 1863	40, 789
Planter, Corn	G. I. Bergen	Galesburgh, Ill	Mar. 7, 1865	46, 649
Planter, Corn	G. I. Bergen	Galesburgh, Ill	Aug. 29, 1865	49, 601
Planter, Corn	E. L. Bergstresser	Hublersburgh, Pa	Sept. 24, 1867	69, 296
Planter, Corn	H. C. Beshler	Berrysburgh, Pa	July 20, 1869	92, 780
Planter, Corn	H. C. Beshler	Berrysburgh, Pa	Oct. 4, 1870	107, 857
Planter, Corn	C. F. Bilhimer	Irwin's Station, Pa	Feb. 14, 1871	111, 808
Planter, Corn	O. Billings	La Grange, Ohio	Dec. 3, 1867	71, 572
Planter, Corn	L. F. Bingham and N. O. Pierce	Chicago, Ill	May 10, 1859	23, 895
Planter, Corn	W. F. Blandin	Macomb, Ill	Dec. 15, 1863	40, 900
Planter, Corn	J. M. Blessing	Jeffersonville, Ohio	Apr. 28, 1868	77, 251
Planter, Corn	W. Blessing	Jeffersonville, Ohio	June 13, 1865	48, 150
Planter, Corn	D. G. Boardman	Albia, Iowa	Mar. 18, 1873	136, 961
Planter, Corn	R. Bocklen	Jersey City, N. J	May 6, 1856	14, 801
Planter, Corn	J. L. Bond	Marshalltown, Iowa	Sept. 6, 1870	107, 156
Planter, Corn	J. H. Bonham	Elizabethtown, Ohio	Dec. 8, 1857	18, 798
Planter, Corn	J. H. Bonham	Elizabethtown, Ohio	May 22, 1860	28, 343
Planter, Corn	R. A. Boulware	Doniphan, Kans	Mar. 21, 1871	112, 894
Planter, Corn	B. Bower	Millersburgh, Ohio	Nov. 27, 1860	30, 713
Planter, Corn	R. M. and W. H. Bowman	London, Ohio	July 30, 1872	129, 924
Planter, Corn	W. C. Bowman	Clarksburgh District, Md	May 16, 1871	114, 908
Planter, Corn	W. H. Boyle	Cazenovia, N. Y	May 9, 1865	47, 614
Planter, Corn	H. Bradt and J. Otis	Schenectady, N. Y	Jan. 30, 1872	123, 080
Planter, Corn	N. Breed	Jeffersonville, Ind	Apr. 6, 1869	88, 604
Planter, Corn	T. R. Brent	Muscatine, Iowa	Aug. 5, 1862	36, 069
Planter, Corn	A. Brightbill	Bethel Township, Pa	Aug. 6, 1861	33, 015
Planter, Corn	A. W. Brinkerhoff	Upper Sandusky, Ohio	May 10, 1859	23, 896
Planter, Corn	A. W. Brinkerhoff	Upper Sandusky, Ohio	Jan. 8, 1861	31, 072
Planter, Corn	A. W. Brinkerhoff	Upper Sandusky, Ohio	Sept. 2, 1862	36, 335
Planter, Corn	J. H. Broad	Lodi, N. Y	Aug. 30, 1864	43, 968
Planter, Corn	T. Brockway	Chatsworth, Ill	Mar. 29, 1870	101, 349
Planter, Corn	J. Broughton	New York, N. Y	May 12, 1857	17, 258
Planter, Corn	E. C. Brown	Crawfordsville, Ind	Sept. 9, 1873	142, 675
Planter, Corn	G. W. Brown	Galesburgh, Ill	Apr. 5, 1864	42, 177
Planter, Corn	G. W. Brown	Galesburgh, Ill	Apr. 5, 1864	42, 178
Planter, Corn	G. W. Brown	Galesburgh, Ill	Apr. 5, 1864	42, 179
Planter, Corn	G. A. Brace	Mechanicsburgh, Ill	Oct. 4, 1853	10, 064
Planter, Corn	A. B. Bruen	Trenton, Mo	Dec. 31, 1872	134, 350
Planter, Corn	J. T. Bryan	Lebanon, Ind	May 2, 1865	47, 514
Planter, Corn	S. B. Buck	Elyria, Ohio	Mar. 8, 1870	100, 495
Planter, Corn	G. Bunch and J. A. Price	Grand River Township, and Breckinridge, Mo.	Jan. 24, 1865	45, 971
Planter, Corn	G. W. Bunker	Saint Anthony, Minn	June 29, 1869	91, 904
Planter, Corn	J. A. Burchard	Beloit, Wis	Oct. 13, 1868	82, 918
Planter, Corn	A. C. and G. F. Burgner	Charleston, Ill	Feb. 4, 1873	135, 521
Planter, Corn	J. Burns	Elyria, Ohio	Apr. 16, 1867	63, 851
Planter, Corn	R. Burns	New York, N Y	July 11, 1865	48, 654
Planter, Corn	T. D. Burrall	Geneva, N. Y	June 26, 1835	....
Planter, Corn	C. Busch	Brooklyn, Iowa	Dec. 17, 1872	134, 032
Planter, Corn	W. R. Butler	Greenbush, Ill	Nov. 20, 1866	59, 742
Planter, Corn	E. M. Butz	Allegheny City, Pa	Feb. 11, 1868	74, 298
Planter, Corn	S. J. Bye	Bluff Point, Ind	July 4, 1871	116, 552
Planter, Corn	W. Caldwell	Bryan, Ohio	July 6, 1869	92, 159
Planter, Corn	H. W. Camp and A. W. Fox	Owego, N. Y	June 25, 1867	66, 002
Planter, Corn	A., W., and J. Campbell	Harrison, Ohio	June 28, 1859	24, 537
Planter, Corn	J. Campbell	Harrison, Ohio	May 8, 1866	54, 642
Planter, Corn	J. Campbell	Harrison, Ohio	Nov. 19, 1872	133, 199
Planter, Corn	J. P. Campbell	Danville, Pa	Nov. 16, 1869	96, 883
Planter, Corn	S. O. Campbell	Leavenworth, Kans	Sept. 1, 1868	81, 597
Planter, Corn	A. C. Carey	Ipswich, Mass	June 1, 1858	20, 467
Planter, Corn	J. H. Carothers	Davidsburgh, Pa	July 26, 1853	9, 882
Planter, Corn	G. J. Carpenter	Adrian, Mich	June 24, 1873	140, 114
Planter, Corn	J. T. Carter and M. B. Williamson	Xenia, Ill	Dec. 20, 1870	110, 341
Planter, Corn	J. Case	La Fayette, Ind	Aug. 18, 1868	81, 066
Planter, Corn	J. Case	Springfield, Ohio	Dec. 9, 1873	145, 396
Planter, Corn	R. and J. L. Cassady	Hardingsville, N. J	Sept. 17, 1867	68, 841
Planter, Corn	A. Cassidy	Putnam, Ohio	Feb. 3, 1863	37, 564
Planter, Corn	W. R. Center	Athens, Ill	Dec. 4, 1860	30, 799
Planter, Corn	J. D. Chambers	Carthage, Mo	Dec. 15, 1868	84, 936
Planter, Corn	W. B. Chambers	Decatur, Ill	Aug. 15, 1871	117, 981
Planter, Corn	J. F. Champlin	Aurora, N. Y	July 24, 1866	56, 667
Planter, Corn	J. B. Chapman	Morrison, Ill	Aug. 24, 1869	94, 077
Planter, Corn	E. C. Chesney	Abingdon, Ill	June 21, 1864	43, 181
Planter, Corn	W. E. Chesney	Abingdon, Ill	Mar. 7, 1865	46, 637

Index of patents issued from the United States Patent Office from 1790 *to* 1873, *inclusive*—Continued.

Invention.	Inventor.	Residence.	Date.	No.
Planter, Corn	Z. T. Clagett	Washington, D. C	July 21, 1868	80, 058
Planter, Corn	B. Clark	Mackinaw, Ill	Sept. 4, 1866	57, 676
Planter, Corn	N. Clark	Rochester, Mass	Oct. 22, 1872	132, 355
Planter, Corn	W. R. Clark	Indianola, Ill	May 26, 1868	78, 262
Planter, Corn	J. Clarridge	Pancoastburgh, Ohio	Apr. 23, 1867	64, 076
Planter, Corn	R. J. Clay	Saint Louis, Mo	May 11, 1858	20, 193
Planter, Corn	J. D. Cochran	Milford, N. H	May 8, 1866	54, 507
Planter, Corn	J. S. Coen	Attica, Ind	Jan. 5, 1869	85, 563
Planter, Corn	J. S. Coen	Attica, Ind	Sept. 21, 1869	95, 085
Planter, Corn	W. Cogswell	Ottawa, Ill	July 10, 1866	56, 183
Planter, Corn	A. Colton	Attica, N. Y	May 28, 1861	32, 412
Planter, Corn	W. Combs	Duquoin, Ill	Jan. 8, 1861	31, 076
Planter, Corn	J. B. Conaty and M. Catt	Indianapolis, Ind	Mar. 28, 1871	113, 018
Planter, Corn	J. Conrad	Centralia, Ill	Nov. 13, 1866	59, 559
Planter, Corn	L. H. Converse	Springfield, Ill	July 16, 1872	129, 320
Planter, Corn	G. T. Cooley	Wooster, Ohio	Feb. 16, 1869	87, 028
Planter, Corn	J. Cooley	Tafton, Wis	Mar. 19, 1861	31, 700
Planter, Corn	J. P. Coonley	Farmington, Mich	Aug. 9, 1859	24, 993
Planter, Corn	A. M. Corbit	Bethlehem, Iowa	Mar. 26, 1867	63, 219
Planter, Corn	R. B. and H. Corson	Henry County, Mo	Dec. 20, 1870	110, 207
Planter, Corn	J. Cosand	Russiaville, Ind	Sept. 20, 1870	107, 457
Planter, Corn	W. H. Cox	Virden, Ill	Oct. 23, 1866	58, 988
Planter, Corn	W. H. Cox	Virden, Ill	Mar. 16, 1869	87, 825
Planter, Corn	W. Craig	Urbana, Ill	Dec. 8, 1863	40, 822
Planter, Corn	W. J. C. Crandall	Steele's, Ind	Jan. 7, 1873	134, 521
Planter, Corn	W. H. Crosby	Parish, N. Y	May 14, 1872	126, 788
Planter, Corn	A. L. Crow	Pennville, Ind	June 1, 1869	90, 640
Planter, Corn	I. Crum	Port Union, Ohio	Dec. 18, 1866	60, 483
Planter, Corn	J. W. Crume	Troy, Mo	Aug. 2, 1870	106, 034
Planter, Corn	W. Daggett, 4th	Cordova, Ill	Mar. 10, 1868	75, 380
Planter, Corn	T. Dale	Russellville, Ky	July 12, 1870	105, 314
Planter, Corn	J. Davis	Allegheny City, Pa	Nov. 14, 1865	50, 915
Planter, Corn	J. A. Davis	Verona, Miss	July 12, 1870	105, 318
Planter, Corn	J. R. Davis	Bloomfield, Iowa	Nov. 17, 1863	40, 610
Planter, Corn	J. R. Davis	Bloomfield, Iowa	May 9, 1865	47, 622
Planter, Corn	J. S. Davis	Arcadia, Ohio	Aug. 17, 1852	21 187
Planter, Corn	J. S. Davis	Ithaca, N. Y	Dec. 23, 1873	145, 728
Planter, Corn	N. C. Davis	Worthington, Ohio	June 14, 1870	104, 280
Planter, Corn	F. Dean	Beloit, Wis	July 4, 1865	48, 528
Planter, Corn	W. A. Dean	New Lexington, Ohio	July 16, 1872	129, 008
Planter, Corn	J. D. Delap	Tyrone Township, Pa	Apr. 18, 1871	113, 747
Planter, Corn	S. G. Dentler	Orangeville, Ill	July 2, 1867	66, 308
Planter, Corn	G. Dickerson	Harveysburgh, Ohio	July 14, 1868	79, 814
Planter, Corn	S. Dixson	Roseville, Ill	Jan. 21, 1873	135, 098
Planter, Corn	J. Doak	Keithsburgh, Ill	Dec. 13, 1864	45, 395
Planter, Corn	S. L. Donnell	Humboldt, Tenn	Mar. 19, 1872	124, 728
Planter, Corn	W. A. Donnell	Greensburgh, Ind	Aug. 13, 1867	67, 735
Planter, Corn	N. Drake	Newton, N. J	Feb. 2, 1858	19, 242
Planter, Corn	W. Drummond	Woodbridge, N. J	July 6, 1858	20, 781
Planter, Corn	T. Duncanson	Buford, Ohio	Jan. 5, 1869	85, 516
Planter, Corn	D. Dutcher	Blue Grass, Iowa	June 26, 1860	28, 847
Planter, Corn	J. W. Eardly	Cascade, Mich	Mar. 9, 1869	87, 651
Planter, Corn	W. A. Eastin	Henderson County, Ky	Jan. 21, 1873	134, 984
Planter, Corn	T. M. Edgar	Paris, Tenn	Oct. 31, 1871	120, 424
Planter, Corn	A. Edwards	Cuba, Mo	Feb. 14, 1871	111, 826
Planter, Corn	R. S. Edwards	Savannah, Mo	Mar. 24, 1868	75, 883
Planter, Corn	Y. I. Edwards	Trenton, Tenn	Jan. 10, 1871	110, 837
Planter, Corn	P. Eidmann	Pekin, Ill	Feb. 11, 1868	74, 326
Planter, Corn	J. Elbertson	Kirksville, Mo	May 12, 1868	77, 807
Planter, Corn	J. Elbertson	Kirksville, Mo	Jan. 5, 1869	85, 652
Planter, Corn	S. Elliott	Washington, Ind	Mar. 15, 1859	23, 235
Planter, Corn	B. H. Elmore	Richmond, Ind	Feb. 19, 1861	31, 448
Planter, Corn	J. H. Elward	Ottawa, Ill	Jan. 5, 1864	41, 176
Planter, Corn	D. J. Ely	Acton, Ind	Dec. 19, 1865	51, 575
Planter, Corn	R. Erdley	Selm's Grove, Pa	Oct. 18, 1870	108, 468
Planter, Corn	J. H. Ernst	Millerstown, Pa	Jan. 8, 1869	91, 106
Planter, Corn	J. W. Fawkes	Decatur, Ill	Aug. 8, 1865	49, 250
Planter, Corn	R. R. Fenner	Paxton, Ill	May 19, 1863	38, 627
Planter, Corn	R. W. Fenwick and R. Bocklen	Brooklyn, N. Y., and Jersey City, N. Y.	Aug. 7, 1855	13, 387
Planter, Corn	J. G. Fetzer	Brunswick, Mo	May 24, 1870	103, 315
Planter, Corn	S. Filson and W. E. Kinert	Bluffton, Ind	Nov. 17, 1868	84, 178
Planter, Corn	W. H. Fish, jr	Scarsdale, N. Y	May 12, 1868	77, 808
Planter, Corn	D. S. Fisher	Cedar Spring, Ind	May 28, 1867	65, 197
Planter, Corn	D. Fitzpatrick and J. Knull	Saint Paris, Ohio	Dec. 21, 1869	98, 157
Planter, Corn	F. G. and E. A. Floyd	Macomb, Ill	Aug. 28, 1860	29, 777
Planter, Corn	M. C. Floyd	Bloomfield, Iowa	Jan. 30, 1866	52, 281
Planter, Corn	W. C. Ford	West Salem, Ohio	Apr. 9, 1861	31, 966
Planter, Corn	J. T. Force	Henry County, Ky	July 11, 1871	111, 827
Planter, Corn	R. Forman	Normal, Ill	July 6, 1869	92, 297
Planter, Corn	C. Foster	Prairie City, Ill	Nov. 24, 1863	40, 681
Planter, Corn	J. S. Fowler	Peoria, Ill	Aug. 14, 1860	29, 578
Planter, Corn	J. M. Foy	Fountain Green, Ill	Apr. 16, 1861	32, 056
Planter, Corn	A. Franklin	Genoa Cross-Roads, Ohio	July 14, 1857	17, 786
Planter, Corn	J. A. Freese	Hanover, Ohio	July 8, 1862	35, 814
Planter, Corn	P. A. Freylinghousen and J. G. Heilman.	Johnstown, Pa	June 21, 1859	24, 452
Planter, Corn	J. P. Fulghum and L. L. Lawrence.	Dublin, Ind	July 1, 1873	140, 493
Planter, Corn	E. C. Gage	Witoka, Minn	Dec. 30, 1873	146, 060
Planter, Corn	T. A. Galt and G. S. Tracy	Sterling, Ill	June 17, 1873	140, 031
Planter, Corn	W. M. Garce	Granville, Ohio	Mar. 6, 1860	27, 358
Planter, Corn	W. C. Gardner	Pokagon, Mich	Feb. 2, 1869	86, 386
Planter, Corn	J. Gilbert	Wyalusing, Wis	Jan. 28, 1868	73, 797
Planter, Corn	R. B. Gilbert	Sutherland Springs, Tex	May 10, 1859	23, 915
Planter, Corn	W. Gilman	Ottawa, Ill	Mar. 15, 1870	100, 748
Planter, Corn	D. H. Gobin	Springfield, Ill	July 2, 1872	128, 613
Planter, Corn	F. A. Goddard	Lexington, Ill	July 31, 1860	29, 433

Index of patents issued from the United States Patent Office from 1790 *to* 1873, *inclusive*—Continued.

Invention.	Inventor.	Residence.	Date.	No.
Planter, Corn	A. J. Going	Clinton, La	May 12, 1868	77, 811
Planter, Corn	J. Golder	Sterling, Ill	Aug. 2, 1864	43, 682
Planter, Corn	W. B. Goodwin	Effingham, Ill	Sept. 29, 1868	82, 515
Planter, Corn	W. B. Goodwin	Kinmundy, Ill	Jan. 5, 1869	85, 581
Planter, Corn	J. M. Gordon and E. Christianson.	Saint Joseph, Mo	Nov. 26, 1867	71, 379
Planter, Corn	H. Gortner	Washport, Ohio	Dec. 14, 1869	97, 906
Planter, Corn	H. Gortner	Washport, Ohio	Nov. 28, 1871	131, 357
Planter, Corn	H. Gortner	Frazeyburgh, Ohio	Nov. 5, 1872	132, 824
Planter, Corn	L. Graham	Plymouth, Ill	Apr. 4, 1871	113, 422
Planter, Corn	J. B. Greely	Summit, Iowa	Oct. 21, 1862	36, 753
Planter, Corn	H. P. Gregg	Cincinnati, Ohio	Mar. 24, 1868	75, 749
Planter, Corn	J. Gross	Decatur, Ill	Jan. 3, 1860	26, 670
Planter, Corn	J. Gross	Decatur, Ill	June 6, 1865	48, 130
Planter, Corn	R. Hackman	Clearfield, Pa	Sept. 3, 1872	131, 097
Planter, Corn	S. H. Hamilton	Deerfield, N. J	June 11, 1872	127, 871
Planter, Corn	J. A. Hamrick	Parnassus, Va	Nov. 9, 1869	96, 586
Planter, Corn	J. W. Harbin	Delaware Station, Ind	Oct. 16, 1860	30, 445
Planter, Corn	E. E. Hardy and W. Dubrul	Joliet, Ill	June 14, 1870	104, 303
Planter, Corn	J. J. Harpel	Lebanon, Pa	Oct. 12, 1869	95, 681
Planter, Corn	S. Harpster	Centre Hall, Pa	Aug. 27, 1867	68, 070
Planter, Corn	W. N. and J. J. Harrison	Hornby, N. Y	June 9, 1868	78, 664
Planter, Corn	S. E. Hartwell	New York, N. Y	May 10, 1859	23, 919
Planter, Corn	C. A. Haskell	Galena, Ill	Dec. 27, 1870	110, 458
Planter, Corn	J. J. S. Hassler	Ripley, Va	Dec. 15, 1857	18, 846
Planter, Corn	P. Hatch	Norwich, Vt	June 22, 1858	20, 639
Planter, Corn	L. L. Haworth	Decatur, Ill	Sept. 19, 1871	119, 142
Planter, Corn	G. D. Haworth	Decatur, Ill	Sept. 3, 1861	33, 196
Planter, Corn	G. D. Haworth	Springfield, Ill	Oct. 18, 1864	44, 725
Planter, Corn	G. D. Haworth	Decatur, Ill	Feb. 22, 1870	100, 032
Planter, Corn	G. D. Haworth	Decatur, Ill	Jan. 14, 1873	134, 747
Planter, Corn	A. Hayes and J. Vancuren	Chenoa, Ill	May 22, 1860	28, 367
Planter, Corn	G. Hayes	Lawrenceburgh, Ind	June 10, 1873	139, 672
Planter, Corn	J. Haynes	Cameron, Ill	Mar. 29, 1859	23, 371
Planter, Corn	A. A. Hazard	New York, N. Y	Aug. 8, 1865	49, 265
Planter, Corn	A. Hearst	Peoria, Ill	Jan. 26, 1869	86, 226
Planter, Corn	A. Hearst	Peoria, Ill	Nov. 7, 1871	120, 643
Planter, Corn	C. W. S. Heaton	Belleville, Ill	Feb. 24, 1863	37, 754
Planter, Corn	G. W. Hendricks	Rushville, Ind	Aug. 6, 1872	130, 220
Planter, Corn	E. E. Henegan	Downsville, Wis	Oct. 7, 1873	143, 450
Planter, Corn	C. W. Henkle	Washington Court-House, Ohio.	Mar. 19, 1867	63, 047
Planter, Corn	C. W. Henkle	Washington Court-House, Ohio.	Feb. 11, 1868	74, 357
Planter, Corn	D., J., and J. F. Herr	Lancaster, Pa	Oct. 9, 1860	30, 316
Planter, Corn	A. E. Herrington and J. D. Richards.	Big Prairie Ronde, Mich	Aug. 25, 1868	81, 502
Planter, Corn	J. A. C. and A. S. Hickman	Summerfield, Ill	Sept. 3, 1861	33, 197
Planter, Corn	W. Hickok	Pharisburgh, Ohio	Oct. 1, 1872	131, 756
Planter, Corn	J. Hildebrand	Taneytown, Md	Dec. 24, 1872	134, 219
Planter, Corn	J. H. Hill and J. T. Hammond	Clinton, Ill	June 23, 1868	79, 120
Planter, Corn	T. M. Hill	Eaton, Ohio	Feb. 6, 1866	52, 413
Planter, Corn	A. F. Hines	Washington, D. C	Oct. 3, 1865	50, 246
Planter, Corn	I. P. Hines	Independence, Iowa	Jan. 5, 1864	41, 120
Planter, Corn	C. Hippensteel	Lee's Cross-Roads, Pa	Apr. 19, 1870	102, 004
Planter, Corn	J. W. Hitchcock and J. K. Deyo	Morrisville, N. Y	Nov. 19, 1867	71, 172
Planter, Corn	J. L. Hoag	Genoa, Ill	Jan. 18, 1859	22, 650
Planter, Corn	W. J. Hobson	Savannah, Mo	Feb. 18, 1867	62, 200
Planter, Corn	A. and E. Hodgson	El Paso, Ill	July 20, 1869	92, 826
Planter, Corn	E. R. Holford	Westford, Wis	Oct. 9, 1866	58, 724
Planter, Corn	H. R. Holland	Wilmington, Va	Oct. 4, 1870	107, 908
Planter, Corn	T. H. Hoskings	Crawfordsville, Ind	Jan. 20, 1844	3, 413
Planter, Corn	G. O. Houck	Springfield, Ohio	Apr. 22, 1873	138, 158
Planter, Corn	W. House	Aurora, Ind	Oct. 21, 1873	143, 905
Planter, Corn	H. J. Howe	Onargo, Ill	May 14, 1861	32, 294
Planter, Corn	M. A. Howell, jr	Ottawa, Ill	Apr. 10, 1860	27, 807
Planter, Corn	W. W. Hubbard	Edinburgh, Ind	Apr. 24, 1866	54, 166
Planter, Corn	W. W. Hubbard	Edinburgh, Ind	July 23, 1867	67, 115
Planter, Corn	D. W. Hughes	New London, Mo	Sept. 8, 1857	18, 148
Planter, Corn	J. Hughes and N. Stonecipher	Cambridge, Ind	Jan. 18, 1859	22, 652
Planter, Corn	N. G. Hughes	Waynesburgh, Pa	Feb. 4, 1868	74, 092
Planter, Corn	D. Humphreys	Cincinnati, Ohio	Apr. 16, 1861	32, 065
Planter, Corn	W. Hunter	Hastings, Minn	Apr. 9, 1867	63, 724
Planter, Corn	W. H. Hunter	Ridgefield, Ill	May 16, 1865	47, 728
Planter, Corn	C. Hutchins	Aubrey, Kans	Oct. 21, 1873	143, 826
Planter, Corn	B. Ingebrightson	Cambridge, Wis	Nov. 7, 1871	120, 649
Planter, Corn	J. Ingels and W. R. Mount	Milton, Ind	Feb. 11, 1873	135, 813
Planter, Corn	H. Ingraham	Naples, N. Y	Oct. 27, 1857	18, 508
Planter, Corn	J. Jackson, jr	Chatsworth, Ill	Aug. 15, 1871	118, 134
Planter, Corn	D. W. Jacoby	Shelbyville, Ill	Sept. 3, 1867	68, 443
Planter, Corn	J. D. Jeffers, J. Sparks, and J. H. Jeffers.	Philadelphia, Pa	July 29, 1856	15, 426
Planter, Corn	L. K. Jenne	Grand Rapids, Mich	Feb. 12, 1861	31, 427
Planter, Corn	J. John	Massillon, Ohio	Nov. 26, 1867	71, 493
Planter, Corn	H. J. Johnson	Saint Peter, Minn	Dec. 24, 1867	72, 501
Planter, Corn	J. Johnson	Naples, Ill	May 29, 1860	28, 490
Planter, Corn	J. A. Johnson	Pendleton, Ind	Aug. 31, 1869	94, 417
Planter, Corn	J. B. Johnson	Rock Island, Ill	July 20, 1869	92, 728
Planter, Corn	J. J. Jones and S. H. Dwight	Decatur, Ill	July 9, 1872	128, 888
Planter, Corn	S. F. Jones	Saint Paul, Ind	Apr. 5, 1864	42, 196
Planter, Corn	S. W. Jones	Bluffton, Ind	Sept. 29, 1868	82, 531
Planter, Corn	H. Jordan	Milton, Ohio	Jan. 3, 1865	45, 721
Planter, Corn	W. H. Karicofe	Harrisonburgh, Va	June 19, 1866	55, 668
Planter, Corn	H. Keeney and C. H. See	Danville and New Florence, Mo.	July 7, 1868	79, 576
Planter, Corn	D. Keethler	Mount Oreb, Ohio	Nov. 23, 1869	97, 200
Planter, Corn	D. Keethler	Mount Oreb, Ohio	July 5, 1870	105, 090
Planter, Corn	C. A. Kellog	Elyria, Ohio	Nov. 24, 1868	84, 426

Index of patents issued from the United States Patent Office from 1790 *to* 1873, *inclusive*—Continued.

Invention.	Inventor.	Residence.	Date.	No.
Planter, Corn	W. G. Kenedy	Greenfield, Ind	Nov. 8, 1864	44, 955
Planter, Corn	D. A. Kershner	Elliottstown, Ill	Apr. 21, 1868	77, 050
Planter, Corn	I. J. Kidd	Young Settlement, Tex	Sept. 8, 1868	81, 911
Planter, Corn	E. W. Kimball	Ottawa, Ill	Jan. 8, 1861	31, 085
Planter, Corn	P. H. Kimball	Prophetstown, Ill	Aug. 1, 1865	49, 121
Planter, Corn	W. H. King	Charleston, Ill	May 10, 1859	23, 980
Planter, Corn	J. M. Kiracofe	Mount Solon, Va	Apr. 19, 1870	102, 017
Planter, Corn	A. Kirlin	New Boston, Ill	Oct. 18, 1859	25, 833
Planter, Corn	M. L. and J. B. Kissell	Springfield, Ohio	Sept. 20, 1870	107, 508
Planter, Corn	J. Klar	Shelbyville, Ill	Dec. 30, 1873	145, 952
Planter, Corn	G. W. Knapp	Corning, N. Y	June 11, 1867	65, 755
Planter, Corn	J. A. Knotzer	Fillmore, Ind	Aug. 1, 1871	117, 646
Planter, Corn	M. Knickerbocker	New Lenox, Ill	Sept. 16, 1873	142, 922
Planter, Corn	J. I. Knight	Philadelphia, Pa	Jan. 10, 1860	26, 819
Planter, Corn	J. Knull and J. P. Pence	Saint Paris, Ohio	Oct. 3, 1871	119, 467
Planter, Corn	H. Koeller	Camp Point, Ill	Nov. 15, 1870	109, 225
Planter, Corn	H. Koeller and W. Uecke	Camp Point, Ill	July 16, 1872	129, 144
Planter, Corn	J. Krebs and A. Johns	Massillon, Ohio	Oct. 1, 1867	69, 449
Planter, Corn	J. L. Kreider	Chesnut Level, Pa	Jan. 16, 1872	122, 770
Planter, Corn	R. Kuschke and P. Merkel	Saint Louis, Mo	May 26, 1857	17, 380
Planter, Corn	E. P. Lacey	Rochester, N. Y	Apr. 8, 1856	14, 631
Planter, Corn	L. Lacey	Bath, Ill	Apr. 29, 1873	138, 259
Planter, Corn	A. Ladd	Saint Lawrence, N. Y	Sept 4, 1866	57, 734
Planter, Corn	D. Ladd	Dearborn, Mich	Nov. 30, 1858	22, 183
Planter, Corn	J. G. La Ponte	Dayton, Ohio	Nov. 25, 1873	144, 989
Planter, Corn	N. C. Lamb	Wyandotte, Mich	Sept. 9, 1873	142, 706
Planter, Corn	R. B. Lanum	Washington, Ohio	May 17, 1864	42, 781
Planter, Corn	H. B. Lansing	Hudson, Mich	Feb. 27, 1866	52, 861
Planter, Corn	H. B. Lansing and H. W. Grenell	Hudson, Mich	Sept. 30, 1862	36, 569
Planter, Corn	L. Larchar	Utica, N. Y	Aug. 27, 1867	68, 091
Planter, Corn	E. B. Lawrence and C. Quick	Lakeville, Ohio	Nov. 10, 1868	83, 979
Planter, Corn	D. P. Leach	Franklin, Ind	Mar. 23, 1869	88, 048
Planter, Corn	J. L. Leas	York Sulphur Springs, Pa	Sept. 22, 1868	82, 422
Planter, Corn	J. L. Leas	Hampton, Pa	Aug. 13, 1872	130, 378
Planter, Corn	J. Lee	Galesburgh, Ill	May 22, 1860	28, 383
Planter, Corn	W. Lees	Germantown, Ohio	Sept. 13, 1859	25, 425
Planter, Corn	J. C. Leffel	Shelbina, Mo	Dec. 1, 1863	40, 760
Planter, Corn	G. W. Lewis	Winchester, Ky	Apr. 4, 1871	113, 717
Planter, Corn	N. A. Lewis	Glen's Falls, N. Y	July 21, 1857	17, 841
Planter, Corn	N. Liddell	La Fayette, N. Y	Apr. 2, 1867	63, 401
Planter, Corn	N. and M. Liddell	La Fayette, N. Y	Mar. 10, 1868	75, 435
Planter, Corn	O. Lippincott	Camden, N. J	Apr. 27, 1858	20, 074
Planter, Corn	W. H. Littel	Prairie du Chien, Wis	July 26, 1870	105, 702
Planter, Corn	H. C. Locke	Somerville, Tenn	Apr. 21, 1868	77, 057
Planter, Corn	C. Long	Paris, Ill	Jan. 1, 1867	60, 914
Planter, Corn	F. Loos	Denver City, Colo	Aug. 9, 1864	43, 782
Planter, Corn	J. T. Lowrey, J. A. Case, and R. Chew.	High Banks, Ind	Mar. 3, 1868	75, 174
Planter, Corn	D. and D. F. Luse	Spring Mills, Pa	Mar. 27, 1866	53, 459
Planter, Corn	J. W. Magers	Reinersville, Ohio	May 3, 1870	102, 564
Planter, Corn	N. Maisonneuve	Kankakee, Ill	May 11, 1869	90, 008
Planter, Corn	S. Malone	Tremont, Ill	Jan. 3, 1854	10, 393
Planter, Corn	W. H. Maple	Chariton, Iowa	Apr. 28, 1863	38, 313
Planter, Corn	T. Marcus	New York, N. Y	Dec. 1, 1857	18, 756
Planter, Corn	A. S. Markham	Monmouth, Ill	Oct. 27, 1863	40, 418
Planter, Corn	A. S. and D. Markham	Monmouth, Ill	Dec. 11, 1860	30, 883
Planter, Corn	T. R. Markillie	Winchester, Ill	Nov. 13, 1855	13, 794
Planter, Corn	A. C. Martin	Waddam's Centre, Ill	Apr. 8, 1873	137, 700
Planter, Corn	A. C. Martin and R. Ferguson	Lena, Ill	Dec. 2, 1873	145, 115
Planter, Corn	J. Matthews	Lincoln, Ill	Nov. 21, 1871	121, 116
Planter, Corn	H. Maxell	Canton, Ohio	Feb. 19, 1867	62, 280
Planter, Corn	J. A. McClure	Mount Carroll, Ill	July 19, 1870	105, 473
Planter, Corn	T. B. McConaughey	Newark, Del	Dec. 4, 1866	60, 218
Planter, Corn	R. McCorkell	Warsaw, Minn	Oct. 25, 1864	44, 810
Planter, Corn	W. H. McCormick	Muncie, Ind	July 6, 1869	92, 200
Planter, Corn	D. McCullough	Oxford Township, Canada	July 27, 1869	92, 984
Planter, Corn	O. C. McCune	Darby Creek, Ohio	Dec. 13, 1859	26, 442
Planter, Corn	O. C. McCune	Pleasant Valley, Ohio	Aug. 2, 1864	43, 699
Planter, Corn	O. C. McCune	Darby Creek, Ohio	Aug. 7, 1866	56, 969
Planter, Corn	I. W. McGaffey	Buffalo, N. Y	June 16, 1857	17, 566
Planter, Corn	J. McGinnis	Frazeysburgh, Ohio	Dec. 10, 1872	133, 789
Planter, Corn	W. McLucas	Reinersville, Ohio	Feb. 4, 1868	74, 112
Planter, Corn	W. McLucas	Reinersville, Ohio	July 20, 1869	92, 861
Planter, Corn	N. Mendenhall	Greensburgh, Ind	Dec. 6, 1870	109, 923
Planter, Corn	N. Mendenhall	Greensburgh, Ind	Nov. 19, 1872	133, 109
Planter, Corn	C. T. Merry and M. A. Dunton	Norwalk, Ohio	Aug. 8, 1871	117, 910
Planter, Corn	C. T. Merry and M. A. Dunton	Norwalk, Ohio	Oct. 15, 1872	132, 306
Planter, Corn	A. Merwin and C. H. Hobart	Padua, Ill	Nov. 3, 1863	40, 492
Planter, Corn	S. P. Metz and M. Rohrer	McDonaldsville, Ohio	Sept. 25, 1866	58, 277
Planter, Corn	W. M. Meyers	New Brunswick, N. J	Sept. 20, 1870	107, 520
Planter, Corn	S. Mickley	York, Pa	Sept. 13, 1870	107, 280
Planter, Corn	G. G. J. Millar	Lockbourne, Ohio	Aug. 20, 1872	130, 653
Planter, Corn	A. Miller	Brockport, N. Y	Aug. 14, 1860	29, 653
Planter, Corn	J. Miller	Bucyrus, Ohio	Mar. 31, 1857	16, 930
Planter, Corn	P. Mills	Ridge Farm, Ill	Oct. 27, 1863	40, 423
Planter, Corn	T. S. Mills	Iberia, Ohio	June 26, 1860	28, 886
Planter, Corn	H. Mitchell	Osborn, Ohio	July 7, 1868	79, 588
Planter, Corn	H. S. Mitchell and C. Search	Hublersburgh, Pa	Aug. 6, 1867	67, 444
Planter, Corn	J. G. Mitchell	Collington, Md	Mar. 29, 1859	23, 382
Planter, Corn	M. Mitchell	Altona, Ill	Aug. 9, 1859	25, 033
Planter, Corn	N. B. Moody	Woodman, Wis	Oct. 3, 1871	119, 470
Planter, Corn	J. Moore	Bushnell, Ill	Feb. 18, 1868	74, 710
Planter, Corn	J. C. Moore	Peoria, Ill	May 29, 1860	28, 498
Planter, Corn	J. C. Moore	Peoria, Ill	July 8, 1862	35, 831
Planter, Corn	Q. R, and P. Moor and E. L. Patrick.	Forest Hill, Ind	Dec. 1, 1868	84, 568
Planter, Corn	W. H. Moore	Blooming Grove, Ind	Feb. 18, 1868	74, 711
Planter, Corn	O. P. Moran	Hainesville, Mo	Nov. 8, 1859	26, 044

Index of patents issued from the United States Patent Office from 1790 to 1873, inclusive—Continued.

Invention.	Inventor.	Residence.	Date.	No.
Planter, Corn	R. W. Moran	Chicago, Ill	Dec. 24, 1867	72, 525
Planter, Corn	L. Morris	Woodbury, Ill	June 26, 1860	28, 891
Planter, Corn	W. Morrison	Carlisle, Pa	Sept. 13, 1859	25, 435
Planter, Corn	W. Morrison	Carlisle, Pa	Apr. 4, 1871	113, 550
Planter, Corn	P. C. Mosier	Homer, Ill	Jan. 26, 1858	19, 198
Planter, Corn	S. Mowry and E. Deppen	Womelsdorf, Pa	Dec. 11, 1860	30, 885
Planter, Corn	D. Moyer	New Hamburgh, Pa	June 5, 1860	28, 595
Planter, Corn	G. W. Moyers	Gordonsville, Va	June 29, 1869	92, 084
Planter, Corn	W. Mull	Rantoul, Ill	Aug. 12, 1873	141, 723
Planter, Corn	J. Y. D. Murphy	Half Moon, Pa	Jan. 22, 1861	31, 186
Planter, Corn	J. Y. D. Murphy	Half Moon, Pa	Mar. 8, 1864	41, 854
Planter, Corn	C. C. Myers	Sterling, Ill	Oct. 16, 1866	58, 876
Planter, Corn	C. K. Myers	Pekin, Ill	Apr. 23, 1861	32, 168
Planter, Corn	D. C. Myers	Richmondale, Ohio	June 12, 1860	28, 684
Planter, Corn	E. J. Myers	Onawa, Iowa	Dec. 5, 1871	121, 649
Planter, Corn	D. Nichols	Onargo, Ill	Jan. 3, 1860	26, 700
Planter, Corn	H. A. Nicholls	Saint Louis, Mo	Sept. 13, 1870	107, 402
Planter, Corn	J. Nitkey	Minneapolis, Minn	Apr. 8, 1873	137, 563
Planter, Corn	A. J. Noe	Mitchell, Ind	Nov. 5, 1872	132, 770
Planter, Corn	R. F. Norwood	Charlotte, N. C	Feb. 28, 1871	112, 171
Planter, Corn	J. Olmsted	Knoxville, Ill	Jan. 19, 1864	41, 345
Planter, Corn	J. Olmsted	Knoxville, Ill	Nov. 1, 1864	44, 920
Planter, Corn	S. Y. Orr	Morning Sun, Iowa	June 29, 1869	91, 961
Planter, Corn	W. F. Osgood	Lowell, Mass	Jan. 12, 1864	41, 263
Planter, Corn	G. Gaddington	Waubeck, Iowa	Dec. 6, 1870	109, 837
Planter, Corn	A. G. Parker	North Gage, N. Y	July 29, 1862	36, 020
Planter, Corn	J. B. Parker	Knobnoster, Mo	Dec. 14, 1869	97, 957
Planter, Corn	T. H. Parker and D. Kellison	Parkersburgh, Ill	Oct. 15, 1867	69, 836
Planter, Corn	E. Parmentier	Clifton, Ill	July 1, 1873	140, 432
Planter, Corn	J. M. Parsons	Charles City, Iowa	Feb. 14, 1871	111, 774
Planter, Corn	B. F. Partridge	Syracuse, N. Y	Jan. 9, 1849	6, 016
Planter, Corn	G. F. Partridge	Adrian, Mich	Sept. 15, 1868	82, 155
Planter, Corn	J. I. Patton	Tiffin, Ohio	July 27, 1869	93, 116
Planter, Corn	E. Peck	Middleport, Ill	June 23, 1863	38, 979
Planter, Corn	G. H. Peek	East Hamburgh, N. Y	Nov. 22, 1870	109, 543
Planter, Corn	L. G. Peel	Webster County, Ga	Aug. 9, 1859	25, 041
Planter, Corn	W. T. Pepper	Rising Sun, Ind	June 16, 1857	17, 582
Planter, Corn	S. M. Perkins	Fort Hill, Ill	Feb. 3, 1857	16, 551
Planter, Corn	L. B. Phelps	Geneva, Ohio	May 18, 1858	20, 297
Planter, Corn	G. W. Phillips and B. C. Richardson.	Oconomowoc, Wis	Feb. 16, 1869	86, 941
Planter, Corn	D. Pierson, J. W. Macy, and J. D. Moore.	Grinnell, Iowa	Feb. 25, 1873	136, 337
Planter, Corn	J. Pies	Sparland, Ill	Dec. 26, 1871	122, 191
Planter, Corn	J. M. Piper	Harrison City, Pa	July 12, 1870	105, 365
Planter, Corn	G. W. Pittman	Winona, Miss	Aug. 9, 1870	106, 202
Planter, Corn	P. Platter	Moore's Hill, Ind	July 5, 1859	24, 703
Planter, Corn	F. W. Poe, jr	Vandalia, Ill	Jan. 3, 1871	110, 786
Planter, Corn	W. R. Pomeroy	Millersburgh, Ohio	Apr. 8, 1862	34, 902
Planter, Corn	B. Porter	Ossian, N. Y	May 24, 1870	103, 499
Planter, Corn	H. G. Porter	Hopkinton, Iowa	Dec. 6, 1870	109, 936
Planter, Corn	N. M. Powers	Kirksville, Mo	Mar. 30, 1869	88, 508
Planter, Corn	F. B. Preston	Fayette, Mo	Feb. 26, 1861	31, 579
Planter, Corn	J. Price	Harrison, Ohio	July 10, 1860	29, 100
Planter, Corn	B. L. Prime	Hamilton, Ohio	Dec. 1, 1857	18, 768
Planter, Corn	W. S. Purdey	Butler, Ind	Sept. 21, 1869	95, 139
Planter, Corn	A. Putnam	Owego, N. Y	Sept. 18, 1866	58, 135
Planter, Corn	J. Rader	Daleville, Ind	July 20, 1869	92, 882
Planter, Corn	J. B. Raines	Tremont, Iowa	Apr. 23, 1867	64, 143
Planter, Corn	P. Raines	London, Ohio	May 1, 1855	12, 783
Planter, Corn	F. A. Ramey and R. R. Cross	Woodstock, Va	Oct. 17, 1871	119, 950
Planter, Corn	B. Randall	Adams, N. Y	Apr. 14, 1868	76, 658
Planter, Corn	J. R. Randall	Camargo, Ill	Feb. 2, 1869	86, 589
Planter, Corn	J. R. Randall	Camargo, Ill	Jan. 3, 1871	110, 788
Planter, Corn	S. G. Randall	Rockford, Ill	Apr. 29, 1856	14, 776
Planter, Corn	J. H. Rankin	Versailles, Mo	Dec. 4, 1869	30, 836
Planter, Corn	W. Rayhill	Pana, Ill	Jan. 21, 1868	73, 649
Planter, Corn	L. M. Reamy	Kokomo, Ind	Mar. 12, 1867	62, 886
Planter, Corn	H. Reed and W. P. Pennewell	Middletown, Mo	Aug. 14, 1866	57, 259
Planter, Corn	S. B. Reeder	Meacham, Ill	Oct. 6, 1868	82, 875
Planter, Corn	J. M. Reeds	Millwood Township, Mo	Feb. 12, 1867	62, 067
Planter, Corn	J. Rice	Prairie Creek, Ind	Nov. 12, 1872	133, 055
Planter, Corn	W. E. Rich	New Providence, Iowa	June 26, 1866	55, 908
Planter, Corn	S. Richardson	Jericho, Vt	June 16, 1857	17, 584
Planter, Corn	J. Richeldorfer	Cridersville, Ohio	Jan. 3, 1871	110, 679
Planter, Corn	J. S. Rickel	Geneseo, Ill	Sept. 25, 1866	58, 295
Planter, Corn	J. S. Rickel	Geneseo, Ill	Mar. 12, 1867	62, 888
Planter, Corn	J. W. Ricketts	Charlestown, Ill	Nov. 23, 1869	97, 229
Planter, Corn	J. J. Rider	Wilton Junction, Iowa	Oct. 13, 1863	40, 281
Planter, Corn	P. Ritchey	Hamilton, Ill	Apr. 16, 1861	32, 084
Planter, Corn	M. Robbins	Cincinnati, Ohio	Feb. 10, 1857	16, 611
Planter, Corn	B. F. Robertson	Cap Au Gris, Mo	May 28, 1867	65, 278
Planter, Corn	P. Rogers	Sharon, Ohio	Mar. 30, 1869	88, 415
Planter, Corn	A. Rubmer	Newport, Ky	June 3, 1873	139, 617
Planter, Corn	C. Ropp	McLean County, Ill	Oct. 4, 1859	25, 575
Planter, Corn	L. Ruf	Rock Falls, Ill	July 16, 1872	129, 595
Planter, Corn	A. Runstetler	Peoria, Ill	Apr. 13, 1869	88, 812
Planter, Corn	D. Ruppart	Ninisilla, Ohio	Feb. 19, 1867	62, 280
Planter, Corn	C. B. Ruth	Doylestown, Pa	Dec. 21, 1869	98, 193
Planter, Corn	H. Ruth	Summerfield, Ill	June 10, 1862	35, 545
Planter, Corn	J. B. Ryder	Wapello, Iowa	Apr. 21, 1863	38, 214
Planter, Corn	W. H. Sabins	Rockford, Ill	Feb. 2, 1869	86, 595
Planter, Corn	W. J. Sager	Milesburgh, Pa	Apr. 18, 1871	113, 936
Planter, Corn	J. M. Sampson	Waynesville, Ill	Nov. 19, 1867	71, 070
Planter, Corn	O. Sampson	Petersburgh, Ill	May 26, 1868	78, 237
Planter, Corn	W. G. Savage	Clinton, Ill	Sept. 25, 1860	30, 158
Planter, Corn	M. Saviers and W. N. Ayres	Wyandot and Bristolville, Ohio.	May 30, 1865	47, 987

Index of patents issued from the United States Patent Office from 1790 *to* 1873, *inclusive*—Continued.

Invention.	Inventor.	Residence.	Date.	No.
Planter, Corn	M. Schnapp and W. J. Hollis	De Witt, Mo	Apr. 16, 1872	125, 847
Planter, Corn	C. Schnepf	Lancaster, Pa	July 28, 1857	17, 898
Planter, Corn	J. Schott	Peoria, Ill	Feb. 21, 1871	111, 979
Planter, Corn	L. Scofield	Watertown, Wis	Feb. 7, 1871	111, 687
Planter, Corn	L. Scofield	Watertown, Wis	Nov. 21, 1871	121, 204
Planter, Corn	A. Seaman	Winnebago County, Ill	May 1, 1860	28, 106
Planter, Corn	J. Seibel	Manlius, Ill	May 30, 1865	47, 990
Planter, Corn	J. Selby	Peoria, Ill	Aug. 30, 1864	44, 019
Planter, Corn	J. Selby	Peoria, Ill	Sept. 15, 1868	82, 162
Planter, Corn	J. Selby	Peoria, Ill	July 13, 1869	92, 659
Planter, Corn	J. Selby	Peoria, Ill	May 14, 1872	126, 751
Planter, Corn	W. G. Selby and J. Bowman	Princeville, Ill	June 4, 1872	127, 648
Planter, Corn	J. P. Selsor	Cherry Box, Mo	June 11, 1867	65, 771
Planter, Corn	H. C. Shafer	Petersburgh, Ind	June 15, 1869	91, 489
Planter, Corn	H. A. Sharp	Harlem, Mo	May 27, 1873	139, 269
Planter, Corn	T. T. Shawcross	Allisonville, Ind	Oct. 2, 1866	58, 557
Planter, Corn	J. Sheer	Lincoln, Ill	Feb. 18, 1873	135, 942
Planter, Corn	D. J. Sheers	Belmont, Wis	Nov. 14, 1871	121, 012
Planter, Corn	A. Shellabarger	Miami County, Ohio	July 20, 1869	92, 889
Planter, Corn	P. Shellenberger	Millerstown, Pa	Aug. 3, 1869	93, 237
Planter, Corn	W. H. Shepherd	College Corner, Ohio	Nov. 12, 1867	70, 753
Planter, Corn	L. Sipe	Keezletown, Va	Nov. 18, 1873	144, 708
Planter, Corn	E. and E. C. Slosson	Vienna, Ill	Apr. 27, 1869	89, 512
Planter, Corn	F. J. Smiley	Marshall, Mich	Dec. 10, 1867	71, 917
Planter, Corn	A. D. Smith	Cincinnati, Ohio	Jan. 14, 1873	134, 942
Planter, Corn	A. F. Smith	Montrose, Iowa	Sept. 30, 1873	143, 262
Planter, Corn	A. N. Smith	Chillicothe, Mo	Sept. 23, 1873	143, 036
Planter, Corn	C. Smith	Nauvoo, Ill	July 31, 1860	29, 412
Planter, Corn	E. F. Smith	Orangeville, Ill	Jan. 15, 1867	61, 273
Planter, Corn	F. J. Smith	Four Corners, Ohio	May 26, 1857	17, 397
Planter, Corn	J. D. Smith	Muskegon, Mich	Feb. 25, 1873	136, 341
Planter, Corn	J. L. Smith	Neoga, Ill	June 26, 1860	28, 943
Planter, Corn	J. L. Smith	Neoga, Ill	Oct. 6, 1863	40, 194
Planter, Corn	J. M. Smith	Galva, Ill	Feb. 14, 1865	46, 398
Planter, Corn	W. T. F. Smith	Lexington, Ill	Dec. 10, 1872	133, 808
Planter, Corn	J. H. Sorey	Xenia, Ill	May 12, 1863	38, 510
Planter, Corn	P. Soule	Windsor, N. Y	Sept. 13, 1870	107, 302
Planter, Corn	A. Springsteen	Oquawka, Ill	Nov. 18, 1873	144, 711
Planter, Corn	A. H. Stark	Nevada, Iowa	May 20, 1873	139, 205
Planter, Corn	A. H. Stark and J. C. Mitchell	Nevada, Iowa	Oct. 24, 1871	120, 340
Planter, Corn	P. S. Stames	Majority Point, Ill	Mar. 22, 1870	101, 176
Planter, Corn	G. W. Starrett	Dublin, Ohio	Sept. 9, 1873	142, 656
Planter, Corn	J. Statz	Cross Plains, Wis	Nov. 4, 1873	144, 233
Planter, Corn	S. Stevens	North Fryeburgh, Me	Nov. 7, 1871	120, 793
Planter, Corn	W. Stinson	Georgetown, Pa	July 3, 1855	13, 185
Planter, Corn	V. D. Stoddard	Muscatine, Iowa	Nov. 22, 1864	45, 192
Planter, Corn	C. S. Stone	Greenwood, Ind	July 30, 1872	130, 082
Planter, Corn	J. Stone and J. T. Archibald	Wapello, Iowa	Apr. 16, 1861	32, 091
Planter, Corn	P. Stover	West Alexandria, Ohio	June 5, 1860	28, 615
Planter, Corn	B. T. Stowell	Quincy, Ill	May 22, 1860	28, 418
Planter, Corn	L. F. Straight	Fairbury, Ill	Sept. 11, 1860	30, 008
Planter, Corn	A. Streeter	Adrian, Mich	May 13, 1862	35, 280
Planter, Corn	L. Study	Plum Hollow, Iowa	Oct. 15, 1867	69, 862
Planter, Corn	S. L. Sweeney	Morrison, Ill	Oct. 13, 1868	83, 105
Planter, Corn	N. Swigart	Richfield, Ohio	Oct. 4, 1870	108, 065
Planter, Corn	D. F. Taft	New Bedford, Mass	Oct. 20, 1868	83, 338
Planter, Corn	D. F. Taft	New Bedford, Mass	May 27, 1873	139, 340
Planter, Corn	J. O. Talmage	Cardington, Ohio	July 16, 1872	129, 619
Planter, Corn	L. B. Tarbox	Colliersville, N. Y	May 3, 1870	102, 618
Planter, Corn	G. Taylor	Richmond, Indiana	Apr. 20, 1858	20, 024
Planter, Corn	R. Taylor and R. Sprague	Prairie City, Ill	June 26, 1860	28, 917
Planter, Corn	C. W. Thiessero	Effingham, Ill	Oct. 27, 1868	83, 567
Planter, Corn	J. C. Thomas	Redpoint, Md	Apr. 25, 1865	47, 468
Planter, Corn	A. G. and A. J. Thompson	Belleville, Ohio	Mar. 1, 1859	23, 126
Planter, Corn	G. Thomson	Springfield, Ill	June 11, 1872	127, 938
Planter, Corn	J. Thompson	Aledo, Ill	July 16, 1872	129, 624
Planter, Corn	J. Thomson and J. Ramsey	Aledo, Ill	Sept. 27, 1864	44, 472
Planter, Corn	G. W. Thrall and W. L. Rayment.	Burlington, Mich	Dec. 29, 1868	85, 346
Planter, Corn	E. A. Thrush	New Kingston, Pa	Feb. 22, 1870	100, 215
Planter, Corn	J. S. Toan	Venice, N. Y	Jan. 13, 1857	16, 410
Planter, Corn	W. S. Todd	Mechanicsville, Iowa	Mar. 3, 1863	37, 834
Planter, Corn	B. F. Tomb	Tiffin, Ohio	June 21, 1870	104, 664
Planter, Corn	W. F. Tunnard	East Baton Rouge Parish, La.	July 27, 1869	93, 140
Planter, Corn	C. G. Udell	Morris, Ill	June 28, 1859	24, 594
Planter, Corn	F. Underwood	South Rutland, N. Y	Sept. 17, 1872	131, 482
Planter, Corn	J. Underwood	Cameron, Ill	Sept. 25, 1860	30, 169
Planter, Corn	H. Upjohn	Richland, Mich	Sept. 20, 1864	44, 355
Planter, Corn	A. Upson	Randolph, Ohio	Apr. 17, 1866	54, 042
Planter, Corn	J. M. E. Valk	Baltimore, Md	Mar. 12, 1872	124, 643
Planter, Corn	A. J. Van Boekel	Uniontown, N. J	Feb. 6, 1866	52, 471
Planter, Corn	J. W. Vandiver	Shelbina, Mo	Oct. 6, 1863	40, 2[illegible]2
Planter, Corn	F. Van Doren	Adrian, Mich	Dec. 1, 1868	84, 594
Planter, Corn	C. Van Honten	Sunbury, Ohio	Sept. 21, 1858	21, 583
Planter, Corn	S. and H. Vannuys	Bethel, Ind	Jan. 21, 1873	135, 178
Planter, Corn	R. M. Varner	Oxford, Miss	Oct. 4, 1859	25, 690
Planter, Corn	G. B. Vaughn	Marshall, Mo	Oct. 25, 1870	108, 656
Planter, Corn	G. J. Vaught	Hanley, Ky	Apr. 18, 1871	113, 819
Planter, Corn	W. E. Vernon	Franklin County, Mo	May 17, 1870	103, 259
Planter, Corn	C. L. Waffle	Sharon, Ohio	Aug. 21, 1860	29, 734
Planter, Corn	D. F. Wagner	West Hanover, Pa	Nov. 23, 1869	97, 139
Planter, Corn	E. M. Walker	Gallatin, Mo	May 1, 1866	54, 451
Planter, Corn	J. F. Walker	Murrayville, Ill	Nov. 3, 1868	83, 677
Planter, Corn	J. C. Walkinshaw	Leavenworth, Kans	Dec. 11, 1866	60, 447
Planter, Corn	W. J. Wallingford	Portland, Ill	Nov. 5, 1872	132, 783

Index of patents issued from the United States Patent Office from 1790 *to* 1873, *inclusive*—Continued.

Invention.	Inventor.	Residence.	Date.	No.
Planter, Corn	H. W. Wansbrough and H. M. Diggins.	Cincinnati, Ohio	Sept. 20, 1864	44, 357
Planter, Corn	J. P. Ware	Montgomery City, Mo	Mar. 23, 1869	88, 244
Planter, Corn	D. R. Warfield	Muscatine, Iowa	July 31, 1866	56, 839
Planter, Corn	J. Waterman	Keokuk, Iowa	May 3, 1864	42, 612
Planter, Corn	Z. D. Waters	Brookville, Md	July 1, 1873	140, 562
Planter, Corn	J. R. Weber	Bourbon, Ind	Dec. 10, 1867	71, 932
Planter, Corn	R. Welch	Worth, Ill	Oct. 1, 1861	33, 410
Planter, Corn	J. D. Wells	Franklin County, Ohio	Apr. 30, 1867	64, 390
Planter, Corn	D. F. Welsh	Nevada, Ohio	Jan. 3, 1871	110, 807
Planter, Corn	J. K. Welter	Springfield, Ill	July 16, 1872	129, 380
Planter, Corn	J. K. Welter	Springfield, Ill	Dec. 24, 1872	134, 336
Planter, Corn	J. E. West	Georgetown, Ky	Feb. 26, 1867	62, 458
Planter, Corn	J. W. West	Hillsborough, Ohio	July 5, 1859	24, 683
Planter, Corn	L. West	Georgetown, Ky	Aug. 30, 1870	106, 898
Planter, Corn	C. L. Westbrook	New York, N. Y	Mar. 7, 1865	46, 738
Planter, Corn	F. F. Westerfield	Fort Dodge, Iowa	Feb. 5, 1867	61, 783
Planter, Corn	C. Whitaker	Davenport, Iowa	Sept 13, 1859	25, 461
Planter, Corn	F. W. White	Worcester, Mass	Aug. 31, 1858	21, 393
Planter, Corn	H. Whitman	Kingsville, Ohio	Aug. 24, 1858	21, 287
Planter, Corn	J. M. Whitmore and J. N. Arvin	Valparaiso, Ind	Jan. 3, 1871	110, 811
Planter, Corn	T. H. Wible	Quincy, Ill	Mar. 31, 1868	76, 282
Planter, Corn	B. Wieland	Orangeville, Ill	July 16, 1867	66, 923
Planter, Corn	L. Wright and O. G. Ewings	Whitewater and La Grange, Wis.	July 16, 1872	129, 383
Planter, Corn	H. Wiley	Frankfort, Ohio	July 5, 1859	24, 687
Planter, Corn	F. L. Wilkens	Saint Mary's, Ohio	Dec. 7, 1869	97, 739
Planter, Corn	M. Wilkinson	Burlington, Mich	May 8, 1866	54, 633
Planter, Corn	W. C. Willey	Princeton, Ill	Jan. 22, 1861	31, 201
Planter, Corn	T. C. Williams	Warrenton, Mo	July 16, 1872	129, 446
Planter, Corn	U. T. Wilson	De Soto City, Miss	Mar. 28, 1871	113, 233
Planter, Corn	W. W. Wilson	Collins Station, Ill	Jan. 14, 1862	34, 176
Planter, Corn	A. Windeck	Peoria, Ill	Dec. 1, 1868	84, 666
Planter, Corn	A. Windeck	Peoria, Ill	Mar. 26, 1872	124, 926
Planter, Corn	A. Windeck and A. Runstetler	Peoria, Ill	Dec. 18, 1866	60, 604
Planter, Corn	L. Winston	Pontiac, Ill	Aug. 30, 1870	106, 976
Planter, Corn	C. Wisdom	Flat Rock, Mich	Nov. 1, 1870	108, 865
Planter, Corn	S. Witt and G. W. Albaugh	Greencastle, Pa	Apr. 29, 1856	14, 785
Planter, Corn	D. and H. Wolf	Lebanon, Pa	Apr. 28, 1860	29, 839
Planter, Corn	D. D. Wood	Paris, Ill	Feb. 16, 1869	87, 018
Planter, Corn	G. H. Wood	Cambridge City, Ind	June 15, 1869	91, 399
Planter, Corn	J. B. Woolsey	Bloomfield, Iowa	Oct. 18, 1864	44, 762
Planter, Corn	E. M. Wright	Wilmington, Ohio	June 13, 1865	48, 229
Planter, Corn	J. J. Wright and J. H. Penny	Harrison, Ohio	May 19, 1868	78, 171
Planter, Corn	L. D. Wyatt	Castleton, Ind	Nov. 17, 1868	84, 075
Planter, Corn	H. M. Wyeth	Bloomfield, Iowa	Sept. 15, 1863	39, 982
Planter, Corn	J. Yoder, J. Gilliford, and E. Gouver.	Juniata, Pa	Oct. 3, 1848	5, 834
Planter, Corn	R. M. Yorks	Schoolcraft, Mich	Nov. 13, 1866	59, 703
Planter, Corn	G. W. N. Yost	Yellow Springs, Ohio	Apr. 3, 1860	27, 764
Planter, Corn and cane	J. M. Fate	Boonesborough, Iowa	Oct. 16, 1866	58, 798
Planter, Corn and cotton	G. C. Boyd	New London, Pa	Apr. 11, 1836	....
Planter, Corn and cotton	J. W. Savage and F. M. Doty	Parmitchie, Miss	May 20, 1873	139, 195
Planter, Corn and cotton	Z. B. Sims	Bonham, Tex	June 3, 1873	139, 620
Planter, Corn and cotton	B. J. Svenson	Manor Station, Tex	Aug. 20, 1872	130, 764
Planter, Corn and cotton-seed	C. C. Garrett	Dayton, Ala	Mar. 12, 1867	62, 834
Planter, Corn and cotton-seed	D. E. Holt	Wilkinson County, Miss	Nov. 2, 1869	96, 323
Planter, Corn and cotton-seed	A. Ingalis	Memphis, Tenn	Sept. 10, 1872	131, 274
Planter, Corn and cotton-seed	R. R. McGregor	Covington, Tenn	Feb. 16, 1869	86, 992
Planter, Corn and cotton-seed	M. L. and R. W. Thornton	Lumpkin, Ga	July 16, 1867	66, 912
Planter, Corn and seed	E. C. Brown	Crawfordsville, Ind	Mar. 24, 1868	75, 850
Planter, Corn and seed	D. D. Franklin	Flora, Ill	July 27, 1869	92, 951
Planter, Corn and seed	J. S. Hersey	Elizabethtown, Pa	May 22, 1866	54, 901
Planter, Corn and seed	W. C. Lockwood	Spring Mills, Mich	July 26, 1870	105, 703
Planter, Corn and seed	F. W. Marriott	Richwood, Ohio	May 26, 1868	78, 303
Planter, Corn and seed	R. H. Mathews	Nebraska City, Nebr	June 6, 1871	115, 755
Planter, Corn and seed	N. Mendenhall	Greensburgh, Ind	Dec. 29, 1868	85, 466
Planter, Corn and seed	M. M. Rutt and A. B. Baer	East Hempfield, Pa	Nov. 26, 1867	71, 540
Planter, Cotton	P. H. Altstatt	Clarke County, Ind	June 20, 1871	116, 003
Planter, Cotton	E. L. Barnett	El Dorado, Ark	June 30, 1868	79, 433
Planter, Cotton	M. Beam	Buffalo, N. C	Feb. 13, 1835	....
Planter, Cotton	M. Beam	Buffalo, N. C	Mar. 30, 1836	....
Planter, Cotton	H. Blair	Glencoss, Md	Aug. 31, 1836	....
Planter, Cotton	H. Blair	United States	Aug. 31, 1836	15
Planter, Cotton	R. E. Bowen	George's Creek, S. C	Aug. 12, 1873	141, 753
Planter, Cotton	J. M. Brooks	Woodbury, Ga	Feb. 11, 1873	135, 767
Planter, Cotton	J. P. Clopton	Terry, Tenn	July 4, 1871	116, 556
Planter, Cotton	W. J. Cox and W. T. Smith	Lilesville, N. C	July 29, 1873	141, 328
Planter, Cotton	J. S. Dickason	Sulphur Well, Tenn	Nov. 22, 1870	109, 499
Planter, Cotton	J. L. A. Edwards	New Orleans, La	Apr. 9, 1867	63, 623
Planter, Cotton	D. C. Ellis and G. N. Deming	Rochester, N. Y	Mar. 14, 1871	112, 567
Planter, Cotton	A. G. W. Foster	Franklin, Ga	Oct. 25, 1870	108, 580
Planter, Cotton	J. Gatling	Murfreesborough, N. C	June 20, 1835	....
Planter, Cotton	W. L. Gebby	New Richland, Ohio	Dec. 4, 1866	60, 170
Planter, Cotton	J. G. Goodman and W. S. McDonald.	Wilson County, Tenn	June 29, 1833	....
Planter, Cotton	R. T. Goodman	Ballsville, Va	Sept. 18, 1835	....
Planter, Cotton	R. G. Hobson	Houlka, Miss	Nov. 7, 1871	120, 742
Planter, Cotton	J. Hughes	New Berne, N. C	Apr. 4, 1871	113, 671
Planter, Cotton	F. E. Moran	Millburn, Ill	Feb. 26, 1867	62, 356
Planter, Cotton	J. H. Nale	Memphis, Tenn	Apr. 30, 1872	126, 320
Planter, Cotton	C. H. Nixon	Polo, Ill	Nov. 18, 1873	144, 784
Planter, Cotton	G. Paterson	Waynesborough, Ga	June 10, 1873	139, 732
Planter, Cotton	J. A. and W. L. D. Pope	Charlotte, N. C	May 7, 1872	126, 487
Planter, Cotton	W. Price	Mount Olive, N. C	Apr. 22, 1873	138, 043
Planter, Cotton	B. A. Ramsay	Trenton, Tenn	Dec. 20, 1870	110, 393
Planter, Cotton	W. E. Rhodes	Darlington Court-House, S. C.	Jan. 28, 1873	135, 290
Planter, Cotton	C. Richmond	Memphis, Tenn	May 12, 1868	77, 761

Index of patents issued from the United States Patent Office from 1790 *to* 1873, *inclusive*—Continued.

Invention.	Inventor.	Residence.	Date.	No.
Planter, Cotton	H. A. Ridley	Jacksonport, Ark	Dec. 12, 1871	121, 815
Planter, Cotton	H. A. Ridley	Jacksonport, Ark	Nov. 12, 1872	132, 984
Planter, Cotton	N. F. Sandelin	New York, N. Y	Mar. 12, 1872	124, 450
Planter, Cotton	J. P. Selsor	Shelbyville, Mo	Feb. 26, 1867	62, 447
Planter, Cotton	P. Seymour	East Bloomfield, N. Y	Oct. 28, 1873	144, 037
Planter, Cotton	O. L. Slater	Wall Hill, Miss	Sept. 2, 1873	142, 355
Planter, Cotton	E. L. Sykes	Okolona, Miss	Jan. 18, 1870	99, 031
Planter, Cotton	T. T. and G. T. Thorne	Whitaker's Station, N. C	Feb. 1, 1870	99, 499
Planter, Cotton	A. R. Wiggs	Iuka, Miss	May 11, 1869	89, 958
Planter, Cotton	J. Wood	Wedowee, Ala	May 28, 1872	127, 399
Planter, Cotton and corn	A. Pennington, jr	Brenham, Tex	Nov. 11, 1873	144, 562
Planter, Cotton and corn	E. Wagner	Westminster, Md	Nov. 1, 1870	108, 947
Planter, Cotton and corn	A. R. Wiggs	Iuka, Miss	Nov. 23, 1869	97, 257
Planter, Cotton-seed	L. Acree	Taliaferro County, Ga	Feb. 14, 1860	27, 091
Planter, Cotton-seed	H. P. Allen	Bowling Green, Ky	Aug. 31, 1858	21, 308
Planter, Cotton-seed	J. P. Allen	Dawson, Ga	Mar. 29, 1870	101, 206
Planter, Cotton-seed	F. M. Bacon	Ripon, Wis	Sept. 26, 1865	50, 089
Planter, Cotton-seed	W. Badger	Memphis, Tenn	Jan. 13, 1857	16, 368
Planter, Cotton-seed	N. E. Badgley	Gadsden, Ala	May 1, 1860	28, 049
Planter, Cotton-seed	N. E. Badgley	New York, N. Y	Nov. 6, 1866	59, 339
Planter, Cotton-seed	P. B. Baker	Wall Hill, Miss	Dec. 13, 1859	26, 404
Planter, Cotton-seed	W. C. Banks	Como Depot, Miss	Aug. 6, 1867	67, 481
Planter, Cotton-seed	C. Battle	Warrenton, Ga	Mar. 13, 1860	27, 421
Planter, Cotton-seed	E. P. Beauchamp	Preston, Ga	Aug. 9, 1859	24, 984
Planter, Cotton-seed	D. I. Beecher	Greenville, Miss	Sept. 2, 1856	15, 640
Planter, Cotton-seed	C. R. Belt	Near Washington, D. C	Oct. 21, 1856	15, 9[illegible]8
Planter, Cotton-seed	E. T. Bastrom	Newnan, Ga	June 29, 1858	20, 694
Planter, Cotton-seed	N. Breed	Jeffersonville, Ind	June 1, 1869	90, 722
Planter, Cotton-seed	A. W. Brian	Ouachita County, Ark	June 15, 1869	91, 206
Planter, Cotton-seed	T. E. C. Brinly	Louisville, Ky	June 13, 1871	115, 929
Planter, Cotton-seed	R. M. Brooks	Greenville, Ga	Dec. 27, 1859	26, 562
Planter, Cotton-seed	A. D. Brown, sr	Columbus, Ga	Feb. 23, 1869	87, 089
Planter, Cotton-seed	F. H. Brown	Chicago, Ill	June 26, 1866	55, 811
Planter, Cotton-seed	F. H. Brown	Auburn, N. Y	Oct. 2, 1866	58, 372
Planter, Cotton-seed	L. B. Brown	Scriven County, Ga	Jan. 15, 1861	31, 107
Planter, Cotton-seed	M. S. Burns	Memphis, Tenn	Oct. 18, 1870	108, 445
Planter, Cotton-seed	A. Carey	Rome, Ga	Feb. 14, 1860	27, 107
Planter, Cotton-seed	F. F. Carroll	Midway, S. C	Apr. 4, 1871	113, 626
Planter, Cotton-seed	Z. Carter	Line Creek, S. C	Dec. 16, 1873	145, 554
Planter, Cotton-seed	J. G. Clark	Middletown, Ohio	Oct. 2, 1866	58, 381
Planter, Cotton-seed	T. T. and H. W. S. Collier	Lavernia, Tex	Sept. 13, 1859	25, 388
Planter, Cotton-seed	J. M. Cooper	Palmyra, Ga	Mar. 7, 1854	10, 595
Planter, Cotton-seed	J. A. Cox	Humboldt, Tenn	June 11, 1867	65, 647
Planter, Cotton-seed	W. W. Croom	Gainesville, Ala	Mar. 29, 1870	101, 232
Planter, Cotton-seed	W. M. Croom	Opelika, Ala	Oct. 22, 1872	132, 388
Planter, Cotton-seed	J. P. Crutcher	Silver Spring, Tenn	Apr. 12, 1859	23, 554
Planter, Cotton-seed	I. T. Donovan and W. J. Fowler	Seguin, Tex	Apr. 27, 1858	20, 049
Planter, Cotton-seed	Z. Doolittle	Perry, Ga	July 10, 1860	29, 065
Planter, Cotton-seed	Z. Doolittle and A. M. Crowder	Houston Factory, Ga	Feb. 18, 1868	74, 518
Planter, Cotton-seed	J. M. Elliott	Winnsborough, S. C	July 20, 1869	92, 804
Planter, Cotton-seed	H. R. Fell and E. Phifer	Trenton, N. J	Nov. 20, 1866	59, 765
Planter, Cotton-seed	D. P. Ferguson	Jonesborough, Ga	Nov. 18, 1873	144, 607
Planter, Cotton-seed	T. C. Garlington	Chambers Court-House, Ala	Aug. 2, 1870	105, 935
Planter, Cotton-seed	C. C. Garrett	Spring Hill, Ala	Mar. 8, 1859	23, 164
Planter, Cotton-seed	W. A. Gates	Mount Comfort, Tenn	Nov. 16, 1852	9, 403
Planter, Cotton-seed	W. Gessner	Cape Girardeau, Mo	July 30, 1872	129, 943
Planter, Cotton-seed	O. L. Gibson	Fort Bend County, Tex	Sept. 11, 1860	29, 964
Planter, Cotton-seed	A. J. Going	Clinton, La	Mar. 10, 1868	75, 410
Planter, Cotton-seed	A. J. Going	Clinton, La	Dec. 29, 1868	85, 379
Planter, Cotton-seed	A. D. Griswold	Rocky Hill, Conn	May 6, 1873	138, 497
Planter, Cotton-seed	F. E. Habersham	Old Church, Va	Dec. 17, 1872	133, 983
Planter, Cotton-seed	J. T. Ham	Senatobia, Miss	Oct. 9, 1860	30, 313
Planter, Cotton-seed	D. Herlong	Sandy Ridge, Ala	Jan. 15, 1861	31, 120
Planter, Cotton-seed	J. L. Horn	Edgecombe County, N. C	Feb. 12, 1856	14, 240
Planter, Cotton-seed	W. A. Horrall	Washington, Ind	Oct. 30, 1866	59, 314
Planter, Cotton-seed	E. J. Hudson	Golconda, Ill	June 14, 1870	104, 157
Planter, Cotton-seed	J. S. Huggins and R. Chapman	Darlington District, S. C	June 1, 1858	20, 432
Planter, Cotton-seed	O. P. Humber	Greenville, N. C	Nov. 16, 1869	96, 919
Planter, Cotton-seed	J. W. Huntley	Lane's Creek, N. C	Aug. 16, 1859	25, 119
Planter, Cotton-seed	G. Jessup	Shortsville, N. Y	Sept. 8, 1868	81, 906
Planter, Cotton-seed	J. M. Jones	Palmyra, N. Y	Jan. 22, 1856	14, 134
Planter, Cotton-seed	H. L. Justice and J. H. Galbreath.	Goodlettsville, Tenn	May 5, 1857	17, 221
Planter, Cotton-seed	C. Kesler and F. Reinhard	Columbus, Tex	Dec. 27, 1859	26, 596
Planter, Cotton-seed	J. C. King	Spring Place, Ga	Nov. 15, 1870	109, 222
Planter, Cotton-seed	L. D. Law	Henderson, Ga	Dec. 22, 1857	18, 913
Planter, Cotton-seed	J. Lytch	Laurinburgh, N. C	May 31, 1870	103, 759
Planter, Cotton-seed	C. W. McClanahan	Victoria, Tex	Nov. 27, 1860	30, 743
Planter, Cotton-seed	A. McDonald	Salem, Miss	June 15, 1858	20, 572
Planter, Cotton-seed	M. McMillian	Caney, Ark	Nov. 30, 1869	97, 424
Planter, Cotton-seed	J. M. Merrymon and W. M. Dunn.	Indianapolis, Ind., and Gurleysville, Ala.	Mar. 5, 1867	62, 664
Planter, Cotton-seed	E. H. Metcalf	Montgomery, Ala	Aug. 13, 1872	130, 521
Planter, Cotton-seed	S. Miller	Washington College, Tenn	Apr. 19, 1853	9, 676
Planter, Cotton-seed	A. H. Morrel	Marlin, Tex	May 15, 1855	12, 867
Planter, Cotton-seed	I. Myers and M. D. Wellman	Pittsburgh, Pa	Jan. 31, 1865	46, 130
Planter, Cotton-seed	R. C. Nash	Somerville, Tenn	Dec. 18, 1860	30, 981
Planter, Cotton-seed	D. B. Neal	Mount Gilead, Ohio	Feb. 23, 1858	19, 438
Planter, Cotton-seed	H. Nicholls	Fairfield, Ky	Aug. 30, 1870	106, 859
Planter, Cotton-seed	A. R. Nixon	Memphis, Tenn	Dec. 21, 1869	98, 183
Planter, Cotton-seed	J. F. Orr	Orville, Ala	Feb. 3, 1857	16, 550
Planter, Cotton-seed	B. Owen	Dayton, Ohio	June 26, 1860	28, 896
Planter, Cotton-seed	B. Owen	Dayton, Ohio	Mar. 20, 1866	53, 377
Planter, Cotton-seed	A. Packham	Prestonville, Ky	May 22, 1866	54, 942
Planter, Cotton-seed	B. and N. Platt	Pana, Ill., and Saint Louis, Mo.	Aug. 1, 1865	49, 150
Planter, Cotton-seed	J. Price	New Harrisburgh, Ohio	Jan. 30, 1866	52, 518
Planter, Cotton-seed	O. Richardson	Boston, Mass	Dec. 31, 1867	72, 907

Index of patents issued from the United States Patent Office from 1790 *to* 1873, *inclusive*—Continued.

Invention.	Inventor.	Residence.	Date.	No.
Planter, Cotton-seed	J. Riggsbee	Chapel Hill, N. C	July 5, 1870	105, 000
Planter, Cotton-seed	T. J. Rogers	Cassville, Ga	Sept. 1, 1857	18, 104
Planter, Cotton-seed	C. A. Rose	Columbia, Ala	Dec. 4, 1860	30, 839
Planter, Cotton-seed	J. Ross	Midway, Ala	Apr. 6, 1858	19, 874
Planter, Cotton-seed	J. L. Russell	Pella, Iowa	July 10, 1866	56, 274
Planter, Cotton-seed	D. H. A. Sanders	Senatobia, Miss	Apr. 6, 1869	88, 670
Planter, Cotton-seed	J. Shearer and M. B. Armstrong	Timbersville, Ill	July 13, 1869	92, 481
Planter, Cotton-seed	N. B. Sherwood	Millville, N. Y	June 5, 1866	55, 372
Planter, Cotton-seed	N. B. Sherwood	Millville, N. Y	June 5, 1866	55, 373
Planter, Cotton-seed	N. B. Sherwood	Millville, N. Y	Mar. 3, 1868	75, 207
Planter, Cotton-seed	F. Sloan	Bolivar, Tenn	Sept. 27, 1870	107, 730
Planter, Cotton-seed	J. L. Slocumb	Raymond, Miss	July 22, 1873	141, 085
Planter, Cotton-seed	A. C. Smith	Roaring Falls, Tenn	Oct. 25, 1870	108, 643
Planter, Cotton-seed	B. Smith	Falkland, N. C	June 2, 1868	78, 614
Planter, Cotton-seed	B. Smith	Falkland, N. C	Feb. 1, 1870	99, 361
Planter, Cotton-seed	B. Smith	Hood Swamp, N. C	July 5, 1870	105, 006
Plantor, Cotton-seed	P. E. Smith	Scotland Neck, N. C	May 23, 1871	115, 123
Planter, Cotton-seed	P. M. Smith and T. T. Collier	Lavernia, Tex	Sept. 13, 1859	25, 449
Planter, Cotton-seed	J. H. Sorey	Flora, Ill	Oct. 31, 1871	120, 544
Planter, Cotton-seed	B. Spencer	Lewisburgh, Pa	Nov. 8, 1864	44, 984
Planter, Cotton-seed	A. V. M. Sprague and R. F. Osgood.	Rochester, N. Y	July 6, 1869	92, 389
Planter, Cotton-seed	J. A. Stewart	Franklin, Ky	July 1, 1856	15, 260
Planter, Cotton-seed	W. A. and J. F. Suddith	Charlestown, Va	June 26, 1860	28, 913
Planter, Cotton-seed	L. B. Sutton	Windsor, N. C	Feb. 25, 1873	136, 342
Planter, Cotton-seed	S. P. Sweeny	Columbia, Tex	Apr. 3, 1860	27, 748
Planter, Cotton-seed	J. C. Tobias	Helena, Ark	Aug. 27, 1867	68, 262
Planter, Cotton-seed	J. Trump	Springfield, Ohio	Apr. 12, 1870	101, 947
Planter, Cotton-seed	W. F. Tunnard	East Baton Rouge Parish, La.	Sept. 7, 1869	94, 671
Planter, Cotton-seed	A. W. Washburn	Yazoo City, Miss	Mar. 26, 1856	14, 529
Planter, Cotton-seed	M. D. Wells	Morgantown, W. Va	Nov. 6, 1866	59, 490
Planter, Cotton-seed	T. W. White	Milledgeville, Ga	Oct. 20, 1857	18, 482
Planter, Cotton-seed	L. T. Wilcox and W. G. Caldwell.	Three Rivers, Mich	July 16, 1867	66, 924
Planter, Cotton-seed	I. Williams and I. W. Bausman.	Allegheny County, Pa	Jan. 23, 1855	12, 294
Planter, Cotton-seed	W. R. Wright	Allendale, S. C	May 28, 1872	127, 4[illegible]0
Planter, Cotton-seed and corn	N. Foster	Palmyra, N. Y	Feb. 11, 1868	74, 334
Planter, Cotton-seed and corn	J. G. B. Gill	Chesnut Grove, S. C	Aug. 31, 1869	94, 301
Planter, Cotton-seed and corn	J. B. Godwin	Williamston, N. C	Mar. 29, 1870	101, 257
Planter, Cotton-seed and corn	A. Runstetler and A. Windeck	Peoria, Ill	Mar. 10, 1868	75, 466
Planter, Cotton-seed and corn	I. A. Towers	Quincy, Fla	Dec. 20, 1870	110, 311
Planter, cultivator, and digger, Potato	C. J. C. Peterson	Port Chester, N. Y	Feb. 2, 1869	86, 442
Planter cultivator, Corn	M. H. Lineback	Greenfield, Ind	Mar. 13, 1866	53, 156
Planter, Double corn	W. McLucas	Rinersville, Ohio	July 14, 1868	79, 998
Planter drill, Seed	J. F. Keller	Hagerstown, Md	July 26, 1870	105, 810
Planter, dropper, and cultivator, Combined seed	C. M. Cornell	Ionia, Mich	July 6, 1869	92, 171
Planter, fertilizer, and plow, Combined seed	H. C. Eves	Orangeville, Pa	Aug. 3, 1869	93, 239
Planter, fertilizer, distributer, cotton-chopper, and cultivator, Combined corn and cotton.	J. F. Tucker	Monticello, Fla	May 31, 1870	103, 688
Planter, Foot corn	G. B. Roe	Paine's Point, Ill	Mar. 17, 1863	37, 922
Planter, Foot corn	N. Silvester	Granger, Ohio	May 20, 1862	35, 330
Planter gearing, Seed	A. Sandoe	Mifflintown, Pa	Apr. 16, 1850	7, 294
Planter, Grain	I. Bogart	Newport, Ind	May 24, 1870	103, 289
Planter, Grain	N. H. Purcell	Avon, N. Y	Nov. 14, 1865	50, 954
Planter, grain-drill, and harrow, Combined corn	H. Smith	Kirksville, Mo	Mar. 2, 1869	87, 376
Planter, Hand	R. Arey	Rock Falls, Ill	Mar. 11, 1873	136, 688
Planter, Hand	J. Jeffcoat	Onawa, Iowa	Mar. 2, 1869	87, 414
Planter, Hand	J. H. Latimer	Crystal Lake, Ill	Dec. 26, 1871	122, 121
Planter, Hand	F. H. Roberts	Wilmington, Ind	Mar. 1, 1864	41, 784
Planter, Hand	N. Safford	Pleasant Valley, Vt	Oct. 2, 1866	58, 486
Planter, Hand	D. B. Seely	Sterling, Ill	Mar. 25, 1873	137, 246
Planter, Hand	S. S. Stults	Cedar Bluffs, Nebr	Jan. 28, 1873	135, 294
Planter, Hand	G. Windle	New Cambria, Mo	Feb. 25, 1873	136, 294
Planter, Hand	Y. F. Wright	Hannahatchee, Ga	Jan. 23, 1872	122, 980
Planter, Hand corn	G. Atkins	Pittsburgh, Pa	June 17, 1856	15, 114
Planter, Hand corn	H. F. Batcheller	Sterling, Ill	Mar. 9, 1858	19, 540
Planter, Hand corn	O. Billings	La Grange, Ohio	May 22, 1866	54, 999
Planter, Hand corn	S. P. Briggs	Saratoga Springs, N. Y	Apr. 30, 1861	32, 178
Planter, Hand corn	D. Broy	Canton, Mo	Apr. 14, 1868	76, 709
Planter, Hand corn	G. Burson	East Palestine, Ohio	Apr. 6, 1869	88, 607
Planter, Hand corn	G. Burson	East Palestine, Ohio	Sept. 20, 1870	107, 594
Planter, Hand corn	G. Burson	East Palestine, Ohio	Sept. 20, 1870	107, 595
Planter, Hand corn	M. Case	Kasoag, N. Y	Aug. 18, 1863	39, 548
Planter, Hand corn	T. A. Chandler	Rockford, Ill	Nov. 25, 1856	16, 135
Planter, Hand corn	J. W. Chappell	Loomisville, Mich	Aug. 6, 1861	32, 975
Planter, Hand corn	D. G. Choppin	Cincinnati, Ohio	Apr. 6, 1858	19, 833
Planter, Hand corn	R. L. Cleveland	Durand, Ill	Sept. 11, 1866	57, 868
Planter, Hand corn	J. W. Coleman	Bridgeville, Mich	June 24, 1873	140, 186
Planter, Hand corn	A. C. L. Davis	Saint Louis, Mo	May 17, 1870	103, 153
Planter, Hand corn	S. L. Denney	Lancaster, Pa	June 3, 1856	15, 035
Planter, Hand corn	W. Douglass	Westport, Mo	July 30, 1861	32, 937
Planter, Hand corn	H. Dyer	Fort Scott, Kans	Aug. 23, 1870	106, 568
Planter, Hand corn	J. Fairbank and E. C. Durfee	Leon, N. Y	Feb. 9, 1858	19, 329
Planter, Hand corn	W. Gilbert	Bardstown, Ky	June 26, 1866	55, 850
Planter, Hand corn	H. Gortner	Nashport, Ohio	Apr. 29, 1873	138, 392
Planter, Hand corn	H. B. Hammon	Bristolville, Ohio	Sept. 9, 1856	15, 696
Planter, Hand corn	H. B. Hammon	Bristolville, Ohio	Mar. 6, 1860	27, 365
Planter, Hand corn	J. M. Harrison	Spartanburgh, Ind	Jan. 24, 1871	111, 202
Planter, Hand corn	E. W. Haven	Brandon, Vt	Oct 12, 1869	95, 683
Planter, Hand corn	F. A. Hawkins	Noblesville, Ind	Jan. 14, 1873	134, 884
Planter, Hand corn	L. O. Hayworth	New Cumberland, Ind	May 18, 1869	90, 169
Planter, Hand corn	H. Hickman	Omaha, Nebr	May 10, 1870	102, 818
Planter, Hand corn	D. H. Howell	Independence, Iowa	Jan. 23, 1866	52, 171
Planter, Hand corn	D. Humphreys	Oskalcosa, Iowa	July 3, 1866	56, 053
Planter, Hand corn	W. Jenks	Alexandria, Va	Mar. 25, 1856	14, 504
Planter, Hand corn	S. G. Jones	Philadelphia, Pa	Dec. 10, 1872	133, 709

Index of patents issued from the United States Patent Office from 1790 *to* 1873, *inclusive*—Continued.

Invention.	Inventor.	Residence.	Date.	No.
Planter, Hand corn	I. A. Keeler	Middleville, Mich	July 19, 1870	105, 578
Planter, Hand corn	C. A. Kellogg	Elyria, Ohio	July 4, 1865	48, 562
Planter, Hand corn	A. C. Kent	Janesville, Wis	Apr. 2, 1872	125, 306
Planter, Hand corn	J. F. Klinglesmith	Hardin County, Ky	Aug. 4, 1868	80, 746
Planter, Hand corn	R. Lambert	Cortland Village, N. Y	Apr. 10, 1866	53, 838
Planter, Hand corn	J. S. Lawson	Disco, Mich	Feb. 11, 1868	74, 231
Planter, Hand corn	W. C. Lowman	Kansas, Ohio	Jan. 8, 1867	61, 076
Planter, Hand corn	C. Martratt	Albany, N. Y	Aug. 26, 1856	15, 616
Planter, Hand corn	P. McCollum	Fayette, Mo	Nov. 22, 1870	109, 437
Planter, Hand corn	D. McKanna	Madison, Wis	June 12, 1866	55, 517
Planter, Hand corn	J. Morris	Auburn, Mo	June 13, 1865	48, 197
Planter, Hand corn	J. H. Morris and T. B. Harrison	Maquoketa, Iowa	Nov. 24, 1868	84, 435
Planter, Hand corn	M. P. Nemmers	Saint Donatus, Iowa	Nov. 25, 1873	144, 919
Planter, Hand corn	M. S. Orton	Galesburgh, Ill	Mar 21, 1865	46, 928
Planter, Hand corn	W. S. Pelham	Kirkville, Iowa	Sept. 14, 1869	94, 910
Planter, Hand corn	L. H. Richards	Rising Sun, Md	Feb. 4, 1873	135, 489
Planter, Hand corn	E. Rogers	Rochester, Ind	Nov. 18, 1873	144, 701
Planter, Hand corn	M. C. Root	Janesville, Wis	Nov. 11, 1873	144, 566
Planter, Hand corn	M. Russell and A. G. Burdick	Mill Rock, Iowa	Nov. 12, 1867	70, 747
Planter, Hand corn	H. Sage	Omaha City, Nebr	June 11, 1872	127, 696
Planter, Hand corn	S. Z. Shores	Towanda, Pa	July 2, 1861	32, 725
Planter, Hand corn	S. S. Smith	North Fairfield, Ohio	June 14, 1864	43, 139
Planter, Hand corn	O. Stoddard	Busti, N. Y	June 26, 1855	13, 151
Planter, Hand corn	W. D. Stroud	Oshkosh, Wis	May 3, 1870	102, 611
Planter, Hand corn	J. O. Talmage	Cardington, Ohio	Nov. 11, 1873	144, 485
Planter, Hand corn	J. D. Tracy and J. F. Platt	Sterling, Ill	Dec. 16, 1873	145, 698
Planter, Hand corn	S. E. Tyler and R. Tattershall	Beloit, Wis	Apr. 3, 1866	53, 752
Planter, Hand corn	J. P. Van Vleck	Rock County, Wis	Feb. 26, 1867	62, 380
Planter, Hand corn	C. A. Wakefield	Pittsfield, Mass	June 5, 1866	55, 399
Planter, Hand corn	J. M. and W. C. Wallis	Milton, Iowa	Oct. 28, 1862	36, 815
Planter, Hand corn	L. Weaver	Canton, Ohio	Jan. 22, 1867	61, 490
Planter, Hand corn	H. B. West and C. A. Kellogg	Elyria, Ohio	Oct. 13, 1863	40, 299
Planter, Hand seed	S. P. Briggs	Saratoga Springs, N. Y	May 26, 1857	17, 362
Planter, Hand seed	J. H. Bruen	Penn Yan, N. Y	Feb. 24, 1857	16, 677
Planter, Hand seed	W. Bullack	Philadelphia, Pa	Nov. 2, 1852	9, 381
Planter, Hand seed	T. Crane	Fort Atkinson, Wis	Apr. 21, 1857	17, 080
Planter, Hand seed	H. V. Davis	Amherst, N. H	Aug. 27, 1867	68, 050
Planter, Hand seed	J. Decker	Sparta, N. J	Apr. 21, 1857	17, 081
Planter, Hand seed	J. Dyson	Philadelphia, Pa	Aug. 17, 1869	93, 690
Planter, Hand seed	P. B. Green	Chicago, Ill	Apr. 21, 1857	17, 089
Planter, Hand seed	J. Haines	West Middleburgh, Ohio	Sept. 8, 1859	18, 145
Planter, Hand seed	J. Heberling	Mount Pleasant, Ohio	June 7, 1870	104, 628
Planter, Hand seed	E. Hopkins	Cincinnati, Ohio	Apr. 29, 1856	14, 767
Planter, Hand seed	D. W. Hughes	New London, Mo	Nov. 20, 1855	13, 820
Planter, Hand seed	J. H. Jones	Rockton, Ill	Aug. 26, 1856	15, 610
Planter, Hand seed	J. H. Jones	Rockford, Ill	Apr. 9, 1867	63, 643
Planter, Hand seed	H. Koeller and W. Uecke	Camp Point, Ill	Feb. 11, 1868	74, 383
Planter, Hand seed	A. T. Large	Tomah, Wis	Nov. 27, 1866	60, 019
Planter, Hand seed	J. H. Latimer	Crystal Lake, Ill	Dec. 4, 1866	60, 200
Planter, Hand seed	J. T. Macomber	Grand Isle, Vt	Apr. 5, 1870	101, 480
Planter, Hand seed	A. C. Miller	Morgantown, Va	July 29, 1856	15, 431
Planter, Hand seed	A. More	Moresville, N. Y	July 13, 1869	92. 632
Planter, Hand seed	G. Sanford	Ellenville, N. Y	June 15, 1852	9, 037
Planter, Hand seed	A. Stickney	Concord, N. H	Dec. 25, 1855	14, 003
Planter, Hand seed	F. Van Doren	Adrian, Mich	Feb. 14, 1860	27, 172
Planter, harrow, and cultivator, Combined	D. D. Stelle	New Brunswick, N. J	July 16, 1867	66, 904
Planter, Hedge	N. L. Griffith	South Plymouth, Ohio	Aug. 1, 1871	117, 622
Planter, Hedge	J. J. Tucker	Eugene, Ill	June 1, 1869	90, 703
Planter, Hedge	J. J. Tucker	Albia, Iowa	Aug. 3, 1869	93, 250
Planter, Hedge	W. Young	Bloomington, Ill	Dec. 22, 1868	85, 195
Planter, hoe, and potato-digger, Combined potato	F. B. Marden	Bangor, Me	Jan. 21, 1868	73, 618
Planter, Potato	J. R. Albertson	Allegheny, Pa	Jan. 26, 1858	19, 178
Planter, Potato	L. A. Aspinwall	Watervliet, N. Y	Nov. 14, 1865	50, 890
Planter, Potato	L. A. Aspinwall	Albany, N. Y	Nov. 30, 1869	97, 339
Planter, Potato	L. A. Aspinwall	Albany, N. Y	Oct. 10, 1871	119, 687
Planter, Potato	J. E. Bendix and M. Deitsch	New York and West Chester, N. Y.	Nov. 12, 1867	70, 786
Planter, Potato	H. Farmer	Pontiac, Mich	Jan. 7, 1868	73, 086
Planter, Potato	S. T. Godfrey	Seaville, N. J	Jan. 11, 1870	98, 687
Planter, Potato	E. E. Hawley	New Haven, Conn	Feb. 9, 1858	19, 294
Planter, Potato	T. Herbert	Philadelphia, Pa	Oct. 25, 1870	108, 591
Planter, Potato	L. J. Holcomb	Nunda, Ill	June 19, 1866	55, 657
Planter, Potato	H. J. Kent	Palmyra, N. Y	Jan. 24, 1871	111, 217
Planter, Potato	G. W. and J. J. Kersey	Beartown, Pa	Sept. 25, 1860	30, 144
Planter, Potato	G. Knowlton	Johnstown, Pa	Mar. 7, 1871	112, 470
Planter, Potato	J. Krehbril	Clarence Centre, N. Y	May 17, 1870	103, 058
Planter, Potato	F. S. McWhorter	Smyrna, Del	Apr. 20, 1858	20, 001
Planter, Potato	J. C. Mills	Rochester, N. Y	Mar. 14, 1871	112, 617
Planter, Potato	J. Moore	Quincy Point, Mass	July 29, 1856	15, 433
Planter, Potato	C. Morgan	Philadelphia, Pa	Feb. 12, 1856	14, 270
Planter, Potato	W. Nevins	Irving, N. Y	Mar. 22, 1864	42, 014
Planter, Potato	A. E. Payne	Jonesville, Mich	Dec. 19, 1871	121, 957
Planter, Potato	A. Randel	Verona, N. Y	Feb. 15, 1848	5, 446
Planter, Potato	J. P. Scudder	Lawrenceville, N. J	June 11, 1867	65, 695
Planter, Potato	N. B. Sherwood	Millville, N. Y	Nov. 24, 1868	84, 309
Planter, Potato	J. C. Stoddard	Worcester, Mass	Mar. 17, 1859	24, 065
Planter, Potato	S. H. Strong	Brunswick, Ohio	Dec. 8, 1857	18, 827
Planter, Potato	C. Svendsen	Blair, Nebr	July 15, 1873	140, 965
Planter, Potato	J. L. True	Benton, Me	Jan. 5, 1869	85, 547
Planter, Potato	J. L. True	Benton, Me	Apr. 16, 1872	125, 705
Planter, Potato	L. Van Wie	Bethlehem, N. Y	June 22, 1869	91, 798
Planter, Potato	H. Wainwright and S. T. Williams	Farmingdale, N. J	Jan. 5, 1858	19, 054
Planter, Potato	F. W. Worstell	Saint Louis, Mo	June 24, 1873	140, 178
Planter, Potato and corn	A. J. Taylor	Manchester, Ind	June 15, 1869	91, 494
Planter, Rice	E. Wagoner	Westminster, Md	Feb. 25, 1868	74, 957
Planter, Seed	M. Adams	Chilmark, Mass	June 8, 1869	90, 911
Planter, Seed	C. C. Aldrich	Faribault, Minn	Sept. 8, 1857	18, 126

Index of patents issued from the United States Patent Office from 1790 *to* 1873, *inclusive*—Continued.

Invention.	Inventor.	Residence.	Date.	No.
Planter, Seed	T. K. Alexander	Decatur, Ill	Feb. 13, 1866	52, 635
Planter, Seed	J. P. Allen	Midville, Ga	Oct. 4, 1859	25, 616
Planter, Seed	J. P. Allen	Dover, Ga	Aug. 14, 1860	29, 555
Planter, Seed	P. Alling	Norwalk, Ohio	July 26, 1870	105, 621
Planter, Seed	A. Anderson	Markham, Canada	Jan. 2, 1855	12, 152
Planter, Seed	C. F. Anderson	Charlestown, N. H	May 31, 1859	24, 185
Planter, Seed	J. Andrews	Winchester, Mass	Dec. 12, 1854	12, 084
Planter, Seed	L. Arnold	Janesville, Wis	Feb. 17, 1857	16, 636
Planter, Seed	M. Atwood	New Sharon, Iowa	Aug. 4, 1868	80, 585
Planter, Seed	L. and S. H. Bachelder	Hampstead, N. H., and Haverhill, Mass.	Apr. 30, 1840	1, 577
Planter, Seed	B. Baker	Hopkinton, Iowa	Aug. 1, 1871	117, 502
Planter, Seed	H. F. Baker	Centreville, Ind	Jan. 5, 1858	19, 010
Planter, Seed	J. C. Baker	Mechanicsburgh, Ohio	Mar. 1, 1859	23, 069
Planter, Seed	S. Baker	Mount Pulaski, Ill	Jan. 19, 1858	19, 122
Planter, Seed	E. F. Ballard	Lincoln, Nebr	Jan. 30, 1872	123, 216
Planter, Seed	G. Banister	Hartford, Vt	June 22, 1869	91, 595
Planter, Seed	W. H. Barber	Wolcottville, Conn	Sept. 4, 1860	29, 853
Planter, Seed	A. J. Barnhart	Schoolcraft, Mich	Feb. 27, 1855	12, 465
Planter, Seed	J. Barnhill	Circleville, Ohio	May 27, 1851	8, 116
Planter, Seed	J. F. Beckwith and A. G. Gage	Alabama, N. Y	Nov. 30, 1858	22, 156
Planter, Seed	J. F. Beckwith and A. G. Gage	South Alabama, N. Y	Jan. 18, 1859	22, 617
Planter, Seed	L. Beemer	Libertyville, N. J	Feb. 3, 1857	16, 522
Planter, Seed	H. Bell	Clinton, Ill	Dec. 13, 1859	26, 456
Planter, Seed	A. Bennett	Rockford, Ill	Mar. 5, 1867	62, 595
Planter, Seed	G. F. Bennett	Mount Olive, N. C	July 17, 1860	29, 134
Planter, Seed	I. C. Benthall	Oakland, Tex	Mar. 1, 1859	23, 071
Planter, Seed	A. Berdan	Macon, Mich	Aug. 10, 1858	21, 112
Planter, Seed	C. W. Billings	South Deerfield, Mass	Mar. 14, 1854	10, 632
Planter, Seed	J. Blackwood	Franklin County, Ohio	Jan. 30, 1855	12, 307
Planter, Seed	W. Blessing	Jeffersonville, Ohio	Dec. 13, 1859	26, 410
Planter, Seed	R. Boeklen	Jersey City, N. J	Feb. 10, 1857	16, 592
Planter, Seed	C. B. and B. S. Borden and A. McLean	West Dresden, N. Y	Apr. 3, 1855	12, 618
Planter, Seed	C. P. Brown	Shortsville, N. Y	Oct. 9, 1866	58, 588
Planter, Seed	G. W. Brown	Tylerville, Ill	Aug. 2, 1853	9, 893
Planter, Seed	G. W. Brown	Galesburgh, Ill	May 8, 1855	12, 811
Planter, Seed	G. W. Brown	Galesburgh, Ill	Feb. 28, 1865	46, 615
Planter, Seed	G. W. Brown	Galesburgh, Ill	Feb. 1, 1870	99, 286
Planter, Seed	J. Brown	Lawn Ridge, Ill	Jan. 30, 1855	12, 308
Planter, Seed	J. A. Brown	Richmond, Ind	Jan. 19, 1858	19, 126
Planter, Seed	Z. B. Brown and M. C. Godard	Granby, Conn	Sept. 13, 1859	25, 380
Planter, Seed	D. Broy	Canton, Mo	Apr. 23, 1861	32, 117
Planter, Seed	J. H. Bruen	Penn Yan, N. Y	May 12, 1857	17, 260
Planter, Seed	J. Bryant	Brooklyn, N. Y	Jan. 4, 1859	22, 484
Planter, Seed	A. Bugbee	Elkhart, Ind	Sept. 19, 1865	50, 065
Planter, Seed	W. Bullock	Red Falls, N. Y	Aug. 1, 1854	11, 464
Planter, Seed	H. Bundel and J. Williams	Dayton, Ohio	Jan. 7, 1868	72, 971
Planter, Seed	M. S. Burdick	Milton, Wis	Dec. 10, 1867	71, 972
Planter, Seed	J. W. Butterick	Farmington, Wis	Jan. 15, 1867	61, 153
Planter, Seed	L. A. Butts	Cuba, N. Y	June 5, 1855	12, 990
Planter, Seed	L. A. Butts	Cuba, N. Y	Feb. 23, 1858	19, 404
Planter, Seed	L. A. Butts	Ripon, Wis	Aug. 6, 1867	67, 409
Planter, Seed	W. Call, jr	Haverstraw, N. Y	June 17, 1873	139, 943
Planter, Seed	J. Campbell	Harrison, Ohio	Nov. 9, 1869	96, 546
Planter, Seed	S. Cannon	New Richmond, Pa	Nov. 12, 1850	7, 770
Planter, Seed	L. R. Carpenter	Lancaster, Ohio	Sept. 27, 1859	25, 556
Planter, Seed	N. R. Carrington	Coldwater, Miss	Mar. 6, 1860	27, 350
Planter, Seed	J. Carroll	La Porte, Ohio	Dec. 8, 1857	18, 802
Planter, Seed	T. Carter	Laurens District, S. C	Apr. 11, 1854	10, 755
Planter, Seed	J. Case	Springfield, Ohio	Jan. 16, 1855	12, 231
Planter, Seed	J. Case	Bloomington, Ill	Dec. 7, 1858	22, 228
Planter, Seed	L. Caswell	Harrison, Me	Aug. 2, 1853	9, 894
Planter, Seed	J. Charlton	Allegheny City, Pa	May 4, 1858	20, 143
Planter, Seed	E. E. Chesney	Abingdon, Ill	Apr. 16, 1867	63, 859
Planter, Seed	E. E. Chesney	Bushnell, Ill	Oct. 24, 1871	120, 241
Planter, Seed	A. Chrisman and M. Whitmer	Sugar Creek, Iowa	Nov. 22, 1864	45, 139
Planter, Seed	W. Clark	Palmyra, Ill	Aug. 9, 1859	24, 991
Planter, Seed	N. Clute and O. W. Marshall	Schenectady, N. Y., and Columbus, Ohio.	Sept. 24, 1867	69, 076
Planter, Seed	L. W. Colver	Louisville, Ky	Feb. 28 1854	10, 565
Planter, Seed	L. W. Colver	Louisville, Ky	May 22, 1855	12, 895
Planter, Seed	L. W. Colver	Louisville, Ky	Dec. 7, 1852	9, 439
Planter, Seed	J. W. Corey	Crawfordsville, Ind	Apr. 10, 1855	15, 672
Planter, Seed	W. M. Cory	Jerseyville, Ill	Oct. 28, 1851	8, 463
Planter, Seed	E. Cox	Point Pleasant, Ohio	July 7, 1863	39, 122
Planter, Seed	G. Crampton	Marshall, Mich	June 28, 1859	24, 541
Planter, Seed	W. Cressler	Shippensburgh, Pa	May 17, 1853	9, 740
Planter, Seed	C. H. Dana	West Lebanon, N. H	Sept. 5, 1854	11, 641
Planter, Seed	C. H. Dana	West Lebanon, N. H	June 5, 1855	12, 995
Planter, Seed	J. H. Dancy	Dancyville, Tenn	Oct. 29, 1872	132, 564
Planter, Seed	L. Daser	Washington, D. C	Sept. 5, 1854	11, 642
Planter, Seed	G. Davis	Mumford, Ala	June 25, 1872	128, 366
Planter, Seed	H. V. Davis	Amherst, N. H	Oct. 24, 1865	50, 657
Planter, Seed	L. H. Davis and S. and M. Pennock.	Kennett's Square, Pa	Nov. 16, 1852	9, 399
Planter, Seed	N. C. Davis	West Jefferson, Ohio	Oct. 25, 1853	10, 169
Planter, Seed	W. Davis	Morgantown, Va	June 6, 1854	11, 032
Planter, Seed	D. Deihl	Hanover, Pa	June 12, 1849	6, 516
Planter, Seed	E. I. Dickey	Chester County, Pa	Jan. 23, 1849	6, 049
Planter, Seed	W. C. Doss	Texana, Tex	Mar. 9, 1858	19, 549
Planter, Seed	J. K. Dugdale	Richmond, Ind	Aug. 4, 1863	39, 386
Planter, Seed	A. B. Earle	Franklin, N. Y	Oct. 24, 1854	11, 829
Planter, Seed	D. Eberly	Strasburgh, Pa	Oct. 8, 1850	7, 698
Planter, Seed	J. Edge	Acquackanonck, N. J	Mar. 26, 1867	63, 147
Planter, Seed	C. R. Edwards	Bowling Green, Ky	Sept. 20, 1870	107, 467
Planter, Seed	J. M. Elliott	Winsborough, S. C	Nov. 8, 1870	108, 988
Planter, Seed	J. W. Ells and J. Charlton	Pittsburgh, Pa	Sept. 8, 1859	18, 140

Index of patents issued from the United States Patent Office from 1790 *to* 1873, *inclusive*—Continued.

Invention.	Inventor.	Residence.	Date.	No.
Planter, Seed	F. E. A. Engelman	Cheektowaga, N. Y	June 8, 1869	91, 002
Planter, Seed	G. T. Enoch and D. Wissinger	Springfield, Ohio	May 9, 1854	10, 881
Planter, Seed	G. M. Evans	Pittsburgh, Pa	Apr. 28, 1857	17, 145
Planter, Seed	G. M. Evans	Pittsburgh, Pa	Oct. 25, 1859	25, 889
Planter, Seed	H. C. Fairchild	Brooklyn, Pa	Aug. 10, 1858	21, 127
Planter, Seed	H. C. Fairchild	Brooklyn, Pa	June 5, 1860	28, 567
Planter, Seed	H. C. Fairchild	Brooklyn, Pa	Nov. 5, 1867	70, 541
Planter, Seed	J. W. Fawkes	Bart Township, Pa	Dec. 17, 1850	7, 837
Planter, Seed	D. S. Fisher	Mauckport, Ind	May 10, 1859	23, 913
Planter, Seed	L. B. Fisher	Cold Water, Mich	Mar. 7, 1854	10, 607
Planter, Seed	L. B. Fisher	Cold Water, Mich	Mar. 28, 1854	10, 694
Planter, Seed	G. Fletcher, sr., and T. Barnes	Greensburgh, Ind	Mar. 26, 1850	7, 213
Planter, Seed	J. Fordyce	Morgantown, Va	Sept. 9, 1856	15, 691
Planter, Seed	N. Foster, G. Jessup, and H. L. and C. P. Brown	Palmyra, N. Y	Nov. 4, 1851	8, 484
Planter, Seed	J. K. Frantz	Goodville, Pa	Dec. 24, 1867	72, 472
Planter, Seed	R. Friday	Crockett, Tex	Feb. 25, 1873	136, 231
Planter, Seed	T. C. Carlington	La Fayette, Ala	Feb. 11, 1873	135, 704
Planter, Seed	I. H. Garretson	Clay, Iowa	Mar. 29, 1853	9, 636
Planter, Seed	C. C. Garrett	Spring Hill, Ala	Feb. 12, 1861	31, 381
Planter, Seed	J. C. Gaston	Reading, Ohio	Mar. 7, 1854	10, 608
Planter, Seed	R. J. Gatling	Murfreesborough, N. C	May 10, 1844	3, 581
Planter, Seed	R. and W. L. Gebby	New Richland, Ohio	Feb. 12, 1856	14, 235
Planter, Seed	W. L. Gebby	New Richland, Ohio	Aug. 14, 1860	29, 581
Planter, Seed	J. Gibbons	Adrian, Mich	Aug. 25, 1840	1, 731
Planter, Seed	W. Y. Gill	Henderson, Ky	Oct. 6, 1857	18, 333
Planter, Seed	R. Gillaspie	New Richmond, Ohio	Mar. 6, 1866	52, 989
Planter, Seed	J. M. Gitchell	Haverhill, N. H	Oct. 20, 1868	83, 151
Planter, Seed	W. W. Golsan	Autaugaville, Ala	Sept. 4, 1860	29, 878
Planter, Seed	F. Goodwin	Astoria, N. Y	Mar. 3, 1857	16, 729
Planter, Seed	A. M. Gould and A. Flanders	Cambria, N. Y	Oct. 6, 1857	18, 334
Planter, Seed	J. Green	Kennett's Square, Pa	June 5, 1860	28, 572
Planter, Seed	J. D. Green	Antrim, Ohio	Mar. 19, 1867	63, 038
Planter, Seed	P. B. Green and E. A. Kenedy	Chicago and Newark, Ill	June 10, 1856	15, 101
Planter, Seed	T. M. Green	Milledgeville, Ga	Aug. 21, 1860	29, 689
Planter, Seed	J. P. Groshon	Yonkers, N. Y	Mar. 19, 1850	7, 187
Planter, Seed	R. B. Ground	Marino Town, Ill	June 29, 1858	20, 709
Planter, Seed	D. Haldeman	Morgantown, Va	Oct. 5, 1852	9, 296
Planter, Seed	G. Hall	Morgantown, Va	June 24, 1856	15, 182
Planter, Seed	J. Hall	Honey Cut, Ala	Oct. 13, 1857	18, 393
Planter, Seed	L. Hariman	Anderson, Ind	Aug. 21, 1860	26, 691
Planter, Seed	W. D. Harrah and B. S. Baldwin	Davenport, Iowa	Aug. 9, 1859	25, 011
Planter, Seed	E. P. Harris	Conneautville, Pa	May 26, 1868	78, 281
Planter, Seed	E. Hart	New Albany, Ind	Aug. 6, 1850	7, 544
Planter, Seed	E. Hart	New Albany, Ind	Oct. 19, 1852	9, 343
Planter, Seed	J. Haselton	Oxford, N. H	May 12, 1857	17, 275
Planter, Seed	W. W. Haupt	Mountain City, Tex	Apr. 27, 1869	89, 404
Planter, Seed	J. D. Havis	Perry, Ga	Oct. 28, 1856	15, 974
Planter, Seed	M. Hayden	Rochester, Mich	May 9, 1865	47, 637
Planter, Seed	A. Hays	Guy's Mills, Pa	June 9, 1868	78, 666
Planter, Seed	T. D. Henson and G. Rohr	Charlestown, Va	Feb. 28, 1854	10, 572
Planter, Seed	G. E. Herrick	Lynn, Mass	July 2, 1867	66, 236
Planter, Seed	I. F. Herrin	San Antonio, Tex	Apr. 13, 1869	88, 870
Planter, Seed	G. Hetrick	Reidsburgh, Pa	July 10, 1860	29, 078
Planter, Seed	J. Hildebrand	East Berlin, Pa	Feb. 10, 1857	16, 590
Planter, Seed	D. Hill	Bartonia, Ind	June 27, 1854	11, 159
Planter, Seed	G. A. Hill and C. Lohnes	Springfield, Ohio	May 14, 1867	64, 670
Planter, Seed	A. F. Hines	Washington, D. C	Sept. 4, 1860	29, 937
Planter, Seed	P. Hinkley	Charleston, Ill	Oct. 20, 1857	18, 450
Planter, Seed	S. M. Hockman	Tom's Brook, Va	Aug. 29, 1854	11, 609
Planter, Seed	C. T. Holman	Conneautville, Pa	June 18, 1867	65, 810
Planter, Seed	D. S. Holman	Conneautville, Pa	Jan. 22, 1867	61, 431
Planter, Seed	S. T. Holly	Rockford, Ill	June 16, 1857	17, 568
Planter, Seed	A. J. Holt	Peru, Ind	Sept. 29, 1868	82, 524
Planter, Seed	P. Horn	Hagerstown, Md	Aug. 23, 1853	9, 955
Planter, Seed	T. B. Houghton	Bloomington, Ill	Feb. 10, 1857	16, 585
Planter, Seed	J. S. Huggins	Timmonsville, S. C	Feb. 28, 1860	27, 291
Planter, Seed	D. H. Hull	Plantsville, Conn	May 21, 1867	64, 877
Planter, Seed	M. J. Hunt and J. H. Haines	Rising Sun, Md	Jan. 5, 1858	19, 026
Planter, Seed	S. C. Hunter	East Hickory, Pa	Nov. 26, 1867	71, 488
Planter, Seed	O. Hyde	Oakland, Cal	Sept. 17, 1872	131, 400
Planter, Seed	S. Ide	East Shelby, N. Y	July 4, 1854	11, 226
Planter, Seed	S. C. Ives	Land of Promise, Va	Aug. 27, 1872	130, 920
Planter, Seed	R. M. Jackson	Penningtonville, Pa	Oct. 5, 1852	9, 298
Planter, Seed	S. Jenkins	Portsmouth, Pa	Sept. 20, 1873	10, 032
Planter, Seed	S. Johnson	Cold Spring, N. Y	Feb. 5, 1861	31, 354
Planter, Seed	W. B. Johnson	Staunton, Va	June 6, 1854	11, 029
Planter, Seed	W. D. Johnson	Raleigh, N. C	Aug. 13, 1867	67, 656
Planter, Seed	J. J. Johnston	Allegheny, Pa	May 4, 1858	20, 158
Planter, Seed	J. F. Keller	Greencastle, Pa	Feb. 14, 1865	46, 364
Planter, Seed	J. F. Keller	Greencastle, Pa	Jan. 29, 1867	61, 544
Planter, Seed	J. F. Keller	Greencastle, Pa	Jan. 29, 1867	61, 545
Planter, Seed	J. M. Kelley	Clinton, Ill	Sept. 15, 1863	39, 931
Planter, Seed	R. Ketcham	South Dansville, N. Y	Mar. 20, 1866	53, 304
Planter, Seed	C. Ketchum	Penn Yan, N. Y	May 12, 1857	17, 305
Planter, Seed	W. Kilburn	Lawrenceville, Pa	June 11, 1842	2, 663
Planter, Seed	W. Kilburn and F. Haines	Lawrenceville and Marietta, Pa.	Dec. 31, 1844	3, 870
Planter, Seed	E. W. Kimball	Ottawa, Ill	Aug. 10, 1858	21, 137
Planter, Seed	J. H. King, jr	Georgetown, D. C	Aug. 29, 1854	11, 611
Planter, Seed	A. Klaus	Belleville, Ill	Oct. 18, 1859	25, 835
Planter, Seed	W. Knowland and K. Collings	Henryville, Ind	June 27, 1871	116, 327
Planter, Seed	A. Kraber	York, Pa	July 27, 1852	9, 151
Planter, Seed	B. Kuhns	Dayton, Ohio	Dec. 29, 1868	85, 455
Planter, Seed	B. Kuhns and M. J. Haines	Dayton, Ohio, and Delaware City, Del.	Sept. 30, 1856	15, 810
Planter, Seed	L. L. Lancaster	Rocky Mount, N. C	Sept. 20, 1859	25, 513
Planter, Seed	J. Landes	Selma, Ohio	Feb. 10, 1857	16, 597

Index of patents issued from the United States Patent Office from 1790 *to* 1873, *inclusive*—Continued.

Invention.	Inventor.	Residence.	Date.	No.
Planter, Seed	J. N. Lane	Bethel, Ky	Apr. 3, 1866	53, 635
Planter, Seed	G. W. Lee	Ercildown, Pa	Nov. 21, 1854	11, 967
Planter, Seed	J. Lee	Galesburgh, Ill	Dec. 8, 1857	18, 821
Planter, Seed	J. H. Lee	Camanche, Iowa	Dec. 13, 1859	26, 439
Planter, Seed	J. S. Lewis	Elkader, Iowa	Sept. 28, 1869	95, 241
Planter, Seed	T. J. and G. F. Lewis	Boston, Mass	June 27, 1840	1, 657
Planter, Seed	T. J. and G. F. Lewis	Boston, Mass	June 27, 1840	1, 661
Planter, Seed	H. C. Locke	Somerville, Tenn	Sept. 21, 1869	95, 121
Planter, Seed	C. O. Luce	Brandon, Vt	Oct. 6, 1857	18, 344
Planter, Seed	I. G. McFarlane	Perry County, Pa	Mar. 14, 1854	10, 655
Planter, Seed	W. A. Mahaffy	Carimona, Minn	Aug. 31, 1858	21, 397
Planter, Seed	J. L. Manlove	Connersville, Ind	Oct. 22, 1867	70, 100
Planter, Seed	E. Marshall	Clinton, N. J	Apr. 11, 1854	10, 753
Planter, Seed	F. M. Marshall	Seguin, Tex	Dec. 28, 1858	22, 438
Planter, Seed	J. W. Masten	Utica, Mich	May 22, 1860	28, 386
Planter, Seed	E. G. Mathews	Newton, Mass	June 8, 1869	91, 143
Planter, Seed	A. Maurer	New Carlisle, Ind	Oct. 11, 1859	25, 746
Planter, Seed	J. M. Maxwell	Cape Elizabeth, Me	Nov. 29, 1864	45, 257
Planter, Seed	T. B. McConaughey	Newark, Del	Mar. 27, 1860	27, 644
Planter, Seed	E. McCormick	Connellsville, Pa	Oct. 16, 1855	13, 683
Planter, Seed	I. W. McGaffey	Syracuse, N. Y	Apr. 3, 1855	12, 641
Planter, Seed	I. W. McGaffey	Chicago, Ill	July 1, 1862	35, 771
Planter, Seed	I. W. McGaffey	Chicago, Ill	July 3, 1866	56, 076
Planter, Seed	J. McKown	Gardstown, Va	June 22, 1858	20, 651
Planter, Seed	J. McKown	Gardstown, Va	May 24, 1859	24, 135
Planter, Seed	J. McLaughlin	Duncannon, Pa	Aug. 28, 1860	29, 803
Planter, Seed	J. B. McMillan	Tipton, Ind	Aug. 9, 1859	25, 026
Planter, Seed	G. A. Meacham	New York, N. Y	June 10, 1856	15, 106
Planter, Seed	G. A. Meacham	New York, N. Y	Mar. 31, 1857	16, 929
Planter, Seed	N. Mendenhall	Greensburgh, Ind	Jan. 25, 1870	99, 217
Planter, Seed	J. T. Mercer	Seneca Township, Ohio	Dec. 27, 1859	26, 605
Planter, Seed	J. Miller	Russellville, Ky	Sept. 26, 1865	50, 213
Planter, Seed	J. C. Miller	Marietta, Pa	Jan. 23, 1849	6, 050
Planter, Seed	J. R. Mills	Bloomfield, Iowa	Aug. 28, 1860	29, 807
Planter, Seed	J. B. Miner	Groton, Conn	June 22, 1869	91, 655
Planter, Seed	T. S. Minniss	Meadville, Pa	May 29, 1855	12, 958
Planter, Seed	E. Moore	Avon, N. Y	Dec. 9, 1856	16, 212
Planter, Seed	H. Moore	Clinay, Mich	Mar. 27, 1855	12, 603
Planter, Seed	L. Moore	Bart Township, Lancaster County, Pa.	Apr. 18, 1848	5, 522
Planter, Seed	E. Morgan	Morgantown, Va	Jan. 16, 1855	12, 256
Planter, Seed	E. Morgan	Morgantown, Va	Mar. 18, 1856	14, 465
Planter, Seed	E. Morse	Walpole, N. H	Mar. 20, 1855	12, 554
Planter, Seed	G. Mottmiller	Columbus, Ohio	Sept. 1, 1843	3, 247
Planter, Seed	W. R. Mozier	Higginsville, Ill	Dec. 17, 1867	72, 318
Planter, Seed	W. G. Murphy	Seguin, Tex	Aug. 16, 1859	25, 134
Planter, Seed	J. Musgrave	New Cumberland, W. Va	Dec. 8, 1868	84, 751
Planter, Seed	E. Myers	Carroll County, Md	June 19, 1849	6, 542
Planter, Seed	D. B. Neal	Mount Gilead, Ohio	Oct. 23, 1855	13, 706
Planter, Seed	D. B. Neal	Mount Gilead, Ohio	Dec. 1, 1857	18, 762
Planter, Seed	H. Nycum	Uniontown, Pa	Dec. 14, 1852	9, 468
Planter, Seed	R. F. Osgood	Rochester, N. Y	Sept. 15, 1868	82, 153
Planter, Seed	B. Owen	Dayton, Ohio	Sept. 7, 1858	21, 440
Planter, Seed	G. Page	Baltimore, Md	May 25, 1840	1, 617
Planter, Seed	E. Parker	Baltimore, Md	June 1, 1858	20, 440
Planter, Seed	J. Peirson	Wilmington, Del	July 5, 1848	5, 655
Planter, Seed	J. Peirson	Wilmington, Del	Dec. 25, 1849	6, 976
Planter, Seed	M. and S. Pennock	East Marlborough, Pa	Mar. 12, 1841	1, 999
Planter, Seed	S. and M. Pennock	Kennett's Square, Pa	Dec. 10, 1850	7, 822
Planter, Seed	H. Perrin and W. Ruddock	Wilmington, Ohio	Sept. 20, 1853	10, 037
Planter, Seed	D. H. Phillips	Greenville, Ill	Mar. 20, 1855	12, 557
Planter, Seed	W. C. Pitts	Austin, Texas	May 15, 1860	28, 304
Planter, Seed	F. Plummer	Manchester, Ind	Jan. 22, 1856	14, 144
Planter, Seed	F. Plummer and G. B. Rollins	Manchester, Ind	Sept. 11, 1855	13, 551
Planter, Seed	W. G. Pollock and J. W. Sener	Fredericksburgh, Va	Oct. 9, 1860	30, 348
Planter, Seed	J. F. Pond	Cleveland, Ohio	Aug. 20, 1861	33, 103
Planter, Seed	L. Pratt	Amherst, N. H	Apr. 25, 1844	3, 562
Planter, Seed	J. A. Preston	Greensborough, Ga	Jan. 16, 1872	122, 736
Planter, Seed	D. R. Prindle	Bethany, New York	Mar. 29, 1859	23, 391
Planter, Seed	J. Putnam	Hamilton, N. Y	Mar. 7, 1846	4, 405
Planter, Seed	B. A. Ramsey	Trenton, Tenn	Oct. 4, 1870	108, 051
Planter, Seed	C. Randall	Palmyra, Ga	Nov. 2, 1852	9, 370
Planter, Seed	S. G. Randall	Rockford, Ill	Feb. 10, 1857	16, 610
Planter, Seed	S. G. Randall	Dixon, Ill	June 2, 1857	17, 450
Planter, Seed	S. G. Randall and J. H. Jones	Rockton, Ill	Aug. 7, 1855	13, 401
Planter, Seed	J. Redhead	Woodville, Miss	Mar. 9, 1858	19, 579
Planter, Seed	W. Redick	Uniontown, Pa	Nov. 18, 1851	8, 532
Planter, Seed	W. Redick	Uniontown, Pa	Aug. 29, 1854	11, 617
Planter, Seed	I. Rexford	Malone, N. Y	Dec. 8, 1868	84, 762
Planter, Seed	I. Reynolds	Republic, Ohio	Mar. 9, 1852	8, 790
Planter, Seed	D. B. Rhodes	Concord, N. Y	Dec. 10, 1850	7, 823
Planter, Seed	A. Richards	Anderson, Tex	May 21, 1872	126, 985
Planter, Seed	J. Robb	Lewistown, Pa	Oct. 12, 1852	9, 333
Planter, Seed	J. S. Robb and S. P. Allison	New Cumberland, W. Va	Jan. 12, 1869	85, 854
Planter, Seed	J. Robinson	Sharpstown, Md	Dec. 1, 1857	18, 772
Planter, Seed	J. Robinson	Sharpstown, Md	Apr. 17, 1860	27, 929
Planter, Seed	T. B. and R. N. Rockwell	Batavia, Ill	Aug. 13, 1861	33, 053
Planter, Seed	A. J. Rogers	Stephentown, N. Y	June 19, 1860	28, 777
Planter, Seed	T. B. Rogers	Wethersfield, Conn	Aug. 9, 1859	25, 047
Planter, Seed	G. Rohr	Charlestown, Va	June 21, 1853	9, 803
Planter, Seed	R. Romaine	Montreal, Canada	Feb. 20, 1855	12, 447
Planter, Seed	J. H. Rose	Versailles, Ill	Aug. 17, 1858	21, 217
Planter, Seed	J. P. Ross	Lewisburgh, Pa	Sept. 25, 1849	6, 743
Planter, Seed	J. P. Ross	Lewisburgh, Pa	Jan. 1, 1851	7, 877
Planter, Seed	J. P. Ross	Lewisburgh, Pa	June 8, 1852	9, 004
Planter, Seed	E. Russell	Coatesville, Pa	May 15, 1860	28, 330
Planter, Seed	T. Russell	Waldoborough, Me	Apr. 13, 1858	19, 953
Planter, Seed	W. J. Saffrey	Bremen, Ohio	July 8, 1873	140, 648

Index of patents issued from the United States Patent Office from 1790 *to* 1873, *inclusive*—Continued.

Invention.	Inventor.	Residence.	Date.	No.
Planter, Seed	B. D. Sanders	Holliday's Cove, Va	June 8, 1852	9, 006
Planter, Seed	C. R. Sargent	Newburyport, Mass	Apr. 25, 1871	114, 047
Planter, Seed	M. Satterlee	Louisa, Ill	July 26, 1853	9, 879
Planter, Seed	B. Saunders	Claverack, N. Y	July 4, 1871	116, 633
Planter, Seed	W. F. Schroeder	La Porte, Ind	Aug. 14, 1860	29, 624
Planter, Seed	J. F. Seaman	Walcott, N. Y	Sept. 30, 1856	15, 822
Planter, Seed	G. M. and S. H. Seward	Guilford, Conn	June 6, 1865	48, 104
Planter, Seed	J. B. Seymour	Pittsburgh, Pa	Sept. 24, 1867	69, 130
Planter, Seed	P. Seymour	East Bloomfield, N. Y	July 24, 1855	13, 326
Planter, Seed	J. Shattuck	Waterloo, N. Y	Jan. 29, 1867	61, 680
Planter, Seed	J. M. Shaw	Water Valley, Miss	Nov. 30, 1869	97, 446
Planter, Seed	J. W. Sherman	Ontario, N. Y	Nov. 6, 1849	6, 853
Planter, Seed	N. C. Sherman and J. Mason	Hazel Green, Wis	Dec. 23, 1856	16, 314
Planter, Seed	J. H. Shireman	East Berlin, Pa	Oct. 21, 1856	15, 955
Planter, Seed	J. Signer and T. N. Shipton	Kishacoquillas Valley, Pa	Dec. 10, 1850	7, 831
Planter, Seed	H. Sloan	Franklin, Ind	June 26, 1860	28, 909
Planter, Seed	D. M. Smith	Springfield, Vt	May 10, 1859	23, 955
Planter, Seed	G. Smith and A. G. Perry	Clyde, Ohio	June 29, 1858	20, 738
Planter, Seed	H. W. Smith	Paradise, Pa	Nov. 4, 1846	4, 833
Planter, Seed	J. D. Smith	Lancaster, Ohio	Oct. 27, 1857	18, 524
Planter, Seed	J. D. Smith	Peoria, Ill	June 24, 1862	35, 713
Planter, Seed	L. Smith	Chester Centre, Mass	Aug. 23, 1870	106, 631
Planter, Seed	T. G. Smith	Canton, Miss	Apr. 13, 1869	88, 989
Planter, Seed	T. H. Smith	Clyde, N. Y	Aug. 25, 1868	81, 548
Planter, Seed	T. H. Smith	Clyde, N. Y	May 24, 1870	103, 382
Planter, Seed	B. M. Snell	Hancock, Md	Mar. 20, 1855	12, 561
Planter, Seed	T. Snow	Social Circle, Ga	Jan. 2, 1872	122, 412
Planter, Seed	J. G. Snyder and J. Young	Wheatfield Township, Pa	Feb. 28, 1854	10, 583
Planter, Seed	W. Sprague	Ellicottsville, N. Y	Mar. 14, 1854	10, 644
Planter, Seed	R. H. Springstead	Wooster, Ohio	July 24, 1849	6, 605
Planter, Seed	J. Stark	Thomasville, Ga	Dec. 17, 1867	72, 333
Planter, Seed	E. M. Stevens, J. B. Crosby, and J. W. Pearson.	Boston and Winchester, Mass.	May 22, 1855	12, 924
Planter, Seed	S. L. Stockstill	Medway, Ohio	May 24, 1859	24, 158
Planter, Seed	S. L. Stockstill and P. H. Humes	Brandt, Ohio	Jan. 16, 1855	12, 260
Planter, Seed	B. T. Stowell and A. Marcellus	Waddam's Grove, Ill	Apr. 13, 1852	8, 877
Planter, Seed	J. L. Strait	Cooksville, Miss	Dec. 28, 1869	98, 313
Planter, Seed	W. H. Stuart	Millington, Md	Oct. 4, 1859	25, 685
Planter, Seed	I. T. Suggs	Greene Hill, Tex	Apr. 22, 1873	138, 105
Planter, Seed	J. F. Tannehill	Staunton, Va	Aug. 14, 1860	29, 636
Planter, Seed	J. H. Thomas and P. P. Mast	Springfield, Ohio	July 27, 1858	21, 034
Planter, Seed	H. Thomason	La Fayette, Ind	Feb. 10, 1857	16, 617
Planter, Seed	J. Thompson	Durhamville, N. Y	Jan. 13, 1857	16, 409
Planter, Seed	S. Thompson	Hopedale, Ohio	Mar. 30, 1858	19, 818
Planter, Seed	D. L. Tilton	Mount Carmel, Ill	Feb. 23, 1858	19, 456
Planter, Seed	F. W. Tilton	Moline, Ill	Mar. 17, 1868	75, 600
Planter, Seed	H. Todd	Oxford, N. H	Dec. 15, 1843	3, 381
Planter, Seed	F. Townsend	Cambria, N. Y	Nov. 2, 1852	9, 372
Planter, Seed	H. M. Tremble	Mattoon, Ill	Oct. 18, 1864	44, 758
Planter, Seed	C. S. Trevitt	Ellicottsville, N. Y	Nov. 2, 1852	9, 373
Planter, Seed	C. J. Turner and M. L. Wilkinson.	Olean, N. Y	Apr. 25, 1871	114, 066
Planter, Seed	J. Urmy	Wilmington, Del	Apr. 6, 1852	8, 866
Planter, Seed	F. Vandoren	Adrian, Mich	Apr. 13, 1852	8, 879
Planter, Seed	T. G. S. Vaniz	Canton, Miss	Aug. 8, 1871	117, 835
Planter, Seed	H. Vermillion	Rising Sun, Md	Nov. 2, 1852	9, 374
Planter, Seed	J. T. and L. J. Wait	Waterloo, S. C	Aug. 15, 1854	11, 537
Planter, Seed	C. A. Wakefield	Essex County, N. Y	Feb. 19, 1850	7, 110
Planter, Seed	C. A. Wakefield	Plainfield, Mass	July 25, 1854	11, 395
Planter, Seed	A. Wales	Pontiac, Ill	June 29, 1858	20, 749
Planter, Seed	L. F. Ward	Marathon, N. Y	Nov. 24, 1857	18, 716
Planter, Seed	L. F. Ward	Marathon, N. Y	Dec. 29, 1857	18, 999
Planter, Seed	M. Ward	Owego, N. Y	Mar. 27, 1855	12, 608
Planter, Seed	S. B. Ward	Auburn, Ind	Jan. 5, 1869	85, 711
Planter, Seed	M. Warner	West Middleburgh, Ohio	Aug. 21, 1860	29, 735
Planter, Seed	D. Warren	Gettysburgh, Pa	June 26, 1860	28, 926
Planter, Seed	S. J. Wasterburg	Altona, Ill	Sept. 27, 1859	25, 595
Planter, Seed	M. Waterbury	Cuba, N. Y	Nov. 21, 1854	11, 980
Planter, Seed	G. Watt	Richmond, Va	Mar. 8, 1859	23, 206
Planter, Seed	M. D. Wells	Morgantown, Va	Dec. 14, 1852	9, 475
Planter, Seed	W. F. West	Haverstraw, N. Y	Jan. 3, 1871	110, 706
Planter, Seed	J. E. White	Clinton, Ill	June 7, 1870	104, 088
Planter, Seed	T. W. White	Milledgeville, Ga	Apr. 16, 1861	32, 098
Planter, Seed	J. Whitehead	Manchester, Va	May 26, 1857	17, 402
Planter, Seed	E. Wicks	Bart, Pa	Feb. 10, 1852	8, 728
Planter, Seed	A. Wieting	Middletown, Pa	Apr. 1, 1851	8, 018
Planter, Seed	J. H. Wiggin	Boston, Mass	Feb. 2, 1858	19, 274
Planter, Seed	S. C. Wilder	Sardinia, Ohio	Mar. 17, 1868	75, 613
Planter, Seed	H. Willard	Vergennes, Vt	Oct. 6, 1857	18, 366
Planter, Seed	W. B. Willis	Charlestown, Va	Jan. 22, 1850	7, 044
Planter, Seed	J. D. Willoughby	Chambersburgh, Pa	June 5, 1849	6, 504
Planter, Seed	J. D. Willoughby	Pleasant Hall, Pa	Jan. 26, 1858	19, 222
Planter, Seed	J. D. Willoughby	Carlisle, Pa	Aug. 3, 1858	21, 102
Planter, Seed	C. B. Winder	North Lewisburgh, Ohio	Nov. 24, 1857	18, 717
Planter, Seed	S. and W. H. Witherow	Gettysburgh, Pa	Jan. 18, 1853	9, 551
Planter, Seed	D. and H. Wolf	Lebanon, Pa	Feb. 15, 1853	9, 589
Planter, Seed	D. and H. Wolf	Lebanon, Pa	Mar. 21, 1854	10, 682
Planter, Seed	L. Woodruff	Ann Arbor, Mich	May 2, 1865	47, 593
Planter, Seed	D. C. Woods	Waxahatchie, Tex	Dec. 21, 1869	98, 136
Planter, Seed	J. Woodward	Haverhill, N. H	July 13, 1852	9, 125
Planter, Seed	A. R. Worth	Nantucket, Mass	Apr. 9, 1867	63, 685
Planter, Seed	W. H. Worth and L. Finlay	Canton, Mo	Dec. 13, 1859	26, 455
Planter, Seed	R. C. Wrenn	Mount Gilead, Ohio	Nov. 29, 1853	10, 278
Planter, Seed	L. R. Wright	Cohoes, N. Y	Apr. 16, 1861	32, 108
Planter, Seed	L. R. Wright	Cohoes, N. Y	Oct. 23, 1866	59, 115
Planter, Seed	R. B. Wright	Vermillion, Ill	Mar. 17, 1868	75, 616
Planter, Seed	H. Wyant	Knox County, Ind	Dec. 9, 1856	16, 198
Planter, Seed	F. G. Wynkoop	Corning, N. Y	Oct. 16, 1855	13, 694
Planter, Seed	E. Young	Fayetteville, Mo	June 26, 1860	28, 936

Index of patents issued from the United States Patent Office from 1790 *to* 1873, *inclusive*—Continued.

Invention.	Inventor.	Residence.	Date.	No.
Planter, Seed and guano	T. W. White	Milledgeville, Ga	June 18, 1867	65, 976
Planter, seed-sower, and cultivator, Corn	D. P. Leach	Franklin, Ind	Oct. 19, 1869	96, 015
Planter, seeder, and cultivator, Combined potato	B. F. Field	Sheboygan Falls, Wis	Sept. 26, 1865	50, 202
Planter, Self-working rotary corn	J. W. Simonton	Taylorsville, Ind	Feb. 15, 1870	99, 792
Planter shield, Corn	R. T. Taylor	Everton, Ind	Oct. 20, 1868	83, 340
Planter, Single-row corn	J. Clarridge	Pancoastburgh, Ohio	Oct. 9, 1866	58, 598
Planter, sower, revolving harrow, and cultivator, Combined.	W. P. Byler	Leavenworth, Kans	Feb. 18, 1868	74, 496
Planter, Sugar-cane	E. Cortés	Sagua la Grande, Cuba	Dec. 18, 1866	60, 620
Planter, Walking	A. W. Dunlevy	Fair Play, Ohio	Oct. 31, 1871	120, 507
Planter, Walking	N. Earlywine	Centerville, Iowa	Oct. 31, 1871	120, 423
Planter, Walking	G. W. Heath	Burlington, Pa	Apr. 9, 1872	125, 568
Planter, Walking	L. D. Noble	Cerro Gordo, Ill	Jan. 30, 1872	123, 121
Planter, Walking	M. W. Stephenson	Pickensville, Ala	Jan. 9, 1872	122, 675
Planters, Actuating the feeding-apparatus of seed	J. K. Dugdale	Richmond, Ind	Apr. 7, 1863	38, 099
Planters, Check-row attachment for corn	W. C. Grimes	Decatur, Ill	Apr. 18, 1871	113, 761
Planters, Cultivating seed	W. Flory and G. A. Grove	Chambersburgh, Pa	Mar. 12, 1850	7, 163
Planters, Device for sowing in seed	W. P. Clements	Ellerslie, Ga	Oct. 7, 1851	8, 408
Planters, Draft-apparatus of seed	J. Mumma	Mount Joy, Pa	Aug. 16, 1853	9, 940
Planters, Drill-teeth in seed	L. Haverstick	Manor Top, Pa	Apr. 2, 1850	7, 251
Planters, Dropper for seed	H. H. Koeller	Camp Point, Ill	Dec. 16, 1873	145, 575
Planters, Gearing for seed	M. J. Hunt	Rising Sun, Md	June 3, 1851	8, 138
Planters, Gearing for seed	J. Pierson	Wilmington, Del	Apr. 9, 1850	7, 268
Planters, Guide for corn and seed	E. C. Brown	Crawfordsville, Ind	Mar. 24, 1868	75, 851
Planters, Marking-attachment to corn	E. Sawyer	Madison, Iowa	Sept. 13, 1870	107, 294
Planters, Seed-distributer for seed	D. and H. Wolf	Lebanon, Pa	June 3, 1851	8, 132
Planter, Seed-roller for seed	E. Wicks	Bart Township, Pa	Mar. 26, 1850	7, 227
Planters, Seed-tube for corn	L. Scofield	Watertown, Wis	Mar. 14, 1871	112, 741
Planters, Seed-tube for corn	L. Scofield	Watertown, Wis	July 11, 1871	116, 998
Planters, Seeding-apparatus for seed	G. S. Gardner	Charlestown, Va	Oct. 8, 1850	7, 699
Planters, Seeding-apparatus for seed	D. Horner	Knox County, Ohio	July 29, 1851	8, 264
Planters, Seeding-apparatus for seed	L. Moore	Bart, Pa	July 2, 1850	7, 479
Planters, Seeding-apparatus for seed	S. and M. Pennock	Kennett's Square, Pa	July 9, 1850	7, 495
Planters, Seeding-apparatus for seed	S. and M. Pennock	Kennett's Square, Pa	Feb. 11, 1851	7, 930
Planters, Seeding-apparatus for seed	G. Rohr	Charlestown, Va	July 16, 1850	7, 513
Planters, Seeding-apparatus for seed	C. C. Van Enery	Victor, N. Y	Oct. 21, 1851	8, 459
Planters, Seeding-roller for seed	A. Palmer	Brockport, N. Y	Sept. 10, 1850	7, 642
Planters, Shoe for seed	A. W. Brinkerhoff	Upper Sandusky, Ohio	July 16, 1861	32, 819
Planters, Slide for seed	R. I. Colvin	Lancaster, Pa	Oct. 1, 1850	7, 680
Planters, Tube for seed	J. C. Haines	Dublin, Ind	Dec. 15, 1857	18, 843
Planting and cultivating machine, Cotton	R. Herbert	Williamson County, Tenn.	Apr. 19, 1828	....
Planting and cultivating machine, Cotton	F. H. Smith	Richmond, Va	Feb. 15, 1826	....
Planting and digging potatoes, plowing corn, &c.	P. Meigs and M. C. Arnold	Madison, Conn	Nov. 3, 1830	....
Planting and fertilizing machine	S. L. Allen	Cinnaminson, N. J	Oct. 25, 1870	108, 672
Planting and preparing seeds, Machine for	Z. Lane	Harrisburgh, N. Y	Sept. 21, 1829	....
Planting barrow, Seed	C. A. Wakefield	Essex County, N. Y	Feb. 19, 1850	7, 109
Planting, Corn	C. R. Belt	Washington, D. C	Jan. 15, 1836	....
Planting corn, &c	H. Todd	Pembroke, N. H	Oct. 1, 1831	....
Planting corn and pease	R. Coffey	Burke County, N. C	Dec. 14, 1830	....
Planting, Cotton	H. Allen	Fayetteville, Tenn	June 16, 1836	....
Planting cotton	C. Ford	Paineville, Va	May 26, 1825	....
Planting, hoeing, and digging potatoes, Combined machine for.	J. C. Clement	Kenduskeag, Me	Dec. 19, 1865	51, 560
Planting Indian corn in rows, Machine for	J. Blecker	Lycoming County, Pa	Mar. 18, 1809	....
Planting-machine	S. L. Allen	Cinnaminson, N. J	Nov. 24, 1868	84, 247
Planting-machine	M. Atwood, jr	Hampstead, N. H	June 24, 1839	1, 188
Planting-machine	J. M. Bright and E. Standt	Bernville, Pa	Aug. 5, 1873	141, 538
Planting-machine	R. Craggs and O. Reynolds	Williamson and Webster, N. Y.	Aug. 14, 1847	5, 237
Planting-machine	J. M. Forrest	Princess Ann Court-House, Va.	June 25, 1839	1, 194
Planting-machine	J. Jones	Newton, N. J	Oct. 11, 1841	2, 303
Planting-machine	M. Nichols	Clearfield, Pa	Oct. 16, 1840	1, 830
Planting-machine	G. Ray	Kinderhook, N. Y	Oct. 23, 1866	59, 071
Planting-machine	J. L. Sater	Cincinnati, Ohio	June 10, 1862	35, 546
Planting-machine	E. Woods	Beloit, Wis	Jan. 10, 1845	3, 879
Planting-machine	R. B. Wright	Vermillion, Ill	Feb. 26, 1867	62, 461
Planting machine, Corn	D. and W. F. Chipman	Mount Carmel, Ill	Apr. 28, 1868	77, 255
Planting machine, Corn	A. Cock	Flushing, N. Y	Mar. 20, 1821	....
Planting machine, Corn, &c	S. W. Cole	Chelsea, Mass	Aug. 25, 1840	1, 729
Planting machine, Corn	M. Lennox, W. Croft, and H. Pitner.	Steubenville, Ohio	Jan. 26, 1829	....
Planting machine, Corn	N. R. and O. G. Merchant	Gilford, N. Y.	Oct. 12, 1849	1, 366
Planting machine, Corn and bean	E. Spooner		Jan. 25, 1799	....
Planting machine, Corn and other seed	D. S. Rockwell	New Canaan, Conn	Mar. 12, 1839	1, 097
Planting machine, Cotton, corn, &c	J. Lobdell	Warren County, Miss	Dec. 26, 1826	....
Planting machine, Cotton-seed	J. Armstrong	Bucyrus, Ohio	Nov. 26, 1867	71, 263
Planting machine, Cotton-seed	Z. N. Morrel	Cameron, Tex	July 5, 1859	24, 652
Planting machine, Cotton-seed	W. Price	Mount Olive, N. C	June 5, 1860	28, 602
Planting machine, Cotton-seed	R. S. Thomas	Bennetsville, S. C	July 30, 1841	2, 205
Planting, Machine for making holes for	E. R. Ensign	East Hartford, Conn	Sept. 3, 1867	68, 424
Planting, Machine for marking ground for	P. McQuaid	Wenona, Ill	Aug. 7, 1866	56, 970
Planting machine, Grain	O. Starr	Richmond, N. Y	Aug. 22, 1828	....
Planting machine, Potato	L. A. Aspinwall	Ireland's Corners, N. Y	Oct 14, 1862	36, 634
Planting machine, Potato	G. J. Bundy	Lyndon, Vt	July 21, 1857	17, 827
Planting machine, Potato	T. Marcus	Washington, D. C	Sept. 6, 1864	44, 103
Planting machine, Potato	J. W. Pelletreau	East Moriches, N. Y	Apr. 6, 1858	19, 869
Planting machine, Potato	J. L. True	Garland, Me	Feb. 7, 1865	46, 281
Planting machine, Potato	T. B. Whyte	Greenwich, N. Y	Feb. 9, 1858	19, 322
Planting machine, Seed	D. Brener	Petersburgh, Tenn	Apr. 4, 1844	3, 525
Planting machine, Seed	W. Buckminster	Framingham, Mass	Oct. 8, 1838	969
Planting machine, Seed	A. H. and L. Robbins, jr	Denmark, N. Y	Aug. 28, 1828	....
Planting machine, Seed	J. Woodward	Haverhill, N. H	Feb. 29, 1848	5, 460
Planting onion-seed, beans, &c	P. Smith and T. H. Arnold	Haddam, Conn	Jan. 20, 1831	....
Planting or sowing, Toothed roller for preparing the earth for.	A. Densmore	Le Roy, N. Y	July 29, 1815	....
Plas er	M. C. Borgia and H. B. Taylor	Philadelphia, Pa	July 17, 1866	56, 353
Plaster	G. E. Mitchell	Lowell, Mass	Sept. 12, 1871	118, 872

Index of patents issued from the United States Patent Office from 1790 *to* 1873, *inclusive*—Continued.

Invention.	Inventor.	Residence.	Date.	No.
Plaster, Adhesive	J. Lynch	Columbia, S. C	Sept. 1, 1868	81, 657
Plaster, Adhesive	J. Melvin	Lowell, Mass	Feb. 3, 1863	37, 604
Plaster, Adhesive	W. H. Sherut and H. H. Day	New York, N. Y	Mar. 26, 1845	3, 965
Plaster, Adhesive compound and	J. Hirsh	Chicago, Ill	Feb. 2, 1869	86, 398
Plaster and seed sower	D. Dick and O. W. Preston, jr	Corning, N. Y	Feb. 11, 1868	74, 320
Plaster, Blister	E. Perkins	Baltimore, Md	Jan. 15, 1830	....
Plaster-crushing mill	E. Kent		Sept. 14, 1804	....
Plaster for walls	B. R. Smith and J. C. Harris	Philadelphia, Pa	Jan. 24, 1871	111, 267
Plaster, Healing	W. Kramer	New York, N. Y	May 10, 1870	102, 836
Plaster, lime, &c	A. and J. Krauss	Upper Milford, Pa	Oct. 16, 1830	....
Plaster, lime, &c., Machine for spreading	G. U. Relyea	Watkins, N. Y	Nov. 3, 1868	83, 816
Plaster, Medical	S. I. Merrill	Falmouth, Me	Nov. 10, 1868	83, 872
Plaster, Medicated	W. S. Bright and J. G. Morey	New Orleans, La	Jan. 8, 1867	61, 045
Plaster-molds, Process of reducing the size of	N. A. Downer	Canandaigua, Mich	Jan. 8, 1869	91, 100
Plaster, Mustard	B. I. Crew	Philadelphia, Pa	July 14, 1868	79, 811
Plaster of Paris, Breaking	P. Geiger	Robeson, Pa	May 12, 1812	....
Plaster of Paris, Mill for breaking	W. Brown	Baltimore, Md	Dec. 8, 1808	....
Plaster on cloth, Machine for spreading adhesive	A. Monson	Milton, Pa	Aug. 8, 1837	341
Plaster-patterns, Manufacture of	J. Semple	Chicago, Ill	Sept. 30, 1873	143, 384
Plaster-sowing machine	A. Bugbee	Elkhart, Ind	Sept. 19, 1865	49, 973
Plaster-spreading apparatus	N. Wood	Portland, Me	Oct. 29, 1872	132, 614
Plaster-spreading gage	J. M. Keep	Bath, Me	Apr. 30, 1850	7, 323
Plaster-spreading machine	W. N. Reed	Arlington, Va	June 4, 1872	127, 517
Plaster, Sticking or adhesive	J. C. Battersby	New York, N. Y	May 16, 1871	114, 750
Plaster, Whooping-cough	F. Hower	Brooklyn, N. Y	Nov. 23, 1869	97, 089
Plasters, Apparatus for making medical	A. D. Richards	Lowell, Mass	Apr. 16, 1867	63, 814
Plasters, Apparatus for spreading medical	A. J. Schafhirt	Washington, D. C	Apr. 4, 1871	113, 453
Plasters, Composition for healing	D. G. Williams	Boston, Mass	July 11, 1871	116, 906
Plasterer's float	L. A. Goodsell	New Haven, Conn	May 27, 1873	139, 383
Plasterer's hawk	T. J. McGeary	Newark, N. J	Oct. 19, 1869	96, 022
Plasterer's head-support	T. T. Wright	Fayetteville, N. C	Apr. 29, 1873	138, 463
Plastering, Corner-bead for	E. F. Rice	Worcester, Mass	Apr. 16, 1872	125, 844
Plastering, Foundation for	J. John	Chicago, Ill	Oct. 3, 1871	119, 615
Plastering-machine	J. L. Coburn	Mineral Point, Wis	Mar. 23, 1869	88, 135
Plastering-machine	I. Hussey	Harveysburgh, Ohio	Mar. 7, 1854	10, 590
Plastering-machine	J. Keene	Washington, D. C	July 9, 1867	66, 664
Plastering-machine	T. McKinley	New York, N. Y	July 27, 1869	92, 985
Plastering-machine	G. Stevens and J. H. Watson	Tawas City, Mich	Dec. 9, 1873	145, 459
Plastering surfaces	J. J. Althause	New York, N. Y	Jan. 7, 1862	34, 099
Plastering walls, Composition for	D. Young	Kokomo, Ind	Aug. 17, 1869	93, 941
Plastic articles, Composition for molding	C. Legg	Malcom, Iowa	Mar. 11, 1873	136, 658
Plastic composition	H. C. Gaskins	Union Vale, N. Y	Oct. 20, 1868	83, 149
Plastic composition	T. B. Gunning	New York, N. Y	Mar. 19, 1872	124, 736
Plastic compound	F. Baschnagel	Beverly, Mass	Mar. 6, 1860	27, 343
Plastic materials, Method of molding	W. B. Gleason	Boston, Mass	Dec. 10, 1867	72, 017
Plastic materials, Mold for	J. J. Wiggin	Cincinnati, Ohio	Dec. 4, 1866	60, 310
Plate	N. D. Stevens	Westbrook, Me	Oct. 12, 1869	95, 742
Plate and tumbler lock, Revolving	L. Jennings	New York, N. Y	Apr. 2, 1850	7, 244
Plate, Anti-slipping	W. B. Gould and W. H. Harris	Boston and Taunton, Mass	Sept. 8, 1868	81, 890
Plate, Army	W. Terneson	Philadelphia, Pa	Feb. 16, 1864	41, 652
Plate, Coated metal	E. Morewood and G. Rogers	Enfield, England	Apr. 6, 1858	19, 866
Plate cover, Dinner	M. M. J. O'Sullivan	New Haven, Conn	Nov. 14, 1871	120, 995
Plate, Dinner	J. K. Andrews	Antrim, Ohio	Nov. 19, 1867	71, 115
Plate handling-fork, Heated	G. W. Hyatt	Auburn, N. Y	Nov. 11, 1856	16, 058
Plate-holder	J. Carlin	New York, N. Y	Jan. 18, 1870	98, 923
Plate-holder	W. T. Stantenborough	Brooklyn, N. Y	Jan. 31, 1871	111, 399
Plate holder, Camera	A. S. Southworth	Boston, Mass	Apr. 10, 1855	12, 700
Plate holder, Focusing	S. W. Burcaw	Allentown, Pa	Oct. 24, 1865	50, 555
Plate holder, Glass	T. Jones	New York, N. Y	Mar. 30, 1869	88, 487
Plate-holder, Non-corrosive metal-coated	J. F. Ryder	Cleveland, Ohio	May 16, 1871	114, 864
Plate, Hollow pressure	J. B. Fontaine	Philadelphia, Pa	Nov. 13, 1866	59, 716
Plate-lifter	D. B. Beaty	Aurora, Ill	Dec. 10, 1867	71, 954
Plate-lifter	C. F. Bosworth	Milford, Conn	Aug. 6, 1867	67, 402
Plate-lifter	H. P. Brooks	Waterbury, Conn	Nov. 29, 1870	109, 715
Plate-lifter	J. A. Burns	New Haven, Conn	Sept. 3, 1867	68, 557
Plate-lifter	E. Gibbs	Richland Centre, Wis	June 14, 1870	104, 298
Plate-lifter	D. E. Roe	Elmira, N. Y	Jan. 7, 1868	73, 199
Plate-lifter	G. O. Roe	Conantsville, Conn	Feb. 18, 1868	74, 598
Plate-lifter	D. M. Skinner	Sandwich Center, N. H	June 11, 1867	65, 616
Plate-lifter	J. M. Smith	Centre Sandwich, N. H	Nov. 26, 1867	71, 336
Plate-lifter	S. J. Talbott	Milford, N. H	Apr. 14, 1868	76, 649
Plate-lifter	D. Welch	Lowell, Mass	Mar. 5, 1867	62, 716
Plate-lifter	J. B. Willett	West Meriden, Conn	June 4, 1867	65, 460
Plate-lifter and bread-toaster	T. D. Keith	Mayville, Wis	Nov. 23, 1869	97, 093
Plate-lifter, Hot	C. L. Merrill	Saint Louis, Mo	Nov. 5, 1872	132, 848
Plate-metal along curved or straight lines, Machine for dividing.	N. Waterman and A. T. Perkins.	Toledo, Ohio	Apr. 30, 1872	126, 357
Plate or salver	H. McManus and J. B. Hatting	New York, N. Y	Dec. 15, 1868	84, 894
Plate, Pie	J. F. Kohler and S. B. Conover	New York, N. Y	July 23, 1867	67, 061
Plate spring, Circular metallic	J. W. Adams	New York, N. Y	May 15, 1855	12, 849
Plate, Table	D. H. Shirley	Boston, Mass	June 12, 1860	28, 693
Plate-warmer	J. C. Palmer	New York, N. Y	Aug. 27, 1867	68, 229
Plates and glass, Composition for cleaning	H. Teats	Ann Arbor, Mich	Apr. 5, 1870	101, 679
Plates and sheets, Manufacturing into	E. J. Ellicott	Baltimore, Md	May 28, 1818	....
Plates, Machine for folding tinned	O. W. Stow	Plantsville, Conn	Dec. 24, 1867	72, 561
Plates or bars, Machine for making holes in horizontal.	B. King	Washington, D. C	May 29, 1817	....
Plates, Wiping and polishing the exterior surface of punched or etched.	S. Conillard, jr	Boston, Mass	Oct. 9, 1830	....
Plated articles, Nickel	J. A. Whitman and T. M. Neal	Auburn, Me	Apr. 16, 1872	125, 808
Plated goods, Finishing the surface of	J. Rogers	Newark, N. J	Nov. 1, 1870	108, 940
Plated metal, Manufacture of	J. D. Grüneberg	Philadelphia, Pa	Dec. 26, 1865	51, 714
Plated ware	J. C. Blackman	West Meriden, Conn	Oct. 29, 1867	70, 156
Plated-ware-burnishing machine	S. A. Chapman	Waterbury, Conn	Feb. 19, 1867	62, 251
Plated-ware handles, Fastening	H. G. Smith and E. M. Pomeroy	West Meriden, Conn	May 8, 1866	54, 620
Platen-gage	H. Byxbe	Williamsport, Pa	Jan. 31, 1872	123, 233
Plating and coating metals	E. E. De Lobstein	Paris, France	Aug. 13, 1872	130, 332
Plating and gilding, Cleaning metallic articles for	G. J. Sturdy and S. W. Young	Providence, R. I	Mar. 2, 1869	87, 442

Index of patents issued from the United States Patent Office from 1790 *to* 1873, *inclusive*—Continued.

Invention.	Inventor.	Residence.	Date.	No.
Plating and polishing compound	M. J. A. Keane	New York, N. Y	Feb. 21, 1871	111, 945
Plating, Compound for silver	E. Hunter	Cleveland, Ohio	May 28, 1867	*65, 084
Plating iron and steel	E. Savage	West Meriden, Conn	Dec. 26, 1865	51, 754
Plating metals	R. G. Pine	Newark, N. J	July 11, 1854	11, 256
Plating metals	H. Tucker	Newton, Mass	Apr. 27, 1869	89, 522
Plating, Nickel	I. Adams, jr	Boston, Mass	July 4, 1871	116, 658
Plating, Nickel	I. Adams, jr	Boston, Mass	Mar. 11, 1873	136, 634
Plating, Nickel	M. G. Farmer	Salem, Mass	July 4, 1871	116, 579
Plating spoons, &c	M. L. Forbes	West Meriden, Conn	Mar. 10, 1868	75, 258
Platform: *See* Boat-builder's platform. Car-platform. Dumping-platform. Elevating-platform. Extension-platform. Ferry-track platform. Harvester-platform. Harvester-binders' platform. Harvester dropping-platform. Harvester dumping-platform. Harvester grain-platform. Ladder-platform. Moving platform. Oyster-platform. Railway-platform. Railway-ferry-track platform. Railway-scale platform. Revolving platform. Scale-platform. Stove-platform. Strainer-platform. Turning platform. Wagon-platform. Window-platform.				
Platform and windlass, Combined	T. I. Burbyte	Fond du Lac, Wis	Nov. 7, 1865	50, 787
Platform-elevator, Hydraulic telescopic	J. Parker and I. Cook	San Francisco, Cal	July 16, 1872	129, 421
Platform-elevator, Safety-catch for	B. Tatham and J. W. Brittin	New York and Brooklyn, N. Y.	Mar. 19, 1872	124, 864
Platform for steamboats and other vessels, Landing	N. W. Wheeler	Brooklyn, N. Y	Apr. 25, 1865	47, 482
Platform spring-coupling	B. T. Parsels and J. L. Hedges	Hanover, N. J	Apr. 4, 1871	113, 335
Platform-supporter	C. E. Flagg	Sherburne, Mass	June 10, 1856	15, 099
Platform-switch	G. M. Guerrero	New Orleans, La	Nov. 5, 1872	132, 721
Platoon-battery	W. Billinghurst and J. Regna	Rochester, N. Y	Sept. 16, 1862	36, 448
Playing-table	H. Seher	New York, N. Y	Jan. 5, 1869	85, 539
Pleating-machine	S. Sterns	New York, N. Y	Dec. 7, 1869	97, 721
Plettro-lira	P. Trajetta	Philadelphia, Pa	July 22, 1833	....
Plier	J. Bounds	Bridgeport, Conn	Jan. 14, 1868	73, 289
Plier	G. W. Moore	Newark, N. J	June 27, 1871	116, 473
Plier	A. M. Olds	New York, N. Y	June 11, 1867	65, 589
Plier	N. Thompson	Brooklyn, N. Y	Oct. 10, 1871	119, 726
Plier	S. Walker	New York, N. Y	Jan. 8, 1867	61, 032
Plier-joint die	G. R. Andrus	East Berlin, Conn	Jan. 5, 1869	85, 503
Plier-making die	S. H. Wood	East Berlin, Conn	Apr. 7, 1868	76, 574
Pliers and scissors, Combined	J. H. Price	Boston, Mass	Oct. 12, 1869	95, 834
Pliers, Cutting	R. Curtis	Springfield, Mass	Oct. 25, 1832	....
Pliers, Cutting	W. L. Truland	Waterford, N. Y	May 18, 1869	90, 136
Pliers, Making	C. W. Sykes	New York, N. Y	Sept. 14, 1858	21, 525
Pliers, Manufacture of	H. Wilkinson	Collinsville, Conn	June 1, 1858	20, 460
Plotting-instrument	S. Barnett	Washington, Ga	July 17, 1860	29, 133
Plotting instrument	J. E. Crupper	Berlin, Ky	June 20, 1871	116, 163
Plotting-instrument	H. S. Herney	Windsor, Conn	Dec. 22, 1863	41, 005
Plotting-instrument	T. Hinkley	Hallowell, Me	Oct. 18, 1853	10, 133
Plotting-instrument	C. R. Iliff	Falmouth, Ky	June 24, 1856	15, 183
Plotting-instrument	C. R. Iliff	Falmouth, Ky	Jan. 12, 1858	19, 091
Plotting-instrument, Surveyor's	W. J. Card	Lancaster, Ohio	Apr. 16, 1842	2, 561
Plow	J. Adams	Eatonton, Ga	Dec. 23, 1856	16, 260
Plow	S. Aland	Rome, N. Y	Apr. 12, 1864	42, 264
Plow	E. Albert	East Germantown, Ind	Mar. 6, 1847	4, 995
Plow	D. M. Allen	Jeffersonville, Ind	Nov. 5, 1872	132, 707
Plow	R. S. Allen	New York, N. Y	Mar. 6, 1866	52, 048
Plow	D. Almy	Tiverton Four Corners, R. I	May 14, 1867	64, 619
Plow	T. G. Andrews and A. Riviere	Barnesville, Ga	Aug. 5, 1873	141, 533
Plow	E. Andrus	Geneva, N. Y	Mar. 31, 1857	16, 901
Plow	A. Anschutz, A. Seidel, and M. Weber.	Livingston County, Mo	Nov. 4, 1873	144, 308
Plow	C. C. Ansley	Americus, Ga	Nov. 30, 1869	97, 337
Plow	D. Anthony, sr	Union Spring, N. Y	May 30, 1846	4, 549
Plow	J. Anthony	Zanesville, Ohio	Feb. 1, 1831	....
Plow	J. Archer	Springfield, Wis	Dec. 27, 1870	110, 417
Plow	W. J. Arrington	Jefferson County, Ga*	July 6, 1869	92, 143
Plow	G. M. Atherton	Friendsville, Ill	Apr. 20, 1869	89, 115
Plow	J. B. Atwater	Chicago, Ill	Mar. 14, 1865	46, 768
Plow	S. Austin	Sempronius, N. Y	Feb. 27, 1824	....
Plow	B. F. Avery	Louisville, Ky	Jan. 8, 1856	14, 044
Plow	B. F. Avery	Louisville, Ky	July 16, 1867	66, 774
Plow	G. D. Avery	Wood County, Va	Dec. 28, 1818	....
Plow	G. D. Avery	Georgetown, D. C	Feb. 15, 1825	....
Plow	G. D. Avery	Georgetown, D. C	Mar. 25, 1825	....
Plow	J. Bader, sr	Perrysburgh, Ohio	Feb. 16, 1869	86, 896
Plow	B. F. Baker	Ballston Spa, N. Y	May 21, 1872	126, 865
Plow	H. F. Baker	Centreville, Ind	Oct. 24, 1854	11, 821
Plow	H. H. Baker	New Market, N. J	Dec. 11, 1860	30, 863
Plow	N. Baker	Penn Township, Mont	Mar. 24, 1835	....
Plow	E. Ball	Greentown, Ohio	Feb. 20, 1845	3, 918
Plow	E. Ball	Greentown, Ohio	Mar. 23, 1852	8, 819
Plow	E. Ball	North Manchester, Ind	Feb. 14, 1865	46, 321
Plow	E. Ball, jr	Canton, Ohio	June 11, 1867	65, 529
Plow	E. Ball, jr	Canton, Ohio	Aug. 17, 1869	93, 853
Plow	J. Ball	Greentown, Ohio	Nov. 8, 1845	4, 263
Plow	J. Ball	Canton, Ohio	Sept. 1, 1868	81, 730
Plow	J. Ball	Canton, Ohio	July 27, 1869	93, 036
Plow	J. Banks	Dadeville, Ala	Dec. 1, 1857	18, 726
Plow	J. Banks	Dadeville, Ala	Jan. 26, 1858	19, 179
Plow	A. Barnaby	Ithaca, N. Y	Sept. 11, 1839	1, 320
Plow	A. P. Barry	Ashland, Miss	July 16, 1872	129, 265
Plow	G. Bartlett	Smithfield, R. I	Feb. 20, 1847	4, 976

Index of patents issued from the United States Patent Office from 1790 *to* 1873, *inclusive*—Continued.

Invention.	Inventor.	Residence.	Date.	No.
Plow	I. W. Bartlett	Otter Creek, Ill	Dec. 18, 1866	60, 461
Plow	A. Barton	Syracuse, N. Y	July 15, 1856	15, 321
Plow	E. Bass	Pachitta, Ga	Dec. 4, 1860	30, 793
Plow	W. J. M. Batchelder and C. Leiber.	Dayton and Harrisburgh, Ohio.	Mar. 13, 1866	53, 100
Plow	N. Batchly	Windsor, N. Y	Dec. 19, 1848	5, 973
Plow	C. Bates	Warsaw, Ill	Aug. 6, 1872	130, 180
Plow	L. M. Bates	Newark, Ohio	Apr. 24, 1866	54, 095
Plow	S. C. Baughn	Calhoun, Ky	Nov. 19, 1872	133, 187
Plow	W. Beach	Philadelphia	June 27, 1827	....
Plow	L. W. Beal	Dixon, Ill	June 16, 1868	78, 856
Plow	Z. M. Beall	Russellville, Ky	June 4, 1861	32, 459
Plow	J. S. Beals	Alabama Centre, N. Y	Apr. 10, 1866	53, 773
Plow	J. S. Beals	Alabama Centre, N. Y	Aug. 27, 1867	68, 152
Plow	H. T. Beam	Palestine, Ill	July 21, 1868	80, 118
Plow	C. A. Beard and E. E. Evans	Zanesville, Ohio	Apr. 18, 1871	113, 960
Plow	S. Beckett	Olive Branch, Ohio	July 3, 1866	55, 984
Plow	J. R. Beggs	New Albany, N. Y	June 9, 1863	38, 803
Plow	C. Beidler	Allentown, Pa	Jan. 21, 1862	34, 191
Plow	C. Beidler	Allentown, Pa	July 17, 1866	56, 350
Plow	J. C. Bell	Lebanon, Ind	Aug. 31, 1869	94, 383
Plow	A. C. Belt	Goresville, Va	Aug. 27, 1867	68, 032
Plow	H. F. Bemendefer and G. Smith	Attica, Ohio	July 2, 1867	66, 284
Plow	E. D. Benjamin	Old Town, Ill	Nov. 24, 1868	84, 252
Plow	A. Benkelmann	Langford, N. Y	Nov. 27, 1860	30, 712
Plow	J. F. Benton	Penn Yan, N. Y	Mar. 9, 1869	87, 533
Plow	J. F. Benton	Penn Yan, N. Y	Aug. 26, 1873	142, 074
Plow	M. Berdan	Maumee City, Ohio	Oct. 13, 1868	83, 030
Plow	C. Bergen	Brooklyn, N. Y	Nov. 11, 1819	....
Plow	C. Bergen	Brooklyn, N. Y	May 16, 1842	2, 626
Plow	E. L. Bergstresser	Hublersburgh, Pa	Sept. 3, 1867	68, 550
Plow	T. F. Bertrand and P. Sanes	Rockford, Ill	Jan. 29, 1867	61, 508
Plow	R. L. and A. C. Betts	Brunswick, N. Y	May 17, 1864	42, 813
Plow	R. W. Biggs	Jacksonville, Fla	May 28, 1867	65, 050
Plow	J. O. Billing	Halcyon Dale, Ga	Dec. 20, 1870	110, 336
Plow	C. Billups	Norfolk, Va	Feb. 20, 1872	123, 858
Plow	W. Black	Anne Arundel County, Md	Oct. 4, 1831	....
Plow	W. Black	Manchester, Pa	Apr. 17, 1858	21, 182
Plow	W. Black	Scott County, Ill	Sept. 14, 1843	3, 259
Plow	W. Blackstone	Shelby County, Ill	Oct. 14, 1873	143, 660
Plow	J. Blanchard	Iowa Falls, Iowa	Apr. 4, 1871	113, 484
Plow	J. Blanchard	East Saginaw, Mich	Nov. 11, 1873	144, 433
Plow	N. Blatchly	Windsor, N. Y	July 20, 1852	9, 129
Plow	S. R. Bliven	McDonough, N. Y	July 27, 1858	20, 984
Plow	C. Blodgett	Morrison, Ill	July 9, 1861	32, 746
Plow	C. A. Blodgett	Columbus, Nebr	June 4, 1872	127, 549
Plow	B. C. Blomsten	Waupaca, Wis	Oct. 11, 1870	108, 095
Plow	E. H. Bloodworth	Thomaston, Ga	Feb. 16, 1858	19, 401
Plow	J. Boatwright	Columbia, S. C	June 11, 1829	....
Plow	A. Boles	Kinder, Ind	Sept. 7, 1869	94, 554
Plow	A. J. Bonander	Rockford, Ill	Sept. 10, 1872	131, 243
Plow	J. C. Bond	Greenwood, S. C	Aug. 14, 1860	29, 564
Plow	A. T. Boon	Galesburgh, Ill	Apr. 3, 1866	53, 559
Plow	T. Borden	Newport, R. I	Jan. 13, 1830	....
Plow	E. Bourne	New Iberia, La	Nov. 18, 1873	144, 653
Plow	H. R. Bowen and L. D. Robnett	New Washington, Ind	Jan. 17, 1871	111, 033
Plow	J. D. Bowen	Roseburgh, Oreg	July 23, 1867	66, 939
Plow	W. B. Bradford	Charlotte, N. C	Apr. 23, 1872	125, 930
Plow	J. Bradley	Owatonna, Minn	Feb. 25, 1868	74, 885
Plow	F. P. Brannan	Richmond, Va	Sept. 24, 1872	131, 596
Plow	H. Briggs	Smithland, Ky	Dec. 16, 1867	71, 966
Plow	C. R. Brinkerhoff	Batavia, N. Y	Oct. 11, 1853	10, 101
Plow	T. E. C. Brinly	Simpsonville, Ky	Apr. 13, 1858	19, 909
Plow	T. E. C. Brinly	Louisville, Ky	Sept. 4, 1860	29, 858
Plow	T. E. C. Brinly	Louisville, Ky	July 3, 1866	55, 999
Plow	T. E. C. Brinly	Louisville, Ky	July 16, 1867	66, 787
Plow	T. E. C. Brinly	Louisville, Ky	Dec. 10, 1867	71, 968
Plow	T. E. C. Brinly	Louisville, Ky	May 18, 1869	90, 232
Plow	T. E. C. Brinly	Louisville, Ky	July 20, 1869	92, 698
Plow	T. E. C. Brinly	Louisville, Ky	Feb. 15, 1870	99, 830
Plow	T. E. C. Brinly	Louisville, Ky	Nov. 15, 1870	109, 290
Plow	T. E. C. Brinly	Louisville, Ky	Nov. 15, 1870	109, 291
Plow	T. E. C. Brinly	Louisville, Ky	June 17, 1873	140, 004
Plow	T. E. C. Brinly and J. G. Dodge	Louisville, Ky	Aug. 12, 1862	36, 136
Plow	W. Britton	Truro, Ill	Apr. 18, 1871	113, 733
Plow	R. H. Brooks	Greenville, Ga	Feb. 14, 1860	27, 099
Plow	R. M. Brooks	Greenville, Ga	Dec. 27, 1859	26, 563
Plow	C. Brown and L. H. Gerth	Peoria, Ill	June 9, 1868	78, 785
Plow	R. A. Brown	Oakland, Miss	Dec. 2, 1873	145, 088
Plow	S. R. Brown and W. McClean	Norfolk, Va	Jan. 19, 1858	19, 125
Plow	T. Brown	New York	Sept. 11, 1829	....
Plow	C. M. Bryan	Wright City, Mo	May 10, 1859	23, 898
Plow	W. Bryant	Davidson County, Tenn	Mar. 31, 1840	1, 527
Plow	J. Buch	East Rushville, Ohio	Dec. 28, 1832	....
Plow	W. Bullock	Jersey City, N. J	July 30, 1845	4, 127
Plow	W. T. Bunn	Humboldt, Tenn	Oct. 18, 1870	108, 444
Plow	L. D. Burch	Sherburne, N. Y	Aug. 14, 1860	29, 567
Plow	L. D. Burch	Sherburne, N. Y	Mar. 12, 1861	31, 654
Plow	W. Burch	North Fairfield, Ohio	Mar. 4, 1873	136, 361
Plow	L. E. Burdin	Paris, Ky	June 28, 1859	24, 536
Plow	L. E. Burdin	Lexington, Ky	Nov. 21, 1871	121, 153
Plow	T. J. Burgess	Kingston, N. Y	Dec. 3, 1872	133, 517
Plow	W. D. Burgess and G. W. Zeigler.	Maumee, Ohio	July 7, 1868	79, 517
Plow	J. M. Burke	Dansville, N. Y	Nov. 9, 1858	22, 013
Plow	P. Burns	Indiana, Pa	Nov. 30, 1869	97, 352
Plow	N. F. Burton	Plymouth, Ill	Oct. 29, 1861	33, 568

Index of patents issued from the United States Patent Office from 1790 *to* 1873, *inclusive*—Continued.

Invention.	Inventor.	Residence.	Date.	No.
Plow	O. F. Burton and L. B. Hoit	New York, N. Y., and Cedar Falls, Iowa	Jan. 9, 1866	51, 917
Plow	W. V. Burton	Orange, Ohio	July 26, 1853	9, 875
Plow	E. T. Bussell	Indianapolis, Ind	July 2, 1872	128, 588
Plow	J. Butler	Huff Township, Ind	Dec. 5, 1871	121, 582
Plow	M. Butler	Vernon, Ind	Oct. 4, 1870	108, 004
Plow	M. Butler	Vernon, Ind	Aug. 20, 1872	130, 697
Plow	J. J. Cadenhead	Macon County, Ala	Mar. 4, 1856	14, 346
Plow	F. M. Caldwell	New York, N. Y	Nov. 23, 1869	97, 162
Plow	J. Campbell	Newtown, Ill	Oct. 13, 1868	83, 036
Plow	W. P. Cannon	Monroe County, Tenn	Mar. 4, 1836	....
Plow	R. Cantleton	Montgomery, Ala	Apr. 9, 1867	63, 610
Plow	S. Canterberry	Holmes County, Miss	Aug. 14, 1860	29, 569
Plow	J. Carothers	Shirleysburgh, Pa	Dec. 10, 1831	....
Plow	A. Carson	Memphis, Tenn	May 14, 1867	64, 747
Plow	T. Carter	Laurens District, S. C	Dec. 16, 1833	....
Plow	E. Cartwright	De Witt, Nebr	Nov. 4, 1873	144, 255
Plow	F. F. Cary	New York, N. Y	May 3, 1864	42, 631
Plow	J. Case	Springfield, Ill	Dec. 8, 1857	18, 803
Plow	V. M. Chafee	Xenia, Ill	Jan. 17, 1860	26, 833
Plow	C. F. Chambers	Huttonsville, Ill	June 6, 1871	115, 701
Plow	J. J. Chandler and P. Ranger	Wilton, Me	Jan. 20, 1836	....
Plow	T. F. Chapin	Walpole, N. H	July 25, 1854	11, 356
Plow	A. B. Chapman	Pittsfield, Mass	Dec. 1, 1863	40, 733
Plow	L. Chapman	Collinsville, Conn	Feb. 6, 1862	123, 330
Plow	R. B. Chenoweth	Baltimore, Md	Nov. 25, 1808	....
Plow	H. S. Chichester	Brunswick, N. Y	Aug. 6, 1861	32, 976
Plow	C. M. Clark	Seward, Nebr	Dec. 16, 1873	145, 627
Plow	E. B. Clark	Tallahassee, Fla	Feb. 14, 1860	27, 109
Plow	E. B. Clark	Tallahassee, Fla	July 17, 1860	29, 139
Plow	G. B. Clarke	Leonardsville, N. Y	May 24, 1864	42, 838
Plow	J. Clark and G. W. N. Yost	Washington, D. C., and Pittsburgh, Pa	Feb. 12, 1856	14, 224
Plow	T. J. and G. M. Clark	Higganum, Conn	Mar. 25, 1873	137, 060
Plow	G. Clezy	Baltimore, Md	Sept. 14, 1843	3, 266
Plow	J. Clifton	Georgetown, Tex	Dec. 27, 1870	110, 550
Plow	S. Cline	Plumstead, Pa	July 17, 1835	....
Plow	J. C. Cloud	May's Landing, N. J	Feb. 6, 1849	6, 100
Plow	H. C. Cloyd	West Alexandria, Ohio	June 7, 1870	103, 982
Plow	P. J. Clute	Schenectady, N. Y	Nov. 16, 1822	....
Plow	J. M. Cobb	Jackson, Tenn	June 26, 1860	28, 836
Plow	J. M. Cobb	Jackson, Tenn	Apr. 22, 1873	138, 131
Plow	W. Cock	Luzerne Township, Pa	Apr. 28, 1826	....
Plow	D. Cockley	Lancaster, Pa	Sept. 28, 1858	21, 598
Plow	R. Coddington and D. McCall	Leonidas and Kalamazoo, Mich.	Oct. 29, 1861	33, 573
Plow	O. Coe	Port Washington, Wis	Feb. 16, 1864	41, 603
Plow	W. Coggeshall	Finley, Ohio	May 27, 1862	35, 415
Plow	G. W. Coles	Canton, Ill	Apr. 14, 1868	76, 714
Plow	G. D. Colton	Galesburgh, Ill	Jan. 10, 1859	22, 029
Plow	W. S. Colwell	Pittsburgh, Pa	Jan. 21, 1868	73, 504
Plow	W. S. Colwell	Allegheny, Pa	Apr. 13, 1869	88, 851
Plow	W. H. Conaway	Dillsborough, Ind	Dec. 3, 1872	133, 631
Plow	J. H. Conklin	Peekskill, N. Y	Feb. 27, 1847	4, 980
Plow	E. S. Cook	Laurel Grove, Va	Aug. 6, 1872	130, 196
Plow	I. Cook and J. T. Bever	Hainesville, Mo	July 5, 1859	24, 617
Plow	S. Cooley	Clarkston, Mich	Nov. 7, 1871	120, 718
Plow	W. Cooley	Bunker Hill, Wis	Feb. 5, 1867	61, 809
Plow	W. G. Coombs	New Gloucester, Me	Aug. 23, 1870	106, 556
Plow	C. W. Cooper	Ogeechee, Ga	Jan. 1, 1856	14, 013
Plow	G. W. Cooper	Palmyra, N. Y	Apr. 16, 1861	32, 052
Plow	J. B. Cooper	Brooklyn, N. Y	Apr. 9, 1861	31, 955
Plow	T. J. Cornell	Decatur, Ill	Feb. 20, 1866	52, 807
Plow	T. R. Cornick	Cap Au Gris, Mo	Nov. 27, 1860	30, 723
Plow	J. Coston	Bowden, Ga	June 6, 1871	115, 710
Plow	T. L. Cotten	Water Valley, Miss	Apr. 15, 1873	137, 894
Plow	T. Cottman	Cincinnati, Ohio	Feb. 13, 1866	52, 539
Plow	G. D. Cotten	Galesburgh, Ill	July 20, 1858	20, 935
Plow	M. C. Cox	Bennettsville, S. C	Nov. 2, 1869	96, 399
Plow	F. Cremer	Elmwood, Ill	Feb. 21, 1871	111, 911
Plow	R. S. Crockett	Rossville, S. C	Nov. 8, 1870	108, 975
Plow	H. F. Cromwell	Cynthiana, Ky	Jan. 3, 1860	27, 655
Plow	J. Cromwell	Cynthiana, Ky	Oct. 9, 1834	....
Plow	C. Crow	Covington, Ind	Dec. 7, 1869	97, 609
Plow	G. Crowl	Sleepy Creek, Va	June 27, 1832	....
Plow	J. M. Cullen	Benton, Miss	May 30, 1846	4, 553
Plow	H. Culver	Dansville, N. Y	Mar. 9, 1869	87, 641
Plow	T. Cuming, jr	Brookhaven, Miss	Oct. 24, 1871	120, 248
Plow	J. G. Cummings	Columbus, Miss	Nov. 24, 1857	18, 682
Plow	W. H. Cummings and H. L. Childs.	Booneshorough, Iowa	June 22, 1869	91, 721
Plow	A. G. and J. R. Cummins	McKinney, Tex	July 6, 1869	92, 277
Plow	S. A. Cummins	Vienna, N. J	Aug. 1, 1871	117, 608
Plow	G. W. Cunningham	Paris, Mo	Sept. 25, 1860	30, 125
Plow	S. Curtis	Fitchburgh, Wis	Feb. 23, 1869	87, 149
Plow	A. K. Dahl	Fox Lake, Wis	Nov. 25, 1873	144, 960
Plow	M. K. Dahl	Waupun, Wis	Mar. 15, 1870	100, 869
Plow	A. A. Dailey	Wilson, N. Y	Sept. 13, 1870	107, 228
Plow	J. Dalkener	Canton, Ohio	Jan. 15, 1836	....
Plow	E. Davidson	Batesville, Ark	May 10, 1859	23, 904
Plow	A. B. Davis	Catahoula Parish, La	Aug. 3, 1869	93, 281
Plow	B. Davis and J. M. Scroggin	La Grange, Ga	Mar. 13, 1860	27, 619
Plow	G. Davis	Sandy Springs, Md	May 26, 1818	....
Plow	G. Davis	Georgetown, D. C	Oct. 1, 1825	....
Plow	J. R. Davis	McKay, Ohio	Nov. 19, 1867	70, 972
Plow	A. Day	Crystal Springs, Miss	Nov. 18, 1870	108, 979
Plow	J. Deats	Middletown, Pa	Apr. 26, 1828	....
Plow	J. Deats	Roxburgh, N. J	Dec. 28, 1831	....

Index of patents issued from the United States Patent Office from 1790 *to* 1873, *inclusive*—Continued.

Invention.	Inventor.	Residence.	Date.	No.
Plow	J. Deats	Roxburgh, N. J	July 31, 1837	327
Plow	J. Deats	Roxburgh, N. J	Nov. 25, 1838	1, 019
Plow	J. Deaver		July 12, 1804	....
Plow	J. Deere	Moline, Ill	Feb. 21, 1865	46, 454
Plow	R. Deighton, jr	Fairweather, Ill	Aug. 15, 1865	49, 389
Plow	N. Demary, jr	Attica, N. Y	Dec. 15, 1857	18, 882
Plow	J. Dement	Dixon, Ill	July 30, 1861	32, 966
Plow	J. Dement	Dixon, Ill	Aug. 30, 1864	43, 978
Plow	S. T. Demise	Red Bank, N. J	Dec. 3, 1867	71, 715
Plow	S. T. Demise	Red Bank, N. J	Oct. 6, 1868	82, 809
Plow	L. B. Dennett	Portland, Me	May 29, 1866	55, 069
Plow	G. W. Depew	Peekskill, N. Y	Mar. 19, 1861	31, 761
Plow	T. J. De Yampert	Shohola, Pa	May 10, 1859	23, 974
Plow	J. M. Dick	Buffalo, N. Y	Feb. 24, 1863	37, 740
Plow	R. Dickie and H. K. Johnston	Bunker Hill, Ill	Aug. 23, 1870	106, 559
Plow	A. Dickson	Hillsborough, N. C	June 22, 1858	20, 633
Plow	A. A. Dickson	Anderson, S. C	Nov. 29, 1859	26, 259
Plow	J. Dickson	Newcastle, Pa	Oct. 19, 1858	21, 824
Plow	J. H. Dickson	Alford, Ind	Aug. 18, 1868	81, 148
Plow	E. Dietsch	Findley, Ohio	Nov. 1, 1870	108, 892
Plow	O. P. Dills	Falmouth, Ky	Sept. 5, 1865	49, 733
Plow	J. Dodge	Adrian, Mich	Jan. 9, 1872	122, 582
Plow	J. G. Dodge	Louisville, Ky	Oct. 8, 1867	69, 643
Plow	A. Doe	Concord, N. H	Jan. 30, 1855	12, 310
Plow	L. Donnell	Jefferson Township, Ohio	July 23, 1833	....
Plow	W. Donnelly	Calverton, N. Y	Aug. 12, 1873	141, 705
Plow	J. M. Dormon	Claiborne Parish, La	May 4, 1869	89, 565
Plow	W. H. H. Doty	Sonora, Ohio	Dec. 26, 1871	122, 162
Plow	J. W. Downs	Bowdon, Ga	July 16, 1872	129, 112
Plow	V. C. Duclos	New Harmony, Ind	Nov. 17, 1868	84, 097
Plow	J. Dudley	Fleming County, Ky	May 8, 1832	....
Plow	J. C. Duncan	Olney, Ill	May 14, 1867	64, 647
Plow	H. B. Durfee	Decatur, Ill	May 18, 1869	90, 246
Plow	H. B. Durfee	Decatur, Ill	Nov. 9, 1869	96, 680
Plow	J. Dutcher	New York, N. Y	Oct. 9, 1839	1, 360
Plow	J. L. Dutton, jr	Cherry Lake, Fla	Jan. 31, 1860	26, 976
Plow	I. Eastwood	Lanark, Ill	Aug. 16, 1870	106, 344
Plow	I. Eastwood	Lanark, Ill	Oct. 11, 1870	108, 243
Plow	H. H. Ebaugh	Hereford, Md	July 9, 1867	66, 477
Plow	J. B. Edgell and J. W. and E. A. Alexander.	Independence, Iowa	Oct. 2, 1866	58, 395
Plow	T. Edmunds	Talcott, Va	Apr. 15, 1871	118, 119
Plow	A. N. Edwards	Greenville, Ala	July 20, 1869	92, 803
Plow	D. Eldred	Monmouth, Ill	Oct. 18, 1859	25, 816
Plow	G. Ellington	Madison County, Ala	Dec. 25, 1824	....
Plow	J. W. Elliott	Locust Mount, Va	Aug. 6, 1872	130, 115
Plow	C. A. Elton	Hillsborough, Ohio	July 16, 1867	66, 691
Plow	J. H. Elward	Polo, Ill	Nov. 3, 1868	83, 768
Plow	G. Emery and A. C. Wilson	Newfield, Me	Oct. 18, 1859	25, 817
Plow	A. F. Eppes	Stony Creek, Va	May 10, 1870	102, 789
Plow	G. Esterly	Heart Prairie, Wis	Feb. 13, 1855	12, 381
Plow	W. A. Estes	China, Me	Sept. 30, 1873	143, 279
Plow	G. Evans	Pittsburgh, Pa	Sept. 18, 1833	....
Plow	J. D. Evans	Pleasant Hill, Ga	Sept. 24, 1867	69, 083
Plow	L. G. Evans	Spring Hill, Ala	June 19, 1855	13, 082
Plow	W. C. Evarts	Danby, N. Y	July 23, 1867	66, 958
Plow	W. P. Everdon	Leavenworth, Ind	Feb. 11, 1868	74, 327
Plow	O. G. Ewings	Heart Prairie, Wis	Aug. 29, 1854	11, 601
Plow	P. Falker	Lanesville, Ill	July 20, 1869	92, 806
Plow	S. A. Fanning	Jacksonville, Ill	Apr. 2, 1872	125, 185
Plow	J. Farlee	Mercer County, Ky	Apr. 21, 1836	....
Plow	A. B. Farquhar	York, Pa	June 18, 1872	127, 970
Plow	N. Faught	Pittsborough, Ind	Jan. 3, 1871	110, 758
Plow	V. Felker	Carmel, Me	Apr. 30, 1861	30, 231
Plow	V. Felker	Carmel, Me	June 27, 1865	48, 387
Plow	R. R. Fenner	Urbana, Ill	Aug. 27, 1872	130, 797
Plow	J. B. Ferguson and S. M. White	Big Lick, Va	Feb. 28, 1871	112, 234
Plow	J. G. Fetzer	Brunswick, Mo	Aug. 11, 1868	80, 859
Plow	J. Fish	Smelser, Wis	Apr. 2, 1867	63, 377
Plow	D. S. Fisher	Cedar Spring, Ind	May 28, 1867	65, 198
Plow	J. Fisher	Middletown, Pa	Aug. 4, 1868	80, 715
Plow	S. Fisher	West Windsor, N. J	Nov. 27, 1860	30, 727
Plow	C. L. Fleischmann	New York, N. Y	Feb. 26, 1867	62, 325
Plow	F. Fogelgesang	Canton, Ohio	Nov. 13, 1866	59, 577
Plow	S. Folton	Huntingdon, Pa	Nov. 1, 1830	....
Plow	G. G. Foreman	Stockton, Ga	June 18, 1872	128, 032
Plow	C. Forster	Lebanon, Pa	Nov. 26, 1867	71, 376
Plow	M. Fort	Talbotton, Ga	Mar. 15, 1870	100, 743
Plow	L. Fosdick	Tiskilwa, Ill	Nov. 3, 1868	93, 703
Plow	J. Fowler	Rahway, N. J	Feb. 6, 1866	52, 407
Plow	J. Fowler	Hartland, Wis	Mar. 26, 1867	63, 151
Plow	J. Fox	Oskaloosa, Iowa	June 25, 1872	128, 295
Plow	W. Frank	Saint Louis, Mo	Aug. 25, 1863	39, 639
Plow	A. Franklin	Springfield, Ohio	Nov. 16, 1869	96, 907
Plow	A. and F. M. Franklin	Springfield, Ohio	Oct. 11, 1870	108, 247
Plow	F. M. Franklin	Springfield, Ohio	June 8, 1869	91, 109
Plow	F. M. Franklin	Springfield, Ohio	Sept. 30, 1870	107, 606
Plow	E. J. Faser	Kansas, Mo	Apr. 23, 1861	32, 129
Plow	L. F. Frazee	Jersey City, N. J	Mar. 21, 1871	112, 913
Plow	E. J. Frazer	Kansas City, Mo	Sept. 25, 1860	30, 136
Plow	O. M. French	Rochester, Pa	June 16, 1868	78, 869
Plow	A. Friberg	Moline, Ill	Oct. 13, 1868	82, 934
Plow	L. C. Frost	Fredericksburgh, Va	Oct. 14, 1873	143, 620
Plow	D. Fulton	Saint Helena, Cal	Sept. 27, 1870	107, 676
Plow	W. Furnas	Ononwa, Iowa	Dec. 3, 1861	33, 834
Plow	H. Gale	Albion, Mich	May 17, 1870	103, 038
Plow	H. Gale	Albion, Mich	Sept. 6, 1870	117, 033
Plow	W. Gallagher	Shullsbury, Wis	Apr. 28, 1868	77, 272

Index of patents issued from the United States Patent Office from 1790 *to* 1873, *inclusive*—Continued.

Invention.	Inventor.	Residence.	Date.	No.
Plow	J. Gallatin	Manchester, Ohio	Apr. 15, 1843	3, 052
Plow	A. Gardner	Cincinnati, Ohio	Oct. 26, 1852	9, 362
Plow	T. C. Garlington	La Fayette, Ala	May 5, 1857	17, 211
Plow	W. Garrell	Trenton, Tenn	Jan. 22, 1861	31, 172
Plow	W. A. Gates	Mount Comfort, Tenn	Dec. 21, 1852	9, 443
Plow	J. Gehr	College of Saint James, Md	Nov. 2, 1858	21, 953
Plow	J. George	Greene, Mo	Apr. 18, 1865	47, 294
Plow	D. Ghormley	Wayne Township, Ohio	Feb. 13, 1835	....
Plow	G. Gibbs	Canton, Ohio	Apr. 30, 1867	64, 213
Plow	G. Gibbs	Canton, Ohio	Dec. 3, 1867	71, 734
Plow	J. Gibbs	Canton, Ohio	June 16, 1836	....
Plow	J. Gibbs	Canton, Ohio	Aug. 15, 1854	11, 523
Plow	L. Gibbs	Canton, Ohio	Mar. 12, 1867	62, 835
Plow	L. Gibbs	Canton, Ohio	June 25, 1867	66, 016
Plow	L. Gibbs	Canton, Ohio	Aug. 6, 1872	130, 292
Plow	M. L. Gibbs	Canton, Ohio	Oct. 25, 1870	108, 699
Plow	M. L. Gibbs	Canton, Ohio	June 20, 1871	116, 048
Plow	R. Gibbs	Brunswick, Mo	Sept. 6, 1870	107, 037
Plow	W. and G. Gibbs and L. P. Wikidal.	Canton, Ohio	Dec. 3, 1867	71, 735
Plow	J. Gibson	Montgomery County, N. Y	May 2, 1822	....
Plow	B. Giger	Sangamon County, Ill	June 24, 1851	8, 170
Plow	J. R. Gilbert	Wootens, Ga	Nov. 30, 1869	97, 388
Plow	P. M. Gilbert	Kewanee, Ill	Jan. 1, 1867	60, 875
Plow	C. G. and G. Gill	Chester, S. C	Mar. 25, 1822	....
Plow	J. L. Gill	Columbus, Ohio	Aug. 28, 1855	13, 493
Plow	H. Gillette	Millville, N. Y	June 4, 1872	127, 475
Plow	J. W. Gilliam	Elkton, Ky	Aug. 10, 1869	93, 430
Plow	J. Gilmer	Monticello, Fla	Sept. 17, 1872	131, 434
Plow	A. Gilmore	Phœnixville, Pa	Dec. 17, 1867	72, 386
Plow	T. Gilson and N. Martin	Port Washington, Wis	June 5, 1866	55, 279
Plow	J. K. Gingrich	North Annville, Pa	Feb. 19, 1861	31, 452
Plow	N. T. Glenn	Watkinsville, Ga	Jan. 23, 1872	122, 885
Plow	C. Glidden	Milwaukee, Wis	Dec. 4, 1866	60, 172
Plow	J. S. Godfrey	Leslie, Mich	Mar. 29, 1870	101, 256
Plow	J. S. Godfrey	Leslie, Mich	July 19, 1870	105, 446
Plow	S. T. Godfrey	Seaville, N. J	June 15, 1869	91, 328
Plow	H. Goldson	Greensborough, Miss	Nov. 18, 1851	8, 524
Plow	E. Goldthait	Fort Wayne, Ind	Nov. 25, 1851	8, 544
Plow	J. H. Gooch	Oxford, N. C	May 8, 1860	28, 169
Plow	C. M. Gordon	La Porte, Ind	Sept. 20, 1870	107, 481
Plow	J. Gorham	Bairdstown, Ga	May 5, 1857	17, 212
Plow	J. Gorham	Bairdstown, Ga	Dec. 6, 1859	26, 349
Plow	J. L. Graham	Carthage, Ill	Dec. 23, 1873	145, 862
Plow	R. A. Graham	New Paris, Ohio	Oct. 4, 1853	10, 069
Plow	C. W. Grant	Iona Island, N. Y	Aug. 20, 1867	67, 976
Plow	L. Green	Great Bend, Pa	July 17, 1860	29, 162
Plow	L. Green	Great Bend, Pa	Mar. 19, 1861	31, 712
Plow	L. Green	Great Bend, Pa	May 19, 1863	38, 581
Plow	L. Green	Great Bend, Pa	Nov. 15, 1864	45, 036
Plow	W. J. Griffies	Marietta, Ga	Mar. 29, 1859	23, 369
Plow	W. Griffin	Bennettsville, S. C	July 3, 1860	28, 981
Plow	C. T. Grimes	Gerrard County, Ky	Sept. 8, 1868	81, 996
Plow	R. B. Ground	Edwardsville, Ill	Apr. 15, 1873	137, 915
Plow	M. Grover	Clyde, Ohio	Oct. 6, 1857	18, 335
Plow	M. Grover	Clyde Village, Ohio	Jan. 7, 1862	34, 092
Plow	J. Hage	Shiloh, Ill	May 12, 1863	38, 478
Plow	J. Hage	Shiloh, Ill	Apr. 2, 1867	63, 381
Plow	D. B. Haight	Perryville, N. Y	Dec. 26, 1848	5, 981
Plow	E. Haiman	Columbus, Ga	Jan. 17, 1871	111, 055
Plow	H. B. Hakes	Worcester, Mass	Sept. 24, 1872	131, 679
Plow	H. G. and E. L. Hall	Putnam, Ohio	May 21, 1867	64, 974
Plow	J. D. Hall	Canton, Ohio	July 2 , 1870	105, 673
Plow	J. M. Hall	Warrenton, Ga	May 18, 1858	20, 269
Plow	J. M. Hall	Warrenton, Ga	Feb. 22, 1859	23, 023
Plow	J. S. Hall	Manchester, Pa	Feb. 7, 1854	10, 505
Plow	J. S. Hall	Manchester, Pa	Aug. 1, 1854	11, 456
Plow	J. S. Hall	West Manchester, Pa	June 2, 1857	17, 430
Plow	J. S. Hall	West Manchester, Pa	Aug. 14, 1860	29, 590
Plow	J. S. Hall	West Manchester, Pa	Aug. 14, 1860	29, 591
Plow	J. S. Hall	Pittsburgh, Pa	Apr. 28, 1868	77, 188
Plow	J. S. Hall	Pittsburgh, Pa	May 21, 1872	126, 952
Plow	J. S. Hall	Pittsburgh, Pa	Nov. 18, 1873	144, 760
Plow	S. Hall	Pittsburgh, Pa	Apr. 25, 1848	5, 529
Plow	T. J. Hall	Tawakana Hills, Tex	Apr. 3, 1855	12, 627
Plow	J. Hanes	Polkville, Ky	Jan. 24, 1865	45, 995
Plow	J. Hapgood	Shrewsbury, Mass	Sept. 12, 1871	118, 932
Plow	J. Hardey	Moline, Ill	Aug. 28, 1860	29, 843
Plow	W. C. Hardin	Bowling Green, Mo	Mar. 24, 1868	75, 901
Plow	T. Harding	La Fayette, Ind	Jan. 17, 1871	111, 056
Plow	W. E. Harding	Bowling Green, Mo	Aug. 13, 1867	67, 756
Plow	D. Harper	Crystal Lake, Ill	Oct. 21, 1862	36, 711
Plow	W. K. Harrell	Clarinda, Iowa	June 18, 1872	128, 141
Plow	H. Harris	Bullitt, Ky	Feb. 24, 1808	....
Plow	I. P. Harris	Byhalia, Miss	July 6, 1858	20, 790
Plow	I. P. Harris	Byhalia, Miss	Nov. 8, 1859	26, 033
Plow	J. Harris	Kansas, Ill	Oct. 23, 1866	59, 010
Plow	I. P. Harris	Byhalia, Miss	Sept. 2, 1856	15, 649
Plow	J. R. Harris	Hazlehurst, Miss	June 25, 1872	128, 390
Plow	J. R. Harris	Hazlehurst, Miss	Nov. 5, 1872	132, 828
Plow	J. R. Harris	Hazlehurst, Miss	Dec. 24, 1872	134, 275
Plow	R. N. Harrison	New York	Dec. 19, 1818	....
Plow	S. Hartpence and J. D. Bowne.	Kingwood, N. J	July 5, 1837	259
Plow	C. Hartzell	Saint Joseph, Mo	June 1, 1869	90, 747
Plow	C. Hartzell	Saint Joseph, Mo	Dec. 31, 1872	134, 377
Plow	W. Haslup	Sidney, Ohio	June 11, 1872	127, 878
Plow	P. Hastings	Dagsborough, Del	Dec. 19, 1833	....
Plow	J. M. Hawley	Holton, Ind	July 2, 1867	66, 335

Index of patents issued from the United States Patent Office from 1790 to 1873, inclusive—Continued.

Invention.	Inventor.	Residence.	Date.	No.
Plow	J. Heckendorn	Elkton, Md	Dec. 23, 1856	16,277
Plow	J. Heckendorn	Reading, Pa	June 8, 1869	91,129
Plow	D. Heiges	Cashtown, Pa	Mar. 8, 1870	100,624
Plow	J. C. Henry	Point Douglass, Minn	Aug. 6, 1867	67,542
Plow	W. Henry	Wyoming, Pa	Mar. 29, 1864	42,086
Plow	W. Hess	Lower Saucon, Pa	Mar. 27, 1835	....
Plow	J. Hibbs	Tullytown, Pa	Nov. 14, 1854	11,935
Plow	B. S. Higgins	Olney, Ill	Oct. 13, 1868	83,061
Plow	D. H. Hill	Union Springs, Ala	June 22, 1869	91,631
Plow	W. Hinds	Little Falls N. Y	Mar. 22, 1864	41,997
Plow	D. Hitchcock	New York	July 16, 1823	....
Plow	R. G. Hobson	Houlka, Miss	Jan. 7, 1873	134,670
Plow	D. Hoke	Byhalia, Miss	Mar. 9, 1858	19,563
Plow	Z. S. Holdridge and H. L. Lawson.	Berne, N. Y	Aug. 9, 1826	....
Plow	W. C. Holmes	Barnesville, Ga	Apr. 12, 1859	23,580
Plow	B. R. Hood	Clinton, N. C	Oct. 11, 1859	25,738
Plow	W. U. Hoover	Daysville, Ky	Dec. 15, 1868	84,946
Plow	S. Horney, jr	Richmond, Ind	Apr. 5, 1853	9,646
Plow	J. T. Hovis	Clintonville, Pa	Aug. 20, 1872	130,717
Plow	W. S. Howell	Alfred, N. Y	Jan. 21, 1868	73,606
Plow	B. C. Hoyt	Port Washington, Wis	Sept. 2, 1856	15,654
Plow	J. S. Huggins	Timmonsville, S. C	May 22, 1860	28,372
Plow	D. W. Hughes	Quincy, Ill	Jan. 21, 1868	73,607
Plow	S. Hulbert	Ogdensburgh, N. Y	Sept. 20, 1853	10,031
Plow	S. Hulbert	Ogdensburgh, N. Y	Sept. 7, 1858	21,423
Plow	S. Hulbert	Ogdensburgh, N. Y	June 2, 1868	78,457
Plow	N. Hull	De Kalb, Miss	May 12, 1842	2,620
Plow	H. A. Hummer	Franklin Township, N. J	Sept. 4, 1866	57,724
Plow	D. F. Humphrey	Saline, Mich	July 29, 1862	36,004
Plow	D. F. Humphrey	Saline, Mich	Feb. 14, 1865	46,362
Plow	G. W. Hunt	Muscatine, Iowa	Mar. 27, 1860	27,634
Plow	G. W. Hunt	Muscatine, Iowa	Sept. 15, 1863	39,929
Plow	J. Y. Hunt	Tunbridge, Vt	Dec. 24, 1824	....
Plow	L. Hunt	Weathersfield, Vt	June 15, 1869	91,445
Plow	L. Hunt	Weathersfield, Vt	Aug. 24, 1869	93,994
Plow	J. T. Hunter and D. L. H. Mitchell.	Forest, Miss	Feb. 25, 1873	136,245
Plow	C. B. Hunting	Clinton, Ill	Sept. 11, 1866	57,909
Plow	W. S. Huntington	Byron, Mich	Jan. 15, 1867	61,203
Plow	A. P. Ingalls	Shelbyville, Ill	Feb. 9, 1869	86,839
Plow	C. B. Ingersoll	Morris, Ill	June 16, 1857	17,577
Plow	W. D. Ivey	Milford, Ga	Apr. 10, 1860	27,808
Plow	W. B. Jackson and J. M. and O. J. Childs.	Utica, N. Y	May 13, 1873	138,895
Plow	J. Jacobs	Maysville, Ky	July 8, 1834	....
Plow	J. Jacobs	Maysville, Ky	July 8, 1834	....
Plow	D. Jacoby	Mendota, Ill	Aug. 14, 1866	57,147
Plow	A. C. Jaques	Leavenworth, Kans	July 14, 1868	79,834
Plow	J. W. Jenkins	Hudson, N. Y	July 6, 1820	....
Plow	J. R. P. Jett	Knoxville, Tenn	Oct. 11, 1870	108,149
Plow	J. E. Jinkins	Milton, Fla	Oct. 22, 1867	69,999
Plow	J. Johnson	Wooster, Ohio	May 30, 1846	4,552
Plow	M. Johnson	Three Rivers, Mich	Dec. 29, 1868	85,451
Plow	R. Johnson	Frederick, Md	Apr. 29, 1862	35,098
Plow	R. Johnson	Lawrence, Kans	Sept. 12, 1871	118,945
Plow	W. H. Johnson	Richmond, Ark	Feb. 14, 1860	27,188
Plow	J. Jones	New Castle, Del	Aug. 10, 1858	21,167
Plow	J. H. and H. P. Jones	Herndon, Ga	Jan. 19, 1869	86,079
Plow	R. Jones	Waynesburgh, Ohio	Feb. 10, 1863	37,626
Plow	W. Jones	Wilson, Minn	Nov. 25, 1862	36,999
Plow	W. T. Jones	Joliet, Ill	Nov. 15, 1859	26,111
Plow	A. C. Judson	Grand Rapids, Ohio	Sept. 7, 1869	94,609
Plow	A. C. Judson	Grand Rapids, Ohio	Feb. 22, 1870	100,042
Plow	C. T. Kee	Chester Court-House, S. C.	July 30, 1872	129,964
Plow	G. W. Keeler	New Haven, Ohio	Oct. 13, 1868	82,959
Plow	H. M. Keith	Commerce, Mich	Oct. 18, 1870	108,488
Plow	H. M. Keith	Commerce, Mich	Nov. 28, 1871	121,382
Plow	E. Kelley	Bainbridge, Ohio	Aug. 26, 1833	....
Plow	I. Kennedy	Ithaca, N. Y	Oct. 2, 1866	58,431
Plow	I. Kennedy	Binghamton, N. Y	May 18, 1869	90,271
Plow	M. Kennedy	Boston, Mass	June 5, 1866	55,312
Plow	H. Killam	Mendon, Mich	Oct. 5, 1869	95,487
Plow	J. Kilmer	Barnerville, N. Y	Mar. 7, 1865	46,755
Plow	J. D. Kincaid	Bowling Green, Mo	Aug. 13, 1867	67,772
Plow	S. B. King	Starkville, Ga	Feb. 7, 1871	111,651
Plow	C. Kinkel	New York, N. Y	Mar. 19, 1867	63,059
Plow	D. J. Kirkman and E. H. Gray.	Winchester, Ill	June 25, 1867	66,031
Plow	J. Klay	Frederick County, Md	Jan. 11, 1812	....
Plow	S. A. Knox	Worcester, Mass	Oct. 14, 1856	15,887
Plow	J. Koffend	Appleton, Wis	May 26, 1868	78,293
Plow	M. Laflin and E. Slosson	Chicago, Ill	June 14, 1870	104,166
Plow	D. M. Lamb	Strathroy, Canada	Dec. 26, 1871	122,261
Plow	J. Lane, jr	Lockport, Ill	Dec. 1, 1857	18,750
Plow	J. Lane	Chicago, Ill	Mar. 31, 1868	76,208
Plow	J. Lane	Chicago, Ill	July 21, 1868	80,189
Plow	J. Lane	Chicago, Ill	July 5, 1870	104,964
Plow	J. Lane	Chicago, Ill	Feb. 14, 1871	111,854
Plow	J. Lane, jr	Chicago, Ill	Apr. 4, 1871	113,436
Plow	M. W. Lane	Hillsborough, Ohio	June 4, 1872	127,613
Plow	B. Langdon	Troy, N. Y	June 22, 1842	2,689
Plow	W. M. Lanham	Noblesville, Ind	May 3, 1870	102,687
Plow	W. M. Lanham	Noblesville, Ind	July 20, 1869	92,845
Plow	W. Lape	Troy, N. Y	June 25, 1861	32,666
Plow	R. K. and J. Laraway	Battle Creek, Mich	Mar. 5, 1867	62,641
Plow	J. S. Lash	Carlisle, Pa	Oct. 20, 1857	18,459
Plow	J. B. Latimer	Stewart County, Ga	Feb. 27, 1872	124,143
Plow	J. L. Laughlin	Peru, Ill	Nov. 11, 1873	144,462

Index of patents issued from the United States Patent Office from 1790 *to* 1873, *inclusive*—Continued.

Invention.	Inventor.	Residence.	Date.	No.
Plow	J. Layman	Lebanon, Ohio	Jan. 2, 1849	5, 999
Plow	S. J. Leach	Tuscaloosa, Ala	Dec. 17, 1867	72, 305
Plow	S. J. Leach	Tuscaloosa, Ala	Oct. 15, 1872	132, 295
Plow	A. Lee	West Chester, N. Y	July 22, 1822	....
Plow	E. D. and Z. W. Lee	Blakely, Ga	Oct. 4, 1859	25, 654
Plow	H. A. Lee	Parkman, Ohio	Oct. 29, 1872	132, 677
Plow	J. Lee	Galesburgh, Ill	Dec. 8, 1857	18, 820
Plow	Z. W. Lee	Blakely, Ga	Aug. 21, 1866	57, 341
Plow	E. D. and L. W. Legg	Speedsville, N. Y	June 16, 1857	17, 579
Plow	J. M. Leonard	Marshall, Mich	Jan. 10, 1871	110, 924
Plow	W. E. Levoy	Cincinnati, Ohio	Aug. 20, 1867	67, 890
Plow	C. C. Lewis	Gainesville, Ala	Aug. 13, 1872	130, 435
Plow	J. W. Lewis	Fetterman, W. Va	Jan. 8, 1867	61, 075
Plow	T. S. and J. A. Lockhart	Wellington, Mo	Nov. 27, 1860	30, 740
Plow	N. Locklin	Sparta, N. Y	Apr. 28, 1836	....
Plow	N. S. Lockwood and J. D. Winn	Dayton, Ohio	June 17, 1856	15, 137
Plow	P. Loeb	Dayton, Ohio	June 17, 1873	140, 053
Plow	I. Long	Terre Haute, Ind	Jan. 22, 1867	61, 470
Plow	J. Long	Leavenworth, Ind	June 15, 1869	91, 354
Plow	J. Love	Cusseta, Tex	Feb. 11, 1873	135, 829
Plow	R. Loveridge	Knox County, Ohio	Mar. 8, 1828	....
Plow	C. M. Lufkin	Acworth, N. H	Aug. 4, 1863	39, 411
Plow	C. M. Lufkin	Claremont, N. H	June 19, 1866	55, 684
Plow	C. B. Magruder	Thomasville, Ga	Oct. 20, 1857	18, 463
Plow	D. H. Maloy	Temperance, Ga	Feb. 26, 1861	31, 549
Plow	A. S. Mann	Saint Louis, Mo	July 8, 1873	140, 716
Plow	H. F. Mann	La Porte, Ind	Apr. 16, 1861	32, 073
Plow	D. A. Manuel	Napa City, Cal	Jan. 12, 1869	85, 746
Plow	S. March	Norfolk, Va	Mar. 26, 1867	63, 276
Plow	T. March	Dallas, Mich	Sept. 10, 1867	68, 635
Plow	T. R. Markillie	Winchester, Ill	June 19, 1860	28, 759
Plow	J. Marr	Simcoe, Canada	Aug. 19, 1873	141, 881
Plow	C. Marsh, 2d	Natchez, Miss	Sept. 16, 1873	142, 800
Plow	G. W. Marsh	Clinton, N. C	Dec. 1, 1868	84, 563
Plow	J. S. Marsh	Lewisburgh, Pa	Oct. 16, 1866	58, 855
Plow	H. Marshall	Atlanta, Ga	June 30, 1868	79, 486
Plow	P. Marshall and J. B. Smith	Lambertsville, N. J	Mar. 3, 1820	....
Plow	H. A. Martin	Jeffersonville, Ind	July 24, 1866	56, 584
Plow	H. D. Martin	Ypsilanti, Mich	Sept. 10, 1861	33, 256
Plow	J. R. Mason	Elgin, Ill	Feb. 11, 1862	34, 371
Plow	B. F. Masters	Middleport, Ill	Mar. 16, 1869	88, 794
Plow	D. Mater	Bellmore, Ind	Oct. 6, 1868	82, 858
Plow	R. G. Matheny and L. R. Barnes	De Kalb, Miss	Nov. 27, 1860	30, 741
Plow	J. T. Mathis and G. W. Harrison.	Kosciusko, Miss	July 5, 1870	105, 103
Plow	D. Matteson and T. P. Williamson.	Stockton, Cal	Mar. 12, 1867	62, 766
Plow	E. G. Matthews	Worcester, Mass	Feb. 23, 1841	1, 993
Plow	E. G. Matthews	South Natick, Mass	Nov. 26, 1867	71, 513
Plow	E. G. Matthews	Boston, Mass	May 19, 1868	77, 996
Plow	E. G. Matthews	Newton, Mass	Mar. 2, 1869	87, 420
Plow	E. G. Matthews	Oakham, Mass	Jan. 30, 1872	123, 272
Plow	T. F. Mattox	Griffin, Ga	Mar. 15, 1870	100, 780
Plow	H. H. May	Galesburgh, Ill	May 2, 1843	3, 069
Plow	H. H. May	Galesburgh, Ill	Apr. 25, 1846	4, 482
Plow	J. M. May	Philadelphia, Pa	May 2, 1846	4, 488
Plow	J. M. May	Philadelphia, Pa	May 2, 1846	4, 493
Plow	T. McConaughy	Burnsville, Ala	Mar. 23, 1858	19, 706
Plow	W. C. McCool	Guthrie Centre, Iowa	Oct. 8, 1872	131, 961
Plow	W. C. McCool	Guthrie Centre, Iowa	Nov. 11, 1873	144, 552
Plow	S. McCormick	Fauquier, Va	Jan. 28, 1826	....
Plow	S. McCormick	Fauquier County, Va	Oct. 22, 1828	....
Plow	S. McCormick	Auburn, Va	Dec. 1, 1837	501
Plow	M. C. McCullers	Herndon, Ga	Apr. 24, 1860	28, 000
Plow	M. C. McCullers	Herndon, Ga	July 17, 1860	29, 184
Plow	J. McKinley	Bethesda, Ohio	Sept. 4, 1866	57, 748
Plow	A. A. McMahen	Oxford, Miss	Nov. 2, 1858	21, 975
Plow	F. W. McMeekin	Morrison's Mill, Fla	Sept. 18, 1866	58, 119
Plow	F. M. McMeekin	Putnam County, Fla	Feb. 13, 1872	123, 717
Plow	F. M. McMeekin	Orange Springs, Fla	Oct. 29, 1872	132, 679
Plow	J. C. McVutt and A. B. Furman	Strattonville, Pa	Sept. 28, 1869	95, 251
Plow	S. Mead	New Haven, Conn	Sept. 15, 1863	39, 943
Plow	J. B. Mell	Riceborough, Ga	Feb. 19, 1856	14, 288
Plow	W. W. Metcalfe	York Springs, Pa	Oct. 17, 1848	5, 855
Plow	J. and E. P. Miles	Bloomingdale, Ind	Jan. 8, 1867	61, 083
Plow	L. Miller and H. Kaller	Camp Point, Ill	Dec. 13, 1864	45, 427
Plow	P. Miller	Washington County, N. Y	Oct. 26, 1818	....
Plow	R. J. Miller	Sherman, Iowa	Jan. 23, 1872	122, 956
Plow	R. M. Miller	Springfield, Ohio	July 16, 1872	129, 418
Plow	T. Miller	Pittsburgh, Pa	Oct. 23, 1829	....
Plow	T. Miller	Pittsburgh, Pa	Feb. 15, 1831	....
Plow	T. Miller	Pittsburgh, Pa	July 17, 1835	....
Plow	T. Miller	Pittsburgh, Pa	July 2, 1836	....
Plow	W. D. Miller	Enon, Ohio	June 15, 1869	91, 472
Plow	M. and S. J. Mims	Starkville, Miss	Dec. 23, 1841	2, 399
Plow	J. G. Miner	Nashville, Tenn	June 6, 1871	115, 629
Plow	J. G. Miner	Nashville, Tenn	Aug. 1, 1871	117, 662
Plow	H. Mitchell	Racine, Wis	Sept. 1, 1863	39, 741
Plow	J. J. Mitchell	Hopkinsville, Ky	Dec. 30, 1873	146, 088
Plow	J. C. Moltrup	Bucyrus, Ohio	May 10, 1859	23, 933
Plow	J. Mooers	Hazleton, Pa	July 1, 1844	3, 644
Plow	A. N. Moore	North Cohocton, N. Y	Dec. 31, 1867	72, 880
Plow	E. Moore	Slabtown, S. C	June 14, 1859	24, 399
Plow	G. Moore	Moline, Ill	Aug. 27, 1867	68, 102
Plow	G. Moore	Moline, Ill	Nov. 26, 1867	71, 507
Plow	G. H. Moore	Rochester, N. Y	Jan. 1, 1861	31, 028
Plow	J. Moore	Cynthiana, Ky	Aug. 17, 1832	....

Index of patents issued from the United States Patent Office from 1790 *to* 1873, *inclusive*—Continued.

Invention.	Inventor.	Residence.	Date.	No.
Plow	J. Moore	Lexington, Ky	July 22, 1843	3, 193
Plow	J. Moore	Bushnell, Ill	Oct. 29, 1867	70, 243
Plow	J. A. Moore	Louisville, Ky	Nov. 6, 1866	59, 437
Plow	J. B. Moore	Wilmington, Del	Nov. 24, 1843	3, 352
Plow	J. Morford	Maysville, Ky	Apr. 8, 1834	....
Plow	J. Morgan and J. B. Harris	Scipio, N. Y	Oct. 11, 1814	....
Plow	D. Morris	Bunker Hill, Ill	Sept. 27, 1870	107, 705
Plow	S. D., D. A., and J. B. Morrison	Fort Madison, Iowa	Aug. 8, 1871	117, 801
Plow	J. Mott	Danville, Cal	May 15, 1866	54, 756
Plow	J. M. and G. W. Moyers	Gordonsville, Va	Apr. 9, 1872	125, 478
Plow	M. Murray	Middletown, Md	Aug. 7, 1813	....
Plow	S. Myers	Marion, Ohio	July 11, 1842	2, 712
Plow	T. Nabers	Elyton, Ala	Apr. 23, 1872	126, 077
Plow	C. Nash	Bacon, Ill	Oct. 22, 1872	132, 406
Plow	J. Nash	Middlebury, Ohio	May 26, 1843	3, 110
Plow	C. O. Nason	Moline, Ill	July 29, 1873	141, 375
Plow	H. W. Neal	Sidney, Ohio	May 26, 1868	78, 390
Plow	H. W. Neal	Sidney, Ohio	Aug. 10, 1869	93, 549
Plow	D. Nelson	Port Washington, Ohio	Sept. 27, 1864	44, 442
Plow	C. Newbold		June 26, 1797	....
Plow	J. R. Nichols	Bastrop, Tex	Feb. 25, 1873	136, 257
Plow	W. Nichols	Floyd County, Ga	May 10, 1859	23, 938
Plow	W. D. Nichols and N. W. Clark	Chicago, Ill	Sept. 24, 1867	69, 117
Plow	S. Nisbet	Toboyne Township, Pa	May 25, 1830	....
Plow	G. Nixon	New York	June 3, 1823	....
Plow	W. Noble	New Haven, Conn	July 27, 1869	92, 992
Plow	H. Nolte	Lincoln, Ill	Sept. 7, 1869	94, 634
Plow	J. B. Norris	Richmond, Va	Dec. 2, 1873	145, 120
Plow	H. Norton	Farmington, Me	Oct. 9, 1855	13, 653
Plow	H. L. Norton	Granville, N. Y	Sept. 23, 1842	2, 784
Plow	J. B. Norton	Philadelphia, Pa	July 17, 1837	273
Plow	C. O'Bryan and H. Kreps	Minerva, Ohio	Aug. 20, 1861	33, 090
Plow	S. Ogle	Adams County, Pa	Feb. 28, 1832	....
Plow	W. Ogle	Frederick, Md	Apr. 6, 1843	3, 034
Plow	J. Oliver	South Bend, Ind	Feb. 21, 1871	111, 965
Plow	J. Oliver	South Bend, Ind	June 18, 1872	128, 061
Plow	J. Oliver	South Bend, Ind	Nov. 18, 1873	144, 785
Plow	S. J. Olmsted	Binghamton, N. Y	Mar. 17, 1863	37, 939
Plow	S. J. Olmsted	Binghamton, N. Y	Sept. 13, 1864	44, 215
Plow	W. O'Neill	Pine Level, Ala	Sept. 13, 1859	25, 436
Plow	W. O'Neill	Pine Level, Ala	Sept. 13, 1859	25, 437
Plow	W. O'Neill	Pine Level, Ala	June 8, 1869	91, 157
Plow	J. Ormiston	Waterford, Ohio	Sept. 5, 1848	5, 750
Plow	J. Ormiston	Centre Township, Ohio	June 2, 1857	17, 476
Plow	O. Osborn	Trumansburgh, N. Y	Feb. 23, 1869	87, 284
Plow	O. Osborn	Trumansburgh, N. Y	Sept. 7, 1869	94, 636
Plow	O. Osborn	Erie, Pa	Nov. 15, 1870	109, 340
Plow	G. Page	Washington, D. C	Aug. 7, 1847	5, 218
Plow	W. F. Pagett	Springfield, Ohio	Sept. 7, 1869	94, 637
Plow	J. Parker	Milroy, Ind	Feb. 26, 1867	62, 497
Plow	W. F. Parker	Troy, Ala	May 2, 1871	114, 334
Plow	J. A. Partlett and J. Thompson	Elmira, N. Y	Apr. 2, 1861	31, 903
Plow	G. W. Parmer	Jefferson Township, Ohio	Aug. 17, 1832	....
Plow	M. Patrick	Scipio, N. Y	June 2, 1813	....
Plow	T. Patterson	Rush, Ill	May 14, 1861	32, 311
Plow	C. Pawling	Gregg Township, Pa	May 21, 1830	....
Plow	H. Peachey	Philadelphia, Pa	July 25, 1834	....
Plow	D. Peacock	Burlington, N. J	Apr. 1, 1807	....
Plow	D. Peacock	New Mills, N. J	May 29, 1817	....
Plow	D. Peacock	Northampton, N. J	Jan. 21, 1822	....
Plow	H. Pease	Enfield, Conn	Aug. 28, 1813	....
Plow	E. Peck	Chicago, Ill	Sept. 15, 1868	82, 157
Plow	J. D. Peck	Triangle, N. Y	Mar. 10, 1868	75, 455
Plow	S. T. Peck	Penfield, Ga	Nov. 29, 1859	26, 289
Plow	M. Penning	Leavenworth City, Kans	Mar. 31, 1868	76, 243
Plow	D. Peters and J. W. Pauley	Keokuk, Iowa	Mar. 12, 1867	62, 881
Plow	P. Pettis	Portage County, Ohio	Feb. 15, 1825	....
Plow	J. P. Pettit	Cold Spring, Ky	Nov. 27, 1860	30, 756
Plow	O. Phelps and G. Morehouse	Lansing, N. Y	Nov. 17, 1821	....
Plow	J. Pierpont	La Harpe, Ill	Oct. 18, 1864	44, 745
Plow	J. Pinkham	New Market, N. H	Jan. 24, 1871	111, 244
Plow	W. C. Pitts	Austin, Tex	May 15, 1860	28, 329
Plow	J. Plank	Carlisle, Pa	June 2, 1836	....
Plow	H. M. Platt	Darien, Conn	June 22, 1858	20, 659
Plow	N. Platt	Saint Louis, Mo	Mar. 21, 1865	46, 937
Plow	F. Poindexter	Franklin, N. C	July 16, 1872	129, 054
Plow	H. A. G. Pomeroy and R. F. Hudson.	Providence, R. I., and Hartford, Conn.	June 12, 1860	28, 687
Plow	J. N. Pond	Wakefield, Va	Nov. 6, 1866	59, 515
Plow	W. R. Pool	Havanna, Ala	Nov. 9, 1869	96, 614
Plow	S. W. Pope	Louisville, Ky	May 10, 1870	102, 860
Plow	S. W. Pope	Louisville, Ky	July 26, 1870	105, 843
Plow	R. D. Porter	Zanesville, Ohio	July 2, 1872	128, 505
Plow	J. C. Potter	Saint Helena, Cal	June 24, 1873	140, 302
Plow	M. Prentiss	Cambridge, N. Y	Oct. 1, 1867	69, 478
Plow	S. Prentiss and G. Flint	De Soto, Mo	Jan. 12, 1869	85, 851
Plow	B. Price	Leesville, Ohio	July 31, 1866	56, 798
Plow	J. Price	Greenfield, Ind	Oct. 30, 1866	59, 262
Plow	W. Price	Wayne County, N. C	June 5, 1860	28, 601
Plow	W. Price	Mount Olive, N. C	Apr. 8, 1873	137, 716
Plow	J. P. Pritchard	Conn Valley, Cal	Jan. 24, 1871	111, 247
Plow	D. Prouty	Dorchester, Mass	Jan. 13, 1847	4, 928
Plow	D. Prouty and J. Mears	Boston, Mass	Mar. 4, 1836	....
Plow	D. Prouty and J. Mears	Boston, Mass	Sept. 15, 1838	922
Plow	D. Prouty and J. Mears	Boston and Dorchester, Mass.	June 16, 1841	2, 132
Plow	T. E. Putnam	Winneconne, Wis	July 16, 1872	129, 296
Plow	J. A. Quick	South Dansville, N. Y	Sept. 18, 1866	58, 177

Index of patents issued from the United States Patent Office from 1790 *to* 1873, *inclusive*—Continued.

Invention.	Inventor.	Residence.	Date.	No.
Plow	T. B. Quigley	Mansfield, Ohio	June 14, 1843	3,137
Plow	W. S. Rabb	Winnsborough, S. C	Apr. 20, 1869	89,242
Plow	W. S. Rabb	Winnsborough, S. C	May 31, 1870	103,777
Plow	Y. Rakestrow	White House, Ohio	Oct. 20, 1868	83,206
Plow	W. Rall	South Bend, Ind	July 16, 1872	129,364
Plow	J. O. Ramage	La Fayette, Ala	Feb. 2, 1858	19,262
Plow	T. Rams	Keokuk, Iowa	Sept. 11, 1866	57,968
Plow	I. N. Rankin	Middletown, Iowa	May 22, 1860	28,407
Plow	T. S. and T. W. Rappleyo	Farmer, N. Y	July 2, 1861	32,714
Plow	G. W. Ream	Canton, Ohio	Oct. 25, 1870	108,730
Plow	W. Reaney	Berzelia, Ga	Oct. 19, 1858	21,846
Plow	J. W. Reed	Careyville, Mo	Jan. 28, 1873	135,289
Plow	R. Reeder and S. D. Ashley	Madisonville, Ohio	Oct. 21, 1829	....
Plow	J. Reedy	Toledo, Iowa	May 14, 1867	64,707
Plow	E. Reese	Eutaw, Ala	Jan. 28, 1873	135,363
Plow	F. Reese	Elyton, Ala	Oct. 13, 1868	82,990
Plow	F. Reese	Wilsonville, Ala	Dec. 10, 1872	133,722
Plow	F. F. Reynolds	Burke County, Ga	Jan. 1, 1867	60,938
Plow	I. Reynolds	Republic, Ohio	Feb. 13, 1855	12,398
Plow	T. H. Reynolds	Rome, Ga	May 16, 1871	114,855
Plow	D. Rhodes	Pawtuxet, R. I	Apr. 19, 1864	42,401
Plow	J. Rhodes	Urbana, Ohio	June 11, 1829	....
Plow	M. G. Rhodes and J. M. Skaggs	Talladega, Ala	May 22, 1860	28,408
Plow	R. Rhodes	Charlotteville, Va	Feb. 20, 1827	....
Plow	R. Rhodes	Albemarle County, Va	Feb. 25, 1823	....
Plow	J. Rich	Troy, N. Y	July 31, 1849	6,611
Plow	J. Rich	Kingsbury, N. Y	July 15, 1856	15,344
Plow	M. Richards and J. Vandegrift	Princeton, Ill	Feb. 5, 1867	61,762
Plow	F. E. Richardson	Hicksford, Va	Nov. 30, 1852	9,433
Plow	L. W. Richardson	Roscoe, Ill	Nov. 15, 1870	109,250
Plow	L. B. Richardson	Carrollton, Ill	July 1, 1873	140,434
Plow	W. Richardson	Hookstown, Md	Sept. 17, 1867	68,901
Plow	C. F. Richter	Columbia, S. C	June 5, 1860	28,605
Plow	W. Richter	Williamsburgh, Ind	Jan. 9, 1849	6,007
Plow	A. Rickard	Schoharie, N. Y	Feb. 21, 1871	111,975
Plow	A. Rickard	Schoharie, N. Y	July 16, 1872	129,057
Plow	M. Rigell	Newton, Ala	Mar. 1, 1870	100,325
Plow	M. Rigell	Newton, Ala	Mar. 1, 1870	100,326
Plow	G. Ringen	Smith City, Mo	Oct. 29, 1867	70,267
Plow	A. W. L. Rivers	Midway, S. C	Nov. 27, 1860	30,762
Plow	A. Riviere	Barnesville, Ga	Nov. 11, 1873	144,477
Plow	J. L. Roberts	Brunswick, Ga	Oct. 23, 1866	59,078
Plow	M. L. Roberts	Smithville, Canada	Feb. 26, 1867	62,367
Plow	S. I. Roberts	Jeffersonville, Pa	Feb. 12, 1845	3,913
Plow	H. H. Robertson	Kingston, Mo	July 10, 1860	29,104
Plow	J Robinson	Hobart, N. Y	Sept. 23, 1862	36,532
Plow	J. G. Robinson	Biddeford, Me	Dec. 4, 1860	30,837
Plow	N. Robinson	Sacket's Harbor, N. Y	Feb. 13, 1835	....
Plow	N. Robinson	Patchogue, N. Y	Sept. 14, 1869	94,841
Plow	A. Roden	Talladega, Ala	Aug. 28, 1860	29,823
Plow	A. Roden	Mumford, Ala	June 27, 1871	116,355
Plow	J. M. Rodman	South Union, Ky	Feb. 26, 1861	31,559
Plow	H. D. Rogers	Grafton, Ohio	Feb. 12, 1861	31,407
Plow	R. Roles	Cary, N. C	May 31, 1870	103,663
Plow	L. Ronat	Carondelet, Mo	Jan. 5, 1869	85,696
Plow	G. W. Roney	Bayley's Mill, Fla	Dec. 27, 1859	26,620
Plow	H. Roney	Dayton, Ohio	Mar. 28, 1871	113,098
Plow	E. Rood	Darien, Wis	Dec. 24, 1872	134,316
Plow	J. Roop	Clinton, Tenn	Apr. 22, 1873	138,197
Plow	M. P. Rose	Napa, Cal	July 25, 1871	117,335
Plow	B. C. Rouse	Morris, Ill	Aug. 27, 1867	68,313
Plow	I. Rulofson	Penn Yan, N. Y	Mar. 1, 1859	23,116
Plow	I. Rulofson and D. De Garmo	Rochester, N. Y	Mar. 12, 1861	31,677
Plow	J. Runnyon	Marshall Township, Mich	Aug. 20, 1872	130,751
Plow	J. Runnyon and G. Ingersoll	Marshall Township and Marshall, Mich.	Aug. 10, 1869	93,558
Plow	G. W. Russell	Rockford, Ill	July 6, 1869	92,371
Plow	G. M. and G. S. Salsbury	Wilson and Clarendon, N. Y.	Aug. 4, 1863	39,425
Plow	A. Sanborn	Saint Johnsbury, Vt	May 28, 1872	127,372
Plow	A. Sanborn	Saint Johnsbury, Vt	Dec. 10, 1872	133,802
Plow	T. Sanford	Redding Ridge, Conn	Feb. 23, 1858	19,455
Plow	W. R. Saunders	Buena Vista, Miss	Mar. 27, 1860	27,651
Plow	W. Savage and H. Davidson	Horse Well, Ky	Aug. 24, 1831	....
Plow	S. D. Sayre	Rockford, Ill	Apr. 19, 1870	102,162
Plow	B. B. Scofield	Andover, Ill	Sept. 28, 1858	21,630
Plow	T. S. Scoville	Elmira, N. Y	Apr. 6, 1858	19,878
Plow	S. F. Seely	Sylvania, Ohio	Apr. 10, 1866	53,891
Plow	G. and J. Seibert	Ashley, Ill	Feb. 26, 1861	31,563
Plow	H. Selick	Lewistown, Pa	Sept. 28, 1869	95,386
Plow	H. Selick	Lewistown, Pa	Feb. 11, 1873	135,850
Plow	E. Seward	Hamilton, Ohio	Aug. 6, 1867	67,456
Plow	B. Seyler	Franklin County, Pa	Oct. 16, 1849	6,788
Plow	L. W. Shaffar	Shelbyville, Ky	Dec. 10, 1861	33,906
Plow	M. R. Shalters	Alliance, Ohio	May 9, 1871	114,611
Plow	J. P. Sharp	Geneseo, N. Y	Dec. 20, 1833	....
Plow	T. Sharp	Nashville, Tenn	Oct. 6, 1857	18,355
Plow	S. Shearer	Big Prairie, Ohio	July 14, 1845	4,112
Plow	W. F. Shedd	Ripley, Ohio	Feb. 26, 1861	31,565
Plow	T. Sheehan	Dunkirk, N. Y	July 5, 1870	105,134
Plow	G. Shelton	Normal, Ill	June 4, 1872	127,030
Plow	J. Shepard	Newport, Me	June 29, 1869	91,972
Plow	J. W. Shipp and C. W. Crenshaw	La Grange, Tenn	June 26, 1860	28,907
Plow	W. T. Shipp, C. J. Peterson, and R. L. McLurd	Brevard Station, N. C	Nov. 18, 1873	144,707
Plow	C. I. Shiver	Camden, S. C	May 22, 1860	28,416
Plow	W. G. Shuart	Orange County, N. Y	Feb. 5, 1822	....

Index of patents issued from the United States Patent Office from 1790 *to* 1873, *inclusive*—Continued.

Invention.	Inventor.	Residence.	Date.	No.
Plow	H. Shultz	Lancaster, Pa	Dec. 17, 1814	....
Plow	A. Shunk, sr	Bucyrus, Ohio	Nov. 19, 1867	71, 234
Plow	J. Sietz	Strasburgh, Pa	Feb. 8, 1813	....
Plow	H. B. Sinclear	Lyndonville, N. Y	Jan. 9, 1849	6, 020
Plow	D. T. Singleton	Eatonton, Ga	Mar. 22, 1870	101, 056
Plow	H. M. Skinner	Rockford, Ill	Sept. 10, 1872	131, 309
Plow	J. B. Skinner	Rockford, Ill	Aug. 14, 1866	57, 200
Plow	J. B. Skinner	Rockford, Ill	Apr. 30, 1867	64, 259
Plow	J. B. Skinner	Rockford, Ill	July 2, 1867	66, 259
Plow	J. B. Skinner	Rockford, Ill	July 2, 1867	66, 260
Plow	S. F. Skinner	Jacksonville, Mo	Jan. 14, 1868	73, 264
Plow	W. W. Skinner	Davenport, Iowa	Dec. 1, 1857	18, 776
Plow	M. G. Slemmons	Cadiz, Ohio	Oct. 9, 1860	30, 357
Plow	L. L. Sloss	Auburn, Ky	Jan. 4, 1870	98, 640
Plow	W. Small	North Argyle, N. Y	Apr. 22, 1839	1, 133
Plow	W. H. Simley	Bentonville, Ark	Sept. 24, 1872	131, 636
Plow	A. Smith	Bloomfield, Mich	May 6, 1844	3, 576
Plow	A. Smith	Macoupin County, Ill	Jan. 16, 1855	12, 241
Plow	A. B. Smith	Rochester, Pa	Nov. 5, 1872	132, 779
Plow	A. C. Smith	Joyner's Depot, N. C	Dec. 27, 1870	110, 506
Plow	D. Smith	Cedar Falls, Iowa	July 28, 1868	80, 314
Plow	D. H. and E. E. Smith	Glenn Springs, S. C	Apr. 3, 1860	27, 745
Plow	D. W. Smith	Dooly County, Ga	Mar. 19, 1861	31, 749
Plow	F. C. Smith	Harper's Ferry, Va	Apr. 25, 1848	5, 526
Plow	G. K. Smith and J. Strasser	Allegheny City, Pa	Jan. 21, 1868	73, 550
Plow	H. Smith	Moline, Ill	Nov. 15, 1870	109, 352
Plow	H. B. Smith	Tremont, Ill	May 21, 1872	127, 110
Plow	J. Smith	Norfolk, Va	Aug. 14, 1860	29, 629
Plow	J. C. Smith	Kingwood, N. J	July 11, 1837	260
Plow	J. M. Smith	Haddam, Conn	Aug. 30, 1870	106, 966
Plow	J. M. Smith	Haddam Neck, Conn	Dec. 5, 1871	121, 676
Plow	M. T. Smith	Keeler, Mich	Nov. 6, 1866	59, 468
Plow	C. W. Snead	Milledgeville, Ga	Oct. 4, 1870	107, 973
Plow	I. Snider	Mount Pleasant, Pa	July 2, 1836	....
Plow	C. Snyder	Middletown, Ill	Sept. 27, 1870	107, 828
Plow	S. W. Soule	New York, N. Y	Sept. 30, 1873	143, 263
Plow	O. Sparks	Shelbina, Mo	Dec. 11, 1860	30, 892
Plow	H. Speihlman and D. Miller	Lancaster, Pa	Mar. 25, 1833	....
Plow	G. Spiehlman	Strasburgh, Pa	Oct. 15, 1867	69, 943
Plow	W. S. Spratt	West Manchester, Pa	Sept. 5, 1865	49, 799
Plow	W. T. Sprouse	Sangamon, Ill	Feb. 15, 1838	604
Plow	W. T. Sprouse	Petersburgh, Ill	Mar. 13, 1849	6, 179
Plow	W. T. Sprouse	Chandlerville, Ill	Aug. 27, 1867	68, 253
Plow	D. Staebler	Lancaster, Pa	Mar. 15, 1834	....
Plow	P. Stahl and T. Diffenbasher	Turbut, Pa	Sept. 18, 1835	....
Plow	S. W. Standart	Bellevue, Ohio	Feb. 18, 1868	74, 729
Plow	H. Stanley	Erie County, Pa	Nov. 10, 1857	18, 609
Plow	T. A. Stansbury	Saybrook, Ill	Feb. 27, 1866	52, 903
Plow	P. H. Starke	Richmond, Va	Aug. 21, 1860	29, 726
Plow	P. H. Starke	Richmond, Va	May 5, 1868	77, 547
Plow	P. H. Starke	Richmond, Va	Dec. 29, 1868	85, 342
Plow	P. H. Starke	Richmond, Va	Aug. 1, 1871	117, 574
Plow	P. H. Starke	Richmond, Va	Aug. 19, 1873	141, 960
Plow	S. S. Starns	Macomb, Ill	Nov. 26, 1867	71, 419
Plow	J. L. Stearns	Mahomet, Ill	Dec. 15, 1868	84, 970
Plow	L. M. Stearns	Cardiff, N. Y	Jan. 4, 1861	32, 489
Plow	C. Steen	Athens, Ohio	Oct. 2, 1866	58, 500
Plow	H. Stein	Mifflinburgh, Pa	July 20, 1869	92, 898
Plow	C. P. Steinmetz	Madison, Wis	June 2, 1868	78, 492
Plow	R. S. Stenton	New York, N. Y	Dec. 8, 1857	18, 825
Plow	J. Stephens	Agency City, Iowa	Sept. 7, 1869	94, 663
Plow	R. L. and E. A. Stevens	Hoboken, N. J	Sept. 13, 1820	....
Plow	J. A. Stewart	Philadelphia, Pa	Oct. 2, 1860	30, 263
Plow	S. M. Stewart	New Harrisburgh, Ohio	Mar. 7, 1871	112, 508
Plow	C. St. John	Keosauqua, Iowa	Oct. 16, 1866	58, 911
Plow	S. L. Stockstill and A. D. Kutez	Medway, Ohio, and Harrisburgh, Pa.	Apr. 15, 1873	137, 870
Plow	A. W. Stoker	Petersburgh, Ill	Sept. 11, 1866	57, 991
Plow	J. T. Story	Magnolia, Ark	June 13, 1871	115, 907
Plow	R. E. Strait	Galesburgh, Mich	Dec. 21, 1869	98, 120
Plow	A. L. W. Stroud	Munford, Ala	Apr. 25, 1871	114, 221
Plow	D. A. Stubblefield and W. H. Luse.	Yazoo County, Miss	Aug. 24, 1869	94, 041
Plow	Z. W. Sturtevant	Dunstable, Mass	Dec. 14, 1869	97, 828
Plow	R. B. Summers and S. Dement.	San José, Ill	Sept. 25, 1866	58, 318
Plow	J. Swan	Scipio, N. Y	July 5, 1814	....
Plow	D. Swartz	Thomas Brook, Va	June 22, 1852	9, 061
Plow	H. H. Sweetland	Centreton, Ohio	Aug. 27, 1872	130, 882
Plow	J. Sweitzer	Mishawaka, Ind	Jan. 22, 1861	31, 194
Plow	C. W. Sykes	Suffield, Conn	Oct. 31, 1865	50, 749
Plow	N. S. Tallmadge	Fond du Lac, Wis	Sept. 1, 1863	39, 761
Plow	E. C. Tavener and O. Nesmith.	Hamilton, Va	May 8, 1855	12, 838
Plow	A. Taylor	New Garden, Ohio	Dec. 19, 1844	3, 864
Plow	C. B. Taylor	Bainbridge, Ohio	Aug. 23, 1833	....
Plow	H. Taylor	Montague, Mass	May 17, 1838	743
Plow	J. V. Taylor	Dixon, Ill	Jan. 3, 1860	26, 718
Plow	A. Teague	Jackson, Tenn	Mar. 10, 1843	2, 998
Plow	A. Teague	Madisonville, Ky	Sept. 21, 1869	95, 163
Plow	J. S. Tefft	Amherst, N. Y	Oct. 17, 1835	....
Plow	J. S. Tefft	Amherst, N. Y	Aug. 25, 1842	2, 762
Plow	M. Tessier	Cairo, Ill	Sept. 7, 1869	94, 667
Plow	B. Thacker, jr	Barnstable, Mass	July 20, 1831	....
Plow	D. K. Thom	Farmington, Tenn	Oct. 20, 1857	18, 475
Plow	G. W. Thompson	Ripley, Ohio	Sept. 4, 1866	57, 796
Plow	J. Thompson	Ripley, Ohio	Apr. 17, 1844	3, 542
Plow	J. T. Thompson	Jackson, Tenn	June 26, 1860	28, 919
Plow	S. R. Thompson	New Market, N. H	May 4, 1869	89, 608
Plow	T. Thompson	Thompsonville, N. C	Jan. 19, 1858	19, 163

Index of patents issued from the United States Patent Office from 1790 *to* 1873, *inclusive*—Continued.

Invention.	Inventor.	Residence.	Date.	No.
Plow	I. P. Tice	New York, N. Y	Sept. 5, 1871	118, 758
Plow	J. Tietz	Baltimore, Md	July 23, 1867	67, 002
Plow	J. M. Tilford	Murfreesborough, Tenn	July 1, 1836	....
Plow	D. L. Tilton	Mount Carmel, Ill	Mar. 23, 1858	19, 725
Plow	J. Tinkler	Warwick Township, Ohio	Mar. 2, 1835	....
Plow	W. D. Titus	Brooklyn, N. Y	Dec. 24, 1867	72, 568
Plow	J. Tomlinson	Racine, Wis	Jan. 26, 1864	41, 407
Plow	J. T. Townsend	Brenham, Tex	Nov. 15, 1859	26, 133
Plow	A. J. Traver	Lisburn, Pa	Mar. 23, 1869	88, 233
Plow	F. Traxler	Salem, Mich	Apr. 16, 1861	32, 094
Plow	M. Turley	Galesburgh, Ill	Feb. 16, 1858	19, 388
Plow	B. W. Tuttle	Galena, Ill	May 14, 1872	126, 656
Plow	W. H. Tyler	Conneautville, Pa	Jan. 1, 1869	90, 895
Plow	C. Urie	Evansville, Ind	Aug. 20, 1872	130, 773
Plow	J. Urie	Evansville, Ind	Jan. 7, 1868	73, 209
Plow	G. Utley	Louisburgh, N. C	Mar. 16, 1858	19, 658
Plow	G. Utley	Chapel Hill, N. C	Feb. 12, 1861	31, 419
Plow	G. Utley	Chapel Hill, N. C	May 26, 1868	78, 339
Plow	A. Vail	Henry, Ill	Aug. 25, 1868	81, 571
Plow	J. Vandegrift	Princeton, Ill	Oct. 15, 1867	69, 867
Plow	W. W. Van Loan	Catskill, N. Y	Feb. 16, 1858	19, 391
Plow	G. B. Vaughan	Peoria, Ill	May 20, 1873	139, 036
Plow	S. O. Vaughan	De Kalb, Ill	Feb. 28, 1860	27, 322
Plow	W. Veber, jr	Shingle Creek, N. Y	Sept. 25, 1866	58, 323
Plow	J. C. Vertrees	Gallatin, Tenn	Sept. 26, 1871	119, 433
Plow	R. Vincent	Rockton, Ill	Dec. 21, 1858	22, 389
Plow	B. Volkmann	New York, N. Y	Nov. 27, 1866	60, 095
Plow	F. Volkman	Hoboken, N. J	June 11, 1867	65, 709
Plow	G. A. Walker	Annville, Pa	Mar. 12, 1861	31, 684
Plow	J. Walker	Bellefontaine, Ohio	June 19, 1847	5, 168
Plow	R. Walker	Washingtonville, Pa	Mar. 10, 1830	....
Plow	R. Walker	Savannah, Mo	Aug. 6, 1872	130, 256
Plow	S. Walker	Kingston, Ga	Dec. 27, 1859	26, 633
Plow	W. Walker	Washingtonville, Pa	Oct. 6, 1835	....
Plow	W. and M. C. Walker	Washingtonville, Pa	Oct. 20, 1843	3, 311
Plow	J. Wallace	Berks County, Pa	Nov. 28, 1865	51, 245
Plow	J. Wallace	Sheridan, Pa	June 25, 1867	66, 109
Plow	J. Wallace	Sheridan, Pa	Feb. 20, 1872	123, 957
Plow	W. R. Walpole	Chicago, Ill	Apr. 2, 1867	63, 586
Plow	N. Warlick	La Fayette, Ala	Apr. 3, 1855	12, 650
Plow	N. Warlick	La Fayette, Ala	Oct. 20, 1857	18, 480
Plow	J. Warren	Warren County, N. Y	July 31, 1849	6, 620
Plow	T. P. Warren	Norfolk, Va	May 12, 1868	77, 855
Plow	W. Warren	Penn Yan, N. Y	July 20, 1858	20, 968
Plow	W. Warren	Penn Yan, N. Y	Sept. 25, 1860	30, 170
Plow	H. Washburn	Pultney, N. Y	Feb. 28, 1871	112, 302
Plow	F. Watson	Harrison, Ill	Oct. 15, 1867	69, 874
Plow	W. Watson	Bishopville, S. C	Mar. 13, 1860	27, 490
Plow	W. M. Watson	Tonica, Ill	Jan. 7, 1873	134, 716
Plow	G. Watt	Gainesville, Ala	Apr. 11, 1842	2, 548
Plow	G. Watt	Richmond, Va	Dec. 9, 1856	16, 218
Plow	G. Watt	Richmond, Va	Feb. 9, 1858	19, 321
Plow	G. Watt	Richmond, Va	July 10, 1866	56, 298
Plow	G. Watt	Richmond, Va	Nov. 26, 1867	71, 560
Plow	G. Watt	Richmond, Va	July 6, 1869	92, 408
Plow	A. Weaber	Lebanon, Pa	Apr. 25, 1871	114, 237
Plow	J. Weaver	Brownsville, Pa	Aug. 17, 1832	....
Plow	J. W. Webb	Cotton Valley, Ala	Nov. 1, 1870	108, 860
Plow	A. P. Webber	Saratoga Township, Ill	Aug. 27, 1872	130, 832
Plow	L. T. Webster	Northfield, Mass	Mar. 8, 1870	100, 696
Plow	L. Weeks and J. S. Trimble	Shelby, Ohio	May 14, 1872	126, 856
Plow	W. B. West	Utica, Wis	Oct. 19, 1869	95, 960
Plow	G. Wharton	Jerseyville, Ill	July 19, 1870	105, 531
Plow	R. J. Wheatley	Saint John, Ill	Feb. 19, 1867	62, 243
Plow	R. J. Wheatley	Duquoin, Ill	Mar. 15, 1870	100, 957
Plow	C. White	Bladensburgh, Md	Feb. 4, 1868	74, 183
Plow	E. J. White	Locke, N. Y	May 13, 1862	35, 272
Plow	L. B. White	Norfolk, Va	July 16, 1872	129, 076
Plow	L. B. White	Norfolk, Va	July 8, 1873	140, 749
Plow	N. Whitehall	Attica, Ind	July 3, 1855	13, 191
Plow	W. Whiteley	Springfield, Ohio	June 2, 1868	78, 501
Plow	T. A. Whities	Bellefontaine, Ohio	Jan. 14, 1832	....
Plow	E. G. Whiting	Racine, Wis	July 11, 1839	1, 232
Plow	E. G. Whiting	Northfield, Minn	Sept. 5, 1865	49, 816
Plow	S. Whitley	Tallahassee, Fla	Dec. 4, 1860	30, 849
Plow	S. W. Whitman	Harpersfield, Ohio	May 7, 1825	....
Plow	J. M. Whitney	Bolton, Mass	Mar. 1, 1859	23, 134
Plow	A. H. Whittick	Clarksville, Ind	Sept. 12, 1871	118, 890
Plow	E. Wiard	Louisville, Ky	Dec. 22, 1868	85, 152
Plow	E. Wiard	Louisville, Ky	May 25, 1869	90, 475
Plow	E. Wiard	Louisville, Ky	Aug. 31, 1869	94, 366
Plow	E. Wiard	Louisville, Ky	Aug. 31, 1869	94, 367
Plow	E. Wiard	Louisville, Ky	July 11, 1871	117, 024
Plow	E. Wiard	Louisville, Ky	Aug. 26, 1873	142, 136
Plow	E. Wiard	Louisville, Ky	Nov. 11, 1873	144, 584
Plow	T. Wiard	Avon, N. Y	Apr. 16, 1842	2, 557
Plow	T. Wiard	Louisville, Ky	Dec. 14, 1858	22, 332
Plow	W. Wiard	Avon, N. Y	Jan. 26, 1828	....
Plow	J. Widman and F. Millica	El Paso, Ill	Feb. 13, 1866	52, 633
Plow	C. Wilkin	Dundas, Ill	Apr. 14, 1868	76, 681
Plow	R. S. Williams	Bairdstown, Ga	June 26, 1868	28, 930
Plow	S. Williams, jr	Hume, N. Y	Mar. 8, 1859	23, 211
Plow	W. B. Williams	Warrenton, N. C	Sept. 13, 1859	25, 463
Plow	W. B. Williams	Warrenton, N. C	Sept. 13, 1859	25, 464
Plow	J. C. Williamson	Washington, Ga	Apr. 6, 1858	19, 886
Plow	J. C. Williamson	Washington, Ga	July 16, 1872	129, 196
Plow	F. R. Willson	Columbus, Ohio	May 7, 1872	126, 507
Plow	J. M. Wilson	Lexington, Miss	June 23, 1868	79, 091

Index of patents issued from the United States Patent Office from 1790 to 1873, inclusive—Continued.

Invention.	Inventor.	Residence.	Date.	No.
Plow	J. S. Wilson	Waynesborough, Ga	May 8, 1860	28, 227
Plow	L. S. Wilson	Wabash, Ind	Sept. 5, 1871	118, 769
Plow	W. H. Wilson	Summerfield, Ohio	Apr. 12, 1859	23, 636
Plow	G. A. Wing	Albany, N. Y	Feb. 2, 1869	86, 615
Plow	T. Winslow	Cleveland, Ohio	Apr. 25, 1865	47, 486
Plow	R. B. Winston	Richmond, Va	Dec. 1, 1857	18, 783
Plow	S. Witherow	Gettysburgh, Pa	Jan. 15, 1836	....
Plow	S. Witherow	Gettysburgh, Pa	Apr. 18, 1846	4, 465
Plow	G. Wolf	Fairfield County, Ohio	Jan. 27, 1832	....
Plow	L. Wolf	Hamburgh, Mo	Jan. 15, 1861	31, 136
Plow	L. Wolf	Saint Louis, Mo	Aug. 5, 1862	36, 122
Plow	C. Wood and G. Brundage	Blooming Grove, N. Y	Nov. 9, 1820	....
Plow	D. Wood and A. Byington	Byron, Ill	May 15, 1860	28, 323
Plow	J. Wood	Scipio, N. Y	July 1, 1814	....
Plow	J. Wood	Poplar Ridge, N. Y	Sept. 1, 1819	....
Plow	B. Woodcock	Mount Pleasant, Pa	Nov. 23, 1836	....
Plow	B. Woodcock	Mount Pleasant, Pa	Nov. 23, 1836	85
Plow	B. Woodcock	Mount Pleasant, Pa	June 14, 1837	233
Plow	B. Woodcock	Mount Pleasant, Pa	Nov. 23, 1837	476
Plow	B. Woodcock	Wheeling, Va	Jan. 31, 1845	3, 898
Plow	B. Woodcock	Williamsburgh, Pa	Aug. 21, 1860	29, 741
Plow	J. Woodward	Haverhill, N. C	Mar. 9, 1852	8, 794
Plow	J. Woolley	Troy, N. Y	Sept. 14, 1822	....
Plow	W. E. Wormell	Germantown, Tenn	Aug. 28, 1860	29, 841
Plow	A. Wright	Allegheny City, Pa	Jan. 21, 1868	73, 685
Plow	A. Wright	Allegheny City, Pa	Dec. 7, 1869	97, 746
Plow	J. S. Wright	Greenwich, N. Y	July 28, 1820	....
Plow	G. D. Wyman	Oshkosh, Wis	Feb. 18, 1873	136, 120
Plow	W. F. Yeager	Starkville, Miss	Dec. 6, 1859	26, 390
Plow	G. W. N. Yost	Nashville, Tenn	Aug. 11, 1863	39, 536
Plow	W. Yost	Goshen, Ohio	Dec. 5, 1871	121, 567
Plow	W. B. Young	Chicago, Ill	Feb. 14, 1865	46, 418
Plow	W. B. Young	Chicago, Ill	Feb. 11, 1868	74, 474
Plow	J. Yuger	Harrisburgh, Pa	June 9, 1830	....
Plow	G. W. Zeigler	Tiffin City, Ohio	May 27, 1856	14, 989
Plow	W. Zeller and R. Lechner	Lebanon County and Berks County, Pa.	Apr. 16, 1867	63, 838
Plow	C. Zimmerman	Collinsville, Ohio	Oct. 2, 1866	58, 534
Plow	W. T. Zollickoffer	Shelbyville, Tenn	Sept. 25, 1860	30, 188
Plow	S. Zook	Armagh Township, Pa	June 18, 1834	....
Plow, Adjustable shovel	T. Donnely and J. B. Cressler	Southampton, Pa	Jan. 28, 1868	73, 699
Plow, Adjustable shovel	G. W. Morter and E. Berry	Hartville, Ohio	Dec. 29, 1868	85, 469
Plow, Adjusting	W. K. Allan	Brownsborough, Ky	Jan. 31, 1844	3, 416
Plow, Agricultural	A. D. Armstrong	Springfield, Ohio	Jan. 15, 1830	....
Plow and action of the mold-board	H. Peachey	Washington, D. C	Apr. 5, 1833	....
Plow and clevis	I. Reynolds	West Liberty, Ohio	Mar. 26, 1850	7, 223
Plow and cotton-scraper	J. V. Greif	Paducah, Ky	May 28, 1867	65, 212
Plow and cultivator	J. Lane	Chicago, Ill	Sept. 15, 1868	82, 130
Plow and cultivator, Combined	S. Huber	Danville, Pa	Aug. 3, 1869	93, 203
Plow and cultivator, Combined	J. M. Landes	Sonders, Pa	Aug. 30, 1870	106, 837
Plow and cultivator, Combined	A. Romann and J. Peterka	Wilton, Iowa	Aug. 27, 1867	68, 114
Plow and cultivator, Combined	J. Wolpert	Louisville, Ky	Mar. 22, 1870	101, 200
Plow and cultivator, Combined	C. J. Woods and J. A. Phillips	Centreville, Ind	Nov. 9, 1869	96, 757
Plow and cultivator, Combined gang	J. A. Mcdaris	Sullivan, Ind	Jan. 24, 1871	111, 228
Plow and cultivator, Combined gang	O. Sampson	Petersburgh, Ill	Feb. 21, 1871	112, 079
Plow and cultivator, Combined gang	W. W. St. John	Saint Louis, Mo	Dec. 18, 1866	60, 589
Plow and cultivator, Combined gang	S. C. Thornton	Macomb, Tex	Aug. 23, 1870	106, 743
Plow and cultivator, Combined shovel	A. Smith	Providence Township, Pa	Sept. 9, 1873	142, 655
Plow and cultivator, Combined sulky	W. G. Crossley	Shullsburgh, Wis	Jan. 7, 1868	72, 981
Plow and cultivator, Convertible	J. Orr and H. H. Martin	Oxford, Ohio	Oct. 27, 1868	83, 404
Plow and cultivator, Convertible	C. L. Reid	Louisville, Ky	Mar. 29, 1870	101, 380
Plow and cultivator, Corn	I. B. Arthur	Sidonsburgh, Pa	Feb. 4, 1868	74, 032
Plow and cultivator, Cotton	A. Runstetler and A. Windeck	Peoria, Ill	Feb. 4, 1868	74, 005
Plow and cultivator fender	J. C. and W. F. Curryer	Thorntown, Ind	July 5, 1870	105, 047
Plow and cultivator, Gang	S. H. Mitchell	El Paso, Ill	Feb. 14, 1865	46, 378
Plow and cultivator, Gang	G. W. Price	Bloomington, Ill	Jan. 29, 1867	61, 566
Plow and cultivator, Shovel	B. F. Ream	Ada, Ohio	Feb. 18, 1868	74, 721
Plow and cultivator, Shovel	E. G. and A. D. Rowell, J. Rice, and S. M. Seeley.	Hartford, Wis	Nov. 1, 1870	108, 834
Plow and cultivator, Steam	P. H. Standish	Martinez, Cal	Mar. 10, 1868	75, 310
Plow and cultivator, Steam-propeller	J. Marquis	San Francisco, Cal	Sept. 29, 1868	82, 538
Plow and grain-drill, Combined corn	C. Hollar	Abington, Ind	July 19, 1870	105, 575
Plow and gun, Combined	C. M. French and W. H. Fancher	Waterloo, N. Y	June 17, 1862	35, 600
Plow and harrow	J. Haessel	Saint Louis, Mo	Oct. 20, 1868	83, 154
Plow and harrow	H. Higley	Canaan, Conn	Dec. 11, 1832	....
Plow and harrow	P. Wonsey	Ogden, N. Y	Feb. 4, 1868	74, 024
Plow and harrow, Combined	J. F. Brancher	Lincoln, Ill	Oct. 4, 1870	107, 862
Plow and harrow, Combined	J. Harsha	Circleville, Ohio	Jan. 7, 1868	73, 002
Plow and harrow, Combined	A. Moore and F. Wendel	Chillicothe, Ohio	Oct. 19, 1869	95, 925
Plow and harrow, Sulky	C. N. Bakewell	Normal, Ill	Mar. 24, 1868	75, 832
Plow and harrow, Sulky	J. E. Cheasebro	Buffalo, N. Y	Oct. 16, 1866	58, 773
Plow and hedge-grader, Ditching	D. Harmon	Coles County, Ill	Apr. 12, 1870	101, 729
Plow and hoe, Combined	D. W. Colburn	Loami, Ill	Sept. 3, 1867	68, 489
Plow and marker, Combined	O. M. Pond	Independence, Iowa	May 2, 1871	114, 472
Plow and marker, Combined corn	D. Wyant	West Lodi, Ohio	Jan. 19, 1869	85, 982
Plow and marker, Corn	G. W. Meixell	Hecktown, Pa	Oct. 22, 1872	132, 405
Plow and marker, Garden	H. Haynsworth	Sumter, S. C	Dec. 7, 1869	97, 508
Plow and planter	G. C. Avery	Waldron, Ind	June 9, 1868	78, 638
Plow and planter	P. Kling	Springfield, Ill	Oct. 4, 1870	108, 002
Plow and planter, Combined	J. D. Marshall	Renick, Mo	Oct. 29, 1867	70, 237
Plow and planter, Combined	M. Shackleford	Montgomery, Ala	Feb. 2, 1869	86, 458
Plow and planter, Combined	I. H. Walker	Newton, Ill	Aug. 4, 1868	80, 686
Plow and planter, Corn	J. H. Hannon	Halifax, N. C	Oct. 18, 1870	108, 476
Plow and planter, Corn	H. C. Osborn	Clarke County, Ohio	Sept. 27, 1870	107, 709
Plow and planter, Corn	C. Sintz	Clarke County, Ohio	Sept. 6, 1870	107, 110
Plow and planter, Cotton	Z. B. Sims	Bonham, Tex	Aug. 31, 1869	94, 349
Plow and planter, Seeding	J. B. and J. C. Watt	Greene County, Ohio	Aug. 30, 1870	106, 896
Plow and rail-cleaner, Snow	L. J. Woodruff	Mohawk, N. Y	May 9, 1871	114, 633

Index of patents issued from the United States Patent Office from 1790 *to* 1873, *inclusive*—Continued.

Invention.	Inventor.	Residence.	Date.	No.
Plow and road-scraper, Combined	T. B. Parker	Linesville, Pa	Oct. 11, 1870	108, 176
Plow and roller, Combined	J. A. Alley	Clifty, Ind	Feb. 11, 1868	74, 267
Plow and scraper, Combined	J. C. Cameron	Madison Station, Miss	Apr. 4, 1871	113, 625
Plow and scraper, Combined	R. Elliot	Plainfield, N. J	Jan. 2, 1866	51, 817
Plow and scraper, Combined	J. Reynolds	Crystal Springs, Miss	Sept. 14, 1869	94, 774
Plow and scraper, Combined cotton	T. P. Warren	Norfolk, Va	June 18, 1867	65, 847
Plow and scraper, Combined subsoil	J. Reynolds	Crystal Springs, Miss	Feb. 22, 1870	100, 188
Plow and scraper, Grading	A. P. Hopkins	Bentleyville, Pa	Jan. 26, 1869	86, 229
Plow and scraper, Railway snow	E. Trenholm	Washington, D. C	June 23, 1863	38, 996
Plow and seed-planter, Combined	W. Crossdale	Hartsville, Pa	Nov. 27, 1849	6, 908
Plow and seeder, Combined	W. C. Bibb	Madison, Ga	Apr. 4, 1871	113, 482
Plow and seeder, Gang	D. Stone	Pittsford, N. Y	Aug. 9, 1864	43, 804
Plow and stock for cultivator, Combined	M. Kennedy	Chicago, Ill	Oct. 3, 1871	119, 618
Plow and subsoiler	G. Robinson	Plymouth, Ohio	Mar. 9, 1869	87, 594
Plow and tobacco-hiller attachment, Sulky	J. L. Spencer	Wellville, Va	July 16, 1867	66, 748
Plow, Angular	J. Lupton	Virginia	July 31, 1817	....
Plow-attachment	W. Bennett	Rushville, Ind	Jan. 7, 1868	73, 158
Plow-attachment	A. E. Cruttenden	Canasaragra, N. Y	Feb. 2, 1869	86, 370
Plow-attachment	W. J. Martin	Catawissa, Pa	Feb. 11, 1868	74, 398
Plow-attachment	J. W. Monical	Mooresville, Ind	Jan. 5, 1869	85, 601
Plow-attachment	M. Roberts	Saint Paul, Minn	Apr. 3, 1866	53, 749
Plow-attachment	H. B. Smith	Eureka, Ill	Feb. 5, 1867	61, 885
Plow-attachment	C. E. Wilson	Palmyra, Me	Sept. 15, 1868	82, 189
Plow-attachment, (double-tree)	L. E. Morey	Vandalia, Ill	Dec. 8, 1868	84, 750
Plow-attachment for cutting stubble	J. T. Hovis	Clintonville, Pa	June 27, 1871	116, 446
Plow-attachments, Making blanks for	O. A. Anthony	Mayfield, N. Y	Aug. 13, 1872	130, 403
Plow, Auburn	W. F. Richardson	Auburn, N. Y	June 28, 1825	....
Plow, Automatic spading	F. L. Cagwin	Joliet, Ill	Oct. 27, 1868	83, 456
Plow, Bar-share	J. and J. Butler	Brandywine and Chester County, Pa.	Mar. 1, 1814	....
Plow, Bar-share	J. Deakyne	Petersburgh, Va	Nov. 18, 1828	....
Plow, Bar-share	M. Huffman	Elizabeth, Tenn	Dec. 28, 1832	....
Plow, Bar-share	A. Mitchell	Washington County, Tenn	Oct. 1, 1830	....
Plow, Bar-share	E. Pugh	Chatham County, N. C	Dec. 24, 1827	....
Plow, Bar-share	E. M. Waggenet	Adair County, Ky	Jan. 12, 1831	....
Plow, Bar-share or fallow	G. Davis	Georgetown, D. C	Mar. 6, 1833	....
Plow-beam	E. Bement	Fostoria, Ohio	Sept. 16, 1862	36, 446
Plow-beam	W. Eddy	Greenwich, N. Y	June 4, 1867	65, 357
Plow-beam	E. T. Ford	Stillwater, N. Y	May 6, 1862	35, 147
Plow-beam	W. Gilman	Ottawa, Ill	July 30, 1867	67, 188
Plow-beam	L. B. Griffith	Honey Brook, Pa	Oct. 4, 1853	10, 085
Plow-beam	J. Houghton	Ogden, N. Y	June 22, 1842	2, 691
Plow-beam	S. H. Miller	Hamilton, Ill	Feb. 14, 1871	111, 764
Plow-beam blanks, Rolled bar for	G. W. Jope and W. Bunton	Pittsburgh, Pa	Dec. 20, 1870	110, 371
Plow-beam-clevis attachment	J. L. Baldwin	Troy, Pa	July 20, 1869	92, 775
Plow beam, Gang	J. W. Sursa	San Francisco, Cal	Sept. 20, 1870	107, 560
Plow-beam handle	M. C. Brelsford	Girard, Ill	Jan. 5, 1864	41, 056
Plow-beam, Iron	A. Ball	Canton, Ohio	Jan. 3, 1871	110, 722
Plow-beam-rolling machine	W. Gilman	Ottawa, Ill	July 26, 1870	105, 668
Plow beams, Attaching colters to	J. H. Hall	Maysville, Ky	Mar. 10, 1868	75, 414
Plow-beams, carriage-shafts, &c., Machine for hewing.	E. G. Matthews	Worcester, Mass	Oct. 22, 1840	1, 836
Plow-beams, Metallic fastening for	L. Gibbs	Canton, Ohio	Apr. 10, 1866	53, 918
Plow, Bogging and ditching	J. Gordon	Copake, N. Y	Jan. 13, 1829	....
Plow-bolt	C. A. Harper	Burlington, N. J	Mar. 4, 1873	136, 508
Plow-brace die	G. Moore	Moline, Ill	July 23, 1867	67, 131
Plow, Bull	B. Johnson	Hickory Grove, Ill	Feb. 20, 1835	....
Plow, Bull	R. Tousley and J. Swan	....	Nov. 9, 1814	....
Plow, Bull	S. Tousley	Onondaga County, N. Y	May 28, 1813	....
Plow, Carpenter's	C. G. Miller	New Britain, Conn	Sept. 17, 1872	131, 367
Plow, Carpenter's	R. B. Rice	Williamsburgh, Mass	Dec. 21, 1869	98, 108
Plow, Carpenter's	H. Vanbuskirk	Vienna, Mich	Nov. 30, 1869	97, 328
Plow, Carriage	H. M. Bullitt	Louisville, Ky	Oct. 7, 1873	143, 434
Plow, Carriage	J. M. Hammitt and H. T. Miller	Toledo, Iowa	Jan. 1, 1867	60, 721
Plow, Carriage	A. Ingalls	Independence, Iowa	May 29, 1866	55, 108
Plow, Carriage	W. L. Jeffries	Lancaster, Ohio	Mar. 30, 1869	88, 486
Plow, Carriage	H. Minuse	Milan, Ohio	Sept. 10, 1867	68, 777
Plow, Carriage	G. F. Willey	Laconia, N. H	Mar. 31, 1868	76, 283
Plow-carrier and cotton-chopper, Combined	F. L. Kirtley	Cleyborne, Tex	May 18, 1869	90, 274
Plow, Cast-iron	T. Altason	Newton Township, N. J	Apr. 16, 1824	....
Plow, Cast-iron	D. Barnard	Washington, Del	Apr. 21, 1830	....
Plow, Cast-iron	D. Chase and A. and J. Gregg	Palmyra, N. Y	Aug. 4, 1824	....
Plow, Cast-iron	J. Dutcher	Durham, N. Y	Feb. 12, 1822	....
Plow, Cast-iron	T. Fairbanks	Saint Johnsbury, Vt	Apr. 19, 1826	....
Plow, Cast-iron	W. Falconner	New York	Sept. 13, 1821	....
Plow, Cast-iron	G. W. Hawkins and H. Emery	Windsor County, Vt	Feb. 16, 1822	....
Plow, Cast-iron	D. Hitchcock	New York	Dec. 29, 1821	....
Plow, Cast-iron	R. McMillen	Middlebury, Ohio	Dec. 14, 1841	2, 389
Plow, Cast-iron	J. Minturn	Urbana, Ohio	June 11, 1829	....
Plow, Cast-iron	J. K. Odell and W. S. Little	Deckertown, N. J	Dec. 6, 1870	109, 929
Plow, Cast-iron	J. J. Schoonmaker and J. Dolson.	Ulster County and Dutchess, N. Y.	Dec. 7, 1822	....
Plow, Cast-iron	J. Seaver and J. Fay	Rockingham, Vt	Dec. 14, 1820	....
Plow, Cast-iron	O. Seely	Pottstown, N. Y	Aug. 27, 1821	....
Plow, Cast-iron	R. L. and E. A. Stevens	Hoboken, N. J	Apr. 23, 1821	....
Plow, Cast-iron	R. Sweeny	Warren County, Ohio	May 18, 1827	....
Plow, Cast-iron	A. Taylor	Green Township, Ohio	Dec. 26, 1833	....
Plow, Cast-iron	R. Winans	Vernon, N. Y	May 11, 1824	....
Plow-castings	F. F. Smith	Collinsville, Conn	May 16, 1865	47, 753
Plow-castings, Chill for	J. Oliver	South Bend, Ind	May 2, 1871	114, 479
Plow-castings, Machine for grinding	J. Gibbs	Canton, Ohio	Oct. 4, 1853	10, 068
Plow-cleaner	G. W. Burr	East Line, N. Y	Dec. 26, 1871	122, 155
Plow-cleaner	C. P. Devereaux	North Newburgh, Mich	July 16, 1867	66, 809
Plow-cleaner	D. D. Gitt	Butler Township, Pa	May 7, 1850	7, 344
Plow-cleaner	J. and E. P. Miles	Pleasant Hill, Ohio	June 29, 1869	91, 957
Plow-cleaner	J. A. B. Patterson	Springfield, Ill	Apr. 20, 1869	89, 237
Plow-cleaner	J. B. Place	Auburn Township, Pa	Nov. 5, 1872	132, 857
Plow-cleaner	J. F. Reasin	Darlington, Md	Apr. 9, 1850	7, 274
Plow-cleaner	D. Warren	Gettysburgh, Pa	Sept. 10, 1850	7, 640

Index of patents issued from the United States Patent Office from 1790 *to* 1873, *inclusive*—Continued.

Invention.	Inventor.	Residence.	Date.	No.
Plow-cleaner	O. F. Warren	Pembroke, N. Y	July 10, 1866	56, 296
Plow-cleaning attachment	T. Terrel	Spring Hill, Ohio	Aug. 14, 1866	57, 216
Plow-clearer	A. C. Black	Kaukauna, Wis	May 25, 1869	90, 488
Plow-clevis	C. Adams	Pittsburgh, Pa	May 22, 1860	28, 337
Plow-clevis	L. W. Alden	Fosterville, N. Y	Jan. 23, 1866	52, 123
Plow-clevis	H. W. Austin and W. Schaw	Kalamazoo, Mich	Sept. 3, 1867	68, 545
Plow-clevis	W. Axford	Carrollton, Ill	Mar. 5, 1872	124, 313
Plow-clevis	J. Ball	Canton, Ohio	Sept. 1, 1868	81, 731
Plow-clevis	J. A. Bilz	Pleasanton, Cal	May 18, 1869	90, 068
Plow-clevis	R. A. Blair and J. B. Reed	New Philadelphia, Ohio	Aug. 20, 1861	33, 066
Plow-clevis	T. E. C. Brinly	Louisville, Ky	May 3, 1870	102, 652
Plow-clevis	J. Brison	Competine, Iowa	Oct. 3, 1871	119, 502
Plow-clevis	A. Carman	Columbus, N. J	June 20, 1838	791
Plow-clevis	A. A. Dailey	Wilson, N. Y	Mar. 5, 1872	124, 339
Plow-clevis	G. P. Darrow	Cincinnati, Ohio	Aug. 21, 1866	57, 436
Plow-clevis	T. Don	Yorktown, Ill	Aug. 31, 1869	94, 296
Plow-clevis	G. Erkson	Hobart, N. Y	Sept. 17, 1850	7, 651
Plow-clevis	P. Gallagher	Chambersburgh, Pa	Nov. 26, 1845	4, 292
Plow-clevis	J. S. Hall	Manchester, Pa	Dec. 4, 1860	30, 873
Plow-clevis	H. Ingraham	Naples, N. Y	Aug. 27, 1867	68, 200
Plow-clevis	A. Kaufman	Davisville, Cal	Mar. 25, 1873	137, 078
Plow-clevis	J. Keech	Waterloo, N. Y	May 5, 1863	38, 394
Plow-clevis	W. Lahman	Wellsville, Ohio	July 8, 1873	140, 715
Plow-clevis	E. G. Matthews	Newton, Mass	Mar. 2, 1869	87, 421
Plow-clevis	A. McCollam	New Orleans, La	July 18, 1871	117, 187
Plow-clevis	C. L. Meech	Preston, Conn	May 28, 1846	4, 544
Plow-clevis	J. Napier	Martin's Ferry, Ohio	Aug. 6, 1872	130, 145
Plow-clevis	J. Newhart	Terre Haute, Ind	Oct. 29, 1867	70, 351
Plow-clevis	F. Reese	Wilsonville, Ala	Dec. 10, 1872	133, 799
Plow-clevis	R. Sandiford	Joliet, Ill	June 2, 1868	78, 542
Plow-clevis	A. Shogren	Sandwich, Ill	July 18, 1865	48, 844
Plow-clevis	D. Stewart	Corinna, Me	Oct. 6, 1868	82, 764
Plow-clevis	E. Stewart	Fort Madison, Iowa	July 16, 1872	129, 614
Plow-clevis	J. B. Stoner	Southampton, Pa	Sept. 10, 1850	7, 638
Plow-clevis	J. H. Sarpley	Greensborough, N. C	Dec. 14, 1869	97, 990
Plow-clevis	J. Van Brocklin	Middleport, N. Y	Mar. 13, 1847	5, 016
Plow-clevis	J. D. Willoughby	Pleasant Hall, Pa	June 2, 1857	17, 462
Plow-clevis	S. Wilt	Hagerstown, Md	Sept. 3, 1846	4, 730
Plow-clevis	E. A. Wright	Fort Madison, Iowa	June 18, 1872	128, 087
Plow-clevis attachment	H. C. Sieverling	Carrollton, Ill	Dec. 13, 1870	110, 080
Plow-clevis attachment	G. Watt	Richmond, Va	July 6, 1869	92, 409
Plow-clevis pin	H. McKiney	Earlham, Iowa	July 16, 1872	129, 355
Plow-clevises, Machine for making	T. Meikle	Louisville, Ky	Mar. 3, 1868	75, 179
Plow-clevises, Spring-link for	J. Ball	Canton, Ohio	Mar. 2, 1869	87, 458
Plow clod-fender	J. and J. W. Higgins	Orth, Ind	Oct. 26, 1869	96, 234
Plow clod-fender	R. A. Kelly	Hope, Ind	July 6, 1869	92, 318
Plow clod-fender	W. B. Kidder	Pike Township, Ind	Aug. 2, 1870	106, 064
Plow-colter	S. Casebeer	Roseburgh, Oreg	Aug. 21, 1866	57, 286
Plow-colter	G. Curkendall	Dixon, Ill	May 17, 1870	103, 148
Plow-colter	J. Custer and C. Rowland	Clinton, Ill	June 12, 1866	55, 472
Plow-colter	C. Furst	Chicago, Ill	Nov. 9, 1869	96, 575
Plow-colter	D. D. Gibson	Springfield, Iowa	Oct. 17, 1871	120, 055
Plow-colter	C. M. Gordon	La Porte, Ind	Apr. 4, 1871	113, 421
Plow-colter	E. C. Hodge	Oneonta, N. Y	Feb. 21, 1871	112, 039
Plow-colter	W. H. Hoefelman	Columbus, Nebr	Aug. 20, 1872	130, 638
Plow-colter	H. S. Hoxie	Adrian, Mich	May 5, 1868	77, 616
Plow-colter	H. S. Hoxie	Raisin Centre, Mich	May 16, 1871	114, 822
Plow-colter	J. Lane	Chicago, Ill	Mar. 30, 1869	88, 309
Plow-colter	J. Pinkham	New Market, N. H	Feb. 20, 1872	123, 838
Plow-colter	F. F. Smith	Collinsville, Conn	Jan. 19, 1869	85, 971
Plow-colter	W. H. Stone	Lebanon, Mich	Apr. 25, 1871	114, 220
Plow-colter	B. W. Tuttle	Galena, Ill	May 14, 1872	126, 655
Plow-colter	S. Way	La Porte, Ind	Jan. 24, 1871	111, 158
Plow-colter	E. Wiard	Louisville, Ky	Aug. 26, 1873	142, 310
Plow colter and share	S. A. Sperry	Ann Arbor Village, Mich	Oct. 27, 1835	....
Plow-colter, Revolving	H. Delano	Mottsville, N. Y	Apr. 1, 1842	2, 529
Plow-colter, Revolving	M. Richards	Princeton, Ill	Apr. 5, 1870	101, 509
Plow-colter, Revolving	M. Sattley	Taylorsville, Ill	Nov. 24, 1868	84, 380
Plow-colters, Apparatus for cleaning	E. C. Bills, jr	Perry, N. Y	Oct. 21, 1856	15, 919
Plow-colters, Attaching	D. W. Hughes	Palmyra, Mo	Aug. 17, 1869	93, 888
Plow, Combined iron and steel	W. Morrisson	Chadd's Ford, Pa	Jan. 28, 1862	34, 262
Plow, Combined single and double shovel	J. W. Nicholson	Indianapolis, Ind	Feb. 2, 1869	86, 577
Plow, Combined subsoil, drill, and side-hill	W. W. Speer	Allegheny City, Pa	Sept. 26, 1871	119, 423
Plow, Combined turn and subsoil	J. C. Gross	Goshen Hill, S. C	Aug. 31, 1869	94, 307
Plow, Common	N. Turbutt	Fredericktown, Md	Sept. 7, 1811	....
Plow, Corn	P. Barnhart	Chillicothe, Ohio	Oct. 23, 1866	58, 973
Plow, Corn	W. H. Bott	York, Pa	Nov. 9, 1869	96, 667
Plow, Corn	W. B. Broadwell	Springfield, Ill	June 3, 1862	35, 434
Plow, Corn	M. C. Buffington	La Harpe, Ill	Sept. 24, 1867	69, 071
Plow, Corn	M. C. Buffington	La Harpe, Ill	Apr. 12, 1870	101, 706
Plow, Corn	A. Canfield	Lyons, Iowa	Aug. 20, 1867	67, 843
Plow, Corn	A. Canfield	Lyons, Iowa	Aug. 20, 1867	67, 844
Plow, Corn	L. H. Castor	Edington, Ill	June 28, 1864	43, 288
Plow, Corn	J. M. Clark	Somerville, Ohio	June 11, 1867	65, 725
Plow, Corn	W. C. Clifton	Elk River Township, Iowa	Oct. 11, 1870	108, 239
Plow, Corn	S. Coats	La Fayette, Wis	June 5, 1849	6, 501
Plow, Corn	S. H. Cox and W. H. Pence	Mattoon, Ill	Jan. 7, 1868	73, 166
Plow, Corn	T. Dillon	Highland, Ohio	Nov. 17, 1868	84, 094
Plow, Corn	N. Etheridge, G. Heath, and I. G. Glynn	Little Falls, N. Y	July 23, 1831	....
Plow, Corn	W. and J. C. French	Keokuk, Iowa	Nov. 22, 1870	109, 403
Plow, Corn	M. B. Goff	Delavan, Wis	June 28, 1870	104, 729
Plow, Corn	L. Guthrie	Waterloo, Ind	Sept. 7, 1869	94, 489
Plow, Corn	T. W. Hammon	Montfort, Wis	Mar. 7, 1865	46, 752
Plow, Corn	H. Harrier	Indianapolis, Ind	Aug. 23, 1870	106, 693
Plow, Corn	J. Hernly	East Hempfield, Pa	Dec. 10, 1838	1, 031
Plow, Corn	J. Hindmarsh	Henry, Ill	Jan. 15, 1867	61, 195
Plow, Corn	R. C. Howard	Lena, Ill	Oct. 23, 1866	59, 022

Index of patents issued from the United States Patent Office from 1790 *to* 1873, *inclusive*—Continued.

Invention.	Inventor.	Residence.	Date.	No.
Plow, Corn	R. J. King	Lancaster, Pa	Mar. 5, 1850	7, 141
Plow, Corn	J. Marsh	Seneca, Ill	Sept. 24, 1867	69, 109
Plow, Corn	A. McCreight	Tranquility, Ohio	Sept. 1, 1868	81, 660
Plow, Corn	A. D. Michener and J. W. Steigmeyer.	Attica, Ohio	June 15, 1869	91, 471
Plow, Corn	S. J. Miller and L. Wright	Economy, Ind	Dec. 22, 1868	85, 118
Plow, Corn	F. G. Mourning	Bascon, Ill	Sept. 13, 1870	107, 400
Plow, Corn	C. J. Paine	Young America, Ill	Oct. 27, 1868	83, 537
Plow, Corn	C. E. Paxson	Salem, Ohio	Nov. 26, 1861	33, 793
Plow, Corn	E. Phillips	Centreville, Ind	Dec. 6, 1870	109, 934
Plow, Corn	J. Preston	Plymouth, Ohio	May 13, 1873	138, 818
Plow, Corn	W. B. Raper	Carthage, Ill	Jan. 11, 1870	98, 708
Plow, Corn	M. Redlinger	Freeport, Ill	July 2, 1867	66, 253
Plow, Corn	B. W. and N. T. Remy	Brookville, Ind	Mar. 28, 1871	113, 206
Plow, Corn	W. C. Rhinehart and R. Gaston.	Oskaloosa, Iowa	Dec. 8, 1868	84, 763
Plow, Corn	J. Snyder	Williamsfield, Ohio	Dec. 10, 1867	72, 104
Plow, Corn	J. R. Thomas	Mifflintown, Pa	Sept. 10, 1867	68, 670
Plow, Corn	J. L. Walter and E. Bushman	White House, Pa	Nov. 1, 1870	108, 857
Plow, Corn	W. S. Weir, jr	Monmouth, Ill	Feb. 7, 1865	46, 285
Plow, Corn	D. Wilde	Washington, Iowa	May 15, 1866	54, 860
Plow, Corn	J. Wolf	Young America, Ill	Apr. 21, 1868	76, 964
Plow, Corn and cotton	J. W. Milroy	Galveston, Ind	Dec. 29, 1868	85, 324
Plow, Corn and cotton cultivating	W. R. Blanchard	Hertford, N. C	Aug. 2, 1870	105, 892
Plow, Corn and potato	W. Notman	Deerfield, Ohio	Nov. 24, 1868	84, 437
Plow, Corn-dressing	E. Bean	Massachusetts	Mar. 4, 1816	....
Plow, Cotton	H. C. Godfrey	Elizabeth City, N. C	Dec. 10, 1872	133, 850
Plow, Cotton	W. W. Graves	Fort Adams, Miss	Nov. 27, 1860	30, 729
Plow, Cotton	T. Guice	Mount Andrew, Ala	Dec. 6, 1870	109, 893
Plow, Cotton, &c	H. W. Pitts	Wilsonville, Ala	Mar. 31, 1836	....
Plow, Cotton	Z. B. Sims	Bonham, Tex	Aug. 31, 1869	94, 350
Plow, Cotton	Z. B. Sims	Bonham, Tex	Aug. 31, 1869	94, 351
Plow, Cotton	T. A. Wainwright	Wilson, N. C	July 7, 1868	79, 706
Plow, Cotton	J. Wiley	Warsaw, N. C	Dec. 20, 1870	110, 411
Plow, Cotton	W. B. Williams	Warrenton, N. C	Aug. 13, 1867	67, 831
Plow, Cotton and corn	D. C. Richardson	Weldon, N. C	Oct. 22, 1867	70, 120
Plow, Cotton-scraping	J. M. Cobb	Jackson, Tenn	Nov. 11, 1873	144, 509
Plow, Cotton-thinning	W. Crichton	Diamond Grove, Va	Oct. 23, 1847	5, 340
Plow, Cotton-thinning	L. B. Joyner	Hilliardston, N. C	Apr. 24, 1860	27, 987
Plow-coupling	G. Owen	Jacksonville, Ill	Feb. 28, 1871	112, 273
Plow-coupling	J. K. Pitts	McLean, Ill	Jan. 28, 1873	135, 360
Plow-coupling	W. Reck	Mendota, Ill	Aug. 31, 1869	94, 339
Plow-coupling	T. L. Thrasher	Paris, Tex	Aug. 19, 1873	141, 964
Plow coupling, Double	G. Owen	Jacksonville, Ill	Feb. 4, 1862	34, 316
Plow, Covering	W. Roberts	Rocheport, Mo	Jan. 15, 1861	31, 130
Plow, Cultivating	W. Bagnall	Otsego, Ohio	Nov. 4, 1873	144, 247
Plow, Cultivating	A. Hughes	Gratiot, Ohio	May 29, 1860	28, 487
Plow, Cultivator	S. B. Forbes	Steubenville, Ohio	Mar. 23, 1869	88, 025
Plow, Cultivator	R. F. and J. V. Guy	Macomb, Ill	Oct. 10, 1868	82, 938
Plow, Cultivator	T. F. Jones	Hicksford, Va	Nov. 29, 1870	109, 629
Plow, cultivator, and marker, Combined	J. Dooley	Saint Paul, Minn	July 13, 1869	92, 590
Plow, cultivator, and planter, Combined	J. Rider	Woodburn, Ill	Mar. 12, 1841	2, 002
Plow, cultivator, and potato-digger, Combined	H. B. Smith	Tremont, Ill	July 27, 1869	93, 154
Plow, Cultivator gang	H. P. McCleave	Tomales, Cal	Nov. 2, 1869	96, 334
Plow, Cultivator gang	T. T. Prosser and M. C. and K. A. Darling.	Chicago, Ill., and Fond du Lac, Wis.	Jan. 3, 1865	45, 750
Plow, cultivator, &c., Shovel	P. A. Ross	Harvey's, Pa	Dec. 10, 1867	72, 085
Plow-cutter	G. F. Pykiet	Fairfield, Ill	Dec. 20, 1870	110, 283
Plow cutter and point	D. H. Jones and W. F. Richardson.	Palmyra, N. Y	July 29, 1823	....
Plow, Digging	E. Peck	Deer Park, N. Y	Nov. 10, 1857	18, 600
Plow, Ditching	E. S. Bartlett	Romulus, N. Y	Sept. 13, 1859	25, 377
Plow, Ditching	J. Brooks	Romulus, N. Y	July 31, 1860	29, 357
Plow, Ditching	T. I. Burhyte	Fond du Lac, Wis	Nov. 21, 1865	51, 014
Plow, Ditching	W. Burton	Lake, Ind	Feb. 13, 1872	123, 613
Plow, Ditching	G. Chamberlin	Olean, N. Y	Sept. 6, 1870	107, 162
Plow, Ditching	W. R. Clark	Indianola, Ill	Sept. 10, 1867	68, 697
Plow, Ditching	E. L. Foreman	Rantoul, Ill	Jan. 26, 1869	86, 294
Plow, Ditching	J. Kelley and W. H. Hennis	Winamac, Ind	Jan. 30, 1872	123, 262
Plow, Ditching	J. Lyon	Harrisburgh, Iowa	July 25, 1854	11, 382
Plow, Ditching	J. T. Miller	Iowa Falls, Iowa	Feb. 19, 1867	62, 215
Plow, Ditching	F. Pimmer	Grand Junction, Tenn	Feb. 12, 1861	31, 399
Plow, Ditching	R. M. Primmer	Vinton, Iowa	Aug. 8, 1871	117, 813
Plow, Ditching	I. S. Sheets	Troy, Ohio	Feb. 16, 1869	86, 947
Plow, Ditching	I. S. Sheets	Troy, Ohio	May 3, 1870	102, 599
Plow, Ditching	D. S. Stafford	Rochester, Ill	July 10, 1845	4, 107
Plow, Ditching	J. C. Tiffany	Coxsackie, N. Y	Apr. 25, 1854	10, 815
Plow, Ditching	W. West	Pecksburgh, Ind	Mar. 16, 1869	87, 892
Plow, Ditching	H. D. Williams	Fair View, Iowa	Nov. 7, 1871	120, 690
Plow, Ditching	J. L. Wilson and J. R. Haworth	Iowa Falls, Iowa	Jan. 28, 1868	73, 658
Plow, Ditching	S. S. Wood	Brooklyn, N. Y	Aug. 16, 1870	106, 443
Plow, Ditching and mole	E. H. Morton	Oxford, Iowa	Jan. 3, 1865	45, 735
Plow, Double	P. and B. Altenderfer	Coots Town, Pa	June 11, 1829	....
Plow, Double	J. Cromwell	Roystertown, Md	Oct. 23, 1816	....
Plow, Double	P. Eichar	Wooster, Ohio	Dec. 5, 1843	3, 372
Plow, Double	A. Peebles	Henry County, Ga	July 20, 1831	....
Plow, Double	A. Smith	Bloomfield, Mich	May 10, 1844	3, 579
Plow, Double	G. W. N. Yost	Nashville, Tenn	Aug. 11, 1863	39, 537
Plow, Double corn	L. Guthrie	Waterloo, Ind	Sept. 7, 1869	94, 490
Plow, Double-mold-board	S. Gregory	Sawpitts, N. Y	Nov. 14, 1838	1, 008
Plow, Double revolving	J. Mott	Danville, Cal	Aug. 14, 1866	57, 171
Plow, Double-shovel	J. Balltiropo	Loudoun County, Va	Sept. 20, 1820	....
Plow, Double-shovel	J. Clarridge	Pancoastville, Ohio	Oct. 9, 1866	58, 597
Plow, Double-shovel	A. J. Craig	Ashmore, Ill	Apr. 28, 1868	77, 260
Plow, Double-shovel	J. M. Eby	Warren, Ill	Apr. 9, 1867	63, 711
Plow, Double-shovel	I. G. Flisher and E. M. Bates	Stark County, Ohio	Apr. 14, 1868	76, 736
Plow, Double-shovel	G. W. Lawbaugh	Geneseo, Ill	Mar. 1, 1870	100, 298
Plow, Double-shovel	G. Perry	Granville, Ill	Apr. 19, 1870	102, 150
Plow, Double-shovel	S. G. Rayl	Agency City, Iowa	Sept. 13, 1870	107, 409

Index of patents issued from the United States Patent Office from 1790 to 1873, inclusive—Continued.

Invention.	Inventor.	Residence.	Date.	No.
Plow, Double-shovel	M. R. Shalters and S. Ray	Alliance, Ohio	Mar. 24, 1868	75, 988
Plow, Double-shovel	L. L. Sloss	South Union, Ky	Aug. 27, 1867	68, 247
Plow, Double-shovel	H. Stephens	Mount Vernon, Ohio	May 7, 1867	64, 592
Plow, Double-shovel	R. P. Van Horne	Gratiot, Ohio	Feb. 14, 1871	111, 793
Plow, Double-shovel	C. I. Voigt	West Salem, Ill	Oct. 5, 1869	95, 541
Plow, Drain	D. Ballard	Texas, Ohio	July 21, 1868	80, 113
Plow, Drain	M. Barrowman	Buffalo, N. Y	June 29, 1858	20, 689
Plow, Drain	R. W. Downman	Georgetown, D. C	June 25, 1867	66, 133
Plow, Drain	C. G. Grabo	Greenfield, Mich	May 26, 1863	38, 677
Plow, Drain	J. Hanon, jr	Taylorsville, Ill	Mar. 27, 1860	27, 630
Plow, Draining	S. Hall	Boston, Mass	July 13, 1810	....
Plow, Draining and ditching	H. B. Smawley	Greensburgh, Ind	Apr. 16, 1867	63, 952
Plow-drawing machine	L. Johnson	Leominster, Mass	Aug. 6, 1816	... .
Plow drill and cotton-thinner	W. Willis	Edgefield District, S. C	May 7, 1822	....
Plow, Expanding	A. W. Wilkins and S. T. Eskridge.	Rome, Ga	Nov. 30, 1869	97, 465
Plow, Expanding double-shovel	E. Wiard	Louisville, Ky	Aug. 10, 1869	93, 507
Plow, Expanding triple-shovel	E. Wiard	Louisville, Ky	Dec. 21, 1869	98, 214
Plow-fastening	D. and S. Swartz	Tom's Brook, Va	Aug. 22, 1854	11, 575
Plow-fastening device	J. Robb	Lewistown, Pa	Oct. 12, 1852	9, 332
Plow-fender	B. F. Neely	Yorktown, Ind	July 25, 1871	117, 317
Plow-fender	S. J. Reed	Middletown, Ohio	Nov. 10, 1868	83, 999
Plow-fender	A. P. Webber	Saratoga Springs, Ill	Dec. 19, 1871	122, 087
Plow fender, Corn	R. Cook	Franklin, Ohio	Mar. 26, 1867	63, 140
Plow fender, Corn	A. B. Thornton	Berlin, Ill	Sept. 28, 1869	95, 287
Plow fender, Corn	L. C. Witt and W. F. Jones	Boston, Mass	Oct. 26, 1869	96, 293
Plow for breaking and cultivating sward-ground	G. Gray	Industry, Me	Sept. 18, 1835	....
Plow for cutting bogs	J. Coffee	Monroe, N. Y	Nov. 13, 1866	59, 557
Plow for cutting potato-roots	H. M. Keith	Commerce, Mich	Sept. 6, 1870	107, 063
Plow for draining and subsoiling	L. Green	Great Bend, Pa	July 23, 1861	32, 866
Plow for excavating ditches	J. Herbert	La Grange, Ind	Apr. 13, 1844	3, 538
Plow for listing and preparing land	J. Weaver	Washington, D. C	May 8, 1823	....
Plow for planting corn	H. Russell	Litchfield, Me	Jan. 16, 1827	....
Plow for planting cotton	B. Murphy	Union District, S. C	Dec. 31, 1827	....
Plow for planting potatoes	W. Price	Goldsborough, N. C	Aug. 29, 1854	11, 616
Plow, Four-wheel	N. B. Norton	Burlington, Wis	Sept. 15, 1868	82, 241
Plow-frame	S. March	Norfolk, Va	Apr. 28, 1868	77, 203
Plow, Furrowing	S. Perry	Troy, N. Y	July 10, 1866	56, 327
Plow, Furrowing	A. P. and C. F. Rohrer and J. H. Blose	Clarke County, Ohio	Dec. 26, 1871	122, 284
Plow-gage	J. Cluckner	Arcadia, Ind	Mar. 22, 1870	100, 984
Plow-gage	H. B. Spedden	Baltimore, Md	Aug. 3, 1869	93, 362
Plow-gage wheel	P. D. Beckwith	Dowagiac, Mich	Jan. 21, 1868	73, 564
Plow, Gang	S. H. Adams	Coultervillle, Ill	Nov. 8, 1864	44, 924
Plow, Gang	J. Alloways	Decatur, Ill	Aug. 2, 1870	105, 879
Plow, Gang	J. H. Andrews	Benicia, Cal	Sept. 1, 1868	81, 724
Plow, Gang	J. H. Andrews	Benicia, Cal	Sept. 6, 1870	106, 982
Plow, Gang	J. and G. Armstrong	Elmira, N. Y	Aug. 6, 1872	130, 177
Plow, Gang	C. Atwood	Lebanon, Ill	Dec. 19, 1865	51, 536
Plow, Gang	C. Atwood	Lebanon, Ill	Apr. 24, 1866	54, 089
Plow, Gang	C. Atwood	Lebanon, Ill	Oct. 18, 1870	108, 311
Plow, Gang	G. C. Avery	Cone's Creek, Ind	Aug. 27, 1867	68, 277
Plow, Gang	W. Battell	Quincy, Ill	Aug. 7, 1866	56, 878
Plow, Gang	R. Baxter	French Camp, Cal	Aug. 6, 1867	67, 483
Plow, Gang	D. Bequeret and E. Dumoulin	Jamestown, Ill	Nov. 20, 1866	59, 813
Plow, Gang	J. C. Bethea	Blakely, Ga	Feb. 5, 1867	61, 796
Plow, Gang	C. Bishop	Norwalk, Ohio	Oct. 12, 1852	9, 314
Plow, Gang	J. F. and W. L. Black	Lancaster, Ill	July 30, 1861	32, 958
Plow, Gang	J. F. and W. L. Black	Lancaster, Ill	Dec. 19, 1865	51, 543
Plow, Gang	J. Blackwood	Madison Township, Ill	Aug. 15, 1871	117, 970
Plow, Gang	S. B. Bowen and A. M. Abbot	Stockton, Cal	Mar. 12, 1872	124, 480
Plow, Gang	W. J. Boyce and G. W. Haines	Maine Prairie, Cal	Nov. 16, 1869	96, 875
Plow, Gang	G. T. Brewer	Prairie du Rocher, Ill	Oct. 29, 1867	70, 159
Plow, Gang	J. C. Brown and G. H. Slimpert	Pinckneyville, Ill	June 6, 1865	48, 049
Plow, Gang	A. H. Burlingame	Sparta, Ill	Aug. 14, 1866	57, 081
Plow, Gang	R. Carson	Meredosia, Ill	Sept. 21, 1869	95, 082
Plow, Gang	G. R. Carter	New York, N. Y	May 18, 1869	90, 237
Plow, Gang	L. Chapman	Collinsville, Conn	June 29, 1869	92, 016
Plow, Gang	L. Chapman	Collinsville, Conn	Apr. 4, 1871	113, 627
Plow, Gang	L. Chapman	Collinsville, Conn	Nov. 7, 1871	120, 572
Plow, Gang	F. G. Charles	Galesburgh, Ill	Feb. 20, 1872	123, 869
Plow, Gang	P. Conrath	Freeburgh, Ill	Aug. 17, 1869	93, 680
Plow, Gang	R. Coreth	West Belleville, Ill	Aug. 26, 1873	142, 084
Plow, Gang	T. J. Cornell	Decatur, Ill	May 15, 1866	54, 693
Plow, Gang	T. J. Cornell	Decatur, Ill	Oct. 2, 1866	58, 384
Plow, Gang	C. W. Corr	Carlinville, Ill	Aug. 1, 1865	49, 087
Plow, Gang	A. T. Covell	San Leandro, Cal	Aug. 6, 1867	67, 501
Plow, Gang	H. Cowing	New Orleans, La	Nov. 26, 1850	7, 795
Plow, Gang	J. and S. Cox	Eugene City, Oreg	Feb. 8, 1870	99, 538
Plow, Gang	M. A. Cravath	Lodi, Ill	Jan. 12, 1858	19, 077
Plow, Gang	F. R. Crothers	Sparta, Ill	Feb. 23, 1864	41, 686
Plow, Gang	F. R. Crothers	Sparta, Ill	June 29, 1869	92, 019
Plow, Gang	M. S. Curtiss	Bradford, Ill	Nov. 19, 1872	133, 206
Plow, Gang	M. S. Curtiss	Bradford, Ill	July 1, 1873	140, 480
Plow, Gang	G. J. Dahl	Stockton, Cal	Mar. 24, 1868	75, 871
Plow, Gang	H. N. Dalton	Pacheco, Cal	Oct. 2, 1866	58, 386
Plow, Gang	F. S. Davenport	Jerseyville, Ill	Feb. 9, 1864	41, 491
Plow, Gang	F. S. Davenport	Jerseyville, Ill	Oct. 9, 1866	58, 612
Plow, Gang	A. Davison	San Leandro, Cal	Mar. 30, 1869	88, 280
Plow, Gang	G. A. Davison	San Leandro, Cal	July 7, 1868	79, 639
Plow, Gang	J. N. Davison and N. Spencer, jr.	Buffalo, Ill	Nov. 3, 1868	83, 698
Plow, Gang	J. W. Donaldson, D. Sheets, and A. C. Miller.	Suisun, Cal	Sept. 4, 1866	57, 688
Plow, Gang	J. H. Douthit	Albany, Oreg	Jan. 22, 1867	61, 408
Plow, Gang	A. P. Durant	Atlanta, Ill	Feb. 6, 1866	52, 401
Plow, Gang	G. R. Duval	Salem, Oreg	Mar. 1, 1870	100, 383
Plow, Gang	C. L. Eastham	Rhodes Point, Ill	June 18, 1867	65, 798

Index of patents issued from the United States Patent Office from 1790 *to* 1873, *inclusive*—Continued.

Invention.	Inventor.	Residence.	Date.	No.
Plow, Gang	J. H. Eckert	Lebanon, Ill	Mar. 27, 1866	53, 424
Plow, Gang	A. Ellison	Marysville, Cal	Aug. 30, 1870	106, 799
Plow, Gang	C. A. Fargo	Soquel, Cal	June 1, 1869	90, 831
Plow, Gang	A. Farrow	Carrollton, Ill	Jan. 28, 1868	73, 707
Plow, Gang	D. M. Fay	Monmouth, Ill	Mar. 19, 1872	124, 802
Plow, Gang	P. M. Flansburgh	Hayward's, Cal	May 10, 1870	102, 801
Plow, Gang	M. Flinn	Saint Louis, Mo	Apr. 14, 1868	76, 735
Plow, Gang	W. Foster	Greenfield, Ind	Dec. 31, 1867	72, 730
Plow, Gang	A. Freeman	Homer, Ill	Apr. 2, 1872	125, 280
Plow, Gang	W. H. Freeman	Bloomfield, Iowa	Jan. 24, 1865	45, 988
Plow, Gang	J. Frye	Springfield, Ill	Mar. 31, 1857	16, 9[illegible]2
Plow, Gang	J. Frye	Mendota, Ill	May 25, 1858	20, 342
Plow, Gang	W. J. Funk	Portland, Oreg	Oct. 19, 1869	95, 997
Plow, Gang	T. E. Gardiner, jr	Bryantown, Md	May 21, 1867	64, 967
Plow, Gang	R. R. Gaskill	El Paso, Ill	Feb. 13, 1866	52, 558
Plow, Gang	W. S. Gatlin and B. R. Hubbard.	Green Top, Mo	Mar. 24, 1868	75, 895
Plow, Gang	C. F. Gay	Albany, Oreg	July 27, 1869	93, 077
Plow, Gang	P. M. Gilbert	Kewanee, Ill	May 30, 1865	47, 942
Plow, Gang	S. I. and G. M. Gillham	Carlisle, Ill	Aug. 27, 1867	68, 065
Plow, Gang	S. J. Gillham, W. C. Taylor, and J. W. Stolle.	Vandalia, Ill	June 27, 1871	116, 428
Plow, Gang	A. L. and B. F. Gilliland	Littleton, Ill	Mar. 6, 1866	52, 990
Plow, Gang	J. H. Glass	McGregor, Iowa	July 25, 1871	117, 404
Plow, Gang	D. H. Gleeson	San Leandro, Cal	Dec. 7, 1869	97, 500
Plow, Gang	E. E. Gore	Phœnix, Oreg	Dec. 28, 1869	98, 370
Plow, Gang	S. Graham	Fennimore, Wis	July 28, 1868	80, 405
Plow, Gang	R. R. Graves	Montgomery, Ala	July 9, 1867	66, 583
Plow, Gang	G. A. Groves	East Clarkson, N. Y	Jan. 7, 1873	134, 540
Plow, Gang	J. Haege	Shiloh, Ill	Dec. 18, 1860	30, 967
Plow, Gang	J. Haege	Shiloh, Ill	Feb. 24, 1863	37, 750
Plow, Gang	J. Haege	Shiloh, Ill	Apr 14, 1863	38, 161
Plow, Gang	G. W. Haines	Maine Prairie, Cal	Mar. 12, 1872	124, 571
Plow, Gang	T. J. Hall	Tawakana Falls, Tex	May 1, 1855	1[illegible], 791
Plow, Gang	T. J. Hall	Bryan, Tex	Apr. 20, 1869	89, 144
Plow, Gang	A. Hammond	Jacksonville, Ill	July 11, 1865	48, 679
Plow, Gang	A. Hammond	Jacksonville, Ill	Mar. 27, 1866	53, 443
Plow, Gang	J. Harris	Santa Clara County, Cal	Nov. 26, 1867	71, 301
Plow, Gang	J. Harris	San Francisco, Cal	Dec. 20, 1870	110, 356
Plow, Gang	M. W. Harris	Des Moines, Iowa	July 2, 1872	128, 482
Plow, Gang	W. Hay and T. B. Freeman	Hillsborough, Oreg	July 11, 1871	116, 956
Plow, Gang	T. S. Heptinstall	Mendota, Ill	Dec. 27, 1859	26, 587
Plow, Gang	T. S. Heptinstall	Mendota, Ill	July 17, 1860	29, 169
Plow, Gang	P. Herbert	Saint Louis, Mo	Aug. 10, 1869	93, 618
Plow, Gang	C. Hess	Lyons City, Iowa	Mar. 10, 1868	75, 268
Plow, Gang	W. F. Higgins and J. Perry	Watsonville, Cal	Apr. 7, 1868	76, 447
Plow, Gang	G. W. Hildreth	Lockport, N. Y	Oct. 13, 1857	18, 397
Plow, Gang	F. A. Hill	Maysville, Cal	Sept. 7, 1869	94, 491
Plow, Gang	L. Holloway	Gilroy, Cal	Mar. 21, 1865	46, 903
Plow, Gang	L. Holloway	San Francisco, Cal	July 27, 1869	93, 068
Plow, Gang	J. D. Hope	Philadelphia, Pa	June 4, 1850	7, 415
Plow, Gang	C. L. Horn, jr., and L. Mancy	Saint Morgan, Ill	Sept. 15, 1868	82, 223
Plow, Gang	A. W. Hoyt	Denver, Ill	Sept. 10, 1872	131, 218
Plow, Gang	H. R. Hine	Hayward's, Cal	Aug. 11, 1863	39, 483
Plow, Gang	H. R. Huie	Hayward's, Cal	Oct. 20, 1868	83, 283
Plow, Gang	H. R. Huie	Hayward's, Cal	Sept. 28, 1869	95, 229
Plow, Gang	J. M. Huie and E. Card	San Francisco, Cal	Oct. 31, 1871	120, 384
Plow, Gang	A. M. Humphrey	Broadhead, Wis	Feb. 18, 1873	136, 062
Plow, Gang	J. B. Hunter	Ashley, Ill	June 19, 1866	55, 660
Plow, Gang	J. B. Hunter	Ashley, Ill	Nov. 9, 1869	96, 599
Plow, Gang	S. Hutchinson	Griggsville, Ill	Aug. 7, 1866	56, 942
Plow, Gang	J. and J. Ingham	San José, Cal	May 28, 1867	65, 087
Plow, Gang	B. Jennings	Gilroy, Cal	Dec. 7, 1869	97, 646
Plow, Gang	E. C. Jones	Pittsburgh, Pa	Dec. 1, 1857	18, 749
Plow, Gang	J. L. Keasor	Laconia, N. H	Mar. 24, 1868	75, 925
Plow, Gang	C. Kewin	San Francisco, Cal	Nov. 12, 1872	132, 910
Plow, Gang	H. Killam and G. Valleau	Scottsville, N. Y	Mar. 30, 1852	8, 844
Plow, Gang	S. L. Kingston and D. Gore	Plain View, Ill	Oct. 6, 1857	18, 343
Plow, Gang	J. D. Kneedler	Collinsville, Ill	Nov. 3, 1868	83, 641
Plow, Gang	W. Kuehn	Lively, Ill	Nov. 17, 1863	40, 629
Plow, Gang	H. P. Kynett	Lisbon, Iowa	Mar. 26, 1867	63, 260
Plow, Gang	J. H. La Boyteaux and C. A. Ashton.	Jacksonville, Ill	July 11, 1865	48, 696
Plow, Gang	J. Lane	Eugene, Ind	Nov. 5, 1872	132, 842
Plow, Gang	L. B. Lathrop	San José, Cal	May 28, 1867	65, 094
Plow, Gang	J. Lee	Galesburgh, Ill	Nov. 24, 1857	18, 698
Plow, Gang	J. T. Legg	Lewis County, Mo	June 25, 1867	66, 155
Plow, Gang	J. W. Lewis	Oregon City, Oreg	May 18, 1869	90, 178
Plow, Gang	J. W. Lewis	Oregon City, Oreg	June 22, 1869	91, 643
Plow, Gang	M. Likes	Mansfield, Ohio	June 27, 1871	116, 332
Plow, Gang	J. J. Lindly	Lebanon, Ill	Aug. 3, 1869	93, 318
Plow, Gang	J. B. Logan	Richview, Ill	Dec. 1, 1868	84, 634
Plow, Gang	D. A. Manuel	Napa, Cal	June 4, 1872	127, 495
Plow, Gang	G. W. Manuel	San Francisco, Cal	May 19, 1868	78, 111
Plow, Gang	G. W. Manuel	Napa, Cal	Apr. 8, 1873	137, 697
Plow, Gang	T. R. Markillie	Winchester, Ill	Sept. 13, 1864	44, 206
Plow, Gang	W. Mason	Independence, Oreg	Jan. 12, 1869	85, 838
Plow, Gang	W. Mason	Independence, Oreg	Nov. 8, 1870	109, 136
Plow, Gang	W. W. Mathews	Yates City, Ill	Feb. 11, 1868	74, 238
Plow, Gang	D. C. Matteson	Stockton, Cal	June 22, 1858	20, 647
Plow, Gang	D. C. Matteson	Stockton, Cal	June 2, 1868	78, 464
Plow, Gang	J. R. McConnell	Marengo, Iowa	Jan. 24, 1871	111, 226
Plow, Gang	J. K. McLennan	Elmira, Ill	Feb. 11, 1873	135, 654
Plow, Gang	F. McFarnahan	Santa Clara, Cal	Dec. 8, 1868	84, 748
Plow, Gang	P. Merkel	Saint Louis, Mo	Nov. 20, 1866	59, 855
Plow, Gang	M. Murphy	Vacaville, Cal	May 4, 1869	89, 788
Plow, Gang	J. Murry	Silveyville, Cal	Mar. 28, 1871	113, 081
Plow, Gang	R. Nation	Chebanse, Ill	Apr. 5, 1864	42, 258

Index of patents issued from the United States Patent Office from 1790 *to* 1873, *inclusive*—Continued.

Invention.	Inventor.	Residence.	Date.	No.
Plow, Gang	W. Nelson	Cacheville, Cal	Mar. 17, 1868	75, 567
Plow, Gang	W. Newlin	Attica, Ind	Oct. 25, 1870	108, 724
Plow, Gang	J. Oler	Eagle Point, Ill	Apr. 25, 1871	114, 033
Plow, Gang	J. S. Padon	Lebanon, Ill	Sept. 1, 1863	39, 743
Plow, Gang	J. S. Padon	Summerfield, Ill	Sept. 12, 1865	49, 911
Plow, Gang	S. E. Parr	Smithville, Ill	Jan. 28, 1873	135, 357
Plow, Gang	W. Parrish	Dayton, Oreg	June 13, 1871	115, 980
Plow, Gang	W. Parrish	Dayton, Oreg	June 25, 1872	128, 245
Plow, Gang	L. M. Patterson	Jordon's Grove, Ill	Mar. 20, 1866	53, 328
Plow, Gang	T. Peppler	Hightstown, N. J	Dec. 20, 1870	110, 281
Plow, Gang	H. L. Perry	Aurora, N. Y	Apr. 30, 1867	64, 358
Plow, Gang	H. L. Perry	Aurora, N. Y	Nov. 19, 1867	71, 053
Plow, Gang	D. Petticrew	Westville, Ohio	Jan. 21, 1868	73, 644
Plow, Gang	J. C. Pfeil	Arenzville, Ill	Jan. 31, 1865	46, 137
Plow, Gang	J. C. Pfeil	Arenzville, Ill	June 25, 1867	66, 039
Plow, Gang	J. F. Porter and A. Norton	Tidioute, Pa	July 14, 1868	79, 917
Plow, Gang	J. Price	San Leandro, Cal	Mar. 22, 1870	101, 038
Plow, Gang	J. L. Purcell	Thompson, Ill	Oct. 18, 1870	108, 516
Plow, Gang	W. B. Quick	Belleville, Ill	June 6, 1871	115, 639
Plow, Gang	G. Quirin and L. Berkel	Smithton, Ill	Nov. 8, 1864	44, 971
Plow, Gang	W. B. Ready	Sacramento, Cal	Dec. 3, 1861	33, 851
Plow, Gang	F. F. Reynolds	Bethany, Ga	Aug. 24, 1869	94, 133
Plow, Gang	W. B. Rice	Oakland, Oreg	June 29, 1869	92, 099
Plow, Gang	D. C. Riggs	Saint Joseph, Mo	July 30, 1867	67, 351
Plow, Gang	L. Roach	Covington, Ky	Mar. 16, 1858	19, 652
Plow, Gang	L. O. Rockwood	Ottawa, Ill	May 21, 1867	64, 906
Plow, Gang	W. T. Rogers	Quincy, Ill	July 24, 1866	56, 613
Plow, Gang	J. L. Runk	Nashville, Ill	Sept. 15, 1863	39, 961
Plow, Gang	J. L. Runk	Nashville, Ill	Feb. 16, 1864	41, 643
Plow, Gang	J. L. Runk, J. H. Brown, and E. M. Morgan.	Nashville, Ill	May 22, 1866	54, 963
Plow, Gang	M. Sattley	Taylorsville, Ill	Feb. 2, 1864	41, 449
Plow, Gang	M. Sattley	Taylorsville, Ill	Dec. 5, 1865	51, 358
Plow, Gang	D. A. Sears	Rockford, Ill	July 19, 1870	105, 600
Plow, Gang	G. and J. Seibert	Ashley, Ill	May 4, 1869	89, 797
Plow, Gang	J. Seibel	Manlius, Ill	May 30, 1865	47, 989
Plow, Gang	E. Sexton	Munson, Mass	Apr. 23, 1867	64, 152
Plow, Gang	T. Short	Fairmount, Ill	Jan. 10, 1865	45, 897
Plow, Gang	J. B. Skinner	Rockford, Ill	July 18, 1865	48, 846
Plow, Gang	J. B. Skinner	Rockford, Ill	Apr. 17, 1866	54, 029
Plow, Gang	A. Smith	Portland, Me	Dec. 1, 1868	84, 652
Plow, Gang	A. Smith and W. P. Watson	Portland, Oreg	Aug. 11, 1868	80, 838
Plow, Gang	A. and T. S. Smith	Troy, Ill	Mar. 4, 1856	14, 373
Plow, Gang	F. P. Smith	Petaluma, Cal	Sept. 15, 1868	82, 165
Plow, Gang	F. P. Smith	Petaluma, Cal	Sept. 15, 1868	82, 166
Plow, Gang	H. Smith	San Lorenzo, Cal	Aug. 22, 1865	49, 564
Plow, Gang	H. B. Smith	Tremont, Ill	Sept. 19, 1871	119, 192
Plow, Gang	H. C. Smith	Ridge Farm, Ill	Jan. 2, 1866	51, 875
Plow, Gang	J. D. Smith	Peoria, Ill	June 26, 1866	55, 921
Plow, Gang	N. Spencer, jr	Eagle Point, Ill	Sept. 27, 1870	107, 829
Plow, Gang	H. P. Stafford	Decatur, Ill	Aug. 13, 1867	67, 814
Plow, Gang	P. H. Standish	Martinez, Cal	Sept. 1, 1868	81, 700
Plow, Gang	G. Steinegger	Highland, Ill	Mar. 3, 1868	75, 069
Plow, Gang	J. Stone	Plattsburgh, Mo	Sept. 20, 1864	44, 351
Plow, Gang	J. W. Sursa	San Leandro, Cal	May 21, 1867	64, 923
Plow, Gang	J. W. Sursa	San Leandro, Cal	Apr. 5, 1870	101, 539
Plow, Gang	J. W. Sursa	San Leandro, Cal	Mar. 7, 1871	112, 394
Plow, Gang	J. Sutter	Saint Louis County, Mo	June 16, 1857	17, 591
Plow, Gang	Z. T. Sweet	Eugene City, Oreg	Jan. 5, 1869	85, 621
Plow, Gang	J. N. Thompson and W. Kenady	Belpassi, Oreg	Oct. 5, 1869	95, 539
Plow, Gang	J. Totten	Adams, Ill	Feb. 2, 1869	86, 472
Plow, Gang	J. E. Travis	Greenville, Ill	May 9, 1865	47, 686
Plow, Gang	J. E. Travis	Greenville, Ill	Feb. 13, 1866	52, 651
Plow, Gang	J. W. Treadway	Crown Point Centre, N. Y	Oct. 11, 1870	108, 214
Plow, Gang	J. Tustin	Portland, Oreg	Feb. 2, 1869	86, 608
Plow, Gang	A. Walker	Claremont, N. H	July 28, 1868	80, 521
Plow, Gang	E. W. Walton	San Leandro, Cal	July 19, 1870	105, 528
Plow, Gang	J. T. Watkins	Santa Clara, Cal	July 14, 1868	79, 880
Plow, Gang	W. M. Watson	Tonica, Ill	May 1, 1866	54, 453
Plow, Gang	H. J. Wattles	Rockford, Ill	Sept. 10, 1867	68, 673
Plow, Gang	S. Way	La Porte, Ind	Jan. 21, 1868	73, 480
Plow, Gang	T. U. Webb	Springfield, Ill	Dec. 29, 1868	85, 496
Plow, Gang	H. Webster	Beetown, Wis	Jan. 31, 1865	46, 164
Plow, Gang	L. T. Webster	Northfield, Mass	Aug. 1, 1871	117, 707
Plow, Gang	S. Welch	Belpassi, Oreg	Feb. 1, 1870	99, 379
Plow, Gang	G. Wharton	Jerseyville, Ill	July 14, 1868	80, 039
Plow, Gang	G. Wharton	Jerseyville, Ill	Apr. 13, 1869	88, 998
Plow, Gang	T. Wiard	East Avon, N. Y	Nov. 24, 1843	3, 356
Plow, Gang	J. C. Wilson	Cedar Hill, Tex	July 3, 1860	29, 027
Plow, Gang	L. Wolf	Saint Louis, Mo	Nov. 24, 1863	40, 721
Plow, Gang	S. A. Worthen	Thompson, Ill	Aug. 1, 1871	117, 711
Plow, Gang	G. W. N. Yost	Cincinnati, Ohio	Apr. 27, 1858	20, 122
Plow, Gang and subsoil	J. L. Bond	Saint Louis, Mo	Jan. 31, 1871	111, 428
Plow, Gang and subsoil	J. L. Bond	Marshalltown, Iowa	Apr. 4, 1871	113, 390
Plow, Gang and subsoil	R. L. Dodge and E. M. Walker	Gallatin, Mo	Oct. 16, 1866	58, 788
Plow, Gang and trench	J. G. Robinson	Springfield, Ill	Mar. 30, 1869	88, 413
Plow, Garden	G. W. Cole	Farmington, Ill	Aug. 3, 1869	93, 277
Plow, Garden	J. M. Crawford	Newcastle, Ky	Sept. 11, 1866	57, 871
Plow, Garden	W. F. Pagett and S. H. Gard	Springfield, Ohio	Aug. 31, 1869	94, 434
Plow, Garden	R. Scott	La Porte, Ind	Sept. 7, 1869	94, 657
Plow, Garden hand	J. Starr	Grand Rapids, Mich	Aug. 11, 1868	80, 842
Plow, Grading	A. P. Hopkins	Bentleyville, Pa	May 28, 1872	127, 239
Plow, Grading and ditching	S. N. Caldwell	Pilot Grove, Ind	Aug. 2, 1870	105. 902
Plow, Grape	R. Hardenbrook	Bath, N. Y	Aug. 27, 1867	68, 190
Plow grinder and polisher	M. Devault	Charleston, Ill	May 3, 1870	102, 508
Plow, Grooving or panel	W. S. Loughborough	Rochester, N. Y	May 3, 1864	42, 585
Plow guard, Shovel	W. J. M. Batchelder	Dayton, Ohio	Mar. 26, 1867	63, 134
Plow, Hand	W. Gowen	Bartlett, Tenn	Aug. 10, 1869	93, 431

Index of patents issued from the United States Patent Office from 1790 *to* 1873, *inclusive*—Continued.

Invention.	Inventor.	Residence.	Date.	No.
Plow, Hand	F. Keefer	Greenfield, Ind	May 14, 1867	64, 771
Plow, Hand	L. McWhinney	Winterset, Iowa	Sept. 28, 1869	95, 367
Plow, Hand	N. Rue	Harrodsburgh, Ky	Mar. 8, 1871	112, 497
Plow, Hand	J. Winecoff	Berlin, Pa	Nov. 24, 1868	84, 331
Plow, Hand	W. B. Winton	Marion, Iowa	Sept. 1, 1868	81, 857
Plow, Hand garden	W. D. Smith	Homerville, Ga	Aug. 22, 1871	118, 290
Plow, Hand garden	J. Von Achen	Bloomfield, Iowa	Mar. 6, 1866	53, 061
Plow-handle	S. J. Olmsted	Binghamton, N. Y	Apr. 5, 1864	42, 220
Plow-handle	J. T. Raftery	El Dara, Ill	Apr. 29, 1873	138, 434
Plow-handle	G. Watt	Richmond, Va	May 7, 1867	64, 464
Plow-handle	W. E. Wyche	Brookville, N. C	Oct. 25, 1870	108, 671
Plow-handles, Device for bending	W. Heylman	Peoria, Ill	May 4, 1869	89, 764
Plow-handles, Forming	G. W. Matthews	York, Pa	May 24, 1859	24, 131
Plow-handles, Machine for bending	B. F. Avery	Louisville, Ky	Jan. 22, 1856	14, 130
Plow-handles, Machine for bending	J. G. Ernst	York, Pa	Aug. 2, 1859	24, 971
Plow-handles, Machine for bending	G. V. Griffith	Fort Wayne, Ind	July 13, 1869	92, 604
Plow-handles, Machine for bending	J. Woodburn and S. F. Smith	Saint Louis, Mo., and Indianapolis, Ind.	Dec. 13, 1870	110, 184
Plow-handles, Machine for forming	W. A. Ellis	Ashtabula, Ohio	July 31, 1866	56, 736
Plow-handles, Machine for forming	G. V. Griffith	Fort Wayne, Ind	Aug. 17, 1869	93, 710
Plow-handles, Machine for forming	E. G. Matthews	Oakham, Mass	Apr. 25, 1871	114, 170
Plow-handles, &c., Machine for polishing	T. Blanchard	Boston, Mass	Feb. 7, 1854	10, 497
Plow-handles, Machine for turning	F. C. Coppage	Terre Haute, Ind	July 23, 1861	32, 910
Plow-handles, Machinery for making	A. B. Farquhar	York, Pa	July 1, 1873	140, 355
Plow-handles, Manufacture of	T. E. C. Brinly	Louisville, Ky	Dec. 24, 1867	72, 596
Plow, harrow, cultivator, and roller, Combined	J. Johnston	Pemberton, Ohio	July 2, 1867	66, 240
Plow, Hilling	G. Notman	Deerfield, Ohio	Mar. 2, 1869	87, 356
Plow, Hill-side	H. B. Abbott	Felicity, Ohio	Mar. 3, 1868	75, 104
Plow, Hill side	H. S. Akins	Speedsville, N. Y	Aug. 31, 1858	21, 306
Plow, Hill-side	W. H. Babbit	Greene County, Pa	Nov. 6, 1847	5, 363
Plow, Hill-side	W. H. Babbit	Waynesburgh, Pa	Aug. 16, 1853	9, 944
Plow, Hill-side	F. G. Bakes	Vevay, Ind	Jan. 1, 1867	60, 820
Plow, Hill-side	I. Brewster	Stamford, N. Y	Nov. 14, 1848	5, 922
Plow, Hill-side	M. L. Chase	Frankfort, Me	July 16, 1850	7, 505
Plow, Hill-side	J. P. Cobbs	Nelson County, Va	Oct. 1, 1830	....
Plow, Hill-side	H. Cox	Peach Bottom, Va	Nov. 7, 1848	5, 909
Plow, Hill-side	S. Dennis, jr	Jasper, N. Y	Mar. 2, 1858	19, 496
Plow, Hill-side	A. Eldred	Little Falls, N. Y	July 24, 1849	6, 606
Plow, Hill-side	R. H. Ewing	Clives, Ohio	Nov. 27, 1860	30, 726
Plow, Hill-side	F. Feldhaus	Baltimore, Md	Oct. 8, 1867	69, 555
Plow, Hill-side	D. Gochnour, jr	Conemaugh, Pa	June 12, 1840	1, 632
Plow, Hill-side	S. Hall	Manchester, Pa	Jan. 4, 1853	9, 519
Plow, Hill-side	A. J. Hardin	Shelby, N. C	Oct. 6, 1857	18, 336
Plow, Hill-side	N. Harrison and J. W. H. Metcalf.	Ridgeville, Va	Oct. 11, 1853	10, 107
Plow, Hill-side	C. H. McCormick	Rockbridge County, Va	June 13, 1831	....
Plow, Hill-side	M. Merk	Rochester, N. Y	July 6, 1858	20, 812
Plow, Hill-side	G. C. Miller and R. Henry	Cincinnati, Ohio	Aug. 21, 1860	29, 708
Plow, Hill-side	D. H. B. Newcomb	Conewango, N. Y	June 21, 1853	9, 801
Plow, Hill-side	D. Robb	Sangamon County, Ill	June 26, 1849	6, 553
Plow, Hill-side	J. Rorabaugh	Luney's Creek, Va	Dec. 3, 1846	4, 870
Plow, Hill-side	A. Sanborn	Glover, Vt	June 18, 1861	32, 587
Plow, Hill-side	H. Sloop	Mount Healthy, Ohio	Jan. 28, 1868	73, 933
Plow, Hill-side	E. J. Smith and H. Griswold	Delhi, N. Y	July 25, 1848	5, 677
Plow, Hill-side	N. Staples	Patrick County, Va	Nov. 1, 1828	....
Plow, Hill-side	I. Teeter	Johnstown, Pa	Oct. 3, 1838	959
Plow, Hill-side	J. W. Thurman	Buchanan, Va	Aug. 28, 1849	6, 677
Plow, Hill-side	J. Trump	Connellsville, Pa	Sept. 9, 1845	4, 186
Plow, Hill-side	E. Van Camp	Readington, N. J	May 10, 1859	23, 964
Plow, Hill-side	J. B. Walder	Belfast, Me	June 21, 1853	9, 808
Plow, Hill-side double-nosed cast-iron	J. Shepherd	De Ruyter, N. Y	Apr. 12, 1826	....
Plow, Hinged-wing hill-side	W. B. Donald	Fairfield, Va	June 29, 1833	....
Plow-holder, Self-acting	J. L. Keasor	Laconia, N. H	Oct. 8, 1867	69, 678
Plow, Ice	J. H. Cutter	Cambridge, Mass	Feb. 20, 1872	123, 873
Plow, Ice	G. B. Gruman	Bridgefield, Conn	Apr. 2, 1872	125, 131
Plow, Iron	J. Boynton	South Coventry, Conn	July 10, 1829	....
Plow-irons, Machine for making	J. M. McGinty and T. Nolan	Moulton, Tex	Mar. 4, 1873	136, 444
Plow-irons, Machine for welding	C. G. Cross	Chicago, Ill	Apr. 30, 1872	126, 188
Plow-jointer	D. Dillenback	Galesburgh, Mich	Nov. 1, 1870	108, 771
Plow-jointer, Adjustable	J. M. Leonard	Marshall City, Mich	Dec. 20, 1870	110, 251
Plow jointer, Prairie	A. J. Spicer	Galesburgh, Mich	Feb. 14, 1871	111, 786
Plow-landside	J. Bacon	Medina, Wis	Apr. 28, 1868	77, 242
Plow-landside	A. Christ	Unity, Ohio	Sept. 18, 1849	6, 724
Plow-landsides, Manufacture of	J. Duff, H. Roberts, and G. D. Nourse.	Peoria, Ill., and Saint Louis, Mo.	Sept. 12, 1871	118, 848
Plow-lays, Forming	J. Lane	Chicago, Ill	Feb. 4, 1868	73, 983
Plow-marking attachment	C. Morris	Stockton Township, N. J	Apr. 2, 1867	63, 551
Plow, Metallic double-shovel	W. W. Love	Athens, Ohio	Oct. 19, 1869	96, 018
Plow mold-board	E. M. Bard	Philadelphia, Pa	Mar. 14, 1854	10, 629
Plow mold-board	G. A. Beard	Washington County, Md	Aug. 28, 1866	57, 463
Plow mold-board	S. H. Dwight and C. Wells	Decatur, Ill., and Pittsburgh, Pa.	May 25, 1869	90, 512
Plow mold-board	I. T. Dyer	Macon, Ga	Apr. 4, 1871	113, 642
Plow mold-board	R. Gaines and M. Scott	Fairfield, Iowa	Feb. 18, 1868	74, 679
Plow mold-board	J. Lane	Chicago, Ill	Oct. 4, 1870	107, 925
Plow mold-board	J. Oliver	South Bend, Ind	Apr. 14, 1868	76, 652
Plow mold board	G. Peacock	Selma, Ala	Aug. 29, 1871	118, 551
Plow mold-board	L. P. Rider	Munson, Ohio	Oct. 30, 1866	59, 267
Plow mold-board	L. P. Rider	Pittsburgh, Pa	Apr. 25, 1871	114, 044
Plow mold-board	T. E. Session and S. A. King	Worcester, Mass	Feb. 1, 1870	99, 516
Plow mold-board	J. Seymour	Coventry, N. Y	Mar. 24, 1868	75, 987
Plow mold-board	R. Smith	....	May 19, 1800	....
Plow mold-board	L. Witherow and D. Pierce	Philadelphia, Pa	Oct. 5, 1839	1, 357
Plow mold-board, Reversible	H. S. Akins	Berkshire, N. Y	June 16, 1857	17, 547
Plow mold-boards, Attaching handles to	C. Williams	Jackson, Miss	June 30, 1868	79, 532
Plow mold-boards, Die for making blanks for	M. W. Watson	Tonica, Ill	Mar. 15, 1870	100, 825
Plow mold-boards, Machine for bending	W. W. Skinner	Davenport, Iowa	Feb. 7, 1860	27, 078
Plow mold-boards, Machinery for swaging	H. H. May	Galesburgh, Ill	June 27, 1846	4, 609

Index of patents issued from the United States Patent Office from 1790 *to* 1873, *inclusive*—Continued.

Invention.	Inventor.	Residence.	Date.	No.
Plow mold-boards, Making	T. A. Chandler	Rockford, Ill	Dec. 4, 1855	13, 865
Plow, Molding	B. F. Avery	Louisville, Ky	Oct. 25, 1859	25, 873
Plow, Mole	J. Adair	Mendota, Ill	Mar. 27, 1860	27, 606
Plow, Mole	S. Adams	Toulon, Ill	Feb. 28, 1860	27, 283
Plow, Mole	W. B. Atkinson	Plymouth, Ill	Sept. 18, 1860	30, 036
Plow, Mole	H. Bagley	Tipton, Iowa	Sept. 18, 1860	30, 041
Plow, Mole	H. F. Baker	Centreville, Ind	Oct. 4, 1859	25, 618
Plow, Mole	M. Bales	Big Plain, Ohio	Feb. 15, 1859	22, 928
Plow, Mole	H. H. Ballard and H. McClure	Mount Pleasant, Iowa	Mar. 26, 1861	31, 771
Plow, Mole	J. R. Barnett	Galesburgh, Ill	July 9, 1872	128, 842
Plow, Mole	A. Bowers, J. H. Griggs, and J. Wilson.	Monmouth, Ill	Nov. 15, 1859	26, 082
Plow, Mole	J. Carrington	Avoca, N. Y	Mar. 29, 1859	23, 348
Plow, Mole	J. Case	Bloomington, Ill	Jan. 25, 1859	22, 701
Plow, Mole	S. A. Clemens	Rockford, Ill	July 7, 1863	39, 118
Plow, Mole	C. U. and J. H. Crandall and H. N. Hawkins.	Cameron, Ill	Aug. 23, 1859	25, 178
Plow, Mole	J. Creamer	Jeffersonville, Ohio	Feb. 19, 1867	62, 116
Plow, Mole	J. Creamer and T. W. Ricards	London, Ohio	Apr. 5, 1859	23, 452
Plow, Mole	A. Defenbaugh	Walnut Run, Ohio	Sept. 14, 1858	21, 491
Plow, Mole	A. Elmer	Shabbona Grove, Ill	Aug. 16, 1859	25, 105
Plow, Mole	J. H. Edward	Ottawa, Fla	Nov. 13, 1860	30, 625
Plow, Mole	A. Gillet	Lyndon, Ill	Jan. 15, 1861	31, 117
Plow, Mole	W. P. Goolman	Dublin, Ind	Mar. 22, 1859	23, 334
Plow, Mole	W. P. Goolman	Dublin, Ind	Dec. 13, 1859	26, 426
Plow, Mole	G. L. Griffin and J. H. Carper	Dallas City, Ill	Feb. 28, 1860	27, 285
Plow, Mole	J. A. Hammer and J. P. Gordon	Lisbon, Iowa	Feb. 5, 1861	31, 313
Plow, Mole	A. Hammond	Jacksonville, Ill	Aug. 16, 1859	25, 114
Plow, Mole	A. Hammond	Jacksonville, Ill	Apr. 10, 1860	27, 796
Plow, Mole	F. E. Hinckley	Galesburgh, Ill	Dec. 6, 1859	26, 355
Plow, Mole	I. Hobson	Stout's Grove, Ill	Sept. 6, 1859	25, 334
Plow, Mole	M. A. Howell, jr	Ottawa, Ill	Feb. 5, 1861	31, 317
Plow, Mole	R. Hussey and U. Thornburgh, sr.	Walnut Run, Ohio	Oct. 4, 1859	25, 649
Plow, Mole	H. R. Jerome	Monroeville, Ohio	Aug. 16, 1859	25, 121
Plow, Mole	S. F. Jones	Saint Paul, Ind	Oct. 25, 1859	25, 902
Plow, Mole	S. F. Jones	Saint Paul, Ind	Aug. 27, 1861	33, 149
Plow, Mole	A. M. Karr	Mount Pleasant, Iowa	July 24, 1860	29, 285
Plow, Mole	J. Lee	Galesburgh, Ill	Aug. 16, 1859	25, 127
Plow, Mole	A. Miller	Mount Pleasant, Iowa	Oct. 18, 1859	25, 845
Plow, Mole	A. Miller	Chicago, Ill	Oct. 23, 1866	59, 049
Plow, Mole	J. Morrison	De Witt, Ill	Oct. 18, 1859	25, 846
Plow, Mole	E. and W. Parish, jr	Galesburgh, Ill	Feb. 21, 1860	27, 233
Plow, Mole	D. F. Robbins and S. Morrison	De Witt, Ill	Apr. 12, 1859	23, 609
Plow, Mole	H. W. Rowland and E. Forbis	Newport, Ohio	Apr. 19, 1859	23, 745
Plow, Mole	R. P. Smith and J. R. Gates	Louisville, Ky	Jan. 7, 1862	34, 074
Plow, Mole	C. W. Stafford.	Burlington, Iowa	July 17, 1860	29, 201
Plow, Mole	O. Sturdevant	Maquon, Ill	Nov. 13, 1860	30, 659
Plow, Mole	A. L. O. Wall, G. Roberts, and M. S. Cartter.	Decatur, Ill	Apr. 3, 1860	27, 751
Plow, Mole	A. Watson	Walnut Run, Ohio	Aug. 2, 1859	24, 969
Plow, Mole	G. Whitcomb	Springfield, Ohio	Nov. 1, 1859	25, 988
Plow or cultivator, Cotton	F. M. Shields	Macon, Miss	July 16, 1867	66, 895
Plow, planter, and cultivator	E. Bourne	New Iberia, La	Sept. 3, 1870	107, 218
Plow, planter, and cultivator, Combined	T. J. Smith	Holly Springs, Miss	Jan. 25, 1870	99, 118
Plow, planter, and cultivator, Combined corn	I. A. Arthur	Sidonsburgh, Pa	Apr. 28, 1868	77, 238
Plow, Planting	A. B. Earle	Colesville, N. Y	Oct. 17, 1848	5, 858
Plow, Planting	B. Hussey	Jonesborough, Tenn	Apr. 1, 1834	....
Plow, Planting	F. Woodward	New Lisbon, N. Y	July 18, 1809	....
Plow plate and point, Shovel	H. Miller	Roadside, Va	Nov. 23, 1869	97, 213
Plow-plates, Cooling and tempering cast-steel	F. F. Smith	Collinsville, Conn	Feb. 2, 1864	41, 464
Plow-plates from molten steel, Making	F. F. Smith	Momence, Ill	Nov. 20, 1860	30, 691
Plow-point	E. Bement	Fostoria, Ohio	Sept. 16, 1862	36, 447
Plow-point	L. D. Burch	Sherburne, N. Y	Oct. 20, 1868	83, 130
Plow-point	E. C. Gero and J. N. Cooley	Kalamazoo, Mich	Mar. 2, 1869	87, 484
Plow-point	M. L. Gibbs	Canton, Ohio	June 25, 1872	128, 221
Plow-point	C. B. Kerr	Columbus, Ind	Apr. 27, 1869	89, 486
Plow-point	H. Kniphals	Davenport, Iowa	Sept. 10, 1867	68, 755
Plow-point	C. W. Saladee and T. Simpson	Newark, Ohio	Mar. 27, 1866	53, 491
Plow-point	D. J. Selden	Mount Vernon, Ohio	May 21, 1867	64, 914
Plow-point	O. O. Storle	Norway, Wis	May 25, 1869	90, 604
Plow-point, Reversible	R. M. Pattills	Cartersville, Ga	Sept. 16, 1873	142, 864
Plow-points, Attaching	G. W. Hildreth	Lockport, N. Y	Apr. 4, 1871	113, 665
Plow-points, Mold for casting	H. J. Brunner	Nazareth, Pa	Apr. 23, 1872	126, 015
Plow, Potato	H. T. Basye	Dyersburgh, Tenn	June 24, 1873	140, 240
Plow, Potato	A. Eldred	Oppenheim, N. Y	July 24, 1846	4, 656
Plow, Potato	A. Leonard	Newell's Run, Ohio	Aug. 23, 1870	106, 702
Plow, Potato	J. P. Stanton	Pedricktown, N. J	Sept. 3, 1867	68, 392
Plow, Potato	M. Stoll	Conestoga Township, Pa	Sept. 8, 1868	82, 043
Plow, Potato	R. P. Terhune and B. J. Romaine.	Hackensack, N. J	Jan. 26, 1869	86, 190
Plow, Potato and corn	C. F. Noftz	Toledo, Ohio	Nov. 10, 1868	83, 992
Plow, potato-planter, and seeder, Combined	S. Shetter	New Cumberland, W. Va	Dec. 31, 1867	72, 922
Plow, Prairie	C. K. Bartlett	Geneseo, Ill	Mar. 28, 1842	2, 515
Plow, Prairie	J. Frye	Springfield, Ill	Mar. 31, 1857	16, 913
Plow, Prairie	M. Turley	Galesburgh, Ill	Dec. 9, 1856	16, 216
Plow-press and drill	T. E. C. Brinly	Simpsonville, Ky	Sept. 21, 1858	21, 547
Plow, Railway	C. Medbury and T. Wyatt	Providence, R. I	Jan. 2, 1866	51, 850
Plow, Railway snow	J. R. Adams	Cisco, Cal	Jan. 14, 1868	73, 276
Plow, Railway snow	J. W. Arnold	Boston, Mass	May 13, 1873	138, 836
Plow, Railway snow	J. K. Babcock	Honeoye Falls, N. Y	Feb. 16, 1858	19, 339
Plow, Railway snow	O. D. Baird	Newark, N. Y	Apr. 22, 1873	138, 116
Plow, Railway snow	P. Boyden	Sandy Creek, N. Y	Apr. 29, 1862	35, 074
Plow, Railway snow	J. C. Carncross	Philadelphia, Pa	Jan. 16, 1866	52, 028
Plow, Railway snow	H. H. Clemons	Oshkosh, Wis	Sept. 17, 1867	68, 845
Plow, Railway snow	T. A. Davies	New York, N. Y	Mar. 11, 1873	136, 709
Plow, Railway snow	W. Davis	Five Corners, N. Y	Mar. 18, 1873	136, 975
Plow, Railway snow	S. L. Denney	Christiana, Pa	Sept. 15, 1863	39, 894

Index of patents issued from the United States Patent Office from 1790 *to* 1873, *inclusive*—Continued.

Invention.	Inventor.	Residence.	Date.	No.
Plow, Railway snow	T. Dougherty	Philadelphia, Pa	Dec. 27, 1870	110, 446
Plow, Railway snow	C. L. Garfield	Albany, N. Y	Aug. 24, 1869	94, 102
Plow, Railway snow	J. A. Gregg	Derry, N. H	June 13, 1846	4, 568
Plow, Railway snow	S. G. Hadley	Cape Vincent, N. Y	Dec. 3, 1867	71, 607
Plow, Railway snow	A. Hall and S. Sturtevant	South Paris, Me	Feb. 28, 1854	10, 568
Plow, Railway snow	H. T. Hartman	Lexington, Va	Feb. 16, 1858	19, 361
Plow, Railway snow	S. W. Hemenway	Lansing, Iowa	Dec. 24, 1872	134, 202
Plow, Railway snow	C. L. Heywood	Boston, Mass	Jan. 2, 1866	51, 829
Plow, Railway snow	A. Hotchkiss	Sharon Valley, Conn	Dec. 22, 1857	18, 903
Plow, Railway snow	J. B. Hulbert and J. Anderson	Hermon, N. Y	Nov. 5, 1872	132, 834
Plow, Railway snow	W. S. Huntington	Andrusville, N. Y	June 21, 1859	24, 463
Plow, Railway snow	B. A. Johnson	North Auburn, Me	Jan. 13, 1863	37, 420
Plow, Railway snow	J. Jones and T. G. Eiswald	Providence, R. I	Sept. 15, 1868	82, 226
Plow, Railway snow	J. S. Munson	Jamaica Plain, Mass	Dec. 10, 1872	133, 792
Plow, Railway snow	W. Rhoads	Baltimore, Md	Apr. 19, 1859	23, 709
Plow, Railway snow	S. Richards	Philadelphia, Pa	May 13, 1856	14, 886
Plow, Railway snow	E. Robbins	Worcester, Mass	Mar. 24, 1868	75, 979
Plow, Railway snow	T. L. Shaw	Omaha, Nebr	Dec. 14, 1869	97, 971
Plow, Railway snow	P. A. Smith	New York, N. Y	Feb. 25, 1873	136, 272
Plow, Railway snow	W. H. Van Gieson	Whitewater, Wis	Feb. 25, 1873	136, 346
Plow, Railway snow	W. Walker	Fort Bridger, Wyo	Feb. 4, 1873	135, 499
Plow, Railway snow	W. Wheelock	Decorah, Iowa	July 8, 1873	140, 663
Plow, Railway snow	R. Wilson	Des Moines, Iowa	Apr. 5, 1870	101, 693
Plow, Railway snow	C. L. Wood	Calais, Me	Nov. 4, 1873	144, 377
Plow-regulator	H. Sprague	Riga, N. Y	Dec. 14, 1852	9, 473
Plow, Reversible	C. F. Barager	Candor, N. Y	June 3, 1862	35, 432
Plow, Reversible	E. C. Hodge	Oneonta, N. Y	Nov. 1, 1870	108, 907
Plow, Reversible	G. W. Howe	Vineland, N. J	July 29, 1873	141, 222
Plow, Reversible	E. Jennings	Candor, N. Y	June 27, 1871	116, 314
Plow, Reversible	J. W. Jones	Thompson, Ill	Aug. 24, 1869	94, 001
Plow, Reversible	G. W. Thompson	Ripley, Ohio	Jan. 3, 1871	110, 692
Plow, Reversible side-hill	C. B. Pettingill	Hebron, Me	Aug. 15, 1871	118, 849
Plow, Reversing hill-side	C. Delano	Livermore, Me	July 5, 1833	....
Plow, Revolving	M. A. and I. A. Cravath	Bloomington, Ill	July 16, 1867	66, 802
Plow, Revolving	W. J. Dawson	Brookfield, Mo	Sept. 21, 1869	95, 005
Plow, Revolving spade	J. W. Milroy	Galveston, Ind	May 12, 1868	77, 830
Plow, Ridge	W. Atwood and D. Hamblet	Cornish, N. H	Dec. 24, 1833	....
Plow, Riding	B. C. Hoyt	Fort Atkinson, Wis	May 23, 1871	115, 057
Plow, Right and left	G. Doller	Fredericktown, Md	Aug. 20, 1827	....
Plow roller-attachment	C. M. Young	Meadville, Pa	June 27, 1871	116, 388
Plow roller-cutter	J. W. Lewis	Oregon City, Oreg	June 22, 1869	91, 644
Plow roller-cutter	R. Newton	Jerseyville, Ill	Aug. 10, 1869	93, 636
Plow roller-cutter	J. H. Sherman	Galesburgh, Ill	July 30, 1867	67, 222
Plow roller-wheel	L. B. Pitcher	Salina, N. Y	Nov. 19, 1867	71, 057
Plow root-cutter	J. W. Baker	Elkton, Md	May 23, 1871	115, 013
Plow, Rotary	H. Berkstresser	Quaker Bottom, Ohio	Sept. 10, 1867	68, 689
Plow, Rotary	W. E. Bleecker	Brooklyn, N. Y	July 16, 1872	129, 266
Plow, Rotary	E. T. Russell	Indianapolis, Ind	Sept. 3, 1867	68, 410
Plow, Rotary	L. H. Colborn	Chicago, Ill	Apr. 3, 1866	53, 577
Plow, Rotary	C. Comstock	Milwaukee, Wis	May 13, 1862	35, 218
Plow, Rotary	C. T. Elliston	Clinton, Mo	Aug 26, 1873	142, 094
Plow, Rotary	J. W. Haggard and G. Bull	Bloomington, Ill	Feb. 13, 1855	12, 387
Plow, Rotary	A. B. Hoffmeyer and J. Schmidt	Copenhagen, Denmark	Dec. 2, 1873	145, 177
Plow, Rotary	N. T. Judd	Washington, D. C	Nov. 29, 1870	109, 741
Plow, Rotary	D. Myers	Chicago, Ill	Feb. 6, 1866	52, 496
Plow, Rotary	J. Young	Jefferson, Me	July 11, 1848	5, 665
Plow, Rotary-cutter	J. C. Pfeil	Argenzville, Ill	Apr. 7, 1868	76, 343
Plow, Rotary-cutter	T. J. Tuthill	Elmira, N. Y	Feb. 6, 1849	6, 091
Plow, Rotary power	J. Tranter, J. Kinsey, and J. M. Carr.	Cincinnati, Ohio	Sept. 20, 1870	107, 639
Plow, Rotary steam	J. T. Wilson	Rochester, N. Y	Aug. 16, 1870	106, 441
Plow, Rotary subsoil	J. R. Morris	Houston, Tex	Mar. 28, 1871	113, 190
Plow, Rotating	J. Hoadley	Zanesville, Ohio	Aug. 23, 1864	43, 913
Plow-scraper	H. Leach	Camanche, Iowa	Aug. 26, 1873	142, 113
Plow, seeder, and roller, Combined	O. B. Cheatham	Henderson, Ky	Jan. 9, 1872	12[illegible], 567
Plow, Seeding	C. Atkinson	Vermont, Ill	Apr. 10, 1860	27, 766
Plow, Seeding	J. Peeler	Tallahassee, Fla	Apr. 24, 1860	28, 009
Plow, Seeding	W. P. Penn	Belleville, Ill	Jan. 15, 1861	31, 129
Plow, Seeding	W. P. Penn	Belleville, Ill	July 30, 1861	32, 924
Plow, Seeding	J. S. Snider	Lancaster, Ohio	Aug. 2, 1859	24, 959
Plow, Self-sharpening	W. Beach	Philadelphia, Pa	Aug. 26, 1823	....
Plow, Self-sharpening	W. Beach	Philadelphia, Pa	Apr. 9, 1824	....
Plow, Self-sharpening	W. Beach	Philadelphia, Pa	Dec. 28, 1832	....
Plow, Self-sharpening	R. B. Chenoweth	Baltimore, Md	Mar. 17, 1834	....
Plow, Self-sharpening	C. and O. Evans	Philadelphia, Pa	Apr. 14, 1825	....
Plow, Self-sharpening	J. B. Norton	Utica, N. Y	Apr. 27, 1832	....
Plow, Self-sharpening	J. Ormiston	Centre, Ohio	Mar. 17, 1838	638
Plow, Self-sharpening	J. W. Post	Baltimore, Md	Oct. 8, 1838	970
Plow, Self-sharpening	I. Snider	Mount Pleasant, Pa	July 29, 1837	295
Plow, Self-sharpening	B. Woodcock	Mount Pleasant, Pa	Jan. 26, 1832	....
Plow, Self-sharpening horizontal	C. H. McCormick	Rockbridge County, Va	Nov. 19, 1833	....
Plowshare	T. K. Bronson	East Avon, N. Y	Apr. 3, 1866	53, 565
Plowshare	B. Carpenter	Somer's Town, N. Y	July 1, 1822	....
Plowshare	G. W. Cooper	Ogeechee, Ga	Dec. 8, 1868	84, 798
Plowshare	J. S. Hall	Pittsburgh, Pa	July 8, 1862	35, 819
Plowshare	B. Harvey	West Bloomfield, N. Y	Dec. 10, 1872	133, 854
Plowshare	J. Lane	Chicago, Ill	Jan. 10, 1871	110, 860
Plowshare	A. Maschka	Chicago, Ill	Nov. 14, 1865	50, 942
Plowshare	W. D. Mendenhall	Farmington, Ill	June 4, 1867	65, 412
Plowshare	L. M. Reed	Troy, Ohio	Jan. 5, 1869	85, 533
Plowshare-blanks, Plate for	W. Parlin	Canton, Ill	July 6, 1869	92, 349
Plowshare, Cast-iron	J. H. Conklin	Peekskill, N. Y	Jan. 13, 1830	....
Plowshare, Chilling	J. Oliver and H. Little	South Bend, Ind	June 30, 1857	17, 694
Plowshare, colter, and mold-board	W. Holt	Buffalo, N. Y	Aug. 27, 1835	....
Plowshare, Extension	G. W. Thorp	Columbus, Kans	Jan. 3, 1871	110, 693
Plowshare forging, bending, and shaping machine	J. S. Hall	Pittsburgh, Pa	Oct. 14, 1862	36, 691
Plowshare, Shifting	J. Wood	Castleton, N. Y	Feb. 1, 1821	....
Plowshares and land-sides, Blank for	D. H. Rowe	Martinsville, Ill	Apr. 23, 1872	125, 909

Index of patents issued from the United States Patent Office from 1790 *to* 1873, *inclusive*—Continued.

Invention.	Inventor.	Residence.	Date.	No.
Plowshares, Apparatus for forming	E. Ball, jr	Canton, Ohio	Feb. 20, 1872	123, 805
Plowshares by rolling out, Making	J. Potts	Montgomery County, Pa	Nov. 7, 1818	....
Plowshares, Casting	J. Hunton	Haversack, N. J	Mar. 24, 1868	75, 919
Plowshares, Hardening edge of	R. L. and E. A. Stevens	Hoboken, N. J	Apr. 23, 1821	....
Plowshares, Machine for manufacturing	D. H. Rowe	Martinsville, Ill	Apr. 23, 1872	125, 908
Plowshares, Manufacture of	J. Lane	Chicago, Ill	Aug. 30, 1870	106, 838
Plowshares, Mold for casting	C. H. Brady	Mount Joy, Pa	July 1, 1862	35, 744
Plowshares to plow-stock, Fastening	W. G. Beckwith	Lowndesborough, Ala	Nov. 2, 1869	96, 383
Plow, Shield	M. Kirkham	Eminence Post-Office, Ind	May 19, 1868	78, 100
Plow shield, Corn	D. F. Brown and E. C. Brown	Champaign, Ill., and Crawfordsville, Ind.	Sept. 21, 1869	95, 079
Plow shield, Corn	J. Fox	Homer, Ind	Feb. 2, 1869	86, 521
Plow, Shifting-shovel	A. Snyder	Packard, Ohio	Dec. 6, 1870	109, 960
Plow, Shovel	I. A. Benedict	West Springfield, Pa	Dec. 13, 1870	109, 999
Plow, Shovel	F. H. Bowlds	Fairfield, Ky	Feb. 28, 1871	112, 212
Plow, Shovel	J. C. Boyd	Milroy, Ind	Oct. 9, 1866	58, 581
Plow, Shovel	J. F. Cameron	Livingston County, Mo	Mar. 6, 1860	27, 347
Plow, Shovel	H. C. Chandler	Erie Township, Ind	Jan. 14, 1868	73, 231
Plow, Shovel	J. C. Daugherty	Bridgeport, Ky	Oct. 30, 1866	59, 191
Plow, Shovel	P. Dennis	Bemus Heights, N. Y	Feb. 23, 1858	19, 412
Plow, Shovel	P. Dennis	Schuylerville, N. Y	June 19, 1866	55, 630
Plow, Shovel	D. Eberly	Waynesville, Ohio	Dec. 22, 1857	18, 894
Plow, Shovel	W. B. Evans	Bracken County, Ky	Oct. 27, 1868	83, 481
Plow, Shovel	C. Ford	Forest City, Ill	May 2, 1865	47, 536
Plow, Shovel	J. H. Forman	Sharon, Ala	Feb. 10, 1852	8, 721
Plow, Shovel	R. J. Gatling	Murfreesborough, N. C	May 29, 1847	5, 130
Plow, Shovel	D. Gilbert	Carbondale, Ill	June 25, 1867	66, 144
Plow, Shovel	F. Goss	Wexford, Pa	Dec. 28, 1869	98, 251
Plow, Shovel	W. R. Harmon	Unionport, Ohio	Sept. 17, 1867	68, 982
Plow, Shovel	S. W. Jackson	Baldville, Ohio	Feb. 19, 1867	62, 272
Plow, Shovel	A. Jennings	West Cairo, Ohio	Aug. 18, 1868	81, 173
Plow, Shovel	G. Jennings	West Cairo, Ohio	Aug. 24, 1869	93, 997
Plow, Shovel	W. Johnson and M. Ranney	Northfield, Ohio	May 6, 1873	138, 656
Plow, Shovel	E. H. Kamerer	Greentown, Ohio	Oct. 19, 1869	96, 010
Plow, Shovel	E. Knepper	Columbus, Ohio	Apr. 6, 1869	88, 723
Plow, Shovel	J. Lattimer	Chattoogaville, Ga	Mar. 16, 1852	8, 802
Plow, Shovel	W. H. Luce	Hampton, Ill	Sept. 18, 1866	58, 112
Plow, Shovel	L. Luppen	Pekin, Ill	Oct. 11, 1870	108, 272
Plow, Shovel	L. Luppen	Pekin, Ill	Oct. 11, 1870	108, 273
Plow, Shovel	L. Luppen	Pekin, Ill	Oct. 11, 1870	108, 274
Plow, Shovel	L. Luppen	Pekin, Ill	Oct. 11, 1870	108, 275
Plow, Shovel	L. Luppen	Pekin, Ill	Oct. 11, 1870	108, 276
Plow, Shovel	B. F. McCollester	California, Mo	Aug. 18, 1868	81, 188
Plow, Shovel	T. Meikle	Louisville, Ky	Dec. 5, 1871	121, 535
Plow, Shovel	J. Meyer	Bloom Township, Ohio	Aug. 18, 1868	81, 189
Plow, Shovel	J. J. Mitchell	Livingston, Ala	Aug. 5, 1873	141, 583
Plow, Shovel	I. Mosher and W. Eddy	Mosherville and Union Village, N. Y.	Mar. 31, 1863	38, 056
Plow, Shovel	W. C. Pagett	Greene County, Ohio	Oct. 17, 1842	2, 818
Plow, Shovel	W. F. Pagett	Stone Bridge, Va	Mar. 30, 1852	8, 842
Plow, Shovel	P. B. Parcell	Ashmore, Ill	Oct. 4, 1870	108, 047
Plow, Shovel	S. W. Pope	Louisville, Ky	Apr. 4, 1871	113, 341
Plow, Shovel	S. Riley	Northcutt's Store, Ky	June 8, 1869	91, 164
Plow, Shovel	G. H. Smith	Des Moines, Iowa	Apr. 9, 1872	125, 494
Plow, Shovel	M. E. Stanger	Wheeling, Ill	Sept. 24, 1867	69, 137
Plow, Shovel	U. T. Stewart	Fayette County, Tenn	Mar. 30, 1869	88, 344
Plow, Shovel	T. Terrell	Springhill, Ohio	May 8, 1866	54, 623
Plow, Shovel	R. Trowbridge	Waterloo, Iowa	July 2, 1867	66, 267
Plow, Shovel	M. D. Wells	Morgantown, Va	May 9, 1846	4, 500
Plow, Shovel	A. Wilcox	Maquoketa, Iowa	May 21, 1867	64, 927
Plow, Shovel	S. H. Wooldridge	Venice, Ill	Dec. 27, 1864	45, 664
Plow, Shovel-cutter	S. Wilson	Darlington, S. C	Feb. 6, 1830	....
Plow, Side-hill	D. C. Day	San José, Cal	July 18, 1871	117, 054
Plow, Side-hill	P. H. Flansburgh	Eden Township, Cal	June 25, 1867	66, 012
Plow, Side-hill	G. W. Leonard	Middle Valley, Pa	June 20, 1871	116, 070
Plow, Side-hill	D. A. Manuel	Napa City, Cal	May 5, 1868	77, 630
Plow, Side-hill	E. McKesson	Philip's Mills, Pa	Jan. 17, 1865	45, 929
Plow, Side-hill	J. Rich, jr	Troy, N. Y	Feb. 10, 1825	....
Plow, Side-hill	I. and H. H. Scoville	Oakland, Cal	May 4, 1869	89, 796
Plow, Side-hill	H. B. Smith	Springfield, Mass	Mar. 7, 1865	46, 716
Plow, Side-hill	N. Vars	New Market, N. J	Jan. 10, 1865	45, 882
Plow, Snow	W. N. Ball	La Porte, Ind	Feb. 14, 1865	46, 322
Plow, Snow	A. L. Bausman	Minneapolis, Minn	Aug. 8, 1865	49, 216
Plow, Snow	G. Beer	Grafton, Wis	Dec. 7, 1869	97, 474
Plow, Snow	T. S. Brown	Greenfield, Mich	Sept. 12, 1871	118, 784
Plow, Snow	R. Bustin	Saint John, New Brunswick.	May 4, 1869	89, 554
Plow, Snow	A. M. Butts	Waterbury, Conn	Oct. 25, 1870	108, 565
Plow, Snow	R. Cassaday and D. Clark	Buffalo, N. Y	Feb. 26, 1861	31, 529
Plow, Snow	T. C. Churchman	Sacramento, Cal	June 14, 1870	104, 113
Plow, Snow	M. A. and I. M. Cravath	Bloomington, Ill	Oct. 26, 1869	96, 205
Plow, Snow	J. N. Drake	Liberty, Mich	July 12, 1870	105, 185
Plow, Snow	A. Dunbar	New York, N. Y	Oct. 18, 1870	108, 338
Plow, Snow	C. L. Ericzon	Salt Lake City, Utah	Oct. 5, 1869	95, 449
Plow, Snow	N. S. Green	Weelaunee Post-Office, Wis	Aug. 18, 1868	81, 081
Plow, Snow	H. Harris	Circleville, Ohio	Nov. 24, 1868	84, 357
Plow, Snow	R. C. Harris	Maple Green, New Brunswick.	Sept. 20, 1870	107, 485
Plow, Snow	R. S. Harris	Dubuque, Iowa	July 23, 1867	67, 049
Plow, Snow	O. Heggem	Chicago, Ill	Mar. 19, 1872	124, 739
Plow, Snow	C. F. Hornbeck and W. J. Carns	Slaterville, N. Y	Dec. 6, 1870	109, 819
Plow, Snow	A. H. Jackson	Bear Valley, Cal	Oct. 19, 1869	96, 006
Plow, Snow	F. L. Knapp	Gosport, N. Y	Feb. 23, 1858	19, 426
Plow, Snow	S. Lewis	Brooklyn, N. Y	July 14, 1868	79, 913
Plow, Snow	C. Lusted	New York, N. Y	Jan. 7, 1868	73, 189
Plow, Snow	R. B. Nevens	Lowell, Mass	Sept. 18, 1866	58, 124
Plow, Snow	A. Nutting	Quincy, Mass	Dec. 10, 1867	71, 902
Plow, Snow	J. H. Pawling	Philadelphia, Pa	Mar. 9, 1858	19, 577

Index of patents issued from the United States Patent Office from 1790 *to* 1873, *inclusive*—Continued.

Invention.	Inventor.	Residence.	Date.	No.
Plow, Snow	G. Place	New York, N. Y	Feb. 25, 1868	74, 844
Plow, Snow	W. A. Plantz	Iowa Falls, Iowa	Jan. 5, 1869	85, 691
Plow, Snow	E. A. Putnam	Oakfield, Wis	Sept. 19, 1865	50, 031
Plow, Snow	S. Richards	Philadelphia, Pa	Apr. 13, 1858	19, 950
Plow, Snow	S. Richards	Philadelphia, Pa	June 11, 1867	65, 605
Plow, Snow	S. Richards	Philadelphia, Pa	Dec. 3, 1867	71, 642
Plow, Snow	W. J. Roberts	Cold Spring, N. Y	Sept. 23, 1873	143, 095
Plow, Snow	J. Sheridan	Saint Louis, Mo	Apr. 22, 1862	35, 066
Plow, Snow	J. Sheridan	Saint Louis, Mo	Apr. 19, 1864	42, 411
Plow, Snow	M. B. Spofford	Warsaw, N. Y	July 19, 1859	24, 829
Plow, Snow	F. J. Steinhauser	Lancaster, Pa	Feb. 19, 1861	31, 491
Plow, Snow	D. D. Stilwell	Philadelphia, Pa	Dec. 22, 1846	4, 898
Plow, Snow	A. Stutzman	Somerset, Pa	May 9, 1871	114, 623
Plow, Snow	B. B. Sweet and J. Noble	Wilmington, Del., and Philadelphia, Pa.	Oct. 14, 1873	143, 731
Plow, Snow	C. W. Tierney	Altoona, Pa	Mar. 16, 1869	87, 989
Plow, Snow	P. Vonlackum	Elba, Minn	Nov. 19, 1867	71, 249
Plow, Snow	W. Y. Warner	Wilmington, Del	May 19, 1868	78, 161
Plow, Snow	J. B. Williams	Madison, Wis	Sept. 20, 1864	44, 360
Plow, Snow	D. L. Winsor	Cambridge, Mass	July 16, 1867	66, 926
Plow, Snow	J. S. Zane	Pleasant Plains, Ill	Nov. 19, 1867	71, 111
Plow, Spade	E. Harris	Princeton, Ill	May 22, 1860	28, 368
Plow, Spade	D. Russell	Drewersburgh, Ind	Feb. 27, 1855	12, 466
Plow spring, Gang	H. N. Dalton	Pacheco, Cal	Oct. 5, 1869	95, 437
Plow, Steam	I. S. Allen, M. P. Browen, and C. W. Moultbrop.	San Francisco, Cal	Nov. 25, 1873	144, 820
Plow, Steam	S. K. Bassett	Galesburgh, Ill	Feb. 8, 1859	22, 848
Plow, Steam	W. Beckett	Kingston, Jamaica	May 31, 1870	103, 704
Plow, Steam	A. Bigelow	Hamilton, Canada	June 19, 1860	28, 732
Plow, Steam	T. H. Burridge	Saint Louis, Mo	July 31, 1860	29, 358
Plow, Steam	J. Curtis	Chicago, Ill	Sept. 6, 1864	44, 077
Plow, Steam	J. C. Delavigne	New Orleans, La	Mar. 31, 1868	76, 060
Plow, Steam	J. W. Evans	New York, N. Y	Oct. 5, 1858	21, 661
Plow, Steam	J. W. Fawkes	Christiana, Pa	Dec. 13, 1859	26, 422
Plow, Steam	J. W. Fawkes	Decatur, Ill	Dec. 10, 1861	33, 882
Plow, Steam	J. Fowler, jr., D. Greig, and R. Noddings.	London and Leeds, England.	Aug. 28, 1866	57, 653
Plow, Steam	J. Fowler, jr., W. Worby, and D. Greig.	Cornhill, Ipswich, and New Cross, England.	Aug. 28, 1866	57, 652
Plow, Steam	J. R. Gray	Fair Play, Wis	Oct. 20, 1857	18, 446
Plow, Steam	N. A. Gray	Cleveland, Ohio	Oct. 27, 1863	40, 403
Plow, Steam	N. A. Gray	Cleveland, Ohio	Apr. 21, 1868	77, 031
Plow, Steam	A. W. Hall	Saint Louis, Mo	Apr. 21, 1863	38, 260
Plow, Steam	J. Hawkins	Wilkins Township, Pa	Oct. 18, 1859	25, 826
Plow, Steam	O. Hyde	Oakland, Cal	Oct. 17, 1871	120, 071
Plow, Steam	C. F. Johnson, jr	Owego N. Y	Sept. 5, 1865	49, 761
Plow, Steam	P. Klingle	Linæon Hill, D. C	Feb. 23, 1858	19, 427
Plow, Steam	J. G. Knapp	Madison, Wis	Nov. 30, 1869	97, 299
Plow, Steam	M. N. Lynn	New Albany, Ind	May 31, 1870	103, 635
Plow, Steam	M. N. Lynn	New Albany, Ind	July 4, 1871	116, 610
Plow, Steam	P. S. McDonald	San Francisco, Cal	June 17, 1873	139, 966
Plow, Steam	A. E. McGaughey	East Minneapolis, Minn	Jan. 14, 1873	134, 913
Plow, Steam	A. E. and S. N. McGaughey	Wastedo, Minn	Nov. 29, 1859	26, 279
Plow, Steam	J. W. McLean	Indianapolis, Ind	Dec. 6, 1859	26, 397
Plow, Steam	W. H. H. Millen	Littleton, N. H	Dec. 11, 1860	30, 884
Plow, Steam	H. Miller	Bellville, Tex	Aug. 15, 1871	118, 143
Plow, Steam	J. H. Northcott	Mechanicsburgh, Ill	June 22, 1869	91, 558
Plow, Steam	E. G. Otis	Yonkers, N. Y	Oct. 20, 1857	18, 468
Plow, Steam	H. E. Paine	Milwaukee, Wis	Aug. 27, 1867	68, 310
Plow, Steam	G. W. Ramsey	New York, N. Y	Feb. 21, 1860	27, 242
Plow, Steam	O. Redmond	Rochester, N. Y	Oct. 23, 1866	59, 073
Plow, Steam	J. Reynolds	New York, N. Y	Dec. 18, 1860	30, 986
Plow, Steam	C. W. Saladee	Jefferson County, Tex.	June 25, 1861	32, 652
Plow, Steam	S. L. Shotwell and S. R. Hicks	Ottawa, Ill., and North Hempstead, N. Y.	May 28, 1861	32, 437
Plow, Steam	G. Simonson	Mount Carmel, Ind	Apr. 24, 1866	54, 224
Plow, Steam	J. R. Smith	Trenton, N. J	Apr. 23, 1861	33, 153
Plow, Steam	D. B. Spencer	Parkersburgh, Va	Mar. 31, 1857	16, 937
Plow, Steam	L. Stewart	San Francisco, Cal	June 15, 1869	91, 383
Plow, Steam	A. L. Savean	Chaptico, Md	Mar. 29, 1870	101, 395
Plow, Steam	A. P. Thayer	Syracuse, N. Y	Nov. 24, 1863	40, 717
Plow, Steam	E. A. Tounley and E. S. Friedrich.	Washington, D. C	Apr. 27, 1869	89, 361
Plow, Steam	S. B. Wilkins	Milton, Pa	June 1, 1869	90, 799
Plow, Steam	G. Willard	New York, N. Y	Mar. 26, 1867	63, 349
Plow, Steam	L. B. Woolfolk	Nashville, Tenn	June 19, 1860	28, 801
Plow, Steam	L. B. Woolfolk	Nashville, Tenn	June 26, 1860	28, 933
Plow, Steam gang	W. H. H. Heydrick	Chestnut Hill, Pa	Mar. 26, 1867	63, 247
Plow-stock	J. A. Byrd	Jackson County, Fla	May 1, 1860	28, 064
Plow-stock	L. G. Peel	Preston, Ga	Mar. 2, 1869	87, 362
Plow-stock, Convertible	E. T. Parker	Berkley, Ala	Jan. 21, 1851	7, 910
Plow-stocks, Double-footed	J. L. Hood	Floyd County, Ga	July 13, 1869	92, 448
Plow, Stubble and subsoil	R. R. Fenner	Urbana, Ill	June 22, 1869	91, 617
Plow stubble-attachment	J. Kinney	London, Canada	Feb. 28, 1871	112, 152
Plow stubble-attachment	A. Osborn and E. Wulzen	Streator, Ill	Jan. 26, 1869	86, 245
Plow stubble-cleaner	J. Lacy and G. Watkins	Bristol, Wis	Oct. 4, 1864	44, 535
Plow stubble-turner	G. B. Smith	Shopiere, Wis	Nov. 21, 1871	121, 207
Plow, Submarine	E. T. Ligon	Demopolis, Ala	Sept. 17, 1867	69, 000
Plow, Submarine	A. Morrison and R. E. Rose	New Orleans, La	July 5, 1870	104, 981
Plow, Subsoil	G. B. Birmingham	Trenton, Tenn	July 1, 1873	140, 460
Plow, Subsoil	T. E. C. Brinly	Louisville, Ky	Sept. 17, 1872	131, 391
Plow, Subsoil	W. H. H. Burnham and S. B. Pierce.	Homer, N. Y	Aug. 27, 1861	33, 131
Plow, Subsoil	J. F. Cameron	Livingston County, Mo	Jan. 15, 1861	31, 108
Plow, Subsoil	L. Carleton	Pomeroy, Ohio	Oct. 18, 1870	108, 329
Plow, Subsoil	T. L. Cotten	Madison County, Miss	July 19, 1870	105, 551
Plow, Subsoil	J. Custer	Corsica, Ohio	Oct. 27, 1868	83, 472
Plow, Subsoil	S. E. Fletcher	Ballstown, Ind	Nov. 19, 1872	133, 216

Index of patents issued from the United States Patent Office from 1790 *to* 1873, *inclusive*—Continued.

Invention.	Inventor.	Residence.	Date.	No.
Plow, Subsoil	C. Garrett and T. Cottman	Cincinnati, Ohio	June 3, 1856	15, 039
Plow, Subsoil	E. Gross	Goshen Hill, S. C	Mar. 13, 1860	27, 626
Plow, Subsoil	C. K. Hartman	Vincennes, Ind	Apr. 14, 1868	76, 627
Plow, Subsoil	J. W. Howard	Greenville, Ala	Mar. 11, 1873	136, 726
Plow, Subsoil	J. H. Johnson	Bentonville, Ark	Oct. 18, 1870	108, 485
Plow, Subsoil	M. R. Jones	Bradford, Wis	Sept. 24, 1867	69, 099
Plow, Subsoil	M. R. Jones	Walworth County, Wis	Jan. 3, 1871	110, 660
Plow, Subsoil	P. Manny	Waddam's Grove, Ill	Apr. 22, 1856	14, 726
Plow, Subsoil	L. V. B. Marting	Tuscaloosa, Ala	Oct. 11, 1870	108, 164
Plow, Subsoil	J. McCollum	Brownsville, Ala	Apr. 23, 1861	32, 139
Plow, Subsoil	J. W. Murfee	Havanna, Ala	June 22, 1869	91, 657
Plow, Subsoil	C. Myers and W. Gummow	Marysville, Cal	Nov. 19, 1872	133, 167
Plow, Subsoil	W. C. Pagett	Xenia, Ohio	Oct. 22, 1850	7, 732
Plow, Subsoil	R. Peet	Castile, N. Y	June 4, 1867	65, 426
Plow, Subsoil	S. T. W. Potter	Scott, N. Y	July 15, 1862	35, 890
Plow, Subsoil	J. B. Pullman	Los Angeles, Cal	Apr. 27, 1869	89, 432
Plow, Subsoil	E. M. Duery	Harris Depot, N. C	July 26, 1870	105, 844
Plow, Subsoil	R. Themar and B. Brothers	Sheboygan, Wis	May 30, 1871	115, 543
Plow, Subsoil	J. R. Turner and J. Jacobs	Fredericktown, Mo	Nov. 11, 1873	144, 487
Plow, Subsoil	S. D. Tuttle	Eaton, Ohio	Mar. 31, 1868	76, 275
Plow, Subsoil	A. L. P. Vairin	Ripley, Miss	July 8, 1873	140, 660
Plow, Subsoil	J. Vaughn and E. Chamness	Miami County and Grant County, Ind.	May 19, 1868	78, 031
Plow, Subsoil	W. Watkins	Joliet, Ill	Feb. 1, 1870	99, 377
Plow, Subsoil	T. G. Wilder	Camden, Miss	Dec. 27, 1870	110, 525
Plow, Subsoil	J. Wood and R. North	Rochester, Wis	Nov. 10, 1857	18, 619
Plow, Subsoil	G. W. N. Yost	Nashville, Tenn	Sept. 15, 1863	40, 003
Plow subsoil-attachment	E. Cutcliffe	East Bethany, N. Y	Sept. 30, 1873	143, 335
Plow subsoil-attachment	C. Hayden	Collinsville, Conn	Mar. 10, 1868	75, 419
Plow subsoil-attachment	R. Johnson	Lawrence, Kans	May 10, 1870	102, 825
Plow subsoil-attachment	R. Johnson	Lawrence, Kans	Feb. 14, 1871	111, 852
Plow subsoil-attachment	J. A. Krake	Alden, N. Y	July 9, 1867	66, 597
Plow subsoil-attachment	J. C. Leonard and J. J. Gobar	Clinton, Mo	July 28, 1868	80, 356
Plow subsoil-attachment	A. F. Roberts	Lexington, Ky	Nov. 1, 1870	108, 939
Plow, Subsoil corn	H. Bacon	Tecumseh, Mich	June 5, 1849	6, 508
Plow, Subsoil gang	C. Myers	Marysville, Cal	Dec. 9, 1873	145, 361
Plow, Sulky	A. Q. Allis	Dayton, Ohio	Feb. 11, 1868	74, 268
Plow, Sulky	J. G. Boyd	Decatur, Tex	Apr. 5, 1870	101, 574
Plow, Sulky	J. Brewer	Albany, Ill	July 4, 1865	48, 512
Plow, Sulky	G. Burket and S. M. Gaskill	Bluffton, Ohio	Jan. 15, 1867	61, 151
Plow, Sulky	D. W. Colburn	Loami, Ill	Apr. 2, 1867	63, 473
Plow, Sulky	J. H. Cole	Vacaville, Cal	Apr. 5, 1870	101, 435
Plow, Sulky	W. B. Cummins	Leon, Iowa	Feb. 13, 1872	123, 616
Plow, Sulky	O. P. Dils	Falmouth, Ky	Mar. 5, 1867	62, 615
Plow, Sulky	I. Donaldson	Toledo, Iowa	Dec. 17, 1867	72, 375
Plow, Sulky	C. A. Edwards	Chatfield, Minn	Dec. 7, 1869	97, 490
Plow, Sulky	M. A. Elliott	Stratford Hollow, N. H	Mar. 7, 1871	112, 434
Plow, Sulky	R. R. Gaskill	Mendota, Ill	Sept. 11, 1866	57, 889
Plow, Sulky	A. Hale	Eureka, Ill	Nov. 6, 1866	59, 390
Plow, Sulky	B. R. Hubbard	Hillsborough, Ill	Sept. 6, 1870	107, 052
Plow, Sulky	D. W. Hughes	Mexico, Mo	July 1, 1873	140, 505
Plow, Sulky	L. Hunt	Weathersfield, Vt	May 5, 1868	77, 617
Plow, Sulky	W. H. Isaacs and G. E. Banner	Terre Haute, Ind., and Newark, N. J.	Jan. 26, 1869	86, 231
Plow, Sulky	J. R. Jackson	Pelahatchee Depot, Miss	Mar. 9, 1869	87, 577
Plow, Sulky	A. Keith	Lisbon, Ill	Nov. 7, 1865	50, 828
Plow, Sulky	G. Knight	Boone, Iowa	Oct. 30, 1866	59, 232
Plow, Sulky	E. Levee	West Point, Iowa	Feb. 18, 1868	74, 554
Plow, Sulky	J. B. Lewis and J. E. Udall	Concord, Ill	Sept. 1, 1868	81, 799
Plow, Sulky	C. H. Littlefield	Turner, Me	Feb. 26, 1867	62, 430
Plow, Sulky	S. M. Lockwood	Chicago, Ill	Mar. 27, 1866	53, 457
Plow, Sulky	H. W. Mason	Hagerstown, Md	Oct. 4, 1870	107, 938
Plow, Sulky	J. R. McConnell	Marengo, Iowa	Sept. 15, 1868	82, 140
Plow, Sulky	E. Meloy and A. R. Stanley	Shullsburgh, Wis	Jan. 31, 1871	111, 366
Plow, Sulky	W. Ough	Orion, Ill	Feb. 18, 1873	136, 003
Plow, Sulky	C. N. Owen	Salem, Ohio	Nov. 5, 1872	132, 772
Plow, Sulky	I. C. Pratt	Morton, Ill	Sept. 5, 1865	49, 833
Plow, Sulky	J. H. Rankin	Versailles, Ill	Nov. 5, 1867	70, 615
Plow, Sulky	J. J. Reed	Polo, Ill	Nov. 13, 1866	59, 654
Plow, Sulky	S. J. Reed	Camden, Ohio	July 21, 1868	80, 221
Plow, Sulky	J. C. Rogers	Alden, N. Y	June 18, 1867	65, 834
Plow, Sulky	J. Root	Hartland, N. Y	Jan. 12, 1869	85, 768
Plow, Sulky	B. Slusser	Sidney, Ohio	July 28, 1868	80, 427
Plow, Sulky	B. Slusser	Sidney, Ohio	Aug. 3, 1869	93, 358
Plow, Sulky	A. R. Stanley and H. W. Ensign	Shullsburgh, Wis	Sept. 1, 1868	81, 701
Plow, Sulky	S. Stout	Tremont, Ill	Sept. 11, 1866	57, 992
Plow, Sulky	B. W. Sutherlen	Freesoil, Minn	Apr. 14, 1868	76, 648
Plow, Sulky	J. L. Van Gorder	Sidney, Ohio	Aug. 2, 1870	106, 003
Plow, Sulky	J. L. Van Gorder	Sidney, Ohio	Aug. 29, 1871	118, 563
Plow, Sulky	J. Warwick	Springborough, Ohio	Jan. 7, 1868	73, 143
Plow, Sulky	I. Wing	Earlville, Iowa	June 25, 1867	66, 200
Plow, Sulky	T. Wolfe	Girard, Ill	Jan. 16, 1866	52, 104
Plow, Sulky	J. Worrell	Clayton, Ind	Dec. 17, 1872	134, 121
Plow, Sulky	J. Worrell and J. H. Rynerson	Clayton, Ind	June 4, 1872	127, 538
Plow, Sulky	P. Young	El Paso, Ill	July 24, 1866	56, 662
Plow sulky-attachment	J. H. Rynerson and J. Worrell	Clayton, Ind	July 26, 1870	105, 849
Plow sulky-attachment	G. M. Smith	Pittsburgh, Ind	May 30, 1871	115, 537
Plow sulky-attachment	J. Worrell and J. H. Rynerson	Clayton, Ind	Mar. 28, 1871	113, 234
Plow sulky-attachment	J. Worrell and J. H. Rynerson	Clayton, Ind	Mar. 28, 1871	113, 235
Plow sulky-attachment	J. Worrell and J. H. Rynerson	Clayton, Ind	Oct. 31, 1871	120, 560
Plow, Sulky gang	I. C. Pratt	Morton, Ill	Mar. 21, 1865	46, 974
Plow, Swivel	F. Holbrook, J. A. Howe, and J. Nourse	Brattleborough, Vt., and Boston, Mass.	May 17, 1870	103, 187
Plow, Swivel	E. G. Mathews	Oakham, Mass	Nov. 1, 1870	108, 919
Plow, Trench	J. G. Robinson	Springfield, Ill	Mar. 30, 1869	88, 414
Plow, Trenching	W. Wise	Washington, D. C	Mar. 9, 1858	19, 597
Plow-truck	J. Clees	Darbyville, Ohio	Mar. 2, 1869	87, 398
Plow-truck	J. Flanagin	Pawnee City, Nebr	June 3, 1873	139, 564

Index of patents issued from the United States Patent Office from 1790 *to* 1873, *inclusive*—Continued.

Invention.	Inventor.	Residence.	Date.	No.
Plow-truck	J. G. Moore	Kingston, Ohio	Mar. 16, 1869	87, 866
Plow, Truck	M. Mickelson	Ashland Mills, Oreg	Aug. 10, 1869	93, 464
Plow, Turf	H. B. Sommers	Greenfield, Ind	June 5, 1847	5, 148
Plow, Twin	N. G. Cryer	Wentworth, N. C	Mar. 24, 1827	....
Plow, Under-drain	J. and E. Nevison	Morgan, Ohio	Nov. 30, 1858	22, 194
Plow, Underground-drain	A. Watson	Walnut Run, Ohio	Feb. 8, 1859	22, 906
Plow-wheel	A. J. Borland	Charlestown, Iowa	Aug. 8, 1871	117, 734
Plow-wheel	A. J. Borland	Donaldson, Iowa	Apr. 30, 1872	126, 255
Plow-wheel	L. C. Bristol	Holly, Mich	June 11, 1872	127, 734
Plow-wheel	E. A. Chubb	Ionia, Mich	Sept. 28, 1869	95, 194
Plow-wheel	G. Dodge	Kalamazoo, Mich	Aug. 6, 1867	67, 513
Plow-wheel	H. Galentine	Greece, N. Y	Nov. 7, 1871	120, 581
Plow-wheel	L. E. Palmer	Le Roy, Pa	Oct. 15, 1867	69, 834
Plow-wheel	M. Potter	Mendon, Mich	Feb. 25, 1873	136, 179
Plow-wheel	E. S. Rice	Paw Paw, Mich	Mar. 3, 1868	75, 054
Plow-wheel	J. B. Webster and R. Baxter	Stockton, Cal	Dec. 3, 1867	71, 827
Plow, Wheel	J. Bingham and O. M. Pond	Waterloo, Iowa	Dec. 2, 1873	145, 147
Plow, Wheel	T. D. Burrall	Geneva, N. Y	Oct. 28, 1843	3, 320
Plow, Wheel	H. C. Carr	Bordentown, N. J	Aug. 23, 1870	106, 548
Plow, Wheel	E. A. Chace	Rosemond, Ill	May 21, 1867	64, 839
Plow, Wheel	J. Cochrane	Indianola, Iowa	May 7, 1872	126, 447
Plow, Wheel	W. J. Connor	Benton, Ill	Sept. 12, 1871	118, 792
Plow, Wheel	R. Coreth	New Braunfels, Tex	Sept. 5, 1871	118, 694
Plow, Wheel	B. J. Crane	Ripon, Wis	July 16, 1872	128, 952
Plow, Wheel	W. G. Crossley	Apple River, Ill	Jan. 2, 1872	122, 442
Plow, Wheel	O. P. Dills	Falmouth, Ky	Sept. 11, 1866	57, 878
Plow, Wheel	J. C. Ferguson	Hydersville, Miss	June 22, 1836	....
Plow, Wheel	H. M. Freeman, J. Lowe, and J. F. Stevens.	Lathrop, Mo	Jan. 14, 1873	134, 878
Plow, Wheel	I. B. Green	Gillespie, Ill	May 20, 1873	139, 059
Plow, Wheel	W. S. Grimshaw	Pittsfield, Ill	Feb. 11, 1873	135, 800
Plow, Wheel	J. D. Harrison	Middletown, Ohio	Sept. 23, 1873	143, 147
Plow, Wheel	F. Hasbrook	Stokes' Mound, Mo	Apr. 29, 1873	138, 329
Plow, Wheel	F. Hasbrook	Stokes' Mound, Mo	Nov. 11, 1873	144, 453
Plow, Wheel	J. H. Janes, H. Tucker, and T. D. Terry.	Knobnoster, Mo	Apr. 8, 1873	137, 611
Plow, Wheel	J. Kay	Salem, Ind	Oct. 27, 1868	83, 507
Plow, Wheel	I. Long	Bucyrus, Ohio	Mar. 9, 1844	3, 465
Plow, Wheel	W. Mason	Monmouth, Oreg	Sept. 3, 1872	131, 063
Plow, Wheel	W. C. McCool	Guthrie Centre, Iowa	May 21, 1872	126, 976
Plow, Wheel	M. A. Melvin	Washington Court-House, Ohio.	Feb. 6, 1872	123, 499
Plow, Wheel	H. W. Neal	Wellsville, Pa	July 16, 1872	129, 161
Plow, Wheel	I. F. Nutting	Palmer, Mass	Nov. 7, 1865	50, 837
Plow, Wheel	C. N. Owen	Salem, Ohio	Feb. 6, 1872	123, 505
Plow, Wheel	W. H. and J. G. Parish	Portland, Oreg	Jan. 30, 1872	123, 194
Plow, Wheel	J. C. Pearl	Mendota, Ill	July 22, 1873	141, 073
Plow, Wheel	J. Pierce	Wyanet, Ill	Dec. 20, 1864	45, 520
Plow, Wheel	T. B. Quigley and H. Hall	Mansfield, Ohio	Oct. 7, 1845	4, 222
Plow, Wheel	L. W. Richardson	Roscoe, Ill	Aug. 19, 1873	141, 951
Plow, Wheel	J. H. and S. Robbins	Bethel, Oreg	Jan. 30, 1872	123, 294
Plow, Wheel	L. Sachse	Monmouth, Oreg	Mar. 11, 1873	136, 674
Plow, Wheel	M. Sattley	Taylorville, Ill	Feb. 4, 1873	135, 594
Plow, Wheel	H. M. Skinner	Rockford, Ill	July 1, 1873	140, 551
Plow, Wheel	C. B. Stevens	Donelson, Iowa	May 20, 1873	139, 032
Plow, Wheel	J. H. Suydam	Atwater, Minn	Dec. 20, 1870	110, 401
Plow, Wheel	J. E. Swallow	Hagerstown, Md	Aug. 9, 1870	106, 294
Plow, Wheel	J. E. Swallow	Hagerstown, Md	Nov. 29, 1870	109, 684
Plow, Wheel	W. P. Sweeny	Marine, Ill	Apr. 15, 1873	137, 975
Plow, Wheel	T. D. Terry, A. Case, and C. Larkin.	Knobnoster, Mo	Sept. 16, 1873	142, 964
Plow, Wheel	G. Tozer	Jackson, Mo	July 9, 1872	128, 829
Plow, Wheel	J. Worrell	Belleville, Ind	Mar. 18, 1873	137, 044
Plow, Wheeled	R. Wilson	Fowler, N. Y	Feb. 14, 1865	46, 412
Plow with double-pointed wings, Cast-iron	W. C. Pancost	Addison, Vt	Jan. 29, 1824	....
Plow, Wrought-iron	J. and H. F. Cromwell	Cynthiana, Ky	Sept. 30, 1841	2, 274
Plows, Adjustable draft-device for	A. J. Jackson	Millersburgh, Ill	Nov. 15, 1870	109, 215
Plows, Adjustable land-side for	G. Heffley, S. Conrad, and J. Wigle.	Berlin, Pa	Mar. 25, 1851	7, 994
Plows, Adjustable mole for Mole	L. Kazar	Leroy, Ill	Aug. 14, 1860	29, 601
Plows, Adjustable share for corn	D. Wolf	North Lebanon, Pa	Apr. 23, 1850	7, 315
Plows, and converting the same into two single plows, Constructing double hill-side.	M. Rich	Ithaca, N. Y	Mar. 24, 1838	653
Plows, Apparatus for cleaning colters of	A. Reeder	Wrightstown, Pa	May 18, 1858	20, 300
Plows, Apparatus for welding together the lay and land-side of.	J. Lane	Chicago, Ill	Aug. 4, 1868	80, 552
Plows, Attaching cast points to steel mold-boards of	A. and S. Peacock	Cincinnati, Ohio	July 10, 1855	13, 228
Plows, Attaching draft to	G. W. Kidwell	Elwood, Ind	Sept. 27, 1870	107, 692
Plows, Attaching handles to	J. L. Laughlin	Peru, Ill	May 23, 1871	115, 069
Plows, Attaching handles to joiners'	C. H. Weigle	York, Pa	May 3, 1870	102, 630
Plows, Attaching points to the shanks of subsoil	J. W. Murfee	Havanna, Ala	Aug. 9, 1870	106, 193
Plows, Attachment for revolving mold-board for	J. S. Godfrey	Leslie, Mich	Mar. 7, 1871	112, 333
Plows, Attachment for revolving mold-board for	J. S. Godfrey	Rochester, Pa	Apr. 25, 1871	114, 002
Plows, Attachment to	J. W. Murfee	Havanna, Ala	Aug. 9, 1870	106, 192
Plows, Bending mold-boards for	B. Pitcher	Peoria, Ill	Apr. 27, 1858	20, 128
Plows by attaching the mold-boards and sheaths, or cast-iron standards together by rivets.	B. F. Jewett	Springfield, Ill	Feb. 12, 1841	1, 976
Plows, Capstan for	H. C. Drew	Stockbridge, Mich	Feb. 19, 1861	31, 446
Plows, Capstan for ditching	E. Forbis	London, Ohio	July 24, 1860	29, 258
Plows, Carriage attachment for	J. Zoeberlein	Baltimore, Md	Nov. 22, 1870	110, 567
Plows, Cast-iron cutter for	J. Swan	Cayuga County, N. Y	Apr. 24, 1824	....
Plows, Clod-fender and pulverizing-attachment for	E. Lanny	Mowrytown, Ohio	May 17, 1870	103, 203
Plows, Combination	B. F. McCarty and J. W. and R. J. Orr.	Florence, Ga	Aug. 10, 1869	93, 543
Plows, Combined	H. Brown	Payson, Ill	Mar. 9, 1844	3, 474
Plows, Combined	S. Cline	New Britain, Pa	Nov. 12, 1830	....
Plows, Combined	S. Cline	Berks County, Pa	Oct. 15, 1836	....
Plows, Combined	S. Cline	Bucks County, Pa	Oct. 15, 1836	53

Index of patents issued from the United States Patent Office from 1790 *to* 1873, *inclusive*—Continued.

Invention.	Inventor.	Residence.	Date.	No.
Plows, Combined	J. Knodle	Bakersville, Md	Apr. 8, 1840	1,543
Plows, Combined	J. and C. Krauser	Reading, Pa	Sept. 5, 1846	4,745
Plows, Combined	A. Leland	Milton, Pa	Jan. 2, 1849	5,998
Plows, Combined	J. Parsons	Dublin, Ind	Dec. 12, 1842	2,879
Plows, Composition-metal for casting	G. K. Smith	Waterloo, Iowa	Oct. 7, 1873	143,539
Plows, Constructing and fastening shares and cutters of.	M. Smith	Tinicum, Pa	Jan. 28, 1840	1,482
Plows, Corn-planting attachment for	J. F. Byland	Walton, Ky	Mar. 22, 1870	100,974
Plows, cultivator-teeth, &c., Plate and bar for construction of.	W. H. Singer	Pittsburgh, Pa	Oct. 25, 1870	108,641
Plows, Cutter-attachment for	J. Long	Leavenworth, Ind	Apr. 8, 1862	34,893
Plows, Cutter-attachment for	T. E. Marable	Petersburgh, Va	Aug. 18, 1868	81,187
Plows, Cutter-stock for swivel	E. G. Mathews	Oakham, Mass	May 17, 1870	103,215
Plows, Device for cleaning	R. Groom	Albany, N. Y	July 27, 1869	92,959
Plows, Device for cleaning weeds from	J. Jameson	Philadelphia, Pa	Aug. 6, 1867	67,550
Plows, Device for operating	J. O. Potter	Rouseville, Pa	Nov. 8, 1870	109,048
Plows, Device for securing clevis to	R. B. Prindle	Coventry, N. Y.	June 14, 1859	24,403
Plows, Device for securing shield to	J. F. Cameron	Bedford, Mo	Jan. 15, 1861	31,109
Plows, Device for securing stock to guide-rod of joiners'.	S. Going	New York, N. Y	July 7, 1857	17,735
Plows, Draft-attachment for	S. H. Dailey	Olcott, N. Y	May 21, 1872	126,934
Plows, Draft-attachment for	N. Westcott	Nelson, N. Y	Nov. 18, 1873	144,811
Plows, Draft-regulator for	M. Prillaman	Tipton, Ind	Dec. 7, 1869	97,690
Plows, Draft-regulator for	M. Prillaman	Tipton, Ind	Jan. 16, 1872	122,853
Plows, Equalizing-attachment for	D. H. King and W. M. Hulse	Palmyra, Ill	July 8, 1873	140,632
Plows, Fastening colters to	A. and A. K. Whittlesy	Springport, N. Y	Oct. 22, 1850	7,736
Plows, Fastening shoe of hill-side	W. L. Chase	Boston, Mass	July 22, 1850	7,518
Plows, Fertilizer-attachment for	J. I. Boswell	Christiansville, Va	Nov. 2, 1869	96,388
Plows for canals, &c., Excavating	J. Richmond	Hudson, N. Y	June 28, 1817	....
Plows from choking, Device for preventing	W. M. Eccles	Saint Louis, Mo	July 26, 1870	105,660
Plows, Gage for constructing mold-boards of	E. Vaughn	Marion, N. Y	Feb. 5, 1861	31,362
Plows, Handle-seat for	E. Wiard	Louisville, Ky	Apr. 19, 1870	102,071
Plows, &c., Hardening and uniting iron and steel in the manufacture of.	W. Howell and N. W. Browning.	Webster City, Iowa	Nov. 24, 1868	84,420
Plows, Inverting hill-side and horizontal	J. W. Jordan	Lexington, Va	Oct. 28, 1835	....
Plows, Inverting hill-side and horizontal	J. W. Jordan	Lexington, Va	Apr. 19, 1839	1,128
Plows, Locomotive machine for propelling	W. P. Miller	Marysville, Cal	May 3, 1859	23,853
Plows, Machine for making	J. Urie	Evansville, Ind	May 3, 1870	102,731
Plows, Machine for making mold-boards for	H. H. Scoville	Syracuse, N. Y	May 29, 1860	28,507
Plows, Making	J. Deere	Moline, Ill	Apr. 2, 1867	63,369
Plows, Making	W. P. Long	Wheatland, Ind	Feb. 19, 1867	62,211
Plows, Manure-rake attachment for	M. Van Duine and J. De Jonge	Zeeland, Mich	Jan. 24, 1871	111,155
Plows, Mode of coupling and draft of	J. Card and G. Newell	Mentor, Ohio	Nov. 9, 1839	1,401
Plows, Mole for drain	T. S. Cox	La Fayette, Ind	Aug. 16, 1859	25,098
Plows, Mole for drain	A. B. Hawkins and J. Puntenney.	Cameron, Ill	June 12, 1860	28,667
Plows, Mole for drain	J. Lane	Lockport, Ill	Jan. 10, 1860	26,771
Plows, Mole for drain	A., E., and C. Marquiss and C. Emerson.	Monticello and Decatur, Ill.	Feb. 19, 1856	14,287
Plows, Mounting the cutters of rotary	P. H. Standish	Martinez, Cal	May 26, 1868	78,400
Plows of malleable cast-iron, Manufacture of	C. Alger	Boston, Mass	Aug. 3, 1838	869
Plows, Operating cutter for steam	O. Hyde	Oakland, Cal	Jan. 24, 1871	111,212
Plows, Operating drain	D. Watson	Newport, Ohio	Jan. 4, 1859	22,522
Plows, Plant-protector attachment to	J. Ahearn	Baltimore, Md	July 20, 1869	92,772
Plows, Plate for making blanks for mold-boards and shares for.	H. Barnes	Franklin Township, Iowa	Aug. 3, 1869	93,162
Plows, Poly-share for	F. Brewster	Burlington, Vt	Jan. 24, 1833	....
Plows, Portable capstan for mole	A. L. O. Wall, G. Roberts, and M. S. Cartter.	Decatur, Ill	Sept. 11, 1860	30,015
Plows, Portable crab for mole	E. Thorn	Selma, Ohio	Oct. 18, 1859	25,855
Plows, Pulverizing-attachment for	A. A. Rhoads and W. Tash	Berlin, Ill	Jan. 25, 1870	99,236
Plows, Regulating draft of	J. M. C. Arnsby	Worcester, Mass	Apr. 18, 1846	4,466
Plows, Revolving mold-board for	J. S. Godfrey	Rochester, Pa	Apr. 18, 1871	113,760
Plows, Riding-attachment for	A. H. Ballagh	Bowensburgh, Ill	Dec. 2, 1873	145,083
Plows, Riding-attachment for	A. E. and A. L. Porter	Lamoille, Ill	Mar. 15, 1870	100,800
Plows, Riding-attachment for gang	L. Dominy	Ottawa, Ill	Sept. 10, 1867	68,718
Plows, Scraper-attachment for	J. P. Swofford	Brownville, Miss	Feb. 11, 1873	135,863
Plows, Securing point to	H. D. Rogers	Grafton, Ohio	Oct. 9, 1860	30,352
Plows, Seed-sower and harrow-attachment for gang	J. B. Webster	Stockton, Cal	Apr. 4, 1868	77,141
Plows, &c., Self-adjusting equalizer for	J. T. Hagerty	Camp Point, Ill	June 29, 1869	92,042
Plows, Sower-attachment for	J. Brown	Southampton, Ill	Dec. 2, 1873	145,148
Plows, Spring-beam for	W. Morrison	Carlisle, Pa	Sept. 17, 1850	7,656
Plows, Steam-carriage for	M. N. Lynn	New Albany, Ind	July 25, 1871	117,307
Plows, Stop for carpenters'	T. Nichols	Vicksburgh, Miss	May 9, 1871	114,590
Plows, stoves, pipes, levees, dams, &c., Finishing iron-work of.	S. P. Townsend	Union County, N. J	May 28, 1867	65,306
Plows, Subsoil-pulverizing attachment for	L. E. Burdin	Paris, Ky	Oct. 4, 1870	108,002
Plows, Sulky-attachment for breaking	A. H. Allison	Charlottesville, Ind	Feb. 14, 1871	111,799
Plows, Swivel-clevis for	J. B. Small, E. G. Matthews, and F. F. Holbrook.	Boston, Oakham, and Boston, Mass.	Apr. 25, 1871	114,212
Plows to traction-engines, Attaching	W. H. H. Heydrick	Chestnut Hill, Pa	Aug. 15, 1871	118,015
Plows to wheeled carriages, Coupling	M. S. Curtiss	Bradford, Ill	Oct. 1, 1867	69,323
Plows, Traction-machine for	S. S. Stuntz	Jamestown, N. Y	May 17, 1870	103,102
Plows, Traction-wheel for rotary	L. S. Fithian	Rahway, N. J	Mar. 28, 1865	47,005
Plows, Varying the motion of	J. Sanford	Sharon, Conn	Aug. 20, 1811	....
Plows, Vibrating-colter for	J. S. Johnston	Rockford, Ill	Apr. 12, 1870	101,739
Plows, Weed-gatherer for	L. M. Doddridge	Mount Pleasant, Ind	Sept. 7, 1869	94,578
Plows, Weed-gathering device for	D. Hills	East Hartford, Conn	Oct. 7, 1844	3,777
Plows, Wheel-gage for	G. S. Deane	Grand Rapids, Mich	June 16, 1868	78,934
Plows, Whiffletree-attachment for	J. B. Ripsom	Kendall, N. Y	Jan. 15, 1867	61,260
Plower	W. Moore and W. Pruett	Kokomo, Ind	May 28, 1872	127,358
Plowing and breaking up ground, Machine for	J. Bryan	Lebanon, Ill	Jan. 8, 1869	90,921
Plowing and harrowing machine	A. Cunningham	Cincinnati, Ohio	Jan. 4, 1870	98,569
Plowing and tilling land by steam, Drum or pulley to prevent ropes from slipping in machinery for.	J. Fowler, jr., R. Burton, D. Greig, and J. Head.	Cornhill, Kingsland, New Cross, and Newcastle-upon-Tyne, England.	July 23, 1861	39,912
Plowing and tilling land by steam, Machinery for	J. Fowler, jr	Leeds, England	Apr. 9, 1861	32,026
Plowing and tilling land, Machinery for	J. Fowler, jr	Leeds, England	Apr. 9, 1861	32,025

Index of patents issued from the United States Patent Office from 1790 *to* 1873, *inclusive*—Continued.

Invention.	Inventor.	Residence.	Date.	No.
Plowing and tilling land, Machinery for	J. Fowler, jr	London, England	July 9, 1861	32, 809
Plowing and tilling land, Machinery for	J. Fowler, jr., and D. Greig	Leeds, England	Apr. 9, 1861	32, 027
Plowing, Device for burying weeds and stubble while.	J. W. Beer and J. B. Wampler	Shelbyville, Ill	Aug. 21, 1866	57, 279
Plowing-in stubble, Device for	J. Kilmer	Cobleskill, N. Y	Dec. 13, 1864	45, 454
Plowing-machine	A. Bondeli	Philadelphia, Pa	Dec. 21, 1869	98, 017
Plowing-machine	J. W. Fawkes	Christiana, Pa	Jan. 26, 1858	19, 189
Plowing-machine	D. D. Foley	Washington, D. C	Jan. 26, 1864	41, 371
Plowing-machine	H. Moeser	Pittsburgh, Pa	Nov. 10, 1857	18, 596
Plowing-machine	J. Pierce	Wyanet, Ill	Nov. 8, 1864	44, 969
Plowing-machine	W. Stoddard	Lowell, Mass	Jan. 26, 1858	19, 215
Plowing-machine, Steam	Z. Rider	Painesville, Ohio	July 18, 1871	117, 113
Plug-cutter	M. Jincks	Wallace, N. Y	Aug. 31, 1869	94, 416
Plug-cutter for wood-workers	D. L. Kniffen	West Camden, N. Y	Aug. 8, 1871	117, 786
Plug-cutting machines, Feed-motion in	S. Ingersoll	New York, N. Y	June 21, 1853	9, 799
Plug, Fusible safety	F. Curtis	Newburyport, Mass	Nov. 24, 1863	40, 677
Plumb and level	S. A. Bostwick	Laconia, N. H	Dec. 17, 1867	72, 159
Plumb and level	S. Sanderson	New Bedford, Mass	Oct. 21, 1873	143, 933
Plumb and level, Combined	F. A. Trout	New Britain, Conn	Apr. 21, 1863	38, 252
Plumb and level indicators, Method of attaching the plumb-line to.	J. L. Rowe	New York, N. Y	Mar. 30, 1858	19, 817
Plumb and square, Reversible	M. V. Nobles	Saint Anthony Falls, Minn	Jan. 19, 1864	41, 344
Plumb-bob	B. F. Chappell	Norwich, Conn	Nov. 13, 1860	30, 612
Plumb-bob	G. B. Cowles	Bridgeport, Conn	May 13, 1873	138, 790
Plumb, square, and level, Combined	A. T. Ward	Marietta, Ohio	July 21, 1868	80, 252
Plumbago-press	J. C. Ford	Cambridgeport, Mass	Mar. 29, 1870	101, 358
Plumbers' clamp	F. R. Langwith	New York, N. Y	July 29, 1856	15, 427
Plumbers' hook blank	B. F. Gladding	Providence, R. I	Aug. 8, 1865	49, 343
Plumbers' joints, Making	J. P. Hayes	Philadelphia, Pa	June 19, 1860	28, 750
Plush, paper, &c., Manufacture of frosted	H. V. Edmund	Norwich, Conn	June 7, 1870	103, 999
Pneumatic and rubber spring, Combined	P. S. Devlan	Jersey City, N. J	Jan. 31, 1871	111, 438
Pneumatic brake and car-starter	C. A. Haskins	Chicago, Ill	May 28, 1872	127, 164
Pneumatic brake for railway-cars	C. R. Peddle	Terre Haute, Ind	Feb. 5, 1867	61, 860
Pneumatic dispatch-apparatus	C. W. Siemens	Westminster, England	July 29, 1873	141, 294
Pneumatic drill	S. Gwynn	New York, N. Y	Oct. 18, 1864	44, 722
Pneumatic drill	S. Gwynn	New York, N. Y	Nov. 1, 1864	44, 862
Pneumatic drill	H. Haupt	Cambridge, Mass	Mar. 7, 1865	46, 668
Pneumatic elevator for water	J. M. Blanchard and W. E. Prall	Washington, D. C	July 15, 1873	140, 879
Pneumatic elevator for wool	J. Penman	Paris, Canada	Mar. 19, 1872	124, 851
Pneumatic engine	J. F. Haskins	Fitchburgh, Mass	May 23, 1871	115, 198
Pneumatic engine	J. R. Remington	Aberfoil, Ala	May 26, 1842	2, 637
Pneumatic engine	R. Spear	New Haven, Conn	Dec. 14, 1869	97, 822
Pneumatic machine	D. Lloyd	Cambria County, Pa	Jan. 7, 1816	......
Pneumatic motor	T. B. Jeffrey	Chicago, Ill	Dec. 5, 1871	121, 626
Pneumatic signal-apparatus	W. Foster, jr	New York, N. Y	Aug. 26, 1873	142, 224
Pneumatic signaling-apparatus	W. Foster, jr	New York, N. Y	Apr. 4, 1871	113, 646
Pneumatic signaling-apparatus	W. Foster, jr	New York, N. Y	Apr. 4, 1871	113, 647
Pneumatic spring	J. Bevan and B. W. Hitchcock	Port Richmond and Flushing, N. Y.	Jan. 31, 1871	111, 303
Pneumatic spring	W. A. Dripps	Fort Wayne, Ind	Oct. 29, 1867	70, 177
Pneumatic spring	W. R. Fee	Cincinnati, Ohio	Mar. 30, 1858	19, 764
Pneumatic spring	J. F. Heyward	Wilmington, Del	July 10, 1855	13, 248
Pneumatic spring	I. W. Hoagland	New Brunswick, N. J	July 16, 1861	32, 848
Pneumatic spring	J. Lewis	New Haven, Conn	Feb. 10, 1847	4, 965
Pneumatic spring	M. F. Maury	Richmond, Va	Apr. 16, 1872	125, 749
Pneumatic spring	L. Sterne	London, England	Feb. 23, 1869	87, 307
Pneumatic spring	D. B. Strope	Fort Wayne, Ind	May 28, 1867	65, 135
Pneumatic spring	E. Ware	Roxbury, Mass	July 6, 1852	9, 107
Pneumatic tube	A. E. Beach	Stratford, Conn	May 7, 1867	64, 400
Pneumatic tube for transporting goods	A. Brisbane	New York, N. Y	June 22, 1869	91, 513
Pneumatic ways for transmission of parcels, &c	E. P. Needham	New York, N. Y	Sept. 6, 1864	44, 107
Pneumo-hydraulic engine	R. Willcox	New York	Nov. 1, 1823	......
Pocket-alarm	I. Goodspeed	Norwich, Conn	Nov. 1, 1859	26, 001
Pocket-alarm	C. W. Simonds	Boscawen, N. H	Dec. 6, 1870	109, 952
Pocket-book	A. L. Adams	Philadelphia, Pa	Oct. 13, 1863	40, 223
Pocket-book	F. Busch	New York, N. Y	Jan. 30, 1872	123, 232
Pocket-book	A. Button	Dunkirk, N. Y	Dec. 21, 1869	98, 145
Pocket-book	S. C. Currie	New York, N. Y	June 8, 1869	90, 997
Pocket-book	J. C. Dickinson and R. Bate	Hudson, Mich	June 10, 1856	15, 068
Pocket-book	J. F. Dubber	Brooklyn, N. Y	Jan. 10, 1865	45, 819
Pocket-book	W. T. Fry	New York, N. Y	Feb. 20, 1866	52, 702
Pocket-book	E. Hassenpflug	Boston, Mass	June 2, 1863	38, 741
Pocket-book	F. Heiles	New York, N. Y	Oct. 12, 1869	95, 797
Pocket-book	W. H. King	Westbury, England	Feb. 18, 1873	135, 990
Pocket-book	J. Lehman	New York, N. Y	Dec. 6, 1870	109, 827
Pocket-book	L. Saarback	Philadelphia, Pa	July 11, 1865	48, 727
Pocket-book	L. Saarback	Philadelphia, Pa	Aug. 8, 1865	49, 307
Pocket-book	L. Wendt	Philadelphia, Pa	Apr. 22, 1873	138, 217
Pocket-book alarm	T. Blodgett and W. S. Weatherby	Belchertown and Granby, Mass.	May 18, 1869	90, 070
Pocket-book alarm	W. Stoddard	Lowell, Mass	Dec. 8, 1857	18, 826
Pocket-book and fan combined	O. Brück	New York, N. Y	Apr. 21, 1868	76, 988
Pocket-book and watch safe	J. Colton	Amherst, Mass	Apr. 18, 1846	4, 462
Pocket-book clasp	J. W. Brown and E. Rice	New York, N. Y	Mar. 27, 1866	53, 520
Pocket-book clasp	J. F. Dubber	Brooklyn, N. Y	Sept. 2, 1873	142, 447
Pocket-book clasp	D. E. Fisk	Springfield, Mass	Nov. 19, 1867	71, 154
Pocket-book clasp	H. Ropes	Brooklyn, N. Y	Mar. 5, 1867	62, 689
Pocket-book, &c., fastening	J. C. Arms	Northampton, Mass	June 16, 1868	78, 914
Pocket-book fastening	J. C. Arms	Northampton, Mass	Sept. 29, 1868	82, 577
Pocket-book fastening	F. Heiles	New York, N. Y	Oct. 28, 1873	144, 094
Pocket-book, Fire-proof	E. W. Glover and G. L. Damon	Malden and Cambridge, Mass.	Oct. 8, 1872	132, 068
Pocket-book guard	C. G. Cahoone and A. W. Robinson.	Providence, R. I	Dec. 26, 1871	122, 223
Pocket-book, porte-monnaie, &c	B. F. Cowan	New York, N. Y	Nov. 14, 1865	50, 909
Pocket-book protector	A. Drummond	Newark, N. J	June 26, 1866	55, 838
Pocket-book safety-attachment	J. Brown, jr	New York, N. Y	May 2, 1871	114, 404
Pocket-book safety-attachment	W. H. Ferguson	Rochester, N. Y	Nov. 19, 1867	70, 982

Index of patents issued from the United States Patent Office from 1790 *to* 1873, *inclusive*—Continued.

Invention.	Inventor.	Residence.	Date.	No.
Pocket-book safety-attachment	A. McNeill	Cincinnati, Ohio	Apr. 21, 1868	77, 068
Pocket-book safety-attachment	S. B. Parker	New York, N. Y	Mar. 26, 1867	63, 291
Pocket-book safety-attachment	T. A. Weber	New York, N. Y	Apr. 2, 1867	63, 590
Pocket-book safety-guard	T. Potter	Jackson, Miss	May 6, 1873	138, 581
Pocket-book safety-hook	F. C. Likar	Baltimore, Md	May 6, 1873	138, 667
Pocket-books, Manufacture of	J. C. Arms	Northampton, Mass	Aug. 16, 1870	106, 450
Pocket-books, &c., Method of securing	O. Cox	Alexandria County, Va	Apr. 6, 1858	19, 832
Pocket-case for railway-schedule	S. E. Allen	Company Shops, N. C	May 25, 1869	90, 418
Pocket-clasp, Safety	A. T. Large	Tomah, Wis	Nov. 6, 1866	59, 415
Pocket-comb	W. J. Thorn	Westbrook, Me	May 17, 1853	9, 732
Pocket-flask	A. T. Becker	Cohoes, N. Y	Jan. 25, 1870	99, 051
Pocket-flask	R. George	New York, N. Y	Mar. 19, 1872	124, 733
Pocket-flask	R. George	New York, N. Y	July 16, 1872	129, 024
Pocket-flask, Self-heating	S. Hughes	Hudson, N. Y	Apr. 25, 1871	114, 143
Pocket for attachment to suspenders, Safety	J. Summers	Winchester, Va	June 29, 1869	92, 123
Pocket for billet-strap	H. B. Smith	Wilmington, Del	Feb. 25, 1868	74, 859
Pocket for garments	D. Harrington	Philadelphia, Pa	Oct. 11, 1841	2, 309
Pocket for garments, Money	E. Schnopp	East New York, N. Y	Apr. 9, 1872	125, 489
Pocket-hook, Safety	P. E. Malmström and P. E. Dummer.	New York, N. Y	Mar. 11, 1873	136, 742
Pocket-implement	J. A. Russ	Springfield, Mass	Mar. 19, 1867	63, 102
Pocket induction-apparatus	C. W. Meyer	New York, N. Y	Apr. 26, 1870	102, 295
Pocket-knife	F. H. Barnard and W. L. Brace	Hartford, Conn	Nov. 23, 1869	97, 154
Pocket-knife	G. H. Gardner	Philadelphia, Pa	July 5, 1864	43, 398
Pocket-knife	G. and J. Kay	Esopus, N. Y	Jan. 1, 1867	60, 905
Pocket-knife	S. Mason	Northfield, Conn	Aug. 26, 1862	36, 321
Pocket-knife	J. E. McBeth	New Orleans, La	June 23, 1868	79, 139
Pocket-knife	T. D. Miller	Kankakee, Ill	Mar. 13, 1866	53, 166
Pocket-knife	R. B. Milliken	Springfield, Vt	Jan. 15, 1867	61, 229
Pocket-knife	J. Mosley	New Haven, Conn	Mar. 24, 1868	75, 955
Pocket-knife	C. H. Palmer	Newark, N. J	Apr. 30, 1867	64, 248
Pocket-knife	J. G. Perry	South Kingston, R. I	Oct. 7, 1862	36, 620
Pocket-knife	H. Staude	Troy, N. Y	Apr. 22, 1873	138, 052
Pocket-knife	H. Twitchell	Naugatuck, Conn	Mar. 22, 1864	42, 031
Pocket-knife	A. G. Waterhouse	San Francisco, Cal	June 22, 1869	91, 693
Pocket-knife and door-fastener	J. Armstrong and O. Dame	Bucyrus and Wyandot County, Ohio.	Feb. 9, 1869	86, 808
Pocket-knife and envelope-opener, Combined	A. S. Pennington	Paterson, N. J	Jan. 23, 1872	122, 960
Pocket-knife and pistol, Combined	A. J. Peavey	South Montville, Me	Sept. 5, 1865	49, 784
Pocket-knife handle	J. W. Ayers	West Meriden, Conn	Apr. 8, 1873	137, 648
Pocket-knife handle	C. Beckham	Alexandria, Va	Sept. 5, 1871	118, 675
Pocket-lock, Burglar-proof	W. J. Scott	Albany, N. Y	Sept. 25, 1860	30, 160
Pocket of wearing-apparel	F. Darling	North Blackstone, Mass	Jan. 29, 1861	31, 227
Pocket of wearing-apparel, Secret	J. C. Howells	Washington, D. C	Aug. 11, 1863	39, 482
Pocket-opening, Fastening	J. W. Davis	Reno, Nev	May 20, 1873	139, 121
Pocket-register of count	P. D. Richards and F. N. Thayer.	New Orleans, La	July 12, 1859	24, 757
Pocket safe or fastening	G. R. McIlroy	Covington, Ky	May 5, 1857	17, 225
Pocket, Safety	C. H. Bagley	Elgin, Ill	July 3, 1866	56, 139
Pocket, Safety	J. A. Bell	Lacon, Ill	Aug. 14, 1866	57, 072
Pocket, Safety	J. T. Chambers	Utica, N. Y	Sept. 18, 1866	58, 062
Pocket, Safety	S. Chittenden	Willimantic, Conn	June 11, 1872	127, 845
Pocket, Safety	J. Colton	New Orleans, La	Dec. 3, 1867	71, 706
Pocket, Safety	S. Engel	Chicago, Ill	Mar. 18, 1873	136, 983
Pocket, Safety	R. W. Fisk	Olney, Ill	July 16, 1867	66, 698
Pocket, Safety	J. Foster	Sandwich, Mass	Apr. 20, 1869	89, 034
Pocket, Safety	S. French	Boston, Mass	June 24, 1862	35, 731
Pocket, Safety	S. French	Boston, Mass	June 9, 1863	38, 860
Pocket, Safety	L. Goodyear	Trumansburgh, N. Y	May 28, 1872	127, 339
Pocket, Safety	G. Hamel	Abington, Pa	Jan. 1, 1867	60, 720
Pocket, Safety	H. Harris	Newark, N. J	June 16, 1857	17, 573
Pocket, Safety	G. G. Hickman	Downingtown, Pa	Oct. 24, 1865	50, 582
Pocket, Safety	W. Hindhaugh	New York, N. Y	July 3, 1866	56, 049
Pocket, Safety	H. L. Hopkins	Morrisville, N. Y	Dec. 2, 1873	145, 178
Pocket, Safety	J. C. Iden	Buckingham Township, Pa	Apr. 28, 1868	77, 192
Pocket, Safety	E. Jones	Grand River, Iowa	June 27, 1871	116, 318
Pocket, Safety	G. A. Mitchell	Turner, Me	July 10, 1866	56, 248
Pocket, Safety	A. Q. Ross	Cleves, Ohio	Jan. 21, 1868	73, 655
Pocket, Safety	F. Russell	Cambridge, Mass	July 9, 1867	66, 638
Pocket, Safety	C. Sanborn	Chichester, N. H	Feb. 4, 1868	74, 008
Pocket, Safety	H. D. Synder	Carbondale, Pa	Oct. 29, 1867	70, 372
Pocket, Safety	P. A. Stecher	New York, N. Y	July 31, 1866	56, 821
Pocket, Safety	A. Wilmot	New Haven, Conn	Nov. 3, 1863	40, 528
Pocket safety-attachment	C. V. Boughton	Titusville, Pa	June 1, 1869	90, 632
Pocket safety-attachment	J. Brosius	Ranch's Gap, Pa	Apr. 7, 1868	76, 297
Pocket safety-attachment	A. W. Browne	Brooklyn, N. Y	Aug. 17, 1869	93, 856
Pocket safety-attachment	H. R. S. Colton	Houghton, Mich	Mar. 15, 1870	100, 728
Pocket safety-attachment	J. Metzendorf	New York, N.Y	Sept. 18, 1866	58, 120
Pocket safety-attachment	F. L. Roell	Northampton, Mass	Mar. 15, 1870	100, 805
Pocket safety-attachment	R. L. Russell	Brooklyn, N. Y	Nov. 18, 1873	144, 702
Pocket safety-attachment	E. Williams	New York, N. Y	Nov. 19, 1867	71, 106
Pocket safety-clasp	D. C. Tuck	Farmington, Me	Dec. 26, 1871	122, 140
Pocket, Safety watch	W. O. Sumner	Brooklyn, N. Y	Aug. 3, 1869	93, 246
Pocket, Seamless	G. Köttgen	Barmen, Prussia	Sept. 11, 1866	58, 035
Pocket, Suspender	J. F. Clark	Baltimore, Md	Nov. 28, 1865	51, 146
Poison, Precautionary attachment for bottles containing.	J. Harrison	Philadelphia, Pa	Jan. 3, 1871	110, 760
Poke	A. E. Cruttenden	Canaseraga, N. Y	May 10, 1870	102, 778
Poke, Animal	A. W. Bishop	York, Ohio	Aug. 1, 1865	49, 071
Poke, Animal	B. E. Blakslee	Medina, Ohio	Mar. 12, 1872	124, 478
Poke, Animal	G. W. Carpender	Jarvis, Ind	Dec. 17, 1867	72, 265
Poke, Animal	H. F. Chapin	Rochester, N. Y	Mar. 14, 1871	112, 546
Poke, Animal	H. F. Chapin	Rochester, N. Y	July 4, 1871	116, 683
Poke, Animal	N. Denny	Saranac, Mich	Aug. 8, 1871	117, 751
Poke, Animal	T. J. Dickerson	Auburn, N. Y	Dec. 10, 1872	133, 765
Poke, Animal	J. Green	Helmick, Ohio	Feb. 13, 1872	123, 692
Poke, Animal	C. Hinman	Akron, Ohio	Jan. 21, 1873	134, 988
Poke, Animal	C. Hinman	Akron, Ohio	June 17, 1873	139, 893

Index of patents issued from the United States Patent Office from 1790 *to* 1873, *inclusive*—Continued.

Invention.	Inventor.	Residence.	Date.	No.
Poke, Animal	J. Hopkins	Akron, Ohio	May 30, 1871	115, 319
Poke, Animal	J. Hopkins	Akron, Ohio	Aug. 5, 1873	141, 438
Poke, Animal	L. Kelley	Saranac, Mich	Feb. 28, 1871	112, 150
Poke, Animal	W. Kelly	Saranac, Mich	June 22, 1869	91, 639
Poke, Animal	W. Kelly	Saranac, Mich	July 19, 1870	105, 579
Poke, Animal	S. C. Leonard	Rushville, N. Y	Jan. 2, 1872	122, 392
Poke, Animal	R. McIntyre	Boston, Mass	June 25, 1872	128, 236
Poke, Animal	F. M. Moulton	Ferrisburgh, Vt	June 13, 1871	115, 884
Poke, Animal	I. N. Peck	Cohocton, N. Y	Oct. 1, 1872	131, 901
Poke, Animal	I. W. Sherwood	Mount Morris, N. Y	May 21, 1872	127, 107
Poke, Animal	J. N. Wallis	Fleming, N. Y	Dec. 10, 1872	133, 815
Poke, Cattle	S. P. Clemons	Dansville, N. Y	May 3, 1870	102, 497
Poke, Cattle	A. H. and M. P. Kennedy	Brunswick, Ohio	Nov. 6, 1866	59, 408
Poke, Cattle	O. and C. H. Sweet	South Glen's Falls, N. Y	Apr. 30, 1872	126, 239
Poke, Horse	M. J. Van Aken	Sturgis, Mich	Feb. 22, 1870	100, 068
Poke, Horse	G. Whitbeck	Phelps, N. Y	Mar. 15, 1870	100, 827
Poke, &c., Horse and cattle	G. W. Bell and G. W. Fulmer	Hinckley, Ohio	Apr. 21, 1868	76, 881
Poke, Horse and cattle	W. Miller	Granger, Ohio	June 14, 1870	104, 185
Poke, Horse and cattle	B. Page	Saint Joseph County, Ind	Nov. 26, 1872	133, 334
Poke, Horse and cattle	N. Silvester	Weymouth, Ohio	June 4, 1867	65, 440
Poke, Horse and cattle	N. Silvester	Weymouth, Ohio	Dec. 10, 1867	71, 916
Poker	J. F. Brener	Plantsville, Conn	Dec. 29, 1868	85, 425
Poker and lid-lifter combined	W. Clegg	Philadelphia, Pa	Feb. 2, 1869	86, 505
Poker and tongs, Combined	J. Blair	Saint Louis, Mo	Nov. 13, 1866	59, 547
Poker, Fire	J. C. Klein	Birmingham, Pa	Sept. 24, 1872	131, 618
Poker, Fire	G. R. Moore	Mount Joy, Pa	Feb. 5, 1856	14, 200
Poker, lifter, and tongs, Combined	P. Sander and J. Manthorn	Philadelphia, Pa	Feb. 6, 1872	123, 349
Poker, tongs, wrench, &c	J. Richards	Washington, D. C	Dec. 18, 1866	60, 556
Pole and hound coupling	F. Bremerman	Indianapolis, Ind	Dec. 1, 1868	84, 531
Pole and post puller	I. Holmes	South New Berlin, N. Y	Sept. 11, 1866	57, 905
Pole and thill coupling	H. L. Taylor	Fredonia, N. Y	Feb. 13, 1866	52, 621
Pole-ascending apparatus	G. Fleming	New York, N. Y	Mar. 8, 1870	100, 613
Pole-attachment	G. N. Compton	Canton, Ohio	Jan. 14, 1868	73, 234
Pole-clamp	H. Haering	New York, N. Y	Jan. 7, 1873	134, 598
Pole-socket	C. W. and J. D. Wilcox	Kingston, R. I	Apr. 26, 1870	102, 456
Pole-straps to neck-yokes, Attaching	N. Irish	Rochester, Minn	Aug. 18, 1868	81, 092
Poles, &c., Evener for wagon	H. F. Willson	Elyria, Ohio	Sept. 4, 1866	57, 806
Police-club	J. L. Rowe	New York, N. Y	Feb. 28, 1860	27, 335
Policeman's rattle	J. McCord	Philadelphia, Pa	Nov. 20, 1855	13, 823
Policeman's rattle	F. P. Smiley, C. K. Becker, and Q. Townsend.	Philadelphia, Pa	Nov. 27, 1866	60, 072
Polisher and grinder, Knife	G. L. Witsil	Philadelphia, Pa	July 11, 1865	48, 755
Polisher, Cylinder	G. Cowing	Seneca Falls, N. Y	Apr. 14, 1863	38, 153
Polisher, Cylindric	E. Stowell	Middlebury, Vt	Aug. 20, 1813	....
Polishing and cleaning metals, &c., Compound for	J. B. Emerson	Rochester, N. Y	Aug. 27, 1872	130, 907
Polishing and cleaning powder	J. W. Bates	Saint Paul, Minn	June 8, 1869	91, 070
Polishing and scouring composition	J. B. Rand	Concord, N. H	Oct. 31, 1871	120, 397
Polishing, Apparatus for holding plates for	A. S. Southworth and J. J. Hawes.	Boston, Mass	June 13, 1846	4, 573
Polishing, Artificial tripoli for	T. J. Platt	Newark, N. J	Jan. 8, 1867	61, 094
Polishing-box	W. A. Morse	Philadelphia, Pa	July 31, 1866	56, 781
Polishing buff and brush plates by revolving cylinder.	W. J. Stone	Washington, D. C	Apr. 30, 1831	....
Polishing-composition	C. J. H. F. Kleemann	New York, N. Y	Feb. 28, 1871	112, 154
Polishing-compound	E. J. Combs	Prattsburgh, N. Y	Nov. 11, 1873	144, 444
Polishing-device	J. C. Morrison	Chicago, Ill	May 20, 1873	139, 178
Polishing-fabric and flexible abrader	J. H. Crane	Charlestown, Mass	Sept. 8, 1868	81, 986
Polishing hard and soft substances, Machine for	B. Green	Hartford, Vt	Mar. 27, 1827	....
Polishing-iron	G. J. Prentiss	Fall River, Mass	Oct. 18, 1859	25, 850
Polishing knives, Composition for	A. W. Stewart	Middletown, Ohio	Dec. 3, 1867	71, 810
Polishing knives, &c., Machinery for	W. Vine	New York	Dec. 5, 1846	4, 875
Polishing-machine	A. Ball	Worcester, Mass	Nov. 10, 1863	40, 546
Polishing-machine	T. Beran	Philadelphia, Pa	Apr. 25, 1871	114, 098
Polishing-machine	J. Gooden	Lockport, N. Y	Nov. 29, 1870	109, 610
Polishing-machine	J. Moore	Gardiner, Me	July 22, 1856	15, 382
Polishing-machine	F. Pederson and R. Olsen	Racine, Wis	July 16, 1872	128, 975
Polishing-machine	P. F. Randolph	Jerseyville, Ill	Sept. 28, 1869	95, 265
Polishing-machine	P. F. Randolph	Jerseyville, Ill	May 2, 1871	114, 341
Polishing-machine	J. Stackhouse	West Pittsburgh, Pa	May 5, 1868	77, 546
Polishing-machine	H. Volkening	New York, N. Y	Oct. 10, 1854	11, 797
Polishing machine, Sheet-metal	H. Todd	Bridgeport, Conn	Mar. 5, 1867	62, 703
Polishing-machine, Traverse-motion for	J. Miller	New York, N. Y	Oct. 22, 1872	132, 481
Polishing metallic surfaces, Preparation for	C. M. Rowley	Chicago, Ill	Mar. 16, 1869	87, 793
Polishing metals, Composition of matter for	A. M. Sawyer	Athol, Mass	Dec. 4, 1866	60, 264
Polishing metals, glass, &c., Compound for	M. C. Bland	Chicago, Ill	July 23, 1872	129, 708
Polishing metals, Paste for	P. McManus and G. W. Latimer	Detroit, Mich	June 7, 1870	103, 909
Polishing plates used in taking likeness, &c., Apparatus for.	J. Johnson	New York	Dec. 14, 1841	2, 391
Polishing-powder	A. Hamilton and J. De Gray	Broad Brook, Conn., and Brooklyn, N. Y.	Oct. 24, 1865	50, 578
Polishing-powder	T. R. Hubbard	Brooklyn, N. Y	July 16, 1872	129, 029
Polishing-powder, Manufacture of	G. Wheelock	Washington, D. C	July 30, 1872	129, 999
Polishing powder, Metal	H. Greenwood	Providence, R. I	Jan. 28, 1818	....
Polishing-tool	H. Cottrell	Newark, N. J	Nov. 18, 1873	144, 744
Polishing-tool	T. J. Mayall	Roxbury, Mass	Jan. 22, 1861	31, 180
Polishing-tools to shafts, Connecting rotary	H. F. Wheeler	Boston, Mass	Sept. 3, 1872	131, 140
Polishing turned articles, Machine for	W. Wadleigh	Sanbornton Bridge, N. H	July 21, 1863	39, 315
Polishing-wheel	A. Broad	Litchfield, Conn	June 6, 1814	....
Polishing-wheel	L. Holtzscheiter	Philadelphia, Pa	Apr. 10, 1866	53, 825
Polishing-wheel	A. M. Sawyer	Athol, Mass	Jan. 5, 1869	85, 538
Polishing-wheel	B. Webb	Unadilla Forks, N. Y	July 25, 1854	11, 396
Polishing-wheel, Elastic	L. Hale	Milford, Mass	Apr. 12, 1859	23, 571
Polishing-wheels, Attachment for balancing	H. K. Jones	Kensington, Conn	July 14, 1868	79, 984
Polishing-wheels, Balancing	H. K. Jones	Kensington, Conn	June 30, 1868	79, 475
Polishing-wheels, Balancing	A. W. Stephenson	Kensington, Conn	June 30, 1868	79, 405
Polishing-wheels, Holder for	E. W. Nichols	Worcester, Mass	Aug. 7, 1860	29, 512
Polygraph	N. Ames	Saugus, Mass	Dec. 12, 1854	12, 048
Polygraph	N. Ames	Saugus, Mass	Dec. 12, 1854	12, 049

Index of patents issued from the United States Patent Office from 1790 to 1873, inclusive—Continued.

Invention.	Inventor.	Residence.	Date.	No.
Pomatum, Putting up	E. Ludde	New York, N. Y	Dec. 6, 1870	109, 917
Poncho	P. W. Neefus	New York, N. Y	Mar. 17, 1863	37, 919
Ponton	H. Carberry and W. Tatham	Washington, D. C	July 19, 1816	....
Ponton	T. J. Mayall	Roxbury, Mass	Oct. 27, 1863	40, 421
Ponton-boat	E. Harrison	New York, N. Y	Nov. 3, 1863	40, 539
Ponton-boat	J. Hegeman	Vischer's Ferry, N. Y	Apr. 23, 1867	64, 103
Ponton-boat	R. Pallett	New York, N. Y	Nov. 24, 1863	40, 706
Ponton-equipage for military and other operations	J. F. Lane	Terre Haute, Ind	June 17, 1840	1, 634
Ponton wagon-boat, Metallic	J. Francis	New York, N. Y	Jan. 19, 1864	41, 288
Pool-ball rack	H. W. Collender	New York, N. Y	July 4, 1871	116, 561
Pool-board	H. W. Collender	New York, N. Y	June 20, 1871	116, 159
Pop-corn balls, Machine for forming	W. N. Thompson	Moline, Ill	July 16, 1872	129, 623
Porcelain, Architectural	W. J. Cheyney and E. F. Dieterichs.	Wallingford and Philadelphia, Pa.	Oct. 1, 1867	69, 315
Porcelain for use in filtering, &c., Porous	A. Vidal	Paris, France	May 18, 1869	90, 324
Porcelain, glass, &c., by the use of flurosilicate, Manufacture of.	T. Cobley and J. C. Coombe	Hahl, Bavaria, and Hoxton, England.	Apr. 28, 1863	38, 286
Porcelain, Manufacture of	W. J. Cheyney	Wallingford, Pa	Apr. 30, 1867	64, 196
Porcelain or pottery to receive designs, &c., Method of preparing the surface of.	J. Cartisser	New York, N. Y	Oct. 24, 1865	50, 561
Porcelain-paste, Apparatus for pulverizing	J. R. Alsing	Newcastle-upon-Tyne, England.	Jan. 25, 1870	99, 128
Porcelain, Plate	W. J. Cheyney and E. F. Dieterichs.	Wallingford and Philadelphia, Pa.	Oct. 1, 1867	69, 316
Porcelain printing-frame	C. L. Lochman	Carlisle, Pa	July 24, 1866	56, 579
Pork, Mode of curing	M. H. Smith	Brooklyn, N. Y	Aug. 14, 1866	57, 203
Porous ware, Manufacture of	B. S. and M. R. Pierce	New Bedford and Mansfield, Mass.	Dec. 27, 1859	26, 613
Port-hole closer	J. V. Murray and C. Borst	Brooklyn and New York, N. Y.	Apr. 5, 1864	42, 215
Port-hole for directing ordnance, Adjustable	P. G. Peltz	United States Navy	May 5, 1863	38, 407
Port-light for vessels	G. C. Gourlay	New York, N. Y	Oct. 23, 1860	30, 519
Port, Ship's	C. E. Marshall	Chicago, Ill	June 21, 1870	104, 474
Port-stopper fastening	J. Carver	Lincolnville, Me	July 22, 1873	141, 113
Portable battery or platoon-gun	C. Stauf and C. J. Steinbach	Saint Louis, Mo	Dec. 24, 1861	34, 017
Portable bath	N. Bartlett	Belvidere, Ill	Sept. 23, 1845	4, 203
Portable bath	J. W. Warren, jr	Boston, Mass	July 31, 1840	1, 710
Portable bellows, Construction of	A. Ewing	New York, N. Y	Oct. 31, 1839	1, 393
Portable boat	R. C. Buchanan	Baltimore, Md	Mar. 31, 1857	16, 904
Portable-boat	R. Knudsen and W. T. Lassoe	Brooklyn, N. Y	June 19, 1866	55, 675
Portable body-battery	S. J. Seeley	Brooklyn, N. Y	Dec. 3, 1861	33, 854
Portable boiler for pitch, &c	L. S. Mills	Brooklyn, N. Y	Mar. 26, 1867	63, 286
Portable box	A. Dreyspring	Montgomery, Ala	Oct. 12, 1858	21, 748
Portable-box alarm	W. Shepard and A. T. Murden	Brooklyn, N. Y	Jan. 9, 1872	122, 540
Portable chair	G. Gardner	Clarksville Post-Office, N. J.	Jan. 17, 1871	111, 053
Portable chair	Z. Lyford	Lowell, Mass	May 13, 1856	14, 877
Portable clamp for school-books	T. H. Dennison	Baltimore, Md	Aug. 3, 1869	93, 287
Portable elevator and conveyor	H. C. Crosby	Chicago, Ill	Nov. 8, 1870	108, 976
Portable evaporator	H. T. Douglas and J. Cooper	Zanesville and Mount Vernon, Ohio.	Sept. 6, 1859	25, 323
Portable evaporator	H. L. Plumb	Hamer, Ohio	Dec. 17, 1867	72, 326
Portable fountain	T. H. Bonham	Elizabethtown, Ohio	May 8, 1866	54, 494
Portable fountain	J. Hegarty	Jersey City, N. J	Oct. 22, 1867	69, 994
Portable fountain and lawn-sprinkler	R. Brusie	New York, N. Y	Mar. 25, 1873	137, 175
Portable furnace	H. Beatty	Massillon, Ohio	Aug. 15, 1871	118, 095
Portable furnace	T. Chadwick	Newton, Iowa	Feb. 22, 1870	100, 117
Portable furnace	B. G. Fitzhugh	Frederick, Md	Oct. 17, 1871	119, 977
Portable furnace	S. Gregory	Galesburgh, Mich	May 31, 1870	103, 736
Portable furnace	J. H. Harper	Pittsburgh, Pa	Dec. 7, 1869	97, 634
Portable furnace	J. P. Hayes	Boston, Mass	Nov. 5, 1850	7, 765
Portable furnace	J. Hulbert, jr	Richmond, Ind	Apr. 5, 1870	101, 621
Portable furnace	W. J. Keep	Troy, N. Y	Feb. 13, 1872	123, 706
Portable furnace	J. Lewis	Derby, Conn	July 8, 1834	....
Portable furnace	H. J. Lingenfelter	Glen, N. Y	Dec. 30, 1873	145, 954
Portable furnace	J. H. Lyon	Pittsburgh, Pa	Apr. 12, 1870	101, 893
Portable furnace	A. Magee	Lawrence, Mass	Nov. 10, 1863	40, 570
Portable furnace	J. B. Marvin	New York, N. Y	Mar. 20, 1860	27, 552
Portable furnace	D. B. Montague	Springfield, Mass	Feb. 28, 1871	112, 164
Portable furnace	J. L. Mott	New York, N. Y	Oct. 19, 1838	983
Portable furnace	M. F. Potter	Charlemont, Mass	Jan. 22, 1850	7, 039
Portable furnace	J. Reynolds	Philadelphia, Pa	Mar. 21, 1871	112, 961
Portable furnace	W. D. Robertson	Knoxville, Tenn	Mar. 21, 1871	112, 965
Portable furnace	M. Saulson	Troy, N. Y	Sept. 3, 1872	131, 030
Portable furnace	T. Stead and J. S. Oram	Cleveland, Ohio	July 30, 1872	130, 081
Portable furnace	C. Van De Mark	Phelps, N. Y	Nov. 29, 1870	109, 780
Portable furnace	G. E. Waring	Stamford, Conn	Mar. 16, 1844	3, 491
Portable furnace and kettle	F. Sapaire and F. X. Jacquot	Wilmington, Del	Nov. 19, 1872	133, 231
Portable furnace and wash-boiler, Combined	H. and T. Humphreville	Lancaster, Pa	June 18, 1872	127, 977
Portable furnace for casting brass, &c	J. Share	Baltimore, Md	Mar. 30, 1813	....
Portable furnace for heating water in bathing-tubs	R. Densmore	Hopewell, N. Y	Nov. 26, 1840	1, 866
Portable heater	A. T. Hutson	Richmond, Ind	Nov. 6, 1866	59, 397
Portable heater	W. Shaw	Albany, N. Y	Nov. 14, 1871	120, 905
Portable house	J. Burnside	Washington, D. C	Oct. 21, 1862	36, 701
Portable house	D. Fitzgerald	New York, N. Y	Mar. 4, 1856	14, 355
Portable house	D. Fitzgerald	New York, N. Y	Mar. 12, 1867	62, 831
Portable house	H. W. Forman	Centralia, Kans	July 23, 1872	129, 805
Portable house	A. E. Hotchkiss	Hubbardsville, N. Y	Oct. 29, 1872	132, 533
Portable house	W. R. Mears	Grafton, Ill	Apr. 6, 1869	88, 657
Portable house	D. N. Skillings	Boston, Mass	Nov. 19, 1861	33, 758
Portable house, Sectional	J. Montgomery	New York, N. Y	Feb. 2, 1869	86, 572
Portable houses, Constructing	D. Fitzgerald	New York, N. Y	May 27, 1856	14, 952
Portable houses, Constructing	D. Fitzgerald	New York, N. Y	May 12, 1857	17, 270
Portable houses, Constructing frame for	J. Morgan	Dundee, Ohio	Jan. 3, 1865	45, 734
Portable houses for emigrants, Constructing	F. S. Barnard	Philadelphia, Pa	Dec. 21, 1839	1, 439
Portable lock	A. D. Perry	New York, N. Y	Sept. 5, 1848	5, 751
Portable lubricator	H. Hammond	Hartford, Conn	Jan. 25, 1870	99, 083
Portable mill	N. Burr	Batavia, Ill	Aug. 7, 1860	29, 462

Index of patents issued from the United States Patent Office from 1790 *to* 1873, *inclusive*—Continued.

Invention.	Inventor.	Residence.	Date.	No.
Portable mill	S. Dodson	Jersey City, N. J	Dec. 31, 1867	72, 822
Portable mill	B. Harnish and R. J. King	Lancaster, Pa	Mar. 9, 1869	87, 667
Portable press	T. L. Chase	Philadelphia, Pa	Nov. 21, 1865	51, 019
Portable refreshment-fountain	A. J. Ohmer	Hamilton, Ohio	Dec. 19, 1865	51, 612
Portable register	C. S. Watson	Philadelphia, Pa	Dec. 27, 1859	26, 639
Portable roller for moving heavy bodies	J. R. Anderson	New York, N. Y	Apr. 18, 1871	113, 830
Portable switch	B. C. Galvin	New York, N. Y	Dec. 17, 1867	72, 184
Portable sprinkler	W. L. Fish	Springfield, Mass	Aug. 30, 1870	106, 802
Porte-monnaie	E. Lindner and C. Hoffman	New York, N. Y	Feb. 12, 1856	14, 246
Porte-monnaie	J. L. Mason	Germantown, Pa	Oct. 14, 1856	15, 891
Porte-monnaie and pocket-book fastening	J. Hewson	Newark, N. J	May 13, 1856	14, 867
Porte-monnaie, Armlet	A. W. Pratt	Pultneyville, N. Y	Mar. 10, 1868	75, 458
Porte-monnaie lock and clasp	D. C. Smith	Tecumseh, Mich	Apr. 14, 1857	17, 057
Porte-monnaie, Machine for manufacturing	B. S. Stedman	West Meriden, Conn	Sept. 14, 1852	9, 264
Porters' box	I. Barman	Portland, Oreg	Sept. 16, 1873	142, 888
Portfolio	R. Arthur	Philadelphia, Pa	May 19, 1857	17, 311
Portfolio	G. and E. Ashworth	Manchester, Great Britain	Sept. 8, 1868	81, 864
Portfolio	G. Harvey	New York, N. Y	Jan. 2, 1872	122, 461
Portfolio	L. Heyl	Philadelphia, Pa	Sept. 6, 1864	44, 094
Portfolio	L. Heyl	Philadelphia, Pa	Mar. 20, 1866	53, 299
Portfolio	C. W. Palmer	Cleveland, Ohio	July 8, 1873	147, 725
Portfolio	W. H. Pratt	Davenport, Iowa	July 16, 1872	129, 589
Portfolio	L. F. Raymond	Bordeaux, France	Dec. 16, 1873	145, 683
Portfolio	J. Shaw	Providence, R. I	Dec. 24, 1850	7, 860
Portfolio	J. Shaw	Providence, R. I	June 17, 1856	15, 150
Portfolio	H. T. Sisson	Providence, R. I	Dec. 29, 1857	18, 994
Portfolio	H. T. Sisson	Providence, R. I	Apr. 5, 1859	23, 506
Portfolio	D. Slote	New York, N. Y	Sept. 17, 1872	131, 473
Portfolio	S. J. H. Smith	Boston, Mass	June 5, 1855	13, 014
Portfolio and writing-tablet	G. C. Hathaway	Plymouth, Mass	Dec. 24, 1861	33, 991
Portfolio, Commercial	S. E. Mandlebaum	Saint Louis, Mo	Feb. 22, 1870	100, 050
Portfolio for filing music, printed matter, &c	J. C. Koch	New York, N. Y	May 21, 1861	32, 374
Portfolio-holder	J. B. Aiken	Franklin, N. H	Sept. 16, 1873	142, 884
Portfolio or music stand	A. Eliaers	Boston, Mass	Sept. 22, 1857	18, 239
Portfolio-stand	L. Dubernet	New York, N. Y	May 9, 1865	47, 624
Post: *See* Arbor or fence post. Clothes-post. Clothes-line post. Composition-post. Fence-post. Fence and trellis post. Gas-lamp post. Gate-post. Guide-post. Hitching-post. Hitching and sign post. Iron post. Iron gate or fence post. Jack-post. Lamp-post. Newel-post. Sewing-machine post. Stone post. Telegraph-post.				
Post and pile driver	O. Hyde	Benicia, Cal	June 15, 1858	20, 563
Post-boring and rail-pointing machine	R. Martin and M. Harner	Taneytown, Md	June 21, 1870	104, 476
Post-boring and rail-pointing machine	J. Young	Middletown, Md	July 10, 1855	13, 243
Post-boring and rail-pointing machine	J. Young and C. I. Grumbine	Fairview and Frederick, Md.	May 25, 1869	90, 622
Post-boring machine	A. Butler	Vermont	Nov. 7, 1811	....
Post-boring machine	W. F. Cline and P. D. Weaver	Bendersville, Pa	Nov. 14, 1871	120, 941
Post-boring machine	J. R. Group	Idaville, Pa	Feb. 7, 1871	111, 532
Post-boring machine	B. F. Mohr	Mifflinburgh, Pa	Mar. 3, 1868	75, 041
Post-driver	I. V. Adair	Romulus, N. Y	Mar. 25, 1873	137, 120
Post-driver	A. B. Clark	Richmond, Ind	Aug. 4, 1868	80, 600
Post-driver	J. Ellenberger	Easton, Ohio	Nov. 17, 1868	84, 100
Post-driver	C. T. Fitch	Harbor Creek, Pa	Nov. 26, 1867	71, 372
Post-driver	D. D. Foley	Washington, D. C	Nov. 27, 1866	59, 990
Post-driver	A. McClara	Whitney's Point, N. Y	Nov. 3, 1868	83, 692
Post-driver	S. McCullough and A. Robins	Buffalo, Ohio	Dec. 24, 1867	72, 518
Post-driver	J. M. Sampson	Waynesville, Ill	Mar. 25, 1856	14, 520
Post-driver	C. H. Williamson and J. H. Allyn.	Whitestown, N. Y	Dec. 31, 1872	134, 412
Post-driving machine	W. Altick	Dayton, Ohio	Aug. 10, 1869	93, 576
Post-driving machine	C. T. Fitch	Harbor Creek, Pa	May 7, 1867	64, 514
Post frost-protector	S. T. Varian	Plainfield, N. J	Feb. 16, 1869	87, 015
Post, Hitching	D. C. Robie	Springfield, Mass	Sept. 17, 1867	69, 027
Post-hole borer	J. Cothron	Illiopolis, Ill	Sept. 22, 1868	82, 293
Post-hole borer	J. K. Miller	New York, N. Y	Mar. 24, 1868	75, 947
Post-hole borer	J. M. Walter and S. Shank	Springfield, Ohio	Sept. 15, 1868	82, 185
Post-hole-boring machine	A. Q. Allis	Dayton, Ohio	Feb. 25, 1868	74, 785
Post-hole-boring machine	J. Corbett	Hanover, Pa	Dec. 21, 1819	....
Post-hole-boring machine	W. I. Hale	Ashley, Ill	Apr. 14, 1868	76, 752
Post-hole-boring machine	W. R. Iles	West Rushville, Ohio	Dec. 24, 1867	72, 497
Post-hole-cutting instrument	N. S. Vance and E. Watkins	Decatur, Ill	Oct. 22, 1867	70, 137
Post-hole digger	W. E. Ball	Bethesda, Ohio	May 14, 1872	126, 773
Post-hole digger	J. Boone	Clintonville, Ky	June 14, 1864	43, 085
Post-hole digger	J. W. Harper	Xenia, Ohio	Feb. 19, 1867	62, 131
Post-hole digger	B. B. Herrick	Decatur, Mich	Mar. 7, 1871	112, 454
Post-hole digger	B. B. Herrick and C. W. Wicker.	Duquoin, Ill	Nov. 23, 1869	97, 086
Post-hole digger	J. Lee	Bolivar, Ohio	June 26, 1860	28, 874
Post-hole digger	A. E. Lindsley	Paw Paw, Mich	Feb. 14, 1871	111, 757
Post-hole digger	A. E. Lindsley	Paw Paw, Mich	May 20, 1873	139, 169
Post-hole digger	E. R. Sumner	Freeport, Ill	Aug. 29, 1871	118, 560
Post-hole digger	E. R. Sumner	Freeport, Ill	Oct. 29, 1872	132, 607
Post-hole digger and ditch	R. L. Jones	Cleveland, Ohio	July 25, 1871	117, 297
Post-hole-digging machine	A. W. Porter	Saint Johnsville, N. Y	Aug. 7, 1860	29, 518
Post-hole-digging machine	J. Thompson and B. B. Herrick.	Edgewood, Ill	Nov. 19, 1867	71, 086
Post-hole-excavating machine	W. K. Johnston	Rock Island, Ill	Mar. 9, 1858	19, 565
Post-hole excavator	H. Sutliff	Waverly, N. Y	Sept. 14, 1869	94, 790
Post-hole-making machine	J. M. Kirkpatrick	Utica, Ohio	Dec. 20, 1870	110, 372
Post-mold for fences, Stone	E. Boston	Gloucester, N. J	Oct. 14, 1815	....
Posts, Setting	P. Burke	Helena, N. Y	Sept. 24, 1861	33, 335
Posts, Setting	U. B. Stribling	Madison, Ind	Sept. 20, 1870	107, 637

Index of patents issued from the United States Patent Office from 1790 *to* 1873, *inclusive*—Continued.

Invention.	Inventor.	Residence.	Date.	No.
Posts, timber, &c., Composition for setting	A. D. Purinton	Dover, Mass	June 9, 1868	78, 691
Post-office box	E. Stockwell and W. H. Taylor.	Stamford, Conn	June 17, 1873	140, 092
Post-office-box front	C. J. Clements	New York, N. Y	June 18, 1872	128, 116
Post-office-box lock	L. Yale, jr	Bernardston, Mass	Oct. 24, 1871	120, 177
Post-office cabinet for Sunday-school and other rooms.	W. M. and A. Clark	Pittsburgh, Pa	Aug. 16, 1870	106, 325
Post-office-delivery box	J. H. Beidler	Lincoln, Ill	Aug. 28, 1866	57, 464
Post-office distributing-table	C. H. Bradley	West Chester, Pa	July 9, 1861	32, 748
Post-office letter-box	C. H. Bradley	West Chester, Pa	Sept. 15, 1863	39, 879
Post-office letter-box	B. C. Davis	Binghamton, N. Y	Dec. 7, 1869	97, 483
Post-office letter-case	A. F. Miller	Galva, Ill	Apr. 9, 1872	125, 666
Post-office, &c., pigeon-hole	H. G. Pearson	New York, N. Y	Nov. 5, 1867	70, 674
Postal card	H. M. Johnston	New York, N. Y	Nov. 18, 1873	144, 677
Postal-card case	C. H. Townsend, T. E. Hughes, and A. A. Keith.	Minneapolis, Minn	Nov. 11, 1873	144, 423
Postal-card machine	J. M. D. Keating	New York, N. Y	Apr. 22, 1873	138, 028
Postal-card screen	L. Conroy	New York, N. Y	Sept. 16, 1873	142, 899
Postal wrapper	E. Reed	Joliet, Ill	Nov. 6, 1866	59, 451
Pot:				
See Annealing-pot. — Glass-making pot.				
Batter-pot. — Glass-melting pot.				
Bean-pot. — Glue-pot.				
Bone burning or charring pot. — Lead-pot.				
— Marking-pot.				
Chamber-pot. — Molder's pot.				
Coffee-pot. — Mucilage-pot.				
Culinary-pot. — Mustard-pot.				
Dinner-pot. — Paint-pot.				
Eel-pot. — Painter's pot.				
Fire-pot. — Salinometer-pot.				
Flower-pot. — Sprinkling-pot.				
Glass-pot. — Sugar-drip pot.				
Glass-house pot. — Tea-pot.				
Glass-maker's pot. — Tea and coffee pot.				
Pot and kettle cover	S. Van Horn	Chicopee, Mass	Aug. 8, 1871	117, 834
Pot and kettle tipping attachment	C. Coester, jr., and W. L. Dewey.	Bridgeport, Conn	Jan. 29, 1867	61, 654
Pot and kettle tilting device	A. J. Redway	Cincinnati, Ohio	Sept. 17, 1872	131, 410
Pot-cover	W. H. Barker	Windsor, Canada	Sept. 12, 1871	118, 896
Pot, kettle, &c., cover	G. Reuben	San Francisco, Cal	Dec. 29, 1868	85, 398
Pot, kettle, &c., elevator	J. Edwards	New York	Mar. 19, 1825	
Potash and leaching ashes, Manufacturing	E. Parce	Lincklaen, N. Y	July 20, 1831	
Potash and leaching ashes, Manufacturing	E. Parce	Pitcher, N. Y	Jan. 23, 1834	
Potash and pearlash	E. Williams	Erie, Pa	Mar. 8, 1836	
Potash and pearlash furnace	E. Craft, jr		May 12, 1804	
Potash and pearlash furnace	E. Ryan		Apr. 29, 1793	
Potash and pearlash, Making	S. Hopkins		July 31, 1790	
Potash and pearlash, Manufacture of	B. F. Jewett	Malone, N. Y	Sept. 4, 1866	57, 728
Potash and pearlash, Manufacturing	J. Latting	New York	Apr. 26, 1816	
Potash and phosphate of lime, Manufacture of	M. B. Manwaring and R. D. Birch.	New York, N. Y., and Philadelphia, Pa.	Mar. 26, 1872	124, 964
Potash and preparing and leaching ashes, Manufacture of.	E. Williams	Harbor Creek, Pa	Aug. 16, 1835	
Potash and soda, Manufacture of the prussiate of	M. Kalbfleisch	Bushwick, N. Y	Jan. 25, 1848	5, 419
Potash and soda, Preparing	G. Thompson	East Tarentum, Pa	July 24, 1855	13, 325
Potash and soda. Preparing hydrated silicates of	J. M. Ordway	Manchester, N. H	May 5, 1863	38, 449
Potash feldspar for obtaining certain salts, Decomposing.	R. A. Tilghman	Philadelphia, Pa	Dec. 4, 1847	5, 384
Potash into sal polychrestum, &c., Separating, &c., sulphate of.	A. McNitt	Geneva, N. Y	June 15, 1805	
Potash, Manufacture of	Hartsuff and French	Aurelius, N. Y	July 17, 1835	
Potash, Manufacture of chromate of	I. Tyson, jr	Baltimore, Md	Oct. 9, 1845	4, 224
Potash, Manufacture of nitrate of	J. C. Pennington	Paterson, N. J	July 22, 1862	35, 946
Potash, Manufacture of pure carbonate of	C. F. Moll	Kenton, Ohio	Aug. 2, 1870	105, 968
Potash, Manufacturing	W. Frobisher		Nov. 17, 1796	
Potash, Manufacturing	T. Power		Sept. 19, 1801	
Potash, Manufacturing	E. Ryan		Nov. 16, 1796	
Potash, Manufacturing	T. H. Sherman	Hastings, N. Y	Feb. 2, 1831	
Potash, Extracting alkali from ashes in the manufacture of.	J. Osborn	Elyria, Ohio	July 18, 1840	1, 697
Potash, Leaching ashes in the manufacture of	J. H. Ward	Randolph, Ohio	July 16, 1841	2, 178
Potash or soda from alkaline silicate, Liberating	F. O. Ward	London, England	Mar. 21, 1865	46, 979
Potash or soda, Making stannate of	J. Young	Manchester, England	Aug. 20, 1850	7, 588
Potash, Retort for the manufacture of prussiate of	D. B. and A. D. Coles	Newark, N. J	Nov. 14, 1865	50, 907
Potash, soda, &c., Manufacture of sulphates of	H. Deacon	Warrington, England	Feb. 20, 1872	123, 875
Potassa and soda, Preparing stannate of	J. Greenwood, J. Mercer, and J. Barnes.	England	Aug. 12, 1846	4, 688
Potassa and soda, Putting up and preserving caustic.	T. C. Taylor	Philadelphia, Pa	Feb. 6, 1866	52, 466
Potassa, Manufacture of bitartrate of	L. Adler	Baltimore, Md	Dec. 16, 1873	145, 607
Potassa, Manufacture of bitartrate of	G. Bourgade	Jersey City, N. J	Aug. 13, 1872	130, 407
Potassa, Preparation of nitrate of	H. S. Osborn	Belvidere, N. J	May 2, 1865	47, 562
Potassa, Retort for the manufacture of prussiate of	J. Clark	Newark, N. J	Aug. 6, 1861	32, 977
Potato and drill machine	S. L. Spencer	Hopewell X Roads, Md	Aug. 14, 1866	57, 207
Potato and pea-nut digger	C. E. Peirce	Centreville, Mich	Mar. 18, 1873	136, 860
Potato and root washer	W. Ellis	Waterville, Me	Feb. 25, 1835	
Potato assorter and sifter	L. Perrine and P. G. Conover	Freehold, N. J	Feb. 16, 1869	87, 001
Potato-assorting apparatus	J. F. Unlish	Webster, N. Y	May 10, 1870	102, 886
Potato-assorting machine	D. M. King	Garrettsville, Ohio	Sept. 30, 1873	143, 357
Potato-assorting machine	B. D. and D. A. Vandeveer and T. Denise.	Freehold, N. J	Oct. 25, 1870	108, 655
Potato-baker	A. Reid	Buffalo, N. Y	Oct. 1, 1867	69, 480
Potato-bug cleaner	J. Corbeil	Lind, Wis	July 28, 1868	80, 458
Potato-bug collector	J. Corbeil	Lind, Wis	Jan. 14, 1873	134, 854
Potato-bug collector	R. P. Main	Oregon, Wis	May 21, 1872	127, 079

Index of patents issued from the United States Patent Office from 1790 *to* 1873, *inclusive*—Continued.

Invention.	Inventor.	Residence.	Date.	No.
Potato-bug destroyer	A. A. Mixer	Hamilton, Ohio	Sept. 30, 1873	143, 245
Potato-bug destroyer	W. H. Myers	Oregon, Wis	Sept. 23, 1873	143, 178
Potato-bug destroyer	J. B. Turney	Inkster, Mich	July 11, 1871	116, 894
Potato-bug exterminator	C. Cole	Chatfield, Minn	Oct. 28, 1873	144, 065
Potato-bugs, Compound for destroying	L. Pagin	Niles, Mich	Mar. 14, 1871	112, 732
Potato-bugs, Machine for destroying	H. Pitchforth and W. Benson	Muscatine, Iowa	June 4, 1867	65, 427
Potato cultivator and digger	W. J. Avery and T. Laberteaux	Marshall Township, Mich.	Feb. 22, 1870	100, 102
Potato cutter and planter	L. J. Newborn	Kinston, N. C	Oct. 14, 1873	143, 707
Potato-cutting machine	A. Williams	Ashfield, Mass	Mar. 12, 1839	1, 101
Potato-digger	G. W. Adams and J. R. Hopper	Rochester, N. Y	Sept. 14, 1869	94, 693
Potato-digger	J. Albaugh	Lyons, N. Y	June 21, 1870	104, 401
Potato-digger	R. L. Allen	New York, N. Y	Jan. 18, 1859	22, 612
Potato-digger	A. Q. Allis	Rochester, N. Y	Oct. 22, 1872	132, 427
Potato-digger	A. Anderson	Markham, Canada	Dec. 29, 1857	18, 009
Potato-digger	W. F. Andrews	Watson, N. Y	June 29, 1869	91, 897
Potato-digger	L. Annis	Quincy, Mich	Dec. 29, 1868	85, 418
Potato-digger	S. E. Anthony	Stillwater, N. Y	Sept. 20, 1870	107, 433
Potato-digger	S. E. Anthony	Stillwater, N. Y	Apr. 25, 1871	114, 088
Potato-digger	P. Antonides	Freehold, N. J	Feb. 23, 1864	41, 739
Potato-digger	P. Antonides	Freehold, N. J	Dec. 31, 1867	72, 773
Potato-digger	L. A. Aspinwall	Watervliet, N. Y	Nov. 14, 1865	50, 889
Potato-digger	L. A. Aspinwall	Albany, N. Y	June 25, 1867	66, 064
Potato-digger	B. Avis	Penn's Grove, N. J	May 14, 1872	126, 619
Potato-digger	J. B. Baker	Syracuse, N. Y	Sept. 12, 1871	118, 895
Potato-digger	T. Baker	Stillwater, N. Y	July 7, 1863	39, 108
Potato-digger	O. W. Baldwin	Greenfield, Ohio	May 18, 1869	90, 065
Potato-digger	O. W. Baldwin and F. H. Pope	Greenfield, Ohio	Apr. 9, 1867	63, 604
Potato-digger	H. G. Barrows	Benton, Me	Jan. 28, 1873	135, 307
Potato-digger	G. C. Bartlett	Paris, N. Y	July 24, 1860	29, 339
Potato-digger	A. L. Bausman	Minneapolis, Minn	Oct. 11, 1870	108, 093
Potato-digger	J. Bawden	Freehold, N. J	May 8, 1860	28, 232
Potato-digger	W. Beaty	Pontiac, Mich	Mar. 16, 1869	87, 901
Potato-digger	J. Belknap	Adrian, Mich	Nov. 16, 1869	96, 770
Potato-digger	D. D. Bell	Wawarsing, N. Y	Dec. 9, 1851	8, 574
Potato-digger	E. Bennett	Nankin, Mich	June 9, 1868	78, 641
Potato-digger	L. Berry	Clyde, N. Y	Sept. 13, 1870	107, 327
Potato-digger	L. L. Bettys	Ontario, N. Y	May 4, 1869	89, 547
Potato-digger	D. Bibbee and W. Rand	Letart Falls, Ohio	July 14, 1870	104, 251
Potato-digger	S. Bigelow and L. C. Bigelow	Newburgh and Kirtland, Ohio.	May 6, 1873	138, 470
Potato-digger	J. H. Billmeyer	Raisin, Mich	May 3, 1870	102, 479
Potato-digger	J. W. Blodgett	Three Rivers, Mich	June 30, 1868	79, 439
Potato-digger	C. Blood	Malta, N. Y	Sept. 1, 1863	39, 712
Potato-digger	J. I. Brinkerhoff	Auburn, N. Y	Mar. 9, 1869	87, 536
Potato-digger	D. D. T. Brown	Mumford, N. Y	June 11, 1872	127, 676
Potato-digger	J. Brown	Hamlin, Mich	Jan. 19, 1869	85, 994
Potato-digger	A. G. Brush	Susquehanna Depot, Pa	Feb. 18, 1873	136, 028
Potato-digger	G. J. Bundy	Lyndon, Vt	July 4, 1854	11, 206
Potato-digger	I. S. Bunnell	Montrose, Pa	Dec. 29, 1857	18, 954
Potato-digger	T. and G. B. Burditt	Dansville, N. Y	May 14, 1867	64, 627
Potato-digger	A. Burhans	Albany, N. Y	Sept. 8, 1868	81, 980
Potato-digger	J. M. Burke	Dansville, N. Y	July 27, 1869	93, 050
Potato-digger	J. W. Burnham and W. Conlon	Middletown Point, N. J	May 26, 1868	78, 257
Potato-digger	W. H. Burridge	Cleveland, Ohio	Sept. 25, 1866	58, 334
Potato-digger	J. Burt	Sturgis, Mich	Sept. 29, 1868	82, 592
Potato-digger	W. F. Caldwell	Oxford, Me	July 17, 1866	56, 366
Potato-digger	C. B. Cannon	Keokuk, Iowa	Dec. 18, 1866	60, 618
Potato-digger	A. S. Capron and D. S. Davis	Grass Lake, Mich	Oct. 11, 1859	25, 721
Potato-digger	H. C. Carr	Bordentown, N. J	Mar. 15, 1870	100, 854
Potato-digger	H. Carrier	Kirtland, Ohio	June 15, 1869	91, 415
Potato-digger	W. H. Chamberlin	Medina, N. Y	Mar. 3, 1868	74, 991
Potato-digger	A. F. Chandler	Winthrop, Me	Apr. 9, 1867	63, 612
Potato-digger	O. M. Chase	Boston, Mass	Feb. 2, 1864	41, 424
Potato-digger	J. C. Clark	Elmira, N. Y	Mar. 3, 1868	75, 123
Potato-digger	L. Cochran	Penn's Grove, N. J	Nov. 8, 1870	109, 111
Potato-digger	W. W. Cole and T. McGhee	Eudora, Kans	July 13, 1869	92, 518
Potato-digger	S. B. Conover	New York, N. Y	June 3, 1862	35, 435
Potato-digger	S. B. Conover	New York, N. Y	Jan. 5, 1864	41, 058
Potato-digger	S B. Conover	New York, N. Y	May 2, 1867	63, 475
Potato-digger	C. Cooney	Columbus, Nebr	May 9, 1871	114, 531
Potato-digger	I. Cooper	Middlefield, Ohio	May 25, 1869	90, 341
Potato-digger	V. P. Corbett	Alexandria County, Va	Nov. 10, 1868	83, 833
Potato-digger	V. P. Corbett	Alexandria County, Va	Aug. 20, 1872	130, 569
Potato-digger	J. W. Corwin	Lebanon, Ohio	Feb. 2, 1869	86, 509
Potato-digger	J. W. Corwin	Lebanon, Ohio	Nov. 22, 1870	109, 389
Potato-digger	W. Cousens	Orono, Me	Oct. 3, 1871	119, 577
Potato-digger	W. J. Cowan	Cortland, N. Y	Oct. 6, 1868	82, 806
Potato-digger	I. Curtis	Des Moines, Iowa	Apr. 6, 1869	88, 613
Potato-digger	C. H. Dana	West Lebanon, N. H	May 16, 1854	10, 926
Potato-digger	J. P. Davison	Rome, N. Y	Aug. 4, 1868	80, 611
Potato-digger	C. Darling	Utica, N. Y	Sept. 13, 1870	107, 229
Potato-digger	J. Davis	Saratoga, N. Y	Jan. 31, 1871	111, 437
Potato-digger	R. G. Dayton	North Granville, N. Y	Mar. 19, 1872	124, 726
Potato-digger	R. G. Dayton	North Granville, N. Y	Nov. 5, 1872	132, 814
Potato-digger	D. De Garmo	Rochester, N. Y	July 24, 1860	29, 252
Potato-digger	D. N. Denman	Milburn, N. J	Sept. 20, 1864	44, 288
Potato-digger	P. Dennis	Stillwater, N. Y	Mar. 3, 1857	16, 722
Potato-digger	W. Dillon	Sonoma, Cal	Oct. 11, 1870	108, 117
Potato-digger	A. Dolloff	Lake Village, N. H	Feb. 27, 1866	52, 830
Potato-digger	E. O. Doud and W. F. Beardsley.	Penfield, N. Y	Apr. 30, 1867	64, 291
Potato-digger	H. M. and W. W. Dowd, jr	Saratoga Springs and North Granville, N. Y.	June 10, 1873	139, 709
Potato-digger	M. F. Drake	Pleasant Ridge, Ohio	Dec. 11, 1866	60, 347
Potato-digger	C. Dunham	Batavia, N. Y	May 22, 1860	28, 356
Potato-digger	R. Dunlap, 1st	South Lyons, Mich	June 29, 1869	92, 024
Potato-digger	W. H. Elliot	New York, N. Y	Jan. 8, 1867	61, 058
Potato-digger	J. Farlow	Harrison, Ind	July 16, 1872	129, 120

Index of patents issued from the United States Patent Office from 1790 *to* 1873, *inclusive*—Continued.

Invention.	Inventor.	Residence.	Date.	No.
Potato-digger	H. Farmer	Pontiac, Mich	July 20, 1869	92, 712
Potato-digger	W. A. Field	Schuylkill Haven, Pa	May 4, 1869	89, 571
Potato-digger	E. Finch	Albion, N. Y	July 12, 1870	105, 324
Potato-digger	L. Finch	Vienna, Va	Apr. 7, 1868	76, 427
Potato-digger	E. T. Ford	Stillwater, N. Y	Aug. 11, 1863	39, 475
Potato-digger	E. T. Ford	Stillwater, N. Y	Oct. 18, 1864	44, 716
Potato-digger	E. T. Ford	Stillwater, N. Y	Sept. 1, 1868	81, 765
Potato-digger	E. T. Ford	Stillwater, N. Y	July 13, 1869	92, 437
Potato-digger	E. T. Ford	Stillwater, N. Y	Oct. 28, 1873	144, 083
Potato-digger	E. G. Ford and J. F. Penquite	Delphos, Ohio	Mar. 6, 1866	52, 984
Potato-digger	H. France	Hinmansville, N. Y	Nov. 17, 1868	84, 056
Potato-digger	J. E. Giles and W. Ferry	Mead's Mills, Mich	Dec. 8, 1868	84, 741
Potato-digger	J. E. Giles and C. S. McRobert	Mead's Mills, Mich	June 25, 1867	66, 145
Potato-digger	A. Gilmore	Fort Atkinson, Wis	Dec. 18, 1866	60, 501
Potato-digger	D. Gorman	Hornellsville, N. Y	Oct. 19, 1869	95, 896
Potato-digger	C. G. Grabo	Detroit, Mich	July 23, 1867	67, 040
Potato-digger	A. Graves	Marcellus Falls, N. Y	Apr. 7, 1868	76, 438
Potato-digger	A. Graves	Marcellus Falls, N. Y	Apr. 28, 1868	77, 279
Potato-digger	J. H. Gray and C. W. Calhoun	Florence Township, Mich	Jan. 14, 1868	73, 319
Potato-digger	B. F. Green	Syracuse, N. Y	Aug. 6, 1872	130, 125
Potato-digger	J. M. Green	West Bloomfield, N. Y	July 16, 1867	66, 704
Potato-digger	W. Green	Holly, Mich	Dec. 7, 1869	97, 630
Potato-digger	I. Griffen	Quaker Springs, N. Y	June 2, 1857	17, 428
Potato-digger	E. V. W. Griffith	Utica, N. Y	Dec. 17, 1867	72, 289
Potato-digger	A. L. Grinnell and J. Z. Williams.	Willet, Wis	June 10, 1856	15, 100
Potato-digger	L. B. Griswold	Pennfield, N. Y	Aug. 9, 1859	25, 010
Potato-digger	I. C. Groom	Albany, N. Y	July 6, 1869	92, 304
Potato-digger	G. W. Haag	Cairo, Pa	May 6, 1873	138, 561
Potato-digger	A. Hadwen	Rochester, N. Y	Mar. 17, 1868	75, 543
Potato-digger	A. M. Hall	Falmouth, Me	Nov. 11, 1868	83, 847
Potato-digger	J. Hall, jr	Temperanceville, Pa	May 10, 1870	102, 812
Potato-digger	J. Hall, jr	Temperanceville, Pa	May 10, 1870	102, 814
Potato-digger	J. Hall, jr., and O. Flanigan	Temperanceville, Pa	May 10, 1870	102, 811
Potato-digger	M. Hall	Livonia, Mich	Apr. 23, 1867	64, 097
Potato-digger	J. E. Hardenbergh	Fultonville, N. Y	Dec. 22, 1857	18, 899
Potato-digger	U. R. Harlow	Farmersville, Cal	Aug. 2, 1870	106, 058
Potato-digger	W. D. Harrell	Moore's Hill, Ind	Mar. 14, 1871	112, 592
Potato-digger	L. W. Harris	Waterville, N. Y	Apr. 6, 1858	19, 849
Potato-digger	D. B. Hart	Mentor, Ohio	Apr. 30, 1867	64, 218
Potato-digger	R. A. Haw	Bucksport, Cal	July 25, 1871	117, 415
Potato-digger	L. Henderson	Manson, N. C	Aug. 23, 1870	106, 583
Potato-digger	M. Henderson	Detroit, Mich	July 21, 1868	80, 070
Potato-digger	T. N. Henderson	Jackson, Mich	May 4, 1869	89, 661
Potato-digger	A. Heulings	Philadelphia, Pa	May 6, 1856	14, 810
Potato-digger	A. Heulings	Philadelphia, Pa	May 28, 1872	127, 345
Potato-digger	S. S. Hickok	Marlborough, N. J	Feb. 17, 1863	37, 691
Potato-digger	I. Hicks	Hartford, Wis	Oct. 3, 1871	119, 461
Potato-digger	J. J. Hill	Xenia, Ohio	Aug. 22, 1865	49, 521
Potato-digger	A. P. Hinz	Dunton, Ill	Sept. 1, 1868	81, 636
Potato-digger	G. N. Hoag	Muscatine, Iowa	Nov. 15, 1870	109, 211
Potato-digger	H. Holcroft and C. S. Smith	Media, Pa	Mar. 24, 1863	39, 964
Potato-digger	J. R. Hopper	Rochester, N. Y	Mar. 30, 1869	88, 295
Potato-digger	J. H. Hotchkin	Prattsburgh, N. Y	July 13, 1869	92, 450
Potato-digger	W. H. Hutchins	Lockport, Ill	Nov. 19, 1867	71, 177
Potato-digger	J. W. Innis	Newburgh, N. Y	Feb. 19, 1867	62, 271
Potato-digger	J. W. Innis	Newburgh, N. Y	Feb. 4, 1873	135, 558
Potato-digger	J. O. Ives	Saint Louis, Mo	Aug. 15, 1865	49, 413
Potato-digger	M. Jacobs	Sturgis, Mich	Dec. 8, 1868	84, 828
Potato-digger	M. Johnson	Three Rivers, Mich	June 30, 1868	79, 473
Potato-digger	M. Johnson	Three Rivers, Mich	May 17, 1870	103, 194
Potato-digger	M. Johnson	Three Rivers, Mich	May 24, 1870	103, 469
Potato-digger	M. Johnson	Three Rivers, Mich	Nov. 28, 1871	121, 379
Potato-digger	M. Johnson	Three Rivers, Mich	Mar. 12, 1872	124, 585
Potato-digger	M. Johnson	Three Rivers, Mich	Apr. 2, 1872	125, 303
Potato-digger	M. Johnson	Three Rivers, Mich	Oct. 22, 1872	132, 364
Potato-digger	L. Johonnott	Burlington, Vt	June 18, 1872	128, 150
Potato-digger	F. Jones	Terre Haute, Ind	Sept. 19, 1871	119, 152
Potato-digger	R. D. Jones and T. Purcell	Rochester, N. Y	Mar. 29, 1864	42, 088
Potato-digger	W. Jones	Saint Louis, Mo	Dec. 8, 1863	40, 842
Potato-digger	W. Joseph	Quincy, Mich	Nov. 15, 1870	109, 217
Potato-digger	A. C. Ketchum	Schenectady, N. Y	Feb. 20, 1844	3, 442
Potato-digger	D. M. and G. E. King	Mantua Station, Ohio	Apr. 15, 1873	137, 931
Potato-digger	S. Kinman	Eureka, Cal	Jan. 12, 1869	85, 833
Potato-digger	G. W. Kintz	West Henrietta, N. Y	June 25, 1867	66, 029
Potato-digger	G. W. Kintz	West Henrietta, N. Y	Mar. 31, 1868	76, 200
Potato-digger	H. J. Kintz	Greece, N. Y	Mar. 12, 1867	62, 858
Potato-digger	G. W. Knight	Camden, N. J	Feb. 8, 1870	99, 575
Potato-digger	M. W. Knox	Sheridan, N. Y	Apr. 5, 1870	101, 474
Potato-digger	M. W. Knox	Sheridan, N. Y	Oct. 18, 1870	108, 490
Potato-digger	G. Koenig and G. Otto	Plymouth, Mich	Feb. 18, 1868	74, 699
Potato-digger	J. P. Lafetra	Shrewsbury, N. J	Apr. 4, 1871	113, 308
Potato-digger	D. M. Lamb	Strathroy, Canada	Dec. 10, 1872	133, 710
Potato-digger	T. Lane	San Francisco, Cal	Dec. 9, 1862	37, 100
Potato-digger	E. S. Lenox	New York, N. Y	Dec. 26, 1865	51, 731
Potato-digger	E. S. Lenox and E. Spaulding	New York, N. Y	Apr. 23, 1867	64, 114
Potato-digger	J. J. Lingley	La Fayette, Ind	Oct. 26, 1869	96, 153
Potato-digger	D. Locke	Geneva, Wis	Sept. 28, 1869	95, 242
Potato-digger	G. C. Love	English Centre, Pa	Feb. 7, 1871	111, 656
Potato-digger	J. M. Lumbard	Decatur, Mich	Apr. 23, 1872	125, 968
Potato-digger	O. E. Mallory	Batavia, N. Y	Feb. 22, 1870	100, 168
Potato-digger	A. Marcellus	Pittsford, N. Y	Feb. 6, 1866	52, 428
Potato-digger	A. Marcellus	Pittsford, N. Y	Sept. 17, 1867	69, 035
Potato-digger	A. A. Marcellus	New York, N. Y	Nov. 20, 1855	13, 822
Potato-digger	P. Marcy	Tunkhannock, Pa	June 21, 1859	24, 474
Potato-digger	P. Marcy	Tunkhannock, Pa	Aug. 20, 1861	33, 093
Potato-digger	G. M. Marks	Half Moon, Pa	Sept. 27, 1870	107, 698
Potato-digger	I. W. McGaffey	Philadelphia, Pa	Nov. 7, 1854	11, 899

Index of patents issued from the United States Patent Office from 1790 to 1873, inclusive—Continued.

Index of patents issued from the United States Patent Office from 1790 *to* 1873, *inclusive*—Continued.

Invention.	Inventor.	Residence.	Date.	No.
Potato-digger	J. W. Woolsey	Niles, Mich	Sept. 9, 1872	36, 434
Potato-digger	S. Woolson	Moodna, N. Y	Aug. 26, 1856	15, 628
Potato-digger	B. P. Wright	San Francisco, Cal	Apr. 7, 1868	76, 371
Potato-digger	J. S. Wright and I. Benedict	Palmyra, N. Y	Feb. 4, 1873	135, 458
Potato-digger	G. W. B. Yocom, R. J. Walker, and E. Sharp.	Arcata, Cal	Nov. 2, 1869	96, 370
Potato-digger	T. C. Zulick	Schuylkill Haven, Pa	Mar. 5, 1861	31, 641
Potato digger and cultivator	M. Johnson	Three Rivers, Mich	Oct. 19, 1869	96, 007
Potato digger and cultivator, Combined	M. Johnson	Three Rivers, Mich	Aug. 3, 1869	93, 307
Potato digger and picker	W. J. Thompson	Normal, Ill	July 6, 1869	92, 402
Potato digger and separator	J. W. Bartlett	Harmar, Ohio	May 30, 1865	47, 916
Potato digger and separator	W. Green	Holly, Mich	Sept. 1, 1868	81, 772
Potato-digger and shovel-plow frame	J. M. Dick	Buffalo, N. Y	Aug. 2, 1870	105, 922
Potato-digger and stone-gatherer	J. T. Foster	New York, N. Y	June 29, 1852	9, 069
Potato-digger and vine-puller	J. W. Newton	Geneva, Wis	Nov. 3, 1868	83, 785
Potato-digger and weeder	G. W. Hall	Triangle, N. Y	Oct. 1, 1867	69, 339
Potato digger, Sweet	R. Darlington and E. L. Watson	Auburn, N. Y	Feb. 15, 1870	99, 859
Potato-digger, Agitator-wheel for	J. Hall, jr	Temperanceville, Pa	May 10, 1870	102, 813
Potato-diggers, Vine-puller for	L. Berry	Clyde, N. Y	Sept. 13, 1870	107, 328
Potato digging and gathering machine	J. B. Parvin	Heightstown, N. J	May 10, 1859	23, 988
Potato-digging fork	H. Doolittle	East Cleveland, Ohio	Oct. 2, 1866	58, 390
Potato-digging machine	J. W. Bartlett	Harmar, Ohio	Feb. 19, 1867	62, 246
Potato-digging machine	D. Clint and I. Lynd	Poestenkill, N. Y	Feb. 5, 1861	31, 299
Potato-digging machine	S. B. Conover and M. Spring	New York, N. Y	Apr. 30, 1861	32, 183
Potato-digging machine	N. Gear	Zanesville, Ohio	Sept. 7, 1858	21, 413
Potato-digging machine	J. Heulings	Philadelphia, Pa	July 21, 1857	17, 856
Potato-digging machine	H. Kewley	Perry, Ohio	July 2, 1867	66, 358
Potato-digging machine	M. Little	Clyde, N. Y	July 20, 1858	20, 949
Potato-digging machine	H. M. Peavey	Swanville, Me	July 10, 1866	56, 260
Potato-digging machine	J. Pettingill	Carroll, N. H	Oct. 14, 1862	36, 669
Potato-digging machine	J. C. Stoddard	Worcester, Mass	Mar. 29, 1859	23, 408
Potato-digging machine	A. Wells	Brooklyn, N. Y	Aug. 17, 1858	21, 225
Potato-digging machine	L. White	Essex, Vt	Aug. 17, 1858	21, 226
Potato-drill	J. Croco	Holmesville, Ohio	Oct. 2, 1866	58, 385
Potato-drill	J. Croco	Holmesville, Ohio	Apr. 23, 1867	64, 080
Potato-dropper	J. Gibbs	Canton, Ohio	June 25, 1872	128, 384
Potatoes, &c., Bin for preserving sweet	H. T. Basye	Dyersburgh, Pa	June 3, 1873	139, 532
Potato-fork	J. S. Patterson	Whitney's Point, N. Y	Oct. 23, 1866	59, 061
Potato-grinding machine	J. King	Florida, N. Y	Mar. 1, 1814	....
Potato-harvesting machine	O. S. and T. C. St. John	Willoughby, Ohio	Apr. 21, 1868	76, 953
Potato-masher	W. Ball	Peru, Mass	Dec. 13, 1864	45, 382
Potato-masher	V. Fountain, jr	Castleton, N. Y	Dec. 11, 1866	60, 355
Potato-masher	C. A. Frederick	San Francisco, Cal	Mar. 14, 1871	112, 577
Potato-masher	A. H. Lorton	New York, N. Y	Nov. 5, 1861	33, 654
Potato-masher	M. H. Monroe	Rochester, N. Y	Jan. 28, 1868	73, 743
Potato masher	A. L. Whitney	Brooklyn, N. Y	May 22, 1866	54, 987
Potato-masher	W. Zeiger	Elmore, Ohio	Mar. 23, 1869	88, 111
Potato masher and strainer	R. Sebille	Baltimore, Md	June 24, 1873	140, 169
Potato-masher, steak-pounder, and ice-breaker, Combined.	H. P. Stichter	Pottsville, Pa	Nov. 28, 1871	121, 434
Potato-probe	J. A. Beal	Waterford, N. Y	Mar. 7, 1871	112, 312
Potato-rot, Compound for treating	L. Reed	Baltimore, Md	July 27, 1858	21, 023
Potato-rot, Mode of preventing	C. Cory	Lima, Ind	June 28, 1864	43, 292
Potato-rot, Treating potatoes to prevent	A. Mallet	Erie, Pa	Mar. 13, 1866	53, 159
Potato-sacking machine	E. Seely	Elkhart, Ind	Dec. 8, 1868	84, 713
Potato-screen	G. Claflin	Miller's Corners, N. Y	July 15, 1873	140, 886
Patato-seedlings, Instrument for cutting	W. P. L. Herr	Brooklyn, N. Y	Apr. 4, 1865	47, 106
Potato-separator	S. Harrison	Saint Michael's, Md	Oct. 13, 1863	40, 258
Potato separator	R. Haviland	North Branch, Md	Apr. 12, 1870	101, 731
Potato-starch, Manufacture of	E. M., J. S., and W. H. Gordon.	Woodstock, N. H	Nov. 28, 1848	5, 940
Potato-vine cutter and potato-digger	T. E. C. Brinly	Louisville, Ky	Aug. 29, 1871	118, 585
Potato-washer	A. Bentley	Honesdale, Pa	May 4, 1852	8, 918
Potato-washer	O. H. Cooke	Morrisville, Vt	May 5, 1868	77, 461
Potato-washer	I. J. Currier and J. Hawse	Wolcott, Vt	July 6, 1869	92, 279
Potato-washer	D. K. Hickok	Morrisville, Vt	Oct. 13, 1868	82, 950
Potato-washer	J. Lawton, S. Hibbert, and J. Rhodes.	Manchester, England	Nov. 17, 1863	40, 631
Potato-washer	E. N. Porter and P. P. Roberts.	Morrisville, Vt	Feb. 11, 1868	74, 420
Potato-washer	L. B. Sherwin	Hyde Park, Vt	Dec. 29, 1868	85, 405
Potato-washer	S. H. Tift	Morrisville, Vt	May 5, 1868	77, 679
Potato-washer	J. H. Williams	East Craftsbury, Vt	July 24, 1866	56, 652
Potato washer and pan, Combined	J. M. Chaplin	Morrisville, Vt	Apr. 7, 1868	76, 399
Potato-washing machine	J. H. Fairchild and S. Richardson.	Jericho, Vt	Apr. 18, 1854	10, 796
Potatoes and carrots, Machine for cutting	J. Clarke	Hampton, Conn	July 7, 1835	....
Potatoes, coal, &c., Machine for assorting	M. D. Dickinson	Pilesgrove, N. J	Sept. 25, 1866	58, 229
Potatoes, Cutting and planting	S. Hutchinson	Rockport, Ind	Oct. 25, 1853	10, 154
Potatoes, Keeping sweet	W. and J. Davis	Jefferson County, Iowa	Sept. 8, 1863	39, 797
Potatoes, Keeping sweet	A. H. Vestal	Cambridge City, Ind	Aug. 16, 1844	3, 708
Potatoes, Machine for covering	H. L. Bennett	Long Branch, N. J	July 31, 1860	29, 351
Potatoes, pumpkins, &c., for cattle or swine by steam, Machine for boiling.	H. E. Potter	Otsego, N. Y	Nov. 20, 1810	....
Potters' clay, Purifying	A. Weber	Womelsdorf, Pa	Sept. 9, 1835	....
Potters' ware, Machine for molding	E. W. Blackmer	McGrawville, N. Y	Nov. 7, 1865	50, 791
Pottery-kiln	G. R. Booth	Hanley, England	Aug. 31, 1852	9, 230
Pottery-machine	S. R. Thompson	Portsmouth, N. H	June 28, 1870	104, 795
Pottery, Machine for making	A. M. Cheeseman	Trenton, N. J	Aug. 24, 1869	93, 966
Pottery, &c., Manufacture of	T. L. and R. Boote	Burslem, England	Nov. 21, 1865	51, 123
Pottery, &c., Manufacture of	P. Marquardt	Buffalo, N. Y	Mar. 1, 1870	100, 304
Pottery, Molding	J. C. Mendall and R. B. Ricketts.	Masonville, Ky	June 30, 1836	....
Pottery-molding machine	A. Keil and J. Tresch	New York, N. Y	Apr. 2, 1867	63, 529
Pottery-molding machine	W. Linton	Baltimore, Md	Feb. 12, 1861	31, 394
Pottery-molding machine	J. Tresch	New York, N. Y	May 5, 1863	38, 430
Pottery-stock, Machine for preparing	S. R. Thompson	Portsmouth, N. H	May 13, 1873	138, 824
Pottery-ware and crucible kiln	J. Dixon	Jersey City, N. J	Mar. 5, 1850	7, 136
Pottery-ware, Machine for making	H. R. Bodine	Falls Township, Ohio	Jan. 30, 1866	52, 261
Pottery-ware, Machine for making	S. R. Thompson	Portsmouth, N. H	Apr. 18, 1871	113, 816

Index of patents issued from the United States Patent Office from 1790 *to* 1873, *inclusive*—Continued.

Invention.	Inventor.	Residence.	Date.	No.
Pottery-ware, Machine for manufacturing	P. Pointon	Baraboo, Wis	Sept. 29, 1857	18, 298
Pottery-ware, Manufacture of	J. Farnam	Stillwater, N. Y	Oct. 25, 1845	4, 242
Pottery-ware, Manufacture of	I. W. Knowles	East Liverpool, Ohio	Oct. 11, 1870	108, 157
Pottery-ware, Molding	R. J. Marcher	New York, N. Y	June 16, 1863	38, 905
Pottery-ware, Molding and finishing	F. Durand	Paris, France	Sept. 13, 1864	44, 270
Pottery-ware, Burning	G. C. Hicks	Baltimore, Md	Mar. 26, 1872	125, 048
Pottery-ware, Safeguard for protecting	B. Jackson	Trenton, N. J	Jan. 31, 1865	46, 109
Pottery-ware, Working clay for	J. Akrill	Williamsburgh, N. Y	Aug. 5, 1851	8, 280
Poudrette-drying apparatus	L. Duvall	Philadelphia, Pa	June 3, 1873	139, 555
Poultice-cloth	M. L. J. Chollet and C. H. E. Hamilton.	Paris, France	Dec. 8, 1868	84, 731
Poultice, Composition for	J. P. Scott	Newport, Ky	Oct. 6, 1863	40, 189
Poultry-fountain	J. S. Orndorff	Virginia City, Nev	Aug. 8, 1871	117, 807
Poultry-hatching apparatus	G. F. Quick	Moorestown, N. J	Oct. 13, 1863	40, 277
Poultry, Machine for curing croup in	T. V. Bush	Clark, Ky	June 29, 1832	....
Pounce-holder	R. Cushman and J. R. Dennis	Pawtucket and Central Falls, R. I.	Nov. 24, 1868	84, 346
Pouncing-machine	J. Rosenkranz	Boston, Mass	Aug. 16, 1870	106, 409
Pounding and chopping block, Spring	F. Köhler and A. J. Alsing	New York, N. Y	Oct. 19, 1869	95, 914
Powder: *See* Abrasive powder. Baking-powder. Blasting-powder. Bronze-powder. Cattle-powder. Condition-powder. Core-powder. Disinfecting-powder. Explosive powder. Fulminating-powder. Gunpowder. Gun and blasting powder. Horse-powder. Horse and cattle powder. Ink-powder. Insect-powder. Lime powder. Metal-powder. Metal-cleaning powder. Percussion-powder. Polishing-powder. Preserving powder. Toilet-powder. Tooth-powder. Welding-powder. Yeast-powder.				
Powder-box	B. F. Brown	Boston, Mass	Aug. 1, 1871	117, 510
Powder-boxes used for domestic purposes, Cap for	H. A. Bartlett	Philadelphia, Pa	Feb. 22, 1870	100, 234
Powder by hydrogen gas and platina, Igniting	A. Jones	Mobile, Ala	Apr. 22, 1834	....
Powder, Exploding compressed	E. Gomez	New York, N. Y	Feb. 8, 1870	99, 665
Powder-flask	J. H. Breckenridge	Meriden, Conn	Feb. 16, 1858	19, 342
Powder-flask	G. A. and G. D. Capewell	Woodbury, Conn	Sept. 24, 1872	131, 497
Powder-flask	G. A., G. D., and J. T. Capewell	Woodbury, Conn	Aug. 15, 1871	118, 104
Powder-flask	J. T. and G. D. Capewell	Woodbury, Conn	Mar. 14, 1871	112, 542
Powder-flask	M. Cilik	Chicago, Ill	Sept. 9, 1873	142, 614
Powder-flask	F. E. Darrow	Bristol, Conn	Dec. 20, 1870	110, 210
Powder-flask	A. Drizel	Omaha, Nebr	Oct. 17, 1871	120, 047
Powder-flask	T. L. Sturtevant	Framingham, Mass	July 2, 1872	128, 673
Powder-flask	T. L. Sturtevant	Framingham, Mass	Apr. 1, 1873	137, 332
Powder-flask	G. W. Whipple	West Acton, Mass	May 1, 1860	28, 134
Powder-flask charger	H. C. Bascom	La Crosse, Wis	Mar. 24, 1868	75, 839
Powder-flask charger	C. C. Dickey	Philadelphia, Pa	Jan. 20, 1866	52, 147
Powder-flask for breech-loading gun	D. Moore	Williamsburgh, N. Y	Oct. 31, 1854	11, 871
Powder-keg	L. Austin and J. E. Hall	Cleveland, Ohio	Feb. 1, 1870	99, 280
Powder-keg	J. B. and D. J. Fleming	Xenia, Ohio	Oct. 6, 1868	82, 818
Powder-keg	C. Green and W. Wilson, jr	Wilmington, Del	July 15, 1873	140, 913
Powder-keg safety-conductor	M. Ward	Mount Carmel, Pa	Jan. 9, 1872	122, 688
Powder-magazine, Fire-proof	R. S. Sanborn	Ripon, Wis	July 30, 1867	67, 220
Powder-mill	G. Keyser		July 27, 1803	....
Powder-mixer	J. Burns	New York, N. Y	Mar. 26, 1867	63, 208
Powder-packing apparatus	K. F. Knowles	New York, N. Y	Aug. 13, 1872	130, 433
Powder pressing and graining machine	P. A. Oliver	New York, N. Y	Mar. 22, 1870	101, 032
Powder-proof lock	W. Hall	Boston, Mass	Aug. 1, 1848	5, 686
Powder-proof lock	E. Kershaw	Boston, Mass	Aug. 1, 1848	5, 685
Powders, Apparatus for dividing	G. P. Allen	Woodbury, Conn	June 27, 1871	116, 253
Powders, Apparatus for dividing	F. Schaeffer	Philadelphia, Pa	Oct. 19, 1869	96, 038
Powders, Instrument for measuring	W. Thomson	Madison, Wis	Oct. 31, 1871	120, 404
Powders, Machine for folding	J. W. Maxwell	Louisville, Ky	May 3, 1870	102, 568
Powders, &c., Putting up	H. Sawyer	Roxbury, Mass	Jan. 5, 1864	41, 097
Powders, Putting up	H. Sawyer	Chelsea, Mass	Sept. 12, 1871	118, 819
Powders, Putting up and using	A. P. Willard	Battle Creek, Mich	Aug. 31, 1869	94, 263
Power: *See* Animal-power. Atmospheric power. Atmospheric motive-power. Churn-power. Dog-power. Fan-power. Foot-power. Hammer-power. Hoisting-power. Hop-power. Horse-power. Human power. Hydraulic-power. Lever-power. Loom-power. Maintaining power. Man-power. Manual power. Mechanical power. Metal expansion and contraction power. Motive-power. Multiplying-power. Pendulum-power. Propelling-power. Railway-power. Ratchet-power. Regulating-power. Screw-power. Sewing-machine power. Spring-power. Steam water-power. Thrashing-power. Tide-power. Transmitting power. Tread-power. Washing-machine power. Water-power. Wave-power. Wind-power. Wind-wheel power. Windlass-crank power.				
Power-accumulating apparatus	J. F. Latimer	Detroit, Mich	Oct. 15, 1867	69, 821
Power-accumulating apparatus	A. C. Platt	Sandusky, Ohio	July 25, 1871	117, 454
Power-accumulating device	G. H. Becker	Memphis, Tenn	Dec. 17, 1867	72, 156
Power and hand drill	H. Woodman	Biddeford, Me	May 25, 1858	20, 385
Power-apparatus	G. W. Vosburgh	Sun Prairie, Wis	Oct. 8, 1872	131, 981
Power by certain fluids, Obtaining	M. J. Brunel	London, England	Mar. 30, 1827	....

Index of patents issued from the United States Patent Office from 1790 *to* 1873, *inclusive*—Continued.

Invention.	Inventor.	Residence.	Date.	No.
Power-generating apparatus	P. Shearer	Reading, Pa	Mar. 5, 1861	31, 631
Power for small machinery	C. L. Johnson	Omaha, Nebr	June 13, 1871	115, 864
Power-indicator	N. P. Bowsher	Ligonier, Ind	May 12, 1868	77, 864
Power-machine	W. Hosford	Washington Township, Ohio.	Dec. 31, 1829	....
Power, Machine for gaining	H. Bickel	Elizabeth City, N. J	June 6, 1865	48, 042
Power-machinery	R. Mitchell	Cynthiana, Ky	Oct. 10, 1829	....
Power, Machinery for accumulating and transmitting.	E. Stevens	Barnet, Vt	Apr. 26, 1859	23, 794
Power, Machinery for gaining	C. Redheffer	Philadelphia, Pa	July 11, 1820	....
Power, Mechanism for transmitting	J. Rankin	Binghamton, N. Y	Oct. 14, 1873	143, 640
Power-meter	J. A. Bradshaw, W. H. Brown, and D. Whithead.	Lowell, Mass	Sept. 13, 1870	107, 331
Power, Obtaining and transmitting	J. J. Gorman	Cincinnati, Ohio	Aug. 9, 1854	43, 768
Power-press	C. J. Beasley	Petersburgh, Va	Apr. 4, 1871	113, 246
Power-press	E. W. Bliss	Brooklyn, N. Y	July 26, 1870	105, 632
Power-press	D. Campbell	Elizabeth, N. J	July 6, 1869	92, 263
Power-press	R. H. Fisher	Beaver Falls, Pa	Jan. 26, 1869	86, 292
Power-press	A. D. Hamlin	Brooklyn, N. Y	Dec. 27, 1870	110, 455
Power-press	C. W. Johnson	Waterbury, Conn	Nov. 7, 1865	50, 826
Power-press	C. Oysten	Little Falls, N. Y	May 22, 1860	28, 401
Power-press	W. F. Parker	Meriden, Conn	Apr. 6, 1869	88, 732
Power-press	C. M. and O. G. Stratton	Greenfield, Mass	July 19, 1870	105, 605
Power-press	J. Weed	Muscatine, Iowa	May 22, 1860	28, 429
Power-press	M. G. Wilder	Meriden, Conn	Aug. 9, 1864	43, 818
Power-press	M. G. Wilder	West Meriden, Conn	June 27, 1871	116, 383
Power-press, Centripetal	E. S. Scripture	Cazenovia, N. Y	Sept. 26, 1834	....
Power-press-feed device	P. E. Austin	New Haven, Conn	Oct. 18, 1870	108, 312
Power-press for punching copper, &c	G. Darracott	Boston, Mass	July 21, 1829	....
Power-press, Hay, &c	J. H. Wittmer	Manor, Pa	May 11, 1869	90, 039
Power-press, Portable	T. B. Wait	Zena, Oreg	June 27, 1871	116, 376
Power-press, Self-acting	E. Davis	Montpelier, Vt	Mar. 23, 1854	10, 960
Power-press, Spring	R. McOmber	Galway, N. Y	Oct. 27, 1829	....
Power-rake	H. N. Tripp	Alfred, Me	Dec. 20, 1853	10, 327
Power-regulator for friction-surfaces	B. Tatham	New York, N. Y	Apr. 12, 1870	101, 940
Power to machinery, Applying	D. Durfey	Fort Seneca, Ohio	Sept. 6, 1859	25, 361
Power to machinery, Applying	G. Page	Keene, N. H	Sept. 28, 1831	....
Power to machinery, Mechanism for applying	L. Curdts	New York, N. Y	Aug. 13, 1867	67, 730
Power-transferring apparatus	E. B. Clark	Cohoes, N. Y	Dec. 23, 1873	145, 721
Power-transmitting apparatus	T. Damon	Thompsonville, Conn	Dec. 12, 1871	121, 854
Preservation of flesh for food	N. B. Marsh	Cincinnati, Ohio	Nov. 30, 1858	22, 185
Preservation of meat and other articles of food	F. Cassel	Cologne, Prussia	Aug. 16, 1870	106, 465
Preservative materials, Impregnating substances with.	E. Sabathé and L. Jourdan	Paris, France	Sept. 11, 1866	58, 036
Preservatory	A. Smith	Perrysburgh, Ohio	Feb. 13, 1866	52, 650
Preserve-can	B. L. Agnew	Indiana, Pa	Oct. 18, 1859	25, 863
Preserve-can	J. K. Chace	Brooklyn, N. Y	Aug. 19, 1873	141, 994
Preserve-can	P. H. Cotton	Demopolis, Ala	Dec. 21, 1858	22, 351
Preserve-can	H. G. Dayton	Maysville, Ky	June 8, 1858	20, 485
Preserve-can	H. S. Fisher	Newburgh, Pa	Aug. 19, 1862	36, 264
Preserve-can	W. Fridley and F. Cornman	Carlisle, Pa	Oct. 25, 1859	25, 894
Preserve-can	N. S. Gilbert	Lockport, N. Y	Jan. 29, 1861	31, 235
Preserve-can	E. W. Gilmore	North Easton, Mass	May 11, 1858	20, 203
Preserve-can	J. L. Gray	Baltimore, Md	Apr. 2, 1867	63, 505
Preserve-can	J. F. Griffen	New York, N. Y	Apr. 5, 1864	42, 186
Preserve-can	W. H. Harn	Carlisle, Pa	May 22, 1860	28, 366
Preserve-can	W. D. Ludlow	New York, N. Y	June 28, 1859	24, 566
Preserve-can	E. Manley	Marion, N. Y	Aug. 3, 1858	21, 078
Preserve-can	J. F. Martin and H. C. Nicholson.	Mount Washington, Ohio	Feb. 15, 1859	22, 962
Preserve-can	F. O. More	Bellefontaine, Ohio	July 12, 1859	24, 751
Preserve-can	S. Morrett	West Pennsborough, Pa	Mar. 29, 1859	23, 384
Preserve-can	C. Newman	Birmingham, Pa	Dec. 20, 1859	26, 515
Preserve-can	P. H. Niles	Boston, Mass	Nov. 28, 1865	51, 264
Preserve-can	J. Roussel, L. Delaugre, and L. L. Robin.	Nantes, France	July 5, 1864	43, 463
Preserve-can	T. Sellers	East Birmingham, Pa	May 22, 1860	28, 413
Preserve-can	C. F. Stilz and J. C. Knoepke	Philadelphia, Pa	Aug. 14, 1866	57, 212
Preserve-can	A. T. Twing, E. Wood, and W. Elderhorst.	Louisburgh and Troy, N. Y	May 22, 1860	28, 424
Preserve-can cover	C. L. Kelling	Mechanicsburgh, Pa	Jan. 24, 1860	26, 907
Preserve can, jar, &c	T. Earle	Smithfield, R. I	Nov. 10, 1863	40, 556
Preserve-can, Self-sealing	R. Arthur	Washington, D. C	Jan. 2, 1855	12, 153
Preserve-can stopper	V. P. Corbett	Washington, D. C	Sept. 20, 1859	25, 499
Preserve-can stopper	W. P. Patton	Harrisburgh, Pa	Jan. 10, 1860	26, 820
Preserve-cans, Apparatus for exhausting air from	D. N. Phelps	San Leandro, Cal	Feb. 6, 1872	123, 418
Preserve-cans, Closing	T. Earle	Valley Falls, R. I	Feb. 2, 1864	41, 425
Preserve-cans, Hermetically-sealing	C. Branwhite	New York, N. Y	Mar. 18, 1856	14, 439
Preserve-cans, Hermetically-sealing	C. E. Russell	Saint Louis, Mo	June 10, 1856	15, 088
Preserve-cans, Sealing	H. S. Fisher	Newburgh, Pa	Nov. 12, 1861	33, 729
Preserve-cans, Sealing	R. W. Lewis	Honesdale, Pa	Feb. 12, 1856	14, 245
Preserve-cans, Sealing	S. Lutz	Philadelphia, Pa	Oct. 23, 1855	13, 707
Preserve-cans, Sealing	W. W. Lyman	West Meriden, Conn	Aug. 31, 1858	21, 348
Preserve-cans, Sealing	H. C. Nicholson and J. Spratt	Cincinnati, Ohio	Jan. 3, 1854	10, 396
Preserve-cans, Sealing	W. W. Paddock	Cincinnati, Ohio	Nov. 6, 1860	30, 585
Preserve-cans, Sealing	A. Taylor	Baltimore, Md	Dec. 7, 1858	22, 247
Preserve-cans, Sealing-device for	W. H. Elliot	Plattsburgh, N. Y	July 17, 1855	13, 291
Preserve-canisters, Sealing	H. Hunt	Brooklyn, N. Y	Sept. 6, 1853	9, 989
Preserve-jar	T. Earle	Valley Falls, Smithfield Township, R. I.	Dec. 27, 1864	45, 594
Preserve-jar	T. Earle	Valley Falls, Smithfield Township, R. I.	Mar. 21, 1865	46, 887
Preserve-jar	N. S. Gilbert	Lockport, N. Y	Dec. 17, 1861	33, 938
Preserve-jar	F. H. Lauterbach	Boston, Mass	June 12, 1866	55, 585
Preserve-jar	T. G. Ootterson	Camden, N. J	Aug. 31, 1869	93, 236
Preserve-jar	P. W. Reid	Birmingham, Pa	July 21, 1863	39, 327
Preserve-jar	F. J. Shefferly	Detroit, Mich	Jan. 7, 1868	73, 202
Preserve-jar	C. F. Spencer	Rochester, N. Y	Feb. 10, 1863	37, 647
Preserve-jar	J. J. Squire	Windsor Locks, Conn	Oct. 18, 1864	44, 752

Index of patents issued from the United States Patent Office from 1790 *to* 1873, *inclusive*—Continued.

Invention.	Inventor.	Residence.	Date.	No.
Preserve-jar	N. Thompson	Brooklyn, N. Y	Mar. 9, 1869	87, 730
Preserve-jar	J. M. Whitall	Philadelphia, Pa	June 18, 1861	32, 594
Preserve-jar	A. Q. Withers	Holly Springs, Miss	Feb. 20, 1872	123, 851
Preserve-jar cap	L. R. Boyd	New York, N. Y	May 11, 1869	89, 845
Preserve-jar cap	S. B. Rowley	Philadelphia, Pa	Feb. 11, 1868	74, 249
Preserve-jar cap	W. Taylor and C. Hodgetts	Brooklyn, N. Y	July 18, 1871	117, 236
Preserve-jar stand	K. E. Ashley	Williamsburgh, N. Y	Oct. 31, 1865	5[illegible], 674
Preserve-jars, Clasp for closing	J. A. Cowles	Chicago, Ill	Apr. 28, 1863	38, 288
Preserve-jars, Composition for sealing	J. Beckley	Cincinnati, Ohio	July 7, 1863	39, 111
Preserve-jars, Sealing	R. M. Dalbey	Mount Washington, Ohio	Nov. 16, 1858	22, 066
Preserve-vessels	J. Adams	Pittsburgh, Pa	May 20, 1862	35, 280
Preserve-vessel, Self-sealing	W. J. Stevenson	New York, N. Y	May 27, 1856	14, 974
Preserves, Packing-apparatus for	C. T. Provost	New York, N. Y	June 7, 1870	103, 927
Preserving and canning meat, fruit, vegetables, &c., Apparatus for.	N. S. Shipley	Baltimore, Md	Jan. 24, 1871	111, 264
Preserving and condensing milk	G. Borden and J. G. Borden	White Plains and South East, N. Y.	Nov. 4, 1873	144, 310
Preserving and curing meat	J. Tilton	New York, N. Y	Oct. 2, 1866	58, 511
Preserving and drying green corn on the cob, Method of.	D. Parker	Shaker Village, N. H	Apr. 10, 1866	53, 861
Preserving and freezing	D. E. Somes	Washington, D. C	Jan. 25, 1870	99, 254
Preserving and freezing fish, meats, &c., Apparatus for.	P. Nunan	Sandusky, Ohio	Oct. 27, 1868	83, 533
Preserving and packing box for perishable articles	B. Yaw	New Concord, Ohio	Nov. 12, 1872	133, 000
Preserving and packing compound for eggs, wood, &c.	J. Connell, jr	Collingwood, Canada	May 20, 1873	139, 116
Preserving and packing meat	C. E. Richardson	Cambridge, Mass	Feb. 11, 1868	74, 247
Preserving and smoking meat	A. Robertson	Seymour, Ind	May 16, 1871	114, 972
Preserving and transporting fish, Apparatus for	W. Davis	Detroit, Mich	Jan. 19, 1869	85, 914
Preserving and transporting fresh meat	W. Dugan	Chicago, Ill	Apr. 4, 1871	113, 505
Preserving animal and vegetable substances	J. Bruckner	New Orleans, La	Aug. 13, 1872	130, 474
Preserving animal and vegetable substances	D. Cooley	New York	Oct. 24, 1814	....
Preserving animal and vegetable substances	J. Coppinger	Beaufort, S. C	Nov. 23, 1809	....
Preserving animal and vegetable substances	J. Gamgee and A. Gamgee	Bayswater, England, and Edinburgh, Scotland.	Nov. 26, 1867	71, 377
Preserving animal and vegetable substances	R. Jones	London, England	Nov. 28, 1865	51, 280
Preserving animal and vegetable substances	L. H. Spear	Braintree, Vt	Mar. 22, 1864	42, 025
Preserving animal and vegetable substances	L. H. Spear	Braintree, Vt	Aug. 20, 1867	67, 921
Preserving animal and vegetable substances	F. Stabler	Baltimore, Md	Nov. 14, 1865	50, 965
Preserving animal and vegetable substances	A. Fryer	Manchester, England	Oct. 8, 1872	132, 002
Preserving animal and vegetable substances, Apparatus for.	E. Piper	Camden, Me	Aug. 5, 1862	36, 107
Preserving animal and vegetable substances during transportation, Case for.	J. G. Staunton	Buffalo, N. Y	Jan. 3, 1865	45, 764
Preserving animal and vegetable substances in transit.	E. R. Morney	McDonough, Del	Sept. 14, 1869	94, 908
Preserving animal and vegetable substances on ship-board.	J. F. Baldwin	Provincetown, Mass	July 6, 1869	92, 246
Preserving animal and vegetable substances, Refrigerating and condensing apparatus for.	E. D. Brainard	Albany, N. Y	Jan. 14, 1868	73, 202
Preserving animal matters, Process and apparatus for.	A. Rocke	New Orleans, La	Feb. 27, 1872	124, 161
Preserving animal substances	E. Daggett and T. Kensett	New York	Jan. 19, 1825	....
Preserving animal substances	O. Lugo	New York, N. Y	Aug. 18, 1868	81, 185
Preserving animal substances	H. Medlock and W. Bailey	London and Wolverhampton, England.	Aug. 1, 1871	117, 660
Preserving apples and other fruit, beets, sweet potatoes, &c., Mode of.	A. Hart	Wharton Township, Pa	Aug. 10, 1829	....
Preserving articles of food	J. McCall and B. G. Sloper	London and Walthamstow, England.	Dec. 9, 1862	37, 137
Preserving beer and other liquids, Apparatus for	T. Byrne	New York, N. Y	Sept. 19, 1865	50, 085
Preserving box, Fruit	O. E. Doolittle	Boston, Mass	June 2, 1868	78, 437
Preserving bread	J. J. A. Mouries	Phalsbourg, France	Sept. 6, 1870	107, 088
Preserving butter	P. Giraud	New York, N. Y	Oct. 15, 1861	33, 482
Preserving butter	M. Johnson		June 30, 1797	....
Preserving butter	J. F. Saiger	Shelby, Ohio	Jan. 25, 1870	99, 240
Preserving butter	H. D. Wetmore	Cleveland, Ohio	Mar. 5, 1861	31, 638
Preserving butter, Can for	W. C. Rentgen	Vicksburgh, Miss	Dec. 5, 1865	51, 352
Preserving butter, Composition for	L. De Corn	Cincinnati, Ohio	Aug. 3, 1852	9, 167
Preserving butter, meat, &c	W. Ross	Day's Store, Pa	Jan. 1, 1867	60, 942
Preserving butter, &c., Vessel for	J. G. Staunton	Buffalo, N. Y	Jan. 3, 1865	45, 763
Preserving-can	A. R. Davis	East Cambridge, Mass	July 16, 1872	129, 214
Preserving-can	O. N. Weaver	Dover, Ky	July 12, 1859	24, 770
Preserving-can for food	A. S. Lyman	New York, N. Y	May 11, 1858	20, 209
Preserving-can for food	A. S. Lyman	New York, N. Y	June 29, 1858	20, 722
Preserving can, Fruit, &c	G. W. Griswold	Logansport, Ind	July 22, 1862	35, 933
Preserving case, Butter, cheese, &c	T. R. Timby	Saratoga, N. Y	Mar. 2, 1869	87, 446
Preserving chopped meat, &c	W. C. Marshall	New York, N. Y	Feb. 23, 1864	41, 712
Preserving coffee	B. T. Babbitt	New York, N. Y	Mar. 24, 1868	75, 828
Preserving coffee	A. Eikerenkotter and F. Silver	Searsville, Cal	Sept. 3, 1867	68, 422
Preserving corn	L. McMurray	Baltimore, Md	June 3, 1873	139, 595
Preserving cranberries	J. W. Campbell	Philadelphia, Pa	July 25, 1871	117, 256
Preserving cranberries	L. Kniffen	Worcester, Mass	Mar. 19, 1872	124, 827
Preserving dead bodies	J. A. Gaussardia	Washington, D. C	Oct. 28, 1856	15, 972
Preserving dead bodies, &c	J. Morgan	Dublin, Ireland	Sept. 27, 1864	44, 495
Preserving dead bodies	G. W. Scollay	Saint Louis, Mo	Oct. 19, 1869	95, 939
Preserving dead bodies	C. C. St. Clair	Washington, D. C	July 23, 1867	67, 145
Preserving eggs	A. M. Blinval	New York, N. Y	Jan. 5, 1864	41, 053
Preserving eggs, &c	C. Boize	New York, N. Y	Dec. 17, 1867	72, 158
Preserving eggs	L. H. Boole	New York, N. Y	Sept. 17, 1867	68, 834
Preserving eggs	J. Brakeley	Bordentown, N. J	Mar. 10, 1868	75, 242
Preserving eggs	A. R. Davis	East Cambridge, Mass	Sept. 24, 1872	131, 507
Preserving eggs	P. Gaughran and L. Sweeney	San Francisco, Cal	Aug. 6, 1867	67, 427
Preserving eggs	J. H. Hall	New York, N. Y	Mar. 9, 1869	87, 562
Preserving eggs	W. Hansford	San Francisco, Cal	Apr. 11, 1865	47, 202
Preserving eggs, &c	W. B. Ingersoll	New York, N. Y	May 31, 1864	42, 948
Preserving eggs	J. Kaye, jr	Setzler's Store, Pa	Nov. 22, 1870	109, 420
Preserving eggs	J. Knapp	Prattsburgh, N. Y	July 12, 1870	105, 342

Index of patents issued from the United States Patent Office from 1790 *to* 1873, *inclusive*—Continued.

Invention.	Inventor.	Residence.	Date.	No.
Preserving eggs	J. Knapp	Prattsburgh, N. Y	Dec. 26, 1871	122, 258
Preserving eggs	A. G. and E. E. Kyle	Newville, Pa	June 12, 1866	55, 502
Preserving eggs	C. A. La Mont	New York, N. Y	Nov. 28, 1865	51, 263
Preserving eggs	J. Longmaid	New York, N. Y	Sept. 21, 1869	95, 028
Preserving eggs	J. K. Marsh	Terre Haute, Ind	Feb. 26, 1867	62, 432
Preserving eggs	W. S. Marsh	Raymond, Wis	Dec. 5, 1871	121, 640
Preserving eggs	N. Patton	Coles County, Ill	Jan. 16, 1866	52, 072
Preserving eggs	R. S. Rhodes and E. Whyte	Chicago, Ill	Dec. 12, 1865	51, 480
Preserving eggs	A. Van Camp	Washington, D. C	Feb. 18, 1868	74, 733
Preserving eggs and other articles	C. Bruner	Marshall, Mo	Nov. 30, 1869	97, 349
Preserving eggs, butter, &c	T. Edmundson	Pipe Creek, Md	Apr. 26, 1827	....
Preserving eggs, Composition for	W. C. Bruson	Chicago, Ill	Apr. 27, 1869	89, 285
Preserving eggs, Composition for	J. T. McKim	Remington, Ind	July 11, 1871	116, 976
Preserving eggs, Composition for	N. Patton	Kansas, Ill	Dec. 29, 1868	85, 328
Preserving eggs, Composition for	P. J. Stouffer	Uniontown, Pa	Apr. 23, 1867	64, 047
Preserving eggs, Compound for	J. B. Patterson	Portage City, Wis	Aug. 18, 1868	81, 104
Preserving eggs, meats, &c	J. Perkins	Newark, N. J	Apr. 19, 1864	42, 396
Preserving fish	R. A. Adams	Cambridge, Mass	Sept. 28, 1869	95, 179
Preserving fish	E. Piper	Camden, Me	Mar. 19, 1861	31, 736
Preserving fish by freezing	S. H. and D. W. Davis	Detroit, Mich	Feb. 28, 1871	112, 129
Preserving fish, meat, &c., Compound for	T. Sim	Charleston, S. C	Jan. 19, 1869	86, 040
Preserving fish and meat	M. Gross	Washington, D. C	Dec. 13, 1859	26, 427
Preserving flour, grain, &c	J. Carlin	Cincinnati, Ohio	Oct. 2, 1866	58, 374
Preserving flowers	E. M. Stigalo	Philadelphia, Pa	Apr. 27, 1869	89, 515
Preserving flowers and other vegetable forms	P. T. Vining	New York, N. Y	June 11, 1867	65, 777
Preserving food	W. A. Keeler	New York, N. Y	July 31, 1860	29, 383
Preserving food for transportation, Apparatus for	M. Brune	New York, N. Y	July 25, 1865	48, 898
Preserving fruit, &c	W. D. Brooks	Baltimore, Md	July 2, 1867	66, 208
Preserving fruit	G. Jáques	Boston, Mass	Sept 27, 1870	107, 689
Preserving fruit	J. K. Jenkins	Kingston, Pa	Dec. 28, 1858	22, 433
Preserving fruit	D. M. Mefford	Norwalk, Ohio	Sept. 22, 1868	82, 429
Preserving fruit	D. M. Mefford	Norwalk, Ohio	Feb. 2, 1869	86, 433
Preserving fruit	E. R. Norny	McDonough, Del	June 22, 1869	91, 557
Preserving fruit, &c	E. C. Roberts	Salem, Mich	Mar. 7, 1865	46, 707
Preserving fruit, &c	N. S. Shaler	Newport, Ky	Oct. 11, 1864	44, 664
Preserving fruit, &c	H. C. Smith	Chicago, Ill	June 21, 1864	43, 232
Preserving fruit	W. H. Trissler	Cleveland, Ohio	June 18, 1867	65, 844
Preserving fruit and vegetables	W. G. Barbee	High Point, N. C	Nov. 18, 1873	144, 592
Preserving fruit and vegetables	J. M. Blanco y Nuño	Havana, Cuba	Sept. 29, 1868	82, 588
Preserving fruit and vegetables	E. G. Holden	Covington, Ky	Nov. 6, 1866	59, 396
Preserving fruit and vegetables	F. C. Hooten	West Chester, Pa	Aug. 8, 1871	117, 884
Preserving fruit and vegetables	E. C. Roberts	Salem, Mich	June 17, 1862	35, 626
Preserving fruit and vegetables and compounds therefor.	E. R. Norny	McDonough, Del	Apr. 27, 1869	89, 330
Preserving fruit, Apparatus for	L. C. Cooley	New York, N. Y	June 3, 1873	139, 547
Preserving fruit, Apparatus for	J. Cope	East Fairfield, Ohio	Nov. 26, 1872	133, 413
Preserving fruit, Apparatus for	W. W. Lyman	Meriden, Conn	May 10, 1870	102, 951
Preserving fruit, Apparatus for	C. C. Williams	Brooklyn, N. Y	Oct. 26, 1869	96, 179
Preserving fruit, Building for	B. M. Nyce	Kingston, Ind	Mar. 19, 1861	31, 734
Preserving fruit from decay, Composition for	G. Jáques	Boston, Mass	Sept. 27, 1870	107, 690
Preserving fruit in jars, &c	S. J. Parker	Ithaca, N. Y	Jan. 26, 1864	41, 392
Preserving fruit, &c., in sealed cans	H. B. Slaughter	Crumpton, Md	Nov. 11, 1862	36, 921
Preserving fruit, meat, &c	E. S. Bartholomew	Westfield, N. Y	Dec. 18, 1866	60, 462
Preserving fruit, meat, &c	V. W. Blanchard	Bridport, Vt	Nov. 16, 1869	96, 871
Preserving fruit, meat, and vegetables	J. B. Hyde	New York, N. Y	Oct. 6, 1868	82, 719
Preserving fruit, meat, &c., Apparatus for	J. G. McMillan	Baltimore, Md	Oct. 15, 1867	69, 924
Preserving fruit, meat, fish, &c	J. G. Staunton	Buffalo, N. Y	Jan. 3, 1865	45, 765
Preserving fruit, vegetables, &c	L. Bradley and T. D. Phillips	Buffalo, N. Y	Feb. 8, 1870	99, 628
Preserving fruit, vegetables, &c	A. T. Jones and L. C. Cooley	New York, N. Y	June 3, 1873	139, 581
Preserving fruit, vegetables, &c	D. M. Mefford	Huron County, Ohio	Dec. 31, 1872	134, 390
Preserving fruit, vegetables, &c., Process and apparatus for.	J. C. Wrenshall	Baltimore, Md	June 21, 1870	104, 680
Preserving green corn	S. J. Carrol	Baltimore, Md	Jan. 22, 1867	61, 316
Preserving green corn	W. L. Gilroy	Philadelphia, Pa	Jan. 11, 1870	98, 758
Preserving green corn	D. Rowe	Baltimore County, Md	June 30, 1857	17, 697
Preserving green corn	I. Winslow	Philadelphia, Pa	May 13, 1862	35, 274
Preserving green corn	I. Winslow	Philadelphia, Pa	May 20, 1862	35, 346
Preserving green corn and other vegetables	G. L. Merrill and O. F. Soule	Syracuse, N. Y	Dec. 16, 1873	145, 581
Preserving grapes and other fruit	E. A. Wible	Georgetown, Cal	June 10, 1862	35, 556
Preserving grated horse-radish	J. G. Lunz	Chicago, Ill	May 7, 1872	126, 557
Preserving hops	J. C. and M. D. Baldwin and R. Brayton.	Watertown, N. Y., Brantford, Canada, and Buffalo, N. Y.	Sept. 25, 1860	30, 113
Preserving-house	H. Blackburn	Bedford County, Pa	Dec. 7, 1869	97, 475
Preserving-house	J. C. Gove	Cleveland, Ohio	Jan. 1, 1867	60, 716
Preserving-house	W. S. Mabbett	Calverton Mills, Md	July 22, 1862	35, 943
Preserving house, Fruit, &c	N. Hellings	Philadelphia, Pa	Oct. 15, 1867	69, 806
Preserving house, Fruit, &c	E. F. Olds	Hudson, Mich	July 18, 1865	48, 833
Preserving house, Fruit	J. Prior	Adrian, Mich	Feb. 1, 1870	99, 349
Preserving-houses, Ice-floor or	T. S. Rankin	Quenemo, Kansas	Mar. 5, 1872	124, 386
Preserving-jar	J. Borden	Bridgeton, N. J	Apr. 13, 1858	19, 964
Preserving jar and can	G. S. G. Spence	Salem, Mass	Nov. 18, 1862	36, 970
Preserving-jar cap	J. Borden	Bridgeton, N. J	Feb. 12, 1867	61, 921
Preserving liquid and other substances	R. d'Heureuse	New York, N. Y	Aug. 3, 1869	93, 182
Preserving malt-liquor, Apparatus for	J. Keane	New York, N. Y	Oct. 12, 1858	21, 761
Preserving meat, &c	W. Davis	Detroit, Mich	June 16, 1868	78, 932
Preserving meat, &c	E. De La Granja	Boston, Mass	Sept. 17, 1867	68, 850
Preserving meat	J. E. Dotch	Washington, D. C	Dec. 1, 1868	84, 481
Preserving meat	J. E. Dotch	Washington, D. C	June 25, 1872	128, 372
Preserving meat	J. E. Dotch and O. Loew	Washington, D. C., and New York, N. Y.	June 25, 1872	128, 371
Preserving meat	H. Gahn	Upsala, Sweden	Jan. 25, 1870	99, 180
Preserving meat, &c	W. Gillespie	Louisa Court-House, Va	Apr. 2, 1872	125, 285
Preserving meat	F. H. Hatch and B. R. Hawley.	New Orleans, La., and Normal, Ill.	Sept. 13, 1870	107, 367
Preserving meat	W. C. Marshall	New York, N. Y	July 12, 1864	43, 516
Preserving meat	H. B. Meech	Fort Edward, N. Y	Oct. 30, 1866	59, 247
Preserving meat, &c	J. F. Meiners	New York, N. Y	June 4, 1872	127, 627

Index of patents issued from the United States Patent Office from 1790 *to* 1873, *inclusive*—Continued.

Invention.	Inventor.	Residence.	Date.	No.
Preserving meat	G. W. Oliver	New York, N. Y	June 19, 1860	28, 812
Preserving meat	M. Perl	Houston, Tex	Oct. 6, 1868	82, 871
Preserving meat	D. E. Somes	Biddeford, Me	Apr. 9, 1861	32, 109
Preserving meat	L. H. Spear	Peekskill, N. Y	Aug. 23, 1870	106, 632
Preserving meat	W. Wiesmann	Bonn, Prussia	Nov. 10, 1868	84, 038
Preserving meat and fruit	J. Hampton	Lexington, Ky	Mar. 12, 1830	....
Preserving meat and other perishable articles, Process and apparatus for.	T. Sim	Charleston, S. C	Dec. 22, 1868	85, 184
Preserving meat and vegetables	A. S. Lyman	New York, N. Y	Apr. 30, 1872	126, 148
Preserving meat and vegetables, Apparatus for	W. Maxwell	Henderson, Tex	Jan. 30, 1872	123, 273
Preserving meat, Apparatus for	C. F. Carr	Norwich, N. Y	Apr. 9, 1872	125, 540
Preserving meat, Composition for	J. J. Harrison	Saint Michael's, Md	July 10, 1866	56, 217
Preserving meat, fish, &c	B. Mosquera	Santiago, Chili	June 25, 1872	128, 320
Preserving meat, fish, &c., and making meat extracts.	T. F. Henley	Pimlico, England	Oct. 1, 1872	131, 820
Preserving meat, fish, &c., Apparatus for	C. F. Pike	Providence, R. I	Dec. 10, 1867	71, 910
Preserving fish, oysters, &c	J. E. Dotch	Washington, D. C	Nov. 8, 1870	108, 983
Preserving meat for food	W. Adamson and C. F. A. Simonin.	Philadelphia, Pa	Apr. 2, 1872	125, 248
Preserving meat for pastry purposes	G. H. Munroe	New York, N. Y	Oct. 12, 1869	95, 715
Preserving meat, fowls, &c	C. Havard and M. X. Harmony	London, England	June 8, 1869	90, 944
Preserving meat, poultry, fish, &c	J. E. Dotch	Washington, D. C	Aug. 3, 1869	93, 183
Preserving meat, poultry, fish, &c	A. Vazquez and J. E. Rosenberg.	Santiago, Chili	Mar. 26, 1872	125, 102
Preserving milk	J. L. Granger	New York, N. Y	Mar. 19, 1836	....
Preserving milk	L. F. Kirchhofer	Union Hill, N. J	Aug. 19, 1873	141, 878
Preserving milk, Apparatus for	N. P. Holmes	Indianapolis, Ind	Sept. 4, 1866	57, 721
Preserving milk, Apparatus for	N. P. Holmes	Indianapolis, Ind	Dec. 4, 1866	60, 190
Preserving milk, Apparatus for	E. Rönning	Brodhead, Wis	Jan. 31, 1871	111, 476
Preserving organic substances	J. G. Staunton	Buffalo, N. Y	Jan. 3, 1865	45, 762
Preserving oysters	E. T. Gilmore	Springfield, Mass	Nov. 20, 1866	59, 833
Preserving potatoes	C. S. Edwards	Rushville, Ill	Dec. 31, 1845	4, 337
Preserving potatoes	J. Mumford	Clarksburgh, Ohio	Oct. 20, 1868	83, 299
Preserving-powder	G. A. Mariner	Chicago, Ill	May 19, 1868	78, 112
Preserving, Preparing fruit and vegetables for	A. Eckert	Trenton, Ohio	Mar. 6, 1866	52, 980
Preserving provisions, Apparatus for	C. Winship	New Haven, Conn	Oct. 11, 1864	44, 683
Preserving provisions, Cooling-room for	D. E. Somes	Biddeford, Me	Sept. 15, 1863	39, 967
Preserving, refrigerating, and transporting perishable articles.	C. F. Pike	Providence, R. I	Dec. 31, 1867	72, 894
Preserving, storing, and transporting fruits, vegetables, and other perishable articles.	J. Rutter	West Chester, Pa	July 9, 1867	66, 666
Preserving sugar-cane	F. O. Darbey	Attakapas, La	Jan. 23, 1872	122, 936
Preserving sweet potatoes	H. T. Basye	Dyersburgh, Tenn	Dec. 13, 1870	110, 108
Preserving sweet potatoes	J. C. Tilton	Pittsburgh, Pa	Mar. 26, 1872	124, 987
Preserving tomatoes	F. G. Ford	Bridgeton, N. J	Oct. 15, 1872	132, 272
Preserving tomatoes, Compound for	D. C. Yates	Big Lick, Va	Sept. 19, 1871	119, 214
Preserving the color in dried fruit	D. Ackart	Schaghticoke, N. Y	Oct. 3, 1871	119, 442
Preserving various articles of food	H. L. Palmer	Stillwater, N. Y	July 19, 1870	105, 484
Preserving vegetable extracts	H. L. Palmer	Stillwater, N. Y	July 19, 1870	105, 485
Preserving vegetables	F. H. Smith	Baltimore, Md	Aug. 17, 1869	93, 841
Preserving vegetables, fruit, &c	M. D. Mefferd	Norwalk, Ohio	Mar. 29, 1870	101, 373
Preserving vegetables in hermetically sealed cans.	I. Winslow	Philadelphia, Pa	Aug. 26, 1862	36, 326
Preserving-vessel	J. B. Wilson	Williamstown, N. J	Dec. 3, 1861	33, 870
Preserving-vessels, Closing	A. Warner	Williamsburgh, N. Y	Mar. 24, 1868	76, 009
Preserving-vessels, Cover for	I. Stratton	Philadelphia, Pa	Feb. 3, 1863	37, 595
Preserving-vessels, Stopper for	J. B. Wilson	Williamstown, N. J	July 9, 1861	32, 805

Press:

See Anti-friction press.
Back-pad press.
Baling-press.
Beater-press.
Book-holding press.
Brick-press.
Brick and tile press.
Bundling-press.
Butter-print press.
Card-press.
Centripetal press.
Cheese-press.
Cheese and fruit press.
Cider-press.
Cider and wine press.
Cigar-mold press.
Cloth-press.
Clothes-press.
Comb-molding press.
Compressing-press.
Compressing and beating press.
Compression and baling press.
Concrete-building-block press.
Copying-press.
Copying and folding press.
Cotton-press.
Crushing-press.
Cutting-press.
Cylinder-press.
Double-cylinder press.
Drop-press.
Embossing-press.
Field-press.
Filter-press.
Filtering-press.
Fly-wheel press.
Foot-press.
Foundry-press.
Fruit-press.
Fruit and lard press.
Fruit-packing press.
Fuel-press.
Furnace-block press.
Garment-press.
Gear-press.
Glass-press.
Glass-ware press.
Glass-ware-molding press.
Gluing-press.
Gunpowder-press.
Harness-pad press.
Hat-press.
Hat-finishing press.
Hay-press.
Hay and cotton press.
Hay and hemp press.
Hay or cotton press.
Heel-press.
Hop-press.
Hop and hay press.
Hydraulic press.
Hydraulic-wheel press.
Hydro-mechanical press.
Hydrostatic press.
Ink-press.
Jeweler's hand-press.
Lard-press.
Lever-press.
Lever-bottom press.
Liquid-removing press.
Lithographic press.
Meat-press.

Index of patents issued from the United States Patent Office from 1790 *to* 1873, *inclusive*—Continued.

Invention.	Inventor.	Residence.	Date.	No.
Press—Continued. *See* Meat or fruit press. Metallic press. Mold-board press. Mop-press. Oil-press. Oil and cider press. Oil or blubber press. Oil, tobacco, and cotton press. Pad-crimp press. Pen-press. Photograph-burnishing press. Photographic-card press. Plumbago-press. Power-press. Printing-press. Propelling-press. Provision-press. Pulley-press. Punching-press. Punching and stamping press. Rafter-press. Revolving-screw press. Rim-press. Rocket-press. Rocket-filling press. Rolling-press. Screw-press. Screw-die press. Sector-press. Self-acting press. Self adjusting press. Shearing and punching press. Stamping-press. Standing press. Steam-press. Steam and hydraulic press. Stencil-printing press. Sugar-press. Sugar-cane press. Tailor's press. Tile-press. Tobacco-press. Toggle-joint press. Upsetting-press. Veneering-press. Water-power press. Wedge-press. Windlass-lever press. Wine-press. Wool-press. Yarn-press.				
Press	J. E. Alford	Franklinville, N. C	Oct. 15, 1872	132, 229
Press	J. P. Arnold	Louisville, Ky	Apr. 3, 1855	12, 615
Press	S. J. Austin	Freeport, Me	Feb. 7, 1865	46, 205
Press	T. and J. P. Bakewell	Pittsburgh, Pa	Jan. 14, 1829	....
Press	H. B. Barber	Scott, N. Y	June 4, 1867	65, 464
Press	E. C. Betts	Huntsville, Ala	Feb. 21, 1860	27, 198
Press	L. Boudreaux	Thibodeaux, La	Apr. 25, 1865	47, 387
Press	T. G. Brooke	Newark, Ohio	June 4, 1872	127, 557
Press	M. D. Cheek	Memphis, Tenn	May 23, 1848	5, 597
Press	G. C. Chesley	Rocky Mount, Va	June 27, 1838	812
Press	J. Christison	New York, N. Y	Sept. 15, 1863	39, 885
Press	J. E. Conor	Brooklyn, N. Y	July 7, 1868	79, 637
Press	J. Cramton	New York	Feb. 18, 1834	....
Press	D. Curtis	Madison, Wis	Mar. 8, 1870	100, 607
Press	R. V. De Guinon and G. W. Barclay.	Hudson, N. J., and Brooklyn, N. Y.	Apr. 15, 1862	34, 996
Press	D. Dick	Meadville, Pa	Oct. 10, 1848	5, 840
Press	D. Dick	Meadville, Pa	Oct. 17, 1848	5, 856
Press	H. Dodge	Albany, N. Y	Apr. 11, 1865	47, 192
Press	S. S Edmonston	New York	Sept. 1, 1813	....
Press	R. Esmond	New York, N. Y	Aug. 26, 1873	142, 220
Press	P. G. Gardner	New York, N. Y	Feb. 22, 1848	5, 453
Press	J. A. McGillirrae	Dyer, Ind	Jan. 10, 1865	45, 844
Press	J. B. Godwin	Washington, D. C	June 17, 1873	140, 033
Press	G. E. Harding	Bath, Me	May 17, 1864	42, 763
Press	G. E. Harding	Bath, Me	Jan. 3, 1865	45, 711
Press	L. W. Harris	Waterville, N. Y	Aug. 28, 1860	29, 783
Press	T. W. Harvey	New York	Feb. 4, 1843	2, 941
Press	H. Henley	New Garden, Ind	June 9, 1868	78, 738
Press	H. F. Hicks	Grand View, Ind	Sept. 20, 1859	25, 546
Press	S. Hitchcock	Vernon, N. Y	June 25, 1813	....
Press	E. Huddleston and B. M. Harrison.	Lawrence, Kans., and Terre Haute, Ind.	Apr. 9, 1861	31, 979
Press	P. C. Ingersoll	Green Point, N. Y	Jan. 24, 1865	46, 051
Press	S. Ingersoll	Brooklyn, N. Y	Mar. 30, 1858	19, 811
Press	A. Jackson	Liberty, Ill	June 24, 1814	3, 637
Press	G. D. Jones	New York, N. Y	June 11, 1867	65, 752
Press	D. King	Aberdeen, Ohio	Mar. 26, 1867	63, 258
Press	R. Kingsley	Springfield, Mass	Jan. 13, 1857	16, 389
Press	J. Huebler	Belleville, Ill	Jan. 20, 1863	37, 475
Press	A. Lasswell	Springfield, Tex	Sept. 24, 1872	131, 619
Press	W. B. Leonard	Matteawan Works, N. Y	Feb. 20, 1843	2, 960
Press	J. Lewis	Wellington, Ohio	Dec. 26, 1865	51, 732
Press	G. Matthewman	Brooklyn, N. Y	Feb. 13, 1866	52, 584
Press	G. Matthewman	Brooklyn, N. Y	Sept. 15, 1868	82, 139
Press	D. McComb	Port Gibson, Miss	Feb. 27, 1849	6, 141
Press	T. H. McCray	Tellico, Tex	Jan. 17, 1860	26, 855
Press	W. J. McDermott	Covington, Tenn	Sept. 28, 1869	95, 249
Press	J. A. McGillerrae	Dyer, Ind	July 24, 1866	56, 588
Press	W. Norman	New York, N. Y	Apr. 3, 1866	53, 657
Press	B. S. Norris	Ripley, Ohio	May 26, 1868	78, 318
Press	A. Norton	Litchfield, Conn	Feb. 11, 1812	....
Press	I. Peck and W. H. H. Glover	Southold and New York, N. Y.	May 19, 1863	38, 599
Press	G. W. Penniston	North Vernon, Ind	Jan. 3, 1860	26, 706
Press	E. B. Platt	Florence, Ind	Apr. 29, 1873	138, 432
Press	W. H. Ragan	Fillmore, Ind	Jan. 30, 1872	123, 125
Press	A. Randel	New York N. Y	Oct. 23, 1860	30, 500
Press	W. Randle	Hadensville Station, Ky	Feb. 13, 1872	123, 644
Press	W. C. Ray	Redington, N. J	Apr. 28, 1863	38, 333
Press	C. O. and J. H. Ritchie	North Madison, Ind	Mar. 31, 1868	76, 101
Press	C. H. Robinson	Bath, Me	Jan. 3, 1865	45, 754
Press	A. Roden	Talladega, Ala	Aug. 21, 1860	29, 719
Press	C. E. Rymes	Somerville, Mass	Aug. 27, 1867	68, 117
Press	U. Schegg	Nauvoo, Ill	Nov. 19, 1867	71, 071
Press	J. Seitz, sr	Bloom, Ohio	Mar. 5, 1861	31, 630
Press	A. C. Semple	Cincinnati, Ohio	June 28, 1853	9, 824

Index of patents issued from the United States Patent Office from 1790 *to* 1873, *inclusive*—Continued.

Invention.	Inventor.	Residence.	Date.	No.
Press	W. I. Tate	Philadelphia, Pa	Oct. 11, 1870	108, 208
Press	E. Thomas	Beverly, Va	Feb. 19, 1861	31, 495
Press	I. H. Trabue	Livingston County, Ky	Aug. 31, 1869	94, 257
Press	C. P. Wagner	New York, N. Y	Sept. 10, 1867	68, 812
Press	R. Wakeman	Port Deposit, Md	Mar. 26, 1861	31, 841
Press	M. Wappich	Sacramento, Cal	Sept. 4, 1860	29, 928
Press	W. E. Warner	Syracuse, N. Y	Apr. 8, 1873	137, 641
Press	T. B. Webster	New York, N. Y	July 25, 1865	49, 046
Press	J. P. White	New York, N. Y	Oct. 25, 1864	44, [illegible]42
Press	J. J. Wise	Baltimore, Md	May 16, 1839	1, 152
Press and strainer, Combined	J. H. Littlefield	Cambridge, Mass	June 16, 1868	78, [illegible]81
Press and strainer, Combined	J. H. Littlefield	Cambridge, Mass	Jan. 5, 1869	85, 530
Press and strainer, Combined	W. D. Medbury	Bangor, Wis	Mar. 29, 1870	101, 294
Press-board, Tailor's	W. Frizzell	Boston, Mass	Apr. 23, 1872	125, [illegible]
Press-board, Tailor's	A. M. Thompson and C. W. Burbank.	Alfred, Me	Sept. 8, 1868	81, 959
Press frame, Brown's standing	R. Hoe	New York	Apr. 22, 1828	....
Press-plate	A. Smith	Cincinnati, Ohio	Feb. 27, 1866	52, 899
Press-spindle adjustment	A. H. Merriman	New Britain, Conn	May 26, 1868	78, 225
Press-tub for pressing oleaginous seeds	F. Follet	Petersburgh, Va	Aug. 14, 1834	....
Presses, Machine for removing molded forms from.	G. Pattern	Chester, Pa	Dec. 24, 1867	72, 533
Presses, Operating	O. T. Earle and T. J. Rider	Norwalk, Conn., and New York, N. Y.	Aug. 26, 1873	142, 093
Presses, Operating	D. L. Miller	Madison, N. J	Aug. 3, 1858	21, 079
Presses, &c., Trip-motion for	A. Hamlin	Brooklyn, N. Y	Sept. 13, 1870	107, 363
Pressing	J. Tinkler	Massillon, Ohio	Aug. 31, 1831	....
Pressing and boring machine	R. Ramsey	Hanover, N. H	July 9, 1808	....
Pressing and molding pliable materials, Machine for.	G. C. Howard	Philadelphia, Pa	June 19, 1866	55, 658
Pressing and stuffing machine	E. S. Holloway	Columbiana, Ohio	July 4, 1871	116, 709
Pressing, hoisting, and lowering machine	P. Freeman	Perth Amboy, N. J	Feb. 20, 1822	....
Pressing, ironing, and smoothing machine	H. Hamill	New York, N. Y	June 6, 1871	115, 604
Pressing-machine	R. P. Cunningham	Pomfret, Conn	May 21, 1808	....
Pressing-machine	L. E. Dennison	Saybrook, Conn	Oct. 1, 1830	....
Pressing-machine	T. Puffer	Paris, N. Y	May 23, 1815	....
Pressing-machine	R. V. W. and J. Thorne	New York	Mar. 26, 1810	....
Pressing-machine, Tailor's	J. W. Thorp	Hillsborough Bridge, N. H	May 12, 1868	77, 934
Pressure alarm, Hydraulic	J. L. Pillsbury and J. S. Shorb	Columbus and Canton, Ohio	May 16, 1871	114, 849
Pressure and gravitation machine	C. Monson	New Haven, Conn	Sept. 26, 1865	50, 151
Pressure and water indicator, Steam	W. C. Grimes	Philadelphia, Pa	Sept. 7, 1858	21, 468
Pressure-bag for separating elaine and stearine	M. H. Shepard	Cincinnati, Ohio	May 26, 1843	3, 109
Pressure engine, Regulated	D. Livermore	Blairsville, Pa	July 24, 1834	....
Pressure-gage	A. Allan	Perth, Scotland	Nov. 15, 1864	45, 010
Pressure-gage	E. H. Ashcroft	Boston, Mass	July 12, 1853	9, 836
Pressure-gage	R. C. Blake	Cincinnati, Ohio	May 30, 1871	115, 420
Pressure-gage	E. Bourdon	Paris, France	Aug. 3, 1852	9, 163
Pressure-gage	J. Brown, jr	New York, N. Y	Jan. 16, 1872	122, 700
Pressure-gage	W. Burnett	Boston, Mass	Aug. 13, 1861	33, 062
Pressure-gage	H. Chwatal	New York, N. Y	Apr. 12, 1870	101, 832
Pressure-gage	B. Crawford	Allegheny City, Pa	Mar. 16, 1852	8, 797
Pressure-gage	G. H. Crosby	East Sommerville, Mass	Dec. 23, 1873	145, 726
Pressure-gage	C. M. Daboll	New London, Conn	Dec. 4, 1860	30, 803
Pressure-gage	J. B. Eads and H. Flad	Saint Louis, Mo	July 25, 1871	117, 394
Pressure-gage	J. B. Eads and H. Flad	Saint Louis, Mo	Aug. 20, 1872	130, 705
Pressure-gage	H. W. Farley	Hannibal, Mo	Jan. 24, 1860	26, 900
Pressure-gage	R. Finnegan and A. F. W Schulte.	New York, N. Y	June 11, 1861	32, 514
Pressure-gage	W. C. Grimes	Philadelphia, Pa	July 6, 1858	20, 848
Pressure-gage	T. C. Hargrave	Boston, Mass	June 11, 1872	127, 874
Pressure-gage	H. J. H. King	Glasgow, Great Britain	Nov. 9, 1869	96, 705
Pressure-gage	J. Leavens	Brooklyn, N. Y	Mar. 19, 1861	31, 721
Pressure-gage	C. Liedke	Sandusky, Ohio	Aug. 9, 1870	106, 180
Pressure-gage	J. W. Maloy	Boston, Mass	Dec. 3, 1867	71, 773
Pressure-gage	J. W. Maloy	Boston, Mass	Apr. 5, 1870	101, 636
Pressure-gage	J. Matthews, jr	New York, N. Y	Aug. 21, 1855	13, 468
Pressure-gage	J. Matthews, jr	New York, N. Y	Apr. 9, 1867	63, 648
Pressure-gage	G. Sewell	Brooklyn, N. Y	Aug. 26, 1873	142, 128
Pressure-gage	J. W. Stiles	New York, N. Y	Apr. 1, 1873	137, 331
Pressure-gage	E. A. Wood	Utica, N. Y	Apr. 26, 1870	102, 350
Pressure gage and alarm	M. Toulmin	New Orleans, La	Sept. 16, 1873	142, 969
Pressure-gage and safety-valve	A. A. Kent	Lyons, Iowa	June 7, 1870	103, 893
Pressure-gage dial	J. Annin	New York, N. Y	July 15, 1873	140, 867
Pressure-gage, Diaphragm	R. C. Robbins	New York, N. Y	July 18, 1865	48, 877
Pressure-gage, Duplex	P. B. Hovey	New London, Conn	June 21, 1870	104, 593
Pressure-gage for hydrostatic presses	T. Harbottle	Brooklyn, N. Y	May 11, 1869	89, 992
Pressure-gage for steam-boilers	E. H. Ashcroft	Boston, Mass	Mar. 6, 1860	27, 341
Pressure-gage for water-backs, stoves, ranges, &c.	J. Anderson	Allegheny, Pa	Aug. 29, 1871	118, 506
Pressure gage, Hot-blast	J. Storer	New York, N. Y	Aug. 3, 1869	93, 366
Pressure-gage, Recording	J. B. Edson	Brooklyn, N. Y	Aug. 16, 1870	106, 345
Pressure-gage, Self-recording	G. P. Clarke and M. B. and J. B. Edson.	New York, N. Y	Aug. 3, 1869	93, 275
Pressure-gages, Adjustment of mercurial	M. E. Campfield	Newark, N. J	Apr. 26, 1870	102, 464
Pressure-indicator	F. T. Riegel	Philadelphia, Pa	Aug. 11, 1868	81, 006
Pressure indicator, Hydrostatic	J. E. Wootten	Philadelphia, Pa	May 14, 1861	32, 329
Pressure-regulator for steam, water, or gas, Balance	E. Osgood	Boston, Mass	Apr. 29, 1862	35, 104
Pressure-regulator for water and steam apparatus	J. C. Hagan	Nashville, Tenn	Nov. 1, 1870	108, 904
Pressure-regulator for water-works	B. Holly	Lockport, N. Y	May 9, 1871	114, 683
Pressure regulator, Hydraulic	T. J. McGowan	Cincinnati, Ohio	Dec. 29, 1868	85, 322
Pressure regulator, Steam	G. H. Corliss	Providence, R. I	Jan. 5, 1869	85, 566
Pressure regulator, Steam and water	N. C. Locke	Salem, Mass	Mar. 28, 1871	113, 069
Pressure regulator, Steam and water supply and	J. C. Hagan	Nashville, Tenn	Nov. 1, 1870	108, 905
Pressure-regulator, Water-pipe	J. Johnson	Lowell, Mass	July 2, 1872	128, 492
Preventing a person while asleep from turning on his back, Apparatus for.	R. T. Sullivan	Needham, Mass	Oct. 22, 1872	132, 500
Price-current sheets, Producing	W. Smith and H. D. Rogers	Cincinnati, Ohio	July 12, 1870	105, 380
Price-indicator	P. Topham	Newark, N. J	Feb. 7, 1860	27, 084
Printer and stamp, Electro-magnetic date-time	J. C. Hinchman	Summit, N. J	July 29, 1873	141, [illegible]2
Printer, Consecutive number	R. M. Evans	Buffalo, N. Y	Nov. 4, 1873	144, 196

Index of patents issued from the United States Patent Office from 1790 *to* 1873, *inclusive*—Continued.

Invention.	Inventor.	Residence.	Date.	No.
Printer's ball-stock	J. Babcock	New Haven, Conn	Mar. 18, 1818	....
Printer's blanket	J. Taylor	Lawrence, Mass	Sept. 5, 1865	49, 804
Printer's blocks, Use of caseine for making	T. J. Denno	London, England	May 21, 1872	127, 032
Printer's bodkin and tweezers, Pocket-case for	W. Quail	New York, N. Y	Oct. 29, 1867	70, 261
Printer's case	A. T. De Puy	New York, N. Y	May 2, 1871	114, 273
Printer's case	W. H. A. Gresham	Atlanta, Ga	Mar. 7, 1871	112, 445
Printer's case-stand	R. Chapman	Dublin, Ireland	Apr. 23, 1872	126, 017
Printer's case-stand	P. S. Hoe	New York, N. Y	July 16, 1872	129, 281
Printer's chair	A. J. Finch	New York, N. Y	Feb. 1, 1870	99, 302
Printer's chase	E. Allen	Norwich, Conn	July 16, 1872	129, 079
Printer's chase	A. M. Blanchard	Saint Louis, Mo	Feb. 20, 1866	52, 662
Printer's chase	L. G. Chaput	New York, N. Y	June 3, 1873	139, 542
Printer's chase	T. A. Clements	Little Rock, Ark	Nov. 1, 1870	108, 759
Printer's chase	T. A. Clements	Little Rock, Ark	Nov. 1, 1870	108, 760
Printer's chase	R. Dick	Buffalo, N. Y	Apr. 14, 1868	76, 722
Printer's chase	J. N. Murray	Chicago, Ill	July 9, 1867	66, 514
Printer's chase	W. O. Stoddard	Champaign, Ill	July 9, 1861	32, 798
Printer's chase	E. A. Warren	Brooklyn, N. Y	May 13, 1873	138, 967
Printer's cloth, &c., Process for cleaning plate	H. M. Baker	Washington, D. C	June 15, 1869	91, 408
Printer's forms, Locking up	E. H. Sprague	Zanesville, Ohio	June 13, 1854	11, 091
Printer's frisket	A. Overend	Philadelphia, Pa	June 13, 1854	11, 072
Printer's furniture	B. B. Blackwell	Jamaica, N. Y	Aug. 29, 1871	118, 425
Printer's furniture	F. T. M. A. Guyon	Saint Brieux, France	Apr. 7, 1868	76, 440
Printer's furniture	H. A. Hempel	Saint Joseph, Mo	Dec. 23, 1873	145, 800
Printer's furniture	A. N. Kellogg and J. J. Schock	Chicago, Ill	Apr. 19, 1870	102, 013
Printer's furniture	H. P. Montague	Belchertown, Mass	Aug. 5, 1873	141, 450
Printer's furniture	R. B. Topham	Washington, D. C	Jan. 3, 1871	110, 695
Printer's furniture	J. F. Uhlhorn	Sacramento, Cal	May 23, 1871	115, 136
Printer's furniture	W. H. Windsor	Little Rock, Ark	May 30, 1871	115, 404
Printer's galley	J. W. Baker	Warsaw, Ind	July 16, 1867	66, 776
Printer's galley	W. G. Blymyer	Findlay, Ohio	Mar. 17, 1868	75, 619
Printer's galley	J. F. Bronson	Waterbury, Conn	Nov. 21, 1871	121, 041
Printer's galley	A. T. De Puy	New York, N. Y	Dec. 4, 1866	60, 151
Printer's galley	A. T. De Puy	New York, N. Y	Aug. 10, 1869	93, 422
Printer's galley	A. T. De Puy	New York, N. Y	July 16, 1872	128, 954
Printer's galley	E. Hutchings	Hartford, Conn	Mar. 23, 1869	88, 041
Printer's galley	C. H. Lawrence	New York, N. Y	July 14, 1868	79, 988
Printer's galley	H. E. Long	Plymouth, Ind	Mar. 24, 1868	75, 935
Printer's galley	P. G. Meek	Bellefonte, Pa	Nov. 19, 1867	71, 200
Printer's galley	W. Quail	New York, N. Y	Oct. 26, 1869	96, 145
Printer's galley	J. W. Skinner	Quincy, Ill	Sept. 24, 1872	131, 713
Printer's galley	J. Snyder	Burlington, Iowa	Jan. 2, 1866	51, 877
Printer's galley	J. Wilson, jr	New York, N. Y	Aug. 24, 1869	94, 053
Printer's galley-rest	J. A. Doyle and J. A. Murphy	Lancaster, Pa	Sept. 30, 1873	143, 228
Printer's galley-rest	J. M. Murphy	Olympia, Wash	Aug. 3, 1869	93, 216
Printer's galleys, Side-stick for	E. B. Hindle	Saint Louis, Mo	July 1, 1873	140, 368
Printer's ink-roller	G. Littleton	Cleveland, Ohio	Apr. 7, 1863	38, 113
Printer's ink-roller	G. Russell and T. B. Hull	New York and Brooklyn, N. Y.	Jan. 1, 1867	60, 792
Printer's ink-rollers, Machine for washing	O. H. Reed and A. L. Carrier	Washington, D. C	Dec. 15, 1868	84, 904
Printer's inking-roller	L. Francis	New York, N. Y	Mar. 8, 1864	41, 887
Printer's inking-roller	C. S. Westcott	Elizabeth, N. J	Dec. 27, 1870	110, 522
Printer's inking-roller from rubber sponge	S. Moulton	Bradford-on-Avon, England.	Sept. 7, 1869	94, 631
Printer's inking-rollers, Composition for	W. Harvey	Portland, Me	Nov. 13, 1866	59, 597
Printer's inking-rollers, Composition for making	L. Francis and F. W. Letmate	New York, N. Y	May 6, 1862	35, 149
Printer's lapping	S. W. Baker	Providence, R. I	May 6, 1862	35, 133
Printer's lead-casting machine	J. I. Sturgis	Milwaukee, Wis	Jan. 28, 1873	135, 384
Printer's lead-rack	O. A. Dearing	San Francisco, Cal	Feb. 18, 1873	135, 894
Printer's leads and rules, Machine for bending	F. C. Smith and H. McCollum	New York and Long Island City, N. Y.	Mar. 5, 1872	124, 295
Printer's leads, Apparatus for casting	W. Filmer	San Francisco, Cal	Sept. 16, 1873	142, 783
Printer's leads, Cutting	J. W. H. Cheney	Hartford, Conn	Aug. 25, 1868	81, 476
Printer's leads, Cutting	W. E. Clark	Boston, Mass	Aug. 18, 1868	81, 140
Printer's leads, Machine for making	K. M. Klees	New York, N. Y	July 19, 1870	105, 463
Printer's leads, Machine for making	I. Schoenberg, jr	New York, N. Y	Mar. 7, 1871	112, 384
Printer's metallic plates, Preparation of	S. W. Lowe	Philadelphia, Pa	Sept. 18, 1855	13, 587
Printer's miter-machine	J. A. Stansbury	Salem, Ohio	Apr. 8, 1873	137, 575
Printer's quoin	C. W. Ames	New York, N. Y	May 27, 1873	139, 351
Printer's quoin	D. Dorrity	Pont-Audemere, France	Aug. 10, 1869	93, 603
Printer's quoin	H. A. Marinoni and F. N. Chandré.	Paris, France	Sept. 22, 1863	40, 081
Printer's rollers, Apparatus for cleaning and preserving.	S. Crump	Brooklyn, N. Y	Oct. 17, 1871	120, 043
Printer's rollers, Composition for	L. K. Bingham	New York, N. Y	Sept. 11, 1866	57, 849
Printer's rollers, Diamond-holder for engraving	J. Hope	Providence, R. I	Nov. 9, 1869	96, 591
Printer's rollers, Recovering the materials of worn out.	J. H. Osgood	Peabody, Mass	Nov. 3, 1868	83, 786
Printer's rolls, Renewing the surface of	C. Sentell	Waterloo, N. Y	June 27, 1865	48, 453
Printer's rule	C. Reuter	Cincinnati, Ohio	Sept. 2, 1873	142, 512
Printer's rules, Machine for cutting and mitering	F. H. Aiken	Franklin, N. H	Feb. 21, 1871	111, 896
Printer's rules, Machine for making	S. D. Tucker	New York, N. Y	Apr. 20, 1869	89, 183
Printer's rules, Machine for mitering	W. E. Cameron and A. A. Dettlaff.	Green Island, N. Y	Sept. 28, 1869	95, 193
Printer's rules, Machine for preparing	J. B. Bancroft	Hopedale, Mass	Dec. 16, 1873	145, 549
Printer's shears	C. E. Fisk	New York, N. Y	May 25, 1869	90, 518
Printer's shooting-stick	B. B. Blackwell	New York, N. Y	Sept. 6, 1870	107, 154
Printer's side-stick and quoin	F. Keehn	Milwaukee, Wis	Dec. 16, 1873	145, 574
Printer's use, Combination-rack for	R. Yeomans	Chicago, Ill	Oct. 24, 1865	50, 650
Printing	B. Beniowski	London, England	Oct. 29, 1850	7, 738
Printing	W. M. Dodge	New York, N. Y	Mar. 6, 1823	....
Printing	C. N. Morris	Cincinnati, Ohio	Oct. 24, 1865	50, 612
Printing, Anastatic	C. F. Baldamus and F. W. Siemens.	Berlin, Prussia	Oct. 25, 1845	4, 239
Printing and drawing checks to prevent counterfeiting.	J. Dainty	Philadelphia, Pa	July 31, 1837	320
Printing and dyeing fabrics	T. Crossley	Bridgeport, Conn	June 10, 1873	139, 705
Printing and dyeing woolen cloth	W. Duncan	Belleville, N. J	Sept. 18, 1835	....

Index of patents issued from the United States Patent Office from 1790 *to* 1873, *inclusive*—Continued.

Invention.	Inventor.	Residence.	Date.	No.
Printing and embossing skirts, Machine for	H. J. Davies	Brooklyn, N. Y	Mar. 25, 1873	137, 184
Printing and finishing, Copper-coated iron-rolls for	A. Paraf	Mulhouse, France	Apr. 23, 1867	64, 135
Printing and mode of preventing counterfeits, Copper-plate.	J. Perkins	Boston, Mass	June 16, 1810	....
Printing and numbering press	G. I. Hill	Buffalo, N. Y	Sept. 7, 1858	21, 418
Printing and ornamenting	J. L. Wells	Philadelphia, Pa	Sept. 12, 1871	118, 992
Printing and registering apparatus, Ticket	J. Dyer	Chicago, Ill	July 29, 1873	141, 336
Printing-apparatus	F. A. De Mey	Brooklyn, N. Y	June 9, 1863	38, 815
Printing-apparatus	J. Warren	Warwick Township, Ohio	May 2, 1835	....
Printing-apparatus	J. M. Wilbur	Cleveland, Ohio	Nov. 19, 1867	71, 103
Printing-apparatus, Attaching blocks to belts of	A. S. Davis	Boston, Mass	Nov. 26, 1861	33, 775
Printing-apparatus, Elastic	J. McDermott	Frederick, Md	Jan. 14, 1868	73, 359
Printing-apparatus, Register-points for	J. F. Shearman	Brooklyn, N. Y	May 28, 1867	65. 127
Printing, Applying dry metallic and colored powder to type.	G. J. Newberry	New York	Feb. 1, 1821	....
Printing bank-notes	A. Tichenor	Newark, N. J	June 7, 1859	24, 341
Printing bank-notes, Apparatus for	J. W. Hayes	Newark, N. J	Oct. 22, 1861	33, 526
Printing bank-notes, &c	A. Sellers	New York, N. Y	Feb. 23, 1864	41, 724
Printing, Biting the figures on steel cylinders for calico.	D. H. Mason and M. W. Baldwin.	Philadelphia, Pa	Oct. 30, 1827	....
Printing-blanket	S. W. Baker	Providence, R. I	Dec. 20, 1859	26, 467
Printing-blocks, Constructing	L. Jarosson	Jersey City, N. J	Nov. 14, 1854	11, 954
Printing-blocks, Constructing molds for making	J. Berry	Roxbury, Mass	Mar. 14, 1854	10, 63[illegible]
Printing-blocks, Construction of	B. Underwood	Brooklyn, N. Y	Jan. 31, 1854	10, 483
Printing-blocks, Making	T. Crossley	Boston, Mass	July 18, 1854	11, 321
Printing books	S. Randall	Warren, R. I	Jan. 4, 1816	....
Printing by means of rollers	B. Allison and W. Elliott	Washington, D. C	Mar. 11, 1819	....
Printing calico	R. Ferguson and J. Clark	Glasgow, Scotland	Apr. 25, 1846	4, 476
Printing, Calico	R. Prince and A. Lovis	Lowell and Boston, Mass	Dec. 11, 1855	13, 915
Printing calico	B. Woodcroft	Manchester, England	Apr. 17, 1847	5, 064
Printing calico and paper, Machine for	L. Beatty	Wilkesbarre, Pa	Dec. 28, 1810	....
Printing calico, Machine for	A. Sibley	Pawtucket, Mass	July 9, 1838	823
Printing, Card-board for	A. R. Davis	East Cambridge, Mass	Mar. 3, 1868	75, 130
Printing checks, Machine for	J. Pollak	Chicago, Ill	July 4, 1865	48, 589
Printing, Cleansing and drying gum-elastic or cloth bands in calico.	J. Hunter	Blockley Township, Pa	Aug. 27, 1850	7, 602
Printing colored figures on hard substances	J. Edwards	New York	Aug. 22, 1816	....
Printing colors on calico, &c	B. Woodcroft	Ardwich, England	Apr. 5, 1838	678
Printing colors on silk, linen, &c., Forming blocks for.	J. Crabtree	New York, N. Y	Dec. 7, 1837	507
Printing colors on textile material	A. Paraf	New York, N. Y	Sept. 21, 1869	95, 040
Printing, Composition for producing elastic forms for.	I. L. Miles	Philadelphia, Pa	Oct. 15, 1867	69, 926
Printing, Composition to be used instead of gum-senegal in calico.	W. Leversidge	Dorchester, Mass	Aug. 8, 1837	344
Printing, Copper-plate	W. Conisbee	Lambeth, Great Britain	July 2, 1872	128, 595
Printing, Cylinder	J. Webster and S. H. Folsom	Boston and Lowell, Mass	Mar. 25, 1856	14, 542
Printing, Cylinder-roller for distributing ink on copper plates for.	A. Maverick	New York	Apr. 17, 1810	....
Printing-cylinders, Attaching blankets to	A. Overend	Philadelphia, Pa	Apr. 30, 1872	126, 228
Printing cylinders, Method of operating the doctors of calico.	J. Baxendale	Fall River, Mass	Mar. 28, 1854	10, 719
Printing, Die-sinking press for obtaining relief-plates for surface.	W. H. Oakes	New York, N. Y	Jan. 5, 1864	41, 088
Printing-disks, Pattern for casting	J. Goldsborough	Philadelphia, Pa	Mar. 5, 1872	124, 350
Printing, Elastic transfer	D. W. Bowdoin	Salem, Mass	July 12, 1870	105, 300
Printing, Electographic	A. W. Thompson	Philadelphia, Pa	Apr. 26, 1845	4, 012
Printing, Etching relief-plates for surface	C. Henry and J. and E. McLoughlin.	Brooklyn, Morrisania, and New York, N. Y.	Nov. 17, 1868	84, 187
Printing fabrics, books, &c., Manufacture of India-rubber blanket or apron for.	C. McBurney	Boston, Mass	June 7, 1859	24, 321
Printing fabrics, Constructing roller or cylinder for.	R. F. Sturges	Birmingham, England	May 5, 1857	17, 239
Printing fabrics	E. A. D. Guichard	Paris, France	June 25, 1872	128, 302
Printing fabrics, Press for	H. H. Townsend	Milton Mills, N. H	Apr. 8, 1873	137, 638
Printing, Facilitating	S. N. Dickenson	Boston, Mass	Nov. 19, 1833	....
Printing-fluid	E. Whitefield	Buffalo, N. Y	June 27, 1865	48, 471
Printing for colors, Plate	J. Enthoffer	Washington, D. C	Nov. 30, 1869	97, 287
Printing-form	E. Edwards	Middlesex County, England.	Dec. 31, 1872	134, 470
Printing, Forming plates for polychromatic	J. Donlevy	New York, N. Y	Jan. 3, 1854	10, 363
Printing-frame	O. H. Burdick	Auburn, N. Y	Apr. 19, 1864	42, 344
Printing-frame	S. F. Conant and H. A. Manley	Skowhegan, Me	Feb. 11, 1868	74, 311
Printing from engraved plates	J. Meyer	Bay Ridge, N. Y	July 5, 1864	43, 421
Printing from engraved plates, Machine for	R. Neale	Clermont County, Ohio	Jan. 9, 1855	12, 213
Printing from engraved plates, Machine for	J. F. Starrett	New York, N. Y	Feb. 19, 1856	14, 295
Printing from engraved plates, Machine for	L. Stewart and J. McClelland	Washington, D. C	Mar. 31, 1857	16, 952
Printing in colors	W. Croome	Brooklyn, N. Y	May 19, 1857	17, 319
Printing in colors	A. D. McKinzie	Philadelphia, Pa	Nov. 6, 1846	4, 838
Printing in colors	J. Rennie	Lodi, N. J	Aug. 9, 1834	....
Printing in colors	H. F. Smart	Worcester, Mass	Nov. 17, 1868	84, 225
Printing in colors for electro-chemical voting apparatus.	A. N. Henderson	Buffalo, N. Y	July 22, 1850	7, 521
Printing in colors, Machine for	T. F. Adams	Philadelphia, Pa	Sept. 17, 1844	3, 744
Printing in colors, Machine for	R. S. Weaver	Maysville, Ky	Oct. 28, 1851	8, 475
Printing in different colors, Machine for	J. R. Wright	Philadelphia, Pa	Aug. 9, 1859	25, 075
Printing-ink roller	E. Pratt	Salem, Mass	Aug. 10, 1858	21, 148
Printing, Inking-apparatus for color	T. L. Bayliss	Richmond, Ind	July 14, 1868	79, 888
Printing, Inking-apparatus for color	L. Biddle	Knoxville, Iowa	Apr. 16, 1872	125, 655
Printing, Inking-apparatus for color	J. Hunt	Richmond, Ind	July 14, 1868	79, 910
Printing, Inking-apparatus for color	T. Moore and P. H. Day	Bloomington, Ill	Apr. 13, 1869	88, 976
Printing, Inking-apparatus for color	I. L. G. Rice	Cambridge, Mass	Apr. 19, 1870	102, 157
Printing, Inking-apparatus for color	I. L. G. Rice	Cambridge, Mass	Oct. 25, 1870	108, 63[illegible]
Printing, Inking apparatus for color	I. L. G. Rice	Cambridge, Mass	Aug 27, 1872	130, 822
Printing, Inking-apparatus for color	L. B. Waterman	Indianapolis, Ind	Mar. 1[illegible], 1868	75, 499
Printing, Inking apparatus for color	G. W. Wood	Richmond, Ind	June 16, 1868	79, 043
Printing, Inking-attachment for chromatic	E. A. Howitt	Whitewater, Wis	May 20, 1873	139, 1[illegible]5
Printing labels on spools, Machine for	G. Hall, jr	South Wilmington, Conn	July 5, 1870	104, 951

Index of patents issued from the United States Patent Office from 1790 *to* 1873, *inclusive*—Continued.

Invention.	Inventor.	Residence.	Date.	No.
Printing long-napped fabrics	W. A. White	Roxbury, Mass	Aug. 15, 1854	11, 540
Printing-machine	M. Bebro	Manchester, Great Britain.	Aug. 29, 1871	118, 509
Printing-machine	W. Bullock	Pittsburgh, Pa	Apr. 14, 1863	38, 200
Printing-machine	W. A. Burt	Detroit, Mich	July 23, 1829	....
Printing-machine	G. W. Cartwright	Mount Pleasant, N. Y	Apr. 29, 1828	....
Printing-machine	J. H. Cooper	Philadelphia, Pa	May 20, 1856	14, 907
Printing-machine	O. T. Eddy	Baltimore, Md	Nov. 12, 1850	7, 771
Printing-machine	J. B. Fairbank	Leon, N. Y	Sept. 17, 1850	7, 652
Printing-machine	S. W. Francis	New York, N. Y	Oct. 27, 1857	18, 504
Printing-machine	R. M. Hoe	New York, N. Y	July 26, 1859	24, 875
Printing-machine	J. M. Jones	Palmyra, N. Y	May 20, 1856	14, 919
Printing-machine	S. D. Tucker	New York, N. Y	July 9, 1867	66, 654
Printing-machine	C. Thurber	Worcester, Mass	Aug. 26, 1843	3, 228
Printing-machine	C. F. Voorhies	Newark, N. J	Apr. 8, 1834	....
Printing machine, Card	J. S. Moody	Cincinnati, Ohio	Dec. 1, 1857	18, 795
Printing machine, Electrical	T. A. Edison	Newark, N. J	Nov. 12, 1872	133, 019
Printing-machine for paper-hangings, cloth, books, &c.	P. Force	Washington, D. C	Aug. 22, 1822	....
Printing machine, Stencil	A. J. Fullam	Springfield, Vt	Oct. 2, 1860	30, 216
Printing machines, Blanket for calico	J. Fallon	Lawrence, Mass	May 19, 1857	17, 322
Printing-machines, Inking-apparatus for	J. K. Lowe	Cleveland, Ohio	Apr. 2, 1867	63, 540
Printing-machines, Movement for doctors of calico	J. Standing	Fall River, Mass	Feb. 5, 1856	14, 214
Printing-machines, Washing blankets of	T. W. Clarke	Manchester, N. H	June 6, 1865	48, 053
Printing-machinery	J. Hatch	Boston, Mass	Mar. 21, 1832	....
Printing-machinery	S. Savage	Fredonia, N. Y	July 16, 1872	129, 176
Printing-machinery, Cloth-feeding apparatus for	E. O. Potter	Pawtucket, R. I	Dec. 29, 1868	85, 475
Printing machinery, Cylindrical-plate	G. F. Lewis and F. D. Stuart	Philadelphia, Pa., and Washington, D. C.	May 12, 1868	77, 896
Printing machinery, Plate	I. T. Robertson	New York, N. Y	Feb. 20, 1872	123, 933
Printing, Making mill and skeleton dies for	T. Paton	Providence, R. I	Oct. 24, 1848	5, 871
Printing names or directions on packages, &c., Machine for.	J. Spencer	Toronto, Canada	Nov. 23, 1858	22, 136
Printing numbers, Machine for	C. L. Sholes	Milwaukee, Wis	Apr. 30, 1867	64, 375
Printing on bottles, Apparatus for	I. L. Miles	Charlestown, Mass	July 24, 1866	56, 593
Printing on and decorating enameled surfaces, Apparatus for.	S. J. Hoggson	New Haven, Conn	Nov. 25, 1873	144, 979
Printing on enameled or vitreous surfaces	S. J. Hoggson	New Haven, Conn	Dec. 26, 1871	122, 170
Printing on fabrics, Machinery for	C. Holliday	Huddersfield, England	May 26, 1868	78, 288
Printing on glass	I. L. Miles	Charlestown, Mass	Jan. 22, 1867	61, 350
Printing on glass, Apparatus for	I. L. Miles	Charlestown, Mass	Mar. 5, 1867	62, 666
Printing on glass, Apparatus for	D. B. Wingate	Natick, Mass	July 7, 1868	79, 713
Printing on glass, porcelain, &c	E. C. Jayne	Philadelphia, Pa	Feb. 6, 1866	52, 418
Printing on paper, calico, &c., Forming raised surfaces for.	G. Woone	London, England	Apr. 2, 1838	670
Printing on solid or inflexible surfaces, Apparatus for.	A. Wilbaux	Paris, France	Sept. 2, 1873	142, 541
Printing on spools, Machine for	G. Hall, jr. and G. W. Averell.	South Wilmington, Conn., and New York, N. Y.	Apr. 26, 1870	102, 257
Printing on tin-foil, Apparatus for	J. Polhemus and C. H. Lilienthal.	Jersey City, N. J., and Yonkers, N. Y.	Nov. 19, 1867	71, 058
Printing on the back or reverse face of bank-notes to prevent counterfeits.	J. Kneass	....	Apr. 28, 1815	....
Printing on uneven surfaces, Apparatus for	A. Leighton	London, England	Mar. 5, 1867	62, 646
Printing on woven fabrics of cotton, linen, &c., Block.	R. Hampson	Manchester, Great Britain	June 7, 1841	2, 116
Printing or painting on surfaces, Composition for.	A. Farrar	Brookline, Mass	Mar. 15, 1870	100, 877
Printing or painting on surfaces, Composition for.	A. Farrar	Brookline, Mass	Apr. 18, 1871	113, 755
Printing, painting, &c., Preparing canvas for	E. Lee	Baltimore, Md	May 12, 1857	17, 308
Printing paper, cloth, &c., Machine for	G. S. Barton	Worcester, Mass	Oct. 12, 1869	95, 758
Printing paper-hangings, Machine for	F. S. Munroe, jr., and T. Mason	Grantville and Boston, Mass.	June 13, 1865	48, 199
Printing paper-hangings, Roller for	T. Van Deventer	New Brunswick, N. J	Dec. 6, 1859	26, 387
Printing paper, leather, &c., Machine for	J. Dixey	....	Jan. 24, 1798	....
Printing paper on both sides	H. Betts	Norwalk, Conn	Sept. 14, 1833	....
Printing, Photo-mechanical	E. Edwards	Middlesex County, England	Dec. 10, 1872	133, 701
Printing, Plate	W. H. Oakes	New York, N. Y	Oct. 23, 1860	30, 495
Printing plate, Relief	B. Day	Hoboken, N. J	Apr. 26, 1864	42, 530
Printing plate, &c., Relief	D. C. Hitchcock and E. B. and E. M. Larchar.	New York, N. Y	Nov. 13, 1860	30, 630
Printing-plates, Removing ink from engraved	J. S. Ives	New York, N. Y	Nov. 11, 1873	144, 544
Printing, Preparation of engraved metal plates for.	J. M. Batchelder and L. L. Smith.	Cambridge, Mass., and New York, N. Y.	Nov. 24, 1857	18, 668
Printing, Preparing fabrics for	J. Coates	Manchester, England	Mar. 21, 1848	5, 480
Printing-press	A. S. Adams	Chelsea, Mass	Mar. 19, 1861	31, 760
Printing-press	S. Adams	Boston, Mass	Sept. 27, 1844	3, 769
Printing-press	S. Adams	Boston, Mass	Mar. 8, 1853	9, 606
Printing-press	J. B. Adt	Baltimore, Md	Oct. 24, 1871	120, 225
Printing-press	E. Allen	Norwich, Conn	Nov. 12, 1867	70, 773
Printing-press	E. Allen	Norwich, Conn	Feb. 4, 1868	73, 943
Printing-press	J. F. Allen and R. W. McGowan	New York, N. Y	Dec. 15, 1863	40, 892
Printing-press	F. J. Austin	New York, N. Y	May 7, 1850	7, 335
Printing-press	G. H. Babcock	Westerly, R. I	Dec. 23, 1856	16, 263
Printing-press	W. H. Babcock	Homer, N. Y	Oct. 23, 1860	30, 452
Printing-press	F. L. Bailey	Boston, Mass	Nov. 25, 1856	16, 109
Printing-press	F. L. Bailey	Boston, Mass	June 23, 1857	17, 615
Printing-press	F. L. Bailey	Boston, Mass	Feb. 21, 1860	27, 197
Printing-press	F. L. Bailey	Boston, Mass	Apr. 30, 1861	32, 172
Printing-press	F. L. Bailey	Boston, Mass	July 5, 1864	43, 383
Printing-press	F. L. Bailey	Boston, Mass	Dec. 13, 1864	45, 461
Printing-press	F. L. Bailey	Boston, Mass	Sept. 12, 1871	118, 835
Printing-press	F. L. Bailey and J. Watson	Boston and Everett, Mass	Feb. 11, 1873	135, 618
Printing-press	S. J. Baird	Staunton, Va	Nov. 24, 1868	84, 403
Printing-press	W. H. Baker and G. J. Hill	Buffalo, N. Y	June 2, 1863	38, 781
Printing-press	W. L. Balch	Boston, Mass	Aug. 22, 1871	118, 182
Printing-press	A. H. Bangle	Brooklyn, Cal	June 7, 1870	103, 961
Printing-press	A. H. Bangle	Brooklyn, Cal	Feb. 14, 1871	111, 805
Printing-press	A. H. Bangle	Brooklyn, Cal	July 16, 1872	129, 084
Printing-press	H. Barth	Cincinnati, Ohio	Oct. 22, 1867	70, 067

Index of patents issued from the United States Patent Office from 1790 *to* 1873, *inclusive*—Continued.

Invention.	Inventor.	Residence.	Date.	No.
Printing-press	H. Barth	Cincinnati, Ohio	Feb. 22, 1870	100, 002
Printing-press	C. F. Batchelder and W. E. Gard	Chicago, Ill	Oct. 29, 1872	132, 517
Printing-press	M. S. Beach	Brooklyn, N. Y	Nov. 9, 1858	22, 010
Printing-press	M. S. Beach	Brooklyn, N. Y	June 26, 1860	28, 823
Printing-press	V. Beaumont	New York, N. Y	Sept. 6, 1853	9, 987
Printing-press	H. Betts	Norwalk, Conn	Apr. 13, 1869	88, 835
Printing-press	H. A. Bills and S. W. Wood	West Winsted, Conn., and Cornwall, N. Y.	Mar. 23, 1858	19, 672
Printing-press	T. W. Bracher	New York, N. Y	Feb. 1, 1870	99, 395
Printing-press	W. Braidwood	Mount Vernon, N. Y	May 5, 1868	77, 577
Printing-press	W. Braidwood	New York, N. Y	July 28, 1868	80, 444
Printing-press	J. B. Brown	Nashville, Tenn	May 24, 1870	103, 422
Printing-press	J. M. Brownson	Brooklyn, N. Y	Mar. 8, 1870	100, 494
Printing-press	W. Bullock	Philadelphia, Pa	Feb. 12, 1867	61, 996
Printing-press	J. S. Burdick	Norwich, N. Y	Mar. 27, 1849	6, 236
Printing-press	J. S. Burdick	New York, N. Y	June 2, 1857	17, 418
Printing-press	T. H. Burridge	Saint Louis, Mo	Dec. 20, 1859	26, 545
Printing-press	H. Camp	Trenton, Ohio	Mar. 4, 1836	....
Printing-press	A. Campbell	Brooklyn, N. Y	July 31, 1866	56, 701
Printing-press	A. Campbell	Brooklyn, N. Y	Mar. 8, 1870	100, 598
Printing-press	A. Campbell	Brooklyn, N. Y	Apr. 25, 1871	114, 105
Printing-press	A. Campbell	Brooklyn, N. Y	Apr. 25, 1871	114, 106
Printing-press	J. A. Campbell	New Orleans, La	Sept. 14, 1858	21, 484
Printing-press	V. M. Chafee	Xenia, Ill	Jan. 17, 1860	26, 834
Printing-press	W. H. Chandler	Winchester, Mass	Mar. 30, 1869	88, 448
Printing-press	W. R. Collier	Washington, D. C	Apr. 11, 1834	....
Printing-press	S. Conillard	Boston, Mass	Oct. 5, 1827	....
Printing-press	R. J. Coons	Greensburgh, Pa	Oct. 17, 1871	120, 040
Printing-press	S. R. Cotton	Green Bay, Wis	Dec. 28, 1858	22, 414
Printing-press	C. B. Cottrell	Westerly, R. I	May 2, 1871	114, 268
Printing-press	C. B. Cottrell	Westerly, R. I	Oct. 17, 1871	120, 041
Printing-press	C. B. Cottrell	Westerly, R. I	July 15, 1873	140, 813
Printing-press	A. H. Cragin, M. and J. H. Buck, and F. A. Temrey.	Lebanon, N. H	Nov. 23, 1852	9, 426
Printing-press	R. Cummings	Newport, Vt	Oct. 27, 1868	83, 471
Printing-press	W. H. Danforth	Salem, Mass	Aug. 5, 1856	15, 477
Printing-press	G. W. Davis	Seneca Falls, N. Y	Mar. 30, 1858	19, 758
Printing-press	M. Davis	New York, N. Y	July 24, 1855	13, 335
Printing-press	M. Davis	New York, N. Y	Nov. 3, 1857	18, 567
Printing-press	G. R. Dean	Mayville, N. Y	May 7, 1861	32, 242
Printing-press	F. O. Degener	New York, N. Y	Jan. 11, 1859	22, 611
Printing-press	F. O. Degener	New York, N. Y	Apr. 24, 1860	27, 973
Printing-press	F. O. Degener	Brooklyn, N. Y	May 28, 1872	127, 316
Printing-press	H. B. Denny	Washington, D. C	Jan. 5, 1869	85, 515
Printing-press	J. Densmore	Blooming Valley, Pa	Nov. 9, 1852	9, 383
Printing-press	T. H. Dodge	Nashua, N. H	Nov. 18, 1851	8, 521
Printing-press	A. Dougherty	Brooklyn, N. Y	Aug. 9, 1859	25, 000
Printing-press	A. Dougherty	Brooklyn, N. Y	Aug. 28, 1866	57, 486
Printing-press	A. A. Dunk	Philadelphia, Pa	Mar. 10, 1868	75, 801
Printing-press	S. Fairlamb	New York	Nov. 4, 1826	....
Printing-press	G. F. Folsom	Roxbury, Mass	Apr. 1, 1856	14, 558
Printing-press	W. H. Forbush	Buffalo, N. Y	Jan. 19, 1869	86, 064
Printing-press	J. H. Frey and W. Heckert	Sharon, Pa	Jan. 15, 1867	61, 186
Printing-press	M. Gally	Rye, N. Y	Nov. 9, 1869	96, 577
Printing-press	M. Gally	Rye, N. Y	Nov. 9, 1869	96, 578
Printing-press	M. Gally	Rye, N. Y	Nov. 9, 1869	96, 579
Printing-press	M. Gally	Rochester, N. Y	Nov. 23, 1869	97, 185
Printing press	M. Gally	Rochester, N. Y	Feb. 8, 1870	99, 551
Printing-press	M. Gally	Rochester, N. Y	Mar. 29, 1870	101, 254
Printing-press	M. Gally	Rochester, N. Y	May 2, 1871	114, 285
Printing press	G. and S. P. Gary	Oshkosh, Wis	Feb. 26, 1861	31, 573
Printing-press	A. Gilman	Troy, N. Y	Aug. 23, 1844	3, 716
Printing-press	W. H. Golding	Chelsea, Mass	Dec. 2, 1873	145, 101
Printing press	H. Goodrich	Stoneham, Mass	Oct. 1, 1861	33, 387
Printing-press	G. P. Gordon	New York, N. Y	Mar. 26, 1850	7, 215
Printing-press	G. P. Gordon	New York, N. Y	Aug. 5, 1851	8, 285
Printing-press	G. P. Gordon	New York, N. Y	Aug. 31, 1852	9, 234
Printing-press	G. P. Gordon	New York, N. Y	June 13, 1854	11, 064
Printing-press	G. P. Gordon	New York, N. Y	Jan. 1, 1856	14, 016
Printing-press	G. P. Gordon	New York, N. Y	Oct. 3, 1857	18, 545
Printing-press	G. P. Gordon	New York, N. Y	July 13, 1858	20, 874
Printing-press	G. P. Gordon	New York, N. Y	Apr. 19, 1859	23, 677
Printing-press	G. P. Gordon	New York, N. Y	July 5, 1859	24, 626
Printing-press	G. P. Gordon	New York, N. Y	Jan. 10, 1860	26, 761
Printing-press	G. P. Gordon	Brooklyn, N. Y	Apr. 23, 1861	32, 130
Printing-press	G. P. Gordon	Brooklyn, N. Y	Jan. 7, 1862	34, 111
Printing-press	G. P. Gordon	Brooklyn, N. Y	Oct. 28, 1862	36, 840
Printing-press	G. P. Gordon	Brooklyn, N. Y	Sept. 29, 1863	40, 099
Printing-press	G. P. Gordon	Brooklyn, N. Y	Mar. 8, 1864	41, 841
Printing-press	G. P. Gordon	Brooklyn, N. Y	June 2, 1864	43, 196
Printing-press	G. P. Gordon	Brooklyn, N. Y	Dec. 18, 1866	60, 504
Printing-press	G. P. Gordon	Rahway, N. J	May 18, 1869	90, 091
Printing-press	G. P. Gordon	Rahway, N. J	Oct. 31, 1871	120, 377
Printing-press	G. P. Gordon and F. O. Degener	New York, N. Y	May 11, 1858	20, 204
Printing-press	G. P. Gordon and F. O. Degener	New York, N. Y	Aug. 9, 1859	25, 008
Printing-press	J. Gordon	Caledonia, N. Y	Oct. 28, 1862	36, 839
Printing-press	J. Gough	London, England	Apr. 25, 1871	114, 130
Printing-press	F. J. Grace	Fort Lee, N. J	Jan. 1, 1867	60, 717
Printing-press	S. T. Guernsey	Montpelier, Vt	Oct. 19, 1852	9, 342
Printing-press	H. M. Hall and G. W. Espey	Moore's Hill, Ind	July 7, 1868	79, 751
Printing-press	C. W. Hawkes	Boston, Mass	June 4, 1850	7, 413
Printing-press	J. R. Hathaway and J. P. Strippel.	Norfolk, Va	Oct. 21, 1851	8, 445
Printing-press	C. G. Havens and F. C. Penfield.	West Meriden, Conn	Oct. 8, 1872	131, 950
Printing-press	C. W. Hawkes	Boston, Mass	Dec. 24, 1850	7, 855
Printing-press	C. W. Hawkes	Boston, Mass	Sept. 14, 1852	9, 259
Printing-press	C. W. Hawkes	Boston, Mass	Dec. 8, 1857	18, 812
Printing-press	G. F. Hebard, G. J. Hill, and S. D. Rockwell.	Buffalo, N. Y	Aug. 7, 1860	29, 547

Index of patents issued from the United States Patent Office from 1790 to 1873, inclusive—Continued.

Invention.	Inventor.	Residence.	Date.	No.
Printing-press	W. Heckert	Newcastle, Pa	Apr. 16, 1872	125, 678
Printing-press	J. Henry	Vevay, Ind	Dec. 1, 1857	18, 744
Printing-press	J. Henry	Jersey City, N. J	Feb. 12, 1867	61, 032
Printing-press	J. Henry	Millburn, N. J	May 9, 1871	114, 557
Printing-press	G. J. Hill and S. Greene	Philadelphia, Pa	Aug. 28, 1866	57, 634
Printing-press	R. M Hoe	New York, N. Y	Apr. 17, 1844	3, 551
Printing-press	R. M. Hoe	New York, N. Y	July 30, 1844	3, 687
Printing-press	R. M. Hoe	New York, N. Y	May 1, 1845	4, 025
Printing-press	R. M. Hoe	New York, N. Y	July 10, 1847	5, 188
Printing-press	R. M. Hoe	New York, N. Y	July 24, 1847	5, 200
Printing-press	R. M. Hoe	New York, N. Y	Apr. 18, 1871	113, 769
Printing-press	R. M. Hoe and S. D. Tucker	New York, N. Y	Dec. 1, 1868	84, 627
Printing-press	R. M. Hoe and S. D. Tucker	New York, N. Y	June 29, 1869	92, 050
Printing-press	R. M. Hoe and S. D. Tucker	New York, N. Y	Nov. 1, 1870	108, 785
Printing-press	R. M. Hoe and S. D. Tucker	West Farms and New York, N. Y.	Sept. 10, 1872	131, 217
Printing-press	J. C. Holbrook and E. H. Thomas.	Brattleborough, Vt	Feb. 7, 1828	....
Printing-press	J. C. Holbrook and E. H. Thomas.	Brattleborough, Vt	Feb. 7, 1828	....
Printing-press	H. Holt	Winchester, Mass	Mar. 17, 1857	16, 837
Printing-press	G. C. Howard	Philadelphia, Pa	June 26, 1860	28, 864
Printing-press	B. Huber	Williamsburgh, N. Y	Nov. 7, 1871	120, 646
Printing-press	B. Huber	Williamsburgh, N. Y	Mar. 19, 1872	124, 822
Printing-press	G. A. Hunt	Newburgh, N. Y	May 20, 1873	139, 156
Printing-press	G. W. Hunt	New York, N. Y	Feb. 28, 1871	112, 249
Printing-press	T. H. Ide	Claremont, N. H	June 20, 1871	116, 060
Printing-press	M. G. Imbach	New York, N. Y	May 1, 1866	54, 357
Printing-press	D. E. James	Utica, N. Y	Nov. 30, 1858	22, 181
Printing-press	W. Johnson	Lambertville, N. J	July 9, 1872	128, 731
Printing-press	J. M. Jones	Palmyra, N. Y	Aug. 11, 1868	80, 865
Printing-press	J. M. Jones	Palmyra, N. Y	Nov. 19, 1872	133, 156
Printing-press	A. Judson	Brooklyn, N. Y	July 30, 1867	67, 198
Printing-press	J. W. Kelberg	Philadelphia, Pa	Dec. 20, 1870	110, 244
Printing-press	A. N. Kellogg	Baraboo, Wis	Jan. 6, 1863	37, 293
Printing-press	S. Kelsey	Erie, Pa	Jan. 2, 1855	12, 183
Printing-press	W. A. Kelsey	Meriden, Conn	May 6, 1873	138, 660
Printing-press	W. A. Kerr	Easton, Pa	June 7, 1870	103, 894
Printing-press	J. L. Kingsley	New York	Apr. 22, 1835	....
Printing-press	J. L. Kingsley	New York, N. Y	Oct. 31, 1839	1, 394
Printing-press	J. L. Kingsley	New York	Jan. 4, 1845	3, 874
Printing-press	S. Kingsley	New York, N. Y	Mar. 2, 1836	....
Printing-press	A. Kinsley	....	Nov. 16, 1796	....
Printing-press	J. C. Kneeland	Troy, N. Y	Feb. 20, 1845	3, 917
Printing-press	M. Laemmel	Bay Ridge, N. Y	Jan. 30, 1872	123, 266
Printing-press	J. W. Latcher	Northville, N. Y	Jan. 31, 1860	26, 999
Printing-press	S. D. Learned	Boston, Mass	May 26, 1857	17, 405
Printing-press	T. Leavitt	Everett, Mass	July 18, 1871	117, 088
Printing-press	J. Lewis	Buffalo, N. Y	Aug. 9, 1853	9, 923
Printing-press	J. Lewis	Prattville, N. Y	Jan. 2, 1855	12, 185
Printing-press	J. C. MacDonald and J. Calverly.	Waddon and Camberwell, England.	June 1, 1869	90, 858
Printing-press	J. C. MacDonald and J. Calverly.	Waddon and Camberwell, England.	Apr. 25, 1871	114, 020
Printing-press	J. M. Marsh	New York, N. Y	Oct. 3, 1848	5, 819
Printing-press	W. W. Marston	New York	Sept. 12, 1846	4, 756
Printing-press	C. C. Maurice	New York, N. Y	May 24, 1870	103, 351
Printing-press	J. W. McDonald	Osgood, Ind	Dec. 24, 1867	72, 660
Printing-press	G. McKay	Boston, Mass	Feb. 15, 1859	22, 968
Printing-press	Z. Mills	Hartford, Conn	Feb. 27, 1813	....
Printing-press	C. Montague	Pittsfield, Mass	Nov. 23, 1852	9, 424
Printing-press	C. Montague	Pittsfield, Mass	Sept. 6, 1853	9, 992
Printing-press	C. Montague	Pittsfield, Mass	Sept. 6, 1853	9, 993
Printing-press	C. Montague	Hartford, Conn	Nov. 9, 1858	22, 027
Printing-press	C. Montague	Hartford, Conn	Dec. 10, 1861	33, 897
Printing-press	C. Montague	Boston, Mass	Dec. 21, 1869	98, 087
Printing-press	C. Montague	Boston, Mass	Dec. 21, 1869	98, 088
Printing-press	R. W. Moran	Saint Louis, Mo	Apr. 18, 1865	47, 319
Printing-press	W. T. Morgans	Youngsville, N. Y	Jan. 25, 1870	99, 101
Printing-press	T. N. Morse	Fairhaven, Mass	Mar. 26, 1872	124, 907
Printing-press	D. Neale	Philadelphia, Pa	Feb. 13, 1834	....
Printing-press	D. Neale	Philadelphia, Pa	Nov. 15, 1825	....
Printing-press	A. and B. Newbury	Windham Centre, N. Y	Sept. 16, 1856	15, 740
Printing-press	F. B. Nichols	Morrisania, N. Y	Aug. 3, 1858	21, 080
Printing-press	J. G. Nicolay	Pittsfield, Ill	Oct. 5, 1852	9, 305
Printing-press	J. G. Northrup	Cortlandville, N. Y	Apr. 1, 1845	3, 985
Printing-press	J. G. Northrup	Syracuse, N. Y	Aug. 9, 1853	9, 925
Printing-press	J. G. Northrup	Syracuse, N. Y	June 12, 1855	13, 069
Printing-press	J. G. Northrup	Cortlandville, N. Y	Sept. 30, 1842	2, 793
Printing-press	T. and A. Parkes	Brooklyn, N. Y	July 29, 1856	15, 437
Printing-press	F. C. Penfield	West Meriden, Conn	May 6, 1873	138, 689
Printing-press	J. G. Peterson	Morgantown, N. C	Sept. 9, 1873	142, 648
Printing-press	D. Phelps	Boston, Mass	Sept. 15, 1826	....
Printing-press	J. N. Phelps	Brooklyn, N. Y	Nov. 2, 1869	96, 480
Printing-press	D. Pierson	Newburyport, Mass	July 16, 1813	....
Printing-press	C. Potter, jr	Westerly, R. I	Mar. 5, 1861	31, 621
Printing-press	C. Potter, jr	Westerly, R. I	Aug. 20, 1867	68, 601
Printing-press	C. Potter, jr	Westerly, R. I	Jan. 7, 1868	73, 195
Printing-press	C. Potter, jr	Plainfield, N. J	Sept. 24, 1872	131, 702
Printing-press	J. E. Priest	Saint Louis, Mo	Aug. 28, 1860	29, 816
Printing-press	G. W. Prouty	Charlestown, Mass	Nov. 28, 1871	121, 256
Printing-press	G. W. Prouty	Charlestown, Mass	Oct. 29, 1872	132, 599
Printing-press	A. Ramage	Philadelphia, Pa	May 28, 1818	....
Printing-press	A. Ramage	Philadelphia, Pa	Nov. 19, 1834	....
Printing-press	H. Redlich	Chicago, Ill	Apr. 25, 1865	47, 454
Printing-press	L. and H. Reese	Philadelphia, Pa	Nov. 24, 1868	84, 440
Printing-press	T. S. Reynolds	Athens, Ga	Apr. 27, 1858	20, 090
Printing-press	J. T. Robertson	New York, N. Y	Apr. 4, 1871	113, 346

Index of patents issued from the United States Patent Office from 1790 *to* 1873, *inclusive*—Continued.

Invention.	Inventor.	Residence.	Date.	No.
Printing-press	J. T. Robertson	New York, N. Y	Aug. 6, 1872	130, 153
Printing-press	L. Rodney	New York, N. Y	Aug. 6, 1867	67, 586
Printing-press	S. P. Ruggles	Boston, Mass	Nov. 10, 1840	1, 851
Printing-press	S. P. Ruggles	Boston, Mass	Jan. 1, 1851	7, 878
Printing-press	S. P. Ruggles	Boston, Mass	Aug. 2, 1853	9, 904
Printing-press	S. P. Ruggles	Boston, Mass	Nov. 16, 1852	9, 410
Printing-press	S. P. Ruggles	Boston, Mass	Feb. 28, 1854	10, 588
Printing-press	S. P. Ruggles	Boston, Mass	May 10, 1859	23, 951
Printing-press	S. Rust	New York	May 13, 1821	....
Printing-press	S. Rust	New York	Apr. 17, 1829	....
Printing-press	J. Sangster	Buffalo, N. Y	June 27, 1865	48, 493
Printing-press	C. G. Sargent and A. Keach	Lowell and Boston, Mass	Dec. 9, 1856	16, 221
Printing-press	J. Saxton	Washington, D. C	Mar. 21, 1845	3, 958
Printing-press	W. and T. Schnebly	Hagerstown, Md	Sept. 7, 1839	1, 315
Printing-press	A. Sherman	New York	Feb. 26, 1831	....
Printing-press	E. H. Smith	Bergen, N. J	Feb. 7, 1871	111, 581
Printing-press	E. H. Smith	New York, N. Y	Mar. 19, 1872	124, 701
Printing-press	J. A. Smith and L. M. Orris	Fond du Lac and Oakfield, Wis.	June 26, 1860	28, 948
Printing-press	P. Smith	New York	Apr. 6, 1822	....
Printing-press	W. W. Smith	New York	Feb. 1, 1842	2, 439
Printing-press	E. E. Sneider	New York, N. Y	Aug. 10, 1858	21, 154
Printing-press	A. O. Stansbury	New York	Apr. 7, 1821	....
Printing-press	R. G. Stuart	Yonkers, N. Y	July 18, 1871	117, 219
Printing-press	C. A. Swett	Boston, Mass	Nov. 27, 1855	13, 857
Printing-press	A. B. Taylor	Newark, N. J	June 26, 1860	28, 915
Printing-press	W. H. R. Toye	Philadelphia, Pa	May 30, 1871	115, 391
Printing-press	E. B. Tripp	Concord, N. H	Oct. 3, 1854	11, 765
Printing-press	E. B. Tripp	New York, N. Y	Sept. 14, 1858	21, 528
Printing-press	S. D. Tucker	New York, N. Y	June 28, 1864	43, 349
Printing-press	S. D. Tucker	New York, N. Y	June 28, 1864	43, 350
Printing-press	S. D. Tucker	New York, N. Y	Dec. 29, 1868	85, 493
Printing-press	S. D. Tucker	New York, N. Y	Mar. 12, 1872	124, 460
Printing-press	O. Tufts	Boston, Mass	July 30, 1831	....
Printing-press	O. Tufts	Boston, Mass	Aug. 22, 1834	....
Printing-press	J. H. Utter	New York, N. Y	Oct. 20, 1857	18, 477
Printing-press	L. T. Verney	Paris, France	Sept. 3, 1867	68, 471
Printing-press	E. A. Warren	New York, N. Y	June 11, 1872	127, 815
Printing-press	R. C. Warwick	New York, N. Y	Jan. 31, 1871	111, 407
Printing-press	J. Watson	Everett, Mass	Dec. 17, 1872	133, 959
Printing-press	C. Wells and H. Barth	Cincinnati, Ohio	Aug. 7, 1860	29, 554
Printing-press	J. J. Wells	Hartford, Conn	Feb. 8, 1819	....
Printing-press	J. J. Wells	Hartford, Conn	June 29, 1829	....
Printing-press	L. T. Wells	Cincinnati, Ohio	Mar. 20, 1855	12, 568
Printing-press	L. T. Wells	Cincinnati, Ohio	Sept. 6, 1859	25, 357
Printing-press	L. T. Wells	Saint Louis, Mo	July 7, 1868	79, 617
Printing-press	O. E. Weston	Roxbury, Mass	Jan. 17, 1860	26, 869
Printing-press	S. Wilcox, jr	Westerly, R. I	Nov. 10, 1857	18, 618
Printing-press	J. A. Wilkinson	Fireplace, N. Y	Jan. 4, 1853	9, 525
Printing-press	D. K. Winder	Cincinnati, Ohio	June 2, 1857	17, 463
Printing-press	D. Wolfe	Dixon, Ohio	Aug. 17, 1858	21, 228
Printing-press	B. O. Woods and W. S. Tuttle	Boston, Mass	Aug. 6, 1867	67, 475
Printing-press	J. Worms	Paris, France	Sept. 23, 1851	8, 386
Printing-press	J. K. Wright	Philadelphia, Pa	June 8, 1869	91, 191
Printing-press	C. D. Wrightington	Fairhaven, Mass	Sept. 20, 1870	107, 583
Printing-press	J. Young	Philadelphia, Pa	May 10, 1853	9, 721
Printing-press and folding-machine, Combined	E. L. Ford	New York, N. Y	Dec. 17, 1872	133, 976
Printing-press and mode of inking type	W. Elliot	New York	Feb. 17, 1813	....
Printing-press apron	C. McBurney	Roxbury, Mass	Jan. 17, 1860	26, 854
Printing-press attachment, Chromatic	E. A. Howitt	Whitewater, Wis	Jan. 7, 1873	134, 602
Printing press, Card	F. L. Bailey	Boston, Mass	June 16, 1857	17, 549
Printing press, Card	W. W. Clarkson	Baltimore, Md	Apr. 27, 1858	20, 039
Printing press, Card	C. E. Emery	Canandaigua, N. Y	June 9, 1857	17, 499
Printing press, Card	T. Harsha	West Union, Ohio	Oct. 23, 1855	13, 703
Printing press, Card	S. Orcutt	Boston, Mass	Aug. 25, 1840	1, 732
Printing press, Card	D. K. Winder	Cincinnati, Ohio	Sept. 18, 1855	13, 576
Printing press, Card	D. K. Winder	Cincinnati, Ohio	Oct. 9, 1855	13, 671
Printing press, Card and ticket	G. E. Peck	Duquoin, Ill	Oct. 21, 1873	143, 780
Printing press, Chromatic	T. H. Burridge and J. M. Kershaw.	Saint Louis, Mo	June 24, 1873	140, 242
Printing press, Chromatic	F. L. Heughes	Rochester, N. Y	Dec. 31, 1872	134, 380
Printing press, Chromatic	G. W. Woodside	Philadelphia, Pa	Jan. 7, 1873	134, 579
Printing-press, Circular	J. P. Sawin and T. B. Wait	Boston, Mass	Feb. 1, 1810	....
Printing press, Color	A. M. and G. H. Babcock	Westerly, R. I	Oct. 31, 1854	11, 853
Printing press, Color	G. W. Wood	Richmond, Ind	Feb. 18, 1868	74, 741
Printing press, Copper and steel plate	E. C. Middleton, E. Nevers, and R. Neale.	Cincinnati and Mount Carmel, Ohio.	Nov. 19, 1850	7, 786
Printing press, Copper and steel plate	J. Perkins	Newburyport, Mass	June 29, 1813	....
Printing press, Copper-plate	T. S. Bates	Boston, Mass	Dec. 16, 1873	145, 610
Printing press, Copper-plate	C. Durand	New York	May 22, 1828	....
Printing-press, Cylinder	B. Beniowski	London, England	Aug. 13, 1850	7, 558
Printing-press, Cylinder	H. C. Chandler	Indianapolis, Ind	Nov. 5, 1867	70, 524
Printing-press, Cylinder	F. O. Degener	New York, N. Y	Nov. 5, 1861	33, 676
Printing-press, Cylinder	J. G. Northrup	Syracuse, N. Y	Nov. 16, 1852	9, 408
Printing-press, Cylindrical	J. P. Sawin and T. B. Wait	Roxbury, Mass	Jan. 28, 1811	....
Printing-press, Cylindrical	C. G. Williams	New York	May 29, 1828	....
Printing, Device for hand	B. Livermore	Hartland, Vt	July 21, 1863	39, 296
Printing press, Double	J. Booth, sr., and J., T., J., and J. Booth, jr.	New York	Sept. 1, 1829	....
Printing-press, (Napier's,) Double	S. Newton	New York, N. Y	Feb. 26, 1833	....
Printing-press, Double-cylinder	R. M. Hoe	New York	May 20, 1842	2, 629
Printing-press, Engraved-plate	P. C. Andrews and J. Shinn	Leverington, Pa	Oct. 21, 1862	36, 759
Printing-press, Engraved-plate	M. C. Gritzner	Washington, D. C	Apr. 14, 1857	17, 036
Printing-press, Faustus	S. Adams	Boston, Mass	May 23, 1832	....
Printing-press feed-board	E. Allen	Norwich, Conn	Nov. 12, 1872	133, 002
Printing-press feed-gage	H. Barth and R. J. Morgan	Cincinnati, Ohio	Apr. 9, 1872	125, 432
Printing-press feed-gage	E. N. Maxwell	Louisville, Ky	Jan. 16, 1872	122, 842
Printing-press feed-gage	C. N. Morris	Cincinnati, Ohio	Dec. 23, 1873	145, 890

Index of patents issued from the United States Patent Office from 1790 *to* 1873, *inclusive*—Continued.

Invention.	Inventor.	Residence.	Date.	No.
Printing-press feed-gage	C. Potter, jr	Westerly, R. I	Jan. 7, 1868	73, 196
Printing-press feed-gage	G. Wilcox	Norwich, Conn	Oct. 14, 1873	143, 652
Printing-press feeder	H. Barth	Cincinnati, Ohio	Oct. 23, 1860	30, 456
Printing-press feeder	F. L. Henghes	Rochester, N. Y	Aug. 12, 1873	141, 713
Printing-press feeder	R. J. Stuart	Yonkers, N. Y	Mar. 25, 1873	137, 156
Printing-press-fly frame	T. H. Mead	Boston, Mass	Mar. 9, 1869	87, 695
Printing-press-fly frame	T. H. Mead	Boston, Mass	May 18, 1869	90, 114
Printing-press-fly frame	A. Overend	Philadelphia, Pa	Apr. 30, 1872	126, 227
Printing-press for cards and bill-heads, Self-feeding	N. Ames	Saugus, Mass	Mar. 29, 1859	23, 421
Printing-press frisket	T. W. M. Castle and J. B. Conner.	Adriance, Ind	June 23, 1868	79, 653
Printing-press gage	A. Campbell	Brooklyn, N. Y	July 26, 1870	105, 638
Printing-press griper	T. J. Plunket	New York, N. Y	Aug 1, 1871	117, 566
Printing-press-griper and gage	H. Barth	Cincinnati, Ohio	Feb. 15, 1870	99, 749
Printing-press guide	A. L. Bevans	Flushing, N. Y	Jan. 31, 1871	111, 304
Printing-press-guide gage	O. H. Reed	Washington, D. C	May 5, 1868	77, 527
Printing press, Hand	F. J. Austin	New York, N. Y	Oct. 8, 1836	41
Printing press, Hand	D. H. Chamberlain	West Roxbury, Mass	Aug. 14, 1855	13, 423
Printing press, Hand	N. L. Chamberlin	West Roxbury, Mass	Mar. 3, 1857	16, 718
Printing press, Hand	F. S. Coburn	Ipswich, Mass	Mar. 17, 1857	16, 861
Printing press, Hand	S. Conillard, jr	Boston	July 14, 1827	....
Printing press, Hand	W. and W. H. Dunkerly, jr	Woonsocket, R. I	Apr. 8, 1873	137, 664
Printing press, Hand	P. Evens, jr	Cincinnati, Ohio	Dec. 23, 1856	16, 270
Printing press, Hand	C. Foster	Cincinnati, Ohio	Oct. 5, 1852	9, 295
Printing press, Hand	C. A. Haskins	New York, N. Y	June 15, 1858	20, 556
Printing press, Hand	J. M. Jones	Palmyra, N. Y	Dec. 22, 1857	18, 907
Printing press, Hand	C. Keniston	Boston, Mass	Mar. 20, 1855	12, 553
Printing press, Hand	H. Moeser	Pittsburgh, Pa	Jan. 20, 1852	8, 674
Printing press, Hand	J. Morse	Canton, Mass	Oct. 27, 1857	18, 527
Printing press, Hand	A. and B. Newbury	Windham Centre, N. Y	July 5, 1859	24, 655
Printing press, Hand	J. N. Phelps	New York, N. Y	Nov. 2, 1858	21, 980
Printing press, Hand	N. Ross	Washington, D. C	Jan. 7, 1873	134, 705
Printing press, Hand	J. Schaffer and E. Spencer	Saint Louis, Mo	Jan. 14, 1862	34, 166
Printing press, Hand	S. I. Smith	New York, N. Y	Nov. 3, 1857	18, 557
Printing press, Hand	H. Underhill	Canandaigua, N. Y	Mar. 28, 1854	10, 717
Printing press, Hand	C. Whipple	Providence, R. I	Nov. 8, 1864	45, 000
Printing press, Hand	D. K. Winder	Cincinnati, Ohio	July 8, 1856	15, 312
Printing-press hand-roller	B. F. Allen	Boston, Mass	Aug. 6, 1872	130, 175
Printing-press ink-fountain	H. B. Allen	Brooklyn, N. Y	Aug. 6, 1872	130, 176
Printing-press ink-fountain	W. V. Wallace	New York, N. Y	July 4, 1871	116, 778
Printing-press inking-apparatus	T. L. Baylies	Richmond, Ind	May 26, 1868	78, 358
Printing-press inking-apparatus	T. L. Baylies	Richmond, Ind	July 7, 1868	79, 722
Printing-press inking-apparatus	F. O. Degener	Brooklyn, E. D., N. Y	Dec. 13, 1870	110, 018
Printing-press inking-apparatus	M. England	New York, N. Y	Dec. 9, 1873	145, 341
Printing-press inking-apparatus	G. K. Farrington and B. S. Potter.	Bloomington, Ill	Oct. 8, 1872	132, 000
Printing-press inking-apparatus	R. M. Hoe	New York, N. Y	July 31, 1847	5, 212
Printing-press inking-apparatus	J. Prince	New York	Dec. 3, 1829	....
Printing-press inking-apparatus	I. L. G. Rice	Cambridge, Mass	Apr 20, 1869	89, 244
Printing-press inking-apparatus	I. L. G. Rice	Cambridge, Mass	June 14, 1870	104, 203
Printing-press inking-apparatus	N. F. Turner	Williamsburgh, N. Y	July 11, 1871	116, 892
Printing-press inking-apparatus	C. Wells	Cincinnati, Ohio	Jan. 26, 1869	86, 263
Printing-press inking-apparatus	G. W. Wood	Richmond, Ind	June 18, 1867	65, 984
Printing-press inking-apparatus	R. Wood	New York	Nov. 4, 1830	....
Printing-press inking-apparatus, Card	D. K. Winder	Cincinnati, Ohio	Oct. 16, 1855	13, 689
Printing-press inking-apparatus, Oscillating	G. W. Prouty	Charlestown, Mass	July 22, 1873	141, 077
Printing-press inking-apparatus, Rotary	E. Allen	Norwich, Conn	Sept. 9, 1873	142, 607
Printing-press inking-roller	E. E. Barrett	Chicago, Ill	July 21, 1857	17, 824
Printing press, Job and card	J. A. Campbell	New Orleans, La	June 28, 1859	24, 538
Printing press, Lithographic	A. G. Shackford	Malden, Mass	Aug. 4, 1868	80, 771
Printing-press, Locomotive-cylinder	C. J. Carr and A. Smith	Belper, England	Sept. 10, 1840	1, 775
Printing-press, Oscillating	C. Potter, jr	Westerly, R. I	June 2, 1857	17, 449
Printing-press paper-gage	J. W. Thyng	Salem, Mass	July 3, 1866	56, 123
Printing press, Plate	G. C. Howard and J. Dainty	Philadelphia, Pa	May 27, 1873	139, 392
Printing-press, Portable	S. W. Lowe	Philadelphia, Pa	July 29, 1856	15, 428
Printing press, Power	I. Adams	Boston, Mass	Oct. 4, 1830	....
Printing press, Power	I. Adams	Boston, Mass	Mar. 2, 1836	....
Printing press, Power	W. H. Danforth	Salem, Mass	June 21, 1853	9, 794
Printing press, Power	J. Morse	Canton, Mass	June 9, 1857	17, 543
Printing press, Power	J. Morse	Canton, Mass	June 7, 1859	24, 357
Printing press, Power	D. Treadwell	Boston, Mass	Mar. 2, 1826	....
Printing press, Railway-ticket	W. H. Forbush	Buffalo, N. Y	Nov. 6, 1866	59, 533
Printing press, Railway-ticket	S. Greene and W. H. Forbush	Philadelphia, Pa., and Buffalo, N. Y.	Aug. 13, 1867	67, 748
Printing-press, Register for Napier's	M. Caton and J. C. Rives	Washington, D. C	Mar. 11, 1835	....
Printing-press register-point	N. Hopkins	New York, N. Y	Nov. 5, 1867	70, 568
Printing-press register-point	R. W. Macgowan	New York, N. Y	Dec. 10, 1867	72, 059
Printing-press register-point	A. Overend	Philadelphia, Pa	Mar. 9, 1869	87, 587
Printing-press registering-apparatus	J. Kirk	Dover, Del	June 19, 1866	55, 774
Printing-press registering-apparatus	G. McKay	Boston, Mass	May 26, 1857	17, 388
Printing-press registering-apparatus	J. W. Richards	Hoboken, N. J	May 10, 1853	9, 711
Printing-press registering-attachment	A. Hilgenreiner	New York, N. Y	Oct. 21, 1873	143, 904
Printing press, Roller	J. Laird	Pittsburgh, Pa	Aug. 16, 1828	....
Printing-press, Rotary	S. H. Bingham	Philadelphia, Pa	May 27, 1873	139, 360
Printing-press, Rotary	W. Braidwood and H. J. Hewitt.	New York, N. Y	July 29, 1873	141, 314
Printing-press, Rotary	W. Braidwood and H. J. Hewitt.	New York, N. Y	July 29, 1873	141, 315
Printing-press, Rotary	C. C. Child	Boston, Mass	Nov. 18, 1873	144, 742
Printing-press, Rotary	C. B. Cottrell	Westerly, R. I	Apr. 1, 1873	137, 347
Printing-press, Rotary	J. C. Davis and W. Miller	Elizabeth, N. J	Mar. 17, 1857	16, 826
Printing-press, Rotary	R. M. Hoe	New York, N. Y	July 24, 1847	5, 199
Printing press, Self-feeding card	N. Ames	Saugus Centre, Mass	Apr. 22, 1862	35, 055
Printing-press sheet-gage	H. Barth	Cincinnati, Ohio	Jan. 25, 18[illegible]0	99, 133
Printing-press sheet-gage	G. B. Cummings	Cambridge, Mass	Jan. 25, 1870	99, 167
Printing press, Stencil	H. W. Rudolf	Louisville, Ky	Aug. 30, 1870	106, 872
Printing-press, Stop-cylinder	H. Barth	Cincinnati, Ohio	May 27, 1873	139, 229
Printing press, Zincographic	G. H. Korff	Hoboken, N. J	May 18, 1858	20, 276

Index of patents issued from the United States Patent Office from 1790 *to* 1873, *inclusive*—Continued.

Invention.	Inventor.	Residence.	Date.	No.
Printing-presses, Adjustable attachment of cams for fliers of.	C. Potter, jr	Plainfield, N. J	Feb. 14, 1871	111, 776
Printing-presses and paper-machines, Apparatus for receiving and transferring to the pile sheets of paper from.	I. Adams	Boston, Mass	Mar. 26, 1850	7, 205
Printing-presses and ruling-machines, Apparatus for feeding paper to.	D. Baldwin	Godwinville, N. J	Jan. 2, 1855	12, 118
Printing-presses, Apparatus for delivering paper from.	C. O. Furbush	Machias, Me	Apr. 25, 1865	47, 411
Printing-presses, Apparatus for delivering sheets from.	T. J. Mayall	Boston, Mass	Dec. 30, 1873	146, 085
Printing-presses, Apparatus for feeding paper to	J. B. Hall	New York, N. Y	Apr. 10, 1855	12, 702
Printing-presses, Apparatus for feeding paper to	W. Kuhlenshmidt and W. Hauff.	New York, N. Y	Apr. 25, 1854	10, 828
Printing-presses, Apparatus for feeding paper to	R. Larter, jr	Newark, N. J	Jan. 10, 1860	26, 772
Printing-presses, Apparatus for feeding paper to hand.	E. Mathers and W. D. Siegfried.	Morgantown, Pa	Apr. 3, 1855	12, 634
Printing-presses, Attaching blankets to cylinders of	W. H. Street	New York, N. Y	May 16, 1854	10, 935
Printing-presses, Attachment for	H. Willson	Chicago, Ill	May 17, 1870	103, 267
Printing-presses, Automatic paper-feeder for	W. Bullock	Newark, N. J	Sept. 21, 1858	21, 591
Printing-presses, Bed-recoil spring for	A. Campbell	Brooklyn, N. Y	July 31, 1866	56, 700
Printing-presses, Checking momentum of	A. B. Taylor	New York, N. Y	Apr. 4, 1846	4, 442
Printing-presses, Driving	C. M. Langley	Lowell, Mass	Aug. 27, 1867	68, 212
Printing-presses, Feed-table for	H. Barth	Cincinnati, Ohio	May 25, 1869	90, 482
Printing-presses, Feeding-apparatus for	T. Hall	Saint Louis, Mo	Aug. 7, 1860	29, 485
Printing-presses, Feeding out paper from	M. S. Beach	Brooklyn, N. Y	Nov. 9, 1858	22, 011
Printing-presses, Feeding paper to	M. S. Beach	Brooklyn, N. Y	Dec. 2, 1856	16, 138
Printing-presses, Feeding paper to	H. W. Dickinson	Rochester, N. Y	Oct. 30, 1855	13, 737
Printing-presses, Feeding paper to	G. H. and S. Ferguson	Malden Bridge, N. Y	July 26, 1859	24, 861
Printing-presses, Feeding paper to	R. M. Hoe	New York, N. Y	Nov. 10, 1857	18, 589
Printing-presses, Feeding paper to	R. M. Hoe	New York, N. Y	Aug. 23, 1859	25, 199
Printing-presses, Feeding paper to	J. Hunt	New York, N. Y	Nov. 1, 1864	44, 917
Printing-presses, Feeding paper to	C. H. Stoddard and E. D. Norton.	Ithaca, N. Y	July 15, 1873	140, 962
Printing-presses, Feeding paper to and from	C. Potter, jr., and C. B. Cottrell	Westerly, R. I	Sept. 20, 1859	25, 524
Printing-presses, Feeding paper to rotary	J. T. Ashley	Brooklyn, N. Y	Nov. 25, 1873	144, 822
Printing-presses, Griper for cylinder	T. E. Manger	New York, N. Y	July 4, 1871	116, 616
Printing-presses, Machine for feeding paper to	D. Baldwin	Godwinville, N. J	Dec. 9, 1856	16, 168
Printing-presses, Machine for feeding paper to	M. S. Beach	Brooklyn, N. Y	Dec. 23, 1856	16, 311
Printing-presses, Machine for feeding paper to	A. Campbell	Newark, N. J	July 24, 1855	13, 333
Printing-presses, Machine for feeding paper to	S. I. Chapman	Charleston, S. C	Jan. 15, 1856	14, 084
Printing-presses, Machine for feeding paper to	A. B. Childs and H. W. Dickinson.	Rochester, N. Y	Feb. 20, 1855	12, 401
Printing-presses, Machine for feeding paper to	W. F. Collier	Worcester, Mass	June 27, 1854	11, 188
Printing-presses, Machine for feeding paper to	I. B. Livingston and M. Waterhouse.	Barnet, Vt	Apr. 24, 1855	12, 761
Printing-presses, Machine for feeding paper to	B. A. Rugg and E. H. Benjamin	Oak Hill, N. Y	Sept. 5, 1854	11, 653
Printing-presses, Machine for feeding sheets of paper to.	D. Babson	Groton, Conn	Sept. 2, 1856	15, 639
Printing-presses, Machine for feeding sheets of paper to.	H. Clark	New Orleans, La	Apr. 25, 1854	10, 824
Printing-presses, Machine for feeding sheets of paper to.	A. H. Rowand	Allegheny City, Pa	July 3, 1855	13, 183
Printing-presses, Machine for wetting and cutting paper for.	M. S. Beach	Brooklyn, N. Y	Aug. 25, 1857	18, 032
Printing-presses, Machinery to feed sheets of paper to.	M. S. Beach	Brooklyn, N. Y	Aug. 9, 1859	24, 982
Printing-presses, Nippers for	C. W. Hawkes	Boston, Mass	Mar. 21, 1854	10, 670
Printing-presses, Operating feed-table of	G. Little	Utica, N. Y	Apr. 25, 1854	10, 827
Printing-presses, Operating fingers of	S. D. Tucker	New York, N. Y	Sept. 6, 1859	25, 356
Printing-presses, Operating fly-frames of	R. M. Hoe	New York, N. Y	Nov. 17, 1857	18, 640
Printing-presses, Paper-feeder for	L. T. Wells	Cincinnati, Ohio	Oct. 19, 1858	21, 859
Printing presses, Self-inking machine for hand	W. J. Spence	New York	Apr. 30, 1834	....
Printing-presses, Straightening toggle-joint in hand	O. Tufts	Boston, Mass	Nov. 7, 1831	....
Printing-presses, Tympan-frame for	D. U. Stoner	Mount Joy, Pa	Mar. 9, 1869	87, 723
Printing-presses, Tympans for	L. T. Wells	Cincinnati, Ohio	May 4, 1858	20, 179
Printing-presses, Tympan-sheet for	J. Gorman	Portland, Me	Dec. 13, 1870	110, 032
Printing, Producing from printed paper new plates for re-.	C. and C. Vogt	New York, N. Y	May 4, 1869	89, 715
Printing, Production of plates for	J. Ramage and T. Nelson	Edinburgh, Scotland	Mar. 17, 1868	75, 700
Printing railway and other tickets, Machine for	R. M. Hoe	New York, N. Y	Mar. 8, 1859	23, 172
Printing railway-tickets, Machine for	G. Bailey	Buffalo, N. Y	Mar. 20, 1860	27, 510
Printing railway-tickets, Machinery for	W. T. Cushing	New York, N. Y	May 29, 1866	55, 201
Printing register, Plate	J. L. Harley	Washington, D. C	Mar. 1, 1870	100, 399
Printing-rollers, Composition for	I. G. Husted	Brooklyn, N. Y	Oct. 11, 1870	108, 145
Printing, Self-acting feeder for calico	A. Boyd	Providence, R. I	Apr. 22, 1845	4, 010
Printing silks and other textile fabrics	L. Prang	Boston, Mass	Apr. 4, 1871	113, 343
Printing, Stencil-plate	S. T. Sanford	Fall River, Mass	May 19, 1857	17, 340
Printing-surfaces, Mode of obtaining curved	W. H. Elliot	Plattsburgh, N. Y	Mar. 15, 1859	23, 236
Printing-surfaces, Process of obtaining	P. Schulze	Brooklyn, N. Y	Dec. 2, 1862	37, 078
Printing textile fabrics, Machine for	J. Melville and J. Burch	Roebank Works and Crag Hall, Great Britain.	Aug. 7, 1855	13, 376
Printing, Typographic	J. Donlevy	New York, N. Y	Apr. 3, 1866	53, 587
Printing-wheel, &c	T. S. Hudson	East Cambridge, Mass	June 5, 1866	55, 300
Printing woolen and other goods, Machine for	T. Crossley	Boston, Mass	June 20, 1854	11, 118
Prints, &c., Process for transferring	H. Loewenberg	New York, N. Y	Nov. 3, 1863	40, 489
Prism, Water	B. Kane	Cincinnati, Ohio	Nov. 26, 1867	71, 309
Prison-alarm apparatus	W. O. Hills	Nottingham, N. H	Aug. 31, 1858	21, 339
Prison-bar	J. A. Marden	Boston, Mass	July 13, 1869	92, 626
Prison-door bolt	B. F. Haugh	Indianapolis, Ind	Aug. 18, 1868	81, 165
Prison-door bolt	T. Lalor	Toronto, Canada	Nov. 4, 1873	144, 210
Prison-door lock	C. E. Felton	Buffalo, N. Y	Aug. 13, 1867	67, 041
Prison-grating	T. F. Baker	Cincinnati, Ohio	May 18, 1869	90, 219
Prison-grating	E. May	Indianapolis, Ind	May 9, 1871	114, 584
Prison, Iron	E. Jacobs	Cincinnati, Ohio	July 24, 1860	29, 282
Prison-lock	S. A. Denio	Boston, Mass	Sept. 13, 1859	25, 394
Prison-lock	H. R. Towne	Stamford, Conn	Feb. 7, 1871	111, 586
Prison-lock attachment	E. Kershaw	Boston, Mass	July 30, 1850	7, 525

Index of patents issued from the United States Patent Office from 1790 to 1873, inclusive—Continued.

Invention.	Inventor.	Residence.	Date.	No.
Prisons, Construction of	I. Hodgson	Indianapolis, Ind	June 16, 1868	78, 966
Prisons, Construction of	E. Jacobs	Cincinnati, Ohio	June 7, 1859	24, 307
Prisons, Construction of	E. May	Indianapolis, Ind	Oct. 4, 1859	25, 662
Prisons, Construction of	E. May	Indianapolis, Ind	Dec. 27, 1870	110, 483
Privy	E. E. Clarke	New Haven, Conn	Oct. 24, 1865	50, 654
Privy	M. M. Duplat	Charleston, S. C	Mar. 16, 1820	....
Privy	D. S. Forney	Wytheville, Va	Oct. 2, 1866	58, 400
Privy	J. Ingram	New York, N. Y	Mar. 17, 1868	75, 549
Privy	F. Riedel	San Francisco, Cal	Feb. 21, 1871	111, 976
Privy	F. Riedel	San Francisco, Cal	May 23, 1871	115, 239
Privy	W. J. Warren	Philadelphia, Pa	Nov. 8, 1870	109, 160
Privy and sink cleaner	R. R. Retowsky	Baltimore, Md	Oct. 7, 1873	143, 420
Privy and sink cleaning apparatus	A. M. Hobbs	Washington, D. C	Apr. 8, 1873	137, 680
Privy-bowls, Detachable flange for	R. Diven	Brooklyn, N. Y	Dec. 4, 1866	60, 153
Privy-indicator	W. C. Duffield	Wheeling, W. Va	Feb. 25, 1873	136, 224
Privy-seat	H. W. Carpenter	Madison County, N. Y	June 18, 1872	128, 111
Privy-seat	W. P. Hastings	Maryville, Tenn	July 16, 1872	129, 223
Privy-seat	E. Mangeon	Paris, France	Oct. 23, 1866	59, 150
Privy-seat	F. Peters and G. Clem	Cincinnati, Ohio	May 18, 1869	90, 298
Privy-seat	F. Reed	Fitchburgh, Mass	Dec. 7, 1869	97, 694
Privy-seat cover	W. Beach	Philadelphia, Pa	Aug. 7, 1866	56, 881
Privy-seat cover	W. Street	New York, N. Y	Oct. 17, 1871	120, 121
Privy, Portable	W. H. Pulver	Schuylersville, N. Y	Oct. 25, 1870	108, 728
Privy-vaults, Apparatus for emptying	J. G. Berger	Nuremberg, Bavaria	Nov. 2, 1869	96, 385
Privies, Apparatus for cleaning	H. C. Bull	New Orleans, La	June 6, 1871	115, 565
Privies, Apparatus for emptying	F. Datichiy	New York, N. Y	Dec. 17, 1850	7, 834
Privies, Signal for	I. H. Doughty	New York, N. Y	Sept. 18, 1849	6, 722
Probangs or instruments for treating diseased orifices.	J. S. and M. F. Lovell	Philadelphia, Pa	July 27, 1869	92, 980
Progressive kiln	H. Dueberg	Baltimore, Md	Sept. 7, 1869	94, 580
Projectile	J. Absterdam	New York, N. Y	Apr. 6, 1869	88, 689
Projectile	H. Bates	New London, Conn	Nov. 10, 1857	18, 568
Projectile	C. C. Brand	Norwich, Conn	May 19, 1857	17, 312
Projectile	C. F. Brown	Warren, R. I	Sept. 22, 1868	82, 284
Projectile	J. G. Butler	Fortress Monroe, Va	Sept. 26, 1871	119, 313
Projectile	W. A. Cobb	Orange, Mass	Mar. 23, 1869	88, 012
Projectile	J. W. Cochran	New York, N. Y	Sept. 25, 1860	30, 123
Projectile	J. Curtis	Chicago, Ill	June 1, 1869	90, 732
Projectile	E. Drake	Stoughton, Mass	Nov. 29, 1870	109, 600
Projectile	E. C. Hancock	New Orleans, La	May 16, 1871	114, 807
Projectile	S. H. Haycock	Ottawa, Canada	June 19, 1866	55, 796
Projectile	S. H. Haycock	Ottawa, Canada	Apr. 1, 1873	137, 442
Projectile	J. H. Helm	Pittsburgh, Pa	Dec. 19, 1871	122, 014
Projectile	J. W. Hill	Jefferson, Iowa	June 22, 1869	91, 633
Projectile	T. Hill	Vallejo, Cal	July 18, 1871	117, 172
Projectile	J. G. Hope	Topeka, Kans	Oct. 4, 1870	107, 909
Projectile	J. G. Hope	Wichita, Kans	Sept. 2, 1873	142, 394
Projectile	C. T. James	Providence, R. I	Feb. 26, 1856	14, 315
Projectile	R. H. Jones	San Francisco, Cal	Mar. 2, 1869	87, 498
Projectile	J. Link	United States Army	May 24, 1870	103, 477
Projectile	T. G. Orwig	New York, N. Y	Feb. 21, 1865	46, 490
Projectile	A. F. Potter	San Francisco, Cal	Sept. 21, 1869	95, 137
Projectile	J. D. Richards	Muscatine, Iowa	Sept. 7, 1869	94, 647
Projectile	J. Rigney	West Point, N. Y	June 11, 1872	127, 796
Projectile	N. Scholfield	Norwich, Conn	Aug. 19, 1856	15, 577
Projectile	M. Shaw	Sandwich, Mass	Mar. 3, 1857	16, 753
Projectile	W. H. Shock	Washington, D. C	May 24, 1870	103, 514
Projectile	C. W. Stafford	Burlington, Iowa	July 7, 1863	39, 180
Projectile	W. J. Von Hammerhuber	Washington, D. C	May 1, 1855	12, 801
Projectile	R. Weir	Philadelphia, Pa	July 1, 1862	35, 795
Projectile, Compound	J. E. Crane and J. Fox	Lowell, Mass	Sept. 11, 1866	57, 870
Projectile, Compound	S. Sawyer	Fitchburgh, Mass	Nov. 13, 1855	13, 799
Projectile, Compound explosive	M. L. Callender	New York, N. Y	Oct. 14, 1862	26, 686
Projectile, Compound explosive	F. A. Markley	Waynesborough, Va	May 13, 1873	138, 907
Projectile, Compound sub-caliber	C. Arick	Saint Clairsville, Ohio	July 28, 1863	39, 369
Projectile, Enameled	J. F. Clew	New York, N. Y	Nov. 15, 1864	45, 022
Projectile, Explosive	T. H. Burrowes	Chicopee, Mass	Oct. 27, 1863	40, 396
Projectile, Explosive	J. M. Connel	Newark, Ohio	Feb. 3, 1863	37, 565
Projectile, Explosive	A. O. H. Hardenstein	Clinton, Miss	Oct. 6, 1868	82, 714
Projectile, Explosive	B. B. Hotchkiss	Sharon, N. Y	May 6, 1862	35, 153
Projectile, Explosive	C. W. Isbell	New York, N. Y	May 13, 1862	35, 277
Projectile, Explosive	J. Jobson	Derby, England	Nov. 9, 1869	96, 595
Projectile, Explosive	J. Johnson	Brooklyn, N. Y	May 12, 1863	38, 491
Projectile, Explosive	J. Long	Shaver's Creek, Pa	Dec. 1, 1868	84, 635
Projectile, Explosive	J. N. Smith	Jersey City, N. J	Dec. 8, 1863	40, 888
Projectile for breech-loading ordnance	W. W. Hubbell	Philadelphia, Pa	May 1, 1860	98, 084
Projectile for fire-arms	W. H. Arnold	Washington, D. C	Apr. 12, 1859	23, 538
Projectile for fire-arms	J. F. Clew	New York, N. Y	Nov. 15, 1864	45, 021
Projectile for fire-arms	R. Cranston and H. Bates	New London, Conn	Aug. 14, 1860	29, 573
Projectile for fire-arms	A. H. Emery	New York, N. Y	Mar. 17, 1863	37, 966
Projectile for fire-arms	J. Holroyd	Washington, D. C	Nov. 1, 1859	25, 967
Projectile for fire-arms	E. Hoyt, jr	Chelsea, Mass	May 1, 1855	12, 795
Projectile for fire-arms	W. W. Hubbell	Philadelphia, Pa	Jan. 24, 1860	26, 904
Projectile for fire-arms	S. E. Jones	Santa Fé, N. Mex	Feb. 20, 1872	123, 828
Projectile for fire-arms	E. Matteson	Brooklyn, N. Y	Nov. 19, 1861	33, 746
Projectile for fire-arms	D. M. Mefford	Cincinnati, Ohio	Jan. 22, 1862	34, 285
Projectile for fire-arms	J. B. Read	Tuscaloosa, Ala	Nov. 24, 1857	18, 707
Projectile for fire-arms	R. Shaler	Madison, Conn	July 16, 1861	32, 844
Projectile for fire-arms	T. Smith	Pittsburgh, Pa	Apr. 22, 1856	14, 742
Projectile for fire-arms	B. Swain	Washington, D. C	Feb. 21, 1860	27, 245
Projectile for fire-arms	W. Taggart	Haverhill, Mass	Nov. 11, 1856	16, 076
Projectile for fire-arms	O. G. Warren	New York, N. Y	Oct. 20, 1863	40, 385
Projectile for fire-arms	C. A. Waterbury	New York, N. Y	Sept. 15, 1863	39, 977
Projectile for fire-arms, Elongated	J. N. Smith	New York, N. Y	Aug. 18, 1863	39, 593
Projectile for fire-arms, Hollow	T. D. Gibson	Wilmington, Del	May 18, 1869	90, 164
Projectile for killing whales	R. Brown	New London, Conn	June 14, 1859	24, 371
Projectile for killing whales	R. Sibley	Norwich, Conn	May 26, 1857	17, 407
Projectile for many-chambered guns	W. McCord	Sing Sing, N. Y	Sept. 15, 1863	39, 940

Index of patents issued from the United States Patent Office from 1790 *to* 1873, *inclusive*—Continued.

Invention.	Inventor.	Residence.	Date.	No.
Projectile for ordnance	S. C. Abbott	Zanesville, Ohio	Jan. 15, 1861	31, 099
Projectile for ordnance	C. F. Brown	Warren, R. I	Oct. 1, 1861	33, 378
Projectile for ordnance	J. G. Butler	Fort Leavenworth, Kans	June 14, 1870	104, 108
Projectile for ordnance	J. G. Butler	United States Army	Feb. 28, 1871	112, 120
Projectile for ordnance	J. W. Cochran	New York, N. Y	Nov. 1, 1859	25, 951
Projectile for ordnance	J. M. Currie	Washington, Iowa	Oct. 16, 1866	58, 783
Projectile for ordnance	J. Gault	Boston, Mass	July 2, 1861	32, 689
Projectile for ordnance	H. Alvorson	Cambridge, Mass	Sept. 29, 1863	40, 142
Projectile for ordnance	A. Hotchkiss	Sharon, Conn	Oct. 16, 1855	13, 679
Projectile for ordnance	B. B. Hotchkiss	New York, N. Y	July 27, 1869	93, 589
Projectile for ordnance	B. B. Hotchkiss	New York, N. Y	Feb. 28, 1871	112, 144
Projectile for ordnance	O. Lugo	New York, N. Y	Sept. 5, 1865	49, 773
Projectile for ordnance, &c	W. Mansfield, J. Morse, and H. H. Mansfield.	Canton, Mass	Apr. 29, 1862	35, 103
Projectile for ordnance	J. B. Read	Tuscaloosa, Ala	Oct. 28, 1856	15, 999
Projectile for ordnance	M. Ritner	Vincennes, Ind	Jan. 6, 1863	37, 361
Projectile for ordnance	D. E. Somes	Washington, D. C	Dec. 15, 1863	40, 958
Projectile for ordnance	C. W. Stafford	Burlington, Iowa	Jan. 27, 1863	37, 557
Projectile for ordnance, Arrow	W. Cousins	New York, N. Y	June 27, 1865	48, 371
Projectile for ordnance, Compound	L. E. Reynolds	Mendon, Ill	Dec. 8, 1863	40, 856
Projectile for ordnance, Explosive	W. E. Brown	Valley Falls, R. I	June 10, 1862	35, 503
Projectile for ordnance, Explosive	L. D. Gerarden	Jersey City, N. J	Feb. 10, 1863	37, 661
Projectile for ordnance, Explosive	W. W. Hanes	Covington, Ky	Aug. 26, 1862	36, 295
Projectile for ordnance, Explosive	J. Johnson	Brooklyn, N. Y	June 7, 1864	43, 029
Projectile for ordnance, Explosive	W. F. Rippon	Providence, R. I	Oct. 28, 1862	36, 858
Projectile for ordnance, Hot	C. T. James	Providence, R. I	Jan. 21, 1862	34, 207
Projectile for ordnance, Rifled	W. Page	New York, N. Y	Nov. 5, 1861	36, 660
Projectile for rifled cannon	T. T. S. Laidley	United States Army	Aug. 4, 1857	17, 935
Projectile for rifled cannon	T. T. S. Laidley	United States Army	Aug. 25, 1857	18, 049
Projectile for rifled cannon	J. M. Sigourney	Watertown, N. Y	Mar. 3, 1857	16, 755
Projectile for rifled fire-arm	C. Sharps	Philadelphia, Pa	July 11, 1865	48, 729
Projectile for rifled ordnance	B. S. Alexander	United States Army	Jan. 7, 1862	34, 091
Projectile for rifled ordnance	J. B. Atwater	Chicago, Ill	Mar. 17, 1863	37, 891
Projectile for rifled ordnance	A. Berney	Jersey City, N. J	July 7, 1863	39, 112
Projectile for rifled ordnance	J. W. Cochran	New York, N. Y	Nov. 8, 1859	26, 016
Projectile for rifled ordnance	J. W. Cochran	New York, N. Y	Nov. 8, 1859	26, 017
Projectile for rifled ordnance	J. W. Cochran	New York, N. Y	Dec. 6, 1859	26, 337
Projectile for rifled ordnance	B. C. Converse	Cincinnati, Ohio	Aug. 16, 1864	43, 835
Projectile for rifled ordnance	J. A. Dahlgren	Philadelphia, Pa	Aug. 6, 18[illegible]	32, 986
Projectile for rifled ordnance	H. E. Dunick	Saint Louis, Mo	July 14, 1863	39, 216
Projectile for rifled ordnance	A. H. Emery	New York, N. Y	Oct. 28, 1862	36, 773
Projectile for rifled ordnance	W. H. Havens	Paterson, N. J	Feb. 25, 1862	34, 493
Projectile for rifled ordnance	J. L. Henry	District of Columbia	July 8, 1862	35, 824
Projectile for rifled ordnance	B. B. Hotchkiss	Sharon, Conn	July 24, 1860	29, 272
Projectile for rifled ordnance	B. B. Hotchkiss	Sharon, Conn	May 14, 1861	32, 293
Projectile for rifled ordnance	B. B. Hotchkiss	New York, N. Y	June 7, 1864	43, 027
Projectile for rifled ordnance	W. W. Hubbell	Philadelphia, Pa	July 8, 1862	35, 825
Projectile for rifled ordnance	J. H. Merrill	Baltimore, Md	Oct. 13, 1857	18, 401
Projectile for rifled ordnance	R. P. Parrott	Cold Spring, N. Y	Aug. 20, 1861	33, 099
Projectile for rifled ordnance	R. P. Parrott	Cold Spring, N. Y	Aug. 20, 1861	33, 100
Projectile for rifled ordnance	B. S. Roberts	United States Army	Mar. 24, 1863	37, 979
Projectile for rifled ordnance	C. H. Sayre	Utica, N. Y	Dec. 24, 1861	34, 031
Projectile for rifled ordnance	C. W. Small	Bangor, Me	Mar. 4, 1862	34, 596
Projectile for rifled ordnance	W. H. Smith	Birmingham, Conn	Dec. 23, 1862	37, 260
Projectile for rifled ordnance	W. H. Smith	Birmingham, Conn	Jan. 5, 1864	41, 127
Projectile for rifled ordnance	B. F. Sturtevant	Boston, Mass	Aug. 5, 1862	36, 116
Projectile for rifled ordnance	I. P. Tice	New York, N. Y	Nov. 12, 1861	33, 718
Projectile for rifled ordnance	W. S. Williams	Canton, Ohio	May 3, 1864	42, 614
Projectile for rifled ordnance, Explosive	C. W. Smith, G. H. Babcock, and B. B. and C. A. Hotchkiss	New York, N. Y., and Sharon, Conn	Apr. 28, 1863	38, 359
Projectile for small-arms	C. Madnell	New Orleans, La	Jan. 9, 1872	122, 620
Projectile for smooth-bored guns	J. L. McConnel	Jacksonville, Ill	July 28, 1857	17, 886
Projectile for smooth-bored ordnance	J. A. Woodbury	Boston, Mass	Dec. 3, 1861	33, 863
Projectile for smooth-bored ordnance, Rotating	T. F. Reilly	New York, N. Y	Mar. 18, 1862	34, 693
Projectile, Hollow	S. Gardiner, jr	New York, N. Y	Nov. 3, 1863	40, 468
Projectile, Incendiary subcaliber	C. W. Stafford	New York, N. Y	Aug. 4, 1863	39, 427
Projectile, Percussion	A. McBurth	Elizabeth, N. J	Aug. 21, 1855	13, 469
Projectile, Percussion	J. Lippincott	Philadelphia, Pa	Mar. 18, 1856	14, 460
Projectile, Rifled	J. McMurtry	Lexington, Ky	Sept. 15, 1863	39, 942
Projectile, Subcaliber	E. A. Dana	Boston, Mass	Nov. 12, 1872	132, 903
Projectile, Subcaliber rifled	J. L. Henry	Kentucky	Dec. 13, 1864	45, 462
Projectile, Submarine explosive	M. L. Callender	New York, N. Y	Aug. 18, 1863	39, 612
Projectile, Submarine explosive	J. D. Willoughby	Washington, D. C	Jan. 5, 1864	41, 112
Projectiles, Alloy for sabots of	T. Taylor	Washington, D. C	Jan. 8, 1867	61, 117
Projectiles, Apparatus for throwing	S. D. Wellison	Pittsfield, Mass	Sept. 8, 1863	39, 858
Projectiles, Application of soft-metal packing to	W. Boekel	Philadelphia, Pa	Sept. 16, 1862	36, 449
Projectiles, Banding and covering	J. Absterdam	New York, N. Y	Nov. 7, 1865	50, 783
Projectiles by pressure, Press for making cylindro-conical hollow.	W. M. B. Hartley	New York, N. Y	Mar. 20, 1855	12, 574
Projectiles, Composition for covering	J. Absterdam	New York, N. Y	Aug. 12, 1862	36, 132
Projectiles, Composition for filling shrapnell and other similar.	A. Berney	Jersey City, N. J	June 17, 1862	35, 659
Projectiles, Composition for packing	P. S. Devlan	Jersey City, N. J	Sept. 15, 1863	39, 897
Projectiles, Directing	S. H. Noble	Prairie du Chien, Wis	Nov. 3, 1863	40, 543
Projectiles, Discharging	A. B. Cooley	Philadelphia, Pa	June 17, 1862	35, 587
Projectiles for fire-arms, Casting	B. A. Mason	Newport, R. I	Jan. 14, 1862	34, 153
Projectiles for ordnance, Attaching sabots to spherical.	G. P. Gauster	New York, N. Y	June 7, 1864	43, 017
Projectiles for ordnance, Mode of preparing	E. Lindner	New York, N. Y	July 30, 1861	32, 949
Projectiles for rifled ordnance, Packing	J. Absterdam	New York, N. Y	Feb. 23, 1864	41, 668
Projectiles for rifled ordnance, Packing	C. Arick	Saint Clairsville, Ohio	Mar. 28, 1865	47, 078
Projectiles for rifled ordnance, Packing	W. Boekel	Philadelphia, Pa	Apr. 12, 1864	42, 268
Projectiles for rifled ordnance, Packing	L. W. Broadwell	New Orleans, La	June 19, 1866	55, 761
Projectiles for rifled ordnance, Packing	J. W. Cochran	New York, N. Y	July 31, 1866	56, 712
Projectiles for rifled ordnance, Packing	B. B. Hotchkiss	New York, N. Y	May 16, 1865	47, 725
Projectiles for rifled ordnance, Packing	B. B. Hotchkiss	New York, N. Y	Oct. 10, 1865	50, 357
Projectiles for rifled ordnance, Packing	A. J. S. Molinard	Baltimore, Md	Apr. 11, 1865	47, 213
Projectiles for rifled ordnance, Packing	W. H. Smith	Birmingham, Conn	Oct. 11, 1864	44, 669

Index of patents issued from the United States Patent Office from 1790 to 1873, inclusive—Continued.

Invention.	Inventor.	Residence.	Date.	No.
Projectiles for rifled ordnance, Packing	W. H. Smith	Birmingham, Conn	May 16, 1865	47, 754
Projectiles for rifled ordnance, Sabot for	M. Ritner	Vincennes, Ind	June 10, 1862	35, 544
Projectiles for rifled ordnance, Sabot for	C. W. Stafford	New York, N. Y	Dec. 20, 1864	45, 567
Projectiles from smooth-bored ordnance, Rotating	T. Goodrem and C. Jackson	Providence, R. I	July 29, 1862	36, 001
Projectiles, Giving rotation to ordnance	H. K. Kenyon	Steubenville, Ohio	Nov. 18, 1862	36, 959
Projectiles, Igniting explosive	E. Pertuiset, A. Mundel, and J. E. A. De Fleron.	Paris, France	May 26, 1868	78, 322
Projectiles, India-rubber wad for	J. N. Bird	Trenton, N. J	Oct. 28, 1862	36, 879
Projectiles, Inflammable composition for filling	E. Harrison	New York, N. Y	Feb. 9, 1864	41, 577
Projectiles, Jacket for subcaliber	W. H. Smith	Birmingham, Conn	Oct. 11, 1864	44, 670
Projectiles, Machine for throwing	A. B. Smith and W. Weaver	Clinton, Pa	Aug. 12, 1856	15, 529
Projectiles, Mold for casting	H. Conant	Hartford, Conn	Jan. 16, 1855	12, 258
Projectiles, Mold for hollow	E. Allen	Worcester, Mass	July 29, 1856	15, 454
Projectiles, Molding chilled-iron	W. Boekel	Philadelphia, Pa	Jan. 6, 1863	37, 272
Projectiles, Sabot for	E. A. Dana	Brookline, Mass	Oct. 31, 1865	50, 692
Projectiles, Sabot for	C. W. Stafford	Burlington, Iowa	July 7, 1863	39, 179
Projectiles, Sabot for feathered	J. P. Woodbury and S. S. Gray.	West Roxbury and Boston, Mass.	July 22, 1862	35, 965
Projectiles, Sabot for ordnance	J. F. Clew	New York, N. Y	Nov. 15, 1864	45, 023
Projector	W. Farini	London, England	June 13, 1871	115, 837
Proof meter and register for alcoholic spirits and other liquids.	W. M. Storm	New York, N. Y	May 7, 1867	64, 456
Propagating-bed, Apparatus for heating	W. Chalmers	Philadelphia, Pa	Feb. 28, 1871	112, 220
Propagating-box	A. D. Manchester	Westport, Mass	Dec. 15, 1868	84, 955
Propagating-box for plants	C. Jillson	Worcester, Mass	Aug. 24, 1869	94, 169
Propagating-box for plants and vines	F. L. Perry	Canandaigua, N. Y	Mar. 27, 1866	53, 477
Propeller. (*See also* Canal-boat.)				
Propeller	J. F. Allen	New York, N. Y	May 8, 1866	54, 481
Propeller	R. Atkin	Brooklyn, N. Y	June 23, 1868	79, 183
Propeller	E. Beard	New Sharon, Me	Oct. 18, 1853	10, 124
Propeller	T. Bell	Bellport, N. Y	Oct. 26, 1869	96, 075
Propeller	J. Busser	Philadelphia, Pa	Jan. 3, 1865	45, 697
Propeller	R. D. Chatterton	Bath, England	Mar. 5, 1867	62, 723
Propeller	W. A. Cobb	Orange, Mass	July 16, 1867	66, 797
Propeller	A. Cotton	Sycamore, Ill	Jan. 25, 1870	99, 063
Propeller	G. R. Comstock	Little Falls, N. Y	Nov. 9, 1858	22, 016
Propeller	J. Covert	New York, N. Y	Feb. 19, 1867	62, 255
Propeller	E. Dalman y Sala	New York, N. Y	Nov. 21, 1865	51, 025
Propeller	C. De Berque	London, England	Jan. 9, 1855	12, 190
Propeller	W. H. Degges	Washington, D. C	Apr. 28, 1863	38, 292
Propeller	A. Duboce	Brooklyn, N. Y	July 24, 1855	13, 336
Propeller	E. Etinge and J. C. Broadhead	Kingston, N. Y	Jan. 10, 1871	110, 908
Propeller	W. Fields	Wilmington, Del	Nov. 19, 1867	71, 153
Propeller	J. D. Ford	Baltimore, Md	Sept. 20, 1870	107, 471
Propeller	R. Forward	Cincinnati, Ohio	Jan. 16, 1872	122, 823
Propeller	F. G. Fowler	Springfield, Ill	May 28, 1867	65, 202
Propeller	H. Fromm	East New York, N. Y	June 4, 1867	65, 370
Propeller	C. F. Gardiner	East Boston, Mass	June 15, 1858	20, 606
Propeller	B. Gilbert	Pittsburgh, Pa	Nov. 15, 1853	10, 229
Propeller	R. Griffiths	London, England	Apr. 7, 1857	17, 016
Propeller	B. Handforth	Chicago, Ill	Jan. 1, 1867	60, 887
Propeller	H. W. Hewett	New York, N. Y	June 7, 1853	9, 673
Propeller	D. H. Heyen	New York, N. Y	June 9, 1868	78, 739
Propeller	J. O. Heyworth	Chicago, Ill	June 29, 1869	91, 845
Propeller	R. Hines and L. Beyer	Washington, D. C	July 16, 1872	129, 028
Propeller	L. and C. Howard	Watkins, N. Y	Aug. 30, 1870	106, 824
Propeller	D. Hughes	Rochester, N. Y	Nov. 15, 1859	26, 106
Propeller	R. Hunter	New York, N. Y	July 7, 1868	79, 657
Propeller	R. Hunter	New York, N. Y	Aug. 11, 1868	80, 960
Propeller	R. Hunter	New York, N. Y	Sept. 13, 1870	107, 376
Propeller	P. M. Hutton and W. Huston	Troy, N. Y., and Wilmington, Del.	June 6, 1848	5, 623
Propeller	F. Jacob	New York, N. Y	Oct. 31, 1865	50, 714
Propeller	A. Johnson	Buffalo, N. Y	Dec. 8, 1857	18, 814
Propeller	W. D. Jones	Poughkeepsie, N. Y	Apr. 17, 1855	12, 752
Propeller	C. Kinzler	New York, N. Y	Apr. 26, 1870	102, 276
Propeller	W. Lawton	Green Point, N. Y	Oct. 1, 1867	69, 451
Propeller	H. Leach	Boston, Mass	Jan. 31, 1854	10, 474
Propeller	H. Link	Little Falls, N. Y	May 8, 1855	12, 823
Propeller	A. C. Loud	San Francisco, Cal	Jan. 19, 1869	86, 086
Propeller	A. C. Loud	San Francisco, Cal	May 4, 1869	89, 777
Propeller	L. H. Markley	Line Lexington, Pa	May 24, 1859	24, 130
Propeller	A. J. Marshall	Warrenton, Va	Mar. 7, 1871	112, 366
Propeller	J. T. Martin	New York, N. Y	Jan. 9, 1867	51, 955
Propeller	G. L. McKay	Linwood Station, Pa	Jan. 30, 1872	123, 274
Propeller	G. Meader	Ottawa, Ill	Jan. 30, 1866	52, 308
Propeller	C. G. Meinhardt	Altoona, Pa	July 28, 1868	80, 421
Propeller	D. E. Merick	Cleveland, Ohio	July 20, 1858	20, 953
Propeller	C. F. Miller and M. Priester	New York, N. Y	June 11, 1872	127, 909
Propeller	S. B. Morey	San Francisco, Cal	Dec. 14, 1869	97, 952
Propeller	N. Nolan	New York, N. Y	Dec. 17, 1867	72, 319
Propeller	C. O'Hara	London, Great Britain	Apr. 22, 1862	35, 035
Propeller	C. M. O'Hara	New York, N. Y	Apr. 14, 1868	76, 651
Propeller	G. Parker	Worcester, Mass	May 24, 1864	42, 870
Propeller	J. Patch	Boston, Mass	Nov. 27, 1849	6, 914
Propeller	F. Peale	Philadelphia, Pa	Feb. 20, 1855	12, 446
Propeller	J. H. Penny and T. B. Rogers	New York, N. Y	June 12, 1855	13, 041
Propeller	O. C. Phelps	New York, N. Y	July 12, 1864	43, 522
Propeller	J. H. Phillips	Saint Louis, Mo	Feb. 5, 1867	61, 861
Propeller	G. R. Pierce	Grand Rapids, Mich	Feb. 27, 1872	124, 009
Propeller	E. S. Renwick	New York, N. Y	Sept. 24, 1861	33, 364
Propeller	H. F. Roberts	Pittsburgh, Pa	Dec. 29, 1868	85, 403
Propeller	H. Rolle	Boston, Mass	July 16, 1867	66, 742
Propeller	T. W. Roys	Southampton, N. Y	June 3, 1862	35, 475
Propeller	J. Salter	Williamsburgh, N. Y	Nov. 29, 1870	109, 673
Propeller	R. Schmidt	Cincinnati, Ohio	Feb. 13, 1872	123, 733
Propeller	C. J. Schveder	Astoria, N. Y	Oct. 9, 1860	30, 355
Propeller	C. Sharps	Philadelphia, Pa	May 25, 1869	90, 590

Index of patents issued from the United States Patent Office from 1790 *to* 1873, *inclusive*—Continued.

Invention.	Inventor.	Residence.	Date.	No.
Propeller	C. Sharps	Philadelphia, Pa	Sept. 21, 1869	95, 150
Propeller	S. P. Snyder and G. W. Cook	Minneapolis, Minn	Mar. 8, 1859	23, 199
Propeller	A. W. Thompson	Philadelphia, Pa	Jan. 21, 1851	7, 907
Propeller	W. Thurber	Olean, N. Y	June 29, 1858	20, 744
Propeller	C. E. Toley	Brooklyn, N. Y	July 30, 1867	67, 376
Propeller	J. Trees	Salem, Pa	Oct. 25, 1853	10, 164
Propeller	W. Van Dusen	Philadelphia, Pa	Nov. 30, 1858	22, 209
Propeller	S. B. Wait	Mariner's Harbor, N. Y	Feb. 19, 1867	62, 238
Propeller	C. R. M. Wall	New York, N. Y	May 17, 1859	24, 069
Propeller	C. Ward	Detroit, Mich	Apr. 6, 1869	88, 595
Propeller	C. T. P. Ware	New York, N. Y	Oct. 4, 1853	10, 091
Propeller	J. White	Mauch Chunk, Pa	June 11, 1829	....
Propeller	A. A. Wilder and W. Gooding	Detroit, Mich	May 15, 1866	54, 801
Propeller	W. L. and T. Winans	London, England, and Baltimore, Md.	Oct. 9, 1866	58, 744
Propeller, Adjustable and reversible	A. B. Cooley	Philadelphia, Pa	May 13, 1862	35, 219
Propeller, Adjustable screw	C. F. Brown	Warren, R. I	July 12, 1853	9, 849
Propeller and fire-extinguisher, Combined	A. Turner	Bronson, Mich	Apr. 2, 1872	125, 232
Propeller and paddle-wheel shaft	W. Peters	Baltimore, Md	Nov. 29, 1859	26, 290
Propeller and steerer for boats, Portable	T. Reece	Philadelphia, Pa	Oct. 23, 1866	59, 074
Propeller and steering-apparatus	H. Ressel	Vienna, Austria	Oct. 11, 1864	44, 696
Propeller-blade	C. H. Burton	Cleveland, Ohio	Apr. 21, 1863	28, 216
Propeller-blade	G. Hibsch	Buffalo, N. Y	May 12, 1857	17, 276
Propeller-blade	W. G. Oliver	Buffalo, N. Y	Apr. 7, 1863	38, 121
Propeller-blade	G. W. Swartz	Buffalo, N. Y	Aug. 4, 1857	17, 943
Propeller blades, Screw	C. C. Gates	Albany, N. Y	Dec. 12, 1865	51, 446
Propeller-blades to their shafts, Means of affixing	L. B. Flanders	Philadelphia, Pa	Aug. 16, 1864	43, 842
Propeller, Buoyant	W. Lansdell	Memphis, Tenn	May 29, 1855	12, 955
Propeller, Buoyant	E. Matteson	Brooklyn, N. Y	Oct. 13, 1868	82, 968
Propeller, Buoyant	J. Montgomery	New York, N. Y	Dec. 21, 1858	22, 373
Propeller, Buoyant spiral	T. J. Wells	New York	Dec. 23, 1841	2, 400
Propeller, Cantering	P. W. Mackenzie	Jersey City, N. J	Aug. 12, 1862	36, 161
Propeller, Centripetal	J. W. Nystrom	Stockholm, Sweden	Mar. 19, 1850	7, 194
Propeller, Chain	J. Neumann	Brooklyn, N. Y	Dec. 10, 1872	133, 794
Propeller, Chain	E. P. Russell	Manlius, N. Y	Sept. 9, 1873	142, 653
Propeller, Child's horse and self	J. H. Nolan	Waterville, N. Y	Mar. 1, 1870	100, 316
Propeller, Conical screw	H. Hubbell	Moyamensing, Pa	July 14, 1845	4, 111
Propeller-cranks, Apparatus for oiling	J. Davis	New York, N. Y	Oct. 8, 1867	69, 638
Propeller, Double-cone marine	J. R. Baylis	Baltimore, Md	Dec. 10, 1861	33, 874
Propeller Duck's-foot	G. Meader	Ottawa, Ill	Sept. 1, 1863	39, 738
Propeller, Duck's-foot	G. Seibert	Hagerstown, Md	Nov. 7, 1848	5, 911
Propeller, Endless-chain	C. C. Bisbee	Rochester, N. Y	Sept. 24, 1872	131, 494
Propeller, Endless-chain	W. W. Bowman	Graves County, Ky	Jan. 26, 1864	41, 360
Propeller, Endless-chain	L. Chevrier	Philadelphia, Pa	June 11, 1836	......
Propeller, Endless-chain	C. F. Fisher	New Orleans, La	Oct. 7, 1851	8, 407
Propeller, Endless-chain	W. Keeler	Towanda, Pa	Sept. 17, 1872	131, 450
Propeller, Endless-chain	A. McDonald	Mattawan, Mich	July 4, 1865	48, 574
Propeller, Endless-chain	A. B. Smith	Yankton, Dak	Oct. 22, 1867	70, 036
Propeller, Endless-chain	T. Teed	Alexandria, Va	Oct. 21, 1873	143, 793
Propeller, Endless-chain	J. W. Whinyates	Liverpool, England	Mar. 25, 1873	137, 269
Propeller-engine	E. D. Ashe	Brompton, England	Apr. 19, 1864	42, 336
Propeller-engine	H. Whitaker	New York, N. Y	July 23, 1872	129, 700
Propeller, Flexible	R. Hunter	Cincinnati, Ohio	May 16, 1871	114, 944
Propeller for boats	J. Hamilton	New York, N. Y	Dec. 28, 1858	22, 422
Propeller for boats	M. Nelson	New York, N. Y	July 13, 1858	20, 889
Propeller for boats	L. C. St. John	Buffalo, N. Y	Aug. 31, 1858	21, 378
Propeller for boats, &c., Oblique-paddle	R. Bulkley	New York, N. Y	Mar. 13, 1844	3, 476
Propeller for boats or carriages	B. Taylor	New York	Apr. 25, 1808	....
Propeller for boats, Spiral	J. Copley	Warrior Marsh, Pa	May 22, 1830	....
Propeller for boats, Steam	A. W. Von Schmidt	San Francisco, Cal	Nov. 13, 1866	59, 686
Propeller for canal-boats	E. Backus	Rochester, N. Y	Aug. 21, 1860	29, 660
Propeller for canal-boats	A. Burbank	Buffalo, N. Y	July 13, 1858	20, 862
Propeller for canal-boats	R. Cartwright	Ithaca, N. Y	July 19, 1859	24, 794
Propeller for canal-boats	S. D. Gilson	Oswego Falls, N. Y	Nov. 17, 1868	84, 108
Propeller for canal-boats	E. Raynale	Birmingham, Mich	Aug. 28, 1866	57, 569
Propeller for canal-boats	O. Plumb	Millport, N. Y	Sept. 23, 1862	36, 526
Propeller for canal-boats	G. W. Swartz	Buffalo, N. Y	June 18, 1857	17, 592
Propeller for canal-boats	G. W. Swartz	Buffalo, N. Y	Mar. 16, 1858	19, 656
Propeller for canal-boats	G. G. Wyland and T. M. Rathmell	Williamsport, Pa	May 2, 1871	114, 382
Propeller for canal-boats, Chain	E. T. Ligon	Demopolis, Ala	June 25, 1872	128, 319
Propeller for canal-navigation	W. F. Tyson	Owingsburgh, Pa	June 21, 1853	9, 810
Propeller for canals	B. Burling	Buffalo, N. Y	Feb. 22, 1859	23, 010
Propeller for canals	N. P. Otis	Yonkers, N. Y	May 9, 1865	47, 657
Propeller for life-boats	M. M. Camp	New Haven, Conn	Dec. 21, 1858	22, 346
Propeller for life-boats	G. W. La Baw	Jersey City, N. J	May 6, 1856	14, 843
Propeller for shallow water, Wave	J. B. Eads	Saint Louis, Mo	Dec. 16, 1862	37, 157
Propeller for small boats, Hand	J. Fehrenbatch	Indianapolis, Ind	Sept. 18, 1866	58, 083
Propeller for spindle-shaped ships	W. L. and T. Winans	London, England	Sept. 4, 1866	57, 836
Propeller for steamships	C. Kinkel and M. Hubbe	New York, N. Y	June 19, 1866	55, 773
Propeller for steam-vessels	R. Sutton	New Castle, Del	Sept. 6, 1870	107, 120
Propeller for vessels	A. Aubert	Nogent-le-Rotrou, France	Sept. 7, 1869	94, 463
Propeller for vessels	L. De Lill	Phœnix, N. Y	Nov. 21, 1871	121, 163
Propeller for vessels	H. Everett	Windsor, Vt	Aug. 7, 1847	5, 219
Propeller for vessels	S. J. Gold	Cornwall, Conn	Mar. 6, 1847	5, 000
Propeller for vessels	C. A. S. Harris	Saint Louis, Mo	Mar. 28, 1871	113, 163
Propeller for vessels	E. C. Hubbard	Green Bay, Wis	Feb. 4, 1873	135, 555
Propeller for vessels	L. W. Langdon	Northampton, Mass	July 16, 1872	129, 571
Propeller for vessels	W. P. McConnell	Washington, D. C	May 16, 1846	4, 529
Propeller for vessels	C. M. O'Hara	London, England	Nov. 5, 1867	70, 460
Propeller for vessels	L. Phleger	Wilmington, Del	Sept. 13, 1845	4, 193
Propeller for vessels	A. Raplin	France	Apr. 3, 1866	53, 762
Propeller for vessels	C. M. Raynale	Birmingham, Mich	Jan. 8, 1867	61, 020
Propeller for vessels	J. E. Smith	New York, N. Y	Mar. 14, 1848	5, 468
Propeller for vessels	H. Stanley	Troy, N. Y	Oct. 9, 1860	30, 360
Propeller for vessels	C. E. Tripler	New York, N. Y	June 27, 1871	116, 513
Propeller for vessels, Chain	T. W. Carter	New Orleans, La	July 1, 1873	140, 468
Propeller for vessels, Form of screw	E. Beard	New Sharon, Me	Apr. 10, 1841	2, 045

Index of patents issued from the United States Patent Office from 1790 *to* 1873, *inclusive*—Continued.

Invention.	Inventor.	Residence.	Date.	No.
Propeller for vessels, Rotary inclined	R. F. Loper	Philadelphia, Pa	Feb. 28, 1844	3, 459
Propeller for vessels, Vibrating	D. Bainbridge	Philadelphia, Pa	Jan. 28, 1873	135, 257
Propeller for vessels, Vibrating	J. D. Fraser	Pictou, Canada	Dec. 9, 1873	145, 342
Propeller, Four-wheel	R. G. Elder	New York, N. Y	Feb. 1, 1870	99, 421
Propeller, Hand	J. G. Ross	New York, N. Y	May 27, 1856	14, 973
Propeller, Hydraulic	J. B. Booth	Portsmouth, Va	June 2, 1863	38, 724
Propeller, Hydraulic	A. Pagenstecher	Valparaiso, Chili	Oct. 4, 1864	44, 584
Propeller, Jointed scull	C. Dann	Rushford, Minn	Mar. 10, 1863	37, 856
Propeller, Marine	W. Aitken	Newark, Ill	Sept. 15, 1863	39, 984
Propeller, Marine	B. T. Babbitt	New York, N. Y	May 5, 1863	38, 368
Propeller, Marine	G. L. Carver	Brazil, South America	Aug. 7, 1860	29, 464
Propeller, Marine	G. B. Cocks	Cincinnati, Ohio	Mar. 28, 1871	113, 143
Propeller, Marine	R. Covington	Washington, D. C	Nov. 15, 1864	45, 127
Propeller, Marine	J. Eaton	Belleville, Canada	Oct. 19, 1858	21, 825
Propeller, Marine	A. E. Harding	Middletown, Ohio	Sept. 11, 1860	29, 969
Propeller, Marine	A. L. Hatch	Addison, N. Y	Apr. 25, 1865	47, 419
Propeller, Marine	H. W. Herbert	Herbertsville, Va	Mar. 27, 1860	27, 629
Propeller, Marine	H. Hirsch	Berlin, Prussia	Aug. 23, 1859	25, 197
Propeller, Marine	L. Holtslander	Oberlin, Ohio	Sept. 6, 1859	65, 336
Propeller, Marine	W. H. Johnson	Springfield, Mass	June 3, 1862	35, 451
Propeller, Marine	E. Lacroix, jr	Rouen, France	Dec. 10, 1861	33, 893
Propeller, Marine	T. Mason	Boston, Mass	Oct. 11, 1864	44, 641
Propeller, Marine	D. D. Porter	United States Navy	June 12, 1860	28, 688
Propeller, Marine	H. D. J. Pratt	Washington, D. C	Nov. 27, 1860	30, 757
Propeller, Marine	J. Reynolds	Providence, R. I	Feb. 14, 1860	27, 156
Propeller, Marine	J. B. Root	Brooklyn, N. Y	May 5, 1863	38, 415
Propeller, Marine	J. Taggart	Roxbury, Mass	Mar. 29, 1859	23, 432
Propeller, Marine	R. C. Vernol and J. T. Martin	New York, N. Y	Jan. 5, 1864	41, 109
Propeller, Marine	M. Wappich	Sacramento, Cal	Oct. 8, 1861	33, 459
Propeller, Marine	E. Webber	Gardiner, Me	May 29, 1860	28, 542
Propeller, Marine	W. H. Willard	Cleveland, Ohio	Sept. 16, 1862	36, 492
Propeller, Marine	W. D Wilson	Newark, N. J	Dec. 19, 1865	51, 640
Propeller, Marine hand	E. C. Brackett	Newton Corner, Mass	June 14, 1859	24, 368
Propeller, Minnow	W. D. Chapman	Theresa, N. Y	July 5, 1870	104, 930
Propeller, Minnow	W. D. Chapman	Theresa, N. Y	May 30, 1871	115, 434
Propeller operating as air-pump and condenser, Submerged.	W. W. Hunter	Gosport, Va	Mar. 12, 1842	2, 487
Propeller, Paddle-wheel	J. May	Columbus, Ga	Sept. 7, 1858	21, 432
Propeller, Reciprocating	M. Depuy	Pittsburgh, Pa	Aug. 1, 1865	49, 095
Propeller, Reciprocating	J. Galt	Philadelphia, Pa	June 28, 1859	24, 552
Propeller, Reciprocating	H. W. Hewet	New York, N. Y	Oct. 9, 1849	6, 782
Propeller, Reversible hydraulic	R. G. Dunbar	Logansport, Ind	Oct. 15, 1872	132, 262
Propeller, Rocking	E. Landis	Baltimore, Md	July 24, 1860	29, 288
Propeller-ropes, Preventing fouling of	R. H. Hudson	Glasgow, Scotland	Aug. 20, 1867	67, 982
Propeller, Screw	C. Arlen and C. Gautschi	New York, N. Y	May 3, 1870	102, 645
Propeller, Screw	A. Aubert	Nogent-le-Rotrou, France	June 25, 1872	128, 263
Propeller, Screw	B. F. Bee	Harwich, Mass	Jan. 4, 1859	22, 525
Propeller, Screw	D. Bell	Buffalo, N. Y	June 25, 1861	32, 609
Propeller, Screw	O. Byrne and J. G. Elliott	New York, N. Y	Oct. 5, 1858	21, 650
Propeller, Screw	J. Cochrane	Wall Township, N. J	June 27, 1871	116, 414
Propeller, Screw	L. H. Colborn	Chicago, Ill	Dec. 5, 1865	51, 295
Propeller, Screw	W. J. Curtis	Holloway, England	May 19, 1863	38, 566
Propeller, Screw	W. E. Davis	Jersey City, N. J	Aug. 7, 1866	56, 907
Propeller, Screw	H. Elliott	Boston, Mass	Aug. 20, 1861	33, 077
Propeller, Screw	J. Ericsson	New York, N. Y	Sept. 9, 1845	4, 181
Propeller, Screw	D. Freed	Philadelphia, Pa	Aug. 19, 1873	141, 871
Propeller, Screw	J. Hartman, jr	Philadelphia, Pa	Apr. 1, 1873	137, 365
Propeller, Screw	L. and C. Howard	Watkins, N. Y	Sept. 3, 1872	131, 099
Propeller, Screw	H. Hirsch	Paris, France	Apr. 26, 1870	102, 399
Propeller, Screw	F. Jacob	New York, N. Y	Jan. 24, 1865	46, 004
Propeller, Screw	A. Jouan	San Francisco, Cal	Jan. 25, 1859	22, 731
Propeller, Screw	J. E. Kennedy	New Orleans, La	May 12, 1868	77, 888
Propeller, Screw	M. Kolb	New York, N. Y	Sept. 19, 1871	119, 155
Propeller, Screw	G. Malo	Dunkirk, France	Nov. 18, 1851	8, 529
Propeller, Screw	J. Montgomery	Memphis, Tenn	Aug. 7, 1847	5, 224
Propeller, Screw	J. Montgomery	Baltimore, Md	Apr. 12, 1859	23, 598
Propeller, Screw	N. A. Patterson	Athens, Tenn	Jan. 21, 1873	135, 149
Propeller, Screw	N. A. Patterson	Cleveland, Tenn	Aug. 26, 1873	142, 269
Propeller, Screw	S. W. Rowell	Rutland, Vt	Oct. 12, 1869	95, 731
Propeller, Screw	J. B. Root	New York, N. Y	May 23, 1865	47, 864
Propeller, Screw	G. Safford	New York, N. Y	Mar. 1, 1859	23, 117
Propeller, Screw	C. G. Toense	Jersey City, N. J	Feb. 22, 1870	100, 216
Propeller, Screw	A. G. Tompkins	New York, N. Y	Mar. 26, 1861	31, 838
Propeller, Screw	H. Vansittart	Richmond, England	May 4, 1869	89, 712
Propeller, Screw	J. J. B. Vergne	Paris, France	June 21, 1859	24, 508
Propeller, Screw	M. M. Wilson	Honey Grove, Tex	Feb. 18, 1873	136, 015
Propeller, Screw	F. Wittram	San Francisco, Cal	Aug. 17, 1869	93, 848
Propeller, Screw	H. Zahn	San Francisco, Cal	Aug. 22, 1871	118, 325
Propeller-screws, Mode of driving	R. C. Jackson	Cleveland, Ohio	Apr. 21, 1863	38, 228
Propeller, Sculler	A. Bond	Philadelphia, Pa	June 19, 1849	6, 533
Propeller-shaft	W. Burtis	New York, N. Y	Nov. 3, 1868	83, 601
Propeller-shaft	A. Jouan	San Francisco, Cal	Nov. 18, 1856	16, 091
Propeller-shaft coupling, Ship	W. L. and T. Winans	London, England	Sept. 4, 1866	57, 835
Propeller-shafts and receiving rudders of stern-propellers, Means of supporting.	J. Beattie	Liverpool, England	Jan. 15, 1856	14, 081
Propeller-shafts, Anti-friction roller for	D. S. Chase	Augusta, Ga	Aug. 14, 1860	29, 570
Propeller-shafts in keels, Inclosing	A. Arnold	Troy, N. Y	Apr. 8, 1856	14, 589
Propeller-shafts, Stern-bearing for	R. E. Campbell	New York, N. Y	Dec. 20, 1864	45, 473
Propeller, Shell	J. Trees	Salem, Pa	May 14, 1850	7, 371
Propeller-shield	J. Carhart	New York, N. Y	July 9, 1861	32, 751
Propeller, Shifting-bucket	J. Busser	Philadelphia, Pa	July 2, 1867	66, 214
Propeller, Shifting-bucket	J. Busser	Philadelphia, Pa	Apr. 14, 1868	76, 711
Propeller, Ship's	J. K. Peters	New York, N. Y	Mar. 1, 1859	23, 112
Propeller, Spiral	B. Woodcroft	Manchester, England	Apr. 4, 1846	4, 441
Propeller, Spring and valve marine	W. Willis	Charleston, S. C	May 2, 1828	....
Propeller, Steam land	G. W. N. Yost	Pittsburgh, Pa	June 3, 1856	15, 050
Propeller, Steamship	A. Doyle	New York, N. Y	Sept. 4, 1866	57, 690
Propeller, Steering	F. G. Fowler	Norwalk, Conn	Jan. 4, 1870	98, 483

Index of patents issued from the United States Patent Office from 1790 *to* 1873, *inclusive*—Continued.

Invention.	Inventor.	Residence.	Date.	No.
Propeller, Steering	F. G. Fowler	Bridgeport, Conn	Oct. 21, 1873	143, 896
Propeller, Steering	H. E. Fessel	Chicago, Ill	Dec. 28, 1858	22, 417
Propeller, Steering	W. Harsen	Green Point, N. Y	Nov. 11, 1873	144, 538
Propeller, Submarine	J. L. Sullivan	Boston, Mass	Mar. 24, 1817	....
Propeller, Submerged	H. Everett	Windsor, Vt	June 9, 1843	3, 129
Propeller, Submerged	R. Vaile	Westbourne Green, England	June 30, 1863	39, 083
Propeller, Submerged	P. Von Schmidt	Washington, D. C	May 30, 1844	3, 606
Propeller, Valve	A. Colton	Le Roy, N. Y	May 25, 1858	20, 332
Propeller, Vibrating	F. Kellsey	Middletown, Conn	Nov. 2, 1852	9, 366
Propeller, Vibrating	C. P. Macowitzky	Corpus Christi, Tex	Feb. 18, 1873	135, 994
Propeller-wheel	C. T. Finlayson	Albany, Oreg	Sept. 28, 1869	95, 335
Propeller-wheel	J. Harrington and F. Caffrey	New Haven, Conn	Mar. 20, 1866	53, 297
Propeller-wheel	R. Johnson	San Francisco, Cal	Apr. 9, 1872	125, 458
Propeller-wheel	J. A. Joyner	New York, N. Y	Dec. 28, 1869	98, 268
Propeller-wheel	T. Tripp	Buffalo, N. Y	Nov. 22, 1859	26, 213
Propeller-wheel	T. Tripp	Chicago, Ill	Apr. 10, 1866	53, 902
Propeller-wheels, Mold-box for casting	W. Lechler and J. K. Schupp	Cleveland, Ohio	Jan. 6, 1863	37, 295
Propellers and chimneys for canal-boats, Arrangement of.	B. M. Smith	Ridgway, N. Y	Apr. 23, 1850	7, 310
Propellers, Application of high-pressure engine to screw.	H. Whittaker	Buffalo, N. Y	Oct. 18, 1853	10, 145
Propellers, Application of steam to	W. Spalding	Port Clinton, Ohio	July 7, 1868	79, 695
Propellers, Applying steam-power to	A. M. Sawyer	Athol, Mass	May 8, 1860	28, 200
Propellers, Arrangement and connection of screw.	P. S. Devlan	Reading, Pa	May 21, 1850	7, 376
Propellers, Attaching and housing	W. Webster	Jefferson County, Wash	June 29, 1858	20, 751
Propellers, Elevating and depressing	S. G. Hill	Muscatine, Iowa	Aug. 5, 1873	141, 558
Propellers for boats and other vessels, Segmental spiral.	J. Laing	Ellicott's Mills, Md	Oct. 22, 1842	2, 825
Propellers for steamboats, &c., Coupling the shafts of submerged.	R. F. Loper	Philadelphia, Pa	Oct. 9, 1844	3, 786
Propellers, Journal for oscillating	M. A. Crooker	New York, N. Y	Oct. 16, 1849	6, 793
Propellers, Manufacture of	J. Sutherland	New York, N. Y	Apr. 4, 1865	47, 137
Propellers, Mechanism for operating screw	F. H. Tobias	New York, N. Y	Dec. 16, 1873	145, 697
Propellers, Mechanism for operating screw	J. Wilcoxen	Russellville, Ill	May 20, 1873	139, 098
Propellers, Molding screw	A. K. Rider	Elizabeth, N. J	Aug. 31, 1869	94, 442
Propellers of machinery to be used in currents	J. Hardie	Victoria, Tex	Nov. 18, 1851	8, 525
Propellers in steam-vessels, Device for regulating the revolution of.	H. Dale	Boston, Mass	Feb. 26, 1867	62, 399
Propellers of steam-vessels, Elevating and depressing.	R. F. Soper	Philadelphia, Pa	Nov. 26, 1845	4, 285
Propellers, Operating screw	A. Lee	Whitemarsh, Pa	Oct. 29, 1872	132, 588
Propellers, Operating screw	A. McDonald	Cumberland, Md	Nov. 25, 1873	145, 000
Propellers, Operating vibrating	T. Spiller and A. Crowhurst	London, England	Nov. 8, 1853	10, 215
Propellers relatively to the draught of water, Apparatus for adjusting.	W. H. Willard	Cleveland, Ohio	June 10, 1862	35, 573
Propellers, Shafting-coupling for	S. Wilmarth, S. L. Hay, and D. N. B. Coffin, jr.	Charlestown, Reading, and Newton, Mass.	Apr. 6, 1858	19, 887
Propellers, Shipping and unshipping	S. R. Parkhurst	New York, N. Y	Nov. 26, 1845	4, 290
Propellers, Shipping and unshipping screw	G. Leach	Leeds, England	Nov. 1, 1870	108, 800
Propellers, Stern bearing for	J. Boiles	New York, N. Y	July 18, 1871	117, 039
Propellers to driving-shafts, Attaching	J. L. Cathcart	Washington, D. C	Apr. 18, 1854	10, 790
Propellers to vessels, Attaching	G. Smith	Philadelphia, Pa	Feb. 27, 1872	124, 170
Propelling and navigating boats, &c	A. G. D. Tuthill	New York	Mar. 22, 1810	....
Propelling and steering apparatus	S. and S. Huse, jr	Chicago, Ill	Dec. 28, 1858	22, 431
Propelling and steering apparatus	R. H. Lecky	McClure, Pa	Sept. 15, 1863	39, 936
Propelling and steering apparatus	M. Lytle	Allegheny, Pa	May 24, 1859	24, 129
Propelling and steering apparatus	J. C. Salomon, jr	Cincinnati, Ohio	Dec. 2, 1851	8, 558
Propelling and steering apparatus for vessels	J. Jopling	Bishopwearmouth, England	Apr. 3, 1866	53, 759
Propelling and steering boats, &c	T. L. Jones	New York	July 10, 1842	2, 727
Propelling and steering canal-boats	S. D. Bennett	Butternuts, N. Y	Jan. 30, 1829	....
Propelling and steering canal-boats	P. Roberts	New York, N. Y	June 18, 1872	128, 069
Propelling-apparatus	J. F. Alexander	New York, N. Y	Nov. 23, 1869	97, 146
Propelling-apparatus	E. Averill	Sacramento, Cal	Dec. 28, 1869	98, 217
Propelling-apparatus	E. S. Barnes	Nebraska City, Nebr	Oct. 6, 1868	82, 787
Propelling-apparatus	J. Bourne	London, England	Dec. 28, 1869	98, 222
Propelling-apparatus	S. D. Carpenter	Madison, Wis	Jan. 12, 1869	85, 789
Propelling-apparatus	J. S. Cunningham	New York, N. Y	Sept. 7, 1869	94, 573
Propelling-apparatus	A. Gemünder	New York, N. Y	Oct. 24, 1865	50, 574
Propelling-apparatus	J. Granger	Zanesville, Ohio	Sept. 29, 1868	82, 516
Propelling-apparatus	R. Hunter	New York, N. Y	Dec. 14, 1869	97, 924
Propelling-apparatus	R. Kearney	Havre de Grace, Newfoundland.	Sept. 13, 1870	107, 384
Propelling-apparatus	E. S. Ligon	Demopolis, Ala	July 5, 1870	105, 098
Propelling-apparatus	D. S. Merrit	Mount Morris, Mich	Apr. 20, 1869	89, 231
Propelling-apparatus	L. Moody	Malden, Mass	June 22, 1869	91, 760
Propelling-apparatus	H. Mulford	Philadelphia, Pa	July 20, 1869	92, 741
Propelling-apparatus	E. Offhaus	Newark, N. J	July 28, 1868	80, 497
Propelling-apparatus	J. Paradis	Brooklyn, N. Y	Aug. 3, 1869	93, 337
Propelling-apparatus	F. R. Pike	New York, N. Y	Aug. 17, 1869	93, 902
Propelling-apparatus	P. Roberts	New York, N. Y	Apr. 27, 1869	89, 435
Propelling-apparatus	C. Schilling	New York, N. Y	Aug. 16, 1870	106, 512
Propelling-apparatus	C. T. Smith	Nyack, N. Y	Mar. 2, 1869	87, 375
Propelling-apparatus	C. Wolf	Washington, D. C	Aug. 3, 1869	93, 384
Propelling-apparatus	A. Zink	Lancaster, Ohio	June 8, 1869	91, 196
Propelling-apparatus for land-conveyance	J. Wolf	Burr Oak, Mich	June 20, 1871	116, 130
Propelling-apparatus, Marine	E. Campbell	Boston, Mass	Sept. 29, 1857	18, 314
Propelling boats	J. S. Anderson	Flintville, Wis	Mar. 25, 1873	137, 164
Propelling boats	A. Bartholomew	Albany, N. Y	Dec. 1, 1818	....
Propelling boats	T. Beach	Wilmington, Ohio	Apr. 2, 1830	....
Propelling boats	A. Beekman	New York, N. Y	Sept. 30, 1873	143, 270
Propelling boats	J. Black and D. W. Jones	Philadelphia, Pa	Nov. 18, 1873	144, 732
Propelling boats	W. Burk	Whitemarsh, Pa	Dec. 2, 1834	....
Propelling boats	C. Dancker	Hoboken, N. J	Feb. 6, 1872	123, 383
Propelling boats, &c	T. Davis	Lawrence, Ind	Mar. 15, 1828	....
Propelling boats	J. P. Durand	New York	Jan. 17, 1818	....
Propelling boats	J. J. Giraud	Baltimore, Md	Apr. 10, 1818	....
Propelling boats	J. Gray	Peekskill, N. Y	Nov. 18, 1818	....
Propelling boats	P. Heaton, M. B. Boyes, and P. Heaton, jr.	Plymouth, Mass	Mar. 4, 1808	....

Index of patents issued from the United States Patent Office from 1790 *to* 1873, *inclusive*—Continued.

Invention.	Inventor.	Residence.	Date.	No.
Propelling boats	J. Heavin	Montgomery County, Va	Apr. 17, 1806	......
Propelling boats	J. Heroux	Saint Paul, Minn	May 4, 1869	89, 663
Propelling boats	G. Hotchkiss	Windsor, N. Y	June 2, 1836	......
Propelling boats	D. Keller	......	May 25, 1795	......
Propelling boats	J. Lyon	Statesburgh	Apr. 3, 1819	......
Propelling boats	M. Mallory	Urbanna, N. Y	Aug. 4, 1841	2, 210
Propelling boats	G. Manny	Montgomery, N. Y	June 8, 1815	......
Propelling boats	P. Noble	Westfield, Mass	Jan. 20, 1836	......
Propelling boats	J. Ring	Ogden, N. Y	Oct. 25, 1832	......
Propelling boats	D. J. Ross	Havre de Grace, Md	Dec. 28, 1869	98, 302
Propelling boats, &c	P. Sear	Boston, Mass	Dec. 30, 1841	2, 410
Propelling boats	J. Shackford	......	Mar. 21, 1799	......
Propelling boats	I. Silliman	Philadelphia, Pa	Dec. 17, 1812	......
Propelling boats	R. Steel	Philadelphia, Pa	June 25, 1872	128, 258
Propelling boats	J. H. Street	Philadelphia, Pa	Dec. 27, 1843	3, 393
Propelling boats	A. Taylor	New York	Nov. 18, 1818	......
Propelling boats	I. Theal	New York	Dec. 24, 1834	......
Propelling boats, &c	P. C. Traver	Rhinebeck, N. Y	Apr. 30, 1842	2, 598
Propelling boats	E. Williams	Marianna, Fla	Apr. 30, 1867	64, 392
Propelling boats	S. W. Wood	Washington, D. C	Apr. 29, 1856	14, 786
Propelling boats across ferries	S. Ross	Troy, N. Y	Oct. 30, 1828	......
Propelling boats and extinguishing fires, Machinery for.	S. Bates and G. Titcomb	Boston, Mass	Mar. 4, 1842	2, 477
Propelling boats and other vessels	G. Erkson	Pittsburgh, Pa	Mar. 26, 1845	3, 967
Propelling boats and vessels	A. Anderson	Philadelphia, Pa	Aug. 21, 1812	......
Propelling boats and vessels	J. J. Giraud	Baltimore, Md	Dec. 12, 1826	......
Propelling boats, Apparatus for	J. Repetti	Philadelphia, Pa	Feb. 7, 1871	111, 681
Propelling boats, &c., Apparatus for	H. Ronalds	Albion, Ill	Dec. 14, 1840	1, 899
Propelling boats, Apparatus for	S. Stillwell and D. T. Wandall	New York	June 3, 1808	......
Propelling boats, Apparatus for	M. Vergnes	New York, N. Y	Sept. 4, 1866	57, 798
Propelling boats by air, Apparatus for	J. Black	Orangeville, Pa	July 24, 1838	850
Propelling boats by condensed and compressed air	J. L. Sullivan	Boston, Mass	Mar. 24, 1817	......
Propelling boats by animal-power	M. Rogers	New York	Feb. 9, 1814	......
Propelling boats by cattle	H. Voight	......	Aug. 10, 1791	......
Propelling boats by endless chain of paddles	A. Bragg	New York	Jan. 24, 1842	2, 434
Propelling boats by horse-wheels	R. Browne and J. Murphy	New York	Nov. 18, 1818	......
Propelling boats, &c., by horses	W. P. Sprague	......	June 19, 1795	......
Propelling boats by horses, Machine for	M. Crafts	Troy, N. Y	Feb. 14, 1809	......
Propelling boats by horses, mules, &c	R. Bowne	New York	July 18, 1816	......
Propelling boats by levers and weights	J. T. Schenck	New York	Oct. 23, 1813	......
Propelling boats, &c., by man-power	A. Renoir	New Orleans, La	July 20, 1831	......
Propelling boats by means of jets of water	T. W. Reiley	McMinn County, Tenn	Jan. 24, 1842	2, 430
Propelling boats by oars	T. G. Pringle	New York, N. Y	June 20, 1871	116, 095
Propelling boats, &c., by oars, &c., Machine for	J. A. Pearce	Louisville, Ohio	Apr. 7, 1807	......
Propelling boats by power of weight	M. Martin	Woodbridge, N. J	July 8, 1816	......
Propelling boats by screws, &c., in still water	J. L. Smith	Charleston, S. C	Sept. 18, 1835	......
Propelling boats by steam	J. Evelett	Boston, Mass	Apr. 23, 1812	......
Propelling boats, &c., by steam, &c	J. Fitch	......	Aug. 26, 1791	......
Propelling boats by steam	J. Rodgers	Albany, N. Y	Sept. 27, 1811	......
Propelling boats, &c., by steam	N. J. Roosevelt	New Jersey	Dec. 1, 1814	......
Propelling boats, &c., by steam	W. Fatham	Washington, D. C	Apr. 18, 1816	......
Propelling boats by steam, Apparatus for	R. Letton	Albany, N. Y	Aug. 29, 1810	......
Propelling boats by steam, Compound wheel for	P. Antrim and S. Bolton	Philadelphia, Pa	Nov. 1, 1810	......
Propelling boats, &c., by steam-engines	J. Starr	......	Apr. 28, 1797	......
Propelling boats by step-wheels	F. Waterman, 3d	Washington, D. C	May 9, 1823	......
Propelling boats by wheels and by animal-power	A. Olney	Canton, Mass	Dec. 11, 1817	......
Propelling boats, Chain-paddle for	A. Pierson	Cumberland County, N. J	Mar. 2, 1811	......
Propelling boats, Elastic floor for	O. Phelps	Lansing, N. Y	Nov. 17, 1821	......
Propelling boats, Elbow-wheel for	R. Bulkley	New York	Apr. 24, 1824	......
Propelling boats for inland navigation	W. Bell	......	Nov. 25, 1803	......
Propelling boats in shallow water, Method of	J. Dougherty	Mount Union, Pa	Aug. 20, 1850	7, 574
Propelling boats, &c., Lever-purchase machine for	S. Kimball	Rhinebeck, N. Y	Apr. 24, 1811	......
Propelling boats, Machine for	T. Kearsing	New York	June 6, 1818	......
Propelling boats, Machine for	J. Martin	......	Nov. 21, 1797	......
Propelling boats, Machine for	W. Wadsworth	Hartford, Conn	Mar. 15, 1808	......
Propelling boats, Machinery for	J. Chittenden and A. W. Delano	Stevenson, N. Y	May 1, 1823	......
Propelling boats, Machinery for	C. Craft	Woodbury, Conn	Mar. 29, 1826	......
Propelling boats, Machinery for	J. J. Giraud	Baltimore, Md	Jan. 3, 1821	......
Propelling boats, Machinery for	E. Savage	Albany, N. Y	May 21, 1872	126, 906
Propelling boats, Machinery for	H. R. Teller	Schenectady, N. Y	Sept. 28, 1825	......
Propelling boats, Mechanism for	P. Rippingham	Virginia City, Mo	July 16, 1872	129, 366
Propelling boats on canals	W. F. Goodwin	Metuchen, N. J	Jan. 11, 1870	98, 760
Propelling boats on canals and in shoal water	L. Burnell	Elyria, N. Y	May 21, 1834	......
Propelling boats on canals by steam	M. Battel	Albany, N. Y	Oct. 14, 1840	1, 820
Propelling boats on canals or rivers	J. Finlay	Baltimore, Md	Apr. 17, 1837	165
Propelling boats or vessels	W. J. Lewis	Virginia	Feb. 3, 1819	......
Propelling boats or vessels	J. Rumsay	......	Aug. 26, 1791	......
Propelling boats or vessels by horses	S. Fuller	Clark County, Ind	Feb. 19, 1811	......
Propelling boats or vessels by steam-engines	R. Fulton	New York	Feb. 9, 1811	......
Propelling boats, &c., Polygonal wheel and chain-paddle for.	T. Elmer and A. Pierson	Bridgetown, N. J	Jan. 31, 1822	......
Propelling boats, Revolving lever for	L. Merchand	Blakely, Ala	July 10, 1822	......
Propelling boats, Screw for	E. P. Fitzpatrick	Mount Morris, N. Y	Nov. 23, 1835	......
Propelling boats, Screw-power applied to	H. William	Wheatley, N. Y	Dec. 30, 1818	......
Propelling boats, Screw-wheel for	J. Widdifield	Philadelphia, Pa	Oct. 11, 1815	......
Propelling boats up rapids, Machinery for	L. Thwing	Dedham, Mass	Aug. 12, 1820	......
Propelling boats, Vertical bucket in	S. Costill	Philadelphia, Pa	Oct. 17, 1827	......
Propelling boats, vessels, &c	D. Fitzgerald	New York	Oct. 9, 1841	2, 292
Propelling boats, Water-wheel for	P. Palmiter, jr	Jamestown, N. Y	Apr. 14, 1831	......
Propelling boats, Water-wheel for	B. Raymond, jr	Beverly, Mass	Nov. 13, 1826	......
Propelling boats, Water-wheel for	A. Waters	Millbury, Mass	Apr. 30, 1831	......
Propelling boats, Wheels of elastic oars for	L. M. Latour	......	Feb. 16, 1813	......
Propelling boats, working mills, &c	W. Shaw	Otsego County, N. Y	June 5, 1812	......
Propelling canal and other boats	H. R. Worthington	New York, N. Y	Feb. 2, 1844	3, 424
Propelling canal and other vessels	J. Jochum	Brooklyn, N. Y	Aug. 22, 1871	118, 244
Propelling canal-boats	A. Ames	Natchez, Miss	Jan. 7, 1873	134, 504
Propelling canal-boats	H. H. Baker	New Market, N. J	June 17, 1873	139, 996
Propelling canal-boats	G. A. Beidler	Philadelphia, Pa	Jan. 2, 1872	122, 429

Index of patents issued from the United States Patent Office from 1790 *to* 1873, *inclusive*—Continued.

Invention.	Inventor.	Residence.	Date.	No.
Propelling canal-boats	N. Bingham and H. Warner	Rochester, N. Y	Dec. 15, 1828	....
Propelling canal-boats	H. Camp	Dunkirk, N. Y	Feb. 16, 1858	19, 346
Propelling canal-boats	J. Echols	Columbus, Ga	Nov. 26, 1845	4, 293
Propelling canal-boats	W. F. Goodwin	Metuchen, N. J	Mar. 22, 1870	101, 004
Propelling canal-boats	J. Hough	Buckingham, Pa	Oct. 15, 1872	132, 288
Propelling canal-boats	C. H. Jenner	Rockport, N. Y	May 28, 1872	127, 401
Propelling canal-boats	A. Johnson	Brooklyn, N. Y	Feb. 18, 1873	136, 064
Propelling canal-boats	W. Leavenworth	New York, N. Y	Apr. 18, 1839	1, 126
Propelling canal-boats	H. B. Meech	Fort Edwards, N. Y	Mar. 7, 1871	112, 369
Propelling canal-boats	W. F. Miller	East Walpole, Mass	June 10, 1873	139, 727
Propelling canal-boats	W. P. More	Moresville, N. Y	Apr. 2, 1872	125, 207
Propelling canal-boats	J. Reid	Catskill, N. Y	Nov. 15, 1870	109, 248
Propelling canal-boats	P. Rippingham	Virginia City, Nev	Dec. 10, 1872	133, 891
Propelling canal-boats	J. Smith and J. Brown	New York	Nov. 19, 1833	....
Propelling canal-boats	O. C. Smith	Ipswich, Mass	Sept. 23, 1873	143, 038
Propelling canal-boats	G. Stackpole	New York, N. Y	Mar. 26, 1867	63, 184
Propelling canal-boats	J. A. Sterling	New York, N. Y	Nov. 26, 1872	133, 495
Propelling canal-boats	J. T. Teasdill	Lockport, N. Y	July 16, 1872	129, 192
Propelling canal-boats	W. W. Virdin	Baltimore, Md	Apr. 18, 1871	113, 951
Propelling canal-boats	E. K. Watson	Shokan, N. Y	Sept. 20, 1870	107, 573
Propelling canal-boats and other vessels	L. Bastel	New York, N. Y	Nov. 4, 1873	144, 248
Propelling canal-boats by lock-paddle wheels	H. L. B. Lewis	Buffalo, N. Y	Nov. 2, 1830	....
Propelling canal-boats by steam	F. A. Dorman	Norfolk, Va	Mar. 7, 1833	....
Propelling canal-boats, Device for	J. B. Calnan	New Haven, Conn	Aug. 23, 1870	106, 547
Propelling canal-boats, Device for	T. K. McDowell	San Francisco, Cal	Apr. 2, 1872	125, 205
Propelling canal-boats, Machinery for	D. W. Horton	Petersburgh, Ind	July 16, 1872	129, 228
Propelling canal-boats, Mode of applying wheels for	T. Jackson	Reading, Pa	Jan. 18, 1839	1, 070
Propelling canal-boats, Water-wheel for	H. Averill	Richland, N. Y	Oct. 1, 1830	....
Propelling engine, Self	N. S. Bean	Manchester, N. H	Mar. 10, 1868	75, 348
Propelling-hoop	J. Faye	Philadelphia, Pa	July 16, 1867	66, 697
Propelling horse-boats	B. Langdon	Troy, N. Y	May 1, 1821	....
Propelling-mechanism	C. Sharps	Philadelphia, Pa	Nov. 22, 1870	109, 458
Propelling pleasure-boats	J. O. Belknap	Mobile, Ala	May 19, 1866	78, 043
Propelling-power	N. B. Baldwin	Chicago, Ill	Nov. 14, 1871	120, 846
Propelling ships	J. Ericsson	New York	Dec. 31, 1844	3, 869
Propelling ships, &c	D. Rudd	Bozrah, Conn	Mar. 23, 1842	2, 509
Propelling ships, Apparatus for	A. E. Harding	Middletown, Ohio	Apr. 5, 1859	23, 463
Propelling ships, boats, &c	E. F. Aldrich	New York	July 30, 1841	2, 204
Propelling ships, &c., Machinery for	A. Hermange	Baltimore, Md	May 31, 1828	....
Propelling ships, Machinery for	A. Hermange	Baltimore, Md	Nov. 26, 1828	....
Propelling ships, Machinery for	A. Hermange	Baltimore, Md	Nov. 26, 1828	....
Propelling small boats	V. Dresser	Leavenworth, Kans	Nov. 2, 1869	96, 404
Propelling steam and other boats	W. Scarborough	Savannah, Ga	Apr. 8, 1835	......
Propelling steam and other vessels	G. H. Mo eau	France	Jan. 26, 1844	3, 414
Propelling steam and other vessels	F. P. Smith	London, England	Nov. 12, 1841	2, 353
Propelling steam and other vessels, Operating-paddle for.	S. Swett, jr	Chelsea, Mass	May 11, 1841	2, 089
Propelling steamboats, &c	T. Cook	New York	May 12, 1842	2, 623
Propelling steamboats	J. Hunt	East Chester, N. Y	Mar. 11, 1812	....
Propelling steamboats, &c	J. McCurdy	Norwich, Conn	Sept. 18, 1833	....
Propelling steamboats, &c	J. Sheffield and J. Ingraham	Erie County, N. Y	May 17, 1832	....
Propelling steamboats and other vessels	A. Planton	Philadelphia, Pa	Apr. 23, 1834	....
Propelling steam-vessels	J. Ericsson	Sweden	Feb. 1, 1838	588
Propelling vessels, Constructing and	W. W. Hunter and B. Harris	United States Navy and Norfolk, Va.	Mar. 12, 1841	2, 004
Propelling vessels	E. F. Aldrich	New York	June 13, 1848	5, 634
Propelling vessels	L. Alexander	New York, N. Y	Mar. 25, 1856	14, 487
Propelling vessels	J. L. Arman	Bordeaux, France	Oct. 11, 1870	108, 230
Propelling vessels	B. T. Babbitt	New York, N. Y	July 14, 1868	79, 937
Propelling vessels	C. W. Cahoon	Portland, Me	Dec. 18, 1866	60, 471
Propelling vessels	H. C. Camp	Saint Paul, Minn	Jan. 14, 1873	134, 730
Propelling vessels	H. Case	Norwalk, Ohio	Aug. 13, 1828	....
Propelling vessels	F. P. Dimpfel	Philadelphia, Pa	Oct. 25, 1853	10, 149
Propelling vessels	G. W. Dow	Brooklyn, N. Y	July 22, 1873	141, 039
Propelling vessels	J. Echols	Columbus, Ga	May 9, 1846	4, 503
Propelling vessels	J. W. Evans	New York, N. Y	July 29, 1873	141, 340
Propelling vessels	E. Fuller	Providence, R. I	Mar. 2, 1827	... .
Propelling vessels	G. W. Fulton	Brazoria, Tex	Aug. 26, 1846	4, 716
Propelling vessels	S. Gerhard	Camden, N. J	Apr. 1, 1873	137, 359
Propelling vessels	J. Ghisi	Genoa, Italy	May 17, 1870	103, 040
Propelling vessels	J. J. Giraud	Baltimore, Md	Dec. 1, 1829	....
Propelling vessels	G. A. Milani	Frankfort, Ind	Oct. 19, 1869	95, 922
Propelling vessels	T. L. Mitchell	Birkenhead, England	July 5, 1853	9, 831
Propelling vessels	J. O'Neil	New York, N. Y	Nov. 11, 1873	144, 474
Propelling vessels	F. Peltier	New York	Oct. 1, 1830	....
Propelling vessels	C. A. Read	Bridgeport, Conn	July 2, 1872	128, 657
Propelling vessels	W. Shepard	Saint Louis, Mo	Aug. 13, 1872	130, 391
Propelling vessels	S. D. Sillman	Jersey City, N. J	June 21, 1870	104, 512
Propelling vessels	R. R. Spedden and D. F. Stafford	Astoria, Oreg	Dec. 10, 1867	72, 106
Propelling vessels	R. R. Stevens	Mokelumne Hill, Cal	Oct. 22, 1867	70, 130
Propelling vessels	O. Tenny	Littleton, Mass	July 21, 1868	80, 239
Propelling vessels	G. Titcomb	Chelsea, Mass	Aug. 13, 1872	130, 455
Propelling vessels	M. K. Wildman	Brooklyn, N. Y	Dec. 26, 1871	122, 301
Propelling vessels	B. Wynkoop	....	Sept. 13, 1794	.. ....
Propelling vessels, Apparatus for	M. A. Crooker	New York, N. Y	June 29, 1852	9, 066
Propelling vessels, Apparatus for	A. D. Owen and J. D. Sherman	Thorntown, Ind., and Paw Paw, Mich.	Nov. 16, 1869	96, 931
Propelling vessels, Apparatus for	B. Phillips	Philadelphia, Pa	Aug. 16, 1828	....
Propelling vessels, Apparatus for	D. Winer	Lockport, N. Y	Aug. 7, 1866	57, 033
Propelling vessels at sea, Machine for	E. Bryan	New York	Dec. 22, 1827	....
Propelling vessels by screw or spiral levers	J. J. Giraud	Baltimore, Md	Apr. 29, 1831	....
Propelling vessels, &c., by animal-power	J. P. Durand	New York	Oct. 3, 1817	....
Propelling vessels by means of continuous streams of water.	R. Schuyler	New York	Mar. 17, 1843	3, 003
Propelling vessels, &c., by pendulum-power	J. Cerneau	New York	Aug. 28, 1817	......
Propelling vessels by reaction	M. W. Ruthven	New York, N. Y	May 22, 1849	6, 468
Propelling vessels by the buoyancy of air in water acting against inclined surfaces.	S. W. Hall	Troy, N. Y	Nov. 6, 1843	3, 332

Index of patents issued from the United States Patent Office from 1790 *to* 1873, *inclusive*—Continued.

Invention.	Inventor.	Residence.	Date.	No.
Propelling vessels by the direct action of steam on the water.	W. J. McIntire	New York, N. Y	Aug. 7, 1855	13, 394
Propelling vessels, Cylinder for	B. Phillips	Philadelphia, Pa	Jan. 23, 1830	....
Propelling vessels, Device for	B. T. Babbitt	New York, N. Y	June 17, 1873	139, 994
Propelling vessels, Device for	S. P. Harbaugh	Allegheny County, Md	July 30, 1872	130, 042
Propelling vessels, Device for	G. R. Moore	Philadelphia, Pa	July 16, 1872	129, 488
Propelling vessels, Endless-chain paddle for	B. Douglas	Albany, N. Y	Sept. 28, 1843	3, 285
Propelling vessels in a calm	S. Rudd	New York	June 14, 1816	....
Propelling vessels in shoal water	J. W. Wetmore	Erie, Pa	June 16, 1857	17, 598
Propelling vessels, &c., Machine for	A. Babcock	....	Dec. 2, 1793	....
Propelling vessels, Machine for	J. J. Giraud	Baltimore, Md	Nov. 22, 1820	....
Propelling vessels, Machine for	C. Stoudinger	....	June 2, 1798	....
Propelling vessels, Machinery for	S. J. Gold	Cornwall, Conn	July 16, 1867	66, 830
Propelling vessels, Mechanism for	L. Murdoch	Marble Hill, Mo	Sept. 23, 1873	143, 091
Propelling vessels, Mechanism for	E. Raynale	Birmingham, Mich	Aug. 15, 1871	118, 151
Propelling vessels, Mechanism for	J. Weich	New York, N. Y	Mar. 4, 1873	136, 475
Propelling vessels on canals	B. O. Mesnil and M. Eyth	Brussels, Belgium, and Stutgart, Wurtemberg.	Feb. 9, 1869	86, 737
Propelling vessels, Sluice for	M. W. Ruthven	Middlesex, England	Apr. 10, 1866	53, 937
Propelling-wheel	C. Comstock and C. Glidden	Milwaukee, Wis	Apr. 30, 1861	32, 182
Propelling-wheel	J. L. Husband	Philadelphia, Pa	Nov. 22, 1859	26, 187
Propelling-wheel	A. C. Loud	San Francisco, Cal	May 3, 1870	102, 747
Propelling-wheel	C. Seymour	La Porte, Ind	Dec. 22, 1868	85, 248
Propelling-wheel	A. A. Wilder	Warsaw, N. Y	Mar. 8, 1836	....
Propelling-wheel for boats	W. Kelly	Pittsburgh, Pa	July 12, 1834	....
Propelling-wheel for boats	W. Kelly	Pittsburgh, Pa	Nov. 17, 1834	....
Propelling-wheel for boats, &c	J. Prince	New York, N. Y	Oct. 2, 1820	....
Propelling-wheel for river and canal boats	H. Fenton	Cleveland, Ohio	July 4, 1865	48, 541
Propelling-wheel for steamboats	J. J. Giraud	Baltimore, Md	Sept. 15, 1826	....
Propelling-wheel for steamships, also employed as wind or water wheel for mills.	J. Hobday and W. I. Cocke	Portsmouth, Va	Mar. 29, 1841	2, 023
Propelling-wheel, Screw	B. M. Smith	Rochester, N. Y	Nov. 20, 1829	....
Propelling-wheel, Submerged	T. Kendall, sr	San Francisco, Cal	Apr. 28, 1857	17, 156
Propulsion, Apparatus for marine	F. W. Harris	Montreal, Canada	Aug. 4, 1863	39, 394
Propulsion by gas-explosion	W. A. Leggo	Montreal, Canada	May 6, 1873	138, 665
Propulsion, Canal	J. L. Nicolai	Chicago Ill	Feb. 6, 1872	123, 504
Propulsion, Device for canal	G. Stackpole	New York, N. Y	Mar. 6, 1866	53, 088
Propulsion for tug-boats, Screw	E. D. Gird	Syracuse, N. Y	Apr. 19, 1870	102, 111
Propulsion, Marine	L. B. Flanders	Cleveland, Ohio	Sept. 18, 1860	30, 057
Propulsion, Marine	B. S. Heath	Philadelphia, Pa	Aug. 15, 1871	118, 127
Propulsion, Marine	A. J. Reynolds	Sturgis, Michigan	Mar. 17, 1868	75, 648
Propulsion, Means of	S. R. Foster	Saint John, Canada	Oct. 29, 1872	132, 568
Propulsion of boats	L. H. Watson	Pittsburgh, Pa	Apr. 30, 1872	126, 168
Propulsion of canal-boats	H. H. Baker	New Market, N. J	Oct. 1, 1872	131, 839
Propulsion of canal-boats	W. C. Bibb	Madison, Ga	Aug. 22, 1871	118, 329
Propulsion of canal-boats	G. W. Bishop	Brooklyn, N. Y	Sept. 24, 1872	131, 652
Propulsion of canal-boats	J. T. Burke	Virginia, Ill	Dec. 12, 1871	121, 749
Propulsion of canal-boats	O. Coogan	Pittsfield, Mass	Oct. 10, 1871	119, 744
Propulsion of canal boats	M. N. Cumminskey	Paterson, N. J	Aug. 8, 1871	117, 869
Propulsion of canal-boats	W. T. Duvall	Georgetown, D. C	June 27, 1871	116, 281
Propulsion of canal-boats	N. T. Edson	New Orleans, La	Oct. 31, 1871	120, 371
Propulsion of canal-boats	P. H. Fontaine	Reidsville, N. C	Oct. 17, 1871	119, 923
Propulsion of canal-boats	H. Fowler	Washington, D. C	Dec. 12, 1871	121, 712
Propulsion of canal-boats	H. Fowler	Washington, D. C	Dec. 12, 1871	121, 713
Propulsion of canal-boats	H. Fowler	Washington, D. C	Dec. 12, 1871	121, 714
Propulsion of canal-boats	I. J. Hilgerd	New York, N. Y	May 14, 1872	126, 701
Propulsion of canal-boats	J. L. Hornig	Chicago, Ill	Dec. 27, 1870	110, 466
Propulsion of canal-boats	J. L. Hornig	Chicago, Ill	May 9, 1871	114, 561
Propulsion of canal-boats	F. M. Mahan	Memphis, Tenn	July 25, 1871	117, 437
Propulsion of canal-boats	J. M. McMaster	Rochester, N. Y	Nov. 14, 1871	120, 823
Propulsion of canal-boats	T. Main	Green Point, N. Y	Jan. 31, 1871	111, 462
Propulsion of canal-boats	M. P. Muller, A. Schwaab, and F. Krenson.	Savannah, Ga	June 11, 1872	127, 911
Propulsion of canal-boats	W. E. Prall and J. D. Defrees	Washington, D. C	Dec. 5, 1871	121, 469
Propulsion of canal-boats	C. Schilling	New York, N. Y	Oct. 10, 1871	119, 792
Propulsion of canal-boats	T. Smith	Green Island, N. Y	June 4, 1872	127, 437
Propulsion of canal-boats	H. W. Trackman	Philadelphia, Pa	Mar. 5, 1872	124, 348
Propulsion of navigable vessels	C. C. Brand and C. C. Moore	Norwich, Conn	July 9, 1872	128, 846
Propulsion of steamboats	A. Fellows	Maquoketa, Iowa	Mar. 28, 1865	47, 003
Propulsion of vessels	G. A. Beidler	Philadelphia, Pa	May 9, 1871	114, 637
Propulsion of vessels	R. Bayman	London, England	May 30, 1871	115, 425
Propulsion of vessels	J. P. Bruce	Brooklyn, N. Y	Oct. 31, 1871	120, 414
Propulsion of vessels	P. Burke	Boston, Mass	May 16, 1871	114, 914
Propulsion of vessels	S. Chase	Portland, Me	Dec. 10, 1872	133, 758
Propulsion of vessels	K. P. Cool	Glen's Falls, N. Y	July 18, 1871	117, 051
Propulsion of vessels	G. B. De Boucherville	Quebec, Canada	Oct. 7, 1873	143, 440
Propulsion of vessels	S. F. Gard	New Orleans, La	Nov. 7, 1871	120, 735
Propulsion of vessels	J. S. Godfrey	Leslie, Mich	Oct. 24, 1871	120, 264
Propulsion of vessels	W. M. Harding	Chelsea, Mass	Aug. 22, 1871	118, 363
Propulsion of vessels	E. Harriott	Plainfield, N. J	Aug. 29, 1871	118, 607
Propulsion of vessels	R. W. Heywood	Baltimore, Md	May 23, 1871	115, 204
Propulsion of vessels	H. Jackson	Brooklyn, N. Y	Sept. 26, 1871	119, 363
Propulsion of vessels	C. P. Macowitzky	Corpus Christi, Tex	Dec. 30, 1873	146, 009
Propulsion of vessels	E. Matteson	Norwich, Conn	Jan. 9, 1872	122, 624
Propulsion of vessels	C. Mickle	Guelph, Canada	Dec. 19, 1871	122, 046
Propulsion of vessels	H. P. Miller	Paducah, Ky	Oct. 1, 1872	131, 770
Propulsion of vessels	H. Niles	Chicago, Ill	Jan. 9, 1872	122, 640
Propulsion of vessels	H. H. Olds	New Haven, Conn	Dec. 12, 1854	12, 067
Propulsion of vessels	T. B. Raymond	Winona, Mich	Jan. 9, 1872	122, 650
Propulsion of vessels	W. Richards	Rochester, N. Y	Sept. 3, 1872	131, 118
Propulsion of vessels	J. S. Stites	Baltimore, Md	June 20, 1871	116, 236
Propulsion of vessels	H. Waterman	Brooklyn, N. Y	Sept. 26, 1871	119 255
Propulsion of vessels	W. Wells	Salem, Mass	Nov. 19, 1872	133, 275
Propulsion of vessels	E. Whitehead	Cincinnati, Ohio	Dec. 12, 1871	121, 738
Propulsion of vessels	A. F. Yardell	San Francisco, Cal	Dec. 1, 1868	84, 602
Propulsion of vessels, Apparatus for	J. Briggs	Louisville, Ky	Oct. 27, 1863	40, 393
Propulsion of vessels, Apparatus for water	J. Watt	Buffalo, N. Y	May 12, 1863	38, 522
Propulsion of vessels by direct action of steam	G. W. Jones	Philadelphia, Pa	Dec. 2, 1873	145, 210

Index of patents issued from the United States Patent Office from 1790 *to* 1873, *inclusive*—Continued.

Invention.	Inventor.	Residence.	Date.	No.
Propulsion of vessels in shoal water, Wheel	O. Olds	Aurora, N. Y	July 11, 1865	48, 715
Propulsion of vessels, Speed in	J. F. Reigart	Washington, D. C	June 30, 1863	39, 067
Propulsion, Screw	E. C. Gregg	Trumansburgh, N. Y	Mar. 7, 1871	112, 335
Propulsion, Wave-power	C. W. Cahoon	Portland, Me	Apr. 24, 1866	54, 107
Propulsion, Wheel for marine	B. Reed	Allegheny City, Pa	Aug. 4, 1863	39, 423
Protector: *See* Bee hive protector. Body-protector. Book-cover protector. Bottle-protector. Breast-protector. Breast-strap protector. Broom-protector. Car-window protector. Carriage-protector. Carriage-top protector. Check-protector. Chest-protector. Collar-edge protector. Corner-protector. Diamond-protector. Dress-protector. Ear-protector. Eave-protector. Eye-protector. Eye and lung protector. Fly-protector. Frost-protector. Furniture-protector. Grape-vine protector. Gun-nipple protector. Hat-protector. Hatchway-protector. Hay and grain protector. Hay-cock protector. Heel-protector. Hoisting-protector. Horse-head protector. Horse-neck protector. Horse-tail protector. Hose-protector. Joist-protector. Key-hole protector. King-bolt protector. Lamp chimney protector. Lamp-shade protector. Omnibus-step protector. Pencil-point protector. Plant-protector. Rope-protector. Safe-protector. Sap-protector. Shaft-protector. Skirt-protector. Spool-protector. Stocking-heel protector. Sun-stroke protector. Tree-protector. Wall-protector.				
Protracting-table	W. T. Steiger	Washington, D. C	Sept. 7, 1838	907
Protractor	C. Gordon	Washington, D. C	Mar. 29, 1859	23, 365
Protractor	J. Leyman	Lenox, Mass	May 25, 1858	20, 356
Protractor and parallel ruler	W. L. Apthorp	Tallahassee, Fla	May 28, 1872	127, 210
Protractor, Compound	W. Rutherford	Athens, Ga	Oct. 15, 1867	69, 844
Protractor, Facilitating	E. Brown	Seneca County, Ohio	Aug. 1, 1833	....
Protractor, Rectangular	N. Goodwin	Hartford, Conn	Dec. 24, 1829	....
Protractor, Universal	E. Williams	Philadelphia, Pa	Jan. 19, 1810	....
Provision-cooler	J. W. Bartlett	New York, N. Y	Apr. 14, 1863	38, 142
Provision-cutter	W. Smith	Cincinnati, Ohio	June 8, 1858	20, 517
Provision-jar	W. Zettle	Cincinnati, Ohio	Aug. 5, 1862	36, 131
Provision-press	F. Potter	Rushville, Ill	June 4, 1872	127, 427
Provisions, Preparing	D. Lardner and J. Davison	New York, N. Y	Aug. 9, 1845	4, 144
Prow	J. Revercomb	Botetourt Springs, Va	June 27, 1854	11, 201
Prunes, Curing	I. Reckhow	Brooklyn, N. Y	May 22, 1860	28, 441
Pruning and hedge shears	E. K. Bigelow	Litchfield, Mich	May 10, 1870	102, 752
Pruning and hedge shears	L. Campbell	Marengo, Mich	Sept. 22, 1868	82, 290
Pruning-hatchet	J. White	Providence, Pa	Jan. 17, 1871	111, 024
Pruning-hook	J. Christy	Clyde, Ohio	Oct. 28, 1873	143, 962
Pruning-hook	W. M. Doty	Woodbridge, N. J	Jan. 4, 1870	98, 575
Pruning-hook	A. Downer	Hammondsville, Ohio	May 3, 1870	102, 513
Pruning-hook	M. Gates	Paw Paw, Mich	June 12, 1866	55, 483
Pruning-hook	L. M. Harris	Mattawan, Mich	Mar. 14, 1865	46, 797
Pruning hook	W. S. McLean	Williamsburgh, Ohio	Jan. 2, 1867	51, 899
Pruning-hook	B. M. Parks	Saint Louis, Mo	Oct. 6, 1868	82, 868
Pruning-hook	D. B. Seeley	Sterling, Ill	July 19, 1870	105, 601
Pruning-hook	J. Stark	Thomasville, Ga	July 7, 1868	79, 700
Pruning-hook	A. Travis	Peekskill, N. Y	Dec. 8, 1863	40, 869
Pruning-hook, Extension	J. Stark	Thomasville, Ga	Mar. 9, 1869	87, 719
Pruning-implement	A. P. Bettersworth	Carlinville, Ill	Oct. 14, 1873	143, 659
Pruning-implement	J. Fasig	Congress, Ohio	Mar. 13, 1860	27, 437
Pruning-implement	A. C. Hulse and J. S. Crum	Palmyra, Ill	Sept. 16, 1873	142, 915
Pruning-implement	J. Reynolds	Washington, D. C	Sept. 24, 1872	131, 703
Pruning-implement	G. F. Waters	Boston, Mass	May 25, 1869	90, 618
Pruning-implement	J. R. Woodworth	Nunda, N. Y	Oct. 26, 1869	96, 294
Pruning-implements, Combined	C. W. Dawson	Paynesville, Mo	Nov. 22, 1870	109, 498
Pruning-instrument	R. Collier	Springfield, Ohio	June 19, 1866	55, 623
Pruning-instrument	W. H. Callings	Town, Mo	Sept. 23, 1873	143, 064
Pruning-instrument	J. Evans and R. H. Seymour	Newark and Bloomfield, N. J.	Sept. 4, 1866	57, 817
Pruning-instrument	J. Schroy	Fortville, Ind	Jan. 12, 1869	85, 861
Pruning-instrument	G. F. Waters	Waterville, Me	Sept. 25, 1866	58, 326
Pruning-knife	H. Alter	Lakeport, Cal	Feb. 1, 1870	99, 390
Pruning-knife	G. G. Belcher	Worcester, Mass	May 10, 1859	23, 975
Pruning-knife	J. S. Crum	Palmyra, Ill	May 20, 1873	139, 119
Pruning-knife	J. Fasig	West Salem, Ohio	May 28, 1872	127, 229
Pruning-knife	F. P. Goodall	Deering, N. H	Oct. 18, 1859	25, 818
Pruning-knife	M. Grover	Clyde, Ohio	June 7, 1870	103, 879
Pruning-knife	A. C. Hulse	Palmyra, Ill	Aug. 26, 1873	142, 236
Pruning-knife	C. McCrea	Leavenworth, Kans	Oct. 9, 1866	58, 659
Pruning-knife	D. Morris	Bartlett, Ohio	Jan. 3, 1872	122, 633
Pruning-knife	L. B. Morris	Hopkinsville, Ky	Dec. 4, 1866	60, 225
Pruning-knife	W. Richard	Clyde, Ohio	June 29, 1869	92, 100
Pruning-knife	W. Richard	Clyde, Ohio	Sept. 20, 1870	107, 542
Pruning-knife	J. Surerus	Newark, N. J	Jan. 30, 1866	52, 337
Pruning-knife	S. O. and P. W. T. Vaughn	De Kalb, Ill	Dec. 12, 1865	51, 497

Index of patents issued from the United States Patent Office from 1790 *to* 1873, *inclusive*—Continued.

Invention.	Inventor.	Residence.	Date.	No.
Pruning knife, hook, and saw	R. W. Porter	Nashua, N. H	Aug. 27, 1867	68, 109
Pruning-saw	P. Eberle	Freetown, Ind	June 13, 1871	115, 834
Pruning-saw	A. Travis	Peekskill, N. Y	July 16, 1872	129, 071
Pruning-shears	G. W. Anesley	Waungo Township, Mich	Aug. 1, 1871	117, 590
Pruning-shears	W. E. Ball	Bethesda, Ohio	Sept. 23, 1873	143, 114
Pruning-shears	W. Baker	Ilion, N. Y	Sept. 16, 1873	142, 886
Pruning-shears	A. Barling	West Chester, Pa	May 30, 1871	115, 415
Pruning-shears	A. Baumann	Philadelphia, Pa	May 29, 1866	55, 043
Pruning-shears	S. J. Beigh and E. F. Beard	Republic, Ohio	Jan. 16, 1872	122, 753
Pruning-shears	S. J. Beigh and E. F. Beard	Republic, Ohio	Oct. 1, 1872	131, 808
Pruning-shears	T. S. Bell	Wapello, Iowa	June 16, 1868	78, 918
Pruning-shears	G. Bergner	Washington, Mo	Apr. 5, 1870	101, 571
Pruning-shears	G. Bergner	Washington, Mo	Aug. 23, 1870	106, 538
Pruning-shears	G. Bergner	Washington, Mo	Nov. 28, 1871	121, 319
Pruning-shears	F. Binder	Baltimore, Md	Oct. 30, 1866	59, 168
Pruning-shears	H. W. Black	Cecilton, Md	July 26, 1870	105, 629
Pruning-shears	T. Borden	Squancum, N. J	Feb. 25, 1873	136, 209
Pruning-shears	G. Brewster	Otsego, Mich	Aug. 6, 1872	130, 274
Pruning-shears	P. Broadbooks	Batavia, N. Y	June 11, 1872	127, 735
Pruning-shears	P. Broadbooks	Batavia, N. Y	Apr. 15, 1873	137, 821
Pruning-shears	R. Bullard	Litchfield, Mich	May 7, 1872	126, 443
Pruning-shears	R. Bullard and G. E. Mills	Litchfield, Mich	Aug. 9, 1870	106, 253
Pruning-shears	J. Calder	Macedon, N. Y	Jan. 24, 1871	111, 106
Pruning-shears	D. Campbell	Elizabeth, N. J	Feb. 19, 1867	62, 180
Pruning-shears	D. Campbell	Elizabeth, N. J	Aug. 4, 1868	80, 595
Pruning-shears	T. Campbell	East Palestine, Ohio	Jan. 30, 1872	123, 234
Pruning-shears	S. P. Carpenter	Milford, Mass	Dec. 10, 1867	71, 975
Pruning-shears	A. B. Chapman	Clyde, Ohio	July 12, 1870	105, 173
Pruning-shears	O. Chase	Rutland, Ohio	Jan. 7, 1873	134, 591
Pruning-shears	W. F. and G. H. Clemmer	Alexandria, Ohio	Sept. 5, 1871	118, 690
Pruning-shears	G. H. Clinton and D. H. Harris	New Haven, Conn	Feb. 5, 1867	61, 806
Pruning-shears	J. Christy	Clyde, Ohio	June 6, 1871	115, 571
Pruning-shears	J. F. Creighton	Placerville, Cal	Oct. 3, 1871	119, 452
Pruning-shears	M. De Groodt	Eaton, N. Y	Mar. 18, 1873	136, 820
Pruning-shears	A. J. Doolittle	Hamden, Conn	Mar. 12, 1867	62, 737
Pruning-shears	J. Evans	San José, Cal	July 16, 1861	32, 827
Pruning-shears	J. Evans	Newark, N. J	July 18, 1865	48, 803
Pruning-shears	C. H. Eggleston	Marshall, Mich	Aug. 1, 1871	117, 525
Pruning-shears	E. D. Gaines and J. A. McMillen	Chicago, Ill	Feb. 14, 1871	111, 738
Pruning-shears	P. M. Gilbert	Camanche, Iowa	Nov. 19, 1872	133, 220
Pruning-shears	I. Grass	Sandusky, Ohio	Sept. 26, 1871	119, 221
Pruning-shears	M. Grover	Clyde, Ohio	Sept. 3, 1872	130, 990
Pruning-shears	R. Hall	Clyde, Ohio	Jan. 9, 1872	122, 603
Pruning-shears	A. L. Hatfield	Clyde, Ohio	May 24, 1870	103, 327
Pruning-shears	V. Hayes and L. Cuatt	Tekonsha and Eckford, Mich.	Oct. 26, 1869	96, 232
Pruning-shears	J. M. Heath	Allegan, Mich	Aug. 13, 1872	130, 495
Pruning-shears	H. C. Hills	Bickney, Mich	July 29, 1873	141, 273
Pruning-shears	W. E. Hughes	Aylmer, Canada	June 6, 1871	115, 735
Pruning-shears	H. H. Ingalsbe and G. A. Prescott.	South Hartford and Sandy Hill, N. Y.	Aug. 8, 1871	117, 888
Pruning-shears	S. Jocelin	New Haven, Conn	July 13, 1809	......
Pruning-shears	S. W. Jones	Bluffton, Ind	July 16, 1867	66, 850
Pruning-shears	P. Keck	Zanesville, Ohio	May 21, 1867	64, 882
Pruning-shears	D. Keethler	Mount Orab, Ohio	May 21, 1872	126, 888
Pruning-shears	C. A. Kellogg	Elyria, Ohio	May 17, 1870	103, 197
Pruning-shears	M. C. Malone	Palmyra, Ill	Apr. 29, 1873	138, 338
Pruning-shears	E. J. Marsters	Shaw's Flat, Cal	Aug. 23, 1870	106, 705
Pruning-shears	W. McCray	Black Oak, Mo	Apr. 29, 1873	138, 340
Pruning-shears	S. A. McFarlane	Grand Rapids, Mich	July 6, 1869	92, 202
Pruning-shears	J. W. Mix	Batavia, N. Y	Sept. 19, 1871	119, 041
Pruning-shears	W. C. Moreau and H. J. Waldo	Batavia and Elba, N. Y	Nov. 5, 1872	132, 850
Pruning-shears	A. R. Moulton	Fall's Branch, Tenn	Jan. 14, 1873	134, 922
Pruning-shears	J. Neff, jr	Pultney, N. Y	May 14, 1867	64, 692
Pruning-shears	W. Richard	Clyde, Ohio	Sept. 7, 1869	94, 514
Pruning-shears	J. G. Rogers	Dowagiac, Mich	July 12, 1870	105, 255
Pruning-shears	J. G. Rogers	Berrien County, Mich	July 1, 1873	140, 388
Pruning-shears	O. L. Samson and J. R. Dill	Crawfordsville, Iowa	Oct. 1, 1872	131, 830
Pruning-shears	D. B. Seely	Portland, Ill	Mar. 2, 1869	87, 436
Pruning-shears	D. B. Seely	Sterling, Ill	June 6, 1871	115, 773
Pruning-shears	J. H. Shehan	Lima, Ind	Dec. 14, 1869	97, 973
Pruning-shears	J. Sill	Montoursville, Pa	Dec. 31, 1867	72, 923
Pruning-shears	F. Smiley	Batavia, N. Y	Mar. 7, 1871	112, 392
Pruning-shears	F. Smiley	Batavia, N. Y	Aug. 29, 1871	118, 492
Pruning-shears	O. Snell	Williamsburgh, Ohio	Mar. 12, 1872	124, 455
Pruning-shears	G. D. Spielman	Lancaster, Ohio	May 28, 1872	127, 279
Pruning-shears	G. Vanfossen	Bethesda, Ohio	July 15, 1873	140, 974
Pruning-shears	W. A. Wallet and I. Harbaugh	West Salem, Ohio	June 13, 1871	115, 995
Pruning-shears	P. R. Walsh and G. C. Eaton	Rochester, N. Y	Jan. 18, 1870	99, 038
Pruning-shears	G. F. Waters	Waterville, Me	Nov. 22, 1864	45, 196
Pruning-shears	H. Wendt	Elizabeth, N. J	Jan. 11, 1870	98, 826
Pruning-shears	F. A. Will	San Francisco, Cal	Sept. 20, 1870	107, 577
Pruning-shears	E. L. Yancey	Batavia, N. Y	June 27, 1871	116, 387
Pruning-shears	E. L. Yancey	Batavia, N. Y	Jan. 30, 1872	123, 321
Pruning-shears	E. L. Yancey	Batavia, N. Y	Dec. 24, 1872	134, 185
Pruning-shears and fruit-picker, Combined	M. Grover	Clyde, Ohio	July 30, 1872	129, 946
Pruning shears and knife	J. Spear and J. A. Hull	Carbondale, Ill	June 2, 1868	78, 615
Pruning-tool	S. Gamwell	Wayland, Mich	Sept. 6, 1870	107, 172
Pruning trees, Implement for	W. W. Harvey	Saltville, Va	Jan. 15, 1856	14, 097
Prussian blue, Manufacture of	J. Clark	Newark, N. J	June 12, 1860	28, 651
Prussian blue, Manufacture of	T. A. Helwig	Minersville, Pa	Apr. 3, 1860	27, 716
Pudding-mold	J. Musgrove	East Newark, N. J	Jan. 5, 1869	85, 605
Puddle-ball squeezer	S. Wiler	Newcastle, Pa	Apr. 15, 1856	14, 701
Puddle-balls, Machine for compressing	J. F. Winslow	Troy, N. Y	Jan. 14, 1862	34, 177
Puddler, Rotary	W. Sellers and G. H. Sellers	Philadelphia, Pa., and Wilmington, Del.	Nov. 11, 1873	144, 416
Puddler's ball-elevator	S. S. Jackman	Lock Haven, Pa	Feb. 12, 1856	14, 242
Puddler's ball-roller	R. M. Bassett	Birmingham, Conn	May 27, 1873	139, 356

Index of patents issued from the United States Patent Office from 1790 *to* 1873, *inclusive*—Continued.

Invention.	Inventor.	Residence.	Date.	No.
Puddler's ball-squeezer	F. Danks	Cincinnati, Ohio	July 29, 1873	141, 207
Puddler's ball-squeezing machine	S. Danks	Cincinnati, Ohio	Nov. 15, 1870	109, 186
Puddlers' balls and other masses of iron, Machine for compressing.	S. S. Jackman	Lock Haven, Pa	July 31, 1855	13, 355
Puddlers' balls, Machine for compressing	J. Dangerfield	New Albany, Ind	Apr. 2, 1872	125, 181
Puddlers' balls, Machinery for rolling	J. F. Winslow and J. Snider	Troy, N. Y	July 5, 1848	5, 660
Puddlers' balls, Pressing out instead of hammering	W. Jones	Haverstraw, N. Y	Apr. 18, 1834	....
Puddlers' balls, Rolling and compressing	J. F. Winslow	Troy, N. Y	Dec. 18, 1847	5, 399
Puddling and boiling furnace	A. and J. Reese	Pittsburgh, Pa	Apr. 23, 1872	125, 987
Puddling and boiling furnace	W. Swindell	Allegheny City, Pa	May 12, 1868	77, 932
Puddling and boiling furnace door and bit	J. G. Haberfield	Wheeling, W. Va	May 27, 1873	139, 244
Puddling and boiling furnaces, Fettlings for	J. C. McAninch and G. Atkins	Youngstown, Ohio	Feb. 18, 1873	136, 083
Puddling and boiling furnaces, Lining or "fix" for	S. Danks	Cincinnati, Ohio	Sept. 10, 1867	68, 715
Puddling and heating furnace	S. Cuddick	Pembroke, Me	Oct. 29, 1867	70, 322
Puddling and heating furnace	O. F. Mayhew	Indianapolis, Ind	July 17, 1866	56, 429
Puddling and other furnace	J. A. Jones, R. Howson, and J. Gjers.	Middlesborough-on-Tees, England.	Apr. 25, 1871	114, 151
Puddling and other furnace	T. E. Purchase	Reading, Pa	Apr. 5, 1870	101, 658
Puddling and other furnace	J. Russell	Washington, D. C	Sept. 19, 1871	119, 182
Puddling and other furnace door	G. E. Reynolds	Philadelphia, Pa	Sept. 24, 1872	131, 562
Puddling and other furnace door	N. L. Sibley and B. Shiverick	Weston and Waltham, Mass	Apr. 10, 1866	53, 894
Puddling and other furnaces, "Fix" for	A. Reese	Pittsburgh, Pa	May 16, 1871	114, 854
Puddling and refining furnace	J. Heatley	Etna, Pa	Apr. 27, 1869	89, 310
Puddling and refining furnace, Iron	J. S. Gustin	New York	Aug. 2, 1842	2, 743
Puddling and reheating furnaces, Combination of ash-trap with.	L. Scofield and E. Cooper	South Trenton, N. J., and New York, N. Y.	Apr. 3, 1849	6, 279
Puddling and reverberatory furnace, Iron and steel	I. Hersey	Fort Edward, N. Y	May 7, 1872	126, 546
Puddling, boiling, and other furnaces, Stack for	W. Swindell	Allegheny City, Pa	Aug. 23, 1870	106, 636
Puddling-furnace	C. D. Baker	Wheeling, W. Va	Feb. 27, 1866	52, 813
Puddling-furnace	S. Bunn	Belleville, Ill	May 9, 1871	114, 641
Puddling-furnace	M. J. Davies	Zanesville, Ohio	Aug. 16, 1864	43, 880
Puddling-furnace	J. Green	Philadelphia, Pa	Feb. 3, 1857	16, 541
Puddling-furnace	J. P. and J. Grove	Montour County, Pa	Apr. 6, 1858	19, 843
Puddling-furnace	W. and J. Groves	Providence, R. I	June 27, 1865	48, 390
Puddling-furnace	D. and J. Hall	Wheeling, W. Va	Nov. 14, 1865	50, 929
Puddling-furnace	D. and J. Hall	West Wheeling, Ohio	Oct 30, 1866	59, 214
Puddling furnace	G. E. Harding	New York, N. Y	July 30, 1872	130, 044
Puddling-furnace	J. Heatley	Harrisburgh, Pa	June 10, 1873	139, 788
Puddling-furnace	W. Jeffries	West Bromwick, England	Oct. 1, 1867	69, 345
Puddling-furnace	C. Jones	Saint Louis, Mo	June 24, 1873	140, 140
Puddling-furnace	T. J Jones	Scranton, Pa	Feb. 11, 1868	74, 374
Puddling-furnace	P. Keenan and E. O. Connor	Chartier Township and West Pittsburgh, Pa.	Nov. 14, 1865	50, 937
Puddling-furnace	E. Lloyd	Pittsburgh, Pa	Feb. 11, 1873	135, 650
Puddling-furnace	H. McDonald	Pittsburgh, Pa	Mar. 9, 1869	87, 578
Puddling-furnace	H. McDonald	Pittsburgh, Pa	July 16, 1872	129, 153
Puddling-furnace	S. McLaughlin	Philadelphia, Pa	Mar. 1, 1870	100, 308
Puddling-furnace	D. Morgan	Pittsburgh, Pa	Jan. 16, 1872	122, 731
Puddling-furnace	J. Neville	Brooklyn, N. Y	May 2, 1871	114, 465
Puddling-furnace	E. Riley	Pontnewynydd, England	Sept. 30, 1873	143, 301
Puddling-furnace	J. B. Robinson	Duncansville, Pa	Aug. 31, 1869	94, 342
Puddling-furnace	H. Ross	Pittsburgh, Pa	Jan. 28, 1868	73, 931
Puddling-furnace	F. M. Ruschhaupt	Philadelphia, Pa	Apr. 19, 1864	42, 446
Puddling-furnace	J. Snyder	Wheeling, W. Va	Oct. 16, 1866	58, 904
Puddling-furnace	J. A. Stearns	Rolla, Mo	July 23, 1872	129, 762
Puddling-furnace	W. Stevenson	Allegheny City, Pa	Jan. 21, 1868	73, 665
Puddling-furnace	G. A. Whipple	West Pittsburgh, Pa	July 10, 1866	56, 333
Puddling-furnace	J. Williams	Montreal, Canada	Oct. 3, 1865	50, 319
Puddling-furnace door	J. S. Rees	Phillipsburgh, N. J	July 8, 1873	147, 730
Puddling-furnace door or frame	P. E. Shear	Saugerties, N. Y	Mar. 9, 1860	87, 438
Puddling-furnace for manufacturing iron with anthracite coal.	T. Cooper	New York, N. Y	Aug. 25, 1840	1, 733
Puddling-furnace for treating iron and steel, Revolving.	S. Danks	Cincinnati, Ohio	Nov. 24, 1868	84, 347
Puddling-furnace, Gas-heated	J. G. Blunt	Leavenworth, Kans	July 16, 1872	129, 268
Puddling furnace, Iron	R. Jenkins	Newark, Ohio	July 4, 1871	116, 598
Puddling furnace, Iron	N. S. Snedeker	Philadelphia, Pa	Aug. 22, 1865	49, 565
Puddling-furnace, Mechanical	R. G. Wood and J. R. Jackson	McKeesport, Pa	July 16, 1872	129, 199
Puddling-furnace, Revolving	L. S. Goodrich	Waverly, Tenn	Dec. 24, 1872	134, 268
Puddling-furnace, Rotary	W. Baynton	Pottsville, Pa	Sept. 5, 1871	118, 674
Puddling-furnace, Rotary	S. Danks	Cincinnati, Ohio	Oct. 26, 1869	96, 206
Puddling-furnace, Rotary	J. Davies	Knoxville, Tenn	May 21, 1872	126, 937
Puddling-furnace, Rotary	L. S. Goodrich	Waverly, Tenn	Dec. 24, 1872	134, 138
Puddling-furnace, Rotary	T. and J. W. Richardson and A. Spencer.	West Hartlepool, Great Britain.	Oct. 15, 1872	132, 180
Puddling-furnace, Rotary	E. Wood	Pittsburgh, Pa	Nov. 15, 1870	109, 284
Puddling-furnace with a steam-generator, Combination of.	G. W. Hawksley and M. Wild	Sheffield, England	May 26, 1868	78, 370
Puddling-furnaces, Apparatus for heating	S. A. Hill and C. F. Thumm	Oil City, Pa	Oct. 12, 1869	95, 686
Puddling-furnaces, Application of hot-blast to	J. Reese	Pittsburgh, Pa	Dec. 6, 1864	45, 343
Puddling-furnaces, Composition for lining	J. Williams	Montreal, Canada	June 14, 1864	43, 168
Puddling-furnaces, Fix for	C. S. Lynch	Boston, Mass	Nov. 1, 1870	108, 802
Puddling-furnaces, Fixing for	H. McDonald	Pittsburgh, Pa	Oct. 17, 1865	50, 483
Puddling-furnaces, &c., in generating steam, Utilizing the waste heat of.	J. Watt	Buffalo, N. Y	May 12, 1863	38, 521
Puddling-furnaces, Lining rotary	S. Danks	Cincinnati, Ohio	Mar. 4, 1873	136, 421
Puddling-furnaces, Lining rotary	G. H. Sellers	Wilmington, Del	July 23, 1872	129, 685
Puddling-furnaces, Shield for	H. McDonald	Pittsburgh, Pa	Dec. 29, 1868	85, 320
Puddling-furnaces, Water-bosh for	J. Stokes and J. Brough	Trenton, N. J	Aug. 4, 1868	80, 573
Puddling, melting, and heating furnace, Rotary	S. Danks	Cincinnati, Ohio	Aug. 26, 1873	142, 152
Puddling, steam-boiler, and other furnaces, Fire-chamber for.	W. F. Beecher	Pittsburgh, Pa	Apr. 4, 1871	113, 481
Puffing-iron	M. Parry	Rochester, N. Y	Feb. 27, 1872	124, 079
Pug-mill	D. H. Gage	Dover, N. H	May 6, 1873	138, 628
Pug-mill	J. A. Hamer	Reading, Pa	Jan. 26, 1858	19, 194
Pug-mill	J. C. McKenzie	Adrian, Mich	Sept. 29, 1868	82, 626
Pug-mill	J. E. Noyes	Washington, D. C	Mar. 22, 1870	101, 154
Pug-mill for mixing clay	C. F. Schlickeysen	Berlin, Prussia	June 24, 1856	15, 197

Index of patents issued from the United States Patent Office from 1790 *to* 1873, *inclusive*—Continued.

Invention.	Inventor.	Residence.	Date.	No.
Pug-mill grinding-attachment	D. H. Cage	Dover, N. H	Mar. 23, 1858	19, 696
Puller:				
See Bean-puller. Cork-puller. Flax-puller. Hoop-pole puller. Stalk and cane puller. Stone-puller. Weed-puller.				
Pulley	A. S. Blake	Waterbury, Conn	Oct. 30, 1866	59, 172
Pulley	J. A. Burnap	Albany, N. Y	Sept. 1, 1868	81, 860
Pulley	J. J. Cowell	Newark, N. J	May 7, 1872	126, 524
Pulley	G. B. Cowles	Bridgeport, Conn	June 14, 1870	104, 119
Pulley	F. Hewitt	Newark, N. J	Apr. 28, 1868	77, 284
Pulley	C. C. Moore	New York, N. Y	Feb. 11, 1873	135, 722
Pulley	R. W. Parker	Woburn, Mass	Oct. 30, 1866	59, 256
Pulley	J. P. Smith	Oshawa, Canada	Aug. 4, 1868	80, 777
Pulley	S. A. Smith	Philadelphia, Pa	Feb. 7, 1871	111, 582
Pulley	J. S. Tibbets	Terre Haute, Ind	Nov. 20, 1866	59, 874
Pulley	M. Ward	Mount Carmel, Pa	Sept. 5, 1871	118, 761
Pulley	T. A. Weston	Birmingham, England	Aug. 6, 1867	67, 470
Pulley	M. K. Whipple	Warren, Mass	Mar. 25, 1873	137, 270
Pulley and block	S. and J. Roebuck	New York, N. Y	Jan. 21, 1868	73, 653
Pulley and cable combined	R. Heneage, G. Milsom, and H. Spendelow.	Buffalo, N. Y	Jan. 29, 1867	61, 554
Pulley and clamp, Combined	M. W. Clark	Worcester, Mass	Apr. 12, 1870	101, 824
Pulley and gearing for machinery	S. Wheeler	Albany, N. Y	Apr. 14, 1868	76, 680
Pulley and spring cord, Calisthenic	J. Wood	New York, N. Y	June 14, 1864	43, 149
Pulley and wheel fastener	A. Newell	Chicago, Ill	Feb. 6, 1872	123, 413
Pulley-attachment for raising weights	G. W. Gregory	Binghamton, N. Y	Aug. 14, 1866	57, 125
Pulley attachment, Loose	C. F. Thayer	Linesville Station, Pa	Apr. 9, 1872	125, 503
Pulley, Band	C. H. Perkins	Providence, R. I	Jan. 21, 1873	135, 151
Pulley, Band	J. Simpson	Decatur, Ga	July 7, 1846	4, 618
Pulley, Band	E. P. West	Jersey City, N. J	Jan 14, 1873	134, 778
Pulley-banding	R. W. Parker	Roxbury, Mass	Feb. 17, 1852	8, 745
Pulley-belting	E. C. Thayer	Providence, R. I	Dec. 13, 1859	26, 451
Pulley-belts. Machine for measuring the strain on.	J. B. Duff and T. W. Keating	New York, N. Y	June 19, 1860	28, 743
Pulley-block	E. A. Barrett	New York, N. Y	Jan. 21, 1873	135, 066
Pulley-block	L. A. Beardsley	South Edmeston, N. Y	July 23, 1861	32, 856
Pulley-block	J. Boylo	New York, N. Y	Sept. 24, 1872	131, 594
Pulley-block	G. C. Brown	Brooklyn, N. Y	May 23, 1871	115, 022
Pulley-block	T. B. Brown	Fairfield, Me	May 7, 1872	126, 376
Pulley-block	J. A. Burnap	Albany, N. Y	Dec. 10, 1867	71, 973
Pulley-block	H. Cherry	Birmingham, England	Nov. 14, 1871	120, 940
Pulley-block	J. C. Cottingham	Philadelphia, Pa	Mar. 5, 1872	124, 253
Pulley-block	A. S. Dickinson	Brooklyn, N. Y	Oct. 7, 1873	143, 441
Pulley-block	J. J. Doyle	Brooklyn, N. Y	Sept. 27, 1864	44, 409
Pulley-block	J. M. Drake	Amityville, N. Y	Mar. 10, 1863	37, 859
Pulley-block	G. A. Ford	Cleveland, Ohio	Aug. 26, 1873	142, 095
Pulley-block	R. P. Fuller	Machias, Me	May 2, 1865	47, 598
Pulley-block	J. L. Hovey	Lockport, N. Y	Mar. 27, 1860	27, 633
Pulley-block	C. H. Knapp	Lawrenceville, Pa	Dec. 28, 1869	98, 275
Pulley-block	S. F. Lewis	San Francisco, Cal	Nov. 8, 1859	26, 038
Pulley-block	P. Luck	Williamsburgh, N. Y	May 2, 1865	47, 558
Pulley-block	R. Marsden	Sheffield, England	July 4, 1871	116, 613
Pulley-block	J. W. Norcross	Boston, Mass	Jan. 7, 1868	73, 030
Pulley-block	J. W. Norcross	Boston, Mass	Apr. 4, 1871	113, 688
Pulley-block	H. F. Shaw	West Roxbury, Mass	June 23, 1868	79, 264
Pulley-block	H. Smith	Providence, R. I	May 23, 1871	115, 248
Pulley-block	S. Van Henneck and T. Allen	New York, N. Y	Apr. 25, 1865	47, 471
Pulley-block	N. C. Whitcomb and W. Paddock.	Oak Hill, N. Y	July 12, 1864	43, 541
Pulley-block, Differential	G. F. and M. Clemens	Springfield and Boston, Mass.	Apr. 23, 1872	125, 882
Pulley-block, Differential	C. Hall	New York, N. Y	Oct. 17, 1871	119, 981
Pulley-block, Differential	C. Hall	New York, N. Y	Feb. 6, 1872	123, 342
Pulley-block, Differential	C. Hall	New York, N. Y	June 11, 1872	127, 689
Pulley-block, Differential	C. Hall and E. Hubner	New York, N. Y	May 7, 1872	126, 391
Pulley-block, Differential	R. A. Hardcastle	Newcastle-upon-Tyne, England.	May 7, 1867	64, 527
Pulley-blocks, Metal	C. W. Gregory	Watertown, N. Y	May 25, 1869	90, 525
Pulley-blocks. Molding and casting the sheaves of.	A. Worrall	New York	June 12, 1816	....
Pulley box, Loose	C. Purdy	Bedford, Ohio	Nov. 26, 1867	71, 535
Pulley-brake, Self-acting	J. Jochum	Brooklyn, N. Y	Nov. 29, 1864	45, 248
Pulley bush, Wood	T. B. Stout	Keyport, N. J	Apr. 6, 1869	88, 751
Pulley case, Sash	C. B. Clark	Buffalo, N. Y	Apr. 23, 1872	126, 019
Pulley casing, Window	N. Thompson	Brooklyn, N. Y	Sept. 19, 1871	119, 199
Pulley, Cast-iron chain	J. Bird	New York, N. Y	Dec. 11, 1866	60, 329
Pulley, Cast-metal	J. A. Evarts	West Meriden, Conn	Nov. 1, 1859	25, 998
Pulley, Clamp	R. Chandler	Springfield, Mass	Apr. 3, 1866	53, 730
Pulley, Clothes-line	G. H. Ryer	New York, N. Y	May 20, 1873	139, 024
Pulley, Clutch	J. Shinn	Leverington, Pa	Sept. 17, 1861	33, 321
Pulley-coupling	J. E. Atwood	Mansfield, Conn	Nov. 29, 1870	109, 709
Pulley-couplings, Clutch for	J. Knickerbacker	Stockport, N. Y	Dec. 6, 1859	26, 369
Pulley, Differential	T. A. Weston	Ridgewood, N. J	Dec. 24, 1872	134, 337
Pulley, Door	J. Reiser	Trenton, N. J	June 8, 1869	91, 163
Pulley, Door	P. J. Tillman	Trenton, N. J	Apr. 6, 1869	88, 592
Pulley, Dynamometer	I. P. Tice	New York, N. Y	Apr. 1, 1873	137, 393
Pulley, Expanding	G. S. Barton	Worcester, Mass	Apr. 18, 1871	113, 723
Pulley, Expanding	J. E. Fales	Worcester, Mass	Apr. 1, 1873	137, 352
Pulley, Expanding	T. H. Savery	Wilmington, Del	June 9, 1868	78, 763
Pulley, Expansible belt	H. A. Hummer and J. H. Stover	Frenchtown, N. J	Dec. 17, 1872	134, 061
Pulley, Extension	W. Onions and I. Bagnall	Saint Louis, Mo	Apr. 18, 1871	113, 789
Pulley-facing machine	T. Blume	Cincinnati, Ohio	Jan. 10, 1860	26, 810
Pulley, Fast and loose	C. H. Brown	Fitchburgh, Mass	Mar. 8, 1864	41, 829
Pulley for band-saws	W. H. Doane	Cincinnati, Ohio	May 21, 1872	127, 033
Pulley for belting	M. Lewis and S. Miller	Greenville, Conn	Mar. 26, 1867	63, 266
Pulley for belts and brakes, Binder	M. C. Bryant	Lowell, Mass	Nov. 13, 1849	6, 864
Pulley for blind-cords, Ratchet	J. B. Holmes, jr	New York, N. Y	Dec. 13, 1859	26, 433
Pulley for fishing-boats, Self-adjusting	W. Woodbury	Chelsea, Mass	Feb. 13, 1866	52, 634
Pulley for hoisting-apparatus	D. L. Miller	Madison, N. J	Mar. 24, 1868	75, 946

Index of patents issued from the United States Patent Office from 1790 *to* 1873, *inclusive*—Continued.

Invention.	Inventor.	Residence.	Date.	No.
Pulley for window-sashes, Spring	D. Bickford	Westerly, R. I	July 13, 1858	20, 857
Pulley for window-sashes, Spring	J. Shopland	Honesdale, Pa	Aug. 12, 1856	15, 528
Pulley, Friction	E. F. Allen	Providence, R. I	Nov. 1, 1870	108, 749
Pulley, Friction	C. Burleigh	Fitchburgh, Mass	July 8, 1862	35, 853
Pulley, Friction	W. Ebbitt	New York, N. Y	Jan. 16, 1872	122, 818
Pulley, Friction	J. Kaceroosky	Bridgeport, Conn	Sept. 17, 1872	131, 449
Pulley, Friction	L. G. Mason	Worcester, Mass	Aug. 21, 1866	57, 352
Pulley, Friction	C. B. Smith	Newark, N. J	Dec. 4, 1866	60, 269
Pulley, Friction	E. Spaulding	Westborough, Mass	May 24, 1859	24, 156
Pulley, Friction	J. Steger	New York, N. Y	Oct. 18, 1870	108, 533
Pulley, Friction	J. Steger	New York, N. Y	July 11, 1871	117, 010
Pulley, Friction-clutch	J. W. Bishop	Stamford, Conn	Oct. 16, 1866	58, 765
Pulley, Friction-clutch	A. B. Clemons	Ansonia, Conn	Dec. 8, 1868	84, 681
Pulley, Friction-clutch	J. D. Crocker	Norwich, Conn	June 3, 1873	139, 456
Pulley, Friction-clutch	D. Harrington	Worcester, Mass	June 16, 1868	78, 961
Pulley, Friction-clutch	C. D. Palmiter	Oswego, N. Y	Nov. 12, 1867	70, 888
Pulley, Friction-clutch	H. K. Smith	Norwich, Conn	July 14, 1868	80, 024
Pulley, Friction-clutch	C. Wright	Newark, N. J	July 14, 1868	80, 045
Pulley, Gate	E. S. Axtell	Macomb, Mich	Sept. 27, 1870	107, 646
Pulley, Gate	D. Bedell	Seneca Falls, N. Y	May 8, 1860	28, 146
Pulley, Grip	A. S. Hallidie	San Francisco, Cal	Feb. 22, 1870	100, 140
Pulley, Griping	A. S. Hallidie	San Francisco, Cal	June 11, 1872	127, 690
Pulley, Grooved	L. Planer	New York, N. Y	Jan. 3, 1860	26, 707
Pulley-hanger	J. Kent	New York, N. Y	June 17, 1873	139, 898
Pulley, Hoisting and transferring	A. D. Manley	Washington, Mich	May 5, 1868	77, 629
Pulley, Loose	E. S. Capen	Worcester, Mass	July 7, 1868	79, 634
Pulley, Loose	W. W. Carey	Lowell, Mass	Dec. 24, 1872	134, 250
Pulley, Loose	J. P. Grosvenor	Lowell, Mass	April 9, 1872	125, 388
Pulley, Loose	D. Harrington	Worcester, Mass	Aug. 11, 1868	80, 948
Pulley, Loose	L. Knight	Salem, N. Y	June 17, 1873	140, 048
Pulley, Loose	N. P. Otis	Yonkers, N. Y	Nov. 21, 1865	51, 077
Pulley, Loose	J. J. Ralya	Cleveland, Ohio	Dec. 31, 1872	134, 487
Pulley, Lubricating	J. H. Gray	Boston, Mass	Sept. 15, 1868	82, 110
Pulley, Lubricating	D. Harrington	Worcester, Mass	July 7, 1868	79, 752
Pulley, Lubricating	J. K. McLanahan	Hollidaysburgh, Pa	Oct. 21, 1873	143, 830
Pulley lubricator, Loose	J. W. Brockway	New York, N. Y	Apr. 26, 1870	102, 215
Pulley lubricator, Loose	F. Keifel	Cincinnati, Ohio	Mar. 12, 1872	124, 589
Pulley lubricator, Loose	C. A. King	Springfield, Mass	Oct. 19, 1869	95, 911
Pulley lubricator, Loose	C. A. King	Springfield, Mass	Oct. 19, 1869	95, 912
Pulley lubricator, Loose	G. M. Morris and J. McCreary	Cohoes, N. Y	Feb. 4, 1868	74, 116
Pulley lubricator, Loose	S. Ustick	Philadelphia, Pa	Feb. 7, 1871	111, 590
Pulley lubricator, Loose	J. P. Wendell and S. P. M. Tasker.	Philadelphia, Pa	May 3, 1870	102, 739
Pulley, Machine	C. S. Hunt	Bridgewater, Mass	July 13, 1858	20, 881
Pulley-mechanism	W. H. Brown	Bangor, Me	Sept. 6, 1870	106, 993
Pulley-molding apparatus	G. L. Scott	Manchester, England	May 18, 1869	90, 126
Pulley-molding apparatus	J. Yocom, jr	Philadelphia, Pa	Apr. 19, 1864	42, 426
Pulley molding machine	T. R. and S. Knowles	Jersey City, N. J	Nov. 12, 1867	70, 864
Pulley press, Power	W. and R. Skene	Louisville, Ky	June 28, 1859	24, 584
Pulley, Sash	J. Andrews	Marlborough, Mass	Jan. 30, 1872	123, 075
Pulley, Sash	J. D. Browne	Cincinnati, Ohio	Sept. 22, 1868	82, 287
Pulley, Sash	H. Cash	Newport, Ky	Aug. 6, 1867	67, 412
Pulley, Sash	C. C. Clark	Buffalo, N. Y	Sept. 21, 1869	94, 945
Pulley, Sash	S. Drum	Allegheny City, Pa	May 21, 1867	64, 957
Pulley, Sash	O. S. Garretson	Buffalo, N. Y	Apr. 12, 1870	101, 854
Pulley, Sash	A. Le Page	Woodhaven, N. Y	Nov. 21, 1871	121, 178
Pulley, Sash	B. Roux	Cincinnati, Ohio	Nov. 16, 1869	96, 969
Pulley, Sash	Z. E. Sawtell	Boston, Mass	May 31, 1870	103, 783
Pulley, Sash	A. P. Seymour, jr., and W. R. Goodrich.	Hecla Works and Whitestown, N. Y.	Aug. 18, 1868	81, 218
Pulley, Sash-rope	J. C. Price	New Philadelphia, Ohio	Sept. 21, 1869	95, 138
Pulley, Self-detaching	J. E. Gustin	Elmira, N. Y	July 7, 1868	79, 653
Pulley, Self-lubricating	W. W. Crane	Auburn, N. Y	Nov. 26, 1872	133, 416
Pulley, Self-oiling	J. Goodrich and H. J. Colburn	Fitchburgh, Mass	May 4, 1869	89, 652
Pulley, Self-oiling	J. Goodrich and H. J. Colburn	Fitchburgh, Mass	Apr. 25, 1871	114, 129
Pulley, Self-stopping	T. C. Richards	New York, N. Y	Mar. 20, 1866	53, 341
Pulley shade-fixture, Rack	W. Johnson	Waterbury, Conn	May 16, 1871	114, 825
Pulley, Sheave	V. Knecht	Cincinnati, Ohio	May 24, 1870	103, 474
Pulley, Sheave	J. B. Vannan and N. P. Cramer	Carbondale, Pa	Aug. 3, 1869	93, 252
Pulley, sheave, and wheel	N. Thompson	Brooklyn, N. Y	Sept. 26, 1871	119, 430
Pulley, Suspension	J. A. Evarts	West Meriden, Conn	Oct. 28, 1873	144, 078
Pulley-suspension hook	D. B. Baker and P. S. Miller	Rollersville, Ohio	Nov. 13, 1866	59, 542
Pulley, Tension	A. B. Nimbs	Buffalo, N. Y	Apr. 18, 1865	47, 323
Pulley, Tight and loose	J. P. Gates	Lincoln, Ill	Dec. 10, 1867	72, 014
Pulley, Tight and loose	J. G. McCormick	Louisville, Ky	Sept. 13, 1870	107, 275
Pulley to ascertain ship's way, &c., Loxodromick	J. S. Donlery	New York	Sept. 21, 1815	....
Pulley-turning machine	G. A. Gray, jr	Hamilton, Ohio	Apr. 29, 1873	138, 394
Pulley-wheels, Mode of constructing wooden	A. Newell	Erie, Pa	May 6, 1873	138, 518
Pulley, Window	J. M. Ford	Brooklyn, N. Y	June 29, 1869	91, 836
Pulley, Window	O. S. Garretson	Buffalo, N. Y	Sept. 17, 1867	68, 868
Pulley, Window-cord	M. C. Ames	Hartford, Conn	Apr. 4, 1865	47, 150
Pulley, Window-cord	J. W. Bliss	Hartford, Conn	Feb. 21, 1854	10, 540
Pulley, Window-sash	O. P. Briggs	Chicago, Ill	Oct. 15, 1872	132, 135
Pulley, Window-sash	O. S. Judd	New Britain, Conn	June 13, 1848	5, 626
Pulleys and journals, Lubricating loose	S. Ustick	Philadelphia, Pa	Sept. 19, 1871	119, 065
Pulleys and wheel-hubs to shafts, Fastening	E. G. Shortt	Carthage, N. Y	June 20, 1871	116, 228
Pulleys and wheel to shafting, Attaching	C. L. Smith	Rahway, N. J	Oct. 8, 1872	132, 0?9
Pulleys, Balancing	E. W. Phelps	Elizabeth, N. J	Nov. 28, 1871	121, 411
Pulleys, &c., Anti-friction bush for	J. Palmer	Saint Catharine's, Canada	July 16, 1872	129, 585
Pulleys, Bearing for loose	W. Campbell	West Philadelphia, Pa	Oct. 24, 1854	11, 848
Pulleys by friction, Tightening	F. Skinner	New Haven, Conn	Aug. 8, 1865	49, 312
Pulleys, Casting metallic boxes or centers of sheaves or trundles for.	L. Larrabee	Nantucket, Mass	Nov. 4, 1811	....
Pulleys, Cog for	W. Shotwell	....	June 24, 1800	....
Pulleys, Compound to increase the friction between belts and.	L. F. Robertson	New York, N. Y	June 14, 1870	104, 3[illegible]6
Pulleys, constructing self-lubricating	W. W. Crane	Auburn, N. Y	June 4, 1867	65, 349
Pulleys, Construction of	A. Warth	Stapleton, N. Y	Apr. 3, 1866	53, 711

Index of patents issued from the United States Patent Office from 1790 *to* 1873, *inclusive*—Continued.

Invention.	Inventor.	Residence.	Date.	No.
Pulleys, Device for locking loose	W. J. Linton	Detroit, Mich	Aug. 25, 1868	81, 384
Pulleys, Device for oiling fast and loose	W. Hamilton	Chicopee, Mass	Mar. 24, 1868	75, 753
Pulleys, Lubricating loose	I. F. Brown	New London, Conn	Apr. 30, 1872	126, 261
Pulleys, Lubricating loose	S. Ristick	Philadelphia, Pa	Sept. 26, 1871	119, 287
Pulleys, Lubricating-sleeve for loose	O. E. Greene	Lawrence, Mass	Apr. 20, 1869	89, 038
Pulleys, Lubrication of clutch	A. Canis and F. Higgins	Manchester, N. H	Mar. 11, 1873	136, 699
Pulleys, Mold for casting	W. D. Rinehart	Pittsburgh, Pa	June 19, 1866	55, 713
Pulleys, Molding	W. Neemes	Pittsburgh, Pa	Mar. 5, 1867	62, 668
Pulleys or sheaves, Bushing and pin for	M. H. Marshall	Gloucester, Mass	July 10, 1834	....
Pulleys, Rubber cover for flat-belt	J. W. Sutton	Akron, Ohio	Apr. 2, 1872	125, 228
Pulleys, Securing	J. W. Reid	New York, N. Y	Nov. 28, 1865	51, 218
Pulleys, Shipper-gear for	A. Bettcley	Boston, Mass	Aug. 23, 1859	25, 169
Pulleys to shafts, Attaching	C. Clareni	New York, N. Y	Sept. 5, 1854	11, 639
Pulleys to shafts, Device for securing	J. H. Buckman	Cincinnati, Ohio	Dec. 21, 1869	98, 144
Pulleys to shafts, Friction-attachment for securing.	H. Cox	Peterborough, Canada	June 10, 1873	139, 704
Pulleys to shafts, Securing	D. K. Overhiser	Williamsport, Pa	Nov. 19, 1872	133, 246
Pulleys to shafts, Securing	W. C. Van Voorhis	New York, N. Y	May 27, 1873	139, 282
Pulleys with shafts, Coupling and uncoupling	P. Yates	Milwaukee, Wis	Apr. 25, 1843	3, 063
Pulmonometer	A. Eckert	Dayton, Ohio	Jan. 10, 1860	26, 754
Pulp-box drier	S. Wheeler and E. Jerome	Albany, N. Y	July 16, 1867	66, 918
Pulp-box, Finishing	S. Wheeler and E. Jerome	Albany, N. Y	July 16, 1867	66, 919
Pulp-dresser	C. Sellers	Philadelphia, Pa	June 6, 1832	....
Pulp-dresser	E. H. Thomas and N. Woodcock.	Brattleborough, Vt	Aug. 11, 1830	....
Pulp-engine or crusher	M. R. Fletcher	Boston, Mass	June 20, 1871	116, 039
Pulp from wood, straw, and other matters, Making.	J. W. Dixon	Philadelphia, Pa	June 26, 1866	55, 836
Pulp-machine	G. Sweetland	New Haven, Conn	Sept. 5, 1848	5, 756
Pulp machine, Wood	S. C. Taft	Mendon, Mass	Apr. 12, 1870	101, 785
Pulp. Making boxes of paper	A. French and C. Frost	Waterbury, Conn	July 1, 1856	15, 228
Pulp, paste, or slip pigments, Preservation of	P. C. Tiemann	New York, N. Y	Sept. 23, 1873	143, 106
Pulp, Preparing and dressing	J. Ames	Springfield, Mass	Sept. 1, 1832	....
Pulp, Process for reducing fibrous substances to	H. B. Meech	Fort Edward, N. Y	Aug. 7, 1866	56, 971
Pulp-screen	G. West	Tyringham, Mass	Aug. 19, 1851	8, [illegible]6
Pulp-strainer	H. H. Olds	New Haven, Conn	Feb. 18, 1873	136, 002
Pulp-strainer	J. Sawyer	Newbury, Vt	Jan. 21, 1832	....
Pulp-washer, Submerged centrifugal	R. R. Sylands	Millburn, N. J	May 25, 1869	90, 472
Pulp-washing machine	F. Goucher	Chester County, Pa	Apr. 12, 1833	....
Pulp, Wood-grinding machine for making	S. B. Zimmer	Elkhart, Ind	Aug. 19, 1873	141, 976
Pulpit, Apparatus for ventilating	J. P. Herron	Huntsville, Ohio	Jan. 12, 1858	19, 089
Pulverized material, Can for	H. Everett	Philadelphia, Pa	Sept. 24, 1872	131, 511
Pulverizer	I. N. Jennings	Danbury, Conn	Jan. 5, 1869	85, 591
Pulverizer, Agricultural	B. F. Stickney	Vistula, Mich	Mar. 1, 1834	....
Pulverizer and amalgamator Quartz	W. H. and I. Scoville	Chicago, Ill	Sept. 25, 1860	30, 162
Pulverizer and corn-marker, Combined	T. B. Jones	Hiawatha, Kans	Dec. 5, 1871	121, 629
Pulverizer and harrow, Earth	J. Lefeber and G. W. Shults	Cambridge City, Ind	Oct. 31, 1871	120, 446
Pulverizer and seed-sower	G. W. Bonham	Henry, Ill	Jan. 7, 1862	34, 049
Pulverizer and seeder, Combined earth	C. Wilson	Stamford, Conn	Sept. 3, 1867	68, 403
Pulverizer, Clay	G. C. Bovey	Cincinnati, Ohio	Aug. 15, 1871	117, 973
Pulverizer, Clay	I. Hersey and J. H. Van Riper.	New York, N. Y	June 9, 1857	17, 505
Pulverizer, Earth	W. Elwell	Gardiner, Me	Aug. 22, 1865	49, 513
Pulverizer, Earth	J. Johnson	Mount Washington, Ohio	Nov. 29, 1864	45, 249
Pulverizer, Earth	J. Prutzman	Hancock County, Ill	Aug. 27, 1867	68, 110
Pulverizer for reducing ligneous substances to powder.	N. Jones	Boston, Mass	July 23, 1816	....
Pulverizer, Ground	E. A. Olleman	Mooresville, Ind	Aug. 16, 1870	106, 397
Pulverizer, Land	T. R. Denby	Carlinville, Ill	Apr. 15, 1873	137, 764
Pulverizer, leveler, and marker	L. Jones	Funk's Grove, Ill	Dec. 15, 1868	85, 010
Pulverizer, Rotary	J. Thompson	Louisville, Ky	Nov. 15, 1870	109, 273
Pulverizer, Soil	C. Berninger	Mier, Ill	Aug. 25, 1868	81, 333
Pulverizer, Soil	D. Osborn	Paoli, Ind	Jan. 9, 1872	122, 643
Pulverizer, Soil	W. Shumard	Richmond, Ind	Oct. 19, 1869	95, 940
Pulverizer, Subsoil	G. S. Newson	Nashville, Tenn	Aug. 9, 1870	106, 195
Pulverizing and crushing vegetable and mineral matter, Machine for.	M. Smith	Pittsburgh, Pa	Oct. 1, 1861	32, 495
Pulverizing and furrowing device	C. Shabley	Brooklyn, N. Y	Dec. 26, 1865	51, 757
Pulverizing and grinding machine	J. J. Webster	Magog, Canada	Jan. 16, 1872	122, 748
Pulverizing and levigating substances, Machine for.	O. E. Noble	Penn Yan, N. Y	Sept. 27, 1864	44, 444
Pulverizing-barrel, Self-discharging	A. B. Paul	San Francisco, Cal	May 10, 1870	102, 857
Pulverizing-chaser	T. Pugh	Chicago, Ill	Dec. 28, 1869	98, 297
Pulverizing-machine	W. Adamson	Philadelphia, Pa	Mar. 29, 1864	42, 060
Pulverizing machine, Mineral	S. and G. E. Mills	New York, N. Y	June 28, 1859	24, 570
Pulverizing machine, Rock, ore, &c	W. F. Goodwin and C. R. Squire	East New York and New York, N. Y.	Oct. 8, 1867	59, 655
Pulverizing machine, Sand, &c	J. G. Savage	South Reading, Mass	Mar. 7, 1865	46, 713
Pulverizing machine, Soil	G. P. De Yo	Groton Township, Ohio	June 27, 1871	116, 278
Pulverizing machine, Soil	L. S. Fithian	Absecom, N. J	Jan. 24, 1865	46, 048
Pulverizing machine, Soil	A. C. Tower	Mendota, Ill	Apr. 29, 1873	138, 301
Pulverizing machine, Wood	A. Tolson	Georgetown, D. C	June [illegible], 1821	....
Pulverizing soil and clods, Roller for	J. Custer	Corsica, Ohio	Mar. 26, 1867	63, 224
Pulverizing soil, Device for	G. Shelton	Normal, Ill	July 25, 1871	117, 473
Pulverizing tailings from gold-washers	J. H. Hanchett	Beloit, Wis	May 23, 1865	47, 818
Pump	W. Adair	Liverpool, England	June 16, 1868	78, 854
Pump	C. C. Alexander	Denver, Colo	Sept. 22, 1863	40, 006
Pump	M. J. Althouse	Waupun, Wis	Jan. 12, 1864	41, 186
Pump	M. J. Althouse	Waupun, Wis	Aug. 7, 1866	56, 871
Pump	M. J. Althouse	Waupun, Wis	Nov. 13, 1866	59, 536
Pump	M. J. Althouse	Waupun, Wis	July 2, 1872	128, 576
Pump	R. H. Andrews	Elizabethtown, Pa	May 8, 1866	54, 483
Pump	A. Arnold	Heidelberg, Pa	Apr. 21, 1863	28, 209
Pump	J. B. Atwater	Brooklyn, N. Y	Mar. 23, 1858	19, 671
Pump	J. E. Atwood	Bucksport, Me	May 8, 1860	28, 140
Pump	S. L. Avery	Norwich, N. Y	Jan. 1, 1867	60, 818
Pump	H. Bachman	Lancaster, Pa	Apr. 2, 1835	[illegible]
Pump	B. C. Bailey	Constitution, Ohio	June 7, 1870	103, 960
Pump	C. Baker	Weymouth, Mass	May 14, 1867	64, 735

Index of patents issued from the United States Patent Office from 1790 *to* 1873, *inclusive*—Continued.

Invention.	Inventor.	Residence.	Date.	No.
Pump	A. Balding	Wheeling, W. Va	June 28, 1870	104, 688
Pump	E. Balding	Memphis, Tenn	Aug. 17, 1869	93, 794
Pump	A. C. Baldwin	Washington, D. C	May 7, 1872	126, 569
Pump	O. Baldwin	Summitville, Iowa	Aug. 16, 1864	43, 826
Pump	J. S. Barden	Providence, R. I	Mar. 3, 1868	75, 112
Pump	J. S. Barden	Providence, R. I	Feb. 8, 1870	99, 521
Pump	J. S. Barden	Providence, R. I	Jan. 7, 1873	134, 626
Pump	A. Barker	Honesdale, Pa	Feb. 17, 1852	8, 733
Pump	W. C. Barker	Ypsilanti, Mich	Nov. 8, 1870	108, 954
Pump	A. B. Barlow	Ripon, Wis	Oct. 23, 1866	58, 972
Pump	E. Barlow	Marietta, Ohio	Feb. 29, 1848	5, 462
Pump	W. Barnes	Maquoketa, Iowa	July 23, 1872	129, 779
Pump	W. A. Barnes	Decatur, Ill	Feb. 6, 1866	52, 377
Pump	W. T. and J. Barnes	Buffalo, N. Y., and Oakville, Canada.	Oct. 14, 1856	15, 878
Pump	N. Barrett	New York, N. Y	June 25, 1861	32, 607
Pump	W. S. Bartle	Newark, N. Y	Jan. 7, 1862	34, 101
Pump	J. R. Bassett	Cincinnati, Ohio	Jan. 3, 1871	110, 622
Pump	A. Batby	New York	May 18, 1830	....
Pump	F. Bauschtleker and B. Vanfleet	Washington, D. C	Oct. 22, 1867	70, 068
Pump	W. D. Baxter	New York, N. Y	Oct. 13, 1868	83, 027
Pump	W. D. Baxter	New York, N. Y	Apr. 8, 1873	137, 527
Pump	J. Bean	Hudson, Mich	Dec. 20, 1864	45, 467
Pump	J. Bean	Hudson, Mich	Feb. 20, 1866	52, 660
Pump	J. Bean	Hudson, Mich	Feb. 19, 1867	62, 247
Pump	J. Bean	Hudson, Mich	Nov. 19, 1867	71, 122
Pump	J. Bean	Hudson, Mich	Feb. 9, 1869	86, 804
Pump	J. Bean	Hudson, Mich	Mar. 26, 1872	125, 008
Pump	J. Bean	Hudson, Mich	Dec. 17, 1872	134, 025
Pump	N. S. Bean	Manchester, Pa	June 12, 1860	28, 644
Pump	A. Beeler and J. B. Christian	Mount Carroll, Ill	Apr. 12, 1859	23, 544
Pump	L. Beemer	Libertyville, N. J	Feb. 6, 1866	52, 485
Pump	S. Belden	Visalia, Cal	Apr. 27, 1869	89, 378
Pump	H. Belfield	Philadelphia, Pa	Apr. 17, 1860	27, 945
Pump	A. Bellingrath	Atlanta, Ga	Jan. 22, 1861	31, 206
Pump	C. F. Bellows	Seneca Falls, N. Y	Feb. 1, 1859	22, 777
Pump	C. Bemis	Mishawaka, Ind	June 11, 1867	65, 718
Pump	W. A. Bemis	Lyndon Centre, Vt	Aug. 9, 1864	43, 755
Pump	S. Benson	Allegheny City, Pa	Dec. 29, 1868	85, 270
Pump	J. A. Bloom	Philadelphia, Pa	Dec. 20, 1865	51, 686
Pump	J. Boley	Baldwinsville, N. Y	July 11, 1865	48, 647
Pump	C. Bollinger	Glen Rock, Pa	Oct. 1, 1861	33, 375
Pump	C. Bollinger	Harrisburgh, Pa	Jan. 5, 1864	41, 133
Pump	C. Bollinger	Harrisburgh, Pa	Jan. 5, 1864	41, 174
Pump	T. Bourke	....	Nov. 16, 1796	....
Pump	W. Boyers	Mount Carroll, Ill	Feb. 9, 1858	19, 286
Pump	H. E. Braunfeld	Philadelphia, Pa	Dec. 12, 1871	121, 843
Pump	J. F. Brickley	Winchester, Ind	Dec. 15, 1868	84, 991
Pump	S. Brillinger	Clarence Centre, N. Y	May 11, 1869	89, 849
Pump	J. Brekenshire	Oswego, N. Y	Aug. 11, 1868	[illegible]
Pump	T. Brooks	Rutland, N. Y	Apr. 13, 1826	....
Pump	A. P. Brown	New York, N. Y	May 2, 1871	114, 403
Pump	W. R. Brown	Cleveland, Ohio	Apr. 5, 1859	23, 443
Pump	G. Bruce	Corydon, Ind	Oct. 15, 1867	69, 758
Pump	J. Bryan	New York, N. Y	Mar. 8, 1870	100, 595
Pump	M. Buchlin	Grafton, N. H	Oct. 25, 1832	....
Pump	J. Budd	Sandy Hill, N. Y	Apr. 29, 1862	35, 077
Pump	J. and D. Budd	Albany, N. Y	Oct. 13, 1863	40, 239
Pump	J. S. Burnham	Yorkville, N. Y	Jan. 13, 1857	16, 373
Pump	J. H. Burns	Clinton Station, N. J	Jan. 31, 1865	46, 013
Pump	A. Burr	Middlesex, Conn	May 1, 1866	54, 289
Pump	F. S. Burt	Mount Pleasant, Iowa	Jan. 13, 1863	37, 283
Pump	L. Button and R. Blake	Waterford, N. Y	Apr. 29, 1862	35, 079
Pump	J. W. Cahill	Madison, Ind	Nov. 13, 1866	59, 708
Pump	J. Camack	Dane, Wis	Sept. 5, 1865	49, 716
Pump	A. S. Cameron	New York, N. Y	May 1, 1866	54, 291
Pump	H. Camp	Rouseville, Pa	Jan. 31, 1871	111, 316
Pump	J. Canfield	Morristown, N. J	June 28, 1809	....
Pump	J. F. Carll	Pleasantville, Pa	May 6, 1873	138, 477
Pump	W. S. Carr	New York, N. Y	Jan. 11, 1870	98, 740
Pump	W. S. Carr	New York, N. Y	Sept. 24, 1872	131, 663
Pump	I. Carter	Champlain, N. Y	June 27, 1854	11, 192
Pump	A. Carver	Little Falls, N. Y	July 11, 1865	48, 707
Pump	A. Carver	Little Falls, N. Y	Apr. 26, 1870	102, 369
Pump	S. Caswell	Harrison, Me	May 19, 1840	1, 610
Pump	F. Catudal	Webster, Mass	Jan. 14, 1873	134, 785
Pump	E. S. Cavnah and D. Yeagley	Bourbon, Ind	Nov. 30, 1869	97, 274
Pump	W. N. Chamberlain	Van Buren, Mich	Nov. 2, 1869	96, 307
Pump	T. Chamberlin and T. E. Garrett	Philadelphia, Pa	Feb. 18, 1868	74, 500
Pump	T. Chambers	Saint Louis, Mo	Nov. 3, 1868	83, 691
Pump	J. B. Christian and A. Beeler	Mount Carroll, Ill	Apr. 5, 1859	23, 447
Pump	D. W. Christman	New York, N. Y	Feb. 8, 1817	....
Pump	M. R. Clapp	New York, N. Y	Mar. 7, 1865	46, 638
Pump	W. H. T. Clark	San Francisco, Cal	May 18, 1869	90, 080
Pump	R. Cochran	Morrison, Ill	June 26, 1866	55, 821
Pump	N. T. Coffin	Knightstown, Ind	Feb. 21, 1871	112, 018
Pump	N. T. Coffin	Knightstown, Ind	Mar. 28, 1871	113, 144
Pump	J. K. Cohick and J. Fesher	Mountville, Pa	Sept. 20, 1864	44, 282
Pump	D. G. Colburn	Wilmington, Vt	Apr. 25, 1846	4, 467
Pump	G. M. Cole	Folsom City, Cal	Sept. 24, 1867	69, 183
Pump	J. W. Cole	Mount Pleasant, Iowa	May 11, 1869	89, 918
Pump	H. T. Coleman	Williamsburgh, N. Y	June 17, 1873	140, 016
Pump	G. Collins and E. Piper	Camden, Me	Jan. 7, 1862	34, 105
Pump	J. F. Collins	New York, N. Y	Jan. 26, 1869	86, 213
Pump	H. Comstock	Seneca Falls, N. Y	Jan. 22, 1867	61, 319
Pump	A. Conant and I. F. Brown	New London, Conn	Nov. 6, 1866	59, 364
Pump	J. Coney	Boston, Mass	Dec. 16, 1856	16, 229

Index of patents issued from the United States Patent Office from 1790 *to* 1873, *inclusive*—Continued.

Invention.	Inventor.	Residence.	Date.	No.
Pump	G.W. Cook and Z. E. B. Nash	Saint Paul, Minn	Mar. 11, 1862	34, 624
Pump	A. Cooley	Springfield, Ill	Nov. 30, 1858	22, 165
Pump	A. Cooley	Chicago, Ill	June 10, 1862	35, 508
Pump	P. Cope	Oakfield, N. Y	June 18, 1872	128, 123
Pump	C. A. Cowell	Newark, N. J	Sept. 21, 1866	95, 088
Pump	G. Cowing	Seneca Falls, N. Y	Mar. 16, 1869	87, 912
Pump	J. P. Cowing	Seneca Falls, N. Y	Oct. 21, 1856	15, 922
Pump	J. C. Crawford	Clintonville, Ill	Apr. 6, 1869	88, 552
Pump	S. B. Crittenden	Cincinnati, Ohio	Oct. 20, 1863	40, 328
Pump	J. E. Cronk	Poughkeepsie, N. Y	Apr. 5, 1859	23, 453
Pump	J. D. Cross	Petaluma, Cal	Dec. 31, 1872	134, 466
Pump	R. W. Crouse	Westminster, Md	July 14, 1868	79, 959
Pump	R. W. Crouse	Westminster, Md	Aug. 17, 1869	93, 810
Pump	E. S. Crowell	Augusta, Me	Aug. 22, 1871	118, 206
Pump	H. G. Crowell	Roxbury, Mass	May 21, 1861	32, 356
Pump	W. H. Culp	Hammondsville, Ohio	Nov. 14, 1865	50, 911
Pump	C. G. Curtis	Springfield, Mass	Mar. 20, 1855	12, 544
Pump	P. C. Curtis	Utica, N. Y	May 22, 1835	....
Pump	L. H. Davis	West Chester, Pa	Apr. 28, 1863	38, 291
Pump	L. H. Davis	Newark, Del	Mar. 16, 1869	87, 828
Pump	J. M. Dearborn	Boston, Mass	May 3, 1817	....
Pump	J. N. Dennisson	Newark, N. J	Oct. 23, 1866	59, 125
Pump	T. De Witt	Detroit, Mich	May 3, 1870	102, 511
Pump	E. N. Dickerson and E. K. Root	New York, N.Y., and Hartford, Conn.	Feb. 5, 1856	14, 186
Pump	I. Dillinbgam	Rockbottom, Mass	Mar. 14, 1871	112, 561
Pump	A. Dixon	Aurora, Ill	Aug. 31, 1869	94, 190
Pump	G. H. Dodge	Camden, N. J	Apr. 1, 1862	34, 819
Pump	L. P. Dodge	Newburgh, N. Y	May 13, 1862	35, 222
Pump	L. P. and W. F. Dodge	Newburgh, N. Y	June 7, 1853	9, 777
Pump	W. F. Dodge	New York, N. Y	Apr. 11, 1865	47, 193
Pump	B. Douglas	Middletown, Conn	Nov. 6, 1860	30, 693
Pump	B. Douglas	Middletown, Conn	Sept. 17, 1861	33, 299
Pump	B. Douglas	Middletown, Conn	July 1, 1862	35, 802
Pump	B. and W. Douglas	Middletown, Conn	Apr. 12, 1859	23, 649
Pump	J. W. Douglas	Middletown, Conn	Apr. 22, 1862	35, 059
Pump	J. W. Douglas	Middletown, Conn	May 5, 1863	38, 444
Pump	J. W. Douglas	Middletown, Conn	July 24, 1866	56, 671
Pump	J. W. Douglas	Middletown, Conn	Feb. 11, 1868	74, 321
Pump	J. W. Douglas	Middletown, Conn	Dec. 14, 1869	97, 767
Pump	J. W. Douglas	Middletown, Conn	Nov. 25, 1873	144, 965
Pump	W. and B. Douglas	Middletown, Conn	Dec. 31, 1842	2, 895
Pump	L. Drescher	Matanzas, Cuba	Sept. 25, 1866	58, 360
Pump	C. H. Dreyer	Nashville, Tenn	Oct. 13, 1868	83, 051
Pump	S. S. Durbon	Lebanon, Ind	July 17, 1866	56, 387
Pump	A. Duvall	Baltimore, Md	Dec. 13, 1864	45, 398
Pump	J. B. Eads	Saint Louis, Mo	July 5, 1870	105, 056
Pump	C. F. Eastlack	Mantua, N. J	June 23, 1868	79, 217
Pump	J. Edson	Boston, Mass	Apr. 4, 1854	10, 746
Pump	J. Edson	Boston, Mass	May 9, 1854	10, 885
Pump	J. Edson	Boston, Mass	Nov. 8, 1859	26, 025
Pump	J. Edson	Boston, Mass	Oct. 21, 1873	143, 751
Pump	J. Edson and P. Noyes	Boston and Lowell, Mass	Aug. 22, 1871	118, 222
Pump	E. Edwards	Westminster, England	June 14, 1870	104, 289
Pump	J. Eldridge	West Buxton, Me	Nov. 21, 1865	51, 028
Pump	E. Elliott	Petaluma, Cal	Apr. 14, 1863	38, 157
Pump	D. S. Evans	Brockway, Mich	Jan. 5, 1869	85, 517
Pump	J. Evans	Whitestown, N. Y	Jan. 6, 1812	....
Pump	J. Evens	Lebanon, Ohio	Nov. 9, 1838	1, 002
Pump	J. Evens	Lebanon, Ohio	May 25, 1840	1, 615
Pump	J. L. Fagan	Anaqua, Tex	Feb. 22, 1859	23, 019
Pump	J. K. Fairbank	Waupun, Wis	Mar. 1, 1864	41, 764
Pump	D. L. Farnam	Philadelphia, Pa	Jan. 24, 1834	....
Pump	G. B. Farnam	Meriden, Conn	July 6, 1858	20, 783
Pump	J. Farnam	Stillwater, N. Y	Nov. 16, 1841	2, 370
Pump	B. W. Felthousen	Milwaukee, Wis	May 24, 1870	103, 435
Pump	A. Fisher	Nashua, N. H	Apr. 7, 1868	76, 318
Pump	E. Fitzgerald	New York, N. Y	Dec. 17, 1861	33, 936
Pump	A. Fitzpatrick	New York, N. Y	June 2, 1863	38, 736
Pump	E. Flanegin and A. B. Smith	Pittsburgh, Pa	Mar. 26, 1867	63, 235
Pump	A. F. Fletcher	Athol, Mass	Sept. 18, 1866	58, 085
Pump	A. F. Fletcher	Athol, Mass	Oct. 16, 1866	58, 801
Pump	A. V. and A. F. Fletcher	Athol, Mass	Oct. 25, 1864	44, 792
Pump	N. T. Fogg	Lewiston, Me	May 22, 1866	54, 881
Pump	D. Foley	Adams, N. Y	Mar. 12, 1872	124, 563
Pump	I. N. Forrester	Bridgeport, Conn	June 8, 1869	90, 940
Pump	I. N. Forrester	Bridgeport, Conn	Dec. 28, 1869	98, 366
Pump	I. N. Forrester	Bridgeport, Conn	Dec. 28, 1869	98, 367
Pump	I. N. Forrester	Bridgeport, Conn	Dec. 28, 1869	98, 368
Pump	I. N. Forrester and J. H. Luddington.	Bridgeport, Conn	Jan. 5, 1869	85, 577
Pump	G. R. Forsyth	Pemberton, Ohio	Dec. 24, 1867	72, 620
Pump	A. D. Foster	Jordan, N. Y	June 13, 1865	48, 165
Pump	T. Foulds, jr	Treverton, Pa	Nov. 5, 1872	132, 718
Pump	B. Frazee	Belleville, N. J	Dec. 5, 1865	51, 302
Pump	W. A. Fry	Worcester, Pa	Oct. 15, 1867	69, 792
Pump	R. M. Fryer	New York, N. Y	Jan. 14, 1868	73, 316
Pump	A. Fuller	Marietta, Pa	Nov. 30, 1869	97, 382
Pump	A. Fuller and F. J. Bray	Buffalo, N. Y	Mar. 26, 1867	63, 152
Pump	G. W. Fulton	Baltimore, Md	May 29, 1849	6, 486
Pump	J. P. Gallagher	Saint Louis, Mo	May 2, 1871	114, 284
Pump	G. W. Gardner and O. Higgins	Napoleon, Ohio	Mar. 31, 1868	76, 072
Pump	J. G. Garretson	Salem, Iowa	Jan. 19, 1864	41, 291
Pump	A. D. Gates	Surrey, Wis	Oct. 26, 1869	96, 099
Pump	G. W. B. Gedney	New York, N. Y	Jan. 6, 1857	16, 366
Pump	A. A. Genung	Painesville, Ohio	July 6, 1858	20, 787
Pump	H. Getty	Brooklyn, N. Y	July 23, 1867	67, 037

Index of patents issued from the United States Patent Office from 1790 *to* 1873, *inclusive*—Continued.

Invention.	Inventor.	Residence.	Date.	No.
Pump	R. Gilliland and R. H. Armstrong.	Hudson, Mich	Mar. 24, 1868	75, 747
Pump	K. Goddard	Philadelphia, Pa	Apr. 11, 1865	47, 200
Pump	J. Goland	Batavia, Ill	Aug. 4, 1863	39, 391
Pump	J. Goodyear	Carlisle Borough, Pa	Apr. 10, 1866	53, 810
Pump	D. J. Gorton	West Eau Claire, Wis	Mar. 15, 1870	100, 885
Pump	D. J. Gorton	Quincy, Ill	July 4, 1871	116, 703
Pump	L. A. Gould	Santa Clara, Cal	Mar. 6, 1866	52, 991
Pump	R. J. Gould	Newark, N. J	Feb. 15, 1870	99, 877
Pump	T. B. Goulding	Columbus, Ga	May 9, 1871	114, 668
Pump	E. Graser	Union City, Pa	Mar. 12, 1872	124, 434
Pump	A. G. Gray	London, England	Sept. 1, 1863	39, 728
Pump	S. H. Gray	Bridgeport, Conn	Aug. 21, 1855	13, 459
Pump	S. H. Gray	Bridgeport, Conn	Sept. 14, 1858	21, 560
Pump	J. Greaves	Utica, N. Y	June 24, 1862	35, 677
Pump	J. Greaves	Utica, N. Y	Oct. 21, 1862	36, 708
Pump	W. W. Green	Cambridge, Pa	Oct. 22, 1872	132, 462
Pump	M. T. Greenleaf	Quincy, Ill	July 16, 1861	32, 831
Pump	C. L. and J. P. Griscom	Mahanoy Plane, Pa	Sept. 28, 1869	95, 222
Pump	J. P. Gruber	New York, N. Y	Aug. 28, 1866	57, 499
Pump	B. F. Gustin	Middletown, Ind	Apr. 19, 1870	102, 199
Pump	E. Halo	Terre Haute, Ind	Jan. 5, 1864	41, 154
Pump	E. J. Hall	Indianapolis, Ind	Nov. 24, 1868	84, 274
Pump	M. P. Hall	Gayville, Ill	Feb. 15, 1870	99, 884
Pump	W. M. Hamilton	Jacksonville, Ill	Oct. 19, 1869	96, 000
Pump	W. M. Hamilton	Jacksonville, Ill	May 31, 1870	103, 738
Pump	E. C. Hammond	Oswego, N. Y	Mar. 29, 1870	101, 261
Pump	J. Hampson and G. Ladue	Newburgh, N. Y	Apr. 18, 1865	47, 299
Pump	T. Hansbrow	Sacramento, Cal	Feb. 5, 1861	31, 314
Pump	T. Hansbrow	Sacramento, Cal	Dec. 15, 1863	40, 927
Pump	T. Hansbrow	Sacramento, Cal	May 15, 1866	54, 719
Pump	T. Hansbrow	Sacramento, Cal	Nov. 24, 1868	81, 355
Pump	A. S. Hanson	Milan, Mich	Sept. 11, 1866	57, 897
Pump	M. Hanstine	Waynesborough, Pa	Dec. 7, 1869	97, 633
Pump	H. A. M. Harris	Philadelphia, Pa	Feb. 20, 1866	52, 709
Pump	H. A. M. Harris	Philadelphia, Pa	Sept. 24, 1867	69, 092
Pump	W. H. Harrison	Philadelphia, Pa	June 23, 1857	17, 625
Pump	R. Hartley	Pittsburgh, Pa	July 10, 1866	56, 218
Pump	W. Hartley	Durand, Ill	Dec. 6, 1870	109, 896
Pump	E. Hartzler	Orville, Ohio	July 26, 1859	24, 868
Pump	P. Harvey	Chicago, Ill	Jan. 24, 1871	111, 120
Pump	A. L. Hatfield	Clyde, Ohio	Oct. 21, 1873	143, 759
Pump	M. C. Hawkins	Edinborough, Pa	Aug. 24, 1869	93, 990
Pump	D. Hayes	New York, N. Y	Mar. 26, 1861	31, 804
Pump	W. M. Henderson	Baltimore, Md	Oct. 4, 1859	25, 642
Pump	W. M. Henderson	Baltimore, Md	Apr. 28, 1863	38, 308
Pump	W. Hendrick	New York	June 12, 1818	....
Pump	D. P. Henry	Windsor, Ill	May 4, 1869	89, 662
Pump	R. Henry	Morrisania, N. Y	June 17, 1862	35, 609
Pump	F. Henshaw	Washington, D. C	Sept. 14, 1858	21, 561
Pump	S. Hewit	Seneca Falls, N. Y	May 19, 1857	17, 327
Pump	S. Hewit	Seneca Falls, N. Y	Oct. 25, 1859	25, 898
Pump	G. Hibberd	Wheeling, W. Va	Oct. 25, 1870	108, 705
Pump	G. Hibsch	Buffalo, N. Y	July 13, 1858	20, 880
Pump	S. C. Higbie	Openheim, N. Y	Mar. 21, 1843	3, 010
Pump	O. Higgins	Napoleon, Ohio	Feb. 7, 1871	111, 645
Pump	B. S. Hill	New York, N. Y	June 27, 1865	48, 401
Pump	J. M. Hirlinger	Red Rock, Pa	Nov. 12, 1867	70, 842
Pump	S. S. Hogle	Lansingburgh, N. Y	May 29, 1841	2, 112
Pump	B. Holly	Seneca Falls, N. Y	June 5, 1849	6, 500
Pump	B. Holly	Seneca Falls, N. Y	July 14, 1857	17, 820
Pump	B. Holly	Lockport, N. Y	Feb. 14, 1860	27, 128
Pump	B. Holly	Lockport, N. Y	July 24, 1860	29, 266
Pump	B. Holly	Lockport, N. Y	July 14, 1863	39, 259
Pump	B. Holly	Lockport, N. Y	Nov. 15, 1864	45, 010
Pump	D. O. Holman	Adams, N. Y	Oct. 11, 1870	108, 264
Pump	D. O. Holman	Adams, N. Y	Oct. 11, 1870	108, 265
Pump	J. Holmes	Saint Clair, Pa	Oct. 2, 1860	30, 221
Pump	C. Hood	Seneca Falls, N. Y	Apr. 17, 1860	27, 905
Pump	W. D. Hooker	Stockton, Cal	Aug. 15, 1865	49, 408
Pump	J. W. Hopkins	New York, N. Y	Jan. 30, 1872	123, 175
Pump	A. Horn	Easton, Pa	Jan. 6, 1809	....
Pump	H. Hosick	Paris, Pa	Jan. 19, 1864	41, 298
Pump	N. Hotz	Brooklyn, N. Y	Feb. 14, 1865	46, 360
Pump	J. G. Hovey	Waverly, Iowa	Nov. 28, 1865	51, 183
Pump	B. J. C. Howe	Syracuse, N. Y	July 12, 1864	43, 502
Pump	B. J. C. Howe	Syracuse, N. Y	Oct. 11, 1864	44, 627
Pump	C. W. Hoyt	South Norwalk, Conn	Oct. 13, 1868	83, 065
Pump	H. S. Huges	Warren County, Va	June 29, 1839	1, 207
Pump	A. J. Hull	Sterling, Ill	Nov. 14, 1871	120, 973
Pump	J. Humphrey	Keene, N. H	Feb. 14, 1871	111, 748
Pump	H. C. Hunt and G. W. Devin	Ottumwa, Iowa	Nov. 4, 1862	36, 837
Pump	J. I. Hurlbutt	Norwalk, Ohio	July 12, 1870	105, 337
Pump	J. Icard	Donaldsonville, La	Nov. 29, 1870	109, 624
Pump	W. L. Jacobs	Lancaster, Pa	Sept. 23, 1842	2, 785
Pump	D. L. Jaques	Hudson, Mich	Oct. 23, 1866	59, 030
Pump	E. A. Jeffery and J. D. Quackenbush.	Corning, N. Y	June 10, 1862	35, 522
Pump	E. S. Jenkins	Brooklyn, N. Y	Jan. 17, 1871	110, 976
Pump	J. Johnson and C. W. Singer	Saginaw, Mich., and Anderson's Store, Va	Mar. 6, 1866	53, 007
Pump	J. B. Johnson	Boston, Mass	Dec. 4, 1860	30, 825
Pump	N. Johnson	Ripon, Wis	Dec. 26, 1865	51, 726
Pump	N. Johnson	Ripon, Wis	Aug. 28, 1866	57, 516
Pump	W. J. Johnson	Newton, Mass	Oct. 23, 1860	30, 480
Pump	T. O. Jones	Galesburgh, Ill	Jan. 31, 1871	111, 347
Pump	W. F. Jones	Syracuse, N. Y	June 22, 1869	91, 750

Index of patents issued from the United States Patent Office from 1790 to 1873, inclusive—Continued.

Invention.	Inventor.	Residence.	Date.	No.
Pump	J. Jonson	Baltimore, Md	Sept. 12, 1871	118, 859
Pump	J. O. Joyce	Cincinnati, Ohio	Mar. 23, 1858	19, 690
Pump	J. O. Joyce	Dayton, Ohio	Nov. 10, 1868	83, 858
Pump	E. B. Jucket	Pawtucket, R. I	Jan. 3, 1865	45, 722
Pump	E. B. Jucket	Roxbury, Mass	Aug. 28, 1866	57, 635
Pump	W. S. Judd	Clanhassen, Minn	Aug. 18, 1863	39, 577
Pump	A. C. Judson	Grand Rapids, Ohio	Dec. 21, 1869	98, 067
Pump	A. Jusberg	Galva, Ill	Oct. 22, 1867	70, 000
Pump	A. B. Keeley and J. S. Bleck	Philadelphia, Pa	May 17, 1859	24, 032
Pump	W. H. Keek	Stockton, Cal	Sept. 1, 1868	81, 648
Pump	A. L. Keeports and G. Palmer	Littlestown, Pa	Nov. 30, 1858	22, 182
Pump	H. M. Keith	Commerce, Mich	July 11, 1865	48, 691
Pump	L. L. Kellogg	Leon Centre, N. Y	Aug. 10, 1869	93, 537
Pump	W. S. Kelly	Schenectady, N. Y	Feb. 25, 1862	34, 505
Pump	W. S. Kelly	Schenectady, N. Y	Sept. 13, 1864	44, 272
Pump	W. S. Kelly	Schenectady, N. Y	Jan. 15, 1867	61, 209
Pump	H. K. Kenyon	Steubenville, Ohio	Nov. 29, 1870	109, 631
Pump	J. T. Kimbel	Vernon, Ind	Nov. 5, 1867	70, 443
Pump	A. Knecht	Ilchester, Md	Oct. 18, 1870	108, 365
Pump	J. Knibbs	Troy, N. Y	May 24, 1864	42, 920
Pump	L. J. Knowles	Worcester, Mass	June 27, 1871	116, 454
Pump	T. C. H. Kraus	Fishkill, N. Y	Aug. 19, 1873	142, 028
Pump	F. Krieg	Baltimore, Md	Feb. 6, 1833	....
Pump	R. M. Lafferty	Three Rivers, Mich	June 21, 1870	104, 604
Pump	R. M. Lafferty	Toledo, Ohio	May 28, 1872	127, 173
Pump	L. M. Laighton	Portsmouth, N. H	Sept. 1, 1817	....
Pump	E. Lamphear	Stepney, Conn	Mar. 22, 1870	101, 138
Pump	J. W. Lane	Newton, N. J	Nov. 19, 1861	33, 765
Pump	S. Lane, jr	Englewood, N. J	Nov. 26, 1872	133, 373
Pump	S. Lane, jr	Englewood, N. J	Oct. 21, 1873	143, 769
Pump	A. C. Lanning	Wilkesbarre, Pa	Apr. 19, 1859	23, 690
Pump	T. J. Lapsley	Nashville, Tenn	Jan. 31, 1871	111, 354
Pump	G. H. Laub	West Lebanon, Ind	Jan. 24, 1871	111, 219
Pump	E. Lawrence	Antrim, N. H	Nov. 8, 1864	44, 963
Pump	E. Lawrence and R. Lafley, 2d	Waterford, N. Y	Mar. 1, 1859	23, 094
Pump	A. D. Laws and J. C. Cooke	Bridgeport, Conn	June 6, 1871	115, 745
Pump	A. Leuchtweiss	Cincinnati, Ohio	Apr. 10, 1867	53, 939
Pump	W. B. Le Van	Philadelphia, Pa	May 16, 1865	47, 737
Pump	C. N. Lewis	Seneca Falls, N. Y	Jan. 1, 1856	14, 024
Pump	D. W. Lewis	Janesville, Wis	July 23, 1861	32, 881
Pump	E. T. Ligon	Richmond, Va	Oct. 14, 1856	15, 888
Pump	G. Lindsay	Petersburgh, Va	May 1, 1860	28, 090
Pump	H. Lindsey	Asheville, N. C	Dec. 22, 1857	18, 916
Pump	H. Lindsey	Asheville, N. C	Dec. 4, 1855	13, 881
Pump	T. Ling	New York, N. Y	Apr. 30, 1867	64, 337
Pump	T. J. Linton	Providence, R. I	Mar. 7, 1865	46, 683
Pump	A. W. Lloyd	Otis, Mass	May 3, 1859	23, 849
Pump	H. Locke	South Boston, Mass	Apr. 30, 1861	32, 196
Pump	F. R. Lockling	Hannibal, Mo	Aug. 20, 1872	130, 647
Pump	D. Loomis, J. Winters, and A. Stark.	Clyde, Ohio	Feb. 28, 1871	112, 157
Pump	G. W. Low	Ravenna, Ohio	Feb. 22, 1870	100, 162
Pump	J. H. Luddington	Bridgeport, Conn	May 4, 1869	89, 672
Pump	J. M. Lunquest	Griffin, Ga	Mar. 1, 1859	23, 100
Pump	R. M. Marchant	London, England	Dec. 20, 1870	110, 380
Pump	C. Markley	New York, N. Y	Nov. 29, 1870	109, 639
Pump	J. Marquis	San Francisco, Cal	Nov. 7, 1871	120, 655
Pump	G. Marshall	New York, N. Y	May 9, 1865	47, 652
Pump	G. Marshall	Brooklyn, N. Y	July 28, 1868	80, 418
Pump	G. W. Martin	Morrisania, N. Y	Apr. 9, 1861	32, 030
Pump	S. G. Mason and C. B. Gill	Rochester, N. Y	Feb. 6, 1866	52, 431
Pump	L. Matthews	Antrim, Ohio	Apr. 24, 1860	27, 996
Pump	J. M. May	Janesville, Wis	Aug. 23, 1859	25, 207
Pump	J. M. May	Janesville, Wis	Aug. 23, 1859	25, 208
Pump	J. M. May	Janesville, Wis	May 7, 1861	32, 251
Pump	J. Mayher	East Hampton, Mass	May 14, 1872	126, 643
Pump	T. Mayhew	New York, N. Y	Jan. 30, 1866	52, 364
Pump	J. McBride	Flint, Mich	Jan. 14, 1868	73, 253
Pump	A. McCarter	Norristown, Pa	Apr. 21, 1863	38, 234
Pump	R. A. McCauley	Baltimore, Md	Aug. 29, 1865	49, 640
Pump	R. A. McCauley	Baltimore, Md	Dec. 26, 1865	51, 736
Pump	R. A. McCauley	Baltimore, Md	Jan. 1, 1867	60, 919
Pump	A. Y. McDonald	Dubuque, Iowa	May 24, 1870	103, 481
Pump	J. H. McGowan	Cincinnati, Ohio	Nov. 22, 1870	109, 534
Pump	J. H. McGowan	Cincinnati, Ohio	Jan. 24, 1871	111, 247
Pump	J. H. McGowan	Cincinnati, Ohio	Jan. 2, 1872	122, 475
Pump	J. H. and T. J. McGowan	Cincinnati, Ohio	Nov. 12, 1867	70, 733
Pump	T. J. McGowan	Cincinnati, Ohio	Oct. 28, 1862	36, 844
Pump	T. J. McGowan	Cincinnati, Ohio	May 26, 1863	38, 689
Pump	T. J. McGowan	Cincinnati, Ohio	Sept. 1, 1868	81, 806
Pump	T. J. McGowan	Cincinnati, Ohio	Feb. 13, 1872	123, 636
Pump	B. F. McKeehan	Clarksburgh, W. Va	Mar. 14, 1871	112, 730
Pump	C. S. McMahan	Centreville, Ind	June 2, 1868	78, 602
Pump	J. L. McPherson	Sacramento City, Cal	July 8, 1862	35, 829
Pump	C. L. Merrill	Watertown, N. Y	Dec. 7, 1869	97, 670
Pump	C. L. Merrill	Watertown, N. Y	Aug. 2, 1870	106, 072
Pump	C. L. Merrill	Watertown, N. Y	Feb. 21, 1871	111, 961
Pump	J. W. Merrill and E. H. Lawrence.	Berlin, Mass	Aug. 13, 1867	67, 666
Pump	M. Metee	Baltimore, Md	May 20, 1830	....
Pump	M. Mettee	Baltimore, Md	Mar. 8, 1833	....
Pump	J. M. P. Metivier	Paris, France	Apr. 7, 1863	38, 117
Pump	T. Metzler	Wooster, Ohio	Oct. 26, 1869	96, 136
Pump	J. Michel	Rochester, N. Y	Mar. 27, 1866	53, 467
Pump	N. Miller	Finley, Ohio	June 18, 1861	32, 577
Pump	O. Miller	Salem, Ohio	Dec. 18, 1866	60, 535
Pump	G. E. Mills	New York, N. Y	July 18, 1865	48, 826
Pump	G. H. Mills	East Boston, Mass	May 22, 1860	28, 439

Index of patents issued from the United States Patent Office from 1790 *to* 1873, *inclusive*—Continued.

Invention.	Inventor.	Residence.	Date.	No.
Pump	J. R. Mills	Bloomfield, Iowa	Oct. 14, 1862	36, 665
Pump	A. Miner	Jordan, N. Y	July 7, 1835	....
Pump	A. T. Mixsell	Oxford, N. J	Sept. 28, 1839	1, 346
Pump	A. T. Mixwell	Oxford, N. J	June 30, 1836	....
Pump	A. Moon	Maquoketa, Iowa	June 11, 1867	65, 684
Pump	J. A. Morrell	Saint Charles, Mo	Jan. 7, 1862	34, 068
Pump	J. A. Morrell	Chicago, Ill	Oct. 29, 1867	70, 214
Pump	J. A. Morrell	New York, N. Y	June 8, 1869	91, 151
Pump	W. Mullally	Saint Paul, Minn	Oct. 28, 1862	36, 825
Pump	W. B. Munger	Hillsdale, Mich	Nov. 1, 1864	44, 879
Pump	A. Munson	Peterborough, N. H	June 3, 1862	35, 465
Pump	J. Munson	San José, Cal	June 14, 1864	43, 124
Pump	J. L. Munson	New Haven, Conn	Apr. 26, 1819	....
Pump	J. Neal and C. W. Emery	Boston, Mass	Jan. 1, 1856	14, 029
Pump	J. P. Nevens	Starks, Me	May 10, 1864	42, 681
Pump	N. Newman	Cincinnati, Ohio	May 6, 1851	8, 078
Pump	J. Nicholson	Allegheny City, Pa	Jan. 22, 1867	61, 447
Pump	T. Nicholson	Washington, D. C	Apr. 29, 1817	....
Pump	T. Nicholson and S. Gum	Genoa, N. Y	Aug. 6, 1813	....
Pump	H. Norman and C. F. Dietrich	New Orleans, La	Feb. 25, 1873	136, 258
Pump	S. B. B. Nowlan	New York, N. Y	Nov. 16, 1869	96, 947
Pump	W. H. Noyes	Franklin, Pa	June 4, 1867	65, 419
Pump	G. F. Nutting	Randolph, Vt	June 23, 1868	79, 143
Pump	J. K. O'Neil	Kingston, N. Y	Apr. 5, 1859	23, 485
Pump	M. S. Orton	Galesburgh, Ill	Mar. 29, 1870	101, 304
Pump	D. C. Owen	Adams County, Ill	May 11, 1869	89, 885
Pump	N. Page, jr	Danvers, Mass	Oct. 8, 1867	69, 578
Pump	A. Palmer	Hudson, Mich	Dec. 8, 1868	84, 754
Pump	G. Palmer	Littlestown, Pa	June 5, 1860	28, 599
Pump	G. Palmer	Littlestown, Pa	July 15, 1862	35, 888
Pump	H. H. Palmer	Rockford, Ill	Apr. 29, 1862	35, 108
Pump	A. N. Parkhurst	Peoria, Ill	Feb. 10, 1863	37, 638
Pump	A. N. Parkhurst	Peoria, Ill	Jan. 16, 1866	52, 070
Pump	A. N. Parkhurst	Knoxville, Ill	Nov. 2, 1869	96, 475
Pump	H. Parks	Athens, N. Y	June 11, 1867	65, 690
Pump	J. S. Patric	Victor, N. Y	May 1, 1866	54, 399
Pump	J. S. Patric	Rochester, N. Y	Feb. 28, 1871	112, 177
Pump	J. Patrick	Pitt County, N. C	Sept. 7, 1869	94, 507
Pump	T. Patterson	New York, N. Y	June 4, 1867	65, 424
Pump	J. Peabody	Dixmont Centre, Me	June 13, 1865	48, 202
Pump	F. S. Pease	Buffalo, N. Y	Dec. 12, 1865	51, 474
Pump	F. S. Pease	Buffalo, N. Y	Jan. 9, 1867	51, 965
Pump	H. Pease	Rockport, N. Y	July 7, 1857	17, 768
Pump	H. Pease	Rockport, N. Y	Sept. 11, 1860	29, 989
Pump	W. Peck	Rockford, Ill	Jan. 25, 1859	22, 743
Pump	W. Peck	Rockford, Ill	Jan. 3, 1860	26, 705
Pump	W. Peck	Rockford, Ill	May 8, 1860	28, 193
Pump	J. B. Pell	New York	Oct. 31, 1829	....
Pump	A. M. Perkins	Springfield, Mass	Apr. 2, 1861	31, 904
Pump	J. Perkins	....	July 9, 1802	....
Pump	J. Perkins	Newburyport, Mass	Mar. 23, 1813	....
Pump	A. Perry and M. C. Hawkins	Edinborough, Pa	Apr. 16, 1867	63, 937
Pump	E. Perry	Baldwinsville, N. Y	Sept. 20, 1864	44, 337
Pump	N. Phelps	Turin, N. Y	Oct. 20, 1831	....
Pump	J. Pike and W. Fisk	Cumberland, R. I	Aug. 13, 1839	1, 285
Pump	J. H. Plank	Pulaski, Iowa	Nov. 24, 1863	40, 708
Pump	J. R. Pomroy and L. J. Walter	Lockport, N. Y	Apr. 27, 1869	89, 338
Pump	J. Poppe	Green Point, N. Y	Sept. 1, 1868	81, 681
Pump	A. F. Porter	Philadelphia, Pa	Apr. 25, 1865	47, 452
Pump	C. Powell	Birmingham, England	Aug. 24, 1869	94, 027
Pump	J. Powers	New York, N. Y	Apr. 5, 1859	23, 489
Pump	G. W. Preston	Clyde, Ohio	May 7, 1872	126, 574
Pump	O. W. Preston, jr	Corning, N. Y	Nov. 30, 1858	22, 291
Pump	A. J. Pritchard	Liverpool, Ohio	May 25, 1869	90, 576
Pump	F. F. Prud'homme	Paris, France	Sept. 22, 1863	40, 062
Pump	W. Purcell	New York	Nov. 1, 1817	....
Pump	A. M. Putnam	Antrim, N. H	May 13, 1862	35, 258
Pump	A. M. Putnam	Antrim, N. H	Aug. 31, 1869	94, 338
Pump	S. S. Putnam	Boston, Mass	June 1, 1858	20, 442
Pump	W. Race	Seneca Falls, N. Y	May 22, 1860	28, 405
Pump	W. Race	Lockport, N. Y	June 30, 1863	39, 065
Pump	R. Ramsden	South Easton, Pa	Apr. 24, 1860	28, 012
Pump	S. G. Randall	Worcester, Mass	Feb. 28, 1860	27, 308
Pump	F. Ransom	Buffalo, N. Y	July 11, 1865	48, 720
Pump	C. D. Rathbone	Belpre, Ohio	July 8, 1873	140, 643
Pump	J. W. Redding	Belleville, Ohio	July 2, 1861	32, 716
Pump	J. Redeslperger	Mansfield, N. J	Oct. 31, 1835	....
Pump	J. Reed	Marshfield, Mass	Aug. 5, 1831	....
Pump	J. Reed	Marshfield, Mass	July 24, 1838	853
Pump	J. Reed	Marshfield, Mass	Apr. 16, 1841	2, 055
Pump	K. Reed	New York, N. Y	Feb. 6, 1866	52, 446
Pump	H. W. Regan	Cressona, Pa	Oct. 12, 1858	21, 801
Pump	A. Richard	Washington, Mo	Feb. 7, 1860	27, 070
Pump	J. W. Reid	New York, N. Y	Jan. 5, 1864	41, 091
Pump	A. Rewerick	San Francisco, Cal	Mar. 13, 1866	53, 185
Pump	A. J. Reynolds	Dayton, Ohio	Mar. 3, 1863	37, 826
Pump	A. J. Reynolds	Sturgis, Mich	May 10, 1864	42, 686
Pump	A. J. Reynolds	Sturgis, Mich	Sept. 24, 1867	69, 250
Pump	G. A. Reynolds and G. H. Babcock.	Mystic Bridge, Conn	Dec. 15, 1863	40, 945
Pump	E. Rhoads, sr	Clyde, Ohio	Oct. 7, 1862	36, 622
Pump	F. Richter	Milwaukee, Wis	June 4, 1867	65, 435
Pump	T. Rider	Valparaiso, Chili	Mar. 28, 1865	47, 075
Pump	J. Roberts	New Madison, Ohio	Oct. 3, 1871	119, 650
Pump	J. Robinson and L. Shaw	Bath, Me	Feb. 1, 1827	....
Pump	W. F. Robinson	Bellaire, Ohio	July 25, 1871	117, 463
Pump	W. W. Robinson	Ripon, Wis	June 4, 1861	32, 485
Pump	J. M. Roebuck and W. R. Reece	Donaldson, Pa	July 3, 1866	56, 101

Index of patents issued from the United States Patent Office from 1790 *to* 1873, *inclusive*—Continued.

Invention.	Inventor.	Residence.	Date.	No.
Pump	H. Roesen	Elkhart, Ind	May 5, 1863	38, 416
Pump	C. Rogers	Allegheny City, Pa	Feb. 11, 1868	74, 431
Pump	J. Ross	Greenville, Mich	Jan. 22, 1867	61, 468
Pump	P. C. Rowe	Boston, Mass	Nov. 8, 1864	44, 978
Pump	M. V. Rowley	Worcester, N. Y	Nov. 8, 1870	109, 057
Pump	M. E. Rudasill	Shelby, N. C	Oct. 9, 1860	30, 354
Pump	A. Russell	Newburyport, Mass	Aug. 6, 1861	33, 008
Pump	E. G. Russell	Ravenna, Ohio	Aug. 26, 1873	142, 123
Pump	S. Russell	Olean, N. Y	Feb. 4, 1826	....
Pump	E. B. Scammon	Scipio, N. Y	June 20, 1816	....
Pump	L. B. Schafer	Baltimore, Md	Mar. 22, 1859	23, 314
Pump	C. Schmidt	Rock Island, Ill	July 6, 1869	92, 373
Pump	A. H. Scholfield and D. C. Sterry	Worcester, Mass	June 23, 1868	79, 150
Pump	J. Scott	Earlville, Iowa	Dec. 28, 1869	98, 422
Pump	J. and J. Seeberger	West Troy, N. Y	May 11, 1869	89, 892
Pump	G. C. Selfridge	North Greenfield, N. Y	Sept. 4, 1860	29, 918
Pump	J. Selser	Williamsport, Pa	May 10, 1859	23, 953
Pump	G. F. and H. F. Shaw	West Roxbury, Mass	Feb. 21, 1871	111, 983
Pump	J. Shaw	Bridgeport, Conn	Aug. 25, 1868	81, 417
Pump	J. W. Sheaffer	Sterling, Ill	May 1, 1866	54, 420
Pump	W. Shearer	Atlanta, Ga	Aug. 21, 1860	29, 723
Pump	W. Shearer	Atlanta, Ga	Nov. 29, 1870	109, 678
Pump	H. A. Sheldon	Middlebury, Vt	Dec. 15, 1857	18, 870
Pump	N. P. Sheldon	San Francisce, Cal	Dec. 20, 1870	110, 297
Pump	N. P. Sheldon	San Francisco, Cal	Jan. 10, 1871	110, 936
Pump	A. Sherman and J. W. Sheaffer	Sterling, Ill	Nov. 15, 1870	109, 349
Pump	G. Shield	Cincinnati, Ohio	Nov. 5, 1867	70, 634
Pump	C. W. Sholes and H. C. Kelly	Morrison, Ill	Dec. 2, 1873	145, 247
Pump	W. Shoup	Saltsburgh, Pa	Dec. 27, 1864	45, 647
Pump	J. A. Sinclair	Woodsfield, Ohio	Apr. 26, 1870	102, 442
Pump	J. A. Sinclair	Woodsfield, Ohio	June 6, 1871	115, 776
Pump	J. A. Sinclair	Woodsfield, Ohio	Sept. 17, 1872	131, 469
Pump	S. B. Sivertson	Chicago, Ill	Dec. 6, 1870	109, 955
Pump	W. N. Slason	South Reading, Mass	Mar. 20, 1860	27, 576
Pump	A. Sluthour	Cleveland, Ohio	Oct. 19, 1869	95, 943
Pump	J. Smart	Easton, Pa	Feb. 3, 1838	589
Pump	J. Smart	Northern Liberties, Pa	Mar. 28, 1854	10, 708
Pump	R. T. Smart and R. T. Smart, jr.	Troy, N. Y	Jan. 30, 1872	123, 203
Pump	J. A. C. J. Smith	Philadelphia, Pa	July 10, 1860	29, 110
Pump	O. Snell	Williamsburgh, Ohio	Dec. 8, 1868	84, 846
Pump	P. M. and O. Snell	Williamsburgh, Ohio	Jan. 28, 1870	104, 786
Pump	P. M. and O. Snell	Williamsburgh, Ohio	Aug. 9, 1870	106, 226
Pump	T. J. Southard	Richmond, Me	July 2, 1861	32, 724
Pump	C. M. Soward and J. A. Dykeman.	Dyer County, Tenn	Dec. 17, 1872	133, 955
Pump	H. Spear	Portland, Me	Nov. 21, 1871	121, 209
Pump	H. Spear	Cape Elizabeth, Me	Nov. 19, 1872	133, 266
Pump	W. P. Squire	Paris, Ill	Nov. 20, 1866	59, 869
Pump	J. A. Stansbury	Baldwinsville, N. Y	July 24, 1866	56, 627
Pump	E. L. Staples	Nashville, Tenn	May 22, 1866	55, 022
Pump	G. Steck and F. Arnold	Hughesville, Pa	Nov. 14, 1871	121, 016
Pump	N. Stedman	Aurora, Ind	Apr. 28, 1863	38, 360
Pump	N. Stedman	Aurora, Ind	Sept. 8, 1863	39, 842
Pump	J. M. Stephenson	Anderson, Ind	May 22, 1860	28, 417
Pump	C. A. Stillman	Westerly, R. I	Dec. 12, 1865	51, 522
Pump	H. M. Stoker	Watson, Ill	Apr. 3, 1866	53, 750
Pump	S. D. Stout	Charleston, Tenn	Oct. 9, 1860	30, 363
Pump	R. E. Strait	West Oneonta, N. Y	Aug. 28, 1866	57, 595
Pump	R. E. Strait	Battle Creek, Mich	Oct. 29, 1867	70, 285
Pump	N. Sutton	New York, N. Y	Nov. 3, 1857	18, 559
Pump	N. Sutton	Detroit, Mich	Apr. 18, 1865	47, 344
Pump	Z. T. Sweet	Tiskilwa, Ill	May 16, 1871	114, 989
Pump	C. O. Sylvester	Attica, Ind	July 9, 1872	128, 921
Pump	J. Tapley	Frankfort, Me	Sept. 19, 1854	11, 704
Pump	C. L. Taverdon and J. Moret	Paris and Sevres, France	Sept. 29, 1868	82, 659
Pump	A. Thayer	Chatham, N. Y	Feb. 1, 1842	2, 438
Pump	D. M. Thomas	Dowagiac, Mich	July 24, 1866	56, 637
Pump	S. Thomas	Boston, Mass	Oct. 10, 1817	....
Pump	W. H. Thomas	Sacramento, Cal	Mar. 29, 1864	42, 128
Pump	J. E. Thorp	New York, N. Y	July 29, 1862	36, 042
Pump	W. P. Tilton	Boston, Mass	May 6, 1817	....
Pump	J. Tomlinson	Racine, Wis	Apr. 19, 1864	42, 417
Pump	A. Tower	New York, N. Y	Feb. 3, 1857	16, 558
Pump	F. W. Tulley and T. Reece	Philadelphia, Pa	July 17, 1866	56, 468
Pump	H. Tyler	Gaines, N. Y	Sept. 10, 1867	68, 810
Pump	H. Tyler	Gaines, N. Y	Oct. 29, 1867	70, 379
Pump	F. J. Underwood	Rock Island, Ill	Oct. 8, 1867	69, 595
Pump	J. Underwood	Mason County, Ill	Jan. 5, 1869	85, 548
Pump	D. Updike	New York	Feb. 19, 1818	....
Pump	C. D. Van Allen	Petersburgh, Pa	July 8, 1841	2, 160
Pump	G. Van Camp	Somerset County, N. J	Sept. 17, 1861	33, 323
Pump	S. Vance	Newburyport Mass	Jan. 30, 1866	52, 346
Pump	H. Van Keuren	Jersey City, N. J	Aug. 16, 1870	106, 433
Pump	J. Vaughn and J. McGee	Galena, Ill	Feb. 18, 1868	74, 638
Pump	C. Verniand and D. J. Lucie	Quincy, Ill	Aug. 20, 1867	68, 014
Pump	W. W. Virdin	Baltimore, Md	Sept. 15, 1863	39, 974
Pump	W. T. Vose	Newtonville, Mass	Mar. 20, 1855	12, 566
Pump	E. Wade	Norwich, Conn	June 12, 1860	28, 704
Pump	T. J. Wadleigh	Sutton, N. H	Nov. 5, 1861	33, 671
Pump	H. Wadsworth	Duxbury, Mass	Apr. 26, 1870	102, 452
Pump	J. F. Walker	Easton, Pa	May 6, 1836	....
Pump	F. Walther	Winchester, Va	May 1, 1845	4, 026
Pump	J. J. Walton	New York, N. Y	Dec. 23, 1873	145, 918
Pump	R. Ward	Newark, N. J	Feb. 27, 1872	124, 024
Pump	S. B. Ward	Drummondtown, Va	Feb. 2, 1869	86, 612
Pump	W. G. Ward	Fayette, N. Y	Dec. 13, 1870	110, 177
Pump	C. Warner	Lexington, Ky	Nov. 10, 1841	2, 350
Pump	C. Warner	Brooklyn, N. Y	July 31, 1860	29, 424

Index of patents issued from the United States Patent Office from 1790 *to* 1873, *inclusive*—Continued.

Invention.	Inventor.	Residence.	Date.	No.
Pump	Z. Waters and S. Bradley	Bloomington, Ill	Dec. 12, 1871	121, 918
Pump	W. E. Watters	East Bend, Ky	May 7, 1861	32, 260
Pump	W. E. Watters	East Bend, Ky	Jan. 3, 1865	45, 799
Pump	J. H. Webster	Saint Louis, Mo	Feb. 28, 1845	3, 933
Pump	J. R. Weisiger	Danville, Ky	Jan. 22, 1867	61, 492
Pump	J. R. Weisiger	Danville, Ky	Mar. 3, 1868	75, 088
Pump	V. Weitz	Cleveland, Ohio	June 18, 1861	32, 592
Pump	J. D. West	New York, N. Y	Sept. 29, 1857	18, 309
Pump	E. C. Wharton	Philadelphia, Pa	Oct. 28, 1873	144, 171
Pump	E. C. Wharton	Philadelphia, Pa	Dec. 23, 1873	145, 921
Pump	N. W. Wheeler	Morristown, N. J	Nov. 14, 1871	120, 841
Pump	W. M. Wheeler	Liberty, Mo	May 15, 1841	2, 097
Pump	J. F. Whipple	Chicago, Ill	May 3, 1870	102, 634
Pump	J. W. Whitaker	Kenosha, Wis	Nov. 19, 1872	133, 278
Pump	J. F. White	Keene, N. H	Apr. 4, 1865	47, 146
Pump	P. B. and H. M. C. White	Dowagiac, Mich	Mar. 3, 1868	75, 228
Pump	R. White and D. Moritz	New York, N. Y	Sept. 24, 1872	131, 644
Pump	S. White	Providence, R. I	Mar. 24, 1868	75, 819
Pump	J. Wilcke	Newark, N. J	May 5, 1868	77, 694
Pump	M. Wilcox	Sacramento, Cal	Aug. 30, 1870	106, 904
Pump	R. Wileman	East Hartford, Conn	July 9, 1829	....
Pump	T. J. Willett	Nunda, N. Y	Dec. 10, 1861	33, 914
Pump	C. Wilson	Bridgeport, Conn	May 28, 1872	127, 284
Pump	C. A. Wilson and W. R. Dunlap	Cincinnati, Ohio	Sept. 29, 1868	82, 670
Pump	J. E. Wilson	Bridgeport, Conn	Apr. 26, 1870	102, 458
Pump	T. C. Wilson	Watkins, N. Y	May 3, 1870	102, 743
Pump	P. Wineman	Joliet, Ill	May 18, 1869	90, 143
Pump	E. B. Winship	Racine, Wis	Dec. 18, 1866	60, 605
Pump	H. E. Wolcott	Elbridge, N. Y	Dec. 13, 1870	110, 182
Pump	T. J. Wolfe	Baltimore, Md	Apr. 10, 1844	3, 529
Pump	J. Wolford and R. Conner	Knightsville, Ind	Aug. 5, 1873	141, 514
Pump	D. S. Wood	Delavan, Wis	May 8, 1866	54, 636
Pump	S. Woodruff and H. B. Beach	Hartford, Conn	Sept. 22, 1868	82, 371
Pump	R. L. Wright	West Nantmeal, Pa	Apr. 12, 1870	101, 960
Pump	W. Wright	Hartford, Conn	Nov. 15, 1859	26, 139
Pump	W. Wright	New York, N. Y	Mar. 8, 1870	100, 702
Pump	H. M. Wyeth	Pulaski, Iowa	Mar. 18, 1862	34, 711
Pump	H. M. Wyeth	Bloomfield, Iowa	Nov. 25, 1862	37, 024
Pump	H. M. Wyeth	Newark, Ohio	July 12, 1870	105, 401
Pump	F. G. Wynkoop	Corning, N. Y	Dec. 29, 1857	19, 003
Pump	F. G. Wynkoop	Corning, N. Y	Dec. 20, 1864	45, 551
Pump	F. Yeiser	Danville, Ky	Jan. 7, 1868	73, 153
Pump	J. H. Young	Saint Louis, Mo	May 17, 1859	24, 073
Pump	S. M. Young and P. Brand	Jacksonville, Ill	Apr. 2, 1872	125, 243
Pump	H. Zeng	Elizabethport, N. J	July 27, 1858	21, 043
Pump	M. W. and J. Zimmerman	Earl Township, Pa	May 3, 1864	42, 617
Pump	O. Zwietusch	Milwaukee, Wis	Nov. 9, 1869	96, 760
Pump, Acid	L. H. Fisher	Oil City, Pa	Dec. 3, 1872	133, 639
Pump, Acid	F. Nichols	New London, Conn	Mar. 18, 1873	136, 934
Pump, Air	G. E. Barker and S. F. Mack	Waverly, N. Y	Dec. 30, 1873	145, 920
Pump, Air	N. H. Borgfeldt	New York, N. Y	Nov. 29, 1864	45, 220
Pump, Air	D. Carpenter	Peekskill, N. Y	Mar. 26, 1867	63, 209
Pump, Air	T. Claxton	Boston, Mass	June 7, 1834	....
Pump, Air	T. Doane	Boston, Mass	Dec. 31, 1867	72, 820
Pump, Air	W. H. Flanigan	Philadelphia, Pa	Mar. 26, 1872	125, 037
Pump, Air	H. Getty	Brooklyn, N. Y	Oct. 1, 1867	69, 423
Pump, Air	C. Goodyear	Philadelphia, Pa	Mar. 30, 1835	....
Pump, Air	C. B. Hardick	Brooklyn, N. Y	Feb. 14, 1871	111, 742
Pump, Air	J. F. Haskins	Fitchburgh, Mass	Nov. 29, 1870	109, 732
Pump, Air	H. H. Hendrick	Cincinnati, Ohio	Apr. 19, 1864	42, 437
Pump, Air	J. W. Hopkins	Brooklyn, N. Y	Jan. 25, 1870	99, 193
Pump, Air	A. S. Lyman	New York, N. Y	Jan. 31, 1865	46, 122
Pump, Air	R. M. Marchant	London, England	Feb. 21, 1871	112, 060
Pump, Air	J. Molyneaux	Bordentown, N. J	Apr. 11, 1865	47, 243
Pump, Air	G. M. Mowbray	Titusville, Pa	Nov. 22, 1864	45, 168
Pump, Air	R. Porter	Melrose, Mass	May 26, 1863	38, 715
Pump, Air	F. Ransom	Buffalo, N. Y	Aug. 8, 1865	49, 301
Pump, Air	E. S. Ritchie	Brookline, Mass	Nov. 19, 1867	71, 218
Pump, Air	D. E. Somes	Washington, D. C	Oct. 5, 1869	95, 613
Pump, Air	J. M. Wightman	Boston, Mass	Nov. 10, 1841	2, 338
Pump, Air	J. H. Wilhelm	Chicago, Ill	Dec. 26, 1865	51, 769
Pump, Air	G. M. Woodward	New York, N. Y	May 30, 1865	48, 009
Pump, Air-compressing	J. N. Dennison	Newark, N. J	Oct. 23, 1866	59, 124
Pump, Air-compressing	B. Holly	Lockport, N. Y	May 22, 1866	54, 905
Pump, Air-compressing	C. W. Wailey	New Orleans, La	Nov. 10, 1868	84, 030
Pump, Air or water	J. Barron	Hampton, Va	Feb. 20, 1819	....
Pump and condenser, Steam air	S. Gibbons	Freedom, Pa	Sept. 2, 1873	142, 333
Pump and cupping-instrument, Stomach and breast	L. B. White	New York	July 22, 1833	....
Pump and cupping-instrument, Stomach and breast	L. B. White	New York	Dec. 23, 1834	....
Pump and engine, Air	T. Beach	Freeport, Pa	Mar. 25, 1873	137, 123
Pump and engine piston	D. Hinman	Brunswick, Ohio	Mar. 7, 1846	4, 396
Pump and engine, Rotary	D. D. Hardy	Cincinnati, Ohio	Dec. 11, 1866	60, 365
Pump and engine, Rotary	D. D. Hardy and J. J. Morris	Cincinnati, Ohio	Dec. 11, 1866	60, 366
Pump and fire-engine	B. T. Babbitt, S. C. Higbee, and P. W. Plantz.	Little Falls and Oppenheim, N. Y.	Oct. 7, 1842	2, 801
Pump and fire-engine	S. Weirick and H. Lathrope	Warsaw, Ind	June 10, 1873	139, 840
Pump and fire-engine, Steam	J. W. Whitaker	Kenosha, Wis	June 13, 1871	115, 918
Pump and fire-engine, Suction	I. Jennings	Philadelphia, Pa	Apr. 1, 1812	....
Pump and fountain	J. J. Sink	Philadelphia, Pa	June 4, 1872	127, 651
Pump and funnel, Combined atmospheric exhaust	G. H. Randall	Forestville, Md	Feb. 25, 1873	136, 264
Pump and gasometer, Compound air	S. Chichester	Poughkeepsie, N. Y	Apr. 20, 1858	19, 931
Pump and measure, Combined	J. M. Brooks and P. Munson	Independence, Iowa	Oct. 15, 1867	69, 756
Pump and nursing-bottle, Combined breast	O. H. Needham	Needham, N. Y	Sept. 19, 1871	119, 044
Pump and other oscillating rods, Protector for	L. Wilson	Middletown, Conn	July 18, 1865	48, 880
Pump and pump-box	J. C. Helme	Newport, R. I	Aug. 1, 1817	....
Pump and reel for artesian wells, Sand	J. A. Fleming	Shamburgh, Pa	Jan. 2, 1872	122, 449
Pump and reservoir, Combined	E. Nickerson	Provincetown, Mass	Nov. 19, 1867	71, 044
Pump and siphon, Combined	J. D. Averill, G. A. Higgins, and T. Gordon.	New York, N. Y., and Shrewsbury, N. J.	Oct. 11, 1870	108, 087

Index of patents issued from the United States Patent Office from 1790 *to* 1873, *inclusive*—Continued.

Index of patents issued from the United States Patent Office from 1790 *to* 1873, *inclusive*—Continued.

Invention.	Inventor.	Residence.	Date.	No.
Pump, Centrifugal	A. W. and J. H. Von Schmidt	Washington, D. C	July 24, 1847	5, 203
Pump, Centrifugal-screw	J. H. White and W. S. Henson	Lima, South America, and Newark, N. J.	Jan. 26, 1869	86, 264
Pump, Chain	A. G. Bradford	Freeport, Ill	June 10, 1862	35, 499
Pump, Chain	A. Coates	Marlborough, Ohio	Aug. 7, 1860	29, 466
Pump, Chain	J. M. Connel	Newark, Ohio	June 21, 1864	43, 253
Pump, Chain	W. Cooper	Ypsilanti, Mich	June 4, 1872	127, 576
Pump, Chain	D. Du Pré	Raleigh, N. C	June 7, 1859	24, 288
Pump, Chain	L. A. Fisher	Chicago, Ill	Dec. 20, 1864	45, 484
Pump, Chain	J. Harrison, jr	New York, N. Y	May 5, 1857	17, 217
Pump, Chain	E. Morris	Burlington, N. J	Feb. 10, 1857	16, 603
Pump, Chain	J. Robingson	New Brighton, Pa	Nov. 4, 1856	16, 024
Pump, Chain	A. Wyckoff	Columbus, Ohio	Apr. 3, 1855	12, 645
Pump, Chinese chain	J. Black	Northumberland, Pa	Aug. 2, 1810	....
Pump, Cog and pendulum	B. Lies	New York, N. Y	Aug. 6, 1812	....
Pump, Common suction	D. Watson	Plymouth, Mass	Mar. 4, 1808	....
Pump, Compound high and low pressure steam	A. J. L. Loretz	Brooklyn, N. Y	May 20, 1873	139, 071
Pump, Compound hydraulic	H. Rodgers	Auburn, N. Y	Oct. 9, 1841	2, 279
Pump, Compound propeller	Y. Shaw	Philadelphia, Pa	Feb. 15, 1870	99, 791
Pump, Compound steam	A. J. L. Loretz	Brooklyn, N. Y	May 20, 1873	139, 070
Pump-coupling	S. H. Gray	Bridgeport, Conn	Apr. 27, 1858	20, 060
Pump, Cream	T. A. Case	Ellington, N. Y	May 25, 1869	90, 424
Pump, Cream	M. A. Richardson	Sherman, N. Y	Sept. 23, 1862	36, 530
Pump-cylinders, Lining for	W. S. Owen	Oskaloosa, Iowa	Dec. 23, 1873	145, 895
Pump, Deep-well	J. A. Bloom	Philadelphia, Pa	Jan. 16, 1866	52, 019
Pump, Deep-well	R. Boeklen	Brooklyn, N. Y	Nov. 21, 1865	51, 007
Pump, Deep-well	A. Carver	Little Falls, N. Y	Aug. 14, 1866	57, 086
Pump, Deep-well	J. J. G. Collins	Philadelphia, Pa	Nov. 28, 1865	51, 148
Pump, Deep-well	R. Cornelius	Philadelphia, Pa	Mar. 13, 1866	53, 117
Pump, Deep-well	R. Cornelius	Philadelphia, Pa	May 1, 1866	54, 300
Pump, Deep-well	R. Cornelius	Philadelphia, Pa	June 26, 1866	55, 822
Pump, Deep-well	B. Crawford	Allegheny, Pa	July 17, 1866	56, 380
Pump, Deep-well	J. H. Davis	Allegheny City, Pa	June 19, 1866	55, 626
Pump, Deep-well	N. Dodge	New York, N. Y	Mar. 12, 1850	7, 161
Pump, Deep-well	N. Dodge	New York, N. Y	June 27, 1865	48, 378
Pump, Deep-well	S. H. Early	Lynchburgh, Va	Mar. 13, 1866	53, 128
Pump, Deep-well	W. H. Genung	Madison, Ohio	July 16, 1872	129, 220
Pump, Deep-well	J. Harrison	Jamestown, N. Y	Apr. 17, 1866	53, 974
Pump, Deep-well	S. E. Hewes	Albany, N. Y	Nov. 14, 1865	50, 932
Pump, Deep-well	W. S. Kelly	Schenectady, N. Y	Apr. 17, 1866	53, 990
Pump, Deep-well	H. K. Kenyon	Steubenville, Ohio	June 28, 1870	104, 855
Pump, Deep-well	T. J. Lovegrove	Philadelphia, Pa	Dec. 19, 1865	51, 602
Pump, Deep-well	W. E. Morrison and W. L. Betts	Funkville, Pa	July 17, 1866	56, 435
Pump, Deep-well	J. Old	Pittsburgh, Pa	Feb. 18, 1862	34, 444
Pump, Deep-well	J. Old	Pittsburgh, Pa	June 6, 1865	48, 089
Pump, Deep-well	J. Sheffield	Pultneyville, N. Y	June 6, 1865	48, 105
Pump, Deep-well	W. B. Snow	Titusville, Pa	Dec. 18, 1866	60, 586
Pump, Deep-well	J. W. Summers	Tarr Farm, Pa	Sept. 18, 1866	58, 150
Pump, Deep-well	T. J. Trapp	Williamsport, Pa	July 11, 1871	117, 019
Pump, Deep-well	J. Warren	Buffalo, N. Y	May 10, 1864	42, 705
Pump, Deep-well	J. T. Whipple	Chicago, Ill	Feb. 18, 1868	74, 738
Pump-device for steam and other enginery	C. E. Blake	San Francisco, Cal	Jan. 26, 1869	86, 273
Pump device, Steam	M. Wilcox	Sacramento, Cal	Mar. 8, 1870	100, 576
Pump, Diaphragm	D. F. Hitchcock	Warren, Mass	Jan. 17, 1854	10, 428
Pump, Diaphragm	J. L. McPherson and J. O. Joice	New Vienna and Cincinnati, Ohio.	Feb. 12, 1856	14, 247
Pump, Diaphragm	J. A. Pease	New York, N. Y	Aug. 22, 1854	11, 559
Pump, Direct-acting hydraulic steam	R. B. Gorsuch	New York, N. Y	Apr. 3, 1855	12, 625
Pump, Direct-acting hydraulic steam	H. R. Worthington	Brooklyn, N. Y	July 31, 1855	13, 370
Pump, Direct acting steam	O. T. Earle	Norwalk, Conn	Aug. 26, 1873	142, 091
Pump, Direct-acting steam	J. A. Reed	Jersey City, N. J	Apr. 28, 1863	38, 334
Pump, Direct-acting steam	W. Sewell and A. S. Cameron	New York, N. Y	May 10, 1864	42, 694
Pump, Double-acting	P. M. Barker	Grand Haven, Mich	June 10, 1873	139, 650
Pump, Double-acting	M. Braun	Cape Vincent, N. Y	July 5, 1870	105, 035
Pump, Double-acting	D. W. Clark	Bridgeport, Conn	Nov. 20, 1855	13, 816
Pump, Double-acting	W. H. Davis	Austin, Ind	July 19, 1859	24, 797
Pump, Double-acting	P. Foley	Nineveh, N. Y	Aug. 4, 1868	80, 617
Pump, Double-acting	T. N. Henderson	Jackson, Mich	Oct. 8, 1872	131, 952
Pump, Double acting	B. J. C. Howe	Syracuse, N. Y	May 17, 1864	42, 773
Pump, Double-acting	W. H. Ivens	Trenton, N. J	Feb. 8, 1870	99, 680
Pump, Double-acting	E. A. Jeffery	Corning, N. Y	June 5, 1855	13, 007
Pump, Double-acting	B. F. Joslyn	Worcester, Mass	Apr. 3, 1855	12, 631
Pump, Double-acting	W. S. Kelly	Schenectady, N. Y	Aug. 23, 1864	43, 917
Pump, Double-acting	F. A. Mack	Niles, Mich	June 22, 1869	91, 648
Pump, Double-acting	K. Nolan	Galesburgh, Ill	Jan. 21, 1873	135, 146
Pump, Double-acting cylinder	P. Mix	Germantown, Pa	Jan. 9, 1835	....
Pump, Double-acting force	A. Bagley	Ypsilanti, Mich	July 12, 1870	105, 162
Pump, Double-acting force	J. Barney	Perry's Mills, N. Y	Mar. 28, 1871	113, 132
Pump, Double-acting force	E. Bellamy	Saint Louis, Mo	Apr. 13, 1858	19, 907
Pump, Double-acting force	W. C. and I. S. Burnham	New York, N. Y	Feb. 6, 1855	12, 342
Pump Double-acting force	D. W. Clark and S. H. Gray	Bridgeport, Conn	Dec. 19, 1854	12, 092
Pump, Double-acting force	D. L. Farnam	New York	Oct. 23, 1837	437
Pump, Double-acting force	J. P. Flanders	Vergennes, Vt	July 15, 1873	140, 819
Pump, Double-acting force	G. Fowler	Northford, Conn	Jan. 30, 1855	12, 312
Pump, Double-acting force	J. H. McGowan, jr	Cincinnati, Ohio	June 27, 1854	11, 169
Pump, Double-acting force	T. J. D. Yampert	Mobile, Ala	Sept. 11, 1855	13, 559
Pump, Double-acting hollow-piston	F. A. Chase	Jordan, N. Y	July 26, 1864	43, 637
Pump, Double-acting submerged	A. J. Reynolds	Sturgis, Mich	Nov. 1, 1864	44, 891
Pump, Double-acting suction and force	J. Farnam	Stillwater, N. Y	Dec. 14, 1841	2, 387
Pump, Double-acting water	J. Farnam	Stillwater, N. Y	Nov. 10, 1841	2, 345
Pump, Double air and water	B. Smallman	Philadelphia, Pa	Mar. 27, 1815	....
Pump, Double-bellows	D. L. Deshon, 2d, and E. L. Webster.	New London, Conn	Sept. 12, 1848	5, 770
Pump, Double-bored pendulum	A. Farra	Montgomery County, Pa	Dec. 5, 1809	....
Pump, Double-cylinder force	W. H. Ivens	Trenton, N. J	Feb. 8, 1870	99, 681
Pump, Double-cylinder suction and force	J. J. Rice	Salina, N. Y	Mar. 10, 1838	625
Pump, Double forcing	W. Baker	Springfield, N. Y	Dec. 31, 1819	....
Pump, Double forcing	W. Beach	Franklin, N. Y	Feb. 19, 1811	....

Index of patents issued from the United States Patent Office from 1790 *to* 1873, *inclusive*—Continued.

Invention.	Inventor.	Residence.	Date.	No.
Pump, Double forcing	C. Colver	Champlain, N. Y	Apr. 1, 1807	....
Pump, Double forcing	W. Douglass	Middletown, Conn	Aug. 20, 1835	....
Pump, Double forcing	T. Ferris	Dutchess County, N. Y	Feb. 20, 1811	....
Pump, Double forcing	S. Gillett	Windsor, Conn	Apr. 8, 1820	....
Pump, Double forcing	L. Gray	Otsego County, N. Y	Mar. 7, 1811	....
Pump, Double forcing	L. Newton	Alexander, N. Y	Mar. 8, 1836	....
Pump, Double forcing	J. G. White	Dryden, N. Y	July 1, 1836	....
Pump, Double piston	S. W. Allen	Geneva, N. Y	July 21, 1831	....
Pump, Draining	W. S. Nelson	Saint Louis, Mo	Jan. 2, 1866	51, 854
Pump-dredge, Submarine	D. Quinn	Chicago, Ill	Apr. 20, 1869	89, 073
Pump, Driven	S. Waite	New Bedford, Mass	Sept. 17, 1867	68, 917
Pump, Driven-well	J. W. Harrington and J. A. Yingling.	Cincinnati and Hamilton, Ohio.	Sept. 30, 1873	143, 345
Pump, Driving	S. H. Rhoades	Clyde, Ohio	Oct. 2, 1866	58, 479
Pump, Dry-bellows	E. Alden	....	Nov. 15, 1803	....
Pump, Elastic-tube	R. Porter and J. D. Bradley	Washington, D. C., and Brattleborough, Vt.	Apr. 17, 1855	12, 753
Pump, Electro-magnetic	C. H. Rudd and G. W. Shawk	Cleveland, Ohio	Sept. 3, 1867	68, 577
Pump, Elliptic-valve	J. Baker	Charlestown, Mass	Jan. 16, 1817	....
Pump, Elliptical rotary	B. Holly	Seneca Falls, N. Y	Feb. 6, 1855	12, 350
Pump, Endless-chain	J. A. and H. A. Pitts	Winthrop and Livermore, Me.	Apr. 28, 1834	....
Pump engine, Steam	J. Sutherland and F. Moakly	East Hampton, Mass	Mar. 24, 1868	75, 998
Pump for agitating the surface of water in wells, Attachment to.	W. D. Mayfield	Elkton, Ky	Dec. 24, 1850	7, 857
Pump for artesian wells, Sand	T. J. Lovegrove	Philadelphia, Pa	Mar. 7, 1865	46, 756
Pump for boxed wells	J. Suggett	Cortland, N. Y	Mar. 29, 1864	42, 126
Pump for continuous discharge	S. D. Lount	Rahtbone, W. Va	July 13, 1869	92, 621
Pump for deep wells, Tubular	C. H. Duncan	Pithole City, Pa	Apr. 20, 1869	89, 134
Pump for diving-bells, Hydropneumatic	G. Williamson	Brooklyn, N. Y	Jan. 1, 1856	14, 039
Pump for elevating meal, grain, &c	J. Ewing and D. Dickey	Oxford, Pa	Oct. 2, 1813	....
Pump for elevating water mixed with mineral substances.	W. Ball	Chicopee, Mass	Dec. 23, 1851	8, 602
Pump for excavating wells, &c., Gravel	J. I. and E. Rice	Salina, N. Y	Aug. 15, 1838	883
Pump for exhausting and sealing cans, Air	W. Y. Gill	Henderson, Ky	Aug. 14, 1860	29, 582
Pump for fire-engines	J. N. Dennisson and R. J. Gould	Newark, N. J	Nov. 26, 1867	71, 287
Pump for fire-engines	H. Gates	Northampton, Mass	June 12, 1835	....
Pump for fire-engines	J. Newman	Baltimore, Md	Apr. 14, 1838	691
Pump for fire-engines, Force	J. N. Dennisson	Newark, N. J	July 13, 1858	20, 867
Pump for fire-engines, Horizontal double-forcing	E. Smith	Palmer, Mass	Nov. 6, 1810	....
Pump for hydraulic presses, Force	W. P. Callahan	Dayton, Ohio	July 14, 1868	79, 949
Pump for liquid freight, Ship's	S. F. Paullin	Baltimore, Md	May 28, 1872	127, 364
Pump for locomotive-engines	C. L. Rice	Milwaukee, Wis	July 17, 1860	29, 195
Pump for locomotive-tenders, &c	A. W. Todd	Chicago, Ill	Dec. 6, 1864	45, 360
Pump for low-pressure steam-engine	J. Vial	Cleveland, Ohio	Sept. 9, 1862	36, 431
Pump for machinery	L. Knap	Brighton, N. Y	May 22, 1826	....
Pump for marine-alarms, Air	S. G. Cabell	Quincy, Ill	June 18, 1867	65, 872
Pump for mines, Portable	W. E. Sidney	Knightsville, Ind	Oct. 14, 1873	143, 724
Pump for oil-wells, Sand	J. F. Carll	Brooklyn, N. Y	Jan. 21, 1868	73, 577
Pump for railway-stations	H. S. Lansdell	New York, N. Y	Feb. 20, 1866	52, 721
Pump for railway-stations	A. W. Todd	Chicago, Ill	July 17, 1866	56, 467
Pump for railway-stations, Steam	W. H. Butler	Chicago, Ill	Mar. 20, 1866	53, 269
Pump for raising and measuring liquids	D. Hunsicker	Hartleton, Pa	Mar. 7, 1846	4, 409
Pump for raising water, Air or bellows	D. M. Miller	....	Mar. 5, 1804	....
Pump for raising water, Spring	W. Finn	Baths, Md	Feb. 13, 1806	....
Pump for ships, mines, &c	J. Stickney	....	Nov. 29, 1799	....
Pump for steam-boilers	A. Judson	Sweden, N. Y	Feb. 23, 1827	....
Pump for steam-boilers, Feed-water	E. W. Ellsworth	East Windsor Hill, Conn	Dec. 9, 1856	16, 206
Pump for steam-engines	J. F. Hamilton	Pittsburgh, Pa	July 24, 1860	29, 271
Pump for steam-engines, Air	T. B. Stillman	New York, N. Y	May 15, 1837	193
Pump for steam-engines, Air	D. A. Woodbury	Rochester, N. N	May 8, 1860	28, 229
Pump for steam-engines, Double air	J. Smallman	Philadelphia, Pa	Aug. 29, 1815	....
Pump for vessels, Alarm	E. Haskell	Dover, N. H	Oct. 24, 1871	120, 194
Pump, Force	W. Beers and W. Raynor	Milan, Ohio	Jan. 11, 1870	98, 374
Pump, Force	B. Brundage	Sumpter, Mich	Aug. 20, 1872	130, 620
Pump, Force	A. Carver	Little Falls, N. Y	June 21, 1864	43, 180
Pump, Force	N. Chapin	Penn Yan, N. Y	Mar. 30, 1836	....
Pump, Force	N. Clute	Schenectady, N. Y	Feb. 9, 1869	86, 645
Pump, Force	W. G. Crutchfield	Dayton, Ohio	Dec. 4, 1866	60, 146
Pump, Force	S. Curroy	Williamson City, Tenn	Jan. 3, 1825	....
Pump, Force	J. De Long	Upper Sandusky, Ohio	Oct. 25, 1864	44, 790
Pump, Force	B. Egbert	Lansing, N. Y	Jan. 23, 1836	....
Pump, Force	J. D. Egbert	Lansing, N. Y	Sept. 22, 1834	....
Pump, Force	J. Eveleth	....	June 13, 1801	....
Pump, Force	G. B. Farnam	New York, N. Y	Jan. 9, 1855	12, 199
Pump, Force	J. Farnam	Stillwater, N. Y	Oct. 22, 1834	....
Pump, Force	S. D. Gilson	Syracuse, N. Y	Nov. 24, 1863	40, 723
Pump, Force	C. B. Guy	Lybrand, Iowa	Aug. 2, 1864	43, 684
Pump, Force	E. B. Harris	Wilmington, Ill	July 24, 1866	56, 556
Pump, Force	F. Kettler	Milwaukee, Wis	July 26, 1859	24, 876
Pump, Force	F. Kettler	Milwaukee, Wis	July 26, 1859	24, 877
Pump, Force	J. L. Kitchin	Fayetteville, N. C	July 2, 1872	128, 635
Pump, Force	B. Sapham	Waterford, N. Y	Nov. 19, 1833	....
Pump, Force	W. W. Lesner	Venice, N. Y	Mar. 12, 1836	....
Pump, Force	J. Paddleford	Lebanon, N. Y	May 6, 1823	....
Pump, Force	R. Poole	Baltimore, Md	Sept. 6, 1859	25, 366
Pump, Force	H. Rogers	Ferrisburgh, Vt	Jan. 30, 1855	12, 326
Pump, Force	J. F. Rogers	Waterford, N. Y	Mar. 30, 1836	....
Pump, Force	O. Snell	Williamsburgh, Ohio	July 29, 1873	141, 394
Pump, Force and suction	A. Bailey	Jefferson, Ohio	May 4, 1838	722
Pump, Force and suction	J. Clark	Westminster, Vt	Sept. 11, 1823	....
Pump, Force and suction	A. Kasslar	Canajoharie, N. Y	Oct. 28, 1837	443
Pump, Force and suction	J. F. Rogers	Waterford, N. Y	Feb. 27, 1833	....
Pump, Force and vacuum	J. O. Joyce	Dayton, Ohio	July 2, 1872	128, 631
Pump, Frictionless	C. V. Card	New Bedford, Mass	Nov. 26, 1836	87
Pump, Frictionless	E. Whitfield	New York, N. Y	June 2, 1836	....
Pump, Gage	J. D. Woodside	Washington, D. C	July 14, 1834	....
Pump, Garden	C. G. Korth	Carlstadt, N. J	Mar. 12, 1872	124, 593

Index of patents issued from the United States Patent Office from 1790 to 1873, inclusive—Continued.

Invention.	Inventor.	Residence.	Date.	No.
Pump, Gas	W. Foster	Brooklyn, N. Y	Sept. 5, 1871	118, 709
Pump, Gas	W. H. Pollard	Seneca Falls, N. Y	Aug. 8, 1871	117, 925
Pump-gear	H. and F. J. L. Blandy	Zanesville, Ohio	July 5, 1864	43, 327
Pump-gearing	J. P. Carr	Mattapoisett, Mass	May 31, 1859	24, 193
Pump, Hand	T. J. Mayall	Boston, Mass	Apr. 16, 1872	125, 824
Pump-handle	I. M. Frazer	Elmer, N. J	Jan. 28, 1873	135, 269
Pump-handle bracket	D. S. Messler	Philadelphia, Pa	May 21, 1872	127, (87
Pump, Horizontal	L. Holland	Belchertown, Mass	Mar. 28, 1810	....
Pump, Horizontal	L. Holland	Belchertown, Mass	Sept. 10, 1811	....
Pump, Hydraulic	E. Squire	Cold Spring, N. Y	Oct. 25, 1864	44, 829
Pump, Hydraulic steam	H. N. Black	Philadelphia, Pa	Mar. 22, 1853	9, 622
Pump, Hydropneumatic force	A. B. Latta	Cincinnati, Ohio	June 27, 1854	11, 165
Pump, Impellent	S. Thayer	....	June 9, 1801	....
Pump into a suction and forcing pump, or vice versa, Converting a lifting.	J. F. Brickley	Winchester, Ind	Mar. 10, 1857	16, 785
Pump, Labor-saving	D. Smith	China, N. Y	June 13, 1831	....
Pump, Lift and force	A. H. Powell	Philadelphia, Pa	Jan. 14, 1873	134, 927
Pump, Lifting	F. Crocker	Titusville, Pa	Oct. 11, 1864	44, 610
Pump, Lifting	F. Raymond and A. Miller	Cleveland, Ohio	May 24, 1864	42, 874
Pump, Liquor	M. Cavanaugh	Philadelphia, Pa	July 26, 1870	105, 776
Pump, Liquor	H. Rodgers	Moravia, N. Y	Sept. 17, 1842	2, 779
Pump-lock, Ship's	A. G. Zeising	Weehawken, N. J	Mar. 15, 1870	100, 833
Pump, Locomotive-engine-feed	P. Cheswell	Manchester, N. H	June 15, 1869	91, 211
Pump-log-boring machine	S. Draper	Camillus, N. Y	Mar. 31, 1828	....
Pump, Measuring	S. H. Wheeler	Dowagiac, Mich	June 5, 1866	55, 433
Pump, Mechanical	W. Shultz	Baltimore, Md	June 24, 1809	....
Pump, Mercury-packed	P. Marcelin	Chambery, France	Aug. 23, 1864	43, 960
Pump, Metallic	J. Beckwith	Saratoga, N. Y	Dec. 27, 1833	....
Pump, Mine	T. M. Fell	New York, N. Y	Apr. 18, 1865	47, 289
Pump, Mining	T. J. Chubb	Williamsburgh, N. Y	Oct. 22, 1872	132, 384
Pump, Mining	G. E. Mills	New York, N. Y	Apr. 22, 1873	138, 037
Pump, Mining	H. Wolf	Saint Louis, Mo	Sept. 21, 1869	95, 061
Pump, Non-corrosive cast-iron	J. A. Rumsey	Seneca Falls, N. Y	May 25, 1869	90, 586
Pump, Oblique	J. Hunt	....	June 7, 1798	....
Pump, Oil	W. H. Elliot	Plattsburgh, N. Y	May 2, 1865	47, 530
Pump, Oil	L. S. Lapham	Providence, R. I	June 14, 1870	104, 168
Pump, Oil	A. P. Odell	Shamburgh, Pa	Dec. 17, 1872	134, 093
Pump, Oil-well	E. H. Ashcroft	Lynn, Mass	Dec. 27, 1864	45, 573
Pump, Oil-well	W. H. Elliot	Plattsburgh, N. Y	Nov. 19, 1861	33, 736
Pump, Oil-well	M. D. Fenner	Rochester, N. Y	June 30, 1868	79, 333
Pump, Oil-well	J. S. Fish	Cleveland, Ohio	Apr. 26, 1870	102, 241
Pump, Oil-well	A. S. Hill	Pleasantville, Pa	June 11, 1872	127, 880
Pump, Oil-well	J. B. Root	New York, N. Y	Apr. 4, 1865	47, 133
Pump, Oil-well	J. B. Root	New York, N. Y	Apr. 11, 1865	47, 224
Pump, Oil-well	T. Rose	Cortlandville, N. Y	Nov. 7, 1865	50, 846
Pump, Oil-well	H. Searl	Rochester, N. Y	July 25, 1865	48, 983
Pump, Oil-well	J. Thomas, jr	Cleveland, Ohio	Sept. 13, 1864	44, 264
Pump, Oil-well	N. Weare	Marietta, Ohio	Oct. 1, 1872	131, 800
Pump, Oil-well	W. S. Wilkinson	Baltimore, Md	Feb. 27, 1866	52, 947
Pump operated by exhaust steam, Steam vacuum	W. Burdon	Brooklyn, N. Y	Jan. 28, 1873	135, 203
Pump or blower, Rotary	A. Brear	Saugatuck, Conn	Dec. 16, 1873	145, 617
Pump or engine for raising water and ejecting air	J. Beckwith	Saratoga, N. Y	Apr. 16, 1831	....
Pump or engine, Rotary	W. P. Maxson	Elmira, N. Y	July 22, 1873	141, 155
Pump or syringe, Hand	W. B. Robins	Shepherd's Bush, England	May 27, 1873	139, 263
Pump or water-elevator, Rotary	T. J. Lovegrove	Philadelphia, Pa	Dec. 12, 1871	121, 793
Pump, Oscillating	E. Cope and I. W. Bragg	Cincinnati, Ohio	Mar. 23, 1858	19, 680
Pump, Oscillating	J. E. Hallett	Greenbush, N. Y	Mar. 20, 1866	53, 373
Pump, Oscillating	W. Painter	Baltimore, Md	Nov. 12, 1872	133, 048
Pump, Oscillating lifting	E. Horsey	Kingston, Canada	Jan. 16, 1872	122, 722
Pump, Package	J. A. Smith	Windham, Conn	Feb. 3, 1830	....
Pump, pan, &c., Vacuum	E. Dunscomb	Boston, Mass	Mar. 12, 1867	62, 739
Pump, Petroleum	J. Sparks	Rouseville, Pa	Aug. 29, 1871	118, 496
Pump-pipe	J. P. Cowing	Seneca Falls, N. Y	Aug. 30, 1864	43, 975
Pump-pipe joint	D. Knowlton	Camden, Me	July 19, 1859	24, 809
Pump-piston	J. D. Alvord	Bridgeport, Conn	Dec. 13, 1870	109, 994
Pump-piston	D. W. Bell	Saint Louis, Mo	Mar. 21, 1871	112, 768
Pump-piston	F. A. Cramblitt	Petroleum Centre, Pa	Aug. 6, 1867	67, 417
Pump-piston	S. Crowell	Syracuse, N. Y	Feb. 23, 1869	87, 148
Pump-piston	W. C. Culbertson	Girard, Pa	Aug. 19, 1873	142, 004
Pump-piston	W. E. Derrick	Jordan, N. Y	Sept. 17, 1867	68, 852
Pump-piston	W. F. Dodge	New York, N. Y	Apr. 4, 1865	47, 095
Pump-piston	A. H. Foe	Strathroy, Canada	Oct. 17, 1871	120, 054
Pump-piston	J. T. Follansbee and G. Doolittle.	Bridgeport, Conn	Mar. 7, 1871	112, 438
Pump-piston	P. Giffard	Paris, France	Jan. 16, 1872	122, 825
Pump-piston	S. P. Gilbert	Racine, Wis	Feb. 4, 1868	73, 969
Pump-piston	C. B. Gill	Weedsport, N. Y	June 4, 1872	127, 594
Pump-piston	C. B. Gill	Rochester, N. Y	Feb. 18, 1873	135, 905
Pump-piston	A. Griffin	Shamburgh, Pa	June 4, 1872	127, 480
Pump-piston	H. S. Hopkins	Boston, Mass	Apr. 14, 1868	76, 635
Pump-piston	J. Humphrey	Keene, N. H	Oct. 17, 1871	120, 070
Pump-piston	D. Johnson	Ashland, Ohio	Oct. 10, 1871	119, 852
Pump-piston	S. G. Mason	Elbridge, N. Y	Aug. 28, 1866	57, 531
Pump-piston	S. G. Mason	Rochester, N. Y	Apr. 14, 1868	76, 643
Pump-piston	S. G. Mason	Rochester, N. Y	Apr. 20, 1869	89, 060
Pump-piston	F. P. Michel	Rochester, N. Y	Apr. 21, 1868	76, 933
Pump-piston	B. H. Naves	Philadelphia, Pa	Dec. 31, 1867	72, 884
Pump piston	W. Newcomb	Baltimore, Md	Nov. 7, 1871	120, 767
Pump piston	G. Page	Washington, D. C	Apr. 17, 1847	5, 071
Pump-piston	J. Patterson	Monroe, N. Y	Nov. 11, 1862	36, 912
Pump-piston	B. and E. Pickering and B. Pickering.	West Milton and Montgomery County, Ohio.	Sept. 4, 1866	57, 760
Pump-piston	R. Poole	Baltimore, Md	Sept. 6, 1859	25, 367
Pump-piston	P. C. Rowe	Boston, Mass	June 6, 1855	48, 098
Pump-piston	O. Salgee	New York, N. Y	May 5, 1868	77, 538
Pump-piston	J. Van Tassel	Toledo, Ohio	Dec. 19, 1871	122, 082
Pump-piston	J. D. Westcott	Waterford, Pa	Aug. 25, 1868	81, 565
Pump-piston	D. S. Wood	Delaware, Wis	July 10, 1866	56, 314

Index of patents issued from the United States Patent Office from 1790 *to* 1873, *inclusive*—Continued.

Invention.	Inventor.	Residence.	Date.	No.
Pump-piston	J. Wood	Franklin, Pa	Aug. 25, 1868	81, 451
Pump-piston	P. Zeiher	Pomeroy, Ohio	June 18, 1872	128, 196
Pump piston, Deep-well	R. N. Allen	Cleveland, Ohio	Sept. 11, 1866	57, 840
Pump piston, Deep-well	J. W. Hoagland	New Brunswick, N. J	July 17, 1866	56, 414
Pump piston, Deep-well	C. Jarecki	Erie, Pa	Nov. 10, 1868	83, 968
Pump piston, Deep-well	E. Y. Kneeland	Buffalo, N. Y	Apr. 3, 1866	53, 630
Pump-piston, Double-acting	C. B. Gill	Rochester, N. Y	July 30, 1867	67, 291
Pump piston, Force	H. H. Cooley	Battle Creek, Mich	Oct. 21, 1862	36, 702
Pump piston and valve, Rotary	J. Gatley	Rome, N. Y	Sept. 12, 1854	11, 666
Pump piston, Steam	L. J. Knowles	Worcester, Mass	July 19, 1870	105, 465
Pump-pistons, engines, &c., Material for	D. Bickford	Boston, Mass	Dec. 24, 1867	72, 590
Pump-pistons, Operating	S. Wood	Worcester, Mass	Feb. 2, 1858	19, 276
Pump-pistons, Working	D. Whittier	Belfast, Me	Apr. 14, 1838	697
Pump-plunger and pump	N. P. Sheldon	San Francisco, Cal	July 15, 1873	140, 795
Pump, Pneumatic	J. A. Bailey	Detroit, Mich	May 11, 1869	89, 909
Pump, Pneumatic	H. Gottfried	New York, N. Y	May 1, 1866	54, 328
Pump, Pneumatic	W. S. Guild	Williamsburgh, N. Y	Nov. 28, 1865	51, 175
Pump, Pneumatic	J. Powell	Cincinnati, Ohio	Oct. 3, 1871	119, 641
Pump, Pneumatic breast	J. P. Stabler	Sandy Spring, Md	June 19, 1834	....
Pump, Pneumatic forcing	P. Upham	Brookfield, Mass	Jan. 6, 1809	....
Pump, Portable	J. A. Cheron	New York, N. Y	July 13, 1869	92, 584
Pump, Portable	W. and B. Douglas	Middletown, Tex	Apr. 6, 1858	19, 834
Pump, Portable	F. Eichler	New Lisbon, Wis	Dec. 15, 1868	85, 002
Pump, Portable	W. T. Vose	Newtonville, Mass	Nov. 15, 1859	26, 135
Pump, Portable sprinkling	W. Servant	Providence, R. I	Sept. 20, 1870	107, 633
Pump, Portable steam	W. Boardman, jr	New York, N. Y	Nov. 6, 1847	5, 352
Pump, Quadruple forcing	I. Ferris	Ellicott, N. Y	Aug. 12, 1828	....
Pump, Quicksilver	M. P. Boss	Bullionville, Nev	Dec. 23, 1873	145, 718
Pump, Railway	J. B. Atwater	Chicago, Ill	Oct. 4, 1864	44, 501
Pump, Reaction	E. Rhodes, jr	Clyde, Ohio	Jan. 14, 1873	134, 770
Pump, Reciprocating	W. D. Andrews	New York, N. Y	June 17, 1862	35, 577
Pump, Reciprocating	D. M. Weston	Boston, Mass	July 3, 1866	56, 129
Pump reel, Sand	H. T. Hunt	Titusville, Pa	May 21, 1872	127, 058
Pump reel, Sand	M. A. Richardson	Sherman, N. Y	Sept. 24, 1872	131, 706
Pump-regulator	S. H. Wheeler	Dowagiac, Mich	Aug. 29, 1871	118, 659
Pump-rod adjustment	B. Densmore	New York, N. Y	Aug. 30, 1870	106, 791
Pump-rod attachment	H. H. Locke	Pleasantville, Pa	Mar. 31, 1870	103, 632
Pump-rod connection for deep wells	H. De Zavala	New York, N. Y	Aug. 23, 1870	106, 673
Pump-rod coupling	I. A. Dewar, D. S. Smith, and R. A. Brashear.	Franklin, Pa	June 8, 1869	90, 935
Pump-rod coupling	H. T. Hunt	Titusville, Pa	Mar. 16, 1869	87, 778
Pump-rods, Connecting	H. F. Purmort	Saginaw City, Mich	Feb 4, 1868	74, 133
Pump, Rotary	C. L. Adancourt	Troy, N. Y	Sept. 22, 1863	40, 008
Pump, Rotary	W. B. Allyn	Boston, Mass	Jan. 21, 1870	104. 403
Pump, Rotary	M. L. Andrew	Cincinnati, Ohio	Apr. 27, 1869	89, 268
Pump, Rotary	M. L. Andrew	Cincinnati, Ohio	Apr. 27, 1869	89, 269
Pump, Rotary	W. D. Andrews	Brook Haven, N. Y	Dec. 20, 1870	110, 330
Pump, Rotary	J. Banks	New York, N. Y	Sept. 23, 1862	36, 500
Pump, Rotary	J. Banks	New York, N. Y	Aug. 18, 1863	39, 540
Pump, Rotary	J. S. Barden	New Haven, Conn	Jan. 26, 1858	19, 180
Pump, Rotary	A. Barker	Honesdale, Pa	Mar. 27, 1855	12, 579
Pump, Rotary	A. Barker	Honesdale, Pa	Mar. 24, 1857	16, 869
Pump, Rotary	J. A. Bazin	Canton, Mass	July 8, 1856	15, 274
Pump, Rotary	J. A. Bazin	Canton, Mass	June 25, 1867	65, 992
Pump, Rotary	P. Bennett	New York, N. Y	Jan. 7, 1851	7, 883
Pump, Rotary	J. D. Birch	Philadelphia, Pa	July 16, 1867	66, 671
Pump, Rotary	J. Broughton	Chicago, Ill	June 10, 1856	15, 059
Pump, Rotary	R. Burdine	Washington, D. C	Apr. 25, 1854	10, 822
Pump, Rotary	L. Burnell	Milwaukee, Wis	Aug. 31, 1858	21, 318
Pump, Rotary	R. Bush	Brooklyn, N. Y	May 10, 1870	102, 768
Pump, Rotary	W. Butterfield	Madison, Wis	Jan. 8, 1867	60, 995
Pump, Rotary	S. D. Carpenter	Madison, Wis	Oct. 10, 1854	11, 776
Pump, Rotary	S. D. Carpenter	Madison, Wis	June 24, 1856	15, 173
Pump, Rotary	L. Chapman	Collinsville, Conn	Sept. 19, 1871	119, 011
Pump, Rotary	L. Chapman	Collinsville, Conn	Mar. 25, 1873	137, 054
Pump, Rotary	L. Chapman	Collinsville, Conn	Mar. 25, 1873	137, 055
Pump, Rotary	L. Chapman	Collinsville, Conn	Mar. 25, 1873	137, 057
Pump, Rotary	L. Chapman	Collinsville, Conn	Mar. 25, 1873	137, 058
Pump, Rotary	M. R. Clapp	Seneca Falls, N. Y	Sept. 21, 1858	21, 550
Pump, Rotary	C. N. Clow	Port Byron, N. Y	July 1, 1856	15, 221
Pump, Rotary	C. N. Clow	Port Byron, N. Y	July 8, 1856	15, 280
Pump, Rotary	T. Crane	Fort Atkinson, Wis	Dec. 25, 1855	13, 979
Pump, Rotary	T. Crane	Fort Atkinson, Wis	Apr. 8, 1856	14, 599
Pump, Rotary	W. H. Davis	Maysville, Ky	Nov. 5, 1850	7, 751
Pump, Rotary	R. Defrees	Washington, D. C	July 31, 1866	56, 850
Pump, Rotary	F. O. Deschamps	Philadelphia, Pa	Mar. 7, 1871	112, 429
Pump, Rotary	J. Doyle and T. A. Martin	Hoboken, N. J., and New York, N. Y.	Dec. 10, 1867	72, 001
Pump, Rotary	J. Doyle and T. A. Martin	Hoboken, N. J., and New York, N. Y.	Nov. 8, 1870	109, 117
Pump, Rotary	C. I. Du Pont, jr	Wilmington, Del	Aug. 6, 1872	130, 113
Pump, Rotary	J. L. Fagan	Anaqua, Tex	Mar. 8, 1859	23, 162
Pump, Rotary	J. J. Flanders	Manchester, N. H	Nov. 1, 1859	25, 999
Pump, Rotary	G. S. Follensbee	Philadelphia, Pa	Apr. 23, 1872	126, 046
Pump, Rotary	J. T. Foster	New York, N. Y	Jan. 4, 1870	98, 482
Pump, Rotary	R. S. Foster	New York, N. Y	Dec. 27, 1864	45, 596
Pump, Rotary	W. and R. Foster	Brooklyn, N. Y	Jan. 9, 1866	51, 936
Pump, Rotary	T. Freeman, jr	Providence, R. I	Apr. 26, 1859	23, 764
Pump, Rotary	C. V. Genung	Duquoin, Ill	Nov. 17, 1868	84, 106
Pump, Rotary	R. Gilbert	Rochester, N. Y	Apr. 7, 1857	16, 974
Pump, Rotary	R. Gilbert	Rochester, N. Y	Mar. 29, 1864	42, 083
Pump, Rotary	J. E. Gillespie	Norwich, Conn	July 22, 1873	141, 000
Pump, Rotary	J. S. Godfrey	Leslie, Mich	May 7, 1872	126, 540
Pump, Rotary	J. Gray	Boston, Mass	Dec. 5, 1854	12, 022
Pump, Rotary	G. W. Griswold	Carbondale, Pa	Mar. 24, 1857	16, 875
Pump, Rotary	R. C. Grover and J. Nickelson	Newton, Mass	Dec. 26, 1865	51, 713
Pump, Rotary	W. H. Guild	Brooklyn, E. D., N. Y	June 17, 1862	35, 604
Pump, Rotary	J. S. Groynne	New York, N. Y	Jan. 14, 1851	7, 901

Index of patents issued from the United States Patent Office from 1790 *to* 1873, *inclusive*—Continued.

Invention.	Inventor.	Residence.	Date.	No.
Pump, Rotary	E. Hale	Hyde Park, N. Y	Aug. 3, 1833	....
Pump, Rotary	E. R. Hale	Hyde Park, N. Y	July 14, 1830	....
Pump, Rotary	I. Hall	Poughkeepsie, N. Y	Apr. 22, 1835	....
Pump, Rotary	D. S. Hamilton	Elmira, N. Y	Nov. 5, 1861	33, 643
Pump, Rotary	D. D. Hardy and J. J. Morris	Cincinnati, Ohio	Mar. 5, 1861	31, 597
Pump, Rotary	C. H. Harrison	San Francisco, Cal	June 21, 1864	43, 204
Pump, Rotary	G. W. Heald and L. D. Cisco	Baldwinsville, N. Y	July 25, 1865	48, 938
Pump, Rotary	G. W. Heald and L. D. Cisco	Baldwinsville, N. Y	Aug. 23, 1870	106, 581
Pump, Rotary	C. H. Hersey	Boston, Mass	July 26, 1859	24, 872
Pump, Rotary	C. H. Hersey	Boston, Mass	Oct. 6, 1868	82, 833
Pump, Rotary	W. Hinds	Little Falls, N. Y	Aug. 16, 1864	43, 849
Pump, Rotary	A. P. Holly	Seneca Falls, N. Y	July 6, 1858	20, 796
Pump, Rotary	C. P. Holmes	Havana, N. Y	Mar. 19, 1872	124, 820
Pump, Rotary	C. P. Holmes	Havana, N. Y	Sept. 9, 1873	142, 696
Pump, Rotary	R. A. Horning	Lenark, Ill	Aug. 2, 1870	106, 061
Pump, Rotary	C. E. Hutson	Saint Louis, Mo	Aug. 29, 1871	118, 455
Pump, Rotary	C. W. Isbell	New York, N. Y	Jan. 11, 1870	98, 772
Pump, Rotary	C. W. Isbell	New York, N. Y	June 6, 1871	115, 613
Pump, Rotary	C. L. Johnson	Little Falls, N. Y	Apr. 9, 1861	32, 029
Pump, Rotary	W. H. Johnson	Springfield, Mass	Mar. 21, 1846	4, 426
Pump, Rotary	D. L. Jones	Nebraska City, Nebr	Apr. 30, 1872	126, 306
Pump, Rotary	J. Jones and A. K. Rider	New York, N. Y., and Poultney, Vt.	May 21, 1861	32, 372
Pump, Rotary	S. W. Kelly	Nashville, Tenn	Nov. 14, 1871	120, 819
Pump, Rotary	F. Kettler	Milwaukee, Wis	June 3, 1862	35, 452
Pump, Rotary	A. H. Knapp	Newton Centre, Mass	Nov. 17, 1863	40, 660
Pump, Rotary	A. H. Knapp	Needham, Mass	July 20, 1869	92, 842
Pump, Rotary	J. C. Lamb and E. Lavens	Middletown, Conn	Dec. 24, 1872	134, 294
Pump, Rotary	E. Lavens and J. C. Lamb	Middletown, Conn	Apr. 19, 1870	102, 136
Pump, Rotary	H. B. Leach	Boston, Mass	May 8, 1866	54, 565
Pump, Rotary	A. Leuchtweiss	Cincinnati, Ohio	June 21, 1870	104, 469
Pump, Rotary	H. Maas	Homestead, Iowa	June 5, 1866	55, 318
Pump, Rotary	H. W. Mather	New York, N. Y	Apr. 19, 1870	102, 140
Pump, Rotary	J. Megaw	Wilmington, Del	July 5, 1859	24, 650
Pump, Rotary	J. J. Morris	Cincinnati, Ohio	May 8, 1866	54, 581
Pump, Rotary	N. E. Nash	Providence, R. I	Aug. 6, 1872	130, 146
Pump, Rotary	J. Naughton	Buffalo, N. Y	Feb. 23, 1869	87, 108
Pump, Rotary	G. W. Nichols	Wheatland, Iowa	June 3, 1862	35, 446
Pump, Rotary	O. Palmer	Buffalo, N. Y	Dec. 29, 1857	18, 986
Pump, Rotary	O. Palmer	Cincinnati, Ohio	Dec. 5, 1865	51, 347
Pump, Rotary	H. Pease	Brockport, N. Y	Oct. 20, 1857	18, 488
Pump, Rotary	H. Pease	Brockport, N. Y	Mar. 31, 1863	38, 058
Pump, Rotary	W. Peirce	New Orleans, La	Mar. 9, 1858	19, 581
Pump, Rotary	E. Perry	Baldwinsville, N. Y	Sept. 30, 1862	36, 588
Pump, Rotary	Peters and Dean	Poughkeepsie, N. Y	Oct. 31, 1835	....
Pump, Rotary	F. B. Pierce	Brockport, N. Y	May 27, 1862	35, 388
Pump, Rotary	J. Poppe	Green Point, N. Y	Dec. 3, 1867	71, 785
Pump, Rotary	G. W. Putnam	South Glen's Falls, N. Y	Mar. 7, 1871	112, 491
Pump, Rotary	R. Ramsden	South Easton, Pa	June 9, 1857	17, 516
Pump, Rotary	F. Rochow	New York, N. Y	Dec. 17, 1861	33, 961
Pump, Rotary	G. W. Rogers	Philadelphia, Pa	Apr. 20, 1869	89, 080
Pump, Rotary	G. W. Rogers	Brooklyn, N. Y	July 23, 1872	129, 864
Pump, Rotary	F. B. Scott	Buffalo, N. Y	Sept. 30, 1862	36, 577
Pump, Rotary	F. B. Shannon	Port Washington, Ohio	Feb. 5, 1867	61, 882
Pump, Rotary	J. H. Shedd and W. Edson	Boston, Mass	Jan. 19, 1864	41, 327
Pump, Rotary	N. P. Sheldon	San Francisco, Cal	Jan. 30, 1872	123, 130
Pump, Rotary	A. Sluthour	Cleveland, Ohio	Apr. 19, 1870	102, 057
Pump, Rotary	H. C. Spalding and G. Stickney	Hartford, Conn	Apr. 20, 1852	8, 893
Pump, Rotary	M. Stannard	New Britain, Conn	Aug. 20, 1861	33, 126
Pump, Rotary	T. Sutton	Norwich, Conn	Jan. 26, 1831	....
Pump, Rotary	T. Swan	Manlius, N. Y	May 31, 1870	103, 684
Pump, Rotary	P. Sweeney	Buffalo, N. Y	Nov 27, 1849	6, 906
Pump, Rotary	B. F. Taber	Buffalo, N. Y	Oct. 22, 1861	33, 550
Pump, Rotary	J. A. Talpey	Somerville, Mass	Mar. 24, 1868	75, 811
Pump, Rotary	H. P. Tenant	East Germantown, Ind	May 23, 1871	115, 254
Pump, Rotary	B. T. Trimmer	Rochester, N. Y	Sept. 28, 1858	21, 632
Pump, Rotary	W. C. Trowbridge	Southeast, N. Y	Jan. 27, 1835	....
Pump, Rotary	P. Ulmboltz	Tremont, Pa	July 4, 1865	48, 604
Pump, Rotary	A. Walker	Claremont, N. H	Nov. 15, 1864	45, 096
Pump, Rotary	D. M. Walker	Cavendish, Vt	Aug. 15, 1835	....
Pump, Rotary	W. H. Ward	Auburn, N. Y	Oct. 3, 1871	119, 482
Pump, Rotary	S. A. West and L. Goodwin	San Francisco, Cal., and Virginia City, Nev.	Oct. 1, 1872	131, 801
Pump, Rotary	I. Williams	Baldwinsville, N. Y	Jan. 17, 1871	111, 026
Pump, Rotary	G. Wingate	Philadelphia, Pa	Jan. 25, 1859	22, 762
Pump, Rotary	C. D. Wright	Fort Atkinson, Wis	Dec. 11, 1855	13, 930
Pump, Rotary	W. A. Young	Charlotte, N. C	Dec. 29, 1857	19, 004
Pump, Rotary air	G. B. Hill	New York, N. Y	July 11, 1865	48, 770
Pump, Rotary forcing	G. W. Billings	Saint Paul, Minn	Nov. 11, 1862	36, 888
Pump, Rotary hydropneumatic	H. M. Paine	Newark, N. J	Mar. 29, 1870	101, 306
Pump, Rotary steam	H. O. Ames	New Orleans, La	Jan. 29, 1867	61, 650
Pump, Rotary water	S. A. Lee	Boston, Mass	Sept. 4, 1841	2, 236
Pump, Rotary wrecking	O. Palmer	Buffalo, N. Y	July 17, 1855	13, 273
Pump, Saliva	F. Davison	Liberty, Va	Jan. 17, 1854	10, 426
Pump, Saliva	A. M. Shurtleff	Boston, Mass	Dec. 3, 1867	71, 799
Pump, Saliva	A. M. Shurtleff	Boston, Mass	Mar. 31, 1868	76, 257
Pump, Sand	J. Benson	Bellaire, Ohio	Oct. 9, 1866	58, 579
Pump, Sand	R. Cornelius	Philadelphia, Pa	May 22, 1866	54, 867
Pump, Sand	J. B. Kibler	Girard, Pa	May 15, 1866	54, 739
Pump, Sand	O. B. Latham	Seneca Falls, N. Y	Dec. 26, 1865	51, 730
Pump, Sand	C. Mather	Manchester, England	Sept. 4, 1866	57, 830
Pump, Sand	B. Morahan	Brooklyn, N. Y	Oct. 2, 1866	58, 457
Pump, Sand	E. A. L. Roberts	Titusville, Pa	May 28, 1867	65, 121
Pump, Sand	J. Smith	Franklin, Pa	June 4, 1872	127, 655
Pump, Sea-motion	B. Lies	New York	Aug. 19, 1811	....
Pump, Segmental	J. Robertson	Brooklyn, N. Y	Feb. 13, 1872	123, 579
Pump, Self adjustable or anchoring	T. Ling	Shelby, Ohio	Mar. 27, 1855	12, 601
Pump, Ship's	T. Bell	Bellport, N. Y	June 18, 1872	128, 060

Index of patents issued from the United States Patent Office from 1790 to 1873, inclusive—Continued.

Invention.	Inventor.	Residence.	Date.	No.
Pump, Ship's	R. Bulkley	New York, N. Y	Dec. 20, 1847	5, 030
Pump, Ship's	G. Clymer		Dec. 22, 1801	....
Pump, Ship's	A. Coates and S. M. Perry	New York and Brooklyn, N. Y.	Sept. 15, 1857	18, 192
Pump, Ship's	D. A. Dunham	Pilatka, Fla	Apr. 4, 1871	113, 641
Pump, Ship's	J. Edson	Boston, Mass	Aug. 15, 1865	49, 393
Pump, Ship's	J. J. Heard	Boston, Mass	July 10, 1855	13, 219
Pump, Ship's	G. Moulton, jr	Bath, Me	May 29, 1866	55, 147
Pump, Ship's, &c	T. Odiorne	Portsmouth, N. H	Aug. 27, 1835	....
Pump, Ship's	S. F. Paullin	Philadelphia, Pa	Dec. 17, 1872	133, 950
Pump, Ship's	S. Pearn	New York, N. Y	Oct. 2, 1855	13, 622
Pump, Ship's	F. Ransom	New York, N. Y	Mar. 19, 1850	7, 198
Pump, Ship's	B. K. Rogers	Jonesport, Me	Dec. 28, 1869	98, 419
Pump, Ship's	T. S. Speakman and N. Hand	Camden, N. J	Nov. 21, 1865	51, 094
Pump, Siphon	J. M. Bois	Salamanca, N. Y	Mar. 18, 1873	136, 809
Pump, Siphon	T. J. Trapp	Williamsport, N. Y	Oct. 25, 1870	108, 652
Pump, Siphon steam	T. O'Rorke	Pittsburgh, Pa	July 25, 1871	117, 450
Pump-spear, Horizontal circular-valve	E. Leman	Boston, Mass	May 23, 1817	....
Pump, Spiral	C. Foss	Perry, Ohio	Aug. 4, 1832	....
Pump-spout	J. H. Gleim	Tipton, Mo	Apr. 19, 1870	101, 995
Pump, Spring or drop valve	E. Kendall	Ashbury, Mass	June 22, 1832	....
Pump-sprinkler	W. F. Class	Cleveland, Ohio	May 15, 1866	54, 685
Pump, Steam	W. Aldrich	Dayton, Ohio	May 20, 1873	139, 103
Pump, Steam	L. H. Allen and W. Barton	Tamaqua, Pa	Feb. 13, 1872	123, 659
Pump, Steam	L. H. Allen and W. Barton	Tamaqua, Pa	Feb. 13, 1872	123, 660
Pump, Steam	L. H. Allen and W. Barton	Tamaqua, Pa	Feb. 13, 1872	123, 661
Pump, Steam	W. Arthur	Newport, Ky	Aug. 27, 1872	130, 890
Pump, Steam	W. Arthur	Newport, Ky	Jan. 21, 1873	135, 061
Pump, Steam	J. B. Atwater	Chicago, Ill	Aug. 1, 1865	49, 063
Pump, Steam	J. S. Barden	New Haven, Conn	Mar. 27, 1860	27, 610
Pump, Steam	J. S. Barden	Providence, R. I	Nov. 24, 1868	84, 251
Pump, Steam	W. Baxter, jr	Newark, N. J	Apr. 18, 1871	113, 725
Pump, Steam	P. Bernard and T. Soper	Whitestown, N. Y	Jan. 16, 1811	....
Pump, Steam	G. F. Blake	Boston, Mass	Oct. 25, 1870	108, 554
Pump, Steam	G. F. Blake	Boston, Mass	Feb. 20, 1872	123, 766
Pump, Steam	A. and F. Brown	New York, N. Y	Mar. 13, 1860	27, 426
Pump, Steam	F. Brown	New York, N. Y	Dec. 5, 1865	51, 291
Pump, Steam	L. and T. E. Button	Waterford, N. Y	Sept. 26, 1871	119, 314
Pump, Steam	A. S. Cameron	New York, N. Y	Dec. 17, 1867	72, 363
Pump, Steam	L. Carricaburu	Havana, Cuba	Nov. 6, 1866	59, 519
Pump, Steam	J. C. and E. H. Dean	Indianapolis, Ind	Nov. 26, 1872	133, 419
Pump, Steam	R. Eickemeyer	Yonkers, N. Y	May 6, 1873	138, 622
Pump, Steam	H. Epping	Pittsburgh, Pa	May 7, 1872	126, 536
Pump, Steam	A. Friedmann	Vienna, Austria	Apr. 6, 1869	88, 620
Pump, Steam	A. Friedmann	Vienna, Austria	Aug. 23, 1870	106, 683
Pump, Steam	J. B. Gardiner	Springfield, Mass	Feb. 4, 1868	74, 075
Pump, Steam	J. B. Gardiner and E. H. Hyde	Springfield, Mass	Dec. 17, 1867	72, 383
Pump, Steam	J. W. Gardner, T. W. Ranson, and E. Martin.	Cleveland, Ohio	Dec. 10, 1872	133, 847
Pump, Steam	L. P. Garner	Ashland, Pa	Aug. 17, 1869	93, 696
Pump, Steam	W. W. Gilbert	New York, N. Y	Apr. 27, 1869	89, 398
Pump, Steam	W. W. Gilbert	New York, N. Y	Nov. 30, 1869	97, 389
Pump, Steam	L. Griscom	Port Carbon, Pa	Mar. 19, 1872	124, 679
Pump, Steam	J. W. Handren	Jersey City, N. J	Jan. 14, 1873	134, 746
Pump, Steam	J. W. Handren	Jersey City, N. J	Feb. 4, 1873	135, 426
Pump, Steam	T. Harrington	Pittsburgh, Pa	Jan. 10, 1871	110, 850
Pump, Steam	W. Harsen	Green Point, N. Y	Feb. 4, 1868	74, 084
Pump, Steam	W. M. Henderson	Philadelphia, Pa	June 18, 1867	65, 911
Pump, Steam	W. L. Horne	Chicago, Ill	July 30, 1872	129, 891
Pump, Steam	I. Illofsky	Pesth, Hungary	Oct. 4, 1864	44, 532
Pump, Steam	J. Jordan	Wyandotte, Kans	Nov. 6, 1866	59, 400
Pump, Steam	W. E. Kelly	New Brunswick, N. J	May 14, 1872	126, 814
Pump, Steam	L. J. Knowles	Worcester, Mass	Apr. 18, 1871	113, 897
Pump, Steam	L. J. Knowles	Worcester, Mass	May 2, 1871	114, 507
Pump, Steam	B. S. Lawson	New York, N. Y	Dec. 23, 1873	145, 806
Pump, Steam	J. A. Lidback	Portland, Me	Dec. 28, 1869	98, 279
Pump, Steam	W. Livingstone	New York, N. Y	May 31, 1870	103, 756
Pump, Steam	G. W. Long	Old Point Comfort, Va	Feb. 16, 1826	....
Pump, Steam	J. R. Maxwell and E. Cope	Cincinnati, Ohio	June 21, 1870	104, 616
Pump, Steam	J. R. Maxwell and E. Cope	Cincinnati, Ohio, and Covington, Ky.	Jan. 30, 1872	123, 185
Pump, Steam	J. R. Maxwell and E. Cope	Cincinnati, Ohio, and Covington, Ky.	Oct. 1, 1872	131, 765
Pump, Steam	J. Mayher	East Hampton, Mass	Oct. 8, 1872	131, 960
Pump, Steam	J. McCloskey	New York, N. Y	Apr. 20, 1869	89, 158
Pump, Steam	G. McFeely	Steubenville, Ohio	Aug. 31, 1869	94, 426
Pump, Steam	J. J. Miller	Chicago, Ill	Aug. 23, 1864	43, 952
Pump, Steam	J. C. Morgan	Alliance, Ohio	Mar. 13, 1866	53, 170
Pump, Steam	J. North	New York, N. Y	Oct. 15, 1872	132, 315
Pump, Steam	J. North	New York, N. Y	Dec. 31, 1872	134, 481
Pump, Steam	J. V. Pangburn	Galesburgh, Ill	Nov. 5, 1872	132, 735
Pump, Steam	A. and G. W. Perry	New Philadelphia and Mahanoy City, Pa.	Dec. 5, 1871	121, 658
Pump, Steam	W. Porter	Brooklyn, N. Y	Apr. 23, 1872	126, 083
Pump, Steam	J. B. Pottmeyer	Pittsburgh, Pa	Feb. 26, 1867	62, 360
Pump, Steam	A. J. Reynolds	Chicago, Ill	Oct. 27, 1868	83, 549
Pump, Steam	G. J. Roberts	Dayton, Ohio	May 14, 1872	126, 834
Pump, Steam	G. J. Roberts	Dayton, Ohio	May 6, 1873	138, 695
Pump, Steam	W. H. Roberts	Mauch Chunk, Pa	Mar. 30, 1869	88, 512
Pump, Steam	L. C. Rodier	Springfield, Mass	Aug. 14, 1866	57, 188
Pump, Steam	G. W. Rogers	Brooklyn, N. Y	June 25, 1872	128, 426
Pump, Steam	R. Schmidt	New York, N. Y	Aug. 6, 1867	67, 592
Pump, Steam	H. C. Sergeant	New York, N. Y	Oct. 7, 1862	36, 623
Pump, Steam	H. C. Sergeant	New York, N. Y	June 22, 1869	91, 676
Pump, Steam	W. Sewell	New York, N. Y	Oct. 28, 1862	36, 802
Pump, Steam	T. Simmons	New York, N. Y	Aug. 13, 1867	67, 812
Pump, Steam	T. and G. T. Snowdon and I. V. Lynn.	Pittsburgh, Pa	Jan. 11, 1870	98, 810

Index of patents issued from the United States Patent Office from 1790 *to* 1873, *inclusive*—Continued.

Invention.	Inventor.	Residence.	Date.	No.
Pump, Steam	S. Stanton	New York, N. Y	Feb. 27, 1872	124, 095
Pump, Steam	C. L. Stevens	Galesburgh, Ill	Oct. 4, 1870	107, 977
Pump, Steam	C. L. Stevens and A. A. Denton	Galesburgh, Ill	Jan. 19, 1869	86, 111
Pump, Steam	C. Swinscoe	Boston, Mass	Mar. 12, 1872	124, 639
Pump, Steam	W. R. Thomas	Catasauqua, Pa	July 14, 1868	80, 030
Pump, Steam	J. E. Thorpe	Erie, Pa	Apr. 23, 1867	64, 166
Pump, Steam	A. W. Todd	Chicago, Ill	July 18, 1865	48, 853
Pump, Steam	L. W. Turrell	Newburgh, N. Y	Jan. 24, 1865	46, 036
Pump, Steam	W. Watts	Newark, N. J	Sept. 8, 1863	39, 852
Pump, Steam	G. M. Weinman	Columbus, Ohio	Dec. 6, 1870	109, 980
Pump, Steam	J. C. Wightman	Newton, Mass	Apr. 27, 1869	89, 534
Pump, Steam	M. Wilcox	Middlebury, Ohio	Apr. 2, 1861	31, 923
Pump, Steam	M. Wilcox	Sacramento, Cal	Aug. 15, 1865	49, 461
Pump, Steam	C. and G. M. Woodward	New York, N. Y	July 22, 1862	35, 966
Pump, Steam and vacuum	A. J. Simmons	Indianapolis, Ind	Oct. 14, 1873	143, 725
Pump, Steam-engine air	A. Hartupee	Pittsburgh, Pa	Feb. 5, 1867	61, 735
Pump, Steam-jet	C. Barnes	Cincinnati, Ohio	Aug. 7, 1866	57, 035
Pump, Steam-jet	A. J. Blakslee and G. C. Williams.	Duquoin, Ill	Dec. 13, 1870	110, 110
Pump, Steam-jet	J. D. Toppin	Newark, N. J	Dec. 17, 1872	134, 018
Pump, Steam siphon	T. O'Rorke	Pittsburgh, Pa	June 18, 1872	127, 984
Pump, Steam vacuum	I. Barnum	New York	Apr. 16, 1834	....
Pump, Steam vacuum	W. Burdon	Brooklyn, N. Y	July 23, 1872	129, 647
Pump, Steam vacuum	W. Burdon	Brooklyn, N. Y	Jan. 28, 1873	135, 198
Pump, Steam vacuum	W. Burdon	Brooklyn, N. Y	Jan. 28, 1873	135, 199
Pump, Steam vacuum	W. Burdon	Brooklyn, N. Y	Jan. 28, 1873	135, 200
Pump, Steam vacuum	W. Burdon	Brooklyn, N. Y	Jan. 28, 1873	135, 201
Pump, Steam vacuum	W. Burdon	Brooklyn, N. Y	Jan. 28, 1873	135, 202
Pump, Steam vacuum	W. Burdon	Brooklyn, N. Y	May 27, 1873	139, 235
Pump, Steam vacuum	W. Burdon	Brooklyn, N. Y	Aug. 5, 1873	141, 540
Pump, Steam vacuum	C. H. Hall	New York, N. Y	Sept. 24, 1872	131, 515
Pump, Steam vacuum	C. H. Hall	New York, N. Y	Sept. 24, 1872	131, 516
Pump, Steam vacuum	C. H. Hall	New York, N. Y	Sept. 24, 1872	131, 517
Pump, Steam vacuum	C. H. Hall	New York, N. Y	Sept. 24, 1872	131, 518
Pump, Steam vacuum	C. H. Hall	New York, N. Y	Sept. 24, 1872	131, 519
Pump, Steam vacuum	C. H. Hall	New York, N. Y	Sept. 24, 1872	131, 520
Pump, Steam vacuum	C. H. Hall	New York, N. Y	Sept. 24, 1872	131, 521
Pump, Steam vacuum	C. H. Hall	New York, N. Y	Sept. 24, 1872	131, 522
Pump, Steam vacuum	C. H. Hall	New York, N. Y	Sept. 24, 1872	131, 523
Pump, Steam vacuum	C. H. Hall	New York, N. Y	Sept. 24, 1872	131, 524
Pump, Steam vacuum	C. H. Hall	New York, N. Y	Sept. 24, 1872	131, 525
Pump, Steam vacuum	C. H. Hall	New York, N. Y	Sept. 24, 1872	131, 526
Pump, Steam vacuum	C. H. Hall	New York, N. Y	Sept. 24, 1872	131, 527
Pump, Steam vacuum	C. H. Hall	New York, N. Y	Sept. 24, 1872	131, 528
Pump, Steam vacuum	C. H. Hall	New York, N. Y	Sept. 24, 1872	131, 529
Pump, Steam vacuum	C. H. Hall	New York, N. Y	Sept. 24, 1872	131, 530
Pump, Steam vacuum	C. H. Hall	New York, N. Y	Sept. 24, 1872	131, 531
Pump, Steam vacuum	C. H. Hall	New York, N. Y	Sept. 24, 1872	131, 532
Pump, Steam vacuum	C. H. Hall	New York, N. Y	Sept. 24, 1872	131, 533
Pump, Steam vacuum	C. H. Hall	New York, N. Y	Sept. 24, 1872	121, 534
Pump, Steam vacuum	C. H. Hall	New York, N. Y	Sept. 24, 1872	131, 535
Pump, Steam vacuum	C. H. Hall	New York, N. Y	Sept. 24, 1872	131, 536
Pump, Steam vacuum	C. H. Hall	New York, N. Y	Sept. 24, 1872	131, 537
Pump, Steam vacuum	C. H. Hall	New York, N. Y	Sept. 24, 1872	131, 538
Pump, Steam vacuum	C. H. Hall	New York, N. Y	Sept. 24, 1872	131, 539
Pump, Steam vacuum	C. H. Hall	New York, N. Y	Sept. 24, 1872	131, 540
Pump, Steam vacuum	C. H. Hall	New York, N. Y	Sept. 24, 1872	131, 541
Pump, Steam vacuum	C. H. Hall	New York, N. Y	Sept. 24, 1872	131, 542
Pump, Steam vacuum	C. H. Hall	New York, N. Y	Sept. 24, 1872	131, 543
Pump, Steam vacuum	J. B. Little	Monmouth, Ill	Mar. 4, 1873	136, 441
Pump, Steam vacuum	J. M. Morehead	Brooklyn, N. Y	Jan. 30, 1872	123, 117
Pump, Steam vacuum	J. M. Morehead	Brooklyn, N. Y	May 21, 1872	126, 898
Pump, Steam vacuum	G. H. Nye	Monmouth, Ill	Nov. 5, 1872	132, 731
Pump, Steam vacuum	G. H. Nye	Monmouth, Ill	Nov. 5, 1872	132, 732
Pump, Steam vacuum	J. H. Pattee	Monmouth, Ill	July 16, 1872	129, 051
Pump, Steam vacuum	J. H. Pattee	Monmouth, Ill	Aug. 6, 1872	130, 309
Pump, Steam vacuum	J. H. Pattee and H. J. Graham	Monmouth, Ill	Jan. 10, 1871	110, 933
Pump, Steam vacuum	J. H. Pattee and G. H. Nye	Monmouth, Ill	May 17, 1870	103, 076
Pump, Steam vacuum	W. E. Prall	Washington, D. C	Jan. 30, 1872	123, 292
Pump, Steam vacuum	W. E. Prall and D. A. Burr	Washington, D. C	Dec. 30, 1873	146, 095
Pump, Steam vacuum	C. L. Riker	Schraalenburgh, N. J	Apr. 9, 1872	125, 410
Pump, Steam vacuum	J. K. Simpson and S. N. Drake	New York, N. Y	Jan. 7, 1873	134, 568
Pump, Steam vacuum	I. P. Tice	New York, N. Y	Feb. 25, 1873	136, 344
Pump, Stock	W. T. Armstrong	Freeland, Ill	Sept. 15, 1868	82, 064
Pump, Stock	W. A. Durrin	Milledgeville, Ill	Feb. 24, 1863	37, 744
Pump, Stock	A. H. Piland and A. H. Turner	Indianapolis, Ind	Aug. 18, 1868	81, 287
Pump, Stock	A. H. Russell	Adrian, Mich	June 23, 1868	79, 146
Pump-stocks, Machine for securing sheet-metal lining in.	N. T. Coffin	Knightstown, Ind	Aug. 2, 1870	105, 912
Pump, Stomach	J. M. Youngblood	Saint Louis, Mo	July 20, 1869	92, 769
Pump-strainer	W. S. Blunt	Brooklyn, N. Y	Aug. 20, 1872	130, 560
Pump, Submerged	A. C. Blethen	Lynn, Mass	Apr. 6, 1869	88, 538
Pump, Submerged	B. F. Brown	Woburn, Mass	Dec. 22, 1868	85, 063
Pump, Submerged	P. H. Cardwell	Knoxville, Tenn	Feb. 1, 1870	99, 403
Pump, Submerged	J. Haney	Yorktown, Ind	July 9, 1872	128, 877
Pump, Submerged	D. G. Hussey	Nantucket, Mass	Aug. 20, 1872	130, 642
Pump, Submerged	H. Lindsey	Asheville, N. C	Sept. 6, 1859	25, 340
Pump, Submerged	J. A. Morrell and J. Brewer	New York, N. Y	Aug. 12, 1873	141, 807
Pump, Submerged	H. M. Stoker	Watson, Ill	June 13, 1865	48, 219
Pump, Submerged	H. M. Stoker	Watson, Ill	June 13, 1865	48, 220
Pump, Submerged	J. H. Williamson	Branchville, N. J	Mar. 7, 1865	49, 764
Pump, Submerged force	A. Balding	Flora, Ill	Aug. 8, 1865	49, 212
Pump, Submerged force	A. T. Hafford	Saint Louis, Mo	May 20, 1873	139, 003
Pump, Submerged force	C. C. Hiatt	Ridgeville, Ind	Jan. 19, 1869	86, 071
Pump, Submerged rotary	D. D. Hardy	Cincinnati, Ohio	Sept. 1, 1868	81, 778
Pump-sucker	J. Weis	Bordentown, N. J	Dec. 18, 1855	13, 967
Pump, Suction	T. C. Barton	Washington, N. Y	Feb. 20, 1836	....
Pump, Suction	A. H. Mathes	Madisonville, Tenn	Dec. 31, 1833	....

Index of patents issued from the United States Patent Office from 1790 *to* 1873, *inclusive*—Continued.

Invention.	Inventor.	Residence.	Date.	No.
Pump, Suction	E. Whiton	Groton, Mass	Oct. 2, 1834	....
Pump, Suction and force	J. Stevens	Newark, N. J	Dec. 20, 1837	523
Pump, Suction and lifting	E. Tolles	Litchfield, Conn	Sept. 28, 1831	... ...
Pump, Supply	J. L. Sullivan	New York	Dec. 28, 1833	
Pump, Test	T. Shaw	Philadelphia, Pa	Oct. 4, 1864	44, 559
Pump to raise water, Force	A. Taylor		Oct. 4, 1804	....
Pump, Tube-well	J. H. Devirs and D. Gravatt	Pleasantville, Pa	May 18, 1869	90, 157
Pump-tubes, Wrench for elevating	J. A. Fleming	Shamburgh, Pa	Sept. 7, 1869	94, 484
Pump, Turbinate force	J. H. A. Gericke	New Orleans, La	May 29, 1866	55, 082
Pump, Universal	A. Kinsley		Oct. 28, 1799	....
Pump used in low-pressure or condensing steam-engines, Air.	C. Reeder	Baltimore, Md	Nov. 15, 1843	3, 334
Pump, Vacuum	E. Dunscomb	Boston, Mass	Mar. 12, 1867	62, 738
Pump-valve chamber	M. C. Taylor	Grass Valley, Cal	Dec. 1, 1868	84, 516
Pump ventilator for ships, mines, &c., Air	R. Robotham		Oct. 10, 1801	....
Pump, Vibrating	S. Davis	Derby, Vt	Feb. 25, 1836	....
Pump, Vibrating	E. Garrette	Wilmington, Del	Oct. 9, 1855	13, 643
Pump, Water	J. Babcock, jr	Edmiston, N. Y	May 6, 1825	....
Pump, Water	A. H. Fitch	Coylersville, N. Y	May 23, 1848	5, 591
Pump, Water	J. F. Flanders	Newburyport, Mass	July 22, 1851	8, 239
Pump, Water	J. W. Hillias	Baltimore, Md	Sept. 16, 1829	....
Pump, Water	D. Hinman	Brunswick, Ohio	Oct. 24, 1846	4, 823
Pump, Water	F. B. Hyatt	Maysville, Ky	June 16, 1846	4, 583
Pump, Water	G. Ketchum	Marshfield, Mich	Dec. 18, 1847	5, 401
Pump, Water	J. B. Read	Tuscaloosa, Ala	Sept. 11, 1849	6, 714
Pump, Water	J. Renfrew	Fayetteville, Pa	June 16, 1846	4, 580
Pump, Water	A. Stiven	New York, N. Y	Dec. 4, 1849	6, 927
Pump, Water	P. Upham	Brookfield, Mass	Mar. 9, 1811	....
Pump, Water	C. Warner	Louisville, Ky	July 8, 1846	4, 619
Pump, Wind-power	H. Bronson	Kent, Conn	Mar. 8, 1820	....
Pump, Wind-wheel	L. D. Parsons	Tremont, N. Y	Oct. 19, 1869	95, 930
Pump, Wind-wheel	E. Pepple	NewCarlislePost-Office, Ind	Apr. 26, 1864	42, 503
Pump with concentric piston, Force	P. M. Kafer	Trenton, N. J	Sept. 7, 1869	94, 611
Pump with double action for engines, Forcing	E. Cady, jr	Chatham, N. Y	Mar. 9, 1808	....
Pumps, Air-chamber for	R. H. Hilton	New Berne, N. C	Nov. 19, 1867	71, 097
Pumps, Air-escape for	H. Smith	Norwalk, Ohio	Mar. 4, 1856	14, 372
Pumps and steam-engines, Construction of cylinders and pistons for.	W. Wells	New York, N. Y	Oct. 12, 1858	21, 789
Pumps, Anti freezing device for	J. G. Hanning	Indianapolis, Ind	May 24, 1870	103, 325
Pumps, Application of power to	N. Underwood	Baltimore County, Md	July 17, 1828	....
Pumps, Arrangement of	L. H. Mosby	Powhatan County, Va	Feb. 15, 1823	....
Pumps, Attaching air-chambers to	C. N. Lewis	Seneca Falls, N. Y	June 23, 1857	17, 653
Pumps, Attachment for direct-acting steam	O. T. Earle	Norwalk, Conn	Aug. 26, 1873	142, 092
Pumps, Basket-attachment for pistons of deep-well	E. D. Brown	Buffalo, N. Y	Jan. 22, 1867	61, 311
Pumps by wind, Operating	H. Moore	Charleston, Mich	Mar. 13, 1855	12, 517
Pumps by wind-wheels, Regulating	J. W. Goodwin and M. C. Hawkins.	Edinborough, Pa	Apr. 8, 1856	14, 626
Pumps, Case for rotary	P. H. and F. M. Roots	Connersville, Ind	Aug. 11, 1868	81, 010
Pumps, &c., Communicating reciprocating motion to	I. Van Olinda	Brooklyn, N. Y	Oct. 16, 1866	58, 918
Pumps, Constructing and working ships'	A. Henshaw and N. Harlow, jr.	Bangor, Mass	Aug. 30, 1811	....
Pumps, Construction of	W. Rhoades	New York	May 16, 1808	....
Pumps, Cylinder for lining wooden	R. W. Wheeler	Bushnell, Ill	July 15, 1873	140, 802
Pumps, Débris-check for	S. O. Blanding	Smithfield, R. I	Nov. 17, 1868	84, 081
Pumps, Device for air-chambers of	J. P. Cowing	Seneca Falls, N. Y	Mar. 20, 1855	12, 542
Pumps, &c., Device for drawing water from and supplying air to air-vessels of.	A. H. Ranch	Bethlehem, Pa	June 5, 1860	28, 604
Pumps, &c., Double-reciprocating split piston-rod for.	J. A. Burnap	Albany, N. Y	July 24, 1855	13, 299
Pumps for ships, Arrangement of means in pendulum.	J. Stever	Bristol, Conn	July 22, 1856	15, 395
Pumps for steam-engines, Method of varying the stroke of feeding.	J. R. Sees	New York, N. Y	Apr. 1, 1856	14, 576
Pumps for wells, &c., Constructing force	T. W. H. Moseley	Paris, Ky	Sept. 30, 1839	1, 351
Pumps, Glass cylinder for	J. Bryan	New York, N. Y	Aug. 27, 1872	130, 896
Pumps, Glass piston and packing for	A. E. Gay	Nashua, N. H	July 8, 1873	140, 697
Pumps, Hand-lever attachment for steam	O. T. Earle	Norwalk, Conn	Sept. 9, 1873	142, 562
Pumps, Handle for ships'	K. Markuson	Gloucester, Mass	Nov. 5, 1867	70, 587
Pumps, Hoisting water by wind	J. Kerr	Maury County, Tenn	Sept. 6, 1833	....
Pumps in deep wells, Apparatus for protecting	F. Armstrong	Pittsburgh, Pa	May 8, 1866	54, 484
Pumps in propelling boats, Application of	M. Ulman and M. Isaacs	Philadelphia, Pa	Mar. 17, 1818	....
Pumps, Instrument for reaming out barrels of ships'	S. H. Sugett	Eden, Me	Oct. 7, 1862	36, 626
Pumps, Lifting-clamp for	W. H. Downing	Pioneer, Pa	Apr. 5, 1870	101, 595
Pumps, Link for connecting handle and pitman of	L. W. and C. Olds	Erie, Pa	Aug. 13, 1872	130, 527
Pumps, Link for connecting handle and valve-rod of	S. Selden and M. Griswold	Erie, Pa	Oct. 4, 1870	107, 965
Pumps, Lubricating pistons of air	A. Barker	Honesdale, Pa	Apr. 3, 1855	12, 617
Pumps, Machine for operating	R. E. Moore	Navasota, Tex	Dec. 7, 1869	97, 672
Pumps, Making suction	D. Johnson	Boston, Mass	Apr. 21, 1824	....
Pumps, Mechanism for operating	J. A. Whipple	Boston, Mass	June 6, 1854	11, 020
Pumps, Mechanism for operating valve of force	J. C. King	New York, N. Y	Sept. 17, 1867	68, 994
Pumps, Metallic mold for casting	H. Post	New York, N. Y	Mar. 25, 1862	34, 774
Pumps, Method of effecting uniform pressure upon the pumping-piston of double-acting steam.	R. B. Gorsuch	New York, N. Y	June 10, 1856	15, 070
Pumps on railways, Operating	G. M. Cole	Folsom City, Cal	Dec. 16, 1862	37, 152
Pumps on railway-stations, Operating	H. H. Call	Rohrerstown, Pa	Dec. 3, 1867	71, 695
Pumps on railway-stations, Operating	W. McVeigh	Boone, Ill	Feb. 9, 1858	19, 304
Pumps, Operating	J. Armstrong	Dobbinsville, N. C	Apr. 16, 1861	32, 038
Pumps, Operating	E. Borton	Morris, Ill	Jan. 28, 1868	73, 771
Pumps, Operating	D. J. Rogers	Magnolia, N. C	Oct. 26, 1858	21, 904
Pumps, Operating ships'	S. R. Deverell	Mount Gambier, South Australia.	Jan. 21, 1873	135, 094
Pumps, Operating ships'	A. Roff	Southport, Conn	June 8, 1869	91, 045
Pumps, Preparing molds for casting double	F. Henshaw	Brookfield, Mass	June 29, 1839	1, 206
Pumps, Pressure-head for siphon and force	D. M. Shapley	Saint Louis, Mo	May 8, 1866	54, 615
Pumps, Protection against freezing in	S. Adams	Townsend, Mass	Mar. 11, 1837	141
Pumps, Regulating discharge of water from	J. Mayher	East Hampton, Mass	Jan. 10, 1871	110, 928
Pumps, Return-drip for	J. B. Stevenson	Bloomington, Ill	Nov. 10, 1868	83, 887
Pumps, Rivet-pocket for oil-well	W. H. Birge	Franklin, Pa	Jan. 21, 1873	134, 970
Pumps, Sand-chamber for well	E. C. Johnson	Williamsport, Pa	Jan. 4, 1870	98, 503

Index of patents issued from the United States Patent Office from 1790 *to* 1873, *inclusive*—Continued.

Invention.	Inventor.	Residence.	Date.	No.
Pumps, Strainer for ship's bilge	S. J. Chapman	Baltimore, Md	Sept. 2, 1873	142, 438
Pumps, Stuffing-box for deep-well	J. B. Pettey and J. Fredricks	Conneaut, Ohio	Apr. 9, 1867	63, 747
Pumps, Sucker-rod joint for	A. M. Williams	Oil City, Pa	Nov. 18, 1873	144, 813
Pumps, Supplying water to air	J. F. Haskins	Fitchburgh, Mass	Aug. 1, 1871	117, 626
Pumps to barrels, Device for attaching	J. F. de Navarro	New York, N. Y	June 16, 1868	78, 865
Pumps to bungs of barrels, Apparatus for attaching.	F. A. Pratt	Hartford, Conn	Sept. 22, 1863	40, 082
Pumps to bungs of barrels, Apparatus for attaching.	F. A. Pratt	Hartford, Conn	Oct. 13, 1863	40, 276
Pumps, Valve-arrangement for	W. Sewell and A. S. Cameron	New York, N. Y	Dec. 27, 1864	45, 644
Pumps, Valvular arrangement for diaphragm	S. A. Clemens	Springfield, Mass	July 11, 1854	11, 300
Pumps, Vetilating-attachment for	C. N. Lewis	Seneca Falls, N. Y	Nov. 17, 1857	18, 660
Pumps, Vulcanized-rubber sack for stock	J. S. Patric	Rochester, N. Y	Oct. 17, 1871	120, 093
Pumps, Water-charger for	T. Dutton and T. Maguire	Port Jervis, N. Y	Oct. 20, 1868	83, 268
Pumps, Water-meter for rotary	E. Keith	Buffalo, N. Y	Sept. 14, 1869	94, 892
Pumps water-tight, Making joints of stone	A. Van Vorhes	Hebardville, Ohio	May 3, 1839	1, 139
Pumps with hydraulic press or ram, Connecting	R. Dillon	New York, N. Y	Feb. 1, 1848	5, 423
Pumps, Working	E. Rigg		July 29, 1794	
Pumps, Working	B. Robbins	Machias, Me	July 19, 1859	24, 822
Pumps, Working	W. Wright	Hartford, Conn	Apr. 7, 1857	17, 009
Pumps, Working ships'	F. E. Boettner	Chicago, Ill	May 12, 1863	38, 460
Pumps, Working ships'	A. Cain	Holyoke, Mass	Jan. 24, 1865	45, 973
Pumps, Working ships'	M. Francis	Boston, Mass	July 7, 1829	
Pumps, Working ships'	D. Gay	Bath, Me	Sept. 8, 1837	373
Pumps, Working ships'	L. Simpkins	Brooklyn, N. Y	June 5, 1866	55, 377
Pumps, Working condenser of steam diaphragm	J. Black and O. Beecher	Philadelphia, Pa	Jan. 25, 1853	9, 553
Pumping and condensing apparatus	G. J. Washburn	Worcester, Mass	Mar. 1, 1864	41, 799
Pumping and measuring apparatus, Liquid	E. F. Wilder	Lowell, Mass	May 7, 1872	126, 604
Pumping and raising water	B. Folger		July 7, 1804	
Pumping-apparatus	D. Stoner	Canton, Ohio	June 14, 1870	104, 374
Pumping-apparatus, Bellows	S. Motte	Paris, France	Sept. 8, 1868	81, 930
Pumping-apparatus for supplying finings and other liquids to casks.	A. M. Dix	Shelton, England	Jan. 26, 1869	86, 287
Pumping-apparatus, Steam	G. H. Corliss	Providence, R. I	June 2, 1857	17, 423
Pumping by the motion of the cars, Device for	R. R. and E. C. Spedden	Astoria, Ga	Sept. 7, 1869	94, 660
Pumping-device	J. Johnson	Saco, Me	Dec. 15, 1868	84, 948
Pumping-engine	R. Allison	Port Carbon, Pa	Sept. 29, 1868	82, 475
Pumping-engine	E. D. and F. W. Eames	Grand Rapids, Mich	May 28, 1872	127, 322
Pumping-engine	C. E. Emery	Brooklyn, N. Y	Nov. 17, 1868	84, 176
Pumping-engine	S. Harrison	Pottsville, Pa	Aug. 5, 1862	36, 089
Pumping-engine	J. L. Lowry	Pittsburgh, Pa	Mar. 10, 1868	75, 284
Pumping-engine	L. C. Rodier	Springfield, Mass	Aug. 23, 1870	106, 727
Pumping-engine	T. Shaw	Philadelphia, Pa	Dec. 15, 1868	84, 912
Pumping-engine	W. J. Stevens	New York, N. Y	Sept. 13, 1864	44, 232
Pumping-engine	H. R. Worthington	Brooklyn, N. Y	July 19, 1859	24, 838
Pumping-engine, Steam	J. S. Barden	New Haven, Conn	Nov. 24, 1857	18, 718
Pumping-engine, Steam	T. E. Blunt	Brookfield, Ohio	Aug. 16, 1870	106, 312
Pumping-engine, Steam	E. Cope	Cincinnati, Ohio	Oct. 26, 1858	21, 873
Pumping-engine, Steam	G. H. and C. P. Deane	Springfield, Mass	May 28, 1872	127, 153
Pumping-engine, Steam	R. J. Gould	Newark, N. J	July 27, 1869	92, 957
Pumping-enging, Steam	W. C. Hicks	Summit, N. J	Oct. 21, 1873	143, 824
Pumping-engine, Steam	L. J. Knowles	Worcester, Mass	May 2, 1871	114, 506
Pumping-engine, Steam	E. D. Leavitt, jr	Cambridge, Mass	July 16, 1872	129, 240
Pumping-engine, Steam	A. J. L. Loretz	Brooklyn, N. Y	Jan. 7, 1873	134, 689
Pumping-engine, Steam	J. H. McConnell	Harmar, Ohio	Dec. 24, 1872	134, 212
Pumping-engine, Steam	G. W. Perry	Shenandoah City, Pa	Sept. 29, 1868	82, 548
Pumping-engine, Steam	W. H. Roberts	Mauch Chunk, Pa	Feb. 15, 1870	99, 953
Pumping-engine, Steam	J. Tufts	Athens, Ohio	Apr. 28, 1868	77, 336
Pumping-engine, Steam	N. W. Wheeler	Morristown, N. J	June 6, 1871	115, 670
Pumping-engine, Steam	H. R. Worthington	New York, N. Y	June 20, 1871	116, 131
Pumping-engine, Steam	W. Wright	Hartford, Conn	Nov. 15, 1859	26, 138
Pumping-engines, Device for securing uniform motion in.	W. H. Roberts	Mauch Chunk, Pa	Mar. 1, 1870	100, 449
Pumping-machine	J. Clark		May 19, 1803	
Pumping-machine	R. Yeamans	Springfield, Mass	May 28, 1810	
Pumping water out of vessels, Method of	A. Kirkwood	Jackson County, Miss	Feb. 27, 1855	12, 458
Pumping water out of vessels by wind	T. Brownell	New York, N. Y	Mar. 23, 1827	
Pumping water, Wind-machine for	F. G. Johnson	Bellwood, Sag Harbor, N.Y	Apr. 3, 1860	27, 758
Punch	M. J. Fitzpatrick and B. Barker	New York, N. Y	Aug. 29, 1865	49, 616
Punch	A. J. Fullam	Springfield, Vt	Apr. 10, 1860	27, 793
Punch	C. B. Holden	Worcester, Mass	Aug. 14, 1866	57, 136
Punch	R. Hughes	Virginia City, Nev	Jan. 8, 1867	61, 069
Punch	R. Humphrey	Unionville, Conn	Oct. 16, 1860	30, 405
Punch	R. J. Mullen	Providence, R. I	Oct. 24, 1871	120, 162
Punch	W. A. Rex	Newville, Ind	Nov. 6, 1866	59, 453
Punch	I. P. Richards	Whitinsville, Mass	June 28, 1870	104, 769
Punch	E. Schlindler and C. H. Metzger	Easton, Pa	Sept. 17, 1867	69, 031
Punch	G. C. Wilder	Lawrence, Kans	Nov. 10, 1868	84, 039
Punch	L. Wolf	Saint Jacob, Ill	July 2, 1867	66, 436
Punch	J. Wright	Middleport, Ohio	June 15, 1869	91, 400
Punch, Adjustable center	E. E. Safford and S. Sawyer	Fitchburgh, Mass	Aug. 22, 1865	49, 553
Punch, Adjustable gang	C. E. Howard	West Bridgewater, Mass	Sept. 13, 1864	44, 191
Punch and awl, Combined	F. P. Pfleghar	Whitneyville, Conn	Jan. 25, 1859	22, 744
Punch and cutter, Leather	H. S. Fickett	Yarmouth, Me	Jan. 14, 1873	134, 794
Punch and cutting-nippers, Hand	J. O. Reilley	Jersey City, N. J	May 20, 1873	139, 079
Punch and die	W. K. Lewis	Boston, Mass	Mar. 7, 1865	46, 681
Punch and die for punching watch-hands	A. L. Dennison	Roxbury, Mass	Aug. 15, 1854	11, 522
Punch and edger combined	G. E. Hernig	Louisville, Ky	Mar. 19, 1872	124, 740
Punch and eyeleting-machine	E. Shaw	Milwaukee, Wis	Mar. 28, 1871	113, 211
Punch and press	W. S. Lord	Crawford, Miss	Feb. 22, 1870	100, 166
Punch and screw-driver, Combined	S. D. Tuttle	Eaton, Ohio	June 1, 1869	90, 704
Punch and screw-driver, Combined band	G. W. Schofield	United States Army	Dec. 26, 1871	122, 286
Punch and shears	D. D. Robinson	Niles, Mich	July 30, 1867	67, 352
Punch and shears, Combined power	A. C. Stanard	Milton, Wis	Sept. 6, 1870	107, 116
Punch, awl, and knife, Combined	H. S. Holmes and D. Williams	South Boston, Mass	May 16, 1871	114, 820
Punch, Belt	J. T. Carson	Greensborough, N. C	Feb. 11, 1868	74, 197
Punch, Belt	M. D. Drake	Scituate, R. I	Jan. 5, 1869	85, 575
Punch, Belt	J. E. Gates	Lowell, Mass	Apr. 30, 1872	126, 141
Punch, Belt	E. Hester	Suffield, Conn	Aug. 4, 1868	80, 734

Index of patents issued from the United States Patent Office from 1790 to 1873, inclusive—Continued.

Invention.	Inventor.	Residence.	Date.	No.
Punch, Belt	J. P. Jubb	Namaha, Mich	Feb. 2, 1869	86, 412
Punch, Belt	J. Mulchahey	Springfield, Mass	May 7, 1867	64, 556
Punch, Belt	A. Simpson	Worcester, Mass	Sept. 23, 1856	15, 795
Punch, Belt	A. M. Southwick	Boston, Mass	Feb. 11, 1873	135, 734
Punch, Belt	D. M. Weston	Boston, Mass	Sept. 1, 1868	81, 717
Punch, Belt	C. D. Wheeler	New York, N. Y	Jan. 24, 1860	26, 940
Punch-block	J. D. and J. M. Filkins	Johnson Township, Ind	Sept. 22, 1863	40, 030
Punch, Brad	R. F. Cook	Potsdam, N. Y	May 24, 1870	103, 429
Punch, Brad	J. Thorndike	North Weare, N. H	June 29, 1858	20, 743
Punch, Canceling	M. E. Berolzheimer	New York, N. Y	Nov. 30, 1869	97, 344
Punch, Car-ticket	R. J. Kellett	San Francisco, Cal	May 28, 1867	65, 090
Punch, Center	M. Bowker	Fitchburgh, Mass	Oct. 17, 1865	50, 442
Punch, Check	J. R. Mesa	Brooklyn, N. Y	Oct. 29, 1872	132, 593
Punch, Conductor's	J. and G. D. Friese	Baltimore, Md	Dec. 8, 1868	84, 739
Punch, Conductor's registering	A. D. Hoffman	Chicago, Ill	Feb. 22, 1870	100, 036
Punch, Conductor's ticket	J. Beck	Philadelphia, Pa	Sept. 10, 1867	68, 687
Punch, Conductor's ticket	W. J. Phelps	Springfield, Mass	Sept. 17, 1867	69, 019
Punch, Drop	S. Andrews	Perth Amboy, N. J	Apr. 13, 1852	8, 868
Punch for attaching buttons by rivets	E. Pincus and G. Rehfuss	Philadelphia, Pa	Oct. 18, 1864	44, 746
Punch for cutting out welts of boots and shoes	J. H. Walker	Worcester, Mass	June 28, 1864	43, 356
Punch for forming clasps	C. D. Flesche	New York, N. Y	Dec. 10, 1867	72, 007
Punch, Gang	J. H. Keating	Marblehead, Mass	Oct. 30, 1866	59, 230
Punch, Gang	B. G. Welch	Danville, Pa	Feb. 18, 1868	74, 646
Punch, Gun-wad	R. Rathbone	New York, N. Y	Dec. 31, 1867	72, 903
Punch, Hand	A. L. Eckert	Newark, N. J	Mar. 20, 1866	53, 282
Punch, Hand	O. C. Ford	Burlington, Conn	Sept. 13, 1870	107, 242
Punch, Hand	J. D. Higgins	Greenville, Conn	July 28, 1868	80, 349
Punch, Hand	W. Nash	Watertown, N. Y	Jan. 3, 1865	45, 738
Punch, Hand	R. Wood	Grand Ledge, Mich	Sept. 27, 1859	25, 599
Punch, Horseshoe	C. Huil	Lockport, N. Y	Oct. 9, 1866	58, 645
Punch, Leather	J. Lyle	Newark, N. J	Nov. 2, 1869	96, 453
Punch, Metal	W. I. Granger	Chicago, Ill	July 6, 1858	20, 846
Punch, Metal hand	S. F. Leach	Chelsea, Mass	Dec. 16, 1873	145, 578
Punch-mixer and egg-beater	T. Fisler	Camden, N. J	Apr. 17, 1866	53, 965
Punch-operating mechanism	L. W. Holmes	Grand Ledge, Mich	Sept. 22, 1868	82, 317
Punch, Paper-collar-button-hole	S. S. Stone	Troy, N. Y	Feb. 7, 1865	46, 279
Punch, Portable	J. J. Lafely	Ottumwa, Iowa	Apr. 22, 1873	138, 201
Punch, Portable metal	S. McKenna	Cincinnati, Ohio	Mar. 21, 1854	10, 681
Punch, Registering ticket	A. D. Hoffman	Chicago, Ill	Jan. 31, 1871	111, 345
Punch, Registering ticket	J. H. Small	Buffalo, N. Y	Jan. 31, 1871	111, 391
Punch, Registering ticket	J. H. Small	Buffalo, N. Y	May 23, 1871	115, 119
Punch, Revolving spring	S. Merrick	Springfield, Mass	Mar. 17, 1838	636
Punch, Revolving spring	S. Merrick	Springfield, Mass	Feb. 8, 1848	5, 435
Punch, rivet, and screw cutter	L. C. Judson	Waterville, N. Y	Nov. 18, 1825	......
Punch, Rotary	A. Bedford	Cold Water, Mich	Aug. 25, 1868	81, 465
Punch, Self-centering	S. Z. Hall	Camden, N. J	May 30, 1867	48, 018
Punch, shears, and iron-shrinker, Combined	D. C. Burdick	Milton, Wis	Mar. 2, 1869	87, 464
Punch, shears, &c., Operating hand	W. B. Mason	Boston, Mass	July 31, 1866	56, 772
Punch, Spring	P. Bauer	Newark, N. J	Mar. 29, 1864	42, 064
Punch, Spring	P. Bauer	Newark, N. J	Aug. 15, 1865	49, 364
Punch, Spring	M. V. Bryant	North Plains, Mich	July 28, 1868	80, 449
Punch, Spring	J. Lyle	Newark, N. J	Mar. 27, 1866	53, 537
Punch, Spring	A. U. Noble	Kalamazoo, Mich	Feb. 11, 1868	74, 241
Punch, Ticket	J. Chapin	Chicopee, Mass	May 4, 1869	89, 828
Punch, Ticket	R. Engels	Philadelphia, Pa	Apr. 5, 1870	101, 598
Punch, Ticket	H. W. Hewet	Cincinnati, Ohio	Feb. 4, 1873	135, 549
Punch, Ticket	W. Hill	Springfield, Mass	June 30, 1868	79, 470
Punch, Ticket	R. J. Kellett	San Francisco, Cal	June 23, 1868	79, 232
Punch, Ticket	C. H. Palmer and T. K. Leslie	Brooklyn, N. Y	Feb. 2, 1869	86, 438
Punch, Ticket	W. J. Phelps	Springfield, Mass	June 30, 1868	79, 498
Punches to stocks, Device for securing	J. Johnson	New York, N. Y	Apr. 2, 1872	125, 302
Punching and eyeleting machine, Shoe	J. Keith	New Bedford, Mass	Dec. 16, 1862	37, 170
Punching and eyeleting machines, Mechanism for operating.	C. H. Morse	Boston, Mass	Feb. 14, 1871	111, 865
Punching and raising disks from metal	C. Diedrichs	Philadelphia, Pa	Jan. 28, 1868	73, 784
Punching and shaping metal, Machine for	G. Haseltine	Washington, D. C	July 28, 1857	17, 876
Punching and shearing apparatus	S. Renfrew	Marseilles, Ill	Mar. 8, 1864	41, 861
Punching and shearing machine	W. Culver	Vineland, N. J	Jan. 25, 1870	99, 065
Punching and shearing machine	S. R. Houser	Sandusky, Ohio	Oct. 7, 1873	143, 510
Punching and shearing machine	W. H. Ivens and W. E. Brooke	Trenton, N. J	Nov. 9, 1869	96, 594
Punching and shearing machine	J. Kissel	Cassopolis, Mich	Oct. 28, 1873	143, 985
Punching and shearing machine	M. A. Lanagan	Brooklyn, N. Y	July 14, 1868	79, 912
Punching and shearing machine	P. C. Perkins	Waterford, N. Y	Mar. 12, 1861	31, 674
Punching and shearing machine, Combined	W. H. Ivens and W. E. Brooks	Trenton, N. J	May 2, 1871	114, 297
Punching and shearing machine, Combined	W. Lyon	Mamaroneck, N. Y	Sept. 16, 1873	142, 926
Punching and shearing machine, Combined	C. Swanson	Newton, Iowa	Dec. 16, 1873	145, 694
Punching and shearing machine, Iron	L. T. Pope	Boston, Mass	Mar. 17, 1838	639
Punching and shearing metal, Machine for	W. H. Ivens and W. E. Brooke	Trenton, N. J	Aug. 24, 1869	94, 117
Punching and shearing metallic plates, Machine for	J. M. Riter	Pittsburgh, Pa	June 25, 1872	128, 250
Punching and shearing tool	J. C. Jordan	Watertown, Wis	Nov. 12, 1867	70, 857
Punching and stamping machine	F. M. Huntington	Chicago, Ill	Sept. 3, 1872	131, 004
Punching and stamping machine, Metal*	J. Van Hagen	Chicago, Ill	Apr. 25, 1871	114, 068
Punching and stamping press	S. P. Ruggles	Boston, Mass	May 3, 1839	23, 864
Punching and upsetting apparatus	S. E. Lockwood	Westbury, New York	Mar. 3, 1868	75, 032
Punching-apparatus	J. F. Allen	New York, N. Y	Feb. 27, 1866	52, 810
Punching-apparatus	F. De Witt	Detroit, Mich	May 3, 1870	102, 510
Punching-apparatus	T. E. Harris	Green Bay, Wis	Aug. 6, 1867	67, 429
Punching-apparatus	R. Kent	Brooklyn, N. Y	Apr. 20, 1869	89, 226
Punching-apparatus	I. Lamplugh	Peoria, Ill	Feb. 18, 1868	74, 700
Punching-apparatus	P. L. Weimer	Lebanon, Pa	Oct. 31, 1865	50, 754
Punching between rollers	R. Montgomery	New York, N. Y	Feb. 5, 1850	7, 073
Punching corrugated metal, Apparatus for	M. J. Montgomery	New York, N. Y	May 8, 1866	54, 580
Punching countersunk holes	J. V. Westlake	Saint Louis, Mo	Dec. 16, 1862	37, 189
Punching designs in sheet-metal, Apparatus for	W. T. Rudd	Amsterdam, Va	July 8, 1851	8, 211
Punching-die	O. C. Burdict	New Haven, Conn	Apr. 10, 1866	53, 781
Punching-dies, Manufacture of	R. J. Mullin	Providence, R. I	Sept. 2, 1873	142, 407
Punching holes in iron or other metallic plates, Machine for.	W. Adams and A. Hammond	Boston, Mass	May 8, 1843	3, 077

* Ordered to be canceled by decree of circuit court of United States northern district of Illinois, October 2, 1873.

Index of patents issued from the United States Patent Office from 1790 *to* 1873, *inclusive*—Continued.

Invention.	Inventor.	Residence.	Date.	No.
Punching holes in leather, Machine for	G. L. Bailey	Portland, Me	Aug. 16, 1859	25, 083
Punching holes in metal, &c., Machine for	J. Sarchet	Philadelphia, Pa	Apr. 2, 1822	....
Punching linchpin holes and cutting off the journals of axles for wagons, &c., Machine for.	S. H. Hartman	Pittsburgh, Pa	Aug. 26, 1862	36, 298
Punching-machine	G. A. Alger	Manchester, N. H	May 22, 1866	54, 997
Punching-machine	R. Baird	Sterling, Scotland	Mar. 18, 1873	136, 954
Punching-machine	H. W. Bill	Cuyahoga Falls, Ohio	July 14, 1863	39, 209
Punching-machine	J. A. Bradshaw and A. F. Colby	Lowell, Mass	July 24, 1860	29, 231
Punching-machine	E. R. Brown and J. Long	Manch Chunk and Packerton, Pa.	May 23, 1871	115, 156
Punching-machine	W. Churchill	Saint Louis, Mo	May 11, 1869	89, 851
Punching-machine	J. Duff	Peoria, Ill	June 27, 1871	116, 420
Punching-machine	D. A. Faulkner	Centreville, Cal	June 27, 1871	116, 288
Punching-machine	E. Heath	Fowlersville, N. Y	May 13, 1856	14, 866
Punching-machine	C. Hughes	New Orleans, La	July 24, 1860	29, 343
Punching-machine	S. Kendall	Kalamazoo, Mich	Apr. 3, 1849	6, 277
Punching-machine	C. Keniston	Somerville, Mass	June 27, 1871	116, 323
Punching-machine	J. M. Laughlin	Boston, Mass	Sept. 2, 1873	142, 403
Punching-machine	G. W. Lewis	Dansville, N. Y	July 4, 1871	116, 726
Punching-machine	J. M. Long	Hamilton, Ohio	Feb. 6, 1872	123, 407
Punching-machine	W. Lyon	New York, N. Y	Nov. 10, 1868	83, 981
Punching-machine	C. S. Moseley	Elgin, Ill	June 28, 1870	104, 755
Punching-machine	S. W. Murray	Milton, Pa	Feb. 25, 1873	136, 172
Punching-machine	F. Nelson	Wyandotte, Mich	Mar. 24, 1868	75, 782
Punching-machine	R. Porter	Washington, D. C	Jan. 29, 1856	14, 166
Punching-machine	H. Powers	Florence, Italy	Feb. 19, 1861	31, 476
Punching-machine	J. C. Rhodes	South Abington, Mass	Mar. 7, 1871	112, 381
Punching-machine	W. T. Richards	Bridgeport, Conn	Oct. 13, 1868	81, 993
Punching-machine	J. F. Sargent and E. Townsend	Boston, Mass	Mar. 29, 1864	42, 146
Punching-machine	W. E. Scott and D. L. Wood	Terre Haute, Ind	Jan. 26, 1869	86, 324
Punching-machine	D. S. Sherman	Lowell, Mass	June 8, 1858	20, 516
Punching-machine	O. Smith	Bridgeton, N. J	May 21, 1872	127, 112
Punching-machine	O. Snow	West Meriden, Conn	Mar. 12, 1867	62, 897
Punching-machine	D. Sprague	Elizabethport, N. J	Mar. 12, 1861	31, 681
Punching-machine	J. Steadman	Pecatonica, Ill	Dec. 19, 1865	51, 630
Punching-machine	C. Steinbach	Lima, Mich	Mar. 3, 1868	75, 068
Punching-machine	W. H. Van Cleve	Ypsilanti, Mich	Oct. 10, 1871	119, 900
Punching-machine	W. Welch	Bridgeport, Conn	Oct. 15, 1867	69, 877
Punching-machine	W. Woiceski	Bloomington, Ill	Aug. 26, 1873	142, 317
Punching-machine, Automatic	J. E. Wiggin	Stoneham, Mass	Oct. 13, 1868	83, 014
Punching machine, Copper	W. Ballard	Boston	Dec. 6, 1827	....
Punching machine, Copper or brass	J. Shugert	Pittsburgh, Pa	July 20, 1831	....
Punching machine feeding-device	W. B. Bement	Philadelphia, Pa	Apr. 1, 1873	137, 407
Punching-machine, Hand	G. C. Miller	Rockford, Ill	May 31, 1870	103, 764
Punching-machine, Hydraulic	J. B. Barnes	Fort Wayne, Ind	Feb. 12, 1867	61 987
Punching-machine, Hydraulic	R. Livingston	Albany, N. Y	Oct. 24, 1871	120, 203
Punching machine, Leather	J. H. Haskell	Baltimore, Md	Oct. 22, 1867	69. 991
Punching machine, Metal	J. Allonas	Mansfield, Ohio	Sept. 14, 1869	94, 857
Punching machine, Metal	J. H. Brown	Grand Ledge, Mich	May 31, 1859	24, 191
Punching machine, Metal	S. H. Brown	Wheeling, Va	Aug. 14, 1839	1, 288
Punching machine, Metal	J. Clark	Haverstraw, N. Y	Aug. 12, 1873	141, 629
Punching machine, Metal	E. Craddock	London, England	Apr. 2, 1872	125. 121
Punching machine, Metal	O. J. Davis and T. W. Stephens	Erie, Pa	Oct. 4, 1853	10, 098
Punching machine, Metal	E. Doty	Jaynesville, Wis	May 10, 1870	102, 924
Punching machine, Metal	E. R. Hollands	Northampton Square, England.	Jan. 24, 1865	46, 056
Punching machine, Metal	H. A. Shipp	London, England	Nov. 2, 1869	96, 492
Punching machine, Metal	N. C. Stiles	Middletown, Conn	Aug. 27, 1872	130, 878
Punching machine, Metal	N. Thompson	Brooklyn, N. Y	Sept. 10, 1872	131, 193
Punching machine, Metal	G. Zender	Caledonia, Minn	Feb. 6, 1872	123, 435
Punching machine, Sheet-metal	M. Stevens	Brooklyn, N. Y	Oct. 7, 1873	143, 4[illegible]
Punching-machine, Steam	J. Sparrow	Portland, Me	Sept. 27, 1859	25, 588
Punching machine, Tube	J. T. Brigden	Hornellsville, N. Y	Sept. 29, 1868	82, 591
Punching-machine with a combination of adjustable gages.	R. S. Tilden	Saint Louis, Mo	Mar. 10, 1849	6, 154
Punching machinery, Metal	D. T. Walker	Brooklyn, E. D., N. Y	Sept. 12, 1865	49, 937
Punching metal	L. C. Bunnel	Otsego, N. Y	Oct. 10, 1834	....
Punching metal	P. Koch	New Haven, Conn	Oct. 4, 1859	25, 653
Punching metallic sheathing, Machine for	A. H. Teeple	New York	Aug. 28, 1846	4, 727
Punching nail-holes, Machine for	C. C. Crosby	Nantucket, Mass	Oct. 9, 1860	30, 374
Punching or cutting iron or steel by machinery	J. Richards	Elbridge, N. Y	Sept. 23, 1831	....
Punching-press	G. H. Corliss and E. Harris	Providence, R. I	Mar. 25, 1856	14, 493
Punching-press	N. C. Stiles	West Meriden, Conn	Jan. 26, 1864	41, 403
Punching-press	M. G. Wilder	West Meriden, Conn	May 28, 1867	65, 143
Punching-presses, Conveyer to transfer blanks from	J. H. Baird	Oakville, Conn	June 8, 1869	90, 914
Punching-press rod	J. V. C. Crate	Waterbury, Vt	Nov. 20, 1866	59, 823
Punching, pressing, cutting, and slitting machine for iron plates, saws, &c.	J. Bennett	Brutus, N. Y	July 25, 1832	....
Punching rivet-holes in hose, Machinery for	J. R. Hague	Pittsburgh, Pa	July 4, 1854	11, 222
Punching sheet-metal	M. Seiferth	Morristown, N. J	Nov. 12, 1867	70, 905
Punching sheets of metal, Machine for	J. M. Riter and L. J. Farquhar	Pittsburgh, Pa	Oct. 9, 1866	58, 676
Punching sheets of metal, Machinery for	S. T. Sanford	Fall River, Mass	Jan. 13, 1852	8, 660
Purchase, Suspended	W. H. Brown	Erie, Pa	Jan. 2, 1855	12, 155
Purifying and refining spirits, Apparatus for	T. Thompson	Baltimore, Md	June 14, 1864	43, 141
Purifying fatty bodies	R. A. Tilghman	Philadelphia, Pa	Oct. 3, 1854	11, 766
Purifying liquids by galvanism	A. Crosse	Broomfield, England	Jan. 6, 1848	5, 409
Purifying spirits and other liquids, Compound for	P. J. Badoux	New York, N. Y	Aug. 27, 1867	68, 028
Pushing-jack	A. Freeman	Peoria, Ill	June 29, 1869	91, 837
Putty, Composition for	J. and W. H. Lucas	Philadelphia, Pa	Aug. 28, 1866	57, 528
Putty-knife	S. W. Gerelds	Worcester, Mass	Jan. 18, 1870	98, 949
Putty-knife and sprig-box, Combined	J. H. Gaches	Atlanta, Ga	Mar. 4, 1873	136, 499
Putty, Manufacture of	G. W. Hatfield	Dayton, Ohio	Oct. 26, 1869	96, 231
Puzzle	J. W. Mueller	Detroit, Mich	Feb. 24, 1863	37, 763
Puzzle, Alphabetical-instruction	E. F. Gilbert	Lyons, N. Y	June 22, 1869	91, 737
Puzzle, Child's	W. B. Rice	Feltonville, Mass	Feb. 9, 1864	41, 536
Pyrite-burning furnace for manufacture of sulphuric acid, &c.	J. Hughes	Edgewater, Mass	June 18, 1867	65, 914
Pyrites, Treating auriferous and argentiferous	L. Solomon	New York, N. Y	Jan. 11, 1859	22, 587

Index of patents issued from the United States Patent Office from 1790 *to* 1873, *inclusive*—Continued.

Invention.	Inventor.	Residence.	Date.	No.
Pyrometer	W. H. Bailey	Albion Works, Salford, Great Britain.	Feb. 28, 1871	112, 106
Pyrometer	E. Brown	Philadelphia, Pa	June 1, 1869	90, 815
Pyrometer	E. Brown	Philadelphia, Pa	May 3, 1870	102, 654
Pyrometer	E. Brown	Philadelphia, Pa	Aug. 27, 1872	130, 894
Pyrometer	E. Brown	Philadelphia, Pa	Aug. 27, 1872	130, 895
Pyrotechnic night-signal	B. F. Coston	Washington, D. C	Apr. 5, 1859	23, 536
Pyrotechnic night-signal	M. J. Coston	Washington, D. C	June 13, 1871	115, 935
Pyrotechnic night-signal	G. A. Lilliendahl	New York, N. Y	Apr. 5, 1859	23, 529
Pyrotechnic signal	H. J. Harris	Shreveport, La	Sept. 29, 1868	82, 619
Pyrotechnic signal	A. Lamarre	Paris, France	Oct. 28, 1873	144, 030
Pyroxile, pyroxyline, &c., for forming useful and ornamental articles, Treating.	L. R. Streeter	Chelsea, Mass	Apr. 20, 1869	89, 254
Pyroxyline and articles therefrom, Manufacture of	S. Deitz and B. P. Wayne	Albany, N. Y	Dec. 17, 1872	133, 969
Pyroxyline, Apparatus for compressing and molding plastic.	J. Brockway	Albany, N. Y	Apr. 18, 1871	113, 735
Pyroxyline articles, Manufacturing	J. W. Hyatt	Albany, N. Y	Apr. 29, 1873	138, 254
Pyroxyline, Process and apparatus for manufacturing.	J. S. and J. W. Hyatt	Albany, N. Y	Nov. 19, 1872	133, 229
Pyroxyline, Treating and molding	J. W., jr., and J. S. Hyatt	Albany, N. Y	July 12, 1870	105, 338

www.ingramcontent.com/pod-product-compliance
Lightning Source LLC
LaVergne TN
LVHW010524100826
845148LV00001B/88

* 9 7 8 1 4 2 5 5 8 6 5 2 2 *